# ANTARCTIC EARTH SCIENCE

*Edited by*

R. L. Oliver and P. R. James
Department of Geology,
University of Adelaide, South Australia
and
J. B. Jago
School of Applied Geology,
South Australian Institute of Technology,
South Australia

CAMBRIDGE UNIVERSITY PRESS
Cambridge
London New York New Rochelle
Melbourne Sydney

CAMBRIDGE UNIVERSITY PRESS
Cambridge, New York, Melbourne, Madrid, Cape Town,
Singapore, São Paulo, Delhi, Tokyo, Mexico City

Cambridge University Press
The Edinburgh Building, Cambridge CB2 8RU, UK

Published in the United States of America by Cambridge University Press, New York

www.cambridge.org
Information on this title: www.cambridge.org/9780521183796



First published 1983
First paperback edition 2011

*A catalogue record for this publication is available from the British Library*

ISBN 978-0-521-25836-4 Hardback
ISBN 978-0-521-18379-6 Paperback

Dedicated to the memory of Sir Douglas Mawson

# CONTENTS

## 5. Scotia Arc and Antarctic Peninsula



## 6. Marine Geology



## 7. Antarctic Resources

## Fourth International Symposium on Antarctic Earth Sciences

Front row from left to right—
P.R. James, J.B. Jago, D. Hahn, C.J.L. Wilson, K.D. Collerson, B.R. Watters, M. Yoshida, P.S. Dahl, D.R. Hunter, W.J. Zinsmeister, O. Gonzalez-Ferran, D.N.B. Skinner, J.M. McLennan, N.R. Kemp, R.D. Hamer, A.B. Ford, P.D. Rowley, R.L. Oliver, R.H.N. Steed, F.J. Davey, C.C. Plummer, J. Splettstoesser, J.F. Lovering, C.G. Gatehouse, V.L. Ivanov, A.G. Dralkin, V.A. Gostin, G.E. Grikurov, K. Yanai, T.S. Laudon, H. Shirahata.

2nd row from left to right—
R.J. Tingey, K. Shiraishi, Y. Yoshida, T.W. Gevers, J.A. Cooper, J.W. Gosgriff, C. Foster, R.A. Askin, C.T. McElroy, K.J. Whitby, J.R. Krynauw, J. Pojeta Jr., J.H. Shergold, M.A. Bradshaw, M.R.A. Thomson, T.H. Jefferson, P.G. Quilty, J.W. Thomson, R.L. Bauer, J.M. Barton Jr., A.B. Moyes, K. Kaminuma, J.C. Dooley, E. Stump, J.C. Behrendt, H.Y. Ling, L.D. McGinnis, B. McKelvey, M. Ogasawara, S. Sato.

3rd row from left to right—
D.J. Ellis, A.J.W. Gleadow, P.R. Vogt, G. Morgan, G.W. Grindley, M.G. Laird, R.H. Findlay, F. Tessensohn, B. Field, J.D. Bradshaw, B.L. Ward, R.G. Warren, W. Buggisch, S. Weaver, G.F. Webers, P.D. Marsh, J. Davidson, P. Barker, I.W.D. Dalziel, B.J.J. Embleton, D.H. Elliot, D.J. Drewry, F.M. Auburn, E. Nelson, G. Kleinschmidt, N.W. Roland, J. Hofmann, P.F. Ciesielski, M.T. Ledbetter.

4th row from left to right—
E.S. Grew, P.C. Grew, C. Craddock, G. Gibson, P.A. Morris, P.J. Barrett, B.C. Walker, A.R. Pyne, S.J. Pyne, I. Clarke, F. Thyssen, E. Godoy, Z. Danya, P.R. Kyle, P.J. Oliver, A. Ikami, H. Kojima, H. Miller, Y. Nakai, Y. Nakai, G. Gurr.

5th row from left to right—
M. Sandiford, R.J.F. Jenkins, C.R. Bentley, G. Delisle, D. Wyborn, M. Sandiford, B. Murrell, C.S.M. Doake, R.J. Adie, H.J. Harrington, D.H. Green, R.W.R. Rutland, W.A. Cassidy, K. Rankama.

6th row from left to right—
A. Cummings, J.G. Ferrigno, A. Pedlar, C.M. Fanning, S. Daly, A. Webb, P. Wellman, J.T. Hunter, L.P. Black, S.L. Harley, J. Foden, J. Scott, Y. Hiroi, Y. Motoyoshi, A.J. Truelove, R.B. Flint.

7th row from left to right—
D.A. Adamson, J.B. Anderson, J. Annexstad, K. Omoto, R.J. Pankhurst, J.L. Smellie, J.A. Crame, P.D. Clarkson, G.W. Farquharson.

8th row from left to right—G. Faure.

9th row from left to right—
J.G. Bockheim, J.W. Collinson, K.S. Taylor, H.T. Brady, J. Pickard, K.J. Hall, M.D. Turner, R.W. Plume, H. Plume, P.N. Webb, J.W. Zawiskie, S.G. Borg.

# PREFACE

This volume comprises the texts of 174 papers, in complete or abstract form, presented at the Fourth International Symposium on Antarctic Earth Sciences in Adelaide, South Australia, during the week 16-20 August, 1982.

Some 200 persons from sixteen countries attended the Symposium. Despite the increased number of earth scientists currently participating in Antarctic research, the attendance at the Adelaide meeting was very similar to that at Madison in 1977. Transport costs to the antipodes were doubtless a deterrent for many.

Five years separated the Adelaide from the Madison meeting, compared with seven years between the previous Antarctic Symposia. The 1982 date for the Adelaide Symposium was chosen to correspond with the centenary of the birth of Sir Douglas Mawson and the Symposium was a fitting enhancement of Mawson centennial commemorative activities in Adelaide. Irrespective, however, of the Mawson centenary, most Symposium participants felt that an interval of five years between symposia is sufficient, and it is likely that this will be adopted henceforth.

The initial proposal in Madison in 1977 to hold the Symposium in Australia was approved at a meeting of Australian Antarctic geologists and geophysicists, convened by R. J. Tingey, in Melbourne, in May 1979, and Adelaide was suggested as a suitable venue. In October 1979, a draft proposal was submitted by R. L. Oliver to the Australian National Committee on Antarctic Research (ANCAR), all members of which approved the proposal with some suggested modifications. The Adelaide Organizing Committee was constituted in June 1980. Approval of the Symposium proposal (again with some suggested modifications) was granted by the SCAR Geology Working Group, meeting in Paris in July 1980.

An international steering committee was proposed by SCAR at its meeting in New Zealand in October 1980. The Steering Committee's responsibilities included approval of the allocation of grants of money to individuals, refereeing of abstracts and approval of the publication procedure.

The Organizing Committee's decision to charge a significant registration fee broke new ground compared with previous Antarctic symposia, but the decision appeared to be generally (if not universally) accepted.

The Symposium sessions were preceded by a five-day field trip to Eyre Peninsula and the Flinders Ranges. Forty delegates participated. Back in Adelaide, 183 papers were presented in four concurrent sessions over five days in the Mawson Laboratories (Geology and Economic Geology Departments) and adjoining Benham Building (Botany Department) of the University of Adelaide. Symposium activities terminated with four post-sessional separate day excursions in the Adelaide region; these also were enjoyed by some forty delegates.

In July 1981, arrangements were made with the Australian Academy of Science to publish the proceedings of the 1982 Antarctic Symposium from camera-ready copy. This decision was made by the Adelaide Organizing Committee, with the approval of the Steering Committee; the decision followed earlier receipt from a number of printing firms of price estimates for complete typesetting and binding which were prohibitive. We were attracted, however, by the scheme offered by Adelaide Executive Services and Gillingham Printers, in Adelaide. This scheme entailed retyping of the author's copy by Adelaide Executive Services' word processor with the inclusion of typesetting codes. The data on the disks was then interfaced via magnetic tape to Gillingham's digitized typesetting system. Though somewhat more expensive than a straight typing or word processor job, the process has resulted in a somewhat more professional product, worthy of being dedicated to Sir Douglas Mawson, and within the Organizing Committee's budget. A rather small print size is the consequence of the combined (i) wish of the Steering Committee to include in this volume all those papers given at the Symposium and (ii) strict overall page limit set by the Australian Academy of Science.

Symposium contributions were grouped into fifteen topics and this grouping has been retained here. The relative popularity of Topic 1, viz. the Precambrian East Antarctic Craton, is to be expected at a Symposium held in Adelaide, reflecting the activity of a number of Australians in that part of Antarctica. In particular, the concerted attack on Enderby Land by a combination of the Bureau of Mineral Resources and the Universities paid dividends. The Mawson Lecture by Professor D. H. Green reflects this Enderby Land effort. The logistic planning and co-operation of the Antarctic Division of the Australian Department of Science contributed greatly to the Enderby Land operation over a number of seasons.

The single topic with the greatest number of papers is that dealing with the East Antarctica-West Antarctica Boundary and the Ross Orogen, including northern Victoria Land. This is due very much to the staging of the joint Australian-New Zealand-United States field project in northern Victoria Land during the 1981-82 field season. The project, involving some fifty earth scientists, produced much new stratigraphic, palaeontological, structural, petrological and geophysical data and material for subsequent laboratory work, as recounted herein. Transport to and from northern Victoria Land via McMurdo was provided by the U.S. Navy Squadron VXE-6. The whole project depended very much on U.S. logistic planning and operation and is a good example of the co-operative spirit which imbues the research programmes in Antarctica.

Work in the Scotia Arc and Antarctic Peninsula has intensified greatly since the concept of plate tectonics became popular. Stratigraphic, structural, palaeontological, and geophysical studies have combined to produce an increasingly plausible story concerning crustal developments in that region.

The area in which probably most development has taken place in the years since the Madison Symposium is that involving a combination of Marine Geology, Glacial Geology and Geomorphology and Cenozoic Tectonics and Climate, onshore and offshore. This is because acquisition of data in many of those fields of investigation involves the application of new technology with regard to drilling and ocean bottom sampling. Geomorphological mapping has taken on a new look with the involvement of Landsat images and satellite radar altimetry.

New crustal structure data, including measurements from Magsat and heat flow measurements represent further manifestation of the wealth of new data from Antarctica being produced by modern techniques.

The topic of Gondwanaland is not as popular as it was a decade ago but new and confirmatory evidence for direct land connections is still forthcoming. The first discovery of marsupials in Antarctica, reviewed herein, was reported just prior to the Adelaide Symposium.

New data on plate tectonics, meteorites, subantarctic islands and Cenozoic igneous activity, considerably updates our knowledge, both descriptive and theoretical, of these disciplines. Work on rates of uplift, recorded by detailed study of hyaloclastites, contributes further interesting data in this specialized field.

At the closing ceremony of the 1977 Madison Symposium, Tore Gjelsvik observed that there was a 'trend away from preliminary descriptive investigations to more detailed, mature and varied scientific studies'. The trend has continued in the research work recorded in this volume. As alluded to above, increasing specialization has facilitated the application of the most up-to-date techniques of investigation, both in the field and in the laboratory. Specialists in diverse fields must associate with workers in corresponding fields, not necessarily involving Antarctica, in order to further their respective specializations. Despite this, there continues to be a bond which unites research workers in Antarctica, a bond strengthened by the rigours of the environment and, at times, the consequent necessity for co-operation and liaison in order to survive. The international nature of this co-operation is a feature of Antarctic programmes. In his foreword to the proceedings of the second Antarctic Symposium, 1970, R. W. Willett refers to that volume as 'another indication of unity and freedom of the SCAR nations working in Antarctica, the international scientific laboratory of mankind'. It would be a pity if this could not be maintained.

R.L. Oliver
Department of Geology
University of Adelaide
South Australia 5000

J.B. Jago
School of Applied Geology
South Australian Institute of Technology
South Australia 5098

P.R. James
Department of Geology
University of Adelaide
South Australia 5000

# ACKNOWLEDGEMENTS

SCAR, the parent body of the Symposium, financed the travel costs of a number of delegates to the Symposium and provided advice and encouragement on several occasions. The Australian Academy of Science also facilitated the attendance at the Symposium of three overseas registrants and gave useful advice; however, as these words are being penned, the Academy's main contribution as publisher, is yet to come.

The International Union of Geological Sciences, the Australian Academy of Technological Sciences and the Geological Society of Australia all contributed financially. The University of Adelaide, as well as helping financially, provided the Symposium venue, with many attendant contributions, referred to below.

Other organizations provided monetary assistance, some very generously; these are listed separately.

The financial state of the Symposium budget was, throughout the preparations for the Symposium, very much the concern of Professor E. A. Rudd; a donation from SANTOS to set the organizational wheels in operation was due to his energy and drive.

Several organizations helped in a variety of other ways. The Antarctic Division of the Commonwealth Department of Science and Technology donated a map of Antarctica for each Symposium delegate and the ANZ Bank donated the Symposium satchels; ANSETT, as the official Symposium carrier, supplied extremely comprehensive printed information concerning conference preparation and also provided a number of other items; provision of these by Bob Birk, Liaison Officer, and the assumption of complete responsibility for the arrangement of accommodation of Symposium delegates by Paula McIntyre eased the work of the Organizing Committee considerably.

In the University of Adelaide, Chairmen Dr. J. B. Jones and Dr. F. A. Smith willingly made available their respective Departments for the holding of the scientific sessions. M. Stojanovic and other staff in the Departments of Geology and Economic Geology worked overtime in the preparation of displays; Evert Bleys and students efficiently organized and operated the slide projection during the presentation of papers; volunteers from the South Australian Field Geology Club assisted during registration and helped serve tea and coffee, ably organized by Helen Oliver. The Symposium Information Centre was manned efficiently by Jan Angus and Pam Hasenohr. Elsewhere in the University, the Bursar's office handled the Symposium accounts, the Staff Club and the Union provided temporary membership to delegates and thus cheap and rapid eating facilities.

B. Daily and M. Fanning helped to lead the pre-Symposium excursion and B. Daily, M. McBriar, V. Gostin and A. Milnes led one or another of the post sessional excursions.

The long haul in preparation for the Symposium was the responsibility of the Adelaide Organizing Committee and the International Steering Committee. The success of the Symposium is adequate tribute to the time consuming and conscientious efforts by the members of these committees and the Adelaide Organizing Committee, in particular. Assisting this latter Committee were two very hard working and able secretaries—Jan Angus as official secretary and Glenice Leonard at the South Australian Institute of Technology, as Jim Jago's helping hand. Fiona McQueen took over from Jan Angus, as part-time Secretary for a few weeks, following the Symposium.

The refereeing of all the papers submitted for publication represents a very considerable aggregate effort. The large number of individuals who contributed to this have been thanked individually.

The drafting of illustrations for the excursion guides was done almost entirely by the drafting staff of the South Australian Department of Mines and Energy, as a contribution initiated by the SADME Chief Geologist, Dr. C. Branch, to the Symposium.

In the preparation of this volume, the assistance of a number of students and other persons in a variety of ways has been necessary, checking figure sizes, segregating manuscripts according to which stage they have reached in the sequence from initial submission to final camera-ready copy, proof reading of text and figure arrangement, etc.

C. G. Gatehouse, A. J. Parker, R. J. Tingey and I. H. Oliver have contributed particularly in the several stages of proof reading and index preparation. R. J. Tingey deserves mention also as one who has assisted throughout the pre-and post-Symposium periods as referee, Steering Committee member and general adviser.

To all the above, to the Symposium contributors themselves, and to Mr. Tonkin, Sir Mark Oliphant and Dr. Law, who spoke at the Symposium opening ceremony, the editors extend their sincere thanks. Only with their participation and the assistance as outlined has the Symposium been a success and the production of this volume made possible.

R.L.O.
J.B.J.
P.R.J.

*Photo by Paul J. Finlay*

**Bust of Sir Douglas Mawson, by John Dowie, located at the North Terrace entrance to the University of Adelaide, unveiled by Dr John Watson, Lord Mayor of Adelaide, on 16 August 1982.**

# SIR DOUGLAS MAWSON, Kt., O.B.E., B.E., D.Sc., F.R.S., F.A.A.

Douglas Mawson, Antarctic explorer, geologist and a leading scientific figure in the Australian Commonwealth, died in Adelaide in 1958 at the age of seventy-six.

In 1982, the 100th anniversary of his birth, numerous commemorative events were held throughout Australia. It was considered fitting to stage the Fourth International Symposium on Antarctic Earth Sciences also in this year.

Mawson was born in England and came to Australia with his parents at the age of two.

In 1901, he graduated with B.E. from the University of Sydney. While at the University, however, influenced by Professor T. W. Edgeworth David, he developed an interest in geology and this was displayed by his undertaking of two years' geological exploration in the New Hebrides. He returned to the geology school of Sydney University and completed a B.Sc. with geology as a major subject, in 1905.

In the same year he was appointed lecturer in Mineralogy and Petrology at the University of Adelaide where he was subsequently to become professor and work until his retirement in 1952.

Mawson's first visit to Antarctica was as a member of Ernest Shackleton's expedition which he was invited to join on the recommendation of Edgeworth David. Late in 1907 the *Nimrod* sailed from Port Lyttelton, New Zealand, and base camp was established on Cape Royds in the Ross Sea area. Mawson's duties were those of physicist and he was chiefly concerned with geomagnetic and auroral studies. He was a member of the party which made the first ascent of Mt Erebus and, together with Edgeworth David and Dr A. F. MacKay, made history by reaching the South Magnetic Pole. The latter involved a man-hauling sledge journey of nearly 1300 miles (2000 kilometres) up to and across the Antarctic plateau and back to base under the severest conditions. On the return journey Professor Edgeworth David handed over the leadership of the party to the younger and fitter Douglas Mawson.

Mawson returned to Adelaide in 1909. He had already decided that he would organise his own expedition to Antarctica and preparations for this were completed in 1911. The Australasian Antarctic Expedition (AAE) sailed in the *Aurora* to set up its Main Base at Cape Denison, Commonwealth Bay (Longitude 142°40′E), in Adélie Land and a subsidiary Western Base, under the leadership of Frank Wild, on the edge of the Shackleton Ice Shelf at Longitude 095°E. During the 1912-13 summer four separate sledge journeys explored the Commonwealth Bay hinterland and five explored the environs of the Western Base, manifesting the aims of AAE which were to concentrate on scientific discovery. The tragedy which befell Mawson's sledge journey east of the Main Base, during which one companion and most of the party's food were lost in a crevasse and the other companion died of exhaustion and starvation, is well known. Mawson eventually reached Base only to find that the *Aurora* had departed only a few hours before to pick up the Western Base party and return to Australia before winter set in. Those members of the expedition who had remained in the expectation that Mawson would return had to spend another winter in Antarctica. The expedition had taken with it the first wireless installation to be established in Antarctica and it was from Cape Denison in February 1913 that the dramatic message of Mawson's ordeal was transmitted to Australia. It was typical of Mawson and the men of the expedition that the high quality of the scientific work was maintained during the second enforced year. The stream of scientific memoirs that thenceforth issued from the Australian Government Press during the succeeding forty years was testimony of this. Mawson's narrative account of the expedition, published under the title 'The Home of the Blizzard', has become one of the classics of polar exploration.

The years following the final return to Australia of members of AAE found Mawson dividing his time between being a geology teacher and the head of a University department, organising the writing up of the results of the scientific work achieved by AAE in Antarctica and subsequently in Australian laboratories and, before long, thinking again of the southern continent and problems associated with it. The 'problems' were mainly those involving ownership and control of the section of the Antarctic coastline 'neighbouring' Australia. Concern among many for the need to protect the animal life of the region, namely seals, penguins and whales, contributed to this feeling of a need for control. But the question of ownership was perhaps of even greater though allied importance in Mawson's eyes at this time; and the claiming by France in 1924 of Adélie Land and the interest of the Norwegian whalers in seeking ownership further west enabled Mawson to stimulate Australia into action. In 1927 Sir Douglas, as a member of the newly established Australian Antarctic Committee, was most outspoken in advocating an Australian expedition to the Antarctic to complete his 1911-14 Antarctic fieldwork and to establish British sovereignty over the 'Australian quadrant'.

The eventual expeditions during the 1929-30 and 1930-31 summers, in the ship *Discovery*, were led by Mawson. In addition to claiming for Australia the Antarctic coastline between Longitude 45° and Longitude ca 140° E, the expedition amassed a wealth of unique geodetic and geological data which contributed significantly to our knowledge of the southern continent. The expedition delineated for the first time the regions, named by Mawson, viz MacRobertson Land and Princess Elizabeth Land. Important oceanographic and marine biological observations and collections supplemented those made on land.

Although Mawson's name is linked inseparably with Antarctica and with his contributions to our knowledge of that continent, more than half Mawson's bibliography of 102 published papers deals with South Australian geology. Included among these are descriptions of Proterozioc glacials, study of mineral deposits, particularly uranium ores, and the establishment of a South Australian stratigraphy.

He continued his work until retirement in 1952.

Honours came to Mawson from many countries and institutions. As a teacher, he gained both the respect and affection of generations of students.

We in Adelaide, particularly those who have worked in Antarctica, are proud to have followed in Mawson's footsteps and to have contributed a little more to the work that he began.

R.L.O.

# SYMPOSIUM OPENING CEREMONY

*Opening address by the then Premier of South Australia, the Honourable David Tonkin, M.P.*

Mr. Chairman, Ladies and Gentlemen,

It was with great pleasure that I accepted your invitation to open this, the Fourth International Symposium on Antarctic Earth Sciences.

Previous Antarctic Symposia have been held at seven-year intervals, the first in 1963 at Cape Town, the second in 1970 in Oslo, and the third in 1977 in Madison, Wisconsin. The fourth has been brought forward by two years to coincide with the centenary of the birth of Sir Douglas Mawson, the Antarctic explorer and pioneer geologist of South Australia. So it is particularly appropriate that the venue for this symposium is the University of Adelaide, where Sir Douglas Mawson held the Chair of Geology and Mineralogy from 1921 to 1952.

Many of you will have just returned from the first of five geological excursions associated with the Symposium. On that excursion your South Australian guides took you to see features of geological interest on Eyre Peninsula and in the Flinders Ranges. You will have observed geological relationships in the older Precambrian rocks of South Australia, which have known counterparts in Antarctica. These are the rocks that in South Australia contain the bulk of our mineral resources, as they are so far known.

This brings me back to Sir Douglas Mawson's contribution. His early work dealt with the geology of the Broken Hill mineral field—a terrain with many similarities to the Eyre Peninsula area you have visited. Then, following his exploratory expeditions in Antarctica, he turned his attentions to the Adelaide Geosyncline, a belt of younger Precambrian sedimentary rocks that were largely the focus of the remainder of the first excursion, and which will be visited again briefly on the later excursions. Here, his experience working under the conditions of the Antarctic ice sheet provided him with a dress-circle insight into glacial processes. He realised the applicability of this knowledge to the interpretation of those rocks in South Australia that are believed to have been deposited by ancient glaciation, about 700 to 800 million years ago. Study of these rocks and that part of Earth history that they represent is still proceeding, but Mawson's contribution provided a foundation on which a great deal of scientific research and mineral exploration, has been based. Our record in this field has attracted world-wide attention, for we have also in South Australia evidence of a younger glaciation as well, about 250 to 300 million years ago. Some of you will visit evidence for this on one of the post-symposium excursions south of Adelaide. These glaciations, together with other palaeoclimatic markers, provide important mileposts in the earth's history.

We are privileged to have the opportunity today to welcome some 200 participants. Many are guests from some sixteen other countries, as well as from each of the other Australian States. Because of the interdisciplinary nature of this earth science symposium, it has been necessary to set up four concurrent sessions to deal with the 183 papers being presented. I hope that this range of topics will ensure that there is something of interest for each one of you.

Whereas Mawson's interests centred mainly on the older rocks in Australia, our link with Antarctica has important implications for the younger sediments that contain so much of our energy resources. This is so because the idea of continental drift, first conceived some sixty years ago, is now widely accepted in a modernised form as the theory of Plate Tectonics. Study of the sea floor of the Southern Ocean, and comparisons of the rocks of Southern Australia and Antarctica have demonstrated that these continents were once joined and started to move apart about forty to fifty million years ago. This is believed to have been one of the last separations that brought about the final breakup of the great southern supercontinent, Gondwanaland. But before the actual separation commenced, large areas of the crust of the Australian and Antarctic continents were subjected to crustal tension. As a result of this, great rift valleys were formed in the vicinity of the future continental breakup, and as these filled with sediments derived from surrounding highlands, so deposits of some of our most important energy resources accumulated in these sedimentary basins—coal in the St. Vincent Basin near Adelaide and in the Polda Basin on Eyre Peninsula, and natural gas in the Bass Strait Basin. So the relationship between the two continents is of great practical significance for Australia.

Last year, the Geological Society of Australia held its fifth Australian Geological Convention in Perth. The first of a series of Earth Science lectures commemorating Sir Douglas Mawson was presented there. During this symposium we will be privileged to hear the second Mawson Lecture, to be presented by Professor D. H. Green of the University of Tasmania, on rocks formed under conditions of extreme temperature and pressure, deep in the crust of Antarctica.

Looking at the programme of papers to be presented, it is obvious that this symposium will be of great interest to all earth scientists, whether they be in universities, government employment, or in private enterprise, exploring for mineral and energy resources.

On behalf of all participants, I would like to thank the organisers for their splendid efforts in bringing you all together for this international exchange of ideas on the earth science aspects of the great continent, Antarctica—once our close neighbour but now almost completely hidden beneath the polar icecap.

It is a tribute to all who have worked under its harsh climatic conditions that such a wealth of information is now available about it.

It is with great pleasure that I welcome you all to South Australia, and, in so doing I declare the Fourth International Symposium on Antarctic Earth Sciences officially open.

*'Mawson the Man' by Sir Mark Oliphant, K.B.E., F.R.S., F.A.A., distinguished nuclear physicist, former Governor of South Australia, a former student and personal friend of Mawson.*

Having agreed to speak today, I became aware of the great gaps in my understanding of a remarkable man, who was my friend. So I wrote to Dr Reg. Sprigg, seeking illumination from a professional geologist who knew Mawson well. At the end of a most helpful tribute he used words which summarized Mawson for him; 'Always in a hurry'. This was so close to my own experience of the man that it must be the text upon which I base my remarks.

Study of the surface of the Earth was the passionate pursuit of three tall handsome men who influenced me profoundly. Patrick Blackett promoted in his laboratory, and in the field, the investigation of the remanent magnetism of rocks, a technique which has proved of immense importance. W. S. Robinson, of the Zinc Corporation, exploited geological information for profit. But it was Mawson who loved the crust and atmosphere of this planet most intensely, who revelled in handling it in all its forms, and who used his great physical energy to explore it. He recorded meticulously every possible aspect, every mood of the elements, which came within his restless and wide-ranging attention.

It is for others to speak of the incredible record of Douglas Mawson in what he described as 'This accursed land'. He was with the ice and snow when I began the course in geology in Adelaide, under Walter Howchin. But when he returned his heroic figure filled us all with the desire to emulate. He exhausted his students as he strode ahead on geological excursions. The class included Alderman, who succeeded him in the Chair, and Dick Thomas, who was to distinguish himself in CSIR, as it then was. Mawson was obviously at home with every piece of rock he held in his hand, whether selected by himself, or presented by a curious student. He treated a morsel of the crust of the Earth as did a dedicated biologist a piece of tissue, or a solid state physicist a flawless crystal. His encyclopaedic knowledge did not make him impatient of the ignorance of his students. He was kind to us all, even when we asked the most idiotic questions. I often wondered how one with that epic lone journey behind him could tolerate the inanities of some of us.

During this period, Professor Kerr Grant invited me to a meeting of a small group of savants who met occasionally in one of the small rooms in the original building of the University. Wood-Jones and Mawson were to discuss the theory of Wegener, according to whom the continents had drifted very great distances in the course of geological time. In particular, Australia, South America, Africa and India had once formed part of a very large land mass, Gondwanaland, which had split up long ago. Wood-Jones spoke first, very much in favour of Wegener's concept, because it helped explain the distribution of animals and plants, which was of great interest to a comparative anatomist. However, his enthusiasm left Mawson quite unmoved. Mawson proceeded to pour ice-cold water over all that Wood-Jones had said, emphasizing the absurdity of any belief that solid rocks could flow like pitch. He was unimpressed by the argument that things which appeared solid under laboratory conditions, might flow slowly under high pressure over geological time. I don't know whether, if ever, he came to terms with the clear evidence for continental drift. He died before plate tectonics became the vogue.

When I returned to Australia I found that the only body coordinating scientific activity, and representing Australia internationally, was an offshoot of ANZAAS, the Australian Research Council. The Council had no fixed abode, Melbourne/Sydney rivalries dictating that it be located alternately in one city and then the other. Attempts to found a more effective national organization, with high standards, had failed because of interstate jealousies, and problems about who would qualify as members. Accordingly, eleven of the twelve Fellows of the Royal Society of London, resident in Australia, decided to take steps to found an Australian Academy of Science. They had all been elected to the Royal Society in recognition of their scientific work, so no-one could question their credentials. Mawson had been a Fellow since 1923. He understood well the nature of a scientific society devoted to excellence. His standing, and his persuasion removed the doubts of many, both scientists and politicians, who were suspicious of the proposed Academy.

This later close association with Mawson, and with his wife, Paquita, confirmed in me a feeling about the man which had haunted me from the moment I met him as a student. He had walked so close to death that there was something fey about him, an other-worldliness which marked him out from other men. His exuberant character belied this feeling, but it was with me always in his presence. Close to the end of his life, he came to Canberra for the Annual General Meeting of the Australian Academy of Science, of which he had been a founder, and stayed in University House. He collapsed, and had great difficulty in breathing. As Hedley Marston and I hauled the immense frame to a sitting position, bolstered with pillows, it all seemed unreal to me, for this man was surely immortal.

I quote from the short and rather unsatisfying Biographical Memoir of the Royal Society, written by Alderman, his student and successor, with the co-operation of his contemporary, C. E. Tilley: 'It was natural that a man with Mawson's heroic record and striking appearance should appeal to students. But these things cannot account for the affection with which they all regarded him. His friendliness was one of his outstanding characteristics; he was one of the most approachable of professors, whose enthusiasm for his subject had a supremely infectious quality. He could view the errors of inexperience with sympathy and tolerance, but could be devastatingly frank in his criticism of what he considered foolish . . . he was much endeared to his colleagues not only by his extraordinary modesty, but by his lusty sense of humour, which also had a most infectious quality.'

## *The International Polar Years and SCAR by Dr. P.G. Law, A.O., C.B.E., Director, from 1948 to 1965, of the Antarctic Division (then of the Department of External Affairs), and the 'father' of the Australian National Antarctic Research Expeditions (ANARE)*

The conquest of Antarctica seems to have taken a long time and to have been plagued by several major frustrations. Yet, if we except Captain Cook's voyage of 1772, the specific exploration of the Antarctic regions extends roughly from 1820 to 1970. Compared with the times taken to open up the other Continents, 150 years is a very short span of time. It could have been shorter if the frustrating interruptions of two World Wars and one World-wide depression had not occurred.

The two major surges in the conquest, those that occurred during the Second Polar Year and the International Geophysical Year, were motivated by scientific curiosity, in contrast to the motivations of territorial expansion and commercial profit that produced other activity throughout the rest of the 150-year period.

### The First Polar Year

The genesis of this scientific motivation, oddly enough, had little to do with the Antarctic. It was the First Polar Year, whose activities were almost entirely confined to northern polar regions. Karl Weyprecht, an Arctic explorer who was a Lieutenant in the Austrian Navy, put forward a proposal to the German Scientific and Medical Association in 1875 that two rings of observing stations be established as close as possible to the Arctic and Antarctic circles to make observations of meteorology, geomagnetism and the aurora. The idea caught on and, in 1880, an International Polar Commission was set up to organize such a program. The 'Year' was to embrace the twelve months beginning 1 August 1882.

In the Northern Hemisphere, about forty stations took part in the program, including twelve specially set up in remote high-latitude areas such as Greenland, northern Finland, Jan Mayen, Spitzbergen, Alaska, Ellesmere Island and northern Siberia. But, in the South there were only two: the French put a station at Cape Horn and the Germans one on South Georgia.

More than twenty volumes of reports were produced, constituting a major advance in meteorology and geomagnetism.

### The Second Polar Year

The Second Polar Year was believed to have been first proposed by the German meteorologist Johannes Georgi in 1927. At an international conference of directors of national weather services held in Copenhagen in 1928, a Commission was appointed to run the Second Polar Year over the twelve months beginning 1 August 1932.

The total plan was most ambitious. Unlike the First Polar Year, it was not to be confined to regions but was to take in the Earth as a whole. The major studies would again be meteorology, geomagnetism and the aurora and special efforts would be made to set up observation posts in the Arctic and Antarctic regions. The Russians proposed observations from villages, farms, schools and factories within the Arctic Circle, and Norway promised observations from whaling fleets in the Southern Ocean.

A number of ideas were mooted that were later to be important components of the IGY—for example, 'international days' for simultaneous launching of weather balloons, radiosondes, rockets, ionospheric soundings and so on.

Unfortunately, the onset of the World Depression severely affected the final outcome. At one stage it was even seriously considered postponing the Year. That it wasn't postponed was largely due to the efforts of the President of the Commission, Professor La Cour, Director of the Danish Meteorological Institute, who fought stubbornly for its retention.

The Year began on 1 August 1932, exactly fifty years after the start of the First Polar Year, with 44 nations taking part. In terms of observations it was highly successful, but the advent of World War II considerably affected the final collection and publication of results. The final bibliography, published in 1951, listed over 300 publications.

### The International Geophysical Year (IGY)

A third Polar Year is believed to have been first mooted at a meeting of the following physicists on 5 April 1950: James Van Allen,

Sydney Chapman, Lloyd Berkner, Wallace Joyce, Fred Singer and Ernest Kestine.

Berkner suggested that it would be a good time for a third polar year, twenty-five years after the second, in 1957-58. This would coincide with a period of major sun-spot activity.

Berkner and Chapman, who became the major protagonists for the scheme, presented a plan for such a year to the Mixed Commission on the Ionosphere in 1950. The Commission supported the proposal and presented it to the International Council of Scientific Unions (ICSU) which, late in 1951, set up a special committee to work on the project. The USSR did not at that time adhere to ICSU but belonged to the International Astronomical Union (IAU) and the World Meteorological Organization (WMO), so the Russians were accorded a special invitation.

When the work began to accumulate towards the end of 1955, Professor Hugh Webster replaced Martyn as Convener and was released by the University of Queensland to work half-time on the organisation. ANCIGY was enlarged in membership to embrace representatives of all the major geophysical sciences and became the responsibility of the Australian Academy of Science, which had been established in February 1954. Professor K. E. Bullen was appointed Chairman. I devoted a large part of my time in 1955 and 1956 to assisting Webster.

In January 1957 Australia established a second Antarctic station, Davis, in time for the IGY.

In the Antarctic regions and the sub-Antarctic islands, eleven nations set up some fifty stations with a fairly uniform geographical distribution. The nations were Argentina, Australia, Belgium, Chile, France, Japan, New Zealand, Norway, the UK, the USA, the USSR.

Co-operative research was conducted in meteorology, geomagnetism, aurora and air-glow, the ionosphere, solar activity, cosmic rays, glaciology, seismology, gravimetry and oceanography. Intensive exploration and mapping was carried out and studies in biology, although not part of the IGY program, were conducted at most of the stations.

Never before in history has there been such a scientific onslaught upon a major region of the Earth. Within two years a Continent that was largely unknown had been cracked wide open. Geographical features were delineated, seismic traverses determined the thickness of the icecap at many places, meteorological variations over the Continent, both in time and space, were recorded and much new information on global circulation patterns was obtained. Geological studies provided new support for the theory of continental drift and new knowledge about the geological structure and history of Antarctica.

Glaciological studies of the vast Antarctic ice-sheet provided for the first time a reasonable understanding of the glaciation epochs of the Earth in the past and of their complex processes.

Linked with the satellite and rocket investigations carried out elsewhere and the discovery of the Van Allen radiation belts surrounding the Earth, polar research in geomagnetism, the aurora, the ionosphere, astronomy, cosmic rays and radio wave propagation painted a fascinating new picture of solar-terrestrial relationships.

Added to all this, the IGY had resulted in a degree of international accord and co-operation never before achieved. In Antarctica this accord was to be perpetuated by the signing of the Antarctic Treaty in 1959 by twelve nations, a result that was directly the result of the international trust and goodwill generated by the IGY.

## The IGC and SCAR

When the IGY finished, CSAGI agreed that an extension of many of the programs was highly desirable, so 1959 was designated The International Geophysical Co-operation year (IGC). Early in 1959 the USA handed over their Wilkes Station in Antarctica to Australia. Also the work of the 'Weather Centre' at Little America was transferred to Melbourne where the International Antarctic Analysis Centre was established.

The IGC was just one extra year. CSAGI, however, had also agreed at the 1957 Paris Conference that there was a need for international scientific co-operation to continue on a long-term basis, particularly in Antarctica, and recommended accordingly to ICSU. ICSU established a permanent committe called The Special Committee on Antarctic Research (SCAR). (The word 'Special' was later changed to 'Scientific'). Its area of concern was the Antarctic, interpreted as being bounded on its northern extremity by the Antarctic Convergence.

SCAR held its first meeting at The Hague in February 1958. Georges Laclavere was elected as its first President. SCAR consisted of one delegate from each of the nations active in Antarctica together with one delegate from WMO, one from ICSU and one from any ICSU union wishing to participate (URSI, IUBS, IUGG, IUGS, IUPS at present). A delegate to SCAR could be accompanied by any number of advisers. The number of nations increased to twelve in 1960 when South Africa established its first Antarctic base. Poland, the Federal Republic of Germany and the Democratic German Republic have recently been admitted to membership.

SCAR has been a remarkable success. Its constitution states: 'SCAR is a scientific committee of ICSU charged with the initiation, promotion and co-ordination of scientific activity in the Antarctic, with a view to framing and reviewing scientific programs of circumpolar scope and significance.'

The work done by SCAR conservation and environmental protection in Antarctica has been particularly notable. SCAR pressed the Treaty Nations, through the National Committees whose delegates constituted the body of SCAR, to take measures for conservation and environmental protection in Antarctica and initiated a series of scientific investigations to provide foundation material upon which such measures could be based. A number of measures have now been recommended by Treaty Consultative Meetings and adopted by the Treaty Nations. Antarctica is now, without doubt, the best environmentally protected continent in the World. Hopefully we can keep it that way.

Looking back, we have come a long way since the First Polar Year, since the amateur, 'one-off' expeditions of the 'Heroic Era' of Antarctic exploration and since Sir Douglas Mawson's recommendations that led to the implementation of conservation measures at Macquarie Island.

Mawson was a great protagonist for international scientific co-operation. He would be delighted, were he alive today, to see the international amity existing between the Treaty Nations and their expeditioners and to note the achievements of SCAR as the co-ordinating body for international scientific research in the Antarctic.

Mainly as a result of representations from the WMO, which considered the program should be world-wide and not merely polar, Chapman proposed the title 'International Geophysical Year' and this was approved by ICSU in October 1952. An organizing committee was then set up, called CSAGI (Comité Spéciale de l'Annee Géophysique Internationale) and Chapman became its first President, Berkner was made Vice-President and Marcelle Nicolet of Belgium the Secretary-General. CSAGI was composed of *scientists* nominated by ICSU and its relevant Unions together with leaders of working groups on disciplines to be studied, members of the Finance and Publications Committees and adjoint secretaries of the regional working groups (of which Antarctica was one). The period constituting the IGY was extended to cover the eighteen months from 1 July 1957 to 31 December 1958. Meetings were held at Amsterdam (October 1952), Brussels (July 1953), Rome (October 1954), Brussels (September 1955) and Barcelona (September 1956). To facilite international co-operation between governments, an Advisory Council for the IGY (ACIGY) was established at the 1955 meeting with two representatives from each participating country and it operated thereafter in parallel with CSAGI.

The 1954 Rome meeting defined the general planetary problems of the Earth with which the IGY was to be involved:—

(a) Problems requiring concurrent synoptic observations at many points by many nations.

(b) Problems of certain geophysical sciences whose solutions require data from other geophysical sciences.

(c) Observations of all major geophysical sciences in relatively inaccessible areas of the Earth that can be occupied because of the extraordinary efforts possible during the IGY.

(d) Epochal observations of slowly varying terrestrial phenomena to establish basic information for subsequent comparison at later epochs.

Two other major IGY principles were determined:

(i) There was to be full and free interchange of all data and results, with no secrecy.

(ii) The IGY was to be non-political; its composition of scientists, not national representatives ensured this.

Arising from definition (c) above, it was decided that the major effort should be devoted to two regions—the Antarctic and Outer

Space. Other areas singled out for special attention were: the Arctic, the Equatorial regions, and three pole-to-pole meridians—longitudes 70-80° W (the Americas), longitude 10° E (Europe and W. Africa) and longitude 140° E (Siberia, Japan, Australia).

A program of 'World Days' (3 per month) for intensive observations in related sciences was worked out and also a number of periodical 'World Meteorological Intervals' (each of ten days) for more intensive meteorological work.

Three 'World Data Centres' were set up to store and organise the collected data:

In USA—selected localities for various disciplines
In USSR—selected localities for various disciplines
In Europe-Australia-Japan—selected localities for various disciplines

A central World Meteorological Data Centre was also set up in Geneva by WMO, and a 'Weather Central, Antarctica' was established at the US base at Little America.

A World Warning Agency was organized to alert stations regarding the onset of solar flares and there was great activity amongst the ICSU members, such as URSI (Union Radio Scientifique Internationale), IUPAP (International Union of Pure and Applied Physics), IUGG (International Union of Geodesy and Geophysics), and in WMO (World Meteorological Organization).

Four CSAGI Antarctic Conferences were held: Paris (July 1955), Brussels (September 1955), Paris (August 1956) and Paris (June 1957). Georges Laclavere was elected President for the Antarctic conferences at the first meeting.

In the final count, sixty-six nations took part in the IGY but, regrettably, the Chinese People's Republic did not participate owing to a dispute about the place of Taiwan in the scheme. Eleven nations operated in the Antarctic.

A vast amount of work was involved in preparing for the IGY but Australia was slow in starting. I remember reading of the ICSU committee set up late in 1951 to work on the IGY proposal and, having a vested interest because of my Antarctic responsibilities, I waited for something to happen here. When, by October 1952, nothing had eventuated, I wrote to Professor Ward, Chairman of the National Research Council, proposing that Australia should set up a national IGY committee. On 14 December 1952, the Council appointed a committee of four—Dr David Martin as Convener, with Dr Richard Wooley, Mr Jack Rayner and myself. But nothing further happened. Eighteen months later, in May 1954, I wrote to David Martyn urgently requesting a meeting and one was eventually held in July 1954. ANCIGY was at last in action. In the meantime, in February 1954, we had established Australia's first permanent Antarctic station—Mawson.

# THE MAWSON LECTURE

The Mawson Lecture was established by the Australian Academy of Science to mark the outstanding contribution to science in Australia by Sir Douglas Mawson. It is given to the earth scientist (normally resident in Australia) who may appear to the Council of the Academy to be most deserving of such honour. The Lecture shall normally be delivered at the time of each convention of the Geological Society of Australia. The Australian Academy of Science, however, agreed to the suggestion that this, the second Mawson Lecture, be delivered at the Fourth International Symposium on Antarctic Earth Sciences.

The honour of being the Second Mawson Lecturer was awarded to Professor D.H. Green, Department of Geology, University of Tasmania and at the time of the award, Chairman of the Australian National Committee for Antarctic Research.

The following is a summary of Professor Green's lecture, which will be published in full at a later date.

## EXPERIMENTAL PETROLOGY: EXPLORATION IN THE LABORATORY AND APPLICATIONS IN ANTARCTICA

D. H. GREEN *Department of Geology, University of Tasmania*

The characteristics of the Earth's Upper Mantle can be modelled by the behaviour of a natural peridotite composition containing 3-4% CaO, $Al_2O_3$ and with

$$\frac{Mg}{Mg+Fe^{++}} \sim 0.9.$$

Peridotites of this composition are among the least refractory of natural mantle samples and have the potential to yield up to 30-35% of picritic or basaltic magma. Experimental studies of petrogenesis and of the volatile contents of basaltic magmas, together with the common presence of amphibole and, less commonly of phlogopite in mantle derived lherzolites, and the presence of $CO_2$-rich fluid inclusions in lherzolite xenoliths, all attest to the importance of the peridotite-C-H-O system in defining the solidus and mineralogy of the upper mantle.

There are three distinctive solidi for peridotite-$H_2O$, ie. (a) anhydrous, (b) water contents $<\sim 0.4\%$ $H_2O$, (c) water contents $>\sim 0.4\%$ water-saturated. At low oxygen fugacities such that carbon is present as graphite or diamond (± carbonates), the solidi for peridotite-C-H-O are very close to those for peridotite-$H_2O$. Solidus (b) has a distinctive shape determined by the water-saturated melting of pargasitic amphibole. This is considered to be the 'normal' mantle solidus and intersection of oceanic or 'steep' geothermal gradients with this solidus predict a low velocity zone (zone of partial ($<1\%$) melting) beginning at depths of 80-90 km beneath ocean basins or areas of continental crust with high heat flow. In areas of low geothermal gradient the lithosphere may be >80-90 km thick and contain a zone to ~90 km with accessary amphibole overlying a zone of garnet peridotite with accessary phlogopite from ~90 km to the low velocity zone at >150 km. If the geothermal gradient is extremely low, then there may be no solidus intersection.

In contrast to these consequences of the 'normal' mantle solidus ($<0.4\%$ $H_2O$), the availability of water contents $>\sim 0.4\%$ may cause water saturated melting of the lithosphere between ~50 km and 90 km along 'oceanic' geothermal gradients. It is suggested that these conditions apply in island arc regions ($H_2O$ release from subducted oceanic crust?) producing basaltic parent magmas with water contents in excess of 1% and with trace-element characteristics of complex source peridotite, ie. refractory characteristics overprinted by enrichment events as evidenced in 'old lithosphere' samples.

Experimental study of mantle and deep crustal processes has proceeded through examples and models drawn directly from natural occurrences and using the evidence of the rocks themselves to constrain the models to be tested. This approach is being applied by Mr N. Ortez and the author to the problem of the petrogenesis of Jurassic Ferrar-Tasmanian dolerites. Studies of chilled margins of sills provide evidence of crystal fractionation (olivine, enstatite, diopside, plagioclase from olivine tholeiite). The chilled margins contain evidence of deep crustal fractionation in the presence of wall-rock dominated by refractory ($An_{88}$) plagioclase. It is inferred that the parent magma for this province is a tholeiitic picrite with low $TiO_2$, low $Na_2O$, and moderate $K_2O$, segregating from residual harzburgite at ~50 km, 1410°C. Crustal melting occurs in the roof zones of some of the deep-crust magma chambers producing distinctive rhyodacite to rhyolite magmas with extrusion temperatures of 1050-1100°C, eg Butcher Ridge complex, Darwin Glacier area, Antarctica.

A second application of experimental petrology to a distinctive Antarctic magma focusses on the Gaussberg olivine leucitite (S. Foley—study in progress). Analogy with a highly potassic, olivine-rich composition from the African rift valley which has previously been studied suggests that high ($CO_3^-$) dissolved in the melt, together with high water contents, and pressures of ~30 kb are appropriate for conditions of genesis—conditions generated beneath the rifted edge of an ancient cratonic block and old, thick lithosphere.

In contrast to these studies of magma genesis applicable to Antarctic rocks, experimental studies of subsolidus equilibria among common mineral assemblages are used to infer conditions of metamorphism, ie. geothermometry and geobarometry. Garnet/clinopyroxene (D. J. Ellis) and garnet/orthopyroxene (S. Harley) equilibria have been calibrated in the laboratory and applied to the Archaean granulites of Enderby Land, Antarctica. Internal consistency of the data can be checked by multiple sampling of diverse bulk compositions from small areas. In Enderby Land the metamorphic maximum (?3100 Ma) attained temperatures of 900-950°C at pressures of 6-10 kb and sub-provinces of different depths can be identified. The region then underwent *isobaric* cooling to temperatures of 650°C by 2500 Ma with preservation of remarkable corona textures preserving evidence of the cooling history. Studies (S. Kuehner) in progress of phenocryst phases of the chilled margins and the metamorphic mineralogy of several suites of mafic dykes will allow evaluation of the post-2500 Ma tectonic history of this 'craton', formed at ~30 km depth but now exposed at the surface and overlying 20-30 km (at least) of continental crust.

# 1

# Precambrian East Antarctic Craton

# THE PRECAMBRIAN GEOLOGICAL EVOLUTION OF THE EAST ANTARCTIC METAMORPHIC SHIELD—A REVIEW

P.R. James, *Department of Geology and Mineralogy, University of Adelaide, Adelaide, S.A., 5000, Australia.*

R.J. Tingey, *Bureau of Mineral Resources, Canberra, Australia.*

*Abstract* Modern geochronological data indicate that the East Antarctic metamorphic shield is composed of a number of Archaean cores preserved among larger areas of younger, mainly Proterozoic age, metamorphic rocks.

Rocks that are definitely older than 3000 Ma are rare and a pre 3000 Ma history has only been revealed for Enderby Land's Napier Complex. High grade metamorphism and associated tectonism at about 2500 Ma has been documented from the Napier Complex,Prince Charles Mountains and Vestfold Hills. After 2500 Ma the Archaean blocks responded to stress by brittle/semi brittle deformation and mafic dyke swarms were emplaced. Evidence for the 2000-1700 Ma episode of crustal evolution—widely recognised in other southern continents—is confined to sparse geochronological data from the southern Prince Charles Mountains and Terre Adelie/King George V Land areas but there is abundant evidence of metamorphic and tectonic activity between 1300 and 900 Ma. Granulite facies metamorphic rocks, commonly accompanied by syntectonic granitoids and charnockites were formed in this interval; they are exposed in the Prince Charles Mountains area in a belt about 600 km wide which continues west through Enderby Land into Dronning Maud Land. East of the Prince Charles Mountains 1300-900 Ma metamorphic rocks probably occur in the Mirny-Denman Glacier area and in Wilkes Land.

Thermal activity at about 500 Ma reset mineral isotopic systems over wide areas of the East Antarctic shield and is locally manifested by granitoid intrusives but does not appear to have been accompanied by folding. It is coeval with the Ross Orogeny in the Transantarctic Mountains and the "Pan African event" in Africa.

Areas of Precambrian metamorphic gneisses mapped in detail reveal broadly similar styles of tectonic evolution. Intense early tectonism has effectively destroyed pre-existing primary structures and stratigraphy, and produced recumbent gneissic piles. Later tectonothermal activity resulted in horizontal shortening of the gneisses, production of large scale non cylindrical upright folds, and, commonly, major granitoid emplacement.

As Douglas Mawson's 1911-14 Australasian Antarctic Expedition made the first significant studies (Stillwell, 1918) on the rocks now collectively referred to as the East Antarctic metamorphic shield, it is fitting that this commemorative symposium should include a review of the present knowledge of these rocks. For our purposes the East Antarctic metamorphic shield is regarded as the area bounded by longitudes 30°W and 150°E (see Figure 1).

The sparse and scattered bedrock outcrops of East Antarctica consist predominantly of metamorphic rocks ranging in grade from greenschist to granulite facies and in age back to the Archaean. These outcrops have, for the most part, only been mapped at regional or reconnaissance scale, as exemplified by studies of Enderby Land and western Kemp Land (Kamenev, 1972; Sheraton et al., 1980; James and Black, 1981) the Prince Charles Mountains (Tingey, 1982a, b; Hofmann, 1982; Federov and Hofmann, 1982) and Dronning Maud Land (Van Autenboer and Loy, 1972). Detailed accounts are only available for a few small areas such as the Windmill Islands (Blight, 1975; Blight and Oliver, 1982) the Bunger Hills (Ravich and Soloviev, 1966; Ravich et al., 1968), the Vestfold Hills and Rauer Islands (Oliver et al., 1982; other authors, this volume), some localities in Enderby Land (this volume), and Lützow-Holm Bay (National Institute for Polar Research, Japan, 1:500,000 scale geological maps).

The first part of this paper is a chronological review of the Precambrian evolution of the East Antarctic shield up to and including Early Palaeozoic events. After this time the shield appears to have acted in a stable manner apart from localised rifting and sedimentary and intrusive activity (Tingey, 1978). The paper's second part is a discussion of the area's Precambrian evolution in the context of concepts, such as orogenic provinces, developed in studies of shield areas elsewhere. The term Archaean refers to the period before 2400 Ma, Early Proterozoic to the period 2400-1540 Ma, Middle Proterozoic to 1540-880 Ma and Late Proterozoic to 880-570 Ma.

## EVOLUTION OF THE EAST ANTARCTIC METAMORPHIC SHIELD—1. CHRONOLOGY

It is evident that some results from earlier isotopic dating studies must now be rejected since subsequent experience has shown early sampling methods to be unsatisfactory and the applied isotope systems to be either susceptible to resetting or inappropriate to the rock types concerned. For example, dates derived from "isochrons" constructed by joining a single analysis point to an assumed $^{87}Sr/^{86}Sr$ initial ratio on the Rb-Sr diagram are suspect since they are dependent on the validity of the initial ratio assumption. In any case isotopic dating only gives information on the behaviour of an isotope system; an age is derived from a geological interpretation of this information. Geochronological studies using more reliably controlled sampling techniques (e.g. Arriens, 1975; Black and James, this volume), and the attempted correlation of isotopic dates and discrete tectonothermal events as displayed in penetrative rock fabrics (see e.g. James and Black, 1981) have led to a deeper understanding of certain areas and provide a basis for critically appraising age data from others.

### Archaean (pre 2400 Ma)

Archaean isotopic dates have been reported from western (see Figure 1) Dronning Maud Land (Halpern, 1970), Enderby Land (Sobotovich et al., 1976; James and Black, 1981), the southern Prince Charles Mountains (Arriens, 1975; Tingey, 1982a) and the Vestfold Hills (Collerson and Arriens, 1979).

Halpern (1970) derived what might be termed a model age of $3060 \pm 80$ Ma for gneiss at Jule Peaks in Dronning Maud Land from the slope of a line joining a single analysis point to an assumed $^{87}Sr/^{86}Sr$ initial ratio a methodology that would not now be acceptable. As Halpern points out, further studies are necessary. Recent South African geochronological research in western Dronning Maud Land is described in this volume by Barton and Copperthwaite, and Krynauw but these authors do not refer to new data on the area's gneissic rocks.

The Napier Complex metamorphic rocks in Enderby Land (Sheraton et al., 1980) have a complex Archaean geological history described, following studies involving a range of geochronological techniques, by James and Black (1981) and other authors in this volume (see also De Paolo et al., 1982). Evidently a major tectonothermal episode at about 3100 Ma involved rocks that had had, even then, a lengthy crustal history, and took place under exceptionally dry, high temperature granulite facies metamorphic conditions (Ellis et al., 1980). A subsequent deformation under similar conditions at about 2900 Ma apparently did not leave a strong isotopic signature, but a third tectonothermal episode between 2500 and 2400 Ma resulted in widespread resetting of isotope systems. Several suites of tholeiitic dykes intersect the Napier Complex rocks (Sheraton and Black, 1981) but they are present only as metamorphosed relics in parts of the younger Rayner Complex nearby.

The southern Prince Charles Mountains geochronological data base is less comprehensive than that relating to the Napier Complex, but Rb-Sr whole rock ages for granitic basement rocks there range back to $2822 \pm 277$ Ma wheras the age of overlying metasediments is constrained by a 2580 Ma Rb-Sr muscovite age obtained from a cross cutting pegmatite (Tingey, 1982, a, b). The granitic basement rocks and metasediments are intersected by mafic dykes but younger granulites to the north are not.

Granulite facies basement gneisses in the Vestfold Hills have yielded late Archaean ages (Arriens, 1975; Collerson and Arriens, 1979), and are intersected by a prominent mafic dyke swarm only seen as metamorphosed relics in outcrops 10 km south where a Rb-Sr whole rock age of $1073 \pm 101$ Ma has been obtained from granulite facies gneisses (Tingey, 1982a).

*—The mafic dyke relationship—*

Mafic dykes that intersect the Napier Complex display compositional similarities with the Vestfold Hills dykes and yield Rb-Sr ages of

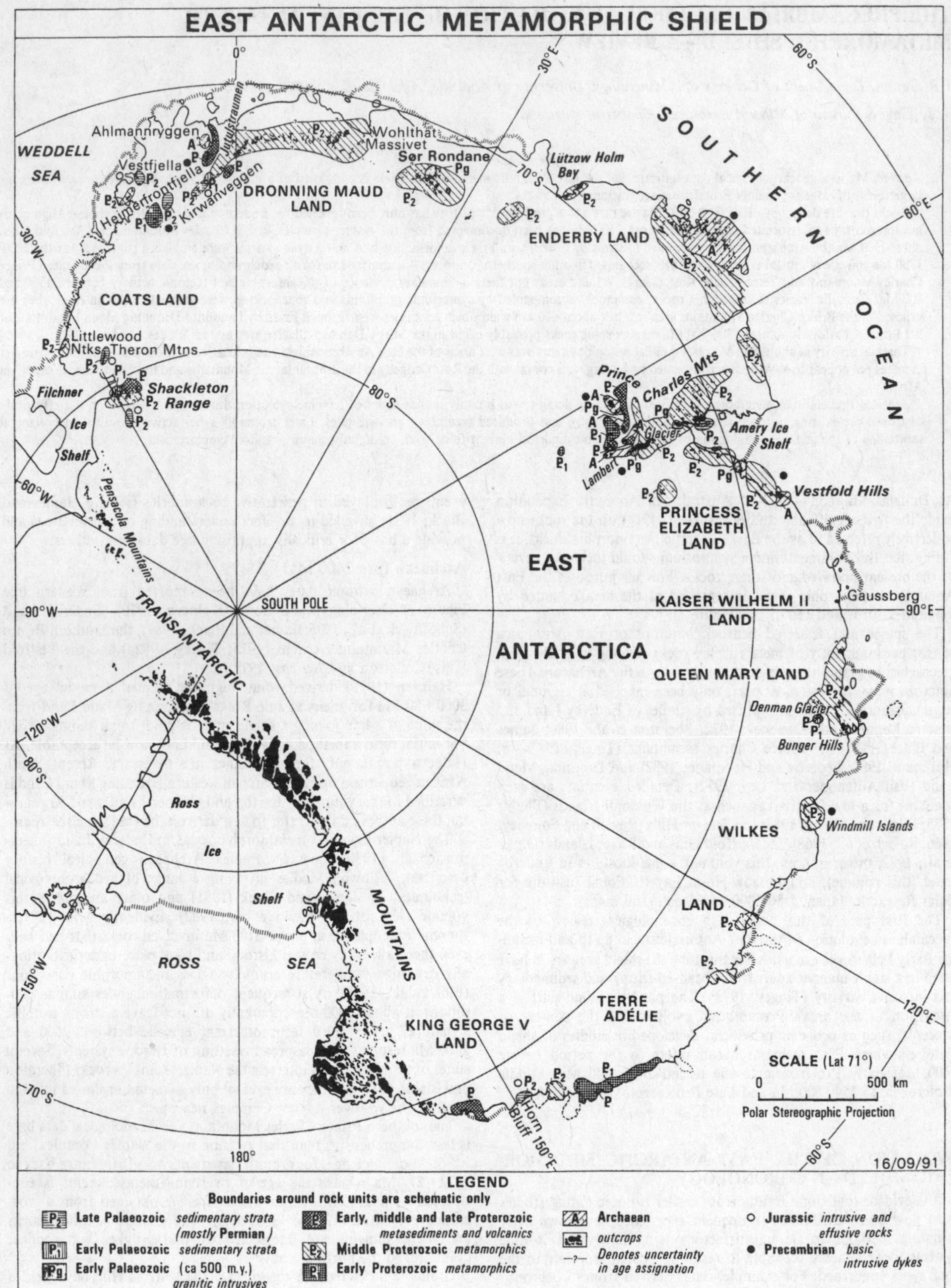


**Figure 1.** Geology of the East Antarctic Metamorphic Shield.

2350 ± 48 Ma for one group and 1190 ± 200 Ma for another (Sheraton and Black, 1981). These ages are bracketed by dates referring to the Napier Complex's latest tectonothermal episode (i.e. about 2400 Ma), and the Rayner Complex age of 987 ± 67 Ma (Grew, 1978), as well as by ages derived from older (dyke intersected) and younger gneissic rocks in both the Vestfold Hills and Prince Charles Mountains areas.

Gneissic basement rocks intersected by mafic dyke swarms elsewhere in East Antarctica may, by analogy with the Napier Complex also have Archaean origins. Thus, basement rocks of possible Archaean age can be tentatively inferred at Commonwealth Bay (142°E) (Stillwell, 1918) the Bunger Hills (101°E), and possibly also at the Windmill Islands (111°E), Heimefrontfjella (10°W) and the Shackleton Range (25°W). Conversely mafic dykes (apart from rare lamprophyres) are unknown from Dronning Maud Land and the metamorphic rocks there are thought likely to have younger origins.

## Proterozoic

*—Early Proterozoic—*

In the East Antarctic metamorphic shield Early Proterozoic rocks are uncommon and evidence for Early Proterozoic tectonothermal activity is sparse. In western Dronning Maud Land these comprise volcanic and sedimentary strata (including jasper bearing types) that are intersected by 1700 Ma mafic intrusives (Bredell, 1982). Other iron rich rocks in the southern Prince Charles Mountains (Tingey, 1982a, b) may also be of Early Proterozoic origin; they are interested by metamorphosed mafic dykes.

In the southern Prince Charles Mountains poorly defined Proterozoic Rb-Sr isotopic dates have been obtained from granitoids which are intersected by metamorphosed mafic dykes, and themselves intrude polymetamorphic schists and gneisses (Tingey, 1982a, b). Early Proterozoic metamorphic rocks may also occur in Adelie Land where Rb-Sr biotite dates of 1543 and 1530 Ma from crosscutting pegmatites (Bellair and Delbos, 1962) define a minimum age for the country rocks. Supporting evidence is provided by Rb-Sr muscovite and biotite dates averaging 1540 Ma for pegmatites in the nearby Commonwealth Bay area (Arriens, unpublished data). The field relationships between the Commonwealth Bay pegmatites and mafic dykes are not known.

*—Middle Proterozoic (1540-880 Ma)—*

Middle Proterozoic supracrustal rocks crop out in the southern Prince Charles Mountains (Tingey, 1982a, b) in the Shackleton Range (Clarkson, 1982) and in Coats Land (Eastin and Faure, 1970). The recumbently folded southern Prince Charles Mountains rocks comprise greenschist facies calcareous phyllites, sandstones and conglomerates which contain clasts of the banded ironstones exposed nearby. They are not intersected by mafic dyke intrusives but are intruded by Cambrian granites; a Middle Proterozoic age is assigned to the metasediments on the basis that recumbently folded high grade rocks in the area have yielded a Rb-Sr date of 891 ± 70 Ma (Tingey, 1982a, b)—a date which is thought to define the time of the folding.

The age of the Turnpike Bluff group in the Shackleton Range is constrained by the age of the local metamorphic complex that it unconformably overlies (1446 ± 60 Ma; Clarkson, 1982). The minimum age is that of the overlying Cambro-Ordovician Blaiklock Glacier Group (Clarkson, 1981). Elsewhere in Coats Land rhyolite porphyries at Littlewood Nunataks have yielded an Rb-Sr whole rock isochron age of 1001 ± 16 Ma (Eastin and Faure, 1970)

Middle Proterozoic metamorphic rocks are widespread in East Antarctica from the Windmill Islands to the Shackleton Range. In the Windmill Islands, Blight and Oliver (1982) have shown that high grade metamorphic rocks which yield Rb-Sr whole rock isochron ages in the range 1100-1400 Ma, are intruded by charnockite (1200 Ma), granite (1100-1140 Ma), and pegmatite and dolerite dykes. The Rb-Sr dates are in broad agreement with earlier K-Ar mineral dates. (For recent geochronological data from the Windmill Islands see Williams et al.; Oliver et al., this volume).

Middle Proterozoic K-Ar dates quoted for the Bunger Hills-Vestfold Hills-Prydz Bay area by Ravich and Krylov (1964) are now not easy to interpret because only limited analytical data were published with them, and because of doubts about the applicability of the K-Ar whole rock method to coarsely crystalline metamorphic rocks. However, high grade metamorphic rocks that give Middle Proterozoic Rb-Sr ages are known in an east west trending 500 km wide belt in the Prydz Bay-Prince Charles Mountains-Mawson area (Tingey, 1982a) and appear to be continuous with the Rayner Complex in Enderby Land (Sheraton et al., 1980), and metamorphic rocks in the Lützow-Holm Bay area and Dronning Maud Land. The edge of this zone is relatively shapr in the Prydz Bay area where Late Proterozoic granulite facies gneisses occur less than 10 km from similar grade Archaean gneisses in the Vestfold Hills, but gradational in the south where the metamorphic grade of Late Proterozoic metamorphic rocks falls from granulite facies to greenschist facies over a wide zone (Tingey, 1982b).

Middle Proterozoic isotopic dates are scarce west of Enderby Land, but Maegoya et al. (1968) report Rb-Sr orthoclase ages of about 1100 Ma from Lützow-Holm Bay, Van Autenboer and Loy (1972) report 950 Ma zircon ages from the Sør Rondane Mountains and a Middle Proterozoic minimum age (1446 Ma) for the Shackleton Range metamorphic complex is noted by Clarkson (1982). The geological situation in western Dronning Maud Land is unclear; Bredell (1982) suggests that andesitic lavas are Late Proterozoic and younger than ca. 1000 Ma intrusive rocks but Wolmarans and Krynauw (1981) imply that these rocks are affected by a ca. 1000 Ma metamorphism.

*—Late Proterozoic (880-570 Ma)—*

Late Proterozoic supracrustal rocks are reported from the Denman Glacier (Ravich et al., 1968), western Dronning Maud Land (Bredell, 1982), and possibly the Shackleton Range and Prince Charles Mountains areas. As discussed above there are doubts about Bredell's (1982) interpretation; in addition the Precambrian age of the Shackleton Range's Turnpike Bluff Group is poorly constrained. The Middle Proterozoic age assigned above to calcareous metasediments in the southern Prince Charles Mountains contrasts with the interpretation of Halpern and Grikurov (1975) who suggested a Late Proterozoic age.

In the Denman Glacier area Cambrian metasediments unconformably overlie greenschist facies metabasaltic rocks for which Ravich et al., (1968) cite a K-Ar whole rock (i.e. minimum) age of 610 Ma. Low grade metamorphic rocks described from Cape Hunter in King George V Land by Stillwell (1918) might also be of Late Proterozoic age. Late Proterozoic low grade metasedimentary rocks are abundant in Oates Land, in the Transantarctic Mountains, and in Marie Byrd Land (Grindley, 1981).

The available Rb-Sr isotopic age data indicate only sparse Late Proterozoic tectonothermal activity in East Antarctica. Early K-Ar whole rock and lead isotope age data range through the whole Late Proterozoic time span but are difficult to interpret.

## Palaeozoic

*—Early Palaeozoic (Cambrian to Silurian)—*

According to Laird (1981) early Palaeozoic Antarctic rocks are much more abundant in the Transantarctic Mountains than in East Antarctica where possible examples are at Mt Sandow in the Denman Glacier area (Ravich et al., 1968), in the Ahlmannryggen area of western Dronning Maud Land (Urfjell Group—Aucamp et al., 1972), and in the Shackleton Range (the Blaiklock Glacier Group—Clarkson, 1982). Mawson's (1940) report of an archaeocyathid-bearing metamorphosed limestone erratic at Commonwealth Bay (142°E) has not been verified and must be treated with suspicion.

By contrast, early Palaeozoic igneous rocks are widespread in East Antarctica though less abundant east of about 80°E compared to areas further west. In the Prince Charles Mountains and Prydz Bay areas Early Palaeozoic granitoids discordantly intrude deformed rocks in which Rb-Sr mineral systems have been reset in early Palaeozoic times (Tingey, 1982a). In Enderby Land minor pegmatites which intrude the Napier Complex have also yielded early Palaeozoic dates; such pegmatites are much more abundant in the adjacent Rayner Complex. Further west, geochronological studies indicate 500 Ma granitoid intrusives in Dronning Maud Land (Van Autenboer and Loy, 1972) and in its Lützow-Holm Bay—Yamato Mountains sector (Yoshida, 1977; Maegoya et al., 1968).

Discordant Early Palaeozoic granitoids like those just described have not been reported in East Antarctica east of about 80°E although granitoids at Cape Webb and Cape Bage (146°E; Ravich et al., 1968) may be examples. Early Palaeozoic K-Ar whole rock dates reported from this eastern sector are of doubtful validity but K-Ar and Rb-Sr mineral isotope studies on Windmill Islands rocks have yielded one Early Palaeozoic date. At Mirny, heterogenous charnock itic rocks collected within a 650 m radius have given rise to a Rb-Sr whole rock isochron age of 502 ± 24 Ma age (McQueen et al., 1972).

Alkaline intrusive rocks are widespread but rare in East Antarctica, the Transantarctic Mountains and possibly Ellsworth Mountains. One example in Enderby Land has yielded an Ordovician age (Sheraton and Black, 1981) and another in the Prince Charles Mountains has given a K-Ar date of $504 \pm 20$ Ma. Lamprophyre dykes in the central Transantarctic Mountains have yielded very similar dates.

Unlike the coeval Ross Orogeny in the Transantarctic Mountains but like the Gondwana-wide Pan African event, the Early Palaeozoic igneous activity and thermal metamorphism in East Antarctica was apparently not accompanied by tectonism. It might have been a manifestation of the internal fracturing of Gondwana in anticipation of eventual break-up, whereas the Ross Orogeny might have involved interaction between the Gondwana Pacific margin and a possible proto Pacific plate. Evidence of other Phanerozoic tectonism is sparse and in particular there are few signs of processes involved in the Mesozoic separation of other continents from Antarctica.

## EVOLUTION OF THE EAST ANTARCTIC METAMORPHIC SHIELD—2. PROCESSES AND COMPARISONS

In considering the crustal evolution of the East Antarctic shield it is appropriate to apply criteria and comparative schemes developed in other areas (see Windley, 1977; Hunter, 1981). The most important criteria for the recognition of Precambrian orogenic provinces are the lithological (and if possible stratigraphic) character and correlation of supracrustal sequences, the identification of the basement to these sequences, and the history of tectonothermal events.

### Supracrustal sequences and basement-cover relationships

The areas of the East Antarctic shield compared on Figure 2 are all composed of middle to upper amphibolite or granulite facies gneisses. These have generally been described as comprising two lithotypes interlayered on all scales (see Yoshida, 1978; Sheraton et al., 1980; Oliver et al., 1982).

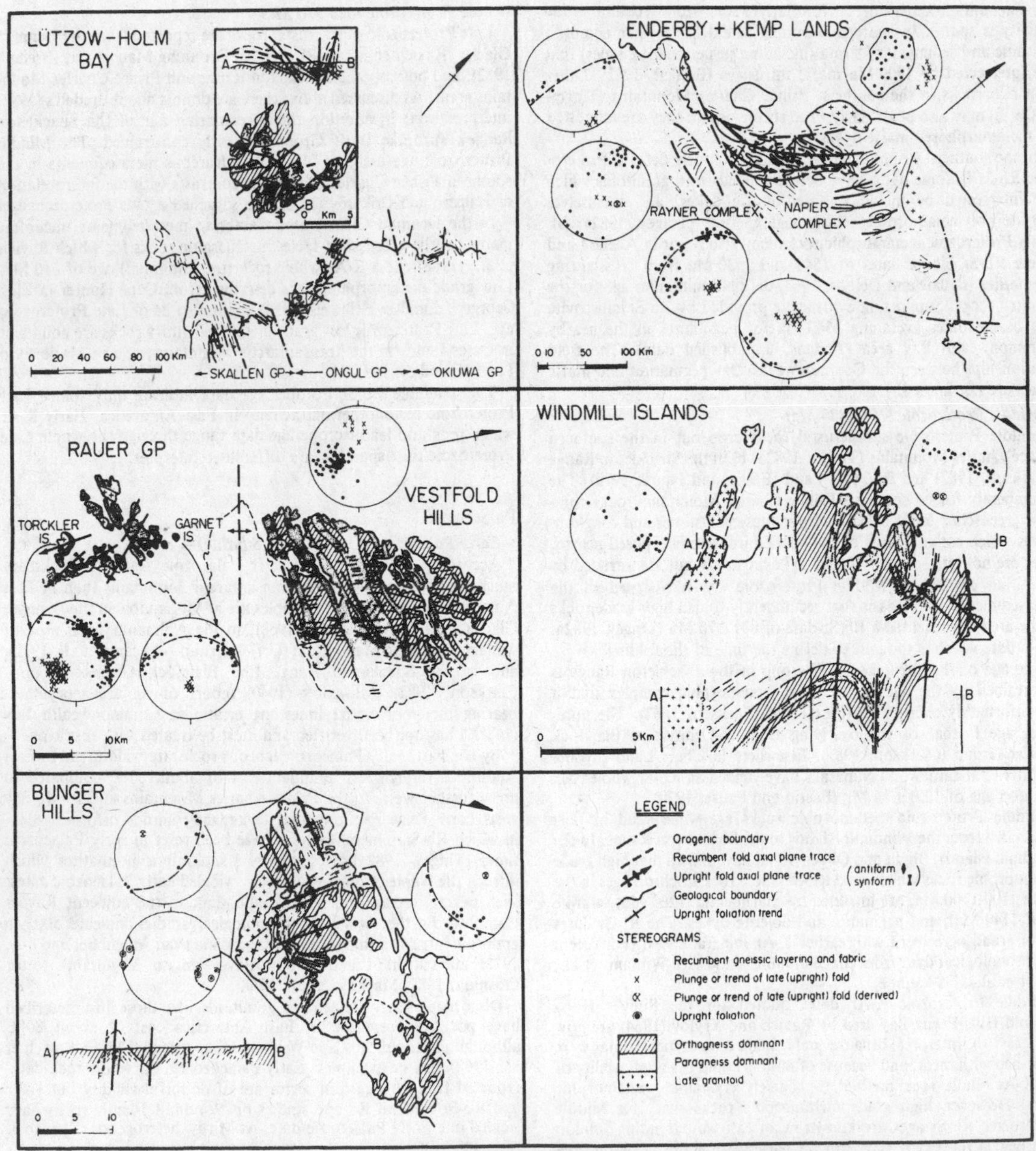


**Figure 2. Comparative geology and tectonics of Lutzow-Holm Bay, Enderby and Kemp Lands, Vestfold and Bunger Hills and the Windmill Islands.**

One lithotype is typically well layered and consists of paragneisses with predominantly psammitic-pelitic compositions now reflected in mineral assemblages with garnet, pyroxene and/or biotite, sillimanite and cordierite. The rocks are interpreted as being derived from greywacke-like sediments and are commonly interleaved with minor quartzites, iron rich metasediments, calc-silicate gneiss, marbles and rocks of magnesian and aluminous composition. Subordinate mafic gneisses also occur. The paragneisses are commonly migmatitic and their layering represents either transposed bedding, metamorphic differentiation or a combination of the two; no original sedimentary features have been identified.

The second lithotype occurs in much the same abundance as the paragneisses and consists of internally poorly layered to massive quartzo-feldspathic gneisses that commonly contain orthopyroxene. They are usually described as orthogneisses of either granitic-granodioritic-tonalitic or charnockitic-enderbitic composition, and are thought to be the metamorphosed representatives of either felsic volcanic or granitic intrusive rocks; primary igneous structures are not preserved.

The relationships between the two lithotypes could be described as reflecting either simple interlayering, or intrusion of lit-par-lit sheets, or the rotation of discordant intrusions, or tectonic interleaving or, most probably, combinations of these processes. The notion that the massive orthogneiss was basement to the now paragneissic supracrustal sequences (see Grikurov et al., 1976) is not now favoured because of local intrusive contacts between the ortho- and paragneisses and the more complex deformation history locally described for the paragneiss (James and Black, 1981). However in the Vestfold Hills finely interlayered mafic-felsic grey orthogneiss may possibly represent basement to nearby paragneisses (Oliver et al., 1982; see also Collerson, this volume). To judge from this example the recognition of relict basement is very difficult, and only possible through detailed structural mapping (preferably supported by isotopic dating).

Similar difficulties are experienced in attempting to determine the protoliths of the high grade metamorphic rocks. The identification in the southern Prince Charles Mountains of at least three Precambrian supracrustal sequences (Tingey, 1982a, b) prompts speculation that they may be equivalents of at least parts of the interlayered middle Proterozoic high grade gneisses in the northern Prince Charles Mountains, though direct evidence is very sparse. To judge from experience elsewhere it is probable that formation of the East Antarctic high grade gneisses involved the metamorphism and intermingling of several supracrustal sequences. This inference can be made from the Prince Charles Mountains example above and from recent studies in the Vestfold Hills (Oliver et al., 1982) and Prydz Bay areas (Sheraton and Collerson, in press).

### Tectonic Evolution

Interlayered ortho- and paragneissic sequences in the East Antarctic shield show patterns of tectonic evolution which are similar in the temporal and geometrical (Figure 2) senses. In most cases it is possible to identify a recumbent gneiss pile produced by early ductile deformation characterised by intense shearing and accompanied by thermal culmination. Later, less intense deformation and slightly lower temperature metamorphism commonly produced upright folds, and was followed by cratonisation, brittle fracture and dyke intrusion.

The early recumbent folds are commonly described in terms of two tectonothermal episodes $D_1$ and $D_2$. $F_1$ folds are rare; they are typically described as rootless recumbent minor isoclines with a layer parallel axial plane LS (lineated schistosity) fabric and, commonly, an axial lineation. The ductile nature of $D_1$ resulted in the obliteration of pre-existing discordances by extreme rotation and flattening, and the characteristic interleaving of the orthogneiss and paragneiss lithotypes (James and Black, 1981).

The $D_1$ ductile deformation was closely followed at least in the Vestfold Hills, and possibly the Ongul Group in Lützow-Holm Bay, by the intrusion of major concordant charnockite bodies. $D_2$, a similar ductile deformation closely followed $D_1$ and resulted in local refolding of the $D_1$ layer parallel fabric. $F_2$ folds are also recumbent to reclined, tight to isoclinal, and they were commonly invaded by axial planar charnockitic or other felsic melts. Axial plane fabrics did not develop during $D_2$, and $F_2$ structures are characterised by granuloblastic polygonal inequidimensional crystal forms. Major $F_2$ fold closures are more easily recognised than the very scarce major $F_1$ folds, and superimposition of $F_1$ and $F_2$ fold phases has typically resulted in fold interference patterns of either type 2 or type 3 in the classification scheme of Ramsay (1967).

The interlayered gneiss sequences illustrated in Figure 2 generally display similar post $F_2$ upright folds ($F_3$). From the Ongul Group of Lützow-Holm Bay to the Windmill Islands recumbent gneisses are folded into major (wavelengths 1-10 km)upright, non-cylindrical open to close folds that rarely plunge more steeply than 30-40° (most of the steeply plunging folds are $F_1$'s and $F_2$'s situated on steeply inclined limbs of $F_3$ folds). The $F_3$ folding was of a ductile type and commonly accompanied by slight retrograde metamorphism and local development of axial plane fabrics that are defined by aligned mica crystals and locally contain a down dip mineral lineation. The $F_3$ upright folds create the dome and basin pattern (type 1 interference pattern of Ramsay, 1967) first recognised in Enderby Land by Soviet geologists (Kamenev, 1972) and elsewhere described in terms of two fold phases by Blight (1975) and Yoshida (1978).

Igneous rocks were locally intruded after the $F_1/F_2$ recumbent folding and before the $F_3$ upright folding. Examples are charnockite and granitic bodies in the Napier Complex, Enderby Land, in the Lützow-Holm Bay area and in the Windmill Islands.

## DISCUSSION

The Archaean Napier Complex in Enderby Land is probably the most thoroughly studied part of the East Antarctic shield and the recumbent folding $D_1$ there is dated at about 3100 Ma, $D_2$ at about 2900 Ma, and the upright $D_3$ folding at 2500 Ma (James and Black, 1981). After this the complex was essentially cratonised although brittle fracturing permitted the emplacement of mafic dyke swarms. Possible correlations with the Prince Charles Mountains have been discussed previously (see also Sheraton et al., 1980) and a similar sequence of events has been deduced in the Vestfold Hills (Oliver et al., 1982; see also Collerson, this volume). Clearly Archaean high grade gneiss sequences are widespread in this general area of Antarctica (see Figure 1); as postulated earlier on the basis of the mafic dyke relationship, other high grade gneisses (e.g. those in the Bunger Hills) may also be of Archaean age.

By contrast, some of the East Antarctic gneiss sequences are much younger than Archaean. For example, although the sequences in Lützow-Holm Bay and the Windmill Islands display structural histories comparable in some respects to those of the Archaean sequences of the Napier Complex and Vestfold Hills, available geochronological data indicate that they are of Proterozoic vintage (see Williams et al., this volume; Yoshida, this volume). High grade metamorphic rocks of Middle Proterozoic age also occur in the Prince Charles Mountains, Enderby Land and, probably Dronning Maud Land (see Figure 1). High grade recumbent mixed paragneissic and orthogneissic sequences are well known from other Archaean continental cratons, (see for example Bridgewater et al., 1974; Coward, 1974, and Fripp, 1981)and from adjacent Proterozoic mobile belts; a similar picture is emerging in the East Antarctic shield. The formation of these younger mobile belts evidently involved reworking of Archaean and Early Proterozoic metamorphic and basement rocks, as well as prograde metamorphism of Proterozoic sediments (Sheraton and Collerson, in press), and the shearing out of discordant dyke swarms into concordant or subconcordant amphibolite pods and schlieren. The nature of the reworking or reactivation orogeny can be more clearly understood by examination of orogenic boundaries such as those between the Napier and Rayner Complexes in Enderby Land, the Vestfold Hills and Prydz Bay areas, and the northern (Proterozoic) and southern (Archaean) Prince Charles Mountains.

The exposed parts of the East Antarctic shield that are of Archaean age contrast with Archaean cratons elsewhere in that greenstone sequences appear to be absent, a banded ironstone bearing sequence in the southern Prince Charles Mountains (over 500 km inland)being the only possible example. In general, greenstone sequences do not occur near (closer than 150 km) to modern continental margins because Mesozoic continental fragmentation took place preferentially in the comparatively thin and more anisotropic crust of the high grade and younger mobile belts that commonly surround Archaean cratons. As most exposures of the East Antarctic shield are within 150 km of the coast, it is perhaps not surprising that greenstone sequences do not crop out, except possibly in the Prince Charles Mountains, the most extensive inland shield exposure.

The East Antarctic shield further contrasts with other Precambrian shield areas in that it is generally not possible to relate the deposition of platform cover sediments to tectonic activity in adjacent mobile belts. One possible exception is in the southern Prince Charles Mountains where the deposition in the south of Middle Proterozoic clastic sediments (now metamorphosed to greenschist facies grade—Tingey, 1982a) might be related to tectonism involved in the formation of high grade metamorphic rocks exposed in the northern Prince Charles Mountains.

## CONCLUSION

The objective of this review has been to show that patterns have emerged from geological research in the East Antarctic Shield area, and other studies in this symposium will doubtless refine this picture. Several areas—for example Enderby Land, the Lützow-Holm Bay area, western Dronning Maud Land and the Shackleton Range—have yielded abundant information following intensive investigations, but others have been covered in reconnaissance style only and need further study. In this latter category we would place the Adelie Land-King George V Land area, the Bunger Hills-Denman Glacier area and Dronning Maud Land.

*Acknowledgements* R.J. Tingey publishes by permission of the Director, Bureau of Mineral Resources, Geology and Geophysics, Canberra, Australia.

## REFERENCES


ADIE, R.J. (ed.), 1972: *Antarctic Geology and Geophysics—Proceedings of the Second International Symposium, Oslo, 1970.* Universitetsforlaget Oslo.

ARRIENS, P.A., 1975: The Precambrian geochronology of Antarctica. Geol. Soc. Aust., 1st Geol. Convent., Abstr., 97-8.

AUCAMP. A.P.H., WOLMARANS, L.G. and NEETHLING D.C., 1972: The Urfjell Group, a deformed ?Early Palaeozoic sedimentary sequence, Kirwanveggen, western Dronning Maud Land; *in* Adie, R.J. (ed.) *Antarctic Geology and Geophysics,* Universitetsforlaget, Oslo, 557-62.

BARTON, J.M., Jnr. and COPPERTHWAITE, Y.E. (this volume): Strontium isotopic studies of some intrusive rocks in the Ahlmannryggen and Annendagstoppane, western Dronning Maud Land, Antarctica.

BELLAIR, P. and DELBOS, L., 1962: Age absolu de la derniére granitisation en Terre Adélie. *C.R. Ac. Sc., Paris, 254.* 1465-6.

BLACK, L.P. and JAMES, P.R. (this volume): Geological evolution of the Napier Complex, Enderby Land, East Antarctica.

BLIGHT, D., 1975: The metamorphic geology of the Windmill Islands and adjacent coastal areas, Antarctica. Unpublished Ph.D. thesis, Univ. of Adelaide.

BLIGHT, D. and OLIVER, R.L., 1982: Aspects of the geologic history of the Windmill Islands, Antarctica; *in* Craddock, C. (ed.) *Antarctic Geoscience.* Univ. Wisconsin Press, Madison, 445-454.

BREDELL, J., 1982: The Precambrian sedimentary-volcanic sequence and associated intrusive rocks of the Ahlmannryggen, western Dronning Maud Land: a new interpretation; *in* Craddock, C. (ed.) *Antarctic Geoscience.* Univ. Wisconsin Press, Madison, 591-7.

BRIDGWATER, D., McGREGOR, V.R. and MYERS, J.S., 1974: A horizontal tectonic regime in the Archaean of Greenland and its implications for early crustal thickening. *Precambrian Res. 1,* 179-97.

CLARKSON, P.D., 1982: Tectonic significance of the Shackleton Range; *in* Craddock, C. (ed.) *Antarctic Geoscience.* Univ. Wisconsin Press, Madison, 835-9.

COLLERSON, K.D. (this volume): Nd and Sr isotope systematics of Archaean and Early Proterozoic granulites in the Vestfold Block, East Antarctica.

COLLERSON, K.D. and ARRIENS, P.A., 1979: Rb-Sr isotope systematics in high grade gneisses for the Vestfold Hills, East Antarctica (abstr.); *Geol. Soc. Aust., J., 26,* 267-8.

COWARD, M.P., 1974: Flat lying structures within the Lewisian basement gneiss complex of NW Scotland. *Proc. Geol. Ass.* 85, 459-72.

DE PAOLO, D.J., MANTON, W.I., GREW, E.S. and HALPERN, M., 1982: Sm-Nd, Rb-Sr and U-Th-Pb systematics of granulite facies from Fyfe Hills, Enderby Land, Antarctica. *Nature, 298,* 614-8.

EASTIN, R. and FAURE, G., 1970: The age of the Littlewood Volcanics of Coats Land, Antarctica; *in* Ohio State Univ., Lab. Isotop. Geol. Geochem., Rep. 5.

ELLIS, D.J., SHERATON, J.W., ENGLAND, R.N. and DALLWITZ, W.B., 1980: Osumilite-sapphirine-quartz granulites from Enderby Land, Antarctica—mineral assemblages and reactions. *Contrib. Mineral. Petrol., 72,* 123-43.

FEDEROV, L.V. and HOFMANN, J., 1982: Structural development of Precambrian rocks in the Mountain fringe of the Lambert Glacier and southern part of the Prince Charles Mountains, East Antarctica (abst.); *in* Craddock, C. (ed.) *Antarctic Geoscience* Univ. Wisconsin Press, Madison, 522.

FRIPP, R.E.P., 1981: The ancient Sand River Gneisses, Limpopo Mobile Belt, South Africa, *in* Glover, J.E. and Groves, D.I. (eds.) *Archaean Geology.* Geol. Soc. Aust., Spec. Pub., 7, 329-35.

GREW, E.S., 1978: Precambrian basement at Molodezhnaya Station, East Antarctica. *Geol. Soc. Am., Bull., 89,* 801-13.

GRIKUROV, G.E., ZNACHKO-YAVORSKY, G.A., KAMENEV, E.N., LOPATIN, B.G. and RAVICH, M.G., 1976: *Geological map of Antarctica scale 1:5,000,000—explanatory notes.* Research Institute of Arctic Geology, Leningrad.

GRINDLEY, G.W., 1981: Precambrian rocks of the Ross sea region. *R. Soc. N.Z., J., 11,* 411-23.

HALPERN M., 1970: Rubidium-strontium date of possibly 3 billion years for a granitic rock from Antarctica. *Science, 164* 977-8.

HALPERN, M. and GRIKUROV, G.E., 1975: Rubidium strontium dates from the southern Prince Charles Mountains, *U.S. Antarct. J.,* 10, (1), 9-15

HOFMANN, J., 1982: Main tectonic features and development of the southern Prince Charles Mountains, East Antarctica. *in* Craddock, C. (ed.) *Antarctic Geoscience* Univ. Wisconsin Press, Madison, 479-488.

HUNTER, D.J. (ed.), 1981: *Precambrian of the southern hemisphere.* Elsevier, Amsterdam.

JAMES, P.R. and BLACK, L.P., 1981: A review of the structural evolution and geochronology of the Archaean Napier Complex of Enderby Land, Australian Antarctic Territory; *in* Glover, J.E. and Groves, D.I. (eds.) *Archaean Geology,* Geol. Soc. Aust. Spec. Pub., 7, 71-83.

KAMENEV, E.N., 1972: Geological Structure of Enderby Land; *in* Adie, R.J. (ed.) *Antarctic Geology and Geophysics,* Universitetsforlaget, Oslo, 579-583.

KRYNAUW, J.R. (this volume): Preliminary report on the geochemistry and petrology of some igneous rocks in the Ahlmannryggen and Giaverryggen, western Dronning Maud Land, Antarctica.

LAIRD, M.G., 1981: Lower Palaeozoic rocks of Antarctica; *in* Holland C.H. (ed.), *Lower Palaeozoic of the Middle East, Eastern and Southern Africa, and Antarctica.* John Wiley and Sons.

MAWSON, D., 1940: Sedimentary rocks. *Australasian Antarctic Expedition 1911-14. Scientific Reports, Series A, 4,* 347-67.

McQUEEN., D.M., SCHARNBERGER, C.K., SCHARON, L. and HALPERN, M., 1972: Cambro-Ordovician palaeomagnetic pole position and Rubidium-Strontium total rock isochron for charnockitic rocks from Mirny Station, East Antarctica. *Earth. Planet. Sci. Lett., 16,* 433-8.

MAEGOYA, T., NOHDA, S. and HAYASE, I., 1968: Rb-Sr dating of the gneissic rocks from the East Coast of Lützow-Holm Bay, Antarctica. *Mem. Coll. Science., Univ. Kyoto, Ser. B, 35,* 131-8.

OLIVER, R.L., JAMES, P.R., COLLERSON, K.D. and RYAN, A.B., 1982: Precambrian geologic relationships in the Vestfold Hills, Antarctica. *in* Craddock, C. (ed.) *Antarctic Geoscience.* Univ. Wisconsin Press, Madison, 435-44.

OLIVER, R.L., COOPER, J.A. and TRUELOVE, A.J. (this volume): Some petrological and zircon geochronological observations of gneisses from Herring Island, Wilkes Land, and Commonwealth Bay, King George V Land.

RAMSAY, J.G., 1967: *Folding and fracturing of rocks:* McGraw-Hill, New York

RAVICH, M.G. and KRYLOV, A.Y., 1964: Absolute ages of rocks from East Antarctica; *in* Adie, R.J. (ed.) *Antarctic Geology.* North Holland, Amsterdam, 570-8.

RAVICH,M.G. and SOLOVIEV, D.S., 1966: *Geologiia i petrologiia tsentral'noi chasti gor zemli korolevy Mod (Geology and petrology of the mountains of central Dronning Maud Land, East Antarctica.) Moscow,* Isdatel'stvo Nedra, 290 pp.

RAVICH, M.G., KLIMOV, L.V. and SOLOVIEV, D.S., 1968: *The Precambrian of East Antarctica.* Israel Program for Scientific Translations, Jerusalem.

SHERATON, J.W., OFFE, L.A., TINGEY, R.J. and ELLIS, D.J., 1980: Enderby Land, Antarctica—an unusual Precambrian high grade metamorphic terrain. *Geol. Soc Aust., J., 27,* 1-18.

SHERATON, J.W. and BLACK, L.P., 1981: Geochemistry and geochronology of Proterozoic tholeiitic dykes of East Antarctica; evidence for mantle metasomatism. *Contrib. Mineral. and Petrol. 78,* 305-17.

SHERATON, J.W. and COLLERSON K.D., in press: Archaean and Proterozoic geological relationships in the Vestfold Hills—Prydz Bay area, Antarctica *BMR J. Aust. Geol. Geophys.*

SOBOTOVICH, E.V., KAMENEV, E.N., KOMARISTYY, A.A. and RUDNIK, V.A., 1976: The oldest rocks in Antarctica (Enderby Land). *Int. Geol. Rev., 18,* 371-88.

STILLWELL, F.L., 1918: The metamorphic rocks of Adelie Land *Australasian Antarctic Expedition 1911-14, Scientific Reports, Ser. A, 3(1).*

TINGEY, R.J. 1978: The exploration and geology of the East Antarctica metamorphic basement shield. Unpublished thesis, Scott Polar Research Institute, Cambridge.

TINGEY, R.J., 1982a: The geologic evolution of the Prince Charles Mountains—an Antarctic Archaean cratonic block; *in* Craddock, C. (ed.) *Antarctic Geoscience* Univ. Wisconsin Press, Madison, 455-64.

TINGEY, R.J., 1982b (map): Geology of the southern Prince Charles Mountains, 1:500,000 scale map. Bureau of Mineral Resources, Canberra.

VAN AUTENBOER, T. and LOY, W., 1972: Recent geological investigations in the Sør Rondane mountains, Belgicafjella and Sverdrupfjella, Dronning Maud Land; *in* Adie R.J. (ed.) *Antarctic Geology and Geophysics,* Universitetsforlaget, Oslo, 563-71.

WILLIAMS, I.S., COMPSTON, W., COLLERSON, K.D., LOVERING, J.F. and ARRIENS, P.A. (this volume): A reassessment of the age of the Windmill metamorphics, Casey area, East Antarctica.

WINDLEY, B.F., 1977: The evolving continents. John Wiley and Sons.

WOLMARANS, L.G. and KRYNAUW, J.R., 1981: *Reconnaissance geological maps of the Ahlmannryggen area, Dronning Maud Land, Antarctica (series of 3—1:250.000 scale)* South African Committee for Antarctic Research, Pretoria.

YOSHIDA, M., 1977: Geology of the Skallen Region, Lützow-Holmbukta, East Antarctica. *Mem. Nat. Inst. Pol. Res., Japan Ser. C, 11.*

YOSHIDA, M. 1978: Tectonics and petrology of charnockites around Lützow-Holmbukta, East Antarctica. *J. Geosci., Osaka City Univ., 21,* 65-152.

# GEOLOGICAL HISTORY OF THE ARCHAEAN NAPIER COMPLEX OF ENDERBY LAND

L.P. Black, *Bureau of Mineral Resources, P.O. Box 378, Canberra, A.C.T., 2601, Australia.*

P.R. James, *Department of Geology and Mineralogy, University of Adelaide, Adelaide, S.A., 5000, Australia.*

*Abstract* Isotopic dating, controlled by structural and geochemical studies, reveals a long complex geological history for the Archaean Napier Complex of Enderby Land. Initial acid igneous crustal formation at 3700-3800 Ma was followed by a sequence of felsic and mafic igneous rocks and metasediments. Intense regional strain ($D_1$) accompanied by granulite-facies metamorphism and granitoid emplacement, rotated into parallelism pre-existing discordances and produced a recumbent gneissic pile at about 3070 Ma. A subsequent ductile decormation ($D_2$) under granulite-facies conditions, and which possibly occurred at about 2900 Ma, was also characterised by tight asymmetric folding, although axial plane foliation was not produced. Cratonisation occurred after a third ductile deformation ($D_3$) (2450-2500 Ma) produced upright non-cylindrical folds under lower grade metamorphism. $D_3$ generally produced no more than weak schistosity. Development of strong $D_3$ schistosity in the Casey Bay area was accompanied by the resetting of Rb-Sr isochrons over at least a cubic metre. Napier Complex monazite and zircon lost Pb during intense granulite-facies tectonism at 1000 Ma in the adjacent Rayner Complex. Post-$D_3$ brittle fractures localised tholeiite dykes at 2350 ± 48 Ma and 1190 ± 200 Ma, minor pegmatites at 520 Ma, and a minor alkaline dyke at 484 Ma.

Until recently, there has been a dearth of geologically well constrained isotopic data from the East Antarctic Shield. However, a series of integrated geochronological-structural studies by the current authors (see below) now provides a firm basis for understanding the evolution of part of this shield, the Napier Complex of Enderby Land (Figure 1). These studies show that not only K-Ar and Rb-Sr mineral systems yield contentious ages, as suggested by Krylov (1972), Elliot (1975), and Sobotovich et al. (1976), but Rb-Sr total rock, U-Pb zircon, and Sm-Nd total rock and mineral systems were in some cases also effectively totally reset under the severe tectonothermal conditions imposed on this terrain. Fortunately, varied development at different localities has allowed the derivation of the chronology of the separate tectonic episodes.

Rb-Sr total rock, U-Pb zircon (both conventional and ion-probe), and Sm-Nd techniques have been used to decipher the involved Late Archaean tectonothermal history. Initial identification of the significance of particular ages was based on the correlation reported by Black et al. (1979) between resetting of Rb-Sr total rock isochrons and the development of penetrative fabrics. Complementary data from other techniques provide convincing evidence that the ages reported below are geologically meaningful and are not merely artifacts resulting from incomplete isotopic resetting. Interpretation of the U-Pb zircon data is based on the premise that morphologically distinct zircons formed under different geological conditions. Hence brown zircons, which have high U (>1000μg/g), are thought to have originally crystallised before or during the onset of the earliest

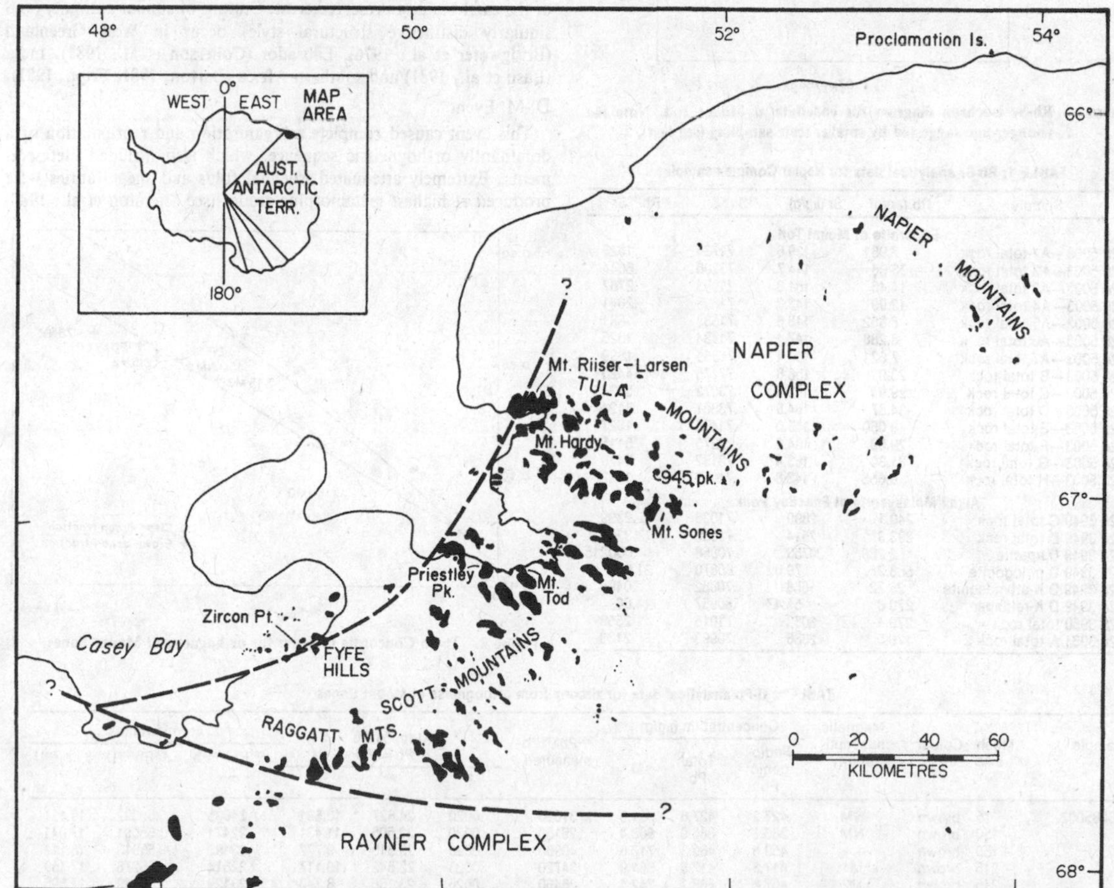


**Figure 1. Map of Enderby Land showing area of Napier Complex and localities discussed in the text.**

recognisable tectonothermal activity, before protracted granulite-facies metamorphism produced the significant U depletion documented by Sheraton and Black (in press). Pale coloured, generally more rounded zircons with low U and higher Th/U ratios are thought to have formed later. This review presents a summary of data being reported elsewhere and pertinent new data.

## ARCHAEAN EVOLUTION

Evolution of the Napier Complex was dominated by a sequence of high grade tectonothermal events in the Late Archaean. These tectonothermal events, which were labelled $D_1$-$M_1$, $D_2$-$M_2$ and $D_3$-$M_3$ on style and overprinting criteria and described in detail by James and Black (1981), produced characteristic major-, minor- and micro-structural features. The relationship between those observed structural features and more detailed geochronological and geochemical systematics is discussed in this article.

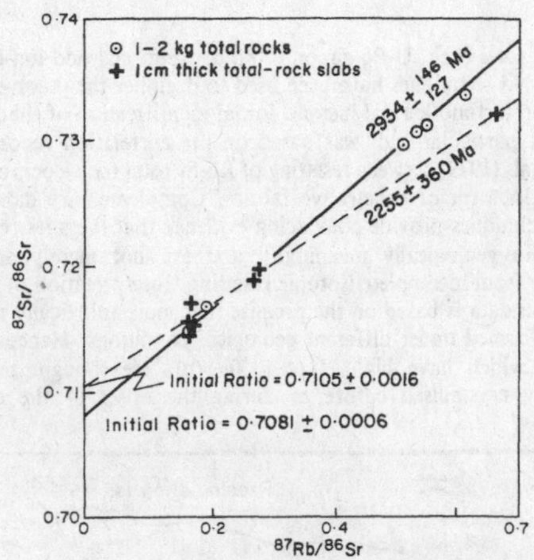


**Figure 2. Rb-Sr isochron diagram for enderbite at Mount Tod. Note the younger age suggested by smaller scale sampling (see text).**

**TABLE 1: Rb-Sr analytical data for Napier Complex samples**

| Sample | Rb (μg/g) | Sr (μg/g) | $^{87}Sr/^{86}Sr$ | $^{87}Rb/^{86}Sr$ |
|---|---|---|---|---|
| **Enderbite at Mount Tod** | | | | |
| 7828 5003—A1 total rock | 7.981 | 139.6 | .71735 | .1625 |
| 7828 5003—A2 total rock | 33.06 | 144.7 | .73205 | .6644 |
| 7828 5003—A3 total rock | 14.45 | 151.0 | .71993 | .2767 |
| 7828 5003—A4 total rock | 12.99 | 142.2 | .71895 | .2641 |
| 7828 5003—A5 total rock | 8.352 | 143.6 | .71539 | .1681 |
| 7828 5003—A6 total rock | 8.286 | 147.4 | .71484 | .1625 |
| 7828 5003—A7 total rock | 7.621 | 134.4 | .71555 | .1639 |
| 7828 5003—B total rock | 23.98 | 156.8 | .72775 | .4423 |
| 7828 5003—C total rock | 29.90 | 162.2 | .73073 | .5334 |
| 7828 5003—D total rock | 34.87 | 164.8 | .73361 | .6124 |
| 7828 5003—E total rock | 8.050 | 143.0 | .71496 | .1627 |
| 7828 5003—F total rock | 29.09 | 164.6 | .72975 | .5115 |
| 7828 5003—G total rock | 31.30 | 165.4 | .73132 | .5479 |
| 7828 5003—H total rock | 9.655 | 143.8 | .71657 | .1942 |
| **Alkali Melasyenite at Priestley Peak** | | | | |
| 7728 3949 C total rock | 240.1 | 2890 | .71026 | .2399 |
| 7728 3949 D total rock | 293.3 | 2514 | .71089 | .3369 |
| 7728 3949 D apatite | 2.286 | 30750 | .70858 | .000215 |
| 7728 3949 D phlogopite | 588.2 | 79.62 | .85670 | 21.65 |
| 7728 3949 D K-arfvedsonite | 26.72 | 1558 | .70882 | .0495 |
| 7728 3949 D K-feldspar | 278.8 | 55.47 | .80957 | 14.66 |
| 7728 3950 total rock | 270.4 | 3053 | .71015 | .2558 |
| 7728 3951 A total rock | 176.0 | 2386 | .70993 | .2130 |

### Early (pre-$D_1$) Evolution

The intensity of the sequence of events, especially $D_1$-$M_1$, imposed a severe tectonothermal veil through which the recognition of a pre-$D_1$ geological history and the style and age of initial crustal formation is almost undecipherable. Until recently, this age could not be rigorously estimated. The 4000 Ma estimate of Sobotovich et al. (1976), based on $^{207}Pb/^{206}Pb$ total rock analyses, was not meaningful for, as the data do not transect the modern end of the terrestrial Pb growth curve, they can not represent a simple secondary isochron. Neither is the reinterpretation of that data by Grew and Manton (1979) valid for it assumes a simple three stage model (inconsistent with the structural data) and an incorrect age of 2500 Ma for initial granulite-facies metamorphism (see below). The $^{207}Pb/^{206}Pb$ data are important, however, in that they demonstrate a prolonged pre-granulite facies crustal history for at least some Napier Complex rocks. High initial $^{87}Sr/^{86}Sr$ ratios $0.724^{+.006}_{-.007}$ for charnockite at Fyfe Hills—Black et al. (in press); $0.717 \pm .021$ for felsic paragneiss at Mt Sones—James and Black (1981); $0.7081 \pm .0006$ for enderbite at Mt Tod—Figure 2, Table 1) at about 3070 Ma, the time of initial granulite-facies metamorphism reported below, complement the $^{207}Pb/^{206}Pb$ total-rock data in indicating an Early Archaean protolith age.

Conventional quantitative data on this issue are currently only provided from an enderbite (78285007) at Mt Sones. Dark brown zircons from this rock define an alignment (Figure 3, Table 2) which meets Concordia at about 2500 Ma and 3700 Ma. The finest (−75 + 45 μm) fraction probably plots off this trend because it could represent a mixed population resulting from the difficulties in resolving brown from pink zircon populations at this grain size. High resolution ion-microprobe U-Pb data from zircon cores (Compston and Williams, 1982; Black and James, 1982) indicate a similar but near concordant and hence considerably better defined protolith age of about 3800 Ma. This requires the Napier Complex to be a remnant of the earth's oldest preserved crust. Cratons of similar antiquity (and similarly distinctive structural style) occur in West Greenland (Bridgwater et al., 1976), Labrador (Collerson et al., 1981), India (Basu et al., 1981) and southern Africa (Barton, 1981; Fripp, 1981).

### $D_1$-$M_1$ Event

This event caused complete reorganisation and reorientation of a dominantly orthogneissic sequence, which also included metasediments. Extremely attenuated isoclinal folds and shear fabrics were produced at highest metamorphic grades (see Sheraton et al., 1980)

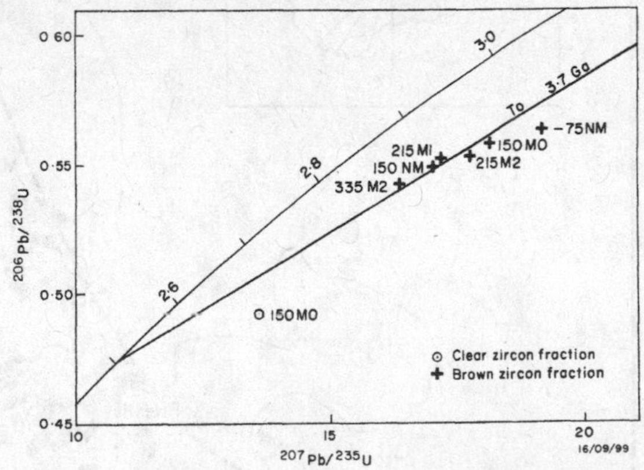


**Figure 3. U-Pb Concordia diagram for orthogneiss at Mount Sones.**

**TABLE 2: U-Pb analytical data for zircons from orthogneiss at Mount Sones**

| Sample | Mesh size | Colour | Magnetic susceptibility | Concentration (μg/g): Radiogenic Pb | Concentration (μg/g): Total Pb | Concentration (μg/g): U | $^{206}Pb/^{204}Pb$ measured | Atom % Pb ($^{206}Pb$ = 100): 204 | Atom % Pb: 207 | Atom % Pb: 208 | Atomic Ratios: $^{207}Pb/^{206}Pb$ | Atomic Ratios: $^{206}Pb/^{238}U$ | Atomic Ratios: $^{207}Pb/^{235}U$ |
|---|---|---|---|---|---|---|---|---|---|---|---|---|---|
| 78285007 | − 75 | brown | NM | 427.2 | 427.6 | 651.5 | 31630 | .0020 | 24.627 | 10.581 | .24605 | .56392 | 19.131 |
| | − 150 | brown | NM | 385.3 | 385.9 | 608.4 | 29150 | .0030 | 22.505 | 11.431 | .22471 | .55001 | 17.041 |
| | − 150 | brown | | 459.8 | 460.3 | 717.6 | 40960 | .0020 | 23.515 | 9.772 | .23493 | .55913 | 18.111 |
| | − 215 | brown | M1 | 417.3 | 417.8 | 661.9 | 34720 | .0025 | 22.542 | 10.117 | .22514 | .55276 | 17.159 |
| | − 215 | brown | M2 | 467.6 | 468.2 | 742.1 | 25460 | .0026 | 23.158 | 8.343 | .23129 | .55489 | 17.696 |
| | − 335 | brown | M2 | 407.4 | 407.9 | 662.6 | 32660 | .0027 | 21.843 | 10.013 | .21813 | .54239 | 16.312 |
| | − 150 | pink | M0 | 162.3 | 162.7 | 239.6 | 9900 | .0064 | 20.124 | 39.830 | .20048 | .49230 | 13.608 |

under very ductile conditions during this long and involved tectonothermal event. Earlier and synchronously formed discordances were rotated into parallelism to produce a recumbent gneissic pile in which all features of initial character or earlier cratonic history were effectively masked. The recumbency, intensity and style of this deformation and its similarity to other high grade Archaean areas (see earlier reference to Greenland—Labrador, southern Africa and India) indicate that this event might result from early (Archaean) plate tectonic activity as it is the most likely lower crustal product of large scale horizontal differential shear between a rigid(?) upper crust and a slowly creeping mantle (see James and Tingey, this volume).

Like the major and minor structures, $D_1$-$M_1$ is also characterised by a distinctive micro-structural style which is described in detail by James and Black (1981). Typical $D_1$-$M_1$ undeformed equilibrium granoblastic polygonal micro-structures from both orthogneissic and paragneissic units are illustrated in Figure 4. Under high grade granulite-facies conditions, stable mineralogies do not contain strong fabric forming elements such as the phyllosilicates which form axial-plane schistosities typical of lower grades. Thus, in the high grade Napier Complex, the preferred orientation of inequidimensional grain aggregates forms the dominant microstructure. These aggregates are axial planar to $F_1$ folds and hence formed during that event.

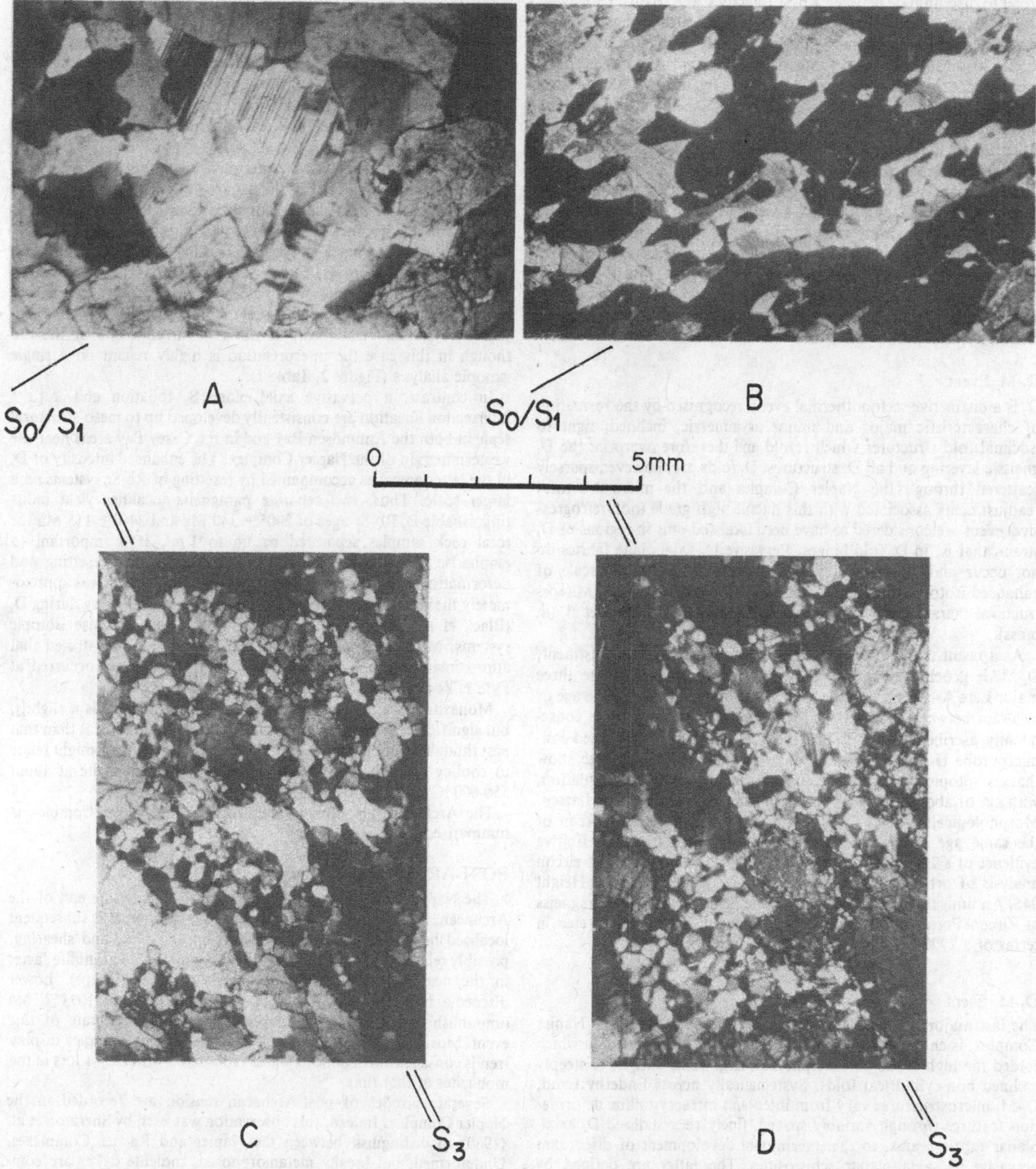


**Figure 4. Typical tectonothermal microstructures (a) $D_1$-$M_1$ in orthogneiss (quartz, plagioclase and orthopyroxene), Mt Hardy. (b) $D_1$-$M_1$ in paragneiss (garnet, sillimanite, quartz) with strong $S_1$ layer parallel fabric, 945 Peak. (c) $D_3$-$M_3$ in orthogneiss—fine "new grain" recrystallisation of quartz and mesoperthitic feldspar, Debenham Pk. (d) $D_3$-$M_3$ in intermediate granulite—two pyroxene—plagioclase assemblage intensely recrystallised, Mt Tod.**

Syntectonic recrystallisation which accompanied the ductile creep deformation would have created microstructural diffusion paths for isotopic migration. The Rb-Sr total rock system, at least on a restricted scale, should thus document the "freezing-in" of the $D_1$-$M_1$ microstructure at the final stage of $D_1$.

The precise significance of U-Pb zircon ages is complicated by the probability that both new growth and isotopic resetting of older grains occurred during $D_1$-$M_1$. However, both types might more reasonably be expected to correlate with the onset of the tectonothermal event when temperatures were high and metamorphic fluid was still abundant. Once annealed, or if newly formed, these zircons would have been relatively resistant to isotopic resetting for the remainder of $D_1$-$M_1$. The precision limits for this event are, however, generally too broad to discriminate between Rb-Sr total rock and U-Pb zircon ages.

On the basis of U-Pb zircon and Rb-Sr total rock evidence, Black and James (1979) and James and Black (1981) postulated an age of about 3050 Ma for $D_1$-$M_1$. Some subsequent U-Pb zircon data approximate the earlier discordia trend (e.g. Fyfe Hills—Black et al., in press; Proclamation Island—Black, unpublished analyses) but no individual localities yield as wide a range of $^{207}Pb/^{206}Pb$ values and consequently they do not define it as precisely. In addition to the Rb-Sr evidence discussed earlier, the regional study of Napier Complex felsic orthogneisses by Sheraton and Black (in press) also provides evidence that initial granulite-facies metamorphism occurred at about 3050 Ma. Indeed, from syntectonic charnockite at Proclamation Island on the northern tip of Enderby Land, those authors were able to provide a precise age estimate for $D_1$ of $3072^{+35}_{-33}$ Ma. $D_1$-$M_1$ was even sufficiently strong to effectively reset the reputedly resistant Sm-Nd total rock system at Fyfe Hills, where a secondary age of 3050 ± 210 ma is recorded (McCulloch and Black, 1982).

### $D_2$-$M_2$ Event

$D_2$ is a distinctive tectonothermal event, recognised by the formation of characteristic major and minor asymmetric, inclined, tight to isoclinal fold structures which refold and therefore overprint the $D_1$ gneissic layering and all $D_1$ structures. $D_2$ folds are, however, sparsely scattered through the Napier Complex and the micro-structural readjustments associated with this ductile high grade (non-retrogressive) event are considered to have been localised only in regions of $D_2$ strain, that is, in $D_2$ fold hinges. Pervasive $D_2$ axial plane fabrics do not occur and, therefore, even in such fold hinges, the scale of enhanced isotopic diffusion is not likely to have been great. Microstructural characteristics of $D_2$ are described in detail in Black et al. (in press).

As a result of the lack of pervasive microstructural readjustment, $D_2$-$M_2$ is geochronologically the most poorly defined of the three major Late Archaean tectonic events. The best evidence for the age of an event between those derived for $D_1$ and $D_3$, and which is consequently ascribed to $D_2$, is provided by currently unpublished ion-microprobe U-Pb zircon analyses by the senior author. These show that an isotopically homogeneous near concordant zircon population, with age of about 2900 Ma, occurs in paragneiss at Mt Riiser-Larsen. Morphologically similar, but isotopically more discordant zircon of the same age occurs in charnockite at Fyfe Hills. Less definitive evidence of a 2900 Ma event is provided by conventional U-Pb zircon analysis of orthogneisses from Mt Tod, Mt Sones and Spot Height 945. An unusual alignment of conventional analyses from paragneiss at Zircon Point in the Casey Bay area might also be interpreted in terms of a 2900 Ma event (Black et al., in prep.).

### $D_3$-$M_3$ Event

The last major Archaean event, which variably affected the Napier Complex, is characterised by weak sub-horizantal compression which folded the high grade gneissic pile into large scale upright to steeply inclined non-cylindrical folds. Systematically across Enderby Land, $D_3$-$M_3$ microstructures vary from inter- and intracrystalline deformation features through variably spaced, finely recrystallised $D_3$ axial-planar fabric seams, to the intermittent development of discordant pervasive $D_3$ axial-planar schistosities. The latter are defined by elongate relics of old ($D_1$-$M_1$) inequidimensional grains and elongate aggregates of recrystallised new ($D_3$-$M_3$) grains (for detail see James and Black, 1981). Examples of such microstructures are illustrated in Figure 4.

Geochronological interpretation is considerably complicated by $D_3$-$M_3$, which significantly but variably reset isotopic systems at most localities. Consequently many zircons, whether or not they originally crystallised at that time, have $^{207}Pb/^{206}Pb$ ages of about 2500 Ma (Black, unpublished analyses), the age originally ascribed by James and Black (1981)—2470 to 2495 Ma—to that event based on data from Mt Hardy. The best examples of fine isotopic detail within near 2500 Ma zircons are provided by distinct but adjacent paragneisses in the Casey Bay area which appear to have slightly different upper intercepts with Concordia (Black et al., in prep.). In addition, that for one of these gneisses, $2485^{+65}_{-30}$ Ma, might also be older than a precise age of $2456^{+8}_{-5}$ Ma for zircon (plotting on a concordant extension of the other gneiss trend) in syn-$D_3$ pegmatite. Because the latter population includes a concordant member, it should unequivocally define a real geological event. The tendency for other zircon alignments to transect Concordia over a small but distinct age range might reflect a real age spread of the $D_3$-$M_3$ event. However, because the concordant analysis appears at the younger extremity of the range, it is more probable that the older intercepts (up to about 2500 Ma) document a small, relatively uniform inherited Pb component which was not completely lost at $D_3$.

During $D_3$ only very weak penetrative microfabric at most was developed over the majority of the Napier Complex. Consequently, at localities such as Fyfe Hills, 1-2 kg sized total rock samples from a cubic metre of charnockite still preserve $D_1$ ages. However, because recrystallisation of quartz and feldspar during $D_3$ produced seams up to 10 cm long (Figure 4), total rock sampling on centimetre scale yields the age of that later event (2463 ± 35 Ma), that is, penetrative $D_3$ microfabric formed on that scale at that time (Black et al., in press). A similar relationship is observed in enderbite at Mt Tod though in this case the interpretation is highly reliant on a single isotopic analysis (Figure 2, Table 1).

In contrast, a pervasive axial plane $S_3$ foliation and a $L_3^{1-2}$ intersection lineation are consistently developed up to meso-structural scale in both the Amundsen Bay and in the Casey Bay areas near the western margin of the Napier Complex. The enhanced intensity of $D_3$ in the latter area was accompanied by resetting of Rb-Sr systems on a larger scale. Thus, two separate paragneiss localities yield indistinguishable $D_3$ Rb-Sr ages of 2405 ± 140 Ma and 2440 ± 115 Ma for total rock samples separated by up to 1 $m^3$. It is important to emphasise this correlation between scale of isotopic resetting and deformation, rather than with metamorphic grade which was approximately the same (700°C, 7 kb) at Fyfe Hills and Casey Bay during $D_3$ (Black et al., in prep.). In addition to resetting of these isotopic systems, McCulloch and Black (1982) have also demonstrated that approximate intermineral equilibration of Nd isotopes occurred at Fyfe Hills at about 2500 Ma.

Monazite in paragneiss from the Casey Bay area yields a slightly, but significantly younger, upper intercept age ($2425^{+16}_{-17}$ Ma) than that best thought to approximate $D_3$ ($2456^{+8}_{-5}$ Ma). The former might relate to cooling through the blocking temperature of monazite at about 550-600°C (Black et al., in prep.).

The Archaean tectonothermal history of the Napier Complex is summarised in Figure 5.

## POST-ARCHAEAN EVOLUTION

The Napier Complex was essentially cratonised by the end of the Archaean, but towards the southwestern margin there was subsequent localised heating, interrelated introduction of fluids, and shearing, possibly related to intense deformation at amphibolite-granulite facies in the nearby Rayner Complex (Black et al., in prep.). Lower discordia intercepts for monazite from this area, of $1073^{+31}_{-36}$ Ma (unpublished data), apparently provide a precise estimate of this event. Most U-Pb zircon data from the entire Napier Complex display trends consistent with minor Pb loss (compared with major loss in the monazites at that time.

Several episodes of post-Archaean tension are recorded in the Napier Complex. Indeed, this observation was used by Sheraton et al. (1980) to distinguish between the Napier and Rayner Complexes. Undeformed, but locally metamorphosed, tholeiite dykes are common in the Napier Complex but are found only as metamorphosed relics in the Rayner Complex which must consequently have undergone a period of later regional deformation. The petrographic subdivision of these mafic dykes by Sheraton et al. (1980) is consistent

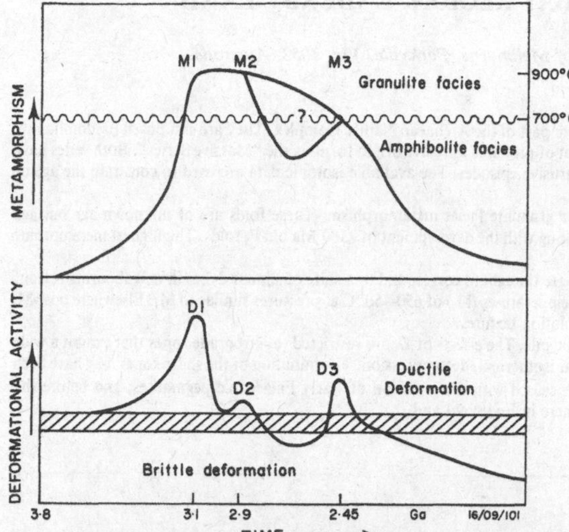


**Figure 5. Diagrammatic representation of the Archaean tectonothermal evolution of the Napier Complex. An alternative interpretation to that shown here is that metamorphic temperatures may have fallen between $M_1$ and $M_2$ rather than remaining relatively constant for about 150 to 200 Ma.**

with geochemical and Rb-Sr isotopic data (Sheraton and Black, 1981), which are in turn consistent with field relations. Metamorphosed quartz tholeiites which crystallised after $D_3$ but while granulite-facies conditions were still prevalent, yield an imprecise age of 2400± 250 Ma. High Mg tholeiites, emplaced at moderate crustal levels (3.1-6.3 kb), define an isochron age of 2350± 48 Ma. A suite of unmetamorphosed quartz tholeiites yield a perfect isochron with age of 1190± 200 Ma. Emplacement of these dyke suites thus appears to have been localised by tension at two different times, one shortly after the last major tectonothermal event ($D_3$) and the other just before major tectonism in the adjoining Rayner Complex. Localised pegmatite intruded along brittle fractures in the Casey Bay area at 522± 10 Ma (Black et al., in prep.). Shortly, but significantly, later at 482± 3 Ma (Table 1, Figure 6) a minor K-rich alkali melasyenite dyke was emplaced at Priestley Peak (Sheraton and England, 1980). This is the youngest rock forming event recognised in the Napier Complex. This age and that of the previously discussed pegmatite are typical "Pan African" ages, so widespread on other continents.

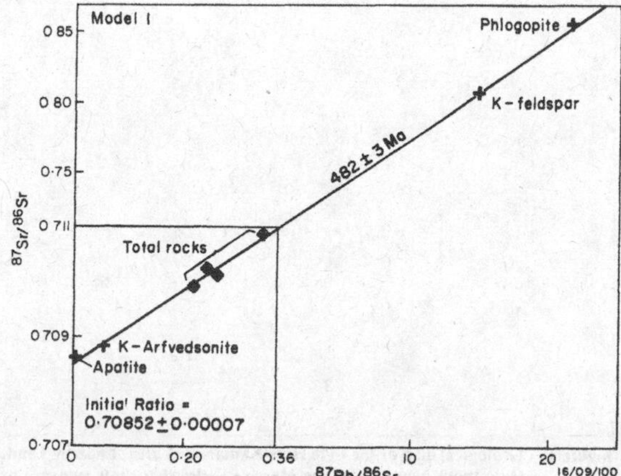


**Figure 6. Rb-Sr isochron diagram for mineral and total rock samples from an alkali melasyenite dyke at Priestly Peak. Note the change in scale.**

*Acknowledgements* Technical assistance was given by M.J. Bower, T.K. Zapasnik, D.B. Guy and N.C. Hyett. The manuscript was critically read by J.W. Sheraton, R.W. Page, J. Ferguson and H. Baadsgaard. The authors gratefully acknowledge the support of R.J. Tingey and S.L. Harley since the inception of this project: also the assistance and logistic support of the Antarctic Division Department of Science and the Environment while they were members of the Australian National Antarctic Research Expeditions during the summers of 1977-78 and 1979-80. Black publishes with the permission of the Director, Bureau of Mineral Resources, Geology and Geophysics, Canberra, Australia.

## REFERENCES


BARTON, J.M. Jr., 1981: The pattern of Archaean crustal evolution in southern Africa as deduced from the evolution of the Limpopo Mobile Belt and the Barberton granite—greenstone terrain; *in* Glover, J.E. and Groves, D.I. (eds.) *Archaean Geology.* Geol. Soc. Aust. Spec. Pub., 7, 21-31.

BASU, A.R., RAY, S.L., SAHA, A.K. and SARKAR, S.N., 1981: Eastern Indian 3800 million year old crust and early mantle differentiation. *Science, 212,* 1502-6.

BLACK, L.P. and JAMES, P.R., 1979: Preliminary isotopic ages from Enderby Land, Antarctica. *Geol. Soc. Aust., J., 26,* 266-7. Abstract.

BLACK, L.P., BELL, T.H., RUBENACH, M.J. and WITHNALL, I.W., 1979: Geochronology of discrete structural metamorphic events in a multiply deformed Precambrian terrain. *Tectonophysics, 54,* 103-37.

BLACK, L.P. and JAMES, P.R., 1982: U-Pb zircon and Rb-Sr systematics of the high grade and multiply deformed Archaean Napier Complex of Enderby Land, Antarctica. Fifth Int. Conf. on Geochronology, Cosmochronology and Isotope Geology, Nikko, Japan, June 27-July 2, 1982, Abstract 29-30.

BLACK, L.P., JAMES, P.R. and HARLEY, S.L., in press: The geochronology, structure and metamorphism of early Archaean rocks at Fyfe Hills, Enderby Land, Antarctica. *Precambrian Res.*

BLACK, L.P., JAMES, P.R. and HARLEY, S.L., in prep.: Geochronology, structure and metamorphism of rocks from the Casey Bay area, Enderby Land, Antarctica: U-Pb zircon and Sr isotopic behaiour in a high grade and multiply deformed terrain.

BRIDGWATER, D., KETO, L., MCGREGOR, V.R. and MYERS, J.S., 1976: The Archaean gneiss complex of Greenland; *in* Escher, A. and Watt, W.S. (eds.) *Geology of Greenland.* Greenland Geological Survey, Copenhagen, 18-75.

COLLERSON, K.D., KERR, A. and COMPSTON, W., 1981: Geochronology and evolution of Late Archaean gneisses in northern Labrador: an example of reworked sialic crust; *in* Glover, J.E. and Groves, D.I. (eds.) *Archaean Geology.* Geol. Soc. Aust., Spec. Pub., 7, 205-22.

COMPSTON, W. and WILLIAMS, I.S., 1982: Protolith ages from inherited zircon cores measured by a high mass resolution ion microprobe. Fifth Int. Conf. on Geochronology, Cosmochronology and Isotope Geology, Nikko, Japan, June 27-July 2, 1982. Abstract 63-4.

ELLIOT, D.H., 1975: Tectonics of Antarctica: A review. *Am. J. Sci., 275-A,* 45-106.

FRIPP, R.E.P., 1981: The ancient Sand River Gneisses, Limpopo Mobile Belt, South Africa; *in* Glover, J.E. and Groves, D.I. (eds.) *Archaean Geology.* Geol. Soc. Aust., Spec. Pub., 7, 329-35.

GREW, E.S. and MANTON, W.I., 1979: Archaean rocks in Antarctica: 2.5 billion year uranium lead ages of pegmatites in Enderby Land. *Science, 206,* 443-5.

JAMES, P.R. and BLACK, L.P., 1981: A review of the structural evolution and geochronology of the Archaean Napier Complex of Enderby Land, Australian Antarctic Territory. *Geol. Soc. Aust., Spec. Pub., 7,* 71-83.

JAMES, P.R. and TINGEY, R.J. (this volume): The geological evolution of the East Antarctic metamorphic shield—A review.

KRYLOV, A. Ya., 1972: Antarctic geochronology; *in* Adie, R.J. (ed.) *Antarctic Geology and Geophysics.* Universitetsforlaget, Oslo, 491-4.

MCCULLOCH, M.T. and BLACK, L.P., 1982: Sm-Nd isotopic systematics of Enderby Land Granulites: evidence for the redistribution of Sm and Nd during metamorphism. Fourth Int. Sym. Antarc. Earth Sci., Adelaide, Australia, August 1982, Abstract 117.

SHERATON, J.W. and ENGLAND, R.N., 1980: Highly potassic mafic dykes from Antarctica. *Geol. Soc. Aust., J., 27,* 129-35.

SHERATON, J.W., OFFE, L.A., TINGEY, R.J. and ELLIS, D.J., 1980: Enderby Land, Antarctica—an unusual Precambrian high grade metamorphic terrain. *Geol. Soc. Aust., J., 27,* 1-18.

SHERATON, J.W. and BLACK, L.P., 1981: Geochemistry and geochronology of Proterozoic tholeiite dykes of East Antarctica: evidence for mantle metasomatism. *Contrib. Mineral. Petrol., 78,* 305-17.

SHERATON, J.W. and BLACK, L.P., in press: Geochemistry of Precambrian gneisses: relevance for the evolution of the East Antarctic Shield. *Lithos.*

SOBOTOVICH, E.V., KAMENEV, Ye. N., KOMARISTYY, A.A. and RUDNIK, V.A., 1976: The oldest rocks of Antarctica (Enderby Land). *Int. Geol. Rev., 18,* 371-88.

# THE GEOLOGY OF THE FYFE HILLS—KHMARA BAY REGION, ENDERBY LAND

M. Sandiford and C.J.L. Wilson, *School of Earth Sciences, University of Melbourne, Parkville, Vic, 3052, Australia.*

*Abstract* The gneisses of the Fyfe Hills-Khmara Bay region in Enderby Land form part of the Archaean Napier Complex. They are composed predominantly of supracrustal rocks forming the "Layered Series", but also include a component of probable intrusive origin forming the "Massive Series". Both series have been affected by four phases of deformation and multiple metamorphic and intrusive episodes. The available isotopic data are used to constrain the age of each of these events.

$F_1$ folds and associated structures indicate intense flattening during prograde granulite facies metamorphism. These folds are of unknown age but age thought to have formed during an episode of continuous metamorphism culminating with the development of 2500 Ma old $F_2$ folds. The highest metamorphic temperatures ($M_1$) of 820-950°C at pressures of 8-11 kb occurred during $D_2$.

Tholeiitic dykes, intruded parallel to the axial surfaces of developing $F_3$ folds, are thought to correspond to a suite of similar dykes in neighbouring regions dated at 2400 ± 250 Ma. The metamorphic assemblages of these dykes indicate temperatures ($M_2$) of 650-750°C at pressures similar to $M_1$. Elsewhere post-$M_1$ isobaric cooling is indicated in the host gneisses by retrograde corona and exsolution textures.

The final deformation phase ($D_4$) occurred during the excavation of the gneissic pile. The effects of $D_4$ are restricted to retrograde zones that exhibit a wide range of amphibolite facies mineral assemblages ($M_3$), indicating progressive, near isothermal, decompression. The initiation of the shear zones may have been coeval with the 1000 Ma Rayner event in nearby crystalline blocks. Tectonism ceased with the intrusion of Early Palaeozoic pegmatites, and before the intrusion of rare mafic alkaline dykes which have been dated at 490 Ma elsewhere in Enderby Land.

The Fyfe Hills area is crucial in the interpretation of the evolution of the Precambrian crystalline rocks of Enderby Land. This is largely the result of the 4.0 ± 0.2 Ga U-Pb age obtained by Sobotovich et al. (1976). Although this age is disputed (Grew and Manton, 1979) it has generated a number of independent isotopic studies (DePaolo et al., 1982; Grew et al., in press; Black and James, this volume; McCulloch and Black, this volume). Much of the data generated by these studies has been collected and interpreted without the constraints imposed by detailed field work. The aim of this contribution is to elucidate the field relationships in the Fyfe Hills-Khmara Bay region and provide a framework for the interpretation of the available isotopic studies. This work is based on a combined total of 14 man-weeks field mapping in the Fyfe Hills and the adjacent islands of Khmara Bay during the 1979/80 Austral summer.

## REGIONAL SETTING OF FYFE HILLS

The geology of Enderby Land has been reviewed by Ravich and Kamenev (1975) and Sheraton et al. (1980). The Fyfe Hills are near the western margin of the older of two metamorphic complexes within Enderby Land; that is, the Archaean Napier Complex which is bounded on its continental side by the Proterozoic Rayner Complex. This bipartite division of the crystalline basement is supported by the restriction of unmetamorphosed Amundsen Dykes to the Napier Complex. These dykes, dated at 1190 Ma (Sheraton and Black, 1981) occur as highly deformed relicts within the Rayner Complex.

Grikurov et al. (1976) divided the Napier Complex into two series; the predominantly orthogneissic Raggatt Series and the predominantly paragneissic Tula Series. Sobotovich et al. (1976) regarded the Fyfe Hills as part of the Raggatt Series, and argued that the 4.0 ± 0.2 Ga age indicates that the Raggatt Series is older than, and hence forms the basement to, the Tula Series. Kamenev (1982) suggested that the layered gneisses of the Fyfe Hills are derived from a primary enderbitic or quartz diorite body by processes of metasomatism and metamorphic differentiation. However both Sandiford and Wilson (in prep.) and DePaolo et al. (1982) have argued that the Fyfe Hills gneisses are derived from predominantly supracrustal precursors and hence are of Tula Series affinities. James and Black (1981) have suggested that the intimate association of the Tula and Taggatt Series in the Amundsen Bay area is the result of tectonic interleaving, but neither isotopic nor field studies have yet been able to successfully distinguish the relative ages of the two series.

## LITHOLOGICAL RELATIONSHIPS

Published geological maps of Enderby Land suggest that the boundary between the Napier and Rayner complexes passes between the Fyfe Hills and the islands of Khmara Bay (Ravich and Kamenev, 1975; Sheraton et al., 1980; James and Black, 1981). However, our mapping has shown that this region (shown in Figure 1) is dominated by high grade granulite facies gneisses with the petrographic, isotopic (DePaolo et al., 1982; Grew et al., in press) and field characteristics of the Napier Complex. The high grade gneisses form 90% of the outcrop, the remaining 10% being retrograde zones composed of amphibolite grade schists, mylonites and pegmatites.

The high grade gneisses form a layered sequence at least 3 km thick. Charnockitic and enderbitic gneisses (felsic granulites) form 60% of

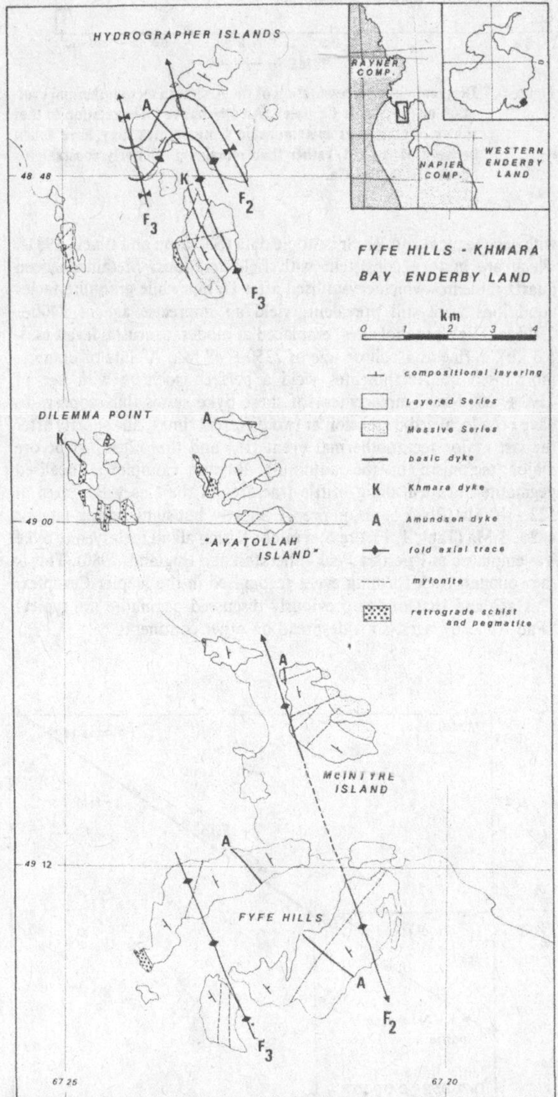


**Figure 1.** **Geological map of the Fyfe Hills-Khmara Bay area, Enderby Land. Only those members of the Massive Series that show evidence of being derived from intrusive sheets are shown.**

outcrop and are interlayered with mafic granulites (15%), ultramafic gneisses (3%), meta-pelites (10%) and minor meta-ironstones and calcsilicates.

Contacts between compositional types are generally concordant and parallel to the bulk compositional layering. However, at a few localities (McIntyre Island and Hydrographer Island) discordant contacts between meta-pelite and relatively massive sheets of charnockite are interpreted as primary intrusive contacts (Sandiford and Wilson, in prep.). The recognition of these discordant contacts allows the division of this sequence into two series: (1) the "Layered Series" derived from a supracrustal sequence composed of high Mg-pelites and Fe-rich sediments together with felsic, mafic and ultramafic volcanic and volcanogenic strata; the occurrence of a celsian-spessartine-Mn clinopyroxene-magnetite paragenesis at Fyfe Hills is particularly interesting as the association of Ba and Mn in chemical precipitates is indicative of submarine exhalative environments; the "Layered Series" comprises at least 70% of outcrop and is characterised by a pronounced layered compositional heterogeneity with individual units rarely exceeding 10 m in thickness and more typically 1-5 m thick: and (2) the "Massive Series" comprising relatively thick (> 50 m) compositionally homogeneous sheet like bodies which appear to be derived from the subconcordant intrusion of intermediate to mafic igneous rocks into the pre-existing Layered Series.

It is tempting to correlate the "Layered" and "Massive Series" with the Tula and Raggatt Series respectively. However, because Sobotovich et al. (1976) suggested the Fyfe Hills rocks are members of the Raggatt Series, we prefer not to use these terms in the present context. Furthermore, until a more detailed picture emerges of the field relationships between orthogneissic and paragneissic sequences throughout the Napier Complex the regional extent of the "Massive/Layered Series" distinction cannot be evaluated.

## STRUCTURAL GEOLOGY

Four generations of structures deform both the Layered and the Massive Series (Figure 1). A set of E-W trending antiformal culminations dominate the structure of the area. The E-W plunging antiform ($F_2$) passing through southern Khmara Bay and northern Fyfe Hills folds a set of earlier mesoscopic folds ($F_1$) and associated structures and is itself folded about an antiform ($F_3$) passing through southern Fyfe Hills. $F_{1-3}$ folds are dissected by retrograde shear zones formed during the final phase of deformation ($D_4$). Metamorphosed basic dykes, the "Khmara dykes", (Sandiford and Wilson, in prep.) were intruded immediately prior to or early in $D_3$ while the generally unmetamorphosed Amundsen Dykes were intruded between $D_3$ and $D_4$.

The $D_1$ event is characterised by mesoscopic isoclinal folding, boudinage and a pervasive gneissic layering. Macroscopic $F_1$ closures are comparatively rare compared with the more central and easterly exposures of the Napier Complex (James and Black, 1981; our own observations, 1979/80). Most $D_1$ structures are associated with uniform medium grained granoblastic polygonal to interlobate textures, although rare $S_1$ flaser textures are preserved in some enderbitic gneisses, suggesting the $D_1$ occurred during prograde granulite facies metamorphism but before metamorphic culmination ($M_1$). Mafic boudins in felsic hosts may be separated in the plane of the compositional layering by up to 20-30 times individual boudin length, suggesting an oblate $D_1$ finite strain ellipse with X:Y:Z> 30:30:1 and indicative of intense subvertical flattening. The $D_1$ mesoscopic geometry and morphology together with the paucity of large scale closures suggests crustal thinning during $D_1$

The $D_2$ event resulted in abundant reclined meso and macroscopic folds. The overall $F_2$ configuration is similar to the "rucked-up" $F_2$ gneissic pile that James and Black (1981) have described in the Amundsen Bay area. Microstructural equilibration of the highest grade assemblages during $D_2$ is apparent in many meta-pelites indicating that metamorphic culmination ($M_1$) occurred during $D_2$ and furthermore suggests a close temporal relationship between $D_1$ and $D_2$. $F_2$ folds are best developed in meta-pelites, and are commonly associated with boudinage. Pegmatites formed in the "pressure shadow" of these boudins have yielded 2.5 Ga zircon ages (Grew et al., in press), a date which we interpret as the age of the $D_2$ event. Late shearing along the axial surfaces of some $F_2$ folds has resulted in recrystallisation of partially hydrated assemblages and may reflect localised reintroduction of $H_2O$ in the waning stages of $D_2$. However the microstructures and assemblages associated with these axial surfaces are reminiscent of $D_3$ microstructures and possibly reflect $D_3$ reworking of suitably oriented $D_2$ structures.

The $D_3$ event produced a series of upright, open to tight meso- and macroscopic E-W trending folds with shallow variously plunging axes. An axial plane schistosity ($S_3$) is locally developed and characterised by fine grained granoblastic textures formed by the recrystallisation and partial hydration of the highest grade assemblages and a near vertical extension lineation ($L_3$).

The "Khmara dykes" form subvertical E-W trending planar bodies approximately parallel to the $S_3$ axial surface; their macroscopic form suggests that they have not been folded. However, examination of the dyke margins revealed folded apophyses with axial planar $S_3$ fabrics. Furthermore the dykes contain the fine-grained granoblastic textures typical of $S_3$ fabrics in the host gneisses. These features suggest to us that the dykes have been intruded either prior to or during the development of the $F_3$ folds in orientations precluding the development of macroscopic folds. Griffin (in prep.) has noted a similar temporal and spatial relationship between metamorphosed tholeiitic dykes and $F_3$ folds in Amundsen Bay. These dykes are petrographically and chemically similar to the $B_2$ tholeiitic dykes of Sheraton and Black (1981) which have been dated at 2.4 ± 0.25 Ga. However, Sheraton and Black (1981) consider the $B_2$ dykes to postdate $D_3$ folding, a relationship which has been used by James and Black (1981) to constrain the age of the $D_3$ event.

Subvertical retrograde zones that truncate all previous structures as well as the Late Proterozoic Amundsen Dykes are the only manifestations of $D_4$. The zones vary from narrow ductile shear zones to zones of amphibolite facies schist and mylonite up to 300 m wide; they locally contain abundant migmatite and pegmatite. Structures within the shear zones are commonly complex, a fact that reflects continuous or episodic reworking during a prolonged history. The mylonitic foliation ($S_4$) formed parallel to the walls of the mylonite zones generally contains a near vertical extension lineation ($L_4$) and indicates that the retrograde zones developed in reponse to differential displacement between rigid blocks of granulite during an essentially vertical tectonic regime.

## METAMORPHIC GEOLOGY

Three dynamic metamorphic/microstructural "events" have been recognised. These are: (1) a prograde granulite facies event ($M_1$) initiated prior to $D_1$ and culminating during $D_2$; (2) a retrograde granulite facies event ($M_2$) coeval with $D_3$; and (3) a retrograde amphibolite facies event ($M_3$) coeval with $D_4$. Superimposed on these metamorphic assemblages are the effects of static re-equilibration to lower grade conditions, namely, a range of corona and exsolution textures and mineralogical zoning. The mineral assemblages diagnostic of each "event" are listed in Table 1 and an evaluation of the conditions of metamorphism is presented below (see also Figure 2).

The stability of mesoperthite, sapphirine-quartz and sub-calcic clinopyroxene testifies to the unusually high temperatures (T) prevailing during $M_1$. Two pyroxene geothermometry (Wood and Banno, 1973; Wells, 1977) yields temperatures in the range 820-950°C. The presence of fine exsolution lamellae in coexisting $M_1$ pyroxenes suggests that these are underestimates, and probably counteracts the overestimation inherent in these thermometric methods (Wood, 1975). Exsolution temperatures in coarsely exsolved pyroxenes range from 770-830°C. The preferred $M_1$ temperature range of 820-950°C is comparable with, although slightly lower than, estimates for the highest grade assemblages eslewhere in the Napier Complex (Ellis, 1980; Grew, 1980).

$M_1$ pressure estimates based on coexisting garnet, orthopyroxene, plagioclase and quartz (Newton and Perkins, 1982) fall in the range 8-10 kb, while Harley's (1981) garnet-orthopyroxene barometer yields pressures in the range 7-10 kb. These estimates are consistent with the absence of garnet in tholeiitic compositions (Green and Ringwood, 1967), the occurrence of sapphirine-quartz and sillimanite-hypersthene assemblages to the exclusion of cordierite (Newton, 1978), and resemble pressure estimates from the Tula Mountains in the central Napier Complex (Ellis, 1980; Grew, 1980). However, a number of lines of evidence indicate that the Fyfe Hills rocks represent deeper crustal sections than are exposed in the Tula Mountains: (i) the restrictions of garnet-clinopyroxene assemblages to Fyfe Hills, albeit to bulk compositions enriched in Mn and/or Ca, (ii) the occurrence of sillimanite and hypersthene in preference to cordierite in coronas

**TABLE 1: Diagnostic mineral assemblages and preferred pressures (P) and temperatures (T) for the metamophic events $M_1$, $M_2$ and $M_3$. Mineral abbreviations: mp = mesoperthite; ap = antiperthite; q = quartz; px = pyroxene (c = clino, o = ortho, s = subcalcic); kf = k-feldspar; pl = plagioclase; hb = hornblende; bt = biotite; gt = garnet; m = muscovite; cc = calcite; cu = cummingtonite; ol = olivine; sp = spinel; se = serpentine; sap = sapphirine; sil = sillimanite; ky = kyanite; gd = gedrite; st = staurolite; cd = cordierite; mg = magnetite; ce = celsian; gru = grunerite; = > mineral reactions or transformations; mp/ap either mp or ap.**

| Rock Type | M1 | | M2 | | M3 (early) | M3 (late) |
|---|---|---|---|---|---|---|
| Felsic granulite | mp/ap, q, opx, cpx | mp = > | kf, pl, q, opx, bt<br>kf + pl | px = > | hb, bt | kf, pl, hb, q, m |
| Mafic granulite | cpx, opx, pl, q | px + pl = > | cpx, opx, gt, pl, hb<br>gt + q | px = > | hb, bt, pl, q, gt, cc<br>hb, bt | hb, bt, pl, q, cc, cu |
| Ultramafic | cpx, opx, ol/pl, sp, hb | | cpx, opx, bt, gt, hb | | | bt, se, cc |
| Metapelite | mp/ap, sap, opx, gt<br>sil, q, sp | sap + q = > | kf, pl, opx, gt, sil<br>sil + opx | sil = > | gt, pl, q, by, ky, gd<br>ky ky + ge + gt = > | gt, pl, q, bt, sil, st, cd, m<br>st + sil + cd |
| Fe- (Mn) meta-sediment | mg, q, opx, cpx, gt,<br>ce, scpx | scpx = > | cpx + opx | | | mg, q, gru, gt |
| T | 820-950 C | | 650-750 C | | 600-700 C | 570-670 C |
| P | 8-11 kb | | 7-10 kb | | 7-9 kb | 3-5 kb |

between $M_1$ sapphirine and quartz in Fyfe Hills, (iii) the absence of osumilite from Fyfe Hills and its replacement by the previously unrecorded higher pressure assemblage sapphirine-hypersthene-quartz ± k-feldspar. In view of this evidence it is believed that 7 kb is an underestimate for Fyfe Hills $M_1$ pressures, and the preferred range is 8-10 kb. The anhydrous nature of $M_1$ assemblages, occurrence of mesoperthite and general scarcity of $M_1$ melt phases suggests extremely dry metamorphism.

A significant temperature reduction between $M_1$ and $M_2$ is indicated by two-feldspar, as opposed to mesoperthite, bearing $M_2$ assemblages and the general scarcity of exsolution lamellae in $M_2$ pyroxenes. $M_2$ garnet-clinopyroxene assemblages in rocks of tholeiitic composition give rise to temperature estimates of between 630 and 700°C (Ellis and Green, 1979) and 700-800 (Ganguly, 1979), while pyroxene solvus thermometry yields temperatures up to 830°C. The discrepancy may be due either to the comparative insensitivity of the pyroxene solvus thermometers at temperatures below 900°C (Davis and Boyd, 1966), or overestimation inherent in these thermometers (Wood, 1975). An $M_2$ temperature range of 650-750°C is preferred. The Newton and Perkins (1982) barometer yields pressures in the range 6-9 kb, which is consistent with the occurrence of garnet in tholeiites (Green and Ringwood, 1967) and sillimanite in aluminous gneisses (Holdaway, 1971). These data suggest that near isobaric cooling of up to 200°C followed the $M_1$ event. Elsewhere isobaric cooling is indicated by the ubiquitous occurrence of sillimanite-hypersthene coronas between $M_1$ sapphirine-quartz assemblages.

The diverse amphibolite facies assemblages formed during $M_3$ are reflecting the influence of a variety of bulk compositions and a range in P-T conditions. Rocks of aluminous composition are used here to establish the prevailing metamorphic conditions as the comparatively narrow stability fields of their constituent phases allow for excellent

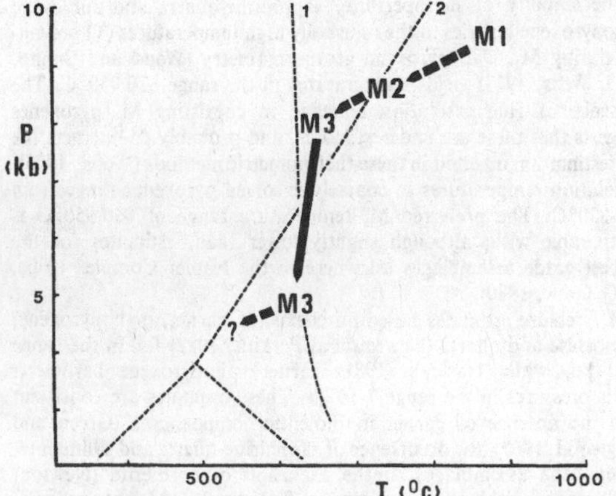


**Figure 2. Pressure (P) temperature (T) evolutionary path of the Fyfe Hills gneisses. The heavy dashed line corresponds to inferred sections of the path while the solid line represents the documented path during $M_3$. Reaction 1 is the wet garnite solidus taken from Kerrick (1972). Reaction 2 is the alumino-silicate phase diagram of Holdaway (1971).**

documentation of this complex metamorphic event. However it should be noted that the application of geobarometric and thermometric techniques to these rocks is complicated by the presence of strongly zoned minerals, in particular garnet and feldspar. The earliest formed $M_3$ assemblages in rocks of aluminous composition include kyanite, gedrite, garnet, plagioclase, quartz and biotite. Garnet-biotite thermometry (Thompson, 1975; Ferry and Spear, 1978) gives garnet core termperatures of 650 ± 50°C and rim temperatures of 570 ± 50°C. The core temperatures are consistent with the field occurrence of migmatites and anatectic pegmatites at $P_{(H_2O)}$ approaching $P_{(total)}$. Kyanite at temperatures of 650°C implies pressures greater than 7 kb (Holdaway, 1971) whie the experimental data of Green and Vernon (1974) suggest the assemblage kyanite + amphibole + quartz in the absence of cordierite is stable only above 9-10 kb. On the other hand Ghent's (1976) garnet-plagioclase-quartz-kyanite barometer yields pressure estimates below 6.5 kb for the earliest formed $M_3$ assemblages, using an ideal solution model for garnet. This estimate is believed to be unrealistically low and can be explained by continued chemical adjustment of the system during subsequent exhumation or to an incorrect activity model for grossular. The preferred conditions for earliest $M_3$ assemblages are pressures of 7-9 kb at 630-660°C.

A succession of reactions in rocks of aluminous composition due to changing metamorphic conditions during the $M_3$ event led to the formation of: (1) epitaxial overgrowths of staurolite on kyanite; (2) sillimanite by replacement of kyanite; and (3) cordierite by reaction between kyanite and/or sillimanite, and garnet and gedrite. Late $M_3$ assemblages which crystallised in the stability field of cordierite give garnet-biotite and garnet-cordierite (Thompson, 1975; Wells, 1979) core temperature estimates of 620 ± 50°C and rim temperatures of 545 ± 50°C, at pressures of 4-5 kb; the Ghent (1976) barometer gives pressure estimates in the range 3-4.5 kb using an ideal solution model for garnet. These data suggest that when late-formed $M_3$ assemblages crystallised pressures were less than 5 kb and temperatures were in the range 620 ± 50°C. This implies near isothermal decompression of 2.5-4 kb during active recrystallisation of $M_3$ assemblages. The rimward zoning common to both early and late formed $M_3$ assemblages is believed to indicate a cooling interval of 60-90°C before closure of the systems, and after the period of isothermal decompression. Interestingly, many of the reactions which took place during $M_3$ (a retrograde event with respect to $M_2$ and $M_1$), were of prograde character, that is they involved an increase in entropy; this is suggestive of either very rapid exhumation or thermal buffering by shear heating during the development of $M_3$ mylonites.

## DISCUSSION OF THE RELEVANT ISOTOPIC DATA

A correlation of the structural, metamorphic and igneous history of the Fyfe Hills area with the relevant isotopic data is presented in Table 2. The $M_1$-$D_2$ event and the intrusion of the Amundsen Dykes are well constrained. The ages of the "Khmara dykes" (and hence the $D_3$-$M_2$ event) and the $D_4$-$M_3$ event are less certain, but can be bracketed by the available data. As yet there are no isotopic data pertaining to the age of $D_1$ and the age of emplacement of the "Massive Series".

Our chronological correlations differ significantly from the conclusions of James and Black (1981). These authors suggest that $D_3$ occurred about 2.5 Ga ago in the waning stages of granulite facies metamorphism and correlate $D_1$ $D_2$ and the metamorphic culmination with what we contend is a poorly resolved isotopic event > 3.0 Ga. We believe that the similarity of the isotopic, structural, and metamorphic

TABLE 2: Correlation of the geologic history of the Fyfe Hills area with the available isotopic data. References: 1. DePaolo et al., 1982; 2. Grew et al. (in prep.); 3. Sheraton & Black (1981); 4. Grew & Manton (1979). (z) — zircon age; (r) — whole rock age. *Age of $B_3$ dykes of Sheraton and Black (1981).

| Event | Age (Ga) | System | |
|---|---|---|---|
| 1. Deposition of Layered Series | 3.4-3.6 | Nd/Sm(r) | 1 |
| 2. Emplacement of Massive Series ? | ? | | |
| 3. $D_1$ | ? | | |
| $M_1$ | 2.5 | Rb/Sr, Nd/Sm(r); U/Pb(z) | 1,2 |
| 4. $D_2$ | 2.5 | U/Pb(z) | 2 |
| 5. Intrusion of Khmara Dykes | 2.4±0.2 | Rb/Sr(r) | 3 |
| $D_3/M_2$ | >2.35* | Rb/Sr(r) | 3 |
| 6. Intrusion of Amundsen Dykes | 1.19 | Rb/Sr(r) | 3 |
| 7. $D_4/M_3$ | ? 1.0-0.5 | U/Pb(z) | 2,4 |
| 8. Mafic alkaline dykes | 0.49 | | 3 |

histories of the Amundsen Bay and Fyfe Hills areas suggests that the 2.5 Ga date corresponds to the $M_1$ and $D_2$ events at both localities. The observations of Griffin (in prep.) and ourselves, that metamorphosed tholeiitic dykes in both areas are in fact pre- or syn-$D_3$ and not, as suggested by previous workers, post-$D_3$ lends some support to this suggestion. However, the poorly resolved age for these dykes of 2.4± 0.25 Ga does not provide conclusive evidence. An absolute minimum age for $D_3$ is provided by the unmetamorphosed 2.35 Ga old $B_3$ dykes of Sheraton and Black (1981).

An important implication of the relative timing of $D_2$ and $D_3$ and hence $M_1$ and $M_2$ pertains to the thermal evolution of the terrain. Our data suggest that $M_1$ temperatures represent a disturbed geotherm that decayed relatively rapidly to a steady state situation in less than 150 Ma. However, this conclusion is not necessarily valid if the scheme of James and Black (1981) is adopted. Accordingly the prolonged isobaric cooling of up to 600 Ma may simply reflect the cooling of the Archaean crust.

The structual and metamorphic relationships suggest that $D_1$ occurred during prograde metamorphism ($M_1$) that culminated in $D_2$. In view of the large heat flux necessary to produce metamorphic temperatures in excess of 900°C followed by near isobaric cooling of up to 200°C, a geologically short time interval between $D_1$ and $D_2$ is envisaged. It is tempting to appeal to the *Massive Series* as the source of heat for $M_1$, implying an age not substantially greater than 2.5 Ga. This suggestion is consistent with the thermal character of $M_1$ and especially post $M_1$ isobaric cooling, but in the absence of rigorous isotopic or field constraints it remains unsubstantiated.

The interval of approximately 1 Ga between deposition and the first apparent tectonism implies remarkably stable crustal conditions for the Archaean. However is is possible that the effects of earlier tectonic events have been obscured by the intense deformation during $D_1$ and $D_2$ and by the pervasive recrystallisation during $M_1$. Indeed such events may be responsible for the isotopic disturbance at 3.1 Ga (James and Black, (1981).

The near isothermal decompression path indicated by $M_3$ assemblages suggests that the exhumation of these deep-seated gneisses was rapid enough to preclude thermal relaxation. However it is possible that the effects of thermal relaxation during an extended exhumation could be buffered by the generation of heat within the mylonites via a shear heating mechanism. The data of Grew et al. (in prep.) suggesting a thermal event at 1.0 Ga may correspond to the initiation of exhumation during the Rayner event. However, until there is a more precise correlation of the 1.0 Ga event with $M_3$ retrogression, the behaviour of the Napier Complex during the Rayner event will remain an enigma. Active tectonism during exhumation appears to have ceased with the intrusion of Early Palaeozoic pegmatites (Grew and Manton, 1979). Mafic alkaline dykes containing richterite, aegirine - augite and baryte and found only on Hydrographer Island almost certainly correspond to the dyke at Priestley Peak described by Sheraton and England (1980) and for which Sheraton and Black (1981) have obtained a 490 Ma age. These dykes show only minor metamorphic alteration and cut the shear zones without apparent displacement.

*Acknowledgements* Logistic support for this work was provided by the Antarctic Division of the Department of Science. The authors are indebted to E.S. Grew and S. Harley for their assistance with field work. S. Kirkby, R.J. Tingey and J.F. Lovering are thanked for their help in initiating and undertaking this work. E.S. Grew is thanked for constructive criticism of the draft manuscript.

## REFERENCES


BLACK, L.P. and JAMES, P.R. (this volume): Geological evolution of the Napier Complex, Enderby Land, East Antarctica.

DAVIS, B.T.C. and BOYD, F.R., 1966: The join $Mg_2Si_2O_6$-$CaMgSi_2O_6$ at 30 kilobars pressure and its application to pyroxenes from kimberlites. *J. Geophys. Res. 71,* 3567-76.

DEPAOLO, D.J., MANTON, W.I., GREW, E.S. and HALPERN, M., 1982: Sm-Nd, Rb-Sr, and U-Th-Pb systematics of granulite facies rocks from Fyfe Hills, Enderby Land, Antarctica. *Nature, 298,* 614-8.

ELLIS, D.J., 1980: Osumilite-sapphirine-quartz granulites from Enderby Land, Antarctica: P-T conditions at metamorphism, implications for garnet-cordierite equilibria and the evolution of the deep crust. *Contrib. Mineral. Petrol., 74,* 201-10.

ELLIS, D.J. and GREEN, D.H., 1979: An experimental study of the effect of Ca upon garnet-clinopyroxene Fe-Mg exchange equilibria. *Contrib. Mineral. Petrol., 71,* 13-22.

FERRY, M. and SPEAR, F.S., 1978: Experimental calibration of the partitioning of Fe and Mg between biotite and garnet. *Contrib. Mineral. Petrol., 66,* 113-7.

GANGULY, J., 1979: Garnet and clinopyroxene solid solutions and geothermometry based on Fe-Mg distribution coefficients. *Geochim. Cosmochim. Acta., 43,* 1021-30.

GHENT, E.D., 1976: Plagioclase-garnet-$Al_2SiO_5$-quartz: a potential geobarometer-geothermometer. *Am. Mineral. 61,* 710-4.

GREEN, D.H. and RINGWOOD, A.E., 1967: An experimental investigation of the gabbro to eclogite transformation and its petrological applications. *Geochim. Cosmochim. Acta, 31,* 767-833.

GREEN, T.H. and VERNON, R.H., 1974: Cordierite breakdown under high-pressure hydrous conditions. *Contrib. Mineral. Petrol. 46,* 215-26.

GREW, E.S., 1980: Sapphirine-quartz association from Archean rocks in Enderby Land, Antarctica. *Am. Mineral., 65,* 821-36.

GREW, E.S. and MANTON, W.I., 1979: Archean rocks in Antarctica: 2.5 billion-year uranium-lead ages of pegmatites in Enderby Land. *Science, 206,* 443-5.

GREW, E.S., MANTON, W.I. and SANDIFORD, M. (in prep.): Geochronologic studies in East Antarctica: age of pegmatites in Casey Bay, Enderby Land.

GRIKUROV, G.E., ZNACHKO-YAVORSKY, G.E., KAMENEV, E.N. and RAVICH, M.G., 1976: *Explanatory notes on the geological map of Antarctica (scale 1:5,000,000).* Research Institute of Arctic Geology, Leningrad.

HARLEY, S.L., 1981: Garnet-orthopyroxene assemblages as pressure-temperature indicators. Unpublished Ph.D. thesis, University of Tasmania, Hobart.

HOLDAWAY, M.J., 1971: Stability of andalusite and the aluminium silicate phase diagram. *Am. J. Sci., 271,* 97-131.

JAMES, P.R. and BLACK, L.P., 1981: A review of the structural evolution and geochronology of the Archaean Napier Complex of Enderby Land, Australian Antarctic Territory. *in* Glover, J.E. and Groves, D.I. (eds.) *Archaean Geology.* Geol. Soc. Aust. Spec. Pub., 7, 71-83.

KAMENEV, E.N., 1982: Antarctica's oldest metamorphic rocks in the Fyfe Hills, Enderby Land; *in* Craddock, C. (ed.) *Antarctic Geoscience,* Univ. Wisconsin Press, Madison, 505-10.

KERRICK, D.M., 1972: Experimental discrimination of muscovite + quartz stability with $P_{H_2O} > P_{tot}$. *Am. J. Sci. 272,* 946-58.

McCULLOCH, M.T. and BLACK, L.P. (this volume): Sm-Nd isotopic systematics of Enderby Land Granulites, evidence for the redistribution of Sm and Nd during metamorphism.

NEWTON, R.C., 1978: Experimental and thermodynamic evidence for the operation of high pressure in Archaean metamorphism, *in* B.F. Windley and S.M. Naqvi (eds.) *Archaean Geochemistry,* Elsevier, Amsterdam, 221-40.

NEWTON, R.C. and PERKINS, D., 1982: Thermodynamic calibration of geobarometers based on the assemblages garnet-plagioclase-orthopyroxene (clinopyroxene)-quartz. *Am. Mineral., 67,* 203-22.

RAVICH, M.G. and KAMENEV, E.N., 1975: *Crystalline basement of the Antarctic Platform.* John Wiley and Sons, New York.

SANDIFORD, M. and WILSON, C.J.L. (in prep.): Structures of the Fyfe Hills-Khmara Bay Region, Enderby Land, Antarctica.

SHERATON, J.W. and ENGLAND, R.N., 1980: Highly potassic mafic dykes from Antarctica *Geol. Soc. Aust., J., 27,* 129-35.

SHERATON, J.W. and BLACK, L.P., 1981: Geochemistry and geochronology of Proterozoic tholeiite dykes of East Antarctica: evidence for mantle metasomatism. *Contrib. Mineral. Petrol., 78,* 305-17.

SHERATON, J.W., OFFE, L.A., TINGEY, R.J. and ELLIS, D.J., 1980: Enderby Land, Antarctica—an unusual Precambrian high grade terrain. *Geol. Soc. Aust. J., 27,* 1-18.

SOBOTOVICH, E.V., KAMENEV, E.N., KOMARISTYY, A.A. and RUDNIK, V.A., 1976: The oldest rocks of Antarctica (Enderby Land). *Int. Geol. Rev., 18,* 71-388.

THOMPSON, A.B., 1975: Mineral reactions in pelitic rocks: (2) calculations of some P-T-X (Fe-Mg) phase relations. *Am. J. Sci. 275,* 425-54.

WELLS, P.R.A., 1977: Pyroxene thermometry in simple and complex systems. *Contrib. Mineral. Petrol., 62,* 129-39.

WELLS, P.R.A., 1979: Chemical and thermal evolution of Archaean sialic crust, southern west Greenland. *J. Petrol., 20,* 187-226.

WOOD, B.J., 1975: The influence of pressure, temperature and bulk composition on the appearance of garnet in orthogneiss—an example from South Harris, Scotland. *Earth Planet. Sci. Lett, 26,* 299-311.

WOOD, B.J. and BANNO, S., 1973: Garnet-orthopyroxene and orthopyroxene-clinopyroxene relationships in simple and complex systems. *Contrib. Mineral. Petrol., 42,* 109-24.

# THE NAPIER AND RAYNER COMPLEXES OF ENDERBY LAND, ANTARCTICA—CONTRASTING STYLES OF METAMORPHISM AND TECTONISM

D.J. Ellis, *Geology Department, University of Tasmania, GPO Box 252C, Hobart, Tasmania 7001, Australia.*

*Abstract* The metamorphic rocks of Enderby Land consist of the Archaean Napier Complex and the Proterozoic Rayner Complex, and are part of the East Antarctic Shield. This paper reviews work on the Napier Complex and presents new data and interpretation for the Rayner Complex. Mineral reactions following the peak conditions of metamorphism of the Napier Complex record the cooling of these granulites to lower temperature at depth. This "near isobaric cooling path" reflects stable tectonic conditions. The Archaean crust was essentially cratonised after the final phase of metamorphism and deformation at ca 2500 Ma. The ca 1000 Ma Rayner Complex represents a major period of hydration, metamorphism and deformation of the margin of the Napier Complex. The P-T path deduced for this reworking is one of "near isothermal uplift" and reflects a fundamental change in tectonic evolution of the craton from being in isostatic equilibrium with the underlying mantle in the Archaean to vertical tectonic movement in the Proterozoic.

The granulite facies metamorphic rocks of the Archaean Napier Complex of Enderby Land have aroused considerable interest amongst petrologists because of the unusual mineral assemblages and exceptionally high P-T conditions of metamorphism inferred for this terrain on a regional scale. The mineral assemblages hypersthene(Hy)-sillimanite(Sill)-quartz(Qtz), sapphirine(Sa)-quartz and spinel(Sp)-quartz occur in pelitic rocks, indicative of metamorphism beyond the P-T stability of coexisting garnet(Ga)-cordierite(Cd) (Sheraton et al., 1980; Ellis et al., 1980). Such features, together with the presence of osumilite instead of cordierite-hypersthene-K feldspar(Kf)-quartz, indicate that the Napier Complex is possibly the highest grade regional granulite terrain exposed at the Earth's surface (7-10 kb, 900-980°C: Ellis, 1980; Sheraton et al., 1980; Grew, 1981; Harley, this volume).

Similar to other Archaean cratons, the 2.5-3.0 Ga Napier Complex is bordered by a younger, approximately 1 Ga mobile zone, the Rayner Complex. This mobile zone has been shown to be retrogressed Napier Complex, deformed and metamorphosed at upper amphibolite to lower granulite facies, but at lower P-T and higher $P_{H_2O}$ conditions than the older high grade complex (Sheraton et al., 1980).

The purpose of this paper is to compare and contrast the styles of metamorphism of these two complexes. Evidence for the P-T path which the Napier Complex followed is reviewed and that for the Rayner Complex determined herein. It is concluded that the development of the Rayner Complex reflects a fundamental change in tectonic stability of the Antarctic Archaean craton.

## GENERAL GEOLOGY

Enderby Land forms part of the Precambrain East Antarctic Shield and has been extensively mapped by both Australian and Soviet geologists (see Sheraton et al., 1980 for references to work prior to 1980). Two metamorphic complexes are recognised—the Archaean Napier Complex and the Proterozoic Rayner Complex (Kamenev, 1972, 1975; Grikurov et al., 1976). This distinction has been further detailed by Sheraton et al., (1980), and the reader is referred to this paper for a description of the geology, petrology and structure of the terrain. More detailed accounts of the petrology, geochemistry, structure and geochronology of the region have been undertaken by other Australian geologists (Ellis and Green, 1979; Ellis, 1980; James and Black, 1981; Sheraton and Black, 1981; Ellis and Green, ms; Sheraton and Black, in prep; Harley, 1981; and this volume).

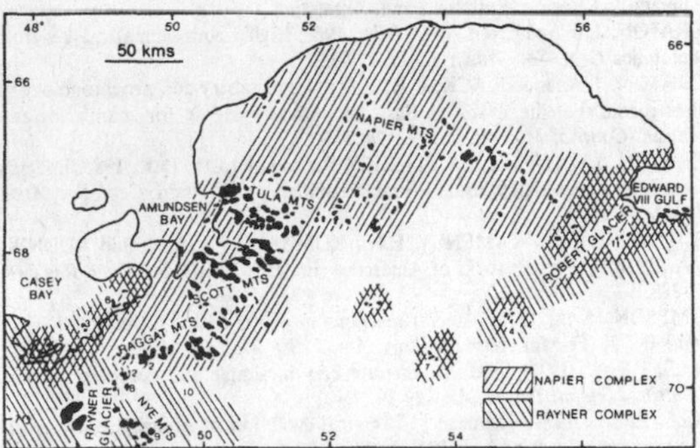


**Figure 1. Metamorphic map of Enderby Land and Western Kemp Land. Localities discussed in the text are—1. Condon Hills, 2. Mt Currie, 3. Demidov Island, 4. Doggers Nunataks, 5. Forefinger Point, 6. Hydrographer Island, 7. Knuckey Peaks, 8. Mt Lira, 9. Mt Robinson, 10. Mt Sibiryakov, 11. Turbulence Bluffs, 12. Mt Yuzhnaya.**

The Napier Complex crops out over an area of 50,000 sq. km and forms the Napier, Tula, Scott and Raggatt Mountains (Figure 1). The Rayner Complex covers most of the area west of the Rayner Glacier, the Nye Mountains and most of Kemp Land and includes the isolated Sandercock and Doggers Nunataks and Knuckey Peaks (Sheraton et al., 1980).

There are several reasons to justify dividing the Precambrian craton of Enderby Land into two distinct complexes (Sheraton et al., 1980). Isotopic dates have given an age of at least 3000 Ma for the Napier Complex (Black and James, 1979). Isotopic resetting, which apparently occurred during discrete tectonothermal events which produced recognisable penetrative fabrics, allowed U/Pb zircon and Rb/Sr age estimates of 3000 ma and 2600 Ma for an early horizontal tectonism and the culmination of dome- and basin-forming events respectively in the Napier Complex (James and Black, 1981). Metatholeiites and high-Mg tholeiites dated at 2350 ± 48 Ma were emplaced during the waning stages of this last Napier Complex metamorphic event, and group one tholeiites (Amundsen dykes) were emplaced at 1190 ± 200 Ma shortly before the major orogeny in the adjoining Rayner Complex (Sheraton and Black, 1981). The fresh Amundsen dykes have been found at almost all major outcrops in the Napier Complex, whereas their chemical equivalents are all metamorphosed in the Rayner Complex. Charnockitic gneiss from the Rayner Complex near Molodezhnaya station has given a Rb-Sr isochron age of 987 ± 60 Ma (Grew, 1978).

On the basis of major and trace element geochemistry as well as further isotopic data on the acid gneisses, Sheraton and Black (in prep.) have shown that the Rayner Complex gneisses are remetamorphosed Napier Complex gneisses of originally igneous derivation.

## PETROLOGY

There are marked difference in metamorphic grade between the two complexes as summarised in Table 1. The petrology of the Napier Complex has received more attention (Dallwitz, 1968; Sheraton et al., 1980; Ellis et al., 1980; Ellis and Green, MS; Grew, 1980; Harley, this volume) than the Rayner Complex (Sheraton et al., 1980; Grew, 1978).

### The Napier Complex

The metamorphic grade of the Napier Complex was beyond the stability of coexisting garnet-cordierite in pelitic rocks. The presence of sapphirine-quartz, hypersthene-sillimanite, spinel-quartz, osumilite-garnet and exceptionally calcic mesoperthite on a regional scale are apparently unique features of the Napier Complex (Sheraton et al., 1980; Ellis et al., 1980; Ellis, 1980). The presence of osumilite and garnet-osumilite in pelitic rocks indicates metamorphism beyond the stability of the chemically equivalent assemblage cordierite-hypersthene-K feldspar-quartz (Ellis et al., 1980). The latter assemblage has hitherto been regarded as diagnostic of intermediate to high pressure granulite terrains, so that the Enderby Land mineral assemblages define a new subfacies of the granulite facies for pelitic rocks.

**TABLE 1: Comparison of some petrological features of the Napier and Rayner Complexes (modified after Sheraton et al., Table 1, 1980)**

| | Napier Complex | Rayner Complex |
|---|---|---|
| Mafic rocks | Opx-Cpx-Pl-(Qtz-Gt) Gt often secondary | Opx-Cpx-Pl-Hbld-(Qtz-Gt) Gt may be replaced by secondary Pl-Opx intergrowth |
| Pelitic rocks | Primary Assemblages | |
| | Os-(Gt) (and in other assemblages) Sa-Qtz, Sp-Qtz, Hy-Sill-Qtz Gt-Sill-Qtz-Kf/Meso Rutile common Gt typically inclusion free | Gt-Cd-Hy-Qtz-(Bi-Pl-Kf) Gt-Cd-Sill-Qtz-(Bi-Pl-Kf) Gt-Bi-Sill-Qtz-Kf-Pl Rare Ky Ilmenite, Bi are Ti-bearing phases Gt may have inclusions of Bi-Sill, or Opx with oxides, feldspar, quartz |
| | Secondary Assemblages | |
| | Gt-Cd-Sill-Qtz Gt-Cd-Hy-Qtz Secondary and zoned rims to primary Opx have less $Al_2O_3$ than primary Opx Fe-Mg $K_D$ increases from core to rim of mineral pairs | Secondary Opx often has higher $Al_2O_3$ than primary Opx Fe-Mg $K_D$ does not markedly increase towards rims |
| Mafic Intrusives | Fresh dolerite dykes abundant | No unmetamorphosed dolerite dykes |
| Acid Intrusives | Bi granite; Opx granite, granodiorite and tonalite Minor pegmatite | Bi granite; Opx-Hbld granite Abundant pegmatite |

Mineral abbreviations—Bi, biotite; Cpx, clinopyroxene; Cd, cordierite; Gt, garnet; Hbld, hornblende; Hy, hypersthene; Kf, K-Feldspar; Ky, kyanite; Meso, mesoperthite; Opx, orthopyroxene; Os, osumilite; Pl, plagioclase; Sa, sapphirine; Sill, sillimanite; Sp, spinel.

A notable feature of these granulites is the clear petrographic distinction between the highest temperature, primary mineral assemblages and the secondary mineral coronas formed during lower temperature reactions after the peak conditions of metamorphism (see Ellis et al., 1980; Sheraton et al., 1980).

A variety of mineral reactions is apparent in these coronas reflecting the instability of sapphirine-quartz at low temperatures (see Ellis et al., 1980; Sheraton et al., 1980) such as:

$$\text{Sa} + \text{Qtz} = \text{Gt} + \text{Cd} + \text{Sill}$$
$$\text{Sa} + \text{Qtz} = \text{Hy} + \text{Sill}$$
$$\text{Sa} + \text{Qtz} = \text{Cd}$$
$$\text{Sa} + \text{Qtz} = \text{Cd} + \text{Sill}$$

The first reaction clearly attests to the retrograde lower temperature recrystallisation of these rocks in the stability field of coexisting garnet-cordierite (see experimental data of Hensen and Green, 1973; Ellis et al., 1980, figure 15). The second reaction is evident in rocks of the northern Scott Mountains whereas the presence of only cordierite coronas is more typical of the Tula Mountains (Sheraton et al., 1980). This was interpreted by Sheraton et al., (1980) as evidence that cooling took place under slightly higher pressure conditions in the Scott Mountains. Also, the presence of primary hypersthene-sillimanite at a number of localities in the northern Scott Mountains-Khmara Bay area is also compatible with higher pressures, lower temperatures or probably both during metamorphism compared with the Tula Mountains. This pressure variation has since been quantitatively confirmed by Harley (this volume).

Mafic rocks usually have the intermediate pressure granulite facies assemblage clinopyroxene (Cpx) + orthopyroxene (Opx) + plagioclase (Plag) ± Qtz ± magnetite (Mt) ± ilmenite (Ilm), although primary garnet is present in some rocks of olivine-normative composition. The distribution of garnet-bearing assemblages in Napier Complex mafic rocks is given by Sheraton et al., (1980).

The metamorphism of a wide range in mafic granulite rock types from one locality in the Napier Complex is described by Ellis and Green (ms). Besides the formation of primary garnet at or near the peak conditions of metamorphism, different generations of secondary garnet formed during the subsequent cooling of these granulites over a temperature range from ~900°C to 640°C with only slightly decreasing pressure (8-10 kb maintained over the cooling interval).

## The Rayner Complex

The Rayner Complex represents a lower grade reworking of the Napier Complex granulites under higher $P_{H_2O}$ conditions (Sheraton et al., 1980). This generally resulted in the complete recrystallisation of the Napier Complex rocks, but relic high pressure mineral assemblages are preserved in rocks which underwent less intense deformation.

The Rayner Complex consists of medium pressure upper amphibolite to granulite facies rocks characterised by the hydrous minerals biotite (Bi) and hornblende (Hbld) and two feldspars instead of calcic mesoperthite in rocks of appropriate composition (Table 1). They are considerably more migmatitic than the Napier Complex, where late granite and pegmatite veins are common (Sheraton et al., 1980).

Evidence that the high pressure conditions of the Napier metamorphism were maintained during early stages of the Rayner Complex metamorphism are found at Forefinger Point, on the western side of the Rayner Glacier, Casey Bay. Here, the high pressure assemblage hypersthene-sillimanite-quartz coexists stably with biotite. By comparison, in the northern Scott Mountains-Khmara Bay area of the higher grade Napier Complex biotite is absent from the hypersthene-sillimanite quartz assemblage.

Apart from the rocks at Forefinger Point, clear petrographic evidence for the retrogression of Napier Complex anhydrous to Rayner Complex hydrous assemblages is rare. An example is seen at the island just east of Demidov Island, Casey Bay, where pyrope rich garnet (Mg/Mg + Fe = 0.5) is corroded and replaced by an intergrowth of cordierite-biotite. Generally, complete recrystallisation has occurred and no Napier Complex minerals have survived this process. The garnet in the Rayner Complex pelitic rocks is much more Fe-rich than that of the Napier Complex.

The presence of K feldspar-sillimanite-quartz in pelitic rocks of the Rayner Complex implies metamorphism beyond the stability of muscovite-quartz, although rare inclusions of muscovite in garnet porphyroblasts from Mount Yuzhnaya provide evidence for an even lower grade history. The assemblages cordierite-hypersthene-K feldspar-quartz and garnet-cordierite (-biotite) are the lower temperature equivalents of the osumilite and sapphirine-quartz assemblages of the Napier Complex (Table 1). Sillimanite is the diagnostic aluminosilicate polymorph, although kyanite (Ky) does occur locally, as in the assemblage Gt-Cd-Bi-Qtz-Ky on the island southwest of Hydrographer Island, Casey Bay.

The assemblages Gt-Bi, Gt-Bi-Sill, Bi-Sill-Cd and Gt-Bi-Opx together with variable amounts of K-feldspar, plagioclase, magnetite and ilmenite occur in pelitic rocks of the Rayner Complex. Rutile is the characteristic Ti-accessory in Napier Complex pelites but is very rare in the Rayner Complex whereas ilmenite and titaniferous biotite are more common. Where there has been intense penetrative deformation, the garnets are often flattened and aligned parallel to the schistosity, suggesting that they are syntectonic. In contrast to garnet of the Napier Complex, that in Rayner Complex pelitic rocks usually contains abundant inclusions of biotite, biotite-sillimanite, biotite-orthopyroxene with or without opaque oxide, feldspar and quartz.

At Mount Sibiryakov in the Nye Mountains, garnet porphyroblasts contain inclusions of biotite and sillimanite, whereas the groundmass consists of biotite-cordierite (-feldspar-quartz). This suggests that during growth of the garnet $P_{H_2O}$-T conditions of metamorphism changed such that the following reaction occurred:

$$\text{Bi} + \text{Sill} + \text{Qtz} \rightarrow \text{Gt} + \text{Cd} + \text{Kf} + H_2O$$

This is one of the most important reactions in high grade pelitic rocks and proceeds from left to right with increase in temperature or decrease in $P_T$ or $P_{H_2O}$ (Figure 2).

Mineral compositions for host and inclusion pairs at garnet cores and towards the rims reveals systematic changes in compositions of minerals to more Fe-rich compositions. The regular shift in tie line between coexisting garnet-biotite, garnet-orthopyroxene and garnet-cordierite pairs is good evidence that equilibrium has been maintained between minerals during P-T conditions of metamorphism. The Gt-Bi $K_D$ ($K_D = Fe/Mg^{Gt}.Mg/Fe^{Bi}$ decreases slightly from core to rim of garnets, suggesting that there was a slight temperature increase during garnet growth in some rocks, however, this is uncertain in view of the fact that $Fe^{3+}$ was not calculated for the biotites.

The assemblage Gt-Cd-Opx-Qtz is present at Mount Lira in the Western Rayner Complex. The garnet is replaced at its margin by a

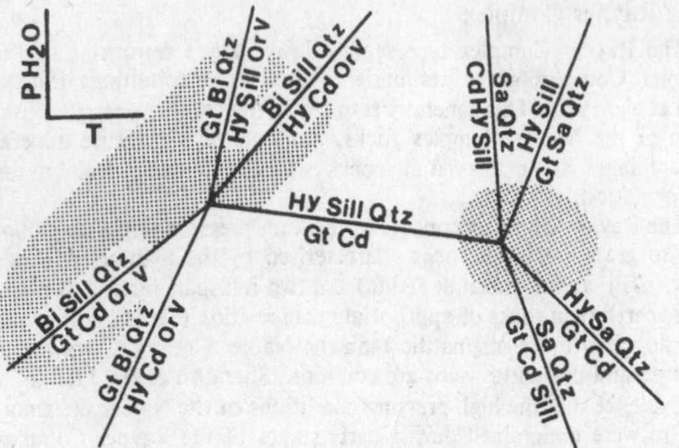


**Figure 2. Schematic representation of the possible univariant reactions in the KFMASH model pelitic system (after Hess, 1969; Hensen and Green, 1973) applicable to the Enderby Land mineral assemblages (ignoring osumilite). The shaded region at high temperature depicts the anhydrous mineral assemblages of the Napier Complex (see Ellis et al., 1980). The shaded region at low temperature depicts the hydrous assemblages of the Rayner Complex.**

symplectite of intergrown cordierite and orthopyroxene, due to the following reaction:

$$Gt + Qtz \rightarrow Cd + Opx$$

This reaction proceeds from left to right with increasing temperature or decreasing pressure. The secondary orthopyroxene in the intergrowth contains a higher amount of $Al_2O_3$ than the primary orthopyroxene (Table 2).

The felsic gneisses consist of layered orthopyroxene-biotite-quartz-feldspar (orthoclase perthite and plagioclase) and in some layers green-brown hornblende. Clinopyroxene occurs in the more melanocratic layers and garnet is locally abundant. Much of this gneiss in the Rayner Glacier area is of probable intrusive origin.

In mafic rocks, Opx-Cpx-Plag assemblages are widespread, but greenish brown hornblende is considerably more abundant than in similar rocks of the Napier Complex. Pyroxene is relatively uncommon in the Molodezhnaya area, although the presence of orthopyroxene in both felsic and mafic gneisses indicates that granulite facies conditions had just been reached.

The high pressure assemblage Gt-Cpx-Opx-Hbld-Plag ± Qtz is present at several localities in the Robert Glacier area of Kemp Land (eg. Turbulence Bluffs) and at Mount Currie and Mount Robinson in the Raggatt and Nye Mountains in western Enderby Land (Sheraton et al., 1980). At Turbulence Bluffs, a medium grained equigranular rock consists of the granulite facies assemblage Gt-Cpx-Plag-Hbld-Qtz-Mt-(Opx). The garnet is replaced by a symplectite of intergrown plagioclase-orthopyroxene, which can be explained by the following reaction:

$$Gt + Cpx + Qtz \rightarrow Plag + Opx$$

which proceeds from left to right with increasing temperature or decreasing pressure. The secondary orthopyroxene has a higher $Al_2O_3$ content than the primary orthopyroxene in this rock (Table 2).

**TABLE 2: Mineral compositions for pelitic and mafic rocks retaining both high and low pressure assemblages from the Rayner Complex**

| | Gt 1A | Opx 1A | Opx 1B | Gt 2A | Opx 2A | Gt 2C | Opx 2C |
|---|---|---|---|---|---|---|---|
| $SiO_2$ | 39.10 | 51.00 | 49.88 | 38.52 | 51.14 | 38.16 | 51.51 |
| $Al_2O_3$ | 22.58 | 4.65 | 5.79 | 21.24 | 1.39 | 20.93 | 1.61 |
| FeO | 25.64 | 21.02 | 20.27 | 26.34 | 28.46 | 26.64 | 27.46 |
| MnO | | | | 1.16 | 0.34 | 2.68 | 0.75 |
| MgO | 11.08 | 24.09 | 23.69 | 5.08 | 17.64 | 3.84 | 18.16 |
| CaO | 1.60 | | | 7.66 | 1.03 | 7.75 | 0.50 |

1. Mt Lira sample 77284665 Gt-Cd-Opx-Plag-Qtz
2. Turbulence Bluffs sample 77283584 Gt-Cpx-Plag-Opx-Hbld-Qtz-Ilm
A Primary mineral
B Corona Opx intergrown with Cd (Mg 86) at Garnet rim
C Corona Opx intergrown with Plag at Garnet rim.
Mineral compositions determined using Si(Li) energy dispersive detector attached to an SEM, University of Tasmania.

## P-T CONDITIONS OF METAMORPHISM

### *The Napier Complex*

Many of the mineral coronas in the Napier Complex pelitic rocks are the result of continuous rather than discontinuous reactions, mainly Mg-Fe and (Mg, Fe) Si = 2Al exchange equilibria taking place in response to decreasing temperature (Ellis et al., 1980; Ellis, 1980). Zoning as well as primary-secondary mineral compositions for Gt-Cd, Gt-Opx and Gt-Cpx pairs in the Napier Complex pelitic and mafic rocks generally have garnets with more Fe-rich rims and the Fe-Mg $K_D$'s increase from cores to rims (or from primary to secondary grains). Also, secondary and rim orthopyroxenes have lower $Al_2O_3$ contents than primary orthopyroxenes (Ellis, 1980; Ellis and Green, ms; Harley, this volume). These features are consistent with decreasing temperature from the peak of metamorphism, and when quantified using experimentally calibrated (Ellis and Green, 1979; Harley, 1981; Harley and Green, 1982; Harley, this volume) or thermodynamically calculated (Ellis, 1980) geothermometers and geobarometers, indicate a considerable cooling interval with only a slight decrease in pressure for these granulites (Table 3). The peak conditions of metamorphism of the Napier Complex were estimated to be 8-10 kb, 900-980°C on a regional scale (Ellis, 1980; Harley, this volume). The formation of different generations of secondary garnet in mafic rocks of the Napier Complex (Ellis and Green, MS) occurred at temperatures down to about 640°C. Thus a temperature interval of cooling of about 300°C can be calculated for these granulites.

### *The Rayner Complex*

The calculated P-T conditions of metamorphism of the Rayner Complex, based on the new data are listed in Table 3. Temperature variations during growth of garnet porphyroblasts can be calculated from the compositions of biotite and orthopyroxene inclusions within garnets as well as those at the margins and in the groundmass of the rocks.

In general, the P-T conditions calculated for the Rayner Complex metamorphism are lower than those of the Napier Complex. Where deformation caused complete recrystallisation then only one P-T estimate can be made, as the minerals are homogeneous in compostition. Such rocks record low pressures of metamorphism. In those cases (eg. Mount Lira, Turbulence Bluffs) where deformation has not resulted in complete recrystallisation to a low pressure assemblage, the rocks have textural evidence for both primary and secondary assemblages. The secondary assemblages formed at lower pressures with nearly constant or only slightly decreasing temperatures compared to the primary assemblages (Table 3). The secondary orthopyroxene contains greater $Al_2O_3$ than the primary orthopyroxene coexisting with garnet. This is the opposite to the compositional trends reported from the Napier Complex orthopyroxenes.

## DISCUSSION

Besides different P-T conditions of metamorphism the P-T paths of metamorphism differ markedly for the two complexes (Figure 3). That of the Napier Complex was one of the near "isobaric cooling"—a large decrease in temperature due to thermal relaxation from a metamorphic peak wih only slight decrease in pressure (Ellis, 1980; Ellis and Green, MS; Harley, this volume). It is believed that the highest temperature mineral assemblages are preserved in the Napier Complex because of the subordinate effect of post-metamorphic deformation following the peak conditions of metamorphism (see Ellis, 1980). This type of P-T cooling-uplift path differs markedly from that evident in many other granulite terrains and by inference suggests that following the peak conditions of metamorphism the Archaean Napier Complex crust was in approximate isostatic equilibrium with the underlying mantle (Ellis, 1980; Harley, this volume).

It is suggested here that the petrologic features of the Rayner Complex can be explained by a P-T path during metamorphism and deformation being one of"near isothermal uplift". That is, uplift was a more important factor than cooling (Figure 3). This change in P-T path implies a major orogenic reworking of the Archaean craton at ~ 1000 Ma, which according to James and Black (1981) had essentially been cratonised at ~ 2500 Ma. It is possible that the whole of the Napier Complex underwent excavation towards the Earth's surface at ~ 1000 Ma, but deformation and introduction of water was restricted to the margins of the Napier Complex, as now defined by the Rayner Complex. Any fluids generated during the early prograde

**TABLE 3: Mineral assemblages and P-T estimates for the Rayner Complex**

| Locality | Sample Number | Assemblage | Core T°C/Pkb | Rim/ Groundmass | Corona/ Intergrowth |
|---|---|---|---|---|---|
| Condom Hills | 77284166 | Gt Bi Sill Kf Plag Qtz |  | $590^{1}/3^{a}$ |  |
|  |  | Cd Bi Sill Kf Plag Qtz |  |  |  |
| Mt Currie | 77284544 | Gt Cpx Plag Qtz | $646^{2}/4.7^{b}$ |  |  |
|  | 77284418 | Gt Opx Plag Kf Qtx |  | $653^{3}/4.4^{c}$ |  |
| Mt Denholm | 77283568 | Gt Bi Sill Plag Kf Qtz | $730^{1}/5.8^{a}$ |  |  |
| Island S.W. of | 77284702 | Gt Bi Cd Ky Qtz Graphite | $551^{1}_{3}$, $534^{4}_{c}/7.5^{e}$ |  |  |
| Hydrographer Is. | 77284486 | Gt Hy Sill Kf Qtz Rutile | $810^{3}_{1}/5.2^{c}$ |  |  |
| Mt Lira | 77284174 | Gt Cd Opx Bi Qtz | $624^{1}_{3}$, $808^{3}_{c}/4.5^{c}$ | $601\text{-}676^{1}$ | $750^{3}/3.3^{c}$ |
|  | 77284665 | Gt Cd Opx Qtz | $779^{3}/5.6^{c}$ |  | $794^{3}/4.6^{c}$ |
| Mt Sibiryakov | 77284608 | Gt Bi Plag Kf Qtz | $560\text{-}619^{1}$ |  |  |
|  | 77284609 | Gt (Bi Sill inclusions) | $513^{1}$ | $560^{1}$ |  |
|  |  | Bi Cd Kf Qtz Op Zircon |  |  |  |
| Mt Underwood | 77283570 | Gt Bi Kf Plag Qtz Ilm Graphite |  | $630^{1}$ |  |
| Nunatak N.E. of Mt Yuzhnaya | 77284573 | Gt Bi Sill Kf Qtz | $502\text{-}591^{1}$ | $585\text{-}602^{1}/3.5^{a}$ |  |
| Mt Yuzhnaya Turbulence Bluffs | 77283583 | Gt Cpx Opx Plag Hbld Qtz Ilm | $716^{2}$, $737^{3}_{3}/11^{c}$, $8.4^{d}$ |  |  |
|  | 77283584 | Gt Cpx Plag Hbld Qtz Ilm | $819^{2}$, $760^{3}/9.5\text{-}11^{c}$ |  | $660^{3}/3.0\text{-}5.0^{c}$, $5.9^{d}$ |

Geothermometers
1. Gt-Bi Thompson (1976)
2. Gt-Cpx Ellis and Green (1979)
3. Gt-Opx Harley (1981)
4. Gt-Cd Thompson (1976)

Geobarometers
a. Gt-Plag-Sill-Qtz Newton and Haselton (1981)
b. Gt-Plag-Cpx-Qtz Perkins and Newton (1981)
c. Gt-Opx Harley and Green (1982)
d. Gt-Plag-Opx-Qtz Perkins and Newton (1981)
c. Gt-Cd Thompson (1976)

history of the Napier Complex were lost from these rocks (see arguments of Sheraton, et al., 1980; Ellis, 1981). If such rocks are non porous and non permeable, then they will have different deformational characteristics from those undergoing extensive rehydration as in the Rayner Complex.

Reasons for the development of a Proterozoic mobile zone in which metamorphic recrystallisation and hydration take place during uplift are not understood. Certainly the contemporaneous and combined effects of hydration and deformation are not coincidental. The source of the introduced water is a considerable problem, as it is in other regions of hydrous retrogression of granulite terrains. Several sources exist:

(1) Water is derived from the mantle, either directly by dehydration reactions or else from water formerly dissolved in basalt magmas crystallising near the base of the crust.
(2) Dehydration of subducted oceanic crust.
(3) Convection of meteoric water from the Earth's surface.
(4) Underplating of sediments beneath the margin of the Napier Complex.
(5) The presence of a significant amount of unrecognised new sediment as part of the Rayner Complex prior to the ~ 1000 Ma metamorphism.

It is not possible to choose from these alternatives, but the underplating of younger, wet sediments beneath the margins of the Napier Complex is a possibility. Underplated sediments would result in dehydration reactions, the liberated water would then react with the overthrust Napier Complex granulites. This mechanism would also explain the inferred marked change in tectonic style from an essentially stabilised craton to one of the vertical tectonics evident in the Rayner mobile zone in which contemporaneous deformation, hydration and metamorphism occurred during uplift along the southern margins of the Napier Complex.

*Acknowledgements* I thank J.W. Sheraton for discussing the conclusions in this paper.

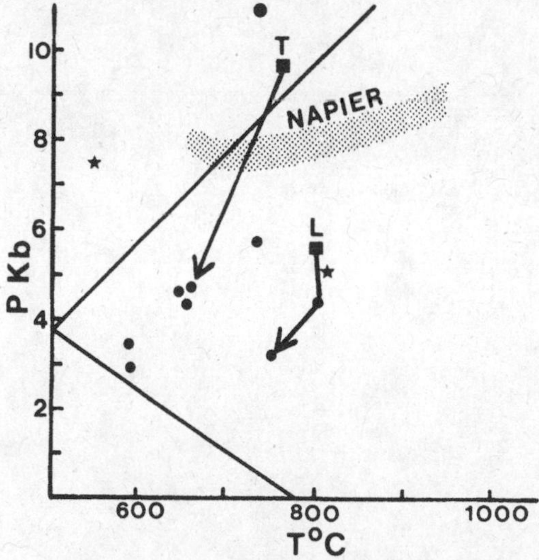


**Figure 3. Summary of the P-T conditions of metamorphism of the Rayner Complex compared to the cooling path of the Napier Complex. The shaded band corresponds to the Scott Mountains and Casey Bay areas of the Napier Complex cooling path (from Harley, Figure 6, this volume). The arrows connect the high presure estimates for the cores of garnets (square symbols) to the low pressure estimates of formation of the symplectite rims replacing garnets at Turbulence Bluffs (T) and Mt Lira (L). The stars are P-T estimates for two samples from the one outcrop from an island S.W. of Hydrographer Island-one kyanite bearing, one sillimanite bearing. Aluminosilicate phase boundaries after Holdaway (1971).**

## REFERENCES

BLACK, L.P. and JAMES, P.R., 1979: Preliminary isotopic ages from Enderby Land, Antarctic. *Geol. Soc. Aust., J. 26,* 266-7.

DALLWITZ, W.B., 1968: Co-existing sapphirine and quartz in granulite from Enderby Land, Antarctica. *Nature, 219,* 476-7.

ELLIS, D.J., 1980: Osumilite-sapphirine-quartz granulites from Enderby Land, Antarctica: P-T conditions of metamorphism, implications for garnet-cordierite equilibria and the evolution of the deep crust. *Contrib. Mineral. Petrol., 74,* 201-10.

ELLIS, D.J. and GREEN, D.H., 1979: An experimental study of the effect of Ca upon garnet-clinopyroxene Fe-Mg exchange equilibria. *Contrib. Mineral. Petrol., 71,* 13-22.

ELLIS, D.J., SHERATON, J.W., ENGLAND, R.N., and DALLWITZ, W.B., 1980: Osumilite-sapphirine quartz granulites from Enderby Land, Antarctica—Mineral assemblages and reactions. *Contrib. Mineral. Petrol., 72,* 123-43.

ELLIS, D.J. and GREEN, D.H., (MS): Garnet-forming reactions in high pressure crustal mafic granulites from Enderby Land, Antarctica—Implications for geothermometry and geobarometry.

GREW, E.S., 1978: Precambrian basement at Molodezhnaya Station, East Antarctica. *Geol. Soc. Am., Bull. 89,* 801-13.

GREW, E.S., 1980: Sapphirine + quartz association from Archaean rocks in Enderby Land, Antarctica. *Am. Mineral., 65,* 821-36.

GREW, E.S., 1981: Granulite-facies metamorphism at Molodezhnaya Station, East Antarctica. *J. Petrol., 22,* 297-336.

GRIKUROV, G.E., ZNACHKO-YAVORSKY, G.A., KAMENEV, E.N. and RAVICH, M.G., 1976: *Explanatory notes to the geological map of Antarctica (scale 1:5 000 000).* Research Institute of Arctic Geology, Leningrad.

HARLEY, S.L., 1981: Garnet-orthopyroxene assemblages as pressure-temperature indicators, an experimental study with applications to granulites from Enderby Land, Antarctica. Unpublished Ph.D thesis, University of Tasmania.

HARLEY, S.L., (this volume): Regional geobarometry-geothermometry and metamorphic evolution of Enderby Land, Antarctica.

HARLEY, S.L. and GREEN, D.H., 1982: Garnet-orthopyroxene barometry for granulites and peridotites. *Nature, 300,* 697-701.

HENSEN, B.J. and GREEN, D.H., 1973: Experimental study of the stability of cordierite and garnet in pelitic compositions at high pressures and temperatures. III Synthesis of experimental data and geological applications. *Contrib. Mineral. Petrol., 38,* 151-66.

HESS, P.C., 1969: The metamorphic paragneses of cordierite in pelitic rocks. *Contrib. Mineral. Petrol., 24,* 191-207.

HOLDAWAY, M.J., 1971: Stability of andalusite and the aluminium silicate phase diagram. *Am. J. Sci., 271,* 97-131.

JAMES, P.R. and BLACK, L.P., 1981: A review of the structural evolution and geochronology of the Archaean Napier Complex of Enderby Land, Australian Antarctic Territory. *Geol. Soc. Aust., Spec. Pub., 7,* 71-83.

KAMENEV, E.N., 1972: Geological Structure of Enderby Land; *in* Adie, R.J. (ed.) *Antarctic Geology and Geophysics,* Universitetsforlaget, Oslo, 579-83.

KAMENEV, E.N., 1975: The geology of Enderby Land (in Russian). *Acad. Sci. USSR, Comm. Antarct. Res. Rep. 14.*

NEWTON, R.C. and HASELTON, H.T., 1981: Thermodynamics of the garnet-plagioclase-$Al_2SiO_5$-quartz geobarometer; *in* Newton, R.C. (ed.) *Thermodynamics of Minerals and Melts.* Advances in Physical Geochemistry, Volume 1, Springer-Verlag, New York, 131-47.

PERKINS, D. III and NEWTON, R.C., 1981: Charnockite geobarometers based on coexisting garnet-pyroxene-plagioclase-quartz. *Nature, 292,* 144-6.

SHERATON, J.W., OFFE, L.O., TINGEY, R.J. and ELLIS, D.J., 1980: Enderby Land, Antarctic—An unusual Precambrian high grade metamorphic terrain. *Geol. Soc. Aust., J., 27,* 305-17.

SHERATON, J.W. and BLACK, L.P., 1981: Geochemistry and geochronology of Proterozoic tholeiite dykes of East Antarctica: Evidence for mantle metasomatism. *Contrib. Mineral. Petrol., 78,* 305-17.

SHERATON, J.W. and BLACK, L.P., (in prep.): Geochemistry of Precambrian gneiss: relevance for the evolution of the East Antarctic Shield.

THOMPSON, A.B., 1976: Mineral reactions in pelitic rocks: II: Calculation of some P-T-X(Fe-Mg) phase relations. *Am. J. Sci., 276,* 425-54.

# REGIONAL GEOBAROMETRY-GEOTHERMOMETRY AND METAMORPHIC EVOLUTION OF ENDERBY LAND, ANTARCTICA

S.L. Harley[1], *Department of Geology, University of Tasmania, GPO Box 252C, Hobart, Tasmania, 7001, Australia.*

[1]*Present Address: Institute fur Kristallographic and Petrographie, E.T.H.-Zentrum, CH-8092, Zurich, Switzerland.*

*Abstract* Experimentally calibrated geothermometers and geobarometers have been used to estimate the temperature and pressure conditions of metamorphism and further evolution of granulites from the Archaean Napier Province, Enderby Land, Antarctica. These granulites display a variety of exsolution, recrystallisation and corona textures, resulting both from cooling and overprinting metamorphism. Compositional data on distinct generations of phases and on zoning patterns in co-existing phases have been used to estimate:

(a) peak metamorphic conditions obtained during the first and second major deformation phases ($D_1$ and $D_2$),
(b) cooling paths from this peak, and
(c) metamorphic conditions prevalent around the time of the third deformation ($D_3$, 2500 Ma).

The granulites were metamorphosed at 900-950°C and 7-10 kb during and subsequent to the major tectonothermal episodes at 3100 Ma-3000 Ma ($D_1$-$D_2$). Rocks presently exposed at the surface in southwestern Enderby Land indicate a regional increase in pressures of this metamorphism towards the Scott Mountains-Casey Bay region, where minimum crustal thicknesses of 30 km were attained. Subsequent to the peak granulite facies event, the Napier Province granulites underwent a prolonged phase of near-isobaric cooling, at depth, to P-T conditions of 5-8 kb and 600-700°C during $D_3$ at 2500 Ma. The near-isobaric pressure-temperature-time (P-T-t) path suggests that the Napier Province acted as a "stable" craton as early as 3000 Ma, and that the major crustal thickening event associated with $D_1$ preceded the thermal peak represented by the earliest recognised metamorphism.

The Napier Complex (Ravich and Kamenev, 1975) is an Archean granulite facies terrain (⅓3000 Ma) of some 100,000 km² area, making up most of Enderby Land, Antarctica. This polymetamorphic province is bounded by the Proterozoic (1100 Ma) Rayner Complex (Ravich and Kamenev, 1975; Sheraton et al., 1980). A schematic summary of the structural-tectonic history is presented in Table 1. The subject of this paper is the relationship between the structural-tectonic history of the Napier Province and the evolution of the metamorphic conditions experienced by the granulites. The aims are to determine the regional variations in pressure-temperature conditions at the metamorphic peak and to establish P-T-t paths for the granulites during the Archaean tectonic episodes (Table 1).

Previous detailed petrological studies of granulites from the Napier Province (Ellis et al., 1980; Ellis, 1980; Grew, 1980) have concentrated mainly on unusual mineral parageneses often found as thin layers within a well-layered granulite sequence, the Tula Series (Ravich and Kamenev, 1975). These studies generally indicate that conditions of 7-9 kb and 900-980°C were attained at the metamorphic peak.

In this study, the more common garnet-orthopyroxene (+ plagioclase + quartz) pelitic and mafic to ultramafic pyroxene granulites which occur both in the Tula Series and in the poorly-layered to massive gneissic group, the Raggatt series (Ravich and Kamenev, 1975), are used to derive estimates of the metamorphic conditions and to determine the Archaean P-T-t path. Many granulite samples exhibit polymetamorphic and multi-deformational textures which must be examined in detail before the significance of any P-T estimates can be fully understood.

**TABLE 1: Generalised deformation-metamorphic history of the Napier Province**

| Event | Age (Ma) | Deformation Features | Grade of Metamorphism & Textures |
|---|---|---|---|
| D1 | 3100-3000 | flat lying isoclinal folds layer-parallel foliation strong elongation lineation symmetric folds with ribbon textures in axial regions. Sillimanite lineations. | Granulite facies conditions. Coarse elongate or polygonal granuloblastic textures. |
| D2 | 3000 | assymmetric tight folds & parasitic folds common. Inclined axial planes, no penetrative foliation. Rodding & sillimanite lineations, boudinage. | Granulite facies conditions. Coarse granuloblastic textures, some pretectonic elongate porphyroblasts. Cooling, granulite facies garnet coronas, exsolutions. |
| D3 | 2500 | upright open to close folds large scale domes & basins weak vertical foliation localised intense deformation and contemporaneous shearing. Mylonite zones, probably reactivated a number of times. Localised deformation. | Granulite to upper Amphibolite facies. Very variable recrystallisation to fine grainsizes. Development of garnet coronas and some biotite and amphibole. |
| D4 | 1100 | Local overprinting by Rayner province deformation. | Amphibolite facies. Biotite, amphibole & cordierite bearing coronas. |

Sources of data: Sheraton et. al. (1980); James & Black (1982); Black et. al (in press); and this study.

A general outline of the primary metamorphic textures, criteria for overprinting relationships and ages, and description of some corona and reaction textures developed in these granulites is given below. This is followed by a summary of the results of the application of various geothermometry-geobarometry techniques to the granulites.

## PETROLOGICAL FEATURES

### Primary Metamorphic Textures and Mineralogy

Typical garnet-orthopyroxene bearing pelitic and siliceous granulites consist of variable modal abundances of garnet, orthopyroxene, quartz and feldspars. Thin orthopyroxene-sillimanite layers also occur in quartzitic gneisses. Normally, the granulites have granuloblastic lobate to polygonal textures and show little preferred orientation of orthopyroxene or quartz, however some well layered samples exhibit strong elongate (ribbon) quartz and garnet-orthopyroxene textures. These textures are attributed to growth of the minerals during $D_1$ (James and Black, 1982) or continued equilibration after $D_1$ or $D_2$ (Table 1). Indeed, relict elongate quartz grains are sometimes recrystallised to finer (0.5-1 mm) equigranular quartz probably during or post-$D_2$. These primary metamorphic textures have been overprinted by a variety of corona and recrystallisation textures attributed both to post-$D_2$ cooling and to $D_3$ deformation.

The broad range in chemistry of co-existing garnets and orthopyroxenes in typical samples is illustrated in Figure 1. The assemblages cover a wide range in composition in terms of $X_{Mg}$ (Mg/Mg + Fe) (Figure 1) and $X_{Ca}$ (Ca/Ca + Mg + Fe), although many of the samples are magnesian ($X_{Mg}^{opx}$ = 70-75, $X_{Mg}^{ga}$ = 50) and low in calcium. Compositions based on core analyses of primary grains yield $K_{D_{Fe\text{-}Mg}}^{ga\text{-}opx}$ values of 1.98 to 3.1, depending on $X_{Ca}^{ga}$ and geographic location. Core analyses

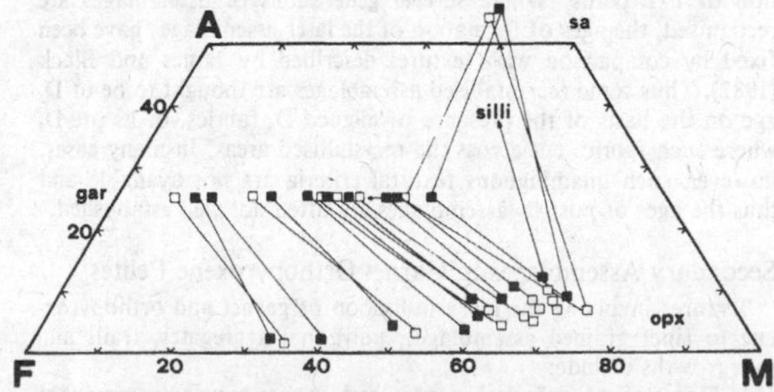


**Figure 1. Partial AFM ($Al_2O_3$-FeO-MgO) diagram showing the range of compositions and zoning patterns in garnets and orthopyroxenes from typical pelitic granulites from the Napier Province. gar = garnet, opx = orthopyroxene, sa = sapphirine, closed symbols = cores of primary grains, open symbols = rims and recrystallised grains.**

of primary metamorphic orthopyroxenes indicate alumina contents of up to 11 wt%.

The mafic and ultramafic pyroxene granulites considered here usually consist of orthopyroxene(opx)-clinopyroxene(cpx) plus plagioclase and ilmenite or magnetite. K-feldspar and quartz may also be present in leucocratic types. Coarse grained (0.5-2 mm) lobate and polygonal granuloblastic primary textures are typical, although elongate ($D_1$) textures may occasionally be preserved. The development of the granuloblastic textures was probably completed in the waning stages of $D_2$ (James and Black, 1982). A number of Fe-rich samples containing inverted primary metamorphic pigeonite have been studied as the compositions of such pigeonites can give minimum temperature estimates for the earliest recorded thermal peak.

Representative compositions of pyroxenes from the mafic granulites are presented in Figure 2. Garnets, formed in secondary coronas in these granulites, are also plotted on this diagram. Tie-lines connect co-existing pyroxene rims, and the less-calcic clinopyroxene cores lie along these tie-lines or within the cpx-opx-garnet three phase triangles in the samples where coronas are developed (Figure 2). Tie-line lengthening and rotation occur as a result of core to rim zoning and exsolution in the pyroxenes.

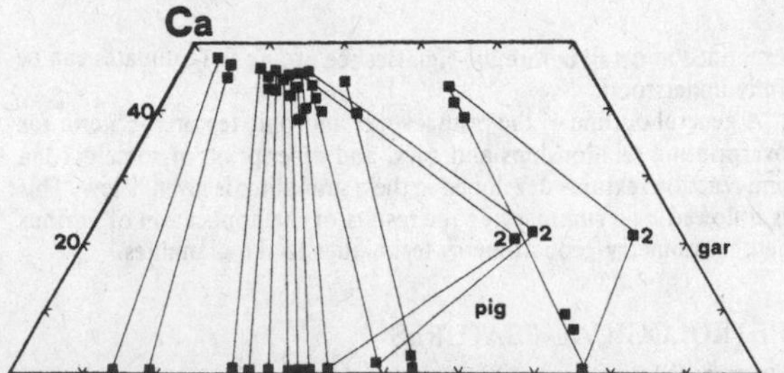


**Figure 2. Ca-Mg-Fe diagram illustrating the range in compositions of pyroxenes and secondary garnets from some mafic pyroxene granulites from the Napier Province. Pyroxene compositions have been corrected for non-quadrilateral components (e.g. NaAl, $CaAl_2$), so that this is a Di-Hd-Fs-En projection for the pyroxenes. pig = reconstructed pigeonite composition, gar = garnet, with the numeral two signifying the secondary development of these garnets.**

## Coronas, Recrystallisation Textures, and Zoning Relationships

Coronas and textures indicating either reaction or simply partial recrystallisation of the early-formed granulite assemblages provide evidence for overprinting metamorphism and information on the P-T history of these rocks. In addition, if the development of the various textures and secondary assemblages can be fitted into an absolute as well as relative time scale then a P-T-t history can be deduced. The textures listed below often involve the formation of distinct generations of minerals, particularly garnet and orthopyroxene. Samples having only minor degrees of recrystallisation and compositional zoning in the co-existing phases have also been used in the construction of P-T paths. Where several generations of assemblages are recognised, the ages of formation of the later assemblages have been fixed by comparison with textures described by James and Black (1982). Thus some recrystallised assemblages are thought to be of $D_3$ age on the basis of the presence of aligned $D_3$ fabrics, or as pre-$D_3$ where such fabrics cut across the recystallised areas. In many cases, however, such unambiguous textural criteria are not available and thus the ages of post-$D_2$ assemblages are often not well established.

### Secondary Assemblages in Garnet-Orthopyroxene Pelites

Textures involving the recrystallisation of garnet and orthopyroxene to finer grained assemblages, polygonal aggregates, trails and overgrowths include:

(1) Fine-grained euhedral garnet and garnet-quartz intergrowths which rim earlier coarse-grained lobate garnet or orthopyroxene and plagioclase.

(2) Kinked and disaggregated coarse orthopyroxene traversed by trails of fine granular euhedral garnet. Such trails may cross kink-band boundaries ($D_3$) and partially rim orthopyroxene.

(3) Partially recrystallised orthopyroxenes which form aggregates with euhedral garnet and polygonal feldspars. These aggregates may rim earlier ragged orthopyroxene or garnet and may also rim the garnet-quartz symplectites (1).

Other corona, rim, and exsolution textures where secondary garnet-orthopyroxene assemblages have developed include:

(4) Garnet exsolution lamellae within orthopyroxene grains and sometimes continuous with garnet rims between these grains.

(5) Thin lamellae garnet rims between polygonal-lobate orthopyroxene grains.

(6) Continuous or granular garnet-quartz coronas between adjacent orthopyroxene and plagioclase.

(7) Fine grained fibrous sillimanite-orthopyroxene-garnet aggregates which occur as coronas or rims on earlier orthopyroxene and sillimanite. These coronas often occur in samples which also show $D_3$ recrystallisation features in quartz-feldspar layers.

(8) Orthopyroxene rims on adjacent garnet and sapphirine.

In the majority of cases where new generations of garnet and orthopyroxene are formed, and in cases where zoning also occurs in the earlier ($D_1$-$D_2$) grains, the following changes in the mineral compositions are observed (Figure 1):

(a) garnet zones to lower $X^{ga}_{Mg}$ from cores to rims, while $X^{ga}_{Ca}$ remains nearly constant. Secondary garnets have the lowest $X^{ga}_{Mg}$ values; and

(b) orthopyroxene zones to higher $X^{opx}_{Mg}$ and lower $X^{opx}_{Al}$ (Al/2) adjacent to rims when occurring with garnet. Orthopyroxene co-existing with sillimanite zones to lower $X^{opx}_{Al}$ at near constant $X^{opx}$. Recrystallised orthopyroxenes usually have the highest $X^{opx}_{Mg}$ and the lowest $X^{opx}_{Al}$ values (see also Ellis et al., 1980).

These zoning patterns and variations between the primary and recrystallised grains lead to an increase in $K_{D_{Fe\text{-}Mg}}^{ga\text{-}opx}$ between cores, rims, and secondary assemblages, consistent with cooling.

### Secondary Assemblages and Textures in Pyroxene Granulites

Some examples of the wide variety of corona textures inferred to have developed in the mafic granulites during $D_3$ are given below:

(1) In Fe-rich pigeonite-quartz-plagioclase samples, pigeonite inversion and subsequent extensive exsolution of calcic pyroxene (Figure 2) is followed by deformation resulting in undulose grains, bent and kinked exsolution lamellae, and disaggregated areas ($D_3$). Secondary garnet occurs along grain boundaries, as finger-like lamellae cutting across earlier exsolutions or as lamellae along kink-band boundaries. The formation of this garnet (post-$D_3$) results in a decrease in alumina contents and an increase in $X^{opx}_{Mg}$ of the exsolved and recrystallised pyroxenes.

(2) Garnet is developed as lobate rims on magnetite and pyroxenes in two-pyroxene + magnetite + plagioclase samples. Pyroxenes zone to more magnesian rims adjacent to this garnet (Figure 2).

(3) In two-pyroxene intermediate granulites, garnet develops as lobate grains which form trails and networks along plagioclase and plagioclase-orthopyroxene grain boundaries, or forms coronas with polygonal plagioclase and clinopyroxene or orthopyroxene.

(4) In quartz-bearing leucocratic two-pyroxene granulites, feldspar rich layers were partially recrystallised to finer aggregates and trails during $D_3$. Garnet forms, along with quartz and secondary pyroxenes, as elongate lens-shaped clots of bead-like and amoeboid grains and coronas on and embaying earlier pyroxenes and opaques. These coronas are often elongate within $S_3$.

## PRESSURE-TEMPERATURE ESTIMATES

### Methods

For any chosen garnet-orthopyroxene metapelite, up to three P-T points have been estimated:

(1) P-T estimates for the metamorphic peak have been obtained using core or averaged core analyses of primary metamorphic garnet and orthopyroxene. This peak may correlate with $D_2$, or with $D_1$ in cases where $D_1$ textures are preserved.

(2) P-T estimates for post-$D_2$ conditions have been obtained from rim analyses of co-existing primary garnet-orthopyroxene pairs.

(3) Minimum P-T data, relating to late post-$D_2$ cooling or to recrystallisation accompanying $D_3$, have been obtained from analyses of recrystallised garnets and orthopyroxenes.

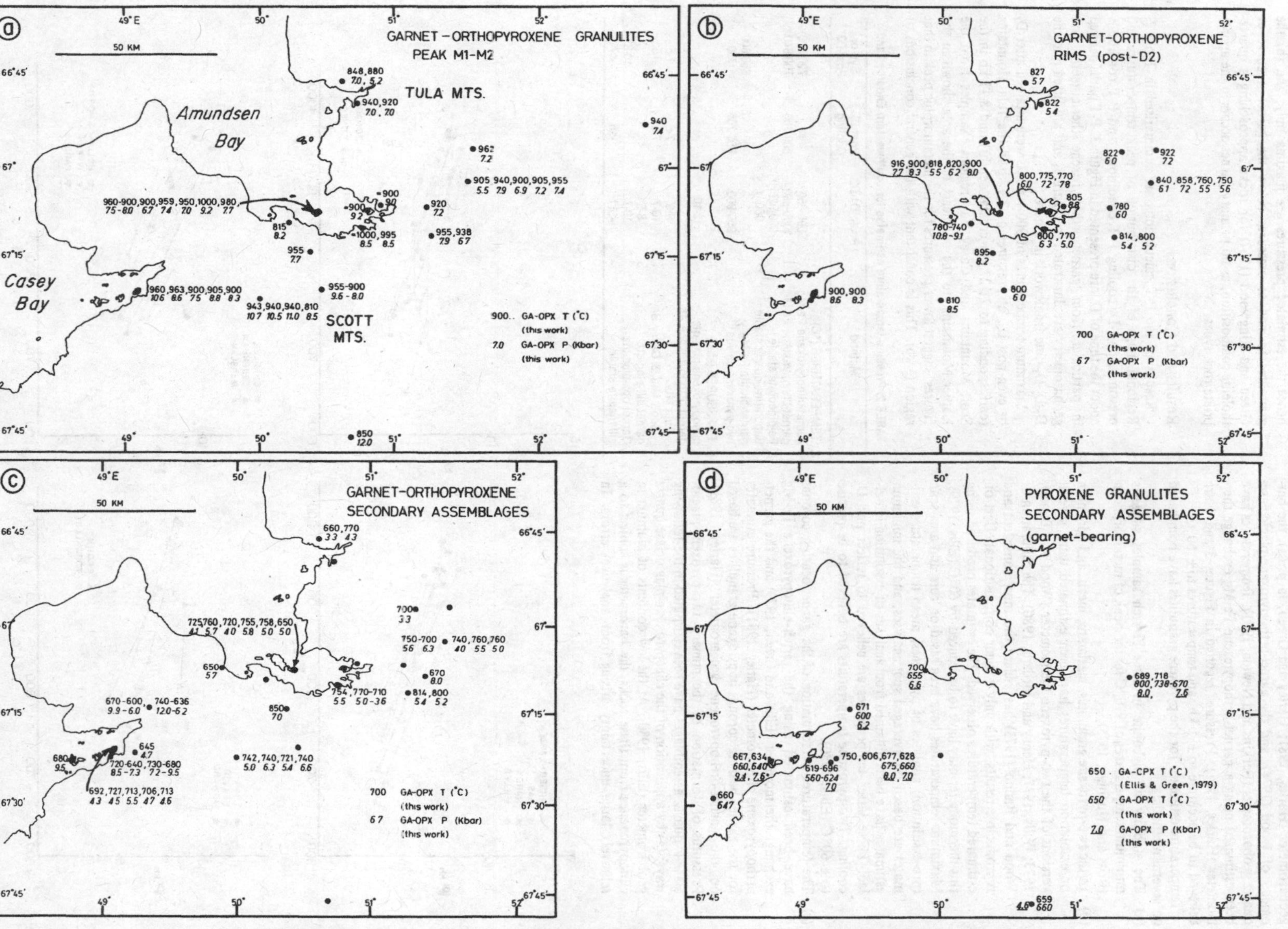


**Figure 3.** Maps of southwestern Enderby Land showing (a) the regional distribution of P-T conditions in the earliest recorded metamorphic peak; (b) during post-$D_2$ cooling; (c and d) recorded by secondary corona and recrystallised assemblages. P-T data based on garnet-orthopyroxene geothermometry and barometry are estimates using the equations of Harley (1981).

Pressures have been estimated using a revised garnet-orthopyroxene geobarometer (Harley, 1981; Harley and Green, in press), the barometer of Wood (1974), and the garnet-orthopyroxene-plagioclase-quartz geobarometer (Perkins and Newton, 1981). Temperatures have been estimated using a garnet-orthopyroxene Fe-Mg exchange thermometer (Harley, 1981). Pressures reported in Figures 3 and 4 are believed to be accurate to ± 1 kb, and temperatures to ± 60°C.

Temperature estimates for the pyroxene granulites have been based on several methods:

(a) The stability of pigeonite in some Fe-rich samples suggests minimum temperatures of 970°C for the peak of metamorphism (Ross and Huebner, 1975).

(b) In the two-pyroxene granulites, maximum, mean, and minimum or exsolution temperatures have been estimated using several versions of the two-pyroxene thermometer (Wood and Banno, 1973; Wells, 1977; Henry and Medaris, 1980). The calibration of Wood and Banno (1973) has yielded the most consistent and reproducible results, with only a minor compositional effect on estimated temperatures observed. The estimates produced using this thermometer may have to be reduced by 60°C (Ellis, 1979). Maximum temperatures have been based on core and area-scan (pre-exsolution) analyses of pyroxenes, mean temperature estimates are based on averaged spot analyses, and the minimum estimates have been obtained from analyses of exsolution lamellae. These latter temperatures are believed to reflect post-$D_2$ cooling. Two-pyroxene temperatures are believed to be precise to ± 60°C (Ellis, 1980).

(c) The temperatures of formation of the garnet-bearing coronas have been estimated using the garnet-clinopyroxene Fe-Mg exchange thermometer (Ellis and Green, 1979) and the garnet-orthopyroxene thermometer (Harley, 1981). Pressure estimates for the garnet-bearing coronas in pyroxene granulites are based on the garnet-orthopyroxene geobarometer (Harley, 1981). Pressures of formation of the primary two-pyroxene-plagioclase + quartz assemblages have been estimated in the following ways: (a) by application of the clinopyroxene-plagioclase-quartz geobarometer (Ellis, 1980). At the low contents of alumina in clinopyroxenes from these rocks, this barometer is subject to a relatively large uncertainty arising from analytical errors. In addition, many of the samples studied do not contain quartz and hence the estimated pressures are maxima only; and (b) the granulites formed at pressures below the garnet-in curve of Green and Ringwood (1967) for rocks approaching a quartz-tholeiite composition, as no primary garnet occurs in the appropriate rock types.

## Results and Conclusions

Results of the P-T calculations are summarised in Table 2. Regional maps of the distribution of peak temperatures and pressures, post-$D_2$ cooling P-T conditions, and P-T conditions about the time of $D_3$ are presented in Figure 3. P-T paths derived in particular from garnet-orthopyroxene thermometry and geobarometry are illustrated in Figure 4, and are fitted into the $D_1$-$D_2$-$D_3$ time framework in Figure 5.

Maximum temperatures of 900-1000°C were attained after $D_1$ or even post-$D_2$, at pressures of 7-9 kb in the Tula Mountains (corresponding to 21-27 km crustal depths) and 8-11 kb in the Scott Mountains and Casey Bay regions. Samples from the Napier Mountains, to the northeast of the area shown in the figures, suggest lower pressures for the metamorphic peak in that region (5 kb). This is consistent with the general conclusions of

**TABLE 2: Pressure-temperature estimates for southwestern Enderby Land**

| Method | Peak, D1-D2 | Post-D2 Cooling | Syn-or Post-D3 |
|---|---|---|---|
| a. TEMPERATURES (°C) | | | |
| garnet-clinopyroxene Fe-Mg | 890-850 | | 720-620 |
| garnet-clinopyroxene Fe-Mg | 1000-860 | 800-700 | 700-640 |
| Fe-pigeonite stable | 960-925 | | |
| Sapphirine-quartz stable | 950 | | |
| garnet-biotite Fe-Mg | | | 650-600 |
| two-pyroxene solvus | 950-900 | 850-750 | 750 |
| b. PRESSURES (kbar) | | | |
| clinopyroxene-plag-qtz | 12-10 | | |
| garnet-orthopyroxene Al: | | | |
| Tula Mts. | 8-6 | 7-5 | 7-5 |
| Scott Mts. & Casey Bay | 10-7 | 7-5 | 9-5 |
| garnet-sill-plag-qtz | 8-6 | | |
| Garnet-opx-plag-qtz | 10-7 | | 10-8 |
| sillimanite stable | <11 | <8 | <7 |

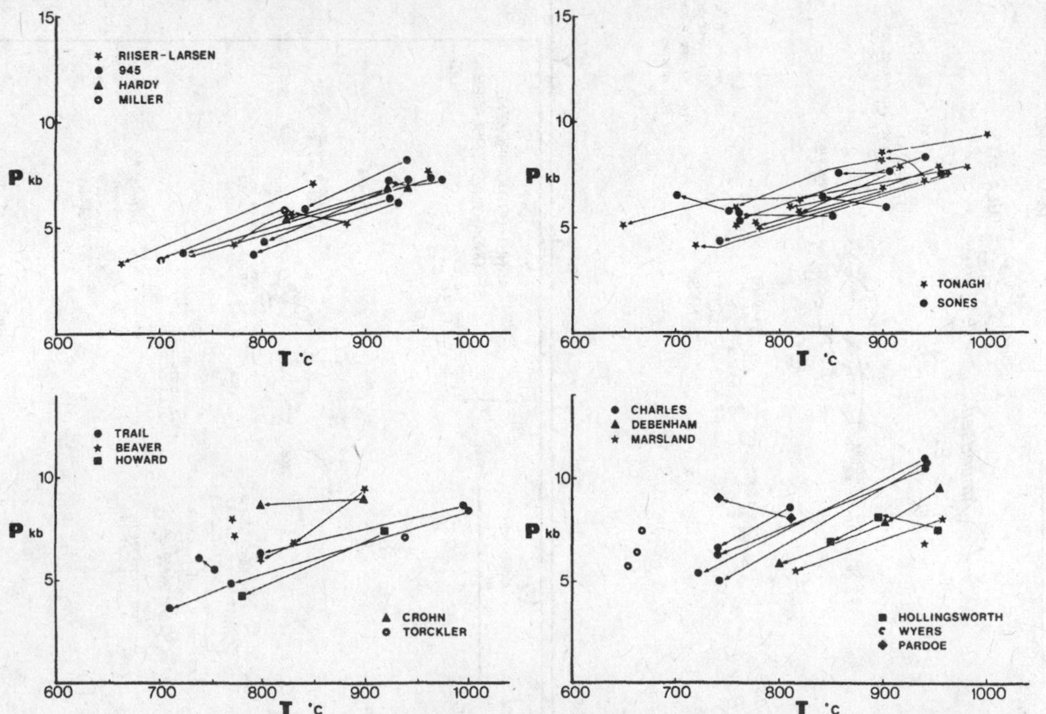


**Figure 4. P-T trajectories for garnet-orthopyroxene bearing granulites. Specific P-T points are based on core (highest T), rim, and recrystallised grain (lowest T) analyses. Points joined by arrowed lines to show the direction of P-T variation with time. Names refer to various localities in Enderby Land.**

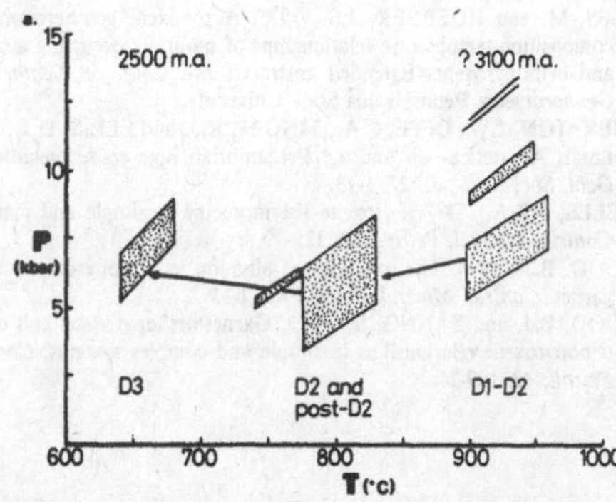


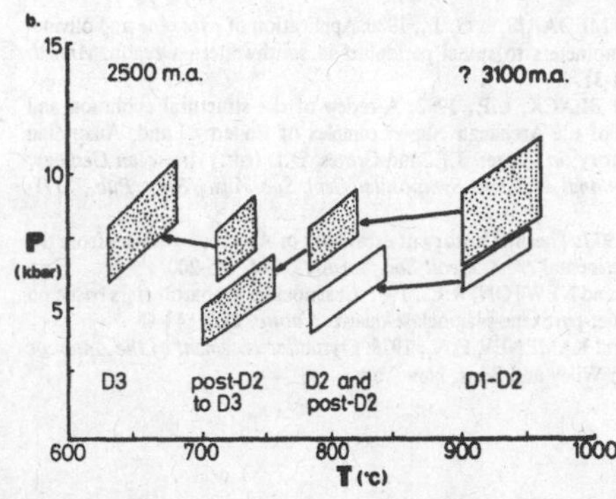


**Figure 5. P-T-t paths for granulites from the Napier Province. Diamond shaped areas result from the intersections of the ga-opx barometer (Harley and Green, 1982) Al-isopleths with the ga-opx thermometer (Harley, 1981) constant-$K_D$ lines. Deformation phases and ages as defined in Table 1. Arrows connect P-T data obtained from samples from adjacent or nearby localities. (a) P-T data for localities from the Tula Mountains and Amundsen Bay. (b) P-T data for localities in the Scott Mountains and Casey Bay areas. Open diamonds are P-T data based on samples from Mount Hollingsworth.**

Sheraton et al. (1980) that pressures of metamorphism increased towards the southwest. Calculated pressures of 8-10 kb at temperatures of 950°C for orthopyroxene-sillimanite parageneses are consistent with recent experimental studies on such assemblages (Annersten and Seifert, 1981).

Secondary recrystallised assemblages related to post-$D_2$ cooling and to $D_3$ deformation and cooling yield P-T estimates of 5-7 kb and 700-750°C. Higher pressures of 6-9 kb at 600-700°C are suggested for the formation of $D_3$ and syn-shearing coronas in the pyroxene granulites. This increase in pressure may have been the result of crustal thickening associated with $D_3$, or may be an artifact of the geothermometry and geobarometry techniques used, as the mineral compositions in the coronas in the pyroxene granulites are outside the range of experimental calibrations.

The P-T-t paths presented here suggest an extremely protracted period of near-isobaric cooling from 950-600°C for the Napier Province granulites subsequent to the major deformation phases, $D_1$ and $D_2$. This kind of post-metamorphic history, previously suggested for the Napier Complex by Ellis (1980), is unusual in comparison to the many terrains where rapid uplift and exhumation have followed the thermal peak (e.g. England and Richardson, 1977; O'Hara, 1977). It is worthwhile to note some important points regarding the P-T-t paths constructed here:

(1) The paths for any particular sample are only well defined at the plotted P-T points. Lines drawn between these points are simply connecting lines and are not necessarily the real P-T trajectories. The absence of significant melting and the continued stability of sillimanite are evidence, however, against large fluctuations in P between the plotted points.

(2) Time-synchroneity of the P-T conditions between samples cannot, in many cases, be unequivocally established. This is especially true for P-T data derived from corona assemblages where no textural evidence for timing in relation to $D_3$ is present, and thus P-T data presented in Figure 3 are only divided into peak, rim and secondary populations.

(3) The schematic P-T-t paths are regarded as a combination of both cooling paths from the thermal peak in $D_2$, and paths generated by the successive overprinting of separate metamorphic and deformation episodes, each producing characteristic mineral assemblages.

(4) Only the average rates of cooling can be ascertained. The relationship between this average rate and the real cooling rates over particular temperature intervals is not known. It is not suggested here that cooling from 950-700°C took place linearly over a 500 Ma time interval, but rather that some coronas formed at the lower temperatures soon after the $D_2$ event, when cooling rates may have been greater. Steady temperatures may have then prevailed until thermal perturbation associated with $D_3$.

The P-T data presented here for granulites from southwestern Enderby Land suggest that this area became an essentially stabilised continental craton following the $D_1$-$D_2$ events at 3000 Ma, and was affected by later adjustments in $D_3$ and by younger events (Table 1) not considered in this work. Minimum crustal thicknesses of 20-30 km were attained after the $D_1$-$D_2$ events and it is likely that greater thicknesses were common as the granulites now exposed at the surface, and underlain by further continental crust, comprise metasediments probably laid down on sialic basement. As the P-T path mapped out by a body of rock reflects its uplift and erosion history (England and Richardson, 1977), it is proposed that the near-isobaric path deduced for these granulites was achieved because the earliest recognised thermal peak post-dated the major deformation and crustal thickening events. A near-isobaric cooling history could only occur if the uplift and erosion resulting from crustal thickening in $D_1$-$D_2$ was largely complete by the time of the thermal maximum.

*Acknowledgements* I thank the officers and expeditioners of the Antarctic Division for their help and support during my field season in Antarctica. This work has benefitted from the encouragement of Prof. D.H. Green and from discussions with Dr L. Black, Dr D. Ellis, Dr B. Hensen, Dr P. James, M. Sandiford, Dr J. Sheraton and Dr R. Tingey. Drs Ellis and Sheraton also provided me with some of the samples which form this work.

## REFERENCES

ANNERSTEN, H. and SEIFERT, F., 1981: Stability of the assemblage orthopyroxene-sillimanite-quartz in the system $MgO$-$FeO$-$Fe_2O_3$-$Al_2O_3$-$SiO_2$-$H_2O$. *Contrib. Mineral. Petrol., 77,* 158-65.

BLACK, L.P. and JAMES, P.R., 1979: Preliminary isotopic ages from Enderby Land, Antarctica (abstr.). *Geol. Soc. Aust. J., 26,* 266-67.

BLACK, L.P., JAMES, P.R. and HARLEY, S.L., 1982: The geology and geochronology of reputedly ancient rocks at Fyfe Hills, Enderby Land, Antarctica. *Precambrian Res.* (in press).

ELLIS, D.J., 1979: Granulites from Enderby Land, Antarctica, application of experimentally determined cation partition data to estimation of pressures and temperatures of metamorphism. Unpublished Ph.D thesis, University of Tasmania. 300pp.

ELLIS, D.J., 1980: Osumilite-sapphirine-quartz granulites from Enderby Land, Antarctica; P-T conditions of metamorphism, implications for garnet-cordierite equilibria and the evolution of the deep crust. *Contrib. Mineral. Petrol., 74,* 201-10.

ELLIS, D.J. and GREEN, D.H., 1979: An experimental study of the effect of Ca upon Garnet-clinopyroxene Fe-Mg exchange equilibria. *Contrib. Mineral. Petrol., 71,* 13-22.

ELLIS, D.J., SHERATON, J.W., ENGLAND, R.N. and DALLWITZ, W.B., 1980: Osumilite-sapphirine-quartz granulites from Enderby Land, Antarctica: mineral assemblages and reactions. *Contrib. Mineral. Petrol., 72,* 123-43.

ENGLAND, P.C. and RICHARDSON, S.W., 1977: The influence of erosion upon the mineral facies of rocks from different metamorphic environments. *Geol. Soc. Lond., J., 134,* 201-13.

GREEN, D.H. and RINGWOOD, A.E., 1967: An experimental investigation of the gabbro to eclogite transformation and its petrological applications. *Geochim. Cosmochim. Acta., 31,* 767-833.

GREW, E.S., 1980: Sapphirine + quartz association from Archaean rocks in Enderby Land, Antarctica. *Am. Mineral., 65,* 821-36.

HARLEY, S.L., 1981: Garnet-orthopyroxene assemblages as pressure-temperature indicators, an experimental study with applications to granulites from Enderby Land, Antarctica. Unpublished Ph.D thesis, University of Tasmania. 366pp.

HARLEY, S.L. and GREEN, D.H., 1982: Garnet-orthopyroxene barometry for granulites and peridotites. *Nature,* (in press).

HENRY, D.J. and MEDARIS, L.G. Jr., 1980: Application of pyroxene and olivine-spinel geothermometers to spinel peridotites in southwestern Oregon. *Am. J. Sci., 280A*, 211-31.

JAMES, P.R. and BLACK, L.P., 1982: A review of the structural evolution and geochronology of the Archaean Napier complex of Enderby Land, Australian Antarctic Territory; *in* Glover, J.E. and Groves, D.I. (eds.) *Archaean Geology: Second International Archaean Symposium. Geol. Soc. Aust., Spec. Pub., 7*, 71-83.

O'HARA, M.J., 1977: Thermal history of excavation of Archaean gneisses from the base of the continental crust. *Geol. Soc. Lond. J., 134*, 185-200.

PERKINS, D.III, and NEWTON, R.C., 1981: Charnockite geobarometers based on co-existing garnet-pyroxene-plagioclase-quartz. *Nature, 292*, 144-46.

RAVICH, M.G. and KAMENEV, E.N., 1975: *Crystalline basement of the Antarctic Platform.* John Wiley and Sons, New York.

ROSS, M. and HUEBNER, J.S., 1975: A pyroxene geothermometer based on composition-temperature relationships of naturally occurring augite, pigeonite, and orthopyroxene. Extended abstracts, *Int. Conf. on Geothermometry and Geobarometry.* Pennsylvania State University.

SHERATON, J.W., OFFE, L.A., TINGEY, R.J. and ELLIS, D.J., 1980: Enderby Land, Antarctica—an unusual Precambrian high-grade metamorphic terrain. *Geol. Soc. Aust., J., 27*, 1-18.

WELLS, P.R.A., 1977: Pyroxene thermometry in simple and complex systems. *Contrib. Mineral. Petrol., 62*, 129-39.

WOOD, B.J., 1974: The solubility of alumina in orthopyroxene co-existing with garnet. *Contrib. Mineral. Petrol., 46*, 1-15.

WOOD, B.J. and BANNO, S., 1973: Garnet-orthopyroxene and orthopyroxene-clinopyroxene relationships in simple and complex systems. *Contrib. Mineral. Petrol., 42*, 109-24.

# Sm-Nd ISOTOPIC SYSTEMATICS OF ENDERBY LAND GRANULITES: EVIDENCE FOR THE REDISTRIBUTION OF Sm AND Nd DURING METAMORPHISM

M.T. McCulloch, *Research School of Earth Sciences, Australian National University, PO Box 4, Canberra, ACT, 2601, Australia.*

L.P. Black, *Bureau of Mineral Resources, PO Box 378, Canberra City, ACT 2601, Australia.*

**Abstract** Measurements of $^{143}Nd/^{144}Nd$ and $^{147}Sm/^{144}Nd$ are reported for whole rocks and mineral separates from granulites of the Napier Complex at Fyfe Hills to (1) determine the primary age of formation of the complex, and (2) evaluate the effects of high grade metamorphism on the Sm-Nd isotopic system. Samples ranging in composition from charnockite to leuconorite and pyroxene bearing gabbros yield a whole rock Sm-Nd isochron age of $3050 \pm 210$ Ma and an initial $^{143}Nd/^{144}Nd$ ratio of $0.50776 \pm 14$ ($\epsilon_{Nd}$(3050 Ma) $= -2.2 \pm 2.0$). The negative $\epsilon_{Nd}$ value and the presence of geologically induced dispersion in the data suggest that the isochron age does not represent the time of primary crystallisation of the complex but instead indicates a time of later redistribution of Sm and Nd and re-equilibration of $^{143}Nd/^{144}Nd$ ratios. This probably occurred during high grade granulite facies metamorphism which has also been dated at ca 3050 Ma by Rb-Sr and U-Pb zircon studies (Black and James, 1979). The Sm-Nd systematics have also been examined in mineral systems. Co-existing plagioclase, clinopyroxene and apatite in two adjacent samples define an approximately linear array corresponding to an age of ca 2500 Ma. This indicates that redistribution of Sm and Nd and re-equilibration of $^{143}Nd/^{144}Nd$ ratios also occurred on an inter-mineral scale during the upper amphibolite to lower granulite facies metamorphism at ca 2500 Ma (Black and James, 1979).

Due to the resetting of the Sm-Nd system on both whole rock and mineral scales, the primary crystallisation age of the complex is not well constrained by the present data, although it is clearly $> 3050$ Ma. If it is assumed that the complex was derived initially from a chondritic reservoir (Nd (T) = 0), evolution of the negative $\epsilon_{Nd}$ value of $-2.2$ with the observed Sm/Nd ratios requires a pre-history of $\leq 300$ Ma. This implies a primary age of $\leq 3350$ Ma. However substantially older primary ages can be inferred if the source reservoirs had (Nd(T) > 0 and/or substantial depletions in the Sm/Nd ratio occurred in whole rocks during the granulite facies metamorphism at 3050 Ma.

# GEOLOGY AND PETROLOGY OF PRINCE OLAV COAST, EAST ANTARCTICA

Y. Hiroi, *Institute of Earth Sciences, Kanazawa University, Kanazawa, 920, Japan.*

K. Shiraishi, *National Institute of Polar Research, 9-10 Kaga I-chome, Itabashi-ku, Tokyo 173, Japan.*

Y. Nakai, *Institute of Geology, Aichi University of Education, Kariya 448, Japan.*

T. Kano, *Department of Geology and Mineralogy, Yamaguchi University, Yamaguchi, Japan.*

S. Yoshikura, *Department of Geology, Kochi University, Kochi, Japan.*

*Abstract* The main rock types exposed in the Prince Olav Coast, East Antarctica, are amphibolite, biotite-hornblende gneiss, garnet-biotite gneiss, and granitic gneiss. The last is most abundant as migmatitic rocks containing xenoliths and schlieren. This gneissic and migmatitic complex has been deformed into a NW-SE trending fold system and cut by small masses of granite and pegmatite. Mylonite zones are found locally, where slightly metamorphosed basaltic rocks occur as discordant masses. Sillimanite + K-feldspar association is present throughout the coast. Metastable kyanite usually occurs as inclusions in garnet and plagioclase in most of the sillimanite-bearing rocks. The Prince Olav Coast is divided into two metamorphic zones A and B. The lower grade zone A to the east is characterised by staurolite and Ca-poor amphiboles. The higher grade zone B to the west differs from zone A by a pleonaste + garnet + aluminium silicates association, the breakdown product of staurolite, and the occurrence of orthopyroxene. Thus, the metamorphic grade of the Prince Olav Coast ranges from upper amphibolite (zone A) to lower granulite facies (zone B). The relation between zones A and B is explained by a gradual increase in metamorphic grade from east to west during a single event, probably 1000 Ma ago. The occasional occurrence of andalusite in rocks closely associated with granite and pegmatite, probably 500 Ma in age, along with local re-equilibration of Mg-Fe distribution between garnet and biotite and of Ca distribution between garnet and plagioclase in sillimanite-quartz-bearing rocks indicates metamorphism contemporaneous with the emplacement of granite and pegmatite under conditions differing from the earlier upper amphibolite to lower granulite facies conditions.

The Prince Olav Coast (lat. 68-69°S, long. 40-45°E), East Antarctica, is underlain by granitic and metamorphic rocks, the latter containing kyanite, staurolite, and Ca-poor amphiboles and belonging to amphibolite facies of intermediate pressure type (Ravich and Kamenev, 1975; Kamenev, 1982). In recent years, geologic and petrographic studies of the bedrock exposures along the coast have been published by Japanese geologists (Yanai and Ishikawa, 1978; Suzuki and Moriwaki, 1979; Suzuki, 1979a, b; Kanisawa et al., 1979; Yoshikura et al., 1979; Nakai et al., 1980, 1981; Kanisawa and Yanai, 1982; Nishida et al., 1982). However, no paper has been published that synthesises present knowledge on the geology and petrology of the Prince Olav Coast. We intend to summarise available data here, especially that concerning physical conditions of metamorphism. We also aim to reveal metamorphic history on the bases of chemical variation of minerals, field relations, and regional geology.

## GEOLOGY

The basement rocks exposed in the Prince Olav Coast consist largely of well layered gneisses, migmatitic rocks containing xenoliths and schlieren, granite and pegmatite.

### Well-layered Gneisses

The well-layered gneisses are divided into two groups

1. Basic to intermediate rocks. These rocks consist of clinopyroxene amphibolite, garnet amphibolite, and biotite-hornblende gneiss, forming layers 10 to several tens of metres thick intercalated with pelitic to psammitic rocks.
2. Pelitic to psammitic rocks. Biotite gneiss, garnet-biotite gneiss, anthophyllite-garnet-biotite gneiss, and various kinds of aluminous biotite gneisses are included in this rock group. In aluminous biotite gneisses, corundum, spinel (pleonaste), kyanite, sillimanite, staurolite, cordierite, and/or garnet occur according to bulk rock composition and metamorphic grade, as will be discussed below. Andalusite occasionally occurs in kyanite-sillimanite bearing rocks closely associated with granite and pegmatite, suggesting local re-equilibration during events following the main metamorphism.

The well-layered gneisses enclose lenses and thin layers of ultrabasic and calcareous rocks. Calcareous rocks usually contain garnet, plagioclase, quartz, and clinopyroxene (including fassite, Kanisawa and Yanai, 1982) with or without epidote. Wollastonite has not been found.

### Migmatitic Rocks

The migmatitic rocks are granitic to granodioritic in composition and often contain muscovite and magnetite as accessory minerals. They commonly enclose xenoliths of well layered basic gneiss and biotite-rich schlieren. The boundary between the migmatitic rocks and biotite gneiss is often ambiguous. Detailed geologic study, however, reveals the migmatitic rocks form concordant to discordant masses of various sizes. The migmatitic rocks are abundant throughout the Prince Olav Coast. Together with the well layered gneisses, they have been deformed into a fold system. The axial planes of the folds trend NW-SE. The migmatitic rocks are of anatectic origin produced during high grade metamorphism.

### Granite and Pegmatite

Small discordant masses (mostly dykes) of white, pink, and reddish brown granite and pegmatite are found throughout the Prince Olav Coast, for example in the area around Lützow-Holm Bay. The granite and pegmatite also have been reported at Molodezhnaya Station (Grew, 1980). They may be about 500 Ma in age.

In addition to the rocks described above, there occur basaltic rocks as dykes in and near the mylonite zones (Akebono Rock and Cape Hinode). The dyke rocks are slightly metamorphosed, the degree of recrystallisation being closely related to the distance from the masses of granite and pegmatite.

## METAMORPHIC GRADE AND ZONES

Regional occurrence of several diagnostic minerals and mineral parageneses is shown in Figure 1. The sillimanite + K-feldspar association is found throughout the Prince Olav Coast. In a silica-deficient rock the corundum + K-feldspar + muscovite association is found. Kyanite occurs as inclusions in garnet and plagioclase in most of the sillimanite-bearing rocks, and is obviously metastable in these rocks. The ubiquitous occurrence of kyanite and sillimanite throughout the Prince Olav Coast indicates the metamorphic facies series of the rocks is of kyanite-sillimanite type.

Staurolite occurs in rocks from the eastern part of the Prince Olav Coast, being included in garnet and plagioclase. It is not in equilibrium with other minerals. On the other hand, a spinel + garnet + aluminium silicates (sillimanite or kyanite) association is found in the western part of the coast. At Cape Hinode a staurolite + spinel + garnet + sillimanite association is found. Thus, the Prince Olav Coast is divided into two metamorphic zones A and B by the following reaction (Figure 1).

1. Staurolite = spinel + garnet + aluminium silicate(s).

Moreover, Ca-poor amphiboles (anthophyllite and cummingtonite) occur in the eastern part of the coast, while orthopyroxene (± quartz) occurs in the western part of the coast (Figure 1). Consequently the following reaction is also inferred to have occurred at or near the isograd between zones A and B.

2. Ca-poor amphiboles = orthopyroxene + quartz.

The occurrence of orthopyroxene in zone B indicates zone B belongs to the granulite facies. Thus, the relation between amphibolite-facies

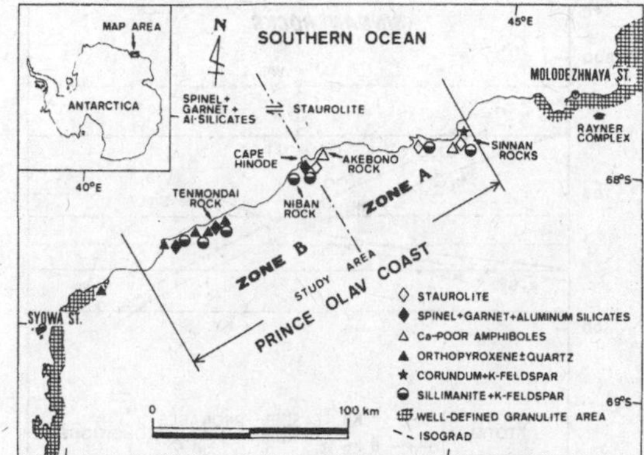


**Figure 1. Map showing regional occurrence of diagnostic minerals and mineral associations and two metamorphic zones A and B. Sources of data are Hiroi and Shiraishi (in prep.), Hiroi et al. (in prep.), Shiraishi (in prep.), and Suzuki (personal communication, 1982).**

rocks (zone A) and granulite-facies rocks (zone B) in the Prince Olav Coast is explained by a gradual increase in metamorphic grade during a single metamorphic event. The following are examples of mineral parageneses in each zone.

*Zone A*

Staurolite ± plagioclase ± garnet
Corundum + K-feldspar + muscovite + plagioclase + biotite ± sillimanite.
Sillimanite + cordierite + garnet + biotite + plagioclase + quartz
Anthophyllite + garnet + biotite + plagioclase + quartz
Cummingtonite + hornblende + biotite + plagioclase ± quartz

*Zone B*

Spinel + garnet + aluminium silicate(s) ± plagioclase
Orthopyroxene + garnet + biotite + plagioclase + quartz
Orthopyroxene + hornblende + biotite + plagioclase ± quartz
Olivine + spinel + orthopyroxene + hornblende + biotite
Spinel + orthopyroxene + hornblende + biotite + plagioclase

## PHYSICAL CONDITIONS OF METAMORPHISM

Electron microprobe analyses of minerals reveal the following two features:

1. By and large, minerals are chemically homogeneous, and compositional variations of minerals among rocks are systematic.
2. Garnet cores are more Ca-rich than rims next to plagioclase, and plagioclase rims next to garnet are more Ca-rich than cores in sillimanite-quartz bearing rocks (Figure 2). Garnet rims next to biotite are more Fe-rich than cores, and biotite next to garnet is more Mg-rich than that in the matrix (Figure 3).

The first feature suggests chemical equilibrium of the rocks during earlier upper amphibolite to lower granulite-facies metamorphism. The second feature suggests local re-equilibration during later events such as emplacement of granite and pegmatite, as will be discussed below.

To estimate physical conditions of metamorphism, the following are utilised in addition to the experimental data (Richardson, 1968; Fonarev and Korolkov, 1980) for reactions (1) and (2) (Figures 4 and 5):

1. Garnet-biotite geothermometer (Thompson, 1976; Goldman and Albee, 1977; Ferry and Spear, 1981).
2. Garnet-plagioclase-sillimanite-quartz geobarometer (Newton and Haselton, 1981).
3. Kyanite-sillimanite relationship (Richardson et al., 1969; Holdaway, 1971).
4. Muscovite breakdown reaction (Chatterjee and Johannes, 1974).
5. Minimum melting of granite (Kerrick, 1972).
6. Garnet-cordierite-sillimanite-quartz geobarometer (Holdaway and Lee, 1977).

### Temperature

Temperatures based on three garnet-biotite geothermometers are listed in Table 1. These thermometers are affected by Ca, Mn, Ti,

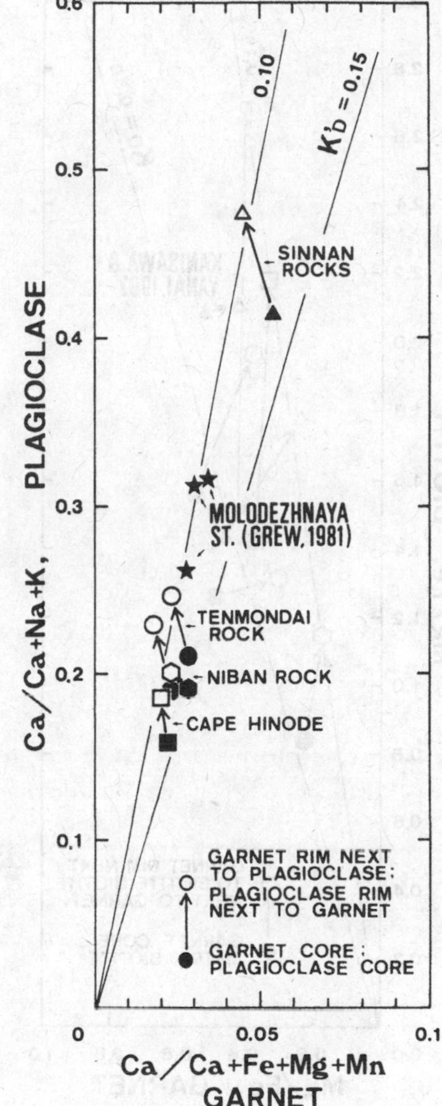


**Figure 2. Effect of polymetamorphism on distribution of Ca between garnet and plagioclase associated with sillimanite and quartz. Lines and slopes ($K_D$) are for reference only.**

$Al^{VI}$, and $Fe^{3+}$. The thermometer of Goldman and Albee (1977) usually gives anomalously low temperatures for high grade rocks, and Ferry and Spear's (1978) thermometer infers significantly higher temperatures for high grade rocks. Thompson's (1976) calibration is based on biotites from amphibolite-facies rocks and includes biotites containing small amounts of $TiO_2$. Temperatures for sp81021401B from Sinnan Rocks in zone A are too high because of high rock oxidation ratio. Temperatures for sp81012409 from Tenmondai Rock in zone B are also too high probably due to high $TiO_2$ content of the biotite. The reasonable temperature for rocks in zone A (upper amphibolite facies) in agreement with other data shown in Figures 4 and 5 is about 680°C, and that for rocks in zone B (lower granulite facies) is about 750°C.

### Total Pressure

Total pressures of metamorphism based on the garnet-plagioclase-sillimanite-quartz geobarometer (Newton and Haselton, 1981) are listed in Table 2. Pressure for zone A rocks is 6-7 kb, and that for zone B rocks 6.5-7.5 kb. As shown in Figure 2, total pressures for rocks in

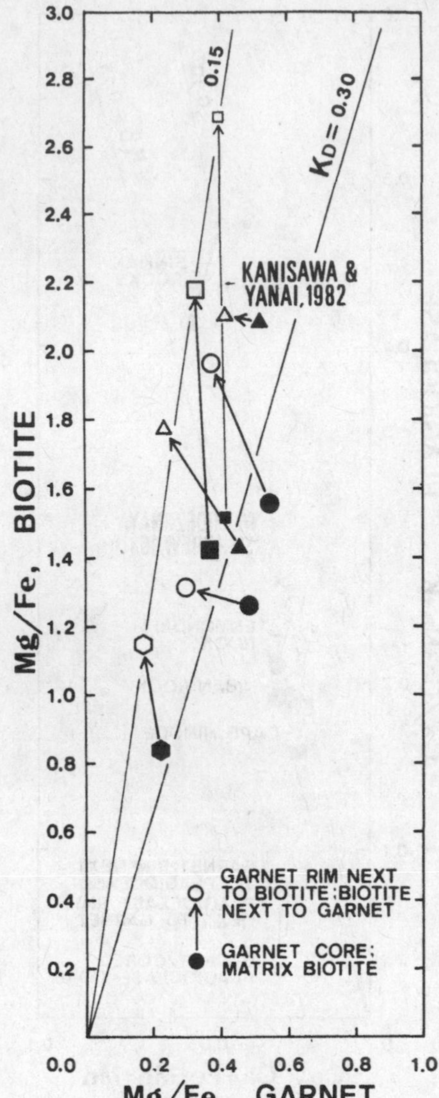


**Figure 3. Effect of polymetamorphism on distribution of Fe and Mg between garnet and biotite. Symbols are the same in Figure 2. Lines and slopes ($K_D$) are for reference only.**

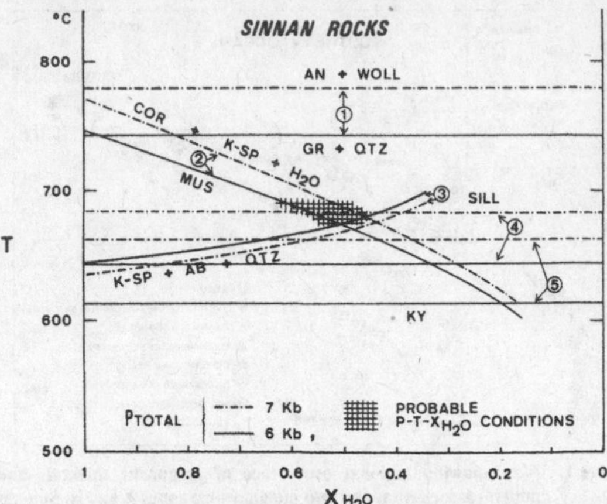


**Figure 4. T-$X_{H_2O}$ diagram at $P_{total}$ © 6 kb and 7kb, showing probable partial pressure of $H_2O$ during upper amphibolite-facies metamorphism at Sinnan Rocks in the eastern most part of the Prince Olav Coast. Curves two are calculated using an equilibrium constant derived from a straight line visually best fitted through the pressure corrected experimental brackets of Chatterjee and Johannes (1974) on a log Kv. 1/T plot. Ideal mixing of $H_2O$ with other gaseous species is assumed. Other sources: curves one from Boettcher (1970), curves three from Kerrick (1972), curves four from Richardson et al. (1969), curves five from Holdaway (1971).**

the Prince Olav Coast are higher than that for rocks at Molodezhnaya Station obtained by Grew (1981).

## Partial Pressure of $H_2O$

Using 6-7 kb total pressure for the rocks at Sinnan Rocks in zone A (see Figure 1), we can estimate partial pressure of $H_2O$ during upper amphibolite-facies metamorphism by the corundum + K-feldspar + muscovite association and minimum melting of granite (Figure 4). The widespread migmatitic rocks are inferred to have been melted during the metamorphism based on the field relation mentioned above and ubiquitous occurrence of muscovite in the rocks. The estimated $X_{H_2O}$ is about 0.5. This value is in good agreement with the experimental data for reactions (1) and (2) by Richardson (1968) and Fonarev and Korolkov (1980), respectively, and with the compositions of garnet-cordierite pairs associated with sillimanite and quartz (Figure 5).

## METAMORPHIC HISTORY

As mentioned above, several lines of evidence such as occasional occurrence of andalusite suggest local re-equilibration of the gneissic and migmatitic complex after the main upper amphibolite to lower granulite-facies metamorphism. Temperatures calculated on garnet rims next to biotite-biotite next to garnet pairs are much lower than those calculated on garnet core-matrix biotite pairs (Table 1). The relatively high temperature for sp81012409 at Tenmondai Rock is in agreement with the occurrence of andalusite in the specimen. Pressure based on the garnet rim next to plagioclase-plagioclase rim next to garnet pair of the specimen is below 4 kb, and is also in harmony with

**TABLE 1: Temperatures based on Mg-Fe distribution between garnet and biotite**

| Locality (Zone) Specimen | Garnet core—matrix biotite | | | | Garnet rim next to biotite—biotite next to garnet | | | |
|---|---|---|---|---|---|---|---|---|
| | ln $K_D^*$ | $T_1$ | $T_2$ | $T_3$ | ln $K_D^*$ | $T_1$ | $T_2$ | $T_3$ |
| Sinnan Rocks (A) | | | | | | | | |
| 81021401B | −1.140 | 745 | 824 | 812 (°C) | −2.018 | 492 | 480 | 525 (°C) |
| 81020910C** | −1.350 | 669 | 716 | 761 | −1.620 | 589 | 505 | 640 |
| Cape Hinode (A-B) | | | | | | | | |
| 74010307 | −1.356 | 667 | 713 | 642 | −1.887 | 522 | 517 | 510 |
| Niban Rock (B) | | | | | | | | |
| 801301 | −1.329 | 675 | 726 | 608 | −1.882 | 523 | 519 | 490 |
| Tenmondai Rock (B) | | | | | | | | |
| 81012409*** | −0.955 | 816 | 941 | 798 | −1.485 | 626 | 657 | 589 |
| 80T127 | −1.040 | 781 | 885 | 731 | −1.655 | 579 | 592 | 553 |

$K_D^* = \frac{(Mg/Fe)\ garnet}{(Mg/Fe)\ biotite}$
$T_1$ Thompson (1976)
$T_2$ Ferry and Spear (1978)
$T_3$ Goldman and Albee (1977)
81010910C** Anthophyllite-garnet-biotite gneiss
81012409*** Andalusite-bearing gneiss

**TABLE 2: Pressures based on garnet and plagioclase coexisting with sillimanite and quartz**

| Locality (Zone) Specimen | Plagioclase* | | Garnet* | | | | | ΔV | T** | P |
|---|---|---|---|---|---|---|---|---|---|---|
| | $^xAn$ | $^\alpha An$ | $^xGr$ | $^xPy$ | $^xAl$ | $^xSp$ | $^\alpha Gr$ | ($cm^3$) | (°C) | (Kb) |
| Sinnan Rocks (A) | | | | | | | | | | |
| 81021401B | 0.415 | 0.608 | 0.054 | 0.289 | 0.630 | 0.027 | 0.070 | 55.8 | 680 | 6.02 |
| Cape Hinode (A-B) | | | | | | | | | | |
| 74010307 | 0.159 | 0.198 | 0.022 | 0.255 | 0.696 | 0.027 | 0.028 | 57.7 | 690 | 6.80 |
| Niban Rock (B) | | | | | | | | | | |
| 801301 | 0.193 | 0.252 | 0.028 | 0.174 | 0.782 | 0.017 | 0.033 | 57.4 | 700 | 6.66 |
| Tenmondai Rock (B) | | | | | | | | | | |
| 81012409 | 0.211 | 0.266 | 0.028 | 0.314 | 0.646 | 0.012 | 0.036 | 57.5 | 750 | 7.47 |
| 80T127 | 0.195 | 0.240 | 0.023 | 0.343 | 0.623 | 0.011 | 0.031 | 57.9 | 750 | 7.22 |

*Core compositions
**Estimated by garnet-biotite geothermometer (Table 1) and other published experimental data (see text).

the occurrence of andalusite. Geologic evidence suggests the re-equilibration contemporaneous with the emplacement of granite and pegmatite, probably 500 Ma ago.

Although we have no geochronologic data of the rocks in the Prince Olav Coast at hand, 1000-1100 Ma ages obtained for the rocks at Molodezhnaya Station (Grew, 1978) and around Syowa Station (Maegoya et al., 1968) suggest the following sequence of events in the Prince Olav Coast.

1. Deposition of original rocks of the gneissic and migmatitic complex.
2. Upper amphibolite to lower granulite-facies metamorphism of kyanite-sillimanite type and folding, 1000 Ma ago (?).
3. Local mylonitisation of the gneissic and migmatitic complex and emplacement of basaltic dykes.
4. Emplacement of granite and pegmatite and related metamorphism under lower pressure conditions, 500 Ma ago (?).

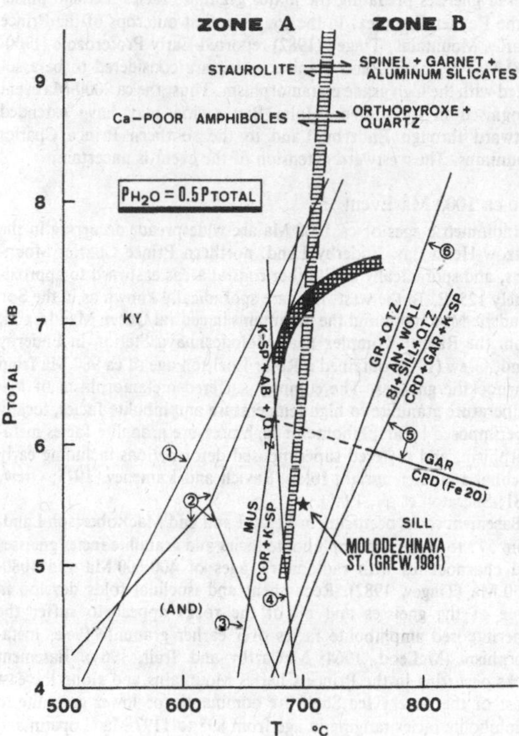


**Figure 5. $P_{total}$ − T diagram showing most probable $P_{total}$ − T conditions of upper amphibolite to lower granulite-facies metamorphism in the Prince Olav Coast. The average Fe/(Fe + Mg) ratio of cordierite from Sinnan Rocks in the most easterly part of the Prince Olav Coast is 0.20. Isogradic reactions are from Richardson (1968) and Fonarev and Korolkov (1980). Other sources: Curve one from Holdaway (1971), curves two from Richardson et al. (1969), curve three from Chatterjee and Johannes (1974), curve four from Kerrick (1972), curves five from Holdaway and Lee (1977), curve six from Boettcher (1970).**

*Acknowledgements* We thank all the members of the Japanese Association of Geologists for Antarctic Research for information on the geology of the Prince Olav Coast and adjacent areas. One of us (Y.H.) gratefully acknowledges warm encouragement and criticism of Professor S. Banno of Kyoto University.

## REFERENCES

BOETTCHER, A.L., 1970: The system $CaO\text{-}Al_2O_3\text{-}SiO_2\text{-}H_2O$ at high pressures and temperatures. *J. Petrol., 11,* 337-79.

CHATTERJEE, N.D. and JOHANNES, W., 1974: Thermal stability and standard thermodynamic properties of synthetic $2M_1$-muscovite, $(KAl_2AlSi_3O_{10}(OH)_2)$. *Contrib. Mineral. Petrol., 48,* 89-114.

FONAREV, V.I. and KOROLKOV, G. Ja., 1980: The assemblage orthopyroxene + cummingtonite + quartz. The low temperature stability limit. *Contrib. Mineral. Petrol., 73,* 413-20.

FERRY, J.M. and SPEAR, F.S., 1978: Experimental calibration of the partitioning of Fe and Mg between biotite and garnet. *Contrib. Mineral. Petrol., 66,* 113-17.

GOLDMAN, D.S. and ALBEE, A.L., 1977: Correlation of Mg/Fe partitioning between garnet and biotite with $^{18}O/^{16}O$ partitioning between quartz and magnetite. *Am. J. Sci., 277,* 750-67.

GREW, E.D., 1978: Precambrian basement at Molodezhnaya Station, East Antarctica. *Geol. Soc. Am., Bull., 89,* 801-13.

GREW, E.D., 1981: Granulite-facies metamorphism at Molodezhnaya Station, East Antarctica. *J. Petrol., 22,* 297-336.

HOLDAWAY, M.J., 1971: Stability of andalusite and aluminium silicate phase diagram. *Am. J. Sci., 271,* 97-131.

HOLDAWAY, M.J. and LEE, S.M., 1977: Fe-Mg cordierite stability in high grade pelitic rocks based on experimental, theoretical, and natural observations. *Contrib. Mineral. Petrol., 63,* 175-98.

KAMENEV, E.N., 1982: Regional metamorphism in Antarctica; *in* Craddock, C. (ed.) *Antarctic Geoscience.* Univ. Wisconsin Press, Madison, 429-33.

KANISAWA, S. and YANAI, K., 1982: Metamorphic rocks of the Cape Hinode district, East Antarctica. *Mem. Nat. Inst. Pol. Res. Japan, Spec. Issue, 21,* 71-85.

KANISAWA, S., YANAI, K. and ISHIKAWA, K., 1979: Major element chemistry of metamorphic rocks of the Cape Hinode district, East Antarctica. *Mem. Nat. Inst. Pol. Res. Japan, Spec. Issue, 14,* 164-71.

KERRICK, D.M., 1972: Experimental determination of muscovite + quartz stability with $P_{H_2O} < P_{total}$. *Am. J. Sci., 272,* 946-58.

MAEGOYA, T., NOHDA, S. and HAYASE, I., 1968: Rb-Sr dating of the gneissic rocks from the east coast of Lützow-Holm Bay, Antarctica. *Mem. Coll. Sci. Univ. Kyoto, Ser. B, 35,* 131-38.

NAKAI, Y., KANO, T. and YOSHIKURA, S., 1980: Geological map of Cape Ryûgû, Antarctica, with explanatory text. *Antarctic Geol. Map Ser.,* Sheet 15, Nat. Inst. Pol. Res. Japan.

NAKAI, Y., KANO, T. and YOSHIKURA, S., 1981: Geological map of Oku-iwa Rock, Antarctica, with explanatory text. *Antarctic Geol. Map Ser.,* Sheet 22, Nat. Inst. Polar Res. Japan.

NEWTON, R.C. and HASELTON, H.J., 1981: Thermodynamics of the garnet-plagioclase-$Al_2SiO_5$-quartz geobarometer; *in* Newton, R.C., Navrotsky, A. and Wood, B.J. (eds.) *Thermodynamics of Minerals and Melts.* Springer-Verlag, New York, 125-45.

NISHIDA, T., YANAI, K. and KOJIMA, H., 1982: Geology of Kasumi Rock, East Antarctica. *Mem. Nat. Inst. Pol. Res. Japan, Spec. Issue, 21,* 1-14.

RAVICH, M.G. and KAMENEV, E.N., 1975: *Crystalline basement of the Antarctic Basement Platform.* John Wiley & Sons, New York, (translated from Russian).

RICHARDSON, S.W., 1968: Staurolite stability in part of the system Fe-Al-Si-O-H. *J. Petrol., 9,* 467-88.

RICHARDSON, S.W., GILBERT, M.C. and BELL, P.M., 1969: Experimental determination of kyanite-andalusite and andalusite-sillimanite equilibria; the aluminium silicate triple point. *Am. J. Sci., 267,* 259-72.

SUZUKI, M., 1979a: Metamorphic and plutonic rocks in the Cape Omega area, East Antarctica. *Mem. Nat. Inst. Pol. Res. Japan, Spec. Issue, 14,* 128-39.

SUZUKI, M., 1979b: On the scapolite-plagioclase equilibria found in the Cape Omega and East Ongul Island areas, East Antarctica. *Mem. Nat. Inst. Pol. Res. Japan, Spec. Issue, 14,* 140-52.

SUZUKI, M. and MORIWAKI, K., 1979: Geological map of Cape Omega, Antarctica, with explanatory text. *Antarctic Geol. Map Ser.,* Sheet 21, Nat. Inst. Pol. Res. Japan.

THOMPSON, A.B., 1976: Mineral reactions in pelitic rocks: II. Calculations of some P-T-X (Fe-Mg) phase relations. *Am. J. Sci., 276,* 425-54.

YANAI, K. and ISHIKAWA, T., 1978: Geological map of Cape Hinode, Antarctica, with explanatory text. *Antarctic Geol. Map Ser.,* Sheet 11, Nat. Inst. Pol. Res., Japan.

YOSHIKURA, S., NAKAI, Y. and KANO, T., 1979: Petrology and geothermometry of the clinopyroxene-garnet rock from Cape Ryûgû, East Antarctica. *Mem. Nat. Inst. Pol. Res. Japan, Spec. Issue, 14,* 172-85.

# TECTONIC SITUATION OF LUTZOW-HOLM BAY IN EAST ANTARCTICA AND ITS SIGNIFICANCE IN GONDWANALAND

M. Yoshida, *Department of Geosciences, Osaka City University, Osaka 558, Japan.*

K. Kizaki, *Department of Oceanography, Ryukyu University, Naha, Japan.*

*Abstract* Three age groups ranging from about 500 to 2000 Ma have been distinguished in the Lutzow-Holm Bay region as reflecting distinctive tectonic-metamorphic events. Similar age groups are recognised elsewhere in east Antarctica and in other Gondwanian regions. ca 2000 Ma ages in East Antarctica occur as sporadic relict ages in the predominant ca 1000 Ma age belt. ca 1000 Ma ages can be traced as a zone from Antarctica to other Gondwanian continents. ca 500 Ma ages are extensively developed along almost all the boundary areas of the Gondwanian continents except at the Antarctica-Australia boundary. The similarity of the tectonic-metamorphic belt, including the Lutzow-Holm Bay region, to the northern part of the Irumide Belt of southern Africa is pointed out. The Napier zone of Enderby Land and the Indian granulite-charnockite belt may be an Archaean mobile belt. The Peninsular Gneiss of Dharwar and the granitic rocks of the Southern Prince Charles Mountains are regarded as Archaean cratonic nuclei.

Reconstruction of Gondwanaland around the Antarctic continent started from early in the 20th century, but substantial correlations of geologic units recognised in the southern continents have been made only since the work of Du Toit (1937), who traced the early Mesozoic Samfrau Geosyncline from Australia to South America through Antarctica and South Africa. However, the correlation of geologic domains in East Antarctica has only been attempted in the last few years. Radiometric dating now allows reconstructions based on the correlation of chronologic zones between continents. The present report attempts to correlate the tectonic-metamorphic events identified in the Lutzow-Holm Bay region with other areas of East Antarctica and discusses the geotectonic situation of these areas in Gondwanaland.

## TECTONIC AND METAMORPHIC EVENTS IN THE AREA AROUND THE LUTZOW-HOLM BAY

The geologic outline of the Lutzow-Holm Bay region, including the Yamato and Belgica Mountains, having been summarised by Yoshida (1979), and Yoshida et al. (1982), along with new Pb-Pb and Rb-Sr isochron dating of the same region (Shirahata, in press), can be tentatively assigned to the superposition of tectonic and metamorphic events. As shown in Table 1, the Lutzow-Holm Bay region suffered superimposed tectonic-metamorphic events of three age groups, viz. ca 1900 Ma (first event), ca 1100 Ma (second event), and ca 500 Ma (third and fourth events). The fifth, fracturing, event is omitted from the succeeding discussions.

**TABLE 1: Tectonic-metamorphic history of the Lutzow-Holm Bay region (after Yoshida, 1979 and Yoshida et al., this volume)**

| | Metamorphism and plutonism | Tectonics |
|---|---|---|
| First event (ca 1900 Ma) | High grade metamorphism or emplacement of original igneous rocks of charnockites | Microfolds in charnockites |
| Second event (ca 1100 Ma) | Intermediate pressure granulite facies metamorphism and formation of charnockites | Earlier recumbent and later isoclinal folds whose axial traces paralleling the modern coast line |
| Third event (ca 560 Ma+) | High amphibolite facies metamorphism | East-west upright open folds |
| Fourth event (ca 460 Ma) | Intrusion of granite—pegmatite and associated low amphibolite facies metamorphism | North-south upright gentle folds and faults |
| Fifth event (Jurassic ?) | | Conjugate set of fractures with accute bisectrix in east-west direction |

## TECTONIC-METAMORPHIC EVENTS IN OTHER AREAS OF EAST ANTARCTICA.

### The ca 2000 Ma Event

Radiometric ages of ca 2000 Ma, corresponding to the age of the first event of the Lutzow-Holm Bay region, have also been reported from Enderby Land and the Southern Prince Charles Mountains (Figure 1). In the Rayner Complex of Enderby Land, Ravich and Kamenev (1975) reported a U-Pb age of 2120 Ma from chevkinite intruding charnockite. Grew (1978) reported a Rb-Sr isochron age of ca 2120 Ma which he considered to be the formation age of the well layered gneisses predating the major granulite facies metamorphism of the Rayner Complex. In the southernmost outcrops of the Prince Charles Mountains, Tingey (1982) reported Early Proterozoic (1700-2000 Ma) pegmatites and granites which are considered to be associated with the high grade metamorphism. Thus, the ca 2000 Ma event recognised in the Lutzow-Holm Bay region may have extended eastward through Enderby Land to the southern Prince Charles Mountains. The westward extension of the event is uncertain.

### The ca 1000 Ma Event

Radiometric ages of ca 1000 Ma are widespread, occurring in the Lutzow-Holm Bay, Enderby Land, northern Prince Charles Mountains, and sporadically in all other coastal areas eastward to approximately 125°E. To the west, they are sporadically known as in the Sør Rondane Mountains and the mountains in central Queen Maud Land. From the Rayner Complex near Molodezhnaya Station in Enderby Land, Grew (1978) obtained a Rb-Sr isochron age of ca 987 Ma from charnockitic gneisses. The complex suffered metamorphism of low temperature granulite to high temperature amphibolite facies, locally superimposed by amphibolite or high-pressure granulite facies metamorphism, and suffered superimposed deformations including early isoclinal and later upright folds (Ravich and Kamenev, 1975; Grew, 1981; Sheraton et al., 1980).

Basement rocks occurring in Kemp Land and MacRobertson Land, from 57° to 64°E, are amphibolite facies and granulite facies gneisses and charnockites with radiometric ages of 400-600 Ma and 1080-1450 Ma (Tingey, 1982). Recumbent and isoclinal folds develop in some of the gneisses and all of the rocks appear to suffer the superimposed amphibolite facies over earlier granulite facies metamorphism (McLeod, 1964; McCarthy and Trail, 1964). Basement rocks occurring in the Prince Charles Mountains and along the east coast of the Amery Ice Shelf are dominantly of lower granulite to amphibolite facies ranging in age from 895 to 1197 Ma (Lopatin and Semenov, 1982; Grew, 1982a; Tingey, 1982). Development of early recumbent isoclinal folds and later mild deformation folding schistosity and localised cataclasis have been described at some places.

In the coastal area east of the Vestfold Hills to about 125°E, 1000 to 1200 Ma ages on whole rocks or minerals, by the K-Ar method are known (Craddock, 1970a). These are granulite facies or superimposed granulite-amphibolite facies gneisses intruded by charnockitic rocks (Webb et al., 1964). Although no isochron age has so far been obtained, there is a possibility that some metamorphism of this area occurred earlier at 1000 Ma.

Metamorphic rocks (Teltet-Vengen Group) occurring in the northern part of the Sør Rondane Mountains gave 935 and 984 Ma U-Pb ages using "detrital" zircons (Pasteels and Michot, 1968). Amphibolite facies rocks predominate, but relict blocks of granulite facies rocks are found suggesting superimposed "epizonal metamorphism" (Van Autenboer and Loy, 1972). Isoclinal folds are developed, which are later crossed by other fold and mylonite zones. Their results appear to indicate that granulite facies metamorphism of ca 1000 Ma occurred.

In central Queen Maud Land, both granulite to amphibolite facies rocks occur; a greenschist facies metamorphism associated with brittle deformation has locally been superimposed on them (Ravich and Soloviev, 1969). K-Ar, Rb-Sr, and Pb-Pb ages of rocks or minerals range generally from 420 to 1750 Ma (Ravich and Soloviev, 1969;

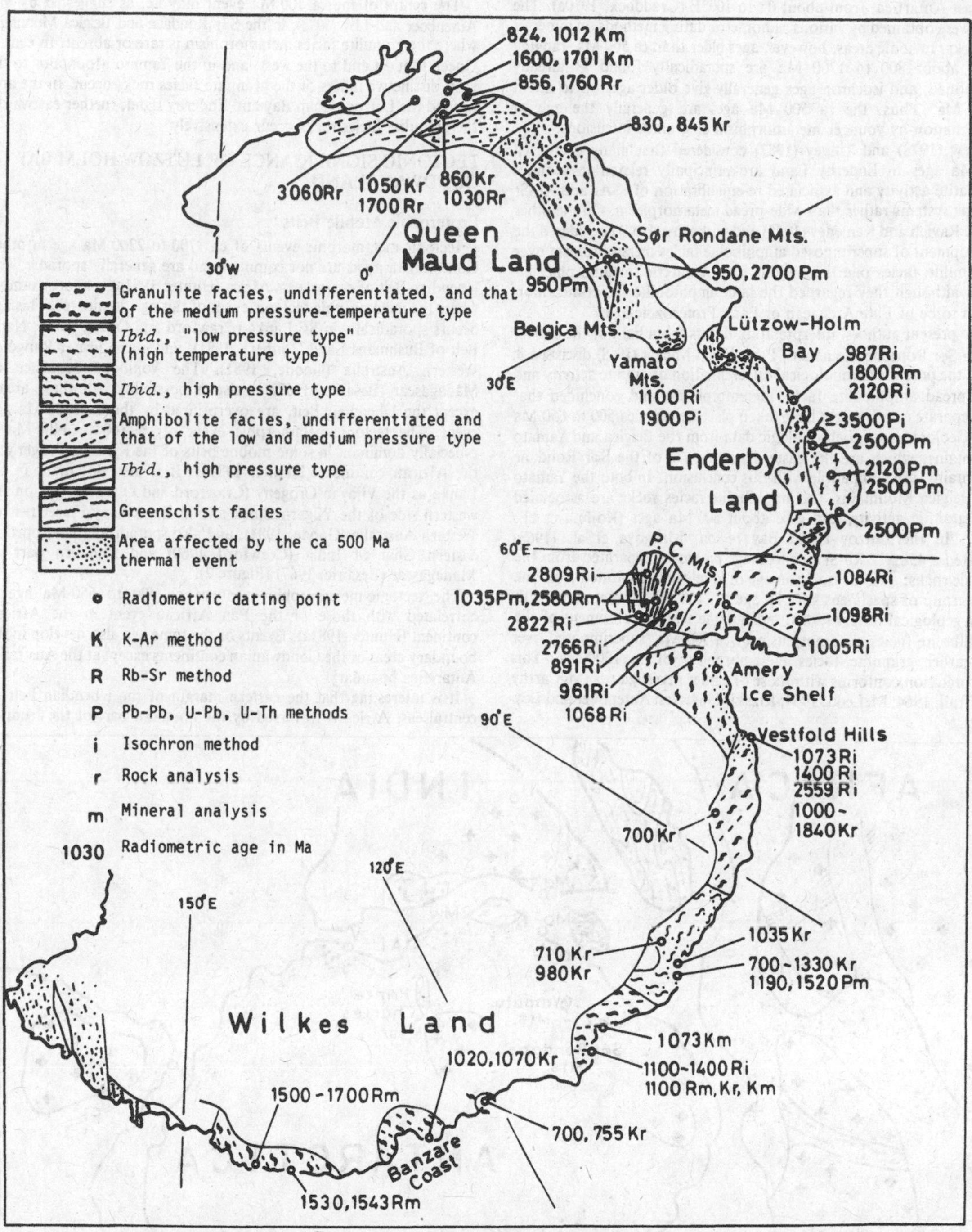


**Figure 1. Metamorphic rocks and selected isotopic ages of East Antarctica. Kinds and distribution of metamorphic rocks are referred to Kamenev and Ravich (1979) with some modifications. Composite symbols indicate superposition of two metamorphisms. Isotopic ages are selected from data after Craddock (1970a), Grew and Manton (1979), and Tingey (1982). P.C.Mt.: Prince Charles Mountains.**

Neethling, 1972), a group of 824 to 1050 Ma among them probably correspond to the widespread ca 1000 Ma event. Ravich and Soloviev (1969) suggested the possibility of a polymetamorphic-folding history in this area.

Thus the ca 1000 Ma tectonic-metamorphic event in the Lutzow-Holm Bay region is considered to extend eastwards at least as far as the east coast of the Amery Ice Shelf through the Rayner Complex of Enderby Land as already pointed out by Tingey (1982) and Grew (1982b) and probably farther eastward to 125°E. Sporadic extension westwards up to about 2°W is possible. Intermediate-pressure granulite facies or the amphibolite-granulite superposition, and some folding tectonics including earlier isoclinal folds, characterise the event of this age group. The superposition is either assigned as monophasial (nearly synchronous with the ca 1000 Ma event) or polyphasial (different from the ca 1000 Ma event).

## The ca 500 Ma Event

Radiometric ages of ca 500 Ma, which correspond to the third and fourth tectonic-metamorphic events in the Lutzow-Holm Bay region, are reported from almost all coastal areas in the Indian Ocean sector of East Antartica, from about 0° to 100°E (Craddock, 1970a). The ages were obtained by various radiometric dating methods on minerals or rocks. In some areas, however, ages older than ca 500 Ma, ranging from about 800 to 1700 Ma are sporadically found as already mentioned, and isochron ages generally give older ages up to about 4000 Ma. Thus, the ca 500 Ma ages are generally the age of rejuvenation by younger metamorphism or granitic intrusions.

Grew (1978) and Tingey (1982) considered that many of the ca 500 Ma ages in Enderby Land are principally related to granite-pegmatite activity and associated re-equilibration of K-Ar and Rb-Sr isotope systems rather than widespread metamorphism. On the other hand, Ravich and Kamenev (1975) and Grikurov (1982) considered the development of superimposed amphibolite facies metamorphism over a granulite facies one throughout East Antarctic metamorphic terrains, although they regarded the later amphibolite facies metamorphism to be of Late Archaean or Early Proterozoic age.

The present authors' interpretation differs from both of the above. In the Sør Rondane Mountains, Pasteels and Michot (1968) discussed in detail the possible chronological differentiation of granite activity and widespread amphibolite facies metamorphism, and concluded that two separate events might have taken place between ca 500 to 650 Ma ago. Geologic and geochronologic data from the Belgica and Yamato Mountains, which are the eastern neighbours of the Sør Rondane Mountains, are compatible with this conclusion. In both the Yamato and Belgica Mountains, the amphibolite facies rocks are associated with granitic activity, and give about 500 Ma ages (Kojima et al., 1982). In the Lutzow-Holm Bay region, Maegoya et al. (1968) reported a 458 Ma Rb-Sr isochron age on biotite separated from the gneissic rocks; a Rb-Sr isochron age of K-feldspar separated from the same group of specimens gave an age of 1100 Ma. These, along with many geological observations, indicate the wide development of the amphibolite facies metamorphism of ca 500 Ma, superimposed over the earlier granulite facies metamorphism of ca 1000 Ma. This interpretation conforms with those of earlier investigators (McCarthy and Trail, 1964; McLeod, 1964; Kizaki, 1964; Van Autenboer and Loy, 1972).

The centre of the ca 500 Ma event may lie, as suggested by Van Aurenboer and Loy (1972), in the Sør Rondane and Belgica Mountains where the granulite facies metamorphism is rare or absent. In central Queen Maud Land to the west, and in the Yamato Mountains to the east, distinctive inliers of the granulite facies rocks occur. In the area around the Lutzow-Holm Bay and Enderby Land, further eastwards, the granulite facies rocks occur extensively.

## TECTONIC SIGNIFICANCE OF LUTZOW-HOLM BAY IN GONDWANALAND

### Proterozoic Mobile Belts

Tectonic-metamorphic events of ca 1700 to 2200 Ma age in other Gondwanian areas are not common and are generally sporadic. The Ubendian Belt of central east Africa (Hunter, 1981a) is a rare example where the metamorphic-tectonic event of this age dominates. This age occurs sporadically in Sri Lanka (Crawford and Oliver, 1969), Natal Belt of Bushmans Land (Joubert, 1981), and in the Fraser Range of Western Australia (Goode, 1981). The Vohibory Sequence of Madagascar (Besairie, 1967) is also a possibility. All these areas, except the Ubendian Belt, are overprinted by the ca 1000 Ma age event. The tectonic-metamorphic event of ca 800 to 1400 Ma is especially dominant in some mobile belts of the Kibaran Orogeny in the African continent (Hunter, 1981a). It is also dominant in Sri Lanka as the Vijayan Orogeny (Crawford and Oliver, 1969), on the western side of the Yilgarn Block and in the Albany-Fraser belt of Western Australia (Goode, 1981), and also sporadically developed in eastern Ghat of India (Crawford 1969) and southern part of Madagascar (Besairie, 1967) (Figure 2).

The tectonic-metamorphic event of ca 400 to 650 Ma age is correlated with those of the Pan African event in the African continent (Hunter, 1981a). Events of the same age also develop in the boundary areas of the Gondwanian continents except at the Australia-Antarctica boundary.

It is interesting that the eastern margin of the Ubendian Belt of central east Africa is traversed by the northern part of the Irumide

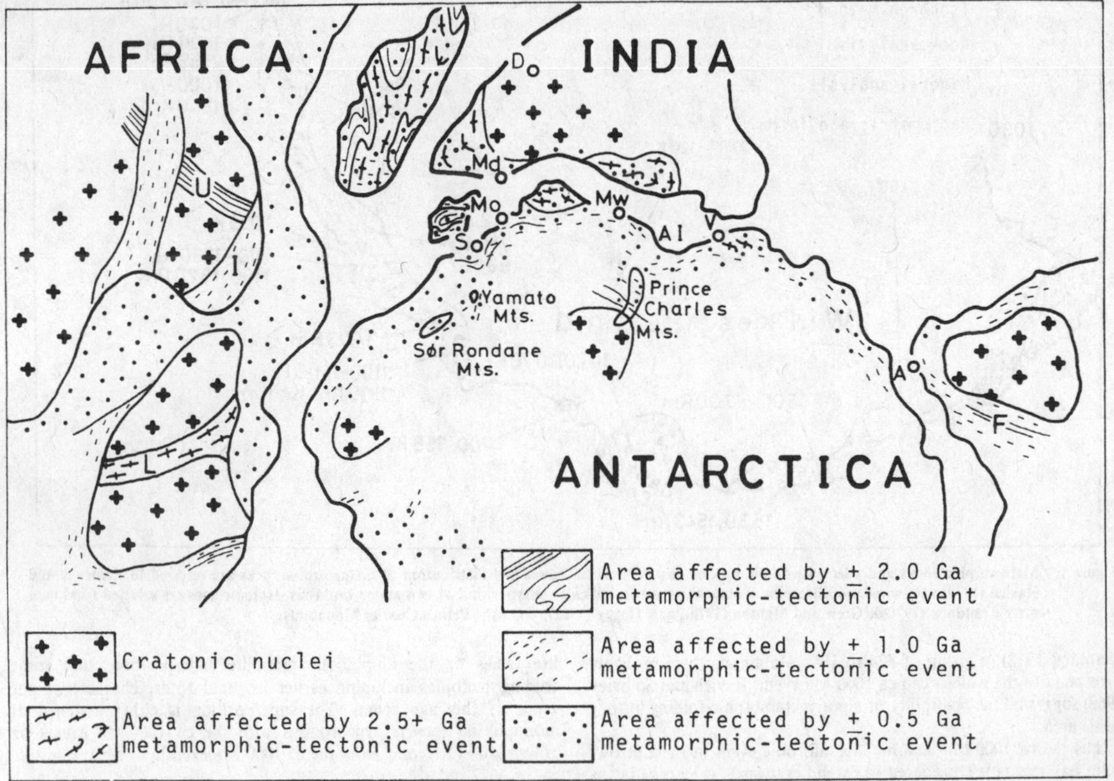


**Figure 2.** **Mobile belts and cratonic nuclei in the Gondwanian regions around Lutzow-Holm Bay. U: Ubendian Belt; I: Irumide Belt; L: Limpopo Belt; F: Fraser Belt; D: Dharwar; Md: Madras; Mo: Molodezhnaya Station; Mw: Mawson Station; S: Syowa Station; AI: Amery Ice Shelf; V: Vestfold Hills; A: Albany. The compilation of geologic data of each continent is based mainly on Hunter (1982b), Besairie (1967), Crawford (1974) and Crawford and Oliver (1969). The base Gondwana reconstruction map is that of Craddock (1982).**

mobile belt of the Kibaran stage, and successively superimposed by the Pan African event. In this area, an older high grade metamorphic event of ca 1970 Ma has been detected, which is superimposed by lower grade metamorphisms with age ranges of 1200 to 1100 Ma and 700 to 400 Ma (Bloomfield, 1981). Thus the northern part of the Irumide Belt is regarded as having similar geotectonic characteristics to the Lutzow-Holm-Rayner Complex-Amery Ice Shelf belt of Antarctica.

## Napier Complex as an Archaean Mobile Belt in Contrast to a Stable Cratonic Nucleus

On the northern side of the ca 1000 Ma mobile belt of East Antarctica, lies the Napier Complex in Enderby Land. Here Archaean rocks were metamorphosed under high temperature granulite facies about 2500 Ma ago and earlier (James and Black, 1981). Similar metamorphic rocks with about 2500 Ma or older ages are found in the main part of the eastern Ghat of India (Ramienger et al., 1978), and the very close resemblance in petrography and geology between the two areas has been reported (Crawford, 1974; Fedorov et al,. 1982).

It is highly probable from the many recent petrologic and geochronologic studies in Enderby Land and in the Vestfold Hills that the high temperature granulite facies metamorphic rocks of ca 2500 to 2700 Ma may continue from the Vestfold Hills to Mysore of India, through northern Orrissa, Enderby Land, and the Madras area as outlined by Crawford (1974) and Grew (1982b). The Androjan and Graphite Sequences of Madagascar (Besairie, 1967) may also be a continuation of this belt. The metamorphic rocks of the area form a belt of about 300 km wide in Enderby Land and can be regarded as a mobile belt, considerable portions of which are supposedly composed of Archaean supracrustal rocks.

In Gondwanaland, a mobile belt with similar tectonic and geochronologic features as the Napier-Madras belt is the Limpopo Belt of southern Africa (Hunter, 1981a). The development of a supracrustal sequence, fold style, granulite facies metamorphism and sporadic occurrence of very old radiometric ages, are all analagous features encountered in both of these belts.

In the Amery Ice Shelf area of Antarctica, Tingey (1982) pointed out that the oldest Archaean rocks occur in rather lower grade metamorphic areas, such as in the southernmost part of the Prince Charles Mountains and that this may form a cratonic block, in contrast to the orogenic belt lying to the north. The observation that the cratonic nucleus is not composed of the highest grade rocks, and that the highest grade rocks occur exclusively in the mobile belts is well documented in the Precambrian of Africa (Tankard et al, 1982). In southern Africa, the ca 2500 to 2700 Ma Limpopo Belt clearly cuts the Archaean cratonic nucleus, although the oldest age in Africa has also been found in this belt. Thus it is possible that cratonic nuclei occur on both sides of the mobile belt. The southernmost part of the Prince Charles Mountains and the Peninsular Gneiss complex around Dharwar could be regarded as stable cratonic nuclei which are divided by the compounded mobile belt ranging in age from ca 2500 Ma to 500 Ma.

## REFERENCES


BESAIRIE, H., 1967: The Precambrian of Madagascar; *in* Rankama, K. (ed.) *The Precambrian, Vol. 3,* Interscience Pub., New York, 133-43.

BLOOMFIELD, K., 1981: The Pan-African event in Malawi and eastern Zambia; *in* Hunter, D.R. (ed.) *Precambrian of the Southern Hemisphere,* Elsevier, Amsterdam, 743-54.

CRADDOCK, C., 1970a: Radiometric age map of Antarctica; *in* Bushnell, V.C. and Craddock, C. (eds.) Geologic maps of Antarctica, *Antarctic Map Folio Ser., Folio 12,*Plate XIX.

CRADDOCK, C., 1970b: Geologic map of Antarctica (entire continent); *in* Bushnell, V.C. and Craddock, C. (eds.) Geologic map of Antarctica, *Antarctic Map Folio Ser., Folio 12,*Plate XX.

CRADDOCK, C., 1982: Antarctica and Gondwanaland; *in* Craddock, C. (ed.) *Antarctic Geoscience,* Univ. Wisconsin Press, Madison, 3-13.

CRAWFORD, A.R., 1969: India, Ceylon, and Pakistan: New age data and comparisons with Australia. *Nature, 223,* 180-4.

CRAWFORD, A.R., 1969: India, Ceylon, and Pakistan: New age data and comparisons with Australia. *Nature, 223,* 180-4.

CRAWFORD, A.R., 1974: Indo-Antarctica, Gondwanaland, and the distortion of a granulite belt. *Tectonophysics., 22,* 141-57.

CRAWFORD, A.R. and OLIVER, R.L., 1969: The Precambrian geochronology of Ceylon. *Geol. Soc. Aust., Spec. Pub., 2,* 283-306.

DU TOIT, A.L., 1937: *Our wandering continents,* Oliver and Boyd, London.

FEDOROV, L.V., RAVICH, M.G. and HOFMANN, J., 1982: Geologic comparison of southeastern Peninsula India and Sri Lanka with a part of East Antarctica (Enderby Land, MacRobertson Land, and Princess Elizabeth Land); *in* Craddock, C. (ed.) *Antarctic Geoscience,* Univ. Wisconsin Press, Madison, 73-8.

GOODE, A.D.T., 1981: Proterozoic geology of western Australia; *in* Hunter, D.R. (ed.) *Precambrian of the Southern Hemisphere,*Elsevier, Amsterdam, 105-204.

GREW, E.S., 1978: Precambrian basement at Molodezhnaya Station, East Antarctica. *Geol. Soc. Am., Bull., 89,* 801-13.

GREW, E.S., 1981: Granulite-facies metamorphism at Molodezhnaya Station, East Antarctica. *J. Petrol., 22,* 297-336.

GREW, E.S., 1982a: Geology of the southern Prince Charles Mountains, East Antarctica; *in* Craddock, C. (ed.) *Antarctic Geoscience,* Univ. Wisconsin Press, Madison, 473-8.

GREW, E.S., 1982b: The Antarctic margins; *in* Nairn, A.E.M. and Stehli, F.G. (eds.) *The ocean basins and margins, 6, The Indian Ocean,*Plenum, New York, 697-755.

GREW, E.S. and MANTON, W.I., 1979: Archaean rocks in Antarctica: 2.5 billion year uranium-lead ages on pegmatites from Enderby Land. *Science, 206,* 443-5.

GRIKUROV, G.E., 1982: Structure of Antarctica and outline of its evolution; *in* Craddock, C. (ed.) *Antarctic Geoscience,* Univ. Wisconsin Press, Madison, 791-804.

HUNTER, D.R., 1981a: Structural province (of Southern Africa); *in* Hunter, D.R. (ed.) *Precambrian of the Southern Hemisphere,* Elsevier, Amsterdam, 397-410.

HUNTER, D.R., 1981b: *Precambrian of the Southern Hemisphere* Elsevier, Amsterdam.

JAMES, P.R. and BLACK, L.P., 1981: A review of the structural evolution and geochronology of the Archaean Napier Complex of Enderby Land, Australian Antarctic territory. *Geol. Soc. Aust., Spec. Pub., 7,* 71-83.

JOUBERT, P., 1981: The Namaqualand metamorphic complex; *in* Hunter, D.R. (ed.) *Precambrian of the Southern Hemisphere,* Elsevier, Amsterdam, 671-704.

KAMENEV, E.N. and RAVICH, M.G., 1979: *Map of metamorphic facies of Antarctica*1:5,000,000. Research Institute of Arctic Geology, Leningrad.

KIZAKI, K., 1964: Tectonics and petrography of the East Ongul Island, Lutzow-Holmbukta, Antarctica. *JARE Sci. Rep., Ser. C, 2,* 1-24.

KOJIMA, H., YANAI, Y. and NISHIDA, T., 1982: Geology of the Belgica Mountains. *Mem. Nat. Inst. Pol. Res. Japan, Spec. Iss., 21,* 32-46.

LOPATIN, B.G. and SEMENOV, V.S., 1982: Amphibolite facies rocks of the southern Prince Charles Mountains, East Antarctica; *in* Craddock, C. (ed.) *Antarctic Geoscience,* Univ. Wisconsin Press, Madison, 465-72.

McCARTHY, W.R. and TRAIL, D.S., 1964: The high grade metamorphic rocks of the MacRobertson Land and Kemp Land coast; *in* Adie, R.J. (ed.) *Antarctic Geology,* North Holland, Amsterdam, 473-81.

McLEOD, E.R., 1964: An outline of the geoogy of the sector from longitude 45° to 80°E, Antarctica. *in* Adie, R.J. (ed.) *Antarctic Geology,* North Holland, Amsterdam, 237-47.

MAEGOYA, T., NOHDA, S. and HAYASE, I., 1968: Rb Sr dating of the gneissic rocks from the east coast of Lutzow-Holm Bay, Antarctica. *Mem. Coll. Sci., Univ. Kyoto, Ser. B, 35,* 131-8.

NEETHLING, D.C., 1972: Age and correlation of the Ritscher Supergroup and other Precambrian rock units, Dronning Maud Land; *in* Adie, R.J. (ed.) *Antarctic Geology and Geophysics,* Universitetsforlaget, Oslo, 547-56.

PASTEELS, P. and MICHOT, J., 1968: Uranium-lead radioactive dating and lead isotope study on sphene and K-feldspar in the Sør Rondane Mountains, Dronning Maud Land, Antarctica. *Eclogae geol. Helv., 63,* 239-54.

RAMIENGER, A.S., RAMAKRISHNAN, M. and VISWANATHA, M.V., 1978: Charnockite-gneiss-complex relationship in southern Karnataka. *Geol. Soc. India, J., 19,* 411-9.

RAVICH, M.G. and KAMENEV, E.N., 1975: *Crystalline basement of the Antarctic Basement Platform.* John Wiley and Sons, New York.

RAVICH, M.G. and SOLOVIEV, D.S., 1969: *Geology and petrology of the mountains of central Queen Maud Land.* Israel program for scientific translations, Jerusalem.

SHERATON, J.W., OFFE, L.A., TINGEY, R.J. and ELLIS, D.J., 1980: Enderby Land, Antarctica—an unusual Precambrian high grade metamorphic terrain. *Geol. Soc. Aust., J., 27,* 1-18.

SHIRAHATA, H. (in press): Lead isotopic compositions in metamorphic rocks from Skarvsnes, East Antarctica (Ms).

TANKARD, A.J., JACKSON, M.P.A., ERIKSSON, K.A., HOBDAY, D.K., HUNTER, D.R. and MINTER, W.E.L., 1982: *Crustal evolution of Southern Africa.* Springer-Verlag, New York.

TINGEY, R.J., 1982: The geologic evolution of the Prince Charles Mountains—An Antarctic Archean cratonic block; *in* Craddock, C. (ed.) *Antarctic Geoscience,* Univ. Wisconsin Press, Madison, 455-64.

VAN AUTENBOER, T. and LOY, W., 1972: Recent geological investigations in the Søx-Rondane Mountains, Belgicafjella and Sverdrupfjella, Dronning Maud Land; *in* Adie, R.J. (ed.) *Antarctic Geology and Geophysics,* Universitetsforlaget, Oslo, 563-71.

WEBB, A.W., McDOUGALL, I. and COOPER, J.A., 1964: Potassium-argon dates from the Vincennes Bay region and Oates Land; *in* Adie, R.J. (ed.) *Antarctic Geology,* North Holland, Amsterdam, 597-600.

YOSHIDA, M., 1979: Tectonics and metamorphism of the region around Lutzow-Holmbukta, East Antactica. *Mem. Nat. Inst. Pol. Res. Japan, Spec. Iss., 14,* 28-40.

YOSHIDA, M., YANAI, K., YOSHIKURA, S., ISHIKAWA, T. and KANISAWA, S., 1982: Tectonics and microstructure of charnockites around Lutzow-Holmbukta, East Antarctica; *in* Craddock, C. (ed.) *Antarctic Geoscience,* Univ. Wisconsin Press, Madison, 511-9.

YOSHIDA, M., KIZAKI, K., KOJIMA, H., SHIRAHATA, H. and SUZUKI, M. (this volume): Tectonic and metamorphic history of the Lutzow-Holm Bay region, East Antarctica.

# SAPPHIRINE-GARNET AND ASSOCIATED PARAGENESES IN ANTARCTICA

E.S. Grew, *Department of Earth and Space Sciences, University of California, Los Angeles, California, 90024, U.S.A.*

*Abstract* In the Schirmacher Hills Queen Maud Land, sapphirine-biotite-garnet granulites (with corundum-plagioclase or sillimante-cordierite) form two lenses up to 1 m thick in polymetamorphic granulite and amphibolite facies gneisses. During later metamorphism, andalusite formed in sillimanite-garnet gneiss. The sapphirine-granulites have low $SiO_2$ and high $Al_2O_3$, MgO and FeO contents and high MgO/FeO ratios. Temperatures and pressures for the appearance of the sapphirine-garnet assemblage in the granulite facies at Schirmacher Hills are estimated to be 750° to 800°C and 5 to 8 kb.

$K_D = (Fe^{2+}/Mg\ (garnet))\ (Mg/Fe^{2+}\ (sapphirine)) \sim 3.5$, for coexisting sapphirine and garnet from the Tula and Scott Mountains, Enderby Land (Napier Complex). The $K_D$'s of the Schirmacher Hills sapphirine-garnet pairs are somewhat higher. Orthopyroxene does not occur in stable association with corundum in either the Napier Complex or the Schirmacher Hills, whereas the assemblage garnet-sapphirine-sillimanite is found in both areas. The orthopyroxene-corundum assemblage may be stable only in very dry rocks at lithostatic pressures above the kyanite-sillimanite univariant curve, well above the metamorphic pressures to which the Napier Complex and Schirmacher Hills rocks were subjected.

The sapphirine-garnet association, known from only a few of the known world sapphirine localities, is well developed in the Archaean Napier Complex of Enderby Land (Grew, 1982a). This paper reports the sapphirine-garnet association from the Schirmacher Hills, Antarctica (Figure 1), which is the first reported locality for sapphirine in Queen Maud Land. The Schirmacher Hills parageneses have several features in common with silica-undersaturated Napier Complex parageneses, but differ in that orthopyroxene is absent. As the sapphirine-garnet association has received little attention in the literature, several petrologic features of the Antarctic sapphirine-garnet rocks, notably Fe-Mg distribution among coexisting sapphirine, garnet and biotite and the significance of the sapphirine-garnet-sillimanite association (which is incompatible with orthopyroxene-corundum) will be considered in the second part of the paper.

Field work in the Schirmacher Hills (and in the nearby Institut Geologii Arktiki Rocks) was done from December 15 to 27, 1973; samples from Enderby Land were collected during the 1977/78 and 1979/80 seasons. Analytical work was done mostly at UCLA from 1978 to 1982; the balance, at the University of Melbourne in 1978.

## GEOLOGY OF THE SCHIRMACHER HILLS

The Schirmacher Hills consist of Precambrian polymetamorphic granulite and amphibolite facies rocks (Ravich and Soloviev, 1966; Ravich and Kamenev, 1972). Garnet-biotite gneiss (locally with hypersthene) and mafic granulites predominate, and sillimanite-garnet gneiss containing lenses (up to 1.5 m thick) of marble and calc-silicate granulite is much less abundant (Figure 1).

## MINERALOGY AND PETROLOGY OF THE SCHIRMACHER HILLS METAMORPHIC ROCKS

### Non-Pelitic Rocks

The metamorphic rocks originally crystallised in the granulite facies and subsequently were altered by at least one amphibolite facies event. Mafic granulites and garnet-biotite gneisses least affected by retrograde metamorphism contain orthopyroxene, clinopyroxene or brown hornblende. Garnet and clinopyroxene are in textural equilibrium in one granulite. Ultramafic rocks include a dunite containing fresh olivine and minor chromite, pyroxene and amphibole. Amphibole-bearing reaction skarns are developed along pegmatite veins cutting the dunite. Marble and calc-silicate granulites contain scapolite, garnet, clinopyroxene, plagioclase, K-feldspar, clinozoisite-epidote, calcite, quartz and secondary amphibole.

### Sillimanite-Garnet Gneiss

Where least affected by the amphibolite facies metamorphism, this gneiss consists of quartz, plagioclase, perthitic K-feldspar, rutile, biotite, garnet and sillimanite. A sample of sillimanite-garnet-gneiss (No. 383B, Figure 1) contains about one modal percent andalusite, which is a relatively rare mineral in the East Antarctic Precambrian Shield. This gneiss (grain size 0.1-2.5 mm) has been considerably altered by an amphibolite facies event, as suggested by absence of K-feldspar, and by textures such as biotite flakes in radiating sprays and plagioclase grains riddled with quartz inclusions. Sillimanite appears both as stout prisms (to 0.5 mm across) and as fine needles. Andalu-

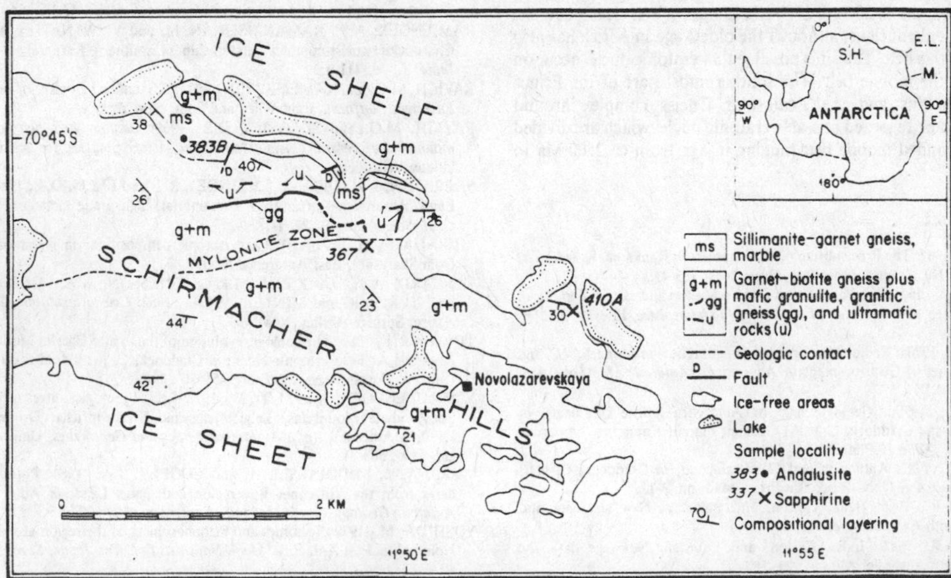


**Figure 1. Geologic sketch map of the Schirmacher Hills around Novolazarevskaya Station (1973 location). Geologic contacts and a mylonite zone are dashed where inferred to pass in areas of no bedrock exposure. Faults: D is downthrown side. Compositional layering: long arm of symbol is parallel to strike; number is dip in degrees. Index map (inset) shows Antarctic sapphirine occurrences: Schirmacher Hills (SH), Enderby Land (EL) and Mawson Station (M).**

**TABLE 1: Compositions of Minerals on Gneiss and Granulite, Schirmacher Hills, Queen Maud Land (Electron Microprobe Analyses)**

| Mineral | And. | Sil. | Garnet | | Biotite | | | Sapphirine[5] | | Crd[6] | Plag[6] |
|---|---|---|---|---|---|---|---|---|---|---|---|
| Sample # | 383B[1] | | 367[2] | 410A[3] | 367 | 410A[4] | 410A[4] | 367 | 410A | 410A | 367 |
| | | | | | weight percent | | | | | | |
| $SiO_2$ | 37.48 | 36.84 | 39.62 | 39.76 | 39.02 | 39.05 | 38.22 | 13.40 | 13.42 | 49.52 | 65.94 |
| $TiO_2$ | 0.03 | 0.03 | 0 | 0 | 2.14 | 2.29 | 2.81 | 0.02 | 0.05 | 0.01 | 0 |
| $Al_2O_3$ | 62.51 | 61.59 | 22.79 | 23.01 | 17.66 | 16.92 | 17.27 | 61.85 | 61.00 | 33.58 | 22.45 |
| $Cr_2O_3$ | 0.03 | 0.10 | 0 | 0.01 | 0.02 | 0.01 | 0 | 0.01 | 0.04 | 0 | 0 |
| FeO[7] | 0.36 | 0.23 | 27.45 | 26.07 | 8.61 | 8.95 | 9.79 | 9.51 | 9.90 | 3.53 | 0.04 |
| MnO | 0 | 0 | 0.24 | 0.24 | 0.01 | 0 | 0.02 | 0.01 | 0.02 | 0 | 0.01 |
| MgO | 0.03 | 0.01 | 10.76 | 11.98 | 19.55 | 19.61 | 18.39 | 15.39 | 15.60 | 11.46 | 0 |
| CaO | — | — | 0.82 | 1.10 | — | — | — | — | — | — | 3.27 |
| $Na_2O$ | 0 | 0 | 0.01 | 0.01 | 0.58 | 0.37 | 0.32 | 0 | 0 | 0.06 | 10.23 |
| $K_2O$ | — | — | — | — | 9.07 | 9.27 | 9.51 | 0.04 | 0 | 0.06 | 0.10 |
| Total | 100.44 | 98.80 | 101.69 | 102.18 | 96.66 | 96.47 | 96.33 | 100.23 | 100.03 | 98.22 | 102.04 |
| Oxygens | 5 | 5 | 12 | 12 | 22 | 22 | 22 | 20 | 20 | 18 | 18 |
| | | | | | Cations per given number of oxygens | | | | | | |
| Si | 1.008 | 1.007 | 2.979 | 2.959 | 5.531 | 5.562 | 5.489 | 1.604 | 1.614 | 4.992 | 2.850 |
| Ti | 0.001 | 0.001 | 0.000 | 0.000 | 0.228 | 0.245 | 0.303 | 0.002 | 0.005 | 0.001 | 0.000 |
| Al | 1.982 | 1.984 | 2.020 | 2.018 | 2.950 | 2.842 | 2.924 | 8.725 | 8.644 | 3.991 | 1.144 |
| Cr | 0.001 | 0.002 | 0.000 | 0.000 | 0.002 | 0.001 | 0.000 | 0.001 | 0.004 | 0 | 0.000 |
| $Fe^{3+}$ | 0.008 | 0.005 | — | — | — | — | — | 0.062 | 0.116 | — | — |
| $Fe^{2+}$ | — | — | 1.726 | 1.622 | 1.021 | 1.066 | 1.176 | 0.891 | 0.879 | 0.298 | 0.001 |
| Mn | 0.000 | 0.000 | 0.015 | 0.015 | 0.002 | 0.000 | 0.002 | 0.001 | 0.002 | 0.000 | — |
| Mg | 0.001 | 0.000 | 1.205 | 1.329 | 4.130 | 4.163 | 3.936 | 2.745 | 2.795 | 1.722 | 0.000 |
| Ca | — | — | 0.066 | 0.088 | — | — | — | — | — | — | 0.152 |
| Na | — | — | — | — | 0.159 | 0.102 | 0.089 | 0.000 | 0.000 | 0.012 | 0.857 |
| K | — | — | — | — | 1.640 | 1.685 | 1.742 | — | 0.000 | 0.008 | 0.005 |
| Total | 3.000 | 2.999 | 8.011 | 8.032 | 15.664 | 15.665 | 15.662 | 14.031 | 14.058 | 11.022 | 5.009 |

1. And. = Andalusite. Sil = Sillimanite. Equivalent $Fe_2O_3$ contents are 0.40 and 0.25 wt % respectively
2. Average of 3 grains. In fourth: FeO = 29.28 wt.%; MgO = 9.50 wt.% and atomic Mg/(Mg + Fe) = 0.366
3. Average of 3 grains. In fourth: FeO = 27.77 wt.%; MgO = l0.5l wt.% and atomic Mg/(Mg + Fe) = 0.403
4. Column 1—Average of two grains. Column 2. Third grain in same section.
5. $Fe^{3+}$ content of sapphirine estimated from stoichiometry (Higgins et al, 1979; Grew, 1980)
6. Crd—Cordierite, Plag—Plagioclase
7. All Fe as FeO

—indicates data not available or not calculated

site generally forms irregular, spongy grains or prisms 0.3 to 1.5 mm across, which are independent of sillimanite.

The andalusite contains more $Fe_2O_3$ and less $Cr_2O_3$ than associated sillimanite (Table 1), a relation characteristic of coexisting andalusite and sillimanite from contact aureoles (Okrusch and Evans, 1970). Andalusite undoubtedly crystallised during an amphibolite facies event. Nonetheless, an equilibrium distribution of $Fe^{3+}$ and Cr was attained between it and prismatic sillimanite, which is probably a relic from the granulite facies assemblage.

## Sapphirine-Garnet-Biotite Granulites

*Petrography.* Sapphirine-garnet-biotite granulites were found in two lenses up to 1 m thick in garnet-biotite gneiss (localities 367 and 410A, Figure 1). Sample 367 is a finely laminated somewhat schistose granulite consisting largely of biotite, grey plagioclase, garnet and dark green plates of sapphirine. Biotite, sapphirine and poikiloblastic, crudely tabular grains of corundum have parallel orientation and grain size ranges mostly from 0.1 to 1.5 mm. The rock appears fresh, although sapphirine is partly replaced by a fine corundum-biotite aggregate. Kyanite and sillimanite are present in trace amounts only. Kyanite prisms, up to 0.4 mm long, typically lie across the foliation and are independent of the rare, fine prisms and needles of sillimanite. Green spinel occurs as cores in sapphirine grains or more rarely as small grains in plagioclase and kyanite. Secondary sericite and zircon are present in small amounts, but no opaques or rutile were found.

Sample 410A is coarser grained (0.2 to about 3.5 mm) and more schistose than 367. Sillimanite prisms, biotite, garnet, sapphirine and cordierite are major constituents; plagioclase and perthitic K-feldspar occur in rare knots and probably constitute no more than a few modal percent. In general, sample 410A is texturally more complex than sample 367. Some sapphirine is enclosed in sillimanite and a selvage of cordierite generally separates sapphirine from plagioclase. Green spinel, corundum and plagioclase occur as inclusions in sapphirine; corundum also is found in biotite, but rarely in contact with cordierite. Cordierite and feldspar are largely fresh, while sapphirine is locally replaced by a fine corundum-biotite aggregate similar to that in 367. The other minerals in 410A, all present in small amounts, are sericite, zircon and kyanite(?).

Sapphirine in both samples is blue in thin section, its optic angle is small ($-2V \sim 10°$ or $20°$) and dispersion of the optic axes is marked.

*Whole-Rock Chemistry* The Schirmacher Hills sapphirine bearing rocks are poor in $SiO_2$ and rich in $Al_2O_3$, total Fe and MgO and have high Mg/Fe ratios (Table 2). They are compositionally similar to silica-unsaturated sapphirine granulites from Forefinger Point, Enderby Land (Sheraton, 1981), Val Codera, Italy (Barker, 1962) and the Anabar Shield, USSR (Lutts and Kopanova, 1968), except that the CaO and $Na_2O$ contents of 367 are relatively high. Moreover, the Schirmacher Hills granulites are relatively low in Cr and Zn, as are some Napier Complex metapelites and Al-Mg-rich granulites (Sheraton, 1980).

The sapphirine-garnet granulites are found, not in the sillimanite-garnet gneiss and marble unit, which probably represents metamorphosed pelitic and calcareous rocks, but in the garnet-biotite gneiss and mafic granulite unit, which may be metamorphosed felsic and mafic volcanic rocks. Assuming isochemical metamorphism, the bulk composition of the granulites suggests that thermally altered mafic igneous rocks, one of the origins Sheraton (1980) proposed for the magnesian Napier Complex rocks, or a bentonite (McKie, 1959), are possible protoliths for the Schirmacher Hills sapphirine-granulites.

*Mineral Chemistry* Microprobe analyses of the ferromagnesian silicates and plagioclase are listed in Table 1. Garnets are relatively magnesian and low in spessartine and grossular. Sapphirine contains a low proportion of $Fe^{3+}$, 6.5% and 11.7% of total Fe in 367 and 410A, respectively (calculated from stoichiometry following Higgins et al.,

**TABLE 2: Composition of Sapphirine-Garnet Granulites from the Schirmacher Hills, Queen Maud Land (X-ray fluorescence, G. Stummer, Analyst)**

| Sample No. | 367[1] | 410A[2] | | 367[1] | 410A[2] |
|---|---|---|---|---|---|
| | weight % | | | parts per million | |
| $SiO_2$ | 48.79 | 41.45 | Nb | 6 | 20 |
| $TiO_2$ | 0.94 | 1.21 | Cr | 34 | 33 |
| $Al_2O_3$ | 19.68 | 23.10 | Ni | 29 | 52 |
| $Fe_2O_3$ | 10.84 | 11.65 | Y | 82 | 79 |
| MnO | 0.054 | 0.046 | Rb | 101 | 356 |
| MgO | 9.97 | 17.28 | Sr | 39 | 3 |
| CaO | 1.40 | 0.27 | Zn | 58 | 42 |
| $Na_2O$ | 5.40 | 0.61 | Zr | 300 | 336 |
| $K_2O$ | 2.60 | 4.15 | | | |
| $P_2O_5$ | 0.098 | 0.008 | | weight % | |
| Total | 99.77 | 99.77 | L.O.I[3] | 0.77 | 1.56 |

All Fe as $Fe_2O_3$

1. Average of analyses on 3 pieces
2. Average of analyses on 2 pieces
3. Loss on ignition

1979; Grew, 1980). $Fe^{3+}$ contents of $Al_2SiO_5$ are also low, about 0.4 weight % $Fe_2O_3$ in sillimanite in 410A and 0.4 weight % $Fe_2O_3$ in kyanite in 367. These low $Fe^{3+}$ contents and absence of a $Fe^{3+}$-bearing oxide imply that the iron in the sapphirine granulites is largely ferrous. Biotites are magnesian, aluminous, and only moderately titanian. A spinel inclusion in sapphirine in sample 410A has a composition (in weight %): $SiO_2$/0.03, $Al_2O_3$/62.89, $TiO_2$ and $Cr_2O_3$/0, MnO/0.03, FeO/26.64, MgO/11.02, Total/100.61 or $Mg_{0.44}Fe_{0.59}Al_{1.98}O_4$. (Zn not detected on energy dispersive scan of elements).

Fe/Mg ratios in garnet and biotite, and $TiO_2$ contents in biotite vary from grain to grain (Table 1). In general, the mineral compositions are similar to those from the Napier Complex sapphirine-bearing rocks, notably in the high Mg-Fe ratios of garnet and biotite, and low $Fe^{3+}$ content of sapphirine (see Ellis et al., 1980; Grew, 1980, 1982a). However, biotites in 367 and 410A are significantly richer in $Al_2O_3$ and poorer in $TiO_2$ by comparison with biotites in Napier Complex metapelites, which contain 13.1 to 14.4 weight % $Al_2O_3$ and 3 to 5.8 weight % $TiO_2$ (Ellis et al., 1980; Grew, 1980, 1982a; and unpublished data on samples 2045G and 2032B). Moreover, the estimated amount of Al in octahedral coordination or $Al^{VI}$, 0.403 to 0.481 per 22 oxygens (from data in Table 1), is greater than $Al^{VI}$ in Napier Complex biotites, which is commonly absent and rarely exceeds 0.1 per 22 oxygens. The low $TiO_2$ contents of the Schirmacher Hills biotites may be due to the abundance of biotite in the rocks, all the $TiO_2$ being accommodated in the biotite and none remaining to form a Ti-saturating phase such as rutile.

*Petrologic Interpretation* On the basis of textures and chemistry, the following mineral assemblages are considered to represent the peak metamorphic conditions: (367) biotite-plagioclase-garnet-sapphirine-corundum; (410A): biotite ± plagioclase ± K-feldspar-garnet-sapphirine-sillimanite-cordierite. Effects of later metamorphic events include modifications to the compositions of garnet and biotite, partial alteration of sapphirine to corundum and biotite and appearance of kyanite. Although there is no evidence of cataclasis in sample 367, the proximity of locality 367 to a mylonite zone (Figure 1) suggests that the appearance of kyanite in 367 is related to the mylonite zone.

Temperatures calculated from garnet and biotite compositions in Table 1 are 553°C in 367 and 604° and 650°C in 410A (for biotite analyses 1 and 2, respectively); and from garnet-cordierite compositions, 683°C in 410A (using calibrations of Thompson, 1976). These estimates appear low and may reflect reequilibration during later metamorphic events, although fluorine in biotite may also be a factor (see below).

The association sapphirine-garnet suggests metamorphic temperatures of at least 730°C and pressures of at least 5 kb, according to a petrogenetic grid constructed according to the Schreinemakers method (Grew, 1982b, Figure 14). The association sapphirine-garnet-corundum may not be stable below pressures of 6 kb at temperatures near 750°C, but the petrogenetic grid involving corundum is less well constrained (Grew, unpublished data). The presence of sillimanite implies that pressures did not exceed 8.9 kb (T = 750°C) or 9.9 kb (T = 800°C) (Holdaway, 1971). In sum, metamorphic temperatures were probably at least 100°C lower than pressures about the same as those which pertained in the Tula Mountains in Enderby Land, for which (Grew, 1980) estimates 900°C and 7 kb.

## PETROLOGY OF ANTARCTIC SAPPHIRINE-GARNET ASSOCIATIONS

### Fe-Mg Partitioning among Sapphirine, Garnet and Biotite

The partitioning of $Fe^{2+}$ and Mg between coexisting sapphirine and garnet from the Tula and Scott Mountains, Enderby Land, is regular (Nos. 3-11, 13, 14, Figure 2), which is consistent with the conclusion that these minerals crystallised in equilibrium. Garnet-sapphirine pairs from granulite facies rocks of Labwor, Uganda (Nixon et al., 1973), and from a crustal xenolith in alkali basalt from the Central Hoggar, Algeria (Leyreloup et al., 1982) lie close to the general Enderby Land trend, while the Casey Bay (Enderby Land) and two Schirmacher Hills pairs (Nos. 1,2,12) have somewhat higher $K_D$'s (where $K_D$ = ($Fe^{2+}$/Mg (garnet)) (Mg/$Fe^{2+}$ (sapphirine) in atomic proportions). The relatively high value for $K_D$ (~ 3.5) suggests that iron-magnesium fractionation between garnet and sapphirine may be sensitive to temperature. Thus the garnet-sapphirine data imply that metamorphic temperatures were uniform within the rock mass now

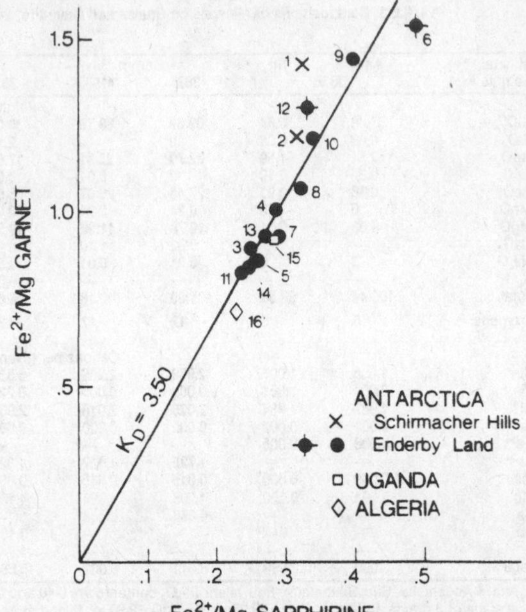


**Figure 2.** **Atomic $Fe^{2+}$-Mg ratios for associated sapphirine and garnet. For garnet: all Fe assumed to be $Fe^{2+}$; sapphirine $Fe^{2+}$ estimated from stoichiometry (Higgins et al., 1979; Grey, 1980). Key to numbers and sources of data (other than this report and my own unpublished data) is as follows: Schirmacher Hills: 1-367 and 2-410A Enderby Land: 3-2018C, Beaver Island; 4-2032B, 5-2032E and 14-76283350 (Ellis et al., 1980), Spot Height 945; 6-2045A, 7-2045G, 8-2050B (Grew, 1978) and 9-2052B, Gage Ridge; 10-2074D, Mount Hardy; 11-2126A, near Mount Torckler (Grew, 1979); 12-2234L (2), Casey Bay (Grew, 1981); and 13-2376C, Mount Hollingsworth. Uganda: 15-PHN-1069, Labwor (Nixon et al., 1973). Algeria: 16-91-1, granulite xenolith from the Central Hoggar (Leyreloup et al., 1982 and A. Leyreloup, pers. comm., 1982). Cross over circle indicates sapphirine contains Be (Grew, 1982); Be in 2945A (no. 6) confirmed with ion microscope. Line for $K_D$ = ($Fe^{2+}$-Mg (garnet)) (Mg-$Fe^{2+}$ (sapphirine)) = 3.5 is for reference only.**

exposed in the Tula and Scott Mountains, an inference also supported by orthopyroxene Al contents (Grew, 1980). Moreover, metamorphic temperatures in Labwor and the Hoggar xenolith were close to the Tula Mountain temperatures. The increased fractionation of iron into the Schirmacher Hills garnets suggests lower peak metamorphic temperatures or reequilibration during a later event.

Partitioning of iron and magnesium between garnet and biotite is a potential geothermometer. Two of the sapphirine-garnet rocks, 2045G and 2032B (Nos. 7 and 4, Figure 2), contain biotite (< 1 modal percent). Biotite in 2045G appears to be in textural equilibrium with garnet and sapphirine, while textures in 2032B are ambiguous. Garnet-biotite temperatures calculated by the method of Thompson (1976) are 563°C (2045G) and 604°C (2032B); corresponding atomic Fe/(Mg + Fe) ratios ($X_{Fe}$) of biotite are 0.142 and 0.174 and fluorine contents, 3.4 and 2.8 weight % F (Grew, unpublished data). The high fluorine contents are consistent with these biotites forming at high temperatures. Secondary biotite ($X_{Fe}$ = 0.17 – 0.19), which occurs sparingly in a sapphirine bearing rock from Mount Hardy in the Tula Mountains (2083A, Grew, 1982a), contains only 1% F. The low temperatures estimated for garnet-biotite in the two sapphirine-garnet rocks may be due to later formation of biotite (Ellis et al., 1980) or to reequilibration of garnet and biotite at lower temperatures after the peak of metamorphism. An alternative explanation is that fluorine and iron in biotite tend to exclude one another (Valley et al., 1982), thereby increasing the fractionation of Fe into coexisting garnet. Thus, temperature calibrations based on low fluorine biotites, such as Thompson's (1976), would give "low" temperatures for pairs involving fluorine rich magnesian biotites.

## Garnet-Sapphirine-Sillimanite and Incompatibility of Orthopyroxene and Corundum

Orthopyroxene is absent in the Schirmacher Hills sapphirine-garnet rocks, while corundum has not previously been reported in silica-undersaturated sapphirine bearing rocks of Enderby Land (e.g. Ravich and Kamenev, 1972, Table 43; Ellis et al., 1980). However, in some Enderby Land rocks corundum occurs with sapphirine in parts of a layer a few metres thick and 3 km in extent at Reference Peak and in a mass 1 x 3 m at Mount Cronus. These rocks characteristically comprise sapphirine-garnet-spinel-corundum. Orthopyroxene is also present in some sections but not in contact with corundum. In a schistose biotite-sapphirine-orthopyroxene-cordierite granulite from Forefinger Point, Enderby Land, corundum is enclosed in aggregates of sapphirine and thereby isolated from orthopyroxene. Noteworthy accessories in this granulite are apatite and wagnerite. Corundum also occurs as tablets along the contact between sapphirine megacrysts and an enclosing sillimanite corona in quartzose granulites (e.g. Casey Bay, Grew, 1981, 1982b, Figure 5, and Mount Hollingsworth); an orthopyroxene corona separates sillimanite from the quartz matrix. In short, there is no evidence for the orthopyroxene-corundum association being stable in either Enderby Land or the Schirmacher Hills at the peak of metamorphism. This assemblage may have been stable in the coronas formed at lower temperatures during cooling (probably isobaric, see Ellis, 1980) of the Napier Complex following the peak conditions. Instead, the assemblage sapphirine-garnet-sillimanite appears to have been stable, for example in sample 410A from the Schirmacher Hills and in a quartz-free granulite from Reference Peak as well as several quartz-bearing granulites (Ellis et al., 1980; Grew, 1979). This assemblage is incompatible with orthopyroxene-corundum and a possible univariant reaction (in the system $MgO-FeO-Al_2O_3-SiO_2$), relating orthopyroxene-corundum and sapphirine-garnet-sillimanite is:

$$8.44(Fe_{0.45}Mg_{1.35}Al_{0.4}Si_{1.8}O_6 + 20.08\ Al_2O_3 =$$
$$3.30(Mg_{2.96}Fe_{0.74}Al_{8.6}Si_{1.7}O_{20}) + Mg_{1.64}Fe_{1.36}Al_2Si_3O_{12} + 6.60Al_2SiO_5$$

(compositions simplified from sample 2126A, see Figure 2 and Grew, 1979). The sapphirine bearing assemblage has lower density (molar volume increase calculated to be 56 $cm^3$ using approach of Grew, 1982a) and probably greater entropy, that is, the reaction has a positive slope and the sapphirine-bearing assemblage is on the high temperature side. The orthopyroxene-corundum association probably requires metamorphic pressures in excess of the kyanite-sillimanite univariant curve and very low water pressures for temperatures of 750°-850°C. These conditions were attained during near isobaric cooling in the Wilson Lake area, Labrador, Canada, where orthopyroxene-corundum and kyanite have formed among the breakdown products of sapphirine-quartz (Morse and Talley, 1971).

*Acknowledgements* I thank P.K. Senko, Yu.M. Zusman and other members of the 18th Soviet Antarctic Expedition for arranging my visit to the Schirmacher Hills and for assistance in the field, and members of the 1977/78 and 1979/80 Australian National Antarctic Research Expeditions for logistic support and assistance to field work in Enderby Land. I also thank A. Leyreloup for unpublished analyses of the Algerian sapphirine and garnet, J.F. Lovering for use of the electron microprobe at the University of Melbourne and G. Stummer for the x-ray fluorescence analyses. An earlier version of this paper was reviewed by an anonymous reviewer. The author appreciates his thoughtful comments, which led to a substantial improvement in the paper. This research was funded by US National Science Foundation grants OPP72-057976 to the University of Wisconsin, Madison and DPP76-80957 and DPP80-19527 to the University of California, Los Angeles.

## REFERENCES

BARKER, F., 1964: Sapphirine bearing rock, Val Codera, Italy. *Am. Mineral., 49,* 146-52.

ELLIS, D.J., 1980: Osumilite-sapphirine-quartz granulites from Enderby Land, Antarctica: P-T conditions of metamorphism, implications for garnet-cordierite equilibria and the evolution of the deep crust. *Contrib. Mineral. Petrol., 74,* 201-10.

ELLIS, D.J., SHERATON, J.W., ENGLAND, R. and DALLWITZ, W.B., 1980: Osumilite-sapphirine-quartz granulites from Enderby Land, Antarctica; mineral assemblages and reactions. *Contrib. Mineral. Petrol., 72,* 123-43.

GREW, E.S., 1978: Osumilite at Gage Ridge, Enderby Land, Antarctica (66°54′S, 51°16′E) (abstract). *Am. Geophys. Union, Trans., 59,* 1216.

GREW, E.S., 1979: Reactions involving sapphirine and sillimanite + orthopyroxene in quartz bearing rocks of the 2.5 b.y. Napier Complex, Enderby Land, East Antarctica (abstract). *Geol. Soc. Am., Abstr. with Prog., 11,* 435-6.

GREW, E.S., 1980: Sapphirine + quartz association for Archean rocks in Enderby Land, Antarctica. *Am. Mineral., 65,* 821-36.

GREW, E.S., 1981: Surinamite, taaffeite and beryllian sapphirine from pegmatites in granulite facies rocks of Casey Bay, Enderby Land, Antarctica. *Am. Mineral., 66,* 1022-33.

GREW, E.S., 1982a: Osumilite in the sapphirine-quartz terrain of Enderby Land, Antarctica: implications for osumilite petrogenesis in the granulite facies. *Am. Mineral., 67,* 762-87.

GREW, E.S., 1982b: Sapphirine, kornerupine and sillimanite + orthopyroxene in the charnockitic region of South India. *Geol. Soc. India, J., 23,* 469-505.

HIGGINS, J.B., RIBBE, P.H. and HERD, R.K., 1979: Sapphirine I. Crystal chemical contributions. *Contrib. Mineral. Petrol., 68,* 349-56.

HOLDAWAY, M.J., 1971: Stability of andalusite and the aluminium silicate phase diagram. *Am. J. Sci., 271,* 97-131.

LEYRELOUP, A., BODINIER, J.L., DUPUY, C. and DOSTAL, J., 1982: Petrology and geochemistry of granulite xenoliths from Central Hoggar (Algeria)—implications for the lower crust. *Contrib. Mineral. Petrol., 79,* 68-75.

LUTTS, B.G. and KOPANEVA, L.N., 1968: A pyrope-sapphirine rock from the Anabar Massif and its conditions of metamorphism. *Dokl. Akad. Nauk SSSR, 179, (5),*1200-02 (English Translation, 1968, in Earth Science Sections, 161-63).

McKIE, D., 1959: Yoderite, a new hydrous magnesium iron aluminosilicate from Mautia Hill, Tanganyika. *Mineral. Mag., 32,* 282-307.

MORSE, S.A. and TALLEY, J.H., 1971: Sapphirine reactions in deep seated granulites near Wilson Lake, central Labrador, Canada. *Earth Planet. Sci. Lett., 10,* 325-8.

NIXON, P.H., REEDMAN, A.J. and BURNS, L.K., 1973: Sapphirine bearing graunulites from Labwor, Uganda. *Mineral. Mag., 39,* 420-8.

OKRUSCH, M. and EVANS, B.W., 1970: Minor element relationships in coexisting andalusite and sillimanite. *Lithos, 3,* 261-8.

RAVICH, M.G. and KAMENEV, Ye.N., 1972: "Kristallicheskiy Fundament Antarkticheskoy Platformy" (in Russian) *Gidrometeoizdat,* Leningrad.

SHERATON, J.W., 1980: Geochemistry of Precambrian metapelites from East Antarctica: secular and metamorphic variations. *BMR J. Aust. Geol. Geophys., 5,* 279-88.

SHERATON, J.W., 1981: Chemical analyses of rocks from East Antarctica. *Aust., Bur. Miner. Resour., Geol. Geophys., Rec. 1981/14.*

THOMPSON, A.B., 1976: Mineral reactions in pelitic rocks: II. Calculations of some P-T-X (Fe-Mg) phase relations. *Am. J. Sci., 276,* 425-54.

VALLEY, J.W., PETERSEN, E.U., ESSENE, E.J. and BOWMAN, J.R., 1982: Fluorphlogopite and fluortremolite in Adirondack marbles and calculated C-O-H-F fluid compositions. *Am. Mineral., 67,* 545-57.

# A REVIEW OF THE TECTONIC AND METAMORPHIC HISTORY OF THE LUTZOW-HOLM BAY REGION, EAST ANTARCTICA

M. Yoshida, *Department of Geosciences, Osaka City University, Osaka 58, Japan.*

M. Suzuki, *Department of Geology and Mineralogy, Hiroshima University, Hiroshima, Japan.*

H. Shirahata, *Department of Mineral Resources Engineering, Muroran Institute of Technology, Muroran 050, Japan.*

H. Kojima, *Institute of Mining Geology, Akita University, Akita 010, Japan.*

K. Kizaki, *Department of Oceanography, Ryukyu University, Naha, Japan.*

*Abstract* Recent studies of the tectonic and metamorphic history of the Lützow-Holm Bay region are critically reviewed and summarised as follows: The earliest tectono-metamorphic episode was a high pressure granulite facies metamorphism probably associated with early recumbent and later isoclinal folding about axes parallel to the present coast lines. A single Pb-Pb isochron age of 1900 Ma is thought to date this event. A subsequent event involved intermediate pressure granulite facies metamorphism and possible but as yet unidentified folding. It is thought to be dated by Rb-Sr isochron ages of 1100 Ma. The third tectono-metamorphic event involved east-west open to close folding and upper amphibolite facies metamorphism. The oldest K-Ar age of 560 Ma is considered to provide a minimum age for this. The fourth event consisted of gentle north-south folding and faulting, syntectonic granitoid intrusion and local low amphibolite facies metamorphism and is thought to be dated by averaged Pb isotope ages of 470 Ma and a Rb-Sr isochron age of 458 Ma. Greenschist facies or slightly lower grade metamorphism occurred subsequently, probably during the latest phase of the fourth event. The fifth event involved fracturing under the influence of a stress field with a north-south maximum extensional axis. It may have occurred during the Jurassic with the stress perhaps reflecting Gondwanaland disruption.

Geological surveys around Lützow-Holm Bay including the Yamato and Belgica Mountains and Prince Olav Coast (Figure 1) since 1959 have led to the geologic, tectonic and metamorphic syntheses recently summarised by Yoshida (1978; 1979a) and Yoshida et al. (1982) as a polyorogenic sequence ranging in age from older than 1100 Ma to younger than 200 Ma. The present article is a critical review and summary of the metamorphic, geochronologic and tectonic sequence in the Lützow-Holm Bay region.

## Former Interpretation of the Polymetamorphic-Tectonic Sequence

Kizaki (1964; 1965) first drew attention to the superposition of the effects of a later amphibolite facies metamorphism upon those of an earlier granulite facies metamorphism and some tectonic differences between the two metamorphisms in the area around Lützow-Holm Bay and the Yamato Mountains.

This polyorogenic view was further developed from tectonic, stratigraphic, metamorphic and geochronologic studies which were recently summarised by Yoshida (1979a) as: 1st stage: early recumbent and later isoclinal folding ($D_1$) and syntectonic high pressure granulite facies metamorphism ($M_1$) of Ongul and Skallen Groups. (The Okuiwa Group formed probably after these events): 2nd stage: intermediate pressure granulite facies metamorphism ($M_2$) under static stress conditions and succeeded by charnockite formation, at 1100 Ma: 3rd stage: east-west open to close folding ($D_2$) and syntectonic, high amphibolite facies metamorphism ($M_3$) at about 500 Ma: 4th stage: north-south gentle and kink folding and faulting ($D_3$), associated with pink granite intrusions and low amphibolite facies metamorphism ($M_4$, thought to be the late phase of $M_3$ metamorphism) at about 400 Ma; local greenschist facies or slightly lower grade metamorphism ($M_5$) related to the pink granite intrusion: 5th stage: fracture system with maximum extension stress axis in the north-south trend ($D_4$); this event is considered tentatively to have occurred less than 200 Ma ago and to be a manifestation of Gondwanaland disruption tectonics.

## Superposed Mineral Parageneses and Element Partitioning Among Mineral Pairs

*Superposition of the amphibolite facies mineral parageneses on granulite facies parageneses* Banno et al. (1964) and Suwa (1968) considered that the mineral parageneses of the metamorphic rocks around Lützow-Holm Bay were principally formed by monometamorphism under equilibrium conditions. Kizaki (1964) described how pyroxene gneiss was altered to hornblende gneiss as a consequence of the emplacement of microcline granite. He explained this as the overprinting of the amphibolite facies metamorphism upon the granulite facies rocks. He later described in the Yamato Mountains the conversion of granulite facies charnockitic rocks into granitic gneiss and migmatite as a result of amphibolite facies metamorphism, which included the intrusion of microcline granite as a very late phase (Kizaki, 1965).

The development of amphibolite facies mineral parageneses from granulite facies rocks throughout the Lützow-Holm Bay region was also pointed out by Yoshida (1975; 1978; 1979b). Yoshida (1978; 1979b) also referred to mineral parageneses composed of plagioclase (pl), orthopyroxene (opx), clinopyroxene (cpx), (plus hornblende (hb) and/or biotite (bi)) in more mafic rocks and quartz (qz), orthoclase (or), sillimanite (sil), garnet (ga) as typical of the intermediate pressure granulite facies metamorphism ($M_2$). The later widespread amphibolite facies parageneses ($M_3$) were described as being typically composed of pl-hb and qz-potash feldspar(kf)-sil-bi-ga; he also showed some overprinting relationships and concluded that the amphibolite facies paragenesis had been imprinted upon the earlier granulite facies paragenesis.

Two different explanations for the dominant mineral parageneses of rocks around Lützow-Holm Bay are:

(1) monometamorphic low granulite facies metamorphism, or
(2) superposition of two different metamorphisms, viz. earlier high granulite facies metamorphism and a later high amphibolite facies one.

A systematic compositional change of biotite in relation to that of orthopyroxene and/or garnet (Banno et al., 1964; Suzuki, 1982) and a progressive metamorphic zoning with increasing grade of metamorphism southwestwards along the Prince Olav Coast (Hiroi et al., this volume, see below) appear to support explanation (1), whereas field and microscopic occurrences of hydrous minerals and the diversity of the garnet-biotite and two pyroxene geothermometry temperatures (see below) support explanation (2).

*Low amphibolite and greenschist facies or slightly lower grade metamorphisms* Yoshida (1978; 1979b) found that a small amount of muscovite and chloritised biotite locally occurs throughout the Lützow-Holm Bay region, especially adjacent to the pink granite and pegmatite. He correlated this event with the occurrence of a qz-kf-pl-sil-mus-bi paragenesis on the Prince Olav Coast because of the common occurrence of muscovite and dominance of the pink granite there. He considered this mineral paragenesis to represent the late phase of the $M_3$ metamorphism ($M_4$ metamorphism, Yoshida, 1979a). Hiroi et al. (this volume) recently clarified the progressive metamorphism from high amphibolite facies to low granulite facies from northeast to southwest along the Prince Olav Coast. Here, muscovite bearing assemblages are regarded as one of the typomorphic parageneses of high amphibolite facies metamorphism (Hiroi et al., this volume; Suzuki, 1979). Late stage granite intrusion has resulted in an andalusite overgrowth in country rocks adjacent to the intrusion and partial re-equilibration of element partitioning among some minerals in rocks from extensive area along the Prince Olav Coast. The re-equilibration has resulted in the lowering of garnet-biotite geothermometry temperatures (method of Thompson, 1966) from about 690°C to 530°C.

From rocks of the west coast of Lützow-Holm Bay (around Botnneset), Yoshida (1975) reported the sporadic development of chlorite, sericite and epidote as secondary minerals and regarded them

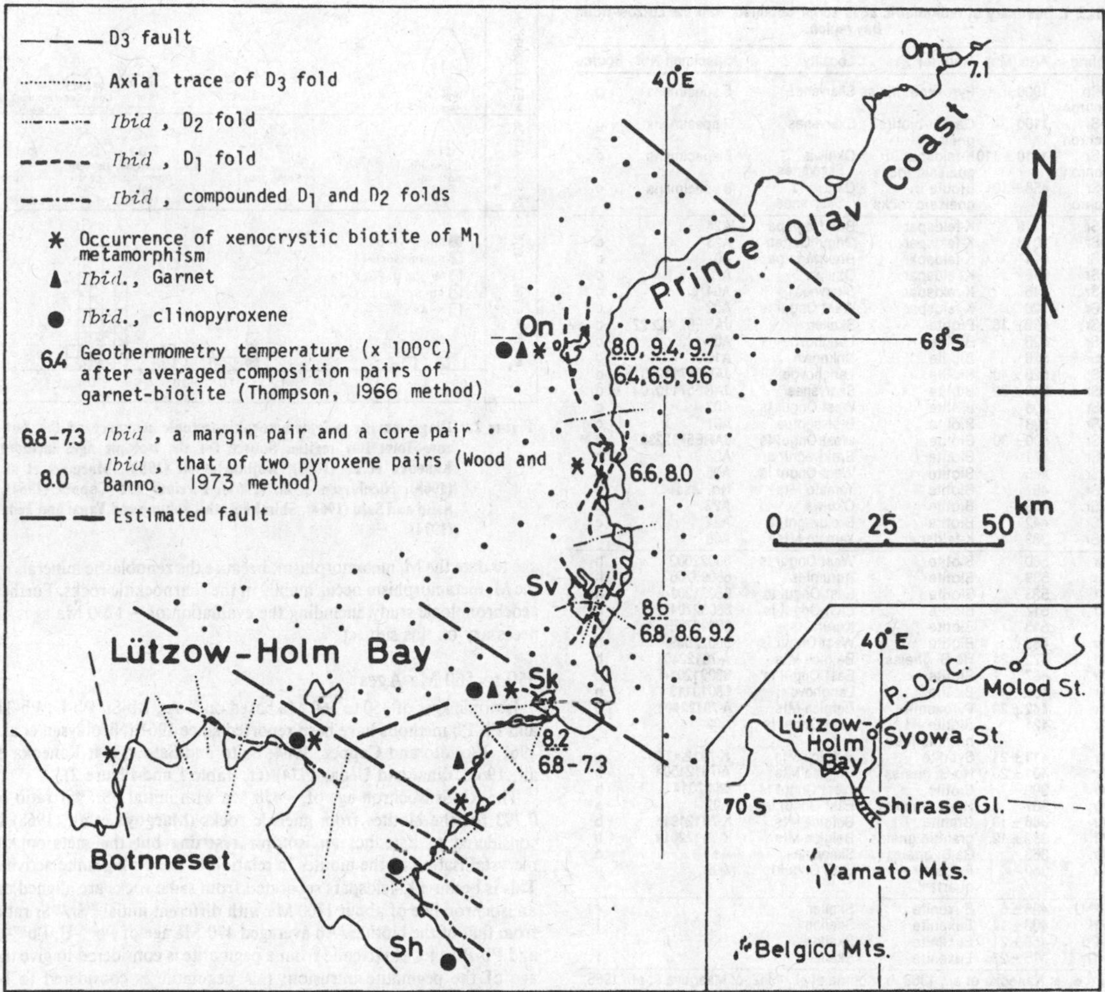


**Figure 1.** Geologic sketch of the Lutzow-Holm Bay region including distribution of xenocrystic minerals of the $M_1$ metamorphism and geothermometry temperatures. Dotted area: Probable distribution of the Ongul Group. Southern blank area: Probable distribution of the Skallen Group. Northern blank area: Probable distribution of the Okuiwa Group. Molod St: Molodezhnaya Station. Om: Cape Omega. On: Ongul Islands where the Syowa Station stands. POC: Prince Olav Coast. Sh: Shirase Glacier. Sk: Skallen. Sv: Skarvsnes. All data are referred to Yoshida (1978, 1979b).

as reflecting lower grades of metamorphism ($M_5$, probably greenschist facies or slightly lower grade conditions). These minerals are invariably dominant in the pink granite and associated rocks. No related deformation is known. Thus, the low grade alteration mineral assemblages are tentatively concluded to reflect later stage alteration by the pink granite activity (Yoshida, 1975). Similar occurrences are widely observed throughout the Lützow-Holm Bay region, the Yamato and Belgica Mountains and Prince Olav Coast (Ohta and Kizaki, 1966; Kojima et al., 1982; Nakai et al., 1980; 1981).

Thus data on alteration or re-equilibration associated with the pink granite activity are accumulating. Some data are not consistent with the previous summary; some revisions, therefore, may be made in future. The low amphibolite facies metamorphism ($M_4$) and the lower grade metamorphism ($M_5$) are possibly cogenetic events.

*Metamorphism prior to the dominant granulite facies metamorphism*
Yoshida (1978) described a widespread occurrence (cf. Figure 1) of xenoblastic clinopyroxene, garnet, quartz and biotite with fine to medium grained polygonal to granular plagioclase or hypersthene in charnockitic rocks and considered that the xenoblastic minerals are relics of a high pressure granulite facies metamorphism ($M_1$) that preceded the dominant intermediate pressure granulite facies metamorphism ($M_2$) now represented by the rocks that contain the xenoblasts.

Yoshida (1978; 1979b) further described relict sodic plagioclase and garnet mantled by calcic plagioclase and plagioclase-hypersthene symplectite in some charnockites and metabasites; these relict minerals were also considered to belong to the $M_1$ metamorphism.

Possibly, however, the xenoblastic and relic mineral parageneses formed under lower temperature conditions in an earlier phase of a single prograde metamorphism. There are thus two possibilities. The xenoblastic and relic minerals:

(1) may represent earlier events ($M_1$) quite distinct from the dominant granulite facies metamorphism ($M_2$), or
(2) may represent the lower grade mineral paragenesis of an earlier phase of prograde $M_2$ metamorphism.

Although decisive data are still insufficient, we prefer the explanation (1) from tectonic relationships (see below).

*On the superposition of element partitioning among pairs of minerals*
Detection of the superposition of metamorphism by element partitioning among pairs of minerals is especially difficult where earlier granulite facies assemblages are intensely overprinted by later amphibolite facies assemblages for the following reasons:

(1) non-equilibrium conditions are generally to be expected from kinetic considerations and inhomogeneity in deformation and material supply,
(2) re-equilibration of (and between) earlier anhydrous minerals with later hydrous ones during the later amphibolite facies metamorphism (often with no microstructural changes of the anhydrous minerals) may disturb the initial equilibrium among anhydrous minerals and further, may result in a "false equilibrium" among them.

Thus, the study of element partitioning among minerals should be made very carefully and involve numerous microprobe analyses of points in a single thin section selected on the basis of detailed

**TABLE 1: Summary of radiometric ages so far obtained from the Lutzow-Holm Bay region**

| Method | Age (Ma) | Material | Locality | Specimen No. | Source |
|---|---|---|---|---|---|
| Pb-Pb isochron | 1900 | Pyroxene gneiss | Skarvsnes | 5 specimens | g |
| Rb-Sr isochron | 1100 | Garnet-biotite gneiss | Skarvsnes | 2 specimens | g |
| Rb-Sr isochron | 1110 ± 110 | K-feldspar in gneissic rocks | Okuiwa-Skarvsnes | 6 specimens | c |
| Rb-Sr isochron | 458 ± 10 | Biotite in gneissic rocks | Okuiwa-Skarvsnes | 8 specimens | c |
| Rb-Sr | 1116 | K-feldspar | Breidvagnipa | A24 | c |
| Rb-Sr | 1013 | K-feldspar | Ongul Galten | A23 | c |
| Rb-Sr | 971 | K-feldspar | Breidvagnipa | A01 | c |
| Rb-Sr | 816 | K-feldspar | Okuiwa | A22 | c |
| Rb-Sr | 745 | K-feldspar | Skarvnes | A04 | c |
| Rb-Sr | 726 | K-feldspar | West Ongul Is | A02 | c |
| Rb-Sr | 530 ± 16 | Biotite | Skallen | JARE57102622 | d |
| Rb-Sr | 526 | Biotite | Langhovde | A09 | c |
| Rb-Sr | 526 | Biotite | Unknown | A10 | c |
| Rb-Sr | 525 ± 40 | Biotite | Langhovde | JARE57112001 | d |
| Rb-Sr | 510 ± 30 | Biotite | Skarvsnes | JARE57110704 | d |
| Rb-Sr | 508 | Biotite | West Ongul Is | A02 | c |
| Rb-Sr | 508 | Biotite | Breivagnipa | A01 | c |
| Rb-Sr | 500 ± 30 | Biotite | West Ongul Is | JARE57122307 | d |
| Rb-Sr | 471 | Biotite | Breidvagnipa | A03 | c |
| Rb-Sr | 465 | Biotite | West Ongul Is | A05 | c |
| Rb-Sr | 457 | Biotite | Yamato Mts | No. less | e |
| Rb-Sr | 448 | Biotite | Okuiwa | A22 | c |
| Rb-Sr | 442 | Biotite | Breidvagnipa | A24 | c |
| Rb-Sr | 383 | K-feldspar | Yamato Mts | A08 | c |
| K-Ar | 560 | Biotite | West Ongul Is | 68022002 | h |
| K-Ar | 539 | Biotite | Kurumi Is | 68090706 | h |
| K-Ar | 533 | Biotite | East Ongul Is | 68091201-2 | h |
| K-Ar | 517 | Biotite | East Ongul Is | 68032704 | h |
| K-Ar | 515 | Biotite | Kurumi Is | 68090706 | h |
| K-Ar | 485 | Biotite | West Ongul Is | 68022609 | h |
| K-Ar | 472 ± 24 | Ho-Bi gneiss | Belgica Mts | A-79122401 | b |
| K-Ar | 467 | Biotite | East Ongul Is | 68091201-1 | h |
| K-Ar | 463 | Biotite | Langhovde | 68013113 | h |
| K-Ar | 442 ± 22 | Pyroxenite | Belgica Mts | A-79122401 | b |
| K-Ar | 421 | Biotite + hornblende | East Ongul Is | A02 | a |
| K-Ar | 411 ± 21 | Syenite | Belgica Mts | K-79122607 | b |
| K-Ar | 401 ± 20 | Ho-Bi gneiss | Belgica Mts | A-79121504 | b |
| K-Ar | 399 | Biotite | West Ongul Is | 68-22014 | h |
| K-Ar | 387 | Ho-Bi gneiss | East Ongul Is | A02 | a |
| K-Ar | 386 ± 19 | Granite | Belgica Mts | K-79121914 | b |
| K-Ar | 382 ± 19 | granitic gneiss | Belgica Mts | K-79122014 | b |
| K-Ar | 363 | Ga-Bi gneiss | Skarvsnes | As | a |
| K-Ar | 350 | Feldspar + quartz | East Ongul Is | A02 | a |
| $Pb^{206}$-U | 485 ± 6 | Euxenite | Skallen | | f |
| $Pb^{207}$ | 468 ± 12 | Euxenite | Skallen | | f |
| Pb-Pb | 458 ± 21 | Euxenite | Skallen | | |
| Pb-Th | 375 ± 27 | Euxenite | Skallen | | f |

Source: a: Kaneoka et al., 1968; b: Kojima et al., 1982; c: Maegoya et al., 1968; d: Nicolaysen et al., 1961; e: Picciotto Coppez., 1964; f: Saito and Sato, 1964; g: Shirahata, in press; h: Yanai and Ueda, 1974.

microstructural relationships. Only preliminary determinations lacking the microstructural considerations of element partitioning among pairs of minerals were made by Yoshida (1979b). The wide range, viz. 643 to 957°C of temperatures calculated from garnet-biotite (except margin pairs) geothermometry (after Thompson's 1966 method) and similar deviation of two pyroxene geothermometry temperatures for the rocks of a restricted area (cf. Figure 1; and Yoshida, 1979b) may reflect nonequilibrium conditions among the pairs of minerals due, perhaps, to superposition of at least two (probably $M_1$, $M_2$ and/or $M_3$) different temperature conditions.

## Geochronology

*Ages between 1900 Ma and 1100 Ma* An Rb-Sr isochron age of 1100 ± 110 Ma with initial $^{87}Sr/^{86}Sr$ ratio of 0.704 for K-feldspars from the gneissic rocks of the east coast of Lützow-Holm Bay was reported by Maegoya et al. (1967). Shirahata (this volume) also reports an Rb-Sr isochron age of 1100 Ma with initial $^{87}Sr/^{86}Sr$ ratio of 0.709 for garnet-biotite gneisses from Skarvsnes on the same coast (Table 1). Their results are interpreted here in terms of a granulite facies metamorphism ($M_2$) at 1100 Ma, as indicated by Yanai and Ueda (1974) and Yoshida (1975; 1979a).

Recently, Shirahata (this volume) obtained a maximum isochron apparent age of 1900 Ma on the basis of a two stage Pb evolution model for charnockitic rocks from Skarvsnes. Also, he presented another calculation of a Pb-Pb isochron age of 1700 Ma by adding analyses of garnet-biotite gneisses to those of the charnockitic rocks and an Rb-Sr age of 1740 Ma for a pyroxene gneiss assuming initial $^{87}Sr/^{86}Sr$ ratio of 0.7037. We prefer tentatively the ~1900 Ma

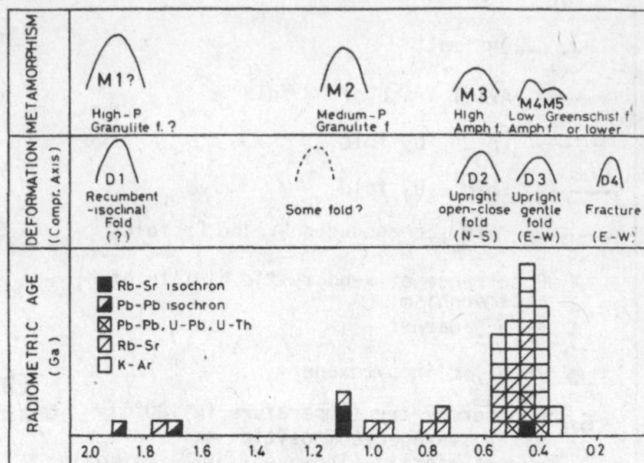


**Figure 2. Diagrammatic polymetamorphic-tectonic sequence of the Lutzow-Holm Bay region. Source for the isotopic ages includes Kaneoka et al. (1968), Kojima et al. (1982), Maegoya et al. (1968), Nicolaysen et al. (1961), Picciotо and Coppez (1964), Saito and Sato (1964), Shirahata, this volume and Yanai and Yeda (1974).**

age to date the $M_1$ metamorphism, because the xenoblastic minerals of the $M_1$ metamorphism occur mainly in the charnockitic rocks. Further geochronologic study, including the evaluation of ~1700 Ma ages, is necessary on this subject.

### 350 to 560 Ma Ages

Isotopic ages of 350 to 560 Ma based on K-Ar, Rb-Sr, Pb-U, Pb-Th and Pb-Pb methods have been reported since 1961 (Nicolaysen et al., 1961; Picciotо and Coppez, 1964; Saito and Sato, 1964; Kaneoka et al., 1968; Yanai and Ueda, 1974) (cf. Table 1 and Figure 2).

The Rb-Sr isochron age of ~458 Ma with initial $^{87}Sr/^{86}Sr$ ratio of 0.793 for the biotites from gneissic rocks (Maegoya et al., 1968) is considered to date not an isotopic resetting but the metasomatic recrystallisation of the biotites in relation to the pink granite activity. This is because K-feldspars separated from same rocks are aligned on an isochron line of about 1100 Ma with different initial $^{87}Sr/^{86}Sr$ ratio from that of the biotites. An averaged 470 Ma age of $Pb^{206}$-U, $Pb^{207}$-U and Pb-Pb ages of euxenite from a pegmatite is considered to give the age of the pegmatite intrusion; this pegmatite is considered to be cogenetic with the pink granite. K-Ar ages are considered to indicate the minimum age of the latest widespread metamorphism ($M_3$) that is 560 Ma and the maximum age of cooling, that is 350 Ma.

## Tectonics associated with Metamorphism

*Tectonics during the $M_1$ metamorphism* Fine grained xenoblasts in charnockitic rocks are thought to have formed coevally with a palimpsestic microfold in the charnockitic rocks, as xenoblastic biotite in them has a preferred lattice orientation parallel to the microfold (Yoshida, 1978). Thus, the microfold and the xenoblast parageneses are both considered to be the oldest products in the Lützow-Holm Bay region (Yoshida, 1979a). This microfold was correlated with the recumbent folds by Yoshida (1978) fom similarity in the trend of microfold lineations with axes of the recumbent folds; this conclusion is tentatively accepted here.

The view that the xenoblastic mineral parageneses were not produced by the $M_1$ metamorphism but were produced during the early phase of the progressive $M_2$ metamorphism (see above) has as yet no support from tectonic relationships and hence cannot be accepted in place of the previous conclusion.

*Dominant granulite facies metamorphism and associated tectonics* The granulite facies metamorphic rocks ($M_2$) in the Lützow-Holm Bay region are characterised by fine to medium grain size (0.3-0.5 mm) and "foam" texture (usage after Smith, 1964, viz. an equigranular polygonal texture as shown by Yoshida, 1978); they consist mainly of opx-cpx-pl-qz-or-(hb)-(bi). As foam texture is generally produced by annealing recrystallisation under static conditions, it is concluded that the metamorphism ($M_2$) is not associated with any deformation (see also Yoshida, 1978; Yoshida et al., 1982). However, the annealing recrystallisation would only have taken place in strained rocks; some folding is expecting to have occurred before $M_2$ and after $M_1$

metamorphisms. In this respect, the time relationships between the recumbent and isoclinal folds and the $M_1$ and $M_2$ metamorphisms should be re-examined carefully in the future because they are not well defined.

## Amphibolite Facies Metamorphisms and Associated Tectonics

Wide development of biotite schistosity parallel to the axial surfaces of east-west open to close folds in the Lützow-Holm Bay region led Yoshida (1978) to conclude that the amphibolite facies metamorphism ($M_3$) was synchronous with that folding. This is considered to strongly support the polymetamorphic view (2) discussed earlier.

Igneous intrusive activity characterised mainly by the pink granites and pegmatites appears to have occurred in conjunction with gentle folding and north-south oriented faulting. The associated metamorphism ($M_4$) which resulted in the local formation of secondary muscovite is thought to have taken place simultaneously (Yoshida, 1978; 1979b). Since the tectonic style of this stage differs significantly from that of $M_3$, this metamorphism could not be regarded as part of $M_3$ as stated by Yoshida (1979b).

## Conclusion

The preferred tectonic-metamorphic sequence in the Lützow-Holm Bay region is schematically shown in Figure 2. Some important revisions and additions to the earlier interpretation, outlined above, are made with regard to radiometric ages and their relationships to metamorphic events later than $M_2$.

## REFERENCES


BANNO, S., TATSUMI, T., OGURA, Y. and KATSURA, T., 1964: Petrographic studies on the rocks from the area around Lützow-Holmbukta; *in* Adie, R.J. (ed.) *Antarctic Geology.* North Holland, Amsterdam, 405-14.

HIROI, Y., SHIRAISHI, K., NAKAI, Y., KANO, T. and YOSHIKURA, A., (this volume): Geology and petrology of Prince Olav Coast, East Antarctica.

KANEOKA, I., OZIMA, M., AYUKAWA, M. and NAGATA, T., 1968: K-Ar ages and palaeomagnetic studies on rocks from the east coast of Lützow-Holm Bay, Antarctica. *Antarctic Rec., 31,* 12-20.

KIZAKI, K., 1964: Tectonics and petrography of the East Ongul Island, Lützow-Holmbukta, Antarctica. *JARE Sci. Rep., Ser. C., 2,* 1-24.

KIZAKI, K., 1965: Geology and petrography of the Yamato Sanmyaku, East Antarctica. *JARE Sci. Rep., Ser. C, 3,* 1-27.

KOJIMA, H., HANAI, Y. and NISHIDA, T., 1982: Geology of the Belgica Mountains. *Mem. Nat. Inst. Pol. Res. Japan, Spec. Issue, 21,* 32-46.

MAEGOYA, T., NOHDA, S. and HAYASE, I., 1968: Rb-Sr dating of the gneissic rocks from the east coast of Lützow-Holm Bay, Antarctica. *Mem. Coll. Sci., Univ. Kyoto, Ser. B, 35,* 131-8.

NAKAI, Y., KANO, T. and YISHIKURA, S., 1980: Explanatory text of geological map of Cape Ryugu, Antarctica. *Antarctic geological map ser.,* Sheet 15, Nat. Inst. Pol. Res., Japan.

NAKAI, Y., KANO, T. and YISHIKURA, S., 1980: Explanatory text of geological map of Oku-iwa Rock, Antarctica. *Antarctic geological map ser.,* Sheet 22, Nat. Inst. Pol. Res., Japan.

NICOLAYSEN, L.O., BURGER, A.J., TATSUMI, T. and AHRENS, L.H., 1961: Age measurements on pegmatites and a basic charnockite lens occurring near Lützow-Holm Bay, Antarctica. *Geochim. Cosmochim. Acta, 22,* 94-8.

OHTA, Y. and KIZAKI, K., 1966: Petrographic studies of potash feldspar from the Yamato Sanmyaku, East Antarctica. *JARE Sci. Rep., Ser. C, 5,* 1-40.

PICCIOTO, E. and COPPEZ, A., 1964: Bibliography of absolute age determinations in Antarctica (addendum); *in* Adie, R.J. (ed.) *Antarctic Geology.* North Holland, Amsterdam, 563-9.

SAITO, N. and SATO, K., 1964: On the age of euxenite from Antarctica; *in* Adie, R.J. (ed.) *Antarctic Geology.* North Holland, Amsterdam, 590-6.

SHIRAHATA, E., (this volume): Lead isotopic compositions in metamorphic rocks from Skarvsnes, East Antarctica.

SMITH, C.S., 1964: Some elementary principles of polycrystalline microstructure. *Metall. Rev., 9,* 1-48.

SUWA, K., 1968: Petrological studies on the metamorphic rocks from Lützow-Holmbukta area, East Antarctica. *23rd Int. Geol. Congr., 4,* 171-87.

SUZUKI, M., 1979: Metamorphic and plutonic rocks in the Cape Omega area, East Antarctica. *Mem. Nat. Inst. Pol. Res. Japan, Spec. Issue, 14,* 128-39.

SUZUKI, M., 1982: On the association of orthopyroxene-garnet-biotite found in the Lützow-Holmbukta region, East Antarctica. *Mem. Nat. Inst. Pol. Res. Japan, Spec. Issue, 21,* 86-102.

THOMPSON, A.B., 1966: Mineral reactions in pelitic rocks, II. Calculation of some P-T-X (Fe-Mg) phase relations. *Am. J. Sci., 276,* 425-54.

WOOD, B.J. and BANNO, S., 1973: Garnet-orthopyroxene and orthopyroxene-clinopyroxene relationships in simple and complex systems. *Contrib. Mineral. Petrol., 42,* 109-24.

YANAI, K. and UEDA, Y., 1974: Absolute ages and geological investigations on the rocks in the area of around Syowa Station, East Antarctica. *Antarctic Rec., 48,* 70-81.

YOSHIDA, M., 1975: Geology of the region around Botnneset, East Antarctica. *Mem. Nat. Inst. Pol. Res. Japan, Ser. C, 8,* 1-43.

YOSHIDA, M., 1978: Tectonics and petrology of charnockites around Lützow-Holmbukta, East Antarctica. *J. Geosci., Osaka City Univ., 21,* 65-152.

YOSHIDA, 1979a: Tectonics and metamorphism of the region around Lützow-Holmbukta, East Antarctica. *Mem. Nat. Inst. Pol. Res. Japan, Spec. Issue, 14,* 28-40.

YOSHIDA, 1979b: Metamorphic conditions of the polymetamorphic Lützow-Holmbukta region, East Antarctica. *J. Geosci., Osaka City Univ., 22,* 97-140.

YOSHIDA, M., YANAI, K., YOSHIKURA, S., ISHIKAWA, T. and KANISAWA, S., 1982: Tectonics and microstructure of charnockites around Lützow-Holmbukta, East Antarctica; *in* Craddock, C. (ed.) *Antarctic Geoscience.* Univ. Wisconsin Press, Madison, 511-9.

# SPINELS IN CALC-SILICATE ROCKS FROM THE COAST OF LUTZOW-HOLM BAY AND SURROUNDING AREAS

H. Matsueda, *Institute of Mining Geology, Mining College, Akita University, 1-1 Tegata-gakuen-cho, Akita 010, Japan.*

Y. Matsumoto, *Department of Mineralogical Science and Geology, Yamaguchi University, 1677-1 Yoshida, Yamaguchi 753, Japan.*

Y. Motoyoshi, *Department of Geology and Mineralogy, Faculty of Science, Hokkaido University, Sapporo 060, Japan.*

*Abstract* Crystalline basement rocks such as paragneisses (garnet gneiss, hornblende gneiss and garnet-biotite gneiss), quartzite, marble and allied rocks, charnockites (pyroxene gneiss), metabasite, granitic rocks (hornblende-biotite granite and garnet-gneissose granite), syenitic gneiss and minor intrusives are exposed along the coast of Lützow-Holm Bay and in surrounding areas. These crystalline rocks are assigned to the Okuiwa, Ongle, Skallen and Belgica Groups.

The marbles and calcareous metamorphic rocks which occur in Kasumi-iwa on the Prince Olav coast, the Skallevikhalsen, Skallen and Breidvognipa regions on the east coast of Lützow-Holm Bay, and the Yamato and Belgica Mountains, are coarse-grained and equigranular, being composed of calcite, dolomite and skarn minerals, such as forsterite, humite, diopside, hastingsitic hornblende, phlogopite and spinel, with apatite, scapolite and corundum as accessories.

The spinels from the Skallevikhalsen region show a variety of colours with the unaided eye, viz. pink, pinkish purple, clear purple, green, blackish green and dark green. The unit cell dimensions of spinels calculated from the powder X-ray diffraction data are $a_0 = 8.09$-$8.13$°A. The quantitative analysis of these minerals was made with an electron probe X-ray micro-analyser. Their compositions are: spinel molecule ($MgAl_2O_4$) 90.3% to 97.3%, hercynite molecule ($Fe^{2+}Al_2O_4$) 2.7% to 7.8%, and gahnite molecule ($ZnAl_2O_4$) 0.0% to 5.9%. Pink spinel contains a high percentage of spinel molecule ($MgAl_2O_4 = 97.3\%$), purplish spinels include a small percentage of hercynite molecule ($Fe^{2+}Al_2O_4 = 3.8$-$4.5\%$ and gahnite molecule ($ZnAl_2O_4 = 4.2$-$5.9\%$), and greenish spinels include a small percentage of hercynite molecule ($Fe^{2+}Al_2O_4 = 5.8$-$7.8\%$). These are identified as spinel, ferro-zincian spinel and ferroan spinel, respectively.

# PETROCHEMICAL STUDY OF METAMORPHIC ROCKS IN THE LUTZOW-HOLM BAY AREA, EAST ANTARCTICA

S. Kanisawa, *Department of Earth Science, College of General Education, Tohuku University, 980 Sendai, Japan.*

K. Yanai, *National Institute of Polar Research, Tokyo 173, Japan.*

*Abstract* The high-grade metamorphic rocks exposed at the Lützow-Holm Bay and Prince Olav Coast regions, East Antarctica can be divided into three lithostratigraphic groups, namely, from southwest to northeast, the Skallen, Ongul and Okuiwa Groups. The first two groups consist of granulite facies rocks, the third of amphibolite facies rocks.

The chemical characteristics of the amphibolite facies Okuiwa Group rocks are quite different from the $K_2O$ rich, charnockitic granulite facies rocks of the Skallen and Ongul Groups in the Lützow-Holm Bay region. For example, most Okuiwa Group rocks have high $Na_2O$, and low $K_2O$ contents, and low $K_2O/Na_2O$ ratios (generally less than 1). By contrast Skallen and Ongul Group rocks have high $K_2O/Na_2O$ ratios. $Fe_2O_3/FeO$ ratios in Okuiwa Group rocks are generally higher than those in Skallen and Ongul Group rocks, a fact that is interpreted as evidence of higher $f_{O_2}$ conditions during the amphibolite facies metamorphism of Okuiwa Group.

The chemical characteristics of the Okuiwa Group rocks indicate derivation either from igneous rocks in the trondhjemite-tonalite suite (Barker, 1979) or sandstones such as those described as the eugeosynclinal clan by Middleton (1960), or the greywacke-sandstones of Pettijohn (1963). The chemical differences between rocks of the Skallen, Ongul, and Okuiwa Groups must be taken into account in considering the tectonic environment of igneous activity, and the composition and provenance of original sediments.

# GEOLOGY AND PETROLOGY OF THE YAMATO MOUNTAINS

K. Shiraishi, *National Institute of Polar Research, Tokyo, Japan.*

M. Asami, *Department of Geological Sciences, Okayama University, Okayama, Japan.*

Y. Ohta, *Norsk Polarinstitutt, Oslo, Norway.*

*Abstract* The basement of the Yamato Mountains consists of high grade metamorphic, syenitic and granitic rocks. Geologic and petrologic observations have led to a two-fold subdivision of the metamorphic rocks: older granulite facies rocks such as two-pyroxene biotite gneiss and associated two-pyroxene amphibolite, orthopyroxene biotite gneiss and calc-silicate gneiss, and younger amphibolite facies rocks such as granitic gneiss, biotite amphibolite and migmatitic gneiss. The contact between the granitic gneiss, and a syenitic rock which is genetically related to the two-pyroxene biotite gneiss, seems to be tectonic. Two-pyroxene phase relations and two-pyroxene geothermometry suggest P-T conditions during the granulite facies metamorphism were generally uniform not only among the two-pyroxene-bearing metamorphic rocks but also between these metamorphic rocks and the two-pyroxene syenite. Probable temperatures around 800°C are indicated. The syenite intrusion is considered to be a plutonic climax of the metamorphism. The younger amphibolite facies metamorphism is characterised by widespread granite and pegmatite intrusion. This activity might correspond to the 500 Ma intrusion of granite and pegmatite which is widely known in East Antarctica, but the age of the older granulite facies metamorphism remains unknown.

The Yamato Mountains (71°30′S, 35°45′E) were first described by Kizaki (1965). Further geological mapping has been carried out by Yoshida and Ando (1971), Shiraishi (1977), Shiraishi et al. (1978, 1982) and Yanai et al. (1982). Kizaki (1965) divided the various metamorphic and plutonic rocks into two major groups: a charnockitic group which crystallised under the granulite facies conditions and has been partly modified by later granitisation which resulted from the emplacement of the second granitic group. This interpretation has been generally accepted, although the geological positions of some lithological units such as the granitic gneiss and a part of the syenite are still controversial.

In this paper, the geology, metamorphism and plutonism of the Yamato Mountains is reviewed on the basis of recent field investigations and petrographic work.

## GEOLOGY

The Yamato Mountains consist of seven massifs, A, B, C, D, E, F and G which range for 50 km from south to north (Figure 1). They are made up of high grade regional metamorphic, syenitic and granitic rocks (Figure 2).

### Metamorphic rocks

The metamorphic rocks can be classified into three groups.

*Two-pyroxene biotite gneiss group*

Two-pyroxene biotite gneiss was described as enderbitic gneiss by Kizaki (1965) and occurs as small masses and xenolithic inclusions in syenitic rocks in Massifs A and G. It is also found in the northwestern part of Massif D, where its intrusion by a large amount of aplitic granite has caused formation of a migmatitic rock. The gneiss is medium-grained and shows a distinct banded structure, defined by an alternation of mafic and felsic layers up to 10 mm thick. In Massif A, calc-silicate gneiss, a few metres in width, is concordantly associated with the two-pyroxene biotite gneiss. Two-pyroxene amphibolite and orthopyroxene biotite gneiss are also found as concordant layers up to 100 cm wide. Concordant association of the calc-silicate gneiss with the two-pyroxene biotite gneiss suggests their sedimentary origin. Their field occurrence suggests that these rocks are the oldest in the Yamato Mountains (Shiraishi et al., 1982).

*Granitic gneiss and biotite amphibolite group*

Granitic gneiss and subordinate biotite amphibolite occupy the central part of the Yamato Mountains. The granitic gneiss shows migmatitic characteristics such as banded, nebulitic and augen structures and is a fine- to medium-grained quartzofeldspathic rock containing abundant pink microcline. Concordant or subconcordant biotite amphibolite layers up to 80 m thick mainly occur with granitic gneiss in Massifs B and C. In both the granitic gneiss and biotite amphibolite, microcline porphyroblasts, up to 2 cm in length, are common especially near the contact of intrusive granitic pegmatite. The direct relationship between the two-pyroxene biotite gneiss and granitic gneiss groups is unknown.

*Migmatitic gneiss and aplitic granite.*

Migmatitic gneiss is found in a transitional zone between two-pyroxene biotite gneiss and granitic gneiss in Massif D. It has agmatitic and nebulitic structures with a leucosome of aplitic granite intruded by hornblende biotite granite, and palaeosomes of two-pyroxene biotite gneiss and metabasic rocks such as two-pyroxene amphibolite and clinopyroxene-biotite-K-feldspar rock. Orthopyroxene in the two-pyroxene biotite gneiss is occasionally altered to radiating aggregates of biotite rods with symplectically intergrown quartz. The nature of the contact between the migmatitic gneiss and the granitic gneiss remains unknown. Foliation in the migmatitic

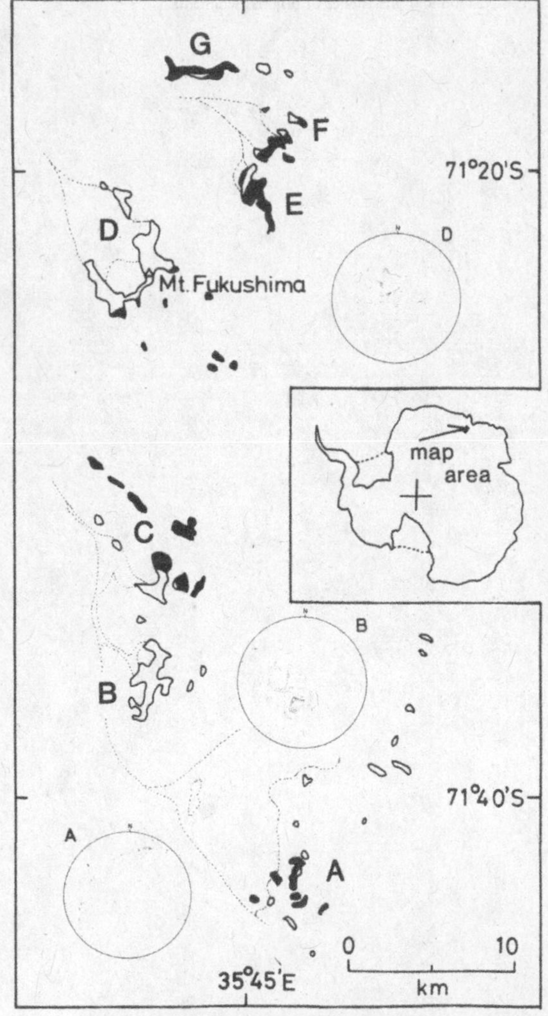


**Figure 1. Location map of the Yamato Mountains. A,B,.........G are tentative names for massifs. Black area shows the distribution of syenitic rocks. Poles to foliation in Massifs A, B and D are also shown (lower hemisphere, contour intervals 1,5,7,10%).**

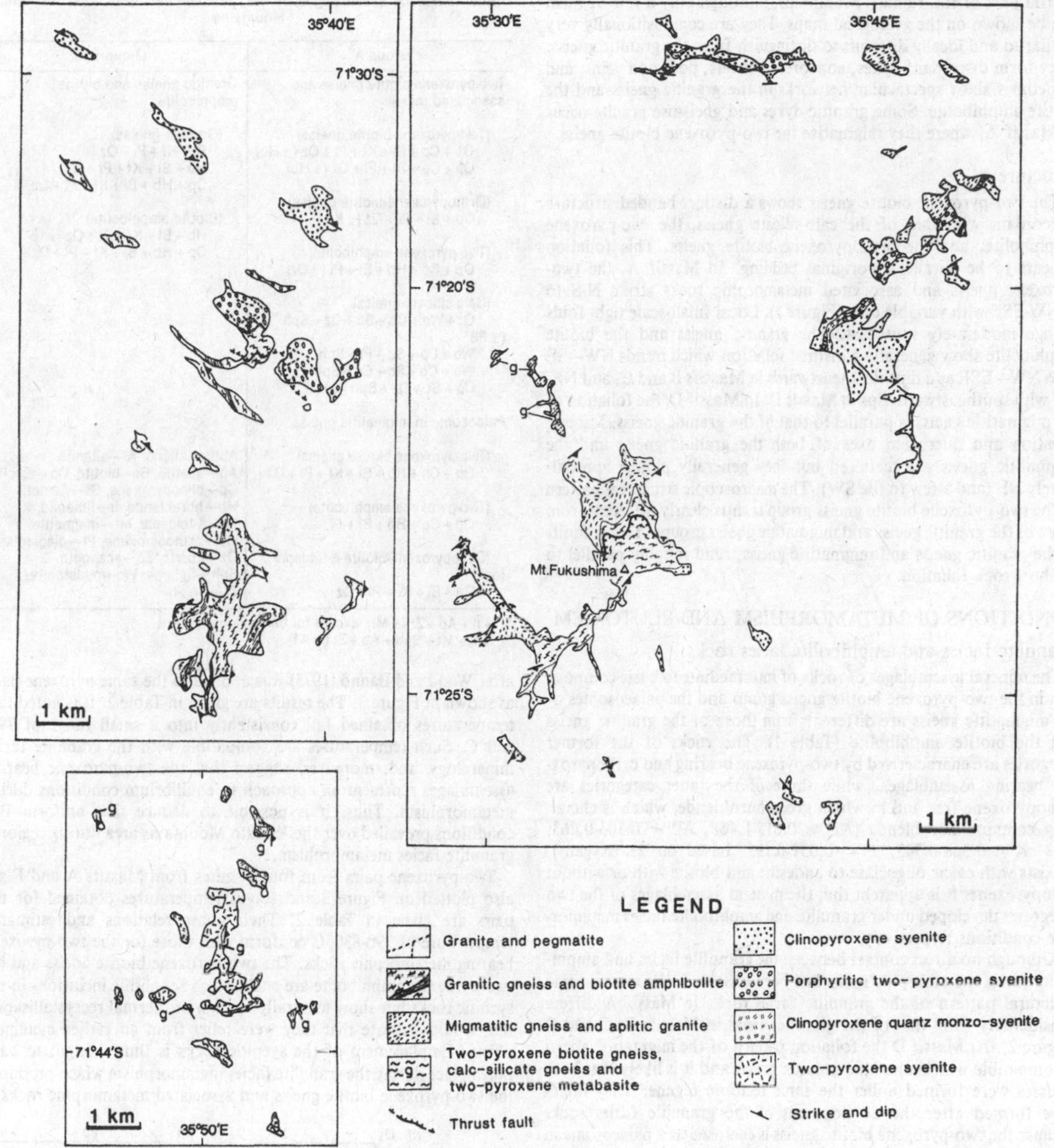


**Figure 2. Geological map of the Yamato Mountains. 1. Granite and pegmatite; 2. Granitic gneiss and biotite amphibolite; 3. Migmatitic gneiss and aplitic granite; 4. Two-pyroxene biotite gneiss, calc-silicate gneiss and two-pyroxene metabasite; 5. Clinopyroxene syenite; 6. Porphyritic two-pyroxene syenite; 7. Clinopyroxene quartz monzo-syenite; 8. Two-pyroxene syenite; 9. Strike and dip; 10. Thrust fault.**

gneiss is locally defined by parallel arrangements of palaeosomes, and basic schlieren in the neosome.

The aplitic granite also occurs alone on the western sides of Massifs E and F and in Massif G as concordant sheets or layers with diffuse contacts with syenitic rocks (see below). These features indicate that the migmatisation by the aplitic granite occurred after the emplacement of the two-pyroxene biotite gneiss and the syenitic rocks.

## Syenitic rocks

Syenitic rocks are widespread in the Yamato Mountains (Figure 2). They could be classified into the following types on the basis of field occurrence and mineral assemblage: (1) two-pyroxene syenite in Massif A, (2) clinopyroxene quartz monzo-syenite in Massif A, (3) porphyritic two-pyroxene syenite in Massifs C, E, F and G, (41 clinopyroxene syenite in Massifs C, D, E, F and G. The two-pyroxene syenite, designated quartz syenitic charnockite by Shiraishi et al. (1982), is characterised by blue-grey to brown-grey feldspars on a fresh surface. Although this rock rarely cuts the two-pyroxene biotite gneiss, the common presence of lens- or ribbon-shaped gneiss inclusions in the syenite indicates a close genetic relationship between the two. The clinopyroxene quartz monzo-syenite intrudes the two-pyroxene syenite in Massif A.

The relationship between the porphyritic two-pyroxene syenite and the clinopyroxene syenite is not clear. Varieties of the clinopyroxene syenite with different mafic mineral associations have gradational contacts. It is probable that the clinopyroxene syenite was derived from the porphyritic two-pyroxene syenite by later granitisation (Ohta and Kizaki, 1966; Shiraishi, 1977). The boundary between granitic gneiss and the clinopyroxene syenite at Massif D, is marked by an intensely deformed narrow zone filled with gneissose pink granite and augen gneiss.

## Pink granite and pegmatite

Pink granite and pegmatite are widely distributed, especially in the central part of the Yamato Mountains, although only a few of them can be shown on the geological maps. They are compositionally very similar to and locally difficult to distinguish from the granitic gneiss. They form discordant dykes, concordant sheets, pods and veins, and sometimes show spectacular networks in the granitic gneiss and the biotite amphibolite. Some granitic dykes and gneissose granite occur in Massif A, where they migmatise the two-pyroxene biotite gneiss.

### Structure

The two-pyroxene biotite gneiss shows a distinct banded structure concordant with that of the calc-silicate gneiss, the two-pyroxene amphibolite, and the orthopyroxene biotite gneiss. This foliation appears to be parallel to original bedding. In Massif A, the two-pyroxene gneiss and associated metamorphic rocks strike N-S to WNW-ESE with variable dips (Figure 1). Local small-scale tight folds plunge moderately southeast. The granitic gneiss and the biotite amphibolite show generally a distinct foliation which trends NW—SE to WNW—ESE and dips northeastwards in Massifs B and C, and NE-SW with southeastward dips in Massif D. In Massif D, the foliation of the migmatitic gneiss is parallel to that of the granitic gneiss. Mineral lineation and microfold axes of both the granitic gneiss and the migmatitic gneiss are scattered but they generally plunge approximately NE (and a few to the SW). The macroscopic structural pattern of the two-pyroxene biotite gneiss group is thus clearly different from those of the granitic gneiss and migmatitic gneiss groups. Thrust faults in the granitic gneiss and migmatitic gneiss trend roughly parallel to the host rock foliation.

## CONDITIONS OF METAMORPHISM AND PLUTONISM

### Granulite facies and amphibolite facies rocks

The mineral assemblages of rocks of intermediate to basic composition in the two-pyroxene biotite gneiss group and the palaeosomes of the migmatitic gneiss are different from those of the granitic gneiss and the biotite amphibolite (Table 1). The rocks of the former categories are characterised by two-pyroxene bearing and orthopyroxene bearing assemblages, while those of the latter categories are orthopyroxene free and in which green hornblende, which is chemically common hornblende ($Al^{IV}$ = 0.615-1.468, $Al^{VI}$ = 0.103-0.263, Na + K = 0.368-0.743, Ti = 0.051-0.195, based on 23 oxygens), coexists with calcic oligoclase to andesine and biotite with or without clinopyroxene. It is apparent that the mineral assemblages of the two categories developed under granulite and amphibolite facies metamorphic conditions respectively.

Although no direct contact between the granulite facies and amphibolite facies metamorphic rocks has been observed, the macroscopic structural pattern of the granulite facies rocks in Massif A differs considerably from that of the granitic gneiss in Massifs B, C and D (Figure 2). At Massif D the foliation pattern of the migmatitic gneiss is compatible with that of the granitic gneiss and it is likely that both gneisses were formed under the same tectonic regime. They might have formed after the development of the granulite facies rocks because the two-pyroxene biotite gneiss is enclosed as a palaeosome in the migmatitic gneiss. Widespread granite and pegmatite intrusion in these amphibolite facies rocks are suggestive of syn- to late-kinematic intrusive activities. The contacts between the granitic gneiss and syenitic rocks in Massifs C and D are tectonic, with a thrust in one place and an intensely deformed narrow zone composed mainly of gneissose granite elsewhere.

### Granulite facies and syenitic rocks

Two-pyroxene assemblages are widely found in metamorphic rocks closely associated with the syenitic rocks in Massifs A and G and in palaeosomes of the migmatitic gneiss in Massif D. Orthopyroxene-clinopyroxene phase relations for eight specimens from small masses and xenolithic inclusions in the syenitic rocks were examined in order to investigate two-pyroxene compatibility. Tie-line relations illustrated in Figure 3 show the minerals to be mostly compatible with each other and suggest that P-T conditions during the granulite facies metamorphism were more or less uniform among the two-pyroxene bearing rocks. Compositions of the two pyroxenes approximate to those of the simple $CaSiO_3$-$MgSiO_3$-$FeSiO_3$ system because of low contents of such other components as $TiO_2$ $Al_2O_3$ MnO and $Na_2O$ in the pyroxenes as shown in Table 2. To investigate further their compatibility and estimation of solvus temperatures, two-pyroxene geothermometry

**TABLE 1: Mineral assemblages of the metamorphic rocks from the Yamato Mountains**

| Group A* | Group B** |
|---|---|
| Two-pyroxene-biotite gneiss and associated rocks | Granitic gneiss and biotite amphibolite |
| (Two-pyroxene-biotite gneiss)<br>Op + Cp + Bi + Kf + Pl + Qz (± Hb)<br>Op + Cp + Bi + Pl + Qz (± Hb) | (Granitic gneiss)<br>Bi + Kf + Pl + Qz<br>Hb + Bi + Kf + Pl + Qz<br>Cp + Hb + Bi + Kf + Pl + Qz |
| (Orthopyroxene-biotite gneiss)<br>Op + Bi + Pl + Qz (± Kf) | (Biotite amphibolite)<br>Hb + Bi + Kf + Pl + Qz<br>Cp + Hb + Bi + Kf + Pl + Qz |
| (Two-pyroxene amphibolite)<br>Op + Cp + Hb + Bi + Pl (± Qz) | |
| (Calc-silicate gneiss)<br>Gt + Wo + Cp + Sc + Qz + Sph (± Pl)<br>Wo + Cp + Sc + Pl + Sph<br>Wo + Cp + Sc + Cc + Sph<br>Cp + Sc + Qz + Sph (± Hb) | |
| Paleosome in migmatitic gneiss | |
| (Two-pyroxene-biotite gneiss)<br>Op + Cp + Hb + Bi + Kf + Pl + Qz | Abbreviations: Al—allanite, Ap—apatite, Bi—biotite, Cc—calcite, Cp—clinopyroxene, Gt—garnet, Hb—hornblende, Il—ilmenite, Kf—K-feldspar, Mt—magnetite, Op—orthopyroxene, Pl—plagioclase, Qz—quartz, Sc—scapolite, Sph—Sphene, Wo-wollastonite, Zr—zircon |
| (Two-pyroxene amphibolite)<br>Op + Cp + Hb + Bi + Pl | |
| (Clinopyroxene-biotite-K-feldspar rock)<br>Cp + Bi + Kf + Pl + Qz | |

* + Il + Ap + Zr (± Mt), except for calc-silicate gneiss
** + Il + Mt + Sph + Ap + Zr (± A1)

after Wood and Banno (1973) was applied to the same pyroxene pairs as shown in Figure 3. The results are given in Table 2. It is noted that temperatures obtained fall consistently into a small range of 790-850°C. Such temperatures are compatible with the granulite facies mineralogy, and, moreover, suggest that the two-pyroxene bearing assemblages represent an approach to equilibrium conditions during metamorphism. Thus, it is possible to deduce that uniform P-T conditions prevailed over the Yamato Mountains area during regional granulite facies metamorphism.

Two-pyroxene pairs from four syenites from Massifs A and F are also plotted on Figure 3 and solvus temperatures obtained for the pairs are given in Table 2. Their phase relations and estimated temperature of 790-830°C conform with those for the two-pyroxene bearing metamorphic rocks. The two-pyroxene biotite gneiss and the two-pyroxene amphibolite are enclosed as xenolithic inclusions in the syenitic rocks but show texturally no sign of thermal recrystallisation that would indicate that they were relict from an earlier metamorphism. Emplacement of the syenitic rocks is thus thought to have taken place during the granulite facies metamorphism which produced the two-pyroxene biotite gneiss and associated metamorphic rocks.

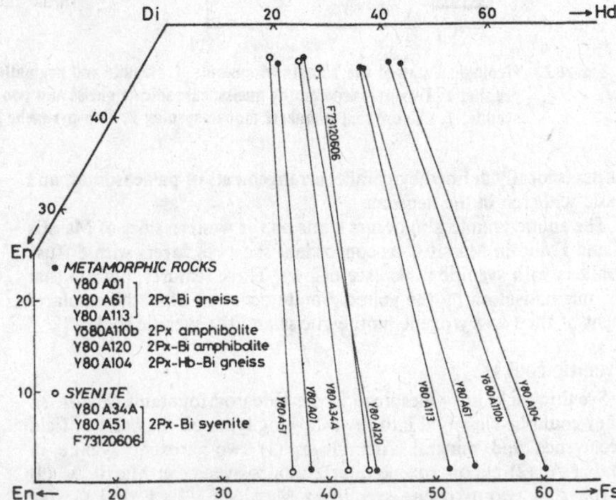


**Figure 3. Two-pyroxene phase relations in the metamorphic rocks and syenite from the Yamato Mountains.**

**TABLE 2: Compositions of coexisting pyroxenes and estimated temperatures for the pyroxene pairs after Wood and Banno (1973)**

| | | $TiO_2$ wt % | $Al_2O_3$ wt % | MnO wt % | $Na_2O$ wt % | Fe/Fe + Mg | $aMg_2Si_2O_6$ | $K = a^{cp}/a^{op}$ | T (°C) |
|---|---|---|---|---|---|---|---|---|---|
| **Metamorphic rocks** | | | | | | | | | |
| Y80A01 | Cp | 0.10 | 0.59 | 0.31 | 0.41 | 0.230 | 0.0415 | 0.115 | 847 |
| | Op | 0.07 | 0.33 | 0.76 | 0.05 | 0.381 | 0.360 | | |
| Y80A61 | Cp | 0.06 | 0.77 | 0.57 | 0.38 | 0.378 | 0.0298 | 0.152 | 804 |
| | Op | 0.05 | 0.32 | 1.61 | 0.00 | 0.536 | 0.197 | | |
| Y80A113 | Cp | 0.15 | 0.77 | 0.48 | 0.31 | 0.372 | 0.0324 | 0.149 | 812 |
| | Op | 0.03 | 0.29 | 1.36 | 0.05 | 0.511 | 0.217 | | |
| Yo80A110b | Cp | 0.09 | 1.06 | 0.21 | 0.30 | 0.421 | 0.0258 | 0.160 | 794 |
| | Op | 0.14 | 0.49 | 0.69 | 0.00 | 0.580 | 0.161 | | |
| Y80A120 | Cp | 0.17 | 1.31 | 0.43 | 0.29 | 0.270 | 0.0359 | 0.123 | 820 |
| | Op | 0.05 | 0.49 | 0.78 | 0.00 | 0.443 | 0.294 | | |
| Y80A104 | Cp | 0.11 | 0.77 | 0.40 | 0.27 | 0.442 | 0.0264 | 0.176 | 797 |
| | Op | 0.11 | 0.27 | 0.70 | 0.00 | 0.601 | 0.150 | | |
| Y80A531 | Cp | 0.15 | 0.92 | 0.51 | 0.33 | 0.454 | 0.0240 | 0.165 | 790 |
| | Op | 0.12 | 0.39 | 1.13 | 0.00 | 0.603 | 0.145 | | |
| D733121010 | Cp | 0.11 | 0.71 | 0.37 | 0.50 | 0.337 | 0.0282 | 0.138 | 806 |
| | Op | 0.09 | 0.56 | 1.11 | 0.04 | 0.505 | 0.205 | | |
| **Syenite** | | | | | | | | | |
| Y80A34A | Cp | 0.21 | 1.24 | 0.20 | 0.46 | 0.265 | 0.0381 | 0.115 | 827 |
| | Op | 0.05 | 0.27 | 0.69 | 0.02 | 0.416 | 0.333 | | |
| Y80A51 | Cp | 0.11 | 0.91 | 0.60 | 0.43 | 0.215 | 0.0323 | 0.0842 | 819 |
| | Op | 0.06 | 0.25 | 0.73 | 0.09 | 0.364 | 0.383 | | |
| F73120606 | Cp | 0.04 | 0.54 | 0.26 | 0.49 | 0.267 | 0.0278 | 0.089 | 794 |
| | Op | 0.06 | 0.38 | 0.78 | 0.04 | 0.423 | 0.313 | | |
| Y80A556 | Cp | 0.16 | 0.75 | 0.43 | 0.42 | 0.387 | 0.0353 | 0.173 | 821 |
| | Op | 0.08 | 0.18 | 1.04 | 0.00 | 0.532 | 0.204 | | |

## Ages of the metamorphism and plutonism

Only a few age determinations have been obtained in the Yamato Mountains. 457 Ma (Rb-Sr) of biotite from the granitic gneiss may be regarded as the age of the last event of the amphibolite facies metamorphism, viz. the granite activity. Younger K-Ar ages of 360 to 400 Ma from whole rock samples of the granitic gneiss and the syenite probably indicate the cooling age of the granitic activity (Picciotto and Coppez, 1964; Yanai et al., 1982). All the above plutonism and metamorphism is thus approximately contemporaneous with the 500 Ma event of many parts of East Antarctica (Yoshida, et al., this volume). The age of the older granulite facies metamorphism remains unknown.

*Acknowledgements* We are much indebted to the members of the Japanese Association of Geologists for Antarctic Research for valuable discussion, and to members of the JARE-21 for their help in the field work.

## REFERENCES

KIZAKI, K., 1965: Geology and petrography of the Yamato Sanmyaku, East Antarctica. *JARE Sci. Rep., Ser. C., 3.*

OHTA, Y. and KIZAKI, K., 1965: Petrographic studies of potash feldspar from the Yamato Sanmyaku, East Antarctica. *JARE Sci. Rep., Ser. C, 5.*

PICCIOTO, E., and COPPEZ, A., 1964: Bibliographie des mesures d'ages absolus en Antartique. (Addendum aout 1963). *Ann. Soc. Geol. Belg., 87,* 115-28.

SHIRAISHI, K., 1977: Geology and petrography of the northern Yamato Mountains, Antarctica. *Mem. Nat. Inst. Pol. Res., Japan, Ser. C, 12.*

SHIRAISHI, K., ASAMI, M. and OHTA, Y., 1982: Plutonic and metamorphic rocks of Massif A in the Yamato Mountains, East Antractica. *Mem. Nat. Inst. Pol. Res., Japan, Spec. Issue 21,* 21-31.

SHIRAISHI, K., KIZAKI, K., YOSHIDA, M. and MATSUMOTO, T., 1978: Mt Fukushima, northern Yamato Mountains. *Antarct. Geol. Map Ser.,* Sheet 27 (1) (with explanatory text 7p, 2pl.), Nat. Inst. Pol. Res., Japan.

WOOD, B.J. and BANNO, S., 1973: Garnet-orthopyroxene and orthopyroxene-clinopyroxene relationships in simple and complex systems. *Contrib. Mineral. Petrol., 42,* 109-24.

YANAI, K., NISHIDA, T., KOJUMA, H., SHIRAISHI, K., ASAMI, M., OHTA, Y., KIZAKI, K. and MATSUMOTO, Y., 1982: Central Yamato Mountains, Massif B and Massif C. *Antarct. Geol. Map Ser.,* Sheet 28 (with explanatory text 10p, 6pl.). Nat. Inst. Pol. Res., Japan.

YOSHIDA, M. and ANDO, H., 1971: Geological surveys in the vicinity of Lützow-Holm Bay and the Yamato Mountains, East Antarctica. *Antarctic. Rec., 39.* 46-54.

YOSHIDA, M., SUZUKI, M., SHIRAHATA, H. and KOJIMA, H. (this volume): Tectonic and metamorphic history of the Lütow-Holm Bay region, East Antarctica.

# GEOLOGY AND PETROLOGY OF THE BELGICA MOUNTAINS

H. Kojima, *Institute of Mining Geology, Mining College, Akita University, Tegata-gakuen-cho, Akita 010, Japan.*

K. Yanai, *National Institute of Polar Research, Tokyo 173, Japan*

T. Nishida, *Institute of Earth Science, Faculty of Education, Saga University, Honjo-machi, Saga 840, Japan.*

*Abstract* The Belgica Mountains (72°18′S-72°43′S lat. and 30°57′E-31°20′E long.) lie about midway between the Yamato Mountains and Sør Rondane Mountains, East Dronning Maud Land, Antarctica. They were visited in December 1979, by a traverse party of the 20th Japanese Antarctic Research Expedition which included three geologists who carried out a geological survey.

The crystalline basement rocks in this area are assigned to the Belgica Group which is divided into the Belgica upper formation and the Belgica lower formation. Various rock types constitute the Belgica Group; they are (1) granitic gneiss, (2) marble and skarn, (3) amphibolite, (4) hornblende-biotite gneiss, (5) augen gneiss, (6) clinopyroxene gneiss, (7) garnet-biotite gneiss, and (8) dyke rocks (basic metadyke, syenite, granodiorite-diorite and pink granite). Granitic gneiss is a main constituent of the Belgica lower formation; hornblende-biotite gneiss occupies a large part of the Belgica upper formation. Marble, skarn and amphibolite are distributed throughout this region, with small amounts of other gneisses. The Belgica Group rocks are slightly to moderately deformed with gentle to open fold structures, which are grouped into two generations.

Whole rock K-Ar measurements of three gneisses and three dyke rocks were carried out. The results of dating range from 382 to 472 Ma suggesting that the metamorphism in these mountains occurred during the Early Palaeozoic time. The mineral assemblages and petrological features of this region reveal that metamorphic grade of the region is amphibolite facies.

These geological features in the Belgica Mountains are more similar to the Teltet-Vengen group of the Sør Rondane Mountains than to the Yamato Mountains, which are characterised by widely distributed charnockitic sequences.

# LEAD ISOTOPIC COMPOSITION IN METAMORPHIC ROCKS FROM SKARVSNES, EAST ANTARCTICA

H. Shirahata, *Department of Mineral Resources Engineering, Muroran Institute of Technology, Muroran 050, Japan.*

***Abstract*** **Bulk chemical and lead isotopic compositions were analysed for pyroxene gneiss, garnet-biotite gneiss, K-feldspar porphyroblastic gneiss and garnet bearing granitic gneiss, all of which belong to the crystalline basement exposed largely in the Skarvsnes area, Lützow-Holm Bay, East Antarctica. Garnet-biotite gneiss is in harmony with pyroxene gneiss in Pb isotope characteristics, though it is closely related to K-feldspar porphyroblastic gneiss in mineral assemblages and bulk chemical characteristic as well as in field relations. The garnet-biotite gneiss is attributed to formation by intermediate pressure granulite facies metamorphism with Rb-Sr age of 1100 Ma. A two stage lead evolution model indicates that the pyroxene gneiss may have formed by high grade granulite facies metamorphism or by mafic igneous activity as old as 1900 Ma. The heterogeneity in Pb isotope systematics of the garnet bearing granitic gneiss implies that the lead has originated from various sialic sources and has not achieved a thorough mixing throughout the later metamorphism. This metamorphic history would be comparable with the sequence of metamorphic events recognised in Lützow-Holm Bay and adjacent regions.**

**The metamorphic terrain of the Lützow-Holm Bay region, East Antarctica has been investigated over a period of 25 years. Metamorphic rocks consisting mainly of K-feldspar porphyroblastic gneiss, garnet-biotite gneiss, marble, hornblende gneiss and pyroxene gneiss, with subordinate quantities of garnet gneiss, metabasite and garnet-bearing granitic gneiss are exposed extensively in the Skarvsnes area on the east coast of Lützow-Holm Bay. The geology and petrography of the metamorphic rocks in this area were first described by Tatsumi**

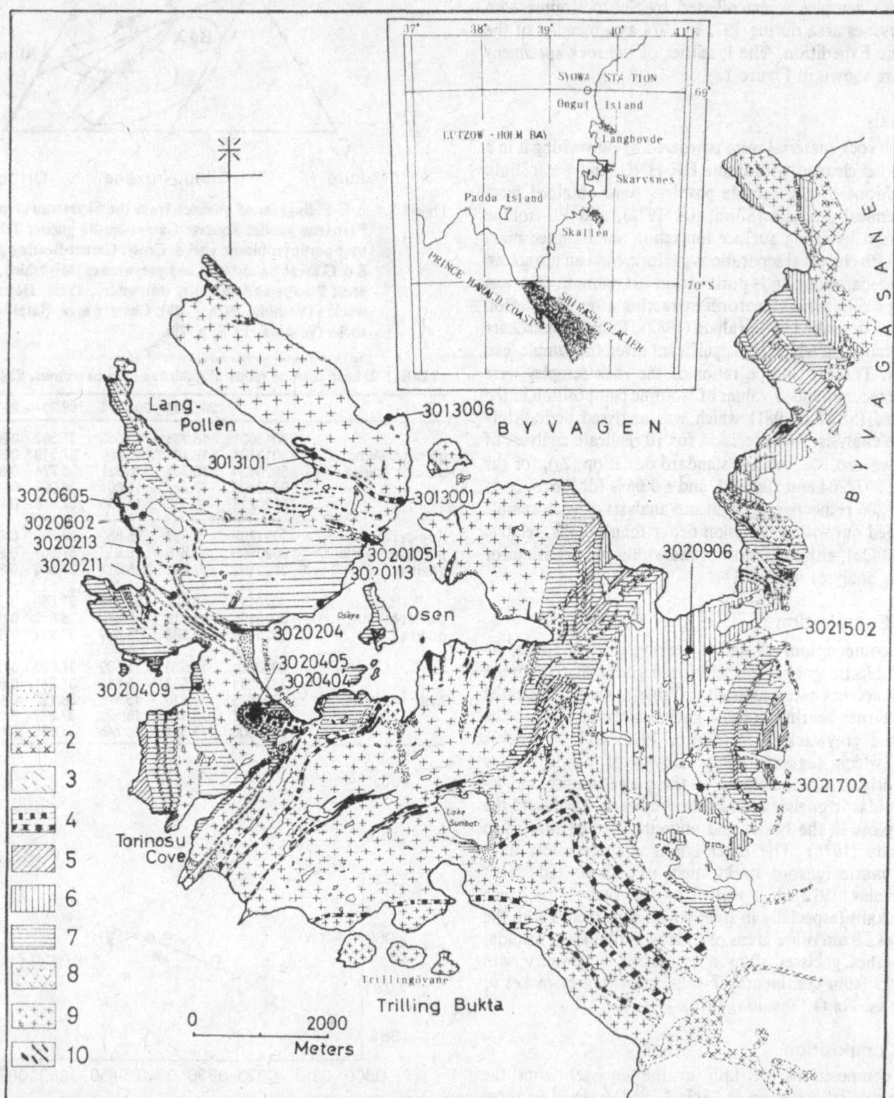


**Figure 1. Geological map of the Skarvsnes area (after Ishikawa et al., 1977). (1) Beach sand and gravel bed; (2) Terrestrial deposits; (3) Pegmatite; (4) Garnet bearing granitic gneiss; (5) Garnet gneiss; (6) Porphyroblastic gneiss; (7) Garnet-biotite gneiss; (8) Hornblende gneiss; (9) Pyroxene gneiss; (10) Metabasite. Sampling localities are marked on the map.**

**TABLE 1: Chemical analyses of gneisses from Skarvsnes, East Antarctica**

| | 1 | 2 | 3 | 4 | 5 | 6 | 7 | 8 | 9 | 10 | 11 | 12 | 13 | 14 |
|---|---|---|---|---|---|---|---|---|---|---|---|---|---|---|
| $SiO_2$ | 76.88 | 78.72 | 75.47 | 75.29 | 70.08 | 71.24 | 70.28 | 64.93 | 71.15 | 69.93 | 46.26 | 49.88 | 47.01 | 49.64 |
| $TiO_2$ | 0.14 | 0.13 | 0.25 | 0.22 | 0.55 | 0.59 | 0.64 | 1.08 | 0.53 | 0.65 | 2.06 | 1.29 | 2.80 | 1.29 |
| $Al_2O_3$ | 12.06 | 11.68 | 12.89 | 13.17 | 14.61 | 14.25 | 14.93 | 16.28 | 14.83 | 14.91 | 16.99 | 17.47 | 16.08 | 17.02 |
| $Fe_2O_3$ | 0.29 | 0.33 | 0.63 | 0.66 | 0.41 | 0.37 | 0.38 | 0.83 | 0.22 | 0.45 | 2.07 | 1.53 | 2.90 | 1.54 |
| FeO | 0.94 | 0.72 | 1.35 | 1.14 | 3.14 | 3.53 | 2.23 | 5.50 | 3.20 | 2.51 | 9.66 | 6.97 | 10.73 | 7.26 |
| MnO | 0.02 | 0.02 | 0.05 | 0.05 | 0.05 | 0.05 | 0.04 | 0.09 | 0.05 | 0.04 | 0.19 | 0.15 | 0.19 | 0.17 |
| MgO | 0.05 | 0.10 | 0.18 | 0.11 | 1.14 | 0.94 | 0.63 | 1.81 | 1.65 | 1.34 | 7.08 | 7.82 | 5.96 | 8.39 |
| CaO | 0.60 | 0.63 | 1.45 | 1.74 | 2.09 | 2.05 | 1.87 | 3.79 | 2.56 | 2.21 | 8.91 | 9.06 | 8.43 | 8.88 |
| $Na_2O$ | 2.60 | 2.87 | 2.40 | 2.75 | 3.20 | 2.95 | 2.97 | 2.95 | 3.13 | 3.61 | 2.84 | 2.97 | 2.97 | 2.85 |
| $K_2O$ | 5.57 | 4.42 | 5.08 | 4.36 | 3.50 | 3.53 | 5.20 | 1.96 | 1.87 | 3.29 | 0.70 | 1.20 | 1.00 | 1.17 |
| $P_2O_5$ | 0.14 | 0.06 | 0.08 | 0.07 | 0.30 | 0.20 | 0.35 | 0.31 | 0.21 | 0.29 | 0.35 | 0.28 | 0.45 | 0.35 |
| $H_2O(+)$ | 0.31 | 0.28 | 0.26 | 0.22 | 0.63 | 0.33 | 0.28 | 0.58 | 0.62 | 0.38 | 2.30 | 0.74 | 0.84 | 0.80 |
| $H_2O(-)$ | 0.13 | 0.04 | 0.10 | 0.06 | 0.15 | 0.11 | 0.10 | 0.15 | 0.15 | 0.08 | 0.19 | 0.14 | 0.17 | 0.17 |
| Total | 99.73 | 100.00 | 100.19 | 99.84 | 99.85 | 100.14 | 99.90 | 100.26 | 100.17 | 99.69 | 99.60 | 99.50 | 99.53 | 99.53 |

| No. | Rock type | Sample |
|---|---|---|
| 1. | Garnet-bearing granitic gneiss | 3013001 |
| 2. | Garnet-bearing granitic gneiss | 3013101 |
| 3. | Garnet-bearing granitic gneiss | 3020404 |
| 4. | Garnet-bearing granitic gneiss | 3020405 |
| 5. | K-feldspar porphyroblastic gneiss | 3020906 |
| 6. | K-feldspar porphyroblastic gneiss | 3021502 |
| 7. | K-feldspar porphyroblastic gneiss | 3021702 |
| 8. | Garnet-biotite gneiss | 3020105 |
| 9. | Garnet-biotite gneiss | 3020113 |
| 10. | Garnet-biotite gneiss | 3020409 |
| 11. | Pyroxene gneiss | 3020204 |
| 12. | Pyroxene gneiss | 3020211 |
| 13. | Pyroxene gneiss | 3020605 |
| 14. | Garnet-bearing pyroxene gneiss | 3021602 |

and Kikuchi (1959a, b); subsequent contributions were summarised and published in the form of a geologic map with an explanatory text book (Ishikawa et al., 1977).

In this present study, both chemical compositions and lead isotope ratios of pyroxene gneiss, garnet-biotite gneiss, K-feldspar porphyroblastic gneiss and garnet-bearing granitic gneiss, all of which are exposed largely in the Skarvsnes area, were analysed to understand their geochemical characteristics and the metamorphic history of the terrain. These rock samples were collected by Shozo Kojima who explored the Skarvsnes area during 1972 to 1974 as a member of the Japanese Antarctic Expedition. The localities of the rock specimens analysed herein are shown in Figure 1.

## Chemical Analysis

1 or 2 kg of each rock material were powdered by pulverising it in a stainless steel mortar cleaned with dilute HF, $HNO_3$ and water. Bulk chemical compositions of the sample powders were obtained by a modified wet chemical analysis (Shirahata, 1972), and Pb isotope ratios were analysed by using surface ionisation solid source mass spectrometry with Pb chemical separations performed in an ultraclean laboratory. The separation and purification of sample Pb was achieved by using a dithizon-chloroform extraction after TED teflon bomb dissolution as described in Shirahata (1982). Total Pb blanks are estimated as less than 2 ng which is insignificant given the sample lead ($\sim 2\mu g$) separated. The Pb isotope ratios of the rock samples were normalised to the recommended values of isotopic composition in the NBS lead standard (SRM no.981) which was analysed periodically during the sample analysis. The precision for 10 replicate analyses of the reference Pb was $\pm 0.1\%$, relative standard deviation ($2\sigma$), for the ratios of 208/204, 207/204 and 206/204, and $\pm 0.06\%$ for the ratios of 208/206 and 207/206 respectively. Duplicate analysis of each sample material was carried out with a precision better than $\pm 0.3\%$, relative standard deviation ($2\sigma$), although most of the within run precision for mass spectrometer analyses was $\pm 0.1\%$.

## Bulk Chemical Composition

Bulk chemical compositions of garnet-bearing granitic gneiss, K-feldspar porphyroblastic gneiss, garnet-biotite gneiss and pyroxene gneiss from the Skarvsnes area are listed in Table 1. As demonstrated in Figure 2, the garnet-bearing granitic gneiss analysed plots in the field of shales and greywackes (Miyashiro, 1973) and greywackes (Yoshida, 1978), which suggests the possibility of pelitic and/or psammitic rock origin. Compositions of the garnet-biotite and K-feldspar porphyroblastic gneisses fall in the greywacke field, and the pyroxene gneiss plots in the basalt and andesite field of Miyashiro (1973) and Yoshida (1978). The latter gneiss could possibly have originated from mafic igneous rocks such as diabase and basalt (Ravich and Kamenev, 1972). It is worthy of note that the pyroxene gneiss is rich in alkalis (especially in $K_2O$) which is in accord with the corresponding rocks from other areas of Lützow-Holm Bay (Yoshida, 1978). The Skarvsnes gneisses show a consistent conformity with metamorphic rocks from continental shields and with greywackes of the world in $K_2O$ vs. $Na_2O$ (Yoshida, 1978).

## Lead Isotopic Composition

The isotopic compositions of lead in the gneisses from the Skarvsnes terrain studied are given in Table 2, and depicted on three Pb isotopic evolution diagrams (Figures 3 to 5). The Pb data plotted on the 207/204 vs. 206/204 diagram in Figure 5 fall to both sides of the primary zero isochron. A line drawn for a suite of the pyroxene

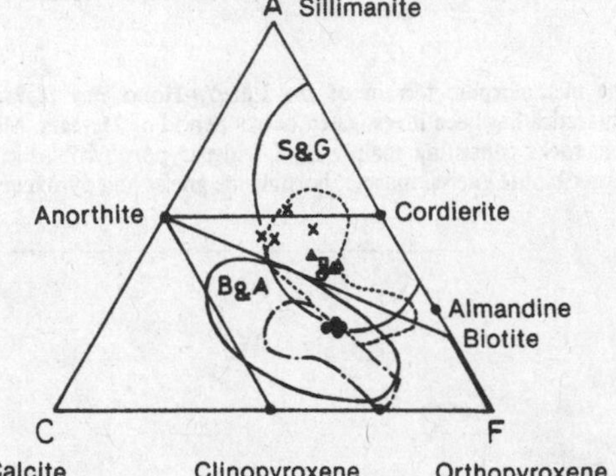


**Figure 2.** A-C-F diagram of gneisses from the Skarvsnes area. Solid circle: Pyroxene gneiss; Square: Garnet-biotite gneiss; Triangle: K-feldspar porphyroblastic gneiss; Cross: Garnet-Bearing granitic gneiss; S & G area: Sandstones and greywackes (Miyashiro, 1973); B & A area: Basalts and andesites (Miyashiro, 1973); Dotted area: Greywackes (Yoshida, 1978, p. 79); Chained area; Basaltic and andesitic rocks (Yoshida, 1978, p. 79).

**TABLE 2: Lead isotope ratios of gneisses from Skarvsnes, East Antarctica**

| | | 206/204 | 207/204 | 208/204 | 207/206 | 208/206 |
|---|---|---|---|---|---|---|
| garnet-bearing granitic gneiss | 3013001 | 18.266 | 15.583 | 37.363 | 0.85299 | 2.0449 |
| | 3013101 | 18.491 | 15.660 | 37.519 | 0.84683 | 2.0295 |
| | 3020404 | 18.364 | 15.621 | 38.774 | 0.85045 | 2.1106 |
| | 3020405 | 17.927 | 15.607 | 38.009 | 0.87063 | 2.1214 |
| | 3020213 | 18.388 | 15.635 | 37.848 | 0.85019 | 2.0579 |
| K-feldspar porphyroblastic gneiss | 3021702 | 18.333 | 15.660 | 38.360 | 0.85453 | 2.0934 |
| | 3020906 | 17.875 | 15.617 | 37.967 | 0.87331 | 2.1237 |
| | 3021502 | 17.797 | 15.612 | 37.859 | 0.87700 | 2.1279 |
| garnet-biotite gneiss | 3020113 | 18.422 | 15.659 | 37.859 | 0.84975 | 2.0543 |
| | 3020105 | 18.919 | 15.686 | 38.572 | 0.82900 | 2.0420 |
| | 3020409 | 18.388 | 15.639 | 37.981 | 0.85037 | 2.0650 |
| pyroxene gneiss | 3020204 | 18.570 | 15.589 | 37.893 | 0.83930 | 2.0399 |
| | 3020211 | 18.567 | 15.614 | 37.971 | 0.84075 | 2.0443 |
| | 3021605 | 18.610 | 15.676 | 38.155 | 0.84220 | 2.0595 |
| | 3020602 | 18.138 | 15.591 | 37.677 | 0.85907 | 2.0758 |
| | 3013006 | 19.524 | 15.745 | 38.865 | 0.80625 | 1.9902 |

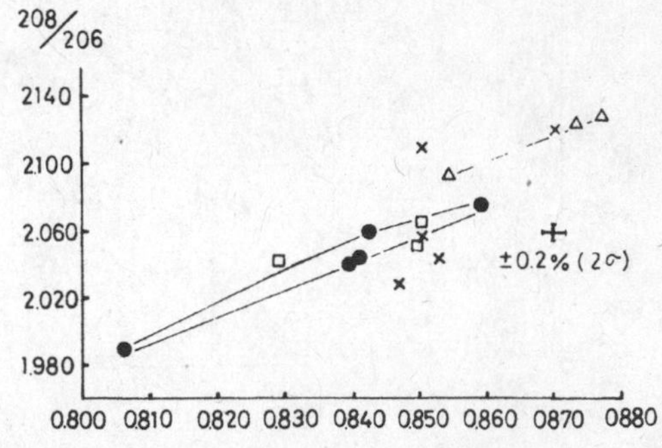


**Figure 3.** 208/206 vs. 207/206 relationship. Each symbol assigned to the four rock types is the same as in Figure 2.

gneiss samples yields a maximum apparent isochron date of 1900 Ma on the basis of the secondary isochron formula. The following parameters were used in the age calculation: $\lambda_1 = 0.155125 \times 10^{-9}yr^{-1}$; $\lambda_2 = 0.98485 \times 10^{-9}yr^{-1}$ (Jaffey et al., 1971), $^{238}U/^{235}U$ = 137.88 (Shields, 1973). The obtained data are based on an assumption that U, Th and Pb in this mafic rock have evolved with a two stage growth history through geologic time, although it may be too simple to fully interpret the Pb isotope system given here. In spite of the possibility of homogenisation of U, Th and Pb in the crystalline basement rocks from Skarvsnes during possibly the last metamorphism under amphibolite facies conditions at ca 460 Ma (Maegoya et al., 1968) or ca 500 Ma ago (Yanai and Ueda, 1974; Yoshida, 1977), no evidence of this younger disturbance is seen in the Pb isotope systematics. Figures 3 and 4 show that Pb isotope variation of the garnet-biotite gneiss is in harmony with that of the pyroxene gneiss. This fact may imply the possibility of homogenisation of the Pb isotope systematics in both gneisses by a regional metamorphic event. If so, an isochron age of 1700 Ma computed on a suite of the garnet-biotite gneiss and pyroxene gneiss would be attributed to the metamorphic event. However, positive evidence, in support of the plausibility of the above possibility, has not yet been seen in the field relations and petrographic characteristics of both metamorphic rocks. There is a significant discrepancy between the 1700 Ma and 1100 Ma obtained by the Rb-Sr whole rock dating on the garnet-biotite gneiss samples. The date of 1100 Ma can be assigned to the intermediate pressure granulite facies metamorphic event (Yoshida et al., this volume) as discussed below. Further geochronologic investigations are required to solve the problem. The garnet-bearing granitic gneiss showing heterogenity in Pb isotopes occurs as a thin layer (Ishikawa et al., 1977) and seems to hae been affected by an amphibolite facies metamorphism (Yoshida, 1978).

## Discussion and Conclusion

Metamorphic basement rocks in the Skarvsnes area of Lützow-Holm Bay appear to have been formed by at least two principal metamorphic events: an earlier granulite facies metamorphism and a later amphibolite facies event (Kizaki, 1964; 1965). Recently, Yoshida (1978; 1979) distinguished three principal episodes in the metamorphic history of the region: high grade granulite facies, followed by intermediate pressure granulite facies and then amphibolite facies metamorphisms. Based on petrographic examinations, the pyroxene gneiss, being most extensively developed in the area, can be attributed to formation by the early high grade metamorphism or original mafic igneous activity. The garnet-biotite gneiss which also occurs widely in the area would be attributed to the result of intermediate pressure granulite facies metamorphism. On the other hand, the K-feldspar porphyroblastic gneiss is distinct from the garnet-biotite gneiss in Pb isotopic characteristics (Figures 3 and 4),

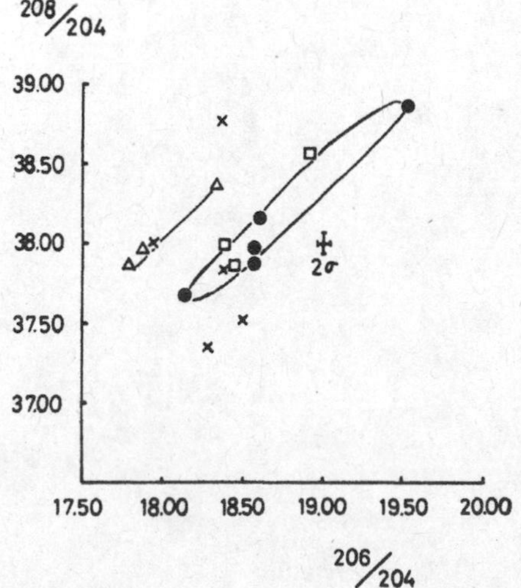


**Figure 4. 208/204 vs. 206/204 relationship. Each symbol assigned to the four rock types is the same as in Figure 2.**

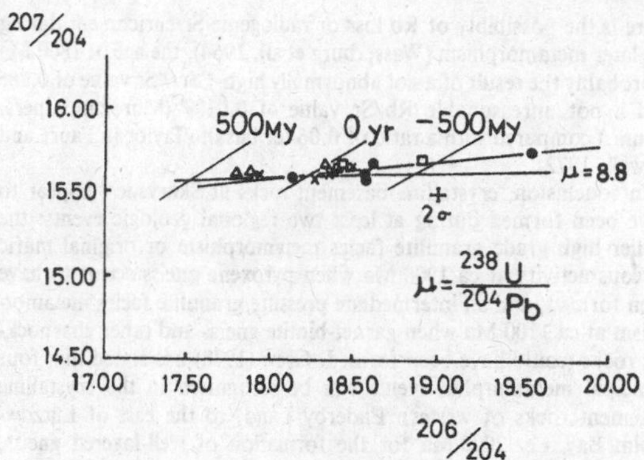


**Figure 5. 207/204 vs. 206/204 diagram. Each symbol assigned to the four rock types is the same as in Figure 2. Linear array is drawn for a suit of pyroxene gneiss samples, whose slope is 0.1170. Single stage growth curve with $\mu$ = 8.8 and the model ages for 500 My, Oyr and − 500 My are also drawn for reference.**

indicating that no homogenisation has occurred in the Pb isotope systems of these rocks through the later metamorphism, although the two rocks are closely related in field occurrence and in modal composition (Ishikawa et al., 1977). There are, however, arguments that most of metamorphic rocks in the Lützow-Holm Bay region have been formed by monometamorphism under low grade granulite facies conditions (Banno et al., 1964; Suwa, 1968).

Murozumi (pers. comm.) has analysed Sr isotope ratios and concentrations of Sr and Rb in some gneissic rocks which were split from each original sample powder used in this study. Sr and Rb concentrations were obtained by isotope dilution mass spectrometry. Duplicate analyses for each sample were done for both isotope ratio and concentration measurements. All $^{87}Sr/^{86}Sr$ ratios were normalised to the $^{87}Sr/^{86}Sr$ value of 0.1197. By using Murozumi's data, Rb-Sr total-rock model dates were computed to be 1080 Ma and 1060 Ma on the garnet-biotite gneiss samples (3020105 and 3020113 respectively), assuming an initial $^{87}Sr/^{86}Sr$ ratio of 0.710, and to be 1800 Ma on the pyroxene gneiss sample (3020204), assuming an initial ratio of 0.7037 which is a a mean ratio of oceanic basaltic rocks (Faure and Powell, 1972). A Rb-Sr whole-rock age of 1180 Ma with an initial strontium ratio of 0.709 was also obtained from a slope of the line drawn for the two garnet-biotite gneiss samples (Figure 6). A decay constant ($\lambda\beta$) of 1.42 x $10^{-11}yr^{-1}$ was used in calculations. Respective total-rock ages obtained here are in relatively good agreement with not only the computed isochron age but also with an isochron age of 1100 ± 100 Ma which is given for K-feldspar separates from gneisses from Lützow-Holm Bay (Maegoya et al., 1968). It is noteworthy that the 1800 Ma seems to be comparable with a maximum estimate of the 1900 Ma obtained by the Pb-Pb method. Although

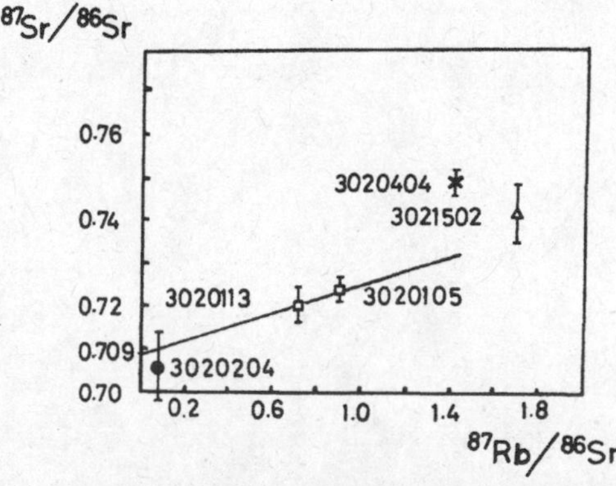


**Figure 6. A plot of $^{87}Sr/^{86}Sr$ ratio vs. $^{87}Rb/^{86}Rb$ ratio of gneissic rocks from the Skarvsnes area. Data plotted come from unpublished analyses by Murozumi. Each symbol is the same as in Figure 2. A linear array is drawn for garnet-biotite gneisses analysed.**

there is the possibility of Rb loss or radiogenic Sr enrichment during the later metamorphism (Wasserburg et al., 1964), the age of 1800 Ma is probably the result of a not abnormally high $^{87}Sr/^{86}Sr$ value of 0.706 and a not unreasonable Rb/Sr value of 0.0309 (Murozumi, pers. comm.) compared with a ratio of 0.06 for basalt (Taylor in Faure and Powell, 1972).

In conclusion, crystalline basement rocks at Skarvsnes, appear to have been formed during at least two regional geologic events: the earlier high grade granulite facies metamorphism or original mafic igneous activity at ca 1900 Ma when pyroxene gneiss seems to have been formed, and an intermediate pressure granulite facies metamorphism at ca 1100 Ma when garnet-biotite gneiss and other charnockitic rocks would have been formed. Grew (1978) has stated that four principal metamorphic events can be recognised in the crystalline basement rocks of western Enderby Land, to the east of Lützow-Holm Bay, i.e. 2000 Ma for the formation of well-layered gneiss, 1000 Ma for granulite facies metamorphism, 500 Ma for amphibolite facies metamorphism and 400 to 500 Ma for secondary mineralisation. The geochronologic data presented here are consistent with this sequence.

*Acknowledgements* I would like to express my sincere thanks to Professor M. Murozumi, Muroran Institute of Technology, for his kind permission to use unpublished Rb-Sr data. I am also indebted to S. Kojima for providing the metamorphic rock specimens available for this study. Lead isotope analyses of the samples were carried out in an ultraclean laboratory of the Institute under direction of Murozumi and his colleague, S. Nakamura, who has helped me in the course of the Pb isotope analyses.

## REFERENCES

BANNO, S., TATSUMI, T., OGURA, Y. and KATSURA, T., 1964: Petrographic studies on the rocks from the area around Lützow-Holmbukta; *in* Adie, R.J. (ed.), *Antarctic Geology,* North Holland, Amsterdam, 405-14.

FAURE, G. and POWELL, J.L., 1972: *Strontium Isotope Geology,* Springer-Verlag, New York.

GREW, E.S., 1978: Precambrian basement at Molodezhnaya Station, East Antarctica. *Geol. Soc. Am., Bull., 89,* 801-13.

ISHIKAWA, T., YANAI, K., MATSUMOTO, Y., KIZAKI, K., KOJIMA, S., TATSUMI, T., KAIKUCHI, T. and YOSHIDA, M., 1977: Skarvsnes. *Antarct. Geol. Map Ser.,* Sheets 6 and 7. Nat. Inst. Pol. Res., Japan.

JAFFREY, A.H., FLYNN, K.F., GLENDENIN, L.E., BENTLY, W.C. and ESSING, A.M., 1971: Precision measurements of half-lives and specific activities of 236U and 238U. *Phys. Rev., C, 4,* 1889-1906.

KIZAKI, K., 1964: Tectonics and petrography of the East Ongle Island, Lützow-Holmbukta, Antarctica. *JARE Sci. Rep., Ser. C, 2,* 1-24.

KIZAKI, K., 1965: Geology and petrography ofthe Yamato Sanmyaku, East Antarctica. *JARE Sci. Rep., Ser. C, 3,* 1-27.

MAEGOYA, M., NOHDA, S. and HAYASE, I., 1968: Rb-Sr dating of the gneissic rocks from the east coast of Lützow-Holm Bay, Antarctica. *Mem. Coll. Sci. Univ. Kyoto, Ser. B., 35,* 131-8.

MIYASHIRO, A., 1973: *Metamorphism and metamorphic belts.* George Allen & Unwin, London.

RAVICH, M.G. and KAMENEV, E.N., 1972: *Kristallicheskii fundament Antarkticheskoi platformy (Crystalline basement of the Antarctic Platform).* Gidrometeoizdat, Leningrad (English translation: 1975, John Wiley, New York).

SHIELDS, W.R., *in* Weast, R.C. and Selby, S.M. (eds.): *Handbook of Chemistry and Physics.* Chemical Rubber Co., Cleveland, Ohio, 1073.

SHIRAHATA, H., 1972: Chemical analysis of the geochemical standards JG-1 and JB-1. *Mem. Muroran Inst. Tech., Sci. Eng., 7,* 831-9 (in Japanese with English abstract).

SHIRAHATA, H., 1982: Lead isotopic composition in the geochemical reference rock JG-1. *J. Japan Assoc. Min. Petr. Econ. Geol., 77,* 100-3.

SUWA, K., 1968: Petrological studies on the metamorphic rocks from Lützow-Holmbukta area, East Antarctica. *23rd Int. Geol. Congr., 4,* 171-87.

TATSUMI, T. and KIKUCHI, T., 1959a: Report of geomorphological and geological studies of the wintering term (1957/58) of the first Japanese Antarctic Research Expedition. Part 1, *Antarctic. Rec., 7,* 1-16.

TATSUMI, T. and KIKUCHI, T., 1959b: Report of geomorphological and geological studies of the wintering term (1957/58) of the first Japanese Antarctic Research Expedition. Part 2, *Antarctic. Rec., 8,* 443-63.

WASSERBURG, G.J., ALBEE, A.L. and LANPHERE, M.A., 1964: Migration of radiogenic strontium during metamorphism. *J. Geophys. Res., 69,* 4395-401.

YANAI, K. and UEDA, Y., 1974: Absolute ages and geological investigations in the area of around Syowa Station, East Antarctica. *Antarctic. Rec., 48,* 70-81.

YOSHIDA, M., 1977: Geology of the Skallen region, Lützow-Holmbukta, East Antarctica. *Mem. Nat. Inst. Pol. Res. Japan, Ser. C, 11,* 1-38.

YOSHIDA, M., 1978: Tectonics and petrology of charnockites around Lützow-Holmbukta, East Antarctica. *J. Geosci., Osaka City Univ., 21,* 65-152.

YOSHIDA, M., 1979: Metamorphic conditions of the polymetamorphic Lützow-Holmbukta region, East Antarctica. *J. Geosci., Osaka City Univ., 22,* 97-140.

YOSHIDA, M., SUZUKI, M., SHIRAHATA, H. and KOJIMA, H. (this volume): A review of the tectonic and metamorphic history of the Lützow-Holm Bay region, East Antarctica.

# Sr-ISOTOPIC STUDIES OF SOME INTRUSIVE ROCKS IN THE AHLMANN RIDGE AND ANNANDAGSTOPPANE, WESTERN QUEEN MAUD LAND, ANTARCTICA

J.M. Barton Jr. and Y.E. Copperthwaite, *Bernard Price Institute of Geophysical Research, University of the Witwatersrand, Johannesburg 2001, South Africa.*

*Abstract* **Rb-Sr whole rock isotopic studies of five suites of mafic igneous rocks from western Queen Maud Land yield ages of about 770 Ma, 1000 Ma, 1430 Ma and 1800 Ma. The ages suggest that igneous activity occurred here intermittently over most of the Proterozoic. These units are also characterised by having high initial $^{87}Sr/^{86}Sr$ ratios, the significance of which is unclear.**

The state of understanding of the geology of western Queen Maud Land south of the SANAE base has been summarised recently by Wolmarans and Kent (1982). This area (Figure 1) may be divided conveniently into five regions: (1) the Ahlmann Ridge; (2) the Borg Massif; (3) the Annandagstoppane; (4) the Sverdrup and Gjelsvik Mountains; and (5) the Kirwan Escarpment. These regions have pronounced northeast trending outcrop patterns and are separated from one another by major crustal breaks along which large drainage glaciers, including the Jululstraumen Glacier, presently flow. The true nature of these crustal breaks is unknown but it is believed that they represent strike-slip faults, movement along which brought the component regions into their present geographic configuration in post-Ross Orogeny times, i.e. more recently than about 480 Ma ago.

The Kirwan Escarpment and Sverdrup and Gjelsvik Mountains comprise migmatites, late-stage granites and both deformed and undeformed mafic dykes and sills. The period of migmatisation appears to have occurred approximately 1050 Ma to 1000 Ma ago, and these rocks have an approximately 480 Ma mineral age veil characteristic of areas affected by the Ross Orogeny (see age data summary of T. Elworthy in Wolmarans and Kent, 1982).

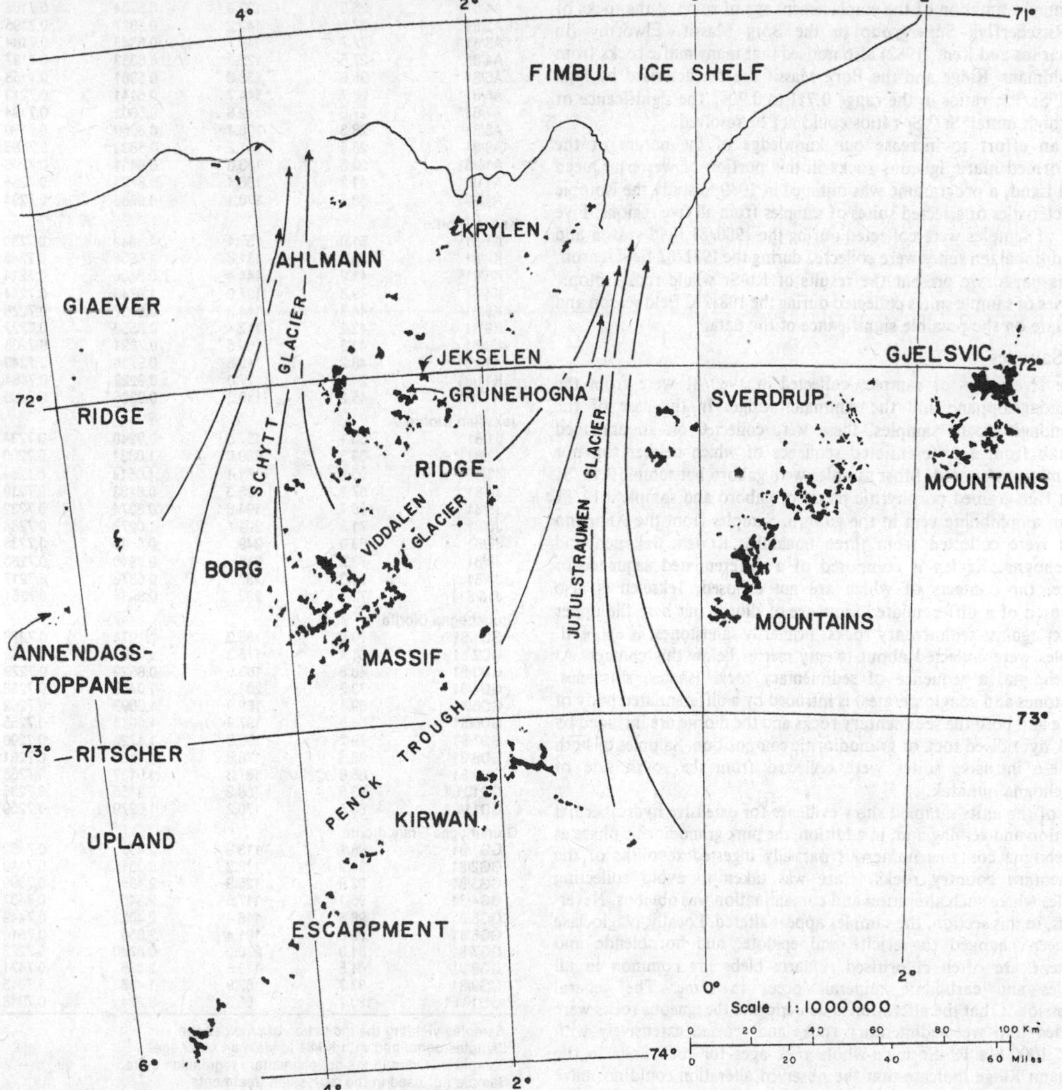


**Figure 1. Outcrop map of western Queen Maud Land, Antarctica, showing the five geographic divisions discussed in the text as well as the four sampling localities.**

The Annandagstoppane are composed of either granite or gabbro, both of which contain a suite of undeformed diabase dykes. The granite is believed to be approximately 2850 Ma old (see age data summary of T. Elworthy in Wolmarans and Kent, 1982).

The Ahlmann Ridge and Borg Massif are composed largely of undeformed fluvial-sedimentary and volcanic rocks of the Ritscherflya Supergroup (Wolmarans and Kent, 1982). The Ritscherflya Supergroup may be divided into groups and formations but limited exposures make the stratigraphic positions of these rock units and correlations among them difficult to establish. The rocks of the Ritscherflya Supergroup are intruded by differentiated sills and other apparently tabular bodies of undeformed but altered diorite, gabbro (sometimes associated with picrite) and granodiorite. In addition, all of these rocks are intruded by diabase dykes.

Very little is known about the age and genetic relationships among the undeformed mafic intrusive rocks of the five regions. Rb-Sr isotopic data for grab-samples from numerous nunataks have been interpreted by Elworthy (in Wolmarans and Kent, 1982) to indicate that periods of intrusion occurred about 1700 Ma and 1000 Ma ago, with a possible intrusive period at about 770 Ma ago. However, data are not sufficiently numerous to allow for the resolution of ages for most units, an exception being that of a sill in the Ahlmann Ridge that yielded a Rb-Sr whole rock isochron age of 1672 ± 79 Ma ($2\sigma$) (T. Elworthy in Wolmarans and Kent, 1982). This age may provide a minimum estimation of the emplacement age of some of the rocks of the Ritscherflya Supergroup in the Borg Massif. Elworthy (in Wolmarans and Kent, 1982) also noticed that many mafic rocks from the Ahlmann Ridge and the Borg Massif are characterised by high initial $^{87}Sr/^{86}Sr$ ratios in the range 0.711 to 0.709. The significance of these high initial $^{87}Sr/^{86}Sr$ ratios could not be resolved.

In an effort to increase our knowledge of the nature of the undeformed mafic igneous rocks in this portion of western Queen Maud Land, a programme was initiated in 1980 to study the isotopic characteristics of selected suites of samples from all five regions. Five suites of samples were collected during the 1980/81 field season and an additional ten suites were collected during the 1981/82 field season. In this paper, we present the results of Rb-Sr whole rock isotopic analyses of sample suites collected during the 1981/82 field season and speculate on the possible significance of the data.

## The Samples

The five suites of samples collected in 1980/81 were from the Annandagstoppane and the Ahlmann Ridge. In the case of the Annandagstoppane samples, these were collected on an unnamed nunatak from a differentiated sequence of which neither top nor bottom were exposed. Most samples were gabbro but sample A11/81 was a fine-grained porphyritic phase of gabbro and sample A12/81 was an amphibolite vein in the gabbro. Samples from the Ahlmann Ridge were collected from three nunataks: Krylen, Jekselen and Grunehogna. Krylen is composed of a differentiated sequence of diorite, the contacts of which are not exposed. Jekselen is also composed of a differentiated sequence of diorite but here the upper contact against sedimentary rocks, primarily sandstones, is exposed. Samples were collected about twenty metres below this contact. At Grunehogna, a sequence of sedimentary rocks (shales, siltstones, sandstones and conglomerates) is intruded by a differentiated body of diorite and both the sedimentary rocks and the diorite are intruded by a pink hybridised rock of granodioritic composition. Samples of both of these intrusive suites were collected from the south side of Grunehogna nunatak.

All of the units sampled show evidence for extensive hydrothermal alteration and veining and, in addition, the pink granodioritic phase at Grunehogna contains numerous partially digested xenoliths of the sedimentary country rocks. Care was taken to avoid collecting samples where such alteration and contamination was obvious. Nevertheless, in this section, the samples appear altered. Locally, plagioclase has been changed to sericite and epidote, and hornblende and pyroxene are often chloritised. Quartz blebs are common in all samples and carbonate minerals occur in some. The general impression is that the alteration is deuteric and the igneous rocks were intruded into wet sedimentary rocks and reacted extensively with them. 1000 Ma Rb-Sr mica-whole rock ages for rock units in the Ahlmann Ridge indicate that the observed alteration could not have been a result of widespread hydrothermal activity associated with the Ross Orogeny (Barton and Copperthwaite, unpublished data).

## Analytical Techniques and Results

The samples were processed and analysed by standard techniques (Barton et al., 1979). The Rb and Sr elemental concentrations are precise to ± 0.5% ($1\sigma$) and the $^{87}Sr/^{86}Sr$ ratios are precise to ± 0.01% ($1\sigma$). Analyses of the N.B.S. standard $SrCO_3$ (SRM-987) yielded an $^{87}Sr/^{86}Sr$ ratio of 0.71022 ± 0.00003 ($1\sigma$). Where appropriate, the data were tested for scatter in excess of the amount expected from experimental uncertainties by a technique modified from that recommended by Brooks et al. (1972). The component of scatter as measured by SUMS/(n-2) (York, 1969) is compared to the appropriate F-variate. If SUMS/(n-2) exceeds the F-variate, the data were assumed to contain excess scatter. Data from units with no excess scatter were regressed by the technique of York (1969). Otherwise, they were regressed by the technique of York (1966). The analytical data are presented in Table 1 and the regression data are presented in Table 2. The data are also plotted on isochron diagrams in Figures 2 through 6.

**TABLE 1: Rb AND Sr ELEMENTAL AND ISOTOPIC DATA**

| Sample | Rb (ppm) | Sr (ppm) | $^{87}Rb/^{86}Sr$ | $^{87}Sr/^{86}Sr$ |
|---|---|---|---|---|
| Annandagstoppane Gabbro | | | | |
| A1/81 | 25.3 | 123.9 | 0.5904 | 0.7188 |
| A2/81* | 27.5 | 142.2 | 0.5627 | 0.7195 |
| A3/81* | 27.7 | 149.7 | 0.5348 | 0.7184 |
| A4/81 | 27.5 | 125.3 | 0.6353 | 0.7197 |
| A5/81 | 28.6 | 138.8 | 0.5961 | 0.7188 |
| A6/81* | 30.7 | 144.7 | 0.6141 | 0.7213 |
| A7/81* | 27.6 | 142.6 | 0.5602 | 0.7194 |
| A8/81 | 28.3 | 135.1 | 0.6059 | 0.7190 |
| A9/81 | 28.5 | 141.2 | 0.5833 | 0.7185 |
| A10/81 | 29.8 | 133.6 | 0.6451 | 0.7206 |
| A11/81 | 41.2 | 136.4 | 0.8731 | 0.7258 |
| A12/81 | 58.7 | 170.1 | 0.9986 | 0.7294 |
| Krylen Diorite | | | | |
| K1/81[1] | 61.0 | 152.1 | 1.1942 | 0.7289 |
| K2/81 | 48.5 | 131.8 | 1.0655 | 0.7245 |
| K3/81[2] | 41.9 | 144.4 | 0.8405 | 0.7214 |
| K4/81 | 49.8 | 134.0 | 1.0748 | 0.7244 |
| K5/81 | 44.7 | 144.8 | 0.8927 | 0.7226 |
| K8/81 | 42.2 | 142.4 | 0.8553 | 0.7222 |
| K9/81 | 43.3 | 169.5 | 0.7391 | 0.7208 |
| K10/81[1] | 48.2 | 143.6 | 0.9716 | 0.7249 |
| K11/81[3] | 24.6 | 77.0 | 0.9229 | 0.7164 |
| K12/81 | 45.2 | 140.3 | 0.9305 | 0.7230 |
| Jekselen Diorite | | | | |
| J1/81** | 63.1 | 197.5 | 0.9240 | 0.7234 |
| J2/81** | 89.3 | 239.0 | 1.0831 | 0.7260 |
| J3/81** | 65.8 | 181.1 | 1.0514 | 0.7251 |
| J4/81** | 57.2 | 198.3 | 0.8283 | 0.7219 |
| J5/81 | 56.7 | 194.8 | 0.8374 | 0.7239 |
| J6/81 | 71.3 | 201. | 1.0279 | 0.7264 |
| J7/81 | 69.0 | 249. | 0.8013 | 0.7236 |
| J8/81 | 73.8 | 215. | 0.9910 | 0.7260 |
| J9/81 | 56.3 | 237. | 0.6873 | 0.7217 |
| J10/81 | 72.1 | 233. | 0.8934 | 0.7251 |
| Grunehögna Diorite | | | | |
| GD1/81 | 70.8 | 185.0 | 1.1074 | 0.7278 |
| GD2/81 | 61.4 | 175.3 | 1.0133 | 0.7259 |
| GD3/81 | 48.8 | 163.5 | 0.8629 | 0.7229 |
| GD4/81 | 73.2 | 203. | 1.0465 | 0.7263 |
| GD5/81 | 59.3 | 169.9 | 1.0093 | 0.7262 |
| GD6/81 | 68.5 | 197.9 | 1.0015 | 0.7255 |
| GD7/81 | 79.7 | 195.9 | 1.1722 | 0.7290 |
| GD8/81 | 63.1 | 178.8 | 1.0165 | 0.7261 |
| GD9/81 | 65.6 | 187.3 | 1.0137 | 0.7258 |
| GD10/81 | 53.5 | 168.9 | 0.9155 | 0.7235 |
| GD11/81 | 60.6 | 170.2 | 1.0279 | 0.7259 |
| Grunehögna Granodiorite | | | | |
| GG1/81 | 96.7 | 113.9 | 2.459 | 0.7452 |
| GG2/81 | 91.9 | 118.2 | 2.251 | 0.7419 |
| GG3/81 | 92.8 | 125.9 | 2.135 | 0.7399 |
| GG4/81 | 95.1 | 117.5 | 2.345 | 0.7437 |
| GG5/81 | 96.6 | 116.6 | 2.402 | 0.7449 |
| GG6/81 | 99.6 | 101.4 | 2.850 | 0.7510 |
| GG7/81 | 81.9 | 270. | 0.8780 | 0.7223 |
| GG8/81 | 101.5 | 113.8 | 2.586 | 0.7471 |
| GG9/81 | 83.7 | 162.9 | 1.4863 | 0.7315 |
| GG10/81 | 120.7 | 56.5 | 6.224 | 0.7988 |

* Samples yielding the indication of an older age.
[1] Samples combined with K9/81 to yield an older age.
[2] Sample included in the supplementary regression suite.
[3] Sample not used in the regression treatments.
** Samples combined to yield one isochron. The unmarked samples in this suite are combined into the other isochron.

**TABLE 2: REGRESSION DATA[1]**

| Unit | Age[2] | Initial $^{87}Sr/^{86}Sr$[2] | S[3] | F[4] |
|---|---|---|---|---|
| Annandagstoppane Gabbro | | | | |
| Suite 1 | 1802 ± 100 | 0.7034 ± 0.0009 | 2.8 | 2.6 |
| Suite 2 | 2518 ± 406 | 0.6990 ± 0.0033 | 0.1 | 3.5 |
| Krylen Diorite | | | | |
| Suite 1 | 767 ± 49 | 0.7128 ± 0.0006 | 0.8 | 2.9 |
| Suite 2 | 806 ± 122 | 0.7122 ± 0.0016 | 6.2 | 2.7 |
| Suite 3 | 1241 ± 61 | 0.7077 ± 0.0008 | 0.05 | 4.4 |
| Jekselen Diorite | | | | |
| Suite 1 | 1079 ± 87 | 0.7091 ± 0.0012 | 2.4 | 3.5 |
| Suite 2 | 984 ± 120 | 0.7122 ± 0.0015 | 4.7 | 2.9 |
| Grunehögna Diorite | 1426 ± 87 | 0.7051 ± 0.0013 | 2.3 | 2.4 |
| Grunehögna Granodiorite | 1008 ± 11 | 0.7097 ± 0.0003 | 2.5 | 2.5 |

[1] Where S > F, data regressed using the technique of York (1966). Otherwise, the data regressed using the technique of York (1969). $\lambda\ ^{87}Rb = 1.42 \times 10^{-11}\ a^{-1}$.
[2] Uncertainties = $2\sigma$.
[3] SUMS/(n-2) from technique of York (1969).
[4] Appropriate F-variate.

## Discussion

The Rb-Sr isotopic data from Annandagstoppane (Figure 2) define two trends: one corresponding to an age of approximately 2520 Ma and the other to an age of about 1800 Ma. The initial $^{87}Sr/^{86}Sr$ ratio of the steeper, four-point trend is probably unrealistically low and by forcing the data through a more reasonable initial $^{87}Sr/^{86}Sr$ ratio of 0.701, a slope indicating an age of about 2250 Ma is obtained. Similarly, the data from Krylen Nunatak (Figure 3) define two trends corresponding to ages of about 1240 Ma and 770 Ma while the data from Jekselen Nunatak (Figure 4) define near parallel trends corresponding to an age of about 1000 Ma. The data from the other two units sampled at Grunehogna yield single trends indicating ages of about 1430 Ma (Figure 5) and 1000 Ma (Figure 6).

Observed structural relationships at Annandagstoppane and Krylen and Jekselen Nunataks neither confirm nor deny the possibility that at each locality two igneous units are exposed and were sampled. If multiple intrusions are present, then significantly older and different younger components were dated at Annandagstoppane and Krylen Nunatak and two coeval phases were dated at Jekselen Nunatak. At Jekselen Nunatak, these phases had significantly different initial $^{87}Sr/^{86}Sr$ ratios reminiscent of the relationship among various phases within the Bushveld igneous complex (Hamilton, 1977, Kruger and Marsh, 1982). On the other hand, the steeper trends of the data from Annandagstoppane and Krylen Nunatak are defined by relatively few points. In must be remembered that rocks at these localities probably were altered deuterically and it is possible that the steeper trends are fortuitous, reflecting local alteration induced Sr-isotopic heterogeneities within the magmas. Clearly, further work is necessary to enable us to choose between these possibilities.

At Grunehogna Nunatak, the two igneous rock units present yield ages that are consistent with their structural relationship. The isochron for the granodiorite is essentially identical to that published by Allsopp and Neethling (1970) for "syenite" at Nils Jorgennutane,

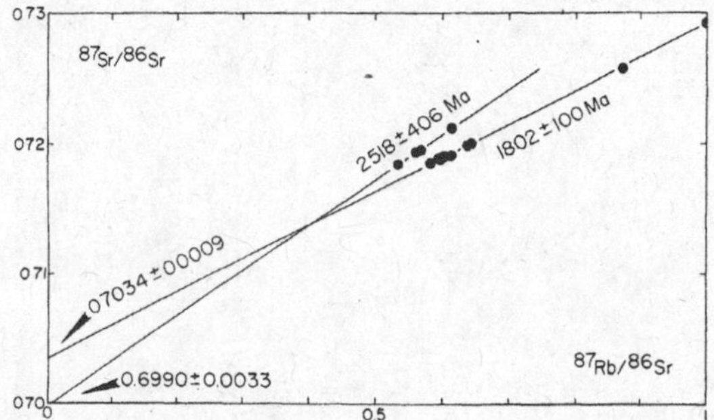


**Figure 2. Rb-Sr isochron diagram showing the data for the gabbro exposed at Annandagstoppane.**

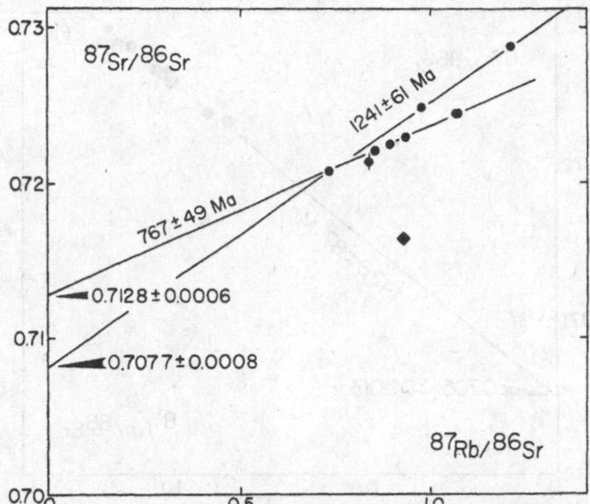


**Figure 3. Rb-Sr isochron diagram showing the data for the diorite exposed on Krylen Nunatak in the Ahlmann Ridge. The diamond indicates the results from sample K11/81 which are not used in the regression treatments. The flagged point is the results from sample K3/81 which plot near the main trend of the data. When combined with the main trend data, the slope of the regression line is increased slightly, corresponding to an age of about 810 Ma.**

twenty kilometres to the north, and we suspect that the syenite and the granodiorite are parts of the same unit. The preservation of an approximately 1430 Ma isochron age for the diorite indicates that alteration associated with the emplacement 1000 Ma ago of the granodiorite did not affect this unit as far as Sr-isotopes are concerned.

The apparent initial $^{87}Sr/^{86}Sr$ ratios relative to emplacement (isochron) ages for the units analysed are too high to indicate that the parental magmas were unaltered derivatives of depleted mantle. In addition, a crude inverse relationship appears to exist between emplacement age and initial $^{87}Sr/^{86}Sr$ ratio (Table 2, Figures 2-6) so that a gabbro at Annandagstoppane was emplaced about 1800 Ma ago with an initial $^{87}Sr/^{86}Sr$ ratio of about .7034; the diorite at Grunehogna was emplaced about 1430 Ma ago with an initial $^{87}Sr/^{86}Sr$ ratio of about 0.7051 and the remaining units were emplaced more recently

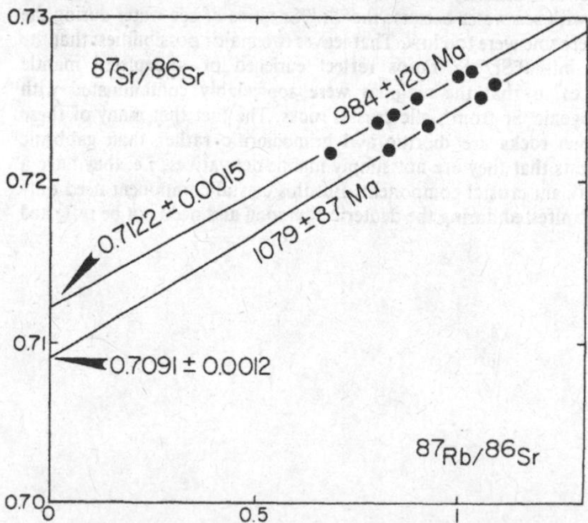


**Figure 4. Rb-Sr isochron diagram showing the data for the diorite exposed on Jekselen Nunatak in the Ahlmann Ridge. Samples J1/81 through J4/81 define the lower trend while samples J5/81 through J10/81 define the upper trend. The two trends may represent separate phases of intrusion with essentially identical ages of emplacement.**

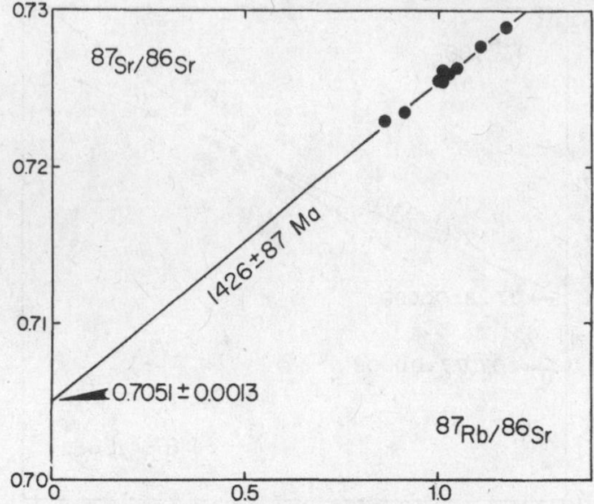


**Figure 5. Rb-Sr isochron diagram showing the data for the diorite exposed on Grunehogna Nunatak in the Ahlmann Range.**

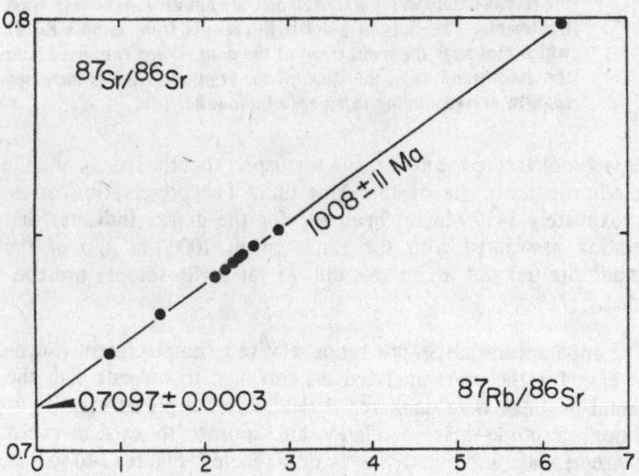


**Figure 6. Rb-Sr isochron diagram showing the data for the granodiorite exposed on Grunehogna Nunatak in the Ahlmann Range.**

than about 1250 Ma ago with initial $^{87}Sr/^{86}Sr$ ratios greater than about 0.708. It is unlikely that the initial $^{87}Sr/^{86}Sr$ ratios can reflect interaction with sea water because the $^{87}Sr/^{86}Sr$ ratios of sea water during the Proterozoic were too low. That leaves two major possibilities: that the high initial $^{87}Sr/^{86}Sr$ ratios reflect enriched or undepleted mantle sources or that the magmas were appreciably contaminated with radiogenic $^{87}Sr$ from sialic crustal rocks. The fact that many of these igneous rocks are dioritic and granodioritic rather than gabbroic suggests that they are not simply mantle derivatives, i.e. they have a significant crustal component. But this crustal component need only be manifested during the deuteric alteration and need not be reflected in the Sr-isotopic composition of the magmas. The problem of choosing between anomalous mantle sources and crustal contamination to explain the high initial $^{87}Sr/^{86}Sr$ ratios has been recognised for some time with regard to certain Mesozoic volcanic rocks in Gondwanaland (see discussion in e.g. Hoefs et al., 1980). We do not know yet which possibility is correct and are undertaking Sm-Nd isotopic studies on these rocks with an aim to deciding. But either way, data from studies of these units of diverse age should enable us to better understand mantle evolution under this portion of Antarctica during most of Proterozoic time.

High initial $^{87}Sr/^{86}Sr$ ratios do not characterise all of the mafic igneous rocks in western Queen Maud Land. The sill in the Alhmann Ridge studies by Elworthy (in Wolmarans and Kent, 1982) has an initial $^{87}Sr/^{86}Sr$ ratio of $0.7008 \pm 0.0012$ ($2\sigma$) for an age of about 1670 Ma, indicating a depleted mantle source. It may be that high initial $^{87}Sr/^{86}Sr$ ratios characterise some but not all mafic igneous rocks exposed in the Ahlmann Ridge and Annandagstoppane.

## Conclusions

The results of age studies reported here have substantiated the observation of Elworthy (in Wolmarans and Kent, 1982) that mafic igneous activity occurred in western Queen Maud Land about 1000 Ma and 770 Ma ago. In addition, these data indicate that other periods of activity occurred about 1430 Ma and 1800 Ma ago. The data also support Elworthy's observation that high initial $^{87}Sr/^{86}Sr$ ratios characterise many of these units.

*Acknowledgement* We wish to thank Dr J. Cooper and Mr J. Krynauw for constructive criticisms of an earlier version of this paper. Financial support for this study came from grants from the Department of Transport of South Africa. Isotopic data reported here are stored in the South African National Isotopic Data Base.

REFERENCES

ALLSOPP, H.L. and NEETHLING, D.C., 1970: Rb-Sr isotopic ages of Precambrian intrusive rocks from Queen Maud Land, Antarctica. *Earth Planet. Sci. Lett., 8,* 66-70.

BARTON, J.M., Jr., FRIPP, R.E.P., HORROCKS, P. and MCLEAN, N., 1979: The geology, age and tectonic setting of the Messina layered Intrusion, Limpopo Mobile Belt, southern Africa. *Am. J. Sci., 279,* 1108-34.

BROOKS, C., HART, S.R. and WENDT, I., 1972: On the realistic use of two error regression treatments as applied to Rb-Sr data. *Rev. Geophys., Space Phys., 10,* 551-77.

HAMILTON, P.J., 1977: Sr-isotope and trace element studies of the Great Dyke and Bushveld mafic phase and their relation to early Proterozoic magma genesis in southern Africa. *J. Petrol., 18,* 24-52.

HOEFS, J., FAURE, G. and ELLIOT, D.H., 1980: Correlation of $\delta^{18}O$ and initial $^{87}Sr/^{86}Sr$ ratios in Kirkpatrick basalt on Mt Falla, Transantarctic Mountains. *Contrib. Mineral. Petrol., 75,* 199-203.

KRUGER, F.J. and MARSH, J.S., 1982: Significance of $^{87}Sr/^{86}Sr$ ratios in the Merensky cyclic unit of the Bushveld Complex. *Nature, 298,* 53-5.

WOLMARANS, L.C. and KENT, L.E., 1982: Geological investigations in western Dronning Maud Land, Antarctica—a synthesis. *CSIR of South Africa, South African National Scientific Programmes Rept.,* (in press).

YORK, D., 1966: Least squares fitting of a straight line. *Can. J. Phys., 44,* 1079-86.

YORK, D., 1969: Least squares fitting of a straight line with correlated errors. *Earth Planet. Sci. Lett., 5,* 320-4

# PRELIMINARY REPORT ON THE GEOCHEMISTRY AND PETROLOGY OF SOME IGNEOUS ROCKS IN THE AHLMANNRYGGEN AND GIAEVERRYGGEN, WESTERN DRONNING MAUD LAND, ANTARCTICA

J. R. Krynauw, *Geology Department, University of Natal, PO Box 375, Pietermaritzburg 3200, South Africa.*

**_Abstract_ Approximately seventy-five percent of the outcrops in the Ahlmannryggen and Giaeverryggen in western Dronning Maud Land comprise ultramafic to felsic rocks of the Borgmassivet Intrusions (Neethling, 1972; Wolmarans et al., in press) and andesitic volcanics of the Straumsnutane Formation (Watters, 1972). Ages vary from approximately 850 to 1,700 Ma (Allsopp and Neethling, 1970; Elworthy, in press).**

**In the west, Annandagstoppane (06°15′W, 72°35′S) and Juletoppane (05°30′W, 72°22′S) consist of gabbro/norites, gabbros and diorites, with minor granopyre and quartz-diorite pegmatoids. Rhythmic feldspar-pyroxene banding, varying in thickness from one to five centimetres, is abundant. Primary igneous minerals are altered, but not as extensively as further east.**

**At Robertskollen (03°15′W, 71°28′S) poorly differentiated diorites overlie an ultramafic suite. Field relationships and geochemistry indicate a genetic relationship between the mafic and ultramafic rocks. Abundant quartz-diorite pegmatoids have invaded the diorites, indicating the presence of large volumes of volatiles during late-stage crystallisation. The pyroxenes in the diorites have been altered extensively to amphiboles.**

**The intrusions at Grunehogna (02°45′W, 72°05′S) comprise gabbros and diorites invading sediments of the Grunehogna and Högronna Formations. A 55 m thick quartz-diorite pegmatoid sill occurs over a large part of the area, and granitoid dykes intrude the mafic rocks. Intrusive relationships and geochemical evidence suggest that the magmas intruded soft, possibly wet, sediments which have supplied some of the volatiles necessary for the large volumes of pegmatoids, and caused the alteration of primary pyroxenes. Low-grade metamorphism, identified by development of tremolite, rare kinks in primary biotite, and curved plagioclase, may be related to movement along the flank of the Penck Trough.**

**The geochemical data of the mafic rocks display typical tholeiitic major and trace element trends. However, the limited data available from the sub-volcanic complex at Jekselen (02°30′W, 72°00′S) plot well away from these trends (e.g. Jekselen rocks are enriched in sodium and magnesium, and depleted in total iron, titanium and calcium), suggesting a separate igneous source region, or extensive metasomatic alteration.**

# PETROLOGY AND ZIRCON GEOCHRONOLOGY OF HERRING ISLAND AND COMMONWEALTH BAY AND EVIDENCE FOR GONDWANA RECONSTRUCTION

R.L. Oliver, J.A.Cooper, A.J. Truelove[1], *Department of Geology and Mineralogy, The University of Adelaide, Adelaide, South Australia, 5000.*

[1]*Present Address: C/- Shell Metals, P.O. Box 5094, Cairns, Queensland, 4870.*

*Abstract* Approximately 1400 Ma metasediments and metavolcanics at Herring Island have been deformed three times and metamorphosed at 850-900°C and 6-7 kb about 1275 Ma ago. Zircons from the metasediment retain a record of a 2500 Ma provenance age. At Cape Denison a very Late Archaean granodiorite gneiss contains zircon which was slightly disturbed isotopically at about 1600 Ma and much more so in the Palaeozoic. This gneiss can be correlated geochronologically and geologically with the Late Archaean Gneisses of Eyre Peninsula, South Australia, whose eastern limit is sharply defined. The match of continental coastlines based on this correlation shows a close resemblance to earlier computer fits and places Herring Island as a continuation of the acutely aligned Fraser Province lineation opposite Albany rather than Esperance. This does not support the linear fitting of the Antarctic Bowers Trough with Tasmanian troughs of similar ages.

Two of the authors (R.L.O. and A.J.T.) made brief visits to Herring Island, one of the Windmill Islands (approx. Lat. 66°30′S, Long. 110°30′E) and Cape Denison, Commonwealth Bay (approx. Lat. 67°5′S, Long. 142°40′E) in January-February 1981. Detailed mapping of the western end of Herring Island (Figure 1) was carried out over four days; material was collected for petrological, geochemical and zircon geochronological study. One hour ashore at Cape Denison permitted the collection of two 30 kg samples of granodiorite but no other geological work could be attempted.

## HERRING ISLAND

### Geological setting

The Windmill Islands consist of schistose to quartzo-feldspathic gneisses which grade from upper amphibolite facies in the north to granulite facies in the south (Blight and Oliver, 1982). Collerson and Reid (1981) recognised three main protolith types in the Windmill Islands area. The oldest is represented by dioritic through tonalitic to granitic layered gneisses of probable igneous parentage, though Blight and Oliver (1982, Figure 54.3) infer that the scattered distribution of some felsic rock analyses on an Ab-An-Or diagram reflect the influence of sedimentary rock parentage. These rocks do not appear to be represented on Herring Island. Interleaved with these orthogneisses are supracrustal paragneisses. The third group is also intrusive and is represented now by leucogneiss.

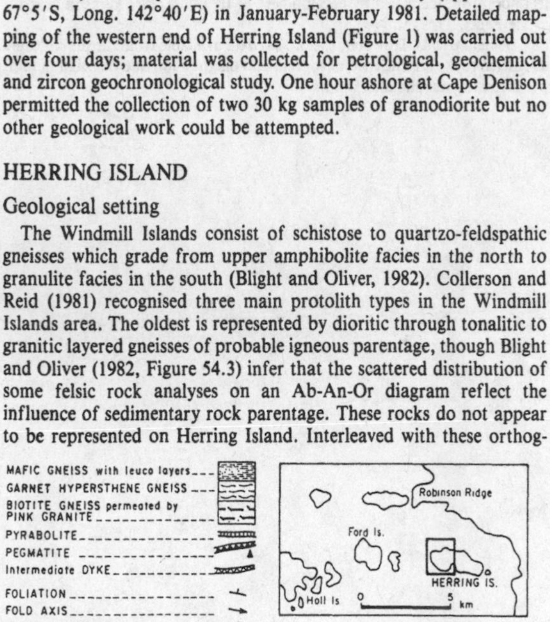

**Figure 1. Geology of western Herring Island**

### Metamorphic Lithologies

A layered sequence of supracrustal units cut by felsic intrusive material is discernible in the western part of Herring Island (Figure 1). The prevalent aluminous hypersthene together with plagioclase in many of the lithologies is indictive of crystallisation in an intermediate pressure granulite facies environment. K/Rb ratios indicate minimal depletion of lithophile elements (Figure 2). A plot of these rocks on ACF and AKF diagrams is shown in Figures 3 and 4. Correlation between actual mineralogy and that predicted from these diagrams is indicative of equilibrium conditions of crystallisation. The presence of biotite in apparent equilibrium with orthopyroxenes is noteworthy. The composition of the biotite is typical of that in granulite facies rocks e.g. high $TiO_2$ at 3.84-5.59%; a relatively high fluorine content of about 1.2% $F_2O$ (cf. 0.85% $F_2O$ of amphibolite facies biotite from Cape Denison (see below)) may be significant in this respect.

Coexisting pyroxenes are salite and hypersthene. Various techniques for determining temperature and pressure from the composition of coexisting pyroxenes (Wood and Banno, 1973; Wells, 1977), garnet-biotite (Goldman and Albee, 1977; Ferry and Spear, 1978) and garnet-orthopyroxene (Wood and Banno, 1973) have been applied to the Herring Island rocks. The results of these are summarised in Figure 5. They indicate a medium pressure relatively high temperature (850-900°C, 6-7 kb) crystallisation environment for the granulite facies event and a geothermal gradient of about 34-45 deg./km.

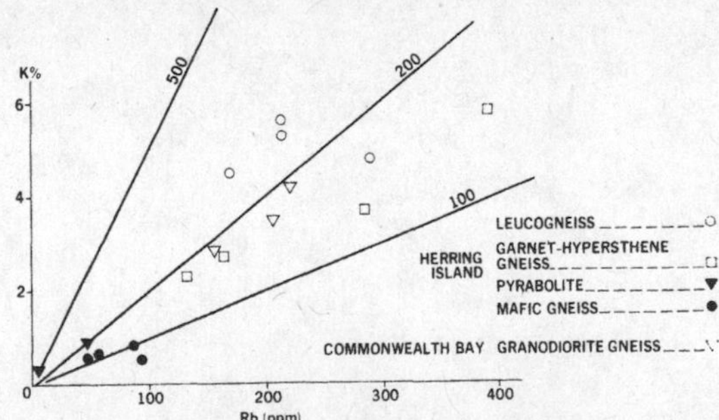


**Figure 2. K-Rb showing minimal Rb depletion.**

### Pre-metamorphic nature of the rocks

Several petrographic features tend to indicate a sedimentary origin for the garnet hypersthene gneiss viz.(a) the high quartz content of some specimens (up to at least 55%) (b) the heterogeneous composition of the unit (e.g. chemical, $SiO_2$ = 56-74%, $Al_2O_3$ = 13-18%, $K_2O$ = 2.7-7%; modal, plagioclase = 0-20%, alkali feldspar = 0-30%, garnet = 0-10%, cordierite = 0-10%) (c) significant rounding of constituent zircon grains. Geochemical support for this conclusion is given by four samples which have discriminant function (Shaw, 1972) values ranging from -0.45 to -3.49. The Niggli plots of Leake (1969), which assume that the elements concerned are relatively immobile during metamorphism, shown in Figure 6, favour a pelitic or semipelitic composition for the garnet hypersthene gneiss and an igneous origin for the pyrabolite ( a concordant pyroxene amphibole rock) and

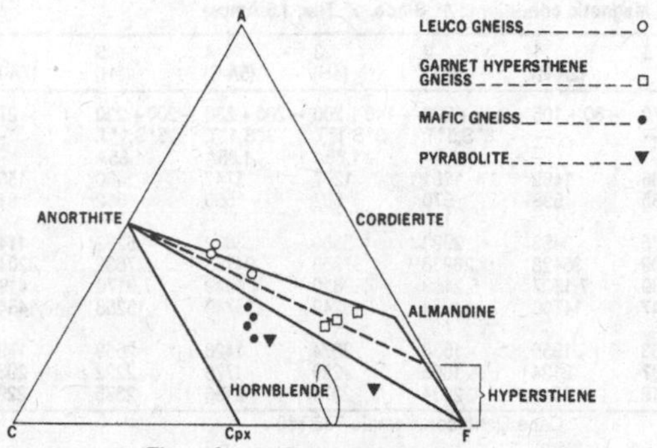


Figure 3. Al-Ca-Fe-, Herring Island.

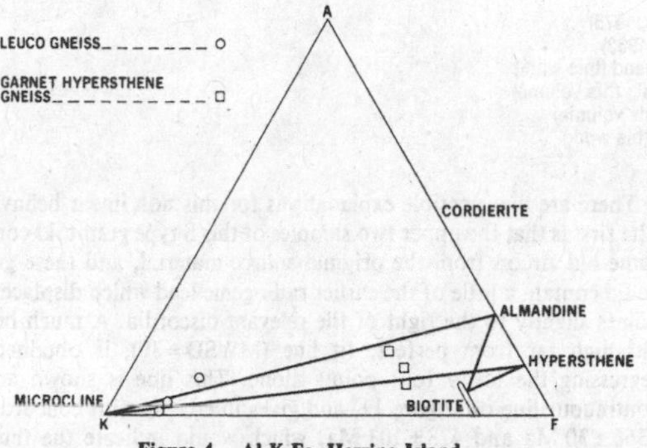


Figure 4. Al-K-Fe, Herring Island.

the mafic gneiss (Figure 6). The mafic gneiss and pyrabolite have tholeiitic affinities (Figures 7A and 7B).

The metamorphic temperature indicated (850-900°C) is probably high enough to cause partial melting of the gneisses and produce magmas of the cross-cutting leucogneiss composition. Plots of the leuco-gneiss on an Ab-An-Or diagram (Figure 8) support the suggestion that it represents a partial melt.

## Geochronology

The garnet hypersthene gneiss (Sample No. 786-T38) was chosen for zircon age study. Six fractions of zircon were analysed isotopically for uranium and lead (Krogh, 1973)—see Table 1. When plotted on a concordia diagram (Wetherill, 1956) they form a highly discordant pattern (Figure 9). Five of the six points define a perfect fit line within experimental error. The sixth point lies well below the straight line defined by the five other points. Unlike the other concentrates this sample had many foreign inclusions within the grains and it is not included in the age calculations.

The regression line through the five linearly related points cuts concordia at 1275 ± 21 Ma and 2529 ± 108 Ma (2 sigma limits). The extremely discordant nature (70%) of these comparatively low uranium zircons, together with the preceding evidence for a sedimentary precursor for the granulite gneiss, suggests that the zircons were present in the original sediment and lost most, but not all, of their radiogenic lead during the granulite facies metamorphism. The lower intercept younger age would then represent the time of that metamorphism, whilst the upper intercept gives a measure of the average of the provenance of the source material of the sediment. The strong linearity of the five points suggests that, for this particular sample, the range of ages of the source material is quite small, and indeed five of the six zircon concentrates actually measured may have all originated in the same rock type.

These results may be compared with previous dating work reported for the Windmill Islands and neighbouring regions (Table 2). The work by Williams et al., (this volume), on Pidgeon Island places the ages of the intrusive tonalites and tonalitic gneisses, not found on Herring Island, at 1477 ± 33 (Rb/Sr total rocks) Ma and a minimum of 1465 ± 34 Ma (TNd CHUR), appropriately earlier than the 1275 ± 21 Ma age derived here for granulite facies metamorphism, and supports Arriens' (1975) earlier conclusion that some isotopic systems remained essentially closed during that severe event. The low resolution, ion microprobe, lead isotope data of Lovering et al. (1981) do not define exact events and individual results are likely to be biased by measurement interferences (Compston and Williams, 1982).

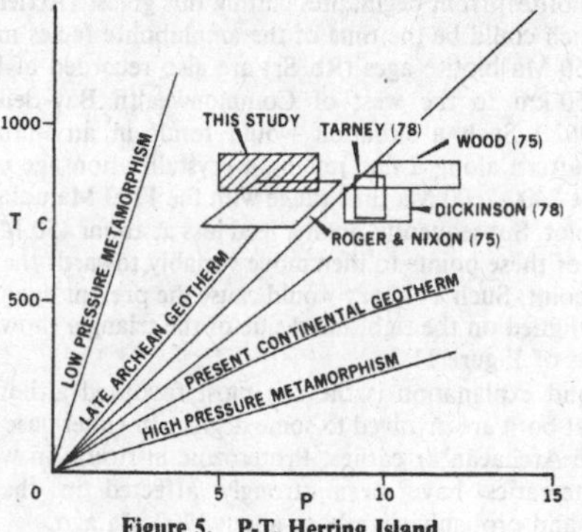


Figure 5. P-T, Herring Island.

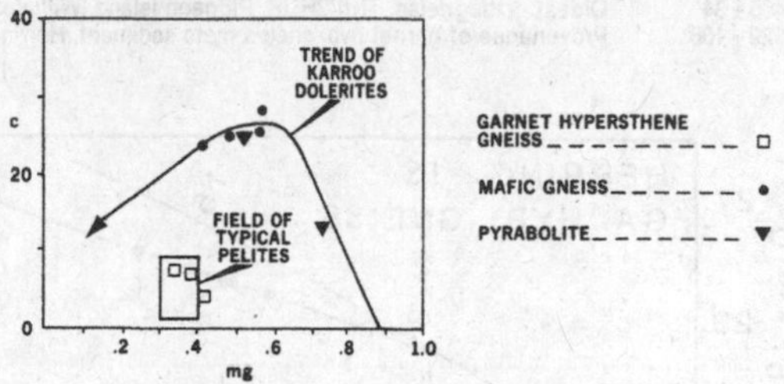


Figure 6. Niggli mg-c, Herring Island.

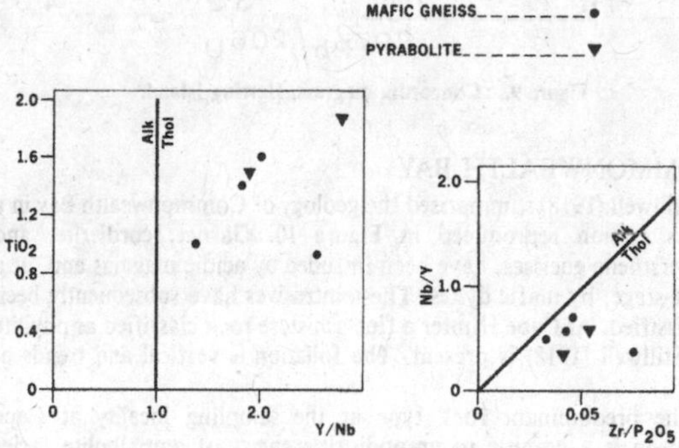


Figure 7. Trace elements, Herring Island mafics.

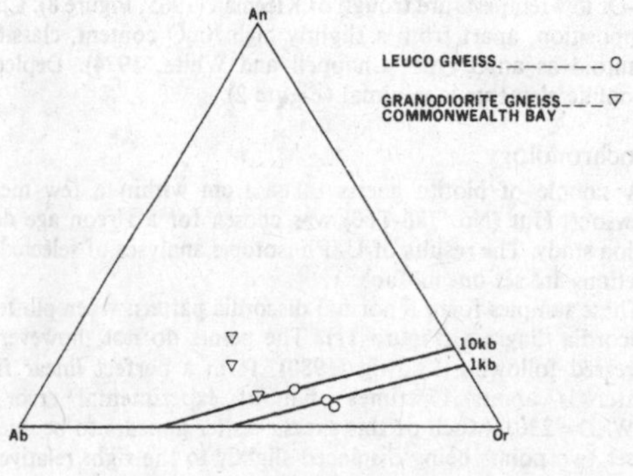


Figure 8. An-Ab-Or of granitoids.

TABLE 1: U/Pb Isotopic measurements and calculations. (NB Magnetic conditions: 4° Slope, 1° Tile, 1.9 Amps)

| Zircon Fraction | 1 | 2 | 3 | 4 | 5 | 6 | 1 (2A-H) | 2 (4A) | 3 (4H) | 4 (5A-C) | 5 (5H) | 6 (7A-H) |
|---|---|---|---|---|---|---|---|---|---|---|---|---|
| Size (B.S.S.) | − 60 + 105 | − 105 + 140 | − 140 + 200 | − 140 + 200 | − 230 + 270 | − 270 | − 60 + 105 | − 140 + 200 | − 140 + 200 | − 200 + 230 | − 200 + 230 | − 270 |
| Magnetic Fraction | — | — | Least Mag. | Most Mag. | — | — | — | 3°S,5°T, 1.3A | 3°S,1°T, 1.85A | 3°S,1°T, 1.85A | 3°S,1°T, 1.85A | — |
| U (ppm) | 427 | 490 | 516 | 582 | 538 | 586 | 1499 | 1881 | 1277 | 1747 | 1400 | 1500 |
| Pb (ppm) | 124 | 127 | 128 | 145 | 132 | 135 | 598 | 570 | 507 | 556 | 552 | 510 |
| Atomic Ratios | | | | | | | | | | | | |
| 206/204 | 626 | 5275 | 10160 | 10260 | 4964 | 7175 | 3453 | 2993 | 5550 | 3002 | 5292 | 4142 |
| 206/238 | .25202 | .24636 | .24183 | .24164 | .23729 | .22409 | .35428 | .26938 | .37959 | .24765 | .37632 | .32013 |
| 207/235 | 3.6012 | 3.4036 | 3.2560 | 3.2824 | 3.1201 | 2.9189 | 7.1807 | 5.2189 | 7.9810 | 4.6949 | 7.9170 | 6.4197 |
| 207/206 | .10364 | .10020 | .09765 | .09852 | .09536 | .09447 | .14700 | .14051 | .15249 | .13749 | .15258 | .14544 |
| Age Estimates (Ma) | | | | | | | | | | | | |
| 206/238 | 1449 | 1420 | 1396 | 1395 | 1373 | 1303 | 1955 | 1538 | 2074 | 1426 | 2059 | 1790 |
| 207/235 | 1550 | 1505 | 1471 | 1477 | 1438 | 1387 | 2134 | 1856 | 2229 | 1776 | 2222 | 2035 |
| 207/206 | 1690 | 1629 | 1580 | 1596 | 1535 | 1518 | 2311 | 2234 | 2374 | 2196 | 2375 | 2293 |
| | Herring Island Sample 786-T38 | | | | | | Cape Dennison Sample 786-T60 | | | | | |

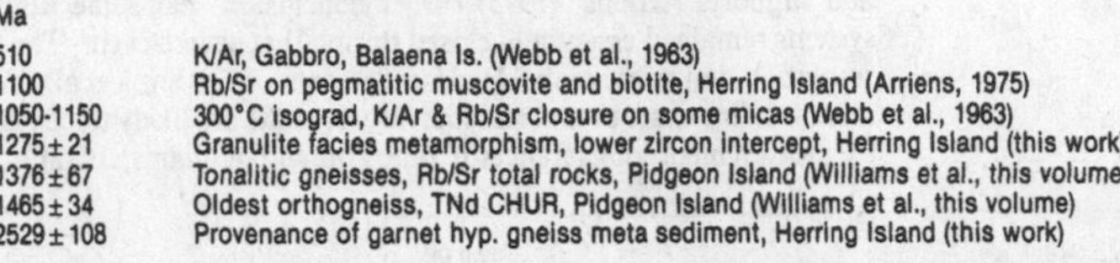

| Ma | |
|---|---|
| 510 | K/Ar, Gabbro, Balaena Is. (Webb et al., 1963) |
| 1100 | Rb/Sr on pegmatitic muscovite and biotite, Herring Island (Arriens, 1975) |
| 1050-1150 | 300°C isograd, K/Ar & Rb/Sr closure on some micas (Webb et al., 1963) |
| 1275±21 | Granulite facies metamorphism, lower zircon intercept, Herring Island (this work) |
| 1376±67 | Tonalitic gneisses, Rb/Sr total rocks, Pidgeon Island (Williams et al., this volume) |
| 1465±34 | Oldest orthogneiss, TNd CHUR, Pidgeon Island (Williams et al., this volume) |
| 2529±108 | Provenance of garnet hyp. gneiss meta sediment, Herring Island (this work) |

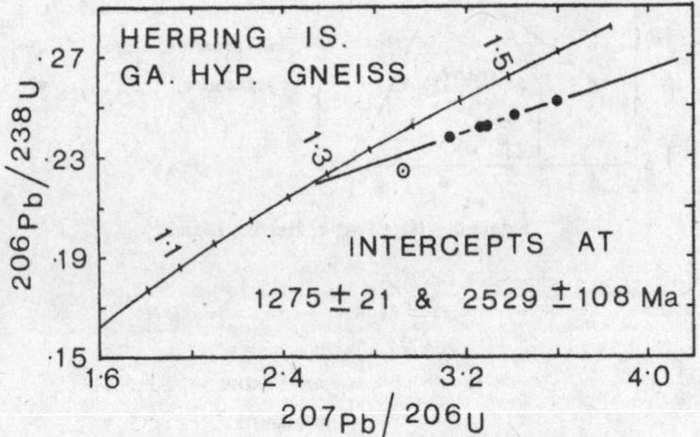


**Figure 9. Concordia diagram, Herring Island.**

## COMMONWEALTH BAY

Stillwell (1918) summarised the geology of Commonwealth Bay in a cross section reproduced in Figure 10. Garnet, cordierite, and hypersthene gneisses, have been intruded by acidic magmas and, at a later stage, by mafic dykes. These intrusives have subsequently been gneissified. At Cape Hunter a fine schistose rock classified as phyllite by Stillwell (1918) is present. The foliation is vertical and trends at 340°.

The predominant rock type at the sampling locality at Cape Denison is a granitic to granodioritic gneiss of amphibolite facies which has been intruded by aplitic and amphibolitic dykes. Other rocks in the region have granulite facies assemblages. An igneous origin is suggested for the granite-granodiorite gneiss on the basis of megacrystic (=phenocrystic) feldspar, a geochemical discriminant function of +0.85-2.98 (Shaw, 1972), and composition near the Ab-An-Or low temperature trough of Kleeman (1965, Figure 8). Chemical composition, apart from a slightly high $Na_2O$ content, classifies the granitoid as an S type (Chappell and White, 1974). Depletion of lithophile elements is minimal (Figure 2).

### Geochronology

A sample of biotite gneiss taken from within a few metres of Mawson's Hut (No. 786-T60) was chosen for a zircon age determination study. The results of U/Pb isotopic analyses of selected zircon fractions are set out in Table 1.

These samples form a normal discordia pattern when plotted on a concordia diagram (Figure 11). The points do not, however, when regressed following Ludwig (1980), form a perfect linear fit. The scatter is about 15 times that of experimental error alone (MWSD = 230). Much of this excess scatter appears to be due to the upper two points being displaced slightly to the right relative to the other four.

There are two possible explanations for this non linear behaviour. The first is that the upper two samples of this S type granitoid contain some old zircon from the original source material, and these grains could contain a little of the earlier radiogenic lead which displaces the points slightly to the right of the relevant discordia. A much better, although far from perfect, fit line (MWSD = 30), is obtained by regressing the lower four points alone. This line is shown as the continuous line on Figure 11, and gives intercepts with concordia at 2366±30 Ma and 428±103 Ma, which would indicate the time of original crystallisation of the granitoid and a time of later lead loss, with some small unknown degree of contamination causing the residual scatter (decay and other isotopic constants used were derived from Steiger and Jager, 1977).

**TABLE 2: Significant dates for the Windmill Islands and neighbouring regions**

| Ma | |
|---|---|
| 510 | K/Ar, Gabbro, Balaena Is. (Webb et al., 1963) |
| 1100 | Rb/Sr on pegmatitic muscovite and biotite, Herring Island (Arriens, 1975) |
| 1050-1150 | 300°C isograd, K/Ar & Rb/Sr closure on some micas (Webb et al., 1963) |
| 1275±21 | Granulite facies metamorphism, lower zircon intercept, Herring Island (this work) |
| 1376±67 | Tonalitic gneisses, Rb/Sr total rocks, Pidgeon Island (Williams et al., this volume) |
| 1465±34 | Oldest orthogneiss, TNd CHUR, Pidgeon Island (Williams et al., this volume) |
| 2529±108 | Provenance of garnet hyp. gneiss metasediment, Herring Island (this work) |

The second explanation for the non linear relationship is that there has been another smaller lead loss event at the time of heating recorded by Rb/Sr ages of about 1600-1700 Ma (muscovites) and 1550 Ma (biotites) from pegmatites cutting this gneiss (Arriens, pers. comm.) which could be the time of the amphibolite facies metamorphism. (1550 Ma biotite ages (Rb/Sr) are also recorded at Dumont D'Urville 50 km to the west of Commonwealth Bay-Bellair and Delbos, 1962.) Such a situation would result in an initial short discordia pattern along a line joining a crystallisation age of somewhere in the 2400-2600 Ma time range with the 1700 Ma point on the concordia plot. Subsequently, greater lead loss at about 450 Ma would cause each of these points to then move variably towards the 450 Ma concordia point. Such a history would cause the present day points to be poorly aligned on the right hand side of the triangle shown by the dashed lines of Figure 11.

The second explanation is the one most favoured although it is possible that both are involved to some degree. In either case we have a very Late Archaean or earliest Proterozoic intrusion in which the zircon systematics have been strongly affected in the Lower Palaeozoic and probably slightly so about 1700 Ma ago.

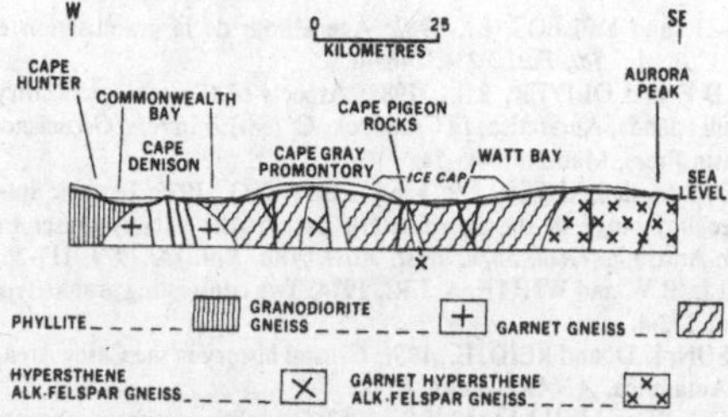


**Figure 10. Commonwealth Bay section.**

## TECTONIC RELATIONSHIPS

The results of this study are relevant to refitting the original Gondwana position of the Australian and the Antarctic continents. Two differing points of view, adapted from Lovering et al. (1981), are essentially summarised in Figure 12. Figure 12A shows the essentially identical computer generated fits of the 1000 fathom isobath (Sproll and Dietz, 1969) and the 500 fathom isobath (Smith and Hallam, 1970). This result differs from the study of reversing sea floor magnetics by Griffith (1974) and the physical matching up of the Early Palaeozoic Bowers Group of Antarctica with either the Dundas Trough or the Adamsfield Trough of Tasmania (Laird et al., 1977; Laird, 1981) as shown in Figure 12B. The latter configuration rotates Antarctica about 400 km to the east of that of Figure 12A.

Lovering et al., (1981) favour the Figure 12B reconstruction with Casey directly opposite Esperance. Contrary to the findings of those authors, this study, and that of Williams et al. (this volume), give evidence that no rocks older than 1500 Ma have been found in the Windmill Islands, although our work reveals that an approximately 2600 Ma provenance has contributed material to some of the sediments. Nevertheless these results of the traditional geochronology

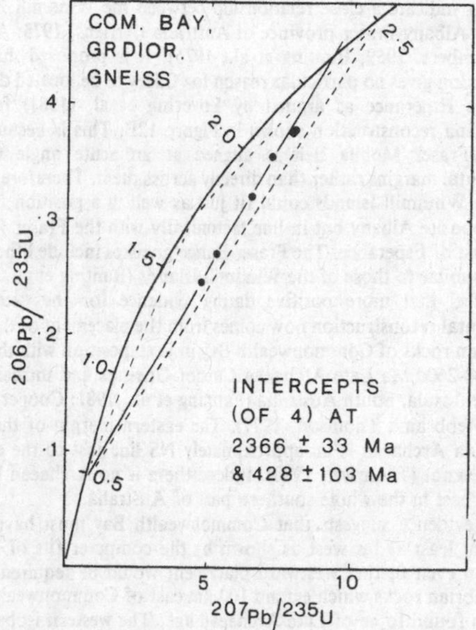


**Figure 11. Concordia diagram, Commonwealth Bay.**

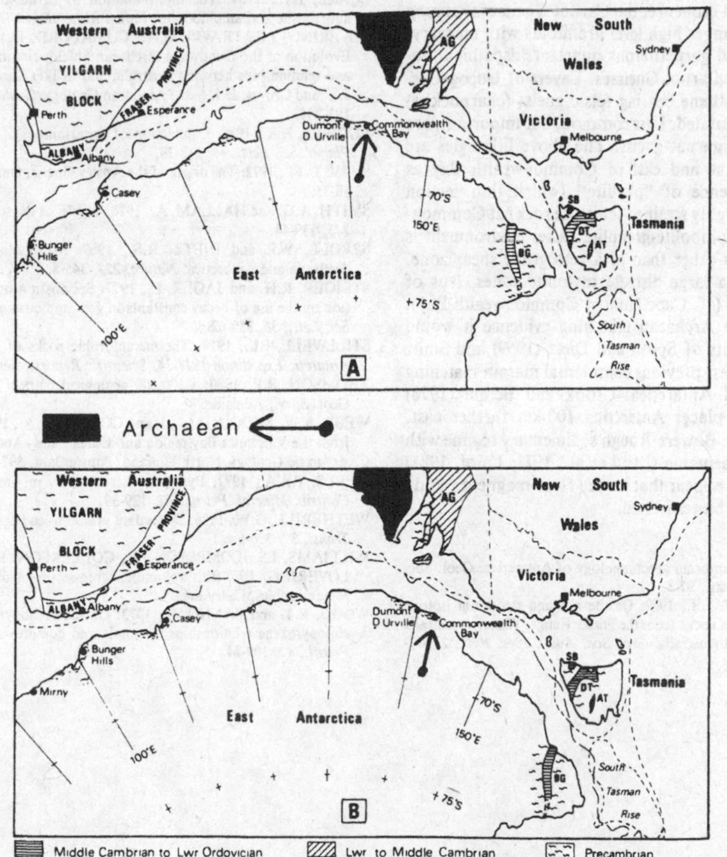


**Figure 12. Continental fit alternatives.**

methods indicate a close relationship between the Windmill Islands and the Albany-Fraser province of Australia (Arriens, 1975; Arriens and Lambert, 1969; Bunting et al., 1975). It is proposed that this observation gives no particular reason for Casey to be located directly opposite Esperance as argued by Lovering et al. (1981) for the Gondwana reconstruction shown in Figure 12B. This is because the Albany-Fraser Mobile Belt is aligned at an acute angle to the continental margins rather than directly across them. Therefore Casey and the Windmill Islands could fit just as well in a position further west opposite Albany but in line tectonically with the Fraser Ranges northeast of Esperance. The Fraser Range gneisses include lithologies not dissimilar to those of the Windmill Islands (Bunting et al., 1975).

We feel that more positive dating evidence for the details of continental reconstruction now comes from the placement of the Late Archaean rocks of Commonwealth Bay in juxtaposition with those of the 2300-2600 Ma Late Archaean Carnot Gneisses and intrusives of Eyre Peninsula, South Australia (Fanning et al., 1981; Cooper et al., 1976; Webb amd Thomson, 1977). The eastern margin of the Eyre Peninsula Archaean is an approximately NS line just to the east of Cape Carnot (Thomson, 1980). Indeed there is no Archaean known further east in the whole southern part of Australia.

This evidence suggests that Commonwealth Bay must have originated at least as far west as shown by the computer fits of Figure 12A. An even further westward placement would be required if the Precambrian rocks which extend 100 km east of Commonwealth Bay are also found to be of Late Archaean age. The western isobaths of Figure 12A would physically allow a movement of Antarctica further westwards but at the expense of increasing the mis-fitting areas in the east. Even so such a misfit would not be as great as that in the Figure 12B proposal.

The geological evidence is in conformity with the geochronological matching of Commonwealth Bay with Eyre Peninsula. The Late Archaean-Early Proterozoic basement of southern Eyre Peninsula, (Sleaford Complex of Thomson, 1980) is composed of a highly metamorphosed supracrustal sequence, the Carnot Gneisses (Fanning et al. 1981), and slightly younger high level granitoids with megacrystic feldspar. A thinly layered garnetiferous quartzofeldspathic lithology is dominant among the Carnot Gneisses. Layers of leucogneiss, biotite garnet gneiss, hypersthene bearing felsic gneiss (charnockite) and basic granulite are intercalated. Less common, but important, are augen gneiss and cordierite garnet gneiss. The above litholigies are similar to those occurring at and east of Commonwealth Bay as described earlier. The presence of "phyllite" (which thin section examination shows to be a highly strained schistose rock) at Commonwealth Bay in an upper amphibolite-granulite facies environment is metamorphically anomalous other than as a retrograde shear zone. On southern Eyre Peninsula large discrete mylonite zones, tens of kilometres long, trend N—S (cf. Cape Hunter, Commonwealth Bay).

On the basis of this late Archaean matching evidence it would appear that the computer fits of Sproll and Dietz (1969) and Smith and Hallam (1970) are the best previous continental margin matching attempts for Australia and Antarctica. Crook and Belbin's(1978) Cretaceous reconstructions places Antarctica 100 km further east. Attempts to linearly align the Bowers Trough sedimentary regime with that of similar troughs in Tasmania (Laird et al., 1977; Laird, 1981) are not compatible. It would appear that the sea floor magnetics study of Griffiths (1974) needs to be reassessed.

## REFERENCES


ARRIENS, P.A., 1975: The Precambrian geochronology of Antarctica: Geol. Soc. Aust., 1st Geol. Convent., Abstr., 97-8.

ARRIENS, P.A. and LAMBERT, I.B., 1969: On the age and strontium isotope geochemistry of granulite facies rocks from the Fraser Range, Western Australia and Musgrave Ranges, Central Australia. *Geol. Soc. Aust., Spec. Pub., 2,* 377-88.

BELLAIR, P. and DELBOS, L., 1962: Age absolu de la granitisation en Terre Adélie. *C.R. Ac. Sc., Paris, 254,* 1465-6.

BLIGHT, D.F. and OLIVER, R.L., 1982: Aspects of the geologic history of the Windmill Islands, Antarctica; *in* Craddock, C. (ed.) *Antarctic Geoscience,* Univ. Wisconsin Press, Madison, 445-54.

BUNTING, J.A., de LAETER, J.R. aand LIBBY, W.G., 1975: Tectonic Subdivision and Geochronology of the northeastern part of the Albany-Fraser Province, Western Australia. *Geol. Surv. West. Aust. Ann. Rpt. for 1974,* 117-26.

CHAPPELL, B.W. and WHITE, A.J.R., 1974: Two contrasting granite types. *Pac. Geol., 8,* 173-4.

COLLERSON, K.D. and REID, E., 1981: Crustal history in the Casey Area, Wilkes Land, Antarctica. *ANARE News, 8,* 1-2.

COMPSTON, W. and WILLIAMS, I.S., 1982: Protolith ages from inherited zircon cores measured by a high mass-resolution ion micro-probe. *Abstr. Fifth Int. Conf. Geochron. Cosmochim. and Iso. Geol.,* Nikko, Japan.

COOPER, J.A., FANNING, C.M., FLOOK, M.M. and OLIVER, R.L., 1976: Archaean and Proterozoic metamorphic rocks on Southern Eyre Peninsula, South Australia. *Geol. Soc. Aust., J., 23,* 287-92.

CROOK, K.A.W. and BELBIN, L., 1978: The southwest Pacific area during the last 90 million years. *Geol. Soc. Aust., J., 25,* 23-40.

FANNING, C.M., OLIVER, R.L. and COOPER, J.A., 1981: The Carnot Gneisses, Southern Eyre Peninsula. *S. Aust., Geol. Surv., Q. Geol. Notes, 80,* 7-12.

FERRY, J.M. and SPEAR, F.S., 1978: Experimental calibration of the partitioning of Fe and Mg between biotite and garnet. *Contrib. Mineral. Petrol., 66,* 113-7.

GOLDMAN, D.S. and ALBEE, A.L., 1977: Correlation of Mg/Fe partitioning between garnet and biotite with $^{18}O/^{16}O$ partitioning between quartz and magnetite. *Am. J. Sci., 277,* 750-67.

GRIFFITHS, J.R., 1974: Revised continental fit of Australia and Antarctica. *Nature, 249,* 336-8.

KLEEMAN, A.W., 1965: The origin of granitic magmas. *Geol. Soc. Aust., J., 12,* 35-52.

KROGH, T.E., 1973: A low contamination method for hydrothermal decomposition of zircon and extraction of U and Pb for isotopic age determination. *Geochim. Cosmochim. Acta, 36,* 485-94.

LAIRD, M.G., 1981: Lower Palaeozoic rocks of the Ross Sea area and their significance in the Gondwana context. *R. Soc. N.Z., J., 11,* 425-38.

LAIRD, M.G., COOPER, R.A. and JAGO, J.B., 1977: New data on the Lower Palaeozoic sequences of northern Victoria Land, Antarctica, and its significance for Australian-Antarctic relationships in the Palaeozoic. *Nature, 265,* 107-10.

LEAKE, B.E., 1969: The discrimination of ortho and paracharnockite rocks, anorthosite and amphibolites. *Indian Mineralogist, 10,* 89-104.

LOVERING, J.F., TRAVIS, G.A., CUMAFORD, D.J. and KELLY, P.R., 1981: Evolution of the Gondwana Archaean Shield; zircon dating by ion microprobe and relationships between Australia and Wilkes Land, (Antarctica); *in* Glover, J.E. and Groves, D.I. (eds.) Archaean Geology. *Geol. Soc. Aust., Spec. Publ., 7,* 193-204.

LUDWIG, K.R., 1980: Calculation of uncertainties of U-Pb isotope data: *Earth Planet. Sci. Lett., 46,* 212-20.

SHAW, D.M., 1972: The origin of the Apsley Gneiss, Ontario. *Can. J. Earth Sci., 9,* 18-35.

SMITH, A.G. and HALLAM, A., 1970: The fit of the southern continents. *Nature, 225,* 139-44.

SPROLL, W.P. and DIETZ, R.S., 1969: Morphological continental drift of Australia and Antarctica. *Nature, 222,* 345-8.

STEIGER, R.H. and JAGER, E., 1977: Submission on Geochronology: Convention on the use of decay constants in geo- and cosmo-chronology. *Earth Planet. Sci. Lett., 36,* 359-62.

STILLWELL, F.L., 1918: The metamorohic rocks of Adélie Land. *Australasian Antarctic Expedition 1911-14, Scientific Reports, Ser. A., 3(1).*

THOMSON, B.P., 1980: 1:100,000 geological map of South Australia. *S. Aust. Geol. Surv.,* Adelaide.

WEBB. A.W., McDOUGALL, I. and COOPER, J.A., 1964: Potassium-argon dates from the Vincennes Bay region and Oates Land, Antarctica; *in* Adie, R.J. (ed.) *Antarctic Geology,* North Holland, Amsterdam, 597-600.

WELLS, P.R.A., 1977: Pyroxene thermometry in simple and complex systems. *Contrib. Mineral. Petrol., 62,* 129-39.

WETHERILL, G.W., 1956: Discordant uranium-lead ages, I. *Am. Geophys. Union, Trans., 37,* 320-6.

WILLIAMS, I.S., COMPSTON, W., COLLERSON, K.D., ARRIENS, P.A. and LOVERING, J.F., (this volume): A reassessment of the age of the Windmill Metamorphics, Casey area.

WOOD, B.J. and BANNO, S., 1973: Garnet-orthopyroxene and orthopyroxene-clinopyroxene relationships in simple and complex systems. *Contrib. Mineral. Petrol., 42,* 109-24.

# MANGANESE-RICH CHEMICAL SEDIMENTS FROM WILKES LAND, ANTARCTICA

I.R. Plimer, *Department of Geology, University of New England, Armidale, NSW, 2351. Australia.*

J.F. Lovering, *Department of Geology, University of Melbourne, Parkville, Victoria 3052, Australia.*

*Abstract*. Three principal facies of chemical sediments have been recognised in a suite of pelitic and felsic metasediments at Clark Peninsula, Wilkes Land, Antarctica which experienced a major metamorphism at ~ 1470 Ma and a later event of retrogression at 1100 Ma. The original age of deposition is unknown but source clastic material derives from sequences of ~ 3000 Ma. A laminated manganese(Mn)-rich chemical metasediment comprising tephroite, normally and reversely zoned garnet, jacobsite, barite and calcite transgressed by late stage ferroan and calcic rhodonite and friedelite represents a sulphate facies. A second facies, probably a temporal or spatial silicate facies equivalent, comprises normal zoned garnet containing magnetite inclusions and a Mn-rich rim, hyalophane with sphene inclusions, barite, rutile, biotite and rare manganoan hedenbergite. Banded iron formation comprising quartz, texturally and compositionally zoned garnet, magnetite, amphibole, K-feldspar, chlorite, fibrolite and pinite represents the most oxidised facies of the chemical sediments. The enclosing quartzofeldspathic metasediments comprise quartz, plagioclase ($An_{22}$), garnet, amphibole, magnetite, fibrolite and chloritised biotite. The Mn content of garnet is highest in a laminated barite-manganese silicate-rich facies and lower in a hyalophane-Mn silicate-amphibole facies which in turn has more Mn in garnet than the banded iron formation. Garnet from the quartzofeldspathic metasediment has the lowest Mn content. Although no abundant sulphide phases are present and there is no significant substitution of zinc(Zn) in olivines, garnets, spinels, pyroxenes or micas, the chemical sediments from Wilkes Land are very similar to chemical sediments (e.g. iron formations, gondites, coticule rock) associated with massive sulphide deposits in Middle Proterozoic sequences of Australia (e.g. Broken Hill, NSW; Pegmont, Qld), South Africa (e.g. Gamsberg, Aggeneys, Rosh Pinah), India (e.g. gondites of Rajasthan), Scandinavia (e.g. Langban, Sweden), USA (Franklin, Sterling Hill, N.J.) and Scotland.

The manganese-rich lens from Clark Peninsula (vicinity of 66°15′S, 110°31′E) in Wilkes Land near Wilkes Station was first collected by Mr. Walter Sullivan of the "New York Times" in 1957 after noting a "vein" of black rock with a metallic sheen enclosed by leucocratic gneiss. Studies by Mason (1959) on Sullivan's samples showed them to be granoblastic tephroite-spessartine-rhodonite-barite rocks enclosed by quartz-andesine-biotite-almandine-cordierite-sillimanite gneiss. During a visit to Casey Station in January 1978 the lens-shaped manganese-rich pod about 1 metre across within gneisses was rediscovered by J.F. Lovering and G.A. Travis. Some little distance to the east an horizon, about 1 metre thick, of banded iron formation was also found within gneiss. Snow cover conceals the relationship between the two occurrences.

The gneiss and manganese-rich horizon from Clark Peninsula are part of the Windmill Metamorphics which comprise a suite of schist, gneiss and migmatite metamorphosed to the upper amphibolite facies in the Clark Peninsula area (Blight and Oliver, 1977). More recently Williams et al. (this volume) have recognized three distinct groups of gneisses within the Windmill Metamorphics, each probably derived from a different protolith. The oldest group is a suite of layered "granitic" orthogneisses with which are interleaved the second group, now paragneisses, occurring as semi-continuous belts of supracrustal rocks. The third, youngest group are intrusive, coarse-grained, leucocratic gneisses. They have reported new whole-rock Rb-Sr and Sm-Nd data, together with ion microprobe U-Pb studies of 30 m areas on individual zircon grains from the gneisses, which indicate the age of the orthogneiss protolith and time of major metamorphism of the sedimentary precursor of the paragneiss to be ~ 1470 Ma. Ion microprobe studies have also shown relict detrital zircons in the paragneisses up to ~ 3000 Ma. Previously published K-Ar and one Rb-Sr age of biotites from metamorphic and intrusive rocks from the Windmill Metamorphics have indicated a later retrogressive event about 1100 Ma (Cameron et al., 1960; Webb et al., 1963).

The manganese-rich lens and banded iron formation horizon discussed here are part of the paragneiss supracrustal sequence of Middle Proterozoic age and metamorphism which contains an Archaean ( ~ 3000 Ma) detrital zircon component.

## PETROGRAPHY

The quartzofeldspathic gneiss of the Windmill Metamorphics enclosing the manganese silicate lens and banded iron formation horizon comprises a granoblastic aggregate of coarse grained deformed quartz, poikiloblastic garnet, deformed oligoclase, pinitised cordierite and slightly chloritised biotite with sulphide inclusions and minor magnetite and sillimanite with traces of apatite.

Banded iron formation comprises essentially deformed quartz with well defined beds about 2 mm thick of poikiloblastic slightly chloritised garnet and magnetite with minor associated hornblende, pinitised cordierite, chloritised biotite and sillimanite and traces of pyrite and chalcopyrite. Garnet cores contain up to 40% inclusions (quartz, biotite and magnetite) with the proportion of the inclusions decreasing towards the rim of the grains.

The manganese silicate-rich lens shows great variation in the relative abundance of constituent phases. Two types have been recognised.

(a) barite-manganese silicate rocks, and

(b) barium feldspar-manganese silicate-amphibole rocks.

The barite-manganese silicate facies (sulphate facies) comprises granoblastic yellowish spessartine (spess) garnet, yellowish grey rhodonite, opaque jacobsite, yellowish grey tephroite and cloudy barite with minor other phases. The following assemblages were observed:

rhodonite > garnet > barite > tephroite > carbonate > rutile

garnet > rhodonite > tephroite > jacobsite > rutile > magnetite > pyrite > ?pyrophanite

jacobsite ~ barite > rhodonite > tephroite > garnet > magnetite > friedelite > apatite > ?retzian

rhodonite > garnet > tephroite > jacobsite > barite > friedelite > zircon

Minor pods and veinlets of friedelite with a rhodonite selvage transgress garnet, rhodonite, tephroite, jacobsite and barite and appear to be controlled by a prominent fracture pattern (Figure 1).

The barium feldspar-manganese silicate-amphibole facies (silicate facies) of the horizon contains abundant granoblastic spessartine, rhodonite, hyalophane, magnetite, hornblende and quartz with minor amounts of biotite, chlorite, manganoan hedenbergite, rutile and spene. Magnetite, garnet and biotite are partially chloritised and hyalophane contains minute inclusion of sphene.

## MINERAL CHEMISTRY

Garnet from the quartzofeldspathic gneiss is compositionally zoned with slightly higher Mn and Ca and lower Fe and Mg in the core as compared to the rim of grains. Of all of the garnets analysed, those from the felsic gneiss have the lowest spessartine component and have an average composition of $spess_{41.4}$ $almandine(alm)_{30.9}$ $grossular(gross)_{19.8}$ $pyrope(pyr)_{7.9}$ (Table 1). Associated biotite is a

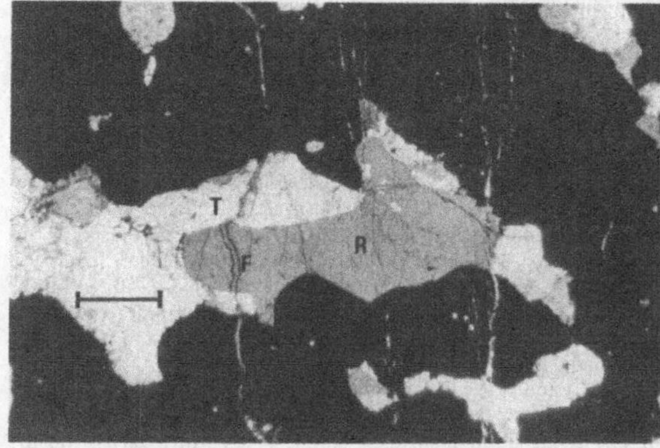


**Figure 1. Subparallel friedelite veinlets (F) transgressing jacobsite (opaque), tephroite (T) and rhodonite (R). Crossed nicols. Bar scale = 0.1 mm.**

**TABLE 1: Representative electron microprobe analyses of abundant phases associated with manganese-rich chemical sediments from Clark Peninsula, Wilkes Land, Antarctica**

| wt % | Olivine | | Pyroxene | | | Feldspar | | Spinel | | | | Friedelite | | Garnet | | | |
|---|---|---|---|---|---|---|---|---|---|---|---|---|---|---|---|---|---|
| oxide | 1 | 2 | 3 | 4 | 5 | 6 | 7 | 8 | 9 | 10 | 11 | 12 | 13 | 14 | 15 | 16 | 17 |
| $SiO_2$ | 29.90 | 29.83 | 54.33 | 47.23 | 46.59 | 65.12 | 58.09 | 0.06 | 0.01 | 0.02 | 0.08 | 34.91 | 34.12 | 37.03 | 37.40 | 36.60 | 37.17 |
| $TiO_2$ | 0.00 | 0.01 | 0.00 | 0.10 | 0.00 | 0.00 | 0.00 | 1.26 | 0.15 | 0.39 | 0.04 | 0.03 | 0.07 | 0.00 | 0.00 | 0.03 | 0.13 |
| $Al_2O_3$ | 0.04 | 0.05 | 1.75 | 0.07 | 0.02 | 21.58 | 20.39 | 0.06 | 0.13 | 0.19 | 0.53 | 0.16 | 0.07 | 20.37 | 20.26 | 17.27 | 19.02 |
| $Fe_2O_3$ | — | — | — | — | — | 0.30 | 0.20 | 66.03 | 66.68 | 84.50 | 73.80 | — | — | 2.65 | 1.21 | — | — |
| FeO | 4.55 | 2.48 | 10.13 | 3.59 | 1.12 | 0.00 | 0.00 | 28.50 | 30.46 | 0.00 | 0.00 | 7.45 | 6.27 | 15.40 | 10.60 | 3.86 | 2.73 |
| MnO | 65.77 | 66.44 | 5.22 | 45.01 | 47.96 | 0.04 | 0.41 | 2.03 | 0.62 | 18.92 | 25.12 | 44.68 | 44.34 | 16.14 | 21.97 | 40.15 | 38.15 |
| MgO | 0.77 | 1.10 | 14.91 | 1.40 | 0.72 | 0.06 | 0.08 | 0.04 | 0.00 | 0.01 | 0.13 | 0.60 | 0.68 | 2.48 | 1.27 | 0.35 | 0.22 |
| CaO | 0.08 | 0.21 | 12.23 | 2.88 | 3.81 | 4.16 | 0.00 | 0.16 | 0.01 | 0.00 | 0.01 | 0.18 | 0.13 | 7.05 | 7.09 | 1.13 | 3.04 |
| $Na_2O$ | 0.01 | 0.07 | 0.27 | 0.01 | 0.24 | 9.24 | 0.52 | 0.02 | 0.11 | 0.27 | 0.19 | 0.06 | 0.06 | 0.06 | 0.09 | 0.04 | 0.13 |
| $K_2O$ | 0.00 | 0.00 | 0.06 | 0.00 | 0.00 | 0.31 | 12.14 | 0.00 | 0.02 | 0.00 | 0.00 | 0.01 | 0.00 | 0.01 | 0.00 | 0.00 | 0.00 |
| $Cr_2O_3$ | 0.00 | 0.00 | 0.00 | 0.06 | 0.00 | 0.00 | 0.03 | 0.07 | 0.08 | 0.05 | 0.00 | 0.01 | 0.03 | 0.16 | 0.03 | 0.03 | 0.06 |
| NIO | 0.10 | 0.03 | 0.09 | 0.28 | 0.00 | 0.00 | 0.00 | 0.06 | 0.11 | 0.01 | 0.00 | 0.00 | 0.01 | 0.00 | 0.05 | 0.00 | 0.06 |
| BaO | 0.12 | 0.16 | 0.00 | 0.00 | 0.08 | 0.22 | 8.97 | 0.00 | 0.12 | 0.14 | 0.00 | n.a. | n.a. | 0.00 | 0.10 | 0.00 | 0.01 |
| TOTAL | 101.34 | 100.39 | 98.41 | 100.63 | 100.54 | 101.03 | 100.83 | 98.30 | 98.51 | 100.49 | 99.80 | 94.14 | 91.34 | 101.07 | 100.08 | 99.45 | 100.72 |
| Structural formulae | | | | | | | | | | | | | | | | | |
| No. of oxygens | 4 | 4 | 6 | 6 | 6 | 32 | 32 | 4 | 4 | 4 | 4 | 20 | 20 | 12 | 12 | 12 | 12 |
| Si | 0.990 | 1.000 | 2.047 | 2.005 | 1.989 | 11.249 | 11.296 | 0.002 | 0.001 | 0.001 | 0.012 | 6.047 | 6.061 | 2.981 | 3.009 | 3.075 | 3.043 |
| Ti | — | — | 0.003 | 0.003 | — | — | — | 0.037 | 0.018 | 0.010 | 0.001 | 0.004 | 0.019 | 0.000 | 0.000 | 0.002 | 0.008 |
| Al | 0.001 | 0.002 | 0.058 | 0.003 | 0.001 | 4.729 | 4.657 | 0.003 | 0.006 | 0.008 | 0.024 | 0.033 | 0.015 | 1.882 | 1.922 | 1.710 | 1.835 |
| Fe | 0.126 | 0.070 | 0.315 | 0.128 | — | — | — | 0.921 | 0.998 | — | — | 1.079 | 0.932 | 1.010 | 0.713 | 0.271 | 0.187 |
| Mn | 1.845 | 1.859 | 0.169 | 1.618 | 1.734 | 0.006 | 0.067 | 0.066 | 0.021 | 0.568 | 0.802 | 6.556 | 6.960 | 1.072 | 1.497 | 2.858 | 2.645 |
| Mg | 0.038 | 0.055 | 0.840 | 0.089 | 0.046 | 0.014 | 0.024 | 0.002 | — | — | 0.007 | 0.155 | 0.180 | 0.290 | 0.153 | 0.043 | 0.026 |
| Ca | 0.003 | 0.008 | 0.478 | 0.131 | 0.174 | 0.758 | — | 0.007 | 0.001 | — | 0.001 | 0.033 | 0.024 | 0.592 | 0.612 | 0.101 | 0.267 |
| Na | — | 0.002 | 0.011 | 0.001 | 0.020 | 3.048 | 0.196 | 0.014 | 0.009 | 0.019 | 0.014 | 0.010 | 0.011 | 0.009 | 0.015 | 0.007 | 0.020 |
| K | — | — | 0.002 | 0.000 | — | 0.067 | 3.000 | — | 0.001 | — | — | 0.001 | — | 0.000 | 0.000 | 0.000 | 0.000 |
| Cr | — | — | 0.001 | 0.002 | — | — | 0.005 | 0.002 | 0.003 | 0.001 | — | 0.002 | 0.006 | 0.010 | 0.002 | 0.002 | 0.004 |
| Ni | 0.002 | — | 0.004 | 0.010 | — | — | — | 0.019 | 0.004 | — | — | — | — | 0.000 | 0.003 | 0.000 | 0.004 |
| Ba | 0.002 | 0.002 | 0.000 | 0.000 | 0.001 | 0.014 | 0.681 | — | 0.002 | 0.002 | — | — | — | 0.000 | 0.003 | 0.000 | 0.000 |
| Fe | — | — | 0.000 | 0.000 | 0.036 | 0.039 | 0.029 | 1.919 | 1.966 | 2.256 | 2.093 | — | — | 0.157 | 0.073 | 0.000 | 0.000 |
| TOTAL | 3.011 | 2.998 | 3.928 | 3.990 | 4.002 | 19.964 | 19.956 | 2.992 | 3.030 | 2.865 | 2.954 | 13.920 | 14.208 | 8.000 | 8.000 | 8.070 | 8.040 |

Sample numbers
1. 7810/33: Tephroite coexisting with garnet, rhodonite and jacobsite
2. 7810/34: Tephroite coexisting with garnet and rhodonite
3. 7810/30: Manganoan hedenbergite coexisting with garnet
4. 7810/32: Rhodonite coexisting with garnet
5. 7810/33: Late stage rhodonite transgressing garnet and jacobsite
6. 7810/36: Plagioclase in quartz-feldspar-garnet-biotite gneiss
7. 7810/30: Hyalophane coexisting with garnet, quartz and hornblende
8. 7810/40A: Magnetite coexisting with quartz and garnet
9. 7810/30: Magnetite coexisting with garnet and hornblende
10. 7810/33: Jacobsite coexisting with barite, garnet and rhodonite
11. 7810/34: Jacobsite coexisting with tephroite, garnet and rhodonite
12. 7810/34: Friedelite transgressing tephroite, rhodonite and spessartine
13. 7810/34: Friedelite transgressing tephroite, rhodonite and spessartine
14. 7810/36: Garnet rim in quartz-feldspar-garnet-biotite gneiss
15. 7810/36: Garnet core of above
16. 7810/32: Garnet rim coexisting with tephroite and rhodonite
17. 7810/32: Garnet core of above.

n.a. = not analysed.
a = includes $ZrO_2$ 0.05%, F 0.00%, Cl 5.97%, $P_2O_5$ 0.03%
b = includes $ZrO_2$ 0.00%, F 0.01%, Cl 5.53%, $P_2O_5$ 0.01%
$Mn^{III}$ and $Mn^{II}$ analysed and calculated as $Mn^{II}$. Total iron analysed as FeO, $Fe_2O_3$ calculated.

titaniferous annitic biotite, the amphibole is a magnesiohornblende and plagioclase varies from $An_{15}$ to $An_{25}$.

In the banded iron formation, the garnet is more manganoan than in the quartzofeldspathic gneiss with a composition of $spess_{42.2}$ $alm_{31.8}$ $gross_{10.9}$ $pyr_{15.1}$. Garnet is also normally zoned (Figure 2) and, in places, is altered to manganoan brunsvigite (chlorite). Amphibole in the banded iron formation is a manganoan magnesio-hornblende and the magnetite varies in composition from magnetite $(mt)_{99.8}$ jacobsite $(jacob)_{0.2}$ to $mt_{94.4}$ $jacob_{6.6}$ (Table 1).

Garnet ($spess_{62.9}$ $alm_{23.1}$ $gross_{7.2}$ $pyr_{6.8}$) from the Ba feldspar-Mn silicate-amphibole facies is normally zoned (Figure 2) and is richer in Mn and poorer in iron (Fe), calcium (Ca) and magnesium (Mg) than garnet from the banded iron formation. Associated hyalophane varies in composition from $celsian_{5.2}$ albite $(ab)_{20.5}$ orthoclase $(or)_{74.3}$ to $celsian_{21.8}$ $ab_{7.6}$ $or_{70.6}$ and contains sphene inclusions which have up to 5.5 wt.% $A1_2O_3$ and 1.7% $Fe_2O_3$. The hyalophane from Wilkes Land is similar in composition to hyalophane from exhalative deposits in other areas e.g. Aberfeldy, Scotland (Fortey and Beddoe-Stephens, 1982), except that the Antarctic hyalophane contains 0.3—0.5% MnO whereas MnO in hyalophane is less than 0.05% in other exhalites. Magnetite in this facies is low in Mn ($mt_{98.0}$ $jacob_{2.0}$) and it coexists with manganoan hedenbergite ($FeSiO_3$ $_{17.4}$ $MnSiO_3$ $_{9.4}$ $MgSiO_3$ $_{46.6}$ $CaSiO_3$ $_{26.5}$) and calcic rhodonite ($MnSiO_3$ $_{83.3}$ $CaSiO_3$ $_{16.1}$ $FeSiO_3$ $_{0.6}$).

The most manganoan garnets ($spess_{82.6}$ $alm_{4.3}$ $gross_{9.8}$ $pyr_{3.3}$) are present in the barite-manganese silicate facies where garnets up to $spess_{89.4}$ $alm_{2.6}$ $gross_{7.5}$ $pyr_{0.5}$ in composition are present. Garnet in this facies is both normally and reversely zoned with Mn, Fe and Mg slightly higher and Ca lower in the rim compared with the core of the reversely zoned grains. Manganiferous garnet is a common mineral associated with metamorphosed Proterozoic exhalative deposits (Table 2). For example, garnet at Broken Hill, NSW coexisting with rhodonite is richer in Mn and poorer in Ca than garnet coexisting with bustamite (Mason, 1973). Because of the lack of Ca minerals and the low Ca content of silicate phases in the Wilkes Land manganese silicate rocks, it is not surprising that spessartine is the dominant garnet component. The jacobsite from the manganese silicate rocks is ferrous and varies in composition from $mt_{38.5}$ $jacob_{61.5}$ to $mt_{18.4}$ $jacob_{81.6}$ as compared with jacobsite from Broken Hill which contains 32 mol.% $Fe_3O_4$ and up to 0.3 wt.% Zn (Segnit, 1977). Tephroite is unzoned, has negligible $Ca_2SiO_4$ and $Mg_2SiO_4$ components and an average composition of $Mn_2SiO_4$ $_{94.3}$ $Fe_2SiO_4$ $_{5.7}$. Tephroite commonly coexists with barite and contains 0.1-0.2% BaO and commonly more BaO than CaO. Compared with tephroite from Broken Hill, NSW (Mason, 1973) and Buritirama (Peters et al., 1977), tephroite from Wilkes Land is almost end member tephroite with very low Fe and Mg. Zincian tephroite ("roepperite") was described from Franklin, New Jersey, but microprobe studies by Hurlbut (1961) showed that roepperite is tephroite with exsolved willemite and that very small amounts of zinc can be included in the tephroite structure. No Zn was detected in tephroite from Wilkes Land.

Rhodonite from Wilkes Land has a far higher $MnSiO_3$ component than rhodonite from Broken Hill, NSW (Mason, 1973; Hodgson, 1975); the Rhetic Alps (Peters et al., 1973) Buritirama, Brazil (Peters et al., 1977) and Bald Knob, USA (Winter et al., 1981) and contains less than 10 mol.% $FeSiO_3$ and $CaSiO_3$ (Figure 3). Late stage rhodonite associated with friedelite is even poorer in $CaSiO_3$ and

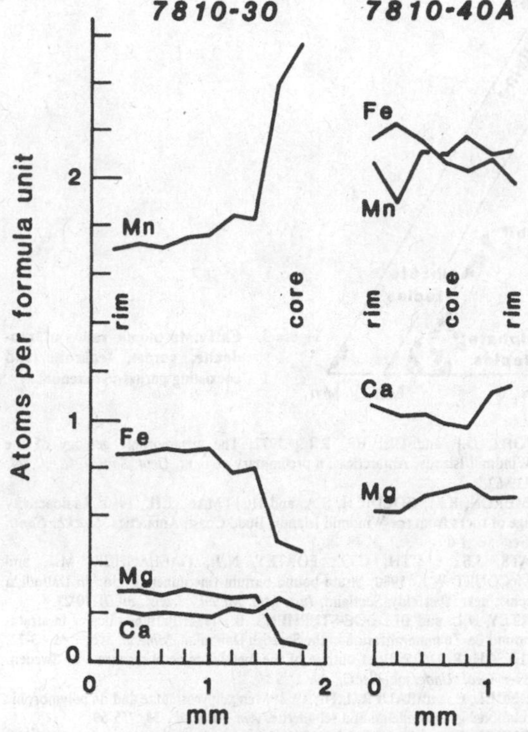


**Figure 2. Atoms/formula unit of Mn, Fe, Mg and Ca from the core to the rim of zoned garnets from 7810/30 (hyalophane-manganese silicate-amphibole facies) and 7810/40A (banded iron formation).**

$FeSiO_3$ (Figure 3) again reflecting the low CaO and FeO content of the Ba-Mn-rich metasediments. Late stage ferroan friedelite identified by a Weissenberg photograph, has $a_o$ = 13.52Å and $c_o$ = 21.48Å. Friedelite is an uncommon late stage phase associated with manganpyrosmalite and schallerite at Franklin and Sterling, New Jersey; Dannemora, Nordmark and Langban, Sweden; Broken Hill, Australia and the Hautes Pyrenees, France (Frondel and Bauer, 1953). Friedelite from Franklin has 0.3-3.4% chlorine(Cl), 0.5-1.5% FeO, 0-1.5% MgO and 0.8-1.4% Zn (Bauer and Benman, 1928) whereas friedelite from Wilkes Land contains 5.5-6% C1, 6.3-7.5% FeO, 0.6-0.7% MgO and no detectable Zn (Table 2).

A rare as yet unidentified phase with Mn, yttrium(Y) and arsenic(As) peaks in the x-ray spectrum may well be retzian ($Mn_2Y(AsO_4)(OH)_4$), especially as the only known occurrence (Nordmarks, Sweden) is from a jacobsite-rich manganiferous horizon (Sjogren, 1897).

## DISCUSSION

Chemical sediments enriched in manganese are usually products of submarine exhalative activity (Peters at al., 1973) and, in many places, are intercalated with banded iron formations and massive sulphide rocks (e.g. Gamsberg, South Africa (Rozendaal, 1980)), are laterally equivalent to barite-rocks, for example Gamsberg, South Africa (Rozendaal, 1980), are laterally equivalent to iron formations, for example central Sweden (Frietsch, 1975), are underlain by iron formations, for example Iberian Pyrite Belt (Strauss and Madel, 1974), and are enriched in sulphides and show both lateral and vertical facies changes to iron formations (e.g. Broken Hill, NSW). In many areas, manganese-rich chemical sediments are lower temperature, more oxidised, distal equivalents of iron formations or massive sulphide deposits of exhalative origin. Manganese-rich chemical sediments and related iron formations are present in greenschist facies sequences (e.g. Iberian Pyrite Belt), amphibolite facies sequences (e.g. Gamsberg, South Africa) and granulite facies rocks (e.g. Broken Hill, NSW: central Sweden) and, although mineral reactions occur during metamorphism, the manganese-rich units essentially act as isochemical units during metamorphism (Mason, 1973; Frondel and Baum, 1974).

Many large submarine exhalative Pb-Zn deposits were deposited in the Middle Proterozoic (e.g. Broken Hill, Mt. Isa and McArthur River, Australia; Gamsberg and Aggeneys, South Africa; Sullivan, Canada) and metamorphosed Mn- and Ba-rich horizons are common facies equivalents of massive sulphide deposits of that age (e.g. Broken Hill, NSW; Gamsberg, South Africa). The Mn-Ba-rich horizon from Wilkes Land has a remarkable mineralogical similarity to other metamorphosed Middle Proterozoic Mn- and Ba-rich exhalites (Table 2). Because of the similarity, and the close association with banded iron formation, the Wilkes Land occurrence is considered here to be an exhalite. Although sulphide phases are very rare in the Wilkes Land exhalites and no Zn was detected in the olivine, pyroxenoid, spinel and garnet analysed, the presence of Mn- Ba-rich assemblages which display very rapid facies changes (oxide, banded iron formations → silicate, Ba silicate-Mn silicate-amphibole → sulphate, barite-manganese silicate) suggests that a Pb-Zn sulphide-rich facies equivalent is a likely possibility in the Windmill Metamorphics. The increase of Mn in garnet from quartzofeldspathic gneiss ($spess_{41}$) to banded iron formation ($spess_{42}$) to Ba silicate-Mn silicate-amphibole facies ($spess_{63}$) and finally to barite-Mn silicate facies ($spess_{83}$) is a very common trend from host rocks to oxide → silicate → sulphate facies of submarine exhalative deposits. The Wilkes Land exhalites contain far less Fe, Ca and base metals than similar metamorphosed chemical sediments at Broken Hill (NSW), Franklin, Sterling Hill, Gamsberg and Langban and Ca-rich carbonate and Fe silicate-sulphide facies might also be expected as time rock equivalents.

These exhalites have experienced the same history of events of deformation and metamorphism as the enclosing quartzofeldspathic gneiss of the Windmill Metamorphics. Blight and Oliver (1977) have established from the gneisses that the Windmill Metamorphics from Clark Peninsula underwent upper amphibolite facies metamorphism. Experimental studies by Peters et al. (1973) showed that at total

**TABLE 2: Comparison of the mineralogy of Wilkes Land Mn-Ba silicate rocks with metamorphosed Proterozoic submarine exhalative ores**

| Locality | Principal Mn and Ba phases |
|---|---|
| Wilkes Land, Antarctica[1] | spessartine, rhodonite, tephroite, jacobsite, friedelite, barite, hyalophane |
| Broken Hill, Australia[2] | rhodonite, spessartine, bustamite, Mn hedenbergite, Mn pyrosmalite, friedelite, jacobsite, Fe tephroite, Mn amphiboles, "sturtite", barite, celsian, hyalophane |
| Pegmont, Australia[3] | Mn fayalite, spessartine, Mn amphibole, Mn clinopyroxene |
| Gamsberg, South Africa[4] | Mn andradite, Mn amphibole; Mn fayalite, Mn hedenbergite, barite |
| Rosh Pinah, South Africa[5] | barite, celsian, witherite, norsethite, benstonite |
| Buritirama, Brazil[6] | rhodonite, spessartine, tephroite, rhodochrosite, Mn hedenbergite, Mn amphiboles, pyroxmangite, braunite, hausmannite, barite |
| Franklin and Sterling Hill, U.S.A.[7] | franklinite, spessartine, tephroite, rhodonite, friedelite, schallerite, manganpyrosmalite, bementite, manganosite, roepperite, rhodochrosite, barite, celsian, hyalophane |
| Bald Knob, U.S.A.[8] | rhodonite, pyroxmagnite, rhodochrosite, Mn amphiboles, tephroite, galaxite, jacobsite, pyrophanite, spessartine |
| Langban, Sweden[9] | spessartine, rhodonite, jacobsite, tephroite, Mn magnetite, Mn amphibole, braunite, hausmannite, barite, celsian |
| Slojdartorp, Sweden[10] | jacobsite, spessartine, rhodonite, dannemourite, piemontite, urbanite, braunite |
| Ultevis, Sweden[11] | hollandite, bixbyite, braunite, spessartine, hausmannite, piemontite, barite, celsian |
| Aberfeldy, Scotland[12] | hyalophane, barite, celsian, Ba muscovite |
| Kodura, India[13] | bixbyite, braunite, hausmannite, hollandite, jacobsite, pyrophanite, pyroxmangite, rhodonite, manganophyllite, rhodonite, spandite, spessartine, tephroite, rhodochrosite, barite, hyalophane |

1. This study. 2. Mason (1973, 1976), Hodgson (1975). 3. Stanton and Vaughan (1979). 4. Rozendaal (1980). 5. Page and Watson (1976). 6. Peters et al. (1977). 7. Frondel and Baum (1974), McSween (1976). 8. Winter et al. (1981). 9. Frietsch (1975), Bostrom et al. (1979), includes Langban, Pajsberg, Harstigen, Jakobsberg, Sjogruvan and Nordmark deposits. 10. Koark (1970). 11. Frietsch (1975). 12. Coats et al (1980). 13. Sivaprakash (1980).

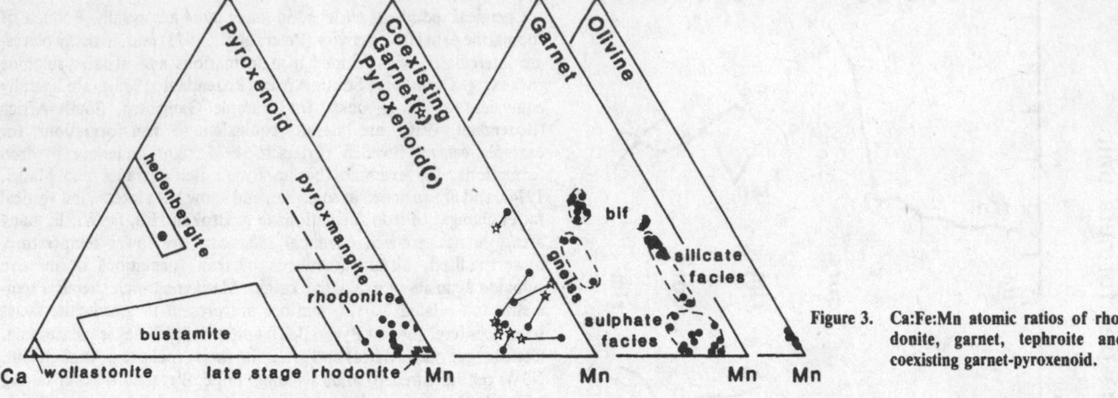


**Figure 3. Ca:Fe:Mn atomic ratios of rhodonite, garnet, tephroite and coexisting garnet-pyroxenoid.**

pressures of 2 kb in the system $MnO$-$SiO_2$-$CO_2$-$H_2O$, the following univariant equilibrium exists at $581 \pm 10°C$:

$$MnCO_3 + MnSiO_3 = Mn_2SiO_4 + CO_2$$

rhodochrosite pyroxmangite tephroite

Because of the $MnCO_3 + SiO_2 = MnSiO_3 + CO_2$ reaction, rhodonite is not commonly found with both quartz and carbonate. If T is constant, the maximum $CaSiO_3$ content of pyroxene coexisting with calcite and quartz is a function of $f_{CO_2}$. Because of the possibility of the reaction (MnFeCa) pyroxenoid + $CO_2$ = ferroan tephroite + $CaCO_3$, tephroite, calcite, rhodonite and bustamite can coexist at any one temperature only at a fixed fugacity of $CO_2$. Therefore the assemblage tephroite—rhodonite (present in the Wilkes Land exhalites) can only exist at high $f_{CO_2}$ and hence tephroite is most commonly recorded in association with rhodonite and calcite.

Calcite is a very minor phase in the Wilkes Land exhalites and most silicate phases are subcalcic, hence the premetamorphic assemblage is likely to have been Mn-Fe oxides + Mn-Fe-Ca carbonate + quartz + barite + ? smectite + ? cymrite. Although the exhalite originally contained quartz, tephroite is only stable in rocks with a low Si/(Mn + Fe) in the presence of carbonate. At temperatures of greater than 550°C at about 5 kb, rhodonite and pyroxmangite have a small miscibility gap (Winter et al., 1981) and, although no pyroxmangite was observed, the very high Mn content of the rhodonite suggests that it derived from pyroxmangite inversion.

In the quartzofeldspathic gneiss and banded iron formation, chlorite, pinite and sulphides are retrograde metamorphic phases and prograde garnet retains its normal zoning from a Mn-rich core to a Mn-poor rim. Such zoning probably derived from magnetite-chlorite-biotite fractionation-depletion during the metamorphic history. In the barite-Mn silicate exhalite, garnet is reversely zoned, possibly as a result of fluid facilitated retrograde reactions with the enclosing rhodonite, tephroite and jacobsite. The presence of friedelite pods and friedelite veinlets lined with secondary rhodonite suggests formation during retrograde metamorphism. These phases are common retrograde phases at Franklin and Broken Hill, NSW. With more detailed sampling, phases such as alabandite, cymrite, Mn carbonates, Mn vesuvianite, Mn axinite, Mn pyrosmalite, bustamite, pyroxmangite, Mn wollastonite, parsettensite, dannemourite, inesite, sturtite, ganophyllite, bannisterite, Mn ilvaite, bementite etc. would be expected as prograde and retrograde species in the exhalites.

*Acknowledgements*. The field collections were made by one of us (JFL) while a member of the Australian National Antarctic Research Expedition summer party relieving Casey Station in January 1978. The logistic and moral support of the Antarctic Division, Department of Science and Technology is gratefully acknowledged. It is also a pleasure to acknowledge and record the assistance and support in the field of G.A. Travis and other members of the Expedition. I.E. Grey carried out the X-ray study of friedelite and P.R. Kelly provided electron microprobe analyses and prepared the specimens for study.

The work was supported in part by a grant to JFL from the Australian Research Grants Committee.

## REFERENCES

BAUER, L.H. and BERMAN, H. 1928: Friedelite, schallerite and related minerals. *Am. Mineral.*, *13*, 341-8.

BOSTROM, K., RYDELL, H. and JOENSUU, O., 1979: Langban—an exhalative sedimentary deposit? *Econ. Geol.*, *74*, 1002-12.

BLIGHT, D.F. and OLIVER, R.L., 1977: The metamorphic geology of the Windmill Islands, Antarctica : a preliminary account. *Geol. Soc. Aust., J.*, *24*, 239-62.

CAMERON, R.L., GOLDICH, S.A. and HOFFMAN, J.H., 1960: Radioactivity age of rocks from the Windmill Islands, Budd Coast, Antarctica. *Stockh. Contr. Geol.*, *6*, 1-6.

COATS, J.S., SMITH, C.G., FORTEY, N.J., GALLAGHER, M.J. and McCOURT, W.J., 1980: Strata-bound barium-zinc mineralization in Dalradian schist near Aberfeldy, Scotland. *Inst. Min. Metall., Trans.*, *89*, B110-22.

FORTEY, N.L. and BEDDOE-STEPHENS, B., 1982: Barium silicates in strata-bound Ba-Zn mineralization in the Scottish Dalradian. *Mineral. Mag.*, *46*, 63-72.

FRIETSCH, R., 1975: Brief outline of the metallic mineral resources of Sweden. *Sver. Geol. Undersok, Ser.C.*, *69*.

FRONDEL, C. and BAUER, L.H., 1953: Manganpyrosmalite and its polymorphic relationship to friedelite and schallerite. *Am. Mineral.*, *38*, 755-60.

FRONDEL, C. and BAUM, J.L., 1974: Structure and mineralogy of the Franklin zinc-iron-manganese deposit, New Jersey. *Econ. Geol.*, *69*, 157-80.

HODGSON, C.J., 1975: The geology and geological development of the Broken Hill lode, in the New Broken Hill Consolidated Mine, Australia. Part II : Mineralogy. *Geol. Soc. Aust., J.*, 33-51.

HURLBUT, C.S., 1961: Tephroite from Franklin, New Jersey. *Am. Mineral.*, *46*, 549-59.

KOARK, H.J., 1970: Zur Geologie des neuendekten Jakobsit-Braunit-Hamatit-Mangansilikaf-Lager Slojdartorp in Nybergetfelde in Zentralschweden. *Geol. Foren. Stock. Forh.*, *92*, 388-401.

MASON, B., 1959: Tephroite from Clark Peninsula, Wilkes Land, Antarctica. *Am. Mineral.*, *44*, 428-30.

MASON, B., 1973: Manganese silicate minerals from Broken Hill, New South Wales. *Geol. Soc. Aust., J.*, *20*, 397-404.

MASON, B., 1976: Famous mineral localities. Broken Hill, Australia. *Mineral* Rec., Feb. 1976, 25-33.

McSWEEN, H.Y., 1976: Manganese-rich ore assemblages from Franklin, New Jersey. *Econ. Geol.*, *71*, 814-7.

PAGE, D.C. and WATSON, M.D., 1976: The Pb-Zn deposits of the Rosh Pinah Mine, South West Africa. *Econ. Geol.*, *71*, 306-27.

PETERS, Tj., SCHWANDER, H. and TROMMSDORFF, V., 1973: Assemblages among tephroite, pyroxmangite, rhodochrosite, quartz : experimental data and occurrences in the Rhetic Alps. *Contrib. Mineral. Petrol.*, *42*, 325-32.

PETERS, Tj., VALARELLI, J.V., COUTINHO, J.M.V., SOMMERAUER, J., and von RAUMER, J., 1977: The manganese deposits of Buritirama (Para, Brazil). *Schweiz, Mineral. Petrog. Mitt.*, *57*, 313-27.

ROZENDAAL, A., 1980: The Gainsberg zinc deposit, South Africa : a banded stratiform base-metal sulfide ore deposit. *Proc. 5th IAGOD Symp.*, 619-33.

SEGNIT, E.R., 1977: Jacobsite and magnetite from Broken Hill, Australia. *Aust. Mineral.*, *9*, 37-9.

SJOGREN, Hj., 1897: Retzian och dess sammansattning. *Geol. Foren. Stock. Furh.*, *19*, 106-12.

SIVAPRAKASH, C., 1980: Mineralogy of manganese deposits of Koduru and Garbham, Andhra Pradesh, India. *Econ. Geol.*, *75*, 1083-1104.

STANTON, R.L. and VAUGHAN, J.P., 1979: Facies of ore formation : A preliminary account of the Pegmont deposit as an example of potential relations between small "iron formations" and stratiform sulphide ores. *Proc. Aust. Inst. Min. Metall.*, *270*, 25-38.

STRAUSS, G.K. and MADEL, J., 1974: Geology of massive sulphide deposits in the Spanish-Portuguese Pyrite Belt. *Geol. Rdsch.*, *63*, 191-211.

WEBB, A.W., McDOUGALL, I. and COOPER, J.A., 1963: Potassium-argon dates from the Vincennes Bay region and Oates Land; in Adie, A.J. (ed.) *Antarctic Geology.* North Holland, Amsterdam, 597-600.

WILLIAMS, I.S., COMPSTON, W., COLLERSON, K.D., ARRIENS, P.A. and LOVERING, J.F., (this volume): A reassessment of the age of the Windmill Metamorphics, Casey Area..

WINTER, G.A., ESSENE, E.J. and PEACOR, D.R., 1981: Carbonates and pyroxenoids from the manganese deposit near Bald Knob, North Carolina..*Am. Mineral.*, *66*, 278-89.

# A REASSESSMENT OF THE AGE OF THE WINDMILL METAMORPHICS, CASEY AREA

I.S. Williams, W. Compston, K.D. Collerson and P.A. Arriens, *Research School of Earth Sciences, A.N.U., P.O. Box 4, Canberra, A.C.T., 2601, Australia.*

J.F. Lovering, *Department of Geology, School of Earth Sciences, University of Melbourne, Parkville, Vic., 3052, Australia.*

**Abstract** Ion microprobe analyses of zircons at high mass-resolution show that the age of the protoliths of orthogneisses from the Windmill Metamorphics, previously suggested to be up to 3500 Ma (Lovering et al., 1981), is not more than approximately 1500 Ma. This is consistent with the age and initial $^{87}Sr/^{86}Sr$ from their whole rock Rb-Sr isochron (1477 ± 73 Ma, 0.7032 ± 0.0004), and with a single Nd model age (1465 ± 34 Ma). The discrepancy probably is due to spectral interferences with $^{207}Pb$ not accounted for at the low mass-resolution used by Lovering et al. A single zircon having an age of approximately 3000 Ma has been observed in an associated paragneiss.

The Windmill Metamorphics near Casey Station, Wilkes Land, consist of polymetamorphic para- and orthogneisses up to hornblende-orthopyroxene granulite grade. Rb-Sr and K-Ar ages determined on these rocks by Cameron et al. (1960), Webb et al. (1963) and Arriens (1974) are in the range 950-1400 Ma. They contrast markedly with $^{207}Pb/^{206}Pb$ apparent ages in the range 1600-3500 Ma measured on zircons by Lovering et al. (1981) using the ARL low-resolution ion microprobe. Lovering et al. proposed that the Windmill Metamorphics contain zircon derived from sialic crust at least 3500 Ma old and, that the Rb-Sr and K-Ar ages record a later thermal event.

In this paper we report Rb-Sr and Sm-Nd total-rock analyses of orthogneiss from the Windmill Metamorphics and reanalysis, using the sensitive high-resolution ion microprobe (SHRIMP) recently constructed at the Australian National University, of some of the same spots on the same zircons analysed by Lovering et al. (1981). We use these data to argue the contrary view that the protolith of at least the orthogneisses cannot be much older than 1470 Ma. The paragneisses do contain an Archaean component.

Ages in this paper are calculated using the constants recommended by Steiger and Jager (1977). The Sm-Nd model age is calculated using $\lambda^{147}Sm = 6.54 \times 10^{-12} yr^{-1}$ (Gupta and MacFarlane, 1970) and the reference values of Jacobsen and Wasserburg (1980). Assigned uncertainties, unless noted otherwise, are 95% confidence limits based on analytical precision.

## GEOLOGY OF THE CASEY AREA

Casey Station and the Windmill Islands are on the Budd Coast of East Antarctica at longitude 110°E. The Windmill Metamorphics (Blight and Oliver, 1977), consisting of interleaved schist, gneiss and migmatite, are the oldest rocks known in the region. The metamorphics are intruded by charnockite (the Ardery Charnockite), K-feldspar megacrystic garnetiferous granite (the Ford Granite) and dolerite and lamprophyre dykes (Figure 1).

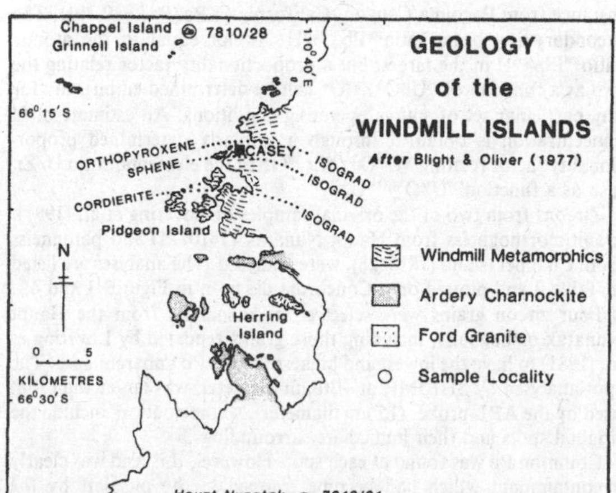


**Figure 1. Sketch geological map of the Windmill Islands, after Blight and Oliver (1977), showing the localities of samples analysed in the present study.**

Geological mapping by Blight and Oliver (1977) showed that in the immediate vicinity of Casey Station, there is a progression in the grade of the Windmill Metamorphics over a distance of about 8 km from almandine amphibolite in the north, through biotite-cordierite-almandine granulite to hornblende-orthopyroxene-plagioclase granulite in the south. The latter extends at least to Haupt Nunatak 30 km further south. The mapping also indicated at least four periods of derformation, two of which produced tight-isoclinal folds. From whole rock major element analyses, Blight and Oliver (1977) concluded that within the Windmill Metamorphics there are both meta-igneous and metasedimentary units. They suggested that the premetamorphic protolith consisted of felsic and mafic volcanics interbedded with sediments ranging from greywacke-type sandstone to shale.

The region was mapped further by K.D.C. during the summer of 1980/81. The gneisses of the Windmill Metamorphics comprise three groups, each probably derived from a different protolith. The oldest group, based on field relationships on Herring and Pidgeon Islands, is a suite of layered "granitic" gneisses ranging from diorite through tonalite to granite. These probably have plutonic igneous precursors, not the sedimentary or volcanic parents suggested by previous workers. The granitic orthogneisses are structurally complex and contain intrafolial folds and strong composite fabrics developed parallel to the axial planes of those folds.

Semi-continuous belts of supracrustal rocks are interleaved with the orthogneisses. In the north of the area, these are dominantly paragneisses derived from pelitic to psammitic precursors and minor calc-silicate rocks. In the south, mafic units metamorphosed to mafic granulite or amphibolite are more abundant. Commonly they are associated with thin layers of quartzite and quartz-magnetite-garnet rocks and may be metamorphosed chert and banded iron formation. In some places, for example on Grinnell and Pidgeon Islands, the layered complex is extensively migmatised and there has been localised generation of granite.

Late in the sequence of events which resulted in the interleaving of the orthogneisses and supracrustal rocks, the layered complex was intruded by coarse grained, leucocratic sheets. Although Oliver (1972) described such rocks on Herring Island, he did not appreciate their intrusive character, preferring to ascribe them to processes of "leuco gneissification".

Blight and Oliver (1977) interpreted the metamorphism of the Windmill Metamorphics as prograde. However there is evidence in the Casey area for retrogression of earlier granulite facies assemblages. For example, in the younger orthogneisses there are disrupted layers of mafic granulite with partially developed amphibolite facies assemblages. This retrogression may have occurred during the emplacement of the parent rocks of the younger leucocratic orthogneisses.

## PREVIOUSLY PUBLISHED GEOCHRONOLOGICAL STUDIES

K-Ar and Rb-Sr ages determined on both the Windmill Metamorphics and the later felsic intrusives are mostly close to 1100 Ma and do not exceed 1400 Ma.

Cameron et al. (1960) dated two samples of the metamorphics by biotite K-Ar, obtaining 950 Ma for gneiss from the Haupt Nunatak and 1120 Ma for gneiss 5 km north of Casey. The average of biotite ages reported for a sample of the Ardery Charnockite, 1090 Ma, is not discernably younger than these.

Samples of metamorphic and intrusive rocks from several of the Windmill Islands were included in the study by Webb et al. (1963). The range in the biotite K-Ar ages determined was small, viz. 1040-1130 Ma, yet significantly greater than the analytical uncertainty. The authors noted that the K-Ar ages of the intrusives are not significantly less than those of the metamorphics and suggested that either emplacement of the intrusives affected the isotopic ages of the pre-existing rocks or reheating caused argon loss over the whole area at a later time.

Rb-Sr ages show a greater range. Arriens (1974) reported total rock ages for suites of gneisses between 1100 and 1400 Ma. Biotite and muscovite Rb-Sr ages of 1100 Ma determined on pegmatites were similar to the youngest gneiss total-rock ages and to the K-Ar ages. Arriens suggested to Blight and Oliver (1977) that 1400 Ma is the age of deposition of the pre-metamorphic protolith and that 1100 Ma is the age of a high grade metamorphic event.

These K-Ar and Rb-Sr ages differ markedly from the zircon $^{207}Pb/^{206}Pb$ apparent ages measured on several gneiss samples and on the Ford Granite by Lovering et al. (1981) using an ARL low-resolution ion microprobe. The range in the zircon ages is considerable, 1600-3500 Ma and all exceed the greatest age determined by the other techniques, 1400 Ma. For all samples, including that of the Ford Granite, the internal range in zircon ages is several hundred million years. There is a general correlation for all samples between increasing apparent age and decreasing relative uranium content.

Lovering et al. (1981) concluded from the range in the zircon ages and from the difference between those ages and the K-Ar and Rb-Sr results, that the metamorphics contain zircon up to 3500 Ma old which has been affected by a later event, probably the ~1100 Ma metamorphism. Their interpretation of the 2100-3100 Ma range in zircon apparent ages from the Ford Granite was that the granite was derived from the partial remelting of Windmill Metamorphics country rock and might well contain relict zircons from those rocks.

## EVIDENCE FOR A LATE PROTEROZOIC PROTOLITH FOR THE ORTHOGNEISSES

### Rb-Sr

Total rock Rb-Sr analyses of nine samples of the early layered tonalitic orthogneisses from Pidgeon Island (Table 1) are plotted on an isochron diagram in Figure 2. The scatter of the data about the isochron is significantly greater than experimental error (MSWD = 58). The pattern of scatter suggests that the samples have a small range of ages but a common initial $^{87}Sr/^{86}Sr$ (Model two, McIntyre et al., 1966). The model two isochron age is 1477 ± 73 Ma, with an initial $^{87}Sr/^{86}Sr$ value of 0.7032 ± 0.0004.

**TABLE 1: Rb-Sr data for tonalitic gneisses from Pidgeon Island**

| Sample No. | Rb (ppm) | Sr (ppm) | $^{87}Rb/^{86}Sr$ (1) | $^{87}Sr/^{86}Sr$ (2) |
|---|---|---|---|---|
| 71-484 | N.M. | N.M. | 0.0910 | 0.7051 |
| 71-498 | 23.2 | 144.8 | 0.4624 | 0.7132 |
| 335-117 | 81.8 | 281.3 | 0.8407 | 0.7214 |
| 71-495 | 40.5 | 127.6 | 0.9176 | 0.7221 |
| 335-113 | 70.6 | 207.9 | 0.9815 | 0.7242 |
| 335-114 | 65.5 | 180.5 | 1.0496 | 0.7262 |
| 71-486 | N.M. | N.M. | 1.1732 | 0.7299 |
| 335-119 | 118.6 | 270.2 | 1.2697 | 0.7285 |
| 335-115 | 107.3 | 198.4 | 1.5656 | 0.7341 |

(1) Coefficient of variation 0.5% (1δ).
(2) Experimental variance in $^{87}Sr/^{86}Sr$ is $1.0 \times 10^{-5}$ (1δ).

This initial $^{87}Sr/^{86}Sr$ is indistinguishable from the average earth $^{87}Sr/^{86}Sr$ 1477 Ma ago, viz. 0.7030, calculated by assuming an initial $^{87}Sr/^{86}Sr$ value of 0.6990 and a present value of 0.7048 for the bulk earth Allègre et al., 1978). As the average $^{87}Rb/^{86}Sr$ of the tonalitic gneiss, 0.93, is an order of magnitude greater than that of the bulk earth, it is highly unlikely that the tonalitic orthogneiss represents crustal material significantly older than 1477 Ma in which the radiogenic Sr has been rehomogenised by metamorphism at that time.

### Sm-Nd

One sample of tonalitic gneiss from the oldest orthogneiss sequence on Pidgeon Island was analysed for total rock Sm-Nd, giving $^{147}Sm/^{144}Nd$ = 0.08483 and $^{143}Nd/^{144}Nd$ = 0.510759 ± 24 (2σ). Its model

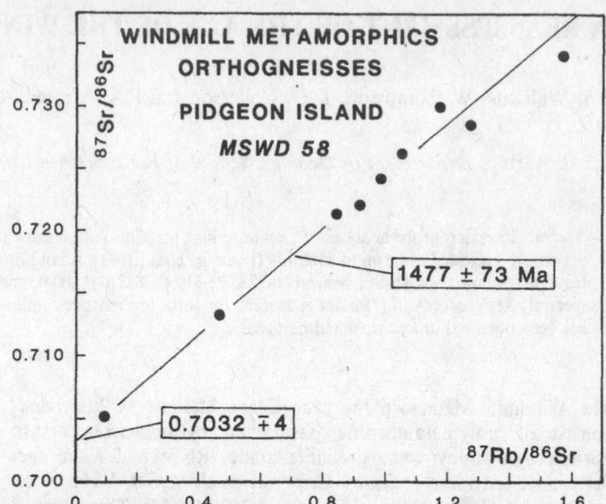


**Figure 2. Isochron diagram of Rb-Sr total rock analyses of samples of Windmill Metamorphics layered tonalitic orthogneisses from Pidgeon Island.**

$T\text{-}T^{Nd}_{CHUR}$ age is 1465 ± 34 Ma. This model age may be interpreted as the time when rare earth elements in the sample were fractionated on extraction of the gneiss's ultimate source materials from a mantle reservoir, assuming that reservoir to have had a chondritic Sm/Nd value. The age is, however, wholly consistent with the Rb-Sr total rock analyses of the Pidgeon Island orthogneiss.

### U-Pb

Some of the same spots on the same zircons from two of the Windmill Metamorphics samples analysed by Lovering et al. (1981) have been analysed on the high-resolution ion microprobe at the Australian National University. In addition to measuring $^{207}Pb/^{206}Pb$ which can be compared directly with the low-resolution ion microprobe data, $^{206}Pb/^{204}Pb$ and Pb/U can also be determined using SHRIMP, making it possible to plot the data on a conventional Concordia diagram.

Zircon U-Pb analysis using SHRIMP has been described by Compston et al. (1982). Briefly, for zircon U-Pb analysis, SHRIMP is operated at a mass resolution of approximtely 7000 (1% valley), which is sufficient to resolve all known major isobaric interferences from Pb, U, ThO and UO. PbH is not resolved, but analyses of standard zircons show its effect is not significant at the 0.1% level. At resolution 7000, no peak stripping is necessary and $^{206}Pb/^{204}Pb$ can be determined directly. The analysed spot was ~30 μm diameter and yielded ~10 counts/sec/ppm of Pb in the zircon. Pb/U is determined via a calibration established using a zircon of known Pb/U, in this instance from Pacoima Canyon, California ($^{206}Pb/^{238}U$ = 0.191). The secondary ion beam ratio $^{206}Pb^+/^{238}U^+$ is not equal to the atomic ratio $^{206}Pb/^{238}U$ in the target, but a proportionality factor relating the two as a function of $^{238}U^{16}O^+/^{238}U^+$ can be determined empirically for any particular set of source operating conditions. An estimate of U concentration is obtained through a similarly determined proportionality factor relating $^{238}U^{16}O^+/^{90}Zr_2{}^{16}O^+$ to the elemental ratio U/Zr, also as a function $^{238}U^{16}O^+/^{238}U^+$.

Zircons from two of the original samples of Lovering et al. (1981), granitic orthogneiss from Haupt Nunatak (7810/21) and paragneiss from Chappel Island (7810/28), were analysed. The analyses are listed in Table 2 and plotted on a Concordia diagram in Figures 3 and 4.

Four zircon grains were selected for reanalysis from the Haupt Nunatak orthogneiss, including those grains reported by Lovering et al. (1981) to have the lowest and highest $^{207}Pb/^{206}Pb$ apparent ages. The spot analysed by SHRIMP, at ~30 μm diameter, was larger than that used by the ARL probe, (15 μm diameter), so the analyses include the original spots and their immediate surroundings.

Common Pb was found at each spot. However, this lead was clearly a contaminant which had become trapped in the pits left by the original analyses. The proportion of common Pb decreased rapidly as the grain was eroded. For the calculation of the radiogenic Pb composition, all the common Pb was assumed to be surficial and to have a composition similar to Broken Hill Pb. The amount of common Pb contributed by the zircons was too small for possible error in this assumption to cause a significant change in the calculated ages.

The $^{207}Pb/^{206}Pb$ apparent ages are all in the range 1147 ± 87 to 1494 ± 125 Ma. There is no evidence of the ages in the range 1729 ± 79 to 3458 ± 98 Ma reported by Lovering et al. (1981). When plotted on a Concordia diagram (Figure 3), the analyses show a broad correlation of decreasing Pb/U with decreasing $^{207}Pb/^{206}Pb$. The line of discordance has Concordia intercepts of $720^{+300}_{-650}$ and $1540^{+780}_{-220}$ Ma. The upper intercept is the primary age of the zircons, which is not significantly older than the ages of 1477 ± 73 and 1465 ± 34 Ma determined by Rb-Sr and Sm-Nd respectively. The lower intercept is loosely defined but discernably not at the Concordia origin, indicating that the event which disturbed the zircons' isotopic systems was not a recent one. The event, however, occurred later than the 1100 Ma metamorphism recorded by K-Ar and Rb-Sr.

**TABLE 2: High-resolution ion microprobe U-Pb analyses of ~30μm spots on zircon from an orthogneiss and paragneiss from the Windmill Metamorphics**

| Grain | Spot | Pb* ppm | U ppm | Atomic Ratios | | | | | Apparent ages (Ma) | | |
|---|---|---|---|---|---|---|---|---|---|---|---|
| | | | | $^{106}Pb†/^{204}Pb$ | $^{207}Pb†/^{206}Pb$ | $^{206}Pb^*/^{238}U$ | $^{207}Pb^*/^{235}U$ | $^{207}Pb^*/^{206}Pb$ | $^{206}Pb^*/^{238}U$ | $^{207}Pb^*/^{235}U$ | $^{207}Pb^*/^{206}Pb$ |
| **7810/21 Granite orthogneiss, Haupt Nunatak** | | | | | | | | | | | |
| 293 | 1 | 180 | 710 | 1215 | 0.0963 | 0.2005 | 2.393 | 0.0866 | 1178 | 1241 | 1352 |
| 300 | 1 | 70 | 450 | 894 | 0.0965 | 0.1613 | 1.851 | 0.0832 | 964 | 1064 | 1274 |
| | 2 | 160 | 960 | 1741 | 0.0858 | 0.1736 | 1.868 | 0.0780 | 1032 | 1070 | 1147 |
| 304 | 3 (Rim) | 170 | 710 | 1004 | 0.0983 | 0.2107 | 2.486 | 0.0856 | 1233 | 1268 | 1329 |
| | 3A | 130 | 560 | 474 | 0.1230 | 0.2049 | 2.635 | 0.0933 | 1202 | 1310 | 1494 |
| | 4 (Core) | 160 | 760 | 228 | 0.1484 | 0.1775 | 2.205 | 0.0901 | 1053 | 1183 | 1428 |
| 308 | 1 | 360 | 1940 | 2006 | 0.0893 | 0.1874 | 2.132 | 0.0825 | 1107 | 1159 | 1257 |
| **7810/28 Paragneiss, Chappel Island** | | | | | | | | | | | |
| 227A | 1 | 110 | 370 | 8668 | 0.0869 | 0.2177 | 2.560 | 0.0853 | 1270 | 1289 | 1322 |
| 231 | 4 | 290 | 1500 | 5263 | 0.0868 | 0.2032 | 2.359 | 0.0842 | 1192 | 1230 | 1297 |
| | 3 (Core) | 50 | 250 | 1281 | 0.0969 | 0.1912 | 2.261 | 0.0858 | 1128 | 1200 | 1334 |
| 246 | 2 | 90 | 360 | 1017 | 0.1067 | 0.2360 | 2.987 | 0.0918 | 1366 | 1404 | 1463 |
| 234 | 1 | 130 | 230 | 1359 | 0.1157 | 0.3146 | 4.610 | 0.1063 | 1763 | 1751 | 1737 |
| 234A | 3 (Rim) | 240 | 890 | 7049 | 0.1130 | 0.2771 | 4.245 | 0.1111 | 1577 | 1683 | 1817 |
| | 2 (Core) | 190 | 380 | 1315 | 0.1871 | 0.4369 | 10.982 | 0.1823 | 2337 | 2522 | 2674 |
| | 1 (Core) | 90 | 180 | 1513 | 0.1937 | 0.4272 | 11.076 | 0.1881 | 2293 | 2530 | 2726 |

†Measured ratios.
*Corrected, assuming a common Pb composition $^{206}Pb/^{204}Pb = 16.0$, $^{207}Pb/^{204}Pb = 15.4$, $^{208}Pb/^{204}Pb = 36.0$.

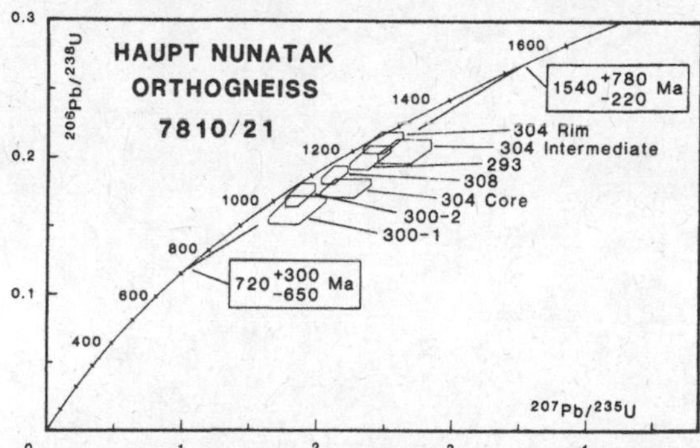


**Figure 3. Concordia diagram of ion microprobe U-Pb analyses of 30 μm spots on zircon crystals from an orthogneiss from Haput Nunatak previously analysed by Lovering et al. (1981). Uncertainties plotted are 2σ estimates of analytical precision.**

Five zircons from the Chappel Island paragneiss were analysed, two of which were not analysed previously. The analyses are plotted in Figure 4. To reduce the contribution of common Pb, some of the analyses were made adjacent to, rather than on, the original spots of Lovering et al. (1981). As with the Haupt Nunatak sample, the reanalyses do not confirm the wide range in $^{207}Pb/^{206}Pb$ apparent age reported by Lovering et al. (1981) (2296 ± 145 to 3090 ± 174 Ma), but instead suggest a zircon age of between 1400 and 1500 Ma. In contrast to the orthogneiss however, the paragneiss does contain significantly older zircon. The analysis of grain 234 is concordant at 1763 ± 103 Ma, and zircon 234A contains a core at least 2726 ± 30 Ma old, rimmed by zircon younger than 1817 ± 27 Ma. If the line of discordance defined by the three analyses of grain 234A is interpreted as a mixing line between two components, the ages of those components

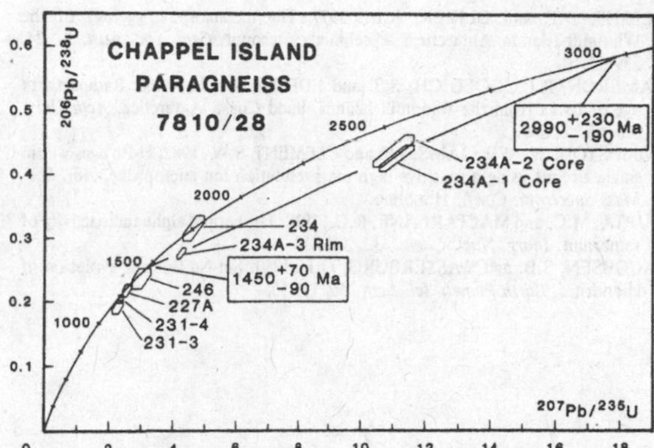


**Figure 4. Concordia diagram of ion microprobe U-Pb analyses of 30 μm spots on zircon crystals from a paragneiss from Chappel Island previously analysed by Lovering et al. (1981). Uncertainties plotted are 2σ estimates of analytical precision.**

are $1450^{+70}_{-90}$ (2σ) Ma (the same age as much of the other zircon in the rock) and $2990^{+230}_{-190}$ (2σ) Ma.

## CONCLUSIONS

No evidence was found to support the suggestion of Lovering et al. (1981) that the orthogneiss from Haupt Nunatak contains zircon derived from a 3500 Ma old terrain. Similarly, reanalyses of zircons from the Chappel Island paragneiss did not confirm the ages originally reported.

Two grains were found in the paragneiss sample, however, which are significantly older than the remainder. These are relict from the protolith of the paragneiss and indicate that it contained detritus ultimately derived from a source at least 2726 ± 30 Ma and probably $2990^{+230}_{-190}$ Ma old.

The primary ages of all the zircons analysed from the orthogneiss and the majority from the paragneiss (including the rim of the grain containing the oldest core) are not discernably different from the Pidgeon Island orthogneiss Rb-Sr and Sm-Nd age of ~ 1470 Ma. We consider this to be the age of the orthogneiss protolith and the time of major metamorphism of the sedimentary precursor of the paragneiss.

The discrepancy between the SHRIMP U-Pb analyses and the Pb data of Lovering et al. (1981) reflects the great difficulty of stripping the effects of isobaric interferences from low-resolution ion microprobe analyses. If common Pb is not subtracted from the total $^{207}Pb$ and $^{206}Pb$, the calculated Pb-Pb age will be too old. Lovering et al. (1981) assumed that common Pb was not present in the areas of zircon selected for analysis, whereas Table 2 (the $^{206}Pb/^{204}Pb$ column) shows that common Pb was found in every zircon. Thus, the previous Pb-Pb ages of Lovering et al. will be too old for this reason alone. However, it is unlikely that the common Pb correction is the sole explanation for Pb-Pb ages as great as 3500 Ma. The SHRIMP analyses contain a contribution from surface common Pb which probably was not present when the original analyses were done, but the $^{207}Pb/^{206}Pb$ apparent ages before common Pb correction are nevertheless not as high as those Lovering et al. (1981) reported. We conclude that the discrepancy arises from the failure of Lovering et al. (1981) to identify and strip all significant isobaric interferences from $^{207}Pb$. The inverse correlation between $^{207}Pb/^{206}Pb$ and U content originally reported may in part be due to the greater proportional effect of the unknown interference(s) on $^{207}Pb$ in samples with lower U and hence lower radiogenic Pb concentrations.

*Acknowledgements* Logistical support for K.D.C. and P.A. in the Windmill Islands was provided by the Australian National Antarctic Research Expedition. We thank Dr M. McCulloch for assistance with the Nd analyses. J.F.L. received financial support from the Australian Research Grants Committee.

## REFERENCES

ALLEGRE, C.J., BEN OTHMAN, D., POLVE, M. and RICHARD, P., 1978: The Nd-Sr isotopic correlation in mantle materials and geodynamic consequences. *Phys. Earth Planet. Int., 19*, 293-306.

ARRIENS, P.A., 1974: The Precambrian geochronology of Antarctica. Geol. Soc. Aust., 1st Geol. Convent., Abstracts 97-8.

BLIGHT, D.F. and OLIVER, R.L., 1977: The metamorphic geology of the Windmill Islands, Antarctica: a preliminary account. *Geol. Soc. Aust. J., 24,* 239-62.

CAMERON, R.L., GOLDICH, S.S. and HOFFMAN, J.H., 1960: Radioactivity age of rocks from the Windmill Islands, Budd Coast, Antarctica. *Acta Univ. Stockh., 6,* 1-6.

COMPSTON, W., WILLIAMS, I.S. and CLEMENT, S.W., 1982: U-Pb ages within single zircons using a sensitive high mass-resolution ion microprobe. *Am. Soc. Mass Spectrom. Conf.,* Honolulu.

GUPTA, M.C. and MACFARLANE, R.D., 1970: The natural alpha radioactivity of samarium. *Inorg. Nucl. Chem., 32,* 3425-32.

JACOBSEN, S.B. and WASSERBURG, G.J., 1980: Sm-Nd isotopic evolution of chondrites. *Earth Planet. Sci. Lett., 50,* 139-55.

LOVERING, J.F., TRAVIS, G.A., COMAFORD, D.J. and KELLY, P.R., 1981: Evolution of the Gondwana Archaean shield: zircon dating by ion microprobe, and relationships between Australia and Wilkes Land; *in* Glover, J.E. and Groves, D.I. (eds.) Archaean Geology. *Geol. Soc. Aust., Spec. Pub., 7,* 193-203.

MCINTYRE, G.A., BROOKS, C., COMPSTON, W. and TUREK, A., 1966: The statistical assessment of Rb-Sr isochrons. *J. Geophys. Res., 71,* 5459-68.

OLIVER, R.L., 1972: Some aspects of Antarctic-Australian geological relationships; *in* Adie, R.J. (ed.) *Antarctic Geology and Geophysics.* Universitetsforlaget, Oslo, 859-64.

STEIGER, R.H. and JAGER, E., 1977: Subcommission on geochronology: convention on the use of decay constants in geo- and cosmochronology. *Earth Planet. Sci. Lett., 36,* 359-62.

WEBB, A.W., MCDOUGALL, I. and COOPER, J.A., 1963: Potassium-argon dates from the Vincennes Bay region and Oates Land; *in* Adie, R.J. (ed.) *Antarctic Geology.* North Holland, Amsterdam, 597-600.

# LITHOLOGICAL AND Sr-Nd ISOTOPIC RELATIONSHIPS IN THE VESTFOLD BLOCK: IMPLICATIONS FOR ARCHAEAN AND PROTEROZOIC CRUSTAL EVOLUTION IN THE EAST ANTARCTIC

K.D. Collerson, *Research School of Earth Sciences, Australian National University, P.O. Box 4, Canberra, A.C.T. 2601, Australia.*

E. Reid, *Department of Geology, University of Tasmania, Box 252C, Hobart, Tasmania 7001, Australia.*

D. Millar, *Research School of Earth Sciences, Australian National University, P.O. Box 4, Canberra, A.C.T. 2601, Australia.*

M.T. McCulloch, *Research School of Earth Sciences, Australian National University, P.O. Box 4, Canberra, A.C.T. 2601, Australia.*

*Abstract* The structural and metamorphic evolution of the Precambrian Vestfold Block, eastern Antarctica (lat. 68°35′S, long 77°85′E) reflects the effects of several periods of granulite facies metamorphism and a complex deformational history which involved a number of periods of folding. The terrain is dominated by E-W striking units of garnetiferous semi-pelitic and pelitic gneiss, banded iron formation, quartzite and marble (the Chelnok supracrustal assemblage) which are tectonically interleaved with compositionally layered units of grey orthogneiss (termed the Mossel gneiss). From field evidence it is clear that the Mossel gneisses formed from metavolcanic protoliths (the Tryne metavolcanics) by the development of incipient trondhjemitic and tonalitic partial melts and by the intrusion of discrete sheets of tonalite and trondhjemite. Cutting this sequence are broad units of medium- to coarse-grained, gabbroic, dioritic, tonalitic, granodioritic and granitic augen gnesisses (the Crooked Lake gneiss), which were emplaced under synkinematic conditions. During the Early and Middle Proterozoic, the Vestfold Block was intruded by several major swarms of dolerite dykes. These are correlated on petrographic, geochemical and isotopic grounds with similar dykes in Enderby Land.

Sr and Nd isotopic data indicate that gneisses in the Vestfold Block experienced at least two granulite facies tectonothermal events during the Archaean (ca 2800-3000 Ma and ca 2400-2500 Ma ago). South of the Vestfold Block, in the Rauer Islands, gneiss and dyke relationships show the effect of reworking (deformation and in situ anatexis) during a Late Proterozoic granulite facies event ca 1100 Ma ago.

In this paper, we review the geological evolution of the Archaean Vestfold Block in the East Antarctic Shield and present preliminary Sr and Nd isotopic data for several of the major lithological units. The relative chronology between units discussed in this paper is largely based on observations made in areas of low finite strain which were recognised during the 1980/1981 field season.

## GEOLOGICAL SETTING AND BACKGROUND

The gneisses in the Vestfold Block comprise a number of distinctive units which can readily be mapped on a regional scale. They have been metamorphosed under granulite facies conditions and exhibit structural and microstructural evidence that they have experienced a complex polyphase deformation history. In an earlier study, Collerson and Arriens (1979) showed that some of the gneisses defined a Rb-Sr isochron equivalent to an age of 2477 ± 44 Ma with an initial $^{87}Sr/^{86}Sr$ ratio ($I_{Sr}$) of 0.7018 ± 3. In view of their extreme depletion in large ion lithophile (LIL) elements this was interpreted to date the last period of granulite facies metamorphism.

In a recent account of the geology and structural evolution of the area, Oliver et al., (1982) subdivided the gneisses into four major units: viz. layered grey gneiss and garnetiferous paragneiss with well-developed composite fabrics, and two groups of younger orthogneisses with simple fabrics. Throughout most of the Vestfold Block, these gneissic units strike east-west and dip moderately to steeply to the north. They are folded about "$F_2$" macroscopic closures, and according to Oliver et al., (1982) and Parker et al., (this volume) show the development of a strong "$S_2$" fabric. This fabric variably overprints intrafolial folds and "$S_1$" axial plane fabrics in the layered gneisses, commonly obliterating them completely. The younger orthogneisses were considered to have been emplaced synkinematically with this period of deformation (viz. "$D_2$").

Within the Vestfold Block, two areas, one near Lake Chelnok and another more extensive area west of Tryne Fjord (Figure 1), appear to have escaped the effects of the "$D_2$" deformation. In these localities, it is possible to resolve structural and lithological relationships which predate the dominant regionally developed "$D_2$" structural elements. As a result, it has been possible to revise the litho-chronological history discussed by Oliver et al., (1982).

### Revised Field Chronology

*Tryne metavolcanics.* The earliest units recognised in the Vestfold Block are a sequence of ultrabasic, basic, intermediate and acidic metavolcanic lithologies (the Tryne metavolcanics). These are commonly associated with minor semicontinuous layers of pelitic and semipelitic gneiss and schist, and with lenses of banded iron-formation. The metavolcanics are generally layered on scales ranging from 10's of centimetres to over one metre, as shown in Figure 2a. The layering is commonly transposed and is paralleled by a strong schistosity (either "$S_1$", or a combination of "$S_1$" and "$S_2$"). Cutting the metavolcanics are differentiated units of coarse-grained pyroxenite (and hornblendite), gabbro and leucogabbro, commonly with compositional layering which is probably of cumulate origin.

West of Tryne Fjord in the core of the macroscopic fold shown in Figure 1, exposures of Tryne metavolcanics appear to have escaped the effects of penetrative "$D_1$" and "$D_2$" deformation. In this area, the metabasic units are relatively massive and exhibit a distinctive fragmental structure defined by a variety of angular to subangular inclusions, up to 50 x 20 cm in size. Two main compositions are present; layered gneiss of crustal origin and ultramafic nodules which include, lherzolite, harzburgite and hornblendite. The latter are clearly of exotic origin and are not the result of boudinage. The nodular-fragmental texture is therefore probably a feature inherited from the volcanic protolith of the unit.

*Mossel gneiss.* Geological relationships are also present in this area which provide information about the origin of the layered grey gneisses in the Vestfold Block. This unit, termed the Mossel gneiss, is composed of a number of different components. These include quartzo-feldspathic orthogneiss, lenses of mafic granulite with a fine compositional layering, boudiné of relatively massive meta-pyroxenite, gabbro and leucogabbro, and discordant pegmatite veins containing cm-scale crystals of blue quartz, plagioclase and orthopyroxene. Two varieties of quartzo-feldspathic orthogneiss can be recognised in most exposures: a variety which is compositionally layered on a mm- to cm-scale and a relatively homogeneous variety. The layered variety displays sharp to diffuse boundaries between different compositions, contains rootless intrafolial folds and commonly exhibits a strong composite fabric. In contrast, the homogeneous orthogneissic component, which constitutes between 10% and 60% of some outcrops, occurs as finely foliated and lineated tonalitic and trondhjemitic sheets that are clearly intrusive into the layered gneiss.

On a regional scale, outcrops of Mossel gneiss preserve morphological differences due to variation in strain. In areas where deformation has been relatively weak, inclusions of mafic granulite appear as agmatitic trains in a felsic matrix. However, in areas of high strain, these mafic horizons are more strongly flattened, as illustrated in Figure 2f.

From field observations in the previously discussed area west of Tryne Fjord, it appears that part of the felsic component of the gneiss formed by melting and deformation of metabasic Tryne metavolcanic protoliths. During the initial stages of this process, mafic granulite partially melted to produce trondhjemitic and tonalitic compositions which net-veined the parent lithology, resulting in the formation of agmatitic structures (Figure 2b). As the amount of felsic melt increased, discordant veins and sheets were formed, commonly with inclusions of "restite", either as diffuse patches or sharply bounded blocks (Figure 2c). With increasing deformation, layered gneisses were formed, like those shown in Figure 2d and e. These gneisses were subsequently cut by sheets of homogeneous tonalitic gneiss and experienced further deformation and/or transposition to produce highly strained Mossel gneisses similar to those shown in Figure 2f.

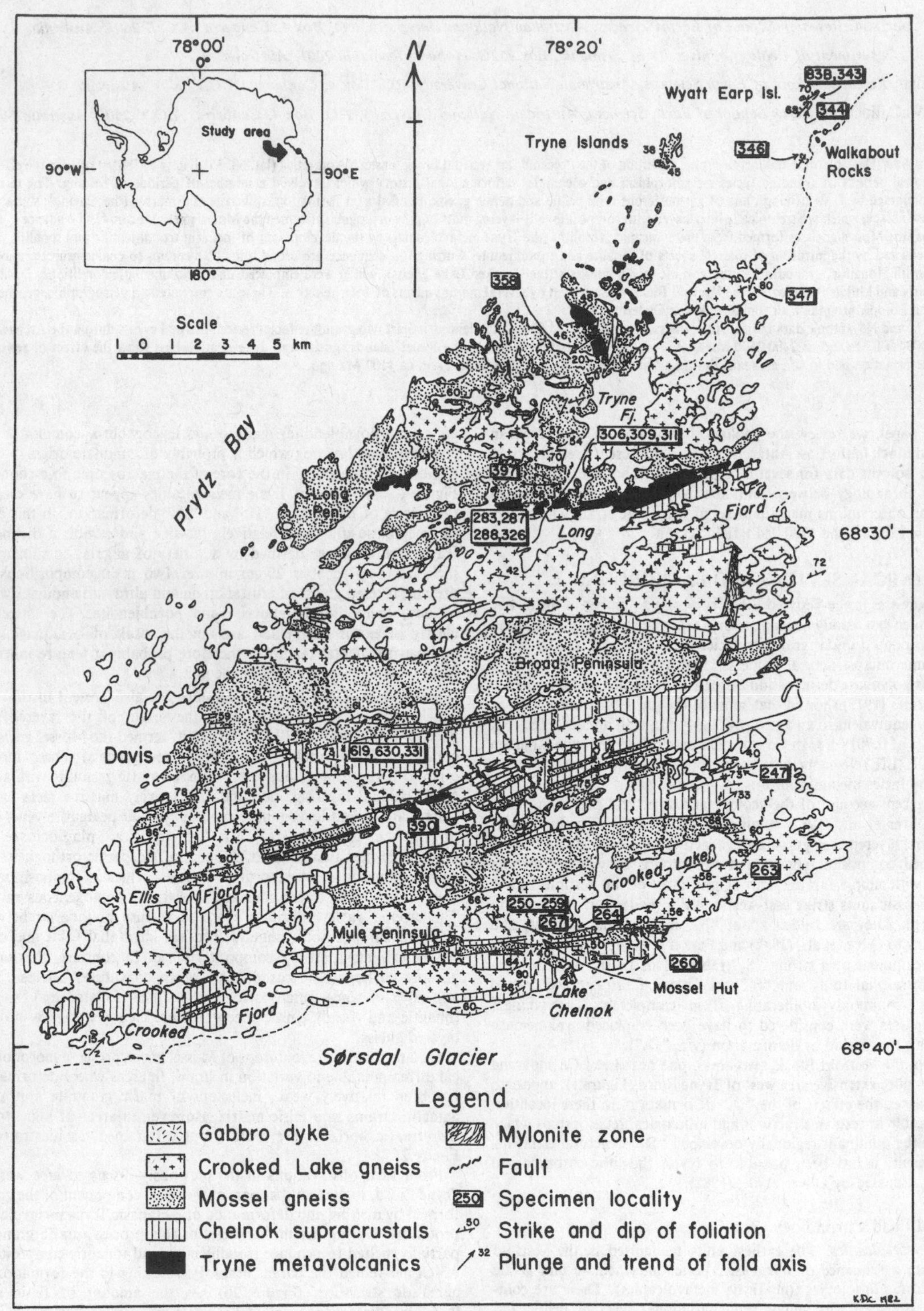


**Figure 1. Geological map of the Vestfold Block based on mapping by Oliver et al. (1982) and the authors. Shown on this map are the locations of specimens analysed in this study.**

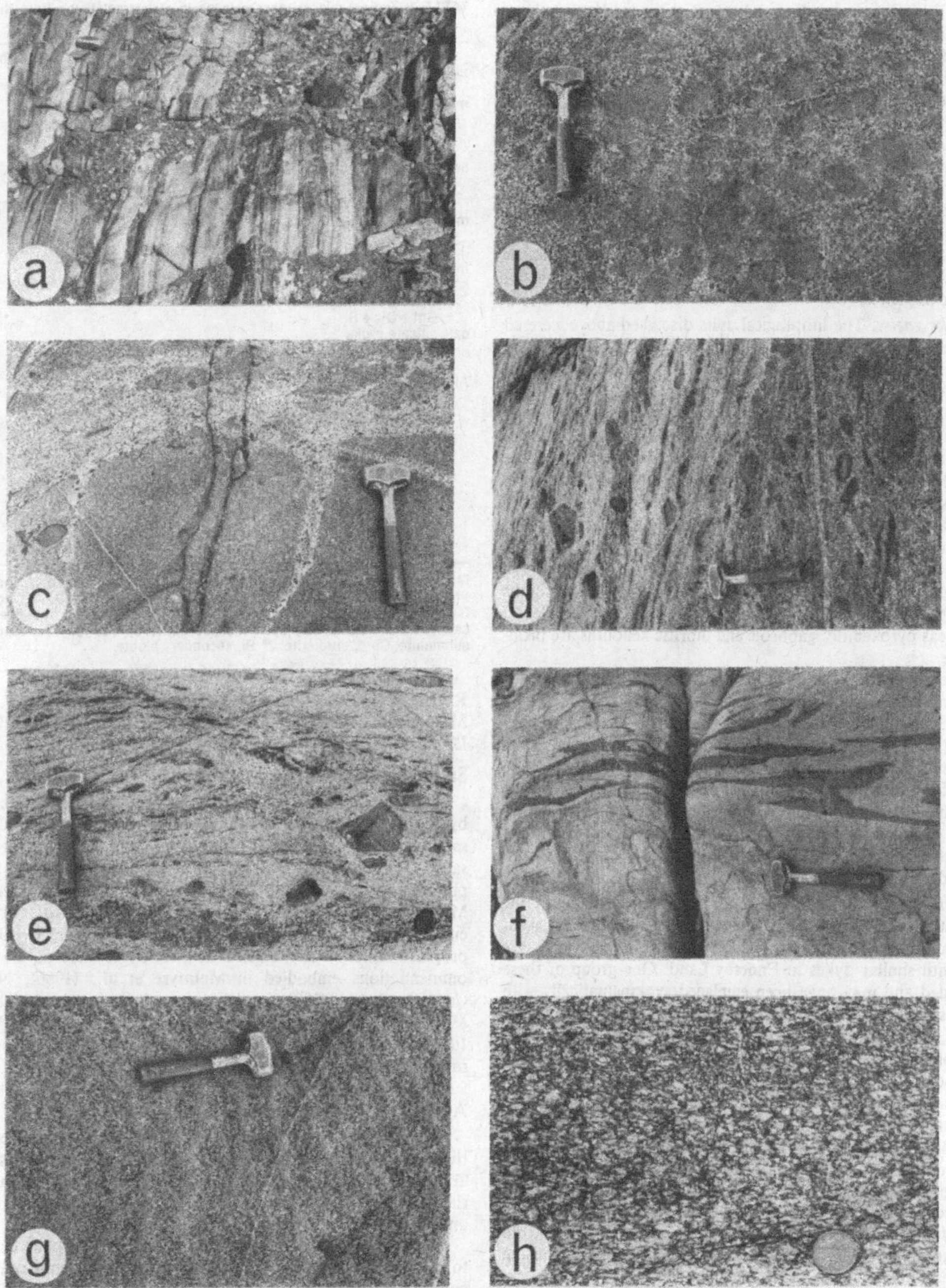


**Figure 2. Plate showing field characteristics of Archaean gneisses from the Vestfold Block. (a) Compositionally layered Tryne metavolcanic sequence near Tryne Crossing. (b) Metabasic Tryne metavolcanic showing incipient partial melts of trondhjemitic-tonalitic composition, northern Long Peninsula. (c) Trondhjemitic-tonalitic veins cutting ultramafic nodule bearing unit of Tryne metavolcanic, northern Long Peninsula. (d and e) Fragments of metabasic "restite" in tonalitic-trondhjemitic Mossel gneiss showing the progressive development of compositional layering, northern Long Peninsula. (f) Relatively homogeneous Mossel gneiss from high-strain domain, showing rootless intrafolial folds defined by disrupted meta-basic layers, Mule Peninsula. (g) Crooked Lake "gneiss" showing cm-scale dioritic and tonalitic to granodioritic cumulus layering. A weak "$S_2$" fabric intersects this layering at a high angle. West of Tryne Fjord.(h) Dioritic Crooked Lake gneiss showing 2 x 1 cm scale plagioclase augen and a simple anastomosing fabric, Crooked Lake.**

*Chelnok supracrustal assemblage.* Rocks belonging to the Chelnok supracrustal assemblage (termed the layered paragneisses by Oliver et al., 1982) form strongly foliated units of variable thickness which are commonly steeply dipping and highly lineated. Contacts with adjacent units of Mossel gneiss are of tectonic origin and presumably reflect folding and transposition during "$D_1$". The assemblage is dominated by metasedimentary compositions, although some belts also contain a minor metavolcanic (mafic granulite) component. The most abundant units are garnetiferous pelitic and semipelitic gneisses. These are generally associated with less abundant compositions such as psammites, quartzites, calc-silicates and banded iron-formations. In rare areas of low strain (eg. near Lake Chelnok, Figure 1) supracrustal units which exhibit cross-bedding and repeated graded layering have been mapped. These structures are believed to have been inherited from the sedimentary protoliths of the sequence.

The pelitic and semipelitic gneisses commonly contain weakly foliated coarse-grained peraluminous garnet-orthopyroxene bearing quartzo-feldspathic pegmatites. These range in width up to 40 cm and discordantly cut the gneisses, enclosing them as residual stringers. They are interpreted as having been formed by high-temperature anatexis during the waning stages of the "$D_2$" deformation.

On the basis of presently available data, it is not possible to resolve whether there is a genetic relationship between the Tryne metavolcanics and the Chelnok supracrustal assemblage, or whether the Chelnok supracrustal assemblage represents a separate sedimentary-volcanic sequence.

*Crooked Lake gneiss.* The lithological units discussed above were all variably affected by macroscopic folding, schistosity development and granulite facies metamorphism during the "$D_2$" tectonothermal event (Oliver et al., 1982). During this event, the complex was also intruded by a major group of apparently related medium to coarse grained orthogneisses, ranging from gabbroic, dioritic and tonalitic compositions to less abundant granodioritic and granitic varieties, termed the Crooked Lake gneiss (Figure 1). They form relatively continuous E-W striking belts, in places exceeding 6 km in width. Contacts are either sharp and concordant to slightly discordant or chaotic, with abundant xenoliths ranging from greater than 50 x 20 m to less than 1 x 0.5 m in size. Within these zones, the xenoliths are generally elongated and are aligned parallel to the foliation in the gneiss. However, the xenoliths vary in composition; some gneissic varieties are clearly of accidental origin, whereas pyroxenitic, gabbroic and dioritic xenoliths are probably cognate.

Unlike the "layered gneisses" which have composite fabrics, outcrops of Crooked Lake gneiss generally exhibit an anastomosing simple "$S_2$" fabric, which commonly forms augen around feldspar megacrysts (Figure 2h). The intensity of this fabric development is, however, quite variable; some domains are clearly dominated by intense L-fabrics (manifest as elongate mosaics of pyroxene, feldspar and sometimes quartz) whereas others are characterised more commonly by L-S and S-fabrics. In areas where "$D_2$" deformation has been relatively weak, gneisses exhibit cm-scale cumulate layering and a weak discordant "$S_2$" schistosity (Figure 2g).

*Dolerite dykes and Proterozoic reworking.* Subsequent to the emplacement of the Crooked Lake gneiss, the gneiss complex was cut by several major swarms of Proterozoic dolerite dykes which are correlated with similar dykes in Enderby Land. One group of these dykes is folded and may have been emplaced synkinematically with the waning stages of the "$D_2$" deformation. However, the majority define N-S, NE-SW striking suites which clearly post-date the last folding deformation recognised by Parker et al. (this volume) in the Vestfold Block.

In the south-western corner of the Vestfold Block on Broad and Mule Peninsulas, the mafic dykes have been extensively recrystallised. Although some of these dykes have foliated (sheared) margins, the majority exhibit fine-grained granoblasic microstructures, commonly containing poikiloblasts of garnet, clinopyroxene and quartz. This metamorphic overprint is believed to be related to the Late Proterozoic granulite facies event which had a major influence on Vestfold Block-type gneiss and dyke relationships in the Rauer Islands, causing deformation as well as in situ anatexis and the formation of charnockitic melts (Sheraton and Collerson, in press).

## Metamorphic Grade

A summary of typical mineral assemblages observed in metamorphic rocks from the Vestfold Block is given in Table 1. The majority of the assemblages, particularly in the relatively reactive supracrustal and metavolcanic compositions, indicate metamorphism under granulite facies conditions. However, evidence for a late stage post-"$D_2$" retrogression under amphibolite facies conditions is present in certain restricted areas. Isotopic data for Mossel gneisses affected by this process are given in Table 2. Further descriptions of the petrology of the different lithological units in the Vestfold Block are given in Collerson et al. (1983), and Sheraton and Collerson (in press).

**TABLE 1: Typical mineral assemblages in major rock-types from the Vestfold Block, Eastern Antarctica**

| |
|---|
| **Tryne metavolcanics (excluding metasedimentary component)** |
| ultramafic granulites |
| Opx + Cpx ± Phl ± Hb ± Pl |
| mafic granulites |
| Opx + Cpx + Pl |
| Opx + Cpx ± Pl ± Hb ± Bi |
| Cpx + Pl ± Hb ± Op |
| intermediate granulites |
| Opx + Cpx + Pl ± Bi ± Qtz |
| **Chelnok supracrustals (excluding metavolcanic component)** |
| meta-pelitic rocks |
| Gnt + Bi + Sp ( ± Cor) + Ksp/Meso ± Qtz ± Pl |
| Gnt + Bi + Si + Ksp |
| meta-psammo-pelitic rocks |
| Gnt + Pl + Ksp + Bi + Qtz |
| Gnt + Opx + Ksp/Meso + Pl + Qtz ± Bi |
| Gnt + Qtz + Bi |
| calc-silicate rocks |
| Ct + Fo + Di ± Phl ± Qtz |
| Sc + Di + Qtz ± Sph |
| Al-Mg-rich rocks |
| Sa + Phl + Sp |
| Cord + Opx + Qtz |
| **Mossel gneiss** |
| Qtz + Pl + Ksp + Opx |
| Qtz + Pl + Ksp + Opx + Hb ± 2°Bi |
| **Crooked Lake gneiss** |
| Plag + Opx + Cpx ± Qtz |
| Qtz + Pl + Ksp + Opx |
| Qtz + Pl + Ksp + Opx + Cpx + Hb |
| Qtz + Pl + Ksp + Fe-amph + Bi |

Mineral abbreviations: Opx, orthopyroxene; Cpx, clinopyroxene; Hb, hornblende; Phl, phlogopite; Bi, biotite; Op, opaque oxides; Qtz, quartz; Pl, plagioclase; Gnt, Garnet; Sp, spinel; Cor, corundum; Ksp, K-feldspar; Meso, mesoperthite; Ct, calcite; Fo, forsterite; Di, diopside; Sc, scapolite; Sph, sphene; Sa, sapphirine; Si, sillimanite; Cord, cordierite; 2° Bi, secondary biotite.

# Rb-Sr AND Sm-Nd GEOCHRONOLOGY OF THE TRYNE METAVOLCANICS, MOSSEL GNEISS AND CROOKED LAKE GNEISS

## Sampling and analytical methods

Samples used in this study were collected using feathers and wedges by K.D.C. and E.J.R. during the 1980-81 field season. Locations of samples are shown in Figure 1. Depending on the average grain size, sample weights ranged from ca 5 to 20 kg. Details of sample preparation and analytical techniques are given in Collerson and McCulloch (this volume). The isotopic data were statistically analysed either by the least squares regression method of York (1969) using a program which was modified at A.N.U. to incorporate the recommendations embodied in McIntyre et al., (1966), or by the Cameron et al., (1981) method. Ages have been calculated using $\lambda^{87} = 1.42 \times 10^{-10}y^{-1}$ for the decay constant of $^{87}Rb$ and $\lambda^{147} = 6.54 \times 10^{-12}y^{-1}$ for the decay constant of $^{147}Sm$. Errors in ages and initial ratios are quoted at the 95% confidence level ($2\sigma$).

## Analytical results

Rb-Sr and Sm-Nd concentrations and isotopic ratios for representative specimens of the Mossel gneiss, Crooked Lake gneiss and the Tryne metavolcanics are given in Tables 2 and 3. Isochron plots of these data are shown in Figure 3. The results of different regression analyses are given in Table 4.

From the concentration data in Table 2, it is apparent that the tonalitic and trondhjemitic Mossel gneisses have significantly lower concentrations of Rb (0.5 to 73.2 ppm) and Sr (219 to 442.7 ppm) than are present in the Crooked Lake gneiss (19.0 to 155.6 ppm Rb and 215 to 1162 ppm Sr). Furthermore, the Mossel gneisses have lower Rb/Sr ratios (Table 2) than the Crooked Lake gneisses, in spite of the fact that the latter group of gneisses is dominated by relatively basic compositions. This depletion in Rb relative to Sr and resulting development of unusually low Rb/Sr ratios is a common feature of felsic granulite facies gneisses in the lower continental crust (cf. Tarney and Windley, 1977). In fact, the mean Rb/Sr for Mossel gneisses from Mule Peninsula (0.0153) is even lower than the mean Rb/Sr reported by Holland and Lambert (1975) for Lewisian granulite facies gneisses (0.02). They are therefore, among the most depleted granulites yet recognised, with a mean range of Rb/Sr less than that estimated by Peterman (1979) for the mantle.

*Sr isotopic data.* Rb and Sr isotopic data for the majority of specimens of Mossel gneiss from Mule Peninsula (Figure 1) exhibit an extremely limited range in $^{87}Rb/^{86}Sr$ (0.0045—0.0526; Table 2). However, one specimen (81-260) collected from the same sequence of gneisses ca 4 km east of the other samples (Figure 2) is less depleted (0.1616) and provides a sufficient spread in $^{87}Rb/^{86}Sr$ to define an isochron. These data, which are plotted in Figure 3a, yield a McIntyre et al., (1966) Model III solution equivalent to an age of 3062 ± 212 Ma, with an initial $^{87}Sr/^{86}Sr$ ratio ($I_{Sr}$) of 0.7012 ± 2. However, an M.S.W.D. of 12.0 reflects excess scatter in the data, which cannot be accounted for by experimental error alone. When the samples with $^{87}Rb/^{86}Sr$ ratios less than 0.0526 are regressed, they yield an isochron which corresponds to an age of 2456 ± 163 Ma with an $I_{Sr}$ of 0.7015 ± 1 (Table 4). As these examples were collected from a single ca. 20 x 5 m exposure, near the nose of a macroscopic "$F_2$" structure (Figure 1), the data demonstrate that Sr isotopic equilibration occurred on an outcrop scale during a tectonothermal event ca 2400-2500 Ma ago.

In Table 2 data are also present for a group of Mossel gneisses which exhibit petrographic evidence of having undergone retrogression to amphibolite facies assemblages. They are slightly younger (2349 ± 50 Ma) and have a significantly higher $I_{Sr}$ viz. 0.7027 ± 1 (Table 4, Figure 3d) than the granulite facies Mossel gneisses. The initial Sr isotopic composition of the retrogressed gneisses, is markedly more radiogenic than expected (assuming 107 Ma of evolution from 2456 Ma to 2349 Ma). This suggests that radiogenic $^{87}Sr$ may have been added during the retrograde event. Local open-system isotopic behaviour is also observed in data for samples of Mossel gneiss from Broad Peninsula, and in regionally collected specimens of Tryne metavolcanics (Table 3). However, these lithologies are not retrogressed and the timing of the disturbance cannot be resolved.

Sr isotopic data for 11 regionally collected specimens of the geologically older Crooked Lake gneiss (Table 4) are poorly correlated (M.S.W.D. = 83.8). When plotted on an isochron diagram (Figure 3b), they yield a McIntyre et al., (1966) Model III solution equivalent to an age of 2454 ± 117 Ma, with a slightly higher $I_{Sr}$ (0.7021 ± 5) than the Mossel gneiss (cf. Table 4). However, when seven samples from a single outcrop are regressed separately, the result is significantly more precise. These data, which are shown in Figure 3c, correspond to an age of 2416 ± 21 Ma with an $I_{Sr}$ of 0.7022 ± 3 (Table 4).

*Nd isotopic results.* Nd isotopic data for Tryne metavolcanics, Mossel gneisses and Crooked Lake gneisses are listed in Table 3. Both groups of felsic gneisses have moderately fractionated REE patterns as is indicated by ΣNd (O) values between −22.5 and −41.0 (Mossel gneisses) and between −22.4 and −32.7 (Crooked Lake gneiss). In

**TABLE 2: Rb-Sr isotopic data for felsic orthogneisses from the Vestfold Block**

| Specimen No. | Rb (ppm) | Sr (ppm) | Rb/Sr | $^{87}Rb/^{86}Sr$(1) | $^{87}Sr/^{86}Sr$(2) |
|---|---|---|---|---|---|
| **Mossel gneisses from Mule Peninsula (granulite facies)** | | | | | |
| 81-252 | 0.490 | 315.4 | 0.0016 | 0.0045 | 0.70156 ± 4 |
| 81-253 | 0.598 | 307.7 | 0.0019 | 0.0056 | 0.70179 ± 4 |
| 81-250 | 2.71 | 316.3 | 0.0086 | 0.0248 | 0.70231 ± 4 |
| 81-251 | 2.86 | 294.7 | 0.0097 | 0.0280 | 0.70252 ± 4 |
| 81-256 | 4.20 | 335.7 | 0.0125 | 0.0361 | 0.70262 ± 4 |
| 81-258 | 4.19 | 325.6 | 0.0129 | 0.0371 | 0.70278 ± 4 |
| 81-257 | 4.41 | 299.1 | 0.0147 | 0.0425 | 0.70305 ± 7 |
| 81-254 | 4.92 | 311.2 | 0.0158 | 0.0456 | 0.70311 ± 4 |
| 81-255 | 5.05 | 316.4 | 0.0160 | 0.0460 | 0.70314 ± 6 |
| 81-259 | 8.07 | 442.7 | 0.0182 | 0.0526 | 0.70335 ± 6 |
| 81-260 | 18.9 | 337.3 | 0.0560 | 0.1616 | 0.70856 ± 2 |
| | | | x̄0.0153 ± 0.0146(3) | | |
| **Mossel gneisses east of Tryne Crossing showing retrogression to amphibolite facies assemblages** | | | | | |
| 81-338 | 2.66 | 379.0 | 0.0070 | 0.0202 | 0.70342 ± 6 |
| 81-335 | 2.46 | 275.7 | 0.0089 | 0.0258 | 0.70346 ± 6 |
| 81-332 | 5.52 | 430.3 | 0.0128 | 0.0371 | 0.70397 ± 6 |
| 81-334 | 2.83 | 218.8 | 0.0129 | 0.0374 | 0.70387 ± 4 |
| 81-339 | 3.58 | 235.7 | 0.0152 | 0.0438 | 0.70436 ± 8 |
| 81-337 | 4.24 | 228.1 | 0.0186 | 0.0536 | 0.70441 ± 2 |
| 81-340 | 5.67 | 303.4 | 0.0187 | 0.0539 | 0.70483 ± 6 |
| 81-336 | 10.0 | 284.2 | 0.0352 | 0.1016 | 0.70622 ± 6 |
| 81-333 | 73.3 | 333.3 | 0.2199 | 0.6362 | 0.72430 ± 2 |
| | | | x̄0.0388 ± 0.0684 | | |
| **Crooked Lake gneisses from different localities in the Vestfold Hills** | | | | | |
| 81-343 | 19.0 | 840.2 | 0.0226 | 0.0656 | 0.70510 ± 6 |
| 81-347 | 20.5 | 834.6 | 0.0246 | 0.0709 | 0.70498 ± 8 |
| 81-288 | 53.1 | 1162.1 | 0.0457 | 0.1319 | 0.70666 ± 4 |
| 81-263 | 35.2 | 734.4 | 0.0479 | 0.1385 | 0.70642 ± 2 |
| 81-247 | 31.3 | 534.4 | 0.0586 | 0.1689 | 0.70794 ± 6 |
| 81-267 | 53.9 | 815.8 | 0.0661 | 0.1908 | 0.70816 ± 6 |
| 81-83B | 67.4 | 812.1 | 0.0830 | 0.2397 | 0.71130 ± 10 |
| 81-264 | 63.8 | 732.7 | 0.0871 | 0.2514 | 0.71046 ± 5 |
| 81-344 | 66.6 | 666.3 | 0.1000 | 0.2887 | 0.71253 ± 4 |
| 81-283 | 81.4 | 590.5 | 0.1378 | 0.3987 | 0.71611 ± 3 |
| 81-287 | 98.8 | 359.5 | 0.2748 | 0.7953 | 0.73045 ± 5 |
| | | | x̄0.0862 ± 0.0711 | | |
| **Crooked Lake gneisses occurring as sheets cutting Tryne metavolcanics at Tryne Crossing** | | | | | |
| 81-302 | 35.7 | 1004.1 | 0.0356 | 0.1025 | 0.70567 ± 4 |
| 81-297 | 40.5 | 980.4 | 0.0413 | 0.1194 | 0.70641 ± 5 |
| 81-303 | 43.1 | 939.7 | 0.0459 | 0.1325 | 0.70671 ± 4 |
| 81-298 | 48.8 | 870.1 | 0.0561 | 0.1621 | 0.70777 ± 6 |
| 81-301 | 54.2 | 843.6 | 0.0642 | 0.1856 | 0.70864 ± 4 |
| 81-299 | 155.6 | 224.0 | 0.6946 | 2.0181 | 0.77298 ± 4 |
| 81-300 | 152.0 | 215.1 | 0.7066 | 2.0539 | 0.77339 ± 6 |
| | | | x̄0.2349 ± 0.3183 | | |

(1)Coefficient of variation 0.5% (2σ).
(2)Errors quoted at 2σ confidence level.
(3)Mean and standard deviation.

**TABLE 3: Sr and Nd isotopic data for granulite facies gneisses from the Vestfold Block**

| Specimen No. | Rb (ppm) | Sr (ppm) | $^{87}Rb/^{86}Sr$ | $^{87}Sr/^{86}Sr$ | Sm (ppm) | Nd (ppm) | $^{147}Sm/^{144}Nd$ | $^{143}Nd/^{144}Nd$* | $\epsilon_{Nd}(0)$+ | $T^{Nd}_{CHUR}$++ (Ma) |
|---|---|---|---|---|---|---|---|---|---|---|
| **Tryne metavolcanics** | | | | | | | | | | |
| 81-358 | 1.25 | 550.1 | 0.0065 | 0.70432 ± 4 | 4.06 | 24.06 | 0.1020 | 0.510174 ± 30 | −32.5 | 2659 |
| 81-390 | 75.4 | 133.4 | 1.6398 | 0.75876 ± 6 | 5.10 | 25.76 | 0.1199 | 0.510620 ± 28 | −23.8 | 2400 |
| 81-309 | 26.5 | 288.5 | 0.2650 | 0.71295 ± 5 | 2.62 | 12.59 | 0.1257 | 0.510768 ± 36 | −20.9 | 2283 |
| 81-326 | 1.40 | 188.1 | 0.0215 | 0.70230 ± 4 | 5.23 | 23.78 | 0.1331 | 0.510878 ± 22 | −18.7 | 2287 |
| 81-306 | 4.11 | 170.8 | 0.0692 | 0.70458 ± 4 | 4.63 | 15.63 | 0.1791 | 0.511585 ± 20 | −4.90 | 2165 |
| 81-311 | 11.4 | 140.2 | 0.2338 | 0.70951 ± 5 | 1.69 | 5.59 | 0.1832 | 0.511858 ± 30 | +0.44 | N.C. |
| **Mossel gneisses** | | | | | | | | | | |
| 81-397 | 56.9 | 99.5 | 1.6614 | 0.76169 ± 3 | 1.27 | 10.73 | 0.0714 | 0.509737 ± 16 | −41.0 | 2540 |
| 81-260 | —See Table 2— | | | | 1.95 | 12.40 | 0.0948 | 0.510182 ± 18 | −32.3 | 2463 |
| 72-630 | 4.81 | 285.6 | 0.0486 | 0.70785 ± 2 | 5.86 | 34.48 | 0.1027 | 0.510186 ± 30 | −32.2 | 2661 |
| 72-619 | 1.33 | 266.7 | 0.0144 | 0.70675 ± 2 | 8.05 | 45.60 | 0.1068 | 0.510239 ± 20 | −31.2 | 2690 |
| 81-331 | 12.7 | 447.6 | 0.0820 | 0.70854 ± 2 | 3.11 | 16.11 | 0.1168 | 0.510467 ± 20 | −26.8 | 2599 |
| 81-253 | —See Table 2— | | | | 5.41 | 26.44 | 0.1238 | 0.510684 ± 26 | −22.5 | 2397 |
| **Crooked Lake gneisses** | | | | | | | | | | |
| 81-283 | —See Table 2— | | | | 11.4 | 73.7 | 0.0937 | 0.510164 ± 12 | −32.7 | 2462 |
| 81-287 | —See Table 2— | | | | 4.15 | 25.8 | 0.0973 | 0.510223 ± 20 | −31.5 | 2462 |
| 81-344 | —See Table 2— | | | | 5.78 | 34.8 | 0.1004 | 0.510242 ± 18 | −31.1 | 2511 |
| 81-267 | —See Table 2— | | | | 7.45 | 42.6 | 0.1058 | 0.510343 ± 12 | −29.2 | 2492 |
| 81-346 | 46.32 | 765.5 | 0.1747 | 0.70833 ± 2 | 7.84 | 42.8 | 0.1108 | 0.510417 ± 20 | −27.7 | 2506 |
| 81-247 | —See Table 2— | | | | 4.41 | 20.98 | 0.1270 | 0.510687 ± 24 | −22.4 | 2498 |

*Normalised to $^{146}Nd/^{142}Nd = 0.636151$ N.C.—not calculated.

$$^{+}\epsilon_{Nd}(0) = \left[\frac{(^{143}Nd/^{144}Nd)\ meas}{(^{143}Nd/^{144}Nd)^{\circ}_{CHUR}} - 1\right] \times 10^4$$

$$^{++}T^{Nd}_{CHUR} = \frac{1}{\lambda}\ln\left[1 + \frac{(^{143}Nd/^{144}Nd)\ meas - (^{143}Nd/^{144}Nd)^{\circ}_{CHUR}}{(^{147}Sm/^{144}Nd)\ meas - (^{147}Sm/^{144}Nd)_{CHUR}}\right]$$

where $(^{143}Nd/^{144}Nd)^{\circ}_{CHUR} = 0.511836$; $(^{147}Sm/^{144}Nd)_{CHUR} = 0.1967$; and $\lambda_{Sm} = 6.54 \times 10^{-12} y^{-1}$.

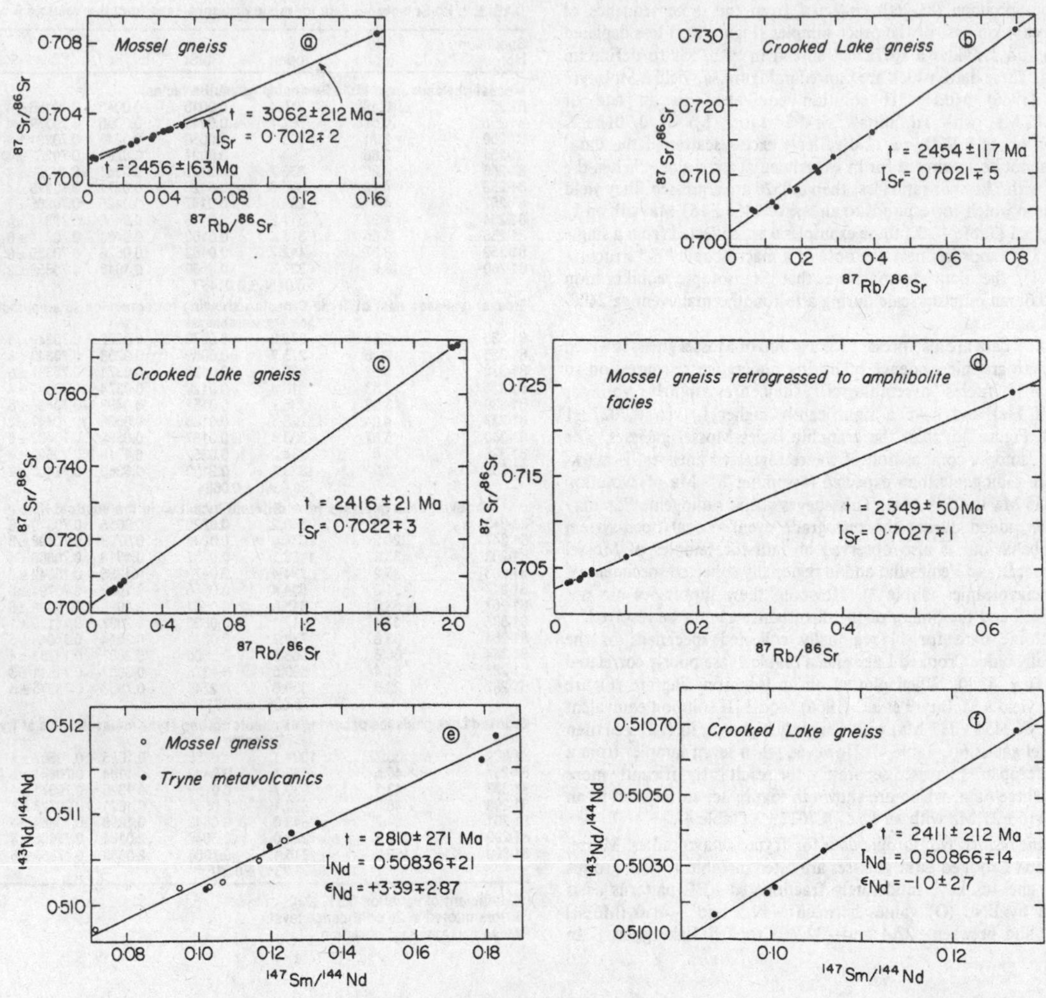


**Figure 3. (a) Rb-Sr evolution diagram for Mossel gneiss from Mule Peninsula. (b) Rb-Sr evolution diagram for samples of regionally collected Crooked Lake gneiss. (c) Rb-Sr evolution diagram for a suite of Crooked Lake gneiss from a single outcrop. (d) Rb-Sr evolution diagram for a suite of Mossel gneiss which shows retrogression to amphibolite facies assemblages. (e) Sm-Nd evolution diagram for samples of Tryne metavolcanics and Mossel gneiss. (f) Sm-Nd evolution diagram for regionally collected samples of Crooked Lake gneiss.**

**TABLE 4: Summary of regression data for felsic and mafic orthogneisses from the Vestfold Block, Antarctica**

| Suite | No. of Samples | MSWD | Age (Ma) | Initial $^{87}$Sr/$^{86}$Sr | Model | $\Sigma_{Nd}$* |
|---|---|---|---|---|---|---|
| **Sr Isotopic Data** | | | | | | |
| Granulite facies Mossel gneisses Mule Peninsula | 11 | 12 | 3062 ± 212 | 0.7012 ∓ 0.0002 | McIntyre et al., III | |
| Granulite facies Mossel gneisses Mule Peninsula | 10 | 2.4 | 2456 ± 163 | 0.7015 ∓ 0.0001 | McIntyre et al., I | |
| Crooked Lake gneisses from different localities | 11 | 83.8 | 2454 ± 117 | 0.7021 ∓ 0.0005 | McIntyre et al., III | |
| Crooked Lake gneisses from single locality near Tryne Crossing | 7 | 3.5 | 2416 ± 21 | 0.7022 ∓ 0.0003 | McIntyre et al., III | |
| Retrogressed Mossel gneisses Tryne Crossing area | 9 | 8.7 | 2349 ± 50 | 0.7027 ∓ 0.0001 | McIntyre et al., III | |
| **Nd Isotopic Data** | | | | | | |
| Mossel gneisses | 6 | 39.5 | 2488 ± 600 | 0.50856 ∓ 0.00038 | McIntyre et al., II | − 1.0 ± 6.1 |
| | | | $2599^{+1160}_{-473}$ | $0.50849^{-0.00075}_{+0.00031}$ | Cameron et al. (F.L.) | $+0.5^{+15.2}_{-5.0}$ |
| Tryne metavolcanics | 6 | 64.4 | 2923 ± 570 | 0.50827 ∓ 0.00049 | McIntyre et al., II | + 4.5 ± 5.0 |
| | | | $2999^{+830}_{-451}$ | $0.50820^{-0.00072}_{+0.00040}$ | Cameron et al., (F.L.) | $+5.1^{+7.3}_{-3.7}$ |
| Mossel gneisses—Tryne metavolcanics | 12 | 50.7 | 2810 ± 271 | 0.50836 ∓ 0.00021 | McIntyre et al., IV | + 3.4 ± 2.9 |
| | | | $2859^{+355}_{-256}$ | $0.50832^{-0.00026}_{+0.00019}$ | Cameron et al., (F.L.) | $+3.8^{+4.0}_{-2.8}$ |
| Crooked Lake gneisses | 6 | 1.94 | 2411 ± 212 | 0.50866 ∓ 0.00014 | McIntyre et al., III | − 1.0 ± 2.7 |

$$^{*}\Sigma_{Nd} = \left[\frac{(^{143}Nd/^{144}Nd)^{T}_{INIT}}{(^{143}Nd/^{144}Nd)^{T}_{CHUR}} - 1\right] \times 10^{4}$$

where $(^{143}Nd/^{144}Nd)^{T}_{CHUR} = (^{143}Nd/^{144}Nd)^{\circ}_{CHUR} - (^{147}Sm/^{144}Nd)^{\circ}_{CHUR}(e^{\lambda t} - 1)$
and $(^{143}Nd/^{144}Nd)^{\circ}_{CHUR} = 0.511836$, $(^{147}Sm/^{144}Nd)^{\circ}_{CHUR} = 0.1967$, and $\lambda^{Sm} = 6.54 \times 10^{-12} y^{-1}$.

contrast, the Tryne metavolcanics exhibit moderately fractionated to almost flat REE patterns (ΣNd (O) − 32.5 to + 0.44). The samples of Mossel gneiss have a much wider range of $T^{Nd}_{CHUR}$ model ages of 2397 to 2690 Ma (mean 2558 Ma) than the specimens of Crooked Lake gneiss; 2462 to 2511 Ma, (mean 2488 Ma). The $T^{Nd}_{CHUR}$ model ages for the Tryne metavolcanics show and even greater range; 2165 to 2659 Ma, (mean 2358 Ma).

Data for the Tryne metavolcanics and Mossel gneisses are poorly correlated (Figure 3e). When they are regressed separately, the two populations yield ages of ca 3000 Ma and 2500—2600 Ma, respectively (Table 4). However, these determinations have large uncertainties due to internal scatter. From field relationships there is evidence that the Mossel gneisses may be genetically related to the Tryne metavolcanics. To assess this possibility both populations have been regressed together. The result of this analysis yields an age of 2810 ± 271 Ma and an initial $^{143}Nd/^{144}Nd$ ratio ($I_{Nd}$) of 0.50836 ± 21 by the McIntyre et al., (1966) method. However, they yield a marginally older result of $2859^{+335}_{-256}$ Ma with an $I_{Nd}$ of $0.50832^{+26}_{-19}$ by the Cameron et al., (1981) free-line method, assuming that local isotopic equilibration occurred 2500 Ma ago. The $I_{Nd}$'s obtained from the 2810 ± 271 Ma and $2859^{+335}_{-256}$ Ma isochrons both lie above the CHUR evolution curve (ΣNd = + 3.4 ± 2.9 and $3.8^{+4.0}_{-2.8}$, respectively). These ages are in excellent agreement with preliminary Cameron et al. regression results for Sr isotopic data for a suite of ten Chelnok supracrustal lithologies, viz. $2774^{+353}_{-278}$, $I_{Sr} = 0.7021^{+15}_{-12}$.

Data for the Crooked Lake gneisses are somewhat better correlated (Figure 3f) and yield an isochron which corresponds to an age of 2411 ± 212 Ma and $I_{Nd}$ of 0.50866 ± 14 (ΣNd = − 1.0 ± 2.7) by the McIntyre et al., (1966) method.

## DISCUSSION

It is significant that there is excellent agreement between the Rb-Sr and Sm-Nd isochron ages for the Crooked Lake gneiss, and the mean $T^{Nd}_{CHUR}$ model age for this unit. However, there is a major discrepancy between the $T^{Nd}_{CHUR}$ model ages for both the Tryne metavolcanics and the Mossel gneisses, and the minimum estimates of the age of these lithologies based on Rb-Sr and Sm-Nd isochrons. Studies by Jacobsen and Wasserburg (1978), Hamilton et al., (1979) and DePaolo (1981) have suggested that Sm and Nd do not fractionate significantly during granulite facies metamorphic events. If this is valid, then the discrepancy between the ages may indicate that the Tryne metavolcanics and the Mossel gneisses were not derived from a primitive CHUR-type mantle, but were derived from a depleted mantle source, with an elevated Sm/Nd ratio. Such a model is supported by the regression results. Alternatively, if fractionation of Sm and Nd does occur under high-grade metamorphic conditions (cf. DePaolo et al., 1982; McCulloch and Black, 1982) then the range of model ages may reflect the influence of such a process, possibly coupled with derivation from a depleted mantle (cf. DePaolo et al., 1982).

### Evidence for crustal pre-history in the Vestfold Block

*Sr isotopic evolution.* Figure 4a shows an $I_{Sr}$-T diagram for the Mossel gneisses and Crooked Lake gneisses. Also shown is the evolution-line of a hypothetical mantle source (Rb/Sr = 0.02) and that of average, 3000 Ma old, continental crust (Rb/Sr = 0.13). In view of the fact that the Mossel gneisses are more depleted than a hypothetical mantle source (Rb/Sr-0.015 cf. 0.02), the strontium isotopic data do not provide any firm constraints on the maximum crustal residence of their protoliths, prior to 3000 Ma. However, the low $I_{Sr}$ ratio exhibited by the ca. 3000 Ma-old Mossel gneisses (0.7012) necessitates that if there was any significant pre-3000 Ma history, then the protoliths of the gneisses must have had an extremely low Rb/Sr. In contrast, the Crooked Lake gneiss has a higher average Rb/Sr than the Mossel gneiss (0.12 cf. 0.015). Extrapolation along this vector from the $I_{Sr}$ of the suite to the mantle growth curve suggests a relatively short crustal residency for the juvenile protoliths of the gneisses. In view of the relatively radiogenic nature of the $I_{Sr}$ of the Crooked Lake gneisses (cf. the $I_{Sr}$ of the Mossel gneisses), partial melting of depleted (low Rb/Sr) pre-existing crust is unlikely to have played a significant role in the genesis of the precursors of the suite.

*Nd isotopic evolution.* $I_{Nd}$-T data for the Tryne metavolcanic-Mosel gneiss assemblage and the Crooked Lake gneisses are illustrated in Figure 4b. Also shown is the evolutionary trajectory for the 3650 Ma-old Uivak-Nulliak suite from Labrador and a number of other

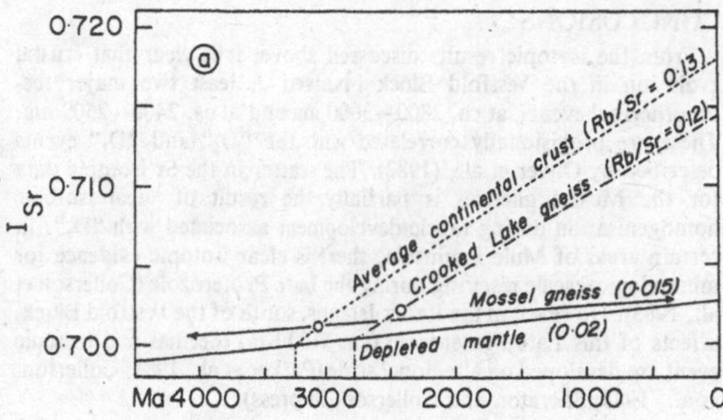


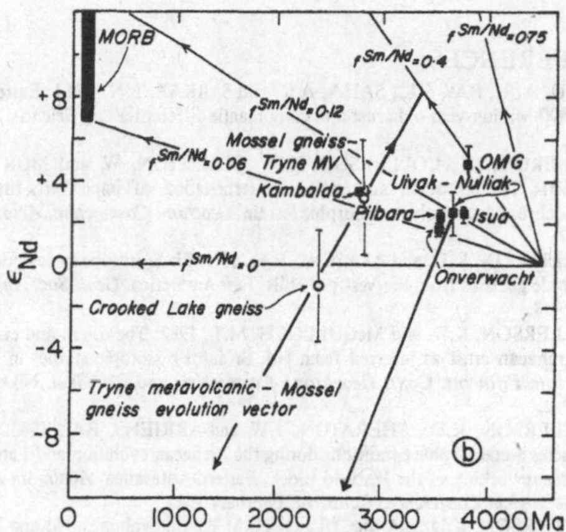


**Figure 4.** (a) Time *versus* initial $^{87}Sr/^{86}Sr$ ratio diagram showing the evolution of the Mossel gneiss and Crooked Lake gneiss. (b) $\Sigma_{Nd}$-T diagram showing evolutionary data for the Tryne metavolcanic—Mossel gneiss assemblage and the Crooked Lake gneiss. The evolutionary vector for the Tryne metavolcanic-Mossel gneiss assemblage was calculated assuming an average $^{147}Sm/^{144}Nd$ = 0.1216. Comparative data are from McCulloch and Compston (1981), Basu et al. (1981), McCulloch (1982) and Collerson and McCulloch (1981). $f^{Sm/Nd}$ is the depletion factor of the Sm/Nd ratio of a suite of rocks of hypothetical mantle composition, relative to that of the primitive mantle (CHUR). It is given by the expression

$$f^{Sm/Nd} = \left[ \frac{(^{147}Sm/^{144}Nd)\ \text{suite}}{(^{147}Sm/^{144}Nd)_{CHUR}} - 1 \right]$$

where $(^{147}Sm/^{144}Nd)_{CHUR} = 0.1967$.

Archaean sequences for comparison. Although there is a relatively large (2σ) error in ΣNd, the data for the Tryne metavolcanic-Mossel gneiss assemblage, nevertheless, show that the suite evolved from a long-lived slightly LREE depleted (high Sm/Nd) source. If the Sm and Nd are not fractionated during metamorphic processes, then the maximum possible pre-history for the protolith of the suite, estimated by extrapolating the average Sm/Nd vector of the suite back in time to the depleted mantle vector through the Older Metamorphic Group (OMG) in India (Basu et al., 1981) is ca 300-400 Ma. However, this scenario would require extremely depleted mantle with ΣNd ~ + 8 at ca 3300 Ma. Mantle of this character has not as yet been confirmed. If the LREE's were enriched by crustal processes prior to 2800 Ma, then a prehistory of up to ca. 500 Ma is also possible. However, in view of the low $I_{Sr}$ and positive ΣNd value of the Tryne-Mossel gneiss suite an extended crustal prehistory before ca. 2800—3000 Ma is considered unlikely.

The Nd isotopic evolution of the Crooked Lake gneisses, depicted in Figure 4, can be explained by; (1) a juvenile addition to the crust derived from a CHUR-type reservoir, (2) a juvenile addition derived from a depleted mantle source but contaminated with older continental crustal melts, or (3) melting of pre-existing crust. However, in the latter scenario, it is difficult to explain the Sr-isotopic data and also the origin of the gabbroic and dioritic members of the suite.

## CONCLUSIONS

From the isotopic results discussed above, it is clear that crustal evolution in the Vestfold Block involved at least two major tectonothermal events, at ca. 2800—3000 ma and at ca. 2400—2500 ma. These are provisionally correlated with the "$D_1$" and "$D_2$" events described by Oliver et al., (1982). The scatter in the Sr isotopic data for the Mossel gneisses is partially the result of local isotopic homogenisation during fabric development associated with "$D_2$". In certain areas of Mule Peninsula, there is clear isotopic evidence for mineralogical scale resetting during the Late Proterozoic (Collerson et al., 1983). However, in the Rauer Islands, south of the Vestfold Block, effects of this Late Proterozoic (ca. 1100 Ma) thermal and tectonic event are developed on a regional scale (Parker et al., 1981; Collerson, et al., 1983; Sheraton and Collerson, in press).

*Acknowledgements.* The authors gratefully acknowledge support by the Australian National Antarctic Research Expeditions (ANARE).

## REFERENCES

BASU, A.R., RAY, S.L., SAHA, A.K. and SARKAR, S.N., 1981: Eastern Indian 3800-Million-year-old crust and early mantle differentiation. *Science, 212,* 1502-6.

CAMERON, M., COLLERSON, K.D., COMPSTON, W. and MORTON. R., 1981: The statistical analysis and interpretation of imperfectly-fitted Rb-Sr isochrons from polymetamorphic terrain. *Geochim. Cosmochim. Acta, 45,* 1087-97.

COLLERSON, K.D. and ARRIENS, P.A., 1979: Rb-Sr isotope systematics in high-grade gneisses from the Vestfold Hills, East Antarctica. *Geol. Soc. Aust., J., 26,* 267-8.

COLLERSON, K.D. and McCULLOCH, M.T., 1982: The origin and evolution of Archaean crust as inferred from Nd, Sr and Pb isotopic studies in Labrador. *Abstr. Fifth Int. Conf. Geochron., Cosmochim. and Iso. Geol.,*Nikko, Japan, 61-2.

COLLERSON, K.D., SHERATON, J.W. and ARRIENS, P.A., 1983: Granulite facies metamorphic conditions during the Archaean evolution and Late Proterozoic reworking of the Vestfold Block, Eastern Antarctica. *Abstr. 6th Australian Geological Conference, Canberra,* February 1983.

COLLERSON, K.D. and McCULLOCH, M.T., (this volume): Nd and Sr isotope geochemistry of leucite bearing lavas from Gaussberg, Eastern Antarctica.

DePAOLO, D.J., 1981: Neodymium isotopes in the Colorado Front Range and crust-mantle evolution in the Proterozoic. *Nature, 291,* 193-6.

DePAOLO, D.J., MANTON, W.I., GREW, E.S. and HALPERN, M., 1982: Sm-Nd, Rb-Sr and U-Th-Pb systematics of granulite facies rocks from Fyfe Hills, Enderby Land, Antarctica. *Nature, 298,* 614-8.

HAMILTON, P.J., EVENSEN, N.M., O'NIONS, R.K. and TARNEY, J., 1979: Sm-Nd systematics of Lewisian gneisses: implications for the origin of granulites. *Nature, 277,* 25-8.

HOLLAND, J.G. and LAMBERT, R. St.J., 1975: The chemistry and origin of the Lewisian gneisses of the Scottish mainland; the Scourie and Inver assemblages and sub-crustal accretion. *Precambrian Res.,* 2, 161-74.

JACOBSEN, S.B. and WASSERBURG, G.J., 1978: Interpretation of Nd, Sr and Pb isotope data from Archaean migmatites in Lofoten-Vesterålen, Norway. *Earth Planet. Sci. Lett., 41,* 245-53.

McCULLOCH, M.T. and COMPSTON, W., 1981: Sm-Nd age of Kambalda and Kanowna greenstones and heterogeneity in the Archaean mantle. *Nature, 294,* 332-7.

McCULLOCH, M.T., 1982: Identifiction of the Earth's earliest differentiates. *Abstr. Fifth Int. Conf. Geochron:, Cosmochim. and Iso. Geol.,* Nikko, Japan, 61-2.

McCULLOCH, M.T. and BLACK, L.P., 1982: Sm-Nd isotopic systematics of Enderby Land granulites: evidence for the redistribution of Sm and Nd during metamorphism. *Fourth Int. Symp. Antarc. Earth Sci., Abstracts,*Adelaide, 117.

McINTYRE, G.A., BROOKS, C., COMPSTON, W. and TUREK, A., 1966: The statistical assessment of Rb-Sr isochrons. *J. Geophys. Res., 71,* 5459-68.

OLIVER, R.L., JAMES, P.R., COLLERSON, K.D. and RYAN, A.B., 1982: Precambrian geologic relationships in the Vestfold Hills, Antarctica; *in* Craddock, C. (ed.) *Antarctic Geoscience.* Univ. Wisconsin Press, Madison, 435-44.

PARKER, A.J., JAMES, P.R., MIELNIK, V. and OLIVER, R.L., (this volume): Metamorphism and fabric development in Archaean gneisses of the Vestfold Hills, Antarctica.

PARKER, A.J., JAMES, P.R., MIELNIK, V. and OLIVER, R.L., 1981: Superposed folding: a classic example in East Antarctica. *Geol. Soc. Aust., Abstr., 3,* 93-4.

PETERMAN, S.E., 1979: Strontium isotope geochemistry of Late Archaean to Late Cretaceous tonalites and trondhjemites; *in* Barker, F. (ed) *Trondhjemites, Dacites and Related Rocks.* Elsevier, New York, 133-47.

SHERATON, J.W. and COLLERSON, K.D. (in press): Archaean and Proterozoic metamorphic relationships in the Vestfold Hills-Prydz Bay area, Antarctica. *BMR J. Aust., Geol. Geophys.*

TARNEY, J. and WINDLEY, B.F., 1977: Geochemistry, thermal gradients and evolution of the lower continental crust. *Geol. Soc. Lond., J., 134,*153-72.

YORK, D., 1969: Least squares fitting of a straight line with correlated errors. *Earth Planet. Sci. Lett., 5,* 320-4.

# STRUCTURE, FABRIC DEVELOPMENT AND METAMORPHISM IN ARCHAEAN GNEISSES OF THE VESTFOLD HILLS, EAST ANTARCTICA

A.J. Parker, *S.A. Department of Mines and Energy, PO Box 151, Eastwood, South Australia, 5063.*

P.R. James and R.L. Oliver, *Department of Geology and Mineralogy, University of Adelaide, South Australia, 5000.*

V. Mielnik, *Mines Administration, 170 Greenhill Road, Parkside, South Australia, 5063.*

*Abstract* Detailed geological mapping at 1:3000 scale has verified the multiphase deformational history of Archaean granulite facies gneisses in the Vestfold Hills and demonstrated the nature and variation of strain, fabric development and metamorphism in selected areas. Early formed fabrics, preserved as"transposed boudins" in a more highly strained/recrystallised matrix, indicate an early history ($D_1$) of high grade metamorphism, migmatisation and probable folding. Concurrent recrystallisation, migmatisation, emplacement of acid/intermediate orthogneisses, pervasive isoclinal folding and interleaving of meta-igneous and metasedimentary rocks were all a function of the subsequent $D_2$ event (Vestfoldian Orogeny). The preservation of various stages in the development of orthogneiss indicates that at least some of the larger masses formed by partial melting in situ during $D_2$. $S_2$ gneissic layering, E—W trending and north dipping, became the principal regional fabric but was locally modified by north and south dipping shear zones associated with monoclinal folding ($D_3$) and numerous N-S trending mylonite zones ($D_{4m}$). Pre- and post-$D_{4m}$ basic dykes have been identified and pseudotachylite-veined microfaults represent a later, much more brittle deformation postdating most basic dykes. The prevalence of hypersthene and plagioclase, together with abundant garnet and red brown biotite, reflect medium load and moderate water pressure granulite facies conditions of metamorphic crystallisation. Temperatures calculated from coexisting mineral pairs are 620-804°C.

Previous investigations in the Vestfold Hills, East Antarctica (Figure 1) have identified the principal lithologies, viz. highly metamorphosed upper amphibolite to granulite facies quartz-feldspar gneisses, garnetiferous gneisses, mafic orthopyroxene-hornblende gneisses and basaltic dykes (Crohn, 1959; Ravich, 1960; McLeod et al., 1966; Oliver et al., 1982). Previous investigations have also recognised the antiquity of the gneisses, viz. 2500 Ma (Arriens, 1975), the Proterozoic age of the basaltic dykes (Harding & McLeod, 1967; Arriens, 1975), and have partly unravelled their complex tectonic evolution (Oliver et al., 1982). More recent work by Collerson et al. (this volume) suggests the possibility of a 3000 Ma age for the Vestfold Hills gneisses.

Oliver et al. (1982) concluded that the gneisses could be subdivided into four major categories: (i) layered grey gneiss with intercalated, sharply bounded felsic and mafic units; (ii) layered garnetiferous paragneiss with subsidiary mafic and submafic units; (iii) acid to intermediate orthogneiss with diffuse layering; and (iv) acid to intermediate homogeneous orthogneiss. These categories have been maintained in this paper although there is some difficulty in distinguishing between diffusely layered orthogneiss, homogenous orthogneiss and felsic end members of layered grey gneiss; an additional lithology, viz. homogeneous garnet gneiss, has also been demarcated.

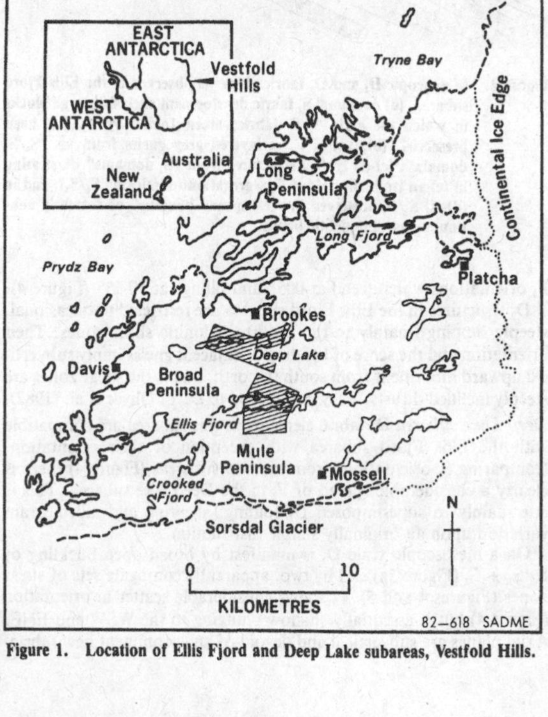

The structural framework constructed by Oliver et al. (1982) demonstrates that rocks of the region have had a complex history. An ENE—WSW and north dipping fabric is superimposed on an earlier tectonothermal fabric and both have been overprinted by regional warping and a complex pattern of shear zones, mylonite zones and fault zones.

The aim of this paper is to furnish more details on the tectonic evolution of the gneisses of the Vestfold Hills with particular reference to fabric development and metamorphic conditions. The data and interpretations presented are the result of several years research by R.L.O. and P.R.J., but are in particular the result of detailed mapping at 1:3000 scale in 1981 by A.J.P. (Parker, 1981; Parker et al., 1981) and V.M. (Mielnik, 1981).



**Figure 1. Location of Ellis Fjord and Deep Lake subareas, Vestfold Hills.**

## Structure

The structural terminology used in this paper follows that of Oliver et al. (1982). Layered grey gneiss and layered garnet paragneiss both record early, pre-$D_2$ histories which may represent more than one tectonic event. However the strong $D_2$ deformation and fabric development has almost completely obscured the early history of those rocks so for the sake of simplicity, all early tectonic events are collectively assigned to $D_1$.

Within the layered gneisses the $S_0 + S_1$ layering is folded by a sequence of tight and isoclinal folds, $F_2$, and universally overprinted by a pervasive new fabric, $S_2$. Clearly where $S_2$ developed in the limbs of isoclinal folds, the observed fabric (layering + schistocity) represents the combined effects of $S_0/S_1 + S_2$.

Modification of $D_2$ structures has occurred largely within discrete E—W trending zones reflecting a transition from wholesale ductile flow to less ductile deformation. These (shear) zones are gently to steeply inclined and both north and south dipping ($D_{3a}$ and $D_{3b}$ respectively). Mylonites are developed along both sets of shear zones. A broad, N—S trending, gently moderately inclined mylonite zone in the eastern Vestfold Hills, the Platcha Mylonite Zone, postdates $D_3$ and is identified as $D_{4m}$. Pseudotachylite veins, $S_{pt}$, overprint all the above structures.

*Ellis Ford subarea* Mesoscopic fabric elements in the subarea are illustrated in Figure 2. Clearly there is an early formed fabric or mineralogical layering folded about $F_2$. That fabric is preserved in a number of ways: in the hinge zones of $F_2$ folds where it is frequently overprinted by a pervasive $S_2$ fabric; and in boudins or dismembered fold hinges either enclosed within a "new" $D_2$ layered gneiss (in which $S_2$ has almost completely overprinted earlier fabrics) or enclosed within homogeneous or diffusely layered orthogneiss.

Of particular importance in the Ellis Fjord subarea is firstly the recognition of macroscopic $F_2$ closures (Figure 3), and secondly the virtual absence of $D_3$ structures. The $D_2$ fold closure is clearly outlined by granulitic amphibolite bands and corroborated by the asymmetry of mesocopic parasitic folds. Both limbs of the structure are parallel, dipping steeply north, and the plunge of the fold parallels mesoscopic

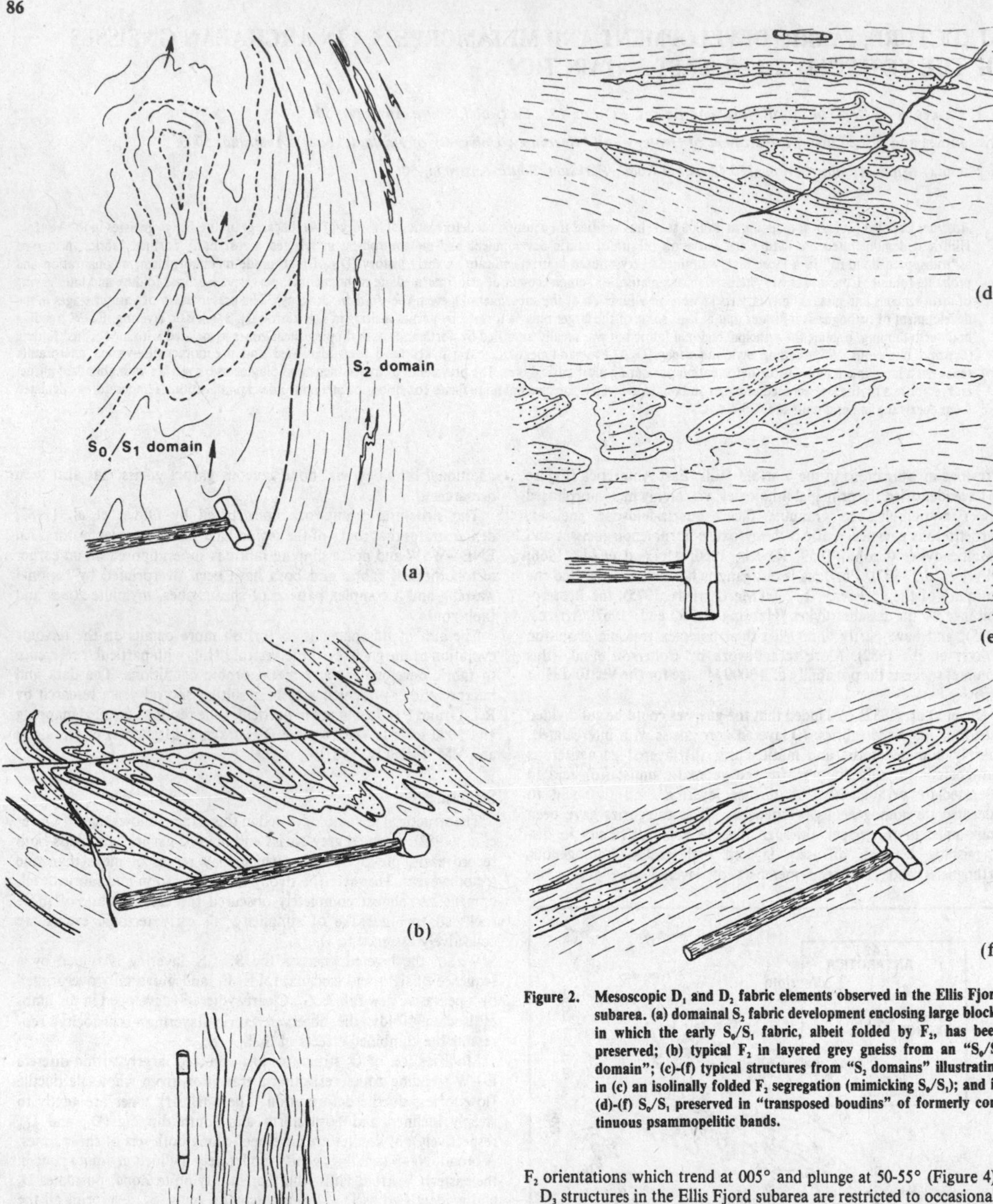


**Figure 2. Mesoscopic $D_1$ and $D_2$ fabric elements observed in the Ellis Fjord subarea. (a) domainal $S_2$ fabric development enclosing large blocks in which the early $S_0/S_1$ fabric, albeit folded by $F_2$, has been preserved; (b) typical $F_2$ in layered grey gneiss from an "$S_0/S_1$ domain"; (c)-(f) typical structures from "$S_2$ domains" illustrating in (c) an isolinally folded $F_2$ segregation (mimicking $S_0/S_1$); and in (d)-(f) $S_0/S_1$ preserved in "transposed boudins" of formerly continuous psammopelitic bands.**

$F_2$ orientations which trend at 005° and plunge at 50-55° (Figure 4).

$D_3$ structures in the Ellis Fjord subarea are restricted to occasional, steeply dipping (mainly to the south), mylonitic shear zones. Their orientation and the sense of bending in adjacent gneiss imply subvertical upward movement from south to north, that is the shear zones are steeply inclined thrusts. They correspond to $D_{3b}$ of Oliver et al. (1982).

*Deep Lake subarea* $D_2$ fabric elements in this subarea are compatible with the Ellis Fjord subarea with exception of their orientation. Comparing $F_2$ orientations from the two subareas (Figure 4) there is clearly a considerable spread of $F_2$ in the Deep lake subarea. This is due mainly to superimposed $D_3$ folding, shearing and allied strain variation upon an originally simple distribution.

On a mesoscopic scale $D_3$ is manifest by broad open buckling of $S_0/S_1$ + $S_2$ (Figure 5a) and by two, apparently conjugate sets of shear zones (Figures 4 and 5). $F_3$ shows considerable scatter in orientation (Figure 4) but is essentially shallow plunging to the WSW and ENE. Axial planes are subvertical and devoid of any prominent new fabric.

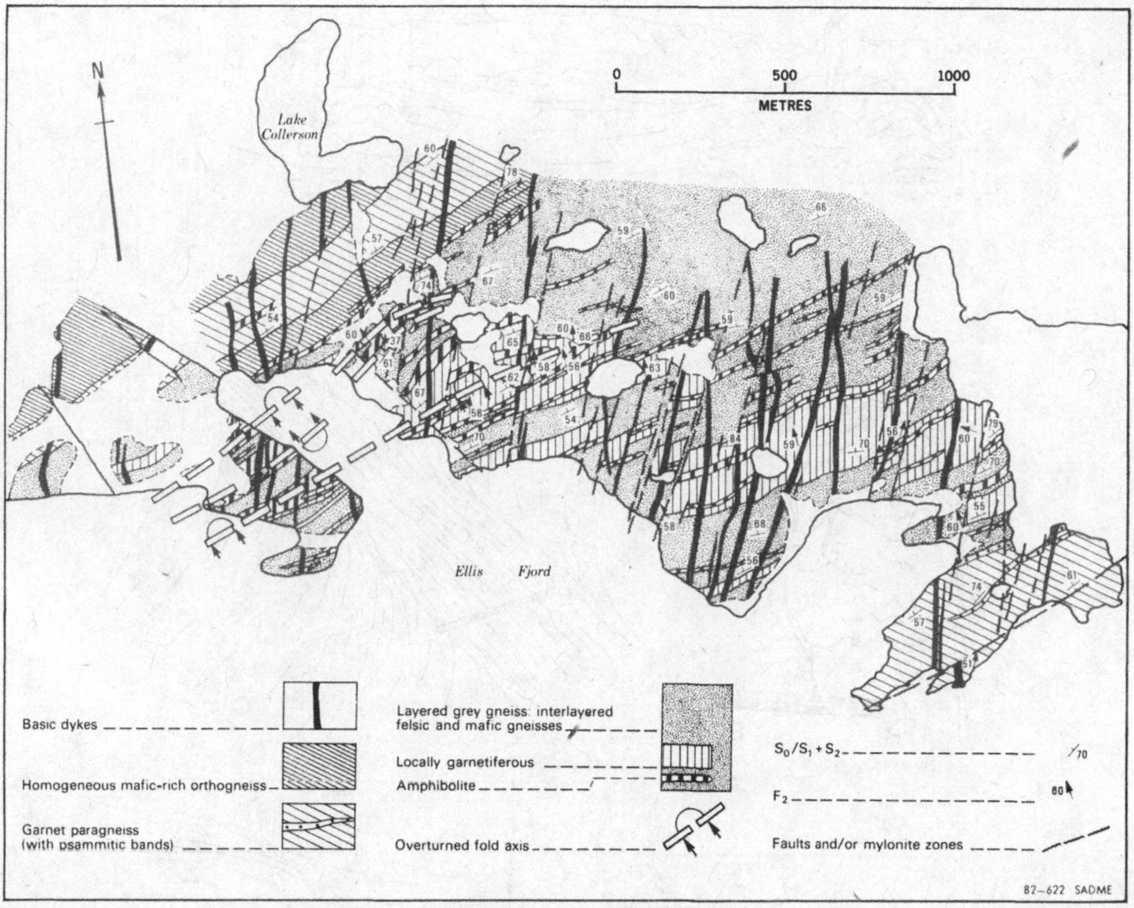

N
Lake Collerson
0
500
1000
METRES
Ellis Fjord
Basic dykes
Homogeneous mafic-rich orthogneiss
Garnet paragneiss (with psammitic bands)
Layered grey gneiss: interlayered felsic and mafic gneisses
Locally garnetiferous
Amphibolite
Overturned fold axis
S0/S1+S2
70
F2
60
Faults and/or mylonite zones
82–622 SADME

**Figure 3. Geological plan of part of the Ellis Fjord subarea.**

The macroscopic expression of $F_3$ is reflected in the shallower $S_0/S_1+S_2$ orientation of this area as documented by Oliver et al. (1982) (Figure 53.10).

Shear zones, often closely associated with $F_3$, and equivalent to $D_{3b}$ of Oliver et al., are common within the Deep Lake subarea and reflect a fundamental difference in post-$D_2$ strain and fabric development between the two cited subareas. The few, rare, shear zones around Ellis Fjord were observed in the northern part of that subarea suggesting an increase in shearing from Ellis Fjord north to Deep Lake.

The dominant shears in the Deep Lake subarea are E—W trending, south dipping and correspond to subvertical reverse fault movements

ELLIS FJORD SUBAREA
DEEP LAKE SUBAREA
124
166
19
96
S0/S1+S2
S0/S1+S2
S3b
Spt
VESTFOLD HILLS
58
65
42
46
F2
F2
F3
S4m
82–619 SADME

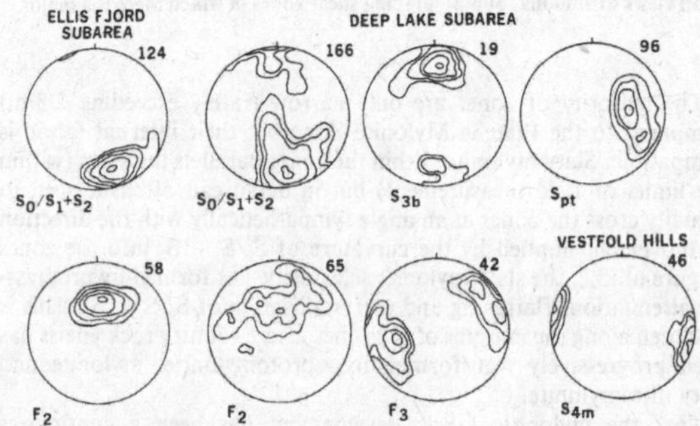

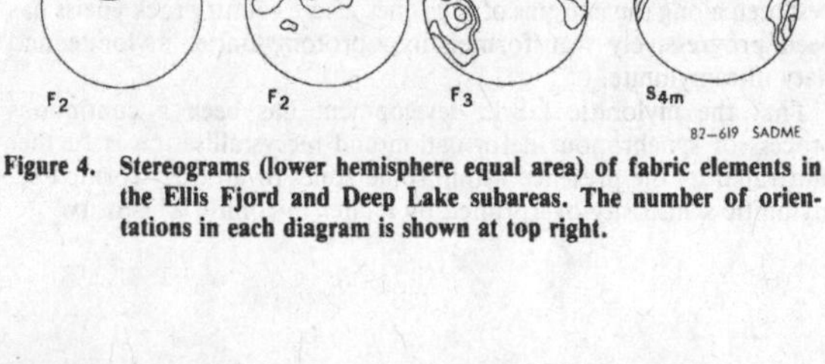

**Figure 4. Stereograms (lower hemisphere equal area) of fabric elements in the Ellis Fjord and Deep Lake subareas. The number of orientations in each diagram is shown at top right.**

from south to north (Figure 5). They can be up to a metre wide and display well developed mylonitic fabrics. Locally, shearing did not proceed fully to produce mylonites but ceased following development of overturned monoclinal folds (Figure 5); these are compatible with the shear zones in orientation, directional fabric, and timing.

While most south dipping shear zones conform to the above criteria, a few vary slightly from the norm. These show a slight splay in trend and record a major component of subhorizontal movement. Nevertheless the sense of movement (very commonly sinistral or left handed along zones oriented north of east and dextral along zones oriented south of east) is directly compatible with orientation of the stress axes which one would infer from the reverse fault movements from south to north, above mentioned, and in fact a component of overthrusting is still evident in most. Steeply inclined lineations within mylonites of these sinistral and dextral shears are a function of the intersection of ($S_0/S_1+S_2$) and $S_{3b}$)—not of elongation.

Shear zones with a pronounced northerly dip are not widely developed but often display a north upward directional fabric. $F_3$ folds associated with one such zone have a pronounced lineation defined by rodding within folded quartz veins. This rodding also is an intersection lineation (defined by the intersection of ($S_0/S_1+S_2$ and $S_{3b}$) rather than a stretching lineation. This suite of shear zones is believed to be conjugate to the south dipping structures.

A further significant difference in strain history and fabric development between the two cited subareas is manifest by the prominence of pseudotachylite-veined faults-microfaults in the Deep Lake subarea. These have been produced relatively late in the tectonic evolution of the region since they postdate discordant, post-$D_3$ pegmatites, $D_4$ mylonite zones (see below), and most dolerite dykes. Furthermore, there is an overall increase in their intensity from south to north across Broad Peninsula. The principal orientations of the pseudotachylites ($S_{pt}$) are recorded in Figure 4: maxima correspond to 000:45°W and

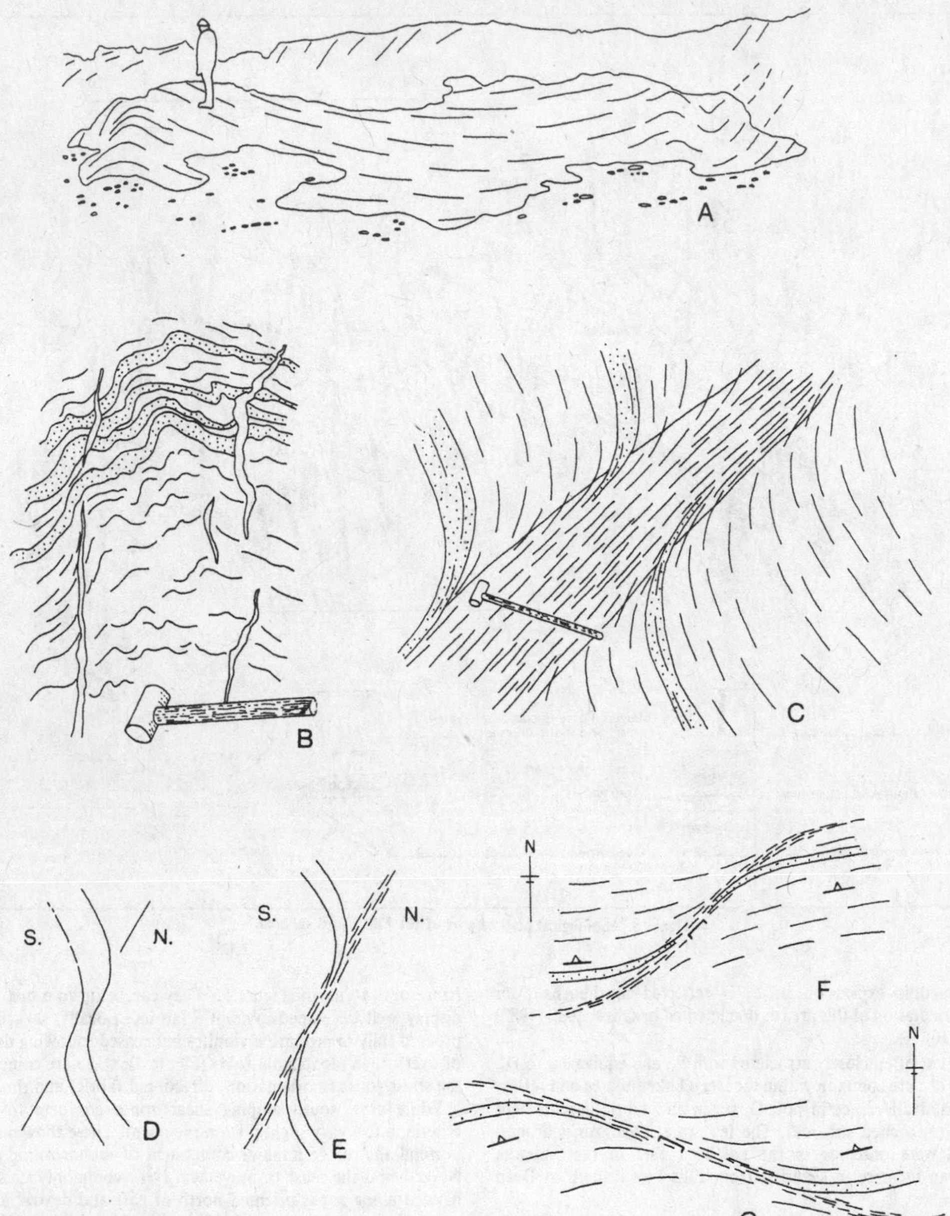


**Figure 5. Mesoscopic $D_3$ structures observed in the Deep Lake subarea. (a) broad, open, monoclinal folding; (b) mesoscopic buckling of migmatite with local pegmatite segregations parallel to $S_3$; (c) mylonite $D_{3b}$ shear zone-cliff section looking west; (d)-(g) schematic diagrams illustrating, (d) monoclinal folding (looking west), (e) typical $D_{3b}$ shear zone (looking west), (f)-(g) plan views of sinuous "south" dipping shear zones in which there is a major strike slip component.**

030:70°WNW. A discussion of pseudotachylite orientation versus directional criteria and relative timing is too complex to present here except to suggest that the N—S striking set may postdate the NNE striking set, and that the former commonly shows at least a component of sinistral strike slip movement.

*Mylonites: Platcha Mylonite Zone* The significance of $D_{4m}$ mylonite zones has been underestimated or even unrecognised in previous studies. The Platcha Mylonite Zone is the only previously recorded zone and while it is the widest known mylonite zone in the Vestfold Hills, it actually represents a major suite of such zones pervasive throughout the Vestfold Hills. Characteristically these mylonites are steeply inclined, N—S trending and, in at least 95% of all zones observed by Parker et al. (1981), they display dextral, subhorizontal displacement (Figure 6).

The majority of zones are only narrow (rarely exceeding 2-3 m) compared to the Platcha Mylonite Zone but their internal fabric is compatible. Slaty mylonite within the zones parallels the sides (within the limits of field measurement) but in detail can often be seen to actually cross the zones at an angle sympathetically with the direction of movement implied by the curvature of $S_0/S_1 + S_2$ into the zones (Figure 6). $S_{4m}$, the slaty mylonite schistosity, has formed by progressive attenuation, flattening and recrystallisation of $S_0/S_1 + S_2$. This is best seen along the margins of the zones where country rock gneiss has been progressively transformed into protomylonite, mylonite and slaty ultramylonite.

That the mylonitic fabric development has been a continuous process of synchronous deformation and recrystallisation is further illustrated by the presence within some zones of an early formed $S_{4m}$ mylonitic schistosity overprinted by a later mylonitic schistosity.

Figure 6. Typical $D_{4m}$ mylonite zone looking vertically down on the outcrop. Note the dextral displacement and progressive development of slaty mylonite from protomylonitic country rock.

A subvertical rodding lineation developed in $S_{4m}$ is an intersection lineation and does not necessarily indicate strain extension. More reliable strain indicators are $F_4$. These confirm subhorizontal, dextral, strike slip shearing.

*Orthogneiss fabric development* Large areas of the early formed gneissic basement have been almost completely reconstituted by processes attributed to $D_2$, leaving only windows of "protolith" in the form of migmatised layered grey gneiss and garnet paragneiss. Most of the areas previously mapped as homogeneous orthogneiss and/or diffusely layered orthogneiss have formed during $D_2$ in response to either intrusion of acid to intermediate magma, and/or in situ partial melting of "protolith". The latter is exemplified in Figure 7 recording examples of the transformation during deformation of layered grey gneiss into diffusely layered orthogneiss. Veining in layered grey gneiss is widespread not only in $F_2$ hinge zones as illustrated, but also as apparent "segregations" within "new" $D_2$ layered gneiss. It represents in situ, "lit-par-lit" migmatisation concurrent with $D_2$ deformation. That is, both the veining and $S_2$ fabric formation are intimately related; $D_2$ providing a source of extra energy for local "melting" and promoting diffusion along and within zones of planar $D_2$ anisotropy (i.e. $S_2$) hence ultimately enhancing the development of the new $S_2$ gneissosity.

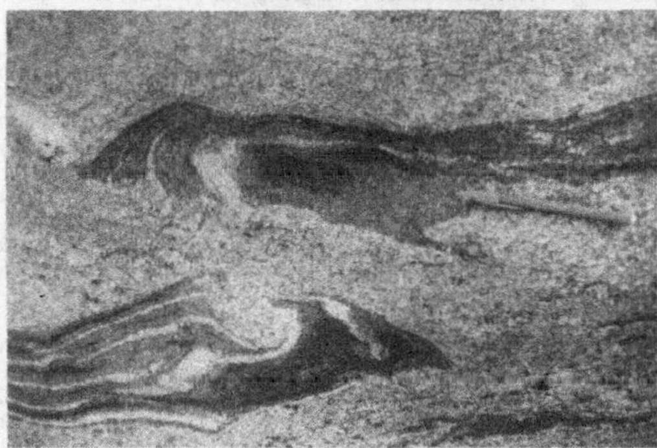

(a)

(b)

Figure 7. Example of $D_2$ orthogneiss veining leading to the formation of diffusely layered orthogneiss. (a) emplacement of orthogneiss along $D_2$ axial planes of folded layered grey gneiss; (b) remnants of layered grey gneiss enclosed within diffuse layered orthogneiss.

In the field both the larger bodies of diffuse layered orthogneiss and the smaller veins paralleling $S_2$ can be recognised by coarse bluish quartz and coarse hornblende or pyroxene.

As noted by Oliver et al. (1982), the orthogneisses have well developed S or LS $D_2$ fabrics. These can vary from simple mineralogical alignment to a segregation schistosity but, particularly in homogeneous orthogneiss, a pronounced augen texture (usually of LS character) is frequently observed. The augen texture is believed to have been, at least in part, produced by the intersection at moderate to high angles of two gneissic fabrics: in this case a strong $S_2$ segregation schistosity superimposed on a relic or "ghost" $S_0/S_1$ gneissosity.

## Basic dyke intrusion

The relative timing of intrusion of the basic dykes so spectacularly developed in the Vestfold Hills is a subject requiring further research for it is clearly more complex than originally conceived (McLeod et al., 1966; Arriens, 1975; Harding and McLeod, 1967). Field relationships clearly distinguish at least two suites: an E—W (to SE—NW) trending suite preceding $D_{4m}$ and a N—S trending suite(s) succeeding or maybe late synchronous with $D_{4m}$ (Figure 8). It has not yet been possible to subdivide the N—S suite although it is believed there is more than one period of dyke emplacement.

## Metamorphism

The prevalence of hypersthene plus plagioclase, together with red brown biotite as an apparent stable phase, suggests crystallisation in a medium load pressure, low to medium water pressure granulite facies environment. The relationship of rock composition to mineral content

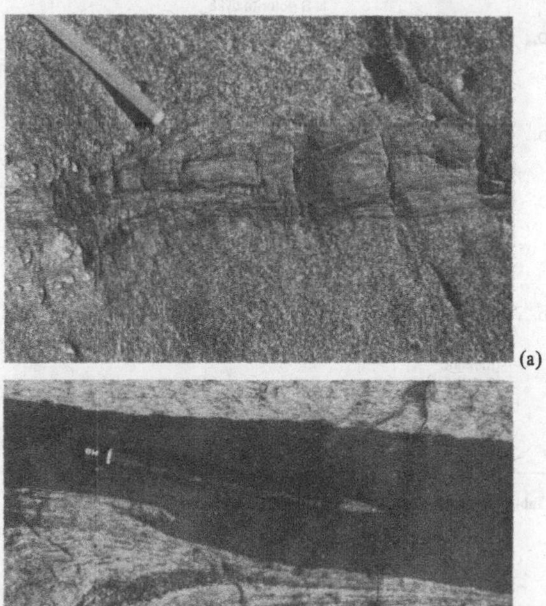

(a)

(b)

Figure 8. The relationship between basic dyke intrusion and $D_{4m}$ mylonites. (a) mylonite cutting a NW-SE trending basic dyke; (b) dolerite dyke intruded along a mylonite zone.

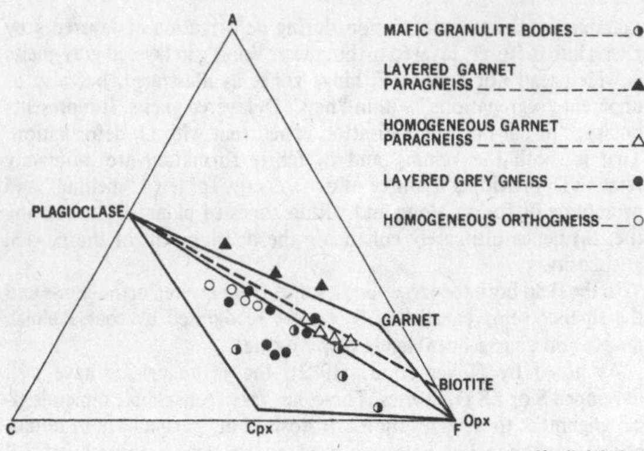


**Figure 9. ACF diagram depicting rock compositions and mineralogies of various lithologies.**

is depicted in Figure 9. The abundance of garnet in preference to cordierite in the layered garnetiferous paragneiss and the homogeneous garnet gneiss, both of which have Mg numbers close to 0.5, further reflects a medium load presure.

Calculated temperatures of metamorphic crystallisation using the composition of coexisting garnet and biotite in two homogeneous paragneisses according to a variety of methods (viz. Perchuk, 1967; Ferry and Spear, 1978; Thompson, 1976; Goldman and Albee, 1977) average 620°C. Using the composition of coexisting clino- and orthopyroxenes in a mafic granulite, according to the methods of Wood and Banno (1973) and Wells (1977), the calculated temperature of crystallisation is 804°C. This discrepancy suggests that the mafic granulite may manifest granulite facies $D_1$ and/or $D_2$ metamorphism

| Event | Fabric Development | Igneous Activity | Metamorphism |
|---|---|---|---|
| | Pseudotachylite veining, $S_{pt}$ | | |
| | | N-S dolerite dyke intrusion | |
| $D_{4m}$ | N-S slaty Mylonite zones, $S_{4m}$ | | |
| | | E-W and SE-NW basic dyke intrusion Widespread pegmatite intrusion | |
| $D_3$ | Monoclinal folding, $F_3$; E-W shear zones (mylonitic thrusts), $S_{3b}$ | Local pegmatite | Amphibolite facies |
| $D_2$ | Isoclinal folding, $F_2$; transportation and boudin formation; formation of secondary gneissic layering or segregation schistosity, $S_2$ | Formation, largely in situ, of diffusely layered orthogneiss and homogeneous orthogneiss. Intense migmatisation | Granulite facies |
| $D_1$ | Probably isoclinal folding; formation of initial gneissic layering, $S_1$ | Migmatisation | Granulite facies |
| | Primary layering, $S_0$ | Acid/intermediate and basic volcanism. Sedimentation | |

**Table 1. Summary of tectonic events in the Vestfold Hills, East Antarctica.**

and the garnet and biotite in the garnetiferous paragneisses may have equilibrated during amphibolite facies $D_3$. In the garnetiferous gneisses, garnet and biotite may not be in chemical equilibrium with earlier formed hypersthene, contrary to textural appearances.

## Conclusions

Detailed geological field studies, including mapping at a scale of 1:3000 in selected areas of the Vestfold Hills, have verified previously published conclusions (Oliver et al., 1982) that Archaean granulite gneisses have undergone at least four phases of deformation (Table 1).

In view of the degree of migmatisation, recrystallisation and reconstitution associated with $D_2$, the age of ca 2500 Ma reported by Arriens (1975) for the gneisses is believed to represent a minimum age for that event. The protoliths of the layered gneisses, therefore, would be considerably older. It is proposed here that the $D_2$ deformation and related processes observed in the Vestfold Hills be known as the Vestfoldian Orogeny.

Metamorphic conditions during evolution of the Vestfold Hills reached a peak during the Vestfoldian Orogeny then subsequently waned during $D_3$ and $D_{4m}$.

*Acknowledgements* The authors acknowledge the logistical support from ANARE and members of ANARE during the 1976/77 season (R.L.O. and P.R.J.) and the 1980/81 season (A.J.P. and V.M.). A.J.P. was supported by A.R.G.C. funds whilst on leave from the South Australian Department of Mines and Energy.

## REFERENCES

ARRIENS, P.A., 1975: The Precambrian geochronology of Antarctica: Geol. Soc. Aust., 1st Geol. Convent., Abstr., 97-8.

COLLERSON, K.D., REID, E., MILLER, D. and MCCULLOCH, M.T. (this volume): Lithological and Sr-Nd relationships in the Vestfold Block: implications for Archaean and Proterozoic crustal evolution in the East Antarctic shield.

CROHN, P.W., 1959: A contribution to the geology and glaciology of the western part of Australian Antarctic Territory. *Aust., Bur. Miner. Resour., Geol. Geophys., Bull.*, 43-4.

FERRY, J.M. and SPEAR, F.S., 1978: Experimental calibration of the partitioning of Fe and Mg between biotite and garnet. *Contrib. Mineral. Petrol., 66,* 113-7.

GOLDMAN, D.S. and ALBEE, A.L., 1977: Correlation of Mg/Fe partitioning between garnet and biotite with $^{18}O/^{16}O$ partitioning between quartz and magnetite. *Am. J. Sci., 277,* 750-67.

HARDING, R.R. and McLEOD, I.R., 1967: Age of dolerite dykes in the Vestfold Hills, Antarctica. *Nature, 215,* 149-51.

McLEOD, I.R., TRAIL, D.S., COOK, P.J. and WALLIS, G.R., 1966: Geological work in Antarctica, January-March, 1965 *Aust., Bur. Miner. Resour., Geol. Geophys., Rec.* 1966/9 (Unpubl).

MIELNIK, V., 1981: The geology of selected areas in the Vestfold Hills, East Antarctica. Unpublished B.Sc. Honours thesis, University of Adelaide, Australia.

OLIVER, R.L., JAMES, P.R., COLLERSON, K.D. and RYAN, A.B., 1982: Precambrian geologic relationships in the Vestfold Hills, Antarctica; *in* Craddock, C. (ed.) *Antarctic Geoscience.* Univ. Wisconsin Press, Madison, 435-44.

PARKER, A.J., 1981: Geological investigations in the Vestfold Hills, Princess Elizabeth Land, East Antarctica: amended field notes, sketches and photographs. Univ. of Adelaide, Report (unpubl).

PARKER, A.J., JAMES, P.R., MIELNIK, V. and OLIVER, R.L., 1981: Superposed folding: a classic example in East Antarctica *Geol. Soc. Aust., Abstr., 3.* 93-4.

PERCHUK, L.L., 1967: The biotite-garnet geothermometer. *Dokl. Akad. Nauk SSR., 177,* 131-4.

RAVICH, M.G., 1960: Gornye porody oazisa Vestfoll'. (Rocks of Vestfold Oasis). *Nauch.—issled. inst. geol. Arktiki, Trudy, 113,* 25-52.

THOMPSON, A.B., 1976: Mineral reactions in pelitic rocks: prediction of P-T-X (Fe-Mg) phase relations. *Am. J. Sci., 276,* 401-24.

WELLS, P.R.A., 1977: Pyroxene thermometry in simple and complex systems. *Contrib. Mineral. Petrol., 62,* 129-39.

WOOD, B.J. and BANNO, S., 1973: Garnet-orthopyroxene and orthopyroxene-clinopyroxene relationships in simple and complex systems. *Contrib. Mineral. Petrol., 42,* 109-24.

# 2

## East Antarctica-West Antarctica Boundary and the Ross Orogen, including Northern Victoria Land

# A REVIEW OF THE ROSS FOLD BELT OF THE TRANSANTARCTIC MOUNTAINS AS A BOUNDARY STRUCTURE BETWEEN EAST ANTARCTICA AND WEST ANTARCTICA

G.E. Grikurov, *Department of Antarctic Geology and Mineral Resources, VNIIOkeangeologiya, Leningrad, 190121, USSR.*

*Abstract* The main features of tectonic evolution of the Transantarctic Mountains are:

(1) Epicratonic nature of the Ross Fold Belt formed on the ancient crystalline basement.

(2) Substantial thickness and rather intensive dislocation of the Riphean complex and predominance of terrigenous formations of the platform and/or miogeosynclinal character.

(3) Very strong inversion of the pre-Riphean infrastructure and its active development at the orogenic (Vendian—Early Palaeozoic) stage, manifested in a wide distribution of synkinematic granitoid batholiths, and volcanic rocks.

(4) Development in the regime of the mobile platform during Middle to Late Palaeozoic and Early Mesozoic time which led to the formation of the platform cover of variable thickness, crowned with the Jurassic trap formation.

The Transantarctic Mountains represent a horst in the flank of the extended rift zone of the complex structure which controls the basins of the Ross and the Weddell Seas and connects subglacial bedrock depressions. The scarp of the Transantarctic horst forms the major lineament of the continent, and adjacent grabens and troughs separate the pre-Riphean (Gondwana) craton of East Antarctica from its Mesozoic (Pacific) geosynclinal-fold fringe on the western margin of the continent. Taken together the above morphostructures trace the ancient structural suture in the Earth's crust emplaced as early as the Early Precambrian around the periphery of the province of Archaean consolidation. At the Riphean-Early Palaeozoic stage a palaeoboundary of the Pacific mobile belt lay along the suture that caused the activation of ancient platform structures and the formation of Ross fold structures in its rear zone. The present boundary separating East Antarctica and West Antarctica is also associated with the area of neotectonic activation inheriting the same structural suture.

# THE EAST ANTARCTICA—WEST ANTARCTICA BOUNDARY BETWEEN THE ICE SHELVES: A REVIEW

C.Craddock, *Department of Geology and Geophysics, University of Wisconsin, Madison, Wisc. 53706, U.S.A.*

*Abstract* East Antarctica (EA) and West Antarctica (WA) show contrasting Phanerozoic tectonic histories. Important problems in Antarctic tectonics include (1) the nature of the EA-WA boundary, and (2) the tectonic behaviour of WA during Cretaceous-Cenozoic time. The Horlick Mountains-Pensacola Mountains-Ellsworth Mountains region is critical to these problems.

The Horlick, Thiel and Pensacola Mountains form a segment of the Transantarctic Mountains chain; deformed (except in the Thiel Mountains) and intruded Proterozoic-Lower Palaeozoic strata, recording two orogenies, are overlain unconformably by the Gondwana sequence (Devonian-Jurassic). To the north, tightly folded metasedimentary rocks in the Stewart Hills yield a mica K-Ar age of 508 Ma. Farther north the Hart Hills reveal deformed Palaeozoic(?) metasedimentary rocks and altered gabbros; granite at nearby Pagano Nunatak gives a 176 Ma K-Ar age. The Whitmore Mountains expose two early Mesozoic granites intruding metasedimentary rocks. The Martin, Nash and Pirrit Hills all display deformed metasedimentary rocks cut by felsic intrusives, with the Nash and Pirrit granites of Triassic age. The Ellsworth Mountains contain a thick, folded sedimentary sequence of Cambrian-Permian age. Haag Nunataks consist of gneisses and schist of 1000 Ma age.

Two tectonic provinces exist, viz. the Transantarctic Mountains (the border of EA) in the south, and the Ellsworth Mountains fold belt (EMFB) to the north. The boundary between them lies north of the Horlick Mountains and probably north of the Hart Hills; its nature is unclear, but it may be a fault. The northwest boundary of the EMFB appears to be a fault zone that truncates the structures. Eight Phanerozoic tectonic events or stages are recognised. Displacement of the EMFB relative to EA has probably occurred, and two models are considered as hypotheses: (1) counter-clockwise rotation with westward translation, and (2) clockwise rotation with eastward translation.

East Antarctica and West Antarctica display strongly contrasting tectonic histories for most of Phanerozoic time. Major deformation and metamorphism last affected East Antarctica during the Cambro-Ordovician Ross orogeny; since then it has behaved as a cohesive shield, accumulating a platform cover and undergoing some vertical displacements. West Antarctica, on the other hand, gives evidence of nearly continuous tectonism—manifest by intrusive and eruptive igneous activity, metamorphism, folding and faulting. A major problem in Antarctic tectonics is the nature of the boundary between these two large and dissimilar provinces.

The triangular area bounded by the Horlick Mountains, the Pensacola Mountains and the Ellsworth Mountains is a critical region in attacking this problem (Thiel, 1961; Craddock, 1971). It is the only segment of the common boundary that contains extensive rock outcrops in both provinces. It holds vital evidence about the evolution of West Antarctica during Mesozoic-Cenozoic time, and about the role of Antarctica in Gondwanaland (Craddock, 1982).

The extensive use of ski-equipped aircraft during IGY allowed the beginning of formal geological projects in 1959 in the U.S. programme. Important advances in Antarctic geology and geochronology followed soon, and the regional geology was summarised in a series of maps (Bushnell and Craddock, 1970). Information about crustal structure can be obtained from certain geophysical measurements, such as seismic refraction profiles (Bentley and Clough, 1972) and airborne magnetic intensity and radio echo soundings (Behrendt et al., 1980; Jankowski and Drewry, 1981). Only a few palaeomagnetic determinations on oriented rock specimens from this region have been made.

## THE TRANSANTARCTIC MOUNTAINS

The Transantarctic Mountains chain extends across the continent from northern Victoria Land on the Pacific coast to western Queen Maud Land on the Atlantic coast. The long chain shows remarkable geological continuity, and it defines one border of the East Antarctic shield. The rocks are divisible into a basement complex and the overlying subhorizontal Gondwana sequence. Basement rocks include deformed and metamorposed Proterozoic and Lower Palaeozoic sedimentary and volcanic rocks, cut by Proterozoic and Cambro-Ordovician batholiths. The Gondwana sequence consists of the Devonian-Triassic Beacon Supergroup and the Jurassic Ferrar Supergroup. Normal faults are common, and uplift of the Transantarctic Mountains probably occurred mainly during Tertiary time (Katz, 1982).

### The Horlick Mountains

The Horlick Mountains (Figure 1) were first visited in December, 1958 (Long, 1959) and parties from Ohio State University worked there during five summers. This work has been summarised by Mirsky (1969). The basement complex in the eastern Horlick Mountains is a large batholith composed mainly of granodiorite and quartz monzonite; it is considered a Ross intrusive of Cambro-Ordovician age. The overlying Beacon strata contain many Devonian and Permian fossils, along with a tillite and coal beds. The Beacon rocks are only gently tilted, but they are cut by a number of faults.

### The Thiel Mountains

A good map and description of the geology of this range are provided by Schmidt and Ford (1969). The oldest rocks exposed here comprise a 100 m sequence of thermally metamorphosed clastic sedimentary beds, considered to be Proterozoic. Stromatolites in a thin marble bed are probably among the oldest known megafossils in Antarctica. These strata are intruded by a thick sill of quartz monzonite porphyry. Discordant Cambro-Ordovician granitic bodies cut the sill. The layered rocks are nearly horizontal.

### The Pensacola Mountains

These interesting and complex mountains are described by Schmidt and Ford (1969).

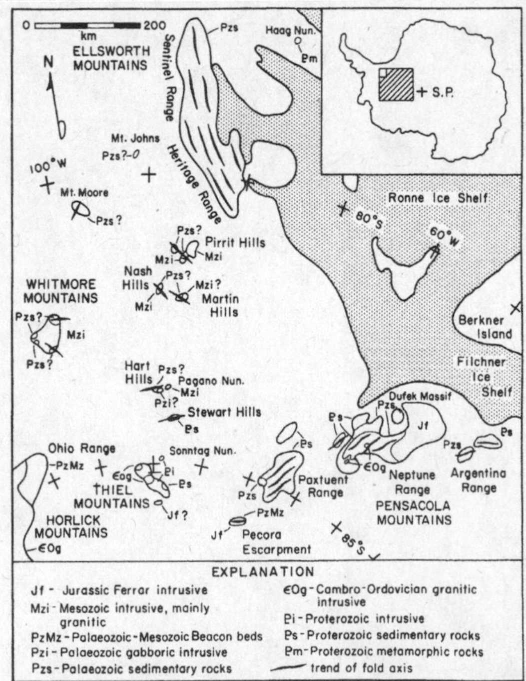


**Figure 1.** **Geologic map, with structural trends, of junction region between East Antarctica (below) and West Antarctica (above). Modified from Craddock (1972) and Craddock (in press). The index map has been rotated a quarter turn so that the Antarctic continent appears in its standard position.**

A very thick sequence of Proterozoic turbidites and lava flows is strongly folded and unconformably overlain by somewhat less deformed Cambrian strata. These Lower Palaeozoic beds in turn are unconformably overlain by slightly deformed Upper Palaeozoic beds. Beacon strata at Pecora Escarpment, however, are horizontal.

The Dufek Massif consists of one of the largest known stratiform gabbro complexes. It is exposed through an area of 8000 km², and geophysical surveys suggest a minimum area of 50,000 km². It is 2,000 m thick locally, and neither top nor bottom are exposed. It appears to postdate all folding and is considered Jurassic in age.

Three cycles of deformation have been recognised—Late Precambrian, Late Cambrian and post-Permian. Most fold axes trend NNE, somewhat oblique to the local trend of the Transantarctic Mountains. Many faults are present, including a large fault inferred along the edge of the Ronne Ice Shelf.

## OUTCROPS NORTH OF THE TRANSANTARCTIC MOUNTAINS

To the north of the segment of the Transantarctic Mountains just described lies a northward trending belt of smaller mountains, hills and nunataks. These exposures display some geologic variety, and the boundary between East Antarctica and West Antarctica is not obvious. Available geophysical data suggest a rugged, irregular subglacial topography and major faults may pass between the scattered outcrop areas. Most major outcrops here have been visited, but there is little published on some areas. The following discussion is based mainly on studies and maps by Craddock (1972, in press).

### The Stewart Hills

Outcrop is limited, but the rocks exposed are mainly grey green and dusky red phyllite, micaceous quartzites and white quartz veins. The rock specimens have not been studied in detail, but the suite appears to be of greenschist facies. All layered rocks obseved are tightly folded; some beds are overturned, and some folds are nearly isoclinal. A well developed foliation dips about 75° northward, and both bedding and foliation trend east northeast. One K-Ar measurement gave a Late Cambrian apparent age.

### The Hart Hills and Pagano Nunatak

The Hart Hills are characterised by low relief, rounded landforms and poor bedrock exposure. Two rock units are about equally abundant. The older unit is a 400 m sequence of folded and metamorphosed clastic metasedimentary rocks. The main rock types are quartz-mica phyllites and subarkosic sandstones. No fossils were found. Bedding and cleavage both trend east northeast; the bedding dips moderately southward, the cleavage moderately northward. The metasedimentary rocks are cut by quartz gabbro intrusions. These igneous rocks are badly altered and locally sheared, and they likely had been emplaced before the sedimentary rocks were folded.

Pagano Nunatak is less than 14 km from the Hart Hills and it rises as a sharp peak about 230 m above the surrounding ice. The bedrock consists of a grey, massive, medium to coarse grained granite cut by small aplite dykes. The rock is fresh, undeformed and unmetamorphosed. A single K-Ar determination on biotite yielded an apparent age of 175 ± 4 Ma.

### The Whitmore Mountains

Bedrock consists of predominant igneous rocks and minor metasedimentary rocks (Webers et al., 1982). Nearly all the exposures consist of a coarse grained, porphyritic Late Triassic granite. A finer, equigranular Early Jurassic granite crosscuts the older granite. The metasedimentary rocks are not abundant and they occur as roof pendants and blocks within the older granite. The largest section of these rocks is about 800 m thick. Rock types include phyllite, slate, greywacke, subarkose and quartzite. Structural trends are poorly defined and probably unreliable; the metasedimentary rocks are likely to have been displaced during emplacement of the granite.

### The Nash Hills

A few rock specimens were collected by the U.S. Ellsworth-Byrd traverse during the 1958/59 season and described by Treves (1959). Most of the bedrock in the outcrops observed is a granitic intrusive, locally a granodiorite. One Rb-Sr and three K-Ar ages all fall in the 172-177 Ma range. The granitic rocks were emplaced into a varied sequence of sedimentary rocks. Types present include grey sandstone with dusky red shale clasts, dusky red siltstone and shale, silty light grey limestone, and light grey pisolitic marble. Although no fossils were found, this suite resembles rocks found in the Cambrian sequence of the Heritage Range. All of the metasedimentary rocks are strongly folded and a good cleavage is common. Both the bedding and the foliation trend northwesterly. Some rocks are spotted hornfels, typical of a contact aureole zone.

### The Martin Hills

The bedrock consists of two distinct units. The more widespread is a sequence of metasedimentary rocks about 300 m thick, and likely a shallow marine facies. It includes sandstones and finer clastic rocks, along with oolitic limestones. These rocks are thermally metamorphosed, and epidote is common. The second unit is fine grained, porphyritic felsite that occurs as dykes and plugs. It is the probable cause of the thermal metamorphism. The layered rocks are strongly folded. Fold axes trend west northwest, but vergence is not clear; cleavage is poorly developed.

### The Pirrit Hills

Most of the Pirrit Hills is underlain by a massive, jointed granite. The rock is pink and very coarse grained with some feldspar crystals 4 cm long. Locally there are flow bands of mafic minerals and elsewhere good pegmatites. This granite intrudes a sequence of clastic metasedimentary rocks. Near the granite contact these rocks are strongly metamorphosed, but farther away they are little affected. Some of the least metamorphosed rocks resemble the Crashsite Quartzite in the Ellsworth Mountains. The layered rocks are strongly folded, and cleavage is well developed. The cleavage trends west northwest and dips steeply northward.

### Mount Moore and Mount Johns

On the Sentinel Range traverse from Byrd Station during the 1957/58 season rock specimens were collected and later described by Anderson (1960). The suite of 13 specimens from Mt Moore includes mainly tan to brown quartzites with some chlorite schists and white quartz and calcite vein material. The strike of bedding is given as northeast, an unusual orientation for the region. The nine rock specimens from Mt Johns are mainly massive, greenish quartzite. From their descriptions they seem similar to the Crashsite Quartzite in the Ellsworth Mountains. The strike of bedding is a little east of north, but no dip is given.

### The Ellsworth Mountains

Anderson (1960) also describes some rocks collected from two nunataks at the western edge of the Sentinel Range.

Craddock et al. (1964) discussed the geology of the Sentinel Range, based on two summers of work, and Craddock (1969) presents a map of the entire chain with a descriptive text. Nearly all the rocks are part of a sequence of sedimentary rocks, with some volcanic rocks, that reaches a total thickness of at least 14,000 m. Middle Cambrian fossils occur low in the sequence, and Permian plants near the top. Unconformities may well exist in the sequence, but none has been clearly demonstrated as yet. The older rocks are quite varied and include boulder conglomerates, lava flows and fossiliferous limestones. The younger beds are mainly green and brown quartzites, a 1000 m glaciomarine tillite and glossopterid-bearing black shales.

All the layered rocks are strongly folded, cleavage is pervasive, and some faults are known. Folds are commonly tight, in places nearly isoclinal. Some axes plunge steeply, and reclined folds occur in the older units. Measurements from throughout the two ranges do not yield a clear signal about vergence direction. Many workers infer but a single, post-Permian cycle of deformation, but others (Hjelle et al., 1982; Yoshida, 1982) argue for additional folding events. The trend of fold axes changes gradually from northwest at the southern end of the mountains to nearly north at the northern end.

### Haag Nunataks

Clarkson and Brook (1977) report on rock specimens collected and some K-Ar age determinations. The suite includes four granodiorite gneisses, one quartzo-feldspathic gneiss, one hornblende schist and one acid vein cutting gneiss—all collected in place. The rocks are considered the result of regional metamorphism to amphibolite facies. Two gneiss specimens yielded concordant apparent ages of about 1,000 Ma on coexisting hornblende and biotite.

## REGIONAL TECTONIC PATTERN

The preceding review of the regional geology provides the basis for consideration of the regional tectonic pattern. This discussion treats the major tectonic provinces, the nature of their boundaries, the major events in their evolution, and the question of relative displacement.

### Tectonic Provinces

The Transantarctic Mountains and the Ellsworth Mountains are two separate and distinct tectonic provinces. Between them lies a rugged highland area; the best portrayal of the bedrock surface is by Jankowski and Drewry (1981). The intervening area is treated separately here.

*The Transantarctic Mountains*

This long mountain chain, probably elevated mainly during Cenozoic time, defines the border of East Antarctica. Locally there is evidence for at least two folding events of Proterozoic age and the Cambro-Ordovician Ross orogen lies upon or Pacificward from these Precambrian orogens. In general, these ancient fold belts trend about parallel to the modern mountain chain. The deformed basement rocks are unconformably overlain by the subhorizontal Beacon-Ferrar Supergroups (Devonian-Jurassic).

*The Ellsworth Mountains*

This province contains Palaeozoic strata of mainly shallow marine facies that are strongly deformed and thicker than those in the Transantarctic Mountains. Although unconformities may exist, the Cambrian-Permian sequence may record continuous deposition near the outer continental shelf. The main folding cycle is considered Triassic in age, but some younger effects may be superposed. Breccia bodies in Cambrian carbonate rocks postdate the major folding, but they are cut by younger faults.

*Intervening Area*

The quartzites and the structure at Mt Johns show that it is almost surely part of the Ellsworth Mountains fold belt (EMFB); the same is true of Mt Moore, and adjacent Mt Woollard, although the northeast strike is anomalous. On the basis of comparable sedimentary rocks, structural trends and early Mesozoic granitic rocks, the Pirrit, Nash and Martin Hills are assigned to the EMFB. The Whitmore Mountains are mainly Triassic-Jurassic granites, with just a few exposures of older sedimentary rocks; they indicate early Mesozoic tectonism and seem part of the EMFB. Pagano Nunatak is fresh, early Mesozoic granite, considered as well to lie within the EMFB. The Hart Hills reveal deformed sedimentary rocks and gabbro bodies, all likely Palaeozoic or older in age. The Stewart Hills consist of folded sedimentary rocks that are probably Lower Palaeozoic or Proterozoic. In both the Hart and Stewart Hills the structural grain roughly parallels the Transantarctic Mountains and vergence is southward.

### Province Boundaries

Rock exposures are sparse and scattered in most of this region, and sharp provincial boundaries are not evident. Nevertheless, knowledge of the bedrock morphology and geology allows some discussion of the position and nature of these important boundaries.

*The Transantarctic Mountains*

The northern boundary is clear along the eastern Ross Ice Shelf where downthrown Beacon rocks occur on the northern side of a major fault. Eastward the boundary becomes indistinct; some linear lows in the subice bedrock surface suggest faults, but the pattern is unclear. The boundary must pass north of the Horlick Mountains, and it probably lies north of the Hart Hills. The Proterozoic beds of the Thiel Mountains appear to be tilted slightly but not folded; they probably lie south of the Ross folds. However, Ross folds are well developed in the Pensacola Mountains. The Ross Orogen may be displaced along a northwest trending fault that lies between the Stewart Hills and the Patuxent Range.

*The Ellsworth Mountains Fold Belt*

The eastern edge of the Ellsworth Mountains is probably a major fault zone, but the nature of the rocks beneath the ice stream and shelf and their relation to the EMFB are unknown. The Proterozoic metamorphic rocks of Haag Nunataks are the only exposures, and they are unlike any known outcrops within hundreds of kilometres. The present northwest edge of the EMFB is defined by a northwest facing steep slope which lies oblique to the structural grain; this truncation of trends and the disappearance of the thick Palaeozoic sequence suggest a fault zone. The Byrd subglacial basin to the northwest, along with its horst-like central ridge (suggestive of the Lomonosov ridge in the Arctic basin) are likely fault-bounded and indicative of an extensional tectonic regime. The southeast dip at Mt Moore may reveal a tilted fault block in this border zone.

The southern edge of the EMFB poses a problem. The bedrock topography is rugged, but it reveals no natural locus for a southern boundary. The fault zone along the southern Ross Ice Shelf may extend eastward past the Horlick Mountains, but it is hard to squeeze a provincial boundary between the Hart Hills and Pagano Nunatak. Indeed, their proximity may show that the early Mesozoic granites here extend into the Transantarctic Mountains province. If so, this relationship agues against (a) a sharp provincial boundary, and (b) a major fault displacement along this boundary.

### Tectonic Evolution

The main events recognised in the tectonic evolution of the region can be summarised as follows:

(1) Proterozoic and Lower Palaeozoic strata were deformed and intruded by granites during th Ross Orogeny. This fold belt trends eastward along the Transantarctic Mountains, probably passes through the Hart and Stewart Hills, and may be offset southeastward to its continuation in the Pensacola Mountains. It is probably not present in the EMFB.

(2) Post-Ross sedimentary rocks (Devonian-Triassic) of shallow marine and terrestrial facies were deposited in the Transantarctic Mountains province. Thicker, probably more distal Devonian-Permian rocks were laid down in the EMFB.

(3) The Ellsworth Orogeny deformed beds as young as Permian in the EMFB. Elsewhere in Antarctica a similar event affects Triassic rocks but not Jurassic rocks. Thus the Ellsworth Orogeny is considered probably Triassic in age.

(4) Early Mesozoic granites, interpreted as post-tectonic, were emplaced in the EMFB.

(5) Intrusive and extrusive mafic igneous rocks of Jurassic age were emplaced along the length of the present Transantarctic Mountains. The Dufek Massif contains the largest of these bodies.

(6) In Late Cretaceous time the Pacific-Antarctic ridge, apparently propagating westward, entered the margin of Gondwanaland and began to separate New Zealand and West Antarctica while Australia and East Antarctica remained joined. Some of the displacement of New Zealand was accommodated by the opening of the Tasman Sea. A complementary eastward displacement of West Antarctica along a fault boundary with East Antarctica might also have occurred. The southern bend in the Antarctic Peninsula and the orientation of the Ellsworth Mountains might have formed at this time.

(7) Any relative movements between West Antarctica and East Antarctica probably stopped in early Tertiary time when the East Antarctica-Australia separation began.

(8) Extensional tectonics seem to have prevailed through much of West Antarctica during late Cenozoic time, as evidenced by the pattern of volcanism. This regime is likely responsible for the subglacial troughs, the extensive tracts in which the bedrock surface lies below sea level, and the truncation of the EMFB.

### Displacement of the EMFB?

The unusual orientation of the Ellsworth Mountains, nearly transverse to the coastal belt on the north and the Transantarctic Mountains on the south, raises the question of possible displacement of the EMFB relative to East Antarctica. If such displacement occurred, the EMFB may have undergone oroclinal bending, province rotation, province translation—or some combination of these. The case for fault truncation along the northwest border is strong, but whether these faults ever underwent major strike-slip displacements is unknown. The stratigraphy and palaeontology of the Ellsworth

Mountains rocks suggest that this sequence was laid down only slightly seaward (north) of the modern Transantarctic Mountains; in overall stratigraphy and structure they most resemble the Pensacola Mountains sequence. Some displacement of the EMFB has probably occurred, and two models are considered here as working hypotheses.

*Counter-clockwise rotation, westward translation*

Schopf (1969) argued that the Ellsworth Mountains have migrated from a position north of the Pensacola Mountains although the present writer is unsure of the sense of his postulated rotation. Dalziel and Elliot (1982) favour a similar reconstruction on the basis of inferred eastward vergence of Ellsworth Mountains structures, palaeomagnetic work by Watts and Bramall (1981), the granites southwest of the Ellsworth Mountains, and the rocks at Haag Nunataks. Such a model places the EMFB closer to the Cape fold belt (CFB) and has appeal because of the similarity in composition and structural style between (1) the Table Mountain Sandstone (TMS) and the Crashsite Quartzite (CQ), and (2) the Dwyka Tillite and the Whiteout Conglomerate. However, this model also encounters certain problems. There are probably no rocks older than Cambrian exposed in the EMFB, although the apparent absence of the Precambrian rocks may only reflect the level of erosion. The EMFB contains a thick Cambrian sequence, but Cambrian time in the CFB is represented by granitic plutons. The basal TMS is Upper Ordovician and locally rests with angular unconformity on Precambrian metasedimentary rocks, but the basal CQ is Upper Cambrian and seems conformable on older Cambrian strata. The Ellsworth Orogeny is probably of Triassic age, but the Cape Orogeny is now regarded as Permian in age (I.W. Halbich, pers. comm., 1981).

*Clockwise rotation, eastward translation*

Cretaceous-Cenozoic tectonic history suggests an alternative model in which the EMFB originally lay farther west and more parallel to the Transantarctic Mountains. Eastward movement of western West Antarctica in Cretaceous time would yield to northward extension in Late Cenozoic time. Some appeals in this model include (1) geologic similarities between Marie Byrd Land and northern Victoria Land, (2) a possible explanation for the southern bend in the Antarctic Peninsula, (3) a means to continue this orogen westward into New Zealand and eastern Australia, (4) an explanation for east-verging late structures in the EMFB, (5) similarity between Lower Palaeozoic volcanic rocks in the EMFB and the Transantarctic Mountains west of the Horlick Mountains, and (6) the major fault between the Transantarctic Mountains and the Ross Ice Shelf. Some problems for this model include (1) whether there is room to accommodate the present width of the EMFB, (2) whether this much rotation could have occurred, and (3) how to explain the Proterozoic metamorphic rocks of Haag Nunataks.

*Acknowledgements* It is a pleasure to acknowledge the financial support of the U.S. National Science Foundation and the logistic support of the U.S. Navy, in particular the aircraft pilots. I thank all my associates in this work for cooperation and discussion, most especially J.J. Anderson, T.W. Bastien, H.J. Meyer, R.H. Rutford, J.F. Splettstoesser, K.B. Sporli and G.F. Webers.

## REFERENCES

ANDERSON, V.H., 1960: The petrography of some rocks from Marie Byrd Land, Antarctica. *Project 825, Report No. 2, Part VIII, IGY Project No. 4.10.* The Ohio State University Research Foundation, Columbus, Ohio, 1-27.

BEHRENDT, J.C., DREWRY, D.J., JANKOWSKI, E. and GRIM, M.S., 1980: Aeromagnetic and radio echo ice sounding measurements show much greater area of the Dufek Intrusion, Antarctica. *Science, 209,* 1014-18.

BENTLEY, C.R. and CLOUGH, J.W., 1972: Antarctic subglacial structure from seismic refraction measurements; *in* Adie, R.J. (ed.) *Antarctic Geology and Geophysics,* Universitetsforlaget, Oslo, 683-91.

BUSHNELL, V.C. and CRADDOCK, C. (eds.), 1970: Geologic Maps of Antarctica. *Antarct. Map Folio Ser.,*Folio 12.

CLARKSON, P.D. and BROOK, M., 1977: Age and position of the Ellsworth Mountains crustal fragment, Antarctica. *Nature, 265,* 615-16.

CRADDOCK, C., 1969: Geology of the Ellsworth Mountains (Sheet 4, Ellsworth Mountains); *in* Bushnell, V.C. and Craddock, C. (eds.) Geologic Maps of Antarctica. *Antarct. Map Folio Ser.,* Folio 12, Pl.IV.

CRADDOCK, C., 1971: The structural relation of East and West Antarctica. *Proc. XXII Int. Geol. Cong. (1964), 4,* 278-92.

CRADDOCK, C., 1972: Antarctic tectonics; *in* Adie, R.J. (ed.) *Antarctic Geology and Geophysics,* Universitetsforlaget, Oslo, 449-55.

CRADDOCK, C., 1982: Antarctica and Gondwanaland (Review Paper); *in* Craddock, C. (ed.) *Antarctic Geoscience,* Univ. Wisconsin Press, Madison, 3-13.

CRADDOCK, C. (compiler), in press: Geologic Map of the Circum-Pacific Region, Antarctic sheet, 1:10,000,000 *Circum-Pacific Map Project,* Am. Assoc. Petrol. Geol., Tulsa.

CRADDOCK, C., ANDERSON, J.J. and WEBERS, G.F., 1964: Geologic outline of the Ellsworth Mountains; *in* Adie, R.J. (ed.) *Antarctic Geology,* North Holland, Amsterdam, 155-70.

DALZIEL, I.W.D. and ELLIOT, D.W., 1982: West Antarctica: problem child of Gondwanaland. *Tectonics, 1,* 3-19.

HJELLE, A., OHTA, Y. and WINSNES, T.S., 1982: Geology and petrology of the Southern Heritage Range, Ellsworth Mountains; *in* Craddock, C. (ed.) *Antarctic Geoscience,* Univ. Wisconsin Press, Madison, 599-608.

JANKOWSKI, E.J. and DREWRY, D.J., 1981: The structure of West Antarctica from geophysical studies. *Nature, 291,* 17-21.

KATZ, H.R., 1982: Post-Beacon tectonics in the region of Amundsen and Scott glaciers, Queen Maud Range, Transantarctic Mountains; *in* Craddock, C. (ed.) *Antarctic Geoscience,* Univ. Wisconsin Press, Madison, 827-34.

LONG, W.E., 1959: Preliminary report of the geology of the central range of the Horlick Mountains, Antarctica. *Report 825-2, Part VII, IGY Project 4.10,* The Ohio State University, Research Foundation, Columbus, Ohio, 2-23.

MIRSKY, A., 1969: Geology of the Ohio Range-Liv Glacier area (Sheet 17); *in* Bushnell, V.C. and Craddock, C. (eds.) Geologic Maps of Antarctica. *Antarct. Map Folio Ser.,*Folio 12, P1. XVI.

SCHMIDT, D.L. and FORD, A.B., 1969: Geology of the Pensacola and Thiel Mountains; *in* Bushnell, V.C. and Craddock, C. (eds.) Geologic Maps of Antarctica. *Antarct. Map Folio Ser.,* Folio 12, P1. V.

SCHOPF, J.M., 1969: Ellsworth Mountains: position in West Antarctica due to sea floor spreading. *Science, 164,* 63-6.

THIEL, E.C., 1961: Antarctica—one continent or two? *Polar Rec., 10, (67),* 335-48.

TREVES, S., 1959: Description of specimens from a nunatak in Ross-Weddell "graben". *Project 825, Report No. 2, Part V, IGY Project No. 4.10.* The Ohio State University Research Foundation, Columbus, Ohio, 2-5.

WEBERS, G.F., CRADDOCK, C., ROGERS, M.A. and ANDERSON, J.J., 1982: Geology of the Whitmore Mountains; *in* Craddock, C. (ed.) *Antarctic Geoscience,* Univ. Wisconsin Press, Madison, 841-47.

WATTS, D.R. and BRAMALL, A.M., 1981: Palaeomagnetic evidence for a displaced terrain in Western Antarctica. *Nature, 293,* 638-42.

YOSHIDA, M., 1982: Superposed deformation and its implication to the geologic history of the Ellsworth Mountains, West Antarctica. *Mem. Nat. Inst. Pol. Res., Japan, Spec. Issue, 21,* 120-71.

# THE PRE-BEACON GEOLOGY OF NORTHERN VICTORIA LAND: A REVIEW

J.D. Bradshaw, *Geology Department, University of Canterbury, Christchurch, New Zealand.*

M.G. Laird, *N.Z. Geological Survey, University of Canterbury, Christchurch, New Zealand.*

*Abstract* A review of our knowledge of the bedrock geology of northern Victoria Land, particularly developments since 1972, reveals important discoveries especially in stratigraphy, sedimentology and structural geology. The stratigraphy of the Bowers Supergroup has been clarified and newly discovered faunas establish a Cambrian age for several formations. The bedrock geology of the Anare Mountains has been mapped and a submarine fan origin for the Robertson Bay Group has been proposed. The lithology and structure of parts of the Wilson Group has been established. Despite the progress in description, controversies have developed in respect of regional interpretation. Problems include the validity of the term Ross Supergroup in northern Victoria Land; the character, importance and duration of the Ross Orogeny and the merits of a separate Silurian Borchgrevink Orogeny. Earlier research had emphasised the existence of three subparallel north—west trending belts underlain by Wilson Group, Bowers Supergroup and Robertson Bay Group from west to east. Recently it has been suggested that parts of the Wilson Group are metamorphosed Robertson Bay Group. If this can be demonstrated it provides major constraints on models for the geological evolution of northern Victoria Land.

We take as our starting point the paper of Dow and Neall (1972) and a review similar in scope to our own by Nathan and Skinner (1972). By that time the reconnaissance geology had been completed and most major map units delineated (Gair et al., 1969). The consensus seems to have been that a late Precambrian metamorphic unit, the Wilson Group, lay to the west, bordered to the east by units of the Ross Supergroup, namely the ?Cambrian Bowers Group and the Precambrian-Cambrian Robertson Bay Group. Both units were considered to have been deformed by the Ordovician Ross Orogeny and to be cut by the plutonic rocks of the Granite Harbour Intrusives. A younger Devono-Carboniferous acid igneous suite, the Admiralty Intrusives and associated Gallipoli Rhyolites, occurred mainly in Robertson Bay rocks but also as far south as Mt Murchison and west to the head of the Canham Glacier. Craddock (1972), while admitting that all units had been affected by the Ross Orogeny, suggested that the more easterly belts, mainly Bowers Group and Robertson Bay Group, had subsequently been involved in a younger Borchgrevink Orogeny characterised by Devonian (Admiralty) granites.

The next phase of geological study was marked by detailed work in the Evans Névé area and in the Bowers Mountains. Studies concentrated on sedimentology, stratigraphy, palaeontology and structure. Cambrian fossils were found initially in the Mariner Glacier area (Laird et al., 1972, 1974) and a regressive sequence of formations of Late Cambrian age overlain with slight unconformity by the marginal marine-fluviatile Camp Ridge Quartzite was described. Detailed facies analysis, palaeocurrent and palaeogeographic analysis was attempted (Andrews and Laird, 1976).

Subsequent work in the Bowers Mountains (Laird et al., 1976; Cooper et al., 1976) identified numerous additional Cambrian fossil localities and led to revised ideas of structure and stratigraphy. In particular an upper Middle and lower Upper Cambrian unit was widely mapped and the supposed basal unit of the Bowers Group, the Carryer Conglomerate, was shown to be one of the younger formations present. In addition many folds were identified (Bradshaw et al. 1982) which reduced the estimated thickness from 25 km to about 10 km (but this does not include the area of thickest Camp Ridge Quartzite). Because of the number of new units mapped a new stratigraphic hierarchy was introduced (Figure 1). Samples from the older units, the Sledgers Group (Molar Formation) and the Robertson Bay Group, contained certain acritarchs which at that time were considered to be restricted to the Vendian or earliest Cambrian (Cooper et al., 1982), consequently considerable breaks in sedimentation were envisaged between the Sledgers and Mariner Groups and also Mariner and Leap Year Groups.

The next major investigations were by members of the GANOVEX expedition summarised in Tessensohn et al. (1981). The contribution

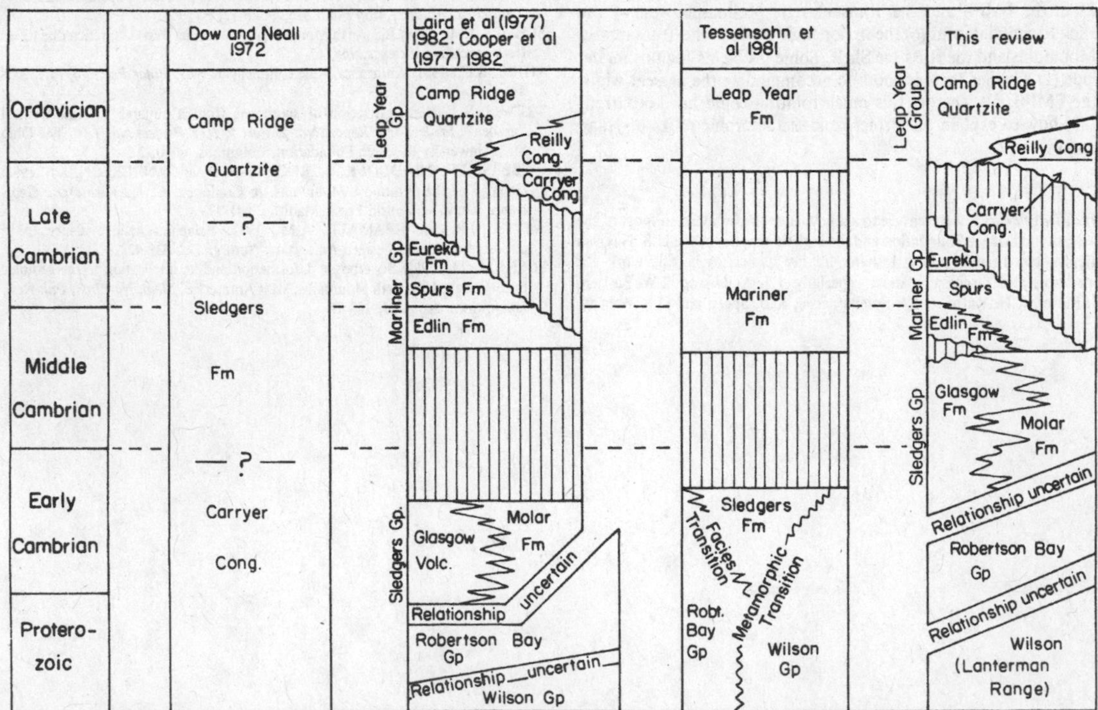


**Figure 1. Development of ideas on the stratigraphic subdivision and relationship of rock units in northern Victoria Land.**

of this expedition was diverse and included the mapping of the Anare Mountains, a previously unexplored area; the first coherent examination of the Robertson Bay Group across the full width of its outcrop, and their recognition of the Group as a submarine fan sequence; and the identification of extrusive volcanics at Litell Rocks, previously mapped as Ferrar Dolerite. The mobility of this expedition allowed its members to examine all the basement units in northern Victoria Land. From this examination they drew the radical conclusions that all deformation and metamorphism related to a single Ross Orogeny spanning the period 530 to 390 Ma, and that there were only one or perhaps two basement rock units present which were exposed at different metamorphic levels due to subsequent differential uplift. Stated briefly they concluded that the Sledgers Group (see Figure 1) was a lateral equivalent of the Robertson Bay Group or that it succeeds the latter conformably; that rocks previously mapped as Wilson Group and Robertson Bay Group in the Morozumi Range, and the metasediments of the Daniels Range (previously Wilson Group) were all Robertson Bay Group; that there is a metamorphic gradation between the Sledgers Group and the Wilson Group in the Lanterman Range. In effect the Wilson Group as a discrete unit was abandoned, and with it the idea of an older Precambrian basement which might represent the edge of the East Antarctic craton.

## STRATIGRAPHIC PROBLEMS

It is not easy to follow the development of ideas on northern Victoria Land. The results of the expeditions of the early seventies were presented at a SCAR meeting in Madison in 1977 and pre-prints were widely distributed. The reports from the GANOVEX expedition were published in 1981 and summarised in Tessensohn et al. (1981). Inevitably these latter reports include comment and criticism of the results presented in 1977 although the latter were not published until 1982. Tessensohn et al. (1981) appear to have misinterpreted some of the material in the pre-prints and this has led to some confusion.

Some difficulties are easily corrected, for example Laird et al. (1982) described the Husky Conglomerate and tentatively correlated it to the Bowers Supergroup or Robertson Bay Group (p.538). Kleinschmidt and Skinner (1981) appear to have assumed that this name was applied to strongly deformed conglomerate within the Wilson Group although the latter had been separately described and figured in Bradshaw et al. (1982) as part of the Wilson Group (Figure 100.4 of Bradshaw et al. is essentially the same as Kleinschmidt and Skinner, Figure 15). The name Husky Conglomerate was never intended for Wilson Group conglomerates. Subsequent work has shown that Husky Conglomerate type rocks are more common than originally thought and occur elsewhere interbedded in the Sledgers Group. We suggest the name is best dropped to avoid further confusion (Laird and Bradshaw, this volume).

Further confusion arises from a different philosophy concerning stratigraphic classification. Tessensohn et al. (1981) are opposed to the revised stratigraphy presented in 1977 (Laird et al., 1982) and in particular object to the status of the Sledgers, Mariner and Leap Year Groups and the Bowers Supergroup, preferring Sledgers, Mariner and Leap Year Formations (Figure 1). Their objection is based on the supposed priority of the Ross Supergroup and they propose a return to a Bowers Group in which all subdivisions are treated as formations or facies. However, if the term Leap Year Formation were to be adopted, the previously widely used Camp Ridge Quartzite, a unit at least 4 km thick and extending 300 km along the strike would be left in limbo as an informal unit, although a type section has been described and measured (LeCouteur and Leitch, 1964; Laird et al., 1974). No type or reference section has been proposed for the Leap Year Formation, and its two main components, the Carryer Conglomerate and the Camp Ridge Quartzite are so dissimilar in detrital composition and geologic setting that it is most unlikely that they are laterally equivalent facies.

The term Ross Supergroup was proposed by Grindley and Warren (1964) to replace the term "Ross System" and was intended to apply to rocks of "late Proterozoic or early Palaeozoic age that were deposited in one or a series of closely linked geosynclinal basins of deposition that formed along the present site of the Transantarctic Mountains" (p.314). Although a number of component groups were named, including the Robertson Bay Group but specifically excluding the Camp Ridge Quartzite, it is clear that the concept of the Supergroup is firmly tied to the concept of a Ross Geosyncline, a term that was already in wide use. Since then the term Ross Supergroup has been used in such widely different ways that Laird and Bradshaw (1982) recommended it be abandoned. In northern Victoria Land it is inappropriate because it implies acceptance of the existence of a Ross Geosyncline and belief in a hypothesis that is widely considered obsolete. It is likely that the rocks of the Bowers Mountains were deposited in basins quite separate from those of southern Victoria Land and the Central Transantarctic Mountains. There is no lithological indication of continuity and as used in northern Victoria Land the term Ross Supergroup embraces rocks which may well post-date the Ross Orogeny as originally defined.

## PROBLEMS OF TECTONIC HISTORY

The status of the Ross Orogeny is closely linked to that of the Ross Geosyncline. The former term was introduced for Late Cambrian to Early Ordovician events in southern Victoria Land that folded and metamorphosed the Ross Supergroup (Gunn and Warren, 1962) and has since been extended in scope. Some geologists, for example Craddock (1972), have listed events as early 1000 Ma as early Ross Orogeny, while Grikurov et al. (1972) used the term to embrace all Riphean to Devonian tectonism in the Transantarctic Mountains. Within this long period other orogenies have been named, for example, Beardmore Orogeny (Grindley and McDougall, 1969) and Borchgrevink Orogeny (Craddock, 1972). Comparison with better understood Mesozoic and Cenozoïc tectonically active areas suggests that although certain zones, particularly plate margins, may be active for long periods, the naming as "orogenies" episodes of activity covering up to 500 Ma is unrealistic if these are conceived of as interruptions to sedimentation. Such long "orogenies" are coeval with sedimentation and are distinct from discrete phases of epizonal folding, plutonism, unconformity and isotopic closure, events which are also termed "orogenies". The Ross Orogeny started as an event of the second type and some authors, for example Skinner (1982) have been at pains to demonstrate intraorogenic sedimentation and to expand the definition to embrace events as young as 500 Ma within the Ross Orogeny. Others have recognised no such constraints and use the term as a label for long periods of geological development.

### Ross Orogeny in Northern Victoria Land

The terms Ross Orogeny and Ross Supergroup were generated during the period of geological reconnaissance prior to 1972. They were widely applied in northern Victorian Land though the evidence of the age of deformation was slight, and subsequently the term Borchgrevink Orogeny was introduced for what was admitted to be a second deformation for many of the rocks involved. Discovery of Middle and Late Cambrian faunas and the determination of isotopic ages for cleavage formation in the Bowers Supergroup poses interesting problems. It seems clear that there was an isotopic event which affected Wilson Group, Robertson Bay Group and the lowest units of the Bowers Supergroup at 520-480 Ma. This event could be interpreted as the Ross Orogeny except that in the Bowers Supergroup basin there appears to be no significant deformation until after the deposition of the Leap Year Group. Recent faunal discoveries compound the problem by eliminating any significant stratigraphic break between the Sledgers and Mariner Groups (Laird and Bradshaw, this volume; Cooper et al., this volume). Mariner and Leap Year rocks give consistently younger ages of cleavage formation (425-400 Ma) and although between 0.5 km and perhaps 1 km of Mariner rocks may have been removed in the northwest prior to Leap Year sedimentation, the discordance is barely perceptible in outcrop except at Reilly Ridge (Bradshaw et al., 1982, figure 100.1). Thus the Bowers Supergroup is essentially concordant throughout the trough and was not strongly deformed during the Ross Orogeny (in the strict sense). Late Cambrian to Early Ordovician activity was restricted to mild tilting or gentle folding near the western margin. (N.B. this margin was also active during Mariner sedimentation, Bradshaw et al., 1982, p.813).

The Leap Year Group is significantly coarser than the Mariner Group and the thick polymict Carryer Conglomerate and the voluminous Camp Ridge Quartzite probably reflect orogeny and regional uplift in an adjacent terrain. The Group was probably deposited in a fault bounded trough developed in an extensional regime. Whether its eventual deformation in the Silurian should be attributed to a "Borchgrevink Orogeny" or simply regarded as an event of local significance is a matter for discussion.

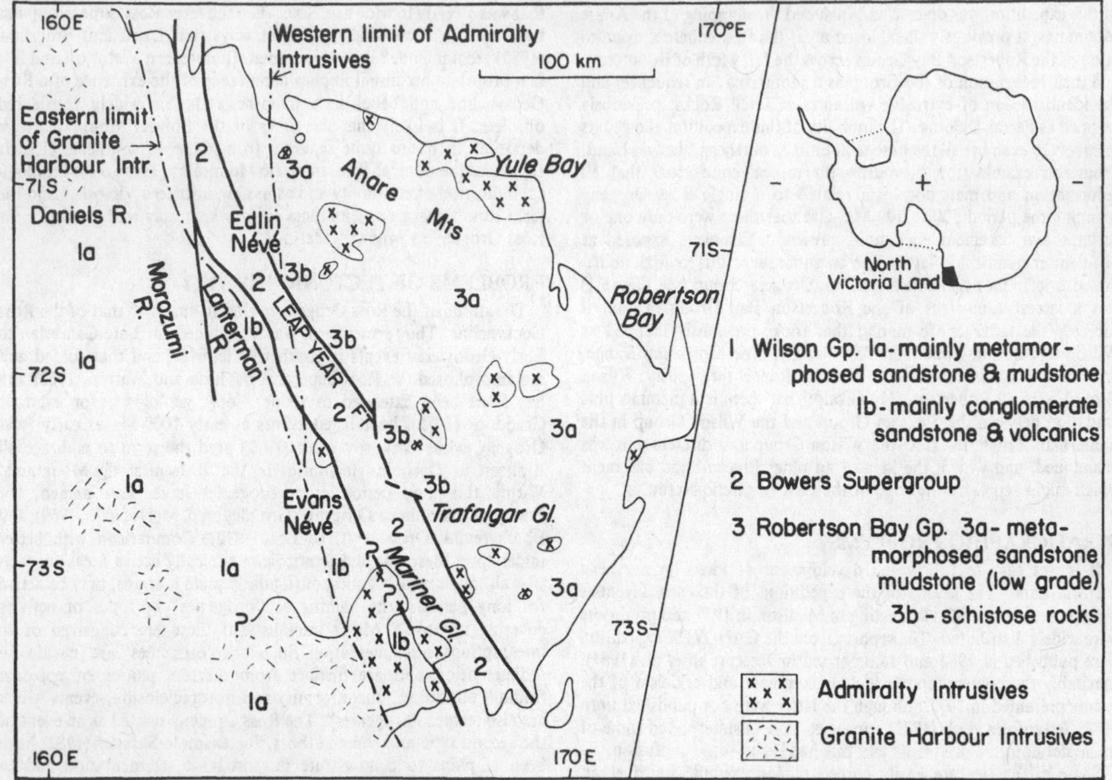


**Figure 2. Much simplified map of the distribution of major rock units in northern Victoria Land.**

West of the Bowers Supergroup rocks Granite Harbour Intrusives with typical Ross Orogeny K-Ar ages (i.e. Late Cambrian—Early Ordovician) are widespread and clearly imply a period a period of elevated temperatures in both granitoids and metasediments. The importance of deformation at this time is as yet unknown.

East of the Bowers Supergroup outcrop the problem is more difficult. The age of sedimentation of the Robertson Bay Group remains uncertain and though many dates in the range 500-475 Ma have been published (Adams et al., 1982) most of the rocks collected were from a zone more structurally complex than typical Robertson Bay rocks (Bradshaw et al., 1982). It is now known that these schistose rocks occur widely in a zone parallel to the Leap Year Fault (Figure 2). Too few typical Robertson Bay rocks have been dated to permit realistic interpretation.

Tessensohn et al. (1981) preferred to extend the Ross Orogeny to cover a period from 530 to 390 Ma. They mistakenly suggest that the Borchgrevink Orogeny was only supported by two dates, seeming to ignore nine dates presented by Adams et al. (1982) although paradoxically the latter are incorporated in their figure 17. As previously pointed out (Bradshaw et al., 1982) the dates do not suggest continuous activity and there is a marked clustering into two groups with a gap of about 30 Ma in the Middle and Late Ordovician. This suggests a significant period of quiescence that may correspond to the time of deposition of the Leap Year Group. We feel that if the term Ross Orogeny is to be used in northern Victoria Land it should be restricted to events of Late Cambrian and Early Ordovician age.

## DISCRIMINATION OF MAJOR BASEMENT UNITS

Tessensohn et al. (1981) clearly believe that the distinction into three subparallel belts (Figure 2) has been overemphasised and the essential unity of the basement has been overlooked. This question is of great importance at the present time. Older ideas of the geosyncline have been largely replaced by more actualistic models based on plate tectonics. Recently these models have in turn been questioned by those (e.g. Davis et al., 1978) who believe that many orogenic belts are a collage of discrete terrains in tectonic contact. Some of the problems of geological history outlined above might be explained if the main belts had only achieved their present relative position by the time of intrusion of the Devonian granites. This type of explanation has been foreshadowed by Harrington (1980) who has suggested the Leap Year Fault represents a fossil plate boundary. The possibility that northern Victoria Land contains two or more distinct terrains would be completely negated if the proposition noted above of Tessensohn et al. (1981) is proved to be correct. If the metasediments of the Morozumi Range and the Daniels Range to the west of the Bowers Mountains are Robertson Bay Group then northern Victoria Land has been a single unit since the Early Cambrian.

Evidence presented for the second view so far is based mainly on field appearances and it is quite clear from the descriptions that the rocks in the Daniels Range (granite, migmatite and complexly deformed metasediments, Tessensohn et al., 1981; Kleinschmidt and Skinner, 1981) are now very different from Robertson Bay Group. The suggestion is that the parent rocks were like Robertson Bay Group and probably were Robertson Bay Group. In this matter the lower grade rocks in the Morozumi Range are particularly important. They do resemble Robertson Bay Group in being an alternating sequence of turbiditic sandstones and mudstone (Wright, 1981; Field and Findlay, this volume) but at present there is no firm evidence for correlation.

It appears that the traditional Wilson Group may contain two suites of metasediments, monotonous metasediments of the Daniels Range type, and a more varied assemblage seen in the Lanterman Range which includes thick conglomerate and amphibolite units. The latter may be informally termed the Lanterman metamorphics. The range of lithologies superficially resembles that of the Sledgers Group and it has been suggested (Tessensohn et al., 1981) that the Lanterman metamorphics are a more metamorphosed equivalent. In detail, however, the parent lithologies appear to have been different, and there is evidence that the Sledgers Group rests unconformably on the Lanterman metamorphics (Laird and Bradshaw, this volume). Finally it has also been argued that the Sledgers Group might be a lateral equivalent of the Robertson Bay Group (Tessensohn et al., 1981). While it is true that some thinly bedded sequences are similar and that turbidites occur in both, the most distinctive lithologies of the Sledgers Group, volcanic breccia, basic lavas, roundstone conglomerate, limestone and carbonate olistostromes are conspicuously absent from the

Robertson Bay Group. The change in character seems too rapid for facies change unless there is considerable telescoping along the Leap Year Fault. In addition the diametrically opposed palaeocurrent pattern (Wright, 1981; Laird et al., 1982) make it unlikely that they shared a common basin of deposition.

## CONCLUSION

Since 1972 considerable advances have been made in understanding the stratigraphy and sedimentology of the Cambro-Ordovician Bowers Supergroup. Current problems seem largely semantic. During the same period a clearer picture has emerged of the Robertson Bay Group structure and sedimentation but the age and tectonic history are still uncertain. The Wilson Group may contain two distinct units but much further work is required on the area west of the Rennick Glacier. The relationship of the Robertson Bay Group east of the Leap Year Fault to the remainder of northern Victoria Land remains a topic of debate.

## REFERENCES


ADAMS, C.J., GABITES, J.E., LAIRD, M.G., WODZICKI, A. and BRADSHAW, J.D., 1982: Potassium-argon geochronology of the Precambrian-Cambrian Wilson and Robertson Bay Groups and Bowers Supergroup, northern Victoria Land, Antarctica; *in* Craddock, C. (ed.) *Antarctic Geoscience*. Univ. Wisconsin Press, Madison, 543-48.

ANDREWS, P.B. and LAIRD, M.G., 1976: Sedimentology of a Late Cambrian regressive sequence (Bowers Group), northern Victoria land, Antarctia. *Sedimt. Geol., 16,* 21-44.

BRADSHAW, J.D., LAIRD, M.G. and WODZICKI, A., 1982: Structural style and tectonic history in northern Victoria Land; *in* Craddock, C. (ed.) *Antarctic Geoscience*. Univ. Wisconsin Press, Madison, 809-16.

COOPER, R.A., JAGO, J.B., MACKINNON, D.I., SIMES, J.E. and BRADDOCK, P.E., 1976: Cambrian fossils from the Bowers Group, northern Victoria Land, Antarctica (Preliminary Note). *N.Z. J. Geol. Geophys., 19,* 283-8.

COOPER, R.A., JAGO, J.B., MACKINNON, D.I., SHERGOLD, J.H. and VIDAL, G., 1982: Late Precambrian and Cambrian fossils from northern Victoria Land and their stratigraphic implications; *in* Craddock, C. (ed.) *Antarctic Geoscience*. Univ. Wisconsin Press, Madison, 629-33.

COOPER, R.A., JAGO, J.B., ROWELL, A.J. and BRADDOCK, P. (this volume): Age and correlation of the Cambro-Ordovician Bowers Supergroup, northern Victoria Land.

CRADDOCK, C., 1972: Antarctic Tectonics; *in* Adie, R.J. (ed.) *Antarctic Geology and Geophysics*. Universitetsforlaget, Oslo, 449-55.

DAVIS, G.A., MONGER, J.W.H. and BURCHFIEL, B.C., 1978: Mesozoic construction of the Cordilleran "collage", central British Columbia to central California; *in* Howell, D.G. and McDougall, K.A. (eds.) *Mesozoic paleogeography of the western United States*. Soc. Econ. Paleont. Mineral. Pacific Section, Los Angeles, 1-32.

DOW, J.A.S. and NEALL, V.E., 1972: Summary of the geology of the lower Rennick Glacier, north Victoria Land; *in* Adie, R.J. (ed.) *Antarctic Geology and Geophysics*. Universitetsforlaget, Oslo, 339-44.

FIELD, B.D. and FINDLAY, R.H. (this volume): The sedimentology of the Robertson Bay Group, northern Victoria Land.

GAIR, H.S., STURM, A., CARRYER, S.J. and GRINDLEY, G.W., 1969: The geology of northern Victoria Land (Sheet 13); *in* Bushnell, V.C. and Craddock, C. (eds.) Geologic maps of Antarctica. *Antarct. Map Folio Ser.,* Folio 12, P1.

GRIKUROV, G.E., RAVICH, M.G. and SOLOVIEV, D.S., 1972: Tectonics of Antarctica; *in* Adie, R.J. (ed.) *Antarctic Geology and Geophysics*. Universitetsforlaget, Oslo, 457-68.

GRINDLEY, G.W. and MCDOUGALL, I., 1969: Age and correlation of the Nimrod Group and other Precambrian rock units in the Central Transantarctic Mountains. *N.Z. J. Geol. Geophys., 12,* 391-411.

GRINDLEY, G.W. and WARREN, G., 1964: Stratigraphic nomenclature and correlation in the western Ross Sea region; *in* Adie, R.J. (ed.) *Antarctic Geology*. North Holland, Amsterdam, 314-33.

GUNN, B.M. and WARREN, G., 1962: Geology of Victoria Land between the Mawson and Mulock Glaciers, Antarctica. *Geol. Surv. N.Z., Bull., 71.*

HARRINGTON, H.J., 1980: Reconstruction of the Australian sector of Gondwanaland from Cambrian to Early Permian. *Fifth Gondwana Symposium, abstracts, Wellington, N.Z.,* 71.

KLEINSCHMIDT, G. and SKINNER, D.N.B., 1981: Deformation styles in the Basement rocks of north Victoria Land, Antarctica. *Geol. Jb. B41,* 155-99.

LAIRD, M.G., ANDREWS, P.B. and KYLE, P.R., 1974: Geology of northern Evans Névé, Victoria Land, Antarctica. *N.Z. J. Geol. Geophys., 17,* 587-601.

LAIRD, M.G., ANDREWS, P.B., KYLE, P.R. and JENNINGS, P., 1972: Late Cambrian fossils and the age of the Ross Orogeny, Antarctica. *Nature 238,* 34-36.

LAIRD, M.G. and BRADSHAW, J.D. (this volume): New data on lower Palaeozoic Bowers Supergroup, northern Victoria Land.

LAIRD, M.G., BRADSHAW, J.D. and WODZICKI, A., 1976: Re-examination of the Bowers Group (Cambrian), northern Victoria Land, Antarctica (Preliminary Note). *N.Z. J. Geol. Geophys., 19,* 275-82.

LAIRD, M.G., BRADSHAW, J.D. and WODZICKI, A., 1982: Stratigraphy of the Upper Precambrian and Lower Palaeozoic Bowers Supergroup, northern Victoria Land, Antarctica; *in* Craddock, C. (ed.) *Antarctic Geoscience*. Univ. Wisconsin Press, Madison, 535-42.

LECOUTEUR, P.C. and LEITCH, E.C., 1964: Preliminary report on the geology of an area southwest of upper Tucker Glacier, northern Victoria Land; *in* Adie, R.J. (ed.) *Antarctic Geology*. North Holland, Amsterdam, 229-36.

NATHAN, S. and SKINNER, D.N.B., 1972: Recent advances in the pre-Permian geology of northern Victoria Land; *in* Adie, R.J. (ed.) *Antarctic Geology and Geophysics*. Universitetsforlaget, Oslo, 333-8.

SKINNER, D.N.B., 1982: Stratigraphy of low grade metasedimentary rocks of the Skelton Group, southern Victoria Land—does Teall Greywacke really exist? *in* Craddock, C. (ed.) *Antarctic Geoscience*. Univ. Wisconsin Press, Madison, 555-63.

TESSENSOHN, F., DUPHORN, K., JORDAN, H., KLEINSCHMIDT, G., SKINNER, D.N.B., VETTER, U., WRIGHT, T.O. and WYBORN, D., 1981: Geological comparison of basement units in north Victoria Land, Antarctica. *Geol. Jb. B41,* 31-88.

WRIGHT, T.O., 1981: Sedimentology of the Robertson Bay Group, north Victoria Land, Antarctica. *Geol. Jb. B41,* 127-38.

# THE SEDIMENTOLOGY OF THE ROBERTSON BAY GROUP, NORTHERN VICTORIA LAND

B.D. Field, *N.Z. Geological Survey (DSIR), University of Canterbury, Christchurch, New Zealand.*

R.H. Findlay[1], *Antarctic Division (DSIR), P.O. Box 13247, Christchurch, New Zealand.*

[1]*Present Address: Geology Department, University of Tasmania, P.O. Box 252C, Hobart, Tasmania 7001, Australia.*

*Abstract* The Robertson Bay Group is a submarine fan deposit of probable Cambrian age. Most of the sandstone beds are turbidites though fluidised sediment flow deposits and debris flow deposits (pebbly mudstones) are also present, particularly in the south. Units of red argillite may be restricted to the southern part of the Group. The sandstone beds occur commonly in either thickening up-sequence or thinning up-sequence units several metres thick. Several subfacies of a submarine fan system are recognised. If the red argillites are pelagic then the Robertson Bay Group in the northeast side of the Millen Range may record an early phase of progradation of part of the fan system. Palaeocurrent direction data indicate sediment transport to the northwest to north. Most of the Group was derived probably from a mainly metamorphic terrain. The trace fossil assemblage suggests a Cambrian rather than Vendian age, but is probably facies controlled as much as age diagnostic. The deformation of the Group, on limited radiometric evidence, may be Early Ordovician.

A zone of schists on the northeast side of the Leap Year Fault is possibly older than the Robertson Bay Group and may form the basement to part of the Group, though in the Millen Range the contact is faulted.

The extent of the Group southwest of the Leap Year Fault is still uncertain. The correlation of the Morozumi Range hornfelses with the Group is questioned and the similarities of the Robertson Bay Group to the southern part of the Molar Formation are discussed.

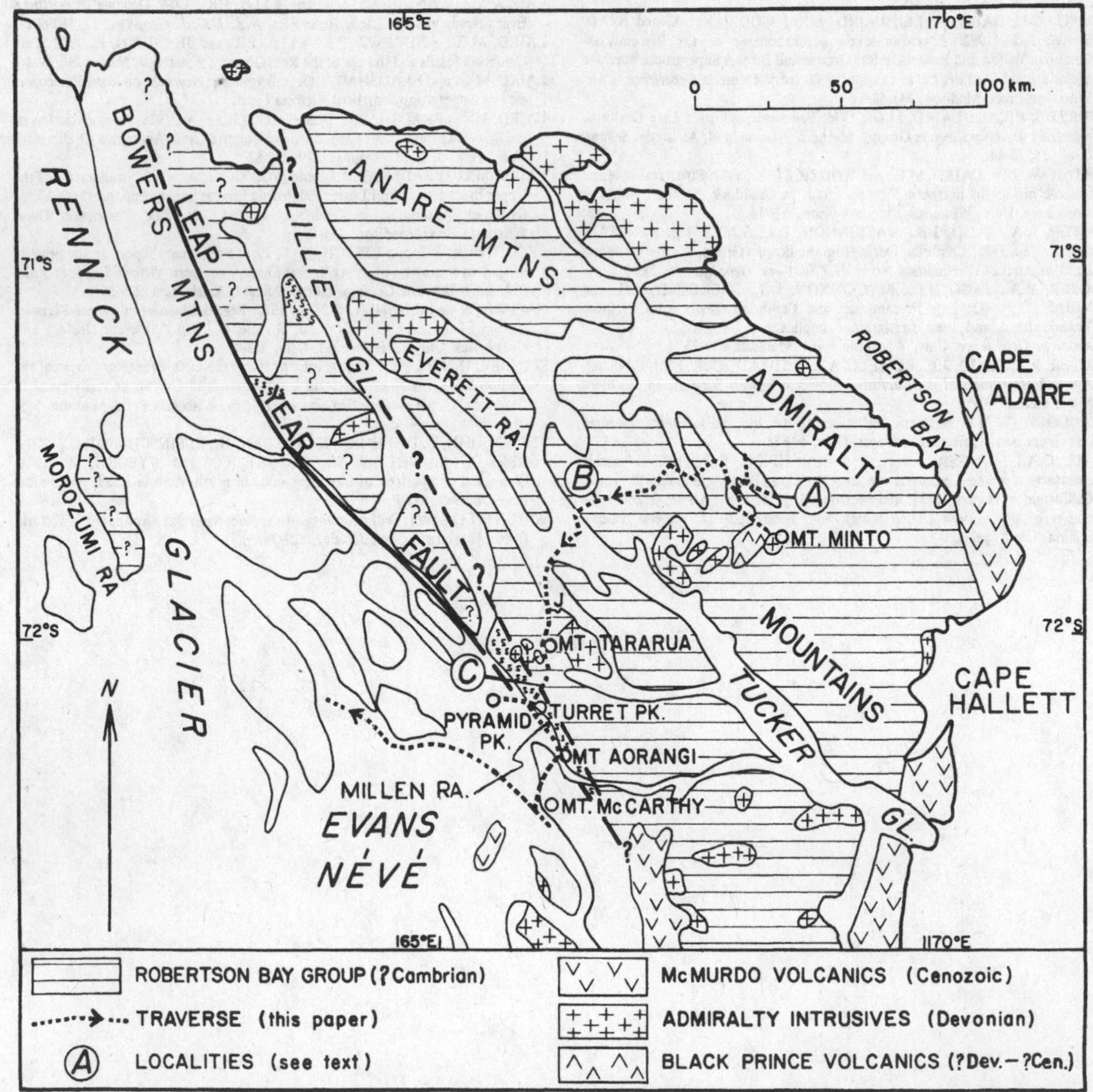


**Figure 1.** **Robertson Bay Group, northern Victoria Land. J = Jutland Glacier; "S"'s = schist (viz. "Millen Range Schist", Findlay and Field, this volume). The fault shown along the Lillie Glacier is proposed by Findlay and Field (this volume). The Black Prince Volcanics are described by Findlay and Field (in press).**

Rastall and Priestley (1921) first described the Robertson Bay Group, at Robertson Bay (Figure 1) as a thick "slate-greywacke series". Harrington et al. (1967) recorded the Robertson Bay Group in the Cape Hallett-Tucker Glacier area as a regularly bedded (cm-3 m), commonly graded, fine to medium grained, quartzose metagreywacke/meta-argillite sequence. Le Couteur and Leitch (1964) found, between the Tucker Glacier and the Evans Névé, some graded sandstone beds which they interpreted as turbidites, some grit beds, a 45 cm thick bed of quartzose conglomerate and some red argillites. They mapped the Pyramid Peak to Mt McCarthy area as Robertson Bay Group. Crowder (1968) and Sturm and Carryer (1970) also interpreted the greywacke sandstone beds as deep water turbidites. Dow and Neall (1974) described quartzo-feldspathic sandstones and siltstones from the eastern Bowers Mountains. Wodzicki et al. (1982) studied the petrology of a schistose zone near the western edge of the Group. Cooper et al. (1982) and Adams et al. (1982) have discussed the age of the Group.

Wright (1981) interpreted all the sandstone beds as turbidites, derived largely from the south to southeast and found them to contain clasts mainly of quartz and rock fragments from a mainly continental source terrain. He reported that the facies characteristics fit those of the middle to outer fan and basin plain environments described by Mutti and Ricchi Lucchi (1978). The trace fossils found were interpreted as suggesting a Late Precambrian age.

The expedition undertaken by the authors in the 1981/2 season covered a large part of the Group which had not been studied before and, being a largely north-south traverse (Figure 1), it was hoped to shed more light on the size and configuration of the sediment body. Sections were measured at 20 localities south of the head of the Shipley Glacier, near Mt Minto. The traverse also crossed the structural grain of the Group (Findlay and Field, this volume).

## SEDIMENTOLOGY

### Bed Types

The sandstone beds are mainly decimetre bedded, with a grain size of mainly fine to medium sand but with a large number of very fine sand beds and a few medium to coarse sand beds. Most can be interpreted readily as turbidites as they display two or more of the Bouma sequence intervals, the most common being a basal graded one, which forms frequently 60-90% of the bed thickness.

Though many beds are so fine grained that it could not be determined whether they are graded or not, a significant proportion of the fine to medium sandstone beds lack obvious grading and appear to be either massive or laminated throughout, commonly with rippled tops. Although some of these beds may have been deposited as traction carpets beneath turbidity currents, we interpret others as fluidised sediment flow deposits. Sand volcanoes were noted on several of these beds.

The sequence exposed in the ridge leading southwest to Turret Peak, in the Millen Range, contains matrix-supported pebbly mudstones and conglomerates that are interpreted as mass flow deposits (Figure 2c). Pebble clasts are generally well rounded, of low to moderate sphericity and 1-5 cm (up to 40 cm) in diameter. They consist dominantly of quartz and fine sandstone with lesser amounts of red chert and one cobble of white, unfossiliferous limestone. Sturm and Carryer (1970) recorded the presence of conglomerate and limestone bands north of Mt Craven (northern Everett Range) but did not describe them.

A type of unit observed only south of the Tucker Glacier consists of 1-2 m thick packets of 1-3 cm beds of sandstone interbedded with 0.1-2 cm beds of mudstone. The sandstones have sharp bases, are massive, or planar or ripple laminated, and usually have sharp rather than gradational tops. They are inferred to be levee or overbank deposits.

Background sediments comprise centimetre to decimetre thick beds of mudstone and decimetre to metre thick units of mudstone with centimetre spaced laminae and thin beds of current bedded siltstone and very fine sandstone. Units of very fine sandstone 5-30 cm thick of 1-2 cm amplitude unimodal trough foresets are found occasionally; they are interpreted as recording periods of particularly strong bottom current activity.

Red argillites were found at two localities. Near Mt Thomson they occur in a sequence containing more than usually green mudstones but

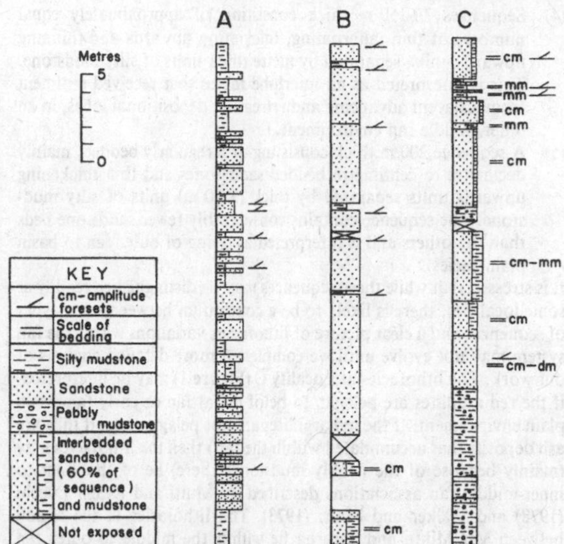


**Figure 2. Measured sections of the Robertson Bay Group. (a) Section of much of the Group (Locality A). (b) Channelised sandstone sequence (Locality B). Note the thinning upwards cycles. (c) Sequence with pebbly mudstones of debris flow origin (Locality C).**

apart from the colours of the argillites, the bedforms in the sequence do not differ markedly from those at other localities. At locality C (Figure 1) they occur in a sequence containing units of pebbly mudstone. No red argillites were seen north of the Jutland Glacier. Red metagreywacke and meta-argillite were noted by Harrington et al. (1967) at Football Mountain near the mouth of the Tucker Glacier, but they may be absent from the northern and western parts of the Group (Wright, 1981). Sturm and Carryer (1970) described a few persistent red argillite beds about 2 m thick, distributed throughout the sequence, but did not state where.

The red argillites could be pelagic sediments or submarine ash deposits (though no interbedded volcanics were found with them). If they do have a pelagic component then pelagic sediments must interfinger with fan sediments in the southern part of the Group.

### Fan Facies

A preliminary study of the up-sequence variations in the thicknesses of the sandstone beds indicates that several subfacies of the fan system can be distinguished. Sandstone beds, interbedded with mudstone, commonly form units in which the beds thicken up-sequence or thin up-sequence. The thickening upwards units are generally up to 25 m thick (average about 6 m) and the thinning upwards units up to 15 m thick (average about 4 m).

As well as the metre scale units there are larger scale sequences of units and beds. Examples of these are:

(1) A sequence consisting of stacked, thinning upwards units each 10-20 m thick of metre thick beds of pebbly mudstone and units of the levee facies described above and red argillites (Figure 2c). This sequence may record the initial advance of the fan system into a possibly pelagic setting (after Rupke, 1977). Alternatively, if the red argillites are not of pelagic origin, it may record deposition in an inner to inner-middle fan environment.

(2) Near Mt Shelton (Locality B, Figure 1) a sequence of metre-bedded (up to 8 m thick) sandstone beds show evidence of fluidisation and indistinct lamination and some contain small pebbles (Figure 2b). The beds occur in three thickening upwards units and comprise the most sand dominated, thickly bedded sequence seen on our traverse. They are interpreted as channel fill mass flow deposits that accumulated probably in an inner to inner-middle fan setting.

(3) A 150 m thick sequence of mainly thickening upwards units but in which the sandstone beds generally thin upwards through the sequence overall. This could record an episode of fan construction in a middle fan environment in which the depocentre of a prograding lobe shifted laterally.

(4) Sequences 30-150 m thick consisting of approximately equal numbers of thin, alternating, thickening upwards and thinning upwards units, separated by metre thick units of silty mudstone. This is interpreted as an interlobe facies that received sediment from adjacent advancing and retreating depositional lobes, in an outer middle fan environment.

(5) A sequence 200 m thick consisting of irregularly bedded, mainly decimetre to centimetre bedded sandstones and thin thickening upwards units separated by thick (5-10 m) units of silty mudstone. The sequence contains considerably fewer sandstone beds than the others and is interpreted as being of outer fan to basin plain facies.

It is stressed that while these sequences may be distinguished readily at some localities, there is likely to be a continuum between some types of sequences and a clear picture of lithofacies variations within the fan system may not evolve until we complete a more detailed analysis of our work. The lithofacies at Locality C (Figure 1) may be interpreted, if the red argillites are pelagic, to be of distal fan or early fan/basin plain environment. If the red argillites are not pelagic (e.g., if they are ash deposits that accumulated within the fan) then the lithofacies may (mainly because of the pebbly mudstones there) be of the inner to inner-middle fan associations described by Mutti and Ricchi Lucchi (1978) and Walker and Mutti (1973). The lithofacies in the region between Mts Minto and Tararua lie within the middle to outer fan associations. Wright (1981) concluded that the sequences exposed in the northern and western parts of the Robertson Bay Group were deposited in middle to outer fan and basin plain settings. Hence there is a trend apparent within the Group for the fan facies to become more proximal towards the south, in a fan system that prograded towards the north. However, the conglomerate and limestone bands recorded by Sturm and Carryer (1970) suggest that the model may not be this simple.

## Palaeocurrents

The current directions recorded on our traverse have been restored allowing for the plunge of the fold structures. As the deformation involved flexural flow, flexural slip and tangential longitudinal strain during progressive inhomogenous shortening, any restoration of current lineations will give approximate directions only. This comment may apply also to the work of Wright (1981).

The dominant palaeocurrent direction is to 320° and is based mainly on flute cast measurements (Figure 3a). This appears also to be the main current direction in the lithofacies near Turret Peak (Locality C, Figure 1). Subsidiary current directions, based on rippled tops of sandstone beds, are to 020° and to 260°. The one to 020° is formed by measurements mainly from the sandstone dominated channelised sequence described in the Mt Shelton area (Locality B, Figure 1) and thus may indicate only the local orientation of the channel axes. The one to 260° is more difficult to interpret, but may reflect the effects of contour currents on the final stages of turbidite deposition. Foresets in mudstone and siltstone units suggest that contour currents flowed to either the southwest or northeast. There appears to be no consistent geographic variation in palaeocurrent directions between Mt Minto and the Millen Range (Figure 3b).

Wright (1981) recorded palaeocurrents of mainly north to northwest trends in the northern and western parts of the Group, and a northwest to northeast trend at Robertson Bay, where fold plunges vary markedly (Kleinschmidt and Skinner, 1981). He was uncertain of the relationships between these palaeocurrent directions and the inferred axis of the depositional basin because he was studying turbidites of outer fan and basin plain environments where current directions can be either transverse or longitudinal to the axis of the basin. By combining Wright's results with those cited here it now seems more likely that the northwest to north trend indicates the main direction of sediment transport for the fan system.

## Petrology

The most noticeable petrological features seen in the field are the high percentage of quartz in the sandstones, the presence of centimetre to decimetre sized brown slightly calcareous concretions in both sandstone and thin siltstone beds and the yellow (?pyritic) and red (?haematitic) staining on a few of the thin siltstone beds.

Preliminary thin section studies suggest that the quartz component of the sand sized clasts is usually over 70% and that much of this is probably clasts of vein quartz. Feldspar is much less common than quartz. Rock fragments of siltstone, chert and probably trachyte occur in minor amounts. Harrington et al. (1967) described quartzose metagreywackes from the southeastern part of the Group. Sturm and Carryer (1970) found the sandstones to be composed mainly of quartz, feldspar and minor rock fragments, with limestone and conglomerate bands present north of Mt Craven.

Wright (1981) inferred, from a preliminary study, that the Group was derived from a continental terrain consisting of medium to high grade rocks with lesser amounts of metasedimentary and volcanic rocks. The preliminary results from our traverse suggest that high grade metamorphic and basic igneous rock fragments may be very rare or absent in the central to southern parts of the Group. The high proportion of probable vein quartz and the apparent lack of acidic igneous rock fragments suggest a metamorphic source terrain for the sediments in this region.

For the area southwest of the proposed (Findlay and Field, this volume) Lillie Glacier Fault, Sturm and Carryer describe, from the King Range (at the head of the Lillie Glacier), numerous unsorted angular grains of feldspar and rock fragments of acid volcanics,

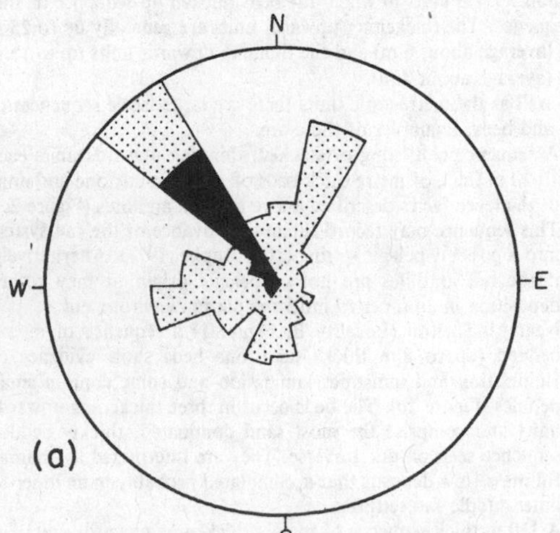


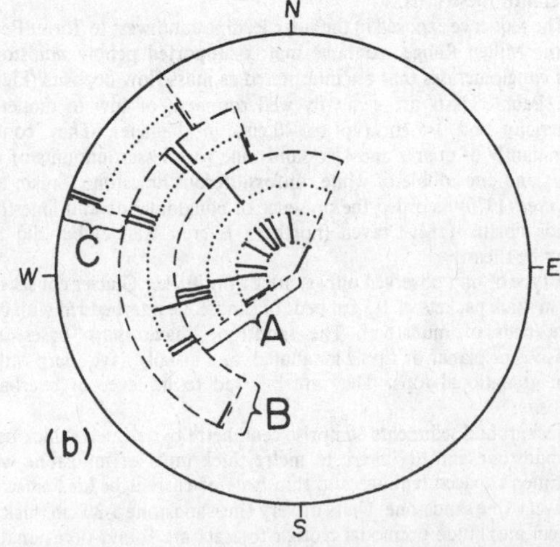


**Figure 3. Palaeocurrent measurements from the Robertson Bay Group, restored allowing for the plunges of the folds (all shown with respect to true north; 20° classes). (a) 72 measurements from 16 localities between Localities A and C (Figure 1). Solid black indicates 13 flute cast measurements from seven localities. (b) Shows lack of geographic variation in palaeocurrent directions. A: Measurements from four sections near Locality A (Figure 1). B: Measurements from two sections near Locality B. C: Measurements from two sections near Locality C.**

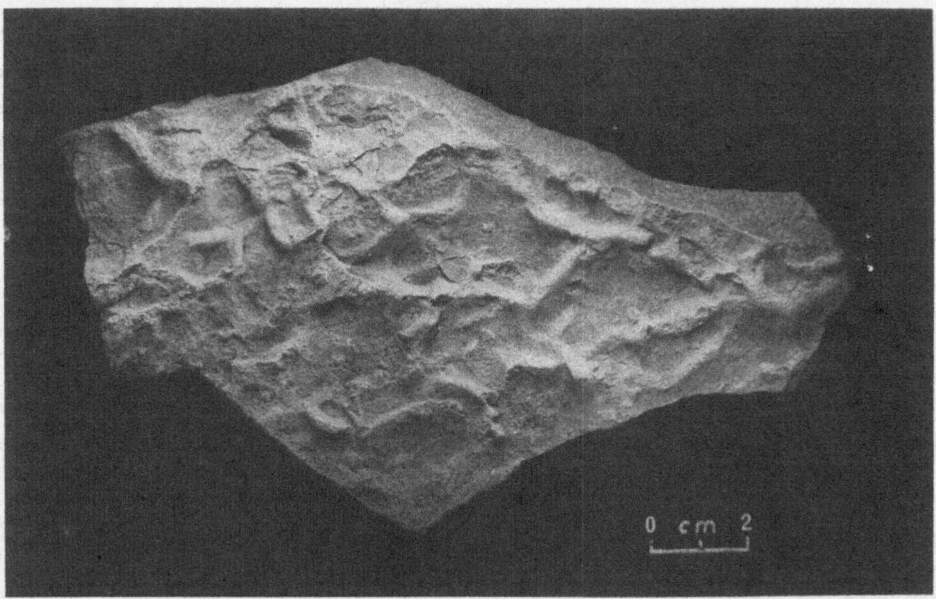


**Figure 4. Trace fossils from Locality C (Figure 1), showing angular regularities of about 60/120°. The repetition of this angle suggests that the organism(s) were starting to develop more efficient feeding patterns and are therefore likely to be of post-Vendian age. (Photo: A. Downing).**

greywacke and quartzite. For the rocks of the eastern Bowers Mountains, Dow and Neall (1974) describe a dominant quartz component, with minor albite grains and very minor rock fragments and Wodzicki et al. (1982) deduce a dominantly granitic provenance. The relationship of these rocks to those northeast of the Lillie Glacier is not well understood (see below and Figure 1).

## PALAEONTOLOGY AND AGE

Although we found no body fossils, 45 samples of trace fossils were collected from several localities. The most common forms are short (generally less than 5 cm long, but up to 15 cm), 1-4 mm diameter burrows parallel to bedding in the mudstones. The traces are simple, being usually unbranched and with straight to open curved nonrepetitive forms. Some poorly preserved burrows 2 mm in diameter and oblique to bedding were seen in two sandstone beds. Sand volcanoes were distinguished as such on the basis of their morphology and the lack of other adjacent biogenic like structures.

The traces resemble some of those recorded commonly from flysch of Cretaceous-Tertiary age as well as some of Precambrian to Cambrian age and are probably facies controlled as much as age controlled. However, Frey and Seilacher (1980) suggest that shallow water and some eurybathic ichnogenera appeared first, in the Late Proterozoic and that deeper water forms lagged in development, being comparably established by the Late Cambrian or Early Ordovician. Thus the fact that trace fossils are fairly common in the Robertson Bay Group argues against assigning a Vendian age to the Group, particularly as some of the traces appear to be more regular than "scribbling traces" (Figure 4). Cooper et al. (this volume) record the presence of Vendian to Middle Cambrian acritarchs in rocks from the southern Bowers Mountains, between the Leap Year Fault and the Lillie Glacier.

Only two samples of undoubted Robertson Bay Group (Adams et al., 1982) have yielded pre-Admiralty Granite K-Ar ages. These ages (491 and 492 Ma) were derived from low grade, thinly bedded "slate" and suggest that the minimum age of the Group is either Late Cambrian or Early Ordovician, depending on the poorly defined absolute age of the Cambrian-Ordovician boundary.

## EXTENT OF THE ROBERTSON BAY GROUP

The northern and eastern boundaries of the Group are unknown (Figure 1). The southwestern boundary is uncertain but may be the proposed Lillie Glacier Fault and the schists described in the Millen Range (Findlay and Field, this volume). The schists display an earlier deformation, possibly representing nappe tectonics, than does the Robertson Bay Group and could suggest that they are a distinct, possibly older, unit in-faulted between the Bowers Supergroup and the Robertson Bay Group. If they are older, then they and possibly also the schists on the northeast side of the Leap Year Fault in the eastern Bowers Mountains (Wodzicki et al., 1982) may form part of the basement for the Group. The southwest boundary of the schists is the Leap Year Fault.

Sturm and Carryer (1970), Wright (1981) and Tessensohn et al. (1981) correlate the hornfelses and greenschists of the Morozumi Range with the Robertson Bay Group (Figure 1). The correlation of virtually unfossiliferous submarine fan deposits across a poorly understood major tectonic feature must rely mainly on radiometric, petrological and geochemical data rather than sedimentology, as fan facies vary considerably even within one fan system. The extensive contact metamorphism caused by the Granite Harbour Intrusives in the Morozumi Range (Dow and Neall, 1974) makes it difficult at present to date the sediments or their age of regional metamorphism radiometrically. Dow and Neall noted the quartzose nature of the rocks but the contact metamorphism makes a more detailed comparison with the Robertson Bay Group difficult. Though Harrington et al. (1967) and Nathan (1976) give geochemical data for the southeastern part of the Robertson Bay Group there appear to be no data available on the geochemistry of the schists of the Morozumi Range. Correlation between the two areas must remain doubtful.

The exposures at Pyramid Peak and at the nunatak north of Mt Watt were examined briefly. It is not clear whether they should be mapped as Robertson Bay Group or Sledgers Group (Molar Formation, see Laird and Bradshaw, this volume). The sequence at Pyramid Peak does not appear to differ greatly from those in known Robertson Bay Group rocks; it even includes red argillites, which appears to be a characteristic of the southern part of the Group. The sequence at the nunatak north of Mt Watt is poorly exposed but could not be distinguished readily from sequences in the Robertson Bay Group. The solution to this problem will depend on the results of more detailed petrological and geochemical studies.

## CONCLUSIONS

Because of the lack of stratigraphic marker horizons and hence structural control, the possibility of repetition by faulting cannot be ruled out and little can be said about the size or shape of the depositional basin of the Robertson Bay Group, except that the submarine fan system within it was probably more than 100 km in length and breadth. The basin suffered downwarping of over 3000 m (based on a minimum thickness estimated near Mt Minto), and that it was fed from the southeast to south relative to its present orientation. Petrological work so far suggests a mainly metamorphic source terrain. The age of the Group is more likely to be Cambrian than Vendian and is probably no younger than Early Ordovician.

The southwestern extent of the Group is not yet well defined. The schists on the northeast side of the Leap Year Fault may be older than the Robertson Bay Group and may therefore form part of the basement to the Group, if not part of its source terrain. Correlation of the Robertson Bay Group with units southwest of the Bowers Tectonic Zone remains questionable.

*Acknowledgements* Our fieldwork was carried out during the International Northern Victoria Land Expedition 1981/82 and was supported by Antarctic Division (N.Z. DSIR). Air support was provided by VXE-6 squadron, U.S. Navy. We thank Dr J. Splettstoesser (NSF) and Messrs P.V. Colbert (ATT) and E. Saxby (Antarctic Division, N.Z. DSIR) for their valuable help with field logistics. We gratefully acknowledge the help of Messrs W. Fowlie and W. Atkinson (Antarctic Division, N.Z. DSIR) for their considerable help and enjoyable company in the field. We thank also Drs Laird, Oliver and van der Linger for their comments on this paper and Mr E.T. Annear for his drafting.

## REFERENCES

ADAMS, C.J., GABITES, J.E., BRADSHAW, J.D., LAIRD, M.G. and WODZICKI, A., 1982: Potassium-argon geochronology of the Precambrian-Cambrian Wilson and Robertson Bay Groups and Bowers Supergroup, northern Victoria Land, Antarctica; *in* Craddock, C. (ed.) *Antarctic Geoscience.* Univ. Wisconsin Press, Madison, 543-8.

COOPER, R.A., JAGO, J.B., MACKINNON, D.I., SHERGOLD, J.H. and VIDAL, G., 1982: Late Precambrian and Cambrian fossils from northern. Victoria Land and their stratigraphic implications; *in* Craddock, C. (ed.) *Antarctic Geoscience.* Univ. Wisconsin Press, Madison, 629-33.

COOPER, R.A., JAGO, J.B., ROWELL, A.J. and BRADDOCK, P., (this volume): Age and correlation of the Cambrian-Ordovician Bowers Supergroup, northern Victoria Land.

CROWDER, D.F., 1968: Geology of part of northern Victoria Land, Antarctica. *U.S. Geol. Surv., Prof. Paper, 600-D,* 95-107.

DOW, J.A.S. and NEALL, V.E., 1974: Geology of the Lower Rennick Glacier, northern Victoria Land, Antarctica. *N.Z. J. Geol. Geophys., 17,* 659-714.

FINDLAY, R.H. and FIELD, B.D., (in press): Black Prince Volcanics, Admiralty Mountains, Antarctica: Reconaissance observations. *N.Z. Ant. Rec.*

FINDLAY, R.H. and FIELD, B.D., (this volume): Tectonic significance of deformations affecting the Robertson Bay Group and associated rocks, northern Victorian Land, Antarctica.

FREY, R.W. and SEILACHER, A., 1980: Uniformity in marine invertebrate ichnology. *Lethaia, 13,* 183-207.

HARRINGTON, H.J., WOOD, B.L., MCKELLAR, I.C. and LENSEN, G.J., 1967: Topography and geology of the Cape Hallet-Tucker Glacier district Antarctica. *N.Z. Geol. Surv., Bull., 80.*

KLEINSCHMIDT, G. and SKINNER, D.N.B., 1981: Deformation styles in the basement rocks of North Victoria Land, Antarctica. *Geol. Jb., B41,* 155-98.

LAIRD, M.G. and BRADSHAW, J.D., (this volume): New data on the Lower Palaeozoic Bowers Supergroup, northern Victoria Land.

LE COUTEUR, P.C. and LEITCH, E.C., 1964: Preliminary report on the geology of an area southwest of upper Tucker Glacier, northern Victoria Land; *in* Adie, R.J. (ed.) *Antarctic Geology.* North Holland, Amsterdam, 229-36.

MUTTI, E. and RICCI LUCCHI, F., 1978: Turbidites of the northern Appenines; Introduction to facies analysis. Translated by T.H. Nilsen, from Le torbiditi dell'Appennino settentrionale; introduzione all' analisi di facies, *Soc. Geol. Ital., Mem.* 1972, 161-99. *Int. Geol. Rev., 20,* 125-66.

NATHAN, S., 1976: Geochemistry of the Greenland Group (Early Ordovician), New Zealand. *N.Z. J. Geol. Geophys., 19,* 683-706.

RASTALL, R.H. and PRIESTLEY, R.E., 1921: The slate-greywacke formation of Robertson Bay: *British Antarctica ("Terra Nova") Expeditions, 1910. Nt. Hist. Rept: Geology,* 121-9.

RUPKE, N.A., 1977: Growth of an ancient deep sea fan. *J. Geol., 85,* 725-44.

STURM, A. and CARRYER, S.J., 1970: Geology of the region between the Matusevitch and Tucker Glacier, northern Victoria Land. *N.Z. J. Geol. Geophys., 13,* 408-35.

TESSENSOHN, F., DUPHORN, K., JORDAN, H., KLEINSCHMIDT, G.F., SKINNER, D.N.B., VETTER, U., WRIGHT, T.O. and WYBORN, D., 1981: Geological comparison of the basement units in North Victoria Land, Antarctica. *Geol. Jb., B41,* 31-88.

WALKER, R.G. and MUTTI, E., 1973: Turbidite facies and facies associations; *in* Middleton, G.V. and Bouma, A.H. (eds.) *Turbidites and Deep Water Sedimentation Short Course, Anaheim, 1973.* Pacific Section, Soc. Econ. Palaeont. Mineral., Los Angeles, California, 119-58.

WODZICKI, A., BRADSHAW, J.D. and LAIRD, M.G., 1982: Petrology of the Wilson and Robertson Bay Groups and Bowers Supergroup, northern Victoria Land, Antarctica; *in* Craddock, C. (ed.) *Antarctic Geoscience.* Univ. Wisconsin Press, Madison, 549-54.

WRIGHT, T.O., 1981: Sedimentology of the Robertson Bay Group, North Victoria Land, Antarctica. *Geol. Jb., B41,* 127-38.

# TECTONIC SIGNIFICANCE OF DEFORMATIONS AFFECTING THE ROBERTSON BAY GROUP AND ASSOCIATED ROCKS, NORTHERN VICTORIA LAND, ANTARCTICA.

R.H. Findlay[1], *Antarctic Division (D.S.I.R.), P.O. Box 13247, Christchurch, New Zealand.*

B.D. Field, *N.Z. Geological Survey (D.S.I.R.), Canterbury University, Christchurch, New Zealand.*

[1]*Present Address: Geology Department, University of Tasmania, G.P.O. Box 252C, Hobart, Australia, 7001.*

*Abstract* Low-grade metagreywacke of the Robertson Bay Group and Chlorite Sub-zone III schists in the Millen Range, formerly incorporated in the Robertson Bay Group, are juxtaposed tectonically by a fault, possibly extending on a trend of 340° along Lillie Glacier, and form two structural domains. Millen Range schists display a possibly early Vendian, syn-metamorphic nappe-like style of deformation seen neither in the Robertson Bay Group to the northeast nor in the Bowers Supergroup to the southwest. Both domains are folded into generally asymmetric, upright to slightly overturned structures with an axial plane cleavage striking 320°. Correlation of folds in both domains is discussed.

Orientation of these folds suggests formation in a dextral wrench regime. Orientation of mapped and inferred faults in northern Victoria Land suggests a Riedel Shear pattern in a dextral transcurrent displacement zone.

The Robertson Bay Group occupies the northeastern part of northern Victoria Land (Figure 1). It is faulted along the Leap Year Fault (Figure 1) against the non-metamorphic, marine to fluvial, Middle Cambrian to Ordovician Bowers Supergroup (Laird and Bradshaw., 1982; Cooper et al., this volume) which forms a northwest trending belt along the eastern side of Rennick Glacier (Figure 1). These rocks overlie unconformably (Laird, pers. comm., 1982) amphibolitic rocks of the Wilson Group, which occupies also the western side of Rennick Glacier.

First described briefly by Rastall and Priestley (1921), the general structure of the Robertson Bay Group was outlined by LeCouteur and Leitch (1964), Harrington et al. (1967), Crowder (1968), Gair (1967), Sturm and Carryer (1970), Kleinschmidt and Skinner (1981), and Bradshaw et al. (1982). These workers recorded the low but variable plunge of fold axes, well-developed axial-plane cleavage, "shear" in fold hinges, open style of folding and lack of regional polyphase deformation. Harrington et al. (1967) suggested that the ubiquitous axial plane cleavage may in places be folded, but described no locality; Bradshaw et al. (1982) made the important observation that between the upper Graveson and Champness Glaciers (Figure 1) there are scattered areas with inverted bedding, and an early metamorphic foliation that predates the more obvious folds and their axial plane cleavage.

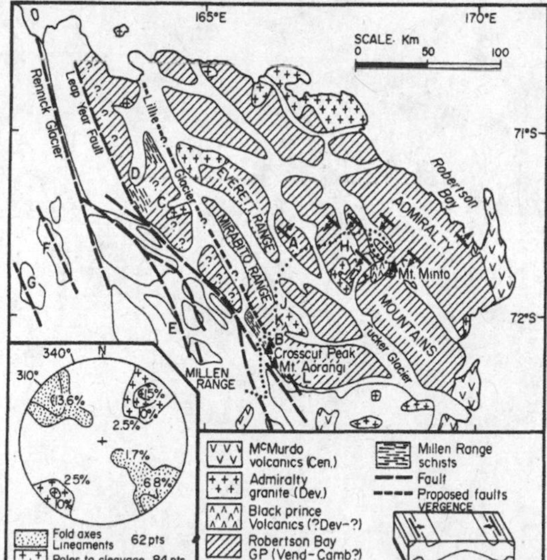


**Figure 1. Locality and sketch geological map to show Robertson Bay Group and supposed Robertson Bay Group. Inset indicates geometries of folds and cleavage in northeastern domain. Localities indicated by letters are: A-Homerun Range; B-Turret Ridge; C-Champness Glacier; D-Graveson Glacier; E-Salamander Range; F-Morozumi Range; G-Helliwell Hills; H-Gadsden Peaks; J-Jutland Glacier; K-Mt Tarurua; L-Wood Glacier. Dotted line indicates sledge route. The Black Prince Volcanics are described by Findlay and Field (1982).**

This paper summarises the results of a detailed study along a section across the structural grain of the Robertson Bay Group (Figure 1) and discusses the tectonic significance of deformations affecting the Robertson Bay Group (northeastern domain) and the previously unknown polydeformed schists of the Millen Range (southwestern domain).

## SUMMARY OF LITHOLOGIES

The Robertson Bay Group consists of an interbedded sequence of low-grade to non-metamorphic (Skinner, pers. comm., 1982) "greywacke" sandstones and siltstones, interbedded on centimetre to 10 m scale, with rare intercalations of red, green, and purple siltstone in the southwest. Wright (1981) and Findlay and Field (this volume) interpret most of the sandstones as turbidites, derived from the south to southeast and deposited in middle fan to basin-plain environments.

Schistose rocks are exposed in a northwest-striking belt in the Millen Range (Figure 1). These rocks are faulted near Turret Peak against siltstones, sandstones, and occasional units of pebbly mudstone which are considered to be Robertson Bay Group (Field and Findlay, this volume). The schists in the Millen Range consist of Chlorite Sub-zone III (Hutton and Turner, 1936) quartzo-feldspathic schist, grey phyllite, graphitic phyllite and chlorite-epidote-actinolite greenschist. The greenschists are exposed clearly in Crosscut Peak and Mount Aorangi where they overlie with a complex, folded discordance, grey phyllites and quartzo-feldspathic schists that are overturned locally.

The greenschist is layered on metre scale and contains occasional metre thick units of very fine-grained, dark quartzite (metachert?). It is deformed internally and on outcrop scale displays a mylonitic texture that intensifies towards the discordant contact with the underlying rocks. We found no evidence suggesting an intrusive contact.

## STRUCTURAL GEOLOGY

### Southwestern Domain (Millen Range)

The Millen Range forms a structural domain distinguished from regions to the northeast by polyphase folding involving schistose rocks.

*$F_1$ folds* These centimetre to 50 m scale (Figure 2) syn-metamorphic folds are tight to isoclinal, and possess an axial plane schistosity ($S_1$) which is the dominant metamorphic fabric. The few axes measured trend northwest and plunge shallowly. Small scale isoclinal folds, thrusting, and the mylonite-like texture in the greenschist at Crosscut Peak are attributed to $F_1$ (see later discussion).

*$F_2$ folds* are open to tight, upright to slightly overturned to the northeast, possess an axial plane crenulation cleavage showing in five thin sections good evidence for solution, but little recrystallisation of mica, along cleavage planes. $F_2$ folds range in scale from centimetre to kilometre half-wavelength. The large, open fold (Figure 3) below Crosscut Peak is included in $F_2$. As these are upright to slightly asymmetric, plunge shallowly and trend at 320°, they were correlated by LeCouteur and Leitch (1964) with the regional northwest-trending folds of the northeastern domain.

Figure 2. 50 m scale isoclinal fold in east face of Crosscut Peak. Telephotograph from locality A in Figure 6.

Figure 3. Telephotograph from northwest (Mt Tararua) of folded discordance (A) in northwest face of Crosscut Peak (incorrectly named Mt Aorangi in Crowder 1968). Note a lower discordance (B) which may be an $F_1$ isoclinal fold. The dark, apparently massive unit above "A" is the greenschist; layered units are quartzo-feldspathic schists and phyllites.

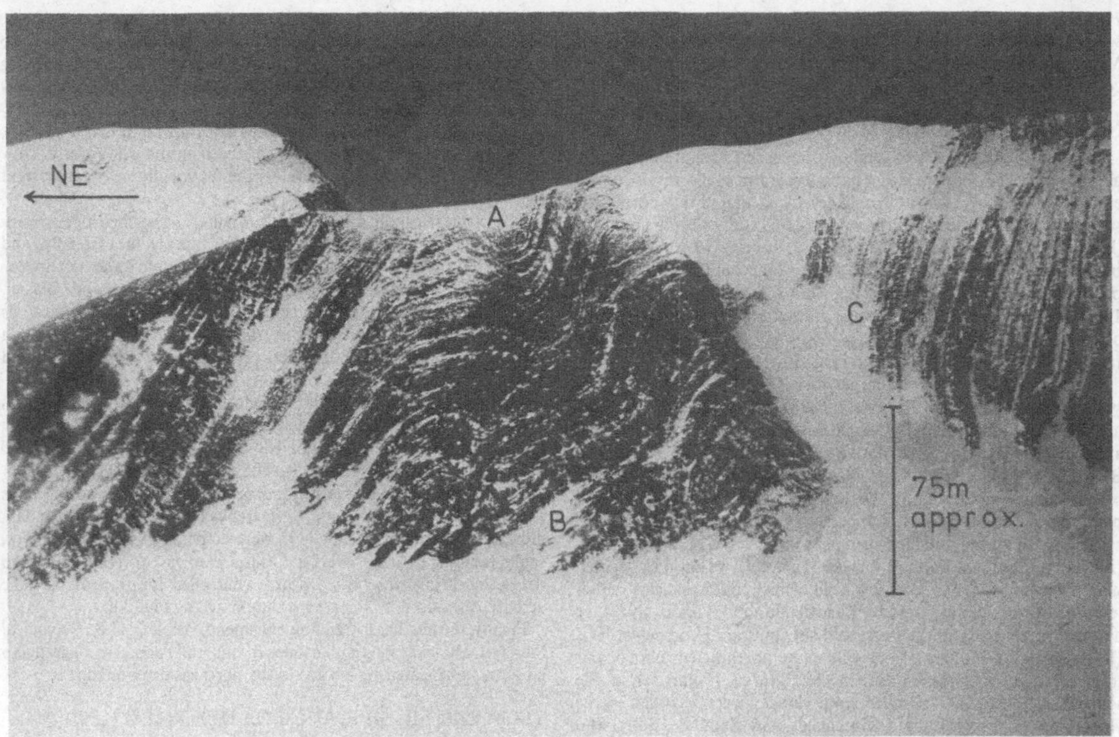


**Figure 4. Multi-hinged folds picked out by snow on northern ridge of unnamed range trending northwest from Mt Faget, Admiralty Mtns. Note single-hinged fold A forms multiply-hinged structure some 150 m below at B. Note also younger sub-coaxial kink-folds in right of photo (C).**

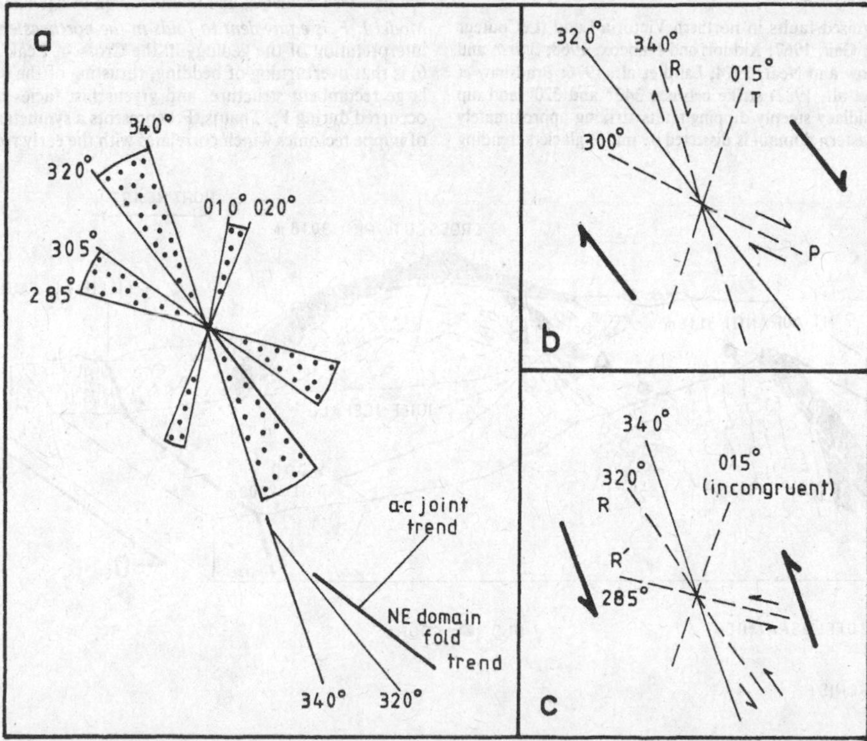


**Figure 5. (a) orientation of folds in northeastern domain, faults and inferred faults in northern Victoria Land. Stippled areas enclose faults and inferred faults. (b) interpretation of fault system in (a) as a dextral Riedel Shear system about displacement zone trending 320° (proposed Lillie Glacier Fault strikes approximately 330°). Note that the 010°-020° trend of Cenozoic-Recent extensional (Jordan, 1981) faults in the Robertson Bay region follows the trend of tensional features associated with this dextral transcurrent system. (c) interpretation as sinistral Riedel Shear system about a sinistral displacement zone trending 340°. Orientation of extensional faults trending 010°-020° is incongruent to R and R' shears, as is the fold trend in the northeastern domain.**

*$F_3$ folds* are associated spatially with the steeply-dipping northwest-striking fault which, in Turret Ridge (Figure 1), separates the Millen Range schists from the Robertson Bay Group. The folds consist of small kinks generally plunging 15°WNW and 50-60°SW.

### Northeastern Domain (Mt Minto to Millen Range)

Folds here are upright to slightly overturned and are commonly asymmetric. They plunge variably but generally shallowly (0-20°) to either 310-320° or 130-140°, and develop a ubiquitous northwest-striking axial plane cleavage, dipping predominantly steeply southwest (Figure 1). Cleavage pre-dates limb-faults (Ramsay, 1974), which may be the "axial plane shears" and "nick-point faults" described by previous workers (Harrington et al., 1967; Kleinschmidt and Skinner, 1981). The two maxima in cleavage orientation (Figure 1) reflect the often strongly-convergent cleavage fans in siltstones, and divergent fans and lenticularity of cleavage in sandstones.

Regional vergences of the larger folds (Figure 1) imply a major synclinorial axis between Robertson Bay (see Kleinschmidt and Skinner, 1981) and the Admiralty Mountains, and a possible anticlinorial axis between the Homerun Range and Gadsden Peaks (Figure 1). Between Mt Minto and Tucker Glacier (Figure 1), the style of folds often changes rapidly along the fold train and up-dip of the axial planes, from chevron folds, sometimes with collapsed hinges (Ramsay, 1974), to multiply-hinged structures with curved or straight limbs (Figure 4). South of Jutland Glacier (Figure 1) we found a tighter, more regular style of chevron fold. This could indicate either increased strain, as suggested by Kleinschmidt and Skinner (1981), or a greater uniformity in thickness and distribution of competent beds (Ramsay, 1974). Plunges of one fold in the northeastern part of our traverse, and of some folds near Mt Tararua (Figure 1) in the southwest, exceed 40°. Similar steep plunges were recorded in the Robertson Bay region by Kleinschmidt and Skinner (1981) who presented evidence for later cross-folding about a northeasterly trend; these structures are discussed more fully later.

### Faulting (Figure 5)

Principal recognised faults in northern Victoria Land (LeCouteur and Leitch, 1964; Gair, 1967; Riddols and Hancox, 1968; Sturm and Carryer, 1970; Dow and Neal, 1974; Laird et al., 1976; Bradshaw et al., 1982; Laird et al., 1982) strike between 340° and 320° and dip steeply, with subsidiary steeply-dipping faults striking approximately 300°. The northeastern domain is dissected by major glaciers trending principally 340° to 320° with a subsidiary trend between 280° and 305° which, as they parallel known faults to the southwest, may also follow important fault zones. Jordan (1981) records steeply-dipping normal faults striking 010°-020° in and east of the Robertson Bay region.

Recorded slip senses of mapped faults (Sturm and Carryer, 1970; Dow and Neal, 1974; Jordan, 1981) are both normal and reverse. Dow and Neal (1974) consider briefly the possibility of transcurrent faulting, but do not discuss it further.

Our proposed Lillie Glacier Fault (Figure 1) cuts the ridges northeast of Turret and Crosscut Peaks, dips steeply to vertically, and would trend along a glacial lineament formed by the northwest-trending Lillie Glacier and the southeast-trending Wood Glacier. Tessensohn has indicated (pers. comm., 1982) that if this fault extends as far northwest as proposed by us, then it is older than the Devonian Admiralty Granite that he considers to extend from the northwest end of the Everett Range across Lillie Glacier (Figure 1).

All mapped faults in northern Victoria Land post-date structures associated with the Ross Orogeny, and three major faults trending 340° cut Beacon Supergroup in Helliwell Hills and the Morozumi and Salamander Ranges (Figure 1). In the Robertson Bay region, Cenozoic volcanics are cut by faults trending 010-020°, and Cenozoic vents and dykes follow the 050° and 340° trends (Jordan 1981). These data indicate that faults trending 340° were active in post-Jurassic times, faults trending 010-020° were active in Cenozoic to Recent times and that pre-existing structural features controlled location of volcanic activity; they do not indicate maximum age of faulting.

Finally, the northeast-trending lineaments visible in U.S. Geological Survey 1:250,000 topographic maps of our study area trend parallel to a-c joints associated with folds in the northeastern domain.

## TECTONIC SIGNIFICANCE OF DEFORMATIONS

### Fold Correlation

There are two interpretations of the significance of folding in the study area depending on fold correlation between northeastern and southwestern domains, and interpretation of age-dating.

*Model 1: $F_2$ is equivalent to folds in the northeastern domain* Our interpretation of the geology in the Crosscut Peak region (Figure 6) is that overturning of bedding, thrusting of the upper limb of a large recumbent structure, and greenschist facies metamorphism occurred during $F_1$. That is, $F_1$ represents a synmetamorphic phase of nappe tectonics which correlates with the early folding recorded

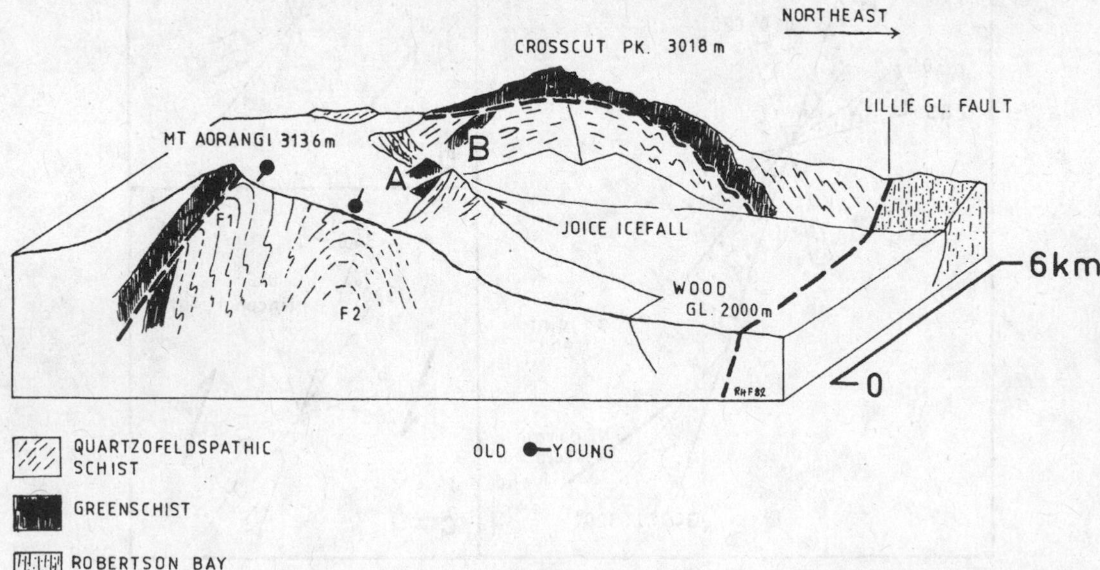


**Figure 6. Sketch block diagram of Crosscut Peak-Mt Aorangi region compiled from telephotographs of east face of Crosscut Peak and field data. Reversals of younging in Mt Aorangi indicates pre-$F_2$ isoclinal folding. Intensely-sheared greenschist at "A" strikes towards massive body interpreted as greenschist at "B" which is cross-cut by greenschist forming summit of Crosscut Peak. Interpretation—the discordance (see also Figure 3) is an $F_1$ thrust; $F_1$ involved syn-metamorphic isoclinal folding and thrusting of upper limb over lower; $F_1$ structures are refolded by $F_2$.**

by Bradshaw et al. (1982) southwest of Lillie Glacier and which also involved inversion of bedding and metamorphic segregation..

The identical geometries of $S_2$ in the southwestern domain and the Cambro-Ordovician (Adams et al., 1982) axial plane cleavage in the northeastern domain suggest that $F_2$ correlates with the northwest trending folds in the northeastern domain and therefore all these structures are Cambro-Ordovician.

The age of $F_1$ is uncertain. Bradshaw et al. (1982) noted that the earliest deformation in the schists southwest of Lillie Glacier, here held equivalent to our $F_1$, had not affected the Husky Conglomerate of the Bowers Supergroup southeast of the Leap Year Fault. As Laird et al. (1982) regarded the Husky Conglomerate as Vendian, Bradshaw et al. (1982) argued that $F_1$ was pre- or early-Vendian. Laird and Bradshaw (1982) now regard Husky Conglomerate as within the lower part of the Middle Cambrian Sledgers Group. Therefore $F_1$ is no younger than pre-Middle Cambrian.

Kleinschmidt and Skinner (1981) suggest that the steep plunges in the Robertson Bay Group are due to cross folding. Other than the steep plunges, we found no structures such as interference folds or later cleavage which would indicate cross folding, and the folds in the northeastern domain do not fold earlier structures. Secondly, the axial planes of the numerous small scale folds in the northeastern domain bifurcate rapidly and fold axes show plunges ranging between 0° and 25° across a few centimetres. This confirms noncylindrical folding in the northeastern domain and indicates that the steep plunges could result from inhomogeneous shortening during progressive deformation.

Finally, if the rocks southwest Fand northeast of Lillie Glacier were deposited in the same submarine fan system (Wright, 1981) then it is surprising that there is no known evidence for an $F_1$ event northeast of the lineament formed by the Lillie and Wood Glaciers. This could imply that the rocks southwest of Lillie Glacier are not part of the Robertson Bay Group, and that the two suites could have been juxtaposed along our Lillie Glacier Fault after $F_1$, before or during $F_2$, and before intrusion of the Devonian Admiralty Granites.

*Model 2: $F_1$ is equivalent to folds in the northeastern domain* At present it is uncertain which of the two axial plane fabrics in the schists southwest of Lillie Glacier were dated by Adams et al. (1982). Our $S_2$ is, on the basis of a study of five thin sections, a solution cleavage accompanied by minimal recrystallisation of mica. The dominant metamorphic fabric is $S_1$, which is an axial plane schistosity formed by profuse growth of phyllosilicates. Therefore, the eleven ages spanning 470-490 Ma from southwest of Lillie Glacier (Adams et al., 1982) could indicate $F_1$ is Cambro-Ordovician and correlates therefore with the dated folds (two ages from the Mirabito Range; Adams et al., 1982) in the northeastern domain. If so, $F_2$ could be synchronous with the Siluro-Devonian deformation recorded in the Bowers Supergroup (Adams et al., 1982; Figure 67.2). There is no record of this later deformation northeast of Lillie Glacier lineament unless the steep plunges of some of the folds in the northeastern domain indicate subcoaxial refolding. However, we consider that the steep plunges are explained satisfactorily by the observed noncylindrical folding.

Although we prefer model 1, model 2 has important implications for the relationship between the Bowers Supergroup, the schistose rocks southwest of Lillie Glacier, and the rocks in our northeastern domain. Adams et al. (1982) record a Cambro-Ordovician isotopic event in the Sledgers Group of the Bowers Supergroup, which is not recorded in the concordantly overlying Mariner and Leap Year Groups. However, there are no structures known in the Sledgers Group that could be related to the Cambro-Ordovician isotopic event. This lack of structures in the Sledgers Group correlatable with our supposed Cambro-Ordovician $F_1$ folds could indicate juxtaposition of Robertson Bay Group and associated schists against Bowers Supergroup along the Leap Year Fault after Early Ordovician times.

### Fault Orientation

Although previous workers in northern Victoria Land mapped extremely long faults separating major stratigraphic units, no one has discussed their tectonic significance. Therefore, it is useful to offer some speculations which we base on the orientation of structural elements and topographic lineaments in northern Victoria Land.

Orientation of folds in the northeastern domain (Figures 1 and 5), with respect to our proposed Lillie Glacier Fault, is consistent with their interpretation as en echelon folds in dextral wrench regime (Hills, 1963; Bishop, 1968; Harding, 1973; Moody, 1973). Although the concept of the Riedel Shear system was developed from small-scale evidence (Cloos, 1928; Riedel, 1929), it is established (Bishop, 1968; Tschalenko, 1970; Tschalenko and Ambraseys, 1970; Freund, 1974) that regional fault systems tend to display geometries consistent with the Riedel Shear pattern. In northern Victoria Land, orientation of known faults and faults inferred along glaciers is consistent with their interpretation as a regional Riedel Shear system in a dextral wrench regime (Figure 5).

## CONCLUSIONS

There are two possible models for correlation of folding in northeastern and southwestern domains. If $F_2$ in the southwestern domain correlates with the regional northwest trending Cambro-Ordovician folds in the northeastern domain, then the two domains may have been juxtaposed tectonically in Late Cambrian to Cambro-Ordovician times. Alternatively if $F_1$ correlates with the regional folds in the northwestern domains, then it is possible that the rocks northeast of the Leap Year Fault were juxtaposed tectonically against the Bowers Supergroup after the Early Ordovician.

Consideration of the geometries of topographical lineaments and mapped faults suggests the possibility of dextral wrench movements across a displacement zone trending approximately 320°.

*Acknowledgements* Field work was carried out during the International northern Victoria Land Expedition (1981-82) and was supported by N.Z. Antarctic Division (DSIR) and VXE-6 Squadron, U.S.A. Navy. Findlay gratefully acknowledges sponsorship by N.Z. Antarctic Division. Grateful thanks also go to Messrs E. Saxby and P.V. Colbert (N.Z. Antarctic Division and ATT) and Dr J. Splettstoesser (NSF) for assistance with field co-ordination. Our field assistants, Messrs W Fowlie and W. Atkinson, are thanked for their most congenial company. Our work has benefitted from discussions with Drs M. Laird (N.Z. Geological Survey), J.D. Bradshaw (University of Canterbury), and C.J. Adams (NZINS). Dr R. Berry (University of Tasmania) is thanked for his valuable critical comments on our structural interpretation of the region.

## REFERENCES

ADAMS, C.J., GABITES, J.E., BRADSHAW, J.D., LAIRD, M.G. and WODZICKI, A., 1982: K/Ar Geochronology of the Precambrian-Cambrian Wilson and Robertson Bay Groups and Bowers Supergroup, northern Victoria Land, Antarctica; *in* Craddock, C. (ed.), *Antarctic Geoscience*, Univ. Wisconsin Press, Madison, 543-48.

BISHOP, D.G., 1968: The geometrical relationships of structural features associated with strike-slip faults in New Zealand. *N.Z. J. Geol. Geophys., 11*, 405-17.

BRADSHAW, J.D., LAIRD, M.G. and WODZICKI, A., 1982: Structural style and tectonic history in northern Victoria Land; *in* Craddock, C. (ed.), *Antarctic Geoscience*, Univ. Wisconsin Press, Madison, 809-16.

CLOOS, H., 1928: Experimente sur innen Tektonik. *Centralb. Mineral. Geol. u, Pal., 1928B*, 609-21.

COOPER, R.A., JAGO, J.B., ROWELL, A.J. and BRADDOCK, P. (this volume): Age and correlation of the Cambrian-Ordovician Bowers Supergroup, northern Victoria Land.

CROWDER, D.F., 1968: Geology of part of northern Victoria Land, Antarctica. *U.S. Geol. Surv. Prof. Pap. 600-D*, D95-107.

DOW, J.S. and NEALL, V.E., 1974: Geology of the lower Rennick Glacier, northern Victoria Land, Antarctica. *N.Z. J. Geol. Geophys., 17*, 659-714.

FIELD, B.D. and FINDLAY, R.H. (this volume): The sedimentology of the Robertson Bay Group, northern Victoria Land.

FINDLAY, R.H. and FIELD, B.D., 1982: Black Prince Volcanics, Admiralty Mountains, Antarctica: Reconnaissance observations. *N.Z. Ant. Rec., 4*, 11-14.

FREUND, R., 1974: Kinematics of transform and transcurrent faults. *Tectonophysics, 21*, 93-134.

GAIR, H.S., 1967: The geology from the upper Rennick Glacier to the coast, northern Victoria Land, Antarctica. *N.Z. J. Geol. Geophys., 10*, 309-44.

HARDING, T.P., 1973: Newport-Inglewood trend, California—an example of wrenching style of deformation. *Am. Ass. Petrol. Geol., Bull., 57*, 97-116.

HARRINGTON, H.J., WOOD, B.L., MCKELLAR, I.C. and LENSEN, G.J., 1967: Topography and geology of the Cape Hallett-Tucker Glacier district, Antarctica. *N.Z. Geol. Surv., Bull., 80*.

HILLS, E.S., 1963: *Elements of Structural Geology*. Methuen & Co. Ltd., London.

HUTTON, C.O. and TURNER, F.J., 1936: Metamorphic zones in NW Otago. *R. Soc. N.Z., Trans, 65*, 405-6.

JORDAN, H., 1981: Tectonic observations in the Hallet Volcanic Province, Antarctica. *Geol. Jb. B41*, 111-25.

KLEINSCHMIDT, G. and SKINNER, D.N., 1981: Deformation styles in the basement rocks of North Victoria Land, Antarctica. *Geol. Jb. B41*. 155-99.

LAIRD, M.G. and BRADSHAW, J.D., 1982: New data on the basement geology of North Victoria Land. *N.Z. Ant. Rec., 4*, 3-10.

LAIRD, M.G., BRADSHAW, J.D. and WODZICKI, A., 1976: Re-examination of the Bowers Group (Cambrian), northern Victoria Land, Antarctica (preliminary note). *N.Z. J. Geol. Geophys., 19*, 275-82.

LAIRD, M.G., BRADSHAW, J.D. and WODZICKI, A., 1982: Stratigraphy of the Upper Precambrian and Lower Palaeozoic Bowers Supergroup, northern Victoria Land, Antarctica; *in* Craddock, C. (ed.), *Antarctic Geoscience*, Univ. Wisconsin Press, Madison, 535-42.

LECOUTEUR, P.C. and LEITCH, E.C., 1964: Preliminary report on the geology of an area southwest of Tucker Glacier, northern Victoria Land; *in* Adie, R.J. (ed.), *Antarctic Geology*, North Holland, Amsterdam, 229-36.

MOODY, J.D., 1973: Petroleum aspects of wrench-fault tectonics. *Am. Ass. Petrol. Geol., Bull., 57*, 449-76.

RAMSAY, J.G., 1974: Development of chevron folds. *Geol. Soc. Am., Bull., 85*, 1741-54.

RASTALL, R.H. and PRIESTLEY, R.E., 1921: The slate-greywacke formation of Robertson Bay, *British Antarctica (Terra Nova) Expedition, 1910*. Natural History Report, Geology, 121-29.

RIDDOLS, B.W. and HANCOX, C.J., 1968: The geology of the upper Mariner Glacier, north Victoria Land, Antarctica. *N.Z. J. Geol. Geophys., 11*, 881-99.

RIEDEL, W., 1929: Zur mechanik geologischer Brucherssheinungen. *Centralb. Mineral. Geol. u. Pal. 1929B*, 354-468.

STURM, A. and CARRYER, S.J., 1970: Geology of the region between the Matusevitch and Tucker Glaciers, north Victoria Land. *N.Z. J. Geol. Geophys., 13*, 408-35.

TSCHALENKO, J.S., 1970: Similarities between shear zones of different magnitudes. *Geol. Soc. Am. Bull., 81*, 1625-40.

TSCHALENKO, J.S. and AMBRAYSEYS, N.N., 1970: Structural analysis of the Dash-e-Bayaz (Iran) earthquake fracture. *Geol. Soc. Am. Bull., 81*, 41-60.

WRIGHT, T.O., 1981: Sedimentology of the Robertson Bay Group, northern Victoria Land. *Geol. Jb. B41*, 127-38.

# GEOLOGY OF THE DANIELS RANGE, NORTHERN VICTORIA LAND, ANTARCTICA: A PRELIMINARY REPORT

C.C. Plummer, *Geology Department, California State University, Sacramento, California, 95819, U.S.A.*

R.S. Babcock, *Geology Department, Western Washington University, Bellingham, Washington, 98225, U.S.A.*

J.W. Sheraton, *Bureau of Mineral Resources, Geology and Geophysics, P.O. Box 378, Canberra City, A.C.T., 2601, Australia.*

C.J.D. Adams, *Institute of Nuclear Sciences, D.S.I.R., Lower Hutt, New Zealand.*

R.L. Oliver, *Department of Geology and Mineralogy, University of Adelaide, Adelaide, S.A., 5000, Australia.*

*Abstract* The Daniels Range is underlain by a variety of intimately intermixed plutonic and metamorphic rocks, which are collectively assigned to the Daniels Range Metamorphic and Intrusive Complex. Metamorphic rocks comprise quartzofeldspathic gneisses and pelitic to psammitic schists with minor marble and calc-silicate intercalations. These underwent several episodes of deformation as indicated by multiple folding. During deformation and associated metamorphism, locally pervasive, mostly subconcordant intrusion of granitic veins took place. Subsequent plutonic activity resulted in the emplacement of a variety of granitic rocks. Foliation within the plutons is parallel to that in the adjoining country rocks, and some of these bodies contain boudinaged and folded synplutonic mafic dykes. The final episode of magmatic activity was emplacement of granitic dykes. Massive, equigranular granitic plutons (of Granite Harbour type) on the western side of the range were probably also emplaced during this post-tectonic episode.

The Daniels Range, part of the USARP Mountains of northern Victoria Land, is a north-south trending range approximately 80 km long with an area of about 2200 km$^2$. It is bounded to the north by the Gressitt Glacier and to the south by the Harlin Glacier; the eastern flank along the Rennick Glacier is generally steep, whereas the more gently inclined western flank merges into the polar plateau. Both flanks have extensive bedrock exposures, principally along ridges jutting out from the glaciated crest of the range. Cliffs and peaks along these ridges commonly have a relief of several hundred metres. The range as a whole appears to be a large fault-block, possibly uplifted to a greater extent along the eastern flank.

This report is based on reconaissance field work carried out in November and December, 1981 by the authors (and by G. Grindley of the New Zealand Geological Survey who took Adams' place for the first two weeks of the field season). Sampling was geared to forthcoming detailed petrological, geochemical, and geochronological programs.

## Previous Work

The Daniels Range was first visited briefly by a New Zealand sledging party (Sturm and Carryer, 1970). Their report states that the entire range was mapped by Gair et al. (1969) as schists, gneisses, and migmatites of the Wilson Group. During the 1979-80 field season, the southern end of the Daniels Range (Thompson and Schroeder Spurs) was investigated in some detail by members of the first German Antarctic North Victoria Land Expedition (Ganovex I). Kleinschmidt (1981), and Kleinschmidt and Skinner (1981) regard the schists and gneisses in this area as the products of a single metamorphic event, although complex interference patterns, indicating several generations of folding, are described. They consider granitic layers in the veined gneisses to have been formed by *in situ* anatexis. K-Ar mica ages of intrusive and country rocks from two localities in the southern Daniels Range are between 471 and 479 Ma (Kreuzer et al., 1981). Similar ages have been obtained from Granite Harbour Intrusives elsewhere in northern Victoria Land.

## Nomenclature of Rock Units

There is some ambiguity regarding names applied to Early Palaeozoic or Precambrian crystalline rock units in northern Victoria Land. Schists, gneisses, and associated granitic rocks in widely separated areas have been collectively referred to as Wilson Group, which Sturm and Carryer (1970) subdivided into Rennick schists and Wilson gneisses. Previously, Gair (1967) defined the Rennick Group as those schists and subordinate marbles that crop out along the western side of the Rennick Glacier. The schists in the Daniels Range conform to Gair's definition. Use of the terms "Wilson Group" and "Wilson gneisses" presents some problems. The Wilson gneisses include both ortho and para-gneisses as well as a wide variety of migmatites. It is felt that the term is too broad to be useful and therefore it is avoided in this paper. Similarly, Wilson Group has been applied to a variety of crystalline metamorphic (and intermixed igneous) rocks at widely separated locations in northern Victoria Land. The implication that they have similar histories and ages is quite premature.

It was found that the role of magmatism was much more important in the formation of the Daniels Range rocks than anticipated. Most of the range is underlain by mixtures of metamorphic and intrusive rocks in complex relationships and the term Daniels Range Metamorphic and Intrusive Complex is therefore introduced for these rocks. Some, if not all, of the plutonic rocks within the complex are time equivalents of the Granite Harbour Intrusives.

## METAMORPHIC ROCKS

Nearly all of the metamorphic rocks were originally sediments, although metadiorites crop out locally and some foliated granitoids were syntectonically recrystallised after solidification of the magma. No metavolcanic rocks were found. Most of these rocks were metamorphosed at amphibolite facies, although locally only greenschist facies conditions were attained. The highest grade rocks are quartzofeldspathic gneisses that most typically crop out in the eastern part of Thompson Spur.

Psammo-pelitic and pelitic schists predominate, and commonly contain layers and lenses of marble and calc-silicate rocks. The schists consist mainly of quartz and biotite, and muscovite is generally also present. Garnet is common in a few areas, and sillimanite occurs in schists and gneisses in the eastern part of Thompson Spur and at the head of Edwards Glacier (Figure 1). Andalusite or cordierite are present in a few rocks. Calc-silicate rocks contain quartz, calcite, epidote, diopside, plagioclase, grossular-rich garnet, and amphibole. The metadiorites are schistose rocks with biotite as the predominant mafic mineral. In some outcrops the schist grades into more massive diorite. The rocks were probably derived from intrusions emplaced early in the plutonic history of the area.

A narrow, north-trending belt of lower grade rocks in the western part of the area (Figure 1), consists mostly of phyllites with a pronounced slaty cleavage. Preliminary thin section study suggests that biotite schist has been retrogressively metamorphosed to phyllite in which white mica crystallised during penetrative deformation.

## Structural Relationships

Throughout most of the area, metamorphic rocks form abundant inclusions within granitic bodies of the Daniels Range Metamorphic and Intrusive Complex. In parts of the complex, they occur as septa and probable roof pendants. Only in the southern part of the range are metamorphic rocks continuously exposed, and even here they are extensively intruded by plutonic dykes and sills, particularly adjacent to major intrusions. The metamorphics have been intensely deformed, as noted by Kleinschmidt and Skinner (1981) in their study of the Thompson Spur area. Although no detailed structural studies were attempted, two generations of tight folds ($F_1$ and $F_2$) and rare open folds ($F_3$) were identified. Transposed layering is common and early folds may have been obliterated at many places. One or two lineations are commonly present. Early concordant granite and quartz veins pre-

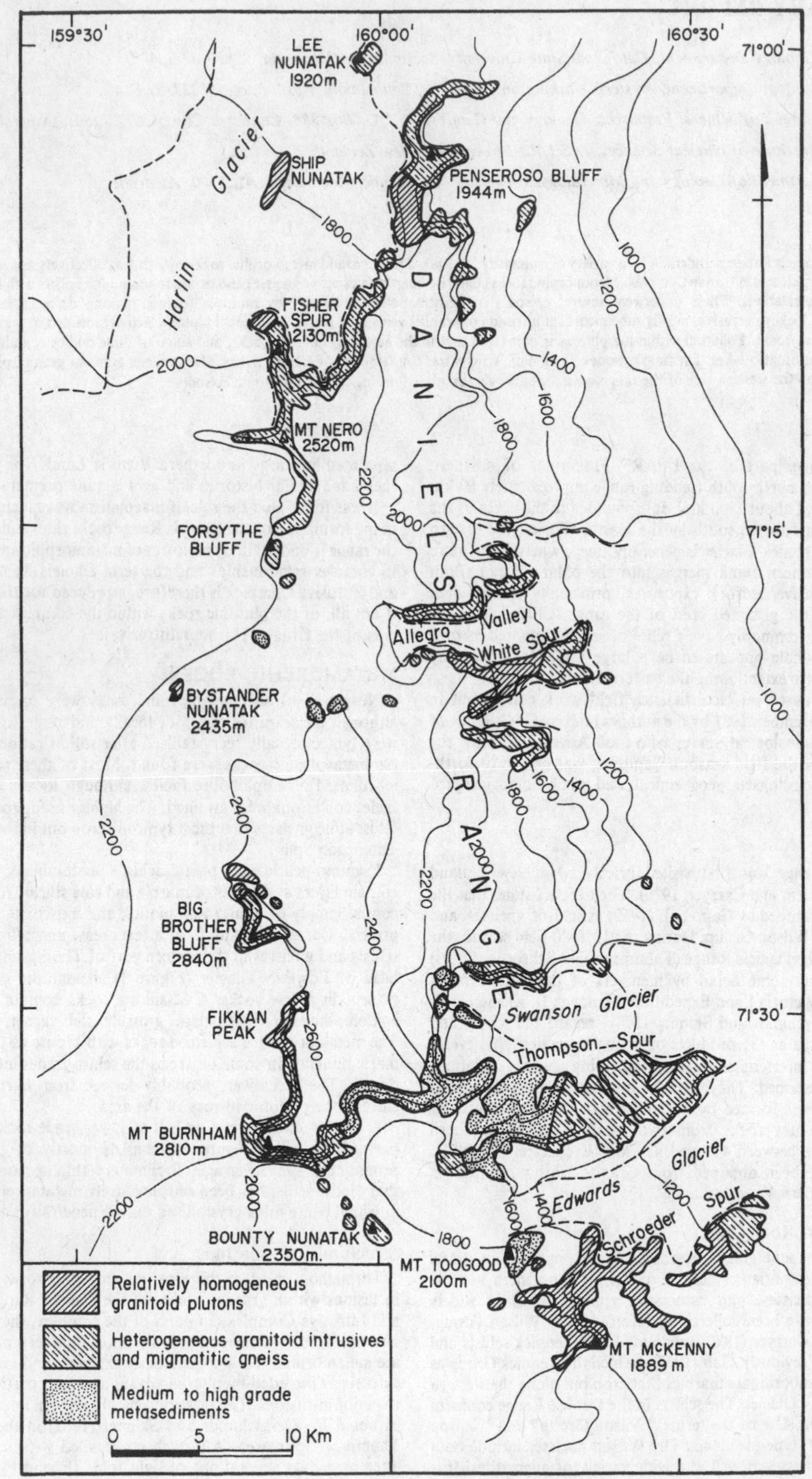


**Figure 1. Geological map of the Daniels Range.**

date $F_1$ and $F_2$ folds, whereas much more extensive granitic intrusions are younger than the folding.

## Breccia Pipe

Of particular interest is a breccia pipe within a roof pendant in the northern part of the area (near Penseroso Bluff). The pipe is approximately 20 m wide and nearly vertical. Rounded quartzite and subordinate schist fragments are enclosed in a dark rock rich in large (1-2 cm) garnet crystals. This dark matrix consists of quartz, garnet, andesine, hypersthene, cordierite, sillimanite, and hercynite. A single crystal of calcite may, like the quartzite fragments, have been derived from underlying country rocks. Some of the hypersthene is poikilitic and appears to have crystallised from a melt. It has been marginally altered to anthophyllite, probably when the material rose to a higher crustal level. The rocks from which the breccia pipe material was derived may possibly represent an extension of East Antarctic Shield rocks which would then form a true basement in this part of the Transantarctic Mountains. Alternatively, they may simply be restite from Rennick schist type rocks at a much lower crustal level than is presently exposed. Radiometric dating may distinguish between these possibilities.

## INTRUSIVE COMPLEX

This complex is composed of a variety of intrusive and intermixed metamorphic rocks. The plutonic rocks are predominantly granite or granodiorite, but include significant more mafic types. They form veins, dykes, sills or sheets, and irregularly shaped plutons, which are commonly markedly heterogeneous. Pegmatite and aplite veins and dykes (mostly tourmaline bearing) are particularly abundant.

The intrusive complex is to a large extent migmatitic in that it is a complex mixture of plutonic and metamorphic rocks. Typical relationships of these two components are (i) concordant veins of granite in schists and gneisses, (ii) highly brecciated metamorphic rocks forming very abundant xenoliths in plutons (Figure 2), and (iii) granitic dykes and veins cross-cutting metamorphic (and older plutonic) rocks (Figure 3).

The earliest plutonic activity took place during regional metamorphism. Most subconcordant granitoids appear to be due to injection of magma, but in some localities field evidence indicates that the leucosome material may locally have been formed by *in situ* anatexis. Subsequently, the most abundant unit of the intrusive complex was emplaced. This granitoid (biotite-muscovite bearing) is characterised by the presence of very abundant (typically 20 to 50%) all-sized

**Figure 2. Inclusion-rich zones within layered granite, Daniels Range.**

**Figure 4. Inclusions of "fruitcake" within layered granite. Daniels Range.**

**Figure 3. Sequences of "fruitcake" grading upwards into and cut by layered granite. Forsythe Bluff, Daniels Range.**

**Figure 5. Transition from "fruitcake" (right) to layered granite (left). Daniels Range.**

**Figure 6. Synplutonic dyke in granite. Schroeder Spur, Daniels Range.**

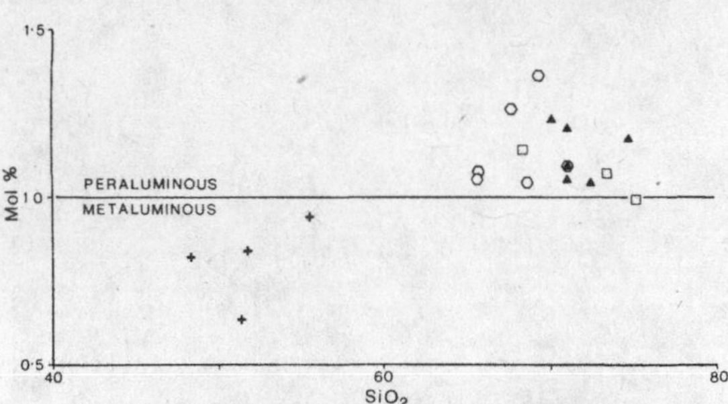


**Figure 7. Mol% $Al_2O_3/CaO + Na_2O + K_2O$ vs $SiO_2$ for various intrusive rocks in the Daniels Range. Squares show post-tectonic aplite dykes; hexagons show "fruitcake" and layered granitoid phases from heterogeneous plutons; triangles show homogenous granitoids from plutonic bodies; crosses show hornblende bearing intrusive bodies.**

fragments of schist and gneiss, many of which are cut by earlier granitic or quartz veins. Owing to its appearance, this unit is referred to as "fruitcake" (Figures 3 and 4). The parallelism of the xenoliths commonly gives the "fruitcake" a pronounced foliation, although mineral grains of the granitic matrix do not necessarily show a preferred orientation. A second type of biotite-muscovite bearing

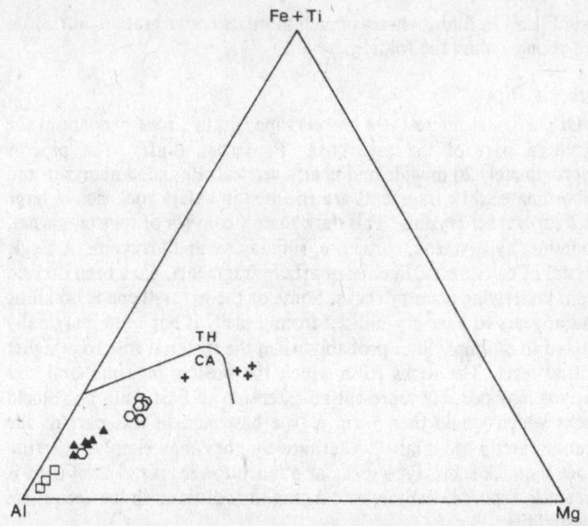


**Figure 8. Jensen (1976) type cation discrimination plot for Daniels Range intrusives. Solid line on diagram separates fields of calc-alkaline (CA) vs. tholeiitic (TH) types.**

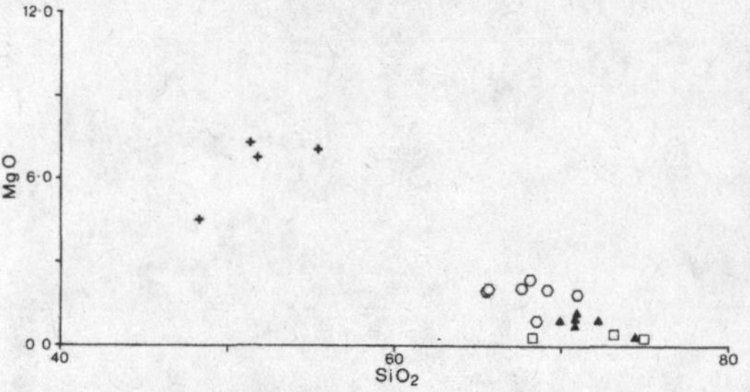


**Figure 9. Silica-magnesia variation diagram for Daniels Range intrusives. Symbols as in Figures 2 and 3.**

granitoid exhibits a marked layering, mainly due to different proportions of biotite. Biotite-rich schlieren may represent restite material. In some places, layered granite appears to grade rather sharply into "fruitcake" (Figure 5), whereas, in others, layered granite clearly intrudes "fruitcake" (Figure 3). Although the layered granite is relatively inclusion free, trains of xenoliths, some of which appear to have been rotated during or after intrusion, are present locally.

In both layered granite and "fruitcake", the foliation appears to be due partly to flow during intrusion, and partly to deformation after intrusion. For example, magmatic flow lines and adjacent inclusions locally show evidence of rotation while the rock was still ductile. Structures within the "fruitcake" indicate that shearing continued

**TABLE 1: Chemical analyses of intrusive rocks from the Daniels Range**

| | 1 | 2 | 3 | 4 | 5 | 6 | 7 | 8 | 9 | 10 | 11 | 12 | 13 | 14 | 15 | 16 | 17 | 18 | 19 |
|---|---|---|---|---|---|---|---|---|---|---|---|---|---|---|---|---|---|---|---|
| $SiO_2$ | 51.75 | 48.29 | 51.29 | 55.41 | 68.48 | 65.51 | 70.92 | 69.08 | 67.57 | 68.08 | 70.85 | 74.55 | 69.88 | 70.92 | 70.85 | 72.28 | 73.24 | 74.98 | 68.22 |
| $Al_2O_3$ | 15.34 | 18.81 | 14.23 | 14.40 | 14.94 | 15.71 | 13.51 | 13.74 | 13.96 | 15.34 | 14.43 | 13.05 | 14.72 | 14.12 | 15.13 | 13.77 | 13.51 | 13.46 | 15.19 |
| $TiO_2$ | 1.19 | 1.21 | 1.63 | 0.70 | 0.28 | 0.69 | 0.55 | 0.59 | 0.59 | 0.27 | 0.35 | 0.20 | 0.57 | 0.59 | 0.42 | 0.47 | 0.08 | 0.06 | 0.03 |
| $Fe_2O_3T$ | 12.48 | 11.87 | 12.78 | 12.00 | 2.35 | 5.42 | 4.02 | 5.79 | 6.40 | 5.61 | 3.03 | 2.53 | 2.75 | 3.61 | 2.03 | 2.53 | 1.26 | 0.76 | 0.63 |
| MnO | 0.14 | 0.15 | 0.15 | 0.23 | 0.04 | 0.07 | 0.05 | 0.06 | 0.05 | 0.05 | 0.04 | 0.07 | 0.04 | 0.05 | 0.04 | 0.04 | 0.06 | 0.02 | 0.02 |
| MgO | 6.74 | 4.51 | 7.28 | 7.04 | 0.88 | 1.96 | 1.84 | 2.01 | 2.04 | 2.36 | 0.96 | 0.25 | 0.86 | 1.13 | 0.69 | 0.87 | 0.38 | 0.27 | 0.27 |
| CaO | 7.02 | 8.70 | 7.98 | 6.77 | 1.84 | 3.61 | 2.05 | 1.48 | 2.21 | 1.11 | 2.09 | 0.82 | 1.11 | 2.31 | 1.22 | 1.26 | 1.36 | 1.36 | 1.04 |
| $Na_2O$ | 1.93 | 3.03 | 2.97 | 0.39 | 2.85 | 3.59 | 2.86 | 1.83 | 2.38 | 1.34 | 3.16 | 2.47 | 2.68 | 2.96 | 2.75 | 3.30 | 3.52 | 2.82 | 1.89 |
| $K_2O$ | 2.21 | 1.93 | 1.56 | 2.15 | 5.86 | 2.29 | 3.88 | 4.00 | 2.92 | 3.50 | 4.28 | 5.37 | 5.09 | 3.80 | 5.37 | 5.15 | 3.93 | 5.99 | 7.91 |
| Totals | 98.80 | 98.50 | 99.87 | 99.09 | 97.52 | 98.85 | 99.68 | 97.10 | 98.02 | 97.66 | 99.29 | 99.54 | 97.70 | 99.49 | 98.50 | 99.67 | 97.34 | 99.72 | 95.18 |

$Fe_2O_3T$ = Total Iron as $Fe_2O_3$

1. Hornblende-plagioclase pegmatite body, cirque south of Allegro Valley
2. Amphibolite dike, northeast of Fisher Spur
3. Hornblende gabbro phase of large pluton, Schroeder Spur
4. Hornblende diorite from chill zone of large pluton above
5. Layered granodiorite pluton cirque south of Allegro Valley
6. More mafic layer in pluton above
7. Heterogeneous granodiorite layer in pluton above (transitional into fruitcake)
8. Fruitcake phase in pluton, south of Penseroso Bluff
9. Fruitcake phase in pluton, northeast of Penseroso Bluff
10. Coarsely crystalline fruitcake phase in pluton, Fisher Spur
11. Homogeneous granodiorite from intrusive sheet, cirque south of Allegro Valley
12. Homogeneous garnet-biotite granite phase of pluton, Thompson Spur
13. Sillimanite-bearing biotite-granodiorite, east of Mt. Burnham
14. Porphyrytic granodiorite pluton, Mt. McKenny
15. Homogeneous biotite-granodiorite pluton, east of Mt. McKenny
16. Homogeneous biotite-granite pluton, Schroeder Spur
17. Post-kinematic aplite dike, cirque south of Allegro Valley
18. Post-kinematic aplite dike, Allegro Valley
19. Post-kinematic aplite dike, Forsyth Bluff

after solidification of the granite matrix. For example, the foliation planes in a schist inclusion have been bent in opposite directions along opposite sides of the inclusion, indicating it was caught in a shear couple.

### Synplutonic Dykes

Whereas emplacement of many of the more mafic (dioritic or basaltic, now amphibolitic) intrusions preceded that of the voluminous granitoids, others post-date the latter. Of particular interest are synplutonic dykes which have been boudinaged and, in some cases, tightly folded (Figure 6). There is no evidence for fracturing of the enclosing granite, but in places, the foliation in the granite has been deformed along with the dykes.

### Post-tectonic Granites

All the previously described intrusive rocks, as well as the metamorphic country rocks, are cut by numerous felsic dykes—most commonly pegmatite or aplite, but also granite or granodiorite—which show no evidence of deformation. Massive, homogeneous equigranular granite, virtually free of xenoliths, crops out in nunataks along the western side of the Daniels Range. It appears to be typical high-level granite and was probably post-tectonically emplaced. It may be genetically related to the post-tectonic dykes, or alternatively may be the higher level equivalent of the layered granite and "fruitcake". Because the post-tectonic dykes are ubiquitous in most of the range it is likely that the K-Ar dates obtained by Kreuzer et al. (1981) define the latest episode of intrusion. If so, it is not at present possible to estimate how much earlier plutonism began.

### Geochemistry of the Intrusive Rocks

Table 1 presents 19 whole-rock major element analyses of intrusive rocks done by atomic absorption spectrophotometry. The results show a clear distinction between minor hornblende-bearing intrusives and the dominant biotite-muscovite-bearing intrusives and the dominant biotite muscovite-bearing intrusives of the Daniels Range. As seen in Figure 7 the biotite-muscovite intrusives are all peraluminous (with one exception) and the hornblende bearing intrusives are metaluminous. Figure 8 indicates that the biotite-muscovite intrusives are calc-alkaline, while the hornblende-bearing intrusives tend to be tholeiitic. The distinct gap in composition between the biotite-muscovite and hornblende-bearing intrusives is also shown by the MgO-$SiO_2$ variation diagram of Figure 9.

The origin of these two different magma types cannot be established from the data available. However, it is possible that the metaluminous hornblende-bearing magmas represent mantle-derived melts that have been emplaced in the lower crust causing secondary melting which has generated the biotite-muscovite-bearing magmas. This is consistent with field relationships which suggest that the peraluminous granitoids have been derived from partial to almost complete melting of pelitic or psammo-pelitic sedimentary source rocks. A more extensive study of the geochemistry of the Rennick schists by Wyborn (pers. comm.) leads to the same conclusion.

## GEOLOGICAL HISTORY

Prior to the Ross Orogeny marine sedimentation took place, although the ages of deposition and subsequent initiation of metamorphism are not at present known. The sediments were intensely deformed and regionally metamorphosed under greenschist to amphibolite facies conditions; there may have been more than one episode of metamorphism. Intrusion of granite and quartz veins took place during this event prior to the formation of tight to isoclinal folds.

Subsequently, the metamorphic rocks were intimately intruded by a variety of granitoids, commonly being thoroughly brecciated. Emplacement of the xenolith-rich "fruitcake" phase was followed by intrusion of less heterogeneous layered granite; in some cases the latter clearly intrudes "fruitcake", but some may represent magma that has separated (perhaps by filter pressing) from "fruitcake" type. Deformation continued to occur during this episode. After these granite bodies had largely crystallised but were still ductile, mafic magmas were emplaced to form synplutonic dykes. Broad folding of the intrusive complex and possibly reverse faulting took place at about this time. Finally, post-tectonic granitoids were emplaced. the magma may well be genetically related to the syntectonic granites and was simply intruded after deformation had ceased, although it is possible that a period of uplift and erosion allowed the rocks to move to a higher crustal level into which a later magma was intruded.

It is not known whether all of the above events occurred during a single continuous orogenic cycle or whether there were significant gaps between the various deformational and intrusive episodes. Detailed radiometric dating may provide some answers.

*Acknowledgements* Funding for the project was provided by National Science Foundation Grant Number DPP-8019956. We benefited from stimulating discussions with G. Grindley, who accompanied us in the early part of the field program, and with many other colleagues in Antarctica and elsewhere. We thank the pilots and crew of U.S. Navy Squadron VXE-6 for their logistic support. The facilities of the Department of Geology and Mineralogy, University of Oxford were kindly made available to Plummer for the preliminary laboratory work. Sheraton publishes with the permission of the Director, Bureau of Mineral Resources.

## REFERENCES

GAIR, H.S., 1967: The geology from the upper Rennick Glacier to the coast, northern Victoria Land, Antarctica. *N.Z. J. Geol. Geophys.*, *10*, 309-44.

JENSEN, L.S., 1976: A new cation plot for classifying subalkalic volcanic rocks. Ontario Dept Mines, Misc. Papers, 66.

KLEINSCHMIDT, G., 1981: Regional metamorphism in the Robertson Bay Group area and in the southern Daniels Range, North Victoria Land, Antarctica—a preliminary comparison. *Geol. Jb.*, *B41*, 201-28.

KLEINSCHMIDT, G. and SKINNER, D.N.B., 1981: Deformation styles in the basement rocks of North Victoria Land, Antarctica. *Geol. Jb.*, *B41*, 155-99.

KREUZER, H., HÖHNDORF, A., LENZ, H., VETTER, U., TESSENSOHN, F., MÜLLER, P., JORDAN, H., HARRE, W. and BESANG, C., 1981: K/Ar and Rb/Sr dating of igneous rocks from North Victoria Land, Antarctica. *Geol. Jb.*, *B41*, 267-73.

STURM, A. and CARRYER, S.J., 1970: Geology of the region between Matusevich and Tucker Glaciers, North Victoria Land, Antarctica. *N.Z. J. Geol. Geophys.*, *13*, 408-35.

# GEOLOGY OF THE DANIELS RANGE INTRUSIVE COMPLEX, NORTHERN VICTORIA LAND, ANTARCTICA

R. S. Babcock, *Geology Department, Western Washington University, Bellingham, Washington, 98225, USA.*

C. C. Plummer, *Geology Department, California State University, Sacramento, California, 95819, USA.*

J. S. Sheraton, *Bureau of Mineral Resources, Geology and Geophysics, PO Box 378, Canberra City, ACT, 2601, Australia.*

C. J. Adams, *Institute of Nuclear Sciences, DSIR, Private Bag, Lower Hutt, New Zealand.*

R. L. Oliver, *Department of Geology and Mineralogy, University of Adelaide, Adelaide, SA, 5000, Australia.*

*Abstract* The Daniels Range of the USARP Mountains in northern Victoria Land is composed primarily of an intrusive complex of S-type granitoids. Field observations indicate that the rocks exposed represent a vertical section of continental crust at least 8 to 10 km thick. Within this section several different structural stages and phases of intrusion can be observed. The initial phase in the lower structural level consists of a granite or granodiorite magma which is initimately comingled with meta-sedimentary xenoliths and xenocrysts of restite and/or wallrock material. This xenolithic magma grades into, and in places, is intruded by a relatively pure granite or granodiorite that is characterised by a distinct compositional layering of biotite-rich versus biotite-poor rock. The layered intrusives grade upwards into more homogeneous granitoids containing only schlieren and/or small clots of biotite-rich material. This phase forms large plutons and sills that constitute an upper structural level that is marked by conspicuous porphyroblasts or megacrysts of K-feldspar. The intrusive complex also includes several generations of pegmatite/aplite dykes as well as tonalite/diorite intrusives that, in places, occur as synplutonic dykes in massive granitic plutons. Distinctive garnet-bearing leuco-granite dykes, sills and plutons are found throughout the complex.

Based on available K-Ar and Rb-Sr dating, the intrusive rocks of the Daniels Range are tentatively assigned to the Granite Harbour Intrusives. However, field evidence suggests that an earlier episode of metamorphism may have occurred. Further geochronologic work to test this hypothesis is underway, along with studies of the distribution of trace elements and isotopes among the various phases of the S-type granitoids.

# TRENDS IN REGIONAL METAMORPHISM AND DEFORMATION IN NORTHERN VICTORIA LAND, ANTARCTICA

G. Kleinschmidt, *Geologisch-Palaontologisches Institut, Technische Hochschule Darmstadt, D-6100, Darmstadt, Federal Republic of Germany.*

*Abstract* The Robertson Bay Group of northern Victoria Land show increases in both metamorphism and deformation from northeast to southwest in the area between Cape Adare and the Bowers Mountains. Metamorphic grade in the area ranges from very low in the east to low in the west, the metamorphic trend being indicated by the appearance of biotite near the western boundary and by an increase in illite crystallinity represented by a change from about 230 to about 190 $Hb_{rel}$. In addition there is evidence from recrystallisation of quartz in syn- or pretectonic quartz veins. Suturing of quartz grains becomes coarser westward and the threshold of true recrystallisation is exceeded only west of the centre of the Robertson Bay Group where quartz veins are common parallel to $S_1$. There is also a progressive change in the deformation from open to tight and finally to complicated folds, and by the tightening of two conjugate sets of cleavage. The phenomena are attributed to a single orogenic event.

Northern Victoria Land consists of four structural units defined in Figure 1 and discussed in detail by Tessensohn et al. (1981). The main rocks in unit 1 (Figure 1) are the slightly metamorphosed turbidites of the Robertson Bay Group, assumed to be of Late Precambrian (Vendian) age. The Bowers (Super-) Group of unit 2 comprises almost unmetamorphosed Cambro-Ordovician sedimentary and volcanic rocks. Unit 4 contains strongly metamorphosed metasediments and migmatites of unknown sedimentary age, known together as the "Wilson Group". The blocks that comprise unit 3 contain rocks associated with the Robertson Bay Group and the "Wilson Group".

Several orogenies are supposed to have been effective in the four units: Precambrian orogenies in units 3 and 4 (Gair et al., 1969; Dow and Neall, 1974), the probable Ordovician Ross Orogeny in units 1-4 and the Mid-Palaeozoic Borchgrevink Orogeny in units 1 and 2 (Bradshaw et al., 1982). By contrast Tessensohn et al. (1981) concluded that northern Victoria Land had undergone only one orogenic event, the Ross Orogeny.

This paper will present new arguments to demonstrate that the Robertson Bay Group records the imprint of only one metamorphism and a single deformational event. Illite crystallinity is used to show the degree of metamorphism and this is compared with mineral assemblages, micro-structures in quartz veins, the formation of certain quartz veins and the development of fold styles and cleavage.

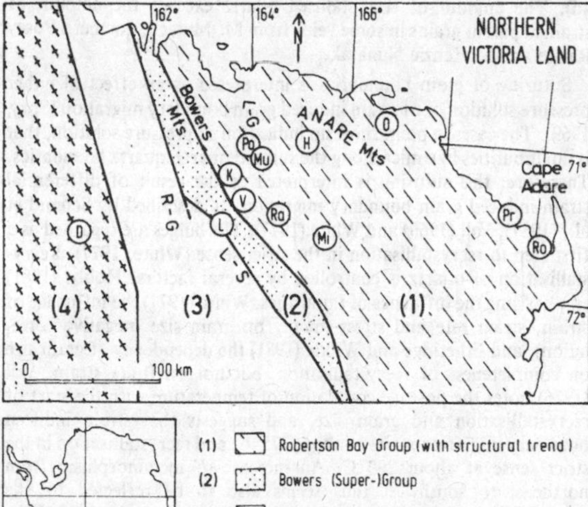


**Figure 1. The four structural units of northern Victoria Land. Key to lettering (the mentioned localities are encircled): D = Daniels Range; H = Mt Hemphill; K = McKenzie Nunatak; L = Lanterman Range, northeastern slope; Mi = Mirabito Range; Mu = Mt Mulach; O = O'Hara Glacier; P = northern Posey Range; Pr = Pressure Bay; Ra = Mt Radspinner; Ro = Robertson Bay; V = Mt Verhage; L-G = Lillie Glacier; L-S = Langerman and Salamander Ranges; R-G = Rennick Glacier.**

## TRENDS OF METAMORPHISM

### Indications from mineral assemblages

In the field Robertson Bay Group rocks between Robertson Bay and the Lillie Glacier appear to be almost unmetamorphosed. They contain minerals which indicate very weak metamorphism: namely newly crystallised chlorite and white mica ("sericite"); clastic biotite is chloritised and plagioclase is albitised. West of the Lillie Glacier higher metamorphic grades are indicated by the sporadic appearance of biotite between Mt Verhage and Ian Peak (Wodzicki et al., 1982) and at Mt Mulach. Wodzicki et al. (1982) have also described patchy occurences of clinozoisite south of Mt Verhage, and observed prehnite and pumpellyite. The presence of biotite (and clinozoisite) might indicate greenschist facies metamorphism.

The appearance of biotite can be interpreted as evidence of a low pressure type of regional metamorphism, since the chlorite zone of the greenschist facies seems to be missing (Miyashiro, 1978; Kleinschmidt, 1981). This interpretation is in agreement with the low pressure succession of andalusite—fibrolite—sillimanite in the southern Daniels Range (Kleinschmidt, 1981).

However, two factors relevant to the "biotite argument" must be taken into consideration: (1) the Robertson Bay Group is intruded by many plutons of the Admiralty Granite, which may have caused contact metamorphism with the nucleation of biotite; (2) according to Frey et al., (1973) biotite can also form at very low grades of metamorphism in rocks with a particular chemical composition related to the content of original glauconite.

### Illite crystallinity

Very low-grade metamorphism (sensu Winkler, 1979) can be subdivided and distinguished from diagenesis and low grade metamorphism by the degree of illite crystallinity (Kübler, 1967; Weber, 1972). The illite crystallinity of 70 Robertson Bay Group specimens has been determined using the relative peak width at half height of the (001) reflection (= $Hb_{rel}$) (Weber, 1972). All values fall within the range of very low grade metamorphism (Figure 2). Plotting the values along a section perpendicular to the structural trend of the Robertson Bay Group area, a gentle decrease of $Hb_{rel}$ (i.e. an increase of crystallinity) from northeast to southwest is indicated (Figure 3). This corresponds, though not closely, with the metamorphic trend suggested by the metamorphic assemblages. Note that the illite crystallinity of slate from the Bowers Group (Sledgers Formation, northwestern slope of the Lanterman Range) is clearly lower (Figures 2 and 3).

### Formation of quartz veins

Quartz veins of different dimensions (thicknesses between < 1 cm to ca 20 cm) occur throughout the Robertson Bay Group, and are especially abundant around Robertson Bay, at McKenzie Nunatak and Mt Verhage. At least three categories of quartz veins are distinguished: (1) quartz veins without a recognisable relation to folds and folding; they cross cut fold structures, run in diverse directions and are commonly irregular; (2) quartz veins running parallel to the bedding ($S_o$); (3) quartz veins following the first cleavage ($S_1$).

Categories (1) and (2) do not show a regular regional distribution. They were observed around Robertson Bay (1,2), O'Hara Glacier (1,2), Mt Hemphill (2), Mirabito Range (1,2) and Mt Mulach (2). By

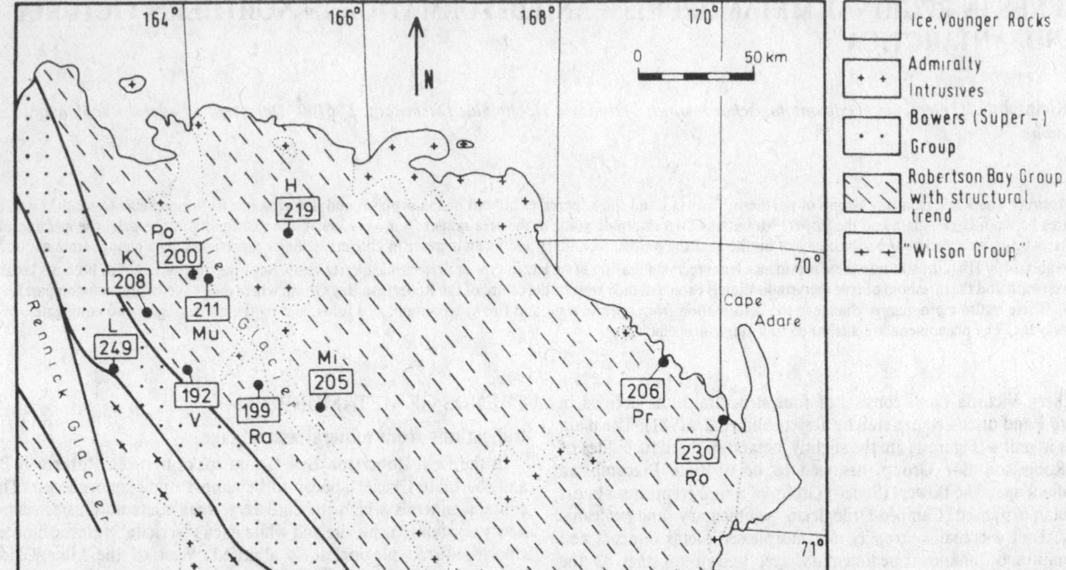


**Figure 2. Means of illite crystallinity ($Hb_{rel}$) in northern Victoria Land. Localities: H = Mt Hemphill; K = McKenzie Nunatak; L = Lanterman Range, northeastern slope; Mi = Mirabito Range; Mu = Mt Mulach; Po = northern Posey Range; Pr = Pressure Bay; Ra = Mt Radspinner; Ro = Robertson Bay; V = Mt Verhage.**

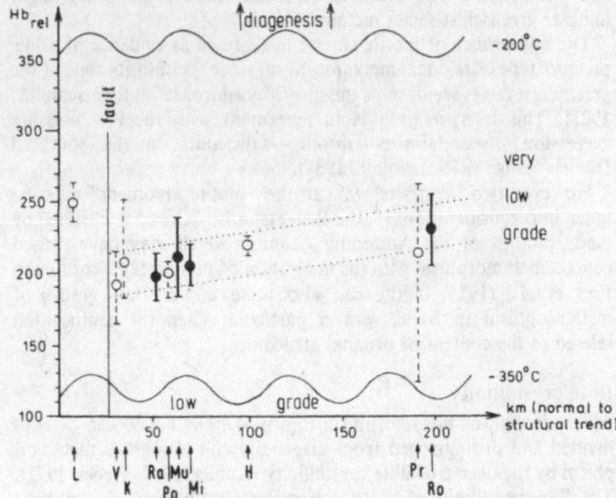


**Figure 3. Illite crystallinity in northern Victoria Land. Correlation of illite crystallinity and Winkler's (1979) metamorphic grades taken from Teichmuller et al. (1979). Arrows show the position of the sampling points on a section perpendicular to the structural trend, starting from the Lanterman Fault (km = 0), crossing the Leap Year Fault ("fault"), and ranging to Robertson Bay (km = 200). Open circles: means of $Hb_{ret}$ statistically uncertain, dashed lines: ranges between minimum and maximum values; full circles with thick lines: means of $Hb_{ret}$ with fiducial limit (95%) (Student-t-test).**

contrast category (3) is restricted to the area west of the Lillie Glacier: quartz veins parallel to $S_1$ are common and predominate at Mt Mulach, northern end of Posey Range, McKenzie Nunatak and Mt Verhage.

From observations in Scotland and the Swiss Alps, Voll (1969) concluded that quartz veins running parallel to $S_1$ form in metapelites which have achieved the phyllitic state of metamorphism. If this interpretation is applied to northern Victoria Land the appearance of quartz veins parallel to $S_1$ is further evidence of increasing metamorphic grade from NE to SW, as indicated by the illite crystallinity.

## Recrystallisation of quartz

This description of quartz recrystallisation is restricted to quartz from quartz veins. The syn- or pre-tectonic quartz veins show a systematic change of the quartz microstructures. Around Robertson Bay a fine suturing of the quartz/quartz boundaries is apparent (Figure 4a). The lobes of the sutures have diameters of up to 0.01 mm. In the Mirabito Range and west of the Lillie Glacier sutures are coarser (diameters of lobes up to 0.1 mm) and the bulges have straight edges. In addition, at Mirabito Range the first, very small, strain free quartz grains form due to recrystallisation within the most strongly deformed primary grains (Figure 4b). The new grains have diameters of 0.02 to 0.025 mm. West of the Lillie Glacier the recrystallised quartz grains increase by grain growth; at Mt Mulach they show diameters of 0.02 to 0.2 mm (Figure 4c), and at northern Posey Range, McKenzie Nunatak and Mt Verhage 0.07 to 0.2 mm (Figure 4d). The amount of recrystallised quartz exceeds the amount of strained parent grains in some veins from Mt Mulach, northern Posey Range and McKenzie Nunatak.

Suturing of grain boundaries is interpreted as an effect of either pressure solution or of strain induced grain boundary migration (Spry, 1969). The cases in point show no indication of pressure solution, that is no impurities like mica along the sutured quartz/quartz boundaries. Therefore, this suturing is interpreted as the result of differential strain-induced grain boundary migration, as described by Schmid et al., (1981), Voll (1969) and White (1977). The bulges are regarded as a first step to recrystallisation in the strict sense (White, 1977). Recrystallisation of quartz is controlled by several factors, Hobbs (1981) emphasising the influence of impurities, White (1977) the influence of strain, strain rate and stress mainly on grain size (negative correlation), and Etheridge and Wilkie (1981) the dependency of grain size on completeness of recrystallisation, but not on finite strain. Voll (1976) notes the positive correlation of temperature with the start of recrystallisation and grain size, and suggests that strain induced boundary migration starts at about 275°C, and recrystallisation in the strict sense at about 290°C. An increase of metamorphism from northeast to southwest thus seems also to be reflected by the recrystallisation phenomena of quartz—namely the exceeding of the threshold temperature at Robertson Bay, but by true recrystallisation (290°C) only at the Mirabito Range and west of the Lillie Glacier. If there were no increase in metamorphism the recrystallisation of quartz grains would be expected to decrease towards the southwest because of the increase of deformation (see below).

Calcite is present in nearly all quartz veins of the Robertson Bay Group. It is strained in the syn- and pre-tectonic veins, but unstrained in the post-tectonic ones, because the threshold for calcite recrystallisation has not been exceeded. This may provide a means for using the the types of quartz vein, as an additional indication that low grade

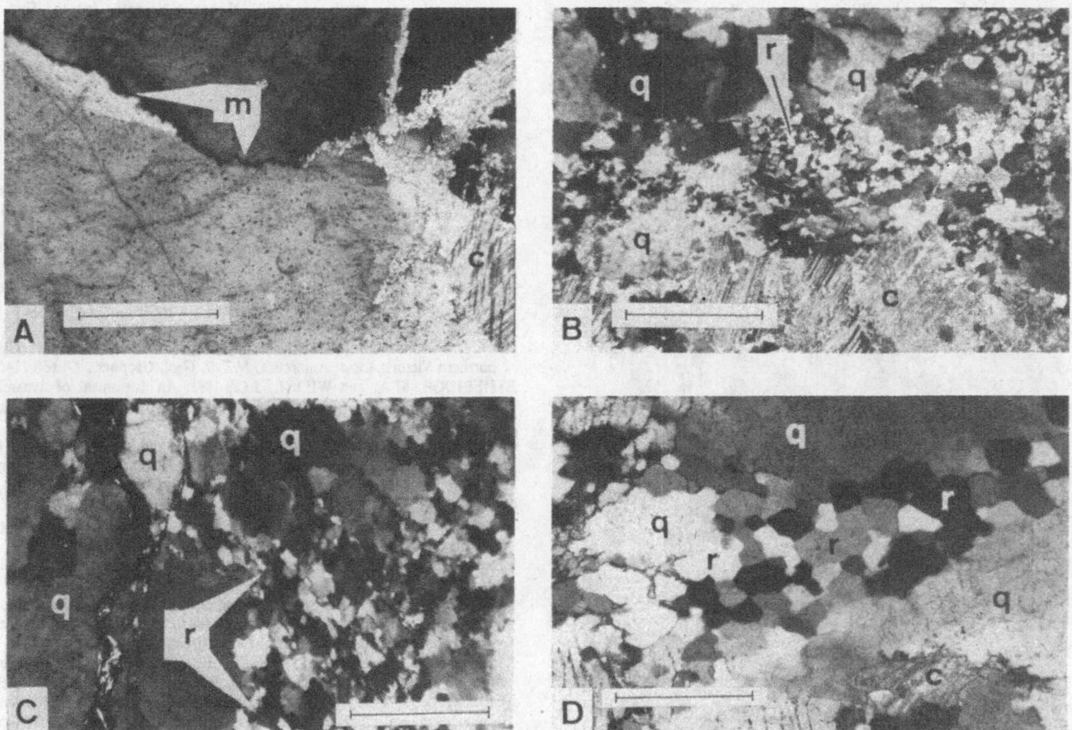


**Figure 4. Photomicrographs. Length of each mark = 0.5 mm. A: strained quartz from quartz vein at Robertson Bay. (m) strain induced boundary migration in quartz, (c) deformed calcite. B: quartz vein from Mirabito Range. (c) deformed calcite; (q) deformed quartz grains, in part very finely recrystallised (r). C: quartz vein from Mt Mulach. (q) strained quartz, (r) recrystallisation of quartz. D: quartz vein from Mt Verhage. Recrystallisation (r) at the boundary of two strained quartz grains (q), (c) deformed calcite (high refraction).**

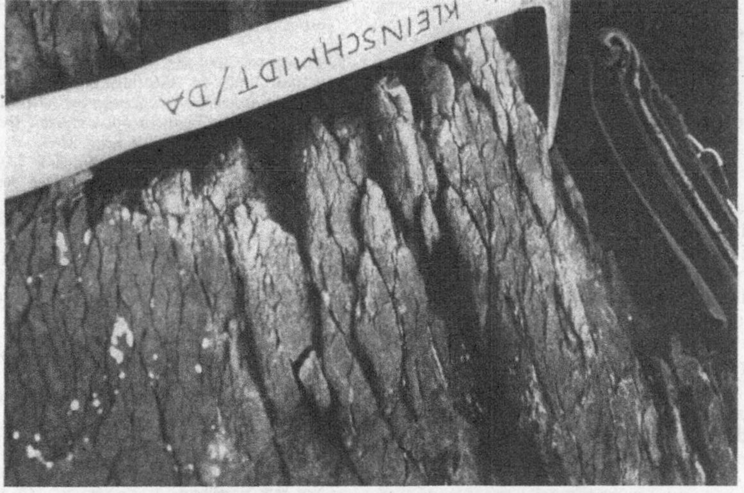


**Figure 5. Two sets of cleavage ($S_1$) in greywacke at Mt Hemphill. Bedding is parallel to handle of hammer.**

metamorphism has not been reached, for the recrystallisation of calcite starts only at 300 to 350°C (Voll, 1980).

## TREND OF DEFORMATION

### Folding

Kleinschmidt and Skinner (1981) have already described the increase in deformation intensity in northern Victoria Land from east to west. Open upright folds characterise the area between Robertson Bay and the central Anare Mountains. In the Anare Mountains the folds are a little tighter. In the Mirabito Range the folds are almost isoclinal and still upright. West of the Lillie Glacier refolding becomes more common and is not restricted to the fine grained rocks. The second deformation has produced drag folds and a second cleavage (e.g. Mt Mulach, McKenzie Nunatak) and leads to the formation of synformal anticlines (Mt Verhage region). In the southern Daniels Range the deformation attains at least heteroaxial triple folding, which can be interpreted as the intensification of the deformation at a deeper level (Kleinschmidt and Skinner, 1981).

### Cleavage

The trend of increasing deformation from northeast to southwest is supported by the parallel development of cleavage. Cleavage can be formed as two sets of cleavage planes, particularly in sandstones or greywackes which are very common in the Robertson Bay Group.

Each set can be described as "rough cleavage" using the terminology of Powell (1979). The two sets of such cleavage planes ($S_1$) intersect at a rather large angle of 25 to 30° in the north-eastern area of the Robertson Bay Group (Robertson Bay, Anare Mountains with Mt Hemphill, Figure 5). The angle tightens to about 10° in the Mirabito Range (Figure 6). West of the Lillie glacier (Mt Mulach) the two sets of $S_1$ are almost indistinguishable and form nearly an axial plane cleavage (see Kleinschmidt and Skinner, 1981, Figure 24).

**Figure 6. Two sets of cleavage ($S_1$) in greywacke from Mirabito Range. Bedding is nearly horizontal.**

## CONCLUSIONS

In the Robertson Bay Group of northern Victoria Land information mainly from illite crystallinity ($Hb_{rel}$ 230 to 192) demonstrate that the effects of a single very low grade metamorphism increase from northeast to southwest. Deformation also increases in the same direction. These parallel trends of one metamorphism and one deformation and the absence of polymetamorphism and polytectonism indicate that only one orogenic event affected the Robertson Bay Group. According to Tessensohn et al. (1981) and age determinations by Adams et al. (1982) this would belong to the Ross Orogeny.

Professor Dr K. Weber, Göttingen, for the first five determinations of illite crystallinity, and last, but not least, to all of Ganovex I and II for discussions and comradeship.

*Acknowledgements* I am grateful to the Bundesantalt für Geowissenschaften und Rohstoffe, Hannover, for the invitation to participate in Ganovex I and II, to the Deutsche Forschungsgemeinschaft for financial support, to the National Science Foundation (U.S.A./McMurdo) for their rescue operation (inclusive of our samples), to Professor Dr K. Weber, Göttingen, for the first five determinations of illite crystallinity, and last, but not least, to all of Ganovex I and II for discussions and comradeship.

## REFERENCES

ADAMS, C.J.D., GABITES, J.E., WODZICKI, A., LAIRD, M.G. and BRADSHAW, J.D., 1982: Potassium-argon geochronology of the Precambrian-Cambrian Wilson and Robertson Bay Groups and Bowers Supergroup, northern Victoria Land, Antarctica; *in* Craddock, C. (ed.) *Antarctic Geoscience,* Univ. Wisconsin Press, Madison, 543-8.

BRADSHAW, J.D., LAIRD, M.G. and WODZICKI, A., 1982: Structural style and tectonic history in northern Victoria Land, Antarctica; *in* Craddock, C. (ed.) *Antarctic Geoscience,* Univ. Wisconsin Press, Madison, 809-16.

DOW, J.A.S. and NEALL, V. E., 1974: Geology of the lower Rennick Glacier, northern Victoria Land, Antarctica. *N.Z. J. Geol. Geophys., 17,* 650-714.

ETHERIDGE, M.A. and WILKIE, J.C., 1981: An assessment of dynamically recrystallised grainsize as a palaeopiezometer in quartz bearing mylonite zones. *Tectonophysics, 78,* 475-508.

FREY, M., HUNZIKER, J.C., ROGGWILLER, P. and SCHINDLER, C., 1973: Progressive neidriggradige Metamorphose glaukonitführender Horizonte in den helvetischen Alpen der Ostschweiz. *Contrib. Mineral. Petrol., 39,* 185-218.

GAIR.H.S., STURM, A., CARRYER, S.J. and GRINDLEY, G.W., 1969: The geology of northern Victoria Land; *in* Bushnell, V.C. and Craddock, C. (eds.) *Geologic maps of Antarctica. Antarct. Map Folio Ser., Folio 12,* Pl XII.

HOBBS, B.E., 1981: The influence of metamorphic environment upon the deformation of minerals. *Tectonophysics, 78,* 335-83.

KLEINSCHMIDT, G., 1981: Regional metamorphism in the Robertson Bay Group area and in the southern Daniels Range, North Victoria Land, Antarctica—a preliminary comparison. *Geol. Jb., B41,* 201-28.

KLEINSCHMIDT, G. and SKINNER, D.N.B., 1981: Deformation styles in the basement rocks of North Victoria Land, Antarctica. *Geol. Jb., B41,* 155-99.

KUBLER, B., 1967: La cristallinité de l'illite et les zones toutà fait supérieures du métamorphisme. *Colloque sur les "Etages tectoniques" Neuchatel.* Neuchâtel, 105-22.

MIYASHIRO, A., 1978: *Metamorphism and metamorphic belts.* 3rd Impr., Allen and Unwin, London.

POWELL, C.McA., 1979: A morphological classification of rock cleavage. *Tectonophysics, 58,* 21-34.

SCHMID, S.M., CASEY, M. and STARKEY, J., 1981: An illustration of the advantages of a complete texture analysis described by the orientation distribution function (ODF) using quartz pole figure data. *Tectonophysics, 78,,* 101-17.

SPRY, A., 1969: *Metamorphic textures.* Pergamon Press, Oxford,

TEICHMULLER, M., TEICHMULLER, R. and WEBER, K., 1979: Inkohlung und Illit-Kristallinität. Vergleichende Untersuchungen im Mesozoikum und Paläozoikum von Westfalen. *Fortschr. Geol. Rheinld. Westf., 27,* 201-76.

TESSENSOHN, F., DUPHORN, K., JORDAN, H., KLEINSCHMIDT, G., SKINNER, D.N.B., VETTER, U., WRIGHT, T.O. and WYBORN, D., 1981: Geological comparison of basement units in North Victoria Land, Antarctica. *Geol. Jb., B41,* 31-88.

VOLL, G., 1969: Klastiche Mineralien aus den Sedimentserien der schottischen Highlands und ihr Schicksal bei aufsteigender Regional-und Kontaktmetamorphose. Habil.-Schrift, Technische Universität Berlin, 206 pp.

VOLL, G., 1976: Recrystallisation of quartz, biotite and feldspars from Erstfield to the Levantine Nappe, Swiss Alps and its geological siginificance, *Schweiz. mineral. geol. Mitt., 56,* 641-7.

VOLL, G., 1980: Deformation, crystallisation and recrystallisation. *International conference on the effect of deformation of rocks, Gottingen 1980.* Göttingen, Appendix 1-9.

WEBER, K., 1972: Notes on determination of illite crystallinity. *N. Jb. Mineral. Mh., 1972,* 267-76.

WHITE, S., 1977: Geological significance of recovery and recrystallisation processes in quartz. *Tectonophysics, 39,* 143-70.

WINKLER, H.G.F., 1979: *Petrogenesis of metamorphic rocks.* 5th ed., Springer, New York/Heidelberg.

WODZICKI, A., BRADSHAW, J.D. and LAIRD, M.G., 1982: Petrology of the Wilson and Robertson Bay Groups and Bowers Supergroup, northern Victoria Land, Antarctica; *in* Craddock, C. (ed.) *Antarctic Geoscience,* Univ. Wisconsin Press, Madison, 549-554.

# NEW DATA ON THE LOWER PALAEOZOIC BOWERS SUPERGROUP, NORTHERN VICTORIA LAND

M.G. Laird, *N.Z. Geological Survey, University of Canterbury, Christchurch, New Zealand.*
J.D. Bradshaw, *Geology Department, University of Canterbury, Christchurch, New Zealand.*

*Abstract* The Lower Palaeozoic Bowers Supergroup occupies a 350 km long belt transecting northern Victoria Land from the Oates Coast to the Ross Sea. To the east the sequence is faulted against the Upper Precambrian or Lower Palaeozoic Robertson Bay Group and in the west faulted and unconformable contacts separate it from metamorphic rocks of probable Precambrian age. Some stratigraphic revision is necessary in the light of new data. The Supergroup, which exceeds 10 km in thickness, consists of three previously defined units: Sledgers Group (oldest), Mariner Group and Leap Year Group, the last two separated by an unconformity. The Sledgers Group is at least 3.5 km thick and contains two formations. The Middle Cambrian Glasgow Formation comprises volcanic breccia, conglomerate, pillow lava and vesicular flow rocks of basic and intermediate composition which interfinger with marine clastic sediments of the Molar Formation. The concordantly overlying Mariner Group (Upper Middle Cambrian to Upper Cambrian) comprises at least 2.5 km of fossiliferous sandstone, calcareous mudstone and limestone forming a regressive sequence dominated by shallow and marginal marine environments. It is truncated by the unconformably overlying mainly fluvial Leap Year Group (?Ordovician), which is exposed in two separate belts of contrasting lithologies. The eastern strip consists of at least 4 km (and perhaps more than 7 km) of quartzose sandstone and minor conglomerate (Camp Ridge Quartzite). In the northwest, the locally deeply-channelled Mariner Group is overlain by up to 800 m of polymict conglomerate and minor sandstone (Carryer Conglomerate), which in turn is overlain by 300 m of quartzose conglomerate and minor sandstone (Reilly Conglomerate). New sedimentological data suggest that the basin of deposition of both the Sledgers and Mariner Groups was bordered by a landmass to the southwest, but that an important source for the Sledgers Group also lay to the northwest. By contrast, the Leap Year Group, although still influenced by a positive area to the southwest, shows dominant transport towards the northwest, indicating a period of tectonic tilting and erosion in post Mariner times.

Further studies of the Bowers Supergroup (Laird, 1981; Laird et al., 1982) which includes all known Early Palaeozoic sedimentary rocks of northern Victoria Land, were carried out during the 1981/82 summer as part of the International Northern Victoria Land Expedition. The sequence, which occupies a belt 20-25 km wide, has now been traced for 350 km from the Oates Coast to the Ross Sea (Figure 1) in a strip transecting northern Victoria Land. It is separated from the Vendian or Lower Palaeozoic Robertson Bay Group (Field and Findlay, this volume) to the northeast by the Leap Year Fault. To the southwest the Bowers Supergroup is flanked by a metasedimentary series of unknown but probably Precambrian age, which in some places reaches amphibolite grade (Wodzicki et al., 1982). This unit has now been traced from the Lanterman Range in the north to the Meander Glacier in the south. In the Houliston Glacier area and south of the Molar Massif the supergroup is overlain unconformably by gently dipping rocks of the Beacon and Ferrar Supergroups of Late Palaeozoic and Early Mesozoic age, and andesitic and rhyolitic volcanics of unknown but probably Mid-Palaeozoic age appear to overlie it in the Lawrence Peaks (Laird and Bradshaw, 1982). The sequence dips steeply and is folded about axes trending northwest-southeast. New fossil discoveries (Cooper et al., this volume) and new outcrop data have resulted in some revision and extension of the earlier accepted stratigraphy (Laird et al., 1982).

## STRATIGRAPHY

The sequence was given supergroup status by Laird et al. (1982), who divided it into four units: the Husky Conglomerate, of unknown age and uncertain affiliation; the Vendian to Lower Cambrian Sledgers Group; the Mariner Group of late Middle Cambrian to late Late Cambrian age; and the Leap Year Group of probable late Late Cambrian or Ordovician age. The presence of an important unconformity was established between the upper two units, and a hiatus was inferred between the Sledgers and Mariner Groups. New data has resulted in a modification of the earlier stratigraphy (Figure 2); the Husky Conglomerate is now considered to be a basal facies of the Sledgers Group and only the unconformity separating the Mariner Group from the Leap Year Group is now recognised.

### Sledgers Group

This group (oldest) is at least 3.5 km thick and has now been examined from Frolov Ridge in the north, to Mt McCarthy in the south (Figure 1). This latter locality had been mapped by previous workers (e.g. Riddolls and Hancox, 1968; Gair et al., 1969) as Robertson Bay Group, but investigation during the 1981/82 field season showed that the rocks are typical of the Sledgers Group. The probable stratigraphic base of the succession was observed at only one locality (Latitude 71°31′S, Longitude 163°12′E), on a ridge extending north from the Lanterman Range and lying west of Reilly Ridge. Here 120 m of well rounded and almost undeformed conglomerate in a greenish matrix, indistinguishable in appearance from the Husky Conglomerate (Laird et al., 1982) the type locality of which lies 5 km along strike to the southeast, rests abruptly against metamorphosed and highly deformed conglomerates belonging to the undifferentiated metamorphics of the Lanterman Range. The contact is undulatory, and the base of the undeformed conglomerate contains large blocks up to 2 m of deformed conglomerate. An unconformity is inferred. The undeformed conglomerate unit has an abrupt, linear, inferred fault contact with overlying alternating sandstone and mudstone of the Molar Formation. Elsewhere, as on a ridge further to the northwest and at the type locality, the unit is bounded by faults. Detailed re-examination of the Husky Conglomerate revealed that although in many outcrops metamorphic clasts predominate, some included breccia units contain a high proportion of mafic volcanic clasts and are indistinguishable from similar breccias in the Sledgers Group. The Husky Conglomerate is thus now considered to be an informal name referring to a basal facies of the Sledgers Group.

The Sledgers Group is divided into two interfingering formations, the Glasgow Formation and Molar Formation (Laird et al., 1982). The Glasgow Formation, (formerly described as the Glasgow Volcanics—Laird et al., 1982), is redefined to include not only flow rocks but closely associated volcanic breccias which occur as sheets and lenses within the sedimentary facies (Molar Formation) of the Sledgers Group. It is at least 2500 mm thick at Mt Glasgow, the type locality. The formation also occupies a considerable thickness at Frolov Ridge, but has variable thickness to the south, being represented by 1000 m of breccia at the northern end of the Molar Massif, but apparently lensing out to the north and south. Thick pillow lavas make up the bulk of the Mt McCarthy massif and crop out for at least 15 km to the south.

The Glasgow Formation consists of porphyritic vesicular basaltic and andesitic (and also subordinate rhyolitic) flow rocks and associated angular, unsorted breccias of similar composition interfingering with the sedimentary Molar Formation. Pillow lavas are common, particularly in the northwest. Most flows appear to be of submarine origin, with the possible exception of a 100 m thick red-brown lava flow in the lower Carryer Glacier, which lacks pillows. The volcanics do not appear to occupy a preferred horizon and particularly in the north, from Frolov Ridge to the Carryer Glacier, they occur throughout the sequence and form the bulk of the Sledgers Group.

The Molar Formation, the type locality of which lies in the western Molar Massif (Laird et al., 1982: Figure 1), interfingers with the Glasgow Formation at many localities and occurs in thick sequences (1350 m+) from east of Frolov Ridge (Jordan, 1981) to Mt McCarthy. The formation consists dominantly of dark mudstone containing thin beds of well-sorted, rippled or wavy-bedded, very fine sandstone, occasionally graded, but more commonly cross-laminated or parallel-laminated throughout. Sole marks are relatively common. Intercalated within the mudstone are thinning and fining upward sequences up to 50 m thick consisting of boulder to granule conglomerate of igneous, sandstone, siltstone, and limestone clasts at the base, and passing upwards into mudstone-dominated units. The succession in the lower Carryer Glacier is somewhat more varied as it includes a laterally persistent bed of limestone 10 m thick, and units of muddy mass-flow deposits (up to 100 m thick), one containing large allochthonous blocks of limestone and volcanics.

Sediments of previous uncertain correlation in the Houliston Glacier (Cooper et al., 1976: Laird et al., 1982) were examined in the 1981/82 field season, and were found to be typical of the Molar Formation, with which they are now correlated. The Molar Formation was originally considered to be of Vendian to Early Cambrian age mainly on the basis of acritarch data (Cooper et al., 1982). However, new macrofossil collections and revision of acritarch age ranges (Cooper et al., this volume) confirm a Middle Cambrian age.

## Mariner Group

This unit, named from the type section at the head of the Mariner Glacier (Andrews and Laird, 1976), crops out intermittently from the head of the Sheehan Glacier to the mouth of the Mariner Glacier. The most continuous and thickest succession through the Mariner Group occurs at the type locality, where the uppermost 1600 m is almost completely exposed. Sporadic exposures at lower horizons suggest that the thickness may exceed 2500 m. The base is not exposed here, but is seen in the Molar Massif and in the Edlin Névé region. Originally inferred to be separated by a time gap from the Sledgers Group, mainly because of the Vendian to Early Cambrian age originally deduced from acritarch collections, it is now considered to have essentially conformable relationships with it. The Group was

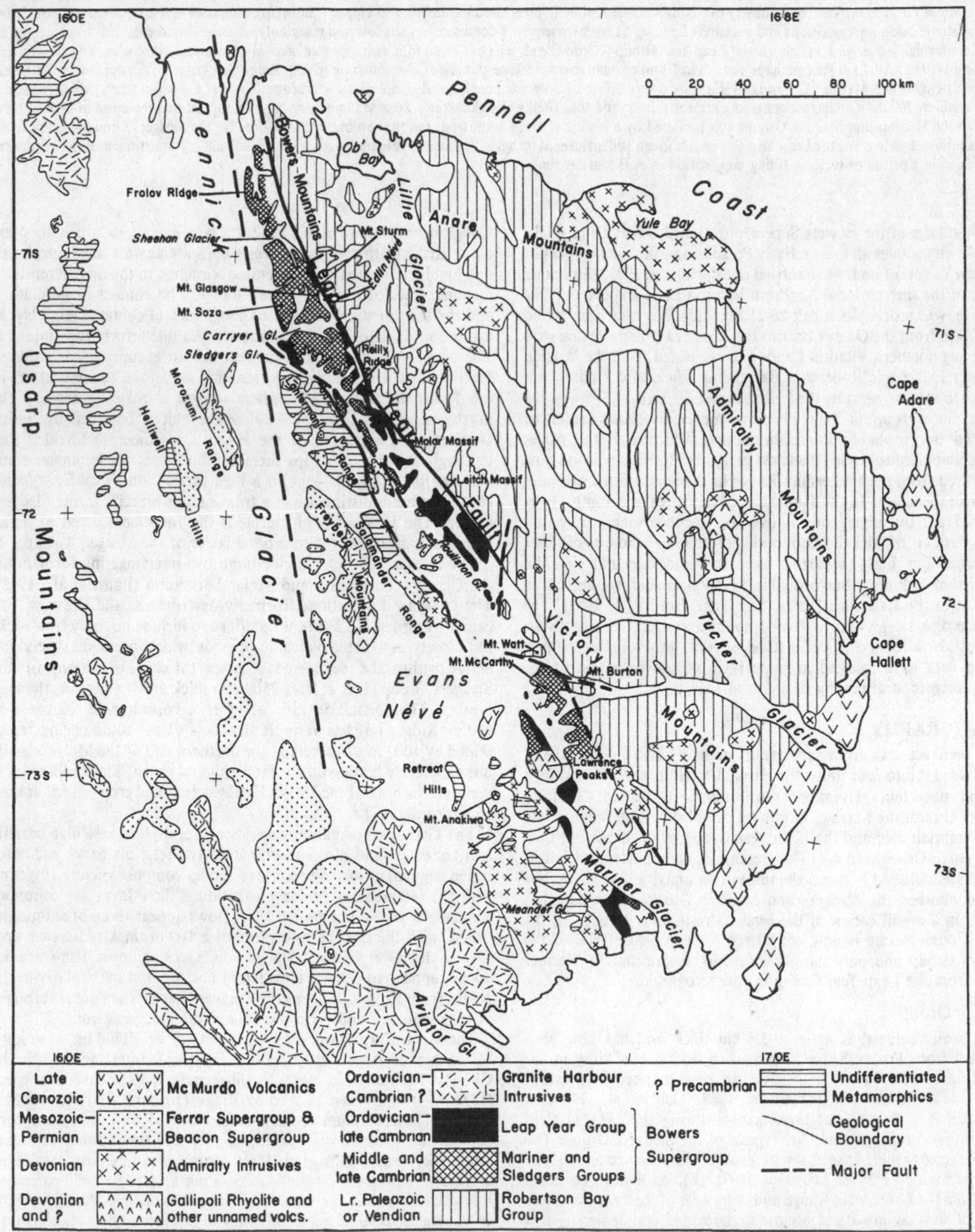


**Figure 1 Geologic map of northern Victoria Land showing the distribution of the Bowers Supergroup and other major rock sequences.**

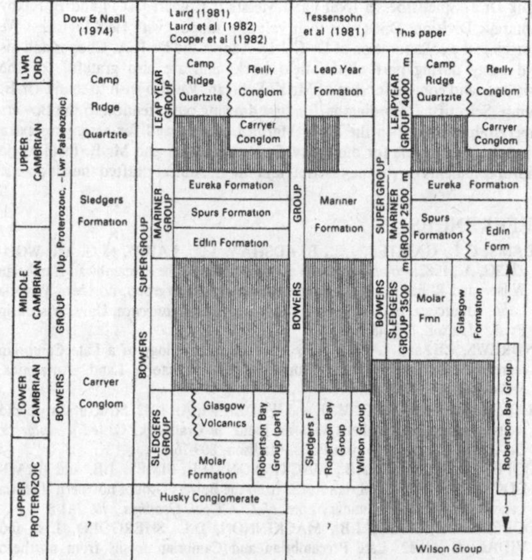


**Figure 2 Lower Palaeozoic stratigraphic units and ages, northern Victoria Land.**

subdivided into three units—the Edlin, Spurs, and Eureka Formations—by Laird et al., (1982).

The Edlin and Spurs Formations are at least in part laterally equivalent, the Edlin Formation resting apparently with sharp contact on the Glasgow Formation on Mt Glasgow, and the Spurs Formation passing gradationally down over a few metres into igneous breccia of the Glasgow Formation in the Molar Massif. The Edlin Formation has been identified only on the northeastern slopes of Mt Glasgow (its type locality), and in a section 15 km to the northwest at the head of the Sheehan Glacier. A sequence of quartzose sandstone occurring in the lower Carryer Glacier, previously correlated with the Edlin Formation, has been re-examined and is now considered to represent a basal part of the Carryer Conglomerate (see below). In the type locality a composite sequence 110 m thick includes thick quartz arenite units, thin tuffs and a volcanic debris flow, overlain by limestone and mudstone of the Spurs Formation. Five metres of red-brown mudstone directly overlying the basal contact on pillow lava of the Glasgow Formation contains probable mudcracks indicating emergence and possibly at least a local hiatus between the two formations. In the succession at the head of the Sheehan Glacier, 260 m of quartzite, tuff, conglomerate, and grey and red mudstone resting on pillow lava of the Glasgow Formation passes upward into limestone correlated with the Spurs Formation.

The Spurs Formation, named from Eureka Spurs at the head of the Mariner Glacier (Laird et al., 1982) is the most widely represented unit of the Mariner Group. It forms the lower part of the Group in the Molar Massif and at the head of the Carryer Glacier, and has continuous exposure for 900 m stratigraphically at the type locality. Scattered outcrops suggest that at least 800 m more may underlie this. It is dominated by fissile mudstone with limestone lenses which locally reach a thickness of 30 m. At Reilly Ridge, a succession otherwise dominated by mudstone contains horizons of channelised breccias, consisting mainly of limestone clasts (some up to 3 m), interpreted as debris flow deposits. Slump folding in the enveloping sediments is relatively common.

At the type locality, fissile mudstone of the Spurs Formation passes gradationally upwards into the Eureka Formation, typified at first by wavy-bedded sandstone, then by mudstone with limestone lenses, and then by brownish grey quartzose sandstone and minor mudstone with numerous trace fossil horizons and mud-cracks, interpreted as intertidal deposits (Andrews and Laird, 1976). The top of the formation is truncated by the Leap Year Group, but the exposed thickness is 700 m. The formation is exposed over a lateral distance of several kilometres in the head of the Mariner Glacier, but does not crop out further north where it was apparently removed by erosion prior to deposition of the Leap Year Group. The Mariner Group is notably fossiliferous, and has an age range of late Middle Cambrian to late Late Cambrian (Cooper et al., 1982).

## Leap Year Group

This group rests on a marked erosion surface which truncates successively older horizons of the Mariner and Sledgers Groups towards the northwest. The Leap Year Group has been divided (Laird et al., 1982), into three dominantly fluvial formations—the Camp Ridge Quartzite, the Carryer Conglomerate, and the Reilly Conglomerate.

The Camp Ridge Quartzite is the most widely represented unit of the Leap Year Group, and has now been traced along strike for a distance of over 300 km, from the head of the Sheehan Glacier to the mouth of the Mariner Glacier. It consists of red-brown or buff-coloured quartzose sandstone, quartzose conglomerate, and minor mudstone. The lithology is relatively uniform throughout the area of outcrop, with no determinable pattern of grain-size variation, except that the quartzose sediments at the mouth of the Mariner Glacier are notably conglomeratic. The basal scoured contact of the formation is seen at the head of the Mariner Glacier and at the head of the Carryer Glacier. The upper limit of the Camp Ridge Quartzite is nowhere seen, but a minimum thickness of 4000 m was examined by the authors in the Leitch Massif. Structural considerations suggest that a stratigraphic thickness of more than 7000 m is not unrealistic.

Sandy, quartzose conglomerate, cropping out on Reilly Ridge and on nunataks to the south and making up the Reilly Conglomerate, may be a lateral, coarser facies of the Camp Ridge quartzite (Laird et al., 1982). Although the basal contact is not well-exposed, the Reilly Conglomerate appears to rest abruptly on Carryer Conglomerate, rather than with rapid gradation as suggested by Laird et al., (1982). No upper stratigraphic limit to the Reilly Conglomerate was seen, but the minimum thickness is 300 m.

The Carryer Conglomerate crops out only in the northwest of the area between the south end of Reilly Ridge and Mt Soza, 15 km north of the mouth of the Carryer Glacier. It is a distinctive polymictic conglomerate with well-rounded clasts locally up to 3 m in diameter. On Reilly Ridge the unit appears to infill large channels up to 200 m deep cut into Spurs Formation of late Middle Cambrian age (Bradshaw et al., 1982), while to the northwest the Mariner Group is missing, and northwards between the Carryer Glacier and Mt Soza the Carryer Conglomerate rests unconformably on progressively older horizons of the Sledgers Group. In this last area the lower 100 m of the succession consists of fine to medium quartz arenite, previously correlated with the lower part of the Edlin Formation (Mariner Group, Laird et al., 1982). However, excellent exposure on a ridge west of Mt Soza (not previously visited) and re-examination of the succession in the lower Carryer Glacier show that the conglomerate and sandstone are interbedded, and the sandstone-dominated facies is best considered part of the Carryer Conglomerate. The formation attains its greatest thickness of approximately 800 m at its type locality, where, however, the top is not seen. Red-brown pebbly sandstone overlying typical Carryer Conglomerate has subsequently been found to represent a local facies of this Formation, rather than the Reilly Conglomerate, as previously inferred (Laird et al., 1982). At Reilly Ridge it has a maximum thickness of 300 m, and wedges-out rapidly to the south.

No shelly fossils are known from the Leap Year Group, but its stratigraphic position, trace fossils and radiometric data (Cooper et al., this volume; Adams et al., 1982) are consistent with a late Late Cambrian to Ordovician age.

## PALAEOGEOGRAPHY

In an earlier paper (Laird et al., 1982) it was inferred that the Bowers Supergroup was deposited in an elongate northwest—southeast trending basin—the Bowers Trough. New data collected during the International Expedition of 1981/82 supports this hypothesis. Palaeocurrent data for the Molar Formation indicate a dominant transport direction towards the southeast, although there is a possible reversal of the general trend in the Mt McCarthy area. Coarser sandstones and conglomerate beds forming part of channelised fining-upward sequences in addition show components from the southwest, and in some instances from the northeast. Palaeocurrent data from the Mariner Group is much sparser; however, trends from the southwest and, in the head of the Mariner Glacier, from the southeast, appear to dominate. Palaeocurrent trends in the Carryer Conglomerate based on cross-stratification and flutes indicate a direction of transport towards the east or northeast. Further data obtained from the Camp Ridge Quartzite confirm that the transport direction was towards the northwest or north-northwest, south of the Molar Massif, swinging round towards the northeast north of this. The exposures newly-examined near the mouth of the Mariner Glacier show a north-northwesterly direction of sediment transport.

Palaeocurrent and palaeoslope data from the Sledgers and Mariner Groups suggest that although most of the finer grained sediments were derived from the northwest, probably along a basin axis, an important source also lay to the southwest along much of the extent of the basin. Debris flow and other channel-fill deposits, together with slump horizons (in the case of the Mariner Group), indicate a significant palaeoslope and probable proximity of a southwestern basin margin. The presence of a northeastern margin is less certain, but scattered palaeocurrent directions derived from flute casts at the base of coarse sandstones in fining upwards channel-fill successions suggest this possibility. The apparent reversal of the dominant southeasterly trend of currents in the Mt McCarthy area may indicate a reversal of axial slope in the basin, suggesting that it had the form of a narrow, canoe-shaped trough. Palaeoslope data was not obtained for volcanic breccia units common in the Glasgow Formation. They are however inferred to have been emplaced by debris flows probably originating on the slopes of volcanoes on the basin margin. Active subsidence of the Bowers Trough continued throughout the deposition of the Sledgers Group, maintaining relatively steep slopes along the basin margin(s), and allowing deposition in excess of 3.5 km of sedimentary and volcanic material. The relatively fine-grained and largely volcanic-free Mariner Group represents a period of slowing subsidence and basin infill, shown by regression culminating in the tidal flat environment of the upper Eureka Formation. Local brief emergence in early Mariner times probably occurred near Mt Glasgow. A locally active basin margin lying to the southwest is however indicated by the debris flows and channelised sediments of the Reilly Ridge area. Evidence for a northeastern margin is absent.

Following deposition of the Mariner Group, the Bowers structural block, incorporating the trough, was uplifted and tilted to the southeast and an excess of 2 km eroded progressively towards the northern end of the block. Block faulting in the late Late Cambrian or Early Ordovician caused relative uplift of the Lanterman block with respect to the Bowers Block, and a northeast-trending valley, infilled with gravels of the Carryer Conglomerate, was eroded in the Mariner Group and in the upper part of the Sledgers Group. Finally, tilting to the northwest of the now downthrown Bowers structural block occurred, and the subsiding graben floor was occupied by mainly northwest flowing river systems, resulting in the deposition of several kilometres of quartzose sands. Although some of the sediments were derived from the metamorphic Lanterman block and its continuation to the south, the dominant source appears to be a granitic, metamorphic, or quartzose terrain lying to the southeast, in the area now occupied by the Ross Sea.

*Acknowledgements* Field work, carried out as part of the International Northern Victoria Land Expedition (1981/82), was supported by Antarctic Division, DSIR, and a VXE-6 Squadron, U.S. Navy. We particularly wish to thank Dr J. Splettstoesser (NSF) and Messrs P. Colbert (ATT) and E. Saxby (Antarctic Division, DSIR) for their valuable support with field logistics. We appreciate the collaboration of Dr C.J.D. Adams and Dr R.A. Cooper and his field party during part of the field work, and are also grateful for the assistance and companionship of Mr K. Sullivan. We also wish to thank Dr E. Stump, Scientific Co-ordinator, for first drawing our attention to the Bowers Supergroup outcrops in the lower Mariner Glacier, and for arranging extra helicopter flying time for our party. Dr R.A. Cooper and Mr B. Field made helpful comments on the manuscript, and Mr E. Annear drafted the diagrams.

## REFERENCES

ADAMS, C.J., GABITES, J.E., BRADSHAW, J.D., LAIRD, M.G. and WODZICKI, A., 1982: Potassium-argon geochronology of the Precambrian-Cambrian Wilson and Robertson Bay Groups and Bowers Supergroup, northern Victoria Land, Antarctica; *in* Craddock, C. (ed.). *Antarctic Geoscience*, Univ. Wisconsin Press, Madison, 543-48.

ANDREWS, P.B. and LAIRD, M.G., 1976: Sedimentology of a Late Cambrian regressive sequence (Bowers Group), northern Victoria Land, Antarctica. *Sedimt. Geol., 16*, 21-44.

BRADSHAW, J.D., LAIRD, M.G. and WODZICKI, A., 1982: Structural style and tectonic history in northern Victoria Land, *in* Craddock, C. (ed.), *Antarctic Geoscience*, Univ. Wisconsin Press, Madison, 809-16.

COOPER, R.A., JAGO, J.B., MACKINNON, D.I., SIMES, J.E. and BRADDOCK, P.E., 1976: Cambrian fossils from the Bowers Group, northern Victoria Land, Antarctica—preliminary note. *N.Z. J. Geol. Geophys., 19*, 283-8.

COOPER, R.A., JAGO, J.B., MACKINNON, D.I., SHERGOLD, J.H. and VIDAL, G., 1982: Late Precambrian and Cambrian fossils from northern Victoria Land and their stratigraphic implications; *in* Craddock, C. (ed.), *Antarctic Geoscience*, Univ. Wisconsin Press, Madison, 629-33.

COOPER, R.A., JAGO, J.B., ROWELL, A.J. and BRADDOCK, P., (this volume): Age and correlation of the Cambrian-Ordovician Bowers Supergroup, northern Victoria Land.

DOW, J.A.S. and NEALL, V.E., 1974: Geology of the Lower Rennick Glacier, North Victoria Land, Antarctica. *N.Z. J. Geol. Geophys., 17*, 659-714

FIELD, B.D. and FINDLAY, R.H., (this volume): The sedimentology of the Robertson Bay Group, northern Victoria Land.

GAIR, H.S., STURM, A., CARRYER, S.J. and GRINDLEY, G.W., 1969: Geology of northern Victoria Land (sheet 13, northern Victoria Land), *in* Bushnell, V.C. and Craddock, C. (eds.). Geologic Maps of Antarctica. *Antarct. Map Folio Ser.*, Folio 12, Pl. XII.

JORDAN, H., 1981: Metasediments and volcanics in the Frolov Ridge, Bowers Mountains, Antarctica. *Geol. Jb., B41*, 139-54.

LAIRD, M.G., 1981: Lower Palaeozoic rocks of Antarctica, *in* Holland, C.H. (ed.). *Lower Palaeozoic of the Middle East, Eastern and Southern Africa, and Antarctica.* John Wiley and Sons Ltd, 257-314.

LAIRD, M.G. and BRADSHAW, J.D., 1982: New data on the basement geology of northern Victoria Land, Antarctica. *N.Z. Antarc. Rec., 4(2).*

LAIRD, M.G., BRADSHAW, J.D. and WODZICKI, A., 1982: Stratigraphy of the Late Precambrian and Early Paleozoic Bowers Supergroup, northern Victoria Land, Antarctica, *in* Craddock, C. (ed.). *Antarctic Geoscience,* Univ. Wisconsin Press, Madison, 525-34.

RIDDOLLS, B.W. and HANCOX, G.T., 1968: The geology of the upper Mariner Glacier region, North Victoria Land, Antarctica.. *N.Z. J. Geol. Geophys., 11*, 881-99.

TESSENSOHN, F., DUPHORN, K., JORDAN, H., KLEINSCHMIDT, G., SKINNER, D.N.B., VETTER, U., WRIGHT, T.O. and WYBORN, D., 1981: Geological comparison of basement units in North Victoria Land, Antarctica. *Geol. Jb., B41*, 31-88.

WODZICKI, A., BRADSHAW, J.D. and LAIRD, M.G., 1982: Petrology of the Wilson and Robertson Bay Groups and Bowers Supergroup, northern Victoria Land, Antarctica, *in* Craddock, C. (ed.). *Antarctic Geoscience*, Univ. Wisconsin Press, Madison, 549-54.

# GEOSYNCLINAL SEDIMENTATION AND ROSS OROGENY IN NORTHERN VICTORIA LAND

F. Tessensohn, *Bundesanstalt fur Geowissenschaften und Rohstoffe, Postfach 51 01 53, 3000 Hannover 51, Federal Republic of Germany.*

**_Abstract_ The pre-Permian basement of northern Victoria Land comprises large areas of almost non-metamorphic sediments besides medium-to high-grade metasediments, migmatites, and granites. It differs from the basement of southern Victoria Land in containing true geosynclinal sediments, especially in the eastern part of the area, roughly between the Rennick Glacier and Ross Sea.**

**The turbidites of the Late Precambrian or Early Cambrian Robertson Bay Group and the clastic sediments and spilitic volcanics can be regarded as typical infill of a geosynclinal basin. These sediments have been tectonised (folded and slightly metamorphosed) in the Ordovician. This is proved by K/Ar dating on phyllites (Adams et al., 1977) and by the minimum age of 479 Ma of an intruding granite in the Morozumi Range (Kreuzer et al., 1981).**

**In the western part of the area the rocks are of higher metamorphic grade (Wilson Group). The parent rocks could well be sediments of a similar character, in some areas equalling the turbidite facies of the Robertson Bay Group, in others more similar to rocks of the Sledgers Formation with their content of conglomerates, amphibolites and marbles. The age of metamorphism of these rocks is similarly Ordovician as proved by K/Ar and Rb/Sr dates on metasediments and intruding granites. It seems clear now, that both areas were affected and tectonised by the same orogeny, and that the whole of northern Victoria Land fits into one geotectonic system, the Ross Orogeny of Ordovician age.**

**However, although both areas comprise similar parent rocks, there is definitely a distinct difference in crustal structure between them. Ordovician syn-orogenic granite intrusions are restricted to the western areas, whereas the turbidite realm in the east contains no such granites. This initially could be explained in terms of higher and lower crustal level or in terms of a continental (in the west) or oceanic (in the east) layer under the sediments.**

**The eastern area, however, was affected by a later large scale intrusive phase in the Devonian. Where does the crustal material for the generation of these granites come from, if the turbidites were initially deposited on an ocean floor?**

**It seems, that the simple model craton in the west (East Antarctic Shield)—ocean in the east (Protopacific) does not fit the facts completely. A model with several original basins and ridges is in better accordance with the field observations.**

# AGE AND CORRELATION OF THE CAMBRIAN-ORDOVICIAN BOWERS SUPERGROUP, NORTHERN VICTORIA LAND

R.A. Cooper, *N.Z. Geological Survey (D.S.I.R), PO Box 30368, Lower Hutt, New Zealand*

J.B. Jago, *School of Applied Geology, South Australian Institute of Technology, PO Box 1, Ingle Farm, S.A. 5098, Australia*

A.J. Rowell, *Department of Geology, University of Kansas, Lawrence, Kansas, 66045, U.S.A.*

P. Braddock, *Queen Charlotte College, Picton, New Zealand*

*Abstract* New evidence bearing on the age and correlation of the Cambrian-Ordovician Bowers Supergroup of northern Victoria Land is outlined. Constraints on the age range of the lower unit, the Sledgers Group, are derived from a variety of fossils including a small assemblage of polymerid trilobites in a limestone clast from conglomerate; an age range within the Middle Cambrian older than the late Middle Cambrian Boomerangian Stage is most probable. Clastic sedimentation infilling the Bowers depositional depression thus commenced in about the early Middle Cambrian, considerably later than the late Precambrian to Early Cambrian age previously inferred. Similarly, the basaltic volcanism represented by the Glasgow Formation is now known to be of Middle Cambrian age rather than Vendian to Early Cambrian. The conformably overlying Mariner Group is now well dated by trilobites and ranges from late Middle Cambrian (Undillan or Boomerangian) to Late Cambrian (late Idamean or early post-Idamean) in age. The youngest unit, the predominantly fluviatile Leap Year Group, contains a trace fossil assemblage consistent with a Late Cambrian to Ordovician age. The Bowers Supergroup, is thought to have formed part of a zone of Cambrian rifting, subsidence, rapid sedimentation and volcanism that extended through Western Tasmania, providing a link between the Antarctic and Australian segments of Gondwanaland in Cambrian time.

Fossil evidence for the age and correlation of units of the Bowers Supergroup has been summarised by Cooper et al. (1976, 1982). The following discussion concentrates on the age implications of new information concerning the known age range of some of the fossils and of new information derived from field mapping and new fossil localities found during the 1981/82 field season. Localities and measured sections are shown in Figures 1 and 2 respectively.

For a general account of the stratigraphy and structure of the Bowers Supergroup and adjacent rock units see Laird et al. (1982), Laird and Bradshaw (this volume). The stratigraphic nomenclature adopted here is that of Laird and Bradshaw (this volume) and is shown in Figure 3 which also shows the ages assigned to Bowers Supergroup units by previous workers.

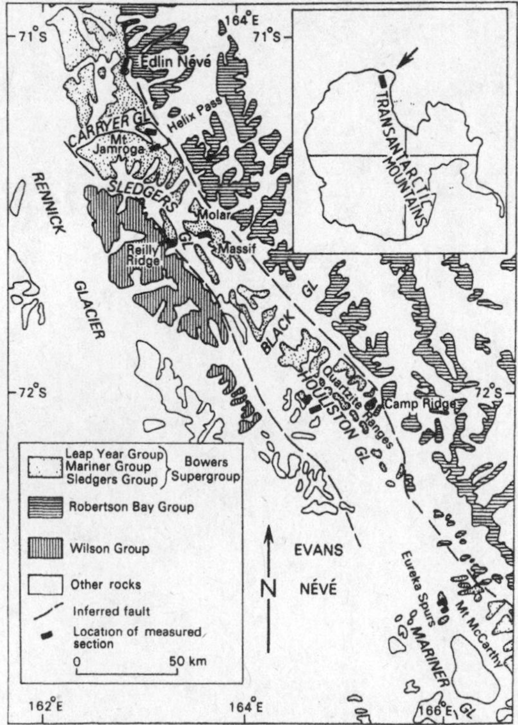


**Figure 1 Simplified basement geology of the Bowers Mountains—Evans Neve area, northern Victoria Land, showing locations of sections represented in Figure 2.**

## MICROFOSSIL AGE RANGES

Organic-walled microfossils were recovered from the insoluble residues of samples of Robertson Bay Group and Sledgers Group (Molar Formation) and reported on by Cooper et al. (1982). A number of acritarch and other taxa were listed and assigned a Vendian to earliest Cambrian age. Although the overall assemblage is characteristic of the Vendian, two fossils in particular were regarded as important in assigning the age, *Bavlinella faveolata* and *Chuaria circularis*. Dr G. Vidal, who is responsible for the acritarch determinations, has kindly provided us with the following new information which leads to a considerably revised assessment of the age range implied by these taxa.

Firstly, the finds reported as *Chuaria circularis* should now be regarded as uncertain. Tasmanitids, which are not found in the Proterozoic, are known to grow as large as *C. circularis* and when found in rocks that have undergone low grade metamorphism become opaque and lose their diagnostic pores. The poorly preserved Antarctic specimens could thus be tasmanitids and be of considerably younger age than previously thought.

Secondly, *Bavlinella faveolata*, in association with *Trachysphaeridium timofeevi* and *T. levis*, is now known from rocks as young as the *Paradoxides paradoxissimus* Zone (middle Middle Cambrian) of Sweden and can no longer be taken as a Vendian indicator.

The net effect of these revisions is that the microfossil assemblages can no longer be regarded as diagnostic of a Vendian to earliest Cambrian age and instead indicate a Vendian to Middle Cambrian age range (Figure 3). The individual localities are discussed further under the appropriate stratigraphic units below.

## AGE OF THE ROBERTSON BAY GROUP

Organic-walled microfossils from rocks mapped as Robertson Bay Group were reported by Cooper et al. (1982) and assigned a Vendian to Early Cambrian age. As outlined above this assemblage should now be regarded as of Vendian to Middle Cambrian age. The fossils come from three localities: the southwestern shoulder of Mt McCarthy, the Leap Year Glacier, and Mt Sturm. The richest assemblage was from Mt McCarthy. However, this region was re-examined in detail by members of the New Zealand Antarctic Research Programme (Events 19 and 20) during the 1981/82 field season and, as suspected by Cooper et al. (1982, p.633), the rocks represent the Molar Formation of the Sledgers Group rather than Robertson Bay Group. From the Leap Year Glacier and Mt Sturm localities, poorly preserved *Bavlinella faveolata* was recovered suggesting an age for Robertson Bay Group rocks in the Bowers Mountains area within the range Vendian to Middle Cambrian.

## AGE OF THE SLEDGERS GROUP

The Vendian to Early Cambrian age assigned to the Molar Formation by Cooper et al. (1982) was based on the presence of a microfossil assemblage recovered from several localities at Molar Massif together with fragmental chitinophosphatic fossils, regarded as "probably inarticulate brachiopods". The microfossils were thought to indicate an age no younger than earliest Cambrian whereas the "probable brachiopods" were taken as indicating a "probable" age of no older than Cambrian. From the above discussion of the revised age ranges of the microfossils it is clear that the fossiliferous beds could be as young as Middle Cambrian. Further evidence bearing on the inferred age of the Molar Formation has arisen from fieldwork carried out during the 1981/82 field season.

Additional chitinophosphatic fossils were collected from several horizons including that which yielded the earlier material in the Molar Massif. The fossils include some characteristic acrotretid brachiopod fragments which indicate a latest Early Cambrian to Middle Cambrian age. The fossiliferous beds are finely fissile dark shales with scattered carbonate concretionary bands and lenses and are now equated with the lower beds of the Spurs Formation (Mariner Group; Figure 2). They pass conformably down into Molar Formation which is distinguished by the presence of thin-bedded to thick-bedded sandstone beds. The contact between the two units in the Molar Massif is taken at the top of the highest band of volcanics in the succession and lies approximately 4 m below the lowest brachiopod horizon.

A more precise control on the younger age limit of the Molar Formation is provided by the late Middle Cambrian *Centropleura*-bearing assemblages of the overlying Spurs Formation in the col southeast of Mt Jamroga discussed below.

From Houliston Glacier poorly preserved ptychagnostid trilobites including *Ptychagnostus* have been recovered, indicating an age no older than Templetonian (middle Middle Cambrian). The fossils are found in cleaved dark shales exposed in the core of a small syncline in a nunatak 5 km south east of Neall Massif. A similar assemblage was recovered from beds exposed in the core of a second syncline in the north east corner of Neall Massif. Assignment of the fossiliferous beds to a formation is, unfortunately, uncertain. Beds underlying the fossiliferous shales are mapped as Molar Formation; they include similar dark thinly bedded shale together with interbedded coarse sandstone and conglomerate units (some of which are turbidites) and are closely similar to the Molar Formation of Molar Massif.

The lower beds of the Spurs Formation, as exposed southwest of Helix Pass and in Molar Massif are comprised of similar thinly bedded shale but without the coarse-grained sandstone and conglomerate of the Molar Formation. With only about 3 m of shale exposed in the core of the Houliston synclines it is uncertain whether the fossiliferous beds represent the basal beds of the Spurs Formation or a shale unit within the Molar Formation as assumed by Cooper et al. (1976). If the latter, then the date can be applied to the Molar Formation itself.

Several polymerid trilobites have been recovered from a limestone block in a conglomerate band in the upper part of the Molar Formation sequence exposed immediately to the east of Neall Massif. Although no specimens have been positively identified to genus two

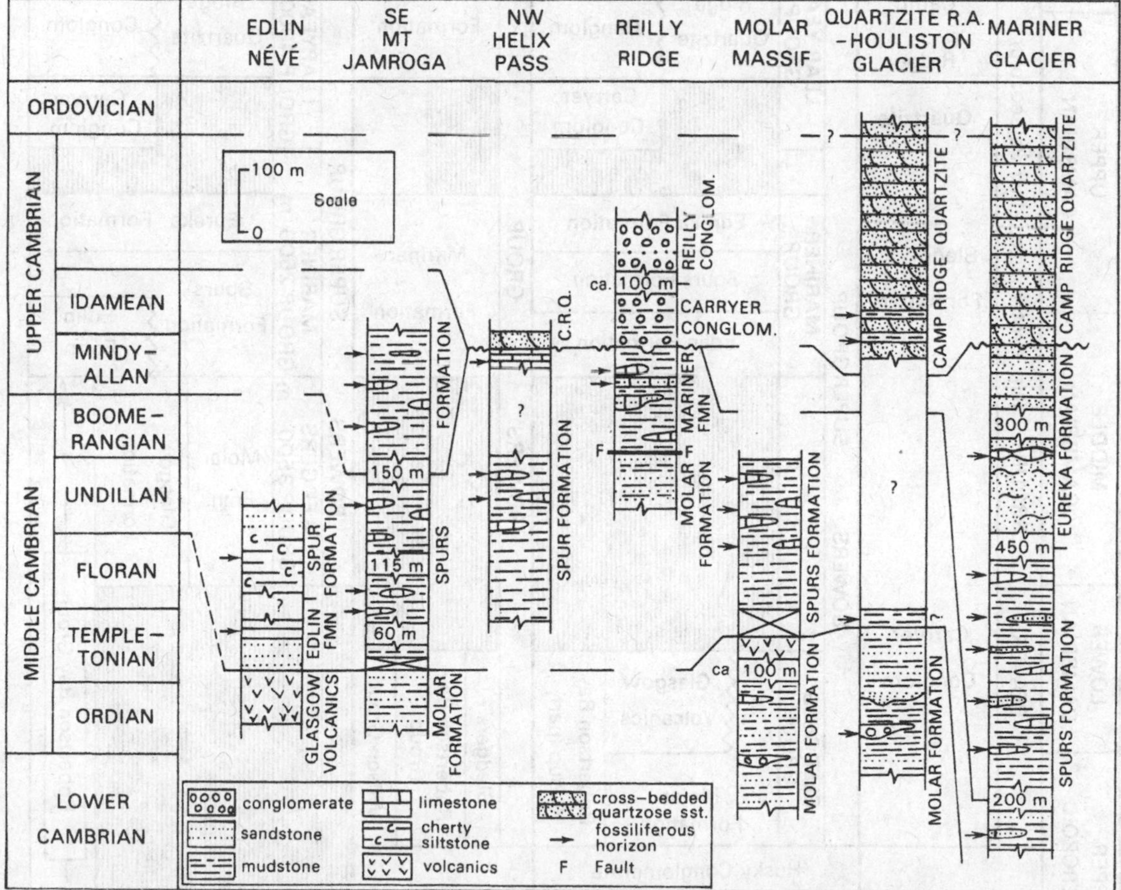


**Figure 2** Correlation of fossiliferous sections from Edlin Neve in the northwest to Mariner Glacier in the southeast. The Edlin Neve section is based on outcrops in the spur 1.5 km to the northeast of Mt Glasgow, and in the southeast trending ridge 3 km northeast of Mt Glasgow. The second section is exposed in the col 5 km southeast of Mt Jamroga. The third section is exposed in low ridges from 3 to 6 km northwest of Helix Pass. The Reilly Ridge section is exposed in the south end of Reilly Ridge. The Molar Massif section is exposed in a ridge 3 km west of point 2210 (U.S. Geological Survey Antarctica Reconnaissance Series Sheet SR 57-58/15, Mt Soza, 1:250 000) and 21 km south of Mt Verhage. The upper part of the sixth column represents the lower part of the Camp Ridge section, the lowest part is based on the outcrops in the two nunataks 7 kms southeast of Neall Massif. The Mariner Glacier section is based on outcrops in Eureka Spurs and in two small spurs to the south.

cranidia represent a form with close affinities to some of the late Middle Cambrian species assigned to the genus *Solenoparia* in Endo and Resser (1937). The remaining forms which include a dolichometopid are consistent with a Middle, rather than Late or Early Cambrian age and the assemblage is here provisionally taken as Middle Cambrian. The assemblage dates the source material of the conglomerate and provides a maximum age for the Molar Formation at this locality.

From the foregoing discussion it is possible to derive general age limits for the Molar Formation. The upper part of the formation is Middle Cambrian, probably late Middle Cambrian but older than Boomerangian; the lower beds are unlikely to be older than early Middle Cambrian. The formation is therefore here regarded as lying within the Middle Cambrian, an age considerably younger than that inferred by Cooper et al. (1982) but in agreement with that originally assigned by Cooper et al. (1976).

The revised age has important implications. Firstly, there is now no need to infer a major hiatus between the Sledgers and Mariner Groups (Laird et al. (1982); Cooper et al. (1982); Figure 3). The second implication concerns the evolution of the Bowers depositional depression; shale and sandstone deposition commenced in the early Middle Cambrian (or thereabouts) rather than in the late Precambrian to earliest Cambrian as previously thought (Bradshaw et al., 1982, p.814). Laird & Bradshaw (this volume) have interpreted a contact in the northeastern Lanterman Range as representing an unconformity between Husky Conglomerate (a basal phase of the Sledgers Group) and underlying metamorphics of the Lanterman Range. On this basis it appears that the Bowers Group rests directly on basement, precluding the presence of a pre-Bowers Group sedimentary succession such as that known in the Nimrod-Byrd Glacier region. The third implication concerns the age of volcanism. The Glasgow Formation (Glasgow Volcanics of Laird et al. (1982), p.538, basaltic breccia sheets and lenses) is a lateral equivalent of the Molar Formation and is therefore now of Middle Cambrian rather than Vendian to Early Cambrian age (Figure 3).

## AGE OF THE MARINER GROUP

A complete section through the Mariner Group is not known and the full succession is compiled by correlating sections exposed at Eureka Spurs and spurs to the south east in the Mariner Glacier area

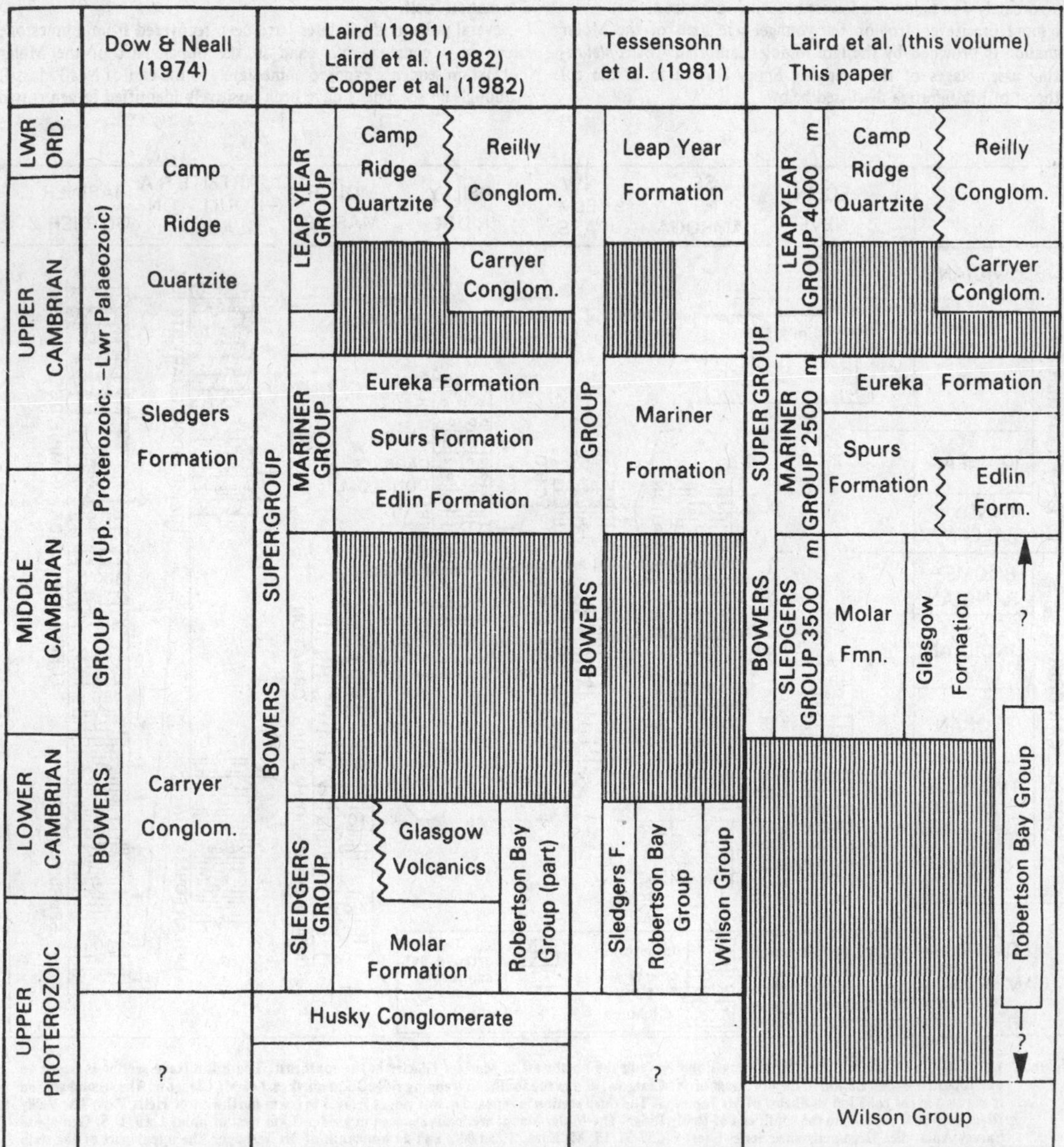


**Figure 3 Ages assigned to units of the Bowers Supergroup and adjacent rocks by previous workers compared with those assigned here.**

with those exposed in the Helix Pass area, 170 km to the northwest (Figures 1 and 2).

New information arising from the recent field work concerning age of the Mariner Group comes from the discovery of the late Middle Cambrian trilobite *Centropleura* in concretionary shales in the lower part of the Spurs Formation in the col section 4 km southwest of Helix Pass (Figure 1). *Centropleura* was found through several tens of metres of shale, the lowest horizon lying approximately 170 m above the contact with the underlying Molar Formation. Age of the base of the Mariner Group is thus taken (Figures 2 and 3) as late Middle Cambrian (Undillan or Boomerangian).

The youngest unit of the Mariner Group, the Eureka Formation contains the youngest marine body fossils of the Bowers Supergroup. The following taxa are present in a thin band of oolitic limestone 500 m from the top of the formation of Eureka Spurs, Mariner Glacier: *Homagnostus* cf. *obesus, Pseudagnostus (Pseudagnostus)* sp. (determinations by J.H. Shergold pers. comm.). Age is late Idamean or possibly early post-Idamean. The fauna enables a reasonably accurate age estimate on the marine regression at the top of the Mariner Group, (Andrews and Laird, 1976), and the passage from marine to intertidal deposition at Eureka Spurs.

From within the Mariner Group, the inarticulate brachiopod *Stilpnotreta* has been recovered from the prominent limestone rib, (Laird et al. (1982), fig. 66.4), in northern Reilly Ridge; present known range of the genus is middle Boomerangian to Mindyallan (Zone of *Cyclagnostus quasivespa*; Henderson and MacKinnon, 1981). Other fossils in the Mariner Group have been described or listed by Shergold et al. (1976), Cooper et al. (1976, 1982).

## AGE OF THE LEAP YEAR GROUP

No new information on the age of this unit is available and a late post-Idamean) Cambrian to Ordovician age, consistent with the presence of the trace fossils, *Daedalus, Arthrophycus*, and radiometric data quoted by Adams et al. (1982) and Laird et al. (1982), is assumed (Cooper et al., 1982; Shergold et al., 1982).

## CORRELATION WITH TASMANIA

The Late Proterozoic to Devonian sequences of southeastern Australia, New Zealand and Antarctica have been correlated and compared by Cooper and Grindley (1982). More specifically, the Bowers Supergroup succession has been compared with the Cambrian and Ordovician of Western Tasmania by Laird et al. (1977). However it is not at all certain that the Bowers depositional "basin" or "trough" was directly aligned with, and formed the southern continuation of, any particular Tasmanian trough. Rather it seems more probable that the Bowers Mountains and Western Tasmania together formed part of a zone of rifting, subsidence, rapid sedimentation and volcanism in Cambrian time (Laird, 1981). This zone, spanning the Antarctic and Australian segments of Gondwanaland, provides a useful link in Cambrian reconstructions of the supercontinent.

*Acknowledgements* We appreciate the co-operation of Dr Laird and his field party. The work formed part of the New Zealand Antarctic Research Program, 1981/82, and was supported by Antarctic Division, Department of Scientific and Industrial Research, New Zealand. J.B. Jago was supported by the Australian Research Grants Committee.

## REFERENCES

ADAMS, C.J.D., GABITES, J.E., WODZICKI, A., LAIRD, M.G. and BRADSHAW, J.D., 1982: Potassium-Argon geochronology of the Precambrian Wilson and Robertson Bay Groups and Bowers Supergroup, northern Victoria Land, Antarctica; *in* Craddock, C., (ed.) *Antarctic Geoscience,* Univ. Wisconsin Press, Madison, 543-48.

ANDREWS, P.B. and LAIRD, M.G., 1976: Sedimentology of a Late Cambrian regressive sequence (Bowers Group), northern Victoria Land, Antarctica. *Sedimt. Geol., 16,* 21-44.

BRADSHAW, J.D., LAIRD, M.G., and WODZICKI, A., 1982: Structural style and tectonic history in northern Victoria Land; *in* Craddock, C. (ed.) *Antarctic Geoscience,* Univ. Wisconsin Press, Madison, 809-16.

COOPER, R.A. and GRINDLEY, G.W. (eds.), 1982: Late Proterozoic to Devonian sequences of southeastern Australia, Antarctica and New Zealand and their correlation. *Geol. Soc. Aust., Spec. Publ., 9.*

COOPER, R.A., JAGO, J.B., MACKINNON, D.I., SIMES, J.E. and BRADDOCK, P.E., 1976: Cambrian fossils from the Bowers Group, northern Victoria Land, Antarctica (preliminary note). *N.Z. J. Geol. Geophys., 19,* 283-8.

COOPER, R.A., JAGO, J.B., MACKINNON, D.I., SHERGOLD, J.H. and VIDAL, G., 1982: Late Precambrian and Cambrian Fossils from northern Victoria Land and their stratigraphic implications; *in* Craddock, C. (ed.) *Antarctic Geoscience*, Univ. Wisconsin Press, Madison, 629-33.

DOW, J.A.S. and NEALL, V.E., 1974: Geology of the lower Rennick Glacier, northern Victoria Land, Antarctica. *N.Z. J. Geol. Geophys., 17,* 659-714.

ENDO, R. and RESSER, C.E., 1937: The Sinian and Cambrian formations and fossils of southern Manchukuo. *Manchur. Sci. Mus., Bull., 1.*

HENDERSON, R.A. and MACKINNON, D.I., 1981: New Cambrian inarticulate Brachiopoda from Australasia and the age of the Tasman Formation. *Alcheringa, 5,* 289-309.

LAIRD, M.G., 1981: Lower Palaeozoic rocks of the Ross Sea area and their significance in the Gondwana context. *R. Soc. N.Z., J., 11,* 425-38.

LAIRD, M.G. and BRADSHAW, J.D. (this volume): New data on the Early Palaeozoic Bowers Supergroup.

LAIRD, M.G., BRADSHAW, J.D. and WODZICKI, A., 1976: Re-examination of the Bowers Group (Cambrian), northern Victoria Land, Antarctica. *N.Z. J. Geol. Geophys., 19,* 275-82.

LAIRD, M.G., BRADSHAW, J.D. and WODZICKI, A., 1982: Stratigraphy of the late Precambrian and early Paleozoic Bowers Supergroup, northern Victoria Land; *in* Craddock, C. (ed.), *Antarctic Geoscience,* Univ. Wisonsin Press, Madison, 525-34.

LAIRD, M.G., COOPER, R.A. and JAGO, J.B., 1977: New data on the Lower Palaeozoic sequence of Northern Victoria Land, Antarctica, and its significance for Australian-Antarctic relations in the Palaeozoic, *Nature, 265,* 107-9.

SHERGOLD, J.H., COOPER, R.A., MACKINNON, D.I. and YOCHELSON, E.L., 1976: Late Cambrian Brachiopoda, Mollusca, and Trilobita from northern Victoria Land, *Palaeontology, 19,* 247-91.

SHERGOLD, J.H., COOPER, R.A., DRUCE, E.C. and WEBBY, B.D., 1982: Synopsis of selected sections at the Cambrian-Ordovician boundary in Australia, New Zealand, and Antarctica; *in* Bassett, M.G. and Dean, W.T., (eds.) *The Cambrian-Ordovician Boundary,* The National Museum of Wales Press, Cardiff, 211-27.

TESSENSOHN, F., DUPHORN, K., JORDAN, K., KLEINSCHMIDT, G., SKINNER, D., VETTER, U., WRIGHT, T.O. and WYBORN, D., 1981: Geological comparison of basement units in North Victoria Land, Antarctica. *Geol. Jb., B41,* 31-88.

# POST-MINDYALLAN LATE CAMBRIAN TRILOBITE FAUNAS FROM ANTARCTICA

J. H. Shergold, *Bureau of Mineral Resources, PO Box 378, Canberra City, ACT 2601, Australia.*

*Abstract* Post-Mindyallan Late Cambrian trilobites are known from northern Victoria Land, and from the Ellsworth Mountains, 2200 km away at the opposite end of the Transantarctic Mountains, in western Antarctica. Six distinct faunal assemblages are currently known, but they represent only a small interval of time: four are of Idamean (Late Dresbachian) age, and the remaining two are possibly immediate post-Idamean (Early Franconian).

Two assemblages, about 200 m apart stratigraphically, occur in the Spurs Formation (Mariner Group) at Eureka Spurs, Mariner Glacier, northern Victoria Land, and in the overlying Eureka Formation at the same locality. The earlier assemblage contains trilobites, brachiopods, and molluscs which indicate a Late Idamean (*Erixanium sentum* Zone?) age, whereas the younger fauna, which contains species of *Olentella, Apheloides*, and possibly an elviniid, may represent either the succeeding *Stigmatoa diloma* Zone or even the *Irvingella tropica* Zone. While neither index fossil had been found, associated agnostids argue for the older zone, while the generic association noted occurs with *Irvingella* in central Kazakhstan, and suggests the younger age.

In the Heritage Range, Ellsworth Mountains, four further Late Cambrian assemblages occur in, or just below, the Minaret Formation. At Yochelson Ridge, an early Idamean *Glyptagnostus reticulatus* Zone assemblage, containing the zone species, aphelaspidines and a pagodiine, is succeeded by an assemblage containing aphelaspidines, a genus similar to *Changshanocephalus* and agnostids of the *Pseudagnostus idalis* group. At Springer Peak, a younger assemblage contains species representative of the *Erixanium sentum* Zone; while at Windy Peak and Pipe Nunatak an assemblage confined to olenid trilobites is considered to post-date the other three faunas, and may have an immediate post-Idamean age. Highly compressed completely articulated trilobite exoskeletons occurring at Inferno Ridge cannot be dated with confidence.

# POST-ROSS OROGENY CRATONISATION OF NORTHERN VICTORIA LAND

G.W. Grindley and P.J. Oliver, *N.Z. Geological Survey, (D.S.I.R.) P.O. Box 30368, Lower Hutt, New Zealand.*

**Abstract** Following the Late Cambrian-Early Ordovician Ross Orogeny and emplacement of the Granite Harbour Intrusives, northern Victoria Land became part of the East Antarctic craton. Non-marine sedimentation in troughs and grabens, rhyolite (ignimbritic) volcanism and fault-folding indicate a typical "Transitional Tectonic Domain" (Tectonic Map of Australia and New Guinea, 1971) during the remainder of the Palaeozoic. The following phases of cratonisation can be recognised.

**Phase I: Silurian-Early Devonian: Uplift of the basement metamorphic infrastructure (Wilson Group) and unroofing of S-type Granite Harbour batholiths. Folding of Middle Cambrian-Lower Ordovician sediments and volcanics of the Bowers Trough along north-west trends.**

**Phase II: Middle-Late Devonian: Rhyolitic volcanism including massive ignimbrites and minor andesites and dacites (Gallipoli Volcanic Complexes). Penecontemporaneous tilting of volcanics up to 90° with uplift of granitic basement and incorporation of granitic breccias and fluvial sediments into a volcano-tectonic depression trending eastnortheast. Intrusion of I-type granitoids (Admiralty Intrusives) and minor geothermal activity.**

**Phase III: Carboniferous-Permian: Uplift of Bowers and Admiralty Mountains and unroofing of Admiralty Intrusives. Deposition of Permian glacigenic sediments followed by Permian and Triassic fluvial sediments (Beacon Supergroup).**

**Phase IV: Jurassic: Subsidence of Rennick Graben and further uplift of Bowers and Admiralty Mountains. Basaltic volcanism and dolerite sill intrusion in intra-continental rift zone.**

**Phase V: Early Cretaceous: Block-faulting and folding of Beacon sequences in the Lanterman Range adjacent to the Bowers Tectonic Zone. Downfaulting of Rennick Graben.**

Northern Victoria Land forms a connecting link between the Early-Mid Palaeozoic orogenic belts of East Antarctica, southeast Australia, New Zealand and West Antarctica (Grindley and Davey, 1982). The triangular block (Figure 1) also contains more exposed rock than most other parts of Antarctica. From the USARP Mountains to the Ross Sea coast (400 km) is an unusually complete section across the Transantarctic Mountains. Fold and fault structures in all basement rock units trend regionally northwest-southeast, being truncated in the southeast by the faulted margin of the Ross Sea Embayment and to the north and northwest by a rifted continental margin facing the Pacific (Southern) Ocean (Gair et al., 1969). The general geological structure of northern Victoria Land is now fairly well known (Figure 1). This paper provides new data on the tectonic history, obtained during palaeomagnetic sampling of Palaeozoic rocks in the 1981/82 season.

## TECTONIC ZONES IN NORTHERN VICTORIA LAND

Three major tectonic zones have long been recognised in the basement rocks (Gair et al., 1969; Bradshaw et al., 1982), comprising two zones of Upper Precambrian metasedimentary rocks (Wilson and Berg Groups in the west and Robertson Bay Group in the east) separated by a northwest trending downfaulted zone of Lower Palaeozoic rocks (Bowers Supergroup). The boundaries between these zones are major faults with a long history of movement dating back to

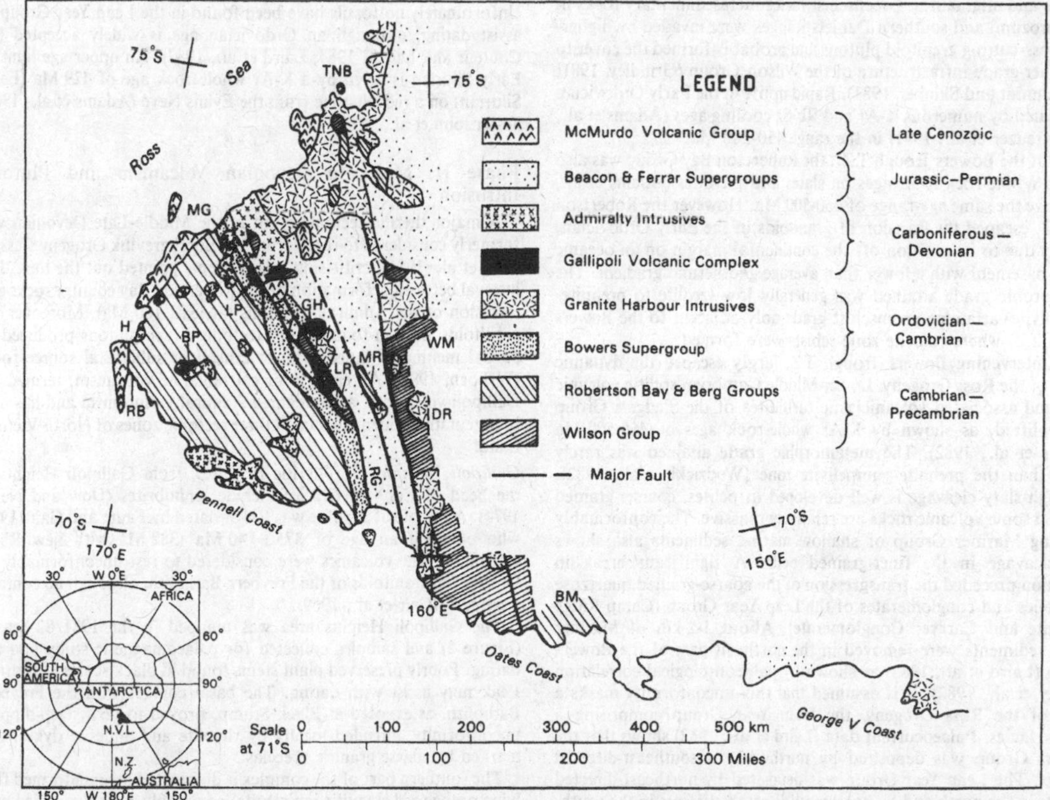


**Figure 1. Generalised geological map of northern Victoria Land showing main structural features and localities mentioned in text. TNB-Terra Nova Bay; LP-Lawrence Peaks; BP-Mt Black Prince; RB-Robertson Bay; MR-Morozumi Range; DR-Daniels Range; MG-Mariner Glacier; GH-Gallipoli Heights; H-Hallett Station; LR-Lanterman Range; WM-Welcome Mountain; RG-Rennick Glacier.**

Cambrian sedimentation. The Lanterman Fault separates the central Bowers Trough Tectonic Zone from the western Rennick Tectonic Zone comprising Upper Precambrian Wilson and Berg Groups and Granite Harbour Intrusives (Ordovician) while the Leap Year Fault separates the Bowers Trough Tectonic Zone from the eastern Robertson Bay Tectonic Zone comprising Upper Precambrian/Cambrian Robertson Bay metagreywackes and Admiralty Intrusives (Upper Devonian). Based on reconnaissance radiometric dating, northern Victoria Land was once considered to contain rocks of two discrete parallel orogens—the Lower Palaeozoic Ross Orogen, southwest of the Lanterman Fault and the Mid-Palaeozoic Borchgrevink Orogen to the northeast (Grindley and Warren, 1964; Gair et al., 1969; Craddock, 1972). Current views include a dominant Ross Orogen (Tessensohn et al., 1981), a wide Ross Orogen split by a synclinorial Bowers Trough Tectonic Zone, deformed in a Mid-Palaeozoic Borchgrevink Orogeny, (Adams, 1981; Bradshaw et al., 1982; Adams et al., 1982), and a dominant Borchgrevink Orogen (Stump, 1981).

## ROSS OROGENY

The Ross Orogeny in northern Victoria Land was a terminal orogeny, completing the accretion of folded metasedimentary Upper Precambrian-Cambrian terrains on to the Pacific margin of the East Antarctic Craton. West of the Bowers Trough Tectonic Zone (T.Z.), the folded metasedimentary Wilson and Berg Groups were intruded by Ordovician granitoid plutons, correlated by radiometric dating with the Granite Harbour Intrusives of the McMurdo Sound region. These granitoids are dominantly of S-type, having originated by melting of the deeper parts of the metasedimentary pile where migmatites commonly grade into anatectic granites (Wyborn, 1981). At least three phases of deformation are recognised in the Wilson Group (Bradshaw et al., 1982; Kleinschmidt and Skinner, 1981), but only the latest can definitely be attributed to the Ross Orogeny. Within the Rennick T.Z., the deeper parts of the Ross Orogen were metamorphosed to amphibolite grade, migmatised, and intruded by a network of granitic veins and dykes to form typical Wilson Group gneisses and migmatites. Greenschist facies metasedimentary rocks in the Morozumi and southern Daniels Ranges were invaded by higher-level cross-cutting granitoid plutons and probably formed the cover to the higher-grade infrastructure of the Wilson Group (Grindley, 1981; Kleinschmidt and Skinner, 1981). Rapid uplift in the Early Ordovician is indicated by numerous K-Ar and Rb-Sr cooling ages (Adams et al., 1982; Kreuzer et al., 1981) in the range 450-500 Ma.

East of the Bowers Trough T.Z., the Robertson Bay Group was also folded. Whole-rock K-Ar ages on slates and phyllites (Adams et al., 1982) give the same age-range of 460-500 Ma. However the Robertson Bay T.Z. escaped the invasion of granitoids in the Early Ordovician, possibly due to its position off the continental margin on an oceanic crustal basement with a lower than average geothermal gradient. The metamorphic grade attained was generally low (zeolite to prehnite-pumpellyite) attaining greenschist grade only adjacent to the Bowers Trough T.Z. where chlorite zone schists were formed.

The intervening Bowers Trough T.Z. largly escaped the dynamic effects of the Ross Orogeny. Lower-Middle Cambrian spilitic volcanic rocks and associated volcaniclastic turbidites of the Sledgers Group were uplifted, as shown by K-Ar whole-rock ages of 460-500 Ma (Adams et al., 1982). The metamorphic grade attained was rarely higher than the prehnite-pumpellyite zone (Wodzicki et al., 1982). Although slaty cleavage is well developed in pelites, coarser-grained beds and some volcanic rocks are relatively massive. The conformably overlying Mariner Group of shallow marine sediments also shows slaty cleavage in the finer-grained beds. A significant break in deposition preceded the transgression of the coarse-grained quartzose sandstones and conglomerates of the Leap Year Group (Camp Ridge Quartzite and Carryer Conglomerate). About 1-2 km of Mariner Group sediments were removed in the northern part of the Bowers Trough (Laird et al., 1982) as shown by palaeontological correlation (Cooper et al., 1982). It is assumed that this unconformity marks a phase of the Ross Orogeny, the Leap Year Group comprising a molasse facies. Palaeocurrent data (Laird et al., 1982) shows that the Sledgers Group was deposited by northwest or southeast-directed currents. The Leap Year Group was deposited by northeast-directed currents in the north and by northwest-directed currents in the south, indicating changes in the palaeogeography following the orogeny. The Carryer Conglomerate (Dow and Neall, 1974) and other coarse clastics on the western margin of the Bowers Trough (Laird et al., 1974) were derived partly from a metasedimentary terrain (Berg Group?) partly from granitoids (Granite Harbour Intrusives) and locally from metamorphic rocks (Wilson Group?) in keeping with uplift and erosion of the Rennick Tectonic Zone at this time.

## POST-ROSS OROGEN CRATONISATION

The process of cratonisation by accretion of island-arc and metasedimentary assemblages to the Precambrian Craton of Eastern Australia (Crook, 1974; Scheibner, 1972) has become the foundation for tectonic zonation on the 1971 Geological Society of Australia *Tectonic Map of Australia and New Guinea*. The term "Transitional Tectonic Domain" applied to cratonisation following major orogeny fits northern Victoria Land following the Ross Orogeny. Phases of cratonisation can be recognised, (see Abstract) which are further described below.

### Phase I: The Borchgrevink Event (Silurian-Early Devonian)

Adams (1981) and Adams et al. (1982) have drawn attention to the concentration of K-Ar cooling ages in the range 380-430 Ma obtained on slates from the Sledgers, Mariner and Leap Year Groups in the Bowers Trough T.Z. Similar ages obtained from the Roberson Bay Group on the Pennel Coast (Ravich and Krylov, 1964) formed the basis for the Borchgrevink Orogeny of Craddock (1972) but have been shown by recent West German studies (Tessensohn et al., 1981) to be reset ages due to Late Devonian Admiralty granitoid intrusion. A more restricted Borchgrevink Event, recognised in the Bowers Trough T.Z. from the radiometric dating evidence cited above, is correlated with strong folding and faulting of Bowers Supergroup rocks, by compression between the Rennick and Robertson Bay of folded terrains during the Late Silurian-Early Devonian. Refolding of the Robertson Bay metagreywackes adjacent to the Bowers Trough T.Z.,described by Bradshaw et al. (1982), may have accompanied closure of the Bowers Trough. Adams (1981) has recognised this event in southeast Australia, western South Island, N.Z., and Marie Byrd Land (see Cooper and Grindley, 1982, for details of correlation). Unfortunately no fossils have been found in the Leap Year Group to assist dating, although an Ordovician age is widely accepted (Le Couteur and Leitch, 1964; Laird et al., 1982). An upper age limit of Early Silurian is given by a K-Ar whole rock age of 429 Ma (Early Silurian) on a red siltstone from the Evans Nevé (Adams et al., 1982; Tessensohn et al., 1981).

### Phase II: Mid-Late Devonian Volcanism and Plutonic Intrusion

A major thermo-tectonic event in the Middle-Late Devonian was formerly considered to result from the Borchgrevink Orogeny. Tessensohn et al. (1981), criticising this concept, pointed out the long time interval between deformation of the Robertson Bay country rocks and intrusion of the Admiralty plutons (at least 100 Ma). Moreover the granitoids are non-foliated post-tectonic I-type plutons produced by partial melting of igneous lower crustal or subcrustal source rocks (Wyborn, 1981). An episode of calc-alkaline volcanism, termed the Gallipoli volcanism, accompanied Admiralty plutonism and has now been tentatively identified in all three tectonic zones of North Victoria Land.

*Gallipoli Volcanics* Porphyritic rhyolites, from Gallipoli Heights at the head of Canham Glacier include ignimbrites (Dow and Neall, 1974). A sample of rhyolite was Rb-Sr dated by Faure and Gair (1970) who obtained an age of 375 ± 40 Ma (382 Ma with new Rb-Sr constants). The volcanics were considered to rest unconformably on Ordovician granitoids of the Freyberg Batholith, although no contacts were seen (Gair et al., 1969).

The Gallipoli Heights area was mapped in the 1981/82 season (Figure 2) and samples collected for palaeomagnetic study and age dating. Poorly preserved plant stems found in black shales at Buttress Peak may assist with dating. The basal contact with the Freyberg Batholith, as exposed at Black Stump, proved to be a steep-dipping unconformity, intruded locally by rhyolite and andesite dykes, and marked by coarse granitic breccias.

The southern part of the complex is dominated by undeformed flat-lying orange-red rhyolitic ignimbrite, commonly strongly welded with eutaxitic texture, and containing corroded quartz and plagioclase phenocrysts in a devitrified felsitic matrix with remnant glass shards. The ignimbrite sheet is at least 250 m thick without an exposed base

This tectonic phase is probably the equivalent of the Bowers Block—Faulting of Tessensohn et al. (1981), postdating the Ross Orogeny.

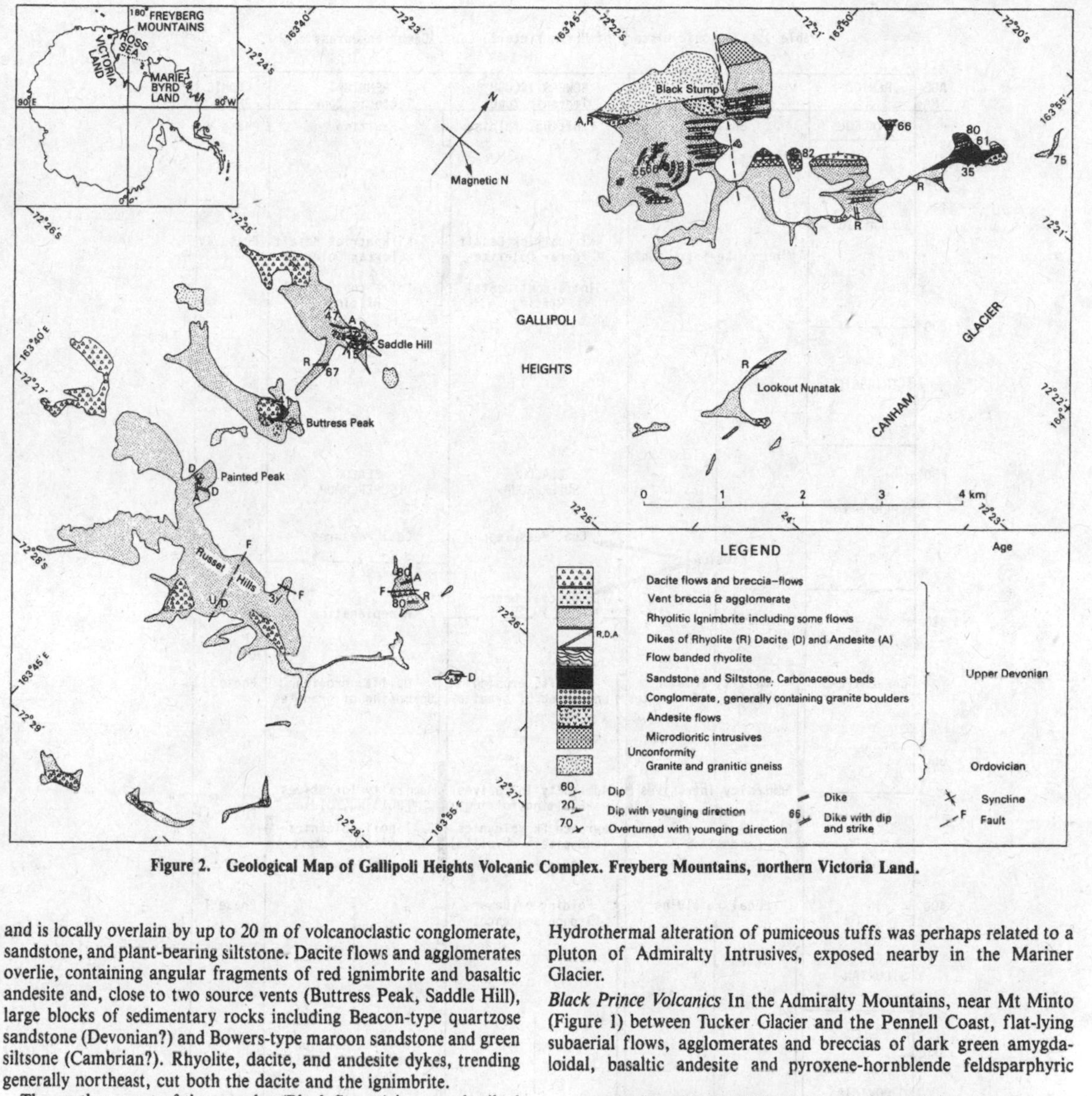

**Figure 2. Geological Map of Gallipoli Heights Volcanic Complex. Freyberg Mountains, northern Victoria Land.**

and is locally overlain by up to 20 m of volcanoclastic conglomerate, sandstone, and plant-bearing siltstone. Dacite flows and agglomerates overlie, containing angular fragments of red ignimbrite and basaltic andesite and, close to two source vents (Buttress Peak, Saddle Hill), large blocks of sedimentary rocks including Beacon-type quartzose sandstone (Devonian?) and Bowers-type maroon sandstone and green siltsone (Cambrian?). Rhyolite, dacite, and andesite dykes, trending generally northeast, cut both the dacite and the ignimbrite.

The northern part of the complex (Black Stump) is a steeply tilted 2 km thick sequence of dark andesite, microdiorite and red ignimbrite in 5-50 m bands containing conspicuous granite breccia-conglomerate layers, both at the base and throughout the lower 500 m of section (Figure 3). These are succeeded by dark andesitic flows and agglomerates, which are capped by ignimbrite and dacite flows containing volcaniclastic sandstone and carbonaceous siltstone layers. Porphyritic red rhyolite and dacite dykes are common, striking northeast subparallel to the regional strike; some may have been intruded after the sequence was tilted. Penecontemporaneous tilting of the sequence probably accompanied volcanism in a northeast-trending graben or rift-zone bordered on the northwest by a fault block range of granite, which provided granite scree to the graben between eruptions.

*Lawrence Peak Volcanics* At the head of Mariner Glacier (Figure 1) the large alpine massif of the Lawrence Peaks, formerly mapped as Robertson Bay Group (Gair et al., 1969) comprises subaerial andesite, rhyolite, and rhyolitic pyroclastics comparable with the Gallipoli Volcanics. In the southeast of the massif, a 100 m sequence of hydrothermal explosion breccia was examined. This sequence is overlain by 150 m of alternating crystal-rich quartzose rhyolite and dark grey feldsparphyric andesite. The volcanics are folded parallel to the Bowers Supergroup rocks to the northwest and dip between 20° and 60°. However, they are uncleaved and less altered (zeolite facies) than the Cambrian Glasgow Volcanics (prehnite-pumpellyite facies). Hydrothermal alteration of pumiceous tuffs was perhaps related to a pluton of Admiralty Intrusives, exposed nearby in the Mariner Glacier.

*Black Prince Volcanics* In the Admiralty Mountains, near Mt Minto (Figure 1) between Tucker Glacier and the Pennell Coast, flat-lying subaerial flows, agglomerates and breccias of dark green amygdaloidal, basaltic andesite and pyroxene-hornblende feldsparphyric

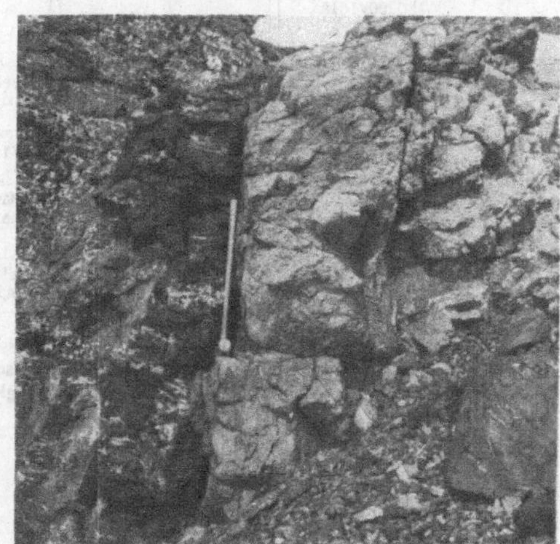

**Figure 3. Vertically dipping granitic breccia (right) overlain by dark fine grained andesite (centre) and rhyolite (left). Northeast side of the Black Stump, Gallipoli Heights.**

Table 1. Tectonic History of North Victoria Land, Cambrian-Jurassic.

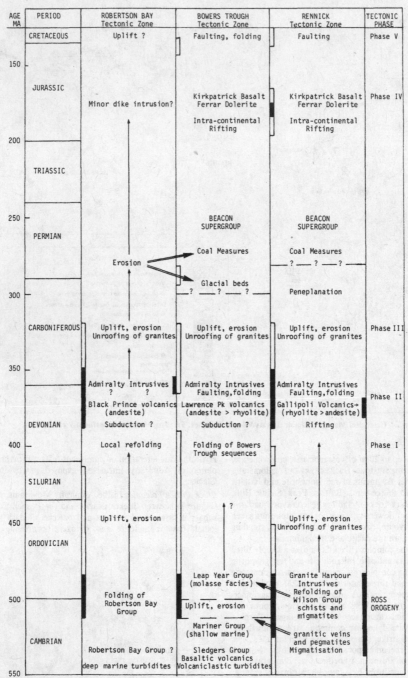

| AGE MA | PERIOD | ROBERTSON BAY Tectonic Zone | BOWERS TROUGH Tectonic Zone | RENNICK Tectonic Zone | TECTONIC PHASE |
|---|---|---|---|---|---|
| | CRETACEOUS | Uplift ? | Faulting, folding | Faulting | Phase V |
| 150 | | | | | |
| | JURASSIC | | | | |
| | | Minor dike intrusion? | Kirkpatrick Basalt Ferrar Dolerite | Kirkpatrick Basalt Ferrar Dolerite | Phase IV |
| | | | Intra-continental Rifting | Intra-continental Rifting | |
| 200 | | | | | |
| | TRIASSIC | | | | |
| 250 | | | BEACON SUPERGROUP | BEACON SUPERGROUP | |
| | PERMIAN | | | | |
| | | | Coal Measures | Coal Measures | |
| | | Erosion | | ? — ? — ? | |
| | | | Glacial beds | Peneplanation | |
| 300 | | | ? — ? — ? | | |
| | CARBONIFEROUS | Uplift, erosion Unroofing of granites | Uplift, erosion Unroofing of granites | Uplift, erosion Unroofing of granites | Phase III |
| 350 | | | | | |
| | | Admiralty Intrusives ? ? | Admiralty Intrusives Faulting,folding | Admiralty Intrusives Faulting,folding | |
| | | Black Prince volcanics (andesite) | Lawrence Pk volcanics (andesite > rhyolite) | Gallipoli Volcanics→ (rhyolite > andesite) | Phase II |
| | DEVONIAN | Subduction ? | Subduction ? | Rifting | |
| 400 | | Local refolding | Folding of Bowers Trough sequences | | Phase I |
| | SILURIAN | | | | |
| | | | ? | | |
| 450 | | Uplift, erosion | | Uplift, erosion Unroofing of granites | |
| | ORDOVICIAN | | | | |
| | | | Leap Year Group (molasse facies) | Granite Harbour Intrusives Refolding of Wilson Group schists and migmatites | |
| 500 | | Folding of Robertson Bay Group | Uplift, erosion | | ROSS OROGENY |
| | | | Mariner Group (shallow marine) | granitic veins and pegmatites | |
| | CAMBRIAN | Robertson Bay Group ? | Sledgers Group | Migmatisation | |
| | | deep marine turbidites | Basaltic volcanics Volcaniclastic turbidites | | |
| 550 | | | | | |

**Table 1. Tectonic History of northern Victoria Land, Cambrian—Jurassic. Range of thermotectonic episodes within each tectonic zone shown by K-Ar (left) and Rb-Sr (right) black boxes. Open boxes indicate uplift episodes identified by K-Ar ages and shown by single arrows. Double arrows represent known influxes of coarse sediment into the Bowers Trough during these uplift episodes. Time scale from Armstrong (1978) with minor modifications.**

andesite or dacite overlie the Robertson Bay Group unconformably (Findlay and Field, 1982). Relationships with the nearby Tucker Granodiorite of Mt Ajax have not been seen. Intrusive (S. Borg, pers. comm.), unconformable (Findlay and Field 1982) or faulted contacts are possible. Petrological composition and structural position suggests correlation with andesites at Lawrence Peaks and Gallipoli Heights (Table 1). Radiometric dating of all these complexes is presently being undertaken.

*Admiralty Intrusives* Numerous I-type biotite-hornblende granite and granodiorite plutons intrude the Robertson Bay T.Z. (Harrington et al., 1967; Crowder, 1968; Wyborn, 1981). Recent K-Ar and Rb-Sr radiometric dating (Kreuzer et al., 1981) has established the age of the Yule Bay and Lillie Glacier plutons as Late Devonian-Early Carboniferous; determinations average 362 ± 5 Ma. Concordance between biotite and hornblende ages indicate rapid cooling. Apparently older ages up to 385 Ma on the Tucker Granodiorite (Gair et al., 1969) suggest some Middle Devonian plutonism but have yet to be confirmed. Some younger ages—for example 300 Ma on the Lillie Glacier pluton—are erroneous due to argon loss.

## POST-CRATONIC COVER AND ITS DEFORMATION

Although cratonisaton was virtually completed following the intrusion of the Admiralty granites, further uplift of the Robertson Bay and Bowers Trough Tectonic Zones ensued during the Carboniferous. Northern Victoria Land remained a highland area during deposition of Beacon Supergroup terrestrial sediments in a major downwarp between Terra Nova Bay and the Oates Coast (Figure 1).

### Phase III: Beacon Supergroup Platform Cover (Carboniferous-Permian)

Glacigene Permo-Carboniferous Beacon Supergroup sediments approximately correlative with the tillites of southern Victoria Land occur deeply downfaulted into the western slopes of the Lanterman Range near the Rennick Glacier (Laird and Bradshaw, 1981). The deposits are infaulted against Wilson Group gneisses and comprise two tillites separated by quartzose sandstones containing dropstrones, and overlain by dark laminated shales also with dropstones, capped by a second unit of quartzose sandstones (Figures 4 and 5). An easterly provenance is indicated by abundant pebbles and cobbles of Robertson Bay/Sledgers metagreywackes, Admiralty granitoids and locally-derived blocks of Wilson Group gneisses. No base to the sequence is seen. Elsewhere in the Lanterman Range and Upper Rennick Glacier, the basal Beacon Beds are quartzose sandstones containing carbonaceous siltstone layers, and thin coal seams with *Glossopteris, Gangamopteris* and other Permian plants. These beds rest on a flat to gently undulating, weathered surface cut into the basement rocks, called the Sub-Beacon Peneplain by Gair (1967) and the Sub-Beacon Surface by Dow and Neall (1974).

**Figure 4. Moderately dipping sequence through Permian glacial deposits, Orr Glacier area, Western Lanterman Range. Tillite (front) underlies black shales (centre) and quartzose sandstone (rear). Morozumi Range in distance.**

### Phase IV: Ferrar Volcanism and Dolerite Sill Intrusion (Jurassic)

In common with Beacon sequences elsewhere, the northern Victoria Land basin was the site of intra-continental rifting in the Early-Middle Jurassic, resulting in the eruption of 2 km of tholeiitic flood basalts in the Upper Rennick Glacier (Gair, 1967) and at Litell Rocks, east of the Morozumi Range (Tessensohn et al., 1981). Basaltic magma invaded the Beacon sequence mainly as dolerite sills but also as cross-cutting dykes. In the northern Lanterman Range near Mt Moody, a thick dolerite sill has invaded the basement ganites (Figure 5). An average age for Ferrar magmatism of 179 ± 7 Ma is given by Kyle et al. (1981). Published ages from northern Victoria Land dolerites range from 139 ± 2 Ma, southern Lanterman Range (Dow and Neall, 1974) to 190 Ma from Horn Bluff (Ravich and Krylov, 1964). Kreuzer et al. (1981) were unable to obtain satisfactory K-Ar ages on the basalts from Litell Rocks; their ages, obviously reset, range from 90 to 120 Ma.

### Phase V: Rennick Faulting (Late Jurassic-Early Cretaceous)

Throughout the Transantarctic Mountains the Beacon Supergroup remains flat-lying or gently warped in large open synclines or monoclines. Block-faulting has been dominant both parallel and normal to the mountain chain (Katz, 1982). In northern Victoria Land, block-faulting has been on a grand scale, producing the Bowers and Admiralty Mountains and the 40 km wide and 2 km deep Rennick Graben (T. Stern, pers. comm). The post-Beacon/Ferrar block-faulting has been termed the Rennick Faulting by Tessensohn et al. (1981).

The Lanterman Range, bounded on the west by the Bowers Fault and on the east by the Lanterman Fault, preserves evidence, in the form of four infaulted and infolded Beacon/Ferrar outliers, for complex post-Jurassic deformation. The glacigene sediments close to the Bowers Fault are folded into two synclines separated by an anticline with dips up to 60° (Figure 4). Further south, Beacon sediments intruded and capped by dolerite form synclinal fault-bounded strips either side of the Hunter Glacier, as noted by Crowder (1968). The intensity of deformation of these Beacon synclines is quite unusual and implies compressional (or strike-slip) movements, rather than simple normal faulting (Figure 5). The age of the Rennick Faulting phase is not precisely known. The determination of Early Cretaceous K-Ar ages (Litell Rocks, Johnstone Glacier) from Ferrar Volcanics and dolerites, either side of the Lanterman Range, suggests some fault-block movements at this time. However, the presence of 3-4 km high mountains indicates neotectonic uplift and reactivation of the cratonic margin in the Late Cenozoic.

*Acknowledgements* The authors are indebted to other members of the North Victoria Land International Expedition for their cooperative attitude and help with logistics and communications especially J. Splettstoesser, E. Stump, G. MacKenzie, E. Saxby and helicopted pilots and crews of Air Development Squadron VXE6. Officers of Antarctic Division, D.S.I.R. supplied field equipment and logistic support. Colleagues at N.Z. Geological Survey, Lower Hutt are thanked for photographic, typing and draughting services, and for comments on the paper.

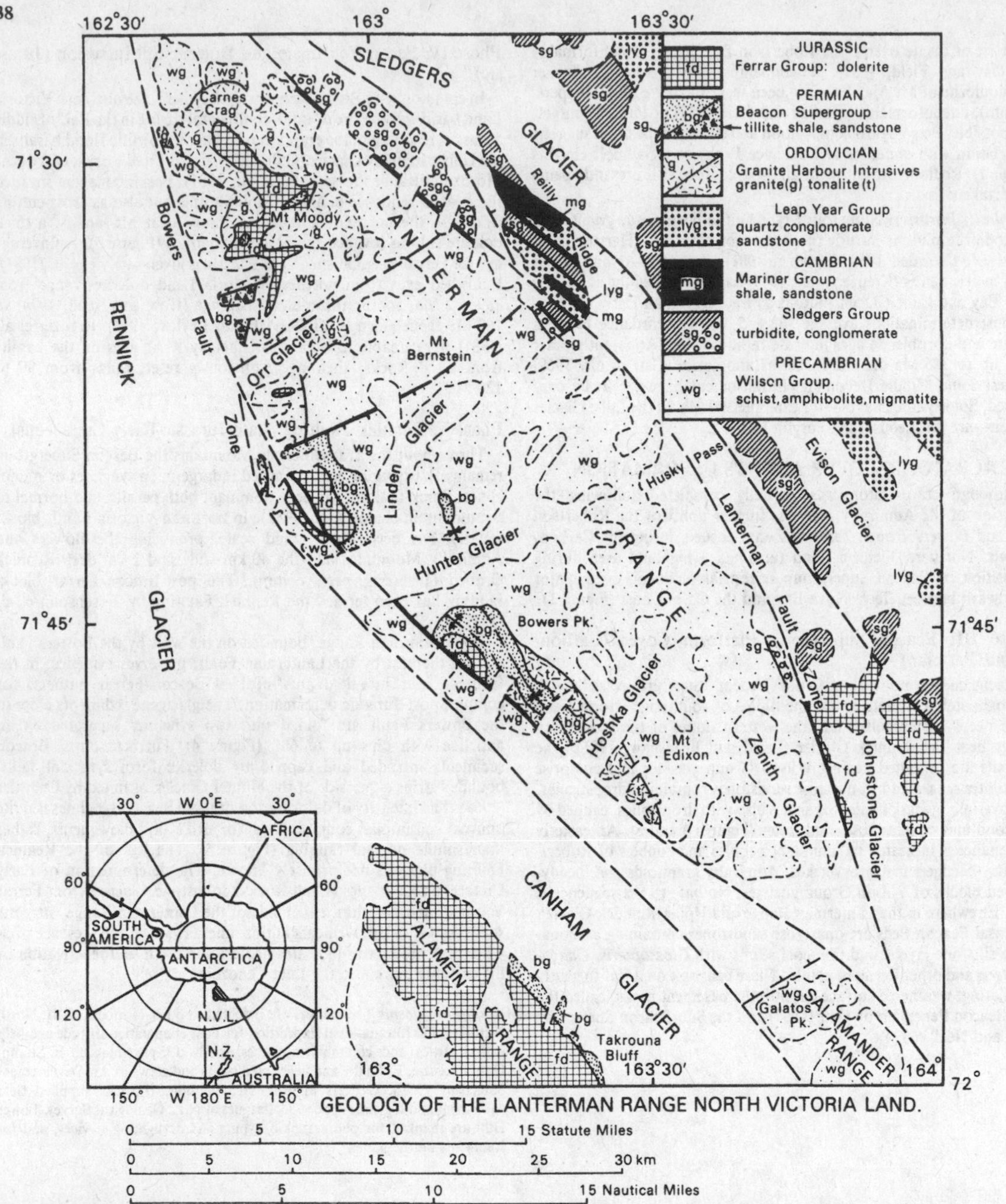


**Figure 5. Geological Map of the Lanterman Range showing infolded and infaulted Beacon sediments and Ferrar dolerites. Geological boundaries east of Lanterman Fault are taken from Bradshaw, et al. (1982).**

## REFERENCES


ADAMS, C.J.D., 1981: Geochronological correlations of Precambrian and Palaeozoic orogens in New Zealand, Marie Byrd Land (West Antarctica), northern Victoria Land (East Antarctica) and Tasmania; *in* Cresswell, M.M. and Vella, P. (eds.) *Gondwana Five,* A.A. Balkema, Rotterdam, 191-8.

ADAMS, C.J.D., GABITES, J.E., WODZICKI, A., BRADSHAW, J.D. and LAIRD, M.G., 1982: Potassium-argon geochronology of the Precambrian-Cambrian Wilson and Robertson Bay Groups and Bowers Supergroup, North Victoria Land, Antarctica; *in* Craddock, C. (ed.) *Antarctic Geoscience.* Univ. Wisconsin Press, Madison, 543-8.

ARMSTRONG, R.L., 1978: Pre-Cenozoic Phanerozoic Time Scale—Computer File of Critical Dates and Consequences of New and In-progress Decay-Constant Revisions. *Amer. Assoc. Pet. Geol., Stud. Geol., 6,* 73-91.

BRADSHAW, J.D., LAIRD, M.G. and WODZICKI, A., 1982: Structural style and tectonic history in northern Victoria Land; *in* Craddock, C. (ed.) *Antarctic Geoscience.* Univ. Wisconsin Press, Madison, 809-16.

COOPER, R.A. and GRINDLEY, G.W. (eds.), 1982: Late Proterozoic to Devonian Sequences of Southeastern Australia, Antarctica and New Zealand and their correlation. *Geol. Soc. Aust., Spec. Pub. 9.*

COOPER, R.A., JAGO, J.B., MACKINNON, D.I., SHERGOLD, J.H. and VIDAL, G., 1982: Late Precambrian and Cambrian fossils from northern Victoria Land and their stratigraphic implications; *in* Craddock, C. (ed.) *Antarctic Geoscience.* Univ. Wisconsin Press, Madison, 629-33.

CRADDOCK, C., 1972: Antarctic Tectonics; *in* Adie, R.J. (ed.) *Antarctic Geology and Geophysics.* Universitetsforlaget, Oslo, 449-68.

CROOK, K.A.W., 1974: Kratonization of West Pacific type geosynclines. *J. Geol., 82,* 24-36.

CROWDER, D.F., 1968: Geology of a part of North Victoria Land, Antarctica. *U.S. Geol. Surv., Prof. Pap. 600D,* D95-107.

DOW, J.A.S. and NEALL, V.E., 1974: Geology of the Lower Rennick Glacier, northern Victoria Land, Antarctica. *N.Z. J. Geol. Geophys., 17,* 659-714.

FAURE, G. and GAIR, H.S., 1970: Age determinations of rocks from northern Victoria Land, Antarctica. *N.Z. J. Geol. Geophys., 13,* 1024-5.

FINDLAY, R.H. and FIELD, B.D., 1982: Reconaissance observations on the Black Prince Volcanics, Admiralty Mountains, Antarctica. *N.Z. Antarc. Rec., 4(2)*, 11-4.

GAIR, H.S., 1967: The Geology from the Upper Rennick glacier to the Coast, northern Victoria Land, Antarctica. *N.Z. J. Geol. Geophys., 10*, 309-44.

GAIR, H.S., STURM, A., CARRYER, S.J. and GRINDLEY, G.W., 1969: The geology of northern Victoria Land (Sheet 13); *in* Bushnell, V.C. and Craddock, C. (eds.) Geologic Maps of Antarctica. *Antarc. Map Folio Ser., Folio 12 Pl XII.*

GRINDLEY, G.W., 1981: Precambrian rocks of the Ross Sea region. *R. Soc. N.Z., J., 11*, 411-23.

GRINDLEY, G.W. and DAVEY, F.J., 1982: The reconstruction of New Zealand, Australia and Antarctica; *in* Craddock, C. (ed.) *Antarctic Geoscience*. Univ. Wisconsin Press, Madison, 15-29.

GRINDLEY, G.W. and WARREN, G., 1964: Stratigraphic Nomenclature and correlation in the western Ross Sea Region, Antarctica; *in* Adie, R.J. (ed.) *Antarctic Geology.* North Holland, Amsterdam, 313-33.

HARRINGTON, H.J., WOOD, B.L., McKELLAR, I.C. and LENSEN, G.J., 1967: Topography and Geology of the Cape Hallett-Tucker Glacier District, Antarctica. *N.Z. Geol. Surv., Bull., 80.*

KATZ, H.R., 1982: Post-Beacon tectonics in the Region of Amundsen and Scott Glaciers, Queen Maud Range, Transantarctic Mountains; *in* Craddock, C. (ed.) *Antarctic Geoscience*. Univ. Wisconsin Press, Madison, 827-34.

KLEINSCHMIDT, G. and SKINNER, D.N.B., 1981: Deformation Styles in the Basement Rocks of North Victoria Land, Antarctica. *Geol. Jb., B41*,155-99.

KRUEZER, H., HÖHNDORF, A., LENZ, H., VETTER, U., TESSENSOHN, F., MÜLLER, P., JORDAN, H., HARRE, W. and BESANG, C., 1981: K/Ar and Rb/Sr Dating of Igneous Rocks from North Victoria Land, Antarctica. *Geol. Jb., B41*, 267-73.

KYLE, P.R., ELLIOT, D.H. and SUTTER, J.F., 1981: Jurassic Ferrar Supergroup tholeiites from the Transantarctic Mountains and their relationship to the initial fragmentation of Gondwana; *in* Cresswell, M.M. and Vella, P. (eds.) *Gondwana Five.* A.A. Balkema, Rotterdam, 283-7.

LAIRD, M.G., ANDREWS, P.B. and KYLE, P.R., 1974: Geology of northern Evans Névé, Victoria Land, Antarctica. *N.Z. J. Geol. Geophys., 17*, 587-601.

LAIRD, M.G. and BRADSHAW, J.D., 1981: Permian tillites of North Victoria Land; *in* Hambrey M.J. and Harland, W.B. (eds.) *Earth's Pre-Pleistocene Glacial Record,*Cambridge University Press, 237-40.

LAIRD, M.G., BRADSHAW, J.D. and WODZICKI, A., 1982: Stratigraphy of the Late Precambrian and Early Paleozoic Bowers Supergroup, northern Victoria Land; *in* Craddock, C. (ed.) *Antarctic Geoscience*. Univ. Wisconsin Press, Madison, 535-42.

LE COUTEUR, P.C. and LEITCH, E.C., 1964: Preliminary report on the geology of an area southwest of the upper Tucker Glacier, northern Victoria Land; *in* Adie, R.J. (ed.) *Antarctic Geology,* North Holland, Amsterdam, 229-36.

RAVICH, M.G. and KRYLOV, A.J., 1964: Absolute ages of rocks from East Antarctica; *in* Adie, R.J. (ed.) *Antarctic Geology,* North Holland, Amsterdam, 579-89.

SCHEIBNER, E., 1972: Tectonic concepts and tectonic mapping. *Geol. Surv. NSW, Rec., 14*, 37-83.

STUMP, E., 1981: Observations on the Ross Orogen, Antarctica; *in* Cresswell, M.M. and Vella, P. (eds.) *Gondwana Five.* A.A. Balkema, Rotterdam, 205-8.

TESSENSOHN, F., DUPHORN, K., JORDAN, H., KLEINSCHMIDT, G., SKINNER. D.N.B., VETTER, U., WRIGHT, T.O. and WYBORN, D., 1981: Geological comparison of Basement units in North Victoria Land, Antarctica. *Geol. Jb., B41*, 31-88.

WODZICKI, A., BRADSHAW, J.D. and LAIRD, M.G., 1982: Petrology of the Wilson and Robertson Bay Groups and Bowers Supergroup, Northern Victoria Land, Antarctica; *in* Craddock, C. (ed.) *Antarctic Geoscience*. Univ. Wisconsin Press, Madison, 549-54.

WYBORN, D., 1981: Granitoids of North Victoria Land, Antarctica—Field and Petrographic Observations. *Geol. Jb. B41*, 229-49.

# GEOCHEMISTRY, PETROGRAPHY, AND GEOCHRONOLOGY OF THE CAMBRO-ORDOVICIAN AND DEVONIAN-CARBONIFEROUS GRANITOIDS OF NORTHERN VICTORIA LAND, ANTARCTICA

U. Vetter, N.W. Roland, H. Kreuzer, A. Höhndorf, H. Lenz and C. Besang, *Federal Institute for Geosciences (BGR), P.O. Box 51 01 53, D-3000 Hannover 51, Federal Republic of Germany.*

*Abstract* West of the Rennick Glacier and in the Lanterman Range, flysch-like sediments of the medium to high grade metamorphic Wilson Group and of the low grade Robertson Bay Group are intruded by Cambro-Ordovician Granite Harbour Intrusives. They are represented mainly by biotite or two-mica granites occasionally containing cordierite, and subordinately by hornblende-biotite tonalites. Tentative Rb-Sr whole-rock (WR) isochrons yield dates of 515 to 480 Ma (initial ratios 0.71 to 0.715). The granites and granodiorites show S-type and the tonalites mostly I-type, characteristics. Similar K-Ar ages for hornblende, muscovite and biotite point to a final quick cooling of high-level intrusions. The range from 490 to 467 Ma suggests more than one pulse of intrusion.

East of the Rennick Glacier the folded Robertson Bay Group metasediments are intruded by the Devonian-Carboniferous Admiralty Intrusives, which are atectonic and strongly discordant. Petrographically nearly all of them show I-type characteristics. Similar biotite and hornblende K-Ar ages suggest high-level quick cooling. The range of K-Ar mineral ages is 367-354 Ma. Multiple intrusions are assumed. A Rb-Sr WR isochron yields a date of 393 ± 20 Ma. Xenoliths of magmatic restite and of metasediments of varying metamorphic grade were found in the Admiralty I-type granitoids. Geochemical analyses and Sr initial ratios of 0.71 to 0.72 demonstrate the influence of metasediments on the composition of the granitoids.

During the first German Antarctic North Victoria Land Expedition (Ganovex I) in 1979/80, two groups of granitoids were confirmed in northern Victoria Land and sampled for more detailed studies; namely

(1) the Cambro-Ordovician Granite Harbour Intrusives (Gunn and Warren, 1962)
(2) the Devonian-Carboniferous Admiralty Intrusives (Grindley and Warren, 1964).

A comprehensive study of the granitoids of northern Victoria Land is essential for the tectonic interpretation of this area, especially of its western part where the boundary between the East Antarctic Shield and the West Antarctic Fold Belts is thought to occur.

## Previous work, outline of geology, and analytical comment

The first study of granitic intrusions related to the Ross Orogeny was carried out by Gunn and Warren (1962). The granites were named after Granite Harbour Bay in southern Victoria Land and radiometric results showed Cambro-Ordovician ages. Granites of Devonian to Carboniferous age which intruded the Vendian to Cambrian Robertson Bay Group schists have been described by Harrington (1958) and Grindley and Warren (1964). These granitoids, the Admiralty Intrusives, are confined to northern Victoria Land. Their pressence is one of the main reasons for postulating a Middle Palaeozoic "Borchgrevink Orogeny" (Sturm and Carryer 1970; Craddock 1970).

Distinguishing the Granite Harbour and Admiralty Intrusives on field evidence proved to be difficult. Hence during Ganovex I and II the Granite Harbour and Admiralty Intrusives of northern Victoria Land were sampled for geochemical, petrological, and radiometric analyses, (Wyborn, 1981; Kreuzer et al., 1981). Radiometric results are calculated with the recommended constants (Steiger and Jäger, 1977). Errors are given as 95% confidence level of analytical precision. For all isochrons the mean squares of weighted deviates are close to 1.

## Granite Harbour Intrusives

These intrusives comprise deep-seated syn- and postkinematic granitoids. In three of the four investigated areas the maximum age of about 480 Ma was determined for muscovite-biotite, and hornblende-biotite pairs. The similarity in ages for minerals of different retentivities from the same rock samples proves that there was a quick final cooling which probably marks the time of postkinematic high-level intrusion (Figure 1). The tail to younger ages, especially a cluster around 472 Ma, including muscovites, we tentatively explain by slightly younger intrusive pulses. But this explanation does not cover the younger ages of Lanterman Range gneisses, especially not the concordant hornblende ages of 460 Ma. Possibly these ages and the much younger hornblende age of 360 Ma are related to the adjacent fault system against the Bowers Graben.

### *Daniels Range and Renirie Rocks*

Daniels Range is situated west of Rennick Glacier (Figure 1). The Thompson Spur/Schroeder Spur area of southern Daniels Range was investigated. Rocks formerly described as Rennick Schists (Sturm and Carryer, 1970) consist of pelitic, quartzofeldspathic and minor carbonate metasediments. They may represent a higher metamorphic equivalent of the Robertson Bay Group flysch-like sediments. Laterally they pass from amphibolite facies to the inhomogeneous migmatitic granitoids and massive granitoids called Wilson Gneiss by Sturm and Carryer (1970).

There is no sharp contact between the metasediments and the S-type granitoid bodies (S- and I-type classification according to the criteria of White and Chappell (1977) and Wyborn (1981). Pegmatite veins, quartz-rich schlieren and obviously synplutonic dykes occur. Rafts of schist and gneiss are common, partly digested, especially in the migmatitic banded zones.

Massive granites in the central part of Schroder Spur and in the western part of Thompson Spur are more homogeneous and more or less devoid of xenoliths. They were interpreted by Wyborn (1981) as anatectic magmas which moved slightly from their source area. Three samples of Schroeder Spur granite poorly define a Rb-Sr isochron of 510 ± 36 Ma with an initial $^{87}Sr/^{86}Sr$ ratio (IR) of 0.7116 ± 0.0023; by including a leucogranite and a pegmatite dyke both of which have considerably larger Rb/Sr ratios, an isochron of 495 ± 10 Ma (IR 0.7125 ± 0.0010) results (Figure 2). The age is still poorly assessed. However, there is no doubt about the comparatively high IR, which is in accordance with the classification of the Schroeder Spur granite as S-type. It is significantly different from the IR of about 0.707/0.708 of the three analysed I-type granitoids from the Schroeder and Thompson Spur areas. On the southwestern tip of Thompson Spur, a different type of granitoid body is intruded with a sharp, steep contact. It shows dark xenoliths in abundance.

Petrographically, Daniels Range and Renirie Rocks comprise different rock types. Three tonalites, 1 granodiorite, 5 adamellites, 1 leucogranite, 1 pegmatite, 3 migmatites, and 2 metasediments were analysed for major and trace elements. In the $Na_2O/K_2O$-diagram most of the migmatites and the metasediments plot in the I-type field (Figure 3). Wyborn (pers comm.) interprets this as a result of the immature character of the sediments. The S-type porphyritic two-mica granite of central Schroeder Spur and intrusive porphyritic adamellites of Thompson Spur and Renirie Rocks are relatively depleted in sodium and plot in the S-type field. CIPW-norms and significantly lower IR separate the three I-type tonalites from all other samples. The tonalites contain either clinopyroxene or < 1% by weight corundum. The S-types have excess $Al_2O_3$ > 1% by weight.

Worth mentioning are the obviously synplutonic I-type tonalite dykes which intruded the migmatites in central Schroeder Spur where they were fragmented by granite tectonic movements. At the contact of these angular blocks pegmatitic reaction rims are developed, showing that the migmatite had not yet cooled at that time. S- and I-type granitoids are partly time equivalent here.

*Morozumi Range*

Morozumi Range is situated east of southern Daniels Range (Figure 1). Most of its outcrops show contact-metamorphosed Robertson Bay Group sediments. Originally, they were mapped as Wilson Group by Dow and Neall (1974). The main granitoid body is found in the northern part of Morozumi Range. It is a biotite granite containing white to pink K feldspar megacrysts. Biotite-rich schistose to gneissic xenoliths, though not common, are randomly distributed throughout the intrusion. Two small granitic outcrops are exposed at Jupiter Amphitheatre, south of the main body, and in Unconformity Valley of eastern central Morozumi Range. All granitic samples show some evidence of strain and recrystallisation.

Well foliated NNW-trending granite dykes cut the higher grade contact metamorphosed Robertson Bay sediments at Graduation Ridge at the northeastern margin of the main granite body (Tessensohn et al., 1981 figure 5). Dow and Neall (1974) postulated a major fault bordering Morozumi Range against the Rennick Glacier (Graben ?) and causing the foliation.

From the main granitoid body seven samples were taken, plus one from Jupiter Amphitheatre, and one from Unconformity Valley. Six of these samples plot in the adamellite field. Included are leucogranites, one from the border facies at the southwestern corner of the main body, close to the contact with Roberston Bay metasediments, and one from Unconformity Valley (Figures 4 and 5). The most mafic sample

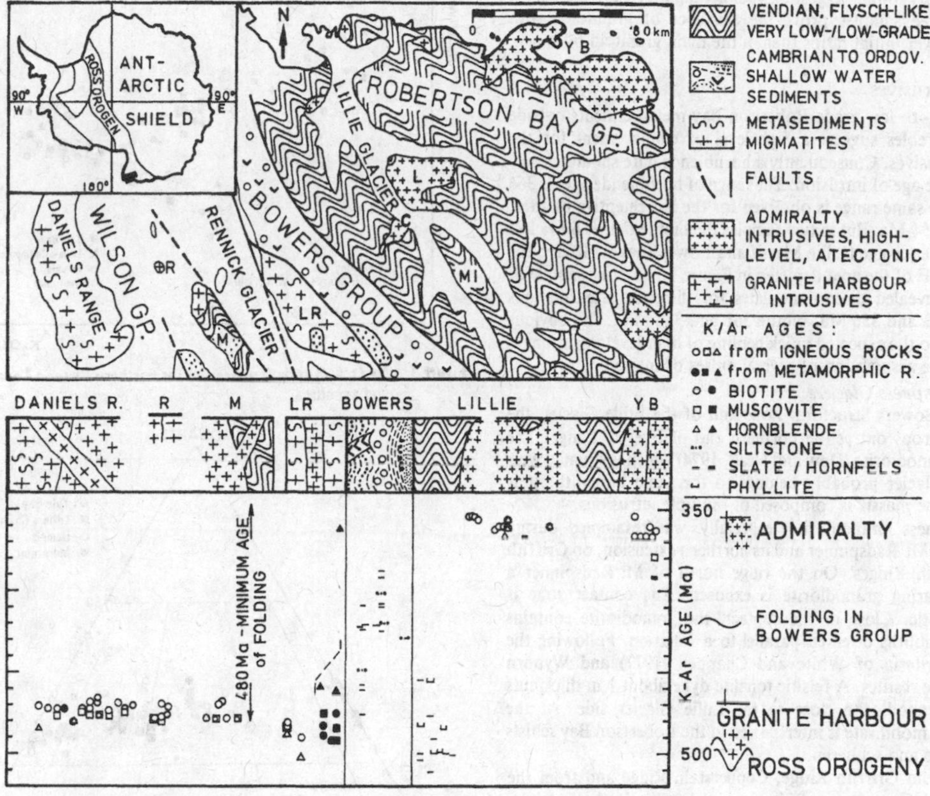


**Figure 1. Age relations along a WSW—ENE section through northern Victoria Land. Daniels, Morozumi (M), and Lanterman (LR) Ranges, Champness (C), Lillie (L), and Yule Bay (YB) granitoids—including Renirie Rocks (R) and Mirabito Range (MI). Data on gneisses, metasediments, and sediments from Adams et al. (1982).**

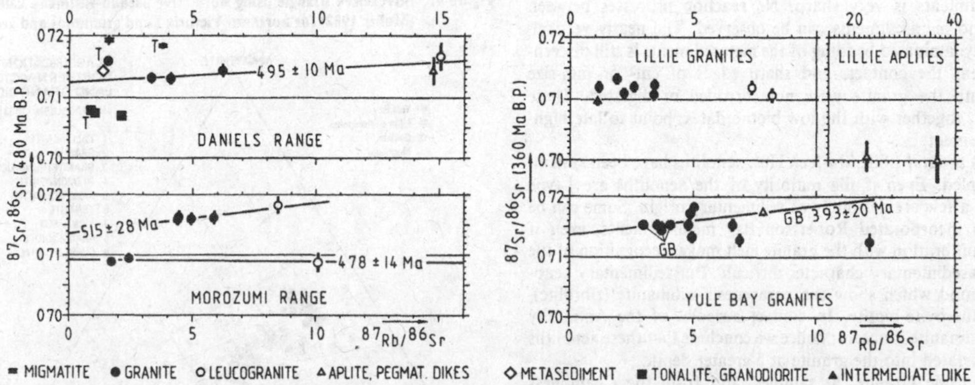


**Figure 2. Left: $^{87}Sr/^{86}Sr$ ratios of Granite Harbour Intrusives at 480 Ma ago. Daniels Range (Schroeder Spur and Thompson Spur (T), and Morozumi Range. This plot shows the reference isochron for 480 Ma and IR = 0.71 (the middle horizontal line) at an enlarged scale compatible with the analytical precision. It shows approximately the initial isotopic compositions and facilitates speculation about possible isochrons. Right: $^{87}Sr/^{86}Sr$ ratios of Admiralty Intrusives 360 Ma ago, ie: at approximately the time of intrusion. GB: Gregory Bluff pluton.**

is a gneissic biotite-rich tonalite from Graduation Ridge. In the $Na_2O$/$K_2O$-diagram (Figure 3) the Morozumi samples are equally distributed in the S- and I-type fields, whereas the CIPW-norm of all samples except the two leucogranites has >1 wt% corundum.

A Rb-Sr whole rock isochron of 515±28 Ma (IR 0.7136±0.0022) is defined by four granites and one leucogranite from the main massif (Figure 2). However, one granite from a spur branching off the main body (Mt Twomey) towards the northwest deviates significantly from the isochron, but plots on the tie-line of the single samples from Jupiter Amphitheatre and Unconformity Valley. Support for a geological meaning of the apparently precise isochron date of these three samples (478±14 Ma, IR 0.7092±0.008) is provided by the observation that south of the main massif the grade of contact metamorphism is markedly lower, suggesting younger high-level intrusions. At least, these three samples prove the existence of granitoids with significantly lower initial ratios than in the main granitoid body.

## Admiralty Intrusives

The very low- to low-grade Robertson Bay metasediments outside the contact aureoles suggest a high level of emplacement for the Admiralty Intrusives. Consequently a hornblende date should closely approximate the age of intrusion. The range of hornblende ages is 366 to 356 Ma. The same range is observed for the less retentive biotites, namely 367 to 354 Ma. But in two samples the biotite date is 6 Ma less than the hornblende age (Lillie Massif and Lower Tucker Glacier, the latter 90 km ESE of the area depicted in Figure 1), and one sample of Lillie Massif revealed discordant dates for different size fractions (hornblende 362 and 356 Ma, biotite 361 and 356 Ma). The obvious contradictions to the expected quick cooling of high-level intrusions is explained by the assumption of several pulses of intrusion.

### *Lillie and Champness Glaciers*

East of the Bowers Structural Zone and of the Lillie Glacier, the Lillie Granite crops out at the northern end of Everett Range. The Champness Granodiorite (Dow and Neall, 1974) which is situated west of the Lillie Glacier probably belongs to the same batholith. It is assumed that the massif is composed of multiple intrusions.

The Champness Granodiorite, especially, was resampled during Ganovex II, on Mt Radspinner and its northern extension, on Griffith and Copperstain Ridges. On the ridge north of Mt Radspinner a hornblende bearing granodiorite is exposed. The contact zone is extremely chaotic. Close to the contact, the granodiorite contains many small xenoliths oriented parallel to a foliation. Following the petrographic criteria of White and Chappell (1977) and Wyborn (1981), these are restites. A felsitic tonalite dyke about 1 m thick cuts through the granodiorite close to the Lillie Glacier side. At the contact, quartz mobilisate is interspersed in the Robertson Bay schists in a lot of veins and schlieren.

The rocks from Griffith Ridge, Copperstain Ridge and from the outcrop west of Copperstain Ridge are all low-biotite hornblende-lacking granitoids. On Griffith Ridge several fine-grained leucocratic schlieren and aplites were observed. The horizontal contact of the Griffith Ridge granite against the overlying steeply folded Robertson Bay metasediments is very sharp. No reaction processes between granitoids and metasediments can be observed. The nearly vertical-dipping finely laminated bedding of the metasediments is still discernible even near the contact, and sharp edges of cm- or mm-size protuding into the granite were not corroded or digested. These observations, together with the low biotite dates, point to late high-level emplacement.

In the area around Lillie Marleen Hut, xenoliths have been systematically sampled. Even if the majority of the xenoliths are I-type restite, quite a few are definitely of sedimentary origin. Some can be explained as incorporated Robertson Bay metasediments, even if advanced equilibration with the granite melt makes recognition of the original metasedimentary character difficult. But sedimentary xenoliths were found which show the paragenesis sillimanite (fibrolite)-corundum-muscovite-biotite. In contact aureoles of the Admiralty Intrusives sillimanite is absent; hence we conclude that these xenoliths were incorportated into the granite at a greater depth.

From the Lillie Granite 20 samples, and from the Champness Granodiorite 10 samples were analysed. In their chemical composition they range from adamellites to granodiorites (Figure 4). Three aplitic samples are classified as alkali feldspar granites by the pseudo-Rittmann calculation. Only the granophyric dykes from the Lillie Granite, the above mentioned felsic dyke from the contact north of Mt Radspinner and four I-type restite xenoliths plot in the granodiorite-tonalite transition zone. Most of the samples clearly belong to the I-type in the $Na_2O$/$K_2O$-diagram though they show CIPW normative corundum. In conflict with the petrographical and geochemical classification of the Lillie intrusives as I-type granitoids, the $^{87}Sr/^{86}Sr$ initial ratios are high (0.71 to 0.713) (Figure 2).

### *Yule Bay*

The Yule Bay Batholith is the largest (1400 km²) of all Admiralty intrusions known so far (Tessensohn et al., 1981). The granitoids are exposed in bluffs along Barnett Glacier, along the coast of Tapsell

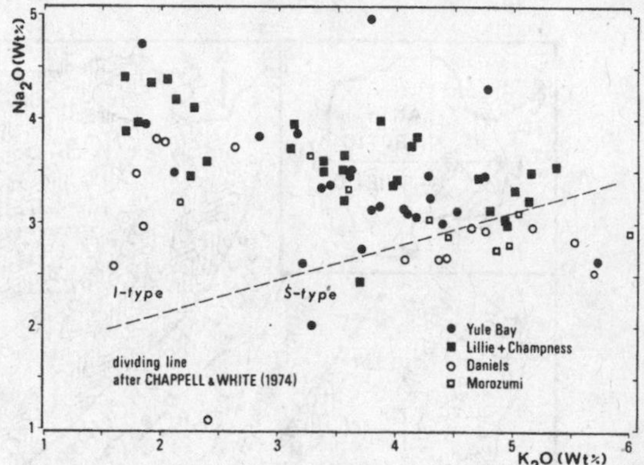


**Figure 3. $Na_2O$/$K_2O$ variation diagram for northern Victoria Land granitoids and xenoliths.**

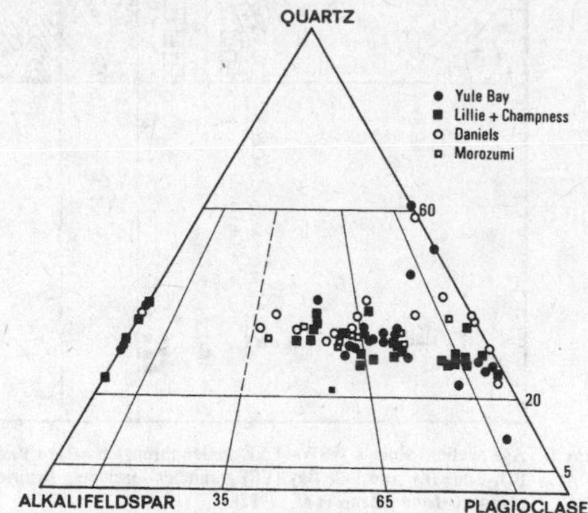


**Figure 4. Streckeisen triangle using normative pseudo-Rittmann calculation (Muller 1982) for northern Victoria Land granitoids and xenoliths.**

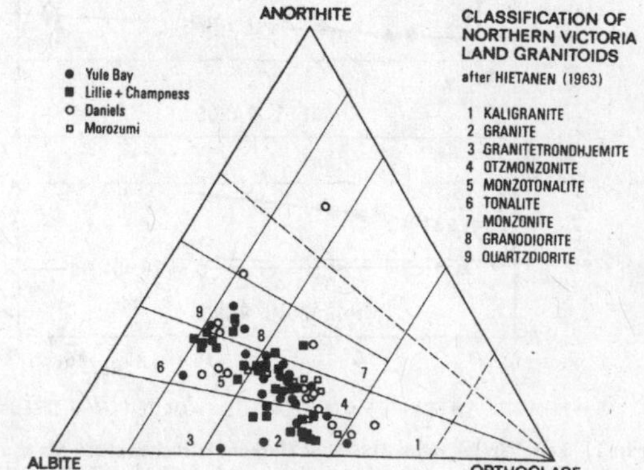


**Figure 5. An-Ab-Or triangle (after Hietanen 1963) for northern Victoria Land granitoids and xenoliths.**

Foreland and on Hughes, Novosan, and Surgeon Islands, further west around Yule Bay and Missen Ridge Foreland including Thala and Nella Islands, and in the Mt Harwood area. Wyborn (1981) distinguished four plutons according to petrographic data.

The Gregory Bluff, Missen Ridge and Ackroyd Point Plutons are all massive, grey homogeneous biotite granites with similar grainsize (2 mm, even-grained). The Tapsell Pluton, covering the easternmost area of Tapsell Foreland, east of Cape Dayman, has a porphyritic texture and a grain size up to 4 mm. Hornblende was encountered in the Missen Ridge and Tapsell Plutons.

Surgeon Island, considered to be part of the Ackroyd Point Pluton by Wyborn (1981), is now considered to be a separate intrusion. It is composed of severely sheared biotite granodiorite. The main direction of the foliation was remeasured during Ganovex II and found to have a NW—SE trend, striking in the direction of Cape Dayman (C. Wilson, pers comm.).

Twenty-five samples were analysed, of which 4 belong to the Tapsell, three each to the Missen Ridge and Surgeon Island Plutons, and the rest to the Ackroyd Point Pluton; six samples are I-type restite xenoliths, three are alkalifeldspathic aplites. Most of the Yule Bay samples plot close to the adamellite-granodiorite boundary, whereas all xenoliths fall in the tonalite field. The three Surgeon Island samples calculate as quartz-rich tonalites (Figure 4). According to the $Na_2O/K_2O$-diagram, the Yule Bay granitoids are typical I-type. Only the three Surgeon Island samples and the three aplite samples fall completely out of this field. Two aplites have a high Na-content whereas one aplite and the Surgeon Island samples have comparatively low sodium (Figure 3). CIPW-norms of the Surgeon Island samples show corundum values above 2.5% by weight which are consistent with the muscovite contents. Wyborn (1981) interpreted this muscovite as secondary, probably formed by the introduction of water into potassium-feldspar during shearing.

It seems that the Yule granitoids have high initial ratios of 0.712 to 0.718 (Figure 2), provided their ages are approximately in the vicinity of 360 Ma as suggested by the consistent pattern of K-Ar biotite ages. This result conflicts with the classification as I-type intrusives. The wide range of initial isotopic compositions proves varying amounts of crustal contamination or even different source rocks for the granitoids. The sampling was too widespread to allow isochron treatment, with the exception of four samples from the Gregory Bluff Pluton. These latter samples yield a date of $393 \pm 20$ Ma (IR $0.7136 \pm 0.0013$) (Figure 2). Similar K-Ar dates, apparently older than the biotite ages, were determined for a muscovite from Surgeon Island and for one of three hornfelses from Unger Island in Yule Bay (Figure 1). However, the meaning of these K-Ar dates is not unequivocal, as both rock samples come from shear zones and might be affected by excess argon.

## Conclusions

The Cambro-Ordovician Granite Harbour Intrusives (about 500 Ma and 480 Ma) are found to the west of the Bowers Graben. They can be considered mainly as anatectic melts derived from arkosic and pelitic sediments. Syn- and postkinematic I-type tonalites and granodiorites are subordinate.

The Devonian-Carboniferous Admiralty Intrusives (cooling age $360 \pm 8$ Ma), situated east of the Rennick Glacier, are, from their petrography, predominantly I-type granitoids. In contrast, Sr initial ratios are high and geochemical trends point to significant contamination by sedimentary material. Only Surgeon Island probably consists of S-type granitoids.

The Ordovician cooling age of 480 Ma of Morozumi Range and the Rb-Sr intrusion age of the granite of about 500 Ma confirms Ross Orogeny folding of Robertson Bay Group metasediments. This was inferred by K-Ar dates on slates east of the Bowers Graben (about 500 Ma, Adams, et al., 1982), and by the hypothesis of Kleinschmidt and Skinner (1981) who assume an increasing degree of deformation of the Robertson Bay Group metasediments in the east to the Wilson Group migmatites in the west, from high to deep crustal level within the same lithofacies. Thus, in northern Victoria Land, a Silurian/Devonian deformation, a proof of the postulated Middle Palaeozoic "Borchgrevink Orogeny", is so far identified only within the Bowers Structural Zone (Figure 1).

*Acknowledgement* The critical comments by N.C.N. Stephenson are gratefully acknowledged.

## REFERENCES


ADAMS, C.J., GABITES, J.E., WODZICKI, A., LARID, M.G. and BRADSHAW, J.D., 1982: Potassium-Argon geochronology of the Precambrian-Cambrian Wilson and Robertson Bay Groups and Bowers Supergroup, Northern Victoria Land, Antarctica; *in* Craddock, C. (ed.) *Antarctic Geoscience*. Univ. Wisconsin Press, Madison, 543-8.

CHAPPELL, B.W. and WHITE, A.J.R.; 1974: Two contrasting granite types. *Pac. Geol., 8*, 173-4.

CRADDOCK, C., 1970: Tectonic map of Antarctica; *in* Bushnell, V.C. & Craddock, C., (eds.) Geologic maps of Antarctica. *Antarct. Map Folio Ser.*, Folio 12, Pl XXIII.

DOW, J.A.S. and NEALL, V.E., 1974: Geology of the Lower Rennick Glacier, northern Victoria Land, Antarctica. *N.Z. J. Geol. Geophys., 17*, 659-714.

GRINDLEY, G.W. and WARREN, G., 1964: Stratigraphic nomenclature and correlation in the western Ross Sea region; *in* Adie, R.J. (ed.) *Antarctic Geology.* North Holland, Amsterdam, 314-33.

GUNN, B.M. and WARREN, G., 1962: Geology of Victoria Land between the Mawson and Mulock Glaciers, Antarctica. *N.Z. Geol. Surv., Bull., 71.*

HARRINGTON, H.J., 1958: Nomenclature of rock units in the Ross Sea region, Antarctica. *Nature, 182*, 290.

HIETANEN, A., 1963: Idaho Batholith Near Pierce and Bungalow, Clearwater County, Idaho. *U.S. Geol. Surv. Prof. Pap., 344-D*, D35-42.

KLEINSCHMIDT, G. and SKINNER, D.N.B., 1981: Deformation in the basement rocks of North Victoria Land, Antarctica. *Geol. Jb., B 41*, 156-99.

KREUZER, H., HÖNDORF, A., LENZ, H., VETTER, U., TESSENSOHN, F., MÜLLER, P., JORDAN, H., HARRE, W. and BESANG, C., 1981: K/Ar and Rb/Sr dating of igneous rocks from North Victoria Land, Antarctica. *Geol. Jb., B 41*, 267-73.

MÜLLER, P., 1982: Von der CIPW-Norm ausgehende Berechnung von Mineralbeständen magmatischer Gesteine in Analogie zu der Modalzusammensetzung plutonischer und vulkanischer Gesteine. *Geol. Jb., D 55*, 3-41.

STRECKEISEN, A., 1976: To each plutonic rock its proper name. *Earth-Sci. Rev., 12*, 1-33.

STEIGER, R.H. and JÄGER, E., 1977: Subcommission on Geochronology: Convention on the use of decay constants in geo- and cosmochronology. *Earth Planet. Sci. Lett, 36*, 359-62.

STURM, A. and CARRYER, S.J., 1970: Geology of the region between Matusevitch and Tucker Glaciers, North Victoria Land, Antarctica. *N.Z. J. Geol. Geophys., 13*, 408-35.

TESSENSOHN, F., DUPHORN, K., JORDAN, H., KLEINSCHMIDT, G., SKINNER, D.N.B., VETTER, U., WRIGHT, T.O. and WYBORN, D., 1981: Geological comparison of basement units in North Victoria Land, Antarctica. *Geol. Jb., B 41*, 31-88.

WHITE, A.J.R. and CHAPPELL, B.W., 1977: Ultrametamorphism and granitoid genesis. *Tectonophysics, 43*, 7-22.

WYBORN, D., 1981: Granitoids of North Victoria Land, Antarctica—field and petrographic observations. *Geol. Jb., B 41*, 229-49.

# CHEMISTRY OF PALAEOZOIC GRANITES OF NORTHERN VICTORIA LAND

D. Wyborn, *Bureau of Mineral Resources, PO Box 378, Canberra City, ACT 2601, Australia*

**_Abstract_ Petrographic data indicate that the Devonian Admirality Intrusives of eastern northern Victoria Land are exclusively I-type (igneous source) whereas the Cambro-Ordovician Granite Harbour Intrusives of western northern Victoria Land are dominantly S-type (sedimentary source) with rare mafic tonalitic I-types. Geochemically, the Admiralty I-types and Granite Harbour S-types are not easily distinguished as the source sediments for the S-types are rich in plagioclase and calcite giving them high calcium and sodium contents. In the Lachlan Fold Belt of SE Australia, S-types are low in calcium and sodium because their source rocks were more mature. However, a peraluminous character is clear for the northern Victoria Land S-types and hornblende is absent.**

**The present sampling of Admirality I-type intrusives has allowed division into two suites. One, high in $Na_2O$, CaO and Sr includes the Gregory Bluffs and Tapsell plutons of the Yule Batholith, the Lillie and Football Granites and the Tucker Granodiorite. The other suite is high in $K_2O$, Rb, Ce and La and includes the Missen Ridge and Ackroyd Point plutons of the Yule Batholith. Strontium and neodymium isotopic ratios of both suites indicate a considerable crustal pre-history for the source rocks, similar to the source rocks for Lachlan Fold Belt I-types.**

# CHEMICAL CONTROL ON STRATIGRAPHIC RELATIONS IN NORTHERN VICTORIA LAND AND SOME POSSIBLE RELATIONS WITH SE AUSTRALIA

D. Wyborn, *Bureau of Mineral Resources, PO Box 378, Canberra City, ACT 2601, Australia*

*Abstract* Northern Victoria Land geology is dominated by a thick monotonous sequence of Late Precambrian to possible Early Cambrian quartz-rich greywackes, the Robertson Bay Group, which was deposited on both sides of a meridional volcanic belt, the Glasgow Volcanics and their clastic apron, the Sledgers Group. The chemistry of the greywackes shows some systematic changes from east to west. In the east, remote from the volcanic belt, the sediments are low in $Na_2O$ and probably represent second-cycle sediments derived from earlier sedimentary sequences to the south; closer to the volcanic belt they increase in $Na_2O$, and particularly Cr, reflecting a detrital input from the volcanic belt. The high Cr is due to the presence of detrital chromite grains within 50 km of the central volcanic belt. Electron probe analyses of these chromites are very similar to those from detrital chromites in the Sledgers Group and microphenocrysts in basalts of the Glasgow Volcanics. To the west of the volcanic belt Robertson Bay Group sediments and similar but higher grade schists from the Daniels Range (Rennick Schist) are higher still in $Na_2O$ and also Sr, but low in Cr. Here they probably represent first-cycle sediments derived from a granitic hinterland.

The composition of northern Victoria Land sediments is appropriate for them being source-rocks for the widespread Ordovician quartz-rich turbidite deposits of the Lachlan Fold Belt, SE Australia. This is supported by a large body of palaeocurrent data in SE Australia indicating a southerly source and a correspondence of turbidite influx in SE Australia with mountain building in northern Victoria Land during the Ross Orogeny.

# RESULTS OF PALAEOMAGNETIC INVESTIGATIONS IN NORTHERN VICTORIA LAND, ANTARCTICA.

G. Delisle, *Federal Institute for Geosciences, P.O. Box 51 01 53, D-3000, Hannover 51, Federal Republic of Germany.*

*Abstract* Palaeomagnetic measurements on material from radiometrically dated rock complexes from the following locations in northern Victoria Land are presented:

(a) Devonian-Carboniferous granites of the Yule Bay area (K-Ar biotite age= 361 ± 1.5 Ma; Kreutzer et al., 1981)

(b) Ferrar Dolerites from Litell Rocks with apparent ages ranging from 90-120 Ma (Kreutzer et al., 1981)

(c) Upper Miocene volcanics from the Hallett Group.

Samples collected during the West German Ganovex I and II expeditions were analysed with the aid of AF and thermal demagnetisation techniques. Evaluation of the Devonian-Carboniferous samples from five sites yielded in four cases a mean magnetisation direction with high inclination values, inferring a near pole position of northern Victoria Land at that time. Measurements on samples from one site gave Dm= 16 and Im= −47, from which a palaeomagnetic pole position at 46.5°S, 7.6°E was derived. From the Cretaceous dolerites a palaeomagnetic pole at 76.9°S, 201°E based on the mean magnetisation direction of three sites (Dm= 315.3; Im= − 84.1; alpha 95 = 7.8) was obtained. This result correlates after reassemblage of Gondwana with an APW curve proposed by Embleton (1981). Upper Tertiary volcanics from ten sites on Adare Peninsula and four sites from Hallett Peninsula were analysed. The Adare sites yielded a palaeomagnetic pole position at 83.6°S and 186.4°E, the Hallett sites a pole position at 81.9°S and 256.8°E. By combining the Hallett Group data with the mean direction of the Cape Hallett lavas published by Turnbull (1959), a mean palaeomagnetic pole position at 89.4°S and 120°E is obtained. This result can be interpreted to imply a geologically younger age for the volcanics of Hallett Peninsula than that of the Adare Peninsula basalts.

A variety of Palaeozoic to Cenozoic rock complexes, well suited for palaeomagnetic research, is exposed in the northern part of northern Victoria Land and was investigated by the West German northern Victoria Land Expedition (Ganovex I and II) in the years 1979-80 and 1981.

The following is a brief summary of the major Phanerozoic tectonic events of the area. The Ross Orogeny of Early Palaeozoic age was accompanied by emplacement of the Granite Harbour Intrusives, which are exposed today in the Lanterman, Morozumi and Daniels Ranges. According to K-Ar age dating (on biotites and muscovites), the complexes from the two latter areas range from 479-467(± 2) Ma (Kreutzer et al., 1981). The Bowers Block faulting during the Devonian and Carboniferous is clearly recognised in the so-called Bowers Structural Zone. It is not clear if the intrusion of the Admiralty Intrusives is synchronous with the block faulting (Tessensohn et al., 1981, 1982). The Admiralty Intrusives consist of granites and granodiorites and were emplaced according to Kreutzer et al. (1981) during the Devonian-Carboniferous transition (around 360 Ma based on K-Ar age dating). After a prolonged period of (glacial?) erosion (Skinner, 1981) and deposition (i.e. Beacon Sandstone), the intrusion of the Ferrar Dolerites during the Mesozoic occurred. The Ferrar Dolerites are commonly thought to be of Jurassic age. The volcanics, which were sampled on Litell Rocks however, yielded mid-Cretaceous ages ranging from 90-120 Ma (Kreutzer et al., 1981). The faulting along the Rennick Glacier, which runs parallel to the trend of the Ross Orogeny, is post-Jurassic in age (Tessensohn et al., 1981). The youngest tectonic event was the onset of the Cenozoic rifting along the Ross Sea, which has been associated with volcanic activity up to the present. The volcanics sampled on the Adare Peninsula are of Late Miocene age (Kreutzer et al., 1981).

## SAMPLING AND SITE SELECTION

The material collected during Ganovex I consists primarily of Tertiary volcanics and a few samples representing each of the Palaeozoic to Mesozoic rock complexes exposed in the area. The intention during Ganovex II was to put emphasis on Ordovician, Devonian-Carboniferous and Jurassic-Cretaceous material. However, due to the sinking of the supply ship servicing Ganovex, the *Gotland II*, on 18th December, 1981 off Yule Bay, the available field time to the author for sample collection was cut to three days and was used to concentrate on the Admiralty Intrusives. A selection of samples as summarised in Table 1, collected during both expeditions is discussed in this paper. Material, which so far has not yielded meaningful results or was available in only insufficient quantities, is excluded from this discussion.

With regard to the Palaeozoic granitic material, it was known from the preliminary investigations on samples from Ganovex I that a considerable proportion of the exposed rock material would prove only marginally suitable for palaeomagnetic studies. However, samples close to or taken from crosscutting features (veins and dykes) promised a greater likelihood of success, because they represent the emplacement of late phase melts, with the material more likely to be reheated and cooled quickly.

**TABLE 1**

| Sample | Lat. | Long. | Age* | Rock Type |
|---|---|---|---|---|
| **Devonian/Carboniferous samples** | | | | |
| Tes 6 | 70.8 S | 167.0 E | 361 ± 1.5 Ma | granite |
| Tes 8 | 71.23S | 164.52E | 362 ± 2 Ma | granite |
| 810Rd.17 | 71.17S | 164.49E | 359-363 ± 4 Ma | granite |
| 810Rd.18 | 71.17S | 164.49E | 359-363 ± 4 Ma | granite |
| Jo 195 | 70.7 S | 166.92E | 320-390 Ma** | spilitic volcanics |
| **Cretaceous samples** | | | | |
| Tes 9 | 71.38S | 162.03E | 90-120 Ma | dolerite |
| Tes 10 | 71.38S | 162.03E | 90-120 Ma | dolerite |
| Tes 12 | 71.45S | 162.03E | 90-120 Ma | dolerite |
| **Tertiary samples** | | | | |
| Tes 4 | 71.3 S | 170.23E | Upper Miocene | basalt |
| Tes 5 | 71.3 S | 170.23E | Upper Miocene | basalt |
| Jo 26a | 72.5 S | 169.67E | Upper Miocene?*** | basalt |
| Jo 26b | 72.5 S | 169.67E | Upper Miocene?*** | basalt |
| Jo 26d | 72.5 E | 169.67E | Upper Miocene?*** | basalt |
| Jo 26e | 72.5 E | 169.67E | Upper Miocene?*** | basalt |
| Jo 157a | 71.58S | 170.34E | Upper Miocene | trachyte |
| Jo 157b | 71.58S | 170.34E | Upper Miocene | trachyte |
| Jo 158 | 71.58S | 170.34E | | basalt |
| Jo 185 | 71.58S | 170.34E | 8.01 ± 0.07 Ma | phonolite |
| Jo 186 | 71.58S | 170.34E | 7.69 ± 0.05 Ma | phonolite |
| Jo 187 | 71.3 S | 170.34E | Upper Miocene | porph. basalt |
| Jo 188 | 71.67S | 170.17E | Upper Miocene | basalt |
| Jo 189 | 71.67S | 170.17E | Upper Miocene | basalt |

*all ages are taken from Kreutzer et. al., 1981.

**whole rock age derived from metamorphosed volcanics

***inferred from adjacent sites; no radiometric age date available.

## SAMPLING TECHNIQUES AND PALAEOMAGNETIC MEASUREMENTS

During both expeditions portable coring apparatuses were used to obtain oriented samples. Core diameter was either 38 mm or 50 mm, maximum length of cores 100 mm. In addition, a number of oriented hand samples, primarily taken from the Hallett volcanics, were collected. The palaeomagnetic measurements were carried out at the BGR laboratory at Grubenhagen. Most samples were magnetically "cleaned" by the Alternating Field (AF) method. A thermal demagnetiser was utilised in the case of a few selected samples. Measurements of the remanent magnetisation of samples were performed with an astatic magnetometer; in the case of the strongly magnetised volcanics, an arrangement of Foerster probes proved sufficient to determine the orientation of magnetic remanence.

## RESULTS

The granitic material is weakly magnetised and shows great scatter in its remanence directions. Alpha-95 angles (A-95) can not be reduced by small values. Only preliminary conclusions concerning the pole position of Antartica at the Devonian-Carboniferous boundary will be drawn. The Cretaceous and Tertiary samples, on the other hand, did yield without exception stable remanence directions with successive demagnetisation steps.

### Devonian-Carboniferous sites

Specimens from the sites Tes 6 and Tes 8 exhibited predominantly positive values of inclination prior to magnetic cleaning; all other specimens yielded negative inclination values suggesting a secondary component of magnetisation parallel to the current earth magnetic field lines. AF-cleaning with 200-300 Oe(16-24kA/m) removed this component. Only in the case of Tes 6, 700 Oe(56kA/m) were required to arrive at a direction of magnetisation of uniform polarity. The removal of the secondary component by thermal cleaning was successful in the case of 810Rd.18 between 300°C and 400°C; other material, notably the one from 810Rd.17, shows this effect gradually in the temperature range between 500°C to 630°C. The removal of this secondary component, which is roughly antiparallel to the primary one, is accompanied by an increase of the magnitude of the mean vector of magnetisation.

Tes 6: Specimens were cored from a granitic outcrop at Birthday Ridge. Based on ten samples, the orientations of the resulting mean magnetisation vector after successive AF-demagnetisation with 400, 700 and 1000 Oe are shown in Figure 1. The minimal change between 700 and 1000 Oe accompanied by a reduction of the A-95 angle to 24° is considered to indicate a stable direction of magnetisation.

Tes 8: This site is located on Mt Dockery. Ten samples were taken from a granitic dyke. Large secondary components of positive polarity with intensities between 20-40 mA/m had to be removed from some samples, while others exhibited from the beginning negative polarities in concurrence with the magnetisation direction obtained during all stages of AF-treatment. The mean direction of magnetisation for each demagnetisation step is shown in Figure 2.

810Rd.17: This and the following site are located on 810 Ridge, a granitic outcrop with crosscutting pegmatite veins of thicknesses ranging from 0.1 m to several metres. The samples were cored from the granite remote from the immediate vicinity to contact zones with veins. The decrease of intensity of magnetisation of five analysed samples with successive demagnetisation steps is shown in Figure 3 as well as the orientation of the magnetisation vectors prior to magnetic cleaning and after treatment in fields of up to 600 Oe. The data points never formed a cluster with a smaller A-95 angle than at this stage (32°). The mean direction of magnetisation, however, did not change significantly after removal of the secondary component at 300 Oe. A palaeo-orientation with a steep inclination value can be inferred.

810Rd.18: Three of the available six samples from this site were cleaned by the AF-method, two samples were thermally demagnetised. The first set readily moved into a position of an easterly directed declination with steep inclination, the data points clustered closely around their mean direction with successive demagnetisation steps (Figure 4). The mean orientation of the thermally cleaned two samples reached this position only after exposure to 600°C. The samples appear to have carried a soft but large antiparallel component of magnetisation of high blocking temperature with respect to the direction obtained after AF-treatment. This is evidenced by the J/Jo-curve (Figure 4), which shows an increase of magnetisation of the samples during removal of the secondary component. No satisfactory explanation for the unexpected rise of the magnetisation at 800-1000 Oe and 400°C, which in the case of the AF-method was not accompanied by a significant change in orientation of the mean magnetisation vector, could be found.

Jo 195: This site is located on Unger Island, where volcanics were apparently metamorphosed by the surrounding Yule Bay Granites (361 to 366±1.5 Ma, see Kreutzer, 1981) No contact between the granite and the volcanics is exposed. It is argued by Tessensohn et al. (1981) that the island is probably part of a volcano-sedimentary roof of the Yule Bay intrusion. Ten samples from this site were magnetically cleaned with increasing demagnetisation steps up to 1000 Oe, six samples up to 2500 Oe. Little change in the orientation of the mean magnetisation vector throughout this process, as depicted in Figure 5, was observed. However, no smaller A-95 angle than 33° was obtained at any stage of AF-treatment. The steep inclination values are in agreement with the near pole position expected for the area at the Devonian-Carboniferous boundary. The observed normal polarity is consistent with the negative inclination determined for site Tes 6, which lies approximately 10 km to the south.

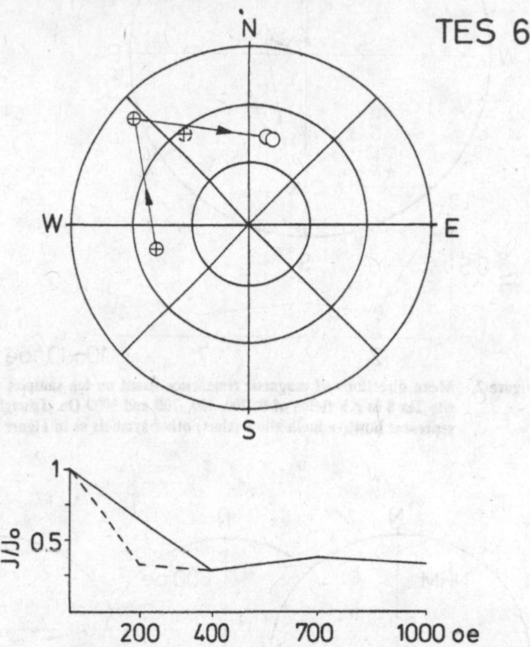


**Figure 1. Successive mean directions of magnetic remanence (as indicated by arrows) based on ten samples of site Tes 6 for AF-fields of 0, 400, 700 and 1000 Oe. Circles with crosses combine inclination values of both polarities; open circles represent negative inclinations; crossed circle with broken lines indicates magnetisation direction at 200 Oe based on three samples. Decrease of magnetisation J of samples shown below stereographic projection.**

*Summary of data*

The measurements on Devonian-Carboniferous samples are summarised in Table 2, where the orientation values of the mean magnetisation vector after the highest demagnetisation step is indicated. At this stage of the investigation, these results are not intended to imply that the given data represent, beyond doubt, primary mean directions of magnetisation. Petrographic-mineralogical studies on the samples and measurements on material from additional sites are needed.

**TABLE 2**

| Site | N | D | I | A-95 | Palaeopole Lat. | Palaeopole Long. |
|---|---|---|---|---|---|---|
| Tes 6 | 10 | 16 | −47 | 24 | 46.5S | 7.6E |
| Tes 8 | 10 | 280 | 81 | 16 | 62.3S | 124.7E |
| 810Rd.17 | 5 | 321 | 77 | 32 | | |
| 810Rd.18 (AF) | 3 | 139 | 72 | 28 | | |
| 810Rd.18 (therm.) | 2 | 177 | 83 | 11 | | |
| Jo 195 | 6 | 137 | −68 | 33 | 35.6S | 135.1E |

### Mesozoic Samples

Doleritic samples were taken from three sites; two on the eastern edge and one at the southern tip of Litell Rocks. This material is thought to belong to the Jurassic Ferrar Dolerite formation. The volcanics collected on Litell Rocks for age dating purposes appear to be altered, probably by hydrothermal action. K-Ar dating on these rocks yielded "rather consistent Mid-Cretaceous apparent ages, ranging from 90-120 Ma"(Kreutzner et al, 1981). Further research is needed to decide, if this material is of Cretaceous or Jurassic age, whose K-Ar age has been affected by the hydrothermal event in the latter case.

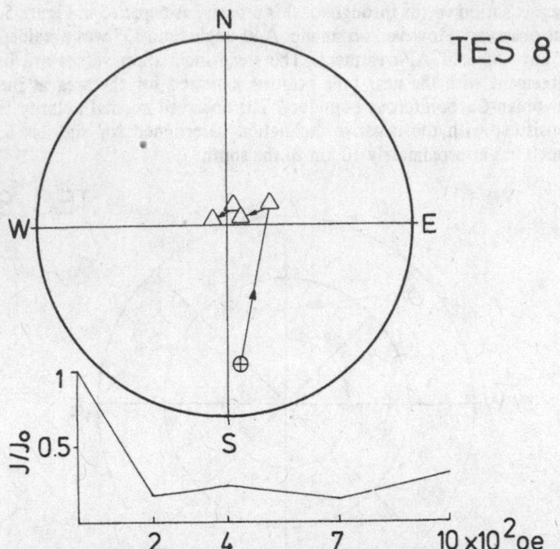


**Figure 2. Mean directions of magnetic remanence based on ten samples of site Tes 8 in AF fields of 0, 200, 400, 700 and 1000 Oe. Triangles represent positive inclination values; other symbols as in Figure 1.**

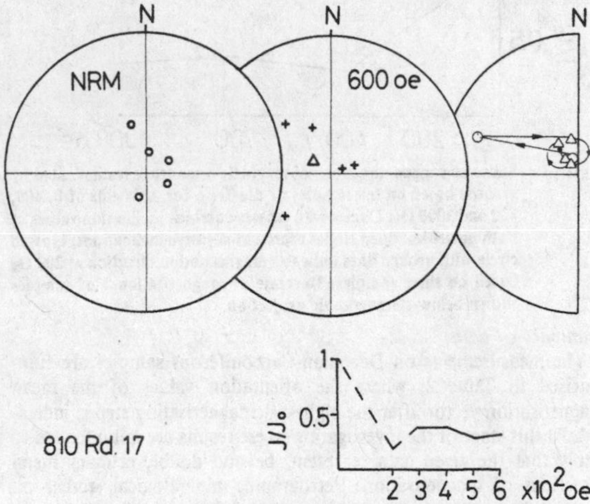


**Figure 3. Direction of magnetic remanence for 5 samples of site 810Rd. 17 in 0 and 600 Oe. Successive mean directions of magnetic remanence shown in the upper right corner. Crosses indicate positive inclination values; otherwise, symbols as in Figure 1.**

All palaeomagnetic samples from the three sites exhibited stable components of NRM for successive AF-demagnetisation steps. The mean directions for the three sites are given in Table 3. A mean of the site mean directions was calculated: Dm = 315.3, Im = −84.1, k = 107, A-95 = 7.8. By defining a common sampling locality at 71.4°S and 162°E, a palaeomagnetic pole at 76.9°S and 201°E has been obtained.

TABLE 3

| Site | AF-field | N | D | I | A-95 |
|---|---|---|---|---|---|
| Tes 9 | 0 | 5 | 237 | −83 | 4 |
| | 200 | 3 | 238 | −86 | 10 |
| | 400 | 3 | 234 | −86 | 5 |
| Tes 10 | 0 | 7 | 316 | −86 | 4 |
| | 200 | 3 | 319 | −84 | 3 |
| | 400 | 5 | 305 | −86 | 3.3 |
| Tes 12 | 0 | 9 | 12 | −81 | 13 |
| | 200 | 3 | 345 | −77 | 15 |
| | 400 | 7 | 335 | −76 | 5.6 |

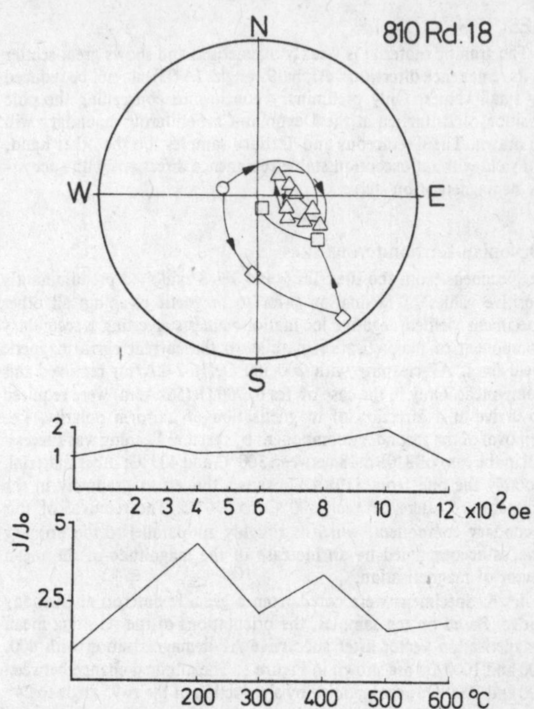


**Figure 4. Mean direction of remanence for three samples for successive AF-field demagnetisation steps (0, 200, 300, 400, 500, 600, 800, 1000 and 1200 Oe). Two samples were thermally demagnetised. Mean directions of remanence are shown for 470K, 570K, 670K and 870K; diamonds indicate negative, squares positive inclination.**

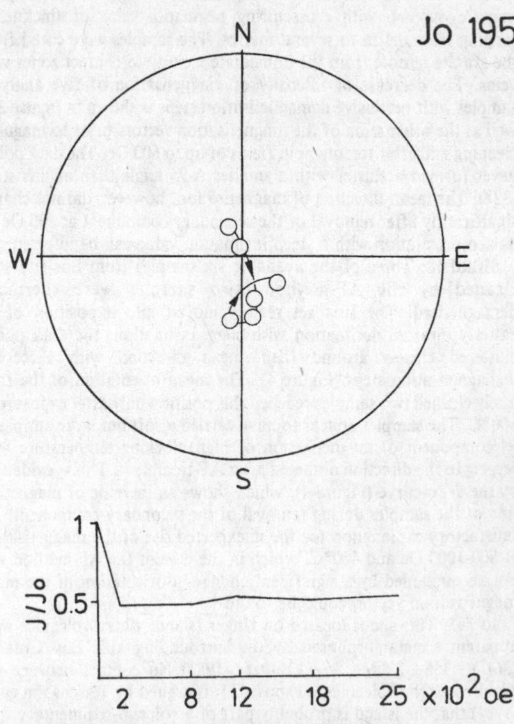


**Figure 5. Site Jo 195 (Unger Island). Mean direction of magnetisation for 0, 200, 800, 1200, 1800 and 2500 Oe are shown; symbols as in previous Figures.**

### Tertiary Samples

All samples, as listed in Table 1, with the exception of Jo 26, were collected on the Adare Peninsula. The samples Jo 26 were taken on the Hallett Peninsula from a sequence of four basaltic flows. A description of the geology and tectonics of the area can be found in Jordan (1981). In agreement with the age dated samples, all the collected material is thought to be of Late Miocene age. All samples were treated by AF-demagnetisation in two steps (200 and 400 Oe), and in all cases a stable direction of magnetisation was observed. The results of all measurements at 400 Oe are listed in Table 4.

**TABLE 4**

| Site | N | D | I | A-95 |
|---|---|---|---|---|
| **Hallett Peninsula** | | | | |
| Jo 26a | 2 | 128 | 78 | 11.1 |
| Jo 26b | 3 | 215 | 81.5 | 7.9 |
| Jo 26d | 3 | 121 | 72 | 7.4 |
| Jo 26e | 2 | 196.5 | 78.6 | 3.9 |
| **Adare Peninsula** | | | | |
| Tes 4 | 6 | 155 | 53 | 5.6 |
| Tes 5 | 9 | 250.9 | 56.4 | 1.7 |
| Jo 157a | 6 | 65 | 82.5 | 4.8 |
| Jo 157b | 2 | 292.5 | 82 | 0.3 |
| Jo 158 | 5 | 349.5 | −73 | 5.7 |
| Jo 185 | 12 | 252.4 | −69.3 | 2.2 |
| Jo 186 | 10 | 297.7 | −82 | 3.0 |
| Jo 187 | 5 | 292.3 | 82.7 | 2.3 |
| Jo 188 | 9 | 316.5 | −82.7 | 13.66 |
| Jo 189 | 3 | 295 | 13.4 | 4.5 |

With the exception of Jo 188, all sites yielded samples, whose magnetic orientation grouped very closely around a mean value as shown by the small A-95 values. The Jo 189 result might be due to one of the frequent changes in polarity of the Earth's magnetic field during the Miocene. Jo 188 and 189 were excluded from further calculations. The mean of the mean directions of the Hallett and Adare sites were calculated:

Hallett Peninsula: Dm = 154.3; Im = 80.3; k = 66; A-95 = 8.6
Adare Peninsula: Dm = 171.8; Im = 83.72; k = 15; A-95 = 12.8

## INTERPRETATION OF RESULTS

The data from the Devonian-Carboniferous material are of poorer quality, as seen from the A-95 angles than the geologically younger material. Their interpretation is accordingly less certain.

Based on the assumption of Antarctica being attached to the southern coast of Australia throughout the Palaeozoic and Mesozoic, a position of northern Victoria Land at intermediate to high latitudes (e.g. Morel and Irving, 1978) during the Devonian-Carboniferous transition is implied by the apparent polar wander path (APW) of Australia (e.g. Embleton, 1981). Two palaeopole positions from the upper Devonian-Carboniferous are known from southeastern Australia: Mulga Downs (Du): 54°S, 96°E (Embleton, 1977); Visean volcanics (Cl): 73°S, 214°E (Luck, 1973). To compare directly the Antarctic palaeomagnetic poles, obtained by Ganovex, with those of Australia, an approximate former site position (fitted by eye) at 45°S,150°E was assumed (see Weissel et al., 1977), for a pre-drift position of Antarctica relative to Australia as derived from marine geophysical data). The corresponding pole of Tes 6 would be located at 69°S, 12.5°E, which is in closer agreement to the Visean volcanics than to the position obtained from the Mulga Downs Group. The other Devonian-Carboniferous sites indicate a near pole position at time of emplacement of the granites. This result is supported by the palaeopole position based on the Mulga Downs material. Given the rather large A-95 angles and the steep inclination values, the calculated mean declininations cannot be considered as accurate.

Figure 6 shows the APW of Australia for the Mesozoic and Cenozoic, as presented by Embleton (1981). For comparison, the palaeopole positions from the dolerites (C), from Adare Peninsula (A) and Hallett Peninsula volcanics (H) are included. By shifting Antarctica back towards Australia as mentioned above, a new palaeopole (C') is obtained. This position is in good agreement with the data from Cretaceous material from Australia. Therefore, the radiometrically determined age range of 90-120 Ma for the Litell dolerites is supported by palaeomagnetic evidence; if this material was emplaced

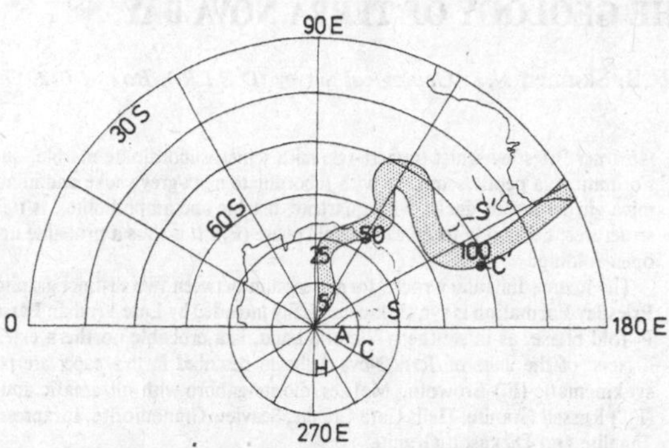


**Figure 6. APW of Australia, redrawn from Embleton (1981). Numbers along path indicate pole positions in Ma before present. S' = S = site locations of Ganovex; C = palaeomagnetic pole of Cretaceous sites; A = palaeomagnetic pole of Adare Peninsula sites; H = Palaeomagnetic pole of Hallett Peninsula sites.**

during the Jurassic, then it must have been effectively magnetically overprinted during the Cretaceous.

The Tertiary material yields palaeomagnetic poles, located consistently to the east of the Australian APW. The calculated pole positions are:

Adare Peninsula: 83.6°S, 186.4°E
Hallett Peninsula: 81.9°S, 256.8°E

The Adare pole position agrees well with a palaeopole derived from lavas from the South Shetland Islands, reported by Blundell (1962, 1966) and Valencio et al. (1979). Later workers combined data from both papers and calculated a pole position of 85°S,191°E. The Hallett Peninsula data listed here and data from Turnbull (1959), who has investigated lavas from Cape Hallett, can be combined to calculate a mean palaeomagnetic pole position. The resulting position of 89.4°S, 120°E by comparison to the APW of Australia implies a geologically younger age of the Hallett Peninsula lavas than that of the Adare Peninsula basalts.

## REFERENCES

BLUNDELL, D.J., 1962: Palaeomagnetic investigations in the Falkland Island Dependencies. *Sci Rep. Br. antarct. Surv., 39*, 1-24.

BLUNDELL, D.J., 1966: Palaeomagnetism of some lavas from the south Sandwich Islands. *Bull. Brit. antarct. Surv., 9*, 61-2.

EMBLETON, B.J.J., 1977: A late Devonian palaeomagnetic pole for the Mulga Downs Group, western New South Wales. *R. Soc. N.S.W., J. Proc., 110*, 25-27.

EMBLETON, B.J.J., 1981: A review of the palaeomagnetism of Australia and Antarctica; *in* McElhinney, M.W., and Valencio, D.A. (eds.) *Palaeoconstruction of the Continents,* American Geophysical Union, Washington, 77-92.

JORDAN, H., 1981: Tectonic observations in the Hallett Volcanic Province, Antarctica. *Geol. Jb. B.41*. 111-25.

KREUTZER, H., HOHNDORF, A., LENA, H., VETTER, U., TESSENSOHN, F., MULLER, P., JORDAN, H., HARRE, W., and BESANG, C., 1981: K-Ar and Rb-Sr dating of igneous rocks from North Victoria Land. *Geol. Jb. B.41*, 267-73.

LUCK, G.R., 1973: Palaeomagnetic results from Palaeozoic rocks of southeast Australia. *Geophys. J., 32*, 35-52.

MOREL, P. and IRVING, E., 1978: Tentative palaeocontinental maps for the early Phanerozoic and Proterozoic. *J. Geol., 86*, 535-61.

SKINNER, D.N.B., 1981: Possible Permian glaciation in North Victoria Land, Antarctica. *Geol. Jb., B.41*, 261-66.

TESSENSOHN, F., DUPHORN, K., JORDAN, H., KLEINSCHMIDT, G., SKINNER, D.N.B., VETTER, U., WRIGHT, T.O. and WYBORN, D., 1981: Geological comparison of basement units in North Victoria Land, Antarctica. *Geol. Jb. B.41*, 31-88.

TURNBULL, G., 1959: Some palaeomagnetic measurements in Antarctica. *Arctic, 12*, 151.

VALENCIO, D.A., MENDIA, J.E. and VILAS, J.F., 1979: Palaeomagnetism and K-Ar age of Mesozoic and Cenozoic igneous rocks from Antarctica. *Earth Planet. Sci. Lett., 45*, 61-8.

WEISSEL, J.J., HAYES, D.E. and HERRON, E.M., 1977: Plate tectonics synthesis: the displacements between Australia, New Zealand and Antarctica since the late Cretaceous. *Mar. Geol., 25*, 231-77.

# THE GEOLOGY OF TERRA NOVA BAY

D.N.B. Skinner, *N.Z. Geological Survey (D.S.I.R.), Box 61-012, Otara, Auckland, New Zealand.*

*Abstract* Priestley schist (quartz-two mica schist, subordinate marble, quartzite and amphibolite), is a higher grade along strike equivalent of the Priestley Formation, a pelitic sequence with subordinate meta-greywacke and minor limestone and quartzite. The Snowy Point paragneiss (quartzo-feldspathic, two mica garnetiferous gneiss with quartzite, marble and amphibolite), is tightly interfolded with Priestley schist ($F_2$) but has incongruent mesofolds to the $F_2$ structures, belonging to an earlier fold phase ($F_1$). It is thus a probable unconformably older unit. A third relatively weak fold phase ($F_3$) caused only broad open folding.

The metasedimentary rocks form a septum between two distinct granitoid associations—the Southern Victoria Land and Terra Nova Batholiths. Because Priestley Formation is syn $F_2$-kinematically intruded by Late Vendian Larsen Granodiorite, Priestley schist and Snowy Point paragneiss are Precambrian, and $F_2$ fold phase, as in southern Victoria Land, is a probable northern extension of the Beardmore Orogeny.

None of the units of Terra Nova Batholith described in this paper are post-kinematic. Instead they are pre-$F_1$ pegmatites; pre-$F_2$ gneiss (Sastrugi Orthogneiss); synkinematic ($F_2$) Browning Mafites (diorite-gabbro with ultramafic and granitic accumulates and differentiates) and Canwe Granodiorite; late kinematic ($F_2$?) Russell Granite, Hells Gate Granite, Seaview Granodiorite, Inexpressible Granite, and Abbott Granite; and late kinematic ($F_3$?) Keinath Granite, Corner Tonalite and Dickason Granite.

The area inland from Terra Nova Bay between Priestley and Campbell Glaciers (Figure 1), is of great historic significance since in the late summer of 1912, Scott's Northern Party with Priestley as geologist, surveyed the region before being forced to winter-over in a snow cave on Inexpressible Island (Smith, 1924; Smith and Priestley, 1921).

The area was next visited briefly in January 1963, the rocks being described as undifferentiated Granite Harbour Intrusives and a metamorphic complex of perhaps two unconformable metasedimentary units (Ricker, 1964; Skinner and Ricker, 1968). However it was clear that the "granites" belonged to a completely different suite from those of the South Victoria Land Batholith which ended along the upper Priestley Glacier. In particular the Terra Nova Batholith lacked the widespread Larsen Granodiorite-Irizar Granite association (Skinner, 1972; also this volume). A belt of contact metamorphosed, but low regional grade semi-pelitic and calcareous sedimentary rocks along the upper Priestley Glacier was named Priestley Formation.

In 1965/66, the Campbell Glacier into Browning Pass and the Priestley Glacier to Corner Glacier were briefly surveyed (Adamson, 1971). The metasedimentary rocks were referred to Priestley Formation and the granites were named in increasing order of age: Northern Foothills Granite with microcline and porphyritic orthoclase varieties (cf. Smith, 1924), Corner Tonalite (Priestley's diorite of the Corner Glacier) and Dickason Granite. An erratic sample believed to be Dickason Granite, has been K-Ar biotite dated at 415± 6 Ma (Nathan, 1971—recalculated age).

This paper is a brief summary of the regional geology after a further season's mapping in 1969/70. It is planned to complete the project during 1982/83.

## METASEDIMENTARY ROCKS

These lie in a northwest-southeast belt from Black Ridge to Gerlache Inlet (Figure 1).

Priestly Formation (Skinner and Ricker, 1968) is a metamorphosed dominantly pelitic, thinly bedded argillite sequence with subordinate quartzo-feldspathic greywacke and minor fine grained limestone and quartzite

Because it is intruded by Larsen Granodiorite which has two Vendian dates in southern Victoria Land (Skinner, this volume), Priestley Formation must be Precambrian. However, its correlation outside Terra Nova Bay is quite problematical. In spite of an earlier correlation with Robertson Bay Group of northern Victoria Land (Nathan and Skinner, 1972), at that time not seen by either of these two authors, it certainly has no resemblance to Robertson Bay Group at the latter's type locality (Wright, 1981; Kleinschmidt and Skinner, 1981). Speculating from the descriptions given by Ravich et al. (1965), there appears to be a possible strong lithological resemblance to Berg Group of the Oates Coast, some 600 km along strike to the northwest.

A large area of apparently upright asymmetrically folded Priestley Formation trends east-southeast from the head of O'Kane Canyon towards the metasedimentary rocks of Corner Glacier (see Figure 3 of Skinner and Ricker, 1968). Southwards from Corner Glacier however, the metasediments are schistose and have undergone medium to high grade regional metamorphism. Until such time as they can be more thoroughly described they are herein informally termed *Priestley schist.* At Cape Sastrugi and Snowy Point and east of Mt Browning, higher grade, strongly deformed paragneisses are herein informally termed *Snowy Point paragneiss.* The contact between the two units has been complexly folded and is best exposed east of Mt Browning. Structural observations suggest that the two units are unconformable, the schist being the younger.

### Snowy Point paragneiss (Figure 2)

Garnetiferous biotite gneisses are known from the Corner Glacier and Browning Pass moraines, from the Mt Browning area and from Cape Sastrugi-Snowy Point (Smith and Priestley, 1921; Ricker, 1964). The rocks are predominantly coarsely foliated, lenticular quartz-microcline-oligoclase-biotite-muscovite paragneisses with layers rich in cordierite, garnet and fibrolite or sillimanite. The lenticles are quartz and boudinaged and folded quartz-garnet-tourmaline pegmatoid layers. Marble and quartzite layers up to 150 mm thick are common and amphibolite lenses form prominent ridges along Gerlache Inlet. In this same area, the paragneisses are strongly migmatised and granite neosome veins parallel the gneiss foliation (Figure 3).

### Priestley Schist (Figure 4)

In their lowest grade between the Boomerang and Corner Glaciers, the lithologies are without doubt schistose Priestley Formation, but this distinction becomes blurred southwards along strike as metamorphic grade and degree of deformation increases.

In spite of extensive recrystallisation, the rocks are recognisably a pelitic greywacke-argillite suite in which bedding laminations, argillite flakes, lenses of medium to coarse sandstone, current bedding, synsedimentary slump folds and load coasts are preserved. 50 mm thick mylonite zones parallel bedding and quartz lenses lie in the cleavage $S_1$. With a more concentric fold style, beds of marble and quartzite are prominent within crenulated, higher grade rocks of Cape Sastrugi to Mt Browning. At their highest grade, the schists are mainly finely fissile with quartz lenticles, but include 10-15 m thick packages of slaty phyllites with 75 mm thick quartzite layers. Close to granite, the schists are extensively migmatised.

The bulk of the rocks are quartz-biotite-muscovite schists with hornblende and garnet rich layers and minor microcline and plagioclase; marble contains scapolite, epidote, diopside and chondrodite. On the Northern Foothills and lower Boomerang Glacier the schists contain sillimanite, however cordierite and andalusite have been described from moraine erratics on Priestley Glacier and Boomerang Glacier-Browning Pass (Smith and Priestley, 1921).

### Tectonism

There is evidence for three deformation events ($F_{1,2,3}$) of which only the younger two affected Priestley schist. North from Browning Pass, Snowy Point paragneiss lies in the core and upper limb (with Priestley schist) of a moderately to gently southeast plunging, northeast closing, $F_2$ recumbent fold. However, between Cape Sastrugi and Corner Glacier, the major $F_2$ structure is a west-verging asymmetric to overturned, gently south plunging synform. This suggests that some fault rotation has occurred. The recumbent $F_2$ fold has also been coaxially deformed into broad open upright, gently southeast plunging folds ($F_3$). Within the recumbent structure, plunges on lineations and many microfolds parallel these $F_2$/$F_3$ axes and the majority of paragneiss mesofolds, like all those in the overlying schist, verge to the east and northeast, appropriate to the $F_2$ fold symmetry.

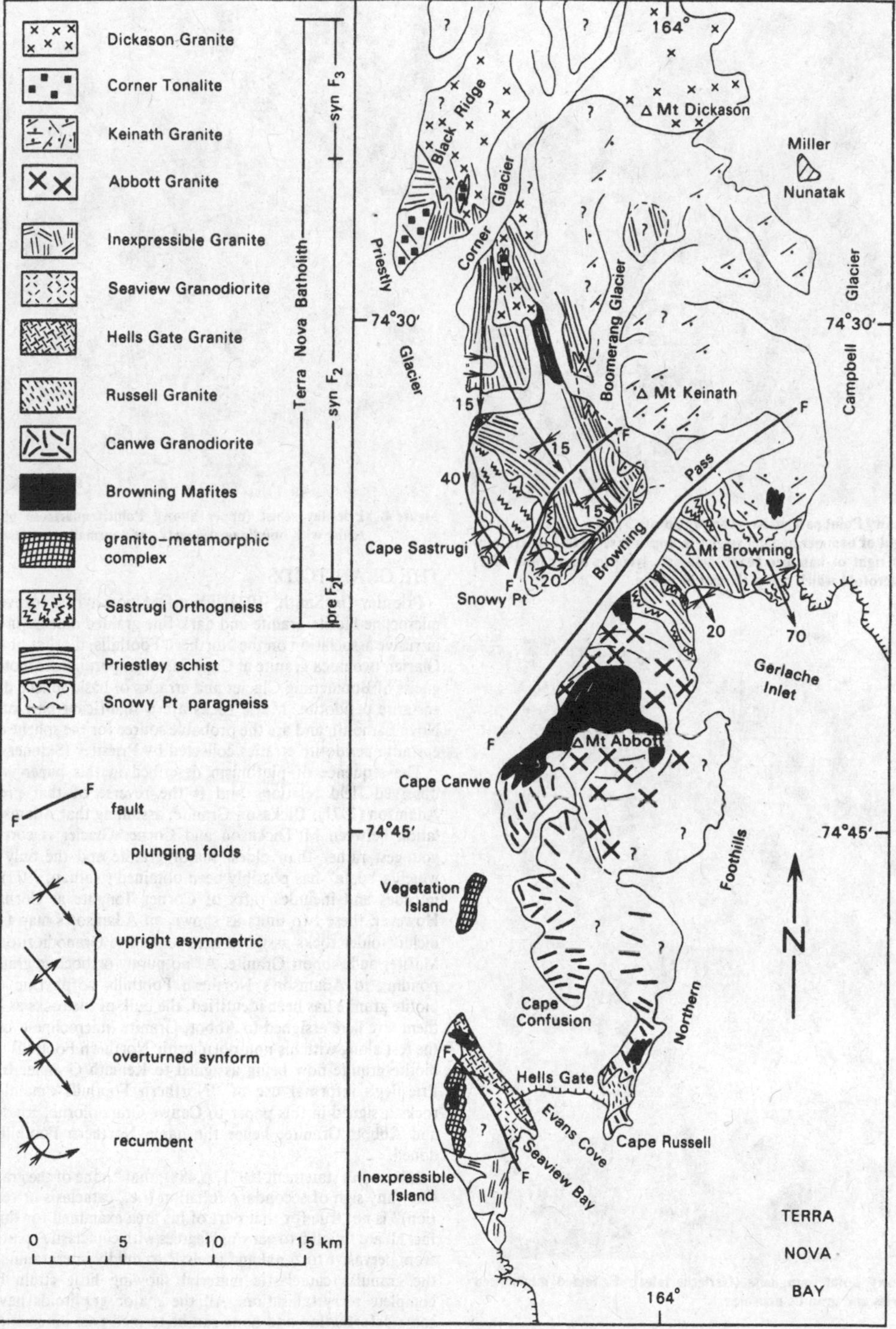


**Figure 1. Generalised regional geological map of Terra Nova Bay region. The orientation of all linear symbols is symbolic of the foliation strike trends; areas with ? have not been surveyed.**

However, several 3 m amplitude folds in Snowy Point paragneiss have the opposite vergence to the west, and are overturned to inverted in form with axes at 35° anticlockwise to $F_2$ fold axes and with folded cleavage. This is good evidence for an earlier fold phase ($F_1$) than the two affecting Priestley schist.

Between Browning Pass and Gerlache Inlet, the first deformation in Priestley schist ($F_2$) has caused tight to isoclinal upright folds with minor chevron fold zones; plunges are low to moderate to the southeast. A major $F_2$ synformal structure is inferred from the mesofold symmetry and Priestley schist-Snowy Point paragneiss distribution. Again fault rotation seems probable along Browning Pass.

Within the Snowy Point paragneiss north of Gerlache Inlet, moderate to subvertical southwest plunging, tight to near isoclinal folds ($F_1$) are refolded by moderate to steep southeast plunging, southwest verging, noncylindrical folds ($F_2$). These have fanned axial cleavage and incongruent vergences of the earlier microfold axes on their limbs. The major structure is a steep southeast plunging antiformal

**Figure 2. Snowy Point paragneiss (at Snowy point): $F_1$ early folds (e.g. lower right of hammer handle) verging in opposite sense to $F_2$ microfolds (to right of hammer head), both sets lying in upper limb of $F_2$ macrofold which closes to right of photo.**

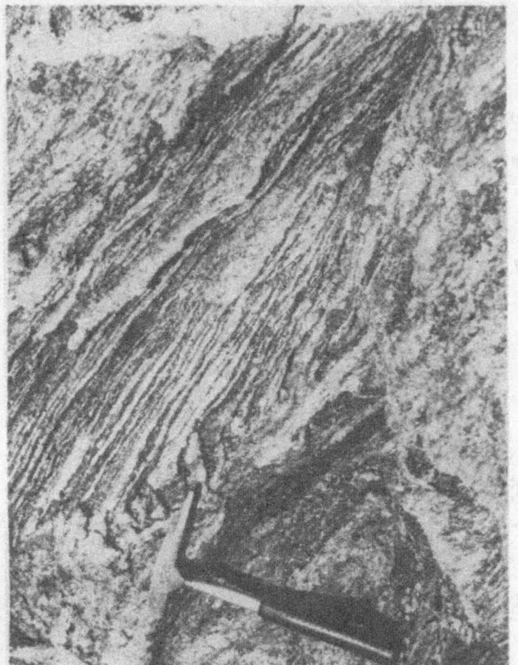

**Figure 3. Snowy point paragneiss (Gerlache Inlet): $F_2$ folded migmatised gneiss and granitic neosome.**

structure ($F_2$) with incongruent $F_1$ mesofold vergences. Small and large scale "mushroom" interference structures also confirm this incongruent two-fold relationship.

$F_3$ structures are not obvious within the steeply dipping Priestley schist south of Browning Pass, but broad open $F_3$ folds are present in the Snowy Point paragneiss of Gerlache Inlet.

Many garnet-tourmaline-quartz "pegmatite" veins in the gneiss terrain have been twice folded and strongly boudinaged, whereas others that are mainly tourmaline bearing, are more simply folded (similar to tourmaline pegmatites cutting Priestley schist). It appears that a period of pegmatite injection preceded $F_1$ folding.

**Figure 4. Priestley schist (upper Snowy Point): quartzose phyllitic mica schist with boudinaged quartz and tourmaline pegmatite veins.**

## THE GRANITOIDS

Priestley (*in* Smith, 1924) identified porphyritic to even grained microcline-biotite granite and dark fine grained diorites in a complex intrusive association on the Northern Foothills, diorites of the Corner Glacier, two mica granite at Cape Sastrugi, porphyritic biotite granitic gneiss of Boomerang Glacier and erratics of basic sphene diorites and enstatite peridotite. Mafic rocks are a significant part of the Terra Nova Batholith and are the probable source for the sphene diorite and enstatite peridotite erratics collected by Priestley (Skinner, 1972).

The sequence of plutonism described in this paper is based on observed field relations and is the reverse of that proposed by Adamson (1971). Dickason Granite, assuming that Adamson's correlation between Mt Dickason and Corner Glacier is correct, is the youngest rather than oldest major granite and the only one from which a "date" has possibly been obtained (Nathan, 1971); it clearly intrudes and includes rafts of Corner Tonalite at Corner Glacier. However, these two units as shown on Adamson's map (1971), also include older rocks assigned here to Canwe Granodiorite, Browning Mafites and Abbott Granite. As no purely orthoclase granite corresponding to Adamson's Northern Foothills porphyritic orthoclase-biotite granite has been identified, the bulk of the rocks as he mapped them, are here assigned to Abbott Granite (microcline + orthoclase), the rest along with his non-porphyritic Northern Foothills microcline-biotite granite now being assigned to Keinath Granite. In addition, Priestley's informal use of "Northern Foothills" mainly included rocks assigned in this paper to Canwe Granodiorite, Russell Granite and Abbott Granite; hence the name Northern Foothills is abandoned.

Adamson's statement (1971, p.488), that "none of the granitic rocks show any sign of secondary foliation (i.e., cataclasis or recrystallisation)" is not true for that part of his area examined for this paper. In fact all are foliated to varying degrees with cataclastic textures ranging from pervasive to zonal and gneissic to mildly intergranular, and with the granular cataclastic material showing little strain because of complete recrystallisation. All the major granitoids have suffered some deformation and none can be termed post-kinematic, although only Sastrugi Orthogneiss and some Browning mafites have been observed actually folded.

### Pre-Kinematic ($F_2$) Intrusives

*Sastrugi Orthogneiss* (Figure 5) is the earliest major granitoid. The main body, some 500 m thick, crops out on Cape Sastrugi as a laminar foliated, strongly cataclasised, biotite-muscovite-quartz-andesine-microcline granodiorite "sill" and folded marginal dykes. The sill lies along the Snowy Point paragneiss—Priestley schist boundary and like them is folded by the $F_2$ recumbent structure. Similar orthogneiss occurs as conformable bands in Priestley schist south of Browning Pass. Several small areas of porphyritic augen gneiss within migmatised Snowy Point paragneiss at Gerlache Inlet are anatectic neosomes derived from the paragneiss and folded with them by $F_2$ noncylindrical folds.

## Synkinematic $F_2$ Intrusives

*The Granito-Metamorphic Complex* (Figure 6) is a portmanteau, large scale mapping term for the western cliffs of Inexpressible Island and Vegetation Island, since it includes many of the other rock types in a complex interrelationship of metasediments, mafic rocks, granites and pegmatites, with various degrees of gneissification.

**Figure 5. Sastrugi Orthogneiss (Mt Browning) cut by very late undeformed microgranite veins.**

**Figure 6. Granito-metamorphic complex on western cliffs of Inexpressible Island. Priestly schist cut by gneissic diorite or Browning Mafites, intruded by sills of Canwe Granodiorite and dykes of the various "granites" of Inexpressible Island.**

*The Browning Mafites* are more or less gneissic, foliated biotite rich, olivine and pyroxene diorites and gabbros in which strong differentiation to granites has occurred (Skinner, 1972). They form a significant part of the Terra Nova Batholith, with a wide age range from early to late synkinematic since they intrude, or are intruded by, the other granitoid units. Many are rafted blocks off the main plutons, whereas others are discrete dykes and plugs. In the Browning Pass pluton, hypersthene-diopside-hornblende monzo-gabbro passes progressively upwards through mafic granodiorite and mafic granite (Figure 7) into porphyritic quartz rich microcline leucogranite with riebeckitic hornblende. The two pyroxene mafic diorite to granodiorite of the Boomerang Pluton is separated by a folded layered diorite sequence from underlying accumulate lenses of biotite bearing pyroxenite, olivine pyroxenite and plagioclase-hornblende-olivine pyroxenite. The rocks have been pervaded by pegmatite veins and have been extensively $K_2O$-$SiO_2$ metasomatised to produce secondary phlogopite, hornblende, biotite, quartz and microcline.

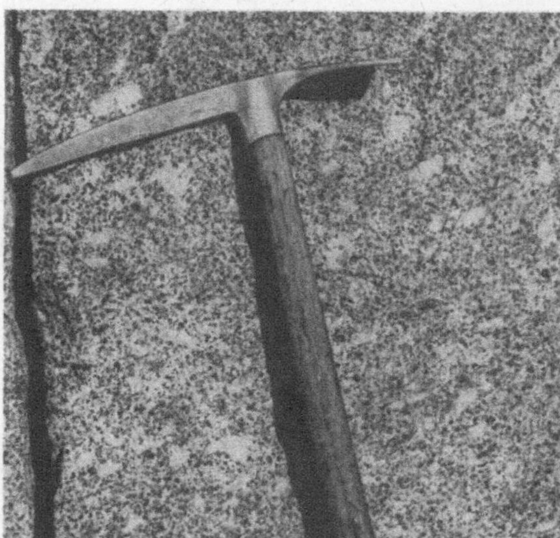

**Figure 7. Mafic granodiorite to granite of differentiated Browning Mafites on ridge to Mt Abbott, Browning Pass.**

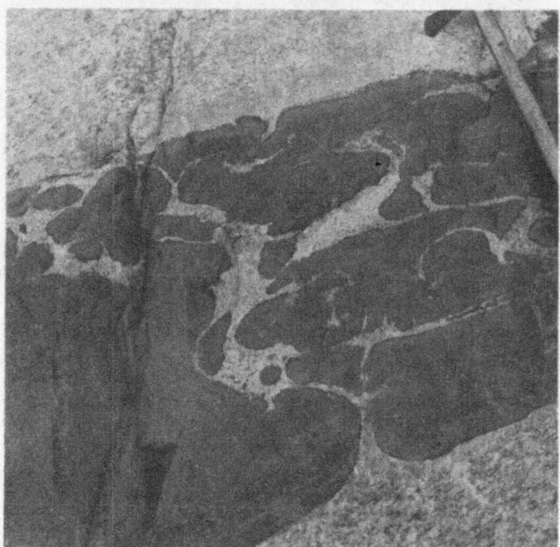

**Figure 8. "Sederhold effect" of microdiorite dyke intruding Canwe Granodiorite; reaction has produced fine grained two mica "granite".**

Biotite microdiorites and pyroxene microgabbros, often with porphyroblastic K-feldspar, are common between Cape Sastrugi. Northern Foothills and Inexpressible Island. Their intrusion into the granitoids has commonly caused palingenesis of the country rock—the so called "Sederholm Effect"—producing a greyish-cream two mica microgranite (Figure 8).

*Canwe Granodiorite* (Figure 8) is a variably foliated to strongly gneissic, zonally cataclastic, yellow-brown to grey biotite and biotite-hornblende quartz diorite to granodiorite, uncommonly with muscovite and orthite but relatively rich in accessory sphene; K-feldspar is concentrated as microcline phenocrysts in zones of porphyritic rocks. Spindle shaped inclusions of amphibolite and migmatised folded rafts of paragneiss are widespread. The main body occurs from Cape

**Figure 9. Russell Granite: coarse porphyritic phase intruding more normal sparsely porphyritic variety.**

Canwe to Cape Russell but dykes and pods are folded with Priestley schist on upper Snowy Point ridge and make up a large proportion of the granito-metamorphic complex on Inexpressible Island.

### Late Kinematic ($F_2$?) Intrusives

*Russell Granite* is a blue-grey, well foliated, often mafic rich biotite-hornblende granite to granodiorite with large perthitic microcline phenocrysts and minor anhedral perthitic orthoclase. Muscovite and remnant clinopyroxene are occasionally present in the least and most mafic rocks respectively; accessories are sphene, apatite and magnetite. Zones of schlieren cause a platy texture and coarsely porphyritic and even grained phases intrude each other and Canwe Granodiorite (Figure 9).

*The Granites of Inexpressible Island* include rafts of Canwe Granodiorite, but otherwise their position in the sequence is uncertain. Along the northeast side of the island, *Hells Gate Granite* is a medium to fine grained, foliated, buff coloured microcline-biotite granite, commonly with muscovite but only minor to sparse hornblende; sphene and orthite are accessory. *Seaview Granodiorite* intrudes Hells Gate Granite. It is a biotite quartz diorite to granodiorite with both perthitic orthoclase and microcline, minor hornblende and relict hypersthene and augite; sphene and orthite are accessory. *Inexpressible Granite* intrudes Seaview Granodiorite and forms most of the southern end of the island. It is a coarse, even grained, pinkish brown, poorly foliated biotite-perthitic microcline quartz poor granite-quartz syenite-quartz monzonite with subordinate hornblende and accessory sphene, orthite and apatite.

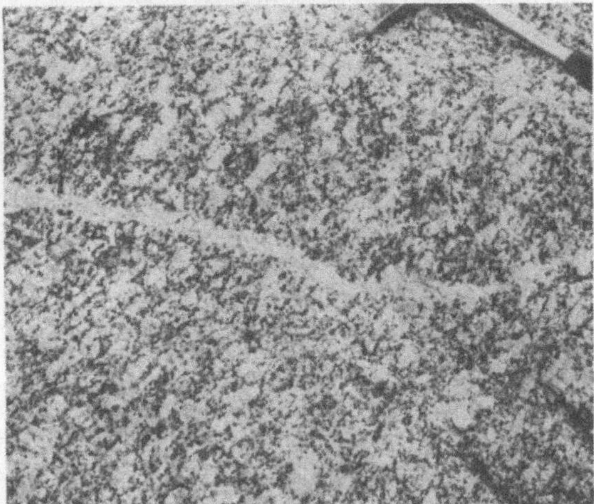

**Figure 10. Abbott Granite, Mt Abbott: subgneissic foliated mafic rich phase near Browning Mafite.**

*Abbott Granite* (Figure 10) forms the central mass of Northern Foothills where it surrounds and intrudes a core of Browning Mafites and intrudes Canwe Granodiorite. It is a foliated, coarse porphyritic leuco-granite with phenocrysts of perthitic microcline and subordinate perthitic orthoclase, in places intergrown and combined in carlsbad twins; plagioclase is andesine and there is sparse biotite and minor hornblende with accessory sphene and apatite.

### Late Kinematic ($F_3$?) Intrusives

*Keinath Granite* extends north from northeast of Mt Browning between the Campbell and Boomerang Glaciers and intrudes Abbott Granite as dykes along Browning Pass. West of Boomerang Glacier and on Mt Browning other dykes cross cut the $F_2$ structures. It is a medium grained, sparsely porphyritic and variably foliated quartz rich plagioclase granite to granodiorite with microcline and biotite and accessory apatite, sphene and magnetite.

*Corner Tonalite* is known with certainty only at the Corner Glacier, on Black Ridge, and as rafts within Dickason Granite. However dykes of similar rock occur on Cape Sastrugi, Cape Canwe and Vegetation Island. It is a variably mafic, foliated hornblende-biotite quartz diorite rich in schist xenoliths and with minor potash feldspar and rare augite; sphene and apatite are accessory.

*Dickason Granite* has been described by Adamson (1971) from Mt Dickason to the Corner Glacier, but contrary to that paper, it clearly intrudes and contains rafts of Corner Tonalite along Black Ridge and south from Corner Glacier and cuts off dykes of Keinath Granite along west Boomerang Glacier. It is a coarsely porphyritic, mafic rich, foliated to subgneissic granite described by Adamson as a biotite-orthoclase granite with oligoclase and subordinate to minor hornblende.

## REGIONAL SIGNIFICANCE

Until the project is finished it may be premature to draw too many conclusions. However, some points do stand out.

The synkinematic intrusion of probable Vendian Larsen Granodiorite into $F_2$ folded Priestley Formation—Priestley schist, immediately implies that these units are Precambrian and that $F_2$, like that in southern Victoria Land, is of Vendian-Earliest Cambrian age. Thus late Beardmore Orogeny (Vendian) has extended into this southern boundary of northern Victoria Land (cf. Skinner, this volume).

Unlike the southern Victoria Land Batholith, no post-kinematic granites such as the Ordovician Irizar Granite (Skinner, this volume) occur in the Terra Nova Batholith, which would limit the age of the $F_3$, presumably Ross Orogeny, deformation to its accepted Cambro-Ordovician age. Instead, the 415 Ma date on the Dickason Granite erratic (assuming this correlation is correct and if it is not a reset age) implies that Ross Orogenic deformation continued in the Terra Nova Bay region into the Devonian. Adamson's (1971) Capsize Adamellite and Recoil Granite, described by him as younger than Dickason Granite, undeformed and post-kinematic, should be of Devonian age and thus equate with the Devonian—Carboniferous Admiralty Intrusives of northern Victoria Land. The so called Devonian Borchgrevink Orogeny of northern Victoria Land may therefore comprise no more than Silurian Ross Orogeny deformation and Devonian post-kinematic granites.

## REFERENCES

ADAMSON, R.G., 1971: Granitic rocks of the Campbell-Priestley divide, northern Victoria Land, Antarctica. *N.Z. J. Geol. Geophys., 14*, 486-503.

KLEINSCHMIDT, G. and SKINNER, D.N.B., 1981: Deformation styles in the basement rocks of northern Victoria Land, Antarctica. *Geol. Jb. B41*, 155-99.

NATHAN, S., 1971: K-Ar dates from the area between the Priestley and Mariner Glaciers, northern Victoria Land, Antarctica. *N.Z. J. Geol. Geophys., 14*, 504-11.

NATHAN, S. and SKINNER, D.N.B., 1972: Recent advances in the pre-Permian geology of northern Victoria Land; *in* Adie, R.J. (ed.) *Antarctic Geology and Geophysics*. Universitetsforlaget, Oslo. 333-7.

RAVICH, M.G., KLYMOV, L.V. and SOLOVIEV, D.S., 1965: Oates Coast and King George V Land; *in The Precambrian of East Antarctica*. Israel programme for scientific translation, Jerusalem, for National Science Foundation, Washington, U.S.A.

RICKER, J., 1964: Outline of the geology between the Mawson and Priestley Glaciers, Victoria Land; *in* Adie, R.J. (ed.) *Antarctic Geology*, North Holland, Amsterdam, 265-75.

SKINNER, D.N.B., 1972: Differentiation source for the mafic and ultramafic rocks ("enstatite peridotites") and some porphyritic granites of Terra Nova Bay, Victoria Land; *in* Adie, R.J. (ed.) *Antarctic Geology and Geophysics*, Universitetsforlaget, Oslo. 299-303.

SKINNER, D.N.B., (this volume): The granites and two orogenies of southern Victoria Land.

SKINNER, D.N.B. and RICKER, J., 1968: The geology of the region between the Mawson and Priestley Glaciers, northern Victoria Land, Antarctica. Part 1—Basement metasedimentary and igneous rocks. *N.Z. J. Geol. Geophys., 11*, 1009-40.

SMITH, W.C., 1924: The plutonic and hypabyssal rocks of southern Victoria Land. *British Antarctica ("Terra Nova") Expeditions, 1910/13 Geol., 1(6)*, 167-227.

SMITH, W.C. and PRIESTLEY, R.E., 1921: The metamorphic rocks of the Terra Nova Bay region. *British Antarctica ("Terra Nova") Expeditions, 1910/13. Geol., 1(5b)*, 145-65.

WRIGHT, T.O., 1981: Sedimentology of the Robertson Bay Group, north Victoria Land, Antarctica. *Geol. Jb. B41*, 127-38.

# THE PETROLOGY AND ORIGIN OF ORBICULAR TONALITE FROM WESTERN TAYLOR VALLEY, SOUTHERN VICTORIA LAND, ANTARCTICA

P.S. Dahl and D.F. Palmer, *Department of Geology, Kent State University, Kent, Ohio, U.S.A. 44242.*

*Abstract* Orbicular tonalite and associated rocks from western Taylor Valley have been studied by field, petrographic, and electron microprobe techniques. Orbicules are biaxially ellipsoidal in shape and typically consist of a diopside-biotite-andesine core surrounded by multiple concentric shells alternately rich in hornblende and plagioclase. Elongate radially-oriented hornblende, plagioclase, and clusters of diopside-biotite are superimposed on the overall concentric structure. The latter clusters give way to hornblende as the tonalite matrix is approached. Microprobe analysis reveals complex mineral-chemical variations from orbicule core to the tonalite matrix, and the bulk composition of the shell region is intermediate relative to compositions of orb core and tonalite matrix. These observations, considered with field evidence, suggest that orbicule shells formed from xenoliths, as a result of incomplete assimilation by and simultaneous chemical exchange with tonalite magma. Chemical exchange occurred by diffusion of components between tonalite magma and partially melted outer parts of xenoliths. Based on calculated bulk chemical analyses, the components $Na_2O$, $Al_2O_3$, and $SiO_2$ are thought to have diffused *from* the tonalite magma to the melting xenolith, whereas FeO, MgO, CaO, $K_2O$ diffused out from the xenolith to the magma. This mass transfer process is postulated on the basis of orbicules representing an intermediate stage of xenolith assimilation. However, when taken nearly to completion, the process also accounts for the common field occurrence of more-advanced orbicules, which have dominantly plagioclase cores (with residual xenolithic diopside ± biotite) and cafemic shells.

During the 1980/81 and 1981/82 austral field seasons, orbicular granitoid and associated rocks from several localities in Taylor Valley were mapped, described, photographed and sampled for subsequent petrographic and chemical analysis. The primary purposes of this research are: (1) to elucidate the mechanism(s) by which orbicular rocks form and (2) to gain new information regarding the processes of magmatic assimilation and chemical diffusion in silicate magmas. Taylor Valley is uniquely suited for such a study because of extensive outcrops in which orbicules from incipient to advanced stages of development are preserved. Previous work on the orbicular rocks in Taylor Valley includes the field and petrographic study of Palmer et al. (1967) and the detailed mapping of Dahl and Palmer (1981).

In this paper we focus our discussion on the petrographic, mineral-chemical and bulk-chemical variations across a single (but representative) orbicular specimen. These data are used to formulate a model for the chemical transfer process involved in orbicule formation. This model is then extended to explain some of the petrographic and chemical characteristics of orbicules in more advanced stages of development. Other orb-forming models, which we consider less favourable in light of all the evidence, will be discussed fully in a later paper.

## FIELD OCCURRENCE—SUMMARY

In Taylor Valley, orbicules and xenoliths are most common along the north wall at a locality 3.5 km west of Rhone Glacier. Here orbicules and xenoliths occur in a tonalitic facies of the Larsen granodiorite in a zone 50 to 100 m from its contact with the Skelton Group metasediments. Orbs and xenoliths typically occur as discontinuous bands, lenses, and irregular pod-shaped zones in the tonalite (Figure 1) although isolated single orbs are commonly found. Xenolith and orbicule zones are parallel to the steeply dipping schlieren in the tonalite, the foliation in the adjacent Skelton metasediments, and the (largely concordant) Larsen-Skelton contact. Field relationships at this locality (henceforth called the Taylor Glacier area) are summarised in Dahl and Palmer (1981).

Orbicule zones in the Taylor Glacier area are characterised by closely-packed (Figure 1) to widely-spaced orbs commonly embedded in a hornblende rich tonalitic matrix that is distinctly coarser-grained than the surrounding tonalite. Orbs within a zone are remarkably well size-sorted, and, although biaxially ellipsoidal in shape, they show no obvious preferred orientation. Orbicule shell geometry and core mineralogy may differ from orb to orb, as is evident in Figure 1.

Orbicules of the Taylor Glacier area typically exhibit cores of diopside ± hornblende + biotite + andesine surrounded by a single monomineralic shell or by multiple shells alternately andesine/labradorite- and hornblende-rich. Within a given shell, radial hornblende, plagioclase, and (less commonly) diopside-biotite intergrowth are superimposed on the overall concentric structure. Orbicule cores are mineralogically identical, and texturally similar, to diopside-biotite hornfelses and gneisses found in place in the nearby Skelton metasediments. Hence, orbicule formation is thought to be the result of incomplete assimilation of metamorphic xenoliths by tonalite magma. Detailed field evidence supporting this origin is presented by Palmer et al. (1967) and Dahl and Palmer (1981).

**Figure 1. Irregular band of closely packed orbicules embedded in the tonalitic facies of the Larsen granodiorite. The outcrop is along the north wall of Taylor Valley, 3.5 km west of Rhone Glacier. Orbicules with cores of plagioclase are especially abundant in this orb band.**

## ORBICULAR SPECIMEN TG81-106D

This specimen is a relatively simple orbicule occurring as a single isolated orb in the tonalite.

### Petrography

A polished thin section of orbicular speciment TG81-106D is shown in Figure 2. Like most orbicules in the Taylor Glacier locality, the xenolithic core contains the assemblage diopside + biotite + andesine. The feldspathic inner shell region ($S_1 + S_2$ in Figure 2) contains the same assemblage, but here intergrowths of finer grained diopside and biotite occur either as irregular patches or radially oriented rods in a "sea" of coarse andesine-labradorite. The outer shell region ($S_3 + S_4$) is dominated by coarse radial hornblende and andesine crystals; biotite is also abundant, but diopside is present in only minor amounts. Quartz and microcline are notably absent in both the core and shells.

Contribution no. 244, Department of Geology, Kent State University, Kent, Ohio, U.S.A. 44242.

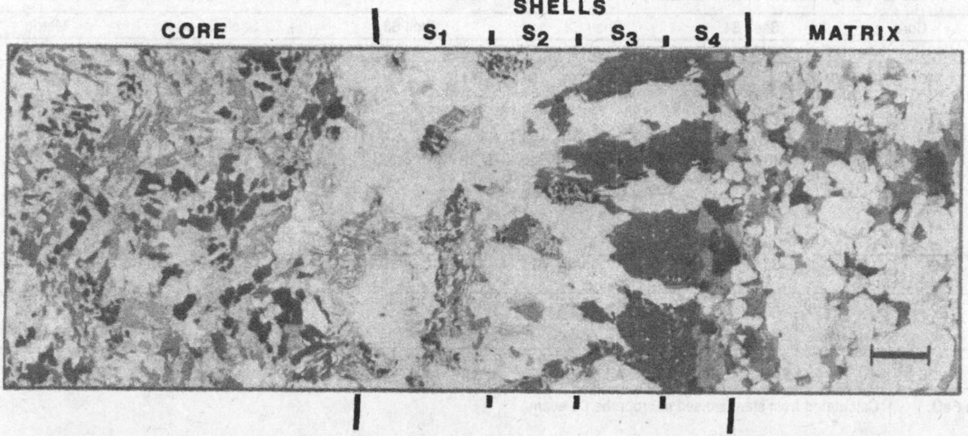


**Figure 2. Photograph of part of polished thin section of orbicular specimen TG81-106D. Core exhibits equigranular texture and contains biotite (dark coloured), diopside (intermediate coloured), and plagioclase (light coloured). Islands of remnant diopside-biotite are enclosed in coarse plagioclase in the inner shells $S_1$ and $S_2$. Shells $S_3$ and $S_4$ predominantly contain coarse, elongate radial hornblende (dark) and plagioclase; biotite is also abundant, particularly as a coarse variety at the extreme outer edge of the orbicule (shell $S_4$). The orbicular matrix is a tonalite; biotite is the dominant ferromagnesian mineral, and the blocky, light coloured mineral is plagioclase. The $S_1$-$S_2$ and $S_3$-$S_4$ boundaries are chosen on the basis of bulk chemical, rather than petrographic, criteria. Scale bar is 2 mm.**

The matrix is a tonalite composed of blocky andesine, biotite, diopside (largely altered to hornblende), and interstitial quartz. Modal analyses of the various zones of orbicule TG81-106D are presented in Table 1.

## Mineral chemistry

Electron microprobe analyses of coexisting minerals in orbicule TG81-106D are given in Table 2. Ferromagnesian minerals show only minor chemical changes from the core through shell S2. However, associated with the marked modal increase in hornblende through shells S3 and S4 is a distinct Fe enrichment trend in all three ferromagnesian minerals. This trend levels asymptotically into the matrix. Mn and Cl parallel the Fe trend, but Mg and F exhibit an antipathetic profile. In any given layer of the orb the mole fraction $X_{Mg}$ decreases in the order diop > hbl > bio. Plagioclase ranges in mean composition from An 36-37 in the core and matrix to An 50 in the shell region. The plagioclase is normally zoned, making it difficult to assign a representative An value for each layer in Table 2. Values reported herein are the mean $\pm$ $1\sigma$. The chemical trends described above in orbicule TG81-106D are observed in all orbicules in the Taylor Glacier locality, regardless of the number of shells.

## Whole-rock chemistry

Bulk chemical analyses of the core, shells, and matrix of orbicule TG81-106D are presented in Table 1. By inspection it is evident that the core is enriched in FeO, MgO, CaO, MnO, $K_2O$, $H_2O$, and F relative to the matrix; the matrix is relatively enriched in $Na_2O$, $Al_2O_3$, $SiO_2$, and Cl. Furthermore, the bulk shell composition ($\Sigma S$ in Table 1) is intermediate to that of core and matrix, for nearly all components and mole fractions.

# ORBICULE GROWTH MECHANISM

## Mass transfer

Implicit in any mass transfer scheme for orbicules of xenolithic origin must be an estimated location for the original xenolith-magma interface. In the following discussion, petrographic and chemical evidence are used to establish this initial boundary in orbicule TG81-106D.

The inner two shells of this orb (Figure 2) resemble partially melted xenolith (adjacent tonalitic magma being the heat source), with diopside-biotite intergrowths occurring as unmelted remnants scattered in a large mass of recrystallised coarse plagioclase. Noteworthy is the deficiency of plagioclase in these intergrowths relative to xenolithic core. This relationship may indicate active feldspar mobilisation and subsequent recrystallisation in the inner shells region. If assimilation of the xenolith occurred by partial melting, as argued above, then the initial xenolith may have extended as far as the last diopside-biotite intergrowth, as marked by the present $S_2$-$S_3$ boundary. This boundary also marks the beginnings of (1) coarse radial hornblende and (2) the Fe-enrichment trend of ferromagnesian minerals. Furthermore, the fact that the bulk shell is chemically intermediate relative to core and matrix (Table 1) indicates that the shell resulted from a near-linear mix of xenolith and tonalite magma. Consistent with such mixing is an initial xenolith-magma interface somewhere within the present shell structure.

Based on the above considerations, the present $S_2$-$S_3$ boundary is selected as the initial interface. Two consequences of this placement

**TABLE 1: Whole-rock and modal analyses of orbicular specimen TG81-106D**

| | Whole-rock analyses (wt. %) | | | | | | |
|---|---|---|---|---|---|---|---|
| | Core | S1 | S2 | S3 | S4 | ΣS* | Matrix |
| $SiO_2$ | 48.27 | 52.04 | 51.85 | 48.65 | 45.71 | 50.0 | 59.68 |
| $TiO_2$ | 1.49 | 0.76 | 0.65 | 1.15 | 2.25 | 1.15 | 1.46 |
| $Al_2O_3$ | 10.98 | 22.00 | 22.17 | 19.53 | 15.00 | 20.00 | 18.15 |
| FeO** | 12.00 | 4.75 | 5.21 | 9.38 | 16.40 | 8.46 | 5.57 |
| MnO | 0.25 | 0.10 | 0.12 | 0.20 | 0.32 | 0.17 | 0.10 |
| MgO | 8.82 | 3.46 | 3.37 | 4.39 | 6.92 | 4.42 | 2.35 |
| CaO | 11.69 | 10.85 | 10.66 | 10.22 | 6.75 | 9.80 | 5.57 |
| $Na_2O$ | 1.79 | 3.99 | 4.05 | 3.43 | 1.90 | 3.43 | 4.12 |
| $K_2O$ | 3.25 | 1.61 | 1.38 | 1.19 | 3.97 | 2.00 | 2.27 |
| $P_2O_5$ | — | 0.25 | 0.27 | 0.22 | — | 0.19 | — |
| F | 0.12 | 0.05 | 0.06 | 0.09 | 0.14 | 0.08 | 0.04 |
| Cl | 0.04 | 0.02 | 0.02 | 0.06 | 0.11 | 0.05 | 0.03 |
| $H_2O+$ | 0.85 | 0.34 | 0.42 | 0.83 | 1.63 | 0.75 | 0.51 |
| | 99.55 | 100.22 | 100.23 | 99.34 | 101.10 | 100.50 | 99.85 |
| O = F, Cl | 0.06 | 0.03 | 0.03 | 0.05 | 0.08 | 0.05 | 0.02 |
| Total | 99.49 | 100.19 | 100.20 | 99.29 | 101.02 | 100.45 | 99.83 |
| $\frac{100mg}{mg+fe}$ | 56.7 | 56.5 | 53.6 | 45.5 | 43.0 | 48.2 | 42.9 |
| $\frac{100f}{f+cl}$ | 86.1 | 79.8 | 83.3 | 75.1 | 69.2 | 74.7 | 70.2 |
| $\frac{100ca}{ca+na}$ | 87.9 | 75.0 | 74.4 | 76.7 | 79.7 | 76.0 | 59.9 |
| Sp.Gr. | 3.067 | 2.806 | 2.814 | 2.905 | 3.006 | 2.870 | 2.759 |
| $V(cm^3)$*** | | 0.552 | 0.334 | 0.331 | 0.322 | 1.539 | |
| | Mineral modes (vol. %) | | | | | | |
| diop | 39.1 | 11.9 | 8.2 | 0.5 | 3.4 | | 3.0 |
| bio | 34.4 | 13.9 | 10.9 | 6.9 | 38.2 | | 18.0 |
| hbl | — | 0.6 | 8.1 | 36.8 | 34.0 | | 2.1 |
| plag | 26.2 | 73.0 | 72.5 | 55.4 | 24.3 | | 57.9 |
| qtz | — | — | — | — | — | | 15.3 |
| sphn | 0.2 | — | — | — | — | | 1.0 |
| apa | — | 0.5 | 0.4 | 0.5 | — | | 1.4 |
| mic | — | — | — | — | — | | 1.4 |

All whole-rock analyses were calculated from mineral modes, average mineral compositions (determined by JEOL electron microprobe) and published mineral specific gravities. $H_2O$ contents of bio and hbl were estimated as 100 wt. % minus the F, Cl-corrected microprobe totals.

*Entire shell structure. **Total Fe as FeO. ***Shell volume used in mass transfer calculations. Abbreviations: diop = diopside, bio = biotite, hbl = hornblende, plag = plagioclase, qtz = quartz, sphn = sphene, apa = apatite, mic = microcline.

**TABLE 2: Average electron microprobe analyses of coexisting diopside, hornblende, and biotite in orbicular specimen TG81-106D**

| | Core | | Shell S1 | | | Shell S2 | | | Shell S3 | | | Shell S4 | | | Matric | | |
|---|---|---|---|---|---|---|---|---|---|---|---|---|---|---|---|---|---|
| | diop | bio | diop | hbl | bio | diop | hbl | bio | diop | hbl | bio | diop | hbl | bio | diop | hbl | bio |
| $SiO_2$ | 51.86 | 36.68 | 51.93 | 44.18 | 36.83 | 52.12 | 43.81 | 36.42 | 51.58 | 43.14 | 36.08 | 51.61 | 43.46 | 36.30 | 51.36 | 43.78 | 36.14 |
| $TiO_2$ | 0.04 | 4.16 | 0.08 | 1.77 | 4.93 | 0.08 | 1.68 | 4.15 | 0.07 | 2.02 | 4.91 | 0.11 | 1.78 | 4.27 | 0.12 | 0.77 | 4.67 |
| $Al_2O_3$ | 0.41 | 14.74 | 1.00 | 10.24 | 15.07 | 0.98 | 10.29 | 15.17 | 0.63 | 10.32 | 14.84 | 0.79 | 10.11 | 14.92 | 0.81 | 9.74 | 15.01 |
| FeO* | 11.52 | 20.09 | 11.57 | 17.19 | 19.54 | 11.26 | 17.77 | 20.63 | 12.64 | 19.26 | 21.60 | 14.76 | 20.11 | 22.76 | 15.10 | 20.76 | 22.56 |
| MnO | 0.39 | 0.25 | 0.41 | 0.29 | 0.25 | 0.49 | 0.38 | 0.30 | 0.63 | 0.42 | 0.30 | 0.68 | 0.45 | 0.34 | 0.68 | 0.43 | 0.32 |
| MgO | 11.89 | 10.95 | 11.90 | 10.27 | 11.15 | 12.14 | 9.87 | 10.68 | 10.75 | 9.03 | 9.72 | 10.39 | 8.43 | 9.26 | 10.43 | 8.47 | 8.86 |
| CaO | 23.04 | — | 23.02 | 12.71 | 0.02 | 23.03 | 12.20 | 0.03 | 23.71 | 11.56 | 0.03 | 22.24 | 11.24 | 0.03 | 21.29 | 12.55 | 0.03 |
| $Na_2O$ | 0.23 | 0.10 | 0.32 | 1.30 | 0.13 | 0.29 | 1.35 | 0.14 | 0.24 | 1.34 | 0.14 | 0.33 | 1.23 | 0.12 | 0.30 | 1.13 | 0.13 |
| $K_2O$ | — | 9.46 | — | 1.09 | 9.58 | — | 1.10 | 9.42 | — | 1.12 | 9.26 | — | 1.09 | 9.34 | — | 0.90 | 9.47 |
| F | — | 0.37 | — | 0.15 | 0.31 | — | 0.17 | 0.40 | — | 0.17 | 0.34 | — | 0.15 | 0.21 | — | 0.13 | 0.20 |
| Cl | — | 0.11 | — | 0.11 | 0.14 | — | 0.12 | 0.10 | — | 0.12 | 0.12 | — | 0.14 | 0.16 | — | 0.16 | 0.15 |
| | 99.40 | 96.91 | 100.23 | 99.31 | 97.96 | 100.39 | 98.74 | 97.44 | 100.25 | 98.50 | 97.34 | 100.91 | 98.19 | 97.71 | 100.09 | 98.82 | 97.54 |
| O = F, Cl | — | 0.18 | — | 0.09 | 0.16 | — | 0.10 | 0.19 | — | 0.10 | 0.17 | — | 0.09 | 0.12 | — | 0.09 | 0.12 |
| Total | 99.40 | 96.73 | 100.23 | 99.22 | 97.80 | 100.39 | 98.64 | 97.25 | 100.25 | 98.40 | 97.17 | 100.91 | 98.10 | 97.59 | 100.09 | 98.73 | 97.42 |
| $\frac{100mg}{mg+fe}$ | 64.8 | 49.3 | 64.7 | 51.5 | 50.4 | 65.8 | 49.8 | 48.0 | 60.3 | 45.5 | 44.5 | 55.7 | 42.8 | 42.1 | 55.2 | 42.1 | 41.2 |
| plag An%** | An 36 ± 3 | | An 50 ± 3 | | | An 50 ± 3 | | | An 50 ± 3 | | | An 42 ± 4 | | | An 37 ± 3 | | |

*Total Fe as FeO. **Calculated from standardised microprobe Ca scans.

are that: (1) xenolith core was converted into what are now shells $S_1 + S_2$ and (2) a thin envelope of tonalite magma was converted into the present shells $S_3 + S_4$. In subsequent discussion, we shall evaluate the mass transfer required to effect the C→ $S_1 + S_2$ and M→ $S_3 + S_4$ conversions, with the initial assumptions that (1) the original xenolith and its immediate magmatic envelope behaved as a closed system chemically and (2) the present tonalite composition closely approxiates that of the magma with which the xenolith reacted.

We consider first the C→ $S_1 + S_2$ conversion. The nature of the mass transfer required to effect this (or any) conversion is dependent on initial and final bulk composition, initial and final rock densities, and the volume change involved in the conversion. For a given oxide component, the equation relating these variables is (after Babcock, 1973):

$$\Delta \text{wt.\%n} = 100\left[K_v X_n^{\beta}\frac{\rho^{\beta}}{\rho^{\alpha}} - X_n^{\alpha}\right]$$

where Δwt.%n = the change in wt.% of oxide n when parent rock $\alpha$ is converted to product rock $\beta$,

$K_v$ = the rock volume ratio $V_\beta/V_\alpha$,

$X_n$ = the weight fraction of oxide n in parent rock $\alpha$ or product rock $\beta$, and

$\tau$ = the density of parent rock $\alpha$ or product rock $\beta$

For a given oxide component in a given conversion, Δwt.%n is a linear function of $K_v$. Thus, for the C→ $S_1 + S_2$ metasomatic convesion, the above equation is usd to generate the family of lines shown in Figure 3. These lines indicate, for any assumed volume change, the weight percent gain or loss of each component required to convert the xenolithic core into shells $S_1 + S_2$. In Figure 3, $K_v$ is expressed as percent volume change, so that $K_v$ values of 0.5, 1, and 2, for example, correspond to percent volume changes of −50%, 0%, and +100%, respectively.

For the metasomatic replacement of core by shells $S_1 + S_2$, we favour a slight volume increase on the basis of several pieces of evidence. First, the diopside/biotite modal ratio drops from 1.14 in the core to 0.86 and 0.75 in shells $S_1$ and $S_2$, respectively. This would suggest that diopside and biotite clusters were not merely separated mechanically during feldspathisation (e.g. see Eskola, 1938). Instead, these minerals must have undergone partial melting at different rates for the modal ratio to change as observed. Extensive partial melting of diopside and biotite would contribute abundant CaO, FeO, MgO and $K_2O$ to the evolving partial melt. Once dissolved into the melt phase these components would be free to diffuse in response to chemical potential ($\mu$) gradients set up by the initial difference in xenolith and magma compositions. Because CaO, FeO, MgO and $K_2O$ are enriched in the xenolith and relatively depleted in the tonalite magma (Table 1), it is likely that diffusion of these components was in the direction of the magma, thus resulting in a loss of these components from shells $S_1$ and $S_2$. According to Figure 3, for these four components to have been lost requires that the C→ $S_1 + S_2$ volume change was less than +17%.

A minimum volume change can be inferred by considering that $\mu_{SiO_2}$ was probably higher in the tonalite magma than in the xenolith during orbicule growth. Both the higher bulk $SiO_2$ content of the tonalite relative to the xenolith (Table 2) and the presence of quartz in the tonalite (notably absent in the xenolith) support this statement. Thus, during orbicule growth, $SiO_2$ was probably diffusing from the magma into the developing shells $S_1$ and $S_2$. For $SiO_2$ to be gained in the C→ $S_1 + S_2$ conversion requires a volume change greater than +1%, according to Figure 3.

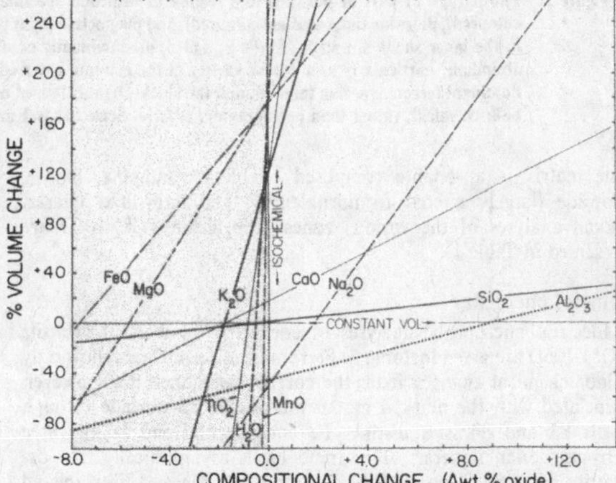


**Figure 3. Compositional variations versus percent volume change for the conversion of xenolith core to orbicular shells $S_1 + S_2$. The stippled and unstippled zones denote the areas of chemical loss and gain, respectively. For example, in any conversion carried out with a volume change less than +177%, the component FeO will be lost.**

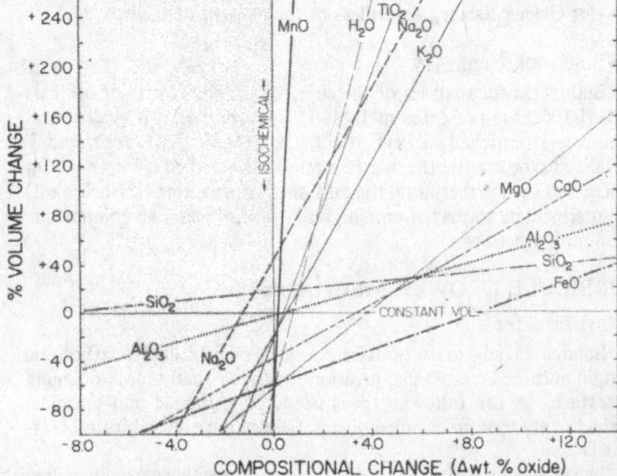


**Figure 4. Compositional variations versus percent volume change for the conversion of tonalite magma to orbicular shells $S_3 + S_4$. Diagram is analogous to Figure 3.**

Thus, the inferred volume change for the C→ $S_1+S_2$ conversion is between +1% and +17%. According to Figure 3, volume changes in this range require that components $Na_2O$, $Al_2O_3$, and $SiO_2$ were added to the xenolith from the magma and that FeO, MgO, $K_2O$, CaO, $TiO_2$, MnO, and $H_2O$ were lost to the magma from the xenolith, resulting in the present shells $S_1+S_2$. Leveson (1963) inferred a remarkably similar chemical migration pattern for the formation of orbicules in the Beartooth Mountains, Montana. Orbicule formation there is though to be the result of granitisation of amphibolites, during which $Na^{+1}$ and $Si^{+4}$ diffused inwards toward the centres of orbicules and $Ca^{+2}$, $Fe^{+2}$, and $Mg^{+2}$ diffused away.

If the original xenolith and its immediate tonalite magma envelope did indeed behave as a closed system chemically, then it follows that the magma envelope must have been: (1) the source of all $Na_2O$, $Al_2O_3$, and $SiO_2$ transferred into the xenolith and (2) the sink for all FeO, MgO, CaO, $K_2O$, MnO, $TiO_2$, and $H_2O$ diffusing out of the xenolith. Qualitative support for this hypothesis is provided by Figure 4, which shows the compositional changes vs volume change required to convert tonalite magma ("matrix" in Table 1) to shells $S_3+S_4$. Growth of cafemic minerals from a tonalite magma is likely to involve a slight volume decrease. Accordingly, we see in Figure 4, that for volume changes in the range −5% to −16%, the components $Na_2O$, $Al_2O_3$, and $SiO_2$ are lost (to developing shells $S_1+S_2$) and the components FeO, MgO, CaO, $K_2O$, MnO, $TiO_2$, and $H_2O$ are gained (from shells $S_1+S_2$, where diopside and biotite have undergone extensive melting).

Our proposed chemical transfer model for orbicule TG81-106D is shown schematically in Figure 5. This model for orbicule TG81-106D is shown schematically in Figure 5. This model involves coupled diffusion of components between the outer edge of the xenolith and its adjacent magmatic envelope to produce the present orbicular shell structure. However, orbicule TG81-106D represents an assimilation reaction that did not to completion. According to our model, continued influx of soda, alumina, and silica and removal of cafemic components (plus $K_2O$ and $H_2$) should result in an orbicule having a plagioclase core and a mafic shell or shells. Indeed, we commonly observe such orbicules in the Taylor Glacier locality; several examples are evident in Figure 1. Most such plagioclase-cored orbs also contain remnant diopside and/or biotite crystals, thus providing clear evidence regarding the assimilative nature of the orb forming process. Simple magmatic accretion, commonly invoked by petrologists as an orb forming process, cannot account for the relict textures exhibited by the cores of these orbs.

Orbicules at this advanced stage, prior to final solidification, must have consisted of a molten feldspathic core surrounded by a solid mafic shell structure all contained in a tonalite magma. As such, the orbicules were highly fragile and could easily have broken due to mutual collisions during times of convective flow of the magma. Evidence for such breakage is the abundance of orbicule hash, namely shell fragments, in the tonalite. These fragments themselves commonly become the nucleation sites for further orbicule development (e.g. see Palmer et al., 1967, Pl 2, Figure 2).

It has been assumed, thus far, that mass transfer in orbicule TG81-106D was restricted to the xenolith and its thin envelope of tonalite that ultimately became shells $S_3+S_4$. A rigorous test of this closed system assumption can be made by determining whether component weight loss from shells $S_1+S_2$ is matched by weight gain in shells $S_3+S_4$, and vice versa. This calculation requires knowledge of the original weights of xenolith and tonalite magma involved in the shell forming reaction. These weights are calculated from present shell volumes (Table 1) in a conical section cut through the orbicule. The height of the cone used is equal to and colinear with the radius of curvature of the orbicular section shown in Figure 2. The sum of $S_1$ and $S_2$ volumes from Table 1 is corrected for an assumed volume change and multiplied by the xenolith specific gravity (Table 1), yielding the weight of xenolith replaced. Assuming a 5% volume increase in the conversion of xenolith to shells $S_1+S_2$, the initial weight of xenolith in the conical section was 2580 mg. Figure 3 gives the component gains and lossess as weight percent of the parent rock. At 5% volume increase, the conversion process involved the gain of 364 mg of $Na_2O+ Al_2O_3+SiO_2$ and the loss of 449 mg of FeO+ MgO+ CaO+ MnO+ $K_2O+ TiO_2+ H_2O$. Similar calculations based on the magma→ $S_3+S_4$ conversion at 5% volume decrease indicate only 249 mg of $Na_2O+ Al_2O_3+ SiO_2$ lost and 299 mg of the order of 30 weight percent of the original now replaced xenolith and magma envelope. Results similar to those given above are obtained for different assumed volume changes within the ranges inferred earlier from Figures 3 and 4.

The obvious discrepancies in mass balance indicate that closed system orbicule formation, as defined earlier, did not occur. (The authors do not believe that analytical error and/or slight mislocation of the initial xenolith-magma interface are sufficient to explain the mass balance discrepancies). Of the 364 mg of $Na_2O+ Al_2O_3+ SiO_2$ gained by the developing inner shells, only 249 mg came from the magma now replaced by shells $S_3$ and $S_4$. The remaining 115 mg must have diffused in from magma as far as several millimetres beyond the present orbicule. Similarly, of the 449 mg of cafemic components (plus $K_2O$ and $H_2O$) lost from the inner shells, 150 mg must have diffused to a similar distance beyond the present orbicule. Chemical gradients in the magma beyond the orbicule were not preserved, however, as evidenced by perfectly flat microprobe profiles of constituent minerals and lack of modal variation in the tonalite. Any such gradients were apparently destroyed during rehomogenisation prior to final solidification.

## Orbicule shell geometry

Orbicule TG81-106D is a simple orbicule. Yet, even the complex multishelled orbicules of the Taylor Glacier locality (Figure 1) exhibit mineral-chemical profiles identical to those of orbicule TG81-106D (Table 2). Thus, to account for the rhythmic layering of these multishelled orbs, it is apparent that we seek some process that is superimposed on the overall mass transfer process outlined in the previous section. We tentatively favour a mechanism of oscillatory nucleation and diffusion controlled growth of shell minerals. Variations in both spacing and number of shells from orb to orb can be qualitatively understood in terms of minor differences in rates of (1) component diffusion, (2) shell mineral nucleation, (3) and shell mineral growth. Rates of these processes, in turn, depend on many variables, including $P_{H_2O}$ and temperature.

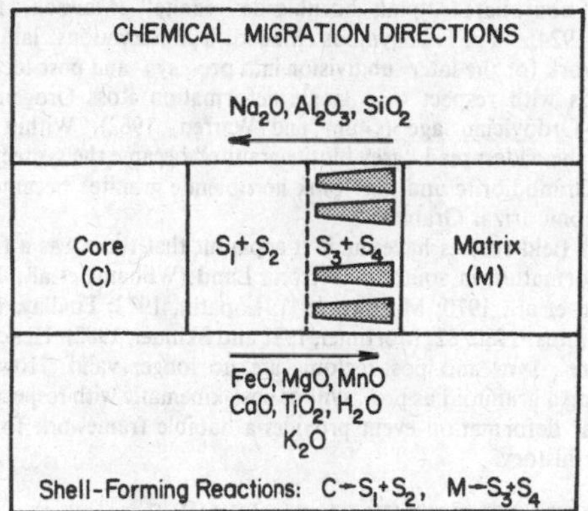


**Figure 5. Chemical concentration (potential?) gradients in orbicular specimen TG81-106D, inferred from mass transfer involved in C→ $S_1+ S_2$ and M→ $S_3+ S_4$ conversions (Figures 3 and 4). Inferred migration directions are consistent with the whole-rock compositional differences between core, bulk shell, and matrix (Table 1). Vertical dashed line marks initial xenolith-magma interface selected for development of the mass transfer model. See text for further explanation.**

*Acknowledgements* We thank Mr Mark Schmidt for field and laboratory assistance and Dr Micheal Holdaway for use of the JEOL electron microprobe at Southern Methodist University. This research was principally supported by grant DPP80-01743 from the U.S. National Science Foundation. One of us (PSD) is also grateful to The Research Corporation for its initial support of this work.

## REFERENCES

BABCOCK, R.S., 1973: Computational models of metasomatic processes. *Lithos, 6*, 279-90.

DAHL, P.S. and PALMER, D.F., 1981: Field study of orbicular granitic rocks in Taylor Valley, Southern Victoria Land. *U.S. Antarct. J., 16(5)*, 47-9.

ESKOLA, P., 1938: On the esboitic crystallisation of orbicular rocks. *J. Geol., 46*, 448-85.

LEVESON, D.J., 1963: Orbicular rocks of the Lonesome Mountain area, Beartooth Mountains, Montana and Wyoming. *Geol. Soc. Am., Bull., 74*, 1015-40.

PALMER, D.F., BRADLEY, J. and PREBBLE, W.M., 1967: Orbicular granodiorite from Taylor Valley, Southern Victoria Land, Antarctica. *Geol. Soc. Am., Bull., 78*, 1423-8.

# THE GRANITES AND TWO OROGENIES OF SOUTHERN VICTORIA LAND

D.N.B. Skinner, *N.Z. Geological Survey (D.S.I.R.), Box 61-012, Otara, Auckland, New Zealand.*

*Abstract* Mawson's grey biotite granite and younger pink hornblende granite were subsequently renamed the syntectonic Larsen Granodiorite and the post-tectonic Irizar Granite with respect to a single pulse Cambro-Ordovician Ross Orogeny. However, more recently three deformation events ($F_1$, $F_2$, $F_3$) have been identified, with considerable time interval between them. Intrusion of Portal Augen Gneiss and other granitoids occurred between $F_1$ and $F_2$. High grade metamorphism and accompanying anatexis to produce Renegar Mafic Gneisses and intrusion of Buddha Diorite, were early synkinematic to $F_2$ and preceded uprising gneiss domes of Chancellor Orthogneiss. These were followed by a second, late synkinematic $F_2$ high grade metamorphism with further anatexis to produce phases of Larsen Granodiorite, namely Olympus Granite Gneiss with anatectic and migmatitic border facies and magmatic Dias Granite (= Carlyon Granodiorite). Rheomorphism or continued injection of Larsen Granodiorite took place during the Early Ordovician third deformation along with intrusion of Theseus and Skelton Granodiorites. High level Crags, Vida and Irizar Granites post dated $F_3$ in the Late Ordovician.

$F_1$ remains undated but is Precambrian in age. $F_1$ and $F_2$ are separated by granitoid intrusion, uplift, erosion and volcanism/sedimentation (Cocks Formation). A zircon U-Pb date on Olympus Granite Gneiss of 588.5 ± 12.5 Ma, and a Rb-Sr whole rock isochron of 568 ± 9 Ma, on Carlyon Granodiorite provide a Vendian order of age for the late second deformation equivalent to the Beardmore Orogeny and imply that Skelton/Koettlitz Groups and pre-Larsen granitoids are Precambrian. The Ross Orogeny is mainly an Ordovician granitoid intrusive thermal event associated with relatively weak tectonism ($F_3$).

There has been less understanding of the basement geology of southern Victoria Land in terms of rock succession and orogenesis than almost any other section of the Transantarctic Mountains, principally because fossiliferous Cambrian is absent (all "possible" Archeocyanthinae have proved conclusively to be calc-silicate rod and mullion tectonites—see Findlay, 1978; Mortimer, 1981). The region is characterised by granitoids (herein named the southern Victoria Land Batholith) with remnants of metasedimentary rocks. To the north, there is a sharp boundary with the different suite comprising the Terra Nova Batholith (Skinner, 1972) and to the south, there is separation from the "granite-less" Byrd Group Cambrian terrain by right lateral displacement on the Byrd Fault (Grindley, 1981).

Early geologists distinguished an older, foliated, often gneissic "grey biotite granite" containing hornblende bearing zones and a younger nonfoliated "pink hornblende granite" (Mawson, 1916; Smith, 1924). The eponymous Mawson's observations laid the groundwork for the later subdivision into pre-, syn- and post-tectonic granitoids with respect to a single deformation Ross Orogeny of Cambro-Ordovician age (Gunn and Warren, 1962). Within this scheme, the widespread "grey biotite granite" became the syntectonic Larsen Granodiorite and the "pink hornblende granite" became the post-tectonic Irizar Granite.

Recent field studies have made it apparent that there was a three-fold deformation in southern Victoria Land (Williams et al., 1971; Smithson et al., 1970; Murphy, 1971; Lopatin, 1972; Findlay, 1978, writt. comms. 1980/82; Mortimer, 1981 and Skinner, 1982). Hence the terms pre-, syn- and post-tectonic are no longer valid. However, classifying a granitoid as pre-, syn- or post-kinematic with respect to a particular deformation event provides a datable framework for the orogenic history.

## THE METASEDIMENTARY SEQUENCES

There has been considerable revision in recent years of the schemes of Grindley and Warren (1964) and Blank et al. (1963), that allows some correlation of the metasedimentary basement rocks in southern Victora Land in spite of differences in metamorphic grade (Lopatin, 1972).

The Skelton Group (generally low grade) consists of an older carbonate-quartzite-mafic volcanic association (Anthill Limestone-greenschist facies) overlain unconformably by a younger calcareous-volcanic-turbidite association (Cocks Formation-greenschist to amphibolite facies) (Skinner, 1982). The unconformable lower part of the Anthill Limestone is more quartzose and less calcareous than the upper part and is a likely correlative of the medium to high grade Koettlitz Group bounding the upper Koettlitz Glacier. These sillimanite bearing quartzose, calcareous schists with amphibolites and calc-silicates, and minor marble and slate are informally designated the Foster schists.

The Koettlitz Group consists, in uncertain stratigraphic order, of the Hobbs Formation (low to medium grade conglomerate, amphibolite, marble, calc-silicate, pillow lava and pelitic and quartzitic schist), Salmon Marble (medium-high grade) and Marshall Formation (high grade quartzo-feldspathic biotite paragneiss with amphibolite and minor quartzite and carbonate) (Findlay, 1978; Mortimer, 1981). From Ferrar Glacier to the Dry Valleys, the Asgaard Formation is a high grade, often migmatised condensed sequence of these same Koettlitz Group lithologies (McKelvey and Webb, 1962; Haskell et al., 1965a; Murphy, 1971). In the north near Larsen Glacier, the Priestley Formation is low to medium grade semipelitic schist with carbonate, quartzite and volcaniclastics (Skinner and Ricker, 1968).

## TECTONISM AND GRANITOIDS

Throughout the region, three deformations ($F_{1,2,3}$) have been identified, of which the last ($F_3$) resulted in rather open, relatively gentle folding but the earlier two produced near isoclinal, overturned and recumbent structures (Skinner, 1982; Findlay, 1978). Fold styles are similar for both Skelton and Koettlitz Groups, but Cocks Formation was deformed by only $F_2$ and $F_3$.

Mafic and felsic dykes and sills, but no granitoids, preceded or accompanied $F_1$. Between $F_1$ and $F_2$ there was sufficient time for the intrusion of basic and felsic sills and dykes and small bosses of granite, for uplift and erosion to expose the granite and for volcanism and sedimentation of Cocks Formation including a basal conglomerate with pebbles of identical granite (Skinner, 1982). This erosion interval has not been recognised in Koettlitz Group. The granite (now orthogneiss) crosscuts $F_1$ structures in Anthill Limestone and has been deformed by $F_2$. At about the same time, the Marshall Formation was intruded by mafic dykes and sills (now amphibolite-metabasite). This preceded migmatisation, and intrusion of a "sill" of granite (Portal Augen Gneiss—Mortimer, 1981) at 35° to the $F_1$ foliation and fold axes which were deformed by $F_2$ to a greater extent than succeeding mafic and granite intrusions.

From the Skelton-Koettlitz divide to the Dry Valley, $F_2$ was accompanied by a characteristic suite of synkinematic granitoids and associated with medium to high grade metamorphism, migmatisation and anatexis. Three main magmas were successively produced and intruded, cooled, and deformed to progressively lesser extents. These were firstly the Renegar Mafic Gneisses (with which may be correlated the Buddha Diorite of Mortimer, 1981); next the Chancellor Orthogneiss (Findlay, 1978; Findlay and Skinner, 1980); and finally the Larsen Granodiorite—McKelvey and Webb's (1962) Olympus Granite Gneiss and Dias Granite (= Carlyon Granodiorite of Haskell et al., 1965b).

There is considerable field evidence that remobilisation (palingenesis) and reintrusion of the Larsen Granodiorite took place during $F_3$ and caused at least some of the stress and local structures. However, $F_3$ was also accompanied by intrusion of the Theseus Granodiorite and Skelton Granodiorite.

A considerable number of high level, unstructured, crosscutting granites and more mafic rocks closely followed $F_3$ and caused some contact retrograde metamorphism.

## THE GRANITOIDS

### Post $F_1$-Pre $F_2$

*Orthogneiss intruding Anthill Limestone* consists of microcline-andesine-hornblende-biotite granite. Microcline contains considerable exsolved albite and has rapakivi textured rims and replacement growth of oligoclase; secondary biotite and green hornblende mantle earlier formed biotite and hornblende. Zircon, orthite and sphene are common accessories. Recrystallisation has produced a matrix of granular quartz-feldspar.

*Portal Augen Gneiss* forms the greater part of Blank et al.'s (1963) now abandoned Garwood Lake Formation, the rest being Marshall Formation and younger granitoids. It is a thinly foliated, mafic poor, fine grained leucocratic granite orthogneiss with oligoclase and minor biotite. Small augen of perthitic orthoclase and of strained granular quartz and quartz-microcline have grain boundaries that have been recrystallised to unstrained quartz-microcline-myrmekite-oligoclase/andesine granules in which secondary biotite, muscovite, chlorite and tourmaline have grown. Zircon-apatite-sphene are accessories.

## Syn $F_2$

*Renegar Mafic Gneisses* are mafic migmatites and anatectic diorites produced by high T-moderate P metamorphism that accompanied the early $F_2$ deformation. At Mt Cocks, Renegar Glacier and Mt Dromedary, the Cocks Formation passes through a migmatite border zone into foliated mafic "diorites" which re-inject through the border zone into Foster schists, and with them the Cocks Formation is folded during $F_2$. Similar rocks occur in Miers, Marshall and Garwood Valleys and include Buddha Diorites (Mortimer, 1981) and intrusives into Portal Augen Gneiss. Such early deformed diorites have also been described by Mawson (1916), Gunn and Warren (1962), Skinner and Ricker (1968), Ghent and Henderson (1968), Lopatin (1972), Grikurov et al. (1982) and Kamenev and Kumeev (1982).

Renegar Mafic Gneisses include fine and coarse grained, foliated and cleaved, mafic rich hornblende-biotite-augite diorite, quartz diorite, tonalite and granodiorite. Folia are predominantly biotite with pyroxene and/or hornblende and wrap around granulated quartz-plagioclase augen. Feldspars are deformed oligoclase/andesine to labradorite with minor perthitic orthoclase. Granular grain boundaries between quartz and feldspar are due to recrystallisation. Coarse sphene is accessory. Green hornblende replaces the augite with granular sphene as a byproduct, and is in turn replaced by a second generation biotite and a late chlorite-epidote(clinozoisite)-quartz retrograde alteration.

*Chancellor Orthogneiss*, a yellowish grey leucocratic, mafic poor, quartz rich granite orthogneiss with its type locality at Chancellor Lakes, pervasively intrudes Renegar Mafic Gneisses and Koettlitz Group from Renegar Glacier to Hobbs Glacier. It forms a series of mantled gneiss domes with the Koettlitz Group drape-folded over them and a swarm of small bosses and dykes intruded into Koettlitz Group and Cocks Formation and folded during $F_2$. In spite of its wide distribution and distinct lithology, it was mapped and referred to as schist by Blank et al. (1963) and Williams et al. (1971). Blattner (1978) and Feary (writt. comm. map 1979) called it biotite flaser gneiss. It has a strained texture of quartz lenticles and rolled oligoclase-perthitic orthoclase augen in a recrystallised granular matrix of quartz, orthoclase, microcline perthite and myrmekite. Small garnets and wispy biotite and muscovite are minor constituents along with accessory zircon and apatite. Minor green hornblende, granular sphene, muscovite and chlorite replace biotite and scattered much larger (2 mm) garnets have grown across the biotite folia. Tourmaline is a rare secondary mineral.

At Chancellor Lakes amphibole rich layers and schlieren (? restite) occur. Pale olive biotite grows in the cleavage of brown hornblende associated with coarse euhedral sphene. Microcline perthite exsolves a high proportion of albite and only remnant cores remain of large calcic plagioclase that have been intensely K-metasomatised to orthoclase.

## Late Syn $F_2$-$F_3$

*Larsen Granodiorite* is a convenient family name for a series of plutons produced by a further high T-moderate P metamorphism accompanying late $F_2$ deformation. They intrude Chancellor Orthogneiss at Chancellor Lakes and Miers Valley and extend from Priestley Glacier to Darwin Glacier. In the Wright Valley, the relationships of plutons to migmatites with mafic restite (amphibolite) and carbonate-quartzite refractory layers and to high grade Asgaard Formation indicate a local partial melt anatectic origin of the igneous rocks (Murphy, 1971; Fikkan, 1968; Smithson et al., 1971 and 1972).

The magmatic schlieren/restite rich border facies is *Olympus Granite Gneiss* and the magmatic inner facies is *Dias Granite*. The degree of gneissosity and augen development has no relationship to the particular facies but is zonally concentrated; for example, parts of Dias Granite on the Dias are more gneissic than Olympus Granite Gneiss east of Bull Pass. In the Larsen Glacier and Blue Glacier areas, slightly higher levels are exposed where the plutons have sharper margins parallel to metasedimentary foliation and migmatitic border zones are narrow. The Dias Granite facies of the Darwin Glacier area is a similarly higher level pluton termed Carlyon Granodiorite (Haskell et al., 1965b). In the Dry Valleys, these rocks have been collectively termed Wright Intrusives (McKelvey and Webb, 1962).

Palingenesis or remobilisation of Dias Granite magma and intrusion through the Olympus Granite Gneiss margins into the country rock, commonly occurred in the plutons of the Larsen Glacier and Blue Glacier—Miers Valley. At Larsen Glacier, dykes of even grained granodiorite (e.g. Mawson's sample) and K-feldspar rich porphyritic granite-syenite represent filter-pressed matrix and phenocryst concentrates respectively of the Dias Granite. This reinjection event seems to be associated with the broad swings of the $F_3$ deformation and may be responsible for some of the stress field of that event. These rocks have no formal name.

The Larsen Granodiorite family are all essentially similar, quartz-K feldspar-plagioclase-biotite granitoids, usually with significant hornblende, and with accessory allanite-sphene-zircon-apatite-Fe oxide. There are however, marked regional differences and a wide range of compositions. The Larsen Glacier rocks are tonalite, quartz monzodiorite, granodiorite and granite (adamellite) whereas in the Dry Valleys, granodiorite is the main rock type with minor granite. The Miers Valley-Chancellor Lakes rocks on the margins of the Blue Glacier pluton include diorite, tonalite, granodiorite and granite (adamellite) and monzodiorite, quartz monzodiorite and quartz syenite. The gneissosity is expressed not only by feldspar and quartz augen, but also by granular, recrystallised cataclastic grain margins to the quartz and feldspar, and growth of secondary biotite and chlorite. Close to the margins a red, usually porphyritic "pseudo Irizar" granite commonly has developed by an increase of hornblende and haematite granules in the K-feldspar. Ferrar (1907) described such rocks at Cathedral Rocks (Ferrar Glacier) as grading into pink (Irizar) hornblende granite.

K-feldspar is predominantly microcline at Larsen Glacier (Skinner and Ricker, 1968) but is slightly albitic, perthitic orthoclase from the Dry Valleys south to Darwin Glacier. Plagioclase is oligoclase/andesine but early formed labrodorite phenocrysts occur in the Miers Valley rocks. Hornblende and biotite are almost ubiquitous, but minor muscovite is reported from Carlyon Granodiorite and occurs in the margins of the Blue Glacier pluton. Pyroxene has previously been reported as augite with "basic segregations" in Olympus Granite Gneiss (McKelvey and Webb, 1962) and as "rare hypersthene" in Carlyon Granodiorite (Haskell et al., 1965b). However, although no Larsen Glacier samples contain pyroxene, almost every sample from the eastern Blue Glacier pluton margins and from Wright Valley between the Dias and east Bull Pass contain significant amounts of colourless augite being replaced by green hornblende; it is least abundant in the K-feldspar rich rocks.

*Theseus Granodiorite* of Wright Valley—part of Wright Intrusives—is the archetypal late kinematic granitoid younger than Larsen Granodiorite. It consists of a dyke swarm and small linear discordant plutons of hydrothermally altered biotite granodiorite with little foliation except marginally where a mild cataclastic fabric is developed. *Skelton Granodiorite* has a cataclastic fabric parallel to $F_3$ folds at Teall Island and Red Dyke Bluff and has been a partial cause of the local $F_3$ deformation.

## Post $F_3$

Undeformed, discordant granites, granodiorites, tonalites and diorites are the youngest members, other than Vanda Porphyries and Lamprophyres, of the Southern Victoria Land Batholith. In the Dry Valleys these are collectively termed Victoria Intrusives (McKelvey and Webb, 1962) and include Vida and Irizar Granites, the former being pale pink to grey with only biotite at its type locality (Lake Vida) and the latter being deep brick red with both biotite and hornblende. There is no Vida Granite in central Wright Valley (Lopatin, 1972) but only northeast of Bull Pass, the extensive sheet of granite above the basement dolerite sill being Dias/Olympus Granite Gneiss (NZARP, 1978; GANOVEX I, 1979; H. Brady—USARP, pers. comm., 1979).

Other post-kinematic units are Delta Diorite (Skelton Glacier), Murray Granite (Mawson Glacier), Crags Granite (Hooper Crags-Skelton Glacier-upper Koettlitz Glacier) and Mt Rich Granite (Darwin Glacier), this last being probably deformed by the Byrd Fault. Crags Granite strongly resembles Hope Granite of the Beardmore and Darwin glaciers and like all the southern granitoids, contains microcline, whereas all granitoids north of Koettlitz Glacier contain perthitic orthoclase.

## DATING OF EVENTS

All dates have been recalculated from the published data by H. Kreuzer (B.G.R.—West Germany), using the Sydney 1976 convention constants.

Many of the early dates were single whole rock determinations and the only unequivocal conclusion, both from K-Ar and Rb-Sr, was that two distinct tectono-magmatic events occurred at 510± 25 Ma and 440± 25 Ma. Problems and backgrounnd to the precise dating of Wright and Victoria Intrusives and the regional tectono-magmatic events are discussed by Faure and Jones (1974).

### Post $F_3$

Post-kinematic intrusive/cooling events took place about late Middle Ordovician. K-Ar biotite dates on Irizar Granite of Mt Falconer (McDougall and Ghent, 1970) range from 459.8± 6 Ma to 469.7± 8 Ma and correspond with an Rb-Sr biotite date of 462± 15 Ma on Irizar Granite at Granite Harbour (Deutsch and Webb, 1964). An Rb-Sr whole rock-feldspar isochron of 460± 7 Ma (Jones and Faure, 1967) and a hornblende-biotite Rb-Sr date of 465± 15 Ma (Deutsch and Webb, 1964), both Vanda Porphyries, are of the same order.

### Late $F_2/F_3$

Dating of the Larsen Granodiorite family would define the late $F_2$/$F_3$ events and partially clarify the enigmatic stratigraphy of southern Victora Land. However, because the family with its three main varieties has a complex cooling and partitioning history during the late $F_2/F_3$ time, the mixing together on one isochron diagram of samples with different consolidation histories would produce considerable scatter and an imprecise "date" such as those obtained by Faure and Jones (1974). Separating their samples by variety and locality, their published data results in "isochrons" of 470± 120 Ma and 490± 110 Ma for Olympus Granite Gneiss, but 497± 38 Ma and 498± 35 Ma for Dias Granite.

These "dates" are also suspect because firstly, deformation and post-cataclastic metamorphism have produced crystalloblastic K-feldspar and biotite so that at best, the dates fix post-intrusive metamorphism. Secondly, the samples all come from the Vanda-Bull Pass area which is pervasively seamed with dykes and a pluton of Theseus Granodiorite. Hence the "dates" may be more appropriately of the order of the $F_3$/Theseus Granodiorite event. Combining the data for the suspect rocks and Theseus Granodiorite produces an "isochron" of 490± 53 Ma. This compares with an Rb-Sr biotite date of 476± 15 Ma on Theseus Granodiorite itself (Deutsch and Webb, 1964), with biotite metamorphic dates on Asgaard Formation of Rb-Sr 486± 15 Ma (Faure and Jones, 1974) and K-Ar 477.7± 8 Ma (McDougall and Ghent, 1970), and with K-Ar biotite dates on pre-Irizar Granite deformed diorites of 477.6± 8 Ma and 490± 8 Ma (McDougall and Ghent, 1970).

The most reasonable age for Olympus Granite Gneiss and hence the late $F_2$ metamorphism/tectonism is the almost concordant zircon U-Pb date of Deutsch and Grögler (1966), the implications of which are discussed by Faure and Jones (1974) who conclude that this could indeed represent the age of magmatism during which the zircons crystallised. The recalculated data produce a maximum age of 620 ± 15 Ma (lower projection fixed at 160 Ma), but 588.5± 12.5 Ma if fixed at 0 Ma. Grikurov et al. (1982) refer to unreferenced ages of 600-630 Ma from "several localities" in the same region. A remarkably consistent Rb-Sr whole rock isochron of 568± 9 Ma (Felder and Faure, 1980) on Carlyon Granodiorite (Dias Granite) is of similar order considering the dates apply to two facies varients of Larsen Granodiorite and the distance separating them.

In any case, all these dates immediately imply that all the rocks older than Larsen Granodiorite, that is Cocks Formation, Skelton and Koettlitz Groups, Chancellor Orthogneiss, Renegar Mafic Gneiss, Portal Augen Gneiss etc., are all of Precambrian (? Late Riphean-Early Vendian) age and that the late $F_2$ deformation extended up from Mid Vendian to earliest Cambrian and must have begun at least rather earlier in the Vendian. Attempts at dating earlier $F_2$ granitoids have not been successful (Blattner, 1978).

The metasedimentary rocks are thus at least equivalent to Beardmore Group if not older rocks and the $F_2$ event corresponds to the Beardmore Orogeny (680-620 Ma), and cooling to 563-576 Ma—Grindley and McDougall (1969), of the Central Transantarctic Mountains.

### $F_1$

There are no data from which the age of the first deformation can be deduced. Perhaps a reasonable estimate of the minimum time for granitoid intrusion, uplift, erosion and the volcanism/sedimentation of Cocks Formation may be comparable with the 30 Ma long Acadian (Devonian) Orogeny of U.S.A. (Naylor, 1971), or the 30 Ma long main episode (510-480 Ma) of the 150 Ma Caledonian Orogeny (McKerrow, 1962). $F_1$ could conceivably be a Riphean event.

**TABLE 1: Granitoid sequence and tectono-magmatic events in South Victoria Land**

| | Darwin | Skelton/Koettlitz/Dry Valleys | | Larsen | Event | |
|---|---|---|---|---|---|---|
| | Hope 450-495<br>Mt. Rich | Crags | Vida | Irizar 486-451 | intrusion | Ross Orogeny |
| Ordovician<br>445-500 Ma | | Skelton | Cathedral rocks 'Dias'. | Theseus 490-499 | deformation<br>$F_3$ | |
| | 515 | | | | | |
| Cambrian<br>500-570<br>± 10 Ma | 540 | | mid Cambrian Bowers Supergroup<br>to the north | | no record; stable shelf carbonate sedimentation to north and south | |
| | Early Cambrian Byrd Group<br>to the south | | | | | |
| | Carlyon 568 | Dias | | | | Beardmore Orogeny |
| Vendian<br>570-675<br>± 25 Ma | | Olympus 588.5<br>Chancellor | | Larsen | Late $F_2$ deformation<br>metamorphism II<br>rising gneiss domes | |
| | | Renegar | Buddha | diorite | $F_2$ deformation<br>metamorphism I | |
| | | Cocks F. (v) | | Priestley F. | sedimentation and volcanism (v)<br>erosion | |
| | | granite Portal | | | intrusion | |
| | | mafic dykes | | | $F_1$ deformation | |
| late<br>Riphean | | mafic and felsic dykes and sills. | | | intrusion | |
| 675-950<br>± 50 Ma | | Anthill Limestone (v)<br>Foster schists | | Salmon Marble<br>Hobbs F. (v)<br>Marshall F.<br>(Asgard F.) | shelf sedimentation<br>and volcanism (v) | |

## CONCLUSION (Table 1)

From the foregoing discussion, the Ross Orogeny is seen mainly as an Ordovician intrusive thermal event accompanying and following the mild $F_3$ deformation and separated by perhaps 25-60 Ma from the late $F_2$ deformation during which Cambrian carbonate shelf sedimentation of the Byrd and Bowers Groups occurred. This $F_2$ deformation event certainly covered the Mid to Late Vendian and this corresponds to the Beardmore Orogeny. It includes at least three major tectonomagmatic pulses (Larsen, Chancellor, Renegar) and could therefore extend back to Early Vendian.

The $F_1$ deformation could be an earlier separate Beardmore pulse of Late Riphean-Early Vendian age, but there are no data from which its age can be deduced. In comparison with other Early Palaeozoic orogenies it may be of the order of 30 Ma older than $F_2$, but an even older pre-Beardmore age is not impossible.

*Acknowledgements* This paper was prepared from part of the work done for GANOVEX during a period as guest scientist of the Alfred Wegener Institut fur Polarforschung, Bremerhaven, at the Bundesanstalt fur Geowissenschaften und Rohstoffe (Hannover).

## REFERENCES

ALLEN, A.D. and GIBSON, G.W., 1962: Geological investigations in southern Victoria Land, Antarctica, part 6—Outline of the geology of Victoria Valley region. *N.Z. Geol. Geophys., 5,* 234-42.

BLANK, H.R., COOPER, R.A., WHEELER, R.H. and WILLIS, I.A.G., 1963: Geology of the Koettlitz—Blue Glacier Region, southern Victoria Land, Antarctica. *R. Soc. N.Z., Trans, Geol., 2 (5),* 79-100.

BLATTNER, P., 1978: Geology and geochemistry of Mt Dromedary Massif, Koettlitz Glacier area, southern Victora Land—preliminary report. *N.Z. Antarc. Rec., 1 (2),* 16-19.

DEUTSCH, S. and GROGLER, N., 1966: Isotopic age of Olympus Granite-gneiss (Victoria Land—Antarctica). *Earth Planet Sci. Lett., 1,* 82-4.

DEUTSCH, S. and WEBB, P.N., 1964: Sr-Rb dating on basement rocks from Victoria Land; evidence for a 700 million year old event; *in* Adie, R.J. (ed.) *Antarctic Geology.* North Holland, Amsterdam. 557-62.

FAURE, G. and JONES, L.M., 1974: Isotopic composition of strontium and geologic history of the basement rocks of Wright Valley, southern Victoria Land, Antarctica. *N.Z. Geol. Geophys., 17,* 611-27.

FELDER, R.P. and FAURE, G., 1980: Rb-Sr age determination of part of the basement complex of the Brown Hills, central Transantarctic Mountains. *U.S. Antarct. J., 15(4),* 16-17.

FERRAR, H.T., 1907: Report on the field geology of the region explored during the "Discovery" Antarctic expedition. 1901-4. *Nat. Antarct. Exped. 1901/04, Nat. Hist., 1,* 1-100.

FIKKAN, P.R., 1968: Granitic rocks in the Dry Valley region of southern Victoria Land, Antarctica. Unpublished M.Sc. thesis—Univ. Wyoming, U.S.A., 119 pp.

FINDLAY, R.H., 1978: Provisional report on the geology of the region between the Renegar and Blue Glaciers, Antarctica. *N.Z. Antarc. Rec., 1,* 39-44.

FINDLAY, R.H. and SKINNER, D.N.B., 1980: Early structural evolution of part of the Gondwana craton: the structural evolution of the Ross Orogen at McMurdo Sound, Antarctica. *Fifth Gondwana Symposium Abstracts* Wellington, New Zealand.

GHENT, E.D. and HENDERSON, R.A., 1968: Geology of the Mt Falconer Pluton, lower Taylor Valley, southern Victoria Land, Antarctica. *N.Z. Geol. Geophys., 11 (4),* 851-80.

GRIKUROV, G.E., KAMENEV, E.N. and KAMENEVA, G.I., 1982: Granitoid complexes in Antarctica; *in* Craddock, C. (ed.) *Antarctic Geoscience.* Univ. Wisconsin Press, Madison, 695-701.

GRINDLEY, G.W., 1981: Precambrian rocks of the Ross Sea region. *R. Soc. N.Z., J., 11,* 411-23.

GRINDLEY, G.W. and McDOUGALL, I., 1969: Age and correlation of the Nimrod Group and other Precambrian rock units in the Central Transantarctic Mountains, Antarctica. *N.Z. J. Geol. Geophys., 12,* 391-411.

GRINDLEY, G.W. and WARREN, G., 1964: Stratigraphic nomenclature and correlation in the western part of the Ross Sea; *in* Adie, R.J. (ed.) *Antarctic Geology.* North Holland, Amsterdam, 314-33.

GUNN, B.M. and WARREN, G., 1962: Geology of Victoria Land between the Mawson and Mulock Glaciers, Antarctica. *N.Z. Geol. Surv. Bull., 71.*

HASKELL, T.R., KENNETT, J.P., PREBBLE, W.M., SMITH, G. and WILLIS, I.A.D., 1965a: The geology of the middle and lower Taylor Valley of southern Victoria Land, Antarctica. *R. Soc. N.Z., Trans, Geol., 2(12).* 169-86.

HASKELL, T.R., KENNETT, J.P. and PREBBLE, W.M., 1965b: Geology of the Brown Hills and Darwin Mountains, southern Victoria Land, Antarctica. *R. Soc. N.Z., Trans, Geol., 2(15),* 231-48.

JONES, L.M. and FAURE, G., 1967: Age of the Vanda Porphyry dykes in the Wright Valley, southern Victoria Land, Antarctica. *Earth Planet Sci. Lett., 3,* 321-4.

KAMENEV, E.N. and KUMEEV, S.S., 1982: Subdivision of granitoids of southern Victoria Land by means of feldspar structural diffractometry; *in* Craddock, C. (ed.) *Antarctic Geoscience.* Univ. Wisconsin Press, Madison, 703-8.

LOPATIN, B.G., 1972: Basement complex of the McMurdo "Oasis", southern Victoria Land; *in* Adie, R.J. (ed.) *Antarctic Geology and Geophysics.* Universitetsforlaget, Oslo, 287-92.

MAWSON, D., 1916: Petrology of rock collections from the mainland of southern Victoria Land. *Rep. Brit. Antarct. Exped. 1907/09, Geol., 2 (13),* 201-37.

McDOUGALL, I. and GHENT, E.D., 1970: K-Ar dates on minerals from the Mt Falconer area, lower Taylor Valley, southern Victoria Land, Antarctica. *N.Z. J. Geol. Geophys., 13,* 1206-9.

McKELVEY, B.C. and WEBB, P.N., 1962: Geological investigations in southern Victoria Land, Antarctica 3. Geology of the Wright Valley. *N.Z. J. Geol. Geophys., 5,* 143-62.

McKERROW, W.S., 1962: The chronology of Caledonian folding in the British Isles. *Nat. Acad. Sci., Proc., 48,* 1905-13.

MORTIMER, G., 1981: Provisional report on the geology of the basement complex between Miers and Salmon Valleys, McMurdo Sound Antarctica. *N.Z. Antarc. Rec., 3 (2),* 1-8.

MURPHY, D.J., 1971: The petrology and deformational history of the basement complex, Wright Valley, Antarctica, with special reference to the origin of the augen gneisses. Unpub. Ph.D. thesis. Univ. Wyoming, U.S.A. 114pp.

NAYLOR, R.S., 1971: Acadian Orogeny: an abrupt and brief event. *Science, 172,* 558-60.

SKINNER, D.N.B., 1972: Differentiation source for the mafic and ultramafic ("enstatite-peridotites") and some porphyritic granites of Terra Nova Bay, Victoria Land; *in* Adie, R.J. (ed.) *Antarctic Geology and Geophysics.* Universitetsforlaget, Oslo, 299-303.

SKINNER, D.N.B., 1982: Stratigraphy and structure of lower grade metasediments of Skelton Group, McMurdo Sound—does Teall greywacke really exist? *in* Craddock, C. (ed.) *Antarctic Geoscience.* Univ. Wisconsin Press, Madison, 555-63.

SKINNER, D.N.B. and RICKER, J., 1968: The geology of the region between the Mawson and Priestley Glaciers, northern Victoria Land, Antarctica. Part 1—Basement metasedimentary and igneous rocks. *N.Z. J. Geol. Geophys., 11,*1009-40.

SMITH, W.C., 1924: The plutonic and hypabyssal rocks of southern Victoria Land. *British Antarctica ("Terra Nova") Expeditions, 1910/13. Nat. Hist. Rep., Geol., 1 (6),* 167-227.

SMITHSON, S.B., FIKKAN, P.R., MURPHY, D.R. and HOUSTON, R.S., 1972: Development of augen gneiss in the ice-free valley area, southern Victoria Land; *in* Adie, R.J. (ed.) *Antarctic Geology and Geophysics.* Universitetsforlaget, Oslo, 293-8.

SMITH, S.B., FIKKAN, P.R. and TOOGOOD, D.J., 1970: Early geologic events in the ice-free valleys, Antarctica. *Geol. Soc. Am., Bull., 81,* 207-10.

SMITHSON, S.B., MURPHY, D.J. and HOUSTON, R.S., 1971: Development of an augen gneiss terrain. *Contrib. Mineral, Petrol., 33,* 184-90.

WILLIAMS, P.F., HOBBS, B.E., VERNON, R.H. and ANDERSON, D.E., 1971: The structural and metamorphic geology of basement rocks in the McMurdo Sound area, Antarctica. *Geol. Soc. Aust., J., 18,* 127-42.

# A RE-INTERPRETATION OF THE BASEMENT GRANITES, McMURDO SOUND, ANTARCTICA

R. H. Findlay, *Geology Department, University of Tasmania, GPO Box 252C, Hobart, Tasmania 7001, Australia.*

*Abstract* Supposed syn-tectonic Larsen Granodiorite consists of two phases, the deformed Dais Phase (Dais Granite and Olympus Granite Gneiss) and the non-deformed Briggs Hills Phase, intruding Dais Phase. Brigs Hills Phase is altered patchily and in some localities is identical in hand specimen to post-tectonic Irizar Granite.

Gneissosity in Dais Phase decreases in intensity into the pluton and represents a flattening fabric generated in the cooling margin by continued forcible injection of magma into the core of the pluton; leucocratic phases adjacent to the margin of the pluton may represent intergranular quartzo-feldspathic phase filter-pressed out from the margin during flattening.

Non-gneissose Briggs Hills Phase represents a continued, possibly lower temperature, intrusion of Larsen Granodiorite, and in places grades into supposed post-tectonic Vida Granite which may represent a minimum melt phase formed during the last stage of intrustion of the regionally extensive Larsen Granodiorite Batholith.

Although anatectic melting is associated with Dais Phase in Wright Valley, leucogneiss and augen gneiss intercalated in Koettlitz Group rocks here do not indicate a gradation from a greywacke terrain to Larsen Granodiorite by 'augenisation'. These rocks were intruded before $F_1$ and $F_2$ deformations in the Koettlitz Group; Dais Phase is, at the earliest, syn-$F_2$.

# PETROLOGY AND GEOCHEMISTRY OF THE QUEEN MAUD BATHOLITH, CENTRAL TRANSANTARCTIC MOUNTAINS, WITH IMPLICATIONS FOR THE ROSS OROGENY

S.G. Borg, *Department of Geology, Arizona State University, Tempe, Arizona 85287, U.S.A.*

*Abstract* A suite of 31 Early Palaeozoic intrusives ranging in composition from granite to diorite from the Scott and Leverett Glacier areas represents a nearly continuous, 190 km long cross section of the Queen Maud Batholith in the central Transantarctic Mountains. Field relations indicate that these granitoids are related to the Cambro-Ordovician Ross Orogeny. The granitoids form a typical calc-alkaline batholith and new data show that chemical variation across this batholith is similar to chemical variation across the major circum-Pacific batholiths. In this study, the Queen Maud Batholith has been examined in terms of the S- and I-type granitoid classification scheme developed in southeastern Australia. Petrographic observations as well as new whole-rock geochemical data indicate that the main portion of the batholith in the middle Scott Glacier area is composed of both I- and S-type plutons whereas granitoids north and east of Leverett Glacier area are exclusively I-type. Thus, an S-I boundary appears to be positioned south of the Leverett Glacier, approximately through the Tapley Mountains. The existence of S-type granitoids suggests the presence of metasedimentary material in the source region and therefore the limit of S-type plutons in a composite batholith (S-I boundary) may be inferred to locate the oceanward boundary of a miogeocline. Elsewhere in the range, S-type granitoids have been described from the Beardmore Glacier area. It is suggested, therefore, that the boundary of the Early Palaeozoic Antarctic miogeocline traverses the Scott Glacier near the Tapley Mountains and runs approximately in a northwesterly direction parallel with the axis of the Queen Maud Batholith to the Beardmore Glacier area. Furthermore, it is suggested that this segment of the Early Palaeozoic Antarctic miogeocline boundary may be continuous with the boundaries inferred in northern Victoria Land and southeastern Australia by the distribution of S- and I-type granitoids.

Granitoid plutons, intruded into Late Precambrian and Cambrian metasedimentary and metavolcanic rocks of the Ross Orogen, make up an extensive batholithic complex which is exposed within the pre-Devonian basement throughout the length of the Transantarctic Mountains. Because granitoids are the products of melting in the lower crust, their mineralogical and chemical characteristics may be used to constrain the composition of their source regions. Furthermore, within the context of plate tectonics, different types of granitic batholiths have been related to different tectonic settings. Thus, an understanding of the batholithic complex in the Transantarctic Mountains is vital to any tectonic reconstruction of Gondwana.

The purpose of this study was to examine the granitoids of the central Transantarctic Mountains on a regional scale and to comment, if possible, on the tectonic environment within which the batholith developed. This is the first such study attempted in the central Transantarctic Mountains. The study area is located in the vicinity of the Scott and Leverett Glaciers in the central Transantarctic Mountains (Figure 1). In this paper, an important petrologic boundary is identified within the batholith which may reflect a major change in lower crustal structure.

The discussion is presented in terms of the S- and I-type model of granitoid genesis (Chappell and White, 1974; White and Chappell, 1977). Use of this framework for a discussion of granites in the Transantarctic Mountains is warranted because of the proximity to,

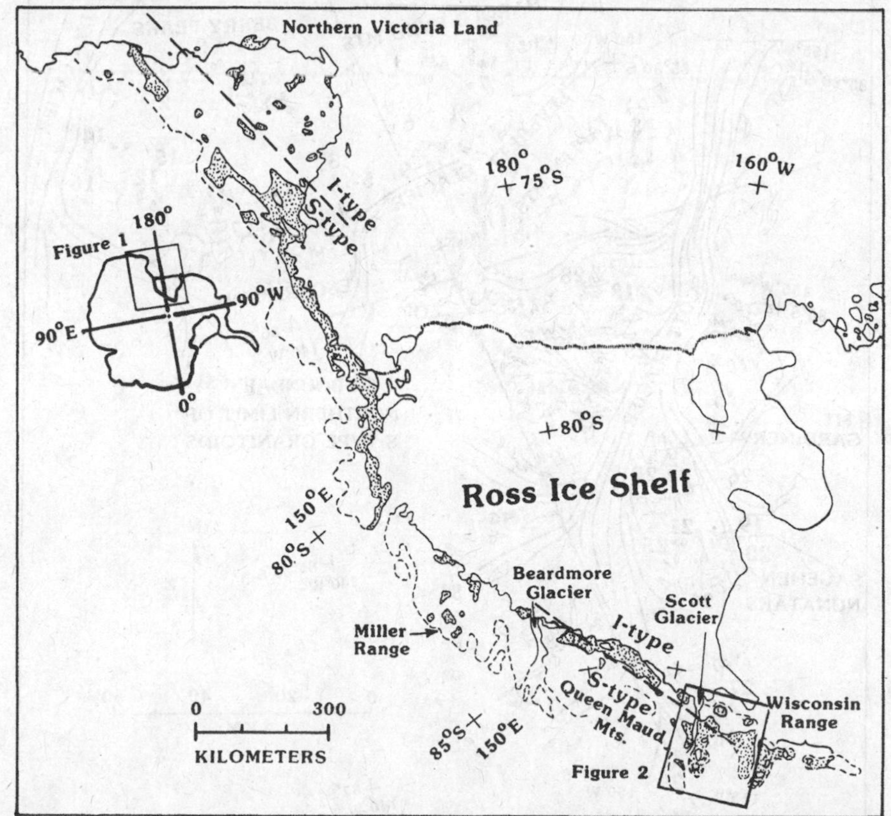


**Figure 1. Location map of the Transantarctic Mountains from northern Victoria Land to the Wisconsin Range showing areas of granitoid outcrop (stippled) and areas of non-granitoid rock outcrop (white encircled area). Note the location of the study area, Figure 2. The S-I boundary separates regions characterised by the presence of peraluminous granitoids (S-type) from regions without peraluminous granitoids (I-type).**

and apparent continuity with (in Gondwana reconstructions) the granitoids of southeastern Australia, where the S-I model was developed. Rock names used here follow Streckheisen (1976).

## GEOLOGIC SETTING

### Previous work

The pre-Devonian granitoids in the Transantarctic Mountains have been referred to as the Granite Harbour Intrusive Complex (Gunn and Warren, 1962). Radiometric dating has shown that the majority of these granitoids are associated with the Cambro-Ordovician Ross Orogeny. However, intrusion and volcanism did occur locally in the Late Precambrian (Faure et al., 1979) during what has been called the Beardmore Orogeny.

In the central Transantarctic Mountains, this intrusive complex crops out nearly continuously from the Beardmore Glacier to the Wisconsin Range (Figure 1). Despite the continuity of exposure, different names have been applied to various portions of the complex. In the Wisconsin Range, Murtaugh (1969) named the intrusive complex the Wisconsin Range Batholith. Earlier, in the Queen Maud Mountains, McGregor (1965) defined the Queen Maud Batholith as comprising post-tectonic intrusives related to the Ross Orogeny although he also described pre-tectonic intrusives related to the Ross Orogeny, the two being distinguished by the absence or presence, respectively, of a tectonic foliation. To eliminate ambiguities in nomenclature, it is suggested here that the term Queen Maud Batholith be extended to cover all intrusives genetically related to the Ross Orogeny, whether foliated or not, which crop out in the central Transantarctic Mountains. As such, granitoids in the Wisconsin Range which are associated with the Ross Orogeny are included in the Queen Maud Batholith.

Work on the Queen Maud Batholith has been mainly reconnaissance mapping, descriptive petrology, and radiometric age determinations (e.g. McGregor, 1965; Murtaugh, 1969; Craddock et al., 1964; Katz and Waterhouse, 1970; Faure et al., 1979; Burgener, 1975), although very little work has been done in the immediate vicinity of the present study. No regional study of mineralogic or chemical characteristics has been previously attempted.

The Queen Maud Batholith is truncated on its northeastern margin by a normal fault which separates the central Transantarctic Mountains from the Ross Sea Depression (Barrett, 1965). Toward the polar plateau the batholith is unconformably overlain by rocks of the Beacon Supergroup of Late Palaeozoic and Mesozoic age, and by the East Antarctic Ice Sheet. The batholith is composed of a calc-alkaline suite of plutons, ranging in composition from granite and quartz monzonite to tonalite, diorite and hornblende gabbro, with quartz monzonite, monzogranite and granodiorite being the most common (McGregor, 1965; Murtaugh, 1969).

Radiometric investigations indicate that the vast majority of granitoids in the central Transantarctic Mountains are associated with the Cambro-Ordovician Ross Orogeny. K-Ar work (Grindley and McDougall, 1969; McDougall and Grindley, 1965; Minshew, 1965; Treves, 1965; Wade et al., 1965) indicates that the granitoids cooled between 450 and 500 Ma while Rb-Sr whole rock isochron and mineral ages fall between 450 and 540 Ma (Craddock et al., 1964; Faure et al., 1979; Gunner, 1974, 1976; Treves, 1965; Gunner and Mattinson, 1975).

Older plutonism, during the so-called Beardmore Orogeny, is suggested from only two localities. One is the Lonely Ridge Granodiorite on the west side of the Nilsen Plateau (to the west of the study area) which yielded a Rb-Sr whole-rock isochron date of 620 ± 13 Ma with an initial $^{87}Sr/^{86}Sr$ of 0.7115 (Faure et al., 1979). The second, in

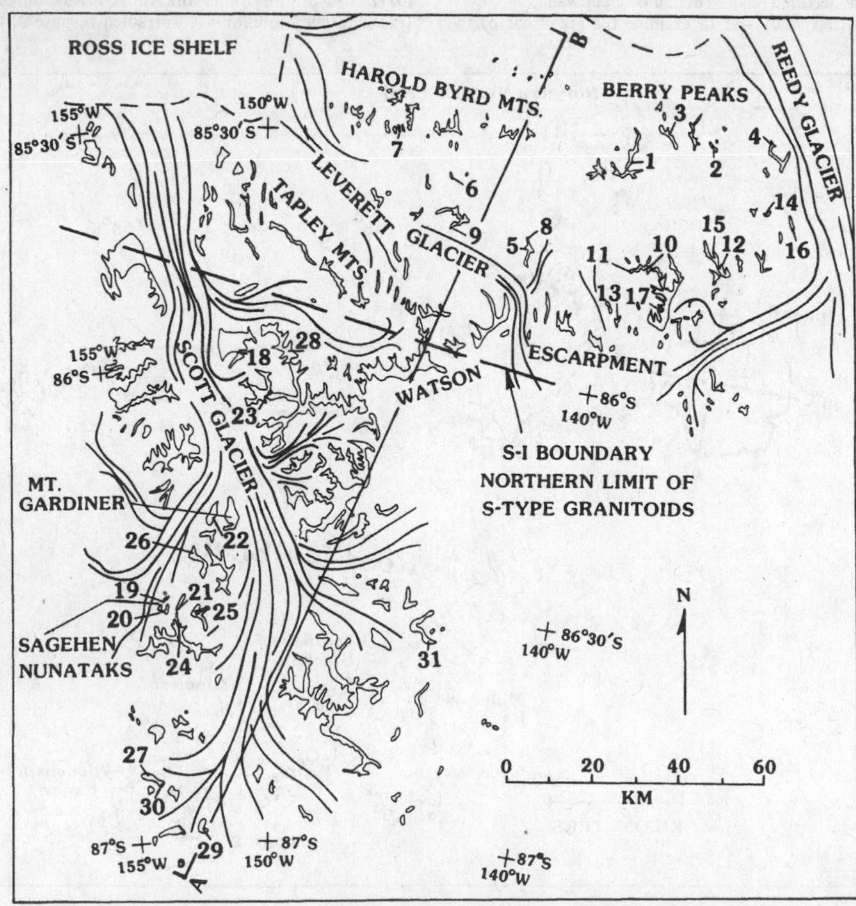


**Figure 2. Sample location map of the study area showing the petrologic boundary (S-I boundary) proposed in this paper. The S-I boundary is defined by the northern limit of the existence of S-type granitoid plutons. Line A-B, approximately perpendicular to the axis of the batholith, defines the traverse of Figure 3.**

the Wisconsin Range, has been called the older granitic suite by Faure et al. (1979). They report a Rb-Sr whole-rock isochron date of 507 ± 23 Ma with initial $^{87}Sr/^{86}Sr$ of 0.7157 and suggest that the data represent isotopic homogenisation of an older granitic terrain during emplacement of the Queen Maud Batholith.

With the present state of radiometric dating in the central Transantarctic Mountains it is clear that extensive plutonism took place during the Cambro-Ordovician Ross Orogeny but it is not clear just what the Beardmore Orogeny really means with respect to granitic magmatism. None of the Beardmore suite granitoids of Faure et al. (1979) has been dated with certainty. Furthermore, it is not clear that the older granitoids, products of the so-called Beardmore Orogeny, are not just an early manifestation of the tectonic regime which was the Ross Orogeny and which produced the bulk of the granitoids in the Transantarctic Mountains.

## Field relations within the study area

The study area, in the vicinity of the Scott Glacier (Figure 2), transects the Queen Maud Batholith at its widest extent perpendicular to its long axis. In the north, the granitoids have intruded Lower Palaeozoic carbonates, volcanics, and clastic sedimentary rocks of the Leverett Formation (Minshew, 1967; Stump et al., 1978). In the south, the granitoids have intruded the Wyatt Formation, a silicic, metavolcanic unit (Minshew, 1967; Borg, 1980) of Late Precambrian age (Faure et al., 1979) as well as the Late Precambrian LaGorce Formation (Minshew, 1967).

Throughout the study area, the plutons display spectacular sharp, discordant contacts. Contact aureoles are narrow and border zones of plutons sometimes contain stoped blocks of country rocks. Contact relations between batholith and country rocks are well exposed at Sagehen Nunataks, Mt Gardiner, and at the eastern end of the Watson Escarpment (Figure 2). The abundance of stoped blocks, narrow contact aureoles, and discordant contacts all suggest a shallow level of emplacement for these granitoids.

McGregor (1965) suggested that the granitoids of the central Transantarctic Mountains were related to the Ross Orogeny and classified them as older pre-tectonic, or younger post-tectonic plutons based on the presence or absence of a cataclastic foliation. Field relations at Sagehen Nunataks (Borg, 1980) show clearly that the foliated granitoids pre-date the unfoliated granitoids, however, the foliation within the deformed plutons is discordant to the regional metamorphic foliation and generally parallel to the pluton margins suggesting that the deformation of the granitoids may be a local feature associated with emplacement of the plutons and not due to regional tectonism (Borg, 1980). Thus, it cannot be argued from a structural point of view that the deformed granitoids are older than the Ross Orogeny. Indeed, field relations suggest that all the granitoids postdate deformation of the rocks of the Ross Orogen. Therefore, despite the lack of radiometric dates all the granitoids discussed here are considered to be genetically related to the Cambro-Ordovician Ross Orogeny.

# PETROLOGY OF THE QUEEN MAUD BATHOLITH

## Sampling and analytical results

The general distribution of granitic rocks in the Transantarctic Mountains is shown on Figure 1. Over 50 samples were collected between the head of the Scott Glacier and the mouth of the Reedy Glacier and examined in hand specimen. Of these, 31 samples which represent the range of texture and composition of plutons throughout the area were selected for whole-rock chemical analyses and examined in thin section. Sample locations are shown on Figure 2. Analytical data are summarised in Table 1. The variation of $K_2O$ in I-type granitoids and $SiO_2$ and $Na_2O/K_2O$ in I- and S-type granitoids along a traverse across the batholith is shown in Figure 3.

## Discussion of data

The Queen Maud Batholith is a typical calc-alkaline batholith (Peacock index ~ 61) composed of plutons of diorite, tonalite, quartz diorite, quartz monzonite, granodiorite, and granite. On the basis of mineralogic and chemical characteristics the granitoids of the Queen Maud Batholith have been divided into two classes, I-type and S-type following Chappell and White (1974) and White et al. (1977). The

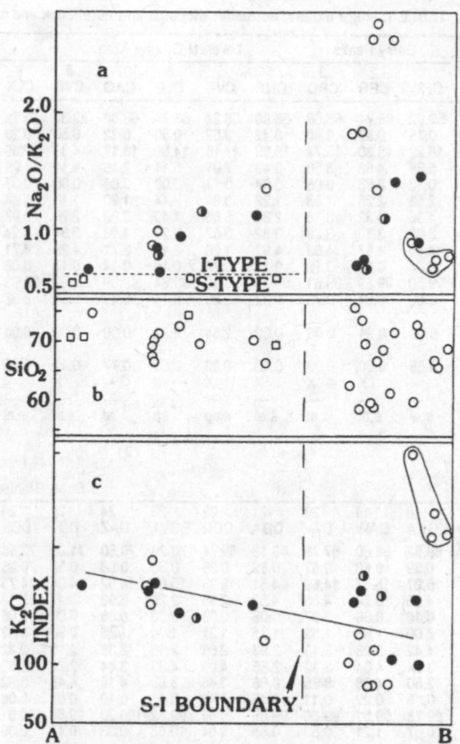


**Figure 3. Variation of selected chemical parameters across the Queen Maud Batholith. Line A-B is shown on Figure 2. Sample locations were projected onto this line, which is approximately perpendicular to the axis of the batholith. (a) Variation in $Na_2O/K_2O$. S-types (squares) are distinguished from I-types (open circles ≤68% $SiO_2$; half filled circles 68%-70% $SiO_2$; filled circles ≥ 70% $SiO_2$) by a $Na_2O/K_2O$ ratio of <0.6 dashed horizontal line). Excluding the Berry Peaks suite (4 encircled I-types), there is a systematic decrease of this ratio from north to south across the batholith and only the high $SiO_2$ I-types approach a $Na_2O/K_2O$ ratio as low as 0.6. The low ratio of the Berry Peaks suite reflects a relatively high $K_2O$ content and suggests an origin different from the other I-types. (b) Variation of $SiO_2$ across the batholith. South of the S-I boundary, only granitoids with $SiO_2$ >66.5% have been found. North of this boundary granitoids range from 58% to 75% $SiO_2$. This shows that the S-I boundary correlates with the southern limit of quartz diorites. Open circles are I-types, squares are S-types. (c) Variation of $K_2O$ Index of Bateman and Dodge (1970) of the I-types across the batholith. $K_2O$ Index = $(1000xK_2O)/(SiO_2 - 45)$. Symbols as in (a). This diagram shows a systematic increase of $K_2O$ from north to south across the batholith in the I-types (excluding the Berry Peaks suite, encircled). A regression line calculated from the $K_2O$ content at 68% $SiO_2$ for each subarea is shown on the graph, emphasising the $K_2O$ trend. This relationship is common in the major circum-Pacific batholiths (Dickinson, 1970; Bateman and Dodge, 1970). The high $K_2O$ Index of the Berry Peaks suite, especially with regard to the lower $SiO_2$ content, indicates a different origin from the other I-types.**

pertinent data are summarised in Table 1 (Samples 1-27 are I-type, 28-31 are S-type).

*I-type Granitoids* I-type granitoids occur as both deformed and undeformed phases indicating that this type of magma was generated throughout the intrusive phase of the Ross Orogeny. The deformed granitoids are medium-grained and weakly porphyritic (pink K-spar phenocrysts). The deformation is seen as a cataclastic foliation composed of fine-grained feldspar, brown biotite, and hornblende (where present) fragments set in a mosaic of polygonally recrystallised quartz. This foliation separates lenses or pods of undeformed granite and individual crystals, generally not more than 2 cm in the long dimension. This style of foliation is uniformly developed within deformed plutons but does not extend beyond the margins as explained above. Sericitic alteration is common in the deformed granitoids but is rarely extensive.

TABLE 1: Major oxides, normative diopside and corundum, and selected minor phases of granitoids of the Leverett and Scott Glacier areas[1]

| | Berry Peaks | | | Leverett Glacier Area | | | | | | Eastern Watson Escarpment | | | | | | |
|---|---|---|---|---|---|---|---|---|---|---|---|---|---|---|---|---|
| | 1 | 2 | 3 | 4 | 5 | 6 | 7 | 8 | 9 | 10 | 11 | 12 | 13 | 14 | 15 | 16 | 17 |
| Sample | CRW | CRR | CRQ | CNF | CVT | CLP | CAO | CVS | CLX | CUS | CVB | CUO | CTQ | CSH | CUL | CSF | CVH |
| $SiO_2$ | 59.92 | 65.44 | 66.68 | 68.30 | 58.27 | 66.74 | 68.50 | 69.32 | 75.72 | 58.86 | 59.48 | 61.12 | 62.58 | 71.16 | 71.45 | 72.60 | 73.70 |
| $TiO_2$ | 0.75 | 0.57 | 0.50 | 0.42 | 0.82 | 0.50 | 0.32 | 0.50 | 0.20 | 0.54 | 0.58 | 0.57 | 0.52 | 0.33 | 0.26 | 0.42 | 0.19 |
| $Al_2O_3$ | 18.26 | 15.30 | 15.74 | 15.69 | 17.16 | 14.59 | 16.17 | 14.38 | 12.05 | 17.43 | 16.08 | 16.86 | 16.54 | 14.15 | 14.44 | 13.21 | 13.52 |
| $Fe_2O_3$ | 5.38 | 4.66 | 3.58 | 2.43 | 7.91 | 5.11 | 2.46 | 4.14 | 1.99 | 7.48 | 7.74 | 6.54 | 5.91 | 3.17 | 2.84 | 2.66 | 1.64 |
| MnO | 0.12 | 0.08 | 0.06 | 0.04 | 0.13 | 0.08 | 0.06 | 0.06 | 0.03 | 0.15 | 0.14 | 0.13 | 0.11 | 0.08 | 0.06 | 0.05 | 0.05 |
| MgO | 2.53 | 2.25 | 1.88 | 1.22 | 3.94 | 2.44 | 1.20 | 1.40 | 0.64 | 3.92 | 4.34 | 3.16 | 3.24 | 1.25 | 1.10 | 0.68 | 0.73 |
| CaO | 4.36 | 3.32 | 3.18 | 2.33 | 6.66 | 4.49 | 2.80 | 2.78 | 0.97 | 7.14 | 7.08 | 6.36 | 5.72 | 2.96 | 2.70 | 1.59 | 1.46 |
| $Na_2O$ | 3.68 | 3.10 | 3.28 | 3.82 | 3.07 | 3.27 | 4.03 | 2.84 | 3.24 | 3.14 | 3.12 | 3.56 | 3.26 | 3.80 | 3.88 | 3.81 | 3.44 |
| $K_2O$ | 4.16 | 4.67 | 4.53 | 4.92 | 1.68 | 2.46 | 3.71 | 4.28 | 4.71 | 1.14 | 1.19 | 1.36 | 1.80 | 2.62 | 2.77 | 4.28 | 4.68 |
| $P_2O_5$ | 0.34 | 0.20 | 0.18 | 0.12 | 0.26 | 0.16 | 0.14 | 0.17 | 0.08 | 0.24 | 0.25 | 0.23 | 0.23 | 0.13 | 0.12 | 0.10 | 0.10 |
| Total | 99.50 | 99.59 | 99.61 | 99.29 | 99.90 | 99.84 | 99.39 | 99.87 | 99.63 | 100.04 | 100.00 | 99.89 | 99.91 | 99.65 | 99.62 | 99.40 | 99.51 |
| LOI | 0.66 | 0.64 | 0.73 | 0.32 | 0.86 | 0.71 | 0.43 | 0.36 | 0.38 | 0.95 | 0.57 | 0.70 | 0.69 | 0.57 | 0.36 | 0.43 | 0.55 |
| Norm DI | 0.00 | 0.91 | 0.00 | 0.00 | 2.64 | 2.72 | 0.00 | 0.00 | 0.00 | 3.04 | 5.76 | 3.28 | 1.37 | 0.02 | 0.00 | 0.77 | 0.00 |
| Norm CO | 0.58 | 0.00 | 0.09 | 0.13 | 0.00 | 0.00 | 0.77 | 0.42 | 0.05 | 0.00 | 0.00 | 0.00 | 0.00 | 0.00 | 0.43 | 0.00 | 0.38 |
| HBLD | — | X | X | X | X | X | — | X | — | X | X | X | X— | — | — | — | — |
| CORD | — | — | — | — | — | — | — | — | — | — | — | — | — | — | — | — | — |
| ACC | s,al | s,al | s,al | s,al | s,ap | ap | al | s,al | s,al | — | — | s,ap | ap | s,ap | — | s | — |

| | Scott Glacier Area | | | | | | | | | | | | | |
|---|---|---|---|---|---|---|---|---|---|---|---|---|---|---|
| | 18 | 19 | 20 | 21 | 22 | 23 | 24 | 25 | 26 | 27 | 28 | 29 | 30 | 31 |
| Sample | EGKA | DAW | DAV | DBA | DCN | EGVA | DAZ | DBT | DCK | DAL | EGSA | DCR | DAO | EHC |
| $SiO_2$ | 66.52 | 66.60 | 67.78 | 69.19 | 69.74 | 70.20 | 70.50 | 71.28 | 72.48 | 75.08 | 69.14 | 70.57 | 70.72 | 74.60 |
| $TiO_2$ | 0.39 | 0.60 | 0.67 | 0.69 | 0.38 | 0.30 | 0.58 | 0.50 | 0.26 | 0.10 | 0.56 | 0.49 | 0.66 | 0.14 |
| $Al_2O_3$ | 15.97 | 15.17 | 14.63 | 14.51 | 15.28 | 15.04 | 14.02 | 14.04 | 14.75 | 13.14 | 14.58 | 14.36 | 14.18 | 13.11 |
| $Fe_2O_3$ | 4.06 | 3.70 | 4.88 | 4.52 | 2.66 | 2.26 | 3.90 | 3.47 | 2.12 | 1.49 | 4.07 | 3.85 | 4.67 | 1.90 |
| MnO | 0.10 | 0.06 | 0.07 | 0.06 | 0.05 | 0.06 | 0.06 | 0.06 | 0.06 | 0.04 | 0.05 | 0.06 | 0.06 | 0.05 |
| MgO | 2.00 | 1.86 | 1.52 | 1.35 | 1.31 | 1.00 | 1.08 | 0.98 | 0.62 | 0.16 | 1.52 | 1.36 | 1.81 | 0.50 |
| CaO | 4.42 | 3.28 | 3.13 | 2.93 | 2.66 | 2.35 | 2.36 | 2.10 | 2.12 | 1.02 | 1.97 | 2.27 | 2.26 | 0.76 |
| $Na_2O$ | 3.54 | 4.04 | 3.12 | 2.85 | 4.00 | 4.20 | 3.44 | 2.72 | 3.62 | 3.14 | 2.80 | 2.34 | 2.29 | 3.07 |
| $K_2O$ | 2.60 | 4.08 | 3.75 | 3.66 | 3.46 | 3.80 | 4.16 | 4.45 | 3.92 | 5.00 | 4.83 | 4.52 | 4.05 | 5.38 |
| $P_2O_5$ | 0.18 | 0.22 | 0.11 | 0.13 | 0.12 | 0.11 | 0.10 | 0.08 | 0.06 | 0.00 | 0.20 | 0.09 | 0.12 | 0.16 |
| Total | 99.78 | 99.61 | 99.66 | 99.89 | 99.66 | 99.32 | 100.20 | 99.68 | 100.01 | 99.17 | 99.72 | 99.91 | 100.82 | 99.67 |
| LOI | 0.67 | 1.21 | 0.63 | 0.55 | 0.64 | 0.63 | 0.33 | 0.41 | 0.39 | 0.50 | 0.76 | 0.67 | 0.83 | 0.71 |
| Norm DI | 0.61 | 2.92 | 0.00 | 0.00 | 0.00 | 0.00 | 0.43 | 0.00 | 0.00 | 0.00 | 0.00 | 0.00 | 0.00 | 0.00 |
| Norm CO | 0.00 | 0.00 | 0.01 | 0.84 | 0.40 | 0.01 | 0.00 | 1.12 | 0.84 | 0.71 | 0.76 | 0.67 | 0.83 | 0.71 |
| HBLD | X | X | X | — | X | X | X | — | — | — | — | — | — | — |
| CORD | — | — | — | — | — | — | — | — | — | — | — | — | — | — |
| ACC | s,al | s,al | s,al z | s,al ap,z | s,al ap,z | s,al | s,z | z | s,al z | — | X (alt) ap | — ap,z | — ap,z | X (alt) — |

Explanation

- 1-4 I-type, Berry Peaks
- 5-9 I-type, Leverett Glacier area and Harold Byrd Mountains
- 10-17 I-type, Eastern Watson Escarpment area
- 18-27 I-type, Scott Glacier area
- 28-31 S-type, Scott Glacier area
- X denotes presence of phase
- (alt) altered
- HBLD = hornblende
- CORD = cordierite
- DI = diopside
- CO = corundum
- ACC = accessory phases
- s = sphene,
- ap = apatite
- al = allanite,
- z = zircon

[1]Analyses are in weight percent on an ignited basis and represent the average of duplicate analyses. X-ray fluorescence (XRF) analyses for all elements (except as noted below) were performed by the author on a Philips Model PW 1410 XRF using sample preparation procedures of Norrish and Chappell (1967). Atomic absorption (AA) analyses for Na and Mg in samples 19-22, 24-27, 29, and 30 were performed by the author on a Varian Tectron Model 1250 AA using sample preparation procedures of Medlin et al (1969). Rock standards used are from Jenkins (1978). Total Fe reported as $Fe_2O_3$. Loss on ignition (LOI) includes loss of $H_2O^+$, $CO_2$, and S, and gain of $O_2$ in oxidising $Fe^{++}$ during ignition of the sample at 1000°C for four hours under oxidising conditions. For the purpose of norm calculations, $Fe^{+++}$ was partitioned to $Fe^{++}$ by the equation: Fe as $FeO = 0.6\ Fe_2O_3$ (total iron).

The undeformed rocks are medium to coarse grained, equigranular and porphyritic (pink K-spar phenocrysts), and have a hypidiomorphic granular texture. Subhedral to euhedral plagioclase, pink K-spar, brown biotite, and hornblende (where present) are set in an anhedral mosaic of quartz and feldspar. Sphene and allanite are common primary accessory phases. Some samples show a parallel alignment of platy and elongate minerals due to magma flowage.

The $SiO_2$ content of these granitoids ranges from 58% to 76% with colour index ranging from 35 to 5. All but the high silica varieties contain significant hornblende and/or sphene. All are metaluminous or have normative corundum less than ~1.1% and have a relatively high $Na_2O/K_2O$ weight ratio (>0.6, Figure 3a). Within each subarea of the study area (see Table 1 and Figure 2) the granitoids form straight lines on Harker diagrams for most major oxides. However, from one subarea to another, position and slope of the line may vary. Potassium shows a linear increase with silica for each area except that in the Leverett and E. Watson Escarpment groups there is a marked increase in $K_2O$ in the high $SiO_2$ granitoids. Because the samples do not represent phases of one pluton but rather are samples of scattered plutons the straight line variation of major oxides with $SiO_2$ for each subarea cannot be interpreted as differentiation paths of a granitic magma, however, they may indicate that the materials for, and processes of magma generation, contamination and differentiation are similar for plutons in each subarea.

With the exception of the Berry Peaks, there is a general increase of $K_2O$ from northeast to southwest across the axis of the batholith toward the craton (Figure 3c). This type of increase has been reported from the Sierra Nevada batholith and has been cited as evidence that the batholith was related to a subduction zone (Bateman and Dodge, 1970; Dickinson, 1970).

The high potassium (and high total alkali) content of the Berry Peaks suite indicate either a source region different from the other I-type granitoids (possibly deeper, or perhaps a source with both igneous and some mica-rich sedimentary rocks) or drastically different magma evolution processes. At present, the only statement to be made with certainty is that these rocks do not appear to fall into the pattern of regional chemical variation which is beginning to emerge from data from other granitoids in the area.

*S-type Granitoids* Four S-type granitoids have been identified. The plutons represent significant volumes of magma and are exposed along some 30 km of ridge and cliff face outcrop. All are undeformed, medium grained rocks exhibiting both equigranular and porphyritic textures. Quartz and plagioclase usually appear as early phases. Potassium feldspar is a late phase which generally occurs as an interstitial poikilitic intergrowth with quartz and rarely as phenocrysts. Biotite is commonly the only essential ferromagnesian phase and varies from bright red to dark brownish-red in colour. Samples 28 and 31 contain cordierite grains which have been almost completely altered to clays. Also present in these rocks are lumps of biotite and quartz similar to features which Clemens and Wall (1981) describe as reaction products after orthopyroxene. Large apatite grains are a common accessory phase. Some muscovite is present but all is arguably secondary based on textural evidence (i.e. developing along twin planes in plagioclase, or developing as a feathery intergrowth with sericitised feldspar).

These granitoids have a relatively low $Na_2O/K_2O$ ratio (wt.% <0.60, Figure 3a) and they are peraluminous with normative corundum in excess of 1.2% (Table 1). Granitoids 29 and 30 contain abundant xenoliths of exclusively metasedimentary material (biotite schist and gneiss).

## IMPLICATIONS FOR THE ROSS OROGENY

Although this study can only be viewed as a preliminary step in the characterisation of the Queen Maud Batholith, the geographical distribution of S- and I-type granitoids in the study area defines a significant petrologic boundary (Figures 2 and 3). S-type plutons have only been found south of the Leverett Glacier, within the core and to the south (the craton side) of the batholith. I-type granitoids occur throughout the area, however, north of the Leverett Glacier only I-type granitoids have been identified. Furthermore, it is significant that no granitoids have been found south of this boundary which contain less than 66.5% $SiO_2$ (Figure 3b). In other words, no quartz diorites, diorites, tonalites, or quartz monzodiorites have been found south of the Leverett Glacier.

This petrologic boundary may reflect a major change in lower crustal structure. Specifically, because generation of S-type granitoids requires the presence of metasedimentary material in the region of partial melting, this boundary is inferred to locate the oceanward margin of a miogeocline at the edge of the Antarctic craton in this region during the Ross Orogeny. Furthermore, the origin of the Queen Maud Batholith may be similar to the origin of the major circum-Pacific batholiths based on the similarity of chemical trends (i.e. $K_2O$) across the batholiths. This analogy infers that a southwestward dipping subduction zone may have existed during the Ross Orogeny and would have been located in the vicinity of the Ross Ice shelf, roughly parallel to the present batholith.

Combining the results of this study with work on granitoids from elsewhere in the Transantarctic Mountains permits a speculation on the extension of the S-I boundary. Initial $^{87}Sr/^{86}Sr$ of 0.7050 from rocks in the Wisconsin Range (Faure et al., 1979) suggests that I-type granites are present there (I-types, $^{87}Sr/^{86}Sr = 0.704$—$0.706$; S-type, $^{87}Sr/^{86}Sr > 0.708$; Chappell and White, 1974; McCulloch and Chappell, 1982). To the northwest of the study area Gunner (1974, 1976) found initial $^{87}Sr/^{86}Sr$ of 0.734 and 0.710 on a pluton from the Miller Range and plutons from the mouth of the Beardmore Glacier respectively. He interpreted this data to indicate extensive contribution of, if not derivation from old metasedimentary material. In the Miller Range, the pluton has intruded pre-Ross Orogen cratonic rocks of the Nimrod Group and may have been derived from similar materials (Gunner, 1976).

The distribution of S-type granitoids inboard on the continent and I-types outboard (Figure 2) in the central Transantarctic Mountains is similar to that in northern Victoria Land, Antarctica (Figure 1; Wyborn, 1981; Stump et al., in press) and southeastern Australia with respect to the margin of Gondwana during the Early Palaeozoic. This suggests that this petrologic boundary was continuous along a 2500 km segment of the Antarctic craton.

*Acknowledgements* Thanks to P. Colbert, M. Sheridan and E. Stump for field collaboration in the upper Scott Glacier area, to E. Stump, P. Lowry, G. Heintz-Stocker, J. Smit, S. Self and P. Colbert for providing samples of granitoids from the middle Scott Glacier and from the area north of the Leverett Glacier and to J. Holloway for valuable discussions relating to this project. I sincerely appreciate the use of the XRF facility of the Center for Solid State Science, Arizona State University, where the analytical work was performed. Robert Gregory, John Holloway and Ed Stump critically reviewed the manuscript. This research was supported by National Science Foundation, Division of Polar Programs, Grant DPP76-82040 and DPP78-20624.

## REFERENCES


BARRETT, P.J., 1965: Geology of the area between the Axel Heiberg and Shackleton Glaciers, Queen Maud Range, Antarctica, Part 2—Beacon Group. *N.Z. J. Geol. Geophys., 8,* 344-70.

BATEMAN, P.C. and DODGE, F.C.W., 1970: Variations of major chemical constituents across the central Sierra Nevada batholith. *Geol. Soc. Am., Bull., 81,* 409-20.

BORG, S.G., 1980: Petrology and geochemistry of the Wyatt Formation and the Queen Maud Batholith, upper Scott Glacier area, Antarctica. Unpublished M.S. thesis, Arizona State University, Tempe, Arizona, U.S.A. 100pp.

BURGENER, J.D., 1975: Petrography of the Queen Maud Batholith, central Transantarctic Mountains, Ross Dependency, Antarctica. Unpublished M.S. thesis, University of Wisconsin, Madison, Wisconsin, U.S.A., 75pp.

CHAPPELL, B.W. and WHITE, A.J.R., 1974: Two contrasting granite types. *Pac. Geol., 8,* 173-4.

CLEMENS, J.D. and WALL, V.J., 1981: Origin and crystallisation of some peraluminous (S-type) granitic magmas. *Can. Mineral., 19,* 111-31.

CRADDOCK, C., GAST, P.W., HANSEN, G.N. and LINDER, H., 1964: Rubidium-strontium ages from Antarctica. *Geol. Soc. Am., Bull., 75,* 237-40.

DICKINSON, W.R., 1970: Relations of andesites, granites, and derivative sandstones to arc-trench tectonics. *Rev. Geophys. Space Phys., 8,* 813-60.

FAURE, G., EASTIN, R., RAY, P.T., McLELLAND, D. and SHULTZ, C., 1979: Geochronology of igneous and metamorphic rocks, central Transantarctic Mountains; *in* Laskar, B. and Raja Rao, C.S. (eds.) *Fourth International Gondwana Symposium, 2,* Hindustan Publishing Corp., Delhi, 805-13.

GRINDLEY, G.W. and McDOUGALL, I., 1969: Age and correlations of the Nimrod Group and other Precambrian rock units in the central Transantarctic Mountains, Antarctica. *N.Z. J. Geol. Geophys., 12,* 391-411.

GUNN, G.M. and WARREN, G., 1962: Geology of Victoria Land between the Mawson and Mullock Glaciers, Antarctica. *N.Z. Geol. Surv., Bull., 71.*

GUNNER, J.D., 1974: Investigations of Lower Palaeozoic granites in the Beardmore Glacier Region. *U.S. Antarct. J., 9,* 76-81.

GUNNER, J., 1976: Isotopic and geochemical studies of the pre-Devonian basement complex, Beardmore Glacier region, Antarctica. *Rep. Inst. polar Stud. Ohio State Univ., 41.*

GUNNER, J. and MATTINSON, J.M., 1975: Rb-Sr and U-Pb isotopic ages of granites in the central Transantarctic Mountains. *Geol. Mag., 112,* 25-31.

JENKINS, L., 1978: Calibration Standards. *X-ray Spectrometry, 7(2),* 99-120.

KATZ, H.R. and WATERHOUSE, B.C., 1970: Geological reconnaissance of the Scott Glacier area, Southern Queen Maud Range, Antarctica. *N.Z. J. Geol. Geophys., 13,* 1030-7.

McCULLOCH, M.T. and CHAPPELL, B.W., 1982: Nd isotopic characteristics of S- and I-type granites. *Earth Planet. Sci. Lett., 58,* 51-64.

McDOUGALL, I. and GRINDLEY, G.W., 1965: Potassium-argon dates on micas from the Nimrod-Beardmore-Axel Heiberg region, Ross Dependency, Antarctica. *N.Z. J. Geol. Geophys., 8,* 304-13.

McGREGOR, V.R., 1965: Geology of the area between the Axel Heiberg and Shackleton Glaciers, Queen Maud Range, Antarctica. *N.Z. J. Geol. Geophys., 8,* 314-43.

MEDLIN, J.H., SUHR, N.H. and BODKIN, J.B., 1969: Atomic absorption analysis of silicates employing $LiBO_2$ fusion. *Atomic Absorption Newsl., 8,* 25-9.

MINSHEW, V.H., 1965: Potassium-argon age from a granite at Mount Wilbur, Queen Maud Mountains, Antarctica. *Science, 150,* 741-3.

MINSHEW, V.H., 1967: Geology of the Scott Glacier and Wisconsin Range areas, central Transantarctic Mountains, Antarctica. Unpublished Ph.D. thesis, The Ohio State University, Columbus, Ohio, U.S.A., 268pp.

MURTAUGH, J.G., 1969: Geology of the Wisconsin Range Batholith, Transantarctic Mountains. *N.Z. J. Geol. Geophys., 12,* 526-50.

NORRISH, K. and CHAPPELL, B.W., 1967: X-ray fluorescence spectrography; *in* Zussman, J. (ed.) *Physical Methods in Determinative Mineralogy.* Academic Press, London, 161-214.

STREICKEISEN, A., 1976: To each plutonic rock its proper name. *Earth Sci. Rev., 12,* 1-33.

STUMP, E., LOWRY, P.H., HEINTZ-STOCKER, G.M. and COLBERT, P.V., 1978: Geological investigations in the Leverett Glacier area. *U.S. Antarct. J., 13(4),* 3-4.

STUMP, E., HOLLOWAY, J.R., BORG, S.G. and LAPHAM, K.E. (in press): Investigations on Lower to Middle Palaeozoic magmatic rocks of northern Victoria Land, Antarctica. *U.S. Antarct. J.*

TREVES, S.B., 1965: Igneous and metamorphic rocks of the Ohio Range, Horlick Mountains, Antarctica; *in* Hadley, J.B. (ed.) Geology and Palaeontology of the Antarctic. *Am. Geophys. Union, Ant. Res. Ser., 6,*117-25.

WADE, F.A., YEATS, V.L., EVERETT, J.R., GREENLEE, D.W., LAPRADE, K.E. and SHENKE, J.C., 1965: The geology of the central Queen Maud Range, Transantarctic Mountains, Antarctica. *Texas Tech. Univ., Antarct. Res. Rep. Ser., 65-1.*

WHITE, A.J.R. and CHAPPELL, B.W., 1977: Ultrametamorphism and granitoid genesis. *Tectonophysics, 43,* 7-22.

WHITE, A.J.R., WILLIAMS, I.S. and CHAPPELL, B.W., 1977: *Geology of the Berridale 1:100,000 Sheet.* Geol. Surv. N.S.W.

WYBORN, D., 1981: Granitoids of North Victoria Land, Antarctica-Field and Petrographic Observations. *Geol. Jb., B41,* 229-49.

# TYPE LOCALITY OF THE ACKERMAN FORMATION, LA GORCE MOUNTAINS, ANTARCTICA

E.Stump, *Department of Geology, Arizona State University, Tempe, Arizona 85287, U.S.A.*

*Abstract* The Ackerman Formation is herein named for the sequence of alternating clastic and volcanic rocks of Late Precambrian(?) age exposed in the La Gorce Mountains, 86°45′S, 147°W, at the head of Scott Glacier. The type locality and only known occurrence of the Ackerman Formation is on Ackerman Ridge, where a steeply dipping 2000 m section crops out. The formation conformably overlies massive porphyries of the Upper Precambrian Wyatt Formation at the northern end of Ackerman Ridge. The upper portion of the formation contacts the La Gorce Formation along a major high angle fault which has produced a shear zone several hundred metres wide in both formations. The Ackerman Formation is characterised by immature sandstones, grey, purple and green shales or phyllites, and massive, silicic porphyries. Some of the clastic rocks contain a minor fraction of calcareous material. In general, the shales and sandstones are thinly bedded and alternating. Two distinctive sandstone units occur in the formation. Each is massive, 2-3 metres thick, and contains sparse rounded cobbles. In the volcanic porphyries phenocrysts of rounded quartz and highly altered plagioclase and biotite occur in varying amounts. In some specimens chlorite completely replaces mafic phenocrysts, which were originally probably orthopyroxene. The groundmass is recrystallised to such an extent that any original textures which might have been present are obliterated. Evidence suggests emplacement of the volcanic units as both ignimbrites and lavas in a marginal marine environment. The entire Ackerman Formation has been slightly metamorphosed. Incipient cleavage is developed in many of the units, increasing in intensity near the high angle fault at the top of the formation.

The first encounter with the Scott Glacier area was by the forward ground party of Byrd's second expedition. Blackburn (1937) saw that most of the Queen Maud Mountains in this area are composed of granitoid intrusions, and noted "black, porphyritic material" on the west side of the headreaches of the glacier. However, it was not until the mid 1960's that the geological framework was established by Minshew (1967). He named the La Gorce Formation for a turbidite sequence of metagreywacke and argillite in the La Gorce Mountains, recently interpreted as the proximal facies of a submarine fan deposit (Smit, 1981).

The "black, porphyritic material" on the west side of Scott Glacier was called the Wyatt Formation by Minshew (1967). It is a dark-grey to black rhyodacite with quartz, plagioclase, biotite and hypersthene phenocrysts. Different portions have either volcanic or intrusive affinities (Stump, 1976). The Wyatt Formation is similar in chemistry and mineralogy to a hypersthene quartz monzonite in the Thiel Mountains 450 km to the east (Ford and Aaron, 1962; Ford, 1964; Ford and Himmelberg, 1976; Borg, 1980)

Although Minshew never saw the two formations in contact he suggested that the Wyatt Formation is younger than the La Gorce on the basis of its topographically higher position in an area where the formations crop out close to each other, but their contact is covered by debris. The La Gorce Formation was correlated with other Upper Precambrian turbidite sequences found in the central Transantarctic Mountains, including the Duncan and Goldie Formations, collectively called the Beardmore Group. The Wyatt Formation was dated at 633 ± 13 Ma (Faure et al., 1968), though more recent dating has produced Ross Orogeny (Cambro-Ordovician) dates or discordant isochrons (Faure et al, 1979). However, it is generally thought that the Wyatt Formation is Precambrian in age. In comparison, dates of 620-670 Ma have been obtained on the quartz monzonite of the Thiel Mountains (Aaron and Ford, 1964; Ford et al., 1963; Faure et al., 1979).

In the course of a traverse down Scott Glacier Katz and Waterhouse (1970) discovered Wyatt Formation at the north end of Ackerman Ridge in the La Gorce Mountains, and an apparently conformable contact with La Gorce Formation. Since sedimentary and volcanic units are interbedded at the boundary, Katz and Waterhouse (1970) described the formations as interfingering. They also suggested that the La Gorce Formation is younger than the Wyatt, though structural complexity prevented a clear answer. In 1971, the present author visited Ackerman Ridge for one day, affirming the conformable contact (Stump, 1976). The Wyatt Formation was interpreted as being stratigraphically above the La Gorce Formation, based on a single set of cross beds found in the zone of interbedded clastic and volcanic units.

## GEOLOGICAL RELATIONSHIPS IN THE LA GORCE MOUNTAINS

During the 1980-1981 field season much of the La Gorce Mountains was mapped in detail (Figure 1) (Stump et al. 1981). The complete results of this work will be presented later, but a brief discussion of the geological relationships is necessary here.

Within the sequence of interbedded volcanic and sedimentary units, tops are undoubtedly toward the south, based on rip-up clasts in the bottom of volcanic flows and on sparse cross-bedding. This is in the opposite direction from a previous interpretation of tops, which was in error (Stump, 1976). The contact on Ackerman Ridge between the uppermost volcanic unit and La Gorce Formation on Ackerman Ridge is a fault which is subparallel to bedding in the La Gorce Formation. Therefore the relative ages of the two sequences cannot be determined at this contact. However, the western ridge of the La Gorce Mountains is composed of a hypabasal phase of the Wyatt Formation which intrudes La Gorce Formation as indicated in Figure 1, thus affirming Minshew's (1967) original suggestion that Wyatt Formation is younger than La Gorce Formation.

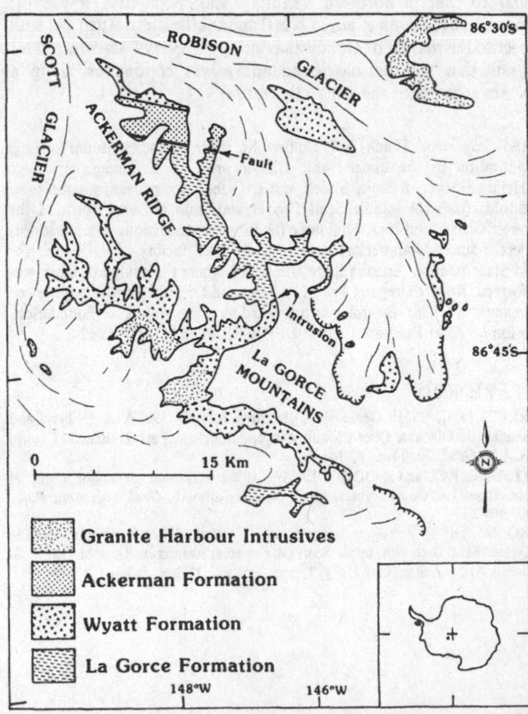


**Figure 1. Geological map of La Gorce Mountains, Antarctica. La Gorce Formation is in fault contact with Ackerman Formation. Ackerman Formation stratigraphically overlies extrusive phase of Wyatt Formation. Hypabyssal phase of Wyatt Formation intrudes La Gorce Formation. Cambro-Ordovician Granite Harbour Intrusives intrude Wyatt Formation.**

## ACKERMAN FORMATION

The sedimentary units interbedded with volcanics on Ackerman Ridge have no affinity with the La Gorce Formation, and indeed are quite distinct. The fact that Wyatt Formation is uniformly massive, aphanitic porphyry wherever it is found, from Nilsen Plateau to Reedy Glacier, warrants distinction of the sequence of interbedded volcanic and sedimentary rocks in the La Gorce Mountains as a separate formation. The name Ackerman Formation is here proposed for these rocks with Ackerman Ridge as the type locality. The type section is presented in Figures 2a and 2b. The section was paced along the main ridge crest using an 85 cm ice axe as a Jacob's staff. Due to a covering of scree and drift from a previous, higher glacial stand, bedrock crops out along only the top 20-50 m of Ackerman Ridge preventing observation of lateral changes in bedding thickness or facies.

Structurally, the Ackerman Formation occurs as a straight section dipping 50°-60°S and striking at 90°-120°. It is truncated by a major fault, so that the top of the formation is unknown. Cleavage which is developed at some levels in the lower portion of the formation, particularly in the argillaceous beds, increases in extent towards the fault. In the upper portion of the formation the shales have been converted to phyllites. Late stage kink folds and small, chevron folds are developed on the cleavage planes of both the volcanic and sedimentary rocks in the vicinity of the fault. Because of the movement and fine-scale recrystallisation suffered by these rocks, bedding is often obscured, particularly in the finer grained sediments.

### Volcanic Units

The volcanic rocks of the Ackerman Formation are massive, dark grey to green porphyries with from 25% to more than 50% phenocrysts. Crystal sizes are up to 5 mm, but generally less than 3 mm. All units contain phenocrysts of rounded or embayed quartz and euhedral plagioclase which is usually twinned and altered to sericite. Units in the lower half of the section contain fairly abundant chlorite pseudomorphs after what appears to have been orthopyroxene phenocrysts, altered biotite is not uncommon, and K-feldspar occurs in one unit. A portion of the phenocrysts (5-50%) in each of the specimens examined in thin section is fragments of crystals. The groundmass of the porphyries is finely recrystallised quartz, feldspar, chlorite and sericite, with nothing of the original texture remaining. On a few exceptional exposures a faint, uneven layering is observable from greater and lesser concentrations of phenocrysts.

The thickness and massiveness of the volcanic rocks, and their fragmented phenocrysts, suggest that most of these rocks were erupted pyroclastically. However, the bottom of the crystal-rich porphyry 198 m from the base of the formation contains pieces of the underlying siltstone that appear to have been caught up in an overriding lava flow. The volcanic units of the Ackerman Formation are similar in appearance and petrology to rocks of the underlying Wyatt Formation.

### Sedimentary Units

The sedimentary rocks of the Ackerman Formation are mainly grey to green, interbedded sandstones and shales. The shales and the matrix of the sandstones have been extensively sericitised; chlorite occurs in some samples. Although these rocks have been recrystallised, the field or hand specimen appearance is generally not metamorphic, and so sandstone and shale are used as descriptive terms throughout most of the report. Specific sandstone names follow the classification of Pettijohn et al.(1973).

Bedding ranges from a few centimetres to about one metre in thickness, and appears to be better developed in the lower portion of the formation. In the upper portion, the units appear rather massive, perhaps due to the overprint of cleavage. Sedimentary structures were either not well developed, or have since been obscured by deformation. A minor fraction (<10%) of the shales display fine lamination, some with ripple-drift laminations. Soft sediment deformation and rip-up clasts are preserved in places. Faint, tabular cross bedding (<5 cm thickness) was observed at a few localities in the sandstones.

A variety of sandstone types occur with wackes more prevalent than arenites. Clasts in samples range from very fine to coarse-grained, subangular to subrounded, and usually are moderately sorted. Quartz is the predominant clast, but volcanic rock fragments and/or plagioclase make up an appreciable fraction in many samples. The matrix of the sandstones is a fine intergrowth of quartz, feldspar, and sericite, sometimes including chlorite. Calcite is present in some units, but only in minor amounts. Some of the sandstones contain a sparse conglomeratic fraction of well rounded pebbles and cobbles of volcanic porphyry and vein quartz, with minor sandstone and shale.

For mapping purposes, two three metre thick units of sandstone, one a quartz arenite, the other a quartzwacke, are distinctive because of their light colouration (located 605 m and 920 m from base of the formation). They are massive, with very faint hints of cross-bedding, and sparse, well rounded cobbles of the rock types mentioned above. Both beds rest on thin shales above volcanic units.

The sedimentary rocks of the Ackerman Formation are distinct from the La Gorce Formation by virtue of differences in colour (green, medium grey, and tan vs. dark grey to black), petrography (varied lithologies vs. uniform greywacke and shale), and sedimentary characteristics (scattered cross-bedding, rip-up clasts, etc, vs. abundant, complete and incomplete Bouma sequences) (Smit, 1981).

Because of the lack of diagnostic criteria, interpretations of the environment of deposition of the Ackerman Formation must be tentative and limited. The narrow width of outcrop along Ackerman Ridge prevents observation of lateral changes in bedding thickness and facies, and the degree of deformation, particularly overprinting by cleavage, has obscured primary sedimentary characteristics.

Throughout the Ackerman Formation there is a general coarsening upwards of the clastic units; however, no specific cycles of sedimentation were recognised. Based on the observations that Wyatt magmatism followed deformation of the La Gorce Formation, and that elsewhere in the range this Late Precambrian deformation resulted in a marked erosion surface (Laird et al., 1971; Schmidt et al., 1965), it is likely that the majority of the extrusive rocks of the Wyatt Formation were erupted subaerially. The calcareous fraction in sedimentary portions of the Ackerman Formation indicates marine deposition. Following this line of reasoning, the Ackerman Formation would appear to have been deposited after a marine transgression subsequent to widespread silicic volcanism.

The detrital components of the Ackerman Formation best fit the model of a clastic shoreline (Elliot, 1978; Reineck and Singh, 1980). The predominance of shales and siltstones in the lower portion of the formation may represent lagoonal facies with the higher, more sandy units representing portions of a barrier island complex and associated near-shore deposits. The conglomeratic fraction in the units could have been derived from actively eroding sea cliffs. No simple cycle of sedimentation related to transgression is apparent since the sequence is disrupted by several thick volcanic units.

Regardless of the precise environment, the Ackerman Formation represents shallow water deposition of sediments around active volcanic islands or at a volcanically active basin margin. Sedimentation began in this area after the extensive volcanism of the Wyatt Formation. Although the volcanic rocks of the Ackerman Formation resemble those of the Wyatt Formation, and are petrologically distinct from the Cambrian silicic volcanics in the region (Stump, 1976), the interbedding of volcanic and sedimentary rocks is characteristic of the widespread Cambrian sequences. As such the Ackerman Formation may represent the initial return of marine conditions to the central Transantarctic Mountains in the latest Precambrian following the folding, uplift, and erosion of the Beardmore Group.

*Acknowledgements* Thanks to Phil Colbert and Terry Stump for assistance on the section measurement of Ackerman Ridge, and to Steve Self and Jerry Smit for stimulating collaboration on other aspects of our work in the La Gorce Mountains. Support from National Science Foundation Grant DPP78-20624.

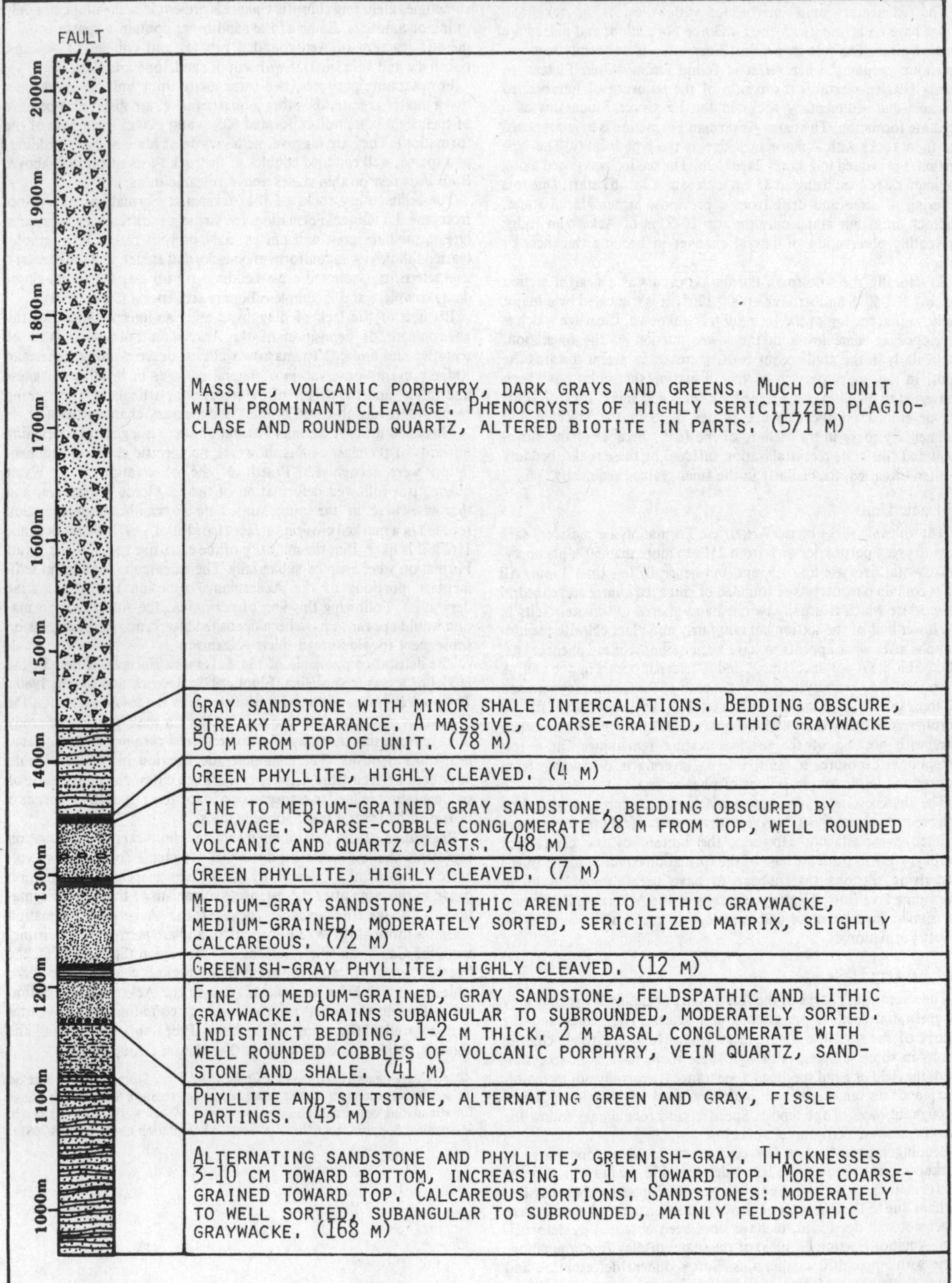

FAULT
2000m
1900m
1800m
1700m
1600m
1500m
1400m
1300m
1200m
1100m
1000m
MASSIVE, VOLCANIC PORPHYRY, DARK GRAYS AND GREENS. MUCH OF UNIT WITH PROMINANT CLEAVAGE. PHENOCRYSTS OF HIGHLY SERICITIZED PLAGIOCLASE AND ROUNDED QUARTZ, ALTERED BIOTITE IN PARTS. (571 M)
GRAY SANDSTONE WITH MINOR SHALE INTERCALATIONS. BEDDING OBSCURE, STREAKY APPEARANCE. A MASSIVE, COARSE-GRAINED, LITHIC GRAYWACKE 50 M FROM TOP OF UNIT. (78 M)
GREEN PHYLLITE, HIGHLY CLEAVED. (4 M)
FINE TO MEDIUM-GRAINED GRAY SANDSTONE, BEDDING OBSCURED BY CLEAVAGE. SPARSE-COBBLE CONGLOMERATE 28 M FROM TOP, WELL ROUNDED VOLCANIC AND QUARTZ CLASTS. (48 M)
GREEN PHYLLITE, HIGHLY CLEAVED. (7 M)
MEDIUM-GRAY SANDSTONE, LITHIC ARENITE TO LITHIC GRAYWACKE, MEDIUM-GRAINED, MODERATELY SORTED, SERICITIZED MATRIX, SLIGHTLY CALCAREOUS. (72 M)
GREENISH-GRAY PHYLLITE, HIGHLY CLEAVED. (12 M)
FINE TO MEDIUM-GRAINED, GRAY SANDSTONE, FELDSPATHIC AND LITHIC GRAYWACKE. GRAINS SUBANGULAR TO SUBROUNDED, MODERATELY SORTED. INDISTINCT BEDDING, 1-2 M THICK. 2 M BASAL CONGLOMERATE WITH WELL ROUNDED COBBLES OF VOLCANIC PORPHYRY, VEIN QUARTZ, SANDSTONE AND SHALE. (41 M)
PHYLLITE AND SILTSTONE, ALTERNATING GREEN AND GRAY, FISSLE PARTINGS. (43 M)
ALTERNATING SANDSTONE AND PHYLLITE, GREENISH-GRAY. THICKNESSES 3-10 CM TOWARD BOTTOM, INCREASING TO 1 M TOWARD TOP. MORE COARSE-GRAINED TOWARD TOP. CALCAREOUS PORTIONS. SANDSTONES: MODERATELY TO WELL SORTED, SUBANGULAR TO SUBROUNDED, MAINLY FELDSPATHIC GRAYWACKE. (168 M)


**Figure 2a. Type section, Ackerman Formation, upper portion.**

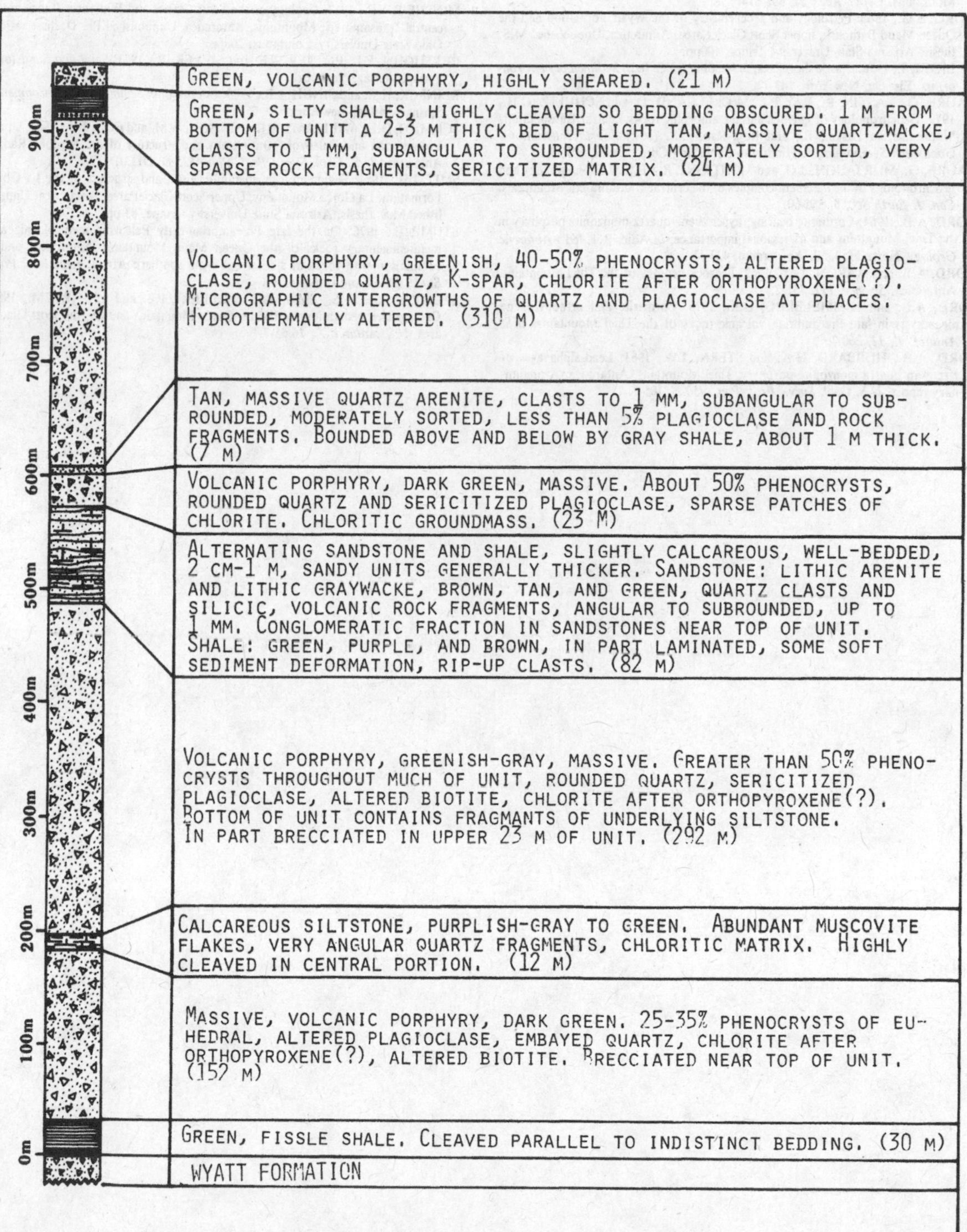

900m
800m
700m
600m
500m
400m
300m
200m
100m
0m
GREEN, VOLCANIC PORPHYRY, HIGHLY SHEARED. (21 M)
GREEN, SILTY SHALES, HIGHLY CLEAVED SO BEDDING OBSCURED. 1 M FROM BOTTOM OF UNIT A 2-3 M THICK BED OF LIGHT TAN, MASSIVE QUARTZWACKE, CLASTS TO 1 MM, SUBANGULAR TO SUBROUNDED, MODERATELY SORTED, VERY SPARSE ROCK FRAGMENTS, SERICITIZED MATRIX. (24 M)
VOLCANIC PORPHYRY, GREENISH, 40-50% PHENOCRYSTS, ALTERED PLAGIOCLASE, ROUNDED QUARTZ, K-SPAR, CHLORITE AFTER ORTHOPYROXENE(?). MICROGRAPHIC INTERGROWTHS OF QUARTZ AND PLAGIOCLASE AT PLACES. HYDROTHERMALLY ALTERED. (310 M)
TAN, MASSIVE QUARTZ ARENITE, CLASTS TO 1 MM, SUBANGULAR TO SUBROUNDED, MODERATELY SORTED, LESS THAN 5% PLAGIOCLASE AND ROCK FRAGMENTS. BOUNDED ABOVE AND BELOW BY GRAY SHALE, ABOUT 1 M THICK. (7 M)
VOLCANIC PORPHYRY, DARK GREEN, MASSIVE. ABOUT 50% PHENOCRYSTS, ROUNDED QUARTZ AND SERICITIZED PLAGIOCLASE, SPARSE PATCHES OF CHLORITE. CHLORITIC GROUNDMASS. (23 M)
ALTERNATING SANDSTONE AND SHALE, SLIGHTLY CALCAREOUS, WELL-BEDDED, 2 CM-1 M, SANDY UNITS GENERALLY THICKER. SANDSTONE: LITHIC ARENITE AND LITHIC GRAYWACKE, BROWN, TAN, AND GREEN, QUARTZ CLASTS AND SILICIC, VOLCANIC ROCK FRAGMENTS, ANGULAR TO SUBROUNDED, UP TO 1 MM. CONGLOMERATIC FRACTION IN SANDSTONES NEAR TOP OF UNIT. SHALE: GREEN, PURPLE, AND BROWN, IN PART LAMINATED, SOME SOFT SEDIMENT DEFORMATION, RIP-UP CLASTS. (82 M)
VOLCANIC PORPHYRY, GREENISH-GRAY, MASSIVE. GREATER THAN 50% PHENOCRYSTS THROUGHOUT MUCH OF UNIT, ROUNDED QUARTZ, SERICITIZED PLAGIOCLASE, ALTERED BIOTITE, CHLORITE AFTER ORTHOPYROXENE(?). BOTTOM OF UNIT CONTAINS FRAGMANTS OF UNDERLYING SILTSTONE. IN PART BRECCIATED IN UPPER 23 M OF UNIT. (292 M)
CALCAREOUS SILTSTONE, PURPLISH-GRAY TO GREEN. ABUNDANT MUSCOVITE FLAKES, VERY ANGULAR QUARTZ FRAGMENTS, CHLORITIC MATRIX. HIGHLY CLEAVED IN CENTRAL PORTION. (12 M)
MASSIVE, VOLCANIC PORPHYRY, DARK GREEN. 25-35% PHENOCRYSTS OF EUHEDRAL, ALTERED PLAGIOCLASE, EMBAYED QUARTZ, CHLORITE AFTER ORTHOPYROXENE(?), ALTERED BIOTITE. BRECCIATED NEAR TOP OF UNIT. (152 M)
GREEN, FISSLE SHALE. CLEAVED PARALLEL TO INDISTINCT BEDDING. (30 M)
WYATT FORMATION


**Figure 2b. Type section, Ackerman Formation, lower portion.**

## REFERENCES

AARON, J.M. and FORD, A.B., 1964: Isotopic age determinations in the Thiel Mountains, Antarctica. *Geol. Soc. Am., Spec. Pap., 76,* 1.

BLACKBURN, Q.A., 1937: The Thorne Glacier period section of the Queen Maud Mountains. *Geogr. Rev., 27,* 598-614.

BORG, S.G., 1980: Petrology and geochemistry of the Wyatt Formation and the Queen Maud Batholith, upper Scott Glacier area, Antarctica. Unpublished M.S. thesis, Arizona State University, Tempe, 100pp.

ELLIOTT, T., 1978: Clastic shorelines; *in* Reading, H.G. (ed.) *Sedimentary environments.* Elsevier, New York, 143-77.

FAURE, G., EASTIN, R., RAY, P.T., McCLELLAND, D. and SCHULTZ, C.H., 1979: Geochronology of igneous and metamorphic rocks, central Transantarctic Mountains; *in* Laskar, B., and Raja Rao, C.S. (eds.) *Fourth International Gondwana Symposium, 2,* Hindustan Publishing Corp., Delhi, 805-13.

FAURE, G., MURTAUGH, J.G. and MONTIGNY, R.J.E., 1968: The geology and geochronology of the basement complex of the central Transantarctic Mountains. *Can. J. Earth Sci., 5,* 555-60.

FORD, A.B., 1964: Cordierite bearing, hypersthene-quartz-monzonite porphyry in the Thiel Mountains and its regional importance; *in* Adie, R.J. (ed.) *Antarctic Geology,* North Holland, Amsterdam, 429-41.

FORD, A.B. and AARON, J.M., 1962: Bedrock geology of the Thiel Mountains, Antarctica. *Science, 137,* 751-2.

FORD, A.B. and HIMMELBERG, G.R., 1976: Cordierite and orthopyroxene megacrysts in late Precambrian volcanic rocks of the Thiel Mountains. *U.S. Antarct. J., 11,* 260-3.

FORD, A.B., HUBBARD, H.A. and STERN, T.W., 1963: Lead-alpha ages of zircon in quartz monzonite porphyry, Thiel Mountains, Antarctica. A preliminary report. *U.S. Geol. Surv., Prof. Pap., 450-E,* 105-7.

KATZ, H.R. and WATERHOUSE B.C., 1970: Geological reconnaissance of the Scott Glacier area, southeastern Queen Maud Range, Antarctica. *N.Z. J. Geol. Geophys., 13,* 1030-7.

LAIRD, M.G., MANSERGH, G.D. and CHAPPELL, J.M.A., 1971: Geology of the central Nimrod Glacier area, Antarctica. *N.Z. J. Geol. Geophys., 14,* 427-68.

MINSHEW, V.E., 1967: Geology of the Scott Glacier and Wisconsin Range areas, central Transantarctic Mountains, Antarctica. Unpublished Ph. D. dissertation, Ohio State University, Columbus. 268pp.

PETTIJOHN, F.J., POTTER, P.E. and SIEVER, R., 1973: *Sand and Sandstone.* Springer-Verlag, New York.

REINECK, H.E. and SINGH, I.B., 1980: *Depositional Sedimentary Environments.* Springer-Verlag, New York.

SCHMIDT, D.L., WILLIAMS, P.L., NELSON, W.M. and EGE, J.R., 1965: Upper Precambrian and Paleozoic stratigraphy and structure of the Neptune Range, Antarctica; *U.S. Geol. Surv., Prof. Pap., 525-D,* D112-D119.

SMIT, J.H., 1981: Sedimentology, metamorphism, and structure of the La Gorce Formation, La Gorce Mountains, Upper Scott Glacier area, Antarctica. Unpublished M.S. thesis, Arizona State University, Tempe, 83 pp.

STUMP, E., 1976: On the late Precambrian-early Paleozoic metavolcanic and metasedimentary rocks of the Queen Maud Mountains, Antarctica, and a comparison with rocks of similar age from southern Africa. *Rep. Inst. Polar Stud., Ohio State Univ., 62.*

STUMP, E., SELF, S., SMIT, J.H., COLBERT, P.B. and STUMP, T.M., 1981: Geological investigations in the La Gorce Mountains and central Scott Glacier area. *U.S. Antarct. J., 16,*55-7.

# THE STRUCTURAL DEVELOPMENT OF SELECTED AREAS IN THE PENSACOLA MOUNTAINS

A. Frischbutter, W. Weber, J. Hofmann, *Bergakademie Freiberg, Sektion Geowissenschaften, 9200 Freiberg, German Democratic Republic.*

H.J. Paech, *Akademie der Wissenschaften der DDR, Zentralinstitut fur Physik der Erde, 150 Potsdam, German Democratic Republic.*

**Abstract** This paper presents the results of field work in the Pensacola Mountains done in conjunction with geologists of the Soviet Antarctic Expedition during the 1978/79 and 1979/80 summer seasons.

The evolution of the western margin of the East Antarctic Platform started with the deposition of a Precambrian flyschoid sequence on a strongly metamorphosed crystalline basement. The flyschoid sequence is overlain discordantly by epicontinental Middle Cambrian limestones which in turn are overlain by acid volcanics of probable Cambrian age. A coarse clastic non-volcanic, partly Upper Devonian red molasse sequence is overlain by incompletely known Carboniferous and Permian platform sediments.

There are several disharmonically superimposed structural levels: a level with slate tectonics; a level of open folding; a fault-fold level and an undeformed level. Each of these levels is characterised by special structural features. The metamorphic processes related to the Beardmore tectogenesis at the Precambrian-Cambrian boundary took place in the range of the 'very low metamorphism' of Winkler (1976). No regional metamorphism is related to the Palaeozoic deformation events. Deformations in association with faults are of different character, at some places reaching the intensity of biotite metamorphism. The synkinematic structure of planes representing the early stages of foliation development start from rupture systems or from grain boundaries in relevant directions.

# A GEOCHRONOLOGICAL INVESTIGATION OF THE SHACKLETON RANGE

R.J. Pankhurst, *British Antarctic Survey, C/- Institute of Geological Sciences, 64 Gray's Inn Road, London WC1X 8NG, U.K.*

P.D. Marsh and P.D. Clarkson, *British Antarctic Survey, Madingley Road, Cambridge, CB3 OET, U.K.*

*Abstract* Rb-Sr radiometric dating has been applied to all major lithological units exposed in the Shackleton Range. This allows the construction of a more complete time scale for the geological evolution of this area than has been previously possible. The oldest rocks identified in the Shackleton Range Metamorphic Complex are high grade ortho- and paragneisses which have disturbed Rb-Sr whole rock systems dating from at least 2,000-2,300 Ma. Two pegmatites from Wedge Ridge have model ages of about 2,700 Ma. A minimum age for the high grade metamorphism of the older part of the basement is given by an isochron of 1,765 ± 15 Ma for granite gneiss in the Read Mountains. Most of the basement rocks, however, show the effects of strong thermal resetting of the isotope systems in Middle to Late Proterozoic times (1,500-1,600 Ma). The younger part of the metamorphic complex consists of supracrustal greenschist to amphibolite grade gneisses and schists, mostly of sedimentary origin. These show a very strong "Pan-African" imprint at 500-600 Ma. We concur with previous authors that this represents the last major metamorphic event in the area, which concluded with a tectonic and metamorphic climax dated by mineral isochrons at 500 ± 5 Ma in the north of the range and the low grade metamorphism of the Turnpike Bluff Group 526 ± 6 Ma ago. Initial $^{87}Sr/^{86}Sr$ ratios constrain the age of deposition of the sediments to a maximum of a out 850 Ma. Thus the main east-west structural trend of the Shackleton Range results from the final Late Cambrian or Ordovician tectonometamorpbism and was succeeded by folding about N-S axes in the extreme west of the range and deposition of the Cambro-Ordovician Blaiklock Glacier Group.

The geology of the Shackleton Range comprises a very varied suite of sedimentary, igneous and metamorphic rocks. Previous work, largely carried out by British geologists (Stephenson, 1966; Clarkson, 1972, 1982) has led to the recognition of late unmetamorphosed sediments (Blaiklock Glacier Group) and low grade greenschists (Turnpike Bluff Group) presumably unconformable on higher grade rocks (Shackleton Range Metamorphic Complex). More detailed investigation of the latter (Grew and Halpern, 1979; Hofmann and Paech, 1980; Clarkson, 1981; Marsh, this volume) has demonstrated the existence of further metamorphosed supracrustal rocks overlying a mainly orthogneissic basement, these two types being repeated in thrust slices or fold cores. The present work has been carried out in close conjunction with the petrological and structural analysis of Marsh (this volume) in an attempt to construct a complete radiometric chronology for the geological evolution of the range. The previous geochronological data of Rex (1972), Grew and Halpern (1979), Grew and Manton (1980) and Hofmann et al. (1980, 1981) have been incorporated in this analysis.

Rb-Sr analyses were carried out at the Institute of Geological Sciences using the same combination of precise X-ray fluorescence and automated thermal ionisation mass spectrometry as for the BAS dating programme on the Antarctic Peninsula (Pankhurst, 1981 and this volume). Approximately 135 whole rock and mineral samples have been analysed from the field collections made in 1968-71 and 1977-78. For obvious logistic reasons, many of the rock samples were rather small (~ 1 kg) compared to the usual requirements for this type of work in polymetamorphic terrains and this factor may account for the tendency towards excess scatter in the data, as reflected in high values for the MSWD statistical parameter. Full details of the data treatment employed are given by Pankhurst (this volume). In this study of the Shackleton Range, only a few sample suites have given well defined isochrons without excess scatter (i.e. MSWD< 3) which

**TABLE 1: Rb-Sr analytical data for Shackleton Range crystalline basement**

| Sample No. | Rb | Sr | $^{87}Rb/^{86}Sr$ | $^{87}Sr/^{86}Sr$ |
|---|---|---|---|---|
| (a) Read Mountains granite | | | | |
| Z.1211.4 | 145 | 227 | 1.859 | 0.75103 |
| Z.1212.1 | 172 | 186 | 2.694 | 0.77223 |
| Z.1212.3 | 193 | 100 | 5.685 | 0.84884 |
| Z.1212.6 | 214 | 133 | 4.724 | 0.82334 |
| Z.1213.1 | 138 | 194 | 2.063 | 0.75674 |
| Z.1214.1 | 138 | 182 | 2.210 | 0.76483 |
| Z.1215.1 | 130 | 186 | 2.032 | 0.76045 |
| Z.1216.1 | 169 | 128 | 3.857 | 0.80242 |
| Z.1221.1 | 93 | 233 | 1.157 | 0.73498 |
| Z.1221.2 | 115 | 206 | 1.617 | 0.74510 |
| Z.1222.1 | 102 | 267 | 1.109 | 0.73592 |
| Z.1224.1 | 119 | 270 | 1.281 | 0.73867 |
| Z.1224.5 | 123 | 269 | 1.326 | 0.73952 |
| Z.1224.6 | 151 | 290 | 1.515 | 0.74469 |
| Z.1225.1 | 111 | 255 | 1.264 | 0.73649 |
| Z.1225.2 | 96 | 381 | 0.731 | 0.72605 |
| Z.1225.2 | 116 | 265 | 1.273 | 0.73365 |
| Z.1226.2 | 154 | 249 | 1.797 | 0.74864 |
| (b) Read Mountains Granodiorite Dykes | | | | |
| Z.1201.4 | 47.5 | 184 | 0.747 | 0.71790 |
| Z.1209.3 | 124 | 1207 | 0.297 | 0.71325 |
| Z.1211.3 | 92.5 | 256 | 1.046 | 0.72323 |
| Z.1214.5 | 34.2 | 280 | 0.353 | 0.71048 |
| Z.1216.4 | 175 | 620 | 0.816 | 0.72705 |
| Z.1222.3 | 40.5 | 185 | 0.634 | 0.71735 |
| Z.1222.4 | 36.8 | 173 | 0.616 | 0.71490 |
| Z.1225.3 | 82.1 | 541 | 0.439 | 0.71774 |
| Z.602.3d | 141 | 801 | 0.509 | 0.71892 |
| Z.602.3e | 156 | 563 | 0.801 | 0.72712 |
| (c) Wedge Ridge gneisses | | | | |
| Z.566.4 | 169 | 63.1 | 7.908 | 0.90722 |
| Z.826.1 | 75.0 | 88.3 | 2.510 | 0.83313 |
| Z.826.2 | 22.3 | 166 | 0.390 | 0.73245 |
| Z.826.4 | 98.2 | 199 | 1.442 | 0.75717 |
| Z.826.5 | 284 | 63.5 | 13.36 | 1.0818 |
| Z.826.6 | 158 | 54.5 | 8.662 | 0.98497 |
| Z.828.1 | 173 | 30.5 | 17.49 | 1.4008 |
| Z.828.2 | 164 | 28.0 | 18.07 | 1.3899 |
| Z.828.3 | 265 | 57.9 | 13.77 | 1.0898 |
| Z.828.4 | 157 | 86.2 | 5.357 | 0.90926 |
| Z.830.3 | 123 | 80.5 | 4.471 | 0.85213 |
| Z.1104.3 | 154 | 152 | 2.975 | 0.81227 |
| Z.1098.6 | 186 | 97.1 | 5.663 | 0.92279 |
| (KF) | 240 | 108 | 6.557 | 0.92900 |
| (bio) | 806 | 29.9 | 84.08 | 1.4860 |
| Z.754.3 | | | | |
| (Musc) | 1120 | 7.92 | 18155 | 433.9 |
| (d) Mount Weston gneisses | | | | |
| Z.1008.3 | 42.0 | 250 | 0.490 | 0.71601 |
| Z.1008.4 | 205 | 728 | 0.817 | 0.72361 |
| Z.1008.5 | 205 | 261 | 2.290 | 0.76933 |
| Z.1008.6 | 139 | 224 | 1.803 | 0.77013 |
| Z.1076.3 | 162 | 159 | 2.976 | 0.78496 |
| Z.1076.4 | 170 | 148 | 3.334 | 0.78763 |
| Z.1076.5 | 135 | 71.0 | 5.591 | 0.83620 |
| Z.1076.6 | 38.5 | 101 | 1.110 | 0.73497 |
| Z.1077.1 | 53.6 | 446 | 0.348 | 0.71944 |
| Z.1077.2 | 70.7 | 500 | 0.410 | 0.71971 |
| Z.1077.6 | 82.3 | 149 | 1.611 | 0.76000 |
| (e) Lagrange Nunataks gneisses | | | | |
| Z.1120.3 | 88.8 | 198 | 1.303 | 0.76545 |
| Z.1120.3 | 25.8 | 100 | 0.753 | 0.74746 |
| Z.1122.1 | 134 | 105 | 3.752 | 0.84321 |
| Z.1122.3 | 146 | 96.6 | 4.440 | 0.87226 |
| Z.1124.1 | 125 | 129 | 2.837 | 0.81285 |
| Z.1131.6 | 114 | 93.2 | 3.571 | 0.84622 |
| Z.1135.2 | 112 | 77.1 | 4.249 | 0.81293 |

Rb/Sr in ppm
Error on $^{87}Rb/^{86}Sr$ generally ~0.5%
Error on $^{87}Sr/^{86}Sr$ 0.01 – 0.03%

may consequently be taken to date a specific event. Otherwise, if the MSWD is less than 30, we have enhanced the errors derived from the regression analysis by a factor of $\sqrt{MSWD}$ in an attempt to allow for minor geological disturbance. In some cases the scatter of data is however so great that it can only loosely be interpreted by comparison with calculated reference isochrons. This can only give rough limits on possible igneous or metamorphic ages, but in favourable circumstances will still distinguish between Precambrian and Palaeozoic rock formation. The initial $^{87}Sr/^{86}Sr$ ratios calculated for linear fits (given in brackets after the corresponding ages) gives a further indication of possible pre-history in the cases of metamorphic resetting.

## RESULTS

### Crystalline Basement

Migmatites and orthogneisses, mostly of acid to intermediate composition but with some pyroxene- and amphibolite-gneisses, occur throughout the range. Early metamorphism is mostly of upper amphibolite facies, locally granulite facies, but in the central and northern areas this has been overprinted, following cataclasis, by metamorphism which affected the cover rocks. The basement has been analysed in three main areas:

*Read Mountains* Here orthogneisses are well developed and relatively little affected by post-supracrustal events. Gneissic igneous bodies derived from intrusive granites were studies from Hatch Plain and between Watts Needle and The Ark (Figure 1). The data for the Hatch Plain area (Figure 2a) yield two true isochrons: 1763 ± 21 Ma (0.704 ± 0.001) for the first five samples listed in Table 1(a) and 1599 ± 38 Ma (0.714 ± 0.001) for the next three. Although all the samples are relatively weakly gneissose granite, the two isochrons correspond to different geographical locality groupings. The first five samples are from a 1.5 km traverse along the southeastern part of the outcrop and the remaining three are from a 1 km traverse further to the west. The older age is interpreted as the age of the original igneous emplacement, its low initial $^{87}Sr/^{86}Sr$ precluding any significant crustal pre-history. It appears to represent a phase of crustal growth by incorporation of material from the mantle. The younger event could either be seen as late anatectic magmatism (because of the high initial $^{87}Sr/^{86}Sr$) or more likely a result of metamorphic resetting. The remaining data for the Read Mountains (Figure 2b) have lower Rb/Sr ratios and are more scattered but reveal the same two trends, for example, three data points lie on a 1820 ± 160 Ma (0.705 ± 0.003) isochron and none lie significantly below a 1550 Ma (0.705) reference line. Data for granodiorite dykes which cut the granite gneiss and the surrounding migmatites are more scattered still, falling between reference lines corresponding to ~1900 and 1300 Ma. The older age limit may reflect assimilation of yet older basement by these magmas, which must be younger than 1760 Ma. Rex (1972) reported a K-Ar whole rock age of 1454 ± 60 Ma (recalculated) for one of these dykes.

*Haskard Highlands* Orthogneisses (mostly quartz-biotite-two feldspar) and pegmatites were analysed from around Wedge Ridge in the Haskard Highlands (Figure 2c). Two samples of pegmatite with Rb-Sr ratios of ca 6 and four of the whole-rock gneisses lie on a moderately good errorchron (MSWD = 18) corresponding to 2700 ± 100 Ma (0.700 ± 0.004). No true isochrons are apparent in the data, even for samples from single localities, but a minimum age is given by muscovite from large books in a pegmatite which has a *measured* $^{87}Sr/^{86}Sr$ ratio of 444 and a model age (assuming an initial value of < 1) of 1700 ± 50 Ma. The whole rock data lie on a perfect isochron indicating reference isochron, although there is a clear tendency to alignment along a ca 1600 Ma reset isochron with an initial $^{87}Sr/^{86}Sr$ ratio of ca 0.8. A further major cause of isotopic disturbance in these rocks is revealed by analysis of mineral separates from one of the gneisses: biotite, K-feldspar and whole rock lie on a perfect isochron indicating total resetting at 504 ± 6 Ma (0.8820 ± 0.0003). High grade *paragneisses* (with garnet, kyanite and sillimanite) from near Mount Weston in the central Haskard Highlands show a similar scatter to the Wedge Ridge whole rock data with a crude alignment close to 1550 Ma (0.707) (Figure 2d).

*Lagrange Nunataks* Acid to intermediate gneisses, containing garnet, from this area which is further from the zone of late N-S folding than the Haskard Highlands, yield slightly more coherent data (Figure 2c). A crude errorchron (MSWD = 43) of 2310 ± 130 Ma includes six of the seven samples, the last having a model age close to 1600 Ma

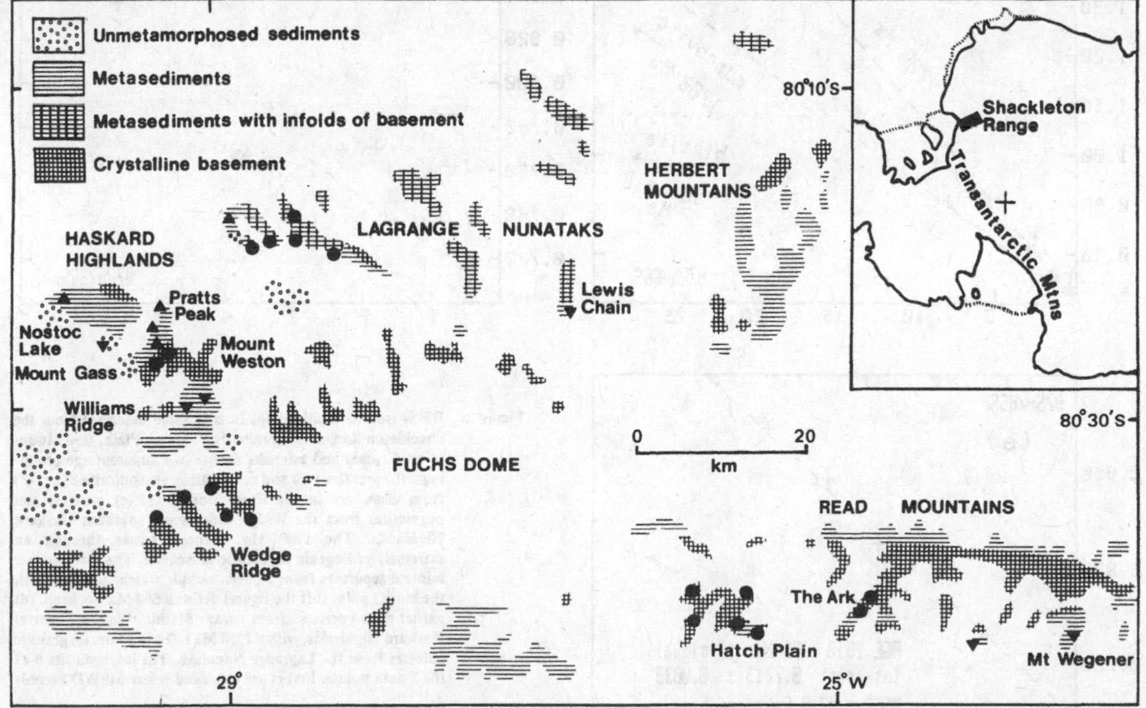


**Figure 1.** Simplified geological sketch map of the Shackleton Range showing outcrop areas, principal place names and sampling sites in the crystalline basement (circles). Nostoc Lake Formation and Mt Gass Formation (triangles), and Williams Ridge Formation, Turnpike Bluff Group and Blaiklock Glacier Group (inverted triangles).

(Figure 2e). The high $^{87}Sr/^{86}Sr$ ratio of 0.722 ± 0.004 indicates a significant crustal pre-history for these rocks, which were either already old basement or sediments derived therefrom by 2300 Ma ago. The Rb-Sr systematics are compatible with a 2700-2800 Ma old ultimate source.

Hofmann et al. (1981) have reported Rb-Sr analyses on six samples of garnet-mica-schist from a single locality in the Herbert Mountains (Figure 1), of which two were taken to define an age of 1384 ± 180 Ma (0.719 ± 0.004) and a further three a reset age of 460 ± 35 Ma (0.728 ± 0.001) (both recalculated with $Rb = 1.42 \times 10^{-11}\ a^{-1}$). Together with a previously published K-Ar age of 1401 ± 28 Ma (Hofmann et al., 1980), it was suggested that this dated a major tectonothermal event in .~1400 Ma in the Shackleton basement complex. It is clear from the disturbance of the whole rock systems observed here that this is probably insufficient evidence upon which to base any precise age of metamorphism.

## Metasediments

The petrology and stratigraphy of these very varied rocks, all of which are seen or inferred to lie unconformably on the crystalline basement are considered by Marsh (this volume) and his nomenclature is used here.

*Nostoc Lake Formation* The northern Haskard Highlands area is one of the most complex, geologically, in the Shackleton Range. High grade garnetiferous gneisses with calc-silicates, marbles, kyanite

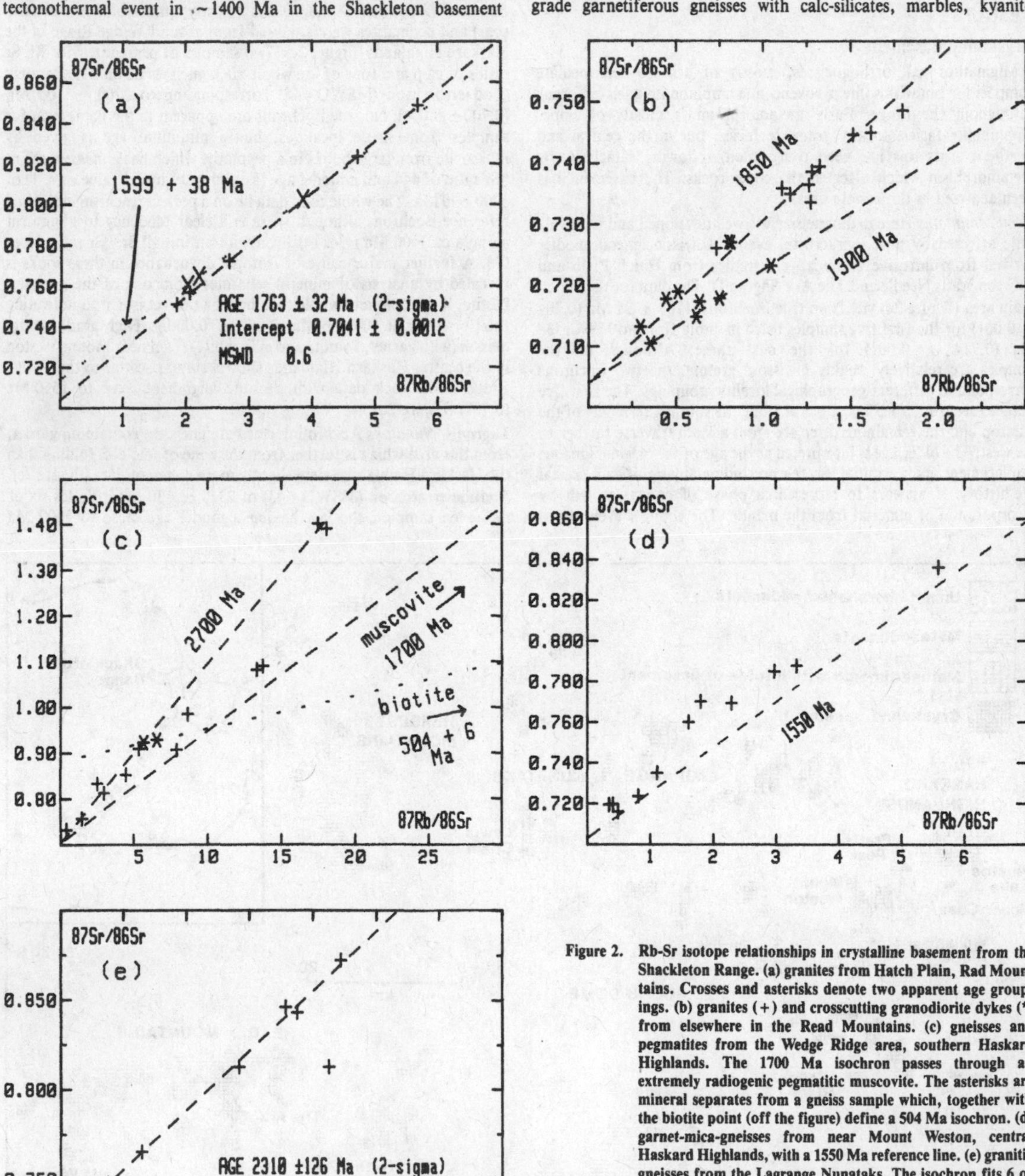


**Figure 2. Rb-Sr isotope relationships in crystalline basement from the Shackleton Range. (a) granites from Hatch Plain, Rad Mountains. Crosses and asterisks denote two apparent age groupings. (b) granites (+) and crosscutting granodiorite dykes (*) from elsewhere in the Read Mountains. (c) gneisses and pegmatites from the Wedge Ridge area, southern Haskard Highlands. The 1700 Ma isochron passes through an extremely radiogenic pegmatitic muscovite. The asterisks are mineral separates from a gneiss sample which, together with the biotite point (off the figure) define a 504 Ma isochron. (d) garnet-mica-gneisses from near Mount Weston, central Haskard Highlands, with a 1550 Ma reference line. (e) granitic gneisses from the Lagrange Nunataks, The isochron fits 6 of the 7 data points. Errors are enhanced where MSWD exceeds 3.**

schists and sheets of granitic and feldspathic augen gneiss are interleaved with migmatitic orthogneisses. Some of the last could be basement rocks equivalent to those described above. Grew and Halpern (1979) report Rb-Sr whole rock isochron ages of 656 ± 66 Ma (0.7078 ± 0.0064) and 583 ± 48 Ma (0.7084 ± 0.0024) for granite gneiss (three points only) and feldspathic augen gneiss respectively. Both were interpreted as related to regional metamorphism at the time of the Beardmore Orogeny. Like the authors, we found the Rb-Sr ratios of the migmatitic gneisses to be uniformly too low to yield useful age information, but hornblende separated from one sample gave a K-Ar age of 531 ± 13 Ma. Grew and Manton (1980) carried out reconnaissance U-Pb zircon dating on four samples from this area. Ages close to 500 Ma were obtained for augen gneiss and migmatites with a possible age of up to 550 Ma for the granite gneiss. These results do not agree very well with the earlier Rb-Sr data. Unless the Rb-Sr isochrons are unreliable (due to mixing of isotopic systems for example) then it would appear that the zircons were isotopically reset during a 500 Ma (?Ross) event. Our new data does not resolve this problem completely—a suite of mainly mesocratic garnet-biotite-hornblende gneisses from the cover rocks gives an errorchron (MSWD = 25) corresponding to 537 ± 36 Ma (0.7086 ± 0.0003) (Figure 3a), although four of our six data points do in fact lie on a single isochron with the feldspathic augen gneiss data of Grew and Halpern (1979), giving 580 ± 10 Ma (0.7077 ± 0.0002). We are inclined to agree with their interpretation of this age as reflecting the early upper amphibolite facies metamorphism apparent in these rocks, whereas later retrogression to middle amphibolite facies, following cataclasis may be recorded by a Rb-Sr biotite age of 519 ± 15 Ma (Grew and Halpern, 1979), a biotite-amphibole-K feldspar-plagioclase whole rock isochron for one of our samples giving 500 ± 5 Ma (0.7085 ± 0.0001) (Figure 3b), and a mineral isochron for a pegmatite clinopyroxene-biotite-apatite body at Pratt's Peak giving 510 ± 5 Ma (0.7082 ± 0.0001). The initial $^{87}Sr/^{86}Sr$ ratios of all these samples indicate a rather immature provenance and coupled with the observed range of Rb/Sr ratio suggest a maximum age of derivation from mantle sources of ~850 Ma.

*Mount Gass Formation* This sequence of quartzite, garnet-mica schists, dolomites and garnet-amphibolite is unconformable on the crystalline basement. It may represent the base of the metasediments in the Nostoc Lake area. 10 samples of schist were analysed (Table 2), but again scatter widely on an isochron diagram (Figure 3e). Individual samples have model ages (assuming an initial $^{87}Sr/^{86}Sr$ ratio of 0.703) between ca 900 Ma and 1500 Ma, with groups of samples cluttering around secondary lines of ca 700-900 Ma but with apparent initial $^{87}Sr/^{86}Sr$ ratios 0.715. The latter ages are probably reasonable minima for sedimentation and an old crustal provenance at ca 700 Ma would then be clearly indicated. Derivation of most of this material from the local crystalline basement is consistent with the Rb-Sr systematics.

*Williams Ridge Formation* These amphibolite grade rocks unconformably overlie the basement and are also derived from shelf facies sediments (limestones, shales and quartzites), although unlike the metasediments referred to above they exhibit evidence for only a single prograde metamorphism. The Rb/Sr results (Figure 4a) are also rather complex. Two samples (one schist, one gneiss) from locality Z.1095 close to a faulted margin of the formation, have relatively radiogenic Sr compositions and give a two point join corresponding to 600 Ma but with a very high initial $^{87}Sr/^{86}Sr$ ratio of 0.742. We suppose that these are tectonically included and mylonitised basement rocks. Four samples of unaltered and lithologically similar Williams Ridge schist, from two separate localities (Z.1090 and Z.1097) define a reasonable "errorchron" (MSWD = 6.8) with an age of 520 ± 24 Ma (0.7134 ± 0.0006) and two further samples containing a significant amount of carbonate fall a little below this line (with initial $^{87}Sr/^{86}Sr$ ratios 520 Ma ago 0.709 and 0.711). A further suite of kyanite-staurolite micaschists from a single locality at Lewis Chain,

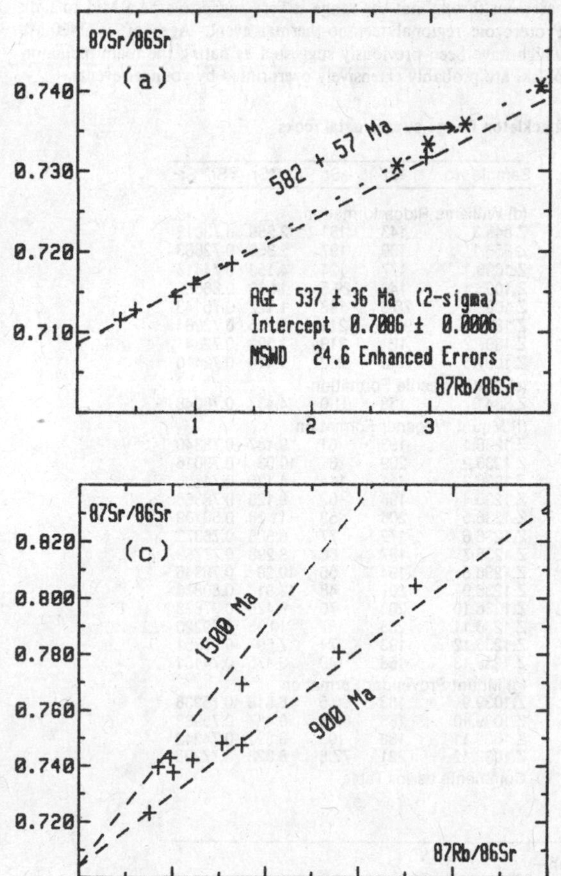


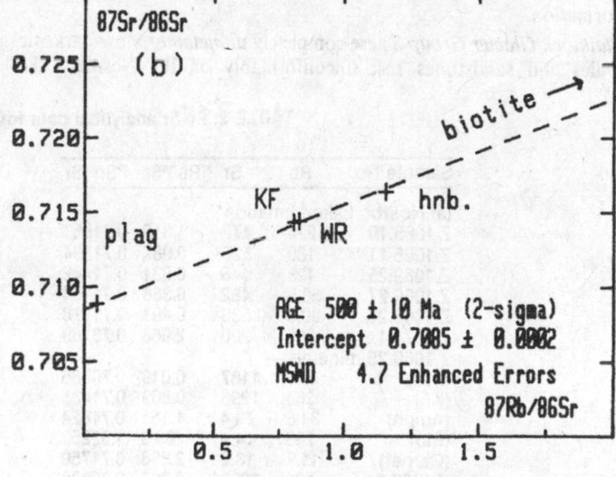


**Figure 3. Rb-Sr isotope relationships in retrogressed high grade metasediment cover rocks from the Shackleton Range. (a) garnetiferous biotite-hornblende gneisses from the Nostoc Lake formation. The dotted line is the 582 Ma recalculated isochron for data given by Grew and Halpern (1979) for augen-gneiss (*). (b) mineral isochron for one sample from the Nostoc Lake formation. (c) garnetiferous mica schists from the Mount Gass formation, with reference isochrons. Errors are enhanced where MSWD exceeds 3.0.**

Lagrange Nunataks, gives a comparable isochron (MSWD = 1.5) with an age of 505 ± 18 Ma (0.7141 ± 0.0001) (Figure 4b). These appear to be lithologically similar to the Williams Ridge schist, but at higher metamorphic grade. In fact, if data for these two are combined, the resultant nine point line is almost a true isochron (MSWD = 3.3) with an age of 512 ± 3 Ma (0.7136 ± 0.0002). The event dated by these isochrons at ca 510 Ma is probably the prograde metamorphism which, both from field observation (Marsh, this volume) and geochronology, appears to equate with the latest one (i.e. post cataclasis) in the Nostoc Lake area. Thus the main part of the Williams Ridge formation could have been deposited after the early (?580 Ma) metamorphism to the north but before Ordovician times. The fine grained schists have similar Rb-Sr systematics to the Turnpike Bluff Group (see below) and may be their stratigraphical equivalent.

*Turnpike Bluff Group* This assemblage of low grade (greenschist) cover rocks occurs only along the southern margin of the Shackleton Range, where exposed contacts with the basement are faulted. Only the Watts Needle Formation, which may be the basal part of a fault repeated succession, is seen in conformable contact. A single sample of purple shale from this formation and a suite of slaty mudstones and siltstones from the Mount Wegener Formation were analysed (Figure 4c). The latter give a good isochron age (MSWD = 1.7) of 526 ± 6 Ma (0.7152 ± 0.0005). Since the micaceous fabric of these samples is predominantly parallel to the bedding and is only partially transposed by the slaty cleavage, it seems reasonable to suppose that the isochron dates a pre-cleavage event, that is diagenesis of the sediments. If so then this formation would appear to be a Cambrian sequence of mature derived sediments, with a source area of perhaps (assuming an average Rb/Sr ratio of ca 2) an ultimate age of less than 700 Ma. The sample from the Watts Needle Formation falls significantly above this isochron, with a model age of 720 Ma (0.715). Since Riphean stromatolites have been reported from the limestones here (Golovanov et al., 1980) a Late Precambrian age is fairly certain and the closest comparison here is with the Mount Gass Formation rather than with the Williams Ridge Formation or the Mount Wegener Formation.

*Blaiklock Glacier Group* These completely *unmetamorphosed* arkosic shales and sandstones rest unconformably on the Nostoc Lake Formation and Mount Gass Formation. Four samples of red shale from the Mount Provender Formation give a rather poor isochron (MSWD = 15) with an age of 475 ± 40 Ma (0.716 ± 0.004), which presumably reflects diagenesis soon after deposition (Figure 4d). In fact three of the data points are on a perfect 482 ± 11 Ma isochron, which is consistent with their unconformable relationship upon schists recording a 500 ± 5 Ma metamorphism. This suggests a Late Cambro-Ordovician age which appears to be too young for the source of Middle Cambrian fossils (Clarkson et al., 1979, for references) in nearby glacial moraine. The true source of these fossils remains enigmatic.

## SUMMARY

This Rb-Sr survey has demonstrated generally complex isotope systematics resulting from polymetamorphic overprinting and/or polygenetic rock formations. Relatively few discrete events have been confidently dated by precise isochrons, but we feel that the overall geological evolution has emerged fairly clearly. There seems little doubt that the history of the Shackleton Range began in Archaean times since pegmatite in the Wedge Ridge gneiss gives a model age of 2700 Ma and many of the whole rock samples both here and in the Lagrange Nunataks contained very radiogenic Sr by 2000-2300 Ma ago. No details of this primary phase have survived later isotopic disturbance, but the probable Archaean lithologies include high grade ortho- and paragneisses. These early basement rocks underwent their main high grade metamorphism prior to the emplacement of granites and granodiorites in the Read Mountains at ca 1760 Ma and possible also ca 1600 Ma. Since the 1760 Ma event represents addition of juvenile material from a subcrustal source it could well be associated with a major orogeny. Although there is no direct petrographic evidence that the Read Mountain granites have been metamorphosed, the strong tendency to reset ages of 1700-1550 Ma in the crystalline basement throughout the range is very suggestive of a Mid to Late Proterozoic regional tectono-thermal event. Ages of ca 1400 Ma which have been previously suggested as dating the main metamorphism are probably extensively overprinted by younger events.

**TABLE 2: Rb-Sr analytical data for Shackleton Range supracrustal rocks**

| Sample No. | Rb | Sr | $^{87}Rb/^{86}Sr$ | $^{87}Sr/^{86}Sr$ |
|---|---|---|---|---|
| (a) Nostoc Lake formation* | | | | |
| Z.1065.10 | 214 | 473 | 1.310 | 0.71851 |
| Z.1065.11 | 180 | 529 | 0.987 | 0.71594 |
| Z.1069.25 | 128 | 448 | 0.831 | 0.71442 |
| Z.1069.27 | 58.5 | 462 | 0.366 | 0.71154 |
| Z.1069.28 | 90.6 | 536 | 0.493 | 0.71278 |
| Z.1074.1 | 205 | 200 | 2.966 | 0.73169 |
| Z.1069.25 minerals:— | | | | |
| (Plag) | 20.1 | 1187 | 0.049 | 0.70885 |
| (KF) | 363 | 1298 | 0.809 | 0.71435 |
| (Amph) | 31.6 | 79.4 | 1.155 | 0.71624 |
| (Bio) | 758 | 14.4 | 169.9 | 1.9225 |
| (Garnet) | 11.7 | 13.2 | 2.558 | 0.71750 |
| Z.1067.1 | 181 | 1963 | 0.267 | 0.70973 |
| (Cpx) | 4.30 | 359 | 0.035 | 0.70855 |
| (Apatite) | 0.4 | 5407 | 0.000 | 0.70825 |
| (Bio) | 328 | 109 | 8.745 | 0.77180 |
| (b) Mount Gass formation | | | | |
| Z.1069.1 | 96.3 | 39.0 | 7.210 | 0.80441 |
| Z.1069.8 | 114 | 169 | 1.959 | 0.74288 |
| Z.1069.9 | 139 | 199 | 2.206 | 0.73770 |
| Z.1069.10 | 120 | 143 | 2.440 | 0.74205 |
| Z.1069.11 | 114 | 193 | 1.704 | 0.73959 |
| Z.1069.16 | 97.0 | 91.8 | 3.069 | 0.74816 |
| Z.1126.8 | 137 | 71.7 | 5.580 | 0.78073 |
| Z.1126.9 | 125 | 104 | 3.492 | 0.76925 |
| Z.1126.11 | 135 | 257 | 1.524 | 0.72318 |
| Z.1127.8 | 190 | 157 | 3.496 | 0.74744 |
| (c) Lewis Chain schists | | | | |
| Z.720.1 | 205 | 99.4 | 5.990 | 0.75752 |
| Z.720.2 | 200 | 94.4 | 6.152 | 0.75866 |
| Z.720.3 | 186 | 78.9 | 6.845 | 0.76333 |
| Z.720.4 | 228 | 88.3 | 7.509 | 0.76793 |
| Z.720.5 | 194 | 111 | 5.085 | 0.75053 |
| (d) Williams Ridge formation | | | | |
| Z.849.3 | 143 | 161 | 2.568 | 0.73013 |
| Z.850.1 | 160 | 197 | 2.350 | 0.72663 |
| Z.1090.1 | 177 | 124 | 4.159 | 0.74415 |
| Z.1095.4 | 146 | 29.5 | 14.60 | 0.86659 |
| Z.1095.5 | 76.6 | 198 | 1.122 | 0.75143 |
| Z.1097.1 | 92.7 | 211 | 1.275 | 0.72264 |
| Z.1097.2 | 121 | 219 | 1.599 | 0.72541 |
| Z.1097.3 | 113 | 232 | 1.416 | 0.72410 |
| (e) Watts Needle Formation | | | | |
| Z.884.3 | 138 | 91.0 | 4.437 | 0.76048 |
| (f) Mount Wegener Formation | | | | |
| Z.1236.1 | 193 | 61 | 9.187 | 0.78340 |
| Z.1236.2 | 209 | 61 | 10.03 | 0.79016 |
| Z.1236.3 | 151 | 114 | 4.125 | 0.74605 |
| Z.1236.4 | 194 | 62 | 9.126 | 0.78368 |
| Z.1236.5 | 206 | 53 | 11.28 | 0.80139 |
| Z.1236.6 | 172 | 77 | 6.506 | 0.76373 |
| Z.1236.7 | 187 | 66 | 8.298 | 0.77789 |
| Z.1236.8 | 194 | 55 | 10.29 | 0.79346 |
| Z.1236.9 | 207 | 48 | 12.61 | 0.80838 |
| Z.1236.10 | 203 | 70 | 8.425 | 0.77838 |
| Z.1236.11 | 203 | 57 | 10.35 | 0.79220 |
| Z.1236.12 | 193 | 74 | 7.597 | 0.77251 |
| Z.1236.13 | 168 | 140 | 3.475 | 0.74131 |
| (g) Mount Provender Formation | | | | |
| Z.1039.9 | 183 | 80.5 | 6.618 | 0.75936 |
| Z.1039.10 | 183 | 82.6 | 6.457 | 0.75988 |
| Z.1039.11 | 138 | 103 | 3.891 | 0.74242 |
| Z.1039.12 | 221 | 77.5 | 8.328 | 0.77298 |

Comments as for Table 1.

*K-Ar data for migmatite sample Z.1060.9 (hornblende):—
K = 1.18%, $^{40}Ar_{rad}$ = 28.29 nl/ng, Atmos. Ar = 8.6%, Age = 531 ± 13 Ma

Deposition of the sedimentary cover rocks may have begun as early as ca 580 Ma with erosion and local sedimentation of material derived from the crystalline basement represented in the Mount Gass Formation and Watts Needle Formation. A continuous upward transition into the Late Precambrian/Early Cambrian sediments of the Nostoc Lake Formation/Williams Ridge Formation/Turnpike Bluff Group is possible. Alternatively, all these sedimentary rocks may have been deposited after about 600 Ma ago, with a gradual increase in low $^{87}Sr/^{86}Sr$ (?volcanic or marine) input with time. There is some evidence that the early metamorphic fabric of the Nostoc Lake Formation may have resulted from a metamorphism ca 580 Ma ago, but little doubt that the major tectonic features (thrusting, nappe formation and mylonitisation) and the late N-S gradational metamorphism reached a climax in Cambro-Ordovician time, that is 500± 5 Ma. Orogenesis was complete before the ?Ordovician deposition of the Blaiklock Glacier Group. Since this last major event is essentially contemporaneous with the Ross orogeny of the Transantarctic Mountains (Faure et al., 1979; Adams et al., 1982), it might be suggested that the E-W trend in the Shackleton Range has been subsequently rotated from parallelism with the margin of the East Antarctic Shield, for example during the break up of Gondwana as has been proposed (Schopf, 1969) for the Ellsworth Mountain block. However, the existence of late N-S fold axes in the extreme west of the Shackleton Range (Clarkson, 1982) and comparison with the geological situation in Queen Maud Land (Marsh, this volume) make the alternative aulacogen hypothesis (Kamenev and Semenov, 1980) equally tenable.

## REFERENCES


ADAMS, C.J.D., GABITES, J. and GRINDLEY, G.W., 1982: Orogenic history of the central Transantarctic Mountains: new K-Ar age data on the Precambrian—Early Paleozoic basement; *in* Craddock, C., (ed.) *Antarctic Geoscience*. Univ. Wisconsin Press, Madison, 917-26.

CLARKSON, P.D., 1972: Geology of the Shackleton Range: a preliminary report. *Bull. Br. antarct. Surv., 31,* 1-15.

CLARKSON, P.D., 1981: Geology of the Shackleton Range: I. The Shackleton Range Metamorphic Complex. *Bull. Br. antarct. Surv., 51,* 257-83.

CLARKSON, P.D., 1982: Tectonic significance of the Shackleton Range; *in* Craddock, C., (ed.) *Antarctic Geoscience*. Univ. Wisconsin Press, Madison, 835-9.

CLARKSON, P.D., HUGHES, C.P. and THOMSON, M.R.A., 1979: Geological significance of a Middle Cambrian fauna from Antarctica. *Nature, 279,* 791-2.

FAURE, G., EASTIN, R., RAY, P.T., MCLELLAND, D. and SCHULZ, C.H., 1979: Geochronology of igneous and metamorphic rocks, central Transantarctic Mountains; *in* Laskav, B. and Raja Rao, C.S. (eds.) *Fourth International Gondwana Symposium, 2.* Hindustan Publishing Company, Delhi, 805-11.

GOLOVANOV, N.P., MIKHAILOV, V.M. and STULIATIN, O.G., 1980:Pervyye diagnostiruemye stromatolite Antarktidy i ikh biostrategrafecheskoe znachenie. (First diagnostic stromatolites from Antarctica and their biostratigraphical significance), *Antarktika, 19,* 152-9.

GREW, E.S. and HALPERN, M., 1979: Rubidium-Strontium dates from the Shackleton Range Metamorphic Complex in the Mount Provender area, Shackleton Range, Antarctica. *J. Geol., 87,* 325-32.

GREW, E.S. and MANTON, W.I., 1980: Uranium-lead ages of zircons from Mount Provender, Shackleton Range, Transantarctic Mountains. *U.S. Antarct. J., 15,* 45-6.

HOFMANN, J., KAISER, G., KLEMM, W. and PAECH, H.J., 1980: K-Ar Alter von Doleriten und Metamorphiten der Shackleton Range und der Whichaway Nunataks, Ostund Sudostumrandung des Filcher-Eisschelfs (Antarktis). *Z. geol. Wiss., 8,* 1227-32.


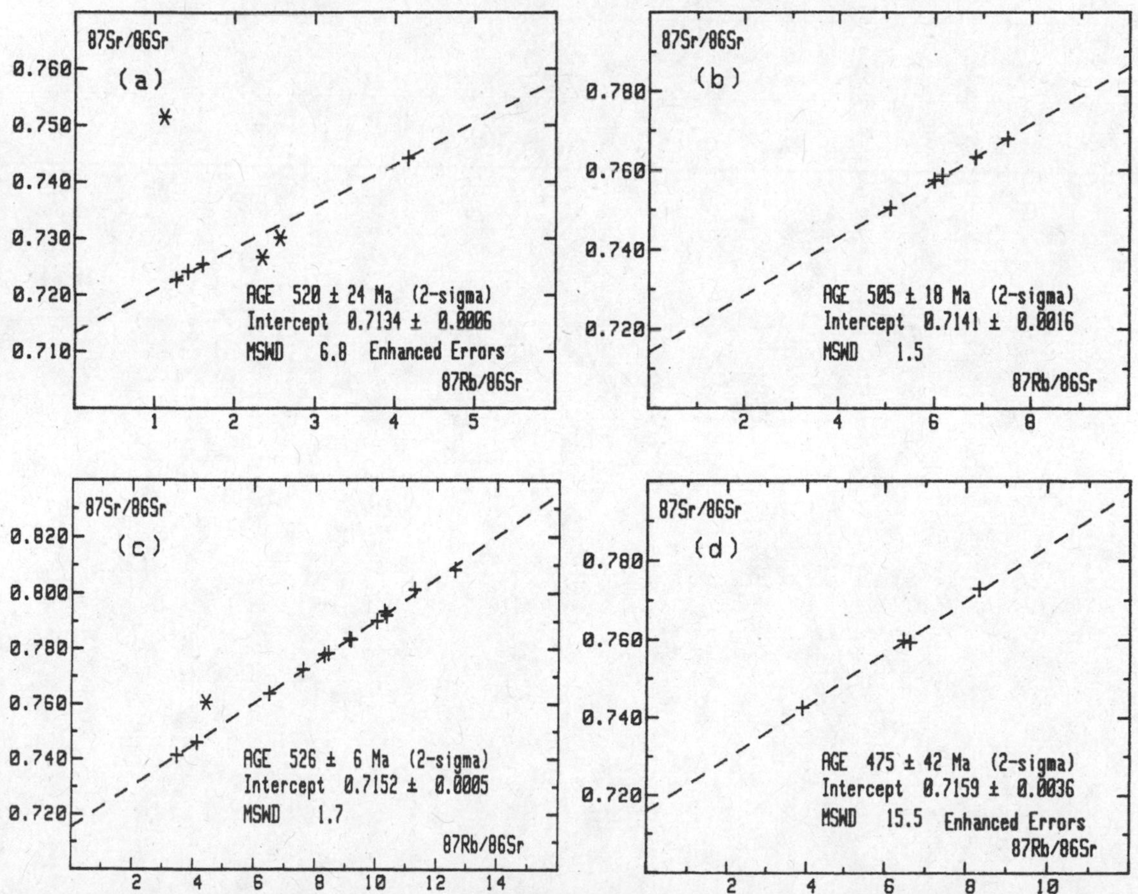


**Figure 4. Rb-Sr isotope relationships in prograde and unmetamorphosed sedimentary rocks from the Shackleton Range. (a) Williams Ridge Formation. The isochron is for four mica schist samples; two of the remaining three samples marked by asterisks contain some carbonate. One extremely radiogenic sample is not shown (see text). (b) schists correlated with the Williams Ridge Formation from Lewis Chain, Lagrange Nunataks. (c) slates of Mount Wegener Formation, Turnpike Bluff Group. The asterisk marks a single analysis from the Watts Needle Formation. (d) shales from the Mount Provender Formation, Blaiklock Glacier Group. Errors are enhanced where MSWD exceeds 3.0.**

HOFMANN, J. and PAECH, H.J., 1980: Zum strukturgeolischen Bau am Westrang der Ostantarktischen Tafel. *Z. geol. Wiss., 8*, 425-37.

HOFMANN, J., PILOT, J., and SCHLICHTING, M., 1981: Das Rb-Sr-Alter von Metamorphiten der Herbert Mountains, Shackleton Range, Antarktica. *Z. geol. Wiss., 9*, 835-42.

KAMENEV, E.N. and SEMENOV, V.S., 1980: Regional metamorphism in Antarctica; *in* Cresswell, M.M. and Vella, P., (eds.) *Gondwana Five.* A.A. Balkema, Rotterdam, 209-15.

MARSH, P.D. (this volume): The Late Precambrian and Early Palaeozoic history of the Shackleton Range.

PANKHURST, R.J. (this volume): Rb-Sr constraints on the age of basement rocks of the Antarctic Peninsula.

REX, D.C., 1972: K-Ar age determinations on volcanic and associated rocks from the Antarctic Peninsula and Dronning Maud Land; *in* Adie, R.J. (ed.) *Antarctic geology and geophysics.* Universitetsforlaget, Oslo, 133-6.

SCHOPF, J.M., 1969: Ellsworth Mountains: Position of west Antarctica due to sea floor spreading. *Science, 164*, 63-6.

STEPHENSON, P.J., 1966: Geology. 1. Theron Mountains, Shackleton Range and Whichaway Nunataks (with a section on palaeomagnetism of the dolerite intrusions, by D.J. Blundell). *Scient. Rep. Transantarctic Exped., 8.*

# TECTONICS AND RELATIONSHIPS BETWEEN STRUCTURAL STAGES IN THE PRECAMBRIAN OF THE SHACKLETON RANGE, WESTERN MARGIN OF THE EAST ANTARCTIC CRATON.

J. Hofmann, *Sektion Geowissenschaften, Bergakademie Freiberg, 9200 Freiberg, D.D.R.*

H.J. Paech, *Zentralinstitut fur Physik der Erde, Akademie der Wissenschaften der DDR, Potsdam, D.D.R.*

*Abstract* This paper presents results of tectonic investigations in two areas of the Shackleton Range (80°20′-80°45′S, 20°00′-31°00′W). In the southern Shackleton Range (Read Mountains, Stephenson Bastion) the Middle to Upper Riphean cover (Turnpike Group) of the East Antarctic Craton, overlies probably the oldest (Archaean) parts of the pre-Middle Riphean basement; the cover rocks were subjected to a saxonotype deformation under the influence of block movements in the basement, and to a low-grade metamorphism. In the northeastern Shackleton Range (Herbert Mountains) the basement rocks are divided into an older complex (Charpentier Series) and a younger one (Herbert Series), possibly of Middle and early-Upper Proterozoic ages respectively. Sedimentation, fold structures and medium grade regional metamorphism of the Herbert Series were controlled by the previously consolidated high grade metamorphic Charpentier Series which, as a rigid basement unit prevented intensive deformation of the Herbert Series. Simultaneously, the Charpentier Series was subjected to retrograde metamorphism and tectonic reworking.

## GEOLOGY AND GEOCHRONOLOGY

The metamorphic rocks of the Shackleton Range were first described by Clarkson (1971, 1972, 1982). According to Soviet Antarctic Expedition (SAE) data (Grikurov, 1980) they represent the western part of the East Antarctic Craton and are divided into three tectono-metamorphic complexes, probably with Archaean (Read Complex), Late Archean to Early-Middle Proterozoic (Provender Complex) and Middle Proterozoic (Skidmore Complex) ages of consolidation. The age of final consolidation of the entire basement is pre-Upper Proterozoic, as the stromatolites in the basal parts of the covering unmetamorphosed to low grade metamorphic Turnpike Group, which in the southern Shackleton Range rests unconformably upon the Read Complex, are of Middle-Upper Riphean age (Golovanov et al., 1979). In the western Shackleton Range, the basement is covered by the Lower Palaeozoic Blaiklock Group (Grikurov and Dibner, 1979; Grikurov et al., 1979) and possibly by Middle Cambrian black shales which are not exposed (Soloviev et al., 1978, 1979).

At present, the tectono-metamorphic development of the basement complexes is not fixed in detail by geochronological data. However Pankhurst et al. (this volume) report the presence of several tectono-metamorphic events in the basement between 2,700 and 1,765 Ma. from areas occupied by members of the Read and Provender Complexes. They confirm the division of the Shackleton Range metamorphic rocks, made by Grikurov (1980). Rb-Sr and K-Ar data from metamorphic and magmatic rocks of the Read and the Herbert Mountains add a few new values to the geochronologic picture concerning the age of consolidation of the Skidmore Complex (Herbert Series) and of younger thermomagnetic events.

From the Herbert Mountains, a series of five garnet bearing mica schists (Herbert Series), taken from a nunatak 4 km SSE of Sumgin Buttress (Figure 9) gave a two point isochron. Indicating an age of 1,414 ± 184 Ma and a three point isochron with 470 ± 36 Ma (Hofmann, 1981). The first isochron is interpreted as the age of pre-Middle Riphean regional metamorphism of the Herbert Series. This interpretation is supported by the occurrence of Middle to Upper Riphean stromatolites in the basal parts of the Turnpike Group (Golovanov et al., 1979). Furthermore, the K-Ar age of a porphyroblastic post-metamorphic granodiorite in the Read Complex (Table 1, 1) is in the same range. Rex (1972) reported a K-Ar age of about 1446 ± 60 Ma for the same rock type in the Read Mountains. Possibly, the 1,400 Ma ages indicate the age of metamorphism of the Skidmore Complex as well as that of the final consolidation connected with granitoid intrusions of the older basement complexes exposed in the Read Mountains.

K-Ar ages of Herbert Series rocks (Table 1, 2-5) indicate reset values between 434 and 268 Ma. They coincide with dolerite ages between 417 and 391 Ma (Table 1, 6-7). The dolerite ages correspond with such, reported from Rex (1972). They indicate a Late Ordovician to Early Devonian dolerite event in the Herbert Mountains. A $^{40}Ar$ versus $^{40}K$ plot of samples 2-7 (Table 1) show, that the metamorphic rocks and dolerites are situated within the range of two isochrons, indicating 402 and 357 Ma respectively (Hofmann et al., 1980). Possibly, the intrusion of the Early Palaeozoic dolerites was connected with a thermal event corresponding to the late stages of the Ross Orogeny which reset the radioactive clocks of the metamorphic rocks on the Herbert Mountains. The 470 ± 36 Ma age of the three point isochron may indicate the same event.

**TABLE 1: K/Ar ages of metamorphic rocks, granitoids and dolerites from the Read and Herbert Mts, Shackleton Range (Hofmann et. al., 1980).**

| | | | |
|---|---|---|---|
| **Read Mts (Read Complex** | | | Ma |
| 1. Porphyroblastic granitoid | Beche Blade | whole rock | 1401 ± 70 |
| **Herbert Mts (Skidmore Complex, Herbert Series)** | | | |
| 2. Amphibolite | Sumgin Buttress | whole rock | 434 ± 35 |
| 3. Fuchsite, quartz-schist | Sumgin Buttress | Fuchsite | 268 ± 21 |
| 4. Granitoid veinlet | unnamed Ntk 4 km south of Sumgin | | 351 ± 28 |
| 5. Amphibolite | Buttress | whole rock | 399 ± 32 |
| 6. Dolerite | Sumgin Buttress | whole rock | 391 ± 31 |
| 7. Dolerite | Charpentier Pyramids | whole rock | 417 ± 33 |

## Read Mountains

The geology of the southern margin of the Shackleton Range (Read Mountains, Stephenson Bastion, Figure 1) is characterised by the occurrence of rocks of the Read Complex, comprising medium to high grade metamorphic rocks partly retrogressed or migmatised, and intruded by acid igneous rocks. Their outcrops form a lens-shaped structure almost 70 km long and 16 km wide. The lens is mantled by formations of the Turnpike Group which are represented by slate, greywacke, sandstone with basal intercalations of Upper Proterozoic stromatolite-bearing limestone (Golovanov et al., 1979).

The contact between the Shackleton Crystalline Complex and the Turnpike Group was observed only at the foot of Mt Wegener (Figure 2) and on the Nicol Crags. Here, the superposition is clearly of sedimentary character. The uppermost part of the Shackleton Crystalline Complex is deeply weathered (about 10 m, locally forming residual soils of pelitic composition of about 1 m thickness). The crust of Proterozoic weathering is overlain by cross-bedded sandstones of the Turnpike Group sediments. In other regions the contact is represented by younger thrust faults.

### *Tectonics of the crystalline basement*

The tectonic structure of the metamorphic rocks in the Read Mountains is characterised by a foliation mainly parallel to the former bedding. The foliation forms an antiform, with a northern limb steeply inclined to the north (Figure 2) and a southern limb with dips to the south (Figure 1). The axis of the antiform trends E-W and is cut at an acute angle by the outer margin of the crystalline nucleus. Fold axes and b-lineations are rarely encountered. These trend between N-S and E-W, but with the NE direction dominating, whilst the plunge varies depending on the inclination of the foliation. Evidently, the formation of these folds precedes the development of the antiform.

Magmatic rocks mostly of acid composition and comprising about 20% of the area occupied by the Read Complex occur predominantly near the axis of the antiform. Their tectonic structure varies considerably in terms of size, contour, character of the contact, internal structure and occurrence of xenoliths and skialiths. The granitoids seem to have intruded before 1,400 Ma (Hofmann et al., 1980) in association with the formation of the antiform.

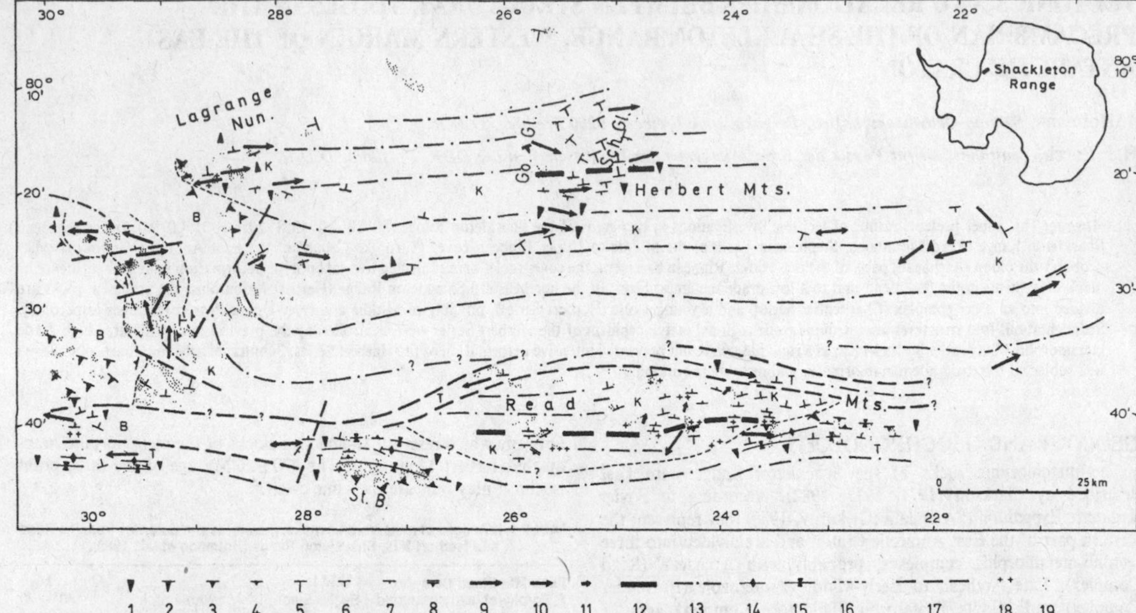


**Figure 1.** **Schematic tectonic map of the Shackleton Range. 1—vergence; 2—4 bedding (2—inclined 1—30°, 3—inclined 31—60°, 4—inclined 61—90°); 5—7 foliation, cleavage (5—inclined 1—30°, 6—inclined 31°—60°, 7—inclined 61—90°); 9—11 fold axis, b-lineation (9—plunge 1—30°, 10—plunge 31—60°, 11—plunge 61—90°); 12—antiform; 13—general tectonic trends; 14—anticline; 15—syncline; 16—geologic boundary; 17—fault; 18—granitoid; 19—outcrops; K—crystalline basement (Pre-Turnpike); T—Turnpike Group; B—Blaiklock Group; St. B.—Stephenson Bastion; Go. Gl.—Gordon Glacier; Sch. Gl.—Schimper Glacier.**

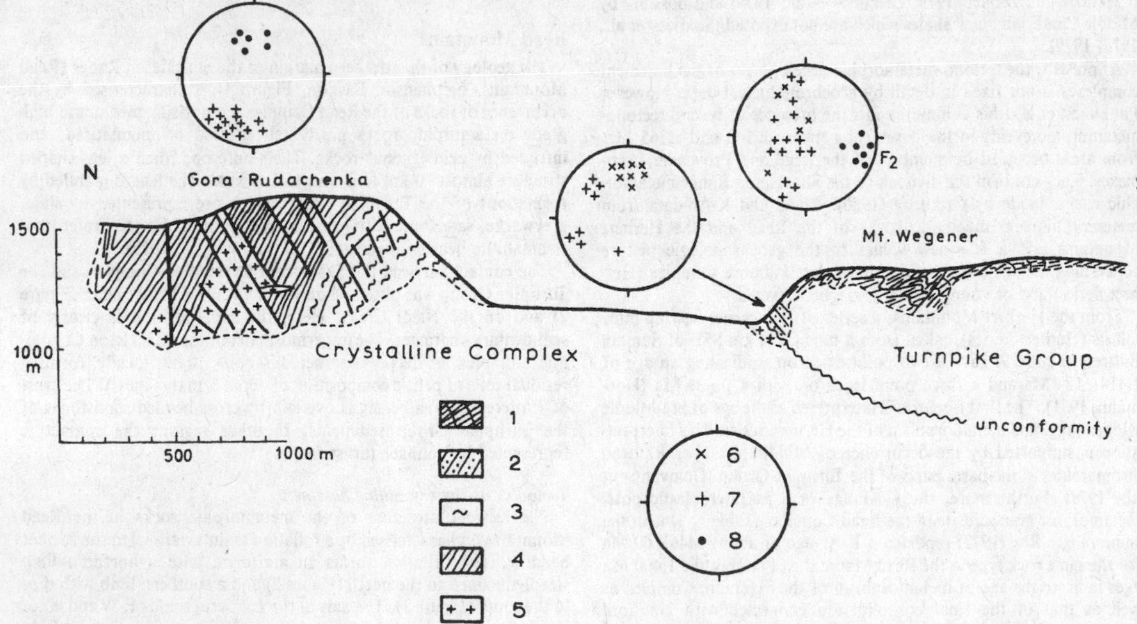


**Figure 2.** **Geologic section through the Read Mountains. 1—slate/phyllite; 2—slates with sandstones and carbonates, only weakly deformed; 3—migmatite/anatectic gneiss; 4—gneiss/schist; 5—granitoid; 6—8 steregraphic plots; 6—bedding surfaces; 7—foliation/cleavage; 8—fold axis/b-lineation.**

*Tectonics of the Turnpike Group*

The rocks of the Read Complex are enframed by sediments of the Turnpike Group which form a narrow faulted northern zone whereas a wider area continues to the south (Figure 1). The structure of the Turnpike Group is distinguished by considerable variation in the intensity of deformation. Flat lying undeformed sediments near the unconformity contrast with the deformed rocks. These range from inclined sediments lacking cleavage to cleaved slates and to doubly folded phyllites. Extreme metamorphic transformation is characterised by the occurrence of biotite indicating a low grade metamorphism.

The structural trend of the Turnpike Group is predominantly E-W but is controlled by the contours of the basement complex (Figure 1). The vergence of folds and the associated cleavage is mainly orientated to the south.

The variation in the tectonic style of the Turnpike Group is caused by the rigidity of the underlying basement. A rigid block of crystalline rocks situated north of the Read Mountains was probably upthrust to the south onto the Turnpike Group. The strongest deformation occurs in the northern limb of the Read anticlinorium, whereas on the southern flank, the Turnpike Group remains in the shadow of this tectonic stress.

The complicated tectonic structure of the Read Mountains can be described as an anticlinorium represented by a southerly verging folded cover of the Turnpike Group and a crystalline core the inner structure of which reflects an antiform whilst its outer surface is characterised by younger updoming (Figure 3). The warping structures are complicated by E-W trending thrust faults and normal faults of different age.

*Palaeotectonic development*

The tectonic structure of the Read anticlinorium and associated metamorphic phenomena reflect a long geologic history (Figure 3). The oldest structures were formed during a first deformation ($D_1$) represented by a relict b-lineation and folds belonging to the Shackleton tectogenesis probably of Lower Proterozoic age. The associated metamorphism has a high grade character. The following deformation ($D_2$) dated at 1400-1450 Ma (Rex, 1972; Hofmann et al., 1980) can be compared to the Nimrod tectogenesis. It is connected with the formation of the antiform, the consequent granitoid intrusions, retrograde metamorphism and anatexis, which recrystallised the rock fabric and obliterated older structures. In this period the formation of fault structures including flat lying overthrusts is common.

The Nimrod tectogenesis is followed by erosion and sedimentation of the Turnpike Group beginning with 80 m of sandstone and carbonate bearing slates and continuing with about 1000 m of an alternation of slates and feldspathic sandstones lacking the sedimentary structures of turbidites. The deformation of the Beardmore tectogenesis ($D_3$) is closely associated with block movements of the basement, with low grade alteration and retrograde metamorphism occurring along faults. Thrust faults inclined towards the centre of the antiform compensate the surface updoming of the crystalline basement. The variable intensity of deformation in the Turnpike Group depends on the position in relation to tectonic pressures associated with local conditions of low grade metamorphism.

## Herbert Mountains

*Series subdivision and metamorphism*

The metamorphic rocks in the Herbert Mountains, exposed between Schimper and Gordon Glaciers are divided into the younger Herbert Series probably belonging to the Skidmore Complex and into the older Charpentier Series considered a reworked member of the Provender Complex.

The Herbert Series (>2900-3000 m) occupies the southern and central parts of the Herbert Mountains (Figure 4). It consists of sequences of quartz-mica schists, biotite feldspar schists and micaceous quartzites, alternating with garnet bearing banded amphibolites. In the Lower Herbert Series (Sumgin Buttress, Shaler Cliffs, Mt Absalom) metacarbonates up to several 10's of m in thickness occur (Figures 8, 9). Relict sedimentary structures are common. The Herbert Series consists of well stratificated psammitic-pelitic sediments with marly and carbonaceous members. The lack of volcanics underlines its more miogeosynclinical (or aulacogene?) character.

The metamorphism of the pelitic-psammitic rock types is characterised by the following associations:

(a) kyanite + garnet + biotite + plagioclase ($An_{9-30}$) + quartz
(b) staurolite + kyanite + garnet + biotite + (plagioclase, $An_{22-30}$ + muscovite) + quartz
(c) kyanite + sillimanite (fibr.) + muscovite + (biotite) + quartz

These associations indicate an almandine-amphibolite facies metamorphism (520°-600°C, ~6.5 kb). The lack of any migmatisation and synkinematic granitoids in the entire Herbert Series is remarkable. Metamorphic mobilisates occur as syntectonic quartz veins, oriented parallel to the foliation and often folded.

The Charpentier Series (>1200 m) which occurs in the northern Herbert Mountains is of polymetamorphic character. Originally consisting of migmatised biotite-plagioclase- and amphibole-plagioclase-gneisses alternating with (?ortho-)amphibolite, it was subjected to a younger retrograde metamorphism and deformation, connected with an alkali-feldspar (microline) blastesis which transformed the gneisses into augen gneisses and blastomylonites. The amphibolites which

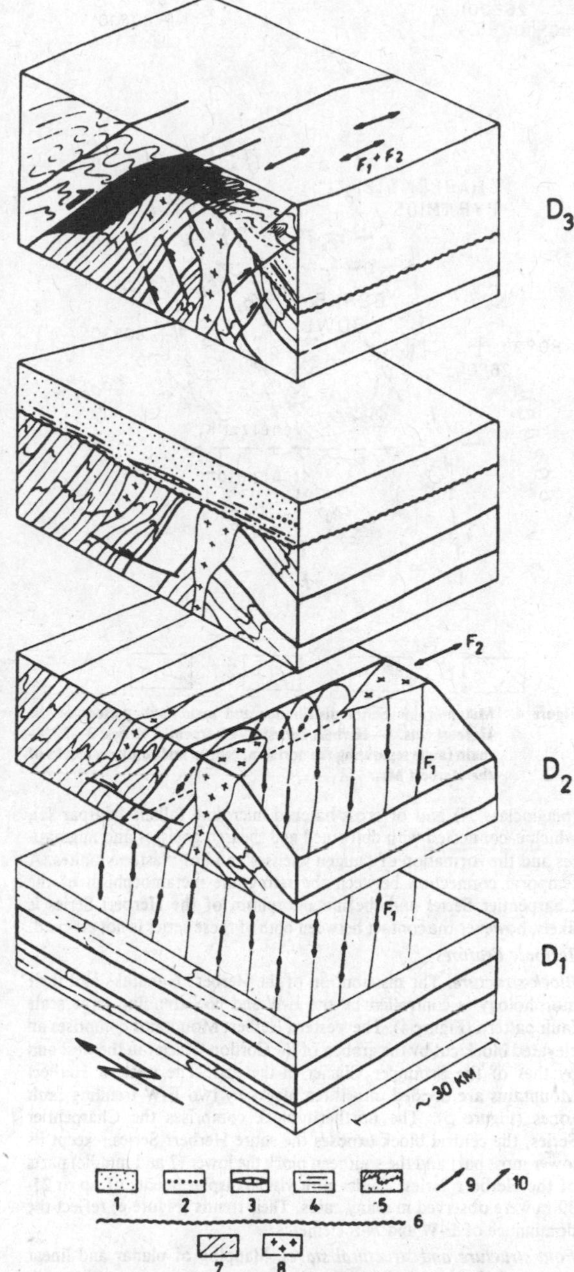


**Figure 3. Paleotectonic development of the Read Mts. 1—psammite; 2—conglomerate; 3—carbonate; 4—siltstone; 5—slate/phyllite; 6—unconformity; 7—gneiss/schist; 8—granitoid.**

represent the palaeosome of the series were subjected to this second metamorphism to a lesser extent. The following typical associations reflect the first metamorphism:

(d) in amphibolites: garnet + clinopyroxene + amphibole + plagioclase ($An_{8-54}$)
(e) in amphibole-biotite- and biotite gneisses: garnet + amphibole + biotite I + plagioclase ($An_{18-26}$) + alkali feldspar I + quartz

The absence of muscovite and presence of alkali feldspar I, clinopyroxene and amphibole indicate an almandine-amphibolite facies metamorphism of the almandine-orthoclase subfacies (675-700°C, P ~ 4 kb).

The retrograde metamorphism is connected with the biotitisation of garnet and amphibole (biotite II), the growth of an albitic plagioclase

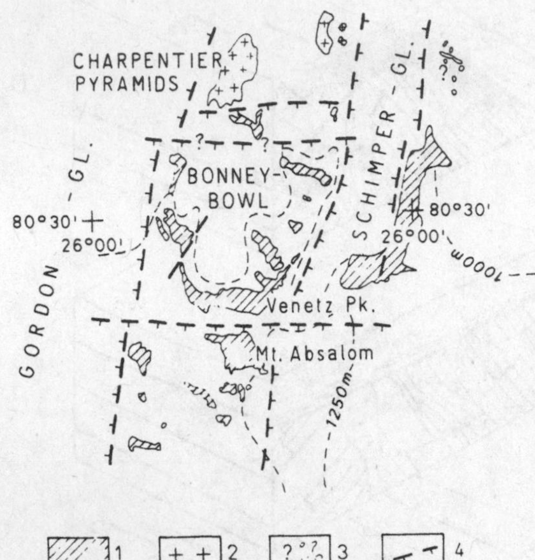


Figure 4. Metamorphic Series distribution and main fault pattern of the Herbert Mts. 1—Herbert Series; 2—Charpentier Series; 3—?; 4—main faults separating the northern, central and southern blocks of the Herbert Mts.

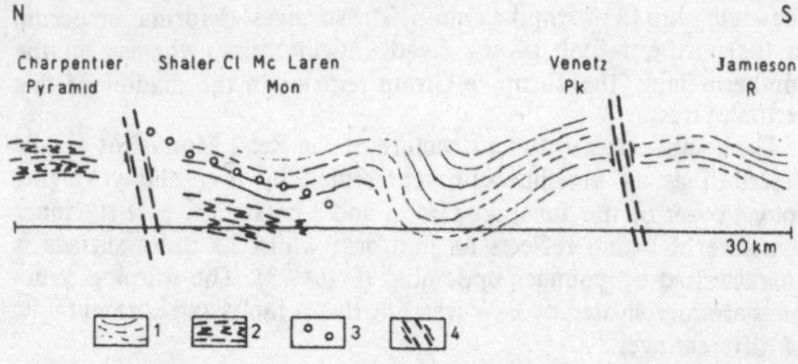


Figure 5. Schematic geologic section through the western Herbert Mts. 1—Herbert Series; 2—Charpentier Series; 3—boundary between upper and lower structural stages; 4—fault zones.

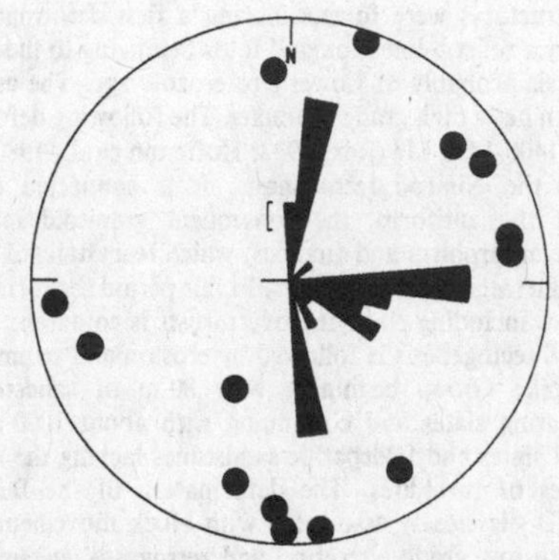


Figure 6. Synoptic representation of minor faults in the central part of the Herbert Mts. Strikes of 35 and stereographic projection of 14 minor faults, mainly from the upper parts of the Herbert Series.

(plagioclase II) and of cross-hatched microline (alkali feldspar II), which is connected with deformed and sheared gneisses and migmatites and the formation of "augen gneisses" and of blastomylonites. A temporal connection between the retrograde metamorphism of the Charpentier Series and the metamorphism of the Herbert Series is likely, however the contact between both of these series is not exposed.

*Tectonic Features*

*Block-structure.* The distribution of the Herbert Nunataks and their morphology is controlled by the E-W and N-S trending large scale fault pattern (Figure 4). The western Herbert Mountains comprises an elevated block cut by the graben of the Gordon Glacier in the west and by that of the Schimper Glacier in the east. The western Herbert Mountains are divided into three blocks by two E-W trending fault zones (Figure 5). The northern block comprises the Charpentier Series, the central block exposes the entire Herbert Series, except its lower-most part and the southern block the lower (? and middle) parts of the Herbert Series. Faults with visible displacements of up to 25-30 m were observed in many cases. Their trends (Figure 6) reflect the dominance of E-W and N-S elements.

*Fold structure and structural stages.* Mapping of planar and linear structural elements and their representation in sterograms (Figure 7) permits the reconstruction of the large scale fold structure and a division of the Herbert metamorphic rocks into two structural stages.

The upper structural stage includes the upper and middle parts of the Herbert Series. The fold structure is characterised by E-W trending linear folds, plunging gently to the east (Figure 5, 7b). The large scale fold structure in the centre is the southerly verging Bonney Bowl syncline adjoining the northern anticline. The wavelength of this fold is about 6-8 km and its amplitude 2-3 km. To the north, the anticline passes into a flat syncline with dips $<$ 15-20° (Figure 5).

The synopsis of planar and linear elements of the central block (except the area north of McLaren Monolith and the northern part of Sumgin Buttress) shows a girdle with a $\pi$-pole plunging 10° to the east, parallel to the orientation of mesocopic fold axes and b-lineations (Figure 7b, c), which is characteristic of cylindrical folds. In the southern block, the Bonney Bowl syncline passes into a flat ENE-WSW trending anticline whose axis plunges gently to the ENE (Figure 7d).

The fold structures of the upper structural stage are characterised by decreasing intensity of folding from the upper to the lower parts of the Herbert Series and pass, in the northern part of the central block (Shaler Cliffs, northern part of Sumgin Buttress, Figure 9) into the homoclinal parts of the lower structural stage. In the southern block and the northern part of the central block, this effect possibly reflects the nearness of the basement, which resisted deformation.

The lower structural stage includes the Charpentier Series and the lowest parts of the Herbert Series and occupies the northern block and the northern part of the central block (Shaler Cliffs, unnamed nunatak north of McLaren Monolith, northern face of Sumgin Buttress, Figures 8, 9) and is characterised by a NE-SW—ENE-WSW trending homocline dipping 15-25° to the SE-SSE (Figure 7a). In the northern central block this homoclinal structure passes to the south into the folded structures of the upper structural stage. The trends of mesoscopic and minor fold axes and of mineral lineations are extremely divergent and have a shallow plunge which varies from the ENE to the SSW (Figure 7a). The internal structure in the lower parts of the Herbert Series and in the Charpentier Series is governed by disharmonic folds of up to several metres, intensive shearing and boudinage of incompetent layers. In the Charpentier Series zones of blastomylonites occur (sub)parallel to the former gneissosity. The high degree of "internal deformation" and the divergences in the trends of linear elements are in striking contrast to the gentle fold structures of the upper structural stage. The transition from the divergent to a strict E-W orientation of the linear elements is restricted to a 300-400 m thick zone (northern face of Sumgin Buttress).

An interpretation of the tectonics in the Herbert Mountains is possible by assuming the deformation and metamorphism of a sedimentary formation (Herbert Series) resting on a rigid basement (?Provender Complex). Lateral compression, which produced a simple large scale fold structure in the higher parts of the Herbert Series (upper structural stage), was prevented by the protective role of the basement in the lower parts and replaced by an intensive intraformational deformation. Simultaneously, the upper part of the Charpentier Series was subjected to an intensive reworking connected with a retrograde metamorphism, blastomylonitisation and an alkali-feldspar blastesis.

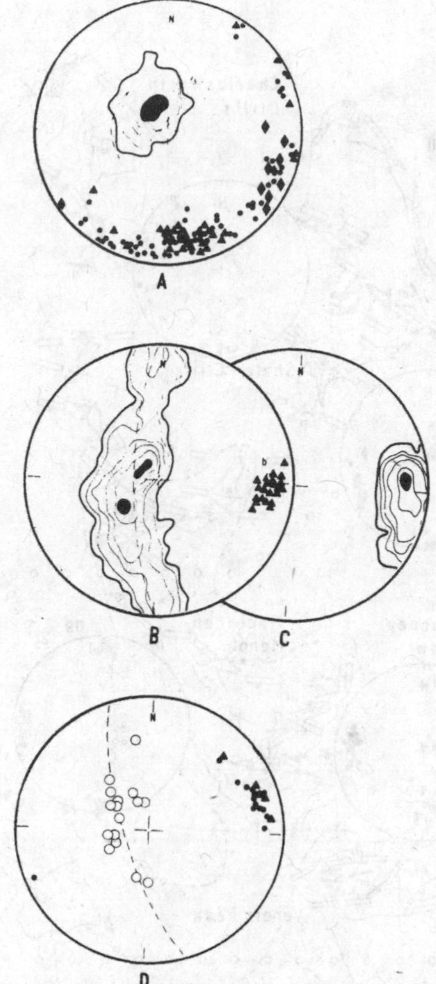


Figure 7. Synoptic representation of planar and linear paracrystalline elements from the Charpentier and the Herbert Series.
A—northern Herbert Mts. (lower structural stage), 75 poles of gneissosity and foliation (contours: 1, 6, 10, 20, 30%); 26 minor fold axes (▲); 75 mineral lineations (●); 9 quartz streaks (◆).
B—central part of the Herbert Mts. (Bonney Bowl area): 221 poles of foliation and bedding (contours: 0.5, 1, 3, 5, 10, 15%); 22 axes of minor folds (▲).
C—the same areas as for B: 102 poles of mineral lineations (contours: 1, 3, 5, 10, 20, 25%).
D—southern part of the Herbert Mts.: 16 poles of foliation (○); 14 mineral lineations (●); 5 axes of minor folds (▲).
B, C and D comprise the upper structural stage.

Figure 8. Northern face of Sumgin Buttress from the North. Lower parts of the Herbert Series consisting of biotite-quartz and biotite-feldspar-quartz schists (right side, light grey) with two metacarbonate layers (m). In the central and left part amphibolites (dark) with a metacarbonate layer (c). Dotted line between biotite-quartz-feldspar schists and amphibolites. Section paraleel to the axial plunge.

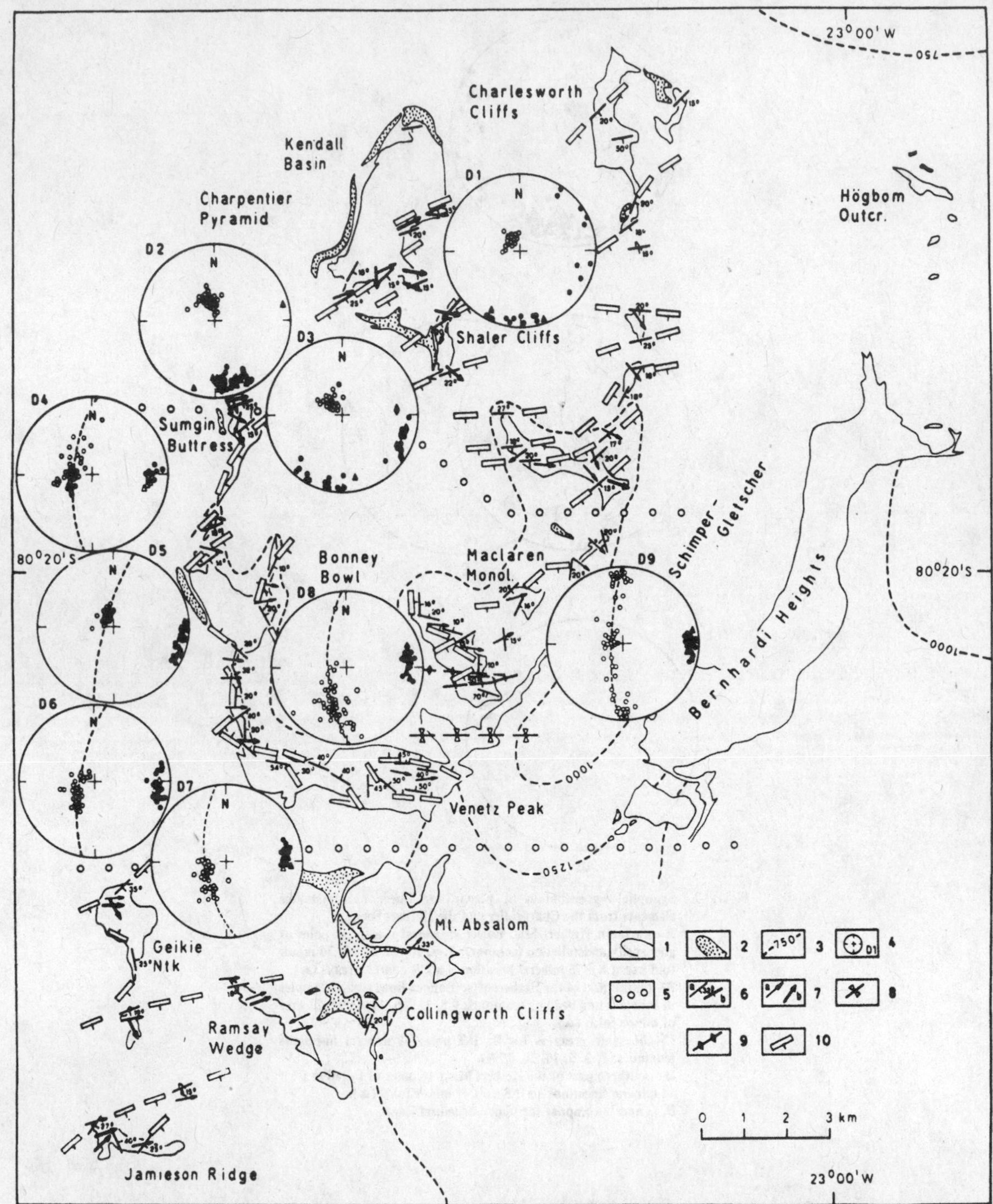


**Figure 9.** **Tectonic Map of the Herbert Mts. (compilation includes field data from S.E. Benovolenskij E.N. Kamenev and V.V. Samsonov), 1—outcrop boundary; 2—moraines; 3—isohypses; 4—stereographic plots—lower hemisphere (○ foliation, bedding, gneissosity; ● —lineations, ▲—fold axes; ◆ —quartz streaks); 5—boundary between fault blocks (comp. fig. 5); 6—planar elements (a—with dips; b—vertical); 7—linear elements (a—lineation; b—fold axes); 8—axes of synclines; 9—axes of anticlines; 10—trends of planar elements. Topographic base: Sheet W80 24/26 1:200 000 British Antarctic Survey.**

*Acknowledgements* H.J. and J.H. carried out field work in the southern and eastern Shackleton Ranges respectively as members of the 1975/76 and 1976/77 Soviet Antarctic Expeditions respectively. The authors wish to thank the field parties and aircraft crews of the 22nd and 23rd SAE, and especially their geological leaders Dr O.G. Suljatin and Dr E.N. Kamenev. The research work was sponsored by the Central Institute of the Earth's Physics, Department of Polar Research of the Academy of Sciences of the German Democratic Republic and the Ministry of Higher and Technical Education of the GDR.

## REFERENCES

CLARKSON, P.D., 1971: Shackleton Range geological survey. *U.S. Antarct. J., 6,* 121-2.

CLARKSON, P.D., 1972: Geology of the Shackleton Range: a preliminary report. *Bull. Br. antarct. Surv., 31,* 1-15.

CLARKSON, P.D., 1982: Tectonic significance of the Shackleton Range, *in* Craddock, C. (ed.) *Antarctic Geoscience.* Univ. Wisconsin Press, Madison, 835-40.

GOLOVANOV, N.P., MIL'STEJN, V.E., MICHAJLOV, V.M. and SULJATIN, O.G., 1979: Stromality i mikrofitolity chrebta Sekltona (Zapadnaja Antarktida). *Dokl. Akad. Nauk SSR, 249, 4,* 977-9.

GRIKUROV, G.E. (ed.) 1980: *Tectonic Map of Antarctica* 1:10,000,000. (Explanatory notes). Research Institute of Arctic Geology, Leningrad.

GRIKUROV, G.E. and DIBNER, A.F., 1979: Vozrast i strukturnoe položenie osadočnych tolso v zapadnojčasti chrebta Šekltona (Antarktida). *Antarktika, 18,* 20-32.

GRIKUROV, G.E., KAMENEV, E.N. and KAMENEVA, G.I., 1979: Epochi granitobrazovanija i glavyne tendencii evoljucii granitoidnych kompleksov v Antarktida. *Antarktika, 18,* 32-44.

HOFMANN, J., KAISER, G., KLEMM, W. and PAECH, H.J., 1980: K-Ar Alter von Doleriten und Metamorphiten der Shackleton Range und der Whichaway Nunataks, West- and Südumrandung des Filchner-Eisschelfes (Antarktika). *Z. geol. Wiss., 8,* 1259-65.

HOFMANN, J., PILOT, H.J. and SCHLICHTING, M., 1981: Das Rb-Sr Alter von Metamorphiten der Herbert Mountains, Shackleton Range, Antarctica. *Zschr. Geol. Wiss., 9, 8,* 835-42.

REX, D.C., 1972: K-Ar Age Determinations on Volcanic and associated Rocks from the Antarctic Peninsula and Dronning Maud Land; *in* Adie, R.J. (ed.): *Antarctic Geology and Geophysics.* Universitetsforlaget, Oslo, 133-6.

PANKHURST, R.J., MARSH, P.D. and CLARKSON, P.D. (this volume): A Geochronological Investigation of the Shackleton Range.

SOLOVIEV, I.A. and GRIKUROV, G.E., 1978: Pervye nachodki srednekembrijskich trilobitov v chrebte Šeklton (Antarktida). *Antarktika, 17,* 187-98.

SOLOVIEV, I.A. and GRIKUROV, G.E., 1979: Novye dannye o rasprostanenii kembrijskich trilobitov v chrebtach Ardžentina i Šeklton. *Antarktika, 18,* 54-73.

# THE LATE PRECAMBRIAN AND EARLY PALAEOZOIC HISTORY OF THE SHACKLETON RANGE, COATS LAND

P.D. Marsh, *British Antarctic Survey, Madingley Road, Cambridge, CB3 OET, U.K.*

*Abstract* Within the range four extensive sequences of metasediments overlie and are interleaved with a Mid-Proterozoic high-grade basement. Two sequences had an active source providing a high proportion of first-cycle sediment. They crop out in the north, at high structural levels and have coarse grained amphibolite facies fabrics cut by retrograde mylonitic fabrics. The metasediments in the south, at lower structural levels, are dominated by more mature sediment and have prograde amphibolite and greenschist facies fabrics. The range is interpreted as a complex of nappes, between a high grade terrain to the north and a foreland to the south, brought together during the events which produced the prograde fabric in the metasediments of the south and the retrograde fabric in those of the north and in the basement. Sedimentation and deformation took place within the same time range as in the Beardmore and Ross Orogens. However the ensialic nature of the Late Precambrian sedimentation, a lack of major plutonism, the present cratonic setting, and the structural trend of the range favour comparison not with these "geosynclinical" orogens of the margin of the East Antarctic craton but with the nearly contemporaneous, intra-Gondwana, "Pan African" mobile belts some of which became the sites of rifting to form the South Atlantic and Indian oceans.

The Shackleton Range (80-81°S, 31-19°W) is composed mainly of amphibolite facies rocks of the Proterozoic-Early Palaeozoic Shackleton Range Metamorphic Complex (Clarkson, 1972; Grew and Halpern, 1979). In the south, this is unconformably overlain by a Late Precambrian quartzite-carbonate sequence and has tectonic contacts with deformed low grade metasediments of the Turnpike Bluff Group (Clarkson, 1972, 1981; Golovanov et al., 1980). In the northwest the complex is overlain by unmetamorphosed, arenaceous strata of the Palaeozoic Blaiklock Glacier Group (Stephenson, 1966; Clarkson et al., 1979; Grikurov and Dibner, 1979). The range is a horst separated from undeformed rocks of the Beacon Supergroup by major glaciers (Stephenson, 1966; Clarkson, 1982).

Recent fieldwork, and a re-examination of previous collections, allows a new map and structural interpretation to be made (Figure 1). This brings together the main features of earlier interpretations, particularly the recognition in the metamorphic complex of the old "basement" and younger "cover" rocks (Clarkson, 1972; Kamenev and Semenov, 1980), widespread blastomylonites (Stephenson, 1966; Grikurov and Dibner, 1979) and major thrusts separating rocks with different thermal histories (Stephenson, 1966). The cross section of the range (Figure 1) is similar to that shown by Hofmann and Paech (1980) but the interpretations differ in both the stratigraphic units recognised and the role played by low angle thrusts and high angle faults. The more detailed interpretation of Hofmann and Paech (this volume) has many features in common with that given here; the most significant differences result from the use of different chronologies.

## Stratigraphy

Within the Shackleton Range Metamorphic Complex (SRMC) in the Haskard Highlands (Figure 1) three formations of metasediments (the Nostoc Lake Formation, the Mount Gass Formation and the Williams Ridge Formation) can be distinguished from older basement gneisses. Sequences correlated with these, for which the same names are used here, are present throughout the northern part of the range where together they approximate equivalents of the Herbert Group of Hofmann and Paech (1980). The Turnpike Bluff Group (TBG) is associated with only the older gneisses of the SRMC (Clarkson, 1982). It has recently been recognised observed basal strata ascribed to the TBG are separated by a zone of high strain from the overlying slates and are structurally overlain by gneiss at Watts Needle. This basal

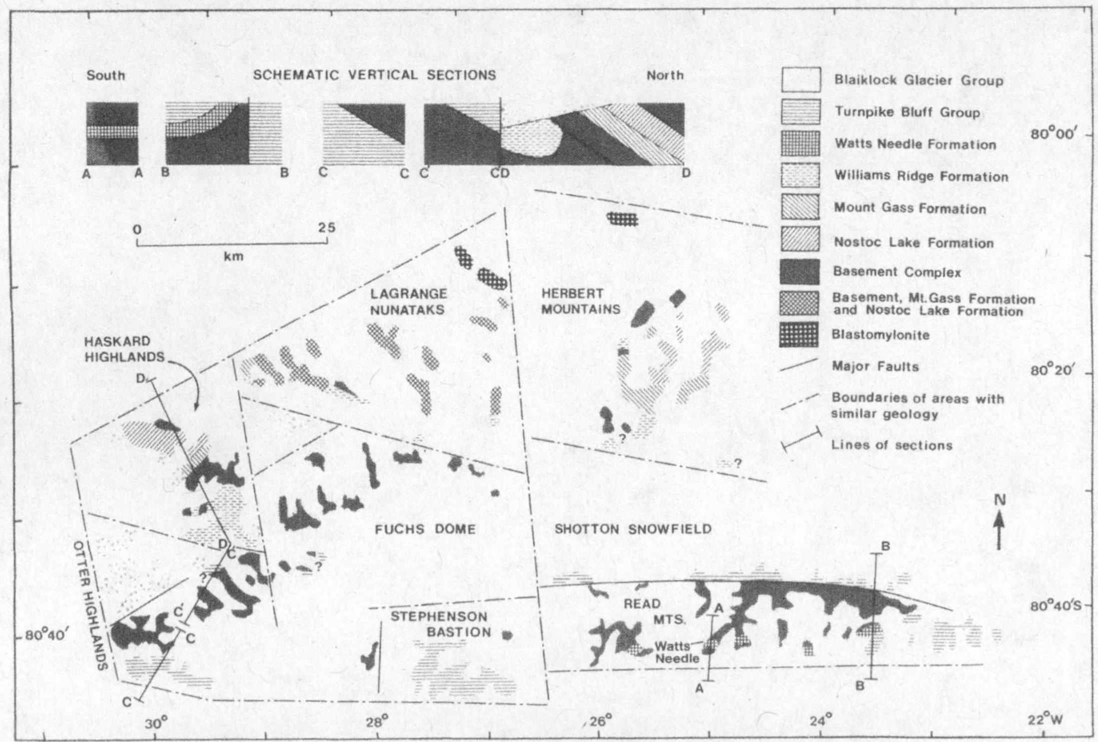


**Figure 1.** Geological sketch map of the central and western parts of the Shackleton Range, with schematic vertical sections.

sequence is here named the Watts Needle Formation. The Blaiklock Glacier Group (BGG) unconformably overlies metasediments of the SRMC. Glacial erratics of shale, siltstone and fine sandstone containing a Middle Cambrian fauna have been ascribed to unexposed strata of the BGG (see Clarkson et al., 1979, for references).

The six main stratigraphic elements of sedimentary origin are summarised in order of descending metamorphic grade.

*Nostoc Lake Formation* In the northern Haskard Highlands three lithological associations recognised by Grew and Halpern (1979) lie in discrete belts. In the northeastern belt marble, siliceous and feldspathic granoblastites, kyanite schist, and garnetiferous gneiss and schist are adjacent to and interleaved with migmatitic orthogneiss regarded as basement. In the central belt the main lithology is richly garnetiferous, mainly mesocratic, biotite—and/or hornblende-bearing gneiss with local calc-silicate layers. Veins and sheets of granitic composition, and isolated balls, schlieren and boudinaged layers of basic (locally ultrabasic) material are present. In the southwestern belt and adjacent to the Mount Gass formation mesocratic garnetiferous gneisses are dominant but marble and subordinate layers of more variable composition (as in the northeast) are present.

Garnetiferous schist and gneiss, with subordinate layers of marble and kyanite schist, in the Herbert Mountains are correlated with this formation. Throughout the Lagrange Nunataks attenuated sequences occur in which garnetiferous gneiss and schist overlie basal rocks with similarities to both the northeastern belt of the Haskard Highlands and to the Mount Gass Formation. Several typical outcrops of the Mount Gass Formation are overlain by the Nostoc Lake Formation or similar rocks and it is therefore uncertain whether the two formations are a single sequence or two similar sequences.

*Mount Gass Formation.* A consistent lithological sequence away from an unconformable base is normally present: muscovite-bearing quartzite (typically green due to the presence of fuchsite), metalimestone, garnetiferous muscovite schist, grey metadolomite and garnetiferous amphibolite, are interlayered on centimetre and metre scales. These reach their maximum abundance in the order quoted.

*Williams Ridge Formation.* In Haskard Highlands cream or grey sandy metalimestone (rarely metadolomite), carbonate mica-schists, and mica-schists are structurally overlain by highly sheared gneiss. A few metres of quartzite are present at the boundary. Similar but higher grade metalimestones in the Lagrange Nunataks, with quartzite above an unconformable contact with gneiss, and a carbonate-dominated sequence in the Herbert Mountains, are correlated with this formation; both are tectonically overlain by gneiss.

*Turnpike Bluff Group.* Clarkson (1972, 1981) described four formations composed of quartzite, quartzose sandstone and siltstone, argillaceous siltstone and, in the east, pebble conglomerates with clasts of quartzite, sandstone, limestone/metalimestone, slate, schist, gneiss and granitic rocks. Metamorphosed quartzite which appears to overlie gneiss near Fuchs Dome may be part of the group, as may garnet-mica-schists in the southern Haskard Highlands, although these also resemble parts of the Williams Ridge Formation. Exposed contacts with gneiss are tectonic. The Watts Needle Formation may or may not be the base.

*Watts Needle Formation.* Gneiss of the SRMC is overlain unconformably by 12 m of quartzite with purple shales which pass upwards into about 60 m thinly bedded sandstones and quartzites with pale grey limestones up to 4 m thick.

*Blaiklock Glacier Group (BGG).* The metamorphic complex is overlain by more than 760 m of red (mainly) and grey, fine to coarse grained sandstones with red and grey shales (Clarkson, 1972). Pebble conglomerates are common near the base and a basal boulder conglomerate is present locally. A great thickness (possibly 5300 m) of grey-green, cross-bedded, feldspathic sandstones with heavy mineral bands, conglomerates and micaceous shales in the Otter Highlands is believed to be higher in the same sequence. The clastic content of the group indicates derivation from rocks similar to those now cropping out in the range.

## Metamorphism

The grain size and metamorphic grade of the metasediment increases northwards. The Watts Needle Formation is lower greenschist facies or less and has little tectonic fabric. The TBG is greenschist facies and has a slaty or spaced cleavage depending on lithology. Biotite is absent in the southern Read Mountains where the cleavage is least well developed but ubiquitous in the northern Read Mountains where the cleavage is most pervasive. It is present in some lithologies at Stephenson Bastion and, particularly in the north of the outcrop, in Otter Highlands. In the Williams Ridge Formation micaceous rocks are schistose; the grade is lowest amphibolite facies in the Haskard Highlands but middle amphibolite facies to the east. In all the above metasediments the dominant foliation is the latest, coarsest grained and highest grade fabric recognised. In the associated basement coarse grained high grade fabrics are cut by finer grained mylonitic fabrics which (where contacts are not faulted) are of similar grade and parallel to the planar fabrics in the metasediments. In the north of the range mylonitic fabrics are found in both basement and metasediments. They are mainly of middle and lower amphibolite facies and are parallel to, and well-developed at, the formation boundaries. Earlier coarser grained and higher grade fabrics were originally middle amphibolite facies in the Mount Gass formation and middle or upper amphibolite facies (with local anatexis) in the Nostoc Lake Formation. Comparisons of grade and grain size suggest that the mylonitic fabrics are broadly equivalent to the similar prograde fabrics in the supracrustal rocks to the south.

## Structure

Except in the northern Haskard Highlands the main folds in the supracrustal rocks have east-west striking axial surfaces and subhorizontal or gently plunging hinges. In the TBG they vary with lithology from open to isoclinal and have northerly dipping axial surfaces to which the cleavage is parallel (Clarkson, 1981, 1982). Local post-cleavage folds have southerly dipping axial surfaces. In the Lagrange Nunataks and Herbert Mountains isoclinal folds with axial surfaces parallel to the dominant foliation, and possibly thrusts, produce repetitions of Mount Gass Formation, Nostoc Lake Formation and basement rocks. These structures, mylonitic fabrics, and the tectonic contact with the Williams Ridge Formation, are deformed by major close folds with mainly southerly dipping axial surfaces. In the northern Haskard Highlands structures similar to those in the Lagrange Nunataks are deformed by major folds with north-south striking axial surfaces and northward plunging hinges which crenulate the mica fabrics.

In the Read Mountains (sections A and B, Figure 1) TBG and basement rocks, possibly in the same nappe, overlie Watts Needle Formation and basement which is in fault contact with higher grade TBG. In the southern Otter and Haskard Highlands (sections C and C′) the mylonitic fabric in the basement and the cleavage in the TBG to the south dip northwards as shown (cf. Stephenson, 1966). An interpretation of the structure of the northern Haskard Highlands prior to the post-foliation folding is shown in section D. The structure of the Lagrange Nunataks and Herbert Mountains appears to be analogous; the Williams Ridge Formation and underlying basement are overthrust by a sequence of interfolded and/or interleaved basement, Mount Gass Formation and Nostoc Lake Formation, which will be referred to as the Lagrange Nappe.

Large areas of the range have characteristic geology. Their boundaries, drawn (Figure 1) with the guidance of sub-ice lineaments seen on Landsat imagery, are probably the trends of major fracture zones; some are major faults. The Lagrange Nappe and Williams Ridge Formation form an ENE trending "northern zone" of amphibolite facies rocks the outcrop of which is offset under the glacier between the Haskard Highlands and the Lagrange Nunataks (and extends eastwards to include outcrops not shown in Figure 1). Variations in metamorphic grade and fabric suggest the recognition of a "central zone" (also offset) to the south of this and a "southern zone" consisting of the Read Mountains south of the major fault and possibly the Stephenson Bastion area. The composite section A-D would thus represent a section across the structure of the whole range.

The range is interpreted as a complex of nappes brought together in an event or events which produced the retrograde fabrics in the basement and the Lagrange Nappe and the prograde fabrics in the Williams Ridge Formation and TBG. The inferred or observed high angle faults forming the "zone" boundaries make it difficult to determine the relative levels of exposure in different areas. However the dip throughout the range is typically to the north and within several areas the metamorphic grade associated with the later structures increases northwards and structurally upwards, supporting the conclusion that the high grade rocks of the Lagrange Nappe are the highest structural level seen and the basement underlying the Watts Needle Formation the lowest.

The north-south trending structures are superimposed on the nappe complex. They are well-developed only in the west of the "northern zone", although flexures with similar trends are detectable in basement rocks of the Fuchs Dome and southern Haskard Highlands. The BGG is restricted in outcrop to this western area. It also has been deformed about the northward plunging axis (Clarkson, 1982), but this may be due to tilting rather than folding.

## Tectonic and Depositional History

Recent geochronological investigations (Pankhurst, et al., this volume, for new data and references) allow the development of the range to be viewed within a more complete time-scale than has previously been possible (Table 1). The metasediments overlie an Early Proterozoic or Late Archaean basement which had undergone major thermal activity in the Mid-Proterozoic. The Nostoc Lake Formation probably underwent its main metamorphism in the Vendian or at about the time of the Vendian-Cambrian boundary. The TBG of the southern Read Mountains underwent diagenesis or metamorphism in the Cambrian. The main metmorphism of the Williams Ridge Formation took place at about the time of the Cambrian-Ordovician boundary and mineral ages in rocks with the mylonitic fabric were also set at this time. A similar but slightly younger age from the overlying BGG probably represents diagenesis in the Early Ordovician. The dating evidence indicates that the east-west trending structures were active, and may have developed, in the Cambrian or earliest Ordovician. If, as appears to be the case, the BGG is less affected by north-south trending structures than the underlying rocks then their initiation must have occurred in the Late Cambrian or Early Ordovician. The most important uncertainty is over whether there was hiatus between the Vendian/Cambrian and the Cambrian/Ordovician activity during which the TBG and/or the Williams Ridge Formation could have been deposited.

In the Lagrange Nappe Late Precambrian sedimentation began under shelf conditions with siliciclastic rocks, carbonates and argillites. Most of the clastic material was derived from the underlying basement or rocks of similar antiquity but the interlayered amphibolites in the Mount Gass Formation suggest the possibility of nearby volcanic activity. Extensive magmatic activity is evidenced within the paragneisses of the Nostoc Lake Formation which were originally immature, first cycle sediments, probably of volcanogenic origin. Their isotopic composition is consistent with their derivation about 850 Ma ago from Late Precambrian mantle-derived magmatic rocks. Extrusive or hypabyssal rocks may be present in the sequence.

The exposed basal rocks in the south were also deposited under shelf conditions. The Watts Needle Formation contains Precambrian (Golovanov et al., 1980) stromatolites and was derived mainly from the basement. The Williams Ridge Formation represents deposition on a shelf with stable margins and the mature arenaceous rocks of the TBG suggest derivations from a stable craton where reworking could occur. In contrast to the metasediments of the Lagrange Nappe there is no indication of a volcanogenic component to the sediment but the strontium isotope ratios of pelites from both formations suggest a source with an average age less than that of the local basement. They are probably younger than the metasediments of the Lagrange Nappe and thus may contain recycled material from the same source and possibly from the terrain represented by the Lagrange Nappe. The minimum age of the TBG is Early-Middle Cambrian and that of the Williams Ridge Formation Late Cambrian. All the evidence points to the BGG being Late Cambrian or Ordovician "post-orogenic" sediment.

Palaeogeographic reconstruction is hindered by uncertainty as to the amount and direction of movement between nappes. The metasediments of the Lagrange Nappe indicate the presence of an active Late Precambrian plate boundary, probably to the north. In the absence of knowledge as to the affinities of the magmatic rocks it is uncertain whether the boundary was extensional or compressive but the deposition of volcanogenic rocks on a sialic shelf hints at the former. The sediments in the south were deposited, probably later, within the craton or on a passive margin. These terrains were sufficiently remote during the Vendian and possibly the Early Cambrian that the event recorded in the Lagrange Nappe had little or no effect on the sediments (if they had been deposited) or basement in the south. By the Early Ordovician they had been brought together and suffered a deformation sufficiently intense to produce the dominant east-west trending structures.

The northward increase in grade, and the dip of the main structures, suggest that the range lies between a foreland to the south and a more extensive high grade terrain to the north. It appears to be part of a Late Precambrian-Early Palaeozoic tectonic belt trending east-west.

## Regional Correlation and Tectonic Significance

The range has geological similarities with both the Transantarctic Mountains and Dronning Maud Land (Clarkson, 1972, 1982; Hofmann and Paech, 1980) which has led to its inclusion by some authors (e.g. Elliot, 1975) in the Beardmore and Ross Orogens of the craton-margin in the Transantarctic Mountains and by others (Kamenev and Semenov, 1980) in intracratonic fold belts developed in aulocogens. The refined stratigraphy and chronology allow further comparisons to be made.

The metasediments of the Lagrange Nappe are broadly similar in their age, lithology, association with volcanic rocks, and (probable) Vendian metamorphism to the Beardmore Group of the central Transantarctic Mountains and the Patuxent Formation of the Pensacola Mountains, although in the latter case the more mature sedimentary types are absent (see Elliot, 1975; Faure et al., 1979; and Grindley, 1981, for reviews). In that it (probably) postdates the Vendian orogeny, is dominated by carbonate rocks and was deformed during the Late Cambrian or Early Ordovician, the Williams Ridge Formation is analogous to the Cambrian rocks of the nearest parts of the Transantarctic Mountains (Laird, 1981, for review). Despite these similarities there are significant differences. In the Shackleton Range the basement underlying the metasediments makes up a high proportion of the outcrop, the metasediments reached amphibolite grade in both Vendian and Cambrian events, volcanic rocks are absent from

**TABLE 1: Supposed sequence of rock units and events**

| Possible Age | North (and north-west) | Centre | South | Possible Time Equivalents in other Regions |
|---|---|---|---|---|
| Early Ordovician | Blaiklock Glacier Group | | | Urfjell Group Neptune Group |
| Cambro-Ordovician boundary | | Major east-west deformation, nappes, metamorphism | | Ross Orogeny |
| | | Williams Ridge Formation | Turnpike Bluff Group | Byrd Group |
| Cambrian or Late Proterozoic | | Probable Vendian or Early Cambrian metamorphism | | Beardmore Orogeny |
| Late Proterozoic | Nostoc Lake fm. Mount Gass fm. | | Watts Needle Formation | Beardmore Group, Patuxent Fm., volcanics in Dronning Maud Land |
| Mid-Proterozoic | | Plutonism and metamorphism | | |
| Early Proterozoic or Late Archaean | Basement complex | Basement complex | Basement complex | East Antarctic shield |

the younger metasediments, the structural trend is east-west, and syn- and post-tectonic acid plutons are absent. In the Transantarctic Mountains south of 80°S basement rocks are restricted in extent and not seen in contact with metasediments, the metasediments only locally exceed the greenschist facies of regional metamorphism, the Cambrian rocks are associated with volcanics, the structural trend is north-south (local) and a Cambro-Ordovician batholith is present.

The structural trend of the range is similar to that within the craton in Dronning Maud Land (Craddock, 1970), where Late Precambrian—Early Palaeozoic ages are also recorded from metamorphic rocks, and high grade gneisses suffered retrogression and cataclasis (e.g. Van Autenboer and Loy, 1972; Juckes, 1972). Of particular significance to correlation with the Lagrange Nappe and the provenance of its sediments is an extensive pre-Middle Proterozoic platform in western Dronning Maud Land which suffered tensional faulting and volcanism at approximately 824-860 Ma (Neethling, 1972). A significant difference between the Shackleton Range and the known geology of Droning Maud Land is the presence of low grade Late Proterozoic sediments, the Turnpike Bluff Group; however, in their mature clastic content these rocks are also atypical of the Transantarctic Mountains.

Many of the differences between the range and the Beardmore and Ross Orogens could be due to the location of the range on the cratonic side of the extrapolated trend of the Transantarctic Mountains; fossiliferous arenaceous, rather than calcareous, Middle Cambrian sediments in the vicinity of the range also suggest a more cratonic environment. However the structural trend suggests that the range is part of a Vendian-Cambrian tectonic belt now situated within the craton, which would have reached the craton margin in the area of the Filchner Ice Shelf.

In the context of a reconstructured Gondwana the range is seen as part of a complex of intra-Gondwana mobile belts of "Pan African" age which reach the "Pacific"margin of the supercontinent in the area which became the site of rifting to form the south Atlantic and Indian Oceans (Tankard, et al., 1982). The Shackleton Range is not sufficiently extensive to determine whether the mobile belt developed in an aulocogen or during the closure of an ocean; the geology of Dronning Maud Land seems to suggest the former. Whatever the origin of the belt the deformation indicates some de-coupling of crustal plates in the area during the development of the Ross Orogen. Thus the oceanic subduction model for the development of the Transantarctic Mountains during this period (Stump, 1976) need not apply beyond their present termination at the Filchner Ice Shelf. This is of significance if the early Gondwana geology of crustal segments is used as an aid to the reconstruction of this part of Gondwana.

*Acknowledgements* Field work during the 1977-78 season was supported by the BAS air unit operating from Halley and also using facilities at Argentine and Soviet stations. I am very grateful to Major Hector Papa and personnel at General Belgrano for the provision of refuelling facilities and for hospitality during the winter of 1978, and to Dr V.N. Masolov and members of the 23rd Societ Antarctic Expedition at Druzhnaya for the supply of fuel in the field and the transport of samples.

## REFERENCES

BRITISH ANTARCTIC SURVEY, 1978: *British Antarctic Survey annual report*, 1977-78. Cambridge, British Antarctic Survey, 72pp.

CLARKSON, P.D., 1972: Geology of the Shackleton Range: a preliminary report. *Bull. Br. antarct. Surv., 31*, 1-15.

CLARKSON, P.D., 1981: Geology of the Shackleton Range: II. The Turnpike Bluff Group. *Bull. Br. antarct. Surv., 52*, 109-24.

CLARKSON, P.D., 1982: Tectonic significance of the Shackleton Range; *in* Craddock, C. (ed.). *Antarctic Geoscience*. Univ. Wisconsin Press, Madison, 835-9.

CLARKSON, P.D., HUGHES C.P. and THOMSON, M.R.A., 1979: Geological significance of a Middle Cambrian fauna from Antarctica. *Nature, 279*, 791-2.

CRADDOCK, C., 1970: Tectonic map of Antarctica; *in* Bushnell, V.C. and Craddock, C. (eds.) Geologic Maps of Antarctica, *Antarct. Map Folio Ser.*, Folio 12, Pl XXIII.

ELLIOT, D.H., 1975: Tectonics of Antarctica: a review. *Am. J. Sci.*, 275-A, 45-106.

FAURE, G., EASTIN, R., RAY, P.T., MCLELLAND, D. and SCHULZ, C.H., 1979: Geochronology of igneous and metamorphic rocks, central Transantarctic Mountains; *in* Laskav, B. and Raja Rao, C.S. (eds.) *Fourth International Gondwana Symposium, 2*, Hindustan Publishing Company, Delhi, 805-11.

GOLOVANOV, N.P., MIKHAILOV, V.M. and STULIATIN, O.G., 1980: Pervyye diagnostiruemye stromatolite Antarktidy i ikh biostrategrafecheskoe znachenie. (First diagnostic stromatolites from Antarctica and their biostratigraphical significance.) *Antarktika, 19*, 152-9.

GREW, E.S. and HALPERN, M., 1979: Rubidium-Strontium dates from the Shackleton Range Metamorphic Complex in the Mount Provender area, Shackleton Range, Antarctica. *J. Geol., 87*, 325-32.

GRIKUROV, G.E. and DIBNER, A.F., 1979: Vozrast i strukturnoe polozheniye osadochnykh tolshch v zapadnoi chasti khebta Shaklton (Antarktida). (Age and structural position of the sedimentary deposits in the western Shackleton Range (Antarctica)). *Antarktika, 18*, 20-31.

GRINDLEY, G.W., 1981: Precambrian rocks of the Ross Sea region. *R. Soc. N.Z., J., 11*, 411-23.

HOFMANN, J. and PAECH, H.J., 1980: Zum strukturgeolischen Bau am Westrand der Ostantarktischen Tafel. (On the geological structure of the western end of the east Antarctic platform.) *Z. geol. Wiss., 8*, 425-37.

JUCKES, L.M., 1972: The geology of north-eastern Heimefrontfjella, Dronning Maud Land. *Scient. Rep. Br. antarct. Surv., 65*.

KAMENEV, E.N. and SEMENOV, V.S., 1980: Regional metamorphism in Antarctica; *in* Cresswell, M.M. and Vella, p. (eds.) *Gondwana Five*, A.A. Balkema, Rotterdam, 209-15.

LAIRD, M.G., 1981: Lower Palaeozoic rocks of the Ross Sea area and their significance in the Gondwana context. *R. Soc. N.Z., J., 11*, 425-38.

NEETHLING, D.C., 1972: Age and correlation of the Ritscher Supergroup and other Precambrian rock units, Dronning Maud Land; *in* Adie, R.J. (ed.) *Antarctic Geology and Geophysics*. Universitetsforlaget, Oslo, 547-56.

PANKHURST, R.J., MARSH, P.D. and CLARKSON, P.D., (this volume): A geochronological investigation of the Shackleton Range.

STEPHENSON, P.J., 1966: Geology. 1. Theron Mountains, Shackleton Range and Whichaway Nunataks (with a section on palaeomagnetism of the dolerite intrusions, by Blundell, D.J.) *Scient. Rep. Transantarctic. Exped., 8*,

STUMP, E., 1976: On the Late Precambrian—Early Palaeozoic metavolcanic and metasedimentary rocks of the Queen Maud Mountains, Antarctica, and a comparison with rocks of similar age from southern Africa. *Rep. Inst. polar Stud. Ohio State Univ., 62*,

TANKARD, A.J., JACKSON, M.P.A., ERIKSSON, K.A., HOBDAY, D.K., HUNTER, D.R. and MINTER, W.E.L., 1982: *Crustal Evolution of Southern Africa*. Springer Verlag, New York, 523pp.

VAN AUTENBOER, T. and LOY, W., 1972: Recent geological investigations in the Sφr-Rondane Mountains, Belgicafjella and Sverdrupfjella, Dronning Maud Land; *in* Adie, R.J. (ed.) *Antarctic Geology and Geophysics*. Universitetsforlaget, Oslo, 563-71.

# STRUCTURE AND OUTLINE OF GEOLOGIC HISTORY OF THE SOUTHERN WEDDELL SEA BASIN

E.N. Kamenev, *Antarctic Division, Department for Marine Geological Exploration, Sevmorgeologia, Leningrad, USSR.*

V.L. Ivanov, *Department of Antarctic Geology and Mineral Resources, VNIIOkeangeologia, Leningrad, USSR.*

*Abstract* The southern Weddell Sea Basin includes parts of the Maudheim Plate, Transantarctic and Ellsworth Mountains fold systems, Antarctic Peninsula fold system and the Weddell Sea Basin proper. In the geologic history of the region there are two major periods recognised: (1) formation of continental crust and (2) formation of sedimentary-volcanic cover. Thick (up to 15 km) sedimentary fill of the Weddell Sea Basin has been forming from the Late Proterozoic to the Cenozoic.

The present report is based on regional geophysical and geological surveys carried out by PGO "Sevmorgeologia" in the Weddell Sea area from the Druzhnaya Base during the austral summers of 1975 to 1982. This work included aeromagnetic, gravity and aerogravity surveys, deep seismic and radio-echo soundings, seismic reflection and marine seismic reflection measurements, as well as geological reconnaissance in mountainous areas. Extra data were obtained by using reports of foreign geologists.

## GEOLOGICAL STRUCTURE

With the exception of the Antarctic Peninsula, the study region is part of an area of pericratonal downwarp occupying a vast area between the East Antarctic craton and the West Antarctic geosynclinal fold system of the Pacific mobile belt. The region covers a variety of structures, such as the Maudheim Plate, fold systems of the Transantarctic and the Ellsworth Mountains, the Weddell Sea Basin proper and a zone of Late Mesozoic folding in the Antarctic Peninsula fold system (Figure 1).

### Maudheim Plate

In the northeast the southern Weddell Sea Basin is contiguous with the epi-Early Proterozoic Maudheim Plate. An important tectonic feature of the structure is the widespread development of a Late Precambrian-Recent volcanogenic-sedimentary platform cover on the Archaean-Early Proterozoic crystalline basement. Late Precambrian, Early Palaeozoic, Middle Palaeozoic, Jurassic and Recent volcanic and sedimentary formations are known to crop out in the Ritscher Upland, in the Kottas, Kraul and Kirwanveggen Mountains (Neethling, 1972; Bredell, 1982). On average the total thickness of the cover formations does not exceed 1-1.5 km; in depressions it reaches 3-5 km. Sills and lavas of Late Proterozoic, Early Palaeozoic and Jurassic trap formations are also present. Locally all the formations underwent weak (open) folding in pre-Jurassic time.

The crystalline basement of the plate is exposed in the Tottanfjella, Kirwanveggen and in the Ritscher Upland. Its metamorphic complexes of the granulite and amphibolite facies differ from similar complexes of the East Antarctic craton in the presence of mineral assemblages typical of high pressure subfacies.

### Transantarctic Mountains Fold System

In the east and southeast the Weddell Sea Basin includes the Late Proterozoic-Early Palaeozoic (Ross) Transantarctic Mountain fold system cutting the Weddell Sea shelf in the Luitpold Coast area. The main tectonic feature of this structure is the intensely folded Precambrian-Early Palaeozoic basement which was peneplained and overlain by the flat lying or gently folded Middle Palaeozoic-Early Mesozoic platform (molassoid in the lower part) Beacon formations. The folded basement exposed in the Shackleton and the Pensacola Mountains and in the Touchdown Hills, includes: (a) pre-Ross metamorphic complexes comparable with the Early Precambrian complexes of the craton, (b) Ross (Late Riphean-Early Palaeozoic) volcanogenic-sedimentary complexes.

The pre-Ross metamorphic complexes build up median massifs of the Ross systems represented in the present erosion section by the Shackleton Range massif. The available geophysical data suggest that a similar massif is buried beneath ice 100 km south of the Shackleton Range (Figure 1). The metamorphic complexes of the Shackleton Range are divided into: (a) an older (presumably Archaean) Provender Complex consisting of granulite and high grade amphibolite facies metamorphic rocks and (b) a younger (probably Early Proterozoic) Skidmore Complex containing metasedimentary-metavolcanogenic rocks of lower amphibolite to high greenschist facies. The metamorphic complexes are cut by small Early Proterozoic intrusions of ultrabasic, basic and silicic igneous rocks and together they underwent a superimposed high pressure metamorphism.

The Ross folded complexes and overlapping Middle Palaeozoic-Early Mesozoic sedimentary and magmatic formations are best known in the Pensacola Mountains (Schmidt and Ford, 1969). Folded Pensacola Mountains structures are divided by regional unconformities into three stages: (1) a stage of isoclinally folded intensely

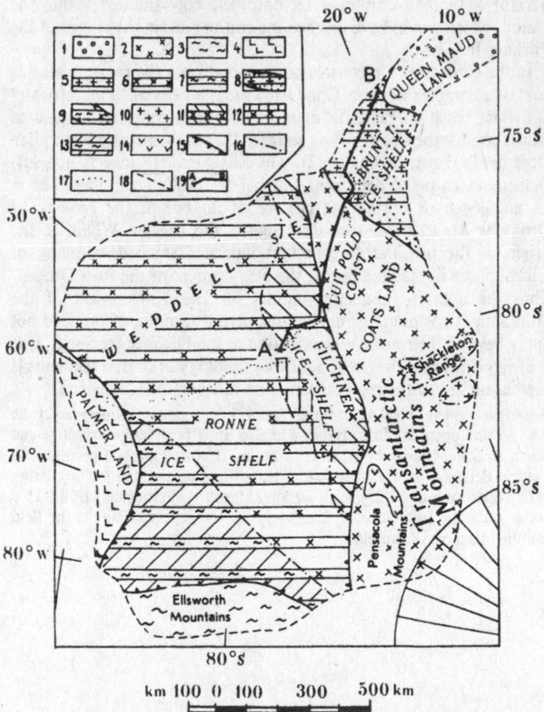


**Figure 1.** **Tectonic provinces in the southern Weddell Sea Basin. (1) epi-Early Proterozoic Maudheim plate; (2) Late Proterozoic-Early Palaeozoic (Ross) Transantarctic Mountains fold system; (3) Middle-Late Palaeozoic Ellworth Mountains fold system; (4) Mesozoic Antarctic Peninsula fold system—zone of Late Mesozoic orogeny. The Weddell Sea Sedimentary Basin: (5) on the early Proterozoic basement; (6) on the Ross basement; (7) on the Gondwanian(?) basement. Major structures: (8) Brunt mega-trough; (9) Ronne syneclise. Minor structures: (10) median masses of fold systems; (11) Filchner trough; (12) Berkner swell; (13) Palmer-Ellsworth monocline; (14) Jurassic differentiated Dufek intrusion; (15) provisional boundaries of structures; (16) provisional boundary of Gondwanian(?) and Ross basements of the Weddell Sea basin; (17) position of generalised geologic section; (18) boundary of the study area; (19) position of the generalised section (see Figure 2).**

schistose volcanogenic-terrigenous sequences of Late Proterozoic-Early Cambrian Patuxent Series exceeding 4000 m in thickness; (2) a stage of open folded volcanogenic-sedimentary and carbonate sequences of the Middle-Late Cambrian Pensacola Series up to 1000 m thick; (3) a stage of terrigenous probably Ordovician-Late Palaeozoic sequences of the Neptune Series and the Gale Suite with a total thickness exceeding 3000 m. These sequences underwent block and gravity folding. In the eastern Neptune Range the Patuxent Series sediments are cut by a pluton of gneissic granites.

In the northeastern and southeastern parts of the Pensacola Mountains there are exposed fragments of the Upper Permian coal bearing platform formation, and in the north there is a gigantic stratiform intrusion of differentiated gabbroids of probable Jurassic age. Geophysical data suggest that the intrusion has an area of about 23,000 km$^2$ and the mean thickness is about 7 km. The Upper Permian platform cover with sills of Jurassic traps is also exposed in the Theron Mountains, in the Whichaway and Faraway Nunataks, in the Otter Peaks, and fragments of the Ordovician-Late Palaeozoic sequences occur in the western Shackleton Mountains, on the eroded surface of metamorphic complexes and folded Late Proterozoic sequences.

As a whole, the Transantarctic Mountains in the Weddell Sea Basin area are part of the intracratonic epi-Early Palaeozoic orogenic belt resting on the heterogeneous base.

## Ellsworth Mountains Fold System

In the west the region includes the Middle-Late Palaeozoic (Gondwanian?) Ellsworth Mountain fold system. It is not yet completely surveyed, but its main tectonic feature is believed to be thick folded Middle-Late Palaeozoic complexes resting on the Ross folded basement and cut by Early Mesozoic intrusions of gabbro and granites.

The Ross folded basement is characterised by intensive folding, a close association of basic effusives and carbonate rocks, and a moderate development of terrigenous rocks (Craddock, 1969). The sequences exceed 5000 m in thickness. Precambrian crystalline metamorphic complexes occur in the Ross folded basement as suggested by their presence in the isolated Haag Nunatak situated northeast of the Ellsworth Mountains in the southern Ronne Ice Shelf (Clarkson and Brook, 1977).

The Middle-Late Palaeozoic folded complexes are divided by a possible regional unconformity into two stages: (a) a stage of Middle Palaeozoic isoclinally folded intensely schistose terrigenous sequences about 3000 m thick, and (b) a stage of Late Palaeozoic open folded terrigenous sequences exceeding 2400 m in thickness (Hjelle et al., 1982).

In the Ellsworth Mountains fold system there are recognised three phases of magmatism: (1) a pre-Middle Cambrian phase represented by quartz keratophyres, rhyolites, alkaline and subalkaline olivine basalts, (2) a Late Palaeozoic phase represented by small plutons and hypabyssal bodies of gabbro and hornblende porphyries, and (3) an Early Mesozoic phase represented by plutons of granites and granodiorites in the Whitmore Mountains and Martin Hills (Webers et al., 1982).

## Zones of Late Mesozoic Orogeny of the Antarctic Peninsula Fold System

In the northwest the Weddell Sea Basin is bounded by the Antarctic Peninsula fold system. It is a part of the Pacific mobile belt. The Late Mesozoic fold zone is adjacent to the Weddell Sea. Folded Middle-Late Jurassic complexes overlie the Precambrian-Palaeozoic folded metamorphic basement and are cut by Late Cretaceous gabbroic to granitic intrusions.

The basement fragments are exposed generally in the northern part of the zone (Black Coast) and are represented by amphibolite facies metamorphic rocks with abundant granite-gneisses.

Middle-Late Jurassic folded complexes (Latady Formation) include terrigenous sedimentary rocks and acid and intermediate volcanics (Rowley and Williams, 1982). The folded structure is characterised by compaction and isoclinal folds combined with less tightly folded forms. Folds of both types have a distinct eastsoutheast vergence. The age of folding is Early Cretaceous. Most of the Late Cretaceous plutons are believed to be roof inliers or apophyses of a single gigantic batholith.

## Weddell Sedimentary Basin

The Weddell Sea Sedimentary Basin is open towards the ocean. It rests on the heterogeneous folded basement (Figure 2). In its northeastern part the basement is believed to be similar to the crystalline Archaean-Early Proterozoic basement of the Maudheim Plate. In the central part of the basin the basement is represented by Ross folded complexes where pre-Ross crystalline basement inliers may also occur, analogous to the adjacent Transantarctic Mountains. In the western part of the basin the basement is probably represented by a folded Gondwanian? one similar to the Ellsworth Mountains, while in the part closest to the Antarctic Peninsula it may be even Late Mesozoic reminiscent of the zone of Late Mesozoic orogeny in Palmer Land.

The complex heterogeneous basement of the basin is between 500 m and 15 km below sea level; and at some localities it is exposed on the sea floor or occurs directly beneath the ice shelves. The basement is mainly overlain by a thick sedimentary cover in which there are recognised the Brunt megatrough (in the northeastern part of the basin) and Ronne syneclise (the southern part) separated by the Luitpold ridge, that is the northern end of the Transantarctic Mountains. The western limb of the Ronne syneclise is represented by the Palmer-Ellsworth monocline, and the Filchner trough and Berkner swell are situated on the eastern limb of the syneclise. The Brunt megatrough may have a similar complex structure.

The structure of the sedimentary cover of the basin is known from seismic measurements only in its eastern part. There, the thickness of sedimentary cover locally reaches 3 km, but generally it ranges from 0

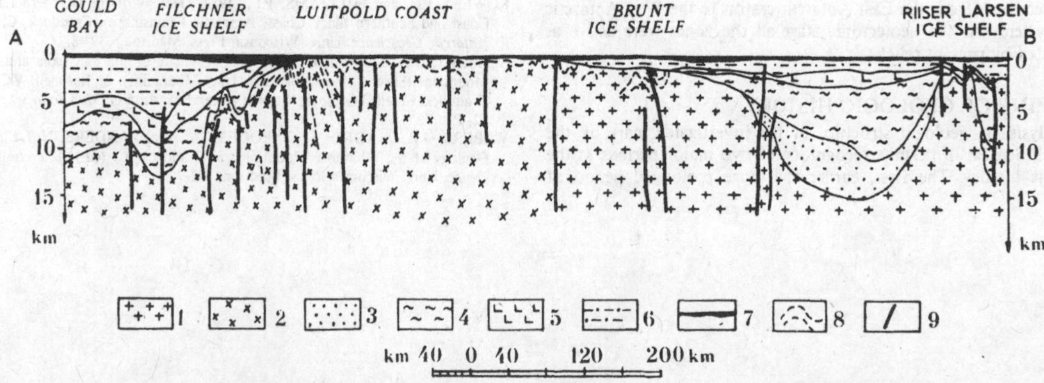


**Figure 2.** Generalised geologic section across the southeastern part of the Weddell Sea. (1) Archaean-Early Proterozoic crystalline basement; (2) Late Proterozoic-Early Palaeozoic (Ross) folded basement; (3) Late Proterozoic-Early Palaeozoic sedimentary cover; (4) Early-Middle Palaeozoic sedimentary cover; (5) Late Palaeozoic-Early Mesozoic sedimentary cover; (6) Late Mesozoic-Cenozoic sedimentary cover; (7) seawater; (8) folded structures in the basement; (9) faults.

to 1.5 km. The cover consists chiefly of Upper Mesozoic and Cenozoic low-density sediments, representing a so-called upper seismological complex. Only in the deepest depressions is there recorded a thin (no more than 1 km) middle seismogeologic complex attributed to the Upper Palaeozoic-Lower Mesozoic.

In a thick (up to 15 km) sedimentary cover of major negative structures there are three distinctly recognised seismogeological complexes viz. the lower, middle and upper complexes attributed to the Lower-Middle Palaeozoic, Upper Palaeozoic-Lower Mesozoic and Upper Mesozoic-Cenozoic, respectively.

In the Ronne syneclise the thickness of sedimentary cover reaches 13.5 km and the folded basement has a complex relief owing to block movements along faults of different rank. Early-Middle Palaeozoic high-density sequences, Late Palaeozoic-Early Mesozoic lower density sequences and loose Late Mesozoic-Cenozoic sequences build up 3-4 km, 2-3 km and 6-8 km of the sedimentary cover section, respectively. Six or seven unconformities are traced throughout the sedimentary cover.

In the western part of the Brunt megatrough an approximately 1 km thick Late Mesozoic-Cenozoic sedimentary cover rests directly on the folded basement. In the central part of the megatrough the sedimentary cover corresponds to those in the Ronne syneclise in seismic characteristics. However, different sedimentary complexes vary in thickness. Two lower complexes, 5-12 km and 1-3 km thick in ascending order, respectively, account for much of the cover. A maximum thickness of the sedimentary fill in the Brunt megatrough reaches 14.5 km and the upper loose sediment complex is 0.7-1.5 km thick. Coarse oblique bedded series occur in the upper complex. On the whole, seven or eight unconformities are recorded throughout the Brunt megatrough sedimentary cover. The lowermost horizons of the cover are thought to be represented by dense Riphean-Early Palaeozoic platform formations exposed in the Maudheim Plate.

Obviously the younger cover rests on the younger basement. Therefore the age of the Palmer-Ellsworth Monocline cover appears to be essentially Late Palaeozoic-Cenozoic, while in the Ronne syneclise areas adjacent to the Antarctic Peninsula the cover is wholly Cenozoic.

### Deep Faults

Zones of ancient deep faults repeatedly activated throughout the geologic history play an important part in the structure and formation of the southern Weddell Sea Basin. The Druzhnaya deep fault zone, about 100-150 km wide, stretches for over 1200 km from the Pensacola Mountains to the Luitpold Coast and furnishes an example of these zones. The Dryzhnaya zone controls the Filchner trough and separates the Transantarctic Mountains fold system from the Ronne syneclise.

Similar zones probably control the Brunt megatrough along the boundary of the Luitpold ridge and the Maudheim Plate, as well as the Ronne syneclise and its boundaries with the fold systems of the Ellsworth Mountains and the Antarctic Peninsula. All the above is supported by geophysical data. The zones of deep faults may have been emplaced at different times and the age of emplacement probably decreases from the East Antarctic craton to the West Antarctic fold system. At the neotectonic stage all the zones were active as suggested by present relief.

## OUTLINE OF GEOLOGIC HISTORY

Analysis of geologic structure of the investigated part of the Weddell Sea Basin permits recognition of two major periods in the geologic history. The first, earlier and more prolonged period of formation of continental crust-craton basement continued throughout Archaean and Early Proterozoic time. This period was accompanied by a gradual increase in volume of the granite-metamorphic layer, and thus, in thickness of craton basement.

The second major period in the history of the region began in the Late Proterozoic and has continued up to now. The formation of a thick sedimentary-volcanic cover and degradation of the craton basement were vital for this period. In the Late Proterozoic there was the development of a system of faults cutting the basement and the formation along them of graben-like troughs and aulacogens (in the position of the Transantarctic and Ellsworth Mountains) accompanied by a simultaneous subsidence of separate marginal parts of the craton (occupied by the Maudheim Plate). Thus an intracratonic trough of considerable size had developed. In the Riphean and Early Cambrian several rises which served as source areas for surrounding troughs were situated in the central part of the present sedimentary basin. In the Early Cambrian the inversion of aulacogens and troughs occurred in the position of the present Transantarctic Mountains, the foredeep moved onto the Maudheim Plate and a fairly prolonged downwarping of the Ronne syneclise and of the Brunt megatrough started; this process continued throughout the Palaeozoic, Mesozoic and Cenozoic. The process was associated with the transformation of basement crystalline metamorphic complexes with consolidation thereof and decrease in volume of the granite-metamorphic layer, partly owing to intrusion of granitic magma into inverted rises.

Downwarping of the Ronne syneclise was uneven in time and space. Epochs of rather intensive downwarping (in Early-Middle Palaeozoic, Early Mesozoic and Cenozoic time) alternated with those of relative stabilisation or even uplift (in Late Palaeozoic and Late Mesozoic time). Inversion of troughs in the position of the Ellsworth Mountains and mountains of the Antarctic Peninsula preceded the epochs of intensive downwarping. During the Late Palaeozoic a rather quiet platform regime took place throughout the basin area accompanied by accumulation of coal measures. In the Early Mesozoic faults activised and trapped magma erupted in the eastern part of the basin. During the Cenozoic crustal block movements intensified again which resulted in the formation of the present sedimentary basin and surrounding mountains.

## REFERENCES

BREDELL, J.H., 1982: The Precambrian sedimentary-volcanic sequence and associated intrusive rocks of the Ahlmannryggen, Western Dronning Maud Land; a new interpretation; *in* Craddock, C. (ed.) *Antarctic Geoscience*. Univ. Wisconsin Press, Madison, 591-7.

CLARKSON, P.D. and BROOK, M., 1977: Age and position of the Ellsworth Mountains crustal fragment, Antarctica. *Nature, 265,* 615-6.

CRADDOCK, C., 1969; Geology of the Ellsworth Mountains (sheet 4, Ellsworth Mountains); *in* Bushnell, V.C. and Craddock, C. (eds.) Geologic Maps of Antarctica. *Antarct. Map Folio Ser.,* Folio 12, P1. IV.

HJELLE, A., OHTA, Y. and WINSNES, T.S., 1982: Geology and petrology of the southern Heritage Range, Ellsworth Mountains; *in* Craddock, C. (ed.) *Antarctic Geoscience*. Univ. Wisconsin Press, Madison, 599-608.

NEETHLING, D.C., 1972: Age and correlation of the Ritscher Supergroup and other Precambrian rock units, Dronning Maud Land; *in* Adie, R.J. (ed.) *Antarctic Geology and Geophysics*. Universitetsforlaget, Oslo, 547-56.

ROWLEY, P.D. and WILLIAMS, P.L., 1982: Geology of the Northern Lassiter Coast and Southern Black Coast, Antarctic Peninsula; *in* Craddock, C. (ed.) *Antarctic Geoscience*. Univ. Wisconsin Press, Madison, 339-48.

SCHMIDT, D.L. and FORD, A.B., 1969: Geology of the Pensacola and Thiel Mountains (Sheet 5, Pensacola and Thiel Mountains); *in* Bushnell, V.C. and Craddock, C. (eds.) Geologic Maps of Antarctica. *Antarct. Map Folio Ser.,* Folio 12, P1. V.

WEBERS, G.F., CRADDOCK, C., ROGERS, M.A. and ANDERSON, J.J., 1982: Geology of the Whitmore Mountains; *in* Craddock, C. (ed.) *Antarctic Geoscience*. Univ. Wisconsin Press, Madison, 841-7.

# MAGNETIC STUDIES OF UPPER CRUSTAL STRUCTURE IN WEST ANTARCTICA AND THE BOUNDARY WITH EAST ANTARCTICA

E.J. Jankowski[1] and D.J. Drewry, *Scott Polar Research Institute, University of Cambridge, Cambridge, CB2 1ER, U.K.*

J.C. Behrendt, *U.S. Geological Survey, Regional Geophysics, Box 25046, MS 964, Federal Center, Denver, Co. 80225, U.S.A.*

*Present Address: British Petroleum Ltd, Britannic House, Moor Lane, London, EC2Y 9BU, U.K.*

*Abstract* The subglacial structure and geology of West Antarctica is poorly known, yet important to the understanding of the tectonic evolution of Gondwanaland. Simultaneous Radio Echo Sounding (RES) and magnetic data have been collected over approximately 1 x $10^6$ kilometres of West Antarctica. Estimates of maximum depth to magnetic basement were made using an automatic computer method and selected anomalies modelled using a two dimensional inversion programme. These studies were combined with subglacial topographic information, derived from RES, and all pertinent geological and geophysical measurements to suggest interpretations of subglacial geology.

Shallow magnetic basement is associated with volcanic outcrop in northern Marie Byrd Land. Basement deepens below the Byrd Subglacial Basin where volcanics are believed to lie beneath a variable, but thin, cover of Mesozoic-Cenozoic sediments. A sinuous ridge cutting the basin is also believed to be of volcanic origin and probably related to Marie Byrd Land volcanics. This deep basin is terminated abruptly by a thick succession of sedimentary strata representing a subglacial extension of the Late Precambrian—Palaeozoic metasedimentary sequence exposed in the Ellsworth Mountains. These mountains and associated subglacial terrain are believed "out of place" in West Antarctica, and evidence of rifting is inferred in the area at the head of the Ronne-Filchner Ice Shelf.

The boundary relationships between East and West Antarctica are complex. Previous suggestions of a major fault separating the two are supported by the abrupt boundary visible in RES records in the vicinity of the Horlick Mountains. Identification of their boundary becomes more problematic towards the Thiel Mountains. Limited geological information and the results of this study suggest that the boundary zone probably passes close to the Thiel Mountains in the area between them and the Stewart Hills, Hart Hills and Pagano Nunatak.

Prior to 1977, airborne magnetometry in West Antarctica, with the exception of the Antarctic Peninsula, had been conducted principally by the University of Wisconsin (Behrendt and Wold, 1963; Behrendt, 1964a, the U.S. Geological Survey (Behrendt et al., 1974) and U.S. Project Magnet (U.S. Naval Oceanographic Office, 1963, 1965). Although aeromagnetic data had high accuracy, they suffered from having no simultaneous ice thickness information and uncertain navigation. During the austral summers 1977/78 and 1978/79 the Scott Polar Research Institute (SPRI) undertook an airborne geophysial programme in West Antarctica combining for the first time Radio Echo Sounding (RES) (by SPRI and the Technical University of Denmark) and magnetometry (by the Applied Physics Laboratory, Johns Hopkins University). The work was carried out in cooperation with the U.S. National Science Foundation. The principal objective was to establish a grid for geophysical investigations of the complex and poorly known structure of West Antarctica and its relationship to the East Antarctic craton.

## DATA REDUCTION

Magnetic studies were based on total field magnetic readings obtained using a Geometrics G803 proton precession magnetometer (precision lnT). The sensor for the magnetometer was mounted at the end of a 3 m long tail stringer. Coefficients compensating for the effect of the aircraft were determined by APL using a vector magnetometer and adopting a technique originally developed by Leliak (1961). During 1978/79 time constraints and hardware problems made the coefficients for this reason unreliable and the data were not compensated. For this reason anomalies occurring on sections of flight where the aircraft was changing altitude or manoeuvring have not been used in further analysis. No diurnal corrections could be applied to the data due to (a) logistic problems in setting up a remote base station; (b) large area covered by survey. This is not considered too serious a limiting factor in a reconnaissance survey of this nature. Residual anomalies were then obtained by removing from the data the effects of the near surface geomagnetic field and its secular variation using a spherical harmonic model, AWC-75 (Peddie and Fabiano, 1976). Because the field data at times suffered from random high frequency noise, of unknown origin, the data were finally cleaned and smoothed using a two stage filtering operation described in Jankowski (1981). Sections of data suffering from excessive noise problems were not used further. RES equipment and methods of reduction of these data are fully described in Jankowski and Drewry (1981). A detailed map of subglacial bedrock topography was prepared using ~ 56,000 RES determinations of elevation and ~ 250 seismically derived elevations (Bentley and Ostenso, 1961; Behrendt, 1964b; Robinson, 1964; American Geographical Society, 1970).

## ANALYSIS OF DATA

As the spacing of flight lines was generally seldom less than 50 km an areal approach to the analysis of the data was virtually impossible. In consequence much emphasis has been placed on interpretation of the magnetic profiles. Simultaneous RES and aeromagnetic measurements allow direct correlation of estimated depth to "magnetic basement" with subglacial topography. Depth to "magnetic basement" was determined using an automatic computer method based on an idea by Werner (1953) adopted for computer analysis by Anderson and Cordell (1973). The method, commonly referred to as Werner Deconvolution, assumes that an observed magnetic field can be referred to a system of magnetically homogenous fault- and dyke-like sources that trend in directions perpendicular to the profile for a distance many times their depth. The magnetism can be in any arbitary direction so long as its direction and intensity are constant. Hartman et al. (1971) showed that by performing Werner Deconvolution on both the total field and calculated horizontal derivative it is possible to identify both dyke- and fault-like bodies.

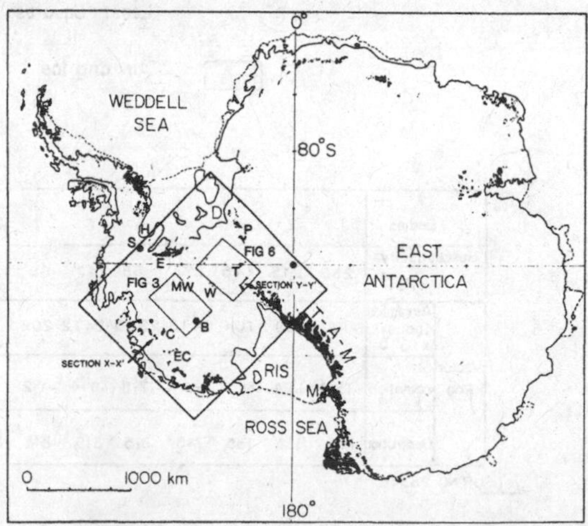


**Figure 1. Antarctic location map showing areas covered by this study (S, Siple Station; B, Byrd Station; M, McMurdo Station; E, Ellsworth Mountains; T, Thiel Mountains; W, Whitmore Mountains, MW, Mounts Woollard and Moore; C, Crary Mountains, EC, Executive Committee Range; P, Pensacola Mountains). The locations of Figures 3 and 6 are shown by boxes. Profiles in Figures 2 and 5 are plotted.**

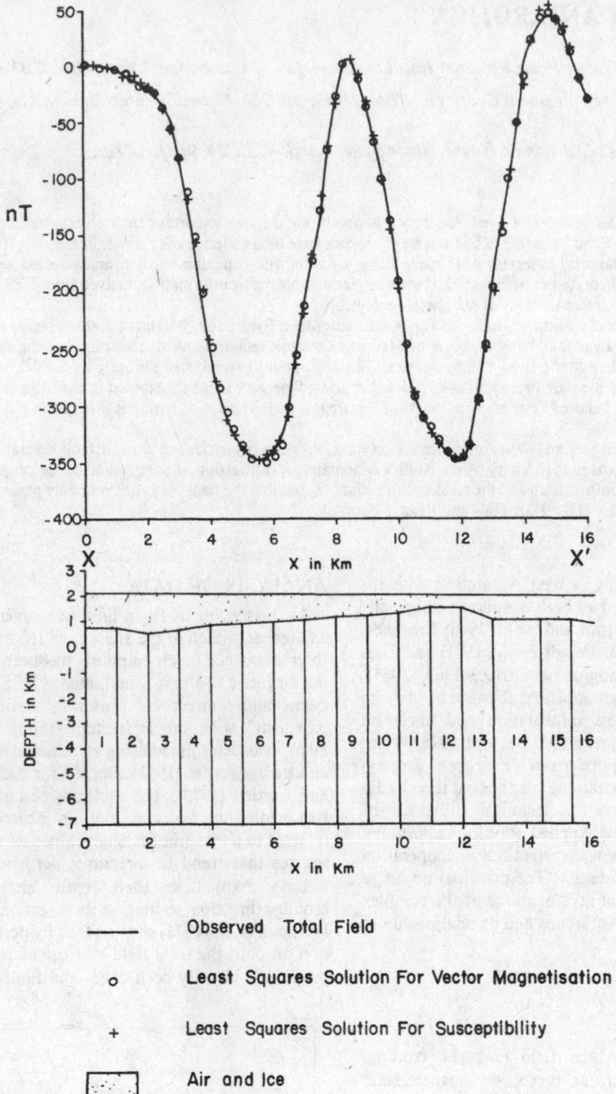


| Bodies | | 1 | 2 | 3 | 4 | 5 | 6 | 7 | 8 | 9 | 10 | 11 | 12 | 13 | 14 | 15 | 16 | Errors |
|---|---|---|---|---|---|---|---|---|---|---|---|---|---|---|---|---|---|---|
| Susceptibility $\times 10^{-4}$ | | 256 | 215 | 445 | 235 | −58 | 42 | −68 | 186 | 320 | 210 | 25 | 19 | −54 | 207 | 343 | 116 | −0.022 ± 8.1 |
| Vector Mag | Amplitude (gauss) $\times 10^{-5}$ | 786 | 1609 | 3104 | 3418 | 2940 | 2412 | 2061 | 1546 | 1139 | 1064 | 990 | 904 | 1050 | 868 | 852 | 842 | −0.026 ± 4.8 |
| | Inclination | −13·4 | −13·4 | −31 7 | −68·4 | −17·5 | −7·4 | −1·2 | 1·7 | 2·2 | 3·0 | 3·8 | 9·0 | 24·8 | 13·6 | 29·0 | 43·8 | |
| | Declination | 135 | 135 | 135 | 315 | 315 | 315 | 315 | 315 | 315 | 315 | 315 | 315 | 315 | 315 | 315 | 135 | |

SUR NO. 28346

**Figure 2. Model of magnetic anomalies across part of the Cenozoic Province near to Crary Mountains (Location shown by profile X-X' in Figure 1). The observed field is shown by continuous line. Inverse solutions for susceptibility are given by crosses and inverse solutions for remanent magnetisations by squares. Triangles represent the residual between observed total field and inverse solution for remanent magnetisation. In 2b the line drawn above the top of the vertical bodies represents the absolute elevation of the measurements. The ice sheet surface has been omitted as it is irrelevant to modelling. All vertical bodies are numbered and correspond to those given in the Table where susceptibilities are in $10^{-4}$ SI units.**

The occurrence of "magnetic basement" at the ice/rock interface, provides an upper limit in modelling the anomalies to provide quantitative estimates of bedrock magnetisation. Selected anomalies were modelled in this way using the 2-D inversion method of Cady (1980). The crust was divided into arbitrary vertical sided, flat-bottomed prisms in a method similar to Behrendt et al. (1981). The models are not unique or geologically realistic but in the absence of *a priori* control provide a first approximation to the magnetic properties of the bedrock.

## GEOLOGICAL DISCUSSION

On the basis of a detailed topographic bedrock map derived from RES and some preliminary magnetic analysis Jankowski and Drewry (1981) divided West Antarctica into seven crude "units". These units are here discussed and refined further drawing upon an extended analysis of the magnetics and all other pertinent geophysical data.

### Cenozoic Volcanic Province

The Cenozoic Volcanic Province is known from exposures to occur in a relatively narrow zone along the Pacific margin of Antarctica. In Marie Byrd Land Cenozoic volcanic outcrops extend some 300 km inland from the coast to the Crary Mountains and Executive Committee Range (Figure 1). Source depths within this area suggest that "magnetic basement" lies mainly close to the ice/rock interface indicating a lack of a thick sedimentary cover (Behrendt and Wold, 1963). Bentley and Chang (1971) described a wide variation in susceptibility for these rocks and this may explain some of the deeper sources in this area. The strongly negative anomaly in the vicinity of the Crary Mountains (Figure 2) suggests an important degree of remanent magnetisation in a direction significantly different from that at present. This is supported by 2-D modelling with both normal and reversed magnetisations required to fit the data (Figure 2). The difference in anomaly amplitudes between those noted by Bentley and Chang (~ 1400 nT) and this study (~ 450-600 nT) again illustrate the variability of the magnetic susceptibility of these rocks.

### Byrd Subglacial Basin

The Byrd Subglacial Basin (Figure 3) is a major feature in West Antarctica and even allowing for complete isostatic recovery (Drewry et al., this volume) still forms a deep submarine connection between the Ross and Amundsen Seas. The basin is characterised by numerous large amplitude 500-1000 nT anomalies, and source depths indicate, on the whole, that "magnetic basement" lies close to the ice/rock interface (Figure 4). This is consistent with seismic refraction data which indicate ~ 0.5 km of sediments overlying ~ 2.5 km of probable volcanic flows and pyroclastics (Bentley and Clough, 1972). On the basis of the magnetic and seismic evidence the floor of the basin would seem to be almost entirely volcanic in origin with the volcanics overlain by a variable but thin amount of Cenozoic sediment.

There is no sharp break in the magnetic field between the Byrd Land volcanic outcrops and the basin to match the steep subglacial topographic boundary. Instead, the field changes gradually to the lower amplitude, longer wavelength anomalies observed over the basin. Taking account of the great distances between magnetometer and source over the basin the anomalies appear similar.

A prominent sinuous ridge (Figure 3) cutting across the basin was thought to be of volcanic origin by Jankowski and Drewry (1981). Further modelling studies support this view and gives susceptibilities and vector magnetisations similar to values obtained near exposed volcanics. Source depths over the ridge vary form the ice/rock interface to ~ 4 km below it and probably reflect a multi-layered volcanic structure to the ridge, bearing in mind the wide variation in susceptibility. The origin of the Byrd Subglacial Basin is subject to much debate, but it is generally thought to have formed in an extensional tectonic regime (Elliot, 1974; Le Masurier and Rex, 1982). This is supported by study of sea floor magnetic anomalies in the South Pacific Ocean (Molnar et al., 1975). Coupled with the lack of a considerable thickness of recent sediment in the deep subglacial troughs in Marie Byrd Basin Land, Le Masurier and Rex (1982) suggest that the entire Byrd Basin has formed largely since the

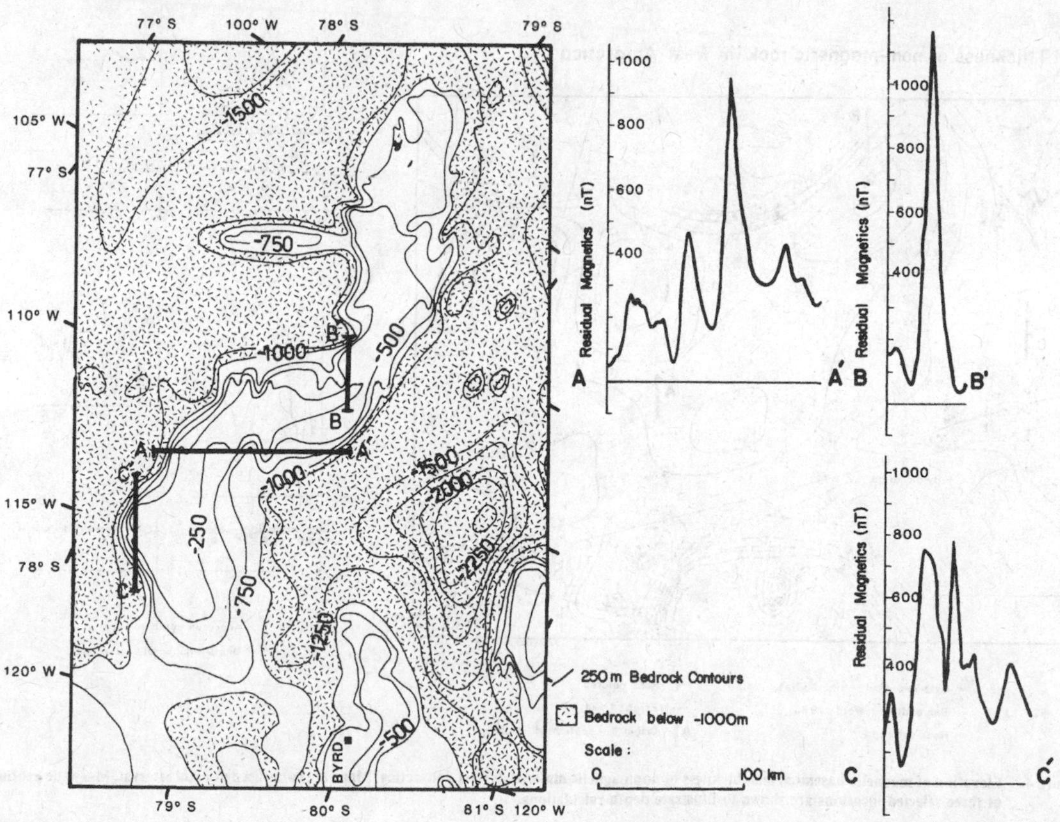


**Figure 3.** **Detailed subglacial topographic map and magnetic profiles in the vicinity of the sinuous ridge which bisects the Byrd Subglacial Basin. Location of Figure 3 is shown in Figure 1.**

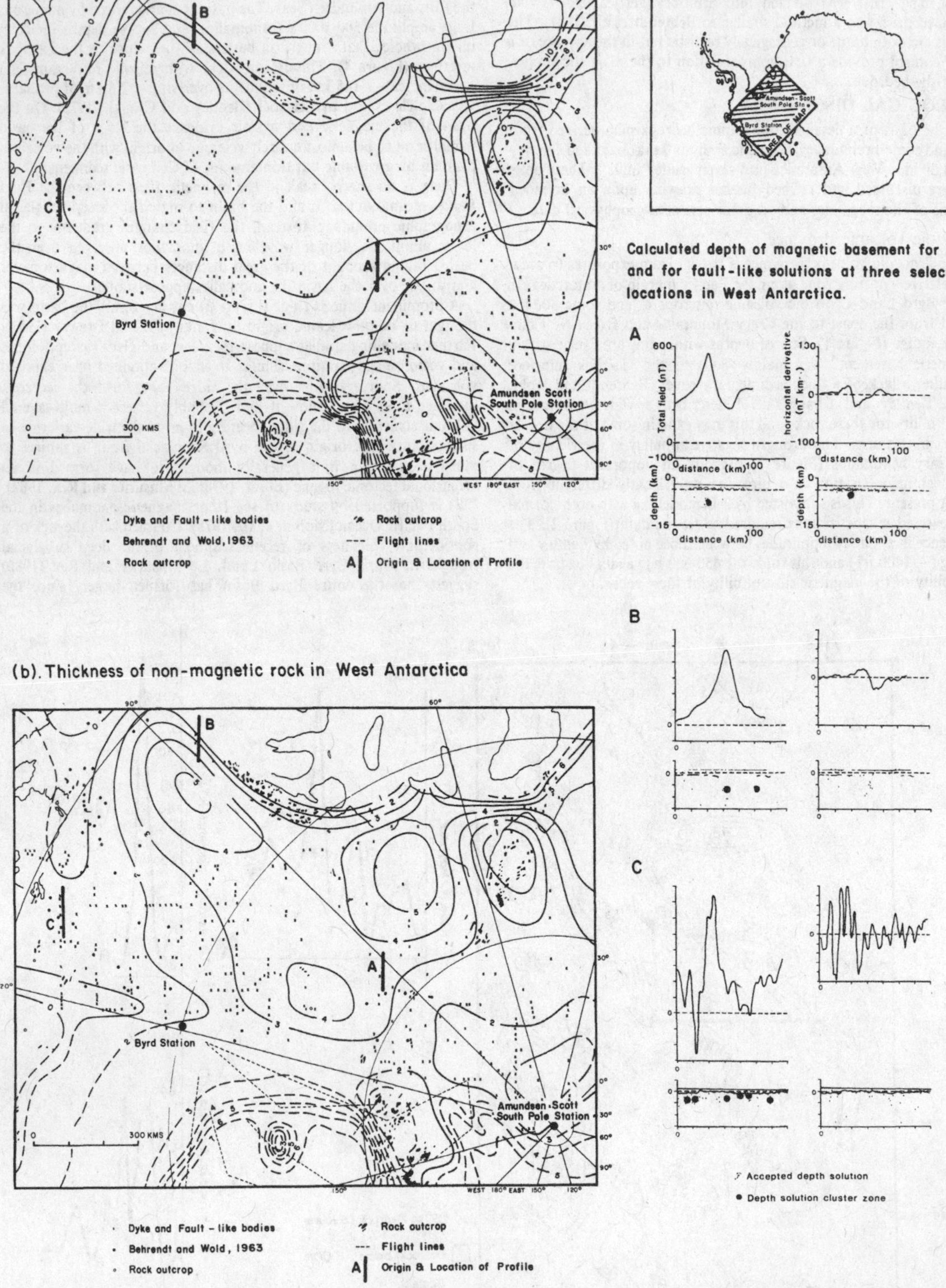


**Figure 4.** Elevation of magnetic basement and thickness of nonmagnetic material in West Antarctica. Maps are contoured at 1 km interval. Magnetic profiles at three selected locations are shown to illustrate depth calculations.

development of the continental ice sheet. Elliot (1974) envisages a tectonic environment in the Middle Mesozoic, somewhat similar to that of back-arc spreading within a continental plate, as responsible for the formation of the Byrd Basin. Some block faulting during the Cenozoic, as seen in the displacement of the very flat prevolcanic erosion surface in Marie Byrd Land is clearly apparent and the dissected appearance of the volcanic outcrops of Toney Mountain and Mount Takahe on RES records does not rule out some vertical displacement. It is difficult, however, to believe that the entire Byrd Basin formed as suggested by Le Masurier and Rex (1982) but, rather developed originally as a result of rifting and major tectonics in the Middle to Late Mesozoic with further rifting and modification during the Cenozoic. Further modelling studies are being carried out on magnetic profiles over the Byrd Subglacial Basin to investigate the origin of this Basin.

Truswell and Drewry (in prep.) have examined palynomorphs from sediments in the Ross Sea, which include a significant component of Late Mesozoic to Early Cenozoic species, with definite marine forms in the Eocene. These suggest a source in sedimentary rocks inland of the Ross Ice Shelf and their provenance is placed in a zone from the Byrd Subglacial Basin through to the Ross Sea embayment, and that this rift zone commenced opening in the Late Mesozoic.

## Ellsworth—Whitmore Mountains

A rugged highland area extending west from the exposed Ellsworth Mountains to the Whitmore Mountains was described by Jankowski and Drewry (1981). Source depths within this area nearly all indicate a thickness of non-magnetic material overlying basement in the range 3.5-8 km (Figure 4). The northern margin of this highland area, abutting the Byrd Subglacial Basin is well defined topographically and magnetically (see Jankowski and Drewry, 1981, Figure 4). Seismic evidence, magnetics and RES all suggest that this margin represents the boundary between two distinctly different geological provinces. The boundary becomes less well defined to the west in the area south of Byrd Station. RES suggests that the boundary follows the edge of the deep subglacial basin around to the west side of the Whitmore and Thiel Mountains.

## Boundary Between East and West Antarctica

The nature of the boundary zone between East and West Antarctica is not well understood and information on the deep structure of the zone is beyond the data available at present. The data presented here, however, enable some conclusions regarding its surface manifestation and possible location to be made. Unlike the western margin of the Transantarctic Mountains, which is a gradual transition the eastern inland margin is in places abrupt (Figure 5). Several workers have reported a large boundary fault between the exposures of East and West Antarctica (Long, 1965; Barrett, 1965; Behrendt et al., 1974). This abrupt margin to the Transantarctic Mountains would appear a logical position for the transition between East and West Antarctica across which the depth to MOHO is believed to decrease by ~ 10 km (Adams, 1972). The cause of the marked Bouguer gravity gradient across the boundary is a matter of debate but is further evidence that the margin of the Transantarctic Mountains observed on RES records near the Horlick Mountains appears consistent with a fault interpretation but the depth to which it may extend is a matter of conjecture.

The precise location of the boundary zone, in the ice covered area between the Horlick Mountains and the southern end of the Pensacola Mountains is difficult to ascertain (Figure 6). The Thiel Mountains, lying roughly halfway between the two, would seem on topographic grounds and geological similarities (Ford and Sumsion, 1971; Craddock et al., 1982) to belong to the same tectonic province as the rest of the Transantarctic Mountains. The Whitmore Mountains lie ~ 250 km to the north of the Thiel Mountains and show geological affinities with the Ellworth Mountains—a separate tectonic province (Craddock et al, 1982).

The scattered exposures of the Stewart Hills, Hart Hills and Pagano Nunatak are more problematic. Pagano Nunatak consists entirely of felsic plutonic rocks and like those of the Whitmore Mountains give isotopic ages in the range 163-190 Ma and are therefore related to the Early Mesozoic Ellsworth Orogen. The Stewart Hills consist of tightly folded metasedimentary rocks with a radiometric age of 508 Ma and, like the Thiel Mountains, retain a record of the Lower Palaeozoic Ross Orogeny (Craddock et al., 1982). The Hart Hills are a problem as their tectonic position is still unknown, consisting of deformed metasediments and probably younger altered gabbroic rocks. The boundary zone between the two different tectonic provinces of the Ellsworth and Transantarctic Mountains must pass through this complex area. There does not appear to be any logical topographic boundary between the two provinces (Jankowski and Drewry, 1981) and the highland area on RES records between the Whitmore and Thiel Mountains may show affinities to either tectonic province or even both. The highland area in the vicinity of the Hart Hills is associated with a pronounced magnetic anomaly of ~ 350 nT (Figure 6). Source depths in this area are all significantly deeper than those surrounding the Thiel Mountains giving a thickness of non-magnetic material in the range of 4.2-9.9 km. The nature of the causal body is unknown. The nearest exposed rocks are the altered gabbros of the Hart Hills although these have very low susceptibilities (Jankowski, 1981). The felsic intrusives of Pagano Nunatak and the granite of the Thiel Mountains, likewise, have low susceptibilities. It is however possible that the Hart Hills are underlain by a larger body of less altered gabbro with much higher susceptibilities (see also Behrendt et al., 1974). Clearly the boundary zone is likely to pass through this area, but until the tectonic position of the Hart Hills and the highland area between them and the Thiel Mountains can be determined more precisely the postion will remain in doubt.

In the area between the Pensacola and Ellsworth Mountains a deep subglacial depression occurs (Jankowski and Drewry, 1981). Source depths are deep in this area (Figure 4) and the field is very quiet suggesting a substantial thickness of sedimentary strata. Crustal thickness is also believed to be thinner than East Antarctica and in the

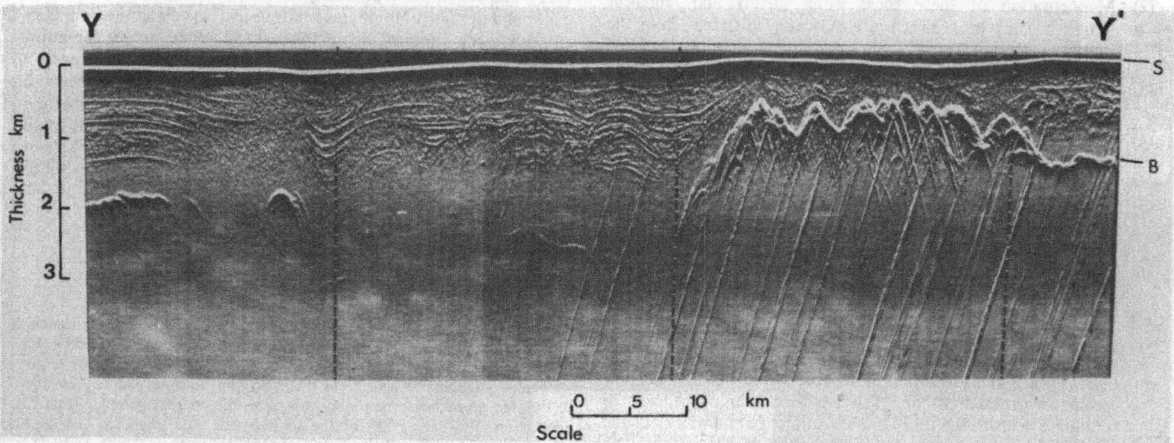


**Figure 5. Radio echo sounding profile over the boundary between East and West Antarctica in the Horlick Mountains area (see section Y-Y' in Figure 1). S represents the ice surface and B the bedrock echo.**

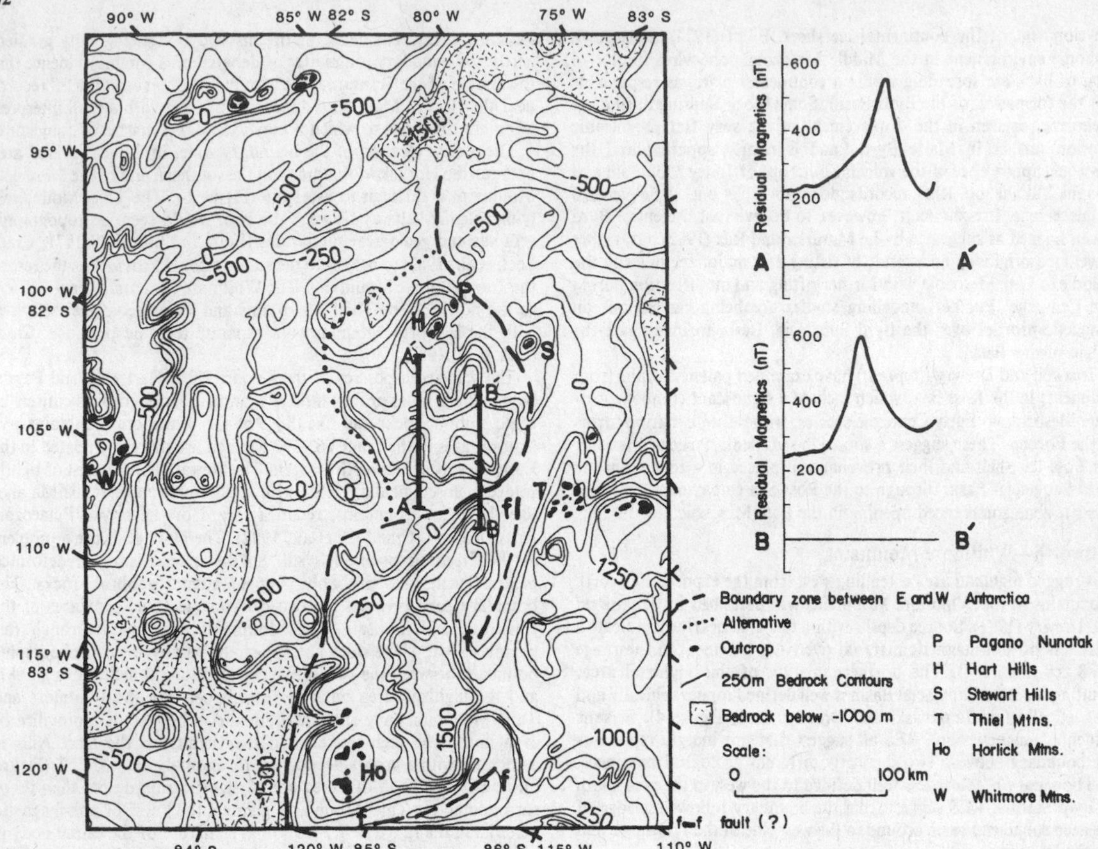


**Figure 6.** **Detailed subglacial topographic map and magnetic profiles in the boundary zone between East and West Antarctica. Location of Figure 6 is shown in Figure 1.**

rest of West Antarctica. This depression is probably related to rifting of the Ellsworth Mountains and marks the continuation of the boundary between East and West Antarctica.

## CONCLUSIONS

The data presented support the view that West Antarctica comprises more than one geological and tectonic province. Some of the main conclusions can be summarised as:

(1) Cenozoic volcanic rocks lie close to the ice/rock interface in a large part of the subglacial areas of Marie Byrd Land.

(2) Magnetisation of these rocks varies greatly and remanent magnetisations in a direction significantly different from the present are important.

(3) There is no clear break in the magnetic field across the boundary between the Cenozoic Volcanic Province and the Byrd Subglacial Basin.

(4) The floor of the basin would seem to be volcanic in origin with the volcanics overlain by a variable, but thin, cover of Cenozoic sediment.

(5) The basin is cut by a prominent sinuous ridge. This ridge too, is believed to be volcanic in origin and related to the Marie Byrd Land volcanics.

(6) The Byrd Basin is a major subglacial feature of West Antarctica and probably represents a tectonic lineament.

(7) The origin of the basin is believed related to extensional tectonics during the break-up and subsequent evolution of Gondwanaland in Late Mesozoic times.

(8) The thick Late Precambrian to Palaeozoic metasedimentary sequence exposed in the Ellsworth Mountains is believed to continue beneath the ice into West Antarctica and to include the scattered outcrops of the Nash, Pirrit, Martin Hills etc. and the Whitmore Mountains.

(9) The northern margin of the Ellsworth—Whitmore Mountains area with the Byrd Subglacial Basin is well defined both topographically and magnetically. The boundary becomes less well defined to the west and south.

(10) The boundary between East and West Antartica is a major problem. The boundary is associated with a steep Bouguer gravity gradient. Suggestions of a major fault between East and West Antarctica are supported by the abrupt boundary to the Transantarctic Mountains on the West Antarctic side. The depth to which a fault zone may extend is a matter of conjecture.

(11) The subglacial topography between the Whitmore Mountains and the Thiel Mountains is extremely rugged and the boundary is more problematic in this area. Limited geological information and the results of this study suggest that the boundary zone passes close to the Thiel Mountains in the area between them and the outcrops of the Stewart and Hart Hills and Pagano Nunatak. The tectonic postion of the Hart Hills could be crucial if the anomalies over the highland area to the west of the exposures are indeed related to a less altered part of the Hart Hills Gabbro.

*Acknowledgements* This work was supported by U.K. Natural Environment Research Council Grant GR3/2291; logistics were generously provided by U.S. National Science Foundation Division of Polar Programmes. C.R. Bentley provided unpublished data from the Wisconsin Traverse and C. Craddock permitted magnetic susceptibility measurements on rock samples from his collection. We thank NSF for the aeromagnetic data, the U.S. Geological Survey for assistance in their reduction and D.H. Elliot and I.W.D. Dalziel for helpful comments.

## REFERENCES

AMERICAN GEOGRAPHICAL SOCIETY, 1970: Antarctica. 1:5,000,000, sheet B.

ADAMS, R.D., 1972: Dispersion wave studies in Antarctica; *in* Adie, R.J. (ed.) *Antarctic Geology and Geophysics*, Universitetsforlaget, Oslo, 473-83.

ANDERSON, W.L. and CORDELL, L., 1973: Digital magnetic interpretation system. *U.S. Geol. Surv., unpub. rep.*

BARRETT, P.J., 1965: Geology of the area between the Axel Heiberg and Shackleton Glaciers, Queen Maud Range, Antarctica. *N.Z. J. Geol. Geophys., 8,* 344-63.

BEHRENDT, J.C., 1964a: Distribution of narrow width magnetic anomalies in Antarctica. *Science, 144,* 993-4.

BEHRENDT, J.C., 1964b: Antarctic Peninsula Traverse geophysical results relating to glaciological and geological studies. *Univ. Wisc. Geophys. Polar Research Cent., Res. Rep. Ser., 64-1.*

BEHRENDT, J.C., DREWRY, D.J., JANKOWSKI, E.J. and GRIM, M.S., 1981: Aeromagnetic and Radio Echo ice-sounding measurements over the Dufek Intrusion, Antarctica. *J. Geophys. Res., 86,* 3014-20.

BEHRENDT, J.C., HENDERSON, J.R., MEISTER, L. and RAMBO, W.L., 1974: Geophysical Investigations of the Pensacola Mountains and adjacent glaciated areas of Antarctica. *U.S. Geol. Surv. Prof. Pap.,*844.

BEHRENDT, J.C. and WOLD, R.J., 1963: Depth to Magnetic "Basement" in West Antarctica. *J. Geophys. Res., 68,* 1145-53.

BENTLEY, C.R. and CHANG, F.K., 1971: Geophysical exploration in Marie Byrd Land, Antarctica. *Am. Geophys. Union Ant. Res. Ser., 16,* 1-38.

BENTLEY, C.R. and CLOUGH, J.W., 1972: Antarctic subglacial structure from seismic refraction measurements; *in* Adie, R.J. (ed.) *Antarctic Geology and Geophysics,* Universitetsforlaget, Oslo, 683-91.

BENTLEY, C.R. and OSTENSO, N.A., 1961: Glacial and subglacial topography of West Antarctica. *J. Glaciol., 3,* 882-911.

CADY, J.W., 1980: Calculation of gravity and magnetic anomalies of finite length right polygonal prisms. *Geophysics, 45, 10,* 1507-12.

CLARKSON, P.D. and BROOK, M., 1977: Age and position of the Ellsworth Mountains crustal fragment, Antarctica. *Nature, 265,* 615-6.

CRADDOCK, C., WEBERS, G.F. and ANDERSON, J.J., 1982: Geology of the Ellsworth Mountains-Thiel Mountains Ridge; *in* Craddock, C. (ed.) *Antarctic Geoscience,* Univ. Wisconsin Press, Madison, 849.

ELLIOT, D.H., 1976: The tectonic setting of the Jurassic Ferrar Group, Antarctica; *in* Gonzalez-Ferran, O. (ed.) *Andean and Antarctic Volcanology Problems,*Int. Assoc. Volcanol. Chem. Earth's Interior, Rome, 357-72.

FORD, A.B. and SUMSION, R.S., 1971: Late Precambrian silicic pyroclastic volcanism in the Thiel Mountains Antarctica. *U.S. Antarc. J., 6, (5),* 185-6.

HARTMAN, R.R., TESKEY, J. and FRIEDBERG, J.L., 1971: A system for rapid digital aeromagnetic interpretation. *Geophysics, 36,* 891-918.

JANKOWSKI, E.J. and DREWRY, D.J., 1981: The structure of West Antarctica from Geophysical Studies. *Nature, 291,* 17-21.

JANKOWSKI, E.J., 1981: Airborne Geophysical investigations of subglacial structure of West Antarctica. Unpublished PhD thesis, University of Cambridge, 293pp.

LELIAK, P., 1961: Identification and evaluation of magnetic field sources of magnetic airborne detector equipped aircraft. IRE Transactions on Aerospace and Navigational Electronics, Sept. 1961.

LEMASURIER, W.E. and REX, D.C., 1982: Volcanic record of glacial history in Marie Byrd Land and Western Ellsworth Land: Revised chronology and evaluation of tectonic interrelationships; *in* Craddock, C. (ed.) *Antarctic Geoscience,* Univ. Wisconsin Press, Madison, 725-34.

LONG, W.E., 1965: Stratigraphy of the Ohio Range Antarctica. *Am. Geophys. Union, Ant. Res. Ser., 6,* 71-116.

MOLNAR, P., ATWTER, T., MAMMERICKX, J. and SMITH, S.M., 1975: Magnetic anomalies, bathymetry and the tectonic evolution of the South Pacific since the Late Cretaceous. *R. Astron. Soc., Geophys. J., 40,* 383-420.

PEDDIE, N.W. and FABIANO, E.D., 1976: Model of Geomagnetic Field for 1975. *J. Geophys. Res., 81,* 2539-42.

ROBINSON, E.S., 1964: Geologic structure of the Transantarctic Mountains and adjacent ice covered areas. Unpub. PhD thesis, Univ. of Wisconsin, Madison.

U.S. NAVAL OCEANOGRAPHIC OFFICE, 1963: Spec. Publ. 66.

U.S. NAVAL OCEANOGRAPHIC OFFICE, 1965: Spec. Publ. 66 Suppl. 1.

WERNER, W., 1953: Interpretation of magnetic anomalies at sheet like bodies. *Sver. Geol. Undersok., Ser. C.C. Arsbok, 43, 6.*

# CRUST AND UPPER MANTLE STUDY OF McMURDO SOUND

L.D. McGinnis, D.D. Wilson, W.J. Burdelik and T.H. Larson, *Department of Geology, Northern Illinois University, DeKalb, Illinois 60115, U.S.A.*

*Abstract.* Crustal thickness is the fundamental distinguishing characteristic between East and West Antarctica. The boundary between the two is located 15 km offshore in McMurdo Sound. The Sound is underlain by as much as 3.5 km of glaciomarine and preglacial sediments resting upon a crystalline basement identical to that in the dry valleys. Basement is underlain, at depths up to six kilometres, by a non-magnetic seismic refractor having a velocity near 6.5 km/s. The refractor is believed to consist of a granulitic facies, petrologically similar to inclusions found on Hut Point Peninsula. A "preliminary" depth to the mantle, calculated from a 200 km-long reversed refraction profile shot between the McMurdo Ice Shelf and the Nordenskjold Ice Tongue, is 25 km under western McMurdo Sound. From gravity data it is shown that the mantle descends abruptly from a depth of 25 km at a point 15 km offshore to a depth over 35 km beneath the Strand Moraine. Ignoring the overlying sediments and water depth, crustal thickness increases from about 21 km beneath McMurdo Sound to over 40 km beneath the Transantarctic Mountains.

It is assumed that the change in crustal thickness occurred during the Ross Orogeny, one of a series of Early Palaeozoic orogenic epochs that created the boundary between East and West Antarctica. The orogenies were episodes of collision between two plates 20 to 25 km thick, comparable to crustal thicknesses associated with present, passive continental margins. All subsequent tectonic events along the Transantarctic Mountains occurred along the boundary of convergence, a consequence of crustal thickening. Complex modification of West Antarctica and the belt of convergence has continued since the Early Palaeozoic orogenies; however, the integrity of the presently remaining core of East Antarctica has not been seriously compromised since cratonisation.

East and West Antarctica are conveniently divided by the Transantarctic Mountains. In addition to forming a major topographic divide, the traditional definition of the boundary is its association with major vertical faulting and intrusive rocks of the Ross Orogen (Craddock, pers. comm.; Elliott, 1975). Recent reports (Molnar et al., 1975) suggest the boundary between East and West Antarctica is not the Transantarctic Mountains, but a Mesozoic-Cenozoic shear zone in the Ross-Weddell Sea trough, along which 500 km of left lateral displacement and subduction has occurred in the past 81 million years. Weissel et al. (1977) find no evidence of a shear zone in their geophysical studies in the Ross Sea.

Hamilton (1963) observed broad, regional doming west of McMurdo Sound whereas Schopf (1969) interprets the Transantarctic Mountains as the boundary of a faulted rift. Grindley and Laird (1969) postulate a major boundary fault along the central Transantarctic Mountains and Kyle and Cole (1974) provide evidence for a northward extension of boundary faulting with the Erebus volcanic province lying at the intersection of two major lineaments. Extensive block faulting in Late Tertiary and Quaternary time was pointed out by Nichols (1970). Calkin and Nichols (1972) suggest that flat-lying Beacon Sandstone of Devonian to Jurassic age lies beneath McMurdo Sound and Ross Island. Since the Beacon Series is found at elevations over 1 km above sea level in the dry valleys, they infer a major fault between the Sound and the mountains. Robinson (1964) interprets a gravity profile across the Transantarctic Mountains to indicate major crustal thickening beneath the mountains. Smithson (1972) gives a similar interpretation.

Explosion seismology, gravity, and aeromagnetics are used in the present study to infer the structural configuration and tectonic history of the boundary between the western Ross Sea and the Transantarctic Mountains. An aeromagnetic study (Pederson et al., 1981) extending from eastern Ross Island to the East Antarctic ice cap did little to resolve the tectonics of the area because of the lack of magnetic anomalies across the boundary. A gravity profile extending from the Strand Moraine in western McMurdo Sound to Ross Island, with stations spaced at 1 km intervals, was established to develop a crustal model beneath the Sound. Thirty-two reversed seismic refraction profiles extending to distances of 42 km and located on sea ice throughout the Sound were used to establish a model of the crust which, in turn, permits quantitative constraints to be placed upon the lower crust configuration as defined by gravity data.

Other field studies, conducted during the 1981-82 field season, include a reversed refraction profile to the mantle and a 13 km line of seismic reflection data. The mantle refraction and reflection data have not been completely interpreted; however, preliminary results are discussed briefly in this paper. Figure 1 illustrates the location of geophysical measurements pertinent to this study.

## DATA ACQUISITION

Seismic refraction data were recorded on an SIE-RS4, 12 channel refraction seismograph and a Texas Instruments, 24 channel DFS-III seismograph. Four and eight hertz vertical seismometers were connected to cables in-line with the shot and were placed on sea ice about two metres thick. Shots up to 900 kg were suspended in sea water 20 to 39 metres below the sea ice.

Sea floor reflection data were obtained with a 12 channel Geometrics Nimbus 1210-F signal enhancement seismograph and sledge hammer blows on the sea ice. Sub-seafloor reflections were obtained using 12-fold, common Depth Point (CDP) techniques and 15 hertz geophones with six geophones per trace. Reflection data were recorded up to 10 s whereas refraction data were recorded up to a maximum of 140 s, which is approximately the time required for a P wave to traverse 200 km in water.

Gravity data were also collected on sea ice with a damped Lacoste-Romberg, model G gravity meter. Observations were made at 1 km intervals between the Strand Moraine on the Victoria Land coast and Hut Point Peninsula on Ross Island.

## INTERPRETATION

The sea floor in McMurdo Sound has velocities ranging from 1.8 to over 3.0 km/s (Figure 2). McGinnis (1981) attributes the high velocities to submarine permafrost; however, recent drilling suggests that the degree of lithification of sea floor sediments can, in many cases, explain the abnormally high velocities. McGinnis and Otis (1979) have shown that velocities as high as 3.0 km/s are found at depths 800 to 900 metres below the sea floor on passive continental margins in a glaciomarine, clastic, depositional environment. Drilling results (Barrett and McKelvey, 1981) indicate sediments on the McMurdo Sound sea floor are also of glaciomarine origin; thus, if the sea floor is not permafrost, it must be assumed that as much as 800 to 900 metres of sediment have been removed from the Sound which, if replaced, would place the land suface in the McMurdo area above sea level. Sediments removed would have been Miocene and younger in age.

Continuous refractors lying below the sea floor in Figure 3 probably correspond with erosional unconformities. Beginning with the sea floor, the higher velocities near 3.0 km/s represent Early or Middle Cenozoic units whereas velocities less than 2.0 km/s are of more recent origin. Barrett and Froggett (1978) have published velocities representative of various units observed in the McMurdo Sound region; however, their data are primarily from observations on hand specimens, including glacially transported erratics, and will be less than velocities observed in situ.

A refractor having a mean velocity of 3.3. km/s lies several hundred metres below the sea floor throughout most of the profile shown in Figure 3. Without further information, such as drilling or high resolution reflection data, any interpretation on the character of the sediments lying below the refractor would be highly speculative. It is our opinion that the refractor marks an erosional unconformity of glacial derivation with the sediment lying below the unconformity being of preglacial age. Wong and Christoffel (1981), in their single channel, marine seismic reflection study, also show an angular unconformity at depths several hundred metres below the sea floor which is believed to mark the boundary between glacial and preglacial sediments. It is assumed that the first refractor corresponds with this angular unconformity and that the sediments lying below are of Mesozoic to Early Tertiary age.

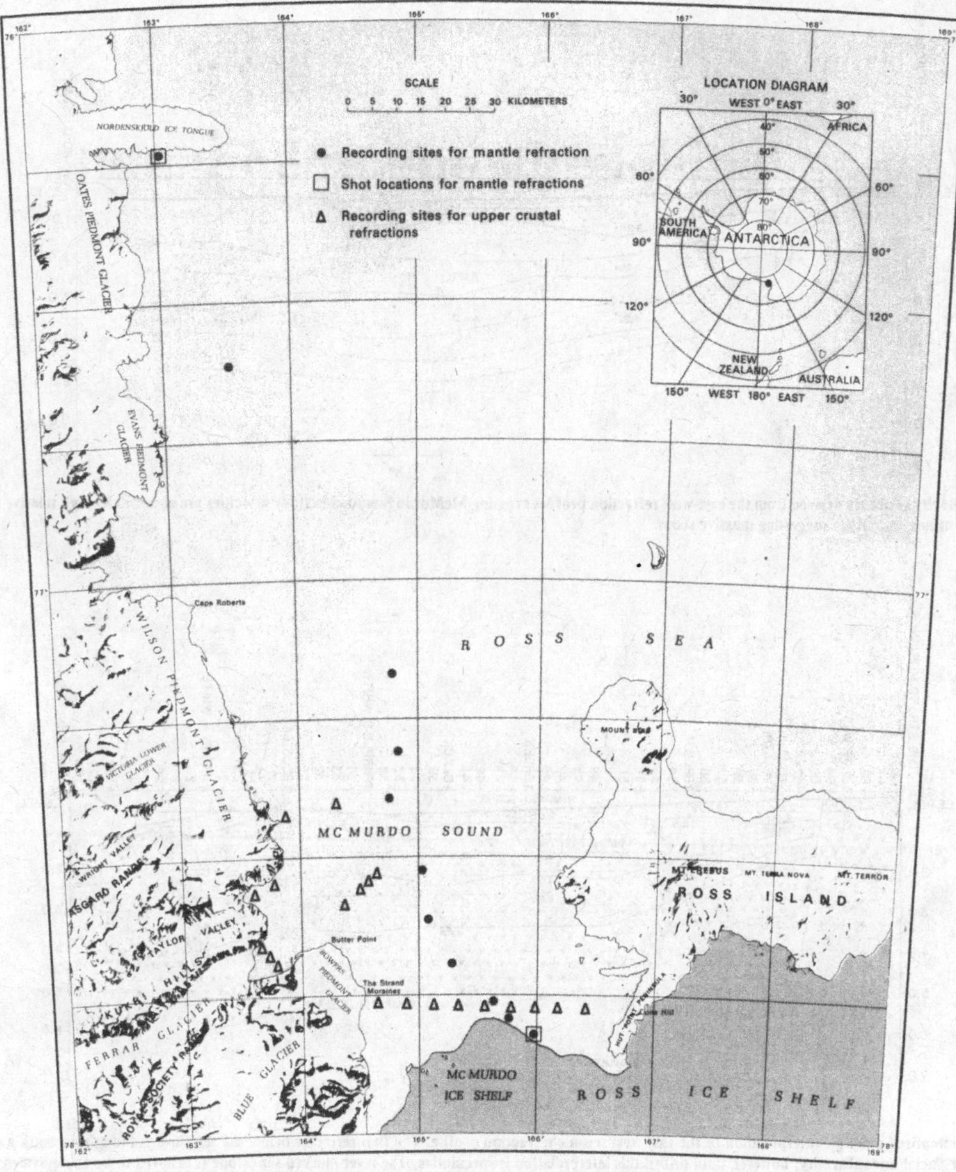


**Figure 1 Location of recording sites of reversed seismic refraction profiles. Triangles represent recording sites where the seismometers were held stationary while shots were moved out. Circles are recording sites where seismometers were moved out and the shot held stationary. Circles are station locations for a mantle refraction experiment conducted in November, 1981. The two squares, one near the McMurdo Ice Shelf and one near the Nordenskjold Ice Tongue, mark the shot locations for the mantle refraction.**

The unit labelled"assorted clastics" in Figure 3 is underlain by a refractor having a mean velocity of 4.0 km/s. The unit is not present near shore but first appears about 8 km offshore where it is only 500 m thick. Beneath Ross Island it attains thicknesses over 2 km. From the velocity data of Barrett and Froggett (1978) we infer that this unit is equivalent in age to the Beacon Sandstone. If the 4.0 km/s section is of Beacon age, then the underlying unit having mean velocities of 5.0 km/s is crystalline basement, and the sub-basement refractor having a mean velocity of 6.5 km/s is a granulite facies that Stuckless and Erickson (1976) described in their study of inclusions on Hut Point Peninsula. From the aeromagnetic study by Pederson et al. (1981) it is evident that the crystalline basement and overlying strata are relatively nonmagnetic and therefore we infer that the Ferrar dolerites are not present in McMurdo Sound.

Depths determined from seismic refraction data are used to provide an approximation of sediment thickness and basement configuration in McMurdo Sound. As stated previously, basement velocity is assumed to be greater than 5 km/s; however, in several cases a refractor having a velocity as low as 4.5 km/s is found at depths where basement must be assumed. Depths to the 5 km/s refractor are contoured in Figure 4.

The basement surface descends from sea level at the coast where Granite Harbour Intrusives outcrop, to 4 km immediatly west of Hut Point Peninsula. Because of the fact that basement lies at a depth greater than 500 m in the mouth of the glaciated Ferrar Valley it is probable that much of the near-coast basement is a glacially scoured surface, at least down to 500 m below sea level. A basement depression northeast of New Harbour is believed to be tectonic in origin because of intermediate velocities found within the depression. The depression is underlain by a 5.3 km/s refractor and half-filled with sediment of 4.2 km/s which is too high for sediment of glacial or glaciomarine origin.

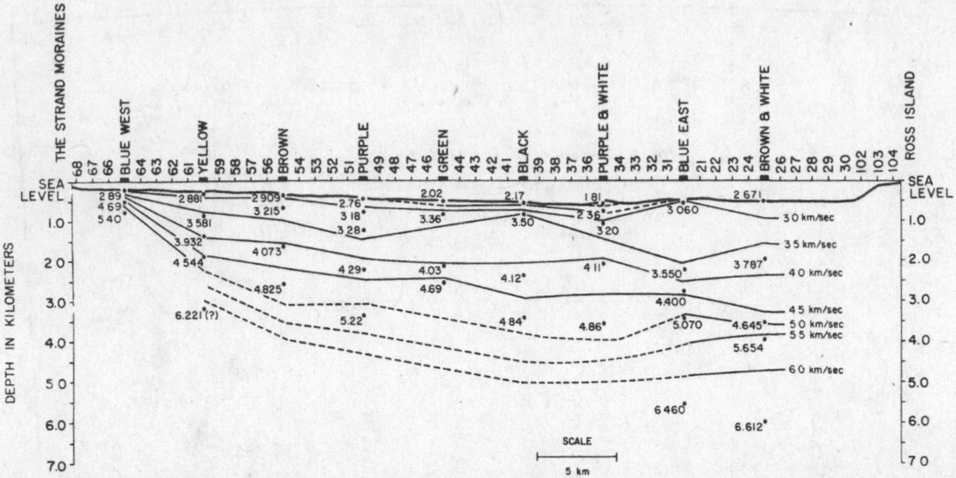


**Figure 2** **Velocity contours derived from the east-west refraction profiles crossing McMurdo Sound. Sea floor velocities are abnormally high relative to most continental shelves suggesting massive scour.**

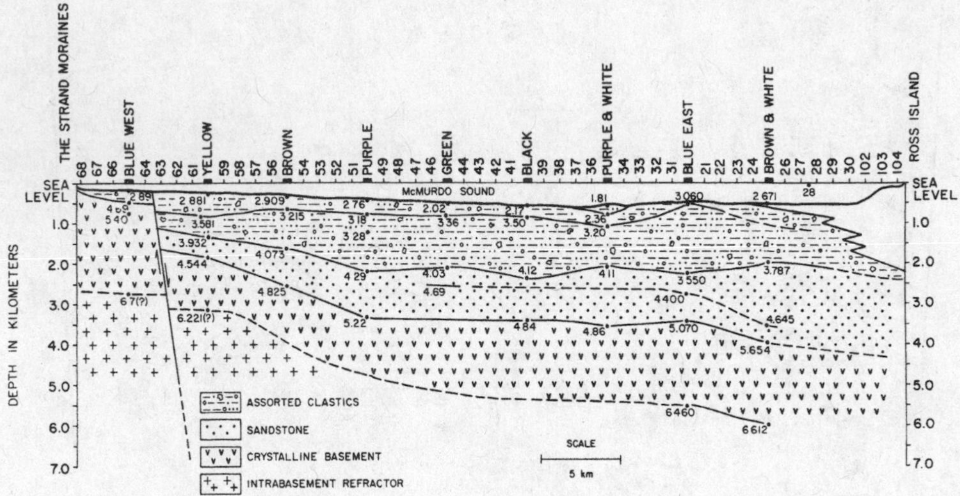


**Figure 3** **Generalised geologic interpretation of the east-west seismic refraction profile. The first refractor below the sea floor is thought to mark a glacial-preglacial unconformity; however, until drilled this interpretation is speculative. The layer marked sandstone is believed to be an equivalent of the Beacon supergroup but of shallow marine origin and without Ferrar intrusives.**

The basement in McMurdo Sound appears to"bottom out" at about 3 km below sea level, at which point it begins to descend again to maximum depths of 4 km near Ross Island. We speculate that about 1 km of the basement depression near Ross Island is due to loading of the island and subsequent deposition of sediments in the submarine depression.

A sediment isopach map (Figure 5) in McMurdo Sound is derived from the difference in elevation of the bathymetric map of McGinnis (1973) and the basement map in Figure 5. Sediments thicken abruptly offshore, eventually reaching 3.5 km near Ross Island. An average thickness in the deeper part of the sound appears to be between 2.5 and 3.0 km. About 1.5 km of sediment are contained in the basement depression northeast of New Harbour. An ubiquitous sub-basement refractor observed throughout McMurdo Sound is probably caused by a granulite facies similar to inclusions of crustal origin found in volcanic vents on Hut Point Peninsula (Stuckless and Erickson, 1976).

A 200 km long refraction profile was shot in the 1981-82 field season to calculate a velocity and depth to the mantle. In this case, the shots were held stationary while the geophone spread was moved away from the shot. Ten shots were recorded with the shots suspended in sea water at the McMurdo Ice Shelf ranging up to 900 kg. The geophone spreads were moved northward toward the Nordenskjold Ice Tongue. Four stations recorded the 6.4 km/s refractor and the last two recorded refractions from the mantle. On the reversed profile, only three shots could be fired near the Nordenskjold Ice Tongue because of dynamite limitations and because large cracks were forming in the sea ice. Recording stations for the three reversed shots were located at sufficiently great distances from the shot so that all recorded the mantle refraction as the first arrival. It is hoped that in future seasons we will be able to get back to the Nordenskjold Ice Tongue area to complete our soundings in order to establish better control on the configuration of the upper crust; however, with the data at hand and

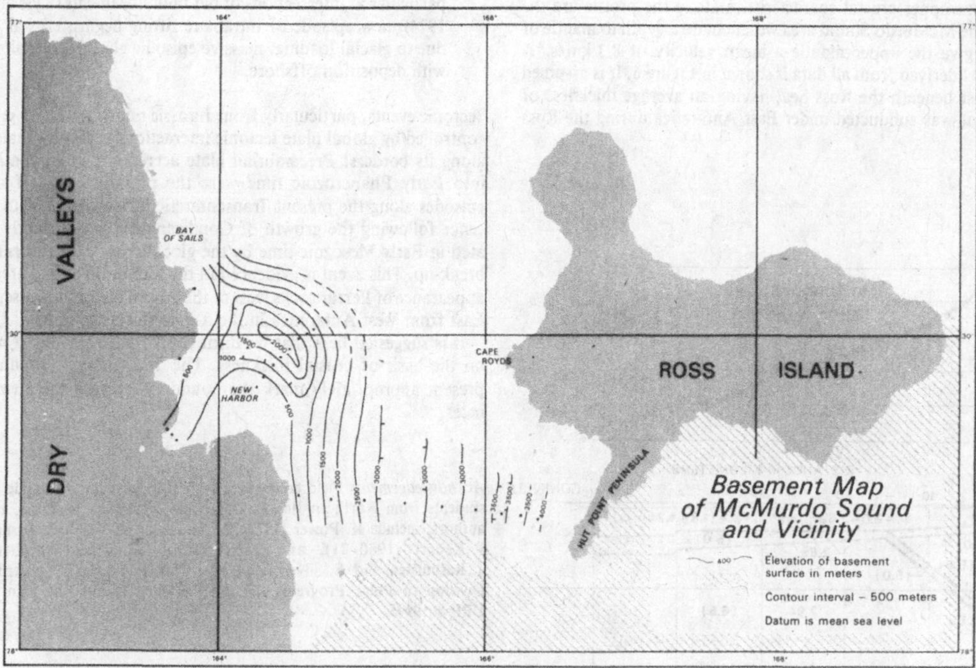


**Figure 4 Configuration of the crystalline basement surface in McMurdo Sound. The depression northeast of New Harbour is believed to be tectonic in origin because it is filled with high velocity, preglacial sediments.**

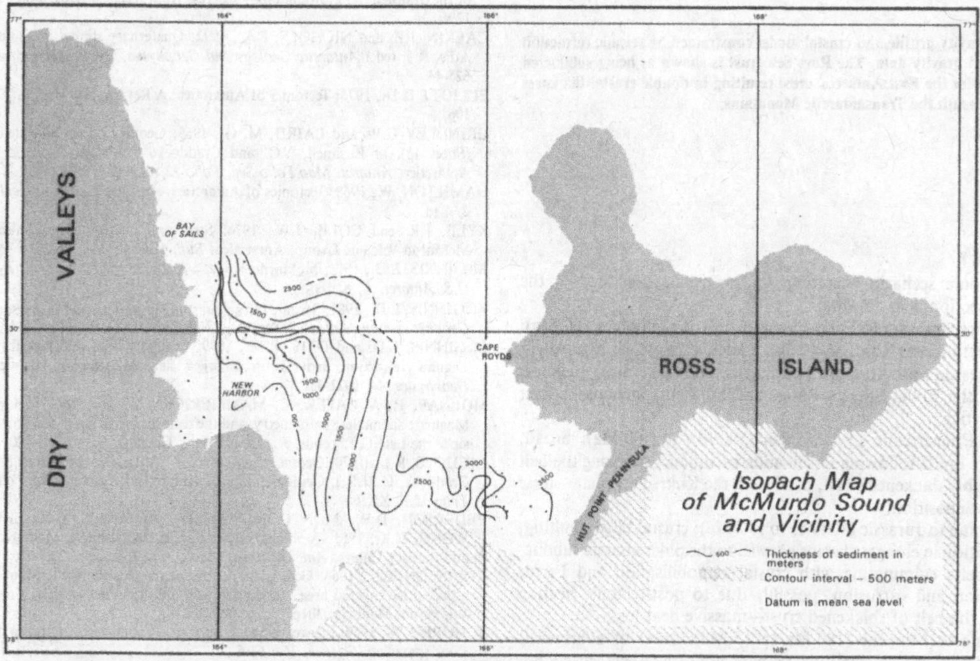


**Figure 5 Sediment thickness in McMurdo Sound. Nearly three kilometres of sediment in the central part of the Sound are thought to result from deposition in a preglacial, elongate basin parallel to the Transantarctic Mountains with deposition possibly beginning during the Ross Orogeny.**

assuming that upper crustal characteristics along the profile are the same as in the McMurdo Sound area we calculate a depth to mantle of 25 km and give the upper mantle a mean velocity of 8.3 km/s. A crustal model derived from all data is shown in Figure 6. It is assumed that the crust beneath the Ross Sea, having an average thickness of about 21 km, was subducted under East Antarctica during the Ross Orogeny.

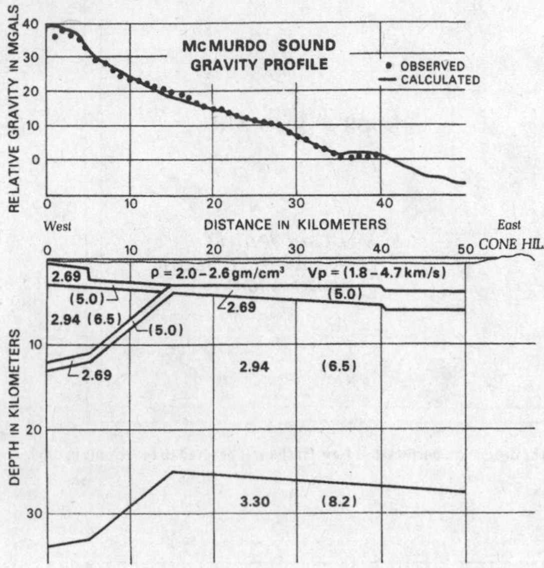


**Figure 6 Gravity profile and crustal model constrained by seismic refraction and gravity data. The Ross Sea crust is shown as being subducted under the East Antarctic crust resulting in double crustal thickness beneath the Transantarctic Mountains.**

## SUMMARY

The tectonic scenario according to the events described in the preceding section is as follows:

(1) Late Precambrian to Early Phanerozoic (one billion to 500 Ma): Subduction of Ross Sea crust under a crust of equivalent thickness in East Antarctica producing a double crustal thickness under the Royal Society Range and adjacent portions of East Antarctica;

(2) Early Phanerozoic to Devonian (500 Ma to 400 Ma): broad, passive uplift following termination of subduction along the belt of doubly thickened crust, producing the Kukri Peneplain—little or no deposition;

(3) Devonian to Jurassic (400 Ma to 180 Ma): crustal block faulting, deposition in elongate basins parallel to the older arcuate subduction belts culminating with crustal remobilisation and Ferrar intrusion and extrusion, possibly due to post-tectonic heating along the belt of thickened crust—massive heat loss;

(4) Jurassic to Miocene (180 Ma to 15 Ma): onset of continuous continental-scale glaciation—perhaps not of ice-sheet variety but certainly along uplifted ranges where climatic and topographic conditions permit deposition of marine and glaciomarine sediments in continental borderlands, on continental shelves, in particular in grabens associated with crustal stretching—uplift of Transantarctic Mountains in post-Cretaceous time associated with the Victoria Orogeny—uplift may be in part due to glacial crustal bending and in part to an episode of reheating;

(5) Miocene—Present (15 Ma—0): onset of ice sheet environment; McMurdo type volcanism along the Transantarctic Mountains, in particular at intersections of old fault lineaments (Kyle and Cole, 1974); new episode of intraplate rifting begins; crustal bending due to glacial loading; massive episodic glacial scour alternating with deposition offshore.

Tectonic events, particularly from Jurassic time on, were in great part controlled by global plate tectonic interactions in Gondwanaland and along its borders. Precambrian plate accretion and its continuation into Early Phanerozoic time were the primary cause of orogenic episodes along the present Transantarctic Mountains. Relative quiescence following the growth of Gondwanaland was abruptly terminated in Early Mesozoic time by the global onset of continental plate break-up. This event is evidenced in the McMurdo Sound area by the appearance of Ferrar rocks west of the crustal discontinuity separating East from West Antarctica in the belt of thickened crust.

It is suggested here that we distinguish East from West Antarctica on the basis of crustal thickness. The Transantarctic Mountains at present appropriately mark the boundary between these two provinces.

*Acknowledgements* Field crews were staffed by graduate and undergraduate students from Northern Illinois University. Field crew members, excluding authors, include K. Power (1978/79), H. Miller (1979/80), T. Fasnacht and P. Rebert (1980/81), and J. Erickson, T. Fasnacht, S. Germanus, J. Rasmussen and S. Silver (1981/82). This study was supported by the Division of Polar Programs, National Science Foundation grant number DPP 8019995.

## REFERENCES

BARRETT, P.J. and FROGETT, P.C., 1978: Densities, porosities, and seismic velocities of some rocks from Victoria Land, Antarctica. *N.Z. J. Geol. Geophys. 21,* 175-87.

BARRETT, P.J. and MCKELVEY, B.C., 1981: Cenozoic glacial and tectonic history of the Transantarctic Mountains—progress from recent drilling. *Polar Rec., 20,* 129.

CALKIN, P.E. and NICHOLS, R.L., 1972: Quaternary studies in Antarctica; *in* Adie, R.J. (ed.) *Antarctic Geology and Geophysics,* Universitetsforlaget, Oslo, 625-44.

ELLIOTT, D.H., 1975: Tectonics of Antarctica: A Review. *Am. J. Sci., 275-A,* 45-106.

GRINDLEY, G.W. and LAIRD, M. G., 1969: Geology of the Shackleton Coast (Sheet 15); *in* Bushnell, V.C. and Craddock, C. (eds.), Geologic maps of Antarctica. *Antarct. Map Folio Ser., Folio 12,* Pl.XV.

HAMILTON, W., 1963: Tectonics of Antarctica. *Am. Assoc. Petrol. Geol., Mem., 2,* 4-15.

KYLE, P.R. and COLE, J.W., 1974: Structural control of volcanism in the McMurdo Volcanic Group, Antarctica. *Bull. Volcan., 38-1,* 16-25.

McGINNIS, L.D., 1973: McMurdo Sound—A Key to the Cenozoic of Antarctica. *U.S. Antarct. J., 8,* 166-9.

McGINNIS, L.D., 1981: Seismic refraction study in western McMurdo Sound. *Am. Geophys. Union, Ant. Res. Ser., 33,* 27-35.

McGINNIS, L.D. and OTIS, R.M., 1979: Compressional velocities from multichannel refraction arrivals on Georges Bank—northwest Atlantic Ocean. *Geophysics, 44,* 1022-33.

MOLNAR, P., ATWATER, T., MAMMERICKX, J. and SMITH, S.M., 1975: Magnetic anomalies, bathymetry, and the tectonic evolution of the South Pacific since the Late Cretaceous. *R. Astron. Soc., Geophys. J., 40,* 383-420.

NICHOLS, R.L., 1970: Geomorphic features of Antarctica; *in* Bushnell, V.C. and Craddock, C. (eds.), Geologic maps of Antarctica. *Antarct. Map Folio Series, Folio 12,*Pl.XXII.

PEDERSON, D.R., MONTGOMERY, L.D., MCGINNIS, C.P., ERVIN and WONG, H.K., 1981: Aeromagnetic survey of Ross Islands, McMurdo Sound, and the Dry Valleys. *Am. Geophys. Union, Ant. Res. Ser., 33,* 7-25.

ROBINSON, E.S., 1964: Geologic structure of the Transantarctic Mountains and adjacent ice covered area, Antarctica. Unpublished Ph.D. thesis, University of Wisconsin, Madison, 301p.

SCHOPF, J.M., 1969: Ellsworth Mountains: position in west Antarctica due to sea floor spreading. *Science, 164,* 63-6.

SMITHSON, S.B., 1972: Gravity interpretation in the Transantarctic Mountains near McMurdo Sound, Antarctica. *Geol. Soc. Am. Bull., 83,* 3437-42.

STUCKLESS, J.S. and ERICKSON, R.L., 1976: Strontium isotopic geochemistry of the volcanic rocks and associated megacrysts and inclusions from Ross Island and vicinity, Antarctica. *Contrib. Mineral. Petrol., 58,* 111-26.

WEISSEL, J.K., HAYES, D.E. and HERRON, E.M., 1977: Plate tectonic synthesis: the displacements between Australia, New Zealand, and Antarctica since the Late Cretaceous. *Mar. Geol., 25,* 231-77.

WONG, H.K. and CHRISTOFFEL, D.A., 1981: A reconnaissance seismic survey of McMurdo Sound and Terra Nova Bay, Ross Sea. *Am. Geophys. Union, Ant. Res. Ser., 33,* 27-35.

# 3

# Beacon Supergroup and Associated Igneous Rocks

# THE BEACON SUPERGROUP OF NORTHERN VICTORIA LAND, ANTARCTICA

B.C. Walker, *Department of Geology and Antarctic Research Centre, Victoria University of Wellington, Wellington, New Zealand.*

*Abstract* The area between Gair Mesa and the Lanterman Range in northern Victoria Land contains up to 300 m of Permian alluvial plain sediments of the Beacon Supergroup (Takrouna Formation). Scattered outcrops rest unconformably on a basement of metasedimentary and plutonic rocks. The Takrouna Formation is divided into two members. Member A occurs only in the northern part of the area, but Member B covers the whole area. Member A was deposited on a northeast dipping palaeoslope, and Member B was deposited on a west dipping palaeoslope. Seven sedimentary facies occurring as two facies associations are recognised within the Takrouna Formation. Facies association 1, considered to be largely flood-deposited, includes the cross-bedded sandstone, ripple laminated sandstone, massive sandstone and mudstone facies. Facies association 2, deposited in shallow channels between flood events, includes the ripple and parallel laminated fine sandstone, tabular cross-bedded sandstone and coal facies. Member A contains both facies associations, whereas Member B contains only facies association 1. The rivers of Member A flowed from the area of the retreating Permian ice sheet. Then uplift of the mountains to the east of the study area reversed the palaeoslope, and the deposition of Member B followed.

During the summer of 1981/82 a two man field party consisting of B.C. Walker and B.C. McKelvey (geologists) spent six weeks investigating the Beacon Supergroup of northern Victoria Land, a coal-bearing alluvial plain sequence. In all, seventeen localities were visited (Figure 1). The Beacon strata occur in widely scattered outcrops, mainly as cliff faces and reach a thickness of 300 m. The strata rest unconformably on a basement of metasedimentary and plutonic rocks. In a few places they occur as rafts within dolerite sills.

Previous investigations of the Beacon Supergroup in northern Victoria Land have been mainly of a reconnaissance nature. Gair (1967) and Skinner and Ricker (1968) worked further south in the upper Rennick Glacier area where they recorded small scattered outcrops of Beacon sediments. Sturm and Carryer (1970) and Dow and Neall (1974) located and briefly described some of the outcrops visited during this study. The latter proposed the name Takrouna Formation for the flat-lying Permian sediments that rest on the sub-Beacon surface.

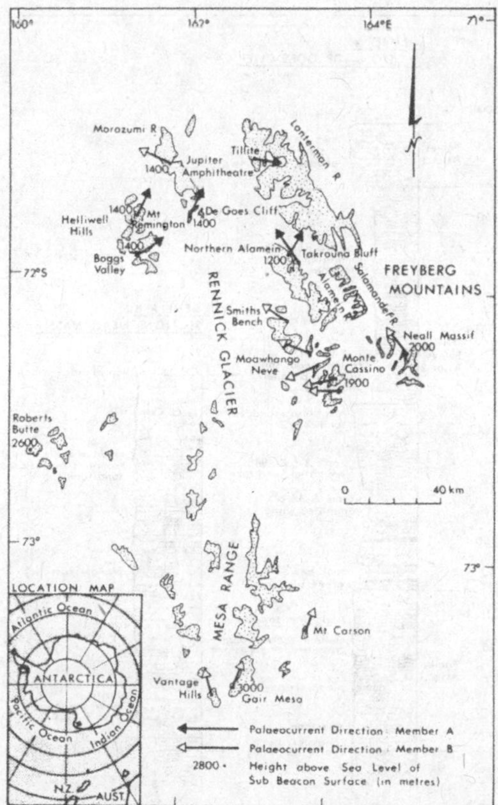


**Figure 1. Palaeocurrent directions for Member A and Member B of the Takrouna Formation, northern Victoria Land, Antarctica.**

## AGE

Leaf impressions of the *Glossopteris* flora were found at a number of localities, indicating a Permian age. They mostly occur in mudstone lenses or in carbonaceous silty claystone believed to be channel fill deposits. Plant fossils were not found at Neall Massif and the outcrops flanking the Moawhango Neve, probably because the strata there are too sandy. However, they are considered Permian in age also because they overlie the same weathered "peneplain" and are similar in gross lithology to the *Glossopteris* bearing localities to the north. The microfloral age of "probable Upper Triassic" by Norris in Sturm and Carryer (1970) from a basal section of sandstone overlying granite in the northern Alamein Range must now be questioned in view of the extensive *Glossopteris* flora in these strata.

## STRUCTURE

The Beacon sequence in northern Victoria Land is either horizontal or tilted up to 10° in blocks of the order of 10 km across. Some faulting is evident within the blocks, for example, in Boggs Valley where the throw is 50 m (Figure 2a). At De Goes Cliff a similar displacement occurs as a monoclinal flexure (Figure 2b). The faulting in Boggs Valley predates dolerite intrusions and at De Goes Cliff occurred during Beacon deposition. An unusual occurrence of isoclinally folded Beacon faulted against the basement complex occurs in the Lanterman Range (Laird and Bradshaw, 1981).

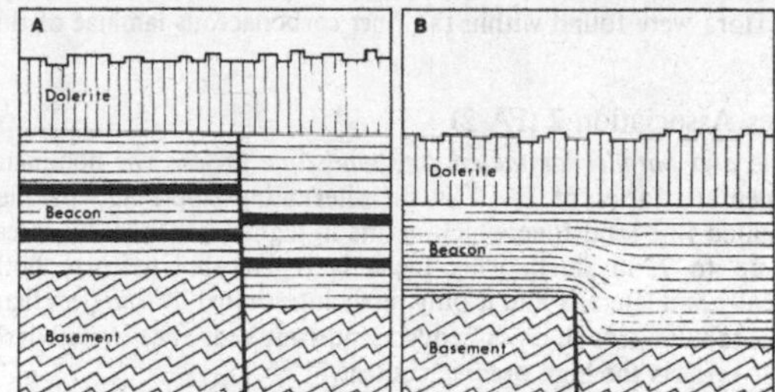


**Figure 2. Diagrammatic sections showing nature and age of faulting: A, Boggs Valley, where faulting began after the Beacon strata were deposited and before dolerite intrusion. B, De Goes Cliff, where the faulting was active during the deposition of the basal sediments.**

## SUB-BEACON SURFACE

The surface on which the Beacon strata were deposited (the sub-Beacon surface) has low relief and is remarkably uniform. At two localities, De Goes Cliff and Boggs Valley, the surface was traced over 5 km and dips 5° west. Local relief is less than a metre except in two places where the surface is displaced 50 m vertically by faults. The low relief is also a feature of the surface further south near the Campbell-Rennick divide (Gair, 1967, figure 14), who in fact referred to the surface as a peneplain. Gair (1967) reported deep weathering of granite to depths of "several hundred feet" near the Mesa Range. The 15 m of weathered granite found in the Freyberg Mountains by Dow and Neall (1974) has the appearance of a massive crumbly arkosic very coarse sandstone. However, we found that further north at De Goes Cliff and Boggs Valley weathering did not extend more than 2 m beneath the sub-Beacon surface. Here the weathered unit consists of clast supported angular blocks of schist in a matrix of coarse sand.

## SEDIMENTARY FACIES OF THE TAKROUNA FORMATION

The Takrouna Formation is divided herein into two members. Member A occurs only in the northern part of the area over a distance of 60 km and at several localities rests directly on the sub-Beacon surface. Member B is more extensive, covering the whole area and in the south also rests on the sub-Beacon surface (Figure 1). Unfortunately nowhere could we find an outcrop containing both members though it is later suggested that Member B followed Member A. Seven sedimentary facies and two facies associations are recognised within the Takrouna Formation (Table 1). Member A contains both facies associations, whereas Member B contains only facies association 1 (Figure 3).

### Facies Association 1 (FA 1)

*Cross-bedded facies.* This facies consists of tabular sheet sandstone beds generally less than 2 m thick composed of up to four sets of trough cross-beds. The troughs are mostly between 1 and 2 m wide and between 0.3 and 0.5 m thick. Occasional single sets of tabular cross-bedding averaging 1 m thick have extensively eroded tops. At Neall Massif tabular cross-beds reach 5 m in thickness. Lower surfaces of cross-bed sets are generally sharp, and in places show evidence of scouring to depths of several centimetres. Grain size and scale of cross-bedding generally remain the same within an individual sandstone bed.

*Ripple laminated facies.* The ripple-laminated facies has similar bed geometry to the cross-bedded sandstone facies. Ripple laminae are frequently picked out by carbonaceous material. This facies is either interbedded within the multistorey sandstones of the cross-bedded facies or, as at De Goes Cliff, forms its own multistorey sandstone bodies.

*Massive sandstone facies.* This facies typically consists of coarse, poorly sorted sandstone overlying a scour surface and containing rounded quartz pebbles, mudstone clasts and plant debris. This facies occurs only beneath the cross-bedded sandstone facies.

*Mudstone facies.* The mudstone facies occurs as thinly laminated carbonaceous mudstone lenses seldom more than 1 m thick and generally less than 0.5 m. This facies extends laterally for up to several hundred metres and occurs between beds of the cross-bedded and ripple laminated sandstone facies. Leaf impressions of the *Glossopteris* flora were found within the finer carbonaceous laminae of this facies.

### Facies Association 2 (FA 2)

*Ripple and parallel laminated fine sandstone facies.* The dominant sedimentary facies of FA 2 is the alternating ripple and parallel laminated fine sandstone which occurs in highly carbonaceous intervals up to 75 m thick. This facies is remarkably uniform both vertically and laterally and is only rarely interrupted by thin (less than 0.5 m) tabular, trough cross-bedded sandstone beds. The overall dark colour reflects the high organic content.

*Tabular cross-bedded sandstone facies.* This facies contains cosets of tabular cross-bedding in fine to medium well sorted sandstone. Set thickness ranges between 0.05 and 0.50 m. Both upper and lower surfaces of the tabular cross-beds are planar and unlike the larger single tabular cross-beds of FA 1 show no evidence of modification.

*Coal facies.* The coal facies contains laterally extensive coal beds, fining upwards trough cross-bedded sandstones, green siltstones and coaly shale. The coal beds have an average thickness of 0.7 m and a maximum of 1.2 m. Sandstone to shale to coal ratios are 3:1:1.

## PALAEOCURRENT ANALYSIS

Over 1000 palaeocurrent measurements were obtained from the Takrouna Formation. Measurements were taken mostly from axes of cross-beds though directions were also obtained for channel scours, planar cross-beds and ripple marks. Difficulty in working on the many steep outcrops and the absence of well exposed platforms meant that in some places only a few palaeocurrent measurements could be obtained. Calculation of vector mean and magnitude was based on vectorial addition of single palaeocurrent readings that were taken up through a section. A summary of the results is presented in Table 2.

Mean directions for Member A (Figure 1) show a consistent trend towards the northeast, except for the Takrouna Bluff section which

**TABLE 1: Description of sedimentary facies and facies associations of the Takrouna Formation, northern Victoria Land.**

| Facies name | Description | % abundance Member A | % abundance Member B |
|---|---|---|---|
| **Facies Association 1** | | | |
| Cross-bedded sandstone facies | Tabular sheet sandstones with up to 4 sets of medium scale trough cross-beds. Single sets of medium to large scale modified tabular cross-bedding (more common Member B) | 42 | 89 |
| Ripple laminated sandstone facies | Tabular sheet sandstones with ripple and climbing ripple lamination. Frequently carbonaceous | 15 | 3 |
| Massive sandstone facies | Massive poorly sorted, medium to coarse sandstone resting on scoured surfaces with lag deposits of quartz pebbles, mudstone clasts, logs and plant stems. (Member B in places has higher relief on scour surfaces and coarser lag deposits) | 6 | 5 |
| Mudstone facies | Thinly laminated carbonaceous mudstone lenses. Leaf impressions | 3 | 3 |
| **Facies Association 2** | | | |
| Ripple and parallel laminated fine sandstone facies | Ripple laminated carbonaceous fine sandstones alternating gradationally with parallel laminated fine sandstones | 25 | — |
| Tabular cross-bedded sandstone facies | Small to medium scale unmodified tabular cross-beds in medium well sorted sandstone | 6 | — |
| Coal facies | Coaly shale, coal beds, green siltstones and fining upwards sandstones | 3 | — |

(Scale of cross-bedding as defined by Conybeare and Crook, 1968, p. 32)

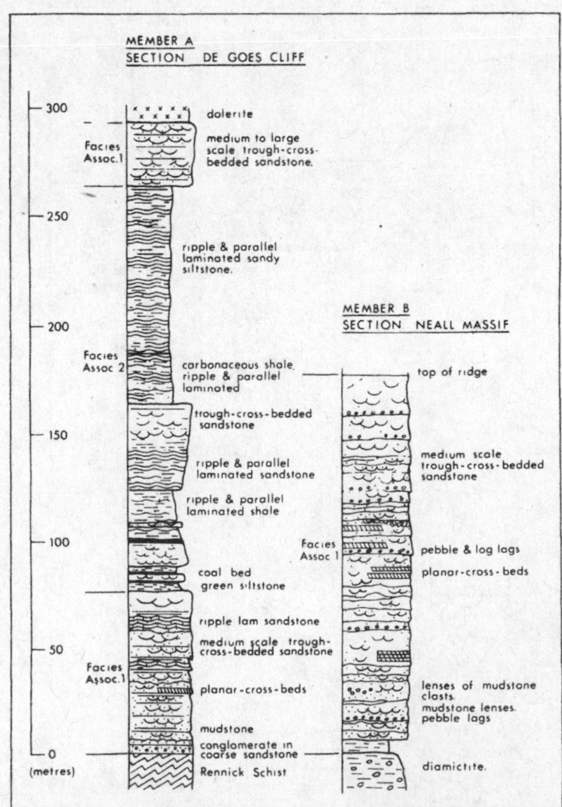


**Figure 3. Representative stratigraphic sections for Members A and B of the Takrouna Formation, northern Victoria Land, Antarctica.**

may have been moved during the intrusion of the 500 m of overlying dolerite. In contrast, mean directions for Member B are towards the west and northwest except for the outcrops in the southern part of the area which have a northerly direction. These localities may represent the overlapping of sediments from a separate sedimentary basin further south.

Vector magnitude ranges between 80 and 90 percent for both members (Table 1). Similar values were obtained for Cretaceous braided stream deposits at Banks Island, Arctic Canada (Miall, 1976). It is suggested here that the river channels of the Takrouna Formation were of low sinuosity and at times braided.

**TABLE 2: Summary of palaeocurrent analysis for the Takrouna Formation, northern Victoria Land. n = number of measurements; Vm = vector mean; L = vector magnitude**

| Location | n | VM (degrees) | L (percent) |
|---|---|---|---|
| **Member A:** | | | |
| De Goes Cliff | 48 | 025 | 86 |
| Mount Remington | 34 | 024 | 80 |
| Boggs Valley | 68 | 056 | 88 |
| Takrouna Bluff | 46 | 322 | 83 |
| North Alamein | 56 | 029 | 83 |
| Lanterman Range | 20 | 099 | 80 |
| **Member B:** | | | |
| Jupiter Amphitheatre | 34 | 296 | 88 |
| Smiths Bench | 38 | 302 | 88 |
| Moawhango Neve 1 | 198 | 292 | 87 |
| Moawhango Neve 2 | 56 | 248 | 88 |
| Monte Cassino 1 | 86 | 258 | 86 |
| Monte Cassino 2 | 116 | 270 | 86 |
| Neall Massif | 135 | 318 | 86 |
| Mount Carson | 6 | 018 | 89 |
| Gair Mesa | 87 | 021 | 89 |
| Vantage Hills | 43 | 343 | 88 |

## PALAEOHYDRAULIC INTERPRETATION

### Facies Association 1

The dominance of tabular sheet sandstone beds consisting of the cross-bedded sandstone and ripple laminated sandstone facies, the absence of fine grained overbank deposits and high vector magnitude values indicate that FA 1 represents the deposits of low sinuosity, braided streams. Rarely exposed channel cross-sections have widths of at least 100 m and depths generally less than 2 m. However, during flood peaks, sheets of water may have extended several hundred metres across the flood plain. The majority of sediments were probably deposited during the high water stage of flood events. This was demonstrated for the Brahmaputra River (Coleman, 1969) and Platte River (Blodgett and Stanley, 1980). The massive sandstone facies indicates that during peak flow of some flood events scouring of underlying sediments occurred and coarse sediment including pebbles, mudstone clasts and plant debris was deposited. This occurred most frequently in Member B and at times resulted in the blocking of channels with subsequent avulsion. As flow became stable dune and linguoid bar migration were the dominant sediment-forming processes. Individual tabular cross-bedded sets average 1 m thick and sandstone beds are seldom greater that 2 m thick suggesting that water depths during high water stage were between 1 and 2 m. However, tabular cross-beds up to 5 m thick in Member B indicate that flood waters may have reached 5 m in depth.

Erosion surfaces on the tops of the tabular cross-beds indicate that as the flood waned and water depth decreased, modification of the bars took place and braid channels formed from the submergence of these bars. During low water stage when water depth was generally less than 0.5 m, considerable quantities of sand were transported by ripple migration through the braid channels.

The mudstone facies was formed when the main flow of water had left the floodplain leaving shallow, still pools of water from which mud would be deposited as mudstone drapes. This also occurred in those channels abandoned by avulsion. Frequent flooding inhibited the development of overbank deposits.

### Facies Association 2

The sediments of FA 2 are fine grained and dominated by the ripple and parallel laminated fine sandstone facies. Blodgett and Stanley (1980) found that ripples in the Platte River formed at depths of 25 to 76 mm and plane bed at depths less than 25 mm. Similar water depths are envisaged for the formation of the ripple and parallel laminated fine sandstone facies. The largest tabular cross-beds in the tabular cross-bedded sandstone facies are 0.5 m thick indicating that the water depth was seldom greater than this.

Channel widths from field observations are indeterminable though the lateral and vertical consistency of the sedimentary facies of this facies association indicates low differentiation of relief within channels. High width to depth ratios are envisaged.

A feature of FA 2 is the occurrence of intervals several tens of metres thick which are made up solely of a single facies. For example, the North Alamein section has a 28 m interval of the tabular cross-bedded sandstone facies, the Boggs Valley section has three, 30 m intervals of the ripple and parallel laminated fine sandstone facies as does 75 m of the De Goes Cliff section. It is evident that the sediment-forming processes of ripple migration, plane bed movement and small scale linguoid bar migration operated uninterrupted for considerable periods of time. These deposits are interpreted as representing normal flow conditions. The rare interbedding of thin (less than 0.5 m) trough cross-bedded tabular sandstones within these deposits may represent deposition from flood events.

Palaeocurrent analysis shows the rivers of FA 2 to be of low sinuosity; however, unlike the rivers of FA 1 braiding is not evident. The small scale sedimentary structures as suggested earlier were formed in shallow channels, which would limit the volume of sediment able to be transported. Channel blocking and resultant avulsion under these conditions could not be expected. As well, the linguoid bars of FA 1 were formed during the high flow stage of a flood event and braiding resulted from the submergence of these bars during low flow. In contrast, FA 2 is considered to have been deposited under normal flow conditions without the major fluctuations in discharge and the topographic differentiation within channels necessary for braiding to occur. A low sinuosity, single or anastomosing channel system is suggested for facies association 2.

The laterally extensive coal beds at De Goes Cliff, that overlie fining-upwards sandstones, represent organic material that has accumulated in rare abandoned channels on the floodplain. The coal-bearing sequence supports the interpretation of a stable environment of deposition for FA 2.

The difference between the palaeohydraulic regimes of facies associations 1 and 2 could result from a change in several variables, including climate, vegetation patterns, sea level, sediment supply and tectonic regime.

Palynological studies in south Victoria Land indicate that the Permian had a cold, temperate, humid climate (Kyle, 1976). In northern Victoria Land, an abundance of plant remains and the widespread occurrence of the *Glossopteris* flora suggest a similar and stable climate. The formation and preservation of cosets of trough and tabular cross-beds and climbing ripple lamination requires an abundant and readily available supply of sediment. This is evident for both facies associations.

It is thus assumed that climate, vegetation patterns and sediment supply remained constant throughout the deposition of the Takrouna Formation. The difference in palaeohydraulic regimes of facies associations 1 and 2 has most likely resulted from fluctuating rates of tectonic uplift. This is supported by the basement faulting at De Goes Cliff which occurred during deposition of the lower part of Member A (Figure 2b).

The sediments of FA 1 were deposited shortly after periods of uplift. The resulting increase in palaeoslope gradient heightened the effects of flood events by increasing the rate of discharge through the channels and over the floodplain enabling large quantities of sediment to be transported.

In contrast, the sediments of FA 2 may represent deposition during periods of tectonic quiescence which enabled streams to return to equilibrium conditions. A low gradient palaeoslope in combination with a well vegetated catchment area considerably lessened the effect of flood events. The result was an increase in stability of the floodplain and channel behaviour.

The overall coarser grain size and larger scale of trough and tabular cross-bedding of Member B suggests deposition on a slightly steeper palaeoslope than FA 1 of Member A. Deposition may have also occurred close to source as evident from the section at Neall Massif where granite pebbles, mudstone clasts up to 0.5 m across, logs and coarse, poorly sorted arkosic sandstones are found.

## DISCUSSION

Members A and B do not occur in the same exposure and although both facies contain *Glossopteris* a greater control on dating is not yet available. Consequently, their relative stratigraphic relationship is not obvious. Both members rest unconformably on basement which could suggest that they are lateral equivalents. However, distinctly different palaeocurrent directions and nature of basement contacts suggest otherwise.

Indirect evidence for their stratigraphic relationship is in the Lanterman Range tillite (Laird and Bradshaw, 1981). The tillite is overlain by several metres of fluvial sandstone that have northeast to east palaeocurrent directions, similar to those obtained for Member A. This indicates deposition on the same palaeoslopes and suggests Member A pre-dates Member B, which was deposited on a west-dipping palaeoslope.

As stated earlier, weathering beneath the sub-Beacon surface was to greater depths in the south of the area than to the north. Correlation with the location of the members (Figure 1) shows that Member A rests on the less weathered basement contact. It is suggested that Member A was probably deposited soon after the sub-Beacon surface was cut while the surface on which Member B was deposited existed as areas of non-deposition, possibly topographic highs that were exposed to the processes of weathering for a longer period of time. This is supported by the outcrops of Member B occurring at distinctly higher topographic levels than the outcrops of Member A (Figure 1).

Uplift of the mountains east of the study area tightly folded the Beacon sediments in the Lanterman Range and may have been responsible for some of the local tilting and faulting of the large blocks further west. The uplift reversed and probably increased the gradient of the palaeoslope and sediments of Member B were deposited by braided streams close to the source, on land that previously existed as topographic highs.

*Acknowledgements* I am grateful to the Antarctic Division of the D.S.I.R., the National Science Foundation and the United States Navy for logistic support in the field. I thank Dr B.C. McKelvey and Dr J.W. Collinson and his U.S.A.R.P. party members for most helpful assistance and companionship in the field. I am also grateful to the Victoria University of Wellington Internal Grants Committee, the University Grants Committee, and Alex Pyne for making VUWAE 26 possible, and to Dr P.J. Barrett for reviewing the manuscript.

## REFERENCES

BLODGETT, R.H. and STANLEY, K.O., 1980: Stratification, Bedforms and Discharge Relations of the Platte Braided River System, Nebraska. *J. Sediment. Petrol.*, *50*, 139-48.

COLEMAN, J.M., 1969: Brahmaputra River: Channel processes and sedimentation. *Sediment. Geol.*, *3*, 129-239.

CONYBEARE, C.E.B. and CROOK, K.A.W., 1968: Manual of Sedimentary Structures. *Aust., Bur. Miner. Geol. Geophys., Bull.*, *102*.

DOW, J.A.S. and NEALL, V.E., 1974: Geology of the Lower Rennick Glacier, northern Victoria Land, Antarctica. *N.Z. J. Geol. Geophys.*, *17*, 659-714.

GAIR, H.S., 1967: The Geology from the Upper Rennick glacier to the coast, northern Victoria Land, Antarctica. *N.Z. J. Geol. Geophys.*, *10*, 309-44.

KYLE, R.A., 1976: Palaeobotanical studies of the Permian and Triassic Victoria Group (Beacon Supergroup) of south Victoria Land, Antarctica. Unpublished Ph.D. thesis, Victoria University of Wellington, New Zealand. 306 pp.

LAIRD, M.G. and BRADSHAW, J.D., 1981: Permian tillites of north Victoria Land, Antarctica; *in* Hambrey, M.J. and Harland, W.B. (eds.) *Earth's pre-Pleistocene glacial record*, Cambridge University Press, 237-40.

MIALL, A.D., 1976: Palaeocurrent and palaeohydrologic analysis of some vertical profiles through a Cretaceous braided stream deposit, Banks Island, Arctic Canada. *Sedimentology*, *23*, 459-83.

SKINNER, D.N.B. and RICKER, J., 1968: The Geology of the region between the Mawson and Priestley Glaciers, northern Victoria Land, Antarctica. *N.Z. J. Geol. Geophys.*, *11*, 1009-40.

STURM, A. and CARRYER, S.J., 1970: Geology of the region between the Matusevich and Tucker Glaciers, northern Victoria Land, Antarctica. *N.Z. J. Geol. Geophys.*, *13*, 408-35.

# TRACE FOSSILS OF THE PERMIAN-TRIASSIC TAKROUNA FORMATION, NORTHERN VICTORIA LAND, ANTARCTICA

J.M. Zawiskie and J.W. Collinson, *Institute of Polar Studies and Department of Geology and Mineralogy, The Ohio State University, Columbus, Ohio, 43210, U.S.A.*

W.R. Hammer, *Geology Department, Augustana College, Rock Island, Illinois, 61201, U.S.A.*

*Abstract* The Takrouna Formation, a sandy braided stream deposit of Permain to Triassic age, contains locally abundant trace fossils. Trace fossils in the channel-form sandstone facies include *Aulichnites, Monocraterion, Skolithos* and horizontal, backfilled epichinial traces. These trace fossils are interpreted as invertebrate dwelling burrows and deposit-feeding trails. A fine-grained facies, interpreted to be of pond origin, lacks all elements of the channel ichnocoenose and is dominated by *in situ* structures interpreted to be reed-like plants and ?root mats. Allochthonous organic detritus was probably the main food resource in the Takrouna ecosystem. The association of *Monocraterion* and *Skolithos*-bearing beds is typically more characteristic of marine environments. Some elements of the Takrouna invertebrate fauna may have been brackish-water species that became adapted to life in fluvial channels.

A fluvial origin for the trace fossil-bearing rocks of the Permian-Triassic Takrouna Formation of northern Victoria Land (Figures 1 and 2) is discussed by Collinson and Kemp (this volume). This report centres on the trace fossils of the Takrouna Formation and has the following objectives: (1) to provide morphologic descriptions of the trace fossils and a discussion of their taxonomic affinities with established ichnotaxa; and (2) to discuss their palaeo-ecologic significance. As emphasised by Ratcliffe and Fagerstrom (1980), there is currently a relative paucity of published research on invertebrate trace fossils from fluvial environments.

## TRACE FOSSIL LOCALITIES

Abundant trace fossils were found at three outcrops of the Takrouna Formation: Moawhango Névé (MNI), Smiths Bench (SB) and patchy exposures near Mount Massell (MM) (Figure 2).

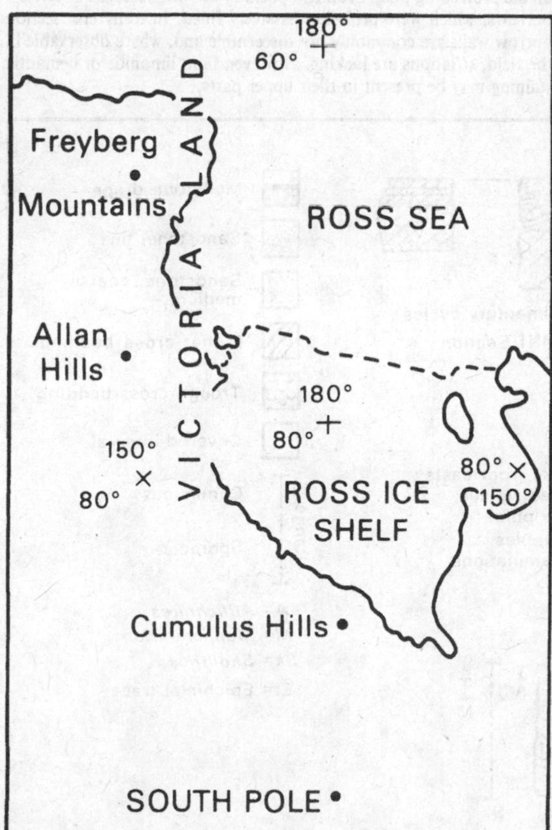


**Figure 1. Map showing location of study area, Freyberg Mountains.**

### MNI and MM Localities

Sedimentary sequences at localities MNI and MM are interpreted by Collinson and Kemp (this volume) as the channel deposits of a sandy braided stream system. Sequences of bedforms and the stratigraphic distribution of trace fossils in the Takrouna channel facies are shown in Figures 3 and 4b. Forms referrable to the following ichnotaxa are present: *Aulichnites, Monocraterion,* and *Skolithos.* An unidentified horizontal epichinial deposit-feeding trace is also present.

### Trace Fossils Description

### Ichnogenus AULICHNITES Fenton and Fenton, 1937

*Aulichnites sp.*, Figure 4g, h, i

Description: Epichinial bilobate trace consisting of two convex ridges separated by median furrow. Ridges may show weak surface crenulation. Trails range from 3 to 7 mm in diameter and are gently

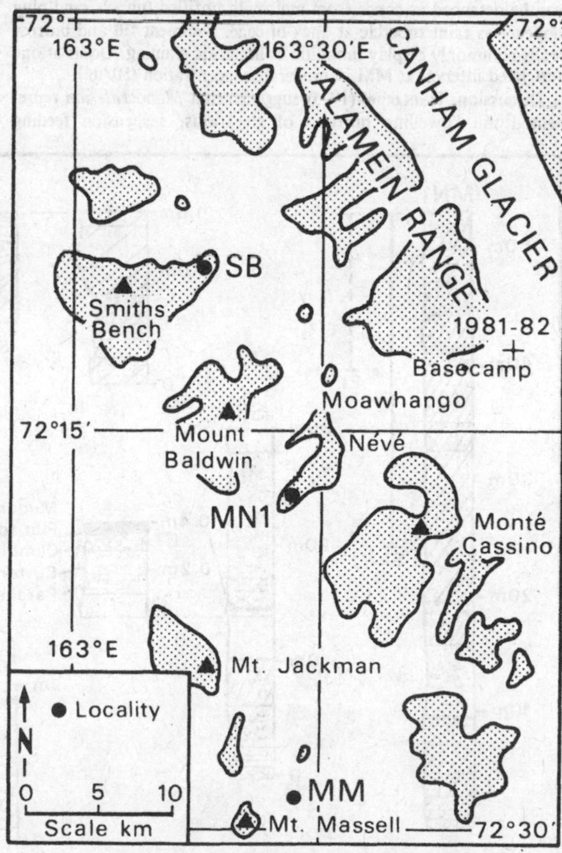


**Figure 2. Detailed map showing location of trace fossil localities, Moawhango Neve (MNI), Mount Massell (MM) and Smiths Bench (SB).**

curved to straight, commonly passing over or under one another. In some instances sharp changes in direction of movement of trace maker are indicated (Figure 4i) as described by Hallam (1970) for *Gyrochorte*. Occur with *Skolithos* on weathered bedding plane at sand-silt interface at MM.

Discussion: Fenton and Fenton (1937) originally suggested that *Aulichnites* represented the burrow of a "shelled and presumably scavenging snail inhabiting sandy shoals." Bandel (1967) described *Aulichnites* from shallow marine rocks in the Pennsylvanian of Kansas, referring to them as "crawling trails, probably of gastropod origin." Osgood (1970) also illustrated "probable gastropod" trails and discussed their variation with respect to different modes of movement. He considered *Aulichnites*- like variants to be burrows produced just beneath the sediment surface, perhaps by gastropods feeding on organic detritus. Gevers and Twomey (1982) likewise considered similar intersecting grooves and ridges in lower Beacon rocks to be gastropod trails.

The Takrouna Formation *Aulichnites* most probably represent the shallow burrows of gastropods grazing on organic detritus in sediments of the channel floor during a period of low flow. Their stratonomic occurrence is also consistent with the horizontal burrowing habit of gastropods, which dictates that their traces should be found on or near the surface (Elders, 1975).

## Ichnogenus MONOCRATERION Torell, 1870

*Monocraterion sp.*, Figure 4e, f

Description: Vertical funnel-topped burrow, ranging from 1 to 3 cm in depth. Funnel tops are circular to elliptical and may be either narrow or broad, the upper opening ranging from 3 to 16 mm in diameter. One specimen constricts from 11 to 3 mm in diameter in a vertical distance of 15 mm. A thin sand-plugged tube (1 to 5 mm in diameter) commonly penetrates the funnel. Inner structure of sediment between inner tube and funnel walls poorly defined. Funnel-fill can be detached as cones from matrix. In unfilled funnels sand plug appears as faint tubercle at apex of cone. Sediment fill and burrow walls commonly display limonitic or hematitic staining. Occur at one restricted interval at MM in moderate concentration (50/m$^2$).

Discussion: Hantzchel (1975) suggested that *Monocraterion* represented the "dwelling burrows of gregarious, suspension feeding worms." In *Monocraterion* a multiple funnel structure may be preserved which was interpreted by Hallam and Swett (1966) as the escape structure of the *Skolithos*- producing organism.

This ichnotaxon has been reported mainly from shallow marine environments, where it is commonly interbedded with *Skolithos*-bearing beds (eg. Westergård, 1931; Hallam and Swett, 1966; Alpert, 1974; Marintsch and Finks, 1982). Pemberton et al. (1982) noted its association with *Skolithos* and *Planolites* in point-bar deposits of channels suggested to be near a marine shoreline. Ratcliffe and Fagerstrom (1980) did not list *Monocraterion* as a potential ichnogenus for the shelter burrows of insects and spiders made in Holocene interior basin floodplains.

## Ichnogenus SKOLITHOS Haldeman, 1840

*Skolithos sp.*, Figure 4a, c, d

Description: Dense concentration of vertical burrows occur in both planar tabular and trough cross-bedded sandstones at localities MNI and MM, but are more commonly preserved and occur in denser concentrations in cycles that include planar tabular cross-beds. These burrows are mainly vertical to steeply inclined and range from 5 to 30 cm in depth. However, in some cases burrows penetrated more than a metre. In one extreme example a single burrow passed through two erosional surfaces, a distance of 2 m. In general original terminations of the burrows are difficult to determine, either because they are poorly preserved or because structures are clearly discernible only on weathered surfaces. Burrow openings are circular to ovoid and may be arranged in chain-like rows (Figure 4c). Their diameters range from 3 to 20 mm with modes of 5, 7, 12 and 14 mm. Some bedding planes exhibit dense crowding of burrows (500/m$^2$), whereas others show only scattered occurrences.

Burrow fills are typically slightly coarser-grained, but may be finer-grained than the surrounding matrix and commonly exhibit epirelief (Figure 4d). In the latter case a shallow central pit may be developed on the protruding plug. Sediment packing was not observed in vertical burrows, which were probably passively filled. In transverse section burrow walls are commonly not discernible and, where observable in the field, striations are lacking. However, faint limonitic or hematitic staining may be present in their upper parts.

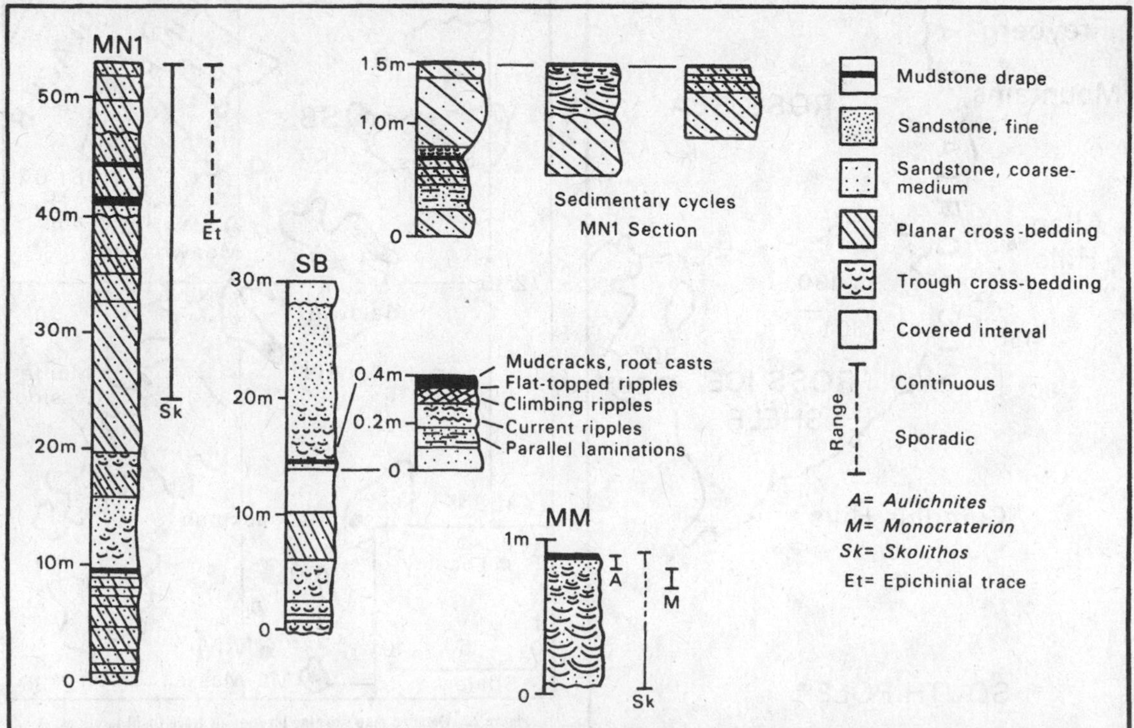


**Figure 3. Stratigraphic sequences and distribution of trace fossils at Moawhango Neve (MNI), Mt Massell (MM) and Smiths Bench (SB).**

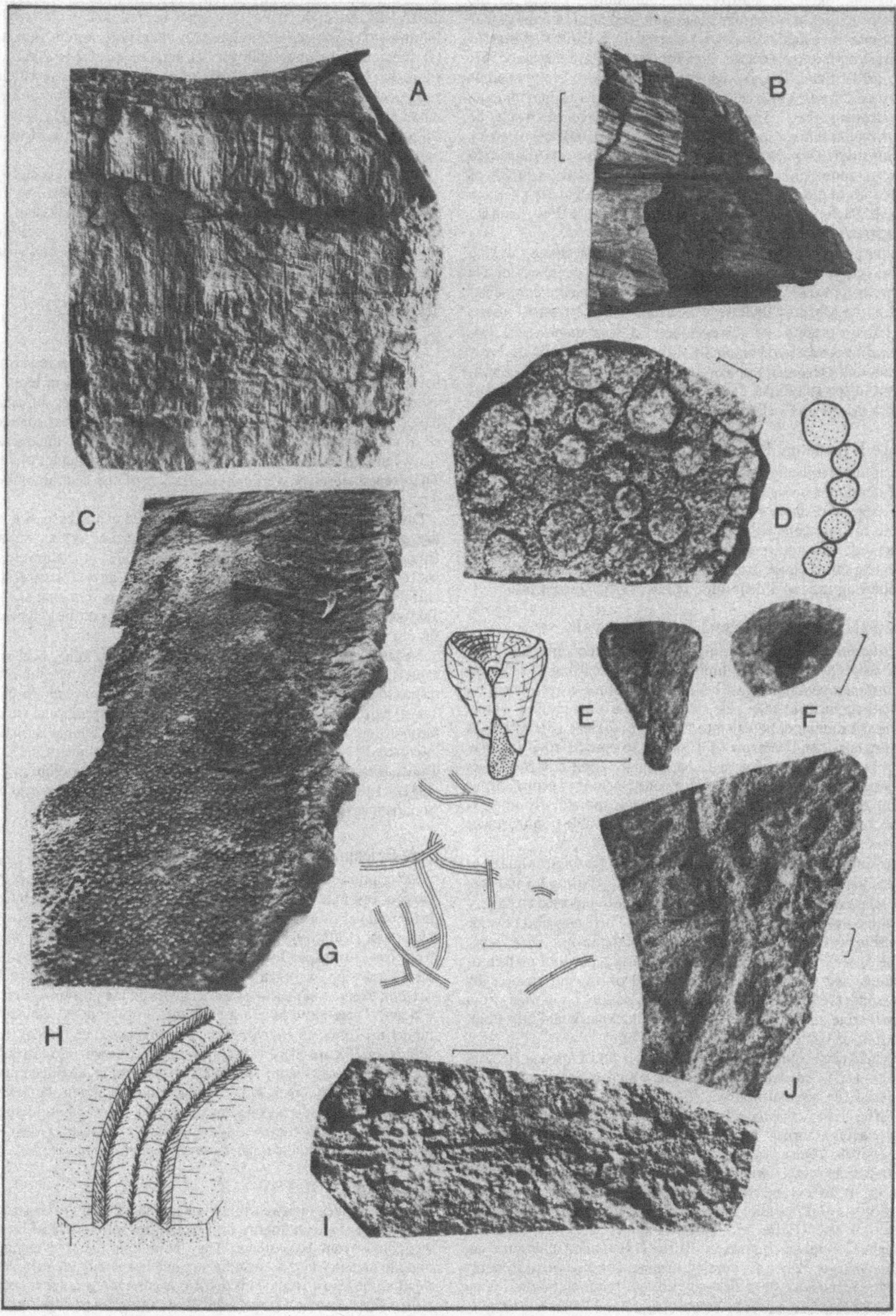


**Figure 4.** (a) *Skolithos sp.* in stacked planar tabular forsets, MNI; (b) channel cycles at MNI, scale .5 m; (c) *Skolithos sp.* in epirelief on trough cross-bedded surface at MM; (d) *Skolithos sp.* in plan view from MNI locality, trace of chain-like row is from right edge of photo, scale 1 cm; (e) *Monocraterion sp.* funnel top and penetrating sand plug from MM, scale 1 cm; (f) *Monocraterion sp.* plan view of unfilled funnel showing tubercle at apex, scale 5 cm; (g) *Aulichnites sp.* trails on bedding plane surface at MM, traced from slide, scale 5 cm; (h) *Aulichnites sp.* sketch showing morphology of Takrouna form; (i) *Aulichnites sp.* from MM with *Skolithos sp.*, note sharp change in orientation of trail, scale 1 cm; (j) Epichinial traces on bedding plane at MNI, scale 1 cm.

Discussion: *Skolithos* burrows are commonly regarded as the dwelling structures of worm-like suspension feeding organisms similar to phoronids or annelids that were formed in a shallow marine to estuarine environment (Fenton and Fenton, 1934; Seilacher, 1967; Alpert, 1974). Ratcliffe and Fagerstrom (1980) noted that numerous spiders and insects produce vertical shelter burrows in Holocene fluvial deposits that, if preserved in the rock record, would be assigned to *Skolithos*. Stanley and Fagerstrom (1974) and Bromley and Asgaard (1979) describe such vertical burrows, attributed to insects, in nonmarine rocks. We observed abundant examples of *Skolithos* in the Lower Feather Sandstone in the Allan Hills (Figure 1), which Barrett and Kohn (1975) identified as a low sinuosity braided stream deposit.

We interpret the Takrouna *Skolithos* to be invertebrate dwelling burrows, but we cannot be sure of the zoological affinities of the trace-maker. However, their association with a *Monocraterion*-bearing bed at the MM locality lessens the likelihood of terrestrial insects as the burrow-makers. This association of *Monocraterion* and *Skolithos*-bearing beds is common in shallow marine rocks of Early Cambrian and Devonian age (Westergård, 1931; Hallam and Swett, 1966; Marintsch and Finks, 1982), which is before the fossil record of terrestrial insects was well established.

### Concave Hypichinial Pits

Description: At the base of some channel cycles concave hypichinial pits are developed on the sole of the lowest bedding plane. These pits have a range of diameters similar to that of the associated *Skolithos*. They are roughly conical, range from 1 to 3 cm in length, and are oriented with their apex pointing up. We interpret these structures as representing failed escape attempts by the *Skolithos*-producing organism, following the rapid deposition of the overlying sediment.

### Horizontal, Convex Epichinial Trace, Figure 4j

Description: Unbranched, straight to sinuous, horizontal, slightly compressed convex epichinial unlined tunnels occurring at sedimentary interfaces, commonly just beneath a mudstone drape. Traces do not intersect; instead, they pass over or under one another. Rarely observed continuing obliquely into bedding plane at angles as low as 15° to the horizontal. Burrow fill is similar to country rock and may exhibit sharp imbricate pattern of backfilling. These chevron ridges represent the surface expression of a cone-in-cone structure. Trails range in diameter from 4 to 16 mm. Occur sporadically in MNI section. Recognition may be a function of favourable bedding plane exposures.

Discussion: These generally poorly preserved traces are similar to the endocone variety of *Scalarituba* described by Marintsch and Finks (1982), who pointed out that *Scalarituba* has been reported from very shallow to deep basinal marine environments. There may also be some relationship with the ichnitaxon *Halapoa* (Martinsson, 1965, Figure 31). The latter is an endocone form exhibiting a chevron pattern of backfilling, and was interpreted as the trail of an epipsammont by Martinsson (1965). Chevron trails have also been reported from shallow marine and lagoonal facies in the Blackhawk and Star Point Formations of Utah (Frey and Howard, 1970).

The ecological counterpart of *Scalarituba* and *Halapoa* in nonmarine rocks is *Scoyenia* (White, 1929), which is Seilacher's (1967) index fossil for nonmarine trace fossil assemblages. The Takrouna traces differ from *Scoyenia* in lacking dense clustered surface wrinkles and by having sharper, more imbricate backfilling (Bromley and Asgaard, 1979, Figure 9a).

An endocene trace similar to traces in the Takrouna Formation occurs in siltstones and fine-grained sandstones of the Fremouw Formation (Lower Triassic) in the central Transantarctic Mountains (Cumulus Hills, Figure 1). These siltstones are interpreted as nonmarine floodplain deposits by Barrett (1969) and Collinson et al. (1981). Whatever their taxonomic assignment these traces probably represent the activity of a deposit-feeding organism grazing along sedimentary interfaces. (Stopftunnel structures of Seilacher, 1963).

### SB Locality

The overall sequence at the SB locality (Figure 3) is similar to those at MNI and MM, but a different facies is represented by a laterally continuous unit at the 14 m level. At its base this upward fining bed consists of massive to parallel laminated medium- to coarse-grained sandstone and is overlain by current ripples followed by type A ripple drift. The ripple drift sequence may be followed by one of the following: (1) draped laminations; (2) planed-off current ripples; or (3) microcurrent ripples with foreset amplitudes of 2 to 3 mm. The entire unit is capped by a 2 to 3 cm drape of silty mudstone with local patches of dessication cracks. This unit is interpreted as recording an initial influx of sediment followed by gradually waning flow and curtailment of sediment supply and finally by an interval of shallow ponding and eventual subaerial exposure.

The trace fossils at locality SB are concentrated in the upper part of the previously described unit at the 14 m level (Figure 3). The traces are dominated by structures interpreted as *in situ* root stalks and a fine network of epichinial ridges comparable to modern root mats. Impressions of glossopterid fruiting bodies (?*Scutum*) and striated stems are also preserved with the traces.

## Trace Fossil Descriptions

### Root Casts, Figure 5b, c

Description: Numerous cylindrical structures are preserved in epirelief on the upper bedding planes of the unit at the 14 m level. The cylinders display up to 5 radiating horizontal ridges of smaller diameter. These ridges may have even finer offshoots that commonly show a complex anastomosing pattern. The central cylinders range from 2 to 28 mm in diameter and generally occur in clumps of 2 to 4. The deepest observed vertical penetration of the central stalk was 5 cm.

Discussion: These structures are interpreted as the base of a reed-like plant, with an essentially horizontal rhizome system radiating from a central stalk. Willows (*Salix*) from an area of shallow ponded water in an intermittent stream in Manti Canyon, central Utah (U.S.A.) display a horizontal root system that is very similar to the Takrouna examples (Figure 5a). The central stalks of the willows also show a similar pattern of close clumping.

Sarjeant (1975) noted that shallow horizontal rooting is generally associated with wet surface soil and a shallow water table as a response to lack of aeration. The functional morphology of the Takrouna root casts is thus consistent with the depositional environment deduced from the stratification sequence. Similar structures, interpreted as roots, have been reported from siltstone in the Lower Triassic Fremouw Formation by Barrett (1969) and Collinson et al. (1981). This siltstone facies, as stated earlier, is interpreted as a floodpain environment by these authors.

### Fine Epichinial Ridges and Grooves, Figure 5e, g

Description and Discussion: In the upper silty mudstone capping the root cast unit is a complex system of slightly compressed intersecting epichinial ridges and grooves. Identical structures are illustrated by Courel et al. (1979) from the Middle Triassic of France and are tentatively identified by these authors as ?roots. In the previously mentioned Manti Canyon stream similar complex patterns are developed on exposed bar surfaces associated with the previously discussed willows. These patterns are a combination of root systems and the tunnel networks of *Heterocerus*, the variegated mud-loving beetle. The latter (Figure 5f) are hollow and are formed just beneath the sediment surface. Figure 5d depicts a root system developed under a rock in the Manti creek-bed. Comparison of ridges in the Takrouna examples with modern examples shown in Figures 5d and f suggests close affinity with the root system, mainly because of their more rectilinear nature and similar pattern of intersection.

## PALAEOCECOLOGY

Recent studies reviewed by Anderson and Sedell (1979) stress the importance of allochthonous organic detritus, primarily leaf litter and wood in stream ecosystems. They point out that the amount of organic material added to some streams is actually greater than the input at the forest floor. This detritus is commonly added in seasonal pulses (Fischer and Lickens, 1973). Large channel-fills containing carbonaceous shale with *Glossopteris* leaves and striated stems and numerous examples of log impressions and coalfield wood attest to the abundance of organic detritus in the Takrouna channel facies.

We envisage episodic influxes of organic detritus to the Takrouna braided channel facies from adjacent wooded highlands, possibly following spring melts. Much of this detritus would be stranded

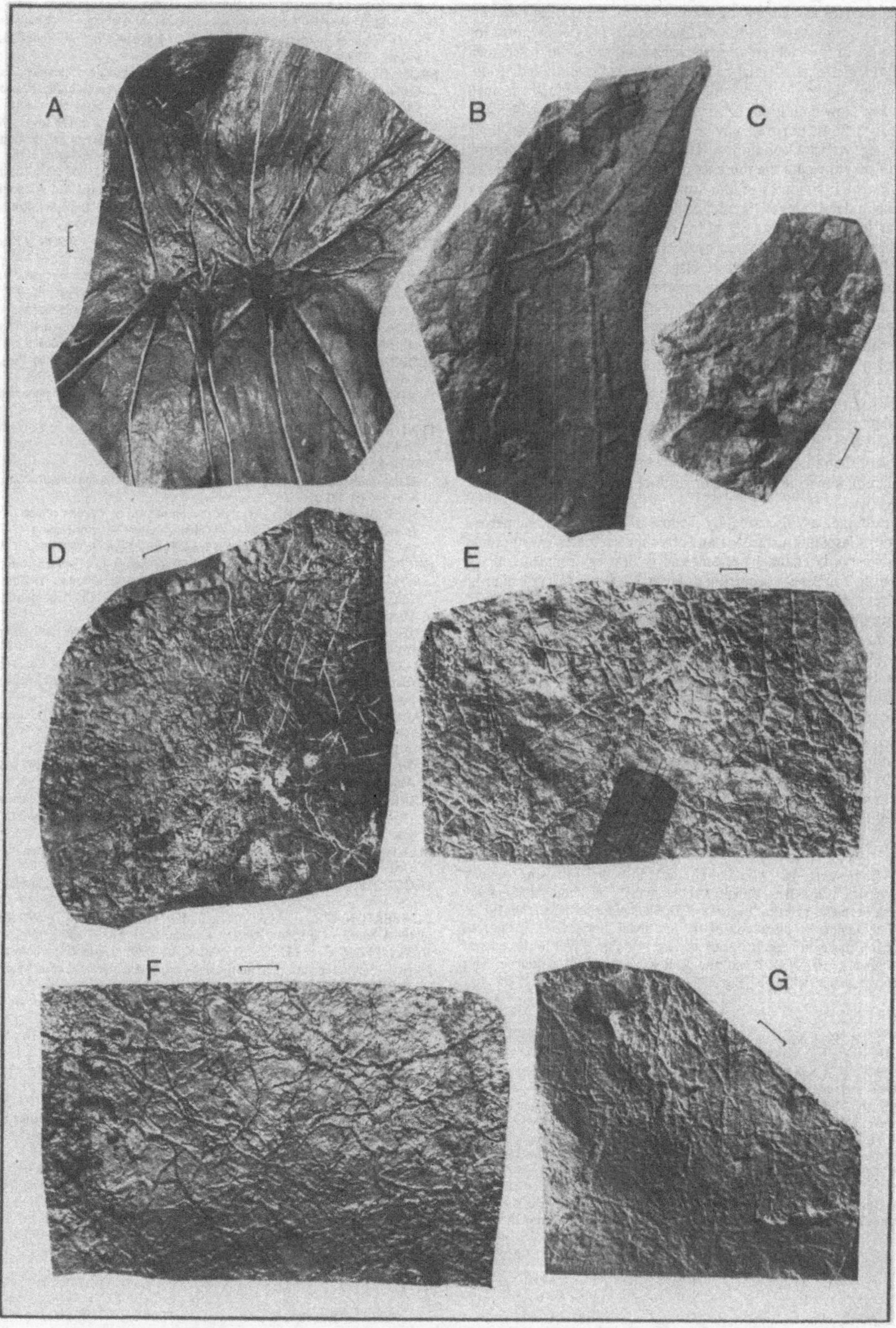


**Figure 5.** **(a) Root system of willow (*Salix sp.*) growing in intermittent stream, Manti Canyon, Utah (U.S.A.), note radiating rhizomes and close clumping of central stalks; (b) root cast preserved in interripple trough at SB, note central stalk and radiating rhizomes; (c) root stalks and rhizomes in epirelief on bedding plane at SB, note close clumping of stalks; (d) root mat under rock in intermittent stream, Manti Canyon, central Utah (U.S.A.); (e) fine epichinial ridges at SB, ?root mat; (f) tunnel system of *Heterocerus sp.* on bar surface, intermittent stream, Manti Canyon, Utah (U.S.A.); (g) fine epichinial ridges, ?root mat, SB locality; all scale bars 1 cm.**

during the subsequent low flow stage forming an "organic sink." No doubt a complex assemblage of micro- and macro-organisms became adapted to exploit these resources. Trace fossils provide evidence for the presence of invertebrate deposit feeders and grazers in the channel facies (*Aulichnites,* epichinial backfilled traces). *Skolithos* and *Monocraterion* are probably the dwelling burrows of invertebrates formed during low flow conditions when sedimentation was negligible. The deepening of some burrows may be a response to a falling water table or freezing surficial conditions. The observed variation in burrow diameter may indicate the presence of different burrowing species or the action of individuals of the same species at different stages of ontogenetic development (Stanley and Fagerstrom, 1974). The pond facies at locality SB lacks all of the elements of the channel facies ichnocoenose and is dominated by local plants.

The Takrouna trace fossil assemblage, although in fluvial rocks, is similar to assemblages previously reported from marine rocks. Pemberton et al. (1982) documented the occurrence of *Skolithos, Monocraterion* and *Planolites* in point-bar deposits and suggested a nearby marine shore-line. According to Gondwana reconstructions (eg. Griffiths, 1974), the nearest Permian marine rocks to known exposures in northern Victoria Land would have been located in Tasmania (Clarke and Banks, 1975). Abundant trace fossils attributed to worm-like organisms are also present in the Tasmanian sequence (Conkin and Conkin, 1968). Although the Takrouna basin was situated more cratonward than the Tasmanian Permian basin (Collinson and Kemp, this volume), a marine influence on its trace fossil assemblage cannot be ruled out.

For example, as discussed by Boltovskoi (1979a, b), numerous burrowing Caspian crustaceans and other invertebrates have become adapted to sandy channel bottoms and littoral environments in the Volga River, commonly occurring as far as 3,000 km from its mouth. These euryhaline species also live in the brackish Caspian Sea. Interestingly, this fauna is found in the main channels and is not present in backwater areas or floodplain water bodies. Raup and Stanley (1978) noted that euryhaline brackish-water species are most commonly derived from marine forms and "are usually present in great abundance because of reduced interspecific competition for ecologic niches."

It is conceivable that brackish-water species may have colonised the Takrouna fluvial system in a fashion similar to the invasion of the Volga by Caspian euryhaline invertebrates. As stressed by Seilacher (1963) invertebrate trace fossils are very poor salinity indicators, their morphology being more closely related to the sedimentary environment.

*Acknowledgements* We would like to thank the following people: J.A. Fagerstrom and B.C. Ratcliffe for comments on photographs of the Takrouna Formation vertical burrows; K.O. Stanley for comments on *Aulichnites;* T.N. Taylor for discussion of the functional morphology of the root casts; R.L. Stuckey for identification of *Salix* sp.; C.L. Vavra for his interest and comments; and N.R. Kemp and B. Roberts for field assistance. This research was supported by NSF grants DPP-8019996 and DPP-8020098.

## REFERENCES

ALPERT, S.P., 1974: Systematic review of the genus *Skolithos. J. Paleontol., 48,* 661-9.

ANDERSON, N.H. and SEDELL, J.R., 1979: Detritus processing by macroinvertebrates in stream ecosystems. *Ann. Rev. Entomol., 24,* 351-77.

BANDEL, K., 1967: Trace fossils from two Upper Pennsylvanian sandstones in Kansas. *Univ. Kansas Paleont. Contrib., Pap, 18,* 1-13.

BARRETT, P.J., 1969: Stratigraphy and petrology of the mainly fluviatile Permian and Triassic Beacon Rocks, Beardmore Glacier Area, Antarctica. *Rep. Inst. polar Stud, Ohio State Univ., 34.*

BARRETT, P.J. and KOHN B.P., 1975: Changing sediment transport directions from Devonian to Triassic in the Beacon Super-Group of South Victoria Land; *in* Campbell, K.S.W. (ed.) *Gondwana Geology,* ANU Press, Canberra, 15-35.

BOLTOVSKOI, M., 1979a: Biogeography of the Volga; *in* Boltovskoi, M. (ed.) *The River Volga and Its Life,* Junk Pub., The Hague, 346-66.

BOLTOVSKOI, M., 1979b: Zoobenthos and other invertebrates living on substrata in the Volga; *in* Boltovskoi M (ed.) *The River Volga and Its Life.* Junk Pub., The Hague, 235-68.

BROMLEY, R. and ASGAARD, U., 1979: Triassic freshwater ichnocoenoses from Carlsberg Fjord, East Greenland. *Palaeogeogr., Palaeoclimatol., Palaeoecol., 28,* 39-80.

CLARKE, M.J. and BANKS, M.R., 1975: The stratigraphy of the lower (Permo-Carboniferous) parts of the Parmeener Super-Group, Tasmania; *in* Campbell, K.S.W. (ed.) *Gondwana Geology,* ANU Press, Canberra, 453-67.

COLLINSON, J.W., STANLEY, K.O. and VAVRA, C.L., 1981: Triassic fluvial depositional systems in the Fremouw Formation, Cumulus Hills, Antarctica; *in* Cresswell, M.M. and Vella, P. (eds.) *Gondwana Five,* A.A. Balkema, Rotterdam, 141-8.

COLLINSON, J.W. and KEMP, N.R., (this volume): Permian-Triassic sedimentary sequence in northern Victoria Land, Antarctica.

CONKIN, J.E. and CONKIN, B.M., 1968: *Scalarituba missouriensis* and its stratigraphic distribution. *Univ. Kansas Paleont. Contrib., Pap., 31,* 1-7.

COUREL, L., DEMATHIEW, G. and GALL, J.C., 1979: Figures sédimentaires et traces d'origine biologique du trias moyen de la bordure oriental du massif central signification sédimentologique et paléoécologique. *Geobios. 12,* 379-97.

ELDERS, C.A., 1975: Experimental Approaches in Neoichnology; *in* Frey, R.W. (ed.) *The Study of Trace Fossils,* Springer-Verlag, New York, 513-36.

FENTON, C.L. and FENTON, M.A., 1934: *Scolithos* is a fossil phoronid. *Pan Amer. Geologist, 61,* 341-8.

FENTON, C.L. and FENTON, M.A., 1937: Burrows and trials from Pennsylvanian rocks of Texas. *Am. Midland Naturalist, 18,* 1079-84.

FISCHER, S.G. and LIKENS, G.W., 1973: Energy flow in Bear Brook, New Hampshire: an integrative approach to stream ecosystem metabolism. *Ecol. Mongr., 43,* 421-39.

FREY, R.W. and HOWARD, J.D., 1970: Comparison of Upper Cretaceous ichnofaunas from siliceous sandstones and chalk, Western Interior Region, U.S.A.; *in* Crimes, T.P. and Harper, J.C. (eds.) Trace Fossils. *Geol, J., Spec. Iss., 3,* 141-66.

GEVERS, T.W. and TWOMEY, A., 1982: Trace fossils and their environment in Devonian (Silurian?) Lower Beacon Strata in the Asgard Range, Victoria Land, Antarctica; *in* Craddock, C. (ed.) *Antarctic Geoscience,* Univ. Wisconsin Press, Madison, 639-47.

GRIFFITHS, J.R., 1974: Revised continental fit of Australia and Antarctica. *Nature, 249,* 336-7.

HALLAM, A., 1970: *Gyrochorte* and other trace fossils, in the Forest Marble (Bathonian) of Dorset, England; *in* Crimes, T.P. and Harper, J.C. (eds.) Trace Fossils. *Geol, J., Spec. Iss., 3,* 189-200.

HALLAM, A. and SWETT, K., 1966: Trace fossils from the Lower Cambrian pipe rocks of the northwest Highlands. *Scot. J. Geol, 2,* 101-6.

HANTZSHEL, W., 1975: Trace fossils and problematica; *in* Teichert, C. (ed.) *Treatise on Invertebrate Paleontology, Part W, suppl. 1.* Geological Society of America and University of Kansas, Lawrence, W1-W269.

MARINTSCH, E.J. and FINKS, R.M., 1982: Lower Devonian ichnofacies at Highland Mills, New York and their gradual replacement across environmental gradients. *J. Paleontol., 56,* 1050-78.

MARTINSSON, A., 1965: Aspects of a Middle Cambrian thanatotope on Öland. *Geol. Foren. Stock., Forh., 87,* 181-230.

OSGOOD, R.G., 1970: Trace fossils of the Cincinnati Area. *Paleont. Americana, 6,* 281-444.

PEMBERTON, G.S., FLACH, P.D. and MOSSOP, G.D., 1982: Trace fossils from the Athabasca Oil Sands, Alberta, Canada. *Science, 217,* 825-7.

RATCLIFFE, B.C. and FAGERSTROM, J.A., 1980: Invertebrate lebensspuren of Holocene floodplains: their morphology, origin and paleoecological significance. *J. Paleontol., 54,* 614-30.

RAUP, D.M. and STANLEY, S.M., 1978: *Principles of Paleontology,* W.H. Freeman, San Francisco.

SARJEANT, W.A.J., 1975: Plant trace fossils; *in* Frey, R.W. (ed.) *The Study of Trace Fossils,* Springer-Verlag, New York, 163-80.

SEILACHER, A., 1963: Lebensspuren und Salinitätsfazies. *Fortschr. Geol. Rheinld. Westf. 10,* 81-94.

SEILACHER, A., 1967: Bathymetry of trace fossils. *Mar. Geol., 5,* 413-28.

STANLEY, K.O. and FAGERSTROM, J.A., 1974: Miocene invertebrate trace fossils from a braided river environment, western Nebraska, U.S.A. *Palaeogeogr., Palaeoclimatol., Palaeoecol., 15,* 63-82.

WESTERGARD, A.H., 1931: *Diplocraterion, Monocraterion,* and *Scolithus* from the Lower Cambrian of Sweden. *Sver. Geol. Undersok. Ser. C, Avh. och Upps., 372* (Arsbok. 25, no. 5), 1-25.

WHITE, C.D., 1929: Flora of the Hermit shale, Grand Canyon, Arizona. *Carnegie Inst. Washington, Publ., 405.*

# PERMIAN-TRIASSIC SEDIMENTARY SEQUENCE IN NORTHERN VICTORIA LAND, ANTARCTICA

J.W. Collinson, *Institute of Polar Studies, Ohio State University, Columbus, Ohio 43210, U.S.A.*

N.R. Kemp, *Tasmanian Museum, Hobart, Tas., 7001, Australia.*

*Abstract* Permian-Triassic rocks in northern Victoria Land were deposited by sandy braided streams in a northwest trending basin that paralleled the present location of the Rennick Glacier. The distribution of facies, composition of sandstones, and palaeocurrent data indicate dispersal toward a northerly and westerly direction from a granitic and metamorphic source terrain. Comparison of this sequence with coeval sequences in the central Transantarctic Mountains, southern Victoria Land and Tasmania suggests deposition in a series of similar, but separate basins along the Pacific Margin of Gondwana. The Takrouna Formation is widespread in the central Rennick Glacier area and along the margin of the Polar Plateau. Regionally, the Takrouna Formation overlies a rugged, locally deeply weathered surface of granitic and metamorphic basement rocks. The maximum preserved thickness of the formation, 300 m, occurs in the Freyberg Mountains, but the upper part has been removed by erosion. Along the margin of the Polar Plateau the formation is less than 50 m thick, resting directly on granitic basement and overlain unconformably by Kirkpatrick Basalt (Jurassic). The formation is characterised by coarse to medium grained feldspathic to quartzose sandstone and by lesser amounts of carbonaceous and noncarbonaceous silty mudstone and coal. Isolated pockets of diamictite, some of which are glacigene, occur at the base. Sedimentary sequences are dominated by large-scale trough and planar tabular cross-bedding and closely resemble the Devonian Battery Point sequence of Quebec, which represents a sandy braided stream environment such as that of the South Saskatchewan River in Canada. Coarse grained, less carbonaceous facies east of the Rennick Glacier change to finer grained coal bearing facies to the west, defining the margin and central axis of the basin.

A thick sequence of Permian-Triassic alluvial sediments accumulated on the cratonic side of a back-arc basin, which extended the length of the Transantarctic Mountains along the Pacific margin of Gondwana. The Transantarctic Mountains belt became part of a foredeep during the Gondwanian Orogeny, which probably began during Late Permian time (Elliot, 1975; Collinson et al., 1981). Permian and/or Triassic sediments along much of this belt are volcanoclastic, reflecting calc-alkaline volcanism along the Pacific margin (Vavra et al., 1981; Vavra and Collinson, 1981). Northern Victoria Land is located at a critical position on the Antarctic continent for geologic comparisons with Australia, according to most Gondwana reconstructions (e.g. Griffiths, 1974). The eastern margin of Australia, particularly Tasmania and the Ross Sea sector of the Transantarctic Mountains formed a continuous belt along the Pacific margin. Comparisons have been made of Upper Precambrian-Lower Palaeozoic geologic relationships (Veevers, 1976; Laird et al., 1977). Although the existence of Jurassic dolerite in Antarctica and Tasmania, but not Australia in general, points to a close relationship, little has been written about the relationships of Permian-Triassic pre-breakup sequences in these areas. The Permian-Triassic of Tasmania is relatively well known (Banks, 1962; Hale, 1962; Banks and Naqvi, 1967), but prior to the 1981/82 field season, only a few field parties have had an opportunity to investigate these rocks in northern Victoria Land (Gair, 1967; Crowder, 1968; Nathan and Schulte, 1968; Sturm and Carryer, 1970; Dow and Neall, 1974; Skinner, 1981). Many of the data presented here are the preliminary results of our participation in the 1981/82 U.S.-New Zealand-Australia joint international expedition, which operated out of a remote C-130 supported field camp between early November and early January on the lower Rennick Glacier (72°13′S, 163°52′E).

## STRATIGRAPHY

Dow and Neall (1974), after recognising the Permian *Glossopteris* flora in a 60 m thick section of sandstone and carbonaceous shale at Takrouna Bluff, introduced the name Takrouna Formation for rocks of the Beacon Supergroup in the lower Rennick Glacier region. Because of the report of an Upper Triassic microflora from a single locality in the Alamein Range (Norris *in* Sturm and Carryer, 1970), the Takrouna Formation has been assigned to both a Permian and Triassic age.

In the upper Rennick Glacier area Gair (1967) reported a relatively thin (10-15 m) arkosic sandstone sequence between granite basement and the Jurassic basalt. Microfloras from Section Peak in the Lichen Hills, supposedly from sediments above the base of the Kirkpatrick Basalt, were assigned a probable Middle to Late Triassic age (Gair et al., 1965; Norris, 1965). We believe that Gair misidentified sills as lava flows on Section Peak and that the entire sedimentary sequence there is older than the Kirkpatrick Basalt. Because no name other than Beacon has been applied to these rocks along the margin of the Polar Plateau and because they are lithologically similar to rocks assigned a Permian-Triassic age in the lower Rennick Glacier area, we are including them in the Takrouna (?) Formation.

During the 1981/82 field season we measured 11 stratigraphic sections at eight localities (Figure 1, localities 1, 3-10, 16) and visited two sections studied by Walker (this volume; localities 14, 15). In addition, we observed sections by helicopter at Takrouna Bluff (locality 11), Lanterman Range (locality 13), Gair Mesa (locality 2), Vulcan Hills, and along the margin of the Polar Plateau from Roberts

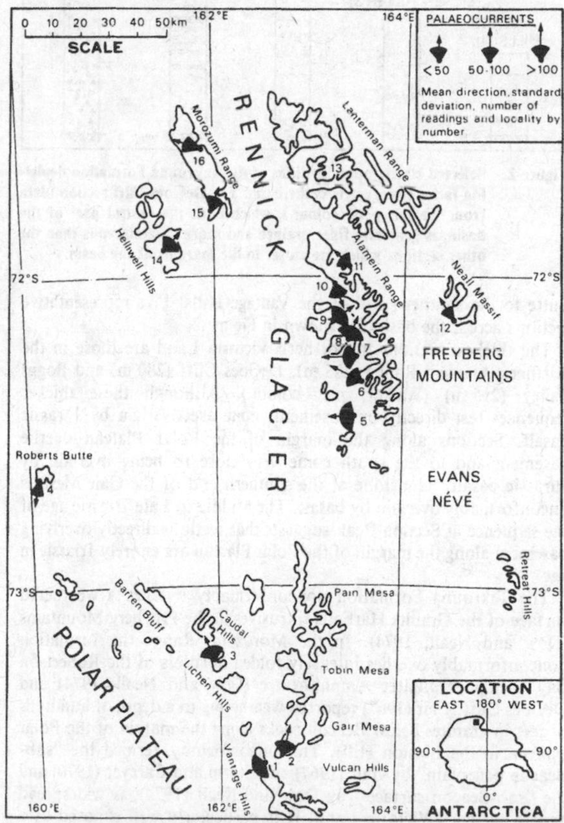


**Figure 1. Map of northern Victoria Land showing localities and palaeocurrent data. Localities are: (1) Vantage Hills, (2) Gair Mesa, (3) Section Peak, (4) Roberts Butte, (5) small ridge east of Mt Massell, (6) Monte Cassino, (7) ridge within Moawhango Neve, (8) Mt Baldwin, (9) Smiths Bench, (10) Alamein Range, (11) Takrouna Bluff, (12) ridge south of Neall Massif, (13) Lanterman Range, (14) Boggs Valley, (15) DeGoes Cliff, and (16) head of Jupiter Amphitheatre. Palaeocurrent data at localities 2, 11 and 14-16 from Walker (this volume).**

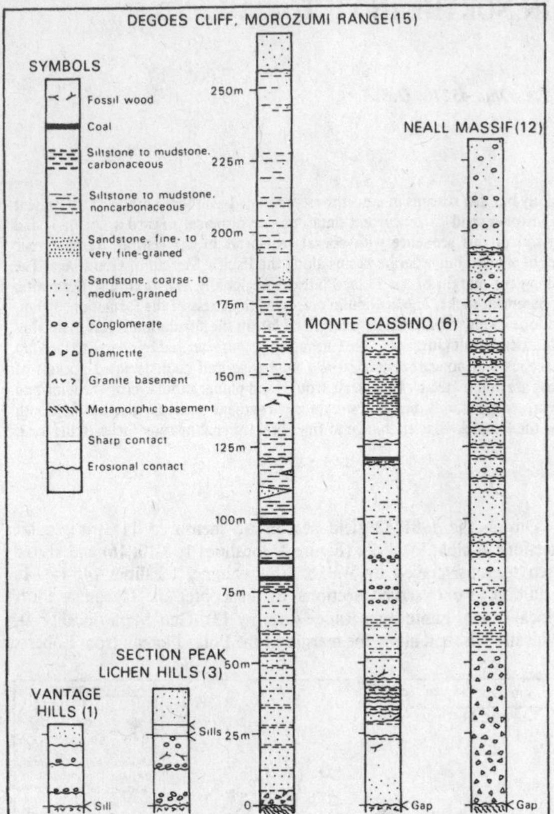


**Figure 2. Selected stratigraphic sections of the Takrouna Formation depicting facies change across the basin. The DeGoes Cliff section (data from Walker, pers. comm.), which is in the central part of the basin, is generally finer grained and more carbonaceous than the other sections which are closer to the margins of the basin.**

Butte to the southern end of the Vantage Hills. Five representative sections across the basin are shown in Figure 2.

The thickest sections in northern Victoria Land are those in the northern Alamein Range (285 m), DeGoes Cliff (280 m) and Boggs Valley (215 m) (Walker, per. comm.). Although these thicker sequences rest directly on basement, none are overlain by Jurassic basalt. Sections along the margin of the Polar Plateau overlie basement and to the south come very close to being overlain by Jurassic basalt. Sandstone at the southern end of the Gair Mesa is unconformably overlain by basalt. The Middle to Late Triassic age of the sequence at Section Peak suggests that sections directly overlying basement along the margin of the Polar Plateau are entirely Triassic in age.

The Takrouna Formation nonconformably overlies a weathered surface of the Granite Harbour Intrusives in the Freyberg Mountains (Dow and Neall, 1974). In the Morozumi Range the formation nonconformably overlies intensely folded hornfels of the Robertson Bay Group at Jupiter Amphitheatre (Dow and Neall, 1974) and DeGoes Cliff. Gair (1967) reported weathering to a depth of hundreds of feet in granites below Beacon rocks along the margin of the Polar Plateau in the Illusion Hills. This unconformity, termed the "sub-Beacon peneplain" by Gair (1967) and Sturm and Carryer (1970) and the "sub-Beacon surface" by Dow and Neall (1974), is widespread across northern Victoria Land. It is particularly well exposed as a prominent bench for 200 km along the margin of the Polar Plateau. Although the term peneplain implies a relatively featureless surface, Sturm and Carryer (1970) suggested relief of up to 100 m on a broad low scale.

The Takrouna Formation is lithologically similar over its regional extent, consisting of crossbedded feldspathic to quartzose sandstone interbedded with carbonaceous or noncarbonaceous silty mudstone and minor coal. No regional members were discerned, partly because of the difficulty in correlating sections. From structural relationships between sections at Monte Cassino (locality 6) and exposures within the Moawhango Névé (locality 7), it appears that in the Freyberg Mountains the lower 200 m of section are more carbonaceous than the rocks above. This observation generally agrees with lithologic relationships in a 290 m thick section measured by Walker (this volume) in the northern Alamein Range. Also, the sequence appears to become finer grained and more carbonaceous westward from the Neall Massif area toward the Morozumi Range and Helliwell Hills (Figure 2). The sequence at Neall Massif contains coarse feldspathic sandstone and granitic conglomerate, indicating a proximal source. This locality is near the eastern limit of Beacon exposures in northern Victoria Land, which we interpret as near the margin of the depositional basin. The westernmost exposures along the margin of the Polar Plateau from Roberts Butte to the Vantage Hills are relatively thin (Figure 2) and consist of coarse feldspathic sandstone, indicating a position near the western margin of the basin. Coals are best developed along the axis of the basin in the Morozumi Range (DeGoes Cliff, Figure 2) and the Helliwell Hills (Boggs Valley and Mt Remington localities of Walker, this volume).

At a few localities diamictite occurs below the Takrouna Formation. McKelvey and Walker (1982) reported more than 150 m of a thick folded sequence of Upper Palaeozoic glacigene strata in the western Lanterman Range. A thin irregular sequence of metasedimentary breccia occurs at the base of the Takrouna Formation at DeGoes Cliff. A similar, but much thicker breccia (56 m), containing angular metasedimentary clasts and rounded cobbles and boulders of aplite, occurs at the base of the Takrouna Formation at Neall Massif. Here, poorly stratified breccia grades upward into carbonaceous mudstone and sandstone (Figure 2). Laterally, however, the basal sandstone channels into the diamictite, concentrating aplite cobbles and boulders along the contact. Glacial features were not observed in these deposits; they probably represent remnants of post-glacial fill in valleys on the unconformity surface.

## SEDIMENTOLOGY

An ideal sedimentary sequence for the coarser grained facies of the Takrouna Formation is shown in Figure 3. The entire sequence is rarely fully developed in one cycle. Sections are dominated by coarse to medium grained, trough crossbedded, channel-form sandstone. Cycles range from less than 1 m to 10 m in thickness. The larger basal channel surfaces exhibit local relief on the order of 0.5 m to 1.5 m. Pebbles of quartz and basement rock and intraformational clasts are concentrated along channel bases and along scour surfaces.

Trough cross-bedding (Figure 4) occurs in sets 0.3 m to 1 m thick; the larger sets generally occur near the base of cycles. Planar tabular

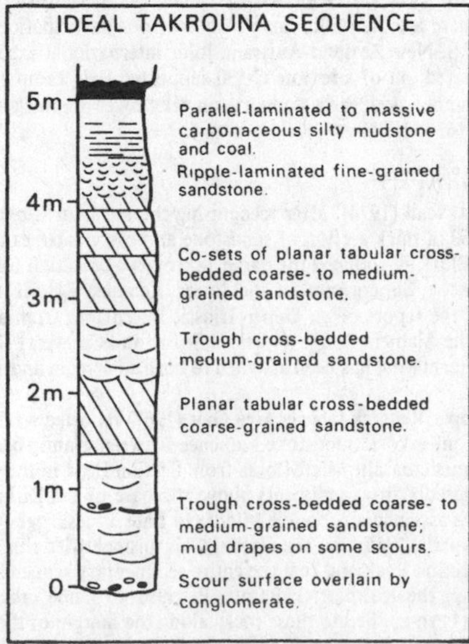


**Figure 3. Composite section of ideal Takrouna sequence, which closely resembles the Battery Point sequence of Cant (1978).**

crossbedding is most common in coarse grained sandstone and therefore is most abundant in sections near the basin margins such as Neall Massif and Vantage Hills. Orientation of planar sets diverge from trough cross-bed directions at these localities by an average of 75° and 95°, respectively. Planar sets are generally 0.3 m to 2 m thick, but may be as thick as 5 m (Figure 5). These occur as solitary sets or as cosets with sets of decreasing thickness upward. At a locality near Moawhango Névé (Figure 1, locality 7) closely crowded vertical tube burrows approximately 1 cm in diameter penetrate as deep as 2 m through the sequence of planar tabular cross-bed sets that have been modified at the tops by trough cross-bedding or ripple cross-lamination (Figure 6, see Zawiskie et al., this volume).

Parallel-laminated fine grained sandstone, in some cases with parting lineation, and silty carbonaceous or greenish-grey mudstone occur at the top of the more complete cycles. Silty mudstone also occurs as thin drapes on scour surfaces (Figure 7) or as lenticular channel fills. Rarely coals occur at the top of cycles. Coalified wood in the form of coaly streaks occurs along scour surfaces. Thicker carbonaceous units represent channel fills. For example, a channel fill at Monte Cassino (Figure 8) is 7 m thick in the centre and approximately 100 m across. It is composed of carbonaceous shale interbedded with 10 cm thick beds of ripple-laminated fine grained carbonaceous sandstone. Coalified calamitid stems and *Glossopteris* leaves occur on parting surfaces. These types of channel fills are most common in the finer grained facies in the Morozumi Range and Helliwell Hills in the central area of the basin.

**Figure 4. Bedding surface at approximately the 125 m level of the Monte Cassino section (Figure 1, locality 6; Figure 2) showing large scale trough cross-beds.**

**Figure 5. Person is standing on a single large scale, 5 m thick, planar tabular cross-bed set at 80 m level of the Neall Massif section (Figure 1, locality 12; Figure 2).**

**Figure 6. Cosets of planar tabular cross-beds showing top truncated by ripple marks on ridge within Moawhango Neve (Figure 1, locality 7).**

Sedimentary cycles in the Takrouna Formation most closely resemble the Battery Point facies model of Cant (1978). In this model large scale trough cross-bedded sandstone represents in-channel deposition by migrating dunes; large planar tabular cross-beds are formed by the progradation of transverse bars. Small scale planar tabular cross-beds are formed by the progradation of transverse bars. Small scale planar tabular cross-beds, ripple-laminated and parallel-laminated sandstone, and mudstone developed on bar tops by vertical accretion. Bloggett and Stanley (1979), in a study of the Platte River system of Nebraska, attributed a similar sequence of bedforms to the formation of large linguoid bars during episodes of high discharge and their subsequent modification during falling and low discharge. A modern analogue for the Battery Point sequence is the South Saskatchewan River, a sandy braided stream that flows from the Rocky Mountains in southern Alberta to southern Saskatchewan and ultimately to Hudson Bay (Cant, 1978; Cant and Walker, 1978).

## DEPOSITIONAL SETTING

The distribution of facies, composition and grain size of sediments, and palaeocurrent data (Figure 1) suggest that the Takrouna Formation was deposited in an elongate northwest trending basin that developed during the latest Carboniferous. This basin approximately conformed to the complex graben feature now occupied by the Rennick Glacier. Probable source terrains were the extensive basement rocks now exposed at the northeast corner of northern Victoria Land and on the East Antarctic craton.

Palaeocurrent directions (Figure 1) generally indicate dispersal in northerly to westerly directions along the axis of the theorised basin. Exceptions are at Roberts Butte (Figure 1, locality 4, mean direction ($\bar{x}$) = 192°, number of readings (n) = 13) and one locality in the Alamein Range (locality 10, $\bar{x}$ = 158°, n = 31). In both cases directions are based on relatively few readings obtained from thin stratigraphic sequences. Elsewhere in the Alamein Range, Walker (this

**Figure 7. Channel form sandstone composed of trough cross-bedded coarse to medium grained sandstone with thin mud drapes on scours on ridge within Moawhango Neve (Figure 1, locality 7).**

**Figure 8. Channel fill of carbonaceous shale and some coal intercalated with ripple-laminated fine grained sandstone at Monte Cassino (Figure 1, locality 6). Channel is up to 7 m thick and approximately 100 m across. It lies within a sequence of trough cross-bedded coarse to medium grained sandstone.**

volume) reports mean directions of 317° (locality 11) and 23°, which are consistent with major trends. Large numbers of trough cross-bed directions sampled through relatively thick sequences remain consistent at each locality (e.g. Monte Cassino, locality 6, $\bar{x}$ = 295°, standard deviation (sd) = 44°, n = 164; Neall Massif, locality 12, $\bar{x}$ = 326°, sd = 26°, n = 77), suggesting that major drainage patterns were stable through time and that aggradation may have been rapid.

The Takrouna Formation, occupying a basin about 200 km wide and having a maximum thickness greater than 300 m, represents a complex river system composed of several streams on a low gradient slope. Because of the piecemeal nature in which stratigraphic sequences are preserved, it is not possible to ascertain whether the Takrouna basin was floored by a relatively flat basement erosion surface such as that exposed along the margin of the Polar Plateau, producing a tabular blanket of sediments, or whether sediments filled in and covered a very irregular topography. The latter hypothesis would help to explain: (1) the sporadic occurrence of basal diamictites, because they were preserved in the deeper valleys, (2) the paucity of well preserved stratigraphic sequences toward the margins of the basin, because thicker sequences are found only where they were deposited in valleys cut into basement, (3) the variation of cross-bed directions from locality to locality whereas directions through thick sequences remain consistent, because streams were controlled by pre-existing drainage cut into basement, and (4) the compositional immaturity and coarse nature of sediments in all sections, because local source areas such as valley walls were important.

## PALAEOGEOGRAPHY

The trend of the northern Victoria Land basin and the thinning of sediments toward the south along the margin of the Polar Plateau suggests that coeval sequences in southern and northern Victoria Land occupied separate structural and depositional basins. Although post-glacial sequences in both areas contain coal measures, those in southern Victoria Land were deposited by meandering rather than braided streams (Barrett and Kohn, 1975). However, the Takrouna Formation is similar to the lower part of the Feather Sandstone (Upper Permian) in southern Victoria Land that overlies the coal measures. Sediments in the Triassic of southern Victoria Land and the central Transantarctic Mountains contain an important volcanogenic component; in northern Victoria Land only those in the Takrouna (?) Formation along the margin of the Polar Plateau do.

The Takrouna Formation is generally very different from most of the Permian in Tasmania, which is primarily marine. The Cygnet Coal Measures (Upper Freshwater Sequence), which marked the end of Permian-Triassic marine conditions in Tasmania (Clarke and Banks, 1975), are similar to finer grained parts of the Takrouna Formation. Also similar to the Takrouna Formation is the Lower Triassic Ross Sandstone in Tasmania, but the latter is noncarbonaceous.

Most Gondwana reconstructions (e.g. Griffiths, 1974) place Tasmania off the Ross Sea coast of nothern Victoria Land. Permian-Triassic rocks in northern Victoria Land and Tasmania were apparently deposited in separate basins, possibly sharing a common source area that is now the rugged basement area of the northeast corner of northern Victoria Land.

*Acknowledgements* We wish to express appreciation to fellow scientists and support crew at the northern Victoria Land camp. We thank Barry L. Roberts for his able assistance in the field and William Hammer and John Zawiskie, with whom we worked side by side. We thank Barrie C. McKelvey and Barry C. Walker for sharing information and showing us localities. David H. Elliot, Charles L. Vavra and Deana Chapman kindly reviewed the manuscript. This research was supported by National Science Foundation grant DPP-80-20098.

## REFERENCES

BARRETT, P.J. and KOHN, B.P., 1975: Changing sediment transport directions from Devonian to Triassic in the Beacon Supergroup of South Victoria Land; *in* Campbell, K.S.W. (ed.) *Gondwana Geology.* ANU Press, Canberra, 15-35.

BANKS, M.R., 1962: Permian System: *Geol. Soc. Aust., J., 9,* 189-215.

BANKS, M.R. and NAQVI, I.H., 1967: Some formations close to the Permo-Triassic boundary in Tasmania. *R. Soc. Tas., Pap. Proc., 110,* 91-105.

BLODGETT, R.H. and STANLEY, K.O., 1980: Stratification, bedforms and discharge relations of the Platte braided river system, Nebraska. *J. Sediment. Petrol., 50,* 139-48.

CANT, D.J., 1978: Development of a facies model for sandy braided river sedimentation comparison of the South Saskatchewan River and the Battery Point Formation; *in* Miall, A.D. (ed.) *Fluvial Sedimentology: Can. Soc. Petrol. Geol., Mem., 5,* 627-39.

CANT, D.J. and WALKER, R.G., 1978: Fluvial processes and facies sequences in the sandy braided South Saskatchewan River, Canada. *Sedimentology, 25,* 625-48.

CLARKE, M.J. and BANKS, M.R., 1975: The stratigraphy of the lower (Permo-Carboniferous) parts of the Parmeener Supergroup, Tasmania; *in* Campbell, K.S.W. (ed.) *Gondwana Geology,* ANU Press, Canberra, 453-67.

COLLINSON, J.W., STANLEY, K.O. and VAVRA, C.L., 1981: Triassic fluvial depositional systems in the Fremouw Formation, Cumulus Hills, Antarctica; *in* Cresswell, M.M. and Vella, P. (eds.) *Gondwana Five.* A.A. Balkema, Rotterdam, 141-8.

CROWDER, D.F., 1968: Geology of a part of northern Victoria Land, Antarctica. *U.S. Geol. Surv., Prof. Pap., 600-D,* 95-107.

DOW, J.A.S. and NEALL, V.E., 1974: Geology of the lower Rennick Glacier, northern Victoria Land, Antarctica. *N.Z. J. Geol. Geophys., 17,* 659-714.

ELLIOT, D.H., 1975: Gondwana basins of Antarctica; *in* Campbell, K.S.W. (ed.) *Gondwana Geology.* ANU Press, Canberra, 493-536.

GAIR, H.S., 1967: The geology from the Upper Rennick Glacier to the coast, northern Victoria Land, Antarctica. *N.Z. J. Geol. Geophys., 10,* 309-44.

GAIR, H.S., NORRIS, G. and RICKER, J., 1965: Early Mesozoic microfloras form Antarctica. *N.Z. J. Geol. Geophys., 8,* 231-5.

GRIFFITHS, J.R., 1974: Revised continental fit of Australia and Antarctica. *Nature, 249,* 336-7.

HALE, G.E., 1962: Triassic system. *Geol. Soc. Aust., J., 9,* 219-31.

LAIRD, M.G., COOPER, R.A. and JAGO, J.B., 1977: New data on the Lower Palaeozoic sequence of northern Victoria Land, Antarctica and its significance for Australian-Antarctic relations in the Palaeozoic. *Nature, 265,* 107-10.

McKELVEY, B.C. and WALKER, B.C., 1982: Late Palaeozoic glacigene strata in northern Victoria Land. Fourth Int. Symp. Antarc. Earth Sci., Adelaide, Australia, August 1982, Abstracts, 123.

NATHAN, S. and SCHULTE, F.J., 1968: Geology and petrology of the Campbell-Aviator divide, northern Victoria Land, Antarctica, Part 1—post-Palaeozoic rocks. *N.Z. J. Geol. Geophys., 11,* 940-75.

NORRIS, G., 1965: Triassic and Jurassic miospores and acritarchs from the Beacon and Ferrar Groups, Victoria Land, Antarctica. *N.Z. J. Geol. Geophys., 8,* 236-77.

SKINNER, D.N.B., 1981: Possible Permian glaciation in northern Victoria Land, Antarctica. *Geol. Jb. B41,* 261-6.

STURM, A. and CARRYER, S.J., 1970: Geology of the region between the Matusevich and Tucker Glaciers, northern Victoria Land, Antarctica. *N.Z. J. Geol. Geophys., 13,* 408-35.

VAVRA, C.L. and COLLINSON, J.W., 1981: Sandstone petrology of the Polarstar Formation (Permian), Ellsworth Mountains. *U.S. Antarct. J., 16,* 15-16.

VAVRA, C.L., STANLEY, K.O. and COLLINSON, J.W., 1981: Provenance and alteration of Triassic Fremouw Formation, central Transantarctic Mountains; *in* Cresswell, M.M. and Vella, P. (eds.) *Gondwana Five.* A.A. Balkema, Rotterdam, 149-53.

VEEVERS, J.J., 1976: Early Phanerozoic events on and alongside the Australian-Antarctic platform. *Geol. Soc. Aust., J., 23,*183-206.

WALKER, B. (this volume): The Beacon Supergroup of northern Victoria Land, Antarctica.

ZAWISKIE, J.M., COLLINSON, J.W. and HAMMER, W.R. (this volume): Trace fossils of the Permian-Triassic Takrouna Formation, northern Victoria Land, Antarctica.

# LATE PALAEOZOIC GLACIGENE STRATA IN NORTHERN VICTORIA LAND

B.C. McKelvey, *Department of Geology, University of New England, Armidale, N.S.W. 2351, Australia.*

B.C. Walker, *Department of Geology and Antarctic Research Centre, Victoria University, Wellington, New Zealand.*

*Abstract* In the western Lanterman Range of northern Victoria Land, adjacent to the Rennick Glacier, an isolated and incomplete folded sequence of Late Palaeozoic glacigene strata occurs downfaulted via steep reverse faults against the ?Precambrian Wilson Group. The Palaeozoic sequence is an equivalent of the Beacon Supergroup Metschel Tillite of southern Victoria Land but in contrast reflects either a fjord or glaciolacustrine setting.

The Palaeozoic strata are of acid plutonic and metasedimentary provenance and are exposed on two spurs about two kilometres apart. On the northern spur the oldest exposed strata consist of at least 70 metres of poorly sorted lensoidal or channel-fill outwash conglomerates, pebbly sandstones and diamictites associated with laminated or cross-bedded coarse sandstones. The coarser rock types are most common at the top and bottom of this interval. A dark mudstone and shale unit containing interbeds of thin sandstone, lensoidal conglomerate and a fine diamictite overlies. This unit thickens eastwards to at least 150 metres and changes rapidly in facies to a monotonous mudstone sequence containing in its uppermost 80 metres thin laterally persistent graded and laminated turbidites sometimes with erosional bases and containing intraformational sandstone clasts. No dropstones are present. More than 100 metres of coarse outwash strata similar to those at the base of the section overlie.

On the southern spur a similar coarse sandstone sequence with lensoidal conglomerates and pebbly sandstones at the base overlies more than 150 metres of fissile or massive mudstones alternating with very coarse tillite sheets and lenses. Most of the mudstone and shale horizons contain sparsely scattered dropstones.

The contact between the folded glacigene sequence in the Lanterman Range and the widespread Permian fluvial Takrouna Formation of northern Victoria Land has not been observed. Coarse metasediment breccias up to 50 metres thick and associated sometimes with minor petromict conglomerate lenses frequently mantle the pre-Permian basement complex beneath the Takrouna Formation. No definite glacial features have been observed in these breccias.

# DEPOSITION OF THE WELLER COAL MEASURES (PERMIAN) IN SOUTH VICTORIA LAND

A.R. Pyne, *Geology Department and Antarctic Research Centre, Victoria University, Wellington, New Zealand.*

**_Abstract_ The Weller Coal Measures crop out along the western edge of the Transantarctic Mountains in southern Victoria Land. This 200 m thick Permian sequence, deposited on the near horizontal Pyramid Erosion Surface, overlies Devonian alluvial strata and remnants of Permo-Carboniferous glacial and glacio-fluvial beds. In some areas at least, coal measure deposition began concurrently with glacial retreat.**

**Three distinctive lithofacies associations have been stratigraphically correlated over 70 km of exposure. Facies A comprises extensive thin beds of sandstone ("sheet sandstone") and carbonaceous shale and coal. The sandstone units "fine upwards" from coarse feldspathic sandstone to fine micaceous sandstone with a corresponding trend in sedimentary structures from large-medium crossbeds through parting lineation to ripple bedding. These sands are an in-channel facies. The sand-shale-coal association can only be considered to represent a single flood event when contacts are gradational and palaeocurrent directions consistent. In the lower coarse sandstone of a "sheet sand", palaeocurrent directions are consistent for single units but highly varied between units. These units therefore are the preserved coarse deposits from several flood events. An indication of meander curve is obtained from progressive palaeocurrent direction changes within single coarse units. This facies was established soon after ice retreat and represents deposition from a high sinuosity river system of low bank relief.**

**Facies B is predominantly shale with thin fine sandstone interbeds. The basal units are strongly lensoidal with moderate relief, plane and ripple bedded and represent levee accretion. The remainder of this facies is finely laminated and ripple bedded lacustrine shale and sandstones.**

**Facies C comprises medium to fine quartzose sandstone and thin gravel lag with minor shale lenses. Large planar crossbedded sand bodies with festoon bedded tops are common. These represent in-channel bar deposits, probably in a low sinuosity river system without any significant overbank facies.**

# FORMATIONAL MAPPING OF THE BEACON SUPERGROUP TYPE AREA WITH SPECIAL REFERENCE TO THE WELLER COAL MEASURES, SOUTH VICTORIA LAND, ANTARCTICA

K.J. Whitby, *McElroy and Associates Pty Ltd, Milsons Point, N.S.W. Australia.*

G. Rose, *Department of Mineral Resources, Sydney, N.S.W. Australia.*

C.T. McElroy, *McElroy and Associates Pty Ltd, Milsons Point, N.S.W. Australia.*

*Abstract* Detailed formational mapping within the Beacon Supergroup has been carried out in the type area, the Beacon Heights region, located between Taylor Glacier and the Ferrar Glacier in southern Victoria Land. The area mapped was approximately 650 square kilometres. The special problems encountered and the non-standard geological mapping techniques employed in this exercise are discussed. Fifteen separate formations and members including extensive exposures of the Ferrar Dolerite have been shown on the map at a scale of 1:50 000. Of special interest is the occurrence of deformed Metschel Tillite in Beacon and Arena Valleys, apart from previously recorded deformation of strata lower in the sequence in the same region. Type sections of the Altar Mountain Formation (230 m thick), Arena Sandstone (358 m), Brawhm Sandstone Member (41 m) and Farnell Sandstone Member (188 m) were selected and measured. At Kennar Valley a 7 m stratigraphic section of part of the Weller Coal Measures was measured and channel samples were taken from three coal seams; at Mt Fleming a coal seam in close proximity to a basic dyke was sampled. On the basis of petrographic analyses, the very high reflectances indicate the coals to be high rank anthracite; however, the volatile content indicates semi-anthracite. Float/sink analyses of individual plies show cumulative floats of the order of 80% at R.D. 1.70 with an ash content (dry basis) of about 10%. The field relationship of the coal with associated intrusives are considered in relation to coal properties. Comments are made on factors to be considered in assessing the possibility of mining coal in Antarctica.

Field investigations were carried out in the summer of 1980/81 in the Quartermain Range (Beacon Heights) area of southern Victoria Land with the principal research objective of furthering the knowledge of the Victoria Group and Taylor Group in the Beacon sequence.

Four basic objectives were established:

(1) to produce, for the first time, a large scale geological map of the Beacon Supergroup of regional extent on a detailed formational basis in the Beacon Heights area,

(2) to establish type sections of incompletely defined stratigraphic units in the Beacon Supergroup,

(3) to study a remarkable exposure of deformed sedimentary rocks on Slump Mountain, near Brawhm Pass, and to further examine the large scale deformation in the Altar Mountain Formation in Arena Valley,

(4) to examine in detail coal occurrences in the Weller Coal Measures and to collect representative samples for analysis.

## GEOLOGICAL MAPPING

### Technique

Standard techniques of regional mapping were not applicable in the Beacon Heights area. For example, only partial stereoscopic cover was available in vertical aerial photography; in many areas ground compass bearings were rendered unreliable by the proximity of magnetic dolerites and the weakness of the magnetic field. Due to the large contour interval on the best available base map, (a preliminary U.S.G.S. 1:100 000 sheet) and the inherent inaccuracy which expectedly and inevitably occurs on such a map, it was necessary to plot geological boundaries by independent positioning. This was achieved by establishing "plotting stations" at twelve strategic positions located with relatively good accuracy using sextant bearings from defined peaks. Thereafter, use was made of a sextant, high resolution binoculars, sections measured by earlier workers and ourselves, trimetragon aerial photography (with vertical and oblique stereoscopy) and hand held photography in conjunction with observations from helicopters and direct field observations to assist in the compilation of a geological map at a scale of 1:25 000 for publication at 1:50 000. The area accurately mapped comprises approximately 200 square kilometres and the total area mapped (including interpolation) covers approximately 650 square kilometres (Figure 1). This was achieved in two periods of field work each of approximately 2 weeks, separated by a week of compilation and review at McMurdo Station. A small part of the completed map is included here as Figure 2. The coloured version of the full map has now been accepted in principle for publication by the New Zealand Geological Survey.

### Stratigraphy

Seventeen separate formations and members including extensive exposures of the Ferrar Dolerite are recognisable in the Beacon Heights area. Detailed lithologic descriptions of most of the formations have been recorded by earlier workers and are not included here. For example, see Barrett et al. (1972) and McKelvey et al. (1977) and references therein. In Figure 3 are tabulated the stratigraphic units observed in the area mapped. A full discussion of all elements of the sequence is in the course of preparation by the present authors, but selected aspects of the stratigraphy merit discussion here.

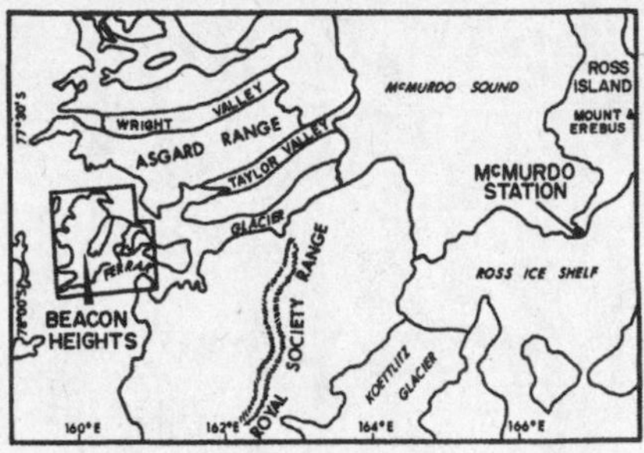


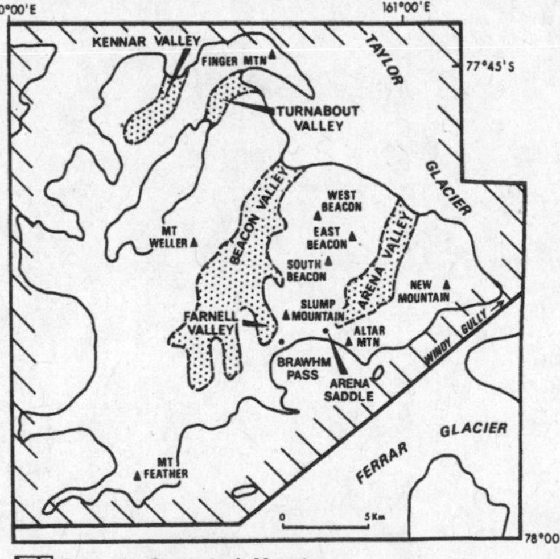


**Figure 1. Locality Map**

*The Farnell—Aztec Problem* The Farnell Sandstone Member is now confirmed as a continuous, mappable unit comprising the top section of the Beacon Heights Orthoquartzite. The type section has now been measured and described on the western side of Beacon Valley approximately 2 km south of Mt Weller. The member is characterised by the flaggy nature of the sandstone outcrops and the presence of thin dark claystone beds throughout, in contrast to the underlying cliff-forming massive sandstones of the West Beacon Sandstone Member. In the Beacon Valley area, the Aztec Siltstone is markedly different in

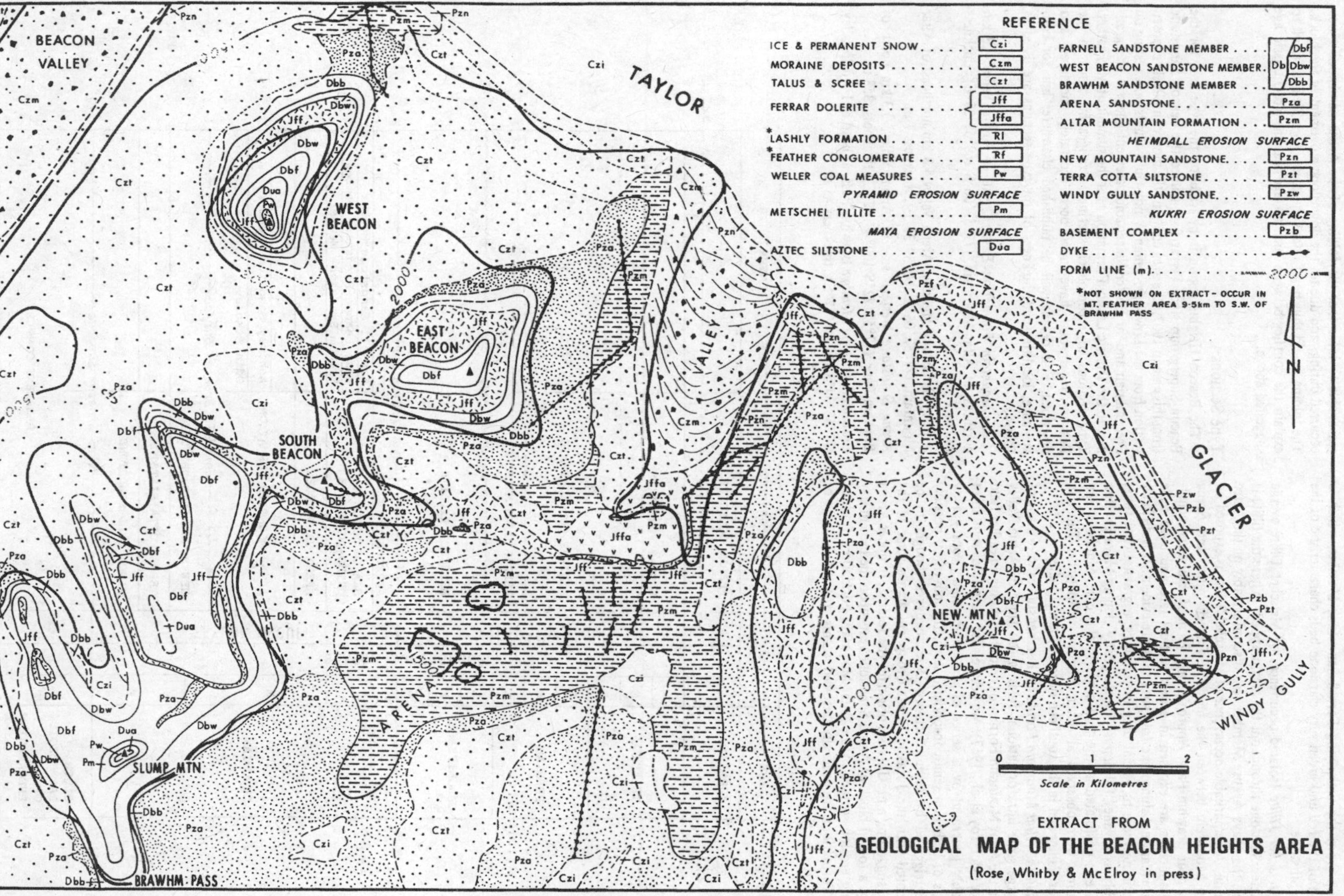


Figure 2. Extract from geological map of the Beacon Heights area.

outcrop from the underlying Farnell Sandstone Member, which is devoid of the red and green silty claystone beds characteristic of the Aztec Siltstone.

*The Metschel Tillite* Isolated occurrences of Metschel Tillite overlie the Aztec Siltstone throughout the area mapped. Where the tillite is absent, the Maya Arkose Member rests disconformably on the Aztec Siltstone. A remarkable occurrence of Metschel Tillite is exposed on Slump Mountain between the heads of Farnell Valley and Arena Valley near Brawhm Pass. Approximately 70 m of sediments including diamictites are exposed in a near vertical face and large scale intraformational slump-like fold structures are evident. It is difficult to attribute these to glacial drag or tectonics; a gravitational slide, possibly as a valley fill deposit on the gently sloping Maya Erosion Surface is postulated as the fold origin in this case. Similar slump structures were observed at other outcrops of Metschel Tillite in Kennar Valley and in the southern ridge of Farnell Valley.

*Folds in the Altar Mountain Formation* Further studies were made of the large scale intraformational folds in the floor of Arena Valley within the Altar Mountain Formation described during an earlier field season (McElroy et al., 1967). We believe that these folds were most probably generated by a series of low angle sub-aqueous gravity slumps of beds in a semi-consolidated state, the movement possibly being triggered by seismic activity. Dolerite dykes thought to be associated with the Jurassic Ferrar Group cut across some of the folds, suggesting a pre-Jurassic origin for the folding. In upper Arena Valley, a folded block of Arena Sandstone rests upon the present day sloping erosion surface of the Altar Mountain Formation. Both of these formations are generally unfolded in this area and it would be reasonable to infer that the folding of the Arena Sandstone block occurred during, and due to, the downslope movement of the block. The folding would thus be of relatively recent origin. It is difficult to explain the conflict between such an age and the pre-Jurassic age suggested above.

## Type Sections

The Beacon Heights region is by tradition the type area for the Beacon Supergroup since Ferrar first named the Beacon Sandstone from this area in 1907. The area contains the most completely exposed section from the Lower Devonian Windy Gully Sandstone, resting on Precambrian and/or Lower Palaeozoic basement gneisses and granites, to the Triassic Lashly Formation near the summit of Mt Feather. Type sections of incompletely defined stratigraphic units in the Beacon Supergroup have now been accurately measured and described. Full details will be published elsewhere, but for convenience, localities and thickness of the type sections are listed as follows:

*Altar Mountain Formation* Eastern wall of Arena Valley 4.3 km northeast of Altar Mountain, (77°52′06″S 160°59′33″E), thickness 230 m.

*Ashtray Sandstone Member.* Arena Valley 4.3 km northeast of Altar Mountain, (77°52′06″S 160°59′33″E), thickness 7 m.

*Arena Sandstone* Western side of Arena Valley, 3.3 km north of Brawhm Pass. (77°52′22″S 160°49′42″E), thickness 358 m.

*Brawhm Sandstone Member* Brawhm Pass, at head of Farnell Valley, (77°53′34″S 160°43′17″E), thickness 41 m.

| AGE | | | | FORMATION | MEMBER |
|---|---|---|---|---|---|
| JURASSIC | | | | FERRAR DOLERITE | |
| TRIASSIC | L | BEACON SUPERGROUP | VICTORIA GROUP | LASHLY FORMATION | |
| | M | | | | |
| | E | | | FEATHER CONGLOMERATE | |
| PERMIAN | L | | | | |
| | M | | | | |
| | E | | | WELLER COAL MEASURES | |
| | | | | *Pyramid Erosion Surface* | |
| CARONIFEROUS | L | | | METSCHEL TILLITE | |
| | M | | | | |
| | E | | | | |
| | | | | *Maya Erosion Surface* | |
| DEVONIAN | L | | TAYLOR GROUP | AZTEC SILTSTONE | |
| | | | | BEACON HEIGHTS ORTHO-QUARTZITE | Farnell Sandstone Member |
| | | | | | West Beacon Sandstone Member |
| | | | | | Brawhm Sandstone Member |
| | M | | | ARENA SANDSTONE | |
| | | | | ALTAR MOUNTAIN FORMATION | Ashtray Sandstone Member |
| | | | | | Odin Arkose |
| | | | | *Heimdall Erosion Surface* | |
| | | | | NEW MOUNTAIN SANDSTONE | |
| | E | | | TERRA COTTA SILTSTONE | |
| | | | | WINDY GULL SANDSTONE | |
| | | | | *Kukri Erosion Surface* | |
| OLDER PALAEOZOIC OR PRECAMBRIAN | | | | BASEMENT COMPLEX | |

**Figure 3. Stratigraphic units, Beacon Heights area.**

*Farnell Sandstone Member* Western side of Beacon Valley, 2.8 km south of Mt Weller, (77°52′21″S 160°31′04″E), thickness 118 m.

## Coal Occurrences and Sampling

In the Beacon Heights area, coal seams within the Permian Weller Coal Measures were sampled at Kennar Valley and in addition, a separate sampling exercise was effected at Mt Fleming. Attempts to locate and sample reported Triassic coals at Shapeless Mountain and Mistake Peak were thwarted by unfavourable weather. A study of the literature had revealed that few coal occurrences in Antarctica have been sampled according to normal coalfield practice. An exhaustive bibliography of coal occurrences is given in McElroy and Rose (in press).

There is a compelling need to acquire a sounder base for discussion on the coal potential of Antarctica, and to initiate a more widespread effort to record and sample coal outcrops throughout the continent. The authors do not condone references to the "vast untapped coal resources" of Antarctica. We emphasise that apart from the enormous practical difficulties of mining and transporting coal to market on an economic basis, so little is known of the purely geological parameters of the coalfields that such statements are unfounded in the light of present knowledge. The factors involved in elevating this level of knowledge relegate any realistic assessment of Antarctic coal resources to the indefinite future. As a small step towards this objective, normal coalfield practice was applied to collect representative samples of coal seams by channel sampling (150 mm x 75 mm cross section) in relation to a detailed measured section of the seam.

## COAL QUALITY

Coal samples were analysed in Australia by two coal testing laboratories using standard analytical techniques. Selected results are summarised in Table 1.

## Kennar Valley Coals

Three coal seams occurring within a 7 m stratigraphic interval of the Weller Coal Measures were sampled. The uppermost seam, units 1/4-1/1, (0.90 m thick) is separated from the middle seam, unit 3/2, (1.49 m thick) by 0.53 m of sandstone and dark grey fossiliferous shale. The lower seam, unit 6/2, comprises only 0.17 m of coal, and occurs 4.9 m below the middle seam. All samples were affected by atmospheric weathering to some degree so analytical results are, at best, only indicative of the true quality of the coals. Results of testing indicate that the coal is semi-anthracite, containing 89% carbon, with a vitrinite reflectance of approximately 3.7% indicative of the extremely high rank of the coal. The coal is uniformly low in volatile matter (approximately 12.0%, dry ash free basis) and has a high specific energy considering the weathered nature of the sample; specific energy of the clean coal (floats at 1.70 RD) is 32.5 MJ/kg (14,000 B.T.U/lb) on a d.a.f. basis. The coals have no coking properties as would be expected of a weathered high rank coal. Raw coal ash analyses in the range 10%—15% are reported in the main coal plies and when washed, a product can be obtained with ash less than 10%. Total sulphur content varies between 6%—7% and comprises mainly organic sulphur.

## Mt Fleming Coals

Two outcrop samples were taken from the same stratigraphic horizon at Mt Fleming; 1.60 m of coal (sample T2-T6) was sampled from one outcrop; a doleritic dyke, approximately 2 m wide, transects the coal measures at the locality. The coal, like that from Kennar Valley, is high rank, probably semi-anthracitic, low volatile, high energy coal which does not exhibit coking properties. Inherent ash is quite high at 27% and when washed at 1.70 RD, a 17% ash product is obtained. Sulphur is high, approximately 1%, and a high vitrinite content of 82.2% was recorded.

## Discussion of Results

The coals tested exhibit very high reflectance of vitrinite indicative of high rank anthracite which would normally contain 4%-5% volatile matter. The coals from Kennar Valley and Mt Fleming contain volatile matter (a.d.b.) in the range 9%-11%. The oxygen:hydrogen ratio is higher than would be expected of a coal with such high reflectance. These results are consistent with the effects of cindering a sub-bituminous coal, and petrographic analysis has indicated the presence of remnant vitrinite structures reminiscent of a sub-bituminous nature. Cindering effects such as pore spaces, now infilled with calcite, were recorded in the petrographic studies.

The samples from Mt Fleming contain a high proportion of sulphur, a deleterious component of potential fuel and coking coals, which occurs primarily as organic sulphur. Relatively minor amounts report as pyritic sulphur or as sulphate. Beneficiation of the coal would not reduce the organic sulphur content and in fact may cause a relative increase in the percentage of total sulphur in the washed product. The samples which were collected were oxidised and most likely affected by igneous activity of the Ferrar Dolerite. Until fresh samples are obtained, preferably by fully cored drilling, the only conclusion possible in relation to the coking potential of these coals is that they are non-coking or that the coking properties have been destroyed by oxidation.

## FUTURE EXPLORATION

In order to properly evaluate the coal resources of Antarctica, co-ordinated and standarised techniques of normal coalfield evaluation practised elsewhere in the world should be adopted. Coalfield exploration can be considered in four categories:

(1) geological mapping—regional and local
(2) geophysical techniques
(3) drilling and exploratory openings
(4) coal analyses.

Our recent work demonstrated that carefully planned and researched regional field mapping can produce meaningful results in a relatively short time. It is hoped that this style of mapping will eventually be extended to cover all likely coal-bearing areas.

The potential for use of ground or airborne geophysical surveys to assist in assessment of coal resources in Antarctica is virtually untested. Completion of comprehensive geological mapping in each "subregional" area should be followed by the drilling of a small number of strategically located fully cored drill holes. This is of the

**TABLE 1: Summary of Coal Analyses**

| | | | | | | Proximate analysis (ADB) | | | | | | |
|---|---|---|---|---|---|---|---|---|---|---|---|---|
| Locality | Sample No. | Thickness (m) | App. Relative Density | Sample State | Yield % | Moist % | Ash % | V.M. % | F.C. % | Total Sulphur % | (ADB) MJ/kg | S.E. |
| Kennar Valley | 1/4 | 0.20 | 1.54 | Raw | 100 | 6.1 | 11.9 | 8.8 | 73.2 | 0.63 | 0 | 25.94 |
| | 1/3 | 0.12 | 1.55 | Raw | 100 | 6.0 | 15.4 | 8.8 | 69.8 | 0.64 | 0 | 24.64 |
| | 1/2 | 0.33 | 1.56 | Raw | 100 | 6.2 | 15.7 | 9.4 | 68.7 | 0.65 | 0 | 25.01 |
| | 1/4 | 0.20 | 1.58 | F1.70 | 93.0 | 5.5 | 12.1 | 9.2 | 73.2 | 0.66 | 0 | 25.87 |
| | 1/3 | 0.12 | 1.63 | F1.70 | 89.9 | 6.7 | 11.8 | 9.2 | 72.3 | 0.65 | 0 | 25.91 |
| | 1/2 | 0.33 | 1.60 | F1.70 | 92.7 | 6.8 | 10.4 | 9.8 | 73.0 | 0.65 | 0 | 26.51 |
| | 1/1-1/4 | 0.90 | 1.59 | F1.70 + fines | 67.2 | 7.2 | 9.6 | 0.6 | 73.6 | 0.68 | 0 | 26.84 |
| | 3/1-1/4 | 1.92 | 1.85 | CF1.70 + fines | 46.9 | 6.7 | 10.3 | 10.7 | 79.0 | 0.72 | 0 | 29.16 |
| | 3/2 | 0.44 | 1.49 | F1.70 | 96.9 | 4.5 | 8.5 | 10.4 | 76.6 | 0.72 | 0 | 27.85 |
| | 6/2 | 0.17 | | CF1.70 + fines | 94.1 | 7.2 | 9.9 | 11.8 | 78.3 | 0.74 | 0 | 29.10 |
| Mt. Fleming | T2-T6 | 1.60 | 1.53 | CF1.70 + fines | 80.5 | 7.9 | 17.3 | 12.2 | 70.5 | 1.01 | 0 | 26.37 |

utmost importance in reaching a better understanding of the geology of the coal measures and in providing invaluable basic information related to coal resource assessment. Having obtained a representative sample by one of the established methods, such as drilling, procedures for analytical studies, as they relate to coalfield exploration of coal seam evaluation, are relatively standardised and can be undertaken by laboratories specialising in these fields.

## COAL MINING CONSIDERATIONS

Exploration of Antarctic coal resources will depend on a satisfactory system of recovery being devised. The mining options available are discussed in more detail by McElroy and Rose (1982). Some thoughts on conventional coal mining techniques as they may apply in Antarctica follow.

The conditions, both climatic and geologic, under which most of the work would be necessary, practically eliminate surface open cut operations from any serious discussion. Underground mining, using a drift or adit opened from or near a seam outcrop, provides a more realistic alternative. The mining operation would probably need to maximise the use of mobile machinery to limit problems of "start up" following freezing and to maintain flexibility. As far as practicable, installations should be developed underground, where some degree of consistent and predictable control over the working environment can be assured. Mine ventilating air could be heated by coal, oil or natural gas rather than refrigerated as in most deep underground coal mines. The geothermal gradient is unlikely to be sufficient to dispense with heating.

Any land based transportation systems would have severe technological and environmental limitations and non-traditional new systems would need to be investigated. Normal beneficiation processes involving coal washing plants would raise significant difficulty simply in maintaining the mobility of the washery fluids in the plant.

Many factors, apart from those related to the geology of the deposits, will ultimately determine the viability of coal mining in Antarctica. Foremost amongst these is the consideration of the energy balance of such an operation; that is, will more energy be expended in winning the coal than will be derived from the final product?

## CONCLUSIONS

It is obvious that without extensive basic exploration, the coal potential of Antarctica will remain essentially unknown. This work must be carried out by geologists trained specifically in coalfield geology, who are familiar with the practices involved. Economic mining of the coal will be dependant upon not only technological, but also environmental, political, sociological and marketing problems. These problems, however, in no way negate the responsibility of the scientist to carry out studies with a view to gaining a proper understanding and evaluation of the coal measures.

*Acknowledgements* As a small party of independent Australians, we wish to express out sincere thanks to the U.S. Antarctic Research Programme, National Science Foundation and to the dedicated pilots of U.S. Navy Squadron VXE-6, for the substantial and indispensable support given in the field. The endorsement of the Australian Antarctic Division is gratefully acknowledged. B. McKelvey and P. Barrett gave us beforehand guidance on location of outcrops and assisted us in many ways, both in advance preparation and critical appraisal of mapping results. We would like to acknowledge out sincere appreciation to B.H.P. Co. Ltd. and Capricorn Coal Management Proprietary Ltd. for providing analytical services for our coal samples through their laboratories at Newcastle and German Creek respectively.

## REFERENCES

BARRETT, P.J., GRINDLEY, G.W. and WEBB, P.N., 1972: The Beacon Supergroup of East Antarctica; *in* Adie, R.J. (ed.) *Antarctic Geology and Geophysics.* Universitetsforlaget, Oslo, 319-32.

McELROY, C.T. and ROSE, G. (in press): Coal Potential in Antarctica; *in* Splettstoesser, J.F. (ed.) *Mineral Resource Potential of Antarctica.* Univ. Texas Press.

McELROY, C.T., ROSE, G. and BRYAN, J.H., 1967: A unique occurrence of deformed sedimentary rocks of the Beacon Group, Antarctica. *U.S. Antarct. J., 2,* 241-4.

McKELVEY, B.C., WEBB, P.N. and KOHN, B.R., 1977: Stratigraphy of the Taylor and Lower Victoria Groups (Beacon Supergroup) between the Mackay Glacier and Boomerang Range, Antarctica. *N.Z. J. Geol. Geophys, 20,* 813-63.

ROSE, G., WHITBY, K.J. and McELROY, C.T. (in press): *Geological Map of the Beacon Heights Area-1:50,000.* N.Z. Geol. Surv., Lower Hutt.

# ENVIRONMENTAL INTERPRETATION OF THE NEW MOUNTAIN SUBGROUP—MARINE OR NON-MARINE? (BEACON SUPERGROUP)

R.W. Plume, *Geology Department, Victoria University, Wellington, New Zealand.*

*Abstract* Rocks of the New Mountain Subgroup (Beacon Supergroup) lie between the Kukri Peneplain and the Heimdall Erosion Surface in south Victoria Land. Environmental interpretation of these rocks has historically been a contentious issue. Recent marine interpretations have been based largely on the character, abundance, and diversity of trace fossils, whereas non-marine interpretations have relied on sedimentological and geochemical evidence.

Many sedimentological features of the sequence strongly support a continental interpretation. Of these, the most important are sub-aerial dessication cracks, red beds, $\delta C^{13}$ values, the palaeocurrent distribution, primary current lineation, cross-bedding style and structure, and stratigraphic and facies relationships. Other features such as ripple marks, ferruginous concretions, and scour-and-fill channels are consistent with a continental interpretation.

Trace fossils are found in all formations of the subgroup. They are particularly plentiful and varied in the upper part of the New Mountain Sandstone. Comparison with sequences similar to the New Mountain Subgroup in South Africa and observations of present day traces in the Indian Ocean have led Gevers and Twomey (1982) to a marginal marine interpretation without due consideration of evidence to the contrary.

## REFERENCE

GEVERS, T. W. and TWOMEY, A., 1982: Trace Fossils and their environment in Devonian (Silurian?) Lower Beacon strata in the Asgard Range, Victoria Land, Antarctica; *in* Craddock, C. (ed.) *Antarctic Geoscience*, Univ. Wisconsin Press, Madison, 639-47.

# ISOTOPIC AND CHEMICAL VARIATIONS IN KIRKPATRICK BASALT GROUP ROCKS FROM SOUTHERN VICTORIA LAND

P.R. Kyle, *Department of Geoscience, New Mexico Institute of Mining and Technology, Socorro, New Mexico, 87801, U.S.A.*

R.J. Pankhurst, *British Antarctic Survey, C/- Institute of Geological Sciences, 64 Gray's Inn Road, London, WCIX 8NG, U.K.*

J.R. Bowman, *Department of Geology and Geophysics, University of Utah, Salt Lake City, Utah, 84112, U.S.A.*

*Abstract* Samples from fifteen Kirkpatrick Basalt Group lava flows at Gorgon Peak, David Glacier, have been analysed for major and trace elements and strontium and oxygen isotope composition. They show many of the unusual features characteristic of Ferrar Supergroup rocks elsewhere, including relative enrichment of lithophile elements such as Rb and high initial $^{87}Sr/^{86}$ ratios (mostly 0.710-0.712). Analyses of mineral separates and leaching experiments show that post-crystallisation alteration has not significantly affected $^{87}Sr/^{86}Sr$ ratios in the basalts. $\delta^{18}O$ values mostly fall in the range +6 to +8 per ml. Consideration of these data, together with results for a high-MgO sill at Painted Cliffs, Beardmore Glacier, show that if the observed variations are due to mixing between a mantle derived magma and a crustal contaminant then the former was nevertheless characterised by an anomalous initial $^{87}Sr/^{86}Sr$ of >0.709. The petrogenesis of the Ferrar Supergroup probably involves some crustal contamination, however, the main features of the geochemistry are believed to reflect a heterogenous mantle source.

Jurassic igneous rocks of the Ferrar Supergroup (FS) occur throughout the 3500 km length of the Transantarctic Mountains. Similar rocks also occur in Tasmania, Australia which was juxtaposed against Antarctica in the Jurassic. The rocks have a tholeiitic character and are predominantly "basaltic", although more acidic compositions occur locally. FS rocks are characterised by an unusual enrichment of lithophile elements and anomalously high Sr-isotope compositions (initial $^{87}Sr/^{86}Sr$ ratios of 0.7089-0.7153) (Compston et al., 1968; Faure et al., 1972, 1974, 1982; Kyle, 1980). The isotopic ratios are high compared to many continental basalts and resemble those found in old crustal material rather than mantle derived magmas (Faure, 1977).

Two end-member models have been proposed to account for the unusual geochemical compositions of the FS, namely crustal contamination or derivation from a grossly heterogenous mantle source. In previous isotopic studies of the Kirkpatrick Basalt Group (KBG) (the extrusive phase of the FS), Faure and co-workers (1972, 1974, 1982) and Hoefs et al., (1980), have interpreted their data in terms of crustal contamination. On the other hand, Brooks and co-workers (1976, 1978) and Kyle (1980) have suggested the isotopic variations are the reflections of a grossly heterogenous mantle. It is extremely important to determine the extent to which each process occurs. If the isotopic data do reflect the mantle source then this has implications to heat flow, continental tectonics, and perhaps to an understanding of the process whereby the supercontinent of Gondwana disintegrated.

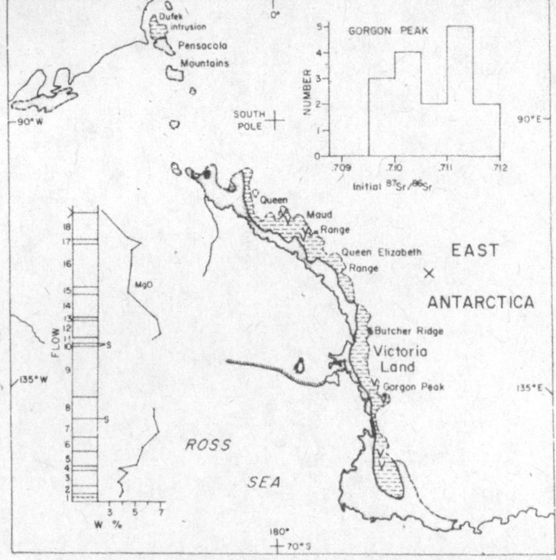


**Figure 1. Distribution of Jurassic Ferrar Supergroup tholeiite intrusions and extrusives in eastern Antarctica and frequency histograms of initial $^{87}Sr/^{86}Sr$ ratios of samples analysed in this study. Insert shows a stratigraphic section of lava flows at Gorgon Peak.**

This paper summarises preliminary results of a comprehensive geochemical study of a suite of KBG samples collected from Gorgon Peak[1] on Griffin Nunatak (Figure 1). An additional nine Ferrar Dolerite Group (the hypabyssal intrusive phase of the FS) samples from locations in southern Victoria Land and the Central Transantarctic Mountains are also included to extend the compositional range observed (Figure 1). Because of space considerations, a detailed discussion of the geochemistry and the analytical data will be presented elsewhere (Kyle, Pankhurst and Bowman, in preparation). The data however are available from the senior author on request.

## MAJOR AND TRACE ELEMENT GEOCHEMISTRY

Major element analyses of Gorgon Peak lavas have a wider range in chemistry than KBG lavas from Storm Peak (Faure et al., 1972) and Mt Falla (Faure et al., 1982) (Figure 2). In general they show higher MgO and lower $SiO_2$ contents, and are thus more basic. Although the chemistry is "andesitic" in character, the normative plagioclase is labradorite and so the samples are classified as tholeiitic basalts. Compared to most continental tholeiites the KBG is enriched in $SiO_2$ and $K_2O$ and depleted in MgO. The trace element data show significant enrichment in the lithophile elements Rb, Ba, and Th. Chondrite-normalised rare earth element (REE) patterns show strong light REE enrichment and weak negative Eu anomalies (Figure 3). Normalised to an assumed primordial mantle composition (Wood, 1979) (Figure 4), the trace element patterns mimic that for an average composition of the crust (Krauskopf, 1979). The trace element geochemistry is quite distinct compared to tholeiitic basalts from mid-ocean ridges; many trace element ratios, for example La/Ba, La/Th, La/Nb and Zr/Nb are more akin to calc-alkali andesites (Gill, 1981). An exception is Sr which is at modest levels typical of oceanic tholeiites.

The sill and dyke samples from southern Victoria Land are chemically similar to the lavas, with one exception. A dolerite sill which caps the sequence of lava flows at Gorgon Peak, is significantly younger (162.8 Ma) than the flows (175.8 Ma) (Kyle et al, 1981) and geochemically unique. The sill is andesitic in composition

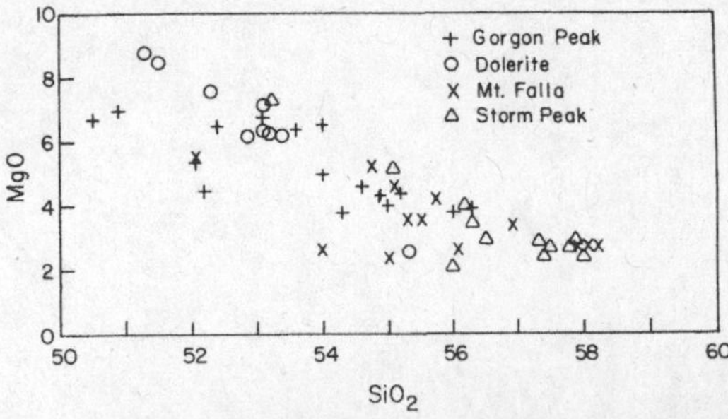


**Figure 2. Major element variation diagram with MgO plotted against $SiO_2$ (both in weight percent) showing the geochemical variability of Kirkpatrick Basalt Group samples from Gorgon Peak, Mt Falla (Faure et. al., 1982) and Storm Peak (Faure et al., 1974). The dolerites are only those analysed in this study.**

[1]Unofficial name, subject to approval by the U.S. Board on Geographic Names.

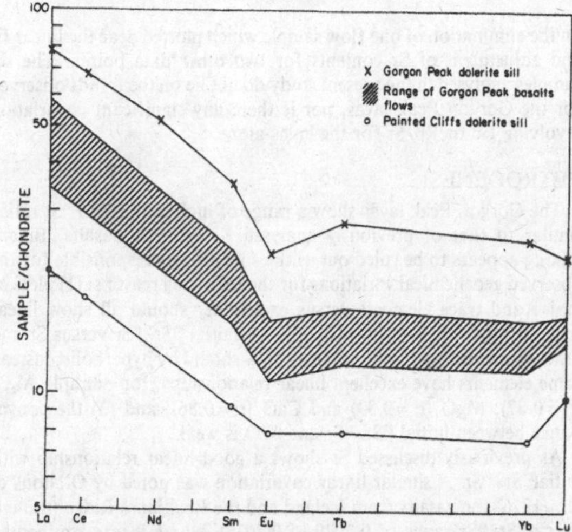


**Figure 3. The range in chrondite-normalised REE abundances in Gorgon Peak Kirkpatrick Basalt Group samples and individual analyses of dolerite sills from Painted Cliffs, Central Transantarctic Mountains and Gorgon Peak.**

(100 An/An + Ab = 44) and richer in FeO and $TiO_2$, and correspondingly depleted in CaO and $Al_2O_3$; the initial $^{87}Sr/^{86}Sr$ ratio of 0.70984 is one of the lowest known in the FS. Two dolerite samples from a sill at Painted Cliffs in the Central Transantarctic Mountains are the most primitive rocks known from the FS. They have Mg numbers (100Mg/Mg + $Fe^{2+}$, atomic) of 69, which is consistent with them being in equilibrium with a typical mantle composition. The sill has usually low $K_2O$ (0.13 wt %) and Rb (3 ppm) concentrations. The samples could be cumulates; however one is from a chilled margin and shows a typical quench texture and only a few pseudomorphs, possibly after olivine. At this time there is no reason to doubt that the samples are representative of fairly primitive, perhaps parental magma for the FS.

## STRONTIUM ISOTOPES

$^{87}Sr/^{86}Sr$ ratios are reported relative to a measured value of 0.70805 for the Einer and Amend standard. The decay constant of $^{87}Rb$ was

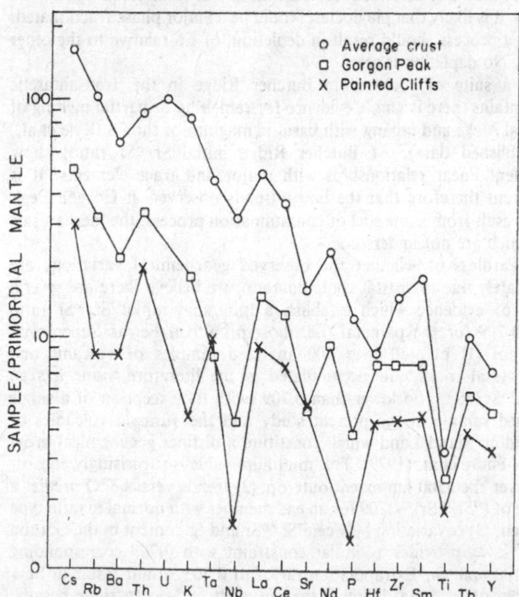


**Figure 4. Abundances of large ion lithophile (LIL) elements in a representative Kirkpatrick Basalt Group sample from Gorgon Peak and a MgO-rich chilled margin of a dolerite sill at Painted Cliffs, Central Transantarctic Mountains. Average composition of the crust from Krauskopf (1979). Normalising factors after Wood (1979).**

taken as $1.42 \times 10^{-11}a^{-1}$. Initial $^{87}Sr/^{86}Sr$ ratios were calculated using ages of 179 Ma for the flows from Gorgon Peak and vicinity and the dolerite sill at Painted Cliff; 162.8 Ma for the sill at Gorgon Peak and 167.3 Ma for the remaining sill and dyke samples (see Kyle et al., 1981 for a discussion of the ages). The initial $^{87}Sr/^{86}Sr$ ratios range from 0.7098 to 0.7115 for the Gorgon Peak lavas with a slightly larger range of 0.7095 to 0.7125 for the dykes and sills (Figure 5). These values are similar to previously published analyses, although the range reported here is not as great. As shown for the Storm Peak and Mount Falla sections (Faure et al., 1974, 1982), there is not a simple progressive increase of isotopic abundances through the flow sequence, but a discontinuous variation between three groups of adjacent flows.

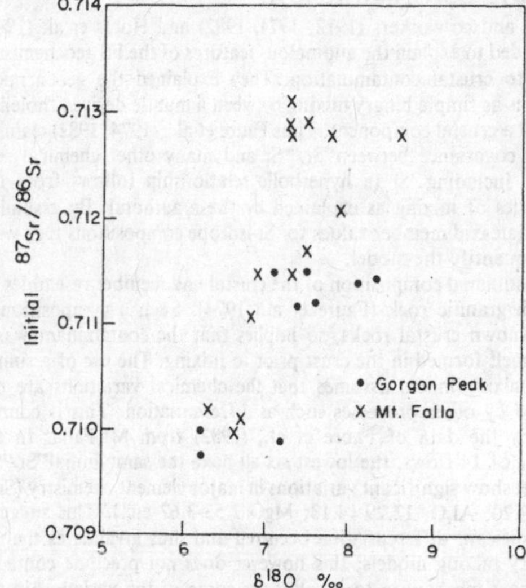


**Figure 5. Plot of initial $^{87}Sr/^{86}Sr$ ratios against $\delta^{18}O$ for Kirkpatrick Basalt Group lava flows from Gorgon Peak and Mt Falla (Hoefs et al., 1980).**

Since the groundmass and interstitial mesostasis in the basalts and dolerites are often altered we have attempted to show that the high initial $^{87}Sr/^{86}Sr$ ratios are not seriously affected by subsolidus processes. This has been done by (1) analysing separates of relatively fresh pyroxene and plagioclase and (2) by leaching minerals and whole rock powders in 6M HCl in a pressure vessel at 120°C, to dissolve low temperature alteration products (O'Nions and Pankhurst, 1976). Of the eight analysed minerals only four have initial $^{87}Sr/^{86}Sr$ ratios which overlap with those of the host rock at the 95% ($2\sigma$) confidence level. Two differ at the 99% ($4\sigma$) confidence level. In the leaching experiments a surprisingly large proportion of the material was removed (up to 84% of the Rb and 40% of the Sr) but the undissolved residues gave initial $^{87}Sr/^{86}Sr$ ratios which in four of the nine cases were statistically indistinguishable, at the 95% confidence level, from those of the unleached powders. Two leached samples from sills and a plagioclase separate from one of the sills differed from the unleached material at the 99.9% confidence level. Overall there is evidence for only slight isotopic disequilibrium within the volcanic rocks and it is concluded that their Sr isotope composition is essentially representative of magmatic liquid. Some dolerite sills show ample evidence of isotopic disequilibrium due possibly to high level contamination as they were emplaced. This will be discussed in more detail elsewhere.

## OXYGEN ISOTOPES

Oxygen isotope data are reported at deviations (per mil) from Standard Mean Ocean Water (SMOW). A value of $+11.8‰$ was obtained from replicate analyses of the Tintic Quartzite, with a coefficient of variation of $2.4‰$ $\delta^{18}O$ (SMOW) values for nine whole rock basalt samples range from $+6.2‰$ to $8.3‰$. Three analyses on one dolerite sill have an average value of $+6.3 \pm 0.1‰$. This range in $\delta^{18}O$ for the basalts is very similar to that reported on a section of KBG rocks from Mt Falla (Hoefs et al., 1980). These values are slightly higher than the range of $+5.0$ to $+7.0‰$ for ultramafic and mafic rock derived from the upper mantle (Hoefs, 1980) and are also greater than the value of $+5.7°/_{00}$ usually accepted for the mantle (Taylor, 1980). The $\delta^{18}O$ values show a weak correlation with the initial $^{87}Sr/^{86}Sr$ ratios ($r^2$ = 0.60) (Figure 5) and $SiO_2$ ($r^2$ = 0.53). For a $\delta^{18}O$ value of $+5.7°/_{00}$ the correlation suggests and initial $^{87}Sr/^{86}Sr$ ratio of 0.7094. This is an important constraint of the possible composition of the primary magma. Enrichment of $\delta^{18}O$ occurs in crustal sediments (due to low temperature exchange with the hydrosphere) and in igneous rocks which are derived from or have interacted with the continental crust, for example granites and andesites. Crustal fractionation, particularly if magnetite is involved, can increase the $\delta^{18}O$ during differentiation.

## CRUSTAL CONTAMINATION

Faure and co-workers (1972, 1974, 1982) and Hoefs et al. (1980) have tended to explain the anomalous features of the FS geochemistry as due to crustal contamination. They explained the geochemical variations as simple binary mixing between a mantle derived tholeiitic melt and a crustal component. Thus Faure et al., (1974, 1982) claimed a linear covariance between$^{87}Sr/^{86}Sr$ and many other chemical parameters, including 'Sr (a hyperbolic relationship follows from the systematics of mixing as explained by these authors). By assuming appropriate end member values for Sr-isotope compositions they were able to quantify the model.

The estimated composition of the crustal end member resembles an iron-rich granitic rock (Faure et al., 1974). Such a composition is unlike known crustal rocks, so implies that the contaminant was a partial melt formed in the crust prior to mixing. The use of a simple binary mixing model assumes that the chemical variations are not modified by other processes such as differentiation. This is contradicted by the data of Faure et al., (1982) from Mt Falla. In the sequence of 14 flows, the lowest six all have the same initial$^{87}Sr/^{86}Sr$ ratio, yet show significant variations in major element chemistry ($SiO_2$ 55.32-58.20; $Al_2O_3$ 12.29-14.13; MgO 2.53-3.67 etc.). This suggests that significant differentiation occurred and thus invalidates the use of binary mixing models; this however does not preclude contamination as a mechanism to account for some of the variation in the isotopic data. Hoefs et al., (1980) demonstrated a crude covariation between initial $^{87}Sr/^{86}Sr$ ratios and $\delta^{18}O$ values in the Mt Falla section. Such positive correlations are common in igneous rock series and are in general believed to represent mixing of mantle-derived igneous rocks with a crustal derived component (Taylor, 1980).

The data summarised here for the Gorgon Peak section show strong covariations between initial$^{87}Sr/^{86}Sr$ ratios and major elements. Positive correlations are noted for $SiO_2$ (correlation coefficient $r^2$= 0.77), $TiO_2$ ($r^2$= 0.81) and negative correlations for $Al_2O_3$ ($r^2$= 0.87), MgO ($r^2$= 0.89) and CaO ($r^2$= 0.86). Trace elements also show excellent positive correlations, particularly for the light rare earth elements, La ($r^2$= 0.89), Ce ($r^2$= 0.87) and Sm ($r^2$= 0.88), and with the Sr content (Figure 6). In the latter case there is slightly better fit to linear covariation ($r^2$= 0.87) than to the hyperbolic covariation expected for binary mixing ($r^2$= 0.85). In this respect it is interesting to note that the observation of a hyperbolic trend by Faure et al., (1974) was based

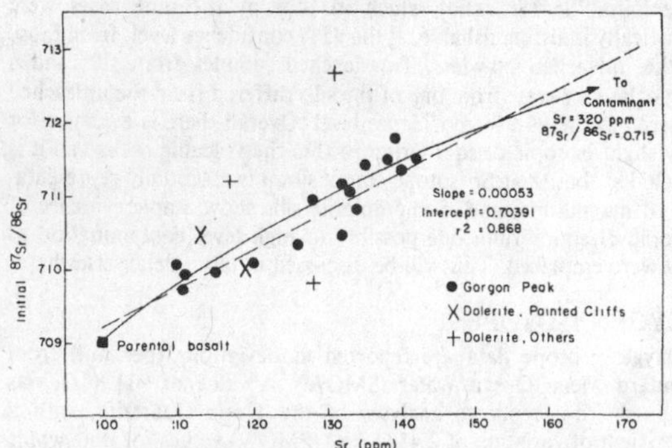


**Figure 6. Plot of initial $^{87}Sr/^{86}Sr$ ratios against Sr (in ppm) for Gorgon Peak samples and selected dolerite samples. Solid lines show the binary mixing trend between a parental basalt ($^{87}Sr/^{86}Sr$ = 0.709, Sr = 100 ppm) and a crustal contaminant ($^{87}Sr/^{86}Sr$ = 0.715, Sr = 320). Dashed line indicates simple linear regression or Gorgon Peak basalts only.**

on the elimination of one flow sample which plotted near the linear fit and adjustment of Sr contents for two other data points. The sill samples analysed in the present study do not lie on the trends observed for the Gorgon Peak lavas, nor is there any significant covariation involving Rb or Rb/Sr for the lavas alone.

## PETROGENESIS

The Gorgon Peak lavas show a range of high initial $^{87}Sr/^{86}Sr$ ratios similar to that of previously analysed Kirkpatrick basalts. Binary mixing appears to be ruled out as the sole process responsible for the observed geochemical variations for the following reasons: (1) plots of major and trace elements versus each other should all show linear relationships—they do not; (2) plots of initial $^{87}Sr/^{86}Sr$ versus Sr and other elements should not be linear, they should by hyperbolic, instead some elements have excellent linear relationships, for example $Al_2O_3$ ($r^2$=0.87), MgO ($r^2$=0.89) and CaO ($r^2$=0.86); and (3) the convariance between initial $^{87}Sr/^{86}Sr$ and $\delta^{18}O$ is weak.

As previously discussed Sr shows a good linear relationship with initial$^{87}Sr/^{86}Sr$. A similar linear covariation was noted by O'Nions et al., (1976) for basalts from Iceland and the Reykjanes Ridge (though over a$^{87}Sr/^{86}Sr$ range of 0.7029 to 0.7035), where it was ascribed to heterogeneity within the mantle source region. We feel that such an explanation, involving old sources with different Rb/Sr ratios, which was dismissed as improbable for the Kirkpatrick basalts by Faure et al., (1982), must still be considered. The main objection, a lack of covariation between initial$^{87}Sr/^{86}Sr$ and the Rb/Sr ratio, would not be valid if Rb has been seriously affected by fractional crystallisation or hydrothermal mobilisation immediately following crystallisation.

It is important to note that it is not possible to reject binary mixing based on the initial$^{87}Sr/^{86}Sr$ versus Sr plot alone. In fact the fit of the initial $^{87}Sr/^{86}Sr$ versus $^1/Sr$ is almost as good as the plot versus Sr. The data appear to fit closely to a mixing line between a basalt end member ($^{87}Sr/^{86}Sr$= 0.709, Sr= 100 ppm) and crustal derived components ($^{87}Sr/^{86}Sr$= 0.715, Sr= 320 ppm) (Figure 6). Our data are restricted to a small interval close to the basaltic end member. In this region the binary mixing line is very close to linear, hence it is critical to use other data to support a binary mixing model.

More complex contamination processes, such as combined assimilation and fractional crystallisation (Taylor, 1980; DePaolo, 1981; James, 1981) can under some circumstances result in linear relationships between$^{87}Sr/^{86}Sr$ versus Sr and other trace elements. If combined assimilation and fractional crystallisation was to affect Gorgon Peak rocks, it is likely that plagioclase would be a major phase fractionated. Such a process should result in depletiom of Eu relative to the other REE. No depletion is seen.

In a suite of rocks from Butcher Ridge in the Transantarctic Mountains there is ample evidence for remelting or partial melting of crustal rocks and mixing with basaltic magmas of the FS (Kyle et al., unpublished data). At Butcher Ridge initial$^{87}Sr/^{86}Sr$ ratios show excellent linear relationships with major and trace elements. It is apparent therefore that the linear trends observed at Gorgon Peak may result from some sort of contamination process, the mechanisms of which are not understood.

Regardless of whether the observed geochemical variations are ultimately due to crustal contamination, we believe there are several lines of evidence which establish a minimum initial$^{87}Sr/^{86}Sr$ ratio of ~ 0.709 for the parental (i.e. most primitive) Ferrar Supergroup magma: (1) of well over 100 analysed samples of volcanic and hypabyssal rocks now accumulated in the literature, none has an initial$^{87}Sr/^{86}Sr$ ratio lower than 0.709 (with the exception of a single leached sample in the present study and the Jurassic volcanics of Dronning Maud Land which constitute a distinct geochemical province—Faure et al. (1979). This minimum value is surprisingly consistent over the 3500 km extent outcrop; (2) trends versus $\delta^{18}O$ predict a value of $(^{87}Sr/^{86}Sr)_i$ ~ .709 for an end member with normal mantle-type oxygen; (3) covariation between$^{87}Sr/^{86}Sr$ and Sr content in the Gorgon Peak lavas provides a similar constraint with 0.709 corresponding to ~ 100 ppm Sr. Extrapolation down to 0.705 would result in Sr = 60 ppm or less, that is lower than in most mid-ocean ridge basalts. Extrapolation of the covariation with the Ce content would likewise rule out initial$^{87}Sr/^{86}Sr$ ratios below about 0.7075 (corresponding to Ce = 0 ppm); (4) unpublished Sr-isotope data from the Dufek Intrusion for the Forrestial Gabbro Group of the Ferrar Supergroup, show an initial$^{87}Sr/^{86}Sr$ ratio of ~ 0.709 for massive basic intrusives

(C.E. Hedge, pers. comm. to P.R. Kyle). The Painted Cliffs still analysed in the present study with an unusually low Rb/Sr ratio of 0.03 also has an initial $^{87}Sr/^{86}Sr$ ratio of 0.7100-0.7105.

Thus we affirm that the problem of the anomalously high $^{87}Sr/^{86}Sr$ ratio of Ferrar Supergroup magmas is not solved by the previously proposed crustal contamination models. Even if they are essentially correct (and we suspect that they may not be the sole explanation of chemical variation), then the predicted primary magma must independently have an elevated initial $^{87}Sr/^{86}Sr$ ratio prior to upper crustal contamination. Any model which requires ultimate derivation of the primary melt in normal, oceanic-type mantle with an $^{87}Sr/^{86}Sr$ ratio of ~ 0.704 or less, would also have to explain this increase by some form of deep lithospheric contamination which did not affect, for example, primary O-isotope compositions. This would apparently require recourse to selective Sr-isotope contamination or zone-refining (e.g. Harris, 1974). We consider it highly improbable that such a process could operate on magmas as they passed through the lower crust in such a way as to raise $^{87}Sr/^{86}Sr$ to an apparently uniform minimum value of ~ 0.709 over the entire length of the Transantarctic Mountains (and Tasmania). We feel compelled to conclude that the sub-crustal mantle throughout this region is genuinely anomalous (cf. Kyle, 1980). This could be a reflection of a high time-integrated Rb/Sr ratio (twice as high as the bulk earth value over the past 2700 Ma for example) or alternatively it may have resulted from a more recent metasomatic introduction of radiogenic Sr.

Trace element analyses of FS samples, including the primitive Painted Cliff sill, have a distinct crustal character (Figure 4). This suggests that mantle contamination may have occurred as a result of a large scale recycling of crustal material back into the mantle by subduction. Such a process would elevate the $\delta^{18}O$ of the mantle source; this is not precluded by our data and would only necessitate increasing the $^{87}Sr/^{86}Sr$ ratio of the mantle to ~ 0.710. Recently White and Hofmann (1982) proposed a similar mechanism to account for the isotopic composition of oceanic basalts.

*Acknowledgements* This work would have been impossible without the field support of Kathy Cashman, Harry Keys and Billy Boy-McIntosh and the U.S. Navy VXE-6 Squadron. We thank G. Faure and M. Logsdon for reviewing the manuscript. This work was supported in part by NSF grants DPP-7721590, DPP-8020002.

## REFERENCES

BROOKS, C. and HART, S.R., 1978: Rb-Sr mantle isochrons and variations in the chemistry of Gondwanaland's lithosphere. *Nature, 271,* 220-3.

BROOKS, C., HART, S.R., HOFMANN, A. and JAMES, D.E., 1976: Ancient lithosphere: its role in young continental volcanism. *Science, 193,* 1086-94.

COMPSTON, W., McDOUGALL, I. and HEIER, K.S., 1968: Geochemical comparison of the Mesozoic basaltic rocks of Antarctica, South Africa, South America and Tasmania. *Geochim. Cosmochim. Acta, 32,* 129-149.

DePAOLO, D.J., 1981: Trace element and isotopic effects of combined wallrock assimilation and fractional crystallisation. *Earth Planet. Sci. Lett., 53,* 189-202.

FAURE, G., 1977: *Principles of Isotope Geology,* Springer-Verlag, New York.

FAURE, G., BOWMAN, J.R., ELLIOT, D.H. and JONES, L.M., 1974: Strontium isotope composition of the Kirkpatrick Basalt, Queen Alexandra Range, Antarctica. *Contrib. Mineral. Petrol., 48,* 153-69.

FAURE, G., HILL, R.I., JONES, L.M. and ELLIOT, D.H., 1972: Isotope composition of strontium and silica content of Mesozoic basalt and dolerite from Antarctic; *in* Adie, R.J. (ed.) *Antarctic Geology and Geophysics,* Universitetsforlaget, Oslo, 617-24.

FAURE, G., BOWMAN, J.R. and ELLIOT, D.H., 1979: The initial $^{87}Sr/^{86}Sr$ ratios of the Kirwan Volcanics of Dronning Maud Land: comparison with the Kirkpatrick Basalt, Transantarctic Mountains. *Chem. Geol., 26,* 77-90.

FAURE, G., PACE, K.K. and ELLIOT, D.H., 1982: Systematic variations of $^{87}Sr/^{86}Sr$ ratios and major element concentrations in the Kirkpatrick Basalt of Mt Falla, Queen Alexandra Range, Transantarctic Mountains; *in* Craddock, C. (ed.) *Antarctic Geoscience,* Univ. Wisconsin Press, 715-23.

GILL, J.B., 1981: *Orogenic Andesites and Plate Tectonics.* Springer-Verlag, New York.

HARRIS, P.G., 1974: Anatexis and other processes within the mantle; *in* Sorensen, H. (ed.) *The Alkaline Rocks.* John Wiley and Sons, New York, 427-36.

HOEFS, J., 1980: *Stable Isotope Geochemistry,* 2nd edition. Springer-Verlag, New York.

HOEFS, J., FAURE, G. and ELLIOT, D.H., 1980: Correlation of $\delta^{18}O$ and initial $^{87}Sr/^{86}Sr$ ratios in Kirkpatrick Basalt on Mt Falla, Transantarctic Mountains. *Contrib. Mineral. Petrol., 75,* 199-203.

JAMES, D.E., 1981: The combined use of oxygen and radiogenic isotopes as indicators of crustal contamination. *Ann. Rev. Earth. Planet. Sci., 9,* 311-44.

KAUSKOPF, K.B., 1979: *Introduction to Geochemistry,* 2nd edition. McGraw-Hill, New York.

KYLE, P.R., 1980: Development of heterogeneities in the subcontinental mantle: evidence from the Ferrar Group, Antarctica. *Contrib. Mineral. Petrol., 73,* 89-104.

KYLE, P.R., ELLIOT, D.H. and SUTTER, J.F., 1981: Jurassic Ferrar Supergroup tholeiites from the Transantarctic Mountains, Antarctica, and their relationship to the initial fragmentation of Gondwana; *in* Cresswell, M.M. and Vella, P. (ed.) *Gondwana Five,* A.A. Balkema, Rotterdam, 283-87.

O'NIONS, R.K. and PANKHURST, R.J., 1976: Sr isotope and rare earth element geochemistry of DSDP Leg 37 basalts. *Earth Planet. Sci. Lett., 31,* 255-61.

O'NIONS, R.K., PANKHURST, R.J. and GRONVOLD, K., 1976: Nature and development of basalt magma sources beneath Iceland and the Reykjanes Ridge. *J. Petrol., 17,* 315-38.

TAYLOR, H.P., 1980: The effects of assimilation of country rocks by magmas on $^{18}O/^{16}O$ and $^{87}Sr/^{86}Sr$ systematics in igneous rocks. *Earth Planet. Sci. Lett., 47,* 243-54.

WHITE, W.M. and HOFMANN, A.W., 1982: Sr and Nd isotope geochemistry of oceanic basalts and mantle evolution, *Nature, 296,* 821-25.

WOOD, D.A., 1979: A variably veined suboceanic upper mantle-Genetic significance for mid-ocean ridge basalts from geochemical evidence. *Geology, 7,* 499-503.

# THE DEPOSITIONAL ENVIRONMENT OF THE LOWER DEVONIAN HORLICK FORMATION, OHIO RANGE

M.A. Bradshaw, *Canterbury Museum, Rolleston Avenue, Christchurch, New Zealand.*

L. McCartan, *U.S. Geological Survey, 1925 Newton Square East, Reston, Virginia, 22092, U.S.A.*

*Abstract* The Lower Devonian Horlick Formation of the Ohio Range is a thin and varied clastic sequence of sandstones and shales that non-conformably overlies Ordovician granitoids. It is in turn overlain disconformably by the Permo-Carboniferous Buckeye Tillite. Exposures along the Ohio Escarpment show rapid variations in thickness of Horlick Formation from 0 to 50 m as a result of pre-Buckeye erosion. Nine lithofacies, repeated several times, are recognised: Lithofacies 1—cross bedded feldspathic sandstone, sometimes with a thin lag gravel; Lithofacies 2—Bioturbated *Pleurothyrella*-rich sandstone; Lithofacies 3—thinly interbedded sandstone and mudstone; Lithofacies 4—ripple and parallel laminated fine sandstone; Lithofacies 5—uniform or upward fining quartz sandstone; Lithofacies 6—bone beds with phosphatised and bryozoa encrusted pebbles, worn fish plates and inarticulate brachiopods; Lithofacies 7—*Nuculites* and *Tentaculites*-rich horizons; Lithofacies 8—spiriferid beds; Lithofacies 9—fluvial feldspathic sandstone.

Most of the sediments and fossils suggest deposition in a shallow marine environment, nearly all shoreface, adjacent to a coastline with fluctuating conditions reflecting minor transgressive and regressive episodes. A coarsening upward motif in the ripple laminated beds indicates a regressive sequence, while a fining upward motif from channel sands to ripple laminated beds represents the onset of a transgressive phase. The phosphatic bone bed horizons represent periods of greatly reduced sedimentation. The appearance of shelf forms in the *Nuculites* and *Tentaculites* horizons suggests that these beds represent the maxima of transgressive cycles.

Connection of the Ohio Range with the Australasian seas is suggested along the Pacific Coast of Gondawana, with convergence of the Subpolar Gyre and Western Boundary Current in the New Zealand region.

The Devonian Horlick Formation of the Ohio Range is unique in that it contains abundant marine fossils (Doumani et al., 1965). Devonian sediments elsewhere in Antarctica are poorly fossiliferous and/or non-marine.

Along the northward facing escarpment of the Ohio Range (Figure 1) Devonian fossils are found at the base of an approximately 1200 m sequence of sediments (Devonian-Permian) that rests non-conformably on Ordovician quartz monzonite and granodiorite (Long, 1965; Treves, 1965). The Horlick Formation is the thinnest of the sequence, reaching a maximum of only 50 m. The thickness of the Devonian beds is highly variable due to pre-Late Carboniferous erosion, and at some localities Late Carboniferous glaciogene Buckeye Tillite (Kyle, 1977; Kemp et al., 1977) rests directly on granitic basement.

The sediments of the Horlick Formation are entirely detrital sandstones with some muddy and shaly intervals. Long (1965) in his initial study of the Horlick Formation concluded that the Devonian sediments had been deposited in nearshore beach and swampy-lagoonal environments. The discovery of psilophyte plant fossils and possible soil horizons in the muddy beds were taken to indicate swamps or lagoons, while unfossiliferous cross-bedded sandstones were regarded as possible fluvial or deltaic deposits.

During the 1979/80 Antarctic field season, a four man joint New Zealand/United States party visited the Ohio Range to make a detailed study of the Horlick Formation as part of the New Zealand Antarctic Research Programme. Except for one locality, body fossils (brachiopods, molluscs, trilobites, crinoids, fish bones) and trace fossils were found to occur throughout the sequence, and an almost entirely marine nature can be demonstrated for the formation. Horizons containing phosphatic pebbles were found for the first time.

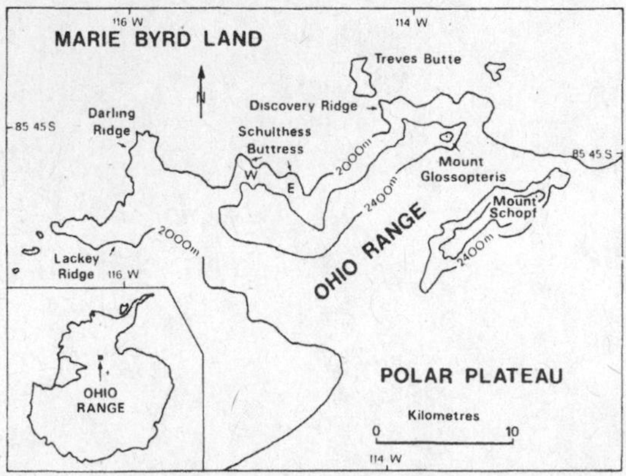


**Figure 1. Locality map of the Ohio Range, Horlick Mountains.**

## The Sediments

The granitic basement below the Horlick Formation is weathered to a depth of 2 m. Basal sediments are coarse, feldspathic, poorly sorted sandstones, sometimes with a thin local lag breccia. The overlying beds are generally more quartzose, better sorted and finer than the basal unit.

The Horlick Formation can be described in terms of nine lithofacies, some of which are distinguished by their fossil content. The most complete section on East Discovery Ridge (Figure 2) contains eight of the nine lithofacies and is regarded as the reference section.

*Lithofacies 1—Cross-bedded feldspathic sandstone.*

Coarse, poorly sorted feldspathic sandstone 2 m to 4 m thick is found at the base of most sections. The feldspathic sandstone strongly resembles the weathered part of the granitic basement in texture and composition (Figure 3), but is distinguished from it by the presence of trilobites. Shale clasts are present locally, particularly at the base of a chiopods *Pleurothyrella* and *Orbiculoidea*, together with phosphatic pebbles, point to a marine origin. Fish spines also occur. Thin, discontinous, ripple laminated grey siltstones yielding psilophyte plant fragments are locally present near the base of the unit (see also Long, 1965, page 81).

*Lithofacies 2—Bioturbated Pleurothyrella-rich beds.*

At most localities, the basal feldspathic sandstone is overlain by a bioturbated, quartz sandstone (Figure 4). Usually four main beds can be recognised, each less than 50 cm thick, separated by thin siltstones or muddy sandstone horizons. *Pleurothyrella* shells are common, particularly at the top of the sand beds. In one of the Darling Ridge sections, the *Pleurothyrella* beds rest directly on the granite with no intervening feldspathic sandstone.

The sandstones are coarse grained, poorly sorted and micaceous, with subrounded quartz grains (with angular overgrowths). Bedding is usually obscure due to intense bioturbation, and both arcuate U-shaped burrows and vertical burrows are present. The cross-section of some of the vertical burrows suggests a bivalve origin, and the presence of internal moulds of conjoined valves of the burrowing bivalve *Palaeosolen* in the same beds, lying parallel to bedding as if exhumed by wave action, suggests that this genus may have been responsible. Other vertical burrows have a distinctive bulbous mud infill perforated by a central sand infilled tube, and are closest in morphology to *Rosselia*. These may represent the fossilised mud-muffs of Polychaete worms (Babin et al., 1971).

The brachiopod *Pleurothyrella* is the most common fossil in these beds, but other brachiopod genera *Orbiculoidea, Lingula,* the molluscs *Modiomorpha* and *Nuculites*, and tribolite pygidia have also been observed. *Pleurothyrella* is a terebratulid that is considered to disarticulate easily (Boucot, et al., 1963). The presence of occasional conjoined valves in this bioturbated sand lithofacies implies little transport, but random orientations suggest that they are not in their life positions.

*Pleurothyrella* shows marked secondary thickening in the posterior part of the shell, has a closed pedicle foramen in the adult stages, and frequently has thick bryozoan overgrowths that are confined to the anterior part of the shell. These features all suggest that during life this brachiopod was unattached, partially buried, and was able to maintain a constant orientation in a high energy environment of shifting sands by weighting of the shell.

The *Pleurothyrella*-rich bioturbated beds were probably deposited in an upper to middle shoreface environment. The concentration of shells at the top of the beds suggests reworking during periodic storms. Similar beds are repeated higher in the succession but never as thick as the lowest occurrence.

*Lithofacies 3—Thinly interbedded sandstone and mudstone.*

The *Pleurothyrella* beds are succeeded by upward coarsening thin alternations of ripple bedded fine sandstone and mudstone 1 m to 2 m thick. Both body fossils—primarily *Burmeisteria* pygidia, *Nuculites* (mollusc) and psilophyte plant fragments in mudstone—and bioturbation structures, are rare. Discontinuous climbing ripples and lenticular bedding with both connected and disconnected flat sand lenses are present, but most common are individual layers of connected sand ripples covered by thin mud drapes. Moderate current action alternating with slack water is necessary to explain both the lenticular bedding and ripple mud drapes (Reineck and Singh, 1975). Thin, cross bedded, coarse to medium grained, poorly sorted sandstones with broken shells and shale clasts along the foresets, and also thin sand beds with convolute bedding, are sometimes interbedded with this lithofacies.

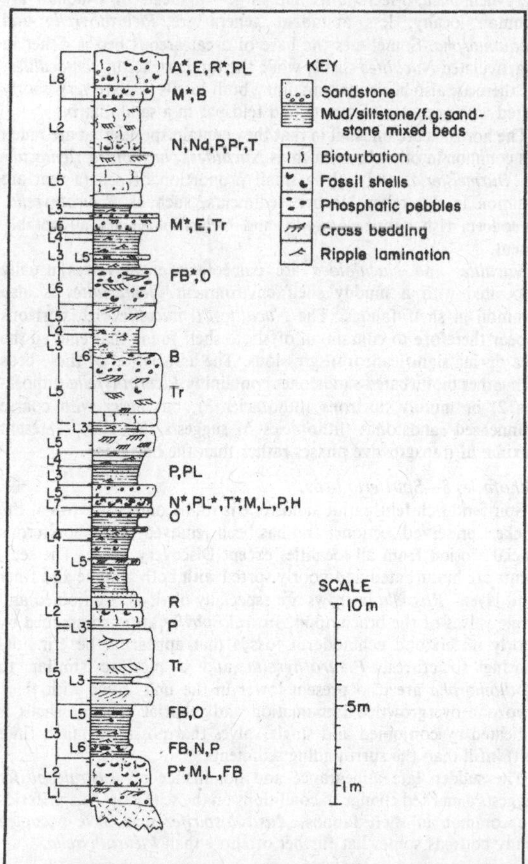


**Figure 2.** **Measured section of the Horlick Formation on East Discovery Ridge. Vertical scale in metres. Lithofacies marked to left. Recognisable fossils indicated where found, and are asterixed where particularly common: A—*Australospirifer*, B—bryozoa, E—echinoderms, FB—fish bone, H—homalonotid, L—*Lingula*, M—*Modiomorpha*, N—*Nuculites*, Nd—*Nuculoidea*, O—*Orbiculoidea*, P—*Pleurothyrella*, Pl—*Plectonotus*, Pr—*Prothyris*, R—*Rosselia*, T—*Tentaculites*, Tr—trackways.**

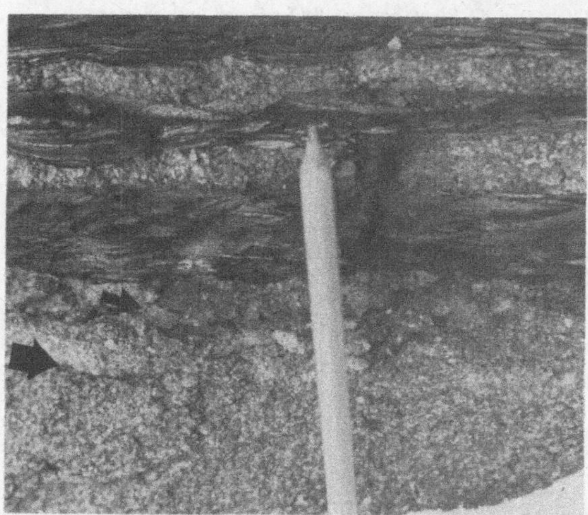

**Figure 3.** **Lower contact of the Horlick Formation (large arrow) on Lackey Ridge. Note that the basal feldsarenites are difficult to distinguish from the granite below. Small arrow points to broken *Orbiculoidea* shell in first sand bed. Darker layers are of laterally inpersistent, ripple laminated siltsone. Pencil for scale.**

**Figure 4.** **Lower units of the Horlick Formation resting unconformably on granite (contact arrowed), Darling Ridge. Cliff is approximately 6 m high. The lowest beds are 1.5 m of very coarse feldsarenite (lithofacies 1). The next unit is of bioturbated sandstone with *Pleurothyrella* shells, separated by thin siltstones and totalling 2.15 m (lithofacies 2). This is in turn followed by 1.60 m of ripple laminated sandstone and siltstone (lithofacies 2). A 75 cm thick very coarse grained feldsarenite bed is present halfway up this muddier unit. A similar succession of lithologies was observed at the base of the Horlick Formation in most of the outcrops along the Ohio Escarpment.**

*Lithofacies 4—Ripple and parallel laminated fine sandstone.*

The thinly interbedded sandstone and mudstone lithofacies (3) usually grades up into interference ripples, well sorted, fine grained feldspathic sandstone with micaceous partings. Body fossils are rare and only isolated shells of the brachiopods *Tanerhynchia* and *Pleurothyrella* were seen. Trace fossils are more common and include infilled circular exit holes less than 1 cm in diameter on bedding surfaces, and narrow collapsed horizontal burrows. Large *Rusophycus* and *Cruziana* occur on the soles of some beds.

In many cases interference rippled sandstone grades up into parallel laminated fine to medium quartz sandstone beds up to 20 cm thick (Figure 5), which probably formed in a beach environment (Reineck and Singh, 1975, page 105).

**Figure 5. Typical coarsening upward regressive motif within the finer beds of the Horlick Formation on West Discovery Ridge. Ripple laminated fine grained sandstone with mud drapes (lithofacies 3) grades up into ripple laminated fine grained feldsarenite (lithofacies 4), and then up into parallel laminated fine grained sandstone, that probably formed in a beach environment. The finer beds are channelled and followed by a very coarse grained ill-sorted cross-bedded sandstone that probably represents the height of the regression. Channel is 30 cm deep.**

*Lithofacies 5—Uniform or upward fining sandstone.*

Relatively thick units (1 m to 4 m) of uniform or upward fining, coarse to medium grained quartz sandstones, notable for their paucity of fossils and medium scale cross-bedding, are also found in the Horlick sequence. A few ribbed brachiopod shells and a very large bivalve similar to *Modiomorpha*, as well as rare large tribolite pygidia, were found. Arthropod trackways and scratch marks are fairly common on bedding planes and were probably made by trilobites. Shale clasts are present locally, particularly at the base of a unit or along foresets. The units have erosional bases and channel down into ripple laminated interbedded sandstone and mudstone or fine grained sandstone. Occasional placoderm fish plates and rare phosphatic pebbles were seen at the base of the unit. All features of this facies suggest deposition in the lower shoreface, adjacent to the thinly interbedded sandstone and mudstone lithofacies (3).

At several localities the sandstones were seen to grade up into ripple laminated fine-grained sandstone (lithofacies 4), and at least one section, into thinly interbedded sand and mudstone (lithofacies 3). This suggests deepening water with the progressive establishment of more distal lithofacies represented by the top of the units (transgressive), after an initial shallowing (progradation) that is represented by the bottom of the unit.

*Lithofacies 6—Bone beds.*

Bone beds, though forming only a small proportion of the total Horlick sequence, are significant horizons that are repeated several times in most sections. The thickness of the bone beds varies from 1 cm to 25 cm, but generally the thicker the unit the lower the phosphate content. The beds can be traced laterally up to several tens of metres but are variable in thickness. The beds are most commonly interbedded with the thinly interbedded sandstone and mudstone horizons (lithofacies 3), and on Darling Ridge a cross bedded sandstone containing phosphatic pebbles passes laterally into the same lithology.

The sediment is characteristically very poorly sorted, very coarse feldspathic sandstone with rounded black phosphatic pebbles up to 14 cm long, some of which contain phosphatised fish debris, and commonly with phosphatised bryozoan overgrowths. In thinner beds, the pebbles are accompanied by worn, phosphatised (secondary) placoderm fish plates and homolonotid pygidia, together with fragmented *Pleurothyrella*, *Orbiculoidea*, and *Lingula* shells.

Thicker beds are graded, with the coarser material at the base often rich in fragmented bony fish plate fragments and teeth free of secondary phosphate. These beds may also contain entire but rare shells of *Pleurothyrella*, *Modiomorpha*, and the bellerophontid mollusc *Plectonotus*.

The concentration of phosphorus in shallow marine environments has been fully discussed by Pevear (1966). In the Ohio Range phosphatic pebbles and phosphatised skeletons can be explained as having formed close to shore in a suitable "nutrient trap". Bryozoan overgrowths on phosphatic pebbles indicate the following depositional history: diagenetic phosphate enrichment of mud and chitonphosphatic skeletons during stillstand, reworking and mechanical enrichment under high energy conditions, bryozoan growth and further phosphatisation during renewed qüiescence, with reworking of the sediment and incorporation of associated biota in a final pulse of sediment discharge.

*Lithofacies 7—Nuculites and Tentaculites-rich horizons.*

Carbonate cemented quartz sandstone and siltstone beds may occur several times in the Horlick Formation. These beds are generally less than 25 cm thick and are dominated by either *Nuculites* or *Tentaculites. Nuculoidea, Plectonotus* and large fragments of tribolites are common locally; less abundant genera are *Pleurothyrella* and *Modiomorpha.* Sometimes the base of a calcareous bed is richer in disarticulated *Nuculites* shells, while the top is richer in *Tentaculites,* but there are also horizons containing both fossils that are very poorly sorted with granules of quartz and feldspar in a sand matrix.

The horizons are unusual in that they contain species that are much less common in other beds, such as *Nuculites, Nuculoidea, Tentaculites, Burmeisteria,* and only a small proportion of genera that are common in the other Horlick sediments, such as *Pleurothyrella.* Placoderm fish debris, *Lingula*, and *Orbiculoidea* are all notably absent.

*Nuculites* and *Nuculoidea* are palaeotaxodont bivalves usually associated with a muddy shelf environment. *Tentaculites* is also common in shelf faunas. The *Nuculites/Tentaculites*-rich horizons appear therefore to contain an offshore shelf fauna that entered the area during significant transgressions. The association of these beds with either bioturbated sandstones containing *Pleurothyrella* (lithofacies 2) or muddy horizons (lithofacies 3), but never with coarse channelised sandstones (lithofacies 5) suggests that they represent maxima of transgressive phases rather than the beginning.

*Lithofacies 8—Spiriferid beds.*

Spiriferid rich feldspathic sandstone is found only at the top of the thickest preserved sections, and has been removed by Carboniferous glacial erosion from all localities except Discovery Ridge. The sediments are bioturbated and poorly sorted with both granule and finer sand layers. *Rosselia* burrows are especially obvious. Conjoined and single valves of the brachiopod *Australospirifer* are accompanied by poorly understood echinoderm fossils that appear to be crinoidal rooting structures. *Pleurothyrella* and a mollusc similar to *Modiomorpha* are also present lower in the unit, some with thick bryozoan overgrowths. Exhumation and reburial of some shells is indicated by conjoined and single valves that possess a much finer (silt) infill than the surrounding sediment.

The sudden late appearance and dominance of *Australospirifer* suggests a marked change in conditions on the sea floor. As spiriferids are common in shelf faunas, *Australospirifer* may have occupied sandy bottoms somewhat further offshore than *Pleurothyrella.*

*Lithofacies 9—Fluvial feldspathic sandstone.*

At West Schulthess Buttress, the entire 10 m section is medium to very coarse, highly feldspathic sandstone. Medium to large scale festoon cross-beds and channels are common, but fossils, both body and trace, are completely absent. The lithology and structures suggests that this is a fluvial sediment. West Schulthess Buttress was apparently the site of the mouth of a river which supplied granitic detritus to the coast in Horlick time.

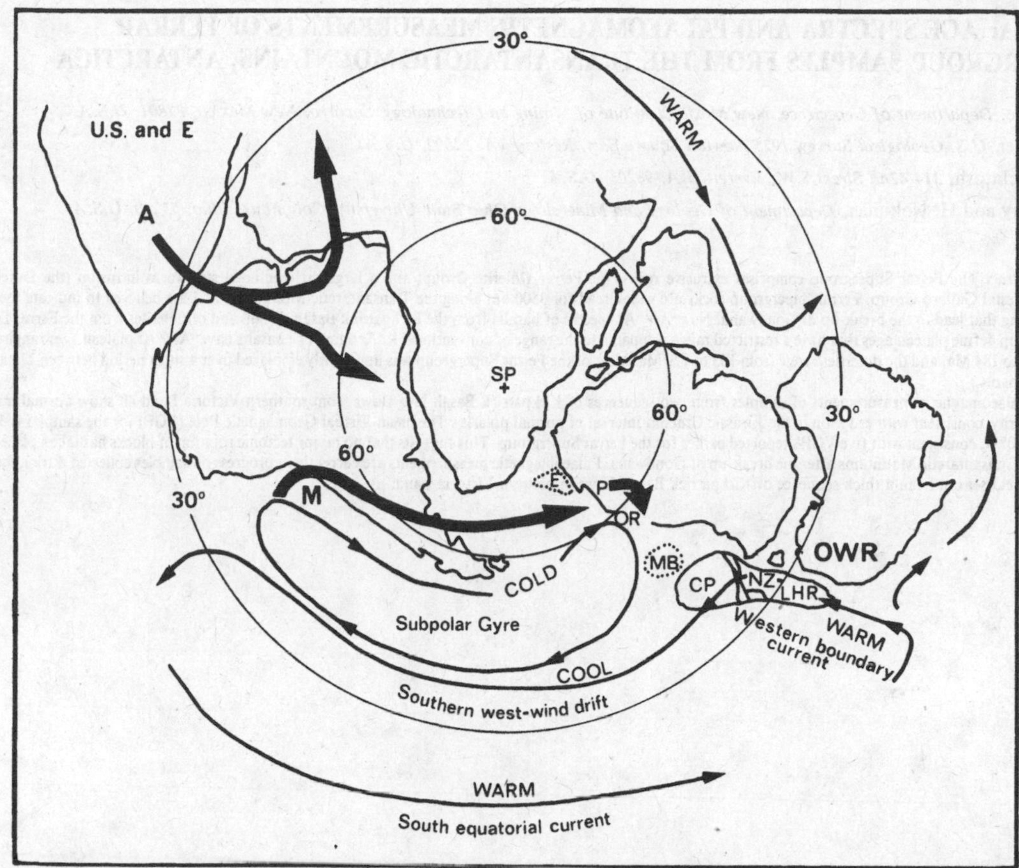


**Figure 6. Polar view of a reconstructed Gondwana during the Early Devonian, showing ocean currents and faunal provinces. Redrawn from Heckel and Witzke, (1979): A—Appalachian faunal province, M—Malvinokaffric faunal province, OWR—Old World Realm, CP—Campbell Plateau, E—Ellsworth Mountains, LHR—Lord Howe Rise, MB—Marie Byrd Land, N.Z.—New Zealand, OR—Ohio Range, P—Pensacola Mountains, SP—Devonian South Pole, U.S. and E—United States and Europe.**

## Conclusions

The lithofacies of the Horlick Formation all record shoreline deposition of granite derived sediment. Feldspar rich beds are concentrated mainly in the one fluvial section and in the basal marine beds.

The alternation of several lithofacies, all of which can be generally ascribed to deposition under varying shoreface and shallow shelf conditions, points to lateral accretion of the deposits. Thus higher and lower energy lithofacies alternate several times in most sections recording minor changes in bottom current regime. These variations may in turn be explained as a series of small transgressive-regressive cyclic events with shelf influence only during the main transgressive phases.

A Middle Devonian reconstruction of Gondwana with oceanic circulation is given by Heckel and Witzke (1979). This has been redrawn as a polar projection in Figure 6. In Early Devonian time the Ohio Range, together with the Pensacola and Ellsworth Mountains, is thought to have occupied the arm of a sea that was probably encroaching from the direction of the then conjoined South America and African sectors of Gondwana, bringing with it a Malvinokaffric fauna (Boucot et al., 1963). Some communication with the seas along the Pacific margin of Gondwana is likely to explain the existence of *Pleurothyrella* and *Tanerhynchia* in New Zealand, the only Australasian country in which the former is found. Conditions were probably less tectonically stable in the Ohio Range than in southern Victoria Land where shallow marine Early Devonian quartzose sandstones indicate much longer reworking of gneiss derived detritus (Bradshaw, 1981). The mixed faunas of New Zealand probably resulted from the convergence of the Early Devonian Subpolar Gyre that carried coldwater Malvinokaffric elements with the Western Boundary Current that carried warmer water Old World elements which dominate the Australian Lower Devonian faunas.

*Acknowledgements* This study was supported by Antarctic Division, New Zealand DSIR, the United States National Science Foundation, the United States Geological Survey, and the Canterbury Museum Trust Board. We appreciated the companionship of Karl Kellogg and Graham Ayres in the field. The Robin S. Allan Memorial Fund contributed towards the preparation of this manuscript.

## REFERENCES

BABIN, C., GLEMAREC, M., TERMIER, H. and TERMIER, G., 1971: Role des Maldanes (Anneliedes Polychetes) dans certains types de bioturbation. *Annals Soc. Geol. Nord, 91,* 203-6.

BOUCOT, A.J., CASTER, K.E., IVES, D. and TALENT, J.A., 1963: Relationships of a new Lower Devonian terebratuloid (Brachiopoda) from Antarctica. *Bull. Am. Paleontol., 207,* 81-151.

BRADSHAW, M.A., 1981: Palaeoenvironmental interpretations and systematics of Devonian trace fossils from the Taylor Group (lower Beacon Supergroup), Antarctica. *N.Z. J. Geol. Geophys., 24,* 615-52.

DOUMANI, G.A., BOARDMAN, R.S., ROWELL, A.J., BOUCOT, A.J., JOHNSON J.G., MCALESTER, A.L., SAUL, J., FISCHER, D.W.and MILES, R.W., 1965: Lower Devonian fauna of the Horlick Formation, Ohio Range, Antarctica. *Am. Geophys. Union, Ant. Res. Ser., 6,* 241-81.

HECKEL, P.H. and WITZKE, B.J., 1979: Devonian world palaeogeography determined from distribution of carbonates and related lithic palaeoclimatic indicators. *Spec. Pap. Palaeontol., 23,* 99-123.

KEMP, E.M., BALME, B.E., HELBY, R.J., KYLE, R.A., PLAYFORD, G. and PRICE, P.L., 1977: Carboniferous palynostatigraphy in Australia and Antarctica: a review. *BMR J. Aust. Geol. Geophys., 2,* 177-208.

KYLE, R.A., 1977: Palynostratigraphy of the Victoria Group of southern Victoria Land, Antarctica. *N.Z. J. Geol. Geophys., 20,* 1081-102.

LONG, W.E., 1965: Stratigraphy of the Ohio Range, Antarctica. *Am. Geophys. Union, Ant. Res. Ser., 6,* 71-116.

PEVEAR, D.R., 1966: The estuarine formation of United States Atlantic coastal plain phosphorite. *Econ. Geol., 61,* 251-6.

REINECK, H.E.and SINGH, I.B., 1975: *Depositional Sedimentary Environments.* Second Edition. Springer-Verlag, Berlin.

TREVES, S.B., 1965: Igneous and metamorphic rocks of the Ohio Range, Horlick Mountains, Antarctica. *Am. Geophys. Union, Ant. Res. Ser., 6,* 117-25.

# $^{40}$Ar/$^{39}$Ar AGE SPECTRA AND PALAEOMAGNETIC MEASUREMENTS OF FERRAR SUPERGROUP SAMPLES FROM THE TRANSANTARCTIC MOUNTAINS, ANTARCTICA

P.R. Kyle, *Department of Geoscience, New Mexico Institute of Mining and Technology, Socorro, New Mexico, 87801, U.S.A.*

J.F. Sutter, *U.S. Geological Survey, 1925 Newton Square East, Reston, VA. 22092, U.S.A.*

W.C. McIntosh, *314 42nd Street S.W., Everett, WA, 98203, U.S.A.*

E. Cherry and H. Noltimier, *Department of Geology and Mineralogy, Ohio State University, Columbus, Ohio, 43210, U.S.A.*

*Abstract* The Ferrar Supergroup comprises extrusive rocks, the Ferrar Dolerite Group; and a large differentiated stratiform intrusion (the Dufek), the Forrestal Gabbro Group. Ferrar Supergroup rocks are widespread for 3500 km along the Transantarctic Mountains, and are believed to indicate the initial rifting that lead to the break-up of Gondwana. New $^{40}$Ar/$^{39}$Ar spectra of basalts from the Kirkpatrick Basalt Group and of dolerites from the Ferrar Dolerite Group define plateau ages that have a restricted range compared to the range of conventional K/Ar ages. The basalts have $^{40}$Ar/$^{39}$Ar plateau ages ranging from 176 to 184 Ma, and the dolerites range from 163 to 179 Ma. Most of the Ferrar Supergroup was apparently emplaced over a short period between 170 and 180 Ma ago.

Palaeomagnetic measurements of samples from two sequences of Kirkpatrick Basalt lava flows from southern Victoria Land all show normal magnetic polarity, consistent with eruption in the Jurassic Graham Interval of normal polarity. The mean Virtual Geomagnetic Pole (VGP) for the samples is 54.9°S, 138.0°W, consistent with five VGP's reported earlier for the Ferrar Supergroup. This suggests that no major tectonic rotation of blocks has taken place within the Transantarctic Mountains after the break-up of Gondwana. Palaeomagnetic measurements are currently in progress on samples collected during the 1981-82 field season from a thick sequence of Kirkpatrick Basalt flows in nothern Victoria Land.

# 4

# West Antarctica

# GEOLOGY AND PLATE TECTONIC SETTING OF THE ORVILLE COAST AND EASTERN ELLSWORTH LAND, ANTARCTICA

P.D. Rowley, *U.S. Geological Survey, Denver, Colorado 80225, U.S.A.*
W.R. Vennum, *Sonama State University, Rohnert Park, California 94928, U.S.A.*
K.S. Kellogg, *U.S. Geological Survey, Denver, Colorado 80225, U.S.A.*
T.S. Laudon, *University of Wisconsin-Oshkosh Wisconsin 54901, U.S.A.*
P.E. Carrara, *U.S. Geological Survey, Denver, Colorado 80225, U.S.A.*
J.M. Boyles, *Cities Service Company, Tulsa, Oklahoma 74102, U.S.A.*
M.R.A. Thomson, *British Antarctic Survey, Madingley Road, Cambridge CB3 OET, U.K.*

*Abstract* Orville Coast and eastern Ellsworth Land are underlain by Middle and Upper Jurassic dark and volcanoclastic shale, siltstone, and sandstone (Latady Formation) and intertongued calc-alkaline silicic to intermediate ash-flow tuff, lava flows, volcanic breccia, and air-fall tuff (Mount Poster Formation). All these rocks were tightly folded and locally thrust faulted with south to southeast vergence, then intruded in Early Cretaceous time by widely spaced calc-alkaline plutons of the Lassiter Coast Intrusive Suite.

The rocks of the southern Antarctic Peninsula and eastern Ellsworth Land formed above a plate of Pacific basin lithosphere being subducted beneath what is now the peninsula. The Jurassic rocks accumulated on a magmatic arc and in its marine back-arc basin. The rocks of the Mount Poster Formation were deposited mostly subaerially along the axis of the magmatic arc along the present topographic crest of the southern Antarctic Peninsula. On the southeastern flank of the arc, where the Latady and Mount Poster Formations intertongue, many rocks likewise are continental deposits. Farther southeast, rocks of the Latady Formation are marine; depositional environments changed progressively southeastward from apparent deltaic and other near-shore environments to open shallow-water. The marine rocks are locally rich in invertebrate fossils. The Cretaceous plutons represent the roots of a younger overlapping magmatic arc.

The Jurassic and Cretaceous igneous rocks are highly evolved, indicating that the magmatic arcs that they represent formed on continental lithosphere at the edge of Gondwanaland. This lithosphere was broken near the site of the back-arc basin during opening of the southern Atlantic Ocean, leaving the crustal sliver that now underlies the peninsula.

Orville Coast, in the southern Antarctic Peninsula, and eastern Ellsworth Land include an area of exposed bedrock spread over about 30,000 km², at about latitude 74°-76°S, and longitude 64°-73°W. Altitudes in the area range from nearly sea level along the Ronne Ice Shelf (Figure 1) to about 2,000 m on the topographic axis of the Antarctic Peninsula and its continuation in eastern Ellsworth Land. Most rock exposures occur on the Atlantic side of the peninsula crestal axis, where there is 5-10% rock exposure. This region was first explored and geologically mapped in reconnaissance during the 1977-78 field season by a U.S. Geological Survey field party consisting of the authors.Some parts of eastern Ellsworth Land, which first had been explored and mapped by field parties led by J.C. Behrendt in 1961/62 and by T.S. Laudon in 1965/66, were re-examined at this time.

The Orville Coast and Eastern Ellsworth Land are underlain by Jurassic sedimentary rocks of the Latady Formation that intertongue with volcanic rocks of the Mount Poster Formation, all of which are intruded by Cretaceous plutons. The geology of the area is similar to that of other nearby areas (Figure 1): the Lassiter Coast and southern Black Coast, mapped during the 1969/70, 1970/71 and 1972/73 seasons (Williams et al., 1972; Rowley and Williams, 1982; Rowley, Kellogg et al. in press): and parts of eastern Ellsworth Land mapped during the 1965/66 season (Laudon et al., 1969; Laudon, 1972). The central Black Coast and nearby Seward Mountains (Singleton, 1980a, 1980b) also contain the same rock units, but in these areas a poorly exposed underlying high-grade metamorphic complex of uncertain age may also be present. Cretaceous plutonic rocks are widespread in the northern parts of this region, but they progressively decrease in abundance southward. Thus plutonic rocks are relatively uncommon in the Orville Coast and eastern Ellsworth Land, and exposed Jurassic rocks have undergone significantly less contact metamorphism here than farther north.

Orville Coast and eastern Ellsworth Land are part of the Pacific-margin belt of magmatism and deformation of Mesozoic and Cenozoic age. This belt extends northward through the Andes of South America, and southward and westward through western Ellsworth Land and Marie Byrd Land and from there northward into the western Pacific area. The belt is characterised by magmatic arcs, that is, arcuate or linear belts of volcanic rocks and underlying plutonic rocks commonly formed along active convergent margins of lithospheric plates. As suggested by its shape, virtually the entire Antarctic Peninsula consists of various, approximately parallel magmatic arcs. They overlap each other and have different ages within the Mesozoic and Cenozoic Eras. All these arcs are thought to have been created by subduction of plates (Farallon, Pacific and perhaps others) of the Pacific Ocean basin under continental lithosphere along the western edge of Gondwanaland before, during and after its initial breakup. Subduction of the Pacific basin ceased in western Ellsworth Land and Marie Byrd Land by Late Cretaceous time, but farther north in the Antarctic Peninsula it continued through most of the Tertiary, and in the northern and central Andes it is still continuing.

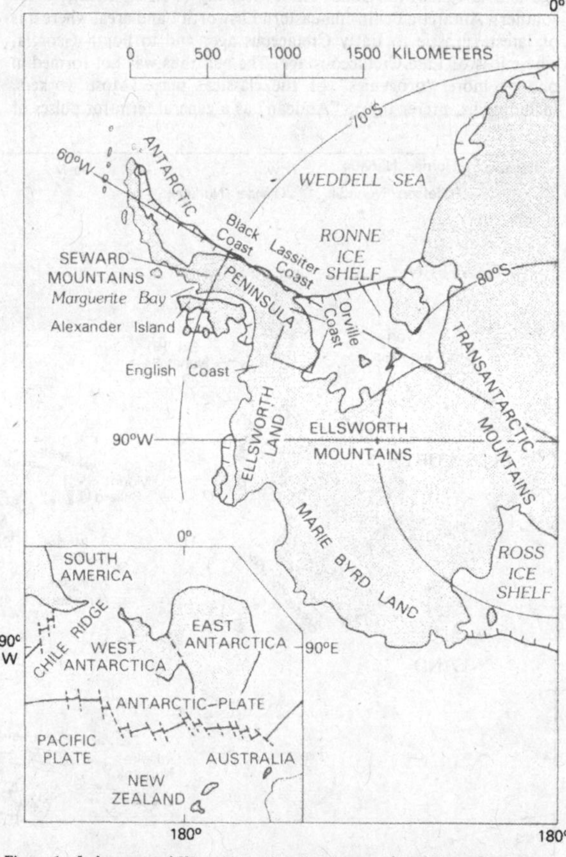


**Figure 1** **Index map of West Antarctica showing location of Orville Coast, eastern Ellsworth Land, and other areas mentioned in the text. Edge of ice shelves shown by hachured lines. Rectangle shows area of Figure 2.**

West Antarctica contains several small plates of continental lithosphere that behaved differently through time and that assumed their present configuration only by the Late Cenozoic (Behrendt, 1964a; Schopf, 1969; Bentley and Clough, 1972; Dalziel and Elliot, 1982). Ellsworth Mountains, western Ellsworth Land and Marie Byrd Land are geologically different from each other and from the Antarctic Peninsula-eastern Ellsworth Land area, thus all these areas may represent different microplates. Palaeomagnetic studies (Scharnberger and Scharon, 1982) indicate that Marie Byrd Land separated from the Antarctic Peninsula-eastern Ellsworth Land-western Ellsworth Land area by Early Cretaceous time. Western Ellsworth Land was joined to the Antarctic Peninsula-eastern Ellsworth Land area at least at this time, although its somewhat different geology and the identification, by geophysics, of a tectonic break in central Ellsworth Land (Behrendt, 1964a; Bentley and Clough, 1972) suggest that this was not always so.

Terminology for deformational and intrusive events in West Antarctica is muddled. The term "Andean" was first applied by Adie (1955) as a stratigraphic name for widespread calc-alkaline plutonic rocks in the Antarctic Peninsula of presumed Late Cretaceous and Early Tertiary age. Isotopic age dating subsequently showed that the plutonic rocks were intruded during several magmatic pulses (Rex, 1976; Pankhurst, 1982) from at least Late Triassic through Tertiary time, and perhaps as early as Late Palaeozoic time (Smellie, 1981). Despite the dating, some workers continue to restrict "Andean" to plutonic rocks of latest Cretaceous and Early Tertiary time in the Antarctic Peninsula, or they use "Andean Orogeny" for deformation as well as plutonism of this age throughout the Pacific margin of West Antarctica, but in Antarctica the Pacific margin belt is characterised primarily by long-lived subduction-related igneous activity. Deformation in the belt is restricted to parts of the northern Antarctic Peninsula and Scotia Arc, where it is of Late Triassic to Early Jurassic age and is called the Gondwanide Orogeny (Dalziel, 1982); to the southern Antarctic Peninsula-eastern Ellsworth Land area, where it is of latest Jurassic to Early Cretaceous age; and to South Georgia, where it is of Late Cretaceous age. The belt thus was not formed in one or more "orogenies" of the classical sense. Most workers, including us, prefer to use "Andean" as a general term for pulses of subduction-related calc-alkaline magmatism and local deformation of post-Gondwanide age (Early Jurassic to Tertiary) as well as for the resulting Pacific margin belt. The products of some Andean magmatic events, such as the Lassiter Coast Intrusive Suite (see below), are worthy of separate names. Gondwanide and possible pre-Gondwanide structures and igneous rocks seem to be due to similar subduction geometrics as Andean ones.

## LATADY FORMATION

The oldest, and by far the most voluminous exposed rocks in Orville Coast and eastern Ellsworth Land consist of intertongued sedimentary and volcanic sequences. Where volcanic rocks dominate, both were mapped together as the Mount Poster Formation; where sedimentary rocks dominate, both were mapped together as the Latady Formation. Intertonguing relations cannot be shown at the scale of the geologic map (Figure 2). The Latady Formation consists of black and grey slate, silstone, and mudstone; subordinate sandstone and coal; and rare conglomerate. The thickness of the Latady Formation is unknown; neither its base or its top is exposed in the field area, and it contains no recognisable regional marker beds; probably it is at least several kilometres thick. The formation was defined by Williams et al. (1972) for exposures in the Latady Mountains of the central Lassiter Coast; mapping and correlation extended it throughout the southern Antarctic Peninsula and eastern Ellsworth Land (Laudon et al., this volume).

On the basis of locally abundant invertebrate marine fossils in the Latady, most rocks of the formation have been assigned a Late Jurassic age (see Thomson, this volume) but in eastern Ellsworth Land, Middle Jurassic faunas are known (Quilty, 1982). Plant fossils in the Latady indicate a generally temperate climate during deposition (Schopf, J.M., written commun., 1976).

Excellent exposures of the Latady Formation in Orville Coast provide a cross section, perpendicular to the strike of the beds and the trend of the arc, of rocks of different depositional environments, most of which were deposited rapidly (Laudon et al., this volume). The Latady was deposited in a back-arc basin on the south to southeastern flank (the craton side) of a magmatic arc whose axis generally coincided with or perhaps lay west of the present topographic axis of

worthy of separate names. Gondwanide and possible pre-Gondwanide

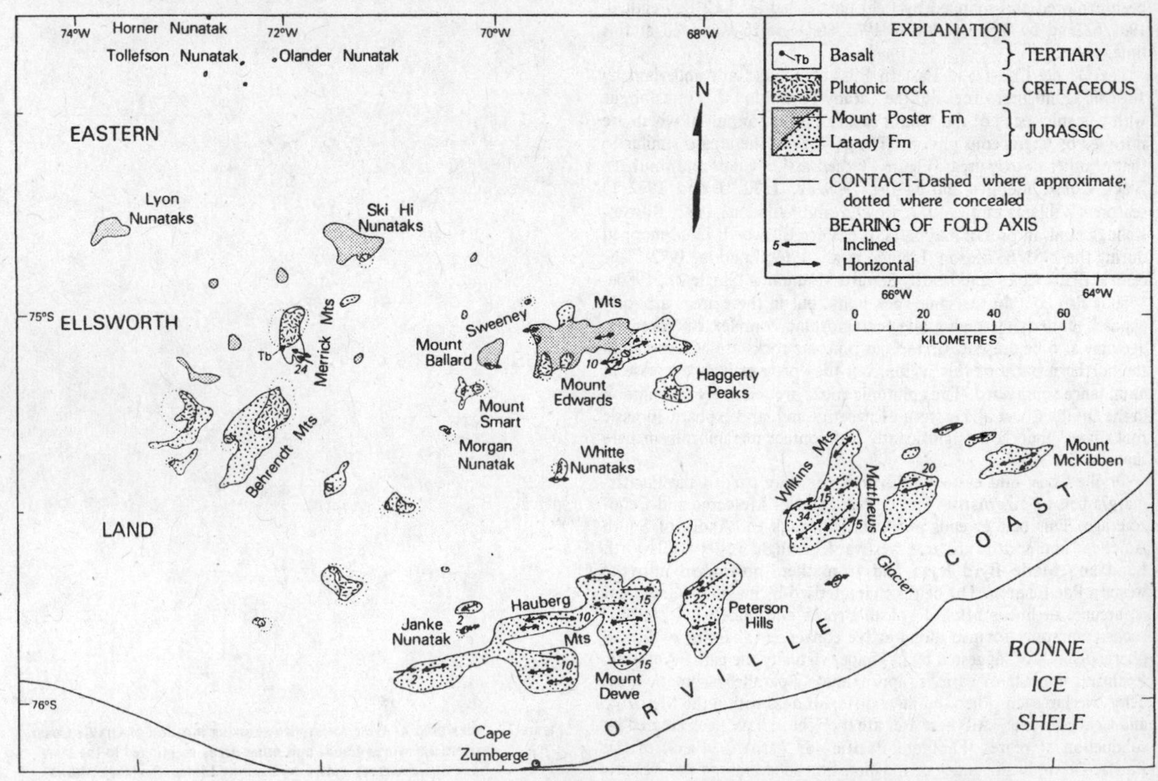


**Figure 2 Geologic map of Orville Coast and eastern Ellsworth Land. Includes mapping of Laudon et al., (1969)**

the southern Antarctic Peninsula and eastern Ellsworth Land. Volcanic rocks of the Mount Poster Formation, accumulated, mainly subaerially, along the axis of this arc. Most of the Latady consists of clastic material that was derived from the contemporaneous volcanic rocks. Southward from the axis, Latady depositional environments change from (1) nonmarine, where the sedimentary rocks intertongue with the volcanic rocks, through (2) shallow-marine, possibly deltaic, to (3) open-marine shallow water farthest south (Rowley, 1978; Thomson et al., 1978).

The Latady Formation in the Merrick and Sweeney Mountains and nearby nunataks contains the best evidence of the mainly nonmarine environment of deposition. Here, mostly subaerial volcanic rocks, especially lava flows, intertongue with beds of Latady, some of which appear to be of lacustrine and fluvial origin, as indicated by abundant plant fossils, numerous thin coal beds, and various sedimentary structures. Marine invertebrate fossils are present also, but are not as abundant as they are farther south; probably the coastline shifted back and forth across this area.

The Hauberg and Wilkins Mountains, Peterson Hills, and areas east of Matthews Glacier are entirely underlain by the Latady Formation, all of sedimentary origin except for rare tuffaceous beds. The stratigraphic sequence is thick and consists mostly of thin to thick bedded mudstone and sandstone. Sandstone is locally trough cross bedded and contains shallow water structures such as ripple marks and root casts. Non-marine(?) carbonaceous shale and fossiliferous marine mudstone are interbedded with the sandstone. The features in the area are suggestive of a low-relief fan-delta complex whose shoreline fluctuated due to progradation and basin subsidence.

The rocks that represent the third type of depositional environment, shallow-water open marine, are known only at several small nunataks farthest to the south. At Cape Zumberge and south of the Wilkins Mountains, black shale containing a latest Tithonian ammonite fauna (Thomson, this volume) that has not been found elsewhere in the field area, are exposed. The rocks of the open-marine environment are finer grained than those farther north-northwest, and probably were deposited farther offshore in deeper water, below effective wave base.

## MOUNT POSTER FORMATION

Flat-lying to folded, silicic to mafic, Mesozoic through Tertiary calc-alkaline volcanic rocks of different magmatic arcs occur throughout the Antarctic Peninsula. Adie (1964) lumped them with the Upper Jurassic Volcanic Group and recognised that they are chemically similar to plutonic rocks in the area. Thomson (1982) redefined them as the Antarctic Peninsula Volcanic Group; Thomson and Pankhurst (this volume) suggest that they range in age from at least Early Jurassic to Tertiary. Definition of formation names within the group is needed where lithology, depositional environment, and age of the volcanic rocks are similar. The first of these names, the Mount Poster Formation, was proposed by Rowley et al. (1982) for Jurassic silicic to intermediate composition volcanic rocks that form the axis and southeastern flank of the southern Antarctic Peninsula and eastern Ellsworth Land. Intertonguing of the Mount Poster Formation is Middle(?) and Late Jurassic; the uppermost rocks in the formation could be as young as Early Cretaceous. The total thickness of the formation is not known, owing to folding and the lack of regional marker beds, but it is at least several kilometres.

The best exposures of the Mount Poster Formation are in the Orville Coast and eastern Ellsworth Land. The formation underlies Horner, Tollefson, Olander, and Sky Hi Nunataks, near the crest of the Antarctic Peninsula-eastern Ellsworth Land area, and it forms much of the Sweeney Mountains and the nunataks west of the Behrendt Mountains, on the flank of the peninsula. The formation also is exposed at the type locality in the southwestern Lassiter Coast (Williams et al., 1972) and in the Sverdup Nunataks of the southwestern Black Coast (Rowley and Williams, 1982; British Antarctic Survey, 1982; Rowley, Kellogg et al., in press). Farther north, similar volcanic rocks in the central Black Coast (Singleton, 1980a), Seward Mountains (Singleton, 1980b) and nearby Gutenko Mountains (British Antarctic Survey, 1982) also may be correlative with the Mount Poster Formation.

Most of the Mount Poster Formation consists of amygdaloidal rhyodacitic to andesitic lava flows and rhyodacitic to dacitic ash-flow tuff. An abundant volcanic rock type in the Horner-Tollefson-Olander Nunataks, Sweeney Mountains, Sky Hi Nunataks, and many other nunataks is massive crystal-rich rhyodacite commonly more than 100 m thick. It contains phenocrysts of sanidine and plagioclase as long as 2 cm and slightly smaller, embayed bipyramidal-morphology quartz, as well as subordinate to minor amounts of small altered ferromagnesian minerals, probably originally mostly hornblende, pyroxene, and Fe-Ti oxides. Broken phenocrysts, collapsed pumice, and other eutaxitic features are visible locally; the volcanic rock is interpreted to be one or more ash-flow sheets. Other unambiguous ash-flow tuffs contain moderate to sparse, locally broken phenocrysts, abundant dark lenticules of collapsed pumice as long as several centimetres, and other eutaxitic features, and they commonly are of similar chemistry and mineralogy to the massive rhyodacite unit. Lava flows contain plagioclase, pyroxene, and hornblende as primary minerals; no olivine was identified in these rocks. Vent breccia, flow breccia, volcanic mudflow breccia, air-fall tuff, sandstone, and siltstone also occur in many places in the formation. Most volcanic and sedimentary rock units are 1 to 10 m thick. Most rocks appear to have been erupted subaerially, but pillow lava flows occur at several localities in the Sweeney and Behrendt Mountains.

## PLUTONIC ROCKS

Upper Triassic to Upper Tertiary calc-alkaline plutons in the Antarctic Peninsula are broadly synchronised with volcanic rocks (Pankhurst, 1982) although plutons and their synchronous volcanic cover are rarely both exposed in individual areas. Plutonism was almost continuous throughout this broad time period, although pulses of especially widespread magmatism, with a peak of 110 to 90 Ma (Middle Cretaceous) are indicated (Rex, 1976; Saunders et al., 1982; Pankhurst, 1982). The plutonic rocks of all pulses seem to be lithologically similar, although Tertiary plutonic rocks are generally more mafic (Saunders et al., 1982). All are one-mica upper-crustal plutonic rocks (cf. Hamilton, 1981) like those in the Sierra Nevada and many other circum-Pacific terrains.

Most exposed plutonic rocks in the southern Antarctic Peninsula and eastern Ellsworth Land range in age from 130 to 95 Ma (Mehnert et al., 1975; Farrar and Rowley, 1980; Pankhurst, 1980). Following usage for intrusive rocks proposed by the North American Commission on Stratigraphic Nomenclature (Henderson et al., 1980), we propose the name Lassiter Coast Intrusive Suite for these rocks because in the Lassiter Coast, the type area, the plutons have been relatively well analysed and studied. This new stratigraphic name, equivalent in rank to a group, refers to calc-alkaline rocks of numerous Lower Cretaceous plutons, some of them named, ranging in composition from gabbro to granite. Fifteen stocks belonging to the Lassiter Coast Intrusive Suite were mapped for the first time in the Sweeney Mountains, in the Morgan, Witte, Sky Hi, and Janke Nunataks, and west of the Behrendt Mountains (Figure 2). Several others were also examined. The stocks have exposed diameters of 1 to 12 km, range in composition from gabbro to granite, and exhibit a range in K-Ar and Rb-Sr ages of 123 to 102 Ma. Some contain small noneconomic copper anomalies that might be deeply-eroded porphyry copper deposits (Rowley, Farrar, et al., in press)

Plutonism probably accompanied eruption of the volcanic rocks of the Mount Poster Formation in the Orville Coast and eastern Ellsworth Land; if so, the plutons would belong to an Intrusive suite of Middle and Late Jurassic age. At the present time, however, all dated plutons are younger than the Mount Poster Formation and are placed in the Lassiter Coast Intrusive Suite. Plutonic rocks contemporaneous with the Mount Poster Formation are most likely to occur where the volcanic rocks are thickest, near or perhaps west of the present crestal axis of the field area; but few nunataks occur near this axis, and no plutonic rocks are exposed in the crestal nunataks we examined.

## CONTACT METAMORPHIC ROCKS

Contact-metamorphic aureoles, locally more than 1 km wide, occur in Jurassic sedimentary and volcanic rocks surrounding all plutons in Orville Coast and eastern Ellsworth Land. The main rock types in these aureoles are resistant black hornfels and slate containing porphyroblasts of andalusite and subordinate cordierite, garnet, biotite and sillimanite. Contact metamorphic rocks are similar to those in the Lassiter Coast and southern Black coast (Kamenev and Orlenko, 1982). The highest metamorphic grade, occurring adjacent to the plutons, is hornblende hornfels facies, whereas farther outward the rocks are of albite-epidote hornfels facies.

## MIOCENE(?) BASANITE

A 10 to 12 m thick sequence of frost-shattered basanite lava flows and flow breccia is exposed in a small nunatak in the southwestern Merrick Mountains (Laudon et al., 1969; Laudon, 1972; Vennum and Laudon, in press). The breccia clasts are set in a palagonitised matrix of glassy ash and lapilli. Halpern (1971) reported a whole-rock K-Ar age of 6 Ma for a sample from this area; the true age of the rock, however, is tentative inasmuch as Halpern did not publish analytical data for the age. The rocks probably are hyaloclastites that erupted partly or entirely under ice, comparable to abundant hyaloclastites of similar composition that are widely exposed in Marie Byrd Land and western Ellsworth Land (LeMasurier and Rex, 1982) and northern Victoria Land (Hamilton, 1972). Alkaline mafic or silicic rocks such as this are associated with extensional tectonics in many parts of the world, including Marie Byrd Land and western Ellsworth Land (LeMasurier and Rex, 1982). Seismic-reflection and gravity data (Behrendt, 1964b) show rough sub-ice topography, including valleys nearly 2,000 m below sea level, in parts of eastern Ellsworth Land; thus fault-block topography may occur as far east as the Orville Coast-eastern Ellsworth Land area. This faulting began after subduction ceased, perhaps as soon as Late Cretaceous time.

## STRUCTURAL GEOLOGY

The Latady and Mount Poster Formations in the Orville Coast and eastern Ellsworth Land are tightly folded. Strikes of fold axes range from mostly west-northwest in the western part of the area to east-northeast in the eastern part (Figure 2). Folds range from open to isoclinal; most are asymmetric or overturned to the south-southeast. Chevron folds are locally well exposed in the Hauberg and Wilkins Mountains. Fold wavelengths range from small crinkles seen in a thin section to broad structures more than a kilometre long. Well-developed axial-plane cleavage formed during folding. Cleavage and most axial planes of folds dip northerly. Most fold axes are horizontal to gently plunging. Some fold axes, however, plunge steeply within several hundred metres of pluton contacts; these represent regional folds that have been refolded during forcible intrusion of the plutons. Several thrust faults that dip gently to the north-northwest were mapped in the Hauberg and Wilkins Mountains. They appear to be of small displacement, with hanging-wall movement toward the south.

The age of the main period of folding and thrusting is latest Jurassic or earliest Cretaceous, between the Middle and Late Jurassic age of the Latady Formation and the Early Cretaceous age of the plutons that cut the folds and thrusts. In the Lassiter Coast and southern Black Coast, extension joints in plutonic rocks are oriented similarly to those in the folded Jurassic rocks, both being perpendicular to fold axes. Some extension joints of this orientation formed during crystallisation of the plutonic rocks, for many mafic dykes, aplites, and pegmatites related to late phases of plutonism were emplaced in the extension joints in plutonic rocks. Thus many plutons are partly syntectonic. In the Orville Coast and eastern Ellsworth Land, in contrast, joints and dykes are randomly oriented in most plutonic rocks, unlike extension joints in the Jurassic rocks, which are perpendicular to the strike of fold axes (Kellogg, 1979). These data suggest that the principal compressive stress that produced the folds continued during and perhaps after plutonism in Lassiter Coast and southern Black Coast, but perhaps not as long in Orville Coast and eastern Ellsworth Land (Kellogg, 1979).

High-angle faults that postdate folding are relatively abundant in Orville Coast and eastern Ellsworth Land. Most strike perpendicular to the axis of the peninsula and appear to have small strike-slip displacements, predominantly in the right-lateral sense. Some faults, however, may be normal faults related to bimodal volcanism and extensional tectonism. On the basis of ages of invertebrate fossils, Thomson (this volume) suggested that a fault may lie between the Wilkins and Hauberg Mountains; this is possible, although it could not be verified by field evidence.

Kellogg (1980), supporting Hamilton (1967), concluded from a palaeomagnetic study of plutons and dykes in the Orville Coast and eastern Ellsworth Land that the oroclinal bend in the southern Antarctic Peninsula (Figure 1) formed after Early Cretaceous plutonism. Kellogg (1980) speculated that strike-slip faulting might have accompanied oroclinal bending. The strike-slip faults we mapped in the field area may have formed in this manner.

## PLATE-TECTONIC SETTING

Most, if not all, of the rocks now exposed in the Antarctic Peninsula and eastern Ellsworth Land formed during Mesozoic and Tertiary time in a magmatic arc environment above a plate of Pacific Ocean lithosphere being subducted beneath what is now the peninsula (Dalziel and Elliot, 1973; Rowley and Williams, 1982; Elliot, this volume). Magmatic arcs have been identified in southern South America (Dalziel, 1981), South Georgia (Tanner et al., 1981), and the northern Antarctic Peninsula (Farquarson, this volume). The oldest known period of magmatic arc development is of pre-Early Jurassic age (Smellie, 1981; Dalziel et al., 1981; Dalziel, 1982; Dalziel, this volume). It may have been roughly axial to (Dalziel et al., 1981) or on the eastern flank of (Smellie, 1981) what is now the northern and central Antarctic Peninsula. Sedimentary rocks of the Trinity Peninsula Formation, of mostly Triassic age (Thomson, 1982; Edwards, 1982) and differently named but roughly correlative sedimentary units, have been related to this magmatic arc (Smellie, 1981). This early arc may have formed on continental lithosphere, based on sparse information cited by Smellie (1981) on the petrologic maturity (that is, how evolved they are; Hamilton, 1981) of its igneous rocks. This information is suspect, however, because no plutonic rocks have yet been confirmed by isotopic dating to be of the assumed age of the arc. This magmatic arc was deformed during Late Triassic to Early Jurassic Gondwanide deformation, coincident with a pulse of calc-alkaline plutonism (Rex, 1976). Initial $^{87}Sr/^{86}Sr$ ratios of 0.706 to 0.707 from these plutons, located in the northern Antarctic Peninsula, demonstrate that continental lithosphere was present at least by this time (Pankhurst, in press), and continued to be present during subsequent Mesozoic and Cenozoic magmatic events. This conclusion is supported by the high petrologic maturity of the igneous rocks, which include granites (Rex, 1976; Pankhurst, 1982).

No evidence of the pre-Jurassic magmatic arc is known in the southern Antarctic Peninsula. The nearest area containing rocks that might predate the Latady Formation is the central Black Coast, where Singleton (1980a) claimed to have found a high grade schist and gneiss complex on which rocks that he correlated with the Latady Formation rest unconformably. If he is correct, the complex would correlate with possible middle crustal (c.f. Hamilton, 1981) metamorphic complexes in the Marguerite Bay farther north in the Antarctic Peninsula. All metamorphic complexes could either represent eroded root zones of the older magmatic arc (Dalziel et al., 1981; Dalziel, 1982; Pankhurst, this volume) or be still older continental lithosphere on which various magmatic arcs were built. Alternatively, the schist and gneiss in the Black Coast and Marguerite Bay area could represent contact phases of Cretaceous plutons (Pankhurst, this volume).

The Mount Poster and Latady Formations represent a Middle to Late Jurassic magmatic arc (Suárez, 1976). The axis of the arc was either along or just west of the present topographic axis of the field area. Although no correlative plutons have been positively identified in the field area, the maturity of the volcanic rocks proves that continental lithosphere underlay this magmatic arc. The southeastern shoreline of the arc trended parallel to the long axis of the present peninsula and through its southeastern flank. A back-arc marine basin formed southeast of the arc during subduction. The Latady Formation was deposited mainly in this basin, but also partly on the southeastern flank of the arc, where continental sedimentary rocks intertongue with mostly subaerially deposited volcanic rocks.

Folds and slaty cleavage of south to southeast vergence as well as south to southeast directed thrusts formed in the southern Antarctic Peninsula and eastern Ellsworth Land during latest Jurassic and Early Cretaceous time. This crustal shortening, following the extensional regime that produced the back-arc basin, could have had several origins. Isostatic uplift of the magmatic arc attendant with crustal thickening during continued subduction could have led to gravitational spreading down a regional slope to the south or southeast. This hypothesis, however, is weakened by the lack of evidence for deep erosion into the Jurassic rocks and initial $^{87}Sr/^{86}Sr$ ratios of plutonic rocks in the northern Antarctic Peninsula (Pankhurst, 1982; Thomson et al., this volume) that suggest a declining, rather than increasing, influence or thickness of continental lithosphere with younger age. Perhaps a better alternative is the uplift and shouldering aside of overlying rocks, with associated gravitational spreading, toward the topographically lower back-arc basin, by emplacement of the batholith that produced the Mount Poster Formation. Alternatively, collision of the magmatic arc and back-arc basin with the craton of West Gondwanaland just before the start of opening of the South Atlantic Ocean could have produced the deformation. Compression also could result from increases in motions of converging plates just before opening of the South Atlantic or from subduction of younger, more buoyant rocks along low-angle subduction zones (Cross and Pilger, 1982).

The opening of the South Atlantic Ocean began in Early Cretaceous time (Sclater et al., 1977). The break separating South Africa and Antarctica was east of the western edge of West Gondwanaland, leaving a sliver of continental lithosphere underlying the Antarctic Peninsula. At about the same time, a second magmatic arc caused by Pacific-margin subduction formed in the southern Antarctic Peninsula-eastern Ellsworth Land area. Calc-alkaline plutons of the Lassiter Coast Intrusive Suite of Early Cretaceous age are the sole evidence for this magmatic arc. No corresponding volcanic ejecta have been identified in the field area, but probably were once present inasmuch as tuff is abundant in Lower Cretaceous sedimentary rocks of Alexander Island (Horne and Thomson, 1972). The maturity of the plutonic rocks proves that this Cretaceous arc formed on continental lithosphere. Either this magmatic arc was broader than the Jurassic arc or its axis was farther south or southeast, for many plutons in the southern Antarctic Peninsula occur well south or southeast of known concentrations of Jurassic arc rocks (Mount Poster Formation). Plutons do not occur in the Hauberg and Wilkins Mountains, so the southern boundary of the arc probably was just north of these ranges. Folding and thrusting apparently had largely ceased before emplacement of plutons in the field area, but southeastward-directed deformation may have continued somewhat longer in the Lassiter and Black Coasts. The paucity of plutons in the Orville Coast and eastern Ellsworth Land, with respect to their distribution in the Lassiter and Black Coasts, may also reflect earlier slowing down or cessation of subduction in the field area. By the time plutonism ended, about Middle Cretaceous time, the southern Antarctic Peninsula and eastern Ellsworth Land had drifted to a position similar to its present latitude (Kellogg and Reynolds, 1978)

The local Miocene(?) basanite in the southwestern Merrick Mountains is perhaps a late product of extensional tectonics. Some of the high-angle faults mapped in the field area might be normal faults related to such a new tectonic regime, which took over from the subduction regime of the Andean belt some time after Middle Cretaceous.

*Acknowledgements* Conversations with D.L. Schmidt, P.L. Williams, and W. Hamilton were invaluable on helping us develop many of the conclusions presented here. We are grateful to M.D. Turner for logistical assistance during the field work. A.M. Kaplan assisted in petrographic studies of the igneous rocks. Technical reviews by W. Hamilton and E.H. Baltz greatly improved the manuscript. We thank R.J. Pankhurst for a reprint of his paper in press and for his editorial comments. Mapping and data reduction were financed by National Science Foundation Grants DPP76-12557, DPP78-24214, and DPP80-07388. Logistic support was provided by the U.S. Antarctic Research Program of the National Science Foundation, and by the U.S. Navy Operation Deep Freeze.

## REFERENCES


ADIE, R.J., 1955: The petrology of Graham Land—II. The Andean granite-gabbro intrusive suite. *Scient. Rep. Falkld. Isl. Depend. Surv., 12.*

ADIE, R.J., 1964: The geochemistry of Graham Land; *in* Adie, R.J. (ed.) *Antarctic Geology.* North Holland, Amsterdam, 541-7.

BEHRENDT, J.C., 1964a: Distribution of narrow-width magnetic anomalies in Antarctica. *Science, 144,* 993-5.

BEHRENDT, J.C., 1964b: Antarctic Peninsula geophysical results relating to glaciological and geological studies. *Univ. Wisc. Geophys. Polar Research Cent., Res. Rep. Ser., 64-1.*

BENTLEY, C.R. and CLOUGH, J.W., 1972: Antarctic subglacial structures from seismic refraction measurements; *in* Adie, R.J., (ed.) *Antarctic Geology and Geophysics.* Universitetsforlaget, Oslo, 683-91.

BRITISH ANTARCTIC SURVEY, 1982: British Antarctic Territory geologic map, northern Palmer Land. *British Antarctic Surv., Series BAS 500G, Sheet 5.*

CROSS, T.A. and PILGER, R.H. Jr., 1982: Controls of subduction geometry, location of magmatic arcs, and tectonics of arc and back-arc regions. *Geol. Soc. Am., Bull., 93,* 545-62.

DALZIEL, I.W.D., 1981: Back-arc extension in the southern Andes—A review and critical appraisal. *R. Soc. Lond., Philos. Trans., Ser. A, 300,* 319-35.

DALZIEL, I.W.D., 1982: The early (pre-Middle Jurassic) history of the Scotia arc region—A review and progress report (Review Paper); *in* Craddock, C. (ed.) *Antarctic Geoscience.* Univ. Wisconsin Press, Madison, 111-26.

DALZIEL, I.W.D., (this volume): The evolution of the Scotia Arc; a Review.

DALZIEL, I.W.D. and ELLIOT, D.H., 1973: The Scotia arc and Antarctic margin; *in* Nairn, A.E.M. and Stehli, F.G. (eds.) *The Ocean Basins and Margins, Vol. The South Atlantic.* Plenum Press, New York, 171-246.

DALZIEL, I.W.D. and ELLIOT, D.H., 1982: West Antarctica—Problem child of Gondwanaland. *Tectonics, 1,* 3-19.

DALZIEL, I.W.D., ELLIOT, D.H., JONES, D.L., THOMSON, J.W., THOMSON, M.R.A., WELLS, N.A. and ZINSMEISTER, W.J., 1981: The geological significance of some Triassic microfossils from the South Orkney Islands, Scotia Ridge. *Geol. Mag., 118,* 15-25.

EDWARDS, C.W., 1982: New Paleontologic evidence of Triassic sedimentation in West Antarctica; *in* Craddock, C. (ed.) *Antarctic Geoscience.* Univ. Wisconsin Press, Madison, 325-30.

ELLIOT, D.H., (this volume): The Mid-Cenozoic active plate margin of the Antarctic Peninsula.

FARQUARSON, G.W. (this volume): Evolution of Late Mesozoic sedimentary basins in northern Antarctic Peninsula.

FARRAR and ROWLEY, P.D., 1980: Potassium-argon ages of Upper Cretaceous plutonic rocks of Orville Coast and eastern Ellsworth land. *U.S. Antarct. J., 15,* 26-8.

HALPERN, 1971: Evidence for Gondwanaland from a review of West Antarctic radiometric ages; *in* Quam, L.O. and Porter, H.D. (eds.) *Research in the Antarctic,* Am. Assoc. Adv. Sci., Washington D.C., 717-30.

HAMILTON, W., 1967: Tectonics of Antarctica. *Tectonophysics, 4,* 555-67.

HAMILTON, W., 1972: The Hallett volcanic province, Antarctica. *U.S. Geol. Surv., Prof. Pap., 456-C.*

HAMILTON, W., 1981: Crustal evolution by arc magmatism. *R. Soc. Lond., Philos. Trans., Ser. A, 301,* 279-91.

HENDERSON, J.B., CALDWELL, W.G.E. and HARRISON, J.E., 1980: North American Commmission on Stratigraphic Nomenclature, Report 8—Amendment of Code concerning terminology for igneous and high grade metamorphic rocks. *Geol. Soc. Am., Bull., 91,* 374-6.

HORNE, R.R. and THOMSON, M.R.A., 1972: Airborne and detrital volcanic material in the Lower Cretaceous sediments of southeastern Alexander Island. *Bull. Br. antarct. Surv., 29,* 103-11.

KAMENEV, E.N. and ORLENKO, E.M., 1982: Metamorphism of sedimentary formations on the Lassiter Coast; *in* Craddock, C. (ed.) *Antarctic Geoscience.* Univ. Wisconsin Press, Madison, 357-61.

KELLOGG, K.S., 1979: Structural geology of Orville Coast and eastern Ellsworth Land. *U.S. Antarct. J., 14,* 19-21.

KELLOGG, K.S., 1980: Paleomagnetic evidence for oroclinal bending of the southern Antarctic Peninsula. *Geol. Soc. Am., Bull., I, 91,* 414-20.

KELLOGG, K.S. and REYNOLDS R.L., 1978: Paleomagnetic results from the Lassiter Coast, Antarctica, and a test for oroclinal bending of the Antarctic Peninsula. *J. Geophys. Res., 83,* 2293-9.

LAUDON, T.S., 1972: Stratigraphy of eastern Ellsworth Land; *in* Adie, R.J. (ed.) *Antarctic Geology and Geophysics.* Universitetsforlaget, Oslo, 215-23.

LAUDON, T.S., LACKEY, L.L., QUILTY, P.G. and OTWAY, P.M., 1969: Geology of eastern Ellsworth Land (Sheet 3, eastern Ellsworth Land); *in* Bushnell, V.C. and Craddock, C. (eds.) Geologic Maps of Antarctica. *Antarct. Map Folio Ser.,* Folio 12, Pl.III.

LAUDON, T.S., THOMSON, M.R.A., WILLIAMS, P.L. and MILLIKEN, K.L. (this volume): The Jurassic Latady Formation, southwestern Antarctic Peninsula.

LeMASURIER, W.E. and REX, D.C., 1982: Volcanic record of Cenozoic glacial history in Marie Byrd Land and western Ellsworth Land—Revised chronology and evaluation of tectonic factors; *in* Craddock, C. (ed.) *Antarctic Geoscience.* Univ. Wisconsin Press, Madison, 725-34.

MEHNERT, H.H., ROWLEY, P.D. and SCHMIDT, D.L., 1975: K-Ar ages of plutonic rocks in the Lassiter Coast area, Antarctica. *U.S. Geol. Surv., J. Res, 3,* 233-6.

PANKHURST, R.J., 1980: Appendix, radiometric dating; *in* Singleton D.G., The geology of the central Black coast, Palmer Land. *Scient. Rep. Br. antarct. Surv., 102,* 49-50.

PANKHURST, R.J., 1982: Rb-Sr geochronology of Graham Land, Antarctica. *Geol. Soc. Lond., J, 139,* 701-12.

PANKHURST, R.J., (this volume): Rb-Sr age determination on possible basement rocks of the Antarctic Peninsula.

QUILTY, P.G., 1982: Bajocian bivalves from Ellsworth Land, Antarctica. *N.Z. J. Geol. Geophys.,* in press.

REX,D.C., 1976: Geochronology in relation to the stratigraphy of the Antarctic Peninsula. *Bull. Br. antarct. Surv., 43,* 49-58.

ROWLEY, P.D., 1978: Geologic studies in Orville Coast and eastern Ellsworth Land, Antarctic Peninsula. *U.S. Antarct. J., 13,* 7-9.

ROWLEY, P.D. and WILLIAMS, P.L., 1982: Geology of the northern Lassiter Coast and southern Black Coast, Antarctic Peninsula; *in* Craddock, C. (ed.) *Antarctic Geoscience.* Univ. Wisconsin Press, Madison, 339-48.

ROWLEY, P.D., SCHMIDT, D.L. and WILLIAMS, P.L., 1982: Mount Poster Formation, southern Antarctic Peninsula and eastern Ellsworth Land. *U.S. Antarct. J., 17,* in press.

ROWLEY, P.D., KELLOGG, K.S., VENNUM, W.R., WAITT, R.B. Jr., and BOYER, S.J., (in press): Geology of the southern Black Coast, Antarctic Peninsula. *U.S. Geol. Surv., Prof. Pap.*

ROWLEY, P.D., FARRAR, E., CARRARA, P.E., VENNUM, W.R. and KELLOGG, K.S., in press: Porphyry-type copper deposits and K-Ar ages of plutonic rocks of the Orville Coast and eastern Ellsworth Land, Antarctica. *U.S. Geol. Surv., Prof. Pap.*

SAUNDERS, A.D., WEAVER, S.D. and TARNEY, J., 1982: The pattern of Antarctic Peninsula plutonism; *in* Craddock, C. (ed.) *Antarctic Geoscience*. Univ. Wisconsin Press, Madison, 305-14.

SCHOPF, J.M., 1969: Ellsworth Mountains—Position in West Antarctica due to sea-floor spreading. *Science, 164*, 63-6.

SCHARNBERG, C.K. and SCHARON, L., 1982: Paleomagnetism of rocks from Graham Land and western Ellsworth Land, Antarctica; *in* Craddock, C. (ed.) *Antarctic Geoscience*. Univ. Wisconsin Press, Madison, 371-5.

SCLATER, J.G. HELLINGER, S. and TAPSCOTT, 1977: The paleobathymetry of the Atlantic Ocean from the Jurassic to the present. *J. Geol., 85*, 509-52.

SINGLETON, D.G., 1980a: The geology of the central Black Coast, Palmer Land. *Scient. Rep. Br. antarct. Surv., 102*.

SINGLETON, D.G., 1980b: Geology of Seaward Mountains, western Palmer Land. *Bull. Br. antarct. Surv., 49*, 81-9.

SMELLIE, J.L., 1981: A complete arc-trench system recognised in Gondwana sequences of the Antarctic Peninsula region. *Geol. Mag., 118*, 139-59.

SUÁREZ, M., 1976: Plate-tectonic model for southern Antarctic Peninsula and its relation to southern Andes. *Geology, 4*, 211-4.

TANNER, P.W.G., STOREY, B.C. and MACDONALD, D.I.M., 1981: Geology of an Upper Jurassic-Lower Cretaceous island-arc assemblage in Hauge Reef, the Pickersgill Islands and adjoining areas of South Georgia. *Bull. Br. antarct. Surv., 53*, 77-117.

THOMSON, M.R.A., 1982: Mesozoic paleogeography of West Antarctica; *in* Craddock, C. (ed.) *Antarctic Geoscience*. Univ. Wisconsin Press, Madison, 331-7.

THOMSON, M.R.A. (this volume): Jurassic ammonite biostratigraphy of the Orville Coast.

THOMSON, M.R.A., LAUDON, T.S. and BOYLES, J.M., 1978: Stratigraphical studies in Orville Coast and eastern Ellsworth Land. *U.S. Antarct. J., 13*, 9-10.

THOMSON. M.R.A. and PANKHURST, R.J. (this volume) Age of Post-Gondwanide calc-alkaline volcanism in the Antarctic Peninsula region.

THOMSON, M.R.A. PANKHURST, R.J. and CLARKSON, P.J. (this volume): The Antarctic Peninsula—a Late Mesozoic-Cenozoic Arc (Review).

VENNUM, W.R., and LAUDON, T.S. (in press): Igneous petrology of the Merrick Mountains, eastern Ellsworth Land, Antarctica. *U.S. Geol. Surv., Prof. Pap.*

WILLIAMS, P.L., SCHMIDT, D.L., PLUMMER, C.C. and BROWN, L.E., 1972: Geology of the Lassiter Coast area, Antarctic Peninsula—Preliminary Report; *in* Adie, R.J. (ed.) *Antarctic Geology and Geophysics*. Universitetsforlaget, Oslo, 143-8.

# GEOLOGY OF PAGANO NUNATAK AND THE HART HILLS

G.F. Webers, *Department of Geology, Macalester College, 1600 Grand Avenue, St. Paul, Minnesota, 55105, U.S.A.*

C.Craddock, *Department of Geology and Geophysics, University of Wisconsin, Madison, Wisconsin, 53706, U.S.A.*

M.A. Rogers, *Exxon Company, Central Plaza, 600 Carondelet, Suite 800, New Orleans, Louisiana, 70130, U.S.A.*

J.J. Anderson, *Department of Geology, Kent State University, Kent, Ohio, 44240, U.S.A.*

*Abstract* The Pagano Nunatak-Hart Hills area (about 83°42'S, 88°30'W) lies near the southern end of a mainly ice-covered rock ridge which extends southward from the Ellsworth Mountains to the Thiel Mountains. This ridge is a key region in understanding the tectonic relationships between East and West Antarctica. Pagano Nunatak is a prominent peak rising about 230 m above the surrounding ice. Almost all bedrock consists of a grey, massive, medium to coarse-grained granite, here called the Pagano Granite. The granite is undeformed, unmetamorphosed, and cut by small aplite dykes. A $^{40}K$-$^{40}Ar$ age determination on biotite from the Pagano Granite yielded an apparent age of 175 ± 4 Ma. The Hart Hills are a series of low, rounded hills aligned in an east-west direction. The exposed bedrock includes almost equal areas of metasedimentary rocks and cross-cutting quartz gabbro intrusions. The metasedimentary unit, here called the Hart Hills Formation, is exposed in a 396 m stratigraphic section; it consists mainly of quartz-mica phyllites and subarkosic sandstones. No fossils were recovered. The quartz gabbro, here called the Johnson Peak Gabbro, is metamorphosed and locally sheared. The metamorphosed quartz gabbro and the metasedimentary rocks in the Hart Hills have not been dated, but they are considered to be older than the nearby undeformed Pagano Granite. Granites exposed in the Whitmore Mountains and in the Nash and Pirrit Hills are similar in age to the Pagano Granite. On this basis, Pagano Nunatak may represent the southernmost exposure of granitic rocks within the Ellsworth orogen. The Hart Hills may be part of this orogen, or of some older terrain.

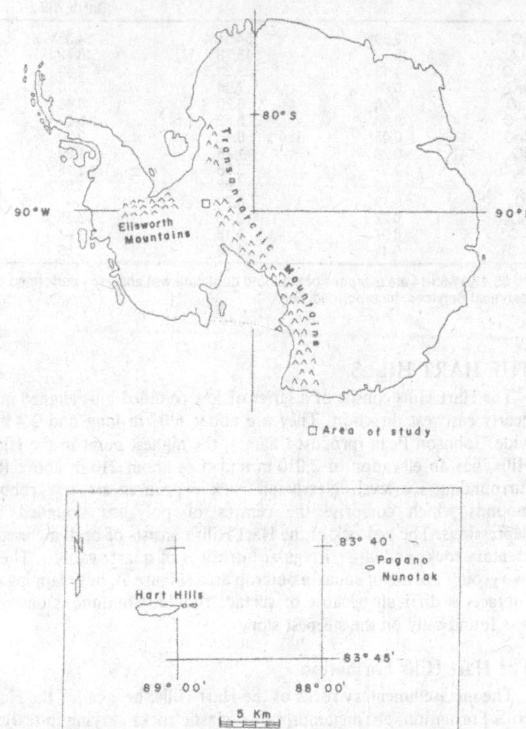


**Figure 1. Map showing location of Pagano Nunatak and Hart Hills.**

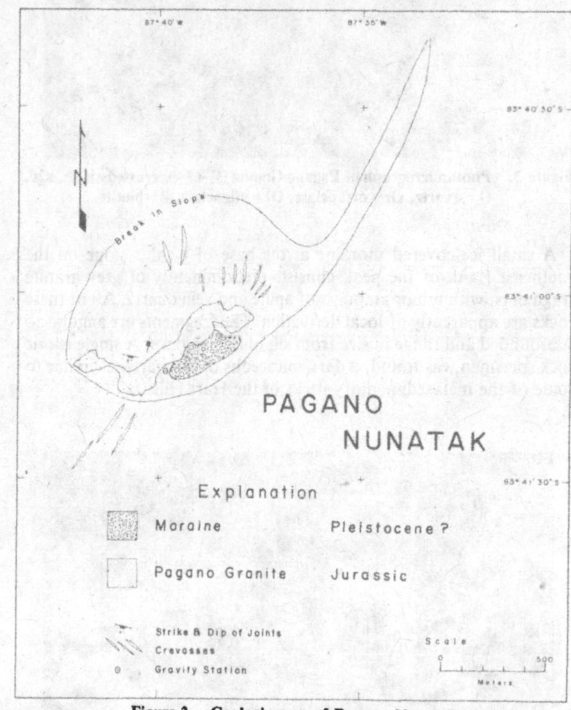


**Figure 2. Geologic map of Pagano Nunatak.**

During the 1964-1965 austral summer a University of Minnesota party conducted a reconnaissance study of the Pagano Nuntak and Hart Hills (Figure 1) as part of a program of geologic exploration in West Antarctica.

## PAGANO NUNATAK

Pagano Nunatak (83°41'S., 87°40'W.) is a single prominent peak that rises about 230 m above the surrounding ice (Figure 2). It extends about 0.8 km in an ENE—WSW direction and is located about 13.6 km ENE of the Hart Hills. The next closest area of rock exposure is the Stewart Hills, located some 35 km SSE. The Thiel Mountains, the nearest part of the Transantarctic Mountains, lie about 150 km to the south.

The exposed bedrock (Figure 3) consists almost entirely of a light grey, massive, medium to coarse-grained granite, here called the Pagano Granite. The Pagano Granite appears fresh and unaltered, both in hand specimen and thin section. It shows no definite evidence of metamorphism or post-emplacement distortion or shearing. Rock texture is seriate to porphyritic, with twinned orthoclase phenocrysts as long as 3 cm commonly with a preferred orientation—comprising as much as 50% of the rock. The rock weathers to rounded pink blocks; disintegration seems to be the dominant and probably sole type of weathering. Two prominent joint sets in the granite are widely spaced; their mean orientations are 1) strike 273°, dip 85°N, and 2) strike 010°, vertical dip. Several aplite dykes (leucogranite) (Figure 4) with an average width of 30 cm were observed in the Pagano Granite. Point-count estimates of mineral percentages in rocks from Pagano Nunatak appear in Table 1, and a chemical analysis of a granite specimen in Table 2.

A $^{40}K$-$^{40}Ar$ age determination by Geochron Laboratories, Inc. on biotite from a specimen of the Pagano Granite yielded an apparent age of 175 ± 4 Ma (Early Jurassic), using constants of $\lambda_\beta = 4.72 \times 10^{-10}$/yr, $\lambda_e = 0.585 \times 10^{-10}$/yr, and $^{40}K/K = 1.22 \times 10^{-4}$g/g.

**Figure 3. Photomicrograph of Pagano Granite (W-65-4), crossed nicols, x20. Q = quartz, Or = orthoclase, O1 = oligoclase, B = biotite.**

A small ice-covered moraine at the base of a talus slope on the southeast flank of the peak consists predominantly of grey granite fragments, with minor amounts of aplite and vein quartz. All of these rocks are apparently of local derivation. The fragments are angular to subrounded and range in size from silt to large blocks. A single exotic rock specimen was found, a dark micaceous orthoquartzite similar to some of the metasedimentary rocks of the Hart Hills.

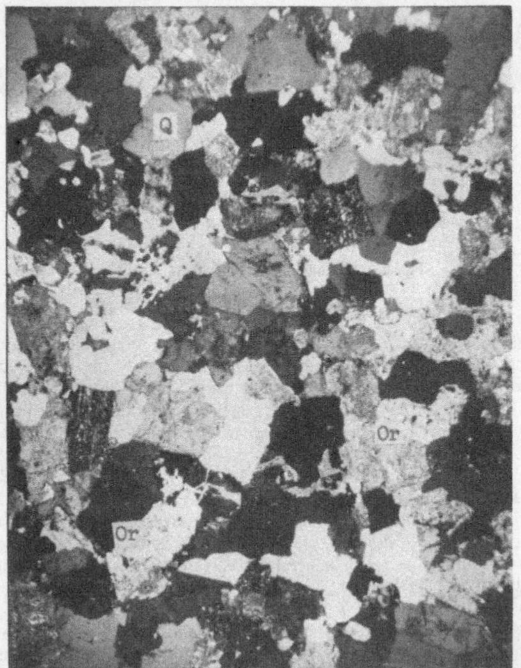


**Figure 4. Photomicrograph of Pagano Nunatak aplite (W-65-1), crossed nicols, x20. Q = quartz, Or = orthoclase, O1 = oligoclase.**

**TABLE 1: Point-count estimates of mineral abundances (in percent) in Pagano Nunatak rock specimens**

| Rock Unit | Aplite Dyke | Pagano Granite | | | | Mean, Pagano Granite |
|---|---|---|---|---|---|---|
| Specimen No. | W-65-1 | W-65-2 | W-65-3 | W-65-4 | W-65-5 | |
| Mineral | | | | | | |
| Quartz | 39% | 32% | 24% | 29% | 30% | 29% |
| Orthoclase | 46 | 37 | 50 | 45 | 44 | 44 |
| Oligoclase | 12 | 21 | 20 | 15 | 20 | 19 |
| Biotite | P | 7 | 3 | 5 | 1 | 4 |
| Chlorite | P | P | 2 | 3 | 2 | 2 |
| Muscovite | 3 | 3 | 1 | 2 | 3 | 2 |
| Hornblende | P | — | — | P | — | P |
| Apatite | P | P | P | P | P | P |
| Magnetite | P | P | P | P | P | P |
| Pyrite | P | — | — | P | — | P |
| Hematite | P | — | — | 1 | — | P |
| Calcite | P | P | P | P | P | P |
| Sericite | P | P | — | — | — | P |

P = present but less than 1%
J. J. Anderson, petrographer

**TABLE 2: Chemical analysis of igneous rocks (in wt %)**

| Oxide | Pagano Granite (W-65-4)* | Johnson Peak Gabbro (W-65-14)* | Average Plutonic Quartz Gabbro (Barth, 1962) |
|---|---|---|---|
| $SiO_2$ | 72.59% | 54.54% | 54.35% |
| $Ai_2O_3$ | 15.72 | 15.78 | 16.72 |
| $Fe_2O_3$ | 1.13 | 3.23 | 2.49 |
| FeO | 0.88 | 7.94 | 7.15 |
| CaO | 1.46 | 6.20 | 6.68 |
| MgO | 0.30 | 3.42 | 0.20 |
| MnO | 0.05 | 0.18 | 0.20 |
| $TiO_2$ | 0.28 | 1.28 | 1.29 |
| $Na_2O$ | 2.51 | 2.08 | 3.15 |
| $K_2O$ | 3.60 | 1.98 | 1.58 |
| $CO_2$ | 0.19 | 0.72 | — |
| $P_2O_5$ | 0.26 | 0.18 | 0.35 |
| $H_2O^-$ | 0.07 | 0.18 | — |
| $H_2O^2$ | 0.41 | 1.66 | 1.85 |

*W-65-4 & W-65-14 are averages of standard duplicate wet analyses performed by Technical Services, Incorporated, Toronto

## THE HART HILLS

The Hart Hills consist of a series of low rounded hills aligned in a nearly eastwest direction. They are about 6.0 km long and 2.4 km wide. Johnson Peak (proposed name), the highest point in the Hart Hills, has an elevation of 2,010 m and rises about 210 m above the surrounding ice level. Nearly all rock exposures are low rubbly mounds which comprise the centres of polygons bounded by depressions. The bedrock of the Hart Hills consists of both metasedimentary rocks and large, irregular intrusives of quartz gabbro. These two groups are about equal in outcrop area (Figure 5), but mapping of contacts is difficult because of surface rubble. Continuous outcrop was found only on the steepest slopes.

### The Hart Hills Formation

The metasedimentary rocks of the Hart Hills, here called the Hart Hills Formation, are metamorphosed clastic rocks varying in texture from lutites to arenites. The best stratigraphic section, about 396 m thick, is exposed on the northwestern flank of the Hart Hills (Figure 5), and it is designated the type section (Figure 6). The fine-grained rocks are foliated, and abundant chlorite indicates a metamorphic rank in the greenschist facies. No fossils have been found in these rocks. Point-count estimates of mineralogic composition of selected rock specimens are given in Table 3.

The Hart Hills Formation can be divided into three members. The uppermost member is composed mainly of thin beds of weathered grey-to-black quartz-mica phyllite. In the upper part of this member the beds are highly (as much as 25% by volume) carbonaceous, and a few thin beds of pyritic, carbonaceous limestone are present. Petrographic study of rock specimens from this member reveals two separate foliations that may record two tectonic events. Most specimens show one conspicuous foliation, but its relation to bedding is unclear except in one specimen where it appears to crosscut bedding at a small angle. One other specimen shows a later foliation crosscutting the main foliation at a large angle. This member has a stratigraphic thickness of about 88 m, and it probably also crops out in small exposures throughout the Hart Hills.

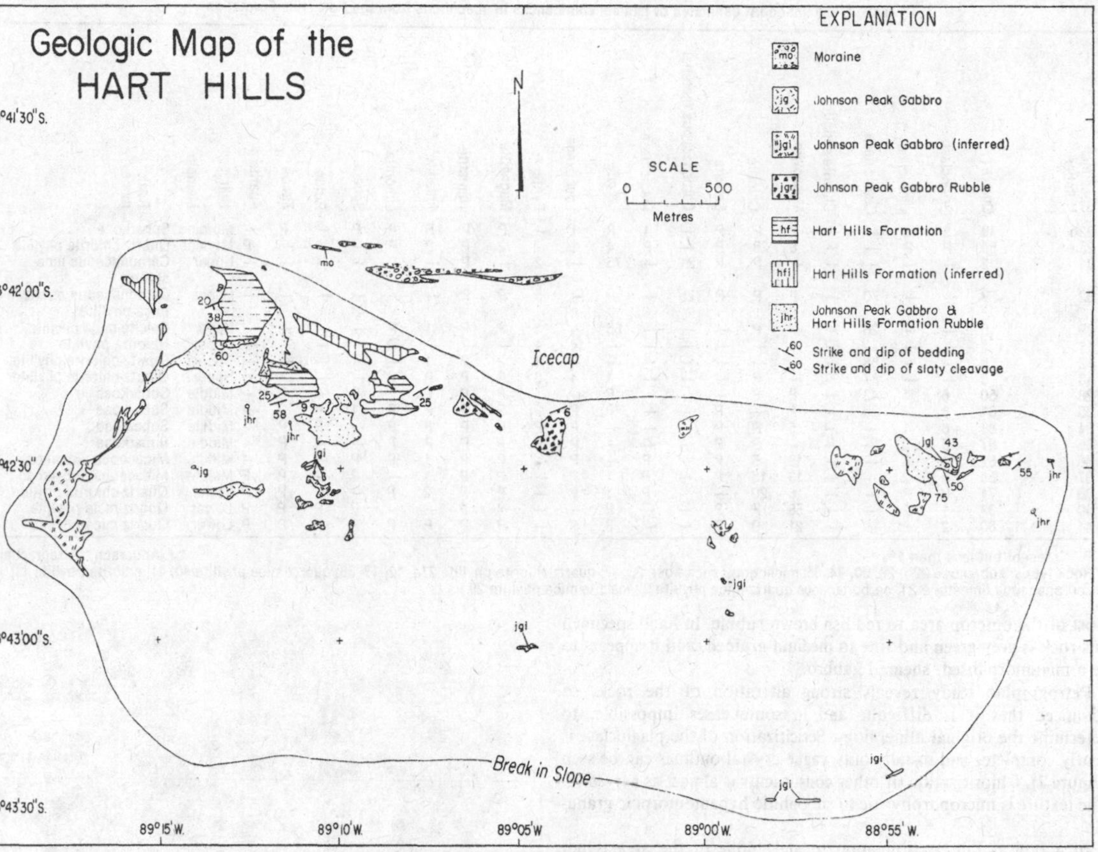

**Figure 5. Geologic map of the Hart Hills.**

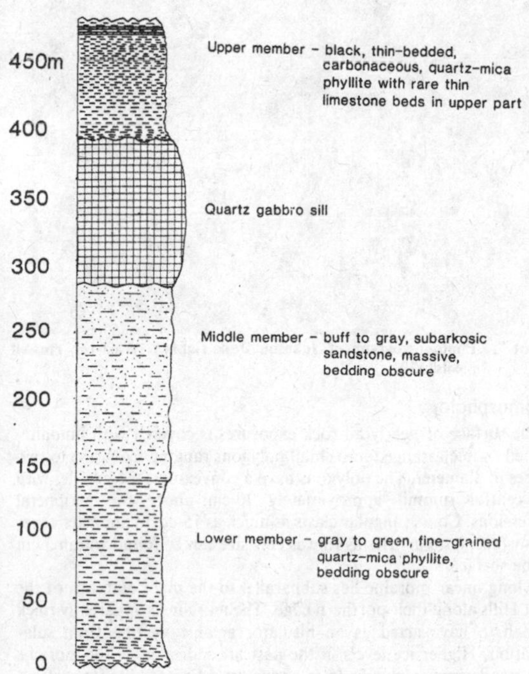


**Figure 6. Type section of the Hart Hills Formation (pre-Jurassic?), northwestern Hart Hills.**

The middle member has a stratigraphic thickness of about 146 m; in the type section it is separated from the upper member by a quartz gabbro sill 107 m thick. These beds consist of metamorphosed buff, siliceous, subarkosic sandstones. Sorting is generally poor, and bedding is massive and obscure. These sandstones contain about 10% potash feldspar. The basal contact of this member is gradational into the phyllites of the lower member. The lower member consists of at least 140 m of greenish-grey, fine-grained quartz-mica phyllite. These rocks are poorly sorted, and bedding is obscure. This member differs from the overlying middle member primarily in its finer grain size, lower feldspar content, and foliated nature. The base of this unit in not exposed in the Hart Hills.

The age of the Hart Hills Formation is unknown. The highly carbonaceous sedimentary rocks of the upper member suggest a Phanerozoic age. The potash feldspar in the middle member of the formation may provide a clue to both the age and the source of the sediments. Granites exposed in the Whitmore Mountains to the west (Webers et al., 1982) and in the Pirrit and Nash Hills to the north (Craddock, 1972a) are similar in age to the Early Jurassic Pagano Granite, but all these granites are probably younger than the Hart Hills Formation. However, granitic plutons exposed in the nearby Transantarctic Mountains are mainly Proterozoic or Cambro-Ordovician in age, for example, quartz monzonite and biotite granitic rocks in the Thiel Mountains (Schmidt and Ford, 1969), the nearest mountain group of the Transantarctic Mountains. The obscure bedding, poor sorting, and high potash feldspar content of the middle member of the Hart Hills Formation suggest derivation from a nearby granitic terrain. Possibly the sedimentary rocks of the Hart Hills resulted from the erosion of a Lower Palaeozoic granitic and metamorphic terrain in the Transantarctic Mountains.

## Johnson Peak Gabbro

The metasedimentary rocks of the Hart Hills are intruded by dykes, sills and irregular bodies of quartz gabbro, here called the Johnson Peak Gabbro. Mechanical and chemical weathering have reduced

**TABLE 3: Point-count estimates of mineral abundances in specimens from the Hart Hills Formation**

| Specimen Number (W-65-) | Quartz | Orthoclase | Microcline | Quartz-mica | Chlorite-sericite | Muscovite | Chlorite | Sericite | Carbonaceous material | Plagioclase | Calcite | Leucoxene | Pyrite | Magnetite | Hematite | Ilmenite | Limonite | Sphene | Rutile | Zircon | Apatite | Unit or Member | Rock name |
|---|---|---|---|---|---|---|---|---|---|---|---|---|---|---|---|---|---|---|---|---|---|---|---|
| 22b | 49 | 3 | 2 | 44 | — | P | P | P | — | 1 | P | 1 | — | P | P | P | P | P | — | P | — | Moraine | Subarkose |
| 23a | 59 | P | P | — | — | 6 | 28 | P | — | P | 4 | 1 | — | 2 | P | P | P | — | — | — | P | Moraine | Quartz-chlorite phyllite |
| 31 | 2 | — | — | — | — | — | P | P | 21 | — | 75 | — | 2 | — | P | — | P | — | — | — | — | Upper | Carbonaceous limestone |
| 32 | P | — | — | 70 | — | P | P | P | 25 | — | — | — | 5 | P | P | — | — | — | — | — | — | Upper | Carbonaceous quartz mica phyllite |
| 29 | 6 | — | — | 26 | — | — | P | — | — | — | 65 | — | 1 | P | P | 1 | P | — | — | — | — | Upper | Calcite-mica phyllite |
| 11 | 3 | — | — | — | 82 | — | — | — | — | — | — | 14 | 1 | — | — | — | — | — | — | — | — | Upper? | Chlorite phyllite |
| 12 | 8 | — | — | 88 | — | — | — | — | — | — | — | P | 1 | 2 | P | 1 | — | — | — | — | — | Upper? | Quartz-chlorite phyllite |
| 13 | 5 | — | — | 87 | — | — | P | — | — | — | 1 | 4 | 3 | P | P | P | P | — | — | — | — | Upper? | Quartz-chlorite phyllite |
| 28 | 50 | 6 | 1 | 43 | — | P | P | — | — | — | P | — | — | P | P | P | P | P | — | P | — | Middle | Subarkose |
| 30 | 61 | 8 | — | 30 | — | P | — | P | — | — | P | P | P | P | P | P | P | 1 | P | P | — | Middle | Subarkose |
| 34 | 88 | 6 | 1 | — | — | 5 | P | P | — | — | — | P | P | P | P | P | P | — | P | P | — | Middle | Subarkose |
| 35 | 87 | 8 | 2 | — | — | — | P | P | — | — | — | P | P | P | P | P | P | — | — | P | — | Middle | Subarkose |
| 36 | 62 | 11 | 3 | — | — | 18 | 5 | P | — | P | — | P | — | P | P | 1 | P | P | — | P | — | Middle | Micaceous subarkose |
| 37 | 56 | 9 | 2 | — | — | 13 | 15 | P | — | P | 1 | P | — | P | P | 1 | — | P | — | P | P | Middle | Micaceous subarkose |
| 38 | 71 | P | P | — | — | 6 | 20 | — | — | P | 1 | P | — | P | P | 2 | P | — | — | P | P | Lower | Quartz-chlorite phyllite |
| 40 | 37 | 1 | 2 | — | — | 56 | P | P | — | — | P | — | — | 4 | P | — | — | P | P | P | P | Lower | Quartz-mica phyllite |
| 41 | 60 | 2 | P | — | — | 21 | 9 | P | — | — | 8 | P | — | P | P | P | P | P | P | P | P | Lower | Quartz-mica phyllite |

P = present but less than 1% J.J. Anderson, petrographer

Rock types: subarkose 22b, 28, 30, 34, 35; micaceous subarkose 36, 37; quartz-chlorite phyllite 23a, 12, 13, 38; quartz-mica phyllite 40, 41; chlorite phyllite 11; carbonaceous limestone 31; carbonaceous quartz-mica phyllite 32, calcite-mica phyllite 29.

most of the outcrop area to reddish brown rubble. In hand specimen this rock is grey-green and fine to medium-grained, and it appears to be a metamorphosed, sheared gabbro.

Petrographic study reveals strong alteration of the rock, so advanced that it is difficult, and in some cases impossible, to determine the original mineralogy. Sericitization of the plagioclase is nearly complete, and usually only vague crystal outlines can be seen (Figure 7). Chloritization of other constituents is almost as pervasive. The texture is microporphyritic to subophitic hypautomorphic-granular.

In a typical thin section approximately 40% of the area is not determinable as specific minerals, but probably contains a mixture of sericite, chlorite, feldspar, and quartz. The average percentages within the remaining area are as follows: quartz 5%, oligoclase 20%, augite 29%, chlorite 4%, and muscovite 2%. Point counts of thin sections of the two least altered rock specimens yield the following average percentages: quartz 4%, quartz-feldspar intergrowths 2%, orthoclase (including sericitic alteration) 19%, (?)oligoclase (including sericitic alteration) 27%, chlorite 18%, augite 25%, amphibole 2%, ilmenite 2%, magnetite 1%.

Chemical analysis of the Johnson Peak Gabbro (Table 2) shows an intermediate silica content, but a high iron, magnesium, and calcium content. It compares closely with the chemical composition of the average plutonic quartz gabbro (Barth, 1962). The rock is difficult to name because of the intense alteration. However, it is classified as a quartz gabbro (locally quartz dolerite) on the basis of its texture, mineralogy, and chemical composition.

## Structural Geology

Slaty cleavage ($S_1$) is found throughout the metasedimentary sequence of the Hart Hills, and it is especially well developed in the black quartz-mica phyllite of the upper member. The strike of the cleavage is nearly parallel to the strike of the bedding, but the dips of the two are quite different (Figure 5). Another foliation, a slip cleavage ($S_2$) at a high angle to the slaty cleavage, has been observed in some specimens of the Hart Hills Formation.

Both slaty cleavage and bedding show considerable variation in strike in the small outcrops along the northeastern margin of the Hart Hills. These small outcrops may be rotated blocks of sedimentary rock engulfed by the quartz gabbro intrusions. Along the northwestern flank of the the Hart Hills the outcrops are larger and exposure more nearly continuous. In this area the bedding is more consistent, with an average strike of 078° and a southerly dip of about 45°. Slaty cleavage has an average strike of 074° with a northerly dip of about 45°. Joints are abundant in the quartz gabbro intrusions, but no consistent pattern was found. These closely spaced joints, together with mechanical weathering, produce the rubble-covered gabbro exposures characteristic of the Hart Hills.

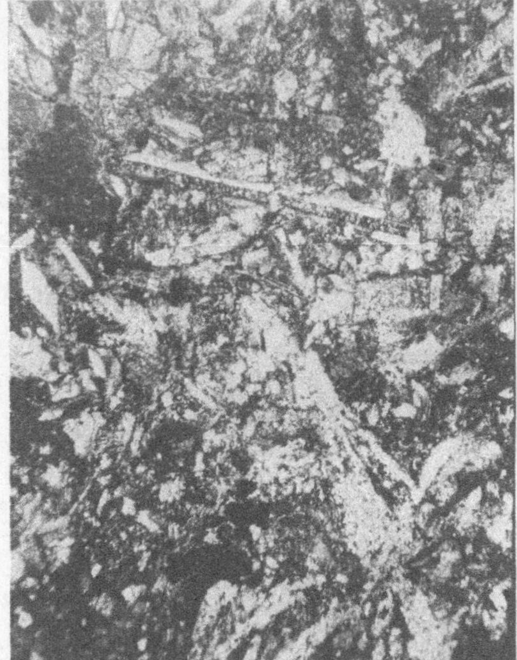

**Figure 7. Photomicrograph of Johnson Peak Gabbro (W-65-14), crossed nicols, x20.**

## Geomorphology

The surface of nearly all rock exposures is covered with limonite-stained rubble arranged into small polygons ranging from one to two metres in diameter. The polygons have a convex-upward profile, with the central summit approximately 30 cm above the peripheral depressions. Coarse angular clasts as much as 15 cm in diameter cover the surface, but particles as fine as clay size can be found within 5 cm of the surface.

A long linear moraine lies subparallel to the main outcrops of the Hart Hills along their northern edge. The moraine is apparently thick enough to have acted as an insulator against melting from solar radiation. Higher ice levels in the past are indicated as the moraine now stands approximately 15 m above the adjacent blue ice surface. Clasts present in the moraine were probably locally derived; they consist of a mixture of phyllite, metamorphosed subarkose, and quartz gabbro fragments similar to the rocks exposed in the Hart Hills. The metasedimentary rocks, however, tend to be much less weathered than in the bedrock exposures.

### Tectonic History

The metasedimentary rocks of the Hart Hills may have been derived from the erosion of granitic terrains in the Transantarctic Mountains, probably after the Cambro-Ordovician Ross Orogeny. The poor sorting, massive appearance, obscure bedding and potash feldspar content of the lower and middle members of the Hart Hills Formation may indicate deposition under terrestrial conditions. The black, pyritic, and in some cases highly carbonaceous sediments of the upper member, however, suggest a subaqueous reducing environment.

These strata were intruded by irregular bodies of quartz gabbro. The time of emplacement of the quartz gabbro is unknown, but these rocks may be related to similar gabbros exposed in the Ellsworth Mountains (Craddock, 1969). The nearly complete sericitisation of the plagioclase and the chloritisation of other constituents in the quartz gabbro indicate that the entire rock complex was metamorphosed by a later tectonic event which probably also deformed the metasedimentary rocks and produced the slaty cleavage. Quartz veins as much as 70 cm wide, crosscutting both the metasedimentary rocks and the quartz gabbro, were probably emplaced during the last stages of this event. These quartz veins are likely related to the Early Jurassic(?) granite of Pagano Nunatak.

## REGIONAL TECTONIC PATTERN

The Pagano Nunatak-Hart Hills area lies near the southern end of a mainly ice-buried ridge which extends southward from the Ellsworth to the Thiel Mountains, the latter a part of the Transantarctic Mountains chain. Several nunataks and groups of hills stand above the ice surface along this ridge. These outcrops include the Stewart Hills, the Hart Hills, Pagano Nunatak, the Whitmore Mountains, the Nash Hills, the Martin Hills, and the Pirrit Hills. This ice-buried ridge was recognised by Bentley and Ostenso (1961) and discussed by Craddock (1966, 1971). The geologic composition of this ridge is important in establishing the regional tectonic pattern and the relationship between East and West Antarctica.

Radiometric dates on granitic rocks from Pagano Nunatak (this paper), the Whitmore Mountains to the west (Webers et al., 1982), and the Nash and Pirrit Hills to the north (Craddock, 1972a) all yield Late Triassic to Early Jurassic apparent ages. The Hart Hills Formation appears to be older than the Pagano Granite, but it is likely to be of Phanerozoic age. The deformation of the Hart Hills Formation and the emplacement of all these granitic rocks were probably parts of an Early Mesozoic tectonic event, the Ellsworth Orogeny of Craddock (1972b).

Within the Hart Hills, in the part where rock exposure is best, both the bedding and the slaty cleavage strike in an ENE direction; the deformation shows a southward vergence. The structual grain in the Hart Hills trends toward Pagano Nunatak, only some 14 km away, so it is likely that both areas are part of the same structural belt. The structural grain in the Hart Hills, however, is roughly parallel to the trend of the Transantarctic Mountains chain to the south.

It is probable that the Pagano Nunatak-Hart Hills area lies within the Early Mesozoic Ellsworth Orogen, and it may represent the southermost exposure of this terrain. However, it is possible that the deformation of the Hart Hills Formation may be pre-Mesozoic in age, and the Hart Hills could be part of an older province such as the Early Paleozoic Ross Orogen.

*Acknowledgements* This work was supported by a grant from the Office of Polar Programs, National Science Foundation, to the University of Minnesota, with C. Craddock the principal investigator. Thanks are also extended to R. Buchanan for help throughout the field session, and to the U.S. Navy for logistic support.

## REFERENCES

BARTH, T.F., 1962: *Theoretical Petrology.* John Wiley and Sons, New York, Second edition.

BENTLEY, C.R. and OSTENSO, N.A., 1961: Glacial and subglacial topography of West Antarctica. *J. Glaciol., 3,* 882-911.

CRADDOCK, C., 1966: The Ellsworth Mountains Fold Belt-a link between East and West Antarctica. *Geol. Soc Am., Spec. Pap., 87,* 37-8.

CRADDOCK, C., 1969: Geology of the Ellsworth Mountains; *in* Bushnell, V.C. and Craddock, C. (eds.) Geologic Maps of Antarctica. *Antarct. Map Folio Series, Folio 12,* Pl. IV.

CRADDOCK, C., 1971: The structural relation of East and West Antarctica. *Proc. XXII Int. Geol. Cong. (1964), 4,* 278-92.

CRADDOCK, C., 1972a: *Geologic Map of Antarctica, 1:5,000,000.* American Geographical Society, New York.

CRADDOCK, C., 1972b: Antarctic tectonics; *in* Adie, R.J. (ed.) *Antarctic Geology and Geophysics.* Universitetsforlaget, Oslo, 449-55.

JOHANNSEN, A., 1939: *A Descriptive Petrology of the Igneous Rocks.* Univ. Chicago Press, Chicago.

SCHMIDT, D.L. and FORD, A.B., 1969: Geology of the Pensacola and Thiel Mountains; *in* Bushnell, V.C. and Craddock, C. (eds.) Geologic Maps of Antarctica. *Antarct. Map Folio Series, Folio 12,* Pl. V.

WEBERS, G.F., CRADDOCK, C., ROGERS, M.A. and ANDERSON, J.J., 1982: Geology of the Whitmore Mountains; *in* Craddock, C. (ed.) *Antarctic Geoscience.* Univ. Wisconsin Press, Madison, 841-7.

# LOW-GRADE METAMORPHISM IN THE HERITAGE RANGE OF THE ELLSWORTH MOUNTAINS, WEST ANTARCTICA

R.L. Bauer[1], *Geology Department, Macalester College, 1600 Grand Avenue, St Paul, Minnesota, 55105, U.S.A.*

[1]*Present Address: Department of Geology, University of Missouri, Columbia, Missouri 65211, U.S.A.*

*Abstract* Mineral assemblages in mafic igneous and volcanoclastic rocks within the Middle to Upper Cambrian Heritage Group indicate these rocks have been recrystallised under pumpellyite-actinolite facies to greenschist facies conditions. Basaltic flows, volcanic breccias, and gabbroic dykes and sills from the Liberty Hills and Springer Peak Formations of the upper Heritage Group contain pumpellyite-actinolite facies assemblages, including: albite, sphene, white mica, chlorite, ± epidote, ± actinolite, ± quartz, ± calcite, and ± pumpellyite. Gabbros in the Drake Icefall Formation of the middle Heritage Group also contain pumpellyite + actinolite bearing assemblages, but clinozoisite replaces epidote in these rocks and pumpellyite has a substantially reduced Fe content. The observed decrease in the Fe content of pumpellyite in the Heritage Group indicates, by comparison with results of other authors, an increase in metamorphic grade with depth. Such an increase in grade is further indicated by the lack of pumpellyite and the presence of greenschist facies mineral assemblages in the lower Heritage Group. Correlation of pumpellyite-actinolite facies to greenschist facies conditions with depth in the Heritage Group is consistent with a major burial metamorphic event in the Ellsworth Mountains ranging from zeolite facies in the Polarstar Formation (Permian) to greenschist facies in the lower Heritage Group. Prehnite-pumpellyite facies assemblages, not reported from the units between the Heritage Group and the Polarstar Formation, may not have developed due to lack of rocks of appropriate composition or as a result of high $CO_2$ in carbonate rich portions of the section.

The Ellsworth Mountains, located between the polar plateau of West Antarctica and the Ronne Ice Shelf, are divided into a southern range, the Heritage, and a northern range, the Sentinel (Figure 1). The mountains comprise a series of Palaeozoic sedimentary rocks ranging in age from Middle Cambrian to Permian (Craddock et al., 1964; Webers and Sporli, this volume, cf. their Figures 2 and 3). Several formations within the lowest stratigraphic unit in the Ellsworth Mountains (the Heritage Group as defined by Webers and Sporli, this volume) contain tuffaceous rocks, intermediate to mafic volcanic flows, and gabbroic dykes and sills. The structure of the range is dominated by a major anticlinorium plunging gently to the northwest, parallel to the long axis of the mountains.

At least one period of low grade regional metamorphism has affected all of the rock units in the Ellsworth Mountains. Most of the units have been reported to contain mineral assemblages in the greenschist facies (Craddock et al., 1964; Hjelle et al., 1978, Yoshida, 1982), excepting the uppermost unit in the section, the Polarstar Formation, which attained a grade no higher than laumontite grade of the zeolite facies (Castle and Craddock, 1975).

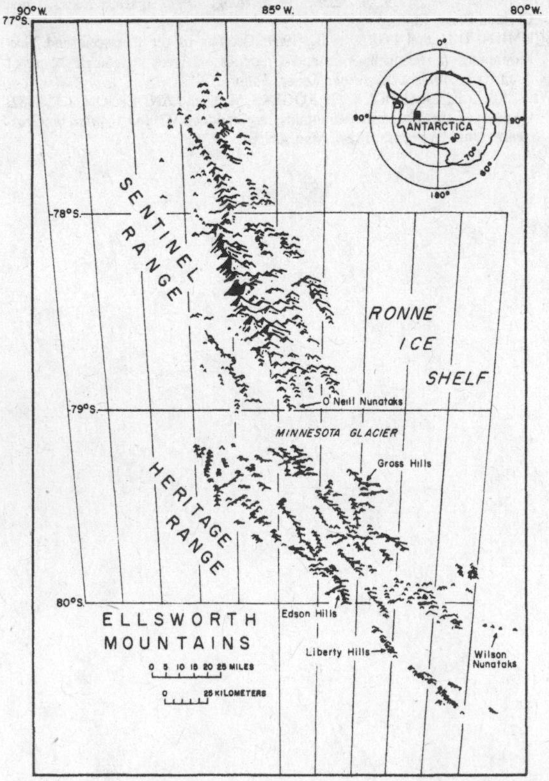

Figure 1. Location and general topography of the Ellsworth Mountains.

Extensive sampling of both sedimentary and igneous rock units within the Heritage Group was conducted during the 1979/80 austral summer as part of a detailed geological study of the Heritage Range of the Ellsworth Mountains (Splettstoesser and Webers, 1980). Selected samples from this expedition and igneous rocks collected during previous University of Minnesota expeditions (1961/62, 1963/64) are being examined to further elucidate the metamorphic history of the range.

This report summarises evidence which indicates that: (1) the upper Heritage Group has been recrystallised in the pumpellyite-actinolite facies, (2) metamorphic grade increases progressively downward in the stratigraphic section to greenschist facies assemblages, and (3) this low grade metamorphic event may locally be superimposed on an older greenschist facies event.

## PETROGRAPHY

The rocks examined are mainly mafic to intermediate lavas and tuffs and gabbroic dykes and sills. Foliated greywackes and argillites from the Edson Hills (Figure 1) were examined to investigate the relationship between deformation and metamorphism. Pumpellyite, which in some samples was difficult to distinguish from clinozoisite, was verified by electron microprobe. Equilibrium mineral assemblages indicative of pumpellyite-actinolite and greenschist facies conditions were observed in areas of a thin section 1 mm or less across, and the assemblages showed no evidence of reaction relationships.

### Pumpellyite-Actinolite Facies Rocks

Pumpellyite + actinolite bearing assemblages were identified in basaltic volcanic flows from the Liberty Hills and adjacent High Nunatak and from gabbroic rocks from the Edson Hills and Gross Hills of the Heritage Range, and in gabbroic rocks from the O'Neill Nunataks in the southern-most Sentinel Range (Figure 1).

*Liberty Hills-High Nunatak volcanic sequence* Basaltic volcanic rocks from the Liberty Hills, belonging to the Liberty Hills Formation of the upper Heritage Group, are plagioclase- and/or augite-phyric flows that are locally amygdaloidal. Similar flows and a volcanic breccia with a basaltic matrix occur in the High Nunatak area just east of the Liberty Hills (W. Vennum, 1981, written communication). All of the rocks contain metamorphic mineral assemblages including albite, white mica, chlorite and sphene, and varying combinations of the minerals epidote, pumpellyite, actinolite, calcite and quartz.

The flows have undergone varying degrees of metamorphic recrystallisation as evidenced by the degree of replacement of augite and plagioclase phenocrysts by metamorphic phases. Samples containing abundant calcite have had all pyroxene replaced by chlorite ± calcite ± quartz within a rim of magnetite ± pyrite that is pseudomorphic after the pyroxene crystal outline. These samples contain no actinolite, minor epidote is the only Ca-Al silicate, and plagioclase phenocrysts are partially altered to white mica ± calcite. Flows with

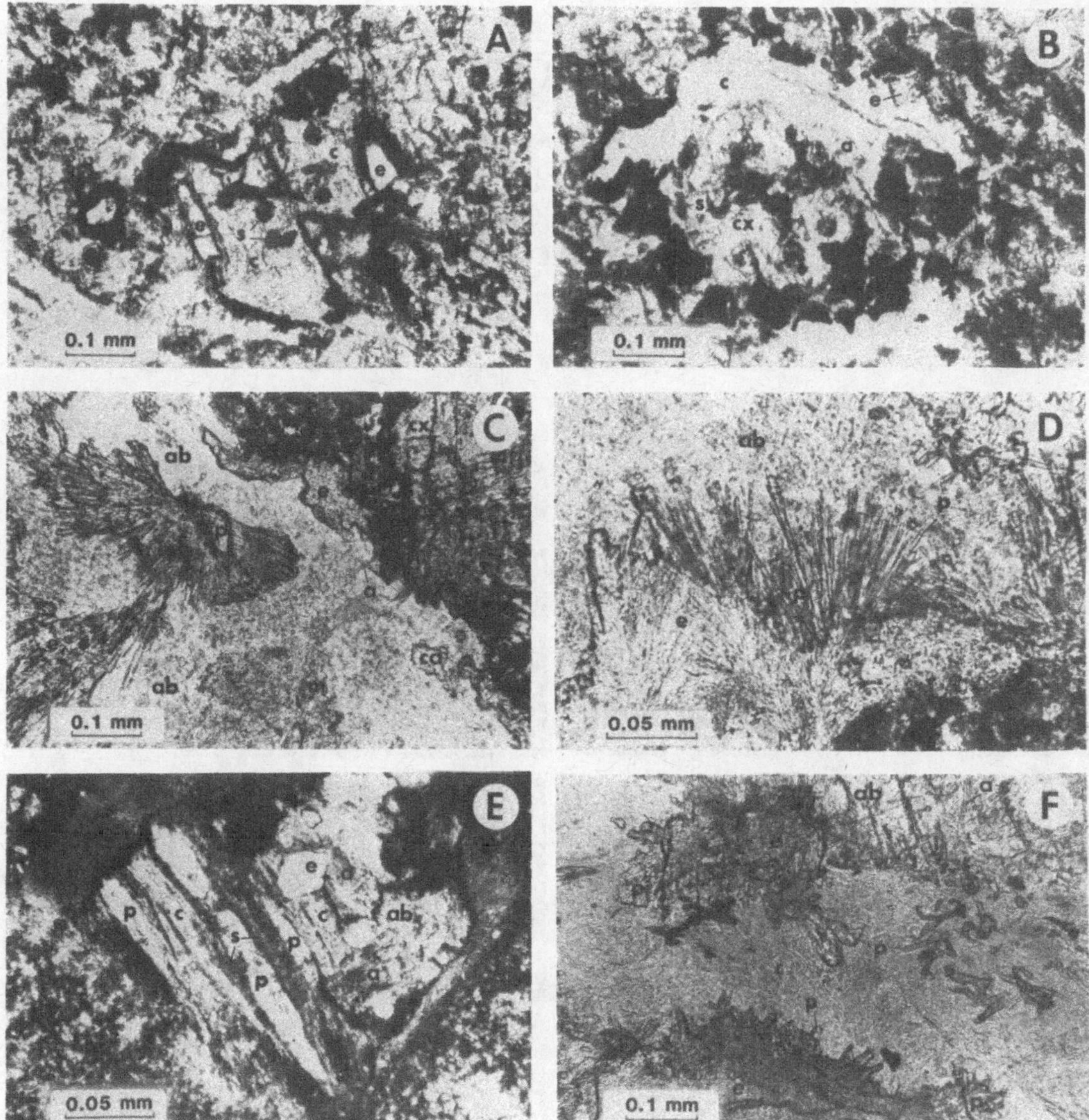


**Figure 2.** **Selected mineral occurrences in the mafic volcanic rocks from the Liberty Hills Formation and Springer Peak Formation of the upper Heritage Group. Symbols used in Figures 2 and 3 are: a = actinolite, ab = albite, c = chlorite, ca = calcite, cx = clinopyroxene, cz = clinozoisite, e = epidote, m = white mica, p = pumpellyite, q = quartz, s = sphene. A. Chlorite, epidote, sphene, magnetite and pyrite after clinopyroxene. The silicates are enclosed in an opaque rim pseudomorphically outlining the replaced clinopyroxene (plane light, sample 79Ve13e). B. Clinopyroxene, partially replaced by chlorite, actinolite, epidote, and opaques (plane light, sample 79Ve13E). C. Radiating acicular mass of pumpellyite in amygdule with albite, epidote and small grains of actinolite, calcite and white mica (plane light, sample 79Ve4A). D. Acicular pumpellyite attached to an epidote substrate with albite in an amygdule (plane light, sample 79Ve4a). E. Spindle shaped lenses of pumpellyite intergrown with chlorite, sphene and actinolite (x-nicols, sample 79Ve3a). F. Epidote with subhedral overgrowths of pumpellyite in an interstitial pod of chlorite in metagabbro (plane light, sample 64-MH-54).**

less abundant calcite contain similar opaque rims pseudomorphic after pyroxene, but the rims enclose varying combinations of epidote, chlorite, actinolite, calcite, sphene (Figure 2A) and, in rare instances, albite or pumpellyite. Replacement of pyroxene phenocrysts is complete in most instances, but selective. Pseudomorphically replaced grains may be adjacent to phenocrysts showing only slight alteration. Remnants of pyroxene within partially formed opaque rims were observed only rarely (Figure 2B). Plagioclase phenocrysts in samples with relatively low modal calcite are partially altered to sericite ± epidote ± pumpellyite. Plagioclase in the basaltic matrix of the High Nunatak breccia is altered to sericite and a fine opaque dust, but many grains have clear albite rims.

Amygdules in the flows are most commonly filled with combinations of epidote, calcite, chlorite, or albite, but pumpellyite, quartz, and white mica were also observed. Fine grained brownish epidote is common as the outer rim phase; coarser grained pale green to greenish yellow epidote and other minerals occur irregularly distributed within the brown epidote rim. Pumpellyite in the amygdules occurs as radiating acicular masses in quartz or albite (Figure 2C) or as acicular bundles attached to an epidote substrate (Figure 2D).

Pumpellyite in contact with actinolite is best observed in the basalt matrix of the High Nunatak volcanic breccia where it occurs as irregular to spindle shaped green pleochroic grains intergrown with chlorite and actinolite. In some cases the spindle shaped grains occur along chlorite cleavage traces (Figure 2E). Actinolite in these rocks as well as in the volcanic flows occurs as small acicular to lath shaped grains in the groundmass and as a direct replacement of the clinopyroxene.

*Gabbro dykes and sills* Gabbros and dolerites from the Springer Peak Formation in the Gross Hills and O'Neill Nunataks, and from the Drake Icefall Formation in the Edson Hills contain pumpellyite + actinolite bearing assemblages.

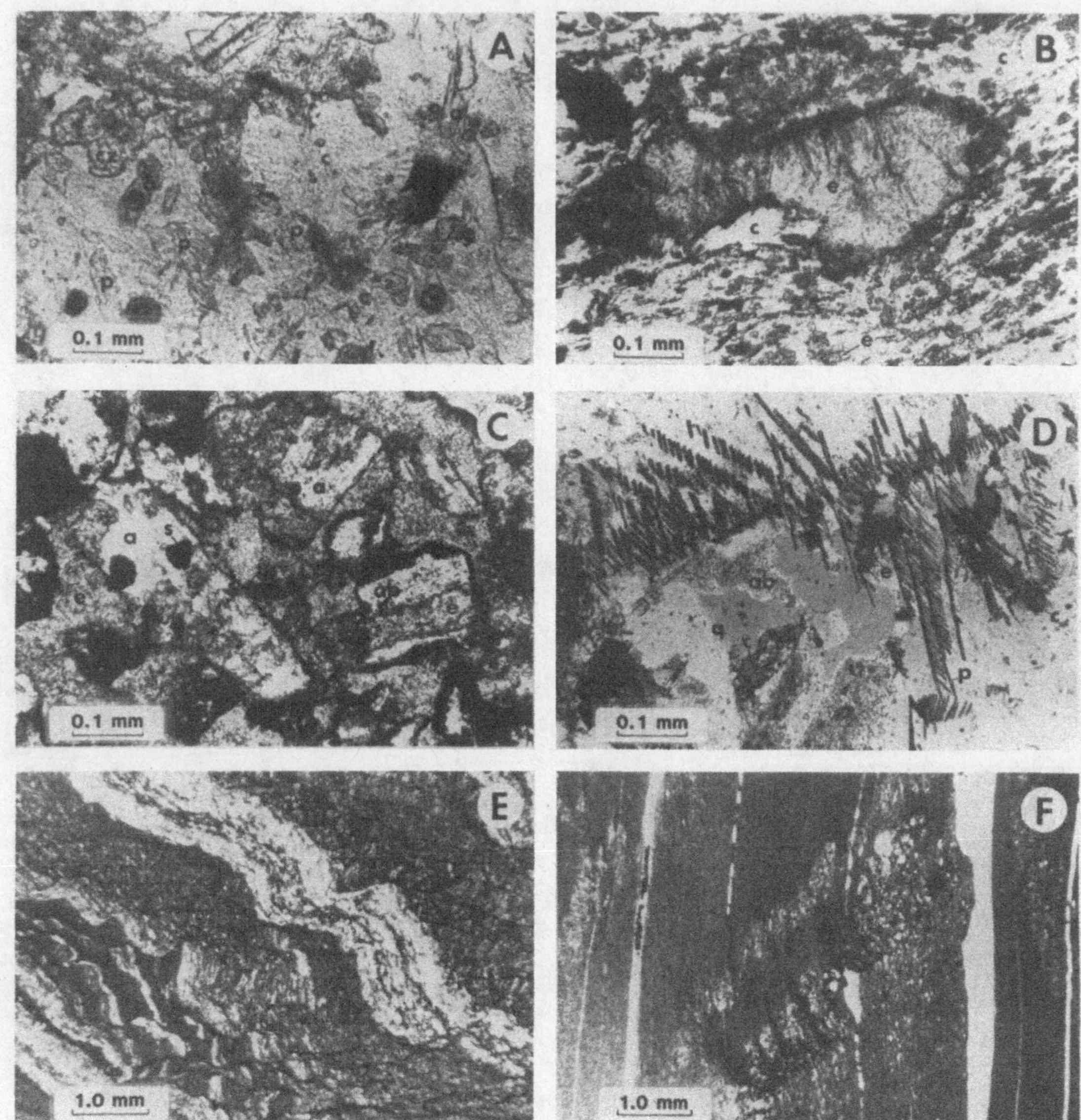


**Figure 3. Selected mineral occurrences in mafic tuffs and gabbros and tectonic fabric in metagreywacke from the lower Heritage Group. A. Clinozoisite and stubby laths and irregular grains of pumpellyite with actinolite in chlorite (plane light, sample W63-226-1900b). B. Elongate cluster of brown radiating acicular epidote mantled with fine grained granular sphene and set in a matrix of chlorite and fine epidote granules (plane light, sample W79-90). C. Epidote cement between albite and actinolite clasts (x-nicols, samples W79-97B). D. Pumpellyite laths with chlorite, epidote and albite in quartz (plane light, sample 63-W-238K). E. $S_1$ foliation parallel to bedding, overprinted by an $S_2$ crenulation cleavage (plane light, sample W79-106). F. $S_3$ kink plane trace overprinting $S_1$ or $S_2$ foliation (x-nicols, sample W79-105).**

Pumpellyite was identified in two samples from the Gross Hills. It occurs with epidote and sericite as an alteration product of plagioclase in both samples. One sample also contains pumpellyite as isolated crystals in chlorite and as lath shaped overgrowths on radiating clusters of epidote in chlorite pods (Figure 2F). The metamorphic mineral assemblage in the latter sample includes pumpellyite, actinolite, epidote, chlorite, sericite, sphene, and albite. Actinolite partially replaces clinopyroxene but also occurs as thin laths and small acicular clusters throughout the sample.

Pumpellyite bearing mineral assemblages in gabbros from O'Neill Nunataks and the Edson Hills differ from those in the Gross Hills only by the occurrence of clinozoisite instead of epidote. Colourless to pale green pumpellyite is abundant as an alteration product of plagioclase in samples from O'Neill Nunataks. Clinozoisite occurs as small clusters of radiating laths in chlorite pods or, less commonly, as more equant grains with a distinct anomalous blue interference colour. Gabbros from the Edson Hills contain stubby laths to equant grains of colourless pumpellyite occurring with clinozoisite and actinolite in chlorite (Figure 3A).

## Greenschist Facies Rocks

*Edson Hills section* The gabbros from the Drake Icefall Formation mark the lowest occurrence of pumpellyite or clinozoisite observed in the Heritage Group in the Edson Hills section. Mafic tuffaceous rocks from the Hyde Glacier Formation and Union Glacier Volcanics contain metamorphic mineral assemblages including chlorite, epidote, white mica, sphene, quartz, ± actinolite, and ± calcite, consistent with greenschist facies metamorphism.

Epidote occurs in several forms: (1) as fine, green pleochroic granules in plagioclase clasts or distributed throughout the tuff matrix, (2) as rounded or elongate clusters of nonpleochroic, light brown, radiating acicular grains (Figure 3B), (3) cementing various clasts and rock fragments (Figure 3C), and (4) as equant, greenish yellow pleochroic grains in veins, in the tuff matrix, or as overgrowths on the brown epidote clusters. Fine grains of chlorite and actinolite occur parallel to the foliation in the tuff matrix and in strain shadows around various clasts. Actinolite also occurs as fine acicular grains in quartz veins and partially or completely replacing clasts of pyroxene or hornblende.

*Wilson Nunataks* Gabbros and dolerites in the Wilson Nunataks intrude clastic rocks interpreted as Heritage Group by Hjelle et al. (1978). Common low grade mineral assemblages in the mafic rocks include chlorite, epidote, white mica, actinolite, quartz, sphene, and ± calcite. Chlorite occurs most commonly as an alteration product of green biotite (see discussion below). Stilpnomelane was observed in the above assemblage without sphene in one iron rich sample containing a deep green to blue-green ferroactinolite.

Pumpellyite occurs in one especially quartz rich sample along with chlorite, epidote, albite, sphene, white mica, and quartz. This pumpellyite is quite unlike that observed in any of the other samples investigated. It has interference colours as high as second order blue, and occurs as prismatic grains arranged en echelon and commonly twinned on {001} with longer prismatic grains (Figure 3D). The resulting pattern is reminiscent of the oakleaf twinning described by Coombs (1953). This occurrence of pumpellyite, indicative of sub-greenschist facies conditions, is anomalous relative to the greenschist facies assemblages observed in all other samples from Wilson Nunataks.

*Biotite bearing greenschist facies assemblages* Several gabbroic samples from the Wilson Nunataks contain green biotite. It occurs along with actinolite as an alteration product of clinopyroxene and as stubby laths partially altered to chlorite and fine sphene granules. Thin lenses of biotite within chlorite pseudomorphs are all that remain of the biotite in some samples.

The alteration of clinopyroxene to biotite may be a result of late stage, deuteric alteration in the dolerites. But the occurrence of biotite after clinopyroxene and the replacement of biotite by chlorite are also consistent with the occurrence of a greenschist facies metamorphic event prior to the chlorite forming (greenschist facies) event described in the previous section. Yoshida (1982, p.146) described similar relationships including: (1) green biotite partially altered to chlorite + leucoxenic matter in "basaltic lavas, tuff breccias, and strongly cleaved dikes" from unspecified locations within the Heritage Group, and (2) biotite and chloritoid partially altered to chlorite in the clastic country rocks from Wilson Nunataks.

A brownish yellow to dark brown micaceous mineral resembling biotite was observed in association with chlorite in a few samples of greywacke from the Edson Hills and in amygdaloidal basalt from the Liberty Hills. The yellow variety is pleochroic from brownish yellow to yellow brown with a birefringence in the range 0.025-0.030. The dark brown variety is pleochroic from dark yellow brown to dark brown and may display no interference colours due to masking by the mineral colour. Both mineral varieties grade into or are interlayered with chlorite and are interpreted as oxidised chlorites similar to those described by Chatterjee (1966) and Brown (1967).

Dark brown biotite, such as that described by Hjelle et al. (1978) as an interstitial constituent in the gabbros of their Middle Horseshoe Formation, was not observed in the samples considered in this study. Green biotite such as that described by Yoshida (1982) was observed only from the Wilson Nunataks.

## DISCUSSION

### Deformation Fabric and Metamorphism

Most of the mafic rocks examined containing pumpellyite-actinolite facies assemblages display little or no deformation fabric, making a correlation of deformation with this event difficult. However, during the course of this investigation several fabric elements affecting metamorphic minerals were observed in the metasedimentary sequence and these will be briefly described.

Three planar fabric elements were observed and are depicted in Figures 3E and 3F. The oldest fabric, $S_1$, is a penetrative foliation parallel to bedding in several samples and is defined by aligned chlorite and white mica. $S_1$ is overprinted by an $S_2$ crenulation cleavage. $S_2$ may occur as a simple spaced cleavage or even a penetrative cleavage in samples not displaying a well-developed $S_1$ foliation. Small scale kinking (Figure 3F), postdating all of the metamorphic minerals, was observed kinking $S_1$ or $S_2$ in some samples.

Distinction of $S_2$ from $S_1$ in thin section was possible only where both foliations were observed together or where porphyroblasts of chlorite elongated in $S_2$ maintain their cleavage traces, defining $S_1$, at a high angle to $S_2$. Epidote clusters, chlorite and actinolite in tuffaceous rocks of the Union Glacier Volcanics are oriented parallel to a strong foliation of uncertain $S_1$-$S_2$ origin. Some epidote clusters in the tuffs, and plagioclase phenocrysts in basalt display strain shadows containing chlorite ± actinolite and calcite, respectively.

The fabric-mineral relationships observed during the course of this study are consistent with a pre-to syn-$S_1$ timing of the metamorphism described here.

### Spatial Variations in Metamorphic Grade

Electron microprobe analyses and variations in the colour of pumpellyite from green in the Liberty Hills Formation to pale green in the Springer Peak Formation to clear in the Drake Icefall Formation indicate a progressive decrease in Fe and an increase in Al content of the pumpellyite (Surdam, 1969). A decrease in the Fe content of pumpellyite has been observed by several workers (Bishop, 1972; Kawachi, 1975; Smith et al., 1982) with increasing metamorphic grade in the pumpellyite-actinolite facies, and such a decrease is predicted by the temperature-composition model of Nakajima et al. (1977) for pumpellyite in the assemblage pumpellyite + chlorite + epidote + actinolite. These observations are consistent with an increase in metamorphic grade downward in the Heritage Group and are further corroborated by the absence of pumpellyite from the above assemblage in rocks of appropriate composition in the lower Heritage Group exposures in the Edson Hills. The low grade mineral assemblage in these rocks is consistent with greenschist facies conditions.

The relationship of the rocks in the Wilson Nunataks to those in the rest of the Heritage Groups is uncertain. The common greenschist facies asemblage in the gabbros from this site is complicated by the identification of pumpellyite from a single sample and by the presence of green biotite possibly indicative of an older greenschist facies event. Another factor complicating this problem is the 935 Ma age reported by Yoshida (1982) from this site.

It is tempting to project a continuous decrease in metamorphic grade from the pumpellyite-actinolite facies in the upper Heritage Group to the zeolite facies in the Polarstar Formation. Such a projection would predict prehnite-pumpellyite facies assemblages in rocks of appropriate compositions in the Minaret Formation, Crashsite Quartzite, or Whiteout Conglomerate. However, the relative lack of mafic rocks in these units and the abundance of carbonates may preclude the formation of the Ca-Al silicates required to distinguish the prehnite-pumpellyite facies.

The correlation of increased metamorphic grade with depth in the stratigraphic section suggests that the pumpellyite-actinolite and greenschist facies assemblages were produced during burial metamorphism within the thick stratigraphic sequence. Recrystallisation at essentially the same metamorphic grades continued during the anticlinorial folding of the Ellsworth Mountains and the generation of the $S_1$ and $S_2$ foliations.

*Acknowledgements* I would like to thank Gerald Webers for his assistance in obtaining the samples considered in this study and for the knowledge imparted during numerous discussions of the geology and stratigraphy of the Ellsworth Mountains. Thanks also to John Splettstoesser and Nicholas Stephenson for critically reading the manuscript. This work was supported by National Science Foundation grant DPP 78-21720 to Macalster College (G.F. Webers, Principal Investigator).

## REFERENCES

BISHOP, D.G., 1972: Progressive metamorphism from prehnite-pumpellyite to greenschist facies in the Dansey Pass area, Otago, New Zealand. *Geol. Soc. Am., Bull., 83,* 3177-98.

BROWN, E.H., 1967: The greenschist facies in part of eastern Otago, New Zealand. *Contrib. Mineral. Petrol., 14,* 259-92.

CASTLE, J.W. and CRADDOCK, C., 1975: Deposition and metamorphism of the Polarstar Formation (Permian), Ellsworth Mountains. *U.S. Antarct. J., 10,* 239-41.

CHATTERJEE, N.D., 1966: On the widespread occurrence of oxidised chlorites in the Pennine Zone of the western Italian Alps. *Beitr. Mineral. Petrogr., 12,* 325-39.

COOMBS, D.S., 1953: The pumpellyite mineral series. *Min. Mag., 30,* 113-35.

CRADDOCK, C., ANDERSON, J.J. and WEBERS, G.F., 1964: Geologic outline of the Ellsworth Mountains; *in* Adie, R.J. (ed.) *Antarctic Geology,* North Holland, Amsterdam, 155-70.

HJELE, A., OHTA, Y. and WINSNES, T.S., 1978: Stratigraphy and igneous petrology of the southern Heritage Range, Ellsworth Mountains, Antarctica. *Nor. Polarinst., Skr., 169,* 5-43.

KAWACHI, Y., 1975: Pumpellyite-actinolite and contiguous facies metamorphism in part of upper Wakatipu district, South Island, New Zealand. *N.Z. J. Geol. Geophys., 18,* 401-41.

NAKAJIMA, T., BANNO, S. and SUZUKE, T., 1977: Reactions leading to the disappearance of pumpellyite in low grade metamorphic rocks of the Sanbagawa metamorphic belt of central Shikoku, Japan. *J. Petrol., 18,* 263-84.

SMITH, R.E., PERDRIX, J.L. and PARKS, T.C., 1982: Burial metamorphism in the Hamersley Basin, Western Australia. *J. Petrol., 23,* 75-102.

SPLETTSTOESSER, J. and WEBERS, G.F., 1980: Geological investigations and logistics in the Ellsworth Mountains, 1979-1980. *U.S. Antarct. J., 15,* 36-9.

SURDAM, R.C., 1969: Electron microprobe study of prehnite and pumpellyite from the Karmutsen Group, Vancouver Island, British Columbia. *Am. Mineral., 54,* 246-66.

WEBERS, G.F. and SPORLI, G.K., (this volume): Palaeontological and stratigraphic investigations in the Ellsworth Mountains, West Antarctica.

YOSHIDA, M., 1982: Superposed deformation and its implications to the geologic history of the Ellsworth Mountains, West Antarctica. *Mem. Nat. Inst. Polar Res. Japan, 21,* 120-71.

# PALAEONTOLOGICAL AND STRATIGRAPHIC INVESTIGATIONS IN THE ELLSWORTH MOUNTAINS, WEST ANTARCTICA

G.F. Webers, *Geology Department, Macalester College, 1600 Grand Ave., St. Paul, Minn., 55105, U.S.A.*

K.B. Sporli, *Geology Department, University of Auckland, Auckland, New Zealand.*

*Abstract* Seventeen fossiliferous localities, ranging in age from Middle Cambrian to Permian, were sampled in the Ellsworth Mountains during the 1979/80 field season. Some of these were known from earlier expeditions. Studies of new trilobite faunas together with stratigraphic and structural field work have permitted a major revision of the lower 7800 m of strata in the Heritage Range of the Ellsworth Mountains. The lowermost stratigraphic unit is the Heritage Group. It is here informally divided into formations. Formal description of these formations is not made here due to lack of space and pending approval of submitted geographic names. The lowest formation of the Heritage Group is the "Union Glacier Volcanics", followed in order by the "Hyde Glacier Formation", "the Drake Icefall Formation", the "Conglomerate Ridge Formation", and finally by three laterally equivalent units: the "Frazier Ridge Formation", the "Springer Peak Formation", and the "Liberty Hills Formation". The Heritage Group ranges in age from Middle to Late Cambrian. The Late Cambrian Minaret Formation overlies the Heritage Group.

The Ellsworth Mountains (Figure 1) comprise a rugged mountain chain located between the polar plateau of West Antarctica and the Ronne Ice Shelf. They trend N55W, are about 370 km long and 80 km wide, and are divided into a northern range, the Sentinel, and a southern range, the Heritage.

During the 1979/80 austral season a total of 42 scientists carried out

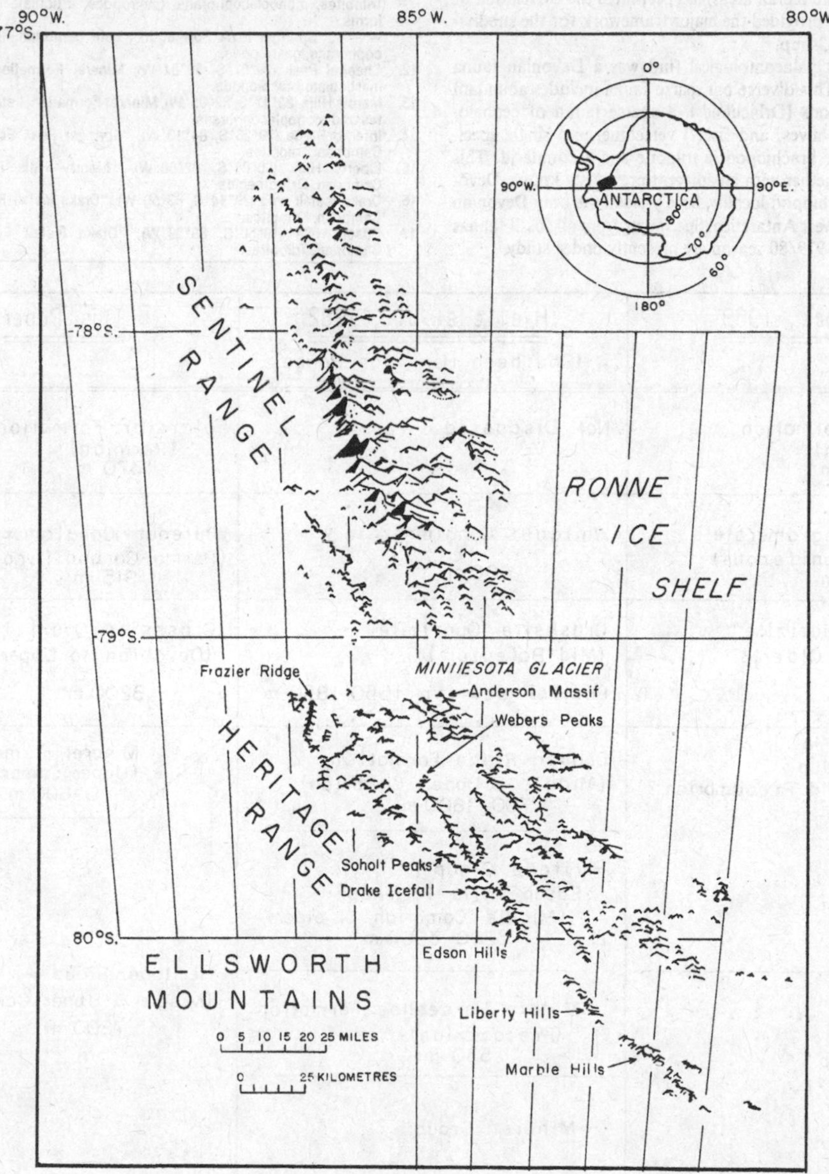


**Figure 1. Location and general topography of the Ellsworth Mountains**

a wide variety of geological investigations in the Ellsworth Mountains (Splettstoesser and Webers, 1980). Some of these scientists were involved in a detailed study of the palaeontology, sedimentology, and stratigraphy of the 13,000 m+ stratigraphic column. The Heritage Range was the primary focus of study because of the relatively low relief and accessibility, and because all of the stratigraphic units, with the exception of the Polarstar Formation, were well exposed there.

Seventeen fossiliferous localities were sampled during the 1979/80 field season (Table 1). Fossils range in age from Middle Cambrian to Permian. The most significant finds were trilobite faunas ranging in age from Middle to Late Cambrian. These finds permitted the correlation of major stratigraphic sections throughout the Heritage Range and the reorganisation of the stratigraphy of the lower 7800 m of the column. Most significant was the discovery of Cambrian trilobites in the Minaret Group, previously thought to be of possible Precambrian age. These trilobites also indicated that the stratigraphy of the Minaret Group and the adjacent clastic sediments are inverted, and that the Minaret Group is correctly placed near the middle of the stratigraphic column rather than at the base. A Middle Cambrian trilobite fauna found in each of two thick stratigraphic sections (Edson Hills and Drake Icefall areas) has permitted the correlation of these sections and has provided the major framework for the subdivision of the Heritage Group.

Another important palaeontological find was a Devonian fauna from Planck Point. This diverse but sparse fauna includes abundant inarticulate brachiopods (Orbiculoids), a sparse fauna of cephalopods, gastropods, bivalves, and fish(?) vertebrae, and single specimens of an articulate brachiopod, a trilobite and a conularid. This Devonian locality, together with an adjacent previously known Devonian inarticulate brachiopod locality, have yielded the only Devonian faunas known from West Antarctica. Specimens from all fossil faunas collected during the 1979/80 season are presently under study.

**TABLE 1: Geographic position, formation, age, and fossils collected from localities in the Ellsworth Mountains, 1979-80**

1. Polarstar Peak (77°32′S, 86°09′W), Polarstar Formation, Permian *Glossopteris* flora.
2. Meyer Hills (79°47′S, 81°06′W), Whiteout Conglomerate, Permo-Carboniferous Archaeocyathids in Cambrian boulders.
3. Planck Point (79°18′S, 85°11′W), Crashsite Quartzite, Devonian. Inarticulate brachiopods, cephalopods, gastropods, bivalves, fish(?) vertebrae, and 1 specimen each of a trilobite, articulate brachiopod, and a conularid.
4. Pipe Peak (79°09′S, 86°15′W), Crashsite Quartzite, Late Cambrian. Trilobites, inarticulate brachiopods, and trace fossils.
5. Windy Peak (79°13′S, 86°04′W), Crashsite Quartzite, Late Cambrian. Trilobites and inarticulate brachiopods.
6. Matney Peak (79°10′S, 86°14′W), Crashsite Quartzite, Late Cambrian. Trilobites and inarticulate brachiopods.
7. Reuther Nunataks (79°10′S, 85°57′W), Crashsite Quartzite, Late Cambrian. Trilobites, inarticulate brachiopods, and trace fossils.
8. Anderson Massif (79°10′S, 84°45′W), Minaret Formation, Late Cambrian. Articulate brachiopods, and an archaeocyathid.
9. Springer Peak (79°24′S, 84°53′W), Minaret Formation, Late Cambrian. Trilobites, monoplacophorans, gastropods, rostroconchs, articulate and inarticulate brachiopods, conodonts, archaeocyathids, pelmatozoans, and algal forms.
10. Bingham Peak (79°26′S, 84°47′W), Minaret Formation, Late Cambrian. Trilobites, monoplacophorans, gastropods, archaeocyathids, pelmatozoans, articulate and inarticulate brachiopods, and algal forms.
11. "Yochelson Ridge" (79°37′S, 84°24′W), Minaret Formation, Late Cambrian. Trilobites, monoplacophorans, gastropods, articulate brachiopods, algal forms.
    ———, "Springer Peak Formation", Late Cambrian. Trilobites, monoplacophorans, gastropods.
12. Chappel Peak (79°57′S, 82°54′W), Minaret Formation, Late Cambrian. Inarticulate brachiopods.
13. Marble Hills (80°17′S, 82°05′W), Minaret Formation, Late Cambrian. Trilobites and monoplacophorans.
14. Inferno Ridge (79°26′S, 84°13′W), "Springer Peak Formation", Middle? Cambrian. Trilobites.
15. Liberty Hills (80°06′S, 82°58′W), "Liberty Hills Formation", Late? Cambrian. Orthothecids.
16. Drake Icefall area (79°44′S, 83°50′W), "Drake Icefall Formation", Middle Cambrian. Trilobites.
17. Edson Hills (79°50′S, 83°39′W), "Drake Icefall Formation", Middle Cambrian. Trilobites.

| Craddock, 1969 | Hjelle et al., 1982<br>(Southern Heritage Range) | This Paper |
|---|---|---|
| Polarstar Formation<br>(Permian)<br>1370 m | Not Discussed | Polarstar Formation<br>(Permian)<br>1370 m |
| Whiteout Conglomerate<br>(Permo-Carboniferous)<br>915 m | Whiteout Conglomerate<br>400 m | Whiteout Conglomerate<br>(Permo-Carboniferous)<br>915 m |
| Crashsite Quartzite<br>(Devonian & Older)<br>3200 m | Crashsite Quartzite<br>(Mid-Paleozoic)<br>Partial section 1550-1850 m | Crashsite Quartzite<br>(Devonian to Upper Cambrian)<br>3200 m |
| Heritage Group<br>(Cambrian and Precambrian ?)<br>6710 m | Dunbar Ridge Formation<br>(Middle & Upper Cambrian)<br>700-1600 m | Minaret Formation<br>(Upper Cambrian)<br>0-600 m |
| | Heritage Group<br>Edson Hills Formation<br>(Middle Cambrian or older)<br>3200-3900 m | Heritage Group<br>(Middle & Upper Cambrian)<br>7200 m |
| Minaret Group<br>(Precambrian ?)<br>975 m | Middle Horseshoe Formation<br>(Precambrian)<br>530 m | |
| | Minaret Group | |

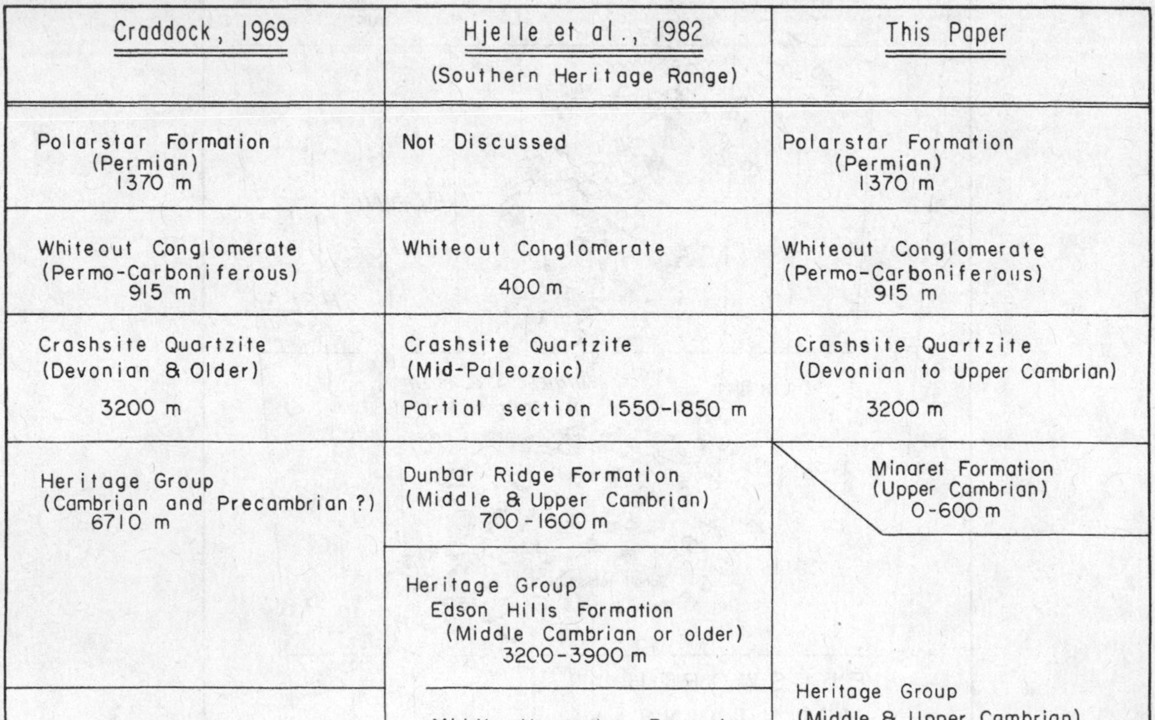

**Figure 2. Generalised geological sections in the Ellsworth Mountains**

The most important changes in the Ellsworth Mountains stratigraphy concern the 7800 m of strata included in the Minaret (600 m) and Heritage Groups (7,200 m).

## The Heritage Group

The Heritage Group is the lowest stratigraphic unit in the Ellsworth Mountains. Craddock (1969) described it and considered it to overlie the Minaret Group. Hjelle et al. (1978) revised the description of the Minaret Group, placing the upper part of the unit in the Heritage Group and describing it as the Middle Horseshoe Formation. The Heritage Group then contained the Middle Horseshoe Formation and a new formation called the Edson Hills Formation. A new formation, the Dunbar Ridge Formation, was erected between the Heritage Group and the Crashsite Quartzite. Generalised geologic sections of Craddock (1969), Hjelle et al. (1978), and this paper are shown in Figure 2.

During the 1979/80 United States Ellsworth Mountains expedition, palaeontological and structural studies indicated that the stratigraphic sections of the Minaret Group and the "lower beds" of the Heritage Group in the Liberty Hills are overturned. This and other detailed stratigraphic investigations conducted during the 1979/80 field season raised major problems with the stratigraphy of Hjelle et al. (1978). It is thus necessary to return to the stratigraphic units of Craddock (1969), reversing the positions of the Minaret and Heritage Groups. The Minaret Group is here downgraded to formational status and the Heritage Group is informally subdivided into new formations. Diagrammatic relationships of the Crashsite Quartzite, Minaret Formation, and the informal units of the Heritage Group are shown in Figure 3. Formal description of formations in the Heritage Group is not made here because of space, pending approval of submitted geographic names, and continuing investigations. The Heritage Group will be formally subdivided in a subsequent publication. A discussion of the informal formations of the Heritage Group follows:

### *The "Union Glacier Volcanics"*

The lowermost unit of the Heritage Group is the "Union Glacier Volcanics". The exposed thickness is about 3000 m and the base is not exposed. About 90 percent of the strata consist of a dark green volcanic diamictite. This material contains slightly deformed basalt clasts as much as 60 cm in diameter, which comprise from 5 to 30 percent of the volume. Some of these clasts have been identified as volcanic bombs. Surrounding the basalt clasts is a tuffaceous matrix which contains abundant highly deformed elongate clasts. The deformed clasts are as large as 30 cm in length, 15 cm in width and 2 cm thick. The volcanic diamictite is considered to be a terrestrial deposit of lahars and/or ashflow tuffs. Deformed clasts are considered to be flattened pumice fragments. Green and maroon tuffaceous argillites are common and grade into the volcanic diamictite.

About 180 m above the lowest exposed beds of the "Union Glacier Volcanics" is a 390 m thick buff conglomerate. The conglomerate is typically clast supported with clasts making up about 70 percent of the material. Clasts average about 3 cm in length and are dominantly composed of grey limestone. Matrix is typically calcareous but occasionally quartzitic. Green and maroon tuffaceous sediments are occasionally interbedded with the conglomerate which is considered to be of fluvial origin.

### *The "Hyde Glacier Formation"*

Above the "Union Glacier Volcanics" is the "Hyde Glacier Formation". It is a clastic unit about 1750 m thick. Rock types include green and maroon tuffaceous argillites, greywackes, polymict conglomerates, and quartzites. The base of this unit is a 70 m thick green polymict conglomerate. Clasts are composed of granite, basalt and quartzite, range in size from 0.5 to 60 cm, and comprise about 75 percent of the conglomerate. Overlying this conglomerate is a green tuffaceous phyllite with lesser amounts of maroon tuffaceous phyllite, green quartzite, and green polymict conglomerate. It is about 1100 m thick with the lower 760 m primarily composed of green and occasionally maroon tuffaceous phyllite, and the upper 340 m primarily composed of green quartzite and lesser amounts of green phyllite and green polymict conglomerate.

Above the green phyllites and quartzites is a 280 m thick sequence of red and buff polymict conglomerates and quartzites. At the base of this unit in the measured section is a 10 m thick cut-and-fill structure. It is cut into alternating bands of red polymict conglomerate and green conglomeratic sandstone. The entire sequence is conformable.

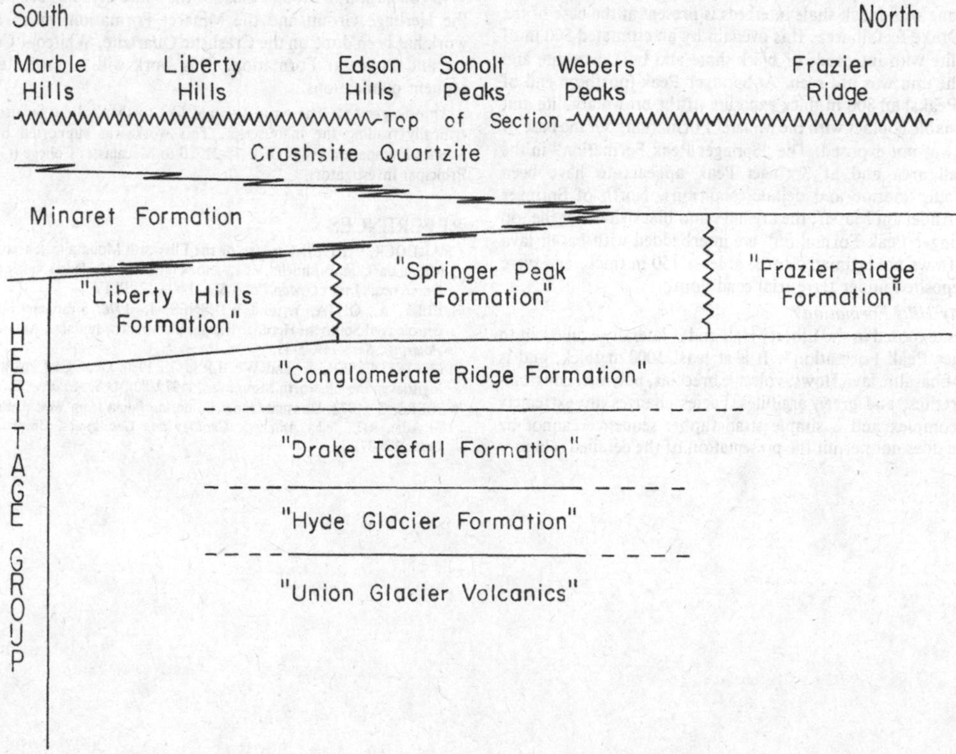


**Figure 3. Diagrammatic cross section through the southwestern Heritage Range showing stratigraphic relationships**

The uppermost unit of the "Hyde Glacier Formation" is primarily composed of green greywacke. Sand-sized material makes up about 80 percent of the unit. Some of the coarser beds are conglomeratic and slate interbeds are present. This unit is about 300 m thick.

The "Hyde Glacier Formation" is considered to have been deposited under deltaic conditions.

*The "Drake Icefall Formation"*

The "Drake Icefall Formation" is a thick unit of black shales and interbedded limestones which conformably overlies the "Hyde Glacier Formation". Total thickness is very difficult to estimate because of intense shearing and parasitic folds within the formation, but it is estimated to be about 800 m thick. In the Drake Icefall area podlike black limestone lenses are common in the lower half of the unit, and a 200 m thick grey limestone occurs at the base. In the Edson Hills the unit is almost entirely composed of black shale. A thin (5 m thick) limestone occurs at the base and black limestone concretions are common in a 20 m thick zone near the base. Both the Edson Hills and the Drake Icefall sections have produced Middle Cambrian trilobite faunas. The "Drake Icefall Formation" is considered to have been deposited under marine conditions.

*The "Conglomerate Ridge Formation."*

This is a 650 m thick clastic unit that conformably overlies the "Drake Icefall Formation". A 150 m thick green argillaceous quartzite occurs at the base. It is overlain by a 450 m thick buff polymict conglomerate. Clasts of predominantly grey limestone and buff quartzite make up about 70 percent of the material. Matrix is dominantly quartz sand. Interbeds of maroon quartzite and green polymict conglomerate are present near the top of the formation. In general appearance, the buff polymict conglomerate of the "Conglomerate Ridge Formation" is similar to the buff polymict conglomerate near the base of the "Union Glacier Volcanics". The "Conglomerate Ridge Formation" could be fluvial and/or shallow marine in origin.

*The "Springer Peak Formation."*

This is estimated to be about 1000 m thick, and conformably overlies the "Conglomerate Ridge Formation" in the area north of the Drake Icefall. Actual thickness is very difficult to estimate because of shearing and abundant parasitic folds within the formation. A 60 m thick limestone with black shale interbeds is present at the base of the unit in the Drake Icefall area. It is overlain by an estimated 500 m of brown argillite with interbeds of black shale and buff quartzite and the top of the unit was not seen. At Springer Peak (northern end of the Webers Peaks) an 800 m thick exposure of the brown argillite unit is in conformable contact with the Minaret Formation, but the base of the argillite was not exposed. The "Springer Peak Formation" in the Drake Icefall area and at Springer Peak appears to have been deposited under marine and deltaic conditions. North of Springer Peak in the Anderson Massif, the argillites and black shales at the top of the "Springer Peak Formation" are interbedded with basalt lava flows. The flows are estimated to be at least 150 m thick, and were probably deposited under terrestrial conditions.

*The "Liberty Hills Formation"*

This is best exposed in the Liberty Hills and is laterally equivalent to the "Springer Peak Formation". It is at least 1000 m thick, and is composed of basaltic lava flows, volcanic breccias, polymict conglomerates, quartzites, and green argillites. Facies changes are extremely rapid and complex and a simple stratigraphic sequence cannot be given. Space does not permit the presentation of the detailed relationships among the rock types. The polymict conglomerate is a clast-supported conglomerate with clasts as much as 60 cm in length. Clasts are composed of granite, vein quartz, quartzite, siltstone, limestone and volcanics. It is as much as 800 m thick in some sections and absent in others. The quartzites are immature and cross-bedded. A basic intrusion as much as 300 m thick is exposed near the top of the formation on the eastern edge of the Liberty Hills. The "Liberty Hills Formation" is in conformable contact with the Minaret Formation in the Liberty Hills. Sediments in the "Liberty Hills Formation" were probably deposited in both fluvial and shallow marine environments.

*The "Frazier Ridge Formation."*

The "Frazier Ridge Formation" is laterally equivalent to the "Liberty Hills Formation" and the "Springer Peak Formation". It is known only from the northwestern Heritage Range. It is primarily a fine-to medium-grained green quartzite with occasional beds of green argillite and black shale. It is estimated to be at least 500 m thick. The base of the formation is not exposed. The Minaret Formation is not present in the northwestern Heritage Range and the "Frazier Ridge Formation" is in conformable contact with the Crashsite Quartzite. The contact with the Crashsite Quartzite is marked by a series of interbedded black shales. The "Frazier Ridge Formation" is considered to be of shallow marine origin.

## The Minaret Formation

The Minaret Formation overlies the Heritage Group and is composed almost entirely of white to grey marble. It reaches its maximum thickness in the Marble Hills where it is at least 600 m thick. It uniformly thins northward to the northern end of the Webers Peaks where it is only 6 m thick; it is not present in the northwestern Heritage Range. The Minaret Formation is primarily composed of white and grey sparry calcite and some dolomite; it is commonly oolitic and oncolitic. Exposures in the Webers Peaks area have produced a well-preserved Late Cambrian fauna of trilobites, monoplacophorans, rostroconchs, gastropods, hyolithids, articulate and inarticulate brachiopods, archaeocyathids, and pelmatozoans (Webers, 1972). The Minaret Formation is considered to have been deposited under shallow marine conditions.

In conclusion, although most of the space here has been devoted to the Heritage Group and the Minaret Formation, much additional work has been done on the Crashsite Quartzite, Whiteout Conglomerate and Polarstar Formation. This work will be presented in subsequent publications.

*Acknowledgements* The writers would like to thank John Splettstoesser for critically reading the manuscript. This work was supported by National Science Foundation grant DPP 78-21720 to Macalaster College (G.F. Webers, Principal Investigator).

## REFERENCES

CRADDOCK, C., 1969: Geology of the Ellsworth Mountains (Sheet 4, Ellsworth Mountains); *in* Bushnell, V.C., and Craddock, C. (eds.) Geologic Maps of Antarctica, *Antarct. Map Folio Ser.*, Folio 12, Pl.IV.

HJELLE, A., OHTA, Y., and WINSNES, T., 1978: Stratigraphy and igneous petrology of Southern Heritage Range, Ellsworth Mountains, Antarctica. *Norsk. Polarinst. Skr., 169,* 5-43.

SPLETTSTOESSER, J., and WEBERS, G., 1980: Geological investigations and logistics in the Ellsworth Mountains, (1979/80). *U.S. Antarct. J., 15,* 36-9.

WEBERS, G., 1972: Unusual Upper Cambrian fauna from West Antarctica, 1972, *in* Adie, R.J., (ed.) *Antarctic Geology and Geophysics.* Universitetsforlaget, Oslo, 235-37.

# PALAEOZOIC CARBONATE ROCKS OF THE ELLSWORTH MOUNTAINS, WEST ANTARCTICA

W. Buggisch, *Geologisch-Palaontologisches Institut, Technischen Hochschule Darmstadt, D-6100 Darmstadt, Federal Republic of Germany.*

*Abstract* The continuous Palaeozoic sequence of the Ellsworth Mountains is over 13000 m thick (Craddock 1969) and contains several carbonate horizons.

The lower beds of the Cambrian Heritage Group contain a conglomerate with limestone pebbles. Microfacies analysis of these clasts indicates the presence of the following types:

(a) Oolitic limestones,
(b) Skeletal and nonskeletal algal limestones,
(c) Intrasparites or lithoclastic grainstones,
(d) Components with cementation of dissolution voids.

The provenance area of these clasts is thought to be a carbonate platform of the Early or Middle Cambrian. A possible source area is not exposed in the Ellsworth Mountains.

Black limestones and nodules with trilobites are found in the Middle Cambrian beds.

In the Upper Heritage Group oncolitic limestones of about 8 m thickness with a well-preserved Late Cambrian fauna are known from the Webers Peaks area (Webers 1972). To the south the thickness increases and it is possible to show that the carbonates of the Marble Hills are of the same age. Therefore the Minaret Formation, previously considered to be Precambrian (?) (Craddock 1969), is of Late Cambrian age. The following microfacies types are found in this limestone horizon: onco- and biosparites, pelmatozoan sparites with rim cement, oolites, mudstones, fenestral textures (type *Stromatactis*) etc., representing a more or less open marine shallow water environment.

The Permo-Carboniferous diamictites (Whiteout Conglomerate) include carbonate pebbles of various microfacies. Reworked fossils (algae, Archaeocyatha) point to a Cambrian age of the fossiliferous clasts.

Considerations concerning the provenance area of the Cambrian carbonate clasts and the facies of the Minaret Formation suggest a shelf. This is in agreement with palaeomagnetic results which record a rotation of the Ellsworth microplate (Watts and Bramall 1981).

# STRUCTURAL AND METAMORPHIC HISTORY OF THE ELLSWORTH MOUNTAINS, WEST ANTARCTICA

M. Yoshida, *Department of Geosciences, Osaka City University, Osaka, 558, Japan.*

*Abstract* A polyorogenic interpretation of the structural development of the Ellsworth Mountains is critically reviewed. Some new data on fold and cleavage structures in the southern and northwestern parts of the Heritage Range are presented and discussed. The data suggest that there are variations in fold style and trend during one folding episode, but there are also fold superposition relationships. These can be directly observed in the field, and can also be inferred from the timing of cleavages in relation to dykes or veins. In the south the first stage main longitudinal folding and cleavage development occurred before the intrusion of the Heritage altered dolerites which have maximum K-Ar ages of ca 400 Ma (Late Silurian). The second stage produced longitudinal minor kink folds and the third produced transverse minor kink folds, both being associated with local spaced cleavages and affecting the Heritage altered dolerites, and hence are considered to be post Late Silurian. The second and third stages in the south can possibly be correlated respectively with second order longitudinal kink folds and gentle second order transverse folds that occur in the northwest.

A 1979-1980 field survey and subsequent laboratory studies resulted in the writer's tentative polyorogenic interpretation of the geology of the Ellsworth Mountains (Yoshida, 1982). This met some criticism (eg. Craddock, pers. comm.), the main criticism being of the writer's interpretation of the superposition of folds and cleavages. The present article reviews the previous interpretations of the geologic development of the Ellsworth Mountains, and discusses the multiple deformation or polyorogenic interpretation, which is based mainly on analysis of fold and cleavage structures.

## REVIEW OF THE GENERAL GEOLOGY AND THE POLYOROGENIC INTERPRETATION

The general geology of the Ellsworth Mountains was described by Craddock et al., (1964) and Craddock (1969), and revisions were made by Hjelle et al. (1978; 1982), Splettstoesser and Webers (1980) and Yoshida (1982). The mountains are composed mainly of clastic rocks, with minor igneous rocks and limestone, ranging in age from Cambrian to Permian, and attaining a thickness of over 13000 metres (Figure 1). The sequence of units, as determined by Splettstoesser and Webers (1980), is: Heritage Group (Middle to Upper Cambrian), Crashsite Quartzite (?pre-Devonian to Devonian), Whiteout Conglomerate (Permo-Carboniferous), and the Polarstar Formation (Permian).

Craddock et al. (1964) and Craddock (1969) concluded that the mountains suffered only one folding episode on an axis paralleling the mountain range. It was associated with synchronous cleavage development and low to very low grade metamorphic recrystallisation occurring during the earliest Mesozoic. This event was termed the Ellsworth Orogeny by Craddock (1972). From stratigraphic, tectonic, and petrologic studies of the southern part of the mountains, Hjelle et al. (1978,1982) suggested the possibility of some tectonic and stratigraphic gaps.

Yoshida (1982) tentatively presented a polyorogenic interpretation and the following sequence of events:-

1) A Precambrian event for which the evidence is a K-Ar age of 935 Ma for a metaclastic rock from the Wilson Nunataks. This might indicate the presence of a fragment of hidden Precambrian basement.
2) Metamorphic recrystallisation under the low grade conditions of Winkler (1974) as indicated by the development of chlorite and sericite in continuous and closely spaced cleavages, longitudinal (axial traces parallel to the mountain range), inclined to reclined and isoclinal folds (fold shape and attitude are referred to that of Fleuty, 1964 and Rickard, 1971, respectively), and longitudinal, nearly horizontal, and close folds, developed after the deposition of the Heritage Group and before the intrusion of the Heritage altered dolerites (early Borchgrevink Orogeny of Bradshaw et al. (1982) or the Ross Orogeny).
3) The intrusion of the Heritage altered dolerites with minimum K-Ar ages of *ca* 400 Ma associated with the annealing recrystallisation of the country rocks resulting in the general hardening of originally cleaved rocks and a local development of chloritoid porphyrblasts.
4) Low grade metamorphic recrystallisation (shown by partial alteration of the chloritoid porphyroblasts into chlorite), and formation of closely spaced cleavage (including cleavage in the Heritage altered dolerites) and longitudinal kink folds (associated with quartz-chlorite veins) occurring after the Crashite Quartzite and before or during the Whiteout Conglomerate. The whole rock K-Ar ages of around 300 Ma obtained for the compact chlorite-sericite phyllite of the Heritage Group are considered to manifest this event and are referred to the late phase of the Borchgrevink Orogeny.
5) Very low grade metamorphic recrystallisation (which affects the Polarstar Formation as well as possibly some of the low grade metamorphites), including formation of roughly spaced cleavages, longitudinal open folds and transverse gentle folds. The K-Ar ages of ca 235 and 254 Ma obtained for strongly schistose and fissile andesitic to basaltic dykes in the Heritage Group are considered to represent the maximum age of these events which are referred to the Early Mesozoic Ellsworth Orogeny of Craddock (1972).

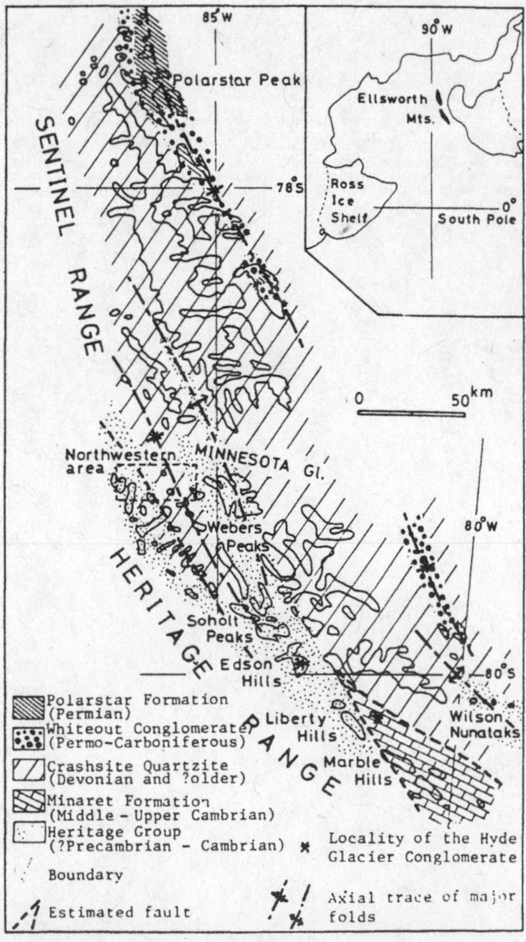


**Figure 1. Geologic outline of the Ellsworth Mountains (from Yoshida, 1982, simplified after Craddock, 1969).**

Craddock (pers. comm.) considers that the different styles of folds, as well as different orientations of cleavages, may reflect different physical characteristics of rocks, but not, as Yoshida (1982) concluded, different times of formation. The following sections provide some new data and discussions of these points, especially with reference to the time relationships of the longitudinal folds and their associated cleavage structures.

## CHARACTERISTICS OF FOLDS AND CLEAVAGES IN THE MARBLE HILLS

The southern part of the Heritage Range is characterised by the abundant development of longitudinal open to tight folds, typified in the Marble Hills (Yoshida, 1982). Minor kink folds of a later generation are also sporadically found. In the major cliff (called the "Fold Cliff" by the field team) at the central-eastern part of the Marble Hills, a full picture of the fold structures was found (Figure 2). Major folds are sub-horizontal, steeply inclined and open to close but inclined to reclined and tight to isoclinal folds of smaller scale are not rare. These folds occur throughout the Marble Hills and continue northwards as far as the central part of the Heritage Range (Yoshida, 1982). The rock cleavage is generally subparallel to, or slightly inclined to, the axial surfaces of all types of folds; intersections between cleavage and bedding are somewhat scattered around the hinge of the folds in the former case, but are generally further scattered in the latter. Equal area stereographic projections of some structural elements observed at the "Fold Cliff" are shown in Figure 3. Mean tectonic axes of the major fold structures, labelled a, b and c respectively, are estimated as follows. Axis b is determined as coincident with the $\beta$ maximum of bedding planes ($\pi$ axis), c is normal to b and nearest to the maximum of cleavage poles, and a is normal to both b and c. Although these estimations of the orthogonal axes follow the rule of Sander (1930), the axes are labelled only for descriptive purposes. It may be noted that lineations are close to the estimated fabric axis a, but scattered about the ac girdle. The considerable scatter of all the planar and linear structures may be explained as a combination of the heterogeneity of distribution of fabric axes during a genetically single folding episode and the superposition of folding; two examples are given below.

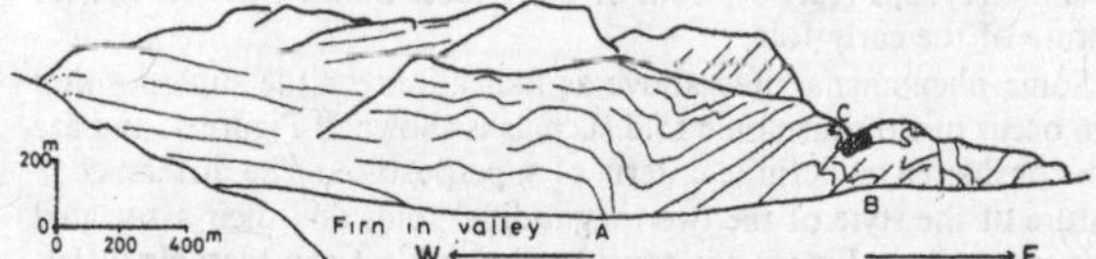


**Figure 2. "Fold cliff" viewed from the eastern slope of Mount Fordel. Dotted part: Dark limestone. Blank: White limestone. Structural data between A and B are plotted on Figure 3. C is the locality of Figure 4. The scale of the sketch is approximate.**

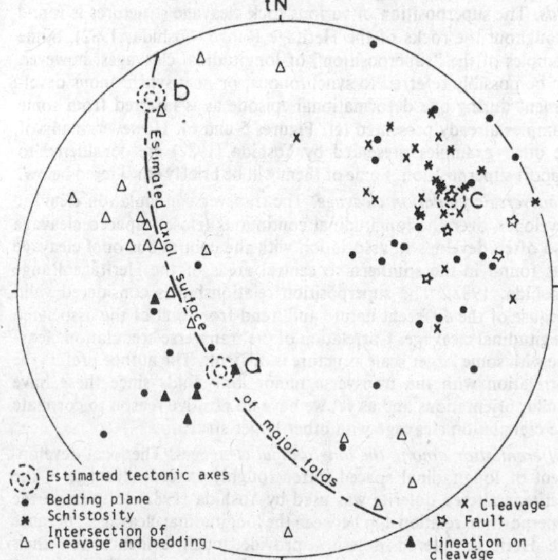


**Figure 3. Equal area stereographic projection (lower hemisphere) of structural data from part of the "Fold cliff", Marble Hills. The location is indicated in the caption to Figure 2.**

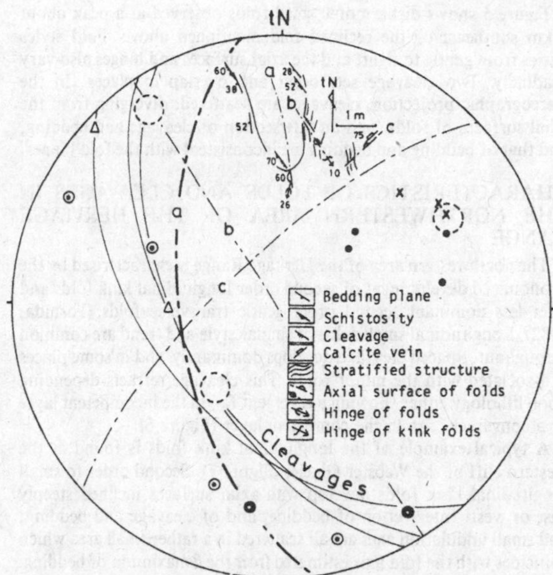


**Figure 4. Equal area stereographic projection (lower hemisphere) and a sketch (horiztontal projection) of a reclined fold at the "Fold cliff", Marble Hills (point MY79121407, the location is indicated in Figure 2). Letters on the great circles are referred to those of the sketch. Great circles denote the axial surface (chain) or the cleavage (solid). Small circles by dashed line are the estimated mean tectonic axes in the "Fold cliff" shown in Figure 3. Other symbols are as in Figure 5.**

Figure 4 is a tight reclined fold observed at the eastern outcrop of the "Fold Cliff". The reclined fold is refolded by a vertical and longitudinal minor kink fold and an inclined and transverse minor kink fold. The stereographic projection shows the association of cleavages slightly inclined to the axial surface of the reclined fold, deviation of intersections of bedding planes from the fold hinge, and inconsistency of the cleavage/bedding intersections with the fold hinge.

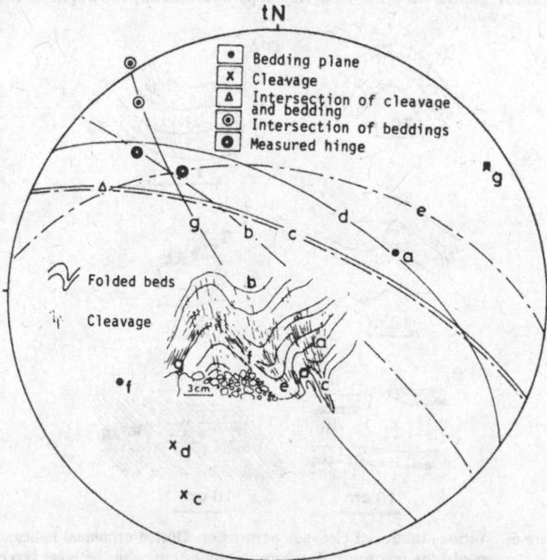


**Figure 5. Equal area stereographic projection (lower hemisphere) of a small disharmonic fold and its profile at Marble Hills (point MY79122201). Great circles denote the axial surface (chain) or the cleavage (solid).**

Figure 5 shows disharmonic small folds observed at a peak about 2 km southeast of the reclined fold mentioned above. Fold styles varies from gentle to tight, and the axial surfaces and hinges also vary gradually. Two cleavage sets occur and overlap in places. In the stereographic projection, cleavages are scattered, diverging from the axial surfaces of folds, and an intersection of cleavage and bedding, and that of bedding and bedding are inconsistent with the fold hinges.

## CHARACTERISTICS OF FOLDS AND CLEAVAGES IN THE NORTHWESTERN AREA OF THE HERITAGE RANGE

The northwestern area of the Heritage Range is characterised by the pronounced development of second order longitudinal kink folds and later less dominant second order gentle transverse folds (Yoshida, 1982). Longitudinal small folds of similar style and trend are common throughout. Spaced cleavage develops dominantly, and in some places is associated with the minor folds. This cleavage refracts depending upon lithology either showing a divergent fan in the incompetent layer or a convergent fan in the competent layer (Figure 6).

A typical example of the longitudinal kink folds is found at the western cliff of the Webster Glacier (Figure 7). Second order to small longitudinal kink folds develop with axial surfaces inclined steeply east or west. Intersection of bedding, and of cleavage and bedding, and small undulation axes are all scattered in a rather small area which coincides with the fold axis estimated from the β maximum of bedding. Cleavages are scattered reflecting either the convergent (sandy layers) or the divergent (muddy layers) fans as suggested also from the inset sketch. In this cliff, no more than one episode of kink folds can be distinguished.

## DISCUSSION AND SUMMARY

*Field evidence for the superposition of folds*

*Superposition of the minor transverse kink folds.* Superposition of the transverse folds over the longitudinal folds is considered valid from data and discussions given by Yoshida (1982); a small-scale example is also seen in Figure 4 where the cleavage structure associated with the reclined fold is refolded by a minor transverse fold.

*Chronology of the longitudinal folds.* The superposition of some longitudinal folds over other longitudinal ones is problematic. Yoshida (1982) noticed the superposition phenomena between longitudinal cleavages and explained it by estimating the superposition-relationships between the longitudinal folds, but presented no direct observations of the refolding. As is seen in Figure 4, a longitudinal minor kink fold with associated spaced cleavage affected the pre-existing continuous cleavage associated with a reclined fold. This type of minor gentle or kink fold develops sporadically throughout the southern to central parts of the Heritage Range (Yoshida, 1982). Refolding can also be estimated either from the systematic change of bedding planes from one side of the fold limbs or from the intersections of bedding planes from the opposing limbs of a fold as shown by Yoshida (1982, Figure 9); both of these cases indicate the curvilinear nature of the early fold.

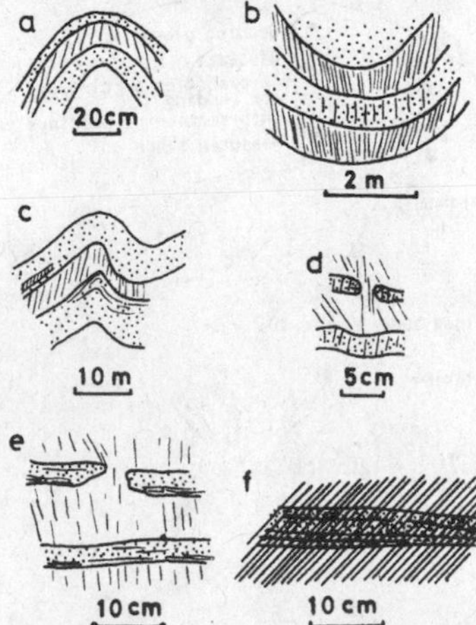


**Figure 6. Various modes of cleavage occurrence. Dotted ornament indicates sandstone (except in d, where it is limestone) and the other layers are shale. a: Near MY79120701, Webster Glacier. b: Point MY79123020, Wilson Nunataks. c: Point MY79120305, Webers Peaks. d: Point MY79120602, Soholt Peak. e: Point MY79120401, Pipe Peak. f: Point MY79120407, Pipe Peak.**

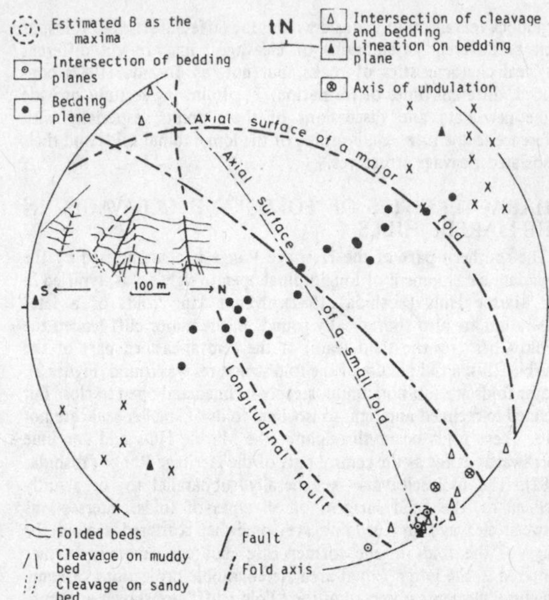


**Figure 7. Equal area stereographic projection (lower hemisphere) and a profile of the longitudinal kink folds at the western cliff of the Webster Glacier (Point MY79120701).**

Some phenomena cited above as indications of the superposition can occur in a disharmonic fold such as is shown in Figure 5, and are not considered as definite criteria of superposition. The difference in nature of the style of the two longitudinal folds and their associated cleavage seen in Figure 4 is considered to reflect considerable difference in either the physical properties of rocks, PT environmental conditions, or the strain rate during each of the folding events, and therefore may be regarded as indicating the superposition of the two folds.

*Superposition of cleavages and its bearing on the superposition of folds.* The superposition of various rock cleavage structures is found throughout the rocks of the Heritage Range (Yoshida, 1982). Some examples of the "superposition" of longitudinal cleavages, however, can be possibly referred to synchronous, or near-synchronous development during one deformational episode as is inferred from some examples already presented (cf. Figures 5 and 6). However, many of the other examples presented by Yoshida (1982) are considered to support superposition. Some of them will be briefly considered below.

*Transverse crenulation cleavage.* The transverse crenulation cleavage developed over the longitudinal continuous (closely spaced cleavage also often develops in association with the continuous one) cleavage was found in the southern to central areas of the Heritage Range (Yoshida, 1982). The superposition relationship is considered valid because of the different nature and trend from that of the associated longitudinal cleavage. Correlation of the transverse crenulation cleavage with some larger scale structure is difficult. The author prefers the correlation with the transverse minor kink folds since these have similar orientations and as yet we have no positive reason to correlate the crenulation cleavage with other larger structures.

*Differentiation among the longitudinal cleavages.* The local development of longitudinal spaced (often roughly spaced) cleavage in the Heritage altered dolerite was used by Yoshida (1982) to indicate the superposition relationship between the longitudinal cleavages, because the Heritage altered dolerites provide much evidence for their intrusion later than the main longitudinal cleavage. The fissile characters of the clastic country rocks is lost in contact with or adjacent to the Heritage altered dolerites. Inclusions of schistose limestone were found in a marginal part of the dolerite mass at the Edson Hills, the periphery of the inclusion being coarsely recrystallised (cf. Yoshida, 1982, Figure 19). The recrystallisation of the pre-existing continuous cleavage is supported also by microscopic observations.

The correlation of the later longitudinal spaced cleavages in the southern area with fold structures is a problem. There is the example of the spaced cleavage associated with a minor longitudinal kink fold superposed over continuous to closely spaced cleavage associated with a reclined fold as mentioned above.

*Correlation of structures between the southern and northwestern areas of the Heritage Range.* Fold and cleavage structures of the southern area and those of the northwestern area differ significantly in various points as described earlier. However they appear to change gradually and may therefore be attributed to one deformational event. The different folds and cleavages may be attributed either to inhomogeneities of the rocks caused by both the stratigraphy and lithology, or to the spatial inhomogeneity of the tectonic force. The former may be eliminated due to the new stratigraphic interpretation of the Heritage Range (that the Marble Hills occupies the upper stratigraphic position of the Heritage Group) and by the abundance of thick conglomerate beds in the southern area where the deformation is more intense than in the northwestern area. As to the latter possibility, some discussions follow, concerning the correlation of longitudinal minor kink folds of the southern area with the longitudinal major kink folds in the northwestern area.

Characteristic chlorite-quartz veins develop throughout the Heritage Range, their chlorite is optically distinctive, and all the veins appear to be cogenetic by their similar petrography and wide distribution (Yoshida, 1982). In the northwestern area, the longitudinal second order kink folds are associated with the chlorite-quartz veins, and in the southern area, the chlorite-quartz veins cut the longitudinal cleavage or develop in the Heritage altered dolerites. Thus it is possible to suggest that the longitudinal second order kink folds of the northwestern area postdate the longitudinal major cleavage formation in the southern area, the latter being correlated to the longitudinal major open to close to tight folds in the southern area. It may be possible to correlate the longitudinal minor kink folds of the southern area with the longitudinal second order kink folds of the northwestern area by their estimated chronologic relationships with the main cleavage and by the similarity in fold style and trend. The longitudinal open folds described by Yoshida (1982) as the latest folds are difficult to discuss at present, because of the paucity of examples.

*A polyorogenic interpretation of the geology of the Ellsworth Mountains*

*Sequence of folding and cleavage development.* From data and discussions above, the simplest sequence of folding and cleavage development in the southern to central area is as follows.

First stage: longitudinal open, close, and tight folds with longitudinal cleavages.

Second stage: longitudinal minor kink folds associated with spaced cleavages.

Third stage: transverse minor gentle folds and minor kink folds associated with local crenulation cleavages.

There is a possibility that the second order longitudinal kink folds and the second order transverse gentle folds in the northwestern area should be included in the second and the third stage structures, respectively. Further chronologic differentiation of the longitudinal folds as made by Yoshida (1982) or of transverse structures may be possible in future.

*Polyorogenic interpretations.* Yoshida (1982) collected various data on folds and cleavages in relation to dykes, veins, stratigraphy, metamorphic recrystallisations, and geochronology, and tentatively presented a polyorgenic summary as mentioned above. Many of his chronologic differentiations and groupings of various geologic phenomena appear possible from the field data presented, and attribution of these events to distinct orogenic phases ascertained in adjacent areas of Antarctica may be probable from the K-Ar age data, although some alternative explanations may be possible. Data and discussions of the present study provided some support on the chronologic differentiation between the longitudinal folds, especially that before and after the Heritage altered dolerites. Further examination of tectonics and metamorphism, including that of the Sentinel Range, are required to make a decisive polyorogenic interpretation.

## REFERENCES

BRADSHAW, J.D., LAIRD, M.G. and WODZICKI, A., 1982: Structural style and tectonic history in northern Victoria Land; *in* Craddock, C. (ed.) *Antarctic Geoscience,* Univ. Wisconsin Press, Madison, 809-16.

CRADDOCK, C., 1969: Geology of the Ellsworth Mountains; *in* Bushnell, V.C. and Craddock, C. (eds.) Geologic maps of Antarctica. *Antarctic Map Folio Ser., Folio 12,* Pl.IV.

CRADDOCK, C., 1972: Tectonics of Antarctica; *in* Adie, R.J. (ed.) *Antarctic Geology and Geophysics,* Universitetsforlaget, Oslo, 449-55.

CRADDOCK, C., ANDERSON, J.J. and WEBERS, G.F., 1964: Geological outline of the Ellsworth Mountains; *in* Adie, R.J. (ed.) *Antarctic Geology,* North Holland, Amsterdam, 155-70.

FLEUTY, M.J., 1964: The description of folds. *Geol. Assoc. Proc., 75,* 461-92.

HJELLE, A., OHTA, Y. and WINSNES, T.S., 1978: Stratigraphy and igneous petrology of southern Heritage Range, Ellsworth Mountains. *Norsk Polarinst., Skr., 169,* 5-43.

HJELLE, A., OHTA, Y. and WINSNES, T.S., 1982: Geology and petrology of the southern Heritage Range, Ellsworth Mountains; *in* Craddock, C. (ed.) *Antarctic Geoscience,* Univ. Wisconsin Press, Madison, 599-608.

RICKARD, M..J., 1971: A classification diagram for fold orientations. *Geol. Mag., 108,* 23-6.

SANDER, B., 1930: *Gefugekunde der Gesteine.* Springer, Vienna.

SPLETTSTOESSER, J. and WEBERS, G.F., 1980: Geological investigations and logistics in the Ellsworth Mountains, 1979-1980. *U.S. Antarct. J., 15,* 36-9.

WINKLER, H.G.F., 1974: *Petrogenesis of metamorphic rocks,* 3rd ed. Springer, New York.

YOSHIDA, M., 1982: Superposed deformation and its implication to the geologic history of the Ellsworth Mountains, West Antarctica. *Mem. Nat. Inst. Pol. Res., Japan, Spec. Issue, 21,* 120-71.

# SUBGLACIAL MORPHOLOGY BETWEEN ELLSWORTH MOUNTAINS AND ANTARCTIC PENINSULA: NEW DATA AND TECTONIC SIGNIFICANCE

C.S.M. Doake, R.D. Crabtree, *British Antarctic Survey, High Cross, Madingley Road, Cambridge CB3 OET, U.K.*

I.W.D. Dalziel, *Lamont-Doherty Geological Observatory, Columbia University, Palisades, New York, 10964, U.S.A.*

*Abstract* Bedrock topography derived from airborne radio echo sounding shows an extensive subglacial rift system between the Ellsworth Mountains and the Antarctic Peninsula. Vinson Massif in the Ellsworth Mountains rises to 5140 m above sea level, the highest peak in Antarctica, while only 45 km to the east the Rutford Ice Stream rests on rock 2000 m below sea level. Rutford Ice Stream occupies a depression several hundred kilometres long and up to 30 km wide. Similar bedrock depressions to the northeast are filled by ice from Carlson Inlet and Evans Ice Stream, both reaching more than 1500 m below sea level. Between the ice streams are steep sided blocks with bedrock elevations around 300 m below sea level.

Although the ice sheet has undoubtedly caused some erosion, the morphology of the troughs suggests they are grabens, perhaps exhumed and emphasised by glacial action but not created by it. This view is supported by the discordant trough across Fowler Peninsula and the contrast with the subdued topography at the head of the Ross Ice Shelf where there is a similar pattern of ice stream flow. The trenches occupied by the ice streams may be the result of NE-SW crustal extension and, together with the blocks, need to be considered when reconstructing the postulated anticlockwise rotation of the Ellsworth Mountains since the Gondwana breakup. Precambrian rocks at Haag Nunataks suggest that a major boundary exists along the Evans Ice Stream separating the Antarctica Peninsula from the rest of Antarctica.

The geological structure of West Antarctica, and its relationship to the Precambrian shield of East Antarctica, together constitute the longest standing major problem in Antarctica earth science. Given the thick ice cover and low ratio of outcrop area to surface area it is hardly surprising that this has proved an intractable problem. The first outline of the sub-ice topography of West Antarctica came from U.S. oversnow traverses in the late 1950's and early 1960's (Behrendt, 1963; Bentley and Ostenso, 1961). Recently, airborne radio echo sounding techniques have greatly expanded the data base of sub-ice topography (Swithinbank, 1977; Jankowski and Drewry, 1981).

Together with geologic and palaeomagnetic data, and scanty data on crustal structure, the sub-ice topography indicates that West Antarctica consists of at least four discrete or semi-discrete continental blocks (Dalziel and Elliot, 1982). In 1981, a joint British Antarctic Survey (BAS)-U.S. Antarctic Research Programme (USARP) project was initiated to shed further light on the tectonics of West Antarctica and its relationship to East Antarctica. We report here the results of the first part of this study. Airborne radio echo sounding was carried out by a BAS Twin Otter aircraft with fuel cached at the site of the 1979-80 USARP camp in the Ellsworth Mountains. The flight lines were jointly planned with combined tectonic and glaciological goals. A number of flights covered the area between the Ellsworth Mountains and the Antarctic Peninsula, complementing sounding carried out in 1975 (Swithinbank,1977). Figure 1 shows these flight lines, other BAS airborne soundings from previous seasons and the route taken by the Antarctic Peninsula Traverse (APT) in 1961-62 when ice thicknesses were found by seismic shooting (Behrendt, 1963).

Navigation in 1981 was controlled by Doppler radar, so that positions are accurate to within 2 km. In 1975 an inertial navigation system was used with comparable accuracy. Ice thickness was measured to about 20 m with a pulsed 60 MHz system. The aircraft altitude was measured by pressure altimetry; the resulting accuracy of bedrock elevations is better than 100 m. Figure 1 shows bedrock topography in eastern Ellsworth Land for the area where ice is grounded. The draft of the ice shelf close to the grounding lines of the ice streams shows that bedrock must go down to 2000 m below sea level (Swithinbank,1977) but there are insufficient data to be able to draw bedrock contours below 1000 m below sea level. In areas with sparse data bedrock contours have been interpolated with the help of Landsat pictures of the ice surface.

## Subglacial Morphology

Between the elevated highlands of the Antarctic Peninsula (higher than 1000 m above sea level) and the Ellsworth Mountains (reaching more than 5000 m above sea level) lies an area of graben- and horst-like structures where bedrock falls to 2000 m below sea level. Three prominent troughs are occupied by glaciers draining the ice sheet—Rutford Ice Stream, Carlson Inlet and Evans Ice Stream. Another deep trough orientated almost perpendicularly to the outlet glaciers splits Fowler Peninsula into two distinct blocks.

The only exposed rocks between Ellsworth Mountains and the Antarctic Peninsula occur at Haag Nunataks, where metamorphic rocks have been radiometrically dated at about 1000 Ma (Clarkson and Brook, 1977). Morphologically the Haag Nunataks block has relief of up to 500 m over distances of a few kilometres (Figure 2b) while the blocks which constitute Fletcher Promontory, the end of Fowler Peninsula, Skytrain Ice Rise and Korff Ice Rise are very smooth (Figure 2a) with roughness scales of tens of metres over tens of kilometres. Another distinguishing feature between Haag Nunataks and the other raised blocks is their elevation; the Haag Nunataks block rises to 1000 m above sea level where the rocks are exposed and has an area above sea level of about 8500 km$^2$. By contrast, the other blocks lie generally between 300 and 500 m below sea level. Their areas at the -1000 m contour are given in Table 1.

**TABLE 1: Area of blocks separating ice streams at the −1000 m elevation contour**

| Block | Area ($10^3$km$^2$) |
|---|---|
| Haag Nunataks | 20 |
| S.E. Fowler Peninsula | 5 |
| Fletcher Promontory | 10 |
| Korff Ice Rise | 6 |
| Skytrain Ice Rise | 6 |

Rutford Ice Stream and Carlson Inlet occupy troughs 30 to 50 km wide while Evans Ice Stream is around 100 km wide. All three troughs are between 200 and 300 km long, although data are sparse both at the inland end where ice is too thick for continuous sounding and at the seaward end where no data exist for water thickness under Ronne Ice Shelf. The trough cutting Fowler Peninsula is about 40 km wide and 100 km long. All the troughs have steep sides ranging up to 34°, which is the limit which can be measured directly by radio echo because of refraction of radio waves in ice (Harrison, 1971). These troughs contrast with the narrower and more sinuous fiord-like structures fringing the Antarctic Peninsula block, which are typically a few kilometres wide, up to 100 km in length and have vertical relief of about 1000 m.

## Tectonic Significance

The deep, wide, steep sided troughs suggest that they are a product of tectonic activity rather than glacial erosion. Although the outlet glaciers have undoubtedly been responsible for emphasising the structures, the horsts and grabens seem too distinctive to have been formed from an unmodified surface. The periods of Antarctic glaciation since the Mesozoic, and the inception of present ice sheet, are not well dated. Build up of ice, starting as local valley glaciation, may have begun during the Early Oligocene (Shackleton and Kennett, 1975) but significant ice cover may not have developed until Mid-Miocene, 12 to 13 Ma ago (Loutit and Kennett, 1980). It is generally agreed that glaciers will preferentially erode along lines of weakness or topographic lows (and therefore tend to occupy pre-existing river drainage networks) until the structural controls are overriden and the size and shape of the trough becomes a function of ice discharge. Haynes (1972) found a relationship between trough size and drainage basin area for outlet glaciers in West Greenland. Applying her

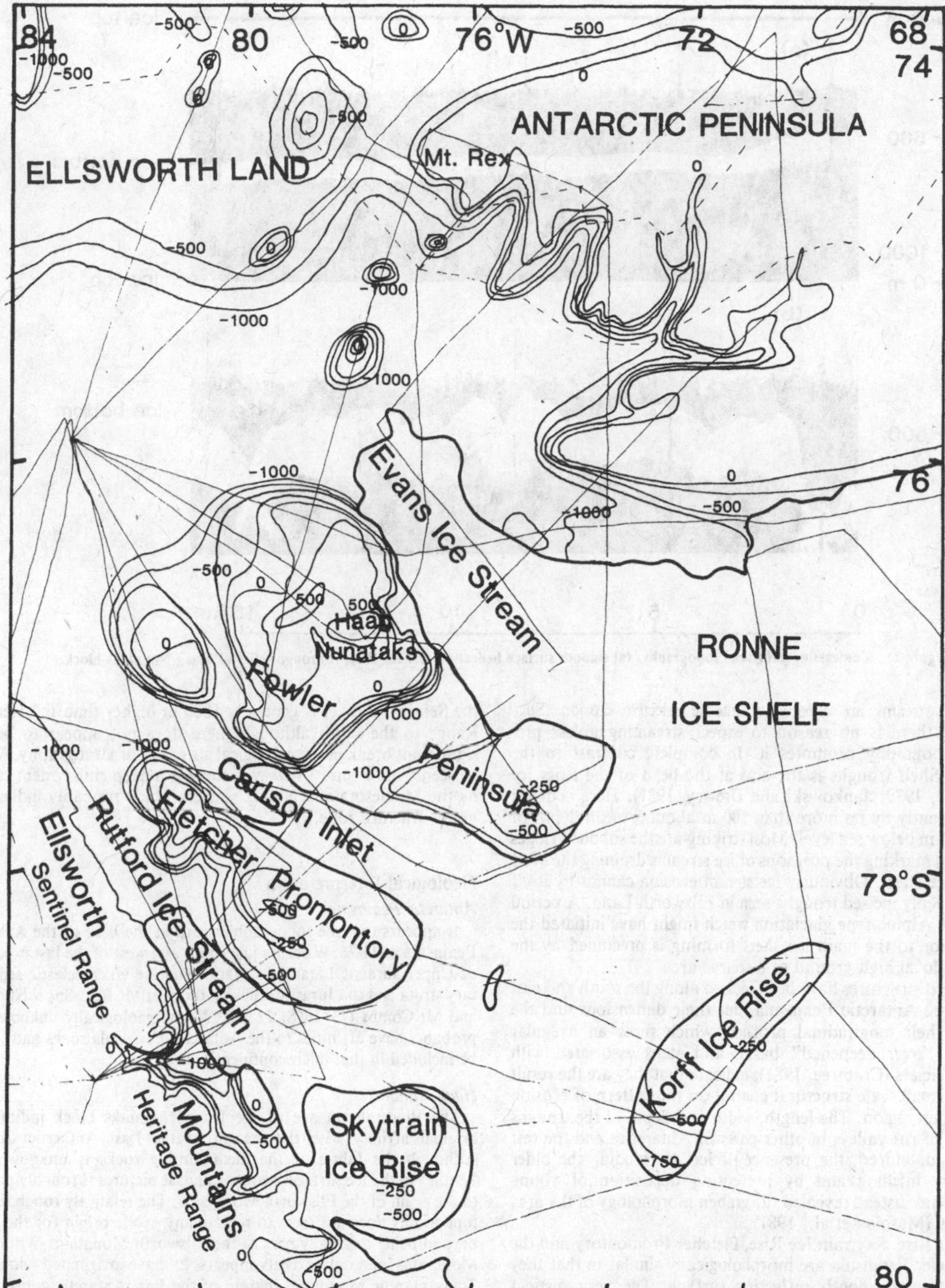


**Figure 1. Bedrock topography (contours at 250 m intervals between 0 and 1000 m below sea level). Radio echo flight lines are marked by thin full lines, seismic traverse by pecked line.**

formula to the cross-sectional areas of Rutford Ice stream, Carlson Inlet and Evans Ice Stream at their grounding lines requires them to have been eroded by ice from a drainage basin of more than 480,000 km², three times their present drainage area (Swithinbank, 1977) which is unrealistically large, even for an expanded ice sheet. Graf (1970) stated that glacial troughs have a depth/width ratio of between 0.24 and 0.45. Sognefiord (Norway) has a ratio of 0.22, Yosemite Valley (U.S.A.) of 0.75 and Clifford Glacier (Antarctic Peninsula) of 0.26 (all authors' estimates). But Rutford Ice Stream has a depth/width ratio of 0.05 and Evans Ice Stream of 0.03 which suggests that they are far too wide in relation to their depths to have been entirely eroded by ice. Lambert Glacier with a similar ratio of about 0.04 has been recognised as a graben structure dating probably from the Late Mesozoic (Fedorov et al., 1982). The discordant trough across Fowler Peninsula gives very strong support against glacial erosion being the process which carved the troughs; there is very little ice flow through this trough now, and there is unlikely to have ever been much flow in the past when the ice sheet may have been a different size. Any increase in outflow would be channelled between the Ellsworth Mountains and the Antarctic Peninsula along present drainage lines, sheltering the Fowler Peninsula trough.

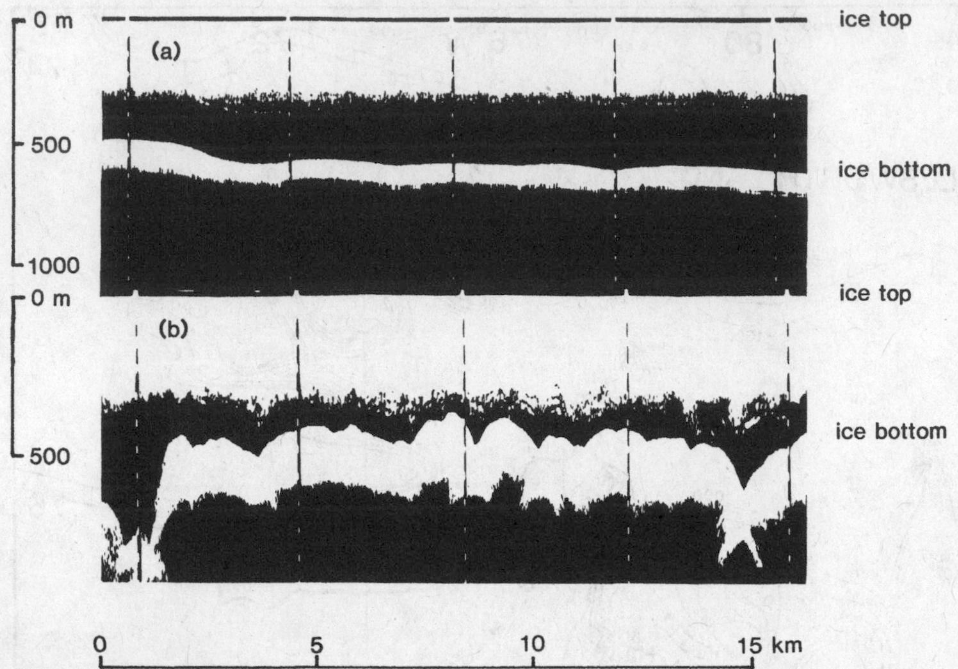


**Figure 2. Contrasting subglacial topography: (a) smooth surface beneath Korff Ice Rise, (b) rough relief of Haag Nunataks block.**

While ice streams are capable of intense selective erosion (Sugden, 1968) there is no reason to expect streaming unless pre-existing topography promotes it. In complete contrast to the Ronne Ice Shelf troughs is the area at the head of the Ross Ice Shelf (Rose, 1979; Jankowski and Drewry, 1981). Here bedrock undulates gently by no more than 500 m about a mean depth of around 750 m below sea level. Most striking are the subdued ridges and troughs marking the positions of ice streams draining the West Antarctic ice sheet. Obviously ice stream erosion cannot by itself form the deeply incised troughs seen in Ellsworth Land. A period of valley or Alpine-type glaciation which might have initiated the troughs prior to the main ice sheet forming is precluded by the absence of local high ground as a source area.

Deep fiord structures have been incised along the south and east coasts of the Antarctic Peninsula, but their dimensions and the nature of their longitudinal profiles, which show an irregular pattern of "over deepened" basins and steps associated with tributary glaciers (Crabtree, 1981), indicate that they are the result of normal small scale structural control on the pattern of erosion in a mountain region. The length, width and depth of the troughs are typical of rift valleys in other parts of Antarctica and the rest of the world; indeed, the presence of ice, by eroding the older sedimentary infilling and by preventing deposition of young sediments, has instead revealed the graben morphology of the area very clearly (Masolov et al., 1981).

Korff Ice Rise, Skytrain Ice Rise, Fletcher Promontory and the end of Fowler Peninsula are morphologically similar in that they all show a very smooth reflecting surface. The very distinct difference between the Haag Nunataks block and the end of the Fowler Peninsula suggests a significant geological or lithological boundary through the trough. Without supporting magnetic, gravity or seismic data it is not possible to determine if there has been considerable lateral movement. Gravity data from the APT traverse (Behrendt, 1964) show a very strong break following the topographic margin along the southern end of the Antarctic Peninsula, supporting the idea of a tectonic boundary through the trough filled by Evans Ice Stream (Kellogg, 1980). On the eastern edge of the Ellsworth Mountains, Vinson Massif, the highest mountain in the Antarctic at 5140 m, lies only 45 km from bedrock more than 2000 m below sea level covered by Rutford Ice stream. The scale of this relief suggests another major tectonic boundary close by. The northern part of the Ellsworth Mountains, the Sentinel Range, is generally 1000 m higher than the Heritage Range to the south although there does not appear to be any significant break in the geological structure or stratigraphy. As can be seen from Figure 1 however, there is a deep embayment caused by the Minnesota and Nimitz glaciers which probably indicates a major fracture zone.

## Geological Interpretation

### *Antarctic Peninsula*

It appears that the topographic scarp at the base of the Antarctic Peninsula at 74-76°W occurs just south and west of the last exposures of Upper Jurassic Latady Formation marine volcaniclastic sedimentary strata and the Jurassic volcanic rocks at Mt Rex. Snow Nunataks and Mt Combs (73°30′S, 77°-79°W) are geologically unknown but probably have affinities to the Antarctic Peninsula rocks and should be included in that microcontinental block.

### *Haag Nunataks*

The Precambrian age of the Haag Nunataks block indicate an original affinity with the margin of the East Antarctic Craton. Although the fabric of the metamorphic rocks is unknown, the "grain" in the ice surface seen on Landsat pictures is roughly parallel to the grain of the Ellsworth Mountains. The relatively rough sub-ice topography does not seem to require any exotic origin for the block beyond being originally part of the Ellsworth Mountains-Whitemore Mountains microplate that appears to have originated along the Transantarctic Mountains margin of the East Antarctic craton. It is worth noting that the Heritage Range with its comparitively low relief and rough topography contains the older and structurally lower part of the Ellsworth section. It is the Sentinel Range that is anomalously high for Palaeozoic strata deformed in the Early Mesozoic.

### *Ice Rises and Peninsulas*

The subdued sub-ice relief of the ice rises and inland ice sheet promontories suggests the possibility that they are cored by Beacon sedimentary strata, particularly given the regional setting and the fact that folding in the Permian (*Glossopteris*-bearing) Polarstar Formation dies out in the Ellsworth Mountains eastwards towards the Rutford Ice Stream (observations made by IWDD and M.R.A. Thomson on Flowers Hills ridge in the Sentinel Range, 1980). Radio Echo profiles taken over presumed Beacon Supergroup strata inland from the Transantarctic Mountains show mesa-like topographic features (Drewry, 1975) but appear to be more rugged than the topography on the "smooth" Ellsworth blocks. This however may be due to different pre-glacial erosion histories, the Transantarctic strata being around 1000 to 2000 m above sea level compared with the -500 m of the Ellsworth strata.

*Ice Streams*

The difference between the fjord-like structures of the base of the Antarctic Peninsula and the trenches occupied by the ice streams suggest that the latter are grabens created by the effects of tectonic extension. While the Rutford, Carlson and Evans streams might reflect NE-SW extension, separation of the Haag Nunataks block from the block to the SE may suggest another direction of extension.

## Conclusions

While there is nothing definitive in the data from the geologic standpoint, it seems that the sub-ice topography and bedrock between Ronne Ice Shelf and the Whitmore Mountains are in keeping with the model of anticlockwise rotation of a crustal block away from the margin of the East Antarctic craton as suggested by the geologic and palaeomagnetic data (Schopf, 1969; Watts and Brammall, 1981; Dalziel and Elliot, 1982).

If the foregoing is correct it means that the major structural boundary of the region is the one between Antarctic Peninsula and Haag Nunataks block, that is, along the Evans Ice Stream. This would be a plate boundary of considerable significance between the Antarctic Peninsula microcontinent along the Pacific margin and the Greater Ellsworth microcontinent. Several theories suggest that the Antarctic Peninsula has rotated clockwise from its original position (Barker and Griffiths, 1977; de Wit, 1977) while the Ellsworth microcontinent has rotated anticlockwise (Watts and Brammall, 1981; Dalziel and Elliott, 1982). In the process of rotation, or else later the Greater Ellsworth microcontinent was subjected to crustal extension on a broad scale, splitting it up into the Haag Nunataks and ice rise blocks, the Ellsworth Mountains block and the Whitmore Mountains block.

An alternative hypothesis would be to consider the Ellsworth Mountains themselves, the Haag Nunataks block and the ice rises to comprise three or more distinct plates of separate origins, that is, a collage of allochthonous terrains. There are no data to refute this idea, but at present it seems unnecessarily complex.

*Acknowledgements* We wish to thank H.E. Thompson for help in collecting radio echo data, and M.R.A. Thomson for discussions on interpretation of the data. This work was supported by National Science Foundation grant DPP 78-20670 to Professor I.W.D. Dalziel by Lamont-Doherty Geology Observatory, Columbia University, U.S.A. The British Antarctic Survey is a component body of the U.K. Natural Environment Research Council.

## REFERENCES

BARKER, P.F. and GRIFFITHS, D.H., 1977: Towards a more certain reconstruction of Gondwanaland. *R. Soc. Lond., Philos. Trans., Ser. B., 279,* 143-59

BENTLEY, C.R. and OSTENSO, N.A., 1961: Glacial amd subglacial topography of West Antarctica. *J. Glaciol., 3,* 882-911.

BEHRENDT, J.C., 1963: Seismic measurements on the ice sheet of the Antarctic Peninsula. *J. Geophys. Res., 68,* 5973-90.

BEHRENDT, J.C., 1964: Crustal geology of Ellsworth Land and the Southern Antarctic Peninsula from Gravity and Magnetic Anomalies. *J. Geophys. Res., 69,* 2047-63.

CLARKSON, P.D. and BROOK, M., 1977: Age and position of the Ellsworth Mountains crustal fragment, Antarctica. *Nature, 265,* 615-6.

CRABTREE, R.D., 1981: Subglacial morphology in northern Palmer Land, Antarctic Peninsula. *Annals Glaciology, 2,* 17-22.

DALZIEL, I.W.D., and ELLIOT, D.H., 1982: West Antarctica: problem child of Gondwanaland. *Tectonics, 1,* 3-19.

DE WIT, M.J., 1977: The evolution of the Scotia arc as a key to the reconstruction of Southwestern Gonwanaland. *Tectonophysics, 37,* 58-81.

DREWRY, D.J., 1975: Radio echo sounding map of Antarctica, (~90°E-180°). *Polar Rec., 17,* 359-74.

FEDEROV, L.V., GRIKUROV, G.E., KURININ, R.G. and MASOLOV, V.N., 1982: Crustal structure of the Lambert Glacier area from geophysical data; *in* Craddock, C. (ed.) *Antarctic Geoscience,* Univ. Wisconsin Press, Madison, 931-6.

GRAF, W.L., 1970: The geomorphology of the glacial valley cross section. *Arctic and Alpine Res., 2,* 303-12.

HARRISON, C.H., 1971: Radio echo records cannot be used for evidence for convection in the Antarctic ice sheet. *Science, 173,* 166-7.

HAYNES, V.M., 1972: The relationship between the drainage areas and sizes of the outlet troughs of the Sukkertoppen ice cap, West Greenland. *Geogr. Annlr., 54A,* 66-75.

JANKOWSKI, E.J., and DREWRY, D. J., 1981: The structure of West Antarctica from geophysical studies. *Nature, 291,* 17-21.

KELLOG, K.S., 1980: Paleomagnetic evidence for oroclinical bending of the southern Antarctic Peninsula. *Geol. Soc. Am., Bull., 91,* 414-20.

LOUTIT, T.S., and KENNETT, J.P., 1980: Polar glacial evolution and global sea level changes. *U.S. Antarct. J., 15,* 99-101.

MASOLOV, V.N., KURININ, R.G. and GRIKUROV, G.E., 1981: Crustal structures and tectonic significance of Antarctic rift zones (from geophysical evidence); *in* Cresswell, M,M, and Vella, P., (eds.) *Gondwana Five,* A.A. Balkema, Rotterdam, 303-9.

ROSE, K.E., 1979: Characteristics of ice flow in Marie Byrd Land, Antarctica. *J. Glaciol., 24,* 63-75.

SCHOPF, J.M., 1969: Ellsworth Mountains: position in West Antarctica due to sea floor spreading. *Science, 164,* 63-6.

SHACKLETON, N.J. AND KENNETT, J.P., 1975: Paleotemperature history of the Cenozoic and the initiation of Antarctic glaciation: oxygen and carbon isotope analysis in DSDP sites 277,279 and 281. *Initial Reports of the Deep Sea Drilling Project, 29,*801-76, U.S. Govt. Printing Office, Washington, DC.

SUGDEN, D.E., 1968: The selectivity of glacial erosion in the Cairngorm Mountains, Scotland. *Trans. Inst. Brit. Geogr., 45,* 79-92.

SWITHINBANK, C.W.M., 1977: Glaciological research in the Antarctic Peninsula. *R. Soc. Lond., Philos. Trans. Ser. B., 279,* 161-83.

WATTS, D.R. and BRAMALL, A.M., 1981: Palaeomagnetic evidence for a displaced terrain in Western Antarctica. *Nature, 293,* 638-41.

# SWANSON FORMATION AND RELATED ROCKS OF MARIE BYRD LAND AND A COMPARISON WITH THE ROBERTSON BAY GROUP OF NORTHERN VICTORIA LAND

J.D. Bradshaw, *Geology Department, University of Canterbury, Christchurch, New Zealand.*

P.B. Andrews, *N.Z. Geological Survey (D.S.I.R), P.O. Box 30368, Lower Hutt, New Zealand.*

B.D. Field, *N.Z. Geological Survey (D.S.I.R), University of Canterbury, Christchurch, New Zealand.*

*Abstract* The Swanson Formation is the oldest unit exposed in the Ford Ranges of Marie Byrd Land; it was folded and metamorphosed in the Middle Ordovician before the intrusion of Late Palaeozoic granitoids. Neither base nor top are exposed but a thickness of several kilometres is likely. The formation is a monotonous quartz-rich metaflysch in which non-graded and graded sandstones, the latter with partial Bouma sequences, alternate with slatey mudstone. Two measured sections totalling 1 km are presented. Palaeocurrent measurements indicate deposition on a surface which dipped to the east but was swept by bottom currents which flowed to the north. Weakly hornfelsed bedded tuffs at Lewisohn Nunatak are distinguished as the Lewisohn Member. They are less deformed than typical Swanson Formation but follow Swanson type rocks concordantly and are correlated with tuffaceous rocks interbedded in the Swanson Formation in the Mackay Mountains 20 km to the north. The name Fosdick metamorphic zone is applied to higher rank Swanson metasediments included in the Fosdick Complex. The latter is unlikely to be older than Middle Ordovician and is not regarded as pre-Swanson Basement. Recent work on the Robertson Bay Group confirms the close sedimentological and compositional similarities. However this is not sufficient ground for considering them to be the same unit. Palaeocurrent patterns are quite different. Their relationship to Devono-Carboniferous granitoids suggests they were in the same orogenic belt and had a similar relationship to Late Palaeozoic plate boundaries.

Marie Byrd Land forms a broad salient lying to the east of the Ross Sea. The Ford Ranges are the best exposed part of that region (Figure 1). The ranges are composed of three main rock types; the low rank metasediments of the Swanson Formation, high rank metasediments and igneous rocks of the Fosdick Complex and granitoids of Late Palaeozoic (345 ± 11 Ma) and Late Mesozoic age (143 ± 4 Ma and 100 ± 10 Ma). Cenozoic volcanics are of minor importance. The results of earlier geological exploration in western Marie Byrd Land are summarised in Wade and Wilbanks (1972) and Wade and Couch (1982). Reconnaissance geological maps have been published (Wade et al., 1977a and b). This paper concentrates on the sedimentology of the Swanson Formation, a thick, unfossiliferous quartzose metaflysch of probable Early Palaeozoic age, which was folded and metamorphosed before the intrusion of the Carboniferous granitoids.

In many Gondwana re-assemblies Marie Byrd Land abuts the Campbell Plateau region of the New Zealand continent and in consequence may be viewed as an extension of the Western Province of New Zealand which has similar granitoids and Early Palaeozoic sedimentary sequences (Cooper, 1979; Cooper et al. 1982; Grindley and Davey, 1982). In addition Marie Byrd Land has been linked with the eastern part of the northern Victoria Land as part of the Borchgrevink Orogen (Craddock, 1972; Wade and Couch, 1982) mainly on the basis of comparable granitoids and similar sedimentary sequences, namely the Swanson Formation and the Robertson Bay Group respectively.

## Swanson Formation

The Swanson Formation includes all sedimentary rocks older than the Late Palaeozoic granites. Two variants are named, the Lewisohn Member for tuffs and tuffaceous sediments at Lewisohn Nunatak and in the Mackay Mountains, and the Fosdick metamorphic zone for the more metamorphosed sediments in the Fosdick Complex. The Swanson Formation is widely exposed in the Denfield, McKay, Clark and Allegheny Mountains (Figure 1), and measured sections of about 0.5 km were logged at Mt Crabtree and Mt Passel. Shorter sections were also taped in the Clark Mountains and at Lewisohn Nunatak. Continuous sections of about 1 km exist but no idea could be formed of the thickness of the full sequence as neither top nor base is exposed. Warner (1945) suggested a thickness of 4.3 km and Lopatin and Orlenko (1972) 12-15 km. In view of the size of the outcrops and scale of folding this latter figure must be speculative. Although there are minor variations there are no distinctive sub-units which could be mapped from outcrop to outcrop.

*Sedimentology* The Swanson Formation is a grey well bedded quartz-rich metaflysch. The sequence is sand-rich (Figure 2) with mudstone interbeds clearly subsidiary. Thick mudstone units are rare. Within the succession coarsening and fining upward sequences can be identified but these are usually only about 15 to 50 m in thickness and occur in unpredictable order. Sandstones are typically sharp based with flame structures, load, flute and groove casts. The overlying sandstones show a range of sedimentary structures (Figure 3) including graded bedding, parallel lamination and ripple scale cross stratification (including spectacular climbing ripples). Some sandstones show amalgamation of several sedimentation units, and massive ungraded sands also occur. Complex patterns of ripple cross lamination are seen in some beds with a sharp distinction between sandstone and overlying mudstone. Consequently, rippled surfaces are exposed at a number of places and suggest reworking of sand by bottom currents. Beds in the upper part of the Mt Passel section (Figure 3b) show deep, scoured troughs (300 mm + )and decimetre scale cross bedding.

Coarse sediments are rare though thick (2 m + ) sandstones sometimes have quartz granules. At Mt Crabtree and Gregor Peak well rounded sandstone pebbles indistinguishable from Swanson Formation occur in an abundant matrix. The pebbles resemble reworked concretions and suggest local cannibalism of partly indurated parts of the sequence.

*Palaeocurrents* Palaeocurrent trends (Figure 4) are distinctly bimodal in all three measured sections. At Mt Passel sole structures clearly indicate that the palaeoslope dipped east and that many currents producing cross beds were flowing downslope. However a majority of measurements at both Mt Passel and Mt Crabtree are strongly oblique or normal to the indicated slope and reflect persistent currents flowing mainly to the north.

*Interpretation* The sandstones appear to have been emplaced mainly as sediment mass flows. Many show partial Bouma sequences and are probably turbidites. Widespread development of climbing ripples suggests that many flows were highly concentrated leading to rapid deposition. In conjunction with the presence of some thick non-graded sandstones they suggest "proximal" situation. The remarkable uniformity of succession through a considerable thickness and large area (no other sediments occur in an area of at least 350 km x 450 km) suggests a submarine fan setting. The fan appears to have sloped east and perhaps northeast in the Clark Mountains. The north flowing current is interpreted as bottom current, probably a contour current (Stow and Lovell, 1979) sufficiently strong to rework the tops of sand bodies to produce new cross-lamination and ripple structures. Laird (1972) also reports bottom currents at a high angle to sole structures in the broadly contemporaneous and lithologically similar Greenland Group of New Zealand. A southerly directed counter current seems to have flowed from time to time. The strong bottom currents may have inhibited the development of an incised channel system on the fan which may explain the irregular coarsening and fining upward cycles.

*Petrography* Petrographic work confirms earlier reports (Passel, 1945; Wade and Couch, 1982). Sandstones are typically medium to fine grained micaceous quartzose litharenites. Quartzitic rock fragments are the principal lithic component with rare acid to intermediate volcanic rock fragments. Sodic plagioclase, detrital muscovite, rutile, and zircon are widespread but not volumetrically significant. The matrix, which is locally abundant, is a mass of quartz, sericite and chlorite. Impure rusty weathering carbonate is particularly common in concretions in finer grained beds. Tuffaceous sandstone interbedded in typical Swanson Formation in the Mackay Mountains are correlated with the Lewisohn Member (see below).

The Swanson Formation is altered to spotted biotite hornfels in

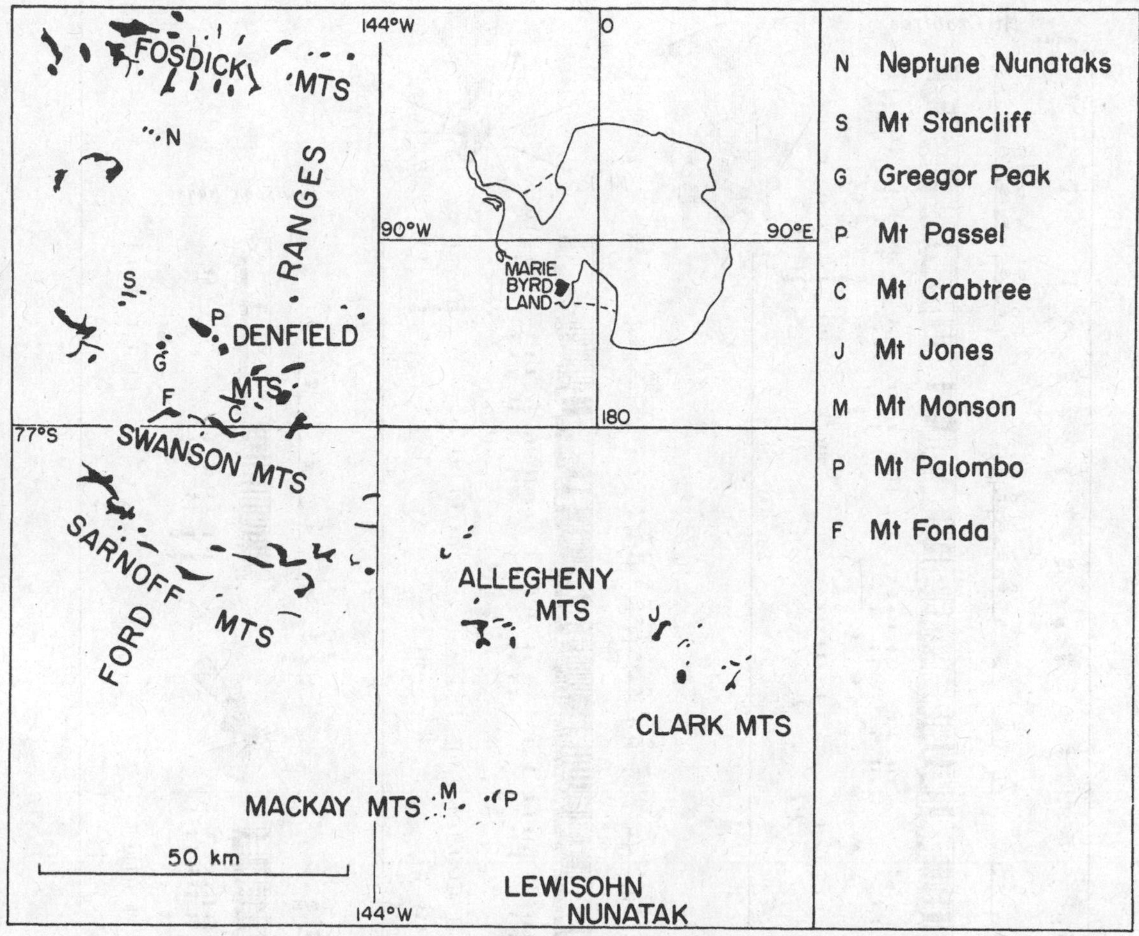


**Figure 1. Distribution of significant outcrop area, Ford Ranges, Marie Byrd Land, Antarctica.**

narrow (< 1 km) aureoles around granites. The concretions are more sensitive to metamorphism and show changes before contact metamorphism can be detected in the normal sediments. With initial warming they become paler and develop a distinctive concentric white and green pattern caused by the growth of zones of acicular tremolite and Fe-Mg chlorite alternating with zones of quartz, porphyroblastic calcic plagioclase, sericite and carbonate. Closer to the granites tremolite is joined by diopside and garnet (probably grossular). Chlorite is less abundant and the concretions develop a uniform pale green colour. The concretions remain distinctive even when isolated as xenoliths in granite.

*Age* No body fossils or microfaunas were recovered from the Swanson Formation. At several localities simple burrows were seen and a single occurrence of *Paleodictyon* (Figure 3f) was found in the Mt Crabtree section. K-Ar whole rock dating of slates (C.J. Adams, pers comm) at 450-470 Ma supports earlier dating by Krylov (1972). If *Paleodictyon* is restricted to the Ordovician and younger rocks (Hantzschell, 1975), a Cambro-Ordovician age of sedimentation for the Swanson Formation followed by mid-Ordovician deformation and uplift is highly probable.

*Lewisohn Member*

The previously unvisited Lewisohn Nunatak (Figure 1) lies 20 km south of the Mackay Mountains and exposes a 175 m section of sediments. The lower part of the sequence consists of alternating fine sandstone and mudstone similar to the Swanson Formation in sedimentology and composition. The upper part (Figure 2) is composed of crystal-lithic tuffs overlain by thinly bedded tuffaceous sandstones and siliceous mudstone. The coarser tuffs have large well formed crystals of augite (commonly altered to actinolite), labradorite (albitised) and fragments of dolerite and aphyric basalt in a matrix of quartz, albite, chlorite, actinolite and phlogopite. The sequence is cut by small basic dykes and is weakly hornfelsed. Probable correlatives of the Lewisohn Member occur within the Swanson Formation in the Mackay Mountains (west face of Mt Monson, east ridge of Mt Palombo). These rocks appear distinctly tuffaceous in the field and are characterised by augen and blebs of chlorite, which appear to be altered by ferromagnesian or basic volcanic grains, granular epidote and scattered magnetite. Wade and Couch (1982) record acid volcanic rocks from the Sarnoff and Allegheny Mountains but these could not be relocated.

*Fosdick Metamorphic Zone*

The Fosdick Complex was examined only on the south side of the Fosdick Mountains and in the Neptune Nunataks. The latter proved to be largely of metamorphic rocks and not granite as previously reported (Wade et al. 1977a). Wilbanks (1972) concluded that the Fosdick Mountains are composed of biotite gneiss, migmatite and nebulite that form an infrastructure antiform below a lower rank Swanson Formation superstructure. Wade and Couch (1982) revived the idea that their Fosdick Complex is an older basement and a possible source of the Swanson Formation. They further compared the Complex to the Wilson Group of northern Victoria Land and suggested a Precambrian age. Our observations of the metasediments support Wilbanks' conclusions. In particular relict sedimentary layering is retained, especially in the Neptune Nunataks, and shows alternating sequences which closely resemble Swanson Formation successions, including the preservation of typical metamorphosed calcareous concretions in metapelites. Lithological similarity and the absence of geochronological evidence of great age suggest that the sedimentary component of the Fosdick Complex is Swanson Formation distinguished as the Fosdick metamorphic zone because of its high metamorphic rank. The Complex is unlikely to be older than Middle Ordovician.

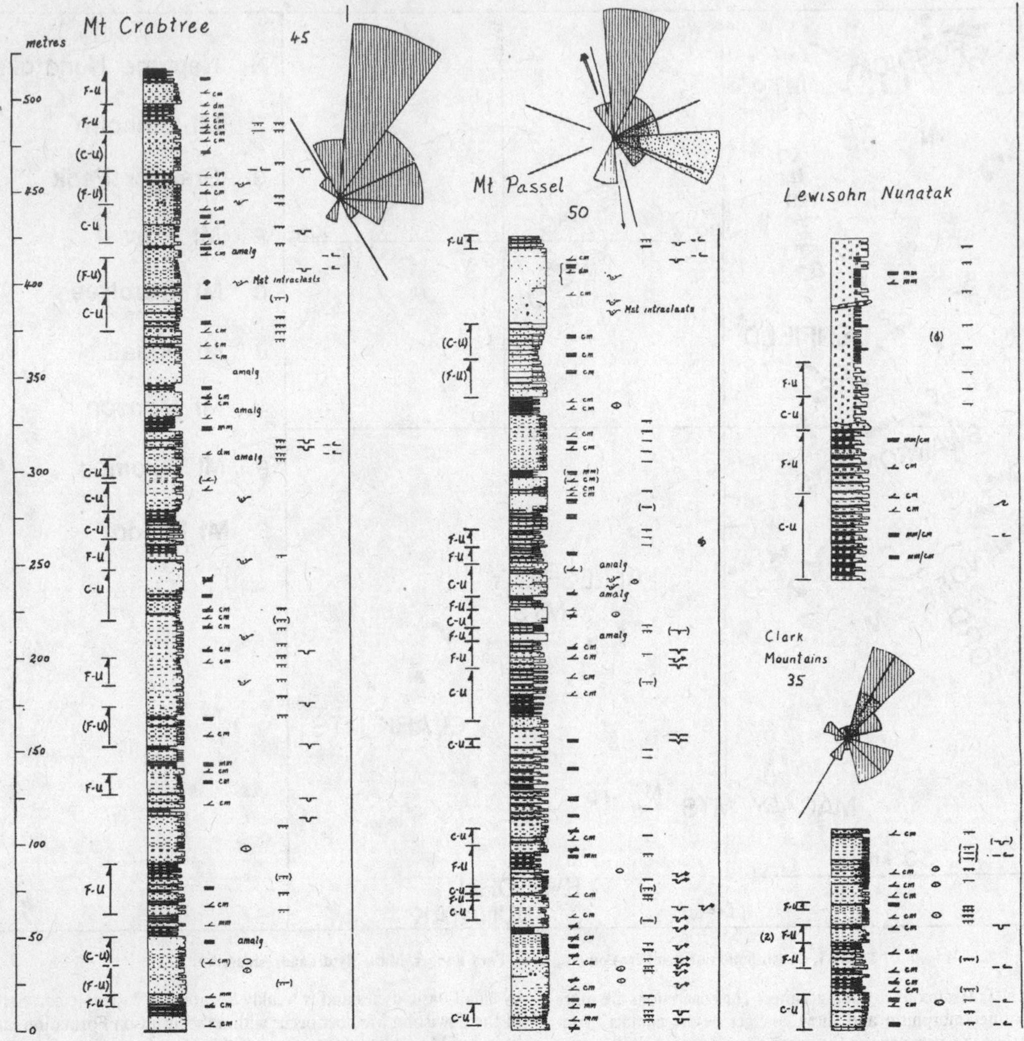


**Figure 2. Measured sections of Swanson Formation, including Lewisohn Member, Marie Byrd Land. Mt Crabtree, north leading ridge, 76°59′S, 145°05′W. Mt Passel, south-west trending spur at 76°52′S, 145°05′W. Clark Mountains, northwest trending ridge, Mt Jones 77°14′S, 142°14′W. Lewisohn Nunatak, north Ridge 77°38′S, 142°48′W.**

### *Structure*

As previously reported (Wade and Couch, 1982) the Swanson Formation is folded into upright or steeply inclined folds. Slatey and fracture cleavage characterise mudstone and sandstone respectively. Data collected in small areas give reasonable $\pi$ diagrams suggesting an axial trend of 290°—315° and highly variable plunge. However a plot of poles to bedding from several folds in the Denfield Mountains show a striking small circle distribution typically associated with conical folds. With a single cone axis direction, adjacent folds must show very different plunge of hinge line, which is consistent with observations. Distribution of fold hinge lines strongly suggests at least two northeast trending sinistral strike slip faults.

### Comparison with Robertson Bay Group

The Swanson Formation and the Robertson Bay Group show strong similarity in outcrop but this is hardly sufficient to suggest a significant relationship. Any comparison should include age, depositional environment, composition and post-depositional history. A Riphean age has been suggested (Iltchenko, 1972) for the Swanson Formation based on acritarchs, but in other instances acritarchs have

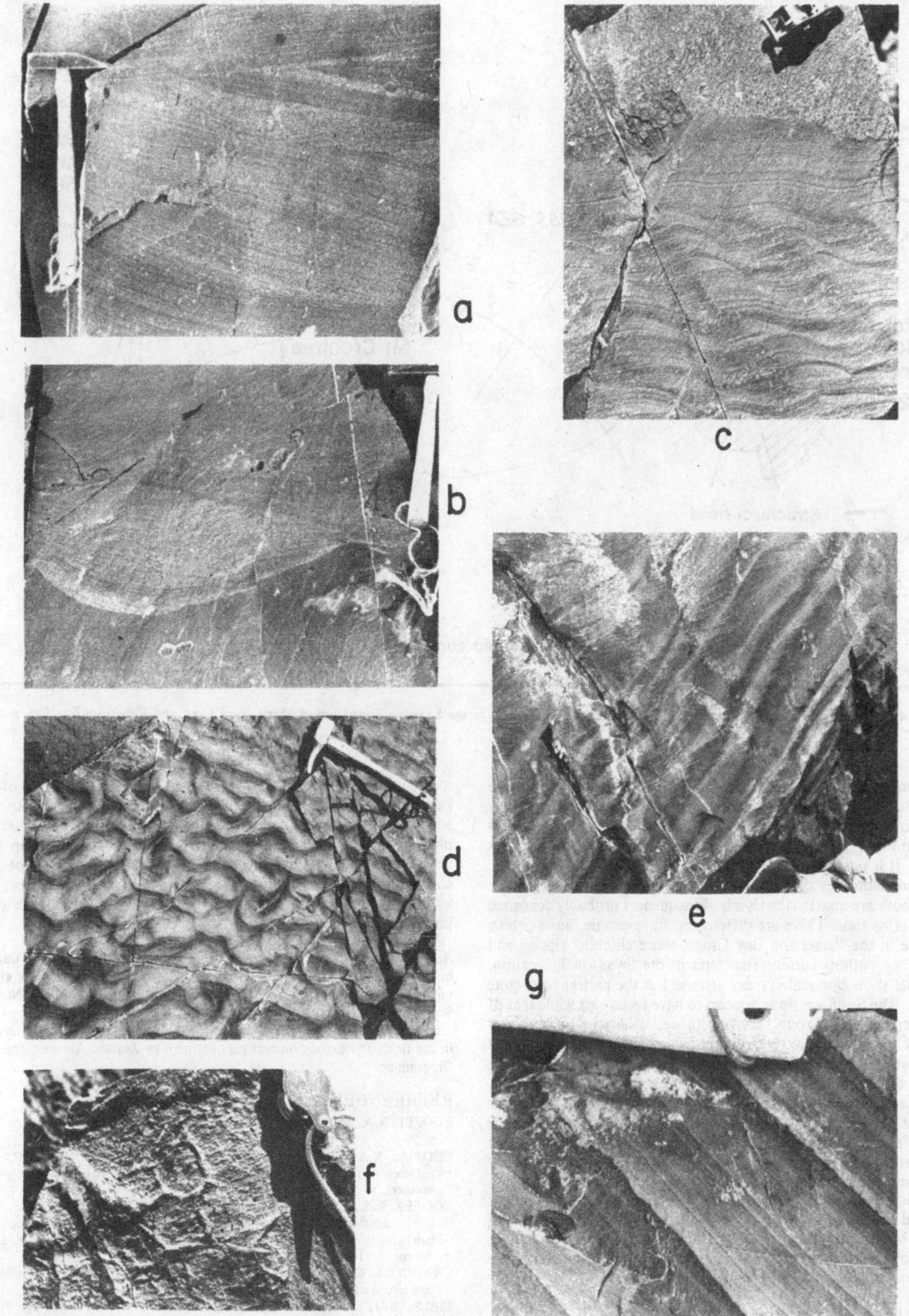


**Figure 3.** **Sedimentary structures and trace fossil, Swanson Formation. a) and b) Cross beds and scour channels in the upper part of the Mt Passel measured section. c) Climbing ripples and laminated fine sandstone overlain by the base of a graded bed, Mt Fonda, western Swanson Mountains. d) Rippled surface in the upper part of Mt Crabtree measured section. e) Climbing ripples in fine sandstone, Mt Stancliff. f) *Paleodictyon,* upper parts Mt Crabtree measured section. g) Thin graded sandstones and laminated siltstones, Mt Crabtree measured section.**

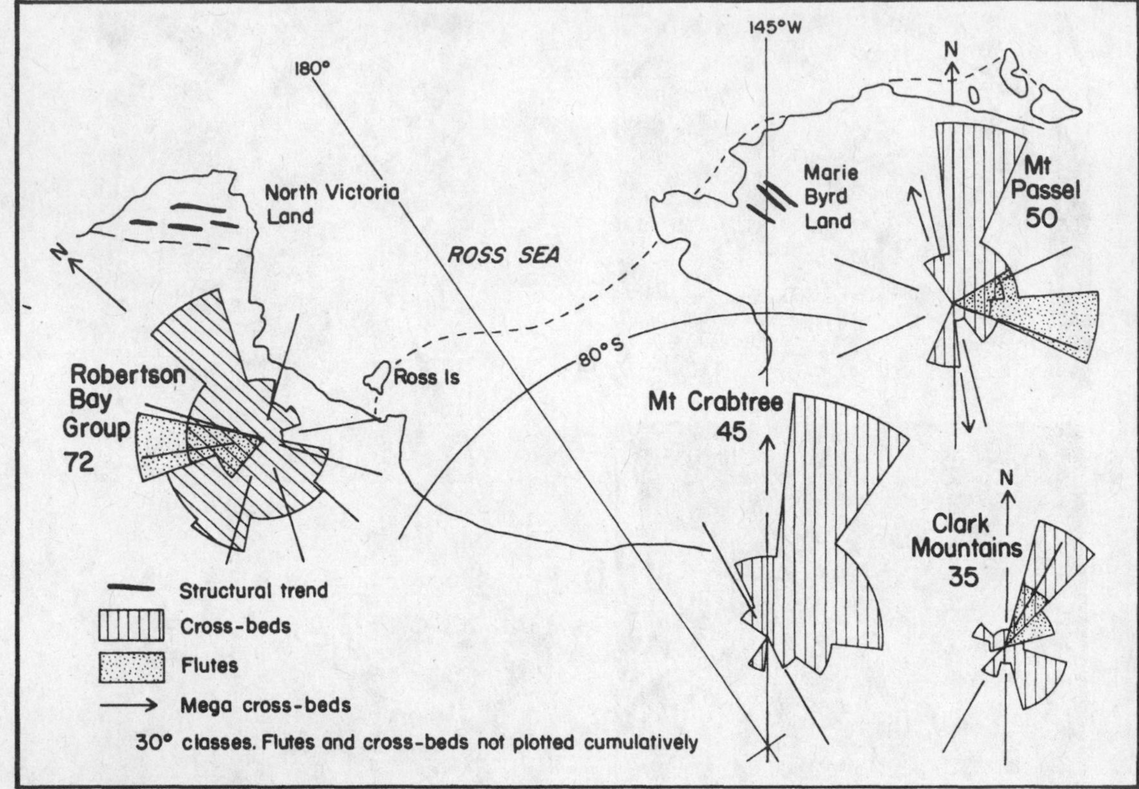


**Figure 4. Relative position, structural trends and palaeocurrent patterns in the Swanson Formation of Marie Byrd Land and Robertson Bay Group of northern Victoria Land.**

proved unreliable (Cooper et al. this volume) and, as noted above, we believe that the Swanson Formation is Phanerozoic and not greatly older than the cleavage age. The Robertson Bay Group is considered to be Cambrian, probably Early Cambrian, by Field and Findlay (this volume). It is possible that the ranges of the two units overlap.

The sedimentology of the two units has much in common, mainly because both are quartz-rich flysch-like sequences probably deposited on submarine fans. There are differences; for example, more pebbly mudstone in the Robertson Bay Group, more climbing ripples and high energy bottom current structures in the Swanson Formation. Both units show bimodal current patterns but the patterns are quite different. The Swanson slope appears to have faced east while that of the Robertson Bay Group faced north west (the true difference is disguised by proximity to the pole, the two palaeoslopes being almost exactly opposite). Both show a north flowing bottom current (Figure 4). The difference in palaeoslope is unlikely to be due to post-Ordovician geotectonic rotation unless it is believed that the similar dominant structural trend is accidental. Bearing in mind that they lie 1400 km apart the difference in palaeocurrent pattern does not rule out a relationship; they could be either different fans in the same basin or different fans on opposite sides of a common source.

The alignment of the two belts, the common fold trend, similar structural history and the presence of the post-tectonic I type Devonian-Carboniferous granites suggests that they may form parts of the same orogenic belt and have a similar relationship to Devonian subduction zones and plate boundaries.

## CONCLUSIONS

The Swanson Formation is a quartzose flysch-like unit probably deposited on a submarine fan. In this respect it resembles the Robertson Bay Group and the two units may overlap in age. In detail it seems impossible to prove that the two are isolated parts of the same unit and there is no evidence to support a general correlation of northern Victoria Land and Marie Byrd Land as suggested by Wade and Couch (1982). In particular we see no similarity between metamorphic rocks of the Fosdick Range and the Wilson Group of the Lanterman and Salamander Ranges (i.e. those closest to the Robertson Bay Group) and note the absence of the distinctive 480-500 Ma plutonic-metamorphic event characteristic of the Wilson terrain. The absence of the voluminous and distinctive lithologies of the Cambrian-Ordovician Bowers Supergroup (Laird and Bradshaw, this volume) from Marie Byrd Land weakens the case for close original links.

*Acknowledgements* We are grateful to Dr D. Shelley (University of Canterbury), Dr G.A. Challis and Mr D. Smale (N.Z. Geological Survey) for examining selected thin sections. Dr C.J. Adams, N.Z. Institute of Nuclear Sciences was a member of the expedition and supplied isotopic ages for cleavage formation. We thank Mr Peter Braddock for his cheerful assistance in the field. The expedition was part of the New Zealand Antartic Research Programme.

## REFERENCES

COOPER, R.A., 1979: Lower Palaeozoic rocks of New Zealand. *R. Soc., N.Z. J.*, 9, 29-84

COOPER, R.A., JAGO, J.B., ROWELL, A.J. and BRADDOCK, P., (this volume): Age and Correlation of the Cambrian-Ordovician Bowers Supergroup, northern Victoria Land.

COOPER, R.A., LANDIS, C.A., LE MESURIER, W.E. and SPEDEN, I.G., 1982: Geologic history and regional patterns in New Zealand: their paleotectonic and paleographic significance; *in* Craddock, C. (ed.) *Antartic Geoscience*. Univ. Wisconsin Press, Madison, 43-53.

CRADDOCK, C., 1972: Antarctic tectonics; *in* Adie,R.J. (ed.) *Antarctic Geology and Geophysics*. Universitetsforlaget, Oslo, 449-55.

FIELD, B.D. and FINDLAY, R.H. (this volume): The Sedimentology of the Robertson Bay Group, northern Victoria Land.

GRINDLEY, G.W. and DAVEY, F.J., 1982: The reconstruction of New Zealand and Antarctica; *in* Craddock, C. (ed.) *Antarctic Geoscience*. Univ. Wisconsin Press, Madison, 15-29.

HANTZSCHELL, W., 1975: Trace fossils and Problematica; *in* Teichert, C. (ed.) *Treatise on Invertebrate Paleontology, part W, suppl. 1.* Geol. Soc. Am. and Univ. Kansas Press, Lawrence.

ILTCHENKO, L.N., 1972: Late Precambrian acritarchs of Antarctica; *in* Adie, R.J. (ed.) *Antarctic Geology and Geophysics*. Universitetsforlaget, Oslo, 599-602.

KRYLOV, A.Y., 1972: Antarctic Geochronology; *in* Adie, R.J. (ed.) *Antarctic Geology and Geophysics.* Universitetsforlaget, Oslo, 491-4.

LAIRD, M.G., 1972: Sedimentology of the Greenland Group in the Paparoa Range, West Coast, South Island. *N.Z. J. Geol. Geophys., 15,* 372-93.

LAIRD, M.G. and BRADSHAW, J.D. (this volume): New data on the Lower Palaeozoic Bowers Supergroup northern Victoria Land.

LOPATIN, B.G., and ORLENKO, E.M. 1972: Outline of the geology of Marie Byrd Land and the Eights Coast; *in* Adie, R.J. (ed.) *Antarctic Geology and Geophysics* Universitetsforlaget, Oslo, 245-50

PASSEL, C.F., 1945: Sedimentary rocks of the southern Edsel Ford Ranges, Marie Byrd Land, Antarctica. *Am. Phil. Soc.,Proc., 89,* 121-31

STOW, D.A.V. and LOVELL, J.P.B., 1979: Contourites: their recognition in modern and ancient sediments. *Earth-Sci. Rev., 14,* 251-91.

WADE, F.A., CATHEY, C.A. and OLDHAM, J.B., 1977a: Reconnaissance geologic map of the Guest Peninsula Quadrangle, Marie Byrd Land, Antarctica, 1:250,000. *U.S. Geological Survey: Antarctic Research Program map series.*

WADE, F.A., CATHEY, C.A. and OLDHAM, J.B., 1977b: Reconnaissance geologic map of the Boyd Glacier Quadrangle, Marie Byrd Land, Antarctica, 1:250,000. *U.S. Geological Survey: Antarctic Research Program map series.*

WADE, F.A. and COUCH, D.R., 1982: The Swanson Formation, Ford Ranges, Marie Byrd Land—Evidence for and against a direct relationship with the Robertson Bay Group, northern Victoria Land; *in* Craddock, C. (ed.) *Antarctic Geoscience.* Univ. Wisconsin Press, Madison, 609-16.

WADE, F.A. and WILBANKS, J.R., 1972: Geology of Marie Byrd and Ellsworth Lands; *in* Adie, R.J. (ed.) *Antarctic Geology and Geophysics.* Universitetsforlaget, Oslo, 207-14.

WARNER, L.A., 1945: Structure and petrography of the southern Edsel Ford Ranges, West Antarctica. *Am. Phil. Soc., Proc., 89,* 78-122.

WILBANKS, J.R., 1972: Geology of the Fosdick Mountains, Marie Byrd Land; *in* Adie, R.J. (ed.) *Antarctic Geology and Geophysics,* Universitetsforlaget, Oslo, 277-84.

# UPPER CAMBRIAN ARCHAEOCYATHANS: NEW MORPHOTYPE

F. and M. Debrenne, *Institut de Paleontologie, 8 Rue de Buffon, 75005 Paris, France.*

G.F. Webers, *Geology Department, Macalaster College, 1600 Grand Avenue, St. Paul, Minnesota 55105, U.S.A.*

*Abstract* A number of cylindrical fossils have been reported as "archaeocyathid-like organisms" by Craddock and Webbers (1964). They occur in the Late Cambrian Minaret Formation of the Heritage Range, Ellsworth Mountains, central West Antarctica. The geology of the Ellsworth Mountains and the stratigraphic column based on the most recent field work have been previously presented (Splettstoesser and Webers, 1980). The Minaret Formation has yielded abundant Late Cambrian trilobites, gastropods, monoplacophorans, and other groups of fossils. All fossil groups are presently under investigation.

A number of specimens of cylindrical fossils that have two "walls" connected by irregularly waved skeletal structures have been reported as "archaeocyathid-like organisms" by Craddock and Webers (1964). Debrenne and Rozanov, examining the important fauna recently collected (1979/80) came to the conclusion, that despite some differences (lack of individualised walls, absence of vesicular tissues and exostructures), these sponge-like forms probably belong to the family Archaeocyathidae confirming Webers' first attribution. A new genus has been proposed, the detailed description and the affinities of which are discussed in a paper presently submitted elsewhere for publication.

The fossils collected during the 1979/80 Ellsworth Mountains expedition provided material demonstrating a new morphotype: branching cups developing into a catenulate pattern. The undifferentiated walls and the exceptional plasticity of the Late Cambrian forms allow the construction of a composite colonial habit. This type is not known in the Early Cambrian ancestors. It is probably because of their plasticity and primitive development, in comparison with the high diversity reached by the porous system of the Early Cambrian Archaeocyathans, that the younger representatives of the phylum have been able to survive in restricted niches into the Late Cambrian.

## REFERENCES

CRADDOCK, C. and WEBERS, G.F., 1964: Fossils from the Ellsworth Mountains, Antarctica. *Nature, 201,* 174-5.

SPLETTSTOESSER, J.F. and WEBERS, G.F., 1980: Geological investigations and logistics in the Ellsworth Mountains, 1979-80. *U.S. Antarct. J., 15, (5)* 36-9.

# 5

# Scotia Arc and Antarctic Peninsula

# THE EVOLUTION OF THE SCOTIA ARC: A REVIEW

I.W.D. Dalziel, *Lamont-Doherty Geological Observatory of Columbia University, Palisades, N.Y. 10964, U.S.A.*

*Abstract* Over the past decade major advances have taken place in our knowledge of the geology and geophysics of the Scotia arc region and our understanding of the evolution of the arc itself. The continental "basement" that was formerly regarded as Precambrian is now known to consist of Palaeozoic, Mesozoic, and possibly even Cenozoic rocks accreted to the South American and Antarctic segments of Gondwanaland as a fore-arc terrain before, during and possibly after break-up, and as the products of early arc magmatism. The Antarctic Peninsula, formerly regarded as essentially a continuation of the entire Andean Cordillera is now known to be a Mesozoic to Cenozoic magmatic arc that, unlike the southernmost Andes, does not incorporate the suture of a closed marginal basin. Thus one can account for differences in tectonic style around the arc.

Recent work on South Georgia and in the South Orkney Islands reinforces the view that the submarine platforms from which they emerge represent fragments of South America and Antarctica that have been transported eastrelative to their respective parent continents. Marine geophysical studies reveal a complex history of Cenozoic tectonics between the South American and Antarctic Plates. A major puzzle lies in the timing and location of the initiation of a westward dipping subduction zone as now seen in the South Sandwich arc.

The original relationship of the Antarctic Peninsula to southern South America remains uncertain. Three possibilities are evaluated.

## HISTORICAL PERSPECTIVE

The Scotia arc is one of the most intriguing geographic features on earth. Clearly similar to the Caribbean arc, and not unlike many of the island arc systems of the western Pacific, the geometrically simple eastward closing loop of the Scotia Ridge joining the Andean Cordillera to the Antarctic Peninsula has generated considerable tectonic speculation. Moreover, the obvious significance of Drake Passage, the youngest and narrowest deep-sea gap between Antarctica and a neighbouring continent, in terms of palaeoclimate and palaeobiogeography has attracted the attention of physical oceanographers, climate modelers, and palaeontologists, as well as geophysicists and geologists.

Understanding of the evolution of the Scotia arc, here taken to mean the entire loop of mountains, submarine ridges and islands between 50°S latitude in the Andes and the base of the Antarctic Peninsula at 70°S (Figure 1), has come from two lines of approach that have only recently started to converge. Geologists have for the most part tended to be impressed by the similarities between the southern Andes, the islands of the Scotia Ridge, and the Antarctic Peninsula (but see Katz, 1973). Hence they have constructed models depicting the disruption of a once continuous mountain chain in the area of Drake Passage and the Scotia Sea (Matthews, 1959; Dalziel and Elliot, 1971 and 1973). Geophysicists studying the history of Cenozoic sea-floor spreading between South America and the Antarctic Peninsula, on the other hand, have tended toward reconstructions of the original connection between the southernmost Andes and the West Antarctic Cordillera as an eastward pointing cusp (see, e.g. Barker and Griffiths, 1972; Hill and Barker, 1980). The present author here reviews the geologic history of the Scotia arc region and re-examine the past relationship of these two segments of the Pacific margin orogen in the light of recent work.

## GEOTECTONIC SETTING

The Scotia arc region is situated on the boundary between two major lithospheric plates, the South American plate and the Antarctic plate (Figure 1). The South American plate is moving slowly left-laterally with respect to the Antarctic plate so that slow subduction of the latter beneath the former is taking place along the western margin of southern South America and the left-lateral motion is distributed across Drake Passage and the Scotia Sea, being concentrated mainly along the North Scotia Ridge and the South Scotia Ridge (Forsyth, 1975). Spreading about a northeast trending spreading centre in Drake Passage has been taking place since approximately 30 Ma ago but is as present extremely slow and the crust generated at this centre is taken to constitute a single plate, the Scotia plate (Barker and Burrell, 1977). At the eastern end of the Scotia arc the South American plate is being subducted beneath the floor of the Scotia Sea at a higher rate, reflecting generation of a north trending back arc spreading centre to the west of the volcanically active South Sandwich Islands. The western side of this centre has been attached to the Scotia plate, the eastern side forms the small, D-shaped, Sandwich plate (Barker, 1972).

The history of plate motion in the area prior to the initiation of Drake Passage spreading is uncertain,partly because of lack of data in the Weddell basin, and partly because some of the record in the southeastern Pacific, and possibly in the Weddell Sea, has been destroyed by subduction (LaBrecque and Barker, 1981; Cande et al., 1982). The geologic histories of the southernmost Andes and the Antarctic Peninsula amply testify to a history of subduction going back into the Early Mesozoic, possibly even the Late Palaoezoic (Dalziel, 1981; Forsythe, 1982; Pankhurst, 1982; Thomson et al., this volume). Evidence of plate interaction before break-up of Gondwanaland comes entirely from the geologic record on land. The history of the Scotia arc region since break-up must have been critically dependent on the relative motion of the South American and Antarctic plates (Norton, 1982). Thus theoretically it should be possible to determine this relative motion through certain "windows" of time and approach an understanding of the evolution of the Scotia arc as Dewey et al. (1973) have done for the western Mediterranean region by deducing the relative motion of the African and Eurasian plates. The situation in the Scotia arc region, however, appears to have been complicated by relative motion of microplates, including one involving the Antarctic Peninsula, during the Late Mesozoic and Cenozoic (Dalziel and Elliot, 1982).

## GEOLOGIC HISTORY

In this section I will summarise the geologic history of the Scotia arc region, mainly as it is known from the on-shore data, without regard for the palaeogeographic relationships of South America, the islands of the North and South Scotia Ridges, and the Antarctic Peninsula. Emphasis will be placed on the tectonic significance of the rocks, field relations and structures.

### Panthalassic margin

I have emphasised before (Dalziel, 1976; Dalziel, 1982), and continue to be impressed by, the profound uniformity beneath volcanic and sedimentary strata of Middle(?) to Late Jurassic or Early Cretaceous age throughout southern South America, the South Orkney Islands, and the Antarctic Peninsula (see Dalziel and Elliot, 1973, Figures 7 and 8). The so-called "basement" rocks of the Scotia arc region beneath this contact appear to represent a subduction complex and fore-arc basin or basins that were located along the Panthalassic margin of Gondwanaland before break-up of the super-continent (Barker, Dalziel et al., 1976; de Wit, 1977; Hyden and Tanner, 1981; Smellie, 1981; Dalziel, 1982; Forsythe, 1982). Strata as old as Pennsylvanian and as young as Upper Triassic were deformed and metamorphosed prior to latest Triassic or Early to Middle Jurassic uplift and erosion. The metamorphic rocks of the Elephant Island subgroup and of Smith Island in the South Shetland Islands (Figure 1) cannot be stratigraphically dated, and, although also representing part of a subduction complex, may in part at least post-date break-up as will be discussed below. Cretaceous radiolaria reported from metacherts from the LeMay Formation on Alexander Island (Anonymous, 1981) underscore this possibility that I expressed at the 1977 SCAR-IUGS Symposium on Antarctic Geoscience when emphasising the lack of clear evidence as to the "older" (i.e. pre-Middle or Late Jurassic) age of some of the highly deformed "basement" rocks of the region (Dalziel, 1982). Nonetheless there is substantial evidence of the subduction of Panthalassic ocean floor beneath Gondwanaland before break-up (see also Pankhurst, 1982).

Uncertainties surround the configuration of the margin of the super-continent, the extent and location of an associated calc-alkaline

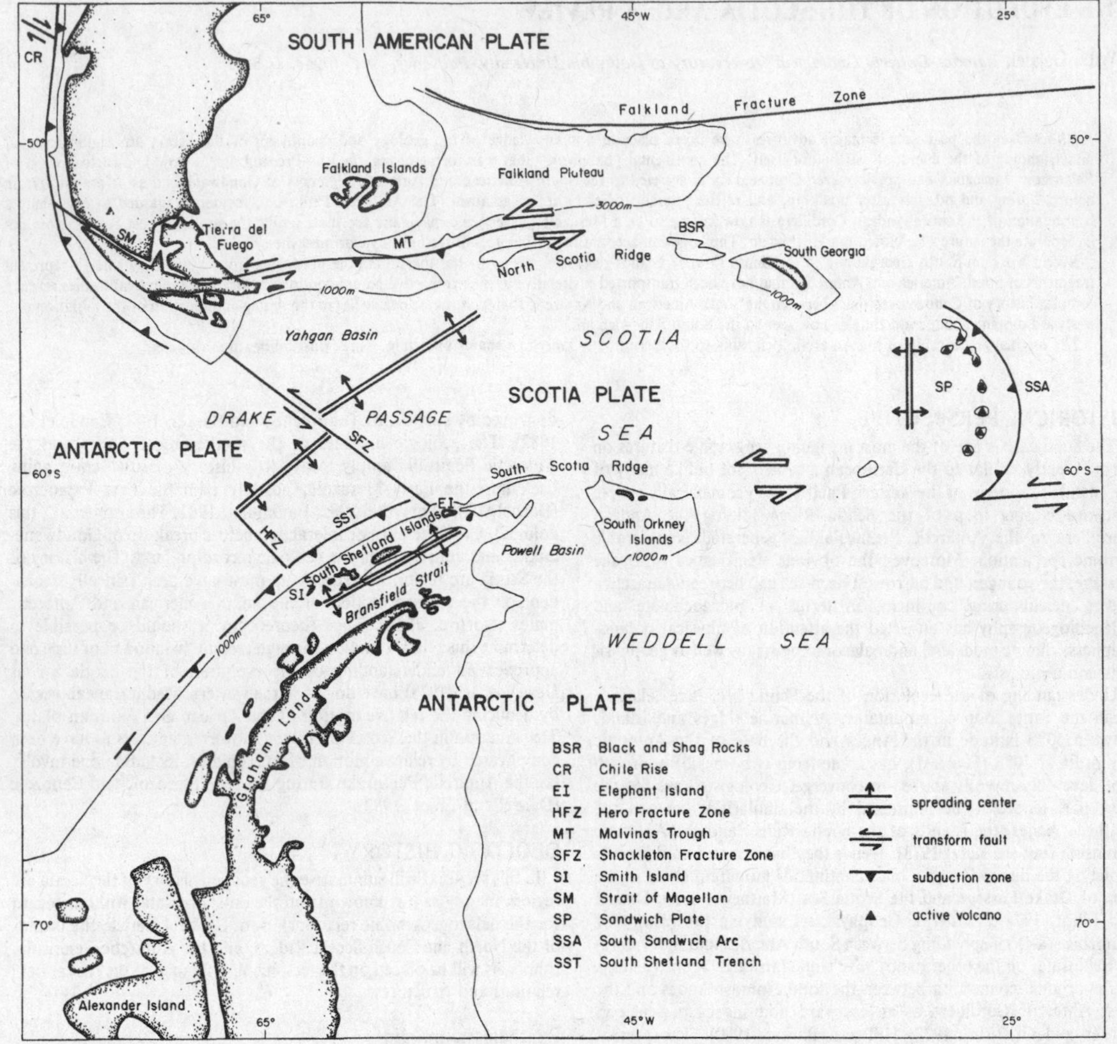


**Figure 1. Present plate setting of the Scotia arc region.**

magmatic arc, and the degrees to which allochthonous material may have been accreted to the Gondwanaland margin in this area. On the latter point there is still, to my mind, no compelling evidence that exotic material was accreted to Gondwanaland in this region other than "offscrapings" (including perhaps, sea mounts, see Forsythe, 1982) in a subduction complex.

## Gondwanaland Break-up

As pointed out previously it seems unlikely that the widespread and profound angular unconformity beneath Middle to Upper Jurassic or Lower Cretaceous strata throughout the South Atlantic region is not genetically related to the incipient Gondwanaland fragmentation (Dalziel, 1982). The unconformity is not confined to the fore-arc terrain along the Panthalassic margin but also occurs to the east on the Falkland Plateau (Barker, Dalziel et al., 1976) and in the foreland fold and thrust belts along the Gondwana craton margin, for example in southern Africa where the Cape fold belt developed in the Early Triassic (Tankard et al., 1982). Anderson (1982) has recently emphasised the potential significance of events like this in recording thermal processes in the mantle.

The lowest strata resting on the uplifted fore-arc terrain in the Scotia arc region are primarily volcanic rocks interlayered with volcanoclastic sandstones and shales indicative of a shallow marine environment. Deposition of these rocks accompanied widespread extension in the latest Jurassic and Early Cretaceous that is particularly well documented in oil company subsurface data from South America (see e.g. Natland et al., 1974; Bruhn et al., 1978). Apart from the South Atlantic basin the most extreme development of this extension occurred along the Pacific margin of South America where an intra-arc basin of uncertain width and depth, but certainly deep and floored in part by mafic quasi-oceanic crust opened up in southernmost South America about the Jurassic-Cretaceous boundary (for review see Dalziel, 1981). It seems clear from the extensive calc-alkaline igneous and volcanoclastic rocks of Late Jurassic and Cretaceous age in the Scotia arc region that subduction related igneous activity both accompanied and outlasted the extensional tectonic regime associated with fragmentation of Gondwanaland (Pankhurst, 1982).

The extent of the so-called "Rocas Verdes" basin in the southernmost Andes towards the south and east into the Weddell basin is uncertain. Clearly, however, the Antarctic Peninsula represents only the Pacific margin magmatic arc terrain of the southern Andes, and does not incorporate a suture marking a closed intra-arc basin (Thomson et al., this volume). Hence the similarities between the geology of the peninsula and southern South America that have been pointed out by some authors (see e.g. Dalziel and Elliot, 1973), can be reconciled with the differences that have been emphasised by others (Katz, 1973; Miller, 1982). To be specific, the peninsula and southern South America share a comparable pre-Gondwanaland break-up geologic history as well as a comparable Late Cretaceous and Cenozoic Pacific margin magmatic arc development. In the case of southern South America, however, the collapse of the marginal basin under compression in the mid-Cretaceous resulted in much more intense deformation than anything seen south of Drake Passage, as well as uplift exposing the arc terrain at a deeper structural level than in the Antarctic Peninsula so that much of the volcanic suprastructure above the Patagonian batholith was eroded.

### Development of the Scotia Ridge

Marine geophysical data bearing on the development of the present Scotia Ridge dates back to approximately 30 Ma and the inception of sea-floor spreading about a northeast oriented spreading centre in Drake Passage. There is still a gap of approximately 70 Ma to understand between this event and the Mid-Cretaceous uplift of the Rocas Verdes marginal basin and penetrative deformation in the southernmost Andes.

Detailed geologic studies on South Georgia (Storey et al., 1977; Tanner et al., 1981) have confirmed the views developed on the basis of a reconnaissance study (Dalziel et al., 1975) that the South Georgia platform likely remained directly attached to the South American continent immediately east of Cape Horn until at least the Mid-Cretaceous phase of Andean orogenesis. Detailed structural studies of the metamorphic complexes in the South Orkney and South Shetland Islands (Dalziel, in preparation) confirm the results outlined in a previous review (Dalziel, 1982) that the South Orkney Islands platform most likely rotated away from the tip of the Antarctic Peninsula in a clockwise sense to open up the Powell Basin (Figure 1) whose bathymetry (Barker, 1972) suggests a Mid-Cenozoic age.

Thus the North and South Scotia Ridges appear to have formed mainly by the eastward translation of microcontinental fragments of southern South America and the Antarctic Peninsula respectively during the Cenozoic even if the exact geometry of the sea-floor spreading and plate movement involved is not yet clear. A study of specimens collected from the Shag Rocks to the west of South Georgia (Tanner, in press) confirms that the platform on which they are situated is also of continental origin, at least in the sense that the rocks appear to be another portion of Pacific (or Panthalassic) margin subduction complex.

The west-dipping subduction zone beneath the South Sandwich Islands has been active for approximately 8 million years (for reviews see Dalziel and Elliot, 1973; Barker and Dalziel, in press). Dredging from topographic highs in the southeastern Scotia Sea indicates, however, that an earlier east-facing arc (or arcs) may have preceded the development of the South Sandwich arc (Barker et al., 1982). It is not clear when a west-dipping subduction zone first developed in the Scotia arc region, but it may date back to the Mid-Cretacerous when strongly northeast vergent structures were formed in the closing Rocas Verdes basin in southern Tierra del Fuego and particularly in South Georgia. One way of interpreting these structures is as a result of the westward subduction of quasi-oceanic marginal basin crust beneath the Pacific margin magmatic arc (Dalziel, 1981; Tanner et al., 1981).

## RECONSTRUCTION

### Premises

It now seems clear following the detailed analysis of southern oceans geophysical data by Norton and Sclater (1979) that the Gondwanaland reconstructions of du Toit (1937) and Smith and Hallam (1970) are essentially correct, except with regard to West Antarctica where extensions of continental crust and motion of microplates in the Late Mesozoic and Cenozoic appear to have distorted the original Panthalassic margin of Gondwanaland. Any modification of the Norton-Sclater reconstruction to remove the overlap between the Antarctic Peninsula and southernmost South America must take into account the geologic history and structure of those two regions as outlined above. Reasons for considering the overlap to be unacceptable are outlined in a recent article by Dalziel and Elliot (1982). As the Antarctic Peninsula geology bears no direct relationship to that of the Transantarctic Mountains and the Gondwana craton, the geologic relationship between the peninsula and southern America is the only real check on reconstructions based on marine geophysical data.

I shall consider three possible relationships, each of which has in one form or another been proposed in the literature. The three reconstructions are based on the following assumptions:

1. South America is kept fixed since its relationship to Africa and Africa's relations to East Antarctica are now well established (Norton and Sclater, 1979).
2. South Georgia is taken to be part of the South American continent and to have been situated south of Isla de los Estados and east of Cape Horn in the location of the present oceanic Yahgan Basin (Katz and Watters, 1966; Dalziel and Elliot, 1973; see Figure 1). No significant relative rotation is involved for reasons discussed in Dalziel et al. (1975) and Tanner (1982).
3. The South Orkney Islands platform is taken to have been part of the Antarctic Peninsula microcontinent and to have been situated south and east of the Elephant Island group in the location now occupied by the present oceanic Powell Basin (Figure 1). Restoration of the South Orkney Islands platform in a counter-clockwise sense is based on the constraints outlined in Dalziel (1982) and set out in more detail in Dalziel (in preparation).
4. While there has clearly been some modification of the shape of the Pacific margin of southernmost South America due to the opening and closing of the Rocas Verdes basin in the latest Jurassic to Mid-Cretaceous, and due to strike-slip faulting along the Strait of Magellan lineament, it is considered unlikely that the entire southwestern extremity of the continent has been distorted. A sharp bend in the Pacific limit of the Gondwana craton takes place at approximately 50°S (Dalziel,1982), and the structural grain of the pre-Late Jurassic basement is northwest trending, even south of the west end of the Strait of Magellan (Forsythe, 1982). Hence the present shape of the Patagonian orocline is retained in the reconstructions. There is palaeomagnetic evidence to support rotation of some crustal blocks (Dalziel et al., 1973; Burns et al., 1980), the data all come from the Cordillera, however, and recent work has shown that the most compelling data in the paper by Dalziel et al. (1973) comes from a syntectonic pluton (M. Suárez, personal communication, 1982). Hence there seems no strong reason now to completely eliminate the Patagonian orocline in reconstructions.
5. With the possible exception of the southern end of the Antarctic Peninsula (south of 72°S) there is no compelling evidence for changing the shape of the Antarctic Peninsula in reconstructions (Dalziel et al., 1973; Kellogg and Reynolds, 1978; Kellogg, 1980).
6. Bransfield Strait is taken to be a Cenozoic spreading centre (for review, see Barker and Dalziel, in press). Hence it is eliminated in the reconstructions.

The reconstructions show the following tectonostratigraphic units of southern South America and the Antarctic Peninsula:

1. Fore-arc terrains: (a) Deep-seated subduction complex: the polyphase deformed metamorphic complex of the South Scotia Ridge—in part pre-Late Jurassic to Early Cretaceous (South Orkney Islands), in part perhaps younger (South Shetland Islands) (Dalziel, 1982; Tanner et al., 1982); also one occurrence on the west coast of southern South America (Forsythe et al., 1982). (b) Fore-arc basin(s): ?Upper Palaeozoic to Triassic sedimentary strata of the Trinity Peninsula and the South Orkney Islands and South Shetland Islands. (c) High level subduction complex: the heterogenous pre-Late Jurassic subduction complex of the Patagonian and Fuegian Andes, and that of Alexander Island that may include rocks as young as Cretaceous. (d) Sedimentary strata of Mesozoic and Cenozoic age.
2. Arc terrain: (a) Granitic plutons: mainly Jurassic to Cretaceous but also Triassic and Tertiary. This includes the southern margins of the South Georgia and South Orkney Islands platforms interpreted to be granitic rocks with mafic intrusives by Simpson and Griffiths (1982) and Harrington et al. (1972) respectively. (b) Silicic and calc-alkaline volcanic rocks: Jurassic to Cretaceous (mainly).
3. Marginal basin terrain: (a) Volcanoclastic infill: Lower Cretaceous turbidites. (b) Mafic igneous floor: uppermost Jurassic to lowermost Cretaceous ophiolitic complexes (gabbro-sheeted dykes-pillow lavas).
4. Back-arc terrain: (a) Crystalline Precambrian rocks (Falkland Islands). (b) Sedimentary strata: Palaeozoic to Triassic (Falkland Islands). (c) Sedimentary strata: Lower Cretaceous to Recent.

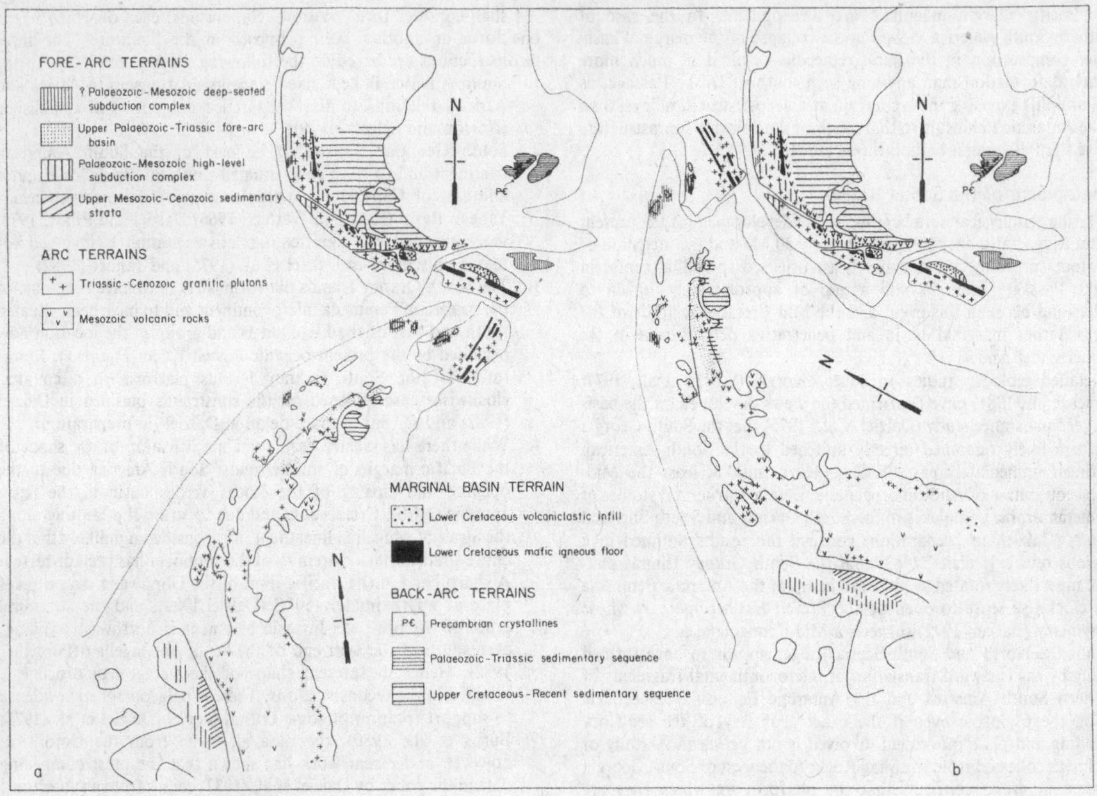

FORE-ARC TERRAINS

? Palaeozoic-Mesozoic deep-seated subduction complex

Upper Palaeozoic-Triassic fore-arc basin

Palaeozoic-Mesozoic high-level subduction complex

Upper Mesozoic-Cenozoic sedimentary strata

ARC TERRAINS

Triassic-Cenozoic granitic plutons

Mesozoic volcanic rocks

MARGINAL BASIN TERRAIN

Lower Cretaceous volcaniclastic infill

Lower Cretaceous mafic igneous floor

BACK-ARC TERRAINS

P€ Precambrian crystallines

Palaeozoic-Triassic sedimentary sequence

Upper Cretaceous-Recent sedimentary sequence

N

P€

N

a

N

P€

N

b

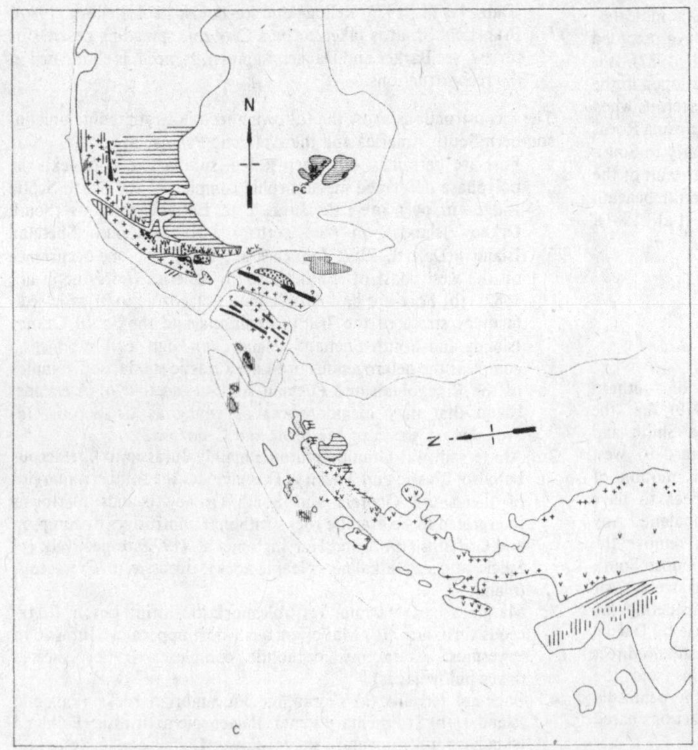


**Figure 2a. "Cuspate" model of South America-Antarctic Peninsula relations.**

**Figure 2b. "Overlapping" model of South America-Antarctic Peninsula relations.**

**Figure 2c. "Continuing" model of South America-Antarctic Peninsula relations (for key, see Figures 2a and 2b).**

### Reconstruction A (Figure 2a)

This is the cuspate reconstruction arrived at by removing the lithosphere created by sea-floor spreading in Drake Passage over the past 30 Ma (Barker and Griffiths,1972; Barker and Burrell, 1977). The South American-Antarctic Peninsula relationship arrived at requires a sharp change in the trends of:

(1) fabrics in the pre-Late Jurassic or Early Cretaceous subduction complexes of southern South America and the South Orkney Islands, and

(2) the trends of mafic intrusions in southern South America and the South Georgia platform on the one hand, and the South Orkneys platform on the other (the presence of intrusions and their trends on the platforms being inferred from the above-mentioned marine geophysical data).

As pointed out by the author at the last SCAR-IUGS Symposium in 1977, such changes do take place along the Pacific margin today, for example, between the Aleutian arc and Kamchatka. Hence the geologic similarities between the two segments of the Pacific margin being considered here may reflect no more than similarity of contemporaneous tectonic processes along different parts of a convergent margin with different orientations (Dalziel, 1982).

The most worrying aspect of this reconstruction is the comparative isolation of South Georgia from the convergent Pacific margin until opening of Drake Passage 30 Ma ago (Dalziel and Elliot, 1973). Given the Late Jurassic to Early Cretaceous magmatic arc-marginal basin setting of the island (Dalziel et al., 1975; Storey et al., 1977; Tanner, 1982), I find this hard to accept.

### Reconstruction B (Figure 2b)

This reconstruction overlapping the Antarctic Peninsula and southern South America is a modified version of that proposed by Harrison et al. (1980) against which the author has argued strongly in the literature (Dalziel, 1980). The reconstruction shown on Figure 2b is modified in that it shows overlap only as far north as the present 50°S latitude in South America as opposed to between 45°S and 40°S in the Harrison et al. version. It is also presented as a possibility only with regard to South America-Antarctic Peninsula relations. As stated above, it is assumed that the Norton and Sclater (1979) fit of Africa to East Antarctica is essentially correct.

Why present this "overlapping" reconstruction as a possibility when the author has argued against it before? Because the author's main geologic argument that such overlap juxtaposed pre-Late Jurassic magmatic arc and fore-arc terrains (Dalziel, 1980) does not seem to be strong for the northern part of the Antarctic Peninsula (Graham Land, see Figure 1) based on new radiometric data (Pankhurst, 1982; Tanner et al., 1982), and also because the possibility of such a reconstruction,again it should be emphasised only for the Antarctic Peninsula and South America and not for Africa-East Antarctica, is emerging as a working hypothesis from study of magmatic anomalies from the Weddell basin (J. LaBrecque, pers. comm., 1982).

The reasonably accurate reconstruction shown on Figure 2b results in a very strange configuration of the Panthalassic margin of Gondwanaland, even with only the northern Antarctic Peninsula overlapping South America to the present 50°S latitude. Once again South Georgia had to be in a suitable location to become part of an arc-marginal basin system by the latest Jurassic. In short there are no geologic data absolutely ruling out this reconstruction, although isotope geochemical data make it an unlikely solution (Pankhurst, 1983). It implies that the Antarctic Peninsula developed as part of an accretionary wedge outboard of South America. Given that the youngest dated rocks of the South American fore-arc are Upper Palaeozoic and those of the Antarctic fore-arc are Triassic this is not unreasonable. Structural trends in the two fore-arc terrains are, however, clearly discordant. Moreover, the reconstruction loses its original raison d'etre if the Norton-Sclater (1979) fit of Africa and East Antarctica, and the likelihood of microplate movement in Antarctica (Dalziel and Elliot, 1982), are both accepted.

### Reconstruction C (Figure 2c)

This is a version of a reconstruction presented independently by Barker, Dalziel et al. (1976) and by de Wit (1977) with the Antarctic Peninsula rotated counterclockwise by over 90° to continue southwestward from the Fuegian Andes. Geologically it is perhaps the most appealing of the models presented here. In terms of fore-arc terrains it presents an acceptable model with the age of rocks involved in the fore-arc becoming younger towards the south and towards the ocean. Trends in the pre-Late Jurassic and Early Cretaceous complexes are slightly discordant between South America and the South Orkney Islands. Such an effect could be original, but closer parallelism could be brought about by further rotation of the peninsula or allowance for the effect of closing the Early Cretaceous marginal basin.

Turning to the arc terrain, this model accounts for the termination of the Patagonian batholith south and east of Cape Horn with only a narrow zone trending across the southern edge of the South Georgia platform (Simpson and Griffiths, 1982). A close relation between the inferred granitic and mafic rocks of the southern South Georgia and South Orkneys platforms is suggested, and the possibility exists of the Rocas Verdes basin in the Andes (now uplifted and destroyed) continuing into the Weddell Sea. If the positive magnetic anomalies on the two platforms are related to mafic igneous rocks, then a distinct possibility is presented of rotation of the peninsula from its "model position" to its present position due to opening of a Rocas Verdes-Weddell basin.

## CONCLUSIONS-MODEL TESTING

The geologic history of the Scotia arc region is now fairly well known despite its remoteness and inaccessibility. The main uncertainties involve the palaeogeographic relationships between South America and the Antarctic Peninsula, firstly before break-up of Gondwanaland, and secondly during the period 100 Ma up to 30 Ma before present.

To my mind the geologic data are most readily explained by a reconstruction with the Antarctic Peninsula continuing southeastward from the Fuegian Andes as in Reconstruction C (Figure 2c). In particular this reconstruction seems to best explain the development of the arc-marginal basin terrain in South Georgia and the fore-arc complexes.

Palaeomagnetism offers the best independent test of models arrived at from marine geophysical or geological standpoints. Collection of critical material may be able to distinguish between Model A on the one hand and Models B and C on the other. The problem even here is that little variation in palaeolatitude is involved and the high latitude makes resolution of palaeo-declination difficult. Nonetheless, mafic dykes in the South Orkney Islands sampled during the course of R.V. Hero Cruise 77-1 (Dalziel et al., 1977) together with material collected by Dr Peter Barker and his colleagues from the South Orkney Islands during an earlier cruise of R.R.S. Shackleton may allow the pre-break-up orientation of the South Orkney Islands platform to be determined, and hence provide a vital clue to the original configuration of the Panthalassic margin in the Scotia arc region.

*Acknowledgements* My work in the Scotia arc region has all been supported by the National Science Foundation, principally by the division of Polar Programs. This paper was prepared with support from NSF grant No. DPP 78 20629. I wish to thank Dr Mort D. Turner, Program Manager for Earth Sciences in the Division of Polar Programs for his continuing help and encouragement.

## REFERENCES

ANDERSON, D.L., 1982: Hotspots, polar wander, Mesozoic convection and the geoid. *Nature, 297,*391-3.

ANONYMOUS, 1981: British Antarctic Survey, Natural Environment Research Council. Ann. Rept. for 1980, Cambridge.

BARKER, P.F., 1972: Magnetic lineations in the Scotia Sea; *in* Adie, R.J., (ed.) *Antarctic Geology and Geophysics,* Universitetsforlaget, Oslo, 17-26.

BARKER, P.F. and BURRELL, J., 1977: The opening of the Drake Passage. *Mar. Geol. 25,* 15-34.

BARKER, P.F. and DALZIEL, I.W.D. et al., 1976: The evolution of the southwestern Atlantic Ocean basin: Leg 36 data. *Initial Reports of Deep Sea Drilling Project, 36,* 993-1014. U.S. Govt Printing Office, Washington, D.C.

BARKER, P.F. and DALZIEL, I.W.D. (in press): Progress in Geodynamics in the Scotia Arc Region. Final Report, Working Group II Interunion Commission on Geodynamics. *Am. Geophys. Union.*

BARKER, P.F. and GRIFFITHS, D.H., 1972: The evolution of the Scotia Ridge and Scotia Sea. *R. Soc. Lond., Philos. Trans., Ser. A., 271,* 151-83.

BARKER, P.F., HILL, I.A., WEAVER, S.D. and PANKHURST, R.J., 1982: The origin of the eastern South Scotia Ridge as an intraoceanic island arc; *in* Craddock, C. (ed.) *Antarctic Geoscience,* Univ. Wisconsin Press, Madison, 203-11.

BRUHN, R.L., STERN, C.R. and de WIT, M.J., 1978: Field and geochemical data bearing on the development of a Mesozoic volcano-tectonic rift zone and back-arc basin in southernmost South America. *Earth Planet. Sci. Lett.*, *41*, 32-46.

BURNS, K.L., RICKARD, M.J., BELBIN, L. and CHAMALAUN, F. 1980: Further palaeomagnetic confirmation of the Magallanes orocline. *Tectonophysics.*, *63*, 75-90.

CANDE, S.C., HERRON, E.M., and HALL, B.R., 1982: The early Cenozoic tectonic history of the southeast Pacific. *Earth Planet. Sci. Lett.*, *57*, 63-74.

DALZIEL, I.W.D., 1976: The margins of the Scotia sea; *in* Burke C.A. and Drake, C.L. (eds.) *The Geology of Continental Margins*, Springer-Verlag, New York, 567-80.

DALZIEL, I.W.D., 1980: Comment on: Mesozoic evolution of the Antarctic Peninsula and the southern Andes by Harrison, C.G.A., Barron, E.J. and Hayes, W.W. *Geology*, 260-1.

DALZIEL, I.W.D., 1981: Back-arc extensions in the southern Andes: a review and critical appraisal. *R. Soc. Lond., Philos. Trans., Ser., A.*, *30*,300-35.

DALZIEL, I.W.D., 1982: The early (pre-Middle Jurassic) history of the Scotia arc region: a review and progress report; *in* Craddock, C. (ed.) *Antarctic Geoscience*, Univ. Wisconsin Press, Madison, 111-26.

DALZIEL, I.W.D. (in prep.): Tectonic evolution of a fore-arc terrain and of the southern Scotia Ridge, Antarctica.

DALZIEL, I.W.D., DOTT, R.H. Jr., WINN, R.D. and BRUHN, R.L., 1975: Tectonic relations of South Georgia to the southernmost Andes. *Geol. Soc. Am. Bull.*, *86*, 1034-40.

DALZIEL. I.W.D., and ELLIOT, D.H., 1971: Evolution of the Scotia arc. *Nature*, *233*, 246-52.

DALZIEL, I.W.D., and ELLIOT, D.H., 1973: The Scotia arc and Antarctic margin; *in* Nairn, A.E.M., and Stehli, F.G. (eds.) *The Ocean Basins and Margins, V. 1, The South Atlantic*, 171-246.

DALZIEL, I.W.D. and ELLIOT, D.H., 1982: West Antarctica: problem child of Gondwanaland. *Tectonics*, *1*, 3-19.

DALZIEL, I.W.D., ELLIOT, D.H., THOMSON, J.W., THOMSON, M.R.A., WELLS, N.A. and ZINSMEISTER, W.J., 1977: Geologic studies in the South Orkney Islands, R/V Hero Cruise 77-1, January 1977. *U.S. Antarct. J.*, *12*, 98-101.

DALZIEL, I.W.D., KLIGFIELD, R., LOWRIE, W. and OPDYKE, N.D., 1973: Palaeomagnetic data from the southernmost Andes and the Antarctandes; *in* Tarling, D.H. and Runcorn, S.K. (eds.) *Implications of Continental Drift for the Earth Sciences, v. 1*, Academic Press, New York, 87-101.

DEWEY, J.F., PITMAN III, W.C., RYAN W.B.F. and BONNIN, J., 1973: Plate tectonics and the evolution of the Alpine system. *Geol. Soc. Am. Bull.*, *84, (10)*, 3137-80.

De WIT, M.J., 1977: The evolution of the Scotia arc as a key to the reconstruction of southwestern Gondwanaland. *Tectonophysics*. *37*, 53-81.

DU TOIT, A.L., 1937: *Our Wandering Continents*, Oliver and Boyd, Edinburgh.

FORSYTH, D.W., 1975: Fault plane solutions and tectonics of the South Atlantic and Scotia Sea. *J. Geophys. Res.*, *80*, 1429-43.

FORSYTHE, R.F., 1982: The Late Palaeozoic-Early Mesozoic evolution of southern South America: a plate tectonic interpretation. *Geol. Soc. Lond., J.*, *139*, 671-82.

FORSYTHE, R.F., MICHAEL, P., and MPODOZIS, C., 1982: Metamorphic complexes of the Estrecho Nelson area: cases for subduction related metamorphism in the southern Andes. *Geol. Soc. Lond., J.*

HARRINGTON, P.K., BARKER, P.F. and GRIFFITHS, D.H. 1972: Crustal structure of the South Orkney Islands area from seismic refraction and magnetic measurement; *in* Adie, R.J., (ed.) *Antarctic Geology and Geophysics*, Universitetsforlaget, Oslo, 27-32.

HARRISON, C.G.A., BARRON, E.J. and HAY, W.W., 1980: Mesozoic evolution of the Antarctic Peninsula and the southern Andes. *Geology*, *7*, 374-8.

HAWKES, D.D., 1962: The structure of the Scotia Arc. *Geol. Mag.*, *99*, 85-91.

HILL, I.A. and BARKER, P.F., 1980: Evidence for Miocene back-arc spreading in the central Scotia Sea. *R. Astron. Soc., Geophys. J.*, *63*, 427-40.

HYDEN, G., and TANNER, P.W.G., 1981: Late Palaeozoic-Early Mesozoic fore-arc basin sedimentary rocks at the Pacific margin in Western Antarctica. *Geol. Rdsch. 70*, 529-41.

KATZ, H.R., 1973: Contrasts in tectonic evolution of orogenic belts in the Southeast Pacific. *R. Soc. N.Z., J.*, *3*, 333-62.

KATZ, H.R., and WATTERS, W.A., 1966: Geologic investigation of the Yahgan Formation, (Upper Mesozoic) and associated igneous rocks of Navarino Island, southern Chile. *N.Z. J. Geol. Geophys.*, *9*, 323-59.

KELLOGG, K.S., 1980: Palaeomagnetic evidence for oroclinal bending of the southern Antarctic Peninsula. *Geol. Soc. Am., Bull.*, *91*, 414-20.

KELLOGG, K.S., and REYNOLDS., R.L., 1978: Paleomagnetic results from the Lassiter Coast, Antarctica, and a test for oroclinal bending of the Antarctic Peninsula. *J. Geophys. Res.*, *83*, 2293-9.

LaBRECQUE, J.L. and BARKER, P.F., 1981: The age of the Weddell Basin. *Nature*, *290*, 489-92.

MATTHEWS, D.H., 1959: Aspects of the geology of the Scotia arc. *Geol. Mag.*, *96*, 425-41.

MILLER, H., 1982: Geologic comparison between the Antarctic Peninsula and southern South America; *in* Craddock C. (ed.) *Antarctic Geoscience*, Univ. Wisconsin, Press, Madison, 127-34.

NATLAND, M.L., GONZALEZ, E. and CANON, A. et al., 1974: A system of stages for correlation of Magallanes basin sediments. *Geol. Soc. Am. Mem.*, *139*.

NORTON, I.O., 1982: Paleomotion between Africa, South America, and Antarctica, and implications for the Antarctic Peninsula; *in* Craddock, C. (ed.) *Antarctic Geoscience*,Univ. Wisconsin Press, Madison, 99-106.

NORTON, I.O. and SCLATER, J.G., 1979: A model for the evolution of the Indian Ocean and the break-up of Gondwanaland. *J. Geophys. Res.*, *84*, 6803-30.

PANKHURST, R.J., 1982: Rb-Sr geochronology of Graham Land, Antarctica. *Geol. Soc. Lond., J.*, *139*, 701-12.

SIMPSON, P. and GRIFFITHS, D.H., 1982: The structure of the South Georgia continental block; *in* Craddock, C. (ed.) *Antarctic Geoscience*, Univ. Wisconsin Press, Madison, 185-92.

SMELLIE, J.L., 1981: A complete arc-trench system recognised in Gondwana sequences of the Antarctic Peninsula region. *Geol. Mag.*, *118*, 139-59.

SMITH, A.G. and HALLAM, A., 1970: The fit of the southern continents. *Nature*, *225*, 139-44.

STOREY, B.C., MAIR, B.F. and BELL, C.M., 1977: The occurrence of Mesozoic oceanic floor and ancient continental crust on South Georgia. *Geol. Mag.*, *114*, 203-8.

TANKARD, A.J., JACKSON, M.P.A., ERIKSSON, K.A., HOBDAY, D.K., HUNTER, D.R. and MINTER, W.E.L., 1982: *Crustal evolution of southern Africa*, Springer-Verlag, New York.

TANNER, P.W.G., 1982: Geologic evolution of South Georgia; *in* Craddock, C. (ed.) *Antarctic Geoscience*, Univ. Wisconsin Press, Madison, 167-76.

TANNER, P.W.G., in press: Geology of Shag Rocks, part of a continental block on the North Scotia Ridge, and possible regional correlations. *Bull. Br. antarct. Surv.*

TANNER, P.W.G., PANKHURST, R.J. and HYDEN, G., 1982: Radiometric evidence for the age of the Scotia metamorphic complex. *Geol. Soc. Lond., J.*, *139*, 683-90.

TANNER, P.W.G., STOREY, B.C. and MACDONALD, D.M., 1982: Geology of an Upper Jurassic-Lower Cretaceous island-arc assemblage in Hauge Reef, the Pickersgill Islands and adjoining areas of South Georgia. *Bull. Br. antarct. Surv.*, *53*, 77-117.

THOMSON, M.R.A., PANKHURST, R.J. and CLARKSON, P.D., (this volume): The Antarctic Peninsula-a late Mesozoic-Cenozoic arc. *Lamont-Doherty Geological Observatory Contribution No. 3418.*

# THE ANTARCTIC PENINSULA—A LATE MESOZOIC-CENOZOIC ARC (REVIEW)

M.R.A. Thomson, R.J. Pankhurst and P.D. Clarkson, *British Antarctic Survey, Madingley Road, Cambridge, CB3 0ET, U.K.*

*Abstract* From at least Early-Middle Jurassic times the Antarctic Peninsula developed as a magmatic arc, related to subduction of Pacific Ocean floor beneath a disintegrating Gondwana supercontinent. The basement to this arc is at least 25 km thick and its exposed parts consist mainly of the deformed metasedimentary rocks and gneisses related to an earlier Late Palaeozoic-Triassic subducted margin. This simple picture is complicated by radiometric dating which suggests that some deformed rocks previously thought to be"basement" accreted in Late Cretaceous times. Enough stratigraphical and facies data now exist to attempt a series of palaeogeographical reconstructions, although these still need to be related geographically to the rest of Gondwana. Geochemical and radiometric studies in the northern Antarctic Peninsula have revealed a westward migration of the site of magma generation across the edge of the continental crust into oceanic mantle and possibly a northward younging of intrusive activity. The latter may in some measure be related to the progressive northward cessation of subduction along the peninsula. Graham Land appears to be divided into two major tectonic segments of which the southern one is the more deeply eroded. Future work needs to address the problems of possible magmatic variations within Palmer Land, and an assessment of the region's uplift history with a view to producing a unified model for the whole of the Antarctic Peninsula.

The Antarctic Peninsula is composed largely of calc-alkaline igneous rocks of Mesozoic-Cenozoic age (Fleming and Thomson, 1979; Thomson, J., 1981; Thomson et al., 1981; Thomson, Harris, Rowley et al., 1982). Like the Andean province of South America, these constitute the remains of a linear magmatic arc related to subduction of Pacific Ocean floor beneath a continental margin during and following the break-up of Gondwana. This activity is superimposed on the pre-existing Pacific margin of the Gondwana subcontinent, and a full interpretation of the later history has to incorporate evidence for the nature and evolution of pre-existing rocks. This paper reviews advances made over the past 10 years in our understanding of the "basement" of the Antarctic Peninsula, the geochemistry and geochronology of the magmatic rocks, and the Upper Jurassic-Lower Cretaceous sedimentary rocks associated with them, which show that the arc had a very complicated history. The simplistic idea that most of the volcanic rocks were erupted during a single Late Jurassic episode is untenable, and the supposition that the"basement" rocks were deformed during the relatively short time span of a Late Triassic-Early Jurassic Gondwanian orogeny needs some modification. Broader features of the geological evolution in terms of palaeogeography, migrating magmatism and its relation to overall tectonic control have also become clearer. In general, only the most recent references are given to save space.

## THE BASEMENT

Prior to the break-up of Gondwana, the Antarctic Peninsula (Figure 1) formed part of its Pacific margin as a continuation of the South American plate (e.g. Dalziel and Elliot, 1982). The pre-Jurassic history of the two regions has been reviewed by Dalziel (1982) who concluded that the uplift and erosion of the"basement" following deformation and metamorphism were causally related to the break-up of Gondwana.

Turbidite-facies sedimentary rocks and low- to high-grade metamorphic rocks assigned to the pre-Jurassic"basement" of the Mesozoic-Cenozoic arc occur throughout the Antarctic Peninsula region, principally in the South Orkney and South Shetland islands, northern Antarctic Peninsula and Alexander Island. The turbidites consist mainly of sandstones and mudstones with minor amounts of conglomerate, quartzite, chert and greenschist (Hyden and Tanner, 1981). One limestone is known and pillow lavas occur locally in northern Graham Land and Alexander Island. A Rb-Sr age of 281 ± 16 Ma probably recording diagenesis of the Hope Bay Formation (Pankhurst, this volume) corroborates the suggestion of Late Palaeozoic sedimentation. However, elsewhere sedimentation is known to have persisted until Triassic times (Thomson, 1975; Edwards, 1980b; Dalziel et al., 1981). In northern Graham Land there is a marked unconformity between arc rocks and the Trinity Peninsula Group of the"basement", and similar relationships have been noted between volcanic rocks and gneiss on the Rhyolite Islands (Davies, in press) and between the Late Jurassic-Early Cretaceous fore-arc Fossil Bluff Formation and the deformed turbidites of the LeMay Formation in eastern Alexander Island (Edwards, 1980a). Thus the recent discovery of Late Cretaceous radiolaria in chert from the latter formation (D.L. Jones, pers. comm. to MRAT) is apparently enigmatic. However, the complicated time relationships between arc rocks and their "basement" can be rationalised if both are considered in an arc-trench setting, similar to that envisaged by Suárez (1976), and well-illustrated in reality by a seismic section across the Palawan trench of Indonesia (Hamilton, 1979, figure 105). The latter demonstrates the possibilities for time-equivalence of fore-arc sedimentation and deformation of more distal parts of the"basement". The Cretaceous chert could represent ponded sediment on the accretionary wedge that became tectonically interleaved with older rocks during subduction.

The metamorphic rocks of the Antarctic Peninsula are diverse but fall into two broad categories: low pressure, high temperature rocks of amphibolite facies, and high pressure, low temperature rocks of blueschist/greenschist facies. In the South Shetland and South Orkney islands both facies are exposed, which led Smellie and Clarkson (1975) to suggest that these constituted a paired metamorphic belt related to pre-Jurassic subduction. Whereas the amphibolite facies terrain has yielded metamorphic ages of up to 200 Ma, and could be a metamorphosed equivalent of the Trinity Peninsula Group, blueschists from Smith Island and Elephant Island appear to be Cretaceous in part (see Pankhurst, this volume). The latter have initial $^{87}Sr/^{86}Sr$ ratios which indicate that sedimentation could have taken place as much as 250 Ma ago but this is considered unlikely (Tanner et al., 1982). In southern Graham Land and Palmer Land the basement outcrops consist of a variety of paragneisses, orthogneisses and migmatites. Some of the orthogneisses are now recognised as early, foliated, Mesozoic igneous rocks (Pankhurst, this volume), but migmatites in eastern Graham Land have yielded mineral ages of ca 250 Ma (Pankhurst, this volume; Rex, 1976) with Rb-Sr whole rock systematics which indicate either an igneous precursor up to 600 Ma old or derivation by anatexis of sedimentary material of ancient provenance. There is as yet no geochronological constraint on the maximum age of most of the gneiss outcrops in Palmer Land.

Smellie (1981) interpreted the turbidites as having been deposited in a pre-Jurassic fore-arc environment although there is an apparent lack of contemporaneous magmatic rocks. Late Palaeozoic subduction of the Pacific plate has however been recorded in an analogous setting in South America (Gonzalez-Bonorino, 1971). The current paucity of evidence may be rectified by further work in Palmer Land where aeromagnetic surveys have shown a concomitant increase in the width of the batholith with the width and crustal thickness of the peninsula (R.G.B. Renner, pers. comm.). This may result from the parallel superposition of the Mesozoic arc on the remnant of an earlier arc.

## PALAEOGEOGRAPHIC DEVELOPMENT

Advances in our understanding of the age of the calc-alkaline volcanic rocks of the Antarctic Peninsula (Thomson and Pankhurst, this volume), and of the origin of the sedimentary rocks associated with them (e.g. Elliot and Wells, 1982; Elliot and Trautman, 1982; Farquharson, this volume; Laudon et al., this volume) allow us to attempt a more detailed interpretation of the Jurassic-Tertiary development of the region (Figure 2) than has previously been possible (Suárez, 1976; Thomson, 1982). Throughout the period of time under consideration the peninsula would have been at a high latitude (mostly south of 60°S) (e.g. Barron et al., 1981; Longshaw, 1981; Smith et al., 1981). However, abundant fossil plant remains in both terrestrial and marine rocks indicate that the arc land surface supported a rich flora dominated by ferns, cycadophytes, ginkgos and conifers in the Jurassic and Early Cretaceous, and by angiosperms and conifers in the Late Cretaceous and Early Tertiary (see Thomson, 1977 for references). Despite the fact that some of these floras, particularly

**Figure 1** Sketch map of the Antarctic Peninsula to show the localities mentioned in the text. To economise on space the Powell Basin has been closed and the South Orkney Islands have been moved closer to the northern end of the peninsula as they would have been in Mesozoic-Early Tertiary times.

those from Alexander Island and the southernmost parts of the peninsula region probably lived beyond 70°S and were subjected to a period of 24-hour darkness each year, many of the constituent species, particularly the cycadophytes, would probably not have been frost resistant. This is an important problem which waits a full palaeobotanical assessment (Jefferson, this volume).

## Late Jurassic

As yet there are not enough data to construct a palaeogeographic map for the earlier part of the Jurassic. However, it is now known that subaerial volcanoes were present in the Jason Peninsula area in Early-Middle Jurassic times, and volcanism may even have begun during the Triassic only a few km to the south (Thomson and Pankhurst, this

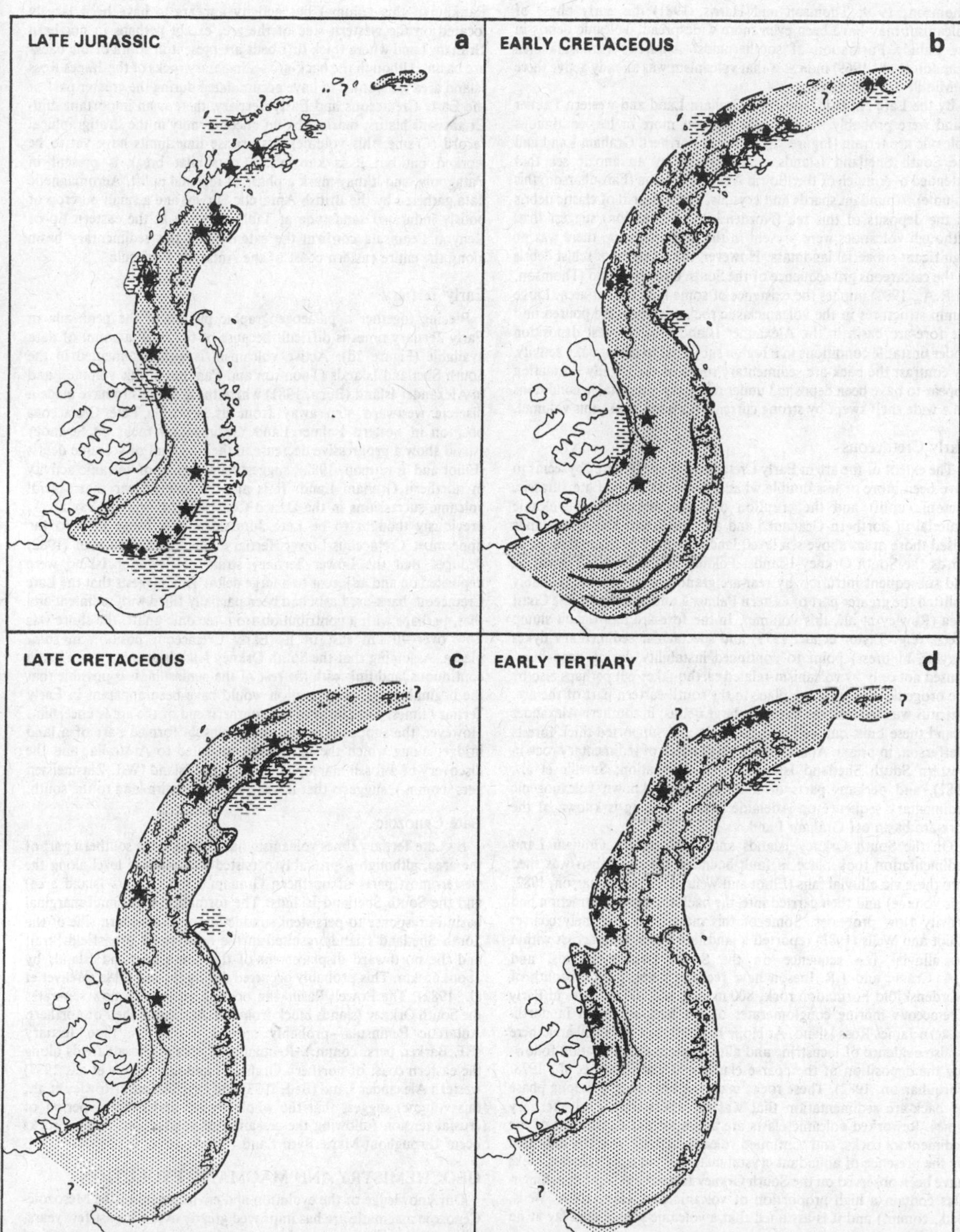


**Figure 2 Sketch maps to illustrate the development of the Antarctic Peninsula as a magmatic arc in Late Jurassic-Early Tertiary times. Bransfield Strait and the Powell Basin have been closed. Fine Stipple = probable land areas; horizontal dashes = known marine sediments; stars = areas of known volcanic activity (diagrammatic only); solid diamonds = terrestrial sediments; thick wavy lines = areas of folding. For full explanation see text.**

volume). Furthermore, judging by the extent of Early-Middle Jurassic plutonism on the eastern coast of Graham Land (Fleming and Thomson, 1979; Thomson and Harris, 1981) this early phase of volcanism may have been even more widespread. Volcanic debris in the Latady Formation of southernmost Antarctic Peninsula area (Laudon et al., 1969) indicates that volcanism was already active there in Middle Jurassic times.

By the Late Jurassic, southern Graham Land and western Palmer Land were probably the site of a narrow, more or less continuous volcanic arc terrain (Figure 2a), whereas northern Graham Land and the South Shetland Islands were covered by an anoxic sea that extended over much of the"South Atlantic" region (Farquharson, this volume). Abundant shards and crystals, and a dearth of clastic debris in the deposits of this sea (Nordenskjöld Formation) suggest that, although volcanoes were present in the northern area, there was no significant subaerial landmass. However, the presence of schist debris in the calcareous grit sequence of the South Orkney Island (Thomson, M.R.A., 1981) implies the existence of some land in that area. Large slump structures in the volcaniclastic rocks, where were poured into the fore-arc basin in the Alexander Island area, suggest deposition under unstable conditions in a region subjected to earthquake activity. By contrast the back-arc sedimentary rocks of the Latady Formation appear to have been deposited under relatively more stable conditions on a wide shelf swept by strong currents (Laudon et al., this volume).

### Early Cretaceous

The extent of the arc in Early Cretaceous times (Figure 2b) seems to have been more or less double what it had been in the Late Jurassic. Tectonic uplift and the eruption of large quantities of volcanic material in northern Graham Land and the South Shetland Islands raised those areas above sea level; land may have been continuous as far as the South Orkney Islands. Folding of the Latady Formation and subsequent intrusion by rear-arc granodioritic plutons probably uplifted the greater part of eastern Palmer Land and the Orville Coast area (Rowley et al., this volume). In the fore-arc basin giant slump sheets (e.g. Taylor et al., 1979) and spectacular sedimentary dykes (Taylor, in press) point to continued instability during deposition, caused not only by volcanism-related earthquakes but perhaps also by the orogenic events taking place in the south-eastern part of the arc. Detritus was fed into the basin via large deltas; in southern Alexander Island these built out well into the basin and supported thick forests (Jefferson, in press). Apart from a small patch of sedimentary rock in western South Shetland Islands (Byers Formation; Smellie et al., 1981), and perhaps parts of a thick, poorly known volcanogenic sedimentary sequence on Adelaide Island, nothing is known of the fore-arc basin off Graham Land.

On the South Orkney Islands and north-eastern Graham Land sedimentation took place in fault-bounded basins. Debris was shed into these via alluvial fans (Elliot and Wells, 1982; Farquharson, 1982, this volume) and then carried into the back-arc basin by traction and gravity flow processes. Some of this material is extremely coarse: Elliot and Wells (1982) reported a landslide block 30 m across within an alluvial fan sequence on the South Orkney Islands, and J.A. Crame and J.R. Ineson have recently mapped an olistolith of Nordenskjöld Formation rock, 800 m long and 200 m thick, in Early Cretaceous marine conglomerates of the back-arc basin in north-eastern James Ross Island. At Hope Bay and the Camp Hill area there is also evidence of lacustrine and alluvial plain sedimentation following the deposition of the coarse clastic material (Elliot et al., 1978; Farquharson, 1982). These rocks mark the beginning of a long phase of back-arc sedimentation that was to persist until Early Tertiary times. Reworked volcanic clasts are an important constituent of the sedimentary rocks, and continued volcanic activity is clearly indicated by the presence of abundant crystal material. No truly volcanic rocks have been observed on the South Orkney Islands but some conglomerates contain a high proportion of volcanic rock clasts (N.A. Wells, pers. comm.) and it is assumed that a volcanic source area lay at no great distance.

### Late Cretaceous

Nothing is known of any possible fore-arc sedimentation during Late Cretaceous times (Figures 2c). Possibly the beginnings of uplift and deformation of the Late Jurassic-Early Cretaceous Fossil Bluff Formation of Alexander Island were taking place at that time and eastern Alexander Island may have been land. Dating of volcanic rocks relies entirely on radiometric evidence (Thomson and Pankhurst, this volume) but activity appears to have been largely located on the western side of the arc, except perhaps in northern Graham Land where thick tuff beds are present in strata of the back-arc basin. Although the back-arc sedimentary rocks of the James Ross Island area are known to have accumulated during the greater part of the Early Cretaceous and Early Tertiary, there is an important mid-Cretaceous hiatus, marked by an unconformity in the stratigraphical record (Crame, this volume). Its precise time limits have yet to be worked out but it is known that a similar break is present in Patagonia, and it may mark a phase of regional uplift. Aeromagnetic data gathered by the British Antarctic Survey, and a small outcrop of poorly indurated sandstone at Table Nunatak at the eastern tip of Kenyon Peninsula confirm the extension of this sedimentary basin along the entire eastern coast of the Antarctic Peninsula

### Early Tertiary

Piecing together a palaeogeographic picture of the peninsula in Early Tertiary times is difficult because of the small amount of data available (Figure 2d). Active volcanism is well-documented in the South Shetland Islands (Thomson and Pankhurst, this volume), and in Alexander Island (Burn, 1981) where the arc seems to have made a discrete westward step away from its probable Late Cretaceous position in western Palmer Land. Sedimentary rocks on Seymour Island show a progressive decrease in their content of volcanic debris (Elliot and Trautman, 1982), suggesting a waning of volcanic activity in northern Graham Land. It is also becoming clear that several volcanic successions in the Danco Coast and Marguerite Bay areas, previously thought to be Late Jurassic in age, range up into the uppermost Cretaceous-Lower Tertiary. Elliot and Trautman (1982) deduced that the Lower Tertiary strata of Seymour Island were deposited on and adjacent to a large delta. This suggests that the Late Cretaceous back-arc basin had been partially filled with sediment and that, perhaps with a contribution from tectonic uplift, the shore was now over 70 km east of its Early Cretaceous position in some places. Assuming that the South Orkney Islands area had once had a continuous land link with the rest of the peninsula, it is possible that the beginnings of its separation would have been apparent in Early Tertiary times. The status of the southern end of the arc is uncertain. However, the supposition that the peninsula formed part of a land bridge, along which the marsupials migrated to Australia, and the discovery of a fossil marsupial on Seymour Island (W.J. Zinsmeister, pers. comm.), suggests that it was continuous with land to the south.

### Late Cenozoic

By Late Tertiary times volcanism had ceased in the southern part of the area, although it probably persisted at a reduced level along the westernmost parts of northern Graham Land (Anvers Island area) and the South Shetland Islands. The formation of a small marginal basin in response to persistent subduction on the western side of the South Shetland Islands resulted in the opening of Bransfield Strait and the northward displacement of the South Shetland Islands by about 65 km. This probably occurred during the last 2 Ma (Weaver et al., 1982). The Powell Basin—an oceanic area which now separates the South Orkney Islands block from the continental shelf of northern Antarctic Peninsula—probably opened during the Late Tertiary (P.F. Barker, pers. comm.). Remnants of alkaline basalt fields along the eastern coast of northern Graham Land (e.g. Baker et al., 1977) western Alexander Land (Bell, 1973) and Orville Coast (Rowley et al., this volume) suggest that the whole region underwent a period of crustal tension following the cessation of subduction. Similar rocks occur throughout Marie Byrd Land.

## GEOCHEMISTRY AND MAGMATIC EVOLUTION

Our knowledge of the evolution and environment of the Mesozoic-Cenozoic magmatic arc has improved greatly over the past few years, with new detailed stratigraphical, geochemical and geochronological studies. Sometimes these have confirmed early, tenuously based ideas, but more often they have necessitated radical revisions, as in the cases of supposed Palaeozoic plutonism (Pankhurst, 1982) and in the timing and nature of volcanism (Thomson and Pankhurst, this volume). One uncertainty concerns the question of how and when volcanic activity commenced. Although the sedimentary rocks deformed during the

Gondwanian event(s) are now thought to be arc-related deposits, there is (with one possible exception considered below) no identification in the Antarctic Peninsula of any corresponding Late Palaeozoic-Triassic magmatism. Plutonic rocks in northern Palmer Land inferred to be of this age by Smellie (1981) are now thought to be much younger. The oldest dated plutonic rock outcrop in the arc is the 209± 3 Ma foliated granite-granodiorite on Cole Peninsula (Pankhurst, 1982). The foliation, which must be older than 175 Ma, could well have developed during the final stage of Late Triassic-Early Jurassic metamorphism recorded by K-Ar dating of the Gondwanian sequences in the South Orkney Islands (Tanner et al., 1982), a timing compatible with the suggestion that the granites were part of the pre-Gondwanian arc. However, Pankhurst (this volume) has proposed that the main orogeny was even older (Early Triassic) on the east coast of the peninsula, so that these granites may simply represent very early plutonic activity in the post-Gondwanian arc. Since the Cole Peninsula granites are chemically similar to late intrusions on the east coast (Hamer, 1978), this alternative is perhaps more likely. A Cretaceous age for some of the blueschist facies sequences in the South Shetland Islands indicates that the subduction history of the Antarctic Peninsula cannot always be separated into unrelated pre- and post-Gondwanian phases, but that it may have been essentially continuous.

By Early-Middle Jurassic times the Antarctic Peninsula was an extremely active magmatic arc. The first major event is recorded by numerous granites and granodiorites on the east coast of Graham Land which give radiometric ages of 165-180 Ma (Rex, 1976; Pankhurst, 1982). Extrusive and intrusive magmatism continued more or less unabated throughout Cretaceous and Early Tertiary times, in response to subduction of Pacific Ocean floor along a trench essentially parallel to the western edge of the present continental shelf (e.g. Weaver et al., 1982). Thus the Mesozoic volcanic rocks increase eastwards in $SiO_2$, $K_2O$ and $K_2O/SiO_2$ as well as in lithophile trace elements and ratios such as Ce/Y. Comparable behaviour is noted in the plutonic rocks (Saunders et al., 1982). Such chemical trends are common in calc-alkaline arcs and are clearly related to depth to Benioff zones, although the causal mechanism is still speculative.

The geochronological evidence shows quite clearly that in Graham Land the principal site of magmatism migrated westwards during the Late Mesozoic (Saunders et al., 1982; Pankhurst, 1982) and it seems that other differences between early and late magmas may be related to this migration. Thus Hamer and Moyes (1982) and Moyes and Hamer (this volume) have studied the occurrence of garnet in the arc products, especially on the east coast, and have shown that whereas some garnets crystallised early from the magmas at depths of about 25 km, others are xenocrysts acquired by assimilation of metamorphic wall-rocks. Both are taken as evidence for thick continental crust underlying the eastern side of the peninsula, even during the early stages of arc formation. A further indication of underlying sialic crust is the progressive decrease in initial $^{87}Sr/^{86}Sr$ ratios from ca 0.707 in the Early-Middle Jurassic plutons to mantle values of $<0.7040$ in the Tertiary ones. Pankhurst (1982) linked this variation with an overall increase in basicity and diversity of composition in the younger magmas and suggested that the migration of the arc took the site of magma generation westwards across the edge of the continental crust into oceanic mantle. The precise nature of the crustal contribution to early magmatism (i.e. contamination or anatexis) should be revealed by more detailed geochemical study of specific igneous complexes in the near future, together with new isotope data. Further work is also necessary to investigate the geochronology of west coast magmatism in northern Graham Land and its relationship to the Late Jurassic-Miocene volcanism of the South Shetland Islands.

With detailed knowledge the other broad feature of the magmatic arc which may become clearer is the interpretation of significant geological variations along it. Differences in the timing of intrusive activity, led Saunders et al. (1982) to suggest a general northward younging of plutons in Graham Land, while Barker (1982) has linked this observation to the sea-floor spreading history in the southern Pacific Ocean. Younging of the oceanic magnetic anomalies away from the coastline testifies to an active spreading centre (the Aluk Ridge) which was gradually consumed by subduction, earlier in the south. Barker proposed that the cessation of igneous activity at any given latitude could be connected to the subduction of very young ocean floor as the offshore Aluk Ridge approached the trench region and he predicted a cut-off point ranging from Early Tertiary near Alexander Island to only 4 Ma ago at Anvers Island. This idea is difficult to test adequately without much closer control on the timing of the youngest magmatism along the whole length of the peninsula, but the ultimate aim must be to unite the available geological and geophysical evidence.

In purely geological terms, lateral variations within Graham Land, are mainly related to the early history of the arc and are preserved in the ocean floor record. Thus the nature of the basement and the timing of igneous climaxes differ between Trinity Peninsula and the arc south of ca 65°S (Table 1). Lateral variations were emphasised by Hawkes (1981) who recognised five segments within Graham Land, separated by the continental extrapolation of oceanic fracture zones (although some are not well-defined from geophysical evidence). He proposed that this "tectonic segmentation" arose from the different nature of oceanic-continental interaction associated with segmented subduction, and also suggested that the subducted fracture zones themselves are sites of more intense mineralisation by deep-mantle fluids. Although we disagree with the details of this model (some of the geological evidence particularly that relating to the basement is now known to be incorrect) it does seem as though the differences observed above (Table 1) probably reflect at least a dual segmentation of Graham Land, with the southern segment possibly representing deeper "pre-volcanic" erosion levels in the core of the peninsula (i.e. the opposite of Hawkes' conclusion about erosion levels). The uplift history of the Antarctic Peninsula is clearly worthy of considerable further study, possibly using fission track dating.

When assessing lateral variations in the Antarctic Peninsula, evidence from Palmer Land will clearly be of great importance. The peninsula is much wider here (generally over 100 km, even excluding Alexander Island), and some aspects of the arc history are clearly quite different from those observed in Graham Land. For instance, the "basement" arc-trench gap and fore-arc sequences are both apparently present within the LeMay Formation on Alexander Island, well to the west of the magmatic arc. The igneous rocks in Palmer Land are geologically similar to those of Graham Land, and Jurassic volcanism (see Thomson and Pankhurst, this volume) and Jurassic-Cretaceous plutonism are represented. So far, owing to the remoteness and relative paucity of outcrop there has been no detailed work in Palmer Land to follow up reconnaissance mapping. In particular there is little geochronological evidence to compare with the now copious data from Graham Land, except for American work on the southern Black Coast and Lassiter Coast. Early-Middle Jurassic plutons occur on the west coast of the mainland (and at Mount Sullivan, Pankhurst, this volume) whereas mid-to-Late Cretaceous plutons abound on the eastern side of the central plateau (Farrar et al., 1982; Singleton, 1980). Tertiary igneous rocks are as yet unknown from Palmer Land itself, activity by this time apparently having rapidly migrated westwards into Alexander Island where it is represented by Early Tertiary basalt-rhyolite volcanic rocks and diorite-granite plutonic complexes, a late member of which has been dated at 46±3 Ma (Pankhurst, 1982).

## CONCLUSIONS

Although reconnaissance geological mapping of Graham Land and northern Palmer Land was essentially complete in the mid-1970's it is only subsequently, with the first detailed stratigraphical and geochronological field and laboratory programme that it has been possible to incorporate observed geology into a reasonable history of the development of the Mesozoic-Cenozoic magmatic arc. This shows a continuous increase in the intensity of calc-alkaline plutonism and volcanism from Early-Middle Jurassic to Early Tertiary times.

**TABLE 1: A geological comparison between northern and southern Graham Land**

| | N. Graham Land | S. Graham Land |
|---|---|---|
| Basement | Trinity Peninsula Group, greywacke, shale | Orthogneiss, migmatite, paragneiss |
| Grade | Greenschist, increasing southwards | Amphibolite |
| Deformation | Open folding | Not known |
| Age of volcanism | Mostly Cretaceous | Mostly Jurassic |
| Age of early plutonism | Mostly Early-Middle Cretaceous | Early-Middle Jurassic |
| Transverse migration of magmatism | Not so well defined as S. Graham Land | Westwards in Late Cretaceous (90 Ma) and Early Tertiary (60 Ma) |

Throughout this period there was a gradual increase in the subaerial growth of the land mass and associated clastic sediments. The linear magmatic arc in Graham Land migrated steadily westward and the chemistry of magmatic products showed the decreasing involvement of pre-existing continental crust. Geochemical and geophysical evidence relate this phase to subduction of Pacific Ocean floor off the west coast of the peninsula. During Tertiary times there was a gradual cessation of magmatism, possibly in response to the progressive northward consumption of very young ocean floor formed at the Aluk Ridge, which was itself eventually consumed. It is possible that segmental differentiation of the peninsula resulted from sectional subduction of the ocean floor between areas separated by major fracture zones.

Further advances in knowledge will probably come only from detailed follow-up studies in areas unexamined since the reconnaissance work, notably north-western Graham Land and its archipelago and the much larger area of Palmer Land. The latter should be particularly crucial to unravelling the mysteries of the pre-Mesozoic subduction history and the nature and timing of the Gondwanian"orogeny". A considerable part of central and southern Palmer Land still awaits primary mapping. Planned work may finally resolve the problems of the geological and geophysical constitution of the Pacific margin of Gondwana in the Late Palaeozoic.

*Acknowledgements* We wish to express our gratitude to our colleagues in the British Antarctic Survey and other Antarctic geologists around the world for open discussions on many occasions.

## REFERENCES


BAKER, P.E., BUCKLEY, F. and REX, D.C., 1977: Cenozoic volcanism in the Antarctic. *R. Soc. Lond., Philos. Trans., Ser.B, 279,* 131-42.

BARKER, P.F., 1982: The Cenozoic subduction history of the Antarctic Peninsula. *Geol. Soc. Lond., J., 139,* 787-801.

BARRON, E.J., HARRISON, C.G.A., SLOAN II, J.L. and HAY, W.W., 1981: Paleogeography, 180 million years ago to the present. *Eclog. geol. Helv., 74,* 443-70.

BELL, C.M., 1973: The geology of Beethoven Peninsula, south-western Alexander Island. *Bull. Br. antarct. Surv., 32,* 75-83.

BURN, R.W., 1981: Early Tertiary calc-alkaline volcanism on Alexander Island. *Bull. Br. antarct. Surv., 53,* 175-93.

CRAME, J.A. (this volume): Cretaceous inoceramid bivalves from Antarctica.

DALZIEL, I.W.D., 1982: The early (pre-Middle Jurassic) history of the Scotia arc region: a review and progress report; *in* Craddock, C. (ed.) *Antarctic Geoscience.* Univ. Wisconsin Press, Madison, 111-26.

DALZIEL, I.W.D., ELLIOT, D.H., JONES, D.L. et al., 1981: The geological significance of some Triassic microfossils from the South Orkney Islands, Scotia Ridge. *Geol. Mag., 118,* 15-25.

DALZIEL, I.W.D. and ELLIOT, D.H., 1982: West Antarctica: problem child of Gondwanaland. *Tectonics, 1,* 3-19.

DAVIES, T.G., (in press): The geology of part of northern Palmer Land. *Scient. Rep. Br. antarct. Surv., 102.*

EDWARDS, C.W., 1980a: New evidence of major faulting on Alexander Island. *Bull. Br. Antarct. Surv., 49,* 15-20.

EDWARDS, C.W., 1980b: Early Mesozoic fossils from central Alexander Island. *Bull. Br. Antarct. Surv., 49,* 33-58.

ELLIOT, D.H. and TRAUTMAN, T.A., 1982: Lower Tertiary strata on Seymour Island, Antarctic Peninsula; *in* Craddock, C. (ed.) *Antarctic Geoscience.* Univ. Wisconsin Press, Madison, 287-97.

ELLIOT, D.H., WATTS, D.R., ALLEY, R.B. and GRACANIN, T.M., 1978: Geologic studies in the northern Antarctic Peninsula, R/V *Hero* cruise 78-1B, February 1978. *U.S. Antarct. J., 13(5),* 12-13.

ELLIOT, D.H. and WELLS, N.A., 1982: Mesozoic alluvial fans of the South Orkney Islands; *in* Craddock, C. (ed.) *Antarctic Geoscience.* Univ. Wisconsin Press, Madison, 235-44.

FARQUHARSON, G.W., 1982: Late Mesozoic sedimentation in the northern Antarctic Peninsula and its relationship to the southern Andes. *Geol. Soc. Lond., J., 139,* 721-7.

FARQUHARSON, G.W., (this volume): Evolution of Late Mesozoic sedimentary basins in the northern Antarctic Peninsula.

FARRAR, E., MCBRIDE, S.L. and ROWLEY, P.D., 1982: Ages and tectonic implications of Andean plutonism in the southern Antarctic Peninsula; *in* Craddock, C. (ed.) *Antarctic Geoscience.* Univ. Wisconsin Press, Madison, 349-56.

FLEMING, E.A. and THOMSON, J.W. (compilers), 1979: Northern Graham Land and South Shetland Islands. *British Antarctic Territory geological map, 1:500,000. BAS 500G Series. Sheet 2.*British Antarctic Survey, Cambridge.

GONZALEZ-BONORINO, F., 1971: Metamorphism of the crystalline basement of southern Chile. *J. Petrol., 12,* 149-75.

HAMER, R.D., 1978: The geochemistry of the igneous rocks of Cape Robinson, Antarctica. Unpublished M.Sc. dissertation. University of Birmingham, U.K.

HAMER, R.D. and MOYES, A.B., 1982: Petrogenesis of garnet-bearing volcanic rocks of the Antarctic Peninsula Volcanic Group from Trinity Peninsula. *Geol. Soc. Lond., J., 139,* 713-20.

HAMILTON, W., 1979: Tectonics of the Indonesian region. *U.S. Geol. Surv., Prof. Pap., 1078.*

HAWKES, D.D., 1981: Tectonic segmentation of the northern Antarctic Peninsula. *Geology, 9,* 220-4.

HYDEN, G. and TANNER, P.W.G., 1981: Late Palaeozoic-early Mesozoic fore-arc basin sedimentary rocks at the Pacific margin in western Antarctica. *Geol. Rdsch., 70,* 529-41.

JEFFERSON, T.H. (this volume): Palaeoclimatic significance of some Mesozoic Antarctic fossil floras.

JEFFERSON, T.H. (in press): The early Cretaceous fossil forests of Alexander Island, Antarctica. *Palaeontology.*

LAUDON, T.S., LACKEY, L.L., QUILTY, P.G. and OTWAY, P.M., 1969: Geology of eastern Ellsworth Land. (Sheet 3, eastern Ellsworth Land); *in* Bushnell, V.C. and Craddock, C. (eds.) Geologic maps of Antarctica. *Antarct. Map Folio Ser., 12,* Pl. III.

LAUDON, T.S., THOMSON, M.R.A., WILLIAMS, P.L. and MILLIKEN, K.L. (this volume): The Jurassic Latady Formation, southwestern Antarctic Peninsula.

LONGSHAW, S.K., 1981: A palaeomagnetic study of the Antarctic Peninsula. Unpublished Ph.D. thesis, University of Birmingham, U.K., 197pp.

MOYES, A.B. and HAMER, R.D. (this volume): Contrasting origins and implications of garnet in rocks of the Antarctic Peninsula.

PANKHURST, R.J., 1982: Rb-Sr geochronology of Graham Land, Antarctica. *Geol. Soc. Lond., J., 139,* 701-11.

PANKHURST, R.J. (this volume): RB-Sr constraints on the ages of basement rocks on the Antarctic Peninsula.

REX, D.C., 1976: Geochronology in relation to the stratigraphy of the Antarctic Peninsula. *Bull. Br. antarct. Surv., 43,* 49-58.

ROWLEY, P.D., VENNUM, W.R., KELLOGG, K.S., LAUDON, T.S., CARRARA, P.E., BOYLES, J.M. and THOMSON, M.R.A. (this volume): Geology and plate tectonic setting of the Orville Coast and eastern Ellsworth Land, Antarctica.

SAUNDERS, A.D., WEAVER, S.D. and TARNEY, J., 1982: The pattern of Antarctica Peninsula plutonism; *in* Craddock, C. (ed.) *Antarctic Geoscience.* Univ. Wisconsin Press, Madison, 305-14.

SINGLETON, D.G., 1980: The geology of the central Black Coast, Palmer Land. *Scient. Rep. Br. antarct. Surv., 102.*

SMELLIE, J.L., 1981: A complete arc-trench system recognized in Gondwana sequences of the Antarctic Peninsula region. *Geol. Mag., 118,* 139-59.

SMELLIE, J.L. and CLARKSON, P.D., 1975: Evidence for pre-Jurassic subduction in western Antarctica. *Nature, 258* 701-2.

SMELLIE, J.L., DAVIES, R.E.S. and THOMSON, M.R.A., 1981: Geology of a Mesozoic intra-arc sequence on Byers Peninsula, Livingston Island, South Shetland Islands. *Bull. Br. antarct. Surv., 50,* 55-76.

SMITH, A.G., HURLEY, A.M. and BRIDEN, J.C. 1981: *Phanerozoic paleocontinental world maps.* Cambridge Univ. Press, Cambridge.

SUAREZ, M., 1976: Plate-tectonic model for southern Antarctic Peninsula and its relation to southern Andes. *Geology, 4,* 211-4.

TANNER, P.W.G., PANKHURST, R.J. and HYDEN, G., 1982: Radiometric evidence for the age of the subduction complex in the South Orkney Islands and South Shetland Islands, West Antarctica. *Geol. Soc. Lond., J., 139,* 683-90.

TAYLOR, B.J. (in press): Sedimentary dykes, pipes and related structures in the Mesozoic sediments of south-eastern Alexander Island. *Bull. Br. Antarct. Surv., 51,* 1-42.

TAYLOR, B.J., THOMSON, M.R.A. and WILLEY, L.E., 1979: The geology of the Ablation Point—Keystone Cliffs area, Alexander Island. *Scient. Rep. Br. Antarct. Surv., 82.*

THOMSON, J.W. (compiler), 1981: Alexander Island. *British Antarctic Territory geological map, 1:500,000. BAS 500G Series, Sheet 4.*British Antarctic Survey, Cambridge.

THOMSON, J.W. and HARRIS, J. (compilers), 1981: Southern Graham Land. *British Antarctic Territory geological map, 1:500,000. BAS 500G Series, Sheet 3.* British Antarctic Survey, Cambridge.

THOMSON, J.W. and HARRIS, J., ROWLEY, P.D. et al. (compilers), 1982: Northern Palmer Land. *British Antarctic Territory geological map, 1:500,000. BAS 500G Series, Sheet 5.*British Antarctic Survey, Cambridge.

THOMSON, M.R.A., 1975: New palaeontological and lithological observations on the Legoupil Formation, north-west Antarctic Peninsula. *Bull. Br. Antarct. Surv., 41 & 42,* 169-85.

THOMSON, M.R.A., 1977: An annotated bibliography of the palaeontology of Lesser Antarctica and the Scotia Ridge. *N.Z. J. Geol. Geophys., 20,* 865-904.

THOMSON, M.R.A., 1981: Late Mesozoic stratigraphy and invertebrate palaeontology of the South Orkney Islands. *Bull. Br. Antarct. Surv., 54,* 65-83.

THOMSON, M.R.A., 1982: Mesozoic palaeogeography of West Antarctica; *in* Craddock, C. (ed.) *Antarctic Geoscience.* Univ. Wisconsin Press, Madison, 331-7.

THOMSON, M.R.A. and PANKHURST, R.J. (this volume): Age of post-Gondwanian calc-alkaline volcanism in the Antarctic Peninsula region.

WEAVER, S.D., SAUNDERS, A.D. and TARNEY, J., 1982: Mesozoic-Cenozoic volcanism in the South Shetland Islands and the Antarctic Peninsula: geochemical nature and plate tectonic significance; *in* Craddock, C. (ed.) *Antarctic Geoscience.* Univ. Wisconsin Press, Madison, 263-73.

# TITHONIAN (UPPERMOST JURASSIC)—BARREMIAN (LOWER CRETACEOUS) SPORES, POLLEN AND MICROPLANKTON FROM THE SOUTH SHETLAND ISLANDS, ANTARCTICA

R.A. Askin, *Geology Department, Colorado School of Mines, Golden, Colorado, 80401. U.S.A.*

*Abstract* Fossil spores, pollen and microplankton have been recovered in samples from five localities on Byers Peninsula, Livingston Island, and two localities on President Head, Snow Island, in the South Shetland Islands, Antarctica. Palynomorph distribution enables subdivision into two broad zones. Assemblage A, represented by samples from the western margin of Byers Peninsula, is of Tithonian-Berriasian-?earliest Valanginian age. Assemblage B, represented by samples from central-southern Byers Peninsula and President Head, is of ?Valanginian-Hauterivian-Barremian age. Similarity of supposed Barremian spore and pollen assemblages of the South Shetlands with those of the Baqueró Formation, southern Argentina, suggest close proximity of the two areas during the Neocomian; while the presence of endemic South Shetland forms suggests environmental differences between the two areas.

Fossil palynomorphs have been recovered from samples collected from two areas in the South Shetland Islands, Antarctica, during the February 1980 expedition, R/V *Hero* cruise 80-2 (Elliot and Askin, 1980). These areas are Byers Peninsula on western Livingston Island, and President Head, eastern Snow Island (Figure 1).

Detailed stratigraphic study of Byers Peninsula sedimentary rocks is difficult. Apart from outcrops of some of the more resistant extrusive and intrusive igneous rocks, the peninsula is mostly covered with surficial periglacial debris, and outcrops of sedimentary rocks are very small and widely scattered. By necessity, much of the geology must be inferred from chips of rock at the surface of float.

Hobbs (1968) initially assigned a Miocene age to the Byers Peninsula sequence, but this was later changed to Late Jurassic—Early Cretaceous with the discovery of ammonites and leaf fossils. Location of invertebrate faunas is shown (as A, B, C) on Figure 1. They have been assigned a Tithonian (A), Berriasian (B) and Valanginian (C) age by González-Ferrán et al. (1970), Tavera (1970) and Covacevich (1976), and more recently with additional collections by Smellie et al. (1980). Plant megafossils from southwest of Cerro Negro were assigned a "Wealden" age by Fuenzalida (1965) and Araya and Hervé (1966), and were later compared by Hernández and Azcárate (1971) to the Barremian Baqueró Formation flora of Santa Cruz Province, Argentina. The sequence of volcanic and volcanogenic rocks on Byers Peninsula has been referred to the Byers Formation (Smellie et al., 1980), consisting of four members: a mudstone member, mixed marine member, volcanic member, and agglomerate member. Fossil plants from one locality in the volcanogenic sequence on President Head, Snow Island, were assigned a Middle Jurassic age by Fuenzalida et al. (1972).

## RESULTS

Localities sampled for palynomorphs are indicated on Figure 1 as circled numbers. These include 10 main localities on Byers Peninsula (BP) and three on President Head (PH). Samples collected at Byers Peninsula were from the marine "mudstone member" (BP1), the "mixed marine member" (BP2-9), and non-marine plant-bearing beds of the "volcanic member" (BP10). Samples from President Head are from marine strata (PH1, 2) and from non-marine plant-bearing beds (PH3). Outcrops of fine-grained sedimentary rocks suitable for palynological study are scarce, and in many cases sediments are too baked by emplacement of adjacent igneous rocks for preservation of palynomorphs. Palynomorphs recovered from five localities (BP2, 4, 6, 9, 10) on Byers Peninsula and two (PH1, 3) on President Head include spores and pollen from land plants, and microplankton (dinoflagellates and acritarchs).

### Palynomorph Preservation and Thermal Metamorphism

Palynomorph preservation is generally poor. Samples from five localities on Byers Peninsula and one on President Head are barren of palynomorphs or contain skeletal grains too poorly preserved to identify. The skeletal grains are black and indicate thermal metamorphism in these localities approaching the destructive stage for palynomorphs, that is, equivalent to meta-anthracite coal rank.

Of the productive localities, BP6 and 10 and, in particular PH1 and 3 are better preserved. The orange colour of President Head and BP6 palynomorphs suggests a thermal maturation equivalent to high-volatile bituminous coal rank, a metamorphic grade relatively low in comparison with other Mesozoic sedimentary rocks in the Antarctic Peninsula area. Other Byers localities are substantially more metamorphosed (palynomorphs dark brown to black), presumably a result of igneous heating. The Byers palynomorphs have also suffered from a moderate to high degree of corrosion and abrasion, and mineral scarring from abundant pyrite. Surface ornamentation, most of the processes, and traces of tabulation (if originally present) have been obliterated on many of the dinoflagellate cysts.

### Palynomorph Composition of Samples

Sample data, and taxonomy of spores, pollen and microplankton species will be included in forthcoming papers. A brief outline of palynomorph distribution and geologic implications is given below.

Species of spores and pollen which consistently and sometimes frequently (>5%) occur on Byers Peninsula and President Head are: *Cyathidites australis* Couper, *C. minor* Couper, *Stereisporites antiquasporites* (Wilson and Webster) Dettmann, *Osmundacidites* spp. including *O. wellmanii* Couper, *Baculatisporites* spp. including *B. comaumensis* (Cookson) Potonié, *Neoraistrickia* spp., *Lycopodiumsporites* spp., *Gleicheniidites circinidites* (Cookson) Dettmann, *Polycingulatisporites* spp., *Contignisporites cooksonii* (Balme) Dettmann, *Callialasporites* spp., mainly *C. trilobatus* (Balme) Dev, *Alisporites* spp. and *Podocarpidites* spp. These species are also long-ranging (throughout much of the Jurassic and Cretaceous or more) in extra-Antarctic regions (e.g. Australia).

Species which are relatively long-ranging elsewhere and rare (only one or very few specimens observed) in the South Shetlands include *Ceratosporites equalis* Cookson and Dettman, *Lycopodiacidites asperatus* Dettmann, *Klukisporites scaberis* (Cookson and Dettmann) Dettmann, *Ischyosporites punctatus* Cookson and Dettmann, *Cibotiumspora juriensis* (Balme) Filatoff, *Cingutriletes clavus* (Balme) Dettmann, *Murospora florida* (Balme) Pocock, *Foraminisporis dailyi* (Cookson and Dettmann) Dettman and *Aequitriradites spinulosus* (Cookson and Dettmann) Cookson and Dettmann. More common but still infrequently occurring species include *Microcachyridites antarcticus* Cookson, *Classopollis* spp. and *Araucariacites australis* Cookson, and the dinoflagellate *Pareodinia ceratophora* Deflandre.

Consistently appearing species in the Byers and President Head samples which have biostratigraphic significance are *Cicatricosisporites australiensis* (Cookson) Potonié, *C. ludbrookii* Dettmann, *Polypodiaceoisporites elegans* Archangelsky and Gamerro, and the dinoflagellate *Oligosphaeridium complex* (White) Davey and Williams. (Note however that samples from BP10 and PH3 are non-marine and contain no dinoflagellates).

The following species are of restricted occurrence on Byers Peninsula and President Head and are of biostratigraphic significance. They are the spores *Cyatheacidites tectifera* Archangelsky and Gamerro, *Appendicisporites* sp., *Concavissimisporites* sp. cf. *C. penolaensis* Dettmann, *Densoisporites velatus* Weyland and Krieger emend. Krasnova, *Biretisporites* sp. A, *Verrucosisporites* sp. A, and *Foveotriletes* sp. A; and the dinoflagellates *"Aptea" attadalica* (Cookson and Eisenack) Davey and Verdier, *Batioladinium micropodum* (Eisenack and Cookson) Brideaux, *Batioladinium* sp. A, cf. *Broomea simplex* Cookson and Eisenack, *Cribroperidinium perforans* (Cookson and Eisenack) Morgan, *Gonyaulacysta helicoidea* (Eisenack and Cookson) Sarjeant, *Hystrichogonyaulax serrata* (Cookson and Eisenack) Stover and Evitt, *Hystrichodinium voigtii* (Alberti emend. Sarjeant) Davey, *Palaeoperidinium cretaceum* Pocock ex Davey, and cf. *Sirmiodiniopsis* sp. A.

The distinctive spore *Cyatheacidites tectifera* predominates the non-marine assemblages of BP10 and PH3 along with the longer-ranging

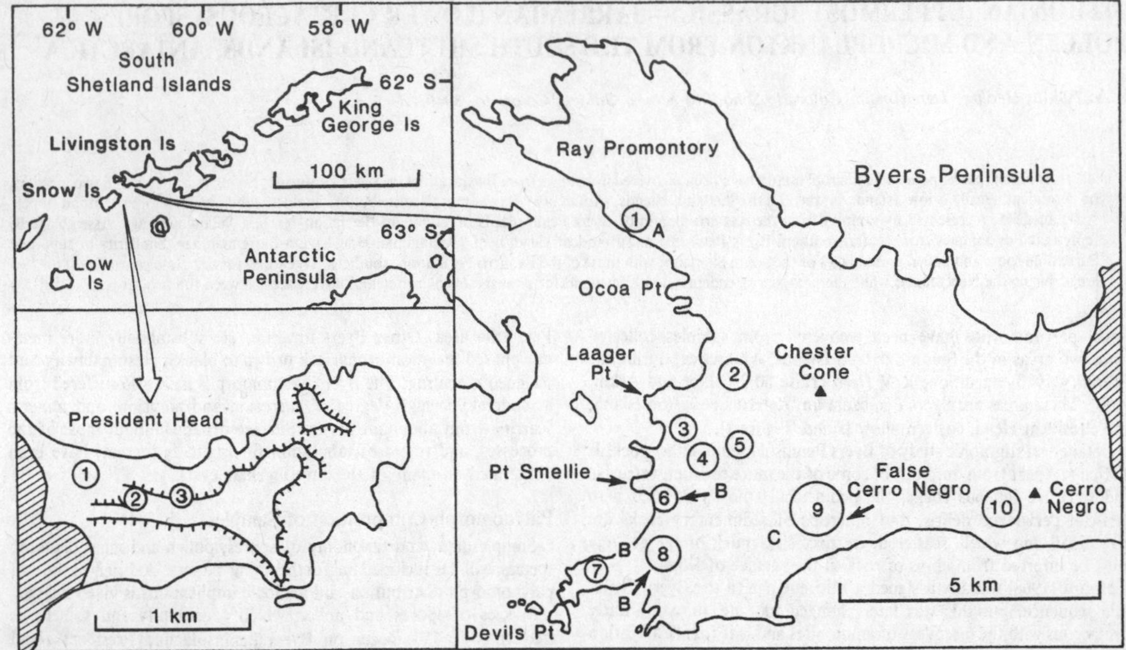


**Figure 1. Locality map for Byers Peninsula, Livingston Island, and President Head, Snow Island. Numbered circles represent main sampling localities.**

*Polypodiaceoisporites elegans*. Very rare specimens of *C. tectifera* occur in marine beds at BP6 and PH1. *Appendicisporites* sp. occurs in BP9 and 10, *Concavissimisporites* sp. cf. *C. penolaensis* in BP6 and PH1, *Densoisporites velatus* in BP9, 10 and PH1; and *Biretisporites* sp. A and *Verrucosisporites* sp. A in BP10 and PH3, with the former occurring very rarely in BP6. *Foveotriletes* sp. A occurs in PH3.

Among the dinoflagellates, specimens with characteristic antapical horns resembling *Broomea simplex* are restricted to BP2, *Batioladinium micropodum* occurs in BP9, a similar form in BP4, and the distinctive *Batioladinium* sp. A occurs in BP6, 9 and PH1. The species cf. *Sirmiodiniopsis* sp. A occurs, often frequently, in BP2, 4 and 9, and very rarely in PH1. *Hystrichodinium voigtii* occurs in BP6 and PH1 with a possible specimen in BP4, and questionable specimens of *Palaeoperidinium cretaceum* occur in BP4 and 9. One specimen of *Hystrichogonyaulax serrata* was found in BP6. *"Aptea" attadalica*, *Cribroperidinium perforans* and *Gonyaulacysta helicoidea* occur in PH1.

## Biostratigraphy

From the distribution of spore, pollen and dinoflagellate species at Byers Peninsula and President Head, 2 broad assemblages or zones can be distinguished: the older represented by samples from BP2 and 4, and the younger by those from BP9 and 10 and the President Head samples. BP6 seems to represent an intermediate position. It is recognised however that dinoflagellates are so poorly preserved in most Byers Peninsula samples that their observed distribution may not be real, and that the somewhat better-preserved assemblage from BP6 at Point Smellie probably reflects a more realistic representation of some species ranges. As far as can be judged from the admittedly sporadic outcrops, and as supported by the similar invertebrate faunas, BP6 is probably close in stratigraphic position to BP4.

*Assemblage A* (BP2, 4 and 6). This assemblage occurs in rocks included in the "mixed marine member" of Smellie et al. (1980) cropping out in the western part of Byers Peninsula south of Ocoa Point. Marine invertebrate faunas in the Point Smellie region were assigned a Berriasian age, and Smellie et al. (1980) suggest their mixed marine member may interfinger with the mudstone member which contains Tithonian ammonites. The mixed marine member may therefore be as old as Tithonian in part.

Spore species suggest correlation of Assemblage A with the Australian *Cicatricosisporites australiensis* Subzone (Burger, 1973; and in Morgan, 1980). Assemblage A includes *C. australiensis* and *C. ludbrookii*, both of which appear at the base of the *C. australiensis* Subzone which Burger (1973) included in the early Neocomian (Berriasian), possibly extending down into the Tithonian. Spore and pollen zones in the basal Cretaceous of Australia have lacked definitive age control throughout. New evidence is provided from a study (Burger, 1982) on dinoflagellates, with associated spores and pollen, from the Carpentaria Basin of northeastern Australia, in which dinoflagellate distribution in Australia and other areas of the world are compared to give more certain age assignments for the Neocomian. A Late Jurassic (Tithonian?)—Berriasian—early Valanginian age for the *C. australiensis* Subzone is suggested by Burger (1982).

BP2 includes dinoflagellates similar to *Broomea simplex* which is of mid-Kimmeridgian—Tithonian age in the Australia—New Guinea region (Cookson and Eisenack, 1958, 1960). Among other dinoflagellates, Assemblage A includes *O. complex*, which is reported from Upper Jurassic rocks of the Papuan Basin (Burger, 1982), and elsewhere appears in lower Neocomian rocks. Unfortunately, no palynomorphs were recovered from the Tithonian "mudstone member" north of Ocoa Point (BP1).

Rocks including Assemblage A are, at least in part, of Berriasian age as suggested both by invertebrate faunas and palynomorphs. Palynomorph evidence suggests these rocks range as old as Tithonian, particularly those northeast of Laager Point, to possibly up into the basal Valanginian. Species characteristic of Assemblage B which make their first rare appearance in Assemblage A are the spores *C. tectifera* and *Biretisporites* sp. A, and the dinoflagellates *Batioladinium* sp. A and *Hystrichodinium voigtii*.

*Assemblage B* (BP9 and 10, PH1 and 3). This assemblage occurs in rocks included in the upper part of the "mixed marine member" (BP9) and from the "volcanic member" (BP10) on Byers Peninsula, and from marine (PH1) and non-marine (PH3) strata on President Head.

Invertebrate fossils in beds believed to stratigraphically underlie those of BP9 were assigned a Valanginian age by Covacevich (1976). Both BP10 and PH3 samples are from the non-marine beds containing plant megafossils dated as Early Cretaceous (Hernández and Azcárate, 1971) for those SW of Cerro Negro (BP10), and Middle Jurassic (Fuenzalida et al., 1972) for those from President Head. Samples from these two areas contain low diversity suites of spores and pollen dominated by *C. tectifera* and *P. elegans*.

Spore and pollen composition, as with the leaf fossils from Cerro Negro, indicates correlation with the Barremian Baqueró Formation assemblage of southern Argentina. Both sequences have many species in common, and *C. tectifera* and *P. elegans* were first described from the Baqueró beds by Archangelsky and Gamerro (1965, 1966).

The principal Australian spore and pollen index species are not recorded in the South Shetlands and do not permit certain comparison with Burger's (1973) *Foraminisporis wonthaggiensis* and *F. asymmetricus* Subzones (which overlie the *C. australiensis* subzone), or with Dettmann and Playford's (1969) *Cyclosporites hughesii* subzone (in part equivalent to the *F. asymmetricus* subzone). The earliest occurrence of *Appendicisporites* in Australia is in the *F. asymmetricus* subzone according to Burger (1976). Morgan (1980) shows *Appendicisporites* in the *F. wonthaggiensis* subzone, although his thesis tables (Morgan, 1978) evidently do not record this occurrence (Truswell, pers. comm. 1982).

The dinoflagellates occurring in Assemblage B, *B. micropodum*, *P. cretaceum*, *"A". attadalica*, *C. perforans* and *G. helicoidea*, all first appear during the *F. wonthaggiensis* subzone equivalent in Morgan's (1980) review of Australian Early and Middle Cretaceous palynostratigraphy. *H. voigtii* appears at the base of Morgan's *Odontochitina operculata* microplankton Zone, and is known in slightly older Neocomian rocks elsewhere. *B. micropodum* and *"A" attadalica* first appear in Burger's (1982) dinoflagellate zone DK3, to which he assigns a Hauterivian age, possibly into early Barremian. Burger (1982) suggests the *F. wonthaggiensis* subzone is "(late) Valanginian, Hauterivian and possibly Barremian", and the *F. asymmetricus* subzone is Barremian to earliest Aptian.

Thus it appears that Assemblage B spans the Hauterivian and Barremian, and part is possibly as old as Valanginian. Strata previously believed to be Middle Jurassic on President Head, Snow Island, appear to be upper Neocomian (Barremian?) in age.

## Palaeogeographic Considerations

Land-plant derived spores and pollen observed in Tithonian and Neocomian strata of the South Shetland Islands include many species common in Australian basins. These species are, however, mostly of wide distribution, stratigraphically as well as geographically. Barremian assemblages from Byers Peninsula and President Head are similar to the southern Argentine Baqueró assemblage. Both areas include restricted species not reported elsewhere, except for recycled specimens of *C. tectifera* recorded in western Weddell Sea piston-core samples (Kemp, 1972; pers. comm. 1981). Similarity of these assemblages suggests their close proximity during the Neocomian. The presence of new, presumably endemic species in the South Shetland Islands is perhaps a consequence of environmental factors caused by, for example, differing proximity to the open "Pacific" ocean, prevailing winds and rainfall, or palaeolatitudinal differences.

*Acknowledgements* I am grateful for the helpful comments of the reviewers, Elizabeth M. Truswell and Dennis Burger. I also wish to thank Captain Lenie and crew of R/V *Hero* for logistic support, fellow members of the 1980 South Shetlands expedition; and Stephen R. Jacobson and David H. Elliot for constructive criticism of the manuscript. This research was supported by U.S. National Science Foundation grant DPP78-21128.

## REFERENCES

ARAYA, R. and HERVÉ, F., 1966: Estudio geomorfológico y geológico en las Islas Shetland del Sur, Antártica. *Publ. Inst. antart. Chil.*, 8, 1-76.

ARCHANGELSKY, S. and GAMERRO, J.C., 1965: Estudio palinológico de la Formación Baqueró (Cretacico); Provincia de Santa Cruz. 1. *Ameghiniana, 4,* 159-67.

ARCHANGELSKY, S. and GAMERRO, J.C., 1966: Estudio palinológico de la Formación Baqueró (Cretacico), Provincia de Santa Cruz. 2. *Ameghiniana, 4,* 201-9.

BURGER, D., 1973: Spore zonation and sedimentary history of the Neocomian, Great Artesian Basin, Queensland. *Geol. Soc. Aust., Spec. Pub., 4,* 87-118.

BURGER, D., 1976: Some Early Cretaceous plant microfossils from Queensland. *Aust., Bur. Miner. Resour., Geol. Geophys., Bull, 160,* 1-22.

BURGER, D., 1982: A basal Cretaceous dinoflagellate suite from northeastern Australia. *Palynology, 6,* 161-92.

COOKSON, I.C. and EISENACK, A., 1958: Microplankton from Australian and New Guinea Upper Mesozoic sediments. *R. Soc. Vic., Proc., 70,* 19-79.

COOKSON, I.C. and EISENACK, A., 1960: Upper Mesozoic microplankton from Australia and New Guinea. *Palaeontology, 2,* 243-61.

COVACEVICH, V.C., 1976: Fauna Valanginiana de Península Byers, Isla Livingston, Artártica. *Revta geol. Chile, 3,* 25-56.

DETTMANN, M.E. and PLAYFORD, G., 1969: Palynology of the Australian Cretaceous, a review; *in* Campbell, K.S.W. (ed.) *Stratigraphy and Palaeontology, Essays in Honour of Dorothy Hill.* ANU Press, Canberra, 174-210.

ELLIOT, D.H. and ASKIN, R.A., 1980: Geologic studies in the South Shetland Islands an at Hope Bay, Antarctic Peninsula, R/V *Hero* cruises 80-1 and 80-2. *U.S. Antarct. J., 15,* 23-24.

FUENZALIDA, H., 1965: Serie sedimentaria con plantas en las Islas Snow y I ivingston. *Soc. Geol. Chile, Res. 10.*

FUENZALIDA, H., ARAYA, R. and HERVE, F., 1972: Middle Jurassic flora from north-eastern Snow Island, South Shetland Islands; *in* Adie F.J. (ed.) *Antarctic Geology and Geophysics.* Universitetsforlaget, Oslo, 93-7.

GONZÁLEZ-FERRÁN, O., KATSUI, Y., and TAVERA, J., 1970: Contribución al conocimiento geológico de la Península Byers de la Isla Livingston, Shetland del Sur, Antártica. *Ser. cient. antart. Chil., 1,* 41-54.

HERNÁNDEZ, P. and AZCÁRATE, V., 1971: Estudio paleobotanico preliminar sobre restos de una tafoflora de la Península Byers (Cerro Negro), Isla Livingston, Islas Shetland del Sur, *Ant'artica. Ser. cient. Inst. antart. Chil., 2,* 15-50.

HOBBS, G.J., 1968: The geology of the South Shetland Islands, IV. The geology of Livingston Island. *Scient. Rep. Br. antarc. Surv., 47.*

KEMP, E.M., 1972: Recycled palynomorphs in continental shelf sediments from Antarctica. *U.S. Antarct. J., 7,* 190-1.

MORGAN, R., 1978: Early and Middle Cretaceous palynostratigraphy of Australia. Unpublished Ph.D Thesis, University of Adelaide, Australia.

MORGAN, R., 1980: Palynostratigraphy of the Australian Early and Middle Cretaceous. *NSW, Geol. Surv., Mem., Palaeont., 18.*

SMELLIE, J.L., DAVIES, R.E.S. and THOMSON, M.R.A., 1980: Geology of a Mesozoic intra-arc sequence on Byers Peninsula, Livingston Island, South Shetland Islands. *Bull. Br. antarc. Surv., 50,* 55-76.

TAVERA, J., 1970: Fauna titoniana-neocomiana de Isla Livingston, Islas Shetland del Sur, *Ant'arctica. Ser. cient. Inst. antart, Chil., 1,* 175-86

# CRETACEOUS INOCERAMID BIVALVES FROM ANTARCTICA

J.A. Crame, *British Antarctic Survey, Madingley Road, Cambridge. CB3 0ET, U.K.*

*Abstract* Inoceramid bivalves provide a useful method for correlating the Cretaceous sedimentary sequences of Antarctica. In the Lower Cretaceous, the cosmopolitan *I. neocomiensis*, *I. concentricus* and *I. anglicus* groups are particularly useful zone fossils. The *I. ovatus* (Berriasian-Valanginian) and *I. heteropterus* (Hauterivian-Albian ?) groups can be used to make circum-Pacific correlations, and *I.* cf. *anomiaeformis* provides a possible link between Hauterivian-Barremian strata of the Scotia Arc and Patagonia. The oldest Upper Cretaceous inoceramids belong to the Turonian *I. lamarcki* group. They are followed by *I. madagascariensis*, a Turonian-Coniacian species which apparently had a widespread distribution in the southern hemisphere. The same is true of the succeeding *I. neocaledonicus* group, which may prove to be particularly useful for correlating Senonian sections in Antarctica, Patagonia, Madagascar and New Caledonia. The youngest Antarctic inoceramids are two 'giant' Campanian species.

It is over 70 years since Henry Woods published a series of papers on *Inoceramus* that was to become a landmark in Cretaceous palaeontology. In two parts of a monograph (Woods, 1911, 1912a) and in a review paper (Woods, 1912b) he described the principal English species and their stratigraphic succession. In so doing he firmly established, for the first time, the marked diversification of the genus through the period: by the Upper Cretaceous each stage was clearly characterised by a variety of species. Subsequent workers found similar successions in many other regions and it became apparent that inoceramids were a useful biostratigraphic tool. At present there are a number of sophisticated correlation schemes based on northern hemisphere Cretaceous inoceramids (e.g. Kauffman, 1977; Pergament, 1978).

Current investigations into the Mesozoic sedimentary formations of Antarctica are revealing the presence of both Lower and Upper Cretaceous inoceramid species. Although much work remains to be done on their taxonomy, these can be arranged into a stratigraphic succession and directly compared with other southern hemisphere inoceramid zonations. The Antarctic data will make an important contribution towards the construction of a detailed Cretaceous inoceramid biostratigraphy for the southern hemisphere. All specimen numbers in this paper refer to the collection of the British Antarctic Survey, Cambridge.

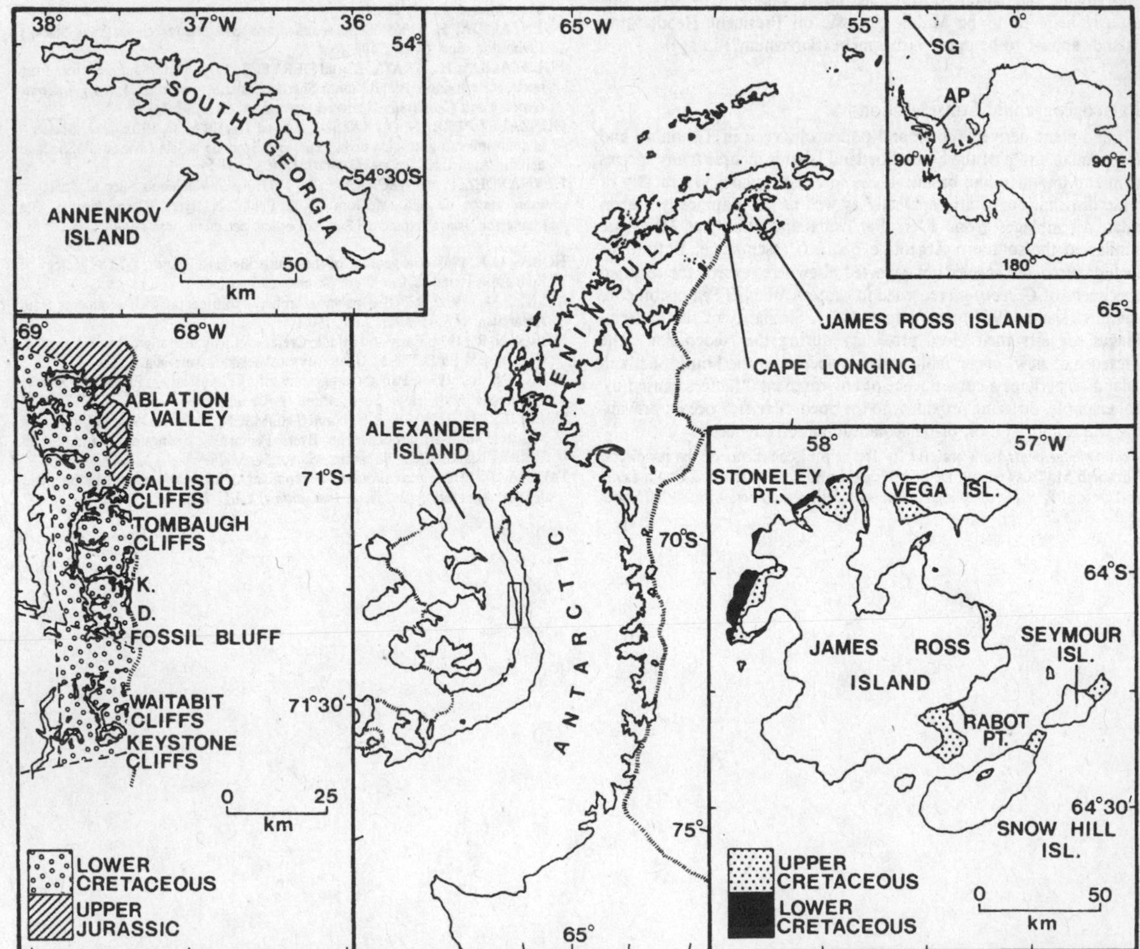


**Figure 1.** **Locality map for the Antarctic Peninsula region. The inset in the top left-hand corner shows the position of Annenkov Island in relation to South Georgia and that in the top right the position of South George (SG) and the Antarctic Peninsula (AP) in relation to the rest of Antarctica. In the inset in the bottom left-hand corner, which is an enlargement of the central east coast of Alexander Island, the Upper Jurassic and Lower Cretaceous comprise the Fossil Bluff Formtion. The inset in the bottom right-hand corner is a sketch geological map of the James Ross Island group.**

## LOWER CRETACEOUS

### Alexander Island

Among the biostratigraphically important fossils obtained from the Fossil Bluff Formation of eastern Alexander Island (Figure 1) are ammonites, belemnites and bivalves (Taylor et al., 1979). The latter have been the subject of a recent review (Crame, 1982a), and only the inoceramids will be discussed further here.

The predominantly Jurassic genus *Retroceramus* is represented in lowermost Cretaceous strata of Callisto and Tombaugh Cliffs (Figures 1 and 2) by *R. everesti* (Oppel), a species known to transgress the Jurassic-Cretaceous boundary at several southern hemisphere localities (Crame, 1982b). At a slightly higher level in the Ablation Valley section the first specimens referrable to *Inoceramus* occur (Figure 2). These are slightly inequivalve forms with a distinctive pyriform outline that can be assigned to the *I. ovatus* Stanton group. This group has a Berriasian-Valanginian age in California and Siberia (Pokhialainen, 1974) and occurs in association with a Berriasian ammonite assemblage in Alexander Island (Taylor et al., 1979). The next inoceramids in the sequence, which occur at localities K and D (Figure 2), belong to the distinctive genus *Anopaea*. Characterised by a high posterior region and a narrow anterior that bears both a lunule and a radial sulcus, this genus is interpreted as having been functionally endobyssate (Crame, 1981). The species occurring at these localities, *A. trapezoidalis* (Thomson and Willey), is a coarsely-ribbed one with no obvious close relatives (Crame, 1981). It has been assigned to an undifferentiated Neocomian age.

The lower levels of the Fossil Bluff section (Figure 2) are distinguished by the presence of an *Inoceramus* resembling *I. neocomiensis* d'Orbigny. The specimens show the rounded-triangular form and narrow, regular ornament that is typical both of this species and close allies such as *I. subneocomiensis* Glazunova from the Aptian of the Volga basin (Glazunova, 1973). Although usually regarded as Aptian, it has to be remembered that there are records of the *I. neocomiensis* group from both the Barremian and Neocomian (e.g. Sornay, 1965; Pokhialainen, 1974). Higher still in the sequence, there are representatives of two further cosmopolitan *Inoceramus* groups. Gryphaeoid specimens from Waitabit Cliffs (Figure 2) can be linked to *I. concentricus* Parkinson, but seem somewhat larger than typical European forms of this species. Nevertheless, some large, broad forms are known from New Zealand and it would seem that the Antarctic material does fall within the range of variation normally accepted for the *concentricus* group (Woods, 1911, 1917). Although a Middle-Upper Albian age is usually assigned to *I. concentricus*, some closely related forms do range into the Cenomanian (e.g. Pergament, 1966; Raine et al., 1981). A small representative of the *I. anglicus* Woods group from Keystone Cliffs (Figure 2) seems closest to *I. anglicus elongatus* Pergament from the Albian of Kamchatka (Pergament, 1965). From higher in the same section a species of *Anopaea*, resembling *A. mandibula* (Mordvilko) from the Albian of Mangishlak (Saveliev, 1962), has been recorded (Figure 2).

The ages suggested by the inoceramid bivalves from the upper levels of the Fossil Bluff Formation agree well with those normally interpreted from the ammonites. Species of *Aconeceras, Theganeceras* and *Acrioceras* at Locality D and Fossil Bluff have Barremian-Aptian affinities, as does *Sanmartinoceras* in the lower levels of Waitabit Cliffs (Thomson, 1981). *Eotetragonites* from higher in Waitabit Cliffs, as well as forms closely resembling *Ptychoceras* and *Hemihoplites* from Keystone Cliffs, are generally taken to be Albian. However, certain heteromorph ammonites from both these localities could be interpreted as being Barremian, or even Neocomian, in age. Their presence so high in the formation is anomalous and has not yet been

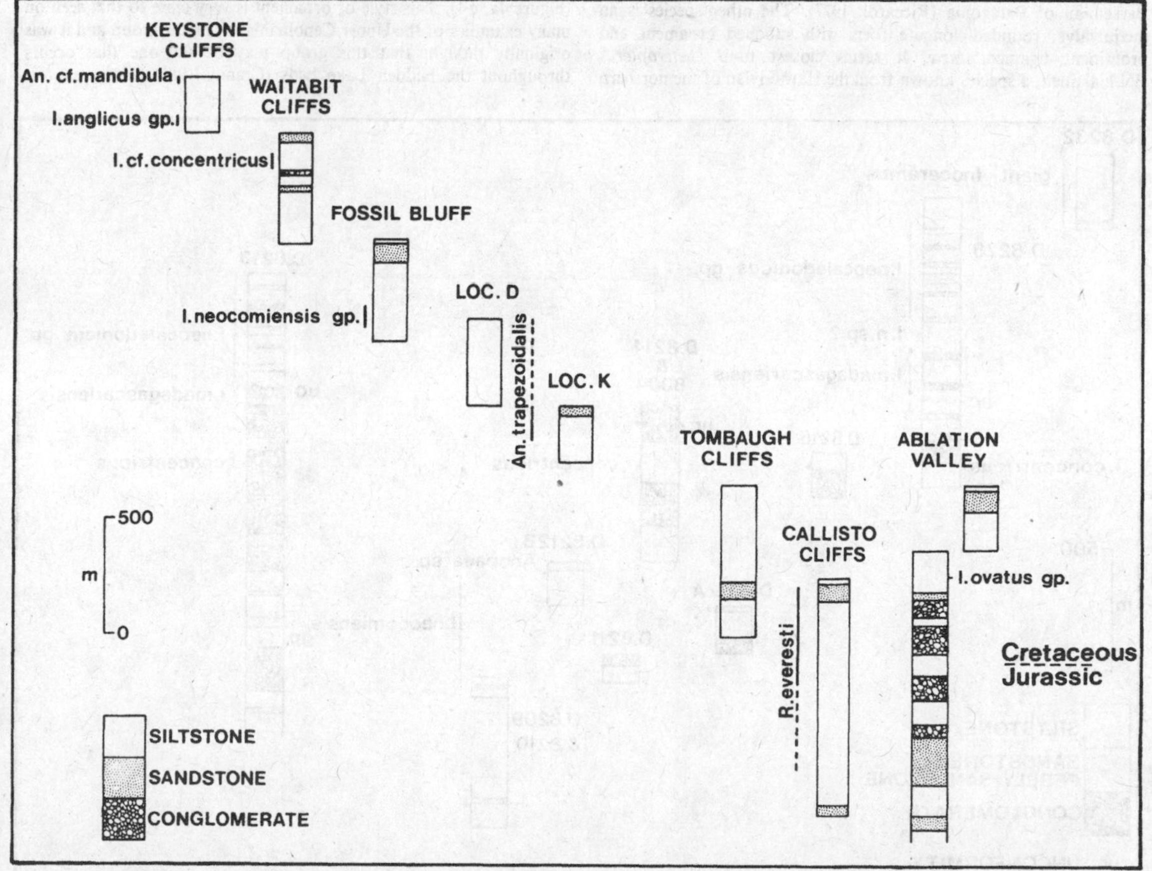


**Figure 2.** Stratigraphic occurrences and ranges of inoceramid bivalves in the Lower Cretaceous strata of the Fossil Bluff Formation, Alexander Island. The approximate position of the Jurassic-Cretaceous boundary is indicated on the right-hand side of the diagram. Generic abbreviations: *An- Anopaea, I- Inoceramus, R- Retroceramus.*

satisfactorily explained (Thomson, 1981). Apart from the *I. ovatus* group, there are no obvious Neocomian inoceramids in the succession.

## James Ross Island

Approximately the lowest 1500 m strata exposed on the northwest coast of James Ross Island is now thought to be Lower Cretaceous in age. Composed of siltstones and fine sandstones with numerous prominent conglomeratic interbeds, they dip southeastwards beneath the extensive Upper Cretaceous outcrops of southern James Ross, Snow Hill and Seymour Islands (Figure 1). The lowest beds in the sequence contain no diagnostic fossils (Figure 3), but at the level of sections D.8211, 8212A and B there are specimens of the *I. neocomiensis* group and small aconeceratid and ancyloceratid ammonites. The *Inoceramus* is significantly smaller than the Alexander Island members of this group but still compares well with European forms such as *I. volgensis* Glazunova from the Aptian of the Volga basin (Glazunova, 1973). Some poorly preserved specimens of *Anopaea* occur at locality D.8212B and at D.8213-15 and 8228 a prominent level can be traced in which members of the *I. concentricus* group are common (Figure 3). These specimens are both smaller and narrower than their Alexander Island counterparts and compare well with many European specimens of *I. concentricus* (e.g. Woods, 1911). The most likely age for these lower beds on James Ross Island is Aptian-Albian.

## Annenkov Island

Pettigrew (1981) divided the sequence of volcaniclastic sediments exposed on Annenkov Island (Figure 1) into a Lower Tuff Member (860 m) and an Upper Breccia Member (1000 m). The former of these units has yielded poorly preserved ammonites with Lower Cretaceous affinities and two species of *Inoceramus*. The lower of these is a small, circular form with irregular ornament that probably corresponds with *I. anomiaeformis* Feruglio, a species known from the Hauterivian-Barremian of Patagonia (Riccardi, 1977). The other species is an inequivalve, rounded-elongate form with subdued ornament and prominent ligament area. It seems closest to *I. heteropterus* Pokhialainen, a species known from the Hauterivian of the northern Pacific region (Pokhialainen, 1974). Studies in progress suggest that the *heteropterus* group may in fact range into the Albian.

## UPPER CRETACEOUS

The lowest Upper Cretaceous strata on James Ross Island are the conglomerate and pebbly sandstone beds exposed in the vicinity of unconformities in sections D.8213, 8214 and 8228 (Figure 3). They pass up into a sequence of sandstones and siltstones exhibiting both small scale slump folds and crossbedding. This unit, the Hidden Lake Beds of Bibby (1966), passes up in turn into the extensive Snow Hill Island Series. A Campanian age for the latter, established from a well preserved ammonite fauna, has been assigned also to all the lower stratigraphic units (Howarth, 1966). However, we now know that some at least of these are Lower Cretaceous and it is likely that some of the earlier Upper Cretaceous stages may also be represented. This was suggested by a preliminary study of the inoceramid bivalves from these lower units (Crame, 1981), although it is apparent that some of the determinations made in this study must now be revised.

The beds associated with the previously mentioned unconformities contain numerous specimens of *I. madagascariensis* Heinz (Figure 3), an Upper Turonian-Lower Coniacian species first recorded from Madagascar (Heinz, 1933). Small and erect in outline, this species is characterised by two distinct orders of concentric ornament (Figure 4g and h). It constitutes part of a form group that includes *I. vancouverensis* Shumard, *I. ernsti* Heinz and some forms of *I. lamarcki* Parkinson (Crame, 1981, p.46). Higher in section D.8228 there are several representatives of a medium sized *Inoceramus* with a prominent umbo that probably belong within a new species (Figure 3). There then follows the first occurrences of the *I. neocaledonicus* Jeannet group, which is common in the Hidden Lake Beds of both sections D.8213 and 8228 (Figure 3). This group is characterised by the fine ring-like ornament that was termed *Anwachsringen* by Heinz (1928a) (Figure 4a, c-f). This style of ornament is very close to that seen on many examples of the Upper Cenomanian *I. pictus* group and it was originally thought that this group may be the one that occurs throughout the Hidden Lake beds (Crame, 1981). However, this

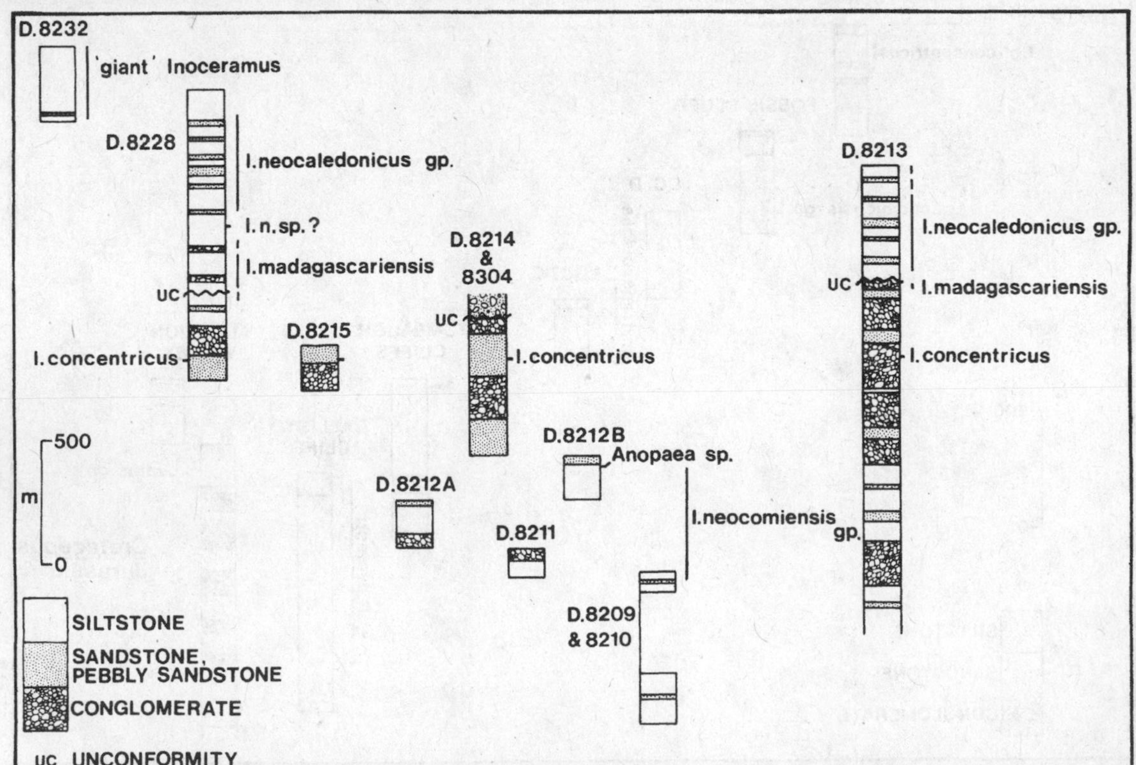


**Figure 3. Stratigraphic occurrences and ranges of inoceramid bivalves in the Cretaceous strata of James Ross Island. All of these localities occur along the strip of coast immediately to the northeast of Stoneley Point (Figure 1).**

identification is now thought to be in error, for the specimens occur above *I. madagascariensis* and are associated with pachydiscid and tetragonitid ammonites with Coniacian-Campanian affinities. They are probably better linked to *I. neocaledonicus* Jeannet, a Senonian species from New Caledonia with clearly defined *Anwachsringen* (Jeannet, 1922). Heinz (1928b) first commented on the similarity between *pictus* and *neocaledonicus*, and his suggestion that the latter should be synonymised in the former was accepted by many subsequent authors. Nevertheless, it is quite likely that this is simply a further example of a similar ornament pattern being developed by two quite separate *Inoceramus* lineages.

The earliest members of the *neocaledonicus* group possess the characteristic ornament only in the juvenile stages (Figure 4e). They are succeeded by types in which the whole valve is covered by *Anwachsringen* and often these are grouped into *Anwachsringreifen* (Figure 4a, c and d). Both erect and obliquely elongated specimens are

**Figure 4. Upper Cretaceous *Inoceramus* from the northwest coast of James Ross Island. (a) incomplete large left valve of the *I. neocaledonicus* Jeannet group (D. 8228.331). (b) a right valve of the "giant" *Inoceramus* from the base of the Snow Hill Island Series (D.8232.89). (c) a suberect right valve of the *I. neocaledonicus* gp. with symmetrical ornament (D.8228.268). (d) an obliquely elongated right valve of the same species group (D.8228.278). (e) a right valve of an early member of the *I. neocaledonicus* gp. (D.8213.98). (f) an erect left valve of the same species group (D.2212.4A). (g) a left valve of *I. madagascariensis* Heinz (5120). (h) a right valve of the same species (5119c). All specimens are internal moulds and x .75, except for (b) which is x .25. Specimens (a), (c), (d) and (f) show well developed *Anwachsringen* and on specimens a, c and d these are grouped into *Anwachsringreifen*. Specimen (e) shows *Anwachsringen* in the early stages only. Faint second order ribs can be seen between strong primaries on (g) and (h).**

common and near the top of the group's range large forms up to 15 cm in length occur (Figure 4a). Some of these large specimens appear to be very similar to *I. andinus* Wilckens from the Senonian of Patagonia (Wilckens, 1907). Other probable members of the group are *"Mytiloides" africanus* Heinz from the Santonian of Madagascar (Heinz, 1933; Sornay, 1964) and certain small forms of *I. australis* Woods (1917, pl. 12, Figures 17 and 19) from the Piripauan stage of New Zealand.

Just above the base of the Snow Hill Island series the *neocaledonicus* group is replaced by the first of two "giant" species of *Inoceramus* (Figure 3). Erect to suberect in outline and bearing only faint concentric ornament, members of this species often attain a shell height of 30 cm (Figure 4b). The second species, which is much more oblique and may reach 80 cm in length, occurs at Rabot Point (Figure 1). Although correlations within the Snow Hill Island Series are uncertain at present, it is likely that this locality represents a considerably higher stratigraphic level.

On the mainland opposite James Ross Island, at Cape Longing (Figure 1), Cretaceous fossils have been collected from a 110 m sandstone sequence. At least some of the inoceramids in this collection belong to the *I. lamarcki* group (Crame, 1981) and indicate a Lower-Middle Turonian age.

**TABLE 1: Principal Antarctic Cretaceous inoceramids**

| Species | Occurrence | Probably age-range |
|---|---|---|
| 'giant' *Inoceramus* sp.1 | James Ross Isl. | Campanian |
| 'giant' *Inoceramus* sp.2 | James Ross Isl. | Campanian |
| *I. neocaledonicus* gp. | James Ross Isl. | Coniacian-Campanian |
| *I. madagascariensis* | James Ross Isl. | U. Turonian-L. Coniacian |
| *I. lamarcki* gp. | Cape Longing | L.-M. Turonian |
| *Anopaea* cf. *mandibula* | Alexander Isl. | Albian |
| *I. anglicus* gp. | Alexander Isl. | Albian |
| *I. concentricus* gp. | James Ross Isl., Alexander Isl. | M-U.Albian |
| *Anopaea* sp. | James Ross Isl. | Aptian-Albian |
| *I. neocomiensis* gp. | James Ross Isl., Alexander Isl. | Barremian-Aptian |
| *I. heretropterus* gp. | Annenkov Isl. | Hauterivian-Albian? |
| *I.* cf. *anomiaeformis* | Annenkov Isl. | Hauterivian-Barremian |
| *Anopaea trapezoidalis* | Alexander Isl. | Neocomian |
| *I. ovatus gp.* | Alexander Isl. | Berriasian |
| *Retroceramus everesti* | Alexander Isl. | Berriasian |

## SUMMARY

The Antarctic Lower Cretaceous inoceramids most useful for stratigraphic correlations are members of the *neocomiensis, concentricus* and *anglicus* groups (Table 1). Although the taxonomy of all three of these groups is in need of revision, it is clear that Antarctic specimens can be closely matched with northern hemisphere counterparts. In New Zealand, a member of either the *neocomiensis* or *anglicus* groups is present in the Albian Motuan stage and is followed by *I. concentricus* in the Ngaterian (Late Albian-Cenomanian) (Raine et al., 1981). The latter species may also be present in South Africa (Heinz, 1930) and Patagonia (Bonarelli and Nágera, 1921). The *ovatus* and *heteropterus* groups (Table 1) are less well known but seem to have essentially circum-Pacific distributions (Pokhialainen, 1974). *Inoceramus* cf. *anomiaeformis* provides a possible palaeontological link between the sedimentary formations of the Scotia Arc and those of southern South America.

The species of *Anopaea* (Table 1) are less useful at present for regional correlations. Nevertheless, it is interesting to note the relative abundance of this genus in the Lower Cretaceous of Antarctica and this may yet be shown to be of some palaeobiogeographical significance.

The earliest Upper Cretaceous inoceramids appear to be members of the Turonian *I. lamarcki* group from Cape Longing. They are followed by the Upper Turonian-Lower Coniacian species *I. madagascariensis* (Table 1), which provides a direct link between Antarctica and Madagascar. Some forms of *I. nukeus* Wellman from the Teratan stage (Coniacian-Santonian) of New Zealand may also belong within this species (Crame, 1981, p.46).

The *I. neocaledonicus* group, although a diverse one, offers considerable potential for correlating Senonian sections in Antarctica with those of Patagonia, Madagascar and New Caledonia. The first giant *Inoceramus* compares closely with *I. expansus* Baily from the Santonian-Campanian of South Africa (Newton, 1909; Kennedy and Klinger, 1975). The second has less obvious affinities, but may be related to one of the giant species known from the Upper Campanian and Maastrichtian of Angola, South Africa and New Zealand.

## REFERENCES

BIBBY, J.S., 1966: The stratigraphy of part of northeast Graham Land and the James Ross Island group. *Scient. Rep., Br. antarc. Surv., 53.*

BONARELLI, G. and NAGERA, J.J., 1921: Observaciones geológicas en las inmediaciones del Lago San Martin (Territoro de Sants Cruz). *Dir. gen. Minas Geol. Hidrol., B. Aires, Boln, 27.*

CRAME, J.A., 1981: The occurrence of *Anopaea* (Bivalvia: Inoceramidae) in the Antarctic Peninsula. *J. molluscan Stud., 47,* 206-19.

CRAME, J.A., 1982a: Late Mesozoic bivalve biostratigraphy of the Antarctic Peninsula region. *Geol. Soc. Lond., J., 139,* in press.

CRAME, J.A., 1982b: Late Jurassic inoceramid bivalves from the Antarctic Peninsula and their stratigraphic use. *Palaeontology, 25,* 555-603.

GLAZUNOVA, A.E., 1973: Palaeontological evidence of the stratigraphic separation of the Cretaceous deposits in the Volga region. Lower Cretaceous. "Nedra", Moscow, 200 pp (in Russian).

HEINZ, R., 1928a:Uber die bisher wenig beachtete Skulptur der Inoceramen-Schale und ihre stratigraphische Bedeutung. Beitrage zur Kenntnis der oberkretazischen Inoceramen IV. *Mitt. miner.-geol. Stlnst. Hamb., 10,* 3-39.

HEINZ, R., 1928b:Uber die Oberkreide-Inoceramen Neu-Seelands und Neu Kaledoniens und ihre Beziehungen zu denen Europas und anderer Gebiete. Beitrage zur Kenntnis der oberkretaziscen Inoceramen VII. *Mitt. miner.-geol. Stlnst. Hamb., 10,* 111-30.

HEINZ, R., 1930:Uber Kreide-Inoceramen der sudafrikánischen Union. *C. r. 15th Int. geol. Congr., Pretoria, 2,* 681-7.

HEINZ, R., 1933: Inoceramen von Madagaskar und ihre Bedentung fur die Kreide-Stratigraphie. *Z. dt. geol. Ges., 85,* 241-59.

HOWARTH, M.K., 1966: Ammonites from the Upper Cretaceous of the James Ross Island group. *Bull. Br. antarct. Surv., 10,* 55-69.

JEANNET, A., 1922: Description d'une espece nouvelle d'Inocerame. *Soc. geol. Fr., Bull., 22,* 251-3.

KAUFMAN, E.G., 1977: Systematic, biostratigraphic and biogeographic relationships between Middle Cretaceous Euro-american and North Pacific Inoceramidae; *in* Matsumoto, T. (ed.) Mid-Cretaceous Events. Hokkaido Symposium, 1976. *Palaeontol. Soc. Japan, Spec. Pap., 21,* 169-212.

KENNEDY, W.J. and KLINGER, H.C., 1975: Cretaceous faunas from Zululand and Natal, South Africa. Introduction, stratigraphy. *Br. Mus. (Nat. Hist.), Bull., Geol., 25,* 263-315.

NEWTON, R.B., 1909: Cretaceous Gastropoda and Pelecypoda from Zululand. *R. Soc. S. Afr., Trans., 1,* 1-106.

PERGAMENT, M.A., 1965: Inocerams and Cretaceous stratigraphy of the Pacific region. *Geol. Inst. Nauk. Mosk., Tr., 118,* 102 pp.

PERGAMENT, M.A., 1966: Zonal stratigraphy and inocerams of the lowermost Upper Cretaceous on the pacific coast of the USSR. *Geol. Inst. Nauk. Most., Tr., 146,* 1-83 (in Russian).

PERGAMENT, M.A., 1978: Upper Cretaceous stratigraphy and inocerams of the northern hemisphere. *Geol. Inst. Nauk. Mosk., Tr., 322,* 214 pp. (in Russian).

PETTIGREW, T.H., 1981: The geology of Annenkov Island. *Bull. Br. antarct. Surv., 53,* 213-54.

POKHIALAINEN, V.P., 1974: Spreading of the Neocomian Pacific Inoceramidae. *Inst. Geol. Geofiz. Sib. Otd., Tr., 80,* 174-87 (in Russian).

RAINE, J.I., SPEDEN, I.G. and STRONG, C.P., 1981: New Zealand; *in* Reyment, R.A. and Bengtson (eds.) *Aspects of Mid-Cretaceous Regional Geology.* Academic Press, London, 221-67.

RICCARDI, A.C., 1977: Berriasian invertebrate fauna from the Springhill Formation of Southern Patagonia. *Neues Jahrb. Geol. Palaeontol., Abh., 155,* 216-52.

SAVELIEV, A.A., 1962: Albian inoceramids of Mangishlak. *Vses. nauchno-issled. geol.-razv. neft Inst., Tr., 196,* 219-54 (in Russian).

SORNAY, J., 1964: Sur quelques nouvelles especes d'Inocérames du Sénonien de Madagascar. *Annls Paléont., Invert., 50,* 167-79.

SORNAY, J., 1965: Les Inocérames de Crétacé Inférieur en France. *Bur. Rech. Géol. miniér., Mém., 34,* 393-7.

TAYLOR, B.J., THOMSON, M.R.A. and WILLEY, L.E., 1979: The geology of the Albation Point—Keystone Cliffs area, Alexander Island. *Scient. Rep. Br. antarct. Surv., 82,* 65 pp.

THOMSON, M.R.A., 1981: Antarctica; *in* Reyment, R.A. and Bengtson, P. (eds.) *Aspects of Mid-Cretaceous Regional Geology.* Academic Press, London, 269-96.

WILCKENS, O., 1907: Die Lamellibranchiaten, Gastropoden etc. der oberen Kreide Sudpatagoniens. *Ber. naturf. Ges. Freiburg. i. B., 15,* 97-166.

WOODS, H., 1911: A monograph of the Cretaceous Lamellibranchia of England, Part 2. *Palaeontogr. Soc. Monogr., 2,* 261-84.

WOODS, H., 1912a: A monograph of the Cretaceous Lamellibranchia of England, Part 3. *Palaeontogr. Soc. Monogr., 2,* 285-340.

WOODS, H., 1912b: The evolution of *Inoceramus* in the Cretaceous period. *Geol. Soc. Lond., Q. J., 68,* 1-20.

WOODS, H., 1917: The Cretaceous faunas of the northeastern part of the South Island of New Zealand. *N.Z. geol. Surv., Palaeont. Bull., 4.*

# CONGLOMERATIC STRATA OF MESOZOIC AGE AT HOPE BAY, NORTHERN ANTARCTIC PENINSULA

D.H. Elliot and T.M. Gracanin, *Institute of Polar Studies, The Ohio State University, Columbus, Ohio 43210, U.S.A.*

*Abstract* Conglomeratic rocks of Mesozoic age are widely but sparsely distributed on the northern Antarctic Peninsula and the South Orkney Islands. At Hope Bay the conglomerates unconformably overlie, but are locally in fault contact with, the Trinity Peninsula Group which constitutes part of the pre-Jurassic Gondwanian basement. The conglomerates and associated sandstones represent sedimentary facies of both mass movements and traction currents. Facies recognised include deposits from debris flows and braided streams, and the succession of facies are interpreted as the evolution of a single fan. The Hope Bay fan is modified by contemporaneous volcanism which is recorded as pyroclastic beds in the middle and upper part of the section. These alluvial fan sediments were probably deposited in a fault bounded basin in the uplifted Gondwanide orogen. The Hope Bay fan is part of a well defined tectonostratigraphic unit that bridges the Gondwanian and Andean events.

Mesozoic conglomeratic rocks, consisting largely of alluvial fan deposits, crop out at widely separated localities from the South Orkney Islands to at least as far south as the Longing Gap region (Figure 1). The rocks unconformably overlie the pre-Jurassic Gondwanian basement terrain (Dalziel, 1982), from which they are clearly derived and locally are overlain by volcanic strata that constitute part of the Andean magmatic arc. The most extensive exposures of Mesozoic conglomerates occur on the South Orkney Islands, where two distinct sequences each over 500 m thick crop out (Elliot and Wells, 1982). The conglomerates are interpreted as alluvial fans that accumulated in fault bounded basins. The age of one sequence is probably Early Cretaceous, but there is no definitive palaeontological data for the other (Thomson, 1981).

The conglomerates at Hope Bay have most recently been described by Bibby (1966) who published a generalised stratigraphic succession, reported on the possible occurrence of tuffs in the section and provided structural information on the unconformity that separates the conglomerates from the underlying greywackes, now designated by Hyden and Tanner (1981) the Trinity Peninsula Group (TPG). The plant remains from the carbonaceous beds have long been regarded as Middle Jurassic (Hallé, 1910) though Schopf (see Elliot, 1975) and Stipanicic and Bonetti (1970) argue for a Late Jurassic age.

Rocks of grossly similar aspect but lacking the plant bearing beds of the Hope Bay section, crop out on Joinville Island (Figure 1) (Elliot, 1967 and unpublished data) and have yielded unidentifiable plant megafossils and poorly preserved plant microfossils assignable only to the Mesozoic (R.A. Askin. pers. comm.). Southwest of Hope Bay, conglomerates crop out at Botany Bay and near Longing Gap (Figure 1). The Botany Bay sequence has been described by Farquharson (in press) and consists of alluvial fan conglomerates, floodplain sandstones and siltstones, and lacustrine mudstones; the age of the beds is uncertain though possibly the same as that at Hope Bay. Thin sequences of conglomerate and finer grained plant bearing sedimentary rocks crop out northeast of Longing Gap and at Tower Peak (Aitkenhead, 1975). At both localities the rocks are unconformable on TPG and at Tower Peak are overlain by volcanic rocks. The plants yield only a Mesozoic age.

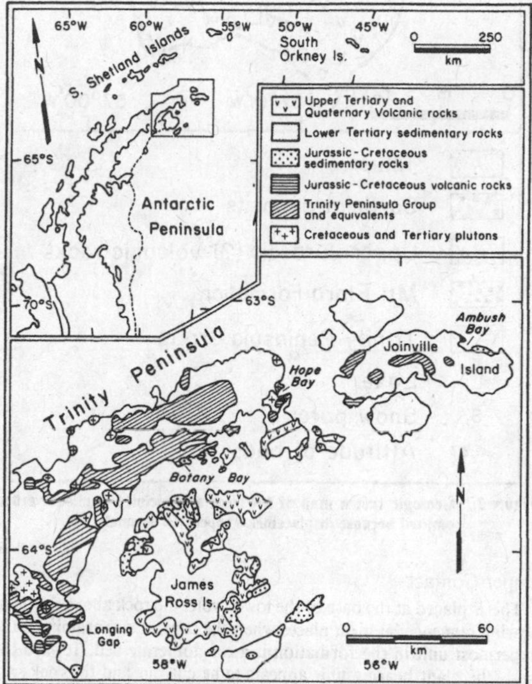


**Figure 1. Location and geologic sketch map for the northern Antarctic Peninsula region.**

## FIELD RELATIONS AND STRATIGRAPHY

The conglomeratic rocks at Hope Bay (Figure 2) are well known, though little published information is available. The measured section (Figure 3) is 366 m thick; an uncertain thickness below the base of the section is not exposed. The section consists of conglomerate in the lower part, becomes finer grained up section, and included carbonaceous sandstones in the top 25 m. Pyroclastic rocks are interbedded in the middle and upper part.

Andersson (1906) referred to these rocks only in terms of their lithology; more recently they have been described as the Mount Flora plant beds (Adie, 1964) or Middle Jurassic beds (Bibby, 1966). The importance of these rocks in establishing the geologic history of the Antarctic Peninsula warrants their formal designation, despite their restricted distribution (the north facing flank of Mt Flora only) and it is here proposed they be named the Mt Flora Formation.

### Lower Contact

A stratigraphic contact is exposed at the west end of the conglomerate cliffs; TPG rocks are overlain by volcanic rocks which in turn are overlain by the massive conglomerates that form the lower part of the northwest buttress of Mt Flora. A fault contact between TPG rocks and volcanic strata is also exposed in the lower slopes beneath that buttress. The rocks forming the buttress and the exposures below it are probably in fault contact with the rest of the conglomerate sequence, but the relations have not been investigated adequately.

Toward the east a contact with TPG can be located within about 5 m (Figure 4). This contact can be interpreted as a fault dipping very steeply to the north and nearly parallel to the cliffs; this would be consistent with the orientation of the fault contact mentioned above.

However, if that TPG outcrop illustrated in Figure 4 is not a fault sliver, and if as previously reported (Bibby, 1966, Figure 3) the unconformity is planar and crosses the base of Mt Flora at a near constant elevation, then the plane of unconformity and the attitude of the conglomerate beds must be discordant; the strike of the conglomerates is slightly north of west whereas the trace of the inferred unconformity is slightly south of west. If the strike of the unconformity is similar to that of the stratigraphically lowest conglomerate beds, then the conglomerates to the east of the small TPG outcrop above the talus slope (Figure 4) project below that outcrop; other orientations of the unconformity also lead to the conclusion that it is nonplanar. Although a rigorous determination of the location and form of the unconformity was not possible, the evidence favours it having significant relief.

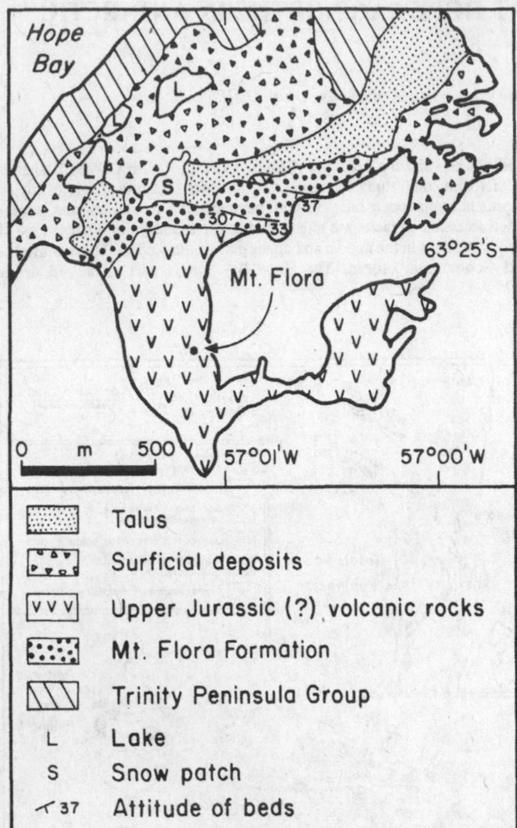


**Figure 2. Geologic sketch map of Mt Flora and vicinity. Faults have been omitted because displacements appear to be small.**

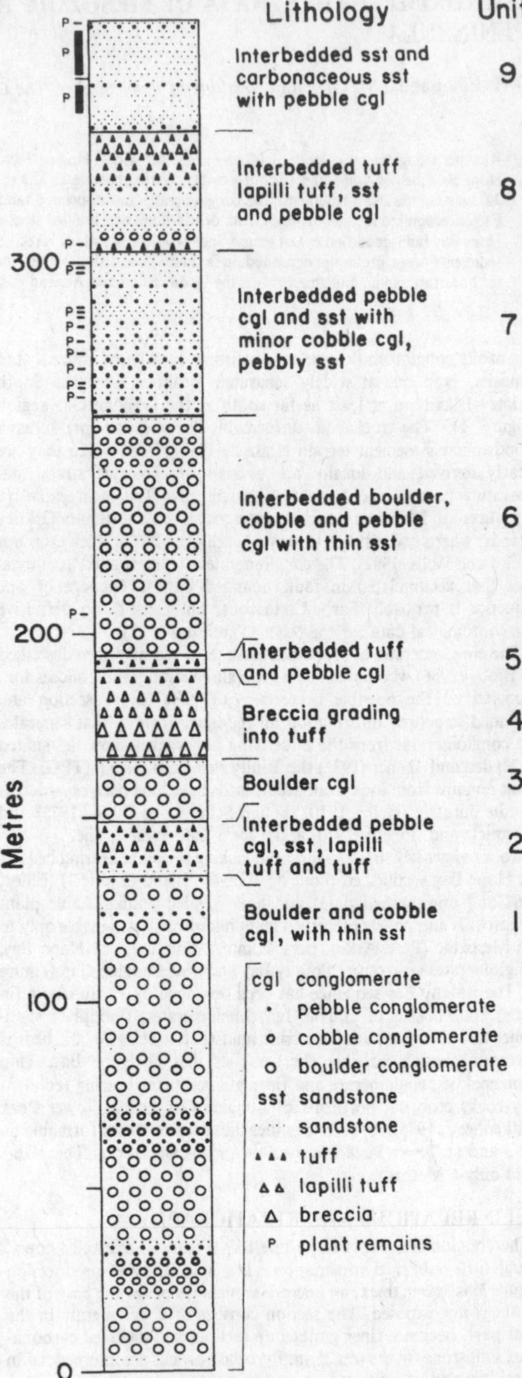


**Figure 3. Composite stratigraphic column, Mt Flora Formation.**

### Upper Contact

This is placed at the base of the lowest volcanic rock above the plant bearing sequence; at most places where the contact was examined, the uppermost unit in the formation is a conglomerate bed. Toward the west, the plant bearing unit appears to be cut out and the rocks are also faulted; the field relations have not been resolved. That an angular discordance exists between the uppermost sedimentary rocks and the volcanic sequence cannot be discounted and is, in fact, regarded as likely.

### Lithology

The lower part of the sequence (unit 1, Figure 3) is a 135 m succession of tabular conglomerate beds 1 to 29 m thick. The conglomerates range up to boulder size and form poorly sorted, densely packed conglomerates with a poor stratification to conglomerates with unsupported clasts. A few beds show inverse grading, passing from pebbly sandstone at the base up into cobble conglomerate. Megaclasts that range up to 2.3 m long occur in a number of beds. The upper parts of some conglomerates pass gradationally through pebbly beds into sandstone over a vertical interval of 2 m or less; in other instances the transition is quite abrupt. The sandstone beds are 10 to 50 cm thick and commonly discontinuous, passing laterally into partings between conglomerate beds. Conglomerate clasts are derived entirely from TPG rocks.

Units 2-6 form a mixed succession of pyroclastic and sedimentary rocks. Unit 2 consists of 15 m of pebble conglomerate, sandstone, lapilli tuff, tuff and sparse accretionary lapilli tuff in beds 25 to 165 m thick, together with one conglomerate-sandstone bed 5 m thick. With the exception of the 5 m bed which is stratified and contains scour surfaces, the conglomerates are clast supported, unstratified and ungraded. Unit 3 is a 17.5 m thick clast-supported unstratified boulder conglomerate with megaclasts up to 1.8 m long. This is overlain by 24 m of breccia grading up into fine tuff; fragments of both volcanic rocks and TPG are present. Four metres of alternating tuff, lapilli tuff and pebble conglomerate form Unit 5. The uppermost unit (Unit 6) is 67 m thick and consists of interbedded boulder, cobble and pebble conglomerate in beds 1 to 18 m thick. The conglomerates are clast supported, ungraded except for one bed near the top, and range from unstratified to poorly stratified. Megaclasts are present in all beds and attain a maximum size of 1.1 m. The conglomerates grade into sandstone beds that are as much as 1.0 m thick and are commonly lens-shaped.

The upper part of the sequence, 104 m thick, consists of thinner bedded cobble and pebble conglomerate and sandstone, and an

**Figure 4. Northeast ridge of Mt Flora. The locations of the measured sections making up the composite column are indicated (...). TPG: outcrops of Trinity Peninsula Group (below dashed line). Eastern limit of conglomerate outcrop at left hand side of photo. The white band and overlying pale rocks in the upper right consist of volcanic rocks.**

interval containing much volcanic material. Unit 7, 41.5 m thick, is formed of interbedded pebble conglomerate and sandstone with minor cobble conglomerate and pebbly sandstone; bed thickness is 1 to 3 m except for a few that are only 20 to 30 cm thick. The conglomerates are better sorted than those lower in the section, most are stratified and grade upward into sandstone that commonly is a third or more of the bed thickness. Cross-bedding is present in a few of the beds. Plant remains occur at the top of most beds. The section passes up into 36 m of beds dominated by tuffs and volcanic sediments, though a 4 m thick pebble conglomerate is interbedded near the base. The conglomerate is stratified, contains lenses of sandstone and passes upward into sandstone. The sandstones occur in beds 5 cm to 1.6 m thick. Volcanic rock clasts are common in the sandstones and conglomerates. Tuff and lapilli tuff beds are as thin as 1 cm, but most are 3 to 60 cm thick. The uppermost unit, unit 9, is 26.6 m thick and is the carbonaceous sandstone unit from which the plant megafossils have been recovered. Sandstone in beds 1 to 40 cm thick are interbedded with thicker beds of pebble conglomerate, sandstone, pebbly sandstone and granulestone. Broad channelling occurs in this unit and both cross-bedding and ripple lamination have been observed in a few sandstones. Plant material is abundant on the upper surfaces of many of the carbonaceous sandstone beds, and one thin coaly bed was located.

## SEDIMENTARY FACIES

The interpretation of the conglomerates and associated finer grained sedimentary rocks as alluvial fan deposits rests on the recognition of facies representing mass movements and traction currents.

### Mass Movement Deposits

*Debris Flow Facies* Debris flows (Bull, 1972) differ from mudflows in the coarser grain size, commonly sand size, matrix. Debris flows

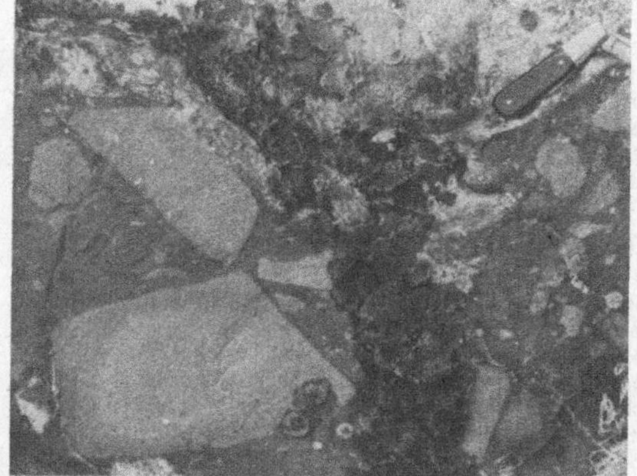

**Figure 5. Matrix-supported debris flow at 8 m in the Mt Flora section. Pocket knife is 8.5 cm long.**

**Figure 6. Clast-supported debris flow at 219 m in the Mt Flora section.**

(Gloppen and Steel, 1981) may yield both matrix and clast supported conglomerates and are distinguished by lack of significant erosion at the base of beds as well as clasts protruding above the upper surface.

Matrix supported debris flow deposits have been identified at Hope Bay (Figure 5). These beds consist of planar sheets of sediment that have very poor to no stratificaion, random orientation of clasts including some which are elongate and vertically oriented, megaclasts up to 1.5 m long, and no grading.

Massive beds of clast supported conglomerate, up to 29 m in thickness, are abundant, particularly in the lower two thirds of the section (Figure 6). These conglomerates, which show little or no erosion into the underlying sandstone or conglomerate, are unstratified and ungraded except at the base of a few beds where inverse grading is present. They contain megaclasts up to 2.3 m long and display in a few cases large clasts at the top of the bed protruding into the overlying finer sediment. The conglomerates commonly grade upward through a poorly stratified interval into sandstone. These sandstones, less than 60 cm thick, are impersistent and when traced laterally are seen to pass into clefts in the conglomerate sequence, or are not represented by any observable feature in outcrop; this probably accounts for the apparently excessive thickness of some conglomerate beds. These features are consistent with the debris flows described by Gloppen and Steel (1981), though most conglomerates of this sort at Hope Bay have subrounded to subangular clasts and appear to be much thicker. They are also similar to those described by Heward (1978) except for the apparent lack of imbrication. Modern debris flows with some of the characteristics of the Hope Bay conglomerates have been recorded by Pierson (1980).

Alternatives considered include sieve deposits (Hooke, 1967) and traction current deposits. Long continued aggradation of coarse clasts on a fan surface with subsequent infiltration of sand could lead to massive, thick, clast supported conglomerate beds, however this seems an unlikely process for such thicknesses of uniform strata. Traction current transport of gravel that includes megaclasts up to 2.3 m, although clearly not impossible in situations of sufficient relief, rainfall and stream gradient, seems unlikely. Such conglomerates might be channel fills, but no evidence has been found to suggest they are so confined, although this may be a functon of the extent of individual exposures. Gravel bars (Rust, 1972) may be structureless and thus a possible interpretation for these rocks. The stacking of such units without any associated facies of the braided stream environment argues against this origin. In some characteristics the Hope Bay conglomerates are similar to those described by Allen (1981) and attributed to traction current deposition as thick gravel sheets.

The mode of deposition cannot be argued with certainty on sedimentological grounds. The preservation in the sequence (Unit 2) of thin-bedded fine grained pyroclastic rocks, showing within the limits of the outcrops no channelling or other evidence of fluvial erosion, suggests an environment in which aggradation was predominant and thus supports the alluvial fan environment proposed here.

## Traction Current Deposits

*Braided Stream Facies* These deposits consist of well sorted, well stratified beds showing rapid lateral variations in grain size (Steel, 1974). Beds of this facies show gradational but rapid changes in grain size which range from sand to pebble size, rarely to cobble size. Bed thickness is up 2.5 m except for one bed 15 m thick which is tentatively assigned to this facies. Some of these beds are clearly erosional at the base and fill broad shallow channels.

**Figure 7. Thin bedded sandstones; 354 m in the Mt Flora section.**

*Sheetflood Facies* This is a facies regarded as typical of the alluvial fan environment (Bull, 1972; Steel, 1974).

Thin sandstones (1 to 10 cm thick) that are laterally continuous over several to tens of metres (Figure 7), and which display normal grading, non-erosive bases, but slight loading into the underlying sediment, are abundant at the top of the Mt Flora section. A few of these sandstones display ripple lamination in the upper part of the bed, but otherwise lack sedimentary structures. The tops of these beds commonly pass to carbonaceous sandstone with abundant plant remains. No root structures were observed in the field, although one in situ tree stump was found. The characteristics of this facies suggest rapid deposition from single "flood" events. Although such beds have been interpreted as crevasse splay deposits, the lateral continuity and vertical stacking of these beds together with the absence of well defined channel fills suggests deposition by sheetfloods. These beds are similar to the thin sheetflood sandstones described by Heward (1978).

*Channel Fills* Deposits that are clearly channel fills occur sparsely between the coarse conglomerate beds or quite commonly in the braided stream setting. At the base of braided stream deposits, channels up to 75 cm deep are filled with pebbly conglomerate.

*Sandstone Interbed Facies* The sandstone beds and lenses at the top of the coarse conglomerates are moderately to well stratified, and display low angle crossbedding. These are traction current deposits probably formed either in the waning stages of the deposition of debris flows, as described by Bluck (1967) or from separate depositional events.

## INTERPRETATION

With the Mt Flora Formation, ignoring the conglomerates and associated volcanic rocks at the west end of the cliffs, the overall progression of sedimentary facies is interpreted as the evolution of an alluvial fan. The predominant facies in the lower half of the sedimentary section is the clast supported debris flow; with this is associated three identified matrix supported debris flow deposits. These coarse clastic beds continue to about 260 m in the section where there is a quite abrupt change to thinner bedded and finer grained rocks that are dominantly braided stream deposits. The lower, debris flow dominated part of the succession probably represents mid to upper fan conditions whereas the overlying rocks represent mid to lower fan conditions. The braided stream environment is followed by both braided stream and sheetflood conditions on the lower part of the fan. The plant bearing sequence (Unit 9) includes braided stream conglomerate and sandstone, however some of the thicker sandstone beds are clearly erosional on a major scale, cutting down 1 to 2 m into the underlying thinner bedded more carbonaceous sandstones (Figure 8).

**Figure 8. Unit 9 of the Mt Flora section, showing large scale channelling into thin bedded sandstones.**

The evidence presented here is taken to support a subaerial, alluvial fan environment which in vertical profile is rather different from the two coarse conglomerate types proposed by Miall (1975). The plant bearing beds have been interpreted as lacustrine deposits (Andersson, 1906; Adie, 1964; Bibby, 1966) at least in part because of the presence of fish vertebrae, beetle elytra and poorly preserved bivalves. However, the coaly bed and tree stump, together with the facies present including coarse conglomerate beds, argue for a depositional environment that includes both high energy channels and low energy settings, is often flooded and possibly swampy, lacks siltstone and mudstone and therefore seems unlikely to be lacustrine. Nevertheless the palaeontology suggests lacustrine conditions may have occurred locally; unfortunately no information is available on the sedimentological setting of those fossils, and no new palaeontological data have been acquired. The abundance of plant debris suggests development of the fan in a humid, rather than arid, climate. The very sparse palaeocurrent data obtained indicate widely dispersed directions of flow and the orientation of the fan cannot be inferred.

The evolution of the fan from coarse proximal deposits to finer grained distal deposits could represent either the retreat of the scarp against which the fan was banked or reduction in source terrain relief with time.

Sedimentary deposition on this alluvial fan was interrupted by contemporaneous volcanism (in fact it may have been preceded by volcanism, but the relationships have not been resolved); all volcanic rocks can be attributed to airfall processes or to reworking of airfall debris by surface waters. Although the larger clasts in the conglomerates are made up of TPG rocks, in beds above the lowest tuffs of the stratigraphic section the finer clasts include volcanic rock fragments and volcanic quartz is present in the sandstones.

## CONCLUSIONS

The rock types present and the sedimentary facies they represent point to the conglomeratic beds at Hope Bay being alluvial fan deposits which can be interpreted in terms of the evolution of a single fan. Conglomeratic rocks such as these require significant local topographic relief and therefore it is inferred they were deposited in fault bounded basins.

The conglomerates at Hope Bay rest on and were derived from the pre-Jurassic Gondwanian basement. They are interbedded with and possibly overlie, volcanic rocks and thus overlap the inception (or the early stages) of volcanism associated with the active plate margin of

Late Mesozoic to Cenozoic age. They constitute part of a well defined tectonostratigraphic unit that extends from the South Orkney Islands to the Longing Gap area and locally bridges the two principal Mesozoic tectonic events of the Antarctic Peninsula.

*Acknowledgements* We wish to thank our colleagues R. Alley, R.A. Askin and D.R. Watts for assistance with the fieldwork. We also wish to thank Captain P. Lenie and the crew of R/V *Hero* for support in the field. This work has been supported by NSF grant DPP-7920194 and DPP-7821102. Contribution 448 of the Institute of Polar Studies.

## REFERENCES

ADIE, R.J., 1964: Geological History; *in* Priestley, R., Adie, R.J. and Robin, G. de Q. (eds.) *Antarctic Research*. Butterworths, London, 118-62.

AITKENHEAD, N., 1975: The geology of the Duse Bay-Larsen Inlet area, northeast Graham Land (with particular reference to the Trinity Peninsula Series). *Scient. Rep. Br. antarc. Surv., 51*.

ALLEN, P.A., 1981: Sediments and processes on a small stream flow dominated, Devonian alluvial fan, Shetland Islands. *Sedimt. Geol., 29*, 31-66.

ANDERSSON, J.G., 1906: On the geology of Graham Land. *Univ. Uppsala., Geol. Inst., Bull., 7*, 19-71.

BIBBY, J.S., 1966: The stratigraphy of part of northeast Graham land and the James Ross Island Group. *Scient. Rep. Br. antarc. Surv., 53*.

BLUCK, B.J., 1967: Deposition of Old Red Sandstone conglomerates in the Clyde area: A study in the significance of bedding. *Scot. J. Geol.,3*, 139-67.

BULL, W.B., 1972: Recognition of alluvial fan deposits in the stratigraphic record; *in* Rigby, J.K. and Hamblin, W.K. (eds.) Recognition of ancient sedimentary environments. *Soc. Econ. Paleont. Mineral., Spec. Publ., 16*, 63-83.

DALZIEL, I.W.D., 1982: The pre-Jurassic history of the Scotia Arc: a review and progress report; *in* Craddock, C. (ed.) *Antarctic Geoscience*. Univ. Wisconsin Press, Madison, 111-26.

ELLIOT, D.H., 1967: The geology of Joinville Island. *Bull. Br. antarct. Surv., 12*, 23-40.

ELLIOT, D.H., 1975: Tectonics of Antarctica: a review. *Am. J. Sci., 275-A*, 45-106.

ELLIOT, D.H. and WELLS, N.A., 1982: Mesozoic alluvial fans of the South *Orkney Islands; in* Craddock, C. (ed.) *Antarctic Geoscience. Univ. Wisconsin* Press, Madison, 235-44.

FARQUHARSON, G.W., in press: Lacustrine deltas in a Mesozoic alluvial sequence from Camp Hill, Antarctica. *Sedimentology.*

GLOPPEN, T.G. and STEEL, R.J., 1981: The deposits, internal structure and geometry in six alluvial fan-fan delta bodies (Devonian-Norway)—a study in the significance of bedding sequence in conglomerates; *in* Ethridge, F.G. and Flores, R.M. (eds.) Recent and ancient nonmarine depositional environments: models *for exploration. Soc. Econ. Paleont. Mineral., Spec. Publ., 31*, 49-69.

HALLÉ, T.G., 1910: The Mesozoic flora of Graham Land. *Wissenschaftliche Ergebnisse der Schwedischen Sudpolarexpedition*, 1901-03, Bd *3*, lief *14*, 1-123.

HEWARD, A.P., 1978: Alluvial fan and lacustrine sediments from the Stephanian A and B (La Magdalena, Cinera—Matallana and Sabero) coalfields, northern Spain. *Sedimentology, 25*, 451-88.

HOOKE, R.L.B., 1967: Processes on arid region alluvial fans. *J. Geol., 75*, 438-60.

HYDEN, G. and TANNER, P.W.G., 1981: Late Palaeozoic—early Mesozoic forearc basin sedimentary rocks at the Pacific margin in Western Antarctica. *Geol. Rdsch., 70*, 529-41.

MIALL, A.D., 1975: Lithofacies types and vertical profile models in braided river deposits: a summary; *in* Miall, A.D. (ed.) Fluvial Sedimentology. *Can. Soc. Petrol. Geol., Mem., 5*, 597-604.

PIERSON, T.C., 1980: Erosion and deposition by debris flows at Mt Thomas, North Canterbury, New Zealand. *Earth Surface Processes, 5*, 227-47.

RUST, B.R., 1972: Structure and process in a braided river. *Sedimentology, 18*, 221-45.

STEEL, R.J., 1974: New Red Sandstone flood plain and piedmont sedimentation in the Hebridean Province, Scotland. *J. Sediment. Petrol., 44*, 336-57.

STIPANICIC, P.N. and BONETTI, M.I.R., 1970: Posiciones estratigráficas y Edades de las principales Floras jurásicas argentinas. II. Floras doggerianas y málmicas. *Ameghiniana, 7*, 101-18.

THOMSON, M.R.A., 1981: Late Mesozoic stratigraphy and invertebrate palaeontology of the South Orkney Islands. *Bull. Br. antarct. Surv., 54*, 65-83.

# THE JURASSIC LATADY FORMATION, SOUTHERN ANTARCTIC PENINSULA

T.S. Laudon, *Geology Department, University of Wisconsin, Oshkosh, Wisc. 54901, U.S.A.*

M.R.A. Thomson, *British Antarctic Survey, Madeingley Road, Cambridge, CB3 0ET, U.K.*

P.L. Williams, *U.S. Geological Survey, Denver, Colorado 80225, U.S.A.*

K.L. Milliken, *The University of Texas, Austin, Texas, U.S.A.*

P.D. Rowley, *U.S. Geological Survey, Denver, Colorado 80225, U.S.A.*

J.M. Boyles, *Cities Service Corporation, Tulsa, Oklahoma, U.S.A.*

*Abstract* The Jurassic Latady Formation, a thick sequence of sandstone, siltstone, shale, and conglomerate with locally interbedded volcanic rocks, and sparse thin coal and limestone, occurs in an area that extends more than 600 km along the Weddell Sea coast of the southern Antarctic Peninsula, and inland as much as 200 km. Most Latady sandstone is lithic sandstone, some is arkosic, and a little is quartzitic. Maximum thickness of the Latady Formation is probably several kilometres; neither the top nor the base of the unit has been definitely identified. Palaeontologic evidence indicates that most of the Latady is of Late Jurassic (Kimmeridgian to Tithonian) age; Middle Jurassic beds occur in the Behrendt Mountains. Most of the Latady was deposited in shallow marine shelf environments. Terrestrial and near shore marine environments are represented in the interior of the southern Antarctic Peninsula. Deeper water, open marine environments are represented in a few areas near the present coast. The Latady Formation was deposited in a back-arc basin located on the southeastern side of a magmatic arc located along the axis and western side of the Antarctic Peninsula. This magmatic arc was built on continental lithosphere during south-southeastward subduction of the southern Pacific plate. Volcanic rocks of the Mount Poster Formation which make up the arc were the principal source of detritus of the Latady Formation. Metamorphic rocks of the pre-Jurassic(?) basement of the arc were an important source of detritus, and plutonic rocks and sedimentary rocks were minor sources.

The Latady Formation of Middle and Late Jurassic age, consists of thick, structurally complex, metamorphosed sequences of marine shale, siltstone, sandstone, and conglomerate with locally interbedded volcanic rocks, coal and limestone. It occurs in an area extending more than 600 km along the Orville, Lassiter and southern Black Coasts of the Weddell Sea in eastern Ellsworth Land and southern Palmer Land (Figure 1), from the vicinity of the Behrendt Mountains in the southwest to the Dana Mountains in the northeast. It probably extends even farther north along the eastern side of the Antarctic Peninsula, for Singleton (1980) correlated the Mount Hill Formation of the central Black Coast with the Latady. Williams et al. (1972) named the Latady Formation from exposures in the Latady Mountains of the Lassiter Coast. The type section is designated along a ridge 4 km north of McLaughlin Peak (Locality 1, Figure 1).

On the north and northwest the Latady Formation intertongues with volcanic rocks of the largely contemporaneous Mt Poster Formation (Rowley et al., 1982) of the Antarctic Peninsula Volcanic Group. The Latady and Mt Poster Formations were folded and thrust faulted towards the Weddell Sea prior to emplacement of Early Cretaceous plutonic and hypabyssal rocks (Laudon et al., 1969; Williams et al., 1972; Rowley and Williams, 1982) of the Lassiter Coast Intrusive Suite (Rowley et al., this volume).

The Latady Formation, the Mount Poster Formation and the Lassiter Coast Intrusive Suite are all products of a Mesozoic magmatic arc (or arcs) associated with south-southeastward subduction of a plate of Pacific Ocean lithosphere beneath what is now the Antarctic Peninsula (Dalziel and Elliot, 1973; Smellie, 1981; Rowley et al., this volume). During accumulation of the Latady and Mount Poster Formations in Middle and Late Jurassic time, the axis of the arc was located along, or slightly northwest of the present topographic axis of the southern Antarctic Peninsula. Most of the Mount Poster Formation accumulated subaerially near the axis of the arc. Most of the Latady Formation was deposited in a marine back-arc basin located southeast of the arc on the craton side. The two units intertongue near the location of the Jurassic shoreline which trended parallel to the long axis of the southern peninsula on its southeastern flank (Thomson et al., 1978; Thomson, 1980; Rowley et al., this volume).

Other stratigraphic units associated with the Mesozoic magmatic arcs of the southern Antarctic Peninsula include the LeMay Formation of Alexander Island, in part of Triassic age (Edwards, 1982), part of which was deposited in deep water on the trench-slope break (Smellie, 1981), and the uncomformably overlying, Upper Jurassic to Lower Cretaceous Fossil Bluff Formation, which was deposited in deltaic environments in a fore-arc basin (Thomson, 1982).

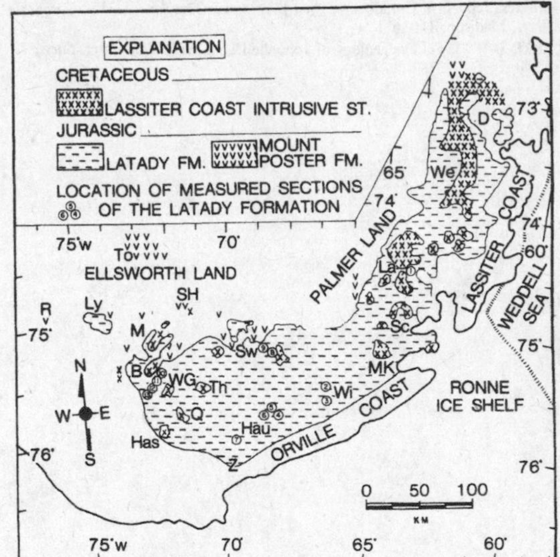


**Figure 1. Generalised geologic map of southern Palmer Land and eastern Ellsworth Land showing locations of geographic features mentioned in text: B = Behrendt Mountains; D = Dana Mountains; Has = Mount Hassage; Hau = Hauberg Mountains; La = Latady Mountains; Ly = Lyon Nunataks; M = Merrick Mountains; MK = Mount McKibben; Q = Quilty Nunataks; R = Mount Rex; Sc = Scaife Mountains; SH = Sky-Hi Nunataks; Sw = Sweeney Mountains; Th = Thomas Mountains; To = Tollefson Nunatak; We = Werner Mountains; WG = Weatherguesser Nunataks; Wi = Wilkins Mountains; Z = Cape Zumberge.**

## STRATIGRAPHY

The most extensive and abundant outcrops of the Latady Formation are in the southern Orville Coast, where the Hauberg and Wilkins Mountains are composed entirely of the Latady. Most outcrops here consist of small, narrow bands occupying ridge crests. In larger outcrops, open, overturned and isoclinal folds and thrust faults are generally evident. Slaty cleavage parallel to fold axes is conspicuous in most outcrops.

In most other parts of its area of exposure Latady outcrops occur near Cretaceous plutons. Contact metamorphic aureoles, locally more than 1 km wide, occur in Latady and Mt Poster rocks surrounding plutons. Latady rocks adjacent to some plutons have been tightly refolded during forceful intrusion.

The Latady Formation consists mostly of interbedded dark coloured sandstone, siltstone and shale. In the northeastern part of the region, including the type area, shale, siltstone and mudstone are the principal rock types present. Sandstone probably makes up less than 10% of the section. Sandstone and siltstone are progressively more abundant to the southwest. Sandstone constitutes approximately 50% of the sections measured in the Wilkins Mountains (Sections 2 and 3, Figure 2) and the Hauberg Mountains (Sections 4, 5, 6 and 7, Figure 3), and more than 50% of the Behrendt Mountains sections (Sections 10, 11, 12 and 13, Figure 4). Terrigenous conglomerate pebbles occur in widely spaced lenses from one pebble to 1 m thick, in many places in the southern Behrendt Mountains and are less abundant in the Hauberg Mountains. Conglomerate found northeast of the Haubergs contains only granules and small pebbles. There seems to be a progressive general increase from northeast to southwest in both the maximum size and average size of terrigenous clastic particles. Intraformational shale-pebble conglomerate containing intraclasts as long as 60 cm are abundant in many Latady outcrops.

Shale, siltstone and sandstone occur commonly in interbedded sequences of thin beds, in gradational cycles coarsening upward and in individual units several metres to tens of metres thick. A few thick-bedded to massive sandstone sequences and some shale sequences are more than 100 m thick. Thick sequences composed entirely of soft, black, fissile shale occur at Cape Zumberge south of the Hauberg Mountains and in nunataks south of Mount McKibben (Figure 1).

Volcanic rocks are interbedded with sedimentary rocks of the Latady Formation at Mount Neuner in the Behrendt Mountains, at several places in the Sweeney Mountains (e.g. Section 9, Figure 4) and in the southern Black Coast. Thin coal seams and lenses and thin limestones occur locally in the northern Lassiter Coast and southern Black Coast.

Planar laminations and gently sloping cross laminations are common in sandy beds within the Latady. Festoon cross laminations, convolute laminations, ripple marks and mud cracks have also been observed. Burrows, tracks and trails and other bioturbation effects are common.

Twelve measured stratigraphic sections of portions of the Latady Formation are presented in Figures 2, 3 and 4. None of the measured stratigraphic sections represents a complete section of the Latady Formation, for neither the stratigraphic top, nor the bottom of the unit have been identified in the field. In section 8 (Figure 3) in the Sweeney Mountains, interbedded sandstone, shale and pyroclastic rocks of the Latady overlie massive dacite of the Mount Poster Formation. At Mount Poster in the Latady Mountains, volcanic rocks of the Mount Poster Formation overlie shale of the Latady Formation. It is doubtful, however, that the Sweeney occurrence represents the oldest part of the Latady, or that the Mount Poster occurrence represents the youngest part. Elsewhere the measured stratigraphic sections begin with the lowest exposed bed and are terminated either at the highest exposed bed or by structural discordance. Most of the measured stratigraphic sections consist of narrow outcrop bands

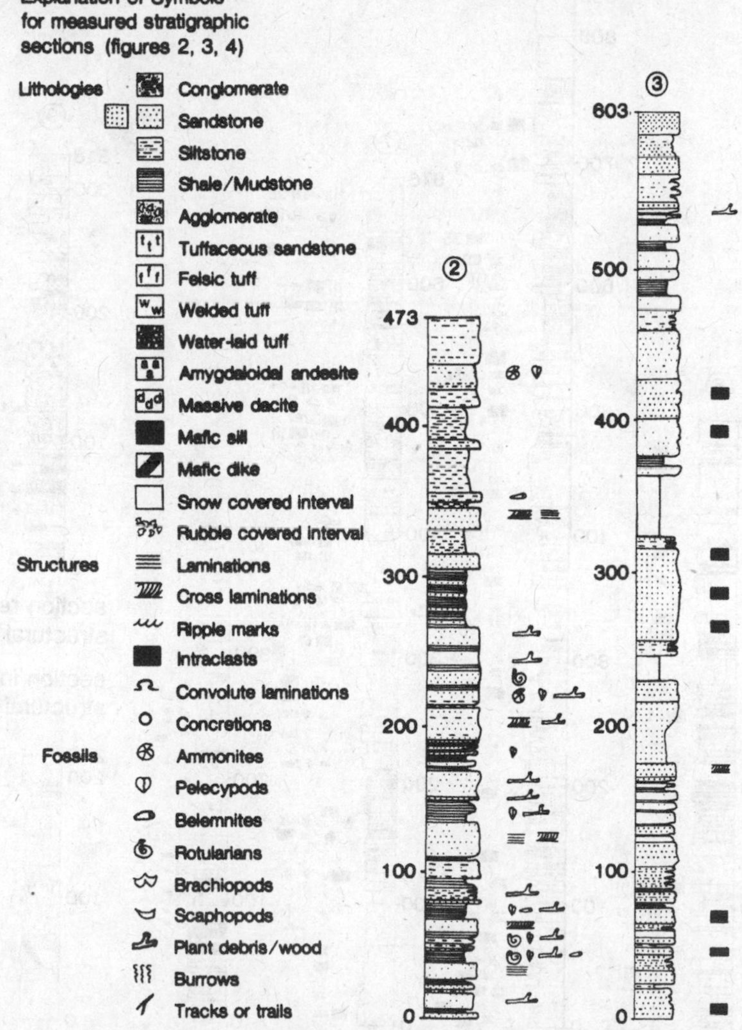


**Figure 2.** **Measured stratigraphic sections of the Latady Formation in the Wilkins Mountains: (2) Northern part of southwestern Wilkins Mountains, 15 km west of Matthews Glacier, measured by Williams and D.L. Schmidt; (3) Southwesternmost Wilkins Mountains, 20 km west of Matthews Glacier, measured by Boyles and Rowley.**

located along ridge crests with many snow covered intervals, so it is possible that some may contain undetected structural complications. For these reasons neither the stratigraphic locations of the measured sections within the Latady, nor the maximum thickness of the unit are known.

Stratigraphic thicknesses in measured sections range between 318 m in section 9 in the Sweeney Mountains and 1675 m in section 13 in the Behrendt Mountains. The greatest measured stratigraphic thickness of the Latady in a continuously exposed section is 830 m in section 4, located south of Mt Leek in the Hauberg Mountains. Based on the lithology, thicknesses, and ages of measured sequences and on plunges of fold axes, we estimate the maximum thickness of the Latady to be several kilometres.

## Age of the Latady Formation

Most of the Latady is barren of fossils; however marine invertebrate fossils are locally very abundant. Stevens (1967) assigned a Late Jurassic, Kimmeridgian(?) age to belemnites, pelecypods and worms from Lyon Nunatak. Quilty (1970, 1972a, 1972b, 1977, 1982 and in press) described faunas of Bajocian, probable Callovian, Oxfordian and Heterian (Kimmeridgian) ages from eastern Ellsworth Land. Imlay and Kaufman (in press) describe Upper Jurassic ammonites and bivalves from the Lassiter Coast from which they define eight biostratigraphic zones ranging in age from Late Kimmeridgian to Late Tithonian. These zones occur in bands which parallel the trend of the Antarctic Peninsula and dip east towards the coastline. Crame (1981) described pelecypod faunas from the Orville Coast to which he assigned Kimmeridgian age. Thomson (this volume) concluded that most of the ammonite faunas from the Orville Coast which he described earlier (1980), and for which he had suggested Kimeridgian ages should probably be placed in the Tithonian. He concluded that faunas from the Hauberg Mountains area favour an Early to earliest Late Tithonian age, that faunas from the Wilkins Mountains area are consistent with a mainly early Late Tithonian age and that a fauna from Cape Zumberge indicates a latest Tithonian age. Kamenev and Orlenko (1982) reported the presence of poorly preserved Late Cretaceous pelecypods at Rivera Peak in the southern Lassiter Coast, but did not describe them.

Thus, with the exception of Middle Jurassic faunas from the southern Behrendt Mountains (Quilty, 1970, 1972a, 1972b, 1982 and

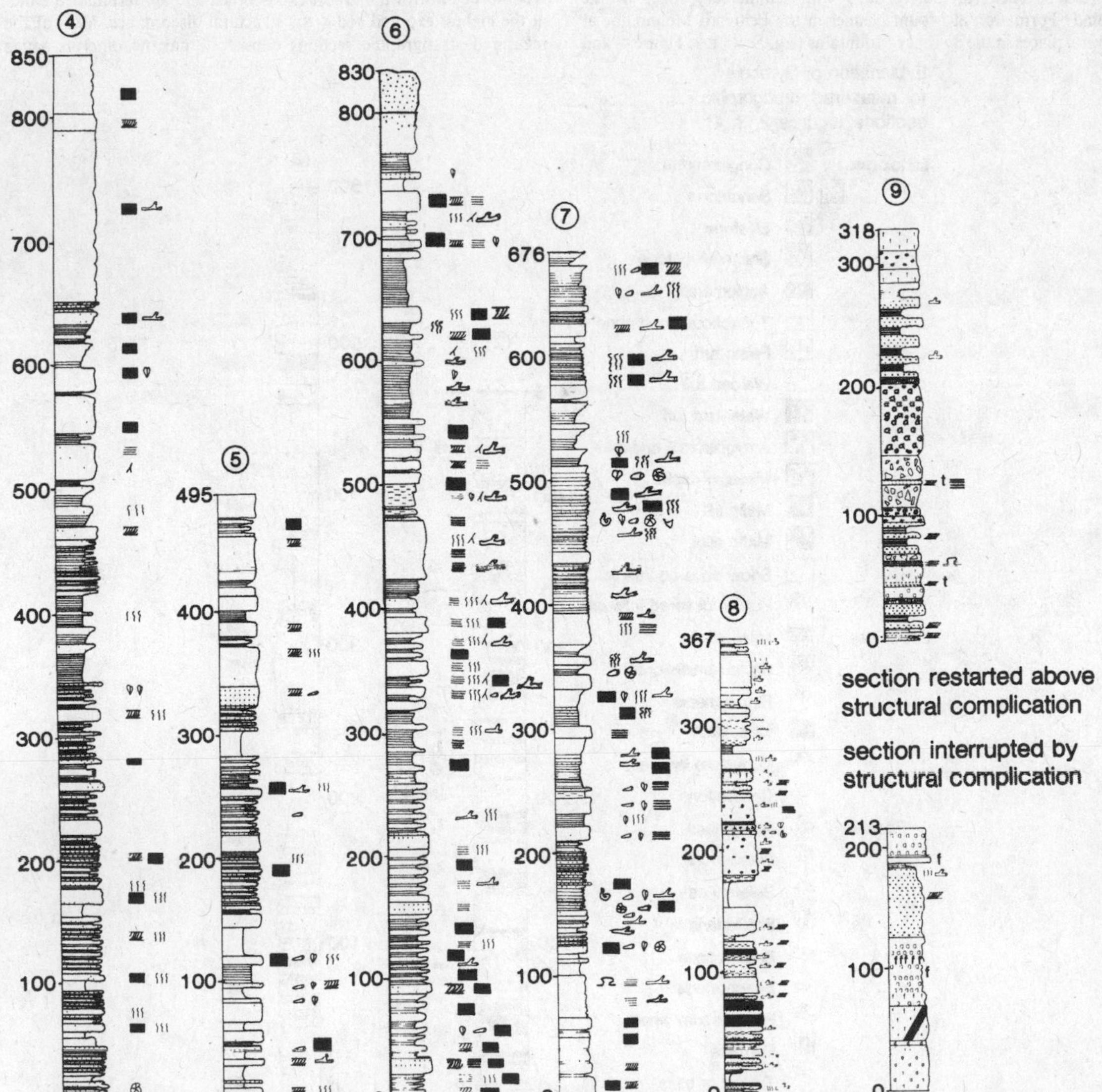


**Figure 3. Measured stratigraphic sections of the Latady Formation in the Hauberg Mountains and the Sweeney Mountains: (4) 5 km south of Mt Leek, measured by Boyles and Rowley; (5) 7 km west of Mt Leek, measured by Boyles, Laudon and Thomson; (6) 14 km north of Mt Dewe, measured by Laudon; (7) 5 km west of Mt Novocin, measured by Laudon, Thomson and Boyles; (8) 6 km southeast of Mt Potter, measured by Laudon and Rowley; (9) 3 km southwest of Mt Potter, measured by Laudon and Rowley.**

in press), all described invertebrate faunas from the Latady Formation are of Late Jurassic age (Kimmeridgian to Tithonian). The faunal evidence indicates that the youngest parts of the Latady Formation are contemporaneous with the oldest parts of the Fossil Bluff Formation of Alexander Island (Crame, 1981; Thomson, this volume, Figure 1).

Small, abraded plant fragments are common in sandy and silty beds of the Latady; fragments as long as 1 m have been observed. Fossil floras from the Lassiter Coast and the Behrendt Mountains include wood, leaves, shoots, pinnules, leaf sheaths and fronds of cycadophytes, conifers, gymnosperms and gynkophytes that bear elements in common with the Middle(?) Jurassic (Halle, 1913) flora from Halle Bay in the northern Antarctic Peninsula (J.M. Schopf, written comunication, 1973).

## PETROLOGY

The following summary is based on petrographic analyses of 227 thin sections of Latady sandstone and conglomerate: 32 are from the Lassiter Coast, 134 are from the Orville Coast and 61 are from eastern Ellsworth Land.

### Textures

Sandstone textures (e.g. Figure 5a) consist essentially of grain mosaics with small amounts of intergranular matrix or cement and reflect the effects of elevated pressures during tectonic deformation. In addition, many show effects of contact metamorphism. Effects of pressure are particularly evident in grain-to-grain relationships between framework components. Only sutured, concave-convex and long contacts between grains are present. Flexible grains are usually deformed. Softer grains are penetrated by harder grains. Pressure shadows and interpenetration of grains by pressure solution are common. Alignment and segregation of framework components, recrystallisation and segregation and streaming of metamorphic sericite, muscovite, biotite and chlorite are common metamorpic effects.

Intergranular material includes cement, phyllosilicate cement, orthomatrix, epimatrix and pseudomatrix. Amounts are less than 5% in most thin sections, although some conglomerates have as much as 10%. Most thin sections contain several kinds of interstitial material.

Most sandstone shows fair sorting of grains, in keeping with its generally well-bedded, frequently laminated character. There are many exceptions to this generalisation, however. Conglomerate especially tends to be essentially unsorted as does much sandstone.

One of the most outstanding general characteristics of Latady sandstone and conglomerate is the fact that most clasts are angular or subangular and that large numbers of grains display extreme angularity. Most sandstone contains sparse subrounded grains and most conglomerate pebbles are well rounded. An interesting and significant exception to this generalisation is the presence in most sandstones of small, well rounded, detrital, heavy mineral grains, particularly zircon and magnetite.

### Composition

*Modal Compositions* Figure 6a shows the compositions of Latady sandstone samples in terms of modal sand sized quartz, feldspar and lithic fragments. The outstanding feature of the modal compositions is their diversity: 11 (5%) are quartzites (Q> 75%), 57 (25%) are arkoses (Q< 75%, F> L) and 159 (70%) are lithic sandstones (Q< 75%, L> F) (classification modified from Folk, 1968).

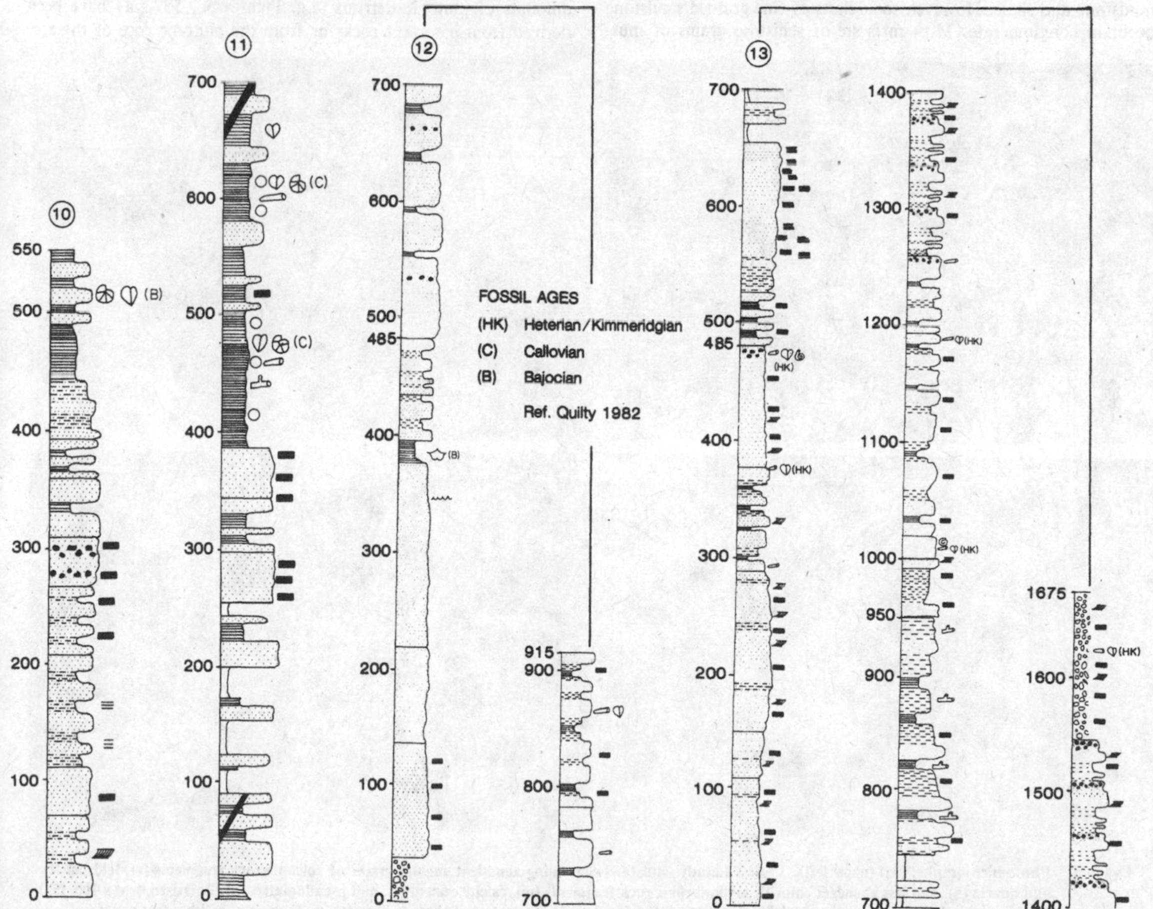


**Figure 4. Measured stratigraphic sections of the Latady Formation in the Behrendt Mountains: (10) 8 km northeast of summit of Mt Hirman, measured by Laudon and P.M. Otway; (11) 4 km northeast of summit of Mt Hirman, measured by Laudon and P.M. Otway; (12) Mt Hirman, northeast of summit, measured by Laudon and P.M. Otway; (13) Mt Hirman, southwest of summit, measured by Laudon and P.M. Otway.**

Figure 6b shows mean modal Q-F-L compositions of Latady sandstone from different areas. Sandstone from the Behrendt Mountains area in the southwestern part of the region has generally higher than average feldspar content and lower than average lithic content, whereas sandstone from the Lassiter Coast area in the northeast has generally higher than average lithic content and lower than average feldspar content. In the northwestern part of the region, sandstone from Lyon Nunataks, the Merrick Mountains, Sky-Hi Nunataks and the southeastern Sweeney Mountains is characterised by considerably higher than average quartz content, whereas sandstone from the northwestern Sweeney Mountains is generally the most feldspathic.

*Framework* Detrital volcanic rock fragments (vrfs), feldspar and quartz are essential framework components of nearly all of the sandstone. Metamorphic rock fragments (mrfs), muscovite, chlorite, skeletal grains, sedimentary rock fragments (srfs), sedimentary intraclasts and a variety of heavy minerals of which zircon and magnetite are the most abundant, are common accessory framework components.

Plagioclase is far more abundant than K-feldspar. Many thin sections contain no K-feldspar. Many grains of albite, the most abundant plagioclase, exhibit chessboard twinning suggestive of formation by replacement of K-feldspar during metamorphism (Starkey, 1959; Moore and Lietz, 1979). Some microcline and some plagioclase containing graphic or perthitic intergrowths are present.

Clear, monocrystalline grains, many containing resorbed bipyramidal crystal faces and bleb inclusions, are the most abundant types of quartz. Some grains of polycrystalline quartz of metamorphic origin also are present.

Vitric grains make up the bulk of vrfs, but felsitic, microlitic and lathwork grains are also present. Most srfs and intraclasts are of mudstone and shale. However, some clasts of fine grained sandstone occur in conglomerate. Most mrfs are of schistose grains of muscovite, chlorite, or biotite; sparse quartzite grains also are present. Rare granule sized clasts of felsic plutonic rock fragments also were found.

Terrigenous conglomerate pebbles consist of volcanic rocks (e.g. Figures 5c, 5d), subordinate metamorphic rocks and minor sedimentary and plutonic rocks (Figure 5d). Laudon (1972, Table 2) reported the following compositions for 164 conglomerate pebbles from the Behrendt Mountains: felsic volcanic, 33%; intermediate and mafic volcanic, 21%, metasedimentary, 30%; unidentified fine grained rocks, 14%; plutonic granodiorite, 1%; and sedimentary (unmetamorphosed greywacke), 1%.

Skeletal grains and plant fragments are not present in most of the thin sections studies, but when present are usually fairly abundant. Detrital magnetite, zircons, muscovite and chlorite are present in most of the slides examined.

## PROVENANCE

Petrologic evidence, as well as stratigraphic relationships and field occurrences support the conclusion that the primary provenance of the Latady Formation was an active magmatic arc (Mount Poster Formation) built on continental lithosphere located above a southeast dipping Benioff zone during subduction of the southern Pacific plate beneath the southern Antarctic Peninsula (Rowley et al., this volume).

Latady sandstone and conglomerate are composed predominantly of volcanogenic material. The vrfs within the Latady Formation (e.g. Figures 5a, 5b and 5c) are of the same rock types as those in the Mount Poster Formation.

Significant amounts of material of metamorphic provenance are considered to represent contributions from the basement upon which the Latady and Mount Poster Formations were deposited. Small amounts of plutonic detritus (e.g. Figures 5c, 5d) may have been derived from basement rocks or from the plutonic core of the arc.

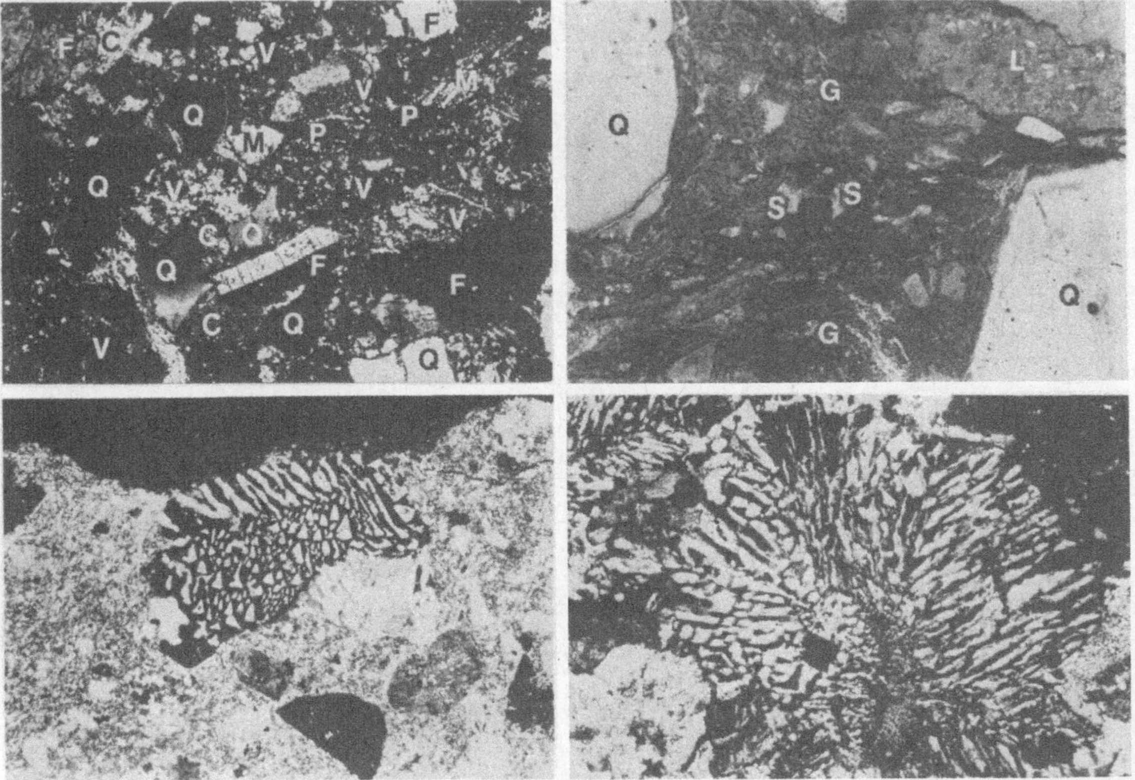


**Figure 5. Photomicrographs: (5a) (upper left), Typical Latady sandstone containing abundant angular grains of volcanic rock fragments (v), feldspar (f), and quartz (q), and less abundant chloritic metamorphic rock fragments (m), calcite cement (c), and pseudomatrix (p). Polarised light x 100. (5b) (upper right), Pebble of felsic, crystal bearing vitric tuff from Latady Formation, Behrendt Mountains. Q-quartz, L-lithic fragments (vrf), S-shards, G-devitrified groundmass. Unpolarised light x 100. (5c) (lower left), Pebble of lithic tuff with sand sized granophyric clast, from Latady Formation, Behrendt Mountains. Polarised light x 100. (5d) (lower right), Granophyric texture in plutonic pebble from Latady Formation, Behrendt Mountains. Polarised light x 100.**

Most srfs are of intraformational origin, although a few pebbles of lithified greywacke are of unknown provenance. Although pre-Latady metamorphic, plutonic and sedimentary rocks have not been observed in the southern Antarctic Peninsula, the basement may be similar to the basement complex described by Singleton (1980) in the Central Black Coast of Palmer Land. A Latady conglomerate pebble of tuff, containing a sand sized clast with granophyric texture (Figure 5c) is particularly interesting from the standpoint of provenance, for it indicates the presence of plutonic rocks either beneath, or on the surface during pyroclastic activity which preceded Latady deposition.

Sandstone compositions are consistent with "transitional" arc provenance of Dickinson and Suczek (1979), where dissection has only partially exposed basement rocks and plutonic core rocks.

The angularity of feldspar and quartz grains, the fact that feldspar and quartz grains are approximately equal in size, the occurrence of both slightly altered and strongly altered feldspars, the absence of carbonate rocks and the presence of the plant fossils all indicate a tectonically active source terrain with rugged topography and temperate climate.

## DEPOSITIONAL ENVIRONMENTS

Latady sandstone and conglomerate accumulated near the source terrain from which it was derived. The preponderance of angular clasts together with the abundance of detrital feldspar and lithic grains imply a relatively short time for the entire sedimentation process (erosion, transportation and deposition), as well as rapid uplift in the source area and rapid subsidence in the basin of deposition.

On the Weddell Sea side of the southern Antarctic Peninsula (Figure 1) there appears to be a general progression in the environments of deposition of Jurassic rocks from (1) terrestrial, near the axis of the peninsula, through (2) shallow, near shore marine, possibly deltaic, to (3) shallow, open marine near the coast at Cape Zumberge and near Mount McKibben. Intertonguing facies suggest that these environments shifted laterally in response to fluctuating rates of uplift, subsidence and sedimentaton during Latady and Mount Poster deposition.

All outcrops in the vicinity of Tollefson Nunatak (Figure 1) near the axis of the Peninsula are composed of subaerially deposited volcanic rocks of the Mount Poster Formation. Outcrops of both the Latady Formation and the Mount Poster Formation occur in a band running parallel to the trend of the peninsula from Mt Rex in the southwest, through Lyon Nunataks, the Merrick Mountains, Sky-Hi Nunataks and the Sweeney Mountains, to the western part of the Latady Mountains on the northeast (Figure 1). Within this area interbedded volcanic rocks and sedimentary rocks occur at several places in the Merrick Mountains, the Sweeney Mountains and the Latady Mountains. Most of the volcanic rocks in this area accumulated subaerially but pillow lava flows and laminated, water-lain tuffs (e.g. Section 9, Figure 3) are present in places in the Sweeney Mountains. Within this band Latady sediments accumulated in both terrestrial and shallow marine environments. Marine fossils are present but are not as abundant as they are in outcrops located farther south and east. The Lassiter Coast floras probably represent deposition in freshwater lakes or streams (J.M. Schopf, pers. comm., 1973). Numerous thin coal beds occur in the Sweeney Mountains and the Latady Mountains. Large plant fossil fragments are particularly abundant in the Sweeney Mountains. Unusually high (for the Latady) quartz content of sandstone from Lyon Nunataks, the Sweeney Mountains, Sky-Hi Nunataks and the southeastern Sweeney Mountains (Figure 6b) may be a result of substantial mineralogic modification of clastic detritus by waves and currents prior to deposition in beach or nearshore settings. In contrast, uncommonly high feldspar content in sandstone from the northwestern Sweeney Mountains may reflect little modification during terrestrial (fluvial or lacustrine) sedimentation. Probably the shoreline shifted back and forth across this area.

With the exception of rare tuffaceous beds and flow rocks, all pre-plutonic rocks in the Behrendt Mountains, the Hauberg Mountains, the Wilkins Mountains and nunataks to the east (Figure 1) are composed of clastic sedimentary rocks of the Latady Formation. The thick sequence of well bedded, frequently laminated and cross laminated, interbedded mudstone, siltstone, sandstone and conglomerate (e.g. Figures 2, 3 and 4) (c.f. heterolithic facies of Johnson, 1978), with locally abundant marine invertebrate fossils, abraded plant fossils, intraformational shale-pebble conglomerate and bioturbation features and less abundant ripple marks, mud cracks, root casts and carbonaceous non-marine(?) shale, accumulated mainly in shallow marine shelf environments regularly affected by waves and currents. Some of this sequence may have been deposited in deltaic environments.

A few small nunataks near the present coastline at Cape Zumberge and south of Mount McKibben are composed entirely of soft, black, fissile shale. A latest Tithonian ammonite fauna was recovered at Cape Zumberge (Thomson 1980, this volume). These rocks represent open marine deposition in shallow water below wave base.

*Acknowledgements* We thank the other members of the field parties to eastern Ellsworth land, the Lassiter Coast and the Orville Coast whose field notes and samples constitute a substantial portion of the data base for the Latady Formation. D.L. Schmidt and P.M. Otway measured stratigraphic sections in the field. S. Weiterman compiled stratigraphic columns from field notes done in five different styles. S. Kowalkowski, C. Seibold, S. Weiterman and C. Purtell assisted in the preparation of maps and columnar sections. This work was supported by grants from the National Science Foundation and the University of Wisconsin-Oshkosh. Logistic support in Antarctica was provided by the National Science Foundation and the U.S. Navy.

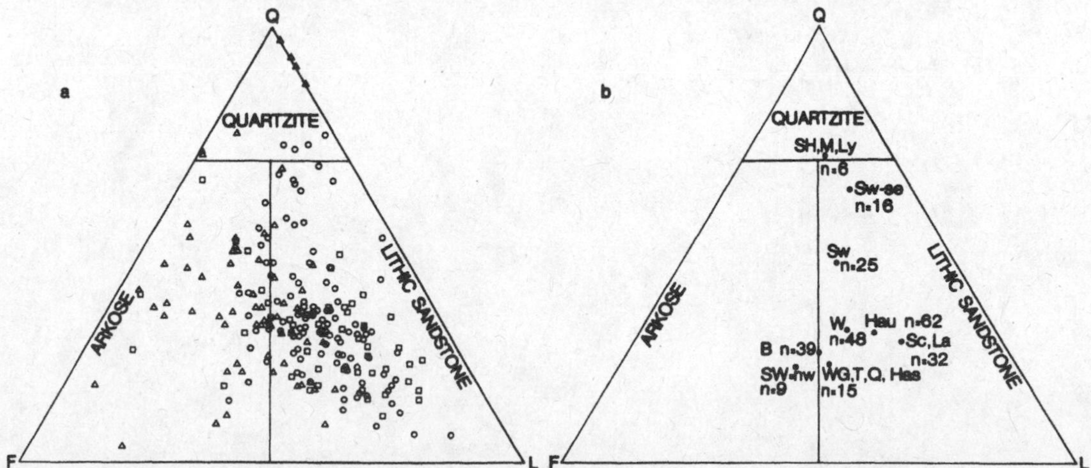


**Figure 6.** **(6a) Modal quartz-feldspar-lithic compositions of Latady sandstones: Δ = from eastern Ellsworth Land (from Laudon, 1972); □ = from the Lassiter Coast (from Williams and Rowley, 1972; Rowley and Williams, 1982); O = from the Orville Coast and eastern Ellsworth Land (new analyses by Milliken); (6b) Mean modal quartz-feldspar-lithic compositions for Latady sandstone from different areas: SH = Sky-Hi Nunataks; M = Merrick Mountains; Ly = Lyon Nunataks; Sw = Sweeney Mountains; W = Wilkins Mountains; Sc = Scaife Mountains; La = Latady Mountains; B = Behrendt Mountains; WG = Weatherguesser Nunataks; T = Thomas Mountains; Q = Quilty Nunataks; Has = Mount Hassage.**

## REFERENCES

CRAME, J.A., 1981: Preliminary bivalve zonation of the Latady Formation. *U.S. Antarct. J., 16,* 8-10.

DALZIEL, I.W.D. and ELLIOT, D.H., 1973: The Scotia arc and Antarctic margin; *in* Nairn, A.E.M. and Stehli, G.F. (eds.) *The Ocean Basins and Margins, Vol. 1, The South Atlantic.* Plenum Press, New York, 171-246.

DICKINSON, W.R. and SUCZEK, C.A., 1979: Plate tectonics and sandstone compositions. *Am. Ass. Petrol. Geol., Bull., 63,* 2164-82.

EDWARDS, C.W., 1982: New paleontologic evidence of Triassic sedimentation in West Antarctica; *in* Craddock, C. (ed.) *Antarctic Geoscience.* Univ. Wisconsin Press, Madison, 325-30.

FOLK, R.L., 1968: *Petrology of Sedimentary Rocks.* Hemphill's, Austin.

HALLE, T.G., 1913: The Mesozoic flora of Graham Land. *Wiss. Ergebn. schwed. Sudpolar edped., 1901-1903, 3, (14),*1-123.

JOHNSON, H.D., 1978: Shallow siliciclastic seas; *in* Reading, H.G. (ed.) *Sedimentary Environments and Facies.* Elsevier, New York, 207-58.

IMLAY, R.W. and KAUFFMAN, E.G. (in press): Dominant Jurassic mollusca from the Lassiter Coast, southern Antarctic Peninsula. *U.S. Geol. Surv. Prof. Pap.*

KAMENEV, E.M. and ORLENKO, E.M., 1982: Metamorphism of sedimentary formations on the Lassiter Coast; *in* Craddock, C. (ed.) *Antarctic Geoscience.* Univ. Wisconsin Press, Madison, 357-61.

LAUDON, T.S., 1972: Stratigraphy of eastern Ellsworth Land; *in* Adie, R.J., (ed.) *Antarctic Geology and Geophysics.* Universitetsforlaget, Oslo, 215-23.

LAUDON, T.S., LACKEY, L.L., QUILTY, P.G. and OTWAY, P.M., 1969: Geology of eastern Ellsworth Land (sheet 3, eastern Ellsworth Land); *in* Bushnell, V.C. and Craddock, C. (eds.) Geological Maps of Antarctica, *Antarct. Map Folio Ser.* Folio 12, P1.III.

MOORE, D.E. and LIETZ, J.G.: 1979: Chessboard twinned albite from Franciscan metaconglomerate of the Ciabolo Range, California. *Am. Mineral., 64,* 329-36.

QUILTY, P.G., 1970: Jurassic ammonites from Ellsworth Land, Antarctica. *J. Paleontol., 44,* 110-6.

QUILTY, P.G., 1972a: Middle Jurassic brachiopods from Ellsworth Land, Antarctica. *N.Z. J. Geol. Geophys., 15,* 147-8.

QUILTY, P.G., 1972b: *Pentacrinites* and (?)*Apiocrinus* rom the Jurassic of Ellsworth Land, Antarctica. *Neues Jahrb. Geol. Palaontol. Mn., 8,* 484-9.

QUILTY, P.G., 1977: Late Jurassic bivalves from Ellsworth Land, Antarctica. Their systematics and palaeogeographic implications. *N.Z. J. Geol. Geophys., 20,* 1033-80.

QUILTY, P.G., 1982: Tectonic and other implications of Middle-Late Jurassic rocks and marine faunas from Ellsworth Land, Antarctica; *in* Craddock, C. (ed.) *Antarctic Geoscience.* Univ. Wisconsin Press, Madison, 669-78.

QUILTY, P.G. (in press): Bajocian bivalves from Ellsworth Land, Antarctica. *N.Z. J. Geol. Geophys.*

ROWLEY, P.D. and WILLIAMS, P.L., 1982: Geology of the northern Lassiter coast and southern Black Coast, Antarctic Peninsula; *in* Craddock, C. (ed.) *Antarctic Geoscience.* Univ. Wisconsin Press, Madison, 339-48.

ROWLEY, P.D., SCHMIDT, D.L. and WILLIAMS, P.L.: 1982: The Mount Poster Formation, southern Antarctic Peninsula. *U.S. Antarct. J., 17.*

ROWLEY, P.D., VENNUM, W.R., KELLOGG, K.S., LAUDON, T.S., CARRARA, P.E., BOYLES, J.M. and THOMSON, M.R.A. (this volume): Geology and plate tectonic setting of the Orville Coast and eastern Ellsworth Land.

SINGLETON, D.G., 1980: The geology of the central Black Coast, Palmer Land. *Scient. Rep. Br. antarc. Surv., 102.*

SMELLIE, J.L., 1981: A complete arc-trench system recognised in Gondwana sequences of the Antarctic Peninsula region. *Geol. Mag., 118,* 139-59.

STARKEY, J., 1959: Chessboard albite from new Brunswick. *Can. Geol. Mag., 96,* 141-5.

STEVENS, G.R., 1967: Upper Jurassic fossils from Ellsworth Land, West Antarctica, and notes on Upper Jurassic biogeography of the South Pacific region. *N.Z. J. Geol. Geophys., 10,* 343-93.

THOMSON, M.R.A., 1980: Late Jurassic ammonite faunas from the Latady Formation, Orville Coast. *U.S. Antarct. J., 15,* 28-30.

THOMSON, M.R.A., 1982: Mesozoic paleogeography of West Antarctica; *in* Craddock, C. (ed.) *Antarctic Geoscience.*Univ. Wisconsin Press, Madison, 331-8.

THOMSON, M.R.A. (this volume): Jurassic ammonite biostratigraphy of the Orville Coast.

THOMSON, M.R.A., LAUDON, T.S. and BOYLES, J.M., 1978: Stratigraphical studies in Orville Coast and Eastern Ellsworth land. *U.S. Antarct. J., 13,* 9-10.

WILLIAMS, P.L. and ROWLEY, P.D., 1972: Composition of Jurassic sandstones, Lassiter Coast, *U.S. Antarct. J., 7,* 145-46.

WILLIAMS, P.L., SCHMIDT, D.L., PLUMMER, C.C. and BROWN, L.E., 1972: Geology of the Lassiter Coast area, Palmer Land. A preliminary report; *in* Adie, R.J., (ed.) *Antarctic Geology and Geophysics.* Universitetsforlaget, Oslo, 143-8.

# LATE JURASSIC AMMONITES FROM THE ORVILLE COAST, ANTARCTICA

M.R.A. Thomson, *British Antarctic Survey, Madingley Road, Cambridge, CB3 OET, U.K.*

*Abstract* Jurassic ammonite faunas in the Latady Formation of the Orville Coast, southernmost Antarctic Peninsula region, are almost entirely composed of perisphinctids and range from Middle to Late Jurassic. The Late Jurassic forms are reviewed and found to be mainly Tithonian, although the possibility that some may range as low as the Kimmeridgian cannot be ruled out. Existing fossil collections were made during the course of a reconnaissance geological survey and, until more detailed field work is done, approximate ages can be assigned on an area basis only. Rocks in the Bean Peaks—Hauberg Mountains—Peterson Hills are the oldest and mainly Early Tithonian to early Late Tithonian, those of the Wilkins Mountains area are early Late Tithonian, and the youngest are those of Cape Zumberge (latest Tithonian).

The Latady Formation is an extensive sequence of back-arc (*sensu lato*) sedimentary rocks, which crop out along the eastern margin of Palmer Land from approximately 72°30′S southward through Lassiter and Orville coasts and into the Behrendt Mountains of eastern Ellsworth Land (Figures 1 and 2; Rowley et al., this volume and references therein). Metasedimentary rocks in northeastern Palmer Land may possibly represent a northward extension of this same sedimentary formation (Taylor et al., 1979). Mainly shallow marine in origin (Laudon et al., this volume), the Latady Formation is abundantly fossiliferous in places. The majority of the fossils are molluscs (bivalves, belemnites and ammonites) and everywhere they indicate a Late Jurassic age for the formation (Stevens, 1967; Quilty, 1977; Thomson, 1980; Crame, 1981), except in the Behrendt Mountains where Middle Jurassic ammonites have been found (Quilty, 1970). In view of this, the suggestion that Late Cretaceous species of the bivalve *Inoceramus* are present at Rivera Peaks (74°48′S, 62°50′W), southern Lassiter Coast (Kamenev and Orlenko, 1982), should be viewed with caution, particularly since the specimens concerned were said to be poorly preserved and have not been described.

Although the Middle Jurassic faunas of southern Behrendt Mountains were recollected in the course of the present investigations, there is little to add at present to the descriptions given by Quilty (1970). Ammonites from the Lassiter Coast have been indentified by Dr R.W. Imlay as *Subdichotomoceras (?), Aulacosphinctoides, Virgatosphinctoides* and *Kossmatia* (Rowley and Williams, 1982), and the present discussion is concerned only with species from the Orville Coast, all of which have Late Jurassic affinities.

Ammonites occur sporadically in the Latady Formation, throughout the whole of the Orville Coast area, but they are most common in the Hauberg and Wilkins Mountains and at Cape Zumberge; areas which were situated at some distance from the palaeocoastline and therefore fully marine in facies. Ammonites are correspondingly rare in the Sweeney Mountains where the facies is nearshore or even lagoonal. However, marine bivalves and belemnites have been found farther north at Sky-Hi Nunataks (74°50′S, 71°30′W) (new data) and at Lyon Nunataks (74°50′S, 73°50′W) (Stevens, 1967); these may represent a short-lived advancement of the sea or possibly a major indentation of the coastline. Only two ammonites have been obtained from the Sweeney Mountains and neither is identifiable more closely than to some form of perisphinctid. All the specimens referred to here came from the area between Bean Peaks and McCaw Ridge, and from Witte Nunataks and Cape Zumberge (Figure 2).

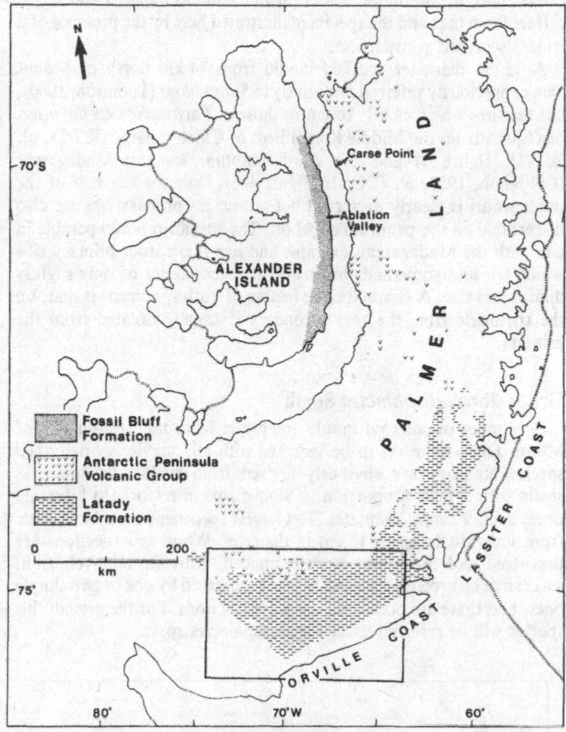


**Figure 1. Location map for the Orville Coast area, showing its position in relation to the volcanic rocks of the Mesozoic arc and the partly contemporaneous fore-arc sedimentary rocks of the Fossil Bluff Formation. The box shows the area of Figure 2.**

## PALAEONTOLOGY

The classification used here follows that outlined by Donovan et al. (1980).

### Family Oppeliidae Bonarelli

The oppeliids are represented by one specimen from an unnamed group of nunataks 11 km south of the Wilkins Mountains; this is the only Late Jurassic ammonite among the whole collection that does not belong to the Superfamily Perisphinctaceae. It is an external mould fragment from a compressed strongly involute ammonite about 35 mm in diameter. It has smooth whorls with a rounded venter and faint falcoid ribs on the ventral half of the outer whorl only. In general morphology it accords well with members of the Subfamily Streblitinae, and the lack of ventral ornament suggests affinities with the genus *Pseudoppelia* (Leanza, 1946) from the Lower Kimmeridgian of Argentina.

### Family Aspidoceratidae Zittel

### Genus *Aspidoceras* Zittel

One incomplete and distorted external mould from northern Wilkins Mountains, and two small ones from a group of nunataks 11 km south of the main mountains, clearly belong to the genus *Aspidoceras*. The best (Figure 3e) is from the latter locality. Although no more than 30 mm in diameter, is shows a close resemblance to *A. euomphalum* Steuer (Steuer, 1897, pl. V, figures 1-4; Leanza, 1980, pl. 8, figures 1a and b) from the Middle Tithonian of Argentina.

### Family Ataxioceratidae Buckman

### Genus *Katroliceras* Spath

Several internal and external moulds in a partly decalcified sandstone from southern Hauberg Mountains (e.g. Figure 3d) are cautiously referred to *Katroliceras*. None is well preserved but all have strongly inflated to depressed whorls, with coarse widely spaced ribs that bifurcate as they pass over the venter; some intercalated ribs are also present. Although no late stage with trifurcate ribs is present, their morphology agrees well with that of smaller examples of *Katroliceras* Spath, and particularly with the earlier stages of *K. pottingeri* from Kenya (Spath, 1927-33, pl. CII, figures 5a and b) and from the *Hybonoticeras hybonotum* zone of Madagascar (Collignon, 1959, pl. CXXI, figure 455).

## Genus *Pachysphinctes* Dietrich

Figure 3c illustrates one of two internal mould fragments from the Hauberg Mountains which may be allied to *Pachysphinctes americanensis* (Leanza, 1980). This specimen came from the western side of Mount Dewe and the second, a slightly smaller specimen, came from a ridge 12 km to the northeast. Coarse primary ribs give rise to two or three stout secondaries which pass over the venter. Several intercalated ribs are present and there is a coarse simple rib where the ornament is interrupted by a constriction. This ornament compares well with that on the middle part of the half outer whorl preserved on the Argentine specimen (Leanza, 1980, pl. 7, figure 1a). The second Antarctic fragment has closer spaced ribbing which is similar to that on the earlier part of the outer whorl on the same example. However, it differs from that and the specimen illustrated here by the presence of a wide discordant constriction.

A 15 cm diameter crushed mould from 14 km north of Mount Dewe, previously referred tentatively to *Subplanites* (Thomson, 1980), has features which closely resemble those of *Pachysphinctes linguiferous* Spath from the Middle Katrol Beds of Cutch (Spath, 1927-33, pl. XCVII, figure 1) and the *H. hybonotum* zone of Madagascar (Collignon, 1959, pl. CXVIII, figure 446). Only the last half of the outer whorl is clearly preserved but prominent primary ribs are also discernible on the penultimate whorl. The specimen is comparable in size with the Madagascan example and has sharp stout primary ribs which tri- and sometimes bifurcate into secondaries of only slightly diminished size. A characteristic feature of both specimens is that, on the trifurcate ribs, the rear secondary is usually isolated from the primary.

## Genus *Torquatisphinctes* Spath

A number of internal mould fragments from the western side of Mount Dewe have yet to be matched with any previously described species but they are obviously derived from evolute perisphinctid shells with a high proportion of simple ribs in relation to bifurcate ones, as in *Torquatisphinctes*. The largest specimens must have come from individuals up to 15 cm in diameter. Whorl cross-sections are depressed and the venter is well rounded. Ribs are relatively thin, successive bifurcate ones usually being separated by one or two simple ones, and there are occasional faint constrictions. For the present this species will be referred to as *Torquatisphinctes* sp. *a*.

A second species (*Torquatisphinctes* sp. *b*) may be represented by some small strongly crushed individuals that occur in baked shales at Witte Nunataks. They have evolute multispiral shells with an ornament of fine regularly spaced radial ribs, some of which bifurcate high on the flank. Constrictions are present on the outer whorl of the largest specimen (ca 45 mm diameter). Attempting a precise identification of such poor material is not warranted but a provisional placing in *Torquatisphinctes* is not unreasonable.

Contrary to expectations, one of the best preserved ammonites (Figure 3b) is one of the most difficult to identify. Its shell is *Discosphinctes*-like, being only moderately evolute with compressed whorls that embrace one another by about one half, yet its ornament more closely resembles that of *Torquatisphinctes* and consists of fine bifurcate ribs with a large number of interspersed simple ones. Trifurcate ribs are present only where the ornament has been disrupted by oblique prorsiradiate constrictions. A specimen from the same bed is more densely ribbed, although along the same general pattern, and has a late-stage with a modified ornament of evenly-spaced sharp ribs, set about two rib-widths apart. This species occurs at the same locality as *Pachysphinctes* cf. *linguiferous* (above) but a few tens of metres lower in the succession. It is previsionally referred to *Torquatisphinctes (?)* sp. *c*.

## Genus *Subdichotomoceras* Spath

Fragments of internal and external moulds from northeastern Bean Peaks, eastern Hauberg Mountains and northwestern Peterson Hills share a common ornament of prominent sharp bifurcate ribs that are set about two rib-widths apart at all stages of growth preserved (e.g. Figure 3f). The shell is evolutely coiled and whorl cross-sections are strongly inflated to depressed from an early stage. Taken as a group they show the general characteristics of *Subdichotomoceras*, notably the species *Perisphinctes biplicatus* Uhlig (1903-10, pls LVII and LIX) from the Spiti Shales of the Himalayas and *S. araucanense* Leanza (1980, pl. 6, figures 1 and 3) from the Middle Tithonian of Argentina.

## Genus *Virgatosphinctes* Uhlig

There are a number of specimens in the collection which may be referred to *Virgatosphinctes;* three of the more readily identifiable forms are noted here. The first is represented by a single well-preserved external/internal mould, from a locality 15 km west of the western end of McCaw Ridge. It has an ornament of fine, dense,

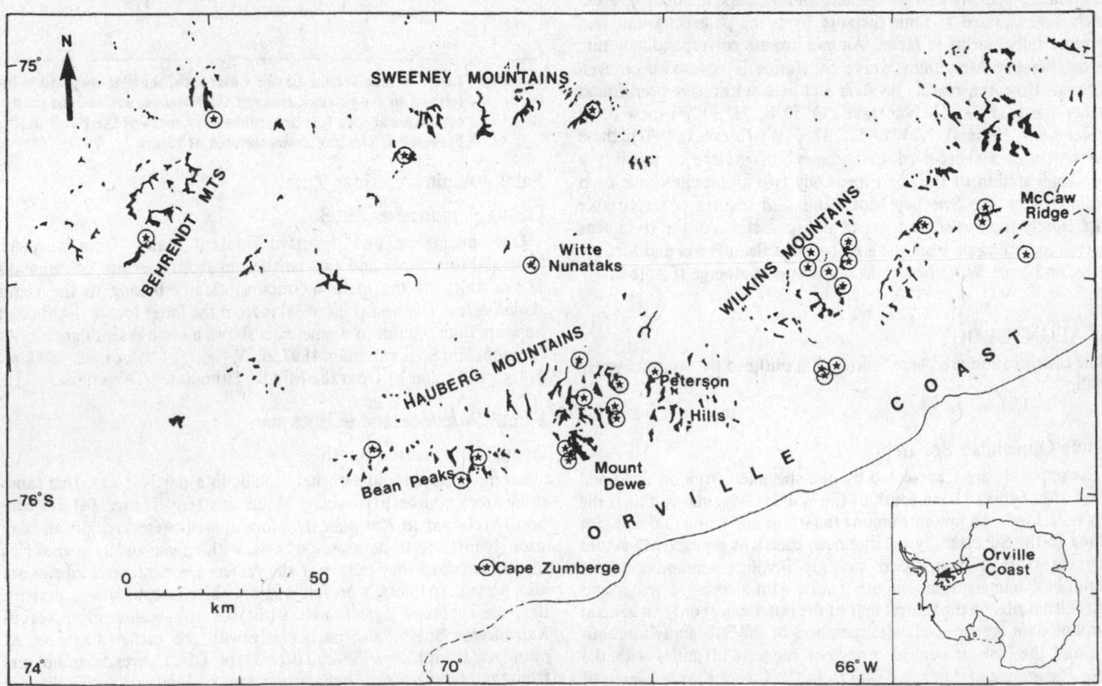


**Figure 2. Schematic outcrop map of the Orville Coast area, showing the main ammonite-bearing localities known (stars).**

Figure 3. Representative ammonites from the Orville Coast. (a) *Virgatosphinctes* aff. *saherense* Spath. (b) *Torquatisphinctes (?)* sp. c.. (c) *Pachysphinctes* aff. *americanensis* Leanza. (d) *Katroliceras* sp. (e) *Aspidoceras* aff. *euomphalum* (Steuer). (f) *Subdichotomoceras* sp. (g) *Kossmatia* aff. *tenuistriata* (Gray). (h) *Kossmatia (?)* sp. nov. (i) *Blanfordiceras* cf. *wallichi* (Gray). (j) *Berriasella (?)* sp. All illustrations are x .75 natural size.

narrowly bifurcate ribs at all stages of growth. Although slightly distorted by tectonism, the shell probably had an original diameter of about 85 mm with whorls whose cross-sections were wider than high. On the inner whorls the venter is flatly arched but on the last preserved part of the outer whorl (just after the last septum) it is more triangular. The moderately evolute shell and fine ornament recall the group of *V. denseplicatus* (Waagen), well described from the Spiti Shales (Uhlig, 1903-1910) and the Lower Umia Group of Cutch (Spath, 1927-33), and also known from Ablation Valley in Alexander Island (Thomson, 1979). Although the present specimen seems to lack the abundant simple ribs present on examples of *denseplicatus* (*s.s.*) from Cutch (Spath, 1927-33) and Madagascar (Collignon, 1960), the species is a variable one and there are very few simple ribs on specimens illustrated by Uhlig from the Spiti Shales (eg. Uhlig, 1903-1910, pl. LV). The lack of simple ribs on the present specimen and the point of bifurcation of the ribs just below the umbilical seam suggest closest affinities to *V. denseplicatus* var. *blakei* (Spath, 1927-33, pl. XC, figure 1; Collignon, 1960, pl. CLIV, figure 619).

The specimen in Figure 3a, from the nunataks 11 km south of the Wilkins Mountains, is a variety of the *denseplicatus* group with more widely spaced ribs, all of which are bifurcate. It compares tolerably well with *V. saherense* from the Lower Umia Group of Cutch (Spath, 1927-33, pl. XCVI, figure 1), which was considered by Spath as being transitional with *V. denseplicatus* var. *blakei*.

Fragments from the western side of Mount Dewe include at least two which probably belong to a third form, *V. frequens* (Oppel). One is a piece of external mould from a whorl flank, showing narrowly branching bi- and trifurcate ribs, typical of those on examples from the Himalayas (Uhlig, 1903-10), Madagascar (Collignon, 1960, pl. CLIV, figures 619, 620) and Alexander Island (Thomson, 1979, pl. Va-d). A second specimen, from a slightly earlier growth stage, has bifurcate and simple ribs only but they are thin and rounded as on the first specimen and the secondaries diverge narrowly.

### Family Neocomitidae Salfeld

### Genus *Kossmatia* Uhlig

Distorted fragments bearing the fine, dense projected ribbing of *Kossmatia* occur at at least four localities in the Wilkins Mountans. On most specimens the ribbing is consistently fine throughout all growth stages (Figure 3g). The ribs curve concavely as they pass over the umbilical rim, flex feebly forwards as they cross the flank, and then pass over the venter in a forward-facing V. Many branch narrowly at or below mid-flank, whereas others remain simple with occasional intercalated ventral ribs. The fragments compare well with the type species, *K. tenuistriata* (Gray) from the Spiti Shales of the Himalayas (cf. Uhlig, 1903-1910, pl. XVI, figures 3a-b). A few fragments seem to be more coarsely ribbed than the example illustrated here and may represent a separate species which is closer to *K. carsensis* (Thomson, 1975, figures 4a-g) from Carse Point, western Palmer Land.

A quite distinct species, which may also belong to *Kossmatia,* is represented by three fragments from an isolated nunatak on the northwestern side of the Wilkins Mountains. The specimens indicate a moderately evolute, multispiral shell having compressed whorls that are only feebly ornamented (Figure 3h). Moderately spaced major ribs bifurcate near the venter and are separated by two or three short intercalated ventral ribs; all cross the venter in a gently projected curve. On the largest specimen, an arcuate whorl fragment from an individual about 60-70 mm in diameter, the dorsal part of the ribbing is reduced so that only the ventral zone of intercalate ribs is clearly visible. The generic placing of these specimens is uncertain but the projected ribbing is *Kossmatia*-like and they are provisionally regarded as a new species of that genus.

### Genus *Blanfordiceras* Cossman

Specimens from Cape Zumberge, typified by the one in Figure 3i, may be referred with little hesitation to the well-known *Blanfordiceras wallichi* (Gray). The shell form, flexuous bifurcate ribs and interspersed simple ribs, and faint ventral groove of the present specimens are all typical of the named species. A closely related form is already known from Alexander Island (Thomson, 1979, pl. VII a-c), and the species proper has been described many times from Madagascar, the Salt Range, the Himalayas and Indonesia (see Helmstaedt, 1969 for a recent synonymy).

### Genus *Berriasella* Uhlig

Associated with *Blanfordiceras* cf. *wallichi* at Cape Zumberge is another berriasellid, whose adult stage is distinguished from the *B.* cf. *wallichi* by its more radial ribs that branch lower on the flank (Figure 3j); also its whorls seem to be relatively higher and more compressed. Earlier stages, however, are more difficult to separate from *Blanfordiceras;* they have a median ventral groove which interrupts the ribbing and the whorl cross section is more inflated. The largest specimen was still septate at a diameter of about 20 cm. Whether these specimens should be regarded as distinct from *Blanfordiceras* is uncertain at present but there is a tolerable likeness to "*Berriasella*" *subprivasensis* Krantz from Ablation Valley, eastern Alexander Island (Thomson, 1979, pl. VII-i), except that the Cape Zumberge specimens show no signs of occasional pairing of major ribs near the umbilical rim.

## CONCLUSIONS

Any biostratigraphical assessment of the Latady Formation is hampered by the patchy distribution of its outcrop (cf. Figure 2 and the geological sketch map in Figure 1), the lack of marker beds, and its complicated structure (Kellogg, 1979). No top or bottom to the formation has been identified and the maximum measured section in any one outcrop is 850 m, whereas the true thickness must be considerably more (Laudon et al., this volume). In the Lassiter Coast area assessments of the bivalve and ammonite faunas suggested that the formation was "late Kimmeridgian to late Tithonian" in age (Rowley and Williams, 1982). A preliminary estimate of the age of the ammonite faunas from the Orville Coast was Kimmeridgian—Tithonian (Thomson, 1980). However, in view of the data presented here, and revisions of Late Jurassic biostratigraphy (eg. Enay, 1973, table 1), this last age assignment needs to be reconsidered.

In Enay's sheme the Middle Katrol Beds of Cutch and the *H. hybonotum* zone of Madagscar, regarded by Spath (1927-33) and Collignon (1959) as Middle Kimmeridgian, are correlated with the lowest part of the Tithonian. Thus most of the faunas from the Orville Coast, originally referred to the Kimmeridgian by Thomson (1980), should probably be placed in the Lower Tithonian. It is still possible that some of the strata in this poorly controlled sequence may be partly Kimmeridgian in age but proof awaits more detailed field work. At present only crude age estimates can be made on an area basis.

The oldest faunas are probably those of the Bean Peaks-Hauberg Mountains-Peterson Hills area, where the presence of *Katroliceras* sp., *Pachysphinctes* aff. *americanensis, Torquatisphinctes* spp. and *Subdichotomoceras* sp. favours an Early to earliest Late Tithonian age (using a twofold rather than a threefold subdivision of the Tithonian). The possibility that the ranges of some of the specimens present may extend down into the Kimmeridgian cannot be ruled out yet.

Two faunas can be recognised in the Wilkins Mountains area and the nunataks 11 km to the south. The first contains *Aspidoceras* sp., *Virgatosphinctes denseplicatus* var. *blakei (?), Kossmatia* aff. *tenuistriata, Kossmatia (?)* sp. nov. and undetermined ataxioceratids, and the second *Pseudoppelia (?)* sp., *Aspidoceras* aff. *euomphalum* and *V.* aff. *saherense.* Both are consistent with a mainly early Late Tithonian age but the former could be slightly younger. The youngest fauna is that at Cape Zumberge, where *Blanfordiceras* cf. *wallichi* and *Berriasella (?)* sp. indicate a latest Tithonian age.

Thus the pattern of an overall younging towards the Weddell Sea, noted in the Lassiter Coast area (Rowley and Williams, 1982), is not so clearly demonstrated here. More significant may be the juxtaposition along strike of Early Tithonian faunas in the Hauberg Mountains area and the Late Tithonian faunas of the Wilkins Mountains, suggesting that the two may be separated by a major fault. In view of this hypothesis it would be useful to investigate further the origin of the marked northeast trending geomorphological break between the Behrendt and Sweeney mountain. Deep north-trending subglacial troughs, which may also be fault-controlled, have been identified on the basis of satellite imagery and geophysical data in the area west of the Behrendt Mountians (cf. Doake et al, this volume, figure 1).

*Acknowledgements* The author is indebted to Dr P.D. Rowley (U.S. Geological Survey) for the opportunity to work on the Orville Coast with a USGS party, funded by National Science Foundation grant DPP 76-12557. He and other members of the party: J.M. Boyles, P.E. Carrara, K.S. Kellogg, T.S. Laudon and W.R. Vennum were responsible for collecting many of the specimens on which this summary is based.

## REFERENCES


COLLIGNON, M., 1959: *Atlas des fossiles caracteristiques de Madagascar (Ammonites). V. Kimmeridgien.* Service géologique, Tananarive.

COLLIGNON, M., 1960: *Atlas des fossiles caracteristiques de Madagascar. VI. Tithonique.* Service géologique, Tananarive.

CRAME, J.A., 1981: Preliminary bivalve zonation of the Latady Formation. *U.S. Antarct. J., 16,* 8-10.

DOAKE, C.S.M., CRABTREE, R.D. and DALZIEL, I.W.D (this volume): Subglacial morphology between Ellsworth Mountains and Antarctic Peninsula: new data and tectonic significance.

DONOVAN, D.T., CALLOMON, J.H. and HOWARTH, M.K., 1980: Classification of the Jurassic ammonites; *in* House, M.R. and Senior, J.R. (eds.) *The Ammonoidea.* Academic Press, London, 101-55.

ENAY, R., 1973: Upper Jurassic (Tithonian) ammonites; *in* Hallam, A. (ed.) *Atlas of palaeobiogeography.* Elsevier Scientific Publishing Company. Amsterdam, 297-307.

HELMSTAEDT, H., 1969: Eine Ammoniten-Fauna aus den Spiti-Schiefern von Muktinath in Nepal. *Zitteliana, 1,* 63-88.

KAMENEV, E.N. and ORLENKO, E.M., 1982: Metamorphism of sedimentary formations on the Lassiter Coast; *in* Craddock, C. (ed.) *Antarctic Geoscience.* Univ. Wisconsin Press, Madison, 357-61.

KELLOGG, K.S., 1979: Structural geology of Orville Coast and eastern Ellsworth Land. *U.S. Antarct. J, 14,* 19-21.

LAUDON, T.S., THOMSON, M.R.A., WILLIAMS, P.L., MILLIKEN, K.L. and BOYLES, J.M. (this volume): The Jurassic Latady Formation, southwestern Antarctic Peninsula.

LEANZA, A.F., 1946: Las oppelias de Chacy-Melehue en el Neuquen: *Streblites (Pseudoppelia) oxynotus* subgen. et sp. nov. *Revta. Asoc. geol. argent., 1,* 63-72.

LEANZA, H.A., 1980: The Lower and Middle Tithonian ammonite fauna from Cerro Lotena, Province of Neuquen, Argentina. *Zitteliana, 5,* 3-49.

QUILTY, P.G., 1970: Jurassic ammonites from Ellsworth Land Antarctica. *J. Paleontol., 44,* 110-6.

QUILTY, P.G., 1977: Late Jurassic bivalves from Ellsworth Land, Antarctica: their systematics and palaeogeographic implications. *N.Z. J. Geol. Geophys., 20,* 1033-80.

ROWLEY, P.D., VENNUM, W.R., KELLOGG, K.S., LAUDON, T.S., CARRARA, P.E., BOYLES, J.M. and THOMSON, M.R.A. (this volume): Geology and plate tectonic setting of the Orville Coast and eastern Ellsworth Land, Antarctica.

ROWLEY, P.D. and WILLIAMS, P.L., 1982: Geology of the northern Lassiter Coast and southern Black Coast, Antarctica; *in* Craddock, C. (ed.) *Antarctic Geoscience.* Univ. Wisconsin Press, Madison, 339-56.

SPATH, L.F., 1927-33: Revision of the Jurassic cephalopod fauna of Kachh (Cutch). *Mem. geol. Surv. India, Palaeont. indica, N.S., 9,* 945pp.

STEUER, A., 1897: Argentinische Jura-Ablagerungen. *Palaont. Abh., N.F., 3,* 129-222.

STEVENS, G.R., 1967: Upper Jurassic fossils from Ellsworth Land, West Antarctica, and notes on Upper Jurassic biogeography of the South Pacific region. *N.Z. J. Geol. Geophys., 10,* 345-93.

TAYLOR, B.J., THOMSON, M.R.A. and WILLEY, L.E., 1979: The geology of the Ablation Point—Keystone Cliffs area, Alexander Island. *Scient. Rep. Br. antarct. Surv., 82.*

THOMSON, M.R.A., 1975: Upper Jurassic Mollusca from Carse Point, Palmer Land. *Bull. Br. antarct. Surv., 41-42,* 31-42.

THOMSON, M.R.A., 1979: Upper Jurassic and Lower Cretaceous ammonite faunas of the Ablation Point area, Alexander Island. *Scient. Rep. Br. antarct. Surv., 97.*

THOMSON, M.R.A., 1980: Late Jurassic ammonite faunas from the Latady Formation, Orville Coast. *U.S. Antarct. J., 15,* 28-30.

UHLIG, V., 1903-1910: The fauna of the Spiti Shales. *Mem. geol. Surv. India, Palaeont. indica, Ser, 15, 4,* 1-395.

# A NEW MARSUPIAL FROM SEYMOUR ISLAND, ANTARCTIC PENINSULA

M.O. Woodburne, *Department of Earth Sciences, University of California, Riverside, Cal. 92521, U.S.A.*

W.J. Zinsmeister, *Institute of Polar Studies, The Ohio State University, Columbus, Ohio 43210, U.S.A.*

*Abstract* Remains of the first fossil land mammal yet found in Antarctica were discovered during the austral summer of 1982. The fossils were recovered from the Upper Eocene La Meseta Formation on Seymour Island, in the northeastern tip of the Antarctic Peninsula. The specimens comprise two jaw fragments and two isolated teeth of an animal that belongs to the extinct South American marsupial family Polydolopidae, and appear to be most closely related to the genus *Polydolops*. Examination of the specimens indicate that at least three individuals were preserved in these deposits, which suggests that a sizable parent population of polydolopids was present on Seymour Island about 40 million years ago. The fossils were recovered from a unit of light grey, finely laminated, crossbedded and loosely consolidated arkosic sandstone, located approximately 510 m stratigraphically above the base of the La Meseta Formation. The marsupial fossils were associated with numerous extinct marine molluscs, sand crab-like arthropods, bones of teleost fish, penguins, and whales, and numerous sharks teeth. Faunal and sedimentological data suggest that the marsupial bearing strata represent a former beach deposit. The discovery of marsupials in Antarctica supports theories that predicted their past presence there and strengthens the proposal that Antarctica served as an intermediate pathway to land mammal dispersal between South America and Australia in the Late Cretaceous or Early Cenozoic. The presence of the South American family Polydolopidae constrains geophysical interpretations as to the connection between Antarctica and South America. The new fossil evidence indicates that either a continuous land connection, or at least a series of closely spaced islands, was present between these two continents in the Paleocene and possibly in the Early to Late Eocene.

South America now contains a land vertebrate fauna dominated by placental mammals, but marsupials are also present, and in the past were more abundant and diverse than they are now (Simpson, 1978; Marshall et al., 1982). On the other hand, Australia is presently dominated by marsupials. In the Neogene (Stirton et al., 1968; Archer and Bartholomai, 1978), and in the late Palaeogene (Tedford et al., 1975), Australia was exclusively populated by marsupials. Australia is now almost completely isolated from other continents of the world, and is thought to have broken away from Antarctica about 56 Ma (Zinsmeister, 1982). Thus, general theories to explain the distribution of the world's marsupials, partially summarised above, have postulated the still earlier presence of marsupials on both Australia and Antarctica at some time from the Late Cretaceous to the Early Cenozoic (e.g. Tedford, 1974).

The specimens discussed here were recovered during the austral summer of 1982, when M.O. Woodburne and W.R. Daily, University of California, Riverside, participated in an investigation of the geology, stratigraphy, invertebrate and vertebrate palaeontology of Seymour Island, Antarctic Peninsula (Figure 1), under the leadership of W.J. Zinsmeister (Woodburne and Zinsmeister, 1982).

All dimensions are metric. Radiometric age assignments have been recalculated, where necessary, to the IUGS constants as presented in Steiger and Jaeger (1977). The classification scheme used here follows Clemens and Marshall (1976). Dental terminology follows that of Paula Couto (1952) and Marshall (1982).

UCR and RV-8200: Specimen number of the fossil mammal collections, Department of Earth Sciences, University of California, Riverside.

## Antarctic Polydolopids

The specimens were collected at RV-8200, on the northwest side of the prominent meseta at the northern end of Seymour Island (Figure 1), from the upper part of the La Meseta Formation (Figure 2). The fossil remains consist of a left mandible, with $P_3$-$M_2$ and alveolus for $M_3$ (UCR 20910; Figure 3), a right mandible fragment with $P_3$-$M_1$ (UCR 20911), and isolated $LM^1$ (UCR 20912), and $LM^2$ (UCR 20913). Differences in occlusal wear suggest that at least three individuals are represented. The specimens appear to pertain to a single species that most closely resembles members of the genus *Polydolops*, as described by Simpson (1948), and Marshall (1982). The Antarctic species differs from *Polydolops*, however, in the absence of $P_2$ (sometimes absent in one species of *Polydolops*; Marshall, 1982), the more elongate proportions of the molars, and in details of their coronal morphology. Until now, the group (Polydolopidae) to which the Antarctic material pertains was considered to have been restricted to South America (e.g. Marshall, 1982). Relationship to any of the other known polydolopid genera appears tenuous on present evidence, and studies now underway may result in the Antarctic species being nominated as a new genus.

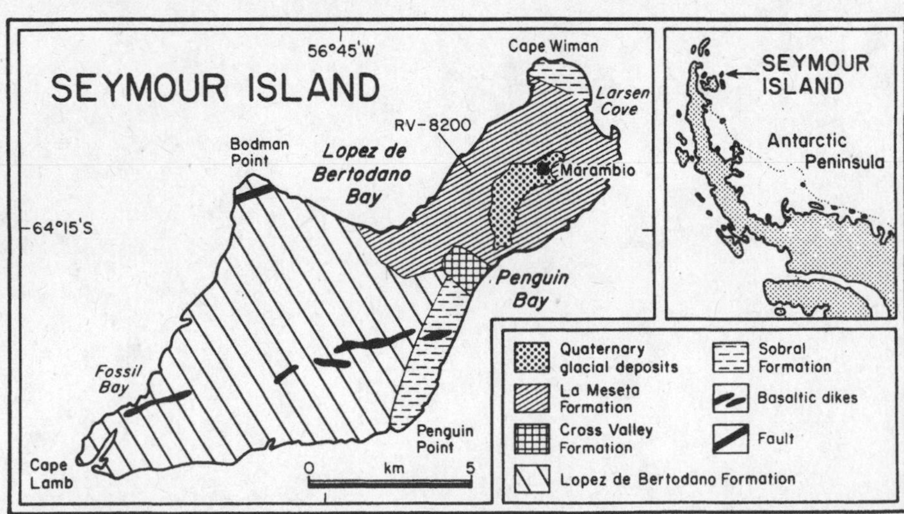


**Figure 1. Map of geographic and geologic features of Seymour Island, Antarctic Peninsula, and approximate location of RV-8200 which produced the fossil marsupial remains reported upon here. The Sobral and Lopez de Bertodano Formations are of Cenomanian and Maestrichtian age respectively. The Cross Valley Formation is thought to be of Paleocene age, and the La Meseta Formation of Eocene age.**

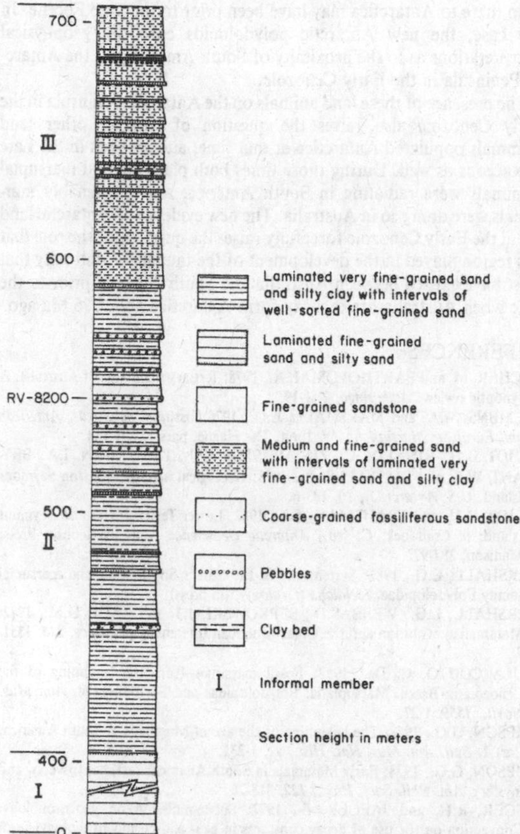


**Figure 2. Stratigraphic column of the La Meseta Formation (Eocene), Seymour Island, Antarctic Peninsula, showing distribution of lithologies, distribution of units I—III, and thickness in metres. The polydolopid remains occur approximately 550 m above the base of the formation at RV-8200.**

## Stratigraphy

The polydolopid material was recovered from a unit of thinly bedded arkosic sandstone of the La Meseta Formation, which comprises the uppermost unit of the thick, nominally marine, sequence that ranges in age from Late Cretaceous to Late Eocene (?or Early Oligocene) on Seymour Island. The locality (RV-8200) occurs approximately 550 m above the base of the La Meseta Formation (Figure 2) in one of the shell banks that characterise Unit II, as defined by Elliot and Trautman (1982). Unit II consists of laminated beds of silty sandstone with interbedded units of coarser-grained, conglomeratic, richly fossiliferous sandstone beds (shell banks). The silty sandstone beds are characterised by finely laminated, wavy bedding, with occasional large, scoured channels. Vertical and obliquely oriented burrows are common, together with varying degrees of bioturbation. Flaser bedding and ripple laminations locally are well developed. Concretionary horizons and scattered wood and coalified plant remains occur throughout these beds.

## Age and Correlation

The marine molluscs associated with the polydolopid are *Antarctodarwinella nordenskjoldi* (Wilckens), *Sthruthioptera camachoi* Zinsmeister, *Lahillia larseni* (Sharman and Newton), *Eutrephoceras argentinae* del Valle and Fourcade, *Aturia* sp., *Eurhomalia antarctica* (Wilckens), *Nielo* n. sp., and a number of undescribed gastropods. These indicate a Late Eocene age for the contemporaneous land mammals (Bartonian in the European and Runangan in the New Zealand marine chronologies; Woodburne and Zinsmeister, 1982).

## Depositional Setting

Many seemingly promising sites in other parts of Seymour Island were prospected unsuccessfully for fossil land mammals in 1982. Prior interpretations of the depositional setting represented by the La Meseta Formation (Elliot et al., 1975; Elliot and Trautman, 1982) suggested that it represented a prograding marine delta, with shell banks reflecting high-energy conditions. However, as outlined below it appears likely, that the site which yielded the fossil mammals represents a former beach deposit. In contrast to many shell banks in the La Meseta Formation, the locality where the marsupial remains were discovered apparently represents a lower-energy facies as indicated by particles of fine grain-size, thin bedding, and a diverse invertebrate fauna, with most species represented by all growth stages.

**Figure 3. Lateral view of UCR 20910, left lower jaw of the polydolopid from Upper Eocene strata of the La Meseta Formation, Seymour Island, Antarctic Peninsula. The teeth shown are the last premolar, which is plagiaulacoid (compressed laterally, and blade-like for cutting), and the first two molars (grinding teeth). Alveoli (holes) for the roots of the last molar are visible behind the second. MOW 8212 is the field number given at the time the fossil was collected on March 5, 1982.**

The presence of many well-preserved delicate thick-shelled species indicates minimal reworking.

One of the most characteristic features of the invertebrate fauna is the large number of nautiloid phragmocones. Although nautiloids are encountered throughout the lower part of the La Meseta Formation, most of those are represented by a single individual. At the mammal locality, most of the nautiloids are small individuals of *Aturia*, but specimens of *Eutrephoceras* are common. Almost all the larger individuals show some degree of breakage. The unusually large number of nautiloids at this locality is believed to be the result of their having been stranded on a beach, similar to the stranding of large numbers of paper argonauts along the beaches of the southwestern Pacific today.

Another curious element in the invertebrate fauna at this locality is the presence of a raninid crab, tentatively referred to the genus *Lyreidus* (R.M. Feldman pers. comm., 1982). This sand crab-like arthropod occurs in large numbers and generally inhabits the swash zone to shallow intertidal region of beaches in temperate regions. In addition, a number of penguin and whale bones, teeth and vertebrae of teleost fish, and teeth of sharks also were collected at this locality, along with the large numbers of mollusc species. This unusual faunal association, together with the fine-grained, laminated cross-bedded sandstone beds, suggests that the polydolopid remains were preserved in a beach setting.

## Discussion

The Antarctic polydolopid species is represented by at least three individuals and was preserved in a former beach deposit of Late Eocene age in northwestern Seymour Island, at the tip of the Antarctic Peninsula. The remains suggest that a larger parent population was present on Seymour Island in the Late Eocene, and that either a land connection, or a series of closely spaced islands, was present between the Antarctic Peninsula and South America in the Early Cenozoic. The present lack of any other kind of mammal in the Antarctic deposits, especially remains of contemporaneous South America placental groups, suggests that an island setting is the more likely. At the same time, the presence of the new marsupials in Antarctica suggests that both inter-island water barriers and the total distance from South America to the Antarctic Peninsula were not great.

The closest taxonomic affinity of the Antarctic species is with extinct South American forms known to range in age from about 55 to 35 Ma (Marshall, 1982). The Antarctic species is about 40 Ma old, and could have dispersed from South America at about that time. On the other hand, evaluation of the closest affinities of the Antarctic form, in conjunction with the fact that South America polydolopids are most abundant and diverse in the Late Paleocene and Early Eocene (Marshall, 1982), suggests that the actual time of dispersal from there to Antarctica may have been prior to the Late Eocene. In any case, the new Antarctic polydolopids constrain geophysical interpretations as to the proximity of South America and the Antarctic Peninsula in the Early Cenozoic.

The presence of these land animals on the Antarctic Peninsula in the Early Cenozoic also raises the question of whether other land mammals populated Antarctica at that time, and perhaps in the Late Cretaceous as well. During those times both placental and marsupial mammals were radiating in South America, and presumably marsupials were doing so in Australia. The new evidence of Antarctic land life in the Early Cenozoic forcefully raises the question of the role that this region played in the development of the faunal discontinuity that must have existed between Australia and South America prior to the time when Australia broke away form Antarctica about 56 Ma ago.

## REFERENCES


ARCHER, M. and BARTHOLOMAI, A., 1978: Tertiary mammals of Australia: A synoptic review. *Alcheringa, 2,* 1-19.

CLEMENS, W.A. and MARSHALL, L.G., 1976:*Fossilium catalogus. American and European Marsupialia.* W. Junk, The Hague, pars 123, 1-114.

ELLIOT, D.H., RINALDI, C., ZINSMEISTER, W.J., TRAUTMAN, T.A., BRYANT, W.A. and DEL VALLE, R., 1975: Geological investigations on Seymour Island. *U.S. Antarct. J., 10,* 183-6.

ELLIOT, D.H. and TRAUTMAN, T.A., 1982: Lower Tertiary strata on Seymour Island; *in* Craddock, C. (ed.) *Antarctic Geoscience,* Univ. Wisconsin Press, Madison, 287-97.

MARSHALL, L.G., 1982: Systematics of the extinct South American marsupial family Polydolopidae. *Fieldiana (Geology),* (in press).

MARSHALL, L.G., WEBB, S.D., SEPKOWSKI, J.J. and RAUP, D.M., 1982: Mammalian evolution and the Great American Interchange. *Science, 215,* 1351-7.

PAULA COUTO, C. De., 1952: Fossil mammals from the beginning of the Cenozoic in Brazil. Marsupialia: Polydolopidae and Borhyaenidae. *Am. Mus. Novit., 1559,* 1-27.

SIMPSON, G.G., 1948: The beginning of the age of Mammals in South America. Part 1. *Bull. Am. Mus. Nat. Hist., 91,* 1-232.

SIMPSON, G.G., 1978: Early Mammals in South America; fact, controversy, and mystery. *Am. Phil. Soc., Proc., 122,* 318-28.

STEGER, R.H. and JAEGER, E., 1977: Subcommission on geochronology: Convention on the use of decay constants in geo- and cosmochronology *Earth Planet Sci. Lett., 36,* 359-62.

STIRTON, R.A., TEDFORD, R.H. and WOODBURNE, M.O., 1968: Australian Tertiary deposits containing terrestrial mammals. *Univ. Calif. Publs. Geol. Sci., 77,* 1-30.

TEDFORD, R.H., 1974: Marsupials and the new paleogeography. *Soc. Econ. Paleont. Mineral., Spec. Publ. 21,* 109-26.

TEDFORD, R.H., BANKS, M.R., KEMP, N.R., MCDOUGALL, I. and SUTHERLAND, F.L., 1975: Recognition of the oldest known fossil marsupials from Australia. *Nature, 255,* 141-2.

WOODBURNE, M.O. and ZINSMEISTER, W.J., 1982: The first fossil land mammal from Antarctica. *Science, 218,* 284-5.

ZINSMEISTER, W.J., 1982: Late Cretaceous—early Tertiary molluscan biogeography of the southern Circum-Pacific. *J. Paleontol., 56,* 84-102.

# EVOLUTION OF LATE MESOZOIC SEDIMENTARY BASINS IN THE NORTHERN ANTARCTIC PENINSULA

G.W. Farquharson, *British Antarctic Survey, Madingley Road, Cambridge, CB3 OET, U.K.*

*Abstract* In Late Jurassic times the Antarctic Peninsula was probably located towards the "southwestern" margin of an anoxic basin that covered the incipient South Atlantic and adjacent areas. Although a volcanic arc was present in the area at that time, it lacked appreciable subaerial relief and alternating radiolaria rich mudstones and ash fall tuffs (Nordenskjöld Formation) accumulated over an extensive area. Uplift in earliest Cretaceous times led to emergence of the magmatic arc and the development of local, fault-bounded alluvial basins which became sites for the accumulation of debris derived from a basement terrain of deformed metasediments. Conglomerates and sandstones were deposited on alluvial fans, whilst finer material accumulated as floodplain and lacustrine deposits. The later phases of terrestrial sedimentation were contemporaneous with and finally gave way to a major episode of calc-alkaline volcanism. To the east of the emergent arc in Cretaceous and Early Tertiary times lay an ensialic back arc basin in which accumulated an extensive and thick (about 5 km) pile of arc derived debris. Petrographic studies indicate that the main phase of arc construction lasted from the Early Cretaceous to Palaeocene and was followed by a period of dissection.

Eastward subduction beneath western Antarctica occurred intermittently from at least Early Mesozoic to Late Tertiary times and resulted in the construction of a complex volcanic arc system. Sedimentary rocks from the northern Antarctic Peninsula enable the Late Mesozoic evolution of that part of the arc to be deduced. Conclusions are based on the recognition and interpretation of three lithostratigraphic units, namely the Late Jurassic Nordenskjöld Formation, and Cretaceous arc terrain and back arc basin deposits.

## NORDENSKJÖLD FORMATION

Fine grained marine deposits of Late Jurassic age are exposed at several localities along the east coast of the Antarctic Peninsula from Joinville Island to Cape Fairweather (Figure 1). These sediments constitute an extensive lithostratigraphic unit, the Nordenskjöld Formation (Farquharson, 1982), and consist of a uniform succession of alternating, thin bedded (0.5-2 cm), radiolaria rich mudstones and tuffs. The tuffs are of ash fall origin, being normally graded, laterally persistent and having non-erosive bases. Sedimentation under quiet water conditions is indicated by the complete absence of wave- or current-induced structures and the preservation of delicate fish remains. Ammonites, belemnites, brachiopods and bivalves (particularly inoceramids) of Kimmeridgian—Tithonian affinities have also been collected. Notable features of the mudstone-tuff sequence include the undisrupted nature of the fine lamination and the lack of trace fossils, both of which can be ascribed to deposition under anoxic conditions that excluded a burrowing infauna.

The problematical group of banded hornfels, described by Aitkenhead (1965) and Elliot (1966) from Trinity Peninsula, consists of a monotonous sequence of alternating argillaceous and quartzofeldspathic laminae that may well be thermally metamorphosed equivalents of the Nordenskjöld Formation. On the South Shetland Islands, the mudstone member of the Byers Formation on Byers

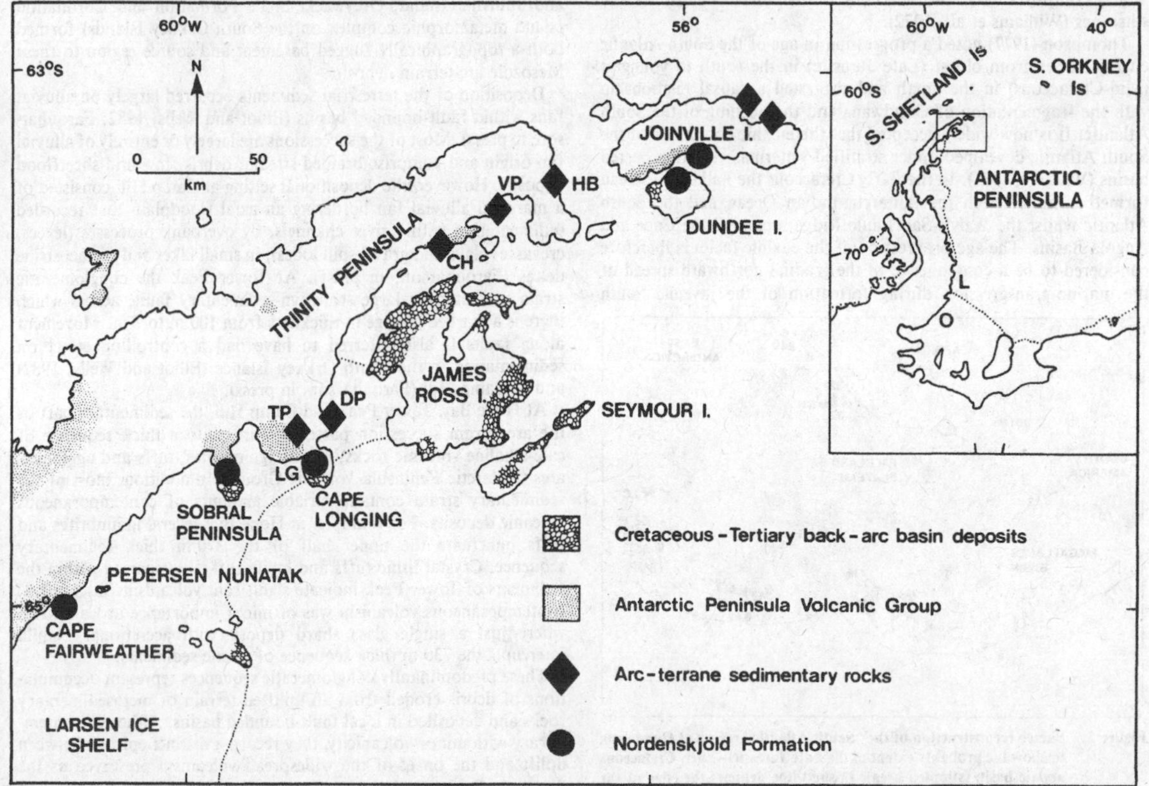


**Figure 1. Location map showing the distribution of the Late Mesozoic lithostratigraphic units. Inset indicates the position of the map at the northern end of the Antarctic Peninsula. CH: Camp Hill, DP: Downham Peak, HB: Hope Bay, L: Lassiter Coast, LG: Longing Gap, O: Orville Coast, TP: Tower Peak, VP: View Point.**

Peninsula is of Early Tithonian age and is lithologically identical to the Nordenskjöld Formation (Smellie et al., 1980). If these three rock units are true correlatives then it is clear that deposition of the mudstone-tuff lithology occurred over much of the northern Antarctic Peninsula. The absence of epiclastic detritus from the Nordenskjöld Formation indicates that its deposition was not coeval with a nearby emergent arc. However, although it lacked appreciable subaerial relief, an island arc was active in the area at that time and regularly contributed ash fall tuffs to the background accumulation of radiolaria rich mudstone.

The interpretation of the Nordenskjöld Formation as a euxinic facies is of regional as well as local significance. Legs 36 and 71 of the Deep Sea Drilling Project (DSDP) have concentrated on unravelling the evolutionary history of the South Atlantic Ocean (Barker et al., 1977; Ludwig et al., 1980). A significant result of this exploration has been the discovery of a widespread Oxfordian-Aptian euxinic facies on the Falkland Plateau (Figure 2). The initial transgression on the Falkland Plateau, associated with the opening of the South Atlantic, occurred in the Late Jurassic bringing to a halt subaerial deposition on a gneissic basement terrain. From Oxfordian to Late Aptian times euxinic conditions prevailed and were recorded by the accumulation of dark, laminated, carbonaceous claystones. Normal, open marine conditions were re-established by Early to Mid Albian times (Thompson, 1977).

Black laminated shales of Early Cretaceous age have been encountered at several other points around the margin of the South Atlantic Ocean (Figure 2). Dark, carbonaceous shales of Neocomian-Aptian age occur on the Mozambique Ridge (DSDP) site 249; Simpson et al., 1974) of Lower Aptian age in the Cape Basin (DSDP site 361; Bolli et al., 1976) and of Upper Aptian-Coniacian age in the Angola Basin (DSDP sites 364 and 365; Bolli et al., 1976). Land occurrences of Late Jurassic-Early Cretaceous carbonaceous shales are widespread in South America (Magallanes Basin, Katz, 1963; Isla de los Estados, Dalziel et al., 1974). A further occurrence of the euxinic facies is the Late Jurassic Latady Formation of the Lassiter Coast (Figure 2) which largely comprises dark, carbonaceous and laminated mudstones and siltstones (Williams et al., 1972).

Thompson (1977) noted a progression in age of the South Atlantic euxinic facies from oldest (Late Jurassic) in the south to youngest (Mid-Cretaceous) in the north and suggested a causal relationship with the fragmentation of Gondwana and the opening of the South Atlantic. It is now widely accepted that the euxinic conditions of the South Atlantic developed under stratified watermasses within barred basins (Weissert, 1981). In the Early Cretaceous the Falkland Plateau formed a sill between the Antarctic-Indian Ocean and the South Atlantic whilst the Walvis-Sao Paulo Ridge separated the Cape and Angola Basins. The age progression of the euxinic facies is therefore considered to be a consequence of the gradual northward spread of the marine transgression during formation of the juvenile South

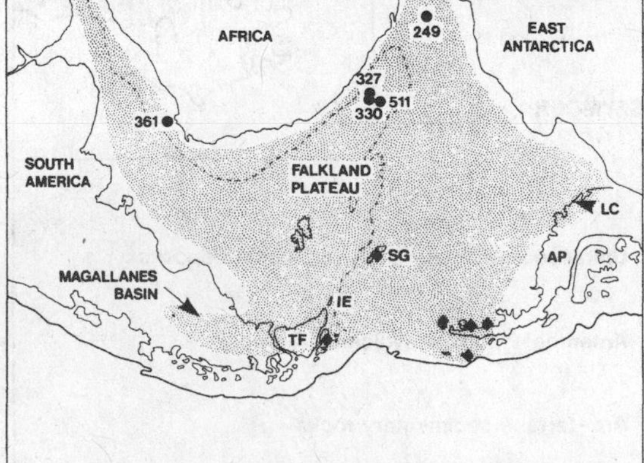


**Figure 2. Sketch reconstruction of the "South Atlantic" region of Gondwana to show the probably extent of the Late Jurassic—Early Cretaceous anoxic basin (stippled area). Dashed line delimits the edge of the South American continental shelf. Solid circles: DSDP sites. Solid diamonds: outcrops of the Nordenskjold Formation and its correlatives. AP: Antarctic Peninsula, IE: Isla de los Estados, LC: Lassiter Coast, SG: South Georgia, TF: Tierra del Fuego.**

Atlantic and the subsequent disruption of the oceanic barriers and commencement of effective oceanic circulation as sea floor spreading advanced.

A similar age progression can be recognised in the euxinic facies of the Scotia arc. Whereas the Nordenskjöld Formation and its correlatives on the South Shetland Islands are of Kimmeridgian-Tithonian age, its correlatives on South Georgia and Tierra del Fuego are dated as Late Jurassic-Early Cretaceous (Farquharson, 1982). However, it seems unlikely that the barred-basin model is applicable to this situation, since, in the absence of an appreciable arc the area would probably be subject to the oceanic influence of the Pacific. The fact that correlatives of the Nordenskjöld Formation crop out in the South Shetland Islands, on the Pacific-ward margin of the island-arc also militates against the restricted basin model. As argued previously (Farquharson, 1982), the Late Jurassic euxinic conditions of the Antarctic Peninsula are likely to have been the result of high nutrient supply and plankton productivity that led to an expanded oxygen depleted layer which intersected much of the shelf area. Whatever the true reasons behind the development of the euxinic conditions it is clear that in Late Jurassic times the Antarctic Peninsula formed part of a widespread anoxic basin that migrated north as the South Atlantic opened.

## ARC-TERRAIN DEPOSITS

A radical change in palaeogeography occurred in the earliest Cretaceous when uplift led to emergence of the magmatic arc, more or less on the site of the present peninsula (Farquharson, 1982) (Figure 3). Preserved arc-terrain deposits associated with this episode include a widespread group of calc-alkaline volcanic rocks (Antarctic Peninsula Volcanic Group) and localised occurrences of non-marine sedimentary rocks. The sedimentary sequences are dominated by conglomeratic strata and crop out on the Antarctic Peninsula (Hope Bay, View Point, Camp Hill, Downham Peak and Tower Peak), Joinville Island and the South Orkney Islands (Figure 1). Metasedimentary rocks (Trinity Peninsula Group on the Antarctic Peninsula and Joinville Island; Greywacke-Shale Formation and Coronation Island metamorphic complex on the South Orkney Islands) formed both a topographically rugged basement and source region to these Mesozoic arc-terrain deposits.

Deposition of the terrestrial sediments occurred largely on alluvial fans within fault-bounded basins (Elliot and Wells, 1982; Farquharson, in press). Most of the successions are largely or entirely of alluvial fan origin and comprise braided stream, debris flow and sheetflood deposits. However, the depositional setting at Camp Hill consisted of a marginal alluvial fan bordering an axial floodplain that recorded sedimentation within river channels, by overbank processes (levees, crevasse splays and gravel spill lobes), in small lakes and on lacustrine deltas (Farquharson, in press). At Tower Peak the conglomeratic strata are cut by a large scale syn-sedimentary fault, across which there is a marked change in thickness from 100 m to 7 m. Movement along faults is also inferred to have had a controlling effect on sedimentation in the South Orkney Islands (Elliot and Wells, 1982) and at Camp Hill (Farquharson, in press).

At Hope Bay, Tower Peak and Camp Hill, the sedimentary part of the arc-terrain succession passes upwards into a thick sequence of calc-alkaline volcanic rocks, mainly ignimbrites, tuffs and agglomerates (Antarctic Peninsula Volcanic Group). In addition, most of the sedimentary strata contain variable amounts of contemporaneous volcanic deposits. For example, at Hope Bay several ignimbrites and tuffs punctuate the upper half of the 370 m thick sedimentary sequence. Crystal lithic tuffs and lapilli tuffs interspersed within the sediments of Tower Peak indicate significant volcanism. In contrast, contemporaneous volcanism was of minor importance at Camp Hill where just a single glass shard deposit with accretionary lapilli interrupts the 730 m thick sequence of clastic sediments.

These predominantly conglomeratic sequences represent accumulations of debris eroded from an uplifted terrain of metasedimentary rocks and deposited in local fault-bounded basins. Although contemporary with minor volcanicity, they record a distinct episode between uplift and the onset of the widespread volcanism preserved as the Antarctic Peninsula Volcanic Group. The only reliable age data on these arc-terrain sediments is supplied by intercalated marine horizons within the alluvial fan deposits of the South Orkney Islands which have yielded gastropods, ammonites and belemnites of Neocomian affinities (Thomson, 1981). Further evidence from indirect stratigraphic relationships and radiometric dating indicate that the lithostratigraphic unit represented by the arc-terrain sediments is of Early Cretaceous age (Farquharson, 1982; Thomson and Pankhurst, this volume).

It is significant that the Late Jurassic-Early Cretaceous evolution of the northern Antarctic Peninsula parallels that of the South Shetland Islands where Late Jurassic marine sediments analogous to the Nordenskjöld Formation give way to terrestrial sedimentary and volcanic rocks of Early Cretaceous age (Smellie, 1980; Smellie et al., 1980).

## BACK-ARC BASIN DEPOSITS

Marine, volcaniclastic sediments of Late Mesozoic age crop out as a discontinuous band along the east coast of the northern Antarctic Peninsula and on adjacent islands (Figure 1). The oldest part of the succession is exposed on the Sobral Peninsula; dinoflagellate cysts and calcareous nannofossils from this locality have yielded a late Hauterivian-Barremian age. Fossils of Albian-Cenomanian age are documented from Dundee Island (Crame, 1980) and of Turonian age from Cape Longing (Crame, 1981). Calcareous nannofossils extracted from material colleced at Pedersen Nunatak indicate a Late Cretaceous (Turonian-Maastrichtian) age for the sequence exposed there. The succession on James Ross Island was thought to have been exclusively Campanian (Bibby, 1966) but recent work has shown that the strata possibly range back as far as the Aptian (J.A. Crame, pers. comm.). On Seymour Island the Cretaceous sequence is overlain, probably unconformably, by Palaeocene-?Oligocene sandstones and siltstones of inferred deltaic and shallow marine origin (Elliot and Trautman, 1982).

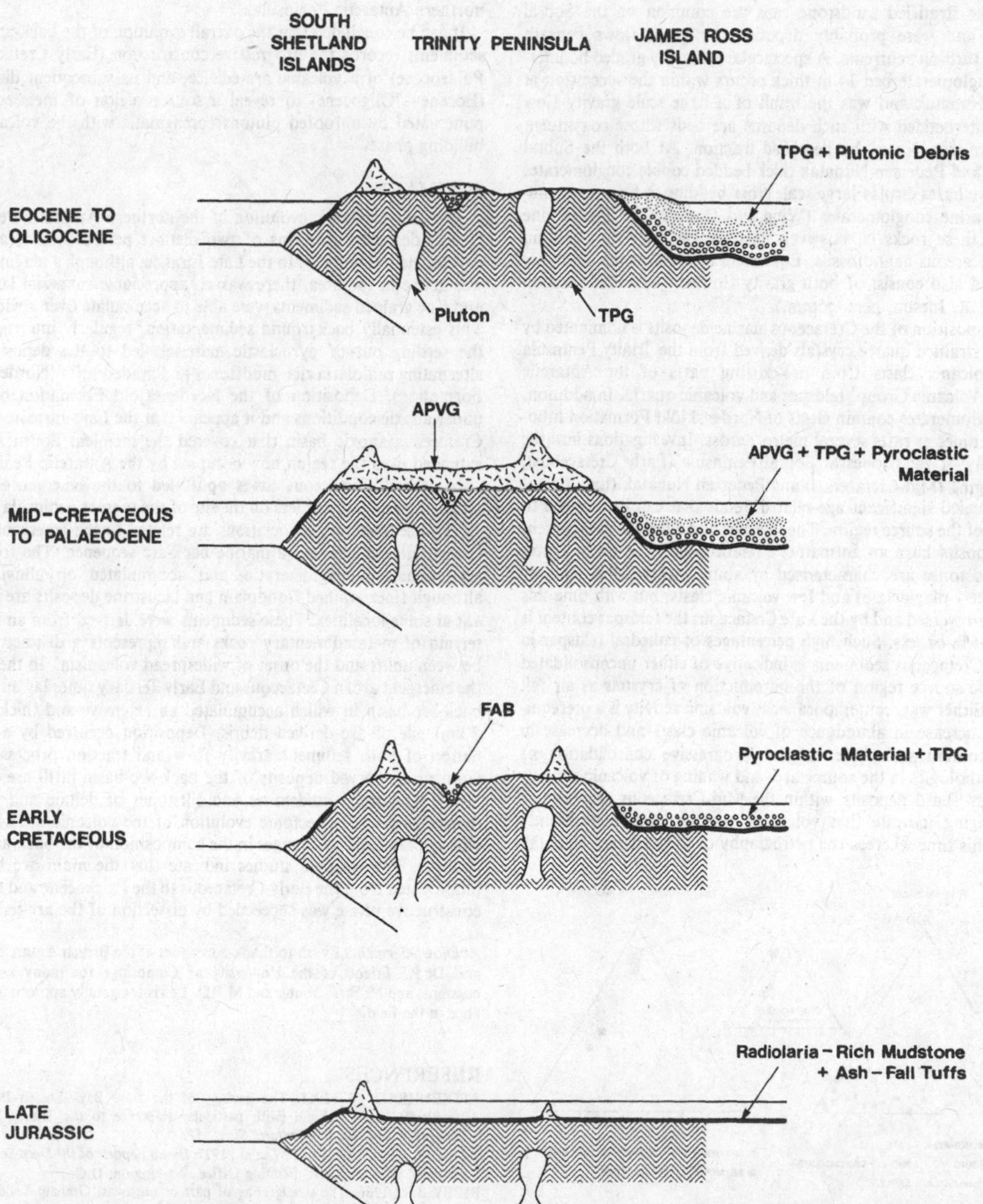


**Figure 3.** **Schematic cross-sections to illustrate the Late Mesozoic evolution of the northern Antarctic Peninsula as discussed in the text. The components of the back-arc basin deposits are listed in order of abundance. Width of cross-section is approximately 300 km. APVG: Antarctic Peninsula Volcanic Group, FAB: fault-bounded alluvial basin, TPG: Trinity Peninsula Group.**

The distribution of these sediments solely to the east (arc-rear) of the Antarctic Peninsula Volcanic Group outcrops, together with palaeocurrent directions towards the southeast (perpendicular to and away from the axis of the peninsula) is clear indication that they accumulated within a back-arc basin as an apron of volcaniclastic material flanking the emergent magmatic arc (Farquharson, 1982, Figure 1). It is probable that these marine Cretaceous sediments represent only part of an extensive ensialic back-arc system that stretched south to the Lassiter and Orville coasts and north to the crustal block to the south of the South Orkney Islands where seismic refraction and magnetic measurements indicate the presence of up to 5 km of sediment of presumed Cretaceous and Tertiary age (Harrington et al., 1972).

Deposition of the proximal back-arc basin sediments of the northern Antarctic Peninsula occurred by both gravity flow and bed load traction mechanisms. Variously graded and imbricated conglomerates with planar stratified sandstone caps are common on the Sobral Peninsula and were probably deposited by grain flows beneath overriding turbidity currents. A spectacular normally graded boulder-pebble conglomerate bed 14 m thick occurs within the succession at Pedersen Nunatak and was the result of a large scale gravity flow process. Interbedded with such deposits are beds whose constituent grains were transported by bed load traction. At both the Sobral Peninsula and Pedersen Nunatak thick bedded cobble conglomerates with erosive bases display large scale cross-bedding, a feature uncommon in marine conglomerates (Winn and Dott, 1977). The marine nature of these rocks is, however, unequivocal since they contain marine calcareous nannofossils. Equivalent strata exposed on James Ross Island also consist of both gravity flow and bed load traction deposits (J.R. Ineson, pers. comm.).

The composition of the Cretaceous marine deposits is dominated by clasts and strained quartz crystals derived from the Trinity Peninsula Group, volcanic clasts from pre-existing parts of the Antarctic Peninsula Volcanic Group, feldspar and volcanic quartz. In addition, many conglomerates contain clasts of Nordenskjöld Formation lithology, sometimes as rafts several metres across. Investigations into the petrography of sandstones at Sobral Peninsula (Early Cretaceous), Cape Longing (Mid-Cretaceous) and Pedersen Nunatak (Late Cretaceous) revealed significant age-related trends that can be ascribed to evolution of the source region. The feldspar and volcanic clast content of the deposits have an antipathetic relationship. The Early Cretaceous sandstones are characterised by abundant euhedral feldspar (55%; largely plagioclase) and few volcanic clasts, but with time this situation is reversed and by the Late Cretaceous the feldspar content is down to 15% or less. Such high percentages of euhedral feldspar in the Early Cretaceous sediments is indicative of either unconsolidated tuffs in the source region of the introduction of crystals as air fall material. Either way, contemporaneous volcanic activity is a prerequisite. The increase in abundance of volcanic clasts and decrease in feldspar content with time suggests progressive consolidation of volcanic lithologies in the source area and waning of volcanic activity. Thick glass shard deposits within the Mid-Cretaceous sequence at Cape Longing indicate that volcanism was at least intermittently active at this time whereas the petrography of the Pedersen Nunatak deposits may suggest that the Late Cretaceous was a period of local volcanic quiescence. The significant effect that these changes in the magmatic arc had on the adjacent sedimentary pile is summarised in the QLF diagram (Figure 4) which shows the progression from feldspar rich Early Cretaceous sandstones to lithic rich Late Cretaceous ones.

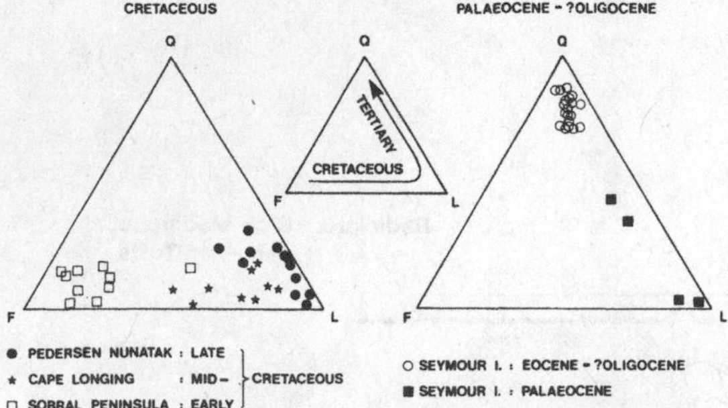


**Figure 4. Quartz (Q), feldspar (F), lithic (L) diagrams for sandstones of various ages from the Late Mesozoic back-arc basin of the northern Antarctic Peninsula. Note the overall trend from feldspathic to lithic to quartzose sandstones. The data points for the Tertiary strata of Seymour Island are from Elliot and Trautman (1982).**

Petrographic studies of the Tertiary marine deposits of Seymour Island (Elliot and Trautman, 1982) enable the evolution of the source region to the back-arc basin to be traced further. These authors recognised that a decrease in the ratio of plagioclase to total feldspar, and of volcanic rock fragments to total detrital grains with age, indicates the decreasing importance of a volcanic source. The older (Palaeocene) sandstones with the highest volcanic clast content also contain air-fall debris. These Tertiary sediments record a predominantly volcanic source region being replaced by a metasedimetary and plutonic source probably similar to the present surface geology of the northern Antarctic Peninsula.

It can be concluded that the overall evolution of the back-arc basin sediments records the progressive construction (Early Cretaceous—Palaeocene) of a volcanic arc edifice, and its subsequent dissection (Eocene—?Oligocene) to reveal a source region of metasediments punctuated by unroofed plutons comagmatic with the volcanic arc building phase.

## SUMMARY

The Late Mesozoic evolution of the northern Antarctic Peninsula can be described in terms of two distinct periods of volcanic arc development (Figure 3). In the Late Jurassic, although a magmatic arc was active in the area, there was no appreciable subaerial landmass and fine grained sediments were able to accumulate over a wide area. This essentially background sedimentation, regularly interrupted by the settling out of pyroclastic material, led to the deposition of alternating radiolaria rich mudstones and graded tuffs (Nordenskjöld Formation). Deposition of the Nordenskjöld Formation occurred under anoxic conditions and it appears that the Late Jurassic— Early Cretaceous anoxic basin that covered the incipient South Atlantic extended over the region now occupied by the Antarctic Peninsula.

In earliest Cretaceous times uplift led to the emergence of the magmatic arc, more or less on the site of the present peninsula (Figure 3). Two sedimentary associations are related to the emergent arc, a terrestrial sequence and a marine back-arc sequence. The former is predominantly conglomeratic, and accumulated on alluvial fans although finer grained floodplain and lacustrine deposits are important at some localities. These sediments were derived from an uplifted terrain of metasedimentary rocks and represents a distinct episode between uplift and the onset of widespread volcanism. To the east of the emergent arc in Cretaceous and Early Tertiary times lay an ensialic back-arc basin in which accumulated an extensive and thick (about 5 km) pile of arc derived debris. Deposition occurred by a combination of both sediment gravity flow and traction processes. The youngest preserved deposits of the back-arc basin infill are Palaeocene—?Oligocene sandstones and siltstones of deltaic and shallow marine origin. The tectonic evolution of the volcanic arc system is closely paralleled by changes in the composition of the back-arc basin deposits. Petrographic studies indicate that the main arc building phase lasted from the Early Cretaceous to the Palaeocene and that this constructive phase was succeeded by dissection of the arc-terrain.

*Acknowledgements* I wish to thank colleagues at the British Antarctic Survey and Dr P.F. Friend of the University of Cambridge for many useful discussions, and N. St J. Young and M.P.D. Lewis for greatly appreciated assistance in the field.

## REFERENCES

AITKENHEAD, N., 1965: The geology of the Duse Bay—Larsen Inlet area, northeast Graham Land (with particular reference to the Trinity Peninsula Series). *Scient. Rep. Br. antarc. Surv., 51.*

BARKER, P.F., DALZIEL, I.W. et al., 1977: *Initial Reports of the Deep Sea Drilling Project, 36.* U.S. Govt. Printing Office, Washington, D.C.

BIBBY, J.S., 1966: The stratigraphy of part of northeast Graham Land and the James Ross Island group. *Scient. Rep. Br. antarct. Surv., 53.*

BOLLI, H.M., RYAN, W.B. et al., 1978: *Initial Reports of the Deep Sea Drilling Project, 40.* U.S. Govt. Printing Office, Washington, D.C.

CRAME, J.A., 1980: The occurrence of the bivalve *Inoceramus concentricus* on Dundee Island, Joinville Island group. *Bull. Br. antarct. Surv., 49,* 283-6.

CRAME, J.A., 1981: Upper Cretaceous inoceramids (Bivalvia) from the James Ross Island group and their stratigraphical significance. *Bull. Br. antarct. Surv., 53,* 29-56.

DALZIEL, I.W., CAMINOS, R., PALMER, K.F., NULLO, F. and CASANOVA, R., 1974: South extremity of Andes: Geology of Isla de los Estados, Argentine Tierra del Fuego. *Am. Assoc. Petrol. Geol., Bull, 58,* 2502-12.

ELLIOT, D.H., 1966: Geology of the Nordenskjöld Coast and a comparison with the northwest Trinity Peninsula, Graham Land. *Bull. Br. antarct. Surv., 10,* 1-43.

ELLIOT, D.H. and TRAUTMAN, T.A., 1982: Lower Tertiary strata on Seymour Island, Antarctic Peninsula; *in* Craddock, C. (ed.) *Antarctic Geoscience.* Univ. Wisconsin Press, Madison, 287-97.

ELLIOT, D.H. and WELLS, N.A., 1982: Mesozoic alluvial fans of the South Orkney Islands; *in* Craddock, C. (ed.) *Antarctic Geoscience.* Univ. Wisconsin Press, Madison, 235-44.

FARQUHARSON, G.W., 1982: Late Mesozoic sedimentation in the northern Antarctic Peninsula and its relationship to the southern Andes. *Geol. Soc. Lond., J., 139,* in press.

FARQUHARSON, G.W. (in press): Lacustrine deltas in a Mesozoic alluvial sequence from Camp Hill, Antarctica. *Sedimentology.*

HARRINGTON, P.K., BARKER, P.F. and GRIFFITHS, D.H., 1972: Crustal structure of the South Orkney Islands area from seismic refraction and magnetic measurements; *in* Adie, R.J. (ed.) *Antarctic Geology and Geophysics,* Universitetsforlaget, Oslo, 27-32.

KATZ, H.R., 1963: Revision of Cretaceous stratigraphy in Patagonian Cordillera of Ultima Esperanza, Magallanes province, Chile. *Am. Assoc. Petrol. Geol., Bull., 47,* 506-24.

LUDWIG, W.J., KRASHENINNIKOV, V. et al., 1980: Tertiary and Cretaceous paleoenvironments in the southwest Atlantic Ocean, preliminary results of Deep Sea Drilling Project Leg 71. *Geol. Soc. Am., Bull., 91,* 655-64.

SIMPSON, E.W., SCHLICH, R. et al., 1974: *Initial Reports of the Deep Sea Drilling Project, 25.* U.S. Govt. Printing Office, Washington, D.C.

SMELLIE, J.L., 1980: The geology of Low Island, South Shetland Islands, and Austin Rocks. *Bull. Br. antarct. Surv., 49,* 235-57.

SMELLIE, J.L., DAVIES, R.E.S. and THOMSON, M.R.A., 1980: Geology of a Mesozoic intra-arc sequence on Byers Peninsula, Livingston Island, South Shetland Islands. *Bull. Br. antarct. Surv., 50,* 55-76.

THOMPSON, R.W., 1977: Mesozoic sedimentation on the eastern Falkland Plateau. *Initial Reports of the Deep Sea Drilling Project, 36,* 877-91. U.S. Govt. Printing Office, Washington, D.C.

THOMSON, M.R.A., 1981: Late Mesozoic stratigraphy and invertebrate palaeontology of the South Shetland Islands. *Bull. Br. antarct. Surv., 54,* 65-83.

THOMSON, M.R.A. and PANKHURST, R.J. (this volume): Age of past-Gondwanian calc-alkaline volcanism in the Antarctic Peninsula region.

WEISSERT, H., 1981: The environment of deposition of black shales in the Early Cretaceous: an ongoing controversy; *in* Warme, J.E., Douglas, R.G. and Winterer, E.L. (eds.) The Deep Sea Drilling Project: a Decade of Progress. *Soc. Econ. Paleont. Mineral., Spec. Publ., 32,* 547-60.

WILLIAMS, P.L. SCHMIDT, D.L., PLUMMER, C.C. and BROWN, L.E., 1972: Geology of the Lassiter Coast area, Antarctic Peninsula: a preliminary report; *in* Adie, R.J. (ed.) *Antarctic Geology and Geophysics.* Universitetsforlaget, Oslo, 143-8.

WINN, R.D. and DOTT, R.H., 1977: Large scale traction produced structures in deep water fan channel conglomerates in southern Chile. *Geology, 5,* 41-4.

# AGE OF POST-GONDWANIAN CALC-ALKALINE VOLCANISM IN THE ANTARCTIC PENINSULA REGION

M.R.A. Thomson and R.J. Pankhurst, *British Antarctic Survey, Madingley Road, Cambridge CB3 OET, U.K.*

*Abstract* After the Gondwanian Orogeny, the Antarctic Peninsula became the site of a magmatic arc that formed in response to subduction of the Pacific beneath a disintegrating Gondwana supercontinent. It was once implied that the calc-alkaline volcanic rocks (Antarctic Peninsula Volcanic Group (APVG)), which resulted from this magmatism, were erupted almost entirely in the Late Jurassic and therefore provided a useful stratigraphic marker unit. However, a review of fossil and sedimentologic data from volcanoclastic parts of the sequence and associated fore- and back-arc deposits, together with radiometric ages from lavas and dykes, clearly indicate that volcanism was active in the area almost continuously from Early Jurassic to Tertiary times (ca 180-60 Ma). Although lithologic and chemical variations have been observed in the volcanic rocks, it is uncertain if any of these have stratigraphic significance. Without fossil or radiometric evidence, no exposure of APVG rocks can be dated more precisely than Jurassic—Tertiary.

Calc-alkaline volcanic rocks are widespread in the Antarctic Peninsula and constitute a major stratigraphic unit in the region (Figure 1). They were erupted after the deformation of Triassic and older rocks by the Gondwanian Orogeny (Dalziel and Elliot, 1973) and during a long phase of magmatic activity, caused by subduction of the Pacific crust beneath a fragmenting Gondwana supercontinent (Weaver et al., 1982). Early investigations suggested that representatives of these rocks in northern Graham Land were Late Jurassic, and the majority of poorly metamorphosed calc-alkaline volcanic rocks in the peninsula were subsequently assigned to a single narrowly-defined stratigraphic unit—the Upper Jurassic Volcanic Group (UJVG). The name was first applied to a sequence of rocks on King George Island (Adie, 1962, table 4) that had been correlated lithologically with the "Upper Jurassic rhyolites and rhyodacites" of northern Graham Land (Hawkes, 1961), and from 1962 onwards the name UJVG appeared frequently in papers on Antarctic Peninsula geology (see references in Adie, 1972). However, with rare exceptions (Thomson, 1972, 1975) there was no local evidence to support the supposed Late Jurassic age of the rocks and recent palaeontologic, petrologic and radiometric evidence suggests that even those to which the name was first applied may be Tertiary (cf. Lucas and Lacey, 1981; Davies, in press). In recognition of these problems, the alternative name Antarctic Peninsula Volcanic Group (APVG) was proposed (Thomson, 1982).

## HOPE BAY

The Late Jurassic age attributed to the volcanic rocks was deduced from field relationships of the strata exposed on the northeast face of Mount Flora, Hope Bay:

Rhyolitic tuffs and ignimbrites > 100 m
Plant bearing sandstones and shales ~ 85 m
Alluvial fan conglomerate sequence ~ 285 m
Unconformity—not exposed
Deformed greywackes of Trinity Peninsula Group

The deformation of the basal Late Palaeozoic-Early Mesozoic Trinity Peninsula Group is thought to have occurred in the Early Mesozoic Gondwanian Orogeny (Dalziel and Elliot, 1973) and fossil plants in the undeformed strata above the conglomerates were dated as Middle Jurassic (Halle, 1913). Although the case has never been argued in print, it was a widely held assumption that the volcanic rocks lying above the plant beds were therefore Late Jurassic. However, tuff beds are present in both the conglomerates and the plant beds, indicating that they were deposited during the volcanic episode, and thus there is no reason to suppose that the overlying volcanic rocks are significantly different in age from the plant beds.

Although the preservation of the plant fossils may be superficially good, all cuticle remains have been destroyed. Thus identification of the plants are based on gross external morphology only, and identifications with supposed extra-Antarctic equivalents are therefore questionable. The Hope Bay flora cannot be regarded as being as precisely dated stratigraphically as is often supposed, and it is significant to the present argument that a reassessment of the flora by Stipanicic and Bonetti (1970) concluded that its age was latest Jurassic or even earliest Cretaceous. Furthermore, elements of the Hope Bay flora appear to be present in the Late Jurassic-Early Cretaceous Springhill Formation of Patagonia (Baldoni, 1979), in conglomerates of probably Early Cretaceous age on the South Orkney Islands (Thomson, 1981), in Early Cretaceous strata on Snow Island (Askin, this volume), in the Late Jurassic-Early Cretaceous Fossil Bluff Formation of Alexander Island (T.H. Jefferson, pers comm), and in the Late Jurassic Latady Formation of Lassiter Coast (Rowley and Williams, 1982). Evidence for the age of the other calc-alkaline volcanic sequences is summarised below. All quoted radiometric ages conform to the decay constants recommended by Steiger and Jäger (1977) and equivalence of absolute and stratigraphic ages follows the time scale compiled by Odin (in press).

## SOUTH SHETLAND ISLANDS

Volcanic rocks in the South Shetland Islands preserve a long record of more a less continuous calc-alkaline activity from Late Jurassic to Oligocene times. This is documented partly by over 60 K-Ar age determinations by RJP, ranging from 132 to 27 Ma (Figure 2), and by Grikurov et al., (1970) and Watts (1982). Most of the rocks are basaltic in composition (low-K tholeiites) with some andesites (Weaver et al., 1982). Occurrences of rhyolite and ignimbrite are uncommon but examples on Byers Peninsula, western Livington Island, have yielded a Rb-Sr whole-rock isochron age of 111 ± 4 Ma (Barremian/Aptian) (Pankhurst, 1982). Low-K tholeiites, similar to those of the South Shetland Islands, also occur on Tower and Two Hummock Islands (Figure 1) and have given K-Ar ages of 35 Ma (Oligocene) and 63 Ma (Paleocene) (Baker et al., 1977).

Radiometric ages are corroborated partly by Late Jurassic and Early Cretaceous fossils from marine beds in the lowest part of the succession (Smellie et al., 1980; Thomson, in press). Although fossil plants in volcanoclastic strata on Snow Island were dated as Middle Jurassic (Fuenzalida et al., 1972), Askin (this volume) has shown from a study of dinocysts in marine beds at the same locality that the sequence is Early Cretaceous (Neocomian).

## NORTHERN GRAHAM LAND

In northern Graham Land and the James Ross Island area, sufficient data now exist to construct a general picture of post-Gondwanian volcanism. From petrologic characteristics of the palaeontologically dated Late Jurassic Nordenskjold Formation of northeast Graham Land, Farquharson (1982) deduced that the Late Jurassic "arc" in that area probably consisted of isolated volcanoes surrounded by anoxic shelf seas, and that a substantial subaerial volcanic landmass did not develop until Early Cretaceous times. Support for this argument comes from the few radiometric age determinations obtained from the volcanic rocks in the area.

An andesite on the Duroch Islands (63°18′S, 57°54′W) has yielded a K-Ar pyroxene age of 88 ± 7 Ma (early Late Cretaceous) (Halpern, 1964). However, since this is known to be a time of intense plutonic activity in Graham Land (Rex, 1976; Pankhurst, 1982) it is possible that thermal loss of argon had occurred in that sample. Older Rb-Sr whole-rock isochron ages of 130 ± 7 Ma (Jurassic/Cretaceous boundary) and 117 ± 4 Ma (Hauterivian/Barremian) were obtained by Pankhurst (1982) for massive crystal tuffs and (?) extrusive porphyries of rhyodacitic and rhyolitic composition, respectively, from near Longing Gap (Figure 3a, b). Some caution is necessary in accepting these ages since Rb-Sr whole-rock systems in acid volcanic rocks are widely considered to be very easily reset without metamorphism. Although the two isochrons from Longing Gap area do have relatively high initial $^{87}Sr/^{86}Sr$ ratios (0.7091 and 0.7153), which would be consistent with such whole-rock scale rehomogenisation, we suggest that they do in fact date crystallisation of the rocks because they clearly cannot both be affected by the same resetting event and the

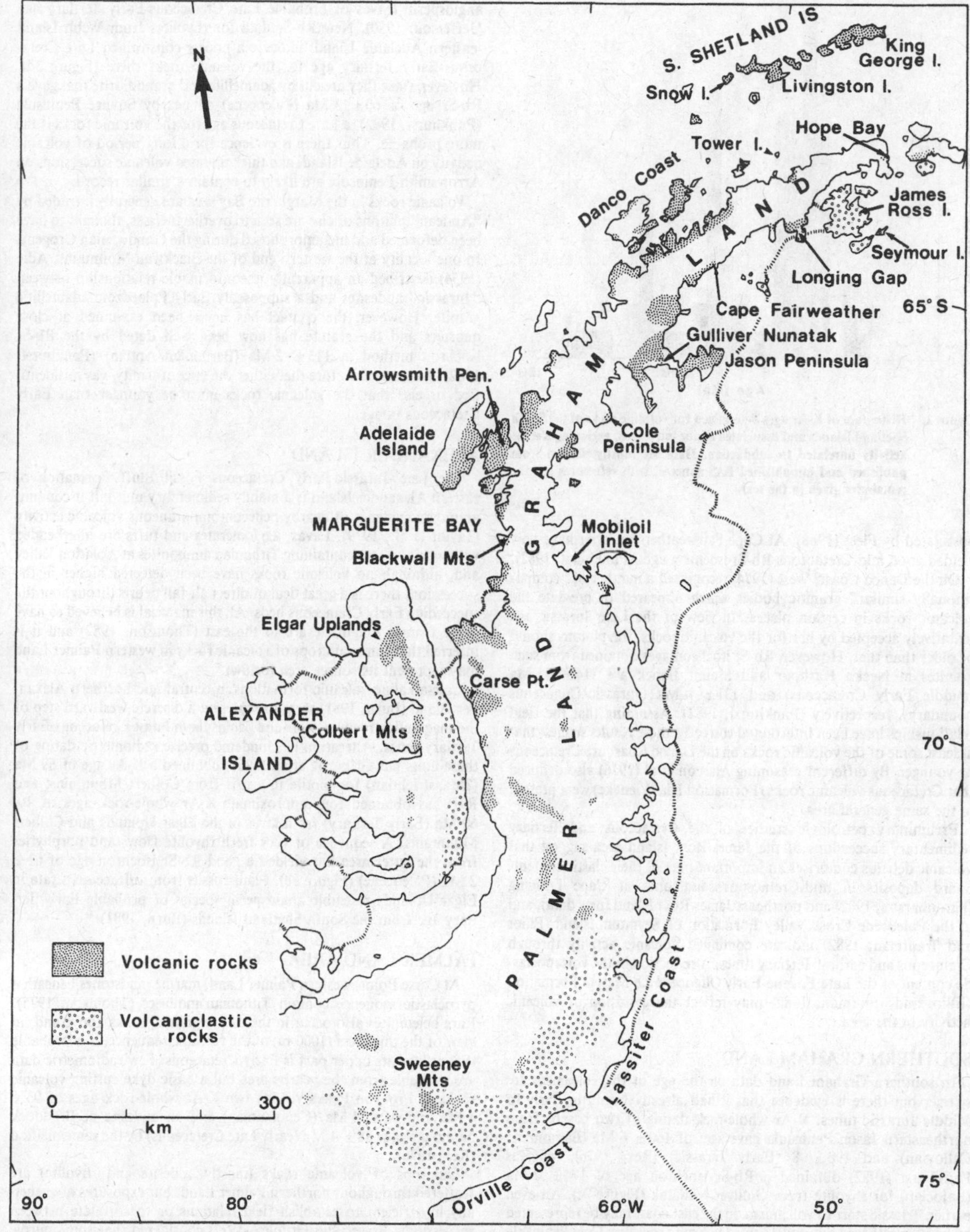


**Figure 1. Sketch map of the Antarctic Peninsula showing the major outcrops of the Antarctic Peninsula Volcanic Group, and associated volcanoclastic sedimentary rocks, together with the principal localities referred to in the text.**

ages obtained are in the logical stratigraphic order. Although most of the samples from Longing Gap show a degree of recrystallisation of the groundmass (in the worst cases with the introduction of some calcite) and slight sericitisation of the feldspar phenocrysts (R.D. Hamer, pers. comm.), the effects are far less marked in the high Rb/Sr samples which give the 117 Ma age than in those which give an older apparent age. This is inconsistent with an interpretation of the isochrons as reflecting subsolidus alteration. Whereas Rb-Sr ages for acid volcanic rocks should strictly be regarded as minima, we feel that in this case they are most likely to be within error of the true times of igneous crystallisation. The rhyolites are lithologically similar to those on Byers Peninsula (above) which give a similar age but have a low initial $^{87}Sr/^{86}Sr$ ratio (0.7051). A genetic link between acid volcanic rocks and adamellitic intrusions of northern Oscar II Coast was

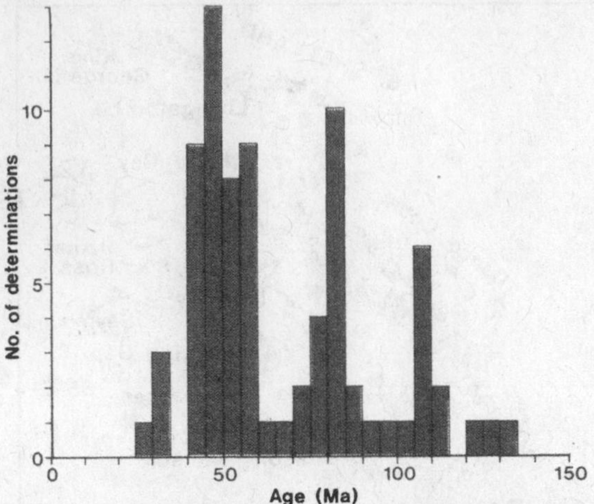


**Figure 2. Histogram of K-Ar ages determined for volcanic rocks of the South Shetland Islands and associated minor intrusions, excluding Recent activity unrelated to subduction. Data are mostly derived from published and unpublished BAS sources, with references for the remainder given in the text.**

postulated by Fleet (1968). At Cape Fairweather the latter have now yielded good mid-Cretaceous Rb-Sr isochron ages (Pankhurst, 1982).

On the Danco Coast, West (1974) recognised a number of "compositionally similar" granitic bodies which appeared to pre-date the volcanic rocks in certain places. In view of the Late Jurassic age tentatively accepted by her for the volcanic rocks, the plutons should be older than that. However, Rb-Sr isochron ages obtained from such granites at Necko Harbour and Mount Banck are 114± 11 Ma (middle Early Cretaceous) and 131± 4 Ma (Jurassic/Cretaceous boundary), respectively (Pankhurst, 1982). Assuming that the field relationships have been interpreted correctly, these results suggest that at least some of the volcanic rocks on the Danco Coast are Cretaceous or younger. By different reasoning Alarcón et al (1976) also deduced that Cretaceous volcanic rocks (Formación Isla Wiencke) were present in the same general area.

Preliminary petrologic studies of the Cretaceous and Tertiary sedimentary successions of the James Ross Island area suggest that volcanic detritus comprises an important part of their clastic content. Shard deposits in mid-Cretaceous sandstones at Cape Longing (Farquharson, 1982) and northeast James Ross Island (new data), and in the Paleocene Cross Valley Formation of Seymour Island (Elliot and Trautman, 1982) indicate continued volcanic activity through Cretaceous and earliest Tertiary times. A reduction in the volcanoclastic content of the Late Eocene-Early Oligocene La Meseta Formation (Elliot and Trautman, 1982) may reflect the cessation of volcanic activity in the area.

## SOUTHERN GRAHAM LAND

In southern Graham Land data on the age of the volcanism are sparser but there is evidence that it had already begun in Early to Middle Jurassic times. K-Ar whole-rock dating of two basalts from northeastern Jason Peninsula gave ages of 160± 6 Ma (Bathonian/Callovian) and 190± 8 (Early Jurassic) (Rex, 1976), whereas Pankhurst (1982) obtained a Rb-Sr isochron age of 174± 2 Ma (Bajocian) for rhyolite from Gulliver Nunatak (Figure 3c). An even earlier, Triassic start to volcanism on the east coast may be represented by badly weathered but undeformed volcanic rocks on Cole Peninsula and farther south, which are apparently cut by granitoids dated by Pankhurst (1982) at 209± 3 Ma (latest Triassic) (A.D. Saunders, pers. comm.).

Volcanic successions on Adelaide Island contain a high proportion of both marine and non-marine sedimentary strata. In the Mount Bouvier area, Late Jurassic (?Kimmeridgian) molluscs occur in volcanoclastic sandstones (Thomson, 1972; Crame, 1982) and a different but stratigraphically undiagnostic Jurassic-Cretaceous fauna is present at Milestone Bluff (MRAT, new data). By contrast, plant bearing shales at Cape Alexendra (southern Adelaide Island) contain angiosperm leaves of probable Late Cretaceous-Early Tertiary age (Jefferson, 1980). New Rb-Sr data for rhyolites from Webb Island (eastern Adelaide Island) indicate a poorly constrained Late Cretaceous-Early Tertiary age for the volcanic rocks there (Figure 3d). However, since they are cut by adamellite and granodiorite that gave a Rb-Sr age of 60± 3 Ma (Paleocene) on nearby Square Peninsula (Pankhurst, 1982), a Late Cretaceous age for the volcanic rocks is the more probable. Thus there is evidence for a long period of volcanic activity on Adelaide Island, and thick undated volcanic successions on Arrowsmith Peninsula are likely to contain a similar record.

Volcanic rocks in the Marguerite Bay area are generally intruded by "Andean" plutons or else are seen to overlie gneisses, thought to have been deformed and metamorphosed during the Gondwanian Orogeny. In one locality at the western end of the Blackwall Mountains, Adie (1954) described an apparently unconformable relationship between "Jurassic" andesites and a supposedly Early Palaeozoic adamellitic granite. However, the contact has never been examined at close quarters and the granite has now been well dated by the Rb-Sr isochron method at 113± 2 Ma (Barremian/Aptian) (Pankhurst, 1982). It seems, therefore that either the unconformity was misidentified or else that the volcanic rocks must be younger than Early Cretaceous in age.

## ALEXANDER ISLAND

The Late Jurassic-Early Cretaceous Fossil Bluff Formation of eastern Alexander Island is a mainly sedimentary unit but it contains abundant evidence of nearby penecontomporaneous volcanic activity (Taylor et al., 1979). Lavas, agglomerates and tuffs are interbedded with marine strata containing Tithonian ammonites at Ablation Valley and, although no volcanic rocks have been detected higher in the succession, there is a great deal of direct air-fall debris throughout the succeeding Early Cretaceous beds. All this material is believed to have come from a magmatic arc to the east (Thomson, 1982) and it is inferred that some outcrops of volcanic rocks in western Palmer Land may represent its source area (below).

Calc-alkaline volcanic formations in central and northern Alexander Island (Burn, 1981) seem to represent a discrete westward step of the magmatic arc away from the peninsula in latest Cretaceous-Early Tertiary times. Alteration has hindered precise radiometric dating of these units but Grikurov et al. (1967) obtained a K-Ar age of 69 Ma (?Maastrichtian) for biotite in a tuff from Colbert Mountains, and RJP has obtained four approximate K-Ar whole-rock ages of 40-60 Ma (Early Tertiary) from lavas in the Elgar Uplands and Colbert Mountains. A sequence of very fresh rhyolite flows and porphyries from the latter area has yielded a good Rb-Sr isochron age of 62± 2 Ma (Paleocene) (Figure 3e). Plant fossils from tuffaceous strata in Elgar Uplands resemble angiosperm species of probable Early Tertiary age from the South Shetland Islands (Burn, 1981).

## PALMER LAND AREA

At Carse Point, western Palmer Land, marine mudstones beneath a pyroclastic sequence contain Tithonian molluscs (Thomson, 1975). Rare belemnites also occur in the coarse grained rocks above and, in view of the thickness (1000 m) of the volcanic sequence, it is probable that at least its upper part is Early Cretaceous. Few radiometric data are available from the nearby area but a basic dyke cutting volcanic rocks at Procyon Peaks yielded two K-Ar whole-rock ages of 93± 4 Ma and 94± 4 Ma (Cenomanian) and an andesite at Braddock Nunataks gave 88 ± 4 Ma (early Late Cretaceous) by the same method (Rex, 1976).

Outcrops of volcanic rocks (mostly andesite and rhyolite) are scattered throughout northern Palmer Land, but exposures are generally insufficient to establish field relations or to correlate between successions. Fraser and Grimley (1972) considered that some purple and black volcanic rocks south of Mobiloil Inlet to be Carboniferous on account of their succession with sediments ascribed to the Trinity Peninsula "Series." However, the correlation and the true age of the latter are both open to doubt and purple tuffs are also common in the dated (Middle Jurassic) volcanic rocks of Gulliver Nunatak (above). A more general distinction has been made between relatively unaltered rocks, previously assigned to the UJVG, and deformed and indurated metavolcanic rocks assigned to the pre-Jurassic metamorphic basement (Davies, 1976). A preliminary Rb-Sr survey of a variety of samples from Davies' traverse of northern Palmer Land has failed to

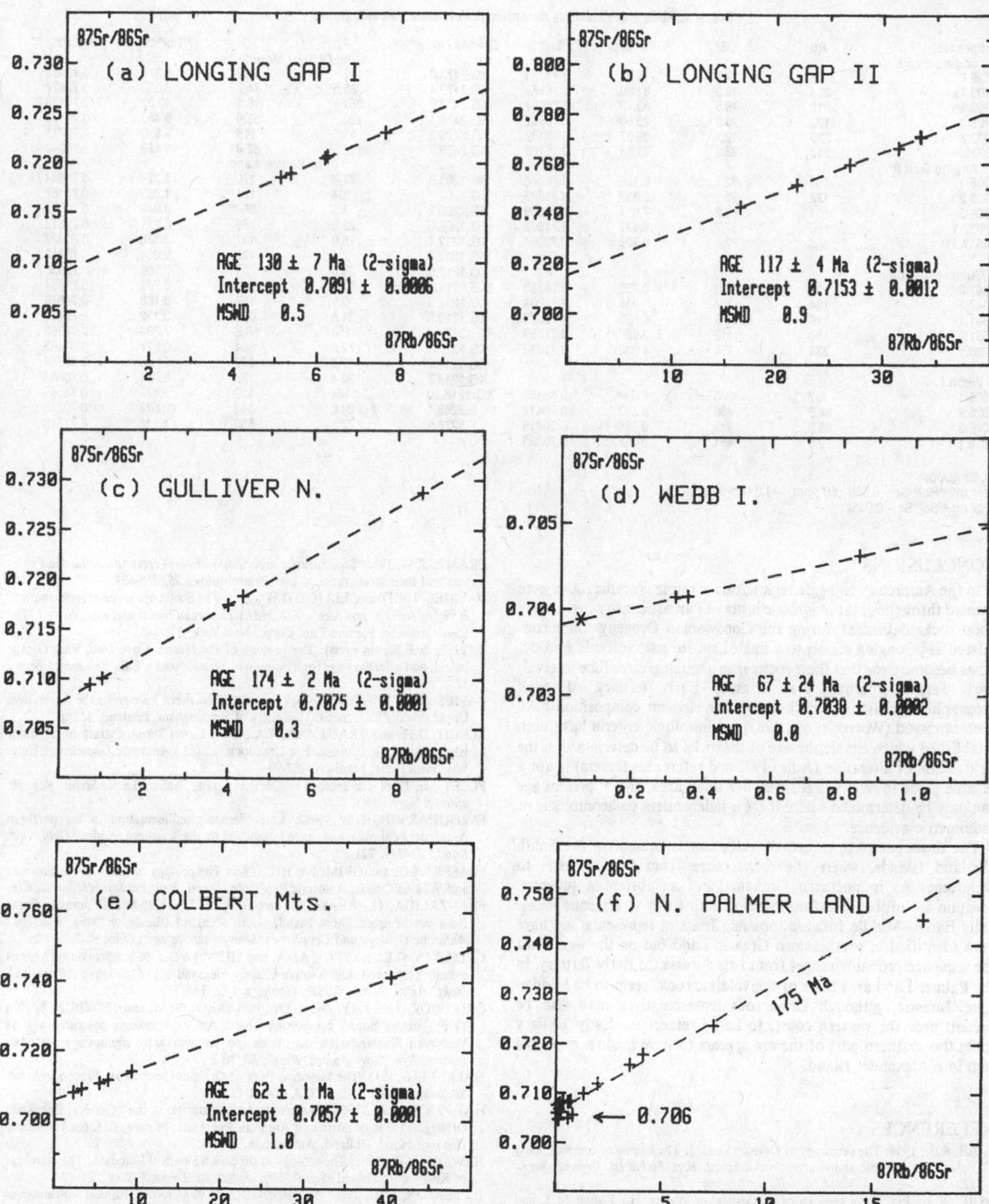


**Figure 3.** Rb-Sr whole-rock isochrons for the Antarctic Peninsula Volcanic Group (see Table 1 for analytical data). (a) dacite from near Hampton Bluff, Longing Gap. (b) rhyolite and ignimbrite from Porphyry Bluff, Longing Gap. (c) dacite and rhyolite from Gulliver Nunatak (one sample from Lyttelton Ridge, 25 km southwest of Gulliver Nunatak). (d) dacite from Webb Island, west of Adelaide Island. (The asterisk is the least radiogenic analysis of a group of isotopically heterogeneous basic volcanic rocks from Lagoon Islands, 20 km to the southwest but this is not included in the isochron fit). (e) rhyolite tuffs and sills from the Colbert Mountains, Alexander Island. (f) altered volcanic and metavolcanic rocks from northern Palmer Land. Errors have been enhanced where MSWD exceeds 3.0.

yield any convincing isochrons for specific localities (Figure 3f) but there is a general trend for most of the data about a ca. 180 Ma reference line. Allowing for isotopic disturbances due to subsequent events, it seems likely that the majority of these rocks, including the metavolcanic ones, are of Early—Middle Jurassic age.

The Middle—Late Jurassic Latady Formation of southeastern Palmer Land and Orville Coast is a mainly marine sequence which has close volcanic associations (Rowley et al., this volume). In the Sweeney Mountains in particular large outcrops of dacitic rock mark the location of an associated volcanic arc, whereas agglomerates, tuffs and lavas are interbedded locally with adjacent marine strata and thin southwards and basinwards. By the far the greater part of the clastic detritus forming the marine sediments is volcanogenic (Laudon et al., this volume).

**TABLE 1: Rb-Sr analytical data for Antarctic Peninsula Volcanic Group**

| Sample No. | Rb | Sr | $^{87}Rb/^{86}Sr$ | $^{87}Sr/^{86}Sr$ |
|---|---|---|---|---|
| (a) Longing Gap I | | | | |
| R.699.4 | 158 | 27.7 | 16.53 | 0.74273 |
| R.903.1a | 214 | 19.9 | 31.40 | 0.76653 |
| R.903.1b | 212 | 18.5 | 33.43 | 0.77097 |
| R.903.2 | 192 | 20.9 | 26.80 | 0.75982 |
| R.903.3 | 220 | 19.1 | 33.37 | 0.77132 |
| D.3926.1 | 212 | 28.2 | 21.86 | 0.75168 |
| (b) Longing Gap II | | | | |
| R.906.1 | 157 | 73.7 | 6.169 | 0.72048 |
| R.906.2 | 172 | 93.1 | 5.357 | 0.71893 |
| R.906.3 | 161 | 61.2 | 7.614 | 0.72309 |
| D.3927.1 | 120 | 67.7 | 5.117 | 0.71852 |
| D.4317.1 | 166 | 77.4 | 6.229 | 0.72067 |
| (c) Gulliver Nunatak | | | | |
| R.347.2 | 102 | 390 | 0.736 | 0.70938 |
| R.347.3 | 124 | 352 | 1.018 | 0.71001 |
| R.351.1 | 216 | 73.3 | 8.569 | 0.72873 |
| R.351.2 | 151 | 110 | 3.986 | 0.71739 |
| R.360.1 | 234 | 156 | 4.333 | 0.71832 |
| (d) Webb I | | | | |
| R.026.1 | 107 | 365 | 0.844 | 0.70463 |
| R.026.2 | 54.5 | 488 | 0.323 | 0.70414 |
| R.026.3 | 55.3 | 449 | 0.356 | 0.70415 |
| R.018.1 | 9.6 | 481 | 0.058 | 0.70388 |
| (e) Colbert Mountains | | | | |
| KG.2473.3 | 132 | 41.2 | 9.277 | 0.73193 |
| KG.2473.4 | 85.5 | 74.7 | 3.313 | 0.70874 |
| KG.2473.6 | 251 | 18.1 | 40.31 | 0.74128 |
| KG.2479.1 | 120 | 53.9 | 6.463 | 0.71136 |
| KG.2479.2 | 91.5 | 115 | 2.310 | 0.70771 |
| KG.2479.3 | 118 | 62.9 | 5.428 | 0.71042 |
| (f) N. Palmer Land | | | | |
| KG.1201.6 | 77.6 | 170 | 1.324 | 0.70984 |
| KG.1202.1 | 104 | 151 | 1.989 | 0.71187 |
| KG.1208.2 | 4.5 | 269 | 0.049 | 0.70592 |
| KG.1208.6 | 22.8 | 286 | 0.231 | 0.70612 |
| KG.1212.1 | 16.0 | 137 | 0.338 | 0.70827 |
| KG.1212.4 | 2.4 | 128 | 0.054 | 0.70752 |
| KG.1212.5 | 25.6 | 704 | 0.105 | 0.70458 |
| KG.1212.7 | 57.9 | 210 | 0.797 | 0.70554 |
| KG.1212.8 | 9.1 | 157 | 0.168 | 0.70670 |
| KG.1212.9 | 31.5 | 690 | 0.132 | 0.70466 |
| KG.1212.10 | 150 | 58.6 | 7.394 | 0.72366 |
| KG.1212.2 | 72.6 | 364 | 0.577 | 0.70487 |
| KG.1218.2 | 17.2 | 270 | 0.184 | 0.70784 |
| KG.1218.7 | 30.4 | 127 | 0.701 | 0.70833 |
| KG.1218.10 | 134 | 112 | 3.472 | 0.71568 |
| E.3268.1 | 28.6 | 143 | 0.578 | 0.70815 |
| E.3278.1 | 77.1 | 650 | 4.113 | 0.71775 |

Rb, Sr in ppm
Error on $^{87}Rb/^{86}Sr$ ~0.5% <50 ppm, ~1.0% >50 ppm
Error on $^{87}Sr/^{86}Sr$ ~0.01%

## CONCLUSIONS

In the Antarctic Peninsula region, calc-alkaline volcanic rocks were erupted throughout the magmatic history of an arc constructed largely upon rocks deformed during the Gondwanian Orogeny. Once considered as belonging mainly to a single Late Jurassic volcanic episode, it has become clear that these rocks span a much greater time interval: Early Jurassic (perhaps even Triassic)—Early Tertiary. Although geographic variations in bulk and minor element composition have been observed (Weaver et al., 1982), no lithologic criteria have been established which enable the age of the rocks to be determined in the field. State of alteration (Adie, 1972 and references therein) is not a reliable guide to relative age, even in a small area, and at present age can only be determined where there is independent palaeontologic or radiometric evidence.

The most complete record of volcanism is present in the South Shetland Islands, where the rocks range from Late Jurassic to Oligocene. As in northern Graham Land a substantial subaerial volcanic arc probably did not develop until Early Cretaceous times. Only Early—Middle Jurassic (possibly Triassic) representatives have been identified in southeastern Graham Land but on the west coast the sequence probably ranges from Late Jurassic to Early Tertiary. In the Palmer Land area most of the volcanic rocks seem to be Middle-Late Jurassic, although Cretaceous representatives may also be present near the western coast. In Late Cretaceous—Early Tertiary times the southern part of the arc appears to have made a westward step into Alexander Island.

## REFERENCES


ADIE, R.J., 1954: The petrology of Graham Land: I. The basement complex; early Palaeozoic plutonic and volcanic rocks. *Scient. Rep. Falkld Isl. Depend. Surv., 11.*

ADIE, R.J., 1962: The geology of Antarctica; *in* Wexler, H., Rubin, M.J. and Caskey, J.E. (eds.) *Antarctic research: the Matthew Maury Memorial Symposium.* American Geophysical Union, Washington, D.C., 26-39.

ADIE, R.J., 1972: Evolution of volcanism in the Antarctic Peninsula; *in* Adie, R.J. (ed.) *Antarctic Geology and Geophysics.* Universitetsforlaget, Olso, 137-41.

ALARCÓN, B., AMBRUS, J., OLCAY, L. and VIERA, C., 1976: Geología del Estrecho de Gerlache entre los parallelos 64° y 65° lat. Sur. Antártica chilena. *Ser. cient. Inst. antart. chil., 4,* 7-51.

ASKIN, R.A.,(this volume): Tithonian—Barremian spores, pollen and microplankton from the South Shetland Islands, Antarctica.

BAKER, P.E., BUCKLEY, F. and REX, D.C., 1977: Cenozoic volcanism in the Antarctic. *R. Soc. Lond., Philos. Trans., Ser. B, 279,* 131-42.

BALDONI, A.M., 1979: Nuevos elementos paleofloristicos de la tafoflora de la Formación Spring Hill, limite Jurásico—Cretácico subsuelo de Argentina y Chile austral. *Ameghiniana, 16,* 103-19.

BURN, R.W., 1981: Early Tertiary calc-alkaline volcanism on Alexander Island. *Bull. Br. antarct. Surv., 53,* 175-93.

CRAME, J.A., 1982: Late Jurassic inoceramid bivalves from the Antarctic Peninsula and their stratigraphical use. *Palaeontology, 25,* 555-603.

DALZIEL, I.W.D. and ELLIOT, D.H., 1973: The Scotia arc and Antarctic margin: *in* Nairn A.E.M. and Stehli, F.G. (eds.) *The ocean basins and margins, V.I, The South Atlantic.* Plenum Pub. Corp., New York, 171-246.

DAVIES, R.E.S., (in press): The geology of the Marian Cove area, King George Island, and a Tertiary age for its supposed Jurassic rocks. *Bull. Br. antarct. Surv., 52,*

DAVIES, T.G., 1976: The geology of part of northern Palmer Land, Antarctica. Unpublished Ph.D. thesis, University of Birmingham, England. 152pp.

ELLIOT, D.H. and TRAUTMAN, T.A., 1982: Lower Tertiary strata on Seymour Island, Antarctic Peninsula; *in* Craddock, C. (ed.) *Antarctic Geoscience.* Univ. Wisconsin Press, Madison, 287-97.

FLEET, M., 1968: The geology of Oscar II Coast, Graham Land. *Scient. Rep. Br. antarct. Surv., 59.*

FARQUHARSON, G.W., 1982: Late Mesozoic sedimentation in the northern Antarctic Peninsula and its relationship to the southern Andes. *Geol. Soc. Lond., J., 139,* 721-8.

FRASER, A.G. and GRIMLEY, P.H., 1972: The geology of parts of the Bowman and Wilkins Coasts, Antarctic Peninsula. *Scient. Rep. Br. antarct. Surv., 67.*

FUENZALIDA, H., ARAYA, R. and HERVÉ, F., 1972: Middle Jurassic flora from northeastern Snow Island, South Shetland Islands; *in* Adie, R.J. (ed.) *Antarctic Geology and Geophysics.* Universitetsforlaget, Oslo, 93-7.

GRIKUROV, G.E., KRYLOV, A.YA. and SILIN, I.YU., 1967: Absolyutnyy vozrast nekotorykh porod dugi Skotiya i Zemli Aleksandra I (Zapadnaya Antarktika). *Dokl. Akad. Nauk. SSSR, Geology, 172,* 168-71.

GRIKUROV, G.E., KRYLOV, A.YA., POLYAKOV, M.M. and TSOVBUN, YA.N., 1970: Vozrast porod v severnoy chasti Antarkticheskogo polyostrova i na Yuzhnykh Shetlandskikh ostrovakh (po dannym kaliy-argonovogo metoda). *Inform. Byul. Sov. Antark. Eksp., 80,* 30-3.

HALLE, T.G., 1913: The Mesozoic flora of Graham Land. *Wiss. Ergebn. schwed. Sudpolarexped., 1901-1903, 3, (14),* 1-123.

HALPERN, M.A., 1964: Cretaceous sedimentation in the "General Bernardo O'Higgins" area of northwest Antarctic Peninsula; *in* Adie, R.J. (ed.) *Antarctic Geology.* North Holland, Amsterdam, 334-51.

HAWKES, D.D., 1961: The geology of the South Shetland Islands: I. The petrology of King George Island. *Scient. Rep. Falkld Isl. Depend. Surv., 26.*

JEFFERSON, T.H., 1980: Angiosperm fossils in supposed Jurassic volcanogenic shales, Antarctica. *Nature, 285,* 157-8.

LAUDON, T.S., THOMSON, M.R.A., WILLIAMS, P.L., ROWLEY, P.D. MILLIKEN, K.L. and BOYLES, J.M. (this volume): The Jurassic Latady Formation, southwestern Antarctic Peninsula.

LUCAS, R.C. and LACEY, W.S., 1981: A permineralised wood of probable early Tertiary age from King George Island, South Shetland Islands. *Bull. Br. antarct. Surv., 53,* 147-51.

ODIN, G.S. (ed.) (in press): *Numerical dating in stratigraphy.* Wylie and Sons, Chichester.

PANKHURST, R.J., 1982: Rb-Sr geochronology of Graham Land, Antarctica. *Geol. Soc. Lond., J., 139,* 701-12.

REX, D., 1976: Geochronology in relation to the stratigraphy of the Antarctic Peninsula. *Bull. Br. antarct. Surv., 43,* 49-58.

ROWLEY, P.D. and WILLIAMS, P.L., 1982: Geology of the northern Lassiter Coast and southern Black Coast, Antarctica; *in* Craddock, C. (ed.) *Antarctic Geoscience.* Univ. Wisconsin Press, Madison, 339-56.

ROWLEY, P.D., KELLOGG, K.S., CARRARA, P.E., et al. (this volume): Geology and plate tectonic setting of the Orville Coast and eastern Ellsworth Land, Antarctica.

SMELLIE, J.L., DAVIES, R.E.S. and THOMSON, M.R.A., 1980: Geology of a Mesozoic intra-arc sequence on Byers Peninsula, Livingston Island, South Shetland Islands. *Bull. Br. antarct. Surv.*, *50*, 55-76.

STEIGER, R.H. and JÄGER, E., 1977: Subcommission on geochronology: convention on the use of decay constants in geo- and cosmochronology. *Earth Planet. Sci. Lett.*, *36*, 359-62.

STIPANICIC, P.N. and BONETTI, M.I.R., 1970: Posiciones estratigráficas y edades de las principales floras jurásicas argentinas. II. Floras doggerianas y malmicas. *Ameghiniana*, *7*, 101-18.

TAYLOR, B.J., THOMSON, M.R.A. and WILLEY, L.E., 1979: The geology of the Ablation Point—Keystone Cliffs area, Alexander Island. *Scient. Rep. Br. antarct. Surv.*, *82*.

THOMSON, M.R.A., 1972: New discoveries of fossils in the Upper Jurassic Volcanic Group of Adelaide Island. *Bull. Br. antarct. Surv.*, *30*, 95-101.

THOMSON, M.R.A., 1975: Upper Jurassic Mollusca from Carse Point, Palmer Land. *Bull. Br. antarct. Surv.*, *41-42*, 31-42.

THOMSON, M.R.A., 1981: Late Mesozoic stratigraphy and invertebrate palaeontology of the South Orkney Islands. *Bull. Br. antarct. Surv.*, *54*, 65-83.

THOMSON, M.R.A., 1982: Mesozoic paleogeography of West Antarctica; *in* Craddock, C. (ed.) *Antarctic Geoscience*. Univ. Wisconsin Press, Madison, 331-7.

THOMSON, M.R.A., (in press): Late Jurassic fossils from Low Island, South Shetland Islands. *Bull. Br. antarct. Surv.*

WATTS, D.R., 1982: Potassium-argon ages and paleomagnetic results from King George Island, South Shetland Islands; *in* Craddock, C. (ed.) *Antarctic Geoscience*. Univ. Wisconsin Press, Madison, 255-61.

WEAVER, S.D., SAUNDERS, A.D. and TARNEY, J., 1982: Mesozoic—Cenozoic volcanism in the South Shetland Islands and the Antarctic Peninsula: geochemical nature and plate tectonic significance; *in* Craddock, C. (ed.) *Antarctic Geoscience*. Univ. Wisconsin Press, Madison, 263-73.

WEST, S.M., 1974: The geology of the Danco Coast, Graham Land. *Scient. Rep. Br. antarct. Surv.*, *84*.

# THE SEAL NUNATAKS: AN ACTIVE VOLCANIC GROUP ON THE LARSEN ICE SHELF, WEST ANTARCTICA

O. Gonzalez-Ferran, *Departamento de Geologia y Geofisica, Universidad de Chile,.Casilla 16, Correo Miramonte, Santiago, Chile.*

*Abstract* Recent investigations carried out in January 1982 confirm that the Seal Nunataks Group (comprising 16 volcanic cones emerging from the Larsen Ice Shelf, about Latitude 65°S and Longitude 60°W), represents an area of active volcanism in Antarctica. Two of these cones, Murdoch Volcano (368 m) and Dallman Volcano (210 m), show active fumaroles and abundant water steam at their parasitic and central craters, respectively. Wide areas of the Larsen Ice shelf around the volcanoes are covered by pyroclastic material, indicating the occurrence of some relatively recent eruption. The absence of a snow and ice cover from the majority of the cones and the abundant melting which has formed small lakes and rivers on the Larsen Ice Shelf, are evidence for present day existence of a high thermal flow in the area. Larsen (as early as 1893) observed volcanic activity at Christensen and Lindenberg volcanoes, belonging to this Group. These volcanic centres morphologically correspond to a series of cones consisting of pyroclastics and lava flows, some of them with well conserved cones which would represent the summit of major volcanoes rooted on the continental shelf, some 300 to 500 metres below the Ice Shelf level on the northwestern sector of the Weddell Sea. These volcanic centres show a clear, N60-70°W striking structural control. Chemically and petrographically they correspond to alkali olivine basalt, indicating volcanism associated with extensive tectonic processes.

The Seal Nunataks were investigated during the 1981/82 Antarctic summer, as part of an extensive geological-geophysical project comprising a number of transverse profiles crossing the Antarctic Peninsula and joining the Bellinghausen and Weddell Seas, as far as latitude 72°S. The project was carried out by investigators of the Depto. de Geologia y Geofisica, Universidad de Chile, with the valuable logistic support of Bell 212 Helicopters and Twin Otter Aeroplanes, specially fitted with skis, granted by the Chilean Air Force. The present work contains the preliminary field observations.

## PREVIOUS STUDIES

The Seal Nunataks, which represent a Volcanic Group of Recent age, have been considered as a part of the James Ross Island volcanic province (Adie, 1972; Baker et al., 1977; Gonzalez-Ferran, 1982). They are located on the Larsen Ice Shelf, around latitude 65°S and longitude 60°W, on the eastern coast of Antarctic Peninsula (Figure 1). They were discovered and visited for the first time by Captain Larsen in December, 1893 (Larsen, 1894), who called them the Seal Islands, at the same time noting their probable volcanic origin and reporting that Lindenberg and Christensen were in full eruptive activity. In 1901-1903, Nordenskjold visited the same region and observed a crater at Christensen without major activity and verified the basaltic character of the effusive rocks (Nordenskjold and Anderson, 1905). After more than 40 years without observations, Adie in 1947 started an extensive regional geological study (Adie, 1964) of the same area. The Seal Nunataks were successively visited by Stoneley (in 1952) and Standring (in 1953) with Fleet's study, carried out between 1962-63, the most complete one (Fleet, 1968). During the same decade, the British Antarctic Survey carried out important geophysical studies (Renner, 1969; Renner, 1980). During November 1978, Argentine investigators undertook geological and glaciological studies. Finally, in January 1982, the author carried out an extensive geological and gravimetrical study of the volcanic centres, observing the existence of two volcanic centres with fumarolic activity from the central cone of Dallman Volcano and from a parasitic crater of Murdoch Volcano.

The volcanic geology of the Seal Nunataks and their tectonic control is described here and a brief consideration of the main petrological and chemical characteristics of the volcanic rocks is given. A detailed study and description is being prepared.

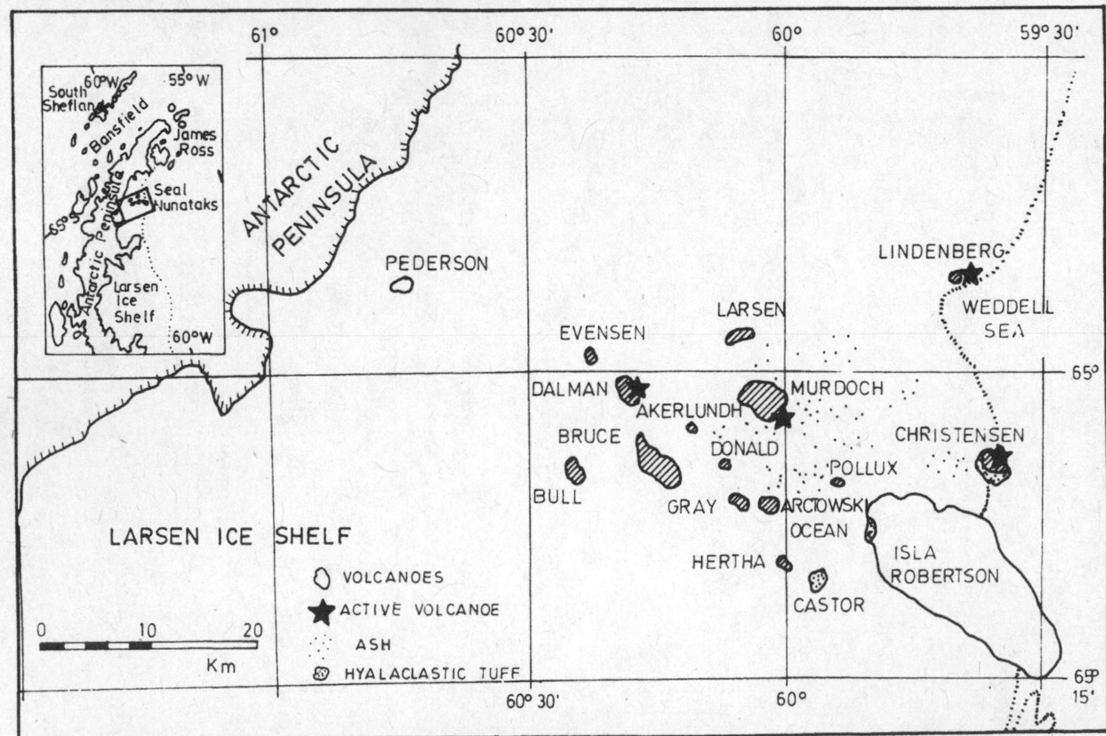


**Figure 1. Map of Seal Nunataks area, showing the locations of the volcanoes, active volcanic centres and hyaloclastite tuff deposits.**

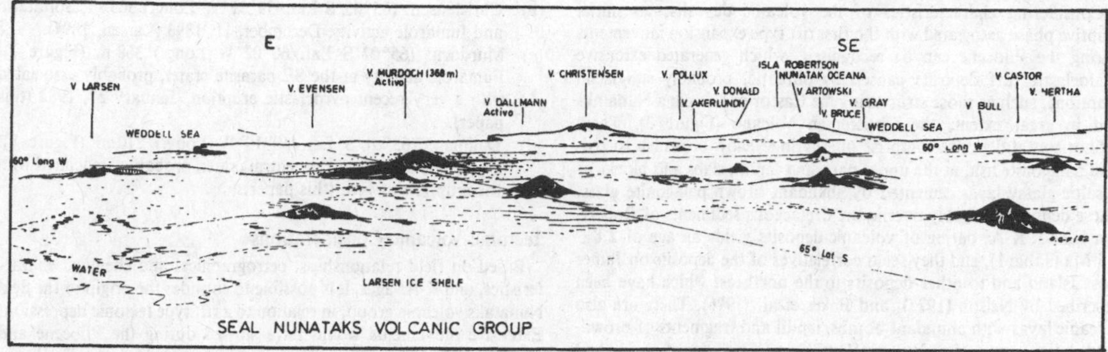


**Figure 2. Panoramic sketch towards the east southeast, showing the general distribution and morphological aspects of the Seal Nunataks on the Larsen Ice Shelf.**

## Volcanic Geology

The Seal Nunataks are located on the northern border of the Larsen Ice Shelf, about 150 km southwest of the James Ross Island (Figure 1). They are composed of at least 16 volcanic islands which overlie basement estimated to be about 500 m below sea level. Their volcanic structure has exceeded the Ice Shelf in an extension of 1,200 km² (Figure 2). A clear NW—SE direction and elongation of the volcanic structures can be observed.

Morphologically, they are mainly volcanic cones, some of them very well conserved, revealing their Recent age, as is the case with Lindenberg Dome, the polygenetic cones Dallman, Donald, Bruce and Bull and the parasite crater of Murdoch. Others such as Larsen, Evensen, Arctowski and Herta, show more amorphous volcanic structures of volcanic agglomerates. The Castor, Oceana and some parts of Christensen comprise tabular beds of hyaloclastic tuff deposits.

Murdoch Volcano (368 m) represents the most prominent feature in the centre of the area. The ice surface of the Shelf extending among these volcanoes is crossed by a number of watercourses and small lakes, and large areas are covered by red and black pyroclastic material. All the volcanoes, except Christensen, Castor and Oceana on Robertson Island, are not covered with ice. Table 1 shows a complete list of the volcanoes and their main characteristics, K-Ar ages and the observed activity.

**TABLE 1: The Seal Nunataks Volcanic Centres.**

| Name | S. Lat. | W. Long. | High over sea level | Historic activity | Ages Ma K/Ar (3) | Refer-ences |
|---|---|---|---|---|---|---|
| Lindenberg | 64°55′30″ | / 59°40′ | — | eruption 11.12.1893 | — | (1) |
| Larsen | 64°57′ | / 60°05′ | 128 | — | 1.5 ± 0.5 | (2) (4) (5) |
| Evenson | 64°59′ | / 60°22′ | 152 | — | 1.4 ± 0.3 | |
| Dallman | 65°01′ | / 60°19′ | 210 | Fumaroles 27.01.1982 | — | |
| Murdoch | 65°02′ | / 60°02′ | 368 | Fumaroles 27.01.1982 | — | |
| Akerlundch | 65°03′ | / 60°11′ | 90 | — | 0.7 ± 0.3 | (4) |
| Bruce | 65°04′30″ | / 60°15′ | 219 | — | 1.5 ± 0.3 | (4) (6) |
| Donald | 65°04′30″ | / 60°07′ | 130 | — | <0.2 | |
| Christensen | 65°05′ | / 59°35′ | 305 | Solfatares 11.12.1893 | 0.7 ± 0.3 | (1) (7) |
| Bull | 65°05′ | / 60°24′ | 230 | — | — | (6) |
| Pollux | 65°05′30″ | / 59°54′ | 85 | — | — | |
| Gray | 65°06′30″ | / 60°05′ | — | — | <0.2 | |
| Arctowski | 65°06′30″ | / 60°02′ | — | — | 1.4 ± 0.3 | |
| Oceana | 65°08′ | / 59°50′ | — | — | 2.8 ± 0.5 | |
| Hertha | 65°09′30″ | / 60°00′ | — | — | — | |
| Castor | 65°10′ | / 59°56′ | — | — | — | |

References:
(1) Larsen (1894).
(2) Baker *et al.* (1977).
(3) Del Valle and Fourcade (letter communication). Age determination (whole rocks) by Institut de Geologia Isotopica. Univ. de B. Aires (unpublished).
(4) Paukhurst (1982).
(5) Saunders (1982).
(6) Fleet (1968).
(7) Nordenskjold and Anderson (1911).
(8) Berminghausen and Neuman Van Padag (1960).

**Table 1. The Seal Nunataks Volcanic Centres location and K-Ar ages in Ma (Whole rock determinations) and historic activity.**

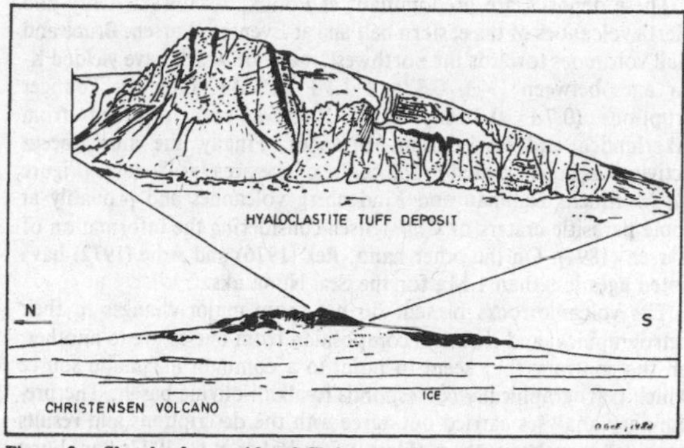


**Figure 3. Sketch of west side of Christensen Volcano. (A) outcrop detail of hyaloclastite tuff deposit.**

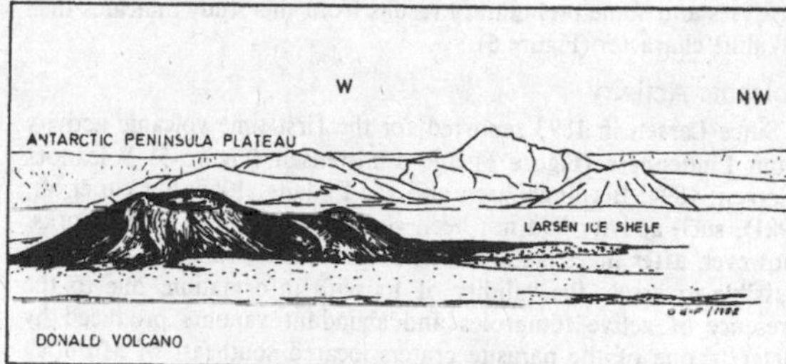


**Figure 4. Donald Volcano, its state of preservation shows that the volcanic structure is very young, probably less than 1 Ma old.**

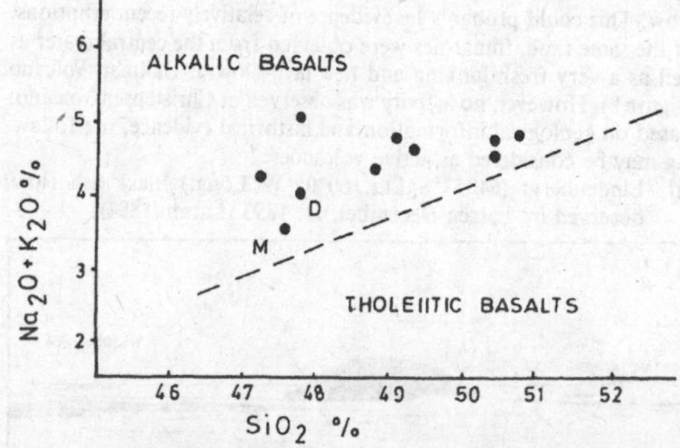


**Figure 5. Total alkalis versus silica diagram, analyses from the Seal Nunataks: (D) Dallman Volcano; (M) Murdoch Volcano. Line separating the tholeiitic and alkaline fields according McDonald and Katsura (1964). Chemistry analyses from Baker et al. (1977) and Gonzalez-Ferran (unpublished).**

Considering characteristics of the volcanic deposits, an initial eruptive phase associated with the first rift type expansion movements during the Pliocene can be recognised, which generated extensive hyaloclastic tuff deposits caused by phreatic, probably subglacial eruptions, such as those structuring the Castor and Oceana Nunataks and in great extent, the Christensen Volcano (Figure 3). These subhorizontal deposits comprise alternating yellow-brownish to reddish palagonite ash, at the upper part, and lapilli, slabs and blocks of basaltic-glassy lavas cemented by abundant brown palagonite glass. These deposits probably overlie the Cretaceous sediments of Robertson Island. K-Ar dating of volcanic deposits yields an age of 2.8 ± 0.5 Ma (Table 1), and they seem correlatives of the deposits on James Ross Island and to other deposits to the northeast which have been described by Nelson (1975), and Baker et al. (1976). There are also volcanic lavas with abundant bombs, lapilli and fragments of brown-reddish lava, some of which contain ultramafic nodules of olivine and pyroxene enclosed fragments of glassy olivine basalt, similar to those found at Bull Volcano (Fleet, 1968) indicating subaerial eruptive phases alternating with phreatic explosions.

These deposits are predominant at Pollux, Arctowski, Gray and Herta volcanoes of the eastern belt and at Evensen, Larsen, Bruce and Bull volcanoes towards the northwest, many of which have yielded K-Ar ages between 1.4 ± 0.3 and 1.5 ± 0.5 Ma (Table 1). Younger eruptions (0.7 ± 0.3 Ma, Table 1) have been recorded from Akerlundch and Christensen Volcanoes. Finally, the most recent activity has concentrated in the centre of the area, at Donald, (Figure 4), Murdoch, Dallman and Lindenberg volcanoes and probably at some parasitic craters of Christensen considering the information of Larsen (1894). On the other hand, Rex (1976) and Adie (1972) have noted ages less than 1 Ma for the Seal Nunataks.

The volcanic rocks present do not show major changes in their petrographical and chemical composition from one centre to another; on the contrary they seem to point to a common magmatic source which petrographically corresponds to alkali-olivine basalt. The preliminary analyses carried out agree with the descriptions and results obtained by other authors (Fleet, 1968; Baker et al. 1977; Pankhurst 1982; Saunders 1982) who identified these rocks as olivine basalts and hawaiites. A variation diagram of total alkalis vs silica for published analyses and some preliminary results from this study indicates their alkaline character (Figure 5).

## Volcanic Activity

Since Larsen in 1893 reported for the first time volcanic activity from Lindenberg (Figure 6) and Christensen (Figure 3) Volcanoes (Larsen, 1894; Berninghausen and van Padang, 1960; Simkin et al., 1981), such activity had not been confirmed by more recent works. However, after investigations carried out during January 1982, it is possible to assess the validity of Larsen's information, due to the presence of active fumaroles and abundant vapours produced by water in one of the parasite craters located southeast of Murdoch Volcano (Figure 7) and about 20 km away from Christensen Volcano. The mentioned pyroclastic cone is only some metres higher than the Ice Shelf and abundant lapilli, fragments and ash of basaltic composition cover wide areas of the ice surface, in parts covered by fresh snow. This could probably be evidence of relatively recent eruptions. At the same time, fumaroles were observed from the central crater as well as a very fresh looking and new lava flow at Dallman Volcano (Figure 8). However, no activity was observed at Christensen Volcano. Based on geological information and historical evidence, the following may be considered as active volcanoes:

(a) Lindenberg: (64°57'S Lat./60°05'W Long.) Black ash flow, observed by Larsen December, 11, 1893 (Larsen 1894).

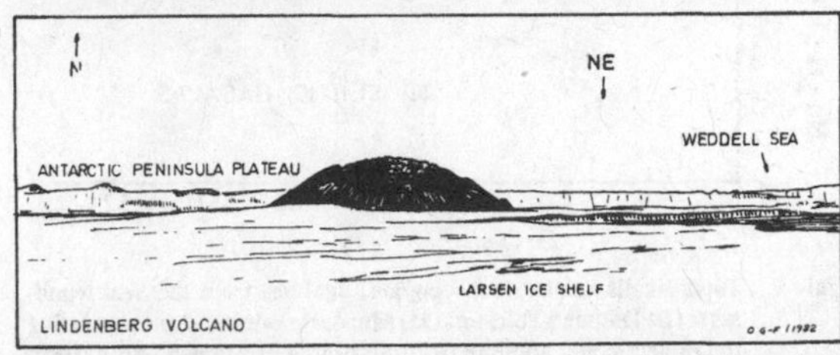


**Figure 6. Sketch showing the south side of the Lindenberg Volcano lava dome.**

(b) Christensen: (65°05'S Lat./59°35'W Long.) 305 m. Solfataric and fumarole activity, December, 11, 1893 (Larsen, 1894).

(c) Murdoch: (65°02'S Lat./60°02'W Long.) 368 m (Figure 7). Fumarole activity at the SE parasite crater, probably associated with a very recent pyroclastic eruption, January 27, 1982 (this paper).

(d) Dallman: (65°01'S Lat./60°19'W Long.) 210 m (Figure 8). Fumarolic activity at the central crater, probably associated with the eruption of lava (this paper).

## Tectonic-Volcanic Considerations

Based on field relationships, petrographical and chemical characteristics, and K-Ar ages, it is possible to consider the origin of the Seal Nunataks volcanic group, in relation to a rift type tectonic depression. Extensive movements would have started during the Pliocene and probably continue at present, indicated by the history of volcanic activity and the fracturing which is affecting the Recent volcanic structures. The presence of Mesozoic rocks at Cape Marsh, Robertson Island, (Fleet, 1966), at the southeastern end of the Seal Nunataks shows the existence of a block which could correspond to the southeastern border of the rift which could extend towards the northeast in the Seymour, Cockburn, Snow Hill and other islands, formed by extensive Cretaceous and Tertiary sedimentary deposits (described by Nelson, 1975 and Adie, 1972) and towards the southwest in the Jason Peninsula Block, where important deposits of calc-alkaline volcanic rocks exist (Saunders, 1982). The Jason Peninsula would be separated by a deep depression of the Antarctic Peninsula, in accord with geophysical information and interpretation by Renner (1980). This southeastern block of the Larsen Rift reaches at least

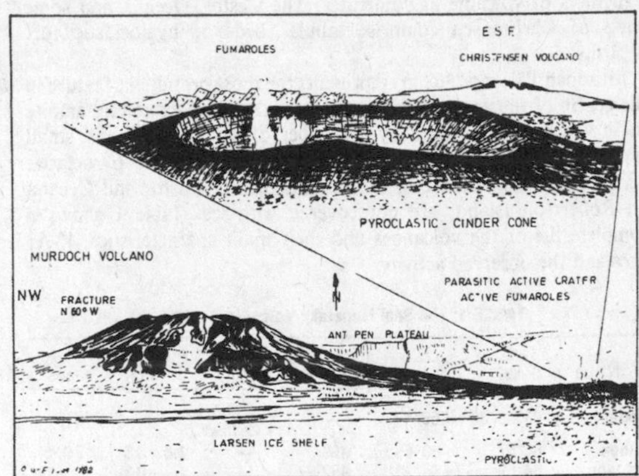


**Figure 7. Murdoch Volcano. The main volcanic structure of the Seal Nunataks. Sketch showing the SSE side of the volcano and the parasitic cinder cone with active fumaroles. Upper sketch, shows detail of this very recent and active parasitic centre; located at the SE slope at the Murdoch volcano, near the Larsen Ice Shelf level.**

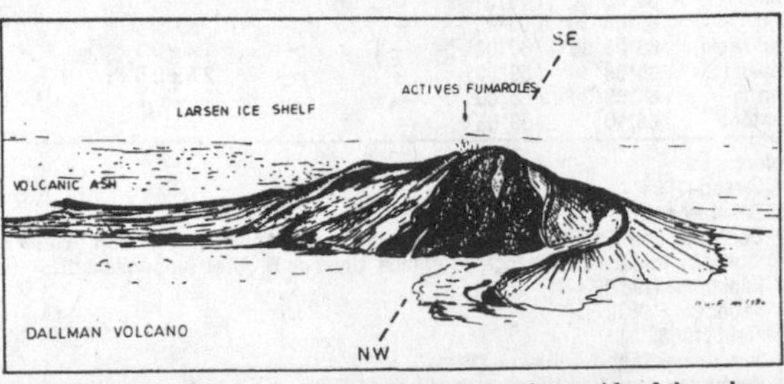


**Figure 8. Dallman Volcano. View showing the northwest side of the volcano and a probably recent basaltic lava flow from the main crater, and active fumaroles at the north side of the summit.**

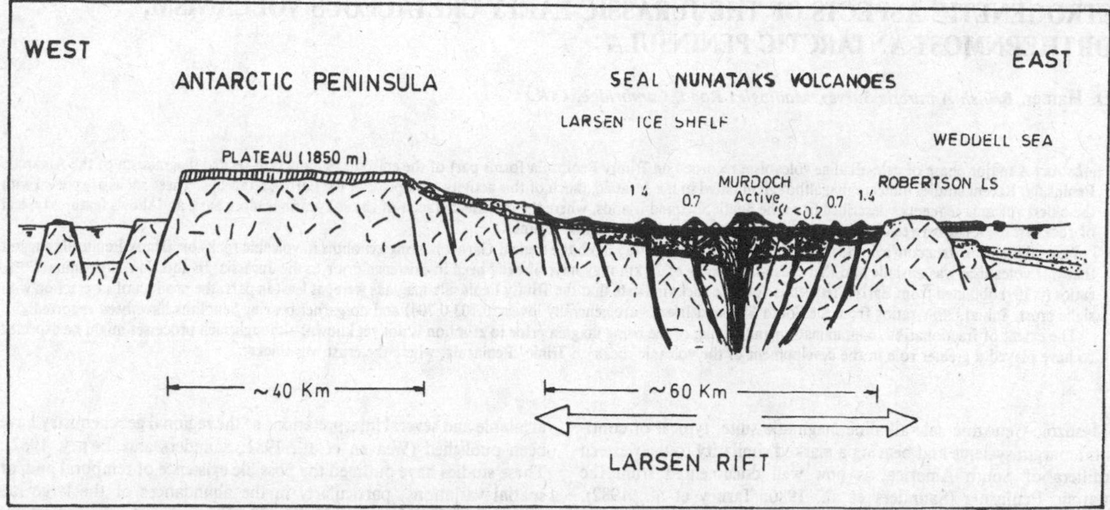


**Figure 9. Sketch showing the interpretation of the Larsen Rift and the open fissure of volcanic activity during Late Pliocene to Recent.**

400 km in length and is separated between 30 and 60 km of the Antarctic Peninsula. In the area of the Seal Nunataks this gap reaches 50 km between Robertson and Pedersen Islands. The cliffs left by the fracture on the eastern slope of the Antarctic Peninsula and by the differential displacement of the blocks, support the hypothesis of the existence of a rift (Figure 9).

The first deposits of volcanic rocks related with the rift system correspond to yellow-brownish hyaloclastic tuffs with abundant palagonite glass, which are very well exposed at Castor and Oceana Nunatak and less well exposed at Christensen Island (Figures 1 and 2); their Pliocene age would be supported by a K-Ar determination of fragments of basaltic lavas from Oceana (Table 1) and which could be correlative with similar deposits described for James Ross Island (Nelson, 1975). On the other hand, morphological evidence from field surveys, supported by radiometric dates (Table 1), show a clear symmetry of the evolution of eruptive activity during the Pleistocene to Recent. A belt of active volcanism with a northeast strike would extend through Lindenberg, Murdoch and Donald volcanoes, these centres providing evidence for historic as well as present day activity, with a high thermal flow leading to fumarolic activity. It is interesting to observe the relative chronological symmetry presented by the volcanic centres as they are located further away on both sides from the active volcanic axis (Figure 9). This symmetry would be evidence for extension processes of the Larsen Rift during the Upper Pliocene-Pleistocene. Its elongate distribution with a WNW strike which reaches the volcanic centres could be assigned to the control due to deep faults of horizontal displacement with the same strike, which could affect the Antarctic Peninsula and which probably is related to the Hero fracture zone noted by Herron and Tucholke (1976) at the subduction segment of the Shetland Islands. Saunders (1982) considered that the Seal Nunataks basalts were quite distinct from the calc-alkaline basalts and andesites of the Antarctic Peninsula and the South Shetland Island, having significantly higher total iron, $TiO_2$ and Nb contents, and lower La/Nb; Zr/Nb; Ba/La and Ba/Nb ratios and suggested that alkaline volcanicity has occurred in response to extensional tectonism. The petrology and chemistry of magmatism associated with the initial stages of back-arc spreading have been discussed by Weaver et al., (1979), Weaver et al., (1982) and Tarney et al., (1981). Finally, the alkaline character of its effusive rocks agrees with the suggested Larsen Rift System.

*Acknowledgements* This study was financed by Grant E 1360-8224 of the Departamento de Desarrollo Cientifico de la Universidad de Chile and the Chilean Air Force. The latter not only supported the cost of the operation in the field but also gave a safe and efficient logistic support. My thanks to the Dean of the Faculty of Physical and Mathematics Sciences and to Prof. E. Kausel, Director of the Departamento de Geologia y Geof'isica de la Universidad de Chile, for their constant support. My gratitude to Dr P. Morris from Dept. of Geology and Geophysics, University of Sydney, who read this manuscript.

## REFERENCES

ADIE, R.J., 1964: Geological History; *in* Priestley, R.E., Adie, R.J. and Robin G de Q. (eds.) *Antarctic Research.* Butterworth, London, 118-62.

ADIE, R.J., 1972: Evolution of volcanism in the Antarctic Peninsula. *in* Adie, R.J. (ed.) *Antarctic Geology and Geophysics,* Universitetsforlaget, Olso, 137-41.

BAKER, P.E., BUCKLEY, F. and REX, D.C., 1977: Cenozoic volcanism in the Antarctic. *R. Soc. Lond., Philos. Trans., Ser. B., 279,*131-42.

BAKER, P.E., GONZALEZ-FERRAN, O. and VERGARA, M., 1976: Geology and geochemistry of Paulet Island and the James Ross Island Volcanic Group; *in* González-Ferrán, O. *Andean and Antarctic Volcanology Problems.* Int. Assoc. Volcanol. Chem. Earth's Interior, Rome, 39-47.

BERNINGHAUSEN, W.H. and VAN PADANG, N., 1960: Catalogue of the active volcanoes of the World including solfatara fields. Part X, Antarctic. *Int. Assoc. Volcanol. Chem. Earth's Interior, Observatorio Vesuviano,* Naples, 1-32.

FLEET, M., 1968: The Geology of the Oscar II Coast, Graham Land, *Scient. Rep. Br. antarct. Surv., 59,* 1-46.

FLEET, M., 1966: Occurrence of fossiliferous Upper Cretaceous sediments at Cape Marsh, Robertson Island. *Bull. Br. antarct. Surv., 8,* 89-91.

GONZALEZ-FERRAN, O., 1982: The Antarctic Cenozoic Volcanic Provinces and their implications in plate tectonic processes; *in* Craddock, C. (ed.) *Antarctic Geoscience.* Univ. Wisconsin Press, Madison, 687-694.

HERRON, E.M. and TUCHOLKE, B.G., 1976: Sea floor magnetic patterns and basement structure in the southeastern Pacific. *Initial Reports of the Deep Sea Drilling Project, 35,* 263-78 U.S. Govt. Printing Office, Washington D.C.

LARSEN, C.A., 1894: The voyage of the "Jason" to the Antarctic regions. *Geogr. J., 4,* 333-44.

NORDENSKJOLD, N.O.G. and ANDERSON, J.G., 1905: *Antarctic.* Hurst and Blackett, London.

NELSON, P.H.H., 1975: The James Ross Island Volcanic Group of northeast Graham Land. *Scient. Rep. Br. antarct. Surv., 54,* 1-62.

PANKHURST, R.J., 1982: Sr-isotope and trace element geochemistry of Cenozoic volcanics from the Scotia arc and the northern Antarctic Peninsula; *in* Craddock, C. (ed.) *Antarctic Geoscience.* Univ. Wisconsin Press, Madison, 229-34.

RENNER, R.G.B., 1980: Gravity and magnetic surveys in Graham Land. *Scient. Rep. Br. antarct. Surv., 77,* 1-99.

RENNER, R.G.B., 1969: Surface elevations on the Larsen Ice Shelf, *Bull. Br. antarct. Surv., 19,* 1-8.

REX, D., 1976: Geochronology in relation to the stratigraphy of the Antarctic Peninsula. *Bull. Br. antarct. Surv., 43,* 59-58.

SAUNDERS, A.D., 1982: Petrology and geochemistry of alkali-basalts from Jason Peninsula, Oscar II Coast, Graham Land. *Bull. Br. antarct. Surv., 55,* 1-9.

SIMKIN, T., SIEBERT, L., MCCLELLAND, L., BRIDGE, D., NEWHALL, C. and LATTER, J.H., 1981: *Volcanoes of the World; a regional directory, gazetter, and chronology of volcanism during the last 10,000 years,* Hutchinson Ross, Stroudburg.

TARNEY, J., SAUNDERS, A.D., MATLEY, D.P., WOOD, D.A. and MARSH, N.G., 1981: Geochemical aspects of back-arc spreading in the Scotia Sea and western Pacific. *R. Soc. Lond., Philos. Trans., Ser. B., 300,* 263-85.

WEAVER, S.D., SAUNDERS, A.D. and TARNEY, J., 1982: Mesozoic-Cenozoic volcanism in the South Shetland Islands and the Antarctic Peninsula: geochemical nature and plate tectonic significance. *In Craddock, C. (ed.) Antarctic Geoscience,* Univ. Wisconsin Press, Madison 263-73.

WEAVER, S.D., SAUNDERS, A.D., PANKHURST, R.J. and TARNEY, J., 1979: A geochemical study of magmatism associated with the initial stages of back-arc spreading: the Quaternary volcanics of Bransfield Strait, from South Shetland Islands. *Contrib. Mineral. Petrol., 68,* 151-69.

# PETROGENETIC ASPECTS OF THE JURASSIC-EARLY CRETACEOUS VOLCANISM, NORTHERNMOST ANTARCTIC PENINSULA

R.D. Hamer, *British Antarctic Survey, Madingley Road, Cambridge, U.K.*

*Abstract* A major phase of calc-alkaline volcanism recorded on Trinity Peninsula forms part of the episodic, subduction-related magmatism of the Antarctic Peninsula. Recent studies indicate that although initiated in the Jurassic, much of this activity took place in the Early Cretaceous. These rocks are coeval with the oldest volcanic sequences described from the South Shetland Islands, where the overall character of the volcanism is more basic and shows features typical of rocks transitional between island-arc tholeiites and normal calc-alkaline regimes.

New whole-rock data confirm the trend of increasing acidity away from the trench. Garnet bearing xenoliths in volcanic rocks on Trinity Peninsula suggest that this volcanism was ensialic and that crustal thickness of 25 km may have already been in existence prior to the Jurassic. In addition, high initial $^{87/86}$Sr ratios (0.709) obtained from Early Cretaceous volcanic rocks indicate that the Trinity Peninsula magmas were, at least in part, the product of interaction with sialic crust. Initial $^{87/86}$Sr ratios from the South Shetland Islands are generally lower (0.703-0.704) and no garnet bearing xenoliths have been reported.

The extent of fractionation, contamination and mixing of the rising magma prior to eruption is not yet known, although such processes might be expected to have played a greater role in the development of the volcanic rocks on Trinity Peninsula, where the crust was thicker.

A Mesozoic-Cenozoic calc-alkaline magmatic suite, typical of continental margin systems and bearing a marked similarity to the western cordillera of South America, is now well documented from the Antarctic Peninsula (Saunders et al., 1980: Tarney et al., 1982). However, the absolute age and correlation of individual rock units, especially within the Antarctic Peninsula Volcanic Group (APVG), is still largely unknown (Thomson and Pankhurst, this volume). Detailed studies on the APVG and associated back-arc sediments at the northern tip of the Antarctic Peninsula (Farquarson, 1982; Pankhurst, 1982), together with the writer's observations, have recently indicated that arc-building activity reached a climax in the Early Cretaceous. This major phase of calc-alkaline volcanism is coeval with the oldest period of activity (Late Jurassic-Early Cretaceous), described from the South Shetland Islands (Smellie, 1979).

A large body of geochemical data relating to the magmatic suite is available and several interpretations of the regional geochemistry have been published (Weaver, et al., 1982; Saunders and Tarney, 1982). These studies have outlined the possible existence of temporal and/or spatial variations, particularly in the abundances of the large ion lithophile (LIL) elements (K. Rb, Th, Pb, etc) and the implications that they have for petrogenetic schemes. Prior to this study, no geochemical data have been available for the APVG of the northern Antarctic Peninsula adjacent to the South Shetland Islands. This region presents an excellent opportunity to investigate any transverse geochemical variations across the magmatic arc.

## GEOLOGICAL SETTING

Although volcanic activity related to the subduction of the Pacific Ocean crust beneath the west of the Antarctic Peninsula is well known in the South Shetland Islands (Smellie, 1979; Weaver, et al., 1982), a

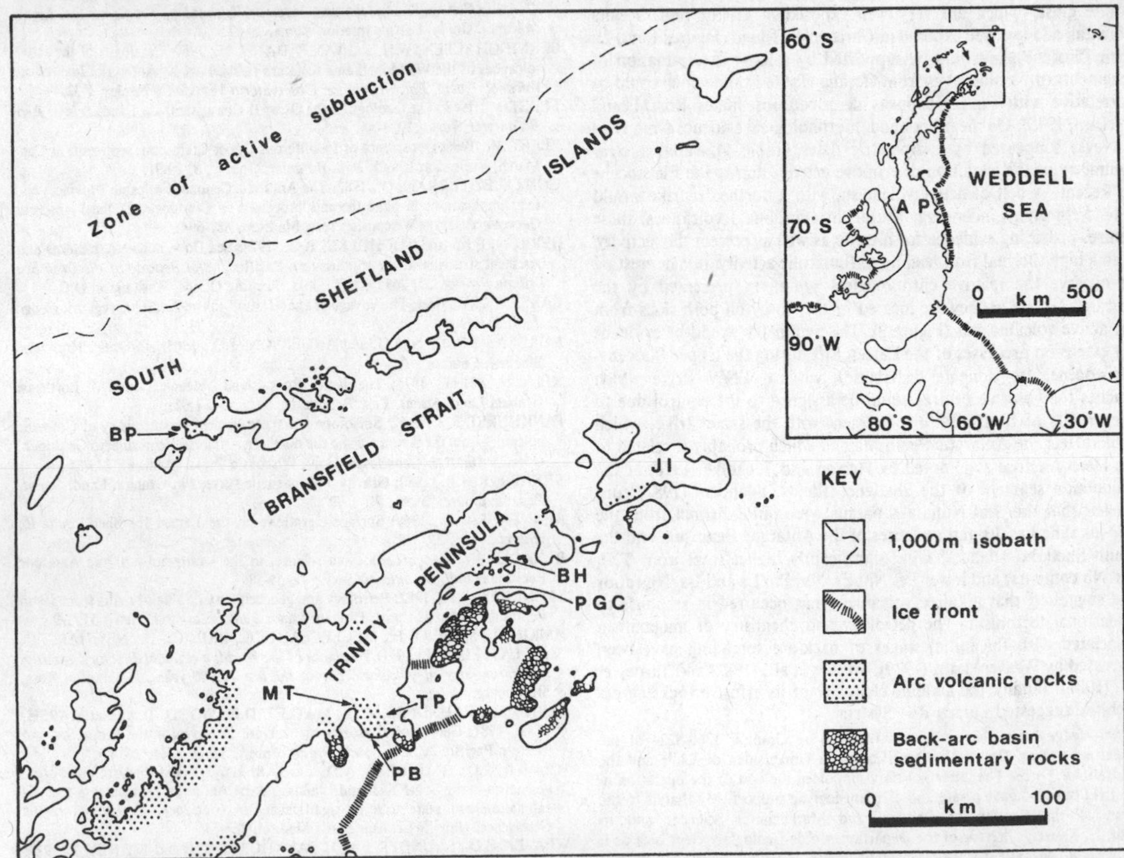


**Figure 1. Sketch map of the northernmost Antarctic Peninsula with Bransfield Strait closed, modified after Smellie (1979). AP-Antarctic Peninsula; BH-Bald Head; BP-Byers Peninsula; JI-Joinville Island; PB-Porphyry Bluff; PGC-Prince Gustav Channel; MT-Mount Tucker and TP-Tower Peak.**

comprehensive account of the equivalent rocks from Trinity Peninsula and Joinville Island (Figure 1) is lacking (Thomson and Pankhurst, this volume). The following comments and comparisons are relevant to this study.

*Age*

The most complete record of Trinity Peninsula volcanism is found in the associated sediments (Cretaceous-Tertiary) present behind the arc. Pyroclastic material occurs throughout the succession, but there is a marked concentration in Lower Cretaceous rocks indicating a peak of volcanic activity during this time. This is supported by a recent radiometric age of 117±4 Ma (Pankhurst, 1982), obtained from rhyodacites at Porphyry Bluff (Figure 1). There is as yet no radiometric evidence for younger calc-alkaline volcanic activity on Trinity Peninsula.

*Stratigraphy*

Correlation between outcrops is severely limited due to extensive snow cover and the diachronous nature of volcanic sucessions. Prior to the intensification of volcanism, an extensive shallow sea existed over much of the region, in which a mudstone-tuff sequence (the Nordenskjöld Formation) was deposited (Farquarson, 1982). Arc terrain deposits of the APVG post-date the Nordenskjöld Formation and comprise large volumes of subaerial, intermediate-acidic pyroclastic material including welded tuffs and rare lava flows. Terrestrial sediments are interbedded with the volcanic rocks at a number of localities. On Byers Peninsula in the South Shetland Islands (Figure 1), a mudstone-tuff association similar to the Nordenskjöld Formation passes diachronously into terrestrial sediments and are volcanic rocks (Smellie, 1979), indicating that the relationship between the two units may be complex.

*Composition*

The predominance of basic material in the South Shetland Islands has been noted by many authors, (e.g. Adie, 1972; Smellie, 1979). Andesites, dacites and rhyolites do occur, but they are subordinate to basalts and basaltic andesites. On Trinity Peninsula basic rocks are generally lacking and only one outcrop of basaltic andesite is known. Rhyolites however, occur throughout the area and andesites comprise much of the succession at Mount Tucker and the outcrops along the coast of the Prince Gustav Channel (Figure 1).

*Petrography*

In general the lavas and hypabyssal intrusions are petrographically similar. Most of the rocks are porphyritic and typically comprise 25-35% phenocrysts. Alteration is common, although its effects appear to be more widespread on Trinity Peninsula than in the South Shetland Islands. It frequently results in replacement of all primary minerals in the matrix (and in some instances the phenocrysts as well), by an assemblage including chlorite, actinolite, epidote, calcite and sericite. In the South Shetland Islands, these assemblages have been attributed to very low pressure metamorphism associated with hydrothermal activity around high level acidic plutons (Smellie, 1979).

Both compositional and regional control of the primary phases is evident. For example, olivine is confined to the basalts and basaltic andesites of the South Shetland Islands, whereas hypersthene and opaque ore are widely developed in the andesites and dacites of both areas. Phenocrysts of garnet and biotite are limited to the andesites and rhyolites of Trinity Peninsula (Hamer and Moyes, 1982). Groundmass pigeonite, commonly present in the lavas and hypabyssal intrusions of the South Shetland Islands, has not been found in the rocks from Trinity Peninsula.

*Bransfield Strait*

This narrow (65 km) marginal basin has opened during the last 4 Ma. Thus in considering the Late Jurassic-Early Cretaceous palaeogeography of the region, the South Shetland Islands have been placed in their former position by closing Bransfield Strait (Figure 1). This reconstruction places the volcanic front no more than 120 km from the trench and indicates that the volcanic arc was at least 150 km in width during the Early Cretaceous.

## GEOCHEMISTRY

Over 100 whole rock analyses have been performed on rocks of the APVG of Trinity Peninsula and Joinville Island using a Philips PW 1400 X-ray fluorescence spectrometer. Major elements were determined on lithium tetraborate/lithium carbonate fusion beads and trace element determinations were carried out on pressed powder discs. All calibrations were effected using international and laboratory standards. Mass absorption corrections for trace elements were made by monitoring the Ag- and W-tube lines or by using mass

**TABLE 1: Representative analyses of Late Jurassic-Early Cretaceous volcanic rocks from the northernmost Antarctic Peninsula**

| | South Shetland Islands | | | | | Trinity Peninsula and Joinville Island | | | | | | |
|---|---|---|---|---|---|---|---|---|---|---|---|---|
| | 1 | 2 | 3 | 4 | 5 | 6 | 7 | 8 | 9 | 10 | 11 | 12 |
| Sample No. | P.726.5 | P.725.2 | P.725.1 | P.848.5 | P.848.15 | R.637.1 | R.624.1 | R.625.1 | R.619.1 | R.653.2 | R.906.1 | R.608.2 |
| $SiO_2$ | 48.27 | 53.20 | 59.50 | 62.09 | 76.84 | 54.65 | 60.98 | 61.09 | 61.83 | 67.80 | 73.62 | 76.30 |
| $TiO_2$ | 1.07 | 1.31 | 1.34 | 0.55 | 0.12 | 0.82 | 0.63 | 0.62 | 0.57 | 0.55 | 0.24 | 0.11 |
| $Al_2O_3$ | 16.54 | 16.15 | 15.50 | 16.34 | 12.88 | 16.07 | 14.70 | 14.69 | 15.32 | 15.65 | 13.39 | 12.07 |
| $Fe_2O_3$* | 10.99 | 10.75 | 8.85 | 5.52 | 1.07 | 9.06 | 6.57 | 6.49 | 6.23 | 3.94 | 0.97 | 1.91 |
| MnO | 0.20 | — | — | 0.26 | 0.02 | 0.15 | 0.12 | 0.13 | 0.12 | 0.06 | 0.01 | 0.04 |
| MgO | 9.58 | 4.55 | 2.11 | 3.60 | 0.26 | 6.47 | 5.13 | 4.65 | 2.70 | 1.07 | 0.40 | 0.45 |
| CaO | 10.49 | 8.79 | 6.55 | 6.45 | 0.56 | 7.92 | 5.87 | 5.98 | 5.65 | 4.03 | 0.66 | 0.47 |
| $Na_2O$ | 2.50 | 3.29 | 3.86 | 3.38 | 3.62 | 2.05 | 2.38 | 2.28 | 2.47 | 2.97 | 2.62 | 2.26 |
| $K_2O$ | 0.24 | 0.50 | 0.54 | 0.89 | 3.83 | 0.68 | 2.01 | 2.01 | 2.39 | 2.66 | 5.99 | 4.90 |
| $P_2O_5$ | 0.22 | 0.22 | 0.32 | 0.22 | 0.01 | 0.18 | 0.14 | 0.14 | 0.14 | 0.13 | 0.06 | 0.03 |
| Total | 100.10 | 98.76 | 98.57 | 99.30 | 99.21 | 98.05 | 98.53 | 98.08 | 97.42 | 98.86 | 97.96 | 98.54 |
| Trace element ppm | | | | | | | | | | | | |
| Ni | 84 | 8 | 3 | 3 | 3 | 29 | 59 | 62 | 13 | >5 | >5 | >5 |
| Cr | 256 | 20 | 7 | 11 | 7 | 205 | 328 | 319 | 91 | 6 | >5 | >5 |
| Rb | 2 | 8 | 31 | 23 | 97 | 13 | 76 | 78 | 88 | 98 | 171 | 142 |
| Ba | 71 | 201 | 352 | 218 | 586 | 334 | 411 | 394 | 503 | 511 | 1169 | 745 |
| Sr | 350 | 457 | 411 | 409 | 47 | 311 | 241 | 237 | 280 | 234 | 79 | 100 |
| Th | <1 | 4 | 5 | 4 | 16 | na | na | na | na | na | na | na |
| Pb | <4 | 7 | 11 | <4 | 11 | na | na | na | na | na | na | na |
| La | 4 | 23 | 29 | 13 | 20 | 22 | 23 | 23 | 27 | 36 | 44 | 23 |
| Ce | 16 | 34 | 49 | 35 | 32 | 34 | 41 | 51 | 45 | 71 | 90 | 55 |
| Y | 18 | 26 | 40 | 18 | 30 | 23 | 22 | 21 | 22 | 26 | 29 | 33 |
| Zr | 69 | 113 | 192 | 140 | 106 | 134 | 132 | 135 | 133 | 214 | 180 | 115 |
| Nb | 3 | 6 | 6 | 4 | 8 | 8 | 8 | 9 | 8 | 10 | 9 | 11 |
| Zn | — | 86 | 119 | 65 | — | na | na | na | na | na | na | na |
| Ga | — | — | — | 15 | — | na | na | na | na | na | na | na |

* Total iron as $Fe_2O_3$
na — not analysed

1. P.726.5: Basalt; Byers Peninsula, South Shetland Islands
2. P.725.2: Basaltic andesite; Byers Peninsula, South Shetland Islands
3. P.725.1: Andesite; Byers Peninsula, South Shetland Islands
4. P.848.5: Andesite; Byers Peninsula, South Shetland Islands
5. P.848.15: Rhyolite; Byers Peninsula, South Shetland Islands.
6. R.637.1: Basaltic andesite; Camp Hill (15 km WSW of Bald Head), Trinity Peninsula
7. R.624.1: Andesite; Bald Head, Trinity Peninsula
8. R.625.1: Andesite; Bald Head, Trinity Peninsula
9. R.619.1: Andesite; Bald Head, Trinity Peninsula
10. R.653.2: Dacite; Joinville Island
11. R.906.1: Rhyolite; Hampton Bluffs (5 km NW of Porphyry Bluff), Trinity Peninsula
12. R.608.2: Rhyolitic welded tuff; Crystal Hill (7 km WSW of Bald Head), Trinity Peninsula

absorption coefficients determined from major element analyses.

Representative analyses of the volcanic rocks from the APVG of Trinity Peninsula are given in Table 1 together with published analyses of the equivalent South Shetland Islands rocks for comparison (Saunders, et al, 1980; Weaver et al., 1982). The Trinity Peninsula rocks display major and trace element chemistries which fall within the range defined for medium to high-K calc-alkaline continental margin magmatic suites (Gill, 1981). They show no Fe-enriched pattern, have high $K_2O$, Rb and Sr contents and low K/Rb and $Na_2O/K_2O$ ratios. These features contrast with those exhibited by the more basic South Shetland Islands rocks, which include a number of Fe-enriched compositions, higher $Al_2O_3$ and $Na_2O$ contents and $Na_2O/K_2O$ and K/Rb ratios and lower $K_2O$ and Rb contents. Although these features suggest affinities with island arc tholeiites, other trace element contents indicate that they are low-K calc-alkaline volcanic rocks (Weaver, et al., 1982).

The new data broadly confirm the patterns of variation outlined by previous workers, although the scatter of data is considerably reduced. This is attributed to the samples having been selected from both a more restricted time range and a limited geographical area. The relative distribution and behaviour of the incompatible elements is summarised in Figure 2. The Trinity Peninsula data exhibit the LIL-element enriched patterns characteristic of calc-alkaline rocks in general. In addition they show a progressive decrease in the abundance of Sr, P and Ti with increasing evolution of the melt. This feature has been noted by Saunders and Tarney (1982), who invoked fractional crystallisation of plagioclase and minor minerals (e.g. apatite), as an explanation. The decrease of Zr at high $SiO_2$ levels is attributed to the removal of zircon from the melt (Saunders et al., 1980).

A detailed assessment of transverse variations between the South Sheltand Islands and Trinity Peninsula is limited by the restricted compositional range of the two areas. The most basic rock present on Trinity Peninsula (Table 1, analysis 6), shows a degree of chemical similarity to the basaltic andesites of the South Shetland Islands, although the Cr and Ni contents are slightly higher, a point discussed later. Similarly, despite the increased scatter at high $SiO_2$ levels, analyses of acidic welded tuffs from the South Shetland Islands overlap with those from Trinity Peninsula. Thus, the differences between the two areas are most apparent in the intermediate compositions, with the andesites of Trinity Peninsula having significantly higher LIL-element contents than those of the South Shetland Islands (Figure 2).

Cr and Ni contents are higher in the rocks farthest from the trench (i.e. Trinity Peninsula), a feature noted from other continental margin arcs and linked to eruption through thicker crust (Gill, 1981). In addition, a distinct group of acidic andesites, rich in Cr, Ni, and MgO, appears to exist within the Trinity Peninsula APVG (Table 1, analyses 7-8). These rocks occur throughout the area, notably Mt Tucker, Tower Peak and Bald Head (Figure 1). Petrographic and chemical characteristics indicate that these rocks are non-accumulative. Their LIL-element concentrations conform to the enriched patterns already described. The Ni and MgO values are slightly lower than those in similar volcanic rocks from the Devonian of Scotland, reported by Thirlwall (1982). These rocks possess features characteristic of rocks transitional between normal calc-alkaline andesites and the primitive mantle-derived andesites (boninites) from modern arcs (Kuroda et al., 1978).

Finally, a striking feature of the APVG of Trinity Peninsula is the frequent occurrence of peraluminous rocks (up to 2.6% corundum normative). This peraluminosity has probably been enhanced by alteration (i.e. alkali loss), but nevertheless there is marked correlation between the paraluminous nature of the host rock and the presence of garnet (Hamer and Moyes, 1982). The significance of this is discussed in detail elsewhere (Moyes and Hamer, this volume).

## ISOTOPE GEOCHEMISTRY

Initial $^{87/86}Sr$ isotope data are now available for a number of APVG rocks at the northern tip of the Antarctic Peninsula (Pankhurst, 1982; and unpublished data). Basalts, basaltic andesites and andesites from the South Shetland Islands, with model ages in the range 125-90 Ma, have yielded initial $^{87/86}Sr$ isotope ratios of 0.703-0.704, whereas rhyolites ($111 \pm 4$ Ma) give a slightly higher ratio of 0.705. The limited data available for Trinity Peninsula indicate significantly higher values of between 0.709-0.715. The possibility that some of these may be reset or affected by hydrothermal alteration soon after crystallisation, could be responsible in part for the high values (R.J. Pankhurst, pers. comm.), but it cannot fully explain the differences between the two areas.

## DISCUSSION

There seems little doubt that magma genesis at convergent margins involves complex, multi-stage processes. Many recent papers (e.g. Hildreth, 1981) have stressed the need for care in the interpretation of regional geochemical variations, especially from sample populations whose age, volume and compositional relationships are poorly

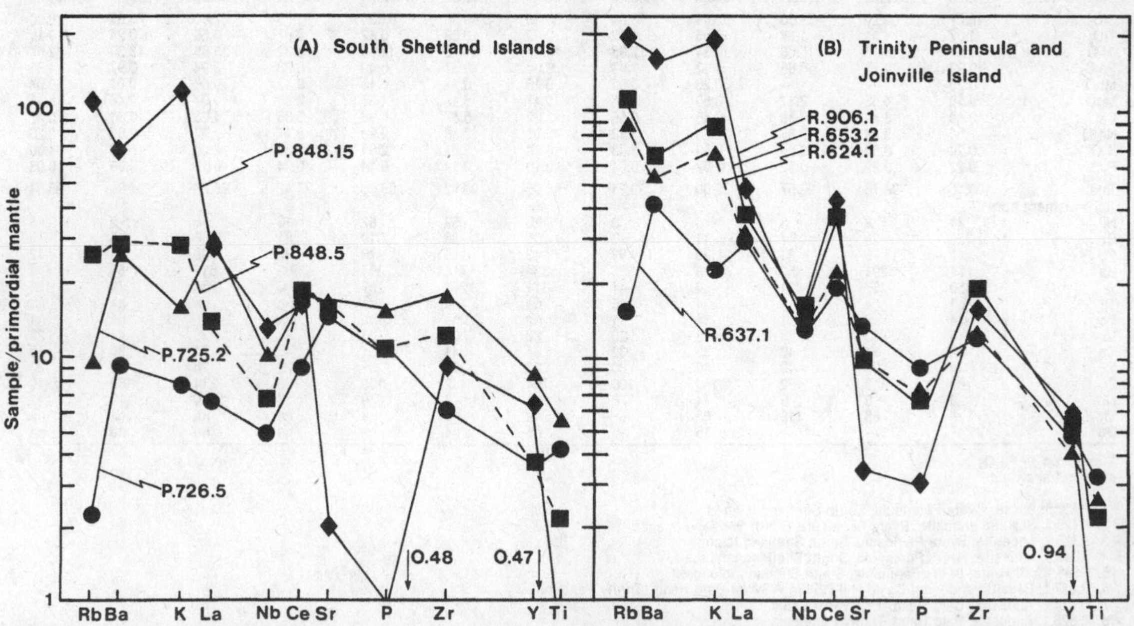


**Figure 2. Multi-element mantle-normalised diagram of selected Late Jurassic-Early Cretaceous calc-alkaline volcanic rocks from a) South Shetland Islands (Saunders et al., 1980; Weaver et al., 1982) and b) Trinity Peninsula and Joinville Island. Mantle normalising values taken from Wood et al., (1979). Rock types and sample localities as in Table 1.**

known. High degrees of data scatter, frequently dismissed as the effects of random weathering and alteration, may hide important systematic variations. Strict control of the sample population is essential if such effects are to be minimised.

Despite the resultant lack of agreement over the extent to which the subducting slab, the overlying mantle wedge and the continental crust are individually involved in the generation of calc-alkaline magmas, there are two points about which there is a consensus:-

(a) The occurrence of mantle asthenosphere at depths of more than 90 km beneath the arc appears to be a prerequisite for andesite generation (Gill, 1981).
(b) Primary andesite magma is unlikely to be produced from one single source region (Wyllie et al., 1976).

The most widely accepted models (Gill, 1981) consider the following mechanisms to be relevant:-

(1) Subduction and dehydration of the oceanic crust at depth as it is metamorphosed to eclogite facies assemblages. This leads to the expulsion of incompatible elements, radiogenic nuclides and silica into the overlying mantle, causing enrichment and the formation of new accessory minerals, which lower the density and viscosity and in some instances result in melting.
(2) Derivation of the primitive arc magmas from enriched mantle by partial melting and subsequent fractional crystallisation of mainly plagioclase, olivine, orthopyroxene, augite and magnetite.
(3) Modification of the rising magmas, through crustal interaction.

Current hypotheses favour the idea that the primary arc magmas of the Antarctic Peninsula are derived from a veined subcontinental mantle, with a varying contribution from the subducting slab (Saunders et al., 1980). This model envisages a similar source for both calc-alkaline and continental alkaline magmas, the latter frequently associated with ensialic arcs and ocurring in extensional regimes behind the arcs. The low HFS-element contents and LIL-element enrichment of the calc-alkaline suite are viewed as primary features resulting from the stabilisation of minor mineral phases and the dehydration of the subducting slab. The transverse variations may reflect an increasing contribution of vein material to melts generated farther from the trench.

The high Cr and Ni bearing rocks of Trinity Peninsula are important in this respect, since high levels of these elements in intermediate volcanic rocks from other arcs have been interpreted as indicating the presence of a significant non-fractionated component, derived from the mantle (Gill, 1981). Fractional crystallisation of a Si-poor assemblage from a more basic magma would be dominated by spinel and calcic-plagioclase, yet high Cr values limit the amount of spinel fractionation and the phenocryst plagioclase is high $SiO_2$ andesine-labradorite (Thirlwall, 1982). The Trinity Peninsula andesites also contain labradorite phenocrysts and have high Cr values suggesting that they too are derived in part, from a primitive source region (i.e. the mantle), although the slightly lower Ni and MgO contents may reflect limited fractional crystallisation of olivine.

An estimate of the relative contribution of mantle material is difficult to make, although initial isotope work for the magmatic arc as a whole provides some useful guidelines. The bulk of the data obtained so far relate to the plutonic rocks. With the exception of a distinct group of Early Jurassic acidic plutons, which have extremely high and constant initial $^{87/86}Sr$ ratios indicative of crustal remelting (Pankhurst, 1982), the majority of the later plutons have initial ratios consistent with the mantle origin. Pankhurst (1982) goes on to suggest that the increasing acidity with passage away from the trench exhibited by these later mantle derived melts reflects greater degrees of differentiation achieved by magmas in response to crustal thickening.

Initial $^{87/86}Sr$ isotope ratios for the volcanic rocks are generally higher, a feature which Pankhurst (1982) attributes to a more complex history. The data available also suggest that the initial ratios for the Trinity Peninsula rocks are higher than those for the South Shetland Islands. Similar high ratios from Andean calc-alkaline volcanic rocks have been discussed recently by several authors, who consider processes involving assimilation of crustal material accompanied by fractionation of plagioclase (James, 1982) and crustal remelting (Hawkesworth et al., 1982) to be important. Saunders et al. (1980) however, maintain that differing degrees of partial melting of a heterogeneous enriched mantle provide an alternative explanation for the observed variations in the Antarctic Peninsula. Clearly, further isotope studies are required for a thorough assessment, because elsewhere in the Antarctic it has been suggested that high ratios in continental tholeiites can be explained without the need to resort to crustal contamination (Kyle, 1980).

Garnet, a rare but potentially important mineral in calc-alkaline magmas (Green, 1977), occurs widely as a minor constituent of the Trinity Peninsula volcanic rocks. Detailed descriptions and discussion of its origin in the APVG have appeared elsewhere (Hamer and Moyes, 1982) Two basic types are present:-

(a) Phenocrysts of igneous garnet which represent the remnant of an early high pressure equilibrium crystallisation phase (5-7 kb).
(b) Microxenoliths and xenocrysts of metamorphic garnet incorporated at low crustal levels. Some of these comprise garnet, biotite, sillimanite, andesine and quartz; they are similar to those from other terrains thought to represent the restite component of partial melting of crustal material (Birch and Gleadow, 1974).

The garnet phenocrysts bear a striking resemblance to those described by Oliver (1956) and Fitton (1972) from the Borrowdale Volcanic Group (BVG). Garnet appears in the BVG andesites at the point where Ca-rich pyroxene disappears, whereas in the Trinity Peninsula andesites garnet and Ca-rich pyroxene coexist. Also, no evidence of a systematic relationship between the composition of the garnet and the composition of the host rock has so far been found on Trinity Peninsula, although the possibility is not discounted (Hamer and Moyes, 1982). There is however, a striking correlation between the presence of garnet and peraluminosity of the host rock. Approximately 15% of orogenic andesites are corundum normative and the presence of peraluminous liquids seems to be a normal consequence of calc-alkaline magmatism (Gill, 1981). A number of models have been proposed including alkali loss during hydrothermal alteration, the assimilation of pelitic crustal material and fractional crystallisation of pyroxene or amphibole without plagioclase (Cawthorn and Brown, 1976). These models are discussed in detail elsewhere (Moyes and Hamer, this volume), but the presence of both diopside-normative (non-garnet bearing) high Cr- and Ni-andesites indicates the possibility of interaction of mantle derived magmas with continental crust. Similar conclusions have been reached by Fitton et al. (1982), who considered that the BVG garnets crystallised at 5-7 kb in response to the assimilation of slates and that resultant magmas were transferred rapidly on the surface with little opportunity for fractional crystallisation.

These conflicting lines of evidence emphasise the complexity of the calc-alkaline magmatic process. Mantle derived magmas may form the largest contribution of material to the arc system, but the influence of pre-existing crustal material can be significant. Rigorous testing of the relative importance of these processes requires more detailed studies.

## SUMMARY

(1) The Trinity Peninsula volcanic rocks form part of a major phase of Late Jurassic-Early Cretaceous ensialic medium to high-K calc-alkaline activity.
(2) New whole rock data confirms the existence of transverse variations in the LIL-elements across the Antarctic Peninsula. The patterns described here are clearly spatial, with LIL-element concentrations increasing away from the trench.
(3) The restricted time range and geographic area of sampling reduce the scatter of data.
(4) Within the APVG of Trinity Peninsula is a group of high Cr and Ni bearing andesites, which indicate derivation of at least part of the magma from mantle sources.
(5) Peraluminous garnet bearing volcanic rocks indicate that pre-existing continental crust may have played a significant role in the modification of the rising magma.
(6) Initial $^{87/86}Sr$ isotope ratios are significantly higher for the Trinity Peninsula volcanic rocks than for the equivalent South Shetland Islands rocks.
(7) Low pressure fractional crystallisation of olivine, pyroxene, plagioclase and magnetite from mantle derived magmas provides the most likely explanation for the variations observed in the South Shetland Islands rocks (Weaver et al., 1982). On Trinity Peninsula,the influence of pre-existing continental crust on at least some of the magmas is indicated.

*Acknowledgements* The author is indebted to the following for their generous assistance and cooperation:- Dr A.D. Saunders and G.F. Marriner, for providing access to the X-ray fluorescence spectrometer at Bedford College; colleagues at BAS, especially A.B. Moyes for helpful discussion and critically reading the manuscript and lastly Mike Sharp, a patient field companion during the 1978-9 austral summer.

## REFERENCES

ADIE, R.J., 1972: Evolution of volcanism in the Antarctic Peninsula; *in* Adie, R.J. (ed.) *Antarctic Geology and Geophysics.* Universitetsforlaget, Oslo, 137-41.

BIRCH, W.D. and GLEADOW, A.J.W., 1974: The genesis of garnet and cordierite in acid volcanic rocks: evidence from the Cerberean Cauldron, Central Victoria, Australia. *Contrib. Mineral. Petrol., 45,* 1-13.

CAWTHORN, R.G. and BROWN, P.A., 1976: A model for the formation of corundum normative calc-alkaline magmas through amphibole fractionation. *J. Geol., 84,* 467-76.

FARQUARSON, G.W., 1982: Late Mesozoic sedimentation in the northern Antarctic Peninsula and its relationship to the southern Andes. *Geol. Soc. Lond., J., 139,* 721-7.

FITTON, J.G., 1972: The genetic significance of alamandine-pyrope phenocrysts in the calc-alkaline Borrowdale Volcanic Group, northern England. *Contrib. Mineral. Petrol., 36,* 231-48.

FITTON, J.G., THIRWALL, M.F. and HUGHES, D.J., 1982: Volcanism in the Caledonian orogenic belt of Britain; *in* Thorpe, R.S. (ed.) *Andesites: orogenic andesites and related rocks.* John Wiley and Sons, Chichester, 611-36.

GILL, J., 1981: *Orogenic andesites and plate tectonics.* Springer-Verlag, New York.

GREEN, T.H., 1977: Garnet in silicic liquids and its possible use as a P-T indicator. *Contrib. Mineral. Petrol., 65,* 59-67.

HAMER, R.D. and MOYES, A.B., 1982: Composition and origin of garnet from the Antarctic Peninsula Volcanic Group of Trinity Peninsula. *Geol. Soc. Lond., J., 139,* 713-20.

HAWKESWORTH, C.J., HAMMILL, M., GLEDHILL, A.R., VAN CALSTEREN, P. and ROGERS, G., 1982: Isotope and trace element evidence for late-stage intra-crustal melting in the High Andes. *Earth Planet. Sci. Lett., 58,* 240-54.

HILDRETH, W., 1981: Gradients in silicic magma chambers: implications for lithosphere magmatism. *J. Geophys. Res., 86,* 10153-92.

JAMES, D.E., 1982: A combined O, Sr, Nd and Pb isotopic and trace element study of crustal contamination in central Andean lavas, I. Local geochemical variations. *Earth Planet. Sci. Lett., 57,* 47-62.

KURODA, N., SHIRAKI, K. and URANO, H., 1978: Boninite as a possible calc-alkaline primary magma. *Bull. Volcanol., 41,* 563-75.

KYLE, P.R., 1980: Development of heterogeneities in the subcontinental mantle: evidence from the Ferrar Group, Antarctica. *Contrib. Mineral. Petrol., 73,* 89-104.

MOYES, A.B. and HAMER, R.D., (this volume): Contrasting origins and implications of garnet in rocks of the Antarctic Peninsula.

OLIVER, R.L., 1956: The origin of garnet in the Borrowdale Volcanic Series and associated rocks, English Lake District. *Geol. Mag., 93,* 121-39.

PANKHURST, R.J., 1982: Rb-Sr geochronology of Graham Land, Antarctica. *Geol. Soc. Lond., J., 139,* 701-711.

SAUNDERS, A.D., TARNEY, J. and WEAVER, S.D., 1980: Transverse geochemical variations across the Antarctic Peninsula; implication for the genesis of calc-alkaline magmas. *Earth Planet. Sci. Lett., 46,* 344-60.

SAUNDERS, A.D. and TARNEY, J., 1982: Igneous activity in the southern Andes and northern Antarctic Peninsula: a review. *Geol. Soc. Lond., J., 139,* 691-700.

SMELLIE, J.L. 1979: Aspects of the geology of the South Shetland Islands. Unpublished Ph.D. thesis, University of Birmingham, United Kingdom, 198 pp.

TARNEY, J., WEAVER, S.D., SAUNDERS, A.D., PANKHURST, R.J. and BARKER, P.F., 1982: Volcanic evolution of the northern Antarctic Peninsula and the Scotia arc; *in* Thorpe, R.S. (ed.) *Andesites: orogenic andesites and related rocks.* John Wiley and Sons, Chichester, 371-400.

THIRLWALL, M.F., 1982: Systematic variation in chemistry and Nd-Sr isotopes across the Caledonian calc-alkaline volcanic arc: implications for source materials. *Earth Planet. Sci. Lett., 58,* 27-50.

THOMSON, M.R.A. and PANKHURST, R.J., (this volume): Age of post-Gondwanian calc-alkaline volcanism in the Antarctic Peninsula region.

WEAVER, S.D., SAUNDERS, A.D. and TARNEY, J., 1982: Mesozoic-Cenozoic volcanism in the South Shetland Islands and the Antarctic Peninsula: geochemical nature and plate tectonic significance; *in* Craddock, C. (ed.) *Antarctic Geoscience.* Univ. Wisconsin Press, Madison, 263-73.

WOOD, D.A., JORON, J.L., TREVIL. M., NORRY, M., and TARNEY, J., 1979: Elemental and Sr isotope variations in basic lavas from Iceland and the surrounding ocean floor. *Contrib. Mineral. Petrol., 70,* 319-39.

WYLLIE, P.J., HUANG. W., STERN. C.R. and MAALOE, S., 1976: Granitic magmas: possible and impossible sources, water contents and crystallisation sequences. *Can. J. Earth. Sci., 13,* 1007-19.

# AN EARLY MIOCENE RIDGE CREST-TRENCH COLLISION ON THE SOUTH SCOTIA RIDGE NEAR 36°W

P.F. Barker, P.L. Barber and E.C. King, *Department of Geological Sciences, Birmingham University, PO Box 363, Birmingham, B15 2TT, England.*

*Abstract* Preliminary interpretation of marine geophysical data and geochemical analyses of dredged rocks has led to the identification of a ridge crest-trench collision zone southeast of the South Orkney Islands. The existence of a series of such collision zones along the South Scotia Ridge had earlier been postulated to explain changes in Scotia Sea evolution. The collision zone takes the form of a double ridge with dissecting trough, aligned northeast-southwest and about 200 km long. Oceanic magnetic lineations in the northern Weddell Sea young northwestward towards the zone, with the youngest (identified as anomaly 6, ca 20 Ma) reaching the central trough. Volcanic rocks dredged from Jane Bank, the northwestern ridge, are chemically very similar to the more siliceous of the low-K tholeiite series which characterises the presently active South Sandwich arc. Jane Bank is interpreted as an *in situ* remnant of the arc and upper fore-arc of an intraoceanic arc produced by subduction of South American oceanic lithosphere (like the present South Sandwich Island). Part of the upper fore-arc and all of an accretionary wedge appear to have been subducted. Unlike ocean floor at collision zones off Chile and the Antarctic Peninsula, the seabed in the trough and on the ridge to the southeast does not conform to the normal oceanic age—depth relationship. The estimated collision time (20 Ma) is close to the time of onset of north-south extension in the Central Scotia Sea. The limited data set described here, however, does not establish the synchroneity of the collision over the entire 200 km of Jane Bank, so an assessment of the extent of any causal relationship between the two events must await the analysis of a more extensive survey.

# THE LARSEN RIFT: AN ACTIVE EXTENSION FRACTURE IN WEST ANTARCTICA

O. Gonzalez-Ferran, *Departamento de Geologia y Geofisica, Universidad de Chile, Casilla 16, Correo Miramonte, Santiago, Chile.*

*Abstract* South of the Antarctic Peninsula a zone of alkali basaltic volcanism, Pleistocene to Recent in age, is still active in some centres represented by Paulet Is., Coley Cone, Seal Nunataks and Argo Point. The volcanism delineates an extension zone 400 km long running parallel to the Antarctic Peninsula, 70-80 km southeast of it. This active fracture affects the continental shelf in this sector of the Weddell Sea, and is here called the Larsen Rift. It is one member of a double rift system, being parallel to the Bransfield Rift, located about 200 km to the northwest, on the other side of the Antarctic Peninsula. Seal Nunataks could represent the intersection with a N60-70°W striking fracture, directly related to the Hero Fracture Zone, which extends northwest of the Shetland trench axis. This fracture affects the continental block of the Peninsula and represents the southern termination of the Bransfield Rift. This Pleistocene to Recent double rift system started as a consequence of the extensive tectonic processes which followed after the ending of subduction activity in the Shetland area during the Miocene. As a conclusion it may be said that at present, the northern part of the Antarctic Peninsula, with its rift system, would be subjected to a "fan type" expansion.

This report aims to analyse the relationships between tectonic and volcanic processes in the northern area of the Antarctic Peninsula and adjacent islands, as a part of the dynamic processes of the Scotia Arc. It is based on published geological and geophysical information and on direct observations by the author.

Evidence from Recent volcanism supports the hypothesis that the fracture zone which controls it corresponds to an extensive rift system. This implies the simultaneous development of a double Rift in the back-arc zone of the Shetland Arc with its evolution starting during the Pliocene, as a consequence of the interruption of the subduction processes (Figure 1).

## GEOLOGICAL EVIDENCE

Only a summary of the evidence from post-Miocene neovolcanic and tectonic events will be given.

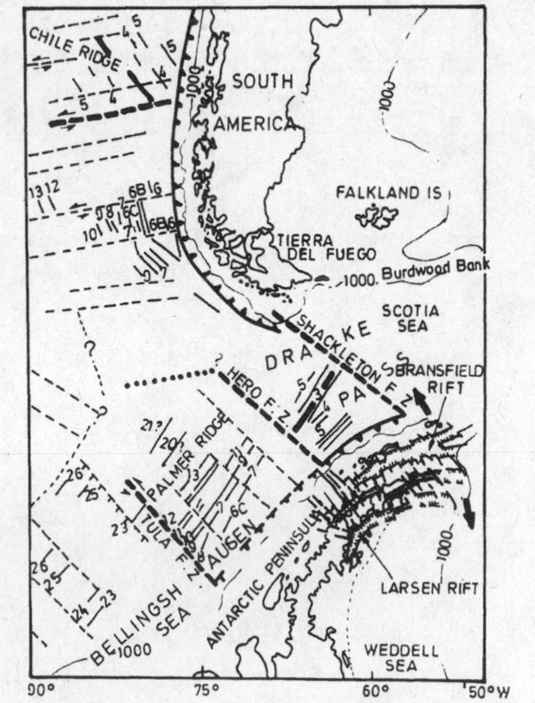

**Figure 1. The Bransfield and Larsen Rift against plate tectonics of the southeastern Pacific Ocean (Modified from Herron and Tucholke, 1976).**

## Larsen Ice Shelf Region—The Larsen Rift

Volcanism of alkali-olivine basalt character, Pliocene-Recent in age, occurs at the following points:

*Paulet Island* (Figure 2) It is a very young volcanic structure, with well conserved craters and alkali-olivine basaltic lava flows, of 0.3 Ma (K/Ar) (Rex, 1976). Probably with some historial eruption, its high thermal flow keeps it free from ice accumulation (Baker et al., 1976; Baker et al., 1977; González-Ferrán and Baker, 1973). The alignment of its craters is northeast (Figure 3) suggesting that this trend is that of the maximum horizontal compression of the tectonic stress, according to the anlysis proposed by Nakamura (1977) and Nakamura and Uyeda (1980).

*Coley Cinder Cone* This is a small volcanic cone on the southern side of the Coley Glacier on James Ross Island and was discovered by the British Antarctic Survey in January 1979. Its intact preservation suggests that it is a very young feature (BAS, 1982).

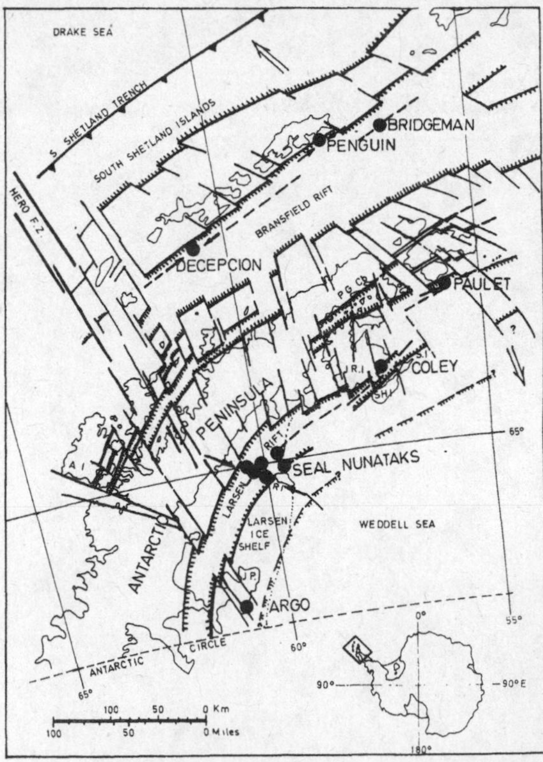


**Figure 2. Sketch map showing the distribution of the Pleistocene-Recent volcanic centre along the Bransfield and Larsen Rifts Region. Segment line, represents the Rift active fracture zone. Continuous line, represents fractures and lineaments which would affect the continental block, causing its enechelon segmentation. (J.R.I.) James Ross Is.; (S.H.I.) Snow Hill Is.; (S.I.) Seymour Is.; (J.P.) Jason Peninsula; (A.I.) Anvers Is., (R.I.) Robertson Is.; (P.G.C.) Principe Gustav Chanel.**

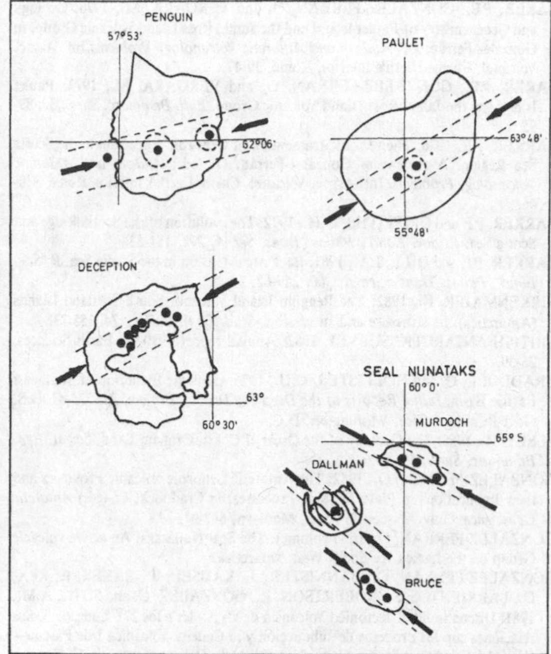


**Figure 3. Sketch of the active volcanoes during the Pleistocene-Recent, showing the distribution and orientation of the monogenetic eruption centres. The arrow, represents the direction of the maximum horizontal compression of the regional stress.**

*Seal Nunataks* A group of 14 volcanoes of Pleistocene-Recent age, four of which are active and have been described by various authors, Fleet (1968), Renner (1980), Saunders (1982), Baker et al. (1977), Pankhurst (1982), González-Ferrán (this volume). Petrologically they have erupted alkali-olivine basalts. Figure 3 shows the northwest trend of the estimated maximum horizontal compression of the tectonic stress.

*Argo Volcanic Cone* (Figure 2) Saunders (1982) reports that due to the state of preservation, he considers that the cone is very young, probably less than 1 Ma old. Its lavas are alkali-olivine basalts.

The alkali character of all this volcanism strongly suggests it is a response to extensive tectonics which affected the southeast margin of the Antarctic Peninsula, reaching its maximum activity during the Pleistocene, as deduced on the basis of K-Ar ages. The symmetry of the ages observed at the Seal Nunataks, suggests a typical rift tectonic process (Figure 4), González-Ferrán (this volume).

## The Bransfield Strait Region—The Bransfield Rift

In this region the alkali character of the volcanism is somewhat less pronounced than in the Larsen Rift. Effectively, Deception Island could represent a transition phase as has been commented by Hawkes (1961) and Baker et al. (1977). Harvey (1974) also notes that within the young volcanic rocks of Deception Island characteristics of calc-alkaline, alkaline and tholeiitic series can be recognised. Baker (op. cit.) considers that Deception Island is probably related to the extension tectonic regime which brought about the opening of Bransfield Strait. Deception Island is still the most active volcano of this region (Figure 2). Another volcanic centre with historic eruption is Penguin Island (Figure 2) characterised by olivine-tholeiites (González-Ferrán and Katsui, 1970; Birkenmajer, 1982). Bridgeman Island is another volcanic centre of Pleistocene age, which is located on the axis of the Bransfield Rift (González-Ferrán and Katsui, 1970; González-Ferrán, 1982).

Deception, Penguin and Bridgeman are clearly related to extension across the Bransfield Rift. This has been noted by different authors who also consider the rift to have generated in response to the termination of subduction in the Shetland area.

# GEOPHYSICAL AND STRUCTURAL EVIDENCE

On the basis of seismic profiles and the interpretation of magnetic lineaments, Herron and Tucholke (1976), Barker and Griffiths (1972), Barker and Hill (1981), Barker (1976) and Craddock and Hollister (1976), have noted the existence of three main oceanic transform faults. Two are considered here: the Shackleton and Hero Fracture Zones (Figure 1). They appear to have incidenced the deep fracturing with similar strike to that which affects the continental block of the Antarctic Peninsula, causing its segmentation. Recently, Hawkes (1981) has suggested the structural segmentation of the whole Antarctic Peninsula, as a result of the segmentation of the oceanic plate and it is implicit in his model that the oceanic fracture zones would generate megafractures in the overlying continental crust during subduction. This appears to be comparable with the active convergence zone between the Nazca Plate and the South America Plate, at latitude 27°S, where a similar effect is observed in the Eastern Fracture zone of the Andean Region; Ojos del Range (González-Ferrán et al., 1981).

# SEQUENCE OF TECTONOVOLCANIC EVENTS AND CONCLUSIONS

In conclusion, the following sequence of events can be deduced:

(1) Late Pre-Cenozoic. An active oceanic subduction regime existed in the Shetland Arc, framed by the Shackleton and Hero Fracture zones. Subduction declined, probably during the Late Miocene.

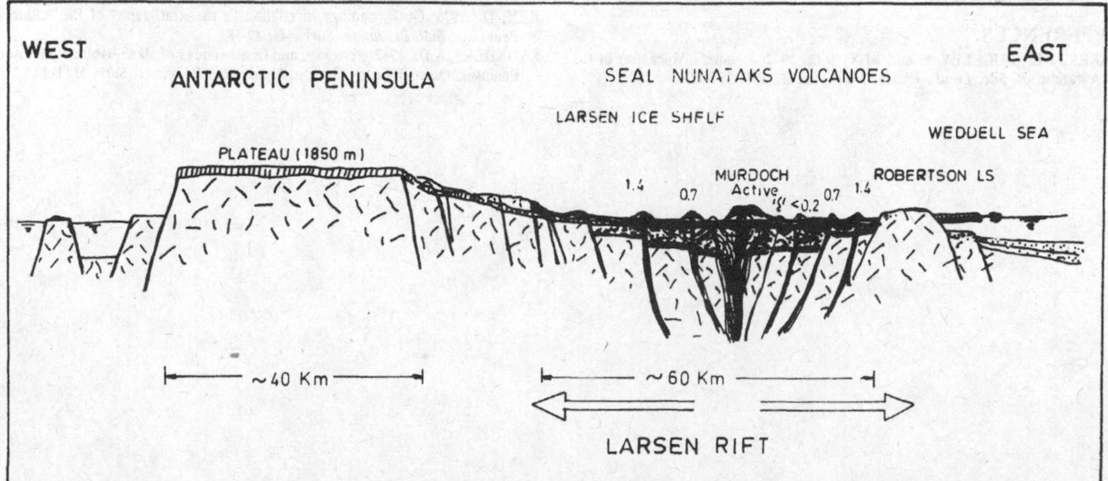


**Figure 4. The Larsen Rift section at the Seal Nunataks latitude, showing the symmetry of the K-Ar ages of the volcanic centres.**

(2) An extensive back-arc tectonic zone developed, possibly during the Pliocene-Pleistocene. It generated two rift structures parallel to the axis of the Shetland Trench, approximately 200 km apart. These rifts are the Bransfield Rift northwest of the Antarctic Peninsula and the Larsen Rift, southeast of the Peninsula.

(3) Development of volcanism during the Pleistocene-Recent, associated with the two rifts: (a) The Bransfield volcanism is expressed by the Deception-Penguin-Bridgeman line, with characteristics which initially show a transition from residual calc-alkaline to alkaline magma. (b) The initial Larsen volcanism involves probably fissure activity, which generated extensive deposits of plateau type palagonitic tuffs (González-Ferrán, this volume), indicating subglacial eruptions. This was followed by a subaerial volcanism which established an active volcanic axis in the Rift during the Pleistocene, marked by the Paulet, Coley, Seal Nunataks and Argo centres (Figure 2). Petrographically, this volcanism involved alkali-olivine basaltic magmas.

(4) The expansion of the Larsen Rift is clearly shown by the symmetry presented by the development of the Pleistocene volcanism in the Seal Nunataks region, according to the K-Ar ages (González-Ferrán, this volume).

(5) Continuation of the southwards influence by oceanic fracturing on the continental strip during the Cenozoic, can explain the segmentation of this region. The Hero and Shackleton fractures limit the Bransfield Rift and also affect the Larsen Rift (Figures 1 and 2).

(6) The direction of maximum horizontal compressive stresses, can be interpreted from alignments of monogenetic eruption centres. The results suggest that the Bransfield Rift is still in an extensive stage, from orientations observed in Penguin and Deception. In the area of Seal Nunataks, the Larsen Rift shows a compressive stress almost perpendicular to the rift, as may be seen from the analysis of the Murdoch, Dallman and other active volcanoes today (Figure 3). In the northern part of the rift, at Paulet Island and Coley Cone the direction suggests that the Larsen Rift also is still in an extensive stage (Figures 2 and 3).

(7) In a final conclusion, it may be said that at present, the northern part of the Antarctic Peninsula with its rifts system, would be subjected to a fan rift type with an axis (rotation point) which should be near Robertson Island.

*Acknowledgments* This study was financed by Grant E 1360-8224 of the Departamento de Desarrollo Científico de la Universidad de Chile and the Chilean Air Force. The latter not only supported the cost of the operation in the field but also gave a safe and efficient logistic support without which those faraway latitudes could not have been reached. My thanks to the Dean of the Faculty of Physical and Mathematical Sciences and to the Director and Colleagues of the Departamento de Geología y Geof'isica de la Universidad de Chile, for their constant support. My gratitude to the colleagues that helped me with their critical reading and discussion of the manuscript.

## REFERENCES

BAKER, P.E., BUCKLEY, F. and REX, D.C., 1977: Cenozoic Volcanism in the Antarctic. *R. Soc. Lond., Philos. Trans., Ser. B., 279,* 131-42.

BAKER, P.E., GONZALEZ-FERRAN, O. and VERGARA, M., 1976: Geology and Geochemistry of Paulet Island and the James Ross Island Volcanic Group. *in* González-Ferrán, O.: *Andean and Antarctic Volcanology Problems.*Int. Assoc. Volcanol. Chem. Earth's Interior, Rome, 39-47.

BAKER, P.E., GONZALEZ-FERRAN, O. and VERGARA, M., 1973: Paulet Island and the James Ross Island Volcanic Group. *Bull. Br. antarc. Surv., 32,* 89-95.

BARKER, P.F., 1976: The tectonic framework of Cenozoic volcanism in the Scotia Sea Region. A review; *in* González-Ferrán, O. (ed.) *Andean and Antarctic Volcanology Problems.* Int. Assoc. Volcanol. Chem. Earth's Interior, Rome, 330-46.

BARKER, P.F. and GRIFFITHS, D.H., 1972: The evolution of the Scotia Ridge and Scotia Sea. *R. Soc. Lond., Philos. Trans. Ser. A. 271,* 151-83.

BARKER, P.F. and HILL, I.A., 1981: Back-arc extension in the Scotia Sea. *R. Soc. Lond., Philos. Trans., Ser. A, 300,* 249-62.

BIRKENMAJER, K., 1982: The Penguin Island Volcano, South Shetland Islands (Antarctica): Its structure and succession. *Stud. Geol. Polon., 74,* 155-73.

BRITISH ANTARCTIC SURVEY, 1982: Annual report 1980-81. Earth Sciences, 25-39.

CRADDOCK, C. and HOLLISTER, C.D., 1976: Geologic Evolution of Southeast Pacific Basin. *Initial Reports of the Deep Sea Drilling Project, 35, 723-43.* U.S. Govt Printing Office, Washington, D.C.

FLEET, M., 1968: The Geology of the Oscar II Coast, Graham Land, *Scient. Rep. Br. antarc. Surv., 59,* 1-46.

GONZALEZ-FERRAN, O., 1982: The Antarctic Cenozoic Volcanic Provinces and their Implications in Plate Tectonic Processes; *in* Craddock, C. (ed.) *Antarctic Geoscience.* Univ. Wisconsin Press, Madison, 687-94.

GONZALEZ-FERRAN, O., (this volume): The Seal Nunataks: An active volcanic Group on the Larsen Ice Shelf. West Antarctica.

GONZALEZ-FERRAN, O., BANNISTER, J., KAUSEL, E., BAKER, P., REX, D., BARRIENTOS, S., ROBERTSON, R., GONZALEZ, C. and SOTO, A.M., 1981: Discontinuidad Tectónico Volcánica de los Andes a los 27° Latitude S. Sus relaciones con los procesos de subducción y la Cadena Volcánica I de Pascua—Ojos del Salado. Informe Final. Unpublished. *Departamento de Geologia y Geofisica,* Universidad de Chile. 1-245.

GONZALEZ-FERRAN, O. and BAKER, P.E., 1973: Isla Paulet y el volcanismo reciente en las islas de Weddell Noroccidental. Antarctica. Terra Australis. *Rev. Geogr. Chile, 22,* 22-43.

GONZALEZ-FERRAN, O. and KATSUI, Y., 1970: Estudio Integral del volcanismo cenozoico superior de las islas Shetland del Sur, Antártica, *Ser. cient. Inst. antart. chil, 1, 2,* 123-74.

HARVEY, M.R., 1974: The geology of the older pyroclastic rocks of Deception Island. Unpublished. Ph.D. Thesis, Leeds University 181pp.

HAWKES, D.D., 1961: The geology of the south Shetland Islands II, Geology and petrology of Deception Island. *Scient. Rep. Falkld Isl. Depend. Surv., 27,* 1-43.

HAWKES, D.D., 1981: Tectonic segmentation of the northern Antarctic Peninsula. *Geology, 9,* 220-4.

HERRON, E.M. and TUCHOLKE, B.G., 1976: Sea floor magnetic patterns and basement structure in the southeastern Pacific. *Initial Reports of the Deep Sea Drilling Projects, 35.* 263-78. U S. Govt. Printing Office, Washington D.C.

NAKAMURA, K., 1977: Volcanoes as possible indicators of tectonic stress orientation principle and proposal. *J. Volcanol. Geotherm. Res. 2,* 1-16.

NAKAMURA, K. and UYEDA, S., 1980: Stress gradient in arc and back-arc regions and plate subduction. *J. Geophys. Res. 85,* 6419.

PANKHURST, R.J., 1982: Sr-isotope and trace element geochemistry of Cenozoic volcanics from the Scotia arc and the northern Antarctic Peninsula; *in* Craddock, C. (ed.) *Antarctic Geoscience.* Univ. Wisconsin Press, Madison, 229-34.

RENNER, R.G.B., 1980: Gravity and magnetic surveys in Graham Land. *Scient. Rep. Br. antarc. Surv., 77,* 1-99.

REX, D., 1976: Geochronology in relation to the stratigraphy of the Antarctic Peninsula. *Bull. Br. antarc. Surv., 43,* 49-58.

SAUNDERS, A.D., 1982: Petrology and Geochemistry of alkali-basalts from Jason Peninsula Oscar II Coast, Graham Land. *Bull. Br. antarc. Surv. 55.* 1-9.

# THE MID-MESOZOIC TO MID-CENOZOIC ACTIVE PLATE MARGIN OF THE ANTARCTIC PENINSULA

D.H. Elliot, *Institute of Polar Studies and Department of Geology and Mineralogy, The Ohio State University, Columbus, Ohio 43210, U.S.A.*

*Abstract* Lithotectonic units of Mid-Mesozoic to Mid-Cenozoic age of the Antarctica Peninsula define an active plate margin, the so-called Andean Orogen. Early to Mid-Mesozoic uplift of the Gondwana plate margin was accompanied by block faulting and deposition of orogenic fanglomerates. The fanglomerates overlap spatially and temporally with the magmatic arc that developed on the Gondwanian Orogen. Magmatic activity was initiated by Mid-Jurassic time and continued into the Tertiary. Upper Jurassic to Mid-Cretaceous fore-arc basin sequences with a volcanic component occur along the Pacific margin. The volcanogenic back-arc sequence along the Weddell flank of the southern Peninsula is Middle to Late Jurassic in age; in contrast, the sequence in the James Ross Island region is Cretaceous to Early Tertiary.

Arc-related deformation was confined to the southern half of the Peninsula and is Early Cretaceous in age in southern Palmer Land, whereas tectonism in the northern Peninsula was confined to major uplift which spanned at least Mid-Cretaceous to Mid-Tertiary time. A major difference in plate tectonic setting is implied. The magmatic history suggests at least intermittent subduction of oceanic crust beneath the Peninsula throughout this time span however, a major break in the subduction history occurred at the time of formation of the Aluk Ridge; relations to sea-floor anomaly patterns can be inferred only for the Cenozoic.

Mid-Mesozoic to Mid-Cenozoic rocks of the Antarctic Peninsula include lithotectonic terrains that define the younger of the two principal events in the tectonic history of the region: (1) the evolution of the Gondwana active plate margin of Late Palaeozoic—Early Mesozoic age (the pre-Jurassic basement); and (2) the Andean mobile belt which was initiated in Middle to Late Jurassic time. The lithotectonic units that can be recognised in Jurassic to Lower Tertiary rocks were formed during Gondwana rifting and dipersal and in response to active plate margin processes. Only on Alexander Island and the northern Antarctic Peninsula are these rocks known to rest unconformably on the pre-Jurassic basement (Edwards, 1982; Elliot and Gracanin, this volume). If microplates that formerly constituted the Pacific margin of Gondwana have been rearranged, then this was accomplished during the latter part of the Mesozoic while the Andean convergent plate margin was active. The purpose here is to review the Andean lithotectonic units and address aspects of the tectonic setting.

## ANDEAN MAGMATIC ARC

The backbone of the Antarctic Peninsula includes widely distributed volcanic and plutonic rocks that represent the Andean magmatic arc (Figures 1 and 2). The volcanogenic sequences consist of both lavas and pyroclastic rocks which range in compostion from calc-alkaline basalt to rhyolite (Weaver et al., 1982). Although the full range of compositions occurs on the west flank of the Peninsula, silicic ash flow tuffs, which locally attain a thickness of 3000 m, appear to dominate on the east flank. Plutonic rocks which have a similar range in composition (Saunders et al., 1982), are exposed throughout the Peninsula region. A K-h relationship has been documented by Saunders et al. (1982) and implies an eastward-dipping subduction zone beneath the Peninsula.

Radiometric dating of the plutonic rocks suggests episodic emplacement, ranging from Early Jurassic to Early Tertiary with the principal culmination between 110 and 90 Ma (Saunders et al., 1982). Dating of the volcanic rocks in the magmatic arc terrain is very poor, but airfall deposits and volcanic detritus that are widespread in the adjacent back-arc and fore-arc basins, suggest inception of Andean magmatism in the Middle Jurassic in eastern Ellsworth Land and in Late Jurassic time elsewhere. During the Early Cretaceous the locus of igneous activity in southern Palmer Land shifted toward the south and east, whereas uppermost Cretaceous-Lower Tertiary volcanic complexes and a Lower Tertiary batholith are located in central and northern Alexander Island (Burn, 1981). A similar, though less marked, migration is apparent to the north where Lower Tertiary igneous rocks are confined to the offshore islands and the South Shetlands (Figure 3).

## ANDEAN BACK-ARC BASINS

Back-arc basin sequences are documented in southern Palmer Land and the James Ross Island area. In southern Palmer Land volcanoclastic sedimentary rocks occur in an arc from eastern Ellsworth Land to the Black Coast (Singleton, 1980; Rowley and Williams, 1982; Thomson, 1982). Shale, siltstone, and sandstone, together with sparse carbonaceous beds and limestone lenses, crop out; in the Orville Coast region a minimum thickness of 830 m has been recorded (Thomson et al., 1978). Near the crest of the Peninsula possible non-marine facies occur and toward the south and southeast deltaic and shelf depositional environments are present. These rocks intertongue with, and are overlain by, andesitic to silicic lavas, ash-flows and ash fall tuffs which on the Black Coast exceed 1200 m in thickness. This represents an expansion of the magmatic arc into the back arc basin. Ammonities indicate a Late Kimmeridgian—Late Tithonian (Late Jurassic) age. To the west in eastern Ellsworth Land a less well-documented sequence of volcanic and sedimentary rocks contains Bajocian to Kimmeridgian (Middle and Upper Jurassic) invertebrate faunas (Quilty, 1977).

Rocks assignable to this lithotectonic unit may crop out on the eastern flank of Palmer Land as far north as the Crabeater Point area where Fraser and Grimley (1972) report volcanoclastic strata that are pre-Cretaceous but younger than the Upper Palaeozoic-Lower Mesozoic Trinity Peninsula Formation (Group).

In Palmer Land, deformation of these arc-related rocks occurred while arc construction was still in progress. Rocks in the Lassiter Coast to Orville Coast part of the basin were strongly deformed into tight and isoclinal folds in Early Cretaceous time, as shown by dated late-Early Cretaceous cross-cutting plutons that lack significant deformational fabrics (Rowley and Williams, 1982). In eastern Ellsworth Land deformation is Early Cretaceous (Halpern, 1967) or older on the basis of dated undeformed dykes that cut isoclinally-folded and thrust-faulted strata (Quilty, 1977). Deformation of equivalent strata on the Black Coast may also have been Late Jurassic or earliest Cretaceous (Singleton, 1980). Structures thoughout this region strike parallel to the coast and verge away from the magmatic arc. The volcanoclastic strata reported by Fraser and Grimley (1972) are structurally complicated and have been cut by thrust faults displaying northeastward movement.

The sequence in the James Ross Island area is similar to a foreland basin, though lacking the normally associated thrusting and folding. The Mesozoic part of the sequence, possibly as much as 5000 m thick, consists of coarse conglomeratic strata at the base, overlain by sandstones and shales, and then poorly consolidated sandstones with intercalated finer grained beds (Bibby, 1966). Volcanic detritus and airfall deposits are now known to be common (J.A. Crame, pers. comm., 1982). The lowest exposed beds are possibly Cenomanian or older (Crame, 1981; Thomson, 1982) and the youngest Mesozoic rocks are Maastrichtian (Rinaldi, 1982). The younger part of the sequence consists of Lower Tertiary marine deltaic sediments that crop out on Seymour Island (Elliot and Trautman, 1982). The ?Middle Cretaceous conglomerates indicate erosion of a terrain of sedimentary rocks, including fossiliferous marine Lower Cretaceous beds, lavas and tuffs, and subordinate metamorphic rocks (Bibby,1966). The conglomerate sequences on Pedersen Nunatak and Sobral Peninsula (Elliot, 1966) also contain much volcanic detritus and may have a similar lithotectonic setting though different age. Other sequences assignable to this unit occur in the Oscar 11 Coast region where a 600 m thick section of clastic sediments is interbedded with and overlain by volcanic rocks (Fleet, 1968), and at Longing Gap where a thin Upper Jurassic sequence contains a significant volcanic component (Bibby, 1966; Farquharson, 1982). Clear evidence of unroofing of plutonic sources is found only in the youngest beds, the Lower Oligocene of Seymour Island.

The limits of this basin are uncertain. Coarse conglomerates on the northwest coast of James Ross Island suggest a location close to the basin edge. Similarly, Sobral Peninsula and Pedersen Nunatak may be close to the margin. To the northeast the basin included Dundee Island where Albian clastic strata crop out (Crame, 1980: Thomson, 1982), and to the southwest Cape Marsh where Upper Cretaceous marine sandstone and shales occur (Fleet, 1966). The basin probably extended south, possibly to include the Cretaceous sediments of Crabeater Point. The Upper Jurassic volcanogenic marine beds at Longing Gap represent part of a back-arc basin, initiated at the time of onset of Andean volcanism. Deepening of the basin occurred in Cretaceous time, but by the Early Tertiary the proximal part of the basin was filled and molasse deposits were laid down.

Deformation in the James Ross Island region is restricted to gentle warping, tilting of Upper Cretaceous strata on Seymour and Snow Hill Islands during earliest Cenozoic time, and tilting of strata along the northwest side of James Ross Island at an uncertain time after the Cretaceous.

## ANDEAN FORE-ARC BASINS

Fore-arc basins located on the Pacific flank of the Peninsula are exposed on Alexander Island and the South Shetland Islands. On the eastern flank of Alexander Island a sequence of uncertain thickness, but possibly more than 5000 m, crops out (Thomson, 1982). The lowest exposed rocks are Kimmeridgian (Late Jurassic) and the section appears to pass without significant break into the Albian (late Early Cretaceous). The lower part of the sequence is predominantly volcanic and passes up into conglomerate, sandstone and shale in which contemporaneous volcanic detritus is abundant (Thomson, 1982). The depositional environment of these rocks is regarded as marine-deltaic, changing to non-marine in the youngest beds. The rocks record

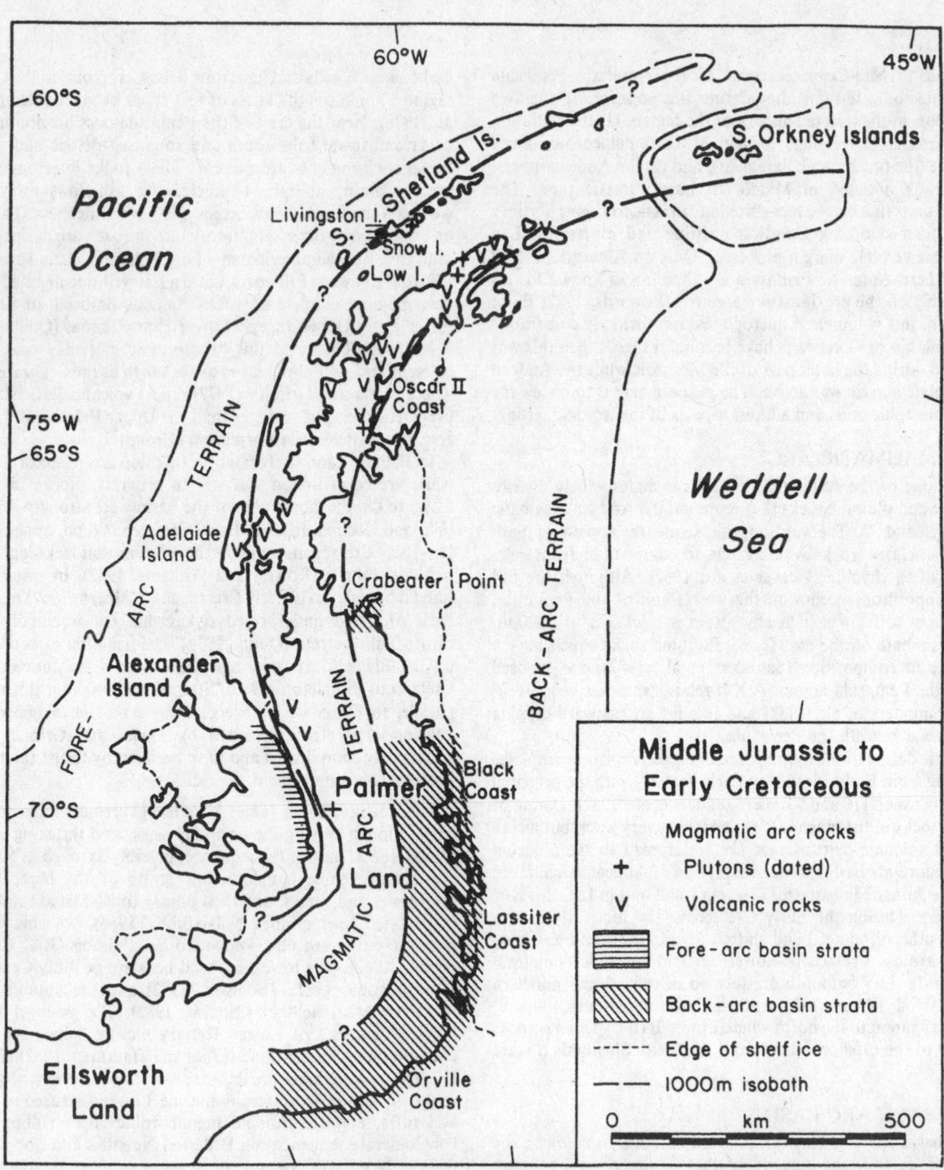


**Figure 1. Sketch map of the Antarctic Peninsula showing Andean lithotectonic units of Middle Jurassic to Early Cretaceous age. Only dated magmatic arc rocks are shown. Note that the boundaries between the arc terrains are generalised and that within the time span indicated may have changed slightly.**

erosion of a volcanic terrain and the unroofing and erosion of crystalline rocks (Elliot, 1974). Adjacent Palmer Land, where thick volcanic rocks crop out and locally overlie thin Middle to Upper Tithonian (Upper Jurassic) marine clastic strata, is regarded as the source terrain. However, there is evidence for a western source for some of the detritus and Thomson (1982) has suggested a setting similar to the Sunda Arc where non-volcanic islands occur in the fore-arc region.

Marine volcanoclastic beds of Late Jurassic to Early Cretaceous age crop out locally on Adelaide Island and in the South Shetland Islands (Thomson, 1982). On Livingston Island shales and sandstone of Tithonian to Valanginian age are flysch-like in the lower part and pass up into (?) shallow water deposits. On Low Island a thin sequence of Upper Jurassic marine volcanoclastic rocks is attributed to turbidite deposition. The Livingston Island sequence passes up into, or is unconformably overlain by (Elliot, unpublished data), subaerially deposited volcanic rocks as young as Barremian (Askin, 1981); similar volcanoclastic strata crop out at President Head, Snow Island and are Early Cretaceous in age (R.A. Askin pers. comm., 1982).

Fore-arc deposits, now exposed, are largely confined to Alexander Island with only scattered outcrops elsewhere. No rocks younger than Mid-Cretaceous have been recorded, but doubtless occur on the continental shelf.

## DISCUSSION

The Late Palaeozoic-Early Mesozoic Gondwana plate margin, which constitutes the pre-Jurassic basement of the Antarctic Peninsula (Dalziel, 1982) was formed and metamorphosed in post-Triassic time. The margin was uplifted and became the source for locally deposited, thick alluvial fans inferred to rest on the eroded Gondwana Orogen in fault block basins (Elliot and Gracanin, this volume). Alluvial fan deposits which form a distinct tectonostratigraphic unit, appear to be confined to the northern Peninsula, where they are probably of Late Jurassic age and the South Orkneys, where the Spence Harbour Conglomerate is probably Early Cretaceous. However, predating (or contemporaneous with) alluvial fan depostion, marine sediments were laid down along the flanks of the Peninsula and underlie or are interbedded with volcanic rocks that mark the

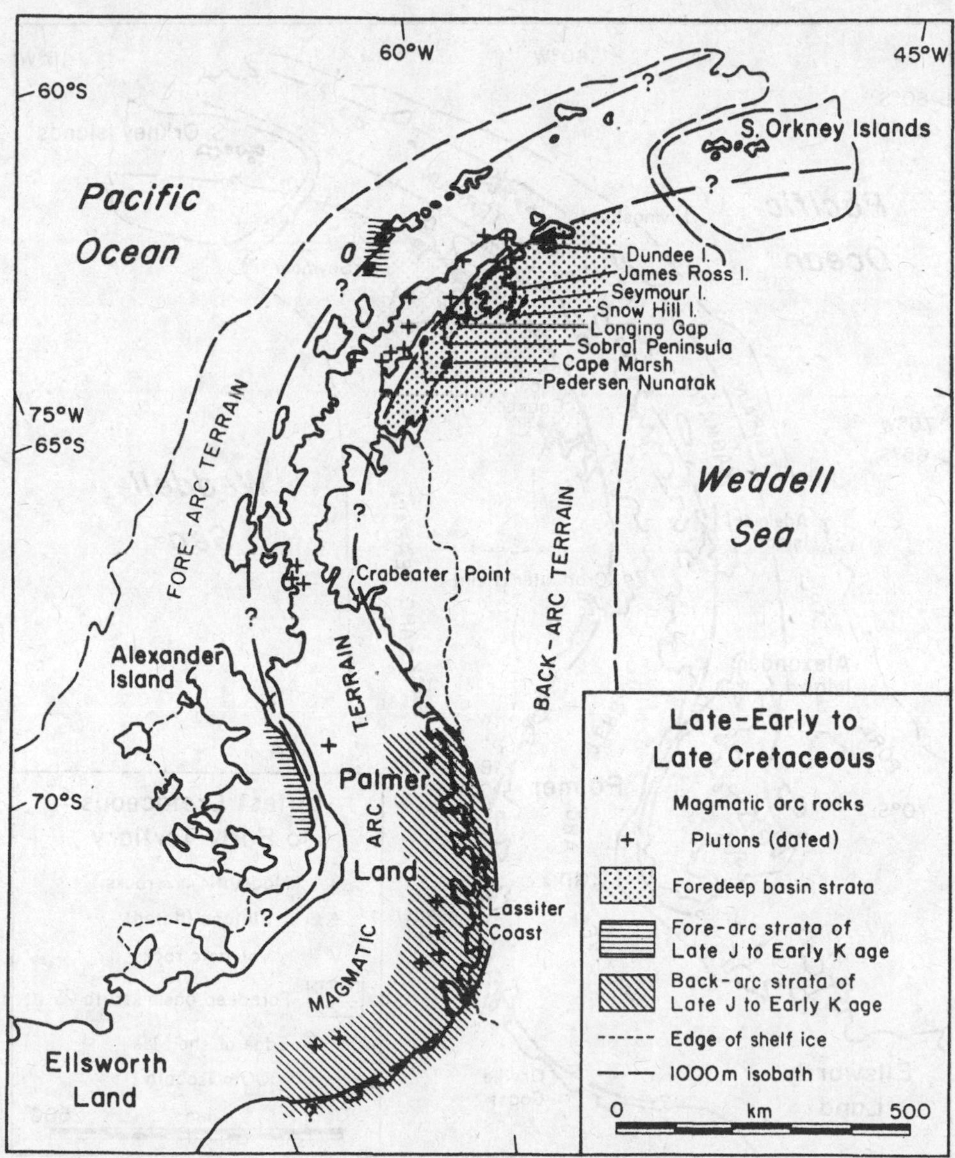


**Figure 2. Sketch map of the Antarctic Peninsula showing Andean lithotectonic units of late-Early Cretaceous to Late Cretaceous age. Only dated magmatic arc rocks are shown. See note for Figure 1.**

onset of Andean magmatism. The magmatic arc was established by late Middle Jurassic time in eastern Ellsworth Land and along the length of the peninsula by the Late Jurassic. This arc appears to have been largely developed on the Gondwana fore-arc terrain though south of about 68°S it partly overlaps magmatic arc rocks of Gondwana age.

In stratigraphic terms, the unconformities separating the pre-Jurassic basement from younger rocks mark a significant event in the geologic history, but in the magmatic history this distinction is unclear because the earliest episode of inferred Andean magmatism, 155-185 Ma, overlaps mineral dates from metamorphic rocks of the Gondwana basement. Undeformed post-tectonic intrusions are common in other plate margins; in the Antarctic Peninsula, presumably, subduction was resumed so rapidly that no clear break exists in the volcanic and plutonic record.

The volcanic record has a marked spatial variation with silicic volcanic rocks dominant on the east flank of the Peninsula. The very great thickness (ca 3000 m) of rhyolitic rocks of probable ash-flow origin suggests comparison with the ignimbrite plateaus identified on the continental flanks of magmatic arcs elsewhere, such as northern Chile.

Fore-arc and back-arc basins flanked the length of the Peninsula from at least Late Jurassic time onwards, and offshore marine strata as old as Middle Jurassic may be present. The back-arc sequence at the base of the Peninsula was folded and thrust away from the arc terrain during the latest Jurassic or earliest Cretaceous, during which time deposition in the Alexander Island fore-arc basin was continuing. The deformation in northeast Palmer Land may be contemporaneous but control is lacking. Farther north the only sign of Cretaceous tectonism lies in the existance of the James Ross basin and the redeposition of Lower Cretaceous marine rocks as clasts in the conglomerates of(?) Middle Cretaceous age in that basin.

For the most part post-Jurassic back-arc deposition is concealed beneath the Ronne Ice Shelf where a thick sedimentary sequence is inferred to be present, and on the continental shelf to the east of the Peninsula. Isolated outcrops at Crabeater Point are part of this back-arc terrain but it is only in the James Ross Island region that any

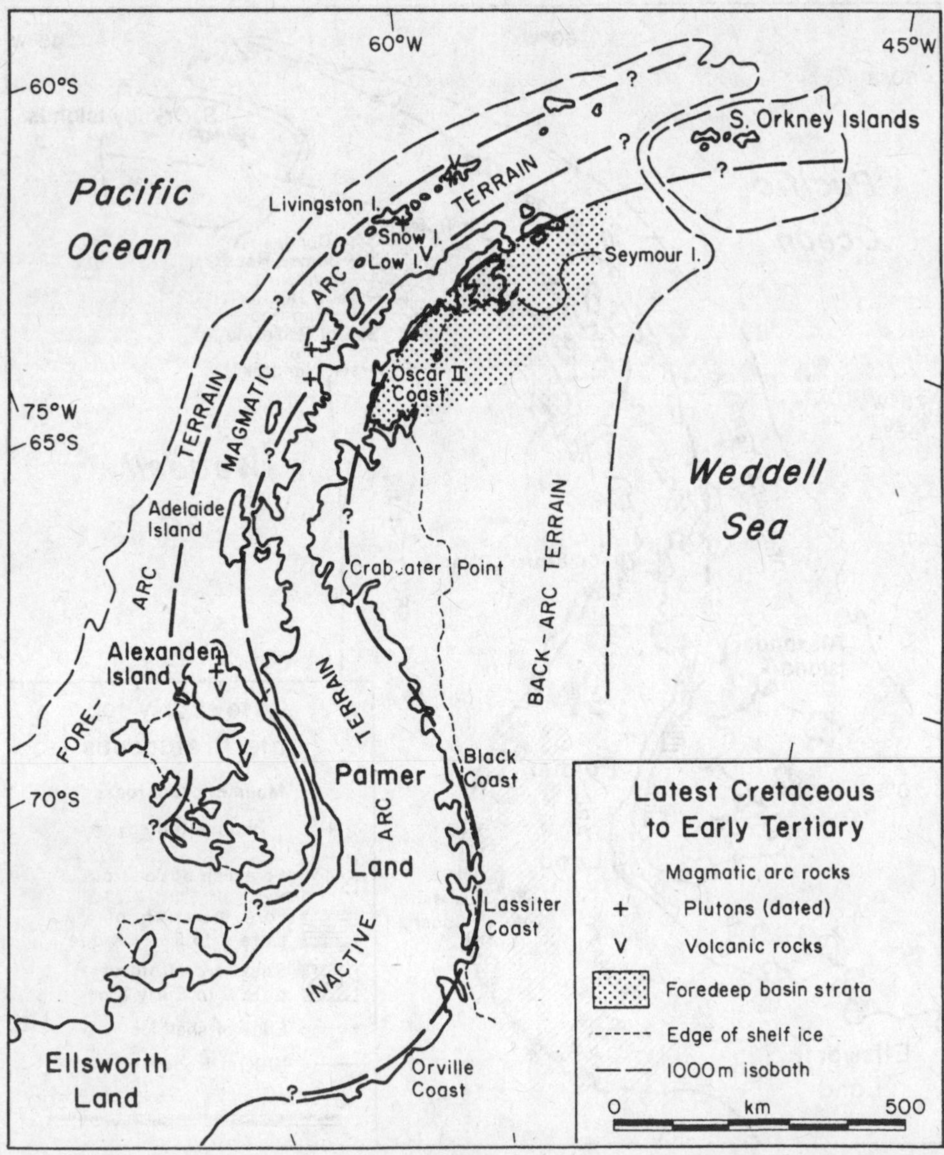


**Figure 3.** **Sketch map of the Antarctic Peninsula showing lithotectonic units of latest Cretaceous to Early Tertiary age. Only dated magmatic arc rocks are shown. See note for Figure 1.**

extensive deposits are exposed, probably as a result of the lava capping and uplift associated with that igneous activity.

Although the post-tectonic Early Cretaceous plutonism of Palmer Land represents a migration of magmatism to the south and east, a more significant event occurred in the latest Cretaceous. This event, the migration of the locus of magmatism toward the north and west during latest Cretaceous-Early Tertiary time, may be attributed to inception of spreading from the Aluk Ridge (Herron and Tucholke, 1976). Sea floor consumption until the Late Cretaceous was related to a spreading system that predates the present Pacific-Antarctic Ridge, possibly the spreading centres that gave rise to the M anomalies of the northwestern Pacific. The oldest anomaly associated with the Aluk Ridge is anomaly 29 (66 Ma). The generation of this ridge probably lead to an hiatus in subduction, reorientation of spreading directions, and a shift in the position of the trench.

The progressive overriding of the Aluk Ridge, inferred to be ca 45 Ma for the sector that lay off Alexander Island and to the west, and 20 Ma for the sector off Adelaide Island to Anvers Island, would have cut off the driving mechanism. Igneous activity continued till ca 40 Ma (Mt Rouen Batholith) in Alexander Island (Burn, 1981) and in the northern Peninsula until ca 35 Ma, but is possibly as young as 20 Ma, on Anvers Island (Gledhill et al., 1982). The extent of volcanic activity in the South Shetlands post ca 40 Ma and pre Bransfield Trough opening (ca 4 Ma) is uncertain (see Elliot and Dupre, this volume).

Tectonism associated with the "Andean" is distinguished by the contrast between the northern Peninsula where structure can be attributed to vertical movements, and Palmer Land where latest Jurassic or earliest Cretaceous deformation consists of strong folding and thrusting away from the arc. Spatially, the different tectonic history and the shift in the location of Early Cretaceous plutonism of Palmer Land coincides with the abrupt widening of the Peninsula. A plate tectonic setting different from that of the northern Peninsula is implied and furthermore the varying orientation of structures in Palmer Land suggests the setting was more complex. Possible explanations include juxtapostion of Palmer Land against continental blocks such that folding and thrusting was related to zones of impingement that changed location and orientation, location (for southern Palmer Land) along a major transform fault with folding and thrusting developed as along the San Andreas Fault in southern California, and a combination of the two. Furthermore the increase in width of the "Pacific" continental shelf suggests either flattening of the Mesozoic subduction zone compared with the northern Peninsula or pre-Aluk Ridge accretion to the continental margin.

In addition to the differences between Palmer Land and the northern Antarctic Peninsula, equal contrasts exist between the latter and southern South America. The northern Peninsula lacks the penetrative deformation and evidence for marginal basin formation and destruction that is the case in Tierra del Fuego (Dalziel, 1981). Even though sea floor data suggest opening of the Drake Passage at ca 29 Ma, a major tectonic break in that region must have existed from Middle Jurassic time onward; certainly the sea floor anomaly patterns for the latest Mesozoic and Early Cenozoic in the southeastern Pacific indicate the presence of a triple junction and hence a complex set of plate boundaries, and thus supports such an argument. Finally, the lack of penetrative deformation in the northern Peninsula suggests that it was backed by oceanic crust of the evolving Weddell Sea, part of which has been documented by LaBrecque and Barker (1981).

*Acknowledgements* I wish to thank K.O. Stanley for discussion of tectonic problems. K.O. Stanley and J.W. Collinson reviewed the manuscript. Funding for preparation of this paper was provided by NSF grant DPP78-21102. Contribution number 446 of the Institute of Polar Studies.

## REFERENCES


ASKIN, R.A., 1981: Jurassic-Cretaceous palynology of Byers Peninsula, Livingston Island, Antarctica. *U.S. Antarct. J., 16,*11-13.

BIBBY, J.S., 1966: The stratigraphy of part of north-east Graham Land and the James Ross Island Group. *Scient. Rep. Br. antarct. Surv., 53.*

BURN, R.W., 1981: Early Tertiary calc alkaline volcanism on Alexander Island. *Bull. Br. antarct. Surv., 53,* 175-93.

CRAME, J.A., 1980: The occurrence of the bivalve *Inoceramus concentricus* on Dundee Island, Joinville Island group. *Bull. Br. antarct. Surv., 49,*283-6.

CRAME, J.A., 1981: Upper Cretaceous inoceramids (bivalvia) from the James Ross Island Group and their stratigraphic significance. *Bull. Br. antarct. Surv., 53,*29-56.

DALZIEL, I.W.D., 1981: Back-arc spreading in the Southern Andes: a review and critical reappraisal. *R. Soc. Lond., Philos. Trans., Ser A, 300,* 319-35.

DALZIEL, I.W.D., 1982: Pre-Jurassic history of the Scotia Arc region; *in* Craddock, C. (ed.) *Antarctic Geoscience.* Univ. Wisconsin Press, Madison, 111-26.

EDWARDS, C.W., 1982: New paleontological evidence of Triassic sedimentation in West Antarctica; *in* Craddock, C. (ed.) *Antarctic Geoscience.* Univ. Wisconsin Press, Madison, 325-30.

ELLIOT, D.H., 1966: Geology of the Nordenskjold Coast and a comparison with northwest Trinity Peninsula, Graham Land. *Bull. Br. antarc. Surv., 10,* 1-43.

ELLIOT, D.H. and DUPRE, D.D., (this volume): The age and geochemistry of some rocks from the South Shetland Islands.

ELLIOT, D.H. and GRACANIN, T.M. (this volume): Conglomeratic strata of Mesozoic age at Hope Bay and Joinville Island, northern Antarctic Peninsula.

ELLIOT, D.H. and TRAUTMAN, T.A., 1982: Lower Tertiary strata on Seymour Island; *in* Craddock, C. (ed.) *Antarctic Geoscience.* Univ. Wisconsin Press, Madison, 287-97.

ELLIOT, M.H., 1974: Stratigraphy and sedimentary petrology of the Ablation Point area, Alexander Island. *Bull. Br. antarc. Surv., 39,*87-113.

FARQUHARSON, G.W., 1982: Late Mesozoic sedimentation in the northern Antarctic Peninsula and its relationship to the southern Andes. *Geol. Soc. Lond., J., 139,* 721-28

FLEET, M., 1966: Occurrence of fossiliferous Upper Cretaceous sediments at Cape Marsh, Robertson Island. *Bull. Br. antarc. Surv., 8,*89-91.

FLEET, M., 1968: The geology of the Oscar II Coast, Graham Land. *Scient. Rep. Br. antarc. Surv., 59.*

FRASER, A.G. and GRIMLEY, P.H., 1972: The geology of parts of the Bowman and Wilkins Coasts, Antarctic Peninsula. *Scient. Rep. Br. antarc. Surv., 67*

GLEDHILL, A., REX, D.C. and TANNER, P.W.G., 1982: K-Ar and Rb-Sr geochronology of igneous and metamorphic rock suites from the Antarctic Peninsula; *in* Craddock, C. (ed.) *Antarctic Geoscience.* Univ. Wisconsin Press, Madison, 315-23.

HALPERN, M., 1967: Rubidium-strontium isotopic age measurements of plutonic igneous rocks in eastern Ellsworth Land and northern Antarctic Peninsula, Antarctica. *J. Geophys. Res. 72,* 5133-42.

HERRON, E.M. and TUCHOLKE, B.E., 1976: Sea floor magnetic patterns and basement structure in the southeastern Pacific. *Initial Reports of the Deep Sea Drilling Project 35,* 263-78. U.S. Govt. Printing Office, Washington, D.C.

LABRECQUE, J.L. and BARKER, P.F., 1981: The age of the Weddell Basin. *Nature, 290,* 489-92.

QUILTY, P.G., 1977: Late Jurassic bivalves from Ellsworth Land, Antarctica: their systematics and palaeogeographic implications. *N.Z. J. Geol. Geophys., 20* 1033-80.

RINALDI, C.A., 1982: The Upper Cretaceous in the James Ross Island Group; *in* Craddock, C. (ed.) *Antarctic Geoscience* Univ. Wisconsin Press, Madison, 281-6.

ROWLEY, P.D. and WILLIAMS, P.L., 1982: Geology of the northern Lassiter Coast and south Black Coast, Antarctic Peninsula; *in* Craddock, C. (ed.) *Antarctic Geoscience.* Univ. Wisconsin Press, Madison, 339-48.

SAUNDERS, A.D., WEAVER, S.D. and TARNEY, J., 1982: The pattern of Antarctic Peninsula plutonism; *in* Craddock, C. (ed.) *Antarctic Geoscience.* Univ. Wisconsin Press, Madison, 305-14.

SINGLETON, D.C., 1980: The geology of the central Black Coast, Palmer Land. *Scient. Rep. Br. antarc. Surv., 102.*

THOMSON, M.R.A., 1982: Mesozoic paleogeography of western Antarctica; *in* Craddock, C. (ed.) *Antarctic Geoscience.* Univ. Wisconsin Press, Madison, 331-7.

THOMSON, M.R.A., LAUDON, T.S. and BOYLES, J.M., 1978: Stratigraphical studies in Orville Coast and eastern Ellsworth Land. *U.S. Antarct. J., 13,* 9-10.

WEAVER, S.D., SAUNDERS, A.D. and TARNEY, J., 1982: Mesozoic-Cenozoic volcanism in the South Shetland Islands and the Antarctic Peninsula: geochemical nature and plate tectonic significance; *in* Craddock, C. (ed.) *Antarctic Geoscience.* Univ. Wisconsin Press, Madison 263-73.

# A GEOCHEMICAL OVERVIEW OF SUBDUCTION-RELATED IGNEOUS ACTIVITY IN THE SOUTH SHETLAND ISLANDS, LESSER ANTARCTICA

J.L. Smellie, *Institute of Geological Sciences, West Mains Road, Edinburgh, EH9 3LA, U.K.*

*Abstract* The South Shetland Islands have been situated on a consuming plate margin for at least 200 Ma but igneous activity probably did not become established there until the earliest Cretaceous. The lavas show some features of island-arc tholeiites but are mainly basalts, basaltic andesites and silica-poor andesites of monotonous low-K, high-alumina calc-alkaline type. Entirely calc-alkaline intermediate magmas were also developed on King George Island during the Early Tertiary. At least five major rock groups have been identified stratigraphically. Despite considerable similarity in major oxides, they differ in dispersed element concentrations. Two groups are shown to be heterogeneous, each containing two chemically distinct members *sensu lato*. All the groups can be related by variable partial melting and/or fractionation of plagioclase, pyroxene, olivine, titaniferous magnetite and apatite.

The plutonic intrusions are a calc-alkaline suite of mainly gabbro, tonalite and granodiorite with element abundances and inter-element ratios closely comparable with the associated, probably cogenetic lavas. They invariably crop out in areas of volcanic rocks altered under conditions ranging up to the prehnite-pumpellyite facies. The alteration probably occurred in fissure-controlled multipass geothermal systems set up above the rising cooling plutons, resulting in leaching and re-precipitation of Si, K, Rb, Ba, Ca and Fe.

The ultimate souce for these arc-front magmas is unclear because lithophile element concentrations depend on a complex interplay of factors but a model is preferred which involves volatile metasomatism on the mantle wedge underlying the island group.

The South Shetland Islands contain a sialic "basement" of strongly deformed low-grade schists and flysch-like strata separated by faults from, but presumably overlain unconformably by, essentially undeformed Jurassic-Quaternary volcanogenic sequences (Figure 1). They are intruded along an axial zone by a Cretaceous-Tertiary plutonic suite and altered hydrothermally under conditions ranging up to the prehnite-pumpellyite facies. The first evidence for volcanic activity located within the islands consists of lavas interbedded with Berriasian marine sediments on western Livingston Island. The subsequent history is dominated by subaerial volcanism and K-Ar radiometric ages have demonstrated an apparent north-easterly migration of the volcanic and plutonic foci between Cretaceous and Tertiary (British Antarctic Survey, unpublished data); although no Late Tertiary ages have yet been obtained, outcrops on southeastern King George Island have been assigned this age by several authors (e.g. Barton, 1965; Birkenmajer, 1980; Weaver et al., 1982). The volcanic sequences are divisible into at least five major rock groups on the basis of age, geographical separation and lithology; their detailed stratigraphy is presented elsewhere (Smellie and others, in preparation; Table 1).

Previous geochemical studies of the igneous activity have, of necessity, used data collected at comparatively few sites scattered throughout the islands (e.g. González-Ferrán and Katsui, 1970; Birkenmajer et al., 1981; Weaver et al., 1982). These authors have suggested that the lavas are a low-K, high-alumina calc-alkaline suite with some tholeiitic characteristics. Mesozoic lavas of Byers Peninsula were shown to be chemically very similar to Tertiary lavas of Fildes Peninsula, with uniformly low lithophile element abundances and inter-element ratios characteristic of magmas evolved in an arc-front position relative to the Antarctic Peninsula. In contrast, it was suggested that lavas of presumed Late Tertiary age on King George Island were enriched in lithophile elements and heralded the opening of Bransfield Strait.

This paper utilises numerous as yet largely unpublished stratigraphical data and a more comprehensive database of geochemical analyses (including the first analyses of the plutonic intrusions) than hitherto was available. The main objectives are to describe and discuss individual rock groups in terms of their major and trace element geochemistry, alteration and stratigraphy. By these means it should be possible to test whether previous conclusions are applicable to the

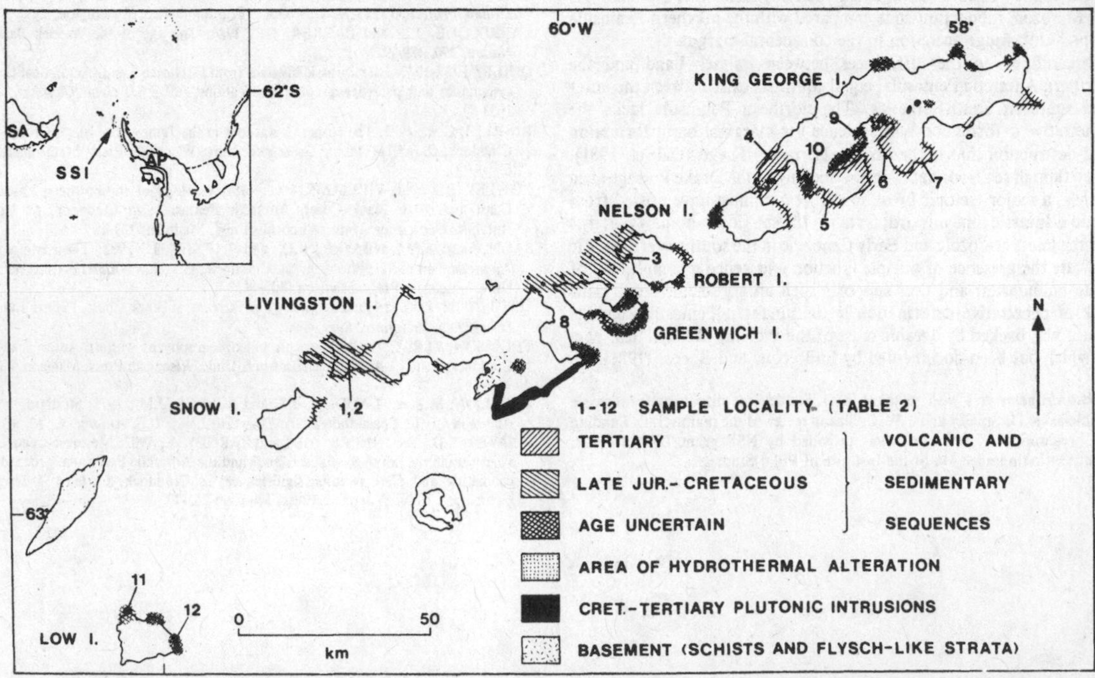


**Figure 1 Geological sketch map of the South Shetland Islands (SSI) and their location (inset) relative to southern South America (SA) and the Antarctic Peninsula (AP). (Abbreviations refer to inset.)**

**TABLE 1: Simplified stratigraphy of the subductiuon-related volcanogenic sequence of the South Shetland Islands**

| Rock Group | Age | Characteristic Lithology | Outcrop Area |
|---|---|---|---|
| 'Late Tertiary' (informal name) | Uncertain, probably mainly late Tertiary extending into the Quaternary | Holocrystalline basalts and basaltic andesites | South-eastern King George Island (? Ternyck Needle, Martins Head, Cinder Spur, Low Head, Cape Melville) |
| Hennequin formation* | Early Tertiary | Glassy andesites | King George Island (east of Admiralty Bay; also Crepin Point, Admiralen Peak, part of Point Thomas, Three Brothers Hill) |
| Fildes formation* | Early Tertiary | Holocrystalline basalts and basaltic andesites | King George Island (west of Admiralty Bay); Eastern Nelson Island |
| Coppermine formation* | Late Cretaceous | Holocrystalline basalts | Northern Robert and Greenwich Islands; north eastern Livingston Island; English and Macfarlane Straits |
| Eyers formation** | Mainly Late Jurassic Cretaceous | Basalts—andesites and rhyolites | Western Livingston Island; Eastern Snow Island |
| Volcanic sequences of uncertain stratigraphical affinities | Late-Jurassic on Low Island***, uncertain elsewhere | Basalts—andesites | Isolated occurrences on all the main islands (see figure 1) |

* Smellie and others, in preparation
** Smellie and others, 1980
*** Smellie, 1980

entire island group. The Quaternary volcanism associated with marginal basin formation has already been the focus of intensive investigation (e.g. Weaver et al., 1979) and is excluded from this discussion.

## Geochemistry

One hundred and sixty-seven rocks were analysed for major oxides and 14 trace elements on a Phillips PW 1450 automatic XRF spectrometer using 46 mm pressed powder pellets; an additional 150 analyses were kindly supplied by A.D. Saunders and S.D. Weaver. In total, the database contains analyses of 282 lavas (104 Mesozoic, 151 Lower Tertiary, 25 Upper Tertiary) and 35 plutonic rocks (Table 2).

*Alteration* Most of the analyses are of holocrystalline lavas sampled in areas away from the axial zone of alteration. For these, the data define coherent trends with comparatively little scatter although slight movement of K may have occurred. However, K, Rb, Ba and possible Ca mobility is evident in lavas with a glassy mesostasis (i.e. principally restricted to the Byers and Hennequin Formations) due to the ease with which patches of acid interstitial glass are readily dissolved by hydrothermal fluids, releasing the most mobile elements to travel in solution for varying distances.

In contrast, metasomatic phenomena are widespread in lavas of the axial zone (e.g. Barton, 1965; Littlefair, 1978). Fifty-five highly altered Mesozoic and Tertiary lavas were analysed in order to examine the chemical effects of the alteration. K, Rb and Ba mobility is ubiquitous and Fe has been mobilized substantially in glassy Hennequin Formation lavas; Ca-migration is comparatively slight, supporting the suggestion that Ca in excess of stoichiometric requirements is largely restricted to fissures (Smellie, 1980). Many of the Mesozoic lavas are enriched in Si, whereas this element was largely stable in the Tertiary lavas. The alteration is envisaged occurring within convective hydrothermal cells set up above rising cooling plutonic intrusions. Fluids in the volcanic pile would be heated at depth and returned to the surface along localised probably structurally controlled pathways which appear to have acted as a locus for precipitation of mobile elements leached from the country rocks. Small amounts of new elements may have been introduced but much of the circulating water was probably meteoric rather than juvenile in origin. Because of the significant chemical effects of this alteration, lavas of the axial zone are excluded from further discussion and from the plots in Figure 2, except where acknowledged.

*Major Oxides* Basalts, basaltic andesites and silica-poor andesites are predominant among the lavas and hypabyssal intrusions. The plutonic intrusions are mainly quartz-gabbros, tonalites and granodiorites and, apart from a small micro-adamellite intrusion on Low Island, more siliceous differentiates are only represented by scarce late-stage veins.

Conventional discriminatory diagrams classify the South Shetland magmas as subalkaline, with total iron (Fe*)/Mg and $Na_2O/K_2O$ ratios mostly similar to island arc tholeiites. On an AFM diagram slight Fe-enrichment is restricted to basaltic compositions, and the more siliceous members show pronounced alkali-enrichment characteristic of calc-alkaline series. Plutonic intrusions on King George Island and Hennequin Formation lavas are entirely calc-alkaline.

The major oxides show regular trends of increasing $K_2O$ and decreasing CaO, $Fe_2O_3$*, MgO and $Al_2O_3$. $Na_2O$ increases in abundance in the basic rocks up to about 55% $SiO_2$ then gradually decreases due to separation of intermediate plagioclase. MnO and $TiO_2$ increase to a maximum in rocks of 54-56% $SiO_2$, falling gradually thereafter corresponding to the incoming of Fe-Ti oxide as a fractionating phase. $P_2O_5$ shows a similar pattern, but peak contents are not reached until ca 60% silica is attained despite the presence of apatite microphenocrysts in lavas with as little as 55% silica. Because apatite is always modally uncommon, its net fractionation effect may have been negligible until enhanced partitioning of P occurred in the more siliceous melts.

The South Shetland magmas are virtually impossible to separate convincingly by major element analysis alone. The Fildes and Byers Formations show division into two sub-groups each using $K_2O$ contents (Figure 2), the Hennequin Formation is slightly enriched in MgO, $Al_2O_3$, $Fe_2O_3$* and CaO, and slightly depleted in $Na_2O$ at equivalent silica and Zr contents, and the basic members of the Byers and Coppermine Formations contain the highest $TiO_2$ and $P_2O_5$.

*Trace Elements* The high contents of La, Ce, Rb, Ba, Sr and Zr resemble other calc-alkaline series in island arcs (as opposed to continental margins: cf. Jakes and White, 1972) and the $Ce_N/Y_N$ (N = normalised with respect to chondrite) ratios suggest slight light rare

**TABLE 2: Representative analyses of South Shetland lavas and plutonic intrusions**

| | 1 | 2 | 3 | 4 | 5 | 6 | 7 | 8 | 9 | 10 | 11 | 12 |
|---|---|---|---|---|---|---|---|---|---|---|---|---|
| $SiO_2$ | 48.3 | 78.0 | 49.8 | 50.7 | 53.1 | 67.5 | 53.7 | 48.2 | 54.1 | 58.2 | 64.0 | 71.4 |
| $TiO_2$ | 1.07 | 0.11 | 0.67 | 0.87 | 0.97 | 0.64 | 0.86 | 0.39 | 0.75 | 0.77 | 0.54 | 0.39 |
| $Al_2O_3$ | 15.5 | 12.4 | 17.1 | 19.9 | 17.2 | 16.2 | 24.0 | 18.7 | 17.3 | 16.9 | 15.2 | 14.4 |
| $Fe_2O_3$ | 2.36 | 0.06 | 2.46 | 2.74 | 2.46 | 0.74 | 0.82 | 2.08 | 2.32 | 2.24 | 1.62 | 0.86 |
| FeO | 7.85 | 0.14 | 6.14 | 6.85 | 6.15 | 1.86 | 2.74 | 5.19 | 5.80 | 5.61 | 4.05 | 2.15 |
| MnO | 0.20 | 0.01 | nd | 0.17 | 0.16 | 0.17 | 0.13 | 0.13 | 0.14 | 0.21 | 0.12 | 0.09 |
| MgO | 9.6 | 0.2 | 9.8 | 3.9 | 5.7 | 0.7 | 1.8 | 8.7 | 6.4 | 4.1 | 1.9 | 1.1 |
| CaO | 10.5 | 0.3 | 11.39 | 11.57 | 8.00 | 2.11 | 10.35 | 13.9 | 8.39 | 7.72 | 4.44 | 2.40 |
| $Ka_aO$ | 2.50 | 2.26 | 2.24 | 3.38 | 3.36 | 5.12 | 4.16 | 8.7 | 3.30 | 3.33 | 4.12 | 4.60 |
| $K_2O$ | 0.24 | 5.48 | 0.34 | 0.32 | 1.00 | 3.81 | 0.66 | 0.20 | 1.41 | 2.18 | 1.82 | 2.52 |
| $P_2O_5$ | 0.22 | 0.01 | 0.12 | 0.12 | 0.22 | 0.14 | 0.20 | 0.04 | 0.17 | 0.21 | 0.12 | 0.05 |
| TOTAL | 98.33 | 98.97 | 100.06 | 100.52 | 98.32 | 98.99 | 99.42 | 99.72 | 100.08 | 101.47 | 97.93 | 99.96 |
| Cr | 256 | — | 360 | — | 20 | 3 | 19 | 150 | 20 | — | — | — |
| Ni | 84 | 6 | 142 | — | 16 | — | 12 | 40 | 11 | — | — | — |
| Zn | — | — | 54 | 69 | 66 | 94 | 99 | 59 | 70 | 86 | 30 | 46 |
| Rb | 2 | 121 | — | — | 14 | 87 | 10 | — | 34 | 38 | 54 | 61 |
| Sr | 350 | 37 | 670 | 609 | 563 | 315 | 10 | 480 | 597 | 719 | 200 | 200 |
| Y | 18 | 15 | 10 | 13 | 23 | 25 | 23 | 7 | 19 | 24 | 36 | 27 |
| Zr | 69 | 102 | 58 | 57 | 93 | 390 | 129 | 21 | 127 | 253 | 140 | 168 |
| Nb | 3 | 9 | — | — | — | 9 | 1 | — | — | 3 | 5 | — |
| Ba | 71 | 665 | 157 | 149 | 333 | 860 | 246 | 62 | 262 | 407 | 319 | 318 |
| La | 4 | 19 | — | — | 16 | 48 | 10 | — | 19 | 29 | 19 | 18 |
| Ce | 16 | 35 | 16 | 17 | 20 | 75 | 21 | — | 33 | 52 | 37 | 34 |
| Pb | 2 | 9 | — | — | — | 10 | 3 | — | — | — | — | 35 |
| Th | 1 | 13 | — | — | — | 18 | 2 | — | — | — | — | 10 |
| Ga | — | — | 20 | 24 | 21 | 19 | 24 | 21 | 24 | 25 | 19 | 15 |

(Major oxides in Wt.%; trace elements in ppm.)
LAVAS: 1–2. Byers Peninsula; 3. Coppermine Peninsula; 4. Fildes Peninsula; 5. Potter Peninsula; 6. Point Hennéquin; 7. Cinder Spur
INTRUSIONS: 8. Half Moon Island; 9. Wegger Peak; 10. Noel Hill; 11. Cape Wallace; 12. North of Cape Hooker (Samples located in Figure 1)

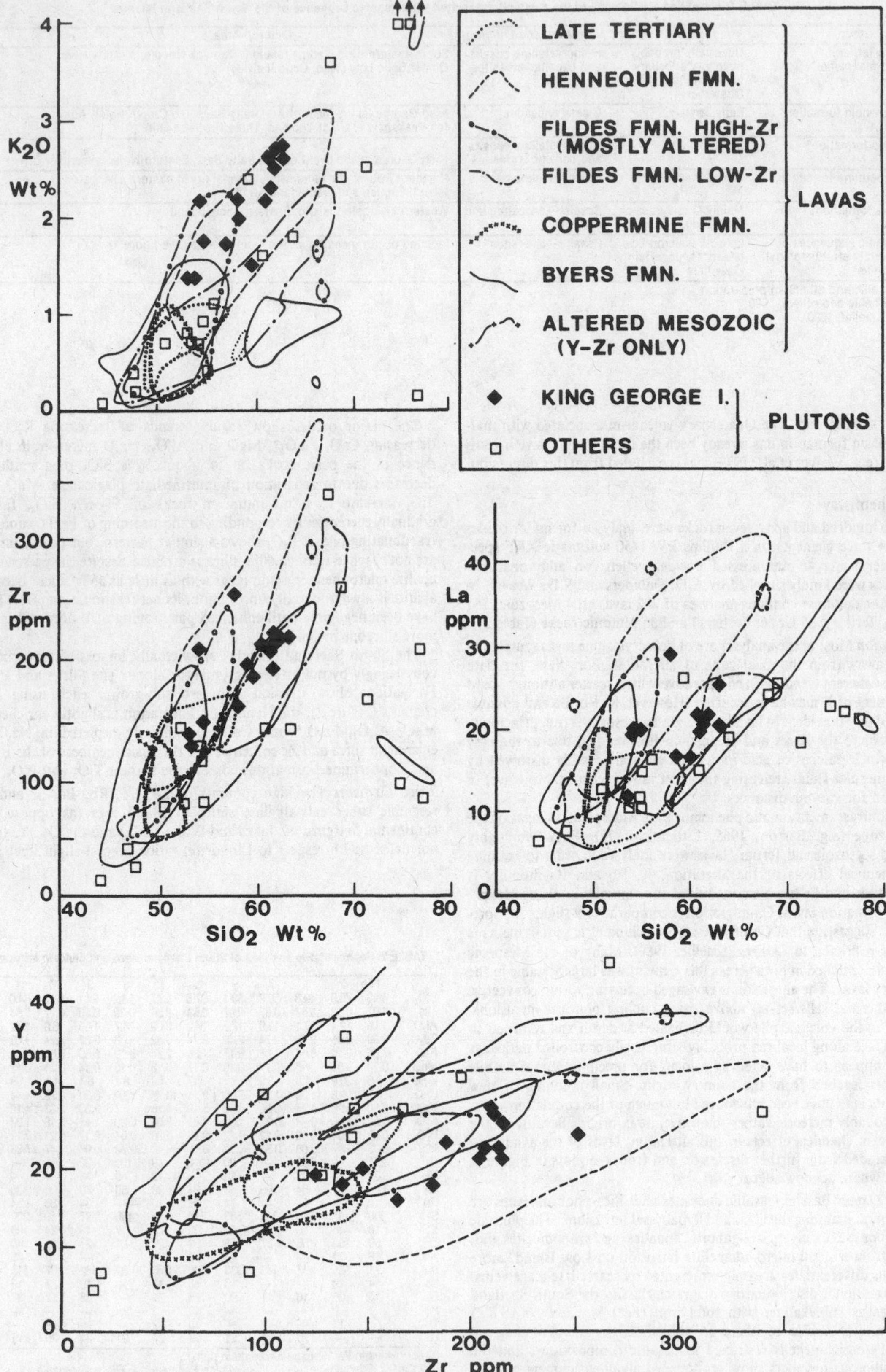


Figure 2 Plots of selected major and trace elements against $SiO_2$ and Zr.

earth element (REE) enrichment atypical of island arc tholeiites. Conversely, tholeiitic characteristics, such as high K/Rb, low Rb/Sr and moderate Cr and Ni, feature in many of the basalts. Rb, Ba, La, Ce, Y and Zr increase smoothly during fractionation. Slight depletion in Ba and Zr occurs at high silica contents (> 67-70%, Figure 2) related to fractionation of intermediate plagioclase and retention by residual zircon (Saunders and others, 1980). Nb, Pb and Th increase more gradually and there is considerable data scatter probably because these elements occur in concentrations close to their detection limits. Zn increases very slightly through the basic rocks but rapidly stablises at 50-100 ppm and may decrease slightly in the most acidic members. Ga is constant at 20-27 ppm, except in lavas from Byers Peninsula which (on the basis of only 10 analyses) contain about 15-20 ppm, decreasing to 5 ppm in the rhyolites. Sr, Cr and Ni behave consistently with fractionation of plagioclase, olivine and clinopyroxene, the phenocryst phases. Moreover, magmas equilibrating with mantle mineral assemblages have Ni/Mg ratios of 0.004-0.006 (Sato, 1977) compared with < 0.001 in South Shetland lavas and it must be assumed that none of these melts are primary magmas but they have been modified by substantial polybaric crystal fractionation prior to eruption. Despite the possible importance of amphibole in island arc magma genesis (Cawthorn and O'Hara, 1976), the lack of covariation between K/Rb and Ce/Y argues against significant amphibole fractionation in South Shetland magmas.

Lavas of the Byers and Fildes Formations are divisible into Zr-rich and Zr-poor members *sensu lato* which, in the former, also corresponds to a division at about 57% silica (Figure 2). Zr, La and Ce contents are comparable between most of the rock groups, but lower in the Byers (high silica) and Fildes (low Zr) Formations. $Y/SiO_2$ ratios are also markedly dissimilar in the two groups of Byers Formation lavas, but are essentially indistinguishable in the other rock groups. Sr contents tend to be lower in the Byers Formation. The Hennequin Formation is distinguished by exceptionally high Sr, Cr and Ni, although lavas at Turret Point differ somewhat in having lower La, Ce and Y. Because of the marked alteration, the affinities and general characteristics of the altered volcanic rocks of possible Mesozoic age (west of Nelson Island) are still uncertain. Element abundances fall within the general data spread for other South Shetland magmas. Interestingly, Y/Zr ratios in altered rocks between eastern Livingston and western Nelson islands are much higher than in the nearby Coppermine Formation (Figure 2), suggesting that the latter may be of restricted occurrence.

Although some separation of the individual rock groups is achieved in plots of the inter-element ratios, much of this variation is attributable to fractionation: Rb/Sr and Ba/Sr ratios are affected because Sr has behaved compatibly even in the basic rocks. Alteration has also reduced the reliability of many ratios involving mobile incompatible elements, and plots of K/Rb and Ba/Rb ratios show little convincing separation. By contrast, K/Ba ratios are clearly higher in the Hennequin Formation (> 33) than any other lavas (< 35). In this respect, the high-Zr lavas of the Fildes Formation show a clear difference from the Hennequin Formation, which they closely resemble in trace element abundances. A most striking feature of the data for the plutonic intrustions is the similarity in trace element contents and inter-element ratios between the King George Island plutons and Hennequin Formation lavas.

No useful separation is achieved in plots of Rb, Ba, Zn, Pb, Th, Nb and Ga. However, the high-silica Byers Formation lavas are slightly depleted in Ba and Ga and, in general terms, this sub-group is unique in consistently showing slight to marked depletion in K, Ba, La, Ce, Y, Zr, Nb and Ga.

Apart from an interesting decoupling of Rb and Ba from K in low-Zr lavas of the Fildes Formation, which have slightly high Rb(and Ba)/Zr ratios compared with other South Shetland lavas but comparable K/Zr, there is little variation in most element/Zr ratios, attesting to a relatively homogeneous source. By contrast, Y/Zr ratios are highly variable and the correlation lines intersect the Y axis (Figure 2). REE analyses carried out by Tarney and other (1982) showed $Ce_N/Yb_N$ ratios ranging from 1.8 to 3.0 in lavas from Byers Peninsula, and significantly higher ratios of 2.0 to 5.8 at Fildes Peninsula due primarily to lower relative Yb contents in the latter. It was suggested that some of the Tertiary lavas were derived by melting at greater depths, leaving some residual garnet, but the generally low $Ce_N/Yb_N$ ratios indicate that garnet played only a minor part in the genesis of most South Shetland magmas.

## Conclusions

The Early Cretaceous-Late Tertiary arc-front magmas of the South Shetland Islands have been divided stratigraphically into at least five major rock groups. The geochemistry broadly supports this division but shows that two, possibly three, are chemically heterogeneous and contain lavas derived from magmas with contrasting trace element abundances. Of these, the high-Zr lavas of the Fildes Formation form isolated outcrops and may deserve separate stratigraphical status, whereas the two magma types in the Byers Formation are apparently interbedded. Only modally dominant plagioclase, pyroxene, olivine, titaniferous magnetite and apatite have had a recognisable effect on the trace element patterns. Additional evidence for fractional crystallisation includes the Cr and Ni contents, which are low in comparison with concentrations expected in mantle-derived primary magmas. There is no significant chemical difference between the volcanic and plutonic rocks, suggesting that both may represent liquid compositions, and their close association geographically and temporally suggests that they are cogenetic.

Most of the differences between the rock groups can be explained by variable partial melting and/or fractional crystallisation. The rocks with the lowest element abundances, probably corresponding to a high degree of partial melting, are of Cretaceous age in the Byers Formation and they include the only undoubted rhyolites found so far. By comparison, the Early Tertiary Hennequin Formation and related plutonic intrustions on King George Island have higher compatible element abundances (Mg, Ca, Fe, Sr, Cr and Ni) than any other South Shetland magmas sampled. Moreover, they are enriched in incompatible elements relative to most lavas of the coeval Fildes Formation. These characteristics were mistakenly assigned by Weaver et al. (1982) to Plio-Pleistocene volcanism connected with the initial stages of rifting open of Bransfield Strait. The absence of basic members in the calc-alkaline Hennequin Formation and subjacent plutons is conspicuous compared with the abundance of these rocks elsewhere in the islands and it appears to indicate a cycle of entirely calc-alkaline intermediate magmatism unique in this area. The development appears to have been temporary, however, because the subsequent Late Tertiary volcanism shows a return to basalt-basaltic andesite-dominated tholeiitic/calc-alkaline compositions characteristic of the South Shetland Islands as a whole.

The ultimate source for these arc-front magmas is unclear because of the extensive masking effects of fractionation. The additional data presented here are closely comparable with data already described by Saunders and Weaver and co-workers, extending many of their conclusions to cover the entire island group, and strengthening the general model of calc-alkaline magma genesis recently proposed for the Antarctic Peninsula region (Saunders et al., 1980). The temporary development of entirely calc-alkaline magmas of intermediate composition on King George Island may simply be due to increased $a_{H_2O}$ (cf. Mysen and Boettcher, 1975) or to an interplay of factors, which are likely to be particularly complex at a major consuming plate boundary.

No regular lateral or transverse variations are evident. Rather, each area is characterised by chemically distinctive magmas derived from a comparatively homogeneous mantle source, to which they can be related by different degrees of fractionation. Minor source heterogeneity is possibly indicated by the decoupling of Rb and Ba from K in Fildes Formation lavas and high K/Ba ratios in the Hennequin Formation, but this may also be due to variable metasomatism of the mantle wedge of lithophile element-enriched fluids expelled from the subducted slab.

*Acknowledgements* This paper is published by permission of the Director, British Antarctic Survey. Sincere thanks are due to many of my colleagues at the British Antarctic Survey for their advice and encouragement and, in particular, to Dr R.J. Pankhurst for critically reading the manuscript.

## REFERENCES

BARTON, C.M., 1965: The geology of the South Shetland Islands. III. The stratigraphy of King George Island. *Scient. Rep. Br. antarct. Surv.*, *44*.

BIRKENMAJER, K., 1980: A revised lithostratigraphic standard for the Tertiary of King George Island, South Shetland Islands (West Antarctica). *Bull. Acad. Polon. Sci., Ser. Sci. Terre*, *26*, 49-57.

BIRKENMAJER, K., NAREBSKI, W., SKUPINSKI, A. and BAKUN-CZUBAROW., 1981: Geochemistry and origin of the Tertiary island-arc calc-alkaline volcanic suite at Admiralty Bay, King George Island (South Shetland Islands, Antarctica). *Stud. Geol. Polon.*, *72*, 7-57.

CAWTHORN, R.G. and O'HARA, M.J., 1976: Amphibole fractionation in calc-alkaline magma genesis. *Am. J. Sci., 276,* 309-29.

GONZÁLEZ-FERRÁN, O. and KATSUI, Y., 1970: Estudio integral del volcanismo cenozoico superior de las Islas Shetland del Sur, Antartica. *Ser. cient. Inst. antart. chil., 1,* 123-74.

JAKES, P. and WHITE, A.J.R., 1972: Major and trace element abundances in volcanic rocks of orogenic areas. *Geol. Soc. Am., Bull., 83,* 29-40.

LITTLEFAIR, M.J., 1978: The "quartz-pyrite" rocks of the South Shetland Islands, western Antarctic Peninsula. *Econ. Geol., 73,* 1184-9.

MYSEN, B. and BOETTCHER, A.L., 1975: Melting of a hydrous mantle: II. Geochemistry of crystals and liquids formed by anatexis of mantle peridotite at high pressures and high temperatures as a function of controlled activities of water, hydrogen, and carbon dioxide. *J. Petrol., 16,* 549-93.

SATO, H., 1977: Nickel content of basaltic magmas: identification of primary magmas and a measure of the degree of olivine fractionation. *Lithos, 10,* 113-20.

SAUNDERS, A.D., TARNEY, J. and WEAVER, S.D., 1980: Transverse geochemical variations across the Antarctic Peninsula: implications for the genesis of calc-alkaline magmas. *Earth Planet. Sci. Lett., 46,* 344-60.

SMELLIE, J.L., 1980: The geology of Low Island, South Shetland Islands, and Austin Rocks. *Bull. Br. antarct. Surv., 49,* 239-57.

SMELLIE, J.L., DAVIES, R.E.S. and THOMSON, M.R.A., 1980: Geology of a Mesozoic intra-arc sequence on Byers Peninsula, Livingston Island, South Shetland Islands. *Bull. Br. antarct. Surv., 50,* 55-76.

SMELLIE, J.L., PANKHURST, R.J., THOMSON, M.R.A. and DAVIES, R.E.S. (in prep.): The geology of the South Shetland Islands. VI. *Scient. Rep. Br. antarct. Surv.*

TARNEY, J., WEAVER, S.D., SAUNDERS, A.D., PANKHURST, R.J. and BARKER, P.F., 1982: Volcanic evolution of the Antarctic Peninsula and the Scotia arc; *in* Thorpe, R.S. (ed.). *Orogenic andesites.* Wiley, London.

WEAVER, S.D., SAUNDERS, A.D., PANKHURST, R.J. and TARNEY, J., 1979: A geochemical study of magmatism associated with initial stages of back-arc spreading: the Quaternary volcanics of Bransfield Strait, from South Shetland Islands. *Contrib. Mineral. Petrol., 68,* 151-69.

WEAVER, S.D., SAUNDERS, A.D. and TARNEY, J., 1982: Mesozoic-Cenozoic volcanism in the South Shetland Islands and the Antarctic Peninsula; geochemical nature and plate tectonic significance; *in* Craddock, C. (ed.). *Antarctic Geoscience.* Univ. Wisconsin Press, Madison, 263-273.

# THE AGE AND GEOCHEMISTRY OF SOME ROCKS FROM THE SOUTH SHETLAND ISLANDS

D. H. Elliot, D. D. Dupre and T. M. Gracanin, *Institute of Polar Studies, Ohio State University, Columbus, Ohio 43210, USA.*

*Abstract* Investigations in the South Shetland Islands have yielded additional data on the age and composition of igneous rocks from Low, Snow, Livingston, Greenwich and King George Islands. Twenty-eight $^{40}Ar/^{39}Ar$ whole rock and mineral age spectra were obtained for 26 rocks. The pluton at False Bay yielded a 38 Ma biotite cooling date. Volcanic and hypabyssal rocks range from 145 to 40 Ma, with one plug, Edinburgh Hill Ca 1 Ma old. The results for Admiralty Bay, King George Island, lead to major revision of the age of Tertiary volcanic strata. Compositionally the rocks are basalts, andesites and subordinate more felsic rocks which include the plutons, and are with one exception, calc-alkaline with low to moderate K and exhibiting characteristics transitional to the tholeiitic series. The exception, Edinburgh Hill, is a tholeiite with some characteristics similar to those of abyssal tholeiites. The older rocks fit well with previously reported data, whereas the young Edinburgh Hill plug is distinct from both the Late Tertiary volcanic rocks of the south coast of King George Island and the active or recently active volcanoes such as Deception Island.

# CONTRASTING ORIGINS AND IMPLICATIONS OF GARNET IN ROCKS OF THE ANTARCTIC PENINSULA

A.B. Moyes and R.D. Hamer, *British Antarctic Survey, Madingley Road, Cambridge CB3 OET, U.K.*

*Abstract* Almandine rich garnet occurs as an accessory mineral in many outcrops on the Antarctic Peninsula. It is found as discrete crystals or in enclaves, within peraluminous calc-alkaline volcanic and plutonic rocks, associated volcaniclastic and terrestrial sediments and in metamorphic rocks. Comparison with published experimental data shows that differences in petrography and chemistry have genetic implications for the origin of garnet and host rock.

Two different types of garnet can be distinguished—primary igneous and metamorphic. Igneous garnet is found in both peraluminous volcanic and plutonic rocks—a high Fe, low Mn variety typical of volcanic rocks is considered a high pressure (5-7 kb) remnant phase and a high Fe, high Mn variety typical of plutonic rocks is a lower pressure (possible as low as 1 kb) mineral.

Metamorphic garnet is distinguished by a higher Mg content, although chemistry varies with host rock and metamorphic grade. Garnet xenocrysts within the calc-alkaline volcanic rocks indicate temperatures of 690-800°C and pressures possibly up to 10 kb, and represent accidental inclusions of lower crust.

It would appear that a considerable thickness of continental crust (at least 25 km) was already developed prior to the initiation of the Mesozoic magmatic arc.

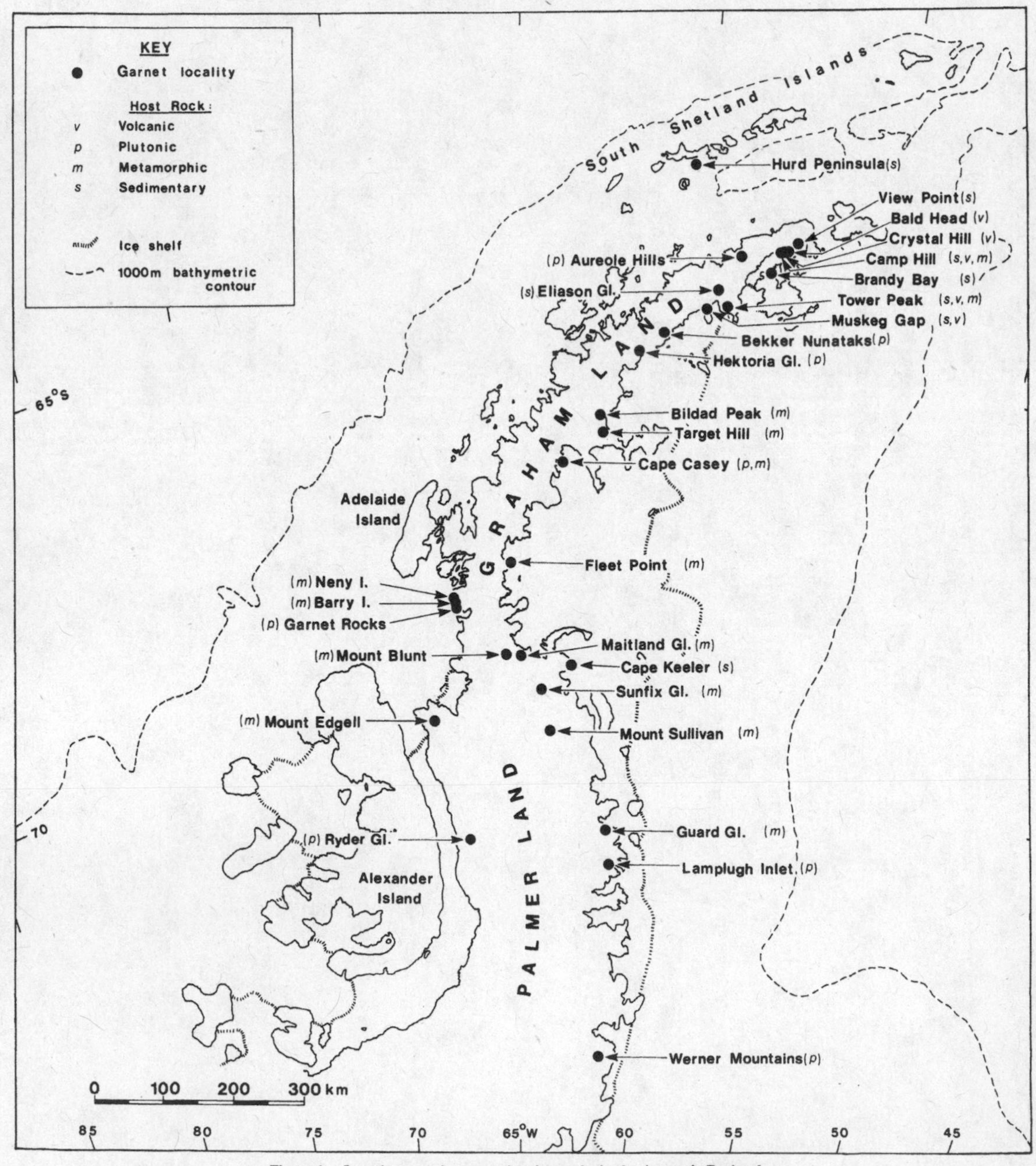


Figure 1. Location map for garnet bearing rocks in the Antarctic Peninsula.

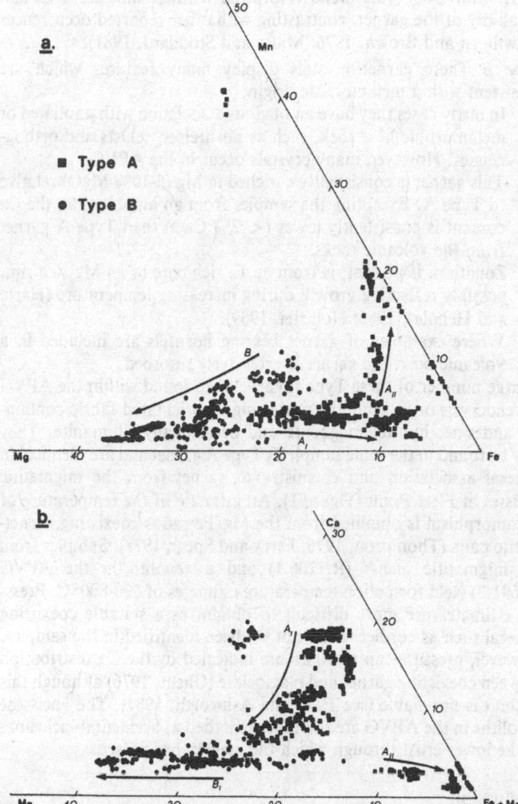


**Figure 2. Triangular variation diagram of garnet chemistry. A. Mg-Fe-Mn, and B. Mg-Ca-Fe+Mn. Garnet analyses plotted from ionic proportions, number of analyses = 685. Arrows indicate zonation from core to rim in igneous garnet from volcanic rocks ($A_I$) and plutonic rocks ($A_{II}$) and in metamorphic garnet from all types (B) and in xenoliths from the APVG ($B_I$).**

**TABLE 1: Representative garnet analyses**

| | 1 | 2 | 3 | 4 | 5 | 6 |
|---|---|---|---|---|---|---|
| $SiO_2$ | 37.52 | 37.01 | 38.58 | 38.98 | 37.92 | 37.81 |
| $TiO_2$ | 0.13 | — | 0.14 | — | — | — |
| $Al_2O_3$ | 21.33 | 20.97 | 22.00 | 21.91 | 21.41 | 21.58 |
| FeO* | 32.97 | 23.35 | 28.95 | 28.10 | 28.21 | 32.88 |
| MnO | 1.75 | 16.87 | 1.26 | 0.44 | 2.88 | 3.22 |
| MgO | 3.69 | 1.34 | 7.64 | 9.00 | 3.46 | 3.70 |
| CaO | 2.74 | 0.74 | 1.74 | 1.48 | 6.39 | 1.72 |
| Total | 100.13 | 100.28 | 100.31 | 99.91 | 100.27 | 100.91 |
| **Number of Ions on the basis of 12 Oxygens** | | | | | | |
| Si | 2.99 | 3.01 | 2.99 | 3.00 | 3.00 | 3.00 |
| Al | 2.01 | 2.01 | 2.01 | 2.00 | 2.00 | 2.02 |
| Fe | 2.20 | 1.59 | 1.88 | 1.82 | 0.41 | 2.18 |
| Mg | 0.44 | 0.16 | 0.88 | 1.04 | 1.87 | 0.44 |
| Mn | 0.12 | 1.16 | 0.08 | 0.03 | 0.19 | 0.22 |
| Ca | 0.23 | 0.06 | 0.15 | 0.12 | 0.54 | 0.15 |
| Ti | 0.01 | — | 0.01 | — | — | — |
| Total | 8.00 | 7.99 | 8.00 | 8.01 | 8.01 | 8.01 |
| **Ionic Proportions (percentage)** | | | | | | |
| Fe | 73.54 | 53.38 | 62.84 | 60.47 | 62.01 | 73.15 |
| Mg | 14.68 | 5.45 | 29.53 | 34.51 | 13.56 | 14.70 |
| Mn | 3.85 | 39.05 | 2.78 | 0.96 | 6.41 | 7.25 |
| Ca | 7.83 | 2.12 | 4.85 | 4.06 | 18.02 | 4.90 |

*Total Fe as FeO
Type A garnet 1-2, Type B garnet 3-6.
1. R.632.2K. Andesite, Camp Hill. 2. D.4702.1. Granite, Bekker Nunataks. 3. R.632.5. Andesite, Camp Hill. 4. R.1206.2. Migmatitic gneiss, Fleet Point. 5. E.1661.5. Amphibolite, Sunfix Glacier. 6. D.4291.1. Contact hornfels, Eliason Glacier.

Garnet is a conspicuous accessory mineral in many rocks of the Antarctic Peninsula (Figure 1). It is prominent in the Mesozoic-Cenozoic calc-alkaline volcanic rocks of northern Graham Land, viz. the Antarctic Peninsula Volcanic Group (APVG) and their chemically equivalent plutonic rocks (the so called Andean Intrusive Suite), particularly on the east coast. It is found in sediments of different ages, such as the Trinity Peninsula Group (?Carboniferous-Triassic) and those associated with the APVG. Garnet is found in many metamorphic rocks, particularly orthogneisses of dominantly Early Mesozoic age (Pankhurst, 1982) in northern Palmer Land.

Previous investigations in the Antarctic Peninsula have been limited to the garnet occurrences in the plutons of the Werner Mountains (Vennum and Meyer, 1979) and in the sediments of the Trinity Peninsula Group at View Point (Hyden and Tanner, 1981). Details of petrography and chemistry of the garnets occurring in the APVG of northern Graham Land have been reported by Hamer and Moyes (1982), and the present study extends the area of investigation to cover much of the Antarctic Peninsula.

Variations in garnet petrography and chemistry reflect different origins and crystallisation histories. By comparison with published experimental data, variables such as temperature and pressure of garnet crystallisation may be constrained, thereby elucidating the petrogenesis of garnet bearing rocks.

## Host Rocks

*Volcanic Rocks* Garnet occurs as discrete crystals or in enclaves within rocks of the APVG in the northern part of Graham Land (Figure 1). It was recorded at Crystal Hill (Adie, 1955) and Camp Hill (Bibby, 1966) and has been found in the AVPG by one of us (RDH) at Bald Head, Tower Peak and Muskeg Gap. Garnet has not been found in equivalent volcanic rocks farther south than Muskeg Gap, but this may be a limitation of sampling. The garnet crystals occur most frequently in corundum-normative andesites, but also in minor rhyolites.

*Plutonic Rocks* Garnet is an accessory mineral at several isolated outcrops of calc-alkaline plutonic rocks, which vary from quartz diorite to granite. It occurs in the plutons at Aureole Hills and Bekker Nunataks (Elliot, 1965, 1966) and at the Hektoria Glacier and Garnet Rocks (R.J. Pankhurst, pers. comm.). Garnet xenocrysts have been reported from the Werner Batholith (Vennum and Meyer, 1979).

*Metamorphic Rocks* Garnet occurs in orthogneisses at Mount Blunt and Maitland Glacier (Fraser and Grimley, 1972), Target Hill and Bildad Peak (Marsh, 1968), in amphibolites at the Sunfix Glacier (Fraser and Grimley, 1972) and in contact metamorphic hornfelses at Eliason Glacier (Aitkenhead, 1965). Garnet bearing migmatitic gneiss has been described from Fleet Point (Hamer and Moyes, 1982) and a minor occurrence of garnet in a cataclastically deformed granite at Barry Island has also been reported (Hoskins, 1963). Garnet bearing metamorphic enclaves have been identified by the authors within both the volcanic rocks at Camp Hill and Tower Peak, and the granodiorite pluton at Cape Casey.

*Sedimentary Rocks* Detrital garnet crystals are often found in non-marine sediments associated with the APVG at Tower Peak (Aitkenhead, 1965), Camp Hill (G.W. Farquharson, pers. comm.) and at Muskeg Gap (RDH). Detrital garnet is also found in marine

**TABLE 2: Representative analyses of garnet host rocks**

| | 1 | 2 | 3 | 4 | 5 | 6 |
|---|---|---|---|---|---|---|
| $SiO_2$ | 56.90 | 61.35 | 78.80 | 65.02 | 66.99 | 76.76 |
| $TiO_2$ | 0.86 | 0.46 | 0.14 | 0.81 | 0.66 | 0.08 |
| $Al_2O_3$ | 16.99 | 13.50 | 13.17 | 16.07 | 15.67 | 12.18 |
| $Fe_2O_3$* | 7.47 | 7.05 | 7.75 | 5.33 | 4.64 | 1.17 |
| MnO | 0.20 | 0.12 | 0.03 | 0.10 | 0.08 | 0.07 |
| MgO | 4.02 | 4.90 | 0.72 | 2.26 | 1.73 | 0.14 |
| CaO | 6.97 | 2.86 | 1.09 | 3.25 | 3.03 | 1.02 |
| $Na_2O$ | 2.17 | 2.41 | 4.46 | 2.94 | 3.34 | 2.69 |
| $K_2O$ | 0.57 | 2.30 | 1.30 | 1.67 | 3.17 | 4.96 |
| $P_2O_5$ | 0.21 | 0.12 | 0.05 | 0.16 | 0.14 | 0.07 |
| Total | 96.36 | 95.07 | 97.51 | 97.61 | 99.45 | 99.14 |
| %C-norm. | 0.66 | 2.26 | 2.54 | 3.99 | 1.58 | 0.70 |

*Total Fe as $Fe_2O_3$
1. R.631.1. Andesite, Camp Hill. 2. R.691.1. Andesite, 5 km S of Tower Peak. 3. R.688.4. Rhyolitic welded tuff, Tower Peak. 4. R.1206.1. Migmatitic gneiss, Fleet Point. 5. R.1102.1. Granodiorite, Cape Casey. 6. R.372.5. Granite, Hektoria Glacier.

sediments, notably the Trinity Peninsula Group at View Point (Hyden and Tanner, 1981) and the Miers Bluff Formation on the Hurd Peninsula, South Shetland Islands (Smellie, 1979). Cretaceous marine sediments on James Ross Island also contain garnet (J.R. Ineson, pers. comm.).

## Chemistry and Petrography

60 individual garnet crystals from 20 localities have been analysed on the energy dispersive electron microprobe at Cambridge University, as plotted in Figure 2. Representative analyses are given in Table 1. A number of host rocks have been analysed for major elements by X-ray fluorescence spectrometry at Bedford College and analyses are given in Table 2.

Petrographic and chemical characteristics of many of the samples have been presented elsewhere (Hamer and Moyes, 1982) but salient features to note are:

(a) Variable degrees of resorption are very common, with the exception of the contact metamorphic garnet and the garnet in the rapidly erupted (ignimbritic) rocks.
(b) Reaction coronas are common only in volcanic host rocks, but may occur occasionally in plutonic rocks. They consist typically of plagioclase or biotite/chlorite.
(c) All the garnets are Fe rich (almandine), differences between samples arising principally through variations in the other major elements, Mg (pyrope), Ca (Grossular) and Mn (spessartine).
(d) Colour is not always diagnostic but generally a deep red-brown colour indicates high Fe, very pale pink indicates high Mg and orange indicates high Ca.
(e) There is a striking positive correlation between the occurrence of garnet and the peraluminous (i.e. corundum/normative) nature of the host rock.
(f) Variations in garnet chemistry with host rock composition have not yet been observed.
(g) Two distinctive groups (Type A and B) can be distinguished both petrographically and chemically. For reasons discussed below, Type A is considered to be magmatic and Type B metamorphic in origin.

## Origin

*Type A* Garnet from both the APVG and plutonic equivalents display many features consistent with a magmatic origin:

(a) The high Fe content concurs with an igneous origin, Fe rich garnet being a liquidus or near-liquidus phase which may crystallise from a melt at pressures as low as 1 kb (Clemens and Wall, 1981).
(b) A higher Mn content in garnet from the plutonic rocks (see Figure 2) is well documented from many other areas (Miller and Stoddard, 1981), in contrast to lower Mn contents from volcanic rocks (Green and Ringwood, 1968).
(c) Zonation, if present, is most commonly from an Mg rich core to an Fe rich rim. This has also been noted from other calc-alkaline areas (Fitton, 1972; Birch and Gleadow, 1974) and may indicate crystallisation under decreasing temperature (Hollister, 1969).
(d) Inclusions are commonly restricted to acicular, skeletal apatites, whose habit is characteristic of crystallisation in the presence of a liquid phase (Wyllie et al., 1962).
(e) In some volcanic rocks, the garnet rim contains inclusions of plagioclase identical to the core composition of large plagioclase phenocrysts in the host rock. A sharp decrease in the Ca content of the garnet has been noted at the point where these inclusions first appear (see Hamer and Moyes, 1982).

By comparison with published experimental data on the stability of almandine garnet, it may be concluded that high Fe, low Mn garnet typical of the APVG equilibrated with melts at pressures of approximately 5-7 kb (Green and Ringwood, 1968; Green, 1976, 1977). This garnet, therefore, is regarded as a remnant from a higher pressure crystallisation phase brought up by the volcanic rocks. The varying degrees of resorption reflect differing rates of magma uprise—euhedral crystals occurring only in the most rapidly erupted, ignimbritic rocks.

The high Fe, high Mn garnet typical of the plutonic rocks indicates equilibration at lower pressures, possibly as low as 1 kb (Clemens and Wall, 1981). There is no evidence to suggest a reaction relationship between biotite and the liquid to produce garnet (Miller and Stoddard, 1981). Moreover, widespread resorption features indicate a certain instability of the garnet, contrasting with other reported occurrences (Cawthorn and Brown, 1976; Miller and Stoddard, 1981).

*Type B* These garnet crystals display many features which are consistent with a metamorphic origin:

(a) In many cases they have an obvious association with a foliated or metamorphic host rock, such as hornfelses, schists and orthogneisses. However, many crystals occur in the APVG.
(b) This garnet is consistently enriched in Mg (5-10% MgO) relative to Type A. Excepting the samples from an amphibolite, the Ca content is consistently lower (< 2% CaO) than Type A garnet from the volcanic rocks.
(c) Zonation, if present, is from an Fe rich core to an Mg rich rim, possibly reflecting growth during increasing temperature (Harte and Hendley, 1966; Hollister, 1969).
(d) Where examples of garnet bearing hornfels are included in a volcanic host, the garnet is extensively resorbed.

A large number of these Type B garnets are found within the APVG as xenocrysts or in xenoliths possessing an orientated fabric containing andesine, biotite, magnetite and occasionally sillimanite. They may be found in the same sample as Type A garnet and are identical in mineral association and chemistry to garnet from the migmatitic gneisses at Fleet Point (Figure 1). An estimate of the temperature of metamorphism is obtained from the Mg/Fe ratios coexisting garnet-biotite pairs (Thompson, 1976; Ferry and Spear, 1978). Samples from the migmatitic gneiss (R.1206.1) and a xenolith in the APVG (R.691.3) yield formation temperature estimates of 690-800°C. Pressure estimates are more difficult to obtain, as a suitable coexisting mineral such as cordierite has not yet been identified in the samples. However, pressures up to 10 kb are indicated by the Ca distribution between coexisting garnet and plagioclase (Ghent, 1976) although this method is unreliable (see Tyler and Ashworth, 1981). The gneissose xenoliths in the APVG are therefore regarded as accidental inclusions of the lower crust through which the magma has passed.

## Discussion

The present outcrop in the Antarctic Peninsula is dominated by calc-alkaline plutonic and volcanic rocks. Magmatic activity is undoubtedly related to melting processes associated with subduction of an easterly migrating Pacific oceanic crust, and may have been essentially continuous since the Late Triassic (Pankhurst, 1982). Garnet is a common accessory mineral in rocks of other calc-alkaline provinces and, in this respect, the Antarctic Peninsula is no exception. Differing origins and crystallisation histories are reflected in varying petrographic and chemical characteristics, although there is often conflicting evidence and no consensus as to whether such garnets are cognate or accidental inclusions (Gill, 1981). Evidence has been presented in this study to show that, for the rocks of the Antarctic Peninsula, sufficient petrographic and chemical differences do exist to allow a tentative separation of igneous (Type A) and metamorphic (Type B) garnets.

All igneous garnets are Fe rich. An antipathetic relationship between Mn and Ca reported by Green (1977) is observed here, with low Mn, high Ca garnet in volcanic rocks and high Mn, low Ca garnet in plutonic types. Comparison with experimental data on the stability of almandine garnet (Green and Ringwood, 1968; Green, 1976, 1977; Clemens and Wall, 1981) suggests that volcanic garnet crystallised at higher pressures (5-7 kb) than that in the plutonic rocks (possibly as low as 1 kb). Despite such differences, all the igneous garnets are found in peraluminous (i.e. corundum-normative) host rocks, a feature noted from many other calc-alkaline provinces (Gill, 1981). With increasing silica content, the rocks of the Antarctic Peninsula also display a linear trend from decreasing diopside- to increasing corundum-normative compositions. Garnet bearing rocks of the Antarctic Peninsula are generally <1% corundum-normative, although some may be >2% (unpublished data), thus displaying dominant I-type with minor S-type granite characteristics (Chappell and White, 1974). Although several mechanisms have been proposed to account for this trend (see Cawthorn and Brown, 1976), only three are considered to have general applications—loss of alkalis through leaching or alteration, assimilation of peletic material and fractional crystallisation of a pyroxene or amphibole.

Alkali loss through leaching or alteration, as evidenced by low totals for the volcanic rocks (see Table 2), may contribute towards producing a corundum-normative composition. However, the linear trend observed in the Antarctic Peninsula is difficult to equate with such a random process. It is felt, therefore, that any effect through alteration served only to accentuate an already pre-existing peraluminous character in the magma.

The high level assimilation of pelitic material by uprising magma may explain the occurrence of "peraluminous" minerals such as garnet, especially in plutonic rocks which are seen to contain abundant enclaves. Indeed the apparent restriction of garnet bearing andesites to continental areas containing pelitic sediments and the high (> 0.705) initial $^{87/86}$Sr ratios of the host lavas, led Gill (1981) to conclude that a peraluminous nature must be due to assimilation. Garnet from the plutons at Bekker Nunataks, Aureole Hills and the Werner Mountains has been ascribed to this origin (Elliot, 1965; Vennum and Meyer, 1979).

Fractional crystallisation of either pyroxene or amphibole has been proposed to explain the observed trends towards a peraluminous nature, but both are very difficult to test. Pyroxenes are abundant in the volcanics, but rare in the plutonics, and amphiboles vice versa. Amphibole fractionation has been favoured by Ujike (1975) and Cawthorn and Brown (1976) because of the linear nature of observed trends and the commonly late stage (euhedral) nature of the garnet in many granitoids. Green (1978) argued, however, that such a mechanism was generally applicable only to I-type (<1% corundum-normative) granites and that many S-type (>1% corundum-normative) granites contain high pressure remnant garnet, characterised by low Mn content and resorption features. Thus, garnet bearing plutonic rocks from the Antarctic Peninsula appear to be more evolved I-types with corresponding garnet compositions (high Mn), whereas the garnet bearing volcanic rocks are less evolved and reflect melting of an S-type source.

There are several lines of evidence which indicate that some form of "crustal basement" may have exerted a considerable influence on the development of the Antarctic Peninsula Mesozoic-Cenozoic magmatic arc.

Firstly, high initial $^{87/86}$Sr ratios (0.706-0.707) from Lower Jurassic acidic plutons have been interpreted as resulting from partial melting within, or at the base of, the crust and not solely to contamination of mantle derived magnas (Pankhurst, 1982). High initial $^{87/86}$Sr ratios (>0.708) from the calc-alkaline volcanics are indicative of a complex origin, involving contamination or multistage remelting. These values are similar to those obtained from Andean rocks, where eruption through thick (40-70 km) crust has been advocated (Francis et al., 1977; Briqueu and Lancelot, 1979) and complex processes such as combined assimilation and fractional crystallisation (James, 1982) or crustal anatexis (Hawkesworth et al., 1982) may have operated. Secondly, gravity surveys within the Antarctic Peninsula have indicated crustal thicknesses at the present day of 32-35 km (Renner, 1980). Utilising data on crustal accretion rates given by Brown (1977) and by comparison with the better known Andes (James, 1971; Saunders et al., 1980), it is suggested that such thicknesses are unlikely to have been derived entirely from mantle sources within the period of approximately 200 Ma since magmatic activity was initiated. Even though these estimates are necessarily crude in their approach, a substantial thickness of pre-existing crustal material is indicated. Thirdly, the Trinity Peninsula Group (?Carboniferous-Triassic) which underlies the APVG, has recently been interpreted as a fore-arc basin sequence with a floor partly of sialic crust and partly of trapped oceanic crust (Hyden and Tanner, 1981).

It seems probable, therefore, that a considerable thickness of crustal material, possibly 25 km or more, was in existence prior to the development of the Mesozoic-Cenozoic magmatic arc. The nature of this "basement" is very difficult to define and much confusion has arisen in the past over its identification (see Pankhurst, this volume). It is probable that it consists partly of crystalline material that formed the Pacific margin of Gondwana and partly of a (probably) Late Palaeozoic magmatic arc system situated on the present position of the Antarctic Peninsula (cf. Smellie, 1981). The presence of garnet bearing xenoliths of lower crust in rocks of the APVG, which bear a striking similarity to migmatitic gneisses in outcrop, indicates that such material may be representative of this "basement". A probable ensialic arc system was established during the Mesozoic in the Antarctic Peninsula, essentially mirroring the situation in South America and the presence of a thick crust in this area must be taken into account when considering variations in chemistry and petrography of the arc products.

*Acknowledgments* The authors are indebted to Drs A.D. Saunders and G.F. Marriner of Bedford College for assistance with the X-ray spectrometry and Dr P. Treloar of Cambridge University for help with the microprobe facilities. All colleagues at BAS are thanked for their advice and criticism during the preparation of the manuscript.

## REFERENCES

ADIE, R.J., 1955: The rocks of Graham Land. Unpublished Ph.D. thesis, University of Cambridge, U.K. 259pp.

AITKENHEAD, N., 1965: The geology of the Duse Eay-Larsen Inlet area, northeast Graham Land (with particular reference to the Trinity Peninsula Series). *Scient. Rep. Br. antarct. Surv., 51.*

BIBBY, J.S., 1966: The stratigraphy of part of northeast Graham Land and the James Ross Island Group. *Scient. Rep. Br. antarct. Surv., 53.*

BIRCH, W.D. and GLEADOW, A.J.W., 1974: The genesis of garnet and cordierite in acid volcanic rocks: evidence from the Cerberean Cauldron, Central Victoria, Australia. *Contrib. Mineral. Petrol., 45,* 1-13.

BRIQUEU, L. and LANCELOT, J.R., 1979: Rb-Sr systematics and crustal contamination models for calc-alkaline igneous rocks. *Earth Planet. Sci. Lett., 43,* 385-96.

BROWN, G.C., 1977: Mantle origin of Cordilleran granites. *Nature, 265,* 21-4.

CAWTHORN, R.G. and BROWN, P.A., 1976: A model for the formation of corundum normative calc-alkaline magmas through amphibole fractionation. *J. Geol., 84,* 467-76.

CHAPPELL, B.W. and WHITE, A.J.R., 1974: Two contrasting granite types. *Pac. Geol., 8,* 173-4.

CLEMENS, J.D. and WALL, V.J., 1981: Origin and crystallisation of some peraluminous (S-type) granitic magmas. *Can. Mineral., 19,* 111-31.

ELLIOT, D.H., 1965: Geology of northwest Trinity Peninsula, Graham Land. *Bull. Br. antarct. Surv., 7,* 1-24.

ELLIOT, D.H., 1966: Geology of the Nordenskjold Coast and a comparison with the northwest Trinity Peninsula, Graham Land. *Bull. Br. antarct. Surv., 10,* 1-43.

FERRY, J.M. and SPEAR, F.S., 1978: Experimental calibration of the partitioning of Fe and Mg between biotite and garnet. *Contrib. Mineral. Petrol., 66,* 113-7.

FITTON, J.G., 1972: The genetic significance of almandine-pyrope phenocrysts in the calc-alkaline Borrowdale Volcanic Group, Northern England. *Contrib. Mineral. Petrol., 36,* 231-48.

FRANCIS, P.W., MOORBATH, S. and THORPE, R.S., 1977: Strontium isotope data for Recent andesites in Ecuador and North Chile. *Earth Planet. Sci. Lett., 37,* 197-202.

FRASER, A.G. and GRIMLEY, P.H., 1972: The geology of parts of the Bowman and Wilkins Coasts, Antarctic Peninsula. *Scient. Rep. Br. antarct. Surv., 67.*

GHENT, E.D., 1976: Plagioclase-garnet-$Al_2SiO_5$-quartz: a potential geobarometer-geothermometer. *Am. Mineral., 61,* 710-4.

GILL, J., 1981: *Orogenic andesites and plate tectonics.* Springer-Verlag, New York.

GREEN, T.H., 1976: Experimental generation of cordierite or garnet bearing granitic liquids from a pelitic composition. *Geology, 4,* 85-8.

GREEN, T.H., 1977: Garnet in silicic liquids and its possible use as a P-T indicator. *Contrib. Mineral. Petrol., 65,* 59-67.

GREEN, T.H., 1978: A model for the formation and crystallisation of corundum-normative calc-alkaline magmas through amphibole fractionation: a discussion. *J. Geol., 86,* 269-72.

GREEN, T.H. and RINGWOOD, A.E., 1968: Origin of garnet phenocrysts in calc-alkaline rocks. *Contrib. Mineral. Petrol., 18,* 163-74.

HAMER, R.D. and MOYES, A.B., 1982: Composition and origin of garnet from the Antarctic Peninsula volcanic Group of Trinity Peninsula. *Geol. Soc. Lond., J., 139,* 713-20.

HARTE, B. and HENDLEY, K.J., 1966: Occurrence of compositionally zoned almandinitic garnets in regionally metamorphosed rocks. *Nature, 210,* 689-92.

HAWKESWORTH, C.J., HAMMILL, M., GLEDHILL, A.R., VAN CALSTEREN, P. and ROGERS, G., 1982: Isotope and trace element evidence for late stage intracrustal melting in the High Andes. *Earth Planet, Sci. Lett., 58,* 240-54.

HENSEN, B.S. and GREEN, D.H., 1973: Experimental study of the stability of cordierite and garnet in pelitic compositions at high pressures and temperatures. III. Synthesis of experimental data and geological applications. *Contrib. Mineral. Petrol., 38,* 151-66.

HOLLISTER, L.S., 1969: Contact metamorphism in the Kwoiek area of British Columbia: an end member of the metamorphic process. *Geol. Soc. Am., Bull., 80,* 2465-93.

HOSKINS, A.K., 1963: The basement complex of Neny Fjord, Graham Land. *Scient. Rep. Brit. antarct. Surv., 43.*

HYDEN, G. and TANNER, P.W.G., 1981: Late Palaeozoic-Early Mesozoic fore-arc basin sedimentary rocks at the Pacific margin in East Antarctica. *Geol. Rdsch., 70,* 529-41.

JAMES, D.E., 1971: Plate tectonic model for the evolution of the Central Andes. *Geol. Soc. Am., Bull., 82,* 3325-46.

JAMES, D.E., 1982: A combined O, Sr, Nd and Pb isotopic and trace element study of crustal contamination in central Andean lavas. I—Local geochemical variations. *Earth Planet. Sci. Lett., 57,* 47-62.

MILLER, C.F. and STODDARD, E.F., 1981: The role of manganese in the paragenesis of magmatic garnet: an example from the Old Woman-Piute Range, California. *J. Geol., 89,*233-46.

MARSH, A.F., 1968: Geology of parts of the Oscar II and Foyn Coasts, Graham Land. Unpublished Ph.D. thesis, University of Birmingham, U.K., 291 pp.
PANKHURST, R.J., 1982: Rb-Sr geochronology of Graham Land, Antarctica. *Geol. Soc. Lond., J., 139,* 701-12.
PANKHURST, R.J. (this volume): Rb-Sr Age determinations on possible basement rocks of the Antarctic Peninsula.
RENNER, R.G.B., 1980: Gravity and magnetic surveys in Graham Land. *Scient. Rep. Br. antarct. Surv., 77.*
SAUNDERS, A.D., TARNEY, J. and WEAVER, S.D., 1980: Transverse geochemical variations across the Antarctic Peninsula: implications for the genesis of calc-alkaline magmas. *Earth Planet. Sci. Lett., 46,* 344-60.
SMELLIE, J.L., 1979: Aspects of the geology of the South Shetland Islands. Unpublished Ph.D. thesis, University of Birmingham, U.K. 198 pp.
SMELLIE, J.L., 1981: A complete arc trench system recognised in Gondwana sequences of the Antarctic Peninsula. *Geol. Mag., 118,* 139-59.
THOMPSON, A.B., 1976: Mineral reactions in pelitic rocks: II. Calculations of some P-T-X (Fe-Mg) phase relations. *Am. J. Sci., 276,* 425-54.
TYLER, I.M. and ASHWORTH, J.R., 1981: Garnet zoning and re-equilibration in the Strontian area, Scotland. *Min. Mag., 44,* 293-300.
UJIKE, O., 1975: Petrogenetic significance of normative corundùm in calc-alkaline volcanic rocks series. *J. Japan. Assoc. Min. Petr. Econ. Geol., 70,* 85-92.
VENNUM, W.R. and MEYER, C.E., 1979: Plutonic garnets from the Werner batholith, Lassiter Coast, Antarctic Peninsula. *Am. Mineral., 64,* 268-73.
WYLLIE, P.J., COX, K.G. and BIGGAR, G.M., 1962: The habit of apatite in synthetic systems and igneous rocks. *J. Petrol., 3,* 238-43.

# BLUESCHIST RELIC CLINOPYROXENES OF SMITH ISLAND (SOUTH SHETLAND ISLANDS): THEIR COMPOSITION, ORIGIN AND SOME TECTONIC IMPLICATIONS

F. Herve and E. Godoy, *Departamento de Geologia y Geofisica, Universidad de Chile, Casilla 13518, Correo 21, Santiago, Chile.*
J. Davidson, *Servicio Nacional de Geologia y Mineria, Casilla 10465, Santiago, Chile.*

*Abstract* Clinopyroxene porphyrocrysts thought to be relics, of igneous origin, were found in mafic blueschists of the pre-Jurassic(?) Smith Island metamorphic complex. Although the blueschists do not preserve mesoscopic relic igneous textures, the composition of the clinopyroxenes (low Ti-augites) and the nature of the associated lithologies, which include possibly metalliferous metacherts, support the idea that the high P-low T metamorphics of the South Shetland Islands are at least in part composed of accreted subalkaline ocean floor or island arc basalts. The geochemical data preclude an origin involving an eclogitic or other high grade metamorphic rock. Because the moderate to high P-low T metamorphic complex of the Southern Coastal Domain in Chile shares common features, a more or less continuous belt of oceanic lithologies accreted to the margin of Gondwana during the Late Palaeozoic and/or Early Mesozoic can be postulated.

The South Shetland Islands comprise predominantly a pluto-volcanic arc and associated sedimentary rocks ranging in age from the Middle(?) Jurassic to the present. A metamorphic complex crops out mainly at the northern and southern ends of the archipelago, in the Elephant Island Group and Smith Island respectively (Figure 1). No depositional contact has been observed between these two different rock units. The South Shetland Islands lie between the Bransfield Marginal Basin to the East and a presently nonsubducting oceanic trench to the west.

Blueschists have been reported as outcrops on Smith Island by Rivano and Cortés (1975, 1976), and Smellie and Clarkson (1975), on Elephant Island (Dalziel, 1976) and as boulders in dredge hauls south of Clarence Island (Tyrrell, 1945). Little is known about the protolith of the blueschists themselves, but their presence has led to the widely accepted inference that subduction processes were active at the time of their metamorphism. Despite younger radiometric ages, the metamorphism is supposed to be of pre-Middle Jurassic age, on the basis of the existence of unmetamorphosed plant bearing sequences of that age both in the South Shetland Islands (Snow Island) and in the Antarctic Peninsula (Hope Bay). Blueschist terrains have also been recognised on Alexander Island, where glaucophane and lawsonite occur in rocks of the Triassic Le May Formation (Edwards, 1982; Hyden and Tanner, 1981a).

During a review study of the samples collected by Rivano and Cortés at Cape Smith (Figure 1), abundant clinopyroxene grains were identified in one of the mafic blueschists. This article reports chemical and petrographic characteristics of these clinopyroxenes which are interpreted as relic. An interpretation of the tectonic setting of the rocks from which they may have been derived is presented.

The microprobe analyses were obtained using a MAC400, with an EDS detector system corrected by a MAGIC IV programme in a KEVEX 7000 system at the Department of Earth Sciences, Colorado University, Boulder.

## PETROGRAPHY

According to Rivano and Cortés (1975, 1976) and Smellie and Clarkson (1975) the outcrops of Cape Smith and nearby islets are mainly composed of mafic blueschists and greenschists, with minor metachert bands. A strong east-west foliation is conspicuous and tight asymmetric microfolding of the foliation implies that the schists are the product of more than one phase of deformation.

Coexisting lawsonite and sodic amphiboles typical of the glaucophane schist facies (Turner, 1981) are reported by Rivano and Cortés (1975, 1976). On the other hand, the coexistence of garnet, epidote and sodic amphiboles would indicate a state transitional

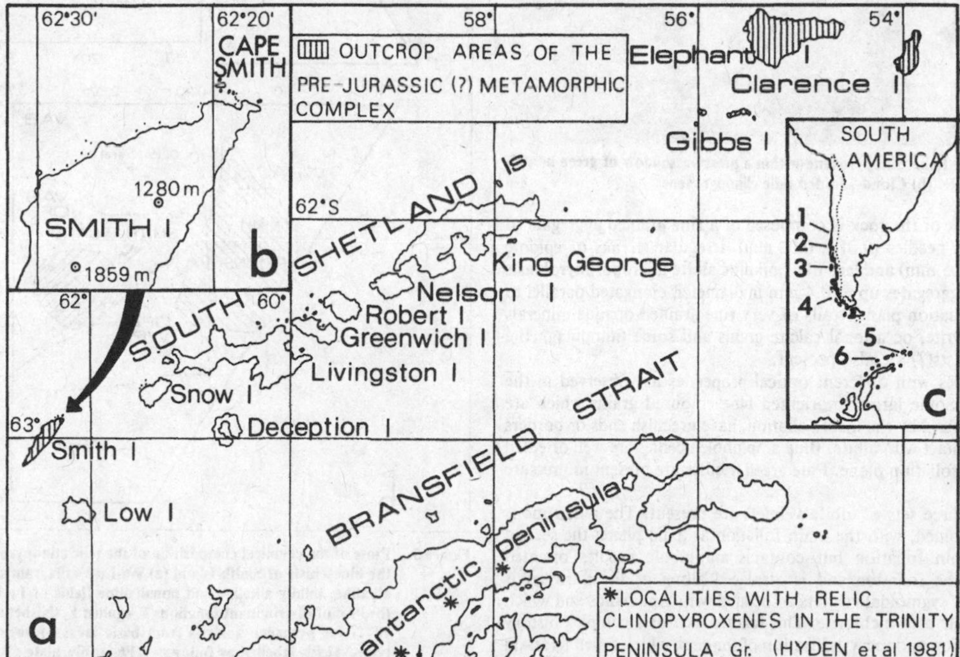


**Figure 1. Location maps slightly modified from Rivano and Cortes (1976). (a) Outcrops of the pre-Jurassic(?) metamorphic complex of the South Shetland Islands. (b) Location of Cape Smith on Smith Island. (c) Localities along the South American Pacific margins with high pressure/low temperature metamorphic minerals.**

between greenschist and glaucophane schist facies (Smellie and Clarkson, 1975).

A bluish-green schist sample (19A of Rivano and Cortés) has 5% pyroxene crystals 0.5 to 2.5 mm long (Figure 2a) with irregular outlines, undulose extinction and, where in the centre of pressure shadows, is associated with a mosaic of albite, chlorite and fibrous pale green amphibole. The pyroxene crystals are clouded by very fine grained opaque minerals and have a brownish colour of variable intensity within a grain or from grain to grain. Some crystals show relative displacement of fragments along fracture planes; others are folded (Figure 2b). Some pale coloured crystals are surrounded by a thin blue amphibole rim.

**Figure 2. (a) Relic clinopyroxene within a pressure shadow of green amphibole. (b) Cloudy, folded relic clinopyroxene.**

The matrix of the rock is composed of a fine grained aggregate of glaucophane needles (0.01 x 0.06 mm), irregular streaks of epidote granules (0.05 mm) and lens like poikilitic albite microporphyroblasts or mosaic aggregates up to 0.4 mm in diameter, elongated parallel to the main foliation plane; trails of very fine grained opaque minerals including pyrite, occasional calcite grains and some titanomagnetite porphyroblasts(?) are also present.

Amphiboles with different optical properties are observed in the same slide. Some larger unoriented blue coloured grains which are wrapped around by the main foliation, have greenish ends or borders where in contact with albite. Blue amphibole needles are well oriented in the main foliation plane. Pale green needles are present in pressure shadows.

At least three sets of albite veinlets are present. The older one is isoclinally folded, with the main foliation as axial plane; the second cuts the main foliation but contains amphibole needles oriented parallel to the foliation and is displaced along it; the third set is composed of sygmoidal veinlets associated with kink bands and which contain a pale brown chlorite. This texture records a complex history of deformation and recrystallisation of the rock which failed to attain mineralogical equilibrium. There is no lawsonite in this relic pyroxene bearing thin section.

A very remarkable rock type from the same locality, however, is composed of poikilitic globular albite porphyroblasts (30%), subhedral lawsonite porphyroblasts (25%) with rare fine (110) polysynthetic twinning, set in an oriented aggregate of pleochroic pale green, highly birefringent stilpnomelane (Mn rich?) laths (35%), which includes also sphene granules, isolated blue amphibole and pyrite automorphs. This rock probably corresponds to a metamorphosed Mn-Fe rich sediment. Epidote-crossite blueschists and epidote pale green amphibole greenschists are also present in the Cape Smith samples.

## CHEMISTRY AND ORIGIN OF THE CLINOPYROXENE CRYSTALS

Microprobe analyses (10) of the clinopyroxene crystals are presented in Table 1. End member calculations according to the method proposed by Cawthorn and Collerson (1974) classify the clinopyroxenes as augite and endiopside in the Wo-En-Fs triangle (Figure 3a).

The metamorphic mineral assemblage of the relic clinopyroxene bearing schists is typical of those developed in rocks of "mafic" or basaltic composition under low grade metamorphic conditions. No chemical analysis of the rock is available. The composition of the clinopyroxenes precludes an eclogitic origin for them, which is another known source for clinopyroxene relics in blueschists. Contact metamorphic(?) pyroxenes were reported from Gibbs Island (de Wit et al., 1977), but their composition also is very different from the ones analysed here (see Figure 3b). The clinopyroxenes which have been analysed are thus thought to be relic igneous.

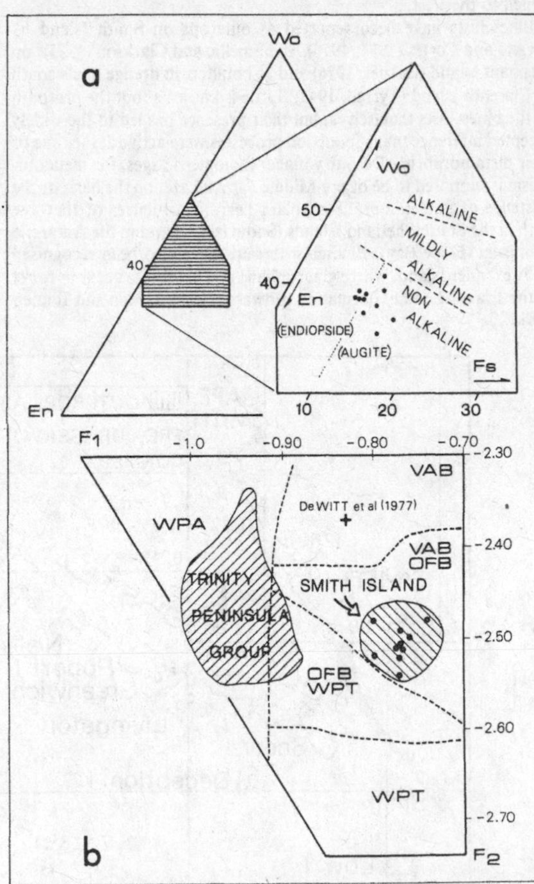


**Figure 3. Plots of the chemical composition of the relic clinopyroxenes from the blueschists of Smith Island (a) Wo-En-Fs diagram showing the alkaline, mildly alkaline and nonalkaline fields of Le Bas (1962). (b) Plot of discriminant functions $F_1$ against $F_2$ (Nisbet and Pearce, 1977) for pyroxene analysis from basic lavas of various magma types. Fields labelled as follows: WPA-within plate alkali basalts; VAB-volcanic arc basalts; OFB-ocean floor basalts; WPT-within tholeiites. Circles: Smith Island relic clinopyroxenes. Hatched field: distribution of relic clinopyroxenes in meta pillow basalts of the Trinity Peninsula Group (after Hyden and Tanner, 1981b).**

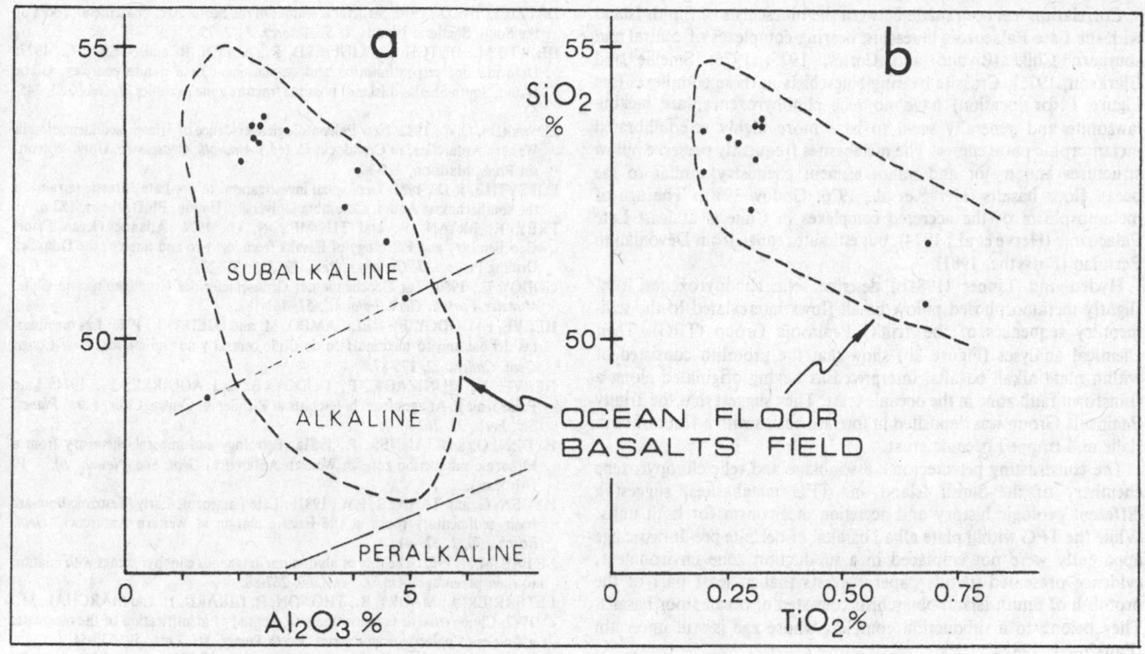


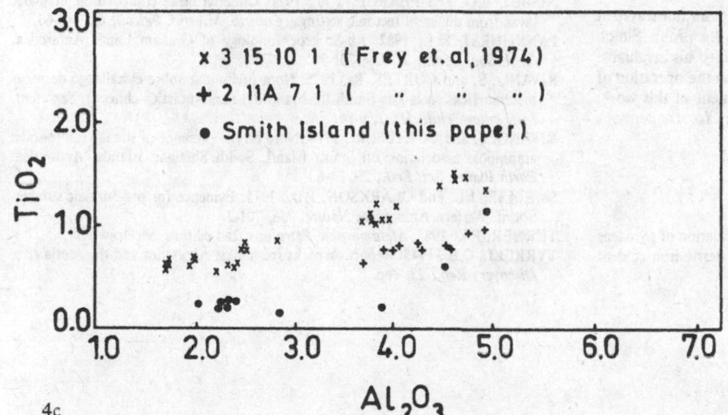


**Figure 4. (a) $Al_2O_3$-$SiO_2$ diagram of Le Bas (1962) modified by Nisbet and Pearce (1977). (b) $TiO_2$-$SiO_2$ diagram of Le Bas (1962) modified by Nisbet and Pearce (1977). (c) $TiO_2$-$Al_2O_3$ diagram (includes data from Frey et al., 1974).**

The pyroxenes plot in the nonalkaline field of the Wo-En-Fs triangle of Le Bas (1962) (Figure 3a) as well as in the ocean floor basalt section of the subalkaline field in the $Al_2O_3$-$SiO_2$ and $TiO_2$-$SiO_2$ diagrams (Figures 4a, 4b). A $TiO_2$-$Al_2O_3$ plot (Figure 4c) shows a positive correlation trend, comparable to those shown by pyroxenes of mid-Atlantic basalts, as presented by Frey et al. (1974).

In the diagram presented by Nisbet and Pearce (1977) to discriminate between pyroxenes of different tectonic settings, the Smith Island relic clinopyroxenes plot in the field of overlap between the ocean floor basalts and the volcanic arc basalts. A within plate alkalic basaltic composition for the original rocks (Figure 3b) is clearly excluded. A considerable scatter was obtained instead on the $TiO_2$-MnO-$Na_2O$ diagram. Following Leterrier et al. (1982), we consider this scatter related to the fact that the concentrations of the last two elements were close to the detection limit of the microprobe. The two unusually high $Na_2O$ values (Table 1) may be due to metamorphic modifications, for example development of glaucophane or jadeite rich submicroscopic domains.

The clinopyroxenes are thought to be relic igneous clinopyroxenes originally crystallised in a "basaltic" magma. Because other relic structures are absent, however, it is impossible to preclude the possibility of a detrital pyroclastic basaltic rock source.

## DISCUSSION

The Smith Island blueschist clinopyroxenes are thus thought to have had an origin as ocean floor or volcanic arc basalts. The associated lithologies suggest an ocean floor origin and the metamorphic mineralogy is generally thought to imply high P-low T gradients such as may develop in subduction zones. Thus, accretion of the oceanic rocks would have occurred after they were transported to great depths in a convergent plate margin.

**TABLE 1: Composition of relic clinopyroxenes from blueschists of Smith Island, South Shetland Islands**

| | | | | | | | | | | |
|---|---|---|---|---|---|---|---|---|---|---|
| $Na_2O$ | 0.52 | 0.51 | 0.39 | 1.55 | 0.76 | n.d. | n.d. | 0.42 | 0.18 | n.d. |
| MgO | 17.71 | 17.79 | 18.26 | 15.47 | 16.09 | 17.55 | 17.93 | 17.37 | 18.00 | 16.26 |
| $Al_2O_3$ | 2.27 | 2.34 | 2.02 | 2.88 | 3.86 | 2.29 | 2.45 | 4.56 | 2.39 | 4.93 |
| $SiO_2$ | 53.44 | 53.39 | 53.10 | 53.59 | 52.20 | 53.64 | 53.77 | 51.61 | 53.67 | 50.67 |
| CaO | 19.05 | 19.06 | 19.20 | 18.65 | 18.82 | 19.15 | 19.62 | 17.13 | 19.81 | 18.69 |
| $TiO_2$ | 0.25 | 0.27 | 0.31 | 0.22 | 0.29 | 0.32 | 0.32 | 0.69 | 0.31 | 1.18 |
| $Cr_2O_3$ | 0.49 | 0.51 | 0.49 | 0.36 | 0.67 | 0.29 | 0.35 | 0.40 | 0.31 | — |
| MnO | 0.44 | 0.06 | 0.15 | 0.12 | 0.17 | — | — | 0.25 | 0.08 | — |
| FeO | 5.78 | 6.20 | 5.65 | 7.25 | 7.31 | 6.51 | 6.16 | 8.54 | 6.18 | 8.28 |
| Total | 99.95 | 100.13 | 99.58 | 100.08 | 100.16 | 99.74 | 100.67 | 100.97 | 100.93 | 100.02 |
| **Numbers of cations on the basis of 24 oxygens** | | | | | | | | | | |
| $Na^+$ | 0.15 | 0.15 | 0.11 | 0.44 | 0.22 | — | — | 0.12 | 0.05 | n.d. |
| $Mg^{2+}$ | 3.85 | 3.87 | 3.99 | 3.38 | 3.52 | 3.82 | 3.87 | 3.77 | 3.88 | 3.57 |
| $Al^{3+}$ | 0.39 | 0.40 | 0.35 | 0.50 | 0.67 | 0.39 | 0.42 | 0.78 | 0.41 | 0.85 |
| $Si^{4+}$ | 7.80 | 7.78 | 7.78 | 7.86 | 7.66 | 7.84 | 7.79 | 7.52 | 7.77 | 7.46 |
| $Ca^{2+}$ | 2.98 | 2.98 | 3.01 | 2.93 | 2.96 | 3.00 | 3.04 | 2.67 | 3.07 | 2.95 |
| $Ti^{4+}$ | 0.03 | 0.03 | 0.03 | 0.02 | 0.03 | 0.03 | 0.04 | 0.08 | 0.03 | 0.13 |
| $Cr^{3+}$ | 0.06 | 0.06 | 0.06 | 0.04 | 0.08 | 0.03 | 0.04 | 0.05 | 0.04 | n.d. |
| $Mn^{2+}$ | 0.05 | 0.01 | 0.02 | 0.02 | 0.02 | — | 0.01 | 0.03 | 0.01 | n.d. |
| $Fe^{2+}$ | 0.71 | 0.76 | 0.69 | 0.89 | 0.90 | 0.80 | 0.75 | 1.04 | 0.75 | 1.02 |
| Total | 16.02 | 16.03 | 16.04 | 16.07 | 16.04 | 15.91 | 15.95 | 16.05 | 16.00 | 15.98 |

Correlation has been made between the blueschists of Smith Island with the Late Palaeozoic blueschist bearing complexes of central and southern Chile (Rivano and Cortés, 1975, 1976; Smellie and Clarkson, 1975). Crossite bearing blueschists in these complexes (see Figure 1 for location) have no relic clinopyroxenes, are lacking lawsonite and generally seem to have more highly re-equilibrated metamorphic parageneses. The metabasites frequently preserve pillow structures and major and minor element chemistry similar to the ocean floor basalts (Hervé et al., 1976; Godoy, 1980). The age of metamorphism of the accreted complexes in Chile is at least Late Palaeozoic (Hervé et al., 1974), but estimates range from Devonian to Permian (Forsythe, 1981).

Hyden and Tanner (1981b) describe relic clinopyroxenes from slightly metamorphosed pillow basalt flows intercalated in the sedimentary sequences of the Trinity Peninsula Group (TPG). Their chemical analyses (Figure 3b) show that the protolith consisted of within plate alkali basalts, interpreted as having originated along a transform fault zone in the oceanic crust. They suggest that the Trinity Peninsula Group was deposited in fore-arc basins with a floor of both sialic and trapped oceanic crust.

The constrasting petrotectonic assemblage and relic clinopyroxene chemistry of the Smith Island and TPG metabasites, suggest a different geologic history and accretion mechanism for both units. While the TPG within plate alkali basalts, of definite pre-Jurassic age apparently were not emplaced in a subduction zone environment, evidence presented in this paper suggests that at least part of the protolith of Smith Island blueschists consisted of ocean floor basalts. They belong to a subduction complex, whose age is still uncertain (Pankhurst, 1982).

*Acknowledgements* This work was carried out while the senior author stayed at Colorado University, Boulder, U.S.A. with a grant of the John Simon Guggenheim Memorial Foundation. S. Rivano kindly provided his invaluable rock samples from Smith Island. Charles Stern instructed on the operation of the microprobe and gave constant support to the development of this work. Instituto Antártico Chileno has provided logistical support for the author's field work in West Antarctica.

## REFERENCES

CAWTHORN, R.G. and COLLERSON, K.D., 1974: The recalculation of pyroxene end member parameters and the estimation of ferrous and ferric iron content from electron microprobe analyses. *Am. Mineral., 59,* 1203-8.

DALZIEL, I.W.D., 1976: Structural studies in the Scotia Arc: "basement" rocks of the South Shetland Islands. *U.S. Antarct. J., 2,* 75-7.

DE WIT, M., DUTCH, S., KIGFIELD, R., ALLEN, R. and STERN, C., 1977: Deformation, serpentinisation and emplacement of a dunite complex, Gibbs Island, South Shetland Island: possible fracture zone tectonics. *J. Geol., 85,* 745-62.

EDWARDS, C.W., 1982: New Paleontological evidence of Triassic sedimentation in Western Antarctica; *in* Craddock, C. (ed.) *Antarctic Geoscience*. Univ. Wisconsin Press, Madison, 325-30.

FORSYTHE, R.D., 1981: Geological investigations of pre-Late Jurassic terrains in the southernmost Andes. Columbia University. Unpub. Ph.D. thesis, 152 p.

FREY, F., BRYAN, W. and THOMPSON, G., 1974: Atlantic Ocean Floor: Geochemistry and Petrology of Basalts from legs two and three of the Deep Sea Drilling Project. *J. Geophys. Res., 79, 35,* 5507-27.

GODOY, E., 1980: Zur Geochemie der Grunschiefer des Grundgebirges in Chile. *Munster. Forsch. Geol. Palaont., 51,* 161-182.

HERVE, F., GODOY, E., DEL CAMPO, M. and OJEDA, J., 1976: Las metabasitas del basamento metamórfico de Chile central y austral. *Actas, First Congr. Geol. Chileno, 2,* 175-87.

HERVE, F., MUNIZAGA, F., GODOY, E. and AGUIRRE, L., 1974: Late Palaeozoic K-Ar ages from blueschists at Pichilemu, Central Chile, *Earth Planet. Sci. Lett., 23,* 262-4.

HYDEN, G. and TANNER, P., 1981a: Petrology and mineral chemistry from a Mesozoic subduction zone in Western Antarctica. *Geol. Soc. Newsl., 10, 5,* 19 (Abstract).

HYDEN, G. and TANNER, P.W., 1981b: Late Palaeozoic-Early Mesozoic Fore-arc basin sedimentary rocks at the Pacific margin in Western Antarctica. *Geol. Rdsch., 70, 1,* 529-41.

LE BAS, M.J., 1962: The role of aluminium in igneous clinopyroxenes with relation to their parentage. *Am. J. Sci., 260,* 267-88.

LETERRIER, J., MAURY, R., THONON, P., GIRARD, D. and MARCHAL, M., 1982: Clinopyroxene composition as a method of identification of the magmatic affinities of palaeovolcanic series. *Earth Planet. Sci. Lett., 59,* 139-54.

NISBET, E.G. and PEARCE, J.A., 1977: Clinopyroxene composition in mafic lavas from different tectonic settings. *Contrib. Mineral. Petrol., 63.* 149-60.

PANKHURST, R.J., 1982: Rb-Sr geochronology of Graham Land, Antarctica. *Geol. Soc. Lond., J., 139,* 701-12.

RIVANO, S. and CORTES, R., 1975: Nota preliminar sobre el hallazgo de rocas metamórficas en la Isla Smith (Shetland del Sur, Antártica chilena). *Ser. cient. Inst. antart. chil., III, 1,* 9-14.

RIVANO, S. and CORTES, R., 1976: Note on the presence of the lawsonite-sodic amphibole association on Smith Island, South Shetland Islands, Antarctica. *Earth Planet. Sci. Lett., 29,* 34-6.

SMELLIE, J.L. and CLARKSON, P.D., 1975: Evidence for pre-Jurassic subduction in Western Antarctica. *Nature, 258,* 701-2.

TURNER, F.J., 1981: *Metamorphic Petrology,* 2nd edition, McGraw Hill.

TYRRELL, C.E., 1945: Report on rocks from West Antarctica and the Scotia Arc. *Discovery Rep., 23,* 76p.

# Rb-Sr CONSTRAINTS ON THE AGES OF BASEMENT ROCKS OF THE ANTARCTIC PENINSULA

R.J. Pankhurst, *British Antarctic Survey, C/- Institute of Geological Sciences, 64 Gray's Inn Road, London WC1X 8NG, U.K.*

*Abstract* Rb-Sr studies have been carried out on samples of the various rock groups in the Antarctic Peninsula region which have been suggested as basement into which the igneous rocks of the Mesozoic-Tertiary arc were emplaced. A Permo-Carboniferous age (281 ± 16 Ma) is confirmed for the Hope Bay Formation of the Trinity Peninsula Group and correlation with other greywacke deposits in the South Shetland and South Orkney Islands seems reasonable. The amphibolite terrain of the Scotia metamorphic complex could be derived from similar material but the blueschist/greenschist terrain seems to be largely of Cretaceous age implying continued ocean floor sediment accretion from pre-Gondwanian times. The orthogneisses of the Neny Fjord area and northeastern Palmer Land are the products of Late Triassic or earliest Jurassic magmatism. The migmatitic gneisses of eastern Graham Land provide the clearest evidence of pre-arc basement, with metamorphism in Early Triassic times (~245 Ma) overprinting a previous crustal history which may extend back into the latest Precambrian.

The exposed geology of the Antarctic Peninsula is dominated by the plutonic and volcanic products of a calc-alkaline magmatic arc which was active at least from Early Jurassic until Tertiary times. Pre-arc crustal basement has been identified largely on the basis of deformation and metamorphism ascribed to the Gondwanian orogeny. Components include metasediments of ocean floor type, low-grade greywacke-shale sequences and, with a southward increase in metamorphic conditions, amphibolite grade para- and ortho-gneisses (Figure 1). Dalziel (1982) has given a comprehensive review of the stratigraphic and radiometric data on these rocks as well as providing a complete bibliography up to 1977.

Radiometric dating is of great importance in understanding the pre-Jurassic evolution of this area since fossil control is poor even in the low grade metasediments. Previous data largely consist of K-Ar dating, although Halpern (1972) and Gledhill et al. (1982) introduced the Rb-Sr method. This paper represents a progress report on a British Antarctic Survey geochronological programme which until now has depended heavily on Rb-Sr whole-rock analyses. Standard analytical techniques were used, with the exception of fully automated mass-spectrometry as employed by Pankhurst (1982). Ages given in the text are followed by the corresponding initial $^{87}Sr/^{86}Sr$ in parentheses.

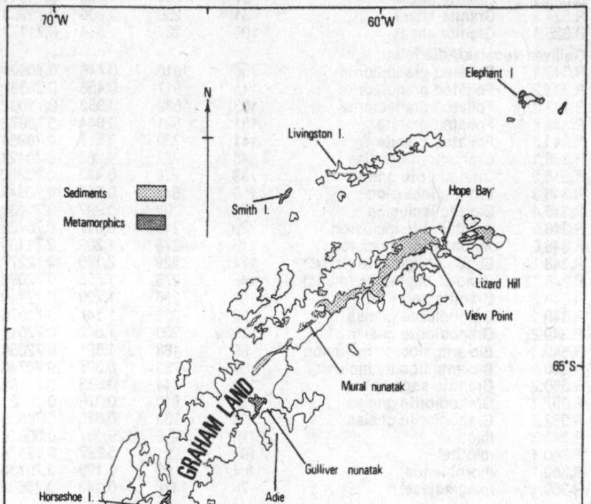

**Figure 1. Sketch map showing the distribution of supposed Pre-Middle Jurassic rocks in the Antarctic Peninsula and place names referred to in the text.**

Isotope constants recommended by Steiger and Jaeger (1977) were adopted and the treatment of raw data follows the general approach of McIntyre et al. (1966), York (1969) and Brooks et al. (1972), goodness of fit to the isochron model being assessed via the MSWD parameter (mean square of weighted deviates). For a perfect isochron, recording a uniquely defined geological event such as igneous crystallisation or diagenesis of fine grained illite bearing sediments, the expected value of MSWD is 1.0 or less. As recommended by Brooks et al. (1972), a cut off value of up to ~2.5 is allowed as being reasonable. Excess scatter, indicated by higher MSWD, is attributed to geological causes—either primary isotopic heterogeneity of the system or secondary disturbance associated with a subsequent event such as metamorphism. For scattered data which nevertheless exhibit good visual collinearity over a wide range of Rb/Sr ratio and for which MSWD is no greater than an order of magnitude too high (say up to ~ 30), we might hope that the non-model effects are relatively minor and that the best-fit line may still be a reasonable estimate of the age of the primary event. Such "errorchrons" (Brooks et al., 1972) have to be interpreted with caution and the excess scatter must clearly be allowed for—in the present case by multiplying all error estimates by the square root of MSWD (York, 1969), yielding a correspondingly increased uncertainty in the age and initial $^{87}Sr/^{86}Sr$ ratio. These are indicated in the Figures as "enhanced errors" although even so they cannot be regarded with the same confidence as true isochron statistics. Where such enhanced errors are too large to be geologically meaningful, or where MSWD exceeds ~30, useful constraints may still be derived by comparison with reference isochrons or by calculation of model ages (assuming a likely initial $^{87}Sr/^{86}Sr$) for individual samples with high measured $^{87}Sr/^{86}Sr$. However, since in these cases *severe* isotopic disturbance is indicated (or even breakdown in the assumption that groups of samples are really cogenetic), such reference or model ages cannot be regarded as proven age *determinations*. At best, they may provide evidence of unexpectedly high crustal residence times for samples with unusually high calculated initial $^{87}Sr/^{86}Sr$ ratios at the time of local rock formation events. This would suggest the involvement of older continental material via erosion and sedimentation, metamorphism or anatexis.

## SCOTIA METAMORPHIC COMPLEX

This term was introduced by Tanner et al. (1982), following a suggestion of Grikurov (1973), for the folded and metamorphosed rocks exposed in the South Orkney and South Shetland Islands. These are often thought to constitute a paired metamorphic belt formed in an oceanic arc-trench gap, with an outer (Pacific) blueschist/greenschist zone of metagreywacke and chert and an inner amphibolite zone of mica schists. By analogy with similar rocks beneath a Jurassic unconformity in Chile, Dalziel (1982) has argued for a Late Palaeozoic-Triassic age for this complex. Attempts at direct dating using the K-Ar method have failed to provide ages older than ~ 100 Ma for the blueschist/greenschist zone and 200 Ma for the amphibolite zone (see Dalziel, 1982 for summary). Grikurov et al. (1970) determined two K-Ar whole rock ages on schists of 220-230 Ma, but this technique is not recommended for low grade metamorphic rocks which could easily contain excess Ar. Whereas the K-Ar ages for metamorphic hornblendes, which average 188 ± 5 Ma (Tanner et al., 1982), may reflect closure during cooling soon after recrystallisation, the younger ages are generally considered to result from thermal overprinting during subsequent arc magmatic activity.

New Rb-Sr data collected as part of the BAS programme were presented by Tanner et al. (1982). These included a whole-rock isochron age of 75 ± 16 Ma (0.7044 ± 0.0001) for the metamorphic recrystallisation of low grade rocks on northern Elephant Island, with compatible data for similar samples from Smith Island. Because Rb-Sr ratios in these rocks are generally very low, even their low initial $^{87}Sr/^{86}Sr$ ratio does not entirely preclude an extended previous existence as sediments, conceivably going back to about 250 Ma if they first formed from mantle with a value of 0.7030. However, making reasonable allowance for their origin as sea floor sediments with an input from sea water Sr as well as oceanic basalt, Tanner et al. (1982) argued that deposition probably did not precede recrystallisation by more than 10-50 Ma. It thus appears that at least some of the blueschist/greenschist facies rocks resulted from Mesozoic subduction are therefore *not* part of the pre-Middle Jurassic basement. This may not apply to all such rocks, however—more radiogenic Sr in low grade schists from Clarence Island itself (F. Hervé, University of Santiago, pers. comm.) would permit sedimentation as long ago as latest Precambrian to Early Palaeozoic times.

Rb-Sr data for the higher grade rocks are more compatible with previous geological interpretation. The K-Ar dates referred to above require pre-Middle Jurassic crystallisation of the quartz-mica schists and amphibolites. Rex (1976) reported a Rb-Sr whole-rock isochron for Signy Island, South Orkney Islands, recalculated by Tanner et al. (1982) to give 281 ± 56 Ma (0.712 ± 0.002), although the scatter of the data suggest a true uncertainty closer to ± 150 Ma. The high initial $^{87}Sr/^{86}Sr$ ratio indicates a possible crustal history extending back into the Palaeozoic, but the provenance of this inner zone of arc-trench sediments may well have included much older continental detritus so that the age of deposition is not closely constrained. Tanner et al. (1982) report data for a locality on southern Elephant Island which give a model age of ~270 Ma but with a *low* initial $^{87}Sr/^{86}Sr$ ratio of 0.704 so that this is effectively a rough maximum age for deposition.

## TRINITY PENINSULA GROUP

Folded greywackes of the Trinity Peninsula Group (TPG) (Table 1) form the observed basement to Jurassic arc rocks in northern Graham Land. Dalziel (1972, 1982) has long argued for their stratigraphical and structural equivalence to similar low grade rocks in the Scotia arc and southern Chile. In the latter environment there is unequivocal palaeontological evidence, indicating a Late Carboniferous-Early Permian age. Plant fragments and spores in the Hope Bay Formation have been assigned a Carboniferous age. On the other hand, Triassic invertebrate fossils have been described from the Legoupil and LeMay Formations (see Dalziel, 1982 for reference).

**TABLE 1: Low grade metasediments of the basement fore-arc**

| | |
|---|---|
| Hope Bay Formation<br>View Point Formation<br>Legoupil Formation | Trinity Peninsula Group<br>(Hyden and Tanner 1981) |
| Miers Bluff Formation (South Shetland Is.) | |
| Greywacke-Shale Formation (South Orkney Is.) | |
| LeMay Formation (Alexander I.) | |

The last three formations have variously been correlated with the Trinity Peninsula Group by different authors (see Dalziel 1982)

Dalziel (1972) mentions a two point Rb-Sr whole rock model age of 242 ± 50 Ma for the Hope Bay Formation. New Rb-Sr data for these rocks are presented here (Table 2, Figure 2a). The samples, of green, red and purple shales and siltstones were collected from Scar Hills along the southern shoreline of Hope Bay. The exposures here are less intensely deformed than the near vertical outcrops of grey-black slate at the Argentine base of Esperanza and only exhibit open flat-lying folds beneath the Jurassic unconformity on Mount Flora. Although the results exhibit a high degree of excess scatter about an isochron model (MSWD = 27), the overall linear array is quite striking and enhancing error estimates to yield MSWD = 1 still gives a reasonably precise errorchron age estimate of 281 ± 16 Ma. Five of the new data points fall on a perfectly good isochron corresponding to 296 ± 4 Ma (0.7063 ± 0.0001). The scatter overall could be due to partly unequilibrated Sr in detrital minerals, especially in the two gritstone samples with the lowest Rb/Sr ratios, but the fairly low initial $^{87}Sr/^{86}Sr$ ratio of 0.7069 ± 0.0003 indicates an immature provenance so that this effect is probably not too serious. On the other hand, some metamorphic rehomogenisation may have occurred since the point with the highest

**TABLE 2: New Rb-Sr Data for Antarctic Peninsula Basement Rocks**

| Rock Unit/ Sample No. | Lithology | Rb | Sr | $^{87}Rb/^{86}Sr$ | $^{87}Sr/^{86}Sr$ |
|---|---|---|---|---|---|
| Trinity Peninsula Group | | | | | |
| (Hope Bay) | | | | | |
| BR.072.1 | Red arkosic grit | 72 | 371 | 0.559 | 0.70950 |
| BR.072.2 | Grey-green grit | 95 | 361 | 0.757 | 0.70950 |
| BR.072.3 | Banded mud/siltstone | 204 | 138 | 4.304 | 0.72332 |
| BR.072.4 | Banded mud/siltstone | 197 | 153 | 3.736 | 0.72218 |
| BR.072.5 | Banded mud/siltstone | 216 | 149 | 4.192 | 0.72405 |
| BR.072.6 | Banded mud/siltstone | 200 | 181 | 3.197 | 0.71979 |
| BR.072.7 | Banded mud/siltstone | 207 | 172 | 3.501 | 0.72080 |
| (Mural Nunatak) | | | | | |
| R.372.1 | Biotite-schist | 77 | 364 | 0.613 | 0.70944 |
| R.372.2 | Biotite-schist | 100 | 255 | 1.132 | 0.71126 |
| R.372.3 | Biotite-schist | 86 | 320 | 0.774 | 0.71009 |
| Miers Bluff Formation (Analysed by M. Halpern)* | | | | | |
| LI.60.6B | Shale with S. cleavage | 211 | 168 | 3.63 | 0.7198 |
| LI.60.6C | Shale with S. cleavage | 173 | 140 | 3.60 | 0.7198 |
| LI.60.6D | Shale with S. cleavage | 210 | 131 | 4.65 | 0.7228 |
| LI.60.6E | Shale with S. cleavage | 199 | 248 | 2.32 | 0.7155 |
| LI.60.6F | Shale with S. cleavage | 234 | 109 | 6.19 | 0.7268 |
| Marguerite Bay Gneisses | | | | | |
| (Roman Four Promontory) | | | | | |
| R.052.1 | Pink granite-gneiss | 174 | 59 | 8.614 | 0.72895 |
| R.052.2 | Pink granite-gneiss | 179 | 58 | 8.911 | 0.72934 |
| R.052.3 | Pink granite-gneiss | 186 | 82 | 6.580 | 0.72177 |
| R.052.4 | Micro-granite gneiss vein | 164 | 65 | 7.281 | 0.72286 |
| R.052.5 | White granodiorite gneiss | 170 | 63 | 7.839 | 0.72666 |
| R.052.6 | White granodiorite gneiss | 163 | 88 | 5.346 | 0.72151 |
| R.052.7 | Late diorite-gneiss vein | 158 | 290 | 1.578 | 0.70913 |
| (Neny Island) | | | | | |
| R.053.1 | Diorite-gneiss | 66 | 741 | 0.255 | 0.70618 |
| R.053.2 | Microgranite dyke | 162 | 146 | 3.215 | 0.71115 |
| R.068.1 | Banded gneiss | 109 | 239 | 1.322 | 0.71705 |
| (Randall Rocks) | | | | | |
| R.054.1 | Granodiorite gneiss | 102 | 432 | 0.684 | 0.70775 |
| (Horseshoe Island) | | | | | |
| R.077.1 | Banded gneiss | 166 | 343 | 1.410 | 0.71225 |
| East Graham Land Gneisses | | | | | |
| Target Hill | | | | | |
| R.320.1 | Biotite-gneiss | 88 | 354 | 0.720 | 0.70888 |
| R.320.2 | Biotite-gneiss | 101 | 357 | 0.821 | 0.70902 |
| R.321.4 | Biotite-gneiss | 68 | 359 | 0.545 | 0.70847 |
| R.322.1 | Biotite-gneiss | 85 | 370 | 0.663 | 0.70878 |
| R.322.2 | Biotite-gneiss | 84 | 373 | 0.655 | 0.70875 |
| R.322.3 | Biotite-gneiss | 84 | 345 | 0.703 | 0.70886 |
| R.322.3 | (plagioclase feldspar) | 14 | 397 | 0.102 | 0.70738 |
| R.324.1 | Granite sheet | 87 | 241 | 1.050 | 0.71041 |
| R.324.2 | Granite sheet | 81 | 259 | 0.900 | 0.70963 |
| R.325.1 | Granite sheet | 109 | 237 | 1.334 | 0.71172 |
| Gulliver Nunatak/Adie Inlet | | | | | |
| R.343.1 | Foliated granodiorite | 156 | 1010 | 0.448 | 0.70901 |
| R.343.2 | Foliated granodiorite | 145 | 917 | 0.456 | 0.70892 |
| R.343.3 | Foliated granodiorite | 158 | 532 | 0.862 | 0.71002 |
| R.343.4 | Foliated granite | 131 | 591 | 0.644 | 0.70972 |
| R.343.5 | Foliated granite | 141 | 789 | 0.517 | 0.70954 |
| R.346.1 | Granodiorite gneiss | 48 | 382 | 0.362 | 0.70421 |
| R.346.2 | Granodiorite gneiss | 139 | 928 | 0.433 | 0.70890 |
| R.346.3 | Dioritic inclusion | 115 | 645 | 0.517 | 0.70941 |
| R.346.4 | Dioritic inclusion | 92 | 292 | 0.907 | 0.71033 |
| R.346.5 | Amphibolite inclusion | 39 | 218 | 0.517 | 0.70765 |
| R.346.6 | Amphibolite inclusion | 115 | 276 | 1.209 | 0.71134 |
| R.348.1A | Banded migmatite (felsic) | 174 | 229 | 2.196 | 0.72277 |
| R.348.1B | Banded migmatitc (mafic) | 148 | 276 | 1.558 | 0.72093 |
| R.348.2 | Granitic segregation | 151 | 256 | 1.709 | 0.72110 |
| R.349.1 | Granodiorite gneiss | 125 | 206 | 1.747 | 0.71976 |
| R.349.2 | Granodiorite gneiss | 129 | 200 | 1.872 | 0.72019 |
| R.349.3 | Bio-amphibolite inclusion | 99 | 188 | 1.521 | 0.72056 |
| R.350.1 | Bio-amphibolite inclusion | 87 | 769 | 0.327 | 0.70746 |
| R.350.2 | Granitic segregation | 134 | 444 | 0.873 | 0.71185 |
| R.352.1 | Granodiorite gneiss | 118 | 910 | 0.376 | 0.70881 |
| R.352.2 | Granodiorite gneiss | 115 | 933 | 0.356 | 0.70885 |
| R.343.3 | (biotite) | 474 | 47.6 | 29.08 | 0.80628 |
| R.350.1 | (biotite) | 164 | 91 | 5.227 | 0.72474 |
| R.350.1 | (hornblende) | 6.4 | 93 | 0.199 | 0.70730 |
| R.350.1 | (plagioclase) | 79 | 1617 | 0.142 | 0.70689 |
| Northern Palmer Land | | | | | |
| R.1909.3 | Leucogranite gneiss | 382 | 7.05 | 163.0 | 1.1216 |
| R.1909.4 | Granite gneiss | 400 | 6.65 | 181.5 | 1.1577 |
| R.1909.5 | Granite gneiss | 377 | 7.44 | 151.9 | 1.0857 |
| R.1909.6 | Leucogranite gneiss | 357 | 7.42 | 145.1 | 1.0699 |
| R.1909.8 | Leucogranite gneiss | 336 | 10.36 | 96.23 | 0.9613 |
| R.1909.9 | Leucogranite gneiss | 321 | 2.57 | 397.7 | 0.7179 |
| R.1911.1 | Granite gneiss | 185 | 81 | 6.646 | 0.72439 |
| R.1911.2 | Granite gneiss | 240 | 45 | 16.19 | 0.74824 |
| R.1911.4 | Granite gneiss | 225 | 55 | 11.92 | 0.73746 |

*Data provided by I. W. D. Dalziel, errors assumed to be 1% on Rb/Sr, 0.05% on $^{87}Sr/^{86}Sr$. All other data analysed at IGS using XRF or ID where appropriate. Errors are 0.5% on Rb/Sr, 0.01% on $^{87}Sr/^{86}Sr$ (1-sigma).

Rb/Sr ratio falls below the best fit line, as do the two data points of Dalziel (1972). It is concluded that the majority of the data represent diagenetic homogenisation of Sr-isotopes during latest Carboniferous or earliest Permian times, in good agreement with previous inferences for this formation. A maximum age for the View Point Formation

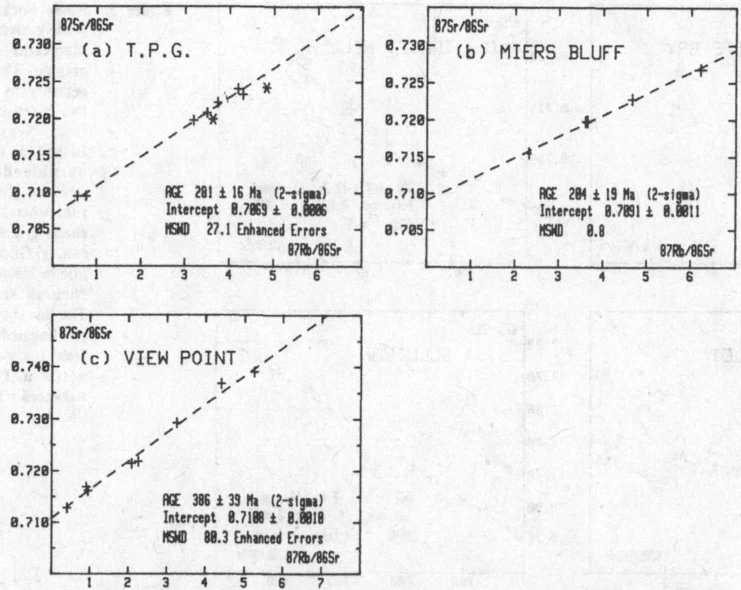


**Figure 2. Rb-Sr isochron relationships for low grade metasediments included in this work. (a) Mudstone/shale siltstone from the Hope Bay Formation of the Trinity Peninsula Group. The two asterisks are data of M. Halpern, supplied by I.W.D. Dalziel. (b) Mudstone/shale from the Miers Bluff Formation, Livingston Island. Data analysed by M. Halpern, supplied by I.W.D. Dalziel. (c) Granophyre cobbles from a conglomerate in the View Point Formation, Trinity Peninsula Group. Material provided by G. Hyden. Errors are enhanced where MSWD exceeds 2.5.**

(TPG) is given by an Early Devonian errorchron of 386 ± 40 Ma (0.7108 ± 0.0010) for a suite of granophyric cobbles in a conglomerate near the base (BAS unpublished data) (Figure 2c). Support is thus provided for Dalziel et al.'s. (1977) correlation, via the Greywacke Shale Formation of the South Orkney Islands with amphibolite grade rocks of the Scotia metamorphic complex. Correlation with the Miers Bluff Formation of the South Shetland Islands (Dalziel, 1972) is not clear however, since that has yielded a possible Mesozoic flora and a five point Triassic-Jurassic Rb-Sr isochron, recalculated by the author to 204 ± 17 Ma (Figure 2b). A rather high initial $^{87}Sr/^{86}Sr$ ratio of 0.709 ± 0.001 and an average $^{87}Sr/^{86}Sr$ ratio of ~ 4 is just about compatible with metamorphic resetting of a Rb-Sr system originally similar to that of the Hope Bay material. Three samples of rather higher grade TPG material at Mural Nunatak, Oscar II Coast (Table 2), lie on another secondary isochron of 244 ± 27 Ma (0.7074 ± 0.0003) which can be interpreted in the same way. The K-Ar data of Rex (1976) and new Rb-Sr data for the east Graham Land gneisses (below) strongly support an age of ~245 Ma for high grade metamorphism in this area.

Previous evidence suggesting a *minimum* age of Devonian for the TPG is represented by K-Ar muscovite and biotite ages of 360-390 Ma (Rex, 1976), supposedly for a granite from Lizard Hill 10 km south of Hope Bay. R.D. Hamer revisited Lizard Hill in 1978 and found the only outcrop to be diorite or granodiorite, dated by Pankhurst (1982) at 92 ± 2 Ma. It then transpired that the original samples, for which no paper records exist, may have come from nearby Last Hill, but in 1981 A. Crame and J. Ineson showed that this is composed of diorite similar to that at Lizard Hill. Although the possibility of Devonian granitic basement in the region would be welcome as a local source for the cobbles in the View Point Formation, this is not backed by any reliable evidence and the K-Ar data should probably be discounted as a valid constraint on the geology of this area.

## MARGUERITE BAY GNEISSES

The pre-volcanic basement rocks of the Neny Fjord area comprise a variety of schists and gneisses (Adie, 1954; Hoskins, 1963). Undeformed igneous rocks of supposed pre-Jurassic age (Adie, 1954) have since been demonstrated to be Cretaceous intrusions (Gledhill et al., 1982; Pankhurst, 1982). Many of the schist outcrops are of extremely limited extent, although they do appear to occur as xenoliths or rafts included in the orthogneiss. The majority of previous age determinations on the orthogneisses have given obviously reset Cretaceous ages, but older ages obtained include a two point model Rb-Sr age of ~200 Ma (Halpern, 1972) and a Rb-Sr whole-rock isochron of 175 ± 7 Ma (0.709 ± 0.001) (Gledhill et al., 1982) for pink granite-gneiss from a minor outcrop in the Debenham Islands (this age would be 171 ± 7 Ma using the decay constant adopted here). Fresh collections for geochronology from Roman Four Promontory, Neny Island and Randall rocks were made by the author working from sea-ice in November 1977. These include a wider range of compositions from granite-gneiss to diorite-gneiss and appear to represent the result of two phases of metamorphism (Hoskins, 1963) of a suite of broadly related but polyphase intrusions. As a whole these do not fit an isochron model (Figure 3a): some of the more acid rocks are distributed close to the isochron given by Gledhill et al. (1982) but others, and all the dioritic gneisses, plot somewhat below this line. Since they interpreted their result as dating the last metamorphism of these rocks it is possible that the steeper reference isochron in Figure 3a of 190 Ma (0.705) is a closer approach to the age of igneous crystallisation. Attempts by the author to date biotite-gneisses from the Debenham Islands and Black Thumb have not been successful, but two samples of migmatitic gneiss, from Neny Island and Horseshoe Island, are shown in Figure 3a. These have very radiogenic Sr for their measured Rb/Sr ratios and may indicate the sparse presence of probable Palaeozoic metasediments in this area (see following section).

## EAST GRAHAM LAND GNEISSES

South of ~66°S on the east coast of the Antarctic Peninsula the TPG is replaced as the observed pre-arc basement by amphibolitic and quartz-feldspar gneisses (Marsh, 1968), many of which have a migmatitic appearance with banded amphibolites intruded by acid material in a *lit-par-lit* fashion. These rocks have been investigated in two areas: Target Hill and Gulliver Nunatak.

Five kilometres west of Target Hill there are outcrops of strongly foliated quartz-biotite gneiss, intruded by undeformed adamellitic granite. Rex (1976) reports K-Ar mica ages of ~170 Ma and a hornblende age of 242 ± 9 Ma from this locality. The granite has since been dated at 180 ± 5 Ma by Rb-Sr whole rock isochron (Pankhurst, 1982). Six samples of the biotite gneiss give a Rb-Sr isochron age of 141 ± 49 Ma, the high error being entirely due to the limited range of Rb/Sr ratios displayed (Figure 3b). Inclusion of a

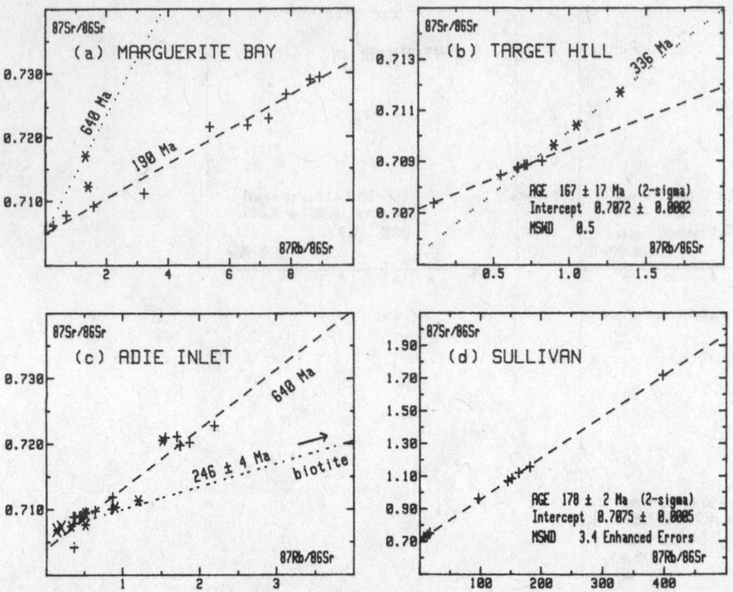


**Figure 3. Rb-Sr isochron relationships for high grade gneisses included in this work. (a) Marguerite Bay Gneisses: diorite to granitic. The best-fit line can only be regarded as a reference isochron. The two asterisks are migmatitic gneisses from Neny Island and Horseshoe Island. (b) Target Hill gneisses: biotite-hornblende-plagioclase gneisses (crosses) give the 167 ± 17 Ma isochron, whereas three samples of granite sheets give an apparent 336 Ma isochron. (c) Gneisses (crosses) and amphibolitic inclusions (asterisks) from Gulliver Nunatak and the head of Adie Inlet. The line shown is a reference isochron. (d) Foliated granite and augen gneiss from the west side of Mount Sullivan, northern Palmer Land. Errors are enhanced where MSWD exceeds 2.5.**

data point for a plagioclase feldspar separate from one of these samples extends the range significantly and results in an age of 167 ± 17 Ma (0.7072 ± 0.0000). These rocks were probably formed as part of the well developed Early-Middle Jurassic arc activity in this area. However, three samples of granitic sheets occurring in banded migmatite at Target Hill itself lie on a line in the Rb-Sr system which would correspond to an age of 336 ± 34 Ma (0.7054 ± 0.0005). Further sampling will be necessary to establish whether or not this is an artefact.

At Gulliver Nunatak and to the west around the head of Adie Inlet, migmatitic granite-gneiss with numerous inclusions of diorite gneiss and amphibolite are structurally overlain by andesites and rhyolites of probable Bajocian age (Pankhurst, 1982, see also Thomson and Pankhurst, this volume). Rex (1976) also reported a K-Ar hornblende age of 248 ± 10 Ma from gneiss in this area. Whole-rock Rb-Sr data for the gneisses and inclusions are shown in Figure 3c, where it is clear that they scatter widely about a reference isochron of ~600 Ma. The xenolithic inclusions also plot about this line and have similar calc-alkaline REE patterns. Individual groups of data points tend to lie along lines of lower slope. For example, at Gulliver Nunatak itself three samples of granite gneiss from a single outcrop define a 98 Ma (0.7088) isochron. Inclusion of two data points for granodioritic gneiss from the same outcrop produces a much poorer isochron (MSWD = 9.8) with an age of 174 ± 30 Ma (0.7080 ± 0.0003). Both these ages correspond to well developed episodes of magmatic activity in the area (Pankhurst 1982) and are likely to represent domain resetting. One group of migmatitic gneisses has much more radiogenic Sr than any analysed Mesozoic rocks from the Antarctic Peninsula—present day $^{87}Sr/^{86}Sr$ ratios ~ 0.720 corresponding to values of 0.710 in Early Jurassic times. If not derived from 600 Ma old igneous parents then these must at least represent remelted sedimentary material with an ancient provenance, unlike any so far analysed (their Rb/Sr ratios are roughly half that of equally radiogenic samples of the TPG or Signy Island schists). A minimum age for rock formation is given by internal isochron data for individual samples—amphibolite and quartzo-feldspathic bands from one of the more radiogenic banded gneisses give 203 ± 24 Ma (0.7164 ± 0.0003), biotite from the pegmatitic phase of granite gneiss at Gulliver Nunatak gives a whole-rock intersection age of 240 ± 4 Ma (0.7070 ± 0.0001) and biotite, hornblende and plagioclase from an amphibolitic inclusion in the gneisses give a mineral isochron of 246 ± 4 Ma (0.7065 ± 0.0001). Together with evidence referred to above and Rex's (1976) K-Ar data this confirms an Early Triassic age for the amphibolite grade metamorphism and migmatisation of these pre-volcanic gneisses.

## NORTHERN PALMER LAND

Exposures in Palmer Land are less continuous than in Graham Land and field relationships are often enigmatic. Davies (in press) recognised a general stratigraphy in which a metamorphic complex (crystalline rocks overlain by metavolcanic and metasedimentary rocks) are followed by a volcano-sedimentary sequence with an early plutonic phase and finally by undeformed ("Andean") intrusive and hypabyssal rocks. These three sequences were tentatively assigned Palaeozoic, Mesozoic and Cretaceous/Tertiary ages on the basis of inferred correlation with the now somewhat outdated stratigraphy previously used in Graham Land. Evidence that the metavolcanic rocks are in part the equivalents of unmetamorphosed Early-Middle Jurassic volcanic rocks in Graham Land is presented elsewhere (Thomson and Pankhurst, this volume). Preliminary results on the crystalline basement are presented here.

The rocks of the metamorphic complex exposed around Mount Sullivan were classed by Davies (in press) as mainly biotite gneisses and amphibolites with augen-gneisses and acid gneisses derived from igneous rocks which intruded the former units. From field observation made in 1980/81 by the author and further north by A. Meneilly (pers. comm.) the augen-gneisses are now thought to be a more foliated variety of the parental granite arising from marginal distension during emplacement. The acid gneisses are almost structureless granite with feldspar megacrysts.

Rb-Sr data for the augen-gneisses and acid gneisses at the west end of Mount Sullivan are shown in Figure 3d. There is a wide range of Rb/Sr ratios, reaching very high values in the megacrystic granite which has Sr contents as low 2.5 ppm. The data define a good isochron giving an age of 177 ± 2 Ma (0.7075 ± 0.0003). It is clear that this essentially dates the primary magmatic episode during which these rocks were emplced and that this coincides with the Early-Middle Jurassic arc magmatism of eastern Graham Land (Pankhurst 1982).

## SUMMARY

Not all rock units previously proposed as pre-volcanic basement in the Antarctic Peninsula yield radiometric data compatible with this assignment. Thus the blueschist terrain of the Scotia metamorphic complex appears to be, at least in part, of Cretaceous age, although the amphibolite terrain which includes terrigenous material seems to be significantly older. A Late Palaeozoic (Permo-Carboniferous) age is however confirmed for the Hope Bay Formation of the Trinity Peninsula Group, which could be a time correlative of the amphibolite terrain. Other formations of the TPG are known to include Triassic fauna and an extended depositional history for these fore-arc sediments seems probable. Similar diversity is apparent in the orthogneisses of Graham Land. Those of the Neny Fjord area are probably derived from Early Jurassic igneous rocks, but earlier migmatitic paragneisses are present elsewhere in Marguerite Bay. The oldest apparent ages so far obtained are for migmatitic gneiss on the east coast of Graham Land where resetting of the Rb-Sr systems during Triassic metamorphism (~245 Ma ago) and later igneous events largely masks a possible latest Precambrian or Early Palaeozoic history. Orthogneisses from north eastern Palmer Land are shown to belong to a mildly deformed igneous suite of Early-Middle Jurassic age, but orthogneisses and paragneisses further south have yet to be investigated. The evidence available so far on orthogneisses has not convincingly revealed the presence of a pre-Gondwanian magmatic arc to match the established Late Palaeozoic-earliest Mesozoic arc-trench and fore-arc basin sediments (Hyden and Tanner, 1981; Smellie, 1981).

*Acknowledgements* Many members of British Antarctic Survey have contributed to the support of geochronological collecting in the field, but for the material considered here S. Artis and R. Atkinson deserve special mention. I.W.D. Dalziel and F. Hervé have kindly made useful unpublished data available. Karen Brotby has one again performed the invaluable task of sample preparation.

## REFERENCES

ADIE, R.J., 1954: The petrology of Graham Land. I. The basement complex; Early Palaeozoic plutonic and volcanic rocks. *Scient. Rep. Falkld Isl. Depend. Surv., 11.*

BROOKS, C., HART, S.R. and WENDT, I., 1972: Realistic use of two error regression treatments as applied to rubidium-strontium data. *Rev. Geophys. Space Phys., 10,* 551-77.

DALZIEL, I.W.D., 1972: Large scale folding in the Scotia arc; *in* Adie, R.J. (ed.) *Antarctic Geology and Geophysics.* Universitetsforlatet, Oslo, 47-55.

DALZIEL, I.W.D., 1982: The early (pre-Middle Jurassic) history of the Scotia arc region: a review and progress report; *in* Craddock, C. (ed.) *Antarctic Geoscience.* Univ. Wisconsin Press, Madison, 111-26.

DALZIEL, I.W.D., ELLIOT, D.H., THOMSON, J.W., THOMSON, M.R.A., WELLS, N.A. and ZINSMEISTER, W.J., 1977: Geologic studies in the South Orkney Islands: R/V Hero Cruise 77-1, January 1977. *U.S. Antarct. J., 12,* 98-101.

DAVIES, T.G., in press: The geology of part of northern Palmer Land. *Scient. Rep. Br. antarct. Surv.*

GLEDHILL, A., REX, D.C. and TANNER, P.W.G., 1982: Rb-Sr and K-Ar geochronology of rocks from the Antarctic Peninsula between Anvers Island and Marguerite Bay; *in* Craddock, C. (ed). *Antarctic Geoscience.* Univ. Wisconsin Press, Madison, 315-23.

GRIKUROV, G.E., 1973: *Geologiya Antarkticheskogo Poluostrova.* (Geology of the Antarctic Peninsula). Acad. Nauk. S.S.S.R., Moscow, 119 pp (English translation: Amerind Publ. Co. Ltd, New Delhi, 140 pp).

GRIKUROV, G.E., KRYLOV, A. Ya., POLYAKOV, M.M. and TSVOBUN, Ya. N., 1970: Vozrast porod v severnoi chasti Antarkticheskogo poluostrova i na Yuzhnykh Shetlandskikh ostrovakh (po danym Kaliy-argonovogo metoda) (Age of rocks from the northern part of the Antarctic Peninsula and the South Shetland Islands (from data of the potassium-argon method)). *Inform. Byul. Sov. Antark. Eksped., 80,* 30-4 (English translation: *Inf. Bul. Sov. Antarct. Exped., 8,* 61-68).

HALPERN, M., 1972: Rb-Sr total rock and mineral ages from the Marguerite Bay area, Kohler Range and Fosdick Mountains; *in* Adie, R.J. (ed.) *Antarctic Geology and Geophysics,* Universitetsforlaget, Oslo, 197-204.

HOSKINS, A.K., 1963: The basement complex of Neny Fjord, Graham Land. *Scient. Rep. Br. antarct. Surv., 43.*

HYDEN, G. and TANNER, P.W.G., 1981: Late Palaeozoic-Early Mesozoic fore-arc basin sedimentary rocks at the Pacific margin in West Antarctica. *Geol. Rdsch., 70,* 529-41.

MCINTYRE, G.A., BROOKS, C., COMPSTON, W. and TUREK, A., 1966: The statistical assessment of Rb-Sr isochrons *J. geophys. Res., 71,* 5459-68.

MARSH, A.F., 1968: Geology of parts of the Oscar II and Foyn coasts. Graham Land. Unpublished Ph.D. thesis, University of Birmingham, England. 291 pp.

PANKHURST, R.J., 1982: Rb-Sr geochronology of Graham Land, Antarctica. *Geol. Soc. Lond., J., 139,* 701-12.

REX, D.C., 1976: Geochronology in relation to the stratigraphy of the Antarctic Peninsula. *Bull. Br. antarct. Surv., 32,* 55-61.

SMELLIE, J.L., 1981: A complete arc-trench system recognised in Gondwana sequences of the Antarctic Peninsula region. *Geol. Mag., 118,* 139-59.

STEIGER, R.H. and JAEGER, E., 1977: Subcommission of geochronology: convention on the use of decay constants in geo- and cosmochronology. *Earth Planet. Sci. Lett., 36,* 359-62.

TANNER, P.W.G., PANKHURST, R.J. and HYDEN, G., 1982: Radiometric evidence for the age of the subduction complex of the South Orkney and South Shetland Islands, West Antarctica. *Geol. Soc. Lond., J., 139,* 683-90.

THOMSON, M.R.A. and PANKHURST, R.J. (this volume): Age of Post-Gondwanide calc-alkaline volcanism in the Antarctic Peninsula region.

YORK, D., 1969: Least square fitting of a straight line with correlated errors. *Earth planet. Sci. Lett., 5,* 320-4.

# PETROLOGY, GEOCHEMISTRY AND ORIGIN OF I- AND S-TYPE 'GRANITES' IN THE SOUTHERNMOST ANDES

E. Nelson, *Geology Department, Colorado School of Mines, Golden, Colorado, 80401, USA.*

D. Elthon, *Geology Department, University of Houston, Houston, Texas, 77004, USA.*

*Abstract* Two intrusive granitic suites, forming outlying portions of the Patagonian batholith, are exposed in the Cordillera Darwin. Herein named the Darwin granite suite and the Beagle tonalite suite, they are easily distinguished on the basis of field relations, petrography, geochemistry and geochronoloy. The 157 Ma old (Rb-Sr whole rock) Darwin suite is penetratively deformed and is intruded by the undeformed Beagle suite (60-85 Ma, K-Ar biotite ages). Quartz and alkali feldspar are more abundant in the Darwin suite than in the Beagle suite. Mafic minerals in the Darwin suite are dominantly biotite, muscovite, and garnet in contrast to hornblende, pyroxene, epidote and biotite in the Beagle suite. $SiO_2$, $K_2O$, Rb and Rb/Sr are generally greater than CaO, $Na_2O$, Sr, Y and Ti/Zr generally lower in the Darwin suite compared to the Beagle suite. The peraluminous Darwin suite and metaluminous Beagle suite, have chemical affinities to I- and S-type granitic suites respectively. The Darwin suite is thus interpreted as being the plutonic, subvolcanic equivalent of the mainly extrusive Upper Jurassic Tobifera Formation, both having been derived from anatexis of metasedimentary rocks of the continental crust. In contrast the Beagle suite was probably derived from partial melting of an igneous source rock (mantle or subducted oceanic crust).

# 6

# Marine Geology

# PROBLEMS IN ANTARCTIC MARINE GEOLOGY: A REVIEW

L.A. Frakes, *Department of Earth Sciences, Monash University, Clayton, Vic., 3168, Australia.*

*Abstract* The slow pace of scientific research in Antarctica has meant that insufficient information is available on climatology and meteorology to permit reliable modelling of past climates from sparse geologic data. Interpretations of information from oxygen isotopes, from sedimentologic data and from palaeogeographic and palaeo-oceanographic information are suspect when the distributions of land and sea and the palaeoclimatic state were greatly different from the present. This particularly is evident with regard to the Eocene and Oligocene glaciations of Antarctica, which apparently are not reflected in oxygen isotope curves or in the sea level record.

The evidence of the earliest Antarctic glaciations is restricted, thus far, to West Antarctica and it is therefore postulated that large Miocene build-ups of ice suggested in isotope studies reflect construction of East Antarctic ice. Large moisture sources (warm oceanic surface waters) existed in the southern Pacific in the Palaeocene-Eocene, contributing to formation of West Antarctic ice sheets. Such sources did not exist in the southern Atlantic and Indian Oceans until the advent of North Atlantic Deep Water in the Middle Miocene; it was at this time that the East Antarctic ice sheets originated.

Around Antarctica, the gap between modern observational science (e.g. physical oceanography and climatology) and interpretative marine geology is enormous, and assumptions about the former may lead geologists to make incorrect judgments about the history of sedimentation in the Southern Ocean. An example of this is that despite admitted ignorance about the significance of the Antarctic Polar Front Zone, or at least a variety of opinion, we select the definition of the zone which relates most closely to climatic conditions and apply this to geologic situations. Similar difficulties plague interpretations of the Subtropical Convergence, the Antarctic Divergence, the significance of Antarctic Bottom Water, and applications of poorly known biogeographic distributions of micro-organisms to ancient settings. In order to understand the history of this region we need rigid and more generally accepted physical-chemical-biological models of modern conditions.

From isotope studies, the Middle Miocene has been suggested as the time of accumulation of Antarctic ice and yet, there is evidence of glaciation at sea level in both the Oligocene and the Eocene. If extreme cooling prevailed in the Early Tertiary, to the extent that large bodies of ice existed at sea-level, would not more elevated areas also have been ice-covered? What then is the significance of large increases in $\delta^{18}O$ in the Mid-Miocene? These questions illustrate some of the uncertainties in current assumptions about the significance of geologic data. There are many other problems of this sort, including the general lack of Mesozoic glaciation in a polar Antarctica while both sea-level fluctuations and isotope variations suggest tangible glaciation at this time. Perhaps it is time re-examine some of our assumptions.

Research in marine earth-science around Antarctica is still very much in the descriptive stage and although many studies now go far beyond this in design, implementation and interpretation, the general lack of activity and the scarcity of information place great limitations on our ability to understand new data and to place them properly in a regional geological framework. First, large tracts of the Antarctic seas remain almost totally unexplored and in these regions we lack the most fundamental knowledge. Second, newly recognised problems in areas which have been surveyed and probed in reconnaissance fashion, have not been pursued in the usual way as in the more equable environments. 25 years after the begining of the International Geophysical Year, the basic data are incomplete, not only in the expected sense of geographic gaps but also, and more importantly, in the sense that most of the large problems of Antarctic marine geology have not even been addressed by specially designed expeditions.

The Antarctic continental margin has been studied by means of multi-channel seismic surveys in the vicinity of the Weddell Sea and by single-channel in the Ross Sea. The latter area has also been drilled by the Deep Sea Drilling Project, but in general we know very little about the deep stratigraphy of the Antarctic margin. The remainder of the continental margin (an area comparable to the margins of the North Atlantic and the Caribbean) is essentially unknown geologically. Yet the Antarctic continental shelf is unique in lying at substantially greater depth than any other margin. The abyssal plains near Antarctica also present rare if not unique features—a great variation in depth from one to the next; a magnetic quiet zone; marginal plateaus with crusts of unknown composition; extensive glacial-marine and contour-current sedimentation; and high latitude manganese nodule fields in part on siliceous ooze substrates. In areas of the circumpolar ridge system which have been examined, the magnetic anomaly patterns are well known and these in themselves present some intriguing problems, such as asymmetric spreading and the highly fractured Antarctic Discordance in the Southeast Indian Ocean and ridgecrest jumping in the southeast Pacific. In any other part of the globe these areas would have been subjects for intensive restudy and data gathering. Other deep regions remain almost completely unstudied: the structurally complex ridge system south of Africa; the region west of Kerguelen Ridge, where Mesozoic seafloor probably exists; magnetic patterns north of the Ross Sea, to name some examples.

Specialist studies based mostly on data obtained during reconnaissance cruises have shed light on some major questions having to do with the unique position of Antarctica as the only polar continent. Naturally, these have been concentrated on the climatic history of Antarctica and its key role in the evolution of global climates. But such research is dependent on a thorough knowledge of the modern polar climate and particularly on the mechanisms of climate control exerted by the oceans. Though Antarctica is a polar continent, most of the high latitude southern hemisphere is a water hemisphere and it is well known that the ocean plays a major part in regulating the thermal regime here. The point is that we may not yet have a sufficient grasp of the present system to be able to extend concepts to past times, even for climates which were similar to the present and certainly not for times when they were markedly different.

## PALAEOCLIMATOLOGY AND PALAEO-OCEANOGRAPHY

As in all geological investigations, the study of Antarctic palaeoclimates is based on incomplete information. We do not have a complete record in the Antarctic region, nor in any other region. The Antarctic situation is more piecemeal than elsewhere, however, because most of the continent and much of the continental margin is inaccessible due to ice cover. We have some information bearing on Cenozoic climates from surficial geology and from drill holes in the McMurdo Sound, Ross Ice Shelf and Dry Valleys Drilling Projects, and efforts have been made to summarise and relate the results to offshore work. However, the integration process is far from being complete, in part because the marine data encompass both nearshore and deep ocean types. Yet, the only valid climatic history will be that which is based on a truly comprehensive synthesis of information from land and sea.

Already there are a few mile-posts in such a historical scheme, mostly based on Antarctic and Subantarctic data (Figure 1). These include (1) the earliest evidence of glaciation in the recent MSST borehole (Webb, this volume) in poorly dated early Tertiary sub-ice hyaloclastites from West Antarctica and in possible ice rafting in Eocene beds from the South Pacific; (2) isotopic evidence for an initial marked cooling of Subantarctic waters near the beginning of the Oligocene, in conjunction with a wealth of other indicators elsewhere on the globe; (3) a large buildup of West Antarctic ice by the end of the Oligocene, as indicated by the thick glacial-marine sequence of the Ross Sea; (4) Subantarctic and global isotopic indications that the Middle Miocene saw the development of most of the Antarctic ice sheets; (5) evidence in Antarctica of an advance of continental ice to the edge of the continental shelf at about the end of the Miocene followed by a rapid retreat and (6) global evidence for the initiation (at or before about 700,000 years ago) of marked temperature cyclicity with a period of about 100,000 years, corresponding to that of eccentricity of the orbit about the sun. Interpretation of all of these reasonably well established events can be questioned on the ground that underlying assumptions could be invalid. More importantly the scientific consequences of these events, both globally and regionally in terms of palaeo-oceanography, are of great significance and therefore the underlying models must be examined carefully.

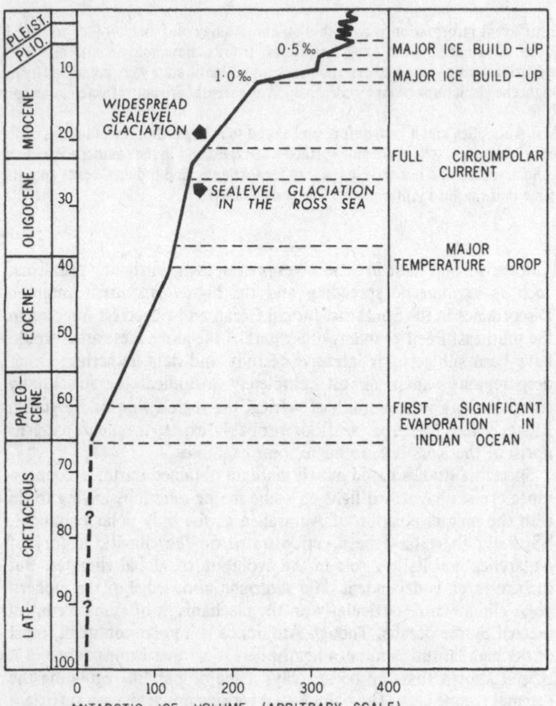


**Figure 1. Speculative ice-volume curve for Antarctica. Right hand column shows significant climatic events suggested by various workers.**

## Initiation of Antarctic Glaciation

Until recently it has been assumed that the glaciation of Antarctica began in the Early Tertiary, coincident with the oldest known glacial deposits or features carved by glacial erosion. Now, however, there is evidence in sea level curves that global ice volumes may have fluctuated throughout Phanerozoic history. Interpretations of such curves is not straightforward, moreover, because of the need to distinguish the glacio-eustatic from the tectono-eustatic component in sea level change, a goal not yet achieved for pre-Middle Cretaceous times. The only Mesozoic sedimentary rocks exposed in Antarctica (Jurassic and Cretaceous rocks of the Antarctic Peninsula region and Triassic non-marine strata of the Trans-Antarctic Mountains) show no sign of glacial activity. Though evidence of Mesozoic glaciation is lacking on a global scale excluding a few vague and unpublished reports from the northern hemisphere, glaciers of limited size may well have been present in Antarctica, especially given its high latitude Mesozoic position. That any such ice bodies would not have extended continent wide is shown by the occurrences of both in situ and reworked microfloras in sediments of the continental shelf (Kemp, 1978; Domack et al., 1980). Thus, while it is possible to say that glaciations had begun probably by Eocene, and certainly by Oligocene time, this represents a minimum date on the earliest Antarctic glaciation and the possibility of at least limited Mesozoic glaciation should be re-examined.

According to data from McMurdo Sound borehole, glaciation was proceeding in the Ross Sea in the Early Eocene, and possibly in the Palaeocene. It is not certain that quartz grains in Pacific sediments acquired their glacial surface textures immediately prior to final deposition—they could be inherited features. Nor is it unequivocably established that Queen Maud Land hyaloclastites originated beneath ice, as opposed to seawater. Late Oligocene-Early Miocene glacial marine sediments of the Ross Sea resting apparently conformably on carbonaceous glauconitic sandstones dated at about 26 Ma (McDougall, 1977), represent the oldest widespread glacial deposits of Antarctica (Hayes and Frakes, 1975).

A related problem in the initiation of Antarctic glaciation derives from the exceptional global warmth revealed by climatic indicators of Early Tertiary and Mesozoic age. Several lines of evidence, including floral assemblages and characteristics and oxygen isotope studies, indicate high latitude warmth and low temperature gradients from equator to pole. Barron et al. (1981) have pointed out the difficulties in establishing and maintaining these characteristics of Mesozoic earth. One might ask further, how was glaciation initiated in such conditions, whether in the Mesozoic or Early Tertiary? It now seems necessary that the Palaeocene was a time of high humidity and precipitation (Figure 1) a condition which contributed to early ice formation in Antarctica (Frakes, 1979).

## Oligocene Cooling

Near the end of the Eocene a sharp global cooling is inferred by many workers (Wolfe and Hopkins, 1967; Frakes and Kemp, 1972; Shackleton and Kennett, 1975). There is no evidence whatever in Antarctica to support this. The best evidence is found in subantarctic oxygen isotope records, which also show a very slow warming through the Oligocene. There is no sign in the isotope curve of either sea level glaciation in the Ross Sea in the Early Eocene or of a marked Middle Oligocene fall of sea level (Vail et al., 1977). It is possible that available isotope records do not adequately cover the Oligocene. If this is not the case, then the Oligocene fall in sea level reflects tectono-eustatics and all changes up to this magnitude throughout the Vail et al. sea level curve may also be of tectonic rather than glacial origin.

The arrival of West Antarctic ice at sea level in the Ross Sea in the Eocene is an event of great importance, signifying truly frigid conditions at low elevations. Previous to this and for an unknown period, ice could have formed in elevated areas while considerably warmer climates affected the coastal region. The event also signifies increased ice volume by Oligocene time, as seismic records show the deposits to blanket the entire Ross Sea floor. The fact that the increased ice is not reflected in an obvious increase in $\delta^{18}O$ in the isotope curve, also suggests poor coverage of the Oligocene.

DSDP site 274 located just outside the Ross Sea, displays evidence of bottom water activity in the Oligocene but not previously (Frakes, 1975). This is not observable at other sites within the Ross Sea, but generation of bottom water by freezing to floating ice shelves in the Oligocene has been suggested as the cause of widespread deepwater erosion giving rise to unconformities in the Tasman Sea (Watkins and Kennett, 1972). Once again the lack of effect on oxygen isotopes is puzzling and additional suites of samples clearly are needed both to clarify the situation and to help define the vertical structure of the Oligocene Antarctic seas.

## Build-up of the Antarctic Ice Sheets

Shackleton and Kennett (1975) suggested on the basis of studies of oxygen isotopes that perhaps half of the ice mass presently on Antarctica accumulated during the Middle Miocene. I make three postulates. First, since the West Antarctic ice sheets constitute about one third of Antarctic ice and since the Ross Sea glacials show pre-Middle Miocene ice build-up in West Antarctica, the Mid-Miocene increase in $\delta^{18}O$ can only represent a major build-up of ice in East Antarctica. The build-up of West Antarctic ice occurred earlier (Eocene and Oligocene), in part because this area was characterised by Mesozoic and Cenozoic mountainous topography and probably because oceanic circulation patterns in the adjacent Pacific provided a convenient source of warm surface water for evaporation and consequent precipitation. Deflection of warm water currents due to the northward passages of Australia-New Guinea probably played a role (Frakes and Kemp, 1972). Ice rafting around East Antarctica apparently began in the Early Miocene but was not widespread until the Middle Miocene (Frakes, 1979). East Antarctic ice did not overtop the Trans-Antarctic Mountains until the Mid-Miocene (Hayes and Frakes, 1975), but declined in volume in the Pliocene.

Second, as East Antarctica was dominantly a shield area, perhaps low-lying and with shallow depressions filled with Mesozoic sediments (Drewry et al., 1979), substantial ice accumulation did not occur there until a similar evaporative source occupied the South Atlantic region in the Middle Miocene. This may have come about through the initial formation of North Atlantic Deep Water (NADW) which upwells in the far South Atlantic (Schnitker, 1980). The Indian Ocean sector may have been the last area to develop a full-fledged ice sheet, owing to its

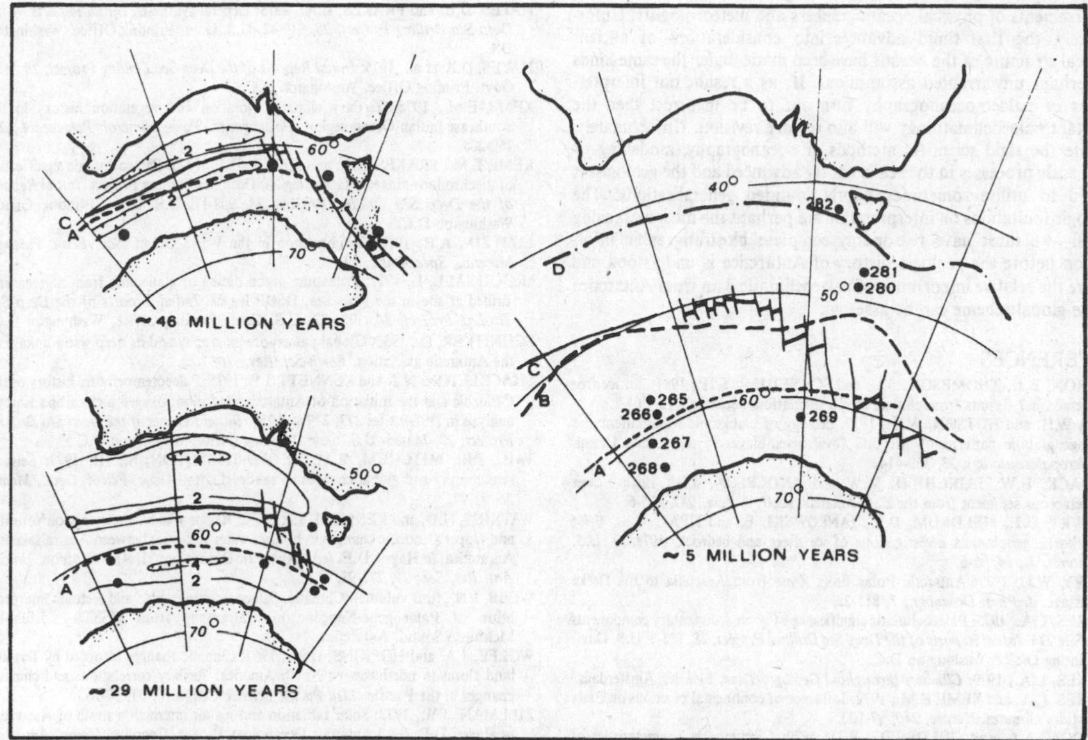


**Figure 2. Palaeogeography and palaeo-oceanography of the southeast Indian Ocean at about 46, 29 and 5 Ma. A= boundary between calcareous ooze (north) and siliceous ooze (south). B= present position of the Antarctic Polar Front Zones. C= southeast Indian Ridge crest. D= Subtropical Convergence. Spiked lines represent approximate depth contours. DSDP sites shown by number on right hand figure.**

remoteness from abundant sources of easily evaporated (warm) surface waters.

Third, the generation by bottom freezing on ice shelves, of bottom waters analogous to modern day Antarctic Bottom Water, first took place in the Ross Sea (Pacific and Indian Ocean sectors) in the Oligocene and later in the Weddell Sea region (Middle Miocene, Atlantic sector). In contrast to other histories of global evolution of water masses, the present one suggests that the full development of the past structure was attained at different times in different oceans, with the final addition of bottom waters. Prior to the Oligocene in the Pacific and Indian Oceans and before the Middle Miocene in the Atlantic, water mass structures lacked bottom water (in the special sense described above) and may therefore have been markedly different from their present configurations. The establishment of the Antarctic Circumpolar Current early in the Miocene had the effect of blocking the southward movement of warm surface waters in the western Pacific, thus diminishing the growth of West Antarctic ice. NADW is and was not similarly affected and growth in the Atlantic-West Indian Ocean sectors thus continued.

## History of the Antarctic Convergence

Leg 28 of the Deep Sea Drilling Project discovered that clayey sediments rich in siliceous microplankton had been deposited in a narrow band around Antarctica in the time interval since the Late Eocene (Hayes et al., 1975). North of this band of siliceous sediments, in many cases consisting of diatom oozes, calcareous sediments were contemporaneously being deposited (Kemp et al., 1975). It is tempting to think of the boundary demarking this separation of sediment types as representing the ancestral position of the modern Antarctic Convergence or Antarctic Polar Front Zone (APFZ), particularly since the boundary at any time roughly parallels lines of latitude as does the APFZ. However, a close look at the APFZ reveals that it is the southern boundary of dominantly foraminiferal sediments, whereas ancient boundaries represent the southern limit of sediments rich in calcareous nanoplankton. In the region south of Australia modern nanoplankton are not prominent components in either the surface waters or the underlying sediment, except in the region north of the Subtropical Convergence, 7-15° of latitude north of the APFZ (Lizitzin, 1972; Zillman, 1972). It thus seems that the high latitude oceanic structure over most of the Tertiary period differed from the modern situation in that there was a single convergence rather than a pair. It was only at the end of the Miocene (5 Ma, Figure 2) that the present system came into being, concurrently with a dramatic expansion outward from Antarctica of the locus of siliceous sedimentation (Kemp et al., 1975).

The significance of the early structure of this part of the Antarctic seas is uncertain, as regards both palaeo-oceanography and palaeoclimatology. There is the problem of estimating tolerances for temperature and other water parameters from assemblages of fossils, many of which are not extant. And of determining which of several factors (temperature, salinity, nutrient availability, etc.) may have controlled their distribution. It is likely that temperature exerts a major control on many modern forms in the Indian Ocean but this is not certain (Be and Hutson, 1977). Further problems of major importance derive from our lack of understanding of southern ocean dynamics in the present system. The APFZ has been defined and mapped on several different criteria over the years, one of the most recent being a salinity (not temperature) minimum at 200 m depth (Gordon and Goldberg, 1970), and is now known to include microscale thermal structures such as rings and meanders (Emery, 1977). There is no general agreement among physical oceanographers as to the dynamic significance of the APFZ; it does not appear to be a "normal" wind generated convergence, as the structure is embedded within the zone of West Wind Drift. With this uncertainty about the oceanographic and meteorological meaning of the APFZ, it is obvious that interpretations of earlier, differing dynamic systems must be tempered with caution.

## CONCLUSIONS

From early, qualitative experiments with the concept of circulation patterns in oceanic surface currents, marine geologists have now moved on to consider three-dimensional structure and flow of water masses. The early work adopted assumptions, particularly about the thermal structure in southern oceans but with only the most guarded endorsements of physical oceanographers and meteorologists. Unfortunately, the first timid advances into considerations of ancient vertical structure of the oceans have been made under the same kinds of perhaps unwarranted assumptions. If, as a result, our interpretations of palaeo-oceanography turn out to be incorrect then the resultant palaeoclimatology will also require revision. Unfortunately, despite the rigid scientific methods of oceanography, modelling of large scale processes in the sea is not far advanced and the geologist is forced to utilise sometimes poorly founded generalisations. The geologic limitations on interpretation are perhaps the most weakening of all—we must have reasonably complete biostratigraphic information before the geologic history of Antarctica is understood and before the relative importance of palaeoclimatic data from Antarctica in the global scheme can be assessed.

## REFERENCES


BARRON, E.J., THOMPSON, S.L. and SCHNEIDER, S.H., 1981: An ice-free Cretaceous? Results from climate model simulations. *Science, 212,* 501-8.

BE, A.W.H. and HUTSON, W.H., 1977: Ecology of planktonic foraminifera and biogeographic patterns of life and fossil assemblages in the Indian Ocean. *Micropalaeontology, 23,* 369-414.

DOMACK, E.W., FAIRCHILD, W.W. and ANDERSON, J.B., 1980: Lower Cretaceous sediment from the East Antarctic shelf. *Nature, 287,* 625-6.

DREWRY, D.J., MELDRUM, D.T., JANKOWSKI, E. and NEAL, C.S., 1979: Airborne geophysical investigations of ice sheet and bedrock, 1978/79. *U.S. Antarct. J., 14,* 95-6.

EMERY, W.J., 1977: Antarctic Polar Front Zone from Australia to the Drake Passage. *J. Phys. Oceanogr., 7,* 811-22.

FRAKES, L.A., 1975: Palaeoclimatic significance of some sedimentary components at Site 274. *Initial Reports of the Deep Sea Drilling Project, 28,* 785-7, U.S. Govt. Printing Office, Washington D.C.

FRAKES, L.A., 1979: *Climates throughout Geologic Time.* Elsevier, Amsterdam.

FRAKES, L.A. and KEMP, E.M., 1972: Influence of continental positions on Early Tertiary climates. *Nature, 240,* 97-100.

GORDON, A.L. and GOLDBERG, R.D., 1970: Circumpolar characteristics of Antarctic waters. *Antarct. Map Folio Ser.,* Folio 13.

HAYES, D.E. and FRAKES, L.A., 1975: General synthesis. *Initial Reports of the Deep Sea Drilling Project, 28,* 919-42. U.S. Govt. Printing Office, Washington D.C.

HAYES, D.E. et al., 1975: *Initial Reports of the Deep Sea Drilling Project, 28.* U.S. Govt. Printing Office, Washington D.C.

KEMP, E.M., 1978: Tertiary climatic evolution and vegetation history in the Southeast Indian Ocean region. *Palaeogeogr., Palaeoclimatol. Palaeoecol., 24,* 169-208.

KEMP, E.M., FRAKES, L.A. and HAYES, D.E., 1975: Palaeoclimatic significance of diachronous biogenic facies, leg 28, Deep Sea Drilling Project. *Initial Reports of the Deep Sea Drilling Project, 28,* 908-17. U.S. Govt. Printing Office, Washington D.C.

LIZITZIN, A.P., 1972: Sedimentaton in the World Ocean. *Soc. Econ. Paleont. Mineral., Spec. Publ., 17.*

McDOUGALL, I., 1977: Potassium-argon dating of glauconite from a greensand drilled at site in the Ross Sea, DSDP leg 28. *Initial Reports of the Deep Sea Drilling Project, 36,* 1071-72. U.S. Govt. Printing Office, Washington D.C.

SCHNITKER, D., 1980: Global palaeo-oceanography and its deep water linkage to the Antarctic glaciation. *Earth-Sci. Rev., 16,* 1-20.

SHACKLETON, N.J. and KENNETT, J.P., 1975: Palaeotemperature history of the Cenozoic and the initiation of Antarctic glaciation: oxygen and carbon isotope analysis in DSDP sites 277, 279 and 281. *Initial Reports of the Deep Sea Drilling Project, 29,* 743-55. U.S. Govt. Printing Office, Washington D.C.

VAIL, P.R., MITCHUM, R.M., Jr. and THOMPSON, S., III, 1977: Seismic stratigraphy and global changes of sea level. *Am. Assoc. Petrol. Geol., Mem., 26,* 83-97.

WATKINS, N.D. and KENNETT, J.P., 1972: Regional sedimentary disconformities and Upper Cenozoic changes in bottom water velocities between Australasia and Antarctica; *in* Hayes, D.E. (ed.) Antarctic Oceanology II, *Am. Geophys. Union, Ant. Res. Ser., 19,* 273-93.

WEBB, P.N., (this volume): Climatic, palaeo-oceanographic and tectonic interpretation of Palaeogene-Neogene biostratigraphy from MSSTS-1 drillhole, McMurdo Sound, Antarctica.

WOLFE, J.A. and HOPKINS, D.M., 1967: Climatic changes recorded by Tertiary land floras in northwestern North America: Tertiary correlations and climatic changes in the Pacific. *11th Pacific Science Congress., 25,* 67-76.

ZILLMAN, J.W., 1972: Solar radiation and sea-air interaction south of Australia; *in* Hayes, D.E. (ed.) Antarctic Oceanology II, *Am. Geophys. Union, Ant. Res. Ser., 19,* 11-40.

# BOTTOM-CURRENT EROSION IN THE SOUTHEAST INDIAN AND SOUTHWEST PACIFIC OCEANS DURING THE LAST 5.4 MILLION YEARS

M.T. Ledbetter, P.F. Ciesielski, N.I. Osborn and E.T. Allison, *Department of Geology, University of Georgia, Athens, Georgia, 30602, U.S.A.*

*Abstract* Hiatus frequency in the sedimentary record in the southeast Indian Ocean during the last 5.4 Ma defines periods of increased velocity of Antarctic Bottom Water (> 4200 m) and Circumpolar Deep Water (< 4200 m). The extent of the erosion by these two bottom-currents varies with time and the disconformities produced by each are not synchronous. Increases in the extent of erosion by Circumpolar Deep Water occurred in the late Gauss Chron (2.96 to 2.47 Ma) and late Matuyama Chron (2.0 to 1.5 Ma). Increased Antarctic Bottom Water erosion occurred in the late Gilbert Chron (3.5-4.0 Ma) with a small increase in the late Matuyama Chron. The upper Matuyama (1.66-0.72 Ma) disconformity is much greater in extent than at present in both deep and shallow areas. Bottom-current erosion has waned extensively since the late Matuyama.

In order to test for palaeoclimatic mechanisms involved in the increased bottom-current activity, the timing of the initiation of increased velocity must be determined. As this record is lost within the scour zone, cores in the marginal, winnowing areas must be examined for evidence of the initiation of scour in the axis of flow. The percent manganese micronodules and the mean particle-size of the noncarbonate silt fraction were used in three cores adjacent to the axis of high velocity Antarctic Bottom Water flow in order to determine the timing of increases in flow. The percent manganese micronodules increases dramatically between 3.0-4.5 Ma and at 1.2-1.3 Ma indicating a reduction in sedimentation rate associated with increased bottom-current velocity. Similarly the silt mean particle-size coarsens at the same times indicating a coarse lag deposit under high velocity currents.

The periods of increased Antarctic Bottom Water velocity correspond to episodes of Antarctic cooling and expansion of ice-sheets/shelves. The sharp decrease in velocity at 3.0-3.2 Ma occurs in response to the decrease in North Atlantic Deep Water production as a result of the freezing over of the Norwegian Sea during the Northern Hemisphere Ice Ages. The widespread shallow and deep disconformities at 2.0-1.5 Ma are in response to cooling of the Ross and Weddell Seas and the steepening of the latitudinal temperature gradient.

Geologic evidence for recent bottom-water activity in the southeast Indian Ocean has been derived, in part, from bottom photographs of current produced features such as sediment free manganese nodules, ripples, and scour marks and sediment lineations (Watkins and Kennett, 1972; Kennett and Watkins, 1976). Other evidence of deep sea current activity in the southeast Indian Ocean has been inferred from hiatuses in the sedimentary record. Regional disconformities resulting from deep sea current erosion or periods of nondeposition have been identified in the Southern Ocean (Watkins and Kennett, 1972; Kennett and Watkins, 1976; Ciesielski et al., 1982). Watkins and Kennett (1972) and Kennett and Watkins (1976) discovered major hiatuses in the Upper Cenozoic sedimentary record in the southeast Indian Ocean. In their reconaissance study of *Eltanin* cores, they found a zone south of the Tasman Basin where middle Gauss through Brunhes Chronozone sediments (t = 3.0 to 0 Ma) are missing. They also found hiatuses in the South Indian and South Australian Basins which they attributed to two major pulses of increased Antarctic Bottom Water (AABW) velocity, during the Matuyama Chron (t = 2.4 to 0.72 Ma) and Brunhes Chron (t = 0.72 to 0 Ma).

Theories have been advanced which relate AABW production with sea ice and climatic conditions, but more research is needed for the interpretation of this complex relationship. Knowledge of past fluctuations in AABW production will lead to a better understanding of palaeoclimatic and palaeo-oceanographic conditions. The purpose of this study is to identify deep sea disconformities produced by AABW and Circumpolar Deep Water (CDW). These hiatuses in the sedimentary record are used to examine the palaeoclimatic influence on increased bottom-water production.

## TEMPORAL AND AREAL DISTRIBUTION OF HIATUSES

The palaeomagnetic polarity of *Eltanin* cores in the southeast Indian Ocean (Figure 1) (Watkins and Kennett, 1977; Osborn et al., 1983) was correlated to the magnetostratigraphic time-scale (LaBrecque et al., 1977) using the diatom zonation of Ciesielski (1983) and the silicoflagellate zonation of Ciesielski (1975). The stratigraphic correlation yielded ages for each core or core segment (e.g. Figure 2). A full description of methods and all of the results for cores shown in Figure 1 are discussed in Osborn et al. (1983).

A method of showing the temporal distribution of hiatuses and variations in hiatus abundance was adapted from Moore et al. (1978). The number of cores with a hiatus is recorded for every half million year interval between 0.5 to 5.0 Ma. This number is expressed as a percentage of the total number of cores represented by either recovered sedimentary material or a hiatus (Figure 3a). The 0 to 0.5 Ma interval was not plotted because of poor biostratigraphic control and the lack of palaeomagnetic events in the Brunhes Chron. If 0.2 Ma or less is missing in the interval, the core is recorded as having sediment present rather than a hiatus. Two peaks of hiatus abundance are evident: one between 5.4 and 4.0 Ma and one between 2.0 and 1.0 Ma; distinct minima in hiatus abundance occur between 3.5 and 2.5 Ma, and 1.0 and 0.5 Ma (Figure 3a). The frequency of hiatuses as a function of water depth (Figure 3b) is used to separate erosion by AABW (> 4200 m) from CDW (< 4200 m). Plots of hiatus frequency for each water mass are shown in Figure 3c. Cores < 4200 m on the continental slope of Antarctica (south of 60°S) were separated from shallow hiatuses (Figure 3d) since AABW may be more responsible for erosion at those sites than CDW.

Four maps depicting the distribution of hiatuses within selected intervals of the magnetostratigraphic time-scale are shown in Figure 4. The magnetostratigraphic time-scale was divided into intervals on the basis of magnetic chron and subchron boundaries. Cores in which the entire time-slice is missing from the sedimentary column are represented along with cores which are missing only the older part of the time-slice or the younger part of the time-slice. Cores which have sediment present during the time-slice are represented on each map and are used to define the extent of zones where hiatuses dominate the sedimentary record (Figure 4). The boundaries of each zone of hiatus occurrence are restricted by bottom topography and water depth range since bottom-currents are likewise restricted. This approach is different from that used by Watkins and Kenett (1972) which grouped widely separated disconformities, often separated by major topographic restrictions.

A minimum of two cores containing hiatuses in a time-slice constitutes a hiatus zone. Hiatus zones with water depths greater than 4200 m occur in the Tasman, Emerald and South Indian Basins. Another deep water hiatus zone occurs east of the Macquarie Ridge where water depths range from about 3800 to 4400 m. A hiatus zone at the junction of the Southeast Indian Rise, the Macquarie Ridge and the Tasman Basin (referred to as the central junction area in this study) contains cores from shallow-water depths ranging from 2650 m on the western side to 4270 m on the eastern side. Two other shallow-water zones occur south of the Emerald Basin and south of the South Indian Basin. Water depths in these zones range from about 2900 to 3200 m.

The temporal pattern of hiatus frequency for each water mass is different for most of the last 5.4 Ma (Figure 3d). Shallow hiatuses are common from 5.4-1.5 Ma with a slight increase in frequency at 2.0-2.5 Ma; deep hiatuses are frequent until 3.5 Ma when the frequency decreased continuously with a slight increase from 2.0-1.5 Ma. The palaeo-oceanographic significance of the temporal and areal distribution of the hiatuses produced by each water mass are discussed for four time-slices from 2.96-0 Ma (Figure 4).

During the period 5.4-3.0 Ma an apparently high hiatus frequency for both shallow and deep disconformities (Figure 3d) makes it difficult to define the areal extent of hiatuses because of the low number of cores with a record of that period. A marked decrease in AABW hiatus frequency beginning at about 3.0 Ma (Figure 3d) results in a better record of hiatus extent in the late Gauss (2.96-2.47 Ma) (Figure 4) because the margins of the erosional zones may be delineated. While AABW erosion decreased at about 3.0 Ma erosion by CDW increased (Figure 3d) to cause an expansion of the shallow hiatus zone in the late Gauss (Figure 4).

During the early Matuyama (2.47-1.66 Ma) the CDW erosional zone expanded (Figure 4). In the early part of this time period AABW scour decreased but expanded in the latter part of the period with the formation of the Emerald Basin scour zone. A scour zone of the Antarctic margin also expanded in the early Matuyama. During the late Matuyama (1.66-0.72 Ma) both AABW and CDW scour zones expanded (Figure 4). The major expansion occurred at about 1.5 Ma and the areal extent of all scour zones is greater during this period than can be demonstrated in any other time in the last 5.4 Ma. A reduction of erosion occurred at about 1.0 Ma in both deep and shallow areas (Figure 4) and the minimum extent of AABW and CDW erosion in the last 5.4 Ma occurred during the Brunhes (0-0.72 Ma).

An analysis of hiatus frequency (Figure 3d) cannot be used to determine the timing of the initiation of bottom-current erosion because the duration of the disconformity is dependent on the depth of erosion. It is possible, however, to determine the initiation of scour by examining cores at the margins of the high velocity current (Ledbetter et al., 1978; Huang and Watkins, 1977). The percent manganese micronodules in cores adjacent to high velocity AABW (Figure 5) were used to identify zones of reduced sedimentation rate which correspond to periods of high velocity flow (Ledbetter and Huang, 1980). Two periods of high velocity AABW were detected during the last 4.6 million years (Figure 5). The earlier period lasted from 4.5 to 3.2 Ma with a short duration decrease at approximately 4.0 Ma; the younger period lasted from 1.5 to 1.0 Ma. The increases in manganese micronodule percent at 4.5 and 1.5 Ma mark the initiation of high velocity AABW.

Because manganese micronodule accumulation rates may not be representative of bottom-current velocity (Immell and Osmond, 1976), the mean particle-size of noncarbonate silt (Ledbetter, 1979) was used to determine fluctuations of bottom-current velocity adjacent to a scour zone. Increases in mean particle-size in cores at the margin of AABW are the result of winnowing at the edge of the erosional zone (Ledbetter, 1981). The earliest increase in AABW velocity occurred at 4.3 Ma with several pulses in the succeeding 1.7 million years (Figure 6); a later increase in velocity occurred at about 1.4 Ma and lasted about 0.4 million years. The periods of both the high manganese micronodule percent and coarse silt mean particle-size are nearly synchronous and both are responding to increases in AABW bottom-current velocity which eroded sediment nearer the axis of flow.

## PALAEOCLIMATIC INFLUENCE ON BOTTOM CIRCULATION

Contrary to what was expected, the timing of hiatuses produced by deep Antarctic Bottom Water (AABW) and shallow Circumpolar Deep Water (CDW) are not synchronous (Figure 3d). Nevertheless, the causes of increased bottom-water circulation may be correlated with palaeoclimatic events (Ledbetter and Ciesielski, this volume). Southern Ocean cooling and the proposed re-establishment of the West Antarctic Ice Sheet in the late Gilbert Chron (Ciesielski and Weaver, 1974; Ciesielski et al., 1982) resulted in increased AABW and CDW activity at 4.3 Ma. Cool Antarctic climatic conditions continued

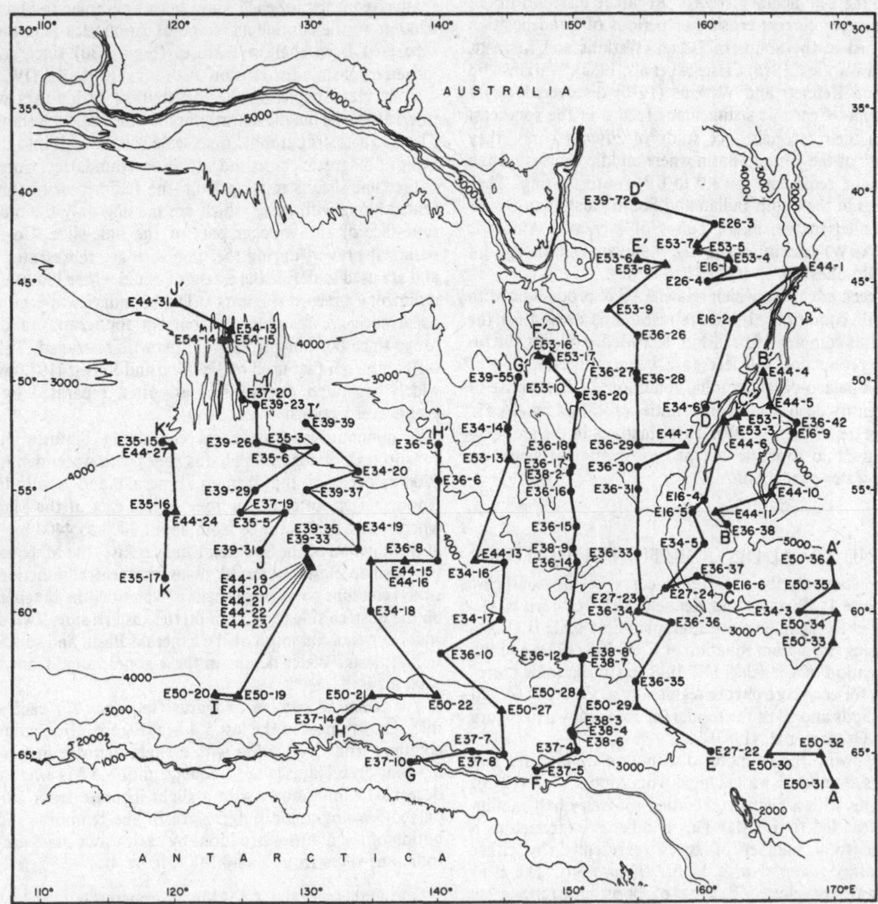


**Figure 1. Location map of cores in the southeast Indian and southwest Pacific Oceans used to define the areal and temporal distribution of disconformities in the sedimentary record of the last 5.4 million years.**

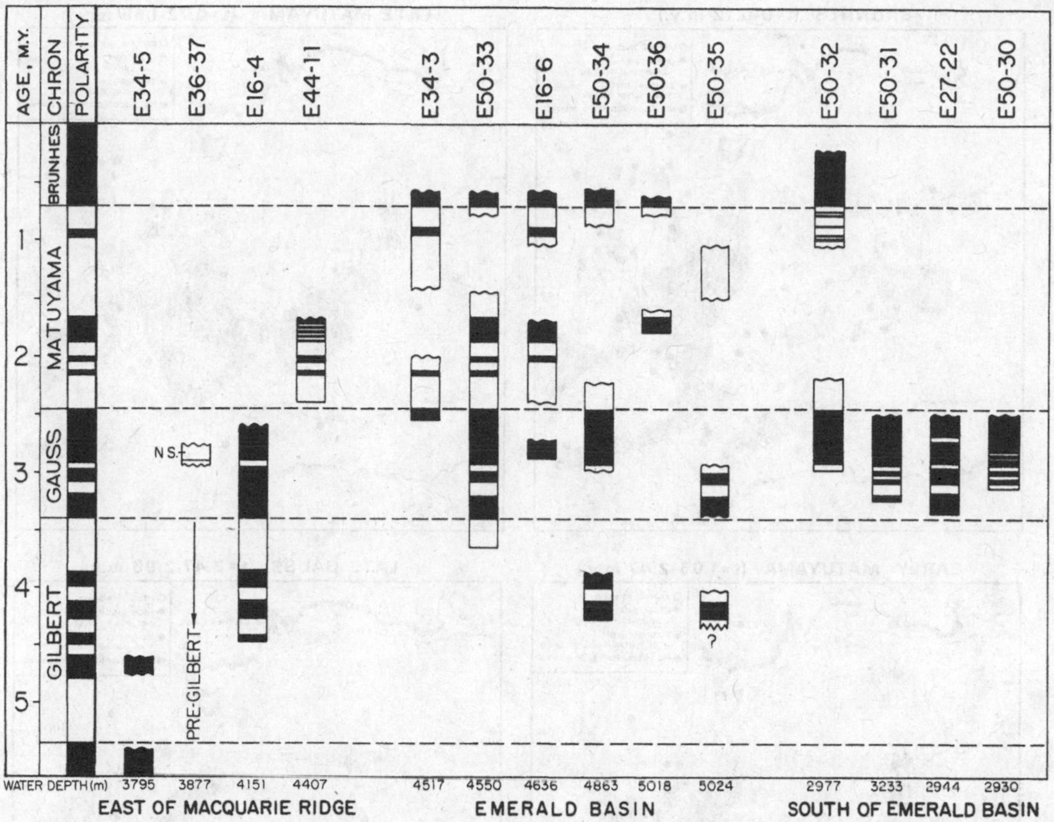

Figure 2. The biostratigraphy and magnetostratigraphy of cores was used to correlate to the magnetostratigraphic time-scale. Examples from cores in Figure 1 show hiatuses formed by both shallow and deep bottom-currents.

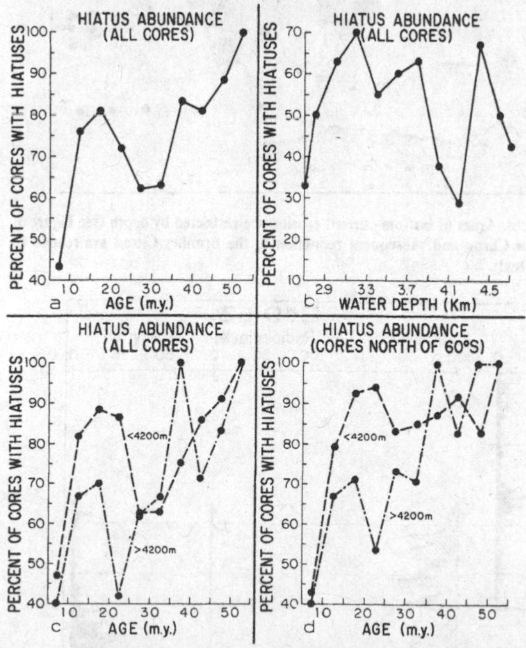


Figure 3. (a) The percent cores with hiatuses in 0.5 million year time-slices indicates periods of erosion. (b) The percent cores with hiatuses was separated into shallow cores (< 4200 m) eroded by Circumpolar Deep Water (CDW) from deep cores (> 4200 m) eroded by Antarctic Bottom Water (AABW). (c) The percent hiatuses for shallow and deep cores indicates periods of erosion by CDW and AABW respectively. (d) Cores on the Antarctic continental slope south of 60°S were excluded from the compilation of shallow diconformaties because AABW and CDW may be more responsible for erosion at those sites.

into the early Gauss Chron and were accompanied by the formation of the Northern Hemisphere Ice Sheet at about 3.2 Ma (Shackleton and Opdyke, 1977) which may have inhibited the production of the Norwegian Sea Overflow Water component of the North Atlantic Deep Water (NADW). This decrease in NADW may have been responsible for a decrease in AABW activity at 3.2 Ma (Figures 3-6) due to the reduction of high salinity NADW from the area where thermohaline-produced AABW is formed (Ledbetter, 1981). The climatic cooling and further expansion of the Antarctic ice sheets during the latest Gauss to Matuyama Chrons (Anderson, 1972) continued to increase atmospheric circulation and CDW activity which produced increased shallow water hiatuses from 2.5-1.5 Ma (Osborn et al., 1983).

Cool climatic conditions in the Southern Ocean and Argentine Patagonia (Mercer, 1976) persisted throughout the early Matuyama Chron and culminated in the late Matuyama Chron. Cool climates in the Southern Hemisphere led to intensified atmospheric circulation which influenced CDW activity throughout the Matuyama Chron. This resulted in a high percentage of cores containing shallow water hiatuses (Figure 3d). Bottom-current activity in the Brunhes and latest Matuyama Chrons (< 1.0 Ma) has been reduced in intensity relative to the expansion in the late Matuyama Chron. This reduction may be due to a change in frequency and intensity of the Northern Hemisphere Ice Sheet (Shackleton and Opdyke, 1976).

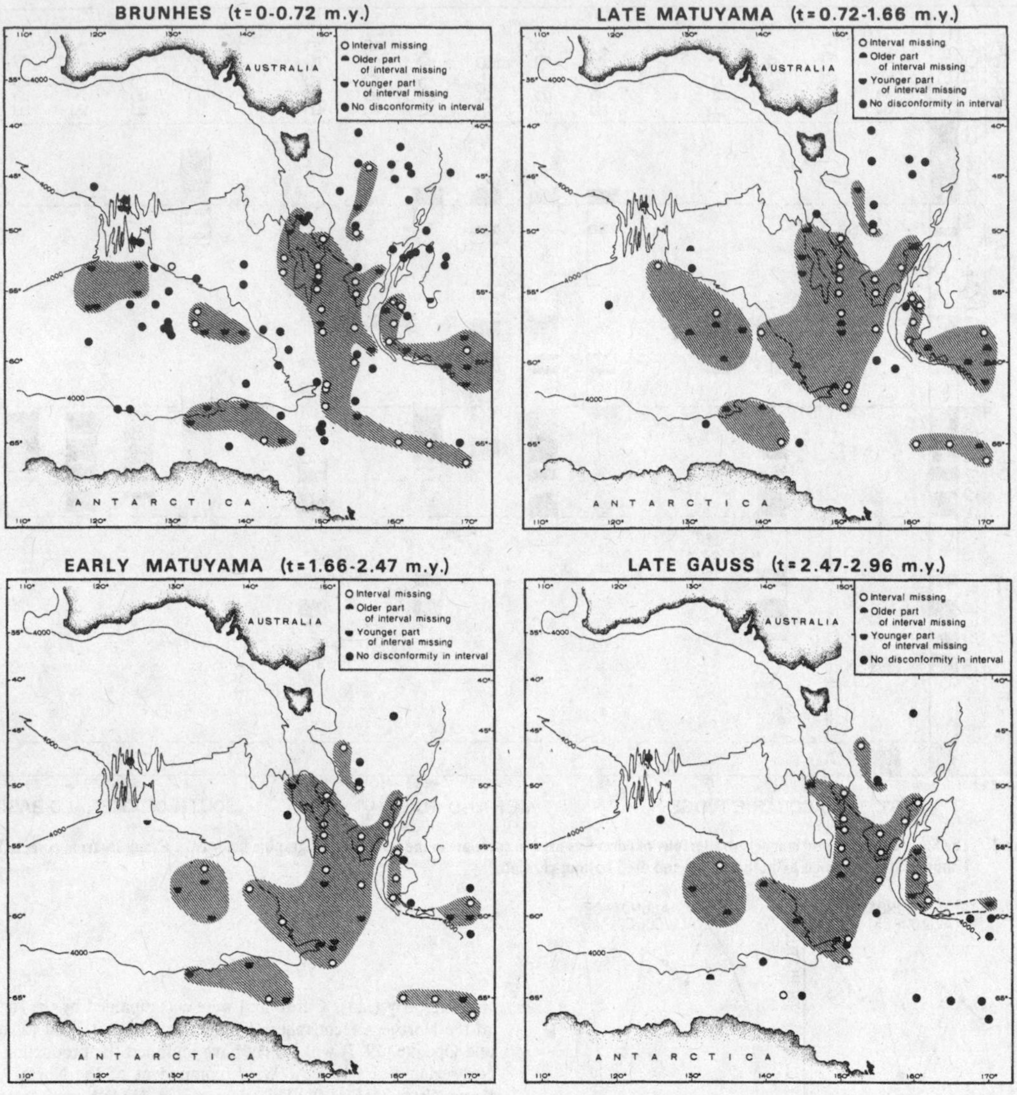

BRUNHES (t=0-0.72 m.y.)

LATE MATUYAMA (t=0.72-1.66 m.y.)

EARLY MATUYAMA (t=1.66-2.47 m.y.)

LATE GAUSS (t=2.47-2.96 m.y.)

**Figure 4. The areal distribution of cores with a hiatus is shown for four time-slices. Areas of bottom-current erosion are restricted by depth (see Figure 3) and topography. The expansion of hiatus zones in the late Matuyama Chron and subsequent reduction in the Brunhes Chron are related to palaeoclimatic changes in the Antarctic or Northern Hemisphere (see text).**

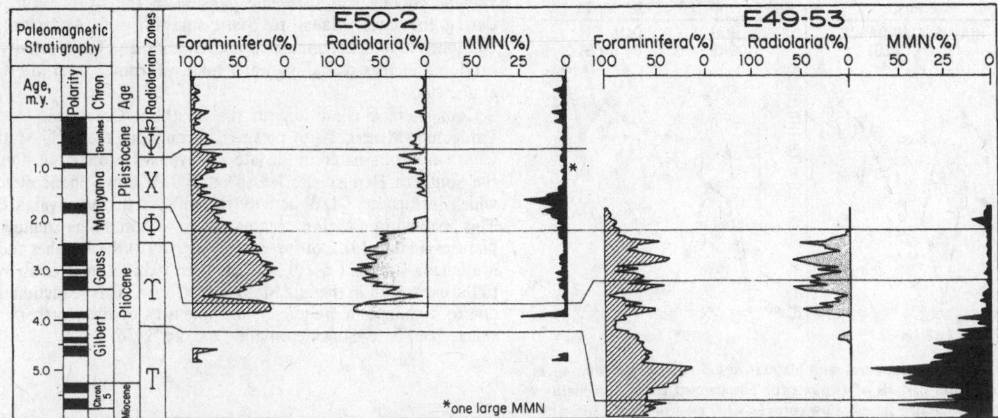


**Figure 5. The percent manganese micronodules in cores adjacent high velocity Antarctic Bottom Water records two periods of increased flow. The older period is from 4.5-3.2 Ma and the younger one from 1.5-1.3 Ma.**

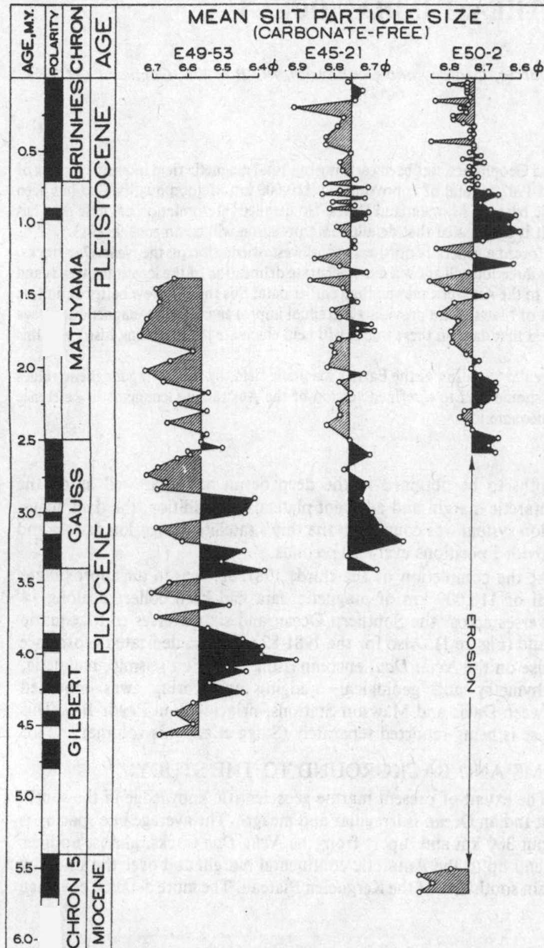


**Figure 6. The mean silt particle-size in cores adjacent an Antarctic Bottom Water erosional zone is used to detect periods of grain-size winnowing which correlate with erosional pulses in areas nearer the axis of flow. The initiation of winnowing in the cores at the margin of the flow may be used to date the initiation of erosion in order to test the palaeoclimatic influence on bottom circulation.**

## REFERENCES


ANDERSON, J.B., 1972: The marine geology of the Weddell Sea. *Florida State Univ., Sedim. Res. Lab., Rep., 35.*

CIESIELSKI, P.F., 1975: Neogene and Oligocene silicoflagellates from cores recovered during Antarctic Leg 28, Deep Sea Drilling Project: Biostratigraphy and palaeoecology. *Initial Reports of the Deep Sea Project, 28,* 625-92. U.S. Govt. Printing Office, Washington D.C.

CIESIELSKI, P.F., 1983: The Neogene diatom biostratigraphy of DSDP, Leg 71, piston cores. *Initial Reports of the Deep Sea Drilling Project, 71.* U.S. Govt. Printing Office, Washington D.C., in press.

CIESIELSKI, P.F. and WEAVER, F.M., 1974: Early Pliocene temperature changes in the Antarctic Seas. *Geology, 2,* 511-5.

CIESIELSKI, P.F., LEDBETTER, M.T. and ELLWOOD, B.B., 1982: The development of Antarctic glaciation and the Neogene paleoenvironment of the Maurice Ewing Bank. *Mar. Geol., 46,* 1-51.

HAUNG, T.C. and WATKINS, N.D., 1977: Antarctic Bottom Water velocity: Contrasts in the associated sediment record between the Brunhes and Matuyama epochs in the South Pacific. *Mar. Geol., 23,* 113-32.

IMMEL, R. and OSMOND, J.K.: Micromanganese nodules in deep sea sediments: Uranium-isotopic evidence for post-depositional origin. *Chem. Geol., 18,* 263-72.

KENNETT, J.P. and WATKINS, N.D., 1976: Regional deep-sea dynamic processes recorded by Late Cenozoic sediments of the southwestern Indian Ocean. *Geol. Soc. Am., Bull., 87,* 321-9.

LABRECQUE, J.L., KENT, D.B. and CANDE, S.C., 1977: Revised magnetic polarity time scale for Late Cretaceous and Cenozoic time. *Geology, 5,* 330-5.

LEDBETTER, M.T., 1979: Fluctuations of Antarctic Bottom Water velocity in the Vema Channel during the last 160,000 years. *Mar. Geol., 33,* 71-89.

LEDBETTER, M.T., 1981: Palaeo-oceanographic significance of bottom-current fluctuations in the Southern Ocean. *Nature, 294,* 554-6.

LEDBETTER, M.T. and CIESIELSKI, P.F. (this volume): Bottom-current erosion in the South Atlantic sector of the Southern Ocean.

LEDBETTER, M.T. and HUANG, T.C., 1980: Reduction of manganese micronodule distribution in the South Pacific during the last three million years. *Mar. Geol., 36,* M26-M28.

LEDBETTER, M.T., WILLIAMS, D.F. and ELLWOOD, B.B., 1978: Late Pliocene climate and southwest Atlantic abyssal circulation. *Nature, 272,* 237-9.

MERCER, J.H., 1976: Glacial history of southernmost South America. *Quat. Res., 6,* 126-66.

MOORE Jr., T.C., VAN ANDEL, Tj.H., SANCETTA, C. and PISAS, N., 1978: Cenozoic hiatuses in pelagic sediments. *Micropaleontology, 24,* 113-38.

OSBORN, N.I., CIESIELSKI, P.F. and LEDBETTER, M.T., 1983: Disconformities and palaeo-oceanography in the southeast Indian Ocean during the last 5.4 million years. *Geol. Soc. Amer., Bull.*

SHACKLETON, N.J. and OPDYKE, N.D., 1976: Oxygen-isotope and paleomagnetic stratigraphy of Pacific core V28-239 Late Pliocene to latest Pleistocene. *Geol. Soc. Am., Mem., 145,* 449-64.

SHACKLETON, N.J. and OPDYKE, N.D., 1977: Oxygen isotope and palaeomagnetic evidence for early Northern Hemisphere glaciation. *Nature, 270,* 216-9.

WATKINS, N.D. and KENNETT, J.P., 1972: Regional sedimentary disconformities and Upper Cenozoic changes in bottom water velocities between Australia and Antarctica. *Am. Geophys. Union, Ant. Res. Ser., 19,* 273-93.

# RECENT BMR MAGNETIC SURVEYS OF THE SOUTHEAST INDIAN OCEAN

H.M.J. Stagg, D.C. Ramsey and R. Whitworth, *Bureau of Mineral Resources, Geology and Geophysics, PO Box 378, Canberra, ACT, 2601, Australia.*

*Abstract* Since late 1979, the Australian Bureau of Mineral Resources, Geology and Geophysics, has been carrying out total magnetic field intensity surveys of the southeast Indian Ocean on the Danish polar supply vessel *Nella Dan*. To mid-1982, a total of approximately 110,000 km of good quality data has been acquired in the area between Melbourne and Hobart and the Australian Antarctic bases of Mawson and Davis. No detailed interpretation of these data has been performed yet as the data are still largely at an early stage of processing. It is anticipated that detailed interpretation will commence in 1983.

The relative positions of southeast Australia and the Mawson-Davis area have forced a generally northeast-southwest orientation on the *Nella Dan* tracks, oblique to the majority of previous ships tracks. Early indications are that this new direction will allow a more accurate delineation of the locations, trends and frequency of fracture zones and transform faults. Most of the lines proceed closer to the Antarctic margin than earlier data; this should allow better definition of the presently poorly defined anomalies close to Antarctica. Tracks into and out of Mawson are providing a gradual improvement in the magnetic coverage of the Enderby Basin between Enderby Land and the Gaussberg Ridge. It is hoped that data on these tracks will help elucidate the spreading history of this largely unknown area.

As a result of the surveys it may be possible to make a revised estimate of the secular variation of the Earth's magnetic field, by comparing magnetic values at line intersections of *Nella Dan* and earlier surveys (principally *Eltanin*). This should lead to a refined version of the Australian Geomagnetic Reference Field, a local improvement to the IGRF in an area where it has been found inadequate.

During the 1979/80 Antarctic summer season, the Australian Bureau of Mineral Resources (BMR) began a moderate Antarctic marine geoscience programme using the *Nella Dan*. It was intended to collect geophysical data over the Antarctic margin, the plateaus surrounding Australian sub-Antarctic Island Territories and in the southeast Indian Ocean. In the initial year magnetic data only were collected using a computer based data acquisition system. Despite teething problems, over 32,000 km of data were collected on four voyages, one to Macquarie Island and three to mainland Antarctica (Stagg et al., 1980). Although excellent coverage was obtained over the Kerguelen Plateau and in the southeast Indian Ocean, only about 1000 km of data were collected over the Antarctic margin because of adverse sea ice conditions.

Improvements were made to the equipment for the second season (1980/81) to increase the scientific value of the programme and the accuracy of the navigational data collected, and to reduce post-processing of data (Tilbury et al., 1980). Installation of BMR's more powerful Raytheon echo-sounder supplemented the existing echo-sounder which could obtain depths to only 3000 m; this allowed water depths to be obtained in the deep ocean basins as well as on the Antarctic margin and adjacent plateaus. In addition, the data acquisition system was coupled to the ship's satellite navigation system and provided positions every 10 seconds.

At the completion of the third, 1981/82, season an approximate total of 110,000 km of magnetic data had been collected along 14 traverses across the Southern Ocean and six traverses to Macquarie Island (Figure 1). Also for the 1981-82 season, a dedicated geoscience cruise on the *Nella Dan*, encompassing reflection seismic, magnetic, bathymetry and geological dredging and coring, was mounted between Davis and Mawson Stations, principally in Prydz Bay. This cruise is being reported separately (Stagg et al., this volume).

## AIMS AND BACKGROUND TO THE STUDY

The extent of present marine geoscientific knowledge in the southeast Indian Ocean is irregular and meagre. The average line spacing is about 300 km and, apart from the *Nella Dan* tracks, almost no lines extend up to the Antarctic continental margin and over the Enderby Basin southwest of the Kerguelen Plateau. The more detailed coverage

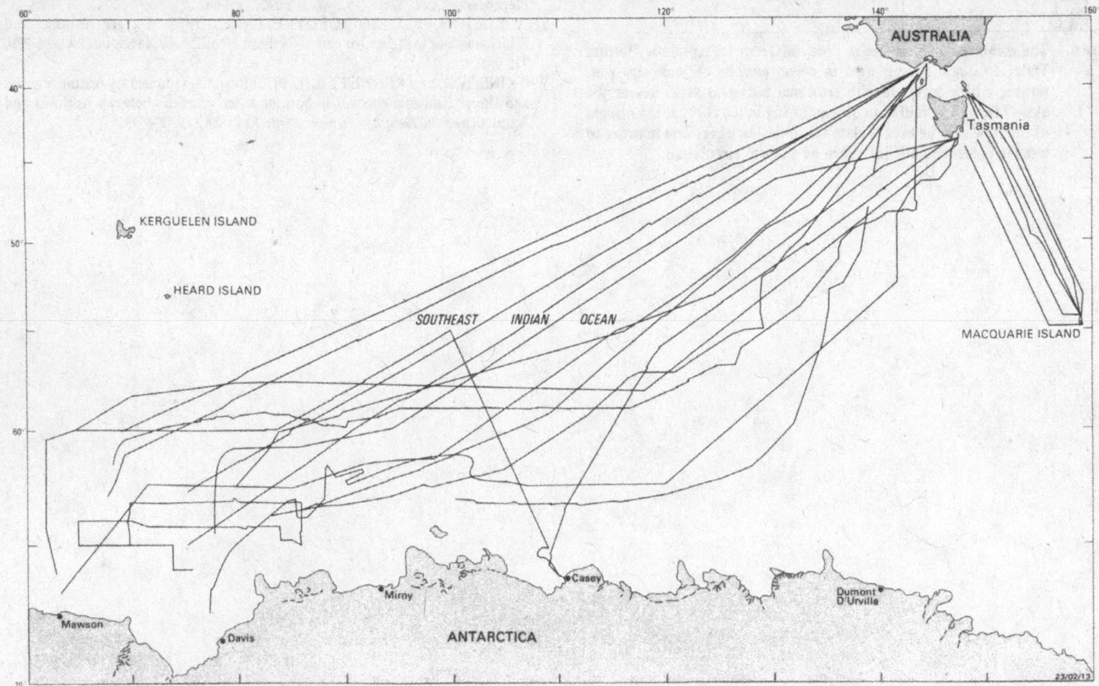


**Figure 1. Tracks of the M.V. *Nella Dan* across the Southern Ocean, 1979/82, on which magnetic data have been recorded (Standard Mercator projection). Tracks close to the Antarctic margin and other ships tracks have been omitted for clarity.**

is over the southeast Indian Ridge, which has attracted the interest of U.S. scientists and over the Kerguelen Plateau which has received some attention, from the French in particular but also from the Americans and Japanese.

The current BMR programme of magnetic measurements on the *Nella Dan* in the southeast Indian Ocean has the following aims:

(a) To contribute to the study of the continental margin of the Australian Antarctic Territory and of plateaus and rises surrounding Australia's Heard and Macquarie Islands. The Heard Island region in particular requires resolution of its tectonic history and possible economic potential. Studies are being oriented to further defining the structural framework and evolution of these regions using the adjacent magnetic anomaly patterns (Tilbury, 1981).

(b) To further define the structural elements and magnetic age of the southeast Indian Ocean region, in order to obtain a better understanding of the evolution of the Australian and Antarctic plates following the breakup of Gondwanaland.

(c) Secular variation studies over the southeast Indian Ocean. Further development of an Australian Geomagnetic Reference Field (AGRF) using estimates of secular variations obtained by comparison at intersections with tracks of earlier surveys, particularly those of the USNS *Eltanin*.

This programme for the present is distinct from the multidisciplinary geoscientific programmes on the *Nella Dan* being carried out on the continental margin of the Australian Antarctic Territory and which will be extended to include the island territories.

## SPECIFIC OBJECTIVES AND METHODS

### Plateaus Surrounding Heard Island and Macquarie Island

The Kerguelen-Gaussberg Plateau is a broad topographic high situated in the south-central Indian Ocean. It is about 2000 km in length, extending from 300 km north of Kerguelen more than 1000 km southeast of Heard Island, to within 800 km of the Australian Antarctic Territory. Studies by Houtz et al. (1977) have shown that much of the plateau appears to be covered by a veneer of sediments up to 1 km thick, with localised accumulations at least 2-3 km thick which are of potential economic interest. The structural framework and evolution of the plateau requires further study to assist in the definition and evaluation of the areas of thicker sedimentary section. In particular the magnetic anomaly patterns adjacent to the plateau margins which would place constraints of any interpretation of its evolution need to be further determined, especially to the southwest of the plateau in the Enderby Basin where there is a paucity of information.

The Macquarie Ridge is a complex feature which forms a narrow topographic high extending southwards some 1400 km from near the south of New Zealand to about latitude 60°S. A comprehensive study of the Macquarie Ridge complex undertaken by Hayes and Talwani (1972) concluded that it evolved as the consequence of interaction at the junction of the Pacific and Indian plates, where the predominant motion has been right-lateral strike-slip at about 4-5 cm/yr. Magnetic and bathymetric data are being collected over this complex feature to add to our knowledge of it, and also over the adjoining Tasman Sea basin where parts of the magnetic anomaly pattern require elucidation. Furthermore, to separate the predominantly north-trending anomalies of the Tasman Sea basin from the predominantly east-trending anomalies of the southeast Indian Ocean, a major tectonic boundary must exist and needs definition.

### Southeast Indian Ocean

Most of the data from the southeast Indian Ocean were collected on the *Eltanin* cruises between 1968 and 1972. Weissel and Hayes (1972) have identified the overall magnetic anomaly pattern, outlined the tectonic framework of the region and discussed the possible evolution of the region from the commencement of seafloor spreading, 55 Ma ago, to the present. However, the *Eltanin* data had a line spacing of about 300 km. While this is sufficient for broad definitions of major features, the definition of finer structures is left in doubt and often is some confusion. An example of the improvement provided by more detailed traversing can be seen in the results of the Project Investigator-1 aeromagnetic survey of the Australian-Antarctic Discordance (Vogt et al., 1979); the data have been processed and displayed at BMR. In that case, the line spacing was about 20 km. The magnetic anomaly pattern and the numerous fracture zones can be identified with confidence. In addition, two oblique sets of fractures not previously apparent are clearly seen to divide the region into a series of wedges. Preliminary studies of the *Nella Dan* data indicate the new traverse direction is going to give a much improved handle on the locations, trends and frequencies of fracture zones and transform faults.

Similarly, the magnetic pattern adjacent to the Antarctic margin and to the Kerguelen-Gaussberg Plateau is very poorly defined. Its definition is essential to an understanding of the tectonics of the margin and plateaus.

### Secular Variation Studies

As a secondary outcome of magnetic traverses across the southeast Indian Ocean, it may be possible to make an estimate of the secular variation in the Earth's magnetic field for this region. This will be determined by comparison of magnetic values at intersections with tracks of earlier surveys, primarily of the *Eltanin*, circa 1970. Because of the rather imprecise navigation used, a statistical comparison using all available intersections will be made.

There are two objectives in this. The first is to study the long wavelength variations in the magnetic field over the southeast Indian Ocean, their variation with time, correlation with depth, and other parameters (vide Vogt et al., 1979). The second is to provide a refined empirical model of the secular variation in the southeast Indian Ocean. This is required for effective integration of data collected on surveys made over a lengthy span of time (10 years or so). The current International Geomagnetic Reference Field (IGRF) had poor control in the Indian Oean region and has been shown to be inadequate; data on the Australian Northwest Shelf could not be accurately integrated, even though surveys were only four years apart (Petkovic, 1974). A preliminary version of an Australian Geomagnetic Reference Field, a modified IGRF, has been developed (Petkovic and Whitworth, 1975; Whitworth and Petkovic, 1977) and is in use by the Marine Geophysics Group at BMR.

## FUTURE WORK

Our present intention is to continue surveying at the present separation of about 100 km until 1985/86, adding four to six lines per year. At the end of this time we will have covered all of the southeast Indian Ocean that can be reached by the *Nella Dan* without adding unduly to the Australia-Antarctic transit time. Beyond 1985/86, future surveying will depend on whether it is justified by the results of the current work. Vogt et al. (1979) have shown the level of improvement that can be achieved with a 20 km line separation; and in the longer term this separation is our general aim. Most of the data acquired so far still remain to be processed. We intend to begin a detailed re-evaluation of the southeast Indian Ocean magnetic anomalies when this processing has been completed, probably in 1983.

*Acknowledgements* These magnetic surveys have been conducted under the auspices of the Australian National Antarctic Research Expeditions (ANARE) as a programme approved by the Antarctic Research Policy Advisory Committee (ARPAC). The survey vessel has been the M.V. *Nella Dan* of the Danish Lauritzen Line, variously under the command of Captain H. Klostermann and Captain J.B. Jensen. This paper is published with permission of the Director, Bureau of Mineral Resources, Australia.

## REFERENCES

HAYES, D.E. and TALWANI, M., 1972: Geophysical investigation of the Macquarie Ridge Complex. *Am. Geophys. Union, Ant. Res. Ser., 19*, 211-34.

HOUTZ, R.E., HAYES, D.E. and MARKL, R.G., 1977: Kerguelen Plateau bathymetry, sediment distribution and crustal structure. *Mar. Geol., 25*, 95-130.

PETKOVIC, J.J., 1974: Magnetic secular variation in the Australian region. *Aust., Bur. Miner. Resour., Geol. Geophys., Rep., 190*, 51-3.

PETKOVIC, J.J. and WHITWORTH, R., 1975: Problems in secular variation in the Australian region (Abstract). *EOS 56*, 547-58.

STAGG, H.M.J., FRASER, A.R. and DULSKI, R.A.W., 1980: Marine magnetic surveys aboard the M.V. *Nella Dan*, 1979/80—operational report. *Aust., Bur. Miner. Resour., Geol. Geophys., Rec., 1980/84.*

STAGG, H.M.J., RAMSAY, D.C. and WHITWORTH, R., (this volume): Preliminary report on a marine geophysical survey between Davis and Mawson Stations, 1982.

TILBURY, L.A., 1981: 1980 Heard Island expedition: marine geophysical operations and preliminary results. *Aust., Bur. Miner. Resour., Geol. Geophys., Rec., 1981/16.*

TILBURY, L.A., WHITWORTH, R., and STAGG, H.M.J., 1980: Operations manual for marine magnetic surveys on the M.V. *Nella Dan*, 1980/81. *Aust., Bur. Miner. Resour., Geol. Geophys., Rec., 1980/84.*

VOGT, P.R., FEDEN, R.H. and MORGAN, G.A., 1979: Project Investigator—1: a joint Australian/U.S. aeromagnetic survey of the Australian-Antarctic Discordance (Abstract). *International Union of Geology and Geophysics*, XVII General Assembly, Canberra, Australia, December 1979.

WEISSEL, J., and HAYES, D.E., 1972: Magnetic anomalies in the southeast Indian Ocean. *Am. Geophys. Union, Ant. Res. Ser., 19*, 165-96.

WHITWORTH, R. and PETKOVIC, J.J., 1977: A method of determining spherical harmonic coefficients for secular variation. Paper given at International Association of Geomagnetism and Aeronomy, Seattle, 1977.

# SEDIMENTARY DYNAMICS OF THE ANTARCTIC CONTINENTAL SHELF

J.B. Anderson, C. Brake, E.W. Domack, N. Myers and J. Singer, *Department of Geology, Rice University, Houston, Texas, 77001, U.S.A.*

*Abstract* Relict basal tills and glacial marine sediments of the Antarctic continental shelf underlie surficial sediments with only minor glacial components. Basal tills, which were deposited by lodgement processes, occur on the inner continental shelf of the Weddell Sea, Ross Sea and off the George V coast. These deposits grade seaward into diamictons, of glacial marine origin, on the outer continental shelf. The ice-rafted component of relict sediments decreases markedly a short distance beyond the shelf break. This abrupt facies boundary reflects the dominance of basal debris transport, and deposition of this debris landward of the calving line. Hence, significant glacial sedimentation on the continental shelf occurs only when the ice sheet is grounded there. Presently, glacial sedimentation on the continental shelf is minimal; siliceous biogenic muds and oozes are the most abundant surficial sediments on the continental shelf, exclusive of the Weddell Sea where terrigenous muds occur. These sediments contain only minor ice-rafted components. Shallow portions of the shelf (above approximately 250 m) are floored by residual sands and gravels which reflect intense bottom current activity. Most of the shelf is deeper than 400 m, so it is below the reach of these currents. Fine sediments entrained by bottom currents have little chance of escaping the continental shelf. This is because the continental shelf is highly irregular, glacial troughs almost completely encircle the East Antarctic continent, and the continental shelf typically slopes toward the continent, not offshore. Given the present lack of meltwater run-off, there is virtually no terrigenous sediment supplied by this mechanism. It is concluded that terrigenous sediment supply to the continental shelf is mostly limited to major glacial advances and to relatively warm periods when meltwater run-off is more pronounced.

During the past decade, considerable attention has been focused on continental shelf sedimentation, the result being important gains in our understanding of those processes which influence sedimentation on continental shelves, except, that is, the Antarctic continental shelf. It is so unique that sedimentation there is totally different from that of other continental shelves.

The most important differences between the Antarctic continental shelf and other continental shelves are:

(1) it is ice covered most of the year, and in many areas throughout the year,

(2) it is quite deep (average depth 450 m), and has considerable topographic relief, due mainly to glacial erosion, and

(3) there is no fluvial input nor is there a wave dominated coastal zone. Glaciers deliver unsorted sediment directly to the deep continental shelf.

It is these unique qualities which inspired a long term research programme aimed at investigating sedimentary dynamics on the Antarctic continental shelf. During the past five years, marine geologic surveys have been conducted in the Weddell Sea, Ross Sea, Bransfield Strait and along the northern Victoria Land coast and the George V coast (Figure 1). Thousands of kilometres of bathymetric profiles, some high resolution seismic reflection data, and hundreds of piston cores and bottom grab samples have been obtained on the continental shelves in these areas. Analyses of these cores and grab samples have led to a number of important observations about sedimentation on the continental shelf. This paper provides a brief

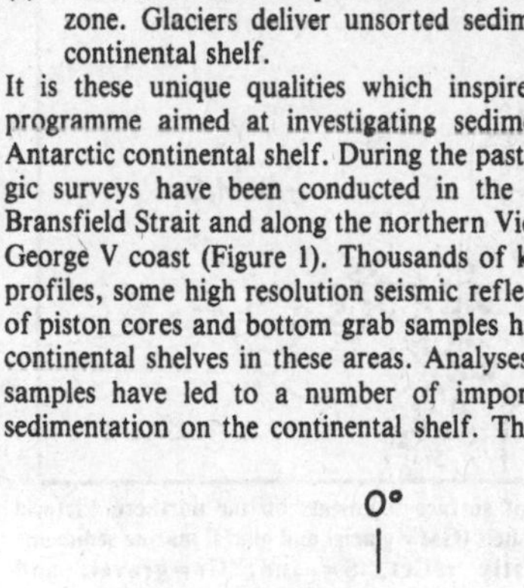

**Figure 1. Locations of continental shelves studied to date. Areas are numbered so as to correspond to Figures 2 to 5 which show surface sediment distribution maps.**

synthesis of these findings. Sediment distribution maps (Figures 2, 3, 4 and 5) provide the primary source of information for this paper.

## RELICT GLACIAL DEPOSITS

Piston cores obtained from the Antarctic continental shelf typically consist of relict glacial sediments overlain by, and in sharp contact with, marine sediments containing only sparse ice-rafted debris. Glacial sediments are almost entirely terrigenous and include both basal tills (sediments deposited by grounded ice) and glacial-marine deposits (sediments derived wholly or in part from ice shelves and/or icebergs). The two are distinguished using a variety of criteria which have been described by Anderson et al., (1980). Basal tills occur on the continental shelves of the Ross Sea, Weddell Sea and George V coast (Anderson et al., 1980; Domack, 1982). They grade seaward into glacial marine sediments which show an increasing marine influence (in the form of marine fossils, stratification, and sorting) in an offshore direction. At the continental slope, the concentration of ice-rafted debris relative to the marine component of glacial marine sediments decreases markedly. Glacial marine sediments of the continental shelf and slope typically contain greater than 50% ice-rafted debris, while continental rise and abyssal sediments normally contain only a few percent ice-rafted debris (Anderson, et al., 1979).

In Antarctica, glacial transport of sediment is in basal debris zones; only in mountainous regions has englacial and supraglacial sediment transport perhaps been significant. The abrupt decrease in ice-rafted debris seaward of the shelf edge implies that these basal debris are deposited near the grounding line of the ice sheet, with little debris making its way to the calving line. This is supported by thermodynamic models for ice shelves (Thomas, 1979; Robin, 1979) which conclude that basal ice melts before reaching the calving line of the ice shelf. This implies that significant quantities of terrigenous sediment are delivered to the sea floor by glaciers only during a major glacial advance. Subglacial sedimentation is primarily by lodgement (Domack et al., 1980).

## SURFICIAL SEDIMENTS

Surficial sediments of the Antarctic continental shelf are strikingly different from the glacial deposits they overlie. The most obvious difference being the greater concentration of marine versus ice-rafted components. In fact, one of the most widespread surficial sediment types, and by far the most abundant based on the thickness of cored units, is siliceous ooze and siliceous mud (Figures 2, 3, 4 and 5). Most of the deep glacial troughs of the continental shelf are presently being filled by these siliceous sediments. They are seldom penetrated by our 6 m corer, and high resolution seismic data indicate that they are several tens of metres thick. These siliceous sediments normally contain less than 30% terrigenous sediment and most of that is in the form of fine silt and clay, not unsorted ice-rafted debris. Normally, the biogenic component of these sediments increases with increasing water depth.

Cores taken within 20 km of the icefront of the George V coast penetrated siliceous ooze, and even those troughs, fiords, and tectonic basins bound by the mountainous coastal regions of northern Victoria

Land and the northern Antarctic Peninsula are floored by siliceous muds and oozes (Figures 3 and 5). Only in Weddell Sea Basins, such as Crary Trough, are siliceous sediments not present (Figure 4.). The reason(s) for their absence there is still problematic.

Shallower banks and ridges of the Antarctic continental shelf are mostly blanketed by sands and gravels (Figures 2, 3, 4 and 5). These sediments are comprised mainly of residual glacial debris and calcareous shell hash which has survived winnowing by currents strong enough to remove up to fine sand-sized particles. They are associated with well-sorted sands which represent traction current deposits. Silts and clays are swept onto the deeper portions of the shelf or across the shelfbreak where they accumulate along with siliceous biogenic sediments.

The continental shelf of northern Victoria Land is the shallowest part of the Antarctic continental shelf surveyed to date; the shelf itself has an average depth of around 200 m. Here the influence of bottom currents is most obvious. Volcanic sands eroded from Adare Peninsula have been transported to the west, across the continental shelf (Brake, 1982) nearly 30 km from their source (Figure 3).

The maximum depth at which strong bottom current activity is recorded in bottom sediments is generally between 200 and 250 m. The current responsible for this sediment erosion and transport is probably the East Wind Drift, although actual bottom current measurements on the continental shelf are too scarce to substantiate this. Most of the Antarctic continental shelf is situated below 400 m, and hence is below the reach of waves and wind driven currents. In the absence of these agents, fine-grained terrigenous and biogenic sediments accumulate.

The irregular glacial topography of the Antarctic continental shelf is of extreme importance in regulating sediment dispersal and gravity related transport. Relief on the shelf is typically sufficient to maintain sediment mass movement. Thus, it is not surprising that slumps, debris flows and turbidites are rather common deposits of the shelf (Figures 2, 3, 4 and 5). In fact, these processes play a key role in redistributing glacial deposits and are even remarkably efficient at sorting these glacial sediments (Wright and Anderson, in press).

Cross-shelf transport of suspended sediment, which is typical of most continental shelves, is the exception rather than the rule in Antarctica. In most areas, the continental shelf slopes toward the continent, not offshore. Also, the continental shelf has numerous glacial troughs which act as sedimentary basins. Most of these troughs are over 1000 m deep. On the East Antarctic continental shelf, large glacial troughs extend subparallel to the coast in an en echelon manner, forming an almost continuous ring of sedimentary basins around the continent (Volokitina, 1975; Vanney and Johnson, 1979). Therefore, only that sediment transported seaward in surface waters and that which is rafted out to sea in icebergs reaches the shelf edge. Piston cores taken at the shelf break—upper slope typically penetrated either outcrops or residual sands and gravels (Figures 2, 3, 4 and 5). These deposits result from strong circumpolar currents impinging on a sediment starved slope.

The widespread and abundant occurrence of siliceous mud and ooze on the continental shelf indicates that the continental shelf, as well as the continental slope, is, in fact, presently receiving very little

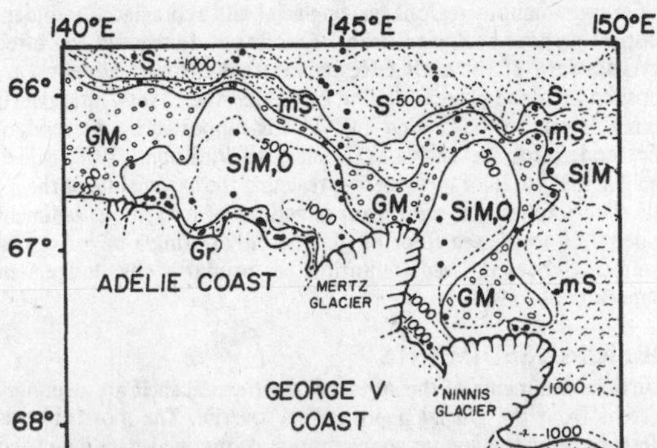


**Figure 2. Surface sediment distribution map for the Adelie-George V continental shelf (GM = glacial and glacial marine sediments which are mostly relict, MS = muddy sand, S = sand, Grb = bioclastic gravel, and SiM,O = siliceous mud and ooze). The 500 m and 1000 m depth contours are also shown. Siliceous muds and oozes occur in depressions at depths below approximately 450 m. They are mostly laminated. Shallower parts of the continental shelf are floored by glacial and glacial marine sediments which are mostly relict. These relict sediments are presently being reworked by bottom currents along the shelf edge. Muddy sands of the outer continental shelf show moderate bottom current winnowing. Sands of the outer continental shelf result from strong, impinging circumpolar currents. Gravels were collected along the ice front at depths shallower than 200 m. These gravels are derived through wave erosion of coastal ice walls. See Figure 1 for area location (Area 2).**

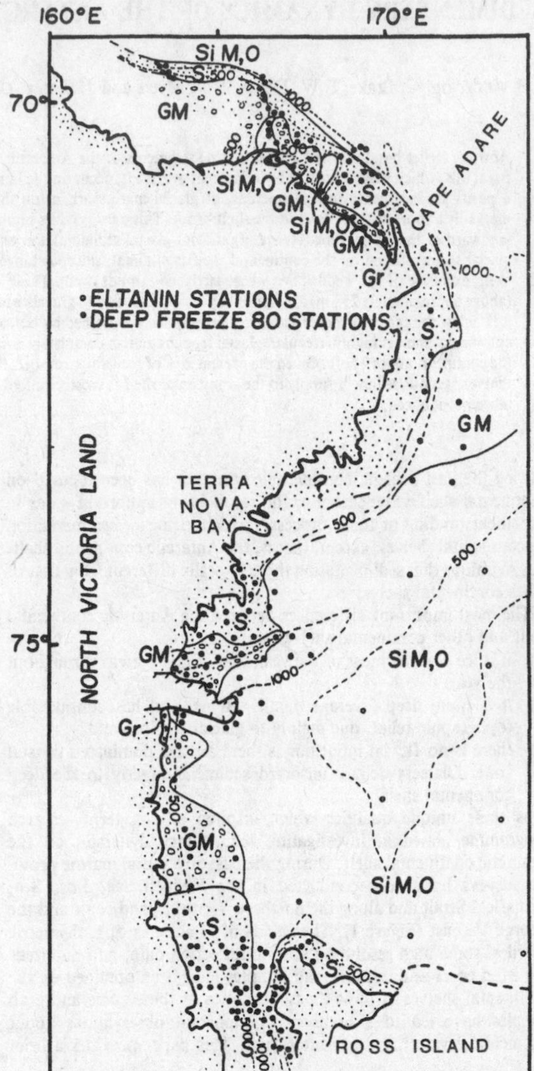


**Figure 3. The distribution of surface sediments on the northern Victoria Land continental shelf (GM = glacial and glacial marine sediments which are mostly relict, S = sand, Gr = gravel, and SiM,O = siliceous mud and ooze). The 500 m and 1000 m depth contours are also shown. Siliceous muds and oozes overlie relict glacial and glacial marine deposits over most of the Ross Sea floor. Siliceous oozes also occur in glacial troughs on the shelf. Relict glacial sediments are exposed at the sea floor along most parts of the coast. Sands of the Terra Nova Bay and Ross Island areas are turbidites and consist of primarily volcanic debris. They grade downward into gravels. Sands of the Cape Adare area are traction current deposits derived from the cape and transported to the west. They too are volcanic. Sands also occur along the outer continental shelf in the northern part of the area and reflect strong bottom current activity in this area, probably impinging circumpolar currents.**

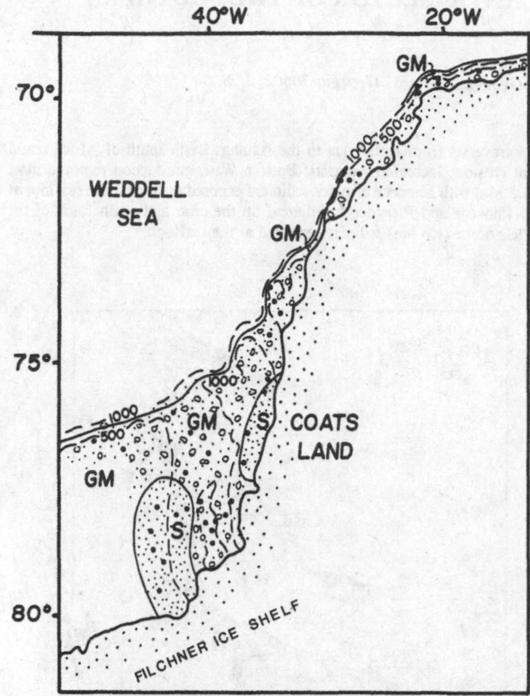


**Figure 4. Distribution of surface sediments on the continental shelf of the Eastern Weddell sea (GM = glacial and glacial marine sediments which are mostly relict, and S = sand). The 500 m and 1000 m depth contours are also shown. Unlike other parts of the continental shelf which have been surveyed to date, the Weddell Sea continental shelf sediments contain only minor siliceous material. Glacial and glacial marine sediments are exposed at the sea floor or are buried beneath a few centimetres of mud. Sands of the Coats Land area are traction current deposits and occur at depths shallower than 300 m. Sands of the Filchner Ice Shelf area include turbidites that were deposited along the steep flank of a large glacial trough, and relict eolian sands that are exposed on the shelf itself.**

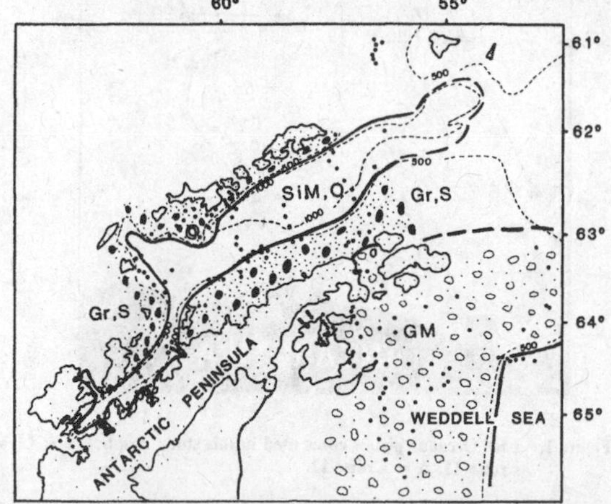


**Figure 5. Sediment distribution in the Northern Antarctic Peninsula area (GM = glacial and glacial marine sediments which are mostly relict, S = sand, Gr = gravel, and SiM,O = siliceous mud and ooze, M = mud). The 500 m and 1000 m depth contours are also shown. Siliceous muds and oozes, which are mostly laminated, occur at depths below approximately 500 m north of the Antarctic Peninsula, but siliceous sediments are absent in the Weddell Sea area. Sands and gravels cover the shallow portions of the continental shelf.**

terrigenous sediment from the continent. Even cores taken in the fiords and bays of the northern Antarctic Peninsula region penetrated siliceous muds and oozes, not meltwater deposits (Figure 3). This is attributed to the present glacial regime of Antarctica, under which virtually no meltwater flows from the continent. Consequently, surface waters contain little or no fine terrigenous sediment. Ice-rafting is the only means by which terrigenous sediment is delivered to the continental margins and this mechanism is pretty much inactive today. From this, one might infer that thick, relict silt deposts that occur on the Antarctic continental slope and rise, which contain very little ice-rafted debris, were deposited at a time when meltwater input was much greater, and hence when the climate was much warmer.

From these observations, we might conclude by stating that significant quantities of terrigenous sediment are delivered to the Antarctic seafloor only during major glacial advances, at which time basal tills and glacial marine sediments are deposited on the shelf, and during significant warming events, when meltwater contributions become significant. The present climatic regime is intermediate between these two climatic extremes, and therefore terrigenous sediment supply to the continental shelf is minor.

*Acknowledgements* Financial support for this research was provided by the National Science Foundation-Division of Polar Programmes (grants DPP77-26407 and DPP-80242) and the American Chemical Society-Petroleum Research Fund (grants PRF-11101-AC2 and PRF-2472-AC2). We wish to thank the officers and crew of the USCGC *Glacier* for their assistance during our marine geologic cruise.

## REFERENCES

ANDERSON, J.B., KURTZ, D.D. and WEAVER, F.M., 1979: Sedimentation on the Antarctic continental slope; *in* Pilkey, O., and Doyle, L. (eds.), Geology of Continental Slopes, *Soc. Econ. Paleont. Mineral., Spec. Publ., 26,* 265-83.

ANDERSON, J.B., KURTZ, D.D., DOMACK, D.W. and BALSHAW, K.M., 1980: Glacial and glacial marine sediments of the Antarctic continental shelf. *J. Geol., 88,* 399-414.

BRAKE, C., 1982: Marine geology of the North Victoria Land continental margin. Unpublished M.A. thesis, Rice University, Houston, Texas, 75pp.

DOMACK, E.W., 1982: Sedimentology of glacial and glacial-marine deposits on the George V-Adélie continental shelf, East Antarctica. *Boreas, 11,* 79-97.

DOMACK, E.W, ANDERSON, J.B. and KURTZ, D.D., 1980: Clast shape as an indicator of transport and depositional mechanisms in glacial-marine sediments: George V continental shelf, Antarctica. *J. Sediment. Petrol., 50,* 813-20.

ROBIN, G., de Q., 1979: Formation, flow, and disintegration of ice shelves. *J. Glaciol., 24,* 259-71.

THOMAS, R.H., 1979: Ice shelves: a review. *J. Glaciol., 24,* 273-86.

VANNEY, J.R. and JOHNSON, G.L., 1979: The sea floor morphology seaward of Terre Adélie (Antartica). *Dtsch. Hydrogr. Zeitschr., 32,* 39-88.

VOLOKITINA, L.P., 1975: Some features of the underwater margin of East Antarctica. *Oceanology, 15,* 323-26.

WRIGHT, R. and ANDERSON, J.B., 1982: The importance of sediment gravity flow to sediment transport and sorting in a modern glacial marine environment: Eastern Weddell Sea, Antarctica. *Geol. Soc. Am.. Bull., 93,* 951-63.

# BOTTOM-CURRENT EROSION IN THE SOUTH ATLANTIC SECTOR OF THE SOUTHERN OCEAN

M.T. Ledbetter and P.F. Ciesielski, *Department of Geology, University of Georgia, Athens, Georgia 30602, U.S.A.*

*Abstract* The magnetostratigraphy and biostratigraphy of 61 piston cores on two traverses from Antarctica to the Agulhas Basin south of Africa reveal disconformities in the sedimentary record which are attributed to bottom-current erosion. Increased Antarctic Bottom Water production removed most sediment at the eastern end of the Weddell Basin younger than Middle Pliocene (2.5 Ma) with Miocene or older sediment exposed at or near the seafloor at some locations. Increases in the velocity of the Circumpolar Deep Water eroded Pliocene and Pleistocene sediment on the crest and south flank of the Southwest Indian Ridge and maintains an active scour zone today with Middle Pleistocene (1.0 Ma) sediment exposed at the seafloor.

Modern bottom-currents in the Southern Ocean maintain disconformities on both abyssal plains (Watkins and Kennett, 1971; Kennett and Watkins, 1976; Ledbetter and Ciesielski, 1982; Osborn et al., 1983) and shallow rises (Ciesielski and Wise, 1977; Ciesielski et al., 1982; Osborn et al., 1983). In addition to modern disconformities at the seafloor, erosion of sediment in the geologic record is responsible for widespread hiatuses in the Neogene sedimentary record (Ciesielski and Wise, 1977; Dingle and Cambden-Smith, 1978; Ciesielski et al., 1982; Osborn et al., 1983). Erosion of the deep sections is attributed to periods of increased Antarctic Bottom Water (AABW) production while erosion of shallow rises is attributed to increased velocity of Circumpolar Deep Water (CDW). With an increasing knowledge of the timing of both erosional and palaeoclimatic events, it is now possible to infer the palaeo-oceanographic response of deep and intermediate circulation to climatic changes. In this paper and Ledbetter et al. (this volume), the history of bottom-current erosion in the Atlantic and Indian sectors, respectively, of the Southern Ocean are examined.

The magnetostratigraphy and biostratigraphy of 61 piston cores from the *Islas Orcadas* collection have been determined. The cores fall in two traverses from near the Queen Maud Land coast of Antarctica to the Agulhas Basin south of Africa (Figure 1). All major physiographic provinces on the traverses are sampled and the depth range (3127-5274 m) of the cores falls within both AABW and CDW. The biostratigraphic age of each core or core segment was determined by using the diatom zonation of Ciesielski (1983) and the silicoflagellate zonation of Ciesielski (1975) as revised (Weaver, 1976; Busen and Wise, 1977). The biostratigraphic zonation was used to assign the palaeomagnetic polarity (determined at 10 cm intervals) to polarity chrons of the established magnetostratigraphic time scale (see Ledbetter and Ciesielski, 1982 for a full discussion of methods). The resulting age determinations reveal disconformities both at the seafloor and within the geologic record (Figure 2).

## MODERN DISCONFORMITIES

The age of sediment at or very near the seafloor reveals that the present velocity of bottom-currents in the South Atlantic sector of the Southern Ocean is sufficiently high to erode or inhibit deposition of sediment in three areas. The most recent sediment from all but the deepest part of the Weddell Basin is Lower Pliocene to Miocene (Figure 2) and forms the Weddell Basin Scour Zone. Erosion of sediments near the crest of the Southwest-Indian Ridge has exposed Lower Pleistocene sediments in the uppermost sediment column. Sediment on the Maud Rise is Lower Pliocene at the seafloor and on the adjacent Astrid Ridge one core recovered Miocene sediment at the seafloor (Figure 2).

Production of AABW within the Weddell Sea has created a dense contour-following bottom-current (Hollister and Elder, 1969) which is currently eroding sediments from an area of the Weddell Basin crossed by our traverses (Figure 1). The area near the Maud Rise and Astrid Ridge where AABW forms by deep vertical convection currents (Gordon, 1978) has the longest duration disconformity with Lower Pliocene and older sediment exposed at or near the seafloor. Scour within the deep, northern Weddell Basin has exposed Upper Pliocene sediment (Figure 2) and is caused by AABW produced in the Weddell Sea which is confined to the deepest part of the basin and flows north reaching the Atlantic-Indian Ridge where it is deflected to the east. The intensification of AABW in the deepest part of the basin causes the erosion of sediment in the Weddell Basin with a northern limit of scour at the base of the ridge (Figure 2). The Weddell Basin, including the area of the Maud Rise, is designated as a scour zone since erosion or winnowing by AABW has removed sediment deposited since the Miocene to Early Pliocene.

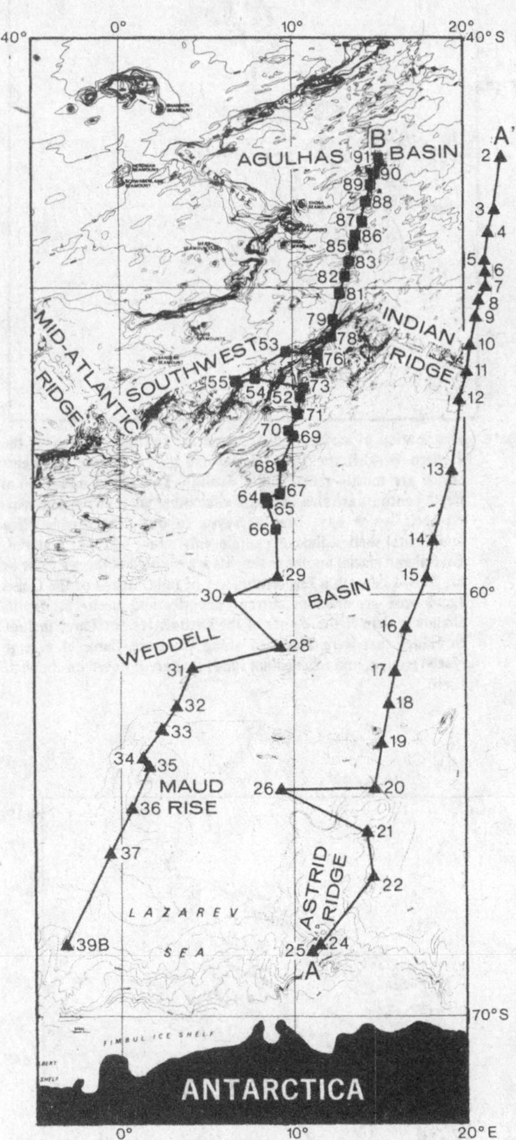


**Figure 1. Islas Orcadas piston cores used in this study. Symbols are: □ = Cruise 11, Δ = Cruise 12.**

A third, large scour zone created by high velocity bottom-currents is found on the crest of the Atlantic-Indian Ridge (Figure 2). This shallow disconformity is separated from the Weddell Basin scour zone by a zone of sedimentation along the south flank of the ridge (Figure 2). The disconformity is maintained by high velocity CDW which has eroded or inhibited deposition at the crest of the ridge leaving Lower to Middle Pleistocene sediments at the seafloor.

## PALAEOCURRENT EROSION

In addition to the disconformities at the present seafloor, three major hiatuses were identified within the sedimentary record (Figure 2). Several short duration disconformities were identified; however, the limited core coverage prohibits the definition of the lateral extent or regional significance of these hiatuses.

Four cores near the crest and south flank of the Southwest Indian Ridge on the western traverse (Figure 1) have a hiatus in the sedimentary record from the Middle Pliocene to Early-Middle Pleistocene (Figure 2). These cores are from water depths less than 3850 m and are presently in the axis of high velocity CDW which has left a disconformity at the seafloor. The scour zone on the Southwest Indian Ridge during the Middle Pliocene to Early Pleistocene represents an increase in the velocity of CDW within an area similar to the present core of high velocity CDW. The temporal fluctuations of percent cores with a hiatus reveals the timing of major disconformities on the Southwest Indian Ridge (Figure 3); a water depth of 4200 m was used to distinguish between AABW and CDW erosion. The CDW velocity began waning by the Olduvai Subchron (1.8 Ma) and decreased sufficiently to allow renewed deposition shortly before the Jaramillo Subchron (1.0 Ma). The initiation of scour cannot be determined since an increase in hiatus frequency may be an artifact of the depth of erosion rather than initiation of scour. However, the high palaeovelocity episode began after the middle Gauss Chron (3.0 Ma) and the hiatus is very similar in time to one identified on the Maurice Ewing Bank at a similar water depth (Ciesielski et al., 1982). Other widespread hiatuses on the Southwest Indian ridge occurred during the late Gilbert Chron (>4.5 Ma) (Figures 2, 3). The major decreases in CDW velocity occurred 4.5 and 3.0 Ma (Figure 3). The timing of these hiatuses are very similar to ones identified on the Maurice Ewing Bank (Ciesielski et al., 1982) and in the southeast Indian Ocean (Osborn et al., 1983).

Buried disconformities in the Weddell Basin (>4200 m) reveal a complex series of hiatuses created by AABW (Figures 2, 3). Major decreases in hiatus frequency occur at 4.5 and 2.0 Ma (Figure 3). the disconformity at 4.5 Ma was more widespread than at present in the Weddell Basin since it extended further north to the site of core 1011-64 (Figure 2). On the Maud Rise and Astrid Ridge, however, sedimentation occurred at this time at the sites of cores 1012-37 and 1012-22, respectively (Figure 2). Because those sites are presently in the Weddell Basin Scour Zone, the bottom-current erosion which was expanded to the north until 4.5 Ma did not extend as far south as at present.

The reduction in AABW scour at 2.0 Ma (Figure 3) is a very complex pattern (Figure 2). Instead of a shrinking of AABW away from the margins of the Weddell Basin as was the case at 4.5 Ma, the reduction in velocity occurred in the deepest (most central) part of the basin. Cores 1012-16, 17 on the eastern traverse and cores 1012-29, 30 on the western traverse record the 2.0 Ma reduction of AABW velocity as a period of deposition while to the north and south the erosion persisted. This relationship suggests that AABW flowed out of the basin as a contour current and the deep return-flow was also a contour current around the Astrid Ridge and Maud Rise.

## PALAEOCLIMATIC EFFECTS ON BOTTOM-CURRENT EROSION

The intensification of bottom circulation which results in disconformities may be caused by climatic deterioration (Kennett and Watkins, 1976; Ciesielski et al., 1982). The exact age of the initial increase in bottom-water velocity cannot be determined, however, since previously deposited sediment may be eroded. Therefore, only the cessation of scour may be dated by the top of the hiatus. The age span of the disconformities in this region encompasses some major climatic coolings which could be responsible for intensification of bottom circulation. The waning of deep and shallow erosion 4.5 Ma (Figure 3) corresponds to an Early Pliocene climatic warming (Hays and Opdyke, 1967; Ciesielski and Weaver, 1974) which may be associated with a marine invasion of Taylor Valley, Antarctica (Webb, 1972). The proposed partial deglaciation of West Antarctica at this time would lead to a reduction of AABW production and the reduced temperature gradient would lessen wind-stress and concomitantly CDW. Therefore, the Late Miocene to Early Pliocene erosion in both shallow and deep sections in the south Atlantic ceased due to a warming of the Antarctic margin.

The increased velocity of both AABW and CDW during the period 3.8-3.2 Ma resulted in widespread erosion. The increased bottom-current velocity occurred at the same time as a major cooling trend in the Southern Ocean (Ciesielski and Weaver, 1974) and in Argentine Patagonia (Mercer, 1976). This cooling has been correlated with the re-establishment of the West Antarctic ice sheet (Ciesielski and Weaver, 1974) and would result in increased production of AABW and an increased wind stress on CDW due to the steepening of the temperature gradient. The reduction of erosion at approximately 2.8-3.2 Ma (Figure 3) corresponds to a late Gauss warming (Anderson, 1972; Fillon, 1975). The reduction of hiatus frequency at ~3.2 Ma occurs at the time Northern Hemisphere glaciations began (Shackleton and Opdyke, 1977) and corresponds to a similar decrease in bottom-water velocity in the Indian Ocean (Ledbetter, 1981).

The third major disconformity in both shallow and deep sections occurred between 2.5 and 1.0 Ma when cooling climatic conditions in the Weddell Sea (Anderson, 1972) and the Ross Sea (Fillon, 1975) resulted in widespread erosion of deep and shallow sections in every sector of the Southern Ocean (e.g. Huang and Watkins, 1977; Ciesielski et al., 1982). The widespread extent of hiatuses and the severity of glacial conditions on Antarctica and in the Argentine Patagonia (Mercer, 1976) at this time suggest an intense cooling of the

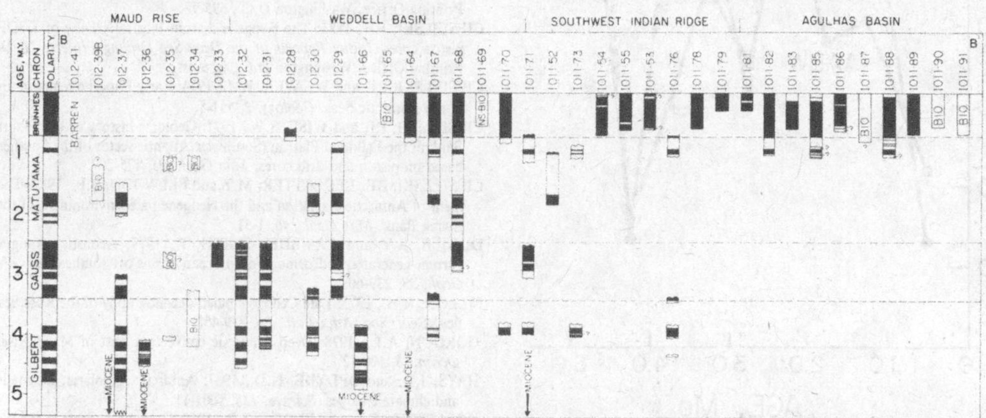


**Figure 2.** **The magnetostratigraphy of cores in Figure 1 is assigned an age based on biostratigraphic correlation to the time-scale. Disconformities at the seafloor and in the geologic record are revealed by hiatuses in the record. Symbols used are: ? in polarity log = no vertical component of magnetisation (therefore no polarity assignment); ? to side of polarity log = age assignment uncertain due to poor preservation or scarcity of microfossils; NS = not sampled for palaeomagnetics due to core disturbance; BIO = biostratigraphic age assignment only, — = disconformity, — = break in core but no disconformity.**

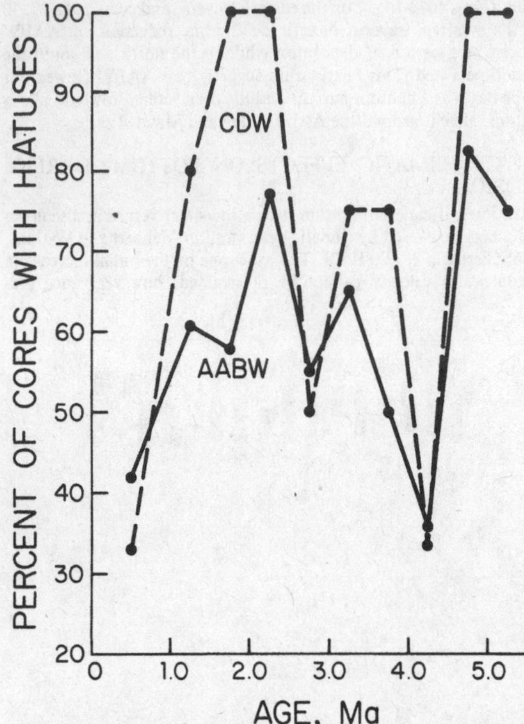


**Figure 3. The percent of core segments with a hiatus was determined for each age interval recovered in the cores. Hiatuses were separated into two groups based on water depth; greater than 4200 m = erosion by Antarctic Bottom Water (AABW) and less than 4200 m = erosion by Circumpolar Deep Water (CDW). The three periods of increased hiatus frequency correspond to three major disconformities found in the area.**

climate which increased AABW production and intensified the wind-driven CDW through steepened temperature gradients. The reduction of bottom-water velocity about 1.0 Ma is a common feature in the Atlantic (Ciesielski et al., 1982), Pacific (Ledbetter and Huang, 1980) and Indian (Osborn et al., 1983) sectors of the Southern Ocean. Since there are no known significant climatic events in the Southern Ocean at this time, the reduction of bottom-water activity may be the result of a change in the Northern Hemisphere glaciations.

## CONCLUSIONS

The magnetostratigraphy and biostratigraphy of piston cores on two traverses between Antarctica and the Agulhas Basin reveal disconformities at the seafloor and in the sedimentary record of both shallow and deep sections. Shallow disconformities at the present seafloor are caused by erosion on the crest of the Southwest Indian Ridge by Circumpolar Deep Water. Disconformities at the seafloor in the Weddell Basin are caused by erosion by Antarctic Bottom Water. Widespread buried disconformities occur at both shallow and deep sites and are attributed to major Antarctic coolings which resulted in increased bottom-current circulation. These events culminated at 4.5-5.0, 3.2-3.8 and 1.5-2.5 Ma.

*Acknowledgements* We thank Dennis Cassidy of the Antarctic Core Storage Facility at Florida State University for help in sampling *Islas Orcadas* cores. Support was received from NSF Grants DPP79-05111 and DPP81-13147.

## REFERENCES

ANDERSON, J.B., 1972: The marine geology of the Weddell Sea. *Florida State Univ., Sedim. Res. Lab., Rep., 35.*

BUSEN, K.E. and WISE, S.W., 1977: Silicoflagellate stratigraphy. *Initial Reports of the Deep Sea Drilling Project, 26,* U.S. Govt. Printing Office, Washington D.C., 697-743.

CIESIELSKI, P.F., 1975: Neogene and Oligocene silicoflagellates from cores recovered during Antarctic Leg 28, Deep Sea Drilling Project: Biostratigraphy and paleocology. *Initial Reports of the Deep Sea Drilling Project, 28.* U.S. Govt. Printing Office, Washington D.C., 695-92.

CIESIELSKI, P.F., 1983: The Neogene diatom biostratigraphy of DSDP, Leg 71, piston cores. *Initial Reports of the Deep Sea Drilling Project, 71.* U.S. Govt. Printing Office, Washington D.C., in press.

CIESIELSKI, P.F. and WEAVER, F.M., 1974: Early Pliocene temperature changes in the Antarctic Seas. *Geology, 2,* 511-5.

CIESIELSKI, P.F. and WISE, S.W., 1977: Geologic history of the Maurice Ewing Bank of the Falkland Plateau (Southwest Atlantic sector of the Southern Ocean) based on piston and drill cores. *Mar. Geol., 25,* 175-207.

CIESIELSKI, P.F., LEDBETTER, M.T. and ELLWOOD, B.B., 1982: The development of Antarctic glaciation and the Neogene paleoenvironment of the Maurice Ewing Bank. *Mar. Geol., 46,* 1-51.

DINGLE, R.V. and CAMBDEN-SMITH, F., 1979: Acoustic stratigraphic and current-generated bedforms in deep ocean basins off southeastern Africa. *Mar. Geol., 33,* 239-60.

FILLON, R.H., 1975: Late Cenozoic paleo-oceanography of the Ross Sea, Antarctica. *Geol. Soc. Am., Bull., 86,* 839-45.

GORDON, A.L., 1978: Deep Antarctic convection west of Maud Rise. *J. Phys. Ocean., 8,* 600-12.

HAYS, J.D. and OPDYKE, N.D., 1967: Antarctic radiolaria, magnetic reversals and climate changes. *Science, 158,* 1001-11.

HOLLISTER, C.D. and ELDER, R.B., 1969: Contour currents in the Weddell Sea. *Deep Sea Res., 16,* 99-101.

HUANG, T.C. and WATKINS, N.D., 1977: Antarctic Bottom Water velocity: contrasts in the associated sediment record between the Brunhes and Matuyama epochs in the South Pacific. *Mar. Geol., 23,* 113-32.

KENNETT, J.P. and WATKINS, N.D., 1976: Regional deep sea dynamic processes recorded by Late Cenozoic sediments of the southwestern Indian Ocean. *Geol. Soc. Am., Bull., 87,* 321-9.

LEDBETTER, M.T., 1981: Palaeo-oceanographic significance of bottom-current fluctuations in the Southern Ocean. *Nature, 294,* 554-56.

LEDBETTER, M.T. and CIESIELSKI, P.F., 1982: Bottom-current erosion along a traverse in the South Atlantic sector of the Southern Ocean. *Mar. Geol., 46,* 329-42.

LEDBETTER, M.T. and HUANG, T.C., 1980: Reduction of manganese micronodule distribution in the South Pacific during the last three million years. *Mar. Geol., 36,* M26-M28.

LEDBETTER, M.T., CIESIELSKI, P.F., OSBORN, N.I. and ALLISON, E.T. (this volume): Bottom-current erosion in the southeast and Indian Pacific Oceans during the last 5.4 Ma.

OSBORN, N.I., CIESIELSKI, P.F. and LEDBETTER, M.T., 1983: Disconformities and palaeo-oceanography in the southeast Indian Ocean during the last 5.4 Ma. *Geol. Soc. Am., Bull,* in press.

SHACKLETON, N.J. and OPDYKE, N.D., 1977: Oxygen isotope and palaeomagnetic evidence for early Northern Hemisphere glaciation. *Nature, 270,* 216-9.

WATKINS, N.D. and KENNETT, J.P., 1971: Antarctic bottom-water: major change in velocity during the Late Cenozoic between Australia and Antarctica. *Science, 173,* 813-8.

WEAVER, F.M., 1976: Late Miocene and Pliocene radiolarian paleobiogeography and biostratigraphy of the Southern Ocean. Unpublished Ph.D. Dissertation, Florida State University, Talahassee, Florida. 175 pp.

WEBB, P.N., 1972: Wright Fjord, Pliocene marine invasion of an Antarctic dry valley. *U.S. Antarct. J., 7,* 225-32.

# MESOZOIC TO HOLOCENE BIOSTRATIGRAPHIC FRAMEWORK FOR THE FALKLAND PLATEAU AND SOUTHEAST ARGENTINE BASIN

I. A. Bassov, *USSR Academy of Sciences, Moscow, USSR.*

P. F. Ciesielski, *Department of Geology, University of Georgia, Athens, Georgia 30602, USA.*

A. M. Gombos, *Exxon Production Research Co., PO Box 2189, Houston, Texas 77001, USA.*

J. A. Jeletzky, *Geological Survey of Canada, Ottawa, Canada.*

I. Kotova, *USSR Academy of Sciences, Moscow, USSR.*

V. A. Krasheninnikov, *USSR Academy of Sciences, Moscow, USSR.*

F. M. Weaver, *Exxon Production Research Co., PO Box 2189, Houston, Texas 77001, USA.*

S. W. Wise, *Florida State University, Tallahassee, Florida, USA.*

**Abstract** Four Deep Sea Drilling Project sites cored continuously on the eastern Falkland Plateau and in the nearby southeast Argentine Basin during *Glomar Challenger* Cruise 71 augment considerably the Jurassic to Holocene record recovered previously during DSDP Leg 36 and provide the basis for an integrated biostratigraphic framework for the region. Principal fossil groups used for correlation are benthic and planktic foraminifera, calcareous nannofossils, diatoms, silicoflagellates, radiolarians, pollen and spores, and benthic and nektonic molluscs. Useful palaeomagnetic data is available for parts of the section sampled by the hydraulic piston corer.

The Jurassic and Lower Cretaceous shallow water (shelf depth) section at DSDP Site 511 can be divided into seven major units on the basis of megafossils (ammonites, belemnites, and bivalves); three distinct palynomorph associations are recognised in this interval, and conventional nannofossil zones are delimited in the Upper Jurassic, Aptian and Albian. Planktonic foraminifera are most useful for picking the base of the Albian (first *Ticinella roberti*) and for subdividing the superjacent deep water Mesozoic strata. All evidence documents a late Tithonian-Neocomian hiatus spanning about 20 million years.

All planktonic groups are represented in a long Middle Eocene section cored continuously at DSDP Site 512. Nine important diatom datums in this interval allow correlation with other Lutetian sections in the South Atlantic. A composite Upper Eocene to Lower Miocene section represented by cores from DSDP Holes 511 and 513A is divided into 12 newly proposed diatom zones. Six new or revised silicoflagellate zones have also been proposed for this interval. A new Plio-Pleistocene diatom zonation consisting of 13 zones has been proposed for two long sequences cored at DSDP Sites 513 and 514.

# GEOLOGICAL IMPLICATIONS OF RECYCLED PALYNOMORPHS IN CONTINENTAL SHELF SEDIMENTS AROUND ANTARCTICA

E.M. Truswell, *Bureau of Mineral Resources, Geology and Geophysics, P.O. Box 378, Canberra, Australia.*

*Abstract* On three areas of the East Antarctic continental shelf, concentrations of recycled palynomorphs in dredge samples collected during the Australasian Antarctic Expedition of 1911-14 are sufficiently high to suggest sedimentary sequences are being eroded nearby. Near the Shackleton Ice Shelf, palynomorphs of Permian, Late Jurassic to Early Cretaceous, and Late Cretaceous to Early Tertiary age occur. On the outer shelf offshore from Cape Carr, Early Cretaceous to Early Tertiary sequences are suggested; the same age span distinguishes sediments west of the Mertz Glacier. On an Australia/Antarctica reconstruction, the last two areas match southern Australian basins where thick Cretaceous rift valley sequences occur. Quantitative analyses of sediments from the Ross Sea, collected by USNS "Eltanin", show a dense concentration of Late Cretaceous and Tertiary palynomorphs in the southeast, coinciding with discharge from ice streams D and E. This may reflect a provenance in the Byrd subglacial basin. Dinoflagellates suggest marine conditions in this area between East and West Antarctica in the Eocene.

Sea floor sediments near Antarctica are rich in recycled spores, pollen, and dinoflagellates. These derive from the erosion of sedimentary sequences either on the Antarctic continent, or on the continental shelf. Potential applications of these organic-walled microfossils are twofold: in a geological sense, they provide information about the position, age, and state of metamorphism of hidden sedimentary rocks; in the botanical sphere, they provide a check list of plant taxa that once grew on Antarctica.

The physical processes controlling distribution of the recycled palynomorphs vary according to the location of the source beds. For sedimentary sequences cropping out on the continent, erosion is by entrainment into the bottom layers of moving ice, followed by meltout and deposition when the ice reaches the sea; this may occur either through a floating ice shelf or through valley glaciers. For sequences cropping out on the continental shelf, erosion is by submarine processes of current action, and, in shallower areas, by gouging by grounded ice.

The presence of recycled palynomorphs in Antarctic muds was noted first by Lizitzin (1960) in offshore East Antarctica, then by Wilson (1968) in the Ross Sea. Subsequently, the present author described assemblages from the West Ice Shelf (Kemp, 1972a) and from the Weddell Sea (Kemp, 1972b). More recently, palynomorphs were examined from a wide span of the Wilkes Land Coast (Truswell, 1982), and from the Ross Sea (Truswell and Drewry, in press). The present account essentially summarises these two investigations.

## SOURCE OF SAMPLES

For areas offshore from East Antarctica, most samples examined were dredged by Sir Douglas Mawson from the S.Y. "Aurora" during the 1911-14 Australasian Antarctic Expedition. Samples were recovered from the driver tubes during the taking of routine depth soundings; they were stored in test tubes labelled with position, date and depth, and lodged in the Australian Museum, Sydney. Sample lithologies, and foraminifera were described by Chapman (1922). Supplementary material from the area between 130° and 150°E was obtained from piston core tops taken during Cruise 37 of USNS "Eltanin". For the Ross Sea, material used in this study came from the top 10 cm of piston cores taken during "Eltanin" Cruises 27 and 32. Samples from this area were dried and weighed before maceration, so that palynomorph numbers per gram of sediment could be determined, following Muller's (1959) method.

## COMPOSITION, DISTRIBUTION AND ORIGIN OF RECYCLED MATERIAL

In three distinct areas in the region offshore from Wilkes Land (Figure 1), concentrations of recycled palynomorphs are sufficiently dense to suggest the presence nearby of eroding sedimentary sequences. These areas, the Shackleton Ice Shelf, and areas offshore from Cape Carr, and west of the Mertz Glacier, are treated separately below.

### Wilkes Land Coastal Region

*1. Shackleton Ice Shelf.* Sample sites in the area west of the Shackleton Ice Shelf are shown in Figure 2, which is redrawn from Mawson's (1942) narrative of the 1911-14 voyages. All samples shown are palynologically productive in varying degrees. Samples numbered 141, 141A, B, and 142, near the areas labelled by Mawson "thick bay ice", and "high grounded shelf ice", yielded 1500 to 2000 palynomorphs per gram of sediment. This is a concentration comparable to that found not far offshore from tropical pollen-laden rivers today (Muller, 1959). The recycled palynomorphs fall into three age groups, namely, Permian, Late Jurassic to Early Cretaceous, and Late Cretaceous to Early Tertiary. Compositional details are given in Truswell (1982, tables 1-3).

The Permian element is most common in samples 25, 29, 30, 31 and 139, and includes taxa previously reported from the Amery Group of the Prince Charles Mountains (Balme and Playford, 1967; Kemp, 1973; Dibner, 1978). There, the entire sequence belongs to Stage 5 of the Australian palynostratigraphic sequence (Kemp et al., 1977), of mid to Late Permian age. In the recycled assemblage there is a hint that Permo-Carboniferous source beds older than that may also have contributed microfossils.

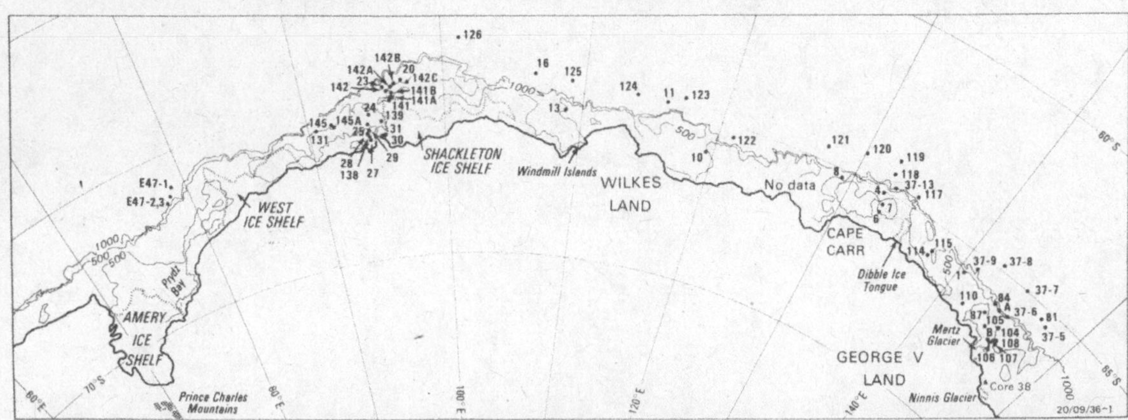


**Figure 1. Coastal East Antarctica, showing position of dredge samples collected by Douglas Mawson. Sample numbers as listed in Chapman (1922). Numbers prefixed 37 are piston core localities from "Eltanin" Cruise 37 (from Truswell, 1982).**

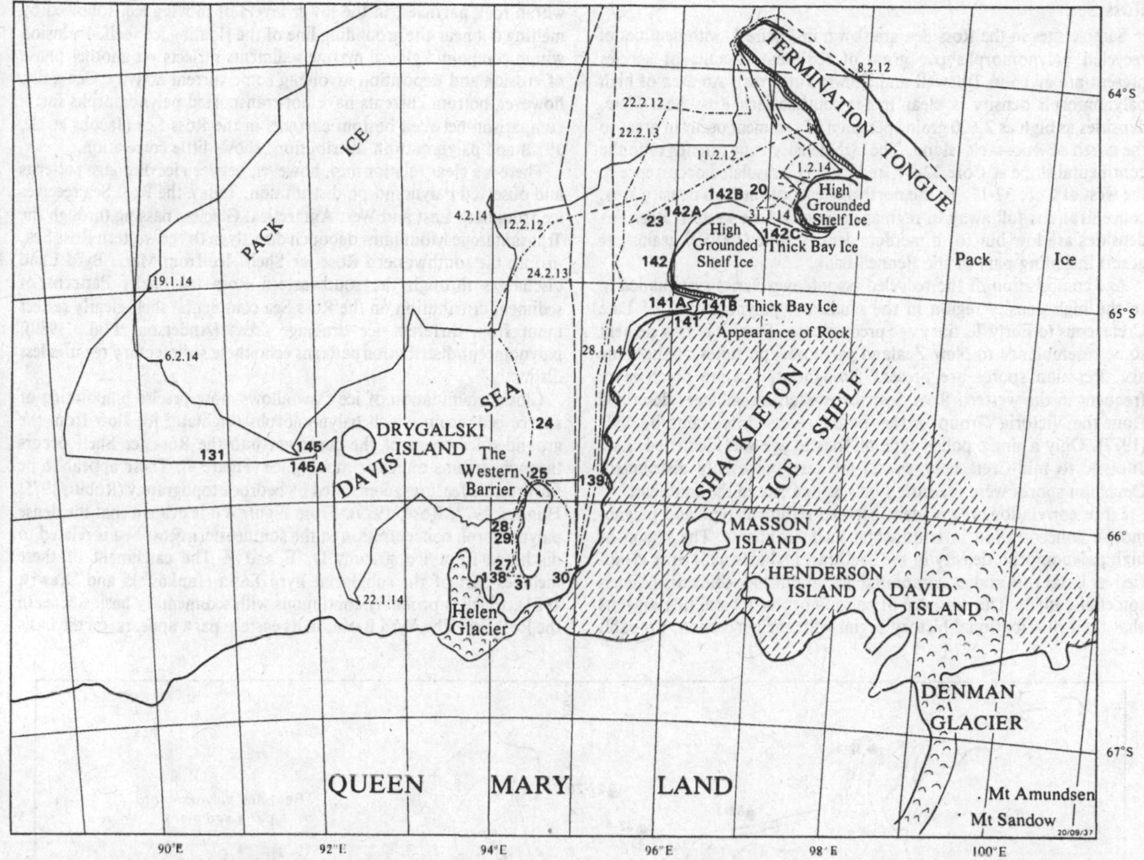


**Figure 2. Shackleton Ice Shelf area, redrawn after Mawson (1942). The routes of the three voyages of the S.Y. "Aurora" are shown, with dates sample stations were occupied. Heavy figures are samples examined in this study, numbered after Chapman (1922).**

The Late Jurassic to Early Cretaceous element is most diverse, and most species are long-ranging within that time interval. However, there are a few which suggest that part of the source beds may be Albian. A single broken dinoflagellate is the only evidence for marine conditions in these rocks. By contrast, dinoflagellates are common in the Late Cretaceous to Early Tertiary element, and may indicate marine source beds of Late Paleocene to Eocene age.

There is no clear indication, either from known geology inland from the Shackleton Shelf, nor from seafloor bathymetry, as to where source beds of the palynomorphs might lie. Dense concentrations in samples from the outer continental shelf suggest that sequences may be eroding there. The Gebco Bathymetric Map (Johnson and Vanney, 1980) showed this area to lie at the western end of a deep, linear trough on the continental shelf parallel to the coastline. Such depressions occur on much of the shelf of East Antarctica. Johnson et al (1982) suggest that they reflect faults associated with rift uplift and subsequently scoured by glacial action. Such action, perhaps during Pleistocene ice advances, may have steepened the walls of the depression in the Shackleton Shelf area, resulting in exposure of Mesozoic and Tertiary sediments and the release of palynomorphs.

The Permian spores possibly reflect deposition of debris from both the Shackleton Ice Shelf and the Helen Glacier. Their source, however, remains unknown. Inland, subglacial topography is poorly defined, but available data do not suggest large sedimentary basins. In the onshore geology, the youngest known rocks are of Riphean to Early Cambrian age (Ravich et al., 1968). There does however, remain some possibility that Palaeozoic rocks are present in the Denman rift, in a manner similar to their occurrence in the Lambert Glacier.

*2. Cape Carr area.* Palynomorph concentrations are high in a region offshore and to the east of Cape Carr, at 64°30′S, between 132° and 135°E. Samples 117 and 120, from the Mawson collection, and "Eltanin" core 37-13 yielded 200-300 palynomorphs per gram. The recycled assemblages at these sites (Truswell, 1982) consist of Early to Mid-Cretaceous and Late Cretaceous to Early Tertiary components. The precise age range of the source beds contributing the microfossils is difficult to determine, but rare spore species suggest that they may be partly Albian to Cenomanian. A late Paleocene to Eocene component was also identified. Only in this interval were there dinoflagellates attesting to marine influence; they form part of a high latitude Eocene flora first reported by Wilson (1967) from boulders at McMurdo Sound. Sparse sampling here makes it difficult to pinpoint a location for source beds. The recycled material may be eroding from strata cropping out on steep slopes of the shelf break, but this is speculative.

*3. Mertz Glacier area.* It was to the west of the Mertz Glacier that the Australasian Antarctic Expedition made its first Antarctic landfall in January, 1912. Samples dredged from the outer continental shelf here yielded pollen and spores predominantly of Early Cretaceous age; Late Cretaceous to Early Tertiary taxa are less common. The Early Cretaceous element resembles an assemblage recovered from a siltstone penetrated by piston coring in the "George V basin", a depression on the continental shelf west of the Ninnis Glacier, during Operation Deep Freeze of 1979. Domack et al (1980) claimed that palynomorphs from the siltstone suggest deposition in a non-marine environment during the Neocomian to Aptian.

The possible presence of intracratonic sedimentary basins which intersect the coast in the vicinity of the Ninnis Glacier has been raised by Steed and Drewry (1982). Using radio echo sounding to describe subglacial topography, they predicted that the Wilkes Basin reaches the coast in the region of the Ninnis Glacier and Cook Ice Shelf, and contains Beacon Supergroup rocks overlain by a younger Mesozoic and Cenozoic cover. Mawson (1940) mentioned lignite fragments in dredges offshore from the Mertz Glacier, which could come from such a sedimentary sequence. Unfortunately, the lignites could not be located in collections held by the Australian Museum, so their age remains unknown.

## Ross Sea

Sample sites in the Ross Sea are shown in Figure 3, with densities of recycled palynomorphs per gram of sediment. Details of species present are given in Truswell and Drewry (in press). An area of high palynomorph density is clear in the southeastern Ross Sea. Here, densities as high as 2,600 grains per gram of sediment occur in areas to the north of Roosevelt Island. The high density zone extends onto the continental slope at Core 32-34, and there is an isolated occurrence to the west at Core 32-17. To the north and east of the high density area, concentrations fall away in regular fashion. In the western Ross Sea densities are low but for a medium density zone (100-500 grains per gram) including part of the Pennell Bank.

Agé composition of the recycled assemblages varies systematically. In the high density region in the southeast, palynomorphs of Late Cretaceous to Early Tertiary age predominate. There is in this element some resemblance to New Zealand palynomorph suites of the same age. Permian spores are present throughout, but are much more frequent in the western Ross Sea; they match assemblages described from the Victoria Group in the Transantarctic Mountains by Kyle (1977). Only a single pollen type of Triassic age was identified. Late Jurassic to mid-Cretaceous spores are rare but evenly distributed. Devonian spores were identified only at one site in the southeast.

Little correlation was evident between sediment type and palynomorph concentration in continental shelf sediments. The region of high palynomorph density in the southeast includes sediments identified as basal tills and as compound glacial marine sediments (Anderson et al., 1980). The presence of palynomorphs in basal tills suggests that their distributional history begins with their inclusion, probably within rock particles, in the lower layers of moving ice, followed by melting out near the grounding line of the floating ice shelf. Inclusion within compound glacial marine sediments reflects yet another phase of erosion and deposition involving some current activity. Generally, however, bottom currents have not transported palynomorphs far; a comparison between bottom currents in the Ross Sea (Jacobs et al., 1970) and palynomorph distributions shows little correlation.

There is a clear relationship, however, between ice drainage patterns and observed palynomorph distribution. Today, the Ross Sea receives ice from both East and West Antarctica. Glaciers passing through the Transantarctic Mountains debouch directly in to the western Ross Sea, or into the southwestern Ross Ice Shelf. Ice from Marie Byrd Land discharges through the southeastern Ross Ice Shelf. Patterns of sediment distribution on the Ross Sea continental shelf clearly reflect input from different ice drainage areas (Anderson et al., 1980); palynomorph distribution patterns echo these sedimentary regimes less distinctly.

Closer examination of ice flow allows more precise pinpointing of source beds from which palynomorphs originate. Ice flow from the grounded ice cover of the continent into the Ross Ice Shelf occurs through streams of faster moving ice (Figure 4). These appear to be fairly stable features, controlled by bedrock topography (Robin, 1975; Hughes, 1977; Rose, 1979). From Figure 4 it is evident that the dense palynomorph concentration in the southeastern Ross Sea is related to discharge from ice streams D, E and F. The catchment of these includes part of the subglacial Byrd Basin (Jankowski and Drewry, 1981), which is probably continuous with sedimentary basins beneath the Ross Sea. The Byrd Basin, in its eastern part, appears, on the basis

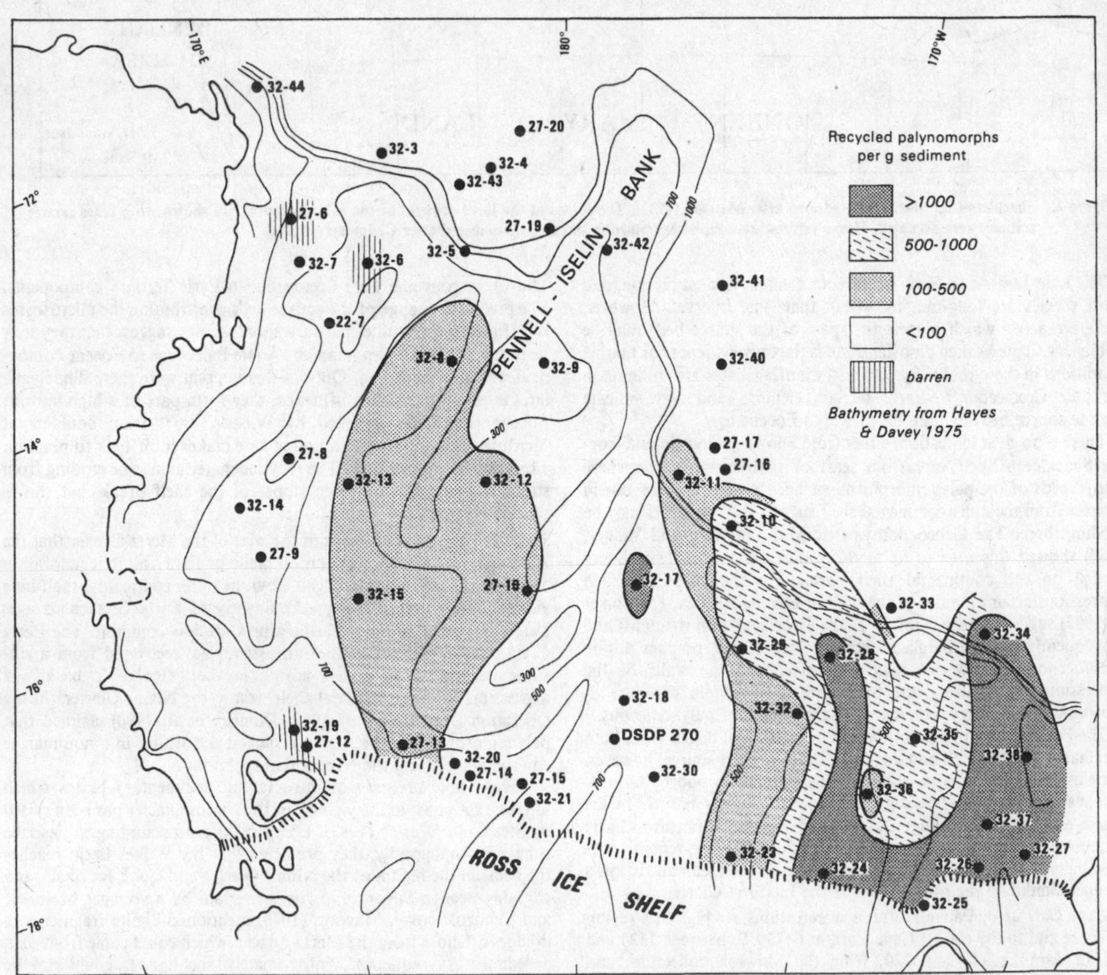


**Figure 3. Palynomorph density in Ross Sea sediments, superimposed on bathymetry. Sample sites are those of "Eltanin" Cruises 27 and 32.**

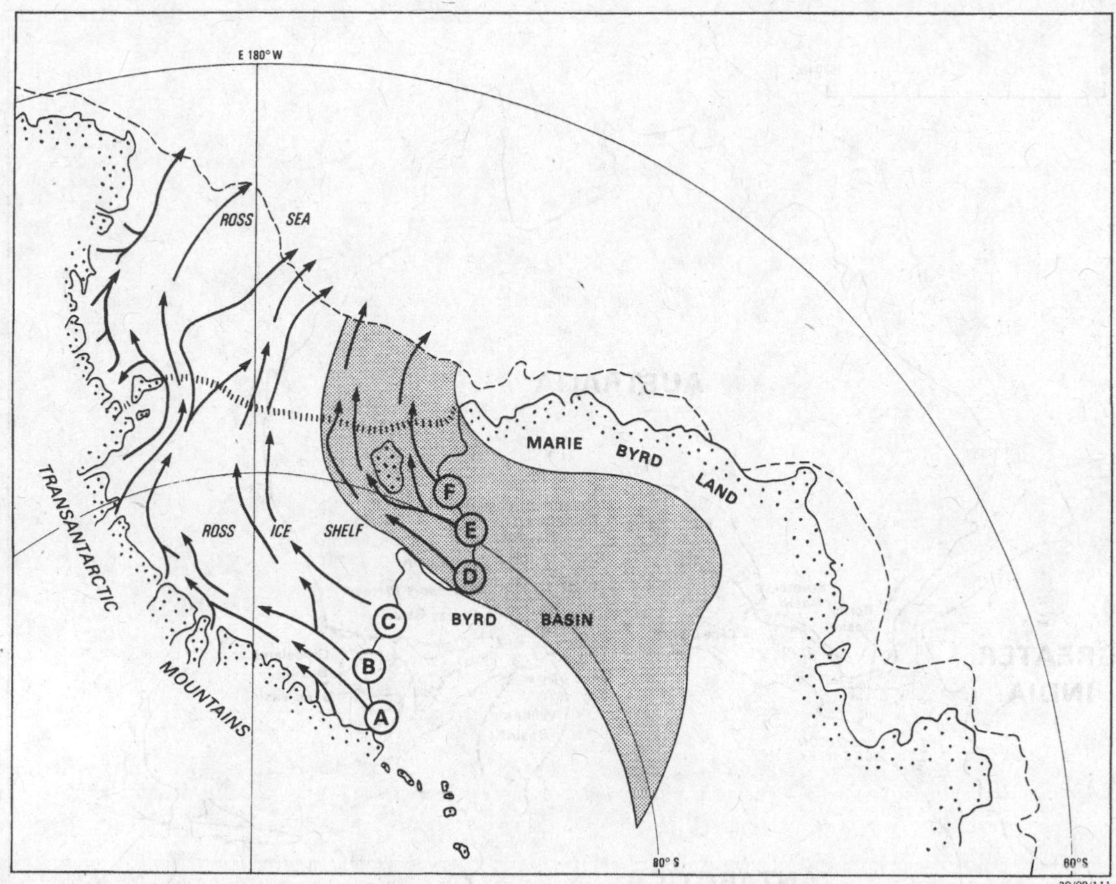


**Figure 4. Ice flow patterns into Ross Ice Shelf and Ross Sea. Circled numbers are ice streams of Hughes (1977): arrows show present ice stream directions and their former extent into the continental shelf during Pleistocene advances (from Stuiver et al., 1981). Hachured area includes zone of high palynomorph density and possible source areas within the catchment of ice streams D, E and F.**

of the composition of the recycled palynomorphs, to include sequences of Late Cretaceous to Early Tertiary clastic rocks. A marine influence in the Eocene is apparent from the dinoflagellate suite.

## IMPLICATIONS FOR CONTINENTAL RECONSTRUCTION

Most of the Antarctic coastline in the region covered by Mawson's dredge sampling was conjugate with southern Australia prior to continental separation in the Late Cretaceous to Paleocene. It is instructive then, to compare the position of sedimentary sequences suggested for Antarctica by palynomorph distributions, with southern Australian sedimentary basins. The Ross Sea region prior to separation abutted the complex area to the south of New Zealand.

In Figure 5, Antarctic localities discussed in this paper are shown on the reconstruction of Crook and Belbin (1978), with Greater India added. On this basis, the sedimentary sequence suspected in the Shackleton Ice Shelf area would have been continuous with depositional areas in Greater India prior to the Early Cretaceous. Subsequently, the area probably faced a widening Indian Ocean. On the present Australian continent, the basin nearest to the area is the Perth Basin. The Shackleton Shelf areas lies too far west to represent a continuation of this basin southward into Antarctica, but there are similarities between sequences suspected in Antarctica on the basis of the recycled palynomorphs, and those from the southern Perth Basin.

Hidden Permian sequnces may correlate with the Sue Coal Measures of the Perth Basin, which are Early to Late Permian (Balme, *in* Kemp et al., 1977). The Late Jurassic to Early Cretaceous sequences hinted at in Antarctica are reminiscent of the dominantly non-marine Yarragadee Formation. The mid-Cretaceous to Early Tertiary sequences of the two "basins", however, differ. Albian palynomorphs from the Shackleton Ice Shelf area are non-marine; in the Perth Basin, facies of this age are mainly shallow marine. Eocene dinoflagellates from the Shackleton Shelf have no clear counterparts in Australia, so Early Tertiary comparisons are obscure.

The offshore Cape Carr region abuts the southern Australian coast in the vicinity of the Ceduna Plateau, in the Great Australian Bight Basin. There, some 9 km of Early and Late Cretaceous sediments are indicated, and have been interpreted as pre-breakup, rift valley fill (Falvey and Mutter, 1981), or as deposits in a newly-formed ocean basin (Cande and Mutter, 1982). The Cretaceous sequences which may, on the basis of the palynomorph data, crop out offshore from Cape Carr, may thus represent the Antarctic part of this depositional sequence. It is interesting to note that outcrop of the Australian counterparts on the continental slope of the Great Australian Bight is also suspected on the basis of seismic data (Talwani et al., 1979).

The Mertz and Ninnis Glacier regions, where both *in situ* and recycled Cretaceous palynomorphs occur, abuts the Beachport Plateau, west of the Otway Basin in Figure 5. The non-marine Antarctic sequences may have correlations within the Otway Group in southeastern Australia.

Reconstructions in the Ross Sea region are more complex than in the Australia/Antarctica conjunction. Most suggest that the Campbell Plateau abutted the Ross Sea continental shelf prior to separation (Crook and Belbin, 1978; Grindley and Davey, 1982); in such configurations the Chatham Rise abuts Marie Byrd Land. This picture accords with Cretaceous palynomorphs from the Ross Sea showing a similarity with those from the Chatham Islands (Mildenhall, 1977). Another point of palaeogeographic interest is that recycled dinoflagellates in the Ross Sea, which probably derive, with pollen and spores, from the Byrd Basin, suggest that a seaway existed between Marie Byrd Land and East Antarctica as long ago as the Eocene.

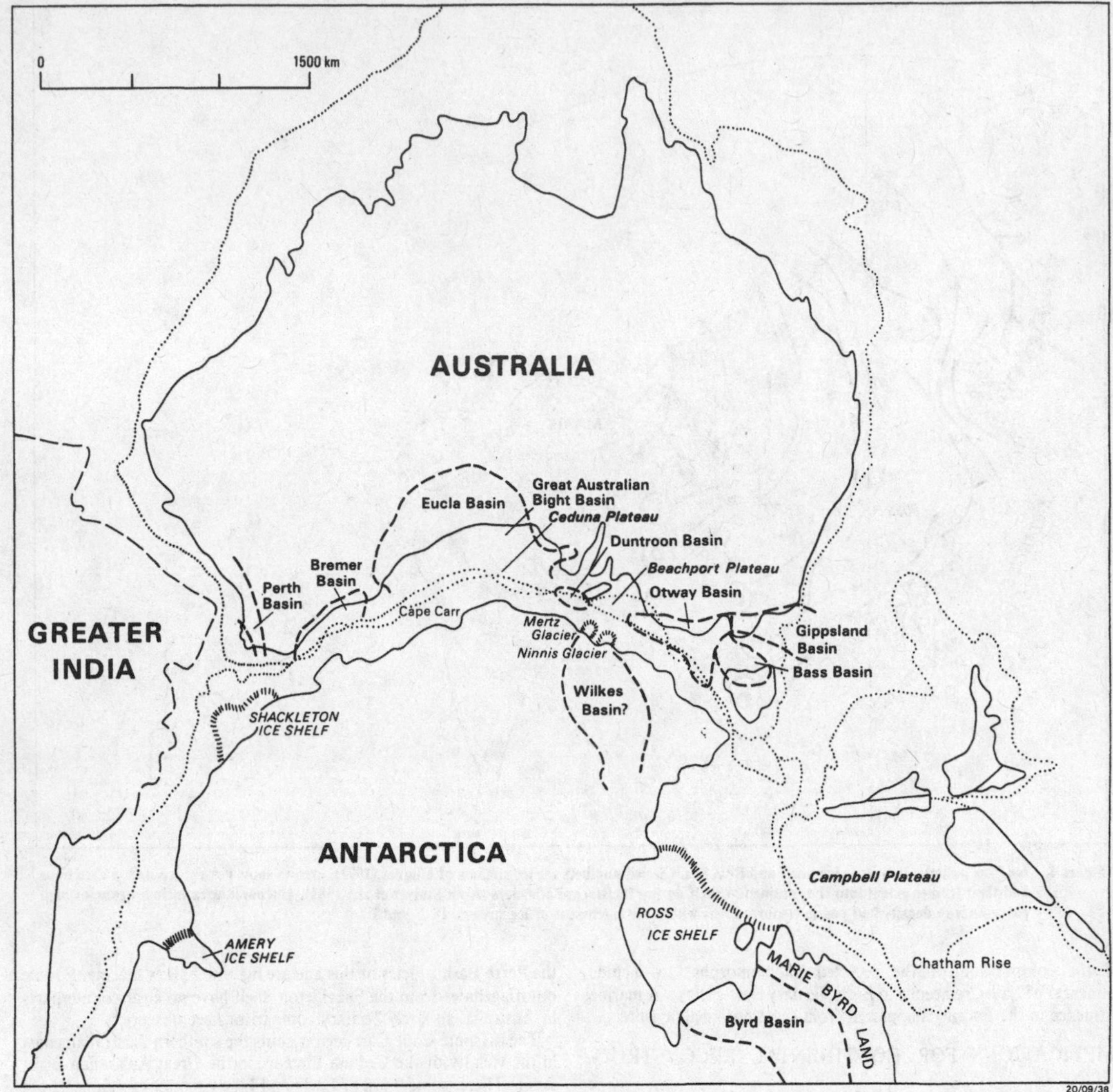


**Figure 5. Continental reconstruction after Crook and Belbin (1978) with Greater India added. Positions of sedimentary basins in southern Australia shown, together with Antarctic localities discussed in text.**

*Acknowledgements* The Mawson samples were provided through courtesy of the Director and Trustees of the Australian Museum; the "Eltanin" cores by the Antarctic Research Facility, Florida State University. Publication is by permission of the Director, Bureau of Mineral Resources.

## REFERENCES

ANDERSON, J.B., KURTZ, D.D., DOMACK, E.W. and BALSHAW, K.M., 1980: Glacial and glacial marine sediments of the Antarctic continental shelf. *J. Geol. 88*, 399-414.

BALME, B.E. and PLAYFORD, G., 1967: Late Permian plant microfossils from the Prince Charles Mountains, Antarctica. *Rev. Micropaleontol., 10*, 179-92.

CANDE, S.C. and MUTTER, J., 1982: A revised identification of the oldest seafloor spreading anomalies between Australia and Antarctica. *Earth Planet. Sci. Lett., 58*, 151-60.

CHAPMAN F., 1922: Seafloor deposits from soundings. *Australasian Antarctic Expedition 1911-14. Scientific Reports Ser. A. 2.* 1-60.

CROOK, K.A.W. and BELBIN, L., 1978: The southwest Pacific area during the last 90 million years. *Geol. Soc. Aust., J., 25*, 23-40.

DIBNER, A.F. 1978: Palynocomplexes and age of the Amery Foundation deposits, East Antarctica. *Pollen et Spores, 20*, 405-22.

DOMACK, E.W., FAIRCHILD, W.W. and ANDERSON, J.B., 1980: Lower Cretaceous sediment from the East Antarctic continental shelf. *Nature, 287*, 625-6.

FALVEY D.A. and MUTTER, J.C., 1981: Regional plate tectonics and the evolution of Australia's passive continental margins. *BMR J. Aust. Geol. Geophys., 6*, 1-29.

GRINDLEY, G.W. and DAVEY, F.J., 1982: The reconstruction of New Zealand, Australia and Antarctica; *in* Craddock, C. (ed.) *Antarctic Geoscience* Univ. Wisconsin Press, Madison, 15-29.

HUGHES, T., 1977: West Antarctic ice streams. *Rev. Geophys. Space Phys., 15*, 1-46.

JACOBS, S.S., AMOS, A.F. and BRUCHHAUSEN, P.M. 1970: Ross Sea oceanography and Antarctic bottom water formation. *Deep Sea Res., 17*, 935-62.

JANKOWSKI, E.J. and DREWRY, D.J., 1981: The structure of West Antarctica from geophysical studies. *Nature, 291*, 17-21.

JOHNSON, G.L. and VANNEY, J.R., 1980: *General Bathymetric chart of the Oceans (GEBCO), 5.18. Polar Stereographic Projection, Scale-1:6,000,000 at 75S lat.* Canadian Hydrographic Service, Ottawa.

JOHNSON, G.L., VANNEY, J.R. and HAYES, D., 1982: The Antarctic continental shelf; *in* Craddock, C. (ed.) *Antarctic Geoscience.* Univ. Wisconsin Press, Madison, 995-1002.

KEMP, E.M., 1972a: Reworked palynomorphs from the West Ice Shelf area, East Antarctica, and their possible geological and palaeoclimatological significance. *Mar. Geol., 13*, 145-57.

KEMP, E.M., 1972b: Recycled palynomorphs in continental shelf sediments from Antarctica. *U.S Antarct. J., 7*, 190-1.

KEMP, E.M., 1973: Permian flora from the Beaver Lake area. Palynological examination of samples. *Aust., Bur. Miner. Resour., Geol. Geophys., Bull., 126,* 7-12.

KEMP. E.M., BALME, B.E., HELBY., R.J., KYLE, R.A., PLAYFORD, G. and PRICE., P.L., 1977: Carboniferous and Permian palynostratigraphy in Australia and Antarctica: a review. *BMR J. Aust. Geol. Geophys., 2,* 177-208.

KYLE, R.A., 1977: Palynostratigraphy of the Victoria Group of south Victoria Land, Antarctica. *N.Z. J. Geol. Geophys., 20,* 1081-1102.

LISITZIN, A.P., 1960: Bottom sediments of the eastern Antarctic and the southern Indian Ocean. *Deep Sea Res., 7,* 89-99.

MAWSON, D., 1940: Sedimentary rocks. *Australasian Antarctic Expedition 1911-14. Scientific Reports Ser. A., 4,* 347-67.

MAWSON, D., 1942: Geographical narrative and cartography. *Australasian Antarctic Expedition 1911-14. Scientific Reports Ser. A., 1,* 1-350.

MILDENHALL., D.C., 1977: Cretaceous palynomorphs from the Waihere Bay Group and Kahmitara Tuff, Chatham Islands, New Zealand. *N.Z. J. Geol Geophys., 20,* 655-72.

MULLER, J., 1959: Palynology of recent Orinoco delta and shelf sediments. *Micropaleontology 5,* 1-32.

RAVICH, M.G. KLIMOV, L.V. and SOLOVIEV, D.S., 1968: The Precambrian of East Antarctica. *Trans. Sci. Res. Institute Geology of the Arctic of State Geological Committee of USSR, 138.* Israel Program for Scientific Translations, Jerusalem.

ROBIN, G. de Q., 1975: Ice shelves and ice flow. *Nature, 253,* 168-72.

ROSE, K.E., 1979: Characteristics of ice flow in Marie Byrd Land, Antarctica. *J. Glaciol., 24,* 63-74.

STEED, R.H.N. and DREWRY, D.J., 1982: Radio-echo sounding investigations of Wilkes Land, Antarctica; *in* Craddock, C. (ed.) *Antarctic Geoscience.* Univ. Wisconsin Press, Madison, 969-75.

STUIVER, M., DENTON, G.H., HUGHES, T.J. and FASTOOK, J.L, 1981: History of the marine ice sheet in West Antarctica during the last glaciation: a working hypothesis; *in* Denton, G.H. and Hughes, T.J. (eds.) *The Last Great Ice Sheets,* John Wiley and Sons, New York, 319-66.

TALWANI, M., MUTTER, J., HOUTZ, R. and KONIG, M., 1979: The crustal structure and evolution of the area underlying the magnetic quiet zone on the margin south of Australia. *Am. Assoc. Petrol. Geol., Mem., 29,* 151-75.

TRUSWELL, E.M., 1982: Palynology of seafloor samples collected by the 1911-14 Australasian Antarctic Expedition: implications for the geology of coastal East Antarctica. *Geol. Soc. Aust. J., 29,* 343-56.

TRUSWELL, E.M. and DREWRY, D.J. (in press): Recycled palynomorphs in surficial sediments of the Ross Sea, Antarctica. *Mar. Geol.*

WILSON, G.J., 1967: Some new species of Lower Tertiary dinoflagellates from McMurdo Sound, Antarctica. *N.Z. J. Bot., 5,* 57-83.

WILSON, G.J., 1968: On the occurrence of fossil microspores, pollen grains and microplankton in bottom sediments of the Ross Sea, Antarctica. *N.Z. J. Mar. Fresh. Res., 2,* 381-9.

# MIDDLE-LATE MIOCENE PALAEOENVIRONMENT OF THE CIRCUM-ANTARCTIC

P.F. Ciesielski and M.T. Ledbetter, *Department of Geology, University of Georgia, Athens, Georgia 30602, USA.*

F.M. Weaver, *Exxon Production Research Co., PO Box 2189, Houston, Texas 77001, USA.*

*Abstract* Upper Middle-Miocene radiolarian, diatom, and silico-flagellate zonal schemes of the circum-Antarctic are correlated to magnetostratigraphy for the first time. Biostratigraphic-magnetostratigraphic correlations are based on analyses of numerous piston cores and hydraulic piston cores from DSDP Hole 512. Siliceous microfossil zones correlated to magnetostratigraphy include the *Eucyrtidium pseudoinflatum, Theocalyptra bicornis spongothorax*, and *Antarctissa conradae* radiolarian zones; *Denticulopsis hustedtii, D. hustedtii/D. lauta*, and *Nitzschia denticuloides* diatom zones; and the *Mesocena diodon, M. diodon/M. circulus*, and *M. circulus* silicoflagellate zones. Some zonal boundaries and datums are correlated to lower latitude biostratigraphic zonations on the basis of the co-occurrence of high and mid latitude calcareous and siliceous microfossils in DSDP Sites 512 and 513.

The developed stratigraphy is used to correlate Southern Ocean drill core sequences, including DSDP Sites 266, 274, 278, 329, 512 and 513. The revised ages and sedimentary characteristics of these cores suggest that prior to the Late Miocene (a) extensive ice shelves were not present along the Antarctic Coast (b) no grounded ice sheet was present in W. Antarctica. Present day W. Antarctica was occupied by an archipelago and W. Antarctic Sea, and (c) most ice-rafted detritus was deposited close to the continent (south of 56°S) by small bergs principally from tide-water glaciers and small ice shelves.

# PALEOCENE RADIOLARIAN BIOSTRATIGRAPHY OF THE SOUTHERN OCEAN

Hsin Y. Ling, *Department of Geology, Northern Illinois University, DeKalb, Illinois 60115, USA.*

**Abstract The Paleocene Epoch, the oldest geochronologic unit within the Cenozoic Era, recently became a centre of controversy among earth scientists. On the one hand, earth scientists view the epoch as a transitional stage for some organic evolution. On the other hand, many new forms made their initial appearance at the beginning of the epoch, the boundary between the Mesozoic and Cenozoic Eras, giving rise to 'the Cretaceous-Tertiary boundary problem', and numerous hypotheses have been proposed to explain this striking world-wide phenomenon. Thus, the understanding of faunal and floral composition of the Paleocene Epoch has become essential.**

**Our knowledge of all aspects of Paleocene siliceous microfossils, radiolarians, silicoflagellates, ebridians, and diatoms, is gradually and steadily advancing, yet it is still far behind that of calcareous forms.**

**To remedy the situation, a biostratigraphic framework for silicoflagellates has been recently proposed, and diatom zonation is being added.**

**Abundant radiolarians have been recovered from the deep-sea sediments of the southern ocean. Although the research is still in progress, the following results are already apparent: (1) the Paleocene section can be divided and characterised by species belonging to genus *Buryella* by their apparent distinct geological ranges and their phylogenic lineages; and (2) some significant faunal differences have been noticed between the assemblage from the southern ocean and the reported occurrences from the northern hemisphere.**

# MARINE GEOLOGY OF THE GEORGE V CONTINENTAL MARGIN: COMBINED RESULTS OF DEEP FREEZE 79 AND THE 1911-14 AUSTRALASIAN EXPEDITION

E.W. Domack and J.B. Anderson, *Department of Geology, Rice University, Houston, Texas, 77251, U.S.A.*

*Abstract* During the Australasian Expedition of 1911-14 geological and glaciological reconnaissance studies were conducted along the coast and in the waters of the George V Coast under the direction of Sir Douglas Mawson. Important accomplishments made during this expedition included mapping of the ice front, observations of sediment laden icebergs, bottom sediment descriptions and geologic mapping of coastal outcrops.

During the United States Deep Freeze 79 expedition we revisited the George V Coast and conducted a detailed marine geologic survey of the area. Mawson, Ninnis and Mertz's sledging track was retraced by helicopter and a new map of the ice front constructed. This map, combined with that of Mawson's party and a 1957 Russian survey, shows significant historical changes in the size and shape of the Mertz and Ninnis ice tongues. We also encountered a number of sediment laden icebergs, including two icebergs with basal debris zones.

The most prominent physiographic feature of the area is the George V Basin, a deep (up to 1300 m) glacial trough oriented subparallel to the coast. Piston cores taken from this depression and elsewhere on the shelf penetrated lodgment till with a mineralogic composition indicative of an easterly source, probably the Cook Ice shelf. At the present time, terrigenous sediment supply to the continental shelf is minimal. Siliceous mud and ooze is the most widespread surficial deposit. Sediment reworking is most pronounced at the shelf edge and upper slope and results from impinging circumpolar currents.

Lower Cretaceous siltstone, containing a rich spore assemblage, identical to that of South Australian deposits of that age, was recovered in one core from the George V Basin. The shallow occurrence of this sediment relative to that of the Australian continental margin, where Lower Cretaceous deposits occur several kilometres subsurface, reflects a very different subsidence and sedimentation history for these two margins.

The 1911-14 Australasian Antarctic Expedition stands as one of the landmarks of Antarctic exploration and research. The accomplishments of this expedition, under the leadership of Sir Douglas Mawson, include: discovery and exploration of King George V Land, location of the south magnetic pole, mapping of two large outlet glaciers (the Mertz and Ninnis Glaciers), geologic mapping of coastal outcrops, establishment of the first profile of a part of the East Antarctic ice sheet and a meteorological account which established the region as the windiest in the world. In addition, sounding data and bottom samples were collected by the expedition vessel, the "S.Y. *Aurora*".

During the austral summer of 1978/79, a marine geologic survey of the George V-Adélie continental shelf was undertaken using the United States Coast Guard Icebreaker *Glacier.* The primary objective of this study was to investigate sedimentation in a grounded ice sheet-continental shelf setting. Our study relied considerably on results of the earlier Australasian Expedition. This paper reports the findings of this more recent expedition to the region.

## GLACIAL SETTING

During the Australasian Expedition, Sir Douglas Mawson and two colleagues, B.E.S. Ninnis and Xavier Mertz, conducted a sledging expedition along the George V Coast. Both Mertz and Ninnis died but their party, along with the Eastern Coastal Party, succeeded in mapping the area between Cape Denison and Cape Wild (Figure 1) with, what we later found to be, considerable accuracy.

The coast of George V Land and eastern Adélie Land is dominated by the edge of an ice covered plateau which rises to a height of over 1219 m at a distance of 64 km inland (Mawson, 1942). The area includes some 390 km of coastline which has a total drainage basin of 145,000 km². Most drainage occurs within a relatively small segment of the coastline through the Mertz and Ninnis Glaciers. Combined results of the Australasian survey, a 1958 Russian survey (Bardin, 1964; Koblents, 1965) and our 1979 survey of the ice front show changes of the order of several tens of kilometres in the configurations of the Mertz and Ninnis ice tongues (Figure 1) between these surveys (Anderson et al., 1980a). It was Mawson (1942) who first recognised the dynamic nature of these glaciers when he commented on the "cracking and groaning" of these outlet glaciers.

Icebergs calved from these glaciers have dimensions of up to several tens of kilometres. Several sediment-laden icebergs were observed by members of the Australasian Expedition. We also observed sediment-laden icebergs in the area, including two basal debris zones (Anderson et al., 1980b). The observed drift of icebergs in the area is to the west, following the East Wind Drift; Mawson (1942) also observed icebergs drifting to the south, onto the shelf, which he attributed to southerly flowing ocean currents. Icebergs grounded on banks shallower than approximately 250 m and become concentrated over these banks.

Most of the coastline of the area consists of sheer rock cliffs and ice walls which are mostly sediment free. A 12 m thick basal debris zone was photographed by Australasian Expedition members at Cape Denison (Figure 1). Glacial drainage along ice walls is probably

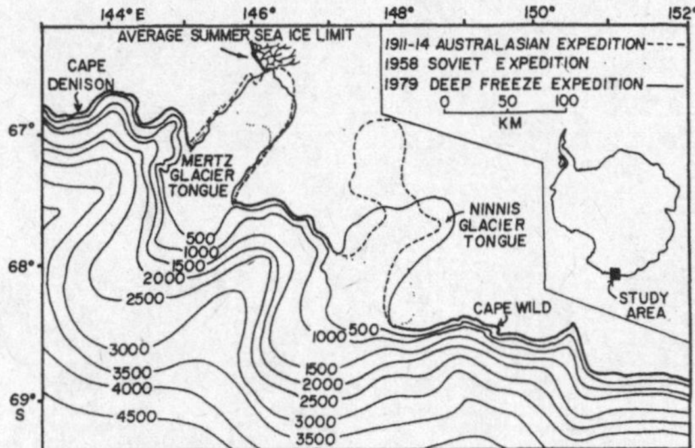


**Figure 1. Glacial setting of the George V Coast and historical changes in the configuration of the Mertz and Ninnis Glacier tongues since 1911. Contours are in metres. Glacial topography from Mawson (1942). Sea ice limits are from Mawson (1942) and conform with recent observations.**

sufficiently slow that wave erosion keeps pace with the rate of advance, resulting in glacial sedimentation near the ice edge (Drewry and Cooper, 1981). A small "dump moraine" was observed by Mawson (1940) some 10 m below the present edge of the ice cap at Cape Denison.

## GEOLOGICAL SETTING

Though the exposed bedrock in the region is limited to ice shrouded coastal cliffs and nunataks, Mawson was able to define a major geologic boundary, coinciding approximately with the Ninnis Glacier, separating a crystalline basement complex of granite, gneiss and high grade metamorphic rocks west of the glacier from dolerites and sandstones exposed east of the Ninnis Glacier (Figure 1). Mawson believed the dolerites and sandstones to be Triassic in age because of their similarity to dolerites and the Ross sandstones of Tasmania. The coastal boundary separating the two terrains most likely corresponds to the western edge of the Wilkes subglacial basin; a sedimentary sequence recognised by Drewry (1976) from radio-echo sounding of the interior. Erratics of arkosic sandstone were collected at Cape Denison (Figure 1) by Mawson (1940). He suggested that outcrops of these rocks occurred to the southeast, under the ice cap.

## MARINE GEOLOGIC RESULTS

### Sea Floor Topography

Bottom soundings acquired by the crew of the *Aurora* showed the continental shelf to be quite deep (up to 822 m deep) and highly irregular. More detailed bathymetric maps were later compiled by Grinnell (1971) and Vanney and Johnson (1979). During Deep Freeze 79, six transects were made onto the continental shelf, allowing for even more detailed mapping of the sea floor (Figure 2). The bottom topography is dominated by the George V Basin, a linear, inner shelf depression which parallels the coast. This depression reaches depths of

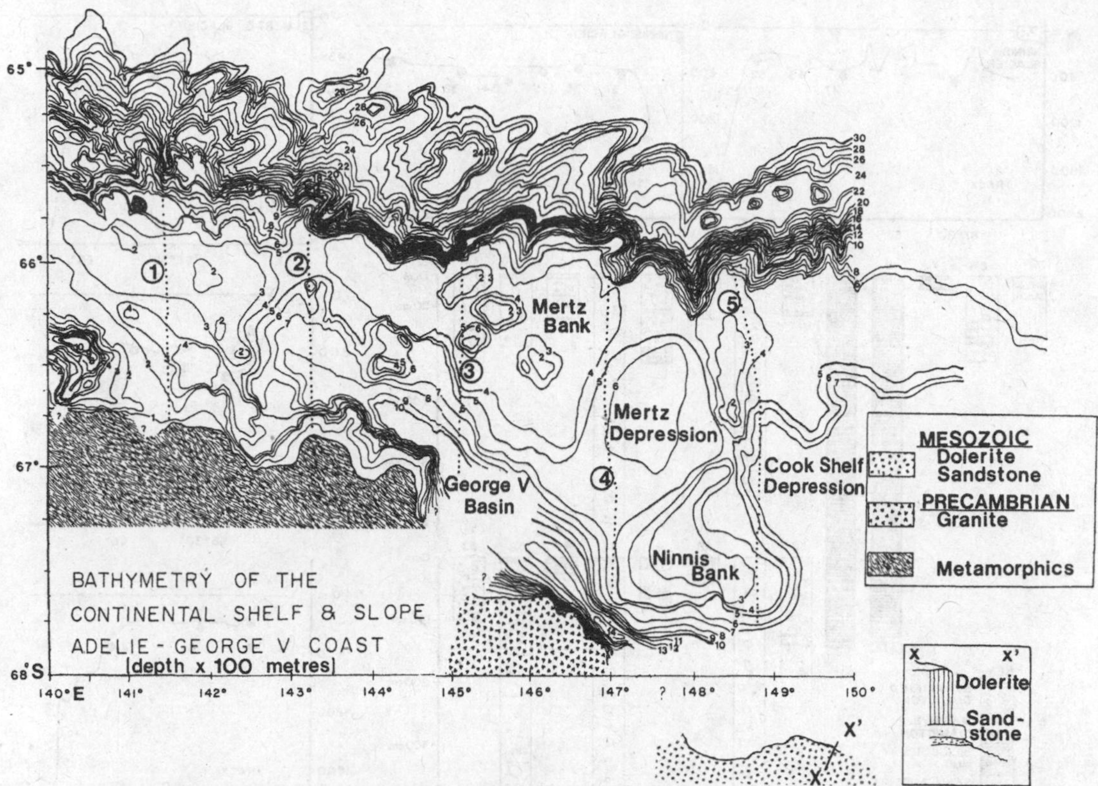


**Figure 2. Geology and bathymetry of the George V and Adélie margin. Geology is from Craddock (1972) and the bathymetry is modified after Vanney and Johnson (1979). Circled numbers indicate shelf profiles (dotted lines) shown in Figure 3. Insert shows geologic section (X-X') measured at Horn Bluff by Mawson (1942).**

over 1400 m just west of the Ninnis Glacier Tongue and apparently is an extension of the much larger Cook Shelf Depression which lies to the east (Figure 2). The George V Basin results in shelf profiles which slope towards the coast (Figure 3). Broad somewhat linear banks, 200 to 400 m deep, are prominent features of the middle and outer portions of the shelf (Figure 2). The two largest of these are given the unofficial names Mertz Bank and Ninnis Bank. A broad 600 m deep depression, which lies between Mertz and Ninnis Banks, extends from the George V Basin to the outer shelf and is given the unofficial name Mertz Depression (Figure 2). The George V Basin most likely represents a tectonic feature (Holtedahl and Holtedahl, 1961; Vanney and Johnson, 1979) but has likely been modified by glacial erosion.

## Continental Shelf Deposits

### *Surficial sediments and modern processes*

Eight samples of sea floor sediment were collected on the George V continental shelf by the crew of the "S.Y. *Aurora*". Chapman (1922) described these sediments and his results demonstrate the profound influence of biological processes upon shelf sedimentation. The sediments are primarily siliceous oozes composed of sponge spicules, diatoms and radiolaria. In general, the terrigenous component of sediments was found to increase above 550 m water depth (Chapman, 1922).

Relict basal tills and glacial marine sediments, discussed in the next section, are overlain by both recent and palimpsest sediments. These surficial sediments are comprised primarily of siliceous biogenic material. Deposition of green, siliceous mud and ooze is occurring at depths of greater than 500 m, while surficial reworking of relict diamicton occurs above this depth (Domack, 1980). Deposition of biogenic sediments occurred soon after recession of glacial ice from the shelf, as marked by sharp contacts between diamicton and overlying siliceous mud and ooze.

Laminated diatomaceous muds have been deposited in other regions (i.e. coastal California) where oxygen depleted (< 0.20 to 0.10 ml/l) water has effectively excluded benthic invertebrates (Calvert, 1964, 1966; Rhoads and Morse, 1971). However, Gordon and Tchernia (1972) have found shelf waters of the George V Basin Adélie Depression to be rather well mixed and near 90% oxygen saturation during the austral summer. The laminated nature of George V-Adélie Shelf sediments implies that sedimentation is sufficiently fast to mask any mixing by benthic organisms.

Up to 40 m of ooze has accumulated in the western end of the George V Basin, where core DF-79-12 penetrated 5.9 m of laminated spicule-diatom ooze before bottoming out in glacial-marine sediment (Figures 3 and 4). Diatom floras from the base of cored oozes are less than 18,000 years old (A. Gombos, pers. comm.). Depositional rates, based on $Pb^{210}$ measurements, are extremely rapid (0.3 cm/yr, K. Cochran, pers. comm.) and, when compared to an average laminae pair thickness of 0.5 cm, suggest annual depositional layering of marine varves.

Deposition of these sediments is most likely related to upwelling of Circumpolar Deep Water (CDW) onto the outer and middle parts of the shelf. This upwelling induces high productivity in ice-free surface waters. The region of most extensive upwelling, as revealed by temperature-salinity profiles, corresponds to the ice-free area described by Mawson (1942); it is here that the thickest accumulations of ooze are found (Figure 4). Extremely strong and persistent offshore winds are a notable feature at Cape Denison (Mawson, 1942; Parish, 1981). These katabatic winds average 19.3 m/s and may play a role in wind forced upwelling of CDW.

The paucity of ice-rafted debris in recent shelf sediments is rather surprising, given the glacial setting of the region and observed debris zones in icebergs (Mawson, 1942; Anderson et al., 1980b). Rates of ice-rafting on the continental shelf are very low and, based on a biogenic sedimentation rate of 0.3 cm/yr (K. Cochran, pers. comm.), an estimate of 0.28 gm/cm²/1000 yr may be given for ice-rafting rates of poorly sorted sand found within these sediments. Where the siliceous ooze is thinner, less than 9 cm in regions of persistent pack ice, ice-rafted material comprises up to 35 to 50% of the surficial sediment. This is likely due to lower rates of biogenic sedimentation. We attribute the minor role of ice rafting to rapid movement of icebergs across the continental shelf, due to strong katabatic winds

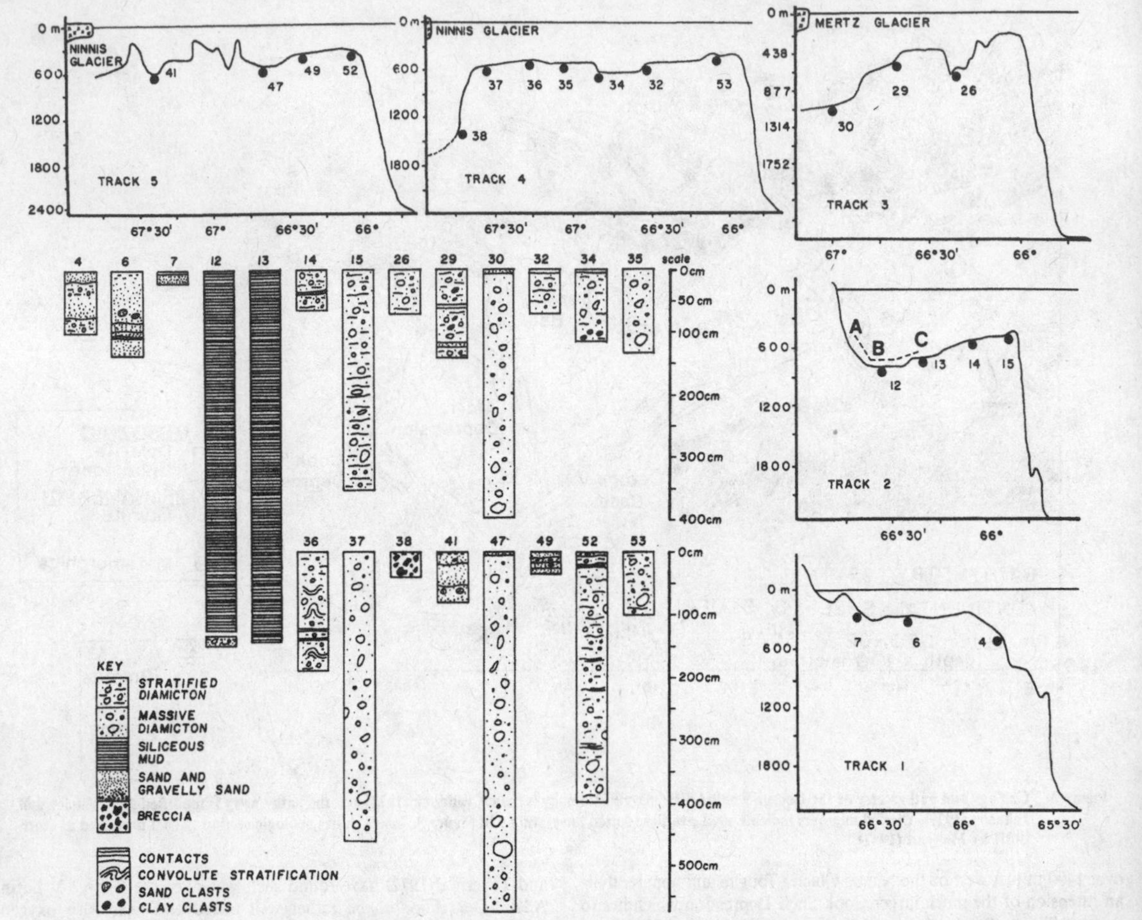

**Figure 3. Slope-shelf bottom profiles and core lithologies for piston cores collected along five transects (see Figure 2 for location of track lines). Line A-B-C on track two refers to location of profile in Figure 4.**

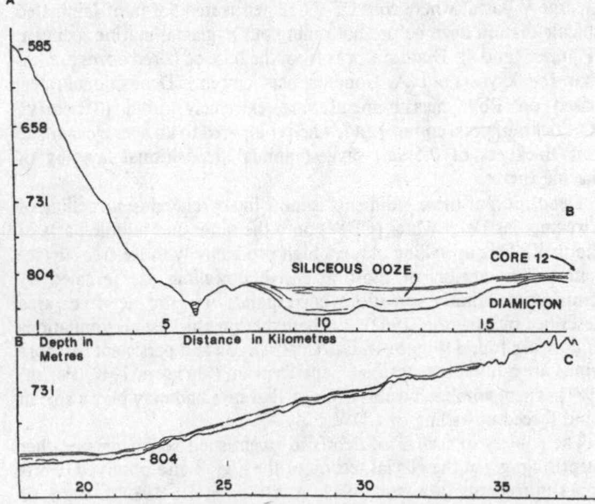


**Figure 4. Distribution of siliceous ooze along ship's track 2 (see Figure 3). Core 12 penetrated 5.9 m of siliceous ooze. The contact between siliceous ooze and underlying glacial and glacial marine diamicton gives a strong reflection on the 12 KHZ bottom profiles.**

and to subzero surface waters. Meltwater deposits are totally lacking in the area.

Given the great depth of the continental shelf, waves and wind driven currents have little influence on sedimentation. In their absence, gravity driven processes and deep sea currents play the most active roles in sediment reworking and transport. Slumping, debris flow transport and turbidity current transport are active on the continental shelf and are attributed to the glacial relief of the shelf. These slumps and flows follow the local relief and hence may move in any direction. These processes are more active in the far western and eastern parts of the shelf. Normally graded sand units, 50 to 100 cm thick, are interbedded with diamictons in cores 4 and 49 (Figure 3). Diamictons in cores 29 and 34 contain deformed clasts of diatomaceous ooze or clay (Figure 3). In some cases, turbidites have apparently flowed toward the coast, down the local slope (e.g. core 6, Figure 3).

Along the shelf break, impinging geostrophic currents sort and rework glacial-marine deposits to produce a thin (approximately 5 cm thick) gravel and sand lag. The cohesive and overcompacted state of the relict glacial-marine deposits implies that organisms play a key role in rendering them more susceptible to bottom current erosion. As seen in Figure 5, sands as coarse as 3.25 $\phi$ have been winnowed by these currents. Flume experiments by Singer (1982) have shown that this degree of sediment erosion and transport is possible, even in overcompacted basal tills, at velocities of less than 20 cm/s provided biological mixing is pronounced.

At least part of the find sand, silt and clay winnowed from relict glacial-marine sediments is transported onto the shelf by impinging currents (Figure 5). Biogenic sediments collected in depressions on the shelf contain up to 20% terrigenous sediment in the find sand to silt-sized range which is believed derived in this manner. Some of the fine sediment winnowed at the shelf break is likely transported along the slope by contour currents and eventually down slope by turbidity currents.

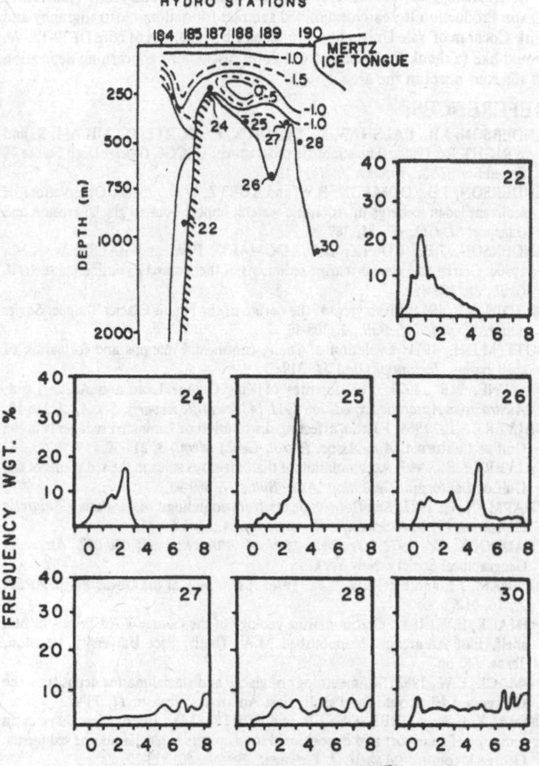


**Figure 5.** **Water column temperature profile and grain-size data for bottom samples collected along track three. Note tongue of relatively warm Circumpolar Deep Water that impinges onto the continental shelf. Bottom sample 22, taken from the upper slope, consists of coarse sand and gravel. Samples 24, 25 and 26 show fining in an onshore direction, away from the influence of this impinging current. Samples 27, 28 and 30, collected on the inner shelf, consist of fine silt and clay and reflect quiescent bottom conditions.**

*Subsurface deposits*

Piston cores were collected along five shelf transects in the area (Figure 3). By far the dominant sediment cored is relict diamicton of glacial and glacial-marine origin. The distinction between glacial-marine deposits and basal till is based on a number of criteria which are outlined in Anderson et al. (1980c). Studies of clast shape and a consideration of unit thickness suggest that lodgment processes were responsible for basal till deposition and that glacial-marine sediments were derived primarily from basal debris zones (Domack et al., 1980a).

Basal tills were recovered on the inner part of the shelf at depths which range from 398 to 1078 m. Their presence provides direct evidence that glacial ice formerly was grounded on the continental shelf (Anderson et al., 1980b; Domack, 1982). Furthermore, heavy minerals and coarse sand lithologies from these sediments indicate that ice flow directions during the latest glacial maximum were to the northwest (Domack, 1982). This is based on the high percentages of dolerite, quartz, pyroxene and garnet comprising these sediments. This mineral assemblage was most likely derived from the Mesozoic sandstones and dolerites exposed east of the Ninnis Glacier (Mawson, 1940; Browne, 1923). Small amounts of minerals derived from Precambrian rocks exposed west of the Ninnis Glacier indicate a minor contribution of ice from this region.

Ice-rafted diamictons or glacial-marine sediments are widespread in the Mertz Depression and on the outer continental shelf, where they exhibit some characteristics of basal till. These sediments were deposited from the melting of basal debris zones seaward of the grounding line of locally grounded ice shelves (Domack, 1980, 1982).

Glacial and glacial-marine diamictons represent at least two periods of deposition during the Quaternary. Piston cores of unlithified till and glacial-marine sediment (Figure 3) contain abundant clasts of lithified diamicton (Domack, 1980, 1982). These clasts contain marine and nonmarine diatoms of Miocene to Recent age (D. Kellogg, pers. comm.). This suggests that a period of glacial reworking, sedimentation and lithification occurred prior to the most recent episode of glacial sedimentation.

*Pre-Quaternary deposits and continental margin development*

The oldest sediment collected on the continental shelf is a non-marine, organic rich siltstone of Early Cretaceous age. This sediment was recovered in piston core DF-79-38 from the George V Basin (Figure 3). It contains a diverse and well preserved spore-pollen assemblage of Aptian age and therefore predates the breakup of Australia and Antarctica (Domack et al., 1980b). Earlier, brown lignites were dredged from the flanks of the George V Basin just west of the Mertz Glacier Tongue (Mawson, 1942). A Cretaceous age for these deposits has also been determined (E. Truswell, pers. comm.).

Figure 6 shows a comparison of the geologic and physiographic settings of the George V margin with its paired margin off southern Australia. The cross-section in Figure 6 is based on best-fit reconstructions (e.g. Sproll and Dietz, 1969; Griffiths, 1974) and structure profiles constructed from seismic data (Denham and Brown, 1976). An interesting comparison centres around the deep physiography and shallow occurrence of the Lower Cretaceous section on the George V margin relative to the southern Australian margin (Figure 6).

Although the exact thickness of the Lower Cretaceous section on the George V margin is unknown, it occurs at least 1400 m below sea level, or 900 m below the surface of the adjacent continental shelf (Figure 6). Lower Cretaceous deposits on the southern Australian margin occur at much greater depths in the subsurface (up to 2 km and more commonly at 3 to 4 km, Denham and Brown, 1976; Falvey and Mutter, 1981). A maximum of 900 m of strata presently separates Lower Cretaceous deposits from Quaternary deposits on the George

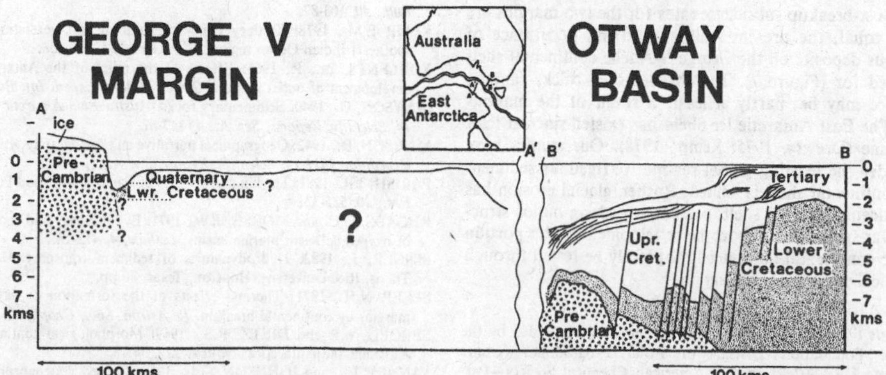


**Figure 6.** **Comparison of continental margin structure and stratigraphy between the George V margin (A-A') and southern Australian margin (B-B'). Geology of the Otway Basin is from Denham and Brown (1976). A fault is inferred on the George V margin where Precambrian granites are juxtaposed next to Cretaceous sediments.**

V continental shelf, while up to 2 km of Upper Cretaceous and Tertiary sediments overlie Lower Cretaceous strata on the southern Australian continental shelf. Water depths average 500 m on the George V continental shelf and are typically much deeper (Figures 2 and 3), which is typical of the Antarctic continental shelf as a whole. The southern Australian continental shelf is much shallower; the shelf break occurs at approximately 200 m. The relationships shown in Figure 6 imply that the George V continental margin has a very different sedimentary and subsidence history from that of its paired margin off southern Australia. Although the stratigraphy of the George V continental shelf is largely unknown, it is interesting to speculate about the reasons for these differences.

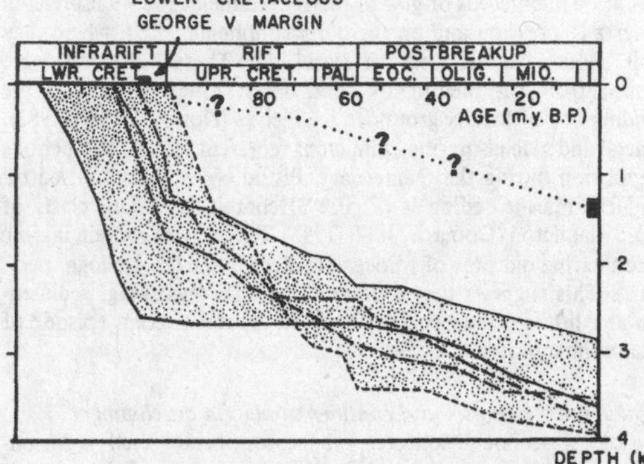


**Figure 7. Subsidence history diagram for Lower Cretaceous, nonmarine sediments recovered in five exploratory wells from Australia's southern margin (from Falvey and Mutter, 1981). If one assumes an equal amount of postbreakup (thermal) subsidence for Lower Cretaceous of the George V margin (dotted line), then significantly less pre-breakup subsidence is implied. Present depths of Lower Cretaceous are on right and individual well histories are shown by dashed and continuous lines.**

A number of subsidence mechanisms are thought to be important in the evolution of passive continental margins. They include, sediment loading and isostatic downwarp (Walcott, 1972), lithosphere cooling (Sleep, 1971), stretching of the continental crust (Bott, 1971) and deep crustal metamorphism (Falvey, 1974). Falvey (1974) and Falvey and Mutter (1981) have recognised the relative importance of these mechanisms during the infra-rift, rift and post-breakup phases in the development of the southern Australian margin. Based on cumulative subsidence histories for exploration wells from Australia's southern margin, they infer that deep crustal metamorphism of greenschist facies rocks to amphibolite resulted in approximately 1500 m of infra-rift (pre-breakup) subsidence (Figure 7). In contrast, the Precambrian upper crustal rocks of the George V-Adélie margin consist of granites, amphibolites and granulite facies metamorphics (Mawson, 1940). Therefore, lower pre-breakup subsidence rates due to this mechanism are implied. If post-breakup subsidence rates for the two margins are assumed to be equal, the present shallow (1407 m) occurrence of Lower Cretaceous deposits on the George V-Adélie continental shelf can be accounted for (Figure 7). The absence of a thick, capping Tertiary sequence may be, partly at least, a result of the margins glacial setting. The East Antarctic Ice Sheet has existed since at least Mid-Miocene time (Drewry, 1975; Kemp, 1978). Our studies have shown that, under the existing glacial regime, terrigenous sedimentation on the continental shelf is limited. Rather, glacial erosion has apparently cut deeply into the shelf, particularly along major structural features. Our speculations about the development of this portion of the Antarctic continental margin will hopefully be tested through marine geophysical surveys of the margin.

*Acknowledgements* Financial support for this research was provided by the National Science Foundation—Division of Polar Programmes (grants DPP77—26407 and DPP-80242) and the American Chemical Society—Petroleum Research Fund (grants PRF-11101-AC2 and PRF-2472-AC2). We wish to thank the officers and crew of the USCGC *Glacier* for their assistance during our marine geologic cruise.

Dr D. Kellogg of the University of Maine at Orono and Andy Gombos of Exxon Production Research examined samples for diatom biostratigraphy and Kirk Cochran of Yale University conducted $Pb^{210}$ analyses of core DF79-12. We would like to thank R. Dunbar for helpful discussions concerning deposition of siliceous oozes in the area.

## REFERENCES

ANDERSON, J.B., BALSHAW, K., DOMACK, E., KURTZ, D., MILAM, R. and WRIGHT, R., 1980a: The scientific programme: USCGC *Glacier* Deep Freeze 79 Expedition: *U.S. Antarct. J., 14,* 142-4.

ANDERSON, J.B., DOMACK, E.W. and KURTZ, D.D., 1980b: Observations of sediment laden icebergs in Antarctic waters: implications to glacial erosion and transport. *J. Glaciol., 25,* 387-96.

ANDERSON, J.B., KURTZ, D.D., DOMACK, E.W. and BALSHAW, K.M., 1980c: Glacial and glacial-marine sediments of the Antarctic continental shelf. *J. Geol., 88,* 399-414.

BARDIN, V.I., 1964: Fresh view of the nature of the Ninnis Glacier Tongue: *Soviet Antarc. Exped. Inf. Bull., 2,* 308-11.

BOTT, M.P.H., 1971: Evolution of young continental margins and formation of shelf basins. *Tectonophysics, 11,* 319-27.

BROWNE, W.R., 1923: The dolerites of King George Land and Adélie Land. *Australasian Antarctic Expedition 1911-14. Scientific Reports, Ser. A., 3,* 246-58.

CALVERT, S.E., 1964: Factors affecting distribution of laminated sediments in the Gulf of California. *Am. Assoc. Petrol. Geol., Mem., 3,* 311-30.

CALVERT, S.E., 1966: Accumulation of diatomaceous silica in the sediments of the Gulf of California. *Geol. Soc. Am., Bull., 77,* 569-96.

CHAPMAN, F., 1922: Sea floor deposits from soundings. *Australasian Antarctic Expedition 1911-14. Scientific Reports, Ser. A., 2,* 1-60.

CRADDOCK, C., 1972: *Geologic Map of Antarctica. 1:5,000,000.* American Geographical Society, New York.

DENHAM, J.I. and BROWN, B.R., 1976: A new look at the Otway Basin. *APEA J., 16,* 91-8.

DOMACK, E.W., 1980: Glacial-marine geology of the George V-Adélie continental shelf, East Antarctica. Unpublished M.A. Thesis, Rice University, Houston, Texas, 142 pp.

DOMACK, E.W., 1982: Sedimentology of glacial and glacial-marine deposits on the George V-Adélie continental shelf, East Antarctica. *Boreas, 11,* 79-97.

DOMACK, E.W., ANDERSON, J.B. and KURTZ, D.D., 1980a: Clast shape as an indicator of transport and depositional mechanisms in glacial-marine sediments: George V continental shelf. *J. Sediment. Petrol., 50,* 813-20.

DOMACK, E.W., FAIRCHILD, W.W. and ANDERSON, J.B., 1980b: Lower Cretaceous sediment from the East Antarctic continental shelf. *Nature, 287,* 625-6.

DREWRY, D.J., 1975: Initiation and growth of the East Antarctic ice sheet. *Geol. Soc. Lond., J., 131,* 255-73.

DREWRY, D.J., 1976: Sedimentary basins of the East Antarctic craton from geophysical evidence. *Tectonophysics, 36,* 301-14.

DREWRY, D.J. and COOPER, A.P.R., 1981: Processes and models of Antarctic glaciomarine sedimentation: *Annals Glaciol., 2,* 117-22.

FALVEY, D.A., 1974: The development of continental margins in plate tectonics theory. *APEA J., 14,* 95-106.

FALVEY, D.A. and MUTTER, J.C., 1981: Regional plate tectonics and the evolution of Australia's passive continental margins. *BMR J. Aust. Geol. Geophys., 6,* 1-29.

GORDON, A.L. and TCHERNIA, P., 1972: Waters of the Continental margin off Adélie Coast, Antarctica; *in* Hayes, D.E. (ed.) Antarctic Oceanography II: The Australian-New Zealand Sector. *Am. Geophys. Union Ant. Res. Ser., 9,* 59-69.

GRIFFITHS, J.R., 1974: Revised continental fit of Australia and Antarctica. *Nature, 249,* 336-38.

GRINNELL, D.V., 1971: Physiography of the continental margin of Antarctica from 125°E to 150°E. *U.S. Antarct. J., 6,* 164-5.

HOLTEDAHL, O. and HOLTEDAHL, H., 1961: On marginal channels along continental borders and the problems of their origins. *Univ. Uppsala, Geol. Inst. Bull., 40,* 103-87.

KEMP, E.M., 1978: Tertiary climatic evolution and vegetation history in the southeast Indian Ocean region. *Mar. Geol., 24,* 169-208.

KOBLENTS, Ya., P., 1965: Effects of the relief of the Antarctic shelf on the development of outlet glaciers: *Soviet, Antarc. Exped. Inf. Bull., 3,* 5-9.

MAWSON, D., 1940: Sedimentary rocks. *Australasian Antarctic Expedition 1911-14, Scientific Reports, Ser. A., 4,* 347-67.

MAWSON, D., 1942: Geographical narrative and cartography. *Australian Antarctic Expedition 1911-14, Scientific Report, Ser. A., 1,* 1-364.

PARISH, T.R., 1981: The katabatic winds of Cape Denison and Port Martin. *Polar. Rec., 20,* 525-32.

RHOADS, D.C. and MORSE, J.W., 1971: Evolutionary and ecologic significance of oxygen deficient marine basins. *Lethaia, 4,* 413-28.

SINGER, J., 1982: Hydrodynamics of sediment transport. Unpublished M.A. Thesis, Rice University, Houston, Texas, 80 pp.

SLEEP, N.H., 1971: Thermal effects of the formation of Atlantic continental margins by continental breakup. *R. Astron. Soc., Geophys J., 24,* 225-50.

SPROLL, W.P. and DIETZ, R.S., 1969: Morphological continental drift fit of Australia and Antarctica. *Nature, 222,* 345-8.

VANNEY, J.R. and JOHNSON, G.L., 1979: The sea floor morphology seaward of Terre Adélie (Antarctica). *Dtsch. Hydrogr. Zeitschr., 32,* 39-88.

WALCOTT, R.I., 1972: Gravity, flexure and the growth of sedimentary basins at a continental edge. *Geol. Soc. Am., Bull., 83,* 1845-8.

# BENTHIC FORAMINIFERA OF McMURDO SOUND

B.L. Ward, *Department of Geology and Antarctic Research Centre, Victoria University, Wellington, New Zealand.*

*Abstract* Sediment samples containing foraminifera have been collected in McMurdo Sound from a range of water depths and physiographic situations, using the sea ice as an operating platform. Splits were treated with rose bengal stain, and tests containing stained protoplasm extracted for further study. Non-stained tests were also separated for comparative purposes. Three areas of varying foraminiferal distribution have been identified as follows: 1) below 560 m there exists an assemblage of agglutinated foraminifera with *Reophax* spp. as the dominant taxa; 2) between 560 and about 210 m there is a mixed assemblage, again with *Reophax* spp. as the dominant agglutinated taxa, and *Trifarina earlandi, Globocassidulina* cf. *subglobosa* and *Cassidulinoides porrectus* as the dominant calcareous taxa; this includes the Granite Harbour area; and 3) the New Harbour area supports and agglutinated population similar to that found below 560 m in the open Sound.

Comparison of living (stained) and dead assemblages from the top 20 mm of five 22 cm-diameter cores indicates that post-mortem alteration of assemblages, specifically, disappearance of calcareous tests, increases progressively with greater water depth until the carbonate compensation depth (CCD) is reached, somewhere between 560 and 850 m. The difference in proportions of calcareous and agglutinated foraminifera in live and dead assemblages increases the difficulty in the ecological interpretation of ancient (dead) assemblages.

McMurdo Sound lies at the southern end of the western Ross Sea between Ross Island on the east and southern Victoria Land on the west. Due to its proximity to McMurdo Station and Scott Base, the Sound is one of the most intensively studied areas of Antarctica. Despite this, surprisingly little is known of the general ecology and physiography of the sea floor. This is exemplified by the fact that no satisfactory bathymetric map of the region was available for the present research.

Previous studies of fossil and Recent foraminifera in and around McMurdo Sound include those by Chapman (1916), Kennett (1968), Webb and Neall (1972), Fillon (1974), Kellogg et al. (1977), Osterman and Kellogg (1979) and Webb and Wrenn (1982). Those dealing with modern faunas did not differentiate between living and dead assemblages or concentrate on McMurdo Sound as an ecological unit.

## Purpose

The objective of this project is to establish distribution patterns for living foraminifera in McMurdo Sound by examining sediment samples from a range of water depths and physiographic situations. This data may be useful in: (1) serving as comparative material for ancient sediments that no longer contain complete foraminiferal assemblages, and thus aid in establishing palaeoenvironmental conditions for those sediments; and (2) providing baseline data for the McMurdo Sound area in the event it is modified by future human activities, such as mineral resource exploitation.

## Collection of Samples

Sediment samples containing foraminifera were collected from a wide range of water depths (8 to 850 m) during three consecutive field seasons. The annual sea ice was used as an operating platform, restricting the area and time of collection, as the ice becomes soft in mid to late December.

The most useful sediment samples were short cores retrieved with a wide diameter (22 cm) gravity sphincter corer designed and built at Victoria University. The cores are undisturbed sections of the sea floor which were photographed and described before being split and preserved in alcohol for transport to New Zealand. The other methods used to obtain sediment samples were a short 5 cm-diameter piston corer, a McIntyre grab and an orange-peel grab. The piston corer did not operate satisfactorily, and the McIntyre grab required a large access hole (2 m) in the sea ice, thus limiting the number of locations that could be sampled in a given time. The orange-peel grab was efficient, but the sample was washed as it was brought to the surface due to the open nature of the grab, and the top 10 cm of sediment was mixed as the grab closed.

Twenty sediment samples collected during the 1981/82 season using the gravity sphincter corer provided the most useful foraminiferal data. Fourteen were intact 22 cm diameter cores, ranging in length from 6 to 56 cm. Six samples were disturbed when the corer fell over as it penetrated the sea floor. This was in areas of coarse substrate, which inhibited penetration of the corer. The longest cores came from Granite Harbour (Cores 14, 15 and 16) and New Harbour (Core 18), where the sea floor sediment is fine-grained and nearly homogeneous, containing only a few scattered pebble- and cobble-sized clasts. Figure 1 shows the sample site locations of cores discussed below.

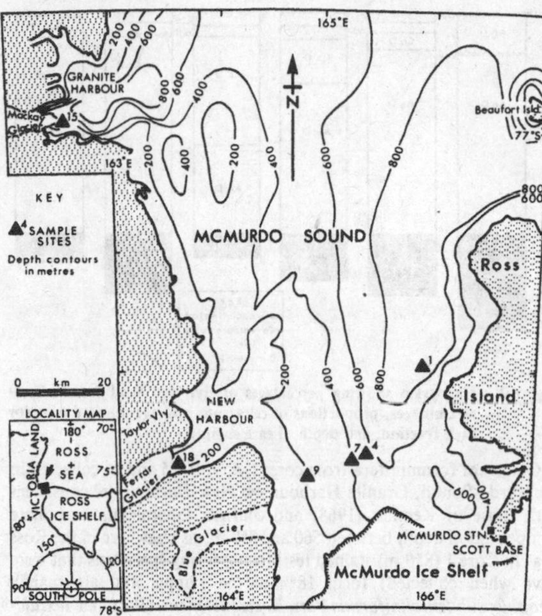


**Figure 1. Map of McMurdo Sound region showing sample site locations for cores 1, 4, 7, 9, 15 and 18.**

## Laboratory Procedures

Portions of the top 20 mm of the sediment samples were washed over a 63 μm brass sieve. The residue retained on the sieve was soaked in a solution of rose bengal stain and ethanol for 45 to 60 minutes (Walton, 1952). The material was rinsed and dried, then floated in carbon tetrachloride ($CCl_4$) to concentrate the biogenic material (Murray, 1979). The floated fractions were split to a size containing approximately 300 to 500 foraminifera tests, which were sorted under the microscope and mounted on faunal slides. Stained tests (tests containing stained protoplasm, and thus considered to have been alive at time of collection) and unstained tests were mounted on separate slides. The heavy residues that sank in the $CCl_4$ were also examined for any foraminifera, which were added to the collections on the slides. Population counts were made from these faunal slides.

## Problems

There is some discussion in the literature (e.g. Walker et al., 1974) concerning the reliability of the rose bengal staining technique. For example, there is some problem in seeing through the test walls of large miliolids and agglutinated forms, necessitating breaking open each individual test to see if it contained stained protoplasm. As the samples used in the present study contained relatively few of these forms, breaking them the examine the interior did not involve much effort. Another problem is the exterior of the tests often being stained. This is easy to differentiate from stained protoplasm under high (x36 to x72) magnification.

A more serious problem involves the staining of bacteria or algae that might be living inside a dead foraminiferal test. If other chambers of the test contain sediment, it can be discounted as being a live test. Also, if an area around a broken part of the test is stained, these are not counted as living forms. Other factors that can contribute to errors being introduced into the calculation of living/dead ratios are: small numbers of tests counted, splitting technique error, and the apparent "patchiness" in the distribution of living populations on the sea floor (Shifflett, 1961).

## FORAMINIFERAL POPULATION COUNTS

### Live: Dead Ratios (See Figure 2)

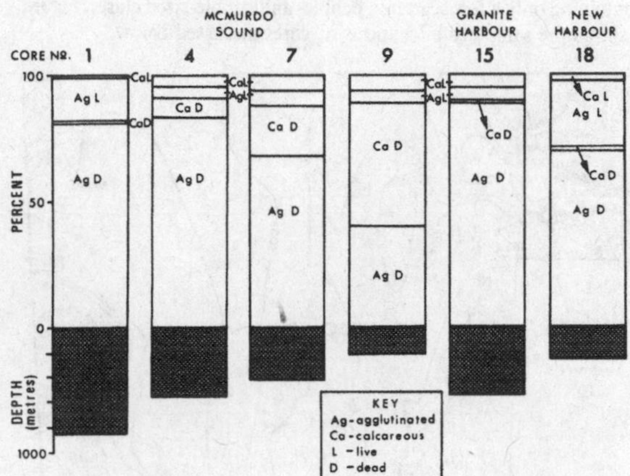


**Figure 2. Histogram showing percentages of live and dead foraminiferal assemblages, proportions of calcareous and agglutinated tests for each fraction, and depth of each sample.**

Counts of foraminifera from cores 1, 4, 7, 9, 15 and 18 collected in McMurdo Sound, Granite Harbour and New Harbour are consistent with those of Kennett (1968) and confirm the calcium carbonate compensation depth between 560 and 850 m, as elsewhere in the Ross Sea. In Core 1 (850 m) stained tests (representing specimens that were alive when collected) form 18% of the sample and are mainly *Reophax subdentaliniformis* Parr, *R. kerguelenensis* Parr and *Textularia antarctica* (Wiesner), with less common *Reophax pilulifer* Brady. Only one stained calcareous test was found, a two-chambered juvenile of *Lenticulina* sp.

Stained calcareous tests form about 4% of the assemblage in Core 4 (560 m) and include, in order of decreasing abundance, *Cassidulinoides porrectus* (Heron-Allen and Earland), *Globocassidulina subglobosa ?* (Brady), *G. crassa* (d'Orbigny), *Trifarina earlandi* (Parr) and *Cibicides lobatulus* (Walker and Jacob). Stained agglutinated tests present are *R. subdentaliniformis, R. kerguelenensis, R. pilulifer* and *Portotrochammina* spp. The numbers of stained agglutinated and calcareous tests are nearly equal, and together form 8% of the total count. Core 7 (420 m) also has approximately equal numbers of stained agglutinated and calcareous tests. Both of these mixed assemblages are within the range of Kennett's (1968) mixed faunal zone and above his CCD.

Core 15, from 550 m in Granite Harbour, has 9.3% stained tests, similar to the above samples, and has 45% calcareous to 55% agglutinated stained tests. The most numerous living (stained) species are the agglutinated *Textularia antarctica* and the calcareous *Fursenkoina* cf. *davisi* (Chapman and Parr). Core 15 is unusual in that the proportion of calcareous tests in the dead assemblage (0.7%) is much smaller than in the living population (45%).

Stained calcareous and agglutinated tests are again present in nearly equal numbers in Core 9 (213 m), 200 m above the top of Kennett's zone of mixed faunas. The stained tests form about 10% of the total count, again similar to the proportions of stained tests found in Cores 4, 7 and 15. The dominant taxa are the calcareous *Trifarina earlandi* and *Ehrenbergina glabra* Heron-Allen and Earland, and the agglutinated *Portotrochammina antarctica* (Parr).

An agglutinated fauna characterises the embayed area of New Harbour. The dominant stained taxa in Core 18 (254 m) are *Reophax subdentaliniformis* and *R. kerguelenensis.* There is an extremely restricted calcareous fauna comparable to that described from Core 1 (850 m in McMurdo Sound). On the basis of depth, this sample should also fall within the zone of mixed faunas described by Kennett (1968). Calcareous species form 6% of the living tests in this core top, compared to the usual 50% in the sample from the open Sound waters.

The evidence from Cores 15 and 18 implies that the waters in Granite Harbour and New Harbour have a lower pH than in the open waters of McMurdo Sound. The very low numbers of calcareous stained tests in the New Harbour sample would indicate that the pH there is lower than in Granite Harbour but similar to that of the deeper Sound waters, that is, those below the CCD.

## SUMMARY

Figure 3 summarises the agglutinated/calcareous test ratios which demonstrate the progressive change of the ratio by loss of calcareous

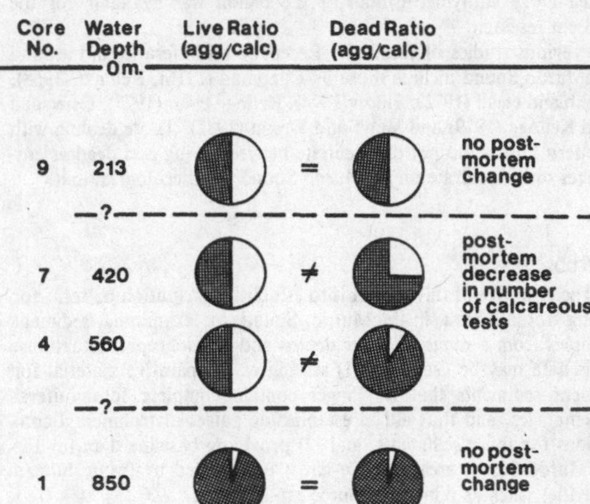


**Figure 3. Summary of amount of post-mortem alteration of foraminiferal assemblages with proximity to the CCD.**

tests as the CCD, between 850 and 560 m in McMurdo Sound, is approached. Well above this zone and below it the post-mortem alteration of assemblages is minimal. It is evident that study of fossil assemblages must involve consideration of how they might differ from the living populations they represent. The removal of calcareous tests by dissolution or other means evidently occurs quite rapidly among non-living foram tests. Some investigation of the palaeo-CCD should go hand-in-hand with study of fossil collections, since the amount of alteration seems to increase with proximity to the CCD.

## CONCLUSIONS

(1) The lower limit of the CCD, as defined by Kennett (1968) lies between 850 and 560 m (supported by the present data), above which living calcareous and agglutinated foraminfera occur in about equal proportions. However, an agglutinated deep water assemblage is also found in 250 m of water in a restricted basin in New Harbour, perhaps due to bottom water of lower pH.

(2) The proportion of calcareous foraminifera in dead assemblages decreases progressively with water depth in the open water of the Sound, clearly a post-depositonal effect. If live and dead assemblages are not distinguished in studies of modern foraminiferal populations, incorrect conclusions on palaeoecology and water depth can be drawn. Furthermore, the difference in proportions of calcareous and agglutinated foraminifera in live and dead assemblages increases the

difficulty in the ecological interpretation of ancient (dead) assemblages.

*Acknowledgements* Fieldwork was supported by Victoria University Antarctic Expeditions 24, 25 and 26. Professor P. Vella provided assistance with foraminifera identifications and critical discussions of the manuscript. P.J. Barrett and A.R. Pyne offered valuable criticism and discussion. Bathymetric data compiled by B.L. Ward and P.J. Barrett.

## REFERENCES

CHAPMAN, F., 1916: Report on the foraminifera and ostracoda from elevated deposits on the shores of the Ross Sea and out of marine muds from soundings in the Ross Sea. *Rep. Brit. Antarct. Exped. 1907/09, Geol., 2 (2, 3)*,27-80.

FILLON, R.H., 1974: Late Cenozoic foraminiferal paleoecology of the Ross Sea, Antarctica. *Micropaleontology, 20*, 129-51.

KELLOGG, T.B., STUIVER, M., KELLOGG, D. and DENTON, G.H., 1977: Marine microfossils on the McMurdo Ice Shelf. *U.S. Antarct., J., 12*, 82-3.

KENNETT, J.P., 1968: The fauna of the Ross Sea-Part 6: Ecology and distribution of foraminifera. *Bull. N.Z. Dep. Scient. ind. Res., 186.*

MURRAY, J.W., 1979: British Nearshore foraminiferids; *in* Kermack, D.M. and Barnes, R.S.K. (eds.) *Synopses of the British Fauna (New Series), 16*, Academic Press, London.

OSTERMAN, L.E. and KELLOGG, T.B., 1979: Recent benthic foraminifera distributions from the Ross Sea, Antarctica. *J. Foramin. Res., 9*, 250-69.

SHIFFLET, E., 1961: Living, dead and total foraminiferal faunas, Heald Bank, Gulf of Mexico. *Micropaleontology, 7*, 45-54.

WALKER, D.A., LINTON, A.E. and SCHAFER, C.T., 1974: Sudan Black B: a superior stain to rose bengal for distinguishing living from non-living foraminifera. *J. Foramin. Res., 4*, 205-15.

WALTON, W.R., 1952: Techniques for recognition of living foraminifera. *Contrib. Cushman Found. Foramin Res., 3*, 56-64.

WEBB, P.N. and NEALL, V.E., 1972: Cretaceous foraminifera in Quaternary deposits from Taylor Valley, Antarctica; *in* Adie, R.J. (ed.) *Antarctic Geology and Geophysics*, Universitetsforlaget, Oslo, 653-7.

WEBB, P.N. and WRENN, J.H. 1982: Late Cenozoic micropaleontology and biostratigraphy of eastern Taylor Valley, Antarctica; *in* Craddock, C., (ed.) *Antarctic Geoscience*, Univ. Wisconsin Press, Madison, 1117-22.

# SILICEOUS BIOSTRATIGRAPHY AND THE ROSS SEA EMBAYMENT

Hsin Y. Ling, *Department of Geology, Northern Illinois University, DeKalb, Illinois 60115, USA.*

*Abstract* The water of the Ross Sea has been characterised by: (1) the unusually low temperature; (2) a rather high salinity; (3) the low to zero productivity due to packed ice in the surface water; and (4) the rather shallow calcium carbonate compensation depth caused by gradual accumulation of $CO_2$.

The development of this water mass is closely related to the tectonic and the glacial history of this part of Antarctica. To evaluate the effects and the timing of oscillation of grounded West Antarctic ice on the Ross Sea sediments and biota, cored sediments along the Pennell Coast margin in north-south direction have been selected and are being analysed for the siliceous microfossil assemblages of radiolarians, silicoflagellates and ebridians.

Such an attempt would bring southern ocean siliceous biostratigraphy inward towards the southern end of the Ross Ice Shelf where subbottom sediments of Neogene age have already been recognised through the attempts of Deep Sea Drilling Project (DSDP), Ross Ice Shelf Project (RISP), and McMurdo Sound Sediments and Tectonic Studies (MSSTS).

# 7

# Antarctic Resources

# MINERAL RESOURCES POTENTIAL IN ANTARCTICA—REVIEW AND PREDICTIONS

J.F. Splettstoesser, *Minnesota Geological Survey, University of Minnesota, 1633 Eustis Street, St. Paul, Minnesota 55108, USA*

**_Abstract_ Minerals, some of economic potential, have been found in Antarctica in considerable variety and in many locations. All are non-commercial on the basis of current market prices. Some, however, merit further study because of their size and possible accessibility in the event that technology and logistics might permit exploitation at some time in the future.**

**Many mineralised occurrences indicate geologic extensions in Antarctica of mineral deposits found in the other southern hemisphere components of Gondwanaland, viz. (1) the widespread coal beds throughout the Transantarctic Mountains, and also in the Beaver Lake area in the Prince Charles Mountains (PCM); (2) Precambrian iron-formation in the PCM, at Mounts Ruker and Stinear; (3) the Dufek intrusion, a layered gabbroic complex including iron, platinum-group metals, vanadium, and other elements; (4) the metallic mineralisation in the Antarctic Peninsula; (5) the hydrocarbon potential of parts of the continental shelf around Antarctica.**

**Coal in Antarctica is not particularly significant as a resource because of the complicated logistics related to its exploitation, and also because coal reserves are relatively abundant in other parts of the world. Banded iron-formations in the Prince Charles Mountains, occurring as jaspilite beds as much as 70 m thick at Mount Ruker, represent a fairly large deposit of iron, including the substantial amounts under the ice sheet that have been inferred from aeromagnetic surveys. These deposits include as much as 58 percent total iron oxides, and are relatively near the coast. The Dufek Intrusion, a stratiform mafic igneous complex of Jurassic age, includes an iron-enriched unit 1.7 m thick in the Forrestal Range.**

**Metallic mineralisation of a variety of elements is known from numerous locations in the Antarctic Peninsula. Most deposits of metallic mineralisation in the Antarctic Peninsula are of hydrothermal origin and are associated with Mesozoic and Tertiary plutons of the Andean Intrusive Suite.**

**The continental shelf, particularly in the Ross Sea and the Weddell Sea, may hold possible oil or gas reserves, but until a geophysical data base is assembled from detailed surveys, any potential is pure speculation.**

**Prospects for any development or exploitation of resources in Antarctica for any of the above, or for any other potential resource, depend in all cases on further exploration, perhaps by commercial firms. To be carried out properly, this exploration phase should be planned with environmental concerns in mind. Until the Antarctic Treaty nations are in accord as to propriety and ownership of resources, controls should be placed on any development. Exploration of hydrocarbons is predicted to receive the most attention in the next decade, with production expected by the year 2000.**

# METALLOGENIC PROVINCES OF ANTARCTICA

P.D. Rowley, *U.S. Geological Survey, Denver, Colorado 80225, U.S.A.*

A.B. Ford, *U.S. Geological Survey, Menlo Park, California 94025, U.S.A.*

P.L. Williams, *U.S. Geological Survey, Denver, Colorado 80225, U.S.A.*

D.E. Pride, *Department of Geology and Mineralogy, Ohio State University, Columbus, Ohio 43210, U.S.A.*

*Abstract* Compilation of information on occurrences of metallic minerals known in Antarctica permits crude definition of metallogenic provinces. No occurrence currently known on the continent is economically exploitable today. The oldest metallogenic province is the East Antarctica iron metallogenic province, which consists of two parts: an iron-formation subprovince of Archaean and Proterozoic age extending from Enderby Land to Wilkes Land and containing widespread jaspilite deposits; and an iron oxide vein subprovince of Archaean to Early Mesozoic age in Queen Maud Land containing veins of magnetite and possible related sulphide minerals that may have been remobilised by hydrothermal processes acting on older iron source terrains. Possible future discoveries in this province are additional syngenetic iron-formation, but gold in mafic volcanic rocks, gold-uranium in conglomerate, and nickel, chromium, and related metals in ultramafic intrusions also may occur in the craton. The Transantarctic metallogenic province consists of two parts: a Ross subprovince of Late Proterozoic (?), Early Palaeozoic, and perhaps Middle and Late Palaeozoic age containing copper, other base metals and precious metals associated with silicic intrusions; and a Ferrar subprovince of Middle Jurassic age that contains iron, copper, cobalt, chromium, nickel , and platinum in stratiform mafic bodies such as the Dufek intrusion. Targets for future discoveries include faulted rocks near the contacts of Lower Palaeozoic intrusive rocks as well as sills of the Middle Jurassic Ferrar Group. The Andean metallogenic province consists of two parts: a broad copper subprovince containing porphyry copper deposits and hydrothermal veins; and a minor narrow iron subprovince of magmatic iron minerals that overlaps the western part of the copper subprovince. The presence of this western iron subprovince argues against Gondwana reconstructions locating the Antarctic Peninsula east or west of southern South America. The most likely future discoveries will be porphyry copper deposits in intrusive rocks beneath their volcanic covers.

Interest in metallic mineral resources of Antarctica and in the possibility of exploitation of these resources have greatly increased during the last decade. During this time, the topic of mineral resources has been the subject of diplomatic conferences and Antarctic Treaty negotiations. Rowley et al. (in press) and Splettstoesser (this volume) have summarised the many reports written on the climatic, logistic, geologic, and political constraints to development of the mineral resources of Antarctica. Most writers have concluded, and we agree, that it is unlikely that any metallic mineral resources on the continent will be economically developed in the foreseeable future.

Despite a lack of mineral deposits of economic potential in Antarctica, the likelihood of finding workable deposits has long been thought to be high because mineral deposits are abundant in other Gondwanaland continents. This report attempts to define metallogenic provinces in Antarctica on the basis of scattered occurrences (see also Kameneva and Grikurov, this volume). Active research on their occurrence, distribution, and genesis will permit proper decisions on their future use.

## GEOLOGIC SETTING

Antarctica can be divided into eastern and western parts. East Antarctica consists of a Precambrian craton and, along its boundary with West Antarctica, occurs the Transantarctic Mountains. By analogy with most other Gondwana cratons, the East Antarctica craton probably consists of Archaean crystalline "nuclei" welded together during Proterozoic and Early Palaeozoic time. Rock exposures, however, are limited mostly to narrow strips along the coastline; thus Archaean nuclei blocks are not well defined except in the Prince Charles Mountains-Enderby Land area (Tingey, 1982a; James and Tingey, this volume).

In contrast to the craton, the Transantarctic Mountains form a linear fold belt (the Ross belt) underlain mostly by Proterozoic and Lower Palaeozoic rocks folded chiefly during a Late Proterozoic Beardmore deformational event and an Early Palaeozoic Ross deformational event. The controversial Middle Palaeozoic Borchgrevink event (Craddock, 1982) in northern Victoria Land and the Late Palaeozoic to Early Mesozoic Weddell event in the Pensacola Mountains also have affected the Transantarctic Mountains. The deformed rocks of the Transantartic Mountains are unconformably overlain by mostly flat-lying unmetamorphosed Lower Palaeozoic to Lower Mesozoic sedimentary rocks of the Beacon Supergroup.

The rocks of West Antarctica are generally younger that those of East Antarctica, and the area is topographically lower than East Antarctica. Continental lithosphere also is much thinner in West Antarctica. West Antarctica may consist of at least four microplates of continental lithosphere that differ from each other in rocks and structural trends (Dalziel and Elliot, 1982). Precambrian rocks have been confirmed only at Haag Nunataks, north of the Ellsworth Mountains. Palaeozoic rocks occur in the Ellsworth Mountains and in widely scattered places in Marie Byrd Land and western Ellsworth Land. Mesozoic and Cenozoic rocks dominate in West Antarctica. Most belong to a Pacific-margin magmatic and deformational belt formed by subduction prior to and during breakup of Gondwanaland. This tectonic feature is the Andean belt (Rowley et al., this volume). It is best exposed in the Antarctic Peninsula and eastern Ellsworth Land, but Andean deformation also overprinted older rocks in western Ellsworth Land and Marie Byrd Land until subduction ceased in these two areas in Late Cretaceous time. The Andean belt consists mostly of scattered magmatic arcs and adjacent associated sedimentary rocks deposited in fore-arc and back-arc basins. Most magmatic arcs formed on continental lithosphere of western Gondwanaland by repeated igneous activity starting in Late Triassic, or perhaps earlier, time and continuing until Late Tertiary time. Deformational and magmatic events occurred locally at various times in this broad span. The Gondwanide deformational event of Late Triassic to Early Jurassic age may be the most widespread (Dalziel, 1982); other deformational events in the belt are more local and have not been named. Different pulses of plutonism are indicated by isotopic ages, and their products may warrant names (Rowley et al., this volume).

## COMPARISONS WITH OTHER GONDWANA CONTINENTS

The location of Antarctica in Gondwanaland reconstructions (Figure 1) has long been used for making predictions (e.g. Wright and Williams, 1974) on what mineral deposits are likely in Antarctica. In these reconstructions, Precambrian rocks in previously adjoining Gondwanaland countries are of special interest because many are metal-rich. That part of the Gondwana craton now constituting eastern South Africa contains a number of large mineral deposits, notably Au-U deposits in conglomerate of the Proterozoic Witwatersrand System and similar rocks; Cr, Ni, Cu, Pt, Fe and V of the Lower Proterozoic Bushveld layered gabbroic intrusion; Archaean and Proterozoic syngenetic sedimentary copper; and Archaean and Proterozoic Fe and Mn in banded iron-formation. Cratonic rocks of India contain Proterozoic and perhaps Archaean banded iron-formation and associated bedded manganese deposits. The Australian craton contains important deposits, including nickel in Archaean ultramafic bodies, epigenetic and syngenetic gold in Archaean mafic volcanic rocks, Archaean and Proterozoic banded iron-formations, and Broken Hill-Mount Isa type syngenetic sedimentary Cu-Pb-Zn-Ag deposits of Proterozoic age.

The Ross deformational belt of the Transantarctic Mountains extends in one direction through central Australia (Craddock, 1982). Its extent in the other direction is not clear but may correlate with the Mozambique Pan-African thermal event (mostly 700 to 500 Ma) that is partly represented along the eastern side of Africa (Kröner, 1977). In central Australia, correlative rocks belong to the western part of the Tasman deformational belt; this part also is known as the Adelaide belt. The rocks of this area are mostly of Late Proterozoic and Early Palaeozoic age that were folded and intruded during the Ordovician Delamerian Orogeny. Mineralisation resulted in

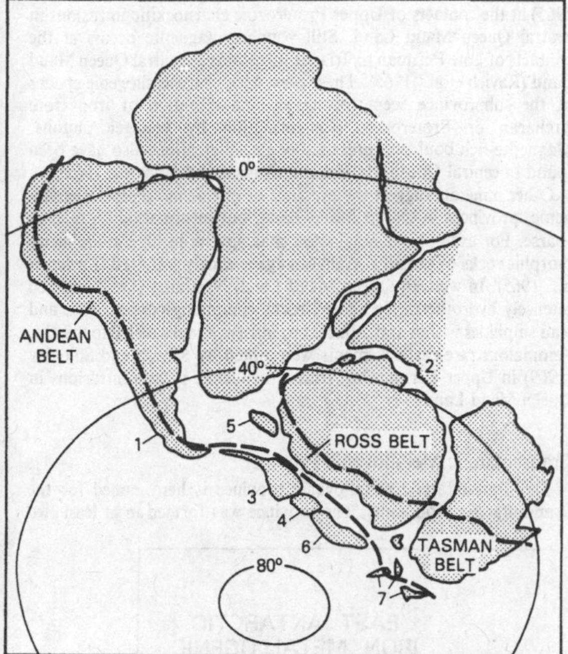


**Figure 1. Gondwanaland reconstruction of the southern continents before initial breakup in Jurassic and Cretaceous. Reconstruction of most of the southern continents is after Norton and Sclater (1979). Reconstruction of West Antarctica and New Zealand is partly after Dalziel and Elliot (1982, figure 5) and Cooper et al. (1982, figure 4.3), respectively. Boundaries of deformational belts, shown by dashed lines, are after Craddock (1982). (1) location where Chile Ridge is subducted under the South American plate, (2) Mahanadi Valley of India, (3) Antarctic Peninsula-eastern Ellsworth Land microplate, (4) western Ellsworth Land microplate, (5) Ellsworth Mountains microplate, (6) Marie Byrd microplate, and (7) various parts of New Zealand.**

numerous, mostly small deposits of Cu and subordinate Au, Pb, Zn, Ag, Ba, Mn, and other metals occurring in veins, stockworks, and replacement bodies generally related to Delamerian igneous activity. Porphyry-Cu deposits and Mississippi Valley type Pb-Zn deposits also occur in the area.

Rocks in more eastern parts of the Tasman belt may correlate with rocks in the eastern Ross belt and with Palaeozoic rocks in West Antarctica. In the central and eastern Tasman belt, deposits of Au, Cu, Ag, As, Pb, Zn, Mo, Bi, Sn, W, Sb, and other metals have been found. Many deposits are syngenetic and are associated with Lower Palaeozoic volcanic rocks. Some deposits are related to intrusive rocks of Ordovician to Permian age. Local Cr, Cu, Pt, and Ni deposits are related to Lower Palaeozoic ultramafic complexes.

The Andean magmatic and deformational belt extends northward from the Antarctic Peninsula through the Andes of South America, and westward from Marie Byrd Land through New Zealand and the western Pacific margin. The belt in the northern and central Andes is one of the richest metal-producing areas in the world. Ericksen (1976) named it the Andean metallogenic province, within which he distinguished five parallel linear subprovinces in which mineralisation is of Mesozoic to Cenozoic age. From west to east, these are characterised by deposits of Fe, Cu, polymetallic base metals-Ag, Sn, and Au. Porphyry-Cu deposits are the most valuable in the province. Hydrothermal vein deposits of Cu, polymetallic base metals-Ag, Sn, and Au are the next most valuable type of ore deposit. Iron occurs as contact metasomatic or hydrothermal magnetite bodies near plutons of mostly Late Cretaceous age. In New Zealand, most mineral deposits are small and contain little more than Au, Sn, and Cu deposits of many ages.

## METALLOGENIC PROVINCES

Locations of metallogenic provinces and subprovinces in Antarctica are shown in Figure 2.

### East Antarctica Iron Metallogenic Province

The East Antarctica iron metallogenic province is here named for iron minerals scattered over most exposed parts of East Antarctica. Lacking information on bedrock from the interior part of the craton, however, we show it only in coastal areas. The province was formed during several metallogenic epochs in Archaean, Proterozoic, and perhaps Phanerozoic time. It seems to consist of two subprovinces, here termed the "iron-formation subprovince" and the "iron oxide vein subprovince".

The iron-formation subprovince extends from western Wilkes Land to western Enderby Land and contains scattered exposures and glacial erratics of banded iron-formation (jaspilite). The deposits of the subprovince are like Superior and Algoma type banded iron-formations found mostly in Precambrian terrains throughout the world. Deposits in other parts of the world span Precambrian time, although their main period of deposition was about 2.0 to 1.8 Ga. Unfortunately, they show enough heterogeneity in ages and lithology (Gole and Klein, 1981) that those from individual districts in India and Australia, which formerly were near the subprovince (Figure 1), probably cannot be correlated with those in Antarctica. Cratonic rocks of the Dharwar Archaean "nucleus" of southern India (Naqvi et al., 1974), however, may correlate with rocks in parts of Enderby Land. Many Indian iron-formation deposits, in the vicinity of this nucleus, are near the graben of the Mahanadi Valley (Figure 1), which may correlate with an apparent graben under the Lambert Glacier and Amery Ice Shelf of Antarctica (Fedorov et al., 1982). Most Australian iron-formation deposits occur in western Australia, apparently formerly adjacent to western Wilkes Land. The Yilgarn Archaean nucleus in southwestern Australia, in particular, contains numerous iron-formation deposits as well as nickel deposits and syngenetic gold deposits.

The thickest exposed iron deposits in the iron formation subprovince occur in the Prince Charles Mountains (Grew, 1982; Hofmann, 1982; Tingey, 1982a,b). The best exposures are at Mount Ruker. In the main sequence here, abundant greenschist-facies jaspilite beds as much as 70 m thick alternate with slate, siltstone, ferruginous quartzite, schist, and metamorphosed gabbro or volcanic rocks (Grew, 1982; Ravich et al., 1982). The main sequence contains laminated type iron-formation and is nearly 400 m thick. It is underlain and overlain by sequences, each more than 300 m thick, in which jaspilite is less abundant than other rock types, especially slate. The age of the deposits is not known with certainty; however, recent Rb-Sr isotopic ages on rocks partly bracketing the iron-formation (Tingey, 1982a,b) suggest that it is Early Proterozoic or Late Archaean. Chemical analyses show that $Fe_2O_3$ exceeds FeO (Ravich et al., 1982), opposite to the ratio in most other large deposits (Gole and Klein, 1981). Based upon their limited exposures, Tingey (pers. comm., 1982) considered the deposits to be of lesser thickness and grade than commercial banded iron-formation in Australia. Hofmann (1982) and Ravich et al. (1982) reported two subparallel aeromagnetic anomalies, each 5 to 10 km wide and containing positive amplitudes of 600 to 3,000 gammas that extend 120 and 180 km west from Mount Ruker under the ice. These anomalies probably represent concealed iron-formation, and they suggest that the deposits compare in size with large deposits of iron-formation in other parts of the world.

Elsewhere in the Prince Charles Mountains, minor faulted jaspilite beds at Mount Stinear and nearby low positive aeromagnetic anomalies might reflect additional iron-bearing beds (Ravich et al., 1982). Tingey (1982b), however, found no banded iron-formation here; he reported that the rocks have Archaean Rb-Sr ages. Farther east, abundant glacial erratics of jaspilite occur in the Vestfold Hills (Ravich et al., 1982). The jaspilite in the boulders is slightly different from that of the Prince Charles Mountains, so it probably was derived from a different up-glacier source (Ravich et al., 1982). Bedrock exposures in the Vestfold Hills contain unspecified volumes of banded iron-formation of apparent Archaean age (Oliver et al., 1982). The nearby Larsemann Hills contain small masses of magnetite of apparent Archaean age in veins and in gneiss (Trail and McLeod, 1969a).

Enderby Land contains many exposures of Archaean magnetite-bearing rocks, none of which has been described in detail. Like Enderby Land, western Wilkes Land also contains outcrops of magnetite-bearing rocks (Lovering and Prescott, 1979), including banded iron-formation near Casey Station (Lovering and Plimer, this volume), that are not well described and that are enclosed in high-grade Precambrian rocks whose age is poorly known. One area, in the Bunger Hills, contain beds of magnetite-bearing schist and gneiss as much as 80 m thick and 100 m long in which magnetite content is as much as 25% (Ravich et al., 1965). Western Wilkes Land also contains scattered films of secondary copper minerals, and near Casey Station (Australia), manganese and barium minerals and secondary copper minerals have been found (McLeod, 1965; Lovering and Plimer, this volume). These minerals suggest the presence of another metallogenic province in the area.

The "iron oxide vein subprovince" occurs in western and central Queen Maud Land, where abundant iron and other metals have been reported. The subprovince boundaries, like those of the iron-formation subprovince, are poorly known because of the lack of bedrock exposures. The craton of eastern South Africa probably lay adjacent to the subprovince (Figure 1) prior to break up of Gondwanaland. This part of Africa contains numerous deposits of Archaean and Proterozoic iron-formation and other base metals and precious metals, and is overprinted by the Mozambique deformational and thermal belt (Kröner, 1977).

Within the subprovince, iron of Archaean (?) age occurs in garnet and quartz-magnetite veins in central Queen Maud Land (Ravich and Soloviev, 1969; Van Autenboer and Loy, 1972). Iron also occurs in garnet and pyroxene-magnetite veins and stockworks (Ravich et al., 1965) at the contacts of Upper Proterozoic charnockitic intrusions in central Queen Maud Land. Still younger magnetite occurs at the contacts of Late Permian to Triassic intrusions in central Queen Maud Land (Ravich et al., 1965). These post-Archaean metallogenic epochs of the subprovince seem to suggest remobilisation of iron from Archaean or Proterozoic iron-formation by younger plutons. Magnetite-rich boulders, presumably glacial erratics., also have been found in central Queen Maud Land (Ravich and Soloviev, 1969).

Other minerals suggest the presence of one or more other metallogenic, provinces in Queen Maud Land, but information on them is sparse. For example, disseminated chalcopyrite in high-grade metamorphic rocks of possible Archaean age is widely scattered (Ravich et al., 1965). In western Queen Maud Land, Neethling (1969) mapped intensely hydrothermally altered rocks containing copper, iron, and lead sulphides within and adjacent to intrusive rocks of 1.0 to 0.8 Ga. Anomalous rare-earth elements were noted by Ravich and Soloviev (1969) in Upper Permian or Triassic nepheline syenite intrusions in Queen Maud Land.

## Transantarctic Metallogenic Province

The Transantarctic metallogenic province is here named for the Transantarctic Mountains. The province was formed in at least two

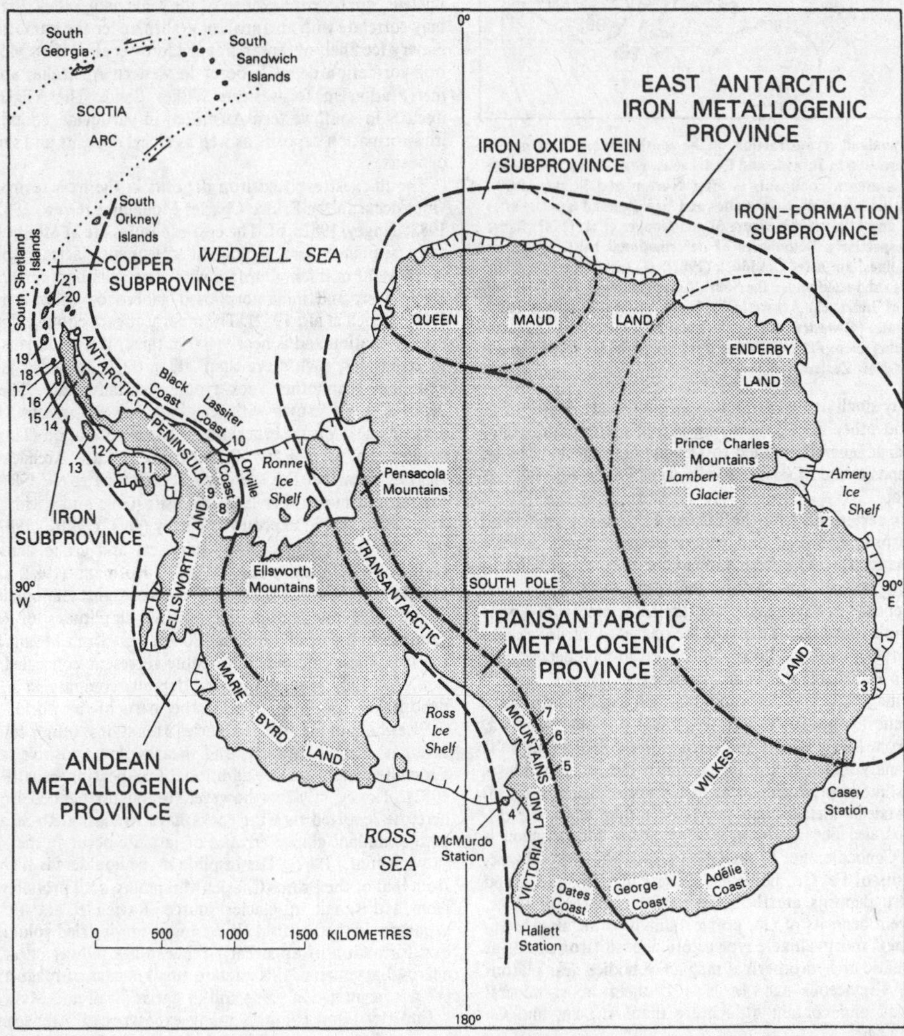


**Figure 2. Map of Antarctica showing metallogenic provinces (dashed line boundaries) and locations of areas referred to in the text. Margins of ice shelves shown by hachured lines. Submarine fracture zones in the Scotia Arc shown by dotted lines. (1) Vestfold Hills, (2) Larsemann Hills, (3) Bunger Hills, (4) Dry Valleys, (5) Warren Range, (6) Darwin Glacier, (7) Byrd Glacier, (8) Merrick Mountains, (9) Sky Hi Nunataks, (10) Copper Nunataks, (11) Alexander Island, (12) Terra Firma Islands, (13) Adelaide Island, (14) Oscar II Coast, (15) Argentine Islands, (16) Anvers Island, (17) Danco and Graham Coasts, (18) Brabant Island, (19) Livingston Island, (20) King George Island, (21) Gibbs Island.**

metallogenic epochs, corresponding to the Early Palaeozoic Ross deformational event and to a second Middle Jurassic event when mafic intrusions (Ferrar Group) were emplaced. These provide the basis for defining two overlapping subprovinces, here named the Ross "subprovince" and the "Ferrar subprovince". Deposits associated with the so-called Borchgrevink deformational event are included in the Ross subprovince.

Metal occurrences in the Ross subprovince are related primarily to plutonic rocks of the Beardmore, Ross, or younger events. The deposits are comparable to those of the eastern (Adelaide) part of the Tasman fold belt (Figure 1). All metal occurrences in the Ross subprovince are small. Minor Precambrian deposits containing molybdenite, pyrite, sphalerite, and arsenopyrite, and local traces of gold and silver, have been noted (Mawson, 1940; Ravich et al., 1965) in the George V Coast and Adélie Coast, which are outside of, but adjacent to and perhaps related to, mineralisation in the province (Figure 2). Minor concentrations of pyrite, chalcopyrite, and arsenopyrite occur in quartz-albite veins in Precambrian phyllite in the Oates Coast (Ravich et al., 1965).

Bornite occurs in the Douglas Conglomerate near a Ross-age pluton (Granite Harbour intrusive rocks) about 30 km south of the mouth of the Byrd Glacier (Stump, pers comm.). Bismuthinite, spodumene, and related bismuth- and lithium-bearing minerals have been noted by Faure (pers comm.) in Ross-age pegmatite dykes and related veins in the Shackleton Limestone south of Byrd Glacier. Minor disseminated copper sulphides and secondary minerals occur in Borchgrevink-age (?) plutonic rocks (Admiralty intrusive rocks) in Copper Cove, near Hallett Station in northern Victoria Land (Harrington et al., 1964). Copper minerals also have been seen near other plutons in the central Transantarctic Mountains (S. Borg and E. Stump, pers comm.). Anomalous amounts of radioactivity occur in pegmatite and plutonic bodies formed during the Ross event in the Dry Valleys area and in alluvial minerals in Devonian parts of the Beacon Supergroup in the Darwin Glacier area (Zeller et al., 1979). Insignificant amounts of tin and rare-earth elements exist in alluvial minerals within Beacon sandstone in Victoria Land (Stewart, 1939; Mawson, 1940) and the Darwin Glacier area (Zeller et al., 1979).

Stratiform mafic sills of the Ferrar Group (Middle Jurassic) occur at many places in the Transantarctic Mountains, and the Ferrar subprovince encompasses the metallic minerals within these sills. These rocks are similar to the Middle Jurassic Karoo Dolerite of southern Africa (Ford and Kistler, 1980), which contains sparse noncommercial Fe, Ni, and Cu. By far the largest and best known of the Ferrar bodies is the gabbroic Dufek intrusion of the Pensacola Mountains, whose possible resources are described by Ford (in press). Cumulus layers of Fe-Ti oxides and scattered copper and iron sulphides are known in the intrusion, but only the top of this mafic body—estimated at more than 7 km thick—is exposed. As in other large layered intrusions, cumulus minerals containing Cr, Ni, Pt-group metals, and perhaps copper may exist in lower parts of the intrusion; and Co, V, Fe, and Cu may exist in upper parts. Hamilton (1964) described trace amounts of Co, Cr, V, Cu, and P from a Ferrar sill in the Dry Valleys.

## Andean Metallogenic Province

The Andean metallogenic province, of Mesozoic and Cenozoic age, primarily contains Cu, but Fe, Mo, Pb, Zn, Ag, and related metals also are present. It is the southward extension of the province of the same name in western South America (Ericksen, 1976), and partly corresponds in dimensions to the Andean magmatic and deformational belt. Most metallic mineral occurrences of the province are related to calc-alkaline plutonism and occur in the Antarctic Peninsula and eastern Ellsworth Land. The occurrences define a "copper subprovince" and a less significant "iron subprovince," plus isolated deposits of several other metals. Present data do not allow the clear separation of subprovinces of different metals as in western South America (Erickson, 1976); metal types in the Antarctic Peninsula appear largely to overlap each other. The most western subprovince (iron) coincides with a belt of Cenozoic plutonic (especially those of trondhjemitic composition; Rowley and Pride, 1982) and volcanic rocks in the islands off the western coast of the Antarctic Peninsula. The most promising of the copper occurences, including most of the porphyry copper deposits, overlaps the iron subprovince, so that the copper subprovince extends across the entire peninsula. Based on sparse chemical analyses of mineral occurrences, neither tin nor gold subprovinces appear to be present in the Antarctic Peninsula-eastern Ellsworth Land region, unlike western South America. Traces of gold, however, have been reported from several places in the Antarctic Peninsula (Knowles, 1945).

Rowley and Pride (1982) and Rowley et al. (in press) summarised the mineral occurrences of the Antarctic Peninsula and eastern Ellsworth Land. Most occurrences contain copper and iron oxides, sulphides, and secondary minerals, but minerals containing Mo, Pb, Zn, Ag, and other elements also are present. Most occurrences are far too small to be of commercial interest; even the larger known deposits do not approach ore grade. Most of the larger deposits belong to the copper subprovince. One exception is chromite in layers and disseminated grains within an ultramafic igneous body on Gibbs Island in the South Shetland Islands (Figure 2); mineralisation apparently is related to an Upper Palaeozoic or Lower Mesozoic subduction complex (de Wit et al, 1977).

The largest known mineral deposit in the Antarctic Peninsula, belonging to the copper subprovince, is on King George Island in the South Shetland Islands. Mapping by Hawkes (1961) and Barton (1965) revealed huge quartz replacement bodies and abundant pyrite and secondary hematite and limonite as well as extensive areas of hydrothermally altered rocks within mostly volcanic rocks of Late Jurassic and Tertiary age. Detailed studies of metallic and alteration minerals by Littlefair (1978) and Cox et al. (1980) found also chalcopyrite, bornite, molybdenite(?), pyrrhotite, marcasite, and magnetite. High copper values occur in analysed samples of pyrite and plutonic rocks. Both low-temperature hot spring type alteration and mineralisation (Littefair, 1978) and high-temperature effects (Cox et al., 1980) probably have taken place during formation of the deposit. Mineralisation and alteration probably are related to nearby plutons, one of which has whole-rock K-Ar ages of 50-46 Ma (Watts, 1982).

Numerous plutons, with probable porphyry copper affinities, occur in the Antarctic Peninsula and eastern Ellsworth Land. Most of the plutonic rocks however, exhibit hypidiomorphic-granular (granitoid) texture, which suggests that they are deeply eroded (Cox and Czamanske, 1981). One of these "plutonic pophryry" deposits comprises widespread altered rocks and copper and molybdenum mineralised rocks of magmatic and hydrothermal origins related to plutons of 59-55 Ma (Rex, 1976) in the Argentine Islands. Hawkes and Littlefair (1981) recognised zones of potassic and propylitic altered rocks within granodiorite and younger quartz monzonite plutonic rocks and in enclosing volcanic rocks. Veinlets of quartz-molybdenite-chalcopyrite and peripheral quartz-magnetite and quartz-pyrite occur within the potassic zone, and disseminated pyrite and magnetite occur within the propylitic zone. Hawkes and Littlefair (1981) interpreted the minerals of the potassic zone to represent the root zone of a porphyry copper-molybdenun deposit. Numerous other porphyry copper deposits occur in the Antarctic Peninsula and surrounding islands (Rowley et al., in press). Hydrothermal polymetallic base-metal veins, containing Cu, Pb, Zn, and other metals, are associated with most of the porphyry-type deposits. However, not enough information exists to define a polymetallic vein subprovince that is distinct from the copper subprovince.

Iron minerals, especially magnetite and pyrite, are abundant in most Antarctic Peninsula-eastern Ellsworth Land mineral deposits. In several other places, however, iron minerals are the most common metallic minerals; when these other occurrences are combined, they define the iron subprovince. The most notable occurrence is on Brabant Island, where magnetite, martite, hematite, and limonite are widespread in Pleistocene mafic lava flows (Vieira et al., 1982). Additional iron occurrences include thin cumulus layers of magmatic magnetite recognised in a Mesozoic or Cenozoic stratiform gabbro in the Argentine Islands (Fraser, 1964, Elliot, 1964).

Metallic mineral occurrences in some islands of the Scotia Arc belong to the Andean metallogenic province but are sparse and apparently economically insignificant. They consist only of the following: (1) pyrite, chalcopyrite, and hematite veins in sedimentary rocks and sills of an Upper Jurassic to Lower Cretaceous island arc sequence on South Georgia (Stone, 1980; Tanner et al., 1981); and (2) minor disseminated sulphide minerals in Palaeozoic(?) schist and a Mesozoic or Cenozoic dyke on southeastern Coronation Island, South Orkney Islands (Thomson, 1974).

Andean subduction left an overprint of calc-alkaline igneous activity and related metals on Palaeozoic rocks in western Ellsworth Land

and Marie Byrd Land. This subduction ceased, however, about 80 Ma ago—much earlier than in the Antarctic Peninsula—and alkaline magmatism and associated block faulting followed (Le Masurier and Rex, 1982). Thus most of the metal deposits of the Andean metallogenic province that might have formed in western Ellsworth Land and Marie Byrd Land have been deeply eroded and are covered by alkaline volcanic rocks. Only widespread but economically insignificant copper "shows" (Wade, 1976) and minor lead and iron suphides have been reported in this huge region. The possibility of finding metallic mineral deposits in these areas seems remote. The Tertiary alkaline igneous rocks do not contain any known metallic mineral occurrences, so the area in which they occur cannot be designated as a separate metallogenic province.

## FUTURE MINERAL DISCOVERIES

The Antarctic Peninsula-eastern Ellsworth Land region of the Andean metallogenic province probably has the greatest likelihood of containing a commercial mineral deposit in Antarctica. Nonetheless, this area does not seem to compare to the metal-rich central and northern Andes. Erickson (1976), in fact, found few metal deposits in the southern part of the Andes, and suggested that the Antarctic Peninsula also may be relatively poor in metallic mineral deposits. Rowley and Pride (1982) agreed with this suggestion based on an appraisal of known mineral occurrences in the peninsula. They suggested that the break between the metal-rich and the metal-poor parts occurs where the Chile Ridge is subducted beneath southern Chile (Figure 1). Compared to the metal-rich part to the north, the metal-poor part to the south experienced a complex subduction history (Herron and Tuchloke, 1976), and the plate being subducted may have been warmer, thinner, softer, lighter, and younger (DeLong and Fox, 1977). Relative to the northern and central Andes, plutons in the province in Antarctica may have been less rich in metals, may have been emplaced at different crustal levels, or may have been more deeply eroded following emplacement.

Exploration for porphyry copper deposits in high-level porphyritic plutonic rocks just beneath their volcanic covers may be the most promising exploration target in the Andean metallogenic province of Antarctica. The topographic axis of the southern Antarctic Peninsula and eastern Ellsworth Land, which appears to be underlain by the axis of a Jurassic magmatic arc (Rowley et al., this volume), is one target. Only the volcanic rocks along this axis have been studied to date, but high-level Jurassic plutons that produced the volcanic rocks also may be exposed locally and, if so, should be investigated. The higher parts of the southern Antarctic Peninsula also are the best places to find porphyry-type deposits of the Lower Cretaceous plutons, apparently because here the rocks are not eroded to as deep a level. Porphyritic plutons in younger volcanic terrains, especially in western parts of the peninsula and its offshore islands, also are worthy of study. The youngest (Tertiary) plutons (Saunders et al., 1982) and the most promising metal deposits found to date occur in the western offshore islands. Furthermore, Care (1980) discovered Tertiary volcanic and plutonic rocks in northeastern Alexander Island that locally contain pyrite, malachite, and molybdenite.

Sillitoe (1972) and later workers have noted that metallic minerals in the Andes Mountains exhibit an eastward progression from Fe deposits near the western coast, to Cu deposits containing some Au and Mo, to Pb-Zn-Ag deposits, to perhaps Sn-Sb-Bi-W-Mo deposits in the east. This eastward progression of mineral deposits bears on Gondwana reconstructions. Specifically, the iron subprovince in the Antarctica Peninsula occurs, as in the Andes, farthest to the west. This similarity argues against any Gondwana reconstruction in which the Antarctic Peninsula is placed east or west of the southern end of South America.

The Antarctic craton undoubtedly contains undiscovered mineral deposits. Banded iron-formation probably is more widespread than now known; future airborne aeromagnetic studies will help define the extent of these rocks. Based on analogy with the formerly nearby Yilgarn nucleus of Australia, Wilkes Land appears to have a geologic environment favorable for nickel deposits in Archaean ultramafic intrusions and gold deposits in Archaean mafic volcanic rocks. Based on analogy with eastern South Africa, Queen Maud Land may contain fossil placer Au-U deposits in Proterozoic conglomerate and may contain magmatic Cr-Ni-Cu-Pt deposits in ultramafic intrusions. The possible occurrence of diamond-bearing kimberlite pipes in Queen Maud Land cannot be ruled out.

Based on analogy with the Tasman belt, the Transantarctic Mountains probably contain larger metal deposits than those found to date. Faulted or otherwise deformed sedimentary rocks (especially carbonates) near the contacts with Lower Palaeozoic (Ross deformational event) or younger plutons may be the most favorable targets for deposits of copper and related metals. Syngenetic deposits containing Au, Cu, Ag, Pb, Zn, and other metals related to deposition of Lower Palaeozoic volcanic and sedimentary rocks and to ultramafic intrusions may also be present. Another likely target in the Transantarctic Mountains is Ferrar sills.

*Acknowledgements* We are grateful to W. Hamilton and D.L. Schmidt for assistance in clarifying concepts on regional tectonics, and to M.D. Turner and G.D. McKenzie for logistic assistance. We thank E. Stump, S.G. Borg and G. Faure for unpublished information on mineral occurrences in the Transantarctic Mountains. Report preparation was financed by National Science Foundation Grant DPP80-07388. Technical reviews of the manuscript by D.L. Schmidt, G.A. Desborough, and G.E. Erickson are greatly appreciated.

## REFERENCES

BARTON, C.M., 1965: The geology of the South Shetland Islands—III. The stratigraphy of King George Island. *Scient. Rep. Br. antarct. Surv., 44.*

CARE, B.W., 1980: The geology of Rothschild Island, northwest Alexander Island. *Bull. Br. antarc. Surv., 50,* 87-112.

COOPER, R.A., LANDIS, C.A., LE MASURIER, W.E. and SPEDEN, I.G., 1982: Geologic history and regional patterns in New Zealand and Western Antarctica—Their paleotectonic and paleogeographic significance; *in* Craddock, C. (ed.) *Antarctic Geoscience.* Univ. Wisconsin Press, Madison, 43-53.

COX, C., CIOCANELEA, R. and PRIDE, D., 1980: Genesis of mineralisation associated with Andean intrusions, northern Antarctic Peninsula region. *U.S. Antarct. J., 15,* 22-3.

COX, D.P. and CZAMANSKE, G.K., 1981: Mineral rich fluid inclusions in the root zone of a porphyry copper system, Ajo, Arizona (abs.). *Abstracts with Programs, 1981. Geol. Soc. Am., 12,* 433.

CRADDOCK, C., 1982: Antarctica and Gondwanaland (Review Paper); *in* Craddock, C. (ed.) *Antarctic Geoscience.* Univ. Wisconsin Press, Madison, 3-13.

DALZIEL, I.W.D., 1982: The early (pre-Middle Jurassic) history of the Scotia Arc region—A review and progress report (Review paper); *in* Craddock, C. (ed.) *Antarctic Geoscience.* Univ. Wisconsin Press, Madison, 111-20.

DALZIEL, I.W.D. and ELLIOT, D.H., 1982: West Antarctica—Problem child of Gondwanaland. *Tectonics, 1,* 3-19.

DELONG, S.E. and FOX, P.J., 1977: Geological consequences of ridge subduction; *in* Talwani, M. and Pitman, W.C. III (eds.) *Island Arcs, Deep Sea Trenches, and Back-Arc Basins, Maurice Ewing Series 1.* American Geophysical Union, Washington D.C., 221-8.

DE WIT, M.J., DUTCH, S., KLIGFIELD, R., ALLEN, R. and STERN, C., 1977: Deformation, serpentisation and emplacement of a dunite complex, Gibbs Island, South Shetland Islands—Possible fracture zone tectonics. *J. Geol., 85,* 745-62.

ELLIOT, D.H., 1964: The petrology of the Argentine Islands. *Scient. Rep. Br. antarc. Surv., 41.*

ERICKSEN, G.E., 1976: Metallogenic provinces of southeastern Pacific region; *in* Halbouty, M.T., Maher, J.C. and Lian, H.M. (eds.) Circum-Pacific Energy and Mineral Resources. *Am. Assoc. Pet. Geol., Mem., 25,* 527-58.

FEDOROV, L.V., RAVICH, M.G. and HOFMANN, J., 1982: Geologic comparison of southeastern peninsula India and Sri Lanka with a part of East Antarctica (Enderby Land, MacRobertson Land, and Princess Elizabeth Land); *in* Craddock, C. (ed.) *Antarctic Geoscience.* Univ. Wisconsin Press, Madison, 73-8.

FORD, A.B. (in press): The Dufek Intrusion and its speculative resources; *in* Splettstoesser, J.F. (ed.) *Mineral Resource Potential of Antarctica.* Univ. Texas Press, Austin.

FORD, A.B. and KISTLER, R.W., 1980: K-Ar age, composition, and origin of Mesozoic mafic rocks related to Ferrar Group, Pensacola Mountains, Antarctica. *N.Z. J. Geol. Geophys., 23,* 371-90.

FRASER, A.G., 1964: Banded gabbros of the Anagram Islands, Graham Land. *Bull. Br. antarc. Surv., 4,* 23-38.

GOLE, M.J. and KLEIN, C., 1981: Banded iron formations through much of Precambrian time. *J. Geol., 89,* 169-83.

GREW, E.S., 1982: Geology of the southern Prince Charles Mountains, East Antarctica; *in* Craddock, C. (ed.) *Antarctic Geoscience.* Univ. Wisconsin Press, Madison, 473-8.

HAMILTON, W., 1964: Diabase sheets differentiated by liquid fractionation, Taylor Glacier region, southern Victoria Land; *in* Adie, R.J. (ed.) *Antarctic Geology.* North Holland, Amsterdam, 442-54.

HARRINGTON, H.J., WOOD, B.L., McKELLAR, I.C. and LENSEN, G.J., 1964: The geology of Cape Hallett-Tucker Glacier district; *in* Adie, R.J. (ed.) *Antarctic Geology.* North Holland, Amsterdam, 220-8.

HAWKES, D.D., 1961: The geology of the South Shetland Islands—I. The petrology of King George Island. *Scient. Rep. Falkld Isl. Depend. Surv., 26.*

HAWKES, D.D. and LITTLEFAIR, M.J., 1981: An occurrence of molybdenum, copper, and iron mineralisation in the Argentine Islands, West Antarctica. *Econ. Geol., 76,* 898-904.

HERRON, E.M. and TUCHOLKE, B.E., 1976: Sea floor magnetic patterns and basement structure in the southeastern Pacific. *Initial Reports of the Deep Sea Drilling Project, 35,* 263-78. U.S. Govt Printing Office, Washington D.C.

HOFMANN, J., 1982: Main tectonic features and development of the southern Prince Charles Mountains, East Antarctica; *in* Craddock, C. (ed.) *Antarctic Geoscience.* Univ. Wisconsin Press, Madison, 479-87.

JAMES, P.R. and TINGEY, R.J. (this volume): The Geological evolution of the East Antarctic metamorphic shield—a review.

KAMENEVA, G.I. and GRIKUROV, G.E. (this volume): Minerogenic Provinces in Antarctica (an attempt of recognition of the basis of formation analysis).

KNOWLES, P.H., 1945: Geology of southern Palmer Peninsula, Antarctica. *Am. Phil. Soc., Proc., 89,* 132-45.

KRÖNER, A., 1977: The Precambrian geotectonic evolution of Africa—Plate accretion versus plate destruction. *Precambrian Res., 4,* 163-213.

LE MASURIER, W.E. and REX, D.C., 1982: Volcanic record of Cenozoic glacial history in Marie Byrd Land and western Ellsworth Land—Revised chronology and evaluation of tectonic factors; *in* Craddock, C. (ed.) *Antarctic Geoscience.* Univ. Wisconsin Press, Madison, 725-34.

LITTLEFAIR, M.J., 1978: The "quartz-pyrite" rocks of the South Shetland Islands, western Antarctic Peninsula. *Econ. Geol., 73,* 1184-9.

LOVERING, J.F. and PLIMER, I.R. (this volume): Manganese-rich chemical sediments from Wilkes Land, Antarctica.

LOVERING, J.F. and PRESCOTT, J.R.V., 1979: *Last of Lands—Antarctica.* Melbourne Univ. Press, Melbourne.

MAWSON, D., 1940: Record of minerals of King George Land, Adélie Land and Queen Maud Land. *Australasian Antarctic Expedition, 1911/14, Scientific Reports, Ser. A., 4,* 371-404.

McLEOD, I.R., 1965: Antarctica—Geology and mineral occurrences; *in* Dew, J.M. (ed.) *Handbook, Australia and New Zealand, Eighth Commonwealth Mining and Metallurgical Congress.* Australasian Institute of Mining and Metallurgy, 165-6.

NAQVI, S.M., DIVAKARA RAO, V. and NARAIN, H., 1974: The protocontinental growth of the Indian shield and the antiquity of its rift valleys. *Precambrian Res., 1,* 345-98.

NEETHLING, D.C., 1969: Geology of the Ahlmann Ridge, western Queen Maud Land; *in* Bushnell, V.C. and Craddock, C. (eds.) Geologic Maps of Antarctica. *Antarc. Map Folio Ser.,* Folio 12, Plate VII.

NORTON, I.O. and SCLATER, J.G., 1979: A model for the evolution of the Indian Ocean and the breakup of Gondwanaland. *J. Geophys. Res., 84,* 6803-30.

OLIVER, R.L., JAMES, P.R., COLLERSON, K.D. and RYAN, A.B., 1982: Precambrian geologic relationships in the Vestfold Hills, Antarctica; *in* Craddock, C. (ed.) *Antarctic Geoscience.* Univ. Wisconsin Press, Madison, 435-44.

RAVICH, M.G. and SOLOVIEV, D.S., 1969: *Geology and Petrology of the Mountains of Central Queen Maud Land (Eastern Antarctica).* Israel Program for Scientific Translations, Jerusalem.

RAVICH, M.G., FEDOROV, L.V. and TARUTIN, O.A., 1982: Precambrian iron deposits of the Prince Charles Mountains (Review Paper); *in* Craddock, C. (ed.) *Antarctic Geoscience.* Univ. Wisconsin Press, Madison, 853-8.

RAVICH, M.G., KLIMOV, L.V. and SOLOVIEV, D.S., 1965: *The Pre-Cambrian of East Antarctica.* Israel Program for Scientific Translations, Jerusalem.

REX, D.C., 1976: Geochronology in relation to the stratigraphy of the Antarctic Peninsula. *Bull. Br. antarc. Surv., 43,* 49-58.

ROWLEY, P.D. and PRIDE, D.E., 1982: Metallic mineral resources of the Antarctic Peninsula (Review Paper); *in* Craddock, C. (ed.) *Antarctic Geoscience.* Univ. Wisconsin Press, Madison, 859-70.

ROWLEY, P.D., WILLIAMS, P.L. and PRIDE, D.E. (in press): Metallic and nonmetallic mineral resources of Antarctica; *in* Splettstoesser, J.F. (ed.) *Mineral Resource Potential of Antarctica.* Univ. Texas Press, Austin.

ROWLEY, P.D., FARRAR, E., CARRARA, P.E., VENNUM, W.R. and KELLOGG, K.S. (this volume): Metallic mineral resources and K-Ar ages of plutonic rocks of the Orville Coast and Eastern Ellsworth Land, Antarctica.

SAUNDERS, A.D., WEAVER, S.D. and TARNEY, J., 1982: The pattern of Antarctic Peninsula plutonism; *in* Craddock, C. (ed.) *Antarctic Geoscience.* Univ. Wisconsin Press, Madison, 305-14.

SILLITOE, R.H., 1972: Relation of metal provinces in western America to subduction of oceanic lithosphere. *Geol. Soc. Am., Bull., 83,* 813-8.

SPLETTSTOESSER, J.F. (this volume): Mineral resources potential in Antarctica—Review and predictions.

STEWART, D., 1939: Petrology of some South Victoria Land rocks: *Am. Mineral., 24,* 155-61.

STONE, P., 1980: The geology of South Georgia—IV. Barff Peninsula and Royal Bay areas. *Scient. Rep. Br. antarc. Surv., 96.*

TANNER, P.W.G., STOREY, B.C. and MacDONALD, D.I.M., 1981: Geology of an Upper Jurassic-Lower Cretaceous island-arc assemblage in Hauge Reef, the Pickersgill Islands and adjoining areas of South Georgia. *Bull. Br. antarc. Surv., 53,* 77-117.

THOMSON, J.W., 1974: The geology of the South Orkney Islands—III. Coronation Island. *Scient. Rep. Br. antarc. Surv., 86.*

TINGEY, R.J., 1982a: The geologic evolution of the Prince Charles Mountains—An Antarctic Archean cratonic block; *in* Craddock, C. (ed.) *Antarctic Geoscience.* Univ. Wisconsin Press, Madison, 455-64.

TINGEY, R.J. (comp.) 1982b: Geology of the southern Prince Charles Mountains, Australian Antarctic Territory. *Aust. Bur. Miner. Resour., Geol. Geophys.*

TRAIL, D.S. and McLEOD, I.R., 1969a: Geology of the Lambert Glacier region; *in* Bushnell, V.C and Craddock, C. (eds.) Geologic Maps of Antarctica, *Antarct Map Folio Ser.,* Folio 12, Plate XI.

TRAIL, D.S. and McLEOD, I.R., 1969b: Geology of Enderby Land; *in* Bushnell, V.C. and Craddock, C. (eds.) Geologic Maps of Antarctica. *Antarct. Map Folio Ser.,* Folio 12, Plate X.

VAN AUTENBOER, T. and LOY, W., 1972: Recent geological investigations in the Sor-Rondane Mountains, Belgicafjella and Sverdrupfjella, Dronning Maud Land; *in* Adie, R.J. (ed.) *Antarctic Geology and Geophysics.* Universitetsforlaget, Oslo, 563-71.

VIEIRA, C., ALARCÓN, B., AMBRUS, J. and OLCAY, L., 1982: Metallic mineralisation in the Gerlache Strait region, Antarctica; *in* Craddock, C. (ed.) *Antarctic Geoscience.* Univ. Wisconsin Press, Madison, 871-6.

WADE, F.A., 1976: Antarctica: An unprospected, unexploited continent—Summary; *in* Halbouty, M.T., Maher, J.C. and Lian, H.M. (eds.) Circum-Pacific Energy and Mineral Resources. *Am. Assoc. Pet. Geol., Mem., 25,* 74-9.

WATTS, D.R., 1982: Potassium-argon ages and paleomagnetic results from King George Island, South Shetland Islands; *in* Craddock, C. (ed.) *Antarctic Geoscience.* Univ. Wisconsin Press, Madison, 255-61.

WRIGHT, N.A. and WILLIAMS, P.L., 1974: Mineral resources of Antarctica. *U.S. Geol. Surv., Circ., 705.*

ZELLER, E., DRESCHHOFF, G., THOSTE, V. and KROPP, W.R., 1979: Radioactivity survey in Antarctica, 1978/79. *U.S. Antarct. J., 14,* 38-9.

# A METALLOGENIC RECONNAISSANCE OF ANTARCTIC MAJOR STRUCTURAL PROVINCES

G.I. Kameneva and G.E. Grikurov, *Department of Antarctic Geology and Mineral Resources, VNIIOkeangeologia, Leningrad, 190121, USSR.*

*Abstract* Three major metallogenic provinces are recognised in Antarctica related to three principal structural provinces. The Gondwanian metallogenic province encompasses the East Antarctic craton; the Circum-Pacific province is associated with the West Antarctic geosynclinal belt, and the "Atlantic" (sub)provinces are related to extensive areas of peri- and epicratonic tectonism. Available data afford but a preliminary assessment of mineral potential of the continent, based to a larger extent on global analogies than on direct Antarctic evidence. The most promising for hard mineral concentrations are the Andean orogen of the West Antarctic geosynclinal belt, and some zones of epicratonic tectonism. The hydrocarbon potential of vast Antarctic sedimentary basins is dubious and requires further investigation. At present only two occurrences of least valuable resources—iron and coal—may be ranked as deposits. Despite favourable structural and minerogenic factors apparently present in some regions, the prospects of discovering major ore and hydrocarbon reserves in Antarctica are highly uncertain.

The problem of Antarctic mineral resources and their possible exploration and exploitation has lately attracted the considerable attention of the international community and is being actively discussed in the Antarctic Treaty framework. In this connection Antarctic geologists are often pressed for answers to questions relevant to assessment of Antarctic mineral potential and outlining the fields of most probable economic interest. In the authors' opinion, only very preliminary answers to these questions can be given at present due to the reconnaissance state of Antarctic geologic exploration and, in particular, extremely meagre evidence directly related to mineral occurrences. Available data suggest only qualitative speculations based on analogies with other continents where, after many decades of detailed geological mapping, drilling, mining, etc. a certain correlation was found between metallogenic and structural features of the Earth's crust. It is legitimate to assume that similar relations exist in Antarctica, and to analyse the extent to which this correlation is confirmed by our present knowledge.

The major structural provinces recognised in Antarctica for the purposes of this paper (Figure 1) basically follow the subdivision suggested previously by Grikurov (1978a, b; 1982), with some modifications arising from latest relevant considerations (Grikurov, this volume; Kadmina et al., this volume). Two principal provinces of indisputable tectonic nature are the East Antarctic craton and the West Antarctic geosynclinal belt; in terms of metallogenic specification they are named respectively "Gondwanian" and "Circum-Pacific" provinces. Around the craton and in the West Antarctic interior there are extensive zones whose structural and, presumably, metallogenic features are determined mainly by multiple tectonic processes affecting the ancient crystalline basement throughout Late Precambrian and Phanerozoic history. These processes are believed to result from the influence on the Antarctic continental landmass of tectonic phenomena related to formation of the surrounding oceanic terrain now occupied by the southern parts of the Atlantic and Indian Oceans; consequently the mobile zones are distinguished as metallogenic provinces (or subprovinces) of Atlantic-Indo-oceanic type and named, for the sake of briefness, "Atlantic" (sub)provinces. The latter fall into two major categories according to prevailing tectonic regimes during Late Precambrian and Phanerozoic time, especially at the latest (Mesozoic-Cenozoic) stage. The present day mountain systems delineate the areas which repeatedly reacted to tectonic impulses by orogenic uplifts and associated igneous activity, and may therefore be considered epiplatform recurrent orogens. On the contrary, evolution of sedimentary basins now constituting the greater part of the "Atlantic" (sub)provinces was dominated by a coilogenic (as opposed to "orogenic" and implying formation of extensive crustal downwarps with thick sedimentary fill. The term was introduced by Spizharsky (1973) and is widely used in the Soviet geological literature) regime closely related to destructive riftogenic processes, and hence these basins may be classified as areas of coilogenic activisation.

## GONDWANIAN METALLOGENIC PROVINCE OF EAST ANTARCTIC CRATON

The most stable part of the craton occupies its western sector between 0° and 35°W; similar cratonic terrains relatively unaffected by superimposed tectonism may also occur in the vast subglacial interior of East Antarctica. They are characterised by the presence of ancient (Precambrian) platform cover formations incorporating large stratiform bodies of mafic igneous rocks. A similar geologic environment in South Africa is favourable for rich Au-U mineralisation of Witwatersrand type, Cu-Co, Pt and chromite ores of the Bushveld type, as well as for diamond bearing kimberlite pipes and related placer deposits. In Antarctica this structural province is too poorly studied to allow reliable metallogenic assessment based on its possible affinities with the South African cratons. However, discovery of rare minor grains of gold in Proterozoic gravelites of this region and a slight increase in concentrations of Ti, Cu and Ni in mafic intrusions may be regarded as promising factors.

In all other exposed areas of East Antarctica the crystalline basement of the craton is affected by orogenic deformation and, consequently, reveals both the relic (Gondwanian) and superimposed ("Atlantic") metallogenic features. The former are preserved in relatively unaltered Lower Precambrian inliers whose structural pattern was essentially formed during the Archaean stages of regional metamorphism and granitisation.

The highest grade Lower Archaean granulite facies complexes are known in Enderby and MacRobertson Lands. Similar crystalline complexes on other continents are characterised by essential concentrations of iron (in magnetite-silicate rocks and ferromagnesian skarns), graphite, phlogopite (in magnesian skarns) and, rarely, manganese. In Antarctica, abundant magnetite-silicate occurrences apparently related to the oldest metamorphic rocks have already been discovered; this confirms that the Lower Archaean crystalline basement complexes may appear productive for metamorphogenic iron ores (Ravich and Kamenev, 1972).

The Upper Archaean high grade amphibolite facies complexes of the crystalline basement also disclose mainly Gondwanian structural and minerogenic affinities. These complexes constitute a great majority of the East Antarctic coastal mountains. Their most characteristic feature is a very high degree of migmatisation which implies that the processes of acid leaching were probably the major metallogenic agents. Such processes are known to be beneficial for concentration of rare, nonferrous and precious metals; during their terminal stages, formation of pegmatites is facilitated with associated enrichment in beryl, muscovite, quartz, rare earths and radioactive metals (Ravich et al., 1965).

Another structural-metallogenic type peculiar to Gondwanian provinces is represented by specific intracratonic mobile zones reminiscent of Lower Precambrian greenstone belts and/or Riphean aulacogens. In Antarctica, the fragments of such systems are exposed in the southern Prince Charles Mountains and the Shackleton Range. Their most conspicuous metallogenic feature is an obvious iron-ore concentration in the Lower Proterozoic structural stage, displayed in the former area by major jaspilite and ferruginous quartzite bodies, and in the latter region—by minor occurrences of similar type; the magnitude of mineralisation in the Prince Charles area ranks it as an iron-ore deposit (Ravich et al., 1978). Some other metallogenic manifestations typical of intracratonic systems on other continents are also observed, such as an increase in polymetallics, cobalt, nickel and chrome content in metamafic dykes and sills, impregnation of breccia zones with disseminated copper sulphides, molybdenite and scheelite, traces of gold and platinoids in quartz veins, metasomatic growths of kyanite, beryl, spodumene, muscovite and quartz in pegmatites, etc. (Ravich et al., 1978). Each of the above listed occurrences is insignificant both in size and degree of mineral concentration; nevertheless,

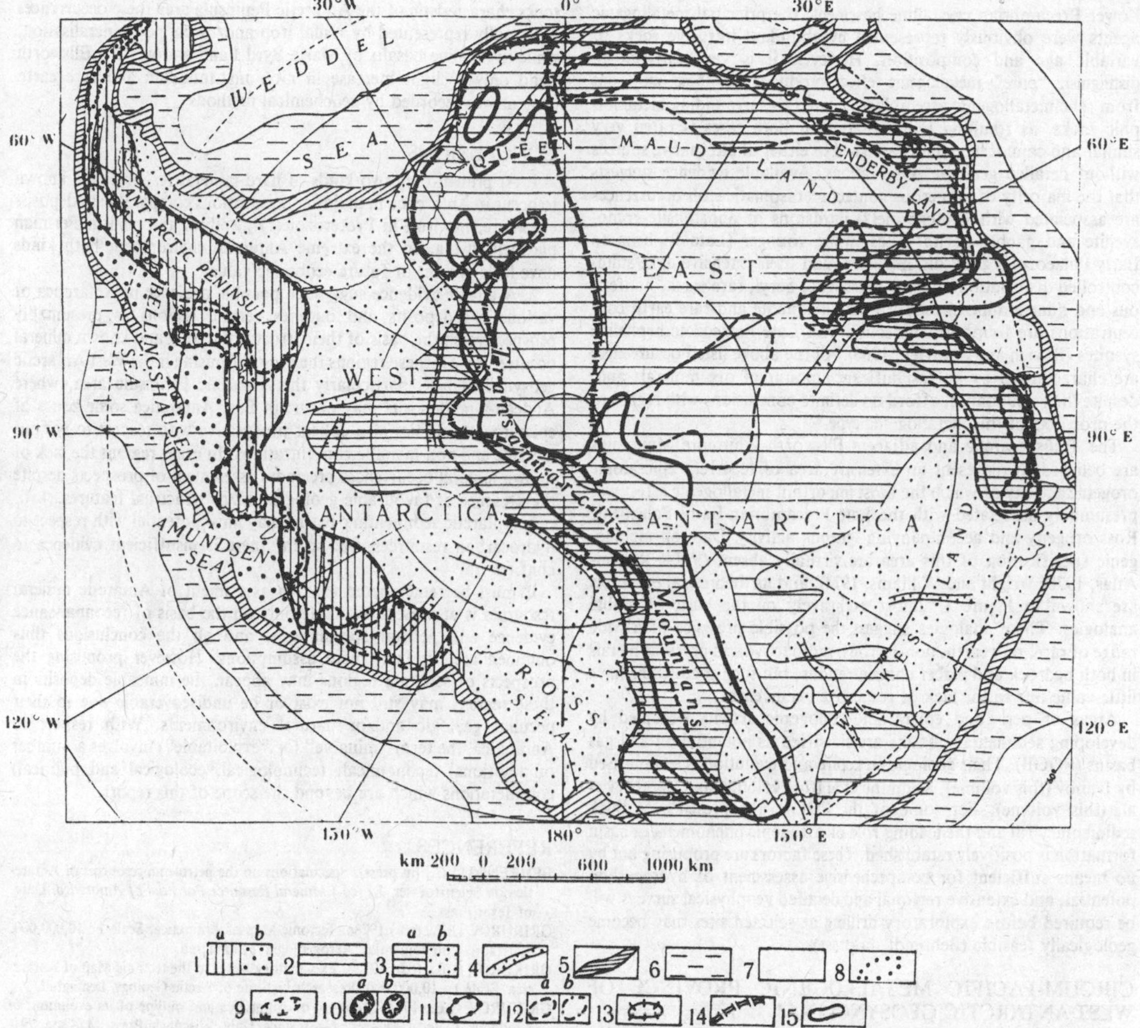


**Figure 1. Structural and metallogenic provinces in Antarctica.** ***Principal subdivisions.*** **1. West Antarctic geosynclinal belt and associated Circum-Pacific minerogenic province, a) Late Mesozoic epigeosynclinal (Andean) orogen, b) Late Mesozoic (?)—Cenozoic foreshelf ("active-margin") sedimentary basins, or potential oil and gas basins (POGB) of Pacific type; 2. East Antarctic craton and associated Gondwanian metallogenic province; 3. areas of peri- and/or epicratonic tectonics and associated "Atlantic" metallogenic (sub)provinces; a) Proterozoic to Phanerozoic epiplatform orogens, b) Mesozoic-Cenozoic "passive-margin" sedimentary basins, or POGB of Atlantic type; 4. major rift zones; 5. continental slope; 6. continental rise; 7. oceanic terrain.** ***Regions preferentially prospective for increased concentrations of hard mineral resources.*** **8. Cr, Ni, Co in eugeosynclinal zone; 9. Cu, Mo, Pb, Zn, Au and Ag in epi-Andean cordillera, essentially composed of Late Mesozoic to Early Cenozoic igneous (mainly intrusive) rocks; 10. Fe, related to a) metamorphogenic mineralisation in the Archaean crystalline basement, b) jaspilites in Proterozoic aulacogens; 11. non-ferrous and precious metals, Mo, W, beryl, mica and piezooptical quartz in zones of polyphase metamorphic and/or hydrothermal-metasomatic reworking of crystalline basement; 12. Nb, Ta, non-ferrous and rare-earth metals, related to; a) intensive Late Riphean to Early Palaeozoic granitoid activity associated with the Ross activisation of the East Antarctic craton and Ross orogeny in the Transantarctic Mountains, b) Late Cenozoic rifting and associated basaltic volcanism; 13. placer deposits of Sn and W in Lower Palaeozoic molasses; 14. Au, platinoids, non-ferrous and radioactive metals in Precambrian platform cover and associated mafic intrusions; 15. coal in Permian and Triassic strata of Phanerozoic platform cover, as well as Fe, Ti and non-ferrous metals in Jurassic tholeiites.** ***Areas preferentially prospective for hydrocarbon resources (POGB):*** **I. Weddell Sea basin; II. Ross Sea basin; III. Prydz Bay basin; IV. Bellingshausen Sea basin; V. Amundsen Sea basin; VI. Victoria Land basin; VII. Wilkes Land basin.**

their obvious localisation in intracratonic mobile zones and a specific set of metallic and/or mineral components usually found in similar structural environments elsewhere, suggests that the Antarctic intracratonic fold systems may be prospective for hard rock mineral resources.

The most typical Gondwanian structural unit in East Antarctica is the Beacon platform cover which includes Permian to Triassic coal measures. Their largest occurrence probably incorporating a few billion tons of moderately metamorphosed Permian coals (Ravich et al., 1978) is known in the Beaver Lake area (northern Prince Charles Mountains). Numerous occurrences are also scattered throughout the Transantarctic Mountains and in the western sector of the craton; their magnitude varies from thin coaly lenses and intercalations to coal beds as much as 5-7 m thick. Evidently, the Beacon strata are highly prospective for coal deposits. The Jurassic tholeiites capping the Beacon cover as thick basaltic piles and piercing it as dykes and sills may be enriched in ferrous and other metals, but so far there are very few known mineral occurrences of this type.

## "ATLANTIC" METALLOGENIC (SUB)PROVINCES IN THE AREAS OF PERI- AND EPICRATONIC TECTONISM

Recognition of this structural metallogenic type is provisional due to uncertainties in definition of the tectonic setting of the respective areas and wide varieties in views on their structural evolution (Grikurov, this volume).

In the zones of epicratonic orogenic activity of the East Antarctic

Lower Precambrian crystalline basement, the principal metallogenic agents were obviously represented by abundant intrusive rocks of variable age and composition. However, it is very difficult to distinguish "pure" metallogenic effect produced by these intrusions from relic metallogenic assemblages related to surrounding metamorphic rocks, as resulting mineralisation in both cases is often very similar and cannot be reliably assigned to either of alternative sources without detailed geologic observations. Available evidence suggests that the majority of numerous contact metasomatic iron occurrences are associated with Proterozoic(?) intrusions of anorthosite-granosyenite and gabbro-norite suites, while younger (Late Riphean to Early Palaeozoic) granitoid intrusions and their extensive dyke suites controlled distribution of some muscovite, beryl, rare earth, nonferrous and quartz mineralisation; titano-magnetite and rare earth concentrations are found in connection with Late Palaeozoic nepheline syenites (Ravich and Soloviev, 1966). All the above listed occurrences are characterised by an insignificant amount of ore minerals and, despite their abundance, afford no definite conclusions with regard to the prospects of this metallogenic type.

The Transantarctic and adjacent Ellsworth-Whitmore Mountains are believed to represent an extensive area of recurrent epicratonic orogenic activity of which the most important metallogenic effect was presumably associated with the Late Precambrian-Early Palaeozoic Ross orogeny and accompanying igneous activity. Data on metallogenic specification of this area are virtually absent (Polar Regions Atlas, 1978; Wright and Williams, 1974), and its theoretical prospects are shown in Figure 1 almost exclusively on the basis of global analogies. These analogies suggest the possible presence of a wide range of rare, rare earth, non-ferrous, radioactive and precious metals in both bedrock and placer concentrations, but such predictions have little value due to the lack of respective occurrences.

Areas of peri- and epicratonic coilogenic activity occupied by developing sedimentary basins are classified as potential oil and gas basins (POGB). Their geologic structure and evolution is summarised by Ivanov (this volume), Kadmina et al (this volume) and Kamenev et al. (this volume). For some of the POGB, the presence of thick sedimentary fill and the leading role of riftogenic phenomena in basin formation is positively established. These factors are promising but by no means sufficient for comprehensive assessment of hydrocarbon potential, and extensive regional and detailed geophysical surveys will be required before exploratory drilling at selected sites may become geologically feasible (Behrendt, in press).

## CIRCUM-PACIFIC METALLOGENIC PROVINCE OF WEST ANTARCTIC GEOSYNCLINAL BELT

In this section only the Andean orogenic zone is considered, as the POGB on its Pacific side are too poorly known to afford even preliminary speculations.

The Andean orogen is characterised by widespread Mesozoic and Cenozoic igneous rocks formed during major deformational and/or orgenic events, as well as in the course of subsequent rifting episodes. Fragments of the Late Palaeozoic (?)—Early Mesozoic eugeosynclinal zone are locally preserved along the Pacific coast. In this zone the presence of dunite-peridotite, gabbro-diabase and gabbro-diorite-plagiogranite intrusive suites and associated volcanic derivatives suggests a possibility of chromite mineralisation accompanied by nickel, cobalt and platinoids; small amounts of some of these metals have been reported from the South Shetland Islands (Rowley and Pride, 1982).

The same authors report numerous metallic occurrences similar to those typical of the mineral-rich South American Andes, i.e. characterised by predominant porphyry-copper mineralisation in connection with hydrothermally altered Mesozoic to Early Cenozoic intrusive rocks synchronous with the major phase of the Andean orogeny. The abundance of such occurrences recorded, despite poor exposure and the reconnaissance nature of geological exploration, suggests a possible presence here of sizeable ore deposits, although a direct analogy with the richest metallogenic Chilean and other South American provinces would be premature.

Some mineral occurrences are apparently related to Upper Cenozoic riftogenic volcanic rocks and associated minor intrusions of mafic to intermediate composition. In calc-alkaline high-alumina basaltic rocks characteristic of the Antarctic Peninsula area these occurrences are usually represented by visual iron and/or copper mineralisation, while in alkaline basalts of Marie Byrd Land and western Ellsworth Land only a slight increase in niobium, tantalum and rare earth content was recorded by geochemical methods.

## CONCLUSIONS

1. At present only two kinds of hard-mineral resources are known to occur in Antarctica in major concentrations comparable to deposits elsewhere, i.e. iron in Proterozoic(?) jaspilites and coal in Permian platform strata. In the extreme Antarctic environment both kinds have little present or future economic value.

2. Available evidence suggests a possible presence in Antarctica of various ore deposits and outlines certain regions as presumably prospective on the basis of their structural setting and known mineral occurrences. Of these regions the most promising is the West Antarctic geosynclinal belt, particularly the Antarctic Peninsula area, where Andean-type deposits may occur; in East Antarctica some zones of epicratonic tectonics also appear prospective with respect to various metallic and non-metallic concentrations. In other regions the lack of known mineral occurrences precludes assessment of prospects despite the presence of favourable geologic and/or structrual features.

3. Antarctic sedimentary basins may have potential with respect to hydrocarbon resources, but so far there is insufficient evidence to confirm it.

It must be remembered that an assessment of Antarctic mineral resources is at present available only on the basis of reconnaissance evidence and general speculations, and all the conclusions thus obtained are, in fact, mere assumptions. However promising the prospects of separate regions may appear, the mineable deposits in these regions may still not exist or be undiscoverable due to their peculiar geologic and/or natural environments. With respect to Antarctica, the term "minable" (or "exploitable") involves a number of additional (economical, technological, ecological and political) considerations which are beyond the scope of this report.

## REFERENCES


BEHRENDT, J.C., (in press): Speculations on the petroleum resources of Antarctica; *in* Splettstoesser, J., (ed.) *Mineral Resource Potential of Antarctica.* Univ. of Texas Press.

GRIKUROV, G.E., (ed.) 1978a: Tectonic Map of Antarctica, Scale 1 : 10,000,000. Kartfabrika ob'edinenija "Aerogeologia", Leningrad.

GRIKUROV, G.E., (ed.) 1978b: Explanatory Notes to the tectonic Map of Antarctica. Scale 1 : 10,000,000. Research Institute of Arctic Geology, Leningrad.

GRIKUROV, G.E., 1982: Structure of Antarctica and outline of its evolution; *in* Craddock, C. (ed.) *Antarctic Geoscience.* Univ. Wisconsin Press, Madison, 791-804.

GRIKUROV, G.E., (this volume): Ross fold belt of the Transantarctic Mountains as a boundary structure between East Antarctica and West Antarctica.

IVANOV, V.L., (this volume): Sedimentary basins of Antarctica and their preliminary structural and morphological classification.

KADMINA, I.N., KURININ, R.G., MASOLOV, V.N. and GRIKUROV, G.E., (this volume): Antarctic crustal structure from geophysical evidence.

KAMENEV, E.N., GRIKUROV, G.E. and IVANOV, V.L., (this volume): Structure and outline of geologic history of southern Weddell Sea basin.

POLAR REGIONS ATLAS, 1978: Central Intelligence Agency, 56-7.

RAVICH, M.G. and KAMENEV, E.N., 1972: *Kristallicheskii fundament Antarkticheskoi platformy (Crystalline basement of the Antarctic platform).* Gidrometeoizdat, Leningrad. (English translation: 1975, John Wiley, New York).

RAVICH, M.G. and SOLOVIEV, D.S., 1966: *Geologiia i petrologiia tsentral'noi chasti gor Zemli Korolevy Mod (Geology and petrology of the mountains of central Dronning Maud Land (eastern Antarctica).* Izdatel'stvo Nedra, Moscow.

RAVICH, M.G., KLIMOV, L.V. and SOLOVIEV, D.S., 1965: *Dokembrii Vostochnoi Antarktidy (The Precambrian of East Antarctica).* Izdatel'stvo Nedra, Moscow.

RAVICH, M.G., SOLOVIEV, D.S. and FEDOROV, L.V., 1978: *Geolocheskoe stroenie Zemli MakRobertsona (Vostochnaya Antarktida) (Geological structure of MacRobertson Land (East Antarctica)).* Gidrometeoizdat, Leningrad.

ROWLEY, P.D. and PRIDE, D.E., 1982: Metallic mineral resources of the Antarctic Peninsula; *in* Craddock, C., (ed.) *Antarctic Geoscience.* Univ. Wisconsin Press, Madison, 859-70.

SPIZHARSKY, T.N., 1973: *Obzornye tectonicheskie karty SSSR (sostavlenie kart i osnovnye voprosy tektoniki) (Reconaissance tectonic maps of the Soviet Union (Compilation of maps and fundamental tectonic problems)).* Izdatel'stvo Nedra, Leningrad.

WRIGHT, N.A. and WILLIAMS, P.L., 1974: Mineral Resources of Antarctica. *U.S. Geol. Surv., Circ., 705.*

# GEOPHYSICAL AND GEOLOGICAL STUDIES RELEVANT TO ASSESSMENT OF THE PETROLEUM RESOURCES OF ANTARCTICA

J.C. Behrendt, *U.S. Geological Survey, MS 964, Federal Center, Denver, Colorado, 80225, U.S.A.*

*Abstract* There are no known petroleum resources in Antarctica and information is lacking to make reliable estimates of undiscovered resources. Only giant (about 700 million tons or 0.5 billion barrels) or more probably supergiant (about 700 million tons or 5 billion barrels) fields would be reasonable to consider as economic in the next few decades in the hostile Antarctic environment. Consideration of locations of known giant oil fields in the world does not make Antarctica appear very prospective. The tectonic history of Antarctica subsequent to the Gondwanaland breakup suggests that West Antarctica is the most likely area to contain petroleum resources although East Antarctica cannot be excluded. Probably only the continental margins (possibly including ice shelf areas) bordering the Ross, Amundsen, Bellingshausen and Weddell Seas and Amery Shelf Ice will be exploitable with present or soon to be developed technology because of the several kilometre thick moving grounded ice sheet covering the rest of Antarctica. Geophysical data are sparse but do suggest the presence of several kilometres of unmetamorphosed sedimentary rock (possibly Cretaceous and Tertiary age) beneath the Ross, Weddell and Bellingshausen margins. Aeromagnetic data indicate significant thicknesses of nonmagnetic sedimentary rock in several areas beneath the grounded ice sheets in East and West Antarctica. Several Deep Sea Drilling Project (DSDP) holes beneath the Ross continental shelf have shown the presence of Tertiary marine and nonmarine sedimentary rocks as old as Oligocene in age overlying early Palaeozoic basement. Shows of gas in the DSDP holes, although provocative, cannot be considered evidence of any hydrocarbon resources.

There are no known petroleum resources in Antarctica (Figure 1). Nonetheless current concern relative to world supplies of oil and gas has turned the attention of geologists, geophysicists, economists, lawyers and diplomats from a number of countries to Antarctica. Exploitation of any metallic minerals that could be mined would be many years in the future (Rowley et al., 1983) even if deposits were to be found that might be economic in other parts of the world. The only mineral commodity with the possibility of exploitation within the next two or three decades is petroleum. Most of Antarctica is covered by a moving ice sheet about 3 km thick. The only areas accessible to presently available or soon to be developed technology are the continental margin possibly including the areas beneath ice shelves. Several previous writers (Wright and Williams, 1974; Zumberge, 1979a and b; Holdgate and Tinker, 1979; Dugger, 1978; Splettstoesser, 1977; Group of Experts, 1977; Rivera, 1977 and Ivanhoe, 1980) have addressed the possibility of petroleum resources on the Antarctic continental margin from various perspectives (geologic, environmental, economic and legal).

In a recent study of world oil resources Nehring (1978) discussed the occurrence of giant fields (0.5 billion barrels (bbl) ~ 70 million tons of recoverable oil) and supergiant fields (5 billion bbl or ~ 700 million tons of recoverable oil). He estimated that a total of four to 10 supergiants containing 30-100 billion tons remain to be discovered in the world. It is likely that nothing smaller than giant and more probably supergiant fields would be economic in the harsh Antarctic environment. Nehring (1978) discussed the "ring of oil" showing the concentration of nearly 85% of known world petroleum resources on a reconstruction of Gondwanaland, from which one would infer that it is unlikely that Antarctica as a whole would be very prospective. Possibly the "ring of oil" is only a reflection of the areas of the world

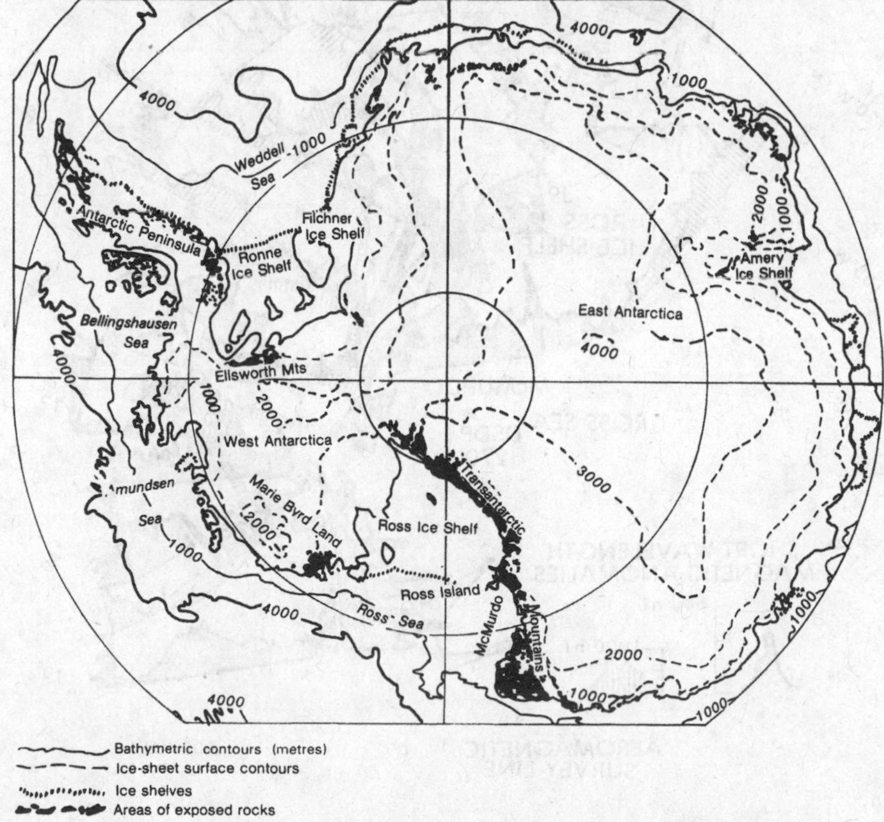


**Figure 1. Index map of Antarctica showing major features discussed in text including continental margin.**

where the most intense exploration has so far taken place.

Consideration of the known geology of Antarctica and inferences from sparse geophysical work suggest the presence of significant thicknesses of sedimentary rock in these areas throughout West Antarctica and several areas in East Antarctica. The Amery Ice Shelf area of East Antarctica might be considered on the basis of the large indentation in the continent suggesting a possible failed rift analogous to the petroleum rich Benue Trough area of West Africa. In this paper I review the available geophysical data and discuss the results of drilling on the continental margin by the Deep Sea Drilling Programme (DSDP).

## GEOPHYSICAL STUDIES

Some geophysical work was carried out in Antarctica during the 1930s and 1940s but the 1950s saw the beginning of systematic seismic reflection, refraction, gravity, land magnetic and aeromagnetic studies. The early seismic reflection work (Bentley, 1964) on the oversnow traverses was primarily directed at measuring ice thickness, with few sub-ice reflection results reported. No modern multichannel seismic reflection data have been collected on the grounded ice sheet or floating ice shelves of Antarctica. The seismic refraction results from the oversnow traverses were biased towards the higher velocities because the seismic velocity of ice is ~ 3.9 km/s, precluding direct obervation of lower velocity sedimentary rocks.

Radio echo ice sounding from the air (Drewry, 1975) has allowed continuous measurements of bedrock topography over large areas of Antarctica, but only recently (Behrendt et al., 1980) have simultaneous aeromagnetic measurements been made, allowing subglacial geologic interpretations. Aeromagnetic data without radio ice thickness measurements have been obtained on a reconnaissance basis mostly on widely spaced profiles throughout large areas of Antarctica. In West Antarctica, Behrendt and Wold (1963) and Behrendt (1964b) reported substantial (> 5 km) thicknesses of non-magnetic presumably sedimentary rocks west of the Ellsworth Mountains. Jankowski and Drewry (1981) also indicated thick sedimentary rocks in this area based on aeromagnetic and radio echo ice sounding. Behrendt et al. (1974) reported thick but undetermined amounts of sedimentary rock between the Pensacola and Ellsworth Mountains.

In the early 1960s widely spaced aeromagnetic profiles were collected across the entire length of Transantarctic Mountains (Behrendt, 1964b). Figure 2 shows the locations of these profiles between 120°W and 135°E in the Ross Ice Shelf area. I reinterpreted the data shown in Figure 2 using the method of Vacquier et al. (1951) to obtain rough estimates of elevation of magnetic basement as shown in Figure 3. For comparison, Figure 4 shows the generalised bedrock elevation of the same area. Beneath the Ross Ice Shelf, and beneath the grounded ice sheet west of the Transantarctic Mountains large areas of rock inferred to be sedimentary have thicknesses of 4-8 km or greater.

The *Eltanin* collected marine magnetic profiles (Hayes and Davey, 1975) over the Ross Sea continental shelf along with single channel seismic reflection profiles but Hayes and Davey did not attempt to calculate depth estimates because temporal variations in the magnetic field relative to the slow ship speed made it difficult to separate temporal from spatial anomalies.

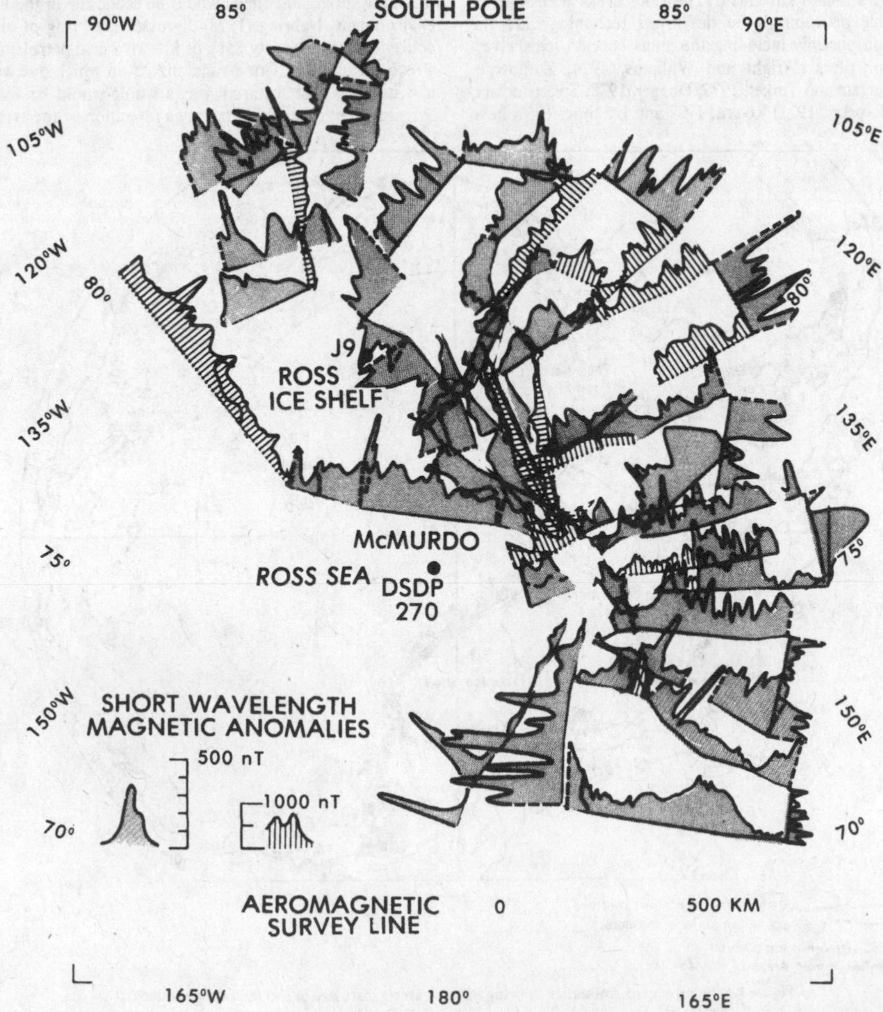


**Figure 2. Magnetic anomalies (1964/64) and track lines over the Ross Ice Shelf and adjacent parts of the Transantarctic Mountains.**

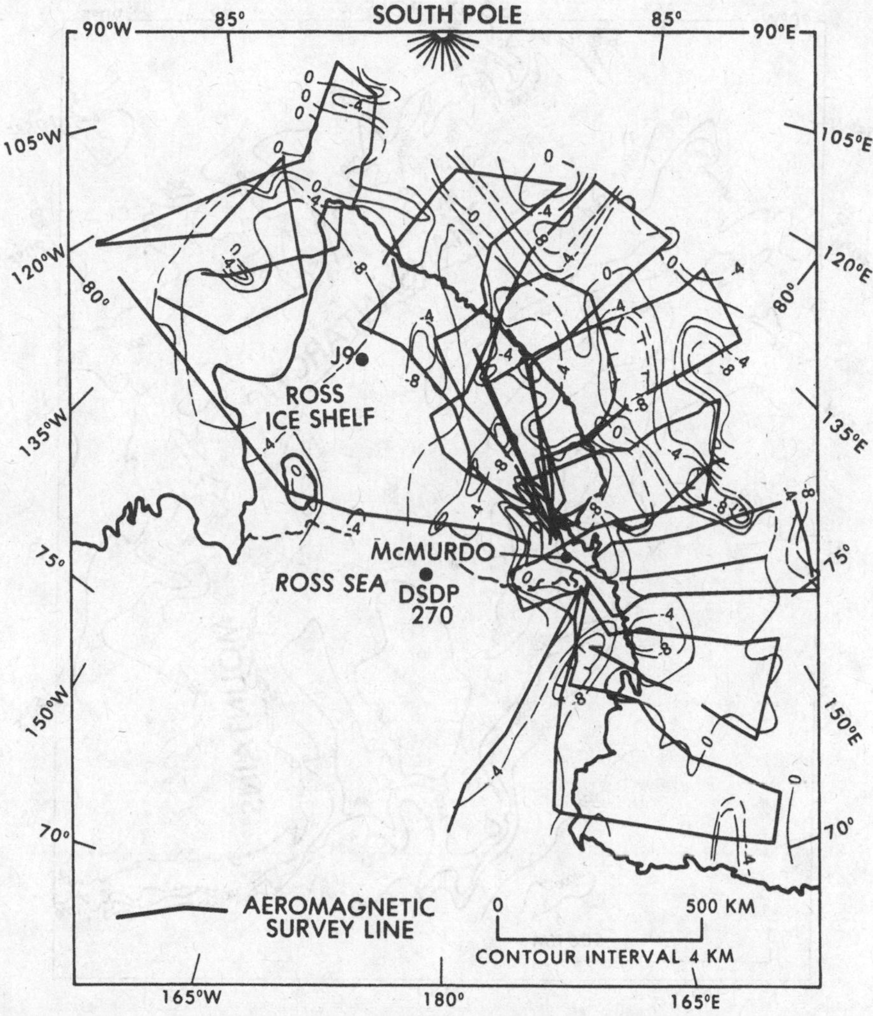


**Figure 3. Contour map showing estimated depth to magnetic basement based on magnetic profiles shown in Figure 2.**

In recent years, geophysicists from the USSR have collected substantial aeromagnetic data in the Filchner and Ronne Ice Shelf areas south of the Weddell Sea (Masolov, 1980). Their data suggest a 12-15 km thickness of sedimentary rock beneath the continental shelf in that area. The British Antarctic Survey (BAS) has also begun a programme of aeromagnetic survey flights over the Ronne Ice Shelf, results of which are consistent with thicknesses of 14 or 15 km of sedimentary rock (G. Renner, pers. comm., 1981). Thus, it appears likely that the area beneath the Ronne Ice Shelf and the continental margin bordering the Weddell Sea to the north may be underlain by a very thick section of sedimentary rock.

The results from magnetic surveys in West Antarctica discussed above suggest that there are several kilometres of sedimentary rock beneath the ice sheet and continental shelves. By analogy to sedimentary basins in other continents and the known geology of West Antarctica, we might expect Cretaceous and Tertiary age rocks to comprise a substantial part of the unexposed sedimentary section. Because a thick section of Palaeozoic and older sedimentary rock is exposed in the Ellsworth Mountains, sedimentary rocks of this age may also underlie the Ronne Ice Shelf (Figure 4). Bibby (1966) reported that a thick succession of Cretaceous sandstone crops at the north end of the Antarctic Peninsula and that a few outcrops of sedimentary rocks of Tertiary age are also found there. Early Cretaceous age rocks do exist beneath the narrow continental shelf of East Antarctica (Domack et al., 1980; Truswell, 1982).

In recent years ships from Norway (Fossum et al., 1982), the Federal Republic of Germany (Hinz and Krause, 1982) and the USSR (G. Grikurov, pers. comm., 1981) have collected multichannel seismic reflection profiles over the continental margin in the Weddell Sea area. In 1976/77 the Norwegian Antarctic Research Expedition (NARE) acquired 16 channel data along lines shown in figure 9 of Fossum et al. (1982). No results from this work have been published except for a profile across the front of the Filchner Ice Shelf (Figure 1) which is shown in Joint Oceanographic Institutions (1981). Reflections dipping westward are consistent with the thick section of sedimentary rock in the western Weddell Sea continental shelf and Ronne Ice Shelf area inferred from aeromagnetic data (Masolov, 1982) discussed above.

In 1978 the Federal Institute for Geosciences and Natural Resources (BGR) of the Federal Republic of Germany collected 5854 km of 48-channel data over the continental shelf between 25°W and 20°E. Hinz and Krause (1982) reported seaward dipping reflectors having seismic velocities >4.5 km/s overlain by sediments up to 3-5.2 km thick having velocities of 1.6-3.6 km/s. These lower velocities seem reasonable for Tertiary or possibly Late Cretaceous age rocks. Hinz and Krause (1982) interpreted the >4.5 km/s velocity seaward dipping reflectors as evidence of volcanic layers rather than sedimentary rock and therefore infers a low petroleum prospectivity for the margin in this area. Neither the 1976 NARE nor the 1978 BGR expeditions penetrated the very heavy pack ice of the Weddell Sea north of the Ronne Ice Shelf where the magnetic data referred to above suggested 12-15 km depth to magnetic basement. The USSR collected ~ 400 km of multichannel seismic reflection data (G.E. Grikurov, pers. comm., 1981) along a track between 15°W and 40°W in 1981, but no results are available.

Davey et al. (1982) and Davey et al. (this volume) have reported on

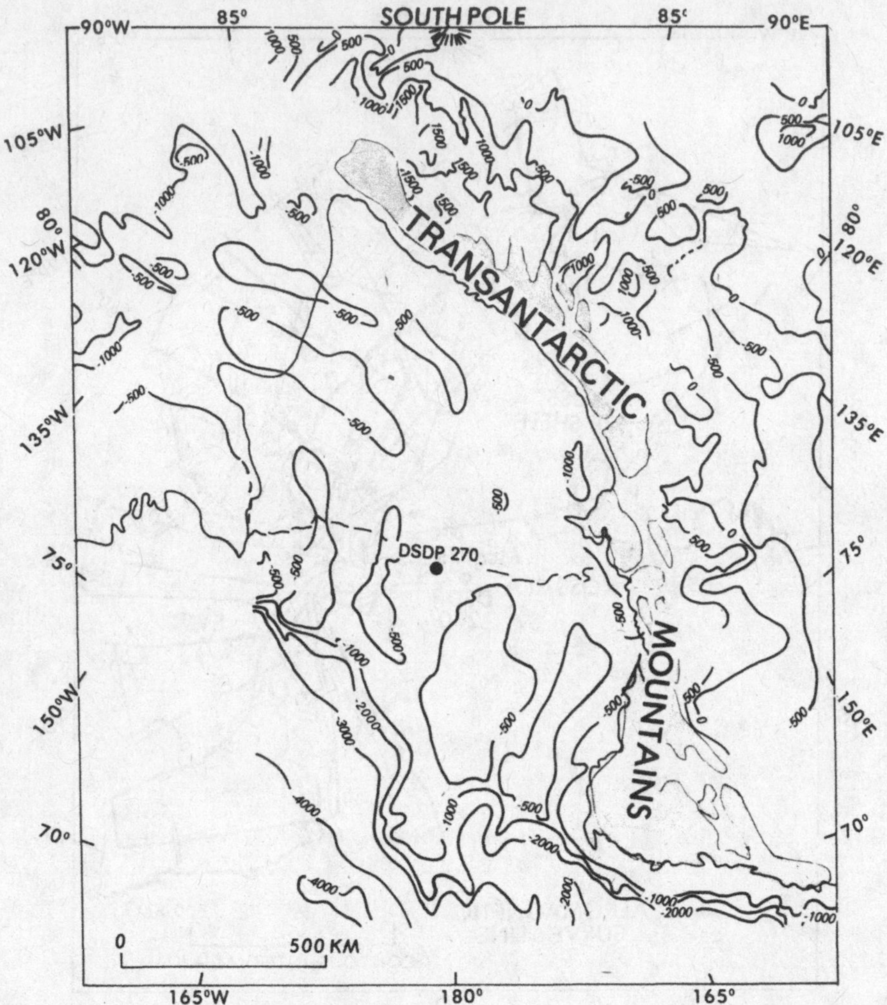


**Figure 4. Bedrock elevation map of area shown in Figures 2 and 3. Contours were generalised from Drewry (1975), Clough and Hansen (1979) and Bentley and Jezek (1981). DSDP Hole 270 is shown.**

sedimentary basins in the Ross Sea based on data from seismic refraction and variable angle reflection measurements using sonobuoys in 1980/81. These results indicate three major basins with sedimentary rock thickness exceeding 4 km in the central trough basin.

In 1980 the BGR acquired 6745 km of 48-channel data over the Ross Sea continental shelf (Fritsch, 1980). He reported that in the east part of the Ross Sea continental shelf two discontinuities were found, which he could correlate with Upper Miocene-Lower Pliocene and Middle Miocene-Upper Miocene contacts recognised in the DSDP core holes in the area. He reported a structural high along about the 180° meridian which divides the Ross Sea into two geologic provinces. Data processing is not completed and a maximum sedimentary rock thickness is not available, but at least several kilometres are suggested.

In 1981/82 the Inst. Francais du Petrol (IFP) using the ship *Explora*, collected about 1500 km of 48-channel data in the Ross Sea area (J. Wannesson, written comm., 1982) but no results from this work are available. In the 1981/82 season *Explora* collected 48-channel reflection data for IFP along the 3000 km of lines between 135°E and 155°E over the East Antarctica continental margin (J. Wannesson written comm., 1982) near Terre Adelie. No results are available yet.

The Japanese ship *Hakurei-Maru* collected more than 3280 km of 12-channel (three fold) reflection data in the Bellingshausen Sea area in 1981 (Kimura, 1982). He reported that maximum sedimentary rock thickness (about 3 s or 3-3.5 km) is greatest in the site of a palaeo-trench lower-slope complex and decreases both seaward and landward. This work is part of a planned three year Antarctic programme by the Japan National Oil Corp.

The Australian Bureau of Mineral Resources (BMR) collected about 5000 km of 6-channel reflection data over the continental margin in the area of the Amery Ice Shelf during 1981/82 between 55°E and 80°E (R.J. Tingey, pers. comm., 1982). No results of this work are available.

## DRILLING STUDIES

The *Glomar Challenger* on Leg 28 drilled a series of holes from December 22, 1972 to February 16, 1973 in the Antarctic area (Shipboard Scientific Party, 1975). Of these, Sites 270, 271, 272 and 273 were drilled beneath the continental shelf at the thinnest part of the sedimentary wedge beneath the Ross Sea Continental Shelf and are most relevant to the question of petroleum resources (the location of site 270 is shown in Figure 4, the others are nearby). The most significant results of this work were summarised in the initial reports (Shipboard Scientific Party, 1975). Results from these four holes (Hayes and Frakes, 1975) showed a Palaeozoic continental basement overlain by a section of Early Oligocene to Late Miocene, Pliocene and Pleistocene rocks. Although small amounts of methane and ethane were reported in parts of the dominantly nonmarine Miocene sedimentary rock (Shipboard Scientific Party, 1975), the authors considered it premature to attach any economic significance to the hydrocarbons. McIver (1975) also analysed samples from the cores of sites 271, 272 and 273. He reported significantly higher amounts of ethane and heavier homologs in these samples than from others collected by DSDP. He suggested this as evidence of local organic diagenesis. These results must be considered equivocal (although provocative) until future drilling takes place.

## SUMMARY

Although no petroleum resources are known in Antarctica and the petroleum industry is not particularly interested at present (Ivanhoe, 1980), economic and political considerations may change this in the next few years, and exploration and exploitation are possible within one or two decades. As noted above, a number of countries are actively carrying out multichannel seismic reflection surveys of the Antarctic continental margin, which are obviously focused on petroleum resource studies. Technology development will probably occur at a more rapid rate than research, exploration and legal developments (Holdgate and Tinker, 1979; Dugger, 1978). By contrast, hard mineral exploitation in Antarctica is probably so much further in the future as not to be seriously addressed other than from a general scientific research interest at this time. The only types of potentially exploitable petroleum resources in Antarctica from economic considerations would be giant or supergiant fields of which probably only four to 10 supergiants remain to be discovered in the world. In this report I have attempted to discuss some of the available information on potential petroleum resources in Antarctica. The points made can be summarised as follows:

(1) West Antarctica is probably the most prospective area of Antarctica because it probably contains large areas of unmetamorphosed sedimentary rock of post-rift age. It is comprised of a number of microplates (e.g. Behrendt, 1964a) that have moved significantly since the breakup of Gondwanaland (Dalziel and Elliot, 1982). East Antarctica, probably contains a number of subglacial sedimentary basins particularly adjacent to high mountain ranges and within the probable failed rift in the Amery Ice Shelf area.

(2) Because of the moving grounded ice sheet several kilometres thick which covers most of Antarctica, the only practical areas for possible exploitation, were petroleum to exist, are the continental margins (possibly including the parts covered by ice shelves) with the most likely areas those bordering the Ross, Amundsen, Bellingshausen and Weddell Seas in West Antarctica and the Amery Ice Shelf in East Antarctica.

(3) The sparse geophysical data suggest that there is a several kilometre thick section of sedimentary rock beneath the Ross and Weddell Sea continental shelves. The Bellingshausen Basin probably contains ± 3 km of sedimentary rock. There is no available information on sedimentary rock thickness beneath the continental shelves bordering the Amundsen Sea and Amery Ice shelf area but recent geophysical cruises can be expected to provide more information soon.

(4) DSDP holes on the Ross Sea continental shelf indicate the presence of rocks from Oligocene to Pleistocene age. Sedimentary rocks of Cretaceous or possibly Jurassic age might be present in the deepest parts of the section indicated by seismic reflection data and depths estimated from aeromagnetic data. Jurassic, Cretaceous and Tertiary sedimentary rocks are probably present beneath the continental shelf and adjacent glaciated areas of East Antarctica based on a number of samples by several investigators.

(5) There is presently no direct information on the petroleum geology beneath Antarctic continental shelves relative to source and reservoir rocks, with the exception of the shows of gas reported in core holes beneath the Ross Sea continental shelf.

*Acknowledgements* I thank P.D. Rowley and K.A. Kvenvolden for helpful discussions. A.B. Ford, C.M. Masters, G.L. Dolton and R.J. Tingey critically reviewed the manuscript.

## REFERENCES

BEHRENDT, J.C., 1964a: Crustal geology of Ellsworth Land and the Southern Antarctic Peninsula from gravity and magnetic anomalies. *J. Geophys. Res., 69,* 2047-63.

BEHRENDT, J.C., 1964b: Distribution of narrow width magnetic anomalies in Antarctica. *Science, 144,* 993-9.

BEHRENDT, J.C. and WOLD, R.J., 1963: Depth to magnetic basement in West Antarctica. *J. Geophys. Res., 68,* 1145-53.

BEHRENDT, J.C., DREWRY, D.J., JANKOWSKI, E. and GRIM, M.S., 1980: Aeromagnetic and radio echo ice sounding measurements show much greater area of the Dufek intrusion, Antarctica. *Science, 209,* 1014-7.

BEHRENDT, J.C., HENDERSON, J.R., MEISTER, L. and RAMBO, W., 1974: Geophysical investigations of the Pensacola Mountains and adjacent glacierised area of Antarctica. *U.S. Geol. Surv., Prof. Pap., 844.*

BENTLEY, C.R., 1964: The structure of Antarctica and its ice cover; *in* Odisbern, H., (ed.) *Research in Geophysics, 2, Solid Earth and Interface Phenomena,* MIT Press, Cambridge, Massachusetts, 335-89.

BENTLEY, C.R. and JEZEK, K.C., 1981: RISS, RISP and RIGGS: Post-IGY glaciological investigations of the Ross Ice Shelf in the U.S. *R. Soc. N.Z., J., 11,* 355-72.

BIBBY, J.S., 1966: The stratigraphy of part of northeast Graham Land and the James Ross Island group. *Scient. Rep. Br. antarc. Surv., 53.*

CLOUGH, J.W. and HANSEN, B.L., 1979: The Ross Ice Shelf Project. *Science 203,* 433-4.

DALZIEL, I.W.D. and ELLIOT, D.H., 1982: West Antarctica—Problem child of Gondwanaland. *Tectonics, 1,* 3-19.

DAVEY, F.J., BENNETT, D.J. and HOUTZ, R.E., 1982: Sedimentary basins of the Ross Sea, Antarctica. *N.Z. J. Geol. Geophys., 25,* 245-55.

DAVEY, F.J., HINZ, K. and SCHROEDER, H., (this volume): Sedimentary basins of the Ross Sea, Antarctica.

DOMACK, E.W., FAIRCHILD, W.W. and ANDERSON, J.B., 1980: Lower Cretaceous sediment from the East Antarctic continental shelf. *Nature, 287,* 622-25.

DREWRY, D.J., 1975: Radio echo sounding map of Antarctica, (90°E-180°). *Polar Rec., 17,* 359-74.

DUGGER, J.A., 1978: Exploiting Antarctic mineral resources—technology, economics and the environment. *Univ. Miami Law Review, 33,* 315-39.

FOSSUM, B.A., MAISEY, G.H. and TORSEN, H.O., 1982: Marine geophysical research in the Weddell Sea during the Norwegian Antarctic Research Expedition 1976-77; *in* Craddock, C. (ed.) *Antarctic Geoscience.* Univ. Wisconsin Press, 397-404.

FRITSCH, J., 1980: Bericht über, Geophysikalische Messungen; *in* Ross Meer Antarktis wahrend der Monate Januar/Februar 1980, Bundesanstalt für Geowissenschaften und Rohstoffe, Hannover, 36 p.

GROUP OF EXPERTS, 1977: Annex 5. *In* Report of the Ninth Consultative Meeting, Antarctic Treaty, London, Foreign and Commonwealth Office, 56-73.

HAYES, D.E. and DAVEY, F.J., 1975: A geophysical study of the Ross Sea, Antarctica. *Initial Reports of the Deep Sea Drilling Project, 28,* 897-908. U.S. Govt. Printing Office, Washington DC.

HAYES, D.E. and FRAKES, L.A., 1975: General synthesis Deep Sea Drilling Project leg 28. *Initial Reports of The Deep Sea Drilling Project, 28,* 919-42. U.S. Govt Printing Office, Washington DC.

HINZ, K. and KRAUSE, W., 1982: The continental margin of Queen Maud Land Antarctica: Seismic sequences, structural elements and geological development, *Geologisches Jahrbuch,* Reihe E., E. Schweizerbartliche Verlagsbuchhandlung, 17-41.

HOLDGATE, M.W. and TINKER, J., 1979: *Oil and other minerals in the Antarctic.* House of Print, London.

IVANHOE, L.F., 1980: Antarctica-operating conditions and petroleum prospects. *Oil and Gas J., 78 (52),*212-20.

JANKOWSKI, E.J. and DREWRY, D.J., 1981: The structure of West Antarctica from Geophysical Studies. *Nature, 291,* 17-21.

JOINT OCEANOGRAPHIC INSTITUTIONS, 1981: Ocean Margin Drilling Programme, Initial Science Plan Appendices.

KIMURA, K., 1982: Geological and geophysical survey in the Bellingshausen Basin off Antarctica, *Antarctic Rec., 75,* 12-24.

MASOLOV, V.N., 1980: Stroeniye magnitoaktivnogo fundamenta yugo-vostochnoiy chasti basseina Morya Ueddella: *Geofizicheskiye Issledovaniya v Antartidye,* Leningrad, 14-28.

McIVER, R. D., 1975: Hydrocarbon gases in canned core samples from Leg 28 sites 271, 272 and 273 Ross Sea. *Initial Reports of the Deep Sea Drilling Project, 28,* 815-9. U.S. Govt Printing Office, Washington DC.

NEHRING, R., 1978: Giant oil fields and world oil resources, *Rand Corporation Report,*R-2284-CIA.

RIVERA, R.C., 1977: Perspectivas de desarrollo de recurso de hidrocarburos en la Antartica; *in* Vicuna, F.O. and Araya, A.S. (eds.) *Desarrolo de la Antartica,* Editorial Universitaria, Institut de Estud. Internac. de la Universitad de Chile, Santiago, 265-71.

ROWLEY, P.D., WILLIAMS, P.C. and PRIDE, D.E., 1983: Metallic and nonmetallic mineral resources of Antarctica; *in* Splettstoesser, J.N. (ed.) *Mineral Resource Potential of Antarctica,* Univ. Texas Press, in press.

SHIPBOARD SCIENTIFIC PARTY, 1975: Initial Reports of the Deep Sea Drilling Project, Part 1 Shipboard Site Reports, 28, 1-369.

SPLETTSTOESSER, J.F., 1977: Offshore development for oil and gas in Antarctica; *in* POAC 77, Fourth international conference on port and ocean engineering under arctic conditions, *Univ. of Newfoundland Memoir,*1-10.

TRUSWELL, E.M., 1982: Palynology of seafloor samples collected by the 1911-14 Australasian Antarctic Expedition: implications for the geology of coastal East Antarctica. *Geol. Soc. Aust., J., 29,* 343-56.

VACQUIER, V., STEENLAND, N.C., HENDERSON, R.G. and ZIETZ, I., 1951: Interpretation of Aeromagnetic maps. *Geol. Soc. Am., Mem., 47.*

WRIGHT, N.A. and WILLIAMS, P.L., 1974: Mineral resources of Antarctica. *U.S. Geol. Surv. Circ., 705.*

ZUMBERGE, J.H., 1979a: Mineral resources and geopolitics in Antarctica, *Am. Scientist, 67,* 68-77.

ZUMBERGE, J.H. (ed.), 1979b: Possible Environmental Effects of Mineral Exploration and Exploitation in Antarctica, *Sci Comm. on Ant. Res.,* University Library, Cambridge, 59 p.

# RADIOMETRIC SURVEY IN NORTHERN VICTORIA LAND

G.A.M. Dreschhoff and E.J. Zeller, *Radiation Physics Laboratory, University of Kansas, Lawrence, Kansas, 66045, U.S.A.*

Wolf-Rudiger Kropp, *Bundesanstalt fur Geowissenschaften und Rohstoffe, D-3000 Hannover 51, Stilleweg 2, Federal Republic of Germany.*

*Abstract* During the 1981/82 field season an airborne radiometric survey was conducted in northern Victoria Land from a remote field camp in the Freyberg Mountains and a total of 3260 km of flight line was completed. The survey included igneous, metamorphic, and sedimentary rocks ranging in age from Precambrian to Late Cenozoic. Gamma-ray data from this region indicate a higher degree of short range variability than has been observed in any of the other four areas surveyed in Antarctica. This undoubtedly reflects the high level of geologic complexity in northern Victoria Land. No extensive concentrations of radioactive elements were detected at any location but the area shows substantial evidence for geochemical mobility of uranium at a number of sites. Since both depletions and enrichments are apparently present the entire region should be viewed as a possible source area for sedimentary deposits in adjacent parts of the continent. Specifically, adjoining areas to the south and west of northern Victoria Land should be examined for potential sedimentary concentrations.

## SPECTROMETER SYSTEM

This airborne survey, flown in a UH-1N helicopter, makes use of a gamma-ray spectrometer and detector array with 8390 cubic centimetre volume (Zeller and Dreschhoff, 1980). The unit has been especially adapted for use in cold regions. The apparatus consists of a 256 multichannel analyser so that gamma-ray spectra can be obtained when desired. In the airborne configuration, the spectrometer is not only highly sensitive as a total count system but also provides selectivity through the availability of four energy windows which measure total count, potassium-40, uranium (Bi-241), and thorium (Tl-208). From these data the ground level abundance of all the natural radioactive elements can be determined.

### Data recording

All data necessary to the interpretation of the spectrometer record are recorded during flight. This includes geological information as well as variations in speed or ground clearance. Fiducial marks are placed on the analog record and matched with the marks placed on the flight line being plotted on the topographic maps. Counts are accumulated over a 1.0 second sampling interval and recorded in analog form by a strip chart recorder with six separate channels. Four channels of the analog chart record the counts from four spectometer windows. The sensitivity in the potassium channel is 41.4 counts/sec. per 1 percent potassium, 3.6 counts/sec. per 1 ppm eU, and 2.3 counts/sec. per 1 ppm eTh. In addition to the evaluation of potential radioactive minerals, the apparatus used for this survey posesses sufficient sensitivity that it can be used for remote sensing of rock type identity by their radioelement signatures.

### Equipment calibration

Compton scattering corrections and sensitivities in terms of count rate per unit concentration have been determined from measurements at the Walker Field calibration pads at Grand Junction, Colorado. In addition, the spectrometer has been flown in an UH-1N (Bell 212) in operational configuration over the Department of Energy dynamic test range at Lake Mead, Arizona.

Two test strips have been established in Antarctica on the Nussbaum Riegel in Taylor Dry Valley (Zeller, et al., 1982) and at the old landing strip on Marble Point (Dreschhoff, et al., in prep.). These test strips are used for calibration at the beginning and end of each survey season. Background measurements are made when starting and concluding each survey flight. During flights, background can be checked constantly whenever the flight path crosses glaciers or thick snow fields exceeding the minimum extent of approximately 1 km width and 3 km length.

### Survey Characteristics of Antarctica

Outcrops are commonly small and surrounded by ice and this necessitates adjustments to the normal flight operation techniques which would be used in other portions of the world. Ground clearances greater than 30 m would often cause portions of the counting area (footprint) to include a substantial percentage of snow or ice. For this reason, we try to hold ground clearance to a range between 15 and 30 m. However, this means that the measurements are heavily influenced by inhomogeneities of surface details. Ratio measurements provide the best means for minimising these factors which can have a major effect on the absolute count rate. In most flight operations in other areas, low ground clearance would require closer flight line spacing to achieve uniform search effectiveness. This does not apply in Antarctica because the small size and isolated nature of the individual outcrops presents the major constraint on survey coverage. The total area covered by the survey is shown in Figure 1.

## METHODS OF DATA INTERPRETATION

Specific attention has been paid to the optimisation of data acquisition and to the interpretation of the measurements. The raw data are digitised and corrected for background and any significant changes in mean ground clearance. In most cases, deviations from mean ground clearance are not necessary and are applied only when clearance exceeds 30 m or drops below 15 m.

### Geologic source of error

Disequilibrium between uranium and radium can occur as a result of recent (< 10,000 years) leaching which may mobilise the uranium and leave behind the long-lived daughter, thorium-230. Whenever this occurs, gamma-ray spectrometry produces an incorrect determination of the uranium concentration. This problem is thought to be less acute in Antarctia than in temperate regions because of the reduction in movement of liquid water in exposed rocks. Radon migration and escape can also result in disequilibrium; it presents serious problems in highly porous media such as soils and alluvium but these are not common in most of the survey areas.

### Outcrop geometry problems

The configuration of the outcrop relative to the flight path of the aircraft can cause significant changes in the absolute values obtained by counting. The extremely rough terrain over which these surveys are conducted make this an important source of error and limits the usefulness of the absolute values obtained in the individual counting channels. However, ratios for example, U/Th or U/K, can be computed and they are relatively insensitive to variations in outcrop clearance, counting geometry, and partial radiation absorption by thin snow cover.

### Uraniferous provinces

In Antarctica, the principle objectives of this type of reconnaissance survey which necessarily uses widely spaced flight lines, is the evaluation of the potential of each area for uranium deposits rather than to search for specific ore occurrences. The techniques used in the evaluation of the data reflect this objective. As observed in other parts of the world, areas containing uranium deposits are generally characterised over tens to hundreds of miles by above average radioactivity, specifically above average thorium abundances. Such areas are termed uraniferous provinces. Ratio measurements of the three radioelements provide a means for recognising count rate anomalies and unusual raidoelement distributions. According to Darnley et al (1977) and Saunders (1979) the three radioelements vary sympathetically with one anouther over a wide range of rock types. If uranium has been removed due to its geochemical mobility relative to thorium, the thorium abundance alone may be a measure of the potential occurrence of uranium deposits in the area.

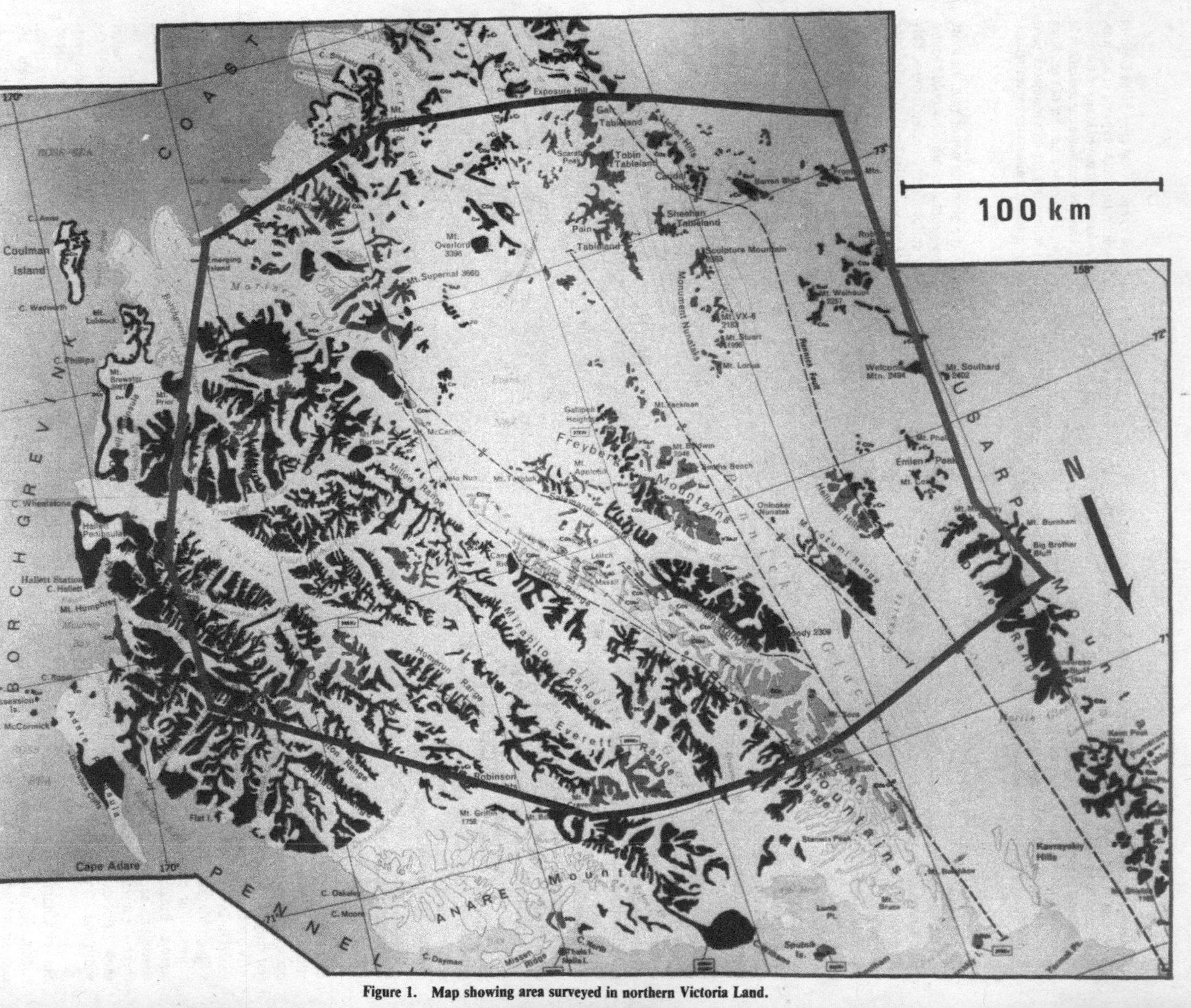


Figure 1. Map showing area surveyed in northern Victoria Land.

## GENERAL GEOLOGY OF THE AREA

### The northeast district

For the purposes of this report, the entire survey area has been divided into two major districts which are separated by the Rennick Fault (see Figure 1). The Robertson Bay Group in the northeastern district attains its maximum development in the coastal mountains. In general, this thick, folded sequence of greywackes and turbidites shows no anomalous concentration of either uranium or thorium. In a few cases small increases in radioactivity were found near contacts with granitic Admiralty Intrusives.

The Bowers Trough is the major strato-tectonic feature that dominates the northeastern district. It intersects the Rennick Fault at a roughly 30 degree angle and can be traced from the Ross Sea at Gauntlet Ridge to the northeastern outcrop at Rennick Bay. The conglomerates of Leap Year Formation which are part of the Bowers group contain concentrations of thorium bearing minerals and make this formation easy to recognise from the aerial survey. These thorium bearing conglomerates occur in the Molar Massif, the Quartzite Ranges and along the Mariner Glacier at Bunker Bluff and at Gauntlet Ridge. Granite Harbour Intrusives occur at several localities between the Rennick Fault and the Bowers Trough and they frequently show large numbers of xenoliths. In this area, individual xenoliths were occasionally found to be substantially more radioactive than the surrounding rock but in most cases the radioactivity was directly related to biotite masses and the aerial extent was always very limited. The Admiralty Intrusives are present on both sides of the Bowers Trough and they are clearly responsible for some of the local concentrations of radioactive minerals which were detected in the contact zones and within the plutons themselves. North of the intrusion at Mt. Dockery, fractures in the contact metamorphic zone were found to show substantial levels of radioactivity.

At Gallipoli Heights, the Gallipoli Rhyolite represents an extrusive phase of the Admiralty Intrusives and the entire area is higher in radioactivity than the surrounding regions. The volcanic ash deposits associated with the extrusives show increased count rates but this is commonly true of extrusive rocks and does not constitute evidence of any active mechanism of concentration that would likely lead to significant deposits of uranium or thorium.

East of the Rennick Glacier, the Kukri Erosion Surface is exposed at a few localities south of the Neall Massif but only minor concentrations of radioactive minerals were detected in the basal Beacon Group conglomerates. At other locations on the Canham Glacier where basal Beacon sediments occur, almost no anomalies could be detected.

The Ferrar Dolerites are widely distributed over much of the area as they are along most of the Transantarctic Mountains but they show no evidence of anomalous radioactivity at any locality where they were monitored. On the other hand, the Tertiary and Quaternary volcanic complex associated with Mt Overlord has a uniformal elevated radioactivity which appears similar to that of the Erebus volcanics of Ross Island. Under the weathering conditions that exist in Antarctica it is highly unlikely that surface geochemical processes could operate to produce significant secondary enrichments of uranium from these volcanics.

### The southwestern district

West of the Rennick Fault, in the Daniels Range, the Wilson Group rocks outcrop in extensive exposures at Thompson and Schroeder Spurs. There is some evidence of small-scale mobilisation of uranium in association with the high grade metamorphic processes that have affected these rocks. Furthermore, near the contact of a Granite Harbour pluton east of Mt Burnham, a small uranium anomaly was detected. In the Morozumi Range, sporadic local increases in radioactivity occur at contact zones but they are of small aerial extent.

Many of the extensive outcrops of the Granite Harbour Intrusives which occur in the Emlen Peaks, the Outback Nunataks, Frontier Mountain, and the Caudal and Lichen Hills, show very large pegmatite bodies containing prominent smoky quartz crystals. Many of these pegmatites have slightly elevated count rates and in a few cases actual low level uranium anomalies were detected.

The Kukri Erosion surface is well displayed in the Caudal and Lichen Hills where large areas have been stripped of the overlying Beacon Group sediments. Overall, the radioactivity of the surface is anomalously high with local enrichments of both thorium and uranium. Thorium enrichments are commonly associated with the outcrop of megacrystic granite which often contain pods of biotite. Anomalous concentrations of uranium are present at a number of localities on the surface where they appear to be associated with weathered zones probably developed at the time of formation of the erosion surface. Although none of these concentrations are high, they do provide very clear evidence of uranium mobility under the conditions that existed at the time of formation of the Kukri Erosion surface.

The basal Beacon Group sediments are well exposed at Mt Bower in the Outback Nunataks, and they show distinctly anomalous concentations of uranium. Although the concentations are clearly much too low to be of any interest as a potential resource, they furnish clear evidence that geochemical processes involving uranium mobility have been active. Examination of basal Beacon Group sediments in the Lichen and Caudal Hills failed to show similar elevated levels of radioactivity.

## RESULTS

In any area, significant deviations of radioelement distribution form normal regional eTh/eU and K/eU values may indicate selective removal of uranium from the "source" rocks and concentration in the "host" rocks. Outcrops in the survey area can be subdivided into source rocks and host rocks by plotting a scatter diagram of the uranium and thorium concentations shown by each outcrop as shown in Figure 2. In this case, adjusted values were used and they are obtained by dividing the uranium and thorium count rates for each outcrop by the respected mean values for the entire survey area. A linear regression line has been fitted to the scatter diagram and the mean deviation is then calculated. Outcrops above the regression line exhibit the characteristics of potential uranium source rocks whereas those below the line have the properties of host rocks. Most outcrop values fall within an envelope of one mean deviation above or below the linear regression line.

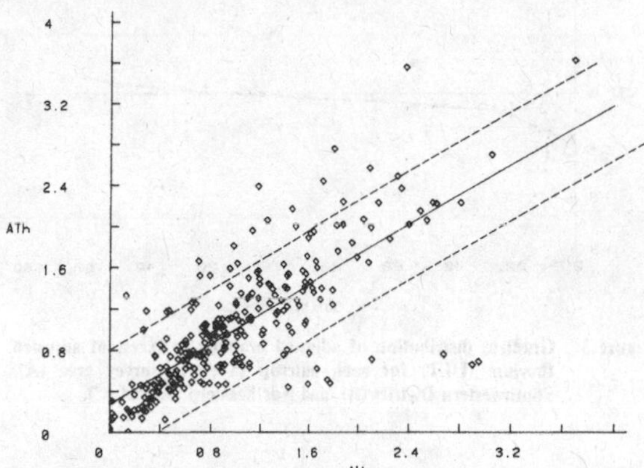


**Figure 2. Scatter diagram of adjusted thorium (ATh) versus adjusted uranium (AU).**

It is convenient to designate as anomalies those points that lie outside an envelope of two mean deviations around the regression line. If we do this, we find that the extreme uranium anomalies in what would be classified as host rocks are associated with the Granite Harbour Intrusives and basal Beacon Group Sediments at Mt Bower in the Outback Nunataks. The extreme thorium anomalies occur in the basement rocks that outcrop on the exposed Kukri Erosion Surface in the Caudal and Lichen Hills and with three outcrops at Mt Dockery in the Everett Range. However, the majority of thorium anomalies are associated with the Admiralty Intrusives, thus they constitute source rocks for uranium.

This conclusion is strongly supported by gradient distribution plots of the data from northern Victoria Land. These graphs are prepared from the standardised values for uranium and thorium that were used to plot the scatter diagrams. The assumption is made that the mean thorium and uranium values for the entire area represent a close approximation of the true abundance of these elements in the crust and that deviations from the mean constitute evidence of geochemical concentration or depletion processes. In northern Victoria Land, the true abundance ratio of thorium to uranium as measured by this survey is approximately 3.8 to 1 which is very close to the world crustal ratio. When standardised, the ratio of thorium to uranium becomes one to one if the measured values from an outcrop are equal to the means for the area. Gradient distribution plots are prepared by subtracting the standardised thorium value for each outcrop from the standardised uranium value. The resultant number is termed the uranium excess over thorium (UET) and is plotted sequentially from the smallest to the largest value.

Figure 3A shows the gradient distribution for all of the survey area in northern Victoria Land. The UET values are zero or close to zero if average amounts of uranium and thorium are present. The spread of the extremes constitute uranium and thorium anomalies. Strong positive values are present almost exclusively in the Granite Harbour Intrusives and the basal Beacon sediments. On the other hand, about 70 percent of the strong negative values are associated with outcrops of the Admiralty Intrusives. Figure 3B and 3C have been included to show the prominent difference in the northeast and southwest districts. This is particularly apparent in the fact that the uranium anomalies are confined to the southwest district though the strength of the thorium anomalies are roughly equal in both districts. Futhermore, the point at which the UET distribution crosses the zero line differs sharply in the two districts. The rocks northeast of the Rennick Glacier are, on the average, enriched in thorium and strongly depleted in uranium. The southwestern district shows the exact opposite trend and the uranium anomalies are much stronger.

In general, the granitic plutons of the Granite Harbour Intrusives and those associated with the Admiralty Intrusives exhibit very distinctive radiation signatures. When compared to the Admiralty Intrusives, we find that the Granite Harbour Intrusives are substantially higher in uranium, about equal in thorium, and always show a higher level of radioelement variability. In this portion of Antarctica, we believe that the rocks of the southwest district and particularly those of the Granite Harbour Intrusives have the characteristics generally associated with a uraniferous province.

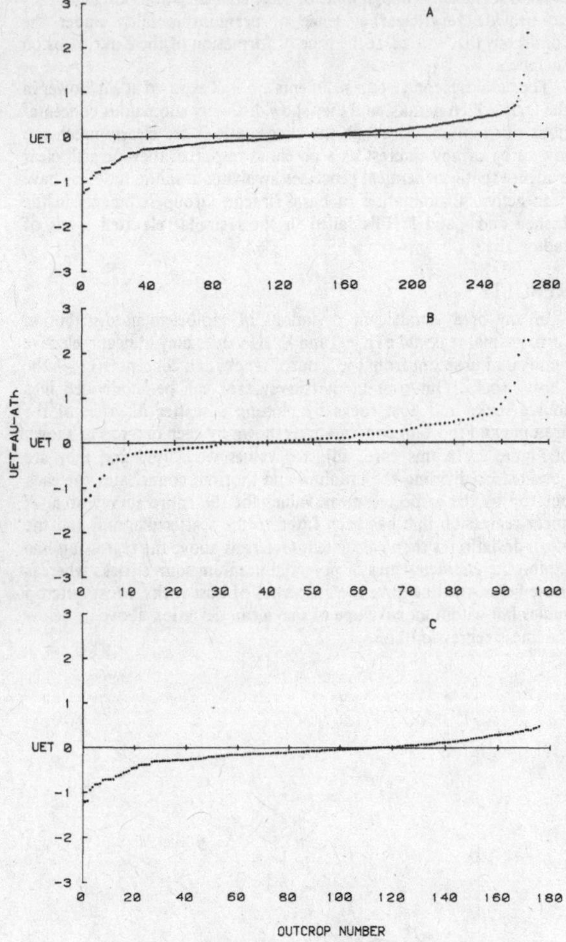


**Figure 3. Gradient distribution of adjusted uranium in excess of adjusted thorium (UET) for each outcrop in entire survey area (A), Southwestern District (B), and Northeastern District (C).**

*Acknowledgements* We wish to express thanks for partial support of this research by the Division of Polar Programs of the National Science Foundation and the Bundesanstalt fur Geowissenschaften und Rohstoffe.

## REFERENCES

DARNLEY, A.G., CHARBONNEAU, B.E. and RICHARDSON, K.A., 1977: Distribution of uranium in rocks as a guide to the recognition of uraniferous regions; *in* Recognition and evaluation of uraniferous areas. Int. Atom. Energy Agency, Vienna, 55-86.

DRESCHHOFF, G.A.M., ZELLER, E.J., BUSCH, K. and BULLA, K. (in prep.): Resource and radioactivity survey in southern Victoria Land, 1980/81.

SAUNDERS, D.F., 1979: Characterisation of uraniferous geochemical provinces by aerial gamma-ray spectrometry *Mining Engineering, 31, 12,*1715-22.

ZELLER, E.J. and DRESCHHOFF, G.A.M., 1980: Evaluation of uranium resources in Antarctica; *in* Uranium evaluation and mining techniques. Int. Atom. Energy Agency, Vienna, 381-90.

ZELLER, E.J., DRESCHHOFF, G.A.M., TESSENSOHN, F. and CRISLER, K., 1982: Resource and radioactivity survey in Antarctica by airborne gamma-ray spectrometry; *in* Craddock, C. (ed.) *Antarctic Geoscience.* Univ. Wisconsin Press, Madison, 877-83.

# RECONNAISSANCE OF MINOR METAL ABUNDANCES AND POSSIBLE RESOURCES OF THE DUFEK INTRUSION, PENSACOLA MOUNTAINS

A.B. Ford, and R.E. Mays, *U.S. Geological Survey, Menlo Park, California 94025, U.S.A.*

J. Haffty, *U.S. Geological Survey, MS 964, Federal Center, Denver, Colorado, 80225, U.S.A.*

B.P. Fabbi[1], *U.S. Geological Survey, National Center, Reston, Virginia, 22092, U.S.A.*

[1] Present address: *Bausch & Lomb Co., 9545 Wentworth St., Sunland, California 91040, U.S.A.*

*Abstract* The Jurassic layered mafic Dufek intrusion (>50,000 km² area, 8-9? km thick) may rival in size South Africa's Bushveld Complex and is far larger than any other known complex of this type. Many similarities with other mafic layered intrusions that contain economically important magmatic deposits make occurrence of resources likely, although none is yet identified in this mostly ice-covered body. Resources such as Cr, Ni, Cu, V, Ti and particularly Pt-group elements (PGE) known in other complexes are considered to be "speculative resources" of the Dufek intrusion, in which PGE would have greatest potential for possible utility. Analyses for these elements show a greater abundance of Cu, V, Ti and PGE in the upper compared with the lower exposed parts of the body, suggesting an enrichment with differentiation. The highest determined contents of Pt (0.0035 ppm) and Rh (0.012 ppm) are in a gabbroic cumulate containing abundant (34%) Fe-Ti oxide minerals near the base of the upper section (Forrestal Range). The maximum Pd abundance (0.044 ppm) found is at 340 m above the base of the section in a gabbroic cumulate containing approximately average amounts (9%) of Fe-Ti oxide minerals. In the lower section (Dufek Massif), Pt, Pd and Rh contents have been found to be only near or below limits of determination. Petrologic comparisons suggest that the chief PGE zone of interest in the Bushveld Complex (Merensky Reef) would occupy an analogous position in the upper part of the concealed section of the Dufek intrusion.

Nearly all major and many minor examples of layered mafic igneous complexes contain economically significant deposits of one or more metals, notably concentrated in lower horizons and associated with early mafic and ultramafic cumulates (Wilson, 1969). Both in size (67,000 km²) and in wealth of resources South Africa's Bushveld Complex is an exceptional example (Willemse, 1969). Of particular economic interest are the major deposits of platinum group elements (PGE) long known and produced in the Bushveld's Merensky Reef and newly discovered (Todd et al., 1979) in the Banded zone of the Stillwater Complex of Montana.

Following its 1957 discovery and early reports of similarity with the Stillwater Complex and of occurrences of ultramafic rock, chromite and copper minerals (Aughenbaugh, 1961; Walker, 1961), the Dufek intrusion has attracted considerable speculation regarding mineral resources (e.g. Runnells, 1970; Wright and Williams, 1974; Lovering and Prescott, 1979). Recent geophysical surveys showing its area to exceed 50,000 km² (Behrendt et al., 1980)—approaching the Bushveld in size and far larger than any other known complex—makes such speculation even more attractive. Only a few percent of the body's area, however, is exposed.

This report presents results of a preliminary survey of abundances and distribution of minor elements (Pt, Pd, Rh, Cu, Cr, Ni, Co, V, Ti and total S) that are of chief interest for study of possible resources of the body. Twenty-four samples of average appearing rock representing all major lithologies and a broad spectrum of the differentiation sequence shown in exposed parts of the intrusion were analysed in this study.

## THE DUFEK INTRUSION

### Geology

*Lithology and stratigraphy* Cumulates of general gabbroic composition make up at least about 90% of exposed parts of the body. They commonly are thinly interlayered with pyroxenitic, anorthositic or, in the upper part, Fe-Ti-oxide-mineral-rich cumulates. The mostly mafic cumulate sequence is concordantly capped by a layer of granophyre 200-300 m thick. Of the total 8-9? km estimated stratigraphic thickness near the southern end of the body (Ford, 1976), 1.8 km of a lower but not basal part makes up Dufek Massif (Ford et al., 1978a) and about 1.7 km of the upper part makes up the Forrestal Range (Ford et al., 1978b). Major unexposed stratigraphic parts of the basal section, estimated from geophysical evidence to be 1.8-3.5 km thick (Behrendt et al., 1974), and a 2-3 km thick intermediate interval inferred beneath Sallee Snowfield between the two main ranges.

*Structure* Primary igneous structures include compositional layering that reflects variable mineral proportions with height, layer-parallel lamination and troughs. Layers range from a few millimetres to 15 m or more in thickness. Major ones of about 1 m or more in thickness typically have sharp bases and gradational tops. Some pyroxenitic layers show no measurable change in thickness for distances of more than 35 km. Cut-and-fill channel-like trough structures are commonly found near the major layers, suggesting an origin related to magma current activity.

Layering attitudes across the body define an overall broad synformal structure, with a northerly trending axis centred in the Forrestal Range. Layered rocks of the gently (5-10°) southeast-dipping limb of Dufek Massif are downdropped northwestward along strands of the Enchanted Valley fault north of the massif.

*Intrusive form* An overall sheetlike form is suggested by the very large ratio (about 5 x 10³:1) of inferred area to thickness. General discordancy is shown by a regional cutting of north-trending fold axes in country rocks across the entire width of the northern Pensacola Mountains. Folded Devonian quartzite is cut at a high angle at the only small exposure of the contact. The intrusive form of a body of such enormous dimensions is doubtless far more complex than that of a simple synformal sheet as apparent in its small area of exposure. At least one large ice-covered lobe, identified by geophysical surveys (Behrendt et al., 1980), extends from the main body southeast of the Forrestal Range, which suggests some similarity with the Bushveld Complex, in which a multiple lobate outline reflects presence of adjoining basin-shaped bodies that are probable intrusive centres (Willemse, 1969). Other lobes may exist but are not identified from the reconnaissance surveys.

*Age and correlation* K-Ar age determinations of 172±4 Ma, near the beginning of the Middle Jurassic, show the body to be closely related in age with the voluminous tholeiitic magmatism of the Ferrar Group elsewhere in the Pensacola Mountains and throughout the Transantarctic Mountains (Ford and Kistler, 1980).

*Tectonic setting* The intrusion lies in a recurrently active mobile belt adjoining the East Antarctic Precambrian craton, in which latest major compressive deformation is dated as probably Triassic (Ford, 1972). Elsewhere along the lengthy belt of Ferrar magmatism, mobility generally ceased by Devonian time. This difference in tectonic setting may in part account for the comparatively immense local scale of Jurassic magmatism in the Pensacola Mountains. This area is also not far from an inferred centre of a radial rift system associated with initial Gondwanaland breaks (Ford and Kistler, 1980).

### Petrology

*Mineralogy* The chief cumulus phases in the lower exposed section (Dufek Massif) are plagioclase and Ca-rich and Ca-poor pyroxenes, with very minor amounts of Fe-Ti oxide minerals only in the highest part. The same minerals are all major cumulus phases in the upper exposed section (Forrestal Range), where Fe-Ti oxide minerals are very abundant and commonly concentrated in magnetite layers a few centimetres to as much as about 1 m thick. Apatite is a ubiquitous phase and Fe-rich olivine a minor one in the upper section. Mg-rich olivine and chromite are only known in rare, metamorphosed xenoliths and have not been found as cumulus minerals.

*Differentiation trends* Stratigraphically upward decrease in An content ($An_{80-50}$) of cumulus plagioclase (Abel et al., 1979) and increase in Fe:Mg ratio of augite-ferroaugite and bronzite-(inverted) pigeonite (Himmelberg and Ford, 1976) show differentiation trends characteristic of layered intrusions of this type (Wager and Brown, 1967). Fe-Ti oxide minerals generally exhibit no systematic variation with height, owing to subsolidus re-equilibriation, except for general upward decrease in $Al_2O_3$ and $V_2O_3$ of ilmeno-magnetite (Himmelberg and Ford, 1977). The differentiation trends continue upward through the exposed sections more or less uniformly until about 1 km from the top, suggesting that the major part of the body here crystallised from a single very large reservoir of magma. Strong reversals in trends of all cumulus minerals at about 1 km from the top are tentatively inferred to result from a major late-stage replenishment either with less fractionated magma from elsewhere within the chamber or with nonfractionated new magma that mixed with Fe-rich residual magma. Normal differentiation trends, leading into the granophyre cap layer, were re-established above the reversal discontinuity.

## Sulphide Mineral Occurrences

Minor amounts of sulphide minerals and efflorescences of chiefly atacamite occur in all lithologies throughout the sequence, including granophyre. Their distribution is highly variable along and across layering. Concentrations of more than a few percent are not known. Walker (1961) reports occurrences of bornite(?), chalcopyrite and pyrite along the northern lower part of Dufek Massif. Sulphide occurrences are much more abundant in the Forrestal Range, most commonly associated with Fe-Ti-oxide-mineral-rich layers. The occurrences appear similar to those in the upper Bushveld Complex in which sulphide precipitation is related to crystallisation of magnetite (von Gruenewaldt, 1976). A preliminary survey of sulphides in polished section shows traces of mainly pyrite and chalcopyrite (Czamanske, G.K., pers. comm., 1981).

# MINOR-ELEMENT INVESTIGATIONS

## Methods

Total-mode mineralogy (cumulus + postcumulus phases) of the 24 samples analysed in this study (Table 1) was determined by counting approximately 1,000 points on stained-rock slabs of 40 $cm^2$ size. (Minerals in granophyre are noncumulus and include large amounts of K-feldspar and quartz, not listed). Stratigraphic heights given in Table 1 are measured from the base of each of the two sections

**TABLE 1: Location and lithology of analysed samples**

| Section | Height (m) | Sample | Lithology | Mode (vol. %)[1] | | | | |
|---|---|---|---|---|---|---|---|---|
| | | | | ap | ox | px | pl | sf |
| Forrestal | 1564 | 316Fa | granophyre | tr | 2 | 12 | 21 | 0 |
| Range | 1483 | 314Fa | granophyre | tr | 2 | 11 | 14 | 0 |
| (upper | 1343 | 159Fa | px-pl-ox cumulate | tr | 12 | 46 | 42 | tr |
| section) | 1320 | 158Fa | pl-px-ox-ap cumulate | 3 | 13 | 38 | 46 | tr |
| | 1149 | 169Fa | pl-px-ox cumulate | tr | 11 | 39 | 50 | tr |
| | 1080? | 269Fb | px-ox-pl cumulate | tr | 18 | 71 | 9 | 2 |
| | 939 | 187Fb | pl cumulate | 0 | 4 | 21 | 70 | 0 |
| | 340 | 301Na | pl-px-ox cumulate | 0 | 9 | 39 | 52 | 0 |
| | 298 | 94Ff | pl cumulate | 0 | 8 | 17 | 75 | 0 |
| | 165 | 98Fa | pl cumulate | 0 | 7 | 12 | 81 | tr |
| | 37 | 78Fa | pl-ox-px cumulate | 0 | 34 | 31 | 35 | tr |
| Dufek | 1798 | 260Fa | pl-px-ox cumulate | 0 | 3 | 30 | 67 | 0 |
| Massif | 1540 | 269Fa | pl-px cumulate | 0 | tr | 32 | 68 | 0 |
| (lower | 1075 | 111Fa | pl-px cumulate | 0 | 0 | 42 | 58 | 0 |
| section) | 1070 | 198Fa | px cumulate | 0 | 0 | 93 | 7 | 0 |
| | 1070 | 226Ff | px cumulate | 0 | 0 | 99 | 1 | 0 |
| | 700 | 52Fb | pl cumulate | 0 | tr | 4 | 96 | 0 |
| | 460 | 281Fm4 | px-pl cumulate | 0 | 0 | 82 | 18 | 0 |
| | 460 | 197Fb | px-pl cumulate | 0 | tr | 76 | 24 | 0 |
| | 380 | 42Fb | pl-px cumulate | 0 | 0 | 33 | 67 | 0 |
| | 300 | 40Fa | pl-px cumulate | 0 | 0 | 40 | 60 | 0 |
| | 230 | 195Fa | pl cumulate | 0 | 0 | 5 | 95 | tr |
| | 200 | 32Ff | pl cumulate | 0 | 0 | 13 | 87 | tr |
| | 20 | 192Fb | pl-px cumulate | 0 | tr | 10 | 90 | tr |

[1]ap = apatite, ox = Fe-Ti oxide minerals, px = pyroxenes, pl = plagioclase, sf = visible sulphides.

The samples were analysed for major elements by the method of Shapiro (1975). Pt, Pd and Rh were determined (Haffty, J., Goss, W.D. and Haubert, A.W.) on 15 g splits using a fire assay preconcentration and emission spectrographic method (Haffty et al., 1977). Lower limits of determination by the method are 0.010 ppm for Pt; 0.004 ppm for Pd; and 0.005 ppm for Rh. Where we used the general term "platinum group elements," or "PGE", in discussing this data it refers only to those 3 elements of the group. Determinations (Mays) for Cu, Ni, Co, Cr and V were by quantitative spectrographic methods having lower detection limits as follows: Cu (1 ppm), Ni (2 ppm), Co (4 ppm), Cr (2 ppm) and V (5 ppm). Spectrographic results are reported to two significant figures, with overall accuracy of ± 15 percent, except for being lower near detection limits. Total S was determined (Fabbi) by X-ray fluorescence analysis having a lower detection limit of 20 ppm.

The degree of statistical inter-element correlation and correlation with stratigraphic height and with mineral modes for the 22 analysed cumulates (granophyre excluded) was determined using the Spearman rank correlation coefficient ($r_s$), a nonparametric method needed (Dixon and Massey, 1957) because of the small population of samples, a probable non-normal distribution, the presence of censored data, and the uncertain thickness of concealed intervals. Linear correlation ($r_s$) with height is an obvious oversimplification, as element variation is known to be nonlinear in other layered intrusions (Wager and Brown, 1967).

## Results

*Platinum group elements* Whereas Rh is determinable in only one sample (78Fa with 0.012 ppm Rh, containing an exceptional amount of Fe-Ti oxide minerals and also the highest determined Pt content of 0.035 ppm), Pt and Pd were each determinable in seven, generally different samples (Table 2). Determinable amounts of Pt and Pd occur

**TABLE 2: Minor-element analyses (ppm, except $TiO_2$ in weight percentage)**

| Section | Sample | Pt | Pd | Rh | S | Cu | V | Co | Cr | Ni | $TiO_2$ |
|---|---|---|---|---|---|---|---|---|---|---|---|
| Forrestal | 316Fa | <0.010 | <0.004 | <0.005 | 130 | 32 | 6 | 7 | 2 | <2 | 1.6 |
| Range | 314Fa | <.010 | <.004 | <.005 | 140 | 36 | 15 | 8 | 2 | <2 | .68 |
| | 159Fa | <.010 | <.004 | <.005 | <20 | 460 | 200 | 90 | <2 | 13 | 4.4 |
| | 158Fa | <.010 | <.004 | <.005 | 320 | 100 | 80 | 70 | 8 | 2 | 5.4 |
| | 169Fa | .025 | <.004 | <.005 | 800 | 460 | 500 | 85 | <2 | 24 | 4.8 |
| | 289Fb | .013 | .006 | <.005 | 4700 | 2000 | 2000 | 100 | 44 | 65 | 7.5 |
| | 187Fb | <.010 | <.004 | <.005 | 90 | 150 | 500 | 38 | 10 | 32 | 1.3 |
| | 301Na | .014 | .044 | <.005 | 170 | 100 | 1000 | 55 | <2 | 50 | 2.9 |
| | 94Ff | <.010 | .011 | <.005 | 100 | 100 | 650 | 60 | 50 | 50 | 2.5 |
| | 98Fa | <.010 | .012 | <.005 | 100 | 90 | 650 | 38 | 65 | 32 | 1.2 |
| | 78Fa | .035 | .006 | .012 | 220 | 100 | 2000 | 120 | 20 | 190 | 8.6 |
| Dufek | 260Fa | <.010 | .010 | <.005 | 130 | 36 | 460 | 70 | 480 | 80 | .87 |
| Massif | 269Fa | <.010 | <.004 | <.005 | 190 | 36 | 170 | 44 | 80 | 55 | .39 |
| | 111Fa | <.010 | <.004 | <.005 | 40 | 12 | 150 | 70 | 320 | 100 | .22 |
| | 198Fa | <.010 | <.004 | <.005 | 110 | 35 | 280 | 85 | 300 | 130 | .36 |
| | 226Ff | .015 | <.004 | <.005 | 90 | 24 | 400 | 90 | 360 | 150 | .46 |
| | 52Fb | <.010 | <.004 | <.005 | 70 | 20 | 65 | 10 | 20 | 12 | .30 |
| | 281Fm4 | <.010 | <.004 | <.005 | 160 | 40 | 320 | 65 | 200 | 110 | .41 |
| | 197Fb | <.010 | <.004 | <.005 | 130 | 40 | 240 | 80 | 240 | 120 | .41 |
| | 42Fb | .015 | <.004 | <.005 | 50 | 24 | 170 | 55 | 55 | 75 | .22 |
| | 40Fa | <.010 | <.004 | <.005 | 100 | 30 | 160 | 55 | 100 | 100 | .28 |
| | 195Fa | .010 | <.004 | <.005 | 60 | 26 | 44 | 8 | 20 | 12 | .06 |
| | 32Ff | <.010 | <.004 | <.005 | 40 | 20 | 55 | 8 | 14 | 12 | .21 |
| | 192Fb | <.010 | .005 | <.005 | 140 | 34 | 55 | 14 | 70 | 38 | .17 |

in five of thirteen (38%) of Dufek Massif cumulates and in six of nine (67%) of Forrestal Range cumulates. Exact average values cannot be calculated owing to the many samples with PGE below limits of determination. Based on limits of determination, average abundances in Dufek Massif cumulates are in the ranges 0.003-0.010 ppm Pt and 0.001-0.004 ppm Pd, and in Forrestal Range cumulates are in the ranges 0.009-0.017 ppm Pt and 0.009-0.011 ppm Pd. Although PGE are not significantly correlated with height (Figure 1), the ranges of their possible averages suggest nearly two-fold enrichment in the upper compared with the lower section. PGE show strongest correlation among elements with V, and with Fe-Ti oxide minerals.

*Copper and sulphur* Cu, but not S, shows strong correlation with height. Except for sample 289Fb, an atypical rock containing two percent visible sulphides, Cu averages 195 ppm in Forrestal Range cumulates and 29 ppm in those of Dufek Massif—and about six-fold increase in the upper section. Except for sample 289Fb, averages for S show an approximate two-fold upward increase, from 101 ppm in Dufek Massif cumulates to 225 ppm in Forrestal Range cumulates. Both elements show marked depletion farther upward into the granophyre, where values approximate those in the Dufek Massif section. In the cumulates, Cu and S are strongly correlated with Fe-Ti oxide mineral abundance (Figure 1).

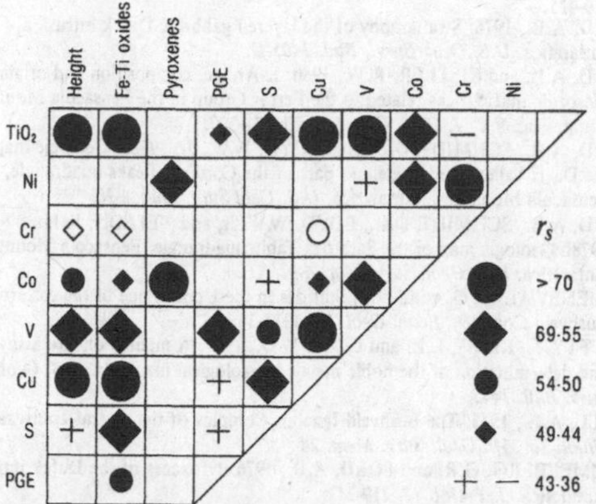


**Figure 1. Correlation of analysed elements in 22 cumulates of the Dufek intrusion. Open symbols and (−) indicate negative corelation, others positive.**

*Titanium, vanadium and cobalt* In decreasing order, $TiO_2$, V, and Co show strong to moderate correlation with height. Average values are 4.3 (weight) percent $TiO_2$, 842 ppmV, and 73 ppm Co in Forrestal Range cumulates compared with 0.34 percent $TiO_2$, 198 ppm V, and 50 ppm Co in Dufek Massif cumulates, which shows upward increases of about ten-fold for $TiO_2$, and about four-fold for V, with Co showing only slight increase. Whereas $TiO_2$ and V, expectedly, show marked correlation with Fe-Ti oxide minerals, Co more strongly correlates with modal pyroxene (Figure 1). Analyses of $TiO_2$ and $V_2O_3$ in Fe-Ti oxide minerals of the intrusion are given in Himmelberg and Ford (1977).

*Chromium and nickel* Cr has slight (−) and Ni no correlation with height. Average values of 22 ppm Cr and 51 ppm Ni in Forrestal Range cumulates compared with 174 ppm Cr and 76 ppm Ni in Dufek Massif cumulates show an approximate eight-fold upward decrease in Cr and only a slight decrease in Ni. Both elements are strongly depleted in the granophyre. Whereas Cr shows no significant correlation with modal minerals, Ni shows strong correlation with pyroxenes (Figure 1).

## SPECULATIVE RESOURCES

At this preliminary stage of a first reconnaissance survey of minor metals there are probably few, if any, types of deposits known in similar bodies that can be arbitrarily excluded as possibly occurring in the Dufek intrusion, in view of (1) the immense area of concealed rock, particularly basal units; (2) the small number of analysed samples, and (3) the lack of field work, sampling and ore microscopic study aimed specifically toward a search for minerals. Accordingly, most or all economic materials known in similar complexes should be considered for this body to be "speculative resources," defined by Brobst and Pratt (1973) as being undiscovered resources that may exist in any unknown district in either a known form or as an as-yet-to-be discovered form of deposit.

The Bushveld Complex is an exceptional example for comparison, with its world's largest reserves of chromite, Pt-group metals and V-rich iron ores and other deposits, of variable economic interest, of Cu, Ni, Pb, Zn, Mo, W, Au, Sn (Willemse, 1969), V, Ag, Bi, As, Co (Hall, 1932), and Ti (Cameron, 1971). Many other complexes, much smaller than the Dufek, contain important resources of one or more of the same metals, including the Stillwater Complex (Page and Dohrenwend, 1973; Todd et al., 1979;) the Sudbury intrusion of Ontario (Cornwall, 1973); the Great Dyke of Zimbabwe (Worst, 1960); the Duluth Complex of Minnesota (Cornwall, 1973); and the La Perouse Layered Gabbro of Alaska (Czamanske et al., 1981). In view of Antarctic operating costs, PGE would be of the greatest interest for possible utility, but their utilisation might benefit recovery of others.

Our results (Table 2) have not yielded indications of PGE or other metal concentrations of present economic significance. However, PGE abundances in other bodies are known to be extremely variable across and along layering (Page et al., 1972), largely owing to their concentration by magmatic currents in thin lenticular layers (Page et al, 1980). They would therefore likely be found only by chance or by very thorough prospecting. Even though known PGE abundances in the Dufek are not high, they are broadly similar to those in equivalent lithologies and stratigraphic positions in layered complexes having high values in basal, in part ultramafic parts (Czamanske et al., 1981; Page et al, 1972). Although complexes such as the Bushveld, Stillwater, Great Dyke and Sudbury occur in Precambrian cratons, the occurrence of potential resources for Ni, Cu and PGE (Czamanske et al., 1981) in the La Perouse Layered Gabbro, of Tertiary age and in a Pacific margin mobile belt, indicates that age and tectonic setting are not sole factors in ore deposition.

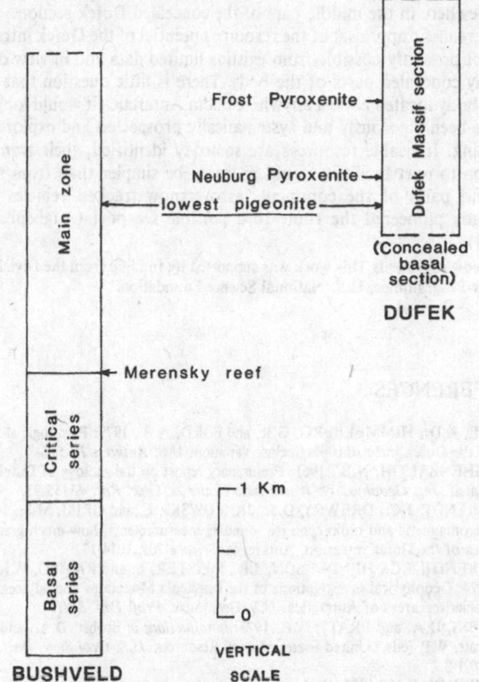


**Figure 2. Possible stratigraphic comparison between lower parts of the Bushveld and Dufek intrusions. Bushveld modified from Wager and Brown (1967), Dufek from Ford (1976).**

Results in Table 2 show stratigraphic variations generally like those in other layered complexes (Wager and Brown, 1967) that can be interpreted in terms of broad differentiation trends. Elements such as Cu, V and Ti having strong correlations with height (Figure 1) or those such as PGE, S and possibly Co with greater average abundance in the upper than lower section are inferred to show enrichment with differentiation, at least in stages represented by exposed rocks. In the upper part of the intrusion, correlations of Figure 1 suggest that future search for possible resources of PGE, V, Cu and Ti should be directed towards horizons with an abundance of Fe-Ti oxide minerals, and that V might be a useful pathfinder element in a search for PGE. An association with sulphide minerals might be expected, in that PGE seems to be collected from magmas by crystallising sulphides (Naldrett and Cabri, 1976). Lacking soils and placers, any search for PGE will be difficult. Deposits of Cu and Ni might also be encountered somewhere in the upper part of the body, as Cu-Ni deposits of possible economic significance are known to occur in a comparable part of the Bushveld Complex (von Gruenewaldt, 1976).

Comparison with other layered complexes (Wilson, 1969), however, indicates that the unexposed basal part of the intrusion should be of greatest potential resource interest, chiefly for PGE and possibly also for Cr, Ni and Cu. The presence of early ultramafic cumulates is unknown, but geophysical data suggest that possibility (Behrendt et al., 1974). If so, occurrences of early chromite might occur. Although cumulus Mg-olivine and chromite are unknown in exposed rocks, the cumulus mineralogy (plagioclase + augite + bronzite) of the lowest exposed rocks is the same as in the Bushveld and Stillwater Complexes above upper limits of Mg-olivine and chromite. In the Bushveld, the lowest occurrence of cumulus pigeonite (co-existing with plagioclase, $An_{60-69}$)is about 3.6 km above the floor and about 1.3 km above the PGE-bearing Merensky Reef (Wager and Brown, 1967). The lowest occurrence of cumulus pigeonite (co-existing with plagioclase, $An_{60-65}$) in the Dufek intrusion is at a height of about 300 m above the concealed basal section and 2.1-3.8 km above the base, using Behrendt's et al. (1974) estimate of the concealed thickness. Thus, if pigeonite appearance is taken as a marker, the Merensky Reef would lie at an equivalent level in the upper part of the concealed basal Dufek section (Figure 2). The chief PGE zone of interest in the Stillwater Complex is about 425 m above the Ultramafic zone (Todd et al. 1979), the top of which is about 1.2 km above the floor (Wager and Brown, 1967). The zone would have an analogous position somewhere in the middle part of the concealed Dufek section.

A realistic appraisal of the resource potential of the Dufek intrusion is not presently possible from existing limited data and in view of the many concealed parts of the body. There is little question that were this body located almost anywhere but in Antarctica it would long ago have been vigorously and systematically prospected and explored by drilling. If usable resources are someday identified, their transportation to port facilities would probably be simpler than from many inland parts of the continent, as oversnow tracked vehicles have already pioneered the route to a possible sea port (Aughenbaugh, 1961).

*Acknowledgements* This work was supported by funding from the Division of Polar Programmes, U.S. National Science Foundation.

## REFERENCES

ABEL, K.D., HIMMELBERG, G.R. and FORD, A.B., 1979: Petrologic studies of the Dufek intrusion: plagioclase variation. *U.S. Antarct. J., 5*, 6-8.

AUGHENBAUGH, N.B., 1961: Preliminary report on the geology of Dufek Massif. *Int. Geophys. Yr. World Data Centre A, Glac. Rpt. 4*, 155-93.

BEHRENDT, J.C., DREWRY, D.J., JANKOWSKI, E. and GRIM, M.S., 1980: Aeromagnetic and radio echo ice sounding measurements show much greater area of the Dufek intrusion, Antarctica. *Science 209*, 1014-17.

BEHRENDT, J.C., HENDERSON, J.R., MEISTER, L. and RAMBO, W.L., 1974: Geophysical investigations of the Pensacola Mountains and adjacent glacierised areas of Antarctica. *U.S. Geol. Surv. Prof. Pap. 844*.

BROBST, D.A. and PRATT, W.P., 1973: *Introduction; in* Brobst, D.A., and Pratt, W.P. (eds.) United States Mineral Resources. *U.S. Geol Surv. Prof. Pap. 820*,1-8.

CAMERON, E.N., 1971: Problems of the eastern Bushveld Complex. *Forschr. Min. 48*, 86-108.

CORNWALL, H.R., 1973: Nickel; *in* Brobst, D.A. and Pratt, W.P. (eds.) United States Mineral Resources. *U.S. Geol Surv. Prof. Pap. 820*, 437-442.

CZAMANSKE, G.K., HAFFTY, J. and NABBS, S.W., 1981: Pt, Pd and Rh analyses and benefication of mineralised mafic rocks from the La Perouse Layered Gabbro, Alaska. *Econ. Geol. 76*, 2001-2011.

DIXON, W.J. and MASSEY, F.J., 1957: *Introduction to Statistical Analysis*, 2nd ed. McGraw-Hill, New York.

FORD, A.B., 1972: The Weddell Orogeny—latest Permian to early Mesozoic deformation at the Weddell Sea margin of the Transantarctic Mountains; *in* Adie, R.J. (ed.) *Antarctic Geology and Geophysics*. Universitetsforlaget, Oslo, 419-425.

FORD, A.B., 1976: Stratigraphy of the layered gabbroic Dufek intrusion, Antarctica. *U.S. Geol. Surv., Bull. 1405-D*.

FORD, A.B. and KISTLER, R.W., 1980: K-Ar age, composition and origin of Mesozoic mafic rocks related to the Ferrar Group in the Pensacola Mountains, Antarctica. *N.Z. J. Geol. Geophys. 23*, 371-390.

FORD, A.B., SCHMIDT, D.L., and BOYD, W.W., Jr, 1978a: Geologic map of the Davis Valley quadrangle and part of the Cordiner Peaks quadrangle, Pensacola Mountains, Antarctica. *U.S. Geol Surv. Map A-10*.

FORD, A.B., SCHMIDT, D.L., BOYD, W.W. Jr, and NELSON, W.H., 1978b:Geologic map of the Saratoga Table quadrangle, Pensacola Mountains, Antarctica. *U.S. Geol. Surv. Map A-9*.

GRUENEWALDT, G. von, 1976: Sulfides in the Upper Zone of the eastern Bushveld Complex. *Econ. Geol. 71*, 1324-36.

HAFFTY, J., RILEY, L.B. and GOSS, W.D., 1977: A manual on fire assaying and determination of the noble metals in geological materials. *U.S. Geol. Surv. Bull. 1445*.

HALL, A.L., 1932: The Bushveld Igneous Complex of the central Transvaal. *Union So. Afr. Geol. Surv. Mem. 28*

HIMMELBERG, G.R. and FORD, A.B., 1976:Pyroxenes of the Dufek intrusion, Antarctica. *J. Petrol. 17*, 219-243.

HIMMELBERG, G.R. and FORD, A.B., 1977: Iron-titanium oxides of the Dufek intrusion, Antarctica. *Am. Mineral. 62*, 623-633.

LOVERING, J.F. and PRESCOTT, J.R.V., 1979: *Last of Lands—Antarctica*. Melbourne Univ. Press, Melbourne.

NALDRETT, A.J. and CABRI, L.J., 1976: Ultramafic and related mafic rocks; their classification and genesis with special reference to the concentration of nickel sulfides and platinum group elements. *Econ. Geol. 71*, 1131-58.

PAGE, N.J. and DOHRENWEND, L.J., 1973: Mineral resource potential of the Stillwater Complex and adjacent rocks in the northern part of the Mount Wood and Mount Douglas quadrangles, southwestern Montana. *U.S. Geol. Surv. Circ. 684*.

PAGE, N.J. MYERS, J.S., HAFFTY, J., SIMON, F.O. and ARUSCAVAGE, P.J., 1980: Platinum, palladium and rhodium in the Fiskenaesset complex, southwestern Greenland, *Econ. Geol. 75*, 907-915.

PAGE, N.J., RILEY, L.B. amd HAFFTY, J., 1972: Vertical and lateral variation of platinum, palladium and rhodium in the Stillwater Complex, Montana. *Econ. Geol. 67*, 915-923.

RUNNELLS, D.D., 1970: Continental drift and economic minerals in Antarctica. *Earth Planet. Sci. Lett. 8*, 400-402.

SHAPIRO, L., 1975: Rapid analysis of silicate, carbonate and phosphate rocks—revised edition. *U.S. Geol. Surv. Bull. 1401*.

TODD, S.G., SCHISSEL, D.J. and IRVINE, T.N., 1979: Lithostratigraphic variations associated with the platinum-rich zone of the Stillwater Complex. *Carnegie Inst. Wash. Yearbook 78*,461-468.

WAGER, L.R. and BROWN, G.M., 1967: *Layered Igneous Rocks*. Freeman, San Francisco.

WALKER, P.T., 1961: Study of some rocks and minerals from the Dufek Massif. Int. *Geophys. Yr. World Data Centre A, Glac. Rep. 4*,195-213.

WILLEMSE, J., 1969: The geology of the Bushveld Igneous Complex, the largest repository of magmatic ore deposits in the world; *in* Wilson, H.D.B. (ed.) Magmatic Ore Deposits, A Symposium. *Econ. Geol. Monogr. 4*, 1-22.

WILSON, H.D.B., 1969 (ed.): Magmatic Ore Deposits, A Symposium. *Econ. Geol. Monogr. 4*.

WORST, B.G., 1960: The Great Dyke of southern Rhodesia. *So. Rhod. Geol Surv Bull. 47*.

WRIGHT, N.A. and WILLIAMS, P.L., 1974: Mineral resources of Antarctica. *U.S. Geol. Surv. Circ. 705*.

# PORPHYRY-TYPE COPPER DEPOSITS AND K-AR AGES OF PLUTONIC ROCKS OF THE ORVILLE COAST AND EASTERN ELLSWORTH LAND, ANTARCTICA

P.D. Rowley, *U.S. Geological Survey, Denver, Colorado, 80225, U.S.A.*

E. Farrar, *Queen's University, Kingston, Ontario K7L 3N6, Canada.*

P.E. Carrara, *U.S. Geological Survey, Denver, Colorado, 80225, U.S.A.*

W.R. Vennum, *Sonoma State University, Rohnert Park, California, 94928, U.S.A.*

K.S. Kellogg, *U.S. Geological Survey, Denver, Colorado, 80225, U.S.A.*

*Abstract* Two weakly mineralised and hydrothermally altered, Lower Cretaceous hypidiomorphic-granular (granitoid) plutons occur in the Orville Coast and eastern Ellsworth Land, Antarctica. The first, the Sky Hi Granodiorite, and its crosscutting dacite porphyry dykes contain disseminated pyrite, magnetite, and chalcopyrite and related secondary minerals, and the same minerals are concentrated in associated quartz veins; average metal additions, however, probably do not exceed 200 ppm (parts per million) copper and 50 ppm zinc. Most plutonic rocks are propylitically altered and are transected by abundant sheared zones that consist of phyllic and subordinate argillic and sparse potassic rocks. The second pluton, the Merrick Mountains Stock, partly consists of a mafic quartz diorite with related crosscutting dykes. These intrusive rocks were transected by numerous randomly oriented shear planes and then intruded by a concentrically zoned quartz diorite and granodiorite and related crosscutting rhyodacite porphyry dykes. Added metallic minerals are like those in the Sky Hi Granodiorite; average metal additions probably do not exceed 100 ppm copper. Large areas of the stock are propylitically altered, and parts are potassically altered; some rocks are transected by sparse sheared zones and altered to phyllic and subordinate argillic rocks.

The Sky Hi and Merrick Mountains copper deposits resemble the Lassiter Coast copper deposit, located 150 km northeast of the Orville Coast. All three deposits appear to represent deeply eroded noneconomic porphyry copper deposits. Isotopic ages show that these plutons and most of the other plutons in the southern Antarctic Peninsula and eastern Ellsworth Land belong to the Lassiter Coast Intrusive Suite, dated at 130 to 95 Ma. Copper mineralisation may be a local characteristic of upper parts of plutons of this suite. The plutons of the suite that are most likely to contain copper deposits are in more axial parts of the Antarctic Peninsula, where erosion has not cut down to as great a depth as it has in coastal areas.

# 8

# Glacial Geology and Geomorphology

# AMERY ICE SHELF TOPOGRAPHY FROM SATELLITE RADAR ALTIMETRY

R.L. Brooks, *Geoscience Research Corporation, Route 4 Box 129, Salisbury, Maryland 21801, U.S.A.*

R.S. Williams, Jr. and Jane G. Ferrigno, *U.S. Geological Survey, 1925 Newton Square East, Reston, Virginia 22092, U.S.A.*

William B. Krabill, *Wallops Flight Centre, National Aeronautics and Space Administration, Wallops Island, Virginia 23337, U.S.A.*

*Abstract* During its operational lifetime, July to October 1978, the Seasat satellite made 43 radar altimeter traverses across the Amery Ice Shelf, East Antarctica. Data were collected from the nominal height of 800 km above the Earth's surface at a sub-satellite traverse spacing of 670 m. The southernmost extent of the satellite areal coverage was 72.06°S latitude which included the terminus of the Lambert Glacier.

An altimeter waveform retracking algorithm was used to determine a total of 6,016 well-distributed surface elevations with a precision of ± 47 cm. Surface contours at 5 m intervals, with supplemental contours at 1 m intervals, have been drawn for the ice shelf. These satellite altimeter-derived contours have been superimposed onto a Landsat mosaic of the ice shelf, resulting in the first merger of data from the two satellite remote sensing systems.

The Seasat radar altimeter was designed to measure the topography of the oceans from a nominal satellite height of 800 km (Townsend, 1980). Analyses have shown that the altimeter also maintained tracking over areas of smooth terrain including deserts, salt flats, ice sheets, and valleys (Brooks, 1981a) (Brooks, 1981b). Studies of the altimeter performance over terrain revealed that, although the onboard altimeter tracker did not respond quickly enough over non-ocean features, the archived waveforms could be retracked to achieve altitude accuracy of ± 1 m.

The altimeter performance over Antarctic ice shelves had not previously been evaluated. Retracking altimeter measurements over the ice shelves could provide the following:

(1) surface elevations to be correlated with ice flow features observed in Landsat imagery. The altimeter-derived elevations would provide additional insight into the ice dynamics in shelf areas.

(2) surface elevations to serve as a reference for future climate studies. Future satellite altimeter studies, performed at regular intervals, could document long-term trends in ice shelf stability, growth or disintegration.

The Amery Ice Shelf in East Antarctica was selected for the evaluation area of the satellite altimeter measurements. The location of the Amery Ice Shelf is shown in Figure 1. This shelf was selected because it is wholly within the Seasat altimeter coverage area (± 72.06° latitude) and because some Amery surface elevations have previously been determined by Australian over-ice traverses (Budd, et al., 1982). In addition, the terminus of the shelf's primary feeder, the Lambert Glacier, is also within the Seasat coverage area and the transition in topography from glacier to shelf may be discerned.

The Amery Ice Shelf evaluation study required:

(1) retracking the Seasat radar altimeter measurements.

(2) generating Amery Ice Shelf surface contours from the retracked altimeter measurements; a contour interval of 5 m would be supplemented by 1 m contours; all the available Seasat altimeter measurements would be utilised.

(3) constructing a Landsat mosaic of the ice shelf.

(4) superimposing the generated surface contours onto the Landsat mosaic of the shelf area.

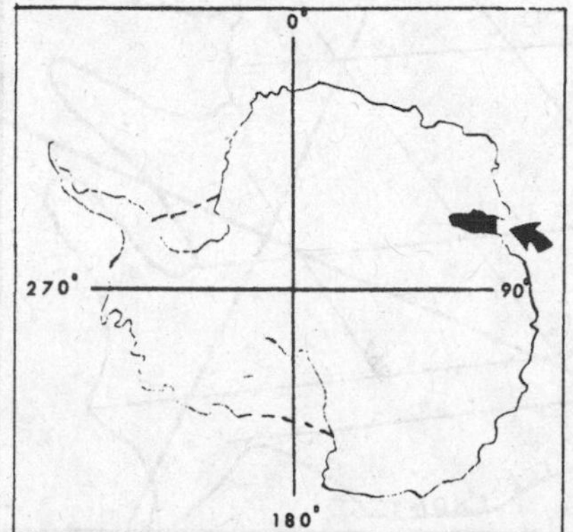


**Figure 1. Location of the Amery Ice Shelf in East Antarctica.**

(5) relating the contours to the imaged features and the Australian surface elevations; some ice shelf topography differences are anticipated in these comparisons due to the Seasat altimeter measurements being aquired in 1978, while the Landsat imagery is from the 1972/74 time period and the Australian surveys were performed during 1968/70.

Additional Seasat radar altimeter design and operational parameters are described by Townsend (1980). The paper by Budd et al. (1980) is an excellent source for further Amery Ice Shelf geophysical information including ice core studies, ice movement surveys, and ice thickness measurements.

## ALTIMETER GROUNDTRACK GEOMETRY

Forty-three Seasat groundtracks traversed the Amery Ice Shelf; their geometry is depicted in Figure 2. The satellite overflew the area in an east-to-west direction. The altimeter groundtracks are shown by the solid lines in Figure 2; gaps in the lines are those areas over which the altimeter did not acquire measurements. The radar altimeter tracker could accommodate neither the abrupt elevation change from the escarpment along the shelf's eastern boundry nor the ice cliff at the ocean-to-shelf transition. After the altimeter reacquired tracking on the ice shelf, it generally maintained a tracking mode across the extent of the shelf. Occasionally the altimeter lost lock, the result of traversing nunataks or encountering rapid surface slope changes in the vicinity of Lambert Glacier.

The measurement rate of the Seasat altimeter was 10 per second, with measurement spacing along each groundtrack at approximately 670 m intervals. A total of 6,016 altimeter-derived surface elevations are available for the Amery Ice Shelf from these 43 altimeter groundtracks.

## DATA PREPARATION

### Altimeter Waveform Retracking

Due to the Seasat altimeter tracking system not being sufficiently responsive over non-ocean surfaces, it was necessary to recompute the altitudes of all the Seasat measurements over the ice shelf. The recomputations were accomplished by repositioning the tracking gate at 50% of peak power on the surface return waveform ramp immediately preceding the peak power point. Previous experience with retracking the Seasat altimeter waveforms over terrain features in the United States has resulted in accuracies of ± 1 m when compared with large-scale maps (Brooks, 1981a).

### Computations of Surface Elevations

A surface elevation corresponding to each altimeter measurement was computed by algebraically subtracting the retracked altitude (A) from the computed satellite height based on orbital computations (H). This surface elevation is with respect to the Seasat orbit reference ellipsoid (a = 6378.137 km, f = 1/298.257) and in order to achieve surface heights referenced to Mean Sea Level (E), the geoid-ellipsoid seperation (G) must be subtracted as $E = H - A - G$ where all units are in metres.

Correlations with GEM10b geoid heights over Antarctic ocean areas and with geoceiver heights over the Antarctic ice sheet have shown that the Seasat orbital computations generally provide H with an accuracy of ± 8 m for the Antarctic area. While the error in H has a long wavelength and may be considered constant for an individual

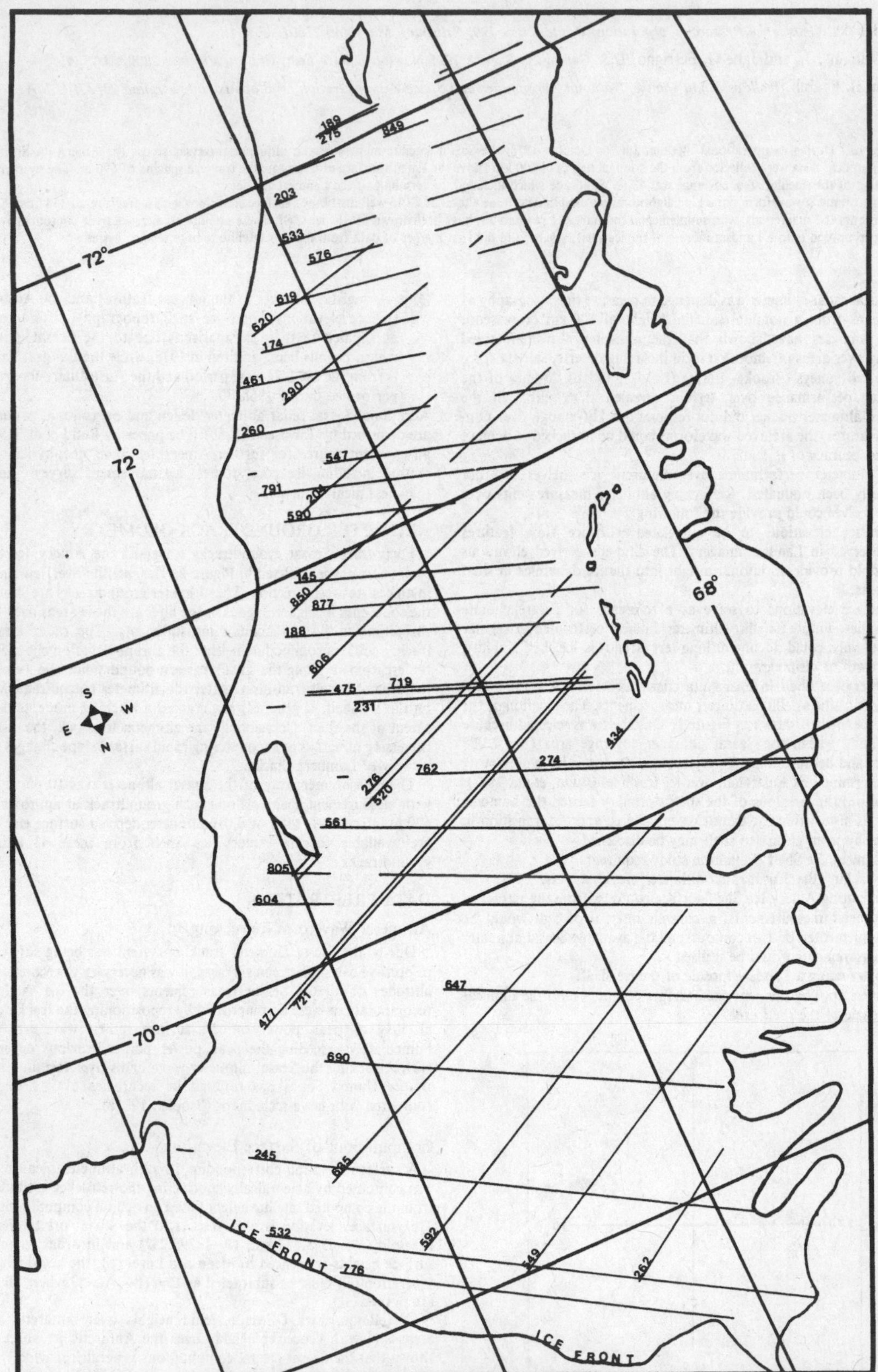


Figure 2. Seasat groundtrack geometry across the Amery Ice Shelf.

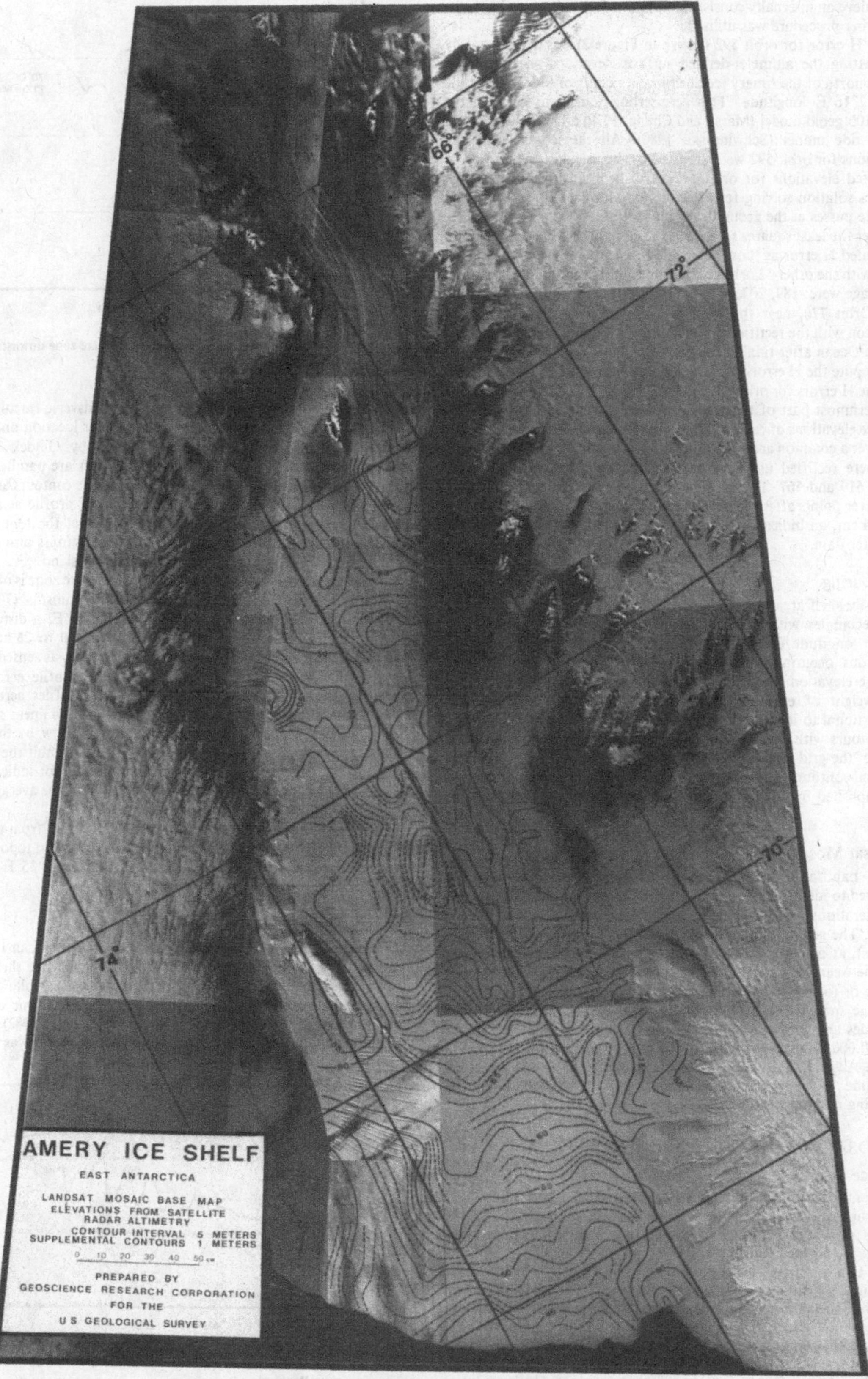


**Figure 3. Amery Ice Shelf mosaic resulting from Landsat imagery and Seasat altimetry.**

groundtrack across the ice shelf area, it changes from orbit to orbit. To achieve an internally consistent data set referenced to sea level, the following procedure was utilised.

The H error for orbit 592 (shown in Figure 2) was determined by zero-setting the altimeter-derived surface elevations on the Indian Ocean north of the Amery Ice Shelf in the vicinity of 68°51′S latitude and 72°16′E longitude. This zero-setting took into account the GEM10b geoid model (Marsh and Chang, 1979) and the Schwiderski ocean tide model (Schwiderski, 1980). All the ice shelf surface elevations for orbit 592 were adjusted by the calculated H error. The corrected elevations for orbit 592 were then constrained in a least squares solution solving for H error value for each of 31 additional satellite passes at the groundtrack intersections.

After the least squares solutions, 11 Seasat orbits remained without computed H error as those orbits did not have groundtrack intersections with the other 32 orbits. The orbits without calculated H error at this stage were: 189, 203, 274, 275, 434, 533, 576, 619, 620, 776, and 849. Orbit 776, near the ice front, was made to agree in surface elevation with the rectified parallel orbit 532 in a common area of the Indian Ocean after tidal corrections. Similarly, orbit 635 was utilised to compute the H error for orbit 434, orbits 762 and 274 were paired; and the H errors for orbits 619, 576, 533, 203, 849, 275 and 189 in the southernmost part of the study area were calculated by forcing the surface elevations of each to agree with the surface elevations of orbit 174 over a common area of the ocean. The surface elevations for orbit 620 were rectified utilising the groundtrack crossover points with orbits 619 and 567. The root-mean-square of fit at the groundtrack crossover points after the computation and application of the H error was 24 cm, an indication of the excellent internal consistency of the altimeter data.

## Contouring

The ice shelf area was prepared for contouring by gridding the area into rectangles with side lengths of 0.125° latitude (13.9 km) and 0.125° longitude (4.3 - 5.0 km, depending upon latitude). All the elevations occuring within a rectangle contributed to a weighted average elevation assigned to the geometric centre of the rectangle. The weight of each elevation within the rectangle was inversely proportional to its distance from the centre.

Contours with 5 m intervals were drawn for the shelf area on the basis of the grid elevation computation described above. Supplemental 1 m contours were drawn where warranted. The gridding was accomplished by computer, while the contouring was performed by hand.

## Landsat Mosaic

The Landsat image archive for the Amery Ice Shelf area was reviewed to identify the optimum images for the mosaic. The imagery considerations were: minimum cloud cover, band selection, and clarity. The selected images were multispectral scanner, band 7 (0.8-1.1 μm), at a scale of 1:500,000. The dates of the selected imagery were between 1972 and 1974. No special processing, such as precise scaling or tone matching, was requested.

Image matching was used to prepare an uncontrolled mosaic. Latitudes and longitudes were assigned to the mosaic by use of the 1:1,000,000 topographic maps of the Amery Ice Shelf printed in 1971 by the Australian Division of National Mapping. The latitude and longitude transferral was accomplished almost exclusively by matching nunatak features.

# LANDSAT-SEASAT CONTOURED MOSAIC

## Surface Features

Five metre contour intervals with supplemental 1 m contours have been superimposed upon the Landsat mosaic; the resultant special map is shown in Figure 3. This synergetic mosaic illustrates many notable ice surface features.

The Landsat imagery clearly delineates the ice front, nunataks, and flow lines of the glacier feeding the ice shelf. The most prominent flow lines are those associated with the Lambert Glacier, the primary feeder of the ice shelf and one of the Earth's largest glaciers. The Lambert flow lines appear upstream from 72°S latitude.

Transverse fractures appear in the ice surface from the northern tip of Gillock Island (70°17′S, 71°48′E) downstream to the ice front, a total distance of approximately 130 km. The transverse fractures are, individually, as long as 40 km. Considering their location on the ice shelf, these fractures are apparently caused by Gillock Island's obstruction of the ice flow. The surface contours are parallel to the transverse fractures only near 69°50′S; usually the contours are more or less orthogonal to the fractures. An altimeter profile across this fractured area, illustrated in Figure 4, shows that the topographic relief across this fractured area is ±5 m. The contours also show a 10 m ice surface depression in the lee of Gillock Island.

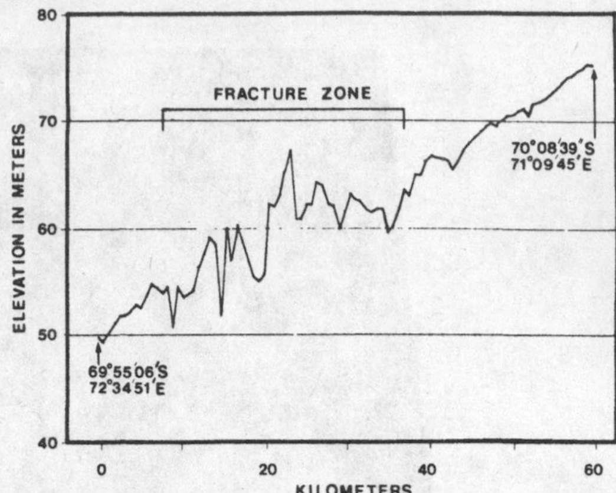


**Figure 4. Altimeter profile across ice surface fracture zone downstream of Gillock Island.**

On the western side of the ice shelf a linear fracture zone is observed to extend from the northern tip of the Jetty Peninsula (70°16′S, 68°53′E) to the ice shelf front at 68°55′S, 72°20′E; a distance of 200 km. This fracture zone's width varies from 10 to 25 km. The topographic relief associated with this fracture zone, as sensed by the Seasat altimeter, is ±4 m. A sample altimeter profile across this fracture zone is in Figure 5; all the altimeter profiles across this fracture zone showed similar relief. The cause of this linear fracture zone is believed to be the obstruction of the ice flow by the Jetty Peninsula. While the individual altimeter profiles exhibit the terrain changes in the fracture area, the final contours do not indicate this surface roughness as the rapid terrain changes tend to be averaged-out in the gridding process prior to contouring.

The ice surface gradually slopes upward from the ice front towards the Lambert Glacier; the average slope is $1.7 \times 10^{-4}$. Two topographic highs appearing at 70°35′S, 70°15′E and 71°05′S, 69°35′E may be due to localised groundings of the ice shelf.

## Comparisons with Australian Expeditions

The Australian expeditions' survey traverses were made in 1968-70. Their primary survey line went from a point near the ice shelf front upstream to near the terminus of the Lambert Glacier, following the approximate centre-line of the shelf. They used electronic distance measuring equipment and theodolites (Budd, et al., 1982). These expeditions established stations T5, G1, G2, G3, and T4 as well as

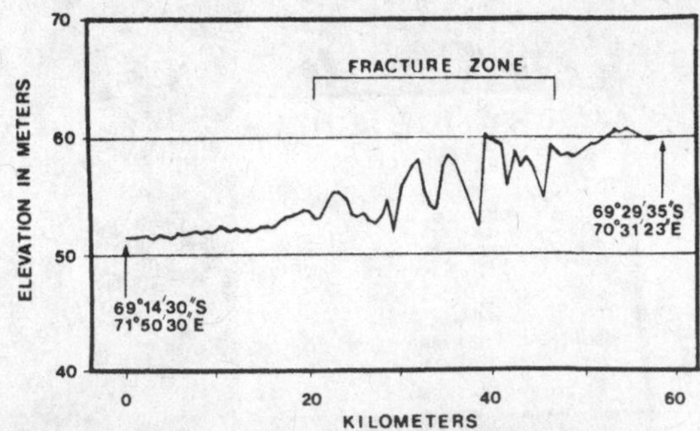


**Figure 5. Altimeter profile across ice surface fracture zone downstream of Jetty Peninsula.**

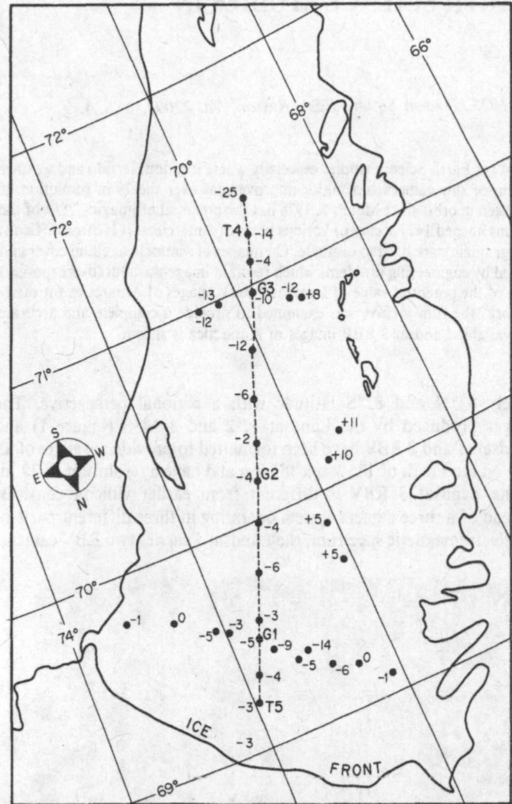


**Figure 6. Surface elevation differences in metres (Seasat altimetry minus Australian surveys).**

ancillary spot elevations on the ice surface. The primary Australian expeditions' survey line is delineated by the dashed line in Figure 6.

The elevation differences in metres (satellite altimeter minus Australian survey) are shown in Figure 6 adjacent to each Australian survey point. The altimeter-derived elevations are generally lower than the survey values from 8 to 10 years earlier, and those differences increase with distance from the ice front. The four survey comparisons in the western portion of the ice shelf between 69°S and 71°S indicated, however, that this area has increased in elevation. A hypothesis is that the Lambert Glacier may be retreating and thus discharging less ice into the shelf area, while the Charybdis Glacier, entering the shelf from the west side, may be surging. The apparent thinning of the main portion of the ice shelf may be related to the ~ 40 km retreat of the ice shelf front between 1955 and 1965, as documented by Budd (1966). Wellman (1982) supports the concept of one or more glaciers in the Amery Ice Shelf drainage basin surging while others are in retreat.

Additional survey elevations at 20 km spacing along the primary Australian traverse line (Budd, et al., 1982) are included in an ice shelf surface centre-line profile in Figure 7. The corresponding profile from the Seasat altimeter measurements is also presented in Figure 7. Although the altimeter-derived profile is consistently lower in elevation, the two profiles agree well in shape from 20 to 220 km from the ice front. The Budd, et al., profile exhibits an abrupt change in slope at 280 km from the ice front, at the shelf grounding line. Beginning with the 280 km point and continuing upstream to the Lambert Glacier, the altimeter's centre-line profile is not as reliable due to the altimeter's losses-of-lock in this area. The altimeter profile increases abruptly 320 km from the ice front at the terminus of the Lambert Glacier.

## SUMMARY

The combined Landsat/Seasat mosaic of the Amery Ice Shelf has produced unprecedented topographic detail for an Antarctic ice shelf. Surface elevations are now available for ice flow features, contributing to ice dynamics studies.

This study has also produced ice shelf reference elevations for future climate studies.

It is anticipated that this combined mosaic will be prepared for a U.S. Geological Survey Special Map. The Special Map will incorporate precise scaling and tone matching for the Landsat images.

*Acknowledgements* The authors gratefully acknowledge the assistance of W.F. Budd of the University of Melbourne, Australia, in providing Amery Ice Shelf survey results from the Australian National Antarctic Research Expedition.

## REFERENCES

BROOKS, R.L., 1981a: Terrain profiling from Seasat Altimetry. *U.S. National Aeronautics and Space Administration CR-156878.*

BROOKS, R.L., 1981b: Ice sheet altimetry. *U.S. National Aeronautics and Space Administration CR-156877.*

BUDD, W., 1966: The dynamics of the Amery Ice Shelf. *J. Glaciol., 6, (45),* 335-57.

BUDD, W., CORRY, M.J. and JACKA, T.H., 1982: Results from the Amery Ice Shelf project. *Annals Glaciology, 3,* 36-41.

MARSH, J.G. and CHANG, E.S., 1979: Global detailed gravimetric geoid. *Mar. Geodesy, 2,* 145-59.

SCHWIDERSKI, E.W., 1980: Ocean tides, part I: Global ocean tidal equations. *Mar. Geodesy, 3,* 161-217.

TOWNSEND, W.F., 1980: An initial assessment of the performance achieved by the Seasat-1 radar altimeter. *IEEE Oceanic Engineering, J., OE:2,* 80-92.

WELLMAN, P., 1982: Surging of Fisher Glacier, Eastern Antarctica: evidence from geomorphology. *J. Glaciol., 28, (98),* 23-8.

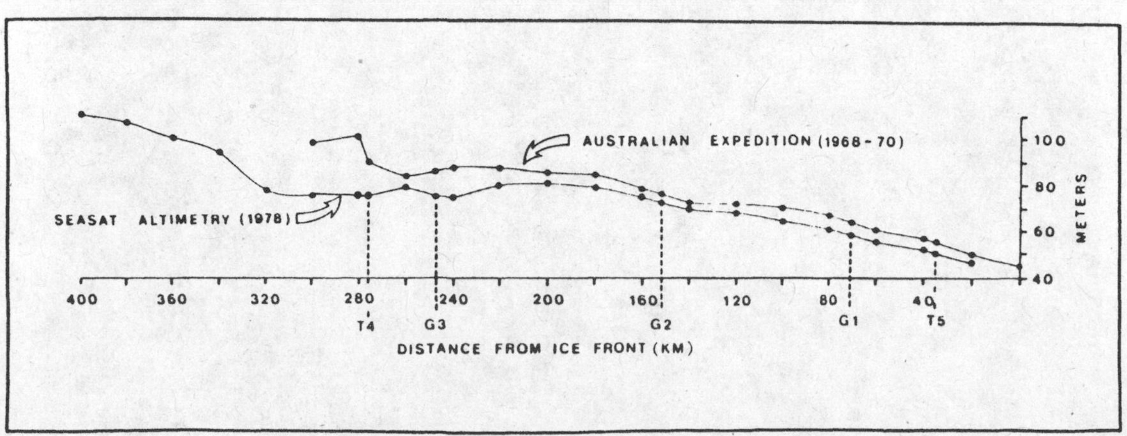


**Figure 7. Surface profiles along the approximate centre-line of the Amery Ice Shelf.**

# EVALUATION OF LANDSAT 3 RBV IMAGES FOR EARTH SCIENCE STUDIES IN ANTARCTICA

J.G. Ferrigno, R.S. Williams Jr. and T.M. Kent, *U.S. Geological Survey, 1925 Newton Square East, Reston, Va. 22092, U.S.A.*

*Abstract* Landsat 3 return beam vidicon (RBV) imagery is a potentially valuable tool for Earth Science studies especially where location, terrain and weather conditions make field logistics difficult, hazardous and costly. The 30 m resolution of this camera is a major improvement over the 79 m resolution of Landsats 1 and 2 RBV and the Landsat multispectral scanner. Landsat 3, which has been in orbit since March 5, 1978 has the potential of imaging 75% of the Antarctic continent. During its four years of operation, the Landsat 3 RBV camera has imaged 1477 scenes of various parts of Antarctica. It is often difficult, however, to determine which of these images are useful because cloud cover and image quality are hard to evaluate. On images of Antarctica, cloud cover and snow cover are easily confused. In addition, the Landsat 3 RBV sensor is handicapped by engineering problems which result in image washout (overexposure) in areas of high reflectance, such as snow or ice covered terrain or deserts. Because of the potential value of Landsat 3 RBV images of Antarctica for many types of earth science studies and the preparation of image base maps for fieldwork, the film archive was examined to provide a complete and accurate evaluation of these images. The location and an assessment of the quality of all available Landsat 3 RBV images of Antarctica is shown.

Landsat 3 was launched on March 5, 1978, carrying the same kind of multispectral scanner (MSS) that was on the previous two Landsat spacecraft, but it also carried a new type of return beam vidicon (RBV) camera. The orbit of Landsat 3 is identical with the previous Landsat satellites: near polar and sun synchronous with an approximate altitude of 920 km. These orbital parameters permit the satellite sensors to repetitively image the Earth every 18 days between approximately 82°N and 82°S latitude with a regional perspective. The images produced by the Landsat 1, 2 and 3 MSS (Figure 1) and Landsats 1 and 2 RBV have been formatted to provide coverage of an area on the Earth of 185 km x 185 km and have a resolution of 79 m.

The Landsat 3 RBV is different from earlier vidicon cameras. Instead of a three camera system operating in three different parts of the electromagnetic spectrum, the Landsat 3 carries two RBV cameras

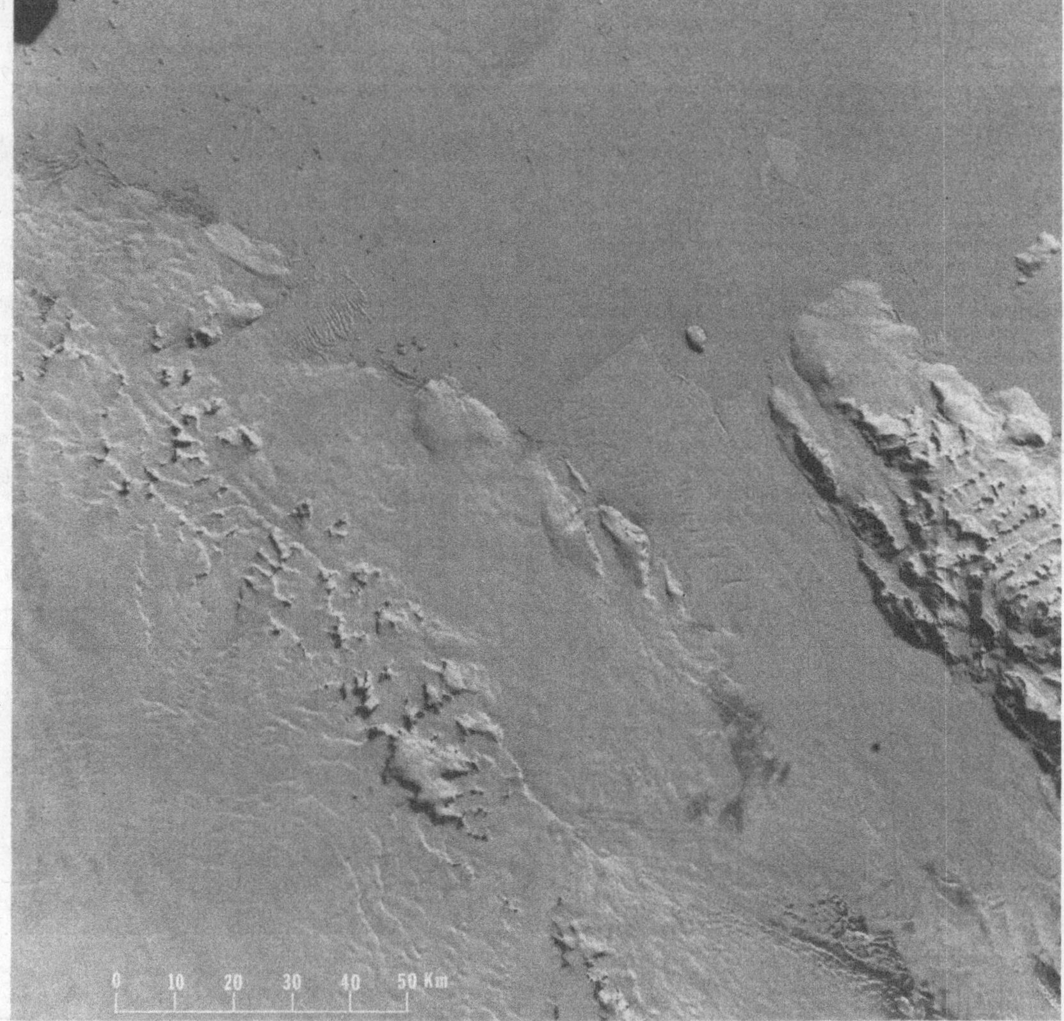


**Figure 1. Landsat 1 MSS image of Rennick Glacier and environs, Oates Coast, northern Victoria Land, Antarctica (Path 72, Row 110; 1460-21103, band 7; 26 October 1973).**

which take simultaneous side by side slightly overlapping images in the 0.51-0.75 micrometer spectral band. The Landsat 3 RBV images are formatted so that four overlapping images (called subscenes A, B, C and D), 99 km on a side, correspond to one of the MSS scenes. The effective focal length of the camera was doubled to 236 mm to increase the spatial resolution to 30 m. The RBV imagery (Figure 2) has better spatial resolution and geometric accuracy than the Landsat MSS image and is especially well suited for many types of geoscience studies.

Landsat has been recognised as an unparalleled tool for studying, inventorying, monitoring and measuring Earth's natural resources from a global point of view (Williams and Carter, 1976). More specifically, the value of Landsat data for glaciological and mapping purposes in Antarctica was recognised quite early (Southard and MacDonald, 1974; MacDonald, 1976 a, b, c; and Swithinbank and Land, 1976). Landsat 3 RBV images are a potentially valuable, cost-effective tool for geologic, glaciologic and many other types of geoscience studies (Williams et al., 1982b). They are especially useful for producing 1:100,000 or smaller scale, geometrically accurate image base maps for field mapping, especially in places like Antarctica where few maps exist at such scales. The increased resolution of Landsat 3 RBV makes it valuable where spectral discrimination is not important.

There have been serious technical problems, however, which have caused the availability of Landsat 3 RBV data of Antarctica to be restricted. Initially, the Landsat 3 data, which were received at NASA's Goddard Space Flight Centre were expected to be transmitted digitally via Domsat to the EROS Data Centre (EDC), Sioux Falls, South Dakota, for data processing and distribution. However, the system for handling the data digitally at Goddard was not ready for operation until February 1979. During this time some, but not all data, were transmitted to EDC as film as the Landsat 1 and 2 data had been. System problems caused a data backlog at Goddard and a data gap at the EROS Data Centre. When the digital image processing system became operational the backlogged data were gradually transmitted to EDC. The MSS data were given preference because these products were more familiar and more in demand. It was not until recently, therefore, that all backlogged RBV data were transmitted to EDC and are now available for users.

As the RBV data became available, it was recognised that other technical problems affected image quality. These problems, which include incorrect shading, corners out of focus, missing and distorted

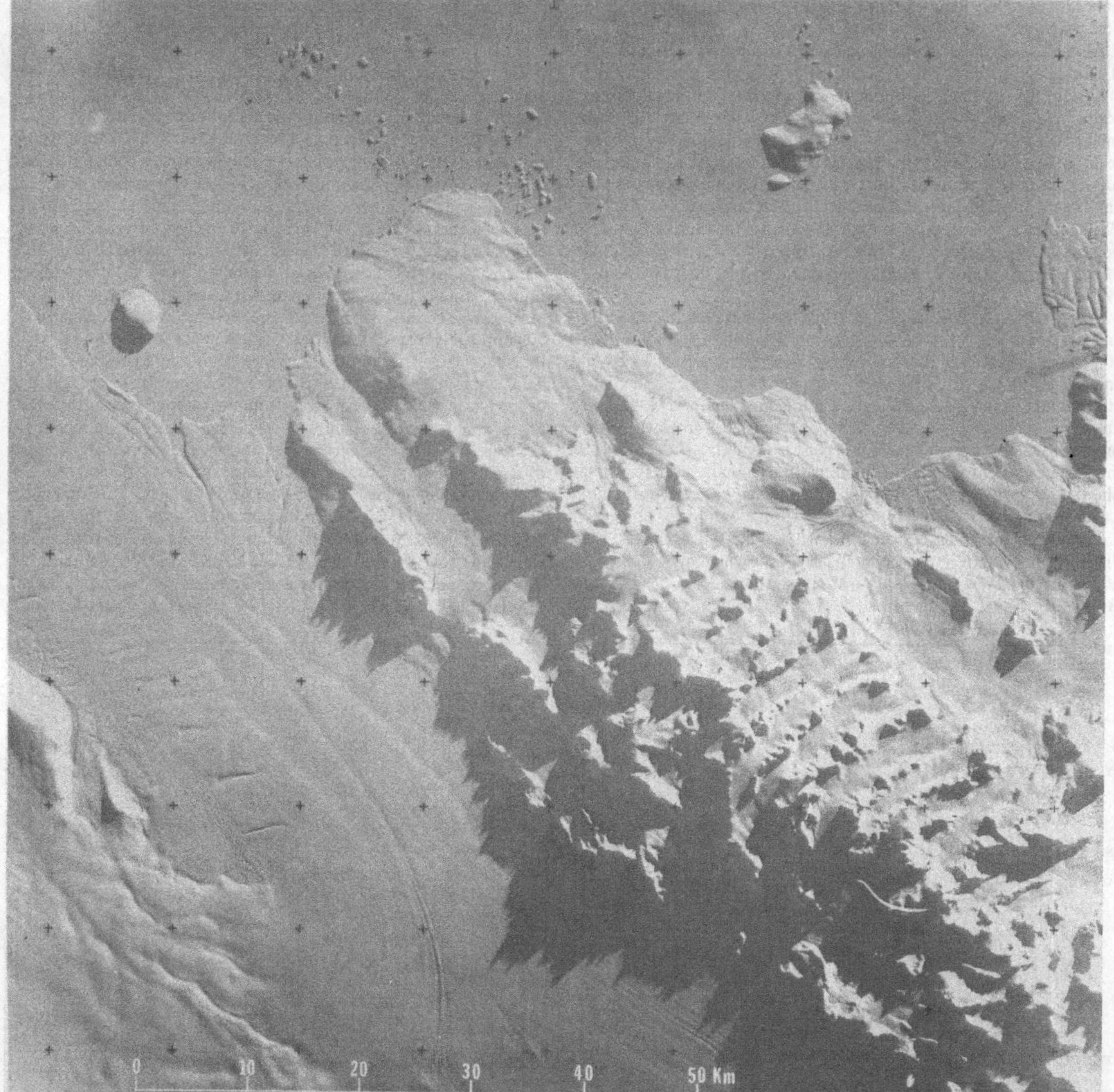


**Figure 2. Landsat 3 RBV image of Rennick Glacier and environs, Oates Coast, northern Victoria Land, Antarctica (Path 70, Row 110; 30927-20382, subscene C; 17 September 1980).**

reseaus, presence of black vertical lines and faceplate contamination, have been catalogued by Clark (1981). The problems have been studied and can in some cases be compensated for by various techniques.

Perhaps the worst problems, however, with imagery of Antarctica are caused by an RBV camera engineering design problem. The RBV sensor becomes oversaturated in areas of high reflectivity, such as deserts or snow or ice covered terrain. As a result, the imagery appears "overexposed" and little or no information can be derived from the data, depending on the severity of the problem. In addition, the sensor is not just temporarily "blinded". If the satellite passes over a large area of high reflectivity, the "blinding" effect is cumulative and many seconds are required for the sensor to recover. This problem cannot be compensated for in the digital image processing phase because the sensor was not able to record any meaningful data.

## EVALUATION OF LANDSAT 3 RBV IMAGERY

In spite of the technical problems some Landsat imagery can be very valuable for studies of Antarctica. Landsat has the potential for imaging about 75% of the Antarctic continent from the coastline to approximately 82°S. Antarctica is covered by all 251 paths of the Landsat orbit, which means that the satellite passes over Antarctica 13 times each day.

Also, in spite of the variations in quality, some of the RBV imagery is excellent (Figure 2). The difficulty lies in determining which images are useful. In searching for optimum Landsat images of Antarctica as part of a large multiyear, multinational project to produce a "Satellite Image Atlas of Glaciers" (Williams and Ferrigno, 1981), it was soon realised how difficult it is to evaluate Landsat imagery of Antarctica properly. Many months of effort were devoted to the preparation of an "Index Map and Table of Optimum Landsat Images of Antarctica" (Williams et al., in press). Landsat MSS images for each of the 2479 nominal scene centres were evaluated. When the new, higher resolution Landsat 3 RBV images became available, they were also evaluated.

When selecting Landsat imagery for most purposes, cloud cover and image quality evaluations and date of acquisition are usually the crucial factors. However, when dealing with Antarctica, cloud cover and image quality are difficult to evaluate. Cloud cover is hard to discriminate in snow covered areas and it is possible for a cloud free image to be classified as completely cloud covered. Moreover, when the quality assessment is made, it is done objectively on the basis of the amount of information contained. In Antarctica, where aerial photographic coverage is often nonexistent, even a very poor quality Landsat image can yield very valuable information.

Because of the potential value of Landsat 3 RBV imagery and the difficulty in determining the usefulness of each image it seemed advisable to review and evaluate all the available RBV imagery of Antarctica. Therefore, in March 1982, after Landsat 3 had been in orbit four years, a computer search of the data base at the EROS Data Centre was made.

The search resulted in an identification of 1477 subscenes. Because there are usually four Landsat 3 RBV subscenes for each nominal scene centre, this figure suggests that only 370 of the 2470 nominal scene centres over the Antarctic Continent had been imaged or less than 1/6 of the total. However, because of image overlap in high latitudes, only 519 scenes precisely located are necessary to cover the entire area of Antarctica imaged by Landsat (Williams et al., 1982a). It is possible therefore that these images could have covered over half the possible coverage area. Unfortunately, however the 1477 subscenes included multiple coverage of several nominal scene centres and only 95 nominal scene centres have been covered with Landsat 3 RBV imagery at least once (Figure 3).

Much of the imagery was not available for inspection anywhere except on the archival 70 mm film roll or 18 cm film chip archived at the EROS Data Centre, Sioux Falls, South Dakota; therefore, it was necessary to travel there and examine individually each of the 1477 subscenes to evaluate the amount of useful information. The result of this review was most disappointing. Because of the previously mentioned technical problems and the lack of availability of the data, not enough had been known about the special problems that Antarctica posed for the RBV sensor. Techniques had not yet been developed to get the maximum amount of useful information from this potentially valuable camera system.

The evaluation revealed that only 11 subscenes or less than 1% were excellent, 164 or 11% contained varying amounts of useful data (Figure 4), and 89% contained little or no information.

Inspection and evaluation of the Landsat 3 RBV imagery of Antarctica resulted in several observations. When comparing four subscenes for any one nominal scene centre, almost invariably subscenes B and D are better quality (have more ground detail visible) than subscenes A and C. This is probably due to a difference in sensor capability because the B and D subscenes are imaged by the same RBV camera; the A and C subscenes are imaged by the other vidicon camera. Usually the quality of subscene C is slightly better than that of A and the quality of subscene D is better than that of B. Sometimes it is noticeable that the lower part of each image is better quality than the upper part. These changes appear due to a direct response to the light conditions. The sun angle of each scene centre changes one degree, decreasing as the spacecraft moves southerly along the orbit. There would be approximately a ½-degree difference in sun angle between one subscene and another and there is apparently enough difference within a subscene to affect the quality of the imagery.

When examining imagery taken sequentially along an orbital path, in December or January when the sun angle is highest, the imagery acquired in Antarctica over the ocean surface appears normal, similar to imagery covering the ocean surface on other parts of the earth. As soon as ice or snow covered surfaces are evident, however, the sensor responds to the increased reflectivity. If only part of the image

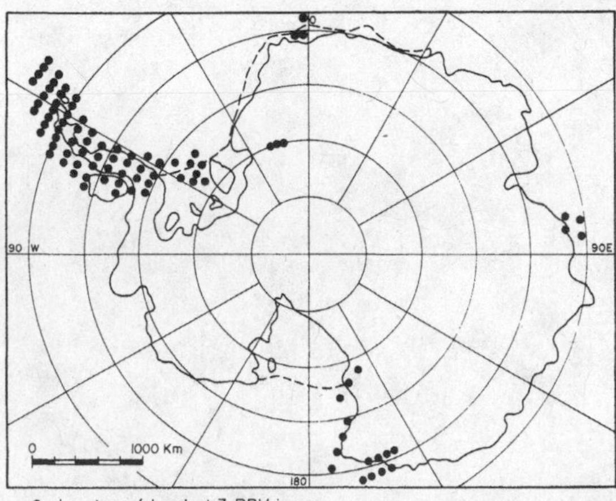


**Figure 3. Index map showing location of Landsat 3 RBV image acquisitions in Antarctica.**

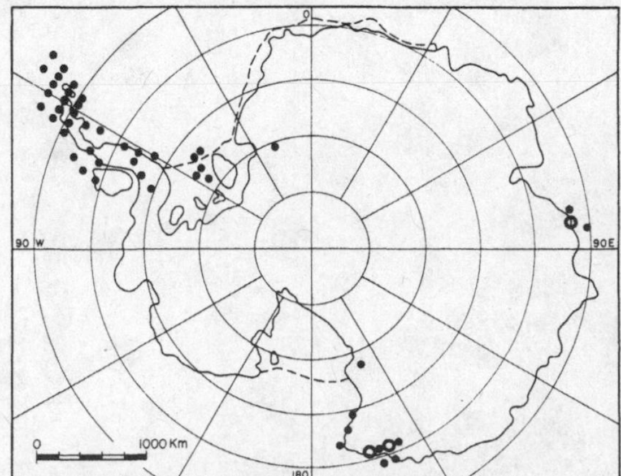


**Figure 4. Index map showing location of Landsat 3 RBV image acquisitions with good or usable information.**

contains snow or ice surfaces, detail is lost in those areas. If most or all of the scene is covered with snow or ice surfaces, the sensor becomes oversaturated, no data are visible on the imagery and even the reseau marks are obliterated.

These conditions persist along the orbital path. However, imagery taken during September, October, February and March—when the sun angle is much lower—are much better. Although there is a problem of long shadows caused by high topographic features, much ground detail is visible and imagery is highly useful.

Another related factor which affects the quality and therefore the usability of the imagery is the shutter setting on the RBV camera. There are five possible settings—2.4, 4.0, 5.6, 8.0 and 12.0 milliseconds. Only the two fastest have been used for the imagery acquired from March 1978 to March 1982 and evaluated for this project. The standard NASA operating procedure for programming the shutter speed has been based on the sun angle at each latitude along the Landsat digital orbital path. Unfortunately, this does not take into consideration the special conditions existing in Antarctica. The high reflectivity has resulted in imagery that is for the most part unusable.

Inspection of the RBV imagery showed that with the fastest shutter speed, 2.4 milliseconds and a sun angle of more than 30°, the only features visible on the Landsat 3 RBV imagery were clouds and land/water boundaries. At about 30° sun angle when the ground features are just beginning to be faintly visible, the shutter speed is automatically changed to 4.0 milliseconds. With this shutter setting and a sun angle of greater than 20° the only visible ground features are land/water boundaries. Below this sun angle ground features gradually become more apparent. From a 10° sun angle down to a 1° sun angle with an optimum around 5 to 7°, the ground features are highly visible, a large amount of data is contained in the imagery, and it is very useful. These observations suggest that it may be possible to acquire usable imagery when the sun angle is 0° (the sun is below the horizon, but illumination is still present) using the three slower Landsat 3 RBV camera shutter settings.

## RECOMMENDATIONS AND CONCLUSIONS

It was worthwhile to evaluate the Landsat 3 RBV imagery not only for the scientific benefits, but for the economic benefits as well. Because of the current U.S. policy to recover operational costs of the Landsat system, the cost to the Landsat data user has been and will be substantially increased. Not only did the cost for data products increase on October 1, 1982, but there will now be a very costly charge for imaging data of any area (special acquisitions) after one time cloud free coverage has been acquired. As a result, Landsat data will continue to be valuable and cost effective for many users, but the selection of data will have to be very careful and the request for new data very selective.

Some guidelines for the acquisiton of new data on Antarctica during the remaining lifetime of the Landsat 3 RBV sensor, based on this study, can be summarised as follows:

(a) For optimum imagery, the shutter setting must be correlated with the sun angle.

(b) For optimum results with a shutter setting of 2.4 milliseconds, exposures should be made when the sun angle is between approximately 15° and 20°.

(c) For optimum results with a shutter setting of 4.0 milliseconds, exposures should be made when the sun angle is approxiamtely 5°.

Research is now being conducted on the best shutter setting at intermediate (5-15°) sun angles and the use of the three slower shutter speeds for sun angles of lower than 5°.

As long as the Landsat 3 RBV has not been oversaturated (overexposed) and the maximum amount of data has been recorded on the image, there is room for further improvement in image quality by digital enhancement of the data.

Because of technical problems with image processing, the first four years of data acquisition have provided relatively few useful Landsat 3 RBV images of Antarctica. In the remaining lifetime of the Landsat 3 RBV sensor, using the guidelines determined from this study, it should be possible to acquire much more useful and valuable image data.

*Acknowledgements* The authors gratefully acknowledge the help of the personnel at the EROS Data Centre, especially Danielle Ehlen, for making it possible to examine the Landsat 3 RBV data archive.

## REFERENCES

CLARK, B.P., 1981: *Landsat 3 return beam vidicon response artifacts: a report on RBV photographic product characteristics and quality coding system.* EROS Data Centre, U.S. Geol. Surv., Sioux Falls, South Dakota.

MACDONALD, W.R., 1976a: Geodetic control in polar regions for accurate mapping with ERTS imagery. *U.S. Geol. Surv., Prof. Pap., 929,* 34-6.

MACDONALD, W.R., 1976b: Antarctic cartography. *U.S. Geol. Surv., Prof. Pap., 929,* 37-43.

MACDONALD, W.R., 1976c: Glaciology in Antarctica. *U.S. Geol. Surv., Prof. Pap., 929,* 194-5.

SOUTHARD, R.B. and MACDONALD, W.R., 1974: The cartographic and scientific applications of ERTS-1 imagery in polar regions. *U.S. Geol. Surv., J. Res., 2,* 385-94.

SWITHINBANK, C.W.M. and LAND, C., 1976; Antarctic mapping from satellite imagery; *in* Pell, R.F. et al. (eds.) *Remote Sensing of the Terrestrial Environment.* Proceedings of the 28th Symposium of the Colston Research Society, Univ. Bristol, England, 212-21.

WILLIAMS, R.S., Jr. and CARTER, W.D., (eds.), 1976: A New Window on Our Planet. *U.S. Geol. Surv., Prof. Pap., 929.*

WILLIAMS, R.S., Jr. and FERRIGNO, J.G., 1981: Satellite image atlas of the Earth's glaciers; *in* Deutsch, M. et al. (eds.) *Satellite Hydrology.* American Water Resources Association, Minneapolis, 173-82.

WILLIAMS, R.S., Jr., FERRIGNO, J.G. and KENT, T.M., (in press): *Index map and table to optimum Landsat images of Antarctica: 1:5,000,000 scale map base.*U.S. Geol. Surv. Open-File Report.

WILLIAMS, R.S., Jr., FERRIGNO, J.G., KENT, T.M. and SCHOONMAKER, J.W., Jr., 1982a: Landsat images and mosaics of Antarctica for mapping and glaciological studies. *Annals Glaciology, 3.*

WILLIAMS, R.S., Jr., MEUNIER, T.K. and FERRIGNO, J.G., 1982b: Delineation of blue-ice areas in Antarctica from satellite imagery; *in Proceedings of the NASA—NSF Workshop on Antarctic Glaciology and Meteorites.*

# THE PROBLEM AND SIGNIFICANCE OF RADIOCARBON GEOCHRONOLOGY IN ANTARCTICA

K. Omoto, *Department of Geography, Tohoku University, Sendai 980, Japan.*

*Abstract* Many radiocarbon dates have been used since the 1960s in order to clarify glacial geohistory of Antarctica, even though problems remain with the basic assumptions of the radiocarbon dating method. Sea water samples and living marine organisms in Antarctica have both yielded anomalously old apparent ages up to 2860 yr B.P. This is a serious problem in considering the precise Holocene geochronology of ice free areas of Antarctica, and is due to low radiocarbon concentrations in surface sea water. These are thought to be caused either by the upwelling of $^{14}C$-deficient deep ocean waters or the dilution of surface seawater by fresh water melting from icebergs, ice shelves, and glaciers. The anomalously old apparent ages make it hazardous to correlate between different areas of Antarctica solely on the basis of the coincidence of dates. Such correlations require the establishment of reservoir correction factors.

Radiocarbon dating techniques have been improved by geochronologists seeking a superior tool. The remarkable progress made in recent years is characterised by the use of micro-computer systems and accelerator techniques. Using these systems very old dates with small uncertainties are now available up to 75,000 yr BP.

Antarctica contains the world's greatest ice sheet, which comprises 84.5% of all modern ice covered areas. Ice free areas at the coast and on inland mountains and nunataks amount to only a few percent of the total area of Antarctica.

Geochronological study of these Antarctic ice free areas is very important, especially for documenting sea level changes, rates of crustal rebound and development of raised beach landforms after deglaciation. But it is in general very difficult to find samples suitable for dating because, due to the severe climate, fossils are very scarce. The most common samples for radiocarbon dating are shells, but whale or seal bones also have been used. Harkness (1979) discussed some aspects of radiocarbon dating in Antarctica and the significance and some problems of Antarctic radiocarbon are discussed in this paper.

## SIGNIFICANCE OF RADIOCARBON DATING

The radiocarbon dating method for determining the age of organic materials in Late Pleistocene and Holocene sediments, has been continuously refined since it was established by W.F. Libby and co-workers in the late 1940s (Table 1).

**TABLE 1: Present radiocarbon dating techniques.**

| Dating system | Tmax (Y BP) | Error (Y) |
|---|---|---|
| Gas proportional counting | 35,000-45,000 | ≦5,000 |
| Liquid scintillation counting | 45,000-65,000 | ≦1,000 |
| Accelerator (mass-spectrograph) | ca. 100,000 | ≦2,500 |

One commonly used radiocarbon dating technique is the "gas proportional counting system" which has the capability of detecting residual levels of radiocarbon corresponding to ages from the present back to 35,000 yr BP. Within the last decade, a new method of dating with a low background liquid scintillation counting system, linked to an online microcomputer system has made it possible to date samples back to 65,000 yr BP with a small uncertainty (Omoto, 1981). Another technique to extend the range of dating has been developed by use of the thermal diffusion method (Grootes et al., 1975), and Stuiver et al. (1978) succeeded in clarifying north American glacial history back to 75,000 yr BP. Modern laboratory techniques of radiocarbon dating are still more accurate and reliable.

Further innovation in radiocarbon dating is promised by the development of a new method in which absolute amounts of radiocarbon are detected by using either a van de Graff accelerator or a cyclotron as a mass spectrometer (Bennett et al., 1977; Muller, 1977). Using this system, we can expect to determine the age of a sample with as little as 5 mg of carbon or a 30 kg ice sample, and open the prospect of extending the range of radiocarbon dating to 100,000 yr BP.

## SOME PROBLEMS RELATED TO RADIOCARBON DATING

### Reservoir Effect

The earth's surface carbon occurs in five major reservoirs: (1) the atmosphere, (2) the ocean, (3) the terrestrial biosphere, (4) the soil, and (5) sediments (Broecker et al., 1980). The sedimentary reservoir contains the vast majority of the carbon and next in importance is the oceanic reservoir.

The radiocarbon content of surface water at different locations in the world's oceans has frequently been measured (for example, Broecker et al., 1961; 1980, Östlund and Stuiver, 1980; Stuiver and Östlund, 1980). All $^{14}C$ measurements around Antarctica have yielded very low $\Delta^{14}C$ values (Figure 1). Modern Antarctic marine organisms have also indicated low radiocarbon concentrations; for example, a seal freshly killed at McMurdo Sound had an apparent age of 1,300 yr BP (Broecker et al., 1961) and a seal which died in 1974, 30 km south of Syowa Station was determined to be 1,455 ± 110 yr BP (Omoto, 1976). Stuiver et al. (1981) have reported radiocarbon dates for freshly killed seals and penguins in the Ross Sea area; a Weddell seal was dated at 1,390 ± 40 yr BP (QL-171), an Emperor penguin was dated at 1,300 ± 50 yr BP (QL-173) and two Adelie penguins were dated at 1,750 ± 70 yr BP (QL-170) and 1,770 ± 50 yr BP (QL-172) respectively. These anomalous apparent ages reflect depletion of $^{14}C$ in the Antarctic oceanic water mass. This depletion is commonly referred to as the "reservoir effect" or "Antarctic reservoir effect"; it necessitates the correction of ages for both modern organisms and fossil materials.

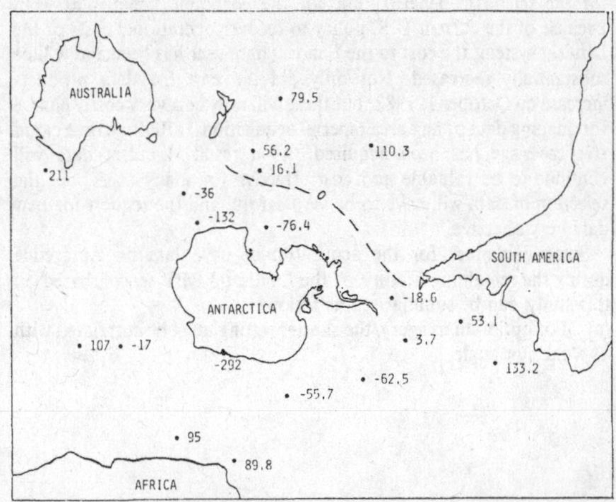


**Figure 1. Radiocarbon concentrations of the surface water around Antarctica. Values are expressed in $\Delta^{14}C$ (°/∘∘). Compiled after Boecker et al. (1961), Omoto (1972), Ostlund and Stuiver (1980) and Stuiver and Ostlund (1980).**

### Sample contamination and pretreatment

Carbon admixture originating before sample processing is called sample contamination and may vary considerably. Old samples are very sensitive to modern contamination because residual $^{14}C$ activity in them is 1,000 to 10,000 times less than that of the modern standard. If, for example, there were a 0.1% addition of modern carbon to a 75,000 yr BP sample, the date would decrease by 19,000 years to 56,000 yr BP. However, careful sample selection and pretreatment to extract contaminants by physical and chemical means can give satisfactory results, even for samples from 50,000 yr BP or older (Grootes, 1978).

It is very difficult to eliminate contamination from samples older than 50,000 yr BP, but contaminants can be reduced sufficiently to permit measurement of the true activities of properly selected samples. Fortunately, most contaminated samples are found in palaeosol, peat and coral samples which are very scarce in Antarctica. A future contamination problem may be caused by nuclear weapons testing in the atmosphere since 1952; this has increased production of $^{14}C$ in the atmosphere to double the natural rates (e.g. Nydal et al., 1979).

### Secular variation of the atmosphere radiocarbon

Radiocarbon dating of samples of known age has revealed that one of the main assumptions, viz. "a constant level of the dynamic equilibrium" is incorrect. This was revealed by dendrochronological studies (e.g. Ralph et al., 1973). A maximum discrepancy at ca 6,500 yr BP is 800-900 years but there is no universal agreement on the exact calibration of radiocarbon ages back to 8,000 yr BP. Beyond the current limit of calibration, the degree of divergence is problematic, although Stuiver (1978) has suggested that the maximum deviation back to 32,000 years is unlikely to exceed 2,000 years. Recently, a high precision calibration curve, derived from the radiocarbon date determinations of 195 decade samples, from AD 1 to 1950, has been produced (Stuiver, 1982). All these studies indicate that there have been secular variations in atmospheric radiocarbon in the past.

## DISCUSSION

Radiocarbon dating of sea-water samples collected from Antarctica have indicated old ages; for example a range of 600 to 1,300 yr BP at McMurdo Sound (Broecker et al., 1961; Marini et al., 1967). In the Syowa Station area Omoto (1972) reports an age of 2860 ± 125 yr BP (N-858) for a sample collected in the Ongul Strait 300 m east of the East Ongul Island and 4 km from the continental ice sheet; sea-water sample collected 50 km northeast of the East Ongul Island was dated at 880 ± 115 yr BP (N-860).

Living marine organisms in Antarctica have also shown old dates (see Tables 2 and 3), while terrestrial water samples and atmospheric samples collected from Syowa Station and its vicinities have indicated "super modern dates" (Omoto, 1972). These are apparently due to greatly enhanced atmospheric $^{14}C$ levels caused by nuclear weapons testings since 1952 AD. It is impossible to reduce the effects of nuclear weapons testings on the ages of live terrestrial samples collected in Antarctica, although the present author has never heard of "super modern dates" on Antarctic living marine organisms.

**TABLE 2: Radiocarbon dates of some modern shell samples in Ross Sea (modified after Stuiver et al., 1981).**

| Sample material | $^{14}C$ Age | Code No. |
|---|---|---|
| *Adamussium colbecki* | 850 ± 50 | QL-98 |
| Do. | 990 ± 50 | QL-996 |
| Mixture of shells | 1,370 ± 50 | QL-77 |
| Do. | 1,290 ± 50 | QL-79 |
| Do. | 1,260 ± 30 | QL-1128 |
| Do. | 1,340 ± 30 | QL-1225 |
| Do. | 1,540 ± 50 | QL-175 |

**TABLE 3: Some modern carbon dates at Syowa Station and its vicinities (compiled after Omoto 1972 and 1976 and Yoshida and Moriwaki 1979 with asterisks).**

| Sample material | Elevation | $^{14}C$ Age ($\Delta^{14}C\%\delta$) | Code No. |
|---|---|---|---|
| Sea water | – 10 m | 2,860 ± 125 (– 292 ± 11) | N-858 |
| Do. | – 10 m | 880 ± 115 (– 101 ± 12) | N-860 |
| Lake water | – 0.5 m | Modern (+ 278 ± 19) | N-859 |
| Do. | – 0.5 m | Modern (+ 253 ± 19) | N-861 |
| Atmospheric $CO_2$ | 10 m a.s.l. | Modern (+ 487 ± 17) | N-922 |
| Do. | Do. | Modern (+ 315 ± 45) | N-923 |
| Crab-eater seal (skin) | 0 m | 1,455 ± 110 | TH-052 |
| *Neoliuccinum eatoni** | – 17 ~ – 35 m | 1,190 ± 90 (– 138 ± 9) | GaK-6789a |
| Do. (shell)* | Do. | 1,300 ± 90 | GaK-6789b |
| *Ophionotus victoriae** | – 92 m | 1,070 ± 90 (– 125 ± 10) | GaK-6790a |
| Do. (shell)* | Do. | 1,210 ± 100 | GaK-6790b |
| *Sterechinus neumayeri** | – 17 m | 1,160 ± 110 (– 134 ± 12) | GaK-6791a |
| Do. (shell)* | Do. | 860 ± 110 | GaK-6791b |
| *Trematomus berunacchii** | – 15 m | 1,160 ± 110 (– 148 ± 9) | GaK-6792 |
| *Zoarcidae** | – 500 m | 1,010 ± 110 (– 118 ± 13) | GaK-6793 |

Note: Sampling sites for N-858 and N-860 are described in the text. Elevation of the lake for N-859 and N-861 is ca. 12 m a.s.l.

Judging from the fact that the age anomalies appear to be greater in sea-water samples collected closer to the continental ice margin, the cause of low $\Delta^{14}C$ concentration could be postulated as due to the dilution of sea waters by fresh water from the submarine melting of icebergs, ice shelves, and glaciers in which carbon dioxide has been sealed for thousands of years (Omoto, 1972). Decrease of salinity by melting of ice in front of a glacier was deduced from oceanographical observations by Wakatsuchi (1982). However, low $\Delta^{14}C$ concentrations are also known to occur in deep oceanic waters (e.g.Östlund and Stuiver, 1980) and upwelling of such water may have caused the anomalous ages reported from the Syowa area.

In Antarctica it is necessary to correct apparent $^{14}C$ ages by using a "reservoir correction". The reservoir correction for radiocarbon dates of *Adamussium colbecki* collected from New Harbour, McMurdo Sound was 850-1,450 years (Stuiver et al., 1976). Yoshida and Moriwaki (1979) derived a similar value of 1,120 years as a reservoir correction around Syowa Station from the mean of determinations on eight living marine organisms collected from various depths between – 15 m and – 500 m (see also Table 3).The present author thus concludes that the reservoir correction in Antarctica appears to range between 800 years and 3,000 years and that it should be determined by radiocarbon dates on living marine organisms similar to the fossil material to be dated.

## CONCLUDING REMARKS

The author has presented the significance of radiocarbon dating, examined some problems related to its basic assumptions, and also discussed the cause of very low radiocarbon concentrations of Antarctic surface water and living marine organisms. As the living marine organisms in Antarctica seldom indicate "modern dates", it is necessary to establish a reservoir correction all over Antarctica before geochronological correlation and interpretation can be discussed with any precision. This has become evident, not because of any new discovery but because of remarkable developments in radiocarbon dating techniques. We need many radiocarbon dates and information on raised beaches, on the behaviour of sea water and ecosystems, sea level changes and crustal uplift.

Some papers still place total reliance on a single radiocarbon date, but one should always bear in mind the difficulty of interpretation. Reliable ages will be derived from plural dates from a single horizon in which a close relationship is established between the death of the sample material and the events whose age is sought.

Professor W. Wolfli (Zurich E.T.H.) and Dr H. Oeschger (Berne University) (pers. comm.) have shown that it is possible to obtain precise dates for the continental ice by using recently developed accelerator techniques and from which we can also expect to obtain information on atmospheric radiocarbon variations. Refined comparison and correlation between radiocarbon dates and tree-ring records can also be expected.

*Acknowledgements* The author would like to express his sincere appreciation to those who encouraged him in, and led him to, the study of Antarctic glacial geomorphology, namely, Emeritus Professor Kasuke Nishimura of Tohoku University, Emeritus Professor Torao Yoshikawa, University of Tokyo and Professor Kou Kusunoki of the National Institute of Polar Research. He also thanks Dr Gordon Thom of the British Embassy, Tokyo, for help in the revising the manuscript.

## REFERENCES

BENNETT, C.L., BEUKENS, R.P., CLOVER, M.R., GOVE, H.E., LIEBERT, R.P., LITHERLAND, A.E., PURSER, K.H. and SONDHEIM, W.E., 1977: Radiocarbon dating using electrostatic accelerators: Negative ions provide the key. *Science, 198,* 508-10.

BROECKER, W.S. and OLSON, E.A., 1961: Lamont radiocarbon measurements VII *Radiocarbon, 3,* 176-204.

BROECKER, W.S., PENG, T.H. and ENGH, R., 1980: Modelling the carbon system. *Radiocarbon, 22,* 565-98.

GROOTES, P.M., 1978: Carbon-14 time scale extended: Comparison of chronologies. *Science, 200,* 11-15.

GROOTES, P.M., MOOK, W.G., VOGEL, J.C., de VRIES, A.E., HARING, A. and KISTEMAKER, J., 1975: Enrichment of radiocarbon for dating samples up to 75,000 years. *Zeit. Naturf., 30A,* 1-14.

HARKNESS, D.D., 1979: Radiocarbon dates from Antarctica. *Bull. Br. antarct. Surv., 47,* 43-59.

MARINI, M.A., ORR, M.F. and COE, E.L., 1967: Surviving Macromolecules in Antarctic seal mummies. *U.S. Antarct. J., 2(5),* 190-91.

MULLER, R.A., 1977: Radioisotope dating with a cyclotron. *Science, 196,* 489-94.

NYDAL, R., LÖVSETH, K. and GULLIKSEN, S., 1979: A survey of Radiocarbon variation in nature since the Test Ban Treaty; *in* Berger, R. and Suess, H.E. (eds.) Radiocarbon Dating, University of California Press, Berkeley, 313-23.

OMOTO, K., 1972: A preliminary report on modern carbon datings at Syowa Station and its neighbourhood, East Antarctica. *Antarctic Rec., 43,* 20-4.

OMOTO, K., 1976: Tohoku University radiocarbon measurements II. *Sci. Repts. Tohoku Univ. 7th Ser. (Geogr)., 26,* 126-50.

OMOTO, K., 1981: Radiocarbon dating by low background liquid scintillation counting system—Carbon-14 time scale extension. *Chiriyo, 20,* 54-5 (in Japanese).

ÖSTLUND, H.G. and STUIVER, M., 1980: GEOSECS Pacific radiocarbon. *Radiocarbon, 22,* 25-53.

RALPH, E.K., MICHEAL, H.N. and HAN, M.C., 1973: Radiocarbon dates and reality. *MASCA Newsletter, 9,* 1-20.

STUIVER, M., 1978: Radiocarbon time scale tested against magnetic and other dating methods. *Nature, 273,* 271-4.

STUIVER, M., 1982: A high precision calibration of the AD radiocarbon time scale. *Radiocarbon, 24,* 1-26.

STUIVER, M. and ÖSTLUND, H.G., 1980: GEOSECS Atlantic radiocarbon. *Radiocarbon, 22,* 1-24.

STUIVER, M., DENTON, G.H. and BORNS, H.W. Jr., 1976: Carbon-14 dates of *Adamussium colbecki* (Mollusca) in marine deposits at New Harbour, Taylor Valley. *U.S. Antarct. J., 11,* 86-8.

STUIVER, M., HEUSSER, C.J. and YANG, In, Che., 1978: North American glacial history extended to 75,000 years ago. *Science, 200,* 16-21.

STUIVER, M., DENTON, G.H., HUGHES, T.J. and FASTOOK, J.L., 1981: History of the marine ice sheet in West Antarctica during the last glaciation: A working hypothesis; *in* Denton, G.H. and Hughes, T.J. (eds.) *The Last Great Ice Sheets.* John Wiley & Sons, New York, 338-45.

WAKATSUCHI, M., 1982: Seasonal variations in water structure under fast ice near Syowa Station, Antarctica, in 1976. *Antarctic Rec., 74,* 85-108.

YOSHIDA, Y. and MORIWAKI, K., 1979: Some consideration on elevated coastal features and their dates around Syowa Station, Antarctica. *Mem. Nat. Inst. Pol. Res. Japan, Spec. Issue, 13,* 220-6.

# *PROVENANCE DATES OF FELDSPAR IN GLACIAL DEPOSITS, SOUTHERN VICTORIA LAND, ANTARCTICA

K.S. Taylor and G. Faure, *Department of Geology and Mineralogy and Institute of Polar Studies, The Ohio State University, Columbus, Ohio, 43210, U.S.A.*

*Abstract* Glacial deposits on Mt. Fleming and at Prospect Mesa in Wright Valley at the west end of Wright Valley contain feldspar that was dated by the Rb-Sr method in order to determine the provenance of these deposits. Two feldspar size fractions extracted from the till on Mt. Fleming and two feldspar fractions from Beacon clasts in the till fit a line on the Rb-Sr isochron diagram whose slope corresponds to an age of 238 ± 4 Ma. The colinearity of these data is probably caused by the presence of varying amounts of alteration products in the feldspar and indicates that most of the feldspar in the till originated from the Beacon sandstone. The fine fractions of feldspar from the Mt. Fleming till were treated to remove these weathering products and they fit a line with a slope corresponding to a date of 499 Ma which is indistinguishable from the age of the granitic basement rocks in southern Victoria Land. This suggests that the feldspar in the Beacon sandstone was originally derived from nearby exposures of the local basement rocks. The apparent absence of Precambrian feldspar suggests that this till was deposited by a local ice cap. Grain-size fractions of feldspar from the Jason glaciomarine diamicton at Prospect Mesa form an isochron whose slope yields a date of 480 ± 95 Ma which is similar to the ages of the igneous rocks in Wright Valley. The provenance date suggests that the feldspar from the Jason diamicton may have been derived from the basement rocks of western Wright Valley. Size fractions of feldspar and one "magnetic" fraction from the Peleus till, situated above the Jason, define a mixing line whose slope is equivalent to a provenance date of 762 ± 90 Ma. The provenance date therefore suggests that the Peleus till may contain a component of feldspar derived from the Precambrian Shield of East Antarctica.

The history of glaciation of Antarctica is recorded in deposits scattered throughout the Transantarctic Mountains. Denton and Hughes (1981) and Mercer (1978) have evaluated and interpreted the accumulated field and laboratory data pertaining to the glaciation of Antarctica. Taylor and Faure (1981), and Faure and Taylor (1981a; this volume) have reported that the Rb/Sr ratios of size fractions of feldspar extracted from till increase with grain size and that such data points form linear arrays on the Rb-Sr isochron diagram. These linear arrays are isochrons in cases where the feldspar was derived from one source. The age of the source is derivable from the slope of such an isochron. In case the feldspar was derived from two sources having different ages, grain-size fractions of feldspar may also form linear arrays on the isochron diagram. However, these are mixing lines whose slopes express the proportion of mixing and have no real time significance. Such "provenance dates" nevertheless have geological value as indicators of the presence of feldspar derived from a second source having a different age.

A K-Ar date of 26 Ma for glauconite (McDougall, 1976) encountered at DSDP site 270 in the Ross Sea beneath a thick deposit of glaciomarine sediment (Barrett, 1975) indicates that glaciation of Antarctica had begun by Late Oligocene time. The East Antarctic ice sheet appears to have reached sufficient size to flow over the Transantarctic Mountains in southern Victoria Land (Denton et al., in press) and in the Beardmore Glacier area (Mayewski and Goldthwait, 1982). Therefore, glacial deposits in the Transantarctic Mountains were apparently deposited both by local ice caps and by the East Antarctic ice sheet. Those deposits that were formed by local ice caps contain feldspar derived from the rocks exposed in the Transantarctic Mountains whereas deposits formed by the ice sheet may also contain rock and mineral particles that originated from the Precambrian Shield of East Antarctica.

The Transantarctic Mountains in southern Victoria Land consist of a crystalline basement complex of Early Palaeozoic age unconformably overlain by quartz arenites and minor siltstone and shale of the Beacon Supergroup (McKelvey and Webb, 1962). Both rock units were intruded by sills of Ferrar Dolerite of Early Jurassic age. Metamorphic rocks of Precambrian age, such as those of the Nimrod Group in the Beardmore Glacier area (Gunner, 1982) are not known from southern Victoria Land. Therefore, glacial deposits derived from local sources contain feldspar derived primarily from the igneous and metamorphic basement rocks whose ages range from 470 Ma to about 560 Ma (Jones and Faure, 1967; Faure et al., 1974, 1979; Felder and Faure, 1980). The Beacon rocks certainly contain less than 5% feldspar on the average and therefore are an important source of feldspar *only* in those glacial deposits that consist largely of ground-up sandstone. Evidence presented by Faure and Taylor (1981a) suggests that the feldspar in the Beacon rocks was originally derived from granitic rocks whose age is about 500 Ma. The available information therefore indicates that the feldspar in the Beacon rocks may have originated from exposures of the underlying basement rocks. The Ferrar Dolerite sills are not significant contributors of feldspar to the glacial deposits perhaps because of their resistance to chemical and mechanical weathering (Faure and Taylor, 1981a).

The objectives of this study are to date feldspar from glacial deposits in and around Wright Valley of southern Victoria Land and to determine their origin and provenance. The deposits chosen for this study are from Mt. Fleming at the western end of Wright Valley and the Peleus till and Jason diamicton (Denton et al., 1983) from Prospect Mesa at the mouth of Bull Pass (Figure 1).

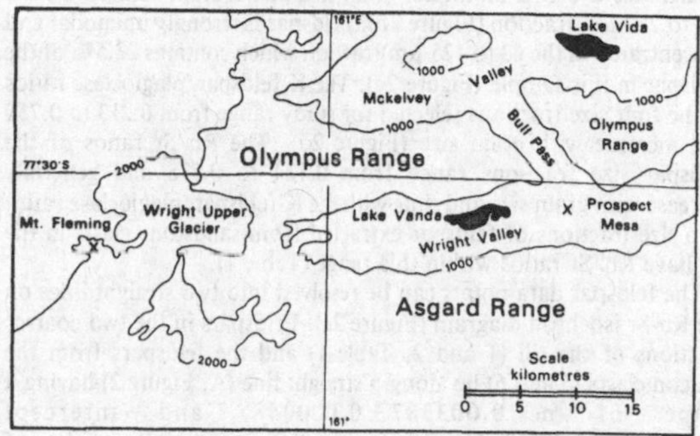


**Figure 1. Map of Wright Valley, southern Victoria Land, showing till deposits on Mt. Fleming and at Prospect Mesa.**

## MOUNT FLEMING TILL

A deposit of till, located on the southwestern flank of Mt. Fleming near the edge of the present East Antarctic ice sheet, rests on sandstones of the Mt. Fleming Formation of the Beacon Supergroup. It is considered to be a basal till because it is massive and highly indurated, but not cemented. A specimen of this till weighing 3087.7 g was disaggregated in water and sieved into size fractions (Figure 2a). Relative abundances of feldspar and quartz in these size fractions were determined in quadruplicate by X-ray diffraction (Figures 2b and c). Feldspar-quartz concentrates were prepared by use of a Frantz Isodynamic Magnetic Separator followed by ultrasonic cleaning in dilute HCl and demineralized water for up to 28 hours. Concentrations of Rb and Sr were determined by X-ray fluorescence as outlined by Faure and Taylor (this volume). The isotopic composition of Sr was measured on a manually operated mass spectrometer (Nuclide Corp., Model 6-60-S, David). The Eimer and Amend $SrCO_3$ isotope standard was analyzed repeatedly and has an average $^{87}Sr/^{86}Sr$ ratio of 0.70811 ± 0.00007 (1σ). The analytical results are compiled in Table 1.

Clasts larger than 4 mm in diameter in the Mt. Fleming till are composed of sandstone (60.6% by weight), grey and greenish siltstone (34.5%), black shale (4.2%), coal (0.4%) and quartz (0.4%). All of these rock types could have originated from the local Beacon rocks.

*Laboratory for Isotope Geology and Geochemistry (Isotopia) Contribution No. 65

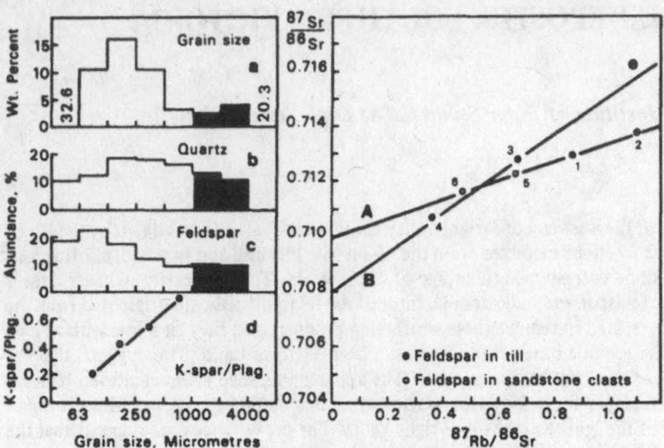


**Figure 2. Granulometry, mineral abundances and Sr isotopic data for feldspar in till and sandstone clasts on Mt Fleming. a) Grain-size distribution; b) and c) Relative abundances of quartz and feldspar; d) Variation of the K-feldspar/plagioclase ratio with grain size; e) Mixing line (A) and tentative isochron (B) yeilding a date of 499 Ma.**

No clasts of igneous or metamorphic rock were found in our sample. The grain-size distribution (Figure 2a) shows a peak between 125 to 250 $\mu$m caused by the presence of quartz grains derived from the well-sorted Beacon sandstones. Grains smaller than 63 $\mu$m make up 32.6% of this sample and give this till an abundant fine-grained matrix. Quartz has a broad unimodal grain-size distribution centred on the 125 to 250 $\mu$m fraction (Figure 2b). Feldspar is strongly unimodal and concentrated in the 63 to 125 $\mu$m fraction which contains 22.5% of the feldspar in this sample (Figure 2c). The K-feldspar/plagioclase ratios of the four size fractions selected for study range from 0.213 to 0.750 and increase with grain size (Figure 2d). The Rb/Sr ratios of the feldspar size fractions range from 0.124 to 0.378 and generally increase with grain size and thus with the K-feldspar/plagioclase ratio. Two size fractions of feldspar extracted from sandstone clasts in the till have Rb/Sr ratios within this range (Table 1).

The feldspar data points can be resolved into two straight lines on the Rb-Sr isochron diagram (Figure 2e). Feldspars in the two coarser fractions of the till (1 and 2, Table 1) and the feldspars from the Beacon clasts (5 and 6) lie along a straight line (A, Figure 2) having a slope of $m = 0.003387 \pm 0.000048$, and intercept $b = 0.71000 \pm 0.00038$. The slope of this line corresponds to a date of $238 \pm 4$ Ma ($\lambda^{87}Rb = 1.42 \times 10^{-11} a^{-1}$) which does not equal the age of any known rock unit in this part of the Transantarctic Mountains. Therefore, the feldspars that form this line could not have been derived from a single source and must be mixtures of two components. The excellent colinearity of these data points establishes a close relationship between the feldspar in the Beacon sandstones and in the till. A possible interpretation of these data is that the feldspar in the till originated predominantly from the Beacon sandstones and that their colinearity results from the presence of varying amounts of alteration products. The feldspars from the till were each cleaned ultrasonically for 12 hours which resulted in the removal of illite, kaolinite and some plagioclase, identified by X-ray diffraction. The feldspars from the sandstone were not cleaned and presumably contain similar alteration products. The sandstone clasts, examined in thin-section, contain plagioclase ($Ab_{63}$ andesine), microcline and some orthoclase. Most feldspar grains are altered to clay and sericite parallel to twinning planes. We conclude that the line formed by feldspars 1, 2, 5 and 6 (Table 1) is a mixing line and not an isochron and that the date derived from its slope has no time significance.

The fine feldspar fractions from the till (3 and 4, Table 1) do not fit the array discussed above. Both samples were cleaned ultrasonically for significantly longer periods of time than samples 1 and 2. A straight line (B, Figure 2) drawn through these points has a slope $m = 0.007108$ and an intercept $b = 0.70805$. The slope of this line corresponds to a date of 499 Ma which is compatible with the known age of the granitic basement rocks in this region (Faure et al., 1974). This result therefore provides additional support to the conclusion expressed previously by Faure and Taylor (1981a) that the feldspar in the Beacon sandstones was derived from igneous and metamorphic rocks similar in age to the crystalline basement upon which the Beacon rocks were deposited.

The final conclusion derivable from the study of till on Mt. Fleming is that no compelling evidence was found for the presence of a component of Precambrian feldspar. The predominance of sandstone clasts and the absence of clasts of igneous and metamorphic rocks strongly suggest that the feldspar in the sandy-silty matrix of the till was derived from the sandstones. This suggestion is further corroborated by the colinearity of feldspars from the sandstone clasts and the till matrix. Based on this evidence, we conclude that the till on Mt. Fleming is composed of locally derived detritus and probably was deposited by an ice cap centred on the Transantarctic Mountains.

## PROSPECT MESA, WRIGHT VALLEY

The sedimentary deposits at Prospect Mesa (Figure 1) have been examined by many investigators (see Denton et al., 1983). The oldest exposed unit is a poorly sorted, unconsolidated and uncemented sandy mud that is stratified in places and contains scattered pebbles. Denton et al. (in press) named this unit the Jason glaciomarine diamicton and concluded that it was deposited in a fjord by floating ice originating from western Wright Valley. The Jason diamicton contains a variety of fossils including marine and non-marine diatoms

**TABLE 1: Analytical results for feldspar in glacial deposits on Mt. Fleming and Prospect Mesa, Wright Valley, southern Victoria Land**

| | Grain Size $\mu$m | | $\frac{Rb}{Sr} = 1\sigma$ | $\frac{^{87}Rb}{^{86}Sr} = 1\sigma$ | $\frac{^{87}Sr}{^{86}Sr} = 1\sigma$ |
|---|---|---|---|---|---|
| | | | Feldspar, Till, Mt. Fleming | | |
| 1 | 500-1000 | | 0.2978 = 0.0085 | 0.8625 = 0.0246 | 0.71294 = 0.00016 |
| 2 | 250-500 | | 0.3785 = 0.0024 | 1.0963 = 0.0069 | 0.71371 = 0.00017 |
| 3 | 125-250 | | 0.2319 = 0.0038 | 0.6716 = 0.0110 | 0.71283 = 0.00012 |
| 4 | 63-125 | | 0.1241 = 0.0011 | 0.3593 = 0.0031 | 0.71061 = 0.00024 |
| | | | Feldspar, Sandstone Clast, Mt. Fleming Till | | |
| 5 | 125-250 | | 0.2273 = 0.0017 | 0.6583 = 0.0049 | 0.71226 = 0.00010 |
| 6 | 63-125 | | 0.1610 = 0.0015 | 0.4662 = 0.0043 | 0.71157 = 0.00013 |
| | | | Feldspar, Jason Diamicton, Prospect Mesa F-80-2 | | |
| 7 | 100-2000 WR | | 0.2882 = 0.0036 | 0.8348 = 0.0104 | 0.71415 = 0.00025 |
| 8 | 1000-2000 | | 0.2370 = 0.0052 | 0.6861 = 0.0150 | 0.71349 = 0.00021 |
| 9 | 500-1000 | | 0.2510 = 0.0029 | 0.7267 = 0.0084 | 0.71326 = 0.0009 |
| 10 | 250-500 | | 0.2290 = 0.0008 | 0.6629 = 0.0023 | 0.71293 = 0.00012 |
| 11 | 125-250 | | 0.2163 = 0.0062 | 0.6264 = 0.0179 | 0.71284 = 0.00014 |
| | | | Feldspar, Peleus Till, Prospect Mesa F-80-30 | | |
| 12 | 500-1000 | (1) | 0.2151 = 0.0017 | 0.6230 = 0.0049 | 0.71247 = 0.00024 |
| | | (2) | | | 0.71210 = 0.00019 |
| 13 | 250-500 | | 0.2080 = 0.0031 | 0.6024 = 0.0089 | 0.71221 = 0.00010 |
| 14 | 125-250 | | 0.2053 = 0.0016 | 0.5945 = 0.0046 | 0.71213 = 0.00018 |
| 15 | 125-250 Mag | | 0.2468 = 0.0022 | 0.7148 = 0.0063 | 0.71336 = 0.00015 |
| | | | F-80-4 | | |
| 16 | 125-250 | | 0.2024 = 0.0045 | 0.5859 = 0.0130 | 0.71187 = 0.00011 |
| | | | Feldspar, Scallop Hill Till, Black Island | | |
| 17 | 125-250 | | 0.0759 = 0.0011 | 0.2196 = 0.0032 | 0.70597 = 0.00007 |

(Truesdale and Kellogg, 1979) and a few foraminifera that may be reworked (Webb, 1972).

The Jason diamicton is overlain by a stratified gravel deposit containing shells of *Chlamys tuftsensis* Turner and an abundant and diverse group of foraminifera (Webb, 1972). The pecten shells and foraminiferal tests are well preserved indicating that the "pecten gravel" was not disturbed after deposition. A massive silty unit overlying the pecten gravel was named the Peleus till by Denton et al. (in press) who concluded that it was deposited by ice advancing from the west. Prospect Mesa is capped by a coarse gravel unit about 2 m thick that is part of a dissected fan originating from Bull Pass.

According to evidence reviewed by Denton et al. (in press), the Jason diamicton was deposited in the time interval from about 15 to 9 Ma (Middle Miocene—early Late Miocene). The Peleus till is younger than the Jason diamicton but underlies the alpine moraines of Bartley Glacier which were deposited between 2.1 Ma and 3.0 Ma (Armstrong, 1978). This means that the Peleus till was deposited between 9 Ma and 2.1 Ma ago during the Pliocene epoch. We have analysed feldspar size fractions from the Jason diamicton and the Peleus till at Prospect Mesa (Figure 3, Table 1).

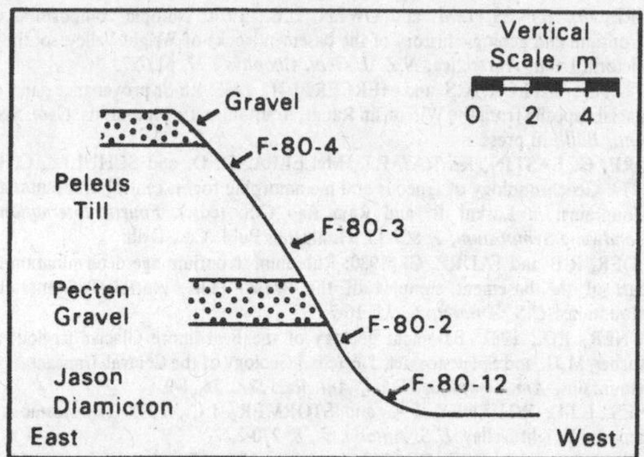


**Figure 3. Cross-section of Prospect Mesa, Wright Valley showing samples taken for study.**

## Jason Diamicton

One sample of sediment from the Jason diamicton weighing 828.0 g (F-80-2, Table 1) was sieved into size fractions and analysed by X-ray diffraction to determine the abundances of quartz and feldspar. The abundance of particle sizes increases with decreasing grain size (Figure 4a) such that particles less than 63 $\mu$m in diameter constitute 34.8% of this sample by weight. Quartz is most highly concentrated in the 250 to 500 $\mu$m fraction whereas feldspar abundances vary only slightly with grain size (Figures 4b,c). The feldspar is composed predominantly of plagioclase (Webb, 1972; Brooks, 1972). Both authors also pictured quartz grains which are generally angular indicating their derivation from igneous and metamorphic rocks and transport by ice. Clasts more than 4 mm in diameter in sample F-80-2 are composed entirely of igneous rocks of dioritic lithology. The Rb/Sr ratios of feldspar range from 0.216 to 0.251 and increase with grain size (Figure 4d).

The feldspar data and one "whole-rock" sample from the 1 to 2 mm fraction of F-80-2 form a linear array of points on the Rb-Sr isochron diagram (Figure 4e). A straight line fitted to these data has a slope $m = 0.00656 \pm 0.00136$ and intercept $b = 0.70869 \pm 0.00096$. The slope of this line corresponds to a date of $460 \pm 95$ Ma. The date indicated by the feldspar from the Jason diamicton is similar to the known ages of the igneous rocks that occur in Wright Valley (Faure et al., 1974; Jones and Faure, 1967). Therefore, the Jason diamicton could have been derived from the basement rocks exposed at the western end of Wright Valley as stated by Denton et al. (in press). We find no evidence for the presence of Precambrian feldspar in the Jason diamicton at Prospect Mesa.

## Peleus Till

Sample F-80-3 (753.9 g) from the Peleus till (Figure 3) has a unimodal grain size distribution (Figure 5a) that is quite unlike that of F-80-2 from the Jason diamicton. The Peleus is deficient in the silt and clay size fractions and is distinctly sandy. Clasts more than 4 mm in diameter are composed almost entirely of igneous rocks with only one small clast (0.9 g) of sandstone. Quartz has a broad unimodal grain-size distribution (125 to 500 $\mu$m, Figure 5b) whereas feldspar is strongly concentrated in the 500 to 1000 $\mu$m fraction. The feldspar in the Peleus till consists mainly of plagioclase much like that of the Jason diamicton. The Rb/Sr ratios of feldspar in the Peleus till (Figure 5d) increase with grain size from 0.205 to 0.215 but are significantly *lower* than those of the Jason diamicton. Because of the small range of Rb/Sr ratios of the feldspar in the Peleus till, we also analysed a "magnetic" fraction in the 125 to 250 $\mu$m size range (Table 1) in order to improve the estimate of the slope on the isochron diagram.

Both the Jason and the Peleus contain high proportions of tan-coloured grains that emerge in the magnetic fraction on the Frantz Isodynamic Separator. A quantitative assessment of the abundances of the mineral components in the 250 to 500 $\mu$m fraction of sample F-80-4 from the Peleus till (Figure 3) indicates that feldspar and quartz make up 23.3% by weight, pyroxene and other ferromagnesian minerals 39.4%, and the tan-coloured fine-grained sediment amounts to 37.3%. This sediment is X-ray amorphous and does not contain detectable clay or oxide minerals. However, some plagioclase and quartz appear in X-ray diffraction scans. These results confirm the work of Webb (1972) and Brooks (1972) who both reported the absence of clay minerals in the sediment at Prospect Mesa.

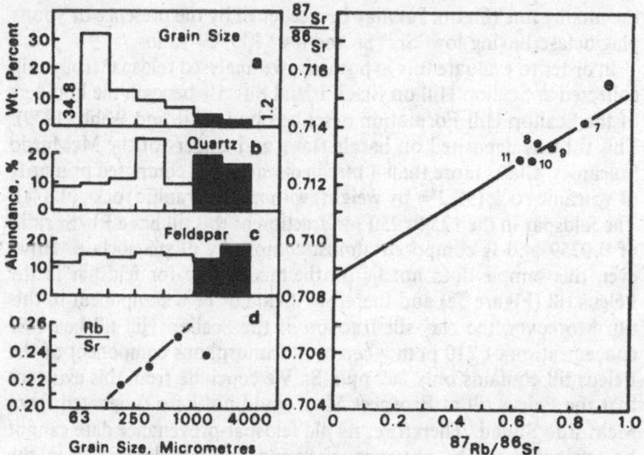


**Figure 4. Granulometry, relative mineral abundances and Sr isotopic data for feldspar and other fractions from sample F-80-2 of the Jason glaciomarine diamicton, Prospect Mesa, Wright Valley. a) Grain-size distribution; b and c) Relative abundances of quartz and feldspar. The shaded portions of the histograms represent lithic grains. d) Variation of Rb/Sr ratios of feldspar with grain size; e) Rb-Sr isochron yielding a date of 460 ± 95 Ma.**

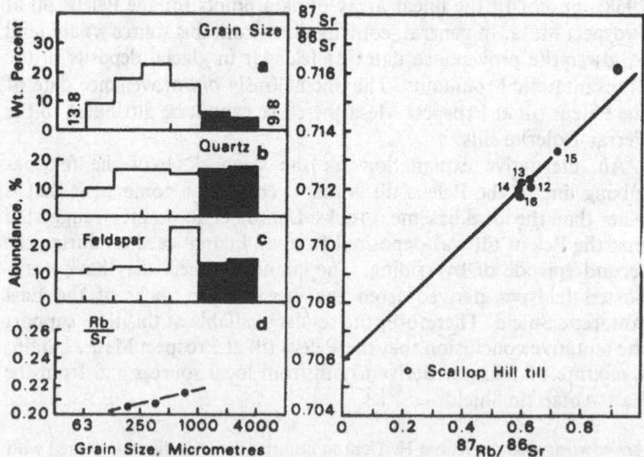


**Figure 5. Granulometry, relative mineral abundances and Sr isotopic data for feldspar and other fractions from sample F-80-3 of the Peleus till, Prospect Mesa, Wright Valley. a) Grain-size distribution; b and c) Relative abundances of quartz and feldspar. The shaded portions of the histograms represent lithic grains. d) Variation of Rb/Sr ratios of feldspar with grain size; e) Rb-Sr mixing line yielding a provenance date of 762 ± 90 Ma that is attributed to the presence of a component of Precambrian feldspar.**

Four size fractions of sample F-80-3 of the Peleus till were analysed for dating by the Rb-Sr method (Table 1). The results form a linear array on the Rb-Sr isochron diagram (Figure 5e) which has a slope of $m = 0.01123 \pm 0.00150$ and an intercept $b = 0.70534 \pm 0.00093$. The slope is equivalent to a date of $762 \pm 90$ Ma. This date is older than the ages of the granitic basement rocks of Wright Valley. A date of $2554 \pm 330$ Ma, reported by Vocke and Hanson (1981) for zircons in the Olympus Granite Gneiss of Victoria Valley, refers to the time of crystallization of zircons of detrital origin and therefore does not represent the age of the crystalline basement rocks in this area.

We have considered the possibility that the Peleus till contains a volcanic component derived from the Tertiary McMurdo Volcanics. This conjecture from the predominance of plagioclase over K-feldspar manifested by the unusually low Rb/Sr ratio, and by the presence of X-ray amorphous sediment in the till. It is supported by Jones et al. (1973) who found water-laid volcanic ash in soil at two locations in the vicinity of Lake Vanda in Wright Valley (Figure 1). Also Nichols (1971) suggested that the pecten shells in the gravel at Prospect Mesa had been transported into Wright Valley by ice advancing from McMurdo Sound. This hypothesis was later disaproved by Webb (1972) and by McSacveney and McSaveney (1972). Nevertheless, we have examined the possibility that the pecten gravel and the Peleus till at Prospect Mesa were originally deposited in McMurdo Sound and were subsequently transported into Wright Valley as a block of frozen sediment without reworking. If this explanation is correct, the slope of the mixing line (Figure 5e) may be steepened by the presence of young plagioclase having low $^{87}Sr/^{86}Sr$ and low $^{87}Rb/^{86}Sr$ ratios.

In order to evaluate this hypothesis, we analysed feldspar from a till collected at Scallop Hill on Black Island directly beneath the boulders of the Scallop Hill Formation described by Leckie and Webb (1979). This till was deposited on basalt flows and cinders of the McMurdo Volcanics. Clasts more than 4 mm in diameter are composed primarily of volcanic rock (99.2% by weight) with minor granitic rocks (0.8%). The feldspar in the 125 to 250 $\mu$m fraction of this till has a Rb/Sr ratio of 0.0759 and is composed almost completely of plagioclase. However, this sample does not lie on the mixing line for feldspar in the Peleus till (Figure 5e) and therefore it cannot be a component in this till. Moreover, the clay-silt fraction of the Scallop Hill till has a Sr concentration of 710 ppm whereas the amorphous component of the Peleus till contains only 260 ppm Sr. We conclude from this evidence that the Peleus till at Prospect Mesa could not have originated from McMurdo Sound. Therefore, its old feldspar-provenance date cannot be attributed to the presence of young feldspar derived from the McMurdo Volcanics. The presence of feldspar derived from basaltic cinder cones within Wright Valley is similarly ruled out because its isotopic and chemical composition are incompatible with those of the Peleus till.

The sills of Ferrar Dolerite (Lower Jurassic) are a potential source of feldspar in glacial deposits of southern Victoria Land. However, feldspars from the Basement Sill in Wright Valley (Compston et al., 1968), do not fit the linear array of data points for the Peleus till at Prospect Mesa. In general, contributions from this source would tend to *lower* the provenance dates of feldspar in glacial deposits in the Transantarctic Mountains. The anomalously *old* provenance date of the Peleus till at Prospect Mesa therefore cannot be attributed to the Ferrar Dolerite sills.

An alternative explanation for the steep slope of the feldspar mixing line of the Peleus till is that it contains a component that is older than the local basement rocks. Denton et al. (in press) suggested that the Peleus till was deposited from an inland ice sheet during the second episode of overriding. The inland ice sheet may have transported feldspar derived from the Precambrian rocks of the East Antarctic Shield. Therefore, the results available at this time support the tentative conclusion that the Peleus till at Prospect Mesa contains a mixture of feldspar derived both from local sources and from the East Antarctic Shield.

*Acknowledgements* George H. Denton inspired these studies and shared with us his understanding of the glacial history of southern Victoria Land. He also collected the sample of till from Mt. Fleming. Teresa M. Mensing, James A. Stewart and Melissa Kallstrom assisted in the laboratory work. We acknowledge the logistical support of squadron VXE-6 of the U.S. Antarctic Support Force. This research was supported by the Division of Polar Programs of the National Science Foundation through grant DPP79-20407.

## REFERENCES

ARMSTRONG, R.L., 1978: K-Ar dating: Late Cenozoic McMurdo Volcanic Group and dry valley glacial history, Victoria Land, Antarctica. *N.Z. J. Geol. Geophys., 21,* 685-98.

BARRETT, P.J., 1975: Textural characteristics of Cenozoic preglacial and glacial sediments at site 270, Ross Sea, Antarctica. *Initial Reports of the Deep Sea Drilling Project, 28,* 757-68. U.S. Govt. Printing Office, Washington, D.C.

BROOKS, H.K., 1972: A fjord deposit in Wright Valley, Antarctica. *U.S. Antarct. J., 7,* 241-3.

COMPSTON, W., McDOUGALL, I. and HEIER, K.S., 1968: Geochemical comparison of the Mesozoic basaltic rocks of Antarctica, South Africa and Tasmania. *Geochim. Cosmochim. Acta, 32,* 129-50.

DENTON, G.H. and HUGHES, T.J. (eds.) 1981: *The Last Great Ice Sheets.* John Wiley and Sons, New York.

DENTON, G.H., PRENTICE, M.O., KELLOGG, D.E. and KELLOGG, T.B. (in press): Origin and early history of the Antarctic ice sheet: Evidence from the Dry Valleys. *Nature.*

FAURE, G. and TAYLOR, K.S., 1980: Interpretation of Rb-Sr dates of feldspar in till on Mt. Tuartara, Byrd Glacier. *U.S. Antarct. J., 15,* 59-60.

FAURE, G. and TAYLOR, K.S., 1981a: Provenance of some glacial deposits in the Transantarctic Mountains based on Rb-Sr dating of feldspars. *Chem. Geol. 32,* 271-90.

FAURE, G. and TAYLOR, K.S. (this volume): Sedimentation in the Ross Embayment: Evidence from RISP core 8 (1977/78).

FAURE, G., JONES, L.M. and OWEN, L.B., 1974: Isotopic composition of strontium and geologic history of the basement rocks of Wright Valley, southern Victoria Land, Antarctica. *N.Z. J. Geol. Geophys., 17,* 611-27.

FAURE, G., TAYLOR, K.S. and MERCER, J.H., 1983: Rb-Sr provenance dates of glacial deposits from the Wisconsin Range, Transantarctic Mountains. *Geol. Soc. Am., Bull.,* in press.

FAURE, G. EASTIN, R., RAY, P.T., McLELLAND, D. and SCHULTZ, C.H., 1979: Geochronology of igneous and metamorphic rocks, central Transantarctic Mountains; *in* Laskar B. and Raja Rao C.S. (eds.). *Fourth International Gondwana Symposium, 2,* 807-13. Hindustan Publ. Co., Delhi.

FELDER, R.P. and FAURE, G., 1980: Rubidium-strontium age determination of part of the basement complex of the Brown Hills, central Transantarctic Mountains. *U.S. Antarct. J., 15,* 16-7.

GUNNER, J.D., 1982: Basement geology of the Beardmore Glacier Region; *in* Turner, M.D. and Splettstoesser, J.F. (eds.) Geology of the Central Transantarctic Mountains. *Am. Geophys. Union, Ant. Res. Ser., 36,* 1-9.

JONES, L.M., WHITNEY, J.A. and STORMER, J.C., 1973: A volcanic ash deposit, Wright Valley. *U.S. Antarct. J., 8,* 270-2.

JONES, L.M. and FAURE, G., 1967: Age of the Vanda Porphyry dykes in Wright Valley, southern Victoria Land, Antarctica. *Earth Planet. Sci. Lett., 3,* 321-4.

LECKIE, R.M. and WEBB, P.N., 1979: Scallop Hill Formation and associated Pliocene marine deposits of southern McMurdo Sound. *U.S. Antarct. J., 14,* 54-6.

MAYEWSKI, P.A. and GOLDTHWAIT, R.P., 1982: Glacial events in the Transantarctic Mountains: A record of the East Antarctic ice sheet; *in* Turner, M.D. and Splettstoesser, J.F., (eds.) Geology of the Central Transantarctic Mountains, *Am. Geophys. Union, Ant. Res. Ser., 36,* (in press).

McDOUGALL, I., 1976: Potassium-argon dating of glauconite from a greensand drilled at site 270 in the Ross Sea, DSDP leg 28. *Initial Reports of the Deep Sea Drilling Project, 36,* 1071. U.S. Govt. Printing Office, Washington, D.C.

McKELVEY, B.C. and WEBB, P.N., 1962: Geological investigations in southern Victoria Land, Antarctica. *N.Z. J. Geol. Geophys., 5,* 143-62.

McAVENEY, M.J. and McAVENEY, E.R., 1972: A reappraisal of the Pecten glacial episode, Wright Valley, Antarctica. *U.S. Antarct. J., 7,* 235-40.

MERCER, J.H., 1978: Glacial development and temperature trends in the Antarctic and in South America; *in* Van Zinderen Bakker, E.M. (ed.) *Antarctic Glacial History and Palaeoenvironment,* 73-93. A.A. Balkema, Rotterdam.

NICHOLS, R.L., 1971: Glacial geology of the Wright Valley, Antarctica; *in* Quam, L.O. and Porter, H.D. (eds.) *Research in the Antarctic,* Am. Assoc. Adv. Sci., Washington, D.C. 293-340

TAYLOR, K.S. and FAURE, G., 1981: Rb-Sr dating of feldspar: A new method to study till. *J. Geol., 89,* 97-107.

TRUESDALE, R.S. and KELLOGG, T.B., 1979: Ross Sea diatoms: Modern assemblage distributions and their relationship to ecologic, oceanographic and sedimentary conditions. *Mar. Micropaleontol., 4,* 13-31.

VOCKE, R.D. and HANSON, G.N., 1981: U-Pb zircon ages and petrogenesis implications for two basement units from Victoria Valley, Antarctica; *in* McGinnis, L.D. (ed.) Dry Valley Drilling Project. *Am. Geophys. Union, Ant. Res. Ser., 33,* 247-55.

WEBB, P.N., 1972: Wright Fjord, Pliocene marine invasion of an Antarctic dry valley. *U.S. Antarct. J., 7,* 227-34.

# USE OF SOILS IN STUDYING THE BEHAVIOUR OF THE McMURDO ICE DOME

J.G. Bockheim, *Department of Soil Science, University of Wisconsin, Madison, Wisc., 53706, U.S.A.*

*Abstract* A chronosequence of soils was examined on a series of four sets of moraines adjacent to the Wright Upper Glacier at Mt Fleming (77°32′S, 160°12′E). Soil development and surface boulder weathering features increase with distance from the glacier. The soils on an ice-cored lateral moraine, an intermediate lateral moraine and an outer lateral moraine are similar to soils on alpine I, II and III moraines in Wright Valley which are < 1.7 to 3.1 ka, about 100 to 140 ka and 2.00 to 3.33 Ma in age, respectively. The soil on ground moraine beyond the lateral moraines is similar to that on the Pliocene or Late Miocene age Peleus till in central Wright Valley. The presence of strongly developed soils on alpine III drift only 40 m from the present margin of the Wright Upper Glacier suggests that it has not thickened substantially in the past 2.00 to 3.33 Ma.

Drewry (1980) recently identified a small ice dome within 100 km of the western edge of the Transantarctic Mountains, just inland of Taylor Glacier. Radio Echo Sounding results suggested a local origin for ice in Taylor and adjacent glaciers. Based on similar $\delta^2H$ ratios of ice at Taylor Glacier snout and at the dome and the occurrence of meteorites on bare ice at the edge of the Transantarctic Mountains at Allan Hills, Drewry (1980) argued that the East Antarctic ice sheet in the McMurdo Sound vicinity has been stable and could not have thickened substantially in 17-33 ka. These results agree with those of Denton et al. (1971) and Stuiver et al. (1981) who reported that the ice sheet in the McMurdo Sound area is now at its maximum height since before the entire Wisconsin (Würm) glaciation, as defined elsewhere in the world.

In view of the difficulty in obtaining suitable materials for radiometric dating, soils have been useful in providing a relative age framework for reconstructing the glacial history of Antarctica (Ugolini and Bull, 1965; Behling, 1971; Everett, 1972; Campbell and Claridge, 1975; Bockheim, 1979). The objective of this paper is to use soils, in conjunction with glacial geology, to show that the local ice dome, termed here the McMurdo ice dome, has not thickened substantially since the Late Pliocene.

## STUDY AREA

Mt Fleming was selected as the study area because it extends into the McMurdo ice dome. Mt Fleming is a 5.8 km ridge at the western edge of the Asgard Range which is bordered on the north by Wright Upper Glacier and on the south by Taylor Glacier (Figure 1). Maximum relief in the area is 760 m. Local bedrock is the Mt Fleming Formation of the Beacon Supergroup (Matz et al., 1972). Lithologies consist of sandstone (conglomeratic in part), carbonaceous siltstones and shales and coals of Permian, Triassic and (?)Jurassic age. Dykes and sills of Ferrar dolerite (Jurassic) intrude the Beacon Supergroup.

Four sets of moraines occur along the southern edge of the Wright Upper Glacier to the east of Airdevronsix Icefalls, including an ice-cored lateral moraine 6 m from the glacier, an intermediate lateral moraine 20 m from the glacier, an outer lateral moraine 40 m from the glacier, and ground moraine 50 m beyond the outermost lateral moraine (Figure 2).

There are no climatic data for the site. The nearest location for which there are short-term climatic data (December 1968-December 1970) is Vanda Station (elevation 94 m), 35 km to the east in Wright Valley. At Vanda Station, mean annual temperature is −19°C and mean annual accumulation of snow is 44 mm (Thompson et al., 1971). Average wind velocity at Vanda Station is 4.7 m/s. Since the study area is at a higher elevation (1900 m) and is closer to the polar plateau, the mean annual temperature is probably lower and the average wind velocity is probably greater. Snowfall likewise is probably greater at Mt Fleming.

## METHODS

In testing the hypothesis that the McMurdo ice dome has not thickened substantially since the Pliocene, soils and surface boulder weathering features were described on the lateral moraines deposited by Wright Upper Glacier and on the outlying ground moraine. To describe surface boulder weathering features, 314 m² circular plots were established on moraine crests. Surface boulders were tallied according to weathering condition, using criteria given by Bockheim (1979). A soil pit was dug to the ice-cemented frost table and samples were collected from each soil horizon. Soil samples were passed through a 2 mm screen in the field and were sent to the University of Wisconsin-Madison for analysis. Analyses included pH, electrical conductivity, and water soluble ions ($Na^+$, $Ca^{2+}$, $Mg^{2+}$, $K^+$, $SO_4^{2-}$, $NO_3^-$, $Cl^-$) on 1:5 soil:water extracts (American Public Health Association et al., 1975); particle size distribution (Day, 1965); and X-ray diffraction analysis of segregated salts and the clay size (< 2 $\mu$m) soil fraction (Jackson, 1956).

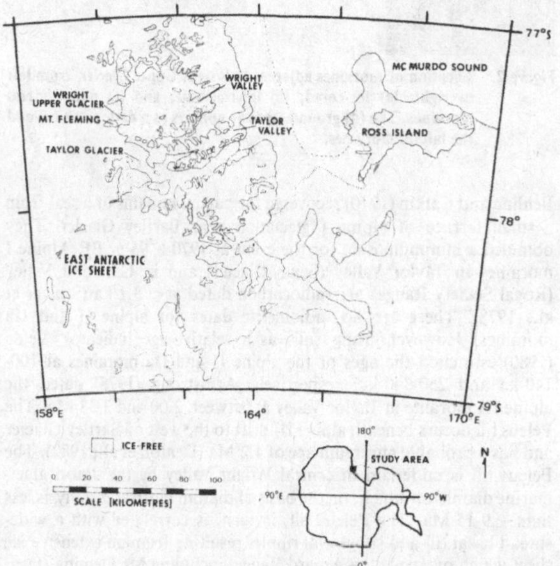


**Figure 1. Location of study area.**

In studying the glacial history of the McMurdo Sound area, soil data are used in conjunction with mapping of glacial geological features by G.H. Denton. In the McMurdo Sound area, soil chronosequences have been identified on moraines deposited by the Ross ice sheet (Bockheim, 1979), Taylor Glacier (Pastor and Bockheim, 1980), and local alpine glaciers (Bockheim, 1978; Leide, 1980). A soil chronosequence is an array of related soils in a geographic area that differ primarily as a result of the soil-forming factor, time (Jenny, 1941; Bockheim, 1980a). In the McMurdo Sound area, soils within a given chronosequence are differentiated on the basis of surface boulder weathering, soil morphology and analytical soil data (e.g. Bockheim, 1979).

In illustrating the utility of soil chronosequences in studying glacial sequences, the glacial history of Wright Valley will be discussed. Soil chronosequences in Wright Valley are confined by either the ice-cored Wright Lower moraine or the ice-cored alpine I moraine and the Peleus till. In eastern Wright Valley, a series of moraines deposited by the Wright Lower Glacier was investigated from the snout of the present glacier to an area 16 km upvalley where the oldest "D moraine" extends over the Pliocene or Late Miocene aged Peleus till (Bockheim, 1979; Denton et al., 1982). Similarly, soils were examined on lateral alpine moraines, including a young, ice-cored alpine I moraine, intermediate aged alpine II and IIa moraines, and the oldest alpine III moraines which overlie the Peleus till to the west of Bartley Glacier (Bockheim, 1978; Leide, 1980; Denton et al., 1982).

Soils within the chronosequences are tied in by absolute ages of the glacial sediments; and ages of sediments for which there are no radiometric dates are estimated using soils as a relative-age indicator. For example, the alpine sequence in Wright Valley is dated as follows.

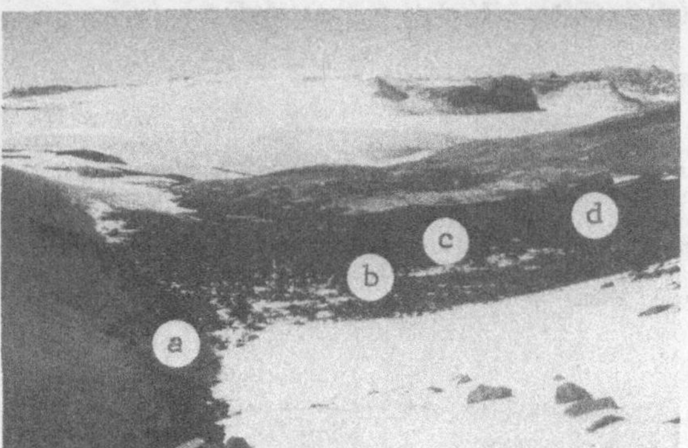


**Figure 2. Location of moraines adjacent to Wright Upper Glacier, from left to right: (a) ice-cored, (b) intermediate, and (c) outer lateral moraines. The (d) ground moraine appears as a dark patch beyond the lateral moraines.**

Behling and Calkin (1970) recovered the partial remains of a seal from a small terrace of alpine I glaciation near Bartley Glacier. They obtained a minimum date for the event at 1970± 95 yr BP. Alpine I moraines in Taylor Valley (Seuss Glacier) and in Garwood Valley (Royal Society Range) are radiocarbon dated at < 3.1 ka (Stuiver et al., 1978). There are no radiometric dates for alpine II and IIa moraines. However, using soils as a relative-age indicator, Leide (1980) estimated the ages of the alpine II and IIa moraines at 100-140 ka and 250-340 ka, respectively. Armstrong (1978) dated the alpine III moraine in Taylor Valley at between 2.00 and 3.33 Ma. The Peleus till occurs beneath alpine III drift to the west of Bartley Glacier and has a probable minimum age of 4.2 Ma (Denton et al., 1982). The Peleus till is underlain in central Wright Valley by the Jason glacimarine diamicton which, on the basis of diatom biostratigraphy, is less than ~ 9-15 Ma. The Peleus till, in turn, is correlated with a widespread basal till and subglacial ripples resulting from an extensive ice sheet which overrode the Asgard Range, including Mt Fleming (Denton et al., 1982).

By establishing soil chronosequences on different parent materials and in different climatic regions, the interdependency of climate and parent material with time can be studied. Previous work (Behling, 1971; Everett, 1972; Campbell and Claridge, 1975; Bockheim, 1979; Pastor and Bockheim, 1980) has indicated that time exerts the greatest influence on soil development in the McMurdo Sound area. Therefore, it is possible to compare the effect of time on soil development at Mt Fleming and in Wright Valley even though the climate and parent materials may differ.

## RESULTS

### Surface Boulder Weathering and Soil Morphology

Although there was no trend in surface boulder frequency on the moraine crests with distance from the Wright Upper Glacier, condition of the boulders did differ. Whereas the ice-cored moraine had 33% sandstone boulders, there were no sandstone boulders on the moraines more distant from the glacier (Table 1). Unlike boulders on the other moraines, 5% of the boulders on the ice-cored moraine were striated. The percentages of boulders which were varnished, pitted, ventifacted and fragmented in situ increased with distance from the Wright Upper Glacier.

Morphological features of the soil likewise bore a relationship with distance from the Wright Upper Glacier. Thickness of solum (i.e. depth of staining) and the depths to which dolerite ghosts and visible salts occurred increased with distance from the glacier (Table 2). Whereas thin (< 1 mm) salt encrustations were visible on surface clasts in the soil on the ice-cored moraine, distinct salt flecks 0.5 mm in diameter were observed to a depth of 15 cm in soils on the intermediate drift. Soils derived from the outer lateral moraine and the ground moraine beyond had weakly cemented salt pans which were 13-14 cm thick. The depth to ice-cemented frost table increased from 7 cm on the ice-cored moraine to 68 cm on the outer lateral moraine. The ice-cemented frost table occurred at 58 cm in the ground moraine.

**TABLE 1: Surface boulder weathering features on moraines adjacent to Wright Upper Glacier, Mt. Fleming.**

| Glaciation | Plot Number | Surface boulder frequency (per 314 m²) | % Sandstone | % Fragmented *in situ* | % Ventifacted | % Pitted | % Striated | % Varnished |
|---|---|---|---|---|---|---|---|---|
| Alpine I (ice-cored, youngest) | 80-24 | 785 | 33 | 1 | 0 | 6 | 5 | 91 |
| Alpine II (intermediate-aged) | 80-23 | 680 | 0 | 2 | 1 | 59 | 0 | 99 |
| Alpine III (oldest) | 80-22 | 850 | 0 | 8 | 3 | 72 | 0 | 100 |

**TABLE 2: Morphology of soils on lateral moraines and ground moraine adjacent to Wright Upper Glacier, Mt. Fleming**

| Glaciation | Profile | Thickness of solum (cm) | Depth of dolerite ghosts (cm) | Depth of visible salts (cm) | Morphologic salt stage† | Depth to ice-cemented frost table (cm) |
|---|---|---|---|---|---|---|
| Alpine I | 80-24 | 0 | 0 | 0 | I | 7 |
| Alpine II | 80-23 | 7 | 7 | 15 | II | 36 |
| Alpine III | 80-22 | 18 | 18 | 18 | IV | 68 |
| Peleus-equivalent | 80-25 | 23 | 23 | 18 | IV | 58 |

† I = salt encrustations beneath clasts; II = salt flecks 0.5 mm diameter; IV = weakly cemented salt pan (Bockheim, 1979).

### Chemical and Physical Properties of Soils

Total water soluble salts to a depth of 40 cm in the soils increased from the ice-cored drift to the ground moraine (Table 3). Salt contents were 16, 1008, 1113 and 1420 mg/cm² on soils from the ice-cored moraine, the intermediate moraine, the outer lateral moraine and the ground moraine beyond, respectively. Electrical conductivity of the horizon of maximum salt enrichment was 0.27, 6.0, 6.6 and 11.0 dS/m in these same soils. Dominant ions in each soil were sodium and sulphate.

Silt (5-2 $\mu$m) and clay (< 2 $\mu$m) contents of the soils increased from the ice-cored moraine to the ground moraine (Table 4). In the first horizon below the desert pavement, silt was 2.7, 3.2, 5.2 and 9.1% in soils on the ice-cored moraine, the intermediate lateral moraine, the outer lateral moraine and the ground moraine beyond, respectively. Maximum clay contents were 2.7, 2.0, 3.6 and 9.9% in the same soils. Coarse fragments ranged between 35 and 66% for all soils. Whereas soils on the lateral moraines were gravelly sands, soils on the ground moraine were gravelly loamy sands.

Based on X-ray diffraction analysis, predominant minerals in the clay-size fraction (< 2 $\mu$m) were mica and quartz, with smaller amounts of chlorite (Table 5). Small amounts of an amphibole mineral were detected in the lower horizon of the soils on the outer lateral moraine and the ground moraine beyond.

In the United States soil taxonomy (U.S. Soil Survey Staff, 1975), the "soil" on the ice-cored moraine would be "non-soil" and the remaining soils would be Pergelic Cryorthents, frigid, sandy-skeletal, micaceous. Using the proposed changes in the United States soil taxonomy by Bockheim (1980b), the "soil" on the ice-cored moraine would be "non-soil" and the remaining soil would be Aridic Pergelic Camborthids, ice-cemented permafrost.

## DISCUSSION

### Surface Boulder and Soil Weathering in Relation to Drift Age

Landscapes and soils appeared to increase in age with distance from the Wright Upper Glacier. Sandstone boulders were present on the ice-cored moraine but appeared to have been reduced to cobbles and smaller clasts on other moraines. Although striations were present on boulders on the ice-cored moraine, they did not occur on the other surfaces, perhaps because of exfoliation and wind abrasion. Pitting, ventifaction and fragmentation increased with distance from the glacier, due primarily to wind erosion and salt weathering.

Soil profiles became more deeply developed with increasing distance from the glacier. Oxidation of iron bearing minerals reported in the Beacon Supergroup (McKelvey and Webb, 1959), such as hypersthene, and release of iron cementing the Beacon sandstone (Matz et al., 1972) have resulted in reddening of the soil profiles to depths ranging from 7 cm in soils on the intermediate lateral moraine to 23 cm in the soils on ground moraine. Dolerite ghosts were not present in soils on the ice-cored drift but occurred at progressively deeper levels in the remaining soils. The ghosts contained an abundance of salts which play an important role in rock comminution.

Morphologic salt stage increased from I (salt encrustations beneath clasts) in the soil on the ice-cored moraine to II (salt flecks) in the soil on the intermediate lateral moraine to IV (weakly cemented salt pan) in the soils on the outer lateral moraine and ground moraine. The salts are primarily sodium and sulphate, as evidenced by soil-water extracts and X-ray diffraction analysis of isolated salts. These salts are believed to have originated from marine aerosols, since they occur in ratios ($NA^+/SO_4^{2-}$ = 0.34-0.87) more similar to that of sea water (1.4) than in igneous rocks (76). However, fractionation of salts following rock weathering cannot be ruled out as an alternative hypothesis. Electrical conductivities remained in excess of 4.5 dS/m at depths exceeding 40 cm, suggesting that some leaching of salts has occurred.

The accumulation of salts has led to particle comminution, as silt contents increase with distance from the glacier. However, these differences, as well as differences in clay contents, may be due to variations in parent materials.

**TABLE 3: Chemistry of 1:5 soil:water extracts of soils adjacent to the Wright Lower Glacier, Mt. Fleming**

| Horizon | Depth (cm) | pH | Electrical conductivity (dS/m) | $Na^+$ | $Ca^{2+}$ | $Mg^{2+}$ | $K^+$ | $Cl^-$ | $NO_3^-$ | $SO_4^{2-}$ | Total salts to 40 cm² (mg/cm²) |
|---|---|---|---|---|---|---|---|---|---|---|---|
| | | | | eq/L | | | | | | | |
| | | | | | **80-24 Alpine I** | | | | | | |
| Cln | 0 – 7 | 7.1 | 0.27 | 1.3 | 0.36 | 0.55 | 0.049 | 0.21 | 0.06 | 1.5 | |
| ice | 10 + | 6.8 | 0.045 | 0.16 | 0.07 | 0.084 | 0.022 | 0.09 | 0.004 | 0.20 | 16.2 |
| | | | | | **80-23 Alpine II** | | | | | | |
| B2 | 0 – 15 | 6.1 | 4.0 | 17 | 15 | 13 | 0.38 | 3.2 | 1.9 | 50 | |
| Clox | 15 – 36 | 6.8 | 6.0 | 54 | 24 | 8.7 | 0.36 | 3.0 | 0.97 | 77 | 1008 |
| | | | | | **80-22 Alpine III** | | | | | | |
| B21 | 0 – 4 | 6.6 | 4.2 | 18 | 28 | 13 | 0.38 | 2.2 | 2.4 | 47 | |
| B22sa | 4 – 18 | 6.7 | 5.0 | 21 | 31 | 18 | 0.46 | 5.2 | 3.3 | 53 | |
| Clox | 18 – 40 | 7.2 | 6.6 | 32 | 29 | 40 | 0.56 | 7.1 | 6.1 | 71 | |
| C2n | 40 – 68 | 7.0 | 4.5 | 18 | 31 | 19 | 0.33 | 6.0 | 2.6 | 51 | 1113 |
| | | | | | **80-25 Peleus-equivalent** | | | | | | |
| B21 | 0 – 5 | 6.3 | 3.8 | 14 | 29 | 7.2 | 0.38 | 2.6 | 2.2 | 38 | |
| B22sa | 5 – 18 | 6.8 | 11.0 | 71 | 24 | 65 | 1.4 | 7.9 | 12 | 91 | |
| Clox | 18 – 26 | 7.0 | 6.6 | 39 | 20 | 27 | 1.2 | 7.1 | 5.0 | 61 | |
| C2n | 26 – 58 | 7.8 | 5.8 | 32 | 26 | 22 | 1.0 | 7.5 | 3.7 | 61 | 1420 |

**TABLE 4: Particle-size distribution of soils adjacent to the Wright Upper Glacier, Mt. Fleming, Antarctica**

| Horizon | Depth (cm) | % Coarse material (>2 mm) | % Sand | % Silt | % Clay | Soil textural class |
|---|---|---|---|---|---|---|
| | | | | **80-24 Alpine I** | | |
| Cln | 0 – 7 | 47 | 94.6 | 2.7 | 2.7 | Gravelly sand |
| | | | | **80-23 Alpine IIa** | | |
| B2 | 0 – 15 | 35 | 94.8 | 3.2 | 2.0 | Gravelly sand |
| Clox | 15 – 36 | 51 | 93.9 | 4.2 | 1.9 | Gravelly sand |
| | | | | **80-22 Alpine III** | | |
| B21 | 0 – 4 | 39 | 92.9 | 5.2 | 1.9 | Gravelly coarse sand |
| C2n | 40 – 68 | 66 | 86.6 | 9.8 | 3.6 | Gravelly loamy sand |
| | | | | **80-18 Peleus-equivalent** | | |
| B21 | 0 – 6 | 46 | 87.5 | 9.1 | 3.4 | Gravelly loamy sand |
| B22sa | 6 – 12 | 55 | 83.2 | 7.4 | 9.4 | Gravelly loamy sand |
| Clox | 30 – 48 | 59 | 84.1 | 6.0 | 9.9 | Gravelly loamy sand |

## Soil Correlation and Relative-Age Dating

Although there are no absolute dates for glacial sediments in the Mt Fleming vicinity, relative-age dates were obtained by comparing the glacial sequence there to those elsewhere in the McMurdo Sound vicinity. Properties of soils at Mt Fleming were compared to alpine soils in Wright and Taylor Valleys.

The soil on the ice-cored moraine at Mt Fleming lacks horizonation and segregated salts (Figure 3a) and is, therefore, similar to soils on alpine I moraines in Taylor and Wright Valleys (Behling, 1971; Everett, 1972; Bockheim, 1978; Leide, 1980). The soil on the intermediate lateral moraine at Mt Fleming is stained to 7 cm and has salt flecks to 15 cm (Figure 3b). This soil is similar to soils on alpine II drift at Meserve Glacier (Behling, 1971; Everett, 1972) and elsewhere in Wright and Taylor Valleys (Bockheim, 1978; Leide, 1980). The soil on the outer lateral moraine at Mt Fleming has deep staining (18 cm), a 14 cm thick salt pan, and dolerite ghosts to 18 cm (Figure 3c). Soils on alpine III drift in Wright Valley show similar strong development (Behling, 1971; Everett, 1972; Bockheim, 1978; Leide, 1980). The soil on the ground moraine is very strongly developed (Figure 3d), and is similar to soils on Peleus till in central Wright Valley (Bockheim, 1978). Therefore, the advances of Wright Upper Glacier at Mt Fleming are here designated as alpine I, II and III to correspond with the alpine sequences in Wright and Taylor Valleys.

**TABLE 5: Distribution of minerals in the clay (<2μm) fraction of soils adjacent to the Wright Upper Glacier, Mt. Fleming, Antarctica†**

| Horizon | Depth (cm) | Chlorite | Mica | Amphibole | Quartz |
|---|---|---|---|---|---|
| | | **80-24 Alpine I** | | | |
| Cln | 0 – 7 | 1 | 3 | — | 3 |
| | | **80-23 Alpine IIa** | | | |
| B2 | 0 – 15 | 1 | 3 | — | 3 |
| Clox | 15 – 36 | 1 | 3 | — | 2 – 3 |
| | | **80-22 Alpine III** | | | |
| B21 | 0 – 4 | 1 | 3 | — | 3 |
| C2n | 40 – 68 | 1 | 2 | 1 | 2 – 3 |
| | | **80-18 Peleus-equivalent** | | | |
| B21 | 0 – 6 | 1 | 3 | — | 3 |
| B22sa | 6 – 12 | 1 | 2 – 3 | 1 | 3 |

†Numbers refer to relative abundance based on X-ray diffraction peak heights (001 reflections): 1 = 1 to 17%; 1 – 2 = 18 to 22%; 2 = 23 to 37%; 2 – 3 = 38 to 42%; 3 = 43 to 57%; 3 – 4 = 58 to 62%; 4 = 63 to 77%; 5 = >77%.

## Soils and Fluctuations of the McMurdo Ice Dome

Drewry (1980) described an ice dome behind the Transantarctic Mountains in the McMurdo Sound area which has not thickened substantially in the past 17-33 ka, substantiating the work of Denton et al. (1971) and Stuiver et al. (1981), who observed that the East Antarctic ice sheet in the McMurdo Sound vicinity is now at its maximum height since before the entire Wisconsin (Würm) glaciation as defined elsewhere in the world. If the McMurdo ice dome has thickened considerably since that time, drift should occur at elevations considerably greater than today. However, soils on alpine III drift, which is 2.00-3.33 Ma, occur within 40 m of the Wright Upper Glacier. Therefore, the McMurdo ice dome does not appear to have thickened substantially since the Pliocene.

## CONCLUSIONS

At Mt Fleming a chronosequence of soils was investigated on a set of three lateral moraines deposited by the Wright Upper Glacier and on an older outlying ground moraine deposited during an extensive glaciation which overrode the Asgard Range. The glacial sequence is, therefore, bounded by an ice-cored moraine adjacent to the present Wright Upper Glacier and a loamy sand till which is equivalent to the Peleus till in central Wright Valley. The lateral moraines are designated alpine I, II and III, because surface boulder weathering, soil morphology and soil analytical properties of the soils on these drifts are similar to alpine I, II and III moraines in Wright Valley. Surface boulder and soil weathering features show a progressive increase in development on the various drifts. The soil on the alpine I moraine is similar to soils on alpine I moraines in Wright and Taylor Valleys which are greater than 1.97 ka and less than 3.1 ka. Soils on alpine II moraines are similar to those on alpine II soils in Wright and Taylor Valleys which are undated but which, based on stratigraphy and relative-age dating, may be 100-140 ka. Soils on alpine III moraines, which are located only 40 m from the present Wright Upper Glacier, are similar to soils on alpine III moraines in Wright and Taylor valleys which are 2.00-3.33 Ma. Finally, soils on the ground moraine beyond the alpine moraines are similar to soils derived from Peleus till in central Wright Valley which is older than 4.2 Ma and younger than 9-15 Ma. These data suggest that the McMurdo ice dome, which

**Figure 3. Soils on moraines adjacent to Wright Upper Glacier, including (a) ice-cored, (b) intermediate (c) outer lateral, and (d) ground moraine.**

supplies the Wright Upper Glacier, has not thickened substantially since the Pliocene.

*Acknowledgements* I am grateful to Dr George H. Denton, University of Maine, for his assistance in all phases of this project. S.C. Wilson assisted in the field work and laboratory analyses were conducted by K. Zuelsdorff. J.E. Leide reviewed the manuscript.

## REFERENCES


AMERICAN PUBLIC HEALTH ASSOCIATION, AMERICAN WATER WORKS ASSOCIATION and WATER POLLUTION CONTROL FEDERATION, 1975: Standard methods for the examination of water and wastewater, 14th edit.

ARMSTRONG, R.L., 1978: K-Ar dating: Late Cenozoic McMurdo Volcanic Group and dry valley glacial history, Victoria Land, Antarctica. *N.Z. J. Geol. Geophys., 21,* 685-98.

BEHLING, R.E., 1971: Pedological development on moraines of the Meserve Glacier, Antarctica. Unpublished Ph.D. thesis, Ohio State Univ., Columbus, Ohio. 216 pp.

BEHLING, R.E. and CALKIN, P.E., 1970: Wright Valley soil studies. *U.S. Antarct. J., 5,* 102-3.

BOCKHEIM, J.G., 1978: Soil weathering sequences in Wright Valley. *U.S. Antarct. J., 13,* 36-9.

BOCKHEIM, J.G., 1979: Relative age and origin of soils in eastern Wright Valley, Antarctica. *Soil Sci., 128,* 142-52.

BOCKHEIM, J.G., 1980a: Solution and use of chronofunctions in studying soil development. *Geoderma, 24,* 71-85.

BOCKHEIM, J.G., 1980b: Properties and classification of some desert soils in coarse textured glacial drift in the Arctic and Antarctic. *Geoderma, 24,* 45-69.

CAMPBELL, I.B. and CLARIDGE, G.G.C., 1975: Morphology and age relationships of Antarctic soils; *in* Suggate, R.P. and Cresswell, M.M. (eds.) *Quaternary Studies.* Royal Soc. New Zealand, Wellington, 83-8.

DAY, P.R., 1965: Particle fractionation and particle sized analysis; *in* Black, C.A. et al. (eds.) *Methods of Soil Analysis,* Part 1, Agron. No. 9. Am. Soc. of Agron., Madison, Wis., 545-56.

DENTON, G.H., ARMSTRONG, R.L. and STUIVER, M., 1971: The Late Cenozoic glacial history of Antarctica; *in* Turekian, K.K. (ed.) *Late Cenozoic Glacial Ages.* Yale Univ. Press, New Haven, Conn., 267-306.

DENTON, G.H., PRENTICE, M.L., KELLOGG, D.E. and KELLOGG, T.B., 1982: Tertiary history of the Antarctic ice sheet: evidence from the dry valleys. *Nature,* in press.

DREWRY, D.J., 1980: Pleistocene bimodal response of Antarctic ice. *Nature, 287,* 214-6.

EVERETT, K.R., 1972: Soils of the Meserve Glacier area, Wright Valley, southern Victoria Land, Antarctica. *Soil Sci., 112,* 425-38.

JACKSON, M.L., 1956: Soil chemical analysis—advanced course. (Third printing, 1967). Published by the author, Dep. of Soil Sci., Univ. Wisconsin, Madison. 893 pp.

JENNY, H., 1941: *Factors of Soil Formation.* McGraw-Hill, New York.

LEIDE, J.E., 1980: Soils and relative-age dating of alpine moraines in Wright and Taylor Valleys, Antarctica. Unpublished M.S. thesis, University of Wisconsin-Madison. 59 pp.

MATZ, D.B., PINET, P.R. and HAYES, M.O., 1972: Stratigraphy and petrology of the Beacon Supergroup, southern Victoria Land; *in* Adie, R.J. (ed.) *Antarctic Geology and Geophysics.* Universitetsforlaget, Oslo, 353-8.

McKELVEY, B.C. and WEBB, P.N., 1959: Geological investigations in southern Victoria Land, Antarctica; part II—Geology of Upper Taylor Glacier region. *N.Z. J. Geol. Geophys., 2,* 718-28.

PASTOR, J. and BOCKHEIM, J.G., 1980: Soil development on moraines of Taylor Glacier, lower Taylor Valley, Antarctica. *Soil Sci. Soc. Am., J., 44,* 341-8.

STUIVER, M., DENTON, G.H., KELLOGG, T.B. and KELLOGG, D.E., 1978: Glacial geologic studies in the McMurdo Sound region. *U.S. Antarct. J., 13,* 44-5.

STUIVER, M., DENTON, G.H., HUGHES, T.J. and FASTOOK, J.L., 1981: Late Würm and Holocene history of the marine ice sheet in Western Antarctica: a working hypothesis; *in* Denton, G.H. and Hughes, T.J. (eds.) *The Last Great Ice Sheets.* Wiley-Interscience, New York, 319-436.

THOMPSON, D.C., CRAIG, R.M.F. and BROMLEY, A.M., 1971: Climate and surface heat balance in an Antarctic dry valley. *N.Z. J. Sci., 14,* 245-51.

UGOLINI, F.C. and BULL, C., 1965: Soil development and glacial events in Antarctica. *Quaternaria, 7,* 251-69.

UNITED STATES SOIL SURVEY STAFF, 1975: Soil taxonomy: a basic system of soil classification for making and interpreting soil surveys. U.S.A. Dep. Agric., Agric. Handbook No. 436. 754 pp.

# A RECONSTRUCTION OF THE QUATERNARY ICE COVER ON MARION ISLAND

K. Hall, *Geography Department, University of Natal, P.O. Box 375, Pietermaritzburg 3200, South Africa.*

*Abstract* Evidence in favour of extensive glaciation on sub-Antarctic Marion Island is presented. Available data indicate three phases of glaciation, each comprising a sequence of stades and interstades. The island stratigraphy is shown and the major glacial deposits briefly described. From the available evidence, particularly the moraines, a number of former glaciers are reconstructed. Temperature fell between 3° and 6.4°C at the glacial maximum, with an equilibrium line altitude depression of 650 m and an accumulation area ratio of 0.6. The interglacials are marked by tectonism and volcanism as a response to isostatic readjustment. Temperatures were similar to present with extensive pedogenesis and plant colonisation. High glacier velocities are indicated to compensate for large ablation rates. Glaciation was initiated by precipitation falling as snow due to the northward shift of the Antarctic Convergence.

Marion Island (46°54′S, 37°45′E) is a 290 km² volcanic complex, rising to a height of 1230 m, situated approximately 2° of latitude north of the Antarctic Convergence (Figure 1). Whilst the present day cover of permanent snow and ice is restricted to a very small area (< 3 km²) above 900 m, there is evidence of several extensive glaciations (Figure 1) in the recent past. Geological, geomorphological and palynological information indicate three periods of glaciation, each comprising a series of stades and interstades. Intervening interglacials were characterised by extensive tectonism, volcanism, pedogenesis and plant colonisation. Whilst a number of specific topics have been dealt with in earlier publications (Hall, 1978a, 1978b, 1979) the aim here is to present some new data and give an overview of the past ca 300,000 years.

## STRATIGRAPHY AND LANDFORMS

At present the island surface exhibits numerous scoria cones, lava flows and spreads of pyroclasts (Verwoerd, 1971). If ice were to advance over the present day surface then initial deposits would certainly be characterised by a high percentage of pyroclasts mixed with fragments of lava bedrock. Observation of the extensive sections along wave-cut cliffs (Figure 2) shows diamicts 1 to 6 m thick, comprising a mixture of reddish pyroclasts and clasts derived from the underlying grey basalts. Going upsection through these diamicts, the percentage of pyroclasts decreases exponentially as that of the basalt clasts increases (Figure 3). Detailed investigation shows some striated basaltic clasts, and a clast a-axis preferred orientation parallel to striations on underlying surfaces. This, taken together with the upward transition into certain glacial sediments, indicates that the diamicts are true tills. As this "pyroclastic till" usually occurs above a lava or palaeosol (Figure 2) it is interpreted as resulting from the first stadial advance of each glacial.

Pyroclast content does not increase in the till sequence beyond the initial pyroclastic till and this is thought to indicate that no further volcanic activity took place at this time. Further pyroclastic till is only found after a period of extensive interglacial volcanism initiated by tectonism caused by isostatic rebound (Hall, 1982). Interglacial age is also indicated by the presence of ca 2 m thick palaeosols and the development of peat. Palynological evidence (Scott and Hall, in press) indicates a climate similar to the present and that sea level fell towards the end of the peat development; suggestive of oceanic depletion as

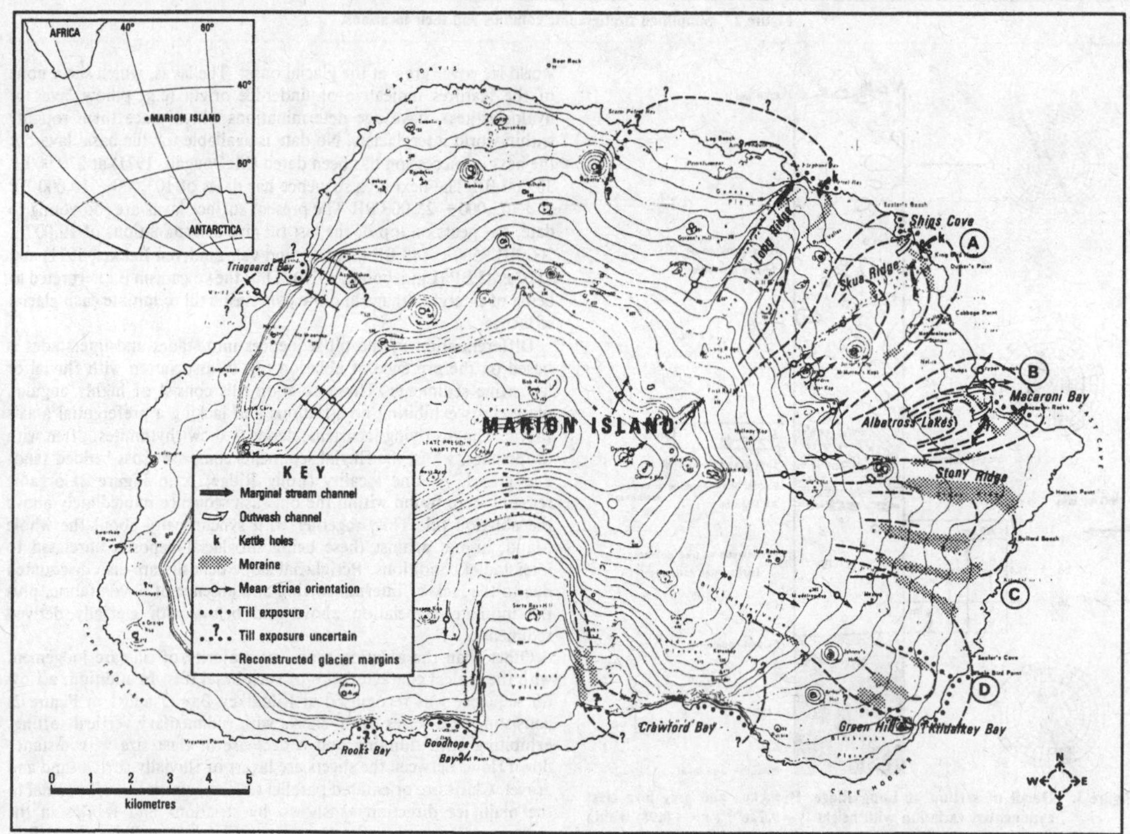


**Figure 1. A reconstruction of some former glacier margins together with a summary of glacio-morphological information.**

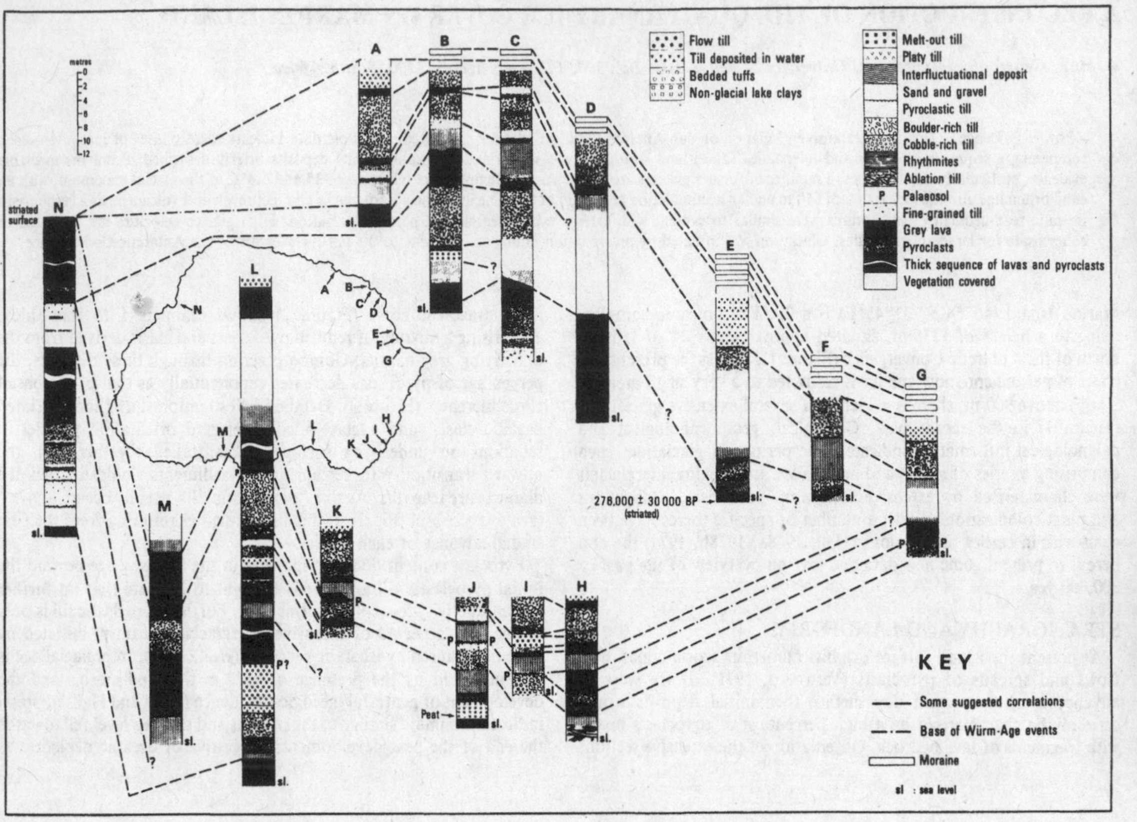

Figure 2. Simplified stratigraphic columns and their locations.

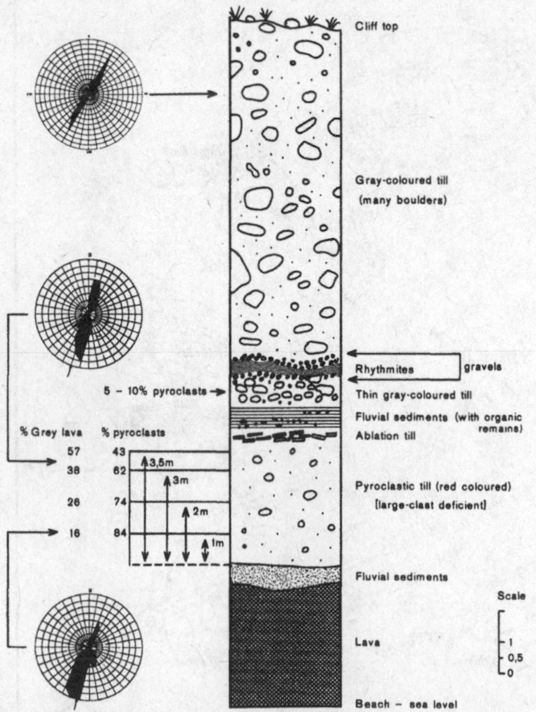


**Figure 3. Detail of section at Long Ridge. Pyroclast and grey lava clast exponential variation with height ($y = 9.22c^{0.5x}$; $r = +0.98$) within the pyroclastic till is shown together with similarity of fabric orientation with the lodgement till above.**

world ice cover grew at the glacial onset. The lavas, which show none of the features indicative of under-ice origin (e.g. pillow lavas or hyaloclastites), have age determinations which place them roughly within world interglacials. No date is available for the basal lava but the next in succession has been dated (McDougall, 1971) at 276,000 ± 30,000 BP. The next lava sequence has dates of 103,000 ± 10,000 BP and 105,000 ± 25,000 BP. The present surface lavas are too young to date, but peats on top of the last till give determinations of 10,000 ± 700 BP, 9500 ± 140 BP (Schalke and van Zinderen Bakker, 1971) and 7120 ± 45 BP (Lindeboom, 1979). Thus the volcanism is interpreted as being of interglacial age and the pyroclastic till to initiate each glacial sequence.

Differentiation of the till sequences into stades and interstades is based on the presence of ablation tills in association with fluvial or lacustrine sediments. These ablation tills consist of highly angular, platy clasts exhibiting no striations and lacking a preferential a-axis fabric. The overlying lacustrine deposits show rhythmites, often with dropstones, whilst the fluvial sediments comprise cross-bedded sands and gravel. At one locality (Long Ridge; A in Figure 2) organic remains were found within the outwash sequence immediately above the ablation till. This, together with synchroneity about the whole island, argues against these being subglacial deposits unrelated to interstadial conditions. Periglacial slope deposits are also discounted due to the lack of internal sorting and absence of a-axis fabric, plus the intimate association above and below with glacially derived sediments.

Other than the ablation type, the majority of tills are lodgement with the typical characteristics of such basal tills. In addition, a flow till sequence was recognised at Kildalkey Bay (J and I in Figure 2) consisting of a series of till sheets with rudimentary vertical sorting, exhibiting both thinning and a decrease in clast size with distance down flow. Between the sheets are layers of fluvially-sorted sand and gravel. Clasts are orientated parallel to flow, which is itself normal to the main ice direction as shown by striations and fabrics in the underlying lodgement till. A melt-out till occurs at Ships Cove (B in Figure 2). This consisting of short "wedges" of subangular debris with poorly developed fabrics and signs of water sorting and imbrication. Clast size increased with height and the wedges are capped with an ablation till and fluvial sediments. Finally, at Goodhope Bay (M in Figure 2) there is a 12 m sequence of alternating beds of water-sorted material, up to gravel size; and beds of non-sorted angular material, of gravel to small-cobble size, with several distinctive beds (0.3 m thick) of silt and clay. This is interpreted as an "interfluctuational" deposit (Miller, 1975) which is a depositional unit resulting from fluvioglacial and glacial processes operating in a subglacial environment near the glacier snout (Kirby, 1969).

Difficulties in differentiating glacially reworked volcanoclastics from in situ volcanic deposits has led to some interpretation problems on Marion Island (Verwoerd, 1971; Gribnitz, 1981). However, there is adequate corroborative information to substantiate a glacial origin for some of the Marion sediments.

Lateral and frontal moraines, of both "push" and "dump" origin (Andrews, 1975), were observed from the present coastline up to an altitude of 250 m. All recorded moraines relate to the most recent glacial event (Würm-Weichselian of N. Hemisphere) but many are affected by postglacial faulting and lava flows which makes estimation of their size impossible. Frontal moraines are conspicuously absent in the Stony Ridge to Kildalkey Bay area (Figure 1) but this is in full accord with the argument that glaciers supported by high precipitation (as here—see below) have large transport values, but little accumulation of frontal moraines, due to continuous melt-water discharge. The survival of end moraine remnants at other localities can be explained by the presence of outflow channels cutting through the moraines (Figure 1). Pollen cores obtained between the moraines suggest interstadial conditions ca 17,000 BP and an end to the glacial ca 12,000 BP (Schalke and van Zinderen Bakker, 1971).

## RECONSTRUCTIONS AND COMPARISONS

Upon the basis of the foregoing information it has been possible to generate an outline of major events during the past ca 300,000 years (Figure 4) and offer a number of palaeocondition reconstructions. The moraines (Figure 1), particularly the frontal, mark successive glacial termini during retreat from their maximum and so allow reconstruction of former glaciers. The sequence of lateral moraines from Stony Ridge to Green Hill (Figure 1) all begin at ca 250 m and thus this altitude can be equated to the approximate position of the former equilibrium line altitude (ELA). Østrem (*in* Andrews, 1975) argues that the climatic snowline would have been situated somewhere between the ELA and 300 m higher (i.e. from 250 to 550 m in this instance). On this basis it was possible to reconstruct temperatures at present day sea level, for glacial maximum conditions, utilising a

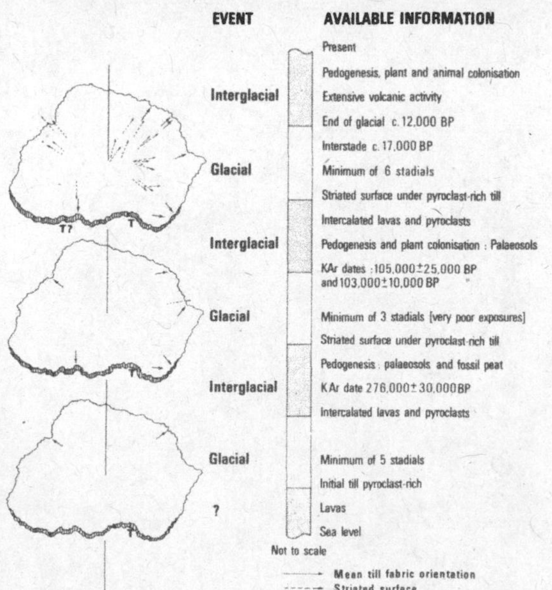


**Figure 4. Synopsis of the history of Marion Island.**

range of mean annual temperatures for the maximum and minimum snowline altitudes. This indicated (Hall, 1979, Table I) that there was a mean annual temperature decrease of 3° to 6.4°C and that temperatures in the former glacier ablation zones were positive for most (66%), if not the whole year. Thus, glaciation of Marion Island is assumed to result from precipitation falling as snow rather than, as at present, rain, with the high annual receipt allowing glacier maintenance despite large ablation rates. Mass transfer from accumulation to ablation zone would have necessitated high ice velocities and the continual ablation would have maintained extensive fluvioglacial outwash systems.

The suggested mean annual decrease in temperature is in good agreement with the 3 to 4°C depression postulated by van Zinderen Bakker (1973) from pollen spectra. It is also within the ranges suggested from various ocean core studies, viz. 2.5 to 3.5°C (Hays, et al., 1976), 3 to 4°C (Prell, et al., 1980), and from other islands, 5°C for Tasmania (Denton and Hughes, 1981) and 3 to 4°C for Macquarie Island (Colhoun and Goede, 1974). The ELA estimation of ca 250 m indicates a fall of 650 m which is similar to that ( 700 m) recently found for Kerguelen by the writer and within the range of successive depressions cited by Porter (1975) for New Zealand.

With the ELA estimation and the presence of lateral and terminal moraines it is possible to delimit and measure the ablation area of several former glaciers from the last glacial maximum. Four reasonably well-defined glaciers (A to D in Figure 1), which equate to roughly 20% of the estimated ice cover, have a combined ablation area of 22.79 km². Steady-state glaciers in equilibrium have accumulation area ratios (AAR) in the order of 0.6 to 0.7 (Andrews, 1975). Assigning a value of O.4 to the known ablation zones a combined accumulation area of 34.19 km² is obtained and thus a total glacier area of 56.98 km². Extrapolation to full island cover indicates a total island ice area in the region of 285 km². The present day island is 290 km² but during glacial times would have been marginally larger due to lower sea levels. However, it is known from moraines and stratigraphy that some glaciers did not extend beyond the present coastline, and others did so only marginally, so an AAR higher than 0.6 would generate too extensive an ice cover.

Such an AAR indicates that the Marion glaciation was a product of high annual snow accumulation rates in conjunction with large ablation rates. The transfer of mass required to maintain the glaciers under these conditions would have necessitated high ice velocities. The steep clast dips ($\bar{x}=23°$) obtained in till fabric analysis, are indicative of just such a flow regime (Andrews and Smith, 1970). The suggested AAR of 0.6 agrees with that found for New Zealand (0.6 ± 0.05) by Porter (1975).

## CONCLUSION

This brief synopsis of events of the past 300,000 years on Marion Island is summarised in Figures 1, 2 and 4. The onset of glaciation is a result of a slight northward shift of the Antarctic Convergence. The consequent decrease in temperatures, caused precipitation to fall as snow. The high annual accumulation offset large ablation rates and so the glaciers had relatively high velocities to compensate. A strong positive relationship exists between precipitation, glacial erosion, glacier transport and proglacial runoff such that the Marion glaciers carried a great deal of debris to their margins and there was extensive fluvioglacial activity. Thus, sequences of tills, fluvioglacial deposits and moraines were built-up.

With an interglacial southward retreat of the Antarctic Convergence temperature rose causing precipitation to revert to rain, raising the ELA above the island summit, and causing rapid deglaciation. Accelerated ice-loss initiated tectonism which, due to location in an active area, caused volcanism. During the warm interglacials there was pedogenesis and plant growth. With reversion to glacial conditions sea level dropped, precipitation fell as snow, the ELA was lowered and glaciers advanced over the interglacial volcanic debris and a further sequence of stadial and interstadial sediments accumulated. The final retreat of ice cover left a sequence of moraines, remnants of which are found on the present island surface. Currently, interglacial tectonism, volcanism, pedogenesis and plant colonisation operate once more.

*Acknowledgements* Sincere thanks are due to Professor van Zinderen Bakker who initiated this work and Professor Visser who spent much time both in the field and discussing the many problems. Logistics were supplied by the

Department of Transport and initial financial support for fieldwork was given by the C.S.I.R. Grateful acknowledgement is made to the University of Natal which, in South Africa, continues to actively support sub-Antarctic glacial geology research. Thanks are due to the University of Natal and the Symposium Committee for financial help towards the costs of attending the Symposium. Mr Bruno Martin redrew all the figures.

## REFERENCES

ANDREWS, J.T., 1975: *Glacial Systems: an Approach to Glaciers and Their Environments.* Duxbury Press, Duxbury, Mass.

ANDREWS, J.T. and SMITH, D.I., 1970: Till fabric analysis: methodology and local and regional variability (with particular reference to the North Yorkshire till cliffs). *Geol. Soc. Lond., Q. J., 125*, 503-42.

COLHOUN, E. and GOEDE, A., 1974: A reconnaissance survey of the glaciation of Macquarie Island. *R. Soc. Tas., Pap. Proc., 108*, 1-14.

DENTON, G.H. and HUGHES, T.J. (eds.) 1981: *The Last Great Ice Sheets.* John Wiley and Sons, New York.

GRIBNITZ, D.J., 1981: Preliminary report on a visit to Marion Island. *Dept. Transport, Pretoria, Unpubl. Rep.*, 7pp.

HALL, K.J., 1978a: Evidence for Quaternary glaciations of Marion Island and some implications; *in* Van Zinderen Bakker, E.M. (ed.), *Antarctic Glacial History and World Palaeoenvironments*, A.A. Balkema, Rotterdam, 137-47.

HALL, K.J., 1978b: Quaternary glacial geology of Marion Island. Unpublished Ph.D. thesis, University of Orange Free State, South Africa, 369pp.

HALL, K.J., 1979: Late glacial ice cover and palaeotemperatures on sub-Antarctic Marion Island. *Palaeogeogr. Palaeoclimatol., Palaeoecol, 29*, 243-59.

HALL, K.J., 1982: Rapid deglaciation as an initiator of volcanic activity: an hypothesis. *Earth Surf. Process. Landform., 7*, 45-51.

HAYS, J.D., LOZANO, J.A., SHACKLETON, N.J. and IRVING, G., 1976: Reconstructions of the Atlantic and Western Indian Ocean sectors of the 18,000 BP Antarctic Ocean. *Geol. Soc. Am., Mem., 145*, 337-72.

KIRBY, R.P., 1969: Till fabric analysis from the Lothians, central Scotland. *Geogr. Ann., Ser. A., 51*, 48-60.

LINDEBOOM, H., 1979: Chemical and microbiological aspects of the nitrogen cycle on Marion Island (sub-Antarctic). Unpublished Ph.D. thesis, University of Groningen, Holland, 138pp.

McDOUGALL, I., 1971: Geochronology; *in* Van Zinderen Bakker, E.M., Winterbottom, J.M. and Dyer, R.A. (eds.) *Marion and Prince Edward Islands.* A.A. Balkema, Cape Town, 72-7.

MILLER, M.M., 1975: The recognition of separate stages of mountain glaciation. *Foundation for Glacier and Environmental Studies, Geology Course Notes 413 and 814.* University of Michigan.

PORTER, S.C., 1975: Equilibrium-line altitudes of Late Quaternary glaciers in the Southern Alps, New Zealand. *Quat. Res., 5*, 27-47.

PRELL, W.L., HUTSON, W.H., WILLIAMS, D.F., BE, A.W.H., GEITZENAVER, K. and MOLFININO, B., 1980: Surface circulation of the Indian Ocean during the last glacial maximum, approximately 18,000 BP. *Quat. Res., 14*, 309-36.

SCHALKE, H.J.W.G. and VAN ZINDEREN BAKKER, E.M., 1971: History of the vegetation; *in* van Zinderen Bakker, E.M., Winterbottom, J.M. and Dyer, R.A. (eds.) *Marion and Prince Edward Islands.* A.A. Balkema, Cape Town, 89-99.

SCOTT, L. and HALL, K.J. (in press): Palynological evidence for preglacial vegetation cover on Marion Island, sub-Antarctic. *Palaeogeogr. Palaeoclimatol. Palaeoecol.*

VAN ZINDEREN BAKKER, E.M., 1973: The glaciation(s) of Marion Island (sub-Antarctic); *in* van Zinderen Bakker, E.M. (ed.) *Palaeoecology of Africa and the surrounding islands and Antarctica, 8.* A.A. Balkema, Cape Town, 161-78.

VERWOERD, W.J., 1971: Geology; *in* van Zinderen Bakker, E.M., Winterbottom, J.M. and Dyer, R.A. (eds.) *Marion and Prince Edward Islands.* A.A. Balkema, Cape Town, 40-62.

# LATE QUATERNARY ICE MOVEMENT ACROSS THE VESTFOLD HILLS, EAST ANTARCTICA

D. Adamson, *School of Biological Sciences, Macquarie University, NSW, 2113, Australia.*

J. Pickard[1], *Antarctic Division, Kingston, Tasmania, 7150, Australia.*

[1]*Present Address: School of Biological Sciences, Macquarie University, N.S.W., 2113, Australia.*

*Abstract* Glacial striations on the Vestfold Hills (68°30′S 78°00′E) record at least three phases of ice movement. During the last major advance of the ice sheet (Vestfold Glaciation) the hills were completely covered but Late Pleistocene fossiliferous sediments survived on Marine Plain. During retreat of the ice sheet a localised lateral advance of the Søorsdal Glacier (Chelnok Glaciation) occurred within the last 2000 years, leaving northerly striations superimposed on the regional west northwest set. Other sets of striae follow minor valleys, particularly in the vicinity of Lichen Lake. These three sets are referred to as the Vestfold, Chelnok and minor valley striae respectively.

Retreat of ice and emergence of the land above sea level is indicated by widespread geomorphological features and by deposits of Holocene marine and non-marine fossils which have been extensively radiocarbon dated. For several thousand years the northern margin of the Sørsdal Glacier has been relatively stable with minor oscillations. The continental ice sheet has retreated in pulses but with an average rate of 1-2 km per 1000 years. Geomorphological and sequential airphoto evidence suggest that the northern margin of the Sørsdal Glacier is grounded and presently retreating and that the edge of the ice sheet is still in active retreat.

The regional warming caused by the large area of exposed rock, combined with sluggish ice flow towards the Vestfold Hills, accounts for the continued expansion of the ice free area, quite apart from any considerations of global warming.

## MOST RECENT PLEISTOCENE GLACIATION

During the last glacial maximum in the terminal Pleistocene from 10,000 to 25,000 years ago, the whole 400 km$^2$ of the Vestfold Hills was covered by ice. Numerous earlier glaciations scoured the region to produce low hills rising only about 160 m above sea level. The direction of ice flow during the last complete glaciation is shown by measurement of some 1600 striae at over 400 sites distributed over the region (Figure 1). At 53 sites, each with more than 10 readings, the mean orientation was 290° with an angular deviation of ± 12°. In this regional glaciation, which we name the Vestfold Glaciation, the ice moves west northwest at right angles to the contours of the present day ice sheet as noted by James and Oliver (1978) but not in the directions claimed by Blandford (1975) using indirect criteria.

Valley orientations were determined by bedrock characteristics not by the orientation of ice flow, as shown by striations of Vestfold direction that occur on ridge tops and valley floors. Vestfold striae also maintain their orientation across the flanks of Ellis, Long and Tryne Fjords, suggesting that these so called fjords were plucked by ice flowing across their long axes rather than along them. They are not true fjords but are merely underwater extensions of the above water valley system. Despite minor deviations of striae around isolated bluffs (as between Lake Stinear and Deep Lake) and a slight rotation to the north in the southwest corner of the Hills, the direction of Vestfold striae is remarkably uniform. Many other glacial features confirm the general orientation and complete coverage of the terminal Pleistocene Vestfold Glaciation. By contrast, striae parallel to the long axis of the Sørsdal Glacier occur only immediately adjacent to its northern margin. These show that it has been closely confined to its trench which is at least 750 m deep (Williams 1981).

## LATE PLEISTOCENE DEPOSITS

The numerous Pleistocene glacial advances must have repeatedly stripped from the Hills sediment that accumulated in the interglacial periods. The former extents of such Pleistocene sediments on the Hills before the Vestfold Glaciation are unknown except for one area. Late Pleistocene sediments have survived on Marine Plain adjacent to Crooked Fjord (Figure 1). We sampled these in 1978/79 and 1979/81 (Pickard and Adamson, in press a). Marine Plain, rising to about 20 m above sea level and occupying broad valleys between bedrock hills, is composed of nearly horizontal deposits of marine diatomite and sandstones containing abundant mollusca. The fossiliferous deposits are exposed in natural escarpments up to 8 m in height. The marine sediments are capped by a till of boulders and cobbles which protects the powdery diatomite from deflation by strong winds and "willy willies" (minor whirlwinds) which occur on the Plain (Pickard, in press a, b).

The diatomaceous deposits of Marine Plain, throughout their exposure, yield a minimum age of 24,000 years (Table 1), the maximum age being indefinite. We consider them to predate 30,000 years and, therefore, to have survived over-riding by the Vestfold Glaciation. They were laid down when at least the western part of the Hills was free of ice but beneath sea level. The upper sediments contain more sand and sandstone beds than the lower sediments, showing that the marine basin became open to the input of large amounts of sand. Present day agencies for the movement of abundant sorted sand in the Hills are primarily water flow and wind. We suggest that the surrounding bedrock areas were partly ice free during the underwater deposition of the sandy beds in the Marine Plain basin. We also conclude from the sequence of sediments that water depth was becoming shallower with time. That these deposits eventually emerged above sea level is shown by the presence of palaeosand wedges in the upper part of the sediments (found by Adamson in 1979; Pickard and Adamson, in press a), which are now being exhumed by deflation. Sand wedges are being formed today on the Plain and elsewhere on the western part of the Hills, some of them one hundred or more metres in length. We suspect that the palaeowedges are Late Pleistocene rather than Holocene in age.

The surface lag of boulders and cobbles on the surface of the fossiliferous sediments was deposited by the most recent glaciation. The permafrosted marine sediments of the Plain have been eroded in the past by meltwater to produce drainage channels which dissect the surface. Present day lakes are sapping the Plain to produce steep escarpments up to 8 m in height.

The Marine Plain deposits show that at periods of glacial retreat during the last 120,000 years ice may also have been withdrawn from at least part of the Hills. Around 120,000 years ago when the sea level and world climate approximated that of the present, the Hills were probably exposed to about their present extent. Such global glacial cycles extended throughout the Pleistocene during which the Hills were repeatedly overridden and at least partly exposed. The present extent of exposure of the region has probably occurred repeatedly although only fleetingly on each occasion. This Quaternary history distinguishes the Vestfold Hills and the Bunger Hills from some of the Victoria Land oases which were not overridden by continental ice during this period (Stuiver et al., 1981).

## HOLOCENE RETREAT

All other fossiliferous deposits on the Hills, which have been dated (Table 1), mark stages in ice retreat during the global warming of the Holocene. The chronology of this retreat, derived from the radiocarbon dates in Table 1 is summarised by grouping dates into broad time zones. Other relics of ice retreat include ubiquitous erratics, linear moraines, thick strews of till, palaeodrainage channels, striae from minor valley glaciation and striae from a lateral surge of the Sørsdal Glacier. Continuing ice retreat is shown by recent collapse of part of the Sørsdal Glacier, by the recessive profile of its northern margin, by recent emergence of rock from ice or snow cover and by major gullies eroded by melt-streams in the ice sheet. Locations of many of these features are shown on Figure 1. Continuing retreat can be explained in part, by local factors, as well as global climatic factors.

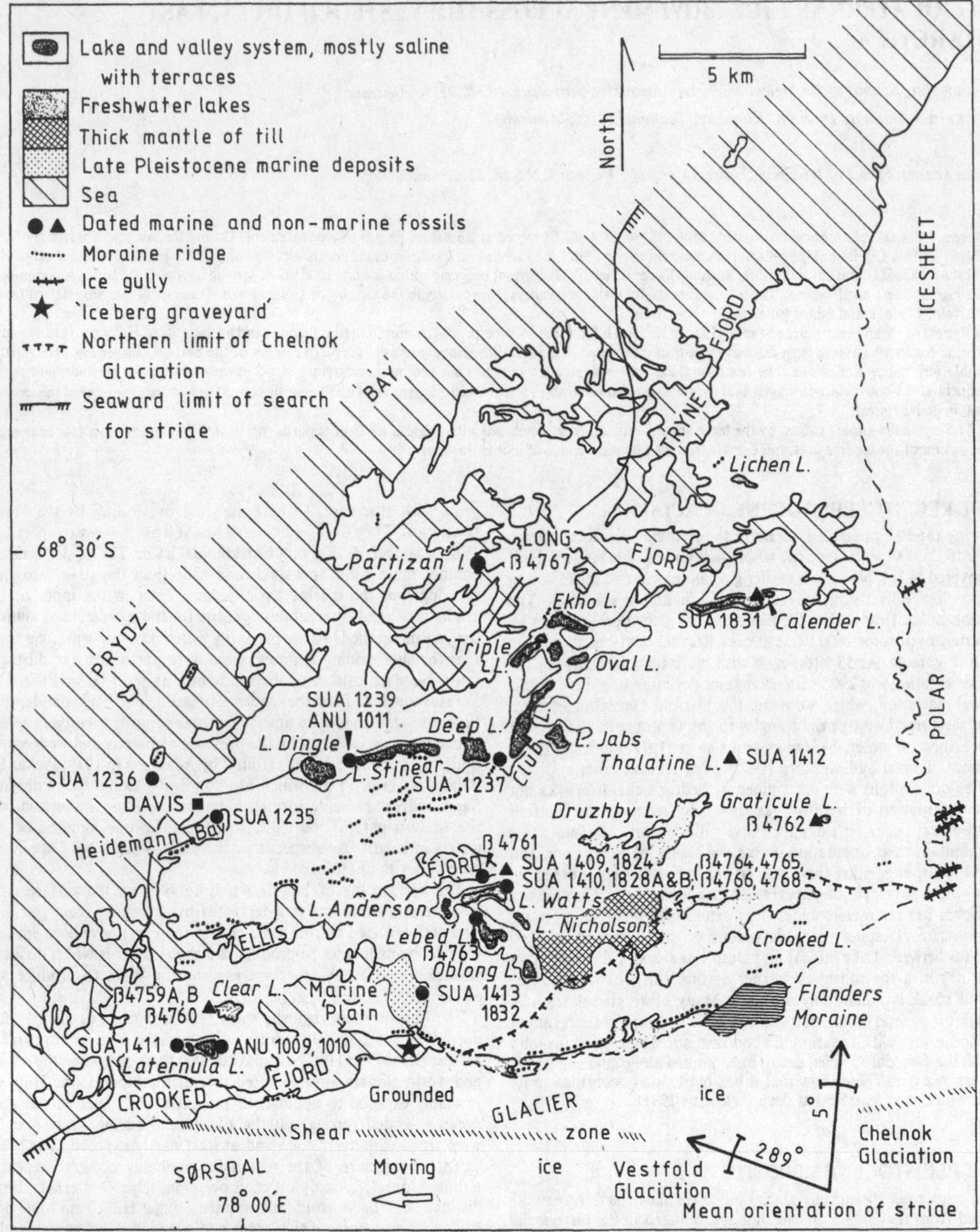


**Figure 1.** **Vestfold Hills showing Late Pleistocene and Holocene features used to reconstruct ice movement. Many examples of thick mantles of till, moraines, lakes, raised beaches and fossil deposits are omitted for clarity.**

From example, there is strong local warming of the atmosphere above the Hills due to the low albedo of the large area of dark coloured rock. Positive feedback exists between the area of bare rock and ice retreat. Retreat will continue so long as the rate of forward movement of the ice sheet is slow. Bedrock topography inland beneath the ice sheet (William 1981) is consistent with sluggish flow of ice westwards towards the Hills.

## Moraines and Till Mantles

The ice retreated in a south easterly direction across the region, on an irregular front roughly parallel to the alignment of linear moraines. Such moraines are found on the southern headland of Heidemann Bay, south of Lake Stinear, on the northern side of Ellis Fjord and elsewhere (Figures 1 and 2). Mantles of till, which are thicker than the general scatter of erratics also occur in well defined areas. The moraines and till mantles mark pauses in ice retreat or minor and temporary readvances. A particularly thick deposit east of the Watts Lake System may be an earlier equivalent of the present day contact zone between the Sørsdal Glacier and the ice sheet in the far southeast corner of the Vestfold Hills. Here, thick deposits of englacial debris are currently accumulating.

## Marine and Non-Marine Deposits Around Lakes

Stages in ice retreat are also recorded by abundant fossil deposits on terraces around lakes which have long attracted attention. Korotkevich (1971) and Johnstone et al. (1973) inferred Holocene ages for such deposits. Dr P. Arriens obtained three radiocarbon dates in 1975 but did not publish them because of uncertainties over reservoir corrections. In 1978/79 the authors began an extensive programme of collecting and radiocarbon dating. This was continued by Pickard in 1980/81. Dates for the Vestfold Hills were first reported at CLIMANZ 1981 (Pickard and Adamson, in press a, b), including dates on modern organisms (Table 1) which suggest reservoir corrections for marine organisms similar to those reported by Stuiver et al. (1981). In 1981 Pickard introduced Zhang Qingsong to a number of sites that had been sampled or dated by us. Parallel sampling of these sites by Zhang yields dates similar to our own. He reports dates for some additional sites Zhang et al. (this volume).

Radiocarbon dating of the fauna shows that three of the lake systems with terraces developed asynchronously (Table 1; Pickard and Adamson, in press a, b). The Watts-Anderson-Lebed-Oblong Lake System (Watts Lake System) appears to be the oldest (Figure 2a). At Lake Watts, marine sediments (SUA 1410) dated at 8260± 110 yr Before Present (BP) are overlain unconformably by thick mats of non-marine blue-green algae (cyanobacteria) dated at 2800± 80 yr BP (SUA 1409). The surface of the cyanobacterial deposit is formed into polygonal units, each about 30 cm in diameter and 10 cm thick and made up of a stack of numerous papery layers. Each stack is an unlithified stromatolite. The level of the prominent terrace around Lake Watts is also marked on cliff faces by lake-full water-level stains which are at least 2000 yr old.

The complex terrace-bordered lake system called the "Valley of Death" (Korotkevich, 1958) includes Lakes Dingle, Stinear, Deep, Club, Jabs, Oval, Triple and Ekho. We refer to the whole group as the Death Valley Lake System. We believe that sea water entered from Long Fjord over a sill at the north eastern end of the System. The ice sheet filled the upper half of the Long Fjord long after the marine entrance to the Watts System was freed of ice. The youngest lake system surrounded by a prominent terrace is the small Laternula Lake System.

At Clear Lake, less than 1 km from Laternula Lake, freshwater mosses are dated 8355± 250 yr BP (Beta 4760). This indicates that the ice and Sørsdal Glacier had retreated from Clear Lake 8000 years ago. However, the Sørsdal Glacier covered Laternula Lake until more recently.

The landscape to the north of Lake Stinear was above sea level and free of ice when marine organisms still inhabited the Death Valley System. This is indicated by the presence of "in situ *Laternula*" shells in an aeolian sand dune draping the terrace of the lake (Pickard, in press).

Allowing reasonable reservoir corrections for radiocarbon dates, by about 7000 years the ice sheet had retreated sufficiently to allow sea water into the Watts Lake System. About 5000 years ago sea water existed through the Death Valley System and at least some of the adjacent land was free of ice and above sea level. Isostatic uplift continued, with approximately stable world sea levels, until these systems were cut off from the sea. The Laternula System was flooded more recently and cut off at about 2000 years ago. A phase of non-marine fossil deposition about 2000 years ago occurred at Lake Watts, possibly related to melt water input from the Crooked Lake drainage system and from Lake Nicholson.

## Chelnok Glaciation

Dramatic events of relatively recent age occurred on the northern margin of the Sørsdal Glacier. Striae oriented at 5° with an angular deviation of ± 10° are superimposed on Vestfold striae (Figure 1) and are therefore younger than the Vestfold striae. Measurements were taken at 87 sites at which both directions occur superimposed. The northerly set indicate an event we name the Chelnok Glaciation. We attribute the Chelnok Glaciation to a lateral northwards ice flow of the margin of the Sørsdal Glacier when rapid retreat of the polar ice sheet left the northern flank of the Glacier relatively unsupported. Flanders Moraine is an extensive thermokarst area of till which mantles stagnant ice. It may be a relic of the ice which flowed northward during the Chelnok Glaciation. Stratigraphy within the stagnant ice may resolve this question. Our studies on Flanders Moraine have concentrated on topography and lake migration (Pickard ms).

## Palaeochannels

Palaeodrainage channels in the southeast corner of the Hills also reflect retreat and lowering of the ice sheet and of the Sørsdal Glacier. As the ice lowered and retreated, streams carrying large volumes of melt water from the ice sheet and the glacier were deprived of their catchment which became diverted into new channels. Palaeodrainage and present day drainage, which we have mapped in the eastern margin and south eastern corner of the Hills, clearly show stream channels and lakes deprived of their former water supply.

## Ice Gullies

Large gullies within the ice sheet, which are formed by melt streams flowing from the blue ice ablation zone, are also indicators of ice retreat (Figure 1). The gullies are up to 1 km long, 200 m wide and over 30 m deep with very steep to vertical N-facing sides and more

**TABLE 1: Radiocarbon dates for the Vestfold Hills. All dates are uncorrected. The dates of modern marine mollusc shells and organic mud suggest reservoir corrections roughly similar to that found elsewhere on the Antarctic coast. The ANU dates were determined by Dr. H. Polach on samples collected in 1972 by Dr. G. W. Johnstone, and are quoted with kind permission.**

| Lab No. | Age yr B.P. | Material | Location |
|---|---|---|---|
| Modern | | | |
| SUA 1235 | 950 ± 110* | Modern *Laternula* shell | Seashore near Davis |
| SUA 1236 | 1310 ± 125 | Modern marine algal sediment (surface 5 cm of sea floor) | Ship's anchorage off Davis, 17 m water depth |
| Marine Plain | | | |
| SUA 1832 | >28,000 | Marine organic sediment | Top 1 m of sediment, Marine Plain |
| SUA 1413 | >24,000 | Marine organic sediment | Marine Plain, 8 m below top of sediment |
| Watts Lake System | | | |
| SUA 1410 | 8260 ± 110* | Marine algal sediment | Watts Lake, E end, slope below terrace |
| SUA 1828A | 7680 ± 120 | Marine worm tubes, $CaCO_3$ | Watts Lake, E end, slope below terrace |
| SUA 1828B | 7380 ± 250 | Algae sieved from SUA 1828A | Watts Lake, E end, slope below terrace |
| Beta 4768 | 7305 ± 130 | Marine scallop shells | Watts Lake, E end, slope below terrace |
| Beta 4765 | 6500 ± 105 | Marine shell | Watts Lake, E end, slope below terrace |
| Beta 4763 | 6452 ± 160 | Algal mud | Lake Lebed |
| Beta 4764 | 6150 ± 95 | Marine serpulid worm tube, *Merceriella enigmatica* | Watts Lake, E end, slope below terrace |
| Beta 4766 | 5795 ± 85 | Marine *Laternula* shell | Watts Lake, E end, slope below terrace |
| SUA 1824 | 4670 ± 190 | Marine algae | Watts Lake, E end, slope below terrace |
| SUA 1409 | 2800 ± 80 | Non-marine stromatolites | Watts Lake, E end, slope below terrace above SUA 1410 |
| Death Valley Lake System | | | |
| SUA 1239 | 5440 ± 110* | Marine *Laternula* shell | Lake Stinear, SW end, flank of terrace |
| SUA 1237 | 5340 ± 90 | Marine worm tubes | Saddle between Club Lake and Deep Lake, flank of terrace |
| ANU 1011 | 4710 ± 70 | Marine *Laternula* shell | Lake Stinear, W end |
| Laternula Lake System | | | |
| ANU 1010 | 3470 ± 80 | Marine *Laternula* shell | Laternula Lake, E end, in lowermost of five layers |
| ANU 1009 | 3270 ± 90 | Marine *Laternula* shell | Laternula Lake, E end, in uppermost layer |
| SUA 1411 | 2410 ± 90* | Marine *Laternula* shell | Between Mud Lake (W end of Laternula Lake) and sea, on terrace |
| Other (arranged in N-S order) | | | |
| Beta 4767 | 7370 ± 95 | Marine shell fragments | Partizan Island |
| SUA 1831 | 1200 ± 170 | Algae | Calender Lake |
| SUA 1412 | 1340 ± 140 | Fresh-water algae | Thalatine Lake |
| Beta 4762 | 405 ± 95 | Fresh-water aquatic moss | Graticule Lake |
| Beta 4761 | 6225 ± 85 | Marine organic sediment | Ellis Fjord S side |
| Beta 4760 | 8355 ± 250 | Fresh-water aquatic moss | Clear Lake W end |
| Beta 4759A | 7310 ± 150 | Fresh-water aquatic moss (Cold HCl treatment) | Clear Lake W end |
| Beta 4759B | 7745 ± 285 | Fresh-water aquatic moss (Full pretreatment of alkali + acid to remove humic acids and some lignin) | Clear Lake W end |

*Ages differ from those listed by Pickard and Adamson (in press b). The revised dates here account for variations in hand-made vials.

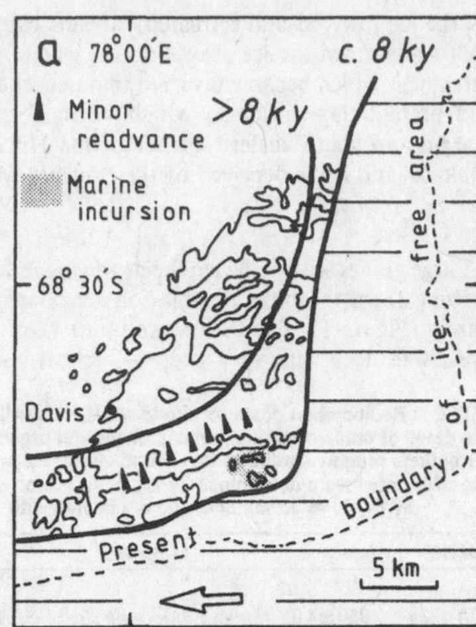


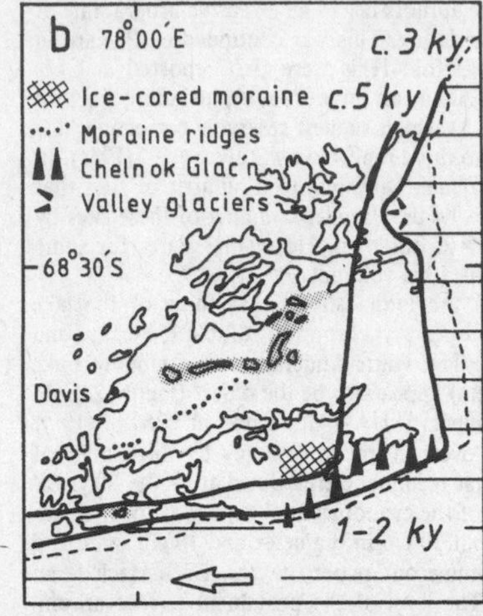


**Figure 2. Reconstructions of Holocene recession of the ice sheet from the Vestfold Hills. Dates are based on radiocarbon measurements which do not take into account reservoir corrections and will be variously up to about 1000 years old in the case of dates based on marine organisms (Table 1). The dates specified are therefore indicative of time zones in the Holocene. See Figure 1 for the location of named places. (a) Before about 8 kyr moraines were left to the east and south of the present location of Davis base, possibly related to still stands or minor readvances of the ice sheet. By about 8 kyr marine organisms were being deposited in the Watts Lake System, the moraines southeast of Davis were exposed and mosses were growing in Clear Lake. (b) By about 5 kyr marine organisms were being deposited in the Death Valley System of Club Lake, etc. The sea water may have entered from the north from Long Fjord. A wasting ice-cored moraine occupied high ground southeast of the Watts Lake System. By about 3 kyr the Laternula Lake System was connected to the sea but Lake Watts contained almost fresh water. Small valley glaciers occupied some mionor valleys with steep gradients in high ground near Lichen Lake. At 1-2 kyr the Chelnok Glaciation surged northward in the southeast corner followed by rapid melting. Recession of the ice sheet and of the northern margin of the Sørsdal Glacier continued to the present, with minor readvances along the northern margin of the Glacier (see moraines in Figure 1).**

gently sloping S-facing sides. The asymmetry is related to differential insolation of the two sides of the gullies. Stratigraphy of the ice in the gully walls shows sediment rich and clean layers of continental ice dipping steeply inland, overlain in some cases by unconformable layers of drift snow and ice dipping toward the coast. The surface of the continental ice at these unconformities is capped by distinct accumulations of till indicating former exposure to ablation.

## Collapse on the Northern Margin of the Sørsdal Glacier

The northern margin of the Sørsdal Glacier is grounded, rather stagnant and shows evidence of recent retreat. The margin is mostly a recessive convex slope seamed by a dense pattern of melt streams whose courses are not disarticulated by ice movement. At the head of Crooked Fjord in contact with the sea, the margin has retreated about 0.9 km over an area of 96 ha in the 20 years between the 1957 and 1979 aerial photography (Figure 1, star). In 1957 this area was an integral part of the glacier showing deep crevasses and entrenched melt streams, but by 1958 the ice had collapsed into isolated bergs, still in the positions they had occupied as ice ridges in 1958. The bergs are now locked in sea ice and have ablated into bizarre pinnacles so that the embayment is known as the "Iceberg Graveyard". The whole northern margin of the Sørsdal Glacier is flanked by numerous fresh moraine ridges from which the ice has withdrawn, probably within the last few centuries (Figure 1).

## Freshly Exposed Rock

Aerial photos and ground inspection also suggest that some rock areas northeast of Thalatine Lake have been freshly exposed during the 20 years between 1958 and 1978. This apparent exposure is not an artefact of differential snow cover, bearing in mind that 1958 was a year of comparatively light snow cover whereas the cover in 1978 was distinctly heavier.

# CONCLUSION

The direction and timing of the last ice retreat across the Vestfold Hills can be generally specified (Figure 2) on the basis of the abundant geomorphic evidence and radiocarbon dates. The dating confirms the Holocene age for ice retreat inferred by Johnstone et al. (1973). Further dating and assessment of reservoir corrections for marine and non-marine materials are in progress and will further resolve the episodic nature of ice movement.

Two striking outcomes of the radiocarbon chronology are the remarkable stability in time of the northern margin of the Sørsdal Glacier and the rapid retreat of the north-south ice cap margin. At the downstream end of the Glacier, even as long ago as 8000 years, the ice boundary was located near Clear Lake, probably within 2 km of its present position. The boundary that has moved rapidly is the north-south contact between continental ice and the Hills and continuing rapid retreat is indicated by the very young date for aquatic moss preserved above the shore line of Graticule Lake (Table 1), about 2.5 km from the present margins of the ice cap. Taken overall, the average rate of eastward retreat of the ice cap is between 1 and 2 km per 1000 years.

It is tempting to interpret the fine detail of ice retreat such as the still stands and the Chelnok Glaciation in terms of wider climatic correlations. However it is also possible that events like the Chelnok Glaciation and its timing can be explained by local factors such as the geometry of the contact between the ice sheet and the Sørsdal Glacier, which induced instability in the glacier itself. Resolving the relative importance of local factors and global climatic events is not yet possible in the Mid to Late Holocene, although it is obvious that the whole process of Holocene ice retreat is dependent on the warm global climatic conditions.

Short of a major global ice advance, we expect that the present day retreat of the ice margins will continue quite apart from any global warming that may occur in the near future. The positive feedback between the area of bare rock and ice retreat will continue to remove ice until the rate of ice recession is balanced by the rate of forward ice movement. The evidence from the marginal ice zone of the Hills suggests that such a balance has not yet been achieved.

*Acknowledgements* We thank the Antarctic Division for transport and logistic support, and many former expeditioners for help and courtesy. The extensive radiocarbon dating programme has been made possible by funds from the Australian Research Grants Scheme. Macquarie University has supported the research with funds, permission to undertake lengthy field work, and in many other ways. We thank Dr Pieter Arriens and Dr Henry Polach of the Australian National University for permission to quote previously unreported radiocarbon dates.

## REFERENCES


BLANDFORD, D.C., 1975: Spatial and temporal patterns of contemporary geomorphic processes in the Vestfold Hills, Antarctica. Unpublished B. Litt Thesis, Univ. New England, Australia.

JAMES, P.R. and OLIVER, R.L., 1978: The Vestfold Hills—geologically speaking. *Aurora*, Spring 1978, 137-40.

JOHNSTONE, G.W., LUGG, D.M. and BROWN, D.A., 1973: The biology of the Vestfold Hills, Antarctica. *ANARE Sci. Rep., Ser. B(1) Zoology, Pub., 123.*

KOROTKEVICH, E.S., 1958: "Antarctic Mummies". *Inform. Byul. Sov. Antark. Eksp., 2,* 89-91.

KOROTKEVICH, E.S., 1971: Quaternary marine deposits and terraces in Antarctica. *Inform. Byul. Sov. Antark. Eksp., 82,* 185-90.

PICKARD, J., in press a: Holocene winds of the Vestfold Hills, Antarctica. *N.Z. J. Geol. Geophys.*

PICKARD, J., in press b: A willy-willy in Antarctica. *Meteoro. Aust.*

PICKARD, J. (ms): Surface lowering of ice-cored moraine by wandering lakes. Submitted to *J. Glaciol.*

PICKARD, J. and ADAMSON, D., in press a: Probable Late Pleistocene marine deposits in the Vestfold Hills, Antarctica 32 ± 5 kyr. CLIMANZ 1981. Climatic history of Australia, New Zealand, Antarctica and surrounding seas. Proceedings of Howman Gap Conference, February 8-13, 1981.

PICKARD, J. and ADAMSON, D., in press b: Holocene marine deposits and ice retreat, Vestfold Hills, Antarctica 7 ± 2 kyr. CLIMANZ 1981. Climatic history of Australia, New Zealand, Antarctica and surrounding seas. Proceedings of Howman Gap Conference, February 8-13, 1981.

STUIVER, M., DENTON, G.H., HUGHES, T.J. and FASTOOK, J.L., 1981: History of the marine ice sheet in west Antarctica during the last glaciation: a working hypothesis; *in* Denton, G.H. and Hughes, T.J. (eds.) *The Last Great Ice Sheets.* Wiley, New York, 319-436.

WILLIAMS, J.W., 1981: Geophysical and surveying work along the Ingrid Christensen Coast near Davis Base, Antarctica, 1980/81 summer. *Aust., Bur. Miner. Resour., Geol. Geophys., Rec.,*1981/59.

ZHANG QINGSONG, XIE YOUYU and LI YUANFANG (this volume): A preliminary study of the evolution of the Vestfold Hills environment, East Antarctica, since the Late Pleistocene.

# PERENNIALLY FROZEN LAKES AT GLACIER/ROCK MARGINS, EAST ANTARCTICA

J. Pickard[1], *Antarctic Division, Kingston, Tasmania, 7150, Australia.*

D.A. Adamson, *School of Biological Sciences, Macquarie University, North Ryde, N.S.W., 2113, Australia.*

[1]*Present Address: School of Biological Sciences, Macquarie University, North Ryde, N.S.W., 2113, Australia.*

*Abstract* Perenially frozen lakes partially bounded by icesheet or glacier ice occur widely in Antarctica. In the Framnes Mountains (67°45′S 62°45′E) at least 15 lakes from 0.02 to 0.85 km² in area occur in embayments on the lee sides of nunataks. In the Vestfold Hills (68°30′S 78°00′E) three lakes up to 2.00 km² in area occur along the margin of the Sørsdal Glacier and low bedrock hills. Lake ice, from three to more than five metres thick differs morphologically from glacier ice and caps water up to 120 m deep. Large boulders "float" on the surface of some lakes because they do not melt through. Lakes form when meltwater at the rock margin becomes deeper than the winter freezing point. Thermo-erosion of glacier ice expands the lake, and as the glacier ice capping the lake ablates, it is replaced from below by refrozen lake water. Diapiric domes occur on smaller lakes, and pressure ridges on large lakes. Pressure ridges are caused by lateral pressure from the icesheet or glacier. Debris deposits on nunatak slopes above lakes include lacustrine sediments as well as scree and till. Strandlines of boulders indicate stages in icesheet lowering, while water marks on rock bluffs indicate stages in lowering of the lakes.

Lakes with frozen surfaces are common in Antarctica. They occur in depressions in the icesheet, against nunataks or ice free land, and fully within ice free land. In this paper we consider only lakes along the rock/ice interface. Although these have been reported from Victoria Land (Lyon, 1979), MacRobertson Land (Crohn, 1959), Dronning Maud Land (Kosenko and Kolobov, 1970) and the Sør Rondane (Autenboer, 1962) they have received little attention. Such lakes are not mentioned in the chapter on glacial lakes in Embleton and King (1975); nor is there a commonly accepted terminology.

We have studied the geomorphology of such lakes in the Framnes Mountains (67°45′S 62°45′E) in MacRobertson Land, and in the Vestfold Hills (68°30′S 78°00′E) in Princess Elizabeth Land (Figures 1 and 2). The aim of this paper is to describe the location of the lakes, the surface morphology and to present a general model for their formation.

In the Framnes Mountains, lakes occur on the lower, down wind sides of nunataks and ranges (Figures 1b, d, e). Close to the ranges ice flow is slow, particularly in the re-entrant bays where the lakes occur. The Vestfold Hills are a 400 km² ice free area of low hills bounded in the south by a small outlet glacier, the Sørsdal (Figure 2). The northern margin of the Sørsdal is stranded on rock and is receding by melt and ablation. Three elongate lakes occur along the junction of the glacier and the hills. Unlike the Framnes Mountains, the glacier here is higher than the adjoining hills and the lakes are not sheltered from winds. During summer, water from the moat around Last Lake overflows into Pineapple Lake. Similarly, water flows from Lake Chelnok into Mossel, Teat and Crooked Lakes and thence into the sea (Tierney, 1975). A small jokulhlaup burst from Lake Chelnok in austral summer 1980/81 two months before the main flow started.

Lakes occur on the northern flanks of two massifs, Mount Riiser-Larsen and Mount Menzies in Enderby Land and the southern Prince Charles Mountains respectively (Figure 1a). In both cases, the lakes are in bays on the ice/rock margin. Such lakes also occur on the edge of the Bunger Hills (Figure 1a).

## DESCRIPTION OF LAKES

The area, ice thickness and water depth of the Framnes Mountains and Vestfold Hills lakes are summarised in Table 1. Area varies considerably, from less than 0.02 km² to almost 2.0 km². Ice thickness ranges from under 3 m to greater than 5 m, near pressure ridges on Lake Chelnok. Typically ice is 3.5-4 m thick, about twice as thick as ice formed each year on freshwater lakes in the Vestfold Hills. Lake depth varies from about 3 m to at least 114 m. Lake Henderson is deepest with depths of 72, 88, 105 and 114 m from widely spaced holes. Pineapple Lake is 58 m deep which indicates that the lake floor is 56 m below sea level. The temperature in Lake Henderson was remarkably constant in January 1981. It was 0.27°C immediately below the ice, and 0.33°C from 6 m below the ice to the bottom at 105 .

Surface features vary between each lake. Microrelief of ablation scallops, and small cracks and hummocks is ubiquitous. The ice is frequently broken by tension cracks about 0.3 m wide which pinch out near the lake margins. Most of the lake surface is not of icesheet or glacier origin. This is clearly shown by flowlines, cryoconite holes and debris bands which stop at the lake edge or close to it. Inferred ice crystals range up to 2 x 0.5 x> 4 m deep on Patterned Lake.

Narrow moats, up to 1 m wide and of unknown depth, melt along the rock margin each summer. Some of these connect with the water below the ice. Others do not as shown by the visible flow or other ice pattens below the melt water. Against the icesheet, there is often a second, wider moat. This is meltwater from the icesheet which accumulates in broad shallow pools on the ice and reduces the surface layers to slurry. All lakes but Bicuspid have domes or ridges of varying sizes which are described below.

A notable feature of lakes in the Framnes Mountains is boulders on the ice surface. These are supra-glacial till boulders resting on lake ice. Their occurrence is a consequence of glacial ice being replaced by lake ice; or boulders falling directly onto the lake. Small erratics melt into the ice forming cryoconite holes but the large boulders "float" on the surface because they are too thick for the necessary heat transfer through them to melt the ice. Approximately 40 boulders occur on Lake Henderson, 20 on Painted Lake, several on Rumdoodle Lake and many more on the icesheet between Painted and Lassitude Lakes. The smallest boulders are approximately 1 x 1.5 x 1.5 m, the largest 6 x 3 x 1.5 m. All perch on ice pedestals about 0.2 m high surrounded by a moat about 0.2 m deep which barely extends beyond the limit of the rock. This moat is caused by heat radiation and wind ablation.

## FORMATION AND MAINTENANCE OF THE LAKES

The lakes are a heat sink and without heat supplied by liquid water, none would form or survive. They are initiated when summer melt water collects in marginal moats and below the surface of glacial ice. Heat stored in the rocks assists and eventually the pool is deep enough that its base is below the winter freezing front. Each summer more water is added and more heat is stored. As underwater thermo-erosion melts the adjoining icesheet or glacier, the lake expands. Initially, the ice above the reservoir of water is primarily of glacial origin. As the glacier surface lowers through ablation and summer melt, the underside of the ice over the pool is replaced by refreezing on the underside (Figure 1e) until the lake is mostly capped by ice of lake origin. At known ablation rates of 0.5 m $y^{-1}$, this would take about 10 years.

As the icesheet lowers, lakes would maintain their position relative to the surface. This would expose till and sediments on the nunatak slopes above the falling lakes. The sediments are both lacustrine, from material washed into the lakes by summer melt, and underwater extensions of scree slopes. Strandlines of boulders above Sonic Lake (Figure 1c, d) probably represent a stillstand in icesheet recession.

If the icesheet expanded, lakes would eventually freeze entirely. Both the source of heat, and ablation would be reduced so that the reservoir of water would refreeze. When the icesheet had risen sufficiently to cover the nunataks, flow over the nunatak would carry away any remaining water.

Bicuspid Lake shows evidence of recent lowering of the lake surface. Water stains on an adjoining rock bluff run from about one to more than 6 m above the lake. Each mark represents a stage in lake lowering. The most recent occurred when ice at the eastern end melted allowing water to drain into Sonic Lake. As a consequence, a large slab of icesheet collapsed into the lake. The 4 m high slab runs the full length of the lake and is separated from the icesheet by a 2 m vertical offset and a 1-3 m wide crevasse. Thermo-erosion of the icesheet had undercut the slab and when the water drained, the unsupported slab collapsed into the lake. The slab and crevasse are visible in 1960 air photos. Subsequently, the surface ice has reformed with no ridges because the flow of ice is past the bay and there is little pressure on the lake.

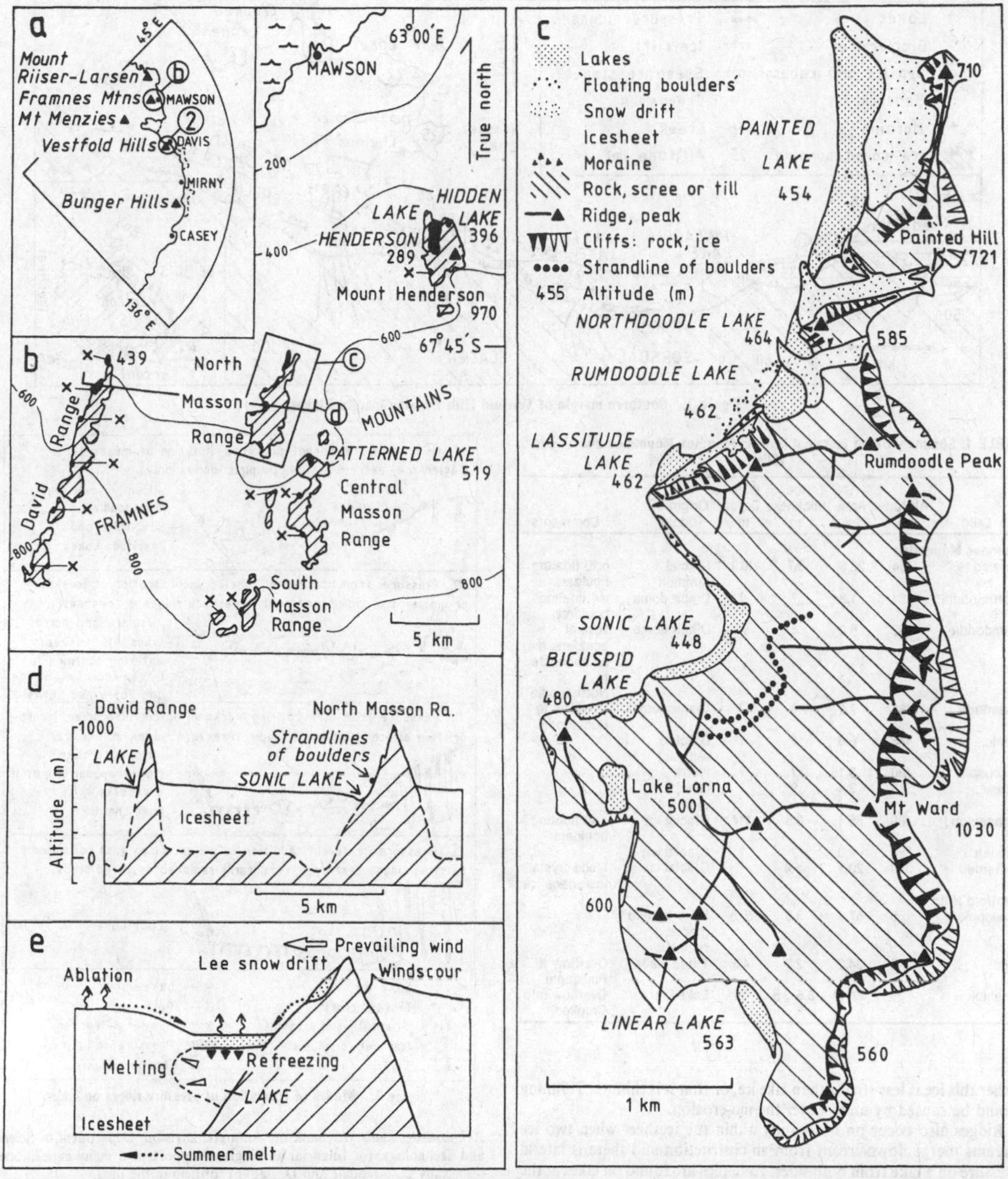


**Figure 1.** **(a) Location map of East Antarctica, (b) Location map of Framnes Mountains. Relevant lakes shown by crosses or by names, altitudes and form lines in m, (c) North Masson Range showing lakes and other features, (d) Section from David Range to North Masson Range showing location of lakes. Ice depths estimated from data on 1:100,000 map of Framnes Mountains, (e) Diagrammatic section of lake showing processes operating.**

## Pressure Ridges

Most lakes have ridges or domes on their surfaces. Some large ridges, up to 100 m wide and 20 m high, are clearly remnants of glacial ice. Pressure ridges are quite distinct from the erosional remnants. They are usually lower, rarely higher than 3 m and have sharper relief when fresh. Domes are restricted to lakes with areas less than 0.05 km² and usually occupy all but the edges. They are probably of diapiric origin (Figure 3a; Bradley and Palmer, 1967). Ridges form in ice of glacial or lake origin, but apparently not in both on the same lake. Three ridges in glacial ice extend up to 70 m onto Lake Henderson.

All ridges examined have several features in common: (a) they are parallel to the glacier margin of the lake, (b) a longitudinal crack caps the ridges, (c) narrow transverse cracks about 0.5 m wide occur, (d) one side of a ridge is steeper regardless of aspect, if melt water laps along that side. The following features were observed on a few ridges: (e) a tunnel running inside the ridge, (f) ice crystals perpendicular to the ridge surface can be followed into the lake ice, (g) cryoconite holes and relict flow lines. The presence of ice crystals and cryoconite holes indicate that the ridges are deformed lake and glacial ice respectively. Together with occurrence of tunnels this indicates that ridges are constructional and not erosional landforms. The parallelism of ridges to the lake margin implies that the ridges formed by external pressure from the icesheet or glacier on the lake. The external pressure sets up stresses in the ice which are partially relieved by deformation of the ice into ridges. Both the source of the pressure and the origin of the buckled ice varies (Figure 3). Buckling of glacial ice indicates that

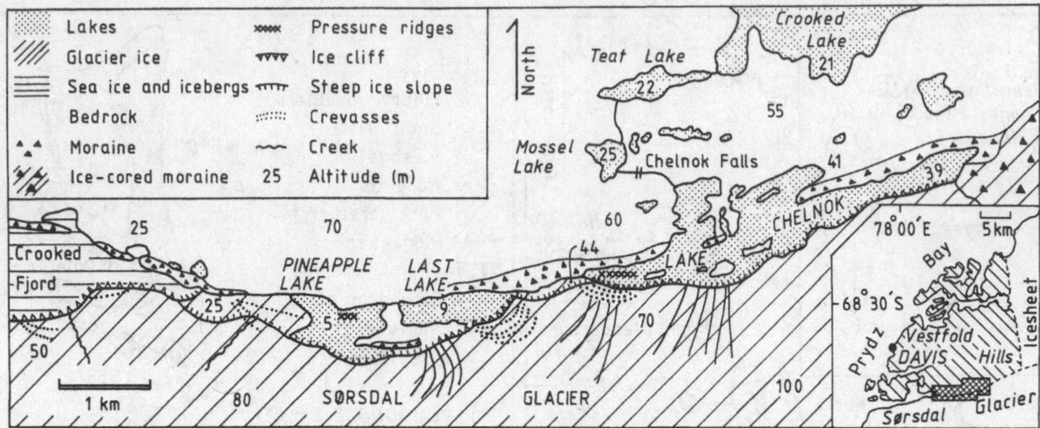


**Figure 2. Southern margin of Vestfold Hills showing marginal lakes.**

**TABLE 1: Summary of ice marginal lakes in Framnes Mountains and Vestfold Hills.**

| Lake | Altitude m | Area ha | Ice thickness m | Water Depth m | Origin of Ridges | Comments |
|---|---|---|---|---|---|---|
| **Framnes Mountains** | | | | | | |
| Painted | 454 | 22.6 | 3.7 | 16.9 | Glacial remnant | c.20 floating boulders |
| Northdoodle | 464 | 1.8 | ? | ? | Diapir dome | c.4 floating boulders |
| Rumdoodle | 462 | 5.0 | 4 | 81 | Diapir dome | several boulders, may be connected to Northdoodle |
| Lassitude | 462 | 2.2 | ? | ? | Diapir dome | Overflow to Rumdoodle |
| Sonic | 448 | 6.6 | ? | ? | Glacial remnant | |
| Biscuspid | 480 | 3.1 | ? | ? | None | |
| Linear | 563 | 3.4 | ? | ? | Glacial remnant | |
| Henderson | 289 | 85.4 | 3.5 | 114 | Glacial ice | c.40 floating boulders |
| Hidden | 396 | 4.0 | ? | ? | Glacial ice | |
| Patterned | 519 | 20.5 | >4 | ? | Glacial ice | Mega-crystals in surface ice |
| **Vestfold Hills** | | | | | | |
| Pineapple | 5 | 57.0 | 3.5 | 57.5 | Lake ice and glacial remnant | |
| Last | 9 | 34.2 | 2.7 | 4.3 | Tilted slabs | Overflow to Pineapple |
| Chelnok | 39 | 197.9 | 3.5->5 | 18 | Lake ice | Overflow into Crooked |

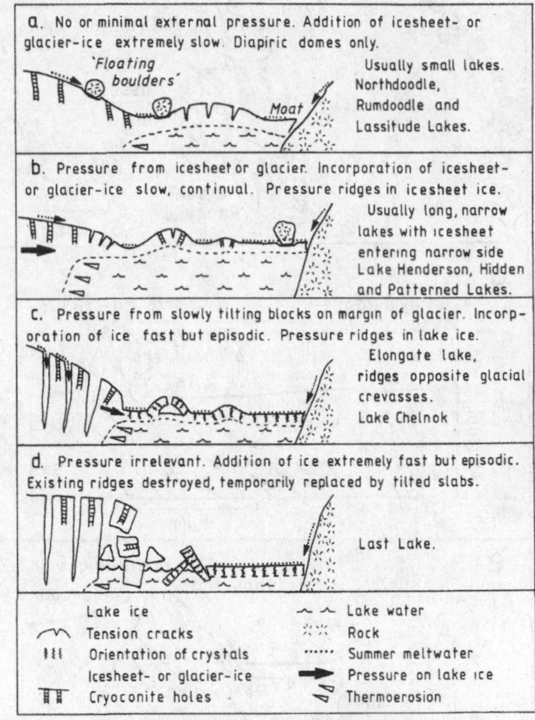


**Figure 3. Modes of formation of pressure ridges on lakes.**

either this ice is less strong than lake ice, or that it is thinner. Thinning would be caused by underwater thermo-erosion.

Ridges also occur on lakes fully within the icesheet when two ice streams merge downstream from an obstruction and there is lateral pressure on a lake from both sides. Examples are found on lakes at the northern end of the David Range (Figure 1b) and at the Chaos Glacier, 50 km west of the Vestfold Hills.

## FUTURE POSSIBILITIES

Henderson Lake, the largest lake in the Framnes Mountains, contains about 80 x 10$^6$m$^3$ of melt water. This represents a considerable heat store. Lake Chelnok in the Vestfold Hills is possibly half that volume but has an annual loss of water and heat which drain into the sea. Energy balance calculations, particularly of lakes with no outlet, would be of interest. They may provide estimates, or at least limits, of the ages and growth rates of the lakes. There is also considerable scope for detailed study of the lake bathymetry, water chemistry, surface ice and growth dynamics.

*Acknowledgements* We thank the Antarctic Division, Department of Science and Technology, for financial and logistic support; and many expeditioners (especially C. Tivendale and D. Paulin) for help in the field.

## REFERENCES

AUTENBOER, T. Van, 1962: Ice mounds and melt phenomena in the Sør-Rondane, Antarctica. *J. Glaciol., 4,* 349-54.

BRADLEY, J. and PALMER, D.F., 1967: Ice covered moraines and ice diapirs, Lake Miers, Victoria Land, Antarctica. *N.Z. J. Geol. Geophys., 10,* 599-623.

CROHN, P.W., 1959: A contribution to the geology and glaciology of the western part of the Australian Antarctic Territory. *ANARE Rep. Ser. A., 3,* 1-103.

EMBLETON, C. and KING, C.A.M., 1975: *Glacial geomorphology.* London, Edward Arnold.

KOSENKO, N.G. and KOLOBOV, D.D., 1970: Exploration of Lake Untersee. *Inform. Byul. Sov. Antark. Eksp., 79,* 33-5.

LYON, G.L., 1979: Some aspects of the ice topography of Trough Lake, southern Victoria Land, Antarctica. *N.Z. J. Geol. Geophys., 22,* 281-4.

TIERNEY, T.J., 1975: An externally draining freshwater system in the Vestfold Hills, Antarctica. *Polar Rec., 17,* 684-5.

# A PRELIMINARY STUDY OF THE EVOLUTION OF THE POST LATE PLEISTOCENE VESTFOLD HILLS ENVIRONMENT, EAST ANTARCTICA

Zhang Qingsong, Xie Youyu and Li Yuanfang, *Institute of Geography, Academia Sinica, Peking, China.*

*Abstract* There were at least two glacial cycles in the Vestfold Hills (68°35′S, 78°00′E) during the Late Quaternary. Much of the present ice-free area was ice-free; some of it was under the sea for a period of unknown duration, ending about 31,000 years BP. Between 30,000 BP, and 10,000 BP, ice covered the presently ice-free area with a minimum thickness of 158 m. The period 6000-5000 BP was one of milder climate with relative sea level higher than at present, so that many valleys were invaded by the sea in which there was abundant plant and animal life. Glacio-isostatic uplift at the rate of 2.5 mm/yr has caused regression and relative sea level has fallen some 15 m during the last 6000 years. A series of young recessional ice-cored shear-moraines may reflect ice fluctuations in sympathy with neoglacial advances and retreats documented for other parts of the world. Ice fronts have been retreating since this period of advances.

The physical geography of the Vestfold Hills has been described by several workers (Crohn 1959; McLeod 1964; Johnstone et al. 1973; Burton and Campbell 1980) but Quaternary strata and geomorphology have only recently attracted particular attention (Pickard and Adamson in press a and b, Zhang in press).

## GLACIAL AND INTERGLACIAL DEPOSITS

There is evidence for at least two Late Quaternary glacial and interglacial episodes (Figure 1).

### Interglacial marine sediments of Late Pleistocene Age

Late Pleistocene fossiliferous marine sediments underlie ground moraine and form a plain 4.5 km² at 20 m a.s.l., opening onto the south shore near the head of Crooked Fjord. Sections through these deposits were formed during the Holocene after deglaciated terrain was invaded by the sea and before glacio-isostatic uplift which was apparently episodic in that some of the sections are in a series of terraces above the eastern shore of Burton Lake. Typical sections show moraine thicknesses of 0.5-2 m; up to 7 m of marine sediments

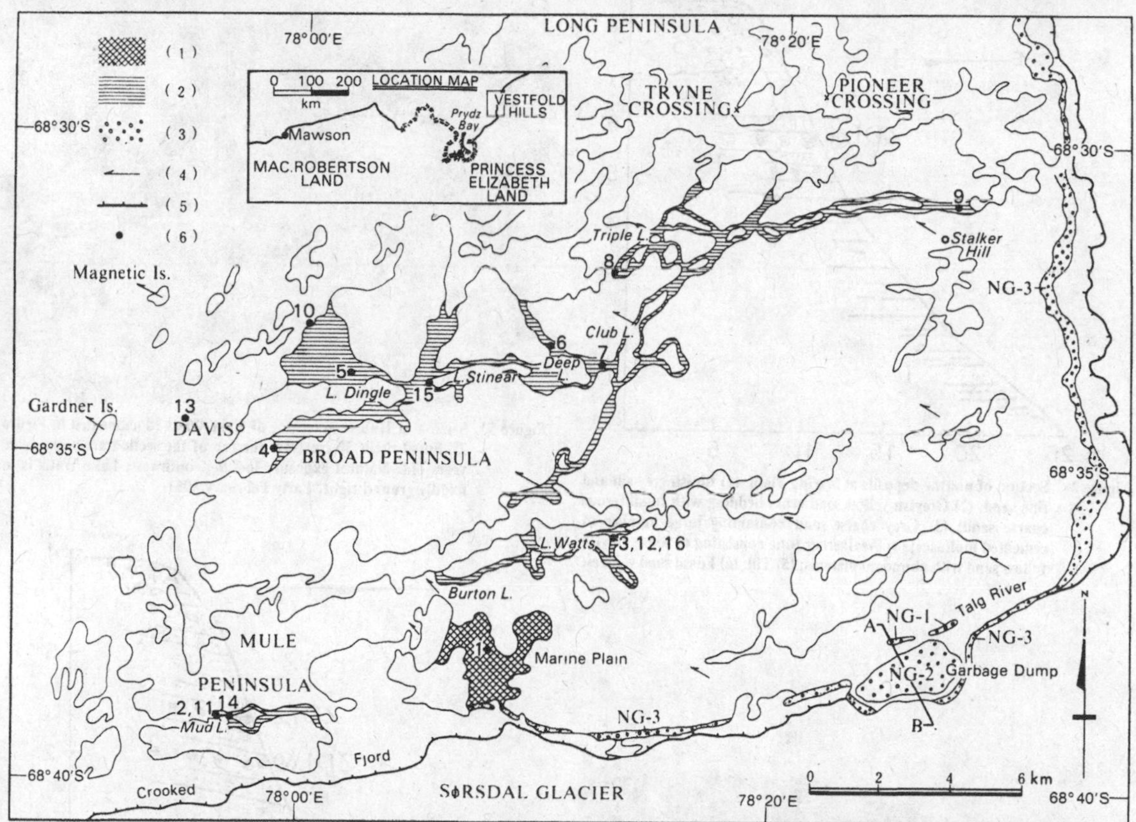


**Figure 1.** **Distribution of glacial and interglacial deposits in the Vestfold Hills. (1) Marine Plain; (2) Terraces and raised beaches; (3) Neoglacial deposits; (4) Orientations of striae; (5) Margin of the ice sheet and Sørsdal Glacier; (6) Sites of ¹⁴C dates. (All radiocarbon dates here are uncorrected. The correction probably will be ~1300 years, see dates for sites 4, 10 and 13.) (1) Marine Plain, Shell, ZDL68, 31,000 ± 474 yrs BP; (2) Terrace, Mud Lake, *Laternula* Shell, ZDL69, 3500 ± 86 yrs BP; (3) Terrace, Lake Watts, Shell, ZDL70, 6100 ± 108 yrs BP; (4) Davis shore, modern *Laternula* Shell, SUA1235, 1295 ± 105 yrs BP (Pickard and Adamson, 1981b); (5) Terrace, Lake Dingle. *Laternula* Shell, ZDL79, 5600 ± 77 yrs BP; (6) Terrace, Deep Lake, *Laternula* Shell, ZDL80, 6632 ± 118 yrs BP; (7) Terrace, Deep and Club Lakes, Worm tubes, SUA1237, 5340 ± 90 yrs BP (Pickard and Adamson, 1981b); (8) Terrace, Triple Lake, *Laternula* Shell, ZDL78, 6141 ± 90 yrs BP; (9) Dry basin near Platcha Hut, algae, ZDL81, 5677 ± 94 yrs BP; (10) Airport beach, modern *Laternula* Shell, ZDL84, 1312 ± 65 yrs BP; (11) Terrace, Mud Lake, *Laternula* Shell ZDL66, 3325 ± 103 yrs BP; (12) Terrace, Lake Watts. Calcareous tufa, ZDL85, 7616 ± 104 yrs BP; (13) Modern marine sediment, 17 m water depth, algal mud, SUA1236, 1310 ± 125 yrs BP (Pickard and Adamson, 1981b); (14) Terrace, Laternula—Mud Lake, *Laternula* Shell, SUA1411, 2800 ± 85 yrs BP (Pickard and Adamson, 1981b); (15) Terrace, Lake Stinear, *Laternula* Shell, SUA1239, 5740 ± 105 yrs BP (Pickard and Adamson, 1981b); (16) Terrace, Lake Watts, marine sediment after acid treatment, SUA1410, 8700 ± 100 yrs BP (Pickard and Adamson, 1981b).**

are exposed in some instances. Shells from amongst fossils collected near the top of the marine section yield a radiocarbon date of 31,000 ± 474 yrs BP (ZDL68) (see Figure 1).

These deposits are mentioned in Pickard and Adamson (in press a) and described in detail by Zhang (in press). In summary, three characteristics should be mentioned: (i) Layers 1 to 4 (Figures 2 and 3) are very rich in fossil diatoms (31 genera and 60 species) including *Coscinodiscus symbolophorus* Grun., *Thalassiosira gravida* Cl., *Diploneis splondida* (Greg.), and *D. sejuncta* (A.S.) Jore. A large number of fossil bivalves (3 genera and 4 species) including *Hiatella* aff. *arctica* Linné, *Nucula oblique* (Lamarek) and *Chlamys (athlopecten)* sp. are found in layers where they are cemented together with $CaCO_3$. A marine enviroment is indicated for the deposition of layers 1- 4; (ii) the exposures indicate from their texture grading (upward coarsening) and fossil content (all organisms in layer 3 are benthic) that marine sediments were deposited in progressively shallowing conditions; and (iii) the moraine deposit (layer 5) has buried a weathered zone (average about 60 cm thick) containing fossil sand wedges (layer 4, containing marine diatoms) indicating a significant interval of subaerial periglacial conditions before a glacial advance.

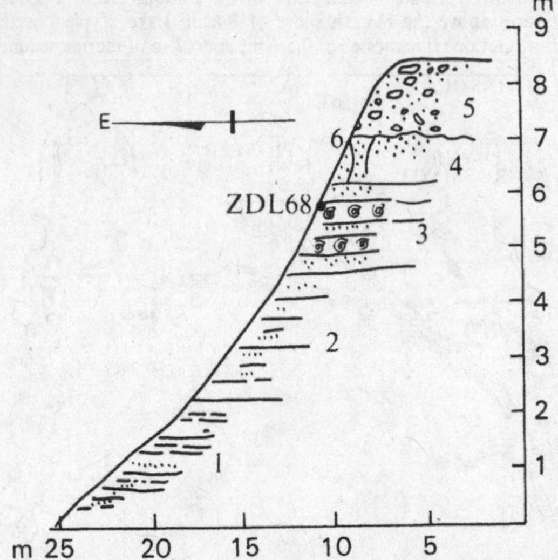


**Figure 2. Section of marine deposits at Marine Plain. (1) Bluish grey silt and fine sand; (2) Greyish yellow sand cross bedding with fossiliferous coarse sand; (3) Grey coarse sand containing large number of cemented molluscs; (4) Weathering zone consisting of loose greyish yellow sand with abundant diatoms; (5) Till; (6) Fossil sand wedges.**

**Figure 3. The section documented in Figure 2, looking south, mid-February 1981.**

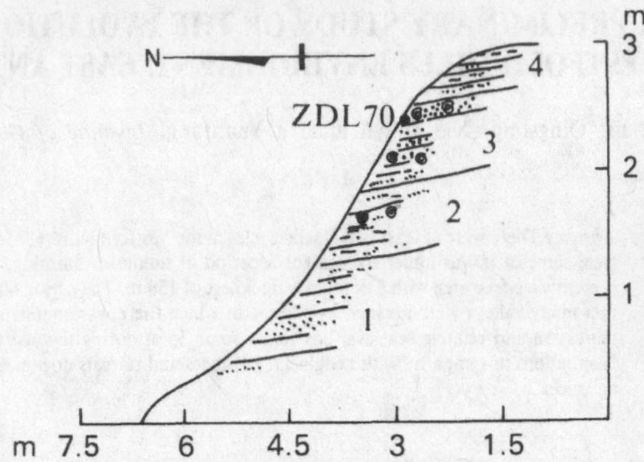


**Figure 4. Postglacial deposits on the terrace of Lake Watts. (1) Grey medium and coarse sand containing worm tubes, *Laternula* and sponges etc.; (2) Greyish yellow and black sand with pebbles containing worm tubes and *Limatula hodysoni* (Smith); (3) Sand and small pebbles with large quantity of worm tubes, *Thracia meridonalis* Smith and *Axinopsida* sp.; (4) Fossil algae.**

**Figure 5. Section in Holocene terrace of Lake Watts (documented in Figure 2). Fossil shells 30 cm from the top of the section are about 6000 years old. Natural exposure looking southwest; Lake Watts is in middle ground right. Early February 1981.**

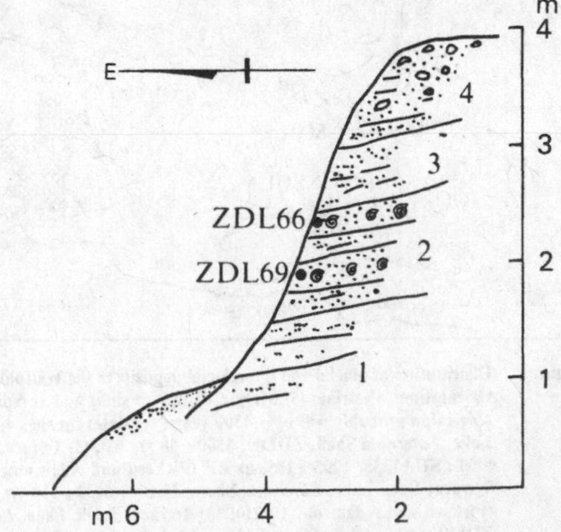


**Figure 6. Postglacial deposits on the terrace of Mud Lake. (1) Grey coarse sand; (2) Cross-bedding of grey medium and fine sand, containing large number of *Laternula elliptica*; (3) Greenish grey sand and small pebbles and fragments; (4) Debris.**

## The latest glacial deposits of the Pleistocene

The ablation till (ground moraine) overlying the marine sediments is less than 31,000 years old and probably correlates with deposits of the last glacial stage as documented from many parts of the world. In the Vestfold Hills ice advanced to cover the highest hills (e.g. Stalker Hill 158 m, a.s.l.). This was a time of lower sea level and hill tops, presently prominent as off-shore islands (e.g. Gardner Island, and Magnetic Island) were ice covered. Both erosional (striae) and depositional (till) evidence testify to this advance which can be extended out to the glacial low sea level "grounding line", which in places was at the edge of the continental shelf (e.g. as far north as about 67°S, see Hughes et al., 1981). Robin (1977) has estimated former ice thicknesses at least several hundred metres thicker than at present for most currently ice-free coastal areas in Eastern Antarctica.

During this glacial phase ice abrasion fashioned abundant roches moutonnées and striae the orientations of which indicate that ice flowed from southeast to northwest except during later stages of thinning and retreat when some ice flow patterns followed the line of valleys draining west. The ice advance that inundated the presently ice-free Vestfold Hills during the last glacial stage has been dubbed "the Vestfold Event" by Pickard (in press) in one paragraph of a paper devoted almost entirely to a discussion of palaeowind directions in "post-glacial" times.

## The post-glacial deposits

Post-glacial deposits are widely distributed in the Vestfold Hills and include depositional terraces around many of the saline lakes and only a very few of the fresh water lakes (e.g. Lake Watts). The terraces are formed mainly of fossiliferous coarse and fine sands mixed with pebbles. A rich assemblage of molluscs, foraminifera, diatoms and sponge remains have been identified and described (Zhang, in press). At some localities (e.g. Lake Watts) a large number of worm tubes were found. Natural sections, one at Lake Watts and the other at Mud Lake, serve to provide information on the age and environment of deposition (Figures 4-7). The sections are similar, showing mainly coarse materials with tilt bedding and characteristics of deposition in shallow water. Fossils are abundant and all are marine. The deposits are rich in benthic animals and plants, e.g. at Lake Watts (Figures 4 and 5) calcareous worm tubes of *Hydroides* sp. and *Sperobis* sp. occupy more than 50% of total sediments. Other fossils such as the six species (from four genera) of bivalves ( *Laternula elliptica* King et Broderip, *L. recta* Reeve, *Adamussium colbecki* (Smith), *Limatula hodysoni* Smith, *Thracia meridonalis* Smith, *Axinopsida* sp.) 76 species (from 44 genera) of foraminifera (*Epistominella patagonica* d'Orbigny, *Cibicides lobatulus* Walker and Jacob, *Triloculina lamellidens* Parr, *Cassidulina erassa* d'Orbigny) and 57 species (from 20 genera) of diatoms (*Cocconeis pinnata* Greg, *Pinularia quadratarea* var. *baitica* Grun, *Trachyneis aspera* (Ehr.) Cl, *Melosira omma* Cl, and *Coscinodiscus biradiatus* Grev, *Grammatophora arcuata* Ehr) are found.

The fossil assemblage at Mud Lake (Figures 6 and 7) is less diverses, 80-90% of the forams being accounted for as *Cassidulina crassa*, the rest of the assemblage being almost entirely dominated by *Laternula elliptica* (see Johnstone et al. 1973, plates 16 and 17). The sediments here are similar to those found in the Lake Watts exposure (Figures 4 and 5) except that they contain rock fragments. As can be seen from the dated horizons (Figure 6) the Mud Lake section is the younger. The lower species diversity in this deposit, compared with the Lake Watts assemblage suggests that water temperatures suffered a net drop between 6100 years BP (ZDL70 Figure 1 caption) and 3500 years BP (ZDL69 Figure 1 caption).

**Figure 7. The section documented in Figure 6; looking southwest, Mud Lake is in middle ground left. Early February 1981.**

## Evidence for Neoglacial activity

### *Deposits*

The youngest tills are found in front and up to 3 km from the present ice edge (Figures 1 and 8). Beyond this zone some bedrock surfaces harbour well-developed lichen covers in marked contrast to the boulders on the young moraines. Here black crustose lichens of small diameter may be found on boulders of the debris-cover that settled after substantial melting of the ice in the recessional debris-mantled ice-cored shear moraines that parallel the comparatively cleaner ice front.

In some places along the front are horizontal bands of debris concentrated in shear zones that dip down towards the glacier base and strike parallel to the front. Thus the most modern moraine deposits in the Vestfold Hills are like those described from Arctic locations (e.g. Boulton 1970) and would be expected to be more obvious along the shear zones most loaded with englacial debris as ablation proceeds, until the stage of shrinkage is brought about by disintegration of the ice core. In some places the englacial debris appears to have been more abundant, most markedly in the area known as the "Garbage Dump", so that part of the series of moraines stand out prominently from the rest.

Evidence for retreat is most marked on the edge of the Søsdal glacier, east of the head of Crooked Fjord and three till types can be distinguished here (Figure 8): (i) about 3 km from the ice front, the outermost moraine crests protrude as a discontinous line above the niveo-aeolian, colluvial and channel deposits of the Talg River. The till consists of greyish-yellow sand and boulders reaching 30-40 m above the Talg River; (ii) there is the blue-grey clay,fine sands and boulders occupying about 3 km² and reaching 70-120 km a.s.l. in the "Garbage Dump" area; and (iii) there is the grey bouldery till of the ice-cored moraines that occupy a 200-500 m zone along the ice front that marks the edge of the ice free area (the Vestfold Hills). These moraines have 10-50 m relief.

### *Striae*

Two generations of striae have been recorded on the bedrock surfaces that border the Søsdal Glacier within the zone where the youngest tills, described immediately above, are found. An older set

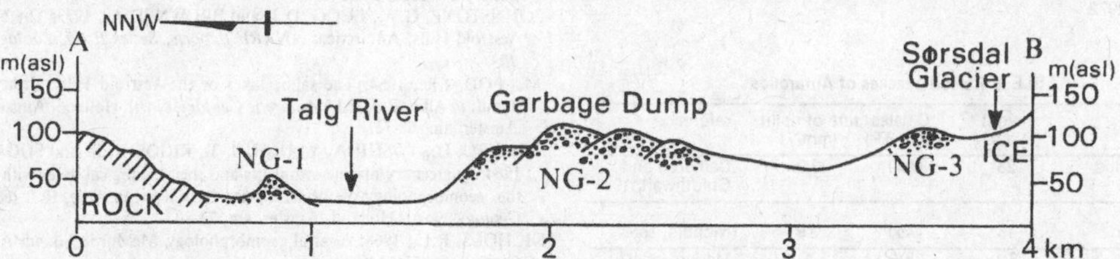


**Figure 8. Neoglacial deposits in the Garbage Dump area. (1) NG-1, early Neoglacial terminae; (2) NG-2, middle Neoglacial terminae; (3) NG-3, late Neoglacial terminae.**

orientated between 250° and 270° is cut across by the other set which is orientated between 330° and 350°. The older set therefore parallels the axis of the Sφrsdal Glacier while the other parallels the dip direction of the shear planes in the ice-cored shear moraines. This suggests that the inland ice reactivated its outlet down the main trough of Crooked Fjord (now beneath the Sφrsdal Glacier) after retreat that followed the Flandrian Transgression. The Sφrsdal Glacier, after establishment of maximum Holocene positions, to which the striations between 250° and 270° refer, suffered thinning so that its northern half sloped north and flow lines developed in that direction.

*Age*

As mentioned above, an ice-free period between 6600 BP and 3500 BP is evident from documentation of marine deposits around many lakes for example, Mud Lake and Lake Watts. The advance of the Sφrsdal Glacier therefore occurred after 3500 BP during the period of Neoglacial advances elsewhere (Denton and Porter, 1970; Clark and Lingle, 1979; Shi Yafeng and Wang Jingtai, 1981). In the case of the outlet of a major ice sheet, ice margin fluctuations are greatly influenced by the dynamics of the inland ice and this may not be entirely a function of climatic change.

A preliminary chronology for glacier fluctuation in the Vestfold Hills during the Late Quaternary is given in Table 1. We have not yet tried to compare it in detail with that established in other parts in the world, however it is interesting that the Vestfold Hills sequence exhibits some similarities to the sequences from McMurdo Sound, Antarctica (Denton et al., 1970; Ward and Webb, 1979) and from East China (Shi Yafeng and Wang Jingtia, 1981). Global-scale climatic change is indicated.

**TABLE 1: Comparison of Glaciation and Interglaciation since the Late Pleistocene**

| Vestfold Hills (this paper) | McMurdo Sound (Denton et al. 1970; Ward & Webb, 1979) | East China (Shi Yafeng & Wang Jingtai, 1981) |
|---|---|---|
| present | present | present |
| neoglaciation (about 2000-3000 years B.P.)* | | neoglaciation (about 2000-3000 years B.P.) |
| Post-Glaciation (3500-6600 years B.P.) | Post-Glaciation (4450-6100 years B.P.) | Optimum climatic stage 5000-6000 years B.P.) |
| Vestfold glaciation** (10000-30000 years B.P.) | Ross glaciation 2 (maximum 18000 years B.P.) | Late Dali glaciation (10000-25000 years B.P.) |
| Interglaciation (31000 years B.P.) | Interglaciation (34800-47000 years B.P.) | Interglaciation 25000-40000 years B.P.) |
| | Ross glaciation 1 | early Dali glaciation (40000-70000 years B.P.) |

** Vestfold Event Pickard in press
* Chelnok Event Pickard in press

**TABLE 2: The height of terraces**

| Position | Altitude of lake Surface (m)* | Height of terraces (m) above lake | above sea | $^{14}C$ dates (years B.P.) |
|---|---|---|---|---|
| Dingle L. | – 10 | 16 | 6 | 5600 ± 77 (ZDL79) |
| Deep L. | – 56 | 60 | 4 | 6632 ± 118 (ZDL80) |
| Triple L. | – 16 | 19 | 3 | 6141 ± 90 (ZDL78) |
| Mud Lake | – 2 | 5 | 3 | 3500 ± 86 (ZDL69) |
| Burton L. | 0 | 15 | 15 | about 6000 |
| Tryne Fjord | | | 15 | ? |
| Pioneer Crossing | | | 10 | ? |

*According to the map of Vestfold Hills, (1:100000) Division of National Mapping of Australia, 1972.

**TABLE 3: Raised beaches of Antarctica**

| Place | height (m a.s.l.) | $^{14}C$ dates (years B.P.) | rate of uplift (mm/y.) | reference |
|---|---|---|---|---|
| Windmill Islands | 23 | 6040 | 3.9 | Cameron and Goldthwait, 1961 |
| McMurdo (Marble Point) | 13 | 4450 | 2.9 | Nichols, 1968 |
| East Ongul Is. | 3-4 | 3840 | 1 | Meguro et al. 1964 |
| Vestfold Hills | 3-15 | 3500-6000 | 1-2.5 | this paper |

## TERRACES AND RAISED BEACHES

Many of the lakes in the Vestfold Hills are surrounded by up to three depositional terraces, some of which can be traced in adjacent valleys. All terraces sampled so far contain marine fossils. The terrace elevations range between 3 and 15 m a.s.l. Relative sea level fall due to glacioisostatic recovery is indicated. This uplift progressively drained the innermost areas of the sea leaving their wider parts isolated as lakes over which evaporation has, since then, caused a steady increase in salinity. Dating of material included in some of the highest terraces (Table 2) indicates that coastal and sub-littoral environments were present at least 6100 years ago at least as far south as Lake Watts. Detailed levelling to obtain elevations above msl and a comprehensive attempt to date all the terraces should yield data on the pattern and rate of glacio-isostatic recovery. Further details and discussion are given by Zhang (in press). In the meantime it can be noted that the general glacio-isostatic recovery rate deduced so far for the Vestfold Hills (Table 3) is of the same order of magnitude as rates for other parts of coastal Antarctica where terraces of similar elevation (normally < 20 m) have been dated.

The oldest terraces date from when the Flandrian transgression was nearing its maximum. According to Clark and Lingle (1979, p.279) "The contribution of the Antarctic ice sheet to the total eustatic sea-level rise is assumed to be 25 m (25% of the assumed total eustatic rise)" between 17,000 and 5000 BP. Eustatic changes since then are thought to have been slight (< 1 m due to glacier melting and steric changes) so that a net fall in sea level has characterised areas affected by glacio-isostatic recovery.

*Acknowledgements* The authors sincerely acknowledge the Antarctic Division, Department of Science and Technology, Australia and the ANARE Wintering party, Davis station, 1981, for their great help and support. We wish to thank Dr J.A. Peterson (Monash University) for his most valuable comments and for help in improving the final manuscript. We are indebted to Mr J. Pickard (Macquarie University) for his assistance in the Vestfold Hills in February 1981, and for his criticisms of an early version of this paper. Radiocarbon dates were provided by Miss Jing Li and Mr Len Daming, of the radiocarbon laboratory, Institute of Geography, Academia Sinica. We also thank Dr Lan Xui and Mr Xue Yaosong, Nanjing Institute of Geology and Palaeontology, Academia Sinica, for identifying bivalves and worm tubes, Dr Li Jiaying, Institute of Geology, Chinese Academy of Geological Sciences, for identifying diatoms, Dr Chen Zhiping, Dr Chen Yongzhong and Miss Zhang Li, Institute of Geography, Academia Sinica, for their comments on the initial manuscript.

## REFERENCES

BOULTON, G.S., 1970: On the origin and transport of englacial debris in Svalbard Glacier. *J. Glaciol., 9, (56),* 213-29.

BURTON, H.R. and CAMPBELL, P.J., 1980: The climate of the Vestfold Hills, Davis station, Antarctica. With a note on its effect on the hydrology of hypersaline Deep Lake. *ANARE Reports, Series D, 129.*

CAMERON, R.L. and GOLDTHWAIT, R.D., 1961: The U.S.-IGY contribution to Antarctic geology, *Union Geodesique et Geophysique Internationale, Association Internationale d'Hydrologie Scientifique. Pub. No. 55,* 7-13.

CLARK, J.A. and LINGLE, C.S., 1979: Predicted relative sea-level changes (18,000 years B.P. to present) caused by late glacial retreat of the Antarctic Ice Sheet. *Quat. Res., 11,*279-97.

CROHN, P.W., 1959: A contribution to the geology and glaciology of the west part of Australian Antarctic Territory, *ANARE Report, Series A., 3.*

DENTON, G.H., ARMSTRONG, R.L. and STUIVER, M., 1970: Late Cenozoic glaciation in Antarctica: the record in the McMurdo Sound region. *U.S. Antarct. J., 5,*15-27

DENTON, G.H. and PORTER, S.C., 1970: Neoglaciation. *Sci Amer., 222,*101-10

HUGHES, T.J., DENTON, G.H., ANDERSEN, B.G., SCHILLING, D.H., FASTOOK, J.L. and LINGLE, C.S., 1981: The last great ice sheets: a global view. *In* Denton, G.H. and Hughes, T.J. (eds.), *The Last Great Ice Sheets.* John Wiley & Sons, New York, 275-318

JOHNSTONE, G.W., LUGG, D.J. and BROWN, D.A., 1973: The biology of the Vestfold Hills, Antarctica. *ANARE Reports, Series B (1), Zoology, Publ. No. 123.*

McLEOD, I.R., 1964: The saline lakes of the Vestfold Hills, Princess Elizabeth Land; *in* Adie, R.J. (ed.) *Antarctic Geology,* North Holland, Amsterdam, 65-72. Amsterdam, 65-72.

MEGURO, H., YOSHIDA, Y., UCHIO, T., KIGOSHI, K. and SUGAWARA, K., 1964: Quaternary marine sediments and their geological dates with reference to the geomorphology of the Kronprins Olav Kyst; *in* Adie, R.J. (ed.) *Antarctic Geology,* North Holland, Amsterdam, 73-80.

NICHOLS, R.L., 1968: Coastal geomorphology, McMurdo Sound, Antarctica. *J. Glaciol., 7, (51),* 449-78

PICKARD, J. (in press): Holocene winds of the Vestfold Hills, Antarctica. *N.Z. J. Geol. Geophys.*

PICKARD, J. and ADAMSON, D. (in press, a): Probable Late-Pleistocene marine deposits in the Vestfold Hills, Antarctica; *in* Bowler, J. (ed.) *Proceedings of the CLIMANZ Conference,* Howmans Gap, Victoria, 1981.

PICKARD, J. and ADAMSON, D. (in press, b): Holocene marine deposits and ice retreat, Vestfold Hills, Antarctica; *in* Bowler, J. (ed.) *Proceedings of the CLIMANZ Conference,* Howmans Gap, Victoria, 1981.

ROBIN, G. de Q., 1977: Ice cores and climatic change. *R. Soc. Lond., Philos. Trans., Ser. B, 280,* 143-68.

SHI YAFENG and WANG JINGTAI, 1981: The fluctuations of climate, glaciers and sea level since Late Pleistocene in China; *in* Allison, I. (ed.), Sea level, Ice and Climatic change. *Int. Assoc. Hydrol. Sci., Publ., 131,* 281-93.

WARD, B.L. and WEBB, P.N., 1979: Investigation of Late Quaternary sediments from Cape Royds—Cape Barne area, Ross Island. *U.S. Antarct. J., 14,* 36-8.

ZHANG QINGSONG, (ed.)(in press): *Late Quaternary studies of the Vestfold Hills, Antarctica*(in Chinese with English abstracts). Science Press, Biejing.

# PERIGLACIAL LANDFORMS IN THE VESTFOLD HILLS, EAST ANTARCTICA: PRELIMINARY OBSERVATIONS AND MEASUREMENTS

Zhang Qingsong, *Institute of Geography, Academia Sinica, Peking, China.*

*Abstract* Ten types of periglacial landforms engendered by frost-wedging, frost heaving and wind are described from the Vestfold Hills, Antarctica. The periglacial landforms developed on surfaces which have been ice-free less than 5000 years. Expansion and contraction of sorted circles was measured between February 1981 and August 1982 at Mossel Lake. The mean annual growth rate of large sorted circles is between 1.2-10 mm/yr, with a maximum of 6-23 mm/yr. Seasonal expansion and contraction of the sorted circles occurred throughout the period of measurement. Expansion occurs in February and March, and contraction during the period November-January. This shows that the development of sorted circles is active in summer and stable in winter.

The Vestfold Hills area (68°35′S, 78°00′E) is a comparatively large (~ 400 km²) ice-free zone on the eastern side of Prydz Bay, on the coast of Princess Elizabeth Land, East Antarctica. It has been described by Law (1959), McLeod (1964) and Johnstone et al. (1973). A reconnaissance survey of periglacial landforms can be found in Blandford (1975).

Meteorological records are available for the period 1957/74, excluding 1965/68, and indicate a mean annual temperature

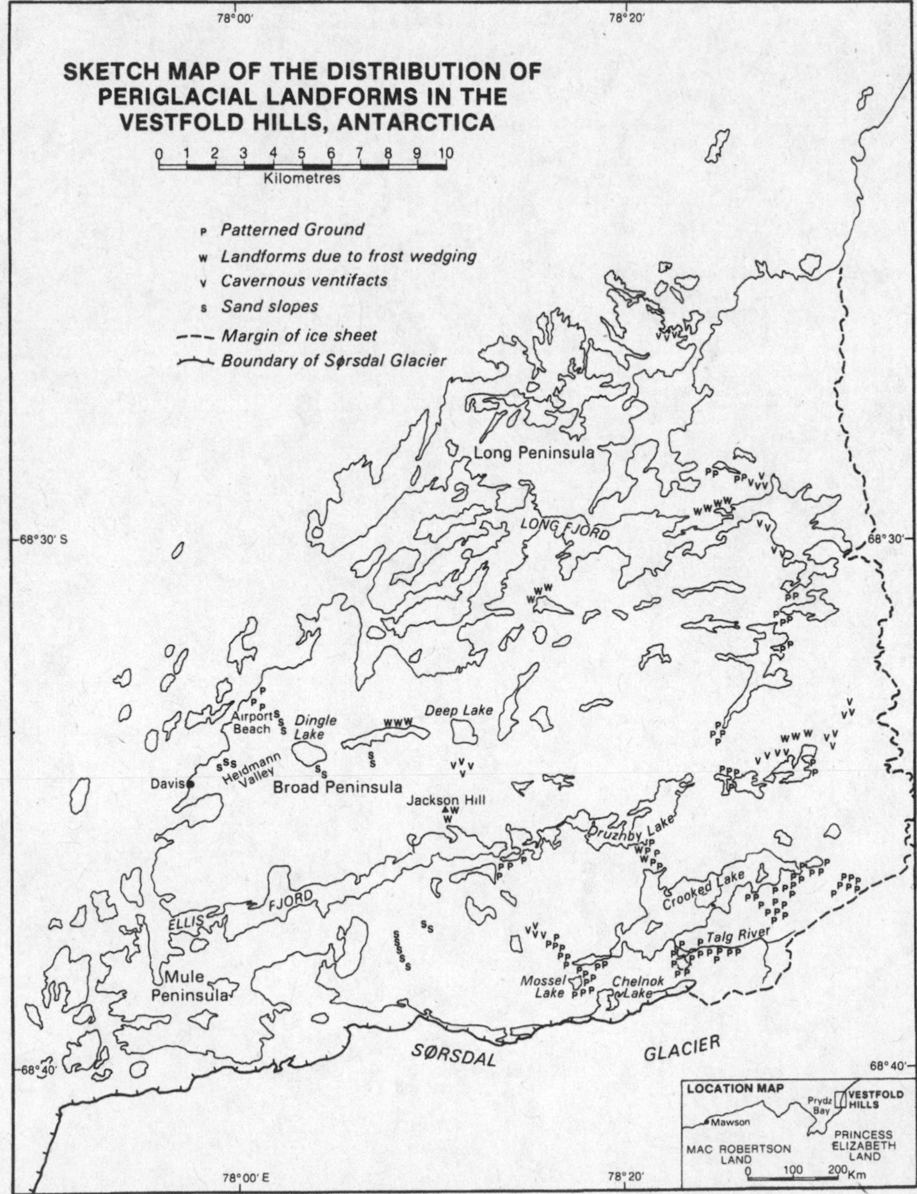


**Figure 1. Sketch map of the distribution of periglacial land forms, Vestfold Hills.**

**Figure 2. Block slope on the east side of Jackson Hill, looking southwest. 20 October 1981.**

**Figure 3. Tors of ice wedged country rock between Druzhby Lake and Crooked Lake: (Facing to West, 18 October 1981).**

**Figure 4. Large sorted circles on the beach of Mossel Lake.**

**Figure 5. Small sorted circles on the youngest moraines Northern side of Chelnok Lake.**

**Figure 6. Large sorted stripe on hillside above Mossel Lake: Mid-September 1981.**

**Figure 7. Debris islands in the bed of Talg River: Mid-February 1981.**

**Figure 8. Cavernous ventifacts developed on the east facing side of an outcrop of the coarse grained layer of the paragneiss, 0.5 km north of the head of Long Fjord.**

**Figure 9. Sand slope and remnant of annual snow patch in the lee of a low northwesterly trending ridge (30 m high) west of Dingle Lake: looking east, early March 1981.**

of – 10.2° (Burton and Campbell, 1980). The only month with a daily mean temperature above freezing (1°C) is January (see Figure 3 of Burton and Campbell, 1980). Even so the Vestfold Hills have a warmer climate than that of certain other coastal locations of similar latitude, because the ice-free surfaces have a comparatively lower albedo and a higher thermal inertia (Burton and Cambell, 1980).

The mean annual wind velocity at Davis is 5.0 m/sec. and the strongest winds (greater than 17 m/sec.) occur mainly in winter and have and overall frequency of less than 4% (Burton and Campbell, 1980). In each month after January the major wind direction shifts successively away from NNE towards ESE, and from June until the end of the year it shifts back again. The movement of drifting snow follows the same pattern as for strong winds, resulting in the accumulation of snow drifts in the lee of topographic irregularities. Annual precipitation totals recorded vary between 70 mm and 140 mm water equivalent (M. Whitehouse 1981, pers. comm.) almost always in the form of snow, but some summer rainfalls have occurred.

## TYPES AND DISTRIBUTION OF PERIGLACIAL LANDFORMS

Landforms due to frost-wedging, frost-heaving, and wind action (Washburn, 1979) can be distinguished (Figure 1).

*Landforms due to frost wedging*

*Block Slopes* Block slopes (e.g. Washburn, 1979, Figure 4.7) are confined to steeper (25°-40°) slopes on fractured gneiss (Figure 2). In some cases the block slope includes erratics from upslope. The blocky mantles are without interstitial fines or ice, and harbour neither long-lying snow drift nor summer melt water. They show no sign of contemporary mass movement and probably accumulated by rock fall from up-slope, possibly aided by minor frost creep.

*Tors* Frost-shattered bedrock ridge tops carry accumulation of blocks frost wedged in situ to resemble single-cycle tors. Some stand 1 m or so above the general surface, for example near Mossel Lake and between Crooked and Druzhby Lakes (Figure 3).

*Landforms due to frost heaving*

*Patterned ground* At least six varieties of patterned ground have been found. They are widely distributed on the east Mule Peninsula close to the ice plateau and in some areas adjacent to fresh water lakes on Broad Peninsula.

*Large sorted circles* These are between 2 and 6 m in diameter and are all found where surface materials are seasonally saturated by summer melt water as on the floors of broad valleys, or on beaches of gentle slope (2-3°) around fresh water lakes. Sand had been blown off the surface of these features but comprises 5-35% (by weight) of the material beneath the surface within the stony borders. In general grain size decreases with circle diameter. Well developed large sorted circles contain secondary polygons between 5 and 15cm in diameter (Figure 4). These features are actively forming (see below).

*Small sorted circles* These range from 0.3 to 2 m in diameter and occur on the young moraine deposits in the southern Vestfold Hills (Figure 5) and therefore are probably younger than the large sorted circles. They are also composed of coarser materials, which probably inhibit the rate of development due to high permeability and low porosity.

*Sorted polygons* These are rare and occur in low-lying areas on the edge of the Talg River bed where there is an abundance of silt and sand. They have irregular, stony borders and are usually saturated, often flooded, during summer. Surfaces may carry dessication cracks forming small non-sorted polygons.

*Large sorted stripes* These occur in gentle slopes (5°-10°) on the lee side of hills below winter snow drifts (Figure 6). Stony stripes are up to several metres apart and tens of metres long. Their distribution appears to be controlled by availability of melt water from up-slope snow drifts and a plentiful supply of frost-shattered stones mantling slopes of suitable gradient.

*Debris islands* These are known from the Talg River bed (Figure 7). They have irregular rocky borders surrounding non-sorted muddy islands often with a block in the centre. They differ from the type of examples quoted elsewhere (e.g. Washburn, 1979, Figures 5.10 and 5.11) and may represent the development of these features on the flatter ground where moisture for ground ice is supplied from a river bed rather than melt water run-off on valley sides. During the summer the fines are subject to diurnal freezing and thawing. The features are actively forming and probably comparatively young because the Talg River bed is amongst the most recently deglaciated parts of the Vestfold Hills (Zhang, et al., this volume).

*Stone pavements* These occur in locations similar to those in which sorted stripes are found but the two are not found together. Pavements were noted at Airport Beach, Heidemann Valley and on some of the terraces surrounding Crooked Lake. The pavements appear to favour locations where stones are flat-sided or facetted and there is a deeper active layer formed in regolith with abundant matrix. Salt in the matrix might favour greater and longer mobility in the active layer thus maximising the opportunity for flatsided stones to distribute themselves on the surface.

*Landforms due to wind action in the periglacial environment*

*Cavernous ventifacts* Outcrops facing east and close to the ice cap north of the Sørsdal Glacier are in the path of the diurnal katabatic winds, which may sustain high velocities for six or eight hours. Such outcrops exhibit well developed ventifacts which, in the coarse-grained intercalations of the layered paragneiss (see Oliver et al. 1982) at the head of Long Fjord take a form reminiscent of cavernous or even honey-comb weathering (Figure 8). The aspect and distribution of outcrops like this strongly suggest that sand, including coarse grades, travels with drifting snow derived from the snow patches that cover dead ice now lying asymmetrically in narrow valleys that now head in the ice-cored moraine at the ice cap edge. Firnification in the snow patch is partly by infiltration congelation and the resultant ice layers are periodically dismembered along with the overlying snow and firn, especially during blizzards. Thus ice particles as well as sand and snow play a role in ventifact formation.

Further downwind (i.e. to the west) similar weathering forms can be found on suitably-placed outcrops of orthogneiss where grain size coarsens. Here cavernous weathering forms include hollows up to 75 cm across and 25 cm deep. Dolerite dykes stand out in high relief in both areas described above, and do not exhibit such weathering features.

*Sand slopes* These are lee-side deposits which have accumulated from residual sand left behind after the melting of snow patches that form during the colder months (Figure 9). Despite their southwesterly aspect, they receive sufficient solar radiation during the months of higher sun angles to be melted then. The orientation and height of the ridges above the sand slopes makes them less efficient "snow fences" than those associated with the permanent snow patches, and also facilitates the receipt of summer solar radiation. The surface of the sand slopes is quite dry but some of the snow-melt water is re-frozen in the lower layers thus accounting for the general stability of these features. The top several centimetres are mobile during summer and may form miniature barchans (< 0.5 m high) migrating downslope. Particle size distribution is dominated by the coarse sand fraction with some fine gravel derived from weathering of nearby bedrock. Estimated thickness of these deposits was always < 10 m.

## MEASUREMENTS OF ACTIVITY OF SORTED CIRCLES

Measurement of seasonal changes in pattered ground is the best way of establishing their active status. In February 1981 eight sets of steel rods (d. = 2 cm) 45 cm long (120 in total) were driven vertically into the active layer, in the centres and on the borders of large sorted circles formed in low lying flat ground near Mossel Lake. Horizontal displacements were monitored by the author until December 1981 and measurements have been continued since then by ANARE expeditioners.

*Results of activity monitoring*

Distinct seasonal activity (Table 1) is evident in the behaviour of the large sorted circles under observation. Expansion takes place in February and March with the expansion upwards of the permafrost and contraction occurs during annual active layer development in November and January. Over winter the circles are stable. Over two seasons mean expansion amounts varied between 3 mm and about 9 mm (Table 1).

Variations of annual rates of expansion of the large sorted circles are indicated by distinct differences between respective radii. The limited data suggest that these differences correlate with size of sorted circle, expansion rate falling exponentially with increasing size. The

TABLE 1: Seasonal expansion and contraction and annual growth rate of large sooted circles

| Site | Diameter (m) | Sand contents of interior of circle (%) | Movement* (mm) Feb 81/Mar 81 Mean | Feb 81/Mar 81 Max. | Mar 81/Nov 81 Mean | Nov 81/Jan 82 Mean | Nov 81/Jan 82 Max. | Jan 82/Feb 82 Mean | Feb 82/Mar 82 Mean | Feb 82/Mar 82 Max. | Mar 82/Aug 82 Mean |
|---|---|---|---|---|---|---|---|---|---|---|---|
| Group G-1 | 3.4 | — | +2.5 | +8 | 0 | −6.4 | −11 | −2 | +6 | +11 | 0 |
| Group B | 2.9 | 45 | +5 | +16 | 0 | −9.4 | −21 | 0 | +7.6 | +12 | 0 |
| Group G-2 | 2.4 | 40 | +9.5 | +19 | 0 | −3 | −7 | +1 | +14 | +21 | 0 |

| Site | net movement (mm) Feb 81-Mar 82 mean | max. |
|---|---|---|
| Group G-1 | 1.2 | 6 |
| Group B | 3.0 | 14 |
| Group G-2 | 10 | 23 |

*Positive movement is expansion; negative is contraction. Distance is taken from the centre to the margin of large sorted circles with nominal accuracy of 1 mm.

most important factor influencing the annual growth rate is the net contraction of the width of the borders (wedges), the wider amongst which seem to be subject to lower rates of narrowing (Table 2). Field evidence indicates that smaller sorted circles normally have wider borders thus, it seems, offering greater scope for narrowing in the process of circle expansion than do the narrow border wedges, tightly filled with pebbles that define the larger sorted circles.

## DISCUSSION

### *The uniformity of periglacial types*

Compared with the range of active periglacial processes and resultant forms to be found in some other periglacial area, the suit of features found in the Vestfold Hills lacks diversity. For example, there are more than 50 types of periglacial landforms engendered by seven periglacial agents on the Tibetan Plateau of China (Cuio Zhijiu, 1982). In the Vestfold Hills only 10 types, engendered by three agents are found, and the majority of features are not well developed. This is in conformity with limitations imposed by the Vestfold environment: comparatively low mean temperatures coupled with comparatively short periods of freeze-thaw activity and widespread dryness in regolith combine to limit the scope for the development of periglacial forms. In addition, saturation of the active layer is in many places due to saline water which inhibits and, close to hypersaline lakes and in deposits that include evaporites, prevents freezing, even at temperatures as low as −19°C, thus eliminating the possibility of periglacial activity there.

TABLE 2: Net contraction of wedges between circles

| Site | (m) | Net contraction (mm) Mean | Maximum | Duration |
|---|---|---|---|---|
| Within Group A | 4.5 | −26 | −29 | February 1981-March 1982 |
| Surrounding Group G-1 | 1.2 | −14 | −1 | February 1981-March 1982 |

*Negative movement is contraction. Distance between pegs set on the borders of wedges is measured monthly with nominal accuracy of 1 mm.

### *Age of the periglacial landforms*

The maximum age for some of the periglacial features (e.g. the various forms of patterned ground) is constrained by the age of the deposits in which they have developed . Thus, for instance, Black and Berg (1963) and Berg and Black (1966) have dated periglacial landforms in Victoria Land as younger than 10,000 years. Large sorted circles and some other types of patterned ground may be found on raised beaches and terraces that have been dated as 5000-6000 years old (Pickard and Adamson, in press; Zhang et al., this volume). The ice-cored moraines probably represent neoglacial advances (Zhang et al., this volume) and small sorted circles on their surfaces date no further back than 2000-3000 years. On those moraines still characterized by substantial ice cores much younger ages are to be expected. The debris islands in the Talg River must have been formed during this interval.

*Acknowledgements* The authors's sincere acknowledgements go to the Antarctic Division of the Department of Science and Technology, Australia for the opportunity to work in the Vestfold Hills. The final manuscript was prepared while the author was a guest of the Antarctic Division and of Monash University during the spring of 1982. I am grateful to Dr J.A. Peterson and staff of the Geography Department of that University for help in preparing the final manuscript. I also thank Associate-Professor Cui Zhijui, Peking University for his suggestions on the composition of this paper and Dr Chen Zhiping, Dr Chen Yong Zhong, Miss Zhang Li, Institute of Geography, Academia Sinica and Mr J. Pickard, Macquarie University for their comments on earlier versions of the manuscript. I thank ANARE expeditioners from the 1981 and 1982 parties for their help with the programme for monitoring the activity of patterned ground.

## REFERENCES

BERG, T.E. and BLACK, R.F., 1966: Preliminary measurements of growth of nonsorted polygons, Victoria Land, Antarctica; *in* Tedrow, J.C.F. (ed.), *Antarctic Soils and Soil Forming Processes,* Am. Geophys. Union, Ant. Res. Ser., 8, 61-108.

BLACK, R.F. and BERG, T.E., 1963: Patterned ground in Antarctica; *in* International Conference on Permafrost, Lafayette, Ind., 1963, Proceedings, Washington, National Academy of Sciences, National Research Council (1966) 121-8.

BLANDFORD, D.C., 1975: Spatial and temporal patterns of contemporary geomorphic processes in the Vestfold Hills, Antarctica. Unpublished B. Lett. thesis, Univ. of New England, Armidale, Australia.

BURTON, H.R. and CAMPBELL, P.J., 1980: The climate of the Vestfold Hills, Davis Station, Antarctica, with a note on its effect on the hydrology of hypersaline Deep Lake. *ANARE Sci. Rep., Ser. D, 129.*

CUI ZHIJIU, 1982: Basic characteristics of periglacial landforms in the QingHai-XiZang (Tibet) Plateau. *Scientia Sinica, Ser. B, 25,* 79-95.

JOHNSTONE, G.W., LUGG, D.J. and BROWN, D.A., 1973: The biology of the Vestfold Hills, Antarctica. *ANARE Sci. Rep., Series B(1), Zoology, 123.*

LAW, P.G., 1959: The Vestfold Hills, *ANARE Sci. Rep., Ser. A(1), 47.*

MCLEOD, I.R., 1964: The saline lakes of the Vestfold Hills, Princess Elizabeth Land; *in* Adie, R.J. (ed.) *Antarctic Geology.* North Holland, Amsterdam, 65-72.

OLIVER, R.L., JAMES, P.R., COLLERSON, K.D. and RYAN, A.B., 1982: Precambrian geologic relationships in the Vestfold Hills, Antarctica; *in* Craddock, C. (ed.) *Antarctic Geoscience.* Univ. Wisconsin Press, Madison, 435-44.

PICKARD, J. and ADAMSON, D. (in press): Holocene marine deposits and ice retreat, Vestfold Hills, Antarctica; *in* Bowler, J. (ed.) *Proceedings of the CLIMANZ Conference,* Howmans Gap, Victoria 1981.

WASHBURN, A.L., 1979: *Geocryology, A Survey of Periglacial processes and Environments.* Edward Arnold Ltd., London.

ZHANG QINGSONG, XIE YOUYU and LI YUANFANG (this volume): A preliminary study on the evolution of Vestfold Hills environment, East Antarctica, since Late Pleistocene.

# SURFACE GEOLOGY AND GEOMORPHOLOGY OF THE LUTZOW-HOLM BAY REGION, ANTARCTICA—AN INTERIM REPORT

Yoshio Yoshida and Kiichi Moriwaki, *National Institute of Polar Research, Tokyo 173, Japan.*
Kiyotaka Sasaki, *Department of Geology, Faculty of Science, Sendai, Japan.*

*Abstract* Surface geological and geomorphological research was carried out in 1981 as a part of a three year project—Synthetic analysis of crustal structure in the Lützow-Holm Bay region, East Antarctica, by geophysical, geological and geomorphological investigatons. Preliminary results of the study are as follows.

In the Yamato Mountains area, 200 km inland from the nearest coast, the surface elevation of the ice sheet has been reduced at least 400 m from its maximum stage of development. A stillstand stage occurred during the retreat. The ice sheet appears to have been stagnant in recent times, viewed from the mode of occurrence of surface moraines. The ice sheet at the maximum stage buried completely whole mountains and exerted effective areal scouring on gentle slopes of the mountains. Wet-based glacial erosion must have been partly responsible for such scouring.

In the coastal area, the former ice sheet extended to the outer margin of the continental shelf and eroded the present ice-free areas and the continental shelf by regional scouring and selective linear erosion. Evidence of wet-based glaciation during that time is distributed in places on ice-free areas. $^{14}C$ ages of elevated beach deposits formed after deglaciation indicate that the ice sheet retreated from the area at least 30,000 years B.P. Stagnation of the ice margin during the retreat appears to have occurred on the margins of the present ice-free areas. The continental shelf is characterised by an inner surface with considerable relief and a smoother outer shelf surface. Topographic features of two troughs and a part of the smoother shelf suggest that Lützow-Holm Bay has been tectonically depressed. This situation caused the concentration of ice streams which led to the formation of a remarkable drowned valley system.

# GLACIAL EROSION AND MORPHOLOGY OF THE EASTERN AND SOUTHEASTERN WEDDELL SEA SHELF

Anders Elverhøi, *Norwegian Research Institute, Box 158, N-1330 Oslo, Lufthavn, Norway.*

George Maisey, *Continental Shelf Institute, Box 1883, N-7001 Trondheim, Norway.*

*Abstract* Sampling (38 samples) of the surface sediments in the eastern and southeastern Weddell Sea show till intermittently overlain by glaciomarine deposits, supporting the concept of Late Wisconsin grounded ice extending to the shelf edge. The 1100 m deep Crary Trough has been covered by grounded ice, indicating increased ice discharge from the Filchner Ice Shelf.

Sparker data show former grounding over the entire shelf and (1) a thin (< 15 m) veneer of sediments above a well defined angular unconformity in most places, and (2) accumulations (100-400 m (two-way travel time) thick) of glaciogenic deposits in the mouth of the Crary Trough, which consist of two units, separated by an angular unconformity. The upper unit, confined to the eastern side of the Crary Trough, is unconformable with the sea floor in its western and southern part.

Tentative dating of the strata unconformably underlying the glaciogenic deposits in the outermost part of the Crary Trough, indicates that grounded ice had extended to the shelf edge by the Mid/Late Tertiary period. This contrasts with the Ross Sea area where glaciers were calving at the head of the embayment.

Glacial conditions have probably prevailed in Antarctica since the Oligocene (Denton et al., 1971; Hayes and Frakes, 1975; Drewry, 1975; Barrett, 1981). The ice cover was initially limited, whereas the period Pliocene-Recent is characterised by extensive ice cover, including ice shelves (Drewry, 1976; Barrett, 1981). Major unconformities in the Ross Sea stratigraphy in the Early Pliocene and the Late Pliocene/Early Pleistocene are assumed to reflect wide scale build-up and expansion to the shelf edge of the Antarctic Ice Sheet (e.g. Drewry, 1976). Repeated seaward advances and retreats probably also occurred later in the Pleistocene, although the frequency and extent of these events are highly disputed (e.g. Drewry, 1976; Drewry, 1979; Kellogg et al., 1979; Thomas, 1979; Stuiver et al., 1981). It should however be noted that the Ross Sea Quaternary sediments show little evidence of glacial erosion (Houtz and Davey, 1973).

Compared with the Ross Sea Embayment, the Late Cenozoic history of the Weddell Sea Embayment is less well known and is so far confined to the Late Wisconsin. During this period, grounded ice probably covered the Weddell Sea Shelf, including the Crary Trough (Elverhøi, 1981). However, as in the case of the Ross Sea, former grounding and seaward advance of adjacent ice masses may also have occurred.

In this paper we use shallow seismic reflection data and sediment sampling to discuss the glacial history of the southeastern and eastern Weddell Sea.

## DATA AND DATA ACQUISITION

Shallow seismic profiling and sediment sampling were conducted on the eastern and southeastern Weddell Sea Shelf (Figures 1a and 1b) during the Norwegian Antarctic Research Expedition (NARE) in 1976/77 and 1978/79. The location of the seismic profiles was partly determined by sea ice conditions. The shallow seismic records were obtained by a 4 kJ sparker and recorded on an analogue recorder after filtration between 40 Hz and 400 Hz. Sediment samples were taken by gravity corer (3 m long), grab (1 m³) and by dredge hauls. Navigation was by dual channel satellite receiver integrated with speed log and gyro.

## SURFACE SEDIMENT DISTRIBUTION

Stiff pebbly mud containing shell fragments is exposed on the sea floor on the outer shelf and in the Crary Trough (Figure 2). Shear strength values measured by cone showed values in the range of 40 to 150 kN/m². Based on its textural and geotechnical properties, the stiff pebbly mud is interpreted as till and/or glacially compacted glaciomarine deposits (Anderson et al., 1980; Orheim and Elverhøi, 1981; Elverhøi and Roaldset, 1982). Soft pebbly mud, or glaciomarine deposits, intermittently overlie the stiff pebbly mud (Figure 2). The sedimentation rate for these deposits is relatively low, 20 to 50 m $Ma^{-1}$, on the outer shelf/upper slope, with a lower rate for the central and inner shelf (Elverhøi and Roaldset, 1982). In general the glaciomarine sediments are homogeneous. Laminations were only observed in cores from station 13 and 214 (Figure 1a).

### Eastern Shelf

Along the eastern shelf, that is the region between SANAE and General Belgrano (Figure 1a, b) only a thin (<15 m) cover of sediments is present above a well defined angular unconformity (Figure 3). On the shelf north of General Belgrano the surface is irregular and acoustically opaque, suggesting exposure of crystalline basement (Figures 4 and 5; profiles 5 and 9). This is confirmed by the high seismic velocity (5.5 km/s) at the sea floor in the area (Haugland pers. comm., 1982). North of Halley, outcrops of sedimentary rocks

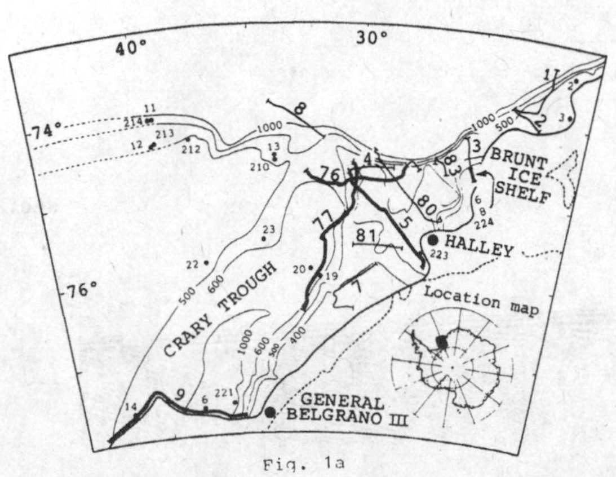


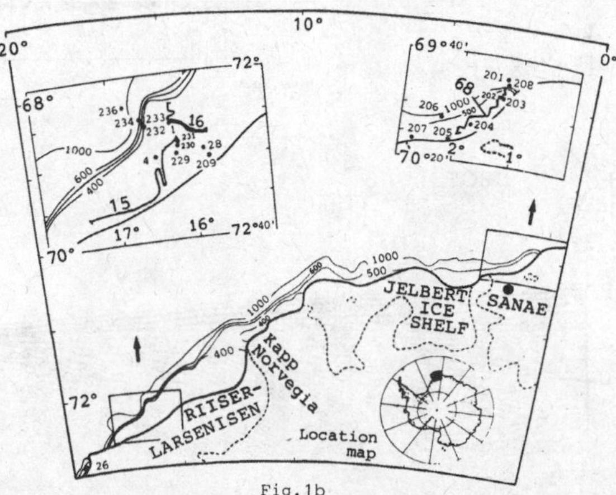


**Figure 1. Bathymetic map of (a) the southeastern and (b) the eastern Weddell Sea Shelf showing the locations of sparker profiles and sediment samples taken by NARE in 1976/77 and 1978/79. Profiles illustrated in Figures 3 and 5 are shown by heavy lines.**

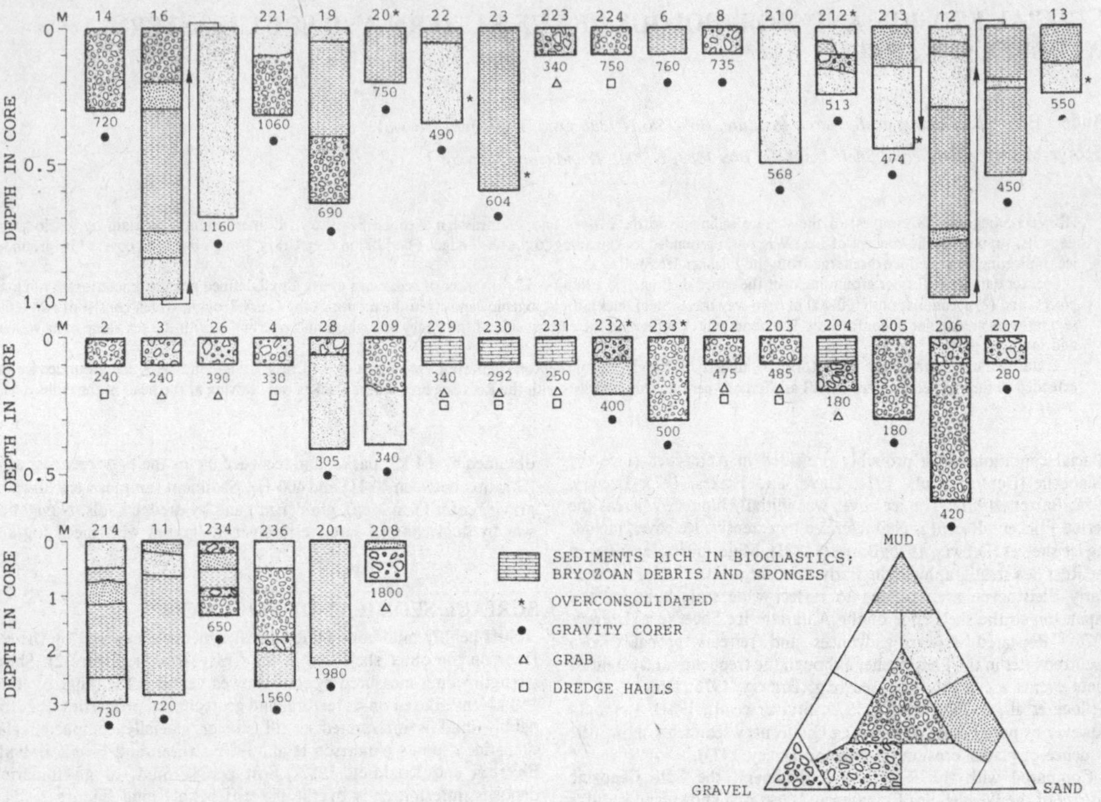

SEDIMENTS RICH IN BIOCLASTICS; BRYOZOAN DEBRIS AND SPONGES
* OVERCONSOLIDATED
● GRAVITY CORER
△ GRAB
□ DREDGE HAULS

MUD
GRAVEL
SAND

DEPTH IN CORE

**Figure 2. Lithology of sediment samples taken at NARE 1976/77 and 1978/79. Station number above and water depth below.**

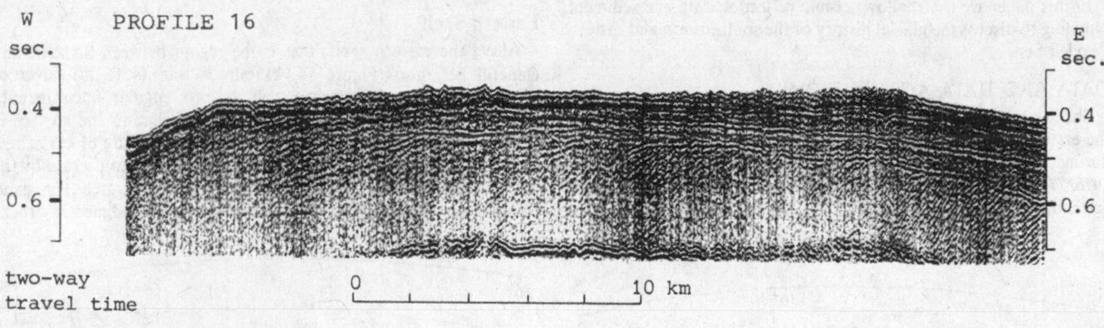


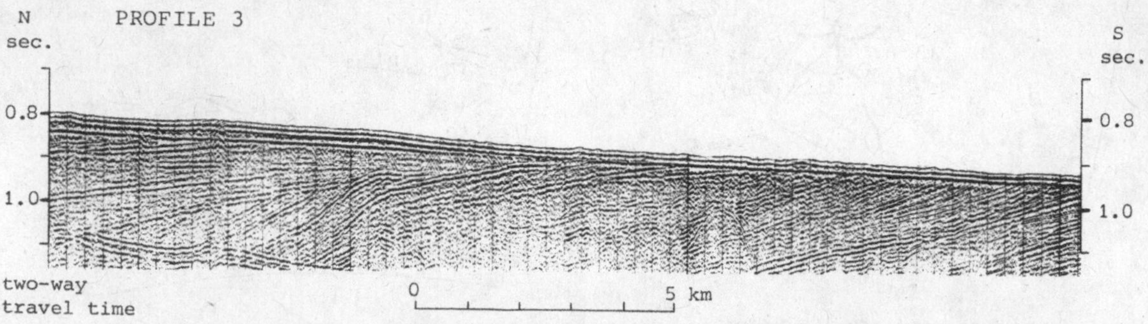


**Figure 3. Sparker profiles outside Riiser-Larsenisen (Profile 16) and north of Halley (Profile 3).**

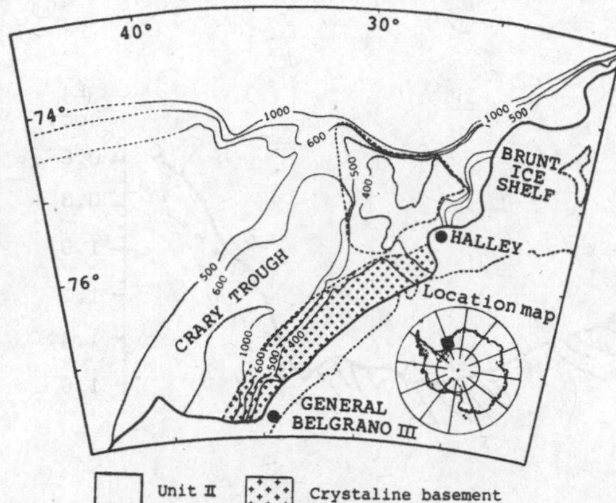


**Figure 4. Bathymetric map of the southeastern Weddell Sea Shelf showing the area of exposure of crystalline basement and the extent of unit II.**

occur on the sea floor (Figure 3, profile 3). In spite of considerable effort we had no success in obtaining in situ material from the area.

Outside Riiser-Larsenisen, close to the ice front, a 200 m (two-way travel time) thick section is present above an angular unconformity (Figure 3, profile 16). The thickness decreases towards the shelf edge, where probably only a thin veneer is preent above the unconformity. To the east, internal truncated reflectors are seen, indicating sediment accretion from westward advancing ice. Sonobuoy measurements using a 4 kJ sparker as the energy source gave for the upper 100 m (two-way travel time) a velocity of 1.8-2.0 km/s, suggesting semi-consolidated deposits (Haugland, pers. comm., 1981).

## Crary Trough

The Crary Trough forms a north-south depression with depths exceeding 500 m relative to the adjacent areas (Figure 1a). In the inner and central parts of the trough only a thin veneer (<15 m) of sediments is present above westward dipping layers (Figure 5, profile 9). Semi-consolidated sediments are present on the western flank of the Crary Trough in front of the Filchner Ice Shelf. The 100-400 m high sill across the outer part of the Crary Trough is mostly formed by an extensive sediment lens, underlain by a well defined angular unconformity (Figure 5, profiles 5, 76 and 77). The lens is made up of two units, with Unit II resting with angular unconformity on Unit I.

Unit II, confined to the eastern part (Figure 4), is a 100-200 m (two-way travel time) thick sequence with few internal reflectors. The eastern boundary of the unit is formed by a 30-50 m high ridge with internal truncated reflectors (Figure 5, profile 5). To the south and the west, internal truncated reflectors are probably also present. To the north, multi-channel seismic data parallel profile 77 (Figure 1a) show the unit to terminate as a ridge (Haugland pers. comm., 1982).

Unit I is a 50-150 m (two-way travel time) thick sequence with seaward dipping internal layers. To the south, east and north, Unit I has a slightly larger areal extent than Unit II and to the west, no outer limits have so far been determined. The compressional seismic velocity in the two units is in the range of 2.4-2.6 km/s (Haugland pers. comm., 1982).

# COMPOSITION OF THE SEMICONSOLIDATED SEDIMENTS

The upper—and sampled—part of the southern Weddell Sea Shelf sediments appears to be exclusively glaciogenic. Whether these are glaciomarine sediments, till and/or glacially compacted glaciomarine sediments, cannot be determined. As concerns Unit II, however, the morphology suggests deposition by ice. The relatively high seismic velocity further indicates a predominance of relatively coarse material and/or compaction by ice. The information on Unit I is too sparse for genetic interpretation, although its close connection with Unit II would tend to suggest a glacial origin.

# THE FORMATION OF THE CRARY TROUGH

The sediment distribution in the Crary Trough reflects at least three major stages: Stage 1: A wide scale erosional phase, predating deposition of Unit I. Stage 2: Deposition and subsequent erosion of Unit II. Stage 3: Deposition and erosion of Unit II.

From the shallow seismic profiles it can be seen that if Units I and II are ignored, the Crary Trough still remains slightly overdeepened. The profile in front of the Filchner Ice Shelf shows also that the Trough is excavated into crystalline rocks. The Crary Trough has further been found to continue at a water depth of 1000 m back to about 83°S (Jankowski and Drewry, 1980), which gives a total length of 900 km. Similar depressions or basins (or also "marginal channels") are a common feature on high latitude shelf areas (e.g. Holtedal and Holtedahl, 1961; Sugden and John, 1976). Their origin is not fully understood, although glacial erosion of a pre-existing relief is suggested as the most probable way of formation. The importance of structural control is demonstrated by the fact that these features are often found along the junction between basement and younger sedimentary rocks (e.g. Sugden and John, 1976). This is also the case for the Crary Trough (Figures 4 and 5; profiles 5 and 9).

The age of Stage 1 is not certain because the underlying strata are undated. According to the multichannel seismic data, Unit I is underlain in the outermost part of the Crary Trough by a reflector extending down the continental slope and rise (Haugland pers. comm., 1982). Sediment thickness above the reflector is 400 m (two-way travel time) on the upper slope decreasing to 300 m (two-way travel time) at the lower slope, with a seismic velocity of 2.1 km/s. Application of the sedimentation rate (20-50 m $Ma^{-1}$) by Elverhøi and Roaldset (1982) for the upper slope/shelf edge gives a Late Tertiary age for the reflector. Even a low sedimentation rate of 10 m $Ma^{-1}$ gives a Mid Tertiary age. The average sedimentation rate is not considered to have been affected by debris flow and turbidity currents reported from the Weddell Sea slope (Anderson et al., 1979). Thus the primary phase of erosion of the Crary Trough probably is synchronous with or postdates the initial phase of Antarctic glaciation. However, this conclusion must remain tentative until more adequate dates are available. Also it is possible that the initial erosion of the Crary Trough could predate the glaciation of the Antarctic.

Concerning Stages 2 and 3 the internal reflection pattern indicates a former wider extension of the deposits indicating an initial period of deposition, followed by erosion. The overdeepened morphology further requires erosion by glaciers. The age of the unconformity between Units I and II may, by analogy with the Ross Sea stratigraphy, correspond to the extensive expansion of the Antarctic Ice Sheet in the Early Pliocene. In the Ross Sea Embayment, 100->400 m of Oligocene-Miocene sediments unconformably underlie the 30-50 m thick horizontal Pliocene-Recent layers (Houtz and Davey, 1973; Barrett, 1975).

Repeated grounding and erosion, at least in the later part of the Pleistocene, is indicated by the sparse sediment cover, by till overlain by glaciomarine deposits and by the angular unconformity. Even with a sedimentation rate as low as 10 m $Ma^{-1}$, the glaciomarine sediments with a maximum thickness of a few metres will not span more than a couple of hundred thousands of years.

# GLACIAL HISTORY OF THE EASTERN WEDDELL SEA SHELF

The sparse sediment cover above an unconformity indicates repeated grounding and expansion to the shelf edge of the adjacent ice shelves, at least during the Pleistocene. The reflection pattern of the semi-consolidated sediments outside Riiser-Larsenisen and Brunt Ice Shelf (Figure 3, profile 16 and Figure 5, profile 5), shows that the advances have, in these areas, been followed by deposition.

# COMPARISON WITH ROSS SEA

Compared to the Ross Sea, the glacial history of the Crary Trough shows the following differences:

(1) During the initial glaciation of the Antarctic, Oligocene-Miocene, grounded ice may have extended to the shelf edge in the Crary Trough, while in the Ross Sea, glaciers were calving at the head of the embayment (Barrett, 1981).

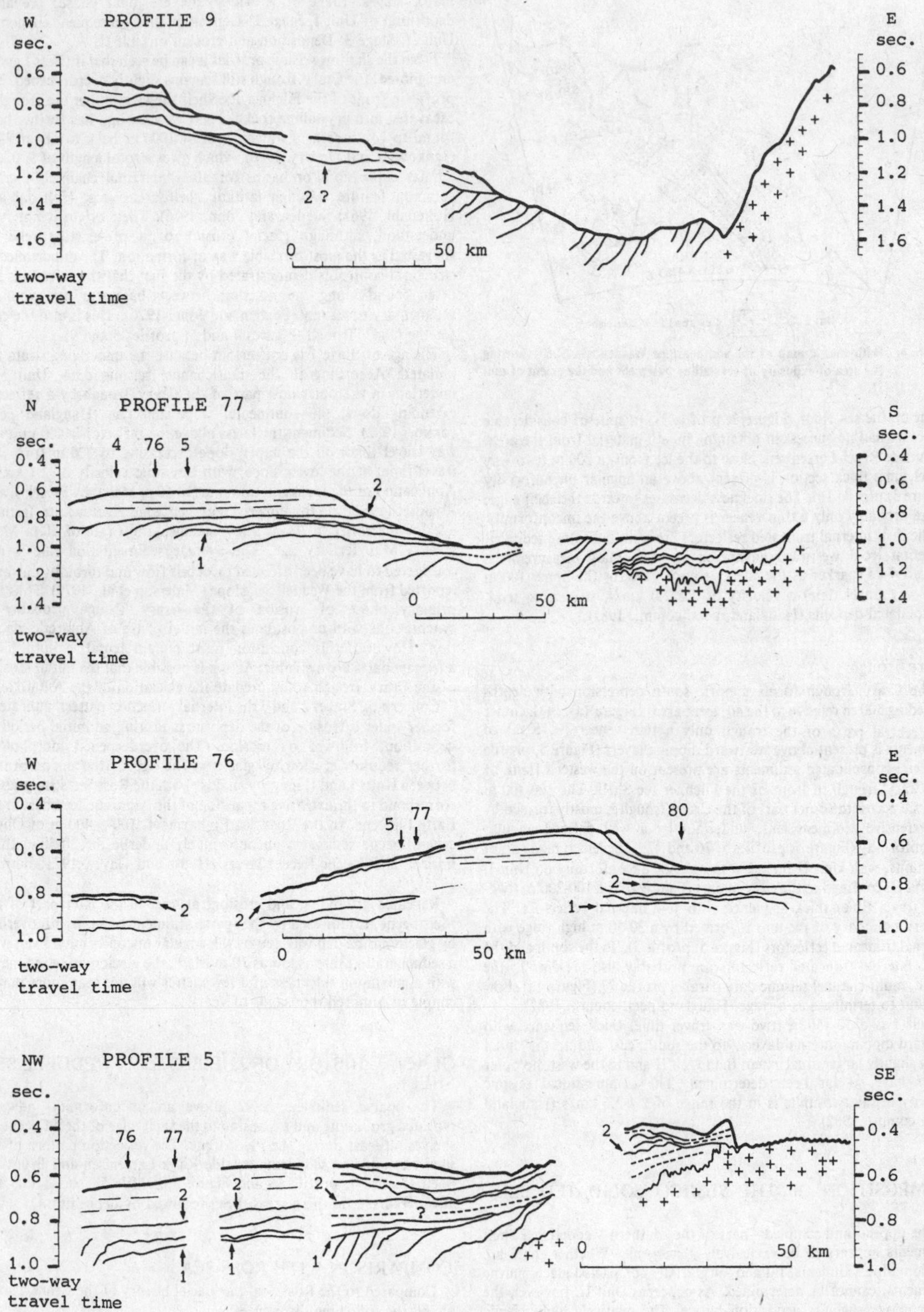


**Figure 5. Line drawings of sparker profiles from outside the front of the Filchner Ice Shelf (Profile 9) and the outer part of the Crary Trough (Profiles 5, 76 and 77). The lower boundaries of units I and II are shown by 1 and 2, respectively.**

(2) In the Ross Sea, there seems little evidence of Pleistocene glacial erosion (Houtz and Davey, 1973), while at least parts of the Filchner-Ronne Ice Shelf were grounded and extended to the shelf edge several times during the Pleistocene, probably as late as Late Wisconsin.

This latter indicates increased ice discharge through the Filchner-Ronne Ice Shelf. Grounding the Crary Trough requires the Filchner Ice Shelf to have been at least 800 m thicker than at present. With the ice shelf advanced 400 km to the edge of the continental shelf this could be achieved either by a modest thickening of the Antarctic Ice Sheet (e.g. Drewry, 1979, 1980) combined with the channelised ice flow, or by a more extensive build-up of the Antarctic Ice Sheet (e.g. Stuiver et al., 1981).

The eastern part of the Filchner-Ronne Ice Shelf drains mostly from the East Antarctic Ice Sheet, while the Ross Ice Sheet is mainly drained from West Antarctica. The differences in glacial history of the Ross Sea and the studied part of the Weddell Sea area could therefore reflect a different behaviour of the two ice masses.

*Acknowledgements* We thank colleagues on NARE in 1976/77 and 1978/79 for their help in the field, and Sven Backström and Leif Rise for their help with the sample analysis. Yngve Kristoffersen and Olav Orheim kindly reviewed the manuscript.

## REFERENCES

ANDERSON, J.B., KURTZ, D.D. and WEAVER, F.M., 1979: Sedimentation on the Antarctic slope. *Soc. Econ. Paleont. Mineral., Spec. Publ., 27,* 265-83.

ANDERSON, J.B., KURTZ, D.D., DOMACK, E.W. and BALSHAW, K.M., 1980: Glacial and glacial marine sediments of the Antarctic continental Shelf. *J. Geol., 88,* 399-414.

BARRETT, P.J., 1975: Textural characteristics of Cenozoic preglacial and glacial sediments at site 270, Ross Sea, Antarctica. *Initial Reports of the Deep Sea Drilling Project, 28,* 757-68. U.S. Govt. Printing Office, Washington, DC.

BARRETT, P.J., 1981: Late Cenozoic glaciomarine sediments of the Ross Sea, Antarctica; *in* Hambrey, M.J. and Harland, W.B. (eds.). *Earth's Pre-Pleistocene glacial record,* Cambridge Univ. Press, 208-11.

DENTON, G.H., ARMSTRONG, R.L. and STUIVER, M., 1971: The late Cenozoic glacial history of Antarctica; *in* Turekian, K.K. (ed.). *Late Cenozoic glacial ages,* Yale Univ. Press, New Haven, 267-306.

DREWRY, D.J., 1975: Initiation and growth of the East Antarctic ice sheet. *Geol. Soc. Lond., J., 131,* 255-73.

DREWRY, D.J., 1976: Deep sea drilling from Glomar Challenger in the Southern Ocean. *Polar Rec., 18,* 47-71.

DREWRY, D.J., 1979: Late Wisconsin reconstruction for the Ross Sea region, Antarctica. *J. Glaciol., 24,* 231-44.

DREWRY, D.J., 1980: Pleistocene bimodal response of Antarctic ice. *Nature, 287,* 214-16.

ELVERHOI, A., 1981: Evidence for a Late Wisconsin glaciation of the Weddell Sea. *Nature, 293,* 641-42.

ELVERHOI, A. and ROALDSET, E., (in press): Glaciomarine sediments and suspended particulate matter, Weddell Sea, Antarctica. *Polar Res., 3.*

HAYES, D.E. and FRAKES, L.A., 1975: General synthesis, Deep Sea Drilling Project Leg 28. *Initial Reports of the Deep Sea Drilling Report, 28,* 919-42. U.S. Govt. Printing Office, Washington, DC.

HOLTEDAHL, O. and HOLTEDAHL, H., 1961: On marginal channels and the problem of their origin. *Geol. Inst. Uppsala, Bull., 49,* 183-87.

HOUTZ, R.E. and DAVEY, F.J., 1973: Seismic profiler and sonobuoy measurements in the Ross Sea, Antarctica. *Geophys. Res., 78,* 3448-68.

JANKOWSKI, E.J. and DREWRY, D.J., 1981: The structure of West Antarctica from geophysical studies. *Nature, 291,*17-21.

KELLOGG, T.B., TRUESDALE, R.S. and OSTERMAN, L.E., 1979: Late Quarternary extent of the West Antarctic ice sheet: new evidence from Ross Sea cores. *Geology, 7,* 249-53.

ORHEIM, O. and ELVERHOI, A., 1981: Model for submarine glacial deposition. *Annals. Glaciology, 2,* 123-28.

STUIVER, M., DENTON, G.H., HUGHES, T.J. and FASTOCK, J.L., 1981: History of the marine ice sheet in west Antarctica during the Last Glaciation: A working hypothesis; *in* Denton, G.H. and Hughes, T.J. (eds.), *The last Great Ice Sheets.* John Wiley & Sons, New York, 319-436.

SUGDEN, D.E. and JOHN, B.S., 1976: *Glaciers and landscape. A geomorphological approach.* Arnold, London.

THOMAS, R.H., 1979: The dynamics of marine ice sheets. *J. Glaciol, 24,* 167-77.

# 9

# Crustal Structure of Antarctica

# CRUSTAL STRUCTURE OF ANTARCTICA FROM GEOPHYSICAL EVIDENCE—A REVIEW

C.R. Bentley, *Geophysical and Polar Research Centre, University of Wisconsin, Madison, Wisc., 53706. U.S.A.*

*Abstract* In many respects, little new has been learned about Antarctic crustal structure in the last decade. Where measured in East Antarctica the thickness of the layer overlying the basement complex is mostly less than half a kilometre, and the crust is about 40 km thick. In West Antarctica, the basement complex typically lies 2 or 3 km below the ice and the crust is only 25 to 30 km thick. Gravity measurements show that the crustal transition is abrupt across the Transantarctic Mountain front. Earlier seismic surface wave studies indicated a "typical continental" structure for East Antarctica; more recent measurements along a somewhat different section suggest a typical Precambrian shield. Scanty body-wave evidence on the upper mantle structure beneath East Antarctica also shows typical sub-shield characteristics.

In West Antarctica, the crust under the Ross Sea embayment, Byrd Subglacial Basin and Weddell Sea embayment is probably only about 25 km thick. The thick sediments, thin crust, and positive gravity belt in the Ross Sea embayment suggest that it may be a major rift zone. Gravity anomalies measured on the Ross Ice Shelf, supported by some seismic refraction measurements of sediment thickness, suggest deep structural control of the submarine and subglacial ridge/trough topography, which could reflect a horst-and-graben structure. That structure disappears abruptly under central West Antarctica, but high seismic velocities typical of deep continental or oceanic crust do occur at shallow depths. Continental palaeomagnetic data suggest at least three microplates in West Antarctica, all separate from East Antarctica.

## THE BASIC PICTURE

In many respects, little new has been learned about Antarctic crustal structure in the last decade. In this review paper, therefore, I will start with the picture as it was known 10 years ago, and then discuss some specific aspects on which further work has been done in the interim. The geographic areas referred to are shown in Figure 1.

Except in the Ross Sea embayment (i.e. the Ross Sea, the Ross Ice Shelf, and the Rockefeller Plateau), the Weddell Sea embayment (i.e. the Weddell Sea and the Filchner and Ronne Ice Shelves), and the Amery Ice Shelf/Enderby Land areas, the seismic evidence on crustal structure is as it was presented in a review by Bentley (1973). The following six paragraphs rest heavily upon that review, with some modifications from more recent information.

Throughout West Antarctica (including the base of the Antarctic Peninsula and the Ross Sea) except along the Ellsworth-Whitmore-Transantarctic Mountains axis, the surface of the basement complex

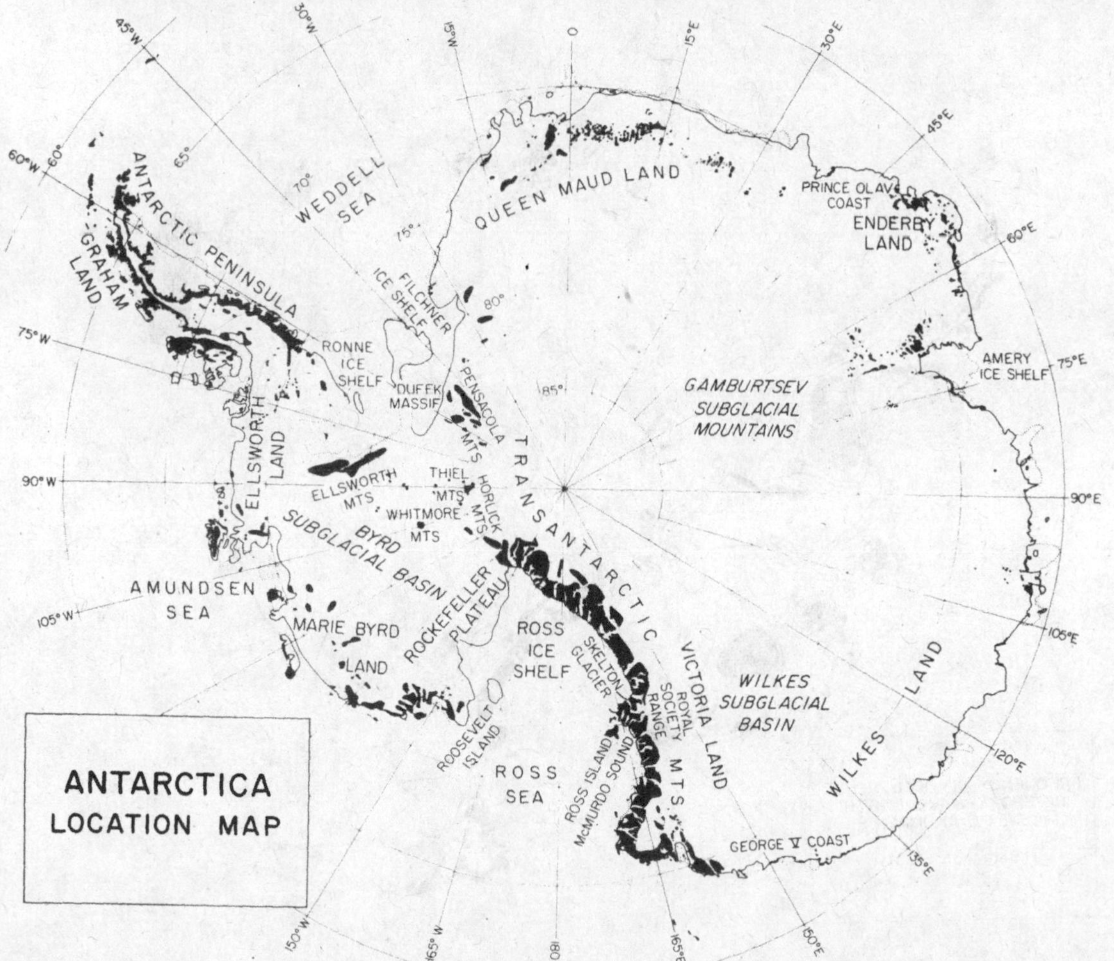


Figure 1. Geographic map of Antarctica, showing features referred to in the text.

as located by seismic refraction shooting, lies at least 2.2 km below sea level, dropping to at least 4 km below sea level beneath the deeper parts of the Byrd Subglacial Basin. In contrast, the surface of the East Antarctic basement complex generally is found close to sea level, and the overlying section is typically no more than a few hundred metres thick. Nowhere next to the Transantarctic Mountains on either side is there evidence for a thick sedimentary column.

At stations within a few hundred kilometres of the Horlick Mountains, on both the East and West Antarctic sides, velocities typical of lower crustal rocks appear at elevations of 1-4 km below sea level. On the West Antarctic side, the high velocities have been observed only in the southwesterly part of the Ellsworth-Whitmore Mountains topographic block. In East Antarctica, similar velocities are found in the Transantarctic Mountains west of Ross Island. Any correlation must, of course, be tentative, but it is interesting to speculate that normally deep-crustal basaltic rocks occur unusually near the surface in the Transantarctic Mountains, and have provided the source of the dolerite intrusions. This interpretation also implies that the Ellsworth-Whitmore block was once an extension of the Transantarctic Mountains, in accord with some palaeographic reconstructions.

The indication that high-velocity (and presumably high-density) rocks are found at shallow depths below the Transantarctic Mountains supports Robinson's (1964) conclusion that the remarkable Bouguer gravity anomaly gradient across those mountains can best be explained by an abrupt change in crustal thickness, rather than by a rapid increase in thickness of the lower-density, upper-crustal rocks.

The thickness of sedimentary rocks appears to increase from the Horlick Mountains towards the South Pole, reaching a probable thickness of more than 1 km under the Pole; no upper limit can be given. A sedimentary basin several kilometres deep could explain the −20 mgal regional free-air gravity anomaly under the South Pole plateau (Bentley, 1968), but in Victoria Land, where the gravity anomalies are even more strongly negative, the single refraction profile shows no sedimentary section. In Queen Maud Land also, the sedimentary section appears generally to be more than half a kilometre thick, and probably considerably less.

The fundamental contrast between East and West Antarctica which is seen from examination of subglacial topography and shallow crustal structure extends through the crust. The mean thickness of the West Antarctic crust as indicated by surface wave dispersion and gravity studies is only about 30 km, compared with about 40 km in East Antarctica. Deep seismic sounding also yields a 40 km thick crust in coastal Queen Maud Land, but one deep seismic reflection profile at 83°S, 70°W indicates a crust only about 25 km thick in the low central regions of West Antarctica. On the basis of average crustal structure, central East Antarctica appears to be "continental" rather than a simple "Precambrian shield", although more recent evidence brings this conclusion into question (see below). The boundary between East Antarctica and the West Antarctic borderland is apparently abrupt.

A low-velocity channel for P-waves with a sharp upper boundary at a depth 60-80 km probably underlies East Antarctica, as it does other continents.

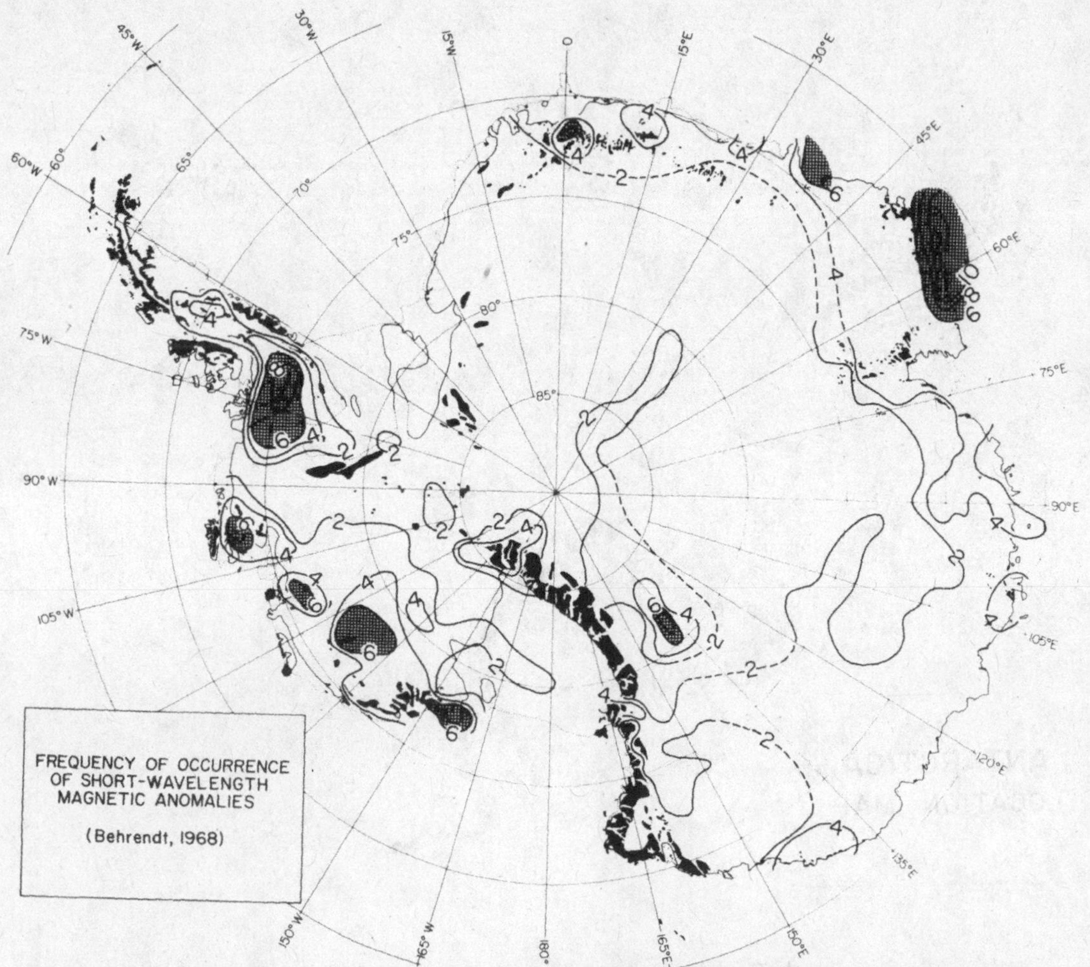


**Figure 2. Frequency of occurrence of short wavelength magnetic anomalies, simplified from Behrendt (1968). Contours show number of anomalies greater than 100 nT per 100 km of flight line, with cross-hatching where that number exceeds six.**

Aeromagnetic investigations a decade ago had been carried out principally in West Antarctica, across the Transantarctic Mountains and in central East Antarctica, with some additional information along the East Antarctic coast. They were summarised by Behrendt (1968).

The distribution and frequency of occurrence of short wavelength anomalies (Figure 2) shows a rather complex distribution of magnetic rock in West Antarctica. There are many anomalies in the regions of exposed volcanic rock in Marie Byrd Land, Ellsworth Land and Palmer Land. Short wavelength anomalies become less common southward into the Byrd Basin, and disappear completely over the Ellsworth-Whitmore block, indicating the presence of a thick sedimentary or metasedimentary section. This is in agreement with the known structure of the Ellsworth Mountains. The magnetics show that the sedimentary rock province extends eastwards beneath the Ronne Ice Shelf.

Many high amplitude magnetic anomalies exist along the coast and interior of East Antarctica. Their sources are probably in the Precambrian to early Palaeozoic orogenic terrain, samples of which outcrop near the coast.

The Transantarctic Mountains show some areas of numerous short wavelength anomalies and others with essentially smooth fields. Some of the anomaly rich areas reflect Late Cenozoic volcanics, particularly in the McMurdo area; others stem from unknown sources. The dolerite sills intruded into the Beacon section surprisingly do not produce observable anomalies, perhaps because they are not sufficiently thick.

The most extensive work on the boundary between East Antarctica and West Antarctica has been in the Pensacola Mountains and adjacent areas (Behrendt et al., 1974). Here, as in the McMurdo Sound area, there is a steep gravity gradient across the mountain front that requires an abrupt change in crustal thickness of at least 5 km. Combined geophysical data on the West Antarctic side of the mountains suggest a thick section of low velocity, low density, non-magnetic, presumably sedimentary rock. Bouguer gravity and magnetic anomalies over the Dufek Massif indicate that the ultrabasic intrusion is between 6 and 9 km thick, and that the Dufek intrusion extends laterally beneath the ice to cover an area of 50,000 km² or more (Behrendt et al., 1980).

The free-air gravity map (Bentley, 1968; Figure 3) shows a mean value near 0 in West Antarctica, implying general isostatic balance. Some areas of regional gravity highs and lows are discussed further below. In East Antarctica the most striking feature is the great "Transantarctic gravity anomaly", a 20-50 mgal negative feature parallel to the Transantarctic Mountains extending almost unbroken across the continent from the Weddell Sea to the George V coast. This indicates either large glacio-isostatic imbalance, an abnormally thick or low-density crust, or a deeper-seated cause in the upper mantle such as that recently proposed by Veevers (see below).

Early Cretaceous rocks from Marie Byrd Land show palaeomagnetic pole positions quite different from those measured on rocks from the Transantarctic Mountains (Scharnberger and Scharon, 1972). Although the latter rocks are somewhat older, the large difference in palaeo-pole position strongly suggests that during the Mesozoic Marie Byrd Land was widely separated from East Antarctica with a zone of plate convergence between them.

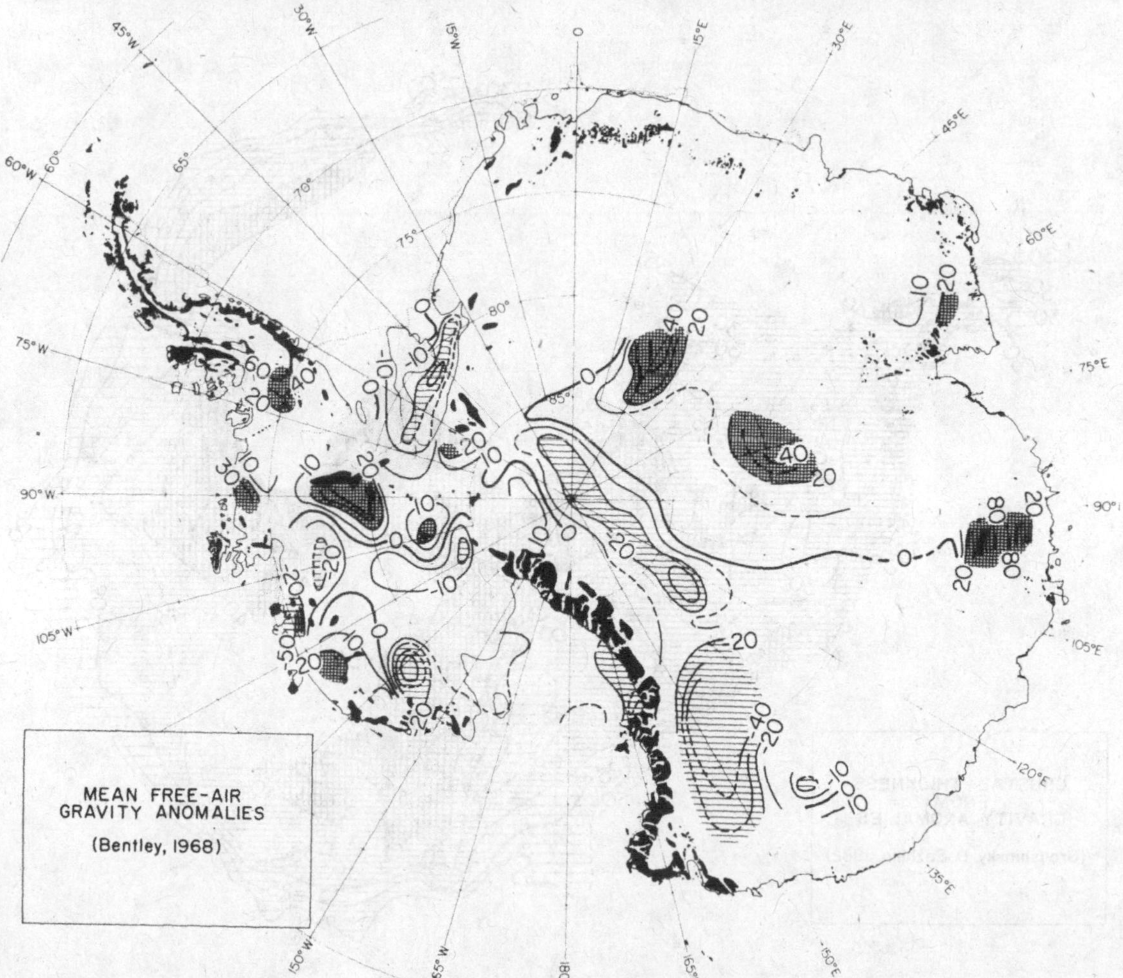


**Figure 3.** Free-air gravity anomalies averaged on 2° x 2° squares, simplified from Bentley (1968). Anomalies greater than 20 mgal in amplitude are indicated by horizontal hachures for negative anomalies and cross-hatching for positive anomalies.

## ADVANCES IN THE LAST DECADE

### East Antarctica

The principal new deep seismic sounding measurements were in the Amery Ice Shelf area (Kurinin and Grikurov, 1982). The soundings generally revealed, as in Dronning Maud Land, typically continental layered crust (Figure 4) with a thickness ranging from 30 to 40 km. The upper crustal layer is about 20 km thick; the seismic velocity within it is somewhat higher than normal for upper crustal layers, consistent with the great age of the East Antarctic crystalline basement rocks.

Central to this area is the Lambert rift, a large and extensive graben extending southward from the coast for almost 700 km (Federov et al., 1982). The crustal thickness within this rift is only about 25 km and the upper crustal layer thins to about 10 km. Magnetic data show a non-magnetic, presumably sedimentary, section some 5 km thick. This rift is the best defined in terms of geophysical evidence of the several rift zones that may exist around Antarctica (Kadmina et al., this volume). A seismic refraction experiment carried out near the Prince Olav Coast in 1979 reconfirmed a 40 km crustal thickness (Ikami et al., this volume).

From analysis of a major new regional gravity survey in Enderby Land, Wellman and Tingey (1982) have shown some success in relating relative gravity highs and lows to density changes in the upper crust. One striking high is interpreted as reflecting a relatively high density upper crust with regional isostatic compensation, in contrast to an earlier interpretation by Koriakine et al. (1970) who attributed the high to the absence of a crustal root and therefore to isostatic

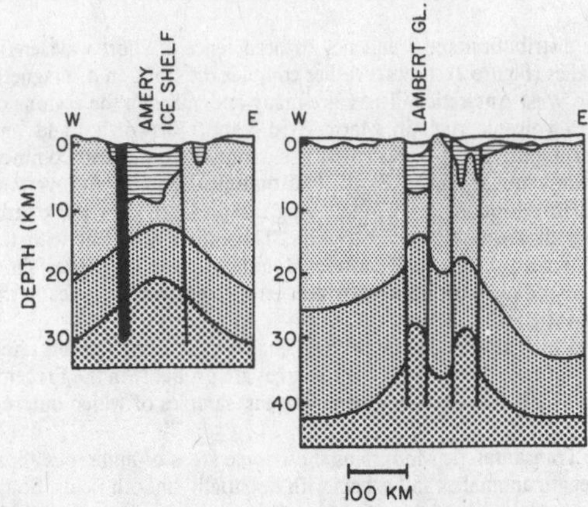


**Figure 4. Crustal section across the Amery Ice Shelf and Lambert Glacier, simplified from Masolov et al. 1981. Horizontal hachures indicate sedimentary rocks; the three intensities of shading mark layers divided by the Conrad and Mohorovicic discontinuities.**

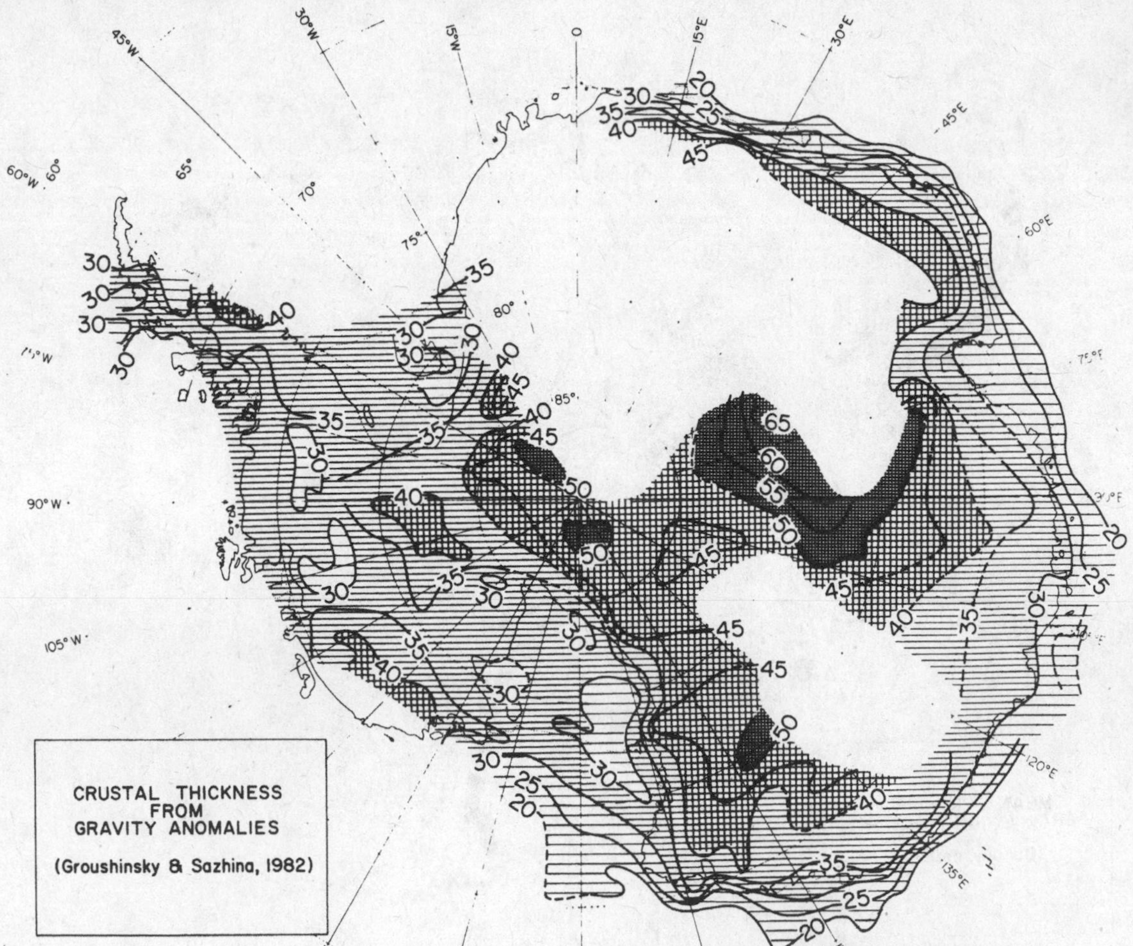

**Figure 5. Crustal thickness (in kilometres) as calculated from gravity anomalies by Groushinsky and Sazhina (1982). Hachuring densities change at thicknesses of 20, 30, 40 and 50 km.**

imbalance. Gravity lows correlate generally with areas of known granite; several regions of high gravity gradients correspond approximately to mapped crustal facies boundaries.

Some new interpretations of continental gravity data have been published in the last 10 years. Groushinsky and Sazhina (1982) suggest that crustal thickness in central East Antarctica in the region of the subglacial Gamburtsev Mountains exceeds 60 km (Figure 5). They also show a thickness greater than 50 km in several places along the "Transantarctic gravity anomaly". Veevers (1982), on the other hand, has suggested that at least that part of the gravity negative that lies in Victoria and Wilkes Lands is underlain by downward convecting asthenosphere.

The seismicity of Antarctica has been reviewed by Kaminuma (1982). Only three earthquakes of magnitude > 4 have been recorded from Antarctic epicentres, all next to major coastal outlet glaciers. The association with glaceirs suggested to Adams (1982) that large-scale movement or cracking in the glacial ice might be the cause. It seems more likely to this reviewer, however, that the earthquakes are associated with subglacial fault zones that determine the position of the outlet glaciers. Kaminuma (1982) concludes that (1) microearthquakes occur at least in the Transantarctic Mountains, although the activity is very low; (2) that the microearthquake activity around Syowa Station on the Prince Olav Coast is less than one per month; (3) that a microearthquake occurs in the McMurdo Sound region every day or two; (4) that seismicity in the volcanic regions is higher than elsewhere in Antarctica; and (5) that the seismicity in West Antarctica is 10 times as great as that in East Antarctica. Kaminuma's review confirms that Antarctica has no large earthquakes and has the lowest seismicity of any continent, a fact that remains a mystery. Interestingly enough, however, a recent compilation by Okal (1981) of earthquakes recorded over a 55 year period shows that despite the quiescence of the continent itself, the total intra-plate seismicity is a high as in other continent-bearing slow moving plates, such as the African plate.

In the only new published report on Antarctic surface wave dispersion Knopoff and Vane (1978/79) examine Rayleigh-wave phase velocities along two paths from Queen Maud Land—to the south pole and to Ross Island, respectively. These measurements clearly show the high velocities that are associated with Precambrian shields (Figure 6), rather than with the average continental structure as previously indicated. However, the paths are not the same (Figure 7)—the new measurements follow paths that avoid the large subglacial mountain ranges of central East Antarctica, which the surface waves studies earlier crossed. The average crustal structure in the two regions may indeed differ significantly.

The high upper mantle velocities implied by Knopoff and Vane's (1978/79) work are supported by the first study of mantle P-wave velocity structure beneath Antarctica from travel times and amplitudes of earthquake waves, recently presented by McMechan (1981). The model he produced (Figure 8), which applies to the grid southeast (Australian) quadrant of Antarctica, contains the prominent velocity increases near depths of 420 and 650 km commonly found elsewhere in the world, and is similar to models for the Canadian and Scandinavian shields. Thus these velocities also are substantially higher than those deduced by Dewart and Toksöz (1965) for central East Antarctica.

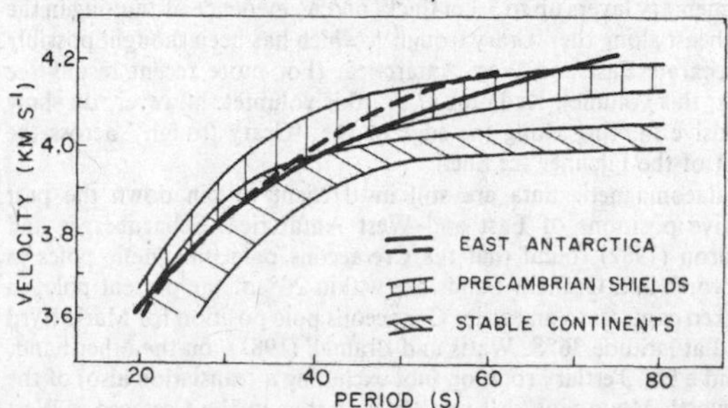


**Figure 6. Rayleigh-wave phase velocities across East Antarctica for the two paths that bound Region B in Figure 7. Ranges covered by dispersion curves for other continental and shield areas are also shown. (From Knopoff and Vane, 1978/79).**

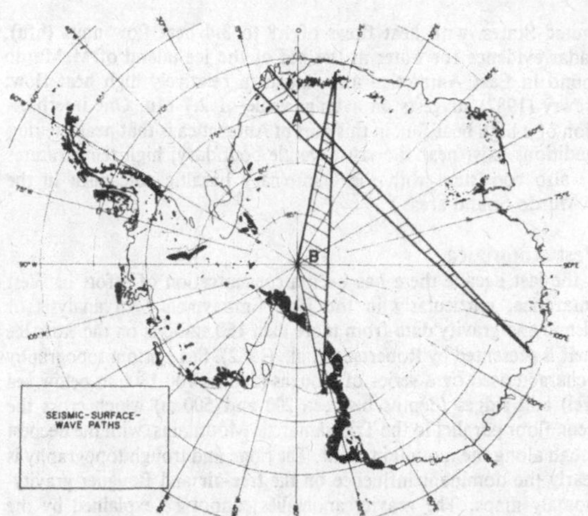


**Figure 7. Surface wave paths across East Antarctica from the studies of Dewart and Toksoz (1965) region A; and Knopoff and Vane (1978/79), region B.**

Detailed study of radar sounding in Wilkes Land has led to some indications of upper crustal structure based on a geological interpretation. Steed and Drewry (1982) conclude that the Wilkes Subglacial Basin and its salients constitute a major intracratonic zone of sedimentation, and that the eastern edge of the basin probably marks the western limit of the orogenic activity responsible for the Transantarctic Mountains. Their study also identifies what is probably a major fault block structure running north-south along 135°E.

Continental heat flow measurements continue to be a rarity—the only good survey is in the McMurdo Sound area (Decker and Bucher, 1982). Heat flow there, both on Ross Island and in the Dry Valleys across the Sound, is much higher than the continental norm (ca 1.6 hfu)—in fact, it is similar to the Basin and Range province in the

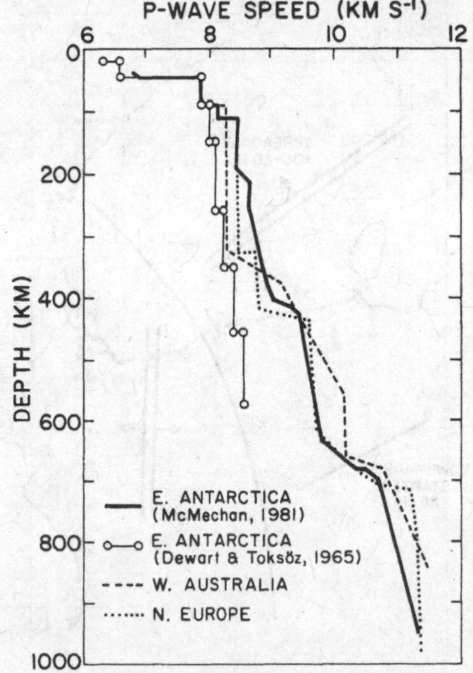


**Figure 8. P-wave speed vs. depth for East Antarctica (solid line), western Australian (dashed line), and northern European (dotted line) body waves, and East Antarctic surface waves. (From McMechan, 1981).**

United States, with heat flows of 1.8 to 3.4 heat flow units (hfu). Radar evidence for water at the bed of the ice inland of McMurdo Sound in East Antarctica also implies a relatively high heat flow; Drewry (1982) suggests an average value of 2.7 hfu. One interpretation of a high heat flux in this part of Antarctica is that near-melting conditions exist near the crust-mantle boundary; high temperatures are also consistent with the Quaternary basaltic volcanism in the McMurdo Sound area.

## West Antarctica

In the last decade there has been a concentration of effort in West Antarctica, particularly in the Ross Embayment. An analysis of seismic and gravity data from more than 150 stations on the Ross Ice Shelf is presented by Robertson et al., (1982). Sea-bottom topography is characterised by a series of troughs (depths 700-1300 m below sea level) and ridges (depths between 200 and 500 m) which cross the ocean floor parallel to the Transantarctic Mountains, with the deepest trough along the mountain front. The ridge and trough topography is clearly the dominant influence on the free-air and Bouguer gravity-anomaly maps. The gravity anomalies cannot be explained by the seismically-derived thicknesses of glacial till; this strongly suggests that tectonic structure and not glaciation is the fundamental determinant of sea-bottom topography.

Interval seismic velocities and acoustic impedances in the layer of sediment at the sea floor match those expected for glacial-marine till. On the basis of seismic refraction measurements, it seems most likely that "basement" beneath the ice shelf, found 1.5 to 2.0 km below sea level, consists of crystalline rocks typical of the Ross or Andean orogenic provinces. On and near Roosevelt Island an intermediate layer 200-700 m thick, intepreted as either Cenozoic volcanic strata or Palaeozoic-Mesozoic sedimentary/metasedimentary rocks, lies between glacial till and crystalline basement.

A combined discussion of the work on the Ross Ice Shelf and that carried out in the Ross Sea has been given by Davey (1981). From the analysis of gravity and topographic data, and by analogy with the seismic sounding in the Weddell embayment, the crust under the Ross Sea embayment is probably about 25 km thick. The thick sediments and possibly thin crust coinciding with the Ross embayment suggests that it may be a major rift zone. The trend of the deeper structures controlling the morphology and gravity anomalies in the Ross Ice Shelf region is consistent with this concept. These structures terminate abruptly along a cross-cutting trend (Figure 9) that runs north from Byrd Glacier, all the way to the edge of the continental shelf, dividing the Ross Sea into two structural regions (Hayes and Davey, 1975). A graben that lies under the Ross Sea west of the boundary and is marked by a pronounced linear positive gravity anomaly is one of the major structural features on the shelf.

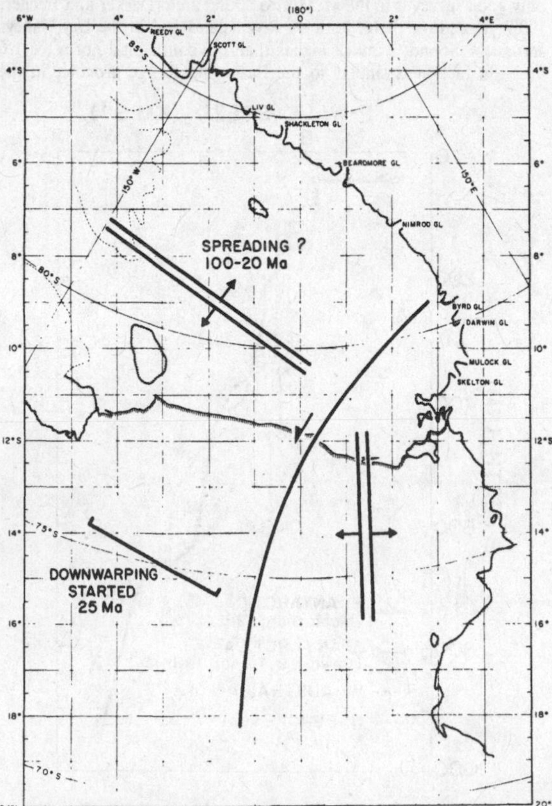


**Figure 9. Structural trends in the Ross Sea embayment, simplified from Davey (1981).**

The faulted margin of the Transantarctic Mountains presumably lies offshore, and there are some gravity data in McMurdo Sound to support this (Sissons, 1980). From seismic shooting McGinnis (1981) found that the basement surface in western McMurdo Sound lies 1.5 to 3 km below sea level and rises gently to the west; the major vertical displacement of 2 to 3 km must lie between the western end of the profile and the coastline. McGinnis et al. (this volume) also find a 25 km depth to Moho in McMurdo Sound.

Radar sounding in the Rockefeller Plateau area by Rose (1982) shows a striking change in subglacial topographic pattern from the ridge-trough structure typical of the Ross Sea embayment to the deep-lying, rugged topography of the central Byrd Subglacial Basin. Rose (1982) attributes the subdued topography and magnetic signatures in the inner Ross Sea embayment to a substantial thickness of Late Mesozoic/Cenozoic sediments extending the Ross Sea sedimentary basin beneath the grounded ice sheet. A pattern of irregular magnetic anomalies over the Byrd Subglacial Basin (Jankowski and Drewry, 1981), together with the earlier seismic refraction studies (Bentley and Clough, 1972), suggest that a sequence of interbedded sedimentary and volcanic strata overlies the basement there.

Jankowski and Drewry (1981) also find that, although the topographic and magnetic boundary between the Byrd Subglacial Basin and the Ellsworth Mountains block becomes less well defined southwestward from the Ellsworth Mountains toward the Whitmore Mountains, probably because of Cenozoic sedimentation, the two mountain ranges are probably part of one crustal block. Modelling of magnetic anomalies suggests the presence of a thick layer of non-magnetic rocks, both in the Transantarctic Mountains, where they presumably correspond to Beacon Supergroup strata and older Precambrian metasediments, and just north of the Thiel Mountains (Drewry et al., 1980).

Groushinsky and Sazhina (1982) interpret regional gravity highs at the base of the Antarctic Peninsula, in the region of the Ellsworth Mountains, and in the mountains of Marie Byrd Land, as indicating crustal thickness as great as 40 km. Bentley and Robertson (1982) have a different interpretation, attributing the first two anomalies to uncompensated crustal thinning and the third to crustal rifting. Bentley and Robertson (1982) further interpret regional negative anomalies in the Whitmore Mountains area and on the northern Rockefeller Plateau as reflecting a thickening of upper crustal rocks at the expense of the lower crust. Another negative along the Amundsen Sea coast they attribute to recent retreat of the ice and to unknown tectonic factors.

In the Weddell Sea, seismic refraction data (Haugland, 1982) show sedimentary layers up to 5 km thick, and no evidence of faulting in the southeast along the "Crary trough", which has been thought possibly to separate East and West Antarctica. (For more recent results see Hinz, this volume). Kadmina et al. (this volume), however, do show extensive faulting along the edge of the "Crary trough" across the front of the Filchner Ice Shelf.

Palaeomagnetic data are still insufficient to pin down the past relative positions of East and West Antarctica. Scharnberger and Scharon (1982) found that the Cretaceous palaeomagnetic poles in Ellsworth and Graham Lands are within 20° of the present pole, in marked contrast to an earlier Cretaceous pole position for Marie Byrd Land at latitude 36°S. Watts and Bramall (1981), on the other hand, found a 90° Tertiary rotation (not excluding a translation also) of the Ellsworth Mountains. All this suggests that in the Cretaceous West Antartica was separated into at least three tectonic units on different lithospheric microplates: Ellsworth Land/Antarctic Peninsula, Marie Byrd Land and the Ellsworth Mountains (i.e. presumably the Ellsworth-Whitmore block).

## CONCLUSION

The increase in knowledge of Antarctic continental crustal structure in the last decade has been regrettably small. The only important advances have been in a very few deep seismic sounding and surface wave dispersion profiles, one seismic body-wave study, aeromagnetic surveys of limited extent, satellite magnetic coverage, and complex geophysical surveys of the Ross Sea and Weddell Sea embayments. The primary future needs are for many more deep seismic soundings not limited to shooting in water, completion of aeromagnetic coverage of the continent, and extension of gravity measurements wherever possible. The development of airborne gravity capability with milligal accuracy (already possible in some countries), and the emplacement of low level magnetic and gravitational satellites in high polar orbits would be extremely valuable. We may hope that the next decade will see extensive application of modern technological advances in geophysics.

*Acknowledgements* This work was supported by the National Science Foundation under grant DPP78-20953. This is contribution No. 402 of the Geophysical and Polar Research Center, University of Wisconsin, Madison, Wisconsin, U.S.A.

## REFERENCES

ADAMS, R.D., 1982: Source properties of the Oates Land Earthquake, October 1974; *in* Craddock, C. (ed.) *Antarctic Geoscience.* Univ. Wisconsin Press, Madison, 955-8.

BEHRENDT, J.C., 1968: Magnetic anomalies of short wavelength. *Antarctic Map Folio Ser., Folio 9,* Pl. VII.

BEHRENDT, J.C., HENDERSON, J.R., MEISTER, L. and RAMBO, W.L., 1974: Geophysical investigations of the Pensacola Mountains and the adjacent glacierised areas of Antarctica. *U.S. Geol. Surv., Prof. Pap., 844.*

BEHRENDT, J.C., DREWRY, D.J., JANKOWSKI, E. and GRIM, M.S., 1980: Aeromagnetic and radio echo ice-sounding measurements show much greater area of the Dufek Intrusion, Antarctica. *Science, 209,* 1014-7.

BENTLEY, C.R., 1968: Free-air gravity anomalies. *Antarctic Map Folio Ser., Folio 9,* plate VIII.

BENTLEY, C.R., 1973: Crustal structure of Antarctica. *Tectonophysics, 20,* 229-40.

BENTLEY, C.R. and CLOUGH, J.W., 1972: Antarctic subglacial structure from seismic refraction measurements; *in* Adie, R.J. (ed.) *Antarctic Geology and Geophysics.* Universitetsforlaget, Oslo, 683-91.

BENTLEY, C.R. and ROBERTSON, J.D., 1982: Isostatic gravity anomalies in West Antarctica; *in* Craddock, C. (ed.) *Antarctic Geoscience.* Univ. Wisconsin Press, Madison, 949-54.

DAVEY, F.J., 1981: Geophysical studies in the Ross Sea region. *R. Soc. N.Z., J., 11,* 465-79.

DECKER, E.R. and BUCHER, G.J., 1982: Geothermal studies in the Ross Island-Dry Valley region (review paper); *in* Craddock, C. (ed.) *Antarctic Geoscience.* Univ. Wisconsin Press, Madison, 887-94.

DEWART, G. and TOKSOZ, M.N., 1965: Crustal structure in East Antarctica from surface wave dispersion. *Geophys. J., 10,* 127-39.

DREWRY, D.J., 1982: Ice flow, bedrock, and geothermal studies from radio-echo sounding inland of McMurdo Sound, Antarctica; *in* Craddock, C. (ed.) *Antarctic Geoscience.* Univ. Wisconsin Press, Madison, 977-83.

DREWRY, D.J., MELDRUM, D.T. and JANKOWSKI, E., 1980: Radio echo and magnetic sounding of the Antarctic Ice Sheet, 1978-79. *Polar Rec., 20,* 43-57.

FEDEROV, L.V., GRIKUROV, G.E., KURININ, R.G. and MASOLOV, V.N., 1982: Crustal structure of the Lambert Glacier area from geophysical data; *in* Craddock, C. (ed.) *Antarctic Geoscience.* Univ. Wisconsin Press, Madison, 931-6.

GROUSHINSKY, N.P. and SAZHINA, N.B., 1982: Some features of Antarctic crustal structure; *in* Craddock, C. (ed.) *Antarctic Geoscience.* Univ. Wisconsin Press, Madison, 907-11.

HAUGLAND, K., 1982: Seismic reconnaissance survey in the Weddell Sea; *in* Craddock, C. (ed.) *Antarctic Geoscience.* Univ. Wisconsin Press, Madison, 405-13.

HAYES, D.E. and DAVEY, F.J., 1975: A geophysical study of the Ross Sea, Antarctica; *Initial Reports of the Deep Sea Drilling Project 28,* 887-907. U.S. Govt. Printing Office, Washington D.C.

HINZ, K. (this volume): Results of geophysical investigations in the Weddell Sea.

IKAMI, A., ITO, K., SHIBUYA, K. and KAMINUMA, K. (this volume): Crustal structure of the Mizuho Plateau, Antarctica, revealed by explosion seismology.

JANKOWSKI, E.J. and DREWRY, D.J., 1981: The structure of West Antarctica from geophysical studies. *Nature, 291,* 17-21.

KADMINA, I.N., KURININ, R.G., MASOLOV, V.N. and GRIKUROV, G.E. (this volume): Antarctic crustal structure from geophysical evidence—a review.

KAMINUMA, K., 1982: Seismicity in Antarctica; *in* Craddock, C. (ed.) *Antarctic Geoscience.* Univ. Wisconsin Press, Madison, 919-23.

KNOPOFF, L. and VANE, G., 1978/79: Age of East Antarctica from surface wave dispersion. *Pure Appl. Geophys., 117,* 806-15.

KORIAKINE, E.D., STROEV, P.A. and FROLOV, A.I., 1970: Resulyaty gravimetricheskikh issledovanii na zemle Enderbi (vostochnai Antarktida) (Results of gravity measurements in Enderby Land (East Antractica)); *in* Fedinski, V.V. (ed.) *Moeskie Gravimetricheskie Issledovanii—sbornik statei, vbipusk 5 (Gravity Measurements at Sea—collection of articles), 5,*83-95.

KURININ, R.G. and GRIKUROV, G.E., 1982: Crustal structure of part of East Antarctica from geophysical data; *in* Craddock, C. (ed.) *Antarctic Geoscience.* Univ. Wisconsin Press, Madison, 895-901.

McGINNIS, L.D., 1981: Seismic refraction study in western McMurdo Sound; *in* McGinnis, L.D. (ed.) Dry Valley Drilling Project. *Am. Geophys. Union, Ant., Res. Ser., 33,* 27-35.

McGINNIS, L.D., WILSON, D.D. and BURDELIK, W.J. (this volume): Crust and upper mantle study in McMurdo Sound.

McMECHAN, G.A., 1981: Mantle P-wave velocity structure beneath Antarctica. *Bull. Seis. Soc. Am., 71,* 1061-74.

MASOLOV, V.N., KURININ, R.G. and GRIKUROV, G.E., 1981: Crustal structures and tectonic significance of Antarctic rift zones (from geophysical evidence); *in* Cresswell, M.M. and Vella, P. (eds.) *Gondwana Five.* A.A. Balkema, Rotterdam, 303-10.

OKAL, E.A., 1981: Intraplate seismicity of Antarctica and tectonic implications. *Earth Planet. Sci. Lett., 52,* 397-409.

ROBERTSON, J.D., BENTLEY, C.R., CLOUGH, J.W. and GREISCHAR, L.L., 1982: Sea-bottom topography and crustal structure below the Ross Ice Shelf, Antarctica; *in* Craddock, C. (ed.) *Antarctic Geoscience.* Univ. Wisconsin Press, Madison, 1083-90.

ROBINSON, E.S., 1964: Geologic structure of the Transantarctic Mountains and adjacent ice covered areas. Unpublished Ph.D. thesis, University of Wisconsin, Madison.

ROSE, K.E., 1982: Radio-echo studies of bedrock in Southern Marie Byrd Land, West Antarctica; *in* Craddock, C. (ed.) *Antarctic Geoscience.* Univ. Wisconsin Press, Madison, 985-92.

SCHARNBERGER, C.K. and SCHARON, L., 1972: Palaeomagnetism and plate tectonics of Antarctica; *in* Adie, R.J. (ed.) *Antarctic Geology and Geophysics.* Universitetsforlaget, Oslo, 843-8.

SCHARNBERGER, C.K. and SCHARON, L., 1982: Paleomagnetism of rocks from Graham Land and Western Ellsworth Land, Antarctica; *in* Craddock, C. (ed.) *Antarctic Geoscience.* Univ. Wisconsin Press, Madison, 371-5.

SISSONS, B.A., 1980: Sea ice gravity survey; *in* Pyne, A. and Waghorn, D.B. (comps.) *Victoria University of Wellington Antarctic Expedition 24 and MSSTS. Immediate Report.* Victoria University, Wellington, 29-32.

STEED, R.H.N. and DREWRY, D.J., 1982: Radio-echo sounding investigations of Wilkes Land, Antarctica; *in* Craddock, C. (ed.) *Antarctic Geoscience.* Univ. Wisconsin Press, Madison, 969-75.

VEEVERS, J.J., 1982: Australian-Antarctic depression from the mid-ocean ridge to adjacent continents. *Nature, 295,*315-7.

WATTS, D.R. and BRAMALL, A.M., 1981: Palaeomagnetic evidence for a displaced terrain in western Antarctica. *Nature, 293,*638-42.

WELLMAN, P. and TINGEY, R.J., 1982: A gravity survey of Enderby and Kemp Lands, Antarctica; *in* Craddock, C. (ed.) *Antarctic Geoscience.* Univ. Wisconsin Press, Madison, 937-40.

# ANTARCTIC CRUSTAL STRUCTURE FROM GEOPHYSICAL EVIDENCE: A REVIEW

I.N. Kadmina, *Antarctic Division, Department for Marine Geological Exploration, Sevmorgeologia, Leningrad. USSR.*

R.G. Kurinin, V.N. Masolov and G.E. Grikurov, *Department of Antarctic Geology and Mineral Resources, VNIIOkeangeologia, Leningrad 190121. USSR.*

*Abstract* The principal features of Antarctic crustal structure are derived from an analysis of three major crustal surfaces whose regional characteristics are obtained from available reconnaissance data, viz. the bedrock surface, the "magnetic" basement surface and the Moho. The relief of the two former surfaces is broadly comparable and indicates the abundance of linear neotectonic features such as rifts/grabens which group into transcontinental systems and are clearly reflected in the behaviour of the Moho and Conrad interfaces. The "magnetic" basement of cratonic terrains and sedimentary basins is believed to consist mainly of Lower Precambrian crystalline basement rocks overlain by upper Precambrian to Lower Palaeozoic sequences whereas in orogenic areas it probably includes both the remobilised ancient infrastructure and surrounding Phanerozoic complexes.

Crustal profiles for selected areas are compiled on the basis of integrated interpretations of gravimetric, magnetic and seismic data to illustrate the major types of Antarctic crustal structure and its reworking by riftogenic processes. It is suggested that the unique intensity of rifting which leads to disintegration of the Antarctic continental landmass may be related, at least partly, to the presence of an ice sheet capable of affecting the tectono-thermal regime at lower crustal levels and a possible mechanism of this influence is discussed.

The geophysical study of Antarctic crustal structure is still at the reconnaissance stage despite considerable international scientific effort. Aerial geophysical surveys have so far been concentrated in a few regions of limited size, and sufficient combinations of methods have not always been employed. A very important contribution to the understanding the subglacial structure has lately been provided by radio-echo sounding surveys of the vast Antarctic interior (Rose, 1982; Steed and Drewry, 1982). These surveys, however, were almost nowhere accompanied by deep crustal studies and, consequently, the correlation of bedrock relief features with crustal structures still depends on global analogies to a greater degree than on integrated interpretation of geophysical data obtained directly in Antarctica. Due to general lack of evidence the crustal models discussed in this paper must be regarded as preliminary, as should certain speculations with respect to particular features of the recent Antarctic crust evolution.

As has been shown previously (Kurinin and Grikurov, 1980; Masolov et al., 1981; Fedorov et al., 1982), the major crustal features are distinctly manifested by three features: viz. (i) in the bedrock topography; (ii) the surface of the "magnetic" basement (base of nonmagnetic rocks) and; (iii) the relief of the Mohorovicic discontinuity. Available geophysical data permit the compilation of very

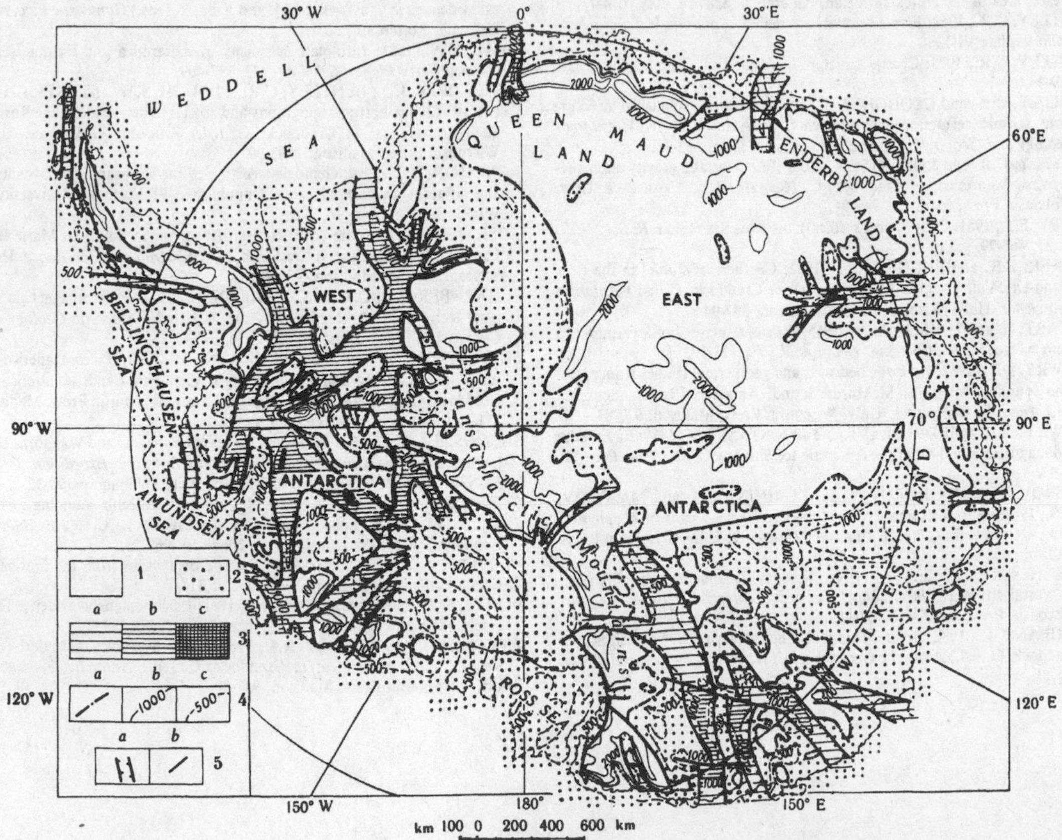


**Figure 1.** Schematic presentation of Antarctic tectonic relief. (1) bedrock elevations above sea level; (2) areal bedrock depressions below sea level without distinct linear grabens; (3) grabens in bedrock surface with depth: (a) from 0 to 1000 m; (b) from 1000 to 2000 m; (c) exceeding 2000 m; (4) contours of bedrock surface in metres: (a) sea level; (b) above sea level; (c) below sea level; (5) major faults reflected in bedrock surface: (a) bounding grabens; (b) undifferentiated.

rough sketches of these critical surfaces for the whole continent and the construction of somewhat more complete crustal profiles across the regions studied in relatively greater detail.

## ANALYSIS OF CRITICAL CRUSTAL LEVELS

The present day morphostructural grain of Antarctica, shown in Figure 1, is compiled on the basis of published data (Hayes and Davey, 1975; Znachko-Yavorskii, 1978; Rose, 1982; Robertson et al., 1982; Steed and Drewry, 1982) and the preliminary results of Soviet work in the Weddell Sea area. Numerous disjunctive features are usually directly reflected in bedrock topography as extensive escarpments, valleys of outlet glaciers, intrashelf trenches, etc., but in some cases their presence is indicated only by discrete dislocations of positive and/or negative forms of tectonic relief and apparent linear correlation of these dislocations over large distances. Bedrock depressions bounded by steep faults and characterised by considerable width and depth are shown as grabens. The fact that tectonic features are clearly seen in the bedrock topography suggests their recent age or even continuing development. This, therefore, indicates the leading role of neotectonic movements in the formation of bedrock relief which thus belongs, in a broad sense, to a "tectonic" rather than erosional or accumulative type.

Figure 2 summarises the results of the quantitative interpretation of aeromagnetic data obtained by Soviet and foreign Antarctic expeditions. The total of 4300 values of depths to magnetic sources (upper boundaries) were calculated in order to derive the position of the "magnetic" basement surface. This is best known in the Lambert Glacier area, in Enderby Land, under the Weddell Sea shelf and in Victoria Land and adjacent parts of the Ross Sea (Kadmina, 1980; Kurinin and Grikurov, 1980, 1982; Masolov, 1980; Behrendt, in press) where its position is confirmed by seismic and gravimetric data.

The geologic nature of the "magnetic" basement surface in various structural provinces probably differs. Within the East Antarctic cratonic terrain and sedimentary basins it is believed to coincide in most cases with the top of Archaean crystalline basement and/or Proterozoic to Lower Palaeozoic supracrustal sequences including both the folded aulacogenic complexes and the relatively undisturbed strata of the ancient platform cover. In fold belts adjacent to sedimentary basins the position of the upper boundaries of magnetic bodies is likely to reflect mainly the level of remobilisation of crystalline infrastructure represented by intrusions of Palaeozoic and Mesozoic age. The folded complexes incorporating these intrusions and, possibly, some of the younger molassic and platform deposits postdating orogenic events may also belong to the "magnetic" layer, but at present they cannot be distinguished from crystalline rocks by magnetic properties. On the whole the surface of the "magnetic" basement outside the sedimentary basins appears identical, within the accuracy of the method, with the exposed and subglacial bedrock surface; in sedimentary basins the positions of these surfaces are, in contrast, entirely different due to the deep subsidence of the "magnetic" basement.

Comparison of Figures 1 and 2 shows that the relief of the "magnetic" basement surface reveals an even greater degree of dissection demonstrated both by the amount of tectonic dislocations

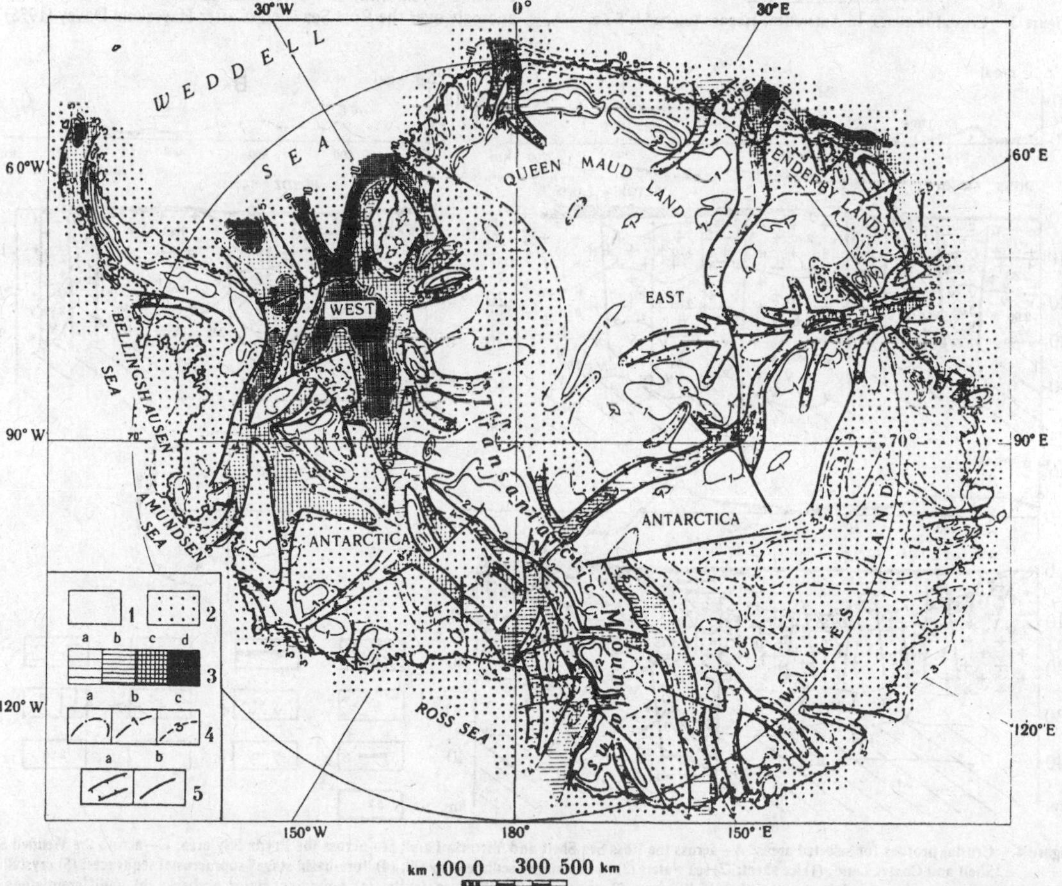


**Figure 2. Schematic presentation of Antarctic "magnetic" basement surface (MBS). (1) MBS elevations above sea level; (2) areal MBS depressions without distinct linear grabens; (3) grabens in MBS with depth: (a) from 0 to 4 km in Victoria Land area and from 0 to 3 km elsewhere; (b) from 4 to 8 km in Victoria Land area and from 3 to 5 km elsewhere; (c) exceeding 8 km in Victoria Land area and from 5 to 10 km elsewhere; (d) exceeding 10 km; (4) contours of MBS in kilometres: (a) sea level; (b) above sea level; (c) below sea level; (5) major faults reflected in MBS: (a) bounding grabens; (b) undifferentiated.**

and by their amplitude; extensive and deep graben-like depressions are particularly conspicuous. Obviously, the "magnetic" basement surface is most critical for the detection of regional crustal block movements which may sometimes be completely masked by subsequent intensive sedimentation.

The morphology of the upper crustal levels is clearly reflected also by the configuration of the Mohorovicic discontinuity (Figure 3). The crustal thickness was determined from regional Bouguer anomalies (Groushinsky et al., 1978); in the absence of gravity data it was

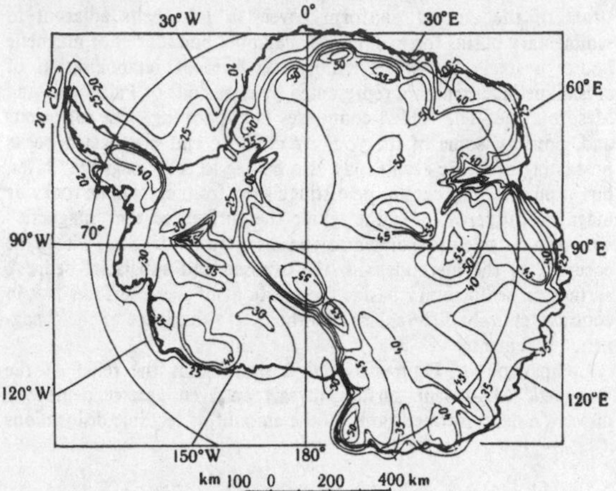


**Figure 3. Crustal thickness in Antarctica. Isopach interval is 5 km.**

deduced from "effective" relief elevations (true bedrock elevations plus thickness of overlying ice recalculated for a density value of 2.67 g/cm³) (Demenitskaya, 1982).

An almost linear zone of sharp gradients in crustal thickness underlying the Transantarctic Mountains escarpment is the most prominent feature of Moho relief. It marks the deep-seated boundary between East and West Antarctica (Grikurov, this volume) and is associated with a transcontinental asymmetric rift graben (or system of grabens) probably connecting the Ross Sea and Weddell Sea basins. The mean value of crustal thickness is in the order of 40-45 km in East Antarctica and 30 km in West Antarctica, with the highest values reaching respectively 55 km (the Transantarctic, Gamburtsev and Queen Maud Land Mountains) and 40 km (the Antarctic Peninsula and Marie Byrd Land) and the lowest values—less than 25 km—related to major rift grabens (the Lambert Glacier graben, the Filchner Ice Shelf graben, etc.).

## CRUSTAL PROFILES ACROSS SELECTED AREAS

Three areas were selected for modelling crustal cross-sections: the southeastern part of the central Weddell Sea basin, the Lambert Glacier—Amery Ice Shelf—Prydz Bay area and the northwestern Ross Sea basin with the adjacent area of Victoria Land (Figure 4). For all these areas relatively abundant geophysical evidence is available including seismic information (Kolmakov et al., 1975; Hayes and Davey, 1975; Wilson et al., 1981; Haugland, 1982; Poselov et al., in preparation) which is a particularly important component in the integrated interpretation of geophysical data.

In the compilation of the crustal profile across the Ross Sea Shelf and the adjacent part of Victoria Land (Figure 4a) new magnetic evidence (Behrendt, in press) was utilised. The Bouguer gravity anomaly over the Ross Sea shelf is after Hayes and Davey (1975) and

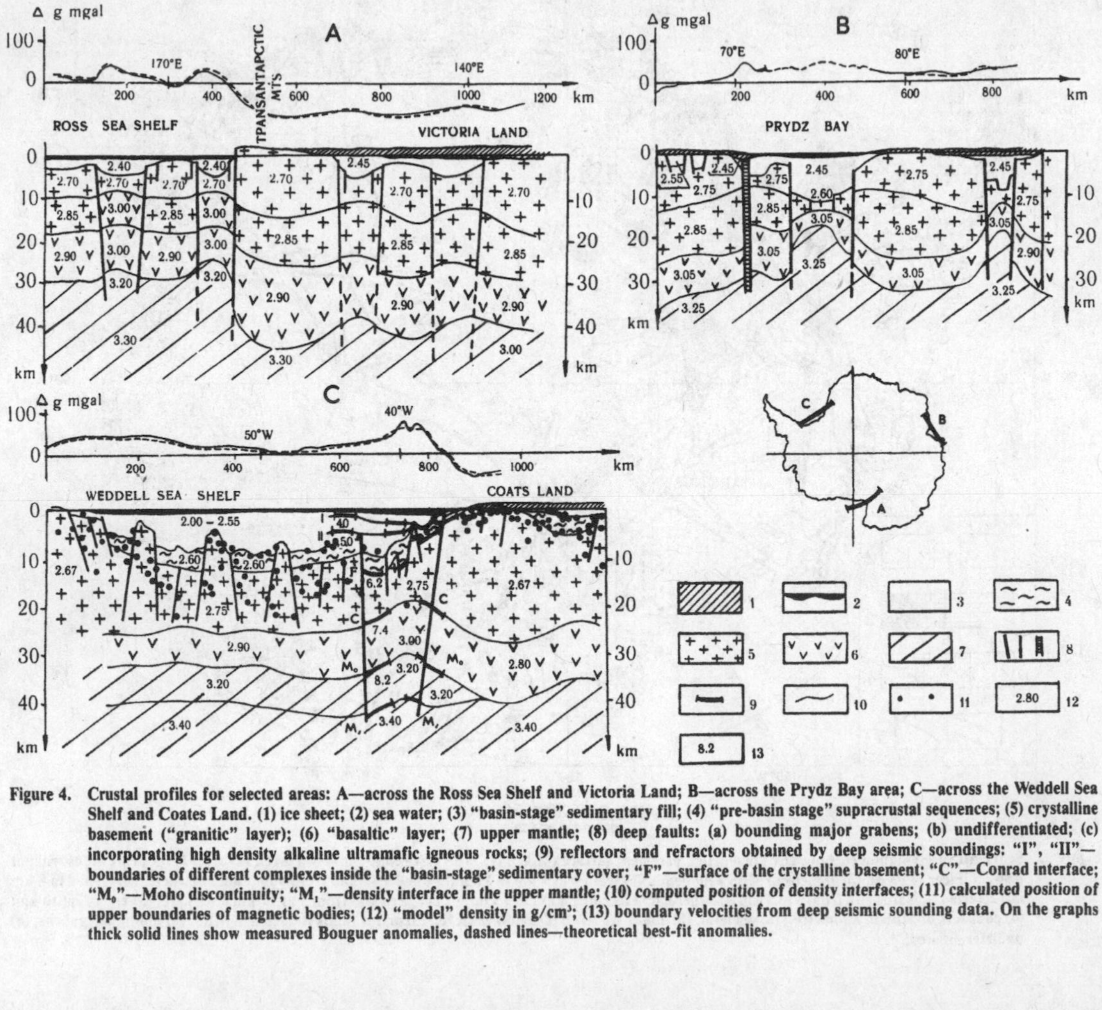

**Figure 4. Crustal profiles for selected areas: A—across the Ross Sea Shelf and Victoria Land; B—across the Prydz Bay area; C—across the Weddell Sea Shelf and Coates Land. (1) ice sheet; (2) sea water; (3) "basin-stage" sedimentary fill; (4) "pre-basin stage" supracrustal sequences; (5) crystalline basement ("granitic" layer); (6) "basaltic" layer; (7) upper mantle; (8) deep faults: (a) bounding major grabens; (b) undifferentiated; (c) incorporating high density alkaline ultramafic igneous rocks; (9) reflectors and refractors obtained by deep seismic soundings: "I", "II"—boundaries of different complexes inside the "basin-stage" sedimentary cover; "F"—surface of the crystalline basement; "C"—Conrad interface; "$M_0$"—Moho discontinuity; "$M_1$"—density interface in the upper mantle; (10) computed position of density interfaces; (11) calculated position of upper boundaries of magnetic bodies; (12) "model" density in g/cm³; (13) boundary velocities from deep seismic sounding data. On the graphs thick solid lines show measured Bouguer anomalies, dashed lines—theoretical best-fit anomalies.**

over the Transantarctic Mountains and Victoria Land after Groushinsky et al. (1978). Crustal densities were chosen mainly by analogy with East Antarctica (Kurinin and Grikurov, 1982) with relevant constraints arising from the interpretation by Hayes and Davey (1975); in contrast to the latter interpretation, however, a combined isostatic effect of the Moho and Conrad interfaces was introduced instead of assuming the compensating influence to be at one interface or the other. The model illustrates the above-mentioned sharp gradient in crustal thickness under the Transantarctic Mountains escarpment and the presence on both sides of isostatically compensated rift-like structures which dissect the continental crust into 100-125 km wide blocks.

The crustal profile across Prydz Bay (Figure 4b) incorporates the results of the statistical evaluation of the calculated parameters of magnetic bodies (Kadmina, 1980), the gravity data interpretation (Kurinin, 1980; Kurinin and Grikurov, 1980) and correlation between seismic velocity and density obtained along the deep seismic sounding profile (Kurinin and Grikurov, 1982). A major crustal feature in this area is the presence of three crustal blocks; the formation of deep faults separating these blocks was apparently related to the evolution of the Lambert Glacier rift zone. The most conspicuous is the western fault bounding the Prydz Bay depression; the latter represents a continuation of the Lambert Glacier rift graben. Geophysical anomalies over this fault and direct geological evidence suggest that it controls intrusion into the upper crust of high density alkaline-ultramafic rocks. The axial graben exceeds 10 km in depth and is associated with compensatory rises of both the Moho and the Conrad interfaces. Under both sides of the axial graben the lower part of the "granitic" layer has an abnormally high density (2.85 g/cm³). To the east of the axial graben there is another graben-like depression of the consolidated crust's surface also underlain by rises of the Moho and Conrad interfaces. The whole Lambert Glacier—Prydz Bay area is characterised by a somewhat thinner crust than the rest of the East Antarctic continental terrain and the upper mantle here has an abnormally low density. This suggests a continuing process of crustal reworking apparently related to rifting.

Preliminary results of deep seismic soundings carried out by the Soviet Antarctic Expedition during the 1979/80 field season in the Weddell Sea area (Figure 4c) are incorporated in an integrated interpretation with other geophysical evidence used in compilation of this crustal profile. The position of a refractor of seismic velocity 6.2 km/s is determined with sufficient certainty and confirmed by the distribution of magnetic sources. This refractor ("F" on Figure 4c) is interpreted as the top of Lower Precambrian crystalline basement which is believed to rise above sea level to the east of the deep seismic sounding profile, but under the Filchner Ice Shelf is downthrown by a series of faults to 14-15 km bsl, thus indicating the presence here of a major graben with thick sedimentary fill. In the latter there are two refractors with seismic velocities 3.9-4.0 km/s ("I") and 5.0 km/s ("II"). The refractors and reflectors with velocities about 7.4 km/s recognised at deeper crustal levels (about 20 km) on both sides of the graben are believed to represent the Conrad interface ("C"). A high velocity boundary (8.2 km/s) at the depth of 34 km is regarded as the Mohorovicic discontinuity ("$M_0$") which seems to rise gently towards the axial part of the graben. The deepest recorded reflector at 41-44 km bsl ("$M_1$") is thought to reflect layering in the upper mantle.

By correlation of the above seismic evidence with Bouguer anomalies, a crustal density section was computed and then extrapolated to the east and west of the deep seismic sounding profile on the basis of gravity data alone. In compilation of the section, the crustal thickness under the Weddell Sea shelf was taken, by analogy with other marginal seas as of the order of 30-33 km, that is similar to the Ross Sea shelf (see Figure 4a). In that case the observed gravity anomaly would require an increase in lower crustal density below the deepest part of the basin, that is from east to west. Within the Filchner Ice Shelf graben the computed position of upper crustal boundaries was corrected in accordance with results of mutlichannel marine seismic profiling carried out by the 26th Soviet Antarctic Expedition during the 1980/81 field season (Poselov et al., in preparation). In particular, a much more detailed configuration of the acoustic basement with seismic velocity 5.8-6.2 km/s was obtained, while its general position remained consistent with that determined by the deep seismic soundings as the "F"-refractor.

The base of the low density upper crustal sedimentary layer (2.0-2.5 g/cm³) in the western part of the crustal profile (Figure 4c) is shown on the basis of unprocessed marine seismic data obtained during the 1981/82 season. Nevertheless, this provisional boundary with seismic velocity of the order of 5.0 km/s appears quite consistent with distribution of the upper boundaries of magnetic bodies; its "mean" position does not contradict the density model, while the presence in this area of such a boundary within a similar range of depths is confirmed by discrete seismic refraction measurements (Haugland, 1982).

The geologic significance of the boundary under consideration is, however, not clear. The suggested model implies that its strongly dissected relief is caused by the presence of several horst-like crystalline basement highs separated by graben-like lows; the latter are partly filled with magnetic rocks with a mean density of 2.6 g/cm³. Multichannel seismic profiling results show that within the Filchner Ice Shelf graben these rocks reveal a distinct internal layering and therefore belong to supracrustal (epicratonic) sequences. Among the rocks which occur extensively in the mountains adjacent to the eastern Weddell Sea the most likely equivalents of such sequences are, in the authors' opinion, the Ross intracratonic folded complexes and/or synchronous subplatform formations of Upper Precambrian to Lower Palaeozoic age. If such correlation is valid, then the layer with a density 2.6 g/cm³, depsite its apparent seismic layering, must be included in a "pre-basin stage" basement rather than in a "basin stage" sedimentary cover; consequently the former will appear identical with the "magnetic" basement of the crust. It is possible, however, that the layer under consideration belongs, in fact, to a much younger stratigraphic level and represents the Upper Palaeozoic(?) or even Mesozoic(?) base of a "basin stage" cover incorporating magnetic (mafic) igneous rocks.

Thus, a major characteristic feature in all selected areas described above is the presence of graben-like crustal downwarps. Their riftogenic nature is suggested by prominent reworking of the overall crustal structure reflected in the considerable increase in the thickness of the sedimentary layer, an abnormally low thicknes of the "granitic" layer and associated compensatory rises of the Conrad and Moho interfaces. By analogy with better studied rift zones elsewhere in the world, the grabens are believed to be underlain by an abnormally low density upper mantle; this is consistent with available gravity data.

## DISCUSSION

A major feature of the Antarctic crustal structure emerging from available evidence is the abundance of rifts and rift-like features, probably unparalleled on other continents. These structural features and related disjunctive dislocations are particularly numerous in West Antarctica and in subglacial parts of Victoria and Wilkes Lands. A linear chain of graben-like crustal depressions can also be traced across East Antarctica from the Lambert Glacier-Enderby Land area through the subglacial Gamburtsev Mountains and towards the central Transantarctic Mountains. The presence of transcontinental systems of crustal fractures often associated with active rift zones indicates that the recent tectonic evolution in Antarctica was dominated by a destructive trend and that rifting was a leading mechanism in disintegration of the Antarctic continental landmass into separate crustal blocks as well as in the formation of vast sedimentary basins (Masolov et al., 1981).

It would be logical to assume that an extreme intensity of riftogenic processes in Antarctica and consequent development of extensive bedrock depressions (occupying as much as 50% of the total area underlain by continental crust) are somehow related to another unique feature of the continent, namely, the presence of a thick ice sheet. A conventional viewpoint implies isostatic compensation of the ice load by a general subsidence of the Antarctic continental mass into the mantle. Such an essentially mechanistic model of compensation may, however, appear oversimplified and the reaction of the mantle may, in fact, be much more complicated. For example, an active thermal regime at the base of the crust seems inevitable due to heating of the upper mantle under the influence of an excessive load. Some authors (e.g. Demenitskaya et al., 1981) believe that heating in the upper lithosphere may be directly caused by a screening effect of the ice cover due to its low thermal conductivity. Whether such an effect primarily causes the heating or just increases it additionally, it would produce a tendency for expansion and, consequently, a considerable built up of pressure with a vertical component directed against the iceload and, presumably, compensating it.

The lateral component of rising pressure would probably be released through the zones of crustal weakness. If sufficient tension is accumulated, such zones are most likely to become the sites of vigorous rifting. It must be pointed out, however, that the ice cover with regard to the riftogenic process may simultaneously perform both creative and conservative functions. The latter effect is probably due to the likely increase in general plasticity of the crust under the higher stress caused by the great ice load; consequently, accumulated tension may be released by deformations before it exceeds the critical level (Volarovich et al., 1972), thus preventing initiation of rifting.

In the light of the above remarks it seems likely that deglaciation of Antarctica would cause a rapid disintegration of its continental crust. The Antarctic ice sheet is known to be just barely stable, particularly in West Antarctica and the process of deglaciation in terms of geologic time would practically occur momentarily. Hence the tension accumulated in the crust, first of all within already "weakened" developing rift zones, would be drastically increased by pressure of the upper mantle and immediately followed by a catastrophic outburst of seismic activity. These phenomena would probably result in intensive crustal movements both in vertical and lateral directions leading to complete disintegration of the Antarctic continental landmass.

## CONCLUSIONS

Antarctica is unique in its abundance of rifts and rift-like crustal structures, the majority of which underlie the vast bedrock depressions; consequently the latter occupy as much as almost half of the continental area surrounding the South Pole. These features are believed to be related, at least to some extent, to the presence of a giant ice sheet responsible for intensification of destructive tectonic processes.

The Antarctic rift zones fall into two major categories—intracratonic symmetric rifts and pericratonic asymmetrical rift systems. The latter delineate to a large extent the developing sedimentary basins of West Antarctica, as well as the configuration of "passive" continental margins facing the Atlantic and Indian Oceans.

Recent riftogenic processes occur mainly in the same regions which were subjected to tectonic fragmentation during a considerable span of preceding geologic history. For some areas (e.g. the Weddell Sea Shelf) the evidence for ensialic riftogenic processes can be traced as far back as Late Precambrian time.

The data presented in this paper suggest that, in the absence of seismic information, a reconnaissance study of crustal structures can be successfully accomplished on the basis of gravity and magnetic data.

## REFERENCES


BEHRENDT, J.C. (in press): Speculations on the petroleum resources of Antarctica; *in* Splettstoesser, J. (ed.) *Mineral Resource Potential of Antarctica.* Univ. of Texas Press.

DEMENITSKAYA, R.M., 1982: Some problems of crustal geodynamics in Antarctica; *in* Craddock, C. (ed.) *Antarctic Geoscience.* Univ. Wisconsin Press, Madison, 903-6.

DEMENITSKAYA, R.M., IVANOV, S.S. and LITVINOV, E.M., 1981: *Estestvennye fizicheskie polia okeanov/Natural physical fields of oceans.* Izdatel'stvo Nedra, Leningrad.

FEDOROV, L.V., GRIKUROV, G.E., KURININ, R.G. and MASOLOV, V.N., 1982: Crustal structure of the Lambert Glacier area from geophysical data; *in* Craddock, C. (ed.) *Antarctic Geoscience.* Univ. Wisconsin Press, Madison, 931-6.

GRIKUROV, G.E. (this volume): Structural history of the Ross orogenic belt and tectonic nature of East-West Antarctica boundary.

GROUSHINSKY, N.P., SAZHINA, N.B., NIKITINA, G.I. and TOROCHKOVA, G.I., 1978: Karta anomaliy sily tiazhesti Antarktidy/Gravity map of Antarctica. Antarktika, 17, 199-203.

HAUGLAND, K., 1982: Seismic reconnaissance survey in the Weddell Sea; *in* Craddock, C. (ed.) *Antarctic Geoscience.* Univ. Wisconsin Press, Madison, 405-13.

HAYES, D.E. and DAVEY, F.J., 1975: A geophysical study of the Ross Sea, Antarctica; *in* Hayes, D.E. and Frakes, L.A. et al. (eds.) *Initial Reports of the Deep Sea Drilling Project, 28.* 887-907, U.S. Govt. Printing Office, Washington, D.C.

KADMINA, I.N., 1980: Struktura magnitoaktivnogo sloya zemnoy kory v raione lednika Lamberta/Structure of magnetic layer of the Earth's crust in the Lambert Glacier; *in "Geofizicheskie issledovania v Antarktide"/Geophysical survey in Antarctica.* Research Institute of Arctic Geology, Leningrad, 44-51.

KOLMAKOV, A.F., MISHEN'KIN, B.P. and SOLOVIEV, D.S., 1975: Glubinnye seismicheskie issledovania v vostochnoy Antarktide/Deep Seismic investigations in East Antarctica. *Inf. Byul. Sov. Antark. Eksp., 91,* 5-15.

KURININ, R.G., 1980: Geologicheskoe istolkovanie gravitatsionnykh anomalii raiona lednika Lamberta (Geologic interpretation of gravity anomalies in Lambert Glacier area). *Trudy SAE, 70,* 134-45.

KURININ, R.G. and GRIKUROV, G.E., 1980: Stroenie riftovoy zony lednika Lamberta (Structure of rift zone of the Lambert Glacier). *Trudy SAE, 70,* 76-86.

KURININ, R.G. and GRIKUROV, G.E., 1982: Crustal structure of part of East Antarctica from geophysical data; *in* Craddock, C. (ed.) *Antarctic Geoscience.* Univ. Wisconsin Press, Madison, 895-901.

MASOLOV, V.N., 1980: Stroenie magnitoaktivnogo fundamenta jugo-vostochnoy chasti basseina moria Weddella/Structure of the magnetic basement in the southeastern part of Weddell Sea basin; *in "Geofizicheskie issledovania v Antarktide"/Geophysical Survey in Antarctica.* Research Institute of Arctic Geology, Leningrad, 14-28.

MASOLOV, V.N., KURININ, R.G. and GRIKUROV, G.E., 1981: Crustal structure and tectonic significance of Antarctic rift zones (from geophysical evidence); *in* Cresswell, M.M. and Vella, P. (eds.) *Gondwana Five.* A.A. Balkema, Rotterdam, 303-9.

POSELOV, V.A., SHELESTOV, F.A. and GRIKUROV, G.E. (in prep.): Predvaritel'nye rezultaty morskikh seismicheskikh issledovanii v more Weddella v sostave 26-oy Sovetskoy Antarkticheskoy Ekspeditsii (Preliminary results of marine seismic sounding in the Weddell Sea during 26th Soviet Antarctic Expedition).

ROBERTSON, J.D., BENTLEY, C.R., CLOUGH, J.W. and GREISCHAR, L.L., 1982: Seabottom topography and crustal structure below the Ross Ice Shelf, Antarctica; *in* Craddock, C. (ed.) *Antarctic Geoscience.* Univ. Wisconsin Press, Madison, 1083-90.

ROSE, K.E., 1982: Radio-echo studies of bedrock in southern Marie Byrd Land, Antarctica; *in* Craddock, C. (ed.) *Antarctic Geoscience.* Univ. Wisconsin Press, Madison, 985-92.

STEED, R.H.N. and DREWRY, D.J., 1982: Radio-echo sounding investigation of Wilkes Land, Antarctica; *in* Craddock, C. (ed.) *Antarctic Geoscience.* Univ. Wisconsin Press, Madison, 969-75.

VOLAROVICH, M.P., TOMASHEVSKAYA, I.S. and KHAMIDULIN, Ya. N., 1972: Svyaz' deformatzionno-prochnostnykh svoystv gornykh porod s istoriey nagruzhenia/Relation of deformation-strength properties of rocks with the history of loading. *Geofiz. sb., 45,* Moscow, 13-22.

WILSON, D.D., McGINNIS, L.D., BURDELIK, W.J. and FASNACHT, T.L., 1981: McMurdo Sound upper crustal geophysics. *U.S. Antarct. J., 16,* 31-3.

ZNACHKO-YAVORSKII, G.A., 1978: Skhema rel'efa Antarktidy bez ledianogo pokrova/Bedrock relief of Antarctica; *in* Grikurov, G.E. (ed.) *Tectonic Map of Antarctica.* Ministry of Geology of the USSR, Moscow, 1980.

# ISOSTATIC DISPLACEMENT OF ANTARCTIC LITHOSPHERE

D.J. Drewry, E.J. Jankowski and R.H.N. Steed, *Scott Polar Research Institute, University of Cambridge, Cambridge CB2 1ER, UK*

*Abstract* Isostatic depression of the Antarctic lithosphere by the weight of the overlying ice has been calculated following determination of ice thickness and sub-glacial bedrock elevation by radio echo sounding. The method adopted assumes that the continent is currently in quasi-isostatic equilibrium and is based upon the deformation function of Brotchie and Silvester (1969) in which downward deflection under an ice load extends to a distance approximately four times the radius of relative crustal stiffness (ca 500 km).

The whole of Antarctica to the continental break-of-slope is considered as a series of blocks, 100 km square, free to move independently. The load on each square is calculated with the response being affected by adjacent squares within an area of 700 km × 700 km. Deformation co-efficients are determined according to the deflection function and representing the contribution of each square to total uplift. In the case of ice unloading account is also taken of areas of bedrock below sea level where ice may be initially replaced by sea water. Iteration of the calculations allows such areas to achieve equilibrium. The amount of uplift varies from about 300 m in West Antarctica to ca 1000 m in central East Antarctica.

Maps and isometric diagrams are produced and discussed of (1) the amount of uplift, and (2) the configuration of continental bedrock following removal of the ice load and full isostatic recovery. The latter, ignoring effects of glacial erosion, is assumed to approximate the preglacial outline of Antarctica, and has implications for morphological correlations between Antarctica and other Gondwana fragments.

# MAGNETIC ANOMALIES OVER ANTARCTICA MEASURED FROM MAGSAT

M.H. Ritzwoller and C.R. Bentley, *Department of Geology and Geophysics, University of Wisconsin, Madison, Wisc., 53706, U.S.A.*

*Abstract* MAGSAT satellite vector magnetic anomaly data have been used to create a scalar magnetic anomaly map over Antarctica and the nearby oceans. The temporal stability of the anomaly features, together with their correlation with well-known oceanic tectonic structures, indicates that the source of the observed anomalies lies within the Earth. Specifically, magnetic anomalies are associated with oceanic basins (negative) and ridges (positive) and with such continental features as tectonic provinces (positive and negative), hypothetical rift features (negative), several mountain ranges (positive and negative), and a subglacial basin(positive). At least two of the largest magnetic anomalies in Antarctica are mirrored by anomalies at the corresponding locations in other Gondwana continents. The map appears to possess information relevant to the large-scale structure and development of Antarctica, but must be used with care due to ambiguities inherent in its interpretation.

It is fortunate and surprising that magnetic fields generated by crustal sources can be isolated in satellite magnetic field data from those generated in the core and those external to the earth. It is surprising because crustal fields (<25 nT) are so much smaller than the core field (30,000-70,000 nT) and external fields (0-2,000 nT), and fortunate because geologic and tectonic features in the deep crust can result in long wave-length magnetic anomalies (Pakiser and Zeitz, 1965; Zeitz et al., 1966; Hall, 1974; Krutikhovskaya and Pashkevich, 1977). Long wave-length magnetic anomalies were first mapped globally from data acquired by the POGO (Polar Orbiting Geophysical Observatories) satellites (Regan et al., 1975) and have been shown to correlate well with upward-continued regional aeromagnetic surveys (Langel et al., 1980). The chief advantage of the MAGSAT data over the POGO data lies in the ability to study vector fields with MAGSAT and in the greater resolution MAGSAT's generally lower orbit (350-560 km compared to 400-1500 km for POGO) affords. For a further description of MAGSAT see Langel, Ousley et al., (1982).

Attempts have been made to isolate anomalies both in the scalar field (Langel, Phillips and Horner, 1982; Coles et al., 1982; Ritzwoller and Bentley, 1982) and in the vector field (Langel, Schnetzler et al., 1982; Coles et al., 1982). Sailor et al., (1982) confirm the general reliability of such attempts and conclude that resolution is possible down to a 250 km spatial wave length in mid-latitudes. Resolution is poorer in higher latitude regions due to the smaller signal to noise ratio that is the consequence of larger external magnetic fields in the auroral regions. Magnetospheric ring-currents in low latitudes generate a long spatial wave-length signature that varies relatively slowly and is easily filtered. However, currents following magnetic field lines in auroral regions have spectra covering a broad band of both temporal and spatial frequencies, thus making filtering difficult. Therefore, only data from passes occurring while field-aligned currents are small can be used in high latitudes, seriously reducing the size of the data set over Antarctica and in the Arctic (Coles et al., 1982). The pronounced radial striping displayed by the anomalies in the Antarctic map of Ritzwoller and Bentley (1982) demonstrates the signal detection problem. In this paper we will present an improved data reduction scheme and a scalar magnetic anomaly map that we believe better approaches the true crustal magnetic field.

The modelling and interpretation of satellite magnetic anomaly maps has a short history and progress is mostly qualitative. On a global scale, anomalies appear to be associated with such large structures as continental shields and platforms, subduction zones (positive), oceanic ridges (positive), and abyssal plains (negative) and appear to be bounded by such "linear" features as sutures, rifts, folded mountains, and age province boundaries (Frey, 1982a). Preliminary regional studies have been performed by Frey (1982b) for Asia, by Hastings (1982) for Africa, by Hinze et al., (1982) for South America, and by Ritzwoller and Bentley (1982) for Antarctica.

Geologic interpretation of the anomalies can only be conducted in the light of the probable mineralogy of the lower crust and upper mantle. Wasilewski et al., (1979) argue that the mantle is probably non-magnetic, so if the Curie isotherm is below the crust the lower-magnetic boundary is the Moho. Moreover, Wasilewski and Mayhew (1982) conclude that, at least for some tectonic settings, the lower crust is the most magnetic crustal layer, and that magnetisation values for lower crustal xenoliths (specifically metabasic rocks of the granulite facies) have values consistent with those inferred from models of long wave-length anomalies. Wasilewski and Fountain (1982) corroborate these findings with a study of the Ivrea Zone in northern Italy, where mafic granulite facies rocks are the only magnetic lithology present and are thick and laterally continuous, thus providing a good candidate for a deep-crustal source of long wave-length magnetic anomalies.

Along with regional mineralisation variations, it is likely that variations in the depth to the Curie isotherm (if above the Moho) will be reflected in the long wave-length anomalies (Mayhew, 1982). Thus, the main sources of long wave-length magnetic anomalies in continental regions are expected to reside above the Moho and the Curie isotherm but principally in the lower crustal layer. It follows that, *ceteris paribus,* regional magnetic anomalies depend inversely on heat flow, and directly on the thickness of the magnetised crust. Therefore, continental highs indicate some combination of a thick lower crust, low heat flow, and higher than normal average magnetic susceptibility in the lower crust, whereas continental lows imply a thin crust, high heat flow, and/or low susceptibilities in the lower crust. It remains uncertain to what degree continental remanent magnetisation may affect long wave-length magnetic anomalies. Galliher and Mayhew (1982) argue that the effect is small.

The base of the oceanic crust, of course, lies much nearer to the surface than that of the continental crust—generally far above the Curie isotherm, except right at the spreading centres. Thus, oceanic magnetic anomalies should reflect regional crustal thicknesses and susceptibility differences, but generally not heat flow. Oceanic basins should be magnetically negative relative to continental regions. Over oceanic spreading centres, however, as has been shown by model calculations, the broad axial region of positive remanent magnetisation results in a positive anomaly, even at satellite elevations.

## Data Reduction

Ritzwoller and Bentley (1982) described a method by which a preliminary map of the crustal scalar magnetic field was produced. The method was designed to filter non-crustal magnetic fields using scalar magnetic field data alone. However, a great deal of information concerning the influence of field-aligned currents is available in the vector magnetic field data; we have now used that information as part of our data selection procedure, which we will now outline.

Only satellite passes over Antarctica that took place during a magnetically quiet time period (planetary magnetic activity index, Kp, no greater than $1^-$ for 6 hours) were considered. Of the approximately 2400 passes over Antarctica between 1st November 1979, and 1st April 1980, 212 met this selection criteria. For these, scalar field values were calculated from the observed vector data, and differences were taken relative to spherical harmonic core field model MGST (4/81), of degree and order 13, created by Langel, et al., (1980). Next a second degree polynomial least squares fit was subtracted from each pass to filter the effects of magnetospheric ring-currents, errors in the core field model, and other errors in measurement (for more details see Langel, Phillips and Horner, 1982). Since the vector data indicate that more than half of the passes remain seriously affected by field aligned currents, passes were accepted for final use only if they satisfied the following three criteria:

(a) the maximum amplitude of the vertical vector anomaly, $\Delta Z$, is less than 25 nT;

(b) the maximum amplitude of the scalar anomaly, $\Delta B$, is less than 20 nT;

(c) $\Delta Z$ and $\Delta B$ are highly correlated.

(For a more detailed discussion of our use of the vector components in data selection, see Ritzwoller, 1982)

The 88 passes selected in this manner appear to provide a reasonable balance between providing a sufficient density of data (Figure 1.) and reducing field-aligned current effects to an acceptable level. Scalar field values calculated from these passes were averaged in

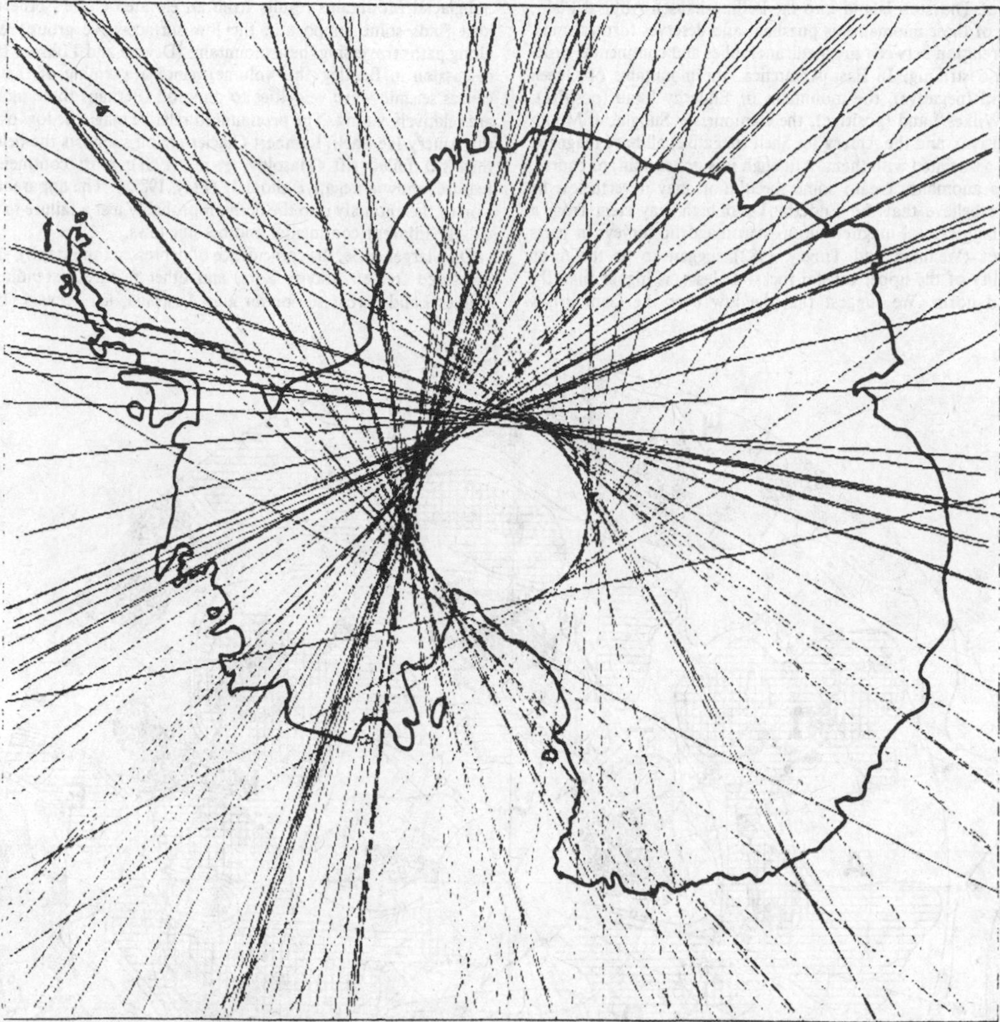

**Figure 1. Flight tracks of the 88 passes used to construct the magnetic anomaly map in Figure 2.**

square bins measuring 330 km on a side, and standard deviations were calculated. Following a check to make sure that the values within the bins were normally distributed, outliers were rejected and the average and standard deviation were recomputed. The averages were then plotted and hand-contoured, yielding the final anomaly (Figure 2.). The average standard deviation is 1.5 nT, leading to an estimate of the standard error in the means of only a few tenths of a nannotesla.

The anomalies on this map have not been reduced to the pole; since all of Antarctica (except the Antarctic Peninsula) is above 60° in geomagnetic latitude, reduction to the pole would have only a small effect. Nor have data been continued to a single elevation before averaging—studies performed elsewhere (R. Sailor, personal communication, 1982) indicate that maps created by simple averaging are "qualitatively and quantatively similar" to those for which passes are first continued to the average elevation (470 km from our data set).

The scalar anomaly map in Figure 1 is a significant improvement over the preliminary map presented in Ritzwoller and Bentley (1982)—we believe it to be a good approximation to the true crustal anomaly field. The data appear internally consistent within the five months of data acquisition (Ritzwoller, 1982), and the map correlates well with the POGO map for Antarctica. Thus, the features are temporally stable, a necessary characteristic of fields produced in the crust.

## Geological Interpretation

The interpretation of MAGSAT data, though still in its infancy, should hold great interest to the Antarctic geoscientist, for here is the first coherent continent-wide data set with information about the Antarctic crust. Furthermore, MAGSAT magnetic anomalies appear to be highly correlated with known Antarctic geologic and tectonic features, especially in oceanic regions where the geology is simplest and best known.

Oceanic magnetic anomalies are almost invariably associated both with basins (negative) and spreading ridges (positive). (All oceanic feature names will be taken from Heezen and Tharp, 1980). Three of the four major oceanic basins surrounding Antarctica (the Weddell, Enderby and Wilkes Abyssal Plains, the exception being the Bellingshausen Abyssal Plain) have negative anomalies associated with them. (Geographic names are indexed in Figure 2.) The most striking conjunction of a positive anomaly with a spreading ridge occurs where the Mid-Indian Ocean Ridge and the East Pacific Ridge meet grid south of the Ross Sea embayment ("grid" directions refer to a Cartesian co-ordinate system laid across the polar map, with grid north parallel to the 0° meridian, grid east parallel to 90°E, etc.). The set of positive anomalies running between 140°E and 120°W, north of 65°S, lies closely over the East Pacific Ridge on the grid west and the Mid-Indian Ocean Ridge on the grid east. There are also highs associated with aseismic volcanic ridges and plateaus, such as the Kerguelen Plateau (about 80°E) and Maud Rise (65°S,0°E), and a relative high in an otherwise pronounced low is associated with the South Sandwich Islands and Trench (60°S, 25°W).

There are, however, some interesting anomalies that do not fit this norm. For example, a positive anomaly runs grid northeast from Maud Rise right into the Enderby Abyssal Plain, and another extends grid west of Thurston Island into the Bellingshausen Abyssal Plain, The cause of these anomalies is puzzling, and deserves further study.

The correlation between magnetic anomalies and continental structures is also striking. In East Antarctica, the mountains of Queen Maud Land (negative), the mountains of Enderby Land (positive), much of Wilkes Land (positive), the Gamburtsev Subglacial Mountains (negative), and the Amery Ice Shelf (negative) all have magnetic anomalies associated with them. Although it is not certain, of course, what these anomalies mean, some speculation may nevertheless be useful. We believe that the Enderby Land high may stem from a relatively high crustal magnetisation—aeromagnetic surveys in parts of the area (Wellman and Tingey, 1982) suggest to us the mean susceptibility of the upper crustal rocks, at least, is higher than the continental norm. We suggest that the low over the Gamburtsev Subglacial Mountains results from an elevated Curie isotherm; this idea finds some support in the low surface-wave group velocities along paths traversing these mountains (Dewart and Töksöz, 1965; see discussion in Bentley, this volume), since a warming of the mantle causes seismic wave velocities to diminish. Perhaps these mountains are relatively young. The pronounced relative magnetic low overlying the Amery Ice Shelf/Lambert Glacier region supports the belief that this is a failed rift (Masolov, et al., 1981); most continental rift features show a negative anomaly (Frey, 1982a). The apparent extension of the anomaly into the ocean is probably just a failure to resolve closely adjacent continental and oceanic lows.

On a larger scale, the coincidence of depressed topography, satellite-measured free-air gravity lows, and other features extending from Wilkes Land across the ocean into Australia, led Veevers (1982) to

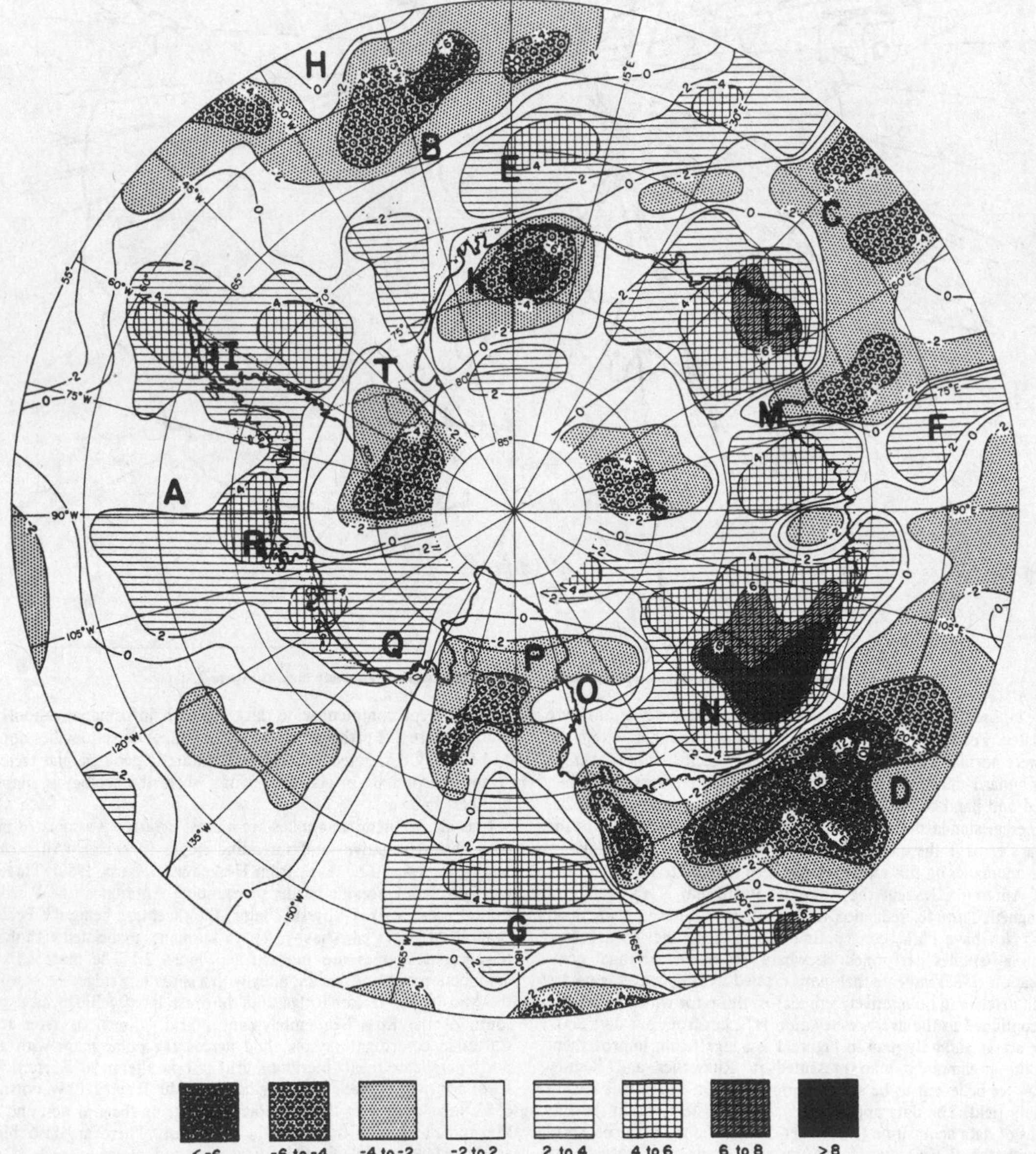


**Figure 2. MAGSAT total field magnetic anomaly map over Antarctica. Units in nT. Average elevation 470 km. Capital letters indicate the approximate location of: A—Bellingshausen Abyssal Plain; B—Weddell Abyssal Plain; C—Enderby Abyssal Plain; D—Wilkes Abyssal Plain; E—Maud Rise; F—Kerguelen Plateau; G—juncture of the Mid-Indian Ocean Ridge and the East Pacific Ridge; H—South Sandwich Islands; I—Antarctic Peninsuls; J—Ellsworth Mountains; K—Queen Maud Land; L—Enderby Land; M—Amery Ice Shelf/Lambert Glacier; N—Wilkes Land; O—Transantarctic Mountains; P—Ross Sea embayment; Q—Marie Byrd Land; R—Thurston Island; S—Gamburtsev Mountains; T—Weddell Sea embayment.**

suggest that the whole vast region is being held down dynamically by downward currents in the mantle. The positive magnetic anomaly in Wilkes Land (and the corresponding one in Australia—see below) is consistent with this suggestion since the convection-convergence zone would be relatively cool.

A noteworthy feature of the anomaly map is the absence of magnetic anomalies over the Transantarctic Mountains. Instead of exhibiting a characteristic anomaly pattern of their own, they mark a distinct boundary zone between largely separate East and West Antarctic anomalies. (It is likely that the negative anomaly that cross-cuts them from the Ross Ice Shelf is, in fact, another case of two separate lows that have not quite been resolved.)

Several anomalies appear in West Antarctica, but their tectonic association is not at all clear. The volcanic province or provinces comprising the Antarctic Peninsula, Thurston Island and Marie Byrd Land all show distinct highs as would be expected, but along the whole region the centres of the highs are inexplicably shifted oceanward. A pronounced low over the Ross Sea may support the concept of a failed rift zone here, but it is not centred over the postulated axis of the rift found from gravity measurements in the grid eastern part of the sea (Hayes and Davey, 1975; Bentley, this volume, Figure 9.).The anomaly does disappear under the Ross Ice Shelf—that is in agreement with the gravity evidence (Davey, 1981).

What may be a mirroring negative anomaly appears in the Weddell Sea embayment between (and partly overlying) the Ellsworth and Pensacola Mountains. However, if this is a rift-zone negative, it is surprising that is does not extend farther grid northward under the Weddell Sea continental shelf.

Comparison of the Gondwana reconstruction of Norton and Sclater (1979) with Figure 2 and the global anomaly map of Langel, Phillips and Horner (1982) shows that the pronounced high in Wilkes Land is mirrored by an even more pronounced high in the Australian shield; there is, in fact, a general similarity between the magnetic appearances of Wikes Land and Australia. Moreover, the low in Queen Maud Land appears to correspond well to lows in southern and southeastern Africa. On the other hand, there is no clear correspondence between the West Antarctic positive anomalies and anything else. It appears, unfortunately, that the MAGSAT map is not yet going to solve the puzzle of where to put the West Antarctic microplates before the breakup of Gondwanaland! For East Antarctica, nevertheless, the Gondwana magnetic reconstruction is very good, implying that the anomaly features in Antarctica were formed prior to breakup. A more exact comparison requires that all data be reduced to the pole.

## Conclusions

Magnetic anomalies mapped in Antarctica from MAGSAT data reflect real features of crustal structures, and many of them can be understood qualitatively. East Antarctic Gondwanian associations appear clear, but West Antarctic associations do not. Further analysis should lead to more quantitative models, and to interpretations of features that are now mysterious.

*Acknowledgements* This work was supported by the National Aeronautics and Space Administration (NASA) under contract NAS5-25977.

## REFERENCES


BENTLEY, C.R. (this volume): Crustal structure of Antarctica from geophysical evidence.

COLES, R.L., HAINES, G.V., JANSEN VAN BEEK, G., NANDI, A. and WALKER, J.K., 1982: Magnetic anomaly maps from 40°N to 83°N derived from MAGSAT satellite data. *Geophys. Res. Lett., 9,* 281-4.

DAVEY, F.J., 1981: Geophysical studies in the Ross Sea region. *R. Soc. N.Z., J., 11,* 465-79.

DEWART, G. and TOKSOZ, M.N., 1965: Crustal structure in East Antarctica from surface wave dispersion. *J. Geophys. 10,* 127-39.

FREY, H.L., 1982a: MAGSAT scalar anomaly distribution: the global perspective. *Geophys. Res. Lett., 9,* 277-90.

FREY, H.L., 1982b: MAGSAT scalar anomalies amd major tectonic boundaries in Asia. *Geophys. Res. Lett., 9,* 299-302.

GALLIHER, S.C. and MAYHEW, M.A., 1982: On the possibility of detecting large-scale crustal remanent magnetisation with MAGSAT vector magnetic anomaly data. *Geophys. Res. Lett., 9,* 325-8.

HALL, D.H., 1974: Long wave-length aeromagnetic anomalies and deep crustal magnetisation in Manitoba and northwestern Canada. *Can. J. Geophys., 40,* 403-30.

HASTINGS, D.A., 1982: Preliminary correlations of MAGSAT anomalies with tectonic features of Africa. *Geophys. Res. Lett., 9,* 303-6.

HAYES, D.E. and DAVEY, F.J., 1975: A geophysical study of the Ross Sea, Antarctica. *Initial Reports of the Deep Sea Drilling Project, 28,* 229-40, U.S. Govt. Printing Office, Washington, D.C.

HEEZEN, B.C. and THARP, M., 1980: *The Floor of the Oceans.* Marie Tharp Oceanographic Cartographer, South Nyack, New York.

HINZE, W.J., VON FRESE, R.B., LONGACRE, M.B., BRAILE L.W., LIDIAK, E.G. and KELLER, G.R., 1982: Regional magnetic and gravity anomalies of South America. *Geophys Res. Lett., 9,* 314-7.

KRUTIKHOVSKAYA, Z.A. and PASHKEVICH, I.K., 1977: Magnetic model for earth's crust under the Ukranian shield. *Can. J. Earth Sci., 14,* 2718-28.

LANGEL, R.A., ESTES, R.H., MEAD, G.D., FABIANO, E.B. and LANCASTER, E.R., 1980: Initial geomagnetic field model from MAGSAT vector data. *Geophys. Res. Lett., 7,*793-6.

LANGEL, R.A., OUSLEY, G., BERBERT, J., MURPHY, J. and SETTLE, M., 1982: The MAGSAT mission. *Geophys. Res. Lett., 9,* 243-5.

LANGEL. R.A., PHILLIPS, J.D. and HORNER, R.J., 1982: Initial scalar magmetic anomaly map from MAGSAT. *Geophys. Res. Lett., 9,* 269-72.

LANGEL, R.A., SCHNETZLER, C.C, PHILLIPS J.D. and HORNER, R.J., 1982: Initial vector anomaly map from MAGSAT. *Geophys. Res. Lett., 9,*273-6.

MASOLOV, V.N., KURININ, R.G. and GRIKUROV, G.E., 1981: Crustal structures and tectonic significance of Antarctic rift zones (from geophysical evidence); *in* Cresswell, M.M. and Vella, P., (eds.) *Gondwana Five.* A.A. Balkema, Rotterdam, 303-10.

MAYHEW, M.A., 1982: Application of satellite magnetic anomaly data to Curie isotherm mapping. *J. Geophys. Res., 87,* 4846-54.

NORTON, I.O. and SCLATER, J.G., 1979: A model for the evolution of the Indian Ocean and the breakup of Gondwanaland. *J. Geophys Res., 84,* 6803-30.

PAKISER, L.C. and ZIETZ, I., 1965: Transcontinental crustal and upper mantle structure. *Rev. Geophys., 3,* 505-20.

REGAN, R.D., CAIN J,C, and DAVIS, W.M., 1975: A global magnetic anomaly map. *J. Geophys Res., 86,* 9567-73.

RITZWOLLER, M.H., 1982: Magnetic anomalies over Antarctica and the surrounding oceans measured by MAGSAT. Unpublished M.S. Thesis, University of Wisconsin, Madison.

RITZWOLLER, M.H. and BENTLEY, C.R., 1982: MAGSAT magnetic anomalies over Antarctica and the surrounding oceans. *Geophys. Res. Lett., 9,* 285-8.

SAILOR, R.V., LAZAREWICZ, A.R. and BRAMMER, R.F., 1982: Spatial resolution and repeatability of MAGSAT crustal anomaly data over the Indian Ocean. *Geophys Res. Lett., 9,* 289-92.

VEEVERS, J.J., 1982: Australian-Antarctic depression from the mid-ocean ridge to adjacent continents. *Nature, 295,* 315-7.

WASILEWSKI, P. and FOUNTAIN, D.M., 1982: The Ivrea Zone as a model for the distribution of magnetisation in the continental crust. *Geophys. Res. Lett., 9,* 333-6.

WASILEWSKI, P. and MAYHEW, M.A., 1982: Crustal xenolith magnetic properties and long wave-length anomaly source parameters. *Geophys Res. Lett., 9,* 329-32.

WASILEWSKI, P., THOMAS, H.H. and MAYHEW, M.A., 1979: The Moho as a magnetic boundary. *Geophys. Res. Lett., 6,* 541-4.

WELLMAN, P. and TINGEY, R.J., 1982: A gravity survey of Enderby and Kemp Lands, Antarctica; *in* Craddock, C. (ed.) *Antarctic Geoscience.* Univ. Wisconsin Press, Madison, 937-40.

ZIETZ, I., KING, E.R., GEDDES, W. and LIDIAK, E.G., 1966: Crustal study of a continental strip from the Atlantic Ocean to the Rocky Mountains. *Geol. Soc. Am., Bull., 77,* 1427-48.

# RESULTS FROM AN AEROMAGNETIC AND RADIO ECHO ICE-SOUNDING SURVEY OVER THE DUFEK INTRUSION, ANTARCTICA

J.C. Behrendt, *U.S. Geological Survey, Denver, Colorado 80225, U.S.A.*

D.J. Drewry, *Scott Polar Research Institute, Cambridge University, Cambridge, CB2 1ER, U.K.*

E. Janowski, *Scott Polar Research Institute, Cambridge University, Cambridge, CB2 1ER, U.K.*

M.S. Grim, *U.S. Geological Survey, Denver, Colorado 80225, U.S.A.*

**Abstract** A combined aeromagnetic and radio echo ice-sounding survey (4200 km of traverse), made in 1978 in Antarctica over the Dufek layered mafic intrusion of Jurassic age, suggests a minimum area of about 50000 km², making it comparable in size with the Bushveld Complex of Africa. Comparisons of the magnetic and subglacial topographic profiles illustrate the usefulness of this combination of methods in studying bedrock geology beneath ice-covered areas. Rocks are exposed in only 3% of the inferred area of intrusion. Magnetic anomalies measured a few hundred metres above outcrops of the intrusion range in peak-to-trough amplitude from ca 50 nT over the lowermost exposed portion of the section in the Dufek Massif to ca 3600 nT over the uppermost part of the section in the Forrestal Range. Theoretical magnetic anomalies, computed from models based on the subice topography fitted to the highest amplitude observed magnetic anomalies, required normal and reversed magnetisations ranging from $10^{-4}$ to $10^{-2}$ emu/cm³, having directions and magnetisations consistent with measurements previously made on oriented samples. This result is interpreted as indicating that the Dufek intrusion cooled through the Curie isotherm during one or more reversals of the earth's magnetic field.

# CRUSTAL STRUCTURE OF THE MIZUHO PLATEAU, ANTARCTICA REVEALED BY EXPLOSION SEISMIC EXPERIMENTS

A. Ikami, *Regional Observation Centre for Earthquake Prediction, Nagoya University, Nagoya 464, Japan.*

K. Ito, *Regional Observation Centre for Earthquake Prediction, Kyoto University, Osaka 569, Japan.*

K. Shibuya, *National Institute of Polar Research, Tokyo 173, Japan.*

K. Kaminuma, *National Institute of Polar Research, Tokyo 173, Japan.*

*Abstract* From 1979 to 1982, explosion seismic experiments were carried out by the 20th and 21st Japanese Antarctic Research Expeditions in the vicinity of Syowa Station and in the northern Mizuho Plateau in East Antarctica. The primary object of these experiments was to investigate a velocity/depth profile of the crust and upper mantle. The biggest explosion was fired in the sea near Syowa Station, and 27 temporary seismic observation stations were set up along a 300 km long profile between Syowa and Mizuho Stations. In addition to this, two big explosions in ice holes were fired at the southern end and the middle point on this profile. Refraction waves from the Conrad and the Moho discontinuities were successfully recorded. The apparent P-wave velocity in the upper crust ranges from 6.0 km/s to 6.4 km/s, suggesting a gradual increase with depth. Apparent velocities of 6.9 km/s for P* and 7.9 km/s for Pn were observed from the biggest shot near Syowa Station, but the true velocities could not be determined. The crustal structure in the northern Mizuho Plateau was determined from travel times, and the depth of the Conrad and Moho discontinuities were defined as about 30 km and about 40 km, respectively.

Explosion seismic experiments carried out by the Japanese Antarctic Research Expeditions (JARE) since 1959 have been mainly concerned with the determination of the P-wave velocity structure of the Antarctic continental ice sheet. Explosion seismic experiments aimed at investigating the crustal structure down to the Moho have not previously been operated by JARE because of numerous difficulties such as instrumentation, the drilling of shot holes in ice, and the use of dynamite at low temperature. The main object of the 1979 experiment was, therefore, to overcome these difficulties, and establish an explosion seismic experiment system. 10 temporary seismic observation stations were set up along a profile 50 km long and 560 kg and 1000 kg explosive charges were fired on the Soya Coast near Syowa Station (Ikami et al., 1980). In 1980 and 1981, 17 explosion seismic experiments were carried out near Syowa Station and in the northern Mizuho Plateau. In this preliminary paper, we discuss the crustal structure as determined from data recorded from the explosion of charges of 1000 kg, 1400 kg and 2918 kg of dynamite.

In the past 30 years, the crustal and the upper mantle structures in Antarctica have been studied by explosion seismology, earthquake seismology, and other geophysical means. Bentley and Clough (1972) compiled many seismic refraction profiles, which provided information about seismic velocities beneath the ice sheet. In 1969, the 14th Soviet Antarctic Expedition carried out the deep seismic sounding near the coast of Queen Maud Land along a 300 km long profile. This was the first time in Antarctica that Pn waves and several intracrustal reflections were successfully recorded. The Soviet Antarctic Research Expedition carried out a second experiment in MacRobertson Land in 1973 and deduced the crustal structure (Kurinin and Grikurov, 1982). In order to reveal the crustal and upper mantle structures, some authors employed surface-wave dispersion (e.g. Evison et al., 1959; 1960; Dewart and Toksöz, 1965; Bentley and Ostenso, 1962) whereas Groushinsky and Sazhina (1982) compiled the gravity data, and Kurinin and Grikurov (1982) analysed magnetic data. Numerous other papers, not referred to, discuss the crustal structure of Antarctica, and Bentley (this volume) gives a comprehensive current review.

## Description of Experiments

Long profiles and large explosions are needed for the study of deep crustal structure. However, in Antarctica, experimental programmes are limited by many logistic conditions including time, manpower, and oversnow traverse transportation. As a consequence only a few profiles have been completed. Following the experience gained by the 20th Japanese Antarctic Research Expedition (JARE-20), the JARE-21 operations were planned as follows: a 300 km long profile from Syowa to Mizuho Stations, with shot points in the sea near Syowa Station (Shot 19), in ice holes at Mizuho Station (Shot 17), and midway between these stations (Shot 18). The ice drilling machinery had been improved since the previous work (Suzuki and Shiraishi, 1982).

Figure 1 shows the locations of shot and observation points in these experiments. As the accuracy of location of these points directly influences the result, an NNSS (Navy Navigation Satellite System) receiver was employed to determine positions more precisely and quickly (see Shibuya et al., 1982). The NNSS also received a precise UTC (Universal Coordinated Time) signal which was used to calibrate clocks, installed in recorders at each observation station (see Shibuya and Kaminuma, 1982). The data recorder, which incorporated a direct analogue recording system, was designed to be driven continuously for 26 days by batteries even at low temperatures. Each recorder had four channels, three for seismic signals and one for the time signal. To improve dynamic range, seismometer signals were fed to these three channels with different amplification constants. Therefore only the vertical component could be observed. A high precision quartz oscillator was included in the timing circuit; it was calibrated more than three times before and after each shot.

The magnetic recording tape was driven at 0.2375 mm/s, an extremely slow speed. Subsequently the signals were replayed at speeds 200 or 400 times faster than that on recording. Each seismogram phase was picked out using the cursor displayed on a digital oscilloscope and hard copies of signals were made at the same time. These were used for constructing record sections.

## Results

Unanticipated troubles, which will be reported in a future paper, happened in the Shot 17 and Shot 18 experiments. In addition, the efficiency of dynamite in ice was found to be very low, and clear seismograms could only be recorded at a few observation stations. However, for Shot 19, the biggest explosion experiment, high quality seismograms were recorded at every station; the Shot 19 record section is shown in Figure 2. The time axis is shifted by a velocity of 6 km/s and the amplitude scale in each seismogram is arbitrary. Initial and subsequent arrivals were picked from each seismogram, and

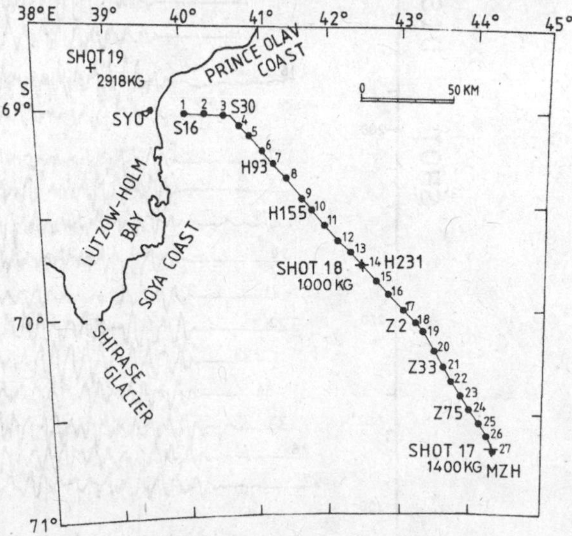


**Figure 1.** **Map showing the profile and the positions of shots and observation stations.**

reduced travel time curves constructed for Shot 17, Shot 18 and Shot 19 (Figure 3). Concerning the observed amplitudes of seismograms it is possible to estimate from the JARE-20 results that, if the charge sizes had been the same, observed amplitude at the same shot distance would have been greater by about one order of magnitude for explosions shot in the sea compared to those in ice holes. In Japan, there have been many explosion experiments, and it is possible to estimate the maximum amplitude of a seismogram and use a moderate amplificaton constant. However, results from Antarctica have either not been published or are otherwise unavailable. From the JARE experiments reported here, it is possible to deduce the relationships between charge size, shot distance and maximum observed amplitude as shown in Figure 4. These data could be useful in the planning of further explosion seismic experiments in Antarctica.

## Structure

Crustal structure could not be constructed with only the travel time curves shown in Figure 3. Making certain assumptions, it is possible to deduce the crustal structure to the Moho as shown in Figure 5. This involved, first, the subtraction of travel times in the continental ice sheet using the ice-sheet elastic wave velocity structures already determined by Chang (1964) and Eto (1971). However, the thickness of the ice sheet along the profile from Syowa to Mizuho Stations was not completely known although Wada et al. (1982) tried to measure it using an ice radar mounted either on an oversnow vehicle, or an aircraft. This work failed to yield results where there was a considerable ice thickness. Bentley (1973) pointed out that the basement in East Antarctica is, on average, nearly at sea level. Shimizu et al. (1972) and Wada et al. (1982) indicated similar results for part of our profile.

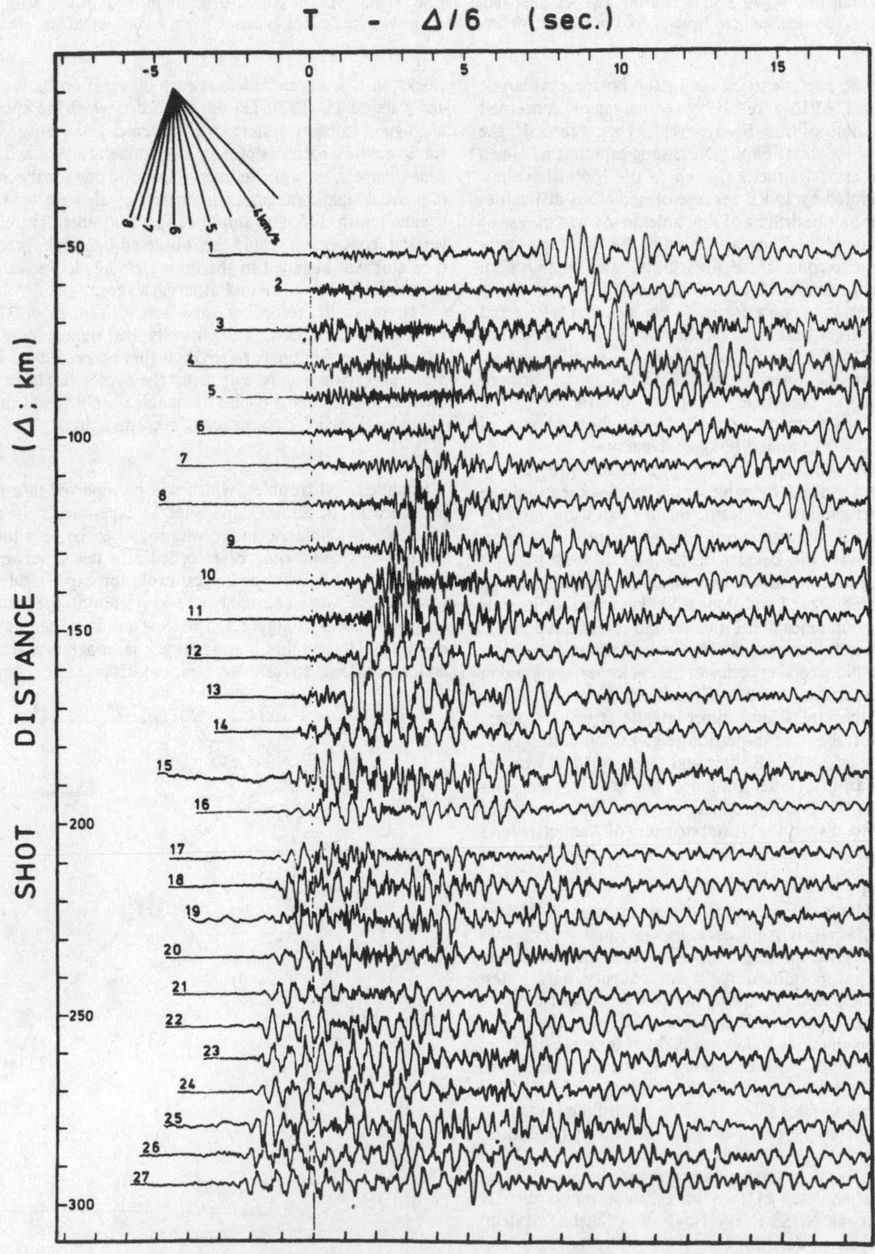


**Figure 2.** **Record section of traces from the temporary observation stations, on a reduced time scale. Numerals attached to traces are station numbers shown in Figure 1.**

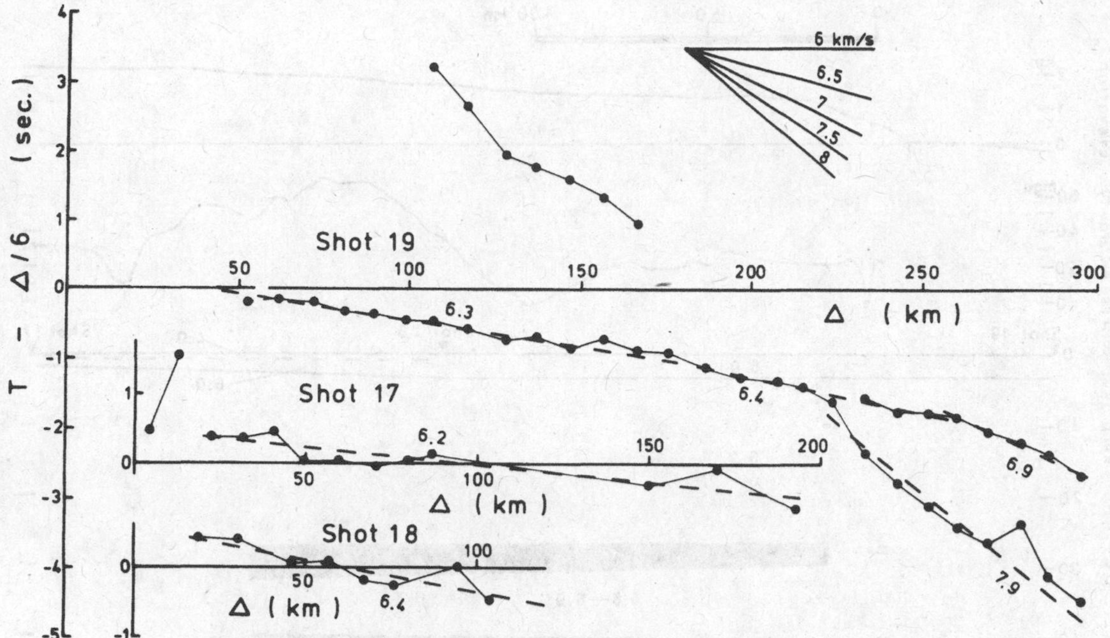


**Figure 3. Travel time plots with a reduction velocity of 6 km/s. Apparent velocities are indicated.**

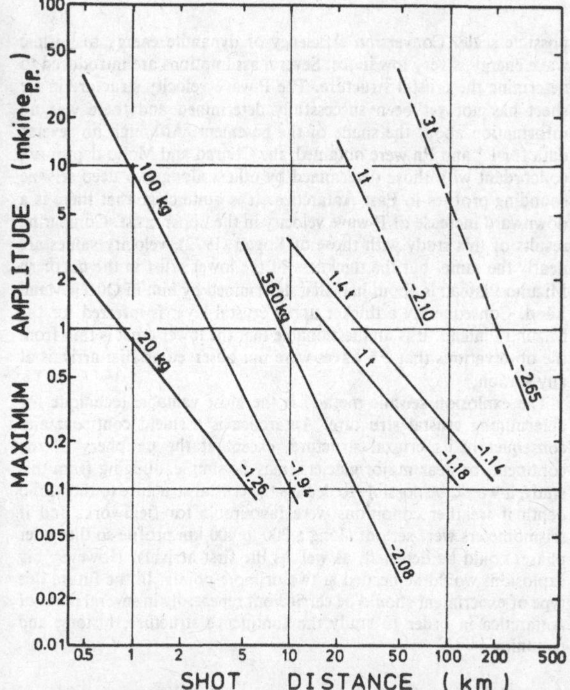


**Figure 4. Relationships between maximum velocity amplitude, epicentral distance and charge size. Dashed line, shots in the sea; solid line, shots in ice holes. Numbers on each line represent charge size and slope.**

Therefore, it was assumed that basement was at sea level. A P-wave velocity structure in the ice sheet was based on the results of Eto (1971), but as that author could not determine a deeper structure, a P-wave velocity of 4.0 km/s from a depth of 300 m to the basement was assumed. Taking both assumptions, time terms and offset distances due to the ice sheet at each point were calculated. These data formed a basis for the construction of travel time curves. The results of JARE-20 work (Ikami et al., 1981) and other explosion experiments by JARE-21 (not reported in this paper) show initial arrivals with apparent velocity about 6.0 km/s near the shot point. The intercept time of these 6.0 km/s curves was zero or nearly equal to zero, which means that any sedimentary layer overlying basement would only be very thin if it existed. This deduction is concordant with Bentley (1973). In Figure 3, a gradual increase in the apparent velocity of the upper crust with shot distance is evident. One interpretation is that this phenomena was due to the inclination of the basement surface. However, in such a case, the basement surface beneath Mizuho Station would be abnormally high. Shots 17 and 18 provided reversed recordings on the southern half of the profile and show an averaged apparent P-wave velocity of 6.3 km/s. Taking an apparent velocity of 6.0 km/s as stated above, we estimate that P-wave velocity of the upper crustal layer increases with depth. A similar result was pointed out by Bentley (1973), who found in the top 15 km of the crust a gradual downward increase in P-wave velocity with depth from 6.0 km/s to 6.3 km/s.

Although it was not possible to recognise any 6.0 km/s branches on the travel time curves shown on Figure 3, we adopted this velocity because it was observed near the shot-points. Judging from the travel time curves observed for Shot 17 and Shot 18, it would be expected that the intercept time for both adopted travel time curves would be about 0.4 s. If it is assumed that this intercept time is due to the ice sheet, the level of the basement may thus be 1 km lower than sea level. True upper crustal velocities of 6.0 km/s and 6.3 km/s were adopted, and an attempt was made to deduce the structure. Arrival time data from the Conrad and the Moho discontinuity is only available from Shot 19, so true velocities beneath these discontinuities must be assumed. Assuming true velocities of 6.8 to 6.9 km/s for P* and 7.9 to 8.0 km/s for Pn, we derive a depth of about 30 km for the Conrad discontinuity and a Moho depth of about 40 km. Crustal and upper mantle velocity structures determined by explosion seismic refraction experiments in South Australia (Muirhead et al., 1977) and in North America (Massé, 1973) are shown for comparison in Figure 6.

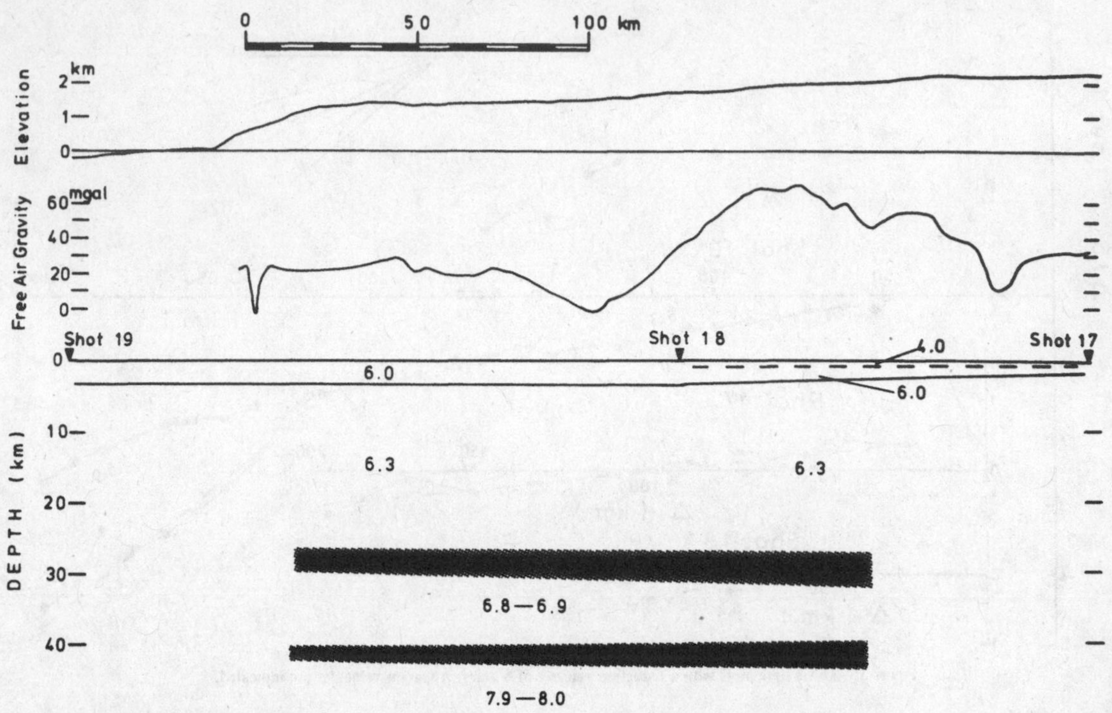


**Figure 5. Crustal structure along the profile shown in Figure 1. The possible range for the boundaries is shown by the hatched area. Basement level between Shot 17 and Shot 18 is indicated by a broken line. Top, surface elevation (Shimizu et al., 1972); middle, free air gravity anomaly (Abe, 1975).**

## DISCUSSION

Bentley (1973) pointed out: (1) that the top of the basement lies near sea level in East Antarctica but typically 2 or 3 km below sea level in West Antarctica; and (2) that throughout much of East Antarctica, the thickness of the layer overlying the basement complex is less than half a kilometre, although one to several kilometres in West Antarctica. From our data, it is difficult to determine whether a surface layer exists or not, because a P-wave structure in the ice sheet is not known and it is not possible to determine the time term and offset distance for the ice sheet. The result of a short profile shows an apparent velocity of 6.0 km/s, and an intercept time nearly equal to zero. This means that the surface layer would be thin if it existed. The observed travel time can be interpreted if the surface layer was replaced by ice of a velocity 4.0 km/s or so just above the basement. It is probable that the velocity structure in the upper crust is smoothly transitional rather than the two-layered structure as shown in Figure 5, but an abrupt change of a P-wave velocity from 6.0 km/s to 6.3 km/s was assumed in order to simplify calculation of the crustal strcture. P-wave velocities by Bentley (1973) resemble those deduced in this study. It is possible to recognise refracted waves from the Conrad and the Moho discontinuities in Antarctica. Previous East Antarctic crustal structure determinations by explosion seismology were published by Kogan (1972) and Kurinin and Grikurov (1982). The present profile is between these profiles. Refraction waves from the Conrad and the Moho discontinuities were recorded from the explosion of Shot 19 but there are no reverse data, so the true velocities cannot be determined. Apparent velocities are 6.9 km/s for P* and 7.9 km/s for Pn. Kogan (1972) showed 6.7 km/s for the average velocity beneath the Conrad discontinuity and a 30 km thick upper crustal layer, and Moho depth about 40 km with 7.9 km/s for Pn. Kurinin and Grikurov (1982) determined 7.1 km/s for P* and 7.7 to 7.9 km/s for Pn with Conrad and Moho depths of 10 km and 30 to 33 km, respectively.

### Conclusion

Inexperience in the operation of a large explosion seismic experiment in Antarctica and rigorous weather conditions both restricted this study. Even so, operations by JARE-21 were on the largest possible scale. Conversion efficiency of dynamite energy to seismic wave energy is very low in ice. Several assumptions are introduced to determine the crustal structure. The P-wave velocity structure in ice sheet has not yet been successfully determined and there was no information about the shape of the basement. Although no reverse data for P* and Pn were obtained, the Conrad and Moho depths are concordant with those determined by others along two deep seismic sounding profiles in East Antarctica. It is quite clear that there is a downward increase of P-wave velocity in the upper crust. Comparing results of this study with those of Kogan (1972), velocity values are nearly the same, but the thickness of the lower crust in the northern Mizuho Plateau is about half that determined by him in Queen Maud Land. Consequently a thicker upper crustal layer is inferred for the Mizuho Plateau. It is unquestionable that the lower crust is thin from the observations that P* waves were not observed as first arrivals at any station.

The explosion seismic method is the most valuable technique for determining crustal structure. Antarctica is a shield continent and consequently its crustal structure, except at the periphery of the continent, or near major glaciers, may be simple. Judging from this study, it would be possible to deduce the crustal structure to the Moho depth if weather conditions were favourable for fieldwork, and if seismometers were set out along a 300 to 400 km profile so that later phases could be detected, as well as the first arrivals. However, big explosions would be needed at two or more points. In the future this type of experiment should be carried out repeatedly in several parts of Antarctica in order to study the continent's structure, history, and tectonics.

*Acknowledgements* We are much obliged to Dr S. Kawaguchi, leader of JARE-21, and all of the wintering members of JARE-21 of whom half took part in these experiments. Support staff at Syowa and Mizuho Stations all participated in the experiments, which could not have been successfully completed without their cooperation. We are also greatly indebted to the members of JARE-22 and the crew of Icebreaker "Fuji". We would like to thank Professor H. Aoki of Nagoya University for critically reading the manuscript. A part of the numerical computation was carried out at the Nagoya University Computing Centre (Problem No. 4001KW2670).

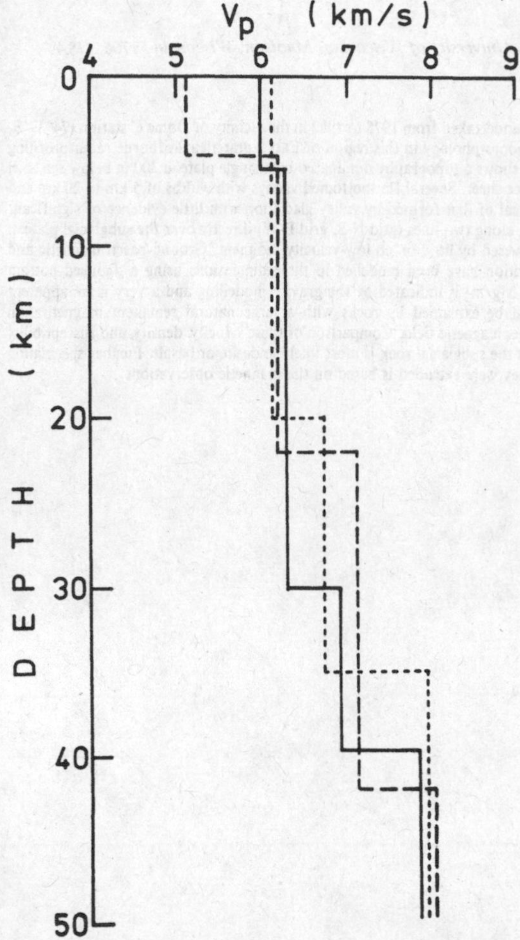


**Figure 6. P-wave velocity models in South Australia (dotted line) by Muirhead et al. (1977). North America (dashed line) by Masse (1973) and the northern Mizuho Plateau (solid line), this study.**

## REFERENCES


ABE, Y., 1975: Gravity data. *JARE Data Reports, 28 (Glaciology),* 114-9.

BENTLEY, C.R. (this volume): Crustal structure of Antarctica from geophysical evidence, a review.

BENTLEY, C.R., 1973: Crustal structure of Antarctica. *Tectonophysics, 20,* 229-40.

BENTLEY, C.R. and OSTENSO, N.A., 1962: On the paper of F.F. Evison, C.E. Ingham, R.H. Orr and J.H. Le Fort, "Thickness of the earth's crust in Antarctica and the surrounding oceans". *R. Astron. Soc., Geophys. J., 6,* 292-8.

BENTLEY, C.R. and CLOUGH, J.W., 1972: Antarctic subglacial structure from seismic refraction measurements; *in* Adie, R.J. (ed.) *Antarctic Geology and Geophysics.* Universitetsforlaget, Oslo, 683-91.

CHANG, FENG-KENG, 1964: Seismic wave studies in northwest Marie Byrd Land, Antarctica. *Bull. Seis. Soc. Am., 54,* 51-65.

DEWART, G. and TOKSOZ, M.N., 1965: Crustal structure in East Antarctica from surface wave dispersion. *R. Astron. Soc., Geophys., J., 10,* 127-39.

ETO, T., 1971: Seismic studies during the JARE South Pole traverse 1968/69. *JARE Sci. Rep., Spec. Iss., 2,* 115-24.

EVISON, F.F., INGHAM, C.E. and ORR, R.H., 1959: Thickness of the earth's crust in Antarctica. *Nature, 183,* 306-08.

EVISON, F.F., INGHAM, C.E., ORR, R.H. and LE FORT, L.H., 1960: Thickness of the earth's crust in Antarctica and the surrounding oceans. *R. Astron. Soc., Geophys. J., 3,* 289-306.

GROUSHINSKY, N.P. and SAZHINA, N.B., 1982: Gravitational field of Antarctica; *in* Craddock, C. (ed.) *Antarctic Geoscience.* Univ. Wisconsin Press, Madison, 913-7.

IKAMI, A., ICHINOSE, Y., HARADA, M. and KAMINUMA, K., 1980: Field operation of explosion seismic experiment in Antarctica. *Antarctic Rec., 70,* 158-82 (in Japanese).

IKAMI, A., KAMINUMA, K. and ICHINOSE, Y., 1981: Upper crustal structure of Soya Coast, Antarctica, revealed by explosion seismology. *Antarctic Rec., 71,* 58-63.

KOGAN, A.L., 1972: Results of deep seismic soundings of the earth's crust in Antarctica; *in* Adie, R.J. (ed.) *Antarctic Geology and Geophysics.* Universitetsforlaget, Oslo, 485-9.

KURININ, R.G. and GRIKUROV, G.E., 1982: Crustal structure of part of East Antarctica from geophysical data; *in* Craddock, C. (ed.) *Antarctic Geoscience,* Univ. Wisconsin Press, Madison, 895-901.

MASSÉ, R.P., 1973: Compressional velocity distribution beneath central and eastern North America. *Bull. Seis. Soc. Am., 63,* 911-35.

MUIRHEAD, K.J., CLEARY, J.R. and FINLAYSON, D.M., 1977: A long range seismic profile in southeastern Australia. *R. Astron. Soc., Geophys. J., 48,* 509-19.

SHIBUYA, K., ITO, K. and KAMINUMA, K., 1982: Utilisation of an NNSS receiver in the explosion seismic experiments on the Prince Olav Coast, East Antarctica, 1. Recovered UTC. *Antarctic Rec., 76,* 63-72.

SHIBUYA, K. and KAMINUMA, K., 1982: Utilisation of an NNSS receiver in the explosion seismic experiments on the Prince Olav Coast, East Antarctica, 2. Positioning. *Antarctic Rec., 76,* 73-88.

SHIMIZU, H., NARUSE, R., OMOTO, K. and YOSHIMURA, A., 1972: Position of stations, surface elevation and thickness of the ice sheet and snow temperature at 10 m depth, in the Mizuho Plateau-West Enderby Land area, East Antarctica, 1969/71. *JARE Data Reports, 17 (Glaciology),*12-37.

SUZUKI, Y. and SHIRAISHI, K. (in press): The drilling system used by JARE-21 and its later improvement. *Mem. Nat. Inst. Pol. Res., Japan.*

WADA, M., YAMANOUCHI, T. and MAE, S., 1981: Glaciological data collected by the Japanese Antarctic Research Expedition from February 1979 to January 1980. *JARE Data Reports, 63 (Glaciology 7),* 1-43.

# SUBGLACIAL GEOLOGY AT DOME C

D.D. Blankenship and C.R. Bentley, *Department of Geology and Geophysics. University of Wisconsin, Madison, Wisconsin 53706, USA.*

*Abstract* The results of both ground-based and airborne geophysical investigations undertaken from 1978 to 1982 in the vicinity of Dome C station (74°39′S, 124°10′E) have been combined to give an indication of subglacial lithology and geomorphology in this region of East Antarctica. Airborne radar profiling (NSF/SPRI/TUD) performed along a 50 km by 50 km grid centred on the station shows a topography dominated by a single plateau 300 m below sea level with a number of small knobs extending upward from it 300 m to 400 m into the ice sheet. Several flat-bottomed valleys with widths of 5 km to 20 km and floors 1000 m below sea level dissect this plateau. The overall morphology is typical of that formed by valley glaciation with little evidence of significant modification after the formation of the ice sheet. Seismic long refraction shooting along two lines (grid N-S, grid E-W) directly over the subglacial plateau imply that it is composed of crystalline bedrock with a velocity of 5.8 km/sec covered by little or no low-velocity sediment. Ground-based magnetic and gravity observations made along a 10 km by 10 km grid also centred on the station have been modelled in three dimensions using a digitised bottom topography based on both airborne and ground radar. A bedrock density of 2.7 Mg/m³ is indicated by the gravity modelling and a very large apparent susceptibility of 0.011 (c.g.s. emu) results from the magnetic models. This can be explained by rocks with a large natural remanent magnetisation (Koenigsberger Q ratio greater than unity) roughly oriented in the direction of the geomagnetic field. Comparison of these velocity, density, and susceptibility determinations with observations available from elsewhere in the world imply that the subglacial rock is most likely andesite or basalt. Further speculation that this portion of the continent was located at a high magnetic latitude when they were extruded is based on the magnetic observations.

# HEAT FLOW MEASUREMENTS IN LUTZOW-HOLM BAY, ANTARCTICA—A PRELIMINARY STUDY

K. Kaminuma, *National Institute of Polar Research, 9-10 Kaga I-chome, Itabashi-ku, Tokyo 173, Japan.*
T. Nagao, *Department of Earth Science, Faculty of Sciences, Chiba University, Chiba, Japan.*

*Abstract* A new technique for measuring temperature gradients on the ocean bottom under the sea ice has been developed, and has been used to measure temperatures at five observation points during the 22nd Japanese Antarctic Research Expedition wintering in Lutzow-Holm Bay, Antarctica, in 1981. The apparatus, used for these measurements, consists of a probe, a connection receptacle, and a pressure vessel. The data from three thermistors in the probe were recorded in IC memories with three channels. The total number of samplings at each channel is 256, at either 30 second or 60 second intervals. The system accuracy of the measurement is 0.01 °C, and the resolution is 0.001 °C. The depths of the sea water at the measurement points are between 500 and 800 m, and the thickness of the sea ice is approximately 1.5 m. Before each measurement, a core of the sediment layer in the sea bottom was collected by a gravity corer. The thermal conductivity of the sediment was obtained from the core. The sea water temperatures at the sea bottom range from -1.30°C to -0.08°C, and those at the top layer of the sediment from −1.24°C to −0.05°C. The heat flows at the individual stations range from 0.86 to 5.43 HFU (1 HFU = $1 \times 10^{-6}$ cal/cm²s). The high heat flow is an unexpected result in the marginal area of the East Antarctic Shield.

In Antarctica the only prior measurements of heat flow have been carried out by Decker and his students in holes which were drilled in McMurdo Sound area during the Dry Valley Drilling Project. The latter was an international project among three countries: Japan, New Zealand and United States, during 1973-1975 (Decker, 1974; Decker et al., 1975; Bucher and Decker, 1976; Decker and Bucher, 1977). They measured vertical temperature distributions in the six holes of which two were located at McMurdo Station, and the other four in the Dry Valley area. The heat flows calculated from these measurements range from 1.5 to 3.4 HFU (1 HFU = $1 \times 10^{-6}$cal/cm²s). The radiogenic heat production from regional core and surface samples was also measured. It ranges from 2.2 to 3.7 HGU (1 HGU = $1 \times 10^{-13}$cal/cm³s) (Decker, 1978).

A new technique to measure sea water and sea bottom sediment temperatures under the sea ice has been developed. The measurements have been carried out from the sea ice in Lutzow-Holm Bay around Syowa Station during the winter season in 1981. Lutzow-Holm Bay is located in the marginal area of the East Antarctic Shield, and is part of the Antarctic continental shelf.

## EQUIPMENT

Figure 1 shows a schematic view of the sensor part of the system which penetrates into the sea bottom. This apparatus consists of three parts: a probe, a connection receptacle and a pressure vessel. Three thermistors are embedded in the probe at points A, B and C as shown in Figure 1. Three probes with two different lengths are used for the measurements. One of these three is 100 cm in length, and the other two are 120 cm in length (different lengths for $L_1$ and $L_2$ are shown in Figure 1). The apparatus weighs 70 kg in air and 50 kg in water. A flow diagram of the system is given in Figure 2. The recording IC memories and the battery for operating both the memory and thermistor are contained in the pressure vessel so that it is usable up to 100 bars pressure. The recording system has three channels. Channel 1 is for recording the temperature at point A, and channels 2 and 3 are for recording temperature differences between A and B, and A and C respectively. The relative accuracy of temperature measurement of the sensor is 0.01°C, and its resolution is 0.001°C. The sampling interval is selected to be either 30 seconds or 60 seconds. The total number of samples for each channel is 256. If the apparatus inclines more than 30 degrees from the vertical at the sea bottom, an alarm signal is recorded. Data retrieval from the IC memory is done after a measurement through the data retrieving circuit in Figure 2. The data can be printed by a digital printer and/or displayed on the light emitting diode screen.

| Probe | Length (mm) | | | | |
|---|---|---|---|---|---|
| No. | L | L 1 | L 2 | L 3 | L 4 |
| 1 | 1680 | 1250 | 1000 | 400 | 400 |
| 2 | 1880 | 1450 | 1200 | 600 | 400 |
| 3 | 1880 | 1450 | 1200 | 500 | 500 |

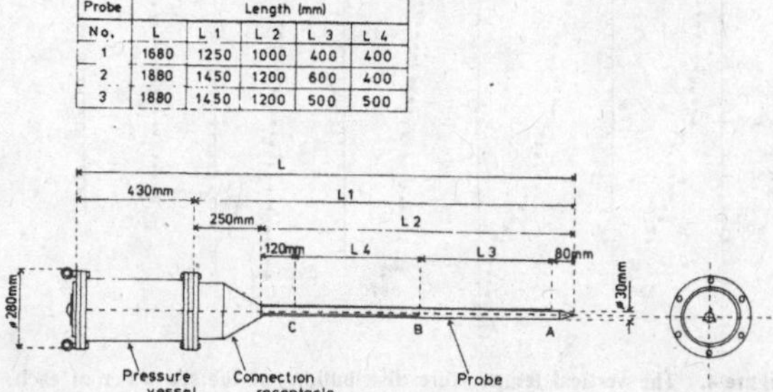


**Figure 1. A schematic diagram of the new apparatus for temperature measurement. Three probes were prepared.**

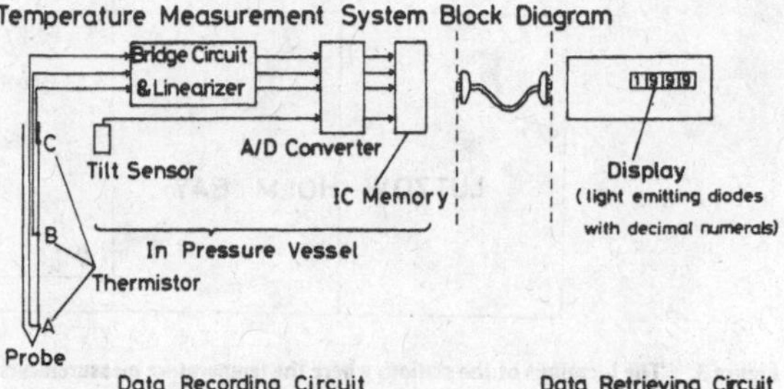


**Figure 2. A block diagram of the new temperature measurement system.**

## OBSERVATIONS

Observations were carried out in November 1981. The mean air temperature at that season around Syowa Station was about −7°C. The thickness of the sea ice in Lutzow-Holm Bay was still more than 1 m, which meant that the sea ice was thick enough for the field operation at that time. The measuring points were selected from the places at which: (1) the depth of the sea was more than 500 m; and (2) a core of sediment was available. The five measurement points are given in Figure 3, and their geographic locations in Table 1. The measurements of sea depth were carried out by using an echo sounder.

The progress of all observations are listed in Table 2. The sampling interval was 30 seconds between every measurement. Sampling started when the magnetic switch which was installed in the pressure vessel was set off. The apparatus was lowered into the sea water through a hole in the sea ice by a winch. Thermistors in the probe were equilibrated for several minutes, when the apparatus reached 100 m above the sea bottom, the point marked "M" in Table 2. Because the water temperature at that depth was nearly constant within one metre vertically, the temperature difference between A and C should be negligible. After the probe penetrated the sea bottom, the apparatus was kept stationary for approximately 20 minutes to sample nearly 40 temperature values. The apparatus was then wound up at a speed of 40 m/minute. All thermistors used for these measurements were re-calibrated on return to Japan. Thus the correction values for all thermistors were determined, and were used to correct the original temperatures. The vertical temperature profiles in sea water at each station are given in Figure 4.

There are two patterns in the vertical temperature distribution as seen in Figure 4. As the depth increases from 0 to 100 m, the temperature increases from approximately −1.7°C to −1.3°C at Stations 3, 4 and 5. However, the temperature remains constant between a depth of 100 m and the sea bottom. On the other hand, at Stations 1 and 2, the gradual temperature increase becomes more steep, and a temperature gradient of 0.006 ~ 0.008°C/m is observed at the depth of 550 m to about 650 m.

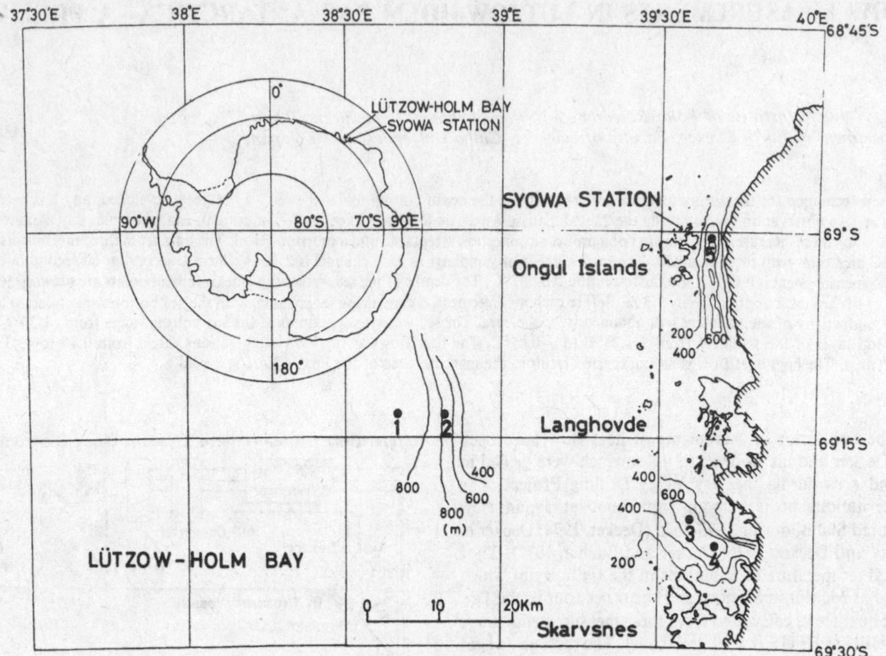


**Figure 3. The locations of the stations where the temperature measurements of the sea water and the sea bottom sediment under the sea ice were carried out.**

**TABLE 1: Temperature measurement stations in the sea bottom**

| Date | Location | Station No. | Latitude | Longitude | Depth by Echo Sounder | Sea ice thickness |
|---|---|---|---|---|---|---|
| Nov. 3, 1981 | Off the coast of Langhovde | 1 | 69°13.4'S | 38°40.5'E | 666 m | 140 cm |
| Nov. 4, 1981 | Off the coast of Langhovde | 2 | 69°13.4'S | 38°49.2'E | 778 m | 140 cm |
| Nov. 8, 1981 | Off the coast of Honnor Glacier | 3 | 69°20.6'S | 39°36.4'E | 668 m | 120 cm |
| Nov. 8, 1981 | Off the coast of Honnor Glacier | 4 | 69°22.5'S | 39°40.9'E | 683 m | 120 cm |
| Nov. 17, 1981 | Ongul Strait | 5 | 69°00.4'S | 39°39.8'E | 537 m | 120 cm |

**TABLE 2: Measurement Schedule**

| Station No. | Probe No. | Started time of measurement | Started time of descent | Duration of winch stop at "M" | Depth of "M" (in wire) | Measurement time of temperature in sediment | End of measurement (Sea level) |
|---|---|---|---|---|---|---|---|
| 1 | 2 | $10^h56^m30^s$ | $11^h04^m15^s$ | $11^h16^m00^s$ ~ $11^h21^m30^s$ | 477 m | $11^h26^m30^s$ ~ $11^h48^m00^s$ | $12^h30^m00^s$ |
| 2 | 2 | $9^h37^m30^s$ | $9^h43^m30^s$ | $9^h58^m30^s$ ~ $10^h05^m30^s$ | 598 m | $10^h10^m30^s$ ~ $10^h30^m00^s$ | $10^h54^m20^s$ |
| 3 | 3 | $13^h40^m30^s$ | $13^h44^m30^s$ | $13^h55^m00^s$ ~ $14^h00^m00^s$ | 447 m | $14^h06^m30^s$ ~ $14^h23^m45^s$ | $14^h44^m10^s$ |
| 4 | 3 | $22^h13^m00^s$ | $22^h19^m00^s$ | $22^h30^m00^s$ ~ $22^h36^m00^s$ | 477 m | $22^h41^m58^s$ ~ $23^h01^m00^s$ | $23^h20^m00^s$ |
| 5 | 2 | $17^h18^m30^s$ | $17^h26^m00^s$ | $17^h37^m00^s$ ~ $17^h43^m00^s$ | 480 m | $17^h49^m00^s$ ~ $18^h06^m00^s$ | $18^h23^m00^s$ |

Temperatures and temperature differences between two of the three thermistors in the sediment at each station are plotted in Figures 5 to 9. Figures 5 and 6 show those at Stations 1 and 2 respectively. Temperatures reach their maximum immediately after penetration of the probe, and become constant thereafter. The temperature differences increase rapidly directly after the penetration, and then decrease gradually afterward. It is apparent that the apparatus was wound up before the temperature differences became constant at these stations. Therefore the temperature difference values for calculating heat flows at these stations were obtained by extrapolation as is shown with arrows in Figures 5 and 6.

As shown in Figures 7 and 8 the temperatures at Stations 3 and 4 increase slightly after the penetration. The temperature differences increase rapidly, and become either constant or decrease slightly. Three temperature difference curves in Figures 7 and 8 show the trend to slightly decrease seven or eight minutes after the penetration. The temperature difference values at the points shown by the arrows in Figures 7 and 8 were used for calculating the heat flows.

At Station 5 in Figure 9, the variation patterns of both the temperature and its difference are different from those of the other four stations. The temperature and its difference are constant from

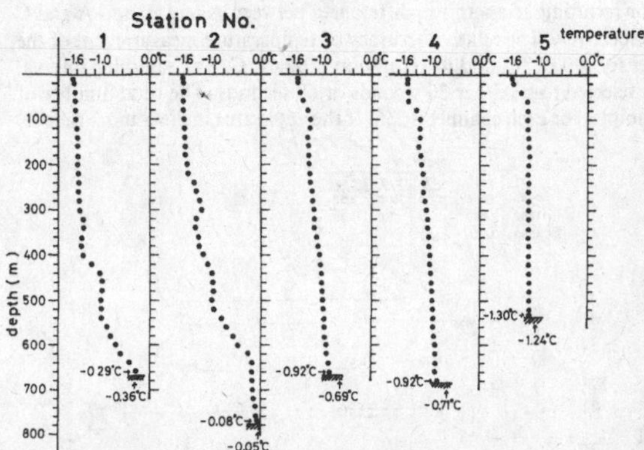


**Figure 4. The vertical temperature distributions of the sea water at each station. The temperatures with arrows are those of 1 m deep of sediment. The temperatures of the sea bottom surface are also given.**

**TABLE 3: Heat Flow**

| Station | No. 1 | No. 2 | No. 3 | No. 4 | No. 5 |
|---|---|---|---|---|---|
| Temperature gradient (°C/m) | 0.028 | 0.092 | 0.188 | 0.209 | 0.055 |
| Thermal conductivity (mcal/cm·s·°C) | 3.07 | 3.07 | 2.53 | 2.60 | 2.53 |
| Heat flow (HFU) | 0.86 | 2.70 | 4.76 | 5.43 | 1.39 |

Station 1 and 2: A-B Value
Station 3, 4, and 5: least squares method.
1 HFU = 1 x $10^{-6}$ cal/cm²·s(= 41.9 mW/m²).

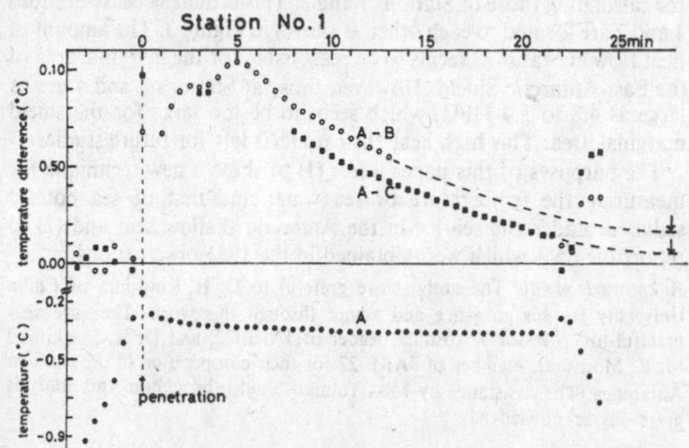


**Figure 5. The temperature and its difference in the sediment at Station 1. Two arrows show the temperature difference values to be used to calculate the temperature gradient.**

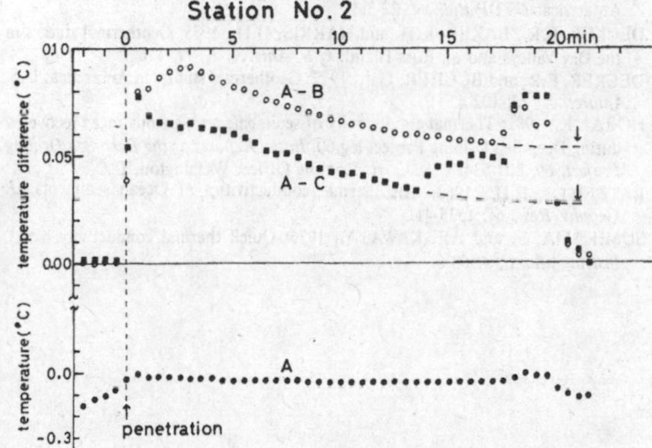


**Figure 6. The temperature and its difference in the sediment at Station 2. Two arrows show the temperature difference values to be used to calculate the temperature gradient.**

seven minutes after the penetration. Therefore these values shown by the arrows in Figure 9 were also used for calculations. The relative vertical temperature distribution in the sediment was estimated from the temperature differences at B and C compared to the temperatures at A which were obtained as mentioned above, and are plotted in Figure 10. The temperature gradient at each station iis determined in Figure 10 and the values are given in Table 3.

## THERMAL CONDUCTIVITY

Before the temperature measurements were carried out at each point, core samples of sediment had been collected using both gravity and piston corers. The sediment was a light grey clay or silt (so called glacial milk). All the core samples were brought back to Japan in frozen states, and were brought to normal temperature on the day before thermal conductivity measurements were made. A quick thermal conductivity meter (QTM method) was used for the measurements (Sumikama and Arakawa, 1976). The measuring instrument is a Shotherm QTM (Showa Denko Company, see Horai, 1972).

Four samples at Stations 2, 3, 4 and 5 were measured for their thermal conductivities. Most of the values of thermal conductivity were in the range of 2.53-3.07 x $10^{-3}$cal/cm s°C. These values were

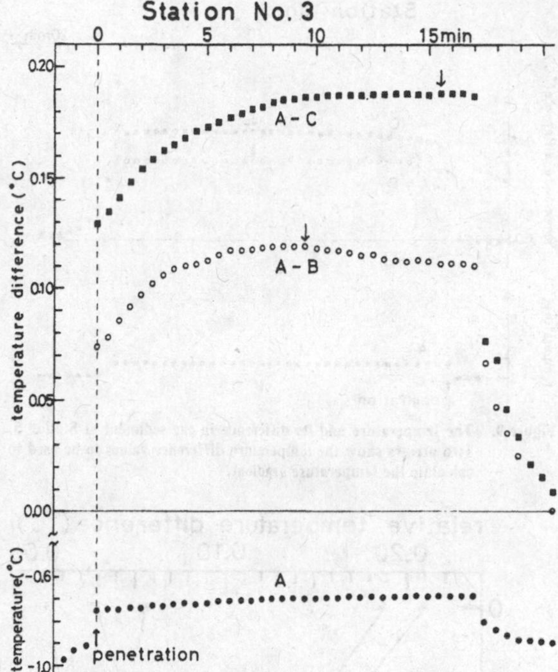


**Figure 7. The temperature and its difference in the sediment at Station 3. Two arrows show the temperature difference values to be used to calculate the temperature gradient.**

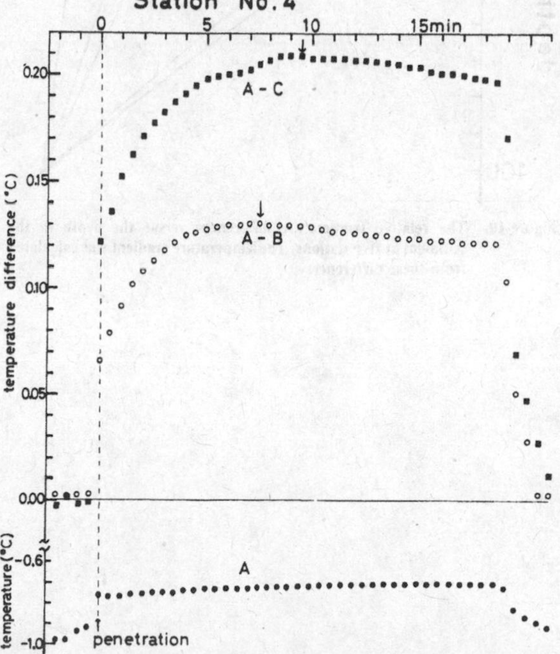


**Figure 8. The temperature and its difference in the sediment at Station 4. Two arrows show the temperature difference values to be used to calculate the temperature gradient.**

corrected for depth and temperature by the empirical formula

$$K_{in} = K\{1-(Tm-T)/400)\}\ \{1+D/183000\}$$

(Ratcliffe, 1960), where K is the value of thermal conductivity measured at Tm°C in the laboratory and $K_{in}$ is the value at T°C and at the water depth D(m). The values of thermal conductivity corrected

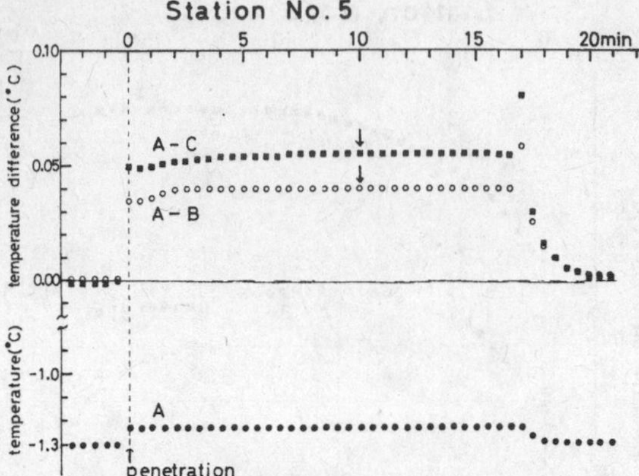


**Figure 9. The temperature and its difference in the sediment at Station 5. Two arrows show the temperature difference values to be used to calculate the temperature gradient.**

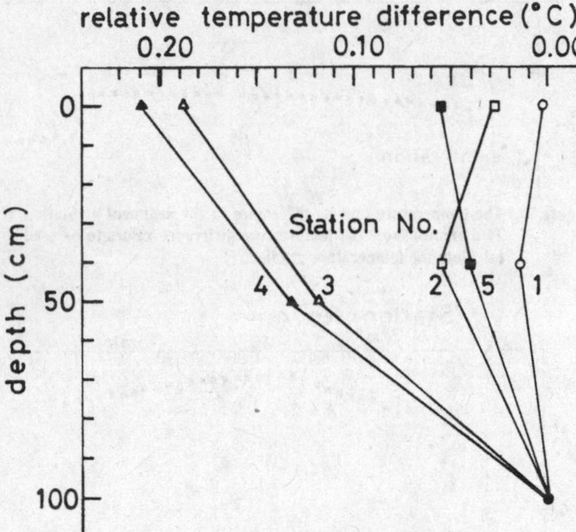


**Figure 10. The relative temperature differences versus the depth in the sediment at five stations. The temperature gradient was calculated from these differences.**

are listed in Table 3. The values of thermal conductivity are reasonable values for ocean sediments with low water content (Ratcliffe, 1960).

## DISCUSSION AND CONCLUSIONS

Using the values for the temperature gradient obtained and thermal conductivity, the heat flows were estimated as shown in Table 3. The heat flows calculated in Table 3 are obtained simply by multiplying the thermal conductivity by the temperature gradient without any correction for the shallow sea. The thermal conductivity at Station 2 is used for calculating those of Stations 1 and 2. This is done because Stations 1 and 2 are located to each other as shown in Figure 3. The amount of heat flow at Station 5 seems to be reasonable for the marginal area of the East Antarctic Shield. However, those at Stations 3 and 4 are as large as 4.8 to 5.4 HFU, which seem to be too large for the shield marginal area. This high heat flow result is left for future studies.

The purposes of this paper were: (1) to show a new technique for measuring the temperature of sea water and that of sea bottom sediment under the sea ice in the Antarctic shallow sea; and (2) to report the data which were obtained in the 1981 observations.

*Acknowledgements* The authors are grateful to Dr H. Kinoshita of Chiba University for his guidance and advice through this work. They are also grateful to Professor Y. Yoshida, leader of JARE-22, and Dr K. Sasaki and Mr K. Moriwaki, member of JARE-22 for their co-operation in the work in Antarctica. The assistance by Miss Yumiko Shudo throughout this study is gratefully acknowledged.

## REFERENCES

BUCHER, G.J. and DECKER, E.R., 1976: Geothermal studies in McMurdo Sound region. *U.S. Antarct. J., 11,* 88-9.

DECKER, E.R., 1974: Preliminary geothermal studies of the Dry Valley Drilling Project holes at McMurdo Station, Lake Vanda, Lake Vida and New Harbour, Antarctica. *DVDP Bull., 4,* 22-3.

DECKER, E.R., BAKER, K.H. and HARRIS, H.H., 1975: Geothermal studies in the Dry Valleys and on Ross Island. *U.S. Antarct. J., 11,* 176.

DECKER, E.R. and BUCHER, G.J., 1977: Geothermal studies in Antarctica. *U.S. Antarct. J., 12,* 102-4.

HORAI, K., 1982: Thermal conductivity of sediments and igneous rocks recovered during Deep Sea Drilling Project leg 60. *Initial Reports of the Deep Sea Drilling Project, 60,* 807-834. U.S. Govt. Printing Office, Washington, D.C.

RATCLIFFE, E.H., 1960: The thermal conductivities of Ocean sediments. *J. Geophy. Res., 64,* 1535-41.

SUMIKAMA, S. and ARAKAWA, Y., 1976: Quick thermal conductivity meter. *Instru. Autom., 4,* 60-6.

# GEOPHYSICAL INVESTIGATION IN THE EASTERN PART OF LUTZOW-HOLM BAY, ANTARCTICA.

K. Kaminuma, and K. Shibuya, *National Institute of Polar Research, Tokyo 173, Japan.*

*Abstract* Almost all gravity stations on the snow free area in the eastern part of Lützow-Holm Bay have negative free air anomalies from − 15 to − 35 mgal. The free air anomaly in the central part of Lützow-Holm Bay is as low as − 90 mgal and increases gradually towards the eastern coast of Lützow-Holm Bay. The Bouguer anomalies in the northern part of Lützow-Holm Bay are about − 15 to − 20 mgal and those in the southern part are about − 30 mgal. From aeromagnetic surveys, the total magnetic intensity in the eastern part of Lützow-Holm Bay is about 44500-45400γ.

A borehole type and a set of water tube tiltmeters were installed at Syowa Station in 1981. An annual change of 12 microradian (maximum) in an E—W direction and 22 microradians (maximum) in a N—S direction were observed from April 1981 to January 1982. The annual change corresponds to the temperature variation of the subsurface at a depth of 1 m at which observation was also started at Syowa Station in 1980.

The Japanese Antarctic Station Syowa is on East Ongul Island in Lützow-Holm Bay, on the coast of the East Antarctic Shield. Many geophysical observations, such as seismological observations, deep seismic soundings, gravity surveys, and aeromagnetic surveys, have been made there since Syowa was established in 1957. Geophysical observations and the surveys were intensified in 1977-1981 because the Japanese Antarctic Research Expedition (JARE) had a special commitment to earth science programmes during those years. This paper reports, and presents preliminary interpretations of, a gravity survey, aeromagnetic surveys, and tiltmeter measurements made in the Lützow-Holm Bay area between 1979 and 1981.

*Gravity anomalies in Lutzow-Holm Bay*

More than 160 gravity stations were established with a LaCoste and Romberg gravimeter (Model G), on the snow free area in the eastern part of the Lützow-Holm Bay (Kaminuma et al., 1980). The gravimeter was mostly transported by oversnow vehicles and snow scooters, but in 1981 it was transported by aeroplane to the southern area of Lützow-Holm Bay and the inland area. At about 50% of the stations, the elevation was determined with an accuracy of better than ± 10 cm, although at some the accuracy was only about ± 1 m. As a general rule at least two gravity measurements were made at each station but at some stations only one measurement was possible because of the time limitations. Taking into account disadvantageous conditions, the accuracy of the measurements for all stations may be less than 100 μgal.

In addition, a gravity survey on the fast ice in the central part of Lützow-Holm Bay was carried out by JARE-22 in October, 1981. An apparent accuracy of only about 5 mgal was achieved because of fast-ice vibrations.

Free air and the Bouguer anomaly maps are being compiled from these data by the Japanese National Institute of Polar Research. The free air anomaly is in the range of − 50 to − 90 mgal in the central part of Lützow-Holm Bay where echo sounding data show a glacier trough about 1000 m deep (Yoshida and Moriwaki, personal communication) and increases gradually towards the east coast. The apparent correspondence between the bathymetric data and a large negative free air anomaly seems reasonable. The free air anomaly over the Ongul Islands ranges from − 10 to − 20 mgal, that over Langhovde and Skarvsnes is − 20 to − 30 mgal and that over Skallen is about − 30 mgal.

**Figure 1. Flight line diagram for the 1500 m elevation aeromagnetic survey in 1981.**

The Bouguer anomaly in the Ongul Islands ranges from − 15 to − 20 mgal, − 2- to − 30 mgal in Langhovde and Skarvsnes and − 30 to − 33 mgal in Skallen. Roughly speaking, the Bouguer anomaly is about − 10 mgal in the northern part of Lützow-Holm Bay and decreases gradually towards the southern part.

*Aeromagnetic Survey*

The first aeromagnetic survey over Lützow-Holm Bay was made using a helicopter in January, 1967, and observations were continued in the following few summer seasons. As a result isomagnetic charts of total magnetic intensity were compiled (Tazima et al., 1972; Kaneko, 1976).

Further, intensive, aeromagnetic surveys were carried out in 1980 and 1981. In 1980, 20 flight lines of three hours each were flown over the eastern and central part of Lützow-Holm Bay, at a flight elevation of 450 m above ground surface.

Figure 1 shows the flight lines of the aeromagnetic survey at 1500 m which was carried out in 1981, with a total flight distance of 1685 km. Surveys at 450 m and 900 m were also carried out, total flight distances being 1120 km and 1325 km respectively. The total magnetic intensity for three different flight elevations of 900, 1500 and 2400 m is given in Figure 2. These surveys were carried out along the longitude of 39°25′E from 68°50′S to 69°40′S latitude. As the survey was done when the K-index was between 0 and 1, and within a few days, the intensity values in Figure 2 are not corrected. These data will provide the basis for an isomagnetic chart soon to be compiled and published by the National Intstitute of Polar Research.

*Tiltmeter measurements*

Two types of tiltmeters were installed in Syowa Station in 1981 for the observation of secular variation of ground tilt after glaciation. One is a borehole type tiltmeter and the other is a water tube tiltmeter. The borehole type, a Kinemetrix Model TM-1B biaxial tiltmeter was installed in a waterproof iron-cased hole of 3 m depth and 15 cm diameter, as shown at the bottom of Figure 3. Two electrical outputs are obtained for the X and Y axes; the first monitors the X and Y DEMOD signals which were used to identify the high frequency motion from DC to approximately 10 Hz; the second reflects the X and Y OUT signals which result from passing Z and Y demodulator signals through low-pass filters having corner frequencies of either 1.6 Hz or 0.008 Hz. The four component signals were recorded with a 12-channel chart recorder in the Data Processing Hut 150 m north of the sensor field. The X-axis corresponds to the E-W component and the Y to the N-S. The scale factor of X and Y OUT outputs is 40 mV per microradian and full scale on the chart has a range of ± 25 mV, corresponding to an attenuation of ± 125 mV or ± 3.125μrad in real scale. The scale factor of X and Y DEMOD outputs is 5 mV per microradian and the full scale is again ± 25 mV.

When the record was over full scale, the position of the pen on the recorder was changed by a bias amplifier. Observations were started in April, 1981 and moderate variations of X and Y OUT were recorded from April to December, 1981 (Figure 4). The underground temperature variations at 1 m and 2 m depth are also shown on Figure 4. Through the observation period the X OUT shows a trend for tilting up of the east side and Y OUT a tilting up of the south side. The

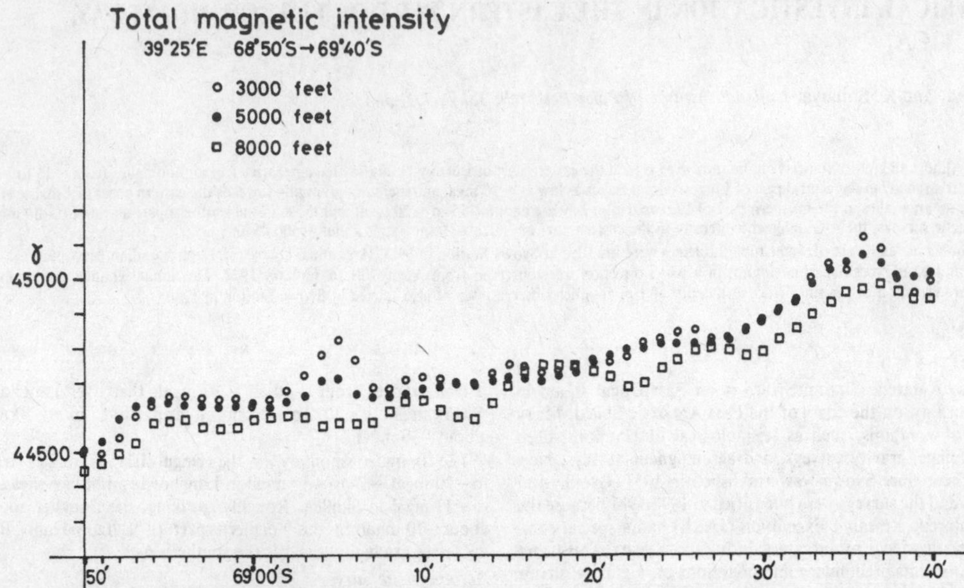


**Figure 2. Uncorrected total magnetic intensity observed at the end of August, 1981, flight elevations of 900, 1500 and 2400 m along the longitude of 39°25′E from 68°50′S to 69°40′S latitude.**

maximum variation in X OUT is 12 μrad, and in Y OUT 22 μrad. The largest variations were observed in the winter season, with tilt values becoming smaller in December, 1981. Figure 4 clearly shows the similarity in trends of annual tilt variation and underground temperature.

Figure 3 also shows the location of water tube tiltmeters installed on granitic gneiss basement rock about 50 m west of the borehole tiltmeter. The sensor of the water tube tiltmeter is similar to the model MF-N2 developed by Shichi et al. (1977). The full scales of A-B and C-D on the chart record ±2.5 V corresponding to ±60 μrad and ±40 μrad respectively in real scale. Observations were started in April 1981 and recorded variations of A-B and C-D which are shown in Figure 5 on which is also plotted air temperature variation at Syowa Station. The variation of both ground tilt and temperature shown in Figure 5 are running means for seven days. The trend of the water-tube tilt variations closely resembles that of the air temperature variations. The large variation of A-B appears to depend on a temperature differential caused by the large vertical difference between the pots and the lowest point of the water tube. This is because the vertical difference of A-B is about 1 m larger than that of C-D.

*Underground Temperature.*

Long term measurements of underground and underwater temperatures using a thermometer with resolution of 0.001°C and relative accuracy of 0.01°C yield important geophysical information. Such a thermometer needs also to be of easy installation and maintenance. A quartz thermometer which satisfies these requirements was developed by Shimamura (1980) and was used near Syowa in 1980 (Shibuya et al., 1982).

Where four quartz sensors were installed at 5 m intervals from 5 m depth to 20 m depth in the borehole. However, the sensor at 15 m was found to be broken after installation.

Figure 6 shows the underground temperatures measured from 17th April, 1980 to 2nd January, 1981; in the upper part the values at intervals of 32 hours are plotted and the lower part shows the plots at 16 hour intervals with an exaggerated temperature scale. The symbols X, Y and Z denote temperatures at depths of 5 m, 10 m and 20 m respectively. At 5 m depth the underground temperature shows a quasi-sinusoidal change in the range from −4.89°C (the maximum was not measured during the observation period) to −10.36°C (at around the 300th day of the year), while at 10 m the temperature ranged from −7.73°C (at around the 150th day) to about −9.31°C (around the 360th day). Although the temperature at 20 m in the upper part of Figure 6 appears to be almost constant, the plot in the

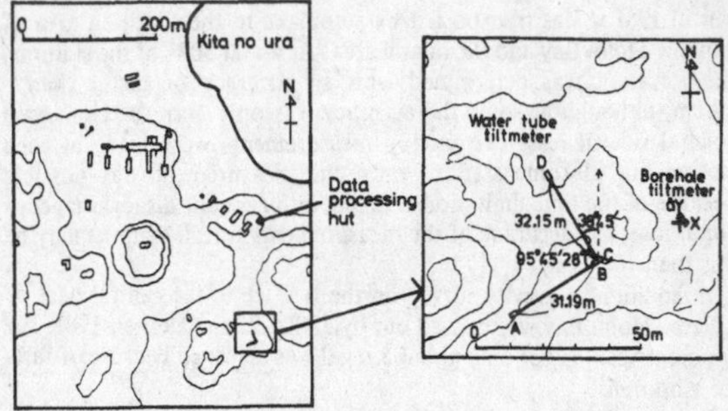


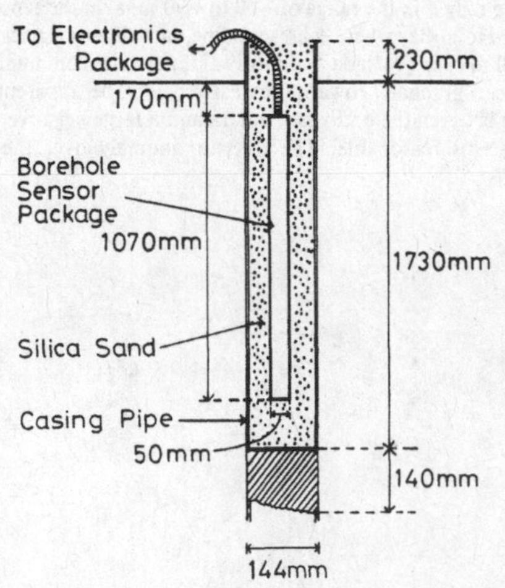


**Figure 3. The locations of the borehole tiltmeter and a set of the water tube tiltmeters, and detail of the borehole tiltmeter installation.**

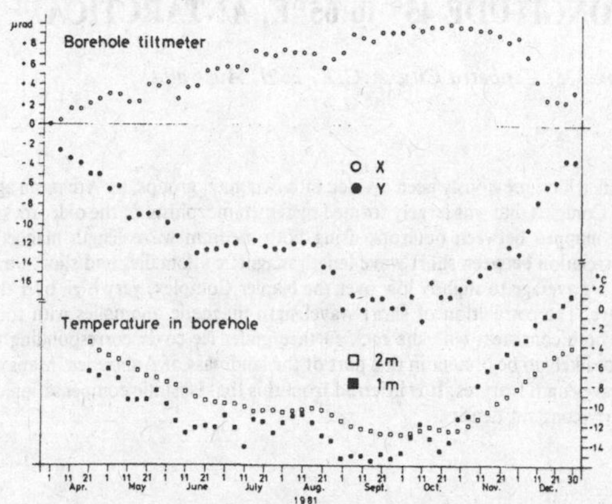


**Figure 4. Ground tilt of X and Y components of the borehole tiltmeter, and ground temperatures at depths of 1 m and 2 m near the Syowa Earth Science Laboratory.**

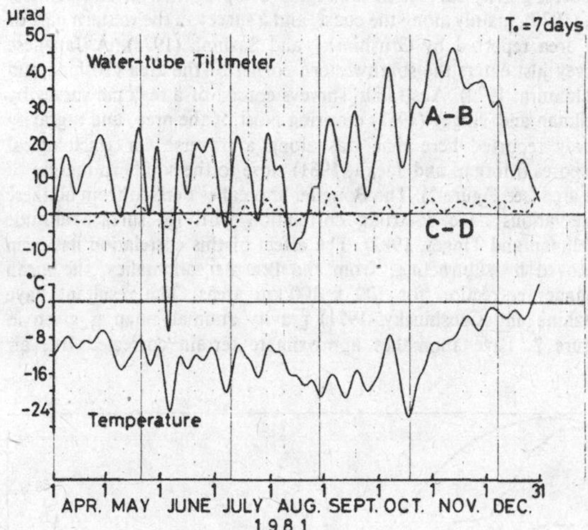


**Figure 5. The tilt variations of the water tube tiltmeters, and the air temperature at Syowa Station.**

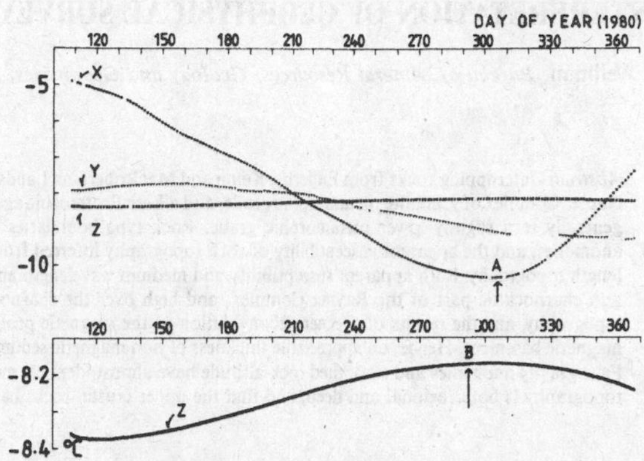


**Figure 6. Ground temperature in April-December, 1980. The plot of temperatures at 20 m depth against an exaggerated temperature scale is given below. "A" indicates the minimum at 5 m depth and "B" the maximum at 20 m.**

lower part clearly shows a sinusoidal change from −8.37°C (at around the 110th day of a year) to −8.17°C (around the 290th day). In comparison with the seasonal variation of air temperature, there is a phase lag of about 90 days at 5 m depth and about 290 days at 20 m depth. Assuming exponential decrease of ground temperature with depth, it was estimated that the effect of the seasonal variation of air temperature was less than 0.01°C at 33.8 m depth.

*Acknowledgements* The authors are grateful to Professors S. Kawaguchi and Y. Yoshida, leaders of JARE-21 and -22 respectively for their kind co-operation in many measurements and observations in Antarctica, and to Mr T. Nagao for his co-ordination of the work in Antarctica. They also thank Professor K. Kusunoki of the National Institute of Polar Research for critically reading the manuscript. The assistance of Miss Yumiko Shudo throughout this study is gratefully acknowledged.

## REFERENCES

KANEKO, H., 1976: Aeromagnetic survey in the vicinity of Prince Olav Coast, Antarctica. *Antarctic Rec.*, *55*, 61-8.

KAMINUMA, K., KUNIMIM, T. and OHTAKI, S., 1980: Gravity survey in Cape Ryugu and Syowa Station, Antarctica. *Antarctic Rec.*, *70*, 149-57.

SHIBUYA, K., NAGAO, T. and KAMINUMA, K., 1982: Measurements of underground and underwater temperatures by quartz thermometers at Syowa Station, East Antarctica. *Antarctic Rec.*, *76*, 89-100.

SHICHI, R., OKUDA, T. and YOSHIDA, S., 1980: A new design of moving float type water-tube tiltmeter. *Geodetic Soc. Japan, J.*, *26*, *(1)*, 1-16.

SHIMAMURA, H., 1980: Precision quartz thermometers for borehole observations. *J. Phys. Earth*, *28*, 243-60.

TAZIMA, M., KAKINUMA, S., YOSHIDA, M., MASUDA, M. and YOSHIMURA, A., 1972: Aeromagnetic survey in the vicinity of Lützow-Holm Bay, Antarctica. *Antarctic Rec.*, *44*, 69-78.

# INTERPRETATION OF GEOPHYSICAL SURVEYS—LONGITUDE 45° to 65°E, ANTARCTICA.

P. Wellman, *Bureau of Mineral Resources, Geology and Geophysics, P.O. Box 378, Canberra City, A.C.T., 2601, Australia.*

*Abstract* Outcropping rocks from Enderby, Kemp and MacRobertson Lands, Antarctica have previously been divided into two main groups, an Archaean age Napier Complex of granulite metamorphic grade, and a Late Proterozoic age Rayner Complex that was largely formed by remetamorphism of the older rocks, generally to a slightly lower metamorphic grade. Rock type boundaries are here mapped between outcrops using both medium wave length magnetic anomalies, and the apparent susceptibility of rock topography inferred from the correlation between short wave length magnetic anomalies and short wave length topography. Both apparent susceptibility and medium wavelength anomalies are average to slightly low over the Napier Complex, very high over the non charnockitic part of the Rayner Complex, and high over the charnockitic part. The correlation of short wavelength magnetic anomalies with rock topography, and the results of Werner deconvolution of the magnetic profiles, are both consistent with the rock surface under ice cover corresponding to magnetic basement. Hence, an appreciable thickness of non magnetic sediments is not likely to be present in this part of the landmass of Antarctica. Maps of Faye gravity anomalies and smoothed rock altitude have almost identical medium wavelength features. It is inferred from this that isostatic compensation of topography is both regional and deep, and that the upper crustal rocks have a nearly constant density.

The part of Antarctica between Mawson and Molodezhnaya bases, from the coast to 69°S latitude, is one of the better exposed parts of the East Antarctic Precambrian shield. Because of this and its relative accessibility, there have been numerous Australian, Soviet and Japanese expeditions to study its geology and geophysics. The relative completeness of these data makes it an excellent area to study the relationships between regional geology (Figure 1) often based on widely spaced outcrops, and geophysical data that covers the area more uniformly. The geophysical data available are gravity, aeromagnetics and sub ice topography.

The exposed rocks are mainly granulite facies metamorphic rocks with minor dolerite dykes, granitic intrusives and pegmatites. Kamenev (1972) distinguished two metamorphic complexes, largely on the basis of metamorphic grade. The older Napier Complex consists of rocks metamorphosed to a high temperature granulite facies; they are exposed as a wide coastal strip in eastern Enderby Land. The younger Rayner Complex comprises rocks of lower metamorphic grade (upper amphibolite to granulite facies), thought to have been largely formed by metamorphism of the Napier Complex rocks. Sheraton et al., (1980) supported this model, and showed that the Napier Complex was metamorphosed to granulite grade from about 3000 Ma in the Archaean, and the Rayner Complex was formed by remetamorphism 1000 Ma ago in the Late Proterozoic.

This paper describes new 1:1,000,000 aeromagnetic, gravity and rock altitude maps of the area (Wellman, 1982), and interprets all available geophysical data in terms of geological structure. The main objective is to map metamorphic facies and rock composition boundaries over the whole area.

## GRAVITY ANOMALIES AND MEAN ROCK ALTITUDE

Soviet gravity surveys in the region comprise that of Koriakine et al., (1970), mainly along the coast, and a survey in the western part of the area reported by Grushinsky and Sazhina (1975). A Japanese survey just enters the southwestern corner of the area (Yoshida and Yoshimura, 1972). Australian surveys consist of a regional survey by Wellman and Tingey (1982) covering most of the area, and a gravity survey reported here that was along a traverse for glaciological purposes (Morgan and Jacka, 1981) close to the southern margin of the area (see Figure 2). The Bouguer anomalies derived from all these observations have a strong correlation with ice surface altitude (Wellman and Tingey, 1982). The effect of this correlation has been removed by subtracting, from the Bouguer anomalies, the mean Bouguer correction for 100 x 100 km areas. The resultant Faye (Sazhina and Grushinsky, 1971) gravity anomaly map is given in Figure 2. Faye anomalies approximate terrain corrected free air

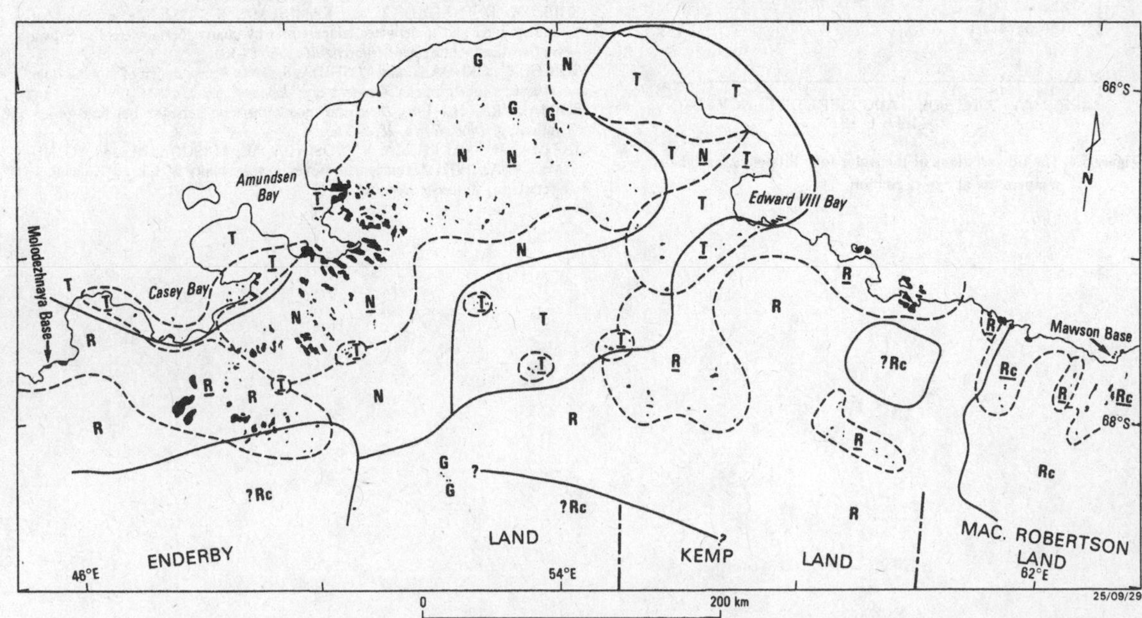


**Figure 1. Distribution of rock types from geological mapping (Sheraton et al., 1980; Trail, 1970) (dashed lines, underlined letters), and from magnetic anomalies (solid lines, non-underlined letters). N, Napier Complex; T, Napier Complex partly remetamorphosed to Rayner Complex; R, Rayner Complex; Rc, charnockites in Rayner Complex; G, outcropping granite.**

anomalies. (An earlier map of this type given by Wellman and Tingey (1982, Figure 3) is incorrect because of a mistake in calculating the Faye corrections.) The main features of the Faye anomaly map are highs in excess of $+1000\mu$ m s$^{-2}$ in northeastern Enderby Land, and along the southern margin of the mapped area in Kemp and eastern Enderby Land.

Ice radar surveys of rock altitude carried out in the region consist of Soviet traverses of widely spaced spot measurements over the western half of the region (Fedorov, 1973; AARI, 1975) and Australian close-spaced traverses mainly over the eastern two thirds of the region (Allison et al, in press; Morgan, in press; Wellman, 1982 in prep.). Figure 3 is a map of the altitude of smoothed rock topography corrected for the isostatic depression by the ice load (from Wellman, in prep.). Rock altitude generally increases from 0 m just inland of the coastline, to a major divide at about 1500 m near the southern margin of the mapped area, but with an isolated high area in northeastern Enderby Land.

The features of the gravity anomaly and rock altitude maps of less than 300 km wavelength are almost identical. This similarity is not due to errors in the Faye anomaly calculation, because free air anomalies are approximately the same as Faye anomalies on average. The similarity is also unlikely to be due to a correlation of rock altitude and rock density, because this would not result in such good correlation of gravity anomaly and topography. Further, in the field no changes in rock types were found that could give rise to the large rock density variation required to explain a 1000 $\mu$m s$^{-2}$ gravity anomaly. The correlation is thought to be caused by isostatic compensation of topography that is both deep and regional, in an area where the upper crustal rocks have nearly constant density.

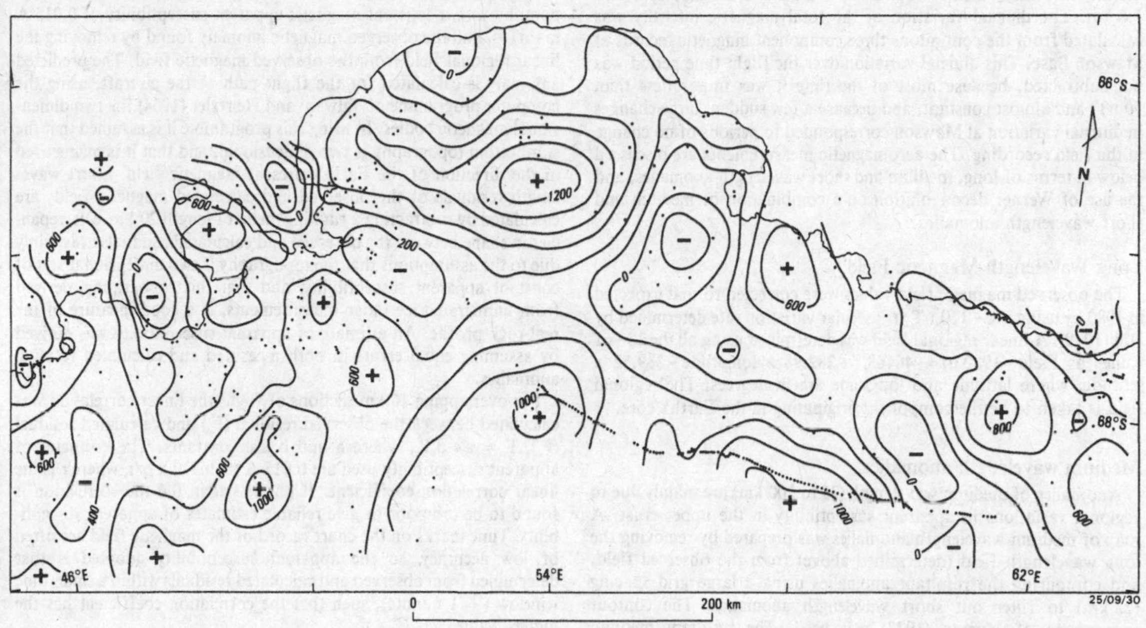


**Figure 2. Faye gravity anomalies. Contour interval 200$\mu$m s$^{-2}$ (= 20 mGal). Densities used are 2.67 t m$^{-3}$ for rock and 0.917 t m$^{-3}$ for ice. Dots show gravity station positions.**

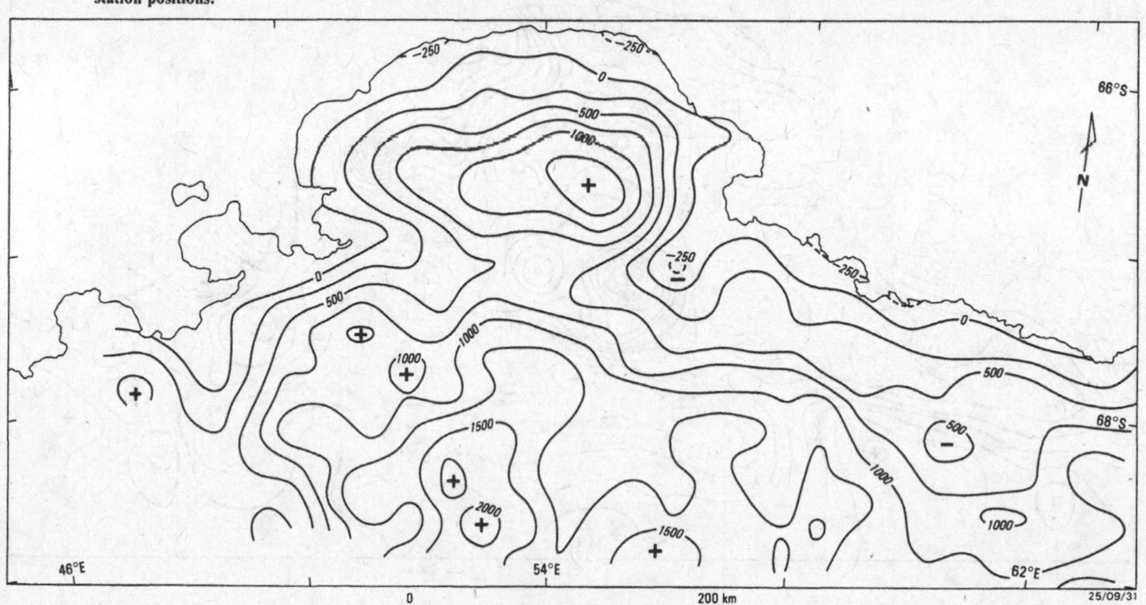


**Figure 3. Smoothed deglacial rock altitude. Contour interval: 250 m. Grid size used in contouring: 22 km.**

## AEROMAGNETIC SURVEY

### Survey Method

The aeromagnetic survey was carried out in early 1977 and early 1980. Magnetic observations were made simultaneously with the radar ice thickness measurements described by Allison et al., (in press) and Morgan et al., (in press). A Pilatus Porter aircraft was flown at about 300 m above the ice surface for flights of about 3 hours duration. Flight line spacing averaged 40 km. Navigation between nunataks was based on dead reckoning, drift side readings and fixes on visible nunataks. Navigational accuracy is thought to be better than 7 km everywhere, and better than 3 km in areas of nunataks.

The total magnetic field was measured using a proton precession magnetometer and electrostatic chart recorder, with the sensor in a towed "bird". The chart record was subsequently "picked" at 0.25 minute intervals, which is equivalent to a ground distance of 0.4 to 1.6 km. The diurnal variation of the total magnetic intensity was calculated from the continuous three component magnetic records at Mawson Base. This diurnal variation over the flight time period was not subtracted, because most of the time it was minor (less than 50 nT) and almost constant, and because a few sudden, large changes in diurnal variation at Mawson corresponded to periods of no change in the field recording. The aeromagnetic measurements are discussed below in terms of long, medium and short wavelength anomalies, and the use of Werner deconvolution on a combination of medium and short wavelength anomalies.

### Long Wavelength Magnetic Field

The observed magnetic field values were corrected to that expected in 1980 by using the $-110$ nT yr$^{-1}$ secular variation rate determined by Hill (1979). A linear regional field was determined using all the 13 961 values as Field (1980.0) = 46468. $+282.98$ x longitude $-389.52$ x latitude, where latitude and longitude are in degrees. The regional field is taken to be the component originating in the Earth's core.

### Medium wavelength anomalies

Anomalies of medium wavelength (25 to 200 km) are mainly due to regional variations in apparent susceptibility in the upper crust. A map of medium wavelength anomalies was prepared by removing the long wavelength field (determined above) from the observed field, and contouring the resultant anomalies using a large grid spacing (22 km) to filter out short wavelength anomalies. The contour programme of Murray (1977) was used. The resultant medium wavelength anomaly map (Figure 4) is dominated by large positive anomalies of up to 800 nT situated at Edward VIII Bay and to 150 km northwest and to 200 km southwest. Other anomalies have amplitudes ranging from $-400$ to $+400$ nT, and many are elongate. A Soviet aeromagnetic residual map of the area (Demenitskaya, 1977) has an almost identical anomaly pattern, but, as that map has only two contour lines, it is more difficult to interpret. The map compiled for the present paper (Figure 4) is interpreted in terms of geology in a later section.

### Short wavelength anomalies due to topography

Magnetic anomalies of wavelengths less than 20 km have been used to obtain estimates of the apparent magnetisation of the sub-ice topography. Figure 5 illustrates, for part of a flight, the measured sub-ice topography, the magnetic anomaly predicted from this for topography with a constant magnetic apparent susceptibility of 0.015 A m$^{-1}$nT$^{-1}$, and the observed magnetic anomaly found by removing the linear regional field from the observed magnetic field. The predicted amomaly is calculated for the flight path of the aircraft, using the computer programme of Talwani and Heirtzler (1964) for two dimensional magnetic bodies. In using this programme it is assumed that the sub-bottom topography is two dimensional, and that it is magnetised in the direction of the Earth's present magnetic field. Short wavelength residuals of the observed and calculated magnetic fields are calculated by subtracting a running mean of length 20 km. Discrepancies in shape between the observed and calculated residuals are mainly due to the assumptions that the topography is two dimensional and of constant apparent susceptibility, and that this topography, derived from unmigrated ice radar measurements, is a good measure of the real rock profile. An estimate of apparent susceptibility was derived by assuming equal errors in both measured and calculated residual anomalies.

For overlapping 16 km sections of flight the linear correlation was calculated between the observed residual ($F_o$) and calculated residual ($F_c$), $F_o = a + b.F_c$, where a and b are constants. The estimates of apparent susceptibility used are 0.015 A m$^{-1}$nT$^{-1}$ x b/r, where r is the linear correlation coefficient. If r is less than 0.6 the correlation is found to be too poor to give reliable estimates of apparent susceptibility. Time marks on the chart record of the magnetic field are often of low accuracy, so the apparent susceptibility adopted is that determined from observed and calculated residuals within a small time window (± 1 minute), such that the correlation coefficient has the highest value.

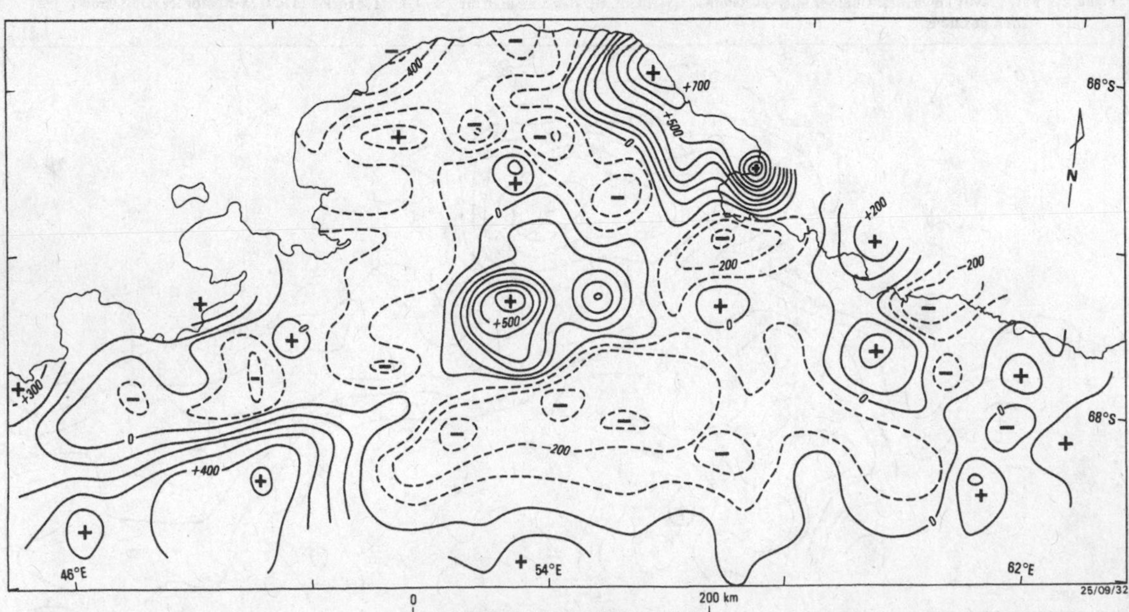


**Figure 4. Total magnetic field intensity anomalies; observed total field intensity minus a long wavelength regional field. Contour interval 100 nT. Grid size used in contouring: 22 km.**

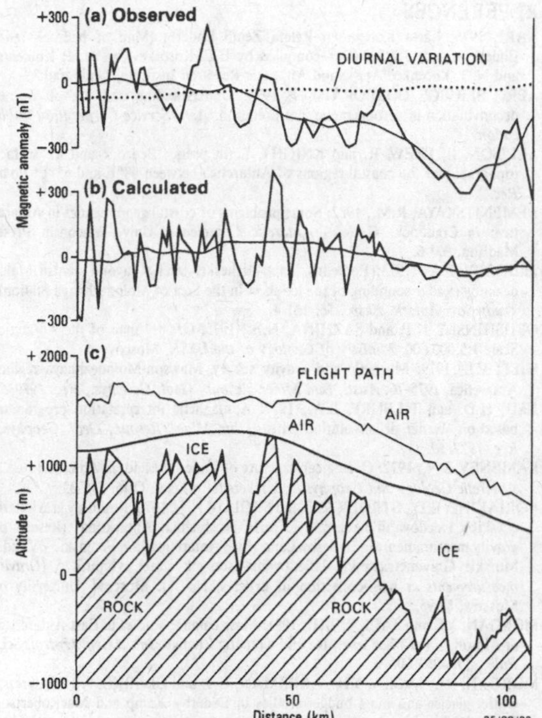


**Figure 5. Results for part of an aeromagnetic ice radar flight, 67.1°S 52.9°E to 66.1°S 52.4°E: (a) observed total magnetic intensity anomaly (after removing the linear regional field), running mean of this anomaly for length 20 km diurnal magnetic variation from Mawson Observatory; (b) magnetic field predicted from distance of rock surface below flight path, and its running mean; (c) altitudes of flight path, ice surface, and rock surface. Note correlation of the observed and calculated residual magnetic anomalies.**

Where the computer could not calculate reliable values because of large timing errors, apparent susceptibilities were estimated by eye. An assumption is made that changes in apparent susceptibility below the base of the topography do not correlate with topography. In general this will not be valid, hence all estimates are thought to have a systematic error causing them to be slightly too high.

Figure 6 is the resultant map of apparent susceptibility of the rock. Note that the contour interval is not linear. The average apparent susceptibility is between 0.015 and 0.030 A $m^{-1}nT^{-1}$; there are small areas above 0.06 A $m^{-1}nT^{-1}$. This map is derived independently of the medium wavelength anomaly map (Figure 4), however there is a good correlation between the maps in the eastern two thirds of the area where control is good. The apparent susceptibility of the rock topography is generally more than 0.02 A $m^{-1}nT^{-1}$ in areas of positive anomaly and less than this in areas of negative anomaly.

## Short wavelength anomalies not due to topography

In most areas the short wavelength anomalies correlate with the sub-ice topography, so it is very difficult to separate non-topographic short wavelength anomalies. However, the nature of these anomalies is known in two areas. In the area of apparent susceptibility less than 0.007 A $m^{-1}nT^{-1}$ there are almost no short wavelength anomalies, so both the non-topographic and topographic anomalies must be very small. The rocks are not sediments because gneiss outcrops. Along the coastal lowland midway between Edward VIII Gulf and Amundsen Bay the short wavelength anomalies have a high amplitude over only minor topography. They are thought to be due to large, short wavelength variations of apparent susceptibility in the upper crust.

## Werner deconvolution

The aeromagnetic profiles were run through a Werner deconvolution computer programme (Werner, 1953; Aero Service Corporation, 1974; Hsu and Tilbury, 1977) to calculate probable depth to the source bodies. For deep magnetic sources no satisfactory results were obtained, possibly because any long wavelength anomalies are badly contaminated by short wavelength anomalies from irregular rock topography, and by not being adequately defined because of the rough digitisation. Results for shallow magnetic sources confirmed that these originated from the level of rock topography.

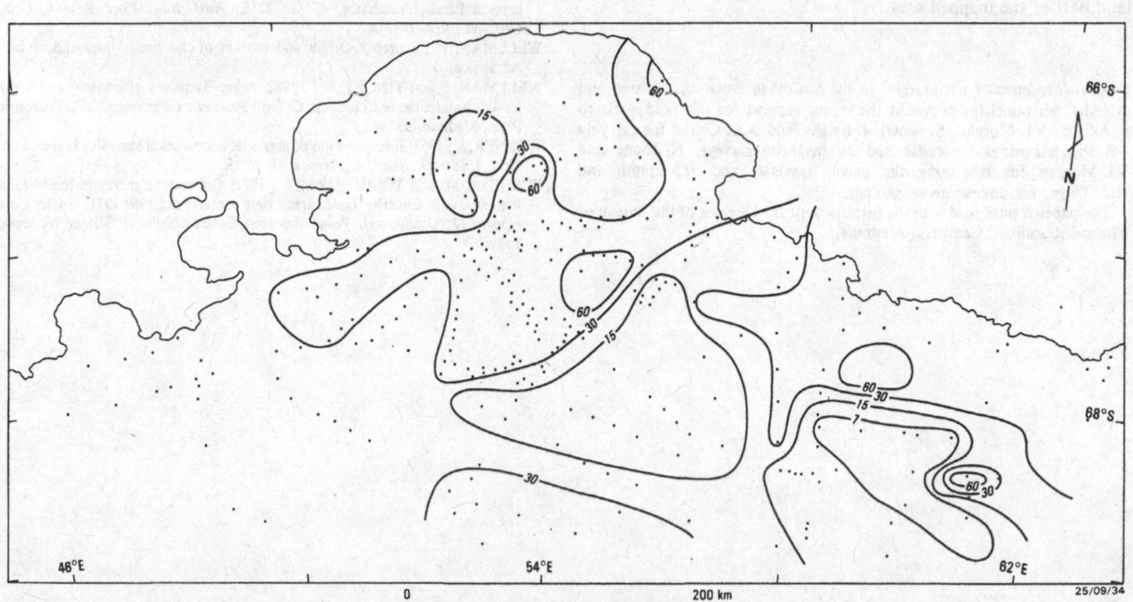


**Figure 6. Apparent susceptibility of rock forming the subglacial topography. Contours at non-uniform intervals. Contour values are in multiples of 0.001 A $m^{-1}nT^{-1}$. Dots show positions of apparent susceptibility estimates.**

## GEOLOGICAL INTERPRETATION OF THE MAGNETIC ANOMALIES

Figure 1 summarises the different rock groups found from geological mapping of the area (Trail, 1979; Sheraton et al., 1980).

The most prominent features of the magnetic anomaly map (Figure 4) are the magnetic highs on and to the southwest and northwest of Edward VIII Bay. These highs lie in the areas mapped by Sheraton et al. (1980) as Napier Complex partly remetamorphosed to Rayner Complex. The magnitude and width of the anomalies is such that they must be caused by bodies of high apparent susceptibility and of considerable thickness. Within these magnetic highs, the rock topography has an apparent susceptibility of over 0.03 A $m^{-1}nT^{-1}$ (Figure 6). A magnetic high in an area of partial remetamorphism occurs in Casey Bay. In all these areas the high apparent susceptibility of the upper crust is interpreted as being caused by partial recrystallisation during partial remetamorphism.

Within the Napier Complex the magnetic anomalies are of small magnitude. The granitic rocks are in areas of magnetic lows. The southwestern part of the area has a slightly higher magnetic anomaly; this area was distinguished by Sheraton et al. (1980, Figure 3) as having orthopyroxene-quartz-plagioclase gneiss rather than orthopyroxene-quartz mesogneiss.

Rocks of the Rayner Complex are thought to underlie the remainder of the area. Most of the exposed rocks are in the areas of magnetic lows. However, in the Mawson area (east of 61°E) rocks are exposed in an area of magnetic high; these rocks differ from most of the other exposed Rayner Complex rocks by being more charnockitic. The rock type causing the magnetic highs along the southern margin of the mapped area is unknown; the few small nunataks within the southwest magnetic high area are similar to those in the magnetic low area to the north.

Figure 1 shows the correlation between previous geological mapping, and the sub-ice boundaries inferred from the magnetic data—using Figures 4 and 6, and, for some marine areas, data of Demenitskaya (1982). Two important features of the map are: (1) Wide, partially remetamorphosed zones are recognised on both the eastern and western sides of the Napier Complex. (2) The Rayner Complex is subdivided into an area of low apparent susceptibility that crops out extensively, and areas of higher apparent susceptibility that, at least in the east, are charnockitic.

Results of the Werner deconvolution, and the generally good correlation of short wavelength anomalies with rock topography, are consistent with there being little or no non-magnetic sediment in the land part of the mapped area.

*Acknowledgements* I am grateful to the Australian Antarctic Division and ANARE personnel for providing the logistic support for this field work, to I. Allison, V.I. Morgan, S. Scherl, I. Knight, and A.L. Clarke for carrying out the hazardous ice radar and aeromagnetic surveys, N. Young and V.I. Morgan for measuring the gravity traverse, and J. Sheraton and R.J. Tingey for discussions on geology.

The paper is published with the permission of the Director of the Bureau of Mineral Resources, Canberra, Australia.

## REFERENCES

AARI, 1975: Karta Korennovo Relefa Zemli Enderbi (Map of bedrock relief Enderby Land) 1:2 500 000—compiled by E.C. Korotkevich, Ya. P. Koblents, and N.G. Kocenko. Arctic and Antarctic Research Institute, Leningrad.

AERO SERVICE CORPORATION, 1974: Understanding and use of Werner deconvolution in aeromagnetic interpretation. *Aero Service Corporation Publication.*

ALLISON, I., FREW, R. and KNIGHT, I. (in press): Bedrock and ice surface topography of the coastal regions of Antarctica between 48°E and 65°E. *Polar Rec.*

DEMENITSKAYA, R.M., 1982: Some problems of crustal geodynamics in Antarctica; *in* Craddock, C. (ed.) *Antarctic Geoscience.* Univ. Wisconsin Press, Madison, 903-6.

FEDOROV, B.A., 1973: Podledniy antarkticheskiy relef v rayone stantsii Molodezhnoy (Radar sounding of the ice sheet in the area of Molodezhnaya Station). *Trudy Sov. Antark. Eksp., 56,* 121-4.

GRUSHINSKY, N.P. and SAZHINA, N.B., 1975: Gravity map of the Antarctic, Scale 1:5,000,000. *Ministry of Geology of the USSR,* Moscow.

HILL, P.J., 1979: Magnetic and gravity survey, Mawson-Molodezhnaya region, Antarctica, 1975-76. *Aust., Bur. Miner. Resour., Geol. Geophys., Rec. 1979/6.*

HSU, H.D. and TILBURY, L.A., 1977: A magnetic interpretation programme based on Werner deconvolution. *Aust., Bur. Miner. Resour., Geol. Geophys., Rec. 1977/50.*

KAMENEV, E.N., 1972: Geological structure of Enderby Land; *in* Adie, R.J. (ed.): *Antarctic Geology and Geophysics,* Universitetforlaget, Oslo, 579-83.

KORIAKINE, E.D., STROEV, P.A. and FROLOV, A.I., 1970: Resultaty gravimetricheskikh issledovanii na zemle Enderbi (vostochnai Antarktida) (Results of gravity measurements in Enderby Land (East Antarctica)); *in* Fedinski, V.V. (ed.) Morskie Gravimetricheskie Issledovanii—sbornik statei, vbipusk 5 (*Gravity measurements at sea*—collection of articles, No. 5). Moscow, University of Moscow, 83-95.

MORGAN, V.I. and JACKA, T.H., 1981: Mass balance studies in East Antarctica; *in* Allison, I. (ed.) Sea level, Ice and Climatic Change, *Int. Assoc. Hydrol. Sci., Publ., 131,* 253-60.

MORGAN, V.I., JACKA, T.H., AKERMAN, G.J. and CLARKE, A.L. (in press): Outlet glacier and mass budget studies in Enderby, Kemp and MacRobertson Lands, Antarctica. *Annals Glaciology.*

MURRAY, A.S., 1977: A guide to the use and operation of programme CONTOR. *Aust., Bur. Miner. Resour., Geol. Geophys., Rec. 1977/17.*

SAZHINA, N. and GRUSHINSKY, N., 1971: *Gravity prospecting.* (transl. A.K. Chatterjee). MIR Publishers, Moscow.

SHERATON, J.W., OFFE, L.A., TINGEY, R.J. and ELLIS, D.J., 1980: Enderby Land, Antarctica—an unusual Precambrian high grade metamorphic terrain. *Geol. Soc. Aust., J., 27,* 1-18.

TALWANI, M. and HEIRTZLER, J.R., 1964: Computation of magnetic anomalies caused by two dimensional structures of arbitrary shape. *Stanford Univ. Publ., Geol. Sci., 9,* 464-80.

TRAIL, D.S., 1970: ANARE 1961 geological traverses on the MacRobertson Land and Kemp Land coast. *Aust., Bur. Miner. Resour., Geol. Geophys., Rep., 135.*

WELLMAN, P., 1982: Geophysical maps of parts of Enderby, Kemp and MacRobertson Lands, Antarctica, 45° to 82°E. *Aust. Bur. Miner. Resour., Geol. Geophys., Rec. 1982/20.*

WELLMAN, P. (in prep.): Origin and erosion of the coastal highland of East Antarctica.

WELLMAN, P. and TINGEY, R.J., 1982: A gravity survey of Enderby and Kemp Lands, Antarctica; *in* Craddock, C. (ed.) *Antarctic Geoscience,* Univ. Wisconsin Press, Madison, 937-40.

WERNER, S., 1953: Interpretation of magnetic anomalies at sheet like bodies. *Sver. Geol. Undersok. Ser. C.C. Arsbok 43, N:06.*

YOSHIDA, M. and YOSHIMURA, A., 1972: Gravimetric survey in the Mizuho Plateau—west Enderby Land area, East Antarctica, 1969-1971. *JARE Data reports 17 (Glaciology),* Polar Research Centre, National Science Museum, Tokyo.

# PRELIMINARY REPORT OF A MARINE GEOPHYSICAL SURVEY BETWEEN DAVIS AND MAWSON STATIONS, 1982

H.M.J. Stagg, D.C. Ramsay and R. Whitworth, *Bureau of Mineral Resources, Geology and Geophysics, PO Box 378, Canberra, ACT, 2601, Australia.*

*Abstract* Between January and March 1982, the Australian Bureau of Mineral Resources carried out a seismic and magnetic survey of the continental shelf, slope and rise between Davis and Mawson Stations in the Australian Antarctic Territory. The survey area includes the potentially economically important Prydz Bay. More than 5000 km of 6-fold digital seismic reflection data and about 8000 km of magnetic and bathymetric data were recorded on lines extending from near the coast out to the 3000 m isobath. Processing of this data is still at an early stage and a detailed interpretation is not yet available. The survey was carried out on the M.V. *Nella Dan* and supported by the Antarctic Division of the Department of Science and Technology.

Preliminary interpretation indicates the existence of a thick sedimentary section beneath Prydz Bay and a minimum of 2-3 km of sediment at the base of the continental slope. Unfortunately, strong water bottom multiples that appear to be a function of hard bottom sediments, preclude an accurate estimate of the sediment thickness in Prydz Bay itself at this time. It is hoped that future computer analysis of the magnetic data will give estimates of the depth to magnetic basement, and hence an estimate of total sediment thickness. Within Prydz Bay, the seismically visible section appears to be little disturbed by faulting, other than in the southwest corner of the bay. The marked angularity of reflectors with the sea floor under much of the bay indicates that little deposition is presently occurring. It is still uncertain whether this angularity is the expression of eroded folded ancient rocks or of the topset beds of an oblique prograded sequence. West of Cape Darnley, the thickness of sedimentary rocks appears to be less than in Prydz Bay, although the shallower shelf depths here, 100-300 m, as against 400-700 m in Prydz Bay, make estimating difficult. Deep gashes, 800-1300 m, on the inner side of the shelf are not obviously structurally controlled and are hard to explain.

During the 1979/80 Antarctic season, the Australian Bureau of Mineral Resources (BMR) began a moderate level Antarctic marine geoscience programme on board the Danish polar supply vessel M.V. *Nella Dan* under charter to the Antarctic Division of the Department of Science and Technology. The aim of the programme was to collect geophysical data over the Antarctic margin, the plateaus surrounding Australia's sub-Antarctic Island Territories and the southeast Indian Ocean but in 1979/80 only digital magnetic data could be collected. Excellent coverage of more than 30,000 km was obtained over the Gaussberg Ridge and in the southeast Indian Ocean but only about 1000 km of data were collected over the Antarctic Margin because of adverse ice conditions (Stagg et al., 1980).

In preparation for the 1980/81 season, *Nella Dan* was extensively modified so that she could undertake major biological and seismic programmes. Deck space was cleared at the stern to allow the handling of trawls and a seismic streamer. The major scientific cruise of the season was primarily devoted to FIBEX—First International BIOMASS Experiment—in January to March 1981. BMR continued their magnetic work and also installed a Raytheon deep echo-sounder (Tilbury et al., 1980). A substantial amount of new data was recorded over the Antarctic margin.

The major scientific cruise of the 1981/82 season was dedicated to geoscience. In addition to underway geophysics conducted by BMR, comprising reflection seismic, magnetic and bathymetric data acquisition, programmes of geological dredging and coring and minor physical oceanography and marine biology were run. The area selected for the survey was between Davis and Mawson Stations in the Australian Antarctic Territory, principally in Prydz Bay (Figure 1). In addition to the obvious logistical reasons for working in this area, it was anticipated from previous work (Fedorov et al., 1982) that Prydz Bay would be a likely target for thick sediment accumulations.

*Nella Dan* left Hobart on December 31, 1981 and arrived at Davis Station on January 16, 1982. After a seismic line had been shot between Davis and Mawson and supplies had been unloaded at both bases, *Nella Dan* commenced the dedicated cruise on January 27. Between January 19 and March 2, 1982 approximately 5000 km of 3-fold and 6-fold seismic data of fair to good quality and 8000 km of magnetic and bathymetric data were recorded between Davis and Mawson, principally in Prydz Bay (Figure 2). In addition, on the return voyage to Hobart, approximately 400 km of single-channel seismic data were recorded across the southern extremity of the Gaussberg Ridge. The abnormally temperate weather and ice free conditions which prevailed for most of the time in Prydz Bay allowed most lines to be run to within 10-30 km of the coast. The northern ends of most lines terminated at 66°S, between 2500-3000 m water depth. West of Prydz Bay, between Cape Darnley and Mawson, the landward end of each line was terminated when icebergs made operations hazardous; generally, this was when the seabed was shallow enough to ground icebergs, that is, about 250-300 m. The large quantity of both seismic and non-seismic data acquired is now in the early stages of processing. Consequently, this paper will be confined to only a brief discussion of the more obvious results. More detailed interpretations should commence in late 1982.

## PREVIOUS WORK

Within the Prydz Bay/Lambert Graben area, previous work has been mostly onshore. The Lambert Graben and Prince Charles Mountains, south of Prydz Bay, have been studied by a number of Russian and Australian expeditions. Fedorov et al. (1982) produced a synthesis and a structural interpretation of all available geophysical and geological data; their paper also provides a comprehensive

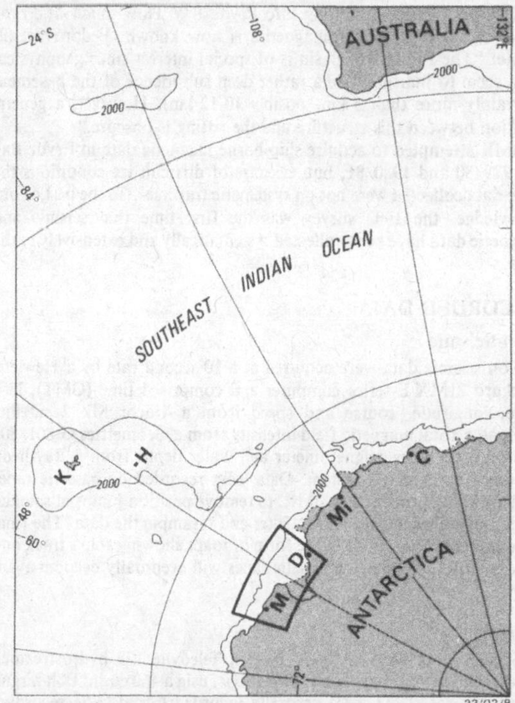


**Figure 1. Locality map showing survey area relative to Australia. Area in box is the area shown in the maps in Figures 2 and 3. Abbreviations: K—Kerguelen Island; H—Heard Island; M—Mawson Station (Australia); D—Davis Station (Australia); Mi—Mirny Station (USSR); C—Casey Station (Australia).**

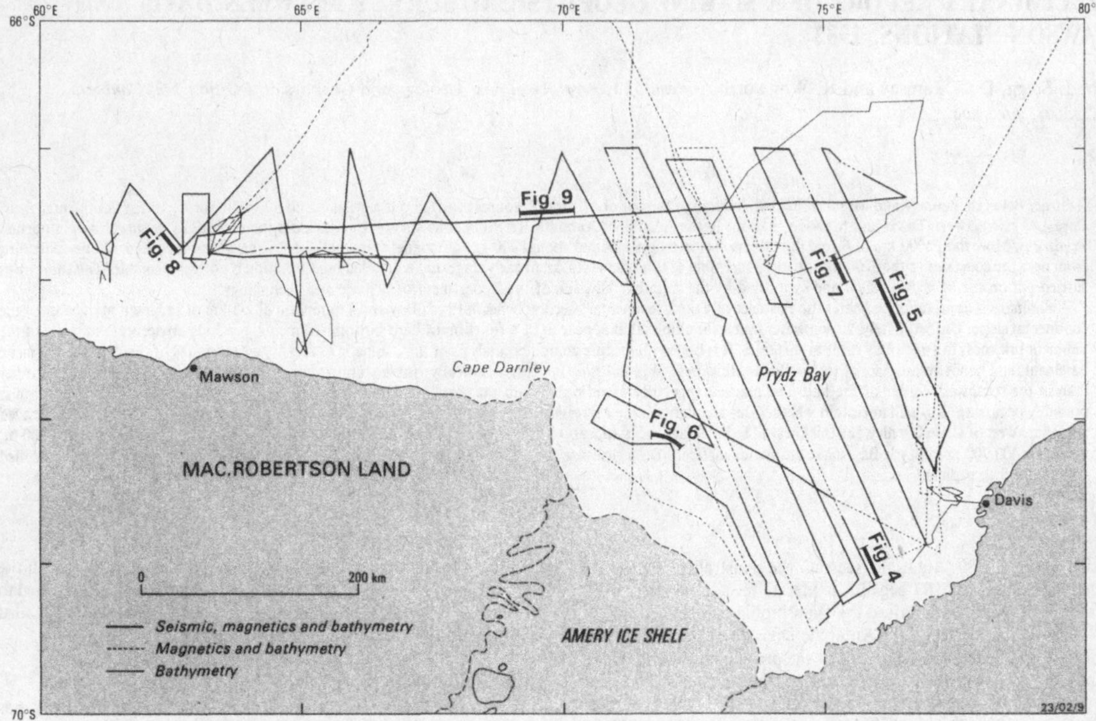


**Figure 2. Tracks of the *Nella Dan* in the 1982 survey. Locations of seismic sections illustrated in Figures 4-9 are shown.**

summary of prior work. The geophysical techniques applied by the Russians in their study include deep seismic sounding on land only, and aeromagnetics extending into Prydz Bay. How much of Prydz Bay was covered by aeromagnetics is now known; Fedorov et al. wrote: "The Prydz Bay Basin is of special interest since geophysical data seem to indicate here a rather deep subsidence of the basement (certainly more than 5 km, posibly 10-12 km). However, a genetic relation between this structure and the rifting is obscure."

BMR attempted to acquire ship-borne magnetic data in Prydz Bay in 1979/80 and 1980/81, but, because of difficult ice conditions the little data collected were not on systematic traverses. To the best of our knowledge, the 1982 survey was the first time that seismic and magnetic data have been collected systematically and extensively in the area.

## RECORDED DATA

### Non-Seismic

Non-seismic data were acquired at a 10 second rate by a Hewlett-Packard 21MX E-series computer and comprised time (GMT), latitude, longitude, course and speed from a Tracor Mk 2 satellite navigator, total magnetic field intensity from a Geometrics G.801/803 proton precession magnetometer and water depth from a Raytheon deep-sea bathymetric system. Data were recorded on cassette tapes and are being processed at BMR, to remove position jumps at satellite fixes and spikes in data and to filter and resample the data. The final data presentation will be in the form of maps showing ship's track and data profiles; summary magnetic tapes will eventually become available.

### Seismic

Seismic data were recorded from a Teledyne 178 hydrostreamer, configured in six 50-metre active sections, usig a 460 cu.in. Bolt airgun as the source. Data were generally recorded 6-fold. The recording system used a second Hewlett-Packard 21MX E series computer and data were recorded on standard 0.5 inch computer magnetic tape in the demultiplexed SEG-Y format. The software was written at BMR. Although the *Nella Dan* is less than an ideal platform for running a seismic system, the quality of the data recorded was fair to good.

## BATHYMETRY

The preliminary bathymetric map (Figure 3) has been compiled primarily from the *Nella Dan* raw data. Only in areas of sparse or no coverage by *Nella Dan*, have contours been guided by using the General Bathymetric Chart of the Oceans (GEBCO) Soundings Sheets 540, 556 and 557. Our experience on the 1982 survey has shown us that the fine detail on the GEBCO charts can frequently not be relied upon, due no doubt to the variety of sources of the data and the inherent variability in quality. It is intended in the final analysis to use the *Nella Dan* data as a base and only use the GEBCO sheets where our coverage is sparse.

The bathymetry of the area surveyed is too complex to allow a complete discussion here; only the major features will be summarised. The continental shelf corresponds quite closely to that described as typical, by Fairbridge (1966), off glaciated land masses. Within Prydz Bay the shelf generally shallows oceanwards from a channel deeper than 800 m near the coast. Most of the bay lies at a depth of 400-700 m with the shallowest area being in the northeast. These depths are about three to five times deeper than the worldwide average of 130 m for nonglaciated shelf depths; the difference is explained as the result of crustal depression due to the ice load. The transverse near-coast channel separates relatively shallow rugged topography landward from the deeper and very smooth seabed of the main part of Prydz Bay. To the west of Cape Darnley the Shelf becomes more complex. The width varies from 60-80 km and at 100-500 m deep, it is distinctly shallower than in Prydz Bay. Also the seabed topography is considerably more rugged, showing numerous banks as well as several major near-coast deeps of which the Nielsen Basin, east of Mawson, is the deepest (> 1300 m). Typically the shelf here can be divided into two regions—an inner shelf, 100-200 m deep which is generally subhorizontal and an outer shelf, deeper than 200 m and sloping with variable gradient to the shelf break.

The shelf break varies greatly in depth and in sharpness of break, being shallowest in the east of Prydz Bay at 350-400 m and deepest west of Cape Darnley to down to 600 m. The break itself varies from gradual, taking place over 2-3 km, to extremely sharp, taking place over a few hundred metres or less. Throughout the survey area the continental slope/rise boundary is 1700-1900 m deep and the rise/

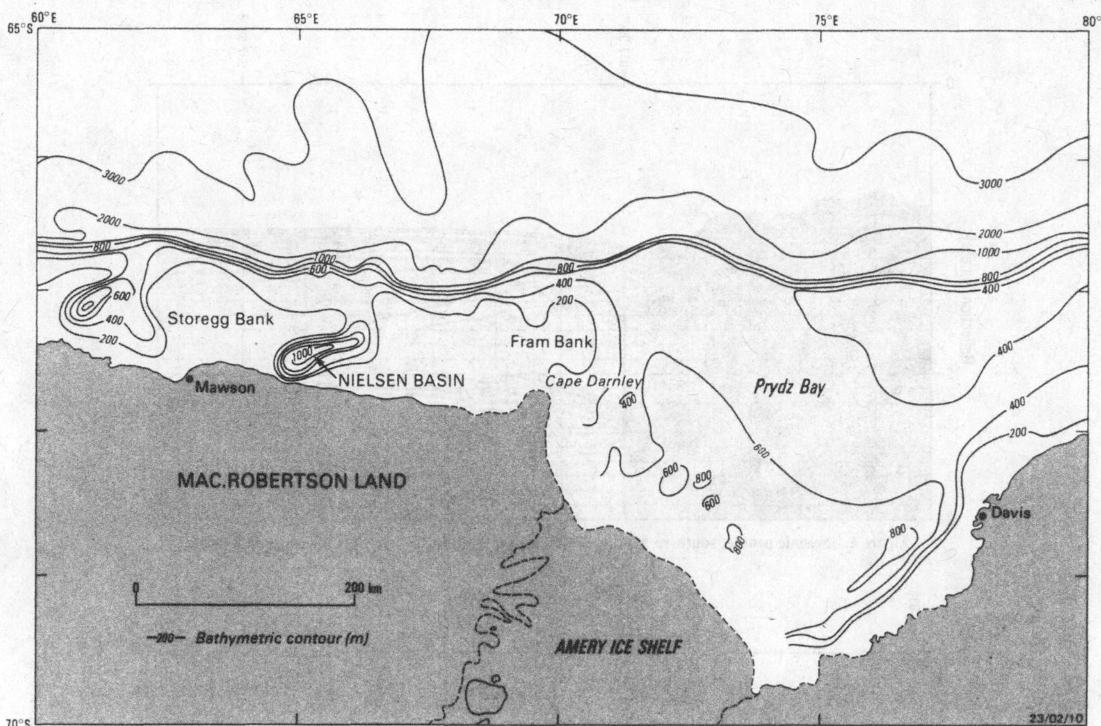


Figure 3. Bathymetry of the survey area. Contours in metres.

abyssal plain boundary is 2600-2800 m deep. Off Prydz Bay the widths of slope and rise are typically 30-50 km and 70-100 km respectively; the greatest widths are attained on the western side of Prydz Bay where the slope/rise boundary is frequently indistinct. West of Cape Darnley the topographic boundaries become obscured by numerous canyons and hills and both slope and rise narrow to a combined width of 40-100 km.

## PRELIMINARY RESULTS

As even simple seismic playback is still incomplete (June, 1982) it is too early for a detailed interpretation to be presented here. Rather it is intended only to present some of the more interesting seismic sections typical of the major structural provinces. The seismic sections illustrated in this paper are single channel and processed for amplitude recovery only.

At this point it is worth commenting on the line orientation and the limits of the survey (Figure 2). Within Prydz Bay, the primary area of interest, lines were oriented so as to be at right angles to what was thought to be the most likely structural trend (i.e. at right angles to the direction of flow of the Lambert Glacier). Lines were taken from as close to the coast as possible north to 66°S at 2500-3000 m water depth, on the lower part of the continental rise. Most time was allocated to this part of the survey. West of Cape Darnley, a zig-zag pattern was adopted to maximise the coverage in the remaining time. In this area the southern ends of lines were limited by the profusion of icebergs which tend to ground in less than 250 m of water.

The nearest stratigraphic drillhole is DSDP Site 268 (63°57′S, 105°9′E), well to the east of the survey area. Consequently we have no direct geological information on the sequences seen in the seismic sections. The dredging carried out during the cruise may help alleviate this problem.

### Prydz Bay

Most of Prydz Bay appears to be underlain by a sedimentary basin. Strong water-bottom multiples preclude any estimate of actual sediment thickness, but the generally quiet magnetics suggest the thickness could be considerable, as Fedorov et al. (1982) believe. It is hoped that future computer processing of the magnetic data, using the technique of Werner Deconvolution (Hsu and Tilbury, 1977) will give estimates of total sediment thickness.

The margins of Prydz Bay are typically underlain by shallow acoustic basement; we do not know if this correponds to true basement. The two most prominent areas of shallow basement are in the northwest, in the vicinity of Cape Darnley, and in the south and southeast near the coast (Figure 4). One of the notable structural features of the bay is the boundary between shallow rugged basement to the south and the gently north-dipping sediments of the main part of the bay. This boundary corresponds to the trough or channel which appears to run the length of the south and southeast margins of the bay, which was referred to in the bathymetry section.

Much of the southern half of the bay is underlain by flat-lying sediments showing little or no sign of faulting or other tectonic disturbance. The sea bed shallows gradually northwards and is mildly though quite distinctly unconformable with the underlying sediments. North of a line approximately through 67°30′S, the northerly dip increases sharply (Figure 5) and the section takes on the appearance of a progradational sequence, generally, though not always, of oblique type.

The only area where faulting is strongly evident is in the extreme southwest of the bay (Figure 6) in the vicinity of the Amery Ice Shelf. On one line only, distinctive tilted fault blocks with a thin sedimentary cover were found. A second line, only 15-30 km to the northeast, strangely enough showed no evidence whatsoever of faulting.

The shelf edge north of Prydz Bay varies considerably in appearance, as noted previously. The illustrated section (Figure 7) is fairly typical, showing strongly dipping reflectors that become parallel to the slope beyond the shelf. A number of strong reflectors can be seen cutting through the shelf edge multiples; unfortunately, these cannot be traced much to the south of the shelf edge.

### Cape Darnley to Mawson

To the west of Cape Darnley most of the shelf is less than 300 m deep. The presence of grounded icebergs allowed only the outer shelf to be surveyed, that area that slopes from about 250 m out to the shelf break; on average, this zone was only 20-30 km wide. Also, the seismic data obtained was badly obscured by water bottom multiples.

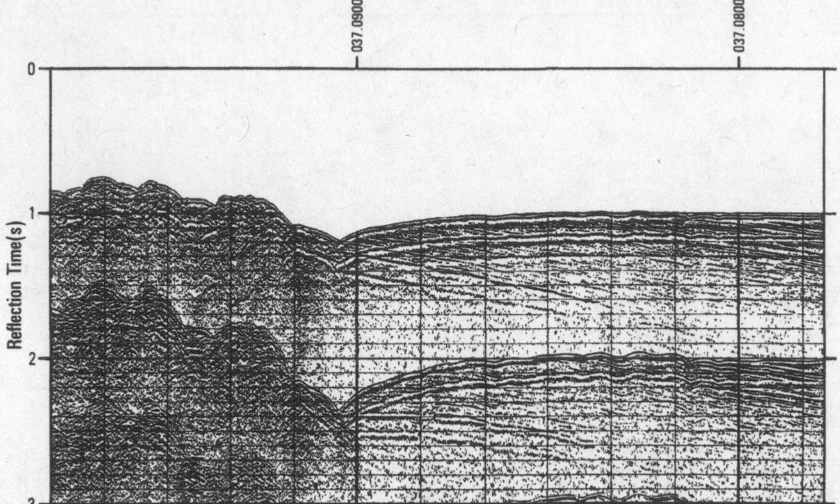


**Figure 4. Seismic profile, southern margin of Prydz Bay, south to the left. Vertical exaggeration 6:1.**

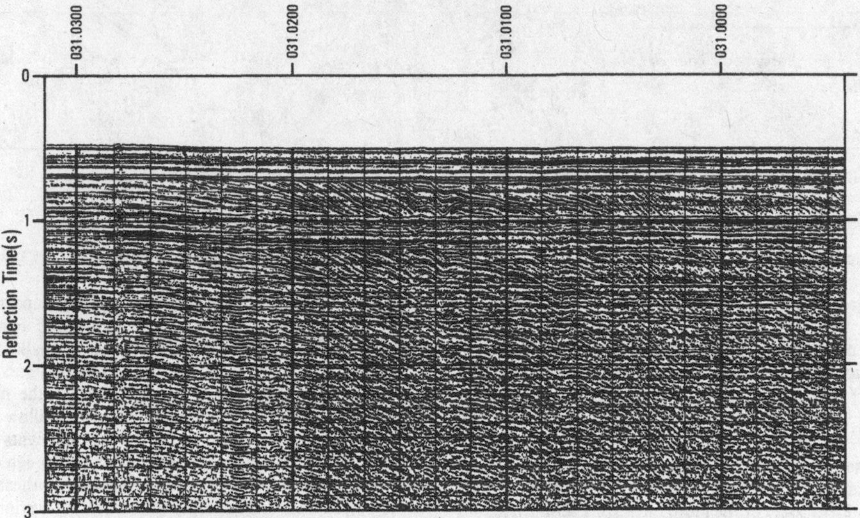


**Figure 5. Seismic profile, northern Prydz Bay, south to the left. Vertical exaggeration 9:1.**

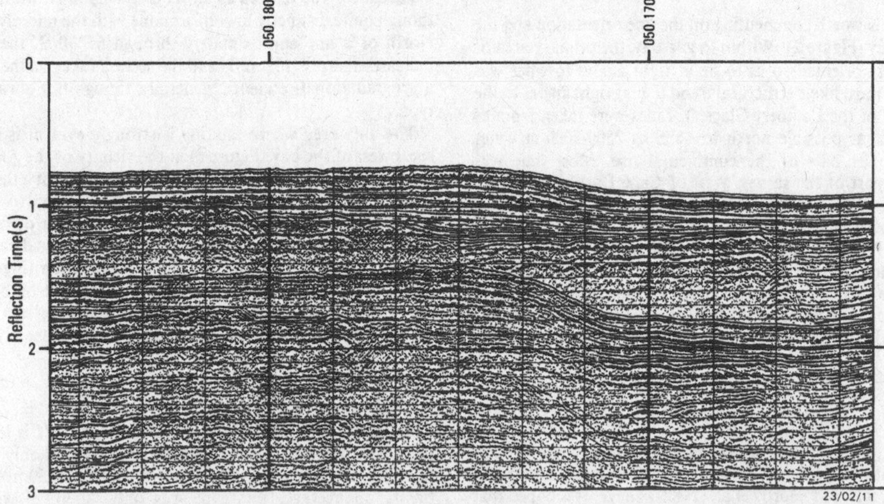


**Figure 6. Seismic profile, southwest Prydz Bay, southeast to the right. Vertical exaggeration 6:1.**

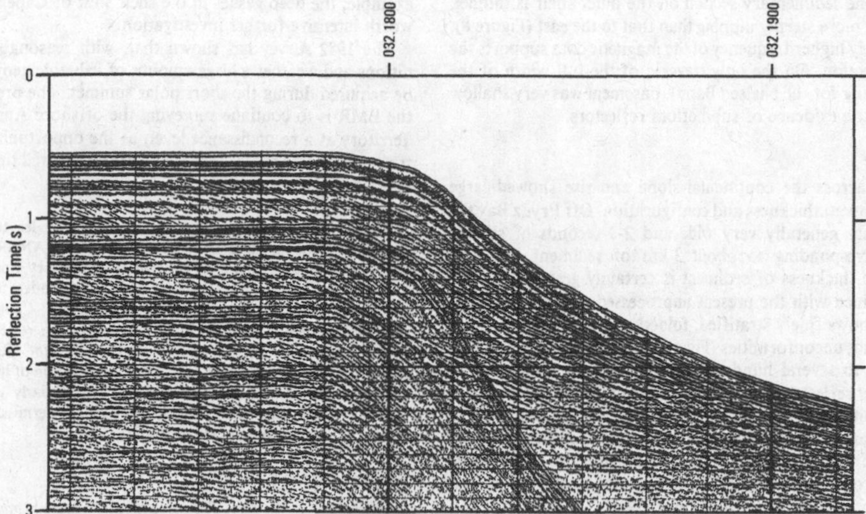


**Figure 7. Seismic profile, shelf edge, northern Prydz Bay, south to the left. Vertical exaggeration 6:1.**

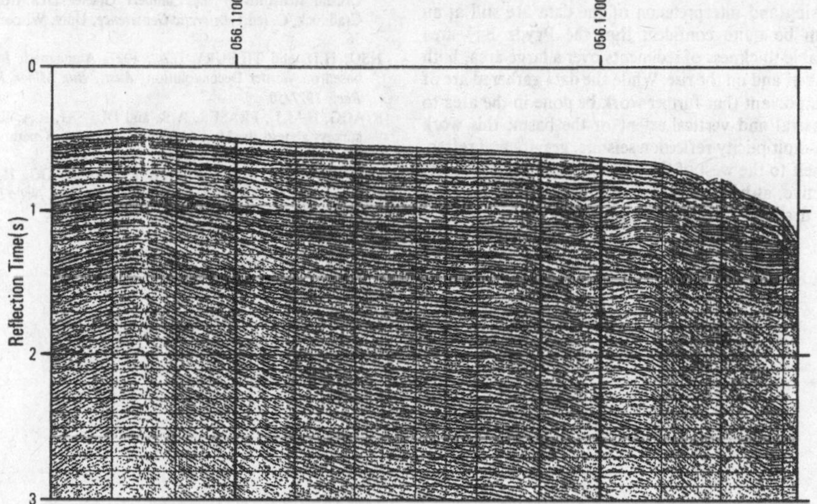


**Figure 8. Seimic profile, continental shelf north of Mawson, southeast to the right. Vertical exaggeration 6:1. The "hump" on the left is due to a course change.**

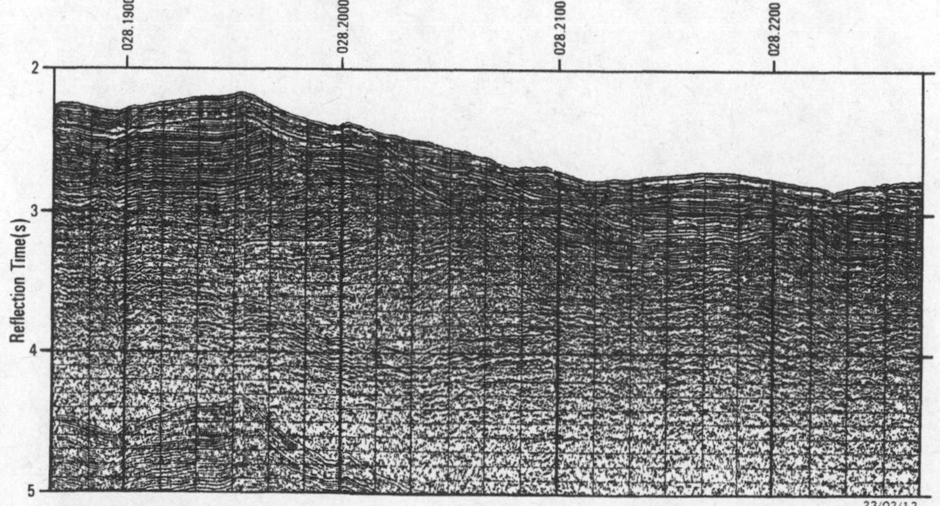


**Figure 9. Seismic profile, continental rise north of Cape Darnley, west to the left. Vertical exaggeration 9:1.**

It appears that the sedimentary section on the outer shelf is thinner, more folded and more steeply dipping than that to the east (Figure 8). The comparatively higher frequency of the magnetic data supports the idea of a thin section. On the only traverse of the full width of the shelf (when looking for the Nielsen Basin), basement was very shallow and there was little evidence of sub-bottom reflectors.

## Slope and Rise

Seismic lines across the continental slope and rise showed large variations in sediment thickness and configuration. Off Prydz Bay the slope and rise are generally very wide and 2-3 seconds of seismic penetration, corresponding to about 3 km of sediment, was not uncommon. The thickness of sediment is certainly greater than the penetration attained with the present unprocessed seismic data. The section mostly shows finely stratified, folded and slumped sediments with many distinct unconformities (Figure 9). Faulting with considerable throws, up to several hundred milliseconds, can sometimes be seen in the deeper reflectors, while minor faulting in the upper part of the section is quite common. The width of the rise and thickness of sediment there, are probably the result of a prolific sediment supply from Prydz Bay. West of Cape Darnley both slope and rise become steeper and narrower, and the seismic data show much thinner and more poorly stratified sediment than further east.

## CONCLUSIONS AND FURTHER WORK

Although processing and interpretation of the data are still at an early stage, we can be quite confident that the Prydz Bay area contains a considerable thickness of sediments over a large area, both on the continental shelf and on the rise. While the data gathered are of good quality, it is important that further work be done in the area to better define the lateral and vertical extent of the basin; this work should include high-multiplicity reflection seismic, gravity and refraction seismic. The shelf to the west of Cape Darnley appears to be less economically attractive, although our limited coverage here makes it more difficult to draw conclusions. Some smaller features, for example, the deep gashes in the shelf west of Cape Darnley, are also worth intensive further investigation.

The 1982 survey has shown that, with reasonably good ice conditions and weather a large amount of valuable geophysical data can be acquired during the short polar summer. The present intention of the BMR is to continue surveying the offshore Australian Antarctic Territory at a reconaissance level, as the opportunity arises, concentrating firstly on those areas which are considered to be more likely to contain sedimentary basins.

*Acknowledgements* The survey was conducted under the auspices of the Australian National Antarctic Research Expedition (ANARE) as a programme approved by the Antarctic Research Policy Advisory Committee (ARPAC). The survey vessel, M.V. *Nella Dan* was under the command of Captain J.B. Jensen of the Danish Lauritzen Line; Captain Jensen and his crew deserve special praise for their handling of the vessel during the survey. Cruise leader was Dr P.G. Quilty, Deputy Director (Research), Antarctic Division. We wish to place on record our appreciation of his efforts and of the assistance given by Antarctic Division staff, particularly Mr Ric Burbury and Mr Mick Webb. This paper is published with the permission of the Director, Bureau of Mineral Resources, Australia.

## REFERENCES

FAIRBRIDGE, R.W., 1966: *The encyclopaedia of oceanography.* Rheinhold, New York.

FEDOROV, L.V., GRIKUROV, G.E., KURININ, R.G. and MASOLOV, V.N., 1982: Crustal structure of the Lambert Glacier area from geophysical data; *in* Craddock, C. (ed.) *Antarctic Geoscience,* Univ. Wisconsin Press, Madison, 931-36.

HSU, H.D. and TILBURY, L.A., 1977: A magnetic interpretation programme based on Werner Deconvolution. *Aust., Bur. Miner. Resour., Geol. Geophys., Rec., 1977/50.*

STAGG, H.M.J., FRASER, A.R. and DULSKI, R.A.W., 1980: Marine magnetic surveys aboard the M.V. *Nella Dan,* 1979/80—Operational Report. *Aust., Bur. Miner. Resour., Geol. Geophys., Rec., 1980/68.*

TILBURY, L.A., WHITWORTH, R. and STAGG, H.M.J., 1980: Operations manual for marine magnetic surveys on the M.V. *Nella Dan,* 1980/81. *Aust., Bur. Miner. Resour., Geol. Geophys., Rec., 1980/84.*

# SEDIMENTARY BASINS OF THE ROSS SEA, ANTARCTICA

F.J. Davey, *Geophysics Division, DSIR, P.O. Box 1320, Wellington, New Zealand.*

K. Hinz and H. Schroeder, *Bundesanstalt fur Geowissenschaften und Rohstoffe, P.O. Box 510153, 3000 Hannover 51, Federal Republic of Germany.*

*Abstract* Seismic refraction and variable angle reflection measurements made using sonobuoys at 98 sites in the Ross Sea define the seismic velocity-depth structure of the sediments and the underlying basement of the Ross Sea continental shelf and margins. 41 new sets of sonobuoy measurements are included in this data set. The data define in greater detail than before two of the three major basins underlying the Ross Sea, that is the Eastern Basin and the Central Trough. New data from the western margin of the Eastern Basin, which detected basement rocks, show that sediment thicknesses were underestimated from the older data, on which basement was not detected. The Central Trough probably extends across the continental shelf edge where sedimentary thicknesses exceeding 4 km are measured. Seismic velocities in excess of 6 km/s were detected at several sonobuoy stations, particularly in the western region suggesting that, perhaps, correlatives of the Jurassic Ferrar Dolerites on land may also occur offshore.

The Ross Sea and the adjacent Ross Ice Shelf cover part of a morphologically depressed region extending across Antarctica from the Ross Sea to the Weddell Sea. This region marks the boundary between the East Antarctic craton and the tectonically younger West Antarctica (Figure 1). It coincides with a zone of thinner continental crust which has been suggested as arising from crustal rifting (e.g., Herron and Tucholke, 1975, Molnar et al., 1975). The western margin of the boundary zone (adjacent to East Antarctica) is formed by the linear mountain chain of the Transantarctic Mountains, along which uplift has been occurring probably since the Jurassic and at an increasing rate in the Cenozoic.

Previous geophysical work in the Ross Sea (Houtz and Meijer, 1970, Houtz and Davey, 1973, Hayes and Davey, 1975, Northey et al., 1975, Wong and Christoffel, 1981, Davey et al., 1982) and on the Ross Ice Shelf (Crary, 1963, Robertson et al., 1982) has shown the existence of thick sedimentary sequences covering a large area in the region. Drillholes in the northern and central Ross Sea (Hayes et al., 1975a,b) demonstrate that these sequences contain sediments of Late Tertiary age. Early Tertiary sediments have recently been found in the McMurdo Sound Sediment and Tectonic Studies (MSSTS)-1 drillhole (Webb, this volume) and as erratics. Houtz and Davey (1973) and Davey et al. (1982) consider that the deeper sediments in the Ross Sea basins would probably be at least Early Tertiary in age possibly as old as Late Cretaceous. Wong and Christoffel (1981) also suggest an Early Tertiary age for the oldest sediments underlying McMurdo Sound.

In this paper we present interpretations of 41 new sets of seismic refraction and variable angle reflection sonobuoy measurements made in the Ross Sea during the 1979/80 season. We combine these with previous geophysical measurements, including 57 sonobuoy measurements reported by Houtz and Davey (1973) and Davey et al. (1982), (Figure 2) to define the broad structure of the sedimentary cover of the Ross Sea continental shelf and margin.

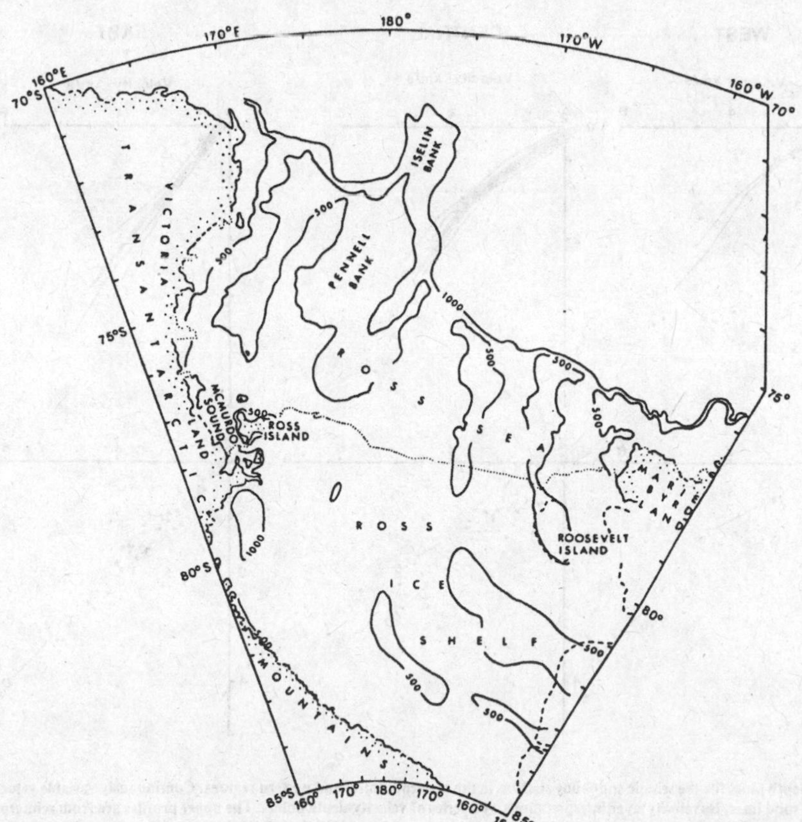


Figure 1. The Ross Sea region of Antarctica showing simplified bathymetry. The dotted line is the northern limit of the Ross Ice Shelf.

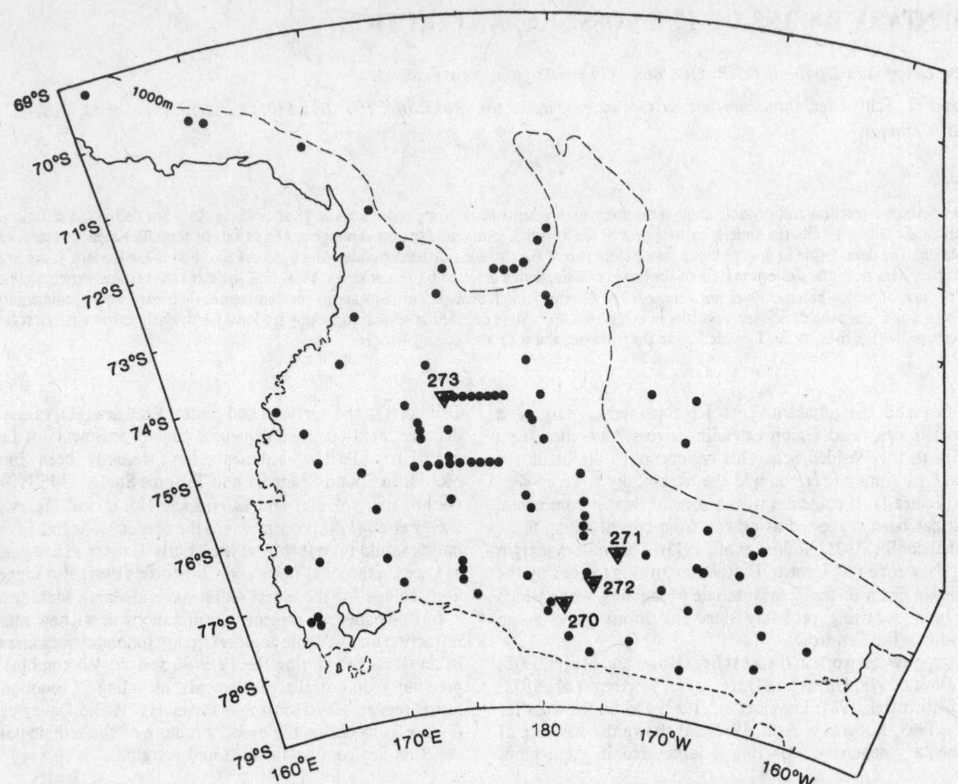


Figure 2. The location of the sonobuoy seismic measurements used in this study (solid circles). DSDP sites 270-273 are marked by open triangles and labelled.

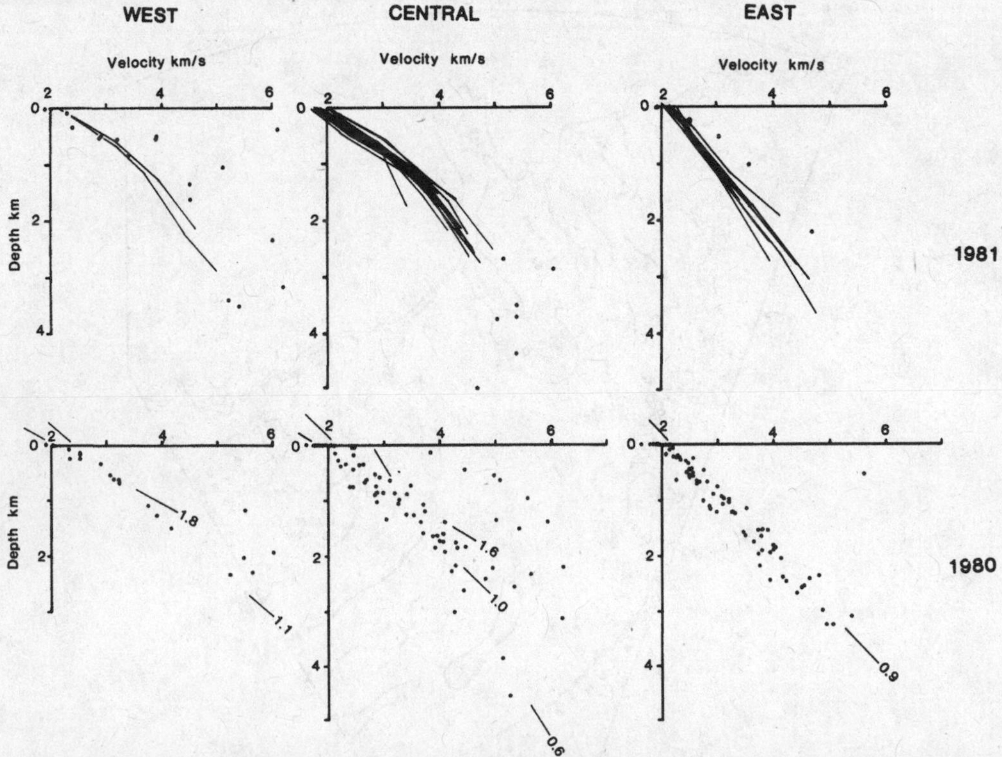


Figure 3. Velocity-depth plots for the seismic sonobuoy stations in the eastern, central and western regions. Continuously variable velocity-depth profiles are shown by solid lines, isovelocity layer interpretations by a series of velocity-depth points. The upper profiles are from reinterpreted Eltanin 52 data and data recorded in 1981 by Davey et al. (1982). The lower plots are from BGR data. Best fit (by eye) linear velocity-depth profiles for sections of the velocity-depth data in the lower plots are shown by solid lines with the gradient marked on them.

**TABLE 1: Seismic Refraction Solutions for the B.G.R. Sonobuoy Measurements in the Ross Sea in 1980**

| Sonobuoy No. | Lat. (S) | Long. | Water depth (m) | $V_1$ / $h_1$ | $V_2$ / $h_2$ | $V_3$ / $h_3$ | $V_4$ / $h_4$ | $V_5$ / $h_5$ | $V_6$ / $h_6$ | $V_7$ / $h_7$ |
|---|---|---|---|---|---|---|---|---|---|---|
| 1 | 69°08.66′ | 161°10.07′ E | 435 | 2.35 | 2.73 | 3.21 | 4.23 | 4.60 | | |
| | | | | 0.41 | 0.20 | 0.71 | 0.48 | | | |
| 2 | 69°54.78′ | 164°58.04′ E | 471 | 2.20 | 2.43 | 2.74 | 2.94 | 3.45 | 4.21 | |
| | | | | 0.14 | 0.36 | 0.50 | 0.33 | 0.82 | | |
| 3 | 70°02.37′ | 165°37.55′ E | 196 | 2.42 | 2.63 | 2.80 | 4.48 | 5.05 | 5.70 | |
| | | | | 0.16 | 0.23 | 0.50 | 0.33 | 0.66 | | |
| 5 | 70°42.11′ | 169°22.50′ E | 326 | 2.79 | 2.96 | 4.58 | | | | |
| | | | | 0.35 | 0.96 | | | | | |
| 6 | 71°42.46′ | 172°01.47′ E | 660 | 2.78 | 3.46 | 4.39 | 5.43 | 6.45 | | |
| | | | | 0.06 | 0.14 | 0.46 | 0.86 | | | |
| 7 | 72°59.73′ | 174°40.35′ E | 348 | 2.14 | 2.68 | 3.52 | 3.88 | 4.26 | 5.25 | 6.73 |
| | | | | 0.67 | 0.58 | 0.42 | 1.33 | 1.52 | 2.35 | |
| 8 | 74°30.14′ | 174°37.72′ E | 522 | 1.89 | 2.13 | 2.87 | 3.41 | 4.09 | 4.46 | |
| | | | | 0.14 | 0.41 | 0.35 | 0.48 | 0.46 | | |
| 9 | 75°24.95′ | 174°35.15′ E | 290 | 1.75 | 2.21 | 2.43 | 2.86 | 3.40 | 3.95 | 4.07 |
| | | | | 0.03 | 0.69 | 0.11 | 0.39 | 0.39 | 0.30 | |
| 10 | 76°42.02′ | 170°01.16′ E | 761 | 2.09 | 2.54 | 3.13 | 3.91 | 5.24 | | |
| | | | | 0.23 | 0.37 | 0.65 | 1.07 | | | |
| 11 | 76°42.74′ | 177°00.01′ E | 304 | 1.81 | 2.19 | 2.93 | 3.26 | 5.41 | 6.22 | |
| | | | | 0.28 | 0.29 | 0.42 | 0.51 | 0.72 | | |
| 12 | 76°42.13′ | 175°08.75′ W | 565 | 2.04 | 2.33 | 2.62 | 3.71 | 4.00 | 4.53 | |
| | | | | 0.29 | 0.38 | 0.86 | 0.35 | 0.67 | | |
| 13 | 76°44.71′ | 167°25.97′ W | 471 | 2.19 | 2.57 | 3.10 | 3.75 | | | |
| | | | | 0.49 | 0.60 | 1.10 | | | | |
| 14 | 77°09.62′ | 165°40.47′ W | 428 | 2.18 | 2.33 | 2.50 | 2.84 | 3.49 | | |
| | | | | 0.25 | 0.35 | 0.52 | 0.50 | | | |
| 15 | 77°08.26′ | 172°00.00′ W | 486 | 1.63 | 2.20 | 2.55 | 2.73 | 3.65 | 4.15 | 4.93 |
| | | | | 0.05 | 0.40 | 0.54 | 0.75 | 0.62 | 0.86 | |
| 16 | 77°07.61′ | 176°38.20′ W | 616 | 1.91 | 2.28 | 2.81 | 3.52 | 3.90 | 4.80 | |
| | | | | 0.24 | 0.43 | 0.45 | 0.44 | 0.77 | | |
| 18 | 77°06.72′ | 178°57.09′ E | 652 | 1.92 | 3.82 | 4.42 | | | | |
| | | | | 0.12 | 0.32 | | | | | |
| 19 | 77°04.43′ | 170°54.15′ E | 761 | 2.35 | 2.65 | 3.48 | | | | |
| | | | | 0.36 | 0.35 | | | | | |
| 20 | 75°59.99′ | 170°01.12′ E | 631 | 2.06 | 2.52 | 3.08 | 3.76 | 4.17 | 5.63 | 6.45 |
| | | | | 0.17 | 0.36 | 0.56 | 0.39 | 0.82 | 1.57 | |
| 21 | 75°59.84′ | 174°28.58′ E | 565 | 2.20 | 2.58 | 3.10 | 3.70 | 4.30 | 4.80 | 6.20 |
| | | | | 0.36 | 0.29 | 0.43 | 0.77 | 0.56 | 0.71 | |
| 22 | 76°00.08′ | 178°56.73′ E | 522 | 1.95 | 2.48 | 2.68 | 2.85 | 2.98 | 3.90 | |
| | | | | 0.45 | 0.16 | 0.12 | 0.15 | 0.95 | | |
| 23 | 75°59.79′ | 177°30.13′ W | 623 | 2.10 | 2.45 | 3.28 | 4.02 | 4.39 | | |
| | | | | 0.30 | 0.90 | 0.62 | 0.86 | | | |
| 24 | 76°05.12′ | 167°19.77′ W | 478 | 2.18 | 2.46 | 3.45 | 3.93 | | | |
| | | | | 0.50 | 1.08 | 0.88 | | | | |
| 25 | 76°47.27′ | 169°40.37′ W | 507 | 1.95 | 2.30 | 2.50 | 2.93 | 3.95 | 4.85 | |
| | | | | 0.20 | 0.24 | 0.44 | 1.07 | 1.04 | | |
| 26 | 77°54.38′ | 173°41.46′ W | 544 | 1.99 | 2.21 | 3.20 | 3.78 | 4.12 | 4.51 | |
| | | | | 0.22 | 0.75 | 0.59 | 0.36 | 0.53 | | |
| 27 | 78°08.23′ | 176°41.00′ W | 602 | 2.05 | 2.30 | 2.48 | 2.75 | 3.00 | 5.14 | |
| | | | | 0.20 | 0.13 | 0.13 | 0.31 | 0.63 | | |
| 29 | 76°40.67′ | 173°17.25′ W | 507 | 1.98 | 2.13 | 2.55 | 2.99 | 3.72 | 4.21 | |
| | | | | 0.10 | 0.47 | 0.64 | 0.76 | 0.49 | | |
| 31 | 74°45.00′ | 179°09.00′ E | 341 | 1.93 | 4.99 | 5.56 | | | | |
| | | | | 0.54 | 0.41 | | | | | |
| 32 | 76°05.40′ | 179°15.60′ E | 413 | 1.85 | 2.65 | 3.08 | 3.30 | | | |
| | | | | 0.63 | 0.29 | 0.30 | | | | |
| 33 | 77°24.00′ | 178°34.80′ W | 638 | 2.16 | 2.58 | 5.61 | | | | |
| | | | | 0.23 | 0.28 | | | | | |
| 34 | 76°57.00′ | 177°00.00′ W | 609 | 1.91 | 2.49 | 3.21 | 3.98 | 4.64 | 5.38 | |
| | | | | 0.53 | 0.49 | 0.78 | 0.61 | 0.69 | | |
| 35 | 76°00.00′ | 175°34.17′ W | 522 | 1.90 | 2.25 | 2.60 | 2.83 | 3.80 | | |
| | | | | 0.64 | 0.05 | 0.48 | 0.73 | | | |
| 36 | 75°35.88′ | 177°40.95′ W | 536 | 1.80 | 2.08 | 3.10 | 3.80 | 5.05 | | |
| | | | | 0.18 | 0.76 | 0.58 | 1.71 | | | |
| 37 | 74°46.26′ | 179°08.53′ E | 341 | 1.76 | 1.94 | 3.04 | 5.00 | | | |
| | | | | 0.96 | 0.38 | 0.01 | | | | |
| 38 | 73°07.34′ | 179°59.95′ E | 547 | 2.13 | 2.86 | 4.06 | 4.94 | 5.62 | | |
| | | | | 0.88 | 0.73 | 0.60 | 0.15 | | | |
| 39 | 74°00.01′ | 179°01.62′ E | 261 | 1.94 | 2.33 | 5.09 | 5.94 | | | |
| | | | | 0.35 | 0.29 | 0.72 | | | | |
| 40 | 74°00.19′ | 173°39.27′ E | 457 | 1.90 | 2.35 | 3.23 | 5.50 | 6.00 | 6.80 | |
| | | | | 0.26 | 0.35 | 0.55 | 0.75 | 0.27 | | |
| 41 | 74°33.55′ | 172°30.25′ E | 529 | 2.04 | 2.33 | 2.91 | 3.25 | 5.48 | | |
| | | | | 0.07 | 0.26 | 0.35 | 1.34 | | | |
| 42 | 75°00.11′ | 174°41.65′ E | 341 | 2.07 | 2.40 | 3.25 | 4.00 | 4.40 | | |
| | | | | 0.73 | 0.29 | 0.68 | 0.92 | | | |
| 43 | 73°10.82′ | 179°53.43′ W | 493 | 1.62 | 2.47 | 3.72 | 4.29 | 5.34 | | |
| | | | | 0.04 | 1.14 | 0.56 | 0.81 | | | |
| 44 | 72°39.32′ | 174°59.97′ E | 399 | 2.30 | 2.47 | 2.86 | 3.69 | 4.29 | 5.12 | |
| | | | | 0.20 | 0.82 | 0.58 | 0.56 | 1.67 | | |
| 45 | 72°16.20′ | 173°07.83′ E | 507 | 1.98 | 2.25 | 3.18 | 3.65 | 4.05 | 4.20 | |
| | | | | 0.37 | 0.50 | 0.52 | 0.33 | 0.56 | | |

V is the seismic velocity in km/s and
h the thickness in km of the isovelocity layers derived from the seismic measurements.

## THE SEISMIC MEASUREMENTS

The seismic records were obtained on board MS Explora during the Bundesanstalt für Geowissenschaften und Rohstoffe (BGR) Ross Sea Expedition early in 1980. Fairfield Industries SB76 sonobuoys were used with a 23 litre airgun array, fired at 50 m intervals, as sound source. The data were digitally recorded and displayed, after amplitude correction for spherical divergence and filtering, as variable area records. These seismic data have been interpreted in terms of isovelocity planar layers (Table 1). Dip corrections have been made where possible from associated seismic reflection records. This method of interpretation tends to underestimate the total sedimentary thickness if the true seismic velocity in the sediments increases continuously with depth, as noted on some data sets in this region (Houtz and Davey, 1973).

Davey et al. (1982) recorded their data using Magnavox SSQ 41B sonobuoys and a two litre airgun fired at 50 m intervals. The data were filtered and recorded as a variable density record. They interpreted their data in terms of a continuous velocity-depth function unless the time distance curve was clearly segmented, in which case an isovelocity plane layer interpretation was used. Some of the BGR records show first arrival data that are clearly formed of straight line segments, indicating isovelocity layers, and this tends to be confirmed by the continuation of the arrival segments as second or later arrivals. This is especially the case for data from the west or north of the central Ross Sea. Other data sets in the region show an apparently curved alignment of the first arrival data, sometimes with a linear initial (or close range) segment to the refracted arrival data.

In Figure 3 the velocity and depth data from all the sonobuoy solutions are plotted. The new BGR data (lower) are plotted separately from the reinterpreted Eltanin data (Houtz and Davey, 1973) and data of Davey et al. (1982) (upper). The data have been split into groups corresponding to the three regions of Houtz and Davey (1973) and the velocity-depth gradients in each region are similar. The sediment velocities tend to lie close to a simple velocity-increase-with-depth curve for each region, with other observed velocities tending to plot well off this curve and on the higher velocity side. These higher

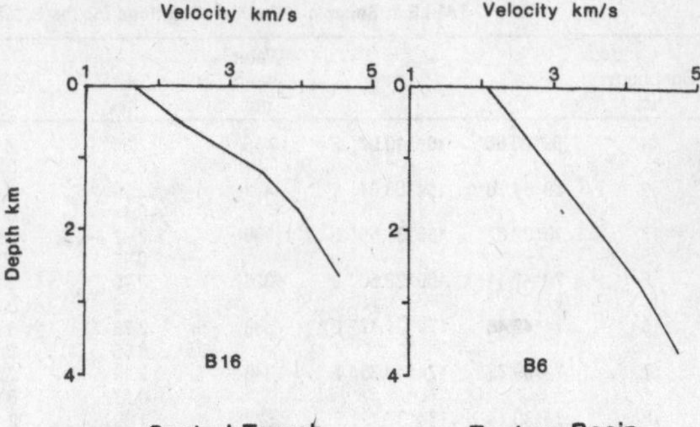


**Figure 4. Typical velocity depth profiles for the Central Trough (Buoy B16) and Eastern Basin (Buoy B6).**

velocities are usually above 5 km/s and are thus presumed to be associated with metamorphic basement. A simple linear velocity depth profile with a low gradient is observed for the data from eastern Ross Sea. A velocity depth profile consisting of several linear segments fits the data from the central Ross Sea with higher gradients in the shallower part of the sedimentary section. In the western Ross Sea the sedimentary section appears to be formed by isovelocity layers. Typical velocity-depth profiles for the Eastern Basin and Central Trough are shown in Figure 4.

Basement velocities appear to be greater than 5 km/s and several apparent velocities greater than 6 km/s were observed. These latter velocities were not well constrained as no dip information for the refractor interfaces was available. They may correspond to updip velocities but their wide distribution and their limited range in values suggest that they may be close to true velocities.

The distribution of sediment thickness in the Ross Sea deduced from the sonobuoy results given above and from data in Houtz and Davey (1973) and Wong and Christoffel (1981) are shown in Figure 5,

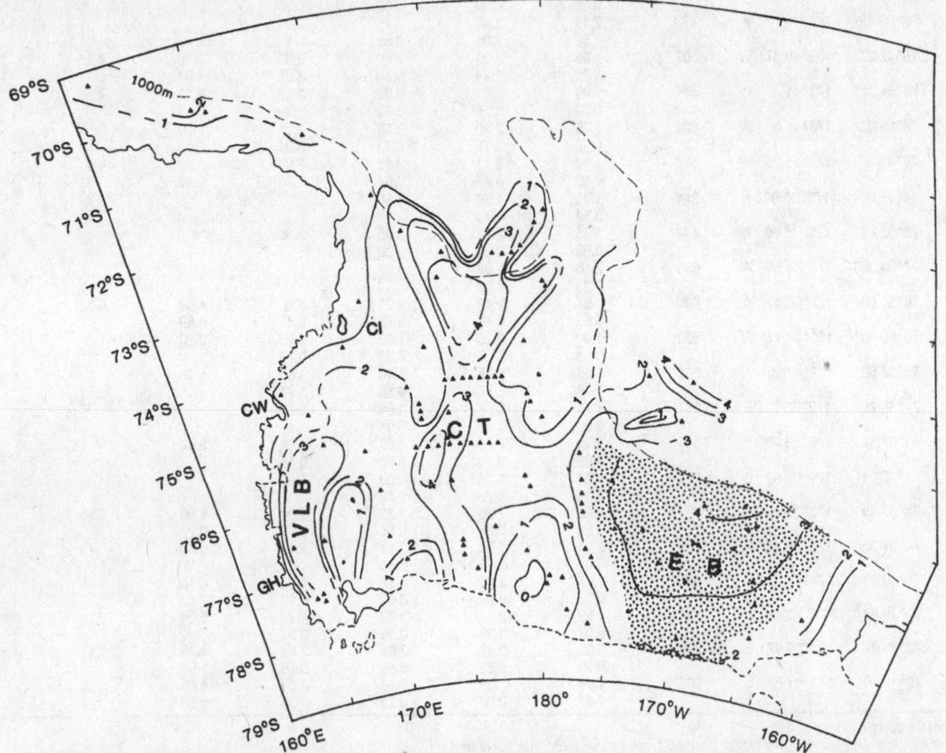


**Figure 5. Isopachs of total sediment thickness (km) for the Ross Sea continental shelf and margin. The location of the sonobuoy stations are marked by solid triangles. Isopachs are dashed where basement velocities were not recorded on the sonobuoy measurements, apart from in the Eastern Basin where the appropriate region is shaded. These isopachs represent minimum sediment thicknesses. VLB—Victoria Land Basin; CT—Central Trough; EB—Eastern Basin; CI—Coulman Island; CW—Cape Washington; GH—Granite Harbour.**

with the location of the sonobuoy measurements. The results delineate more precisely the three major basins underlying the Ross Sea in which sedimentary strata are in excess of 3 km thick. These are the Eastern Basin, the Central Trough and the Victoria Land Basin. In places the isopachs depict minimum thicknesses as basement rocks were not detected on the seismic measurements. Velocities over 5 km/s in the case of a layered velocity interpretation, or velocities of the deep refractor below the rocks giving the fastest arrivals on the curved time-distance plots of Davey et al. (1982), are assumed to be associated with basement rocks. In the Eastern Basin where no basement velocities were detected the contours within the shaded region are interpreted minimum sediment thicknesses.

In Figure 6 the locations of sonobuoy measurements in which basement velocities were detected are shown along with the smallest basement velocity detected at each station. The lack of observed basement velocities in the eastern Ross Sea is striking, and perhaps indicates a much greater depth to basement. Basement velocities are varied and lie in the range 4.7-6.8 km/s, the lower values typical of the metamorphic rocks found around the Ross Sea. Seismic velocities in excess of 6 km/s were recorded at 13 sonobuoy stations which are delineated by the diamond line in Figure 6. In several cases these velocities will correspond to sub-basement rock layers. Rocks with a seismic velocity of greater than 6 km/s thus occur widely under the western Ross Sea. Their high velocity suggests they may be offshore correlatives of the Jurassic Ferrar Dolerites which occur extensively in the Transantarctic Mountains.

## THE SEDIMENTARY BASINS

### The Eastern Basin

The Eastern Basin underlies most of the Ross Sea to the east of 180° and extends from the continental slope southwards under the Ross Ice Shelf (Davey, 1981). The sonobuoy refraction results presented here are consistent with the depth and shape of the basin defined by Davey et al. (1982). No basement velocities were definitely detected on the sonobuoy data in the deeper part of the basin. Metamorphic basement must lie at a greater depth. The new data define the western margin of the basin more precisely as basement rocks are detected down to depths of over 3 km in this region. Sonobuoy stations, 33, 12 and 34 are close to Deep Sea Drilling Project (DSDP) sites 270, 271 and 272 respectively and show similar velocities over the upper few hundred metres of sediments to the measurements obtained on the drillcore by Hayes et al. (1975a).

Seismic reflection data show that the sedimentary layers dip gently towards the centre of the basin and towards the shelf edge where the dip steepens sharply. The section is thought to be building outwards as a pro-grading sedimentary wedge. Around the margin of the basin, where the sedimentary layers dip significantly, there is a marked angular unconformity of Late Miocene to Early Pliocene age overlain by a thin cover of sediments built up in low north-south trending banks. Broad folds striking approximately north-south lie along both east and west margins of the basin (Houtz and Davey, 1973).

DSDP data (Hayes et al., 1975a) along the western margin of this basin show thin subaerial or shallow water sediments of Oligocene age directly overlies metamorphic basement. These sediments are overlain by Upper Oligocene and Miocene marine glacial sediments which are interpreted as indication of contemporary depression of the region and the onset of continental glaciation. The thick sediments detected in the basin are probably Oligocene or younger in age. The cause of the downwarping of the basin is not clear. It may have arisen from crustal loading by these Upper Tertiary glacial sediments since the onset of continental glaciation (Davey et al., 1982) or alternatively may be the result of tectonic subsidence. As basement rocks have not been detected on the seismic data, an unknown thickness of older sediments is presumed to exist beneath this young section.

### The Central Trough

The Central Trough trends approximately north-south through the central Ross Sea along 175°E from the Ross Ice Shelf to the continental shelf edge where the new data show at least 4.5 km of sediments. It may continue southwards under the Ross Ice Shelf. The single channel seismic reflection data yield no evidence of a deep

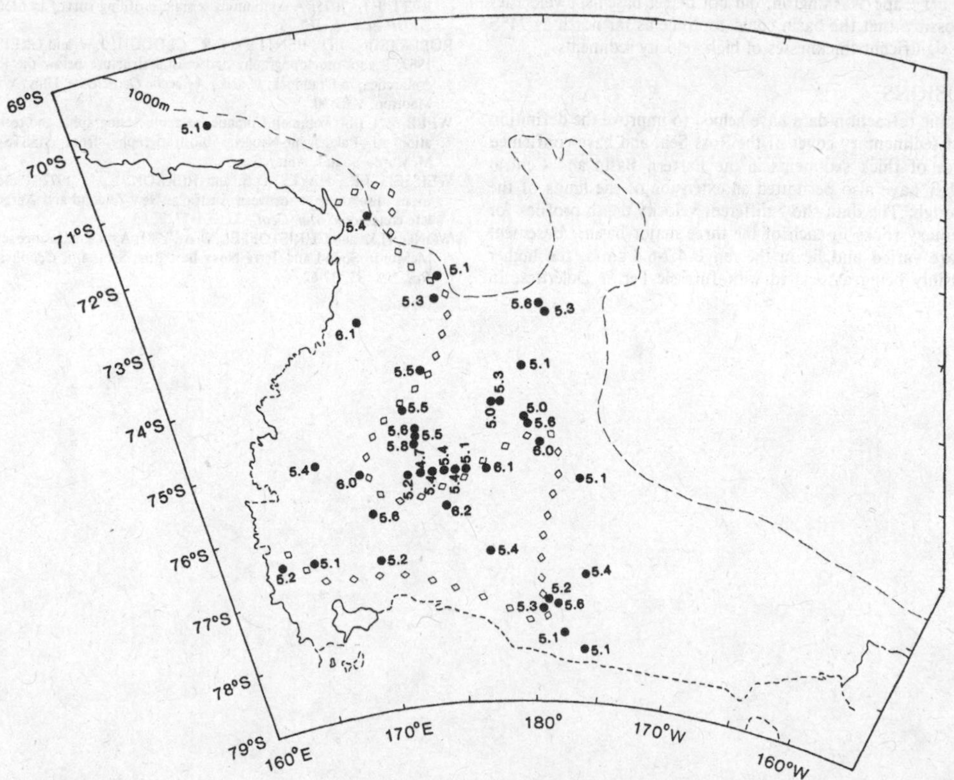


**Figure 6. The location of sonobuoy measurements which recorded basement velocities and the basement velocities measured. The region where basement or sub-basement velocities are greater than 6 km/s lies to the west of the diamond line.**

trough (Houtz and Davey, 1973), but BGR multichannel seismic data shows the presence of one that is thought to contain pre-Oligocene sediments. Its delineation in this paper is based primarily on sonobuoy seismic refraction measurements. The trough appears to be irregular in width and depth, but this may be an artifact of the limited data coverage. The basement ridge along its eastern margin, along 180°, is broken by a depression in excess of 2 km deep at 76°S. At its northern end the trough widens to cover most of the western Ross Sea margin and is apparently continuous with the sedimentary basin (Houtz and Davey, 1973) underlying the slope between Iselin Bank and the western Ross Sea Shelf edge. A postulated basement high (Houtz and Davey, 1973) at the eastern end of the western Ross Sea continental shelf edge may split the northern end of the Central Trough into two, with a minor basin to the east lying alongside the Iselin Bank (Figure 5). The sonobuoy data indicates that the graben would be primarily in the oldest sediments.

The age of the deeper sediments is unknown. Drilling results at DSDP site 273 showed sediments of possible Early Miocene age at a depth of 365 m (Hayes et al., 1975b). Sonobuoy eight lies close to site 273 and shows low seismic velocities, 1.9-2.1 km/s, for the upper sediments, similar to those obtained on drillcore (Hayes et al., 1975b). The seismic reflection data, which extend to a depth of about 1000 m, do not show any offsets which might correspond to a graben structure at depth (Houtz and Davey, 1973, Figure 9). The sediments in the deeper part of the trough are thus probably much older than Miocene in age. Davey (1981) postulates an Early Tertiary age, suggesting that the graben formed as a failed arm of the spreading centre which abutted the continental margin of the western Ross Sea at about 55 Ma (Weissel et al., 1977).

### The Victoria Land Basin

The Victoria Land Basin lies along the Transantarctic Mountains from Coulman Island to McMurdo Sound, being apparently deepest off the Granite Harbour-Cape Washington region (Figure 5). We have no new data in this region. However we note that the isopachs of Davey et al. (1982) show the basin closing off at its northern end of the basin at about 75°S. The two sonobuoy stations immediately to the north, off Cape Washington, did not detect basement velocities and it is possible that the basin could continue as far north as 74°S containing significant thicknesses of high velocity sediments.

## CONCLUSIONS

New seismic refraction data have helped to improve the definition of the total sedimentary cover of the Ross Sea, and have confirmed the presence of thick sediments in the Eastern Basin and Central Trough. They have also permitted an extension of the limits of the Central Trough. The data show different velocity depth profiles for the sedimentary rocks in each of the three major basins. Basement velocities are varied and lie in the range 4.7-6.8 km/s, the higher values possibly being associated with Jurassic Ferrar Dolerites. In addition the tectonic history of each basin appears to be different; this may be the reason for differences between the characteristic sediment velocity structures of each basin. The Eastern Basin, or the upper few kilometres of it, appears to result from downwarping of the crust caused by the accumulation of Late Cenozoic marine glacial sediments in a pre-existing trough, which perhaps contains Late Cretaceous-Early Tertiary sediments. The Central Trough possibly formed as a "failed" rift in the Eocene spreading episode in the southwest Pacific Ocean. The Victoria Land Basin appears to be intimately related to the uplift of the Transantarctic Mountains and may thus show greatest development in the Late Cenozoic.

## REFERENCES


CRARY, A.P., 1963: Marine sediment thickness in the eastern Ross Sea area, Antarctica. *Geol. Soc. Am., Bull., 72,* 787-90.

DAVEY, F.J., 1981: Geophysical studies in the Ross Sea region. *R. Soc. N.Z. J., 11,* 465-79.

DAVEY, F.J., BENNETT, D.J. and HOUTZ, R.E., 1982: Sedimentary basins of the Ross Sea, Antarctica. *N.Z. J. Geol. Geophys., 25,* 245-55.

HAYES, D.E. and DAVEY, F.J., 1975: Geophysical study of the Ross Sea, Antarctica. *Initial reports of the Deep Sea Drilling Project 28,* 887-908. U.S. Govt. Printing Office, Washington, D.C..

HAYES, D.E., FRAKES, L.A., BARRETT, P.J., BURNS, D.A., CHEN, P., FORD, A.B., KANEPS, A.G., KEMP, E.M., MCCOLLUM, D.W., PIPER, D.J., WALL, R.E. and WEBB, P.N., 1975a: Sites 270, 271, 272. *Initial reports of the Deep Sea Drilling Project, 28,* 211-234. U.S. Govt. Printing Office, Washington, D.C.

HAYES, D.E., FRAKES, L.A., BARRETT, P.J., BURNS, D.A., CHEN, P., FORD, A.B., KANEPS, A.G., KEMP, E.M., MCCOLLUM, D.W., PIPER, D.J., WALL, R.E. and WEBB, P.N., 1975b: Site 273. *Initial reports of the Deep Sea Drilling Project, 28,* 335-43. U.S. Govt. Printing Office, Washington, D.C.

HERRON, E.M. and TUCHOLKE, B.E., 1975: Seafloor magnetic pattern and basement structure in the southeastern Pacific. *Initial reports of the Deep Sea Drilling Project, 35,* 263-78. U.S. Govt. Printing Office, Washington, D.C.

HOUTZ, R.E. and DAVEY, F.J., 1973: Seismic profiler and sonobuoy measurements in the Ross Sea, Antarctica. *J. Geophys. Res., 78,* 3448-68.

HOUTZ, R.E. and MEIJER, R., 1970: Structure of the Ross Sea Shelf from profiler data. *J. Geophys. Res., 75,* 6592-7.

MOLNAR, P., ATWATER, T., MAMMERICKX, J. and SMITH, S.M., 1975: Magnetic anomalies, bathymetry and tectonic evolution of the South Pacific since the Late Cretaceous. *R. Astron. Soc., Geophys. J., 49,* 383-420.

NORTHEY, D.J., BROWN, C., CHRISTOFFEL, D.A., WONG, H.K. and BARRETT, P.J., 1975: A continuous seismic profiling survey in McMurdo Sound. *DVDP Bull., 5,* 167-79.

ROBERTSON, J.D., BENTLEY, C.R., CLOUGH, J.W. and GREISCHER, L.L., 1982: Seabottom topography and crustal structure below the Ross Ice Shelf, Antarctica; *in* Craddock, C. (ed.) *Antarctic Geoscience.* Univ. Wisconsin Press, Madison, 1083-90.

WEBB, P.N. (this volume): Climatic, palaeo-oceanographic and tectonic interpretation of Palaeogene-Neogene biostratigraphy from MSSTS-1 drill hole, McMurdo Sound, Antarctica.

WEISSEL, J.K., HAYES, D.E. and HERRON, E.M., 1977: Plate tectonic synthesis: displacements between Australia, New Zealand and Antarctica since the late cretaceous. *Mar. Geol., 25,* 231-77.

WONG, H.K. and CHRISTOFFEL, D.A., 1981: A reconnaissance seismic survey of McMurdo Sound and Terra Nova Bay, Ross Sea. *Am. Geophys. Union, Ant. Res. Ser., 33,* 37-62.

# SEDIMENTARY BASINS OF ANTARCTICA AND THEIR PRELIMINARY STRUCTURAL AND MORPHOLOGICAL CLASSIFICATION

V.L. Ivanov, *Department of Antarctic Geology and Mineral Resources, VNIIOkeangeologia, Leningrad, USSR.*

*Abstract* The knowledge of sedimentary basins of Antarctica is important to the understanding of the evolution of the entire polar segment of the Earth. These basins occupy major bedrock depressions under marginal seas or ice. Hence, direct information is difficult to obtain and geophysical studies are of considerable importance. A preliminary structural and morphological classification of the basins and some ideas about their genesis are proposed. Sedimentary basins of East Antarctica are developed on an ancient stable platform and fall into intracratonic (Wilkes and Victoria Land) or pericratonic (Prydz Bay—Amery Ice Shelf) types. The basins known from West Antarctica are confined to a frontal zone of a geosynclinal fold belt (Bellingshausen and Amundsen Seas) and those of boundary type situated in a craton-geosynclinal belt transition area (Weddell and Ross Seas). The study of the latter is of prime importance because they give clues to the relationship and pattern of transition between the Gondwanian and Pacific structures in Antarctica.

## EAST ANTARCTIC BASINS

The structure of the subglacial Wilkes and Victoria Land basins seems to be very simple, although this may be due to the paucity of data available (Figure 1). Tectonically, these basins occupying vast bed-rock depressions up to 500-1000 m deep exemplify simple platform structures of the syneclise type, that is areas of long-term and persistent downwarping and accumulation of sediments directly on Precambrian crystalline basement. Some geophysical evidence (Drewry, 1976; Steed and Drewry, 1982) implies subsidence of the folded basement surface to 5-8 km from bedrock elevations of about 1 km. Radio-echo sounding investigations of the Wilkes Land basin established steep gradients of the bedrock surface suggesting the presence of superimposed active graben-type structures. The occurrence at the base of the section of the Beacon, possibly coal measures, not subject to subsequent tectonic dislocations and conformably overlain by "coilogenic" (as distinct from "orogenic" and implying formation of extensive crustal downwarps with thick sedimentary fill; the term was introduced by Spizharsky (1973) and is widely used in the Soviet geological literature.) Meso-Cenozoic low density sediments has been already suggested (Grikurov, 1978; Steed and Drewry, 1982). Recent calculations show that without ice load the entire area of the basins discussed would be about 500-1000 m a.s.l. (Figure 2). Therefore, land above sea level at the site of the present basins can be inferred for Early Cenozoic time (prior to glaciation) and hence the predominance of Palaeozoic-Mesozoic sediments in their section. If we consider the Victoria Land basin, in particular its eastern part, as a compensatory deep adjacent to the epiplatform orogenic belt of the Transantarctic Mountains, then the section could contain older molasse-type sequences due to the Ross Orogeny, and young (Andean) molasse sequences resulting from the uplift of the Transantarctic Mountains.

The Prydz Bay basin is a pericontinental riftogenic sedimentary basin formed at the junction of old continental and young oceanic structures due to the destruction of the Antarctic platform because of the formation of surrounding oceanic basins. Morphologically the Prydz Bay basin is a shallow depression on the Sodruzhestvo Sea shelf stretching north-eastwards for more than 200 km with a width of over 100 km. The depth to the magnetic basement is 10-12 km. A basement rise, which separates the depression from the Lambert Glacier graben, strikes meridianally far inland.

No direct evidence has yet been obtained for a sedimentary infilling of the Prydz Bay basin; however some assumptions can be based on geological and geophysical data resulting from Soviet studies in the Lambert Glacier area (Ravich et al., 1978; Federov et al., 1982). The Lambert Glacier graben as an inland extension of the Prydz Bay basin is a typical continental rift dissecting the ancient arch-like bulge of the Archaean crystalline basement. The rift is clearly reflected in the bedrock topography in the form of branching step-wise grabens with a maximum depth of the surface down to 1.5 km b.s.l., as well as in the consolidated crustal relief where maximum submergence reaches 8 km. The crustal thicknesses are respectively 22-25 km and 40 km in the axial graben area and on the flanks. A major deep fault cuts undeformed Permian deposits in the Beaver Lake area and is healed by Early Cretaceous (130-110 Ma) alkali-ultrabasic intrusions. This time is postulated for the initiation of rift and its sedimentary infilling (Fedorov et al., 1982). Recently a rifting of the Earth's crust was shown to be preceded by a long-term (one or two geologic periods) stage of a "trough" development of rift structure corresponding to the existence of a large river valley with a subsequent shallow marine transgression (Johnstone, 1981). Such a section is marked by the presence of coal measures at the base. In this case, Permian coal measures resting on the Early Precambrian basement in the Beaver Lake area may represent basal beds accumulated at the beginning of the "trough" stage which continued until the Late Mesozoic and only then was it succeeded by a "riftogenic" stage. The Lambert Glacier riftogenic structure seems to evolve in a north-south direction, that is inland from the ocean. This fact is responsible for the great depth and width of the graben in the north and higher mean density of sediments (2.45 g/cm³ to the north and 2.30 g/cm³ to the south). Thus, one may suggest that sedimentation in the Sodruzhestvo Sea shelf began not later than the close of the Palaeozoic and a wide spectrum of "pre-rift" facies including coal, deltaic, and shallow marine facies may be inferred to occur under Mesozoic molasse sediments associated with later stages of riftogenic development.

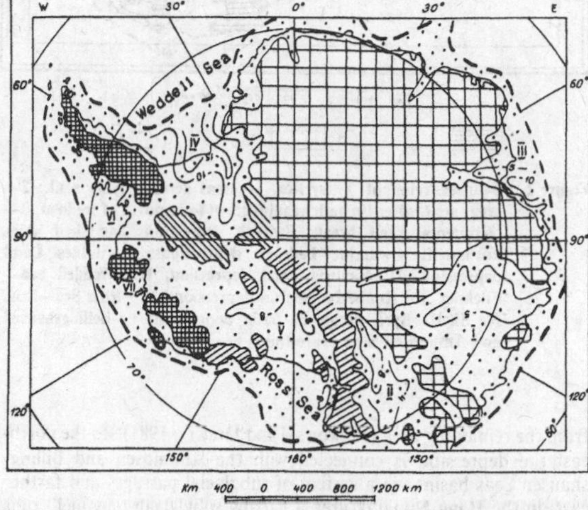


**Figure 1. Sedimentary basins of Antarctica. 1—areas occupied by sedimentary basins; 2—depth to the magnetic basement, in km; 3-6—bedrock elevations composed of hetrogeneous folded basement, including: 3—of Precambrian consolidation; 4—of Ross consolidation; 5—of Late Palaeozoic consolidation; 6—of Mesozoic consolidation; 7—continental shelf break; 8—sedimentary basins: I—Wilkes Land basin, II—Victoria Land basin, III—Prydz Bay—Amery Ice Shelf basin, IV—Weddell Sea basin, V—Ross Sea basin, VI—Bellingshausen basin, VII—Amundsen Sea basin.**

## WEST ANTARCTIC BASINS

Recent Soviet and international studies (Jankowski and Drewry, 1981; Doake et al., this volume; Drewry, 1982) of subglacial morphology in West Antarctica allow the subdivision of the formerly recognised single gigantic subglacial depression into several separate morphological elements. The largest bedrock depression encompasses the Ross Sea shelf, the Ross Ice Shelf and the subglacial Byrd basin, the latter differing greatly in botton topography and field pattern

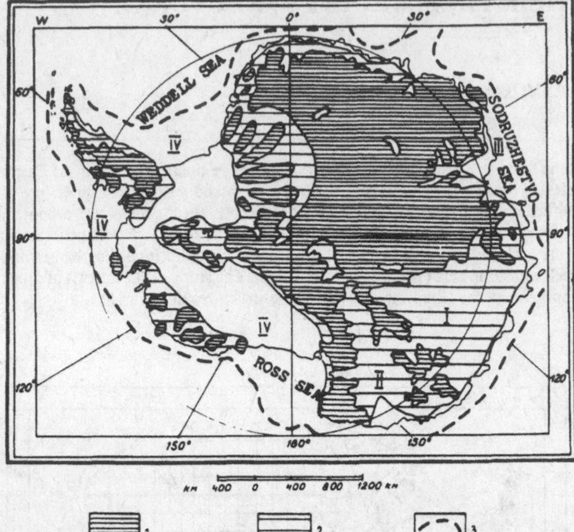


**Figure 2. Bedrock relief of Antarctica. 1—areas actually lying a.s.l.; 2—areas a.s.l. after isostatic readjustment to removal of ice load; 3—continental shelf break. Bedrock depressions and shelf areas unshaded; 4—major bedrock depressions; I—Wilkes Land depression, II—Victoria Land depression, III—Weddell Sea—Filchner and Ronne Ice Shelves depressions, IV—Ross Sea—Ross Ice Shelf—Byrd subglacial basin depression, V—Bellingshausen and Amundsen Seas depression.**

from the remaining area (Jankowski and Drewry, 1981). To the north-east the depression is connected with the Amundsen and Billingshausen Seas basins via a system of subglacial passages and farther east, in the Haag Nunatak area a narrow subglacial "channel" runs towards the Ronne Ice Shelf (Doake et al., op. cit) which is part of the Weddell Sea basin. In both cases, calculations of isostatic readjustment after the removal of the ice show that these connections would be severed. To the south-east, there is a continuous barrier separating the Byrd and Weddell Seas basins in the triangle formed by the Ellsworth—Whitmore—Thiel Mountains.

Therefore, three major negative morphostructures corresponding to main sedimentary basins can be distinguished in West Antarctica. They are: (i) Weddell Sea basin, including the Ronne and Filchner Ice Shelves; (ii) Ross Sea basin, including the Ross Ice Shelf and the subglacial Byrd basin joining it to the east; and (iii) Bellingshausen and Amundsen Seas basins.

## Weddell Sea Basin

The Weddell Sea basin has an area of 1.4 x $10^6$ $km^2$ and opens oceanwards in a northerly direction. The greatest depths are known around the inner periphery from Ellsworth Land to the Pensacola Mountains and farther north-east under the Filchner Ice Shelf. This zone with depths 1250-1500 m b.s.l., follows the strike of a large rift zone. Apparently the active Cenozoic downwarping is sediment starved there due to the presence of ice shelves. Over much of the basin a depth to the bedrock surface is 250-400 m b.s.l.

The structural position and peculiar features of the Weddell Sea sedimentary basin suggest that its evolution followed two "programmes". The basin is underlain by a thin continental crust, 30-35 km thick (25 km in the Filchner Glacier riftogenic graben), and represents a typical marginal system between an ancient platform and Mesozoic geosynclinal fold belt. Therefore, the platform side of the basin should be considered as a pericratonic downwarping area, while the opposite side—as a fore-deep in the sense of classical terminology. On the other hand, since the close of the Mesozoic the structure has been affected by the emplacement of the southern segment of the Atlantic Ocean, and hence it is a passive continental margin of the "Atlantic" type with all the inherent features (aseismicity, absence of young volcanism, weak anomalous mosaic magnetic field, thin continental crust, etc.)

The mountainous rim of the basin is built mainly of platform structures (Figure 1), viz. to the east (western extremity of Queen Maud Land, Coats Land) it is a plate resulting from the Precambrian stabilisation, and to the southeast (Pensacola Mountains), south and south-west (Ellsworth Mountains) it is the result of the epi-Ross intracratonic orogen. Only the western side of the basin consists of geosynclinal-folded rocks of the Antarctic Peninsula. The boundary between the latter structures and those of the Ellsworth Mountains lies under the glaciers situated in Ellsworth Land. Doake et al. (op cit) suggest that a subglacial extension of the Ellsworth Mountains is bounded by steep tectonic escarpments with the bedrock surface downthrown for several thousand metres. Geophysical data suggest the emplacement of the basin on the Archaean crystalline basement. However, unlike the East Antarctic craton proper, which was stabilised as an Early Precambrian shield, the area discussed was developed during the Late Precambrian-Palaeozoic under an active platform regime with accumulation of thick Ross and Beacon strata. Repeated intracratonic folding was due to the existence of an adjacent geosynclinal belt. The development of the Ross and epi-Ross orogenic belts was accompanied by the initiation of compensatory troughs, that is sedimentary basins inherited by recent depressions which suggests an old—even Ross—age of the lower horizons of the undeformed sedimentary cover. Airborne magnetic data mainly outline the surface of the crystalline basement; its maximum depth of 15 km refers to the eastern open part of the Weddell Sea basin and the entire central closed area. The postion of the folded basement surface is inferred from the magnetic, seismic, and gravity data. Its depth of 10-12 km is established on the platform side and in the axial part of the Weddell Sea basin. On the geosynclinal side it never exceeds 5 km. Magnetic data suggest that in the Antarctic Peninsula this basement comes to the surface as a remobilised Mesozoic infrastructure which allows a comparison of the area with the pre-Andean molasse basins.

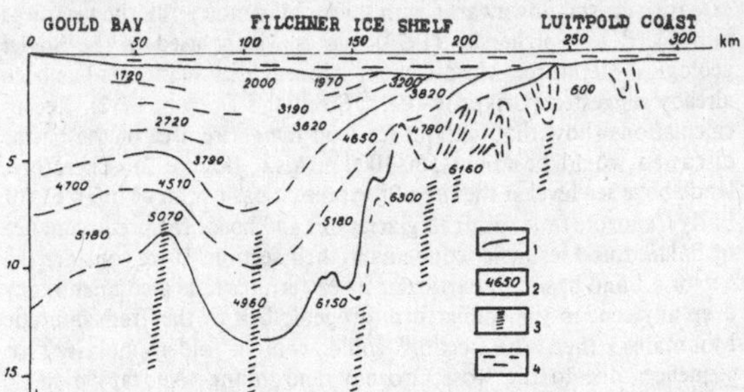


**Figure 3. Schematic seismic section across the southeastern Weddell Sea 1—boundaries of major seismic complexes; 2—seismic velocities in m/sec; 3—deep faults; 4—sea water.**

As a whole, the basin is asymmetric with a deeply submerged platform side. One can assure the importance of rifting for the initation of this part of the basin. A complex system of graben-like troughs about 1100 km long with a surface depth of 1.5 km is distinguished under the Filchner Glacier. Gravity data imply that the Moho surface rises to 35-25 km beneath these structures. However, the surface of the consolidated crust indicates only a successive step-wise submergance from platform to trough with an amplitude of 10-12 km. A western flank of the riftogenic graben has not been recognised as yet. The multichannel seismic profile suggests a successive lowering of the basement surface (velocity 5.8—6.3 km/s) down to 12-13 km (Figure 3). The dissected topography of the surface is not step-wise but wavy in outline. Seismic horizons are not horizontal and conformable with respect to the basement relief; they replace each other in the bottom surface from the basement rise in Luitpold Coast towards the centre of the basin. Seismic discontinuities within a sedimentary sequence are not easily discernible; however, three sequences can be tentatively distingushed, that is 4.9-5.1 km/s; 4.5-4.7 km/s; and 1.7-2.0 to 3.2-3.8 km/s (Poselov et al., in prep.; Hinz, 1978; Haugland et al., 1982). Beds with velocity 3 km/s assigned to young unconsolidated sediments do not exceed 2-3 km in thickness and are not ubiquitously distributed. Normally lithified rocks of a wide age range including Palaeozoic, dominate the section. At present we can only speculate about the basin genesis using published data of our foreign colleagues.

## Ross Sea Basin

Unlike the Weddell Sea basin that of the Ross Sea belongs to an active continental margin of the "Pacific" type. The basin is rimmed by a belt of Late Mesozoic-Cenozoic volcanics including active volcanoes, along the western edge of the open part of the shelf (Borchgrevink Coast—Mount Melbourne—McMurdo Station) and to the north-east (Marie Byrd Land). Recent air-borne magnetic and echo-sounding surveys (Jankowski and Drewry, 1981) suggest that a belt of young volcanics occupies the entire subglacial Byrd Depression encompassing the basin to the east as well. The history of the Ross Sea sedimentary basin as a separate structure is short as compared to that of the Weddell Sea. Its initiation was associated with the emplacement of the Mesozoic fold belt and synchronous uplift of the Transantarctic Mountains dated as Late Cretaceous—Early Cenozoic. The presence of reworked Upper Cretaceous foraminifers in Palaeogene shelf sections suggest the age (Webb, this volume ). The sedimentary cover is also thinner than that of the Weddell Sea. Recent seismic surveys (Davey et al., this volume) revealed three submeridianal sedimentary "troughs" on the shelf, where the maximum thickness of sediments is 3-4 km. The DSDP and MSSTS-1 borehole data in McMurdo Sound show that the Cenozoic section is represented by marine and glacial deposits incorporating volcanoclastic material and marked by deep erosional hiatuses implying repeated drainage of the basin (Webb, this volume). Along the foothills of the Transantarctic Mountains there are some vestiges of rifting such as crustal thinning to 25 km (35 km and over most of the shelf) (McGinnis et al., this volume), lower bottom and magnetic basement relief, gravity highs, etc. Consequently, processes of riftogenic destruction of the continental crust in Antarctica are obvious in all sedimentary basins both in East and West Antarctica irrespective of their original tectonic position (Figure 4).

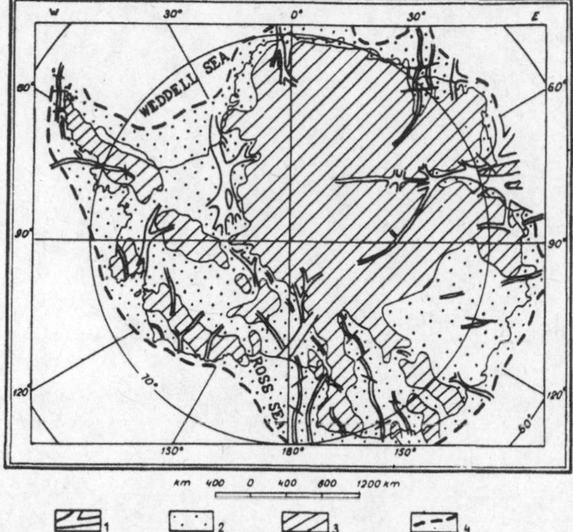


**Figure 4. Distribution of major rift zones in Antarctica. 1—rift zones; 2—sedimentary cover; 3—consolidated crust; 4—continental shelf break.**

## Bellingshausen and Amundsen Sea Basins

Rather small shelf basins of the Bellingshausen and Amundsen Seas are situated on the frontal (inner) flank of the West Antarctic Mesozoic system. However, no direct evidence for their structure is available. One can suggest a Late Cretaceous—Early Cenozoic sedimentary cover resting on folded Palaeozoic-Mesozoic formations.

## REFERENCES

DAVEY, F.J., HINZ, K. and SCHRODER, H., (this volume): Sedimentary basins of the Ross Sea, Antarctica.

DOAKE, C.S.M., CRABTREE, R.D. and DALZIEL, I.W.D., (this volume): Subglacial morphology between Ellsworth Mountains and Antarctic Peninsula.

DREWRY, D.J., 1976: Sedimentary basin of the east Antarctic craton from geophysical evidence. *Tectonophysics, 36*, 301-14.

DREWRY, D.J., 1982: Antarctica unveiled. *New Scientist, 95*, 246-51.

FEDOROV, L.V., GRIKUROV, G.E., KURININ, R.G. and MASOLOV, V.N., 1982: Crustal structure of the Lambert Glacier area from geophysical data; *in* Craddock, C. (ed.) *Antarctic Geoscience*. Univ. Wisconsin Press, Madison, 931-6.

GRIKUROV, G.E. (ed.), 1978: Tectonic Map of Antarctica, Scale 1:10,000,000. Kartfabrika ob'edinenija "Aerogeologia", Leningrad.

GRIKUROV, G.E., KADMINA, I.N., KAMENEV, E.N., KURININ, R.G., MASOLOV, V.N. and SHULYATIN, O.G., 1980: Tecktonicheskoe stroenie basseina moria Weddella (Tectonic structure of the Weddell Sea basin). *Geofizicheskiye issledovanya y Antarktidye (Geophysical Survey in Antarctica).* Research Institute of Arctic Geology, Leningrad, 29-43.

HAUGLAND, K., 1982: Seismic reconnaissance survey in the Weddell Sea; *in* Craddock, C. (ed.) *Antarctic Geoscience*. Univ. Wisconsin Press, Madison, 405-13.

HINZ, K., 1978: Bericht über geophysikalische Untersuchungen im Weddell Meer und am ostantarktischen Kontinentalrand mit M/S Explora. Bundesanstalt für Geowissenschaften und Rohstoffe (BGR), Hannover. Archiv 79617.

JANKOWSKI, E.J. and DREWRY, D.J., 1981: The structure of West Antarctica from geophysical studies. *Nature, 291*, 17-21.

JOHNSTONE, M.H., 1981: The importance of continental fragmentation history to petroleum accumulation; *in* Cresswell, M.M. and Vella, P., (eds) *Gondwana Five*, A.A. Balkema, Rotterdam, 329-34.

McGINNIS, L.D., WILSON, D.D. and BURDELIK, W.J., (this volume): Crust and upper mantle study in McMurdo Sound.

POSELOV, V.A., SHELESTOV, F.A., and GRIKUROV, G.E., (in prep): Predvaritel'nye rezultaty morskikh seismicheskikh issledovanii v more Weddella v sostave 26oy Sovetskoy Antarkticheskoy Ekspeditsii (Preliminary results of marine seismic sounding in the Weddell Sea during 26th Soviet Antarctic Expedition).

RAVICH, M.G., SOLOVIEV, G.S. and FEDOROV, L.V., 1978: *Geologicheskoe stroenie Zemel' Mak-Robertsona i Printsessy Elizavety v Vostochnoy Antarktide (Geological structure of MacRobertson Land and Princess Elizabeth Land, East Antarctica)*. Izdatel'stvo Nedra, Leningrad.

SPIZHARSKY, T.N., 1973: *Obzornye tektonicheskie karty SSSR (sostavlenie kart i osnovnye voprosy tektoniki) Reconnaissance tectonic maps of the Soviet Union (Compilation of maps and fundamental tectonic problems).* Izdatel'stvo Nedra, Leningrad.

STEED, R.H.N. and DREWRY, D.J., 1982: Radio-echo sounding investigation of Wilkes Land, Antarctica; *in* Craddock, C. (ed.) *Antarctic Geoscience*. Univ. Wisconsin Press, Madison, 969-75.

WEBB, P.N., (this volume): Climatic, palaeo-oceanographic and tectonic interpretation of Paleogene-Neogene biostratigraphy from MSSTS-1 drillhole, McMurdo Sound, Antarctica.

# 10

# Cenozoic Tectonics and Climatic Record—Onshore and Offshore Evidence

# A REVIEW OF LATE CRETACEOUS-CENOZOIC STRATIGRAPHY, TECTONICS, PALAEONTOLOGY AND CLIMATE IN THE ROSS SECTOR

P.N. Webb, *Institute of Polar Studies, Ohio State University, Columbus, Ohio 43210. U.S.A.*

*Abstract* The Ross Sector contains a varied and significant record of Late Cretaceous and Cenozoic geological history. The separation of the New Zealand subcontinent from Antarctica in the Late Cretaceous, the separation of Australia in the Palaeogene, the uplift of the Transantarctic Mountains in the Cretaceous and the submergence of the Ross embayment, also in the Cretaceous, controlled or at least strongly influenced the major preglacial and glacial events within and beyond this part of Antarctica. Lateral and vertical tectonic events influenced the fragmentation of Late Mesozoic basins, the development of more localised Cenozoic basins, the formation of seaways within and across Antarctica, the evolution of oceanic circulation patterns and biotic migration routes within and around Antarctica, the growth of fault-related volcanic centres, the provision of ice-nucleation centres and physiographic restriction and channelling of ice sheets and shelves.

While there are no known outcrops of Cretaceous rocks in the Ross Sea area, they almost certainly exist within the continental shelf succession. Recycled Cretaceous foraminifera, radiolaria and palynomorphs occur in Cenozoic sediments. The Palaeogene is still poorly documented, due largely to a lack of exploration by drilling techniques. Diverse but poorly preserved Late Paleocene to Early-Middle Eocene foraminifera occur in the thick glaciomarine successions just offshore in western McMurdo Sound. Transported blocks of fossiliferous Eocene sediments occur in the Quaternary deposits of southern McMurdo Sound and are probably of quite local origin. Late Oligocene glaciomarine sediments are known from both western McMurdo Sound and the south-central Ross Sea. No terrestrial Palaeogene rocks are yet documented from the Transantarctic Mountains but are very likely to be present. Preglacial subaerial and shallow water marine sediments occur in the south-central Ross Sea.

Neogene sediments are widely distributed across the Ross Sea continental shelf and in scattered areas of the Transantarctic Mountains. Early and Middle Miocene glaciomarine sediments of the continental shelf consist of deep water pebbly mudstones. Late Miocene sediments are unknown in continental shelf successions but do occur within valley basins or palaeofiords in the eastern Transantarctic Mountains. Early Miocene volcanic centres are present in the southern Transantarctic Mountains. Middle and Late Miocene volcanic activity was concentrated in and north of the McMurdo Sound area. The Pliocene is represented by often richly fossiliferous coarse clastic marine successions in palaeofiords, on the shelf area of western McMurdo Sound and on the flanks of offshore volcanic islands. More biogenic and finer grained Pliocene sediments are distributed across the Ross Sea continental shelf. Pliocene volcanic centres are common and generally have the same distribution patterns as that for Late Miocene volcanics. A blanket of Quaternary marine sediment is widely distributed across the floor of the Ross Sea. Holocene ages have been established for littoral deposits which crop out around the periphery of McMurdo Sound. Terrestrial Quaternary deposits blanket valley floors and upland areas of the Transantarctic Mountains.

The Transantarctic Mountains provide abundant materials for precise absolute dating but the continuity of record is poor. The Ross Sea continental shelf provides the most complete Late Cretaceous-Cenozoic record but improvement of the present data base will only come with further stratigraphic drilling.

# *SEDIMENTATION IN THE ROSS EMBAYMENT: EVIDENCE FROM RISP CORE 8 (1977/78)

G. Faure and K.S. Taylor, *Department of Geology and Mineralogy and Institute of Polar Studies, The Ohio State University, Columbus, Ohio 43210, U.S.A.*

*Abstract* Sediment recovered at site J9 (88°22′S, 168°38′W) from beneath the Ross Ice Shelf was deposited from floating ice in Mid-Miocene time but contains some older reworked material. Twenty sediment samples (< 150 $\mu$m, non-carbonate fractions) taken at regular interals along core 8 (1977/78) have nearly constant concentrations of Rb, Sr and $^{87}Sr/^{86}Sr$ ratios. Representative averages are Rb = 118.8 ± 4.6 ppm, Sr = 103.6 ± 2.5 ppm and $^{87}Sr/^{86}Sr = 0.72331 \pm 0.00007$. No significant variations occur at the boudary between the "upper" and "lower" units of the sediment in this core.

The sediment from Ross Ice Shelf Project (RISP) core 8 (1977/78) is colinear on a Sr-isotope mixing diagram with sediment from *USNS Eltanin* piston cores 32-16, 25 and 36 taken in the Ross Sea. The colinearity of 50 data points indicates that these samples are mixtures of the same two sediment components in varying proportions. The sediment under the Ross Ice Shelf represents one of these components in nearly pure form. The other component is volcanogenic detritus of calc-alkaline composition derived from Mesozoic volcanics in West Antarctica.

Two feldspar size fractions of bulk sample PNW-23 (1977/78) form a line on the Rb-Sr isochron diagram whose slope yields a date of 174 ± 75 Ma and an initial $^{87}Sr/^{86}Sr$ ratio of 0.7155 ± 0.0023. This date is in excellent agreement with previous age determinations of granitic rocks in the Whitmore Mountains of West Antarctica. Such rocks probably are the source of most of the sediment cored at site J9. However, a study of undifferentiated feldspar concentrates prepared from core 8 (1977/78) indicates that these feldspars are mixtures and contain a small and variable component of grains that are older than 174 Ma. The older component could have originated either from the 190 ± 8 Ma old Mt Seelig granite of the Whitmore Mountains or from 500 Ma old granitic basement rocks of the Transantarctic Mountains.

In this report we present new measurements of the Rb and Sr concentrations of the detrital noncarbonate fractions of sediment from core 8 (1977/78) taken at site J9 (88°22′S), 168°38′W) on the Ross Ice Shelf shown in Figure 1 (Webb, 1978, 1979). This study continues our previous work on piston cores from the Ross Sea (Shaffer and Faure, 1976; Kovach and Faure, 1977a, b, 1978a). These studies demonstrated that the detrital component of sediment deposited in the Ross Sea forms linear mixing arrays in coordinates of the $^{87}Sr/^{86}Sr$ ratios and the reciprocals of their Sr concentrations. The two components are volcanogenic detritus characterised by low $^{87}Sr/^{86}Sr$ ratios and high Sr concentrations and sialic detritus derived from igneous and metamorphic rocks of West Antarctica having elevated $^{87}Sr/^{86}Sr$ ratios and low Sr concentrations. The results for cores 32-16, 32-36 and 32-25 taken by the *USNS Eltanin* indicate that all three cores are mixtures of the same components in varying proportions (Kovach and Faure, 1978a). In addition, we demonstrated that the abundance of the volcanogenic detritus increases toward West Antarctica and that it was probably derived from calc-alkaline volcanics of Mesozoic age described by LeMasurier and Wade (1977).

The objectives of the present study are to relate the sediment in RISP core 8 (1977/78) to that previously studied in the Ross Sea and to determine its provenance by dating detrital feldspar by the Rb-Sr method.

## DESCRIPTION AND AGE OF THE SEDIMENT

The sediment in 11 gravity cores recovered at RISP site J9 during 1977/78 consists of pebble-granule diatomaceous mud and has been subdivided into an "upper" and a "lower" unit (Webb et al., 1979).

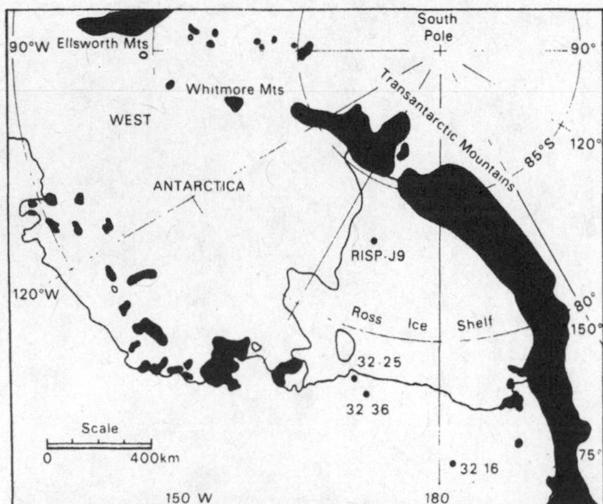


**Figure 1. Location of core samples in the Ross Sea and at RISP site J9 on the Ross Ice Shelf.**

The upper unit varies in thickness from 9 to 30 cm and is light olive grey (5Y 5/2) whereas the lower unit is olive grey (5Y 3/2). The boundary between them is marked by a thin (< 1 cm) brown orange iron rich layer containing iron micronodules that are concentrated into laminae. The sediment contains angular or subrounded pebbles, granules and coarse sandsized particles composed primarily of igneous and metamorphic rocks. The sediment also contains angular to subrounded sediment fragments composed in part of diatomaceous ooze. Most of the sediment clasts occur at the top of the core. Blocks of indurated sediment were also observed with underwater television cameras (Webb et al., 1979).

The sediment recovered at site J9 contains microfossils including diatoms, silicoflagellates and calcareous benthic foraminifera. Brady and Martin (1979) identified approximately 50 species of planktonic diatoms and dated this flora as late Middle Miocene in age (about 14 Ma). Webb et al. (1979) assigned an Early to Middle Miocene age to benthic foraminifera which occur only in the lower unit. The entire deposit was apparently formed beneath floating ice that originated primarily from centres of glaciation in West Antarctica. Pollen grains examined by Brady and Martin (1979) suggest the existence of vegetation of low diversity in favourable habitats along the coast and between glaciers. The pollen data are compatible with a Middle Miocene age deduced from the diatoms and foraminifera. However, Wrenn and Beckman (1982) found a Late Eocene dinocyst flora of low diversity which indicates that a reworked sediment component of Palaeogene age is present at this site. The assignment of a Middle Miocene age to the sediment in the RISP cores has been challenged by Kellogg and Kellogg (1981) on the grounds that the diatom populations contain not only Miocene but also Late Pleistocene species which are distributed throughout the cored sediment. Therefore, they concluded that the sediment at site J9 consists of Miocene and older sediment that was reworked in Late Pleistocene time.

The age of the sediment at site J9 may also be determined from measurements of the abundance of cosmogenic $^{10}Be(T_{0.5} = 1.6 \times 10^6 a)$, formed in the atmosphere by interactions with cosmic rays. Yiou et al. (1981) reported $^{10}Be$ concentrations of $< 10^7$ atoms/g for sediment samples taken from the upper and lower units. These values are more than one order of magnitude less than those of recently deposited marine sediment. This result indicates that the sediment at site J9 is not of Quaternary age but was last in communication with the atmosphere several million years ago. Yiou et al. (1981) also concluded "....that the upper 10 cm of sediment has not been mixed...." which presumably means that it is not a mixture of Miocene and Quaternary sediment. Therefore, the $^{10}Be$ data appear to support the interpretation that the sediment at site J9 is of Miocene age. However, the low $^{10}Be$ concentration may also be caused by a low rate of deposition from the atmosphere typical of high latitudes and by the presence of an ice cover over the site.

## RISP CORE 8 (1977/78)

Twenty samples of 2 cm³ each, taken at regular intervals from core 8 (1977/78), were disaggregated in demineralised water. Coarse particles

*Laboratory for Isotope Geology and Geochemistry (Isotopia), Contribution Number 64.

larger than 150 μm were removed by sieving. The fine fractions were leached with purified 2N HC1 to remove calcium carbonate. The Rb and Sr concentrations of the residues were determined by X-ray fluorescence on a Bausch and Lomb (Diano) Model XRD-6 air-path fluorescence spectrometer equipped with a Mo-target X-ray tube and a LiF (220) diffracting crystal. Calibrations were based on the rock standards of the U.S. Geological Survey (Flanagan, 1973) and matrix corrections were made by means of the Mo K-alpha Compton-scattered peak (Reynolds, 1963). Instrument drift was corrected by means of a monitor analysed at hourly intervals. The reproducibility is ± 1.48 ppm for Sr and ± 2.39 ppm Rb based on duplicate analyses of all samples in this suite.

Strontium was separated from the noncarbonate samples by cation exchange chromatography and isotope ratios were measured on two solid source mass spectrometers. Samples 1-10 (Table 1) were analysed on an automated instrument (Nuclide Corp. Model 12-90-S, called Goliath) whereas samples 11-20 were analysed on a manually operated instrument (Nuclide Corp. Model 6-60-S, called David). The analytical precision of Goliath, expressed as one standard deviation of the mean of all recorded mass scans is ± 0.00008 on the average, whereas that of David is ± 0.00024. The Eimer and Amend $SrCO_3$ isotope standard yielded $^{87}Sr/^{86}Sr$ ratios of 0.70816 ± 0.00067 (one measurement, Goliath) and 0.70811 ± 0.00007(average of eight measurements, David). These results indicate that there is no detectable difference in the $^{87}Sr/^{86}Sr$ ratios measured on the two instruments.

**TABLE 1: Analytical Data for <150μm Fractions of Non-Carbonate Sediment, Core 8 (1977/78), Site J9, Ross Ice Shelf**

| Sample Number | Depth cm | Sr ppm | Rb ppm | $^{87}Sr^{*}/^{86}Sr$ ± 1σ |
|---|---|---|---|---|
| 1/20 | 1.5-2.0 | 102.7 | 121.6 | 0.72358 ± 0.00008 |
| 2/20 | 4.5-6.5 | — | — | 0.72270 ± 0.00024 |
| 3/20 | 8.0-9.0 | 106.9 | 120.4 | 0.72326 ± 0.00007 |
| 4/20 | 10.0-11.0 | 106.1 | 119.5 | 0.72345 ± 0.00007 |
| 5/20 | 14.0-15.0 | 105.9 | 115.8 | 0.72352 ± 0.00007 |
| 6/20 | 19.5-20.5 | 106.9 | 116.5 | 0.72333 ± 0.00008 |
| 7/20 | 24.5-25.5 | 104.3 | 116.3 | 0.72347 ± 0.00007 |
| 8/20 | 29.0-30.0 | 106.2 | 117.4 | 0.72338 ± 0.00005 |
| 9/20 | 34.5-35.5 | 102.6 | 121.3 | 0.72366 ± 0.00006 |
| 10/20 | 39.5-40.5 | 102.6 | 116.6 | 0.72323 ± 0.00003 |
| 11/20 | 44.5-45.5 | 102.1 | 114.0 | 0.72354 ± 0.00048 |
| 12/20 | 49.5-50.5 | 104.1 | 113.1 | 0.72353 ± 0.00041 |
| 13/20 | 54.5-55.5 | 103.9 | 114.5 | 0.72368 ± 0.00030 |
| 14/20 | 59.5-60.5 | 105.4 | 116.0 | 0.72376 ± 0.00040 |
| 15/20 | 64.5-65.5 | 103.7 | 116.0 | 0.72343 ± 0.00024 |
| 16/20 | 70.0-71.0 | 102.2 | 117.2 | 0.72263 ± 0.00019 |
| 17/20 | 74.5-75.5 | 102.5 | 118.7 | 0.72273 ± 0.00008 |
| 18/20 | 80.0-81.0 | 103.0 | 129.6 | 0.72347 ± 0.00009 |
| 19/20 | 84.5-85.5 | 99.2 | 129.7 | 0.72327 ± 0.00019 |
| 20/20 | 89.0-90.0 | 97.3 | 122.8 | 0.72352 ± 0.00011 |

*All $^{87}Sr/^{86}Sr$ ratios were corrected for isotope fractionation to a value of 0.1194 for the $^{86}Sr/^{88}Sr$ ratio.

The concentrations of Rb and Sr vary only between narrow limits down the core and have average values of Sr = 103.6 ± 2.5 ppm, Rb = 118.8 ± 4.6 ppm and Rb/Sr = 1.148 ± 0.064. The only significant deviation occurs below a depth of 75 cm where the Rb content of the sediment rises from 117.2 ± 0.6 ppm up to 129.7 ppm at 85 cm. The Sr concentration decreases slightly below 80 cm to 97.3 ppm. A plot of these data versus depth (Figure 2) emphasises the lack of stratigraphic variation of Rb and Sr concentrations in this core. The homogeneity of the sediment at site J9 contrasts sharply with the systematic variations of Rb and Sr concentrations reported by Kovach and Faure (1977a, b) for *USNS Eltanin* Cores 32-16, 32-25 and 32-36 from the Ross Sea.

The range of $^{87}Sr/^{86}Sr$ ratios in RISP core 8 (1977/78) from 0.72263 to 0.72376 is significant because it exceeds the analytical precision of the measurements. The greatest variation occurs between depths of 65 and 80 cm where the $^{87}Sr/^{86}Sr$ ratio decreases to 0.72268. A low value of 0.72270 was also recorded for sediment between 4.5 and 6.5 cm. The $^{87}Sr/^{86}Sr$ ratios of sediment in this core are significantly higher (0.72331 ± 0.00007) and more constant than those of the piston cores from the Ross Sea analysed previously by Kovach and Faure (1977a, b). No significant changes take place in the concentrations of Rb and Sr and the $^{87}Sr/^{86}Sr$ ratios at the boundary between the upper and lower units. The apparent homogeneity of the sediment supports the conclusion of Webb et al. (1979) that the upper unit formed by in situ alteration of sediment similar to that preserved in the lower unit. The significance of the very small variations of Rb and Sr concentrations

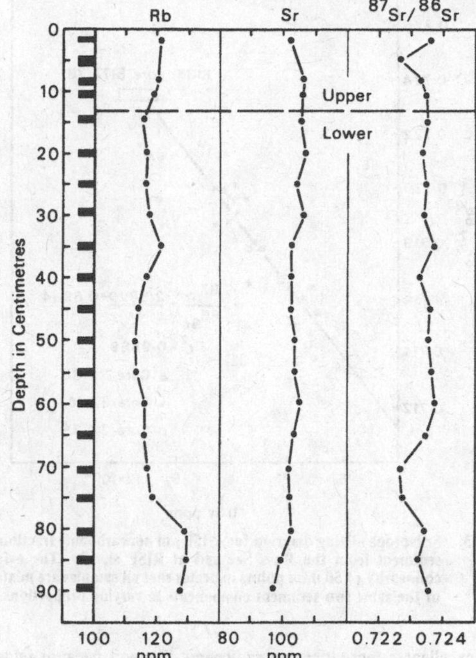


**Figure 2. Profile of Rb and Sr concentrations and $^{87}Sr/^{86}Sr$ ratios of <150 μm noncarbonate fractions in core 8 (1977/78) at RISP site J9. Note the absence of variation of these parameters at the boundary between the upper and lower units.**

and $^{87}Sr/^{86}Sr$ ratios in core 8 (1977/78) is not apparent to us at this time.

These data now permit us to relate the sediment deposited under the Ross Ice Shelf at site J9 to sediment in the Ross Sea. When the $^{87}Sr/^{86}Sr$ ratios of sediment from RISP core (1977/78) are plotted versus the reciprocals of their Sr concentrations (Figure 3), all of the data points are clustered into a small area on the Sr-isotope mixing diagram that appears to be colinear with sediment from the Ross Sea analysed by Kovach and Faure (1977a, b; 1978a). A least-squares regression of all 50 data points for *USNS Eltanin* cores 32-16, 32-25 and 32-36 with the data for core 8 (1977/78) of RISP yields the mixing equation:

$$\frac{^{87}Sr}{^{86}Sr} = \frac{2.5829}{Sr} + 0.6984$$

and a correlation coefficient $r^2 = 0.9589$. The colinearity of this set of data points indicates that these sediment samples are, to a good approximation, mixtures of the same two sediment components in varying proportions.

## PROVENANCE OF FELDSPAR IN CORE 8 (1977/78)

The position of sediment data points from RISP core 8 at one end of the mixing array (Figure 3) indicates that they represent one of the two end-member components in nearly pure form. Webb et al. (1979) suggested that the sediment was derived from West Antarctica because pebbles of Ferrar Dolerite and of sedimentary rocks of the Beacon Supergroup, which occur only in the Transantarctic Mountains, appear to be absent. However, Brady and Martin (1979) stated that rounded sand grains typical of the Beacon Sandstones and grains of coal are present, thereby implying that the Transantarctic Mountains may have been a source of some of the sediment deposited at site J9. Knowledge of the provenance of the sediment is important because this information is needed for the reconstruction of events in the development of glacial conditions in Antarctica during the Miocene. The glaciation of Antarctica may have begun with the formation of ice caps on high plateaus of the Transantarctic Mountains (Mercer, 1968; Faure et al., 1983) and on mountainous terrain in West Antarctica. A mixed provenance for sediment at site J9 therefore cannot be ruled out.

The provenance of feldspar grains in glacial deposits can be determined by the Rb-Sr method of dating (Taylor and Faure, 1981). Dating is facilitated by the fact that both K-feldspar and plagioclase

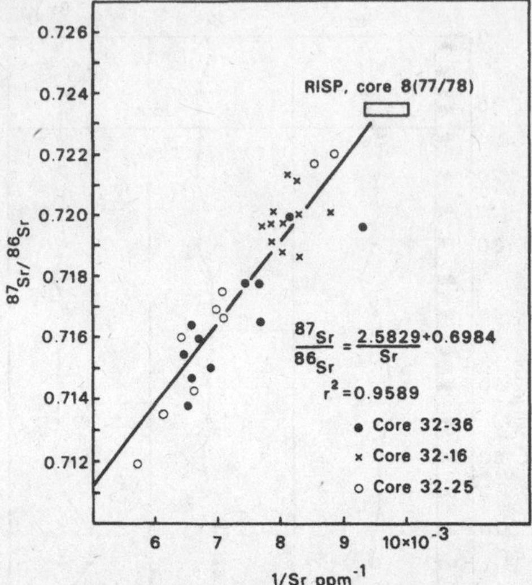


**Figure 3.** Sr-isotope mixing diagram for <150 μm noncarbonate fractions of sediment from the Ross Sea and at RISP site J9. The evident colinearity of 50 data points indicates that all samples are mixtures of the same two sediment components in varying proportions.

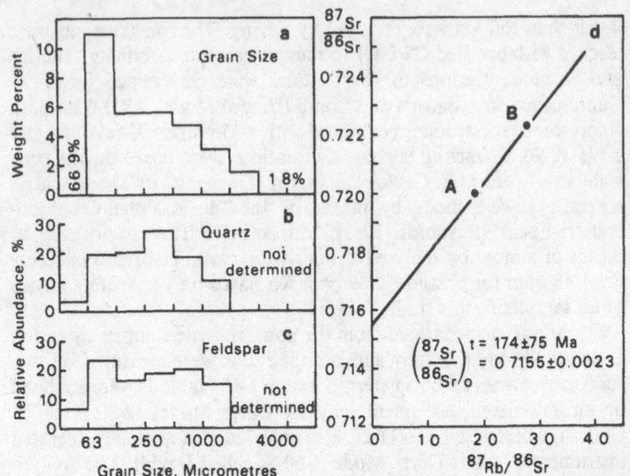


**Figure 4.** (a) Grain size distribution of bulk sample PNW-23 recovered at RISP site J9 in 1977/78; (b) Relative abundance of quartz in PNW-23 based on X-ray diffraction. The histogram shows that about 27% of the quartz in the sample occurs in the 250 to 500 μm grain size fraction; (c) Relative abundance of feldspar in PNW-23. The bimodal distribution is caused by the presence of plagioclase in the 63 to 125 μm fraction and by both K-feldspar and plagioclase in the 250 to 1000 μm fraction; (d) Two point "isochron" formed by feldspars of different grain size in PNW-23. Sample A is from 63 to 125 μm whereas sample B is from 125 to 1000 μm.

have excellent retentivities for radiogenic $^{87}Sr$ and because feldspar grains in different grain size fractions tend to have different Rb-Sr ratios (Faure and Taylor, 1981). Therefore, size fractions of feldspar separated from glacial deposits in many cases form linear arrays on the Rb-Sr isochron diagram whose slopes can be used to calculate dates. If all grains were derived from a source of uniform age, the feldspars indicate the age of the rocks in that source area and the line is an isochron. When the feldspar grains were derived from two sources having different ages, they yield an intermediate "provenance date" which has no time significance but reflects the proportion of mixing of grains derived from the two source regions (Faure and Taylor, 1981). This method of study was used to determine the provenance of feldspar in sediment collected at RISP site J9.

## Sample PNW-23

A bulk sample of sediment collected at site J9 (1977/78) weighing 64.43 g (dry) and labelled PNW-23 (Foster CDZ No. 2, 31 December) was disaggregated in demineralised water and sieved into size fractions. The relative abundances of quartz and feldspar in these size fractions were estimated from X-ray diffraction scans. Feldspar-quartz concentrates were prepared from the 63 to 125 μm (Sample A) and from the 125 to 1000 μm fractions (Sample B) by use of a Frantz Isodynamic Separator followed by ultrasonic cleaning of the grains in purified HC1 and demineralised water. Concentrations of Rb and Sr and $^{87}Sr/^{86}Sr$ ratios were determined as before (Table 2).

The results in Figure 4a indicate that the sediment is poorly sorted but that more than 65% by weight is composed of grains 63 μm or smaller. Quartz has a broad unimodal distribution centred between 125 μm and 500 μm (Figure 4b). The distribution of feldspar is bimodal (Figure 4c). An abundance peak in the 63 to 125 μm fraction is caused by dominant plagioclase whereas the coarse fraction (125 to 1000 μm) contains a mixture of both plagioclase and K-feldspar. In addition, the sediment contains illite (muscovite), chlorite and mixed-layer clay with d-spacings ranging from 10.0 to 16.0 Å.

**TABLE 2: Analytical Data for Feldspar-Quartz Concentrates of Sediment Collected at RISP Site J9, Ross Ice Shelf**

| Sample | Interval cm | $\frac{Rb}{Sr} \pm \sigma$ | $\frac{^{87}Rb}{^{86}Sr} \pm 1\sigma$ | $\frac{^{87}Sr^*}{^{86}Sr} \pm 1\sigma$ |
|---|---|---|---|---|
| | | **PNW-23** | | |
| A | bulk | 0.632 ± 0.011 | 1.831 ± 0.030 | 0.72005 ± 0.00065 |
| B | bulk | 0.946 ± 0.017 | 2.743 ± 0.050 | 0.72231 ± 0.00037 |
| | | **Core 8 (1977/78)** | | |
| F-1 | 1.5-15.0 | 0.944 ± 0.018 | 2.738 ± 0.053 | 0.72310 ± 0.00014 |
| F-2 | 19.5-40.5 | 0.897 ± 0.013 | 2.600 ± 0.039 | 0.72114 ± 0.00013 |
| F-3 | 44.5-65.5 | 0.907 ± 0.003 | 2.629 ± 0.007 | 0.72165 ± 0.00013 |
| F-4 | 70.0-90.0 | 0.940 ± 0.013 | 2.752 ± 0.037 | 0.72231 ± 0.00009 |
| Mag (1-4) | 1.5-90.0 | 0.582 ± 0.006 | 1.686 ± 0.016 | 0.71882 ± 0.00019 |

*All $^{87}Sr/^{86}Sr$ ratios were corrected for isotope fractionation to a value of 0.1194 for the $^{86}Sr/^{88}Sr$ ratio.

The two feldspar concentrates have been plotted on the Rb-Sr isochron diagram (Figure 4d). Sample B (125 to 1000 μm) has a higher Rb/Sr ratio than Sample A (63 to 125 μm) consistent with the fact that sample B contains more K-feldspar than does Sample A. The two points were used to draw a straight line whose slope yields a date of 174 ± 75 Ma ($\lambda^{87}Rb = 1.42 \times 10^{-11}a^{-1}$). The initial $^{87}Sr/^{86}Sr$ ratio is 0.7155 ± 0.0023. This date is in excellent agreement with a Rb-Sr mineral isochron date of 173 Ma and an initial $^{87}Sr/^{86}Sr$ ratio of 0.7148 for coarse and fine grained facies of granitic rocks at Mt Chapman in the Whitmore Mountains (Kovach and Faure, 1978b). It is also in good agreement with a K-Ar date of 176 ± 5 Ma for biotite from the Linck Granite of the Whitmore Mountains (Webers et al., 1982) but is less than the age of the Mt Seelig granite for which they reported a K-Ar biotite date of 190 ± 8 Ma.

These results permit the conclusion that the feldspar in sample PNW-23 could have originated from source areas in West Antarctica underlain by granitic rocks of Early Jurassic age and having an elevated initial $^{87}Sr/^{86}Sr$ ratio like the granites exposed on Mt Chapman in the Whitmore Mountains. In this case, the line in Figure 4d is interpreted as an isochron whose slope yields the age of the rocks whence the feldspar was derived.

## Core 8 (1977/78)

Grains ranging in size from 150 to 1000 μm were separated from core 8 (1977/78) and combined into four composites labelled F-1 through F-4 as shown in Table 2. Feldspar-quartz concentrates were prepared from these composites as before. The magnetic fractions of all four composites were combined to form sample Mag (1-4). All five resulting samples were analysed for Rb and Sr concentrations and $^{87}Sr/^{86}Sr$ ratios. The results are listed in Table 2.

The four feldspar concentrates taken from different depths in core 8 form a linear array on the isochron diagram (Figure 5) with a slope m = 0.009848, intercept b = 0.6956, and correlation coefficient $r^2 = 0.8880$. This line cannot be an isochron because the value of the intercept is less than the $^{87}Sr/^{86}Sr$ ratio of terrestrial Sr at any time in the history of the Earth, except during its formation 4.6 Ma ago (Faure 1977). Consequently, the line fitted to the feldspar points is a mixing line and the feldspars are mixtures derived from two or possibly more sources having different ages.

The simplest interpretation that can be made is that the feldspars are mixtures in varying proportions of grains derived from granitic rocks of Early Jurassic age (175 Ma) with feldspar derived from older rocks. In order to illustrate this interpretation, a reference line having

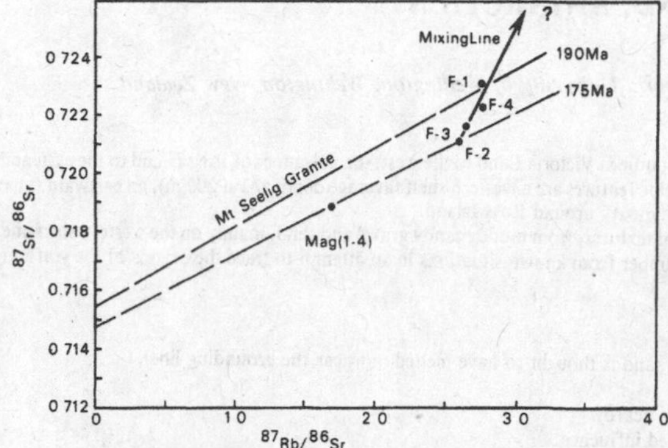


**Figure 5. Interpretation of data for feldspars and magnetic fraction of sediment from RISP core 8 (1977/78). The feldspars are mixtures of two components. The older component could be the Mt Seelig granite in the Whitmore Mountains or it could have been derived from the granitic basement rocks of the Transantarctic Mountains whose ages range from 510 to 540 Ma.**

a slope corresponding to a date of 175 Ma has been drawn on Figure 5 through F-2 and Mag (1-4), both of which fit this line. The feldspar composites then represent mixtures of grains like those of F-2 with varying amounts of an older component. The older component could have originated from granitic rocks like the Mt Seelig granite whose age is 190 ± 8 Ma (Webers et al., 1982). However, the data do not exclude the possibility that the older feldspar component originated from the granitic gneisses of the Transantarctic Mountains. The ages of granitic basement rocks from the Transantarctic Mountains range from about 510 to 540 Ma (Faure et al., 1979). The extension of the mixing line in Figure 5 to a 500 Ma isochron indicates that feldspar derived from such a source should have had a Rb/Sr ratio of about 2.6 and an $^{87}Sr/^{86}Sr$ ratio of 0.77 in order to generate the observed mixing array. K-feldspar having such chemical and isotopic compositions do indeed exist in the granitic basement rocks of the Transantarctic Mountains. However, the abundance of feldspar grains derived from this region cannot be more than a few percent.

## SUMMARY

This study of sediment taken from beneath the Ross Ice Shelf indicates that the Rb and Sr concentrations and the $^{87}Sr/^{86}Sr$ ratios of $<150$ $\mu$m noncarbonate fractions from core 8 (1977/78) vary very little down the core. In addition, the sediment is colinear on a Sr-isotope mixing diagram with similar sediment from piston cores in the Ross Sea. We conclude from these results that the sediment at RISP site J9 and in *USNS Eltanin* piston cores 32-16, 25 and 36 consists of mixtures in varying proportions of the same two sediment components. The sediment at site J9 may be representative of one of the two components that was presumably derived from igneous and metamorpic rocks of West Antarctica. Previous studies have suggested that the other sediment component is volcanogenic detritus derived from calc-alkaline lavas of Mesozoic age in West Antarctica.

Age determinations of feldspar size fractions of bulk sample PNW-23 yield a date of 174 ± 75 Ma that agrees with the age of granitic rocks that form Mt Chapman in the Whitmore Mountains. Unsieved feldspar composites extracted from core 8 (1977/78) indicate the presence of a small percentage of older feldspar derived either from rocks like the Mt Seelig granite (190 ± 8 Ma) or from the granitic basement rocks of the Transantarctic Mountains (510 to 540 Ma).

*Acknowledgements* We thank P.N. Webb and T.B. Kellogg for helpful comments in the course of this study. The sediment samples were provided by D.S. Cassidy from collections curated in the Antarctic Research Facility of the Department of Geology, The Florida State University, Tallahassee. This research was supported by the Division of Polar Programmes of the National Science Foundation through grant DPP-7920407.

## REFERENCES


BRADY, H. and MARTIN, H., 1979: Ross Sea region in the Middle Miocene: A glimpse into the past. *Science, 203,* 437-38.

FAURE, G., 1977: *Principles of Isotope Geology.* Wiley and Sons, New York.

FAURE, G., EASTIN, R., RAY, P.T., McLELLAND, D. and SHULTZ, C.H., 1979: Geochronology of igneous and metamorphic rocks, central Transantarctic Mountains; *in* Laskar, B. and Raja Rao, C.S. (eds.) *Fourth International Gondwana Symposium, 2,* Hindustan Publishing Corp., Delhi, 805-13.

FAURE, G. and TAYLOR, K.S., 1981: Provenance of some glacial deposits in the Transantarctic Mountains based on Rb-Sr dating of feldspars. *Chem. Geol., 32,* 271-90.

FAURE, G., TAYLOR, K.S. and MERCER, J.H., 1983: Rb-Sr provenance dates of feldspar in glacial deposits of the Wisconsin Range, Transantarctic Mountains. *Geol. Soc. Am., Bull.,* in press.

FLANAGAN, F.J., 1973: 1972 values for international geochemical reference samples. *Geochim. Cosmochim. Acta, 37,* 1189-200.

KELLOGG, T.B. and KELLOGG, D.E., 1981: Pleistocene sediments beneath the Ross Ice Shelf. *Nature, 293,* 130-33.

KOVACH, J. and FAURE, G., 1977a: Sources and abundance of volcanogenic sediment in piston cores from the Ross Sea, Antarctica. *N.Z. J. Geol. Geophys., 20,* 1017-26.

KOVACH, J. and FAURE, G., 1977b: Strontium isotopic study of sediment from the Ross Sea. *U.S. Antarct. J., 12,* 77-8.

KOVACH, J. and FAURE, G., 1978a: Use of strontium isotopes to study mixing of sediment derived from different sources: The Ross Sea, Antarctica; *in* Zartman, R.E. (ed.) Short papers of the Fourth International Conference Geochronology, Cosmochronology, Isotope Geology. *U.S. Geol. Surv. Open-File Rep., 78-701,* 230-32.

KOVACH, J. and FAURE, G., 1978b: Rubidium-strontium geochronology of granitic rocks from Mt Chapman, Whitmore Mountains, West Antarctica. *U.S. Antarct. J., 13,* 17-18.

LEMASURIER, W.E. and WADE, F.A., 1977: Volcanic history in Marie Byrd Land; *in* Gonzalez-Ferran, O. (ed.) *Andean and Antarctic Volcanology Problems.* Int. Assoc. Volcanol. Chem. Earth's Interior, Rome, 398-424.

MERCER, J.H., 1968: Glacial geology of the Reedy Glacier area, Antarctica. *Geol. Soc. Am., Bull., 79,* 471-86.

REYNOLDS, R.C., 1963: Matrix corrections in trace element analyses by X-ray fluorescence. *Am. Mineral., 48,* 1133-43.

SHAFFER, N.R. and FAURE, G., 1976: Regional variation of $^{87}Sr/^{86}Sr$ ratios and mineral compositions of sediment from the Ross Sea, Antarctica. *Geol. Soc. Am., Bull., 87,* 1491-500.

TAYLOR, K.S. and FAURE, G., 1981: Rb-Sr dating of detrital feldspar: A new method to study till. *J. Geol., 89,* 97-107.

WEBB, P.N., 1978: Initial report on geological materials collected at RISP site J9, 1977/78. RISP Technical Report 78-1, Ross Ice Shelf Project Management Office, University of Nebraska, Lincoln, 46p.

WEBB, P.N., 1979: Initial report on geological materials collected at RISP site J9, 1978/79. RISP Technical Report 79-1, Ross Ice Shelf Project Management Office, University of Nebraska, Lincoln, 127p.

WEBB, P.N., RONAN, T.E., Jr, LIPPS, J.H. and DELACA, T.E., 1979: The glaciomarine sediments from beneath the southern Ross Ice Shelf, Antarctica. *Science, 203,* 435-37.

WEBERS, G.F., CRADDOCK, C., ROGERS, M.A. and ANDERSON, J.A., 1982: Geology of the Whitmore Mountains; *in* Craddock, C. (ed.) *Antarctic Geoscience,* Univ. Wisconsin Press, Madison, 841-47.

WRENN, J.H. and BECKMAN, S.W., 1982: Maceral, total organic carbon and palynological analyses of Ross Ice Shelf Project site J9 cores. *Science, 216,* 187-89.

YIOU, F., RAISBECK, G.M. and BERNAS, R., 1981: The age of sediments beneath the Ross Ice Shelf as implied by cosmogenic $^{10}$Be concentrations. *EOS, 62,* 297.

# MODERN SEDIMENTATION IN McMURDO SOUND, ANTARCTICA

P.J. Barrett, A.R. Pyne and B.L. Ward, *Department of Geology, Victoria University of Wellington, Wellington, New Zealand.*

*Abstract* McMurdo Sound is 50 km wide, lying between the glaciated coast of southern Victoria Land to the west, the volcanoes of Ross Island to the east, and bounded to the south by the McMurdo Ice Shelf. The Sound's main physiographic features are a western shelf (average depth about 200 m), an eastward slope of about 1° and an elongate basin 900 m below sea level, which is part of a "moat" around Ross Island.

More than 60 samples taken from the floor of the Sound show a wide range of textures, from muddy sandy gravel and sand, mainly on the western shelf and slope, to mud in the deep basins. The samples are compared with a similar number from known situations in an attempt to trace the origins of the seafloor sediment.

The main sources identified are:

(1) Coastal sand blown by wind onto the sea ice.
(2) Basal glacial debris (but only the fine fraction, as the gravel and coarse sand is thought to have melted out near the grounding line).
(3) Silt size diatom debris.
(4) Supraglacial debris from the McMurdo Ice Shelf and ice-cored moraines nearby.

Gravity and bottom currents may be active locally, but appear to be of limited influence.

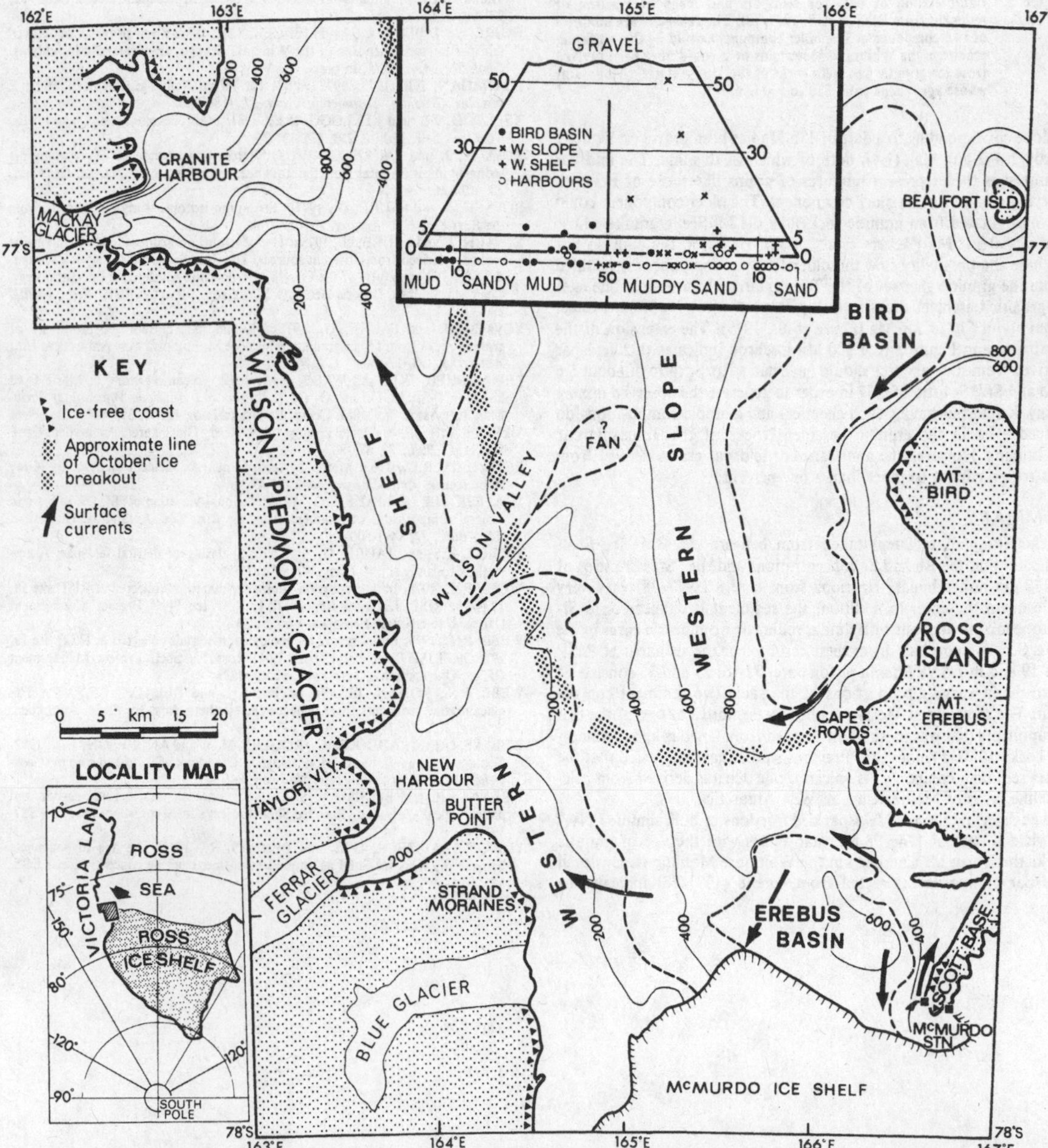


**Figure 1** Map of McMurdo Sound, showing bathymetry, surface currents and major physiographic divisions (names informal). Bathymetry is simplified from a compilation by Ward, Barrett and Pyne in Barrett (1982). Inset shows texture of sea floor samples located in Figure 2.

This paper examines nearshore marine sedimentation in a glaciated region by relating the character of the seafloor sediment in McMurdo Sound (Figure 1) to known sediment sources and processes there. Modern sediment on the floor of the Sound is highly varied in texture, ranging from muddy sandy gravel to mud, with many samples containing a high proportion of sand. Therefore, we have attempted to follow sediment transport paths by comparing the textures of likely sources with those of the modern seafloor sediment.

Grainsize distribution was determined for about 60 samples from all known possible sources, including glacial debris (basal and supraglacial), beach and shallow sub-tidal sand, and wind blown sand collected from the sea ice, and for a similar number of seafloor samples. Forty-six of the latter were collected by orange peel grab, sampling the top 100 mm, and 20 by sphincter or box corer, from which the top 30 mm was taken. Size distribution was determined by sieving at 0.5 phi intervals in the sand range, and by pipette or Sedigraph in the mud range. Sample size was normally 20 g, except for some gravelly samples, when more than a kilo was required for a representative analysis.

The data have been presented as frequency curves because most of the seafloor sediments are polymodal and the modes are well-developed. We think this feature of sediment texture can be used to trace sediment in McMurdo Sound from source to sink and in some cases to estimate proportions of sediment from different sources. Survival of textural modes from entrainment to deposition is unusual in near-shore marine environments, but has come about here through two aspects of the McMurdo Sound sediment transport system:

(1) Passive transport of sediment by floating ice.
(2) Lack of textural modification by waves or in most places by bottom currents.

There are as yet no reliable sedimentation rates for McMurdo Sound, as no cores have been dated or direct measurements made. However, the present sedimentary regime has been operating for probably the last 5000 years, after the sea reached its present level and the Sound

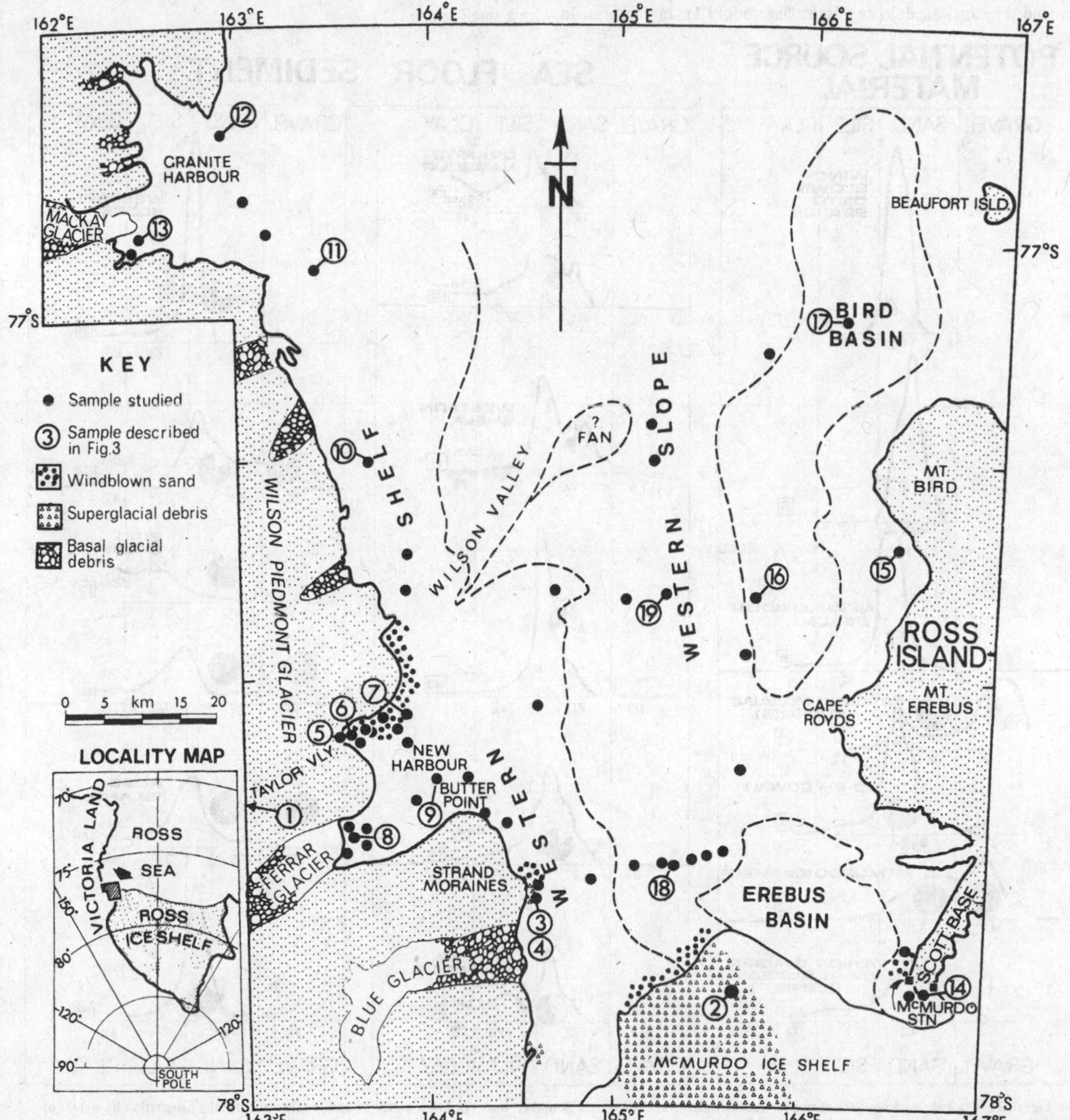


**Figure 2** Map of McMurdo Sound, showing likely sediment sources, together with locations of typical samples and of sea floor sediment samples. Numbers refer to size frequency curves in Figure 3.

was cleared of glacial ice (Stuiver et al., 1981). Widespread net deposition of sediment today is shown in bottom photographs and cores by features such as dead shells (Bullivant, 1967, Plate 15b) and spicule mats (our observations) partially buried by fine sediment, and Bullivant (1967, p.60) also noted that for station A538 "the type of bryozoans present and the appearance of the substrate suggests a relatively high sedimentation rate". Bentley (1979) calculated that wind blown sand was accumulating in northern New Harbour at about 1 mm/year. Even if the rate elsewhere is a small fraction of this our samples from the upper 100 mm of the seafloor should be reasonably representative of the present sedimentary regime.

## MAIN FEATURES OF McMURDO SOUND

In its geological setting McMurdo Sound lies between two major provinces. The Victoria Land coast to the west is part of a Transantarctic Mountains province with a basement of granitic and metamorphic rocks (Late Precambrian to Early Palaeozoic), overlain further inland by more than 2 km of Beacon sandstone (Devonian to Triassic) intruded by Jurassic dolerite sills totalling about 1 km in thickness. To the east the Sound is bounded by Ross Island, part of a Late Cenozoic basaltic volcanic province. Similar volcanic piles are found to the south, and are almost the sole source of debris on the northward-moving McMurdo Ice Shelf. This simple setting offers potential for an independent line of evidence regarding the source of the seafloor sediment.

The Sound itself can be divided into four areas:

(1) The steep slopes (6°) off the Ross Island coast.
(2) The Erebus and Bird basins, relatively flat-floored at depths of about 600 m and 900 m respectively.
(3) The Western Slope, essentially a planar surface rising at 1° to the west from 850 m to 250 m.
(4) The Western Shelf, a broad platform with an average depth of 200 m in the south, but somewhat greater in the north.

This geometry has most probably developed over the last five million years with the growth of Ross Island and its concomitant depression of the crust (McGinnis, 1973). The Sound also has some smaller scale features of more recent origin, such as the submarine valley (Wilson Valley, Figure 1) that from detailed bathymetry appears to terminate in a small fan.

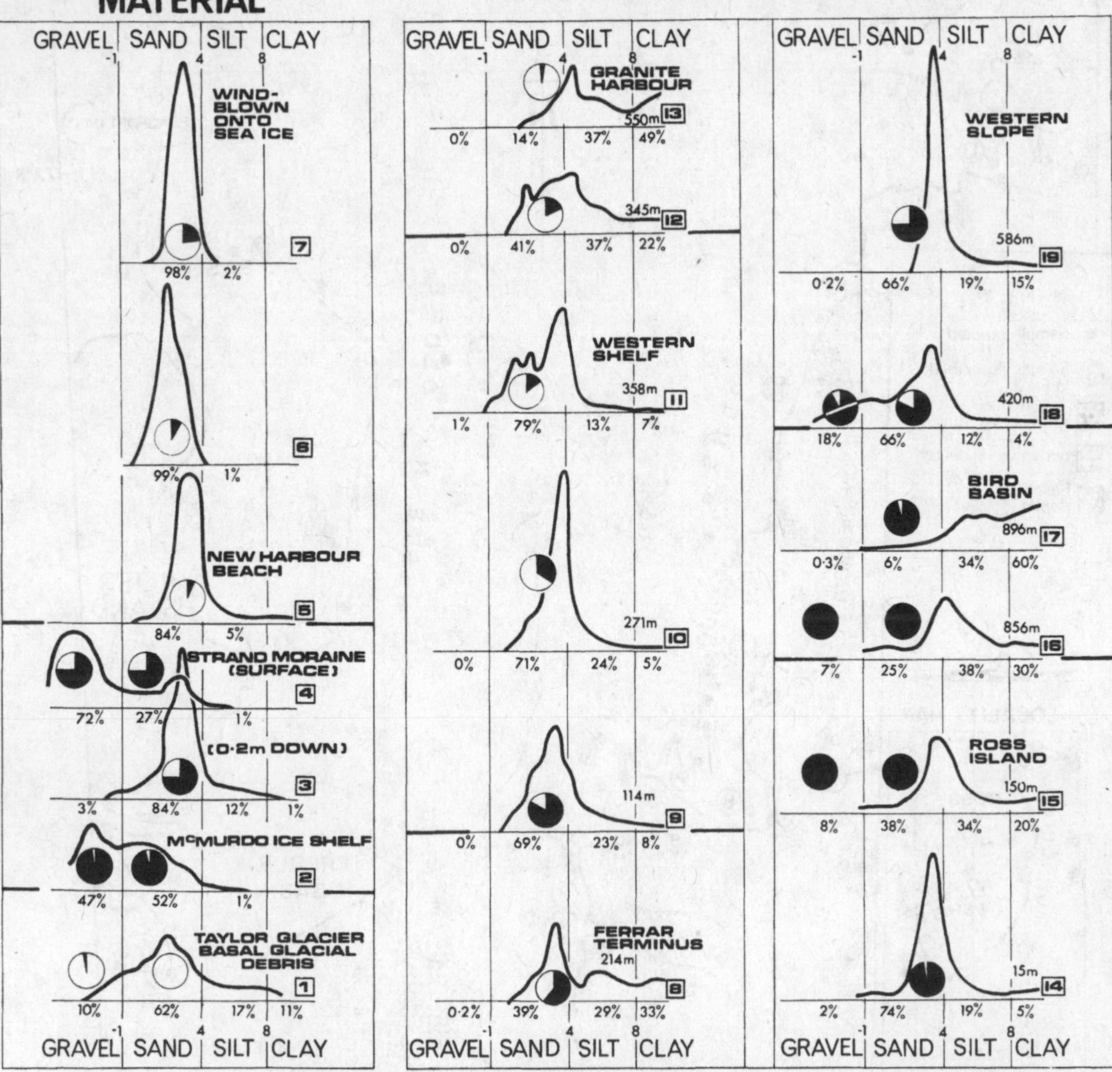


**Figure 3** Typical size frequency curves for sediment from the McMurdo Sound region. The black sector in each circle indicates the proportion by weight of Late Cenozoic basaltic material in the sand or gravel fraction. The frequency curves were drawn from histograms with 0.5 phi class intervals (10 percent vertical scale = 0.5 phi horizontal scale).

The climate of the region is harsh, with mean monthly temperature at sea level ranging from − 5°C in January to − 30°C in August (Keys, 1981), and snow and ice cover is extensive. Nevertheless, about a third of the coastline of the Sound is exposed rock or gravel beach. The Sound is completely frozen over from April to October, when the ice (1-3 m thick) breaks out from the centre, leaving a rim about 20 km wide around the southern and western border (Figure 1). This rim itself normally breaks out in January or February, leaving the Sound ice-free for two or three months before freezing begins again. The prevailing wind over most of the Sound is from the south southeast, with velocities recorded in excess of 14 m/sec (Keys, 1981). However, katabatic winds from the Polar Plateau flow down the valleys to the Victoria Land coast, forcing a westerly component on near-shore winds.

Water circulation in McMurdo Sound is mainly clockwise, the flow coming south past Ross Island and flowing north along the Victoria Land coast, though a considerable flow continues south around Ross Island beneath the McMurdo Ice Shelf (Heath, 1977). Velocities are low on the western side of the Sound (a maximum of 0.12 m/sec in the water column and less than 0.02 m/sec near the seafloor, Barrett, 1982), but are much higher off McMurdo Station where the Sound has been constricted, and 0.25 m/sec has been recorded near the seafloor (Carter et al., 1981).

## SOURCES AND PROCESSES

Sediment on the floor of McMurdo Sound must all come ultimately from the surrounding landscape or through biological precipitation of silica or carbonate. The terrigenous material may be introduced and distributed by a variety of processes, including rafting by glacier, shelf and sea ice, wind, currents in the watercolumn, and gravity currents, whereas the biogenic material forms mainly within the Sound itself.

Glacial debris is an obvious sediment source for a polar setting. Almost all of this debris is carried on and near the glacier surface, having fallen or been blown from rock faces above, or within a few metres of the glacier sole, after incorporation from the rock or debris-covered floor. The glaciers of the McMurdo Sound region carry very little supraglacial debris. For example, the surface of Taylor Glacier in the lower 10 km has a few small patches of sand and scattered rocks up to boulder size (Robinson, 1979), but on average they would amount to a layer only a fraction of a millimetre thick. In contrast, the basal debris layer in Taylor Glacier, which is exposed for about 5 km around the glacier terminus, is about 4 m thick. This layer consists mainly of ice containing between 20 and 43% debris by volume, but there are also some laminae and lenses of debris-free ice (Robinson, 1979, Appendix 10). The thickness of the basal debris layer on an ice-free basis is estimated at about 1 m, roughly $10^3$ times that of the average for supraglacial debris.

Robinson (in press) has calculated from heat flow, frictional heat, surface temperature and ice thickness that the sole of lower Taylor Glacier is at melting point over at least half of its area, thus accounting for the observed similarity in texture between basal debris in Taylor Glacier and temperate or "wet-based" glaciers. The critical ice thickness above which basal melting occurs in ice around McMurdo Sound (assuming accumulation or ablation is low) is about 450 m. Both the Mackay and Ferrar Glaciers exceed this substantially over almost all of their length and to their grounding lines, where they rapidly thin (Calkin, 1974), and so should contain a well-developed basal debris layer. Both glaciers terminate in floating ice tongues calving several kilometres seaward of the grounding line, by which time they will have lost most or all of their debris through basal melting by sea water, if it is proceeding at a rate near the 1.4 m/year calculated by Jacobs et al. (1981) for the Erebus Glacier Tongue across the Sound. However, basal debris may survive longer in bergs from coastal piedmont glaciers, such as the Wilson, that calve directly into the Sound. Several areas are thick enough to be wet-based (shown in Figure 2) and also they may still contain some debris entrained several thousand years ago, when the ice was thicker.

No means has yet been found to sample debris from the base of glaciers calving into McMurdo Sound, but because they are moving over a plutonic terrain similar to that beneath lower Taylor Glacier, we have taken the grainsize distribution of Taylor basal debris (Figure 3, Sample 1), which shows little lateral variation, to represent all basal debris entering McMurdo Sound. The debris has ten percent gravel (subrounded and with some striated pebbles), a broad sand mode, and about 12 per cent clay.

Supraglacial debris is important in two settings in McMurdo Sound. A large area of the McMurdo Ice Shelf is covered by coarse angular basaltic debris (Sample 2) frozen in beneath and moved up through the ice shelf as the surface has ablated (Debenham, 1948, Swithinbank, 1970). The other setting is found at the Strand Moraines, where a blanket of debris only 0.3 m thick covers hummocky ice up to 180 m thick. The debris is gravelly on the surface but sandy beneath (Samples 3 and 4), and has a complex origin; although three quarters is basaltic, the remainder includes a wide range of rock types from the mountains to the west, and some of the stones are subrounded and striated, indicating a period of transport at the base of a glacier. Bergs carrying supraglacial debris similar to that from both settings have been recognised along the western margin of McMurdo Sound, showing them to be potential sources of seafloor sediments.

Beach and shallow subtidal sand is another potential source of seafloor sediment where it is frozen into the sea ice and later rafted out to be deposited as the ice melts. Sand can be incorporated into the ice not only by freezing of the sea surfaces but also by the formation of "anchor ice", which grows as platelets on the seafloor close to the coast to depths of 33 m (Dayton et al., 1969). This sand (Sample 5) is not as well sorted as that from beaches in warmer climates, due to the limited effect of sea ice on wave action, and the higher viscosity of cold water.

The extensive areas of wind blown sand in the sea ice near ice-free parts of the coastline, and our observations of sand grains on the sea ice many kilometres from the coast suggest that the wind may be important in carrying sand offshore. Sand might not travel far when the Sound is ice-free, but could travel tens or even hundreds of kilometres on the surface of drifting sea ice before melting releases it. Bentley's study (1979) of wind blown sand in northern New Harbour showed it to be fine and well sorted and becoming finer with distance from source (Samples 6 and 7). About ten percent of the sand is basaltic. Local bedrock is granitic and metamorphic, but is mantled with moraine containing basaltic debris (in places up to 50 percent, P. Robinson, pers. comm.) transported across the Sound from Ross Island by grounded Ross Sea Ice (Stuiver et al., 1981).

Sediment gravity flows might also distribute sediment in McMurdo Sound, and Kurtz and Anderson (1979) have made a strong case for the importance of gravity flows on the Antarctic Continental Shelf. Distinguishing gravity flow from glacial deposits on textural grounds is difficult, but the muddy character of the basin floors in McMurdo Sound, in comparison to the sandy margins, seems to us incompatible with the widespread gravity flow deposition suggested by Myers (1982), at least for modern seafloor sediment. Nevertheless, the fan-like feature at the mouth of Wilson Valley does point to local gravity flow activity.

The living fauna on the floor of McMurdo Sound is abundant and varied, comprising mainly polychaete worms, bryozoans, sponges and echinoderms (Bullivant, 1967), but their hard parts provide only a small mass contribution to samples from the top 100 mm, normally less than two percent of the sand and gravel fractions we processed. However, high phytoplankton activity has resulted in a much higher biogenic content for the silt fraction, between 30 and 50 percent in the east (Carter et al., 1981) and 10 to 30 percent in the west (Alloway, 1982), where the ice persists for longer. The biogenic material consists largely of diatoms or diatom debris, much of it as fecal pellets.

The main features of sediment from identifiable sources and processes in McMurdo Sound are summarised in Table 1. Ice-rafted sediment may be slightly modified as it passes through the water column, but bottom currents may have a significant effect, where they exceed the entrainment velocity for silt (about 0.25 m/sec, 1 m above the seafloor (Nowell et al., 1981), by inhibiting the accumulation of silt and clay.

## SEA FLOOR SEDIMENTS AND THEIR ORIGINS

Key features of seafloor sediment in McMurdo Sound are the well developed fine to very fine sand modes, and the relatively high clay content considering the lack of clay in most potential source sediment (Figure 3). Samples from near the seaward margins of Ferrar and Mackay Glaciers (such as 8 and 13) are mainly terrigenous silt and clay, consistent with our earlier argument that the basal debris is

**TABLE 1: Features of debris accumulating in McMurdo Sound from different sources and processes**

| Source | Process | Features of debris as released into the water column |
|---|---|---|
| Basal glacial debris | Ice rafting | Broad size distribution with 10% gravel, medium sand mode and 30% mud.[1] |
| Supraglacial debris | Ice rafting | Gravel and sand with very little mud. Virtually all basaltic and angular from McMurdo Ice Shelf; Mixture of angular and rounded, basaltic and plutonic from Victoria Land coast. |
| Beach and subtidal sand | Ice rafting | Moderately sorted sand, little mud.[2] |
| Exposed rock and beach | Wind blown onto sea ice and rafted out | Well sorted fine to very fine sand, very little mud.[2] |
| Varied | Gravity currents | Texture and composition depend on source. |
| Biogenic debris | Settling | Mainly diatomaceous and of silt size. |

Notes: [1]Virtually all from Victoria Land, therefore lacking basalt.
[2]Mainly plutonic but 10 to 20% basalt in southwest. Basalt around Ross Island.

melting out before the ice is reaching the calving line. The gravel and sand is presumably accumulating close to the grounding line, the fine fraction being transported out in suspension.

Samples from the Western Shelf and Slope all have the distinctive well-formed fine-very find sand mode, and indeed in samples such as 10 and 19 it comprises more than 70 percent of the entire sediment. Such a mode cannot result from current sorting because current velocities on the Western Shelf (0 to 0.12 m/sec) are too low to entrain sediment. The most likely origin, considering the range of possible sources, is coastal sand wind blown onto the sea ice and released or washed off during the annual ice break-out, though some may be ice-rafted beach or shallow subtidal sand. The mode can also be seen in samples close to Ross Island (such as 14 and 15) but diminishes northward and is absent in Sample 17 from northern Bird Basin.

The other obvious feature of the textural spectrum is the coarse sand and gravel component of samples such as 16 and 18 from the southern end of Bird Basin and the Western Slope. The dominance and angularity of the basaltic material suggest it is supraglacial debris off bergs from the McMurdo Ice Shelf immediately south. Smaller amounts of coarse debris also occur in samples from the Western Slope and Shelf as far north as Granite Harbour (Sample 11). This inferred pattern of distribution for berg debris follows the observed distribution of ice bergs along the Victoria Land coast, probably a consequence of the surface currents (Figure 1) and prevailing winds.

Clay content of sediment in McMurdo Sound is greatest at the distal end of Bird Basin (Sample 17), and declines to the south and west. It is consistently low (less than five percent on either side of Erebus Basin between Samples 14 and 18, perhaps because of relatively strong bottom currents in this most constricted part of the Sound.

## CONCLUSIONS

Sediment is accumulating on the floor of McMurdo Sound today mainly from the following sources (in decreasing order of importance):

(1) Coastal sand blown by wind onto the sea ice (though there may be some contribution from ice rafting of beach and shallow subtidal sand).
(2) Basal glacial debris melted out close to the grounding line, the fine fraction being transported out into the Sound in suspension.
(3) Diatomaceous debris produced in the Sound itself.
(4) Supraglacial debris from the McMurdo Ice Shelf and ice-cored moraines in the southwest corner of the Sound.

*Acknowledgements* We are pleased to acknowledge support for the seafloor sampling programme from the University Grants Committee, the VUW Internal Research Committee and the N.Z. Lottery Board. Antarctic Division, DSIR, provided the necessary logistic support. We are also grateful to Dr J. Anderson and D. Kurtz, Rice University, for splits of their samples from the floor of McMurdo Sound. The typescript was prepared by Val Hibbert and diagrams by E.F. Hardy.

## REFERENCES

ALLOWAY, B.V., 1982: Radiolarian and sedimentological studies of recent sediments on the floor of McMurdo Sound, Antarctica. Unpublished BSc(Hons) thesis, Victoria University of Wellington, New Zealand, 69 pp.

BARRETT, P.J., 1982: Immediate Report of VUWAE 26, 1981-82. Antarctic Research Centre, Victoria University of Wellington, New Zealand.

BENTLEY, P.N., 1979: Characteristics and distribution of wind-blow sediment, western McMurdo Sound, Antarctica. Unpublished BSc(Hons) thesis, Victoria University of Wellington, New Zealand, 46 pp.

BULLIVANT, J.S., 1967: Ecology of the Ross Sea benthos. *Bull. N.Z. Dep. Scient. ind. Res., 176*, 49-75.

CALKIN, P.E., 1974: Subglacial geomorphology surrounding the ice-free valleys of southern Victoria Land, Antarctica. *J. Glaciol., 13*, 415-29.

CARTER, L., MITCHELL, J.S. and DAY, N.J., 1981: Suspended sediment beneath permanent and seasonal ice, Ross Ice Shelf, Antarctica. *N.Z. J. Geol. Geophys., 24*, 249-62.

DAYTON, P.K., ROBILLIARD, G.A. and DEVRIES, A.L., 1969: Anchor ice formation in McMurdo Sound, Antarctica, and its biological effects. *Science, 163*, 273-4.

DEBENHAM, F., 1948: The problem of the Great Ross Barrier. *Geogr. J., 112*, 196-218.

HEATH, R.A., 1977: Circulation across the ice shelf edge in McMurdo Sound, Antarctica; *in* Dunbar, M.J. (ed.). *Polar Oceans.* Arctic Institute of North America, Calgary, 129-49.

JACOBS, S.S., HUPPERT, H.E., HOLDSWORTH, G. and DREWRY D.J., 1981: Thermohaline steps induced by melting of the Erebus Glacier Tongue. *J. Geophys. Res., 86*, 6547-55.

KEYS, J.R., 1981: Air temperature, wind, precipitation and atmospheric humidity in the McMurdo region. *Antarctic Data Series No. 9*, Antarctic Research Centre, Victoria University of Wellington, New Zealand. 57 pp.

KURTZ, D.D. and ANDERSON, J.B., 1979: Recognition and sedimentologic description of recent debris flow deposits from the Ross and Weddell Seas, Antarctica. *J. Sediment. Petrol., 49*, 1159-70.

McGINNIS, L.D., 1973: McMurdo Sound—a key to the Cenozoic of Antarctica. *U.S. Antarct. J., 8*, 166-79.

MYERS, N.C., 1982: Marine geology of the western Ross Sea: implications for Antarctic glacial history. Unpublished M.S. thesis, Rice University, Houston. 234 pp.

NOWELL, R.M., JUMARS, P.A. and ECKMAN, J.E., 1981: Effects of biological activity on entrainment of marine sediments. *Mar. Geol., 42*, 133-53.

ROBINSON, P.H., 1979: An investigation into the processes of entrainment, transportation and deposition of debris in polar ice, with special reference to the Taylor Glacier, Antarctica. Unpublished PH.D thesis, Victoria University of Wellington, New Zealand, 235pp.

ROBINSON, P.H., in press: Ice dynamics and thermal regime of Taylor Glacier, south Victoria Land, Antarctica. *J. Glaciol.*

STUIVER, M., DENTON, G.H., HUGHES, T.J. and FASTOK, J.I., 1981: History of the marine ice sheet in West Antarctica during the last glaciation: a working hypothesis; *in* Denton, G.H., and Hughes, T.J. (eds.). *The Last Great Ice Sheets.* Wiley-Interscience, New York, 319-436.

SWITHINBANK, C., 1970: Ice movement in the McMurdo Sound area of Antarctica. *International Symposium on Antarctic Glaciological Exploration (ISAGE)*, Hanover, NH, September 1968, International Association of Scientific Hydrology and SCAR Publ. 86, 472-87.

# A REVISED HISTORY OF GLACIAL SEDIMENTATION IN THE ROSS SEA REGION

M.L. Savage and P.F. Ciesielski, *Department of Geology, University of Georgia, Athens, Georgia 30602. U.S.A.*

*Abstract* DSDP drill cores from the Ross Sea of Antarctica contain a lengthy stratigraphic record (Oligocene-Holocene) of glacial marine sedimentation in close proximity to the Antarctic continent; previous palaeo environmental interpretations of this important record have been hindered by scanty biostratigraphic information. New diatom, silicoflagellate, and radiolarian zonations are used in a micropalaeontologic study of Miocene-Holocene sediments from Ross Sea Site 273 (western Ross Sea), Site 272 (eastern Ross Sea) and Site 274 (Ross Sea continental rise). Miocene sediments of Site 272 and 273 represent only a fragmentary record of the Early and Middle Miocene. Extremely high sediment accumulation rates (> 150 m/Ma) and the sedimentology and micropalaeontology of these sites are indicative of an open Ross Sea (no large ice shelves) with sedimentation dominated by rapidly calving, debris-laden outlet glaciers. By 16 Ma ago small ice shelves began forming along the western margin of the Ross Sea. The Ross Sea regional disconformity represents a minimum of 10 Ma; bracketing ages are about 14.7 to 4.0 Ma in the western Ross Sea (Site 273) and about 14.0 to< 0.6 Ma in the eastern Ross Sea (Site 272). The earliest evidence for an extensive ice shelf is found at continental rise Site 274. A major increase in sand size ice-rafted detritus (IRD) and the initial occurrence of dropstones in Core 12 suggests Ross Ice Shelf formation by 10.0 to 8.5 Ma. Initial formation of the West Antarctic Ice Sheet (WAIS) must have, therefore, occurred after 10 Ma.

Tertiary glacial changes in Antarctica have played an influential role in the global evolution of Tertiary climate and oceanography (Kennett, 1977; Ciesielski et al., 1982). Sediments recovered in Deep Sea Drilling Project (DSDP) cores from the Ross Sea region (Figure 1) are particularly important to reconstructions of Antarctic glacial conditions because of their lengthy stratigraphic sequence dating to the Oligocene; nowhere else is this record available so close to Antarctica. The Ross Sea sites, therefore, provide the earliest record showing the initial affects of Antarctic glacial conditions from which interpretations of the Tertiary climatic and glacial developments can be made. In addition, the Ross Sea sites, located between East and West Antarctica, provide good evidence for glacial conditions in both areas.

Since the original studies of the Ross Sea DSDP holes (Hayes et al., 1975) siliceous microfossil biostratigraphic zonal schemes have been greatly improved. This study re-examines the diatom contents of DSDP Sites 272, 273 and the Upper Miocene of Site 274 and correlates the Neogene sedimentary sequences of these holes to the new diatom zonations of Weaver and Gombos (1981) and Ciesielski (in press). This new chronology is utilized to reinterpret the Neogene palaeoenvironment of the Ross Sea; including the timing of the formation of extensive ice shelves in the Ross Sea, the development of the West Antarctic Ice Sheet (WAIS), and the formation of the regional Ross Sea disconformity.

## RESULTS AND DISCUSSIONS

### Site 272 Sediment Ages (Figure 2)

*Unit 2: Miocene* Results of this study indicate that Site 272 Miocene sediments are older than determined by McCollum (1975). Core 3, Section 1 through Core 16, Section 2 (Unit 2A), represents a very short period (about 14.1 to about 13.8 Ma) of the Middle Miocene, the upper *Nitzschia grossepunctata* and possible lower *N. denticuloides* Zone. Core 47 through Core 16, Section 3 (Units 2B and 2C) mid-to-upper Lower Miocene with Core 16, Section 3 to Core 19, Section 6 containing diatoms indicative of the lower *Nitzschia maleinterpretaria* Zone (about 18.19-18.34 Ma) and Core 21, Section 1 through Core 47 containing a diatom assemblage equivalent to the *Coscinodiscus rhombicus* Zone (about 18.34-19.23 Ma). An Early to

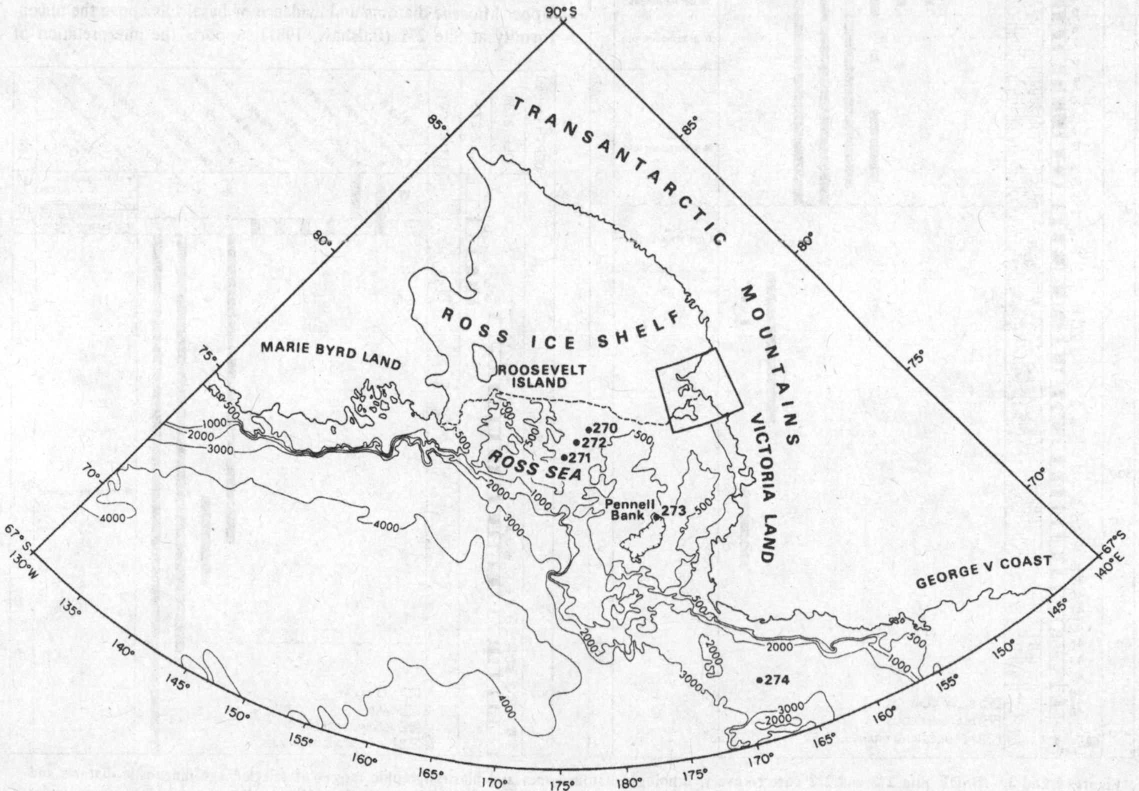


**Figure 1. Locations of DSDP Ross Sea Sites 270-274. Sites 272 and 273 are re-examined herein.**

Middle Miocene unconformity occurs between Core 16, Section 2 and 3; this hiatus spans more than 4 Ma. This erosional and/or non depositional event occurred sometime between 18.2 and 14.1 Ma and possibly is correlative with the erosional event that caused the Early Miocene unconformity at Site 273.

*Unit 1: Pleistocene—Holocene* Unit 1 consists of Brunhes Chronozone sediment (< 0.65 Ma) with diatoms indicative of the *Coscinodiscus lentiginosus* Zone.

This is in agreement with Hayes et al., (1975), McCollum (1975), Kellogg et al., (1979 a, b), Osterman and Kellogg (1979) and Truesdale and Kellogg (1979) who also gave Unit 1 a Brunhes age. The postulated Matuyama disconformity of Fillon (1975), therefore, does not exist at this site.

## Site 272 Depositional and Erosional History (Figure 4)

*Unit 2 Miocene* Since the site is the closest to West Antarctica, its sedimentary record should reflect the history of the West Antarctic Ice Sheet. Unit 2B lacks any bedding for the most part; however, it contains several shell fragments consisting mostly of thin-shelled bivalves and some small gastropods, and shows evidence of bioturbation. The lack of any significant bedding and the high sedimentation rates seem to indicate the absence of ice shelves and their associated currents.

Miocene sediment accumulation rates were extremely high at Sites 272 and 273 inferring an open Ross Sea, the presence of land-based ice at sea level, extremely rapid calving of outlet glaciers and probable wet-based glacial conditions. Below the Early to Middle Miocene unconformity at Site 272 sediment was deposited at a rate of 275 m/Ma between about 18.2-19.2 Ma. Rare pebbles and granules appear to have originated in West Antarctica. Sediment above the Early to Middle Miocene unconformity (about 14.1-13.8 Ma) was deposited at greatly increased sedimentation rates in excess of 420 m/Ma indicating increased calving from West Antarctica.

The only bedding observable in Unit 2A sediments consists of thin diatom-rich beds from the upper portion of the unit which occur at about 15 metre intervals (Hayes et al., 1975), implying periodic blooms of diatoms in the largely ice free Ross Sea during the deposition of Unit 2A sediments (middle Middle Miocene).

The first common dolerite pebbles in Core 15A indicate transport by ice through the Transantarctic Mountains. Perhaps this marks the time when the growing East Antarctic ice sheet first topped the Transantarctic Mountains to introduce this new supply of debris. Sporadic bedding within Unit 2A (Cores 6-8A) may have resulted from bottom currents produced by fringing ice shelves. The source of the majority of ice-rafted pebbles is inferred to come from southern Marie Byrd Land; the incursion of dolerite pebbles in uppermost Unit 2A sediments (Cores 4, 5 and 6) during the middle Middle Miocene suggests an additional source from the Transantarctic Mountains (Barrett, 1975). Ice shelves and outlet glaciers fringing the western portion of the Ross Sea must have been extensive enough at this time to raft pebbles across the Ross Sea to Site 272.

Variations in the abundance of ice-rafted granules and pebbles represent fluctuations in the extent of calving outlet glaciers and the possible presence of fringing ice shelves. Clasts 10 mm or larger are relatively common in Unit 2A sediments but are rare in Units 2B and 2C (Hayes et al., 1975) which indicates more intense West Antarctic glacial conditions during the deposition of Unit 2A sediments. The abundance of granules and pebbles is also seen to vary within Units 2A and 2B from three to four per metre to as many as 60 per metre in many places over distances of tens of centimetres (Hayes et al., 1975). Rapid fluctuations in calving and ice shelf conditions, therefore, occurred during the depositing of Lower and lower Middle Miocene sediments.

*Unit 1: Pleistocene—Holcene* The presence of reworked Middle and Upper Miocene diatoms and evidence of basal tills above the unconformity at Site 272 (Balshaw, 1981) supports the interpretation of

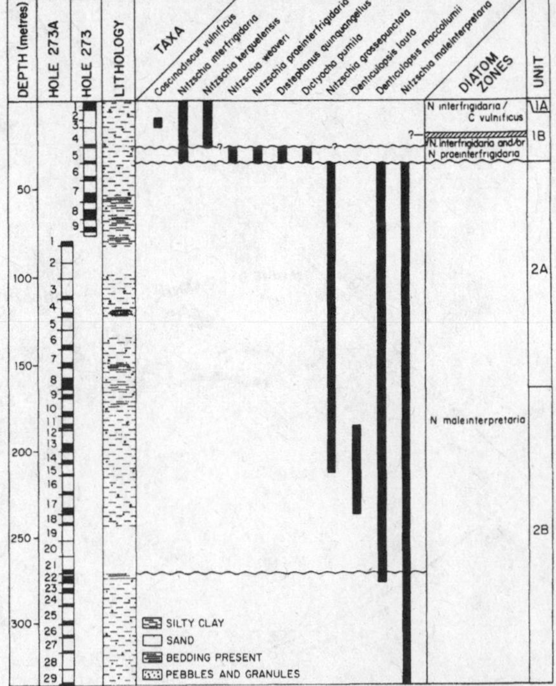


**Figures 2 and 3.** DSDP Site 272 and 273 core recovery, lithology, diatom zones and biostratigraphic ranges of selected age diagnostic diatoms and silicoflagellates. Although no age diagnostic species were found below Core 38 of Hole 272 (Unit 2C), the diatom assemblage is very similar to Unit 2B and also appears to be Early Miocene.

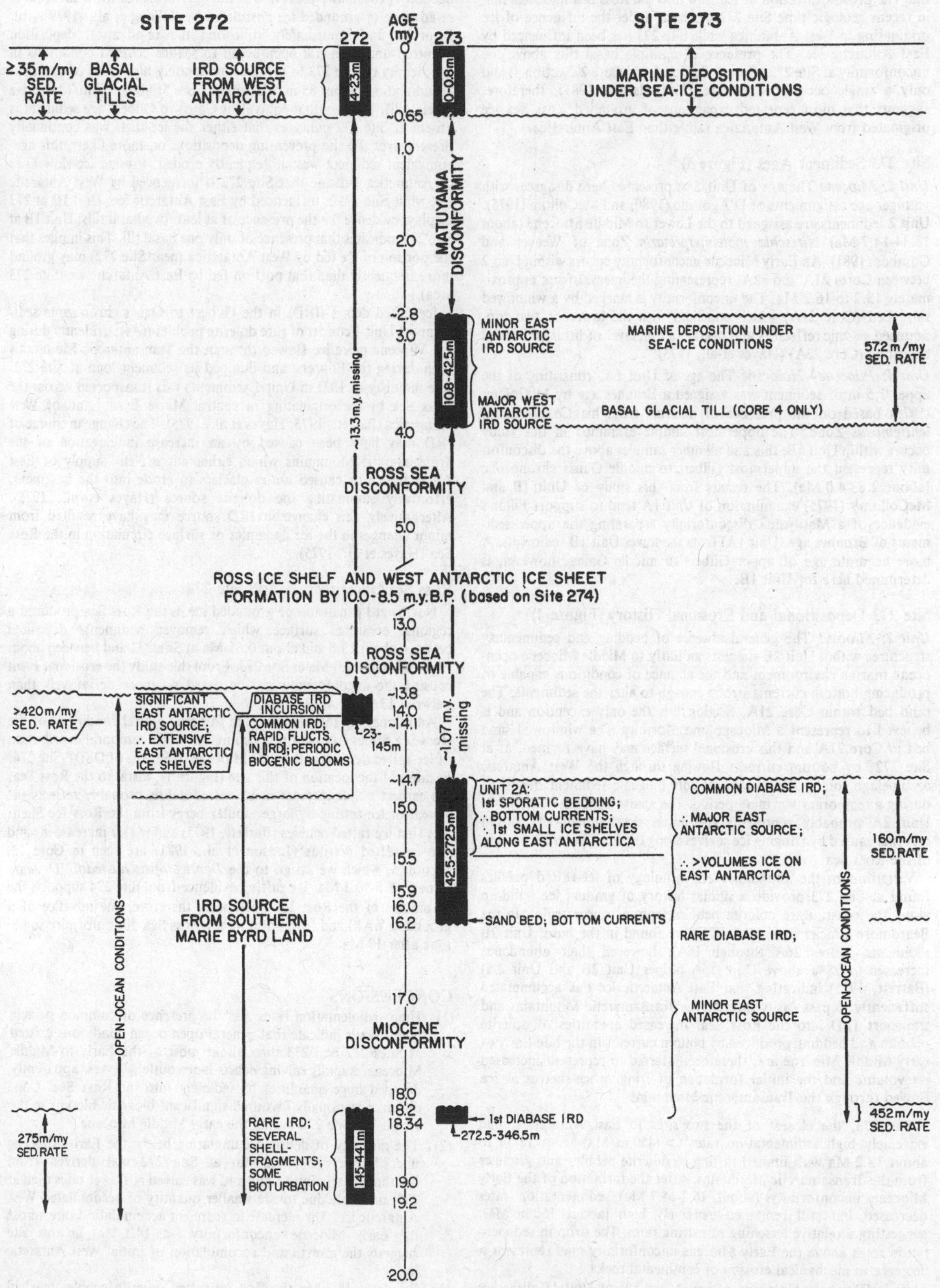


**Figure 4.** Summary of the Neogene palaeoenvironment of the Ross Sea based largely on the sedimentology and micropalaeontology of DSDP Sites 272 and 273.

Kellogg et al., (1979 a, b) that Unit 1B was deposited as a basal glacial till. The present direction of ice flow into the Ross Sea indicates that in recent geologic time Site 272 has been under the influence of ice originating in West Antarctica while Site 273 has been influenced by East Antarctic ice. The presence of multiple basal tills above the unconformity at Site 272 (Core 1, Section 5 and Core 2, Section 4) and only a single occurrence at Site 273 (Balshaw, 1981), therefore, suggests that most repeated expansions of grounded Ross Sea ice originated from West Antarctica rather than East Antarctica.

### Site 273 Sediment Ages (Figure 3)

*Unit 2: Miocene* The age of Unit 2 as presented here disagrees with younger age assignments of D'Agostino (1980) and McCollum (1975). Unit 2 sediments are assigned to the Lower to Middle Miocene (about 18.34-14.7 Ma) *Nitzschia maleinterpretaria* Zone of Weaver and Gombos (1981). An Early Miocene unconformity occurs within Unit 2 between Cores 21A and 22A, representing the interval from approximately 18.2 to 16.2 Ma. The unconformity is marked by a winnowed sand bed (Core 21A) which coincides with the lower of two pronounced seismic reflectors and overlies a thin layer of lithified bedded claystone (Core 22A) (Hayes et al., 1975).

*Unit 1: Pliocene—Holocene* The age of Unit 1A, consisting of the upper 1.5 m of sediment was assigned a Brunhes age by McCollum (1975) based on the presence of diatoms from his *Coscinodiscus lentiginosus* Zone. The uppermost sample examined in this study occurs within Unit 1B; this and all other samples above the disconformity represent the uppermost Gilbert to middle Gauss chronozone (about 2.83-4.0 Ma). The results from this study of Unit 1B and McCollum's (1975) examination of Unit 1A tend to support Fillon's evidence of a Matuyama disconformity separating the upper sediments of Brunhes age (Unit 1A) from the lower Unit 1B sediments. A more accurate age of upper Gilbert to middle Gauss, however, is determined here for Unit 1B.

### Site 273 Depositional and Erosional History (Figure 4)

*Unit 2: Miocene* The general absence of bedding and sedimentary structures within Unit 2B suggests an Early to Middle Miocene open-ocean marine environment and the absence of conditions capable of producing bottom currents strong enough to alter the sediments. The sand bed within Core 21A, Section 1 is the only exception and is believed to represent a Miocene unconformity. The winnowed sand bed in Core 21A and this erosional surface may have formed, as at Site 272, by bottom currents flowing through the West Antarctic archipelago, or by the erosion of soft biogenic sediment deposited during a temporary warming period. The sparse bedding observed in Unit 2A probably represents the initial development of bottom currents caused by fringing ice shelves along the East Antarctic margin of the Ross Sea.

Variations in the abundance and lithology of ice-rafted pebbles found at Site 273 provide a similar history of gradual ice build-up along the coast. Rare dolerite pebbles from a source region in the Beardmore Glacier area (82°-86°S) are found in the basal Unit 2B sediments (Cores 26A through 16A); however, their abundance increases to 28% above Core 16A (upper Unit 2B and Unit 2A) (Barrett, 1975) indicating that East Antarctic ice has accumulated sufficiently to pass easily through the Transantarctic Mountains and transport IRD into the Ross Sea. Increased quantities of dolerite pebbles and bedding produced by bottom currents in the late Early to early Middle Miocene are, therefore, inferred to represent increased ice volume and the initial formation of fringing ice-shelves as ice flowed through the Transantarctic Mountains.

Site 273, the closest of the two sites to East Antarctica, had extremely high sedimentation rates (> 450 m/Ma) from 18.34 to about 18.2 Ma with limited rafting of dolerite pebbles and granules from the Transantarctic Mountains. After the formation of the Early Miocene unconformity (about 16.2-14.7 Ma) sedimentation rates decreased but still remained extremely high (about 160 m/Ma), suggesting a relative lessening of calving rates. The drop in sedimentation rates above the Early Miocene unconformity may represent a decrease in mechanical erosion of continental rocks.

*Unit 1: Pliocene—Holocene* Although our age of Unit 1B disagrees with the Brunhes age of Kellogg et al., (1979b), their suggested mode of deposition as a basal glacial till appears to be correct, in part. A single basal till was found above the unconformity at Site 273 (Core 4, Section 1) (Balshaw, 1981) and is the only evidence for a subsequent re-advance of grounded ice postulated by Kellogg et al., (1979 a, b). Prior to and immediately following the re-advance, deposition occurred under general open-ocean to sea-ice covered conditions in the vicinity of Site 273 to produce moderately high rates of sediment accumulation (about 85 m/Ma within Core 5; about 70 m/Ma above glacial till). The fact that equivalent Gauss to Gilbert age sediment is absent at Site 272 indicates that either the ice-shelf was continually present over the site preventing deposition, or, more likely, this age-equivalent sediment was subsequently eroded. Present ice-flow lines of Antarctica indicate that Site 272 is influenced by West Antarctic ice, while Site 273 is influenced by East Antarctic ice. Unit 1B at 272 displays evidence for the presence of at least two basal tills; Unit 1B at Site 273 indicates that presence of only one basal till. This implies that the portion of ice fed by West Antarctica (near Site 272) may ground more frequently than that portion fed by East Antarctic ice (Site 273 area).

Ice-rafted debris (IRD) in the Gilbert to Gauss chronozone sediments of Unit 1 consist of rare dolerite pebbles (6%); evidently during the Miocene more ice flowed through the Transantarctic Mountains than during the Pliocene and dumped its sediment load at Site 273. The majority of IRD in Unit 1 sediments was transported across the Ross Sea by ice originating in central Marie Byrd Land of West Antarctica (Barrett, 1975; Hayes et al., 1975). The change in source of IRD may have been caused by an increase in elevation of the Transantarctic Mountains which either cut off the supply of East Antarctic ice or caused outlet glaciers to erode into the basement, effectively eliminating the dolerite source (Hayes et al., 1975). Alternatively, this change in IRD source may have resulted from minor changes in the ice dynamics or surface circulation in the Ross Sea (Hayes et al., 1975).

### Ross Sea Unconformity and Site 274

Northward expansion of grounded ice in the Ross Sea produced a regional erosional surface which removed sediments deposited between about 13.8 and about 0.65 Ma at Site 272 and between about 14.7 and about 4.0 Ma at Site 273. From this study the erosional event causing the unconformity can be dated no more definitively than between 13.8 and 4.0 Ma.

Additional insight into the Late Miocene glacial conditions of the Ross Sea comes from our re-examination of the temporal distribution of ice-rafted detritus in the Late Miocene sediments of DSDP Site 274. Because of the location of this site (Figure 1), distal to the Ross Sea, the earliest major increase in ice-rafted debris probably reflects the advent of ice-rafting by large tabular bergs from the Ross Ice Shelf. The first ice-rafted pebbles (Barrett, 1975) and major increase in sand size ice-rafted detritus (Judson et al., 1981) are seen in Core 12, Section 6, which we assign to the *Denticulopsis hustedtii/ D. lauta* Zone (~8.8-10.3 Ma. Ice rafting evidence from Site 274 supports the formation of the Ross Ice Shelf and, therefore, the advance of a grounded WAIS and formation of the Ross Sea disconformity sometime after 10 Ma.

## CONCLUSIONS

(1) High sedimentation rates and the presence of common pelagic microfossils indicate that general open-ocean conditions existed at Sites 272 and 273 throughout most of the Early to Middle Miocene. Rapidly calving debris-laden outlet glaciers apparently dumped huge quantities of sediment into the Ross Sea. Conditions occasionally favoured significant biogenic blooms in the vicinity of Site 272 during the early Middle Miocene.

(2) The majority of debris accumulating below the Early Miocene unconformity (> 18.2 Ma) at Site 272 was derived from Southern Marie Byrd Land and was calved at slower rates than at 272, probably due to the smaller quantity of accumulated West Antarctic ice. The increase in sediment accumulation rate across the Early Miocene unconformity (< 14.1 Ma) at this site suggests the growth and accumulation of initial West Antarctic ice.

(3) By about 18.3 Ma the first ice-rafted dolerite pebbles reached Site 273 (Core 27A), indicating an increased source of ice from the Transantarctic Mountains.

(4) East Antarctic ice had accumulated further, by about 15.9 Ma, to allow abundant ice flow through the Transantarctic Mountains. This less restricted flow of ice deposited abundent dolerite pebbles in Site 273 Cores 15A through six.

(5) Initial development of small fringing ice shelves along the East Antarctic coast had begun by about 15.5 Ma. The first evidence of significant ice-shelf produced bottom currents was indicated at this point by sporadic bedding within Unit 2A at Site 273.

(6) By about 13.8 Ma large volumes of ice had accumulated on East Antarctica. The incursion of dolerite pebbles within Site 272 Cores 4, 5 and 6 (about 13.8 Ma) indicate that the debris had been rafted across the Ross Sea to this site..

(7) The Ross Sea regional unconformity probably represents the removal of more than 10 Ma sedimentary record by the expansion of a grounded Ross Ice Shelf; at Site 272 the hiatus spans about 13.8-0.65 Ma while at Site 273 it represents the interval from about 14.7 to 4.0 Ma. Sedimentological evidence from these two sites, therefore, indicates the formation of a Ross Ice Shelf and West Antarctic Ice Sheet of present day dimensions at some time after 13.8 Ma. Ice-rafting evidence from continental rise Site 274 suggests that the Ross Ice Shelf actually was formed by 10.0-8.5 Ma which implies an initial West Antarctic Ice Sheet formation sometime after 10.0 Ma.

(8) Since the retreat of the grounded ice sheet, Gilbert to Gauss and Brunhes-age sediments have been preserved at Sites 273 and 272, respectively. At least two subsequent grounding of the ice sheet were recorded in DSDP Site 272 sediments, while evidence for only one was recorded at Site 273.

*Acknowledgements* Drs Michael T. Ledbetter and Peter J. Barrett are thanked for a critical review of the manuscript. Drafting of the text figures was by Ms Barbara Daniel. Financial support for this research was provided by NSF Grant DPP-811347 to M.T. Ledbetter and P.F. Ciesielski.

## REFERENCES

BALSHAW, K.M., 1981: Antarctic glacial chronology reflected in the Oligocene through Pliocene sedimentary sections in the Ross Sea. Unpublished Ph.D Thesis, Rice University, Houston, Texas. 140pp.

BARRETT, P.J., 1975: Characteristics of pebbles from Cenozoic marine glacial sediments in the Ross Sea (DSDP Sites 270-274) and the south Indian Ocean (Site 268). *Initial Reports of the Deep Sea Drilling Project 28,* 769-84. U.S. Govt. Printing Office, Washington, D.C.

CIESIELSKI, P.F. (in press): Neogene diatom stratigraphy of Deep Sea Drilling Project Leg 71 sediments. *Initial Reports of the Deep Sea Drilling Project 71.* U.S. Govt. Printing Office, Washington, D.C.

CIESIELSKI, P.F., LEDBETTER, M.T. and ELLWOOD, B.B., 1982: The development of Antarctic glaciation and the Neogene paleoenvironment of the Maurice Ewing Bank. *Mar. Geol., 46,* 1-51.

D'AGOSTINO, A., 1980: Foraminiferal biostratigraphy, paleoecology and systematics of DSDP Site 273, Ross Sea, Antarctica. Unpublished M.S. Thesis, Northern Illinois University, Dekalb, Illinois. 124pp.

FILLON, R.H., 1975: Late Cenozoic Paleo-Oceanography of the Ross Sea, Antarctica. *Geol. Soc. Amer., Bull., 86,* 839-45.

HAYES, D.E., FRAKES, L.A., BARRETT, P.T., BURNS, D.A., CHEN, P.H., FORD, A.B., KANEPS, A.G., KEMP, E.M., McCOLLUM, D.W., PIPER, D.J., WALL, R.E. and WEBB, P.N., 1975: *Initial Reports of the Deep Sea Drilling Project, 28.* U.S. Govt. Printing Office, Washington, D.C.

JUDSON, M., EHRLICH, R., WILLIAMS, D., and CIESIELSKI, P.F., 1981: Early Miocene to Pleistocene fluctuations in ice-rafted quartz from DSDP Site 274: Evidence from Fourier grain shape analysis. *Geol. Soc. Am. Abstr. Programs, 13,* 481.

KELLOGG, T.B., OSTERMAN, L.E. and STUIVER, M., 1979a: Late Quaternary sedimentology and benthic foraminiferal paleoecology of the Ross Sea, Antarctica. *J. Foram. Res., 9,* 322-35.

KELLOGG, T.B., TRUESDALE, R.S. and OSTERMAN, L.E., 1979b: Late Quaternary extent of the West Antarctic Ice Sheet: new evidence from Ross Sea cores. *Geology, 7,* 249-53.

KENNETT, J.P., 1977: Cenozoic evolution of Antarctic glaciation, the Circum-Antarctic Ocean and their impact on global palaeoceanography. *J. Geophys. Res., 82,* 3843-60.

McCOLLUM, D.W., 1975: Diatom stratigraphy of the Southern Ocean, *Initial Reports of the Deep Sea Drilling Project, 28,* 515-71. U.S. Govt. Printing Office, Washington, D.C.

OSTERMAN, L.E. and KELLOGG, T.B., 1979: Recent benthic foraminiferal distributions from the Ross Sea, Antarctica: Relation to ecologic and oceanographic conditions. *J. Foram. Res., 9,* 250-69.

TRUESDALE, R.S. and KELLOGG, T.B., 1979: Ross Sea diatoms: modern assemblage distributions and their relationship to ecologic, oceanographic and sedimentary conditions. *Mar. Micropaleontol., 4,* 13-31.

WEAVER, F.M. and GOMBOS, A.M., Jr., 1981: Southern high latitude diatom biostratigraphy; in Warme, J.E., Douglas, R.C. and Winterer, E.L. (eds.). The Deep Sea Drilling Project: A Decade of Progress. *Soc. Econ. Paleont. Mineral., Spec. Publ., 32,* 445-70.

# CLIMATIC, PALAEO-OCEANOGRAPHIC AND TECTONIC INTERPRETATION OF PALAEOGENE-NEOGENE BIOSTRATIGRAPHY FROM MSSTS-1 DRILLHOLE, McMURDO SOUND, ANTARCTICA

P.N. Webb, *Institute of Polar Studies, Ohio State University, Columbus, Ohio, 43210 U.S.A.*

**_Abstract_ McMurdo Sound Sedimentary and Tectonic Study (MSSTS) drillhole No.1 penetrated 230 m of diamictite, sandstone and mudstone. The succession extends from 196 to 426 m below sea level. Sedimentary analysis suggests that these sediments were deposited under glaciomarine conditions. Foraminiferal studies indicate that the succession spans parts of Palaeogene and Neogene, that a six-fold zonal subdivision is possible and that zone boundaries mark the positions of significant regional disconformities. The interval between the bottom of the drillhole (230 m) and 141 m is dated as Late Paleocene to Early/Middle Eocene; 141 m and 115 m is dated as Late Oligocene-Early Miocene; 115 and 32 m is poorly fossiliferous but probably Miocene; 32 and 20 m is Pliocene; and 20-0 m is poorly fossiliferous and probably Pleistocene-Recent.**

**The structural block on which the MSSTS-1 succession is located underwent a history of uplift and truncation through much of the Cenozoic. The adjacent Transantarctic Mountain block was also elevated during the same period but at a much faster rate. Sediment derived from the Transantarctic Mountains was transported over the MSSTS-1 area and into the deeper parts of the western Ross Sea. Sediment deformation in Palaeogene marine sediments suggests active slumping immediately seaward of the Transantarctic Mountains.**

**Three of the five faunal and sedimentary disconformities coincide with seismic reflectors S, A and K. These are calibrated and along with intervening sedimentary units may be traced laterally into the deeper parts of the southwestern Ross Sea. The association of Late Paleocene to Early-Middle Eocene foraminifera and glaciomarine sediments low in the MSSTS-1 succession indicates that the adjacent Transantarctic Mountains had attained a significant elevation and were the site of significant ice development as early as 45-55 Ma. Effects of glaciation apparently reached central areas of the Ross Sea later in the Palaeogene. The record at DSDP Site 270 points to a non-glacial climate near sea level at about 26 Ma with deposition of glaciomarine sedimentation commencing in the latest Oligocene (about 25 Ma). The foraminiferal record at MSSTS-1 points to the existence of cool temperate marine circulation close to the significantly colder terrestrial climate of the Transantarctic Mountains.**

**Recycled Late Cretaceous foraminifera are quite common within the Palaeogene succession (230 to 141 m). The presence of _Globotruncana_ spp. (double keeled forms), _Rugoglobigerina, Hedbergella, Bolivinoides_ suggests that a non-glacial temperate marine climate characterised the Ross Sea in the latest Cretaceous. Initiation of glaciation in the western Ross Sea is presumably an Early Cenozoic event.**

# THE AGE OF SEDIMENTS AT THE J-9 DRILLING SITE, ROSS ICE SHELF

H.T. Brady, *Honorary Associate, School of Biological Sciences, Macquarie University, North Ryde, NSW 2113, Australia*

*Abstract* The sedimentary cover of the Ross Sea at the J-9 RISP drill site was reported to be Middle Miocene by Webb and Brady (1978). Kellogg et al. (1981) describe the sediments as Pleistocene with reworked Pliocene and Miocene fossils. Brady contends that wrong stratigraphic ages were ascribed by Kellogg et al to some taxa and he suspects that other taxa were wrongly identified by those authors. This paper sets out the parameters of this important debate and the arguments on either side.

# CLIMATIC AND TECTONIC IMPLICATIONS FOR THE LATE CENOZOIC IN THE McMURDO SOUND REGION FROM OFFSHORE DRILLING (MSSTS-1)

P.J. Barrett, *Geology Department and Antarctic Research Centre, Victoria University, Wellington, New Zealand.*

B.C. McKelvey, *Department of Geology, University of New England, Armidale, NSW 2351, Australia.*

*Abstract* MSSTS-1 was drilled in 195 m of water 12 km northwest of Butter Point in McMurdo Sound, reaching 226 m sub-bottom before being terminated by sea-ice movement. The entire cored sequence is of marine glacial origin, mainly mud and sand in varying porportions with scattered clasts up to boulder size. The upper 11 m is soft and probably of Recent age; below this the core is semilithified, and has been dated tentatively as Pliocene from 18 to 32 m and Middle Miocene below 54 m. Foraminifera clearly indicate a Middle Miocene age for the interval 118 to 186 m.

The fine-grained nature of the Miocene section and the scattered clasts, many of which are striated, indicates sedimentation of ice-rafted glacial debris generally in water more than 100 m deep. A significant exception is the shoaling sequence leading up to beach sand from 200 to 190 m sub-bottom. The observed supply of sediment requires ice both more extensive and warmer than at present, though the clay mineralogy indicates weathering on land like the present. Also there appears to have been no vegetation in the nearby mountains, judging from the absence of contemporaneous palynomorphs.

The core below 18 m sub-bottom it surprisingly firm, and this is reflected in the seismic velocities. The sandstones show rapid and irregular velocity increases with depth, reaching 5.1 $km/s^{-1}$, probably due to silica cementation. The mudstones remain at about 2.2 $km/s^{-1}$ throughout the hole, high for such a shallow depth. One explanation is that the cored sequence has been buried by 1 km or more of younger strata, since removed, implying sinking and uplift of that order since the Middle Miocene.

# FISSION TRACK GEOCHRONOLOGY OF GRANITOIDS AND UPLIFT HISTORY OF THE TRANSANTARCTIC MOUNTAINS, VICTORIA LAND, ANTARCTICA

A.J.W. Gleadow, *Department of Geology, University of Melbourne, Parkville, Victoria 3052, Australia.*

*Abstract* Apatite fission track ages of basement samples from Wright and Victoria Valleys increase systematically with sample elevation. The ages observed so far all post-date the emplacement of Ferrar Dolerites and range from 157 ± 7 to 68 ± 4 Ma. These ages reflect the times at which different samples cooled below a temperature of about 100°C and began to retain stable fission tracks. The rate at which the apatite ages change with elevation is controlled by the uplift rate and offsets in the pattern give information on faulting.

The apatite age gradient indicates a steady, slow uplift over this period at about 15 m/Ma. Taken in conjunction with the present geothermal gradient in this area, the youngest ages indicate a greatly increased rate of uplift during the Cenozoic, and give a reliable older limit to the onset of the Victoria Orogeny. The results indicate that nearly 4 km of uplift has occurred during the Cenozoic with a minimum average uplift rate of about 55 m/Ma, assuming the thermal gradient has remained constant over that time. Later onset of rapid uplift would lead to correspondingly greater Cenozoic uplift rates.

The pattern of sphene and zircon fission track ages is quite different to that for apatites and is controlled largely by the proximity of dolerite sheets in the Dry Valleys area. These minerals record the time of cooling below temperatures of about 250 ± 50 and 200 ± 50°C respectively and their fission track ages increase with distance from the dolerite contacts. Preliminary fission track results will also be presented for granitic rocks from northern Victoria Land.

# 11

# Antarctica in Gondwanaland

# STRUCTURAL INTERPRETATIONS OF WILKES LAND, ANTARCTICA

R.H.N. Steed[1], *Scott Polar Research Institute, University of Cambridge, Cambridge CB2 1ER, U.K.*

[1]*Present Address: British Petroleum Limited, Britannic House, Moor Lane, London EC2Y 9BU, U.K.*

*Abstract* Radio echo sounding data collected in East Antarctica have been compiled into a bedrock topography map, and reveal areas of separate structural character and distinct N-S structural symmetry. The outcrop of Beacon sediments and Ferrar Dolerite at Horn Bluff appear to have an association with plateau features within the northern Wilkes Basin; and distinct landforms associated with Beacon strata have been identified from radio echo data as occurring along the eastern flank of the basin. The terrain to the west of the basin has rough and faulted forms and satellite magnetic data indicates that the western basin edge is also the boundary of the Precambrian craton. It is concluded that the Wilkes Subglacial Basin is a sedimentary basin developed by tensional growth and subsidence similar to the development of the North Sea Basin. A simple model is presented indicating the tectonic evolution.

Current structural interpretation of the Wilkes Land region of Antarctica draw on geological observations made at a small number of rock outcrops on the periphery of the continent, but incorporate little information from the subglacial continental interior (Craddock, 1972; Elliot, 1975). A further stage of interpretation is now attempted using information from airborne radio echo soundings (RES) which provide a source of data on subglacial topography. Subglacial topography maps have been previously published (Drewry, 1975; Steed and Drewry, 1982). Additional RES data have now been compiled and a bedrock map for Wilkes Land is presented (Figure 1); a provisional form of this data was included in Johnson and Vanney (1980). The grid of flightlines is shown in Figure 2 and descriptions of the methods of data collection and reduction can be found in Steed (1980), Steed and Drewry (1982), and Drewry et al. (1982).

To aid further discussion the mapped area, shown in Figure 2, is divided into primary regions, delineated by selected 250 m below sea level (bsl) contours. Of all the regions marked only the Transantarctic Mountains and the Wilkes Basin are recognised by the Place Names Commission, but the Aurora Basin has already been identified and named by Drewry (1975, 1976). Other regions have been given sample labels.

## TRANSANTARCTIC MOUNTAINS

Only the western flank of the Transantarctic Mountains is shown in Figure 2, and the survey grid is too coarse to adequately resolve the topography. Two major breaks, however, are apparent in the mountain chain. These are large depressions which correspond to the Byrd and David Glaciers. The extent to which these breaches in the

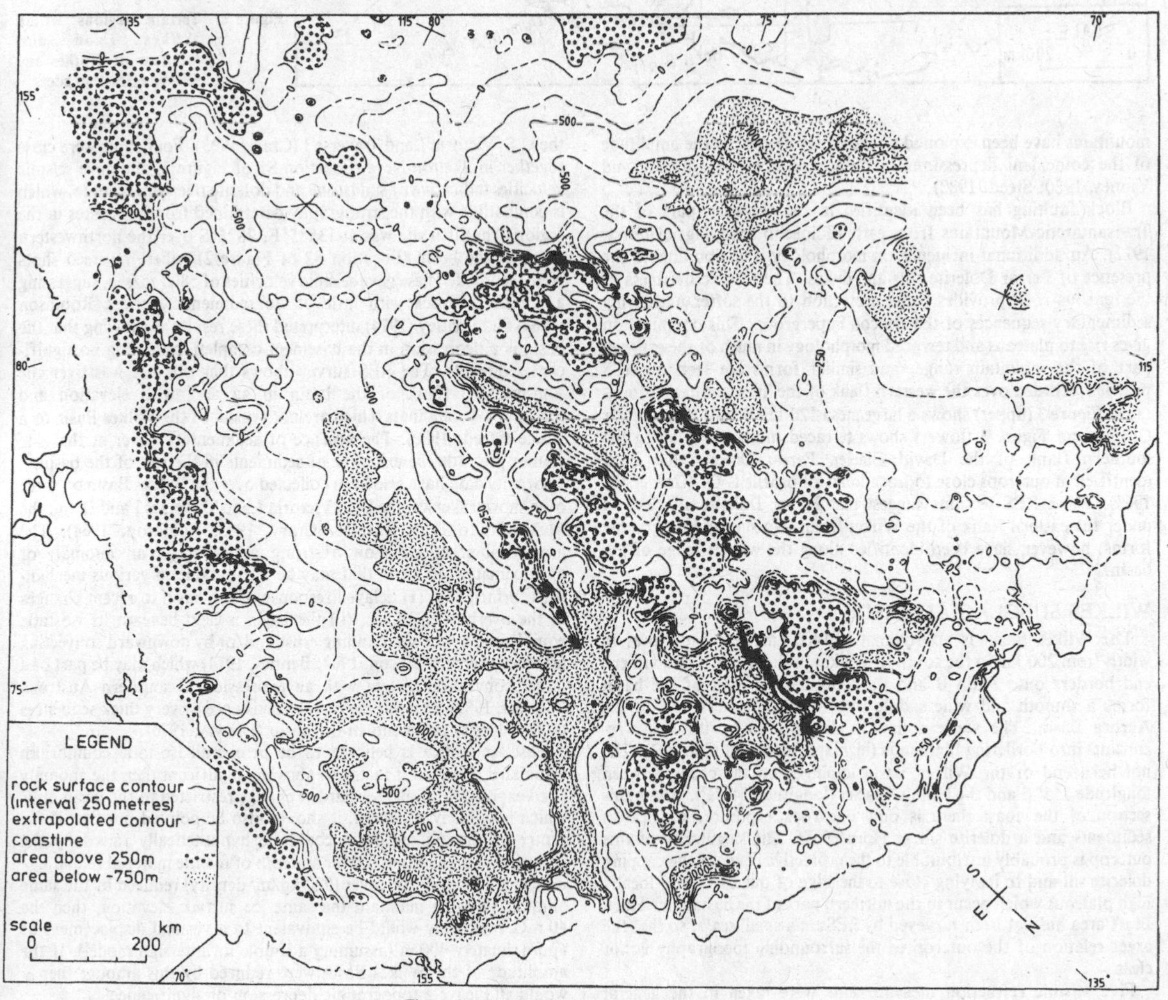


**Figure 1. Bedrock topography in Wilkes Land.**

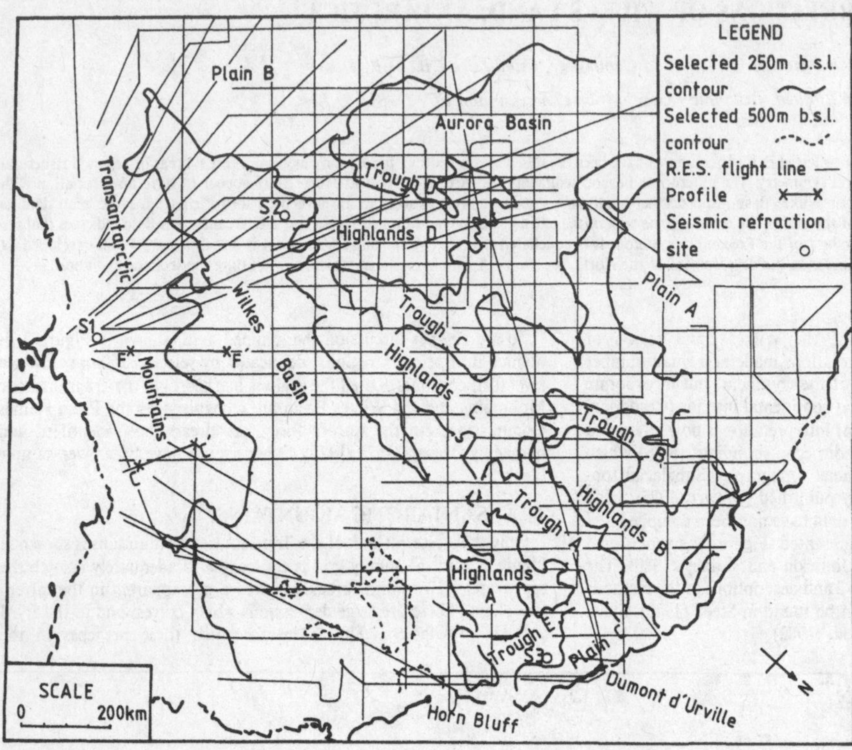


**Figure 2. Terrain regions within Wilkes Land and locations of profiles and seismic refraction sites.**

mountains have been exploited by the ice is reflected by the amplitude of the coincident depressions in the ice surface (see Johnson and Vanney, 1980; Steed, 1980).

Block faulting has been identified in the southern part of the Transantarctic Mountains from early radioecho soundings (Drewry, 1972). An additional influence on morphology is attributable to the presence of Ferrar Dolerite sills and dykes. The relative hardness of the igneous rock provides some protection to the softer underlying sedimentary sequences of the Beacon Supergroup. This combination gives rise to plateaus and terraced morphology in much of the exposed part of the mountain range, and similar forms are seen on RES profiles recorded over the western flank of the Transantarctic Mountains. Figure 3 (upper) shows a large mesa 120 km WNW of the Ferrar Glacier and Figure 3 (lower) shows terraced morphology along the southern flank of the David Glacier. Ferrar Dolerite has been identified at outcrops close to Location C of profile C-C′ (Gair et al., 1969). These RES sections suggest that Ferrar Dolerite sills extend under the eastern flank of the Wilkes Basin. No similar topographic forms, however, have been identified along the western edge of the basin.

## WILKES SUBGLACIAL BASIN

The Wilkes Basin is approximately 1400 km long and varies in width from 200 km in the south to 600 km in the north. The southern end borders onto Plain B and the southwestern limit of the basin forms a smooth and wide saddle with the southeastern end of the Aurora Basin. The western limit is discontinuous with many re-entrants into bordering highlands (highland blocks A, C and D). The northern end of the Wilkes Basin impinges on the coast between longitude 155°E and the Ninnis Glacier (longitude 147°E). Along this section of the coast there is only one rock exposure, of Beacon sediments and a dolerite sill at Horn Bluff. The singularity of this outcrop is probably attributable to the protective cover of the capping dolerite sill and to its lying close to the edge of one of several locally high plateaus which occur in the northern part of the basin. The Horn Bluff area has not been surveyed by RES on a small scale, so that the exact relation of the outcrop to the surrounding topography is not clear.

Five seismic refraction measurements were taken in the general region of the Wilkes Basin and the Transantarctic Mountains during the U.S. Victoria Land Traverse I (Crary, 1963). Four shots were close together in Skelton Névé (Location S1 of Figure 2) and gave seismic velocities indicative of sandstone and dolerite overlying granite, which is compatible with the stratigraphy determined from exposures in the region. The fifth site was at 135°5′E, 78°5′S over the northwestern flank of the basin (Location S2 of Figure 2), where reversed shots were recorded. These gave seismic velocities of 5820 m/sec, suggesting a granitic bedrock with little or no sedimentary cover. Robinson (1964) and Bentley (1974) interpreted these results as showing that the basin is a depression in the basement complex containing no significant sediments. The RES survey shows that this site was over the southwestern flank of the basin in an area with elevation and topographic roughness which are not typical of the Wilkes Basin as a whole (Steed, 1980). The absence of sedimentary cover at that site cannot preclude the existence of sediments in the rest of the basin.

Gravity data have also been collected over the Wilkes Basin on three over-snow traverses (the U.S. Victoria Land Traverses I and II and the McMurdo to Pole Traverse, Crary, 1963; Robinson, 1964). The gravity observations show a strong negative free air anomaly of approximately 40 mGal that may be attributable to various mechanisms, principally: (1) delayed response of the basin to recent changes in the overburden of ice, (ii) the basin is held beneath its isostatic equilibrium by the surrounding crust and/or by downward convecting asthenosphere (Robinson, 1964; Bentley, 1974) which may be part of a larger convection system with an expression in southern Australia (Veevers, 1982), and (iii) the possible existence of very thick sequences of low density sediments in the basin (Drewry, 1976).

East Antarctica is believed to be in overall isostatic equilibrium (Woollard, 1962) so if factor (i) above is significant then the anomaly is an expression of an overburden change restricted to the basin only, which is unlikely. However, it should also be pointed out that if the centre of the Wilkes Basin could be hypothetically raised by the inclusion of an additional layer at depth of mantle material (3,300 kg/$m^3$ density) and the ice cover (920 kg/$m^3$ density) reduced by the same amount so as to maintain the same ice surface elevation, then the 40 mGal anomaly would be equivalent to a vertical displacement of approximately 400 m (assuming a simple infinite slab model). If the amplitude of the Wilkes Basin were reduced by this amount then it would still leave a topographic depression of significance.

The second mechanism mentioned above refers to a convection

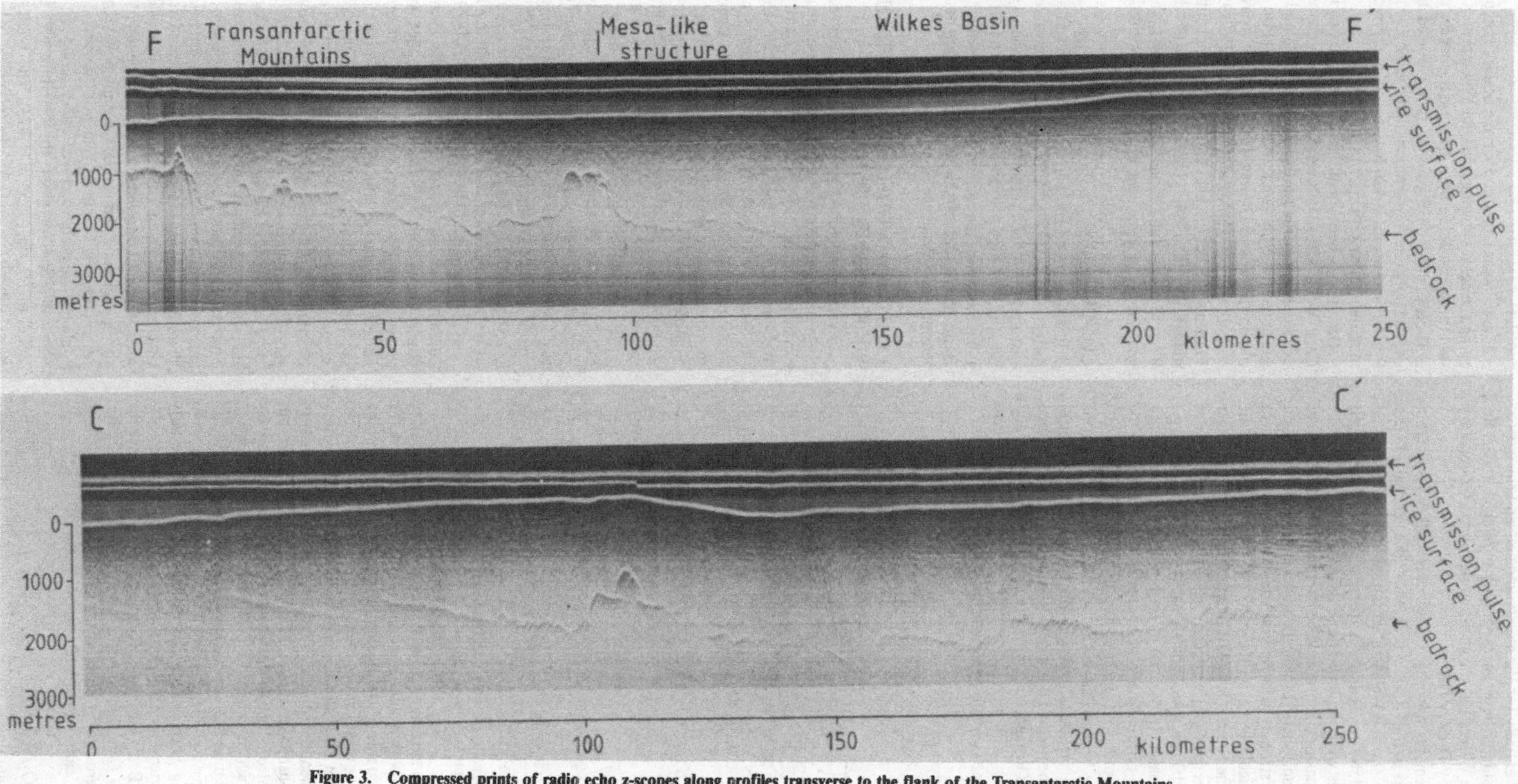


**Figure 3. Compressed prints of radio echo z-scopes along profiles transverse to the flank of the Transantarctic Mountains.**

system that is on a larger scale than the Wilkes Basin. Although there is good evidence for the existence of such a system, it cannot be expected to account entirely for the gravity anomaly and the basinal depression. It is considered here that the gravity anomaly is in part the manifestation of the Wilkes Basin containing significant sedimentary infill, as suggested by Drewry (1976).

Magnetic measurements have been made over Wilkes Land from the POGO satellite as part of a wider programme (Regan et al., 1975). Figure 4 shows the short wavelength variations over Wilkes Land (wavelengths over 1500 km removed), (pers. comm. from Langel: reduction procedures were similar to those described by Regan et al., 1975). How these measurements should be interpreted is not clear, but in the southern hemisphere continents there appears to be good correlation between strongly positive field strength and known shield areas. This is particularly so in the case of Australia where the boundary between the Precambrian shield in the west and the younger orogenic areas in the east (with extensive platform cover) compares closely with the POGO magnetic anomaly pattern (positive in the west and negative in the east). In the case of East Antarctica a boundary between the strongly positive magnetic anomalies in the west and the slightly negative anomalies in the east runs approximately along the western flank of the Wilkes Basin. This boundary may prove to be related to a major structural change similar to that in central Australia. It should be pointed out here that a more recent and refined version of the POGO magnetic data is presented by Ritzwoller and Bentley (this volume). Although their version has a less well defined boundary between the two anomalies, it does confirm the pattern.

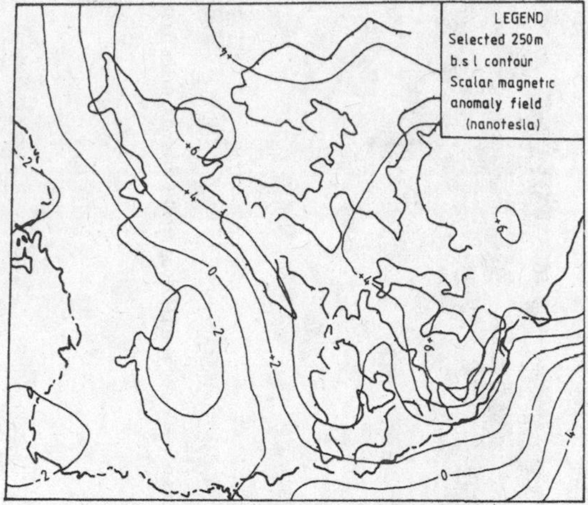


**Figure 4. Magnetic anomaly map from POGO satellite data (by permission of Langel).**

The hypothesis, therefore, that the Wilkes Basin is a sedimentary basin is more in line with available observations. The existence in the Transantarctic Mountains of the Beacon Supergroup, gently dipping towards the basin, the mesas and terracing on the basin flank, and the Beacon strata at Horn Bluff all add weight to this model. Sedimentary erratics have also been identified at Cape Dennison and in Commonwealth Bay which must have originated from strata further inland, but probably to the west of the basin. These erratics include a relatively unlithified arkose which might be post-Beacon in age (Mawson, 1940). Samples of lignite have been dredged up off the Mertz Glacier (ibid.), and may have come from either inland of the glacier or from further east, having been transported by icebergs in the westerly flowing currents. Truswell (1982) examined palynomorphs from offshore of the Wilkes Basin (near Mertz Glacier) and found evidence for Mesozoic and Early Cenozoic source rocks in this vicinity. Whether these derive from erosion of sediments on the continental shelf or from inland basins, cannot be unambiguously decided.

## HIGHLAND BLOCKS

An irregular mountainous region lies to the west of the Wilkes subglacial Basin and shows an approximate north—south orientation. In Figure 2 it is shown divided into highland blocks A, B, C and D. The region is bounded at the southern and northern ends by plains close to sea level and has many deep troughs between the blocks. Large amplitude faults are apparent in many places from the RES profiles.

Plain C near Dumont d'Urville Station is exposed at several points along the coast showing crystalline basement. One seismic refraction shot (unreversed) has been made on the border of Plain C and Trough E (Location S3 on Figure 2). It is reported that the measurement showed only one subsurface boundary, the ice-rock interface, giving a P-wave velocity of 4,920 m/sec and interpreted as being basement material (Imbert, 1953).

Granite has been identified at the outcrops in the Cape Bage and Cape Webb areas, between the Ninnis and Mertz Glaciers. These exposures are at the northern end of Highland Block A. It has been suggested that this block is the manifestation of a massive granitic pluton (Steed and Drewry, 1980). Gravity measurements and seismic reflection shots were made during the 1958-1959 Expédition Polaire Francaise Traverse which crossed over the southwestern flank of the massif (Tardi, 1964). The gravity profile does not allow exact interpretation, but a model incorporating granitic plutons of simple geometry is in agreement with the observations (Steed, 1980).

## CONCLUSION

It is suggested here that the Wilkes Basin is a sedimentary basin of different structural style from the Transantarctic Mountains to the east and the older Precambrian area to the west. It is also proposed that the local high plateaus in the northern part of the basin are either fault blocks or monoclinal structures, similar to those seen in the Basin and Range of the western U.S.A., both being indicative of tension and subsidence.

## PROPOSED MODEL FOR THE EVOLUTION OF THE WILKES BASIN

Several correlations have been made between the major tectonic units of southeastern Australia and those identified in the Transantarctic Mountains (e.g. Craddock, 1970; Laird et al., 1977; Grindley and Davey, 1982). These have the limitation that they do not consider the Wilkes Basin and any relative movement between the two continents prior to separation in the early Cenozoic, although large movements have been proposed by several authors, principally Harrington et al. (1973). A simple evolutionary model for Wilkes Land is proposed here, based upon additional subglacial topographic information and other geophysical and geological data.

In pre-Riphean times, prior to the deposition of the Adelaide Supergroup in Australia and the Nimrod and Wilson Groups in Antarctica, it is envisaged that the western part of Australia, comprising Gawler, Pilbara and Yilgarn Blocks, was juxtaposed with the western part of Wilkes Land (see Figure 5a). This configuration brings together: (i) the Windmill Islands with the Australian Albany Granites as favoured by Oliver (1972), (ii) the crystalline basement at and to the west of Dumont d'Urville with the Gawler Block, both of which contain charnockites, and (iii) brings the Tyennan Block and future Rocky Cape Group of Tasmania close to the Gawler Block, an affinity suggested by Harrington et al. (1973). The approximate line of the eastern boundary of the older Precambrian craton of the Antarctic continent lies along the western edge of the present Wilkes Basin, as suggested from the POGO magnetic data. This configuration provides a framework for the eastward continental growth in Late Precambrian—Early Palaeozoic times represented by the Beardmore, Ross, Byrd and Bowers Group in Antarctica, the Dundas Group in Tasmania, and the Adelaide Supergroup in southern Australia (see Figure 5b); up to this point this is essentially the same configuration as that proposed by Craddock (1972). The development of the Wilkes Basin, therefore, is envisaged as having commenced during the closing stages of cratonisation, or later (Figure 5c). The basin developed from an area of tension between the older continental platform to the west and the younger metasedimentary groups now seen in the Transantarctic Mountains. Strike-slip movement of northern Victoria Land against southern Australia occurred either along the Gambier—Gippsland Fracture Zones or along the Gambier and Beaconsfield Fracture Zones of Harrington et al. (1973). The Wilkes Basin expanded with block subsidence and experience substantial sedimentary infilling. The four regional highs within the basin (outlined by 500 m below sea level contours in Figure 2) and the Horn Bluff outcrop are now considered the expression of tension and not to be eroded anticlines,

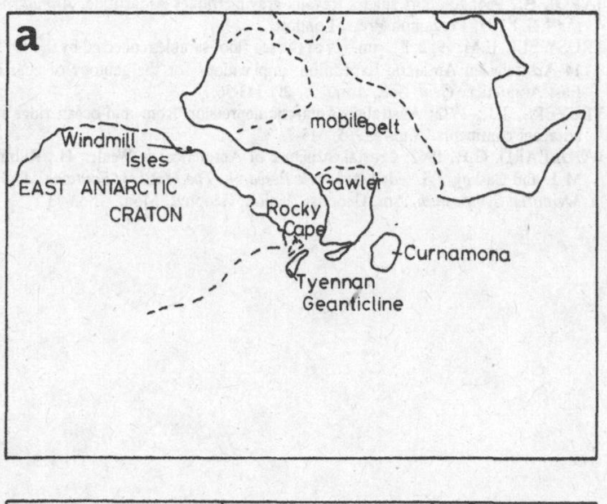


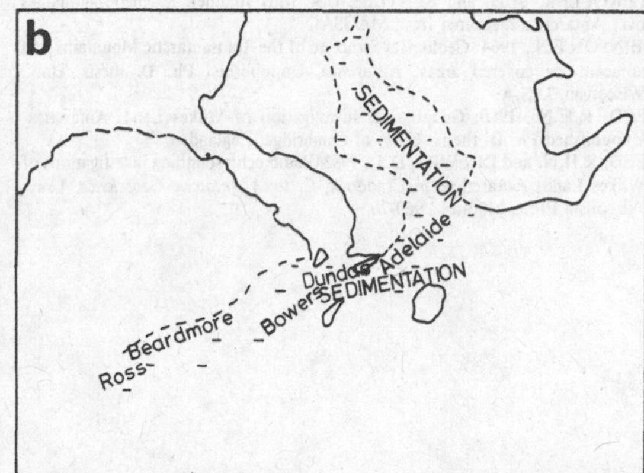


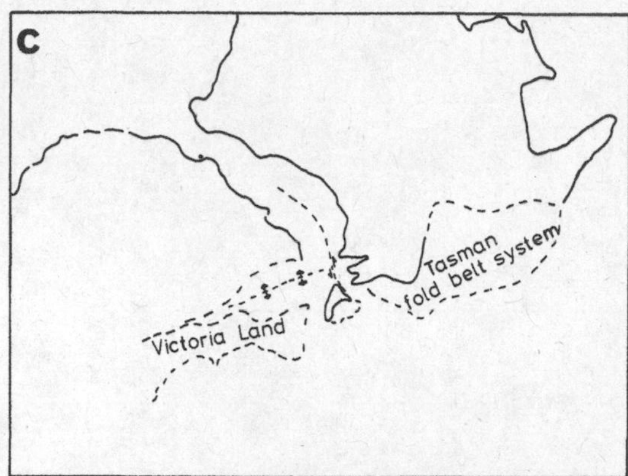


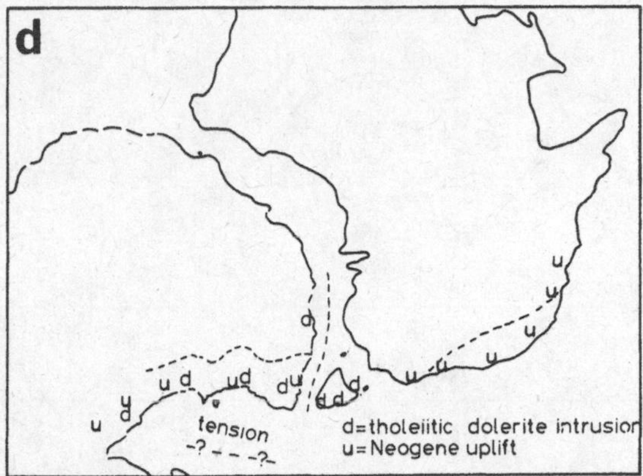


**Figure 5. a) Proposed palaeogeographical configuration of Antarctic and Australian cratons at the end of the Carpentarian Era (Mid Proterozoic). b) Proposed configuration of passive continental growth along the eastern margin of the Australian and Antarctic plates at the end of the Precambrian-Early Palaeozoic. c) Proposed configuration of active continental growth from Mid Palaeozoic to Mesozoic times. d) Occurrence of dolerite intrusions in the Jurassic (prior to separation) and epeirogenic uplift during the Neogene.**

as suggested by Steed and Drewry (1982). This model of the Wilkes Basin is similar to that of the North Sea Basin.

This model is a preliminary hypothesis and requires more work. It is not clear whether the basin opened continuously or in stages and when it may have occurred. It should be pointed out, however, that igneous intrusion would be expected in such a tensional regime and of the several intrusive sequences identified in Victoria Land and Wilkes Land the Jurassic Ferrar Dolerites have the most extensive association with the basin and occur in Tasmania. Furthermore, the fresh appearance of the graben structures (troughs A, B, C and D) which all have north—south trends, suggests a young age and a close affinity to the tensional activity that formed the Wilkes Basin.

*Acknowledgements* Previously unpublished data presented here were collected in 1974-1975 and 1977-1978 during a joint U.S. National Science Foundation, Scott Polar Research Institute, and Technical University of Denmark Antarctic Radio Echo Sounding Programme. Logistic support was generously provided by the U.S.N.S.F., the U.S. Navy Antarctic Support Force, and the U.S. Air Development Squadron VXE-6. The S.P.R.I. project was funded under a U.K.N.E.R.C. grant. I wish to thank the VXE-6 aircrew, members of the S.P.R.I. Radio Echo Research Group, and especially D.J. Drewry for discussion and comments on Antarctic geological interpretations.

## REFERENCES

BENTLEY, C.R., 1974: Crustal structure of Antarctica; *in* Mueller, S. (ed.) *The Structure of the Earth's Crust,* Elsevier, Amsterdam, 229-40.

CRADDOCK, C., 1970: Tectonic map of Antarctica; *in* Bushnell, V.C., and Craddock, C. (eds.) Geologic maps of Antarctica. *Antarct. Map Folio Ser.,* Folio 12, Pl.XXI.

CRADDOCK, C., 1972: Antarctic Tectonics; *in* Adie, R.J. (ed.) *Antarctic Geology and Geophysics,* Universitetsforlaget, Oslo, 449-55.

CRARY, A.P., 1963: Results of the United States Traverses in East Antarctica, 1958-61. *I.G.Y. Glaciol. Rep., 7.*

DREWRY, D.J., 1972: Subglacial morphology between the Transantarctic Mountains and the South Pole; *in* Adie, R.J. (ed.) *Antarctic Geology and Geophysics,* Universitetsforlaget, Oslo, 693-703.

DREWRY, D.J., 1975: Radio echo sounding map of Antarctica. *Polar Rec., 17,* 359-74.

DREWRY, D.J., 1976: Sedimentary basins of the East Antarctic craton from geophysical evidence. *Tectonophysics, 3,* 301-14.

DREWRY, D.J., JORDAN, S.R. and JANKOWSKI, E.J., 1982: Measured properties of the Antarctic ice sheet: surface configuration, ice thickness, volume and bedrock characteristics. *Annals Glaciology, 3,* 83-91.

ELLIOT, D.H., 1975: Tectonics of Antarctica: a review. *Am. J. Sci., 275,* 45-106.

GAIR, H.S., STURM, A., CARRYER, S.J. and GRINDLEY, G.W., 1969: The geology of northern Victoria Land (Sheet 13); *in* Bushnell, V.C. and Craddock, C. (eds.) Geologic maps of Antarctica. *Antarct. Map Folio Ser.,* Folio 12, Pl. XII.

GRINDLEY, G.W. and DAVEY, F.J., 1982: The reconstruction of New Zealand, Australia and Antarctica; *in* Craddock, C. (ed.) *Antarctic Geoscience.* Univ. Wisconsin Press, Madison, 15-29.

HARRINGTON, H.J., BURNS, K.L. and THOMPSON, B.R., 1973: Gambier—Beaconsfield and Gambier—Sorell Fracture Zones and the movement of plates in the Australia—Antarctica—New Zealand region. *Nature Phys. Sci., 245,* 109-12.

IMBERT, V.B., 1953: Sondages Seismiques en Terre Adélie. *Annls. Geophys., 9,* no.1, 85-92.

JOHNSON, L. and VANNEY, J.R., 1980: *General bathymetric chart of the oceans (GEBCO), 5.18 Polar stereographic projection-scale 1:6,000,000 at 75S lat.* Canadian Hydrographic Service, Ottawa.

LAIRD, M.G., COOPER, R.A. and JAGO, J.B., 1977: New data on the Lower Palaeozoic sequence of northern Victoria Land, Antarctica, and its significance for Australia-Antarctic relationships in the Palaeozoic. *Nature, 265,* 107-10.

MAWSON, D., 1940: Sedimentary Rocks. *Australasian Antarctic Expedition 1911-14, Scientific Reports, Ser. A, 4,* 347-67.

OLIVER, R.L., 1972: Some aspects of Antarctic-Australian geological relationships; *in* Adie, R.J. (ed.) *Antarctic Geology and Geophysics.* Universitetsforlaget, Oslo, 859-64.

REGAN, R.D., CAIN, J.C. and DAVIS, W.M., 1975: A global magnetic anomaly map. *J. Geophys. Res., 80,* 794-802.

RITZWOLLER, M.H. and BENTLEY, C.R. (this volume): Magnetic anomalies over Antarctica measured from MAGSAT.

ROBINSON E.S., 1964: Geological Structure of the Transantarctic Mountains and adjacent ice covered areas, Antarctica. Unpublished Ph. D. thesis, Univ. Wisconsin, U.S.A.

STEED, R.H.N., 1980: Geophysical Investigation of Wilkes Land, Antarctica. Unpublished Ph. D. thesis, Univ. of Cambridge, England.

STEED, R.H.N. and DREWRY, D.J., 1982: Radio echo sounding investigations of Wilkes Land, Antarctica; *in* Craddock, C. (ed.) *Antarctic Geoscience*, Univ. Wisconsin Press, Madison, 969-76.

TARDI, P., 1964: Rapport sur les travaux gravimétriques Antartique. *Annuals of the I.G.Y., 31,* Pergamon Press, London.

TRUSWELL, E.M. 1982: Palynology of the sea floor samples collected by the 1911-14 Australasian Antarctic Expedition: implications for the geology of coastal East Antarctica. *Geol. Soc. Aust., J., 29,* 343-56.

VEEVERS, J.J., 1982: Australian-Antarctic depression from mid-ocean ridge to adjacent continents. *Nature, 295,* 315-7.

WOOLLARD, G.P., 1962: Crustal Structure of Antarctica; *in* Wexler, H., Rubin, M.J. and Caskey, J.S. (eds.) *Antarctic Research: The Matthew Fontaine Maury Memorial Symposium,* Am. Geophys. Union, Geophys. Mon., 7, 53-73.

# PALAEOMAGNETISM OF CRETACEOUS VOLCANIC ROCKS FROM MARIE BYRD LAND, ANTARCTICA

G.W. Grindley and P.J. Oliver, *N.Z. Geological Survey, Lower Hutt, New Zealand.*

*Abstract* Palaeomagnetic determinations from 26 widely separated sites (42 oriented blocks) from Upper Cretaceous rhyolitic volcanics and mafic dyke swarms in the Ruppert/Hobbs Coast sector of Marie Byrd Land define a mean direction of stable magnetisation of I = −84° D = 223° ($\alpha_{95}$ = 4.5°) giving a palaeomagnetic pole at 65°S 118°W ($\alpha_{95}$ = 9°). The sampled rocks post-date the emplacement of Lower Cretaceous calc-alkaline granite plutons and are associated with the early stages of rifting of the South Pacific oceanic basin. All sampled rocks are normally magnetised and were emplaced during the long Late Cretaceous epoch of normal polarity as indicated by radiometric dates (90-110 Ma). No evidence for significant post-emplacement remagnetisation or tectonic disturbance was found. The Cretaceous palaeopole position for Marie Byrd Land lies significantly eastward of the Apparent Polar Wander Path (APWP) for Australia and New Zealand, reconstructed by reversal of sea-floor magnetic anomalies on to Antarctica. This might indicate minor extension and/or dextral rotation along a plate boundary with East Antarctica. This palaeomagnetic study does not support speculations that Marie Byrd Land is an exotic microplate that drifted into its present position in the Late Cretaceous or Early Tertiary. Indeed, the evidence is consistent with Marie Byrd Land being an eastward extension of New Zealand and the Campbell Plateau.

The geologic evolution of Marie Byrd Land, one of the principle continental blocks or microplates comprising West Antarctica has been the subject of much speculation. A conservative view is that Marie Byrd Land is the eastward extension of the Campbell Plateau of New Zealand, and an important link in the Late Palaeozoic-Mesozoic circum-Pacific island arc system joining New Zealand with the Antarctic Peninsula and South America (Grindley and Davey, 1982). A more mobilistic view is that Marie Byrd Land is an exotic microplate that drifted into its present position and was accreted on to East Antarctica during the Late Mesozoic (Scharnberger and Scharon, 1972) or Late Palaeozoic-Early Mesozoic (Wade and Couch, 1982).

Marie Byrd Land is also of interest in global plate reconstructions during the Late Mesozoic and Cenozoic. The nature, position and displacements along or across a plate boundary between East and West Antarctica have been controversial topics (see Stock and Molnar, 1982). Reliable palaeomagnetic data from Marie Byrd Land could provide decisive on-site information on the past position and orientation of one of the more problematic blocks in the Gondwana jigsaw.

## GEOLOGICAL SETTING (Figure 1)

### Basement Complex (Palaeozoic-?Precambrian)

*Gneisses* The oldest rocks are quartzofeldspathic gneisses, strongly retrograded to dark wavy-banded schist and phyllonites exposed below the rhyolites of Mt Petras. Model Rb-Sr total-rock ages from Navarette Peak (286 Ma) and Wallace Rock (283 Ma) indicate a Late Palaeozoic deformation (Halpern and Wade, 1979). These infaulted slivers probably indicate the presence of an underlying Palaeozoic or Precambrian basement gneiss complex.

*Layered Gabbros* Layered cumulate gabbro, olivine gabbro with minor noritic and anorthositic layers form a folded basement in the Hobbs Coast area (Cape Burks, Lynch Point). Metagabbros and amphibolites outcropping near younger granites (Holms Bluff, Mt Grey, Mt Giles, Peden Cliffs, Peacock Peak) are probably correlative. Rb-Sr dating has been inconclusive (Spörli and Craddock, 1981).

### Ruppert Coast Metavolcanics (Upper Palaeozoic)

The Ruppert Coast is underlain by a suite of northwest striking, strongly folded, cleaved, calc-alkaline metavolcanic rocks ranging from basaltic andesite to andesite and rhyodacite. Flows, tuffs, agglomerates and breccias are typical. At Wilkins Nunatak, they overlie quartzofeldspathic sandstone and carbonaceous siltstone considered to be the source of plant bearing erratics discovered at Milan Rock at the head of Land Glacier (Grindley et al., 1980). Radiometric dating has so far proved inconclusive, but the plants provide a maximum Late Devonian (Frasnian) age for the inception of volcanism.

### Granitoid Complexes (Upper Palaeozoic and Mesozoic)

*Upper Palaeozoic Granitoids* Foliated quartz monzonite, granodiorite, quartz diorite and syenite at Mt Grey, Dee Nunatak, Mt Giles and Mt Steinfeld include remnants of Upper Palaeozoic felsic plutons (Spörli and Craddock, 1981). They intrude layered gabbro at Mt Giles and Peden Cliffs and are intruded by dykes of granodiorite and quartz porphyry. The foliation strikes east-west in the Mt Grey area. Preliminary Rb-Sr dating yielded a minimum isochron age of 154 ± 35 Ma for metagabbro and quartz monzonite from Mt Giles and granodiorite from Mt Grey (Metcalfe et al., 1978). An Rb-Sr model age of 113 Ma from Mt Grey indicates resetting by Cretaceous plutonism (Halpern and Wade, 1979).

*Upper Mesozoic Granitoids* High-level felsic intrusives ranging from biotite-hornblende tonalite and granodiorite (Hobbs Coast) to quartz monzonite and quartz syenite (Ruppert Coast) and two-mica garnet-bearing granite (Mt Shirley) form most coastal exposures. Aplitic dykes invade the country rocks producing contact metamorphism and inhomogeneous hybrid rocks. Metavolcanic and metaplutonic enclaves and xenoliths are common (Spörli and Craddock, 1981). Rb-Sr dates on biotites, assuming an initial $^{87}Sr/^{86}Sr$ ratio of 0.705, range from 128 Ma to 75 Ma. A whole rock Rb-Sr isochron of 94 ± 12 Ma from Billey Bluff combined with a biotite age of 102 ± 3 Ma probably dates uplift and cooling of the pluton (Halpern and Wade, 1979).

### Rhyolitic Volcanics (Upper Cretaceous)

Mt Petras, in the McCuddin Mountains, 150 km inland from the Hobbs Coast is an accumulation of viscous rhyodacite flows and minor near-source ignimbrites, capped by Miocene hyaloclastites. The thick flows show a low-dipping platy jointing dipping gently away from an eastwest trending central spine, probably representing the

**TABLE 1: Potassium-Argon Age Determinations on Cretaceous Volcanic Rocks, Marie Byrd Land, Antarctica (supplied by Dr C.J. Adams, D.S.I.R., Lower Hutt)**

| INS 'R' | Field | Rock type | Locality | Lat. S | Long. W | K (wt %) | $^{40}Ar$ (radiogenic) nl/g | $^{40}Ar$ (radiogenic) % total | Age (Ma) |
|---|---|---|---|---|---|---|---|---|---|
| 7677Tr | RC1A | Andesite dyke | Mt Pearson | 75°54′ | 140°58′ | 0.184 | 0.807 | 30.1 | 109 ± 9 |
| | | | | | | | 0.827 | 31.2 | 112 ± 9 |
| 7668 Tr | RC7A | Camptonite dyke | Bailey nun. | 75°39′ | 140°05′ | 2.894 | 10.491 | 91.5 | 91 ± 2 |
| 7682 Tr | RC23B | Dolerite dyke | Lewis Bluff | 73°52′ | 140°30′ | 0.924 | 3.856 | 87.4 | 104 ± 3 |
| 7670 Tr | RC42A | Qtz dolerite dyke | Lambert nun. | 75°24′ | 137°54′ | 1.628 | 6.125 | 69.6 | 94 ± 3 |
| 7669 Bi | HC8 | Microdiorite dyke | Patton Bluff | 75°14′ | 133°42′ | 1.473 | 5.846 | 62.0 | 99 ± 4 |
| 7673 Tr | HC17 | Microdiorite dyke | Mt Prince | 74°58′ | 134°10′ | 0.684 | 2.496 | 75.4 | 91 ± 3 |
| 7671 Hb | MP7A | Rhyodacite flows | Navarette Pk | 75°55′ | 128°46′ | 0.393 | 1.631 | 73.6 | 104 ± 3 |
| 7672 Tr | CB7A | Andesite dyke | Cape Burks | 74°45′ | 136°50′ | 1.005 | 3.997 | 67.0 | 99 ± 3 |
| | | | | | | | 3.821 | 68.3 | 95 ± 3 |

hb = hornblende; bi = biotite; Tr = Total rock. Errors are one standard deviation
Decay constants $^{40}K$:$\lambda B = 0.472 \times 10^{-9}$/yr; $\lambda e = 0.584 \times 10^{-10}$/yr; $^{40}K/K = 0.119$ atomic %

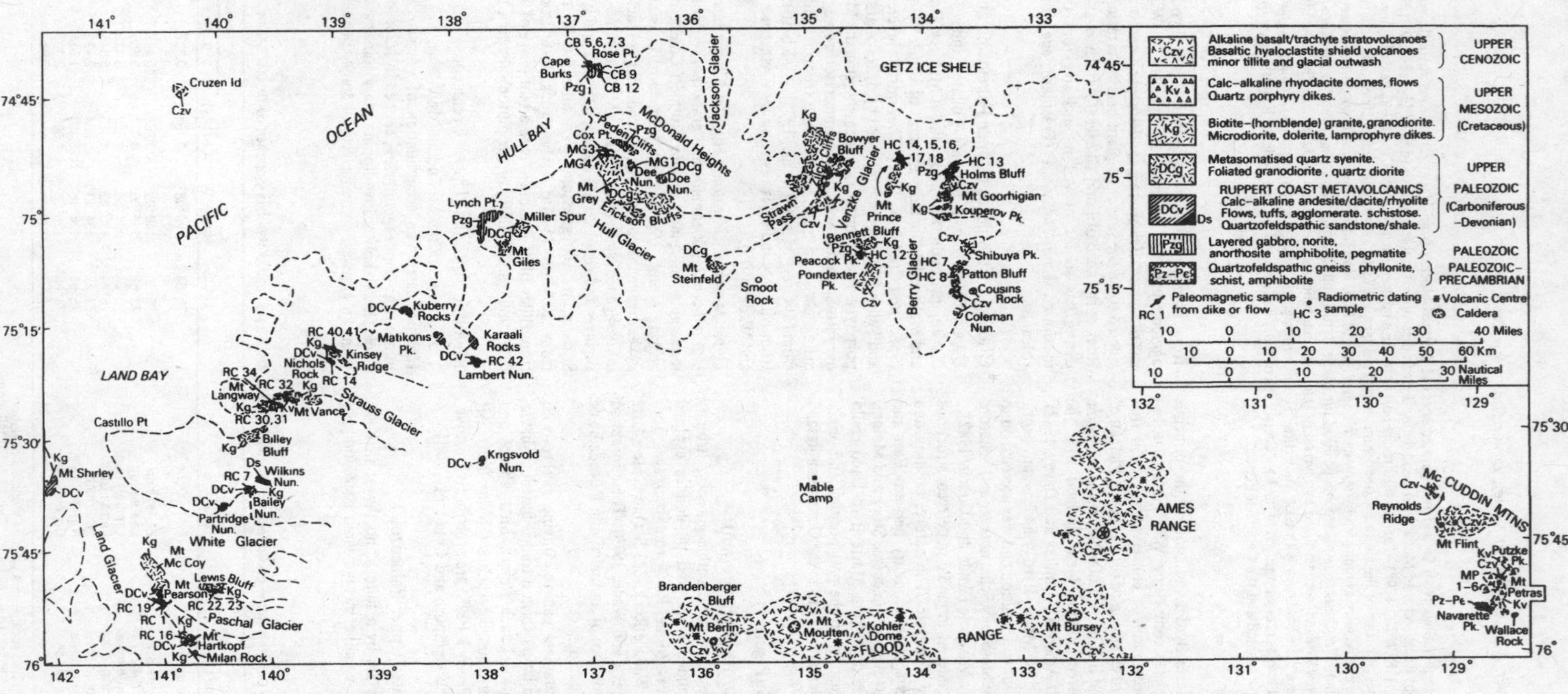


Figure 1. Generalised geology of the Ruppert-Hobbs Coast sector of Marie Byrd Land showing location of palaeomagnetic sites and radiometric dating samples.

axis of a volcanic dome. Calc-alkaline affinities are suggested by chemical analyses and hornblende and biotite phenocrysts (Lemasurier and Wade, 1976). Plagioclase separates have been K-Ar dated at 81±6 Ma (Mt Petras) and 88±4 Ma (Mt Galla) (LeMasurier and Wade, 1976), while hornblendes have recently been K-Ar dated at 104±3 Ma (Navarette Park) bv C.J. Adams (Table 1).

On the Hobbs Coast, quartz porphyry dykes intrude gabbro at Cape Burks and Lynch Point and quartz syenite at Cox's Point and Mt Grey. On the Ruppert Coast, pink porphyritic rhyolite dykes and flows intrude the Cretaceous granites of Bailey Nunatak, Mt Langway and Mt Vance (Figure 1). Rhyolite from Bailey Nunatak gave a K-Ar whole rock age of 98±3 Ma (Spörli and Craddock, 1981).

### Mafic Dyke Swarms (Cretaceous)

The Ruppert and Hobbs Coast basement rocks are cut by mafic dyke swarms that increase in abundance coastwards. The mean strike of dykes is east-west but the range is from NW to NE. Dips are generally within 20° from vertical. The most common rock-type is a medium to fine grained microdiorite/andesite; more alkaline, trachy-andesite dykes with skeletal oligoclase microlites intrude the alkalic granite of Mt Langway and Mt Vance. Camptonitic lamprophyres and dolerite dykes, 3-15 m wide, intruding metavolcanics are also present.

The mafic dyke swarms post-date the Cretaceous granitoids and rhyolites. A single K-Ar date of 113±3 Ma has been determined on a mafic dyke cutting orbicular granite on Billey Bluff (Spörli and Craddock, 1981). Preliminary K-Ar dating of some of our palaeomagnetic samples by C.J. Adams (Table 1) gives ages ranging from 110 to 90 Ma, averaging about 100 Ma in good agreement with the cooling ages on the granites.

## PALAEOMAGNETIC MEASUREMENTS

In late 1977, 42 blocks were collected from 26 separate sites and oriented in the field by sun compass and magnetic compass. Because of the highly magnetic gabbroic basement rocks on the Hobbs Coast, deviations of the magnetic compass up to 10° were recorded by comparison with suncompass readings. Magnetic declinations varied across the region from 68°E in the east (Mt Petras) to 76°E in the west (Mt Pearson).

The oriented blocks were cored in the laboratory to yield 119 specimens, and 10, representative of the main rock types and localities, were subjected to step-wise demagnetisation in a non-tumbling AF demagnetiser, and measured on a Schonstedt-type spinner magnetometer. The changes in magnetic direction and intensity at each step of the demagnetisation of the pilot specimens are shown in the stereographic projections, relative intensity plots and Zijderveld-type diagrams of vector components (Figure 2). The step-wise demagnetisation showed that all the rock types behave in a similar manner, containing a dominant single-component primary TRM magnetism. Some specimens showed a very weak and magnetically soft deviation from the primary direction, that was removed in fields higher than 5 mT.

A cleaning field of 15 mT was chosen for the remaining specimens to determine their primary magnetic directions. To ensure that this was stable, they were also measured at 10 mT and 20 mT. All magnetisations were found to be of normal polarity consistent with emplacement in the Late Cretaceous episode of normal polarity (Helsley and Steiner, 1969). The site results are summarised in Table 2.

The dispersion of the magnetic directions of the sites after cleaning is slightly greater than the dispersion of the NRM directions, possibly due to the removal of the small secondary component seen to be removed in some of the Zijderveld-diagrams. The mean directions for the NRM are D= 226° I= −85.5° compared with the cleaned directions at 15 mT of D= 223° I= −84.2°. The grouping is only slightly poorer (K=52 to K=40) and then $\alpha_{95}$ has increased from 4.0° to 4.5°. It is possible that the small decrease in grouping is due to some unknown instrumental error. Because of the good grouping af palaeomagnetic directions and the absence of reliable evidence for post-Cretaceous tectonic tilting, no tectonic corrections were made.

### Pole Position

The palaeomagnetic pole position, based on mean locality coordinates of 75°S and 223°E (137°W), is at 66°S latitude and 241°E (119°W) longitude with polar errors of dp= 8.7° and dm= −8.8° and a co-latitude of 11.4°. This sector of Marie Byrd Land, therefore,

**TABLE 2: Palaeomagnetic data for Cretaceous Volcanic Rocks and Dykes, Marie Byrd Land**

| | SITE | | | NRM | | CLEANED AT 15 mT | | | | |
|---|---|---|---|---|---|---|---|---|---|---|
| No. | Lat.S | Long.W | N | D | I | D | I | k | $\alpha_{95}$ | R |
| MP1 | 75°53′ | 128°51′ | 6 | 191 | -66 | 191 | -65 | 81 | 7.5 | 5.9 |
| MP2 | 75°53′ | 128°51′ | 3 | 214 | -71 | 206 | -71 | 5269 | 1.7 | 3.0 |
| MP3 | 75°53′ | 128°51′ | 4 | 206 | -81 | 210 | -81 | 42 | 14.3 | 3.9 |
| MP4 | 75°53′ | 128°51′ | 3 | 222 | -76 | 212 | -70 | 1453 | 3.2 | 3.0 |
| MP5 | 75°53′ | 128°51′ | 5 | 253 | -84 | 243 | -82 | 21 | 17.3 | 4.8 |
| MP6 | 75°53′ | 128°51′ | 6 | 258 | -77 | 259 | -76 | 173 | 5.1 | 5.9 |
| MG3 | 74°57′ | 136°40′ | 6 | 177 | -83 | 192 | -74 | 8 | 26.0 | 5.3 |
| MG4 | 74°57′ | 136°40′ | 6 | 254 | -80 | 257 | -82 | 70 | 8.1 | 5.9 |
| CB5 | 74°45′ | 136°50′ | 5 | 172 | -69 | 163 | -68 | 22 | 16.9 | 4.8 |
| CB6 | 74°45′ | 136°50′ | 7 | 261 | -87 | 247 | -87 | 54 | 8.3 | 6.9 |
| CB7 | 74°45′ | 136°50′ | 8 | 113 | -78 | 138 | -81 | 150 | 4.5 | 8.0 |
| HC14 | 74°58′ | 134°10′ | 3 | 091 | -85 | 094 | -84 | 263 | 7.6 | 3.0 |
| HC15 | 74°58′ | 134°10′ | 3 | 316 | -87 | 283 | -86 | 781 | 4.4 | 3.0 |
| HC16 | 74°58′ | 134°10′ | 4 | 340 | -79 | 341 | -75 | 1072 | 2.8 | 4.0 |
| HC17 | 74°58′ | 134°10′ | 3 | 250 | -87 | 271 | -87 | 2894 | 2.3 | 3.0 |
| HC18 | 74°58′ | 134°10′ | 4 | 223 | -82 | 177 | -81 | 520 | 4.0 | 4.0 |
| RC1 | 75°55′ | 140°58′ | 7 | 240 | -81 | 233 | -79 | 68 | 7.4 | 6.9 |
| RC16 | 75°59′ | 140°45′ | 3 | 038 | -75 | 029 | -77 | 174 | 9.4 | 3.0 |
| RC19 | 75°53′ | 140°59′ | 3 | 153 | -77 | 148 | -76 | 374 | 6.4 | 3.0 |
| RC22 | 73°52′ | 140°30′ | 6 | 230 | -78 | 227 | -69 | 47 | 9.9 | 5.9 |
| RC23 | 73°52′ | 140°30′ | 5 | 075 | -86 | 099 | -86 | 108 | 7.4 | 5.0 |
| RC32 | 75°27′ | 139°38′ | 5 | 323 | -71 | 325 | -65 | 120 | 7.0 | 5.0 |
| RC34 | 75°29′ | 139°45′ | 3 | 351 | -82 | 335 | -81 | 1310 | 3.4 | 3.0 |
| RC40 | 75°22′ | 139°10′ | 4 | 220 | -80 | 237 | -79 | 32 | 16.6 | 3.9 |
| RC41 | 75°22′ | 139°10′ | 3 | 300 | -78 | 299 | -76 | 5050 | 1.7 | 3.0 |
| RC42 | 75°24′ | 137°54′ | 4 | 227 | -89 | 185 | -88 | 6921 | 1.1 | 4.0 |
| Mean of 26 sites: | | | | 226 | -86 | 223 | -84 | 40 | 4.5 | 25.4 |

N = No of samples per site
D = Declination of magnetic vector (in degrees)
I = Inclination of magnetic vector (in degrees)
k = Fisher's precision parameter
$\alpha_{95}$ = Radius of 95% cone of confidence of mean value
R = Resultant magnitude of sum of unit vectors

lay in high latitudes during the Late Cretaceous as it does today. The Marie Byrd Land data alone indicates the South Pole to be northeast of the Ruppert-Hobbs Coast sector and east of the Campbell Plateau-Chatham Rise sectors of New Zealand in their pre-drift positions (Grindley and Davey, 1982).

## TECTONIC INTERPRETATION

### Regional Structure

The palaeomagnetic samples were collected from:

(1) Rhyolitic to rhyodacitic flows and quartz prophyry dykes intruding Lower Cretaceous granitoids.
(2) Mafic to intermediate dykes forming east—west striking swarms cutting all basement rocks including the Cretaceous rhyolites.

A mature erosion surface, cut across the basement during the Late Cretaceous and Early Cenozoic (LeMasurier and Wade, 1976), is preserved as remnants on the coastal nunataks and disappears below Pliocene and Miocene hyaloclastites. The low slopes (<10°) on the "peneplain" remnants indicate only epeirogenic uplift or simple block-faulting. Along the coast, a slight northward monoclinal tilt is apparent but elsewhere dips are variable.

The dyke swarms cut across the steep dipping NW-SE structural trends in the basement metavolcanics and the NE-SW folds in the layered gabbros. Locally they may be intruded along cleavage or joint surfaces in these older rocks. The east-west dykes trend sub-parallel to the Marie Byrd Land continental margin and reflect an extensional tectonic regime prior to the northward drift of the Campbell Plateau (New Zealand) starting near the end of the Cretaceous (Anomaly 34-80 Ma). Similar east-west dyke swarms in New Zealand (Figure 3) are dated at 80-100 Ma (Grindley et al., 1977).

### Polar Wander Path for Australia and New Zealand (Figure 3)

The Cretaceous-Cenozoic polar wander path for Australia has been established (McElhinny et al., 1974; Embleton, 1981) from studies of Eastern Australian volcanic rocks and laterites. By removing the relative motion between Australia and East Antarctica, a polar wander path can be constructed for Antarctica. This polar wander path (Figure 3) follows a circuitous course from the entrance to the Ross Sea in the early Late Cretaceous to the present South Pole via the Weddell Sea and Dronning Maud Land (McElhinny et al., 1974; Grindley et al., 1981). The Late Cretaceous-Cenozoic polar wander path for New Zealand has also been determined (Grindley et al., 1977;

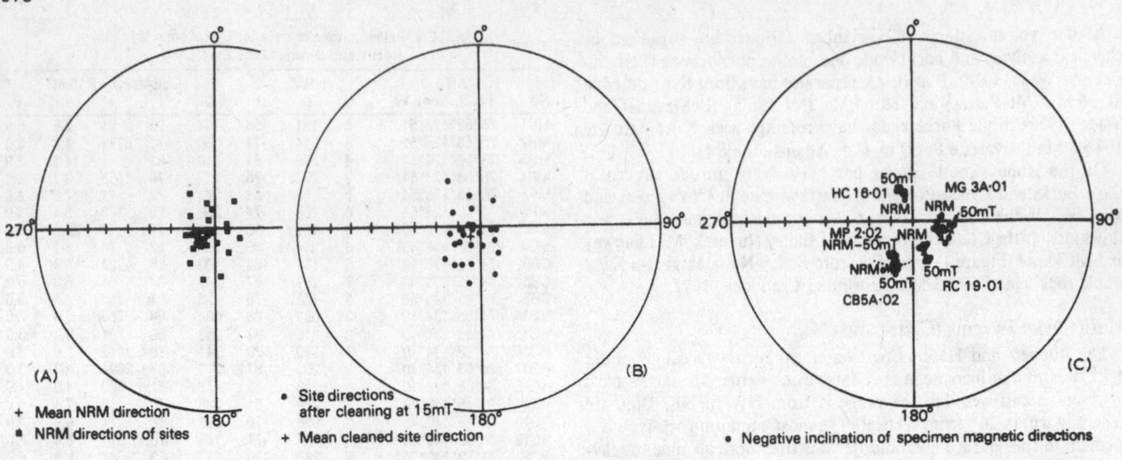

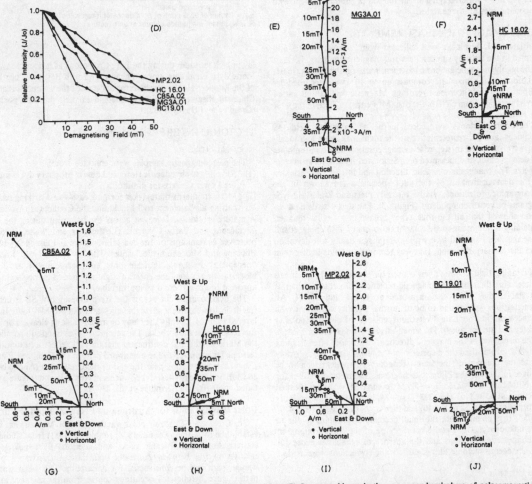


**Figure 2.** (A) Stereographic projection on upper hemisphere of NRM directions; (B) Stereographic projection on upper hemisphere of palaeomagnetic directions after cleaning at 15 mT; (C) Stereographic projections of typical directions on progressive Af demagnetisation from 0 to 50 mT demagnetising fields; (D) Relative intensity plot of pilot specimens on Af demangetisation; (E-J) Zijderveld-type demagnetisation projections of palaeomagnetic directions on progressive Af demagnetisation.

Oliver et al., 1979; Grindley and Oliver, 1979). By reversing the northward drift inferred from sea-floor magnetic anomalies, these palaeopoles and the APWP can be reconstructed back to Antarctica and matched with the Australian APWP (Grindley et al, 1981). Any discrepancies are within the tolerances of the palaeopole positions as shown by their $\alpha_{95}$ circles of confidence (Figure 3).

## West Antarctica Palaeomagnetic Data

Palaeomagnetic data for West Antarctica have mainly been obtained from the Cretaceous Andean Intrusive Suite and from Cenozoic and Jurassic volcanic rocks from the Antarctic Peninsula.

*Antarctic Peninsula* Most palaeopoles (mean age 95 Ma) from the Andean Intrusive Suite plot near the present South Pole (Figure 3) close to the New Zealand 75 Ma pole on the combined APWP and some 1800 km south of the 95 Ma poles from Australia and New Zealand at the entrance to the Ross Sea (Kellogg and Reynolds 1974; Reynolds and Kellogg, 1974). The discrepancy with the Australian-New Zealand 95 Ma poles may be resolved by assuming a younger age (75-80 Ma) for the plutons or by sinistral rotation of the central section of the Antarctic Peninsula, a rotating compatible with its S-shaped configuration.

Eocene palaeopoles (40 to 50 Ma) from the South Shetland Islands

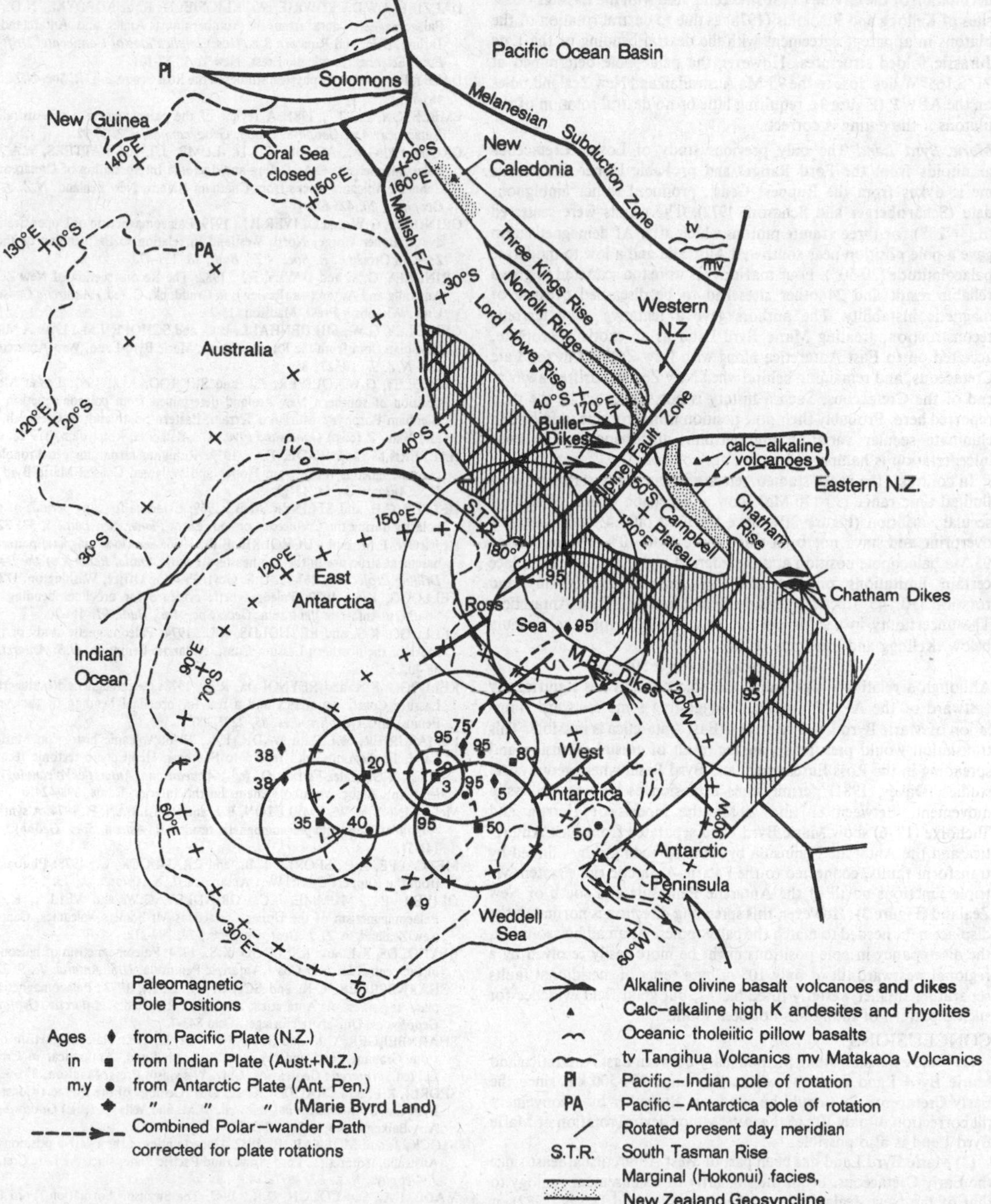


**Figure 3. Reconstruction of Antarctica, Australia and New Zealand at Anomaly 32 (75 Ma) time, showing Late Cretaceous to Cenozoic palaeomagnetic poles reconstructed back to Antarctica and the combined Apparent Polar Wander Path (APWP) inferred from the data. Circles surrounding palaeopoles are $\alpha_{95}$ circles of confidence. Lambert equal-area projection centred at 60°S, 160°E. Latitude/longitude grids refer to present positions of continental blocks.**

at the northern tip of the Antarctic Peninsula plot close to the Australian-New Zealand APWP and indicate little relative rotation or displacement with respect to East Antarctica over the past 50 Ma (Dalziel et al., 1973; Watts, 1982). Other results from the central Peninsula and from Western Ellsworth Land are more scattered, but still suggest high latitudes during the Late Cretaceous and Early Tertiary (Scharnberger and Scharon, 1982).

Kellogg (1980) reported palaeomagnetic data from plutons of the Andean Intrusive Suite (mean age 95 Ma) cutting east-west trending folded Jurassic sediments on the Orville Coast at the base of the Antarctic Peninsula. Kellogg interpreted the 51° difference in mean declination of the Orville Coast sites compared with the Lassiter Coast sites of Kellogg and Reynolds (1978) as due to dextral rotation of the plutons in apparent agreement with the dextral bending of the Late Jurassic folded structures. However, the palaeopole determined at 71°S 165°W lies close to the 95 Ma Australian and New Zealand poles on the APWP (Figure 3), requiring little or no dextral rotation of the plutons if the dating is correct.

*Marie Byrd Land* The only previous study of Lower Cretaceous granitoids from the Ford Ranges and probable Upper Cretaceous mafic dykes from the Ruppert Coast, produced rather ambiguous data (Scharnberger and Scharon, 1972). The results were scattered ($\alpha_{95} = 28°$) for three granite plutons which after Af demagnetisation gave a pole position near southwest Australia and a low to moderate palaeolatitude (20-50°). Four mafic dykes were too scattered to give a reliable result and 24 other sites had to be discarded because of magnetic instability. The authors gave a tentative plate tectonic reconstruction, treating Marie Byrd Land as an exotic microplate, accreted on to East Antarctica along with New Zealand in the Late Cretaceous, and remaining behind when New Zealand drifted away at end of the Cretaceous. Such a history is not supported by the data reported here. Probably their pole position is in error due to failure to eliminate secular variation and tectonic disturbance, while their interpretation is hampered by poor age control.

In contrast, the rocks studied here are magnetically stable, cover a limited time range (90-110 Ma) show a dispersion typical of normal secular variation (Figure 2B), have acquired only a weak magnetic overprint and have not been tilted more than 10°. The calculated 95 Ma palaeopole position and the inferred high palaeolatitude place certain limitations on possible displacement (200-500 km) and rotation (10°-45°) between Marie Byrd Land and East Antarctica. The uncertainty in declination can be calculated from the formula below (Kellogg and Reynolds, 1978);

$$\delta_{95} = \sin^{-1}(\sin\alpha_{95}/\cos I) = \pm 50°$$

Although a relatively high value, the pole position lies significantly eastward of the APWP (Figure 3) suggesting some eastward translation of Marie Byrd Land relative to East Antarctica is possible. This translation would presumably be the result of crustal extension and spreading in the Ross Embayment and Byrd Basin where geophysical studies (Davey, 1981) permit some extension as well as strike-slip movement. Between 80 and 20 Ma the models of Herron and Tucholke (1976) show Marie Byrd Land separated from East Antarctica and the Antarctic Peninsula by short spreading ridges linked by transform faults, connected to the Pacific-Antarctic ridge system via triple junctions north of the Antarctic Peninsula and south of New Zealand (Figure 3). However, this spreading direction is normal to the displacement needed to match the palaeopoles. As an ad hoc solution, the discrepancy in pole positions might be more easily resolved by a regional westward tilt of only 10° or by a series of meridional faults separating smaller westerly tilted blocks, but good field evidence for such a pattern of tilting has not been found.

## CONCLUSIONS

(1) Plate extension along the boundary between East Antarctica and Marie Byrd Land could be within the range 200-500 km since the Early Cretaceous, but could be ruled out altogether by a convenient tilt correction of only 10° to the west. Minor dextral rotation of Marie Byrd Land is also possible.

(2) Marie Byrd Land has been part of West Antarctica at least since the Early Cretaceous. From the similarity of its basement geology to that of the New Zealand Geanticline (Grindley and Davey, 1982), it has probably been there since the Early Palaeozoic.

(3) More evidence defining its pre-Cretaceous position in Gondwana will come from further studies of basement geology and geochronology coupled with aeromagnetic surveys to trace the large magnetic anomalies (Stokes Anomaly System) related to the Late Palaeozoic island arc volcanic/plutonic belt from New Zealand and the Campbell Plateau into Marie Byrd Land and Ellsworth Land.

*Acknowledgements* G.W.G. is indebted to the late Dr F.A. Wade for the invitation to participate in the 1977/78 United States expedition to Marie Byrd Land. Field work was greatly facilitated by the support provided by J. Wilbanks (Scientific Leader) and W. LeMasurier (Project Leader) and the helicopter pilots and crews of Squadron VXE6. Thanks are due to colleagues at New Zealand Geological Survey for photography, typing, draughting and comments on the manuscript. J. Sukroo assisted with the palaeomagnetic measurements.

## REFERENCES

DALZIEL, I.W.D., LOWRIE, W., KLIGFIELD, R. and OPDYKE, N.D., 1973: Palaeomagnetic data from the southernmost Andes and Antarctandes; *in* Tarling, D.H. and Runcorn S.K. (eds.) *Implications of Continental Drift to the Earth Sciences.* Academic Press, New York, 87-101.

DAVEY, F.J., 1981: Geophysical studies in the Ross Sea region. *R. Soc. N.Z., J., 11,* 465-79.

EMBLETON, B.J.J. , 1981: A review of the paleomagnetism of Australia and Antarctica. *Am. Geophys. Union, Geodynamic Ser. 2,* 77-92.

GRINDLEY, G.W., ADAMS, C.J.D., LUMB, J.T. and WATTERS, W.A., 1977: Palaeomagnetism, K-Ar Dating and Tectonic Interpretation of Cretaceous and Cenozoic Volcanic Rocks from Chatham Islands, New Zealand. *N.Z. J. Geol. Geophys., 20,* 425-67.

GRINDLEY, G.W. and OLIVER P.J., 1979: Palaeomagnetism of Upper Cretaceous dykes, Buller Gorge, North Westland in relation to the bending of the New Zealand Orocline. *R. Soc. N.Z., Bull., 18,* 131-47.

GRINDLEY, G.W. and DAVEY, F.J., 1982: The Reconstruction of New Zealand, Australia and Antarctica (Review); *in* Craddock, C. (ed.) *Antarctic Geoscience.* Univ. Wisconsin Press, Madison, 15-29.

GRINDLEY, G.W., MILDENHALL, D.C. and SCHOPF, J.M., 1980: A Mid-Late Devonian flora from the Ruppert Coast, Marie Byrd Land, West Antarctica. *R. Soc. N.Z. J., 10,* 271-85.

GRINDLEY, G.W., OLIVER, P.J. and SUKROO, J.C., 1981: Lower Mesozoic position of southern New Zealand determined from palaeomagnetism of the Glenham Porphyry, Murihiku Terrain, Eastern Southland; *in* Cresswell, M.M. and Vella, P. (eds.) *Gondwana Five,* A.A.Balkema, Rotterdam, 319-26.

HALPERN,M. and WADE, F.A., 1979: Rubidium-strontium geochronology of plutonic igneous rocks from Hobbs and Walgreen Coasts, Marie Byrd Land. *U.S. Antarct. J., 14,* 18-19.

HELSEY, C.E. and STEINER, M.B., 1969: Evidence for long periods of normal polarity during the Cretaceous period. *Earth Planet. Sci. Lett., 5,* 325-32.

HERRON, E.M. and TUCHOLKE, B.E., 1976: Sea-floor magnetic patterns and basement structure in the southeastern Pacific. *Initial Reports of the Deep Sea Drilling Project 35,* 263-75. U.S. Govt. Printing Office, Washington, D.C.

KELLOGG, K.S., 1980: Paleomagnetic evidence for oroclinal bending of the Southern Antarctic Peninsula. *Geol. Soc. Am., Bull., 91,* 414-20.

KELLOGG, K.S. and REYNOLDS, R.L., 1974: Paleomagnetic study of igneous rocks of the northern Lassiter Coast, Antarctic Peninsula. *U.S. Antarct. J., 9,* 38-40.

KELLOGG, K.S. and REYNOLDS, R.L., 1978: Paleomagnetic Results from the Lassiter Coast, Antarctica and a test for oroclinal bending of the Antarctic Peninsula. *J.Geophys. Res. 83,* 2293-300.

LEMASURIER, W.E. and WADE, F.A., 1976: Volcanic history in Marie Byrd Land: Implications with regard to Southern Hemisphere tectonic reconstructions; *in* Gonzalez-Ferran, O. (ed.) *Andean and Antarctic Volcanology Problems.* Int. Assoc. Volcanol. Chem. Earth's Interior, Rome, 398-424.

McELHINNY, M.W., EMBLETON, B.J. and WELLMAN, P., 1974: A synthesis of Australian Cenozoic palaeomagnetic results. *R. Astron. Soc., Geophys. J., 36,* 141-51.

METCALFE, A.P., SPORLI, K.B., and CRADDOCK, C., 1978: Plutonic rocks from the Ruppert Coast, West Antarctica. *U.S. Antarct. J., 13,* 5-6.

OLIVER, P.J., MUMMÉ, T.C., GRINDLEY, G.W. and VELLA, P., 1979: Palaeomagnetism of the Upper Cretaceous Mt Somers Volcanics, Canterbury, New Zealand. *N.Z. J. Geol. Geophys., 22,* 199-212.

REYNOLDS, R.L. and KELLOGG, K.S., 1974: Paleomagnetism of igneous rocks of the central Lassiter Coast,Antarctic Peninsula. *U.S. Antarct. J., 9,* 227-8.

SCHARNBERGER, C.K. and SCHARON, Le R., 1972: Palaeomagnetism and plate tectonics of Antarctica, *in* Adie, R.J. (ed.) *Antarctic Geology and Geophysics.* Universitetsforlaget, Oslo 843-7.

SHARNBERGER, C.K. and SCHARON, Le R., 1982: Paleomagnetism of rocks from Graham Land and Western Ellsworth Land, Antarctica. *in* Craddock, C., (ed.) *Antarctic Geoscience.* Univ. Wisconsin Press, Madison, 371-6.

SPORLI, K.B. and CRADDOCK, C., 1981: Geology of the Ruppert Coast, Marie Byrd Land, Antarctica; *in* Cresswell, M.M. and Vella, P. (eds.) *Gondwana Five,* A.A.Balkema, Rotterdam, 243-50.

STOCK, J. and MOLNAR, P., 1982: Uncertainties in the relative positions of the Australia, Antarctica, Lord Howe, and Pacific Plates since the Late Cretaceous. *J. Geophys. Res., 87,* 4697-714.

WADE, F.A., and COUCH, D.R., 1982: The Swanson Formation, Ford Ranges, Marie Byrd Land—evidence for direct relationship with Robertson Bay Group, northern Victoria Land; *in* Craddock, C. (ed.) *Antarctic Geoscience,* Univ. Wisconsin Press, Madison, 609-16.

WATTS, D.R., 1982: Potassium-Argon Ages and Paleomagnetic Results from King George Island, South Shetland Islands; *in* Craddock, C. (ed.) *Antarctic Geoscience,* Univ. Wisconsin Press, Madison, 255-61.

# THE POSITION OF ANTARCTICA WITHIN GONDWANA IN THE LIGHT OF PALAEOZOIC OROGENIC DEVELOPMENT

H. Miller, *Geologisch-Palaontologisches Institut, Westfalische Wilhelms-Universitat, Munster, Federal Republic of Germany.*

**Abstract** Westward progressing rejuvenation and consolidation of the Pacific edge of Gondwana from the shield regions in the direction of a Proto-Pacific ocean eventually terminated in the Antarctic Peninsula's Late Triassic to Early Jurassic metamorphism and deformation as the latest stage of pre-Andean basement evolution. Gondwana reconstructions that place the Antarctic Peninsula west of the southern tip of South America, are confirmed by the joint pre-Andean history of the Gondwana-Pacific border. Palaeozoic and Mesozoic sediment-filled troughs around the southern Atlantic in southern South America and central west Antarctica are interpreted as aulacogens accompanying early stages of Atlantic rifting.

The fit of Antarctica with neighbouring continents is a basic problem in the reconstruction of the pre-drift Gondwana configuration. While the fit of South America to Africa and of Australia and New Zealand to Antarctica is well established, the relative position of Antarctica in relation to Madagascar, India and South America remains controversial, particularly, the pre-Jurassic relative position of the Antarctic Peninsula to the southern tip of South America. Four main reconstructions have been proposed: (1) Peninsula to the east of Tierra del Fuego (Smith and Hallam 1970, Craddock 1975), (2) Peninsula to the south of Tierra del Fuego (Suárez 1976), (3) Peninsula to the west of Tierra del Fuego (Harrison et al. 1979, Quilty, 1982) and (4) Peninsula twisted alongside South Africa or East Antarctica (De Wit 1977, Dalziel 1982).

In order to reconstruct Gondwanaland, as well as to evaluate the paths of its fragments by palaeomagnetic research, knowledge of the pre-drift geological facts and correlations is the most important key. The Pacific margin of Gondwana in the Palaeozoic is remarkably continuous and shows nearly the same orogenic history from central South America through Antarctica into Australia and New Zealand. This large belt was called the "Samfrau geosyncline" by Du Toit (1937). Antarctic research permits a good definition of Palaeozoic orogenic periods (Craddock 1975; Grikurov et al. 1980). When dealing with the Palaeozoic evolution of Australia and New Zealand, the present author has used Brown et al. (1968), Plumb (1979) and Suggate et al. (1978). However, the Palaeozoic orogenic development of South America was, until a few years ago, known to only a few local specialists. In the Lower Palaeozoic of Argentina Aceñolaza and Toselli (1976) contributed to the classification of the orogenic periods. Near the recent coastline Hervé et al. (1981) and Miller (1979) proposed different models of orogenic development.

## OROGENIES AT THE PACIFIC EDGE OF GONDWANA

In Figure 1 orogenic events at the circum-Pacific edge of former Gondwana are summarised. Their coincidence is striking.

### 1. Late Precambrian events

Only the latest Precambrian will be considered here. In South America an orogenic cycle around 650 Ma was named the Brazilian cycle by Cordani et al. (1973). Its effects are concentrated in Brazil, but there are some hints of it in the southern parts of the Sierras Pampeanas of Argentina and in the North Patagonian Massif. In Antarctica, the Beardmore Orogeny has been dated at 600 to 630 Ma (Adams et al. 1982). In central Australia, the same orogeny is called the Petermann Ranges Orogeny.

### 2. Orogenies of Middle Cambrian to Early Ordovician age.

In South America the Palaeozoic age of a strong orogeny has been recognised only by the discovery of Cambrian trace fossils in rocks formerly considered as Precambrian (Aceñolaza 1973). This orogenic cycle climaxing in the Middle Cambrian was named the "Pampean Cycle" by Aceñolaza and Toselli (1976) in northwest Argentina, and was compared with events of similar age in Argentina and the coastal regions along the Atlantic by Aceñolaza and Miller (1982). There is evidence that magmatism and tectono-metamorphic overprint belonging to this orogeny continued until the Ordovician (Knüver and Miller 1981).

In the Transantarctic Mountains, the Ross Orogeny folded Late Precambrian to Cambrian sediments. Though its influence varies along the Transantarctic Mountains (Stump, 1981), it can be recognised as disconformities, thermal and magmatic events and folding phases of Middle Cambrian to Early Ordovician age throughout the mountain chain (Laird and Bradshaw 1982). Many sedimentological and metamorphic details of the Robertson Bay and Wilson Groups in northern Victoria Land (Tessensohn 1982) can be compared with nearly identical units in northwestern Argentina (Puncoviscana Formation and its metamorphic equivalents). Probably folding of this age also occurred in Marie Byrd Land (Katz 1982). In Australia and Tasmania, the Delamerian and Tyennan Orogenies occurred along the Pacific border of eastern Australia. From New Zealand the "Haupiri disturbance" of Late Cambrian age should be mentioned.

### 3. Late Silurian to Middle Devonian orogenic events

In central and southern South America the Famatinian cycle has been defined by Aceñolaza and Toselli (1976) as an orogenic cycle of Early to Middle Palaeozoic age. Sedimentation begins with the Upper Cambrian molasse series of the Late Pampean cycle (El Mesón Group) and finishes with a Devonian molasse. In northern South America, some poorly known tectonic phases may be attributed to this cycle according to Zeil (1979). In the Chilean coast range its existence is under discussion (Miller 1979).

In Antarctica, the Borchgrevink Orogeny of Silurian/Devonian age is known. Grikurov et al. (1980) consider it as "eo-Gondwanian" in the Gondwanian cycle. Tessensohn et al. (1981) deny its existence in northern Victoria Land, while Katz (1982) relates magmatic events in Marie Byrd Land to it. In Australia and Tasmania the Bowning and Tabberabberan Orogenies indicate strong tectonic activity at this time. In New Zealand the Tuhua Orogeny finishes the Lower Palaeozoic sedimentary cycle.

### 4. Orogenies of Carboniferous age

In South America, this cycle under the names "Variscan", "Hercynian" or "Gondwanian" has been considered as the only Palaeozoic orogeny, though divided into several phases (Dalmayrac et al. 1980). The present author prefers to use the name "Variscan" exclusively to

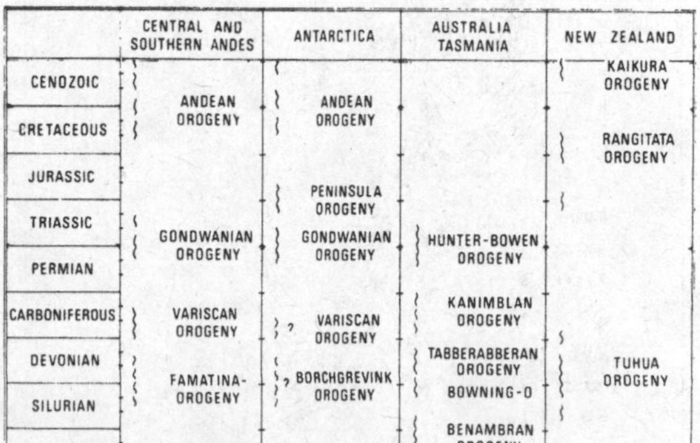

| | CENTRAL AND SOUTHERN ANDES | ANTARCTICA | AUSTRALIA TASMANIA | NEW ZEALAND |
|---|---|---|---|---|
| CENOZOIC | ANDEAN OROGENY | ANDEAN OROGENY | | KAIKURA OROGENY |
| CRETACEOUS | | | | RANGITATA OROGENY |
| JURASSIC | | PENINSULA OROGENY | | |
| TRIASSIC | GONDWANIAN OROGENY | GONDWANIAN OROGENY | HUNTER-BOWEN OROGENY | |
| PERMIAN | | | | |
| CARBONIFEROUS | VARISCAN OROGENY | ? VARISCAN OROGENY | KANIMBLAN OROGENY | |
| DEVONIAN | | | TABBERABBERAN OROGENY | TUHUA OROGENY |
| SILURIAN | FAMATINA-OROGENY | ? BORCHGREVINK OROGENY | BOWNING-O | |
| ORDOVICIAN | | | BENAMBRAN OROGENY | |
| CAMBRIAN | PAMPA-OROGENY | ROSS OROGENY | DELAMERIAN/TYENNAN-O | HAUPIRI DISTURBANCE |
| LATEST PRECAMBRIAN | ? BRAZILIAN OROGENY | BEARDMORE OROGENY | PETERMANN-RANGES-O | |

**Figure 1. Comparative compilation of pre-Andean orogenies on the Pacific edge of Gondwana.**

define orogenic events developed in the Carboniferous which are typical Variscan foldings as defined by Stille (1928). Metamorphism and folding of this age have been observed or assumed at many points in the central and southern Andes (Munizaga et al. 1973, Miller 1979, Davidson et al. 1981). From Peru and Venezuela, Zeil (1979) reports data on deformational phases of Early Carboniferous age; from Bolivia Carboniferous metamorphic dates are known (Cordani et al. 1980). Flysch and molasse sequences (Sierra de Languiñeo and Paganzo Groups) as well as numerous granitic intrusions prove the existence of Variscan orogenic events in the basement of the Andes.

From Antarctica data are very scarce. Grikurov et al. (1980) mention an Early Gondwanian tectonic event as synonymous with the Borchgrevink Orogeny at about 325 Ma referring to the synkinematic and postkinematic granites. From the Antarctic Peninsula Gledhill et al. (1982) give hints of possible Variscan or older events. Hjelle et al. (1982) assigned folding and metamorphism in the Ellsworth Mountains to the Carboniferous. In Australia, tectonic activity documented for the Kanimblan Orogeny has been rather weak.

## 5. Permo-Triassic orogenies

The term "Gondwanian" has been defined by Du Toit (1927, p.105) as follows: "at the close of the Permian a lengthy arc of compression (the Gondwanides) came into being". The application of the term should therefore be restricted to orogenic events near the Permian/Triassic boundary. In South America, in the Chilean Andes, folding and metamorphism occur from the Late Permian to the Early or Middle Triassic. Folding phases of this age were named "Late Variscan" (Miller 1979), before it was realised that they are rather well defined by the term "Gondwanian". From Peru and Venezuela Zeil (1979) refers to folding at the end of the Palaeozoic. In Antarctica orogenic activity at the Permo-Triassic boundary around the Weddell Sea region has been called Weddell Orogeny (Ford 1972) or Ellsworth Orogeny (Craddock 1975). In eastern Australia, the Hunter-Bowen Orogeny is known as the last orogenic phase of the continent.

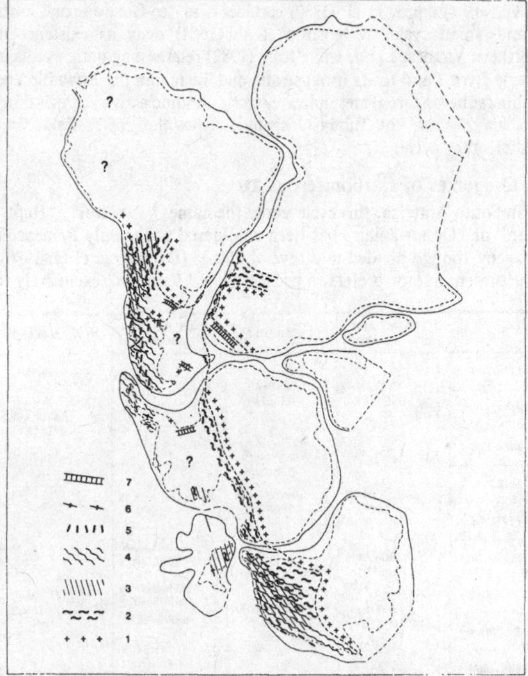


**Figure 2. Gondwana reconstruction based on pre-Andean orogenic development of the border between the continent and the Proto-Pacific Ocean. 1: Precambrian basement. 2: Middle Cambrian to Early Ordovician orogenies. 3: Late Silurian to Middle Devonian orogenies. 4: Carboniferous orogenies. 5: Permo-Triassic orogenies. 6: Late Triassic to Early Jurasic orogeny. 7: Palaeozoic aulacogens.**

## 6. Younger orogenies.

In the folded and metamorphosed basement of the Antarctic Peninsula, fossils of Triassic age confirm the geochronologically known "young" age of the last pre-Andean orogeny on the Penisula (Gledhill et al. 1982). The existence of folding, metamorphism and plutonism at the Triassic/Jurassic boundary is so essential for the reconstruction of Gondwana that the present author believes it to be appropriate to give the name "Peninsula Orogeny" to this event. In New Zealand a metamorphic event of that time has been named "Rangitata I" by Bradshaw et al (1981). The Cretaceous to Recent circum-Pacific Andean Orogeny is not discussed herein.

## THE CONTINUITY OF THE OROGENIC PROCESS AT THE PACIFIC EDGE OF GONDWANA

As demonstrated above, at the Pacific border of Gondwana, from the Late Precambrian until the breaking up of the supercontinent, orogenic activities occurred almost continuously. Orogenic cycles overlap one another at many places, and only a general migration of activity from the Precambrian cratons towards the Pacific can be recognised (Figure 2). The obvious continuity of orogenic events has led several authors to arrange the orogenic events in only one or two groups (e.g. Ross stage: 1000-480 Ma, Gondwanian stage: 480-180 Ma; Grikurov et al. 1980). However, a more individual delineation and description of orogenic units brings advantages for correlating them from one continent to another and for recognising the general laws of orogenic developments at this persistently active continental border.

The comparison of the orogenies of the Pacific/Gondwana edge (Figure 1) shows clearly that the times of orogenic climaxes are not conformable with the classic model elaborated by Stille (1928) according to data from Laurasia. One might assume that this difference is based on the dissimilarity of geotectonic positions. The European geosynclines were intra-continental troughs, developing only small basins of oceanic crust, if at all. Opening and shutting of these troughs developed independently of the geotectonic events occurring on the Pacific/Gondwana border, which were exclusively controlled by almost continuous spreading processes of varying directions within the Proto-Pacific.

In all Gondwana continents a consolidation process is observed which advances towards the Pacific Ocean. There is only one Antarctica/South America fit which fulfills this fact in a reconstructed Gondwana map: placing the Antarctic Peninsula with its very young basement consolidation age (Peninsula Orogeny) on the western side of the southern tip of South America (Figure 2). For different reasons, this reconstruction has been proposed by many authors during the last decade; it not only fits well the pre-Andean orogenic history, but also agrees with recent theories on the South Andean geotectonic setting (Harrison et al. 1979, Suárez 1976, Miller 1981).

## SOUTH ATLANTIC AULACOGENS AND THE SUBGLACIAL BASINS OF CENTRAL WEST ANTARCTICA

The geology of some smaller mountain chains is not consistent with this pattern of a pacificwards consolidation, e.g. the Sierra de la Ventana, the Silurian/Devonian strata of Sierra Grande, the Palaeozoic rocks of the Falkland Islands, the Cape Range in South Africa and the Ellsworth Mountains in West Antarctica. These areas differ from the orogens bordering the Pacific by the scarcity of magmatic rocks, the maturity of their sediments, the uncomplicated deformation and the weakness of metamorphism. Harrington (1970) considered the Sierra de la Ventana as an aulacogenic chain, a term which could be used for describing the other mentioned chains. The present author regards them as aulacogens that accompanied an early Atlantic rifting. It was not until the Mesozoic that the Gondwana continent started to finally split up.

Comparisons of Marie Byrd Land and Western Ellsworth Land with the formerly neighbouring Australia and New Zealand (Katz 1982, Grindley and Davey 1982) demonstrate the integration of at least the Pacific region of West Antarctica into the orogenic pattern described above. A glance at the formerly close lying Patagonia shows that there, perpendicular to the South Atlantic Ocean, several unfolded basins were filled with sediments during the Mesozoic and Cenozoic, i.e. the Salado, Colorado, Chubut and San Jorge Basins.

The gravimetric pattern of the area between the Ellsworth Mountains and Marie Byrd Land (Bentley and Robertson 1982, Groushinsky and Sazhina 1982a) indicates the possible presence of Mesozoic sedimentary basins, separated by horstlike consolidated areas perpendicular to a weakly developed rift zone as a prolongation of the Atlantic Ocean. Geophysical data presented by Jankowski and Drewry (1981) prove the possibility of Mesozoic or Palaeozoic sedimentary basins below the ice.

The crustal thickness of West Antarctica corresponds generally to continental conditions (Groushinsky and Sazhina 1982b); the morphological furrows between the Ross and Weddell Seas (Rose 1982) can partly be explained by glacial erosion, partly by the lack of sedimentary filling since the beginning of the glaciation and partly by recent rifting.

Prior to the commencement of major sea floor spreading, Antarctica was shifted away from South America to the southeast along transform faults, at the same time minor spreading centres were active in the Scotia Arc; similar centres may have also been active in the Weddell Sea.

*Acknowledgements* I am grateful to G.F. Aceñolaza, F. Hervé, E. Godoy and A. Toselli, who helped to form my ideas about South American orogenies by numerous discussions, and to A. Willner for critically reviewing the manuscript.

## REFERENCES

ACENOLAZA, F.G., 1973: Sobre la presencia de *Oldhamia* sp. en la formación Puncoviscana de Cuesta Muñano, Provincia de Salta, Républic Argentina. *Revta. Asoc. geol. argent., 28,* 56-80.

ACENOLAZA, F.G. and MILLER, H., 1982: Early Palaeozoic orogeny in southern South America. *Precambrian Res., 17,* 133-46.

ACENOLAZA, F.G. and TOSELLI, A., 1976: Consideraciones estratigráficas y tectónicas sobre el Palaeozoico inferior del Noroeste Argentino. *Mem. Seg. Congr. Latinoamer. Geol. 1973, 3,* 755-64.

ADAMS, C.J.D., GABITES, J.E. and GRINDLEY, G.W., 1982: Orogenic history of the Central Transantarctic Mountains: New K-Ar age data on the Precambrian-Lower Paleozoic basement; *in* Craddock, C. (ed.) *Antarctic Geoscience.* Univ. Wisconsin Press, Madison, 817-26.

BENTLEY, C.R. and ROBERTSON, J.D., 1982: Isostatic gravity anomalies in West Antarctica; *in* Craddock, C. (ed.) *Antarctic Geoscience.* Univ. Wisconsin Press, Madison, 949-54.

BRADSHAW, J.D., ADAMS, C.J. and ANDREWS, P.B., 1981: Carboniferous to Cretaceous on the Pacific margin of Gondwana. The Rangitata phase of New Zealand; *in* Cresswell, M.M. and Vella, P. (eds.) *Gondwana Five.* A.A. Balkema, Rotterdam, 217-21.

BROWN, D.A., CAMPBELL, K.S.W. and CROOK, K.A.W., 1968: *The Geological evolution of Australia and New Zealand.* Pergamon Press, Oxford.

CORDANI, U.G., AMARAL, G. and KAWASHITA, K., 1973: The Precambrian evolution of South America. *Geol Rdsch., 62,* 309-17.

CORDANI, U.G., KAWASHITA, K. and CORTEZ, G., 1980: Chronology of the tectonomagmatic events of the Cordillera Real of Bolivia. *26th. Int. Geol. Congr., Abstr, 1,* 32.

CRADDOCK, C., 1975: Tectonic evolution of the Pacific margin of Gondwanaland; *in* Campbell, K.S.W. (ed.) *Gondwana Geology.* A.N.U. Press, Canberra, 609-18.

DALMAYRAC, B., LAUBACHER, G., MAROCCO, R., MARTINEZ, C. and TOMASI, P., 1980: La chaîne hercynienne d'Amérique du Sud. Structure et évolution d'un orogène intracratonique. *Geol. Rdsch., 69,* 1-21.

DALZIEL, I.W.D., 1982: The early (pre-Middle Jurassic) history of the Scotia Arc region: A review and progress report; *in* Craddock. C. (ed.) *Antarctic Geoscience.* Univ. Wisconsin Press, Madison, 111-26.

DAVIDSON, J., MPODOZIS, C. and RIVANO, S., 1981: Evidencias de tectogénesis del Devónico superior-Carbonifero inferior al oeste de Augusta Victoria, Antofagasta, Chile. *Revta. geol. Chile, 12,* 79-86.

DE WIT, M.J., 1977: The evolution of the Scotia Arc as a key to the reconstruction of southwestern Gondwanaland. *Tectonophysics, 37,* 53-81.

DU TOIT, A.L., 1927: A geological comparison of South America with South Africa. *Publ. Carnegie Inst, 381,* Washington DC.

DU TOIT, A.L., 1937: *Our wandering continents.* Oliver and Boyd, Edinburgh and London.

FORD, A.B., 1972: Weddell Orogeny—Latest Permian to Early Mesozoic deformation at the Weddell Sea margin of the Transantarctic Mountains; *in* Adie, R. (ed.) *Antarctic Geology and Geophysics.* Universitetsforlaget, Oslo, 419-25.

GLEDHILL, A., REX, D.C. and TANNER, P.W.G., 1982: Rb-Sr and K-Ar geochronology of rocks from the Antarctic Peninsula between Anvers Island and Marguerite Bay; *in* Craddock, C. (ed.) *Antarctic Geoscience.* Univ. Wisconsin Press, Madison, 315-23.

GRIKUROV, G.E., ZNACHKO-YAVORSKY, G.A., KAMENEV, E.N. and KURININ, R.G., 1980: *Explanatory notes to the Tectonic Map of Antarctica (scale 1:10 000 000).* Res. Inst. Geol. Arctic, Leningrad.

GRINDLEY, G.W. and DAVEY, F.J., 1981: The reconstruction of New Zealand, Australia and Antarctica; *in* Craddock, C. (ed.) *Antarctic Geoscience.* Univ. Wisconsin Press, Madison. 15-29.

GROUSHINSKY, N.P. and SAZHINA, N.B., 1982a: Some features of Antarctic crustal structure; *in* Craddock, C. (ed.) *Antarctic Geoscience* Univ. Wisconsin Press, Madison, 907-11.

GROUSHINSKY, N.P. and SAZHINA, N.B., 1982b: Gravitational field of Antarctica; *in* Craddock, C. (ed.) *Antarctic Geoscience.* Univ. Wisconsin Press, Madison, 913-7.

HARRINGTON, H.J., 1970: Las Sierras Australes de Buenos Aires, República Argentina: Cadena aulacogénica. *Revta. Asoc. geol. argent., 25,* 151-81.

HARRISON, C.G.A., BARRON, E.J. and HAY, W.W., 1979: Mesozoic evolution of the Antarctic Peninsula and the southern Andes. *Geology, 7,* 374-8.

HERVÉ, F., DAVIDSON, J., GODOY, E., MPODOZIS, C. and COVACEVICH, V., 1981: The late Palaeozoic in Chile: Stratigraphy, structure and possible tectonic framework. *An. Acad. brasil. Cienc., 53,* 361-73.

HJELLE, A., OHTA, Y. and WINSNES, T.S., 1982: Geology and petrology of the southern Heritage Range, Ellsworth Mountains; *in* Craddock, C. (ed.) *Antarctic Geoscience.* Univ. Wisconsin Press, Madison, 599-608.

JANKOWSKI, E.J. and DREWRY, D.J., 1981: The structure of West Antarctica from geophysical studies. *Nature, 291,* 17-21.

KATZ, H.R., 1982: West Antarctica and New Zealand: A geological test of the model of continental split; *in* Craddock, C. (ed.) *Antarctic Geoscience.* Univ. Wisconsin Press, Madison, 31-41.

KNÜVER, M. and MILLER, H., 1981: Ages of metamorphic and deformational events in the Sierra de Ancasti (Pampean Ranges; Argentina) *Geol. Rdsch., 70,* 1020-9.

LAIRD, M.G. and BRADSHAW, J.D., 1982: Uppermost Proterozoic and Lower Palaeozoic geology of the Transantarctic Mountains; *in* Craddock, C. (ed.) *Antarctic Geoscience.* Univ. Wisconsin Press, Madison, 525-33.

MILLER, H., 1979: Das Grundgebirge der Anden in Chonos-Archipel, Region Aisén, Chile. *Geol. Rdsch., 68,* 428-56.

MILLER, H., 1981: Pre-Andean orogenies of southern South America in the context of Gondwana; *in* Cresswell, M.M. and Vella, P. (eds.) *Gondwana Five* A.A. Balkema, Rotterdam, 237-42.

MUNIZAGA, F., AGUIRRE, L. and HERVÉ, F., 1973: Rb/Sr ages of rocks from the Chilean metamorphic basement. *Earth Planet. Sci. Lett., 18,* 87-92.

PLUMB, K.A., 1979: The tectonic evolution of Australia. *Earth-Sci. Rev., 14,* 205-49.

QUILTY, P.G., 1982: Tectonic and other implications of Middle-Upper Jurassic rocks and marine faunas from Ellsworth Land, Antarctica; *in* Craddock, C. (ed.) *Antarctic Geoscience.*Univ. Wisconsin Press, Madison, 669-78.

ROSE, K.E., 1982: Radio-echo studies of bedrock in southern Marie Byrd Land, West Antarctica; *in* Craddock, C. (ed.) *Antarctic Geoscience.* Univ. Wisconsin Press, Madison, 985-92.

SMITH, A. and HALLAM, A., 1970: The fit of the southern continents. *Nature, 225,* 139-44.

STILLE, H., 1928: Zur Einführung in die Phasen der paläozoischen Gebirgsbildung. *Z. dt. geol. Ges., 80,* 1-25.

STUMP, E., 1981: Observations on the Ross Orogen, Antarctica; *in* Cresswell, M.M. and Vella, P. (eds.) *Gondwana Five,* A.A. Balkema, Rotterdam, 237-42.

SUÁREZ, M., 1976: Plate-tectonic model for southern Antarctic Peninsula and its relation to southern Andes. *Geology, 4,* 211-4.

SUGGATE, R.P., STEVENS, G.R. and TE PUNGA, M.T., (eds.), 1978: *The Geology of New Zealand.* Government Printer, Wellington, 2 vols.

TESSENSOHN, F., 1982: Significance of late Precambrian turbidite sequences bordering the East Antarctic shield. *Geol. Rdsch., 71,* 361-9.

TESSENSOHN, F., DUPHORN, K., JORDAN, H.,, KLEINSCHMIDT, G., SKINNER D.N.B., VETTER, U., WRIGHT, T.O. and WYBORN, D., 1981: Geological comparison of basement units in North Victoria Land, Antarctica. *Geol. Jb. B41,* 31-88.

ZEIL, W., 1979: *The Andes. A geological review.* Borntraeger, Berlin.

# AUSTRALIA-ANTARCTIC BREAKUP: APATITE FISSION TRACK EVIDENCE FROM SOUTH AND WESTERN AUSTRALIA AND EAST ANTARCTICA

K.U. Ferguson, P.R. Kelly, A.J.W. Gleadow and J.F. Lovering, *Department of Geology, University of Melbourne, Parkville, Victoria 3052, Australia.*

*Abstract* Apatite fission track ages from granitic components of the Archaean Yilgarn Block in the south of Western Australia indicate a widespread, low temperature 500 Ma 'Pan African' tectonic event in the central shield area with a localised higher temperature along the western margin as recorded by sphene fission track ages. A transition from these relatively old apatite ages in the central shield to younger ages of approximately 260 Ma in the west is correlated with gradual cooling from the 500 Ma event. The youngest age adjacent to the bounding faults of the Perth Basin may be related to a thermal pulse in the early stages of formation of the basin. However the retention of Permo-Triassic apparent apatite ages along the western margin indicates that high heat flow, uplift and erosion were not sufficiently intense to completely reset the apatite ages after this time. However track length studies show that the southwestern margin of the Yilgarn Block was affected by a mild thermal event during the Early Cretaceous which was probably associated with the onset of sea floor spreading in that region.

The southern continental margin of Western Australia shows apatite ages which are similar or slightly older than those of the formerly adjacent east Antarctic coast, but are significantly older than those found in southeastern Australia where both uplift and thermal disturbance associated with the onset of the Antarctica/Australia rift in the Early Cretaceous were apparently more intense. However track length studies on apatites from the southern continental margin of Western Australia imply a relatively recent (earliest Tertiary) low temperature thermal pulse which is consistent with the established Paleocene-Eocene timing of the onset of sea floor spreading between Australia and Antarctica.

# BREAKUP OF GONDWANA: APATITE FISSION TRACK DATING EVIDENCE FROM SOUTHEAST AUSTRALIA

M.E. Moore, A.J.W. Gleadow and J.F. Lovering, *Department of Geology, University of Melbourne, Parkville, Victoria 3052, Australia.*

**_Abstract_ Processes related to continental rifting and breakup in the Tasman Sea have profoundly affected fission track ages of apatites from Palaeozoic granitic rocks within about 150 km of the continental margin in southeastern Australia. Apatite ages decrease from 250-360 Ma at a distance of 100 km inland to 80-126 Ma along the coast, the youngest ages correlating with the initiation of seafloor spreading in the Tasman Sea.**

**Near-coastal samples yield ages of less than 200 Ma, which are believed to have been either partially or completely reset during development of the Tasman rift system. At this time increased heatflow, independently evidenced by alkaline magmatic activity, resulted in heating of the continental margin. In addition, accelerated uplift and erosion along the flanks of the developing rift have emphasised the thermally-induced apatite age pattern by exposing basement rocks from initially deeper levels in the crust.**

**This interpretation is complemented and strongly supported by measured distributions of confined track lengths in apatite. These indicate that coastal apatites reached temperatures as high as 100-125°C 70-80 Ma ago, rapidly diminishing to temperatures of less than 30°C 100 km inland. An inferred depth of erosion of 1-1.5 km in near-coastal regions points to maximum transient palaeothermal gradients of 65-110°C/km.**

# A GONDWANA RECONSTRUCTION BETWEEN ANTARCTICA AND SOUTH AFRICA

J. Hofmann and W. Weber, *Bergakademie Freiberg, Sektion Geowissenschaften, 9200 Freiberg, German Democratic Republic.*

*Abstract* Recent geological data from the western part of the East Antarctic Platform and from West Antarctica are used to test the Gondwana reconstruction model of Barker and Griffiths (1977) with special reference to the relationship between Antarctica and South Africa. Archaean rocks in Queen Maud Land are interpreted as a continuation of the Kaapvaal Craton in South Africa. Early to Middle Proterozoic metamorphic rocks in the Shackleton Range, the Haag Nunataks, and the Cape Meredith gneisses of the Falkland Islands are considered to be comparable with the South African Namaqua-Natal Belt. Middle to Late Proterozoic platform development was accompanied by volcanism in both Africa and Antarctica. Late Proterozoic to Early Palaeozoic Ross System (exposed in the Pensacola Mountains, with a miogeosynclinal element in the Argentina Range) is correlated with the Late Proterozoic Malmesbury Belt in South Africa. The present block structure of West Antarctica is discussed in relation to transform fault systems in the southeast Pacific and the southwest Atlantic Oceans. In Antarctica, initial Gondwana breakup was marked by Silurian to Devonian dolerite intrusive activity and related basement thermal reactivation. Final separation of continental Antarctica and Africa took place along overlapping arms of the Early to Middle Jurassic Dufek and Mozambique Triple Junctions.

## A GEOLOGIC COMPARISON BETWEEN SOUTH AFRICA AND ANTARCTICA

The Gondwana reconstruction (Figure 1) proposed by Barker and Griffiths (1977) includes the Falkland Plateau, the continental character of which was postulated by Barker et al. (1974). The Ellsworth Mountains are positioned along the east coast of the modern Weddell Sea as proposed by Schopf (1969), and Clarkson and Brook (1977). These arrangements eliminated the overlaps that were a feature of previously postulated reconstructions. However, Barker and Griffiths (1977) did not give a geological example of their model. In this contribution, which is based upon a review of geological data from Antarctica, the authors discuss geological connections between Antarctica and South Africa, and interpret the block structure of West Antarctica in the light of continental drift.

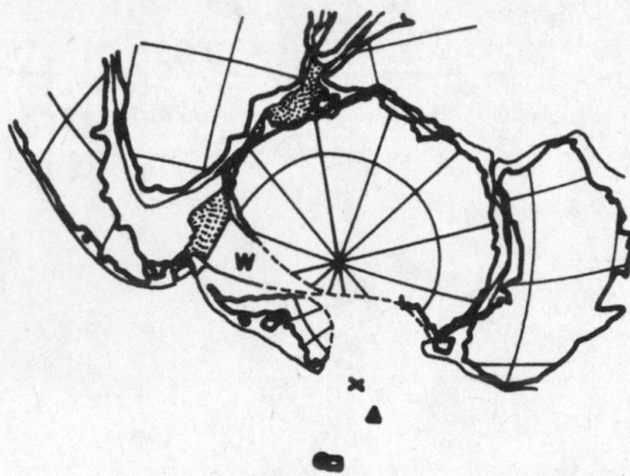


**Figure 1. Gondwana reconstruction "model A" by Barker and Griffiths (1977). Pole of projection (triangle) is mid-Jurassic East Antarctic palaeopole. Jurassic palaeopoles from Australia x, Africa o, South America □, Weddell Sea W, overlaps stippled.**

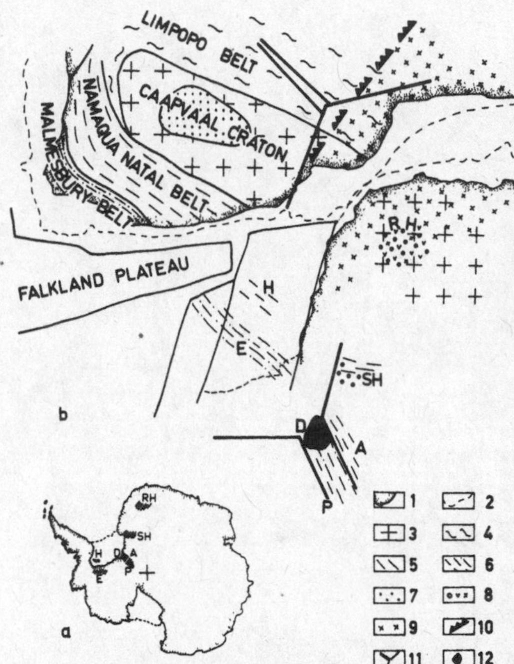


**Figure 2. Predrift situation of Africa/Antarctica. (a) Antarctic geological complexes relevant to the geological Africa/Antarctica reassembly. (b) Precambrian structural provinces in South Africa/Antarctica: (1) modern shore line; (2) modern shelf margin; (3) Archaean cratons; (4) Late Archaean/Early Proterozoic mobile belts; (5) Early to Middle Proterozoic mobile belts; (6) Late Proterozoic/Early Palaeozoic mobile belts; (7) Late Archaean/Early Proterozoic platform cover of the Kaapvaal Craton; (8) Middle/Late Proterozoic platform cover of the East Antarctic Craton (circles = terrigenous sediments; oblique crests and arrows = volcanites; (9) Mozambique Belt; (10) Mozambique Front; (11) triple junctions, Mesozoic; (12) Dufek Intrusion (Lower Jurassic); (P) Pensacola Mountains; (A) Argentina Range; (D) Dufek Massif; (SH) Shackleton Range; (E) Ellsworth Mountains; (H) Haag Nunataks; (RH) Ritscher Highland.**

### Crystalline basement rocks in East Antarctica and South Africa

*Archaean.* The Archaean of Queen Maud Land (Figure 2) is divided into the older Humboldt and the younger Insel Complexes (Grikurov, 1980). Ravich and Grikurov (1977) proposed that metamorphism and deformation of the Humboldt Complex (granulite facies rocks, migmatites, charnockites, nebulitic granites, granite gneisses, marbles and quartzites) took place 3200-3000 Ma ago and that the Insel Complex (biotite gneisses with genetically related migmatites and plagiogranites, orthoamphibolites, marbles, and quartzites) formed between 3000 and 2500 Ma ago. Present knowledge of the Archaean of Queen Maud Land mainly comprises petrographic and petrological data and a few isotopic dates. Nevertheless the exposed rock types appear to correspond to those described from the Archaean Kaapvaal Craton of South Africa (Condie, 1981). This overall lithological similarity leads us to propose that the Archaean rocks in central Queen Maud Land were once connected to those in the Kaapvaal Craton.

*Early to Middle Proterozoic.* Early to Middle Proterozoic metamorphics (which probably contain relict Archaean blocks) occur in the Shackleton Range at the western margin of the East Antarctic Platform (Grikurov, 1980). The youngest age of metamorphism in the upper structural stage of the Shackleton metamorphics is 1414 Ma (Hofmann et al., 1981); late kinematic granodiorites which intrude the lower (Archaean) structural stage have yielded a date of 1400 Ma (Rex, 1972). It is proposed that the Shackleton metamorphic complex be correlated with metamorphics in the Haag Nunataks at the western margin of the Ronne Ice Shelf (Figure 2); the Haag Nunataks gave K/Ar ages of 1018-1031 Ma (Clarkson and Brook, 1977). These ages are also close to ages between 953 and 977 Ma reported for the Cape

Meredith gneisses in the Falkland Islands by Rex and Tanner (1982). It is considered that the Shackleton Range, Haag Nunataks, and Cape Meredith rocks were once joined in a belt that continued into the Middle Proterozoic Namaqua-Natal Belt south of the Kaapvaal Craton. Tectono-metamorphic development of the Namaqua-Natal belt terminated with the emplacement of postkinematic granites which have given isotopic ages between 1100 and 1000 Ma (Joubert, 1981; Matthews, 1981).

## Precambrian platform cover

The Precambrian platform cover sequences in East Antarctica and South Africa appear to have few lithofacial features in common probably because the existence of distinct intracratonic basins makes strict formational comparison difficult. However Proterozoic to Early Palaeozoic volcanism during platform cover development allows some comparisons to be made. The Middle to Upper Proterozoic platform cover in Queen Maud Land (Ritscher Supergroup) is lithologically similar in its lower parts (Ahlmannrygg Group) to the Waterberg Group in South Africa (Neethling, 1972). The Upper Ritscher Supergroup is possibly the equivalent of the Koras Group in the northern Cape Province of South Africa. Equivalents of the Late Archaean/Early Proterozoic Witwatersrand, Ventersdorp and Transvaal Systems, which overlie the Kaapvaal Craton have not been recognised in Queen Maud Land and other areas of East Antarctica.

Evidence for correlation of the Ahlmannrygg Group with the Waterberg Group and the upper Ritscher Supergroup with the Koras Group is provided by isotopic ages for interbedded volcanic rocks and associated intrusives (Table 1). Thus, basal volcanics in the Ahlmannrygg Group (1760-1650 Ma, Neethling, 1972; Bredell, 1982) correspond with those of the Waterberg Group (1790 ± 70 Ma, Salop, 1977). Volcanics of the South African Koras Group (1080 ± 70 Ma, Salop, 1977; 1280 ± 50 Ma, Joubert, 1981) are comparable in age to the Jutul Volcanics of the upper Ritscher Supergroup (1036 ± 1007 Ma, Neethling, 1972) and the Littlewood volcanics in Coats Land (1036-976 Ma, Ravich and Grikurov, 1977; 1050-850 Ma, Grikurov, 1980) at the Weddell Sea margin of the East Antarctic Platform.

## Late Proterozoic—Early Palaeozoic mobile belts

Late Proterozoic—Early Palaeozoic developments at the Gondwana margin can be traced from East Antarctica into Africa more distinctly than earlier phenomena. Miogeosynclinal and/or thick platform formations were deposited along the western margins of the East Antarctic and Kalahari platforms with eugeosynclinal troughs containing turbiditic sediments and ophiolites further to the west. The intensity of deformation decreases towards the cratons, with vergences directed towards the platforms or larger median massifs. Calc-alkaline magmatism accompanied the tectono-metamorphic events. During the Late Proterozoic/Early Palaeozoic geotectonic developments, the Gondwana continental margin which appears to have resembled a modern island arc, migrated to the west.

In the Shackleton Range, the Riphean Turnpike Group with 2000-2500 m of nonturbiditic clastic sediments (conglomerates, greywackes, sandstones, pelites) covers the pre-Riphean basement. The age of sedimentation of the middle Turnpike Group is 900 Ma and that of its "germanotype" deformation under the influence of basement block movements 600 Ma (Pankhurst et al., this volume). The Turnpike Group is possibly overlain by Middle Cambrian black shales found only in moraines (Soloviev and Grikurov, 1979). In the Argentina Range, a Riphean miogeosynclinal sequence more than 4000 m thick consists of greywackes, argillites, and volcanics, deformed in a "germanotype" style. The deformation took place during latest Riphean to Early Palaeozoic times (Grikurov, 1980). In the Pensacola Mountains, as in the entire Ross system, eugeosynclinal development began during the Late Proterozoic. The flyschoid Patuxent Formation, which consists of > 10,000 m of rhythmically bedded psammitic-pelitic turbiditic sediments, intruded by dolerites which comprise up to 25% of the total thickness, is isoclinally folded with vergences directed to the East Antarctic Platform. The Middle or Late Cambrian miogeosynclinal Nelson Formation (700 m) comprises shallow-marine sediments and volcanics and unconformably overlies the Patuxent Formation (Elliot, 1975). Late Proterozoic to Early Palaeozoic sequences are also exposed in the Ellsworth Mountains (Craddock et al., 1964); they were last folded in Late Palaeozoic/Early Mesozoic times (Ford, 1972a). Despite their present position, the Ellsworth Mountains are considered to be an atypical geotectonic element of the Transantarctic Mountains Ross System, comparable possibly with the Dnepr-Donezk aulacogen of the East European Platform.

Assuming the pre-drift position of the Ellsworth Mountains as proposed by Schopf (1969) it is postulated that the Late Proterozoic to Early Palaeozoic Ross System mobile belt in Antarctica can be traced into the Malmesbury Belt of South Africa (Figure 2b). However, because of the large distances involved, this correlation is not yet completely satisfactory (Figure 3). The Malmesbury Group (> 6500 m of pelites, greywackes, sandstones, quartzites, carbonates and basic volcanics) shows flyschoid character only partially in the Tigerberg Formation and seems to be more of mio- than of eugeosynclinal origin (Salop, 1977). The age of deformation is 900-800 Ma viz. that of synorogenic granites in the Cape Fold Belt 1000 Ma (Machens, 1968). The higher Kango Group with 2000-3000 m of psammitic and pelitic sediments, conglomerates and limestones seems to be a lithofacial equivalent of the Nelson Formation in the Pensacola Mountains. The Kango Group is covered by equivalents of the Late Riphean to Early Cambrian Nama Series (Salop, 1977). The Koras Group which consists of 2500 m of conglomerates, psammites and pelites with basic and acid volcanics appears to have a similar palaeogeographic position to the Blaiklock Glacier Group in the Shackleton Range (Clarkson, 1982). The age of acid volcanics in the Koras Group is reported to be 1080 ± 70 Ma (Salop, 1977). The authors have previously related this age to that of the Jutul Volcanics in Queen Maud Land and other volcanic rocks at the Weddell Sea margin of East Antarctica.

The Malmesbury Belt and similar rocks in the region of the Pensacola Mountains are thus considered to represent different troughs in a Late Proterozoic/Early Palaeozoic geosynclinal belt on the passive "western" continental margin of Gondwana.

# THE BLOCK STRUCTURE OF WEST ANTARCTICA

West Antarctica, the "Ross-Weddell Province", is separated from the Antarctic Peninsula by the Ellsworth Fault (Ford, 1972a, b; Figure 4F-B) which we regard as a continuation of the southern Eltanin Fracture Zone of Weissel et al. (1977). The northern continuation of

**TABLE 1: Isotopic ages from volcanic rocks and temporally-related intrusions from the Proterozoic to Lower Palaeozoic cover sequences in East Antarctica and South Africa.**

| | South Africa | Antarctica | |
|---|---|---|---|
| | Kalahari Craton | Coats Ld., Shackleton Range | Ritscher Highlands |
| Palaeozoic | Bremen Syenite 506 ± 10 Ma<br>Kuboos Granite 550 ± 30 Ma | | |
| Upper Proterozoic $PR_3$ | Nama Group | Turnpike Group | Trollkjellrygg Gr. 860-824 Ma |
| | Koras Group 1080 ± 70 Ma | Littlewood Volcanics 1036-840 Ma | Jutul Volcanics 1036-1007 Ma<br>Ytstenut Intrusion 1030 ± 70 Ma |
| Middle Proterozoic $PR_2$ | Leeufontein Syenites 1420 ± 70 Ma | Shackleton Diorites 1446-1400 Ma | |
| Lower Proterozoic $PR_1$ | Waterberg Group 1790 ± 70 Ma | | Krylen Intrusion 1700 ± 30 Ma<br>Ahlmannrygg Group 1760-1600 Ma |

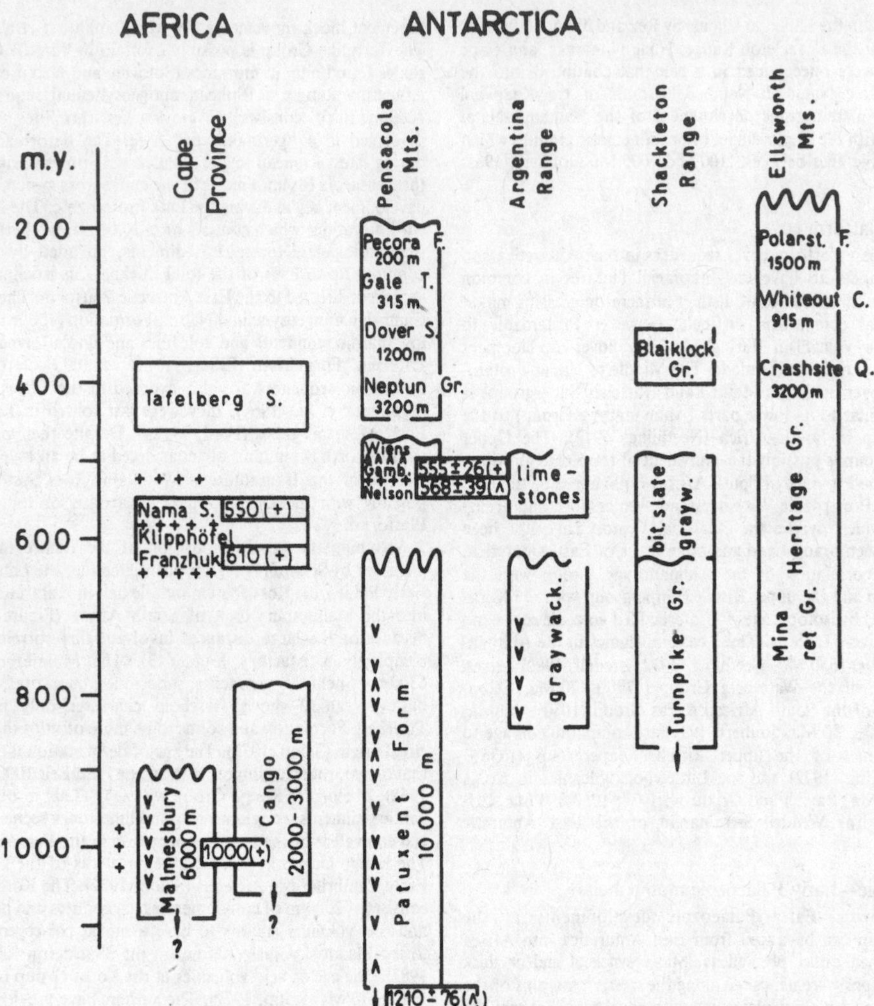


**Figure 3. Scheme of Upper Proterozoic to Early Palaeozoic stratigraphic sequences in Antarctica and South Africa.**

the Ellsworth Fault into the Weddell Sea continental margin region is marked by isobath kinks and seamounts (Vinogradov and Zivago, 1974; Korotkevich, et al., 1975). The boundary of the Ross-Weddell Province against the Ross System and the margin of the East Antarctic Platform is marked by the Cenozoic Filchner-Ross Sea Rift Zones (Masolov, et al., 1981).

In Figure 4 the Ross-Weddell Province is divided into several block-structure units. The individual blocks, irrespective of their inner fault patterns, are bounded by fundamental fault zones which are partly marked by Late Mesozoic to Cenozoic rift zones (Grikurov, 1980; Masolov et al., 1981). The Ellsworth Mountains—Weddell Sea Block occupies the region between a poorly defined fault zone south of the Byrd Basin (Figure 4A) and the Weddell Sea continental margin (Figure 4B); the Western Marie Byrd Land-Ross Shelf Block has a largely uninvestigated fault boundary against the Ross System (Figure 4C); the Marie Byrd Block is bounded by faults D and E (Figure 4) and the Eights Coast Block lies between faults E and F.

In the adopted reconstruction model the pre-drift position of the Ellsworth Mountains-Weddell Sea Block is adjacent to the modern Weddell Sea margin of East Antarctica (Figure 2b). It follows that the western Marie Byrd Land-Ross Shelf Block and the Marie Byrd Land Block must have been located between the present day 100° and 110°W meridians which now enclose the Byrd Basin (Figure 4).

Little is known about the palaeotectonic development of the Ross-Weddell Province since the Gondwana breakup. Independently, Ford and Kistler (1980) and Hofmann and Weber (1982) suggested that development of a triple junction accompanied intrusion of the Early Jurassic Dufek Massif (Figures 2b, 4). It is also possible that the Dufek Triple Junction was connected with the Mozambique Triple Junction (Reeves, 1978) where Fitch (1972) reported ages betwen 195 and 190 Ma for the main phase of volcanism. The age difference of about 25 to 20 Ma between the two triple junctions could be explained by the north to south opening of the Malagasy Rift (Fairbridge, 1978).

After fragmentation of the Gondwana supercontinental mass the Ross-Weddell Province was affected by the development of transform fault systems in the opening southeast Atlantic ocean, and of the West Antarctic plate. Maps given in Weissel et al. (1977, Figure 7) show some correspondence between the main transform faults in the West Pacific plate and the fault zones, which separate block structures discussed previously. The Dufek Triple Junction was incorporated into the pattern of these fault zones (Figure 5) during transform movements in the Ross-Weddell province. Evidence of Early Tertiary block movements, directed towards the Pacific, is provided by the palaeomagnetic data of Kellog (1980).

Development of the Ross-Weddell province block pattern probably occurred in two stages:

(a) differential movement of the entire Ross-Weddell province against the West Antarctic plate (Figure 5a, b);
(b) separation of the Ross-Weddell province into an "Atlantic" and a "Pacific" part due to formation of the Byrd Basin, accompanied by crustal thinning (Grikurov, 1980, Figure 5b, c).

During the first stage rotational movements of the East Antarctic

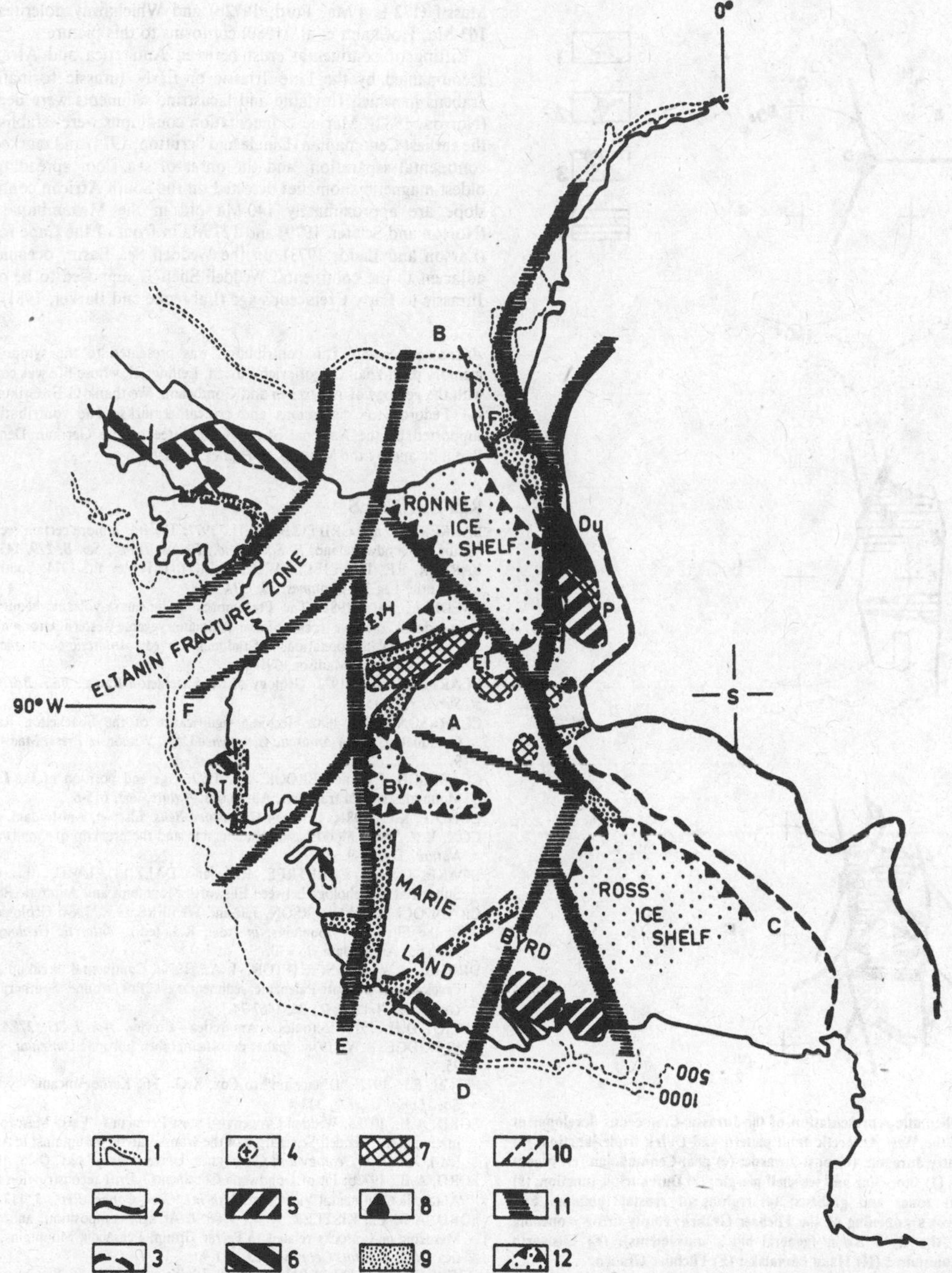


**Figure 4.** **Schematic tectonic map (post-Jurassic fault blocks) of West Antarctica. Constructed after the Tectonic Map of Antarctica (Grikurov, 1980) and the Subglacial Basement Map of Antarctica (Korotkevich et al., 1975). (1) shore line, ice shelves and isobaths; (2) margin of the East Antarctic crystalline basement; (3) margin of the Ross System; (4) pre-Late Proterozoic Precambrian units; (5) Late Proterozoic and Early Palaeozoic consolidation during the Ross Orogeny (650-450 Ma); (6) Late Proterozoic and Palaeozoic units, reworked during Mesozoic orogenies; (7) Palaeozoic platform cover subjected to Mesozoic deformation; (8) Dufek Massif; (9) (?Late Mesozoic) to Cenozoic sediments in grabens and rift zones; (10) rift zones and grabens; (11) fault zones; (12) regions of crustal thinning; (A-F) see text. (H) Haag Nunataks; (El) Ellsworth Mountains; (P) Pensacola and Patuxent Mountains; (T) Thurston Island; (Bi) Byrd Basin; (Fi) Filchner Graben; (Du) Dufek Massif.**

Platform and the Ross-Weddell province possibly caused opening of the Filchner Graben and related structures (Masolov et al., 1981). The "splitting off" of the Ross-Weddell province in the second stage was related to the formation of the Byrd Basin, which Kellog (1980) supposes to be an inactive spreading centre. A migrated former arm of the Dufek triple junction may be incorporated into this basin (Figure 5b, c). The suggested counterclockwise block rotation of the Ross-Weddell province is compatible with results of bedrock morphology investigations in parts of the Eights Coast block and the Eltanin Fracture Zone reported by Doake et al. (this volume).

## TEMPORAL RECONSTRUCTION OF GONDWANA FRAGMENTATION

Gondwana fragmentation possibly started in the latest Precambrian or the earliest Palaeozoic. The Pan-African event (Mozambique Belt, Figure 2b, 6) between 600 and 440 Ma corresponds to reset ages between 640 and 420 Ma reported from metamorphic rocks of the crystalline basement in Queen Maud Land by Ravich et al. (1965), and in the Shackleton Range by Grew and Halpern (1979) and Grew and Manton (1980). These ages possibly indicate a Late Precambrian to

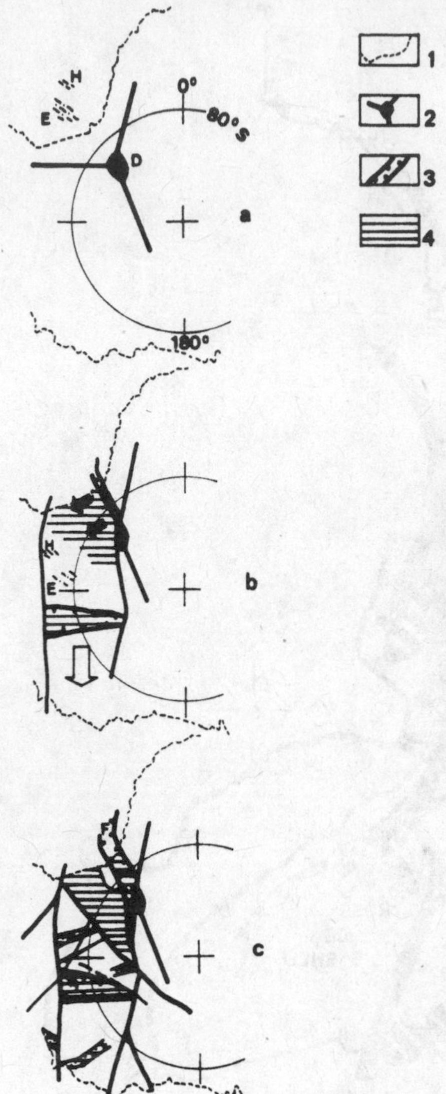


**Figure 5. Schematic representation of the Jurassic-Cretaceous development of the West Antarctic fault pattern and Dufek triple junction: (a) Early Jurassic; (b) post-Jurassic; (c) post-Cenomanian (cf. Figure 4); (1) shore line and ice shelf margin; (2) Dufek triple junction; (3) rift zones and grabens; (4) regions of crustal thinning. Full arrows = opening of the Filchner Graben; empty arrow = opening of the Byrd Basin (general block movements). (E) Ellsworth Mountains; (H) Haag Nunataks; (F) Filchner Graben.**

Early Palaeozoic crustal weakening along a trend later followed by the Malagasy Rift which separated Antarctica and Peninsular India from Africa.

In the Shackleton Range the intrusion of Late Ordovician to Devonian dolerites between 400 and 360 Ma and an associated basement thermal reactiviation (Clarkson 1972; Hofmann et al., 1980) provide some evidence in support of initial Early Palaeozoic Gondwana breakup as proposed by Fairbridge (1978).

The East Antarctica-South Africa breakup was heralded by the Early Jurassic Ferrar dolerites and Kirkpatrick basalts in Antarctica and the penecontemporaneous South African Karroo dolerites (Cox, 1978; Kyle et al., 1981). The Early Jurassic intrusion of the Dufek Massif (172 ± 4 Ma, Ford, 1972b) and Whichaway dolerites (171-163 Ma, Hofmann et al., 1980) conforms to this picture.

Rifting of continental crust between Antarctica and Africa was accompanied by the Late Triassic or Early Jurassic formation of grabens in which fluviatile and lacustrine sediments were deposited (Norton, 1982). Marine sedimentation conditions were established in the earliest Cenomanian (Dingle and Scrutton, 1974) and marked final continental separation, and the onset of sea floor spreading. The oldest magnetic anomalies detected on the South African continental slope are approximately 140 Ma old in the Mozambique Basin (Norton and Sclater, 1979) and 127 Ma in front of the Cape fold belt (Larson and Ladd, 1973). In the Weddell Sea Basin, oceanic crust adjacent to the continental Weddell Shelf is supposed to be of Late Jurassic to Early Cretaceous age (Labreque and Barker, 1981).

*Acknowledgements* This contribution was presented to the symposium in memory of Mikhail Grigorievich Ravich, Leningrad, whose life was connected with the geology of Antarctica and Gondwana. We thank G.E. Grikurov and L.F. Fedorov for discussions and critical remarks. The contribution was supported by the National SCAR Committee of the German Democratic Republic and by the Ministry of Higher Education.

## REFERENCES

BARKER, P.F. and GRIFFITHS, D.H., 1977: Towards a more certain reconstruction of Gondwanaland. *R. Soc. Lond., Philos. Trans., Ser. B, 279,* 143-59.

BARKER, P.F., DALZIEL, I.W.D., ELLIOT, D.H., et al., 1974: Southwestern Atlantic, Leg 36. *Geotimes, 19, (11),* 16-8.

BREDELL, J.H., 1982: The Precambrian sedimentary-volcanic sequence and associated intrusive rocks of the Ahlmannryggen, Western Dronning Maud Land: a new interpretation, *in* Craddock, C. (ed.) *Antarctic Geoscience,* Univ. Wisconsin Press, Madison, 591-7.

CLARKSON, P.D., 1972: Geology of the Shackleton Range. *Bull. Brit. antarct. Surv., 31,* 1-15.

CLARKSON, P.D., 1982: Tectonic significance of the Shackelton Range; *in* Craddock, C. (ed.) *Antarctic Geoscience* Univ. Wisconsin Press, Madison, 835-9.

CLARKSON, P.D. and BROOK, M., 1977: Age and position of the Ellsworth Mountains crustal fragment, Antarctica. *Nature, 265,* 615-6.

CONDIE, K.C., 1981: *Archaean Greenstone Belts.* Elsevier, Amsterdam.

COX, K.C., 1978: Flood basalts, subduction and the breakup of Gondwanaland. *Nature, 274,* 47-9.

DOAKE, C.S.M., CRABTREE, R.D. and DALZIEL, I.W.D. (this volume): Subglacial Morphology between Ellsworth Mountains and Antarctic Peninsula.

CRADDOCK, C., ANDERSON, J.J. and WEBERS, G.F., 1964: Geologic outline of the Ellsworth Mountains; *in* Adie, R.J. (ed.). *Antarctic Geology.* North Holland, Amsterdam, 155-70.

DINGLE, R.V. and SCRUTTON, K.A., 1974: Continental breakup and the development of post-Paleozoic sedimentary basins around Southern Africa. *Geol. Soc. Am., Bull., 85,* 1467-74.

ELLIOT, D.H., 1975: Tectonics of Antarctica—a review. *Am. J. Sci., 275A,* 45-106.

FAIRBRIDGE, R.W., 1978: Spaltet der Rheingraben Europa? *Umschau, 78(3),* 69-75.

FITCH, F.J., 1972: "Discussion" to Cox, K.G.; The Karoo Volcanic Cycle. *Geol. Soc. Lond., J., 128,* 333-4.

FORD, A.B., 1972a: Weddell Orogeny—Latest Permian to Early Mesozoic deformation at the Weddell-Sea margin of the Transantarctic Mountains; *in* Adie, R.J. (ed.), *Antarctic Geology and Geophysics,* Universitetsforlaget, Oslo, 419-25.

FORD, A.B., 1972b: Fit of Gondwana Continents—Drift reconstruction from the Antarctic Continental Viewpoint. *24th Int. Geol. Congr., Rep., 3,* 113-21.

FORD, A.B. and KISTLER, R.W., 1980: K-Ar age, composition, and origin of Mesozoic mafic rocks related to Ferrar Group, Pensacola Mountains, Antarctica. *N.Z. J. Geol. Geophys., 23,* 371-90.

GREW, E.S. and HALPERN, M., 1979: Rb/Sr dates from the Shackleton Range metamorphic complex in the Mount Provender area, Shackleton Range, Antarctica. *J. Geol., 78,* 325-32.

GREW, E.S. and MANTON, W., 1980: Uranium-lead ages of zircon from the Mt Provender, Shackleton Range, Transantarctic Mountains. *U.S. Antarct. J., 15(5),* 9-15.

GRIKUROV, G.E., (ed.) 1980: *Tektoniceska ja Karta Antarktidy, 1:10,000,000.* Minist. Geologii SSR, NIIGA, Moskva.

HOFMANN, J. and WEBER, W., 1982: Versuch einer Gondwana-Rekonstruction im Bereich Antarktika—Afrika. *Z. Geol. Wiss. 10,* 457-72.

HOFMANN, J., PILOT, J., and SCHLICHTING, M., 1981: Das Rb/Sr-Alter von Metamorphiten der Herbert Mountains, Shackleton Range. *Z. Geol. Wiss., 9,* 835-42.

HOFMANN, J., KAISER, G., KLEMM, W. and PAECH, H.-J., 1980: K/Ar-Alter von doleriten und Metamorphiten der Shackleton Range und der Whichaway Nuntaks. *Z. Geol. Wiss., 8,* 1227-32.

JOUBERT, P., 1981: The Namaqualand metamorphic complex; *in* Hunter, D.R. (ed.) *Precambrian of the Southern Hemisphere.* Elsevier, Amsterdam, 617-703.

KELLOGG, K.S., 1980: Paleomagnetic evidence for oroclinal bending of the southern Antarctic Peninsula. *Geol. Soc. Am., Bull., 91,* 414-20.

KOROTKEVICH, E.S., KOBLENZ, J.P. and KOSENKO, N.G., 1975: *Karta korennogo rel'efa Antarktidy 1:10,000,000.* Soiusmorniiprojekt, Leningrad.

KYLE, P.R., ELLIOT, D.H. and SUTTER, J.F., 1981: Jurassic Ferrar Supergroup tholeiites from the Transantarctic Mountains and their relationship to the initial fragmentation of Gondwana; *in* Cresswell, M.M. and Vella, P. (eds.) *Gondwana Five*, A.A. Balkema, Rotterdam, 283-7.

LABRECQUE, J.L. and BARKER, P., 1981: The age of the Weddell Basin. *Nature, 290*, 489-92.

LARSON, R.L. and LADD, J.W., 1973: Evidence fo the opening of the South Atlantic in the Early Cretaceous. *Nature, 246*, 209-12.

MACHENS, E., 1968: Das Prakambrium von Afrika; *in* Looze, F., *Prakambrium, II. Teil—Sudliche Halbkugel.* Enke, Stuttgart.

MASOLOV, V.N., KURININ, R.G. and GRIKUROV, G.E., 1981: Crustal structures and tectonic significance of Antarctic rift zones; *in* Cresswell, M.M. and Vella, P. (eds.) *Gondwana Five.* A.A. Balkema, Rotterdam, 303-9.

MATTHEWS, P.E., 1981: Eastern or Natal sector of the Namaqua—Natal mobile belt in Southern Africa; *in* Hunter, D.R. (ed.) *Precambrian of the Southern Hemisphere.* Elsevier, Amsterdam, 705-14.

NEETHLING, C.D., 1972: Age and correlation of the Ritscher Supergroup and other Precambrian rock units, Dronning Maud Land; *in* Adie, R.J. (ed.) *Antarctic Geology and Geophysics.* Universitetsforlaget, Oslo, 547-56.

NORTON, N.I., 1982: Paleomotion between Africa, South America, and Antarctica and Implications for the Antarctic Peninsula; *in* Craddock, C. (ed.) *Antarctic Geoscience*. Univ. Wisconsin Press, Madison, 99-106.

NORTON, O.I. and SCLATER, J.G., 1979: A model for the evolution of the Indian Ocean and the breakup of Gondwanaland. *J. Geophys. Res., 84*, 6803-30.

PANKHURST, R.J., MARSH, P.D. and CLARKSON, P.D. (this volume): A Geochronological Investigation of the Shackleton Range.

RAVICH, M.G. and GRIKUROV, G.E., 1977: *Geologiceskaja Karta Antarktidy 1:5,000,000.* Naucho-Issled. Inst. Geol. Arktiki i Antarktiki, Leningrad.

RAVICH, M.G., KLIMOV, L.V. and SOLOVIEV, D.S., 1965: Dokembrij Vostoonoj Antarktidy. *Nauchi Issled. Inst. Geol. Arktiki i Antarktiki, Trudy, 138.*

REEVES, C.V., 1978: A failed Gondwana spreading axis in southern Africa. *Nature, 273*, 222-3.

REX, D.C., 1972: K/Ar age determinations on volcanic and associated rocks from the Antarctic Peninsula and Dronning Maud Land; *in* Adie, R.J. (ed.) *Antarctic Geology and Geophysics.* Universitetsforlaget, Oslo, 133-6.

REX, D.C. and TANNER, P.W.G., 1982: Precambrian age of gneisses of Cape Meredith in the Falkland Islands; *in* Craddock, C. (ed.), *Antarctic Geoscience.* Univ. Wisconsin Press, Madison, 107-8.

SALOP, L.I., 1977: *Periodizacija i korreljacija dokembrija juznych materikov—Dokembrij Afriki.* Nedra, Leningrad.

SCHOPF, J.M., 1969: Ellsworth Mountains: Position in West Antarctica due to sea floor spreading. *Science, 164*, 63-6.

SOLOVIEV, I.A. and GRIKUROV, G.E., 1979: Novye dannye o rasprostranenii kembrijskich trilovitov chrebtach Ardzentina i Sekltona. *Antarktika., 18*, 54-73.

VINOGRADOV, O.A. and ZIVAGO, A.V., 1974: *Batimetriceskaja Karta Antarktidy 1:15,000,000.* Glavn. Upr. Geod. i Kartogr. Sovjet., Minist. SSR, Moskva.

WEISSEL, J.K., HAYES, D.E. and HERRON, E.M., 1977: Plate tectonic synthesis: the displacements between Australia, New Zealand and Antarctica since the Late Cretaceous. *Mar. Geol., 25*, 231-77.

# THE LABYRINTHODONT AMPHIBIANS OF THE EARLIEST TRIASSIC FROM ANTARCTICA, TASMANIA AND SOUTH AFRICA

J.W. Cosgriff, Jr., *Department of Biological Sciences, Wayne State University, Detroit, Michigan, 48202, U.S.A.*

W.R. Hammer, *Department of Geology, Augustana College, Rock Island, Illinois, U.S.A.*

*Abstract* The remarkable and extensively documented correspondence between the reptilian assemblages of the lowermost Triassic of Antarctica and South Africa is now seen to extend to the labyrinthodont assemblages associated with them. Representatives of four families—Brachyopidae, Capitosauridae, Lydekkerinidae and Rhytidosteidae—comprise the total known labyrinthodont assemblage from the Fremouw Formation of Antarctica as well as the major part of that from the *Lystrosaurus* Zone of South Africa. These families plus one additional—Indobrachyopidae—comprise the labyrinthodont assemblage from the Knocklofty Sandstones and Shales of Tasmania. This taxonomic correspondence is of palaeobiogeographical significance as the sites of the three noted assemblages were much closer together on Pangaean geography than they are at present and as they were, during the earliest Triassic, the most southerly of all sites from around the world which have produced tetrapod assemblages of this interval. The Capitosauridae, Lydekkerinidae and Rhytidosteidae all seem to be evolutionary derivatives of the Upper Permian Rhinesuchidae; Brachyopidae are a separate lineage tracing back to the Lower Permian. Each of the four families was adapted to a particular ecologic niche. Although each is represented in a number of labyrinthodont assemblages from the south temperate, equatorial and far northern portions of Pangaea, not one of these assemblages shares a close taxonomic correspondence on the family level with the set of assemblages from the far south. The four families, here termed the Austral Complex, together with related Lower Triassic families found in other areas, seems to have replaced or supplanted a very different labyrinthodont complex of the Upper Permian.

The Subclass Labyrinthodontia, an extinct group of tetrapod vertebrates, ranged from Late Devonian to Early Triassic. They form a component of the Class Amphibia, a placement confirmed by a variety of morphologic features in adult skeletons plus a few fossils of larval individuals which exhibit the characteristics of tadpoles. The remaining Amphibia are distributed among the Subclasses Lissamphibia (living and extinct frogs, salamanders and apodans) and Lepospondyli (another extinct group). The Labyrinthodontia contains three orders—Ichthyostegalia, Anthracosauria and Temnospondyli—of which the last is the most important in terms of diversity and representation in the fossil record and the only one to survive the Permian. For a complete listing of all fossil and recent amphibian genera and species see Carroll and Winer (1977).

The Labyrinthodontia are a central group in the history of vertebrate animals, both for their evolutionary position and for their abundance and variety in terrestrial assemblages of the Late Palaeozoic and Early Mesozoic. Regarding evolution, they span a threshold as they were the first tetrapods, having evolved from certain crossopterygian fish and as some of their primitive representatives variously contain the ancestry of all other tetrapod groups; lissamphibians and lepospondyls originated separately from certain labyrinthodonts; reptiles originated from others; and from reptiles, in turn, came birds and mammals. Regarding abundance and variety, they are a prominent and diverse element in most of the fossil assemblages in which they are found. Indeed, in many of those from the Pennsylvanian and Triassic, they outnumber contemporary reptiles as individuals and as species.

They ranged in maximum adult body length from small species (10 to 20 cm) up to very large species (5 m or more). Variety in the adaptive particulars of the skeleton, particularly those of the skull and dentition, indicate that they occupied many trophic levels in the food chains reflected in the fossil assemblages of which they form components. The characterisations of the Triassic families which follow provide an example of this adaptive diversity.

The beginning of the Triassic witnessed the last major evolutionary radiation of the Labyrinthodontia, an expansion involving only the Order Temnospondyli. A number of new families of this order appear at this time. Succeeding intervals of the Triassic document a gradual and progressive reduction in diversity of the Temnospondyli until, by the end of the period, they verged on extinction.

## THE AUSTRAL COMPLEX OF THE TEMNOSPONDYLI

Four temnospondyl families, Brachyopidae, Capitosauridae, Lydekkerinidae and Rhytidosteidae, predominate in the temnospondyl assemblages of three stratigraphic units deposited in the most southerly reaches of the Pangaean continent and are here referred to, compositely, as the Austral Complex. These units, the Lower Member of the Fremouw Formation of East Antarctica, the *Lystrosaurus* Zone of southern Africa and the Poets Road Member of the Knocklofty Sandstones and Shales of Tasmania, were all formed under terrestrial conditions. They are reasonably close to each other in age within the earlier part of the Early Triassic (Anderson and Cruickshank, 1978). Accounts of the Antarctic material are provided by Colbert and Cosgriff (1974), of the African material by Chernin (1978), Kitching (1978) and Cosgriff and Zawiskie (1979) and of the Tasmanian material by Cosgriff (1974) and Banks et al. (1978).

The genera and species of the four families in the three units include both documented and undocumented forms. Taking the families in order, the representatives are: for the Brachyopidae, *Austrobrachyops jenseni* in Antarctica; undescribed material in Africa and *Blinasaurus townrowi* in Tasmania; for the Capitosauridae, generically and specifically indeterminate material in Antarctica, *Kestrosaurus dreyeri* in Africa and undescribed material in Tasmania; for the Lydekkerinidae, *Cryobatrachus kitchingi* in Antarctica, *Lydekkerina huxleyi* and *Limnoiketes paludinatans* in Africa and *Chomatobatrachus halei* in Tasmania; and, for the Rhytidosteidae, unreported material in Antarctica, *Pneumatostega potamia* and *Rhytidosteus capensis* in Africa and *Deltasaurus kimberleyensis* in Tasmania.

The families of the Austral Complex are, by no means, limited to the three southern units but it is only in each of these that they all occur together and compose the bulk of the assemblage. All variously occur in stratigraphic units of the earliest Triassic in the south temperate, equatorial, north temperate and boreal regions of Pangaea. Such units include the Arcadia Formation of Queensland, the Blina Shale of Western Australia, the Middle Sakamena Group of Malagasy, the Panchet Formation of India, the Vetluga Series of European USSR, the Wordy Creek Formation of Greenland and the Sticky Keep Formation of Svalbard. A listing of the taxonomic composition of the temnospondyl assemblage in each of these geographically removed units is not pertinent to the present discussion. It suffices to note that not one of them contains the four families together and, further, that most of them contain representatives of other families not present in the southern units. The unity of the Austral Complex implies the operation of special factors at the southern end of Pangaea such as close similarity in climatic-ecologic conditions, open migration routes for the dispersal of the fauna, proximity of all three areas to the evolutionary centre that produced the complex or the various combinations of these. Certainly, East Antarctica, southern Africa and Tasmania were much closer to each other at the beginning of the Mesozoic than they are at present.

Although the Austral Complex is defined by the four families, rare specimens pertaining to other families are reported from two of the units. The *Lystrosaurus* Zone has produced *Uranocentrodon senekalensis* of the Rhinesuchidae and *Micropholis stowi*, a possible member of the Dissorophidae. The Poets Road Member has produced *Derwentia warreni* of the Indobrachyopidae. Rhinesuchids and dissorophids are Permian families for the most part. The former, as discussed below, appears to contain the ancestry of much of the Austral Complex. The latter is not closely related to any other Triassic family. Indobrachyopids are closely related to rhytidosteids and, in common with them, are limited to the earliest Triassic.

### Palaeoecology

Most temnospondyls of the earliest Triassic were adapted for existence in bodies of fresh water and along surrounding banks, bars and flats. Adaptive features of the cranium, particularly skull shape and size, jaw muscle attachment areas and dentition, vary markedly among the four families. This indicates that each was adapted to a particular diet and, thus, occupied a particular niche in the fresh water aquatic regime of the Pangaean biome. In total, as interpreted here, they seem to have exploited most of the available resources and to have been a fully developed and balanced complex. As space does not permit detailed considerations of the adaptive morphologies of the families and as these are provided elsewhere, only a brief characterisation of each, sketching its salient features, will be attempted. For lengthier discussions, see Cosgriff (1974) for brachyopids and lydekkerinids, Chernin and Cruickshank (1975) for capitosaurids and Cosgriff and Zawiskie (1979) for rhytidosteids.

*Brachyopidae* (Figure 1a)

These were animals with round (as viewed from above), deep and massive skulls. They were small to medium in size for Triassic temnospondyls with one of the larger skulls from the late Early Triassic measuring 24 cm in midline length; the largest of those from the earliest Triassic are only half of this. The skulls exhibit prominent and extensive origin areas for the muscles closing the jaws, indicative of a powerful bite. Functionally coupled with this musculature are the very large tusks seated on the palatal surfaces internal to the marginal tooth row. Obviously, brachyopids were well equipped predators. A reasonable surmise is that they preyed on large osteichthyans such as lungfish, on smaller labyrinthodonts, and perhaps, on small reptiles. They may have waited in ambush along the water's edge or lurked on pond bottoms.

*Capitosauridae* (Figure 1b)

Members of this family were more streamlined, at least in skull shape, than the brachyopids. The skull, viewed dorsally, is bullet shaped with a comparatively long snout portion. They reached the largest adult size of all of the temnospondyls of the earliest Triassic. According to Chernin (1978, figure 4), the holotypic skull of *K. dreyeri*, is about 60 cm in midline length. Some complete capitosaurid skulls from the Middle and Late Triassic are considerably larger than this and fragments from the Antarctic and Tasmanian units are from only slightly smaller skulls. Skull shape shows that they were more active predators than brachyopids although their palatal tusks and marginal teeth are relatively smaller. Chernin and Cruickshank (1975) reasonably characterise them as crocodile/alligator analogues and as midwater feeders. They probably occupied a position close to the top of the food chain, preying on large fish and small temnospondyls. In taking the latter, they may have relied on seizing and then drowning as do modern crocodiles rather than crushing and stabbing as did the brachyopids.

*Lydekkerinidae* (Figure 1c)

Lydekkerinid species were quite similar to the capitosaurids in most adaptive features of the skull but differed in being far smaller animals as full adults. The largest reported skull, a specimen from the earliest Triassic of the USSR, measures ~17 cm in midline length but all reported skulls of this interval in the Austral Complex are considerably smaller. Cosgriff (1974) interprets the family as a group of small surface swimming species that were principally insectivorous in diet. This is based mainly on the relatively small teeth and on the attachment areas for muscles serving to open the jaws. The latter imply a rapid snap action that would have been functional in seizing small organisms from the water's surface.

*Rhytidosteidae* (Figure 1d)

Of all of the temnospondyls of the Austral Complex, these were the most specialised for aquatic existence. They were medium sized with skulls ranging up to 35 cm. Analysis of surface bone from skulls indicates a highly developed cutaneous respiratory system used in obtaining oxygen from water. The skulls were effective cut-waters for rapid swimming as they were triangular in shape with pointed snouts. The rhytidosteid mouth unequivocally denotes a piscivorous diet as the palatal surfaces and inner surfaces of the lower jaw carry extensive pavements of small denticles used in securing slippery prey. These pavements are very similar to those of living North American garpikes. Rhytidosteids, from all considerations, must have been active predators on a variety of fish at mid water levels.

**Figure 1. Reconstructions of representatives of Triassic temnospondyl families: (a) Brachyopidae (*Blinasaurus townrowi*); (b) Capitosauridae (composite); (c) Lydekkerinidae (*Chromatobatrachus halei*); (d) Rhytidosteidae (*Deltasaurus kimberleyensis*). Not to scale.**

## ORIGIN OF THE AUSTRAL COMPLEX

Considering the terrestrial fossil record of the Upper Permian on a world basis, it seems quite likely that the Austral Complex had an autochthonous origin in the southern portion of Pangaea. Three of the families, Capitosauridae, Lydekkerinidae and Rhytidosteidae are clearly, on available evidence, phyletic derivatives of a single Late Permian family, Rhinesuchidae. The remaining family, Brachyopidae, was already in existence in the Late Permian. The fossil record of both of these evolutionary sources, Rhinesuchidae and early Brachyopidae, is confined to areas proximal to the three Triassic units containing the Austral Complex.

The Rhinesuchidae appear to have been endemic to the southern portion of Africa as no species assignable to the family with certainty are reported outside of the Beaufort Series. They range through the entire Upper Permian portion of this series, namely the *Tapinocephalus, Cistecephalus* and *Daptocephalus* Zones. As Kitching (1978) points out, they are found widely through the areal extent of each of these zones but are rare as individuals and limited in taxonomic diversity throughout. Indeed, most of the described species are referrable to the type genus, *Rhinesuchus*.

As compared with the spectrum of specialised types spanned by their various descendants, rhinesuchids were an adaptively generalised and ecologically restricted group. Throughout their stratigraphic range, they exhibit evolutionary stasis with little morphologic change and virtually no adaptive radiation. Somewhere around the Permian-Triassic boundary, evolutionary divergences, prompted by unknown environmental changes, broke up the unity of the basic stock and initiated lineages leading to new families. Rhinesuchid derivatives include not only the families considered here but also certain other families found in the earliest part of the Triassic in areas to the north of the Austral Complex. These are the Benthosuchidae and Trematosauridae (considered capitosaur relatives by Shishkin, 1980) and the Indobrachyopidae (a collateral group to the Rhytidosteidae according to Cosgriff and Zawiskie, 1979).

The capitosaurids are the least modified derivatives of the rhinesuchids (Romer, 1947). Although rhinesuchids were medium sized temnospondyls (skull length up to 40 cm) in contrast to the much larger capitosaurids, these families share most adaptive features of the skull. Rhinesuchids differ mainly in exhibiting more generalised states of the features. It is notable that the one rhinesuchid from the *Lystrosaurus* Zone, *U. senekalensis* is intermediate in many respects between Permian members of this family and the capitosaurids. This intermediacy has caused *U. senekalensis* to be placed in its own family, Uranocentrodontidae, in some classifications.

Lydekkerinids, although very small temnospondyls, are very little advanced over rhinesuchids in most particulars of skull construction. They differ principally in certain modifications associated with dietary specialisation in the attachment areas for jaw musculature. Romer (1947) considered them rhinesuchid derivatives along with the capitosaurids. A possible ancestor for the lydekkerinids among the rhinesuchids of the *Daptocephalus* Zone is *Muchocephalus muchos* (Cosgriff, 1974).

The Rhytidosteidae also appears to stem from the Rhinesuchidae (Cosgriff and Zawiskie, 1979) although their skull and dentition specialisations associated with a narrowly piscivorous diet are more extreme than those observed in capitosaurids and lydekkerinids. Underlying these specialisations, however, is a basic construction which is clearly rhinesuchid.

The Brachyopidae are the only component of the Austral Complex whose source lies outside of the Rhinesuchidae. They are a highly evolved group with a lineage tracing back to the Trimerorhachoidea of the Early Permian (Olson and Lammers, 1975). The earliest true brachyopids are two genera from the Upper Permian Newcastle Coal Measures of New South Wales. Thus, an austral origin for this family as well seems plausible. However, a closely related family, the Dvinosauridae, is present in the Upper Permian of the USSR.

Negative evidence in support of the present evolutionary and palaeobiogeographical construct is provided by the very different taxonomic composition of the labyrinthodont assemblage in the Middle and Late Permian of the northern portions of Pangaea. The most extensive record of this assemblage lies in the Kazanian (Guadalupian) and Tatarian (Dzhulfian) of the USSR. The most striking difference of this record from that of southern portions of Pangaea is that it contains members of the Order Anthracosauria as well as members of the Order Temnospondyli. First, regarding the Temnospondyli, these are quite varied but are all members of rather primitive families such as the Trimerorhachidae, Dvinosauridae, Dissorophidae, Intasuchidae, Archegosauridae and Melosauridae. All of these are quite removed in phylogeny from the Rhinesuchidae although, as mentioned, the Dvinosauridae are closely related to the Brachyopidae. The Anthracosauria are almost equally varied with four families, Chroniosuchidae, Kotlassidae, Lanthosuchidae and Nycteroletoridae. As an aside to the present discussion, it is worthy of mention that the Class Reptilia had its origin among earlier members of the Anthracosauria during the Carboniferous.

Thus, the labyrinthodontia were far more diverse during the Late Permian in the northern portions of Pangaea. But, in spite of this diversity, not one of its elements, on the family level, is present in the Late Permian of the far south. Further, none of the families survived into the Triassic with the possible exception of the Dissorophidae (if *M. stowi* of the *Lystrosaurus* Zone is a member of this family).

The conclusion from this brief review of labyrinthodont taxonomy and distribution during the Late Permian and Early Triassic, is that a series of major events occurred close to the boundary between the periods. One of these events was the almost total extinction of the Late Permian assemblage which is best represented in the Kazanian and Dzhulfian of the USSR. A second is the evolutionary radiation within the Rhinesuchidae which produced most of the families of the Early Triassic. And, third is the deployment over the Pangaean world of these rhinesuchid-derived families plus the Brachyopidae. These events may have been sequential but stratigraphic control is not yet fine enough for documentation. On the other hand, they may have been nearly synchronous if competitive replacement was involved, that is, if the new families of the Triassic invaded and usurped the niches of the older families and, thereby, brought about their extinctions. In any event, the area of origin for the new set of families, here termed the Austral Complex, appears to have been the far south of Pangaea.

*Acknowledgements* The present was supported by National Science Foundation Grant DPP79-26279 to the Department of Biological Sciences, Wayne State University and by National Science Foundation Grant DPP80-19996 to the Department of Biological Sciences, Wayne State University and Augustana College. Figures 1a, 1b, 1c and 1d were executed by Ms Meslissa Putt; 1a, 1c and 1d are based closely on the painting by Ms Jennifer Rofe which is reproduced in Banks et al. (1978, p. 154).

## REFERENCES

ANDERSON, J.M. and CRUICKSHANK, A.R.I., 1978: The biostratigraphy of the Permian and Triassic. Pt. 5. A review of the classificaton and distribution of Permo-Triassic Tetrapods. *Palaeontol. Africana, 21,* 15-44.

BANKS, M.R., COSGRIFF, J.W. and KEMP, N.R., 1978; A Tasmanian Triassic Stream Community. *Aust. Nat. Hist., 19,* 150-57.

CARROLL, R.L., 1977: Patterns of amphibian evolution: an extended example of the incompleteness of the fossil record; *in* Hallam, A. (ed.) *Patterns of Evolution.* Elsevier, Amsterdam, 405-37.

CARROLL, R.L. and WINER, L., 1977: Appendix to accompany Chapter 13. Patterns of amphibian evolution: an extended example of incompleteness of the fossil record. Classification of amphibians and list of genera and species known as fossils, 1-13. (This work refers to Carroll, 1977 and was distributed separately).

CHERNIN, S., 1978: Three capitosaurs from the Triassic of South Africa: *Parotosuchus africanus* (Broom, 1909); *Kestrosaurus dreyeri* (Haughton, 1925); and *Parotosuchus dirus* sp. nov. *Palaeontol. Africana, 21,* 79-100.

CHERNIN, S. and CRUICKSHANK, A.R.I., 1975: The Myth of the Bottom-dwelling Capitosaur Amphibians. *S. Afr. J. Sci., 74,* 111-12.

COLBERT, E.H. and COSGRIFF, J.W., 1974: Labyrinthodont amphibians from Antarctica. *Am. Mus. Novit., 2552,* 1-30.

COSGRIFF, J.W., 1974: Lower Triassic temnospondyli of Tasmania. *Geol. Soc. Am., Sec. Pap., 149.*

COSGRIFF, J.W. and ZAWISKIE, J.M., 1979: A new species of the Rhytidosteidae from the *Lystrosaurus* Zone and a review of the Rhytidosteoidae. *Palaeontol. Africana, 22,* 1-27.

KITCHING, J.W., 1978: The stratigraphic distribution and occurrence of South African fossil Amphibia in the Beaufort Beds. *Palaeontol. Africana, 21,* 101-12.

OLSON, E.C. and LAMMERS, G.E., 1976: A new brachyopoid amphibian; *in* Churcher, C.S. (ed.) Athlon. Essays on Palaeontology in Honor of Loris Shano Russell. *Roy. Ont. Mus. Life Sci., Misc. Publ.,*45-57.

ROMER, A.S., 1947: A review of the Labyrinthodontia. *Bull. Mus. Comp. Zool., Harv., 99,* 1-352.

SHISHKIN, M.A., 1980: The Luzocephalidae, a new Triassic labyrinthodont family. *Palaeont. Jour., 1,* 88-101. (Translated from *Palaeont. Zhur., 1,* 104-19).

# PALAEOCLIMATIC SIGNIFICANCE OF SOME MESOZOIC ANTARCTIC FOSSIL FLORAS

T.H. Jefferson, *Institute of Polar Studies and Department of Botany, Ohio State University, Columbus, Ohio, 43210, U.S.A.*

*Abstract* The Early Cretaceous floras of Alexander Island are discussed as indicators of high-latitude palaeoclimate in Antarctica. Forests of trees with large growth rings (up to 9.5 mm) and diverse plant communities grew at supposed palaeolatitudes of 70° to 80°S. Relatively large, entire-margined leaves suggest susceptibility to low temperatures. Complete dormancy, however, which would be necessary to survive the dark polar winter, usually requires low temperatures. Wood anatomy and growth patterns compare most closely with austral warm-temperate rain forests and suggest that there was abundant light throughout a long growing season. High palaeolatitude forests of Permo-Triassic age from the Transantarctic Mountains show similar features of growth and leaf flora composition. The palaeobotanical evidence is apparently inconsistent with light availability in a polar environment even if high temperatures were maintained throughout the winter. This "inconsistency" may be a result of a global warming during the Mesozoic combined with highly specialised floras unique to Mesozoic polar regions, or as a result of large discrepancies between the magnetic and rotational poles during the period. The possibility also exists that there was a reduced Mesozoic obliquity of the Earth, which allowed equable light conditions and temperate floras to extend much further poleward.

Diverse plant communities, including forests of arborescent gymnosperms, were present at high palaeolatitudes; on Alexander Island (Early Cretaceous), on the western margins of East Antarctica (Permian and Triassic), and in West Antarctica (Jurassic to Early Tertiary). Abundant structurally preserved wood is an important feature of the Antarctic fossil floras. Leaf physiognomy and floral diversity within these and other floras are discussed herein as indicators of palaeoclimate. Also, an analysis of secondary xylem, often from trees preserved in growth position, was undertaken using techniques developed for the study of growth, and growth-climate relationships, in living trees. Although all palaeoclimatic interpretation involves a considerable amount of speculation, a rigorous investigation of growth in fossil plants is an important new source of information which can lead to a more quantified assessment of palaeoclimate.

## Previous work on high latitude palaeoclimate

*Northern Hemisphere fossil floras.* Arctic fossil floras occur at palaeolatitudes of 65° to 85° N in Alaska (Wolfe, 1980), Spitsbergen (Schweitzer 1974), and Canada (Donn, in press). These comprise warm-temperate and sub-tropical leaf forms requiring long, warm growing seasons, and frost-free winters; conditions which could not prevail within the polar circles today.

*Southern Hemisphere fossil floras.* The palaeoclimatic significance of Early Cretaceous floras of Victoria, Australia (palaeolatitude 65° to 85°S) was discussed by Douglas and Williams (1982). Many leaf forms are similar to those in the Alexander Island flora. Jurassic-Cretaceous and Triassic floras from New Zealand and Australia (Arber, 1917; Retallack, 1977) probably grew at similar palaeolatitudes. In Antarctica the Hope Bay flora (Halle, 1913) and elements of the Permo-Triassic floras from the Transantarctic Mountains are of interest (Schopf, 1970; Plumstead, 1962).

*Fossil vertebrates.* Fossil reptiles and amphibians are found at numerous high palaeolatitude sites; in the Cretaceous of the arctic (XuQinqi, 1980, Donn, in press), in the Lower Cretaceous of Australia (Douglas and Williams, 1982) and the Permo-Triassic of Gondwana (Cosgriff and Hammer, this volume). Most of these forms could not have survived long polar winters with little heat input. Their presence at high palaeolatitudes is an inconsistency similar to that represented by the fossil floras.

## Early Cretaceous floras from Alexander Island

*Location, age and palaeolatitude.* Fossil plants are abundant within the non-marine Barremian-Albian (Taylor et al., 1979) upper part of the Fossil Bluff Formation of southeastern Alexander Island (Figure 1). Palaeocontinental reconstructions consistently place southern Alexander Island at 70° to 80°S between 100 Ma and 120 Ma (Figure 2).

*Sedimentary and preservational environment.* The non-marine facies of the Fossil Bluff Formation were deposited on a fluvially dominated delta-top (Jefferson, 1981). All of the sediment had a high volcanoclastic component. Wood was preserved in porous volcanoclastic sandstone and large numbers of standing trees were often silicified in growth position as fossil forests (Jefferson, 1982b). Leaves were preserved as compressions in fine sandstones and siltstones but burial metamorphism led to their disruptive mineralisation (Jefferson, 1982a). Cell outlines are therefore poorly preserved in cuticles and conclusions about evolutionary relationships must be limited.

### *Leaf Fossil Assemblages*

Because of the absence of well preserved cuticle, and in view of some of the problems in conventional fossil taxonomy, particularly with reference to plant material (Hughes and Moody-Stuart, 1969), leaf forms in the fossil flora were compared with rather than referred to, existing taxa. This was done using the Cf system of Hughes and Moody-Stuart (1969). "CfA" indicates only minor quantitative differences of a character, "CfB" indicates also one qualitative difference, and "CfC" indicates more than one qualitative difference. The most common and important forms are illustrated in Figure 5, where authors of the comparison taxa are given.

*Floral composition and general palaeoecology.* Detailed sedimentology, fragmentation and size sorting of leaves, and assemblage composition, determined using quadrat sampling (Scott, 1977), were used to assess the general palaeoecology of 50 very productive plant beds.

Ferns including CfA *Cladophlebis oblonga,* CfB *Phlebopteris dunkeri,* and CfB *Almatus bifarius* dominated flood plain-crevasse splay sedimentary sequences (Jefferson 1981, 1982a). This suggests that fern communities dominated areas frequently inundated by water and sediment, though CfB *Ginkgo huttoni* and CfB *G. digitata* indicate local development of ginkgophytes. Several pteridophyte assemblages, composed entirely of delicate ferns of *"Coniopteris"* type, however, are found in finely laminated siltstones. These beds are interpreted as being derived from restricted or pioneer communities developed on poorly drained lake margins.

Diverse conifer-cycad-fern assemblages also occur in lacustrine silts and fine sands. The size range and non-fragmented nature of the leaves suggests input of plant material from mature communities on dryer sites close to the lake margins. The podocarp-like conifer leaf CfB *Pagiophyllum insigne* is dominant, though several other types of conifer shoots also occur (notably CfA *Elatocladus conferta*). The cycadophytes CfA *Taeniopteris daintreei* and two other *Taeniopteris* forms are very well represented, and many of the large-pinnuled ferns and "ginkgophytes" are also abundant. Bryophytes occur in several of the assemblages.

*Palaeoclimatic implications of the leaf floras*

In virtually all the assemblages studied, dominant leaves were large, entire margined, and probably from non-deciduous plants. The flora is dominated by "notophyls" (leaves 7.6-12.7 cm long) and 76% of the taxa have entire margins. These features are characteristic of the "tropical rain forest" classification of Wolfe (1971) and similar to those found in Australian warm-temperate rain forests today. Although, for the reasons stated above, such a precise classification is not valid, these features do suggest warm, equable conditions.

The dominant conifer shoots (CfB *Pagiophyllum insigne* ) have broad and often fleshy leaves (Figure 5 (1-2)). Short, acicular leaves are not well represented. Cycadophyte leaves probably reached over 30 cm long and 5 cm wide and also possessed entire margins (Figure 5 (4)). Leaves of the fern CfB *Phlebopteris dunkeri* must have reached at least 20 cm long and 24 cm wide and individual segments are over 1 cm across and 12 cm long (Figure 5 (9-10)). The entire-margined leaf of CfB *Hausmania dichotoma* reached over 7 cm by 6 cm (Figure 5 (7-8)). The abundance of large ginkgophyte leaves is also notable. These features of the flora lead to the following conclusions:

1) Unless all the plants (even those similar in form to extant non-deciduous leaves) were deciduous, the extensive lamina area of large leaves with entire margins means that they are unlikely to survive frosts. Temperatures throughout the year can rarely have fallen below 0°C.

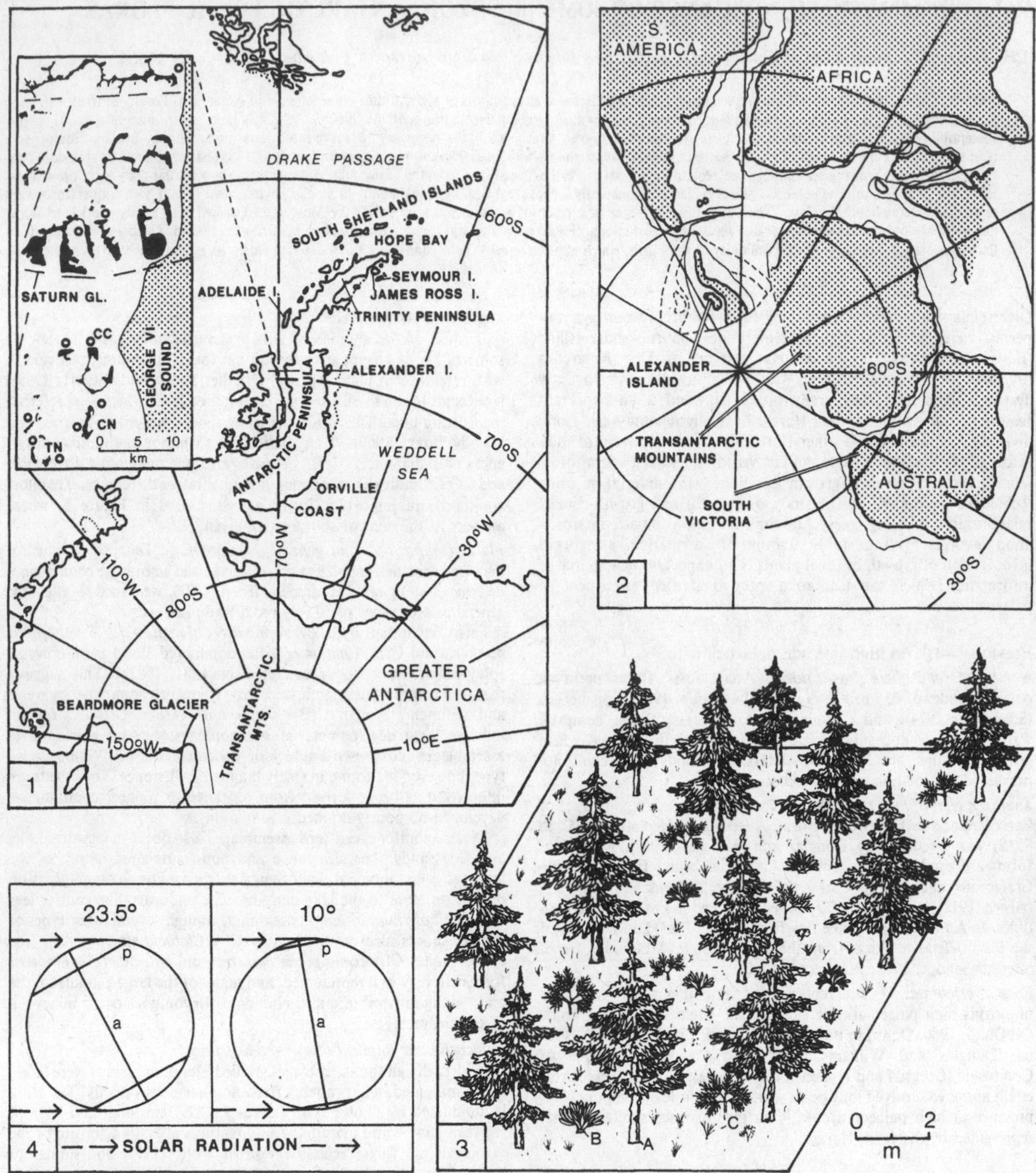


**Figure 1.** Map of the Antarctic Peninsula and Western Greater Antarctica showing localities mentioned in the text. Inset shows the southern part of the Fossil Bluff Formation outcrop. CC = Corner Cliffs, CN = Coal Nunatak, TN = Titan Nunataks. O = location of sections with important fossil plant beds and fossil forests.

**Figure 2.** Reconstruction of the southern continents combining Smith et al. (1981) and the "three plate" Lesser Antarctica model of DeWit (1977). The possible range of positions of the Antarctic Peninsula is shown in dashed lines. Virtually all published reconstructions place Alexander Island at an Early Cretaceous palaeolatitude of 70°-80°S. Important sites of fossil floras are arrowed.

**Figure 3.** Artist's reconstruction of an Early Cretaceous forest in Alexander Island. Tree density is to scale and based on the distribution of stumps in the fossil forest floor at Coal Nunatak. A = Podocarp-like trees. B = cycadophytes (eg. "Taeniopteris"). C = General fern-like undergrowth.

**Figure 4.** The position of polar circles (p) with the Earth's axis (a) at 23.5° to the axis of the ecliptic, as at present; and at 10° to the axis of the ecliptic, as suggested for the Early Cretaceous.

**Figure 5.** **FOSSIL LEAVES, LOWER CRETACEOUS OF ALEXANDER ISLAND. (1)-(2) CfB *Pagiophyllum insigne* Kendal. Specimen KG.2815.164. Part and counterpart. (1) x 0.6; (2) x 1.7. (3) CfB *Bellarinea barklyi* (McCoy) Florin. Specimen KG. 2816.91, x 0.6. (4)-(5) CfA *Taeniopteris daintreei* McCoy. (4) Specimen KG.2815.158c, x 0.6. (5) Specimen KG.2815.179, x 0.6. (6) CfA *Cladophlebis oblonga* Halle. Specimen KG.2815.169b, (7)-(8) CfB *Hausmannia dichotoma* Dunker. Specimen KG.2821.64, part and counterpart, x 0.6. (9)-(10) CfB *Phlebopteris dunkeri* Schenk. Specimen KG.2816.88. (9) x 0.4, (10) x 1.3. (11) ?CfC *Gonatosorous nathorsti* Raciborski ("Coniopteris type" fern). Specimen KG.2814.22, x 0.6. (12) CfB *Ginkgo huttoni* Sternberg. Specimen KG.2815.105a, x 0.6. All specimens housed at British Antarctic Survey, Cambridge, England.**

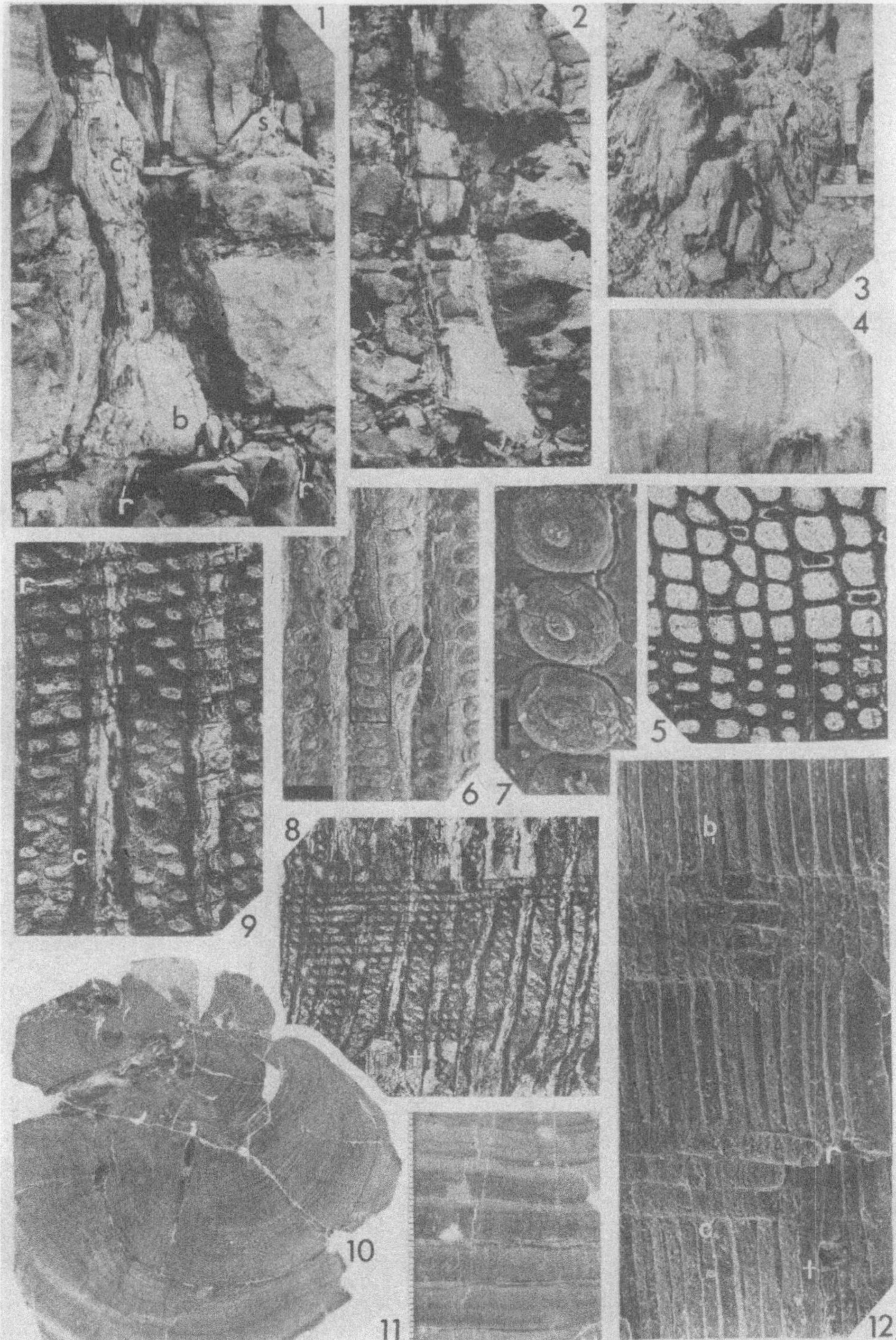


Figure 6. FOSSIL WOOD (1)-(9) Early Cretaceous wood (Cf. *Circoporoxylon*) from Alexander Island (See Figure 1 for localities). (1) Silicified trees in growth position, east Titan Nunatak. Hammer is 40 cm long. Tree height > 2 m. Roots (r) penentrated underlying sand bed. Bulbous base (b) and central section (c) suggest survival after inundation by water and sediment. The stump (s), which apparently grew at the same time as the tree to the left, indicates rapid colonisation of new surfaces and high density tree cover. (2) Silicified tree in growth position, Corner Cliffs. Hammer (arrowed) is 35 cm long. Tree height > 4 m. (3) Silicified tree stump from the fossil forest floor at the top of Coal Nunatak. The hammer is 40 cm long. (4) Large growth rings, KG.1704.10, x 0.6. (5) Growth ring boundary showing late-wood (bottom) and large early wood cells of the next year's growth. KG.2814.254, x 60. (6)-(7) Scanning electron micrograph of borded bits on tracheid radial walls, KG.2817.15, (6) x 150, scale bar 40$\mu$ m, (7) detail x 600, scale bar 10$\mu$. (8)-(9— Transmitted light photomicrographs of tracheids (t), medullary rays (r), and cross-field pits (c) in radial section KG.1702.2b, (8) x 60, (9) x 150. (10)-(12) Triassic wood, Transantarctic Mountains. (10) Acetate peel of cross section through basal portion of a tree trunk, CB 365.4 x 0.4 (scale in mm). (11) Large growth rings up to 8 mm across, CB 353, x.0.6 (scale in mm). (12) Scanning electron micrograph showing broken radial surface with tracheids (t), lightly bordered pits (b), medullary rays (r), and cross-field pitting (c) with up to 10 pits per cross-field, CB 467 x 100. (4)-(9) Specimens housed as in Plate 1. (10)-(12) Specimens housed in Orton Museum, Ohio State University, Columbus, Ohio.

2) The above suggests that not all plants became completely dormant during the winter. Since complete dormancy is usually controlled by low temperatures (Fritts 1976), it is likely that at least some light for photosynthesis must have been available throughout the winter to support respiration if evergreen plants were to survive.

3) The area was humid; water stress cannot have been an important factor in an environment which supported a diverse flora containing large-leaved pteridophytes and cycadophytes but devoid of leaves with highly developed xeromorphic adaptions.

4) The diverstiy, abundance and leaf size of the flora suggest that temperatures, water availability, and equability of light were all at levels to those found in a warm-temperate climate today.

## Fossil Forests

*Distribution.* Silicified wood was found at all localities where detailed sections in non-marine rocks were studied (Jefferson, 1982b). Forest floors exposed on extensive bedding planes at the tops of nunataks, and standing trees exposed in cliff sections (Figure 6 (1-3)) indicate that forests with trees up to 0.5 m in diameter covered much of the area (Figure 3). Since many of the sedimentary units which would have favoured silicification of wood (Jefferson, 1981) do, in fact contain *in situ* fossil trees, it is likely that this forest cover persisted for much of the late Early Cretaceous.

*Preservation and anatomy.* The wood was silicified in two stages, early cell wall impregnation followed by cell lumen fill. Scanning electron microscopy has revealed fine detail of inter-tracheid and tracheid-ray pitting. Both forms of the wood (distinguished by the distribution of parenchyma cells) can be compared most closely to the form genus *Circoporoxylon* Krausel (1949). Although the wood cannot be directly compared with any living genus, similarities in anatomical structure suggest that the general physiology of the trees was similar to that of living podocarps. (Figure 6 (4-7)).

*Growth rings*

Three features of the well preserved growth rings were used as palaeoclimateic evidence: (1) morphology of annual rings, (2) ring widths, and (3) variability (sensitivity) of growth. The well defined late wood (small cells with thick walls at the end of each annual ring) and abrupt late wood-early wood boundaries in all trees (Figure 6 (4)) indicate a seasonal climate. Mean ring widths for trees (up to 2.5 mm) and very large rings (up to 9.55 mm), even in the outer part of large trees, indicate growth rates which are exceptional even by the standards of modern mid-latitude trees. Annual ring width is very variable. This is measured by "mean sensitivity" (MS);

$$\frac{1}{n-1}\sum_{t=1}^{t=n-1}\left|\frac{2\,(x_{t+1}-x_t)}{x_{t+1}+x_t}\right|$$

(x= ring width, t= year number of ring, n= number of rings in sequence). MSs in living trees range from 0 to 1.0. Conventionally, "complacent" trees are those with MSs less than 0.3 and "sensitive" trees those with MSs greater than 0.3. Moisture availability is usually the "limiting factor" producing sensitive growth patterns. Bristlecone pines in semi-arid southwestern U.S.A., for example, have very high MSs (commonly 0.4—0.6) because of fluctuations in moisture levels. These moisture sensitive trees are also very slow-growing. In humid, warm-temperate forests, however, most trees are complacent because there is no single factor which limits growth.

Values for Alexander Island fossil trees in *in situ* forests were consistent and high, at 0.4 to 0.45, suggesting that the trees were growing in an environment in which a single climatic variable, limiting to growth, was liable to fluctuation. Trees were clearly capable, however, of very rapid growth given optimum conditions. This, together with considerations of sedimentology and the morphology of leaf fossils, suggests that moisture was readily available.

*Analogies with Living Forest Communities*

The growth patterns of the fossil trees were used to draw broad analogies with living trees to provide further palaeoclimatic evidence. The most relevant forest environments for comparison are those growing at high latitudes or those in which trees show similar features of growth patterns.

*Boreal forests.* Even in the fastest growing species, growth rates in arctic Norway (Ording, 1941), Alaska (Giddings, 1943), and Canada (Drew, 1975) are very low (the largest ring width from Norway north of 60° was 2.05 mm, and the largest from the North American arctic was 2.02). In all cases trees were found to be complacent; MSs were consistently below 0.2.

Examination of data from many sub-arctic and arctic North American sites substantiated these findings. Growth rates are consistently low close to the northern tree line at 69°N (rings rarely in excess of 1 mm) but rates increase southwards into temperate regions. South of 55°N, annual increments may be up to 7 mm in very favourable years. This is due to increased light availability as well as temperature; many interior Canadian sites with high growth rates have lower mean monthly temperatures (even during the growing season) than arctic Alaskan sites with low growth rates (UNESCO 1979).

*Warm-temperate rain forests.* Several species in warm-temperate rain forests show high sensitivity but are capable of rapid growth. Light availability during the winter and spring is apparently the factor limiting growth in Douglas fir of Washington (Brubaker, 1980) and in the podocarp genus *Phyllocladus* of New Zealand and Tasmania (LaMarche, pers. comm.). Tree-ring data from 19 *Phyllocladus* sites were analysed to obtain comparative statistics. The trees show high MS (usually between 0.35 and 0.45 but up to 0.66) together with high growth rates. In *P. trichomanoides* mean ring widths are up to 2.28 mm, and 20 year means are up to 3.80 mm, and in *P. glaucus* values are 1.86 mm and 2.77 mm respectively. These trees are capable of 6 mm of lateral annual growth and 4 mm rings are common; *P. aspleniifolius* from Tasmania is slower growing but 20 year mean ring widths are up to 2.33 mm.

*Discussion.* Recent work on light utilization by temperate forest trees (Jarvis and Leverenz, in press) and calculations of high latitude light levels and light-tree growth relationships by G.T. Creber (pers. comm.) suggest that high growth rates would be theoretically possible over a short (possibly only 90 day) high latitude growing season. It is therefore possible that Early Cretaceous trees growing at high latitudes were adapted to utilise light efficiently enough over three summer months to sustain rapid growth, such efficient utilisation requires relatively high spring and summer temperatures quite different from those possible in polar regions today. Shading within the forests would also have been an important factor in view of the low angle of incidence of light even during the growing season.

Although growth patterns in the Early Cretaceous trees are remarkable for their combination of high sensitivity and high growth rates, they are similar to those of podocarps in warm-temperate austral forests. In such forests the growing season is more than five months long and large increments of growth depend on high light levels during the late winter-early spring period (levels impossible at high latitudes). Furthermore the anatomy of the fossil trees compares more closely with these podocarps than with any other extant trees and this supports a broad comparison of growth patterns.

## Other Antarctic fossil floras (Permian to Tertiary)

*Beacon Supergroup*

Fossil plants occur throughout the Permian and Triassic sequences of the Transantarctic Mountains and particularly in the Buckley (Permian) and Fremouw (Triassic) Formations in the Beardmore Glacier area (Figure 1). Palaeocontinental reconstructions indicate palaeolatitudes of approximately 80°S.

*Leaf Floras.* These are dominated by "glossopterids," generally large, entire-margined forms. Although evidence from their distribution in seasonal varves suggests that they were deciduous (Taylor, 1981), the size and form of leaves indicate a warm, frost-free growing season.

*Fossil wood.* This occurs as drifted logs up to 22 m long (Barrett 1969), as trees in growth position (J. Collinson pers. comm.) and as an important component of silicified "peat" (Schopf 1970). The "araucaroid" wood has been referred to *Dadoxylon* (e.g. Krausel, in Plumstead 1962) though a more complete classification is under way. Well defined growth rings (Figure 6, 10-11) indicate a seasonal climate and individual rings up to 8 mm across suggest warm temperatures and high levels of light, probably over a long growing season. High variability of ring widths (MS= 0.3-0.45) may again indicate warm-temperate conditions.

*Jurassic to Tertiary flora*

Jurassic floras from Carapace Nunatak (Plumstead, 1962), on the Orville Coast (J.W. Schopf, unpub. ms) and at Hope Bay (Halle 1913), together with Early Tertiary leaf fossils in Lesser Antarctica

(Thomson and Burn, 1977), and Late Cretaceous-Tertiary fossil wood from James and Ross Island and Seymour Island (J.A. Crame and W. Zinsmeister, pers. comm.) and the South Shetland Islands (Orlando 1964) suggest that temperate conditions persisted throughout the Mesozoic and into the Early Tertiary.

## Conclusions

The Late Permian-Early Tertiary floras, which occurred in Antarctica at supposed palaeolatitudes of 65° to 80°S, indicate equability of light, long growing seasons, high temperatures, and frost-free winters. This is apparently inconsistent with polar latitudes, since no light or heat energy would have been available for several months during the winter, and growing seasons would have been of three months or less. There are several possible explanations for this palaeoclimatic enigma.

*1) Global warming.* Mesozoic polar regions, and particularly coastal areas, may have been kept warm by globally high sea temperatures, for which there is $O^{18}$ evidence (Lowenstam, 1964). The problem of light availability (especially in late winter-early spring), however, still remains. It is also difficult to envisage a frost-free winter during periods when there would be no incoming solar radiation even given the mechanisms by which heat could have been trapped by cloud cover or transported more rapidly poleward, as postulated by Barron et al. (1981) to explain an ice-free Cretaceous.

*2) Over-reliance on uniformitarianism.* There is no close modern analogue for polar climates with warm winters. Plants adapted to such conditions may have been highly specialised organisms unrepresented in modern floras; an unquestioned reliance on uniformitarian principles is inadvisable. The combination of a global warming and misinterpretation of Mesozoic plants may be sufficient to explain the apparent inconsistency between palaeobotanical and palaeolatitudinal evidence. The morphological and anatomical similarities between many of the fossils and living forms, however, suggest similar physiological requirements and mitigate against the existence of highly adapted high-latitude plants unique to the Mesozoic. Further modelling of possible Mesozoic climates, and more research on Mesozoic plant anatomy and on the physiology of living plants with reference to light, temperature and growth, is necessary.

*3) Inaccurate palaeolatitudes.* Palaeomagnetic data is used to reconstruct palaeocontinental positions relative to the magnetic pole rather than to the geographic pole. If, during the Mesozoic, these two poles were not closely related, as suggested by Donn (in press), the "polar regions" in such reconstructions would have no palaeogeographic or palaeoclimatic significance. The palaeobotanical evidence does support this suggestion. It is, however, difficult to envisage any mechanism for producing a di-polar magnetic field in the Earth which does not involve the Earth's rotation; a close association between the axis of rotation and the magnetic poles would be expected when averaged out over several million years.

*4) Changes in axial obliquity.* The Earth's spin axis is usually assumed to have been at the same angle to its axis of rotation around the sun throughout geological time. Recent theories (e.g. Macdonald, 1964, Williams, 1972) contend that this may not be the case. Reduced axial obliquity during the Mesozoic would result in smaller polar circles and explain the occurrence of "mid-latitude" vegetation at high latitudes (Figure 4). Although small-scale changes of obliquity (e.g. the Milankovich cycle) are widely accepted, there is at present no astronomical theory to support a larger, and much slower, axial change. The palaeobotanical and other palaeontological evidence, however, does suggest that the possibility of major changes in axial obliquity should be seriously considered and a subject of further research.

*Acknowledgements* A Research Studentship from the Natural Environmental Research Council (U.K.) and a Harkness Fellowship of the Commonwealth Fund (U.S.A.) are gratefully acknowledged. The work depended on the field support and research facilities of the British Antarctic Survey and the research facilities and collections of the Department of Earth Sciences of Cambridge (U.K.), Ohio State University Institute of Polar Studies, and Universtiy of Arizona Laboratory of Tree-Ring Research. I thank N.F. Hughes, M.R.A. Thomson, G.T. Creber and L. Cranwell-Smith for their helpful advice, encouragement and discussions.

## REFERENCES

ARBER, E.N., 1917: The earlier Mesozoic floras of New Zealand. *N.Z. Geol. Surv., Palaeont. Bull., 6.*

BARRETT, P.J., 1969: Stratigraphy and petrology of the mainly fluviatile Permian and Triassic Beacon rocks, Beardmore Glacier area, Antarctica. *Rep. Inst. Polar Stud., Ohio State Univ., 34.*

BARRON, E.J., SLOAN, J.L. and HARRISON, C.G.A., 1981: The potential significance of land-sea distribution and surface albedo variations as a climate forcing mechanism 180 m.y. to present. *Palaeogeogr., Palaeoclimatol., Palaeoecol., 30,* 17-40.

BRUBAKER, L.B., 1980: Spatial patterns of tree growth anomalies in the Pacific northwest. *Ecol., 61,* 798-807.

COSGRIFF, J.W. and HAMMER, W.R. (this volume): The Labyrinthodont amphibians of the earliest Triassic from Antarctica, and Tasmania and South Africa.

DEWIT, M.J., 1977: The evolution of the Scotia Arc as a key to the reconstruction of Gondwanaland. *Tectonophysics, 37,* 53-81.

DONN, W.L. (in press): The enigma of high-latitude paleoclimate; *in* Symposium volume *"Paleogeography and Climate"* Geol. Soc. Am., annual meeting, Cincinnati, 1981.

DOUGLAS, J. and WILLIAMS, G.E., 1982: Southern polar forests; the Early Cretaceous floras of Victoria and their palaeoclimatic significance. *Palaeogeogr., Palaeoclimatol., Palaeoecol.,* (in press).

DREW, L.G., (ed.), 1975: Tree ring chronologies of Western America , VI, Western Canada and Mexico. *Chronology series 1,* Lab. Tree Ring Research, University of Arizona, Tucson, Arizona.

FRITTS, H.C., 1976: *Tree Rings and Climate.* Academic Press, London.

GIDDINGS, J.L., 1943: Some climatic aspects of tree growth in Alaska. *Tree Ring Bull. 9,* 25-32.

HALLE, T.G., 1913: The Mesozoic floras of Graham Land. *Wiss. Ergebn. schwed. Sudpolarexped., 1901-03, 3., 14,* 1-23.

HUGHES, N.F. and MOODY-STUART, J.C., 1969: A method of stratigraphic correlation using early Cretaceous miospores. *Palaeontology, 12,* 84-91.

JARVIS, P.G. and LEVERENZ, J.W., (in press): Physiological processes related to photosynthetic productivity of a specific ecosystem: Temperate, deciduous-evergreen forests; *in Encyclopedia of Plant Physiology, 12,D. Physiological Plant Ecology: Productivity and Ecosystem Process.*

JEFFERSON, T.H., 1981: Palaeobotanical contributions to the geology of Alexander Island, Antarctica. Unpublished Ph.D. Thesis, University of Cambridge.

JEFFERSON, T.H., 1982a: Preservation of leaf fossils in volcaniclastic rocks from the Lower Cretaceous of Alexander Island, Antarctica. *Geol. Mag., 119,* 291-300.

JEFFERSON, T.H., 1982b: The Early Cretaceous fossil forests of Alexander Island, Antarctica. *Palaeontology, 25,* 681-708.

KRAUSEL, R., 1949: Die fossilen Koniferholzer. *Palaeontographica B, 89,* 83-203.

LOWENSTAM, H.A., 1964: Palaeotemperatures of the Permian and Cretaceous periods; *in:* Nairn, A.E.M. (ed.) *Problems in Palaeoclimatology.* Interscience, New York, 227-52.

MACDONALD, G.J.F., 1964: Tidal friction. *Rev. Geophysics, 2,* 467-514.

ORDING, A., 1941: Arringanalyser pa gron og furu. *Medd. Det norske Skogsforsoksvseen, 25.*

ORLANDO, H.A., 1964: The fossil flora of the surroundings of Ardley Peninsula, 25 de Mayo Island (King George Island), South Shetland Islands; *in* Adie, R.J. (ed.) *Antarctic Geology.* North Holland, Amsterdam, 629-37.

PLUMSTEAD, E.P., 1962: Geology 2. Fossil floras of Antarctica. *Scient. Rep. Transantarctic Exped., 9.*

RETALLACK, G.J., 1977: Reconstructing Triassic vegetation of eastern Australasia: a new approach for the biostratigraphy of Gondwanaland. *Alcheringa, 1,* 247-77.

SCHOPF, J.W., 1970: Petrified peat from a Permian coal bed in Antarctica. *Science, 169,* 274-7.

SCHWEITZER, H.J., 1974: Die tertiaren Koniferen Spitzbergens. *Palaeontographica, B, 149,* 1-89.

SCOTT, A.C., 1977: A review of the ecology of upper Carboniferous plant assemblages with new data from Strathclyde. *Palaeontology, 20,* 447-75.

SMITH, A.G., HURLEY, A.M. and BRIDEN, J.C., 1981: *Phanerozoic palaeocontinental world maps.* Cambridge University Press.

TAYLOR, B.J., THOMSON, M.R.A. and WILLEY, L.E., 1979: The geology of the Ablation Point to Keystone Cliffs area, Alexander Island. *Scient. Rep. Br. antarct. Surv., 82.*

TAYLOR, T.N., 1981: *Paleobotany.* McGraw-Hill, New York.

THOMSON, M.R.A. and BURN, R.W., 1977: Angiosperm fossils from latitude 70° South. *Nature, 269,* 139-41.

UNESCO, 1979: *Climatic Atlas of North and Central America.* UNESCO, WMO, Cartographica, Hungary.

WILLIAMS, G.E., 1972: Geological evidence relating to the origin and secular rotation of the solar system. *Mod. Geol., 3.* 165-81.

WOLFE, J.A., 1971: Tertiary climatic flucuations and methods of analysis of Tertiary floras. *Palaeogeogr., Palaeoclimatol., Palaeoecol., 9,* 27-57.

WOLFE, J.A., 1980: Tertiary climates and floristic relationships at high latitudes in the Northern Hemisphere. *Palaeogeogr., Palaeoclimatol., Palaeoecol., 30,* 313-23.

XU QUINQI, 1980: Climatic variation and the obliquity. *Vertebrata Pal. Asiatica, 18,* 334-43. (Chinese with English summary).

# THE RECONSTRUCTION OF LESSER ANTARCTICA WITHIN GONDWANA

P.D. Clarkson, *British Antarctic Survey, Madingley Road, Cambridge, CB3 OET, UK.*

*Abstract* A model for the reconstruction of Lesser Antarctica within Gondwana is proposed. Five main plates comprise the region: Antarctic Peninsula, Ellsworth Mountains, Thurston Island, eastern and western Marie Byrd Land. The existence of other micro-continental fragments in the Weddell embayment is discussed.

The Ellsworth Mountains plate is restored to a position off present-day Coats Land, behind the Antarctic Peninsula. Clockwise rotation and northward movement of the Antarctic Peninsula was followed by anticlockwise rotation and translation of the Ellsworth Mountains plate coupled with failed rifting between the craton and Berkner Island-Pensacola Mountains. The Ellsworth Mountains plate probably comprises several microplates divided by major crustal fractures which allowed relative movement without separation; the Rutford Ice Stream may mark the site of one such fracture between the Ellsworth Mountains and the Haag Nunataks microplate.

The Thurston Island and Marie Byrd Land plates are restored to positions adjacent to the Transantarctic Mountains. Initial separation of these plates from the Transantarctic Mountains would have resulted in the formation of new oceanic crust now buried beneath thick sedimentary piles, similar to other deep subglacial areas of Lesser Antarctica. The alkaline volcanism of Marie Byrd Land, part of a discontinuous belt of alkaline volcanic rocks extending from James Ross Island to Victoria Land, is probably due to extensional tectonics related to continued movement of these plates.

# 12

# Plate Tectonics

# SECULAR MOTION OF THE SOUTH MAGNETIC POLE

P.M. McGregor, A.J. McEwin, J.C. Dooley, *Bureau of Mineral Resources, Geology and Geophysics, PO Box 378, Canberra City, ACT 2601, Australia.*

*Abstract* The Earth's magnetic poles are those places where a freely-suspended magnetic needle would stand vertically—i.e. where the angle of dip is 90°. Douglas Mawson and his two companions from Shackleton's "Nimrod" expedition were the first and only persons to stand at the south magnetic pole (SMP) which they reached on 16 January 1909. They fixed its position (72.4°S, 155.3°E, in Victoria Land) by direct measurements, but since then it has been determined mainly from analyses of global magnetic observatory and survey results. Thus, from the 1980 analysis, the SMP is estimated now to be about 1000 km to the NW of Mawson's location, some 100 km out to sea from the French station Dumont d'Urville. This motion of the magnetic (or dip) poles contrasts with that of the geomagnetic (or dipole) poles which have been stationary for at least 150 years. Such motion is a manifestation of the general secular variation of the Earth's internal magnetic field, and in common with other features is erratic and not accurately predictable. Because secular variations are thought to be due to localised current systems near the surface of the Earth's core, it is to be expected that the north and south magnetic poles have independent characteristics—e.g. they are not antipodal. Although their motions are not synchronous their average rates and directions during this century have been similar: 10-15 km per annum (kma⁻¹) north to northwesterly.

## HISTORICAL BACKGROUND

In 1840-41 Ross made the first of his research voyages in "Erebus" and "Terror" aimed at reaching the magnetic pole. After discovering the Ross Sea, the Ice Shelf, Mt. Erebus, and McMurdo Sound he found his way to the west barred by the Transantarctic Mountains. Unable to reach a safe place to winter he abandoned the quest at 76°S, 164°E where his magnetic observations indicated that the pole was 250 km to the west (Ross, 1847). On 16 January 1909 Douglas Mawson, Edgeworth David and Alistair Mackay located the south magnetic pole, ending a search begun seventy years earlier by Dumont d'Urville, Charles Wilkes and James Clark Ross. Mawson and his companions were members of Shackelton's British Antarctic Expedition 1907-1909. To reach the pole they manhauled more than 2000 km in 109 days (David, 1909)—one of the most remarkable efforts in Antarctic exploration, considering that the pole was much farther than they had planned for, and that a general geological survey of the coast of Victoria Land was part of Shackleton's instructions to them. Mawson placed the pole at 72.4°S, 155.3°E, in Victoria Land. Figure 1 includes the pole positions determined by Ross and Mawson, and the estimated position in 1980. It shows that in 140 years the pole has moved more than a thousand kilometres in a north to northwesterly direction. The purpose of this paper is to review some of the information on this secular motion of the south magnetic pole.

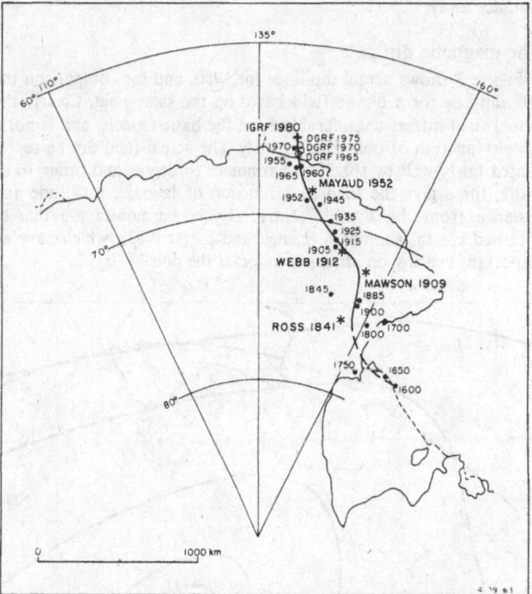


Figure 1. Path of the south magnetic dip pole 1600-1980.

## MAGNETIC POLES OF THE EARTH

The Earth has four pairs of magnetic poles, each important in a different aspect of geomagnetism. To avoid confusion each type of pole is described briefly. Figure 2 shows the positions of the southern poles.

### Axial dipole pole

About 97% of the Earth's field would be accounted for if it was due to a bar magnet at the centre, aligned along the rotation axis and with its north pole south. In such a dipole world the field lines are parallel to the geographic meridians (i.e. the magnetic declination (D) is zero everywhere), and the magnetic and geographic poles coincide. In the southern hemisphere the field lines in fact point upwards at an angle (the dip or inclination, I) which varies from $-90°$ at the pole to 0° at the equator. One can navigate in this world to the extent that the compass points directly to the pole, and the dip angle is a function of magnetic latitude $\phi$ ($\tan I = 2 \tan \phi$), which is the same as the geographic latitude in this case.

This result is used in palaeomagnetism to fix the latitude of rocks at the time of their formation. It is also used to estimate the position of the real dip pole from observations of D and I made not too far away. Finally the rate-of-change of I indicates the distance from the pole; near the poles the angular distance equals twice the change in I—a fact used by Mawson in the last stages of his exploit.

### Inclined dipole (geomagnetic) pole

A better model of the actual field is obtained if the centred dipole is tilted about 11.5° to give a south pole at latitude 78.5°S, longitude 110°E; this is usually called the geomagnetic pole. The field is important in upper atmosphere physics, because away from the surface it controls the plasma radiated by the Sun. For example the

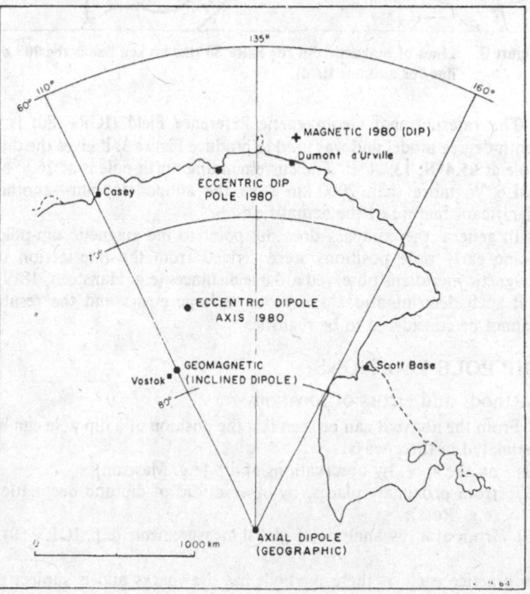


Figure 2. South magnetic poles.

auroras are more or less centred on this pole. The Soviet station Vostok was placed near the geomagnetic pole to utilise these properties. The main difference between this and the axial field is that the magnetic and geographic meridians no longer coincide—there is a geomagnetic declination, which is a simple mathematical function of position, but the relation between polar distance and inclination is the same (in geomagnetic co-ordinates).

### Eccentric dipole pole

An even better model results if the dipole is displaced 488 km towards latitude 20.4°N, 147.3°E, a point just north of the Mariana Islands. From a transformation based on MAGSAT data given by Wallis et al (1982), the axis of this eccentric dipole intersected the Antarctic region at 75.1°S, 119.2°E in 1980. Because of the eccentricity, the magnetic field is not vertical at this point. Using the eccentric dipole model, the position of its pole has been calculated at 67.1°S, 129.0°E, i.e. on the coast of Antarctica at Porpoise Bay. This is a much better approximation to the actual dip pole, but is still some distance away.

### The magnetic dip pole

Figure 3 shows actual dip-lines for 1980, and for comparison the 70° dip-line for a dipole field based on the same pole. Clearly the actual field differs considerably from the basic models, and is not a simple function of pole-distance only. The actual field can be represented fairly well by spherical harmonics to degree and order 10 or more; these give the field as a function of latitude, longitude and distance from the Earth's centre. Higher harmonics are due to localised crustal anomalies (Langel and Estes, 1982), which have an important bearing on attempts to locate the pole.

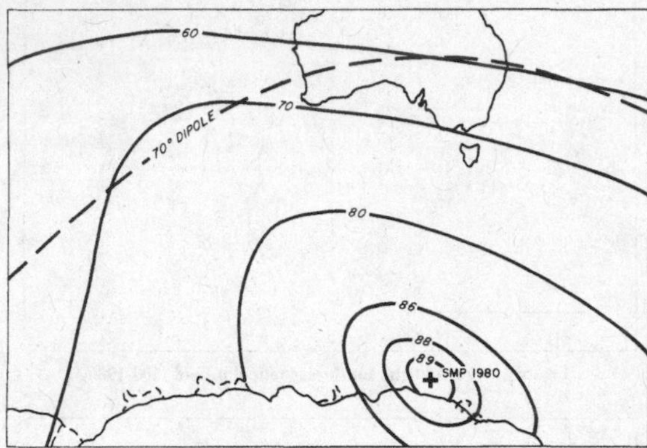


**Figure 3. Lines of magnetic dip for IGRF 80 (the broken line is the 70° dip line for a dipole field).**

The International Geomagnetic Reference Field (IGRF 80) is a tenth-degree model and was used to produce Figure 3: it gives the dip-pole at 65.4°S, 139.4°E. The corresponding north pole is at 76.8°N, 101.6°W, more than 2000 km from the antipodal point—another significant feature of the actual field.

In general the compass does not point to the magnetic dip-pole. Some early pole positions were derived from the intersection of magnetic meridians observed at large distances (e.g. Hansteen, 1819), but such determinations are prone to large errors and the results cannot be considered to be realistic.

## DIP POLE POSITIONS

### Methods and errors of locations

From the above it can be seen that the position of a dip pole can be estimated in three ways:

(1) on the spot, by observations of dip (e.g. Mawson);
(2) from proximate places, by observations of dip and declination (e.g. Ross);
(3) from afar, by analysis of global measurements (e.g. IGRF 80).

In practice each of these methods has drawbacks and is subject to error.

*Direct observations.* Apart from the physical and logistical factors faced by Mawson, the problems with direct observations are:

1) the effect of external magnetic fields; ionospheric electric currents and their concomitant magnetic fields at the Earth's surface cause the point where I = 90° to appear to move cyclically each day; Figure 4 was obtained from the results for January 1978 at the magnetic observatory at Dumont d'Urville; it shows that the apparent path of the dip pole on quiet days is roughly circular with a radius of 10-15 km; on magnetically disturbed days the area covered is expanded greatly.
2) the absence of a horizontal field to direct the compass; as the pole is approached the compass becomes increasingly sluggish and very prone to errors due to pivot-friction and stray fields.

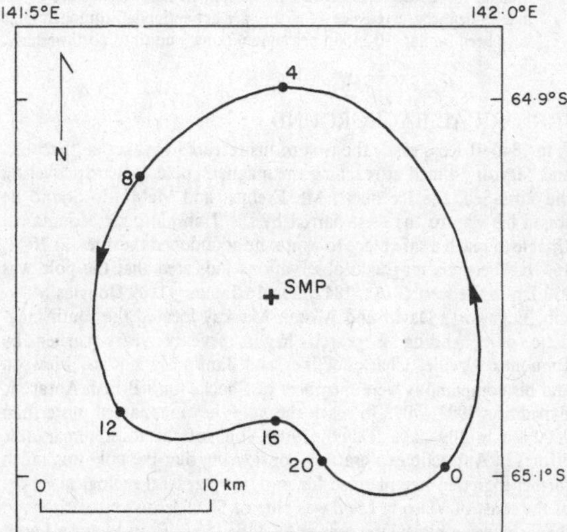


**Figure 4. Apparent daily path of south magnetic pole on quiet days. January 1978.**

Mawson was fully aware of these matters, and realised the impracticability of trying to observe I = 90° exactly. On the evening before their dash to the pole, he said: "..... in order to accurately locate the mean position, possibly a month of continuous observations would be needed.....". He had just observed a dip of 89°48′ and estimated they needed to travel another 20 km the next day. They laid out the compass direction, and marched in a straight line by placing as route markers their sledge, tripods and tent every 3 km. Over the last 8 km they carried only a camera and flagpole to record their achievement—an act of no mean daring.

*Proximate observations.* In determining the pole distance using nearby observations, errors arise from local magnetic anomalies and from the fact that the field is not truly dipolar. Inaccuracies increase rapidly as the pole distance increases. For example, Ross's observations in the Ross Sea, at distances from 200 to 500 km, are remarkably consistent—they give pole positions within 30 km of each other. When he moved away and observed off Adelie Land at a distance of 1200 km, the result differed (in longitude) by about 300 km. This latter result was due to the non-dipolar factor rather than local anomalies; Ross was a skilled and careful observer, and was very sceptical about observations made on islands or rock outcrops (Ross, 1847). The effect of local anomalies is amply demonstrated by French observations in Adelie Land. At Port Martin, Mayaud (1953) observed differences in declination of more than 47° over a 5 km radius. Currently the observatory results from Dumont d'Urville give a pole position about 200 km to the north—northeast whereas the analytical position is about 100 km to the west of north. We consider the observatory result to be affected by local crustal fields.

*Analytical determinations.* Spherical harmonic (SHA) models yield the point where I = ±90°. The models are derived from global observations and their reliability depends critically on the distribution of the observations. In this respect the southern hemisphere has always been deficient with respect to the northern, and this is reflected in the accuracy of south pole determinations. The recent results from the MAGSAT satellite should overcome this deficiency but the models for Antarctica have not yet been tested. The analytical results are theoretically free from external field effects which have been removed from the data (at least in recent models).

## Results

The results presented here are not all those available. They are those we have obtained from original sources, re-examined or re-calculated, and which are based on the three methods described above. We have not used very early positions derived from the intersection of distant magnetic meridians because the method is too inaccurate. Our polar path is shown in Figure 1. The only truly direct observation is that made by Mawson in 1909; but approaches within 100 km on thick ice were made by Webb in 1912 (Webb, 1925) and Mayaud in 1952 (Mayaud, 1953). We have included their results in the "direct" category.

Proximate locations were made by Dumont d'Urville and Wilkes in 1840. They concentrated their efforts in the region predicted by Gauss; that is, around latitude 66°S, longitude 146°E. This turned out to be a poor estimate and they were too far distant to obtain good locations. Ross also had been instructed to search in this region. But when he arrived at Hobart in August 1840 and learned of the explorations and results of Dumont d'Urville and Wilkes he selected a much more easterly meridian (170°E), on which to endeavour to reach the magnetic pole (Ross, 1847). Thus was the way paved for the first reliable location of the south magnetic pole. Table 1 gives details of the experimental determinations we have used.

**TABLE 1**
**Observed positions of the south magnetic pole**

| Year | Observer | Latitude | Longitude |
|---|---|---|---|
| | | °S | °E |
| 1841 | Ross⁺ | 75.5 | 154.8 |
| 1909 | Mawson | 72.4 | 155.3 |
| 1912 | Webb | 71.2 | 150.8 |
| 1952 | Mayaud | 68.1 | 143.0 |

⁺ Mean of 4 observations in the Ross Sea

The analytical positions have been derived from eighteen spherical harmonic models included in a comprehensive list by Barraclough (1978). They were chosen to meet our criteria of "goodness" (see below), and to provide points about every fifty years from 1600 to 1845, and more frequently afterwards. Table 2 gives details of the models.

**TABLE 2**

| Model | Degree | Coordinates of poles: Geomagnetic Lat | Geomagnetic Long | Dip Lat | Dip Long | Equatorial Intensity Ho |
|---|---|---|---|---|---|---|
| | | °S | °E | °S | °E | nT |
| 1600/07 | 4 | 84.6 | 146.2 | 77.5 | 173.9 | 36050 |
| 1650/07 | 4 | 83.2 | 143.7 | 77.5 | 170.7 | 35270 |
| 1700/07 | 4 | 81.5 | 132.9 | 74.6 | 163.1 | 34419 |
| 1750/05 | 5 | 80.0 | 127.1 | 78.1 | 163.3 | 33700 |
| 1800.03 | 5 | 79.3 | 121.0 | 75.3 | 159.8 | 32910 |
| 1845/02 | 6 | 78.7 | 115.7 | 74.2 | 149.1 | 32187 |
| 1885/12 | 7 | 78.6 | 112.2 | 74.1 | 156.0 | 31635 |
| 1900/02 | 6 | 78.6 | 111.2 | 74.3 | 155.7 | 31412 |
| 1905/01 | 6 | 78.6 | 110.0 | 71.3 | 148.2 | 31423 |
| 1915/01 | 6 | 78.6 | 110.1 | 70.9 | 148.1 | 31176 |
| 1925/01 | 6 | 78.6 | 110.1 | 70.4 | 147.4 | 30892 |
| 1935/01 | 6 | 78.6 | 110.1 | 69.8 | 146.2 | 30662 |
| 1945/12 | 10 | 78.5 | 111.4 | 69.0 | 144.1 | 30581 |
| 1952/01 | 6 | 78.6 | 111.3 | 68.9 | 142.0 | 30644 |
| 1955/13 | 10 | 78.5 | 110.9 | 66.8 | 138.5 | 30518 |
| 1960/15 | 10 | 78.5 | 110.5 | 66.9 | 140.8 | 30413 |
| 1965/09 | 10 | 78.6 | 110.2 | 66.9 | 140.1 | 30388 |
| 1970/01 | 10 | 78.6 | 109.8 | 65.9 | 139.2 | 30206 |
| IGRF/80 | 10 | 78.8 | 109.2 | 65.4 | 139.4 | 29988 |

The strength and pole-position of the inclined dipole field can be calculated from the first three coefficients of an SHA model. Modern (more reliable) results show that the field strength has been decreasing steadily, and that the geomagnetic poles have been fixed in position (Vestine, 1967). Thus we consider a model to be "good" if its intensity lies on the trend line, and its poles are within 100 km of the fixed position. Figure 5 shows how our models stand; all models are remarkably good for intensity, but before 1840, pole positions are unreliable. This is indicated by the broken line on the polar path of Figure 1.

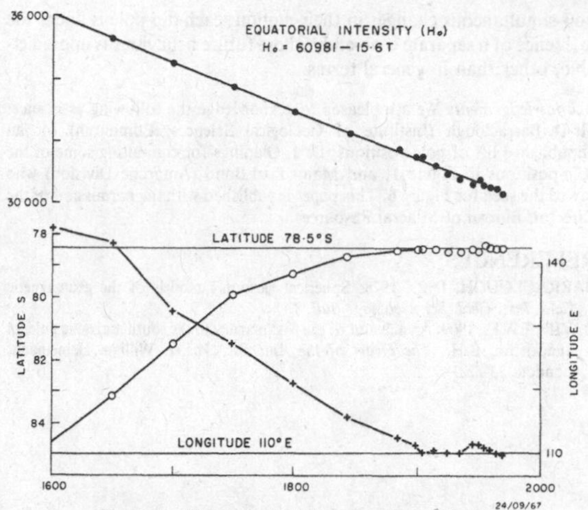


**Figure 5. Geomagnetic field parameters 1600-1980.**

## Discussion

Observational data show that the south magnetic pole has moved more than 1200 km north—northwesterly at a variable rate in the last 140 years. Results for the north magnetic pole are similar (Serson, 1980)—since Ross visited it in 1831 it has moved about 800 km in roughly the same direction and at about the same average rate, although the changes in rate have not occurred in phase at both poles. On the other hand, the global data show that the geomagnetic poles have remained fixed. Clearly more than one internal magnetic field is needed to explain these disparate results.

Figure 6 displays non-dipole fields near the south polar regions obtained from two of the SHA models, one for 1905 (near the time of Mawson's location) and one for 1980; the pole positions are shown for comparison. At each epoch a localised cell is situated near the pole, and the poles have followed the cells; this suggests that the motion of the poles is due to changes in the non-dipolar components of the internal field.

The non-dipolar fields probably arise from a small number of localised electric currents near the surface of the Earth's core, superimposed on a general current system which produces the main field. These may also be invoked to explain other features of the secular variation including the non-antipodality of the dip poles and

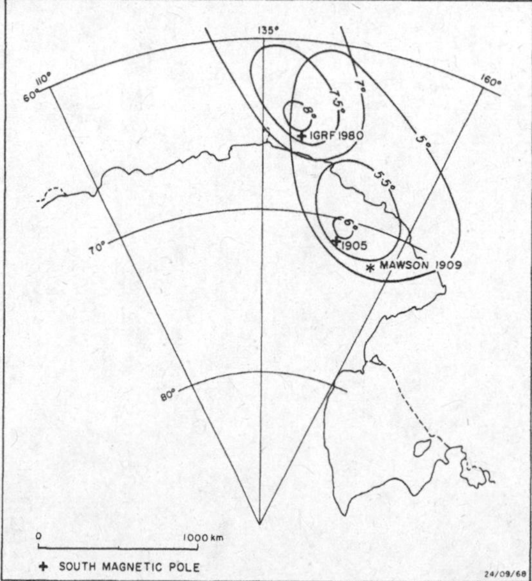


**Figure 6. Non-dipole part of the dip near the magnetic poles in 1905 and 1980.**

non-simultaneous changes in their motion; each dip pole is under the influence of a separate core-eddy whose future behaviour is unpredictable, other than in general terms.

*Acknowledgements* We are pleased to acknowledge the following assistance: Mr D. Barraclough (Institute of Geological Sciences, Edinburgh) for an unpublished list of pole-positions; Dr J. Giddings for computing some of the pole-positions in Figure 1, and Major Fred Bond (Antarctic Division) who sowed the seed for Figure 6. This paper is published with the permission of the Director, Bureau of Mineral Resources.

## REFERENCES

BARRACLOUGH, D.R., 1978: Spherical harmonic models of the geomagnetic field. *Inst. Geol. Sci. Geomag. Bull.* 8.

DAVID, T.W.E., 1909: An account of the first journey to the south magnetic pole; *in* Shackleton, E.H., *The Heart of the Antarctic, Vol II.* William Heinemann, London, 73-222.

HANSTEEN, C., 1819: *Untersuchungen uber den magnetismus der Erde.* J. Lehmann, and Chr. Groudhl, Christiania.

LANGEL, R.A. and ESTES, R.H., 1982: A geomagnetic field spectrum. *Geophys. Res. Lett. 9*,250-3.

MAYAUD, P.N., 1953: Le pole magnetique sud en 1952 et les deplacements compares des poles nord et sud de 1842 a 1952 *Annls. Geophys. 9*,266-76.

ROSS, J.C., 1847: *A Voyage of Discovery and Research in the Southern and Antarctic Regions During the Years 1839-1843.* John Murray, London.

SERSON, P.H., 1980: Tracking the north magnetic pole *GEOS (Winter)* 15-17.

VESTINE, E.H., 1967: Main geomagnetic field; *in* Matsushita, S and Campbell, W.H., (eds.) *Physics of Geomagnetic Phenomena.* Academic Press, New York, 181-234.

WALLIS, D.D., BURROWS, J.R., HUGHES, J.T. and WILSON, M.D., 1982: Eccentric dipole coordinates for MAGSAT data presentation and analysis of external current effects. *Geophys. Res. Lett.*, *9*,353-6.

WEBB, E.N., 1925: Field survey and reduction of magnetograph curves; *in* Terrestrial Magnetism. *Australasian Antarctic Expedition 1911-1914, Scientific Reports,* Ser. B1.

# THE AMERICAN-ANTARCTIC RIDGE/BOUVET TRIPLE JUNCTION

L.A. Lawver, *Massachusetts Institute of Technology, 54-812, Cambridge, Mass. 02139, USA.*

H.J.B. Dick, *Woods Hole Oceanographic Institution, Woods Hole, Mass. 02543, USA.*

*Abstract* A recent survey of the ridge was made between the Conrad fracture zone at 56°S, 4°W and a spreading centre at 60°S, 18.5°W. Five transform faults south of the Conrad fracture zone and six spreading axes were located. Prior to this cruise the plate boundary between these two points was generalised on the basis of seismicity as consisting of two or three spreading centres offset by two or three transform faults. The Bullard fracture zone at 58°S was shown to be a strikingly linear feature approximately 560 kilometres long with a maximum transform fault depth of nearly 6400 m. On the basis of new bathymetric and magnetic data we have revised the pole of rotation for the Antarctic and the South American plates.

With a better pole of rotation for Antarctica-South America we were then able to model the three plate interaction at the Bouvet triple junction. With the instantaneous plate velocities for the three plates at the triple junction and the poles of rotation we were able to conclude that the Bouvet triple junction migrates slowly southeastward but because it is stable over a long period it episodically jumps northward creating new transform faults and spreading centres on the American-Antarctica Ridge and on the Southwest Indian Ridge.

A careful compilation of the available bathymetric data of the two ridges south of the Bouvet triple junction should enable us to calculate the palaeopoles of the three plates by comparing calculated versus measured transform fault orientations. If we can determine when the American-Antarctic ridge assumed its present orientation we may be able to reconstruct the opening of Drake Passage.

# PROJECT INVESTIGATOR-I: EVOLUTION OF THE AUSTRALIA-ANTARCTIC DISCORDANCE DEDUCED FROM A DETAILED AEROMAGNETIC STUDY

P.R. Vogt and N.Z. Cherkis, *Code 5110 Naval Research Laboratory, Washington D.C. 20375, U.S.A.*

G.A. Morgan, *Air Systems Engineering Group, Defence Research Centre, P.O. Box 2151, GPO Adelaide, South Australia 5001.*

*Abstract* A detailed aeromagnetic study over the Australia-Antarctic Discordance (AAD) and adjacent portions of the Southeast Indian Ridge (SEIR) exposes the kinematic details of plate boundary evolution over the last 25 Ma. The present crenulated geometry of the AAD developed from a combination of continuous asymmetric spreading along most of the boundary (north flank faster by 5 mm/year), and propagating rifts which abruptly changed the offsets of the transforms bounding the AAD. Rift propagation may have been initiated by a change in rotation pole 7 to 4 Ma, and also by asthenosphere flow towards the AAD along adjacent spreading centres. Continuous asymmetric spreading in the AAD area has been episodic, with maximum asymmetry (ca. 10-15 mm/year) at 2-3 Ma intervals.

The "Australia-Antarctic Discordance" (AAD) is a zone of anomalously rough, complex and deep topography along the Southeast Indian Ridge (SEIR) which bounds the Indo-Australian and Antarctic plates (Figure 1A; Weissel and Hayes, 1974; Hayes, 1976; Anderson et al., 1980). Asymmetries in spreading rate, magnetic anomaly amplitude, basement roughness and depth anomaly also locally characterise the AAD (zone "B") and adjacent zones A (west) and C (east). The unusual features of the AAD area may attest to a mantle cold spot or sink (Weissel and Hayes, 1974) towards which dissimilar asthenosphere types are flowing along the plate boundary (Vogt and Johnson, 1973). The asymmetries may reflect relative motions between the plate boundary and subjacent temperature anomalies (Hayes, 1976). Veevers (1982) suggests that mantle processes responsible for the AAD have existed in the area of the Antarctica-Australia break-up as early as the Late Carboniferous long before the actual break-up (100 Ma, Cande and Mutter, 1982). Logistics and high sea states have limited shipboard investigations of the AAD, a situation which could in part be remedied by the 1978 detailed aeromagnetic survey described herein. Space limits us to a summary of magnetic anomaly patterns, crustal isochrons, and implications for the evolution of the geometry of the AAD plate boundary by asymmetric spreading and propagating rifts. Other topics (plate kinematics, total opening rate, depth anomalies, and magnetic anomaly amplitudes) will be discussed in a more complete treatment (in preparation).

## Survey Techniques and Initial Processing

Survey design, instrumentation, navigation and initial processing of the data set were discussed by Morgan et al. (1979). Figure 1B shows the residual magnetic anomalies, plotted along the flight tracks. We first identified the magnetic lineations, both peaks and troughs, in terms of the geomagnetic reversal time scale (La Brecque et al., 1977). The 4500 anomaly picks (Figure 2) give a high-resolution picture of crustal isochrons and their disruption by fracture zones (FZ) and propagating rifts. Reliable anomaly identifications could be made over about 50% of the AAD; another 25% is questionable and the rest conjecture. The flights extended to crust locally as old as anomaly 7(25.5 Ma) in the north and anomaly 6(19.5 Ma) in the south. The interpreted lineation patterns on the two flanks are similar but there are some mismatches. We cannot yet decide between anomaly misidentification on one or both flanks, vs. subsequent intraplate deformation as explanations for the lack of accurate match.

## CRUSTAL ISOCHRON MAP

Using computed model profiles and Figure 2, we next constructed a crustal isochron chart at 0.5 Ma contour interval (Figure 3). The major fracture zones and several propagating rifts are well constrained; the unnumbered fractures, particularly within the AAD, are speculative. Note the change in transform trend from 000°-010°T to 012°-017°T at about 4 to 7 Ma. This previously unsuspected change in plate rotation pole, describing motion between the Indian and Antarctic plates, correlates with the highest total opening rate (90-100 mm/year). The lineation trends show that the spreading axis adjusted itself to the new transform trend by turning clockwise by several degrees. This is particularly clear between FZ 2 and 3 (Figures 2, 3).

Our identification of the present plate boundary, straightforward except in some parts of the AAD, is generally consistent with the earthquake epicentre pattern as well as with the locations of dredges recovering basalt glass along the plate boundary postulated from pre-Investigator data by Anderson et al. (1980). However, Figure 3 explains why Anderson et al. failed to recover fresh basalt at site no. 8, which we find lies on 1.2 Ma crust, well south of the actual axis in sector B1.

To first order the present plate boundary consists of three northern segments we call B1, B3, and B5, separated by two southern segments, B2 and B4. This crenulated shape of the plate boundary is unusual for the Mid-Oceanic Ridge (MOR), where the axis is generally offset progressively in a staircase patern. As recognised by Weissel and Hayes (1972), the present AAD plate boundary does not replicate the line of break-up between Antarctica and Australia. The crenulated form appears to have evolved mainly in the Late Tertiary, becoming progressively more exaggerated over time. At 10 Ma the axis already had its present geometry (Figure 3), although the amplitude of the crenulations was only about half that of the present boundary. At 20 Ma the axis had only a gentle hint of its present configuration; FZ 4 did not exist, while FZ's 3, 5 and 6 had offsets only a third their present value.

Exactly how did the plate boundary evolve to its present shape? There are three possible processes: (1) small but resolvable northward jumps of spreading axes B1, B3 and B4, or southward jumps of the remainder of the plate boundary; (2) propagating rifts, which progressively transfer crust from one flank to the other and cause sudden changes in offset of any major FZ which terminates rift propagation (Hey and Wilson, 1982); or (3) asymmetric spreading, continuous down to the 10 km resolution of sea-surface magnetic anomaly data. We failed to find conclusive evidence in our data for isolated jumps of entire ridge segments although such jumps may have occurred in parts of the AAD where the anomaly pattern is confused. This leaves continuous asymmetric spreading and propagating rifts, and we elaborate on these two processes below.

## CONTINUOUS ASYMMETRIC SPREADING

When more lithosphere is accreted per unit time to one of two diverging plates, the spreading is said to be asymmetric. Weissel and Hayes (1972) showed that significant spreading asymmetries have characterised portions of the SEIR south of Australia for long periods of time. Net asymmetric spreading can be accomplished by jumps of finite segments of spreading or by continuously propagating rifts. Such discontinuous processes, discussed in the following section, can be recognised on magnetic profiles by omission or replication or linear anomalies. However, in some areas, such as south of Australia (Hayes, 1976), asymmetric spreading appears continuous at least down to the 10 km resolution limit imposed by the magnetic anomalies.

Most examples of asymmetric spreading along the MOR support a fluid-dynamic hypothesis (Stein et al., 1977) which predicts that the faster-spreading flank belongs to whichever plate is more nearly stationary over the mantle. However, south of Australia the rapidly moving northern flank (Minster and Jordan, 1978) is spreading faster.

We examined our detailed isochron map (Figure 3) for asymmetric spreading by contouring local asymmetry (north flank rate minus south flank rate) as a function of age and distance along the ridge (Figure 4) and by plotting offsets across six major transform faults as

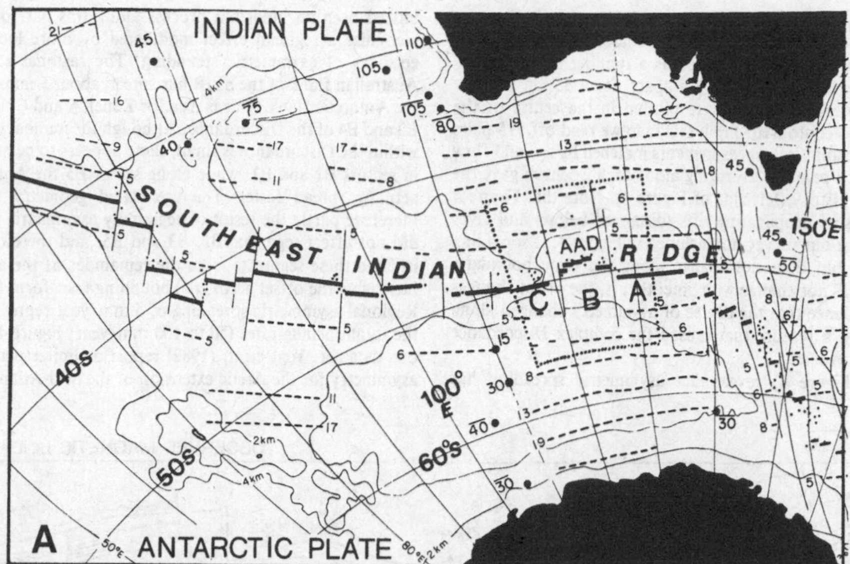


**Figure 1A.** Location of Investigator-I aeromagnetic survey (dotted box) in area of Australia—Antarctica Discordance (AAD). Zones A, B and C from Weissel and Hayes (1972). Base chart, showing DSDP basement ages (Ma), generalised magnetic lineations (dashed) and earthquakes after Veevers and McElhinney(1976).

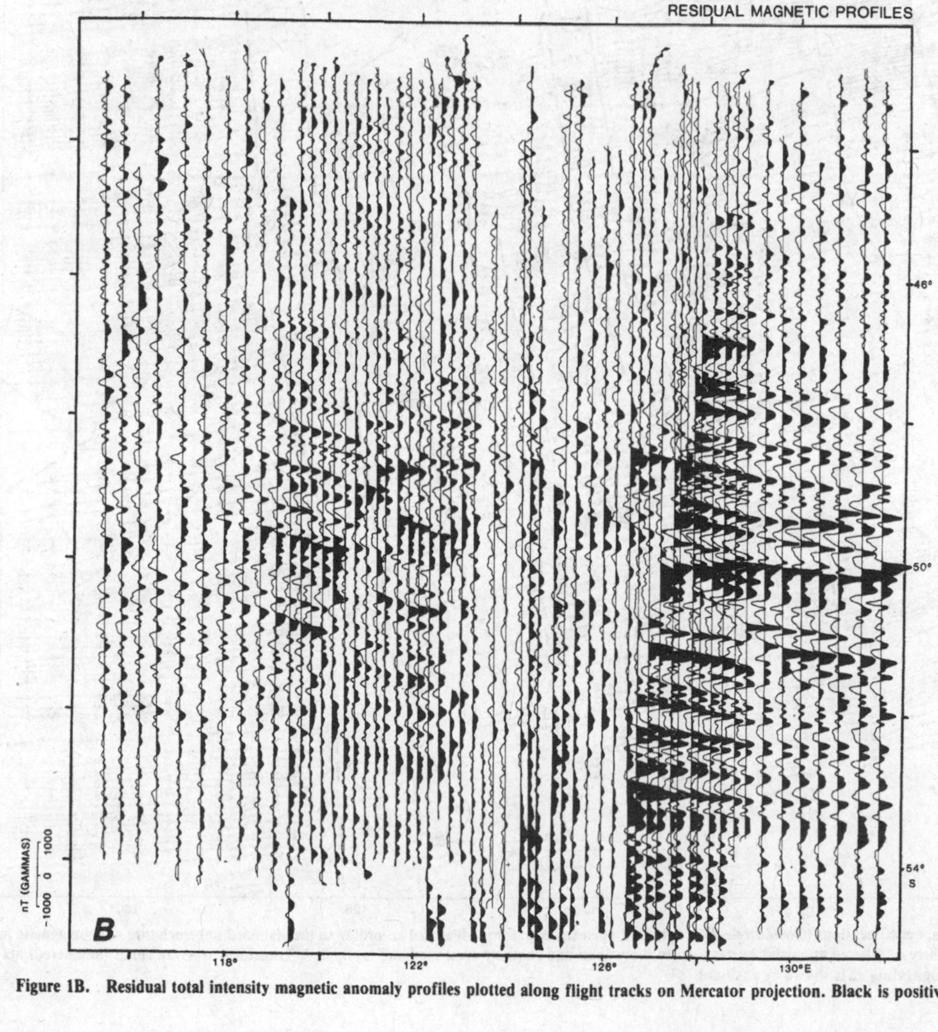

RESIDUAL MAGNETIC PROFILES

nT (GAMMAS) −1000 0 1000

118° 122° 126° 130°E

46° 50° 54° S

B

**Figure 1B.** Residual total intensity magnetic anomaly profiles plotted along flight tracks on Mercator projection. Black is positive.

a function of age at 0.5 Ma intervals (Figure 5). To prepare Figure 4 we first "differentiated" Figure 3 to make a smooth contour map of spreading rate. The latter was measured as a running average every 0.5 Ma, along flow lines roughly 50 km apart, for a 55 km window (about 1 to 2 Ma). These values were plotted in the centre of the 55 km windows and contoured. From this map we read off 515 pairs of high quality spreading-rate measurements matched by age and flow lines. Summing each pair and averaging along each isochron gave the total opening rate history (left side of Figure 4). Note that Figure 5 includes the effects of propagating rifts (discussed below) and gives results from parts of our survey not included in Figure 4. If spreading asymmetry occurs but does not change along the plate boundary, transform offset will not change with time, that is, the offset changes in Figure 5 are evidence that the degree of asymmetry changed along the plate boundary, a feature that caused the complex Discordance plate boundary to evolve.

The contours (Figure 4) reveal that asymmetric spreading has indeed been an inportant process along this part of the SEIR. We recognise a regional effect modulated by more local, short-period episodes of asymmetric spreading. The regional effect is that the Australian flank of the SEIR has spread about 5 mm/year faster than the Antarctic flank. This is true for Zones A and C, as well as sectors B2 and B4 of the Discordance. Although the number of measurements within the Discordance is small, there appears to be no net asymmetry in sectors B1 and B3, while along sector B5 the Antarctic flank has actually spread faster. The crenulated geometry of the AAD is therefore partly the result of regionally asymmetric spreading which did not affect segments B1, B3 and B5, and therefore progressively isolated these segments from the remainder of the accretion axis by increasing the offset across the bounding transform faults (Figure 5). Regional asymmetric spreading of 5 mm/year represents 5 to 10% of the total opening rates (50 to 100 mm/year; Figure 4) obtained from our data set. Vogt et al. (1982) report a similar percent of regional asymmetry for the Arctic extension of the Mid-Atlantic Ridge, where

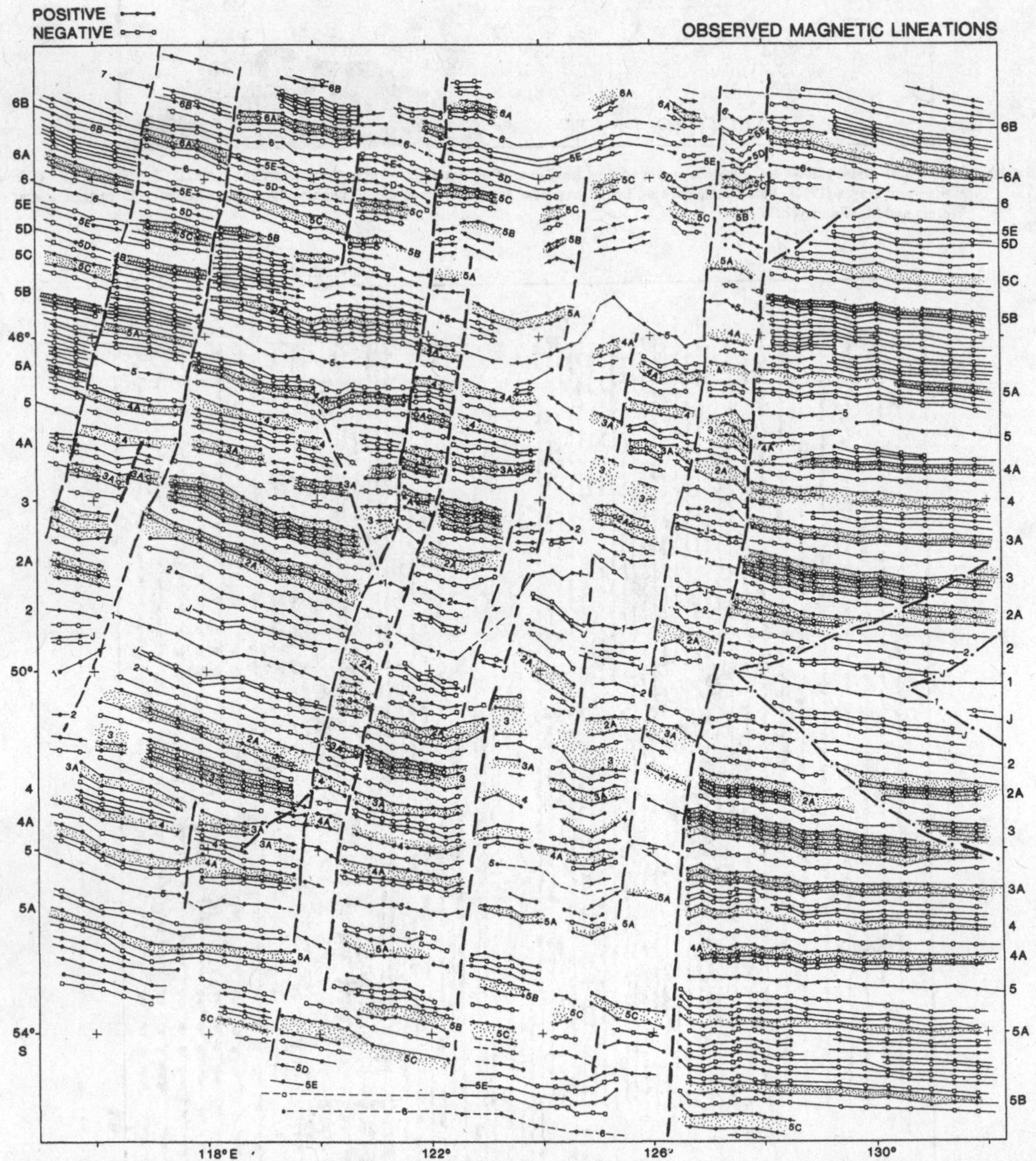


**Figure 2.** **Magnetic lineations (closed circles for highs and open squares, lows) identified according to the standard nomenclature of geomagnetic reversals. Where numbered anomalies correspond to more than one anomaly crest the band is stippled. Transform fracture zones (numbered) are dotted; propagating rifts shown by dash-dot.**

the total opening rate has been much lower (10 to 40 mm/year). However, with respect to absolute plate motions the sense of asymmetry is reversed—in the Arctic the faster-spreading flank is on the stationary plate.

Superposed on the regional spreading asymmetry are episodes with amplitudes ± 5 to 15 mm/year and periods of 2 to 3 Ma (Figure 4). Some episodes simultaneously affected long segments of plate boundary (Zones A and C), but there is little or no correlation between asymmetry and total opening rate. If the data of Figure 4 are simply averaged by time interval, without regard to zone or sector, asymmetry maxima of about + 10 mm/year are found at 12, 8 and 5 Ma, while minima of 0 to + 5 mm/year are located at 14, 10, 6 and 0 Ma. Work is in progress to relate these asymmetry episodes to other measurable parameters, such as depth anomaly and topographic structure.

Two examples of local asymmetry anomalies are the strong high at 220 km and 6 Ma (Figure 4)and the low at 0-2 Ma between Propagator II and the axis (Figures 3, 4). (Propagating rifts are discussed in the following section). The first case seems to reflect a clockwise reorientation of the spreading axis in reponse to a new rotation pole (new transform trend) which became established about 7 to 4 Ma; the asymmetric spreading occurred just "behind" (west) of Propagator I. In the second example the spreading axis "ahead" (west of Propagator II appears to have shifted northward, so as to reduce the offset between the old and the new ridge axis

## PROPAGATING RIFTS

The spoors of four propagating rifts (or "pseudofaults") are evident as diagonal disruptions in the isochron pattern (Figure 3). (Propagating rifts are defined by Hey and Wilson, 1982). Propagator I originated within zone C about 8 Ma, propagated eastward at 27 m/year and terminated at FZ 3 around 4 Ma, increasing the offset of this transform by 60 km (Figure 5). Propagators II and III originated west of the surveyed area sometime before 5 and 2 Ma, respectively, These two rifts have been propagating west at 70 to 90 mm/year.Both involve left-lateral offsets of the plate boundary and, in the future, will reduce the offset of the major right-lateral transform (FZ 6) that forms the boundary between zones A and B. Propagator IV formed about 20 Ma in zone A, advanced west at 32 mm/year and reached FZ 6 about 14 Ma, increasing its offset by 45 km. Our tracks did not extend far enough south to map the trace of this propagator on the Antarctic flank of the ridge. Our data density is insufficient to

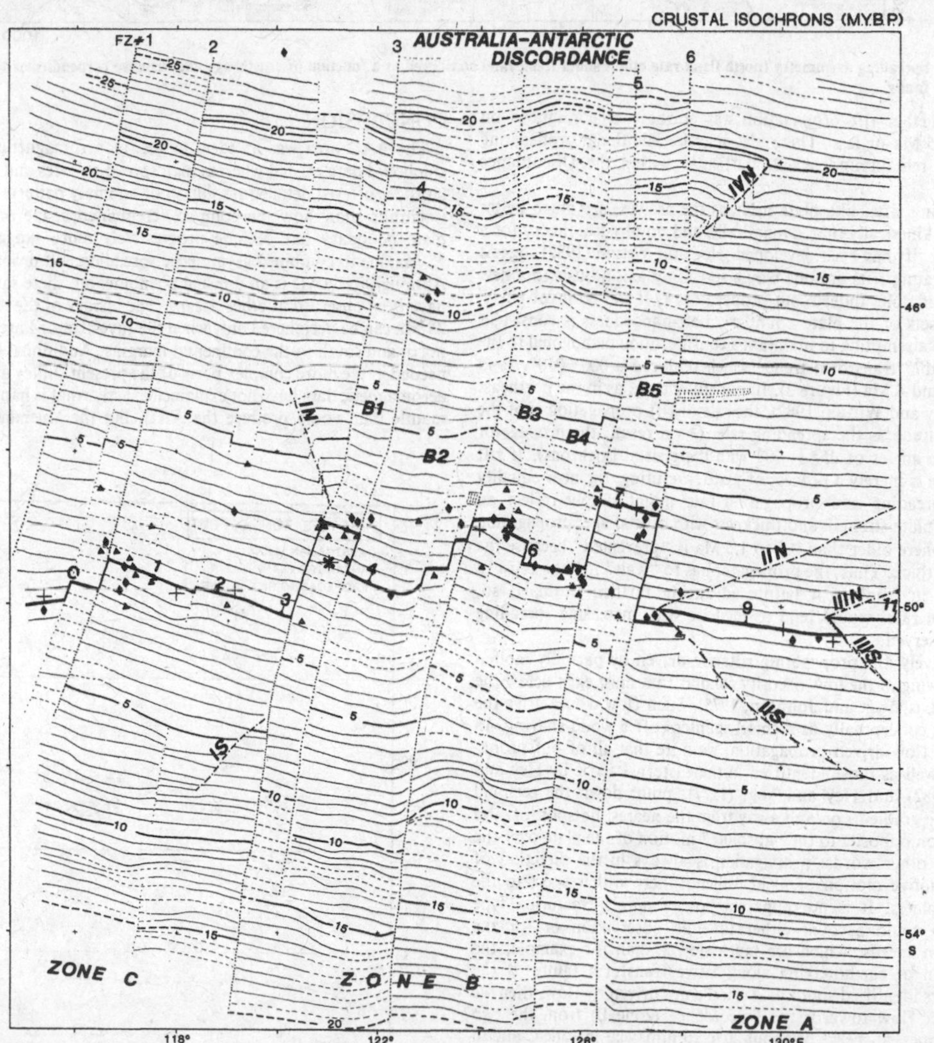


**Figure 3.** **Contours of crustal age at 0.5 Ma interval, interpreted from Figure 2 and the LaBrecque et al. (1977) reversal time scale. Heavier contours (labelled) show multiples of 5 Ma crustal age. FZ and propagating rift designations as in Figure 3. Zone B (Discordance) is subdivided into sectors B1 to B5; propagating rifts identified by Roman numerals. Uncertain contours are dashed. Thick lines show present active plate boundary. Crosses show where basalt glass was dredged by Anderson et al. (1980); no glass was found at site 8 (asterisk). Distances eastward perpendicular to fracture zones, measured from small FZ west of FZ 2, indicated in units of 100 km. Earthquake epicentres shown as open triangles (pre-1961 events of magnitides < 4) or closed diamonds (magnitudes4).**

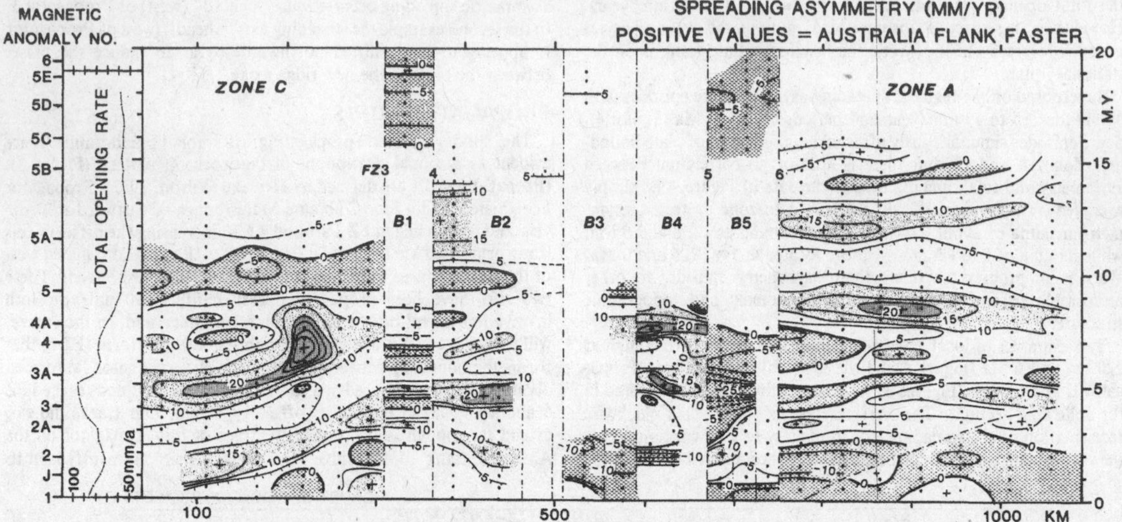


**Figure 4. Spreading asymmetry (north flank rate minus south flank rate) contoured as a function of crustal age and distance perpendicular to transform faults.**

establish whether rift propagation was episodic or continuous on scales of 0.5 Ma or less. There also may be additional propagating rifts which remain undetected, particular within the Discordance itself.

Propagating rifts, like plate motions themselves, are much better understood kinetically than dynamically, and no physical theory for a propagating rift has been developed. Hey and Wilson (1982) suggest that propagating rifts are one way a spreading axis adjusts to a new plate rotation pole. Indeed, our propagators I, II and III involve left-handed offsets of the plate boundary. Left-handedness is consistent with a ridge attempting to maintain a configuration orthogonal to the new 012°-017° transform trend, which replaced a 000°-010° trend between 7 and 4 Ma (Figure 3). In our study area, as in the northeast Pacific (Hey and Wilson, 1982), the rate of rift propagation is of the same magnitude as the spreading rate. (As a result, the propagator traces make angles of 45° ± 20° with the crustal isochrons). If rift propagation is merely a process of plate fracturing, the near equality between spreading and propagation rate might simply reflect an increase of plate strength and thickness with crustal age: propagation into lithosphere older than 0.5 to 1.5 Ma is "prevented" because the plate is too thick. Thus, the propagator has to "wait" for new crust to be formed in front of it before advancing further. If this is so, propagation rates would tend to increase with increasing spreading rates as observed.

Alternatively, the propagating rifts are driven by partially molten mantle flowing in the low-viscosity conduit that must exist below the spreading axis (Vogt and Johnson, 1975). Such flow would drive the propagator conceptually as a chisel splinters off a wood shaving. In support of flow-driven propagation, we note that all of our propagators, as well as those identified in the eastern Pacific by Hey and Wilson (1982) and Hey and Vogt (1977), point down the regional topographic gradient, or/and away from the nearest hotspot, or/and in a direction opposite to the "absolute" motion of the plate over the mentle. In other words, propagation is always in the direction of inferred relative motion of axial asthenosphere with respect to the overlying plates. It is not apparent how a purely fracturing type mechanism could account either for this observation or for the propagation speeds, which are much slower than for example, the propagation of earthquaking along active transform faults. If the propagators identified south of Australia are driven by flow, then we would infer flow towards the Discordance (zone B) from the two adjacent zones, at speeds of about 20 to 20 mm/year in zone C and in zone A before 10 Ma, and 70 to 90 mm/year in zone A after 10 Ma. Such flow speeds are comparable in magnitude to what might be expected from the topographic gradients and other considerations (e.g. Vogt and Johnson, 1973, 1975). The faster propagationm of II and III could reflect the relatively higher melt temperatures (Anderson et al., 1980) inferred from basalt chemistry.

## CONCLUSION

The AAD area was an ideal target for aeromagnetic work: little shipboard data existed whereas high spreading rates and geomagnetic latitudes offered detailed resolution of anomaly patterns. The crustal isochrons show how the complex, crenulated shape of the present plate boundary has become progressively more exaggerated since 25 Ma due to combined asymmetric spreading and propagating rifts. Although the AAD is an unusual phenomenon, these same processes have been observed in other oceans. The history of the AAD prior to 25 Ma can be deciphered only when the Investigator-I area is extended north and south to the continental margins. Additional lines are also needed in the more complex parts of the present survey area. Detailed aeromagnetic data have not explained exactly what is happening in the mantle that would produce the AAD, but the kinematic and other

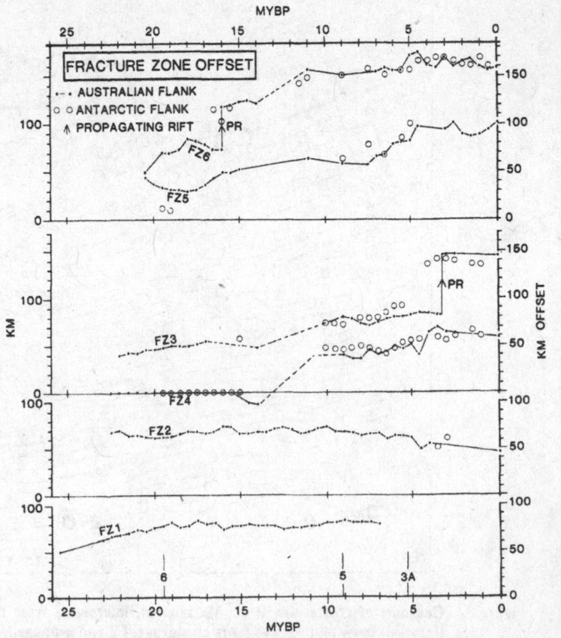


**Figure 5. Offset across transform faults (FZ) No's 1 to 6, vs. crustal age; constructed from Figure 3. Dots show measurements on north flank; open circles denote south flank. Abrupt increases in offset were caused when propagating rifts (PR) intercepted FZ.**

constraints described here (and more completely elsewhere) sharply constrain possible physical models.

*Acknowledgements* The authors wish to acknowledge the tireless efforts of the officers and crews of the Royal Australian Air Force, Squadron 10, the NRL Flight Detachment, and the ground support groups at Edinburgh and Pearce RAAF bases. The concept of a joint U.S./Australian experiment was the brain child of Code 5110 Branch Head, H.S. Fleming, and one of us (G.M.). Special thanks to Wing Commander R. Tayles and Flight Lieutenant P. Jabornicky (both RAAF), and Captain R. Featherstone, USN. R. Feden, L. Kovacs, J. Ostrander, J. Eskinzes, W. Cherry. N. Giuliano, D. Clammons and J. Bright (all NRL) contributed a great deal of time to the project in the field, and I. Jewett, C. Rockelli, J. Watkins and T. Williamson assisted with manuscript and illustrations. This research was supported in part by the Office of Naval Research.

## REFERENCES

ANDERSON, R.N., SPARIOSU, D.J., WEISSEL, J.K. and HAYES, D.E., 1980: The interrelation between variations in magnetic anomaly and amplitudes and basalt magnetisation and chemistry along the Southeast Indian Ridge. *J. Geophys. Res., 85,* 3883-98.

CANDE, S.C. and MUTTER, J.C., 1982: A revised identification of the oldest seaflor spreading anomalies between Australia and Antarctica. *Earth Planet. Sci. Lett., 58,* 151-60.

HAYES, D.E., 1976: Nature and implications of asymmetric seafloor spreading—"Different rates for different plates". *Geol. Soc. Am., Bull., 87,* 994-1002.

HEY, R. and VOGT, P., 1977: Spreading centre jumps and subaxial asthenosphere flow near the Galapagos hotspot. *Tectonophysics, 37,* 41-52.

HEY, R.N. and WILSON, D.S., 1982: Propagating rift explanation for the tectonic evolution of the northeast Pacific—a pseudomovie. *Earth Planet Sci. Lett., 58,* 167-88.

LABRECQUE, J.J., KENT, D.V. and THUNNELL, R.C., 1977: Revised magnetic polarity timescale for Late Cretaceous and Cenozoic time. *Geology, 5,* 330-35.

MINSTER, J.B. and JORDAN, T.H., 1978: Present-day plate motions. *J. Geophys. Res., 83.* 5331-51.

MORGAN, G.A., FLEMING, H.S. and FEDEN, R.H., 1979: Project Investigator-I: A cooperative U.S./Australian airborne aeromagnetic study south of Australia; *in*Proc. 13th. Internat. Symp. Remote Sensing of the Environment, Environ. Res. Inst. of Michigan, Ann Arbor, 1439-44.

STEIN, S., MELOSH, H.J. and MINSTER, J.B., 1977: Ridge migration and asymmetric seafloor spreading. *Earth Planet Sci. Lett., 36,* 51-62.

VEEVERS, J.J., 1982: Australian-Antarctic depression from the mid-ocean ridge to adjacent continents. *Nature, 295,* 315-7.

VEEVERS, J.J. and McELHINNEY, M.W., 1976: The separation of Australia from the other continents. *Earth-Sci. Rev., 12,* 139-59.

VOGT, P.R. and JOHNSON, G.L., 1973: A longitudinal seismic reflection profile of the Reykjanes Ridge, 2, Implications for the mantle hotspot hypothesis. *Earth Planet. Sci., Lett., 18,* 49-58.

VOGT, P.R. and JOHNSON, G.L., 1975: Transform faults and longitudinal flow below the Mid-Oceanic Ridge. *J. Geophys. Res., 80,* 1399-428.

VOGT, P.R., KOVACS, L.C., BERNERO, C. and SRIVASTAVA, S.P., 1982: Asymmetric geophysical signatures in the Greenland—Norwegian and southern Labrador Seas and the Eurasia Basin. *Tectonophysics, 89,* 95-150.

WEISSEL, J.K. and HAYES, D.E., 1972: Magnetic anomalies in the southeast Indian Ocean. *Am. Geophys. Union., Ant. Res. Ser., 19,* 165-96.

WEISSEL, J.K. and HAYES, D.E., 1974: The Australian-Antarctic Discordance: New results and implications. *J. Geophys. Res., 179,* 2579-87.

# PROJECT INVESTIGATOR-I: BATHYMETRIC ANALYSIS OF THE AUSTRALIA-ANTARCTIC DISCORDANCE

N.Z. Cherkis and P.R. Vogt, *Code 5110, Naval Research Laboratory, Washington DC 20375, USA.*

G. A. Morgan, *Air Systems Engineering Group, Defence Research Centre, Box 2150, GPO Adelaide, South Australia, 5001.*

**_Abstract_ The available bathymetric data from the area covered by detailed aeromagnetics (114° to 131°E, 42° to 55°S) has been recontoured using magnetic and SEASAT altimetric (geoid height) anomalies as a guide. The Australia-Antarctic Discordance is very sparsely sounded, with only about 162000 line-km of bathymetric data within the study area. Most of the lines are oriented north-south and are thus better suited for studies of spreading and subsidence history and rift-type topography than for delineation of fracture zone topography.**

**Typically 50 to 150 km separate adjacent sounding lines, and unsurveyed polygons exceeding 60000 km² in area exist between the tracks in this inaccessible and stormy region.**

**In analysing the bathymetry we will present Neogene age-depth curves for both the Australian and Antarctic flanks of the Southeast Indian Ridge (SEIR), and for all three zones: the 'typical' SEIR west of the Discordance, the Discordance, and the magnetic high-amplitude zone east of the Discordance. Topographic asymmetry on the two flanks will be related to magnetic amplitude asymmetry and spreading rate asymmetry. The observed asymmetries will be compared to similar phenomena observed along the slow-spreading accreting plate boundary north of Iceland and in the Arctic. Wavelength and amplitude of topographic relief will be related where possible to spreading rate, province, and magnetic anomaly amplitude.**

# 13

## Antarctic Meteorites

# MEASUREMENTS OF THE TRIANGULATION NETWORK AT THE ALLAN HILLS METEORITE ICEFIELD

J.O. Annexstad, *Johnson Space Centre, NASA, Houston, Texas 77058, U.S.A.*

L. Schultz, *Max-Planck-Institut für Chemie, Mainz, Germany.*

*Abstract* In 1978 a triangulation network marked by stakes set into drilled holes was established across the Allan Hills icefield where large concentrations of meteorites have accumulated. The interior angles of the network's 18 constituent triangles were remeasured in 1979 and 1981 and ablation measurements conducted annually from 1979. The results show, within the limits of error, that the Allan Hills icefield moves in an easterly direction. The rate of ablation averages less than 5 cm per year, is not constant from year to year and varies widely from station to station. Although the icefield is limited in extent and exhibits a low flow rate, the discovery of a large number of meteorite fragments found there shows that it has been an efficient collector of specimens.

The first meteorite found in Antarctica was discovered in 1912 by the western sledging party of the Mawson 1911-1914 expedition based at Commonwealth Bay (142°E) (Bayly and Stillwell, 1923). Other single specimens were found near the Lazarev Coast in 1961, the Thiel Mountains in 1964 and the Pensacola Mountains in 1964 (Cassidy, 1979). Since 1969, Japanese and American expeditions have discovered large concentrations of specimens at two icefields nearly 3000 km apart. The Japanese discoveries were on a large (4000 km$^2$) blue ice area called the Meteorite Ice Field near the Queen Fabiola (Yamato) Mountains at about 72°S,36°E (Yoshida et al., 1971). The American finds have been primarily from a limited (100 km$^2$) ice field west of the Allan Hills (159°30′E, 76°45′S). These two regions have now yielded nearly 5000 fragments of meteorites, which is a significant increase in the world's supply.

The accumulation of meteorites in these two areas has prompted some workers to propose a mechanism of concentration (Nagata, 1978; Yanai, et al., 1978; Cassidy, et al., 1977). The process is not completely understood but in general the theory predicts that meteorites entrapped in ice flowing from the interior of Antarctica will emerge and remain for long periods of time on the surface of stagnant blue ice fields found near the coast.

In an attempt to provide a quantitative measure of the ice flow, triangulation networks were established on the Meteorite Ice Field (Naruse, et al., 1972) and the Allan Hills Ice Field (Annexstad and Nishio, 1979). This paper discusses briefly the measurements obtained from the Allan Hills Ice Field triangulation network.

## Experiment

In December 1978, a 20 station triangulation network which extends 13 km onto the eastern Antarctic ice sheet was established westward from the Allan Hills. The net is shown in Figure 1; baseline stations 1 and 2 are located on bedrock. The meteorite finds from the 1976, 1977 and 1978 summer seasons are represented by dots on the exposed blue ice area.

Stations 3 to 20 are marked by three metre poles set into 50-80 cm deep holes drilled into the ice or firn surface. The holes were filled with fibreglass insulation to provide for easy pole removal and to reduce melting from solar radiation.

Wild T2 theodolites were used to measure the horizontal interior angles of the 18 constituent triangles and the vertical angles between stations. A horizontal angle was measured four times using face R and face L at the graduated circle readings of 0° and 90° respectively while a vertical angle was measured twice using face R and face L pointings. Observational errors were generally less than 25 seconds as a double angle difference on horizontal pointings and 20 seconds on vertical differences. The initial measurements of the network were conducted in December 1978, and remeasurements completed in December 1979 and 1981.

The slope differences from stations 1 to 2 and 19 to 20 were measured using an optical distance meter (SDM-1C) in 1978 and 1979. In 1981, the distance measurements were conducted at those same locations and along one side of every interior triangle using an infrared distance meter (Distomat Wild DI4L). A baseline from station 1 to station 2 of 1553.41 ± .03 m on a true bearing of approximately 011° has been adopted as the standard for this network.

From 1978 on, the heights of the poles at each station on firn and ice were measured annually to determine the rate of surface accumulation or ablation. Since the blue ice surface is heavily sun cupped the poles were measured on two sides (east and west) normal to the prevailing wind direction and the results averaged.

## Data and Discussion

### *Ablation*

The ablation/accumulation measurements were normalised to the annual rate using $A = A_m \frac{365}{N}$ where N is the number of days between two measurements and $A_m$ is the difference between two measurements.

The data for the three seasons are listed in Table 1. Ablation is represented by negative values and accumulation by positive values.

The data show that the ice surface ablates much faster than the firn. Station 8, which is on firn, and station 13, which is on ice, are similarly exposed to the prevailing katabatic winds, but the average ablation at 8 is less than half that measured at 13. At some stations the firn exhibited accumulation instead of ablation during one of the three years.

The rate of ablation varies from station to station and, at most stations, from year to year. The most significant yearly difference was from the 1979 to the 1980 season when the average values dropped to about half of those measured from 1978 to 1979. The initial measurement in the summer of 1979/80 was conducted in early November when surface temperatures are commonly around −30°C. Normally, the initial measurements are taken in early December when temperatures are a little warmer (−20°C). This change from one year to the next could be attributed to a higher ablation rate occurring in the month of December when the ice field was warmer. A comparison of the data by season as listed in Table 1 shows that although the ablation rates vary from year to year the variation is not correlated with the measurement month.

The rate from one year to the next does remain fairly constant at some stations as shown by the data at 6, 13 and 18. Station 6 is on firn and 13 and 18 are located some distance away on an ice surface.

### *Horizontal movement*

Station 10, as seen on Figure 1, is near the main meteorite find area and on the lower limb of a step like feature which has been described as a monocline (Yanai, et al., 1978). The crest of the monocline or step feature is fairly close to stations 12 and 13 with a difference in height of nearly 100 m and a 10% grade between the lower and upper limbs. This sharp change in elevation has been attributed to the existence of a sub-glacial peak in the vicinity of station 13, but attempts to locate the feature by radar depth measurements have been unsuccessful (Kovacs, 1980).

The triangulation network straddles a small ice valley which extends westward from stations 13 to 16 on the northern leg and 12 to 17 on the southern. The northern ridge is smooth and slightly lower in elevation than the southern ridge which has a rough or pinnacled surface with many small crevasses. The general crevasse pattern is parallel to the southern ridge of the ice valley, that is roughly east-west.

Figure 2, a plot of the horizontal movement of stations 10, 11, 12, 13, 14 and 15 from 1978 to 1981 includes an error bar at each station showing the uncertainty in direction of the surface movement. The magnitude of movement over the three year period is listed under each vector along with the calculated error. The positions of the stations or vector tails are in proportion to the location of the surrounding stations. For illustration purposes the magnitude of movement is plotted to a much larger scale than the station locations. Although these stations are located in diverse parts of the icefield, their general direction of horizontal movement is similar and the magnitude of the vectors decreases from station to station as the line progresses from west to east.

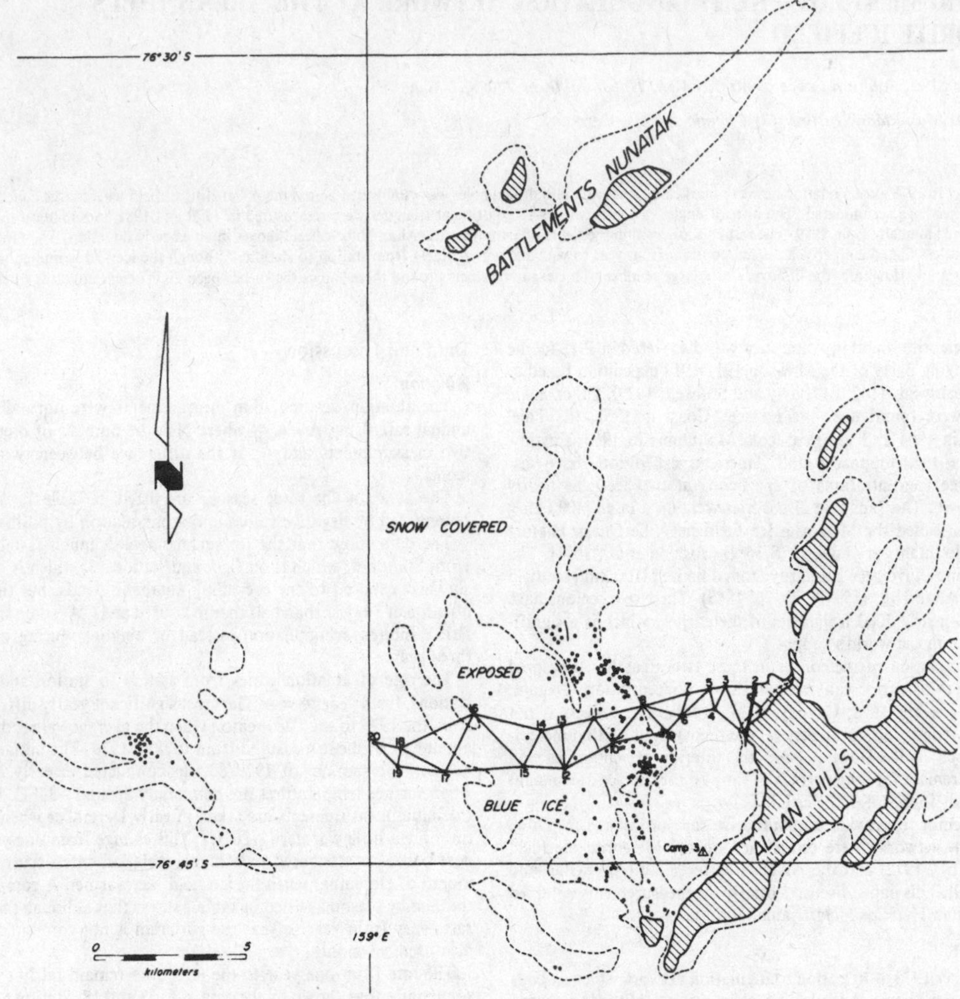


**Figure 1. The triangulation network at the Allan Hills icefield extending westward across the meteorite discovery area.**

Nishio and Annexstad (1980) reported values for the horizontal velocity and azimuth of the triangulation network stations after one year of movement (1978 to 1979) while the data presented in Figure 2 represent the motion and direction of the same stations after three years (1978 to 1981). A comparison of the data from the two surveys shows that the average yearly horizontal velocity measured over three years is about half that reported by Nishio and Annexstad. The measurement errors after one year of movement averaged 20% greater than the measurement errors after three years of movement.

Although the time between surveys was extended, the change in position of some stations relative to the base line is subject to measurement errors that approach 100% of the calculated values. At the stations represented in Figure 2, straight line motion from 1978 to 1981 has been assumed so that a direction within uncertainty limits can be noted from the plots and some idea of the magnitude of the movement of the ice surface can be deduced. At many other stations, the uncertainty limits are too great to discern a clear indication of vector magnitude and direction.

An analysis of the horizontal movement data calculated for each station in the network shows within the limits of error the following results for a three year period. Reference is made to Figure 1 for station locations.

*Stations 3, 4, 5, 6, 7* This area of firn covered ice extends from the Allan Hills outcrop westward toward the East Antarctic ice sheet and is essentially motionless along the surface horizontal plane. The measurement errors are too large, relative to the indicated movement, to permit reasonable confidence in the direction and magnitude of the vectors.

*Stations 8, 9* These stations on firn (8) and ice (9) are located near the border of the transition line between the two surfaces. Minimal movement is detected here, as seen in stations located to the east, with an ill defined direction at 8 but a direction at 9 which appears to be similar to that seen at the more western stations illustrated in Figure 2.

*Stations 11, 13, 14, 16* These stations, beginning with 11 on the lower step of the monocline, are on the northern edge of the triangular network. The horizontal movement direction is slightly to the south of east and the vector magnitude increases gradually to the west as shown in Figure 2.

*Stations 10, 12, 15, 17, 18, 19, 20* The horizontal direction is east at all these points and the magnitude of the flow at each point along this section also increases from east to west. It appears that the maximum flow of the blue ice is about 1 m per year at the far western stations and gradually approaches zero as the Allan Hills are reached.

It has been reported (Nishio, et al., in press) that the ice has probably come from a depth source some distance west of the Allan Hills which indicates that the surface ice is old. The existence of meteorites with very old terrestrial ages ($10^5$ to $10^6$ years, Evans, et al., 1979) seem to corroborate this premise. As more meteorites are dated and more measurements on the ice movement in the area are gained some estimate of the age of the icefield can probably be attempted. For the present, it seems that, in keeping with Drewry's (1980), estimates the ice sheet has probably not thickened appreciably since at least 17,000 years ago and perhaps as long ago as 33,000 years.

**TABLE 1: Ablation (−) and accumulation (+) of the surface at each station of the triangulation network on the Allan Hills ice field. Listed values are normalised to the annual rate.**

| Station | Surface | 1978 to 1979 cm/yr | 1979 to 1980 cm/yr | 1980 to 1981 cm/yr | Average cm/yr |
|---|---|---|---|---|---|
| 3 | firn | −2.2 | +1.7 | −1.9 | −0.8 |
| 4 | firn | −2.3 | −0.8 | +17.4 | N.A. |
| 5 | firn | −1.1 | −2.0 | +1.3 | −0.6 |
| 6 | firn | −1.7 | −2.3 | −2.7 | −2.2 |
| 7 | firn | −2.6 | −2.6 | +2.5 | −0.9 |
| 8 | firn | −3.8 | −1.6 | −1.8 | −2.4 |
| 9 | ice | −4.2 | −1.9 | −3.2 | −3.1 |
| 10 | ice | −6.5 | −3.5 | −5.5 | −5.2 |
| 11 | ice | −5.6 | −2.5 | −5.6 | −4.6 |
| 12 | ice | −4.5 | −2.8 | −4.9 | −4.1 |
| 13 | ice | −5.6 | −6.2 | −5.9 | −5.9 |
| 14 | ice | −7.1 | −2.5 | −4.1 | −4.6 |
| 15 | ice | −5.6 | −2.0 | −4.7 | −4.1 |
| 16 | ice | −4.1 | −2.9 | −1.1 | −2.7 |
| 17 | ice | −5.1 | −3.3 | −2.8 | −3.7 |
| 18 | ice | −4.5 | −4.0 | −3.3 | −3.9 |
| 19 | firn | −3.0 | −1.0 | −1.9 | −2.0 |
| 20 | firn | −1.7 | −1.2 | −1.6 | −1.5 |

Initial measurement were conducted each year during December for 1980 when the measurements was made in November. The final seasonal measurement was conducted each year in January.

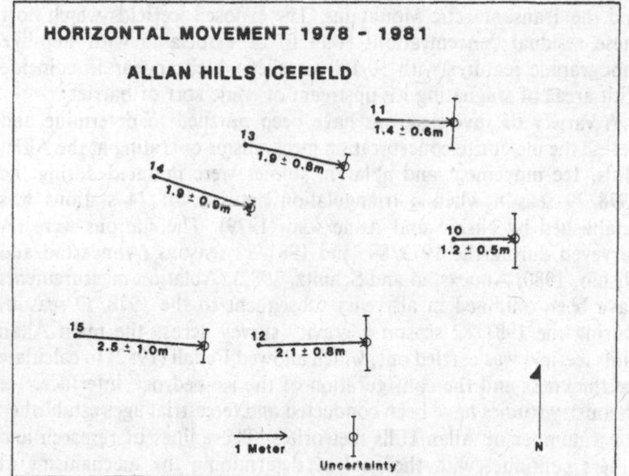


**Figure 2. Vector directions and magnitudes of the surface horizontal movement at stations 10 to 15, Allan Hills icefield. The error bars represent the uncertainty in position of the vector directions.**

## Conclusions

The rate of ablation of the icefield near the Allan Hills ranges from an average of 1.5 cm per year in the firn area to 4.2 cm per year in the ice area. Although the measurements vary from year to year, they average about 5±1 cm annually in the area of high meteorite concentrations. This compares favourably with the values reported by Japanese researchers measuring ablation rates in the Meteorite Ice Field (Naruse, 1979).

There seems to be little correlation between the rate of ablation and the months of the year when measurements are conducted. It has been suggested that blue ice ablation is the result of katabatic winds blowing across the surface and scraping away the ice (Nagata, 1978). From the general similarity of ablation rates at stations in both protected and exposed positions along this network, it is assumed that sublimation also plays an important role in the process.

The stations located close to the Allan Hills show little if any horizontal movement over the span of three years. As the main ice sheet is approached, on a westward track along the network, the movement vectors increase in magnitude. The direction of the vectors is also difficult to define precisely but it appears to be easterly throughout much of the network. We conclude therefore, that the ice flowing from the East Antarctic ice sheet slows down as the Allan Hills are approached and becomes essentially stagnant in the area of the triangulation network.

Until more data are available on mass balance and the micro climate in the area it can be said that the Allan Hills ice field though limited in extent appears to be a rather efficient concentrator of meteorite specimens. In addition, it appears to be essentially stagnant with a yearly flow of 1 m or less and an average rate of ablation around 4 cm per year.

*Acknowledgements* This research was supported by funds from the National Aeronautics and Space Administration, the Deutsche Forschungsgemeinschaft, and the National Science Foundation.

## REFERENCES

ANNEXSTAD, J.O. and NISHIO, F., 1979: Glaciological studies in Allan Hills, 1978-79. *U.S. Antarct. J., 14, 5,* 87-88.

BAYLY, P.G.W. and STILLWELL, F.L., 1923: The Adelie Land meteorite. *Australasian Antarctic Expedition 1911-14. Scientific Reports, Ser. A, 4, Geology.*

CASSIDY, W.A., OLSEN, E. and YANAI, K., 1977: Antarctica: A deep freeze storehouse for meteorites. *Science 198,* 727-31.

CASSIDY, W.A., 1979: Antarctic meteorites. *EOS, 60,* 175-7.

DREWRY, D.J., 1980: Pleistocene bimodal response of Antarctic ice. *Nature, 287,* 214-6.

EVANS, J.C., RANCITELLI, L.A. and REEVES J.H., 1979: $^{26}$Al content of Antarctic meteorites: Implications for terrestrial ages and bombardment history. *Proc. Lunar Planet. Sci. Conf., 10,* 1061-72.

KOVACS, A., 1980: Radio-echo sounding in the Allan Hills, Antarctica, in support of the meteorite field programme. *Cold Regions Res. Eng. Lab. Spec. Rep., 80-23,* 9p.

NAGATA, T., 1978: A possible mechanism of concentration of meteorites within the meteorite ice field in Antarctica; *in* Nagata, T. (ed.) *Proc. Second Symp. on Yamato Meteorites,*70-92. Nat. Inst. Polar Res., Tokyo.

NARUSE, R., 1979: Dynamical features of the Meteorite Ice Field, Antarctica; *in* Nagata, T. (ed.) *Proc. of the Third Symp. on Antarctic Meteorites,*19-24. Nat. Inst. Polar Res., Tokyo.

NARUSE, R., YOSHIMURA, A. and SHIMIZU, H., 1972: Installation of a triangulation chain and a traverse survey line on the ice sheet in the Mizuho plateau—west Enderby Land area, East Antarctica, 1969-70. *JARE Data Report 17 (Glaciology),* 111-3.

NISHIO, F. and ANNEXSTAD, J.O., 1980: Studies on the ice flow in the bare ice area near the Allan Hills in Victoria Land, Antarctica; *in* Nagata, T. (ed.) *Proc. Fifth Symp. on Antarctic Meteorites,* 1-13. Nat. Inst. Polar Res., Tokyo.

NISHIO, F., AZUMA, N., HIGASHI, A. and ANNEXSTAD, J.O., (in press): Structural studies of bare ice near the Allan Hills, Victoria Land, Antarctica: A mechanism of meteorite concentration. *Annals Glaciology, 3.*

YANAI, K., CASSIDY W.A., FUNAKI, M. and GLASS, B.P., 1978: Meteorite recoveries in Antarctica during field season 1977-78. *Proc. Lunar Planet, Sci. Conf., 9,* 977-87.

YOSHIDA, M., ANDO, H., OMOTO, K., NARUSE, R. and AGETA Y., 1971: Discovery of meteorites near Yamato Mountains, East Antarctica. *Antarctic Rec., 39,* 62-5.

# RESULTS OF METEORITE SEARCH AND RECOVERY ACTIVITIES IN THE VICINITY OF THE ALLAN HILLS, ANTARCTICA, DEC. 1981—JAN. 1982

J.W. Schutt[1] and W.A. Cassidy, *Department of Geology and Planetary Sciences, University of Pittsburgh, Pittsburgh, Penn. 15260, U.S.A.*

G. Crozaz, *Department of Earth and Planetary Sciences, Washington University, St. Louis, Missouri, U.S.A.*

R.F. Fudali, *National Museum of Natural History, Smithsonian Institution, Washington, D.C., U.S.A.*

U.B. Marvin, *Smithsonian Astrophysical Observatory, Cambridge, Massachusetts, U.S.A.*

[1]*Present Address: Salisbury and Dietz, Inc., S. 1815 Lewis, Spokane, Washington 99204, U.S.A.*

*Abstract* The Antarctic Search for Meteorites (ANSMET) Expedition spent part of the 1981/82 austral summer working on exposed ice patches in the vicinity of the Allan Hills (76°45′S;159°40′E), southern Victoria Land. This was the sixth season this region had been visited and meteorites recovered. Three separate icefields were searched either systematically or by reconnaissance and 378 tentatively identified meteorite specimens were recovered. In earlier field seasons little emphasis was placed on accurate mapping of the positions of the meteorites; in recent years, increasing attention has been given to this aspect. During the 1981/82 season most localities were determined by using simple surveying methods. Detailed location maps currently being produced will prove useful in examining questions regarding the distribution and relationship of fragments of individual meteorites, meteorite distributions on the icefields, terrestrial age distributions, and how these may relate to the mechanisms which concentrate meteorites on the Antarctic ice cap.

The exposed icefields in the vicinity of the Allan Hills (76°45′S;159°40′E), southern Victoria Land (Figure 1) are some of the most important sites of meteorite concentrations in Antarctica. Other sites are known, such as the Reckling Moraine, Elephant Moraine areas, and the Darwin Glacier region (Cassidy, 1979, 1980; Cassidy and Annexstad, 1981), but only the Yamato Mountains icefields have yielded more meteorite specimens than the Allan Hills icefields. The Allan Hills icefields were visited by the authors, as members of the Antarctic Search for Meteorites (ANSMET) project, for a sixth austral summer in December 1981 and January 1982.

Meteorites were first discovered on the Allan Hills icefields during the 1976/77 field season, the first of a three year joint U.S.-Japanese effort. 45 fragments representing nine individual meteorites were found (Cassidy, 1977). Expeditions to the area during the subsequent four austral summers recovered approximately 660 additional specimens (Score et al., 1981, 1982). The 1981/82 season was the most successful to date—378 tentatively identified meteorite specimens were recovered.

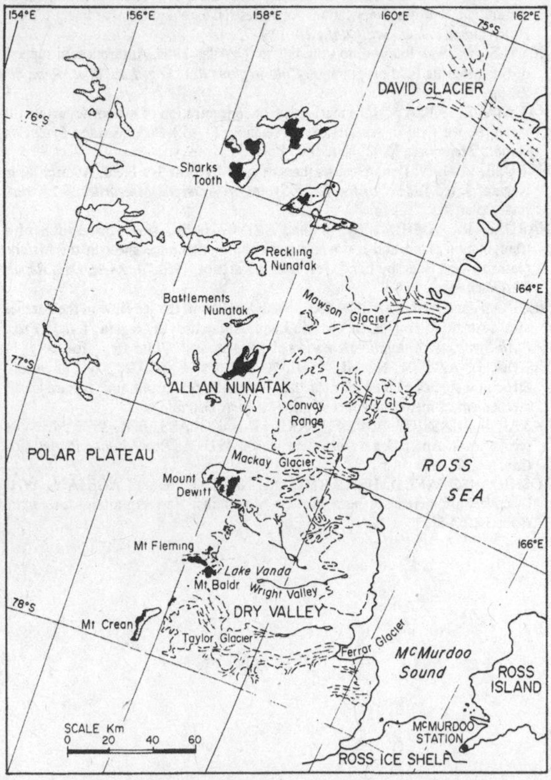

**Figure 1 Location map of the Allan Hills (After Nishio and Annexstad, 1979) area in southern Victoria Land and the bare ice areas on the ice sheet (dotted areas). Allan Nunatak is now officially Allan Hills.**

Meteorite concentration sites in southern Victoria Land occur in ablation zones at the interface between the East Antarctic Ice Sheet and the Transantarctic Mountains. The exposed icefields which host these residual concentrations seem to be associated with step-like topographic features with 50-100 m relief which appear to coincide with areas of stagnating ice upstream of some sort of barrier.

A variety of investigations have been pursued to determine and define the meteorite concentration mechanisms operating at the Allan Hills. Ice movement and ablation studies were initiated during the 1978/79 season when a triangulation network of 24 stations was established by Nishio and Annexstad (1979). The stations were resurveyed during the 1979/80, and 1981/82 seasons (Annexstad and Nishio, 1980; Annexstad and Schultz, 1982). Ablation measurements have been obtained in all years subsequent to the 1978/79 season. During the 1981/82 season a gravity survey across the main Allan Hills icefield was carried out, which allowed Fudali (1982) to calculate ice thickness and the configuration of the ice-bedrock interface. Ice chemistry studies have been conducted and terrestrial ages established for a number of Allan Hills meteorites. These lines of research and others continue, with the goal of determining the mechanisms of meteorite concentration and their relation to Antarctic glaciology. Current concepts regarding the formation and significance of residual concentrations of meteorites in Antarctica have recently been discussed by Cassidy and Rancitelli (1982), Cassidy (this volume), and Annexstad and Schultz (this volume).

Terminology describing Antarctic meteorites and their occurrence is somewhat confusing. This is because a meteorite may either fall to the ice intact, break apart in the atmosphere and fall as a shower, or fragment upon impact or at a later time. For example, many meteorite specimens recovered from the Allan Hills icefields have been paired with specimens found during the same or different years, and it is estimated that the approximately 1100 specimens recovered from this area may represent only about 100-200 meteorite falls. Until final pairing determinations are made, specimens recovered from Antarctica should be considered as fragments of meteorites. However, in reference to meteoritic material from Antarctica, the term "meteorite" is often used to describe any single specimen, and in this broad sense is used here. These may be specimens which have a complete or nearly complete fusion crust or fragments with little or no fusion crust.

## SEARCH, RECOVERY AND SURVEYING METHODS

Effective search methods and strategy have evolved through experience, accompanied by adaptation of equipment. Helicopter reconnaissance searches are made of icefields which are within flying range of McMurdo or a remote camp, such as the northern Victoria Land

Camp. Meteorites as small as 2 cm in diameter can be seen from the helicopter during reconnaissance searches. Ground checks are occasionally made if abundant terrestrial rocks occur on the ice surface. When icefields are out of helicopter range, snowmobiles must be used for reconnaissance. Helicopter reconnaissance resulted in the initial discovery of meteorites on the Allan Hills icefields.

During the first five seasons of work at the Allan Hills icefields the searches were more or less on a random basis. Detailed foot searches of the area were made in the 1977/78 and 1978/79 seasons. Small areas with abundant surficial terrestrial rocks are still searched on foot, but because ice-cleated snowmobiles can cover a great deal of area more rapidly, these are now used almost exclusively for detailed searches in clear areas.

During the 1981/82 season, following reconnaissance snowmobile traverses, the most promising areas were systematically searched in detail on arbitrary grids with search lines 30 to 50 metres apart.

Increasing attention has been given to accurately documenting the locations from which the meteorites were recovered; this was not done prior to the 1981/82 season. One of the major tasks of the past season was to survey the positions of the recovered meteorites so that detailed location maps could be produced. A triangulation network of 24 stations established by Nishio and Annexstad (1979) was used for determining meteorite positions using simple field surveying methods. When the triangulation network was not practically available for reference, map locations were established and plotted, with somewhat lower accuracy, on a U.S. Geological Survey Convoy Range 1:250,000 scale quadrangle by measuring relative angles between three landmarks.

## RESULTS OF THE 1981/82 SEASON

Four separate bare ice areas comprise the Allan Hills icefields (Figure 2). The Main Icefield is located immediately west of the Allan Hills; it comprises about 75 km of blue ice and extends nearly 22 km to the NNW from Peak 2230, near the southernmost tip of the Allan Hills, and is approximately 7.5 km at its widest point. The Main Icefield has received the most attention because of its large concentration of meteorites. The 1981/82 search of this area yielded 286 meteorite specimens, bringing the total number of specimens recovered from there to approximately 1020.

Three western icefields appear to be upstream, in the sense of ice flow, of the Main Icefield. The Allan Hills near Western Icefield is approximately 18 km west of Peak 2330. This icefield trends NW—SE and is about 15 km long and a maximum of 3 km wide, with about 14 km² of ice exposed. The Near Western Icefield was visited briefly by helicopter during the 1977/78 season when 25 meteorites or fragments were recovered and again in the 1978/79 season when five specimens were found (Cassidy, 1978, 1979). This last season, a portion of the Near Western Icefield was searched systematically, yielding 78 specimens probably representing a maximum of 24 individual meteorites. At least 52 fragments recovered from this icefield are similar to fragments recovered in previous years from the same area and are probably from a single individual.

The Allan Hills Middle Western Icefield, located some 31 km WSW of Peak 2330, generally trends parallel to the Near Western Icefield and is nearly 17 km long and 3.5 km wide. Small separate ice patches make up the northwest portion of the approximately 30 km² icefield. One helicopter reconnaissance search of this icefield in 1978/79 yielded three meteorites. Reconnaissance of nearly two-thirds the length of the Middle Western Icefield in 1981/82 resulted in recovery of 14 fragments representing 11 individual meteorites. Several other specimens were found but were not collected due to poor weather and insufficient time.

The Allan Hills Far Western Icefield, a vast area of greater than 100 km² of exposed ice trending NW—SE, is over 40 km long and from 2-8 km wide. Its southeast end is located some 70 km WSW of Peak 2230. Brief helicopter reconnaissance of this area during the 1978/79 season did not reveal any meteorites. Planned reconnaissance traverses of the Far Western Icefield were not carried out because of insufficient time.

Of the 378 meteorite specimens recovered in the 1981/82 season, three are carbonaceous chondrites, nine are achondrites, two are probably stony irons (mesosiderites), two are irons, and the remaining 362 are tentatively identified as ordinary chondrites (Table 1). There will probably be some surprises when these "chondrites" are subjected to

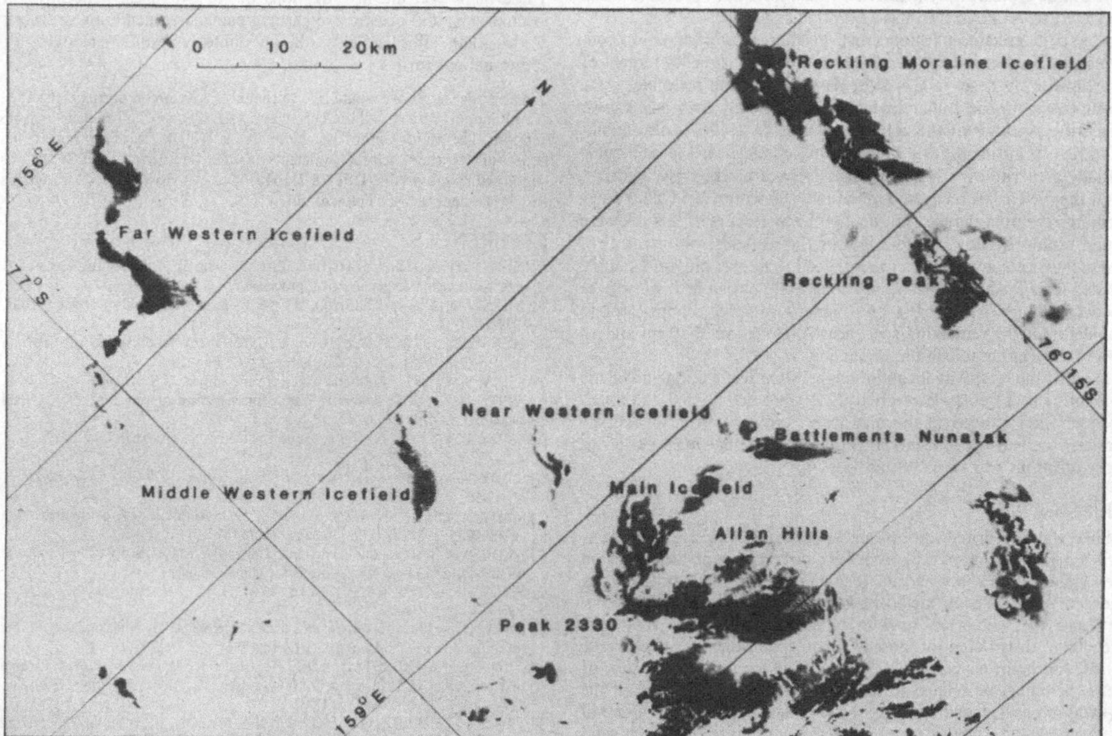


**Figure 2** Satellite photo (Near-infrared-Band 7) of Allan Hills-Reckling Peak region, southern Victoria Land. Bare ice areas are grey and bedrock areas are dark grey and black. (Photo courtesy of U.S. Geological Survey).

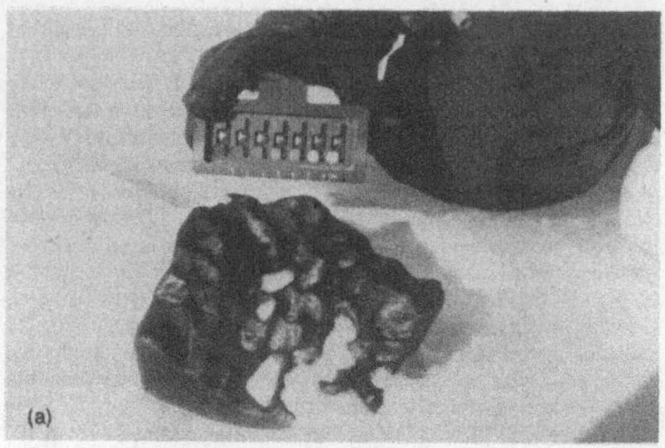


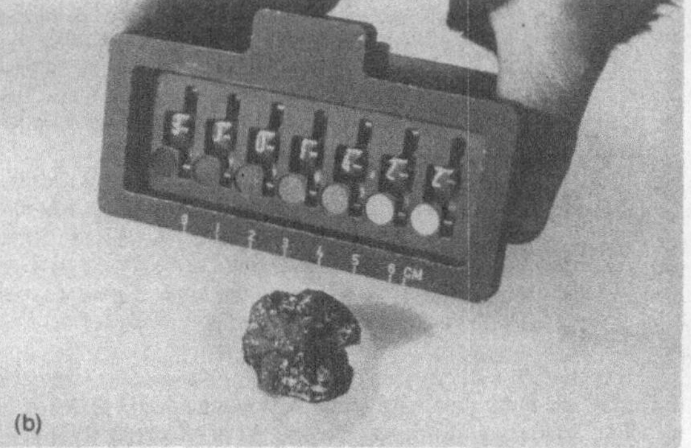


**Figure 3 Allan Hills meteorites: a) Iron meteorite (ALHA-81013) showing unusual cubic form. b) Achondrite breccia (ALHA-81005) with remnants of light coloured fusion crust.**

**TABLE 1: Preliminary tabulation and tentative identification of meteorite specimens from the Allan Hills icefields, 1981-1982**

| | Ordinary Chondrites | Carbonaceous Chondrites | Achondrites | Irons | Stony Irons | Total |
|---|---|---|---|---|---|---|
| Main Icefield | 275 | 2 | 8 | — | 1 | 286 |
| Near Western Icefield | 75 | — | — | 2 | 1 | 78 |
| Middle Western Icefield | 12 | 1 | 1 | — | — | 14 |
| Total | 362 | 3 | 9 | 2 | 2 | 378 |

laboratory examination. The largest specimen (ALHA-81013: Figure 3a) is an unusally shaped iron meteorite weighing approximately 19 kilograms. Although ablated on one face, it is otherwise cubic in form, probably reflecting the crystallography of a hexahedrite; it may be a remnant of a large single crystal. The most unusual meteorite of the 1981/82 collection is an achondritic breccia (ALHA-81005: Figure 3b). Unlike typical stony meteorites which have dark brown to black fusion crust, ALHA-81005 has a very thin and somewhat transparent, frothy, light green-tan fusion crust. Petrographic examination confirms that this is indeed a unique meteorite. It consists largely of anorthositic clasts set in a brown, glassy matrix and resembles some specimens from the Lunar highlands more closely than any known meteorite. Another of the achondrites, ALHA-81001, is also anomalous. It was mistaken for a carbonaceous chondrite in the field but is actually a eucrite consisting almost entirely of black, glassy matrix.

Of the meteorites recovered in the 1981/82 season 295 (78%) are ≤ 4 cm in maximum dimension. Most of the meteorites of this size were found along the down-wind edges of the icefields, on ice or firn, where they were apparently concentrated by the ever-present katabitic winds. They are sometimes found completely buried in firn, or over snow-filled crevasses, or even trapped in wind scoops around boulders. Of the remaining specimens 70 (19%) are 5-10 cm and 13 (3%) are 11 cm in maximum dimension.

Most of the meteorite localities on the Main Icefield, and all of the locations on the Near Western Icefield were determined. Although none of the locations of the meteorites recovered from the Middle Western Icefield were documented, it is hoped that markers left at those positions will be intact next season.

## DISCUSSION

Surveyed meteorite locations are an important aspect of field data. Detailed location maps will record the distribution of meteorites on the fields of occurrence and the relation of specific meteorites to one another. Within the few meteorite-bearing icefields in southern Victoria Land the meteorites tend to be abundant in some areas and irregularly distributed or absent in others. Other than katabatic winds, mechanisms controlling the occurrence and distribution of meteorites must be related to ice sheet dynamics. Knowledge of the distribution of meteorites in conjunction with their terrestrial ages will provide clues to rates of meteorite accumulation, ice dynamics, and other possible factors controlling meteorite concentrating mechanisms.

Fragments of meteorites that have broken on impact or at a later time, may be paired with other nearby specimens recovered during different field seasons. If the locations of these fragments are documented each season, it will be easier to establish pairings.

Preliminary meteorite location maps with field sample numbers have been completed. The Allan Hills Main Icefield Meteorite Location Map will show the locations of most of the meteorites recovered in the 1979/80, 1980/81, and the 1981/82 seasons, and will be published shortly after the meteorite processing has been completed and generic numbers assigned. The Allan Hills Near Western Icefield Meteorite Locations Map will be published after a detailed search of the Near Western Icefield has been carried out in the 1982/83 season.

## SUMMARY

Success of the 1981/82 activities in areas which had previously been searched in detail can be attributed to systematic grid searching. The surveying methods used to obtain meteorite locations proved to be adequate and practical. The resulting maps will show meteorite distributions and their terrestrial ages across the fields of occurrence. These maps will aid investigations of ice dynamics, concentration mechanisms, and questions regarding paired fragments. Future work at the Allan Hills icefields will undoubtedly result in recovery of additional meteorites.

*Acknowledgements* We would like to thank ITT-Antarctic Services, the U.S. Naval Support Force, and the U.S. Navy VXE-6 Squadron for their logistics support. The senior author wishes to thank William Salisbury and James Ridenour for their helpful comments on the manuscript. This work was supported by grant NSF/DPP-78-21104 from the Division of Polar Programs of the National Science Foundation.

## REFERENCES

ANNEXSTAD, J.O. and SCHULTZ, L. (this volume): Antarctic meteorites and their relationship to glaciological processes.

ANNEXSTAD, J.O. and NISHIO, F., 1980: Glaciological studies in Allan Hills, 1979/80. *U.S. Antarct. J., 15,* 65-6.

ANNEXSTAD, J.O. and SCHULTZ, L., 1982: Triangulation survey of the Allan Hills icefields 1981/82. *U.S. Antarct. J., 17,* (in press).

CASSIDY, W.A., 1977: Antarctic search for meteorites. *U.S. Antarct. J., 12,* 96-8.

CASSIDY, W.A., 1978: Antarctic search for meteorites during the 1977/78 field season. *U.S. Antarct. J., 13,* 39-40.

CASSIDY, W.A., 1979: Antarctic search for meteorites (ANSMET 1978/79). *U.S. Antarct. J., 14,* 41-2.

CASSIDY, W.A., 1980: Antarctic search for meteorites, 1979/80. *U.S. Antarct. J., 15,* 49-50.

CASSIDY, W.A. and ANNEXSTAD, J.O., 1981: Antarctic search for meteorites (ANSMET), 1980/81. *U.S. Antarct. J., 16,* 61-2.

CASSIDY, W.A. (this volume): The remarkably low surface density of meteorites at Allan Hills and implications in this for climate change.

CASSIDY, W.A. and RANCITELLI, L.A., 1982: Antarctic meteorites. *Am. Scientist, 70,* 156-64.

FUDALI, R.F., 1982: Gravity measurements across Allan Hills main meteorite collecting area. *U.S. Antarct. J., 17,* (in press).

NISHIO, F. and ANNEXSTAD, J.O., 1979: Glaciological survey in the bare ice area near the Allan Hills in Victoria Land, Antarctica. *Mem. Nat. Inst. Inst. Pol. Res., Japan, Spec. Issue, 15,* 13-23.

SCORE, R., SCHWARZ, C.M., KING, T.V.V., MASON, B., BOGARD, D.D. and GABEL, E.M., 1981: Antarctic meteorite descriptions, 1976-1977-1978-1979. *Antarc. Meteorite News., 4,(1).* 144 pp.

SCORE, R., SCHWARZ, C.M., MASON, B. and BOGARD, D., 1982: Antarctic Meteorite Descriptions, 1980: *Antarc. Meteorite News., 5 (1),* 55 pp.

# THE REMARKABLY LOW SURFACE DENSITY OF METEORITES AT ALLAN HILLS AND IMPLICATIONS IN THIS FOR CLIMATE CHANGE

William A. Cassidy, *Department of Geology and Planetary Science, University of Pittsburgh, Penn., 15260, U.S.A.*

*Abstract* Using a conservative calculation to estimate the expected number of meteorite specimens at Allan Hills, it becomes clear that this meteorite concentration site has been present for only a very small fraction of the lifetime of the East Antarctic ice sheet. Climatic changes probably control the formation and removal of meteorite concentration sites. The last major change to a warmer climate established conditions that produced the present concentration of meteorites at Allan Hills. This probably occurred around 100,000 years ago.

## Age of the Allan Hills Meteorite Concentration Surface

If we knew how fast meteorites were accumulating at Allan Hills we could estimate how long the area has functioned as a collecting surface. Minimum estimates of the modern rate of fall of meteorites onto the earth's surface (Brown, 1960; Hawkins, 1960) are approximately 1 $km^{-2}Ma^{-1}$. Adjusted for seasonal and diurnal variations in recovery rates (Brown, 1960), this becomes 3 $km^{-2}Ma^{-1}$. Adjusted for diurnal variations in the meteoroid influx rate instead of in meteorite recovery rates (Hughes, 1980), this becomes 6 $km^{-2}Ma^{-1}$. Millard (1963) has suggested additional upward revisions due to geographic and sociological factors that would further increase the estimate by the multipliers 7 and 2, respectively. Millard's factors, however, are of uncertain applicability because they probably change with mass for an individual meteorite fall, and with size of strewnfield and total number of fragments in a shower. If we assume, conservatively, the minimum estimate for modern meteorite falls to the earth's surface and further assume that this infall rate has been constant, we can calculate the minimum number of falls expected to occur directly onto a blue ice area in Antarctica over a given period. If a corrected estimate of infall rate were used, that is, using corrections due to diurnal and seasonal effects, the number of fragments per unit area in a given time period would be larger by a factor of six. Note that these are falls directly onto the Allan Hills ice surface.

Two other processes that are peculiar to the ice sheet operate to further affect the surface density of meteorite fragments on a blue-ice area which is in a condition of stagnant flow (Whillans and Cassidy, in press). These processes are: (1) transport of meteorites from their fall sites in the snow accumulation zone to a surface storage site in the ablation zone, and (2) lateral concentration within the ablation zone by compressive ice flow. Transport into the area is estimated at least to double the surface concentrations in the ablation zone, while compressive ice flow is expected to crowd the exposed meteorite fragments down toward the barrier so that in a narrow zone before the barrier the surface concentration would be higher by another factor of 2 to 3. Compressive ice flow, however, would not change the total number of fragments at the surface storage site. Remembering now that the original minimum estimate (1 $km^{-2}Ma^{-1}$) was for the number of falls reaching the earth's surface, one must introduce another factor to provide for fragmentation in the air and upon striking the surface. Ten is suggested as a conservative estimate of the average number of fragments representing a single fall. The effects of these factors on estimates of meteorite fragments expected that Allan Hills over a variety of accumulation periods can be examined in Table 1.

When the Allan Hills meteorite concentration site was first discovered it seemed astonishing that such a high concentration of meteorites should be present anywhere on earth, and a feeling developed that very great lengths of time for accumulation must have been involved. After six field seasons the number of specimens recovered there is approximately 1,000, and a sense of astonishment still prevails. Examination of Table 1 suggests, however, that the meteorite concentration at Allan Hills in fact is surprisingly low: 9,000 fragments are expected to have accumulated there over 1 Ma.

Not considered in the foregoing is the fact that all rocks carried by a glacier eventually are deposited in a terminal moraine. Perhaps, therefore, a terminal moraine exists and contains many more meteorites than have yet been discovered at the Allan Hills site; this would infer a greater age for the site as a collecting surface. Ice directly in front of the Allan Hills barrier is covered in snow, and this may conceal a terminal moraine. There is no evidence to support this, however: there are some small areas of blue ice within a few hundred metres of the barrier, and these do not have meteorites on them; there is an extensive embayment of stagnant blue ice enclosed on three sides by Allan Hills outcrops which does not have meteorites on its surface; and there is a small ice cored moraine within a few hundred metres of Allan Hills that does not contain meteorites. The inferences to be drawn are that a terminal moraine concentration does not exist, and that the Allan Hills have been acting as a concentration site for meteorites for a period of time that is closer to 100,000 years than to 1,000,000 years.

This kind of age estimate cannot be made for the Yamato Mountains site. It is estimated that the surface concentration of meteoritic material at the Yamato Mountains is much lower than that at Allan Hills and suggests this is because the Yamato Mountains is only a partial barrier while the Allan Hills is an absolute barrier to ice flow. This implies that over the years, in proportion to the surface areas at each site, the total number of meteorites detained temporarily at the Yamato Mountains partial barrier, and later discharged downstream could be less, or as great or greater than the number retained permanently at Allan Hills.

Meteorites with measured terrestrial ages of about 700,000 years have been found at both the Allan Hills and the Yamato Mountains sites, and many specimens have been found to have terrestrial ages in the hundreds of thousands of years. This is not inconsistent with an age of 100,000 years for the Allan Hills collecting surface if they have spent most of their terrestrial residence times embedded in the ice, being transported towards a site that had not yet, but was to become, a barrier to ice movement.

## Implications for Climate Changes

Recognition that one of the two known major sites of meteorite concentration on the Antarctic ice sheet has a young age suggests that such concentration sites can be relatively transitory features. Mechanisms that could change ice flow characteristics enough to erase existing meteorite accumulation sites or to create new ones could operate as a result of climate change.

Whillans (1981) has analysed the effect of global climatic warming on a continental ice sheet flow regime. For the 50% increase in precipitation and 8°C warming of the atmosphere assumed by

**TABLE 1: Number of meteorite falls and resulting fragmants expected during different time periods on the 75 km² area of blue ice at Allan Hills (Millard's corrections not used). Estimated age ot the East Antarctic ice cap is 14 Ma (Kennett, 1977).**

| (1) Time Period (Ma) | (2) Minimum estimate of Falls ($km^{-2}$) | (3) Falls corrected for seasonal and diurnal variations (x6) ($km^{-2}$) | (4) Col. 3 corrected for transport into the ablation area (x2) ($km^{-2}$) | (5) Fragments at 10/fall ($km^{-2}$) | (6) Total fragments in 75 km² at Allan Hills |
|---|---|---|---|---|---|
| 1 | 1 | 6 | 12 | 120 | 9000 |
| 5 | 5 | 30 | 60 | 600 | 45000 |
| 10 | 10 | 60 | 120 | 1200 | 90000 |
| 14 | 14 | 84 | 168 | 1680 | 126000 |

Whillans, there would be an initial thickening of the ice sheet in the zone of accumulation, followed by a progressive thinning which eventually would carry the surface elevation below its prewarming level. For the example given by Whillans, the elevation of the ice surface would reach a maximum 240 m higher in 10,000 years, would then decrease and cross its prewarming equilibrium elevation in about 30,000 years, and by 60,000 years would be approaching a new equilibrium level 140 m lower relative to its prewarming elevation.

Whillans' model applies only to the zone of accumulation, whereas meteorite concentrations occur within ablation zones located farther downstream; it seems likely, however, that changes in thickness induced in the accumulation zone would propagate into the ablation zone. The boundary between accumulation and ablation, however, also might migrate as a result of climate change. Migration of the equilibrium line will result from interactions between two influences having opposite sign, that is, the tendency of the equilibrium line to drift downstream because of generally increased precipitation rates over the ice cap, and an opposite tendency for it to migrate upstream because of higher ablation rates in warmer air that would tend to increase the size of the ablation zone. It is difficult to say which of these, if either, would prevail. Certainly if an area that had been an ablation zone during a cold regime became a snow accumulation site during warmer weather, the ice surface would rise and any concentration of meteorites lying on it would be swept away as the barrier was overridden. This is illustrated schematically in Figure 1. In like manner, if the equilibrium line remained approximately stationary, balanced between two opposing influences, the eventual pulse of thicker ice, migrating toward the edge of the continent, should swell the ice thickness even in the ablation zone and flush out the meteorite concentration. This is illustrated schematically in Figure 2. In the third case, in which the equilibrium line migrates upstream because the effect of increased ablation rate is greater than that of increased precipitation, there is a possibility that an advancing pulse of thicker ice would be substantially, or even completely, erased so that a pre-existing meteorite concentration not only would remain but would continue to be augmented. Either of the first two possibilities would help explain the low numbers of meteorite fragments at Allan Hills; the last would not. According to this discussion, then, if the ablation rate becomes much more effective than the accumulation rate as a result of climatic warming, ablation zones will be permanent features of the ice sheet, areas of stagnant flow will persist for great lengths of time, and meteorite concentrations at such sites will represent very long term accumulations. On the other hand, if the ablation rate is only marginally more effective, or less effective, than the snow accumulation rate, then meteorite concentration sites will be only transitory features of the ice cap, recurring repeatedly at the same locations only to be flushed out later. This is the favoured explanation.

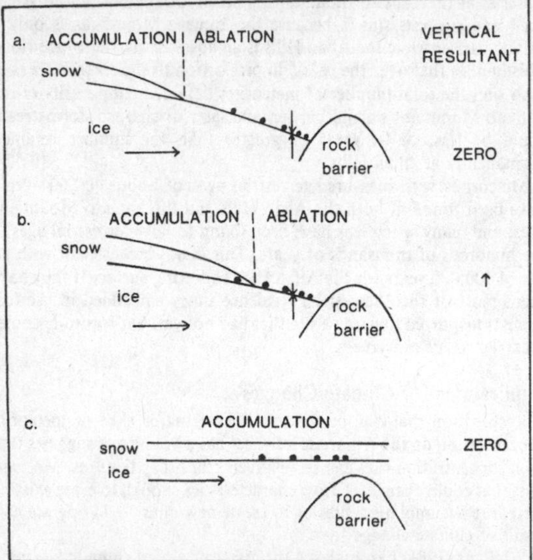


**Figure 1. Possible stages in the removal of a meteorite concentration as a result of climate warming. (a) A meteorite concentration exposed on the surface in a zone of stagnant ice flow behind a barrier. The ablation rate equals the upward vector of arriving ice. (b) Climate warming has resulted in migration of the equilibrium line toward the barrier. The rate of ablation has decreased while the upward vector of arriving ice has increased or remained the same. The ablation surface rises and the meteorite accumulation is partially covered. (c) The equilibrium line has passed the barrier, the former ablation surface has risen above the barrier, and the meteorite concentration is carried away. Because of lower ice viscosity due to a warming effect that lags behind the increase in accumulation rate, the ice surface will later drop below the level shown in (a), and the meteorite accumulation process will be re-established behind the re-exposed rock barrier.**

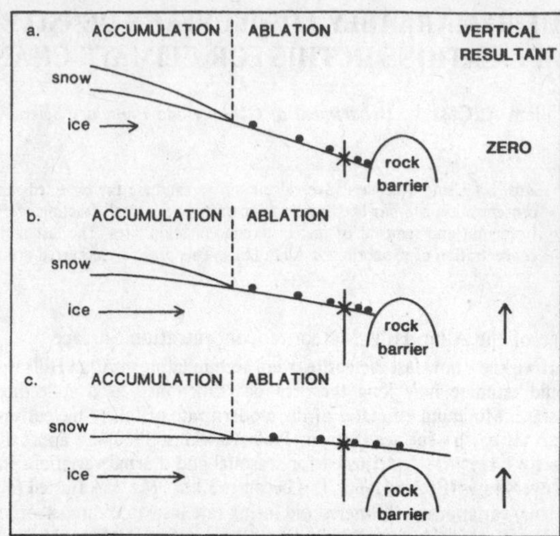


**Figure 2. Possible stages in the removal of a meteorite concentration as a result of climate warming. (a) A meteorite concentration exposed on the surface in a zone of stagnant ice behind a barrier. The ablation rate equals the upward vector of arriving ice. (b) Climate warming has resulted in a thickening of the ice cap in the accumulation zone, but vigorous ablation effects due to higher temperature have kept the equilibrium line from advancing. A pulse of thicker ice travels downstream, however, as a result of faster accumulation in the interior, and this effectively increases the upward vector in the meteorite concentration zone. (c) The barrier is overwhelmed and the meteorite concentration is carried away. Later the ice level will drop below the barrier once again and meteorite accumulation will recur.**

Would the Allan Hills and Yamato sites lend themselves to such an intermittent accumulation and flushing-out process? It can be seen that with a total variation in ice-surface elevation of 380 m over a 50,000 year period, ridge crests such as Allan Hills and the Yamato Mountains alternately could be overridden by flowing ice and exposed as nunataks. In such circumstances it is easy to see that these sites would only be transitory concentration zones for meteorites, having a duration about equal to the length of the period of warmer climate that had caused the ice to thin plus the length of a succeeding colder period. Interestingly, a change to a warmer climate would carry out the functions, first, of flushing out meteorites accumulated during the preceding colder period and, second, of re-establishing those conditions that must lead to formation of a new meteorite accumulation.

## Fossil Meteorite Concentrations

Because meteorite concentration sites may be rather transitory, it seems quite likely that we are seeing only the most recent of them; sites like those at the Allan Hills and the Yamato Mountains may have existed on the Antarctic ice sheet in many places and at many times in the past. Some of these may have had absolute barriers to ice flow for much longer periods than the Allan Hills site, and have accumulated much higher surface concentrations of meteorites. It is not a trivial question, therefore, to ask what has become of these earlier concentrations. In each case illustrated in Figures 1 and 2, the accumulated meteorites are carried away, possibly to be stranded elsewhere but probably to be carried away to the edge of the ice sheet. Very large ablation zones exist around the borders of the Antarctic continent. Figure 3 suggests the re-emergence of a fossil meteorite accumulation at one of these.

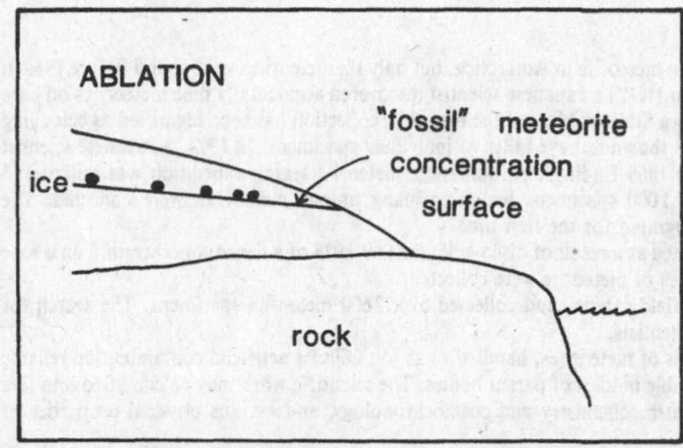


**Figure 3. Eventual re-emergence of a transported "fossil" meteorite concentration in an ablation zone downstream from the site where it originally had accumulated.**

## Summary

Implicit in the foregoing has been the idea that the origin, duration, and eventual burial or removal of a meteorite concentration depends primarily on climatic conditions. Suggested is the conclusion that the removal of a concentration of meteorites specifically requires a climatic warming. The suspected relatively young age of the Allan Hills meteorite concentration site means that the Allan Hills would not have been acting as a barrier to ice flow for too long a period. That it is present today suggests that the last major climatic change to higher temperatures may have occurred around 100,000 years ago, and that we are still within the colder period that succeeded it.

*Acknowledgement* This work was supported by grant NSF/DPP 78-21104.

## REFERENCES

BROWN, H., 1960: The density and mass distribution of meteoritic bodies in the neighbourhood of the earth's orbit. *J. Geophys. Res., 65,* 1679-83.

HAWKINS, G.S., 1960: Asteroidal fragments. *Astron. J., 65,* 318-22.

HUGHES, D.W., 1980: On the mass distribution of meteorites and their influx rate; *in* Halliday, I. and McIntosh B.A. (eds.) *Solid Particles in the Solar System.* Reidel, Boston, 207-10.

KENNETT, J.P., 1977: Cenozoic evolution of Antarctic glaciation, the circum-Antarctic ocean, and their impact on global palaeoceanography. *J. Geophys. Res. 82,* 3843-60.

MILLARD, H.T. Jr., 1963: The rate of arrival of meteorites at the surface of the earth. *J. Geophys. Res., 68,* 4297-303.

WHILLANS, I., 1981: Reaction of the accumulation zone portions of glaciers to climatic change. *J. Geophys. Res.,.86,* 4274-82.

WHILLANS, I., and CASSIDY, W. (in press): Catch a falling star: meteorites and very old ice. *Science.*

# ANTARCTIC METEORITES

K. Yanai, *National Institute of Polar Research, Tokyo 173, Japan*

*Abstract* In December 1912, an Australasian Expedition party first discovered one meteorite in Antarctica, but only six meteorites were found before 1969 in four localities throughout the continent. After the International Geophysical Year (IGY) a Japanese scientist discovered accidentally nine meteorites on bare ice (blue ice) adjacent to the Yamato Mountains, about 300 km southwest of Syowa Station. Most of the meteorite collection has been identified as belonging to five types after examination petrologically and chemically, most of which were shown to have fallen as individual specimens. In 1974, a Japanese scientist also found twelve specimens in the same area, and in a broader area than the 1969 findings. An Antarctic meteorite search expedition was initiated in November to December 1974 and November 1975 to January 1976. About 1000 specimens including many unique meteorites were collected. The accumulation and concentration of the meteorites on the bare ice field was recognised for the first time.

A Japan-US joint program for collecting meteorites in Victoria Land was initiated as a result of discoveries in 1969-1975 of a dense concentration on a bare ice field adjacent to the Yamato Mountains. From 1976-1978 over 600 new pieces of meteorite were collected.

An oversnow traverse party visited the Yamato Mountains in the 1979-1980 field season, and collected over 3600 meteorite specimens. The search for meteorites is now being continued independently mainly by Japanese and US scientists.

Since the Antarctic meteorite collections cover a wide variety of different kinds of meteorites, handled so as to minimise artificial contamination reliable petrological, mineralogical and physical data of these meteorites will lead to possible models of parent bodies. The scientific work may be classified into five different fields, namely the Antarctic field work, mineralogy and petrology, cosmochemistry and cosmochronology, and various physical properties of meteorites.

# SIMILARITIES AND DIFFERENCES BETWEEN THE YAMATO MOUNTAINS AND VICTORIA LAND METEORITE CONCENTRATIONS

W.A. Cassidy, *Department of Geology and Planetary Sciences, University of Pittsburg, Penn, 15260, USA*

*Abstract* The types of meteorites falling in Antarctica are presumed to reflect the abundance of the various types of meteorites in space. A mature concentration of meteorites is defined therefore as a concentration that has been accumulating long enough, or has accumulated over a large enough area, so that its members reflect the true abundance of the various types in space. The maturity of a residual concentration of Antarctic meteorites can be tested in two ways: by comparison with the worldwide falls ratio and by degree of convergence between it and other Antarctic meteorite concentrations. In the first case we assume the modern worldwide falls ratio to be correct, not only for the Present but for the Past as well. In the second case this assumption is not necessary. In this case we assume only that, given enough time or a large enough collecting surface, the same ratio of types will be approached. For the Antarctic meteorites there is evidence of long accumulation times, therefore it should be possible to deduce the correct ratio of types by finding convergence on the same value at two or more major accumulation sites. When this has been done the Present falls ratio can be tested for consistency with the (Present + Past) cumulative ratio to determine if the nature of the meteorite flux at the earth has changed during time. When degree of maturity of a given meteorite accumulation has been established it will have value as a measure of the relative length of time the given ice conditions that produced it have prevailed. The Yamato Mountains and Victoria Land accumulations are discussed in the light of these concepts.

# 14

# Subantarctic Islands

# VOLCANIC EVOLUTION OF HEARD AND McDONALD ISLANDS, SOUTHERN INDIAN OCEAN

I. Clarke[1], *Department of Earth Sciences, Monash University, Clayton, Vic, 3168. Australia.*

I. McDougall, *Research School of Earth Sciences, Australian National University, P.O. Box 4, Canberra, A.C.T., 2601. Australia.*

D.J. Whitford, *Division of Mineralogy, CSIRO, North Ryde, N.S.W., 2113. Australia.*

[1]*Present Address: Geological Survey of New South Wales, Department of Mineral Resources, CAGA Centre, 8-18 Bent Street, Sydney, NSW, 2000. Australia.*

*Abstract* Heard Island and the McDonald Islands were formed by volcanic activity on the Kerguelen Plateau which is believed to have an oceanic crustal origin. Heard Island was built up by three main phases of volcanic activity. The extrusive products of the earliest phase are apparently not exposed and are thought to have been removed by erosion; however, intrusive equivalents occur as mafic bodies in the limestone basement of the island. The intrusions were probably associated with uplift no earlier than the Late Miocene. A second phase of sporadic basaltic volcanism accompanied by glaciation and shallow marine sedimentation commenced in Late Miocene to Early Pliocene times and gave rise to the Drygalski Formation. The final phase produced the youthful volcanic morphology of the island. K-Ar ages are consistent with the final phase commencing no more than 1 Ma ago with the construction of the Big Ben basalt volcano, and they indicate that trachyte and trachyandesite erupted on Laurens Peninsula within the last 10,000 years. The latest volcanism of the final phase was the development of basalt scoria cones and lava fields around the coast, and Mawson Peak on Big Ben. The McDonald Islands are the denuded remains of phonolite volcanism which, according to K-Ar dating, occurred within the last 100,000 years. $^{87}Sr/^{86}Sr$ ratios of Heard Island rocks (0.7047-0.7058) are similar to those of rocks from Iles Kerguelen and the Ninetyeast Ridge. This observation, together with the youthful age of volcanism on Heard Island, is consistent with a model in which all three features were derived from the same mantle source as part of a hot-spot trace.

Two alternative origins have been proposed for the Kerguelen Plateau in the southern Indian Ocean (Figure 1). Arguments supporting a continental origin are based on the thick sediment cover on the plateau (Schlich et al., 1971; Watkins et al., 1974) and the apparent contiguity of the Kerguelen Plateau and the Broken Ridge, another possible continental fragment (e.g. Norton and Molnar, 1977). However, the geophysical properties of the crust of the plateau (Houtz et al., 1977) and the geology, geochemistry and geophysics of Iles Kerguelen (e.g. Dosso and Murthy, 1980) suggest more strongly an oceanic crustal origin. With this in mind, a study is in progress to investigate the nature and origin of Heard and McDonald Islands (Figure 2), the other two exposed edifices on the Kerguelen Plateau.

The bulk of our current knowledge of Heard Island geology is due to Lambeth (1952). Subsequently, Stephenson (1964) reviewed and augmented Lambeth's work, and analysed some igneous rocks (Stephenson, 1972). Both Lambeth's and Stephenson's collections were available for the present study, and one of the writers visited Heard and McDonald Islands in 1980 (Clarke, 1982), and had access to Lambeth's field notebook.

This paper outlines the volcanic evolution of Heard and McDonald Islands by reviewing their geology and petrology, reporting K-Ar ages and Sr isotope compositions, and using age constraints resulting from recent micropalaeontological studies (Quilty et al., this volume). A need for clarification of the age of volcanism on Heard Island was pointed out by Luyendyk and Rennick (1977) who realised that it would constrain models of Indian Ocean crust formation.

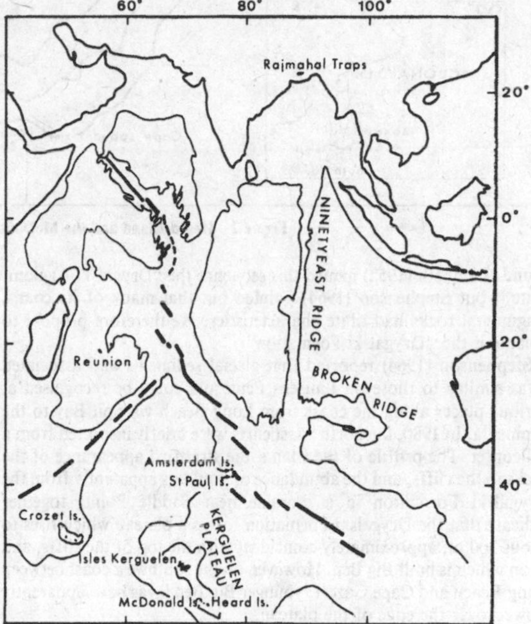


**Figure 1 Location map Indian Ocean. The spreading Mid-Indian Ocean Ridge is shown by thick broken line segments. Isobaths are 4000 m and 1000 m. Modified from Luyendyk and Rennick (1977).**

## GEOLOGY

### Heard Island

This section integrates the geology of Heard Island as outlined by Lambeth (1952, and notebook) and Stephenson (1964) with observations made in 1980 (Clarke, 1982). Three geological units have been recognised, each representing a phase of volcanic activity:
Uppermost: Lavas
Intermediate: Drygalski Formation
Lowermost: Limestone and mafic intrusions

*Limestone, and Mafic Intrusions* Cliffs on the south coast of Laurens Peninsula and behind First Beach expose a unit of pelagic limestone with minor chert, intruded by concordant dolerite sills and a larger gabbro body. The limestone has been strongly indurated by the intrusions. The dolerite sills range from a few centimetres to several metres in thickness and appear to be at least as voluminous as the limestone unit they intrude. A sill at First Beach contains abundant wehrlite xenoliths up to 10 cm in diameter and concentrated towards its base. Also at First Beach, a gabbro intrusion over 30 m thick has a concordant lower contact, although lateral contacts are scree covered. The limestone and sills are folded about a northwest-southeast axis on Laurens Peninsula and at First Beach they dip 25° to the northeast. At both locations the limestone and intrusions are truncated by an erosional surface, and therefore we infer that the intrusions are the products of the earliest phase of igneous activity exposed on Heard Island.

Limestone and chert clasts in moraines east of First Beach and in beach gravels at Spit Bay indicate that the limestone unit must also crop out on the northeast coast of Heard Island but no such indications have been found either from the south side of Big Ben, or from the north coast of the Laurens Peninsula.

*Drygalski Formation* The limestone and mafic intrusions are unconformably overlain by a sequence of clastic deposits and basaltic lavas; similar rocks also crop out on the northeast coast of Laurens Peninsula and Mt Drygalski. The sequence is extremely irregular and comprises subhorizontal conglomerate, sandstone, mudstone, and basalt flows, with basalt and trachyte intrusions. The mudstone is commonly finely laminated in the manner of glacial varves. The sandstone and conglomerate contain clasts that range from rounded to angular, and from a few millimetres to a metre across, although very few are larger than a few centimetres. The conglomerate ranges from matrix to clast supported. Basalt is by far the most abundant clast type, trachyte clasts are rare, and no limestone clasts have been

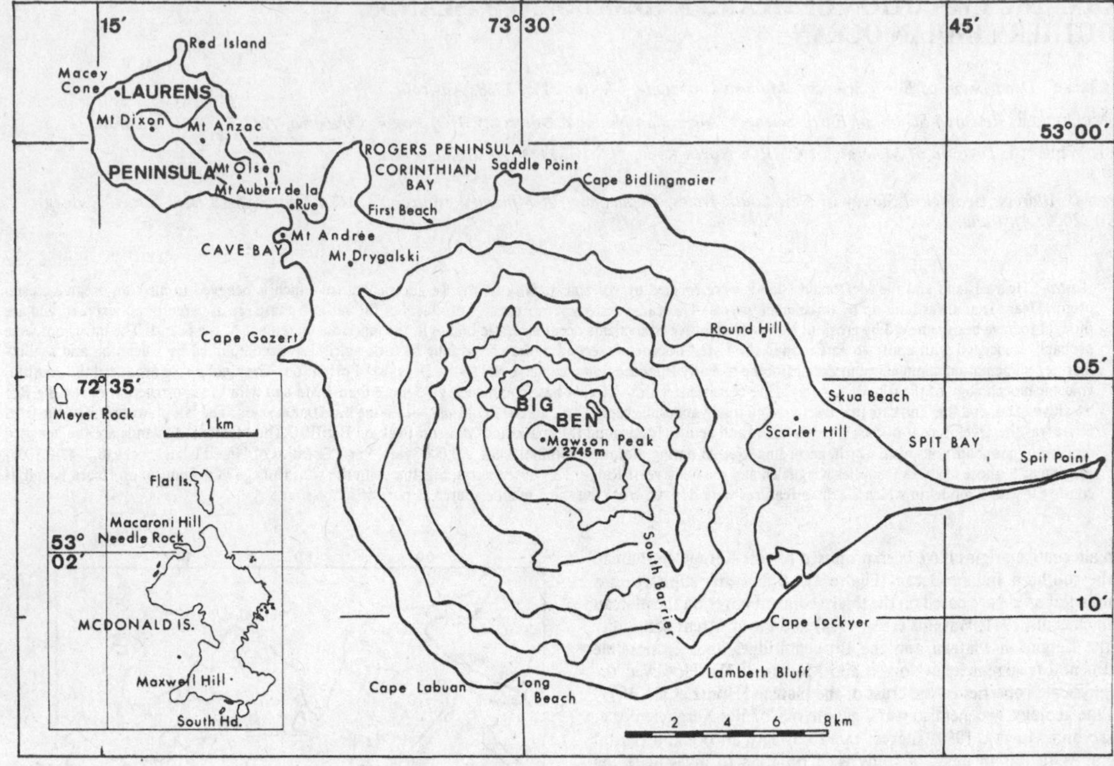


**Figure 2 Heard Island and the McDonald Islands (inset). Contour interval is 400 m.**

found. Lambeth (1952) named this sequence the "Drygalski Agglomerate", but Stephenson (1964) pointed out that many of its coarse fragmental rocks had tillite characteristics. We therefore propose to rename it the "Drygalski Formation".

Stephenson (1964) reported that glacial sediments and associated lavas similar to those on Laurens Peninsula could be recognised at various places along the coast from Long Beach via Spit Bay to the peninsula. In 1980, the north coast cliffs were briefly inspected from a helicopter. The profile of the island, the stratified appearance of the rocks in the cliffs, and the abundance of boulders apparently from the Drygalski Formation in a moraine near Saddle Point, together indicate that the Drygalski Formation forms a plateau which rises to about 300 m, approximately coincident with the top of the cliffs, and upon which is built Big Ben. However, on the southwest coast between Long Beach and Cape Gazert, younger Big Ben lavas have apparently flowed over the edge of the plateau.

*Lavas* The bulk of Heard Island consists of lavas of the Big Ben volcano. At Long Beach, Stephenson (1964) noted that basal lavas of Big Ben had well developed pillow structures which he believed to be the result of submarine extrusion. The pillows were not observed in 1980, but a suggested alternative explanation is that they could have resulted from subglacial eruption. Most sampling of Big Ben has been from peripheral moraines, and it appears that the volcano is entirely basaltic.

The northwest part of Heard Island is a separate volcanic centre containing more differentiated lavas. A cluster of hills around Cave Bay are the remnants of a trachyte volcano that has been breached by the sea. On Laurens Peninsula, the Drygalski Formation is overlain by the denuded volcanoes Mt Olsen, and Mt Anzac. The basal flows of Mt Olsen are trachyandesite and the same lava type caps the Mt Aubert de la Rue razorback. Mt Anzac and the bulk of Mt Olsen are composed of trachyte, and trachyte sills and stocks in the Drygalski Formation exposed on the northeast coast of Laurens Peninsula may also be products of this volcanism.

Mt Dixon makes up more than half of the area of Laurens Peninsula. It is a trachyandesite cone built upon a trachyte basement that is probably related to the Mt Anzac and Mt Olsen volcanism. The uppermost flows show well developed ropey structure and levees, and appear to be very young.

Around the Heard Island coast there are numerous examples of relatively minor, young, parasitic basaltic volcanism. Mt Andree is a remnant of a basalt cone which laps onto the Cave Bay trachyte. Rogers Peninsula is built up of basalt scoria overlain by basalt flows and four scoria cones. Lava plains and scoria cones, some well denuded, are developed at Red Island, Macey Cone, Saddle Point, Cape Bidlingmaier, Round Hill, and Scarlet Hill. There are several small cones at South Barrier, and small lava fields occur at Cape Labuan, the east end of Long Beach, Lambeth Bluff, and the south end of Skua Beach. Recent scoriaceous basalt makes up Mt Aubert de la Rue and caps Mt Drygalski. On the summit of Big Ben, Mawson Peak marks the volcano's latest vent.

## McDonald Islands

The McDonald Islands consist of the eroded products of phonolite volcanism. The northern part of the main island is a plateau composed of laminated tuff which, in coastal cliff exposures, shows cross stratification. Clasts in the tuff are generally rounded and most are of phonolite; clasts of white chalky limestone are also abundant, and there are in addition a few chert clasts. No basalt clasts were seen. We infer that the tuff was probably produced by erosion of a phonolite volcano atop a pelagic sediment basement.

On the east coast of McDonald Island the tuff is intruded by several phonolite dykes. The southern part of the island, Maxwell Hill, consists of a phonolite dome which has intruded and partly uplifted the tuff. South Head, Macaroni Hill, and Meyer Rock appear similar to Maxwell Hill, and Needle Rock is a dyke, but none of these have been sampled. The intrusions generally have well developed columnar jointing. Flat Island consists of a series of thin phonolite flows overlying a massive, columnar jointed, volcanic basement.

## PETROLOGY

### Petrography

The igneous rocks of Heard and McDonald Islands show petrographic characteristics that are exclusively of alkaline affinity. Their mineral assemblages are summarised in Table 1. The dolerite sills are

**TABLE 1: Petrography of igneous rocks from Heard and McDonald Islands**

| | Phenocrysts | Groundmass |
|---|---|---|
| Dolerite | (cpx), (pl), (ol) | pl, cpx, ox, (an) |
| Gabbro | | pl, cpx, ox, (alk), (bi), (ap), (an), (amph) |
| Basalt | ol, cpx, pl, ox, (ks) | pl, cpx, ox, gl, (ol), (alk) |
| Trachyandesite | cpx, ks, pl, ox, bi, (ol) | pl, cpx, ox, alk |
| Trachyte | alk, cpx, ox, (pl), (ks), (bi) | alk, cpx, ox, (ol) |
| Phonolite | alk, ks, cpx, bi, ox, (ol), (sph) | alk, cpx, gl, ne, lc, ox |

Minerals are listed in roughly decreasing order of abundance with minor minerals in parenthesis. ol = olivine, cpx = clinopyroxene, pl = plagioclase, ox = Fe-Ti oxide, alk = alkali feldspar, ks = kaersutite, bi = biotite, ap = apatite, an = analcime, amph = other amphibole, sph = sphene, ne = nepheline, gl = glass, lc = leucite.

aphyric to weakly porphyritic, fine grained, and composed primarily of plagioclase, clinopyroxene and oxides. The gabbro intrusion exhibits decreasing grainsize towards the margin of the body, and in addition to the dolerite assemblage, contains minor alkali feldspar, biotite, amphibole, analcime and apatite. All the mafic intrusions are hydrothermally altered, probably as a result of interaction with the wet carbonate sediments during intrusion.

Basalts from the Drygalski Formation are petrographically indistinguishable from those of the uppermost lavas apart from some alteration in lower members. They are generally strongly porphyritic in olivine and clinopyroxene, porphyritic to a lesser degree in plagioclase and oxides and contain abundant mono- and poly-mineralic glomerophenocrysts. Phenocrystic olivine is usually volumetrically dominant over clinopyroxene, which however, forms larger crystals. The phenocrysts are usually subhedral to euhedral, and only a few show signs of resorption. Some basalts are rich in glass at the expense of groundmass plagioclase. No modal feldspathoids such as nepheline or leucite have been detected.

Trachyandesites from the Laurens Peninsula are characterised by phenocrysts of kaersutite and less abundant titaniferous biotite, both partially or entirely replaced by an oxide-dominated assemblage. The trachytes are composed mainly of alkali feldspar, and some contain groundmass fayalitic olivine.

The McDonald Islands phonolitic rocks are petrographically extremely diverse. The phonolite intrusions and flows, and some clasts are strongly porphyritic, and phenocrysts may include alkali feldspar, kaersutite, titaniferous biotite, clinopyroxene, oxides and sphene. In contrast to those in the Heard Island trachyandesites, kaersutite and biotite phenocrysts in the phonolites are commonly euhedral and free from alteration. Magnesian olivine is a rare xenocrystic phase derived from fragmentation of more mafic rocks. Two groundmass assemblages can be recognised in the porphyritic phonolites: alkali feldspar + sodic clinopyroxene + oxides; and glass + alkali feldspar + leucite. A distinctive phenocryst-poor phonolite is a common clast type in the tuff, and comprises rare alkali feldspar and sodic clinopyroxene phenocrysts set in a groundmass of the same minerals, plus nepheline, and oxides.

*Xenoliths* Four types of xenoliths have been recognised in Heard Island basaltic rocks:

(1) crustal xenoliths: chert, siltstone, quartzite or limestone derived from underlying oceanic sediments
(2) gabbro: olivine and clinopyroxene cumulates with intercumulus plagioclase; oxides may be cumulus or intercumulus phases
(3) wehrlite: olivine and clinopyroxene cumulates with little or no plagioclase
(4) lherzolite: assemblages of olivine, orthopyroxene, chrome diopside and spinel derived from the upper mantle.

A trachyandesite boulder from Mt Olsen contains xenoliths of a rock composed of 40% kaersutite phenocrysts and less abundant phenocrysts of titanium-poor clinopyroxene, oxides and apatite, and rare olivine, in a groundmass of titaniferous clinopyroxene and oxides, and rare plagioclase. These xenoliths are interpreted to be cumulates from a magma undergoing fractionation to trachyandesite.

## Geochemistry

The basalts, dolerites and gabbros are all chemically indistinguishable, regardless of age (Figure 3). They possess the usual chemical properties of alkaline rocks, namely, high alkali, $TiO_2$, incompatible trace element contents, and light rare earth element enrichment; they are also moderately enriched in potassium relative to sodium (Stephenson, 1972; Collerson et al., in prep.). Compositional variation among the mafic rocks is due to removal and accumulation of olivine, clinopyroxene and plagioclase (Stephenson, 1972).

On chemical variation diagrams such as Figure 3, McDonald Islands phonolites mostly plot to one side of the Heard Island basalt-trachyandesite-trachyte lineage, but they merge compositionally with the trachyandesites. This suggests that the phonolites and trachytes

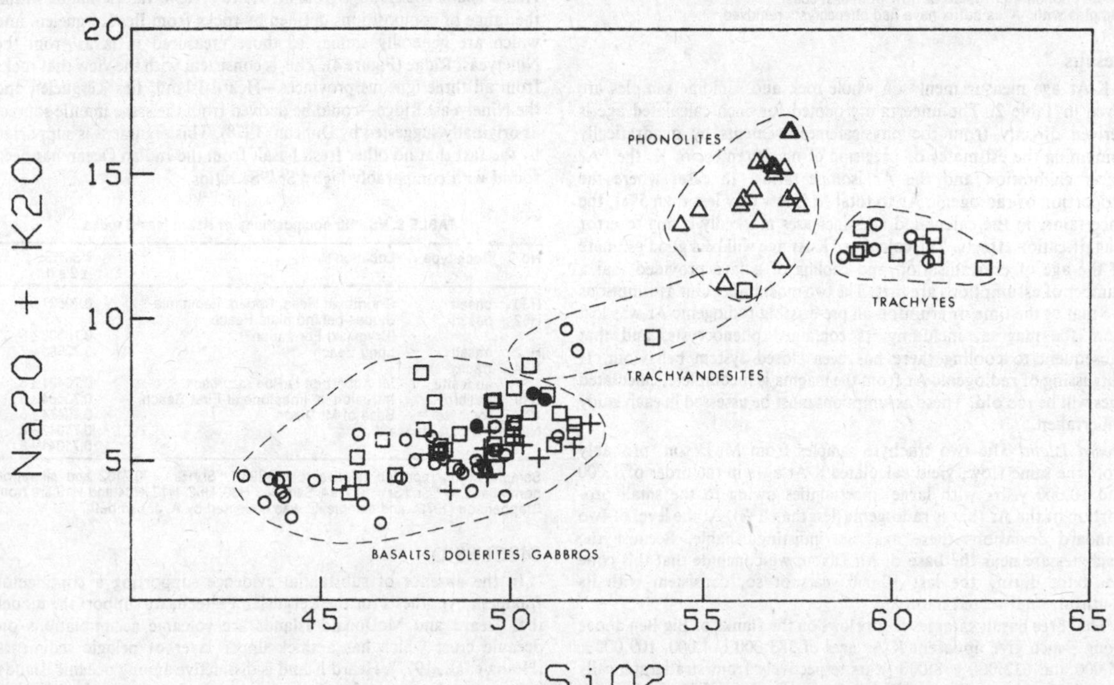


**Figure 3** $Na_2O + K_2O$ **vs.** $SiO_2$ **plot. Volcanic rocks from the Drygalski Formation and uppermost lavas are shown by squares (1980 expedition) and circles (Stephenson, 1972); mafic intrusions in the limestone by crosses (1980) and spots (1972); phonolites from McDonald Islands by triangles.**

could have both evolved via magmas of trachyandesite composition, but by different *fractionation mechanisms*. Abundant euhedral and unresorbed kaersutite phenocrysts in the phonolites prompt speculation that kaersutite fractionation may have been more dominant in magma chambers beneath the McDonald Islands. A more detailed account of magmatic differentiation will be presented elsewhere.

## POTASSIUM-ARGON DATING

### Method

Six samples of "uppermost" lavas from Heard Island were selected for whole rock K-Ar dating. Unfortunately the older rocks, that is the mafic intrusions and lavas and clasts in the Drygalski Formation, were considered to be too altered for reliable K-Ar dating. From McDonald Island three samples of phonolite were selected for K-Ar dating of sanidine phenocrysts. Potassium was measured using a IL443 flame photometer on solutions containing a lithium internal standard and sodium buffer as recommended by Cooper (1963). Argon was measured by isotope dilution as outlined by McDougall and Schmincke (1976). The decay constants recommended by Steiger and Jäger (1977) were used.

**TABLE 2: K-Ar age measurements on (a) whole-rock samples from Heard Island. (b) sanidine separates from McDonald Island.**

| Monash No. | ANU No. | K (wt. %) | Rad.$^{40}$Ar ($10^{-12}$mol/g) | 100 Rad. $^{40}$Ar / Total $^{40}$Ar | Calculated age ($10^3$yrs ± ls.d.) |
|---|---|---|---|---|---|
| (a) 65015 | 81-632 | 4.947,4.989 | .012 | .2 | 1 ± 2 |
| 65017 | 81-633 | 4.951,4.981 | .082 | .7 | 10 ± 4 |
| 65060 | 81-634 | 2.745,2.760 | 1.244 | 26.9 | 261 ± 4 |
| | | | 1.272 | 29.0 | 266 ± 4 |
| 65081 | 81-648 | 2.543,2.549 | 1.677 | 17.2 | 380 ± 7 |
| | | | 1.703 | 26.0 | 386 ± 5 |
| 65081-A | 81-648-A | 2.942,2.947 | .667 | 11.2 | 131 ± 4 |
| 65082 | 81-649 | 2.894,2.905 | .459 | 3.0 | 91 ± 8 |
| | | | .633 | 10.9 | 126 ± 4 |
| 65083 | 81-650 | 1.409,1.409 | 1.544 | 38.4 | 632 ± 8 |
| 65083-A | 81-650-A | 1.624,1628 | .413 | 13.0 | 147 ± 4 |
| (b) 65110 | 81-635 | 6.810,6.845 | .854 | 8.3 | 72 ± 2 |
| 65118 | 81-636 | 7.847,7.990 | 1.079 | 17.4 | 79 ± 3 |
| 65119 | 81-637 | 8.031,8.043 | .505 | 10.0 | 36 ± 3 |

65015 & 6501: trachyte flow from base of Mt Dixon.
65060: basalt dyke from south coast of Laurens Peninsula.
65081, 85082 & 65083: basalt flows from Big Ben above Long Beach.
65110: phonolite dome—Maxwell Hill.
65118: phonolite dyke from east coast.
65119: phonolite intrusion or flow from east coast.
Samples with '-A' as suffix have had phenocrysts removed.

### Results

K-Ar age measurements on whole rock and sanidine samples are given in Table 2. The uncertainty quoted for each calculated age is derived directly from the physical measurements by quadratically combining the estimates of precision of measurement of K, the $^{38}$Ar tracer calibration and the Ar isotope ratios. In cases where the proportion of radiogenic Ar to total Ar is low (say less than 5%), the uncertainty in the calculated ages increases markedly owing to error magnification effects. The calculated K-Ar age will be a good estimate of the age of crystallisation and cooling of a lava provided that a number of assumptions are met. The two most important assumptions are that at the time of eruption all pre-existing radiogenic Ar was lost from the magma, including its contained phenocrysts, and that subsequent to cooling there has been closed system behaviour. If outgassing of radiogenic Ar from the magma is incomplete, calculated ages will be too old. These assumptions must be assessed in each study undertaken.

*Heard Island* The two trachyte samples from Mt Dixon, probably from the same flow, yield calculated K-Ar ages in the order of 1,000 and 10,000 years with large uncertainties owing to the small proportion of the Ar that is radiogenic (less than 1%). At the level of two standard deviations these ages are indistinguishable. Because the trachytes are near the base of Mt Dixon, we conclude that this cone was built during the last 10,000 years or so, consistent with its youthful, undissected nature.

The three basalt samples from flows on the flanks of Big Ben above Long Beach give apparent K-Ar ages of 383,000 ± 4,000, 109,000 ± 25,000 and 632,000 ± 8,000 years respectively from stratigraphically youngest to oldest. These calculated ages are all significantly different from one another, and the apparent age for the middle sample is inconsistent with the stratigraphic order. Clearly not all the apparent ages can be correct. From the general youthful, undissected appearance of Big Ben volcano and the close proximity of the three samples to one another in the lava sequence, it is concluded that at least one of the apparent ages is anomalously old. The three samples contain 10-25% of olivine and clinopyroxene phenocrysts, and in view of the inconsistent ages it was decided to separate phenocrysts from the two samples giving the oldest apparent ages, and then measure the phenocryst-poor remains. These samples, identified as 65081-A and 65083-A in Table 2, yield much younger apparent ages of 131,000 ± 4,000 and 147,000 ± 4,000 yers. These results are nearly concordant with the average apparent age of 109,000 ± 25,000 years for the third whole rock sample. The three results together yield a mean age of 129,000 ± 19,000 years. The evidence strongly suggests that the phenocrysts contain significant amounts of radiogenic Ar (excess$^{40}$Ar), presumably incorporated when the crystals formed intratellurically. However, as not all of the phenocryst material was removed from the two samples, and the third sample contained an appreciable proportion of phenocrysts (10%), we believe that the mean age of 129,000 ± 19,000 years should be regarded as a maximum age for these lavas. A basalt dyke from Laurens Peninsula yielded an average K-Ar age of 263,000 ± 4,000 years. Because this rock is strongly porphyritic, this value must also be considered to be a maximum age in view of the likelihood of excess $^{40}$Ar in the phenocrysts.

*McDonald Islands* The three sanidine separates from McDonald Island phonolites gave K-Ar ages ranging from 36,000 ± 3,000 years to 79,000 ± 3,000 years. These results provide strong evidence that the McDonald Islands evolved within the last 100,000 years.

## STRONTIUM ISOTOPES

### Method

$^{87}$Sr/$^{86}$Sr ratios were determined in six rocks from Heard Island. Strontium was separated using standard cation exchange techniques after dissolution with HF/$HC10_4$. Isotope ratios were measured on an AVCO Series 900A 35 cm radius, thermal ionisation, solid-source mass spectrometer. Techniques for strontium separation and mass spectrometry were similar to those described by Riley et al. (1980).

### Results

Results are presented in Table 3. Measured $^{87}$Sr/$^{86}$Sr ratios in the Heard Island rocks range from 0.7047 to 0.7058, values that lie within the range of compositions defined by rocks from Iles Kerguelen, and which are generally similar to those measured in lavas from the Ninetyeast Ridge (Figure 4). This is consistent with the view that rocks from all three igneous provinces—Heard Island, Iles Kerguelen and the Ninetyeast Ridge—could be derived from the same mantle source, as originally suggested by Duncan (1978). This argument is supported by the fact that no other fresh basalt from the Indian Ocean has been found with comparably high $^{87}$Sr/$^{86}$Sr ratios.

**TABLE 3: $^{87}$Sr/$^{86}$Sr compositions of Heard Island rocks.**

| No. | Rock type | Location | $^{87}$Sr/$^{86}$Sr ± 2.s.d. |
|---|---|---|---|
| H30 | basalt | Corinthian Head, Rogers Peninsula | 0.70537 ± 6 |
| H62 | basalt | Slopes behind Skua Beach (Drygalski Formation?) | 0.70503 ± 6 |
| H12 | basalt | Long Beach | 0.70583 ± 7 |
| H34 | trachy-andesite | Mt Aubert de la Rue razorback | 0.70481 ± 8 |
| H19 | gabbro | Intrusion in limestone at First Beach | 0.70564 ± 6 |
| 47 | trachyte | Base of Mt Dixon | 0.70474 ± 6 |
| NBS987 $SrCO_3$ standard | | | 0.71042 ± 8 |
| | | | 0.71041 ± 6 |

Sample ratios reported relative to NBS987 $^{87}$Sr/$^{86}$Sr = 0.7102 and all ratios normalised to $^{87}$Sr/$^{86}$Sr = 0.1194. Samples H30, H62, H12, H34 and H19 are from Stephenson (1972), and sample 47 was collected by A. J. Lambeth.

## DISCUSSION

In the absence of substantial evidence supporting a continental fragment hypothesis for the Kerguelen Plateau, we support the model that Heard and McDonald Islands are volcanic accumulations on oceanic crust which has a thick upper layer of pelagic sediments (Houtz et al., 1977). Heard Island is distinctive among oceanic islands because the basement for the volcanic pile that forms the island is exposed, namely, the limestone for which Quilty et al. (this volume) have determined a Middle Eocene to Early Oligocene age. Extrusive

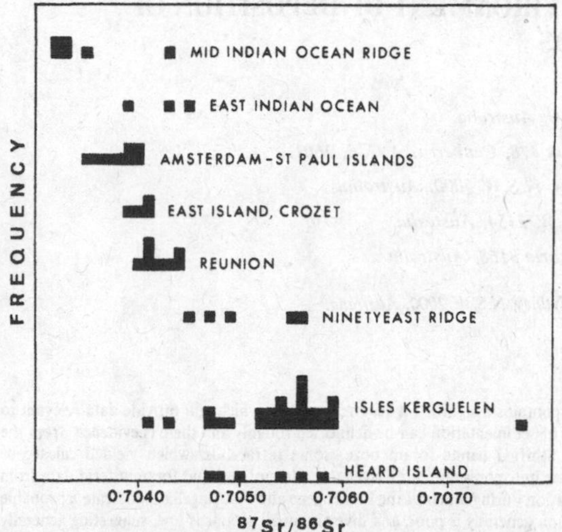


**Figure 4 Frequency histogram of $^{87}Sr/^{86}Sr$ in rocks from the Indian Ocean. Data from Dosso et al. (1979, Table 2 only), Hedge et al. (1973), Subbarao and Hedge (1973), Whitford (1975), Whitford and Duncan (1978) and Table 3.**

products of the earliest phase of Heard Island volcanism have not been located and are presumed to have been removed by erosion. Their intrusive equivalents are present as mafic intrusions in the limestone but are unfortunately too altered for K-Ar dating.

Geochemical similarity between the mafic intrusions and the overlying basalts in the Drygalski Formation and uppermost lavas (Figure 3 and Table 3) is evidence that all three phases of volcanism on Heard Island are related to a single partial melting event. It is suggested that the intrusions are the subvolcanic products of volcanism that was associated with uplift of part of the Kerguelen Plateau, in response to an increased rising geothermal gradient in the lithosphere as it passed over a mantle hot-spot. During uplift the area changed from a depositional regime to an erosional one, and a brief hiatus in the volcanism would have allowed removal of the volcanic and sedimentary overburden to expose the intrusions prior to deposition of the Drygalski Formation. Since a maximum age of Late Miocene to Early Pliocene has been inferred for the Drygalski Formation by Quilty et al. (this volume) a maximum age of Late Miocene is proposed also for the earlier volcanism and uplift. There is no evidence to support Luyendyk and Rennick's (1977) suggestion that volcanism on Heard Island commenced before that on Iles Kerguelen, where the oldest dated rocks are Late Eocene to Early Oligocene (Giret and Lameyre, this volume).

The Drygalski Formation was produced during a phase of sporadic basaltic volcanism, shallow marine sedimentation and glaciation. A final phase of volcanism was marked by an increase in the rate of basalt eruption which led to the build up of Big Ben. From the available K-Ar ages, Big Ben probably began to form a few hundred thousand years ago. Its youthful geomorphology and the similarity with other oceanic volcanoes (McDougall and Duncan, 1980) suggest that it is highly unlikely that Big Ben has a history extending beyond 1 Ma. Within the last 10,000 years trachyte and trachyandesite were erupted from three central vents directly over the Drygalski Formation on Laurens Peninsula. Most recently, volcanic activity has produced numerous small basalt lava fields and scoria cones around the coast of Heard Island, and Mawson Peak on Big Ben.

Phonolite volcanism within the last 100,000 years gave rise to the McDonald Islands. The phonolites are compositionally distinct from the Heard Island trachytes, but both may have evolved via magmas of similar trachyandesite composition.

The relatively young age of volcanism on Heard Island, and the relatively high $^{87}Sr/^{86}Sr$ ratios measured in some products of this volcanism (Figure 4) are together consistent with Duncan's (1978) model in which Heard Island is the latest manifestation of a hot-spot that also produced the Ninetyeast Ridge and Iles Kerguelen. If Duncan's (1978) model is valid, then an important corollary concerns the nature of the underlying mantle. The Ninetyeast Ridge and the Kerguelen Plateau are now separated by the southest Indian Ocean spreading ridge. That the distinctive Sr isotopic signature of the Ninetyeast Ridge and the Kerguelen Plateau is preserved across a spreading centre lends support to the view that oceanic island or hot-spot basalts are derived from deep in the mantle, perhaps from plumes (Hofmann and White, 1982), rather than from shallow depths (Anderson, 1982).

*Acknowledgements* The support of the Antarctic Division, Department of Science, and the Division of National Mapping, Department of National Development, for the 1980 expedition to Heard Island is gratefully acknowledged. I. Clarke thanks I.A. Nicholls and P.J. Stephenson for their encouragement in the preparation of the geology and petrology sections of this paper. D.J. Whitford thanks B.L. Gulson and M. Korsch for valuable assistance and advice. I. Clarke's contribution is published with permission of the Director, Geological Survey of NSW.

## REFERENCES

ANDERSON, D., 1982: Isotopic evolution of the mantle: a model. *Earth Planet. Sci. Lett., 57,* 13-24.

CLARKE, I., 1982: Report on geology; *in* Veenstra, C. and Manning, J. (eds.) *Expedition to the Australian Territory of Heard and McDonald Islands 1980,* Tech Rep.31, Division of National Mapping, Canberra, 46-51.

COLLERSON, K.D., CLARKE, I. and McCULLOCH, M. (in prep.): Sr and Nd isotope and incompatible element geochemistry of Heard and McDonald Islands.

COOPER, J.A., 1963: The flame photometric determination of potassium in geological materials used for potassium argon dating. *Geochim. Cosmochim. Acta, 27,* 525-46.

DOSSO, L. and MURTHY, V.R., 1980: A Nd isotope study of the Kerguelen Islands: inferences on enriched oceanic mantle sources. *Earth Planet Sci. Lett., 48,* 268-76.

DOSSO, L., VIDAL, P., CANTAGREL, J.M., LAMEYRE, J., MAROT, A. and ZIMINE, S., 1979: *"Kerguelen: continental fragment or oceanic island?": petrology and isotopic geochemistry evidence.*Earth Planet. Sci. Lett., 43, 46-60.

DUNCAN, R.A., 1978: Geochronology of basaltic rocks from the Ninetyeast Ridge and continental dispersion in the eastern Indian Ocean. *J. Volcanol., Geotherm. Res., 4,* 283-305.

GIRET, A. and LAMEYRE, J., 1982 (this volume): A study of Kerguelen plutonism: petrology and geochronology.

HEDGE, C.E., WATKINS, N.D., HILDRETH, R.A. and DOERING, W.P., 1973:$^{87}Sr/^{86}Sr$ ratios in basalts from islands in the Indian Ocean. *Earth Planet. Sci. Lett., 21,* 29-34.

HOFMANN, A.W. and WHITE, W.M., 1982: Mantle plumes from ancient oceanic crust. *Earth Planet. Sci. Lett., 57,* 421-36.

HOUTZ, R.E., HAYES, D.E. and MARKL, R.G., 1977: Kerguelen Plateau bathymetry, sediment distribution and crustal structure. *Mar. Geol., 25,* 95-130.

LAMBETH, A.J., 1952: A geological account of Heard Island. *Roy. Soc. NSW, J. Proc., 86,* 14-19.

LUYENDYK, B.P. and RENNICK. W., 1977: Tectonic history of aseismic ridges in the eastern Indian Ocean. *Geol. Soc. Am., Bull., 88,* 1347-56.

McDOUGALL, I. and DUNCAN, R.A., 1980: Linear volcanic chains—recording plate motions? *Tectonophys., 63,* 275-95.

McDOUGALL, I. and SCHIMNKE, H.-U., 1976: Geochronology of Gran Canaria, Canary Islands: age of shield building volcanism and other magmatic phases. *Bull. Volcan., 40,* 1-21.

NORTON, I. and MOLNAR, P., 1977: Implications of a revised fit between Australia and Antarctica for the evolution of the eastern Indian Ocean. *Nature, 267,* 338-40.

QUILTY, P.G., SHAFIK, S., McMINN, A., BRADY, H. and CLARKE, I., (this volume): Microfossil evidence for the age and environment of deposition of sediments on Heard and McDonald Islands.

RILEY, G.H., BINNS, R.A. and CRAVEN, S.J., 1980: Rb-Sr chronology of micas at Jabiluka. *Proc. Int. Uranium Symp. on the Pine Creek Geosyncline,* Int. Atom. Energy Agency, Vienna, 457-68.

SCHLICH, R., DELTEIL, J.R., MOULIN, J., PATRIAT, P. and GUILLAUME, R., 1971: Mise en evidence d'une sedimentation de marge continentale sur le plateau de Kerguelen/Heard. *C. R. Ac. Sci., Paris, 272,* 2060-63.

STEIGER, R.H. and JÄGER, E., 1977: Subcommission on geochronology: convention on the use of decay constants in geo- and cosmochronology. *Earth Planet. Sci. Lett., 36,* 359-62.

STEPHENSON, P.J., 1964: Some geological observations on Heard Island; *in* Adie, R.J. (ed.), *Antarctic Geology.* North-Holland, Amsterdam, 14-23.

STEPHENSON, P.J., 1972: Geochemistry of some Heard Island igneous rocks; *in*Adie, R.J., (ed.), *Antarctic Geology and Geophysics.* Universitetsforlaget, Oslo, 793-801.

SUBBARAO, K.V. and HEDGE, C.E., 1973: K, Rb, Sr and $^{87}Sr/^{86}Sr$ in rocks from the Mid-Indian Oceanic Ridge. *Earth Planet. Sci. Lett., 18,* 223-8.

WATKINS, N.D., GUNN, B.M., NOUGIER, J. and BAKSI, A.K., 1974: Kerguelen: continental fragment or oceanic island? *Geol. Soc. Am., Bull., 85,* 201-12.

WHITFORD, D.J., 1975: Strontium isotopic studies of the volcanic rocks of the Sunda arc, Indonesia, and their petrogenetic implications. *Geochim. Cosmochim. Acta, 39,*1287-1302.

WHITFORD, D.J. and DUNCAN, R.A., 1978: Origin of the Nintyeast Ridge: trace element and Sr isotope evidence. *Carnegie Inst. Washington Yrbk, 77,* 606-13.

# MICROFOSSIL EVIDENCE FOR THE AGE AND ENVIRONMENT OF DEPOSITION OF SEDIMENTS OF HEARD AND McDONALD ISLANDS

P.G. Quilty, *Antarctic Division, Channel Highway, Kingston, Tasmania 7150, Australia.*

S. Shafik, *Bureau of Mineral Resources, Geology and Geophysics, P.O. Box 378, Canberra, A.C.T., 2601.*

A. McMinn, *New South Wales Geological Survey, 36 George Street, Sydney, N.S.W. 2000, Australia.*

H. Brady, *Diatom Research Associates, 8 Elberta Avenue, Castle Hill, N.S.W. 2154, Australia.*

I. Clarke[1], *Department of Earth Science, Monash University, Clayton, Victoria 3168, Australia.*

[1]*Present Address: New South Wales Geological Survey, CAGA Centre, Bent Street, Sydney, N.S.W. 2000, Australia.*

*Abstract* Heard Island is dominated by the volcanic edifice of Big Ben but also contains a significant sediment record, which will provide data relevant to unravelling the history of the Gaussberg-Kerguelen Ridge. So far, three intervals of sedimentation can be delineated roughly and there is evidence, from the McDonald Islands, of a fourth. Laurens Peninsula (Heard Island) contains lithified nanno-foram ooze, some tuffaceous, which yielded calcareous nannofossils and foraminiferids. An argillaceous limestone has also provided some miospores and dinoflagellates. Nannofossil and foraminiferid data from beach pebbles and outcrop samples indicate roughly continuous marine sedimentation during much of the Palaeogene and dinoflagellates indicate a probable Early Oligocene to Middle Miocene age for the argillaceous limestone. Preservation generally is poor, and diversity in all groups is low, suggesting generally low surface water temperatures although there are exceptions. Keeled globorotalids are absent and discoasters uncommon. Nannofossil evidence suggests that waters in the Middle-Late Eocene were generally warmer than in the Oligocene. Dark siltstone from Cape Lockyer (south end Heard Island) has yielded a flora of miospores and marine benthic diatoms. All are consistent with a Late Miocene-Pliocene age. Diatoms suggest deposition in water deeper than 120-140 m and the island's vegetation consisted of low scrubby vegetation in a cold wet environment. Several samples of the Drygalski Formation from the southern side of the Laurens Peninsula contain Neogene foraminiferids, Palaeogene nannofossil taxa recycled from the Laurens Peninsula limestone and a monospecific diatom flora, probably from a freshwater source. The nannofossils include some of Late Paleocene-Early Eocene age, an age older than recorded from the carbonate samples of Laurens Peninsula. McDonald Islands, some 43 km west of Laurens Peninsula, are composed of phonolite instrusions and flows, and tuffs that contain a few clasts of lithified ooze, one of which has yielded a diverse discoaster flora of latest Early Eocene-earliest Middle Eocene age.

Heard and McDonald Islands lie in the southern Indian Ocean (Figure 1) and are administered by Australia which operated a station there between 1947 and 1955. Other visits have been sporadic and most geological studies have been undertaken on an *ad hoc* basis. The islands are unoccupied at present.

The islands are two of three areas (with Iles Kerguelen) which are the only subaerial expressions of an underlying ridge, the Gaussberg-Kerguelen Ridge (Figure 1) of debated origin. The question of an oceanic or microcontinental genesis for the ridge is at present unresolved despite the statements of Anderson (1982) and any new data from elements of the ridge may contribute to a solution to the problem. The problem is discussed further by Clarke et al (this volume). Any proposed origin must take into account the apparent 'twinned' or 'paired' relationship between Gaussberg-Kerguelen and Broken Ridges (Quilty, 1973). Consequent on an understanding of the development of the Indian Ocean circulation patterns, are an understanding of the evolution of the krill ecosystem, and even the development of Antarctic glaciation. An account of the geology of the islands is presented by Clarke et al (this volume).

Heard Island consists of a major volcano (Big Ben whose apex, Mawson Peak, is 2745 m high and displays minor sporadic activity), which has intruded through and built upon older sedimentary rocks (described below) which can be regarded as basement to the island, and probably representative of extensive equivalent units elsewhere on the Gaussberg-Kerguelen Ridge. Heard Island was said to be a "typical "Oceanic" Island" (Stephenson, 1964) because of the association of volcanic rocks and marine sediments. In one respect it is not truly typical of "oceanic" islands because it sits on the Gaussberg-Kerguelen Ridge which is a distinctly atypical platform for oceanic islands. Such islands "typically" rise from ocean floor to sea surface as individual clusters and their life generally is short and consists of essentially one episode, related often to a hot spot origin.

McDonald Islands consist of three separate islands, the largest being McDonald Island which has a maximum elevation of 186 m.

Palaeontological studies on rocks from the entire Gaussberg-Kerguelen Ridge are few and consist of reference to Palaeogene *Globigerina* and *Guembelina* (Glaessner *in* Lambeth, 1952), the bivalve *Chlamys (Zygochlamys) heardensis* (Fleming, 1957), both from Heard Island, undated limestone fragments from Iles Kerguelen (Girod and Nougier, 1972) and mid-Cretaceous and Neogene sediments northeast of Heard Island (Kaharoeddin et al., 1973; Quilty, 1973). Kaharoeddin et al. also recorded Eocene sediments.

## RESULTS

Fossiliferous sediments recorded here come from many locations representing three discrete sequences on Heard Island and a limestone clast in tuffs on McDonald Island. Samples are held by Professor P.J. Stephenson, James Cook University and I. Clarke.

### Limestone from Laurens Peninsula

The most indurated sedimentary rocks on Heard Island are the limestone and associated lithologies of Laurens Peninsula and First Beach. These are the rocks from which Glaessner (in Lambeth, 1952) recorded *Globigerina* and *Guembelina*. They are the oldest rocks known on the island and form the basement on which contemporaneous and later volcanism occurred. Samples examined include about 30 (all from Laurens Peninsula) collected by Professor P.J. Stephenson in 1963 and seven collected by I. Clarke in 1980.

The rocks generally are pale nanno oozes with planktic foraminiferids in various proportions, sporadically rich enough to be termed a nanno/foram ooze. The rocks appear to be well bedded and there are horizons rich in volcanic debris, one an argillaceous limestone. Stylolites and associated concentrations of insoluble residue are common in some samples, but generally this phenomenon is not obvious on the material so far examined.

Most samples are highly indurated so that foraminiferids cannot be extracted and can be identified only roughly from thin sections. Many have yielded calcareous nannofossils but preservation generally is poor due to diagenetic overgrowth. The dark argillaceous limestone noted above has yielded palynomorphs. Radiolaria appear as "ghosts" in thin section but have not been studied as yet.

The foraminiferid fauna varies greatly from sample to sample but generally is dominated by thin or thick walled "globigerinid" species, sometimes with *Guembelina* and very rarely with anything identifiable as a globorotalid. Only one keeled globorotalid was observed.

Calcareous nannofossils are known only from rare *in situ* material from limestone at Laurens Peninsula. One sample (Monash University 65047) yielded four positively and one tentatively identified species which can be assigned an age of mid Eocene-mid Oligocene and to an open ocean environment (Figure 2). Monash University 65097 is the argillaceous limestone from outcrop at First Beach Corinthian Bay. The nannofossil flora includes *Chiasmolithus oamaruensis* and *Reticulofenestra umbilica* of Late Eocene-mid Oligocene age. *Cyclicargolithus reticulatus* is tentatively identified and if correct, would restrict the age to the Late Eocene. This sample also yielded a palynomorph flora, the only one obtained from the Palaeogene rocks in the northern part of the island. Present are the dinoflagellates, *Operculodinium* sp., *Spiniferites* sp., *Impagidinium maculatum, Batiacasphaera micropapillata, Cymatiosphaera* sp. and *Nematosphaeropsis* sp. Miospores present include: *Gleicheniidites* sp., *Cyathidites australis* and *C. minor.*

Stover (1977) indicated that *Batiacasphaera micropapillata* had a range, on the Blake Plateau, between the Middle Oligocene and the Early Miocene. There are few other documented occurrences of this species, and its total range therefore remains uncertain. Wilson and Clowes (1980) recorded a minimum age of Middle Miocene for the genus *Batiacasphaera*. Most other genera and species from the Heard Island assemblage are long ranging. Stover (1977) suggested that the early Middle Oligocene was characterised by assemblages of long ranging species. Partridge (pers. comm.) also has a zone for the Early-Middle Oligocene in the Gippsland Basin in which long ranging forms, particularly *Operculodinium* spp., are dominant and there is an absence of diagnostic species. The assemblage from Heard Island probably correlates with this interval.

The dinoflagellates indicate an Early Oligocene-Middle Miocene age and the dominance of *Operculodinium* (68%) and the low diversity also are consistent with the indications of cool open marine conditions suggested by the nannofossils. Perhaps surprising is the absence of evidence of *Nothofagus* among the miospores, a genus well represented in coeval miospore floras from most "nearby" localities, (e.g. southern Australia), although the open ocean associations of other microfossil groups may suggest that the flora is an oceanic island one where *Nothofagus* may not be necessarily expected. Beach pebbles of Laurens Peninsula limestone also contain calcareous nannofossils of mid-Late Eocene age and one (Monash University 65107) includes several species of *Discoaster* attesting to warmer water periods.

Generally the limestone from Laurens Peninsula outcrop and beach pebbles, yields evidence of Palaeogene (mid Eocene-mid Oligocene) age, usually deposited from cool open ocean conditions but with less common intervals of warmer oceanic conditions (Figure 2).

### *Further Evidence of Palaeogene Sedimentation*

As will be discussed below, the Drygalski Formation contains Neogene foraminiferids and diatoms. Cursory examination showed that calcareous nannofossils are present so it was examined in more detail. All nannofossils appear to be recycled from Palaeogene sources and include evidence of sedimentation older than that suggested by analysis of the known Laurens Peninsula limestone. Because of extensive reworking of older calcareous nannofossils into younger sediments, mixing of floras must be expected so that, while good age data may still be obtained because of the presence of

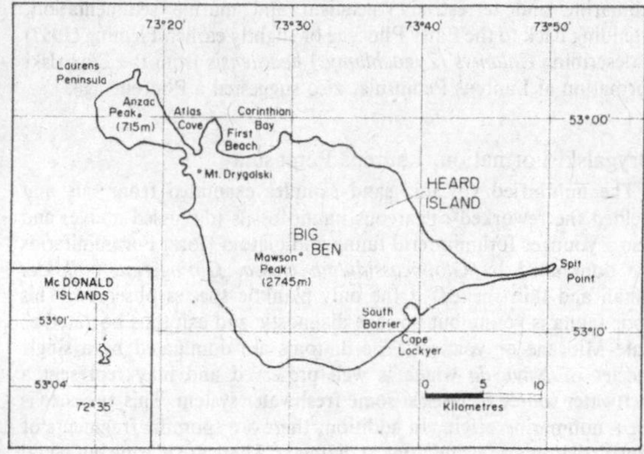


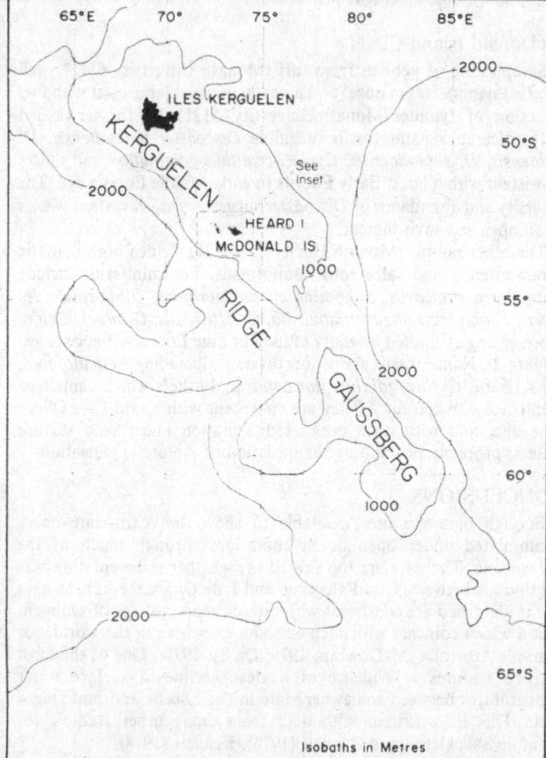


**Figure 1. Gaussberg-Kerguelen Ridge and Heard and McDonald Islands showing localities mentioned in the text.**

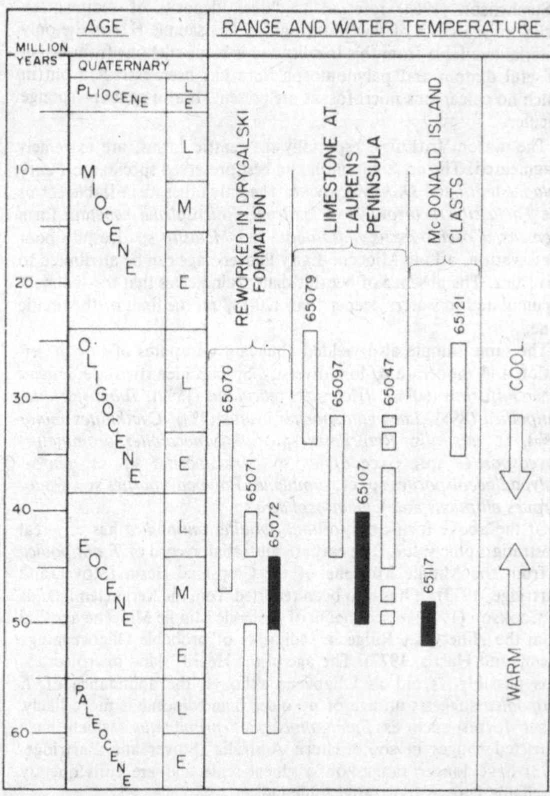


**Figure 2. Summary of results. Specimen numbers (e.g. 65070) refer to specimens in the Department of Earth Science, Monash University. Open bars indicate cool surface water, closed bars, warm water.**

diagnostic species, some environmental aspects cannot be so well delineated.

The Drygalski Formation sediments are represented by three volcanic sand samples, taken one metre apart from the south coast of Laurens Peninsula (Monash University 65070-65072) and a contained limestone clast (Monash University 65079) from the same area. Preservation of the nannofossils is poor, the specimens suffering from overgrowths, dissolution effects and breakage.

Nannofossils from the volcanic sand samples all contain evidence of Late Paleocene to Early Eocene age in the form of *Discoaster multiradiatus, Discoasteroides* spp. and *Cyclococcolithus gammation*. The nannofossil taxa from one sample (Monash University 65072) include *Cyclicargolithus reticulatus, C. floridanus, Coccolithus eopelagicus, Reticulofenestra umbilica, R. scissura, R. scrippsae, Markalius inversus, Discoaster saipanensis, D. sublodoensis, D. lodoensis, D. barbadiensis, D. multiradiatus, D. binodosus, D. tani, D. deflandrei, Cyclococcolithus robustus, Coccolithus pelagicus, Prinsius bisculus, Neococcolithes* sp., *Zygrhablithus bijugatus, Triquetrorhabdus inversus, Chiasmolithus solitus, C. grandis, Discoasteroides bramlettei,*and *Fasciculithus* spp. which are the result of mixing of probably three assemblages varying in age. A Late Eocene age (based on *Chiasmolithus oamaruensis* and *Discoaster saipanensis* in 65071) is also indicated amongst the reworked specimens and many forms range up well into the Oligocene showing that sedimentation may have continued into the Oligocene.

Generally, deposition occurred in an open ocean environment. Distance from shore and water temperature indications are unreliable although those intervals indicated by diversified *Discoaster* may represent "warmer" water periods.

The limestone clast (65079) from the Drygalski Formation contains abundant *Chiasmolithus* spp. (mostly represented by rims resembling those of *C. altus* and *C. oamaruensis*), *Cyclicargolithus floridanus* and *Reticulofenestra scissura* of mid to Late Oligocene age.

### Siltstone from Cape Lockyer

Stephenson (1964) referred to "thick deposits of sedimentary rocks" in sea cliffs of the South Barrier. His sample H13 is the only material available from this locality and is black siltstone from which a useful diatom and palynomorph flora has been extracted but in which no calcareous microfossils are present. It also contains sponge spicules.

The diatom frustules, especially the centric forms, are extremely fragmented. The most common and best preserved species are *Denticula hustedtii* and *D. h.* var. *ovata*. The only other identifiable forms are *Thalassiosira torokina, T. oestrupi, Rhizosolenia hebetata* form *hiemalis, Coscinodiscus marginatus* and *Melosira* sp. Despite poor preservation, a Late Miocene-Early Pliocene age can be attributed to this flora. The absence of benthic diatoms indicates that the sediment accumulated in waters deeper than 120-140 m, the limit of the photic zone.

The same sample also yielded abundant miospores of fair preservation and moderate to low diversity. Species identified are *Triletes tuberculiformis* (60%), *Tricolpites reticulatus* (15%), *Tubulifloridites antipodica* (8%), *Laevigatosporites ovatus* (7%), *Cyathidites minor* (6%), *C. australis, Retitriletes* sp., *Spinizonocolpites prominatus, Foveosporites* sp., *Cycadopites* sp., *Tricolporites* sp. cf. *leuros, Polypodiaceoisporites* sp. cf. *tumulatus, Polypodiisporites* sp., *Podocarpites ellipticus* and *Cymatiosphaera* sp.

Of the above forms only *Tubulifloridites antipodica* has any real biostratigraphic value. The earliest Australian record of *T. antipodica* is from the Middle Miocene of the Gippsland Basin (Stover and Partridge, 1973). It has also been reported from the Kerguelen Islands by Cookson (1947) in sediments of probable Middle Miocene age and from the Ninetyeast Ridge in sediments of probable Oligocene age (Kemp and Harris, 1977). The age of the Heard Island microflora is thus possibly as old as Oligocene although the abundance of *T. antipodica* suggests an age of no older than Miocene is more likely. Other forms such as *Spinizonocolpites prominatus,* which have restricted ranges in southeastern Australia (Stover and Partridge, 1973) have longer ranges on a global scale and are consequently unreliable here as biostratigraphic indices.

There is a close similarity between the microfloral compositions of the samples from Heard Island, Kerguelen Islands and Ninetyeast Ridge. Of the 14 forms of spores and pollen from Heard Island 10 were also identified at site 254, Ninetyeast Ridge (Kemp and Harris, 1977) and nine from the Kerguelen Islands (Cookson, 1947). The assemblages from Heard Island and the Kerguelen Islands are particularly similar. Both are dominated by pteridophyte spores (and by *T. tuberculiformis* in particular) and they each have a similar and impoverished angiosperm microfloral component. This angiosperm component lacks forms typical of a continental province (Bliss, 1979) such as Centrolepidaceae, Proteaceae, Restionaceae, Myrtaceae and *Nothofagus*. The absence of these forms together with the dominance of forms of pteridophyte affinity suggests a parent vegetation not dissimilar to the fern bush vegetation of some modern sub-antarctic islands.

Stephenson (1964) included these sediments in his "glacial sediments and interbedded lavas" which included the Drygalski Formation (defined in Clarke et al this volume) and all volcanic sediments and debris associated with the episode of the volcanism culminating in the formation of Big Ben and ultimately the island itself.

The Late Miocene-Early Pliocene age suggests that Big Ben and its associated volcanism are part of a sequence of events, including submarine and terrestrial volcanism and marine sedimentation, extending back to the Early Pliocene or slightly earlier. Fleming (1957) in describing *Chlamys (Zygochlamys) heardensis* from the Drygalski Formation at Laurens Peninsula, also suggested a Pliocene age.

### Drygalski Formation, Laurens Peninsula

The unlithified volcanic sand samples examined from this unit yielded the reworked calcareous nannofossils (discussed above) and also a younger foraminiferid fauna and diatom flora. Foraminiferids are dominated by *Globocassidulina crassa. Globigerina bulloides* (small and thin shelled) is the only planktic species observed. This poor fauna is young but not age diagnostic and can thus be dated as Late Miocene or younger. The diatoms are dominated by a single species of *Navicula* which is well preserved and may represent a meltwater source or at least some freshwater system. This evidence is for a nonmarine origin. In addition, there are sporadic fragments of centric diatoms a few microns in diameter. These are marine but could easily be windblown fragments.

### McDonald Island Clasts

Samples are of pebbles from tuff (Monash University 65117, well lithified nanno/foram ooze) and a partly lithified large clast with two directions of stylolites (Monash University 65121). The former yielded only calcareous nannofossils including *Discoaster barbadiensis, D. lodoensis, D. sublodoensis, Cyclococcolithus gammation* and others consistent with a latest Early Eocene to early Middle Eocene age. The diversity and abundance of *Discoaster* suggest warmish surface waters in an open sea environment.

The other sample (Monash University 65121) yielded both planktic foraminiferids and calcareous nannofossils. Foraminiferids include *Catapsydrax unicavus, Subbotina* cf. *angiporoides, Globorotaloides suteri, Globigerina angustiumbilicata, G. officinalis, G. praebulloides* representing a "limited diversity" fauna of Late Eocene-Oligocene age (Figure 2). Nannofossils are of low diversity, including *Reticulofenestra scissura, Cyclicargolithus floridanus* and others which cannot be identified with certainty. They are consistent with a mid-Late Oligocene age, and with open ocean sedimentation under cold surface waters, probably post dating the initiation of Antarctic glaciation.

## CONCLUSIONS

Enough data are now available to show that carbonate oozes accumulated under open ocean conditions through much of the Palaeogene. The data are too few to say whether sedimentation was continuous between Late Paleocene and Late Oligocene but the ages so far identified are consistent with intermittent episodes of sedimentation which coincide with such episodes elsewhere in the world, for example Australia (McGowran, 1979; Quilty, 1977). One of the most marked features is evidence of a clear decline in surface water temperatures between somewhere late in the Eocene and mid-Oligocene. This is consistent with data from many other studies, for example Shackleton and Kennett (1975), Kennett (1978).

The Drygalski Formation formed, at least in part, during the Late Miocene-Early Pliocene and is thus not simply "Pleistocene". The suggestion of a glacial origin in part is consistent with evidence of a marked increase in intensity of Antarctic glaciation in the Middle Miocene (see also Shackleton and Kennett, 1975). Palaeogene sediments are open ocean carbonate, whereas the Neogene deposits are volcanic-glacial. There is little evidence of the buildup of Heard Island by volcanism earlier than Late Miocene. The age of formation of the rocks comprising McDonald Islands is not revealed by this study.

*Acknowledgements* McMinn and Shafik publish with the permission of the Secretary of the New South Wales Department of Mineral Resources and of the Director, Bureau of Mineral Resources, respectively; John Cox and Lesley-Manson, Antarctic Division, drafted the figures and typed the paper respectively. Peter Keage of the Antarctic Division provided valuable information on the Islands.

## REFERENCES

ANDERSON, D.L., 1982: Isotopic evolution of the mantle: a model. *Earth Planet. Sci. Lett., 57,* 13-24.

BLISS, L.C., 1979: Vascular plant vegetation of the Southern Circumpolar Region in relation to Antarctic, alpine and Arctic vegetation. *Can J. Bot., 57,* (20), 2167-78.

CLARKE, I., McDOUGALL and WHITFORD, D. (this volume): Volcanic evolution of Heard and McDonald Islands, Southern Indian Ocean.

COOKSON, I.E., 1947: Plant microfossils from the lignites of Kerguelen Archipelago. *BANZ Ant. Res. Exped. 1929-1931. Rept. Ser. A., 5, 2(8),* 127-42.

FLEMING, C.A., 1957: A new species of fossil *Chlamys* from the Drygalski Agglomerate of Heard Island, Indian Ocean. *Geol. Soc. Aust., J., 4,* 13-19.

GIROD, M. and NOUGIER, J., 1972: Volcanism of Sub-Antarctic Islands; *in* Adie, R.J. (ed.) *Antarctic Geology and Geophysics.* Universitetsforlaget, Oslo, 777-88.

KAHAROEDDIN, F.A., WEAVER, F.M. and WISE, S.W., 1973: Cretaceous and Paleogene cores from the Kerguelen Plateau, southern ocean. *U.S. Antarct. J., 8,* 197-8.

KEMP, E.M. and HARRIS, W.K., 1977: The palynology of Early Tertiary sediments, Ninetyeast Ridge, Indian Ocean. *Spec. Pap. Palaeontology, 19,* 1-69.

KENNETT, J.P., 1978: The development of planktonic biogeography in the Southern Ocean during the Cenozoic. *Mar. Micropalaeontol., 3,* 301-45.

LAMBETH, A.J., 1952: A geological account of Heard Island *R. Soc. NSW., J. Proc., 86,* 14-9.

McGOWRAN, B., 1979: The Tertiary of Australia: Foraminiferal overview. *Mar. Micropalaeontol., 4,* 235-64.

QUILTY, P.G., 1973: Cenomanian-Turonian and Neogene sediments from northeast of Kerguelen Ridge, Indian Ocean. *Geol. Soc. Aust., J., 20,* 361-71.

QUILTY, P.G., 1977: Cenozoic sedimentation cycles in Western Australia. *Geology, 5,* 336-40.

SHACKLETON, N.J. and KENNETT, J.P., 1975: Paleotemperature history of the Cenozoic and the initiation of Antarctic glaciation: oxygen and carbon isotope analyses in DSDP sites 277, 279 and 281. *Initial Reports of the Deep Sea Drilling Project, 29,* 743-56. U.S. Govt. Printing Office, Washington, DC.

STEPHENSON, P.J., 1964: Some geological observations on Heard Island; *in* Adie, R.J. (ed.) *Antarctic Geology,* North Holland, Amsterdam, 14-23.

STEPHENSON, P.J., 1972: Geochemistry of some Heard Island igneous rocks; *in* Adie, R.J. (ed.) *Antarctic Geology and Geophysics.* Universitetsforlaget, Oslo, 793-801.

STOVER, L.E., 1977: Oligocene and Early Miocene dinoflagellates from Atlantic corehole 5/5B, Blake Plateau *in* Elsik, W.C. (ed.), Contributions of Stratigraphic Palynology, Volume 1, Cenozoic Palynology. *Amer. Assoc. Stratigraphic Palynologists, Contr. Ser., 5A,* 66-89.

STOVER, L.E. and PARTRIDGE, A.D., 1973: Tertiary and Late Cretaceous spores and pollen from the Gippsland Basin, southeastern Australia. *R. Soc. Vic., Proc., 85,*237-86.

WILSON, G.J. and CLOWES, C.D., 1980: A concise catalogue of organic-walled fossil dinoflagellate genera. *N.Z. Geol. Surv. Rept., 92.*

# CHRONO-SPATIAL EVOLUTION OF THE VOLCANIC ACTIVITY IN SOUTHEASTERN KERGUELEN (T.A.A.F.)

J. Nougier and D. Pawlowski, *Laboratoire de Géologie, 33 rue Louis Pastuer, 84000 Avignon, France.*

J. M. Cantagrel, *Laboratoire de Géologie, 5 rue Kessler, 63031 Clermont-Ferrand Cédex, France.*

*Abstract* In southeastern Kerguelen alkali basalt sequences were erupted from 25 to 5 Ma. The alkali basalts include (a) basalts without nepheline which flowed between 21 and 18 Ma at a mean rate of one metre thickness in 4000 years, overlain by very recent (0.25 Ma) differentiated bodies of trachy-phonolites; (b) in the southeast, olivine-alkali basalts, nepheline-basalts and basanites that erupted between 22 to 20 Ma ago, at a similar mean flow rate, followed by comagmatic extrusions between 10 to 7 Ma.

Petrochemical evidence supports the flood basalt eruption process in related areas during the Miocene period as consisting of (a) simultaneous eruptions generated in separate magma chambers, and (b) differentiated felsic lavas erupted at the same time as the nepheline basalts, and suggests a correlation of the distribution of alkali magmatism on Kerguelen with the orientation of transform faults in the surrounding ocean crust.

During the last 15 years considerable improvements have been made concerning our knowledge of the submarine geophysical and topographic features of the southeastern Indian Ocean (Schlich, 1975; Luyendyk et al., 1977; Houtz, 1977) as well as the volcanic features of Kerguelen (Nougier, 1970; 1972 a-b; Watkins et al., 1974; Marot and Zimine, 1976; Dosso, 1977; Dosso et al., 1979, 1980; Giret et al., 1981).

The Kerguelen-Heard plateau surface is cut by troughs filled with Cenomanian sediments (90 Ma), with an Eocene seismic reflector interpreted as a peneplenation phase. Consequently we suggest that the oceanic basement floor structure is about 100 Ma old, but that the Broken Ridge-Kerguelen plateau break occurred about 55 Ma ago. The oldest magnetic anomalies buried by the basaltic slopes are identified as Anomaly 18—that is 45 Ma.

The subaerial plateau (flood) basalts consist of a strongly zeolitized basal unit of transitional theoleiites, overlain by an upper alkali unit containing basanites and hawaiites. The upper part of the basal unit could not be located. The upper unit occurs mainly on the top of plateaus and in the whole of the southeast province.

Petrological and isotopic studies recently carried out on intrusive plutonic bodies are in agreement with an origin by oceanic crust differentiation. These plutonic complexes and the new ones discovered during the past two years (Giret et al., 1981) have chemical equivalents in the differentiated volcanic suites.

During the 1977 field season we have carried out some detailed studies in the southeastern province (Figure 1), where two alkali flood basalt units occur together with a huge recent strombolian volcano (Mt Ross, 1840 m) and two kinds of differentiated extrusions; an older phonolitic group and a recent trachy-phonolitic group. The whole area is a good example of magmatism and its evolution with time according to the Kerguelen model. Chemical and chronological data also provide information about the basaltic activity rate, and allow correlation of the age of the differentiated basalts with their parental magma and put some limits on the ages of features such as lignite beds and Pecten layers.

## Field Work

We consider that the southeastern province is made of two main units (Figure 2). The Gallieni unit, and the Southeast unit (Table 1) which latter comprises both the Jeanne d'Arc and Ronarc'h peninsulas.

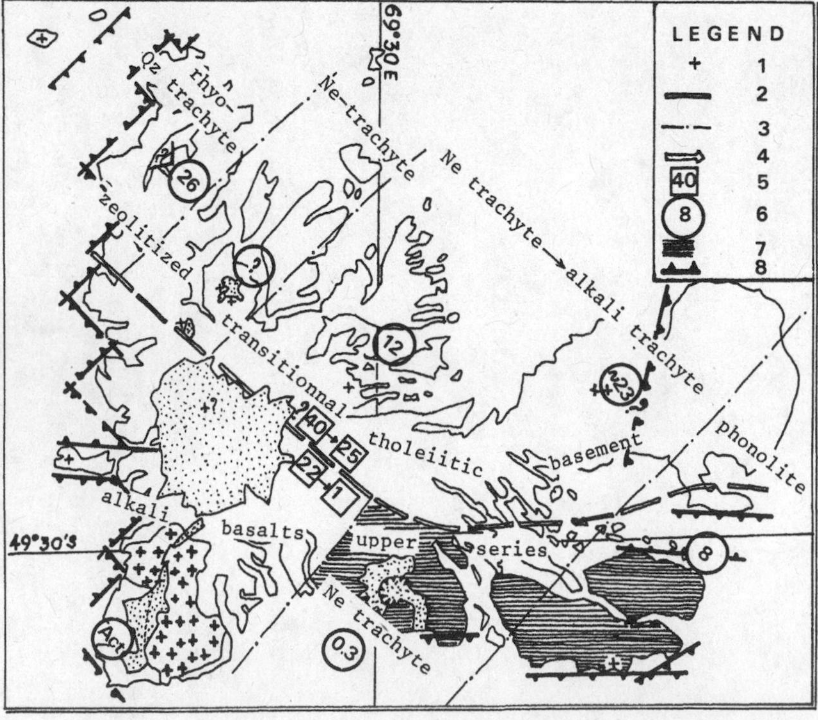


**Figure 1.** Schematic chrono-spatial relationship on Kerguelen archipelago. *Legend:* 1: plutonic bodies or intrusions; 2: limit between basaltic fields; 3: limit between differentiated intrusive lavas series; 4: direction of chronomigration; 5: K-Ar age of basaltic fields; 6: K-Ar age of differentiated lavas; 7: southeastern province; 8: possible fault-system in relationship with peri-insular platform structure.

TABLE 1: Modal analyses of alkali basalts from Gallieni and Southeast units.

Branca Section (base → top)

| N° | Plag. phen | Plag. micr | Cpx phen | Cpx micr | Ol phen | Ol micr | Op | Gdm | Vac |
|---|---|---|---|---|---|---|---|---|---|
| 53 | — | 50,5 | — | 14.6 | — | 6,2 | 14.4 | 14.3 | — |
| 60 | 9,2 | 30,8 | 5,5 | 10,2 | 6,4 | 2,4 | 16.2 | 14.3 | 3.8 |
| (total) | 40 | | 15.7 | | 8.8 | | | | |
| 69 | 26.5 | 27.7 | — | 15.2 | 2 | 7 | 10.4 | 10.6 | 0.6 |
| (total) | 54.2 | | | | 9 | | | | |

South-East Section (base → top)

| N° | Plag. phen | Plag. micr | CPX phen | CPX micr | Ol phen | Ol micr | Horn | Biot | Neph | Apat | Op | Gdm | Vac |
|---|---|---|---|---|---|---|---|---|---|---|---|---|---|
| 125 | — | — | — | 2.9 | — | — | 10.7 | — | 2.8 | — | 13.1 | 70.4 | — |
| 31 | 10 | 25 | 2.8 | 14.5 | 4.1 | 0.2 | — | — | — | — | 29 | 14.3 | — |
| (total) | 35 | | 17.2 | | 4.3 | | | | | | | | |
| 134 | — | 42.9 | — | 12.5 | 14.6 | 4.6 | — | — | — | — | 16.8 | 6.9 | 1.6 |
| (total) | | | | | 19.2 | | | | | | | | |
| 42 | 0.4 | 23.9 | 14 | 5.3 | 13.4 | 5.1 | — | — | — | — | 13.1 | 20.7 | 0.7 |
| (total) | 24.3 | | 19.3 | | 18.5 | | | | | | | | |
| 44 dyke | — | 0,1 | 7,1 | 29,7 | — | 1,2 | 0,5 | 5,4 | — | 0.7 | 13.8 | 36.4 | 3.6 |
| (total) | | | 36.8 | | | | | | | | | | |

*The Gallieni unit* This is morphologically controlled by the shape of the Ross volcano. All peripheral flood basalts are deeply dissected into characteristic pyramidal shaped mountains nearly 900 m high. They are cut in the wide tectono-glacial valleys by trachy-phonolitic sills or needle like extrusions.

The Ross volcano itself is set on a basement of basalt and is 15 km in diameter with a crater 4 km in diameter. Associated with this main volcano are numerous recent scoriaceous cones set within a radius of less than 20 km.

Basaltic flows are 3 to 10 m thick, sub-horizontal and slightly zeolitized, and typically interbedded with red cinders or pyroclasts. New K-Ar data (Table 2) show these strata to have been erupted at wide intervals and indicate cyclic activity possibly starting with a gas-rich magma erupted first as tephra, followed by flow with its scoriaceous surface weathered during inactive periods. The gentle dip towards the SE implies that the flood basalts flowed from the west. A detailed study has been carried out on Pyramide Branca (altitude 905 m)where the basal flows are dated at 21 Ma and upper lavas are 17.5 Ma old (Table 2). However, 32 km to the west, upper lavas from Sentine Mountain have been dated at only 5.9 Ma (Nougier, 1972). Mt

TABLE 2: Geochronological data for lava series from southeastern province on Kerguelen archipelago. Analyst CNRS—LA 10, Clermont-Ferrand. For location of samples see map Figure 2.

| Sample N° | Location | Nature | K% | $Ar^{+}_{40}$ | $A^{tm}$ Ar% | Age (Ma) |
|---|---|---|---|---|---|---|
| K24 (R7855) | Ross Volc. extrusion Doigt-Ste-Anne | Ne-trachyte | 4.47 | 0.012 | 99.4 | 0.04 ± 0.02 |
| K12 (R7852) | Ross Volcano Table Beaulieu | Ne-trachyte | 4.99 | 0.056 | 93.2 | 0.20 ± 0.05 |
| K18 (R7854) | Ross Volcano Table du cratere | Ne-trachyte | 5.44 | 0.080 | 90.5 | 0.20 ± 0.05 |
| K30 (R7853) | Ross Volcano Table Boisguehenneuc | Ne-trachyte | 5.27 | 0.112 | 89.0 | 0.30 ± 0.08 |
| K120 (R7857) | Mt Thomson extrusion Ronarc'h peninsula | phonolite | 5.91 | 2.69 | 26.8 | 6.6 ± 0.2 |
| K150 (R7849) | Lava Flow alt 655 Ronarc'h peninsula | nephelinite | 3.63 | 1.86 | 19.9 | 7.4 ± 0.2 |
| K107 (R7856) | Oreilles de chat extrusion Ronarc'h peninsula | phonolite | 5.35 | 3.13 | 16.6 | 8.4 ± 0.2 |
| K84 (R7858) | Mt Tizard extrusion Jeanne d'Arc peninsula | phonolite | 4.98 | 3.55 | 33.7 | 10.2 ± 0.3 |
| K52 (R7847) | Mt Branca alt 850 Gallieni unit | trachy basalt | 1.97 | 2.42 | 36.8 | 17.7 ± 0.5 |
| K138 (R7850) | Ronarc'h peninsula see level lava flow | alk. basalt | 1.59 | 2.21 | 42.3 | 20.0 ± 0.6 |
| K72 (R7848) | Mt Branca see-level Gallieni unit | Ol.alk.basalt | 0.945 | 1.38 | 31.8 | 21.0 ± 0.6 |
| K34 (R7851) | Jeanne d'Arc peninsula alt 460m | alk.basalt | 1.20 | 1.83 | 28.9 | 21.9 ± 0.5 |

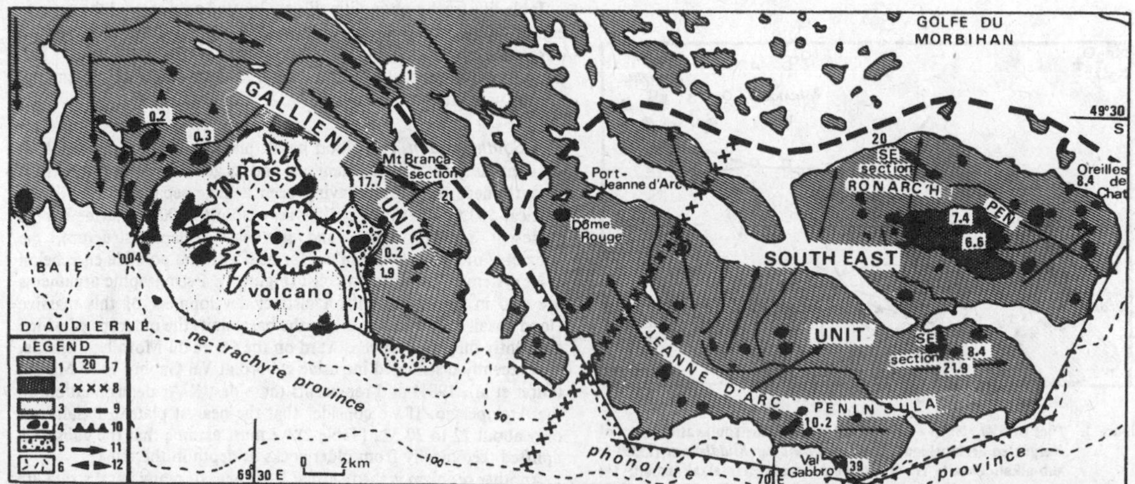


Figure 2. Schematic volcanological map of southeastern Kerguelen province (after J. Nougier, 1970). *Legend:* 1: flood alkali basalt plateau; 2: basanites and nephelinites of Ronarc'h pen.; 3: Pliocene Ross volcano; 4: differentiated extrusion; 5: moraine deposits; 6: glaciers; 7: K-Ar dates; 8: theoretical limit of Ne-trachytes and phonolites; 9: theoretical limit of Gallieni and Southeast units; 10: faults; 11: main morphological flow of flood basalt; 12: petrochemical section of flood basalts.

TABLE 3: Representative chemical analyses from southeastern province of Kerguelen archipelago. For K/Ar data see Table 2

| | Alkali basalts series | | | | | differentiated lavas | | | | | basalts averages | |
|---|---|---|---|---|---|---|---|---|---|---|---|---|
| | South-East unit | | | Gallieni unit | | South-East unit | | | Gallieni unit | | Gall unit | SE unit |
| | K 150 | K 138 | K 34 | K 47 | K 72 | K 84 | K 107 | K 120 | K 30 | K 24 | CS13B* | CS14B** |
| $SiO_2$ | 46.50 | 46.80 | 47.60 | 46.00 | 45.40 | 56.65 | 54.55 | 51.70 | 58.05 | 61.40 | 48.10 | 46.50 |
| $Al_2O_3$ | 16.40 | 16.90 | 15.20 | 14.50 | 16.50 | 21.55 | 21.45 | 20.10 | 18.45 | 18.50 | 15.60 | 15.30 |
| $Fe_2O_3$ | 4.99 | 5.70 | 7.62 | 7.26 | 4.68 | 2.23 | 2.29 | 2.71 | 2.59 | 3.28 | 6.32 | 6.81 |
| FeO | 5.50 | 7.39 | 4.49 | 4.27 | 8.40 | .60 | 1.45 | .80 | 3.84 | 1.01 | 5.95 | 5.11 |
| MgO | 4.40 | 4.60 | 7.30 | 10.40 | 6.90 | .18 | .47 | .66 | .54 | .03 | 8.22 | 5.85 |
| CaO | 6.20 | 6.10 | 10.00 | 9.00 | 8.30 | .85 | 1.10 | 1.70 | 2.25 | .70 | 8.27 | 8.49 |
| $Na_2O$ | 5.80 | 3.50 | 2.60 | 2.70 | 2.80 | 9.10 | 9.20 | 9.00 | 5.65 | 8.00 | 2.75 | 3.62 |
| $K_2O$ | 4.10 | 1.85 | 1.70 | 1.00 | 1.10 | 6.10 | 6.50 | 6.60 | 6.85 | 5.60 | 1.69 | 2.35 |
| $TiO_2$ | 3.00 | 3.50 | 3.30 | 2.20 | 3.20 | .15 | .50 | .50 | .40 | .00 | 3.10 | 3.13 |
| MnO | .20 | .15 | .17 | .16 | .19 | .14 | .13 | .12 | .18 | .18 | .17 | .18 |
| $H_2O^+$ | 1.61 | 1.54 | .35 | 1.78 | 1.00 | 1.29 | .87 | 4.15 | .01 | .85 | 1.26 | 1.78 |
| $H_2O^-$ | .45 | 1.11 | .37 | 1.16 | .69 | .19 | .16 | .97 | .20 | .28 | .77 | .84 |
| TOTAL | 99.15 | 99.14 | 100.70 | 100.43 | 99.16 | 99.03 | 99.07 | 99.01 | 99.01 | 99.83 | 99.70 | 99.90 |
| Or | 24.98 | 11.34 | 10.06 | 6.07 | 6.68 | 36.99 | 39.38 | 41.58 | 41.01 | 33.56 | 10.00 | 13.90 |
| Ab | 11.88 | 30.66 | 21.98 | 23.41 | 24.28 | 32.34 | 19.23 | 8.95 | 33.84 | 50.01 | 23.24 | 22.84 |
| An | 6.76 | 25.80 | 24.75 | 25.09 | 29.92 | — | — | — | 4.75 | — | 25.19 | 18.52 |
| Ne | 20.92 | — | — | — | — | 25.09 | 30.63 | 33.52 | 7.86 | 7.94 | — | 4.20 |
| Di | 19.63 | 4.45 | 19.35 | 16.12 | 9.92 | 1.00 | 4.62 | 5.53 | 5.64 | .99 | 12.39 | 18.36 |
| Hy | — | 9.83 | 5.92 | 5.43 | 6.47 | — | — | — | — | — | 11.05 | — |
| Mo | 7.45 | 8.57 | 5.45 | 8.11 | 6.96 | 2.00 | 1.68 | — | 3.80 | 3.12 | 9.16 | 7.98 |
| Il | 5.87 | 6.89 | 6.27 | 4.29 | 6.24 | .29 | .97 | 1.01 | 0.77 | — | 5.89 | 5.95 |
| He | — | — | 3.86 | 1.86 | — | .84 | — | — | — | — | — | 1.31 |
| Aeg | — | — | — | — | — | .19 | 3.43 | 8.33 | — | 3.39 | — | — |
| Woll | — | — | — | — | — | 1.27 | — | .90 | — | .99 | — | — |
| Ol | 2.51 | 2.46 | 2.36 | 9.64 | 9.54 | — | .06 | — | 2.34 | — | 3.25 | 4.29 |
| D.I. | 57.8 | 42.0 | 32.0 | 29.5 | 30.9 | 94.4 | 89.2 | 84.0 | 82.7 | 91.5 | 33.1 | 42.0 |
| S.I. | 18.1 | 20.5 | 31.8 | 41.8 | 29.5 | 1.0 | 2.4 | 3.4 | 2.8 | .2 | 33.8 | 25.3 |

*average of 4 analyses. **average of 16 analyses. Norms are calculated free of water.

(1) Alkali basalt series
- (a) Southeast Unit:
  - K 150: Ronarc'h pen. alt. 655 m, nephelinite
  - K 138: Ronarc'h pen. sea level, alk-basalt
  - K 34: Jeanne d'Arc pen. alt. 460 m, alk-basalt.
- (b) Gallieni Unit:
  - K 47: Mt Branca alt. 905 m, Ol-alk-basalt
  - K 72: Mt Branca sea level, Ol-alk-basalt.

(2) Differentiated lavas
- (a) K 84: Jeanne d'Arc pen., Mt Tizard, phonolite
  - K 107: Ronarc'h pen., Oreilles de Chat, phonolite
  - K 120: Ronarc'h pen., Mt Thomson alt. 930 m, phonolite.
- (b) Gallieni Unit:
  - K 30: Table de Boisguehenneuc, Ne-trachyte
  - K 24: Doigt-de-Ste-Anne, Ne-trachyte.

(3) Basalt averages
- CS13B: Gallieni unit, 4 analyses, alk-basalt with normative Hy.
- CS14B: Southeast unit, 16 analyses, Ne-alk-basalt.

All data from CNRS—L.A. 10, Clermont-Ferrand. Norms are calculated free of water, computer program written by J-L. Larice, Faculte des Sciences d'Avignon.

Ross volcano is built up of pyroclastics laced and strengthened by feeder dykes. The age of Ross activity is about 2 Ma and the strombolian parasitic cones are 1 Ma old (Nougier, 1972).

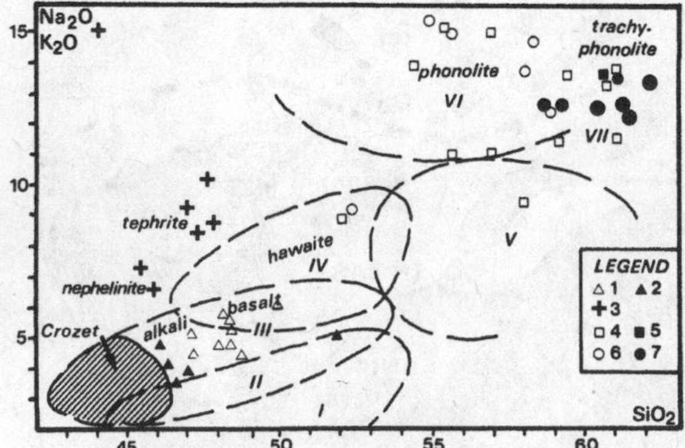


Figure 3. Plot of $SiO_2/Na_2O+K_2O$ for all lavas of the southeastern part of Kerguelen archipelago. *Nomenclature after Middlemost (1980)* 1: sub-alkali basalt; II: transitional basalt; III: alkali basalt; IV: trachy-basalt; V: trachy-andesite; VI: phonolite; VII: trachyte. *Legend: Southeast unit:* 1: alkali basalt; 3: tephrite; 4: phonolite (data from Nougier, 1970); 6: phonolite, new data. *Gallieni unit:* 2: alkali basalt; trachy-phonolite (data from Nougier, 1970); 7: trachy-phonolite, new data.

Trachy-phonolitic extrusions occur only at lower topographic levels and result from Ross volcano activity and their age is from 1.9 Ma (Nougier, 1972) to 0.04 Ma (Doigt de Ste Anne, Table 2).

Table-like trachy-phonolitic sills are located only around the periphery of the Ross volcano. At least three 80 m thick series can be observed, from 100-250 m, 600-700 m and over 1000 m altitude. The middle series is dated 0.3-0.2 Ma (Table 2) and is probably contemporaneous with late Ross activity.

*The Southeast unit* This flood basalt plateau is not in morphological continuity with the Gallieni one and its general dip to the north is in agreement with there having been a huge separate unit. Today, vertical 500 to 800 m high cliffs very close to 100 m marine depths represent a major south and southeast collapse structure, the geometry of which is not in conformity with the possible en echelon fault system of the western coast (Figure 1). Petrographic arguments are also in agreement with a distinct development of this massive flood basalt unit which overlaps the basal unit, the outcrops of which are gently dipping southeastward on the Golfe du Morbihan islands.

A recently discovered intrusive gabbro at Val Gabbro is 39 Ma old (Giret et al., 1981) and represents the oldest K-Ar determination on the Archipelago. If we consider that the nearest plateau basalts are only about 22 to 20 Ma (Table 2) we must assume that the gabbro is uplifted tectonically from older rocks at depth in the crust.

Another problem is the fact that K-Ar ages of many basal strata are younger than those of upper flows. An explanation for such apparent contradiction is afforded by field observations and may be interpreted as argon loss due to hydrothermal activity associated with zeolitic facies development. A characteristic example comes from basal strata

(K 138, age 20 ± 0.6 Ma) and a 460 m high lava flow (K 34, age 21.9 ± 0.6 Ma only) in the southeastern unit (Table 2).

The upper volcanism located on Ronarc'h peninsula consists of nephelinite flows, dated at 7 Ma at the top of the sequence, associated with a tephritic plug and synchronous to late phonolitic extrusions.

Differentiated extrusions (see Table 3) are of two main types that differ both in age and in petrology and are termed types (a) and (b).

(a) Old trachy-phonolites like Dôme Rouge near Port-Jeanne-d'Arc Station, whose estimated age may be near 25-23 Ma. The structure is a widely rooted cumulate dome, deduced from magneto-telluric profiles, and surrounded by younger basaltic flows overlapping a deltaic type conglomerate (Nougier, 1965) and intercalated with lignite deposits described by Cookson (1947). The topographic position of these lignite beds throughout the whole archipelago is commonly about 50 to 150 m above sea level and interpreted as a general climatic and plant-life high established during a long period of inactivity near 20 Ma.

(b) Phonolitic spines, cupolas, domes, sugar-loaves or ring-dykes are located on the northeastern side of the unit. In spite of the density of extrusions, they correspond to only 4.3% of surface outcrops, but many large phonolitic structures are as yet barely exposed by erosion.

The spreading index (height of the extrusion: basal surface radius), defined by Girod (1971) varies between 4 and 0.3 and MT profiles indicate that feeders are relatively deep and narrow. All data are in agreement with extrusives from 10.2 to 6.6 Ma in age (Table 2), being ejected through a 20 Ma basaltic basement. A K-Ar age from Oreilles de Chat is 8.4 Ma, in agreement with crushed marine fossils described by Fletcher (1938) which indicate an age of at least Upper Miocene.

## Petrography

All basaltic lavas are typical alkali olivine basalts with a trend toward basanite in the upper series of the Southeastern unit. Differentiated lavas are nepheline normative trachytes in the Gallieni unit, and phonolites in the Southeastern unit.

*The Gallieni basaltic unit* This was studied in the section at Mt Branca, where it is particularly homogeneous chemically, despite its textural variation being doleritic, poikilitic and porphyritic in flows and more glassy in dykes. The basal lava (Table 1) is mainly plagioclase-rich and poor in olivine, whereas higher in the section, olivine phenocrysts are predominant. The latter may reach 2-3 mm in size, and are corroded and iddingsitized. Clinopyroxenes are colourless titaniferous augites and are often twinned. The plagioclase is labradorite An 57-60. Opaques (titanomagnetite) constitute 12 to 14% of the rock. The lavas are often choritized, and include secondary calcite and silica. Zeolites are more abundant at the top of the section and are in places associated with interstitial analcime. Geochronological K-Ar measurements indicate a 21 Ma age for the basal flow and 17.7 Ma for the top. This yields a theoretical eruption rate of about 1 m thickness of lava flow in 3650 years (1/3650).

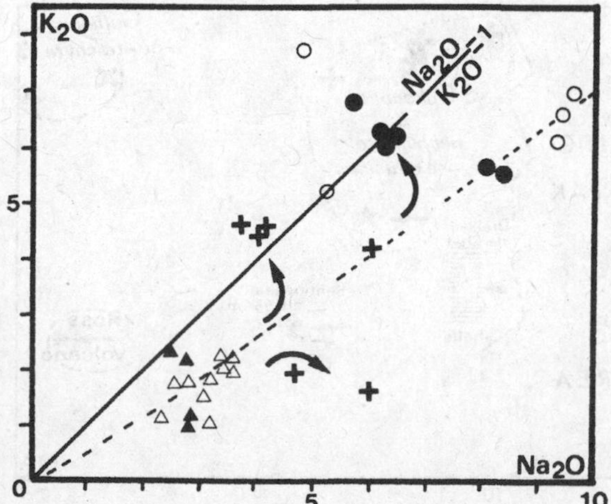


**Figure 4. Plot of $Na_2O/K_2O$ for new analyses of volcanic rocks from the southeastern part of Kerguelen. Same legend as Figure 3.**

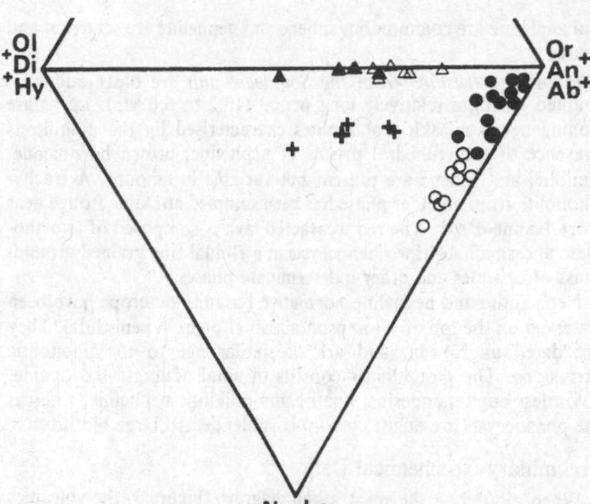


**Figure 5. Plot of (Ol + Di + Hy) – (Or + An + Ab) – (Ne + Le) for new analyses of southeastern Kerguelen province. *Legend: Southeast unit:* 1: alkali basalt; 3: tephrite; 5: phonolite. *Gallieni unit:* 2: alkali basalt; 4: ne-trachyte.**

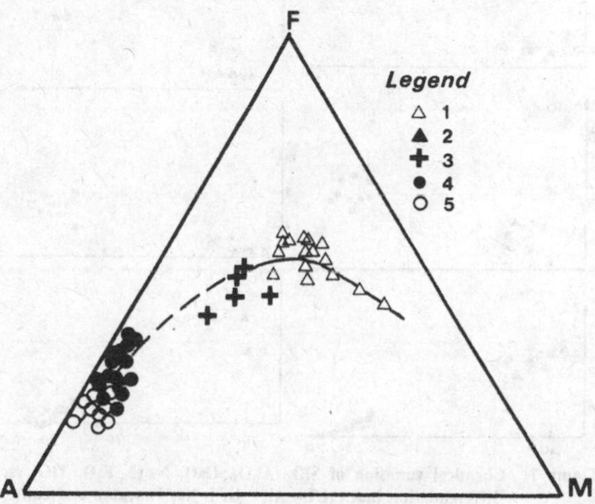


**Figure 6. AFM diagram. Same legend for Figure 5.**

*The Southeast basaltic unit* This is about 700 m thick. Basalts are mainly porphyritic. The olivine in places has an iddingsite core and plagioclase is more calcic (An 60-70) than in the Gallieni lavas. Fresh biotite and hornblende with apatite inclusions or aegirine-rich rims spotted with dark inclusions are common (Table 1).

A particular aspect is the occurrence during the last phonolitic extrusions near 7.4 Ma, of nephelinite flows and probably basanitic plugs, all located around Mt Thomson massif on Ronarc'h peninsula. A typical plug with cone sheet structure has been discovered on the eastern flank of Mt Thompson near an altitude of 500 m. The lava is a massive but glass-rich tephrite with fluidal texture, containing pyroxene and brown hornblende. The alkali content is over 8.37% and the Differentiation Index (DI) is 51. Moreover, the presence of nepheline—or analcime—bearing basanites, many rich in biotite, together with limburgites (Edwards, 1938), suggests alkaline magmatism abnormally rich in potassium.

Differentiated lavas are of two types, similar in composition but with very distinct mineralogy (Figure 3).

*Trachy-phonolites from the Galliene unit* are the youngest differentiated lavas (0.3 to 0.4 Ma, Table 2). They are fresh, with a fluidal texture and fine-grained. Sanidine phenocrysts occur as sheaves or star-shaped clusters. Rare clinopyroxene is titaniferous augite, and brown hornblende and apatite were noted. In some lavas, biotite occurs, rimmed by clinopyroxene granules, and zeolites and interstitial analcime are common but sphene and nepheline are scattered and scarce.

*Phonolitic extrusions from the Southeast unit* are older and were erupted during a relatively long period (10.2 to 6.6 Ma). Lavas are homogeneous miaskitic phonolites characterised by the ubiquitous presence of tiny euhedral prisms of nepheline; brown hornblende, sanidine and sphene are present but variable in amount. A trachyphonolite from an older phase has been sampled at Dôme Rouge near Port-Jeanne-d'Arc. The red weathered lava is composed of anorthoclase and small aegirine phenocrysts in a fluidal fine grained groundmass of opaques and other indeterminate phases.

Nephelinite and nepheline normative basanite outcrops have been observed on the top of Thompson massif (Ronarc'h peninsula). They are dated at 7.4 Ma, and are of similar age to the phonolitic extrusions. The groundmass consists of small iddingsitised olivine, colourless augite, andesine, apatite and poikilitic nepheline, whereas the phenocrysts are augite, brown hornblende and large biotite.

## Preliminary Geochemical Data

When plotted on the silica/alkali diagram (Figure 3) the volcanics of southeastern Kerguelen range from alkali basalt to phonolite and form a widely differentiated alkaline series, possibly with a gap

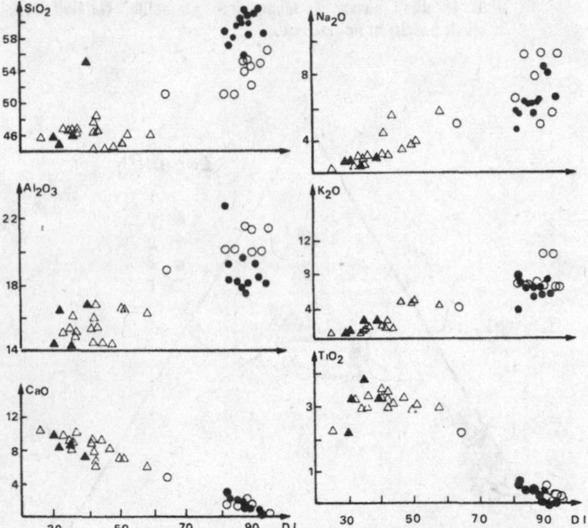


**Figure 7. Chemical variation of $SiO_2$, $Al_2O_3$, CaO, $Na_2O$, $K_2O$, $TiO_2$ vs. Differentiation Index (DI = Ab + Or + Ne; Thornton and Tuttle, 1960).** ***Legend:*** **open triangle: alk-basalt and tephrite of Southeast unit; black triangle: alk-basalt of Gallieni unit; open circle: phonolite of Southeast unit; black circle: Ne-trachyte of Gallieni unit.**

between 49 to 53% $SiO_2$. The Gallieni unit tends to be lower in alkali than the Southeastern unit though there is some overlap. In this unit, the mafic volcanics display a wide range of $Na_2O + K_2O$ and on Middlemost's (1980) classification, they range from alkali basalt to tephrite. Conversely, the phonolitic lavas range relatively widely in their silica contents (from 54 to 60%) but little in their $Na_2O + K_2O$ contents (15-16%). All lavas are miaskitic (agpaitic ratio : 0.75).

These chemical variations can be compared with the well-described alkali rock series of the Tristan da Cunha group (Baker et al., 1964), Gough Island (le Maitre, 1962), St Helena Island (Baker, 1969), the Anjouan Island in the Comoro Archipelago (Flower, 1973), Moheli Island (Strong, 1972), Mayotte Island (Nougier et al., 1981) and closer localities such as Heard Island (Stephenson, 1973) or Possession Island (Chevallier et al., in press). For example, typical volcanics from Crozet Island within the dashed area of Figure 3, show an average which is more basic than at Kerguelen where cumulate rocks are generally lacking. On Heard Island the lack of undersaturated alkali volcanics (nephelinite, phonolite) is apparent and the $Na_2O/K_2O$ ratio is close to unity. In southeastern Kerguelen values are more scattered with various alkali trends, viz. a nepheline-trachyte ($Na_2O/K_2O \sim 1$) in the Gallieni unit and phonolitic ($Na_2O/K_2O \sim 1.5$) and tephritic ($1 > Na_2O/K_2O > 3$) trends in the southeastern unit (Figure 4).

These trends are particularly well displayed on the (olivine + pyroxene)-(felspars)-(feldspathoids) diagram (Figure 5), where they converge towards the felsic pole. The (ol. + pyr.)—(feldsp.) side corresponds to the alkali-basalt suite, the feldsp.-feldspathoid side is the trachy-phonolitic trend which corresponds to increasing nepheline, and the third median line corresponds to tephritic intermediate magma.

The AFM diagram (Figure 6) shows a typical iron-enrichment trend terminating with nepheline-trachytes and phonolites. The close approach to the Fe-Alk side reflects an efficient Fe/Mg fractionation, as on St Helena Island. The same variations have been clearly observed on Possession Island in spite of the scarcity of differentiated lavas. On the Comoro Archipelago, very undersaturated silica trends are recorded. For example, on Mayotte Island, extrusive miaskitic phonolites (DI = 80-85) cross-cut alkali basalts, basanite basement (DI = 25-40) and nephelinites (DI = 30-38). On Moheli the most mafic volcanics are melanephelinite possibly parental to the phonolites, as proposed on Gough Island, in spite of the lack of similar undersaturated rocks. On the whole, the Kerguelen southeastern province geochemical evolution resembles most of the trends on Anjouan or St Helena Islands, where well differentiated volcanic rock suites also occur.

Variations of some major oxides plotted against DI are shown graphically in Figure 7. The differences between the differentiation trends of the Gallieni and southeast units are well shown by $SiO_2$, $Al_2O_3$, $Na_2O$ and $TiO_2$ but are not well displayed for $K_2O$ and CaO. It is suggested that these variations reflect the presence of different parental magmas, both having undergone similar differentiation processes.

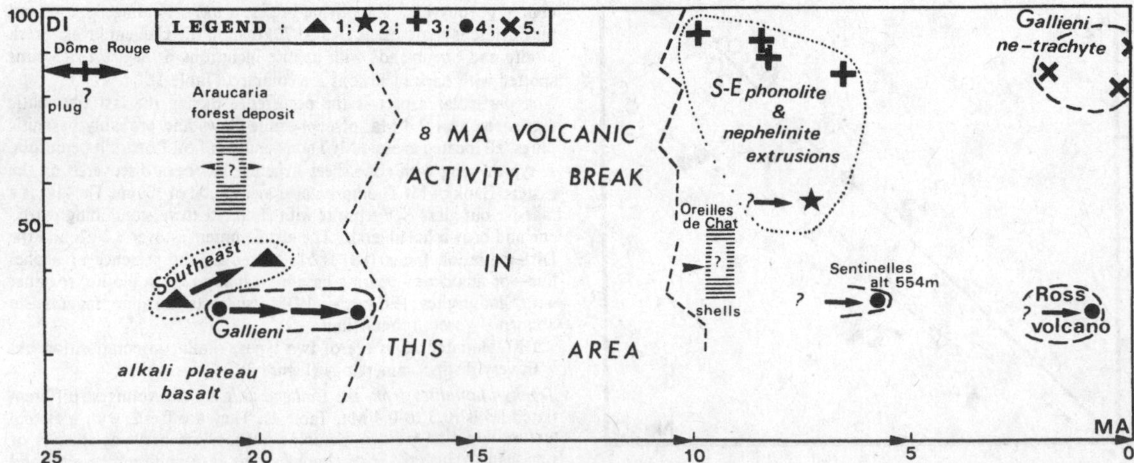


**Figure 8. Differentiation evolution diagram: age (in Ma) vs. DI.** ***Legend: Southeast unit:*** **1: alk-basalt plateau; 2: nephelinite; 3: phonolite;** ***Gallieni unit:*** **4: alk-basalt plateau and Ross volcano; 5: Ne-trachyte.**

It is emphasised that these highly undersaturated volcanics occur only in the southeastern part of the Archipelago. Although co-magmatic plutonic complexes have not yet been found equivalent to those found with the nepheline trachytes or phonolites, this is possibly because erosion has not yet cut deeply enough to expose them.

## Chronospatial Evolution

Postulated schemes of magmatic differentiation need to take account of the time relations. In Figure 8, the Differentiation Index is plotted against age.

The following observations seem pertinent:

(a) the main alkali basalt activity seems to be restricted to a relatively short period (23 to 17 Ma), with some local recurrences (near 5 Ma or 2 Ma for Ross activity). Defined above is a"theoretical eruption rate" of 1/3650 for the Gallieni unit, which is of the same order (1/4500) as for the southeast where however the base is less well-known, but greater than the rate of near 1/6000 for the Courbet peninsula unit (Mt du Château). These rates seem very low when compared with those of present day active volcanoes such as Piton des Neiges (1/1600 to 1/950) and Piton de la Fournaise (1/250) on Reunion Island (Chevallier, 1979) or Amsterdam Island (1/600). In general, Kerguelen's activity rate is 10 to 40 times lower than present day active strato-volcanoes.

(b) Episodes of rythmic volcanic activity include long periods of quiescence (about 10000 years on average), revealed by the occurrence of basal breccias, angular unconformities, conglomerate and lignite development.

(c) Furthermore, in the southeastern province, we note that between about 8 Ma and 18 Ma there is a complex cessation of volcanic activity (Figure 8) broken by the extrusion of phonolitic and nephelinitic lavas. We suggest that this new phase of activity is not related to a former flood basalt phase, although the nepheline trachytes of the Gallieni unit may be related to the Ross basaltic phase which is contemporaneous. No relatively simple relationship between the DI and the stratigraphic position of lavas (Baker, 1969), has been observed on Kerguelen.

(d) Watkins et al., (1974) have suggested a geographic migration of differentiating magmatism from oversaturated and saturated lavas located in the northwest of the archipelago to undersaturated lavas located in the southeast. If we consider the age of these extrusive lavas that erupted on a 25 to 40 Ma basaltic basement; rhyolites and quartz trachytes are the oldest (26 Ma and perhaps older), alkali trachytes from the centre unit are 12 Ma and the southeast phonolites are 8 Ma. So, in a general sense, intrusive and extrusive saturated rocks are located along the western coast, intermediate series occur throughout the largest central part of the archipelago, and undersaturated lavas are restricted to the far east. These spatial geochemical variations may parallel regional ocean floor magnetic anomalies, with their limits defined by the principal transform fault direction (Figure 1).

(e) In detail, no simple scheme of magmatic chronospatial migration is apparent. For instance, the extrusive differentiated lavas erupted since 8 Ma are located along an east to west direction (Jeanne d'Arc, Gallieni to Rallier du Baty units) and are not in agreement with the above direction.

To conclude, chronospatial parameters are necessary to understand volcanological processes. On a local scale new data have been determined. With regard to the whole archipelago, complex trends suggest that although it is premature to relate subaerial volcanics to oceanic floor features, some of the regional variation suggests that there is a connection; this must be studied with a better knowledge of the peri-insular margin and nearby oceanic regions.

*Acknowledgements* We gratefully acknowledge the support of the Mission Recherche of the Territoire des Terres Australes et Antarctiques Francaises, both for the field word and for the laboratory studies.

## REFERENCES


BAKER, I., 1969: Petrology of the volcanic rocks of Saint Helena Island, South Atlantic, *Geol. Soc. Am. Bull., 80,* 1283-1310.

BAKER, P.E., GASS, I., HARRIS, P. and LE MAITRE, R., 1964: The volcanological report of the Royal Society Expedition to Tristan da Cunha, 1962. *R. Soc. Lond., Philos. Trans., Ser. A., 256,* 439-578.

CHEVALLIER, L., 1979: Structures et évolution du volcan Piton des Neiges, île de la Réunion, leurs relations avec les structures du bassin des Mascareignes. Thèse 3° cycle, Grenoble, 187p.

CHEVALLIER, L., NOUGIER, J. and CANTAGREL, J-M. (this volume): Volcanology of Possession Island Crozet archipelago.

COOKSON, I.C., 1947: Plants microfossils from the lignites of Kerguelen archipelago. *BANZ Ant. Res. Exped. 1929/1931. Rept. Ser. A., 2(8),* 129-42.

DOSSO, L. 1977: Données isotopiques (Sr, Pb) sur le volcanisme et le plutonisme des îles Kerguelen. Le problème de la contamination crustale. Thèse 3° cycle, Rennes, 35p.

DOSSO, L., VIDAL, P., CANTAGREL, J.M., LAMEYRE, J., MAROT, A. and ZIMINE, S., 1979: Kerguelen, continental fragment or oceanic island ? Petrology and isotopic geochemistry evidence. *Earth Planet. Sci. Lett., 43,* 46-60.

DOSSO, L. and RAMA MURTHY, V., 1980: A Nd isotopic study of the Kerguelen Islands: inferences on enriched oceanic mantle sources. *Earth Planet. Sci. Lett., 48,* 268-76.

EDWARDS, A.B., 1938: Tertiary lavas from the Kerguelen archipelago. *BANZ Ant. Res. Exped. 1929-1932. Rept. Sèr. A., 2(5),*72-100.

FLETCHER, H.O., 1938: Marine tertiary fossils from Kerguelen Island. *BANZ Ant. Res. Exped. 1929-1931. Rept. Ser. A., 2(6),*101-16.

FLOWER, M.F.J., 1973: Petrology of volcanic rocks from Anjouan, Comores archipelago. *Bull. Volcan., 36,* 238-50.

GIRET, A., CANTAGREL, J-M. and NOUGIER, J., 1982: Nouvelles données sur les complexes volcano-plutoniques des îles Kerguelen (TAAF). *C.R.Ac.Sc., Paris, 293.* 191-4.

GIROD, M., 1971: Le massif volcanique de l'Atakor (Hoggar, Sahara algérien) étude pétrographique, structurale et volcanologique, *CNRS-CRZA, 12,* 158p.

HOUTZ, R., HAYES, D. and MARKL, R., 1977: Kerguelen plateau bathymetry, sediment distribution and crustal structure. *Mar. Geol., 25,* 95-130.

LE MAITRE, R.W., 1962: Petrology of volcanic rocks, Gough Island, South Atlantic. *Geol. Soc. Am., Bull., 73,* 1309-40.

LUENDYK, B. and RENNICK, W., 1977: Tectonic history of aseismic ridges in the eastern Indian Ocean. *Geol. Soc. Am., Bull., 88,* 1347-56.

MAROT, A. and ZIMINE, S., 1976: Les complexes annulaires de syénites et granites alcalins dans la péninsule Rallier du Baty, îles Kerguelen. Thèse 3° cycle Univ. Paris VI., t, 1: 131 p.; t.2; 47p. and *CNFRA,* (1981), *49.* 23-114.

MIDDLEMOST, E.A.K., 1980: A contribution to the nomenclature and classification of volcanic rocks. *Geol. Mag., 117,* 51-7.

NOUGIER, J., 1965: Etude des formations de Port-Jeanne d'Arc (archipel de Kerguelen), *CNFRA, 11,* 31-55

NOUGIER, J., 1970: Contribution à l'étude géologique et géomorphologique des îles Kerguelen (TAAF), *CNFRA, 27.*

NOUGIER, J., 1972a: Geochronology of the volcanic activity in îles Kerguelen; *in* Adie, R.J. (ed.) *Antarctic Geology and Geophysics.* Universitetsforlaget, Oslo, 803-8.

NOUGIER, J., 1972b: Volcanic associations in îles Kerguelen *in* Adie, R.J. (ed.) *Antarctic Geology and Geophysics.* Universitetsforlaget, Oslo, 809-13.

NOUGIER, J., 1972c: Aspects de morpho-tectonique glaciaire aux îles Kerguelen (TAAF). *Rev. Geogr. Phys. et Géol.* Dyn, Paris, (2), *XIV-5,* 499-506.

NOUGIER, J., CANTAGREL, J-M., WATELET, P. and VATIN-PERIGNON, N., 1981: Volcanologie de l'île Mayotte (archipel des Comores). *Soc. gèol* .Fr. C.R. Som., 4, 139.

SCHLICH, R., 1975: Structure et âge de l'Océan Indien occidental. *Soc. Geol. Fr., Mem. Hors Ser., 6.*

STEPHENSON, P.J., 1972: Geochemistry of some Heard Island igneous rocks *in* Adie R.J. (ed.) *Antarctic Geology and Geophysics.* Universitetsforlaget, Oslo, 793-801.

STRONG, D.F., 1972: Petrology of volcanic rocks from Anjouan, Comores archipelago. *Bull. Volcan., 36,* 238-50.

THORNTON, C.P. and TUTTLE, O.F., 1960: Chemistry of igneous rocks. I. Differentiation index. *Am. J. Sci., 258,* 664-84.

WATKINS, N., GUNN, B., NOUGIER, J. and BASKI, A., 1974: Kerguelen: continental fragment or oceanic island ? *Geol. Soc. Am. Bull., 85,* 201-12.

# A STUDY OF KERGUELEN PLUTONISM: PETROLOGY, GEOCHRONOLOGY AND GEOLOGICAL IMPLICATIONS.

A. Giret and J. Lameyre, *Laboratoire de Pétrologie, Université Pierre et Marie Curie, 4 Place Jussieu, F-75230, Paris Cedex 05, France.*

*Abstract* In the Kerguelen Archipelago, studies of plutonic rocks since 1973 provide evidence of several important stages of geological history. There are two kinds of plutonic complexes differing both in their structural setting and their petrological characteristics.

The first type occurs as sills and massifs sited in the older plateau basalt series (Nougier, 1970). It contains very abundant weakly differentiated rocks reminiscent of tholeiitic or transitional types. J.M. Cantagrel has obtained a 39 Ma K-Ar date for two such intrusive complexes; an even older age for the surrounding basalts is inferred.

The second type occurs as volcano-plutonic complexes mainly made up of differentiated alkaline rocks. The rock-type distribution defines two petrological provinces which have a NNW-SSE boundary; the western province is silica (Si)-saturated to Si-oversaturated, while the eastern one is Si-undersaturated. The alkaline magmatism, which appears at present to be in the process of extinction began 26 Ma ago; it was responsible for the metamorphism and isotopic resetting of the plateau basalts.

The linear emplacement of the felsic ring-complexes which parallel the western coast of the main island, and some N-S normal faults which have been observed recently suggest the existence of a rift structure in the open sea to the west, which is consistent with the scarce bathymetric data from that area.

The subaerial part of Iles Kerguelen is built up of mainly flood basalts of which the oldest K-Ar age determined is 26 Ma. Several plutonic complexes intrude the basaltic pile, and the largest exposure of these is found on the Rallier du Baty Peninsula. Here, five intersecting ring complexes can be delineated, and their structure, setting and origin have been discussed by Lameyre et al., (1976, 1981), Marot and Zimine (1976), Dosso et al. (1979), and Dosso and Murthy (1980). A geological map has been compiled by Giret et al. (1980). A systematic

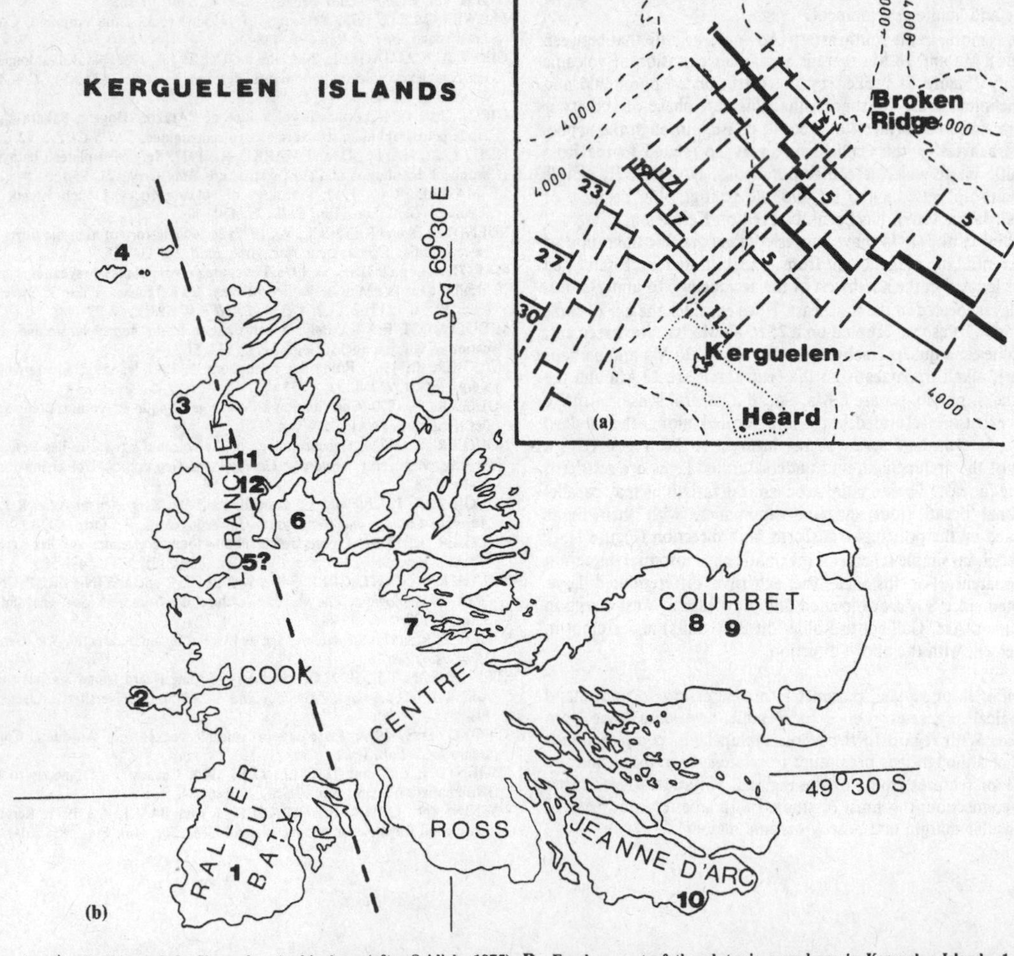


Figure 1. A—Position of the Kerguelen Archipelago (after Schlich, 1975). B—Emplacement of the plutonic complexes in Kerguelen Islands. 1—Ring complexes of the Rallier du Baty Peninsula. 2—Ring complex of Ile de l'Ouest. 3—Ring complex of Mont Lacroix and Cap d'Aiguillon. 4—Ring aomplex of Iles Nuageuses. 5—Plutonic complex of Mont Guynemer (sampled in moraines). 6—Plutonic complex of the G'eographie P'eninsul- 7—Plutonic complex of the Monts Ballons. 8—Plutonic complex of the Montagnes Vertes. 9—Gabbroic complex of the Monts Mamelles. 10—Gabbroic complex of the Val Gabbro, Jeanne d'Arc Peninsula. 11—Basic sills of Vallee du Thermometre. 12—The Righi plutonic complex. The dashed line separates the two alkaline provinces, the western silica-oversaturated one and the eastern silica-undersaturated one (see the text).

study of the other plutonic complexes originally reported by Nougier (1970) as well as some recently discovered ones has been underway since 1977 (Figure 1.) This work has yielded some important petrological, structural and geochronological evidence which has bearing on the magmatic history of Iles Kerguelen.

## Petrology

The plutonic complexes on Iles Kerguelen can be divided on the basis of mineralogy and geochemistry into three groups: (i) tholeiitic to transitional; (ii) silica-oversaturated alkaline; (iii) silica-undersaturated alkaline.

### *Tholeiitic to transitional complexes*

Gabbroic sills, recently described on the Loranchet Peninsula by Giret et al., (1981), are abundant in the northern part of the archipelago. They are generally a few metres to some tens of metres thick, but some are laccolithic, with maximum thicknesses greater than 200 m and have uplifted the overlying basalts and accumulated. The sills are composed of porphyritic olivine-bearing microgabbros and show evidence of gravitational plagioclase accumulation towards their bases.

The Monts Mamelles on the Courbet Peninsula and the Val Gabbro on the Jeanne d'Arc Peninsula are mafic intrusions mineralogically similar to the gabbroic sills. They occur as small massifs with marginal breccias made of basaltic and gabbroic blocks in a basaltic matrix which are indicative of a diapiric emplacement. The Mamelles gabbros display weak horizontal mineralogical layering. The Val Gabbro exhibits well defined subvertical centimetre-scale layering which probably represents several magma injections. In both these complexes, the gabbros are traversed by small syenitic and granitic dykes which represent either a residual liquid from the same parent magma (Giret and Lameyre, 1980) or a differentiated magma of later magmatism. (see below)

The Righi intrusion is a porphyritic microgranite rich in quartz and albite and poor in potassic feldspar. The rocks are mineralogically and geochemically similar to some tholeiitic plagiogranites and granophyric rocks from Iceland (Jauzein, 1981). Minor tholeiitic gabbros are present in the margin of the granite intrusion, a feature also found in Iceland. A diapiric mode of intrusion is indicated by abundant basaltic breccias in the contact zone, by the absence of basaltic inclusions in the intrusion, and not least by the def..mation of a 500 m wide zone of basalts around the intrusion.

Most of the mafic rocks of this group of complexes are augite-rich and contain minor olivine. Evidence of fractional crystallisation processes is found in either horizontal or vertical layering. Differentiated rocks are either alkaline (Mamelles and Val Gabbro) or tholeiitic (Righi). The granitic intrusion and the margins of the mafic intrusions are characterised by a low grade metamorphic development of minerals such as chlorite, albite, epidote and zeolites.

On chemical variation diagrams the mafic rocks (Table 1) are heterogenous, and this is considered to be the result of cumulus processes as evidenced by layering. However, all the rocks of the tholeiitic to transitional complexes are characterised by low potassium and relatively high iron contents. The slightly higher aluminium content of the silicic rocks of these complexes on Iles Kerguelen compared to those of Iceland is a reflection of greater plagioclase abundance.

### *Silica-oversaturated alkaline complexes*

This group consists entirely of ring complexes (Figure 2) which vary in diameter from a few kilometres to 15 km, and which are exposed at different erosion levels. The basaltic Mont Lacroix (Loranchet Peninsula) is traversed by a concentric and radial dyke system characteristic of the hypovolcanic part of a ring complex. On Iles Nuageuses, and particularly on Ile de Croy one can observe the roof of a ring complex and its relationship with associated volcanism, as manifested by trachytic plugs (Figure 3.). The western part of Ile de l'Ouest comprises the upper part of a plutonic body disposed in ring structures whereas numerous associated dykes intrude the flood basalts of the eastern part (Figure 2.). On Rallier du Baty Peninsula, one can observe sections at gradually different levels from the porphyritic roof of an intrusion in the north to isogranular ring dyke sections in the south.

Such differences in the erosion levels are related to isostatic movements rather than the age or the emplacement depth of the intrusions (Marot and Zimine, 1976; Nougier, in press). This view is supported by the discovery of submarine pillow basalts between units of subacrial columnar basalts, which indicates that a subsidence of the subaerial basalt surface to below sea level was followed by an uplift of more than 300 m. After considering these movements, and examining some meridional normal faults along the west coast, Giret and Lameyre (in preparation) propose the existence of a western rift structure, the eastern side of which has been intruded by plutonic ring complexes; this is in conformity with a 1000-2000 m uplift inferred

**ROCK ASSOCIATIONS OF ALKALIC AND SILICA UNDERSATURATED CHARACTERS—WHOLE ROCK ANALYSES**

| | VAL GABBRO K90 | K91 | 7802 | 7812 | 7808 | 7806 | 7813 |
|---|---|---|---|---|---|---|---|
| $SiO_2$ | 40.55 | 42.41 | 47.61 | 44.92 | 57.01 | 57.64 | 58.38 |
| $TiO_2$ | 2.95 | 1.51 | 0.73 | 1.70 | 1.45 | 1.33 | 1.02 |
| $Al_2O_3$ | 18.02 | 7.07 | 14.86 | 12.92 | 16.57 | 17.27 | 18.83 |
| $Fe_2O_3$ | 10.00 | 6.89 | 3.63 | 4.58 | 5.43 | 5.43 | 4.08 |
| FeO | 6.61 | 6.69 | 5.24 | 7.19 | 2.97 | 2.15 | 2.19 |
| MnO | 0.20 | 0.20 | 0.20 | 0.12 | 0.30 | 0.02 | 0.10 |
| MgO | 6.55 | 14.93 | 9.64 | 11.10 | 2.02 | 1.23 | 0.69 |
| Ca0 | 12.70 | 12.86 | 14.28 | 14.26 | 3.65 | 3.14 | 3.37 |
| $Na_2O$ | 1.22 | 0.31 | 2.48 | 1.43 | 4.65 | 4.57 | 4.51 |
| $K_2O$ | 0.33 | 0.09 | 0.38 | 0.54 | 3.85 | 4.16 | 5.15 |
| $P_2O_5$ | 0.17 | 0.01 | 0.22 | 0.05 | 0.20 | 0.20 | 0.08 |
| $H_2O^+$ | 1.44 | 5.60 | 1.09 | 1.39 | 1.31 | 0.93 | 0.83 |
| $H_2O^-$ | 0.14 | 0.46 | | 0.11 | 0.21 | 0.18 | 0.09 |
| TOTAL | 100.88 | 99.03 | 100.39 | 100.31 | 99.62 | 98.25 | 99.32 |

| | MONTS MAMMELLES 77134 | 77102c | 77103a | 77100 | 77132 | 77102b | 77103d |
|---|---|---|---|---|---|---|---|
| $SiO_2$ | 38.31 | 43.48 | 45.19 | 47.42 | 51.52 | 56.94 | 61.53 |
| $TiO_2$ | 4.23 | 3.28 | 1.55 | 1.49 | 2.05 | 1.56 | 0.73 |
| $Al_22O_3$ | 14.20 | 17.99 | 19.54 | 18.93 | 16.26 | 17.04 | 17.07 |
| $Fe_2O_3$ | 9.44 | 6.01 | 5.71 | 5.32 | 8.46 | 5.36 | 3.93 |
| FeO | 8.72 | 7.61 | 6.59 | 4.05 | 3.12 | 2.99 | 2.19 |
| MnO | + | 0.10 | 0.11 | 0.11 | 0.13 | 0.08 | 0.08 |
| MgO | 7.78 | 6.33 | 5.04 | 4.92 | 2.05 | 1.66 | 1.43 |
| CaO | 13.88 | 10.31 | 12.19 | 8.86 | 4.15 | 3.27 | 1.65 |
| $Na_2O$ | 1.50 | 2.63 | 2.02 | 2.88 | 3.86 | 5.28 | 4.02 |
| $K_2O$ | 0.45 | 0.63 | 0.72 | 0.45 | 3.10 | 3.92 | 3.81 |
| $P_2O_5$ | 0.27 | 0.26 | 0.08 | 0.05 | 0.54 | + | 0.11 |
| $H_2O^+$ | 1.19 | 1.18 | 1.07 | 3.39 | 3.16 | 1.34 | 0.58 |
| $H_2O^{\prime}$ | 0.06 | 0.15 | 0.13 | 0.53 | 0.63 | 0.41 | 0.24 |
| TOTAL | 100.03 | 99.96 | 99.94 | 98.40 | 99.03 | 99.85 | 97.37 |

| | PENINSULE LORANCHET 8103 | 8109 | 8122 | 8125 | 81103 | CN55 | 8135 | 8113 | 8101 |
|---|---|---|---|---|---|---|---|---|---|
| $SiO_2$ | 46.99 | 49.09 | 49.02 | 49.46 | 45.54 | 45.88 | 69.63 | 66.76 | 46.53 |
| $TiO_2$ | 2.90 | 1.73 | 1.77 | 1.68 | 1.35 | 1.43 | 0.42 | 0.40 | 1.85 |
| $Al_2O_3$ | 14.79 | 20.32 | 19.61 | 19.85 | 12.89 | 15.45 | 15.30 | 14.49 | 15.29 |
| $Fe_2O_3$ | 5.96 | 3.78 | 2.88 | 2.66 | 4.93 | 4.68 | 2.31 | 2.06 | 6.27 |
| FeO | 5.98 | 4.80 | 5.91 | 6.39 | 6.24 | 6.50 | 0.99 | 1.25 | 5.55 |
| MnO | 0.13 | 0.12 | 0.12 | 0.13 | 0.15 | 0.19 | 0.05 | 0.05 | 0.20 |
| MgO | 6.10 | 3.67 | 3.54 | 3.44 | 11.52 | 7.56 | 1.15 | 0.49 | 5.55 |
| CaO | 9.51 | 11.50 | 10.48 | 10.88 | 8.64 | 9.63 | 2.48 | 4.19 | 11.16 |
| $Na_2O$ | 2.59 | 2.92 | 3.02 | 3.06 | 1.83 | 2.86 | 4.68 | 3.66 | 2.29 |
| $K_2O$ | 0.74 | 0.44 | 0.59 | 0.56 | 0.42 | 0.22 | 1.57 | 1.25 | 0.27 |
| $P_2O_5$ | 0.39 | 0.22 | 0.25 | 0.24 | 0.13 | 0.12 | 0.11 | 0.12 | 0.18 |
| P.F. | 3.71 | 1.20 | 1.87 | 1.16 | 5.87 | 4.61 | 1.31 | 3.74 | 3.90 |
| TOTAL | 100.46 | 100.32 | 99.72 | 100.22 | 100.21 | 99.85 | 100.11 | 98.60 | 99.66 |

**Table 1. Tholeiitic and transitional series, whole rock analyses. *Val Gabbro*: K90 = gabbro; K91 = melagabbro; 7802 = layered gabbro; 7872 = gabbro; 7808 = micromonzonitic gabbro; 7806 = quartz microsyenite; 7813 = quartz microsyenite. *Monts Mamelles*: 77134, 77102c, 77103a, 77100 = gabbros; 77132 = monzonite quartz syenite; 77102b = quartz syenite; 77103d = quartz poor granite. *Peninsule Loranchet*: 8103 = tectonic uplifted gabbro in Righi, 81109, 81122,91125 = coarse-grained gabbros; 81103 = Gabbro inclusion in an alkaline trachytic dyke of Lac Athena; CN55 = sill of gabbro in the Vallée du Thermometre; 8135,8113 = quartz + albite rich microgranite of the Righi intrusion; 8101 = flood basalt near Righi.**

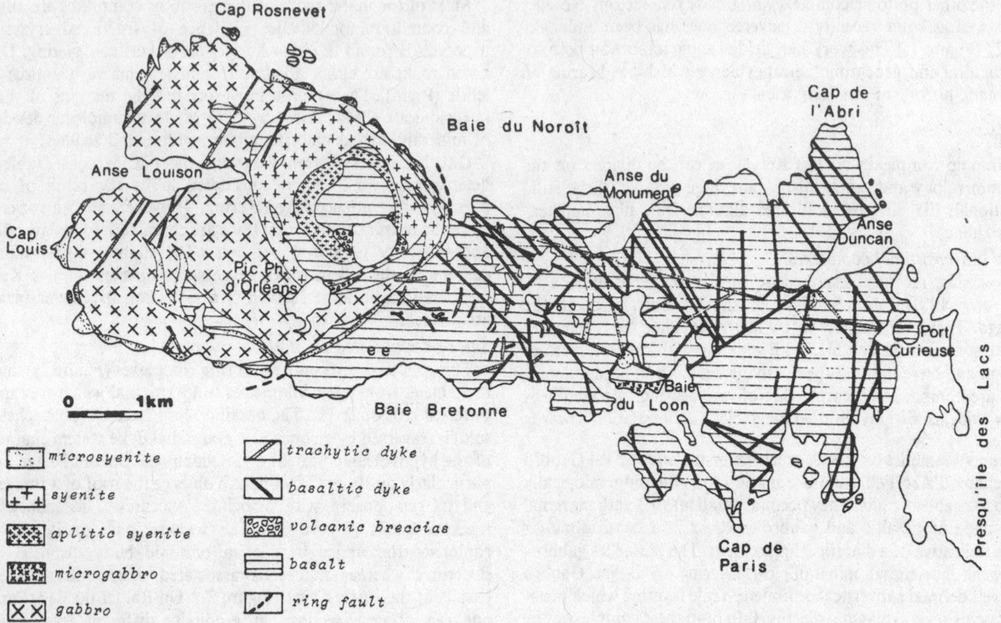


**Figure 2. Geological map of Ile de l'Ouest.**

from geothermal and mineralogical equilibria studies, and which has exposed these deep intrusions.

In these complexes, the intrusion of gabbroic magma was followed by several ring intrusions of silicic magma associated with cauldrom subsidence. The gabbros range from biotite-rich types to alkali feldspar-rich types. They are iron-rich, and as a reflection of their mineralogy, they are also high in potassium. Differentiated rocks include monzonitic syenites which are commonly quartz-bearing, abundant quartz-syenites (nordmarkites described by Nougier and Lameyre, 1973), fayalite-bearing syenites, and rare alkaline granites. The rock types constitute a differentiation series which is characteristically alkaline, with $K_2O$ increasing with differentiation to become equal with $Na_2O$ (on a weight percent basis) in the syenitic rocks.

*Silica-undersaturated alkaline complexes*

These are larger than 1 km in diameter and resemble interlocked pipes rather than ring dykes. As with the previous group, these complexes vary in their erosion levels. The complex on the Geographie Peninsula appears as a criss-cross of dykes which is particularly dense

**ROCK ASSOCIATIONS OF ALKALIC AND SILICA OVERSATURATED CHARACTERS**

| | ILE DE L'OUEST | | | | | | | | | CAP D'AIGUILLON |
|---|---|---|---|---|---|---|---|---|---|---|
| | 8082 | 8025 | 80143 | 8024 | 80144 | 80147 | 80108 | 8099 | 80102 | CN88 |
| $SiO_2$ | 43.87 | 45.42 | 48.17 | 49.36 | 59.53 | 62.58 | 64.55 | 69.43 | 72.56 | 51.93 |
| $TiO_2$ | 3.58 | 3.36 | 2.94 | 1.96 | 0.72 | 0.18 | 0.37 | 0.16 | 0.04 | 3.25 |
| $Al_2O_3$ | 16.91 | 16.81 | 15.08 | 20.14 | 17.46 | 16.95 | 16.36 | 14.36 | 13.30 | 12.53 |
| $Fe_2O_3$ | 4.25 | 4.20 | 3.59 | 2.69 | 4.12 | 3.77 | 2.80 | 2.11 | 2.44 | 4.65 |
| FeO | 9.08 | 8.24 | 6.46 | 4.80 | 3.28 | 1.75 | 1.89 | 1.05 | 0.83 | 7.66 |
| MnO | 0.15 | 0.23 | 0.14 | 0.12 | 0.13 | 0.19 | 0.13 | 0.14 | 0.07 | 0.19 |
| MgO | 6.02 | 3.90 | 5.17 | 2.51 | + | 0.14 | 0.10 | + | + | 4.00 |
| CaO | 11.08 | 10.04 | 6.00 | 9.24 | 0.66 | 0.56 | 0.79 | 0.16 | + | 6.89 |
| $Na_2O$ | 2.41 | 3.25 | 3.91 | 3.43 | 5.74 | 6.44 | 5.36 | 4.52 | 4.10 | 3.61 |
| $K_2O$ | 0.97 | 1.42 | 3.50 | 2.08 | 5.73 | 5.04 | 5.77 | 5.99 | 5.46 | 1.94 |
| $P_2O_5$ | 0.14 | 1.08 | 0.71 | 0.72 | 0.15 | + | 0.04 | + | 0.13 | 0.31 |
| P.F. | 0.60 | 1.14 | 2.76 | 2.41 | 0.74 | 1.14 | 0.62 | 0.66 | 0.79 | 1.65 |
| TOTAL | 99.06 | 99.10 | 98.43 | 99.46 | 98.26 | 98.74 | 98.78 | 98.58 | 99.72 | 99.46 |

| | PENINSULE RALLIER DU BATY—CENTRE SUD | | | | | | | | | |
|---|---|---|---|---|---|---|---|---|---|---|
| | PJ10 | PJ4b | POS13 | LIE1b | LUT5 | VEL1 | AS1 | 2F11 | 2F17 | CHAS1 |
| $SiO_2$ | 43.27 | 44.78 | 43.73 | 59.08 | 62.22 | 63.45 | 61.42 | 69.68 | 75.48 | 74.28 |
| $TiO_2$ | 1.86 | 3.60 | 3.60 | 1.00 | 0.60 | 0.58 | 1.13 | 0.47 | 0.29 | 0.26 |
| $Al_2O_3$ | 9.91 | 18.50 | 13.65 | 17.64 | 17.53 | 17.08 | 15.44 | 14.37 | 11.58 | 12.77 |
| $Fe_2O_3$ | 6.78 | 4.02 | 3.97 | 2.97 | 2.97 | 2.51 | 4.28 | 2.38 | 3.57 | 1.45 |
| FeO | 9.85 | 6.83 | 8.74 | 2.77 | 2.34 | 2.46 | 2.95 | 1.44 | 0.36 | 1.35 |
| MnO | + | 0.10 | 0.10 | 0.09 | 0.06 | 0.09 | 0.04 | 0.04 | + | 0.03 |
| MgO | 15.81 | 5.08 | 11.60 | 0.62 | 0.08 | 0.28 | 0.48 | + | + | + |
| CaO | 7.81 | 10.87 | 9.37 | 2.64 | 0.74 | 0.59 | 0.73 | 0.02 | + | 0.02 |
| $Na_2O$ | 1.24 | 2.74 | 2.46 | 5.45 | 6.26 | 5.91 | 6.03 | 5.54 | 3.87 | 4.03 |
| $K_2O$ | 0.84 | 1.39 | 2.13 | 6.65 | 5.32 | 5.79 | 5.73 | 4.18 | 4.07 | 4.20 |
| $P_2O_5$ | 0.18 | + | + | 0.16 | 0.07 | 0.09 | 0.02 | 0.09 | + | 0.09 |
| $H_2O^+$ | 2.63 | 1.98 | 1.19 | 0.62 | 0.55 | 0.65 | 1.05 | 0.67 | 0.55 | 1.19 |
| $H_2O^-$ | 0.06 | 0.25 | 0.26 | 0.22 | 0.18 | 0.20 | 0.33 | 0.18 | 0.18 | 0.16 |
| TOTAL | 100.24 | 100.14 | 100.80 | 99.91 | 98.92 | 99.68 | 99.63 | 99.06 | 99.95 | 99.83 |

**Table 2. Alkaline and silica oversaturated series, whole rock analyses. *Ile de l'Ouest*: 8082 = biotite poor gabbro; 8025 = biotite gabbro; 80143, 8024 = biotite rich gabbros; 80144 = syenite; 80147 = fayalite bearing syenite; 80108,8099 = quartz syenites; 80102 = alkaline granite. *Rallier du Baty*: PJ10 = biotite poor gabbro; PJ4b = biotite bearing gabbro; POS13 = biotite rich gabbro; LIEUT1b, LUT 5, VEL1, AS1 = quartz syenite; 2F11 = quartz rich syenite, 2F17, CHAS1 = alkaline granite.**

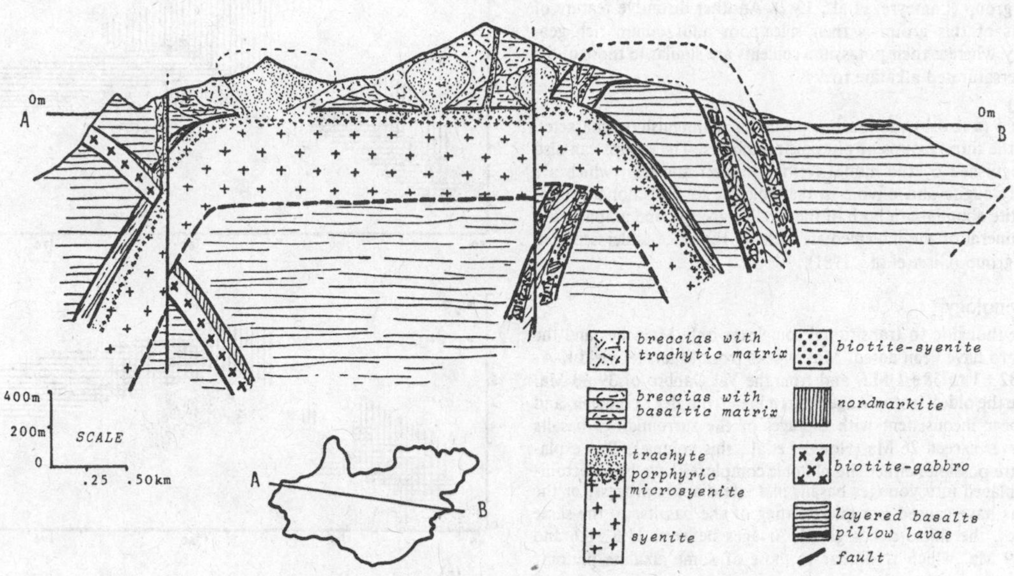


**Figure 3. Geological cross section of Ile de Croy (Iles Nuageuses).**

at the ice-covered summit. On Monts Ballons, dykes are also abundant and they have a radical and concentric distribution around a complex of four distinctive xenolith-rich intrusions. In the Montagnes Vertes complex the erosion level is deeper, and one can observe a progressive compositional variation from the marginal mafic rocks to inner nepheline-bearing syenites (Giret and Lameyre, 1980). The Mont Ross volcano (Nougier et al., this volume) probably represents the subaerial part of a volcano-plutonic complex (Giret et al., 1981).

The early intrusions are mafic, being biotite-rich gabbros and biotite and amphibole bearing monzogabbros. These were followed by intrusions of nepheline-bearing monzonites and amphibole-rich monzosyenites. The latest intrusions are nepheline-syenites in which the resorption of amphibole gives rise to a feldspathoidal liquid (Giret, 1979). Hence, a major feature of this group of complexes is continuous magmatic evolution which produced a compositional variation from basic to syenitic rocks, with no particular lack of

**ROCK ASSOCIATIONS OF ALKALIC AND SILICA UNDERSATURATED CHARACTERS—WHOLE ROCK ANALYSES**

| | MONTAGNES VERTES | | | | | | |
|---|---|---|---|---|---|---|---|
| | MV9 | 7738 | 7718 | 7745 | 7705 | 7741 | 7714 |
| $SiO_2$ | 45.07 | 47.51 | 44.34 | 50.00 | 46.53 | 55.53 | 61.19 |
| $TiO_2$ | 2.85 | 2.86 | 3.23 | 1.83 | 2.66 | 1.07 | 0.73 |
| $Al_2O_3$ | 18.37 | 17.93 | 14.77 | 16.79 | 16.97 | 19.37 | 18.44 |
| $Fe_2O_3$ | 7.46 | 4.15 | 7.77 | 4.06 | 7.82 | 4.42 | 2.92 |
| FeO | 44.74 | 7.10 | 5.59 | 5.65 | 4.58 | 2.49 | 0.81 |
| MnO | 0.11 | 0.19 | 0.19 | 0.13 | 0.21 | 0.15 | 0.06 |
| MgO | 3.65 | 2.73 | 5.03 | 2.50 | 2.94 | 0.79 | 0.38 |
| CaO | 10.89 | 8.79 | 9.77 | 5.36 | 8.06 | 2.67 | 1.01 |
| $Na_2O$ | 3.39 | 4.46 | 3.91 | 4.53 | 5.16 | 5.99 | 5.34 |
| $K_2O$ | 1.25 | 2.13 | 1.91 | 2.10 | 3.10 | 5.21 | 4.36 |
| $P_2O_5$ | 0.27 | 0.08 | 0.36 | 0.52 | 1.28 | 0.21 | 0.29 |
| $H_2O^+$ | 1.39 | 1.43 | 1.58 | 3.61 | 1.42 | 1.56 | 1.33 |
| $H_2O^-$ | 0.50 | 0.20 | 0.38 | 0.47 | 1.28 | 0.26 | 0.64 |
| TOTAL | 99.94 | 99.56 | 98.83 | 97.96 | 100.89 | 99.72 | 97.50 |

| | PRESQU'ILE DE GEOGRAPHIE | | | | | | |
|---|---|---|---|---|---|---|---|
| | CN14 | CN44b | CN60 | 8159 | 8150 | 8190 | 81100 |
| $SiO_2$ | 44.55 | 47.34 | 52.22 | 55.43 | 59.88 | 63.19 | 65.99 |
| $TiO_2$ | 2.94 | 2.70 | 2.22 | 0.06 | 0.59 | 0.08 | + |
| $Al_2O_3$ | 15.83 | 17.45 | 17.46 | 19.97 | 19.44 | 17.11 | 17.32 |
| $Fe_2O_3$ | 4.12 | 3.48 | 2.15 | 2.54 | 1.24 | 2.10 | 1.91 |
| FeO | 6.73 | 6.02 | 6.19 | 1.91 | 3.08 | 1.80 | 1.35 |
| MnO | 0.16 | 0.17 | 0.13 | 0.14 | 0.11 | 0.13 | 0.10 |
| MgO | 3.98 | 3.92 | 3.51 | 0.10 | 0.42 | + | + |
| CaO | 9.25 | 8.36 | 6.28 | 0.56 | 2.07 | 0.85 | 0.26 |
| $Na_2O$ | 4.24 | 3.49 | 4.14 | 7.26 | 5.82 | 6.50 | 7.13 |
| $K_2O$ | 2.95 | 3.97 | 3.92 | 6.27 | 6.44 | 5.12 | 4.92 |
| $P_2O_5$ | 1.23 | 0.63 | 0.64 | 0.01 | 0.10 | 0.02 | + |
| P.F. | 3.72 | 1.89 | 0.63 | 5.31 | 0.64 | 1.56 | 0.83 |
| TOTAL | 99.70 | 99.42 | 99.49 | 99.56 | 99.83 | 98.46 | 99.81 |

| | MONTS BALLONS | | | | | | | | | | PEN LORANCHET | | |
|---|---|---|---|---|---|---|---|---|---|---|---|---|---|
| | 7869b | 7851 | 7860 | 7856 | 7852 | 7831c | 7846 | 7833 | 7857 | 7861 | 81104 | 81112 | 8129 |
| $SiO_2$ | 44.96 | 43.47 | 45.33 | 46.72 | 47.72 | 46.79 | 55.62 | 57.58 | 59.63 | 59.25 | 48.15 | 45.65 | 58.44 |
| $TiO_2$ | 2.67 | 3.27 | 3.02 | 3.09 | 2.66 | 2.35 | 0.61 | 0.09 | 0.69 | 0.51 | 3.37 | 2.71 | 0.44 |
| $Al_2O_3$ | 18.22 | 14.82 | 15.52 | 15.56 | 18.62 | 16.86 | 20.42 | 22.80 | 18.68 | 19.07 | 12.54 | 15.53 | 19.20 |
| $Fe_2O_3$ | 6.30 | 9.24 | 7.51 | 7.50 | 6.28 | 5.53 | 2.92 | 1.35 | 2.11 | 2.02 | 10.11 | 6.05 | 1.78 |
| FeO | 3.57 | 3.69 | 3.43 | 3.42 | 2.34 | 3.76 | 1.96 | 1.70 | 3.61 | 3.18 | 5.44 | 6.02 | 1.30 |
| MnO | 0.15 | 0.21 | 0.14 | 0.17 | 0.17 | 0.17 | 0.13 | 0.10 | 0.19 | 0.13 | 0.21 | 0.18 | 0.05 |
| MgO | 3.83 | 6.15 | 5.91 | 5.80 | 3.12 | 3.36 | 0.65 | + | 0.56 | 0.37 | 4.42 | 4.80 | 1.20 |
| CaO | 9.94 | 9.46 | 8.74 | 8.08 | 7.61 | 6.75 | 2.22 | 0.71 | 1.86 | 1.66 | 5.76 | 5.92 | 1.32 |
| $Na_2O$ | 3.95 | 3.17 | 3.50 | 4.12 | 5.03 | 4.57 | 6.74 | 8.33 | 5.52 | 5.51 | 4.05 | 3.67 | 3.40 |
| $K_2O$ | 2.17 | 2.90 | 2.66 | 3.04 | 3.55 | 3.86 | 5.78 | 6.07 | 6.06 | 6.78 | 1.98 | 3.05 | 10.15 |
| $P_2O_5$ | 1.09 | 1.05 | 0.76 | 0.74 | 0.98 | 0.72 | 0.22 | + | 0.16 | 0.19 | 0.47 | 0.74 | 0.09 |
| P.F. | 1.35 | 0.83 | 1.38 | 1.40 | 1.32 | 3.31 | 2.68 | 1.20 | 1.18 | 0.83 | 1.64 | 4.61 | 2.73 |
| TOTAL | 98.20 | 98.26 | 97.90 | 99.64 | 99.40 | 98.03 | 99.95 | 99.93 | 100.25 | 99.50 | 98.14 | 98.93 | 100.10 |

**Table 3. Alkaline and silica-undersaturated series, whole rock analyses.** ***Montagnes Vertes*****: MV9,7738,7718 = gabbros; 7745 = amphibole bearing monzonites; 7705 = biotite and amphibole bearing monzogabbro; 7741 = amphibole rich nephelinitic monzosyenite; 7714 = syenite.** ***Monts Ballons*****: 7869b,7851,7860,7856 = amphibole rich gabbros; 7852,7831c = nepheline bearing monzonitic gabbros; 7846 = acmite bearing syenite; 7833 = amphibole rich Ne syenite; 7857, 7861 = Ne syenite.** ***Loranchet Peninsula*****: 81104 = gabbroic dyke near Lac Athena; 81122 = dyke of gabbro in the Vallée du Thermometre; 8129 = inclusions of nepheline rich syenite in a basaltic dyke near the Righi intrusion.**

intermediate compositions as is the case for the silica-oversaturated alkaline group (Lameyre, et al., 1982). Another distintive feature of the rocks of this group is their silica-poor and sodium rich geochemistry, whereas their potassium contents are similar to those of the silica-oversaturated alkaline rocks.

*Summary*

Figure 4 (a and b) summarises the major distinguishing characteristics of the three groups of plutonic complexes. The groups can also be distinguished by the modal ferromagnesian minerals, which are essentially augite and olivine in the tholeiitic to transitional group, with biotite abundant in both of the alkaline groups, and amphibole a major mineral of the intermediate rocks in the silica-undersaturated alkaline group (Giret et al., 1981).

## Geochronology

Of the tholeiitic to transitional complexes only Mamelles and the Val Gabbro have been dated. Mafic rocks from Mamelles yield K-Ar ages of $32 \pm 1$ to $38 \pm 1$ Ma, and from the Val Gabbro of $39 \pm 3$ Ma. These are the oldest isotopic ages determined from Iles Kerguelen, and they appear inconsistent with the ages of the surroundings basalts which never exceed 26 Ma (Nougier et al., this volume). Two explanations are possible; either the plutonic complexes have been tectonically emplaced into younger basalts in a late diapiric process, or the intrusions have caused isotopic resetting of the basalts. In the same complexes, the silicic dykes give K-Ar ages between $23.6 \pm 0.6$ and $27.9 \pm 0.9$ Ma, which is similar to those of some alkaline plutonic complexes.

The alkaline plutonic complexes are much younger than the tholeiitic to transitional ones, and are of similar age to the alkaline volcanic complexes (Nougier et al., this volume). Of the silica-oversaturated group, the Montagnes Vertes complex was intruded between $22.7 \pm 0.8$ and $26 \pm 3$ Ma ago, and is the oldest alkaline plutonic complex of the archipelago. In Monts Ballons, the intrusions were emplaced between $12.6 \pm 0.7$ and $17 \pm 1$ Ma ago. Intrusion of the silica-undersaturated alkaline complexes began somewhat later than that of the silica-oversaturated ones. In the Ile d'Ouest complex, the gabbros are $16.6 \pm 0.9$ Ma old and the syenites are $12.4 \pm 0.7$ Ma old. The five successive intrusions on the Rallier du Baty Peninsula commenced in the same time interval with the southern gabbros ($15 \pm 0.4$ Ma), and continued much later with the northern syenites emplaced at $4.8 \pm 1.0$ Ma (Lameyre et al., 1976), that is, at about the same time as the alkaline Mont Ross volcano.

It appears that the intrusions of both types of alkaline complexes took place simultaneously, with a probable focus of silica-undersaturated magmatism in the Courbet Peninsula. In contrast, the intrusion of the tholeiitic to transitional gabbros is the oldest magmatic event known on the Iles Kerguelen, although this appears inconsistent with the maximum K-Ar age determined on the basalts. To attempt a solution to this geochronological paradox, new K-Ar ages have been determined for some of the basalts surrounding the plutonic complexes.

On the Courbet Peninsula, two basaltic mountains have been sampled for K-Ar dating of their bases and summits. On Mont Crozier, the K-Ar age of basalt bedrock at the base (80 m above sea level) is $23 \pm 1$ Ma, whereas that at the summit (980 m) is $25.8 \pm 0.7$ Ma. A similar apparent inversion has been obtained from Monts du Chateau where the base (110 m) gives a K-Ar age of $23.0 \pm 0.9$ Ma and the summit (630 m) $24.5 \pm 0.7$ Ma. It is noteworthy that the oldest K-Ar age in each case is similar to the ages determined by Nougier (1970) for the ancient flood basalts. A larger geochronological inversion has been observed from the basaltic pile of Anse de Duncan (Ile de l'Ouest) where rocks at the base (10 m) have a K-Ar age of $11.8 \pm 0.6$ Ma and those at the summit (250 m) $21.8 \pm 0.8$ Ma. These three examples show an incontrovertible isotope resetting, and we note that in each case the younger age of the base is similar to that of the later plutonic event in the area. This explanation is supported by K-Ar ages from Monts Ballons where $12.6 \pm 0.7$ Ma nepheline syenites are nearly contemporaneous with the surrounding (13.8 Ma) basalts which are in turn younger than the intruded amphibole-rich monzogabbros ($17 \pm 1$ Ma).

The above results show that the alkaline plutonism was responsible for the K-Ar isotopic resetting of the intruded basalts. The intrusions were probably also the cause of zeolitisation of the flood basalts, which is well developed near the trachytic dykes and at the base of the

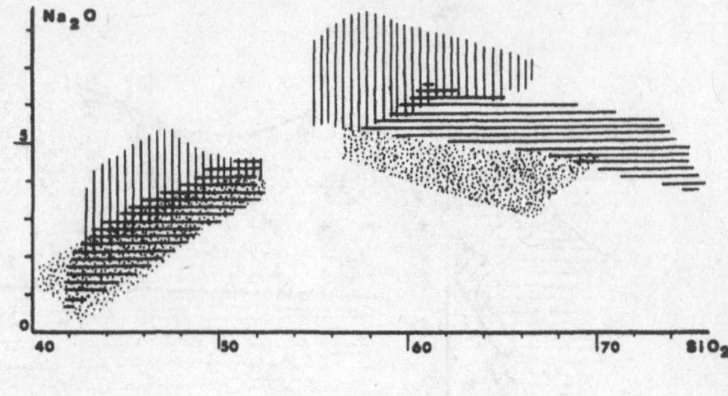


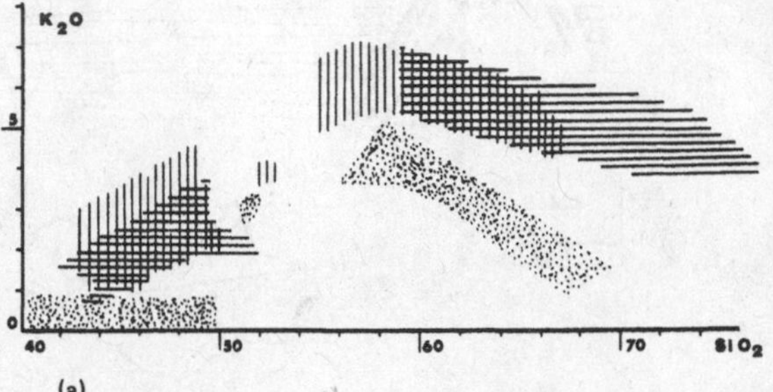


(a)

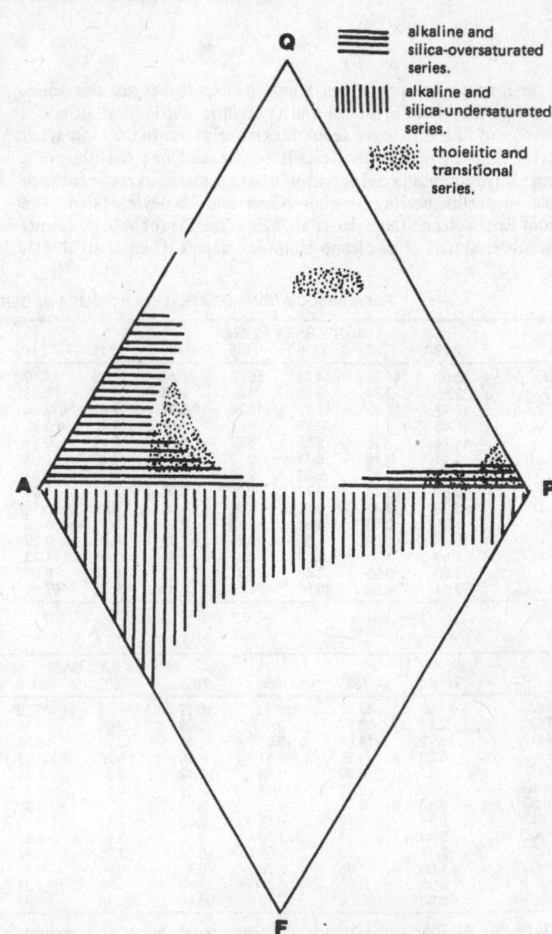


(b)

**Figure 4. The main characteristics of the different plutonic series. (a) $Na_2O/SiO_2$ and $K_2O/SiO_2$ (wt.%) diagrams. (b) The QAPF (IUGS) modal representation.**

basalt flows. The effects of a low-grade metamorphism can also be seen in the granitic rocks of the Righi intrusion and in the brecciated fine grained margins of the tholeiitic to transitional gabbros; it is manifested by the development of minerals such as chlorite, epidote, zeolites and albite.

Care is therefore necessary when considering the isotopic age of basalts. Nevertheless, this study implies the existence of basalts at least as old as the 38 and 39 Ma old tholeiitic to transitional gabbros. These results are consistent with palaeomagnetic data (Schlich, 1975) which locate Anomaly 11 (33-34 Ma) near the northern boundary of the Kerguelen Plateau.

## Conclusions

Two main plutonic episodes can be recognised on Ile Kerguelen: the first, comprising tholeiitic to transitional intrusions, began at least 40 Ma ago, whereas the second comprising alkaline intrusions began only 26 Ma ago. The second event was responsible for K-Ar isotopic resetting of the oldest basalts from which original dates will be difficult to obtain.

The later alkaline magmatism produced two different types of plutonic complexes, although they are synchronous. The first comprises large ring complexes of silica-oversaturated rocks which define a western magmatic province. The second consists of silica-undersaturated rocks which make up central and eastern provinces. The boundary between the provinces of the two types of alkaline complex trends NNW-SSE and traverses the elongated western part of Kerguelen (Figure 1.). The boundary appears to be parallel to a western rift structure that also parallels a linear zone which separated two areas of oceanic crust with different spreading directions (Schlich, 1975). The linear zone is bordered by magnetic anomaly 19 (47 Ma) in the north and anomaly 25 (63 Ma) in the south; there is thus some evidence of the tectonic control of Kerguelen magmatism.

*Acknowledgements* We are indebted to Drs I. Black (CNRS) and I. Clarke (Monash University) for their aid and comments in preparing the manuscript. The K-Ar dates have been determined by Dr J.M. Cantagrel (CNRS). Research facilities were obtained from the Mission de Recherche des TAAF, from CNRS through its grant to LA 298. Participation in the 4th International Symposium on Antarctic Geology has been possible thanks to the aid of SCAR and CNFRA.

## REFERENCES

DOSSO, L., VIDAL, P., CANTAGREL, J.M., LAMEYRE, J., MAROT, A. and ZIMINE, S., 1979: Kerguelen, continental fragment or oceanic island ? Petrology and isotopic geochemistry evidence. *Earth Planet. Sci. Lett., 43*,46-60

DOSSO, L. and MURTHY, V.R., 1980: An Nd isotopic study of the Kerguelen Islands: inferences on enriched oceanic mantle sources. *Earth Planet Sci. Lett., 48*, 268-76.

GIRET, A., 1979: Genese de roches feldspathoidiques par la déstabilisation des amphiboles: massif des Montagnes Vertes, Kerguelen (TAAF). *C.R. Ac. Sci., Paris, 289*,379-82.

GIRET, A. and LAMEYRE, J., 1980: Mise en place et = volution magmatique des complexes plutoniques de la caldéra de Courbet,Île de Kerguelen (TAAF). *Soc. Geol. Fr. Bull., 22*, 437-46

GIRET, A., BONIN, B. and LEGER, J.M., 1980: Amphibole compositional trends in oversaturated and undersaturated alkaline plutonic ring complexes. *Can. Mineral., 18*, 481-95.

GIRET, A., CANTAGREL J.M. and NOUGIER, J., 1981: Nouvelles données sur les complexes volcano-plutoniques des Iles.Kerguelen (TAAF). *C.R. Ac. Sci., Paris, 293*, 191-3.

GIRET, A., LAMEYRE, J., MAROT, A. and ZIMINE, S., 1980: Carte géologique au 1/50000 de la péninsule Rallier du Baty. *CNFRA, 45*, notice 12p.

JAUZEIN, P., 1981: Les granitoides des séries tholéiitiques sur les exemples de l'Islande, de la Corse et de la Brévenne. Thése 3°cycle, Université Pierre et Marie Curie (Paris VI), 230p.

LAMEYRE, J., MAROT, A., ZIMINE, S., CANTAGREL, J.M., DOSSO, L. and VIDAL, P., 1976: Chronological evolution of the Kerguelen Islands syenite-granite ring complex. *Nature, 263*, 306-7.

LAMEYRE, J., MAROT, A., ZIMINE, S., CANTAGREL, J.M., DOSSO, L., VIDAL, P., GIRET, A., JORON, J.L., TREUIL, M., and HOTTIN, G., 1981: Etude géologique du complexe plutonique de la Péninsule Rallier du Baty, Iles Kerguelen. *CNFRA, 49*.

LAMEYRE, J., BLACK, R., BONIN, B., GIRET, A. and PLATEVOET, B., 1982: Coupures dans les suites magmatiques. Le Dalygap représente-t-il la composition de liquides magmatiques primaires? *Reun. Ann. Sci. Terre, Summ.*, 354.

MAROT, A. and ZIMINE, S., 1976: Les complexes annulaires de syénites et de granites alcalins dans la péninsula Rallier du Baty, Iles Kerguelen, TAAF. Thése 3° cycle, Université Pierre et Marie Curie (Paris VI), 2 Vol.

NOUGIER, J., 1970: Contribution á l'étude géologique et géomorphologique des Iles Kerguelen, TAAF. *CNFRA, 27*, 1, 440p, 2, 246p.

NOUGIER, J. and LAMEYRE, J., 1973: Les nordmarkites des Iles Kerguelen (TAAF) dans leur cadre structural. Problème de leur origine et de celle de certaines roches en domaine océanique. *Soc. geol. Fr., Bull., 15*, 306-11.

NOUGIER, J., CANTAGREL, J.M. AND PAWLOWSKI, D. (this volume): Chrono-spatial evolution of the volcanic activity from southeastern Kerguelen (TAAF).

NOUGIER, J., (in press): Volcanology of French sub-antarctic islands. *in* Volcanological Atlas of Antarctica.

SCHLICH, R., 1975: Structure etâge de l'Océan Indien occidental. *CNFRA, 38*,

# VOLCANOLOGY OF POSSESSION ISLAND, CROZET ARCHIPELAGO (TAAF)

L. Chevallier and J. Nougier, *Laboratoire de Geologie, 33 rue Louis Pasteur, 84000 Avignon, France.*

J.M. Cantagrel, *Laboratoire de Geologie, 5 rue Kessler, 63031 Clermont-Ferrand Cedex, France.*

*Abstract* This paper summarises the evolution of the composite oceanic volcano of Possession Island and includes four studies:

(a) *volcano-tectonism:* five phases in three structural cycles are characterised by ringfeeders and radial and linear-rift systems produced by tectonics and pressures exerted from the magma chamber.

(b) *petrography:* includes an alkali suite of ankaramitic-olivine-basalts and feldspar basalts and underlying basanites and phonolites; mineral assemblages of the eruptive rocks are correlated with structural data and activity of the magma chamber.

(c) *geochemistry:* 34 representative samples of distinct phases specify the evolution of Na-alkali magma via fractional crystallisation.

(d) *geochronology:* three cycles, the first being the longest and dated at over 8 Ma, whereas the second (very active) has been going on for 0.3 Ma; the third cycle continues to be present. Correlations with the eastern island of the archipelago indicate that: both volcanoes belong to the same magmatic province; there is no chrono-spatial evolution but rather simultaneous build up; and the activity is connected to the structure and to oceanic floor motion.

The Crozet Archipelago lies on an oceanic plateau located at the boundary between magnetic anomalies 30 and 31 of Upper Cretaceous oceanic floor derived from the western and eastern Indian ridges (Schlich, 1975, Goslin et al., 1981, Figure 1A). The Crozet plateau, as outlined by the 2000 m isobath consists of a broad western platform from which emerge the île aux Cochons and the remnant islands of the Pingouins and Apôtres, and a similar eastern platform including the two islands of Possession and Est (Figure 1B).

## Chronostratigraphic and structural evolution of Possession Island

Geological mapping (Chevallier, 1981) and a volcano-structural study (Chevallier and Nougier, 1981) were recently carried out on Possession Island. Five major volcanic phases were recognised.

*Phase I formations* (Figure 2) outcrop at the western end of the island on Rocher des Moines. They consist of a pile of palagonitized hyaloclastites with fossiliferous beds rich in mussels and pectens alternating with detrital materials (Figure 3). Rocks of this series are too altered to be dated isotopically. They appear to reflect shallow submarine volcanic activity uplifted by recent tectonic movements of phase V.

*Phase II* corresponds to the development of a subaerial strato-volcano that now comprises the island basement (IIa); the central part is cut by dyke system (IIb). The succession comprises a sequence of thick flows apparently derived from the west, with interbedded detrital material increasingly abundant towards the top. Using K/Ar methods, the middle part of the series has been dated at 8.1 ± 0.6 Ma and a lava flow from near the summit has yielded an age of 2.7 ± 0.8 Ma (Figure 3 and Table 1). Because of the lack of feeder dykes within the succession, we can deduce that the lavas were derived from an upper cone in the central part of the strato-volcano now located at the west side of the island. The intrusive system (IIb) which intersects the volcano is composed of linked ring dykes. The geometric centre of this structure, presumably corresponding to the volcano centre, is located offshore, to the west of the island (Figure 2). Two kinds of intrusion

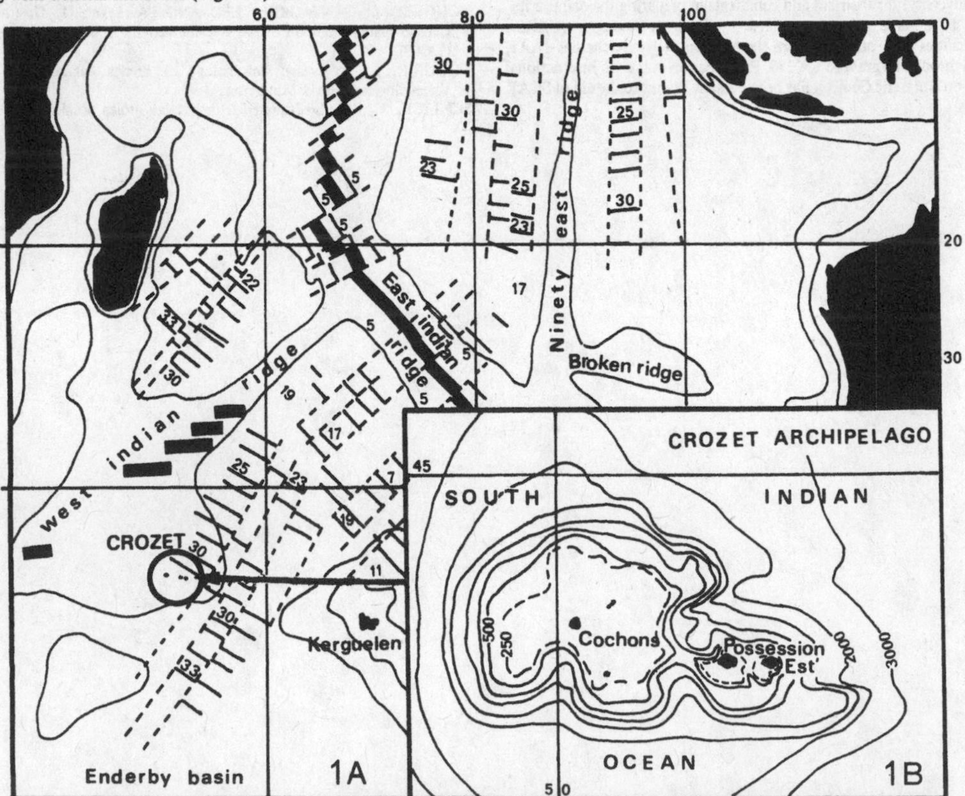


**Figure 1A. Fracture zones and magnetic lineations of the southwestern part of the Indian Ocean after Schlich (1975); Goslin et al. (1981).**

**Figure 1B. Bathymetric sketch map of the Crozet Archipelago.**

**TABLE 1: Geochronological data of lava series from Possession Island**

| Sample No | Nature | K % | $Ar_{40}$ | Atm At % | Age (Ma) |
|---|---|---|---|---|---|
| 1 (R 9417) | ankaramite (phase IV) | 1.30 | 0.0479 | 84.0 | 0.53 ± 0.09 |
| 2 (R 9418) | ankaramite (phase IV) | 0.769 | 0.0349 | 87.4 | 0.65 ± 0.15 |
| 3 (R 9251) | ankaramite (phase IV) | 0.753 | 0.0376 | 86.4 | 0.70 ± 0.15 |
| 4 (R 9419) | Ol. basalt (phase III) | 1.08 | 0.0541 | 83.4 | 0.72 ± 0.11 |
| 5 (R 9253) | Feldspar basalt (ph. III) | 1.70 | 0.120 | 85.7 | 0.80 ± 0.20 |
| 6 (R 9252) | phonolite (phase III) | 4.64 | 0.332 | 50.5 | 1.03 ± 0.04 |
| 7 (R 9255) | basanite (phase $II_b$) | 2.31 | 0.206 | 89.9 | 1.30 ± 0.4 |
| 8 (R 9416) | picritic basalt (phase $II_a$) | 0.218 | 0.0409 | 95.2 | 2.70 ± 0.8 |
| 9 (R 9286) | Ol. basalt (phase $II_a$) | 0.513 | 0.290 | 70.5 | 8.10 ± 0.6 |

Analyst : CNRS, L.A. 10; Clermont-Ferrand.
For location of samples, see map Fig. 2. Samples 3 to 9 are also analysed for geochemistry (see Table 2).

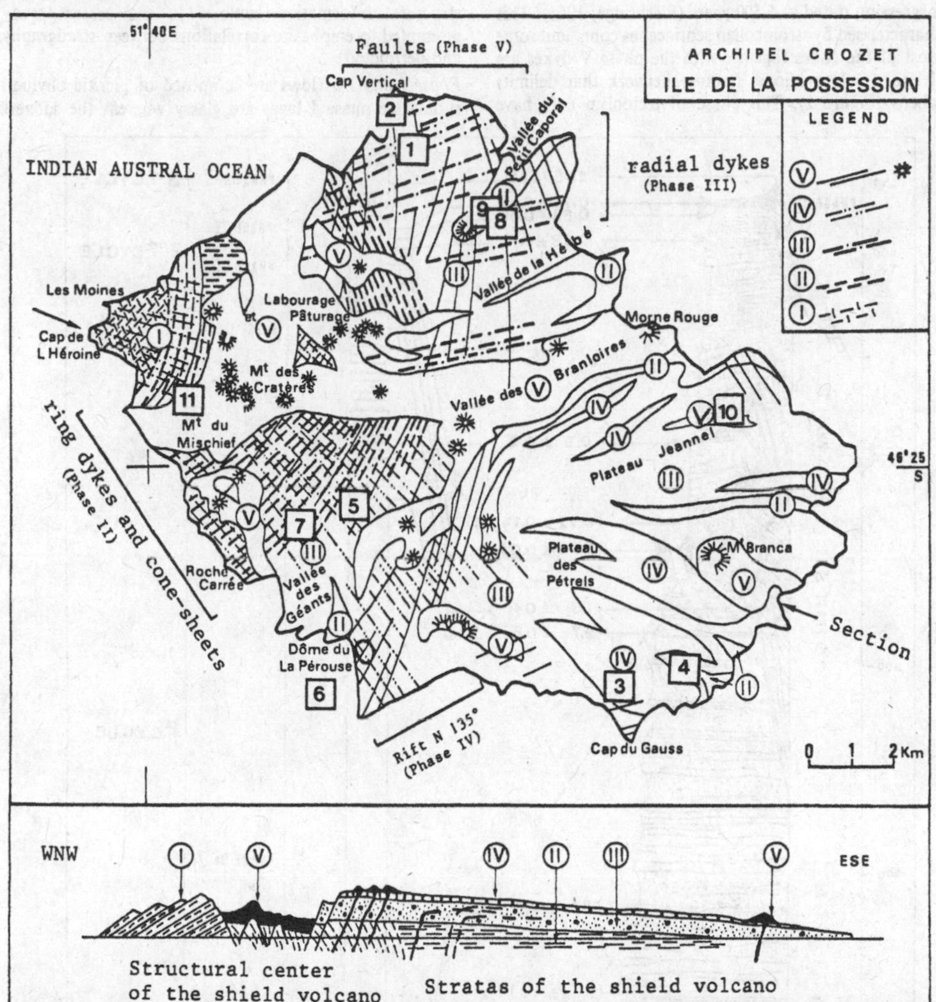


**Figure 2.** **Schematic volcano-structural map of Possession Island and section WNW-ESE (from Chevallier and Nougier, 1981, Chevallier, 1982). I: Moines' series; II: strato-volcano, ring-dykes; IV: plateau basalts and rift N135; V: Strombolian volcanism and volcano-tectonic faults and strombolian volcanoes. 1 to 11: location of samples quoted on Tables 1-2 and Figure 3.**

can be distinguished, the oldest and most numerous being dipping outward ring dykes dated at 1.3 ± 0.4 Ma; these are cut by less numerous dipping inward cone sheets (Figure 3). According to the numerical models of stress distribution over a magma chamber (Anderson, 1936) and to geological studies samples in Scotland (Clough et al., 1909) or Reunion Island (Chevallier and Vatin Perignon, 1982), these basaltic ring intrusions appear to be caused by decreasing magma pressure and consequent cauldron subsidences inside the volcanic pile. In Possession Island some normal faults connected with conic intrusions and thick detrital deposits in the upper part of the succession are compatible with the subsidence theory and with a quiescent period towards the end of the phase II. The size of the Possession ring structure (diameter, thickness of injected zone, geometry of dykes) indicates that the magmatic chamber had subsided to a depth greater than 5 km.

*Phase III* rocks unconformably overlie the phase II formations the lowest representative being a basal conglomerate 10 to 50 m thick and composed of coarse torrential deposits. The upper part of the succession is composed of detrital material interbedded with flows and sills that become more numerous towards the top (Figure 3). The lavas were fed from a huge radial dyke system converging towards a volcanic centre now located in the western part of the island (Figure 2). The earlier intrusion of the La Pérouse plug in the south of the island has been dated at 1.03 ± 0.4 Ma. Two sills of the series have been dated at 0.8 ± 0.2 Ma and 0.72 ± 0.11 Ma.

*Phase IV* is due to activity along a rift zone that trends 135°. Two samples have been dated at 0.7 ± 0.15 and 0.53 ± 0.09 Ma. The intrusive zone is more than 5 km wide and 12 km long with, on average, one dyke every 10 m. A few of these intrusions have poured out to form vast 10 m thick plateau forming flows. In the western part (Rocher des Moines) there is a second rift direction aligned at 105°. The Possession rift extends beyond the structural limits of the volcano and corresponds to a major extensional zone within the oceanic crust. Its origin may be connected with the old structures of the Crozet basin (Late Cretaceous magnetic anomalies 31-32). Phase IV was followed by a glacial epoch which carved wide U-shape valleys such Branloires, Hébé and Géants Valleys.

*Phase V* is represented by recent or near recent volcanism prior to the Flandrian transgression dated at 5,500 years (Fairbridge, 1961). This volcanism is characterised by strombolian scoriaceous cones and some lava flows. Most of the cones together with the phase V dykes are aligned parallel to a submeridional fracture network that delimits horst and grabens (Figure 2). This phase of tectonism may have caused the tilting of the volcano complex that led to the development of the coastal cliffs of Est Island, and appears to be linked to dextral motions along submeridional faults, now recorded by a conjugate 70-80° network of fractures and dykes, folding of older intrusions, rotation of the Moines' block hyaloclastites (phase I), stripe carbonate deposits and flattened dykes. The deformation is not related to local stresses within the volcanic structure but it is a regional phenomena which can be observed also on the other islands of the Crozet Archipelago. It probably results from motions of the oceanic floor.

## Petrography of Possession Island lavas

Previous petrographic studies have been carried out by Reinisch (1908), Tyrrell (1937) and Lacroix (1940) without any stratigraphic or structural information context. In the present study we have attempted to emphasise correlations between stratigraphy, structure, and petrology.

*Phase I and IIa* flows are composed of picritic olivine basalt. The submarine phase I lavas are glassy whereas the subaerial lavas of

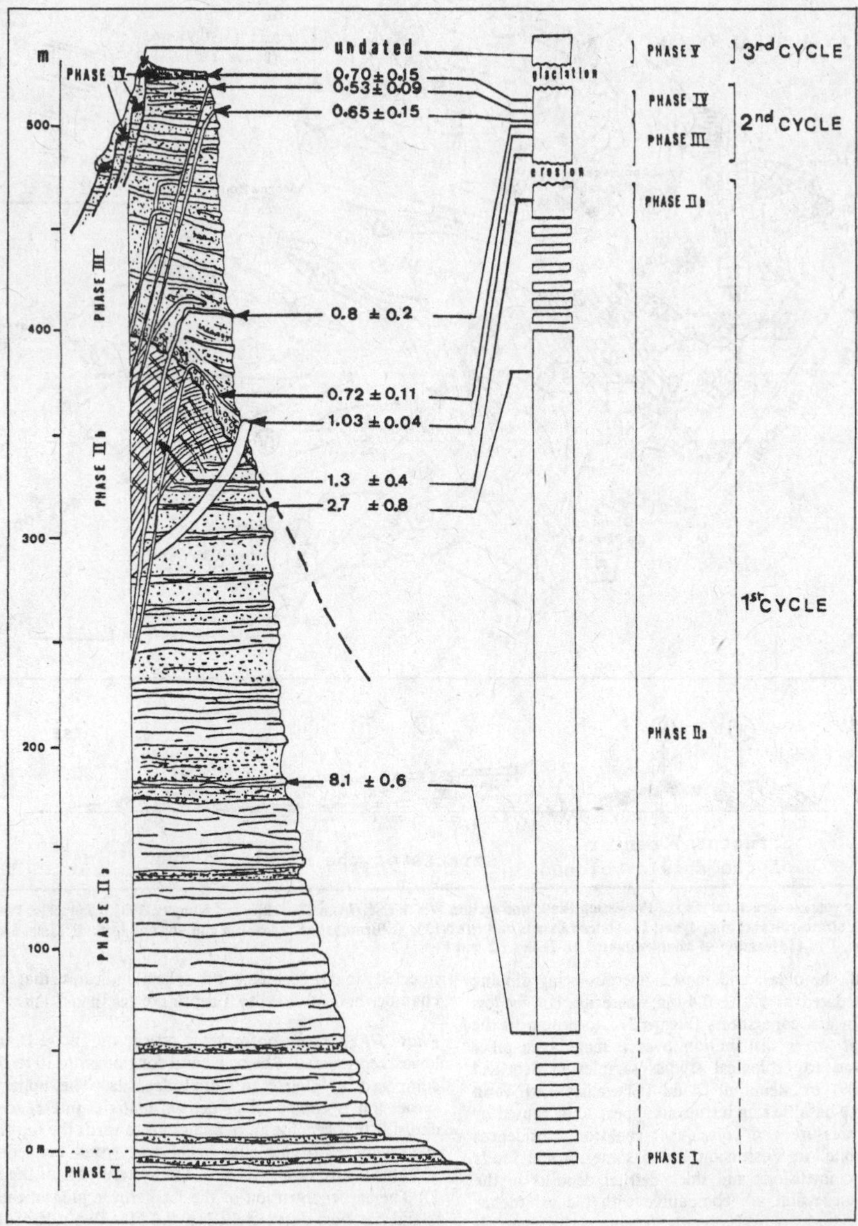


**Figure 3. Chronostratigraphic log of Possession Island sequence. For geochronological data, see Table 1.**

phase II are porphyritic. Phenocryst phases include augite (2-3 mm length) some of it titaniferous, olivine (1-2 mm) commonly iddingsitised, titanomagnetite (square section 0.2 to 1 mm) and some scarce labradorite. The groundmass is dominated by an angular feldspathic framework defining a fluidal texture with clinopyroxenes, olivine, apatite and magnetite.

*The dykes of phase IIb* are basanites with a sparsely porphyritic texture. Phenocrysts include augite that is commonly altered to calcite and chlorite, iddingsitised olivine and some largely sericitised plagioclase, ($An_{60}$). The groundmass is composed of strongly sericitised microliths of plagioclase, altered clinopyroxene, brown hornblende and needle-like dark magnetite. Like the feldspars, nepheline is strongly altered and hardly recognisable. A light coloured felsic tristanite dyke was observed in this intrusive system. The rock has a fluidal texture defined by feldspar phenocrysts and sericite. Intrusion of the ring structure system was accompanied by hydrothermal alteration that affected the whole strato-volcano. Calcite, chlorite and zeolite (probably heulandite) are abundant as amygdales and other cavity fillings. Rare epidote (zoisite and pistacite) has developed within the groundmass and in cavities. Alteration is very strong in the intrusive zone, in the centre of the volcanic edifice in the western half of the island. Pyroxene has been altered to calcite and chlorite and olivine is strongly iddingsitised. The felsic ground mass and nepheline are hardly recognisable and almost entirely sericitised. Hydrothermal quartz has developed inside feeder dykes zone.

*Phases III and IV* exhibit an inverted petrological sequence: differentiated lavas were erupted first and cumulate phenocryst-rich lavas last. The intrusion "La Pérouse" plug, comprises early nepheline phonolite of phase III. A fluidal texture is defined by feldspar (sanidine and anorthoclase) automorphic microliths and microphenocrysts. Nepheline occurs as large polycrystal patches or as square automorph crystals. Aegirine augite occurs as microliths and microphenocrysts. The first lava flows are porphyritic felsic basalts with a fluidal texture defined by plagioclase ($An_{40}$) phenocrysts, brown hornblende and augite. The plagioclase rich groundmass includes some clinopyroxene microliths. Ankaramitic olivine basalts in the middle part of the sequence contain augite, olivine and labradorite ($An_{65}$) phenocrysts. The microlitic groundmass is composed of plagioclase and clinopyroxene scattered through a dark glassy matrix. Ankaramitic flows in the upper part of the series are characterised by huge zoned augite grains (up to 1 cm), smaller olivine (1-2 mm) and less abundant labradorite. The groundmass comprises 50-60% of the lava with the same minerals together with magnetite. Some gabbros occur in the rift dykes and consist of augite or diallage (partially replaced by brown hornblende), hypersthene, biotite and altered plagioclase.

*Phase V* comprises a differentiated suite ranging from felsic basalt flows to intrusive phonolites. The felsic basalts have a porphyritic texture with phenocrysts of augite (some of them zoned aegirine augites) and plagioclase ($An_{40-50}$). Whereas the phonolites have a fluidal texture with phenocrysts including zoned sanidine and anorthoclase. Nepheline occurs in polycrystal aggregates and is locally replaced by haüyne. Other phenocrysts include aegyrine augite, aegyrine, sphene and brown hornblende.

*Petrographic summary* Possession Island is dominated by a suite of alkali basalts and their derivatives, typical of many oceanic islands. The differentiation series ranges from cumulate lavas with olivine and clinopyroxene to small volumes of dyke-basanites of phase IIb and phonolite of phases III and V.

## Major element geochemistry

34 rocks have been analysed and nine representative analyses from different volcanic series are listed in table 2. The silica/alkali diagram (Figure 4) illustrates the alkalic characteristics of the Possession Island lavas. According to Middlemost's classification (1980) olivine and picritic basalts of phases I and IIa are included within the alkali basalt field. The phase IIb intrusive basanitic system plots in the undersaturated alkali lava field. The ankaramitic olivine basalts (basalts), felsic basalts (trachy-basalts) represent phase III. A phonolite plots in the trachytic field owing to secondary hydrothermal albitisation (Sample 6, Table 2). Phase IV ankaramites and gabbros plot in the transitional basalt field and the differentiated suite of phase V (trachy-basalt, tristanite, phonolite) stands out clarly on the diagram, total alkalis ranging from 12 to 13%. The AFM diagram (Figure 5) is consistent with the usual differentiation trends, especially within the limits outlined by Borley (1974) for various islands of the Indian ocean. The $Na_2O$-$K_2O$-CaO diagram (Figure 6) highlights the homogeneity of the alkali lavas. In phase V, the excess of $Na_2O$ is reflected by the presence of both aegirine augite and even aegirine. The tetrahedron diagram from Yoder and Tilley (1962) clearly shows several magmatic phases discriminated by normative nepheline (Figure 7).

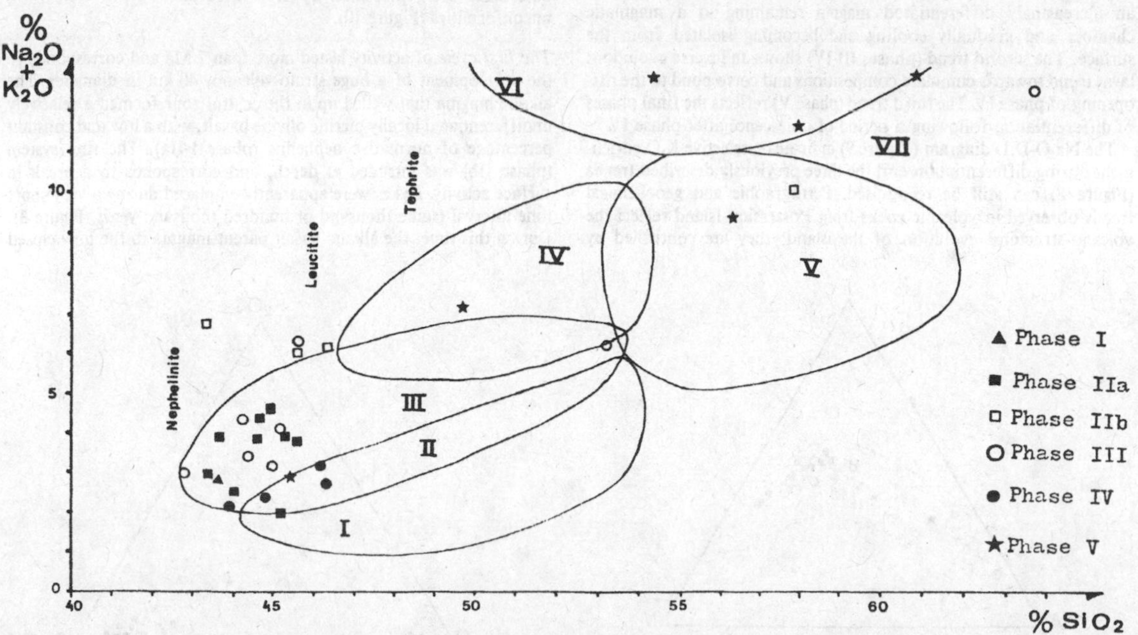


**Figure 4.** Alkali/Silica diagram of Possession lavas. I to VII: areas after Middlemost (1980) nomenclature; I: sub-alkali basalts; II: transitional basalts; III: alkali basalts; IV: trachy-basalts (s.e.)—if $Na_2O/K_2O > 1$: hawaiite—if $Na_2O/K_2O < 1$: trachy-basalts (s.s.); V: trachy-andesite (s.l.)—if $Na_2O/K_2O > 1.5$: benmoreite, if $Na_2O/K_2O < 1.5$: tristanite; VI: phonolite; VII: trachyte.

TABLE 2: Characteristic representative chemical analyses from Possession Island. For location of samples; see map, Fig. 2.

| | Phase IIa | | Phase IIb | | Phase III | Phase IV | | Phase V | |
|---|---|---|---|---|---|---|---|---|---|
| | 9 | 8 | 7 | 6 | 5 | 4 | 3 | 10 | 11 |
| $SiO_2$ | 42.00 | 43.85 | 43.50 | 63.00 | 40.80 | 44.50 | 45.90 | 57.50 | 48.95 |
| $Al_2O_3$ | 12.90 | 10.05 | 14.50 | 19.10 | 13.80 | 15.10 | 13.00 | 19.50 | 17.65 |
| $Fe_2O_3$ | 9.85 | 5.22 | 5.33 | 2.50 | 7.05 | 4.79 | 4.41 | 3.58 | 5.56 |
| FeO | 2.75 | 5.21 | 6.37 | — | 5.36 | 7.67 | 6.66 | 1.01 | 3.91 |
| MgO | 10.80 | 15.00 | 4.85 | .40 | 8.10 | 8.40 | 11.50 | 1.10 | 4.70 |
| CaO | 12.40 | 13.50 | 10.30 | 1.10 | 9.80 | 11.50 | 12.20 | 3.50 | 7.70 |
| $Na_2O$ | 2.30 | 1.50 | 3.00 | 6.60 | 2.10 | 3.00 | 2.00 | 6.80 | 4.60 |
| $K_2O$ | .55 | .35 | 2.60 | 5.70 | 1.90 | 1.40 | 1.20 | 4.80 | 3.00 |
| $TiO_2$ | 2.90 | 2.20 | 3.10 | .40 | 3.20 | 3.00 | 2.50 | 1.30 | 2.50 |
| $MnO_2$ | .20 | .16 | .19 | .15 | .18 | .20 | .18 | .23 | .19 |
| $H_2O^+$ | 2.04 | 1.62 | 4.99 | .90 | 4.19 | .24 | .37 | .26 | .17 |
| $H_2O^-$ | .73 | .35 | .72 | .10 | 2.58 | .21 | .10 | .02 | .13 |
| TOTAL | 99.42 | 99.01 | 99.45 | 99.95 | 99.06 | 100.01 | 100.02 | 99.60 | 99.06 |
| Or | 3.37 | 2.13 | 16.41 | 34.07 | 12.18 | 8.32 | 7.13 | 28.59 | 17.97 |
| Ab | 11.09 | 9.24 | 11.62 | 56.37 | 12.30 | 10.42 | 12.37 | 41.17 | 23.54 |
| An | 24.02 | 20.23 | 19.61 | 5.52 | 24.46 | 23.67 | 23.02 | 8.52 | 18.84 |
| Na | 4.89 | 2.07 | 8.36 | — | 3.76 | 8.16 | 2.50 | 9.05 | 8.58 |
| Di | 30.82 | 38.15 | 27.70 | — | 21.95 | 26.81 | 29.87 | 5.98 | 15.44 |
| Hy | — | — | — | .64 | — | — | — | — | — |
| Mt | 1.14 | 7.80 | 8.24 | — | 9.30 | 6.98 | 6.42 | .24 | 6.04 |
| Il | 5.70 | 4.31 | 6.28 | .32 | 6.59 | 5.73 | 4.77 | 2.49 | 4.81 |
| He | 9.40 | — | — | 2.53 | 1.23 | — | — | 3.44 | 1.46 |
| Ol | 9.57 | 16.07 | 1.78 | .26 | 8.25 | 9.94 | 13.92 | — | 3.33 |
| D.I. | 19.3 | 13.4 | 36.4 | 90.4 | 28.2 | 26.9 | 22.0 | 78.8 | 50.1 |
| S.I. | 42.7 | 56.0 | 22.4 | 2.7 | 34.0 | 33.9 | 45.4 | 6.5 | 22.2 |

Sample 9 : Ol-basalt; Petit Caporal valley, right side; K/Ar age 8.1 ± 0.6 Ma
8 : picritic basalt; Petit Caporal valley, top of the unit; K/Ar age 2.7 ± 0.8 Ma
7 : basanite, Geants valley ring-dykes; K/Ar age 1.3 ± 0.4 Ma
6 : miaskite phonolite; La Perouse plug; K/Ar age 1.03 ± 0.04 Ma
5 : feldspar-basalt; Geants valley, left side; K/Ar age 0.8 ± 0.2 Ma
4 : Ol-ankaramitic basalt; Cirque de Noel; K/Ar age 0.72 ± 0.11 Ma
3 : ankaramite; Petrels plateau; K/Ar age 0.70 ± 0.15 Ma
10 : miaskite phonolite; Jeannel plateau dyke
11 : trachy-basalt; Mont des Crateras flow

All data from CNRS (L.A. 10) Clermont-Ferrand. Norms are calculated free of water
Computer program written by J-L Larice; Faculte des Sciences Avignon

Samples defining field A are parallel to the olivine-diopside line whereas those from field B (intrusive ring dyke system) record a significant nepheline increase in the basanites. In field C (phases III-IV) nepheline is less abundant and but it is at increased levels again in field D (Phase V). The MgO differentiation index (D.I.) diagram (Figure 8) shows a negative correlation in agreement with differentiation through fractional crystallisation processes. Three parallel magmatic trends are apparent: the first one corresponds to the flows of phases I-IIa and the late intrusive system of phase IIb and suggests an increasingly differentiated magma remaining in a magmatic chamber and gradually cooling and becoming isolated from the surface. The second trend (phases III-IV) shows an inverse evolution; lavas trend towards cumulate compositions and correspond to the rift-opening of phase IV. The third trend (phase V) reflects the final phases of differentiation following a period of quiescence after phase IV.

The $Na_2O$-D.I. diagram (Figure 9) indicates a selective $K_2O$ enrichment during differentiation and the three previously described trends (Figure 8) can still be recognised. Petrographic and geochemical trends observed in volcanic rocks from Possession Island reflect the volcano-structural evolution of the island; they are controlled by nepheline, K-feldspar and titanomagnetite fractionation; picritic lavas and ankaramites are cumulate types. Starting with a Na-alkali basalt parent, magmatic evolution appears to have been controlled by opening and closing of the magma chamber.

## Evolutionary model of Possession Island volcano

From chronostratigraphic, structural and petrologic data we have established three main cycles of volcanic activity. Two periods of quiescence are represented by erosive surfaces which now form unconformities (Figure 10).

*The first cycle* of activity lasted more than 7 Ma and corresponds to the development of a huge strato-volcano, 40 km in diameter. The alkalic magma that welled up in the central zone formed a relatively undifferentiated locally picritic olivine basalt, with a low and constant percentage of normative nepheline (phase I-IIa). The ring system (phase IIb) was intruded at depth, and corresponds to a break in surface activity. Dykes were apparently emplaced during a very short time interval (some thousand or hundred thousand years, Figure 3). During this time, the alkalic basalt parent magma in the now closed

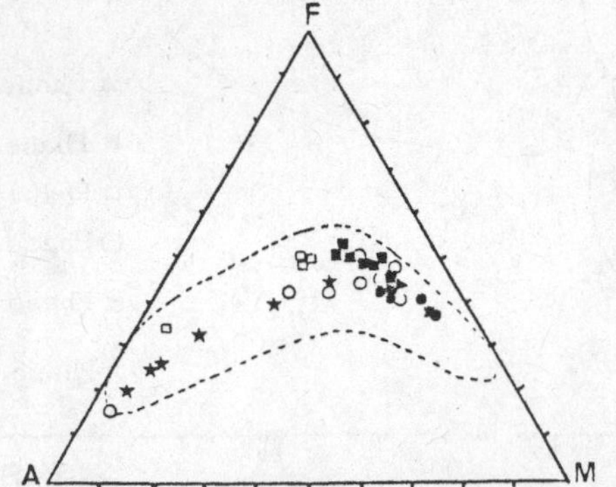


**Figure 5. AFM diagram of Possession Island lavas. Pecked line: alkali suite of differentiated lavas from Indian Ocean volcanoes after Borley (1974).**

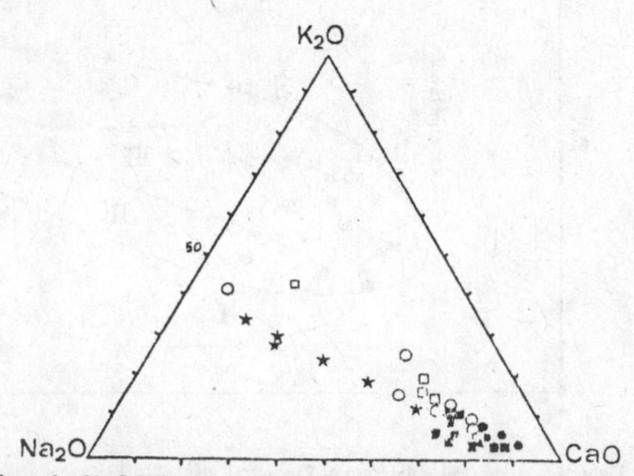


**Figure 6. $Na_2O$-$K_2O$-CaO diagram of Possession Island lavas. Same symbols as Figure 4.**

magma chamber at more than 5 km in depth was perhaps differentiating along a basanite trend. The intrusive activity metamorphosed the whole volcano (epidote zeolite facies).

*The second cycle* (phases III-IV) extended over about 500,000 years and was preceded by a period of intense erosion over 270,000 years while a new alkalic parent magma was apparently differentiating at depth. When volcanic activity was renewed, lavas were erupted along radial fractures and became less differentiated but more abundant with the time. The maximum activity took place during phase IV (170,000 years) with opening of the 135° trending rift giving access to a deep primary magma (ankaramites and gabbros). This observation is consistent with a model in which magma was drawn from differewnt levels of a magma chamber with an increase in the rate of rifting.

A second period of quiescence after phase IV corresponds to a glacial episode. During this time, magmatic differentiation may have occurred, with tectonic movements at Est, Cochons and Pingouins Islands perhaps being related to the reactivation of volcanism on Possession Island. Lavas which reached the surface or remain trapped along fissures include trachy-basalts, tristanites and intrusive phonolites. The small amount of lavas, is due to the fact that *the third cycle* (phase V) did not last longer than a few thousand years.

## Volcanological correlations inside the Crozet Archipelago

Geological units of Est Island described by Gunn et al., 1970, 1972; Cantagrel et al., 1980; LaMeyre and Nougier, 1982, show many stratigraphic and chronological similarities to those of Possession Island. Three main activity cycles separated by two periods of quiescence can be recognised on Est Island and are apparently synchronous to those observed on Possession Island (Figure 11). The first cycle on Est Island corresponds to the development of a primitive volcano in which a gabbroic plutonic complex was intruded at 8.75 Ma. A younger monzo-dioritic dyke system may be related to the observed epidote-zeolite facies metamorphism of the host-rock basement. This cycle could be equivalent to phases I and II, described on Possession Island although a stratified gabbroic complex was not observed in the Moines series (phase I) there. K/Ar ages range from about 8 Ma to 1.3 Ma implying about 7 Ma for the first cycle of volcanism although the oldest rocks observed on Possesson and Est Islands could not be dated. The other cycles lasted only about 1 Ma in which time most of the subaerial part of the volcano was formed. The second cycle observed on Est Island starts with a basal conglomerate unconformably overlying first cycle rocks. The upper part comprises a very thick sequence of flows and agglomerates. A dyke system trending 110° intrudes the series and may be the equivalent to the 135° dyke system observed on the Possession Island. This dyke direction which is due to a recent (age 0.7 Ma) volcanic rift may be correlated with the 120° direction of magnetic anomaly 31 identified on the northeast side of the Crozet plateau (Figure 1A). This anomaly

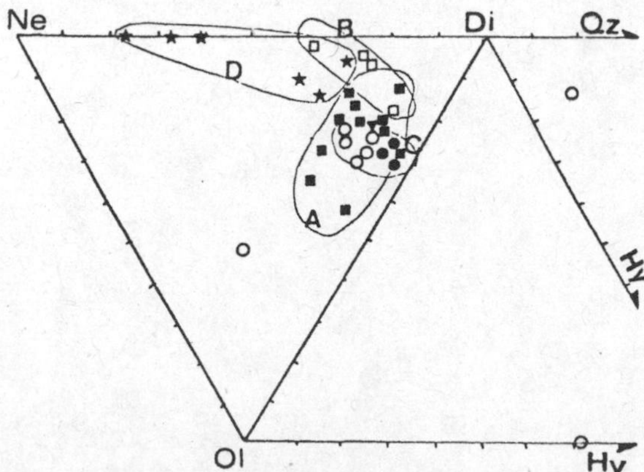


**Figure 7. Tetrahedron diagram after Yoder and Tilley (1962), of Possession Island lavas. Field A: phases I and IIa; build up of the strato-volcano. Field B: phase IIb; intrusive ring system. Field C: phases III and IV; volcano-detritic sequence and plateau flood basalt associated with sills, radial dykes and rift system. Field D: phase V; strombolian volcanism.**

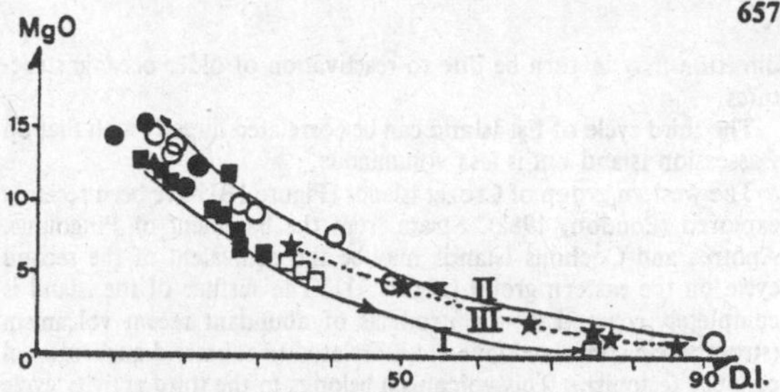


**Figure 8. MgO differentiation index of Possession Island lavas (D.I. = Ab + Or + Ne) after Thornton and Tuttle (1960). Symbols explanation, see Figure 4. I, II, III: magmatic trends; explanation in text. Arrows indicate the chronological volcanic process.**

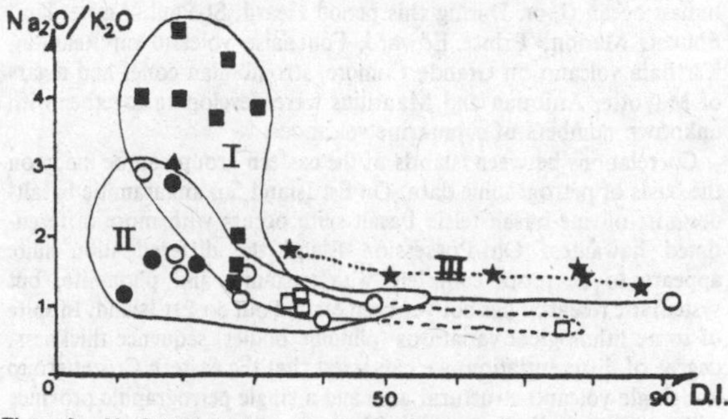


**Figure 9. $Na_2O/K_2O_2$/D.I. I, II, III: magmatic trends; explanation in text.**

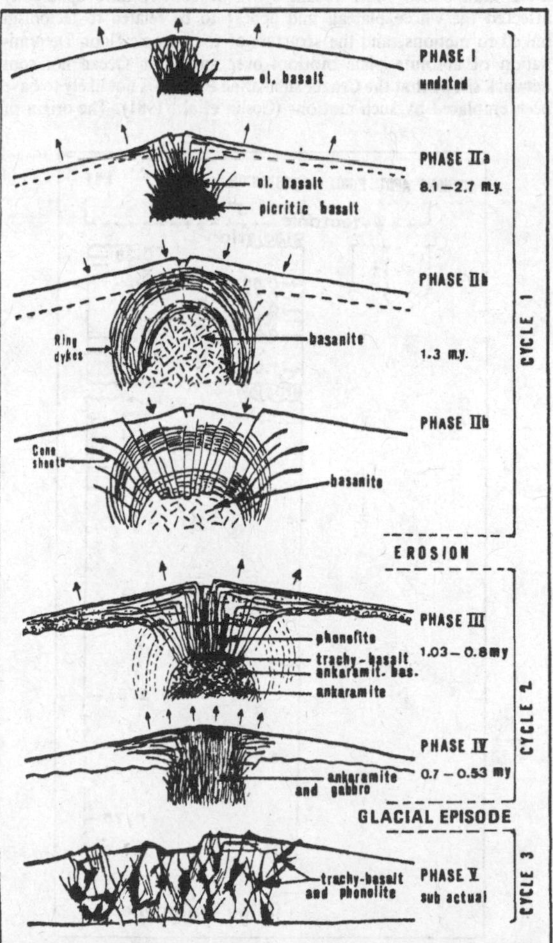


**Figure 10. Magmato-structural evolution of Possession volcanoes.**

direction may in turn be due to reactivation of older oceanic structures.

The third cycle of Est Island can be correlated directly with that on Possession Island but is less voluminous.

The western group of Crozet islands (Figure 1B) have been recently explored (Boudon, 1982). Strata from the basement of Pingouins, Apôtres and Cochons Islands may be the equivalent of the second cycle on the eastern group (Figure 11). The surface of the island is completely covered by the products of abundant recent volcanism (strombolian cones and lava flows) related to submeridianal regional collapse tectonism. This volcanism belongs to the third activity cycle of the Crozet plateau and active submeridianal tectonic structures observed on all islands can perhaps be correlated with the regional pattern of the ocean floor. The recent resurgence of volcanism is characteristic of the last ten thousand years and affects the whole Indian ocean floor. During this period Heard, St-Paul, Amsterdam, Bouvet, Marion, Prince Edward, Fournaise volcano on Reunion, Karthala volcano on Grande Comore, strombolian cones and maars of Mayotte, Anjouan and Mauritius were developing together with unknown numbers of submarine volcanoes.

Correlations between islands of the eastern group can be made on the basis of petrographic data. On Est Island, an ankaramitic basalt-oceanite-olivine-basalt-felsic basalt suite occurs with more differentiated hawaiites. On Possession Island the differentiation suite appears to be more complete with tristanite and phonolite, but systematic research has not yet been carried out on Est Island. In spite of some lithological variations (plutonic bodies, sequence thickness, degree of differentiation) we can assert that the eastern Crozet group is a single volcano-structural area and a single petrographic province with a parental alkalic magma. Rhythmic activity has taken place for more than 8 Ma. More recent cycles of activity have apparently affected the whole plateau and appear to be related to tectonism linked to motions, and the structure of the oceanic floor. Determination of absolute plate motions over an Indian Ocean hot spot network shows that the Crozet submarine Plateau is not likely to have been emplaced by such motions (Goslin et al., 1981). The origin of Crozet Island's volcanism, as with many Indian Ocean volcanoes, may be related to the structure of the oceanic floor.

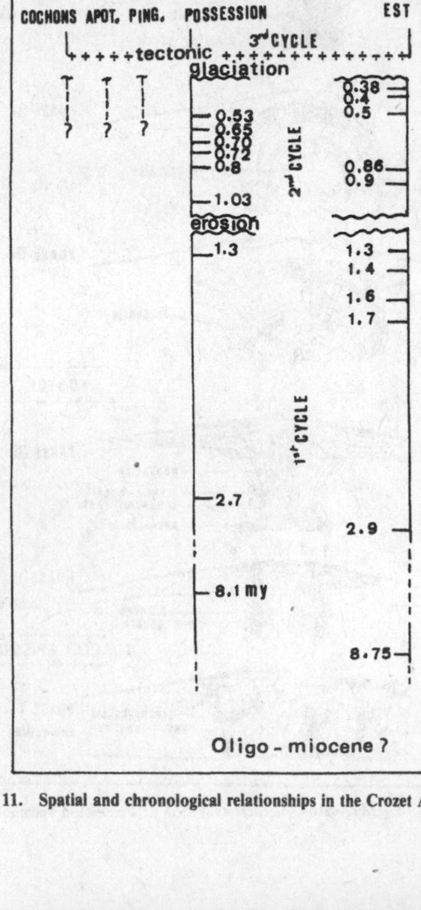

**Figure 11.** **Spatial and chronological relationships in the Crozet Archipelago.**

*Acknowledgements* The field work (February—April, 1980) and laboratory work were supported by Mission Recherche of the Territoire des Terres Australes et Antarctiques Francaises. This support is gratefully acknowledged.

## REFERENCES


ANDERSON, E.M., 1936: The dynamics of the formation of cone-sheets, ring-dykes and cauldron subsidence. *R. Soc. Edinb., Proc., 56,* 128-57.

BORLEY, G.D., 1974: Oceanic islands *in* Sorensen, H. (ed.) *Alkaline rocks,* Wiley Interscience Pub., London, 311-30.

BOUDON, G., 1982: Premiere reconnaissance volcanolgique du groupe occidental de l'archipel Crozet. *C.R. Ac. Sc., Paris,* in press.

CANTAGREL, J.M., LAMEYRE J. and NOUGIER, J., 1980: Volcanologie et géochronologie d'une île volcanique, île de l'Est (archipel Crozet). *Int. Geol. Congr. Abstr., 26, 1,* 27.

CHEVALLIER, L., 1981: Carte géologique de l'île de la Possession avec notice. *CNFRA, 50.*

CHEVALLIER, L. and NOUGIER, J., 1981: Premiere étude volcano-structurale de l'île de la Possession, îles Crozet (TAAF), Ocèan Indian Austral. *C.R. Ac. Sc. Paris, 292,* 363-8.

CHEVALLIER, L. and VATIN PERIGNON, N., 1982: Volcano-structural evolution of Piton des Neiges, Reunion Island, Indian Ocean. *Bull. Volcan.*, in press.

CLOUGH, C.T., MAUFE, H.B. and BAILEY, E.B., 1909: The cauldron subsidence of Glencoe and the associated igneous phenomena. *Geol. Soc. Lond., Q. J., 65,* 611-74.

FAIRBRIDGE, R.W., 1961: Eustatic changes in Sea Level. *Phys. Chem. Earth, 4,* 99-185.

GOSLIN, J., RECQ, M. and SCHLICH, R., 1981: Mise en place et évolution des plateaux sousmarins de Madagascar et de Crozet. *Soc. geol. Fr., Bull., 23,* 609-18.

GUNN, B.M., COY-YLL, R., WATKINS, N.D., ABRANSON, C.E. and NOUGIER, J., 1970: Geochemistry of an Oceanite-Ankaramite-Basalt suite from East Island, Crozet Archipelago. *Contrib. Mineral. and Petrol., 28,* 319-39.

GUNN, B.M., ABRANSON, C.E., WATKINS, N.D. and NOUGIER, J., 1972: Petrology and Geochemistry of isles Crozet; a summary; *in* Adie, R.J. (ed.) *Antarctic Geology and Geophysics,* Universitetsforlaget, Oslo, 825-9.

LACROIX, A., 1940: Les laves des volcans inactifs des Archipels Marion et Crozet. *Mem. Mus. Hist. Nat., Paris, 14,* 47-62.

LAMEYRE, J. and NOUGIER, J., 1982: Geology of Ile de l'Est, Crozet Archipelago (TAAF); *in* Craddock, C. (ed.) *Antarctic Geoscience,* Univ. Wisconsin Press, Madison, 767-70.

MIDDLEMOST, E.A.K., 1980: A contribution to the nomenclature and classification of volcanic rocks. *Geol. Mag., 117,* 51-7.

REINISCH, 1908: Gesteine von der Possession Insel (Crozet gruppe). *Dt. Sudpolar Exped., 2,* H.T., 4. 235-43.

SCHLICH, R., 1975: Structure et âge de l'Océan Indien Occidental. *Soc. geol. Fr., Mem. Hors Ser., 6.*

THORNTON, C.P. and TUTTLE, O.F., 1960: Chemistry of igneous rocks. I. Differentiation Index. *Am. J. Sci., 258,* 664-84.

TYRRELL, G.W., 1937: The petrology of Possession Island. *Rep. BANZ Ant. Res. Exped. 1929-31. Rept. Ser. A., 5, 2(4),* 57-68.

YODER, H.S. and TILLEY, C.E., 1962: Origin of basaltic magmas. *J. Petrol., 3,* 342-532.

# THE MACQUARIE ISLAND OPHIOLITE COMPLEX: MAJOR AND TRACE ELEMENT GEOCHEMISTRY OF THE LAVAS AND DYKES

B.J. Griffin, *Electron Optical Centre, University of Adelaide, GPO Box 498, Adelaide, South Australia 5001, Australia.*

R. Varne, *Geology Department, University of Tasmania, GPO Box 252, Hobart, Tasmania 7001, Australia.*

*Abstract* Macquarie Island is unique amongst subantarctic islands in that it is part of the Mid-Tertiary oceanic lithosphere from a major ocean basin. The rocks were probably created at the Indian-Australian-Pacific spreading ridge.

The lavas and dykes are usually porphyritic, carry plagioclase ($An_{87-80}$) as a dominant phenocryst phase with less abundant olivine ($Fo_{89-85}$), chrome spinel and rare clinopyroxene ($Ca_{45}Mg_{50}Fe_5$—$Ca_{38}Mg_{50}Fe_{12}$) phenocrysts. Major and trace element abundance of rocks little affected by alteration show systematic variations. These variations include a range of CIPW compositions from *Q*- to *Ne*-normative which correlate with a petrographic range from typical ocean floor basalt mineralogy through to alkali olivine basalts. Hygromagmatophile elements, particularly Nb (20-60 ppm) and the light rare earth elements, are relatively enriched in the petrographically and CIPW normative alkaline rocks.

Statistical analysis of the chemical data distinguishes three groups which geochemically may be termed the major and compatible trace elements, moderately incompatible trace elements and extremely incompatible or hygromagmatophile trace elements. Shallow level crystal fractionation processes involving the massive and layered gabbro sequences account for most of the major and trace element variations in these samples but are not sufficient to explain the variations in hygromagmatophile elements. These latter variations could result from either the operation of dynamic melting processes or heterogeneity in the source mantle.

# PETROLOGY OF IGNEOUS ROCKS ON CAMPBELL ISLAND

P. Morris, *Department of Geology and Geophysics, University of Sydney, NSW 2006, Australia.*

*Abstract* Igneous rocks of Campbell Island comprise a suite of alkaline volcanic flows and high-level intrusions ranging from alkali olivine basalt to rhyolite. Flows are represented by rare alkali olivine basalt, abundant hawaiite, rare trachyte and rhyolite, whereas intrusive rocks comprise hawaiite, mugearite, benmoreite and trachyte. No chronologic distinction can be made between flow and intrusive rocks, which have been emplaced during approximately 5 Ma of the Late Miocene. A gabbro intrusion pre-dates volcanism at 16 Ma.

Major element chemistry and least-squares modelling of the flow and high-level intrusive sequence suggests a process of first mafic then felsic dominated fractional crystallisation, although poor agreement for trace element modelling could indicate derivation of the suite by fractionation of more than one parent magma.

# 15

## Cenozoic Igneous Activity

# RATES OF UPLIFT AND THE SCALE OF ICE LEVEL INSTABILITIES RECORDED BY VOLCANIC ROCKS IN MARIE BYRD LAND, WEST ANTARCTICA

W.E. LeMasurier, *Geology Department, University of Colorado, Denver, Colorado, 80202, U.S.A.*

D.C. Rex, *Department of Earth Sciences, University of Leeds, Leeds, LS2 9JT, U.K.*

*Abstract* The Hobbs Coast region of Marie Byrd Land includes two major chains of trachytic shield volcanoes and a large number of small satellitic centres of dominantly basaltic volcanism. Basaltic and trachytic hyaloclastites, associated in this region with tillites and glacially striated granitic basement rocks, record the existence of glacial ice during the past volcanic episodes. The ages of these hyaloclastites range from 25 Ma to 0.4 Ma, with the majority being Pliocene and Pleistocene.

The only reliable structural datum in this region is a flat pre-volcanic erosion surface. It is well exposed in a number of isolated nunataks, at elevations that range from roughly 600 m to 2700 m above sea level and from 200 m to 700 m above the present level of the ice sheet. Much of the hyaloclastite is found on the erosion surface, with the oldest deposits generally found in the highest basement rocks. In these blocks, the erosion surface has been elevated tectonically, while in the adjacent areas it has either remained near sea level where it presumably originated, or been tectonically depressed. We calculate that the rates of uplift and of subsidence have been between 105 and 138 m/Ma. This slow rate implies that hyaloclastites younger than 1 Ma old owe their present exposure above ice level almost entirely to the lowering of the ice surface. In the Hobbs Coast region, Pleistocene hyaloclastites stand up to 2000 m above present ice levels. The results to date suggest that it may be possible to infer how long these large scale fluctuations in ice level have been taking place.

The Cenozoic volcanoes in Marie Byrd Land comprise an alkaline volcanic province that lies on one flank of a major intracontinental rift system (LeMasurier, 1978; Cooper et al., 1982). The effects of contemporaneous block faulting and volcanism are clearly displayed throughout the province by displacements of a pre-volcanic erosion surface, by the rectilinear outlines of basement nunataks, and by the north-south and east-west alignments of linear volcanic chains (Figures 1 and 2). Another apparent manifestation of tectonic uplift is the exposure of hyaloclastite sections at elevations up to as much as 2000 m above the level of the continental ice sheet. Hyaloclastites have been formed by subglacial eruptions throughout the past 27 Ma. The exposure of these deposits above present ice level must have been caused either by tectonic uplift, recession of the ice, or a combination of both (LeMasurier and Rex, 1982). Some of the youngest hyaloclastites stand more than 2000 m above ice level, and would require rates of uplift of the order of 1 cm/year to produce this elevation by tectonic processes alone. Such a rate seems unreasonable for an aseismic region, and leads one to suspect that substantial changes in ice level have been at least as important as tectonic uplift in producing the extensive exposures of hyaloslastite in Marie Byrd Land.

The purpose of this paper is to describe the chronology of glacial and tectonic events recorded by hyaloclastites, subaerial lavas, and displacements of the pre-volcanic erosion surface in this region. The presentation of new data will focus mainly on the problem of ice level fluctuations.

## Characteristics of the Erosion Surface

The key structural datum in Marie Byrd Land is a very flat pre-volcanic erosion surface. It is exposed as isolated block faulted

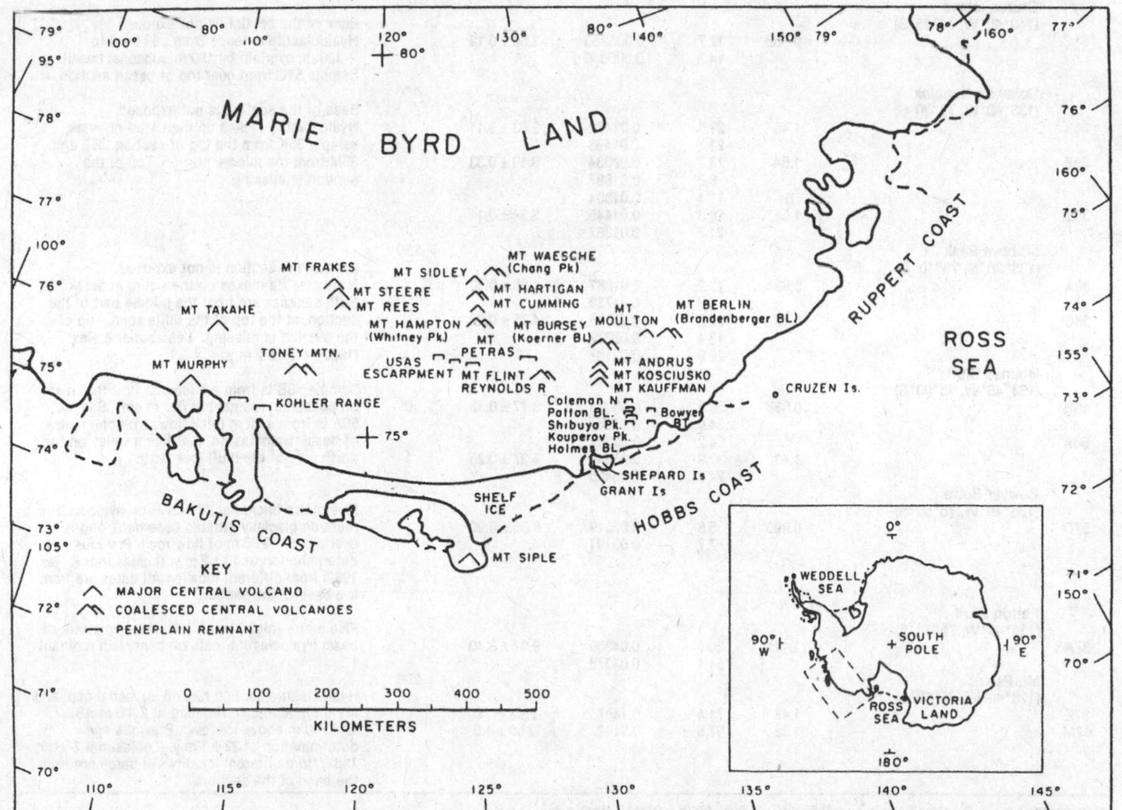


**Figure 1. Index map of Marie Byrd Land, showing localities mentioned in the text and on other figures.**

remnants along the coast of Marie Byrd Land and adjacent regions to the east, from 70°W to 140°W, and at elevations ranging from roughly 200 m to 2700 m above sea level (LeMasurier and Rex, 1982; Laudon, 1972; Rutford et al., 1968). In the Hobbs Coast region of Marie Byrd Land it is exceptionally well exposed (Figure 2), and is overlain by lavas and hyaloclastites in a large number of nunatak exposures.

The configuration of the surface is clearly not controlled by bedrock structure, because it truncates granite and/or highly deformed gneisses in nearly all exposures. It is believed to be Early Tertiary in age because the youngest rocks cut by the surface are those *in an 80 Ma old rhyodacite* dome at Mt Petras (LeMasurier and Rex, 1982), and the oldest rocks that post-date the surface are 23-25 Ma old hyaloclastites, also found at Mt Petras (Table 1). Similar age relations for the surface are found at many other localities along the coast of Marie Byrd Land and eastward, though none constrain the age as precisely as the relations at Mt Petras.

In order to evaluate its usefulness as a structural datum, it is important to determine how much erosional relief exists on the surface and whether there is indeed only one erosion surface, or a series of levels. Perhaps the earliest description of the erosion surface was by Craddock et al. (1964), in which they noted a maximum erosional relief of up to 50 m (one contour interval), but generally less than 15 m in the Jones Mountains (94°00′W, 73°30′S) exposures roughly 450 km east of the region shown in Figure 1. Based on a reconnaissance visit in 1967, LeMasurier and Rex (1982) noted a maximum relief of roughly 400 m on the erosion surface at Mt Petras. However, this estimate was recently revised to less than 100 m after a

**TABLE 1: Chronology and Thicknesses of Hyaloclastite Sections**

| Sample No. | Locality | K-Ar age data* %K | %$^{40}$Ar rad. | Vol. $^{40}$Ar rad. scc/g $\times 10^{-5}$ | Age (m.a.) | Thickness (metres) | Comments |
|---|---|---|---|---|---|---|---|
| | Shepard Island (132°30′W, 74°25′S) | | | | | 500 | Base of the section is not exposed. Hyaloclastite extends from sea level to +500 m. 8E is from +100 m level, 10D is from the +300 m level of the hyaloclastite section. 9F is a late subaerial trachyte flow at sea level. Top of the section is missing. |
| 8E | | 1.14 | 5.5 | 0.00266 | 0.6 ± 0.1 | | |
| 9F | | 4.31 | 8.1 | 0.007289 | 0.42 ± 0.06 | | |
| | | | 7.7 | 0.006437 | | | |
| | | | 7.7 | 0.007440 | | | |
| 10D | | 1.29 | 3.8 | 0.00307 | 0.6 ± 0.1 | | |
| | Grant Island (131°50′W, 74°25′S) | | | | | | |
| 11E | | 3.21 | 7.8 | 0.00935 | 0.7 ± 0.1 | UNK | Hyaloclastite from a phreatomagmatic tuff cone at +500 m elev. |
| | Brandenberger Bluff (136°00′W, 75°57′S) | | | | | 300–400 | Base of the section is not exposed. Samples 3C and 3D are from the base of the section, at ice level. 4C is from the top of the section. Top of the section is missing. |
| 3C | | 3.92 | 55.9 | 0.03935 | 2.58 ± 0.10 | | |
| | | | 38.1 | 0.03949 | | | |
| 3D | | 3.79 | 44.3 | 0.03914 | 2.23 ± 0.26 | | |
| | | | 61.5 | 0.02961 | | | |
| | | | 64.0 | 0.03373 | | | |
| 4C | | 3.88 | 30.8 | 0.04118 | 2.70 ± 0.10 | | |
| | | | 61.6 | 0.04045 | | | |
| | Cruzen Island (140°40′W, 74°45′S) | | | | | 100 | Base of the section is not exposed. Hyaloclastite extends from sea level to +100 m, overlain by 150 m subaerial basalt. Sample 51C from near top of basalt section. |
| 51C | | 0.533 | 12.7 | 0.005280 | 2.68 ± 0.13 | | |
| | | | 14.3 | 0.005840 | | | |
| | Coleman Nunatak (133°40′W, 75°20′S) | | | | | 100 | Base of the section is not exposed. Hyaloclastite makes up the entire nunatak, sample 35K from the top of section, 35E and 35M from the middle portion. Top of the section is missing. |
| 35K | | 1.45 | 21.6 | 0.01473 | 2.63 ± 0.11 | | |
| | | | 21.7 | 0.01498 | | | |
| 35E | | 1.54 | 13.7 | 0.02034 | 3.19 ± 0.33 | | |
| | | | 6.8 | 0.01887 | | | |
| | | | 11.4 | 0.01804 | | | |
| 35M | | 1.55 | 16.3 | 0.01445 | 2.34 ± 0.11 | | |
| | | | 21.2 | 0.01367 | | | |
| | Shibuya Peak (133°35′N, 75°10′S) | | | | | 250 | Base of the section is not exposed. Hyaloclastite makes up the entire nunatak. Both samples are from the middle part of the section, at the top of the tillite zone. Top of the section is missing. LeMasurier & Rex (1982), 4.4 ± 0.2 m.y. |
| 38A | | 0.994 | 13.2 | 0.01867 | 4.66 ± 0.50 | | |
| | | | 4.8 | 0.01739 | | | |
| 38C | | 1.13 | 29.8 | 0.02068 | 4.75 ± 0.20 | | |
| | | | 18.4 | 0.02072 | | | |
| | | | 29.8 | 0.02137 | | | |
| | Holmes Bluff (133°45′W, 75°00′S) | | | | | | Sample 48B is from a subaerial flow that rests on peneplain remnant at 600 m elev. Sample 50A is from a 20 m thick flow, with one metre of basal hyaloclastite, that fills a valley on the north side of the bluff (see text). |
| 48B | | 0.938 | 50.7 | 0.03073 | 8.17 ± 0.33 | 0 | |
| | | | 34.0 | 0.02851 | | | |
| 50A | | | 54.2 | 0.03035 | | | |
| | | 1.41 | 32.9 | 0.03440 | 6.27 ± 0.25 | 1 | |
| | | | 24.6 | 0.03426 | | | |
| | Bowyer Butte (135°40′W, 75°00′S) | | | | | 5 | Five metre thickness of basalt on hyaloclastite rests on glacially striated basement, and is overlain by 10-20 m of flow rock. Previous age determination of 13 ± 2 m.y. (LeMasurier & Rex, 1982) from different locality. All dates are from the base of the section. |
| 57D | | 0.900 | 5.5 | 0.03519 | 9.56 ± 0.90 | | |
| | | | 7.7 | 0.03191 | | | |
| | Patton Bluff (133°40′W, 75°13′S) | | | | | 1 | Five metre thick basalt flow with one metre of basal hyaloclastite rests on peneplain remnant. |
| 37A | | 1.12 | 50.5 | 0.04300 | 9.97 ± 0.40 | | |
| | | | 54.1 | 0.04378 | | | |
| | Mt. Petras (128°40′W, 75°52′S) | | | | | 200 | Hyaloclastite section has no subaerial cap, and rests on peneplain remnant at 2700 m ASL, and 700 m above ice level. Previous age determination of 22 ± 1 m.y. (LeMasurier & Rex, 1982) from different locality. All dates are from the base of the section. |
| 67E | | 1.47 | 71.4 | 0.1457 | 25.3 ± 1.0 | | |
| 67M | | 1.33 | 67.8 | 0.1192 | 23.0 ± 1.0 | | |

*Constants: $\lambda_\beta = 4.72 \times 10^{-10}$ yr$^{-1}$; $\lambda_e = 0.58 \times 10^{-10}$ yr$^{-1}$; $^{40}$K/K = 0.0119 atomic %.
All dated materials are cogenetic basalt lenses or nodules in hyaloclastite, or feldspars separated from these nodules, unless otherwise noted under comments.

more thorough investigation of Mt Petras during the 1977 season (LeMasurier et al., 1981). Other localities along the Hobbs Coast and Ruppert Coast are consistent with an estimated erosional relief of less than 100 m, and probably well under 50 m in most places, based on visual estimates where the contour interval on base maps is 200 m (e.g. Figure 2).

Evidence for multiple surfaces was sought, particularly in nunataks where there is high relief and good physiographic exposure, but no evidence was found for more than one level. The erosion surface can be reasonably interpreted in each nunatak as a remnant of an Early Tertiary peneplain, that originated at or near sea level, and was subsequently disrupted by block faulting. The surface has not, however, been tilted significantly.

There is very little evidence for later Tertiary erosional modification or dissection of the peneplain, and the evidence that does exist is ambiguous. The best example of a possible valley cut into the peneplain is found at the north end of Holmes Bluff (Figure 1), where an east-west trending valley floor is preserved at the base of a 200 m cliff, beneath an Upper Miocene laval flow. It is very likely that this valley is entirely structural in origin, however, because it parallels a major fault system and the cliff boundary of the nunatak is the only valley wall.

In summary, the available evidence suggests that all exposures of the erosion surface are remnants of a single Late Cretaceous-Early Tertiary peneplain, with generally less than 50 m of erosional relief, that has not been significantly modified except by vertical fault displacements.

## Chronology and Occurrence of Hyaloclastites

Data on the ages and thicknesses of 19 hyaloclastite sections in the Hobbs Coast regions are presented in Table 1, in chronological order. Previously published ages for the Mt Petras, Bowyer Butte and Shibuya Peak sections are listed under "comments". None of the other localities have been previously described. All of the dated materials were fresh holocrystalline nodules interbedded in the hyaloclastite. The precautions taken to avoid anomalous ages have been described previously (LeMasurier and Rex, 1982). Multiple samples from the same localities have yielded confirming dates (e.g. Shepard Island, Brandenberger Bluff, Shibuya Peak, Mt Petras). We have no reservations about the new dates presented here, but it should be noted that dates used in Figures 3, 4 and 5 from localities that were not revisited in 1977 (e.g. Mt Steere, Mt Frakes, Mt Murphy, Toney Mountain, Mt Takahe, Kohler Range) are comparatively sparse. See Figure 1 for the locations of sample sites mentioned above.

**Figure 2. View westward along the Hobbs Coast to Mt Prince (foreground), Bowyer Butte, (middle ground) and Hagey Ridge (background). Nunataks are composed predominantly of granitic rocks of uncertain age, truncated by the Early Tertiary erosion surface (see text), and overlain at Bowyer Butte by a thin veneer of Late Miocene basaltic hyaloclastite. The rectilinear outlines of each nunatak, and the lateral discontinuity of the erosion surface, illustrate the effects of vertical displacements along the N-S and E-W system of block faults. Rock exposures at Mt Prince extend a maximum of 10 km north-south. Mt Prince to Hasgey Ridge is 20 km. Cliffs at all three localities are 350-400 m high. U.S. Navy photograph TMA 1621, F33, no.30, 2/6/65 available through U.S. Geological Survey, Reston, Virginia.**

Hyaloclastites occur through Marie Byrd Land as subhorizontally stratified deposits with locally well developed, large scale, foreset bedding. The thickness of each deposit is therefore estimated mainly by determining the topographic relief from the base to the top of the section, using USGS 1:250,000 topographic reconnaissance maps and interpolating within the 200 m contour interval. The thickness of the deposit can generally be regarded as a minimum value for ice thickness at the time of eruption, if the total original thickness can be determined (LeMasurier, 1972a). Hyaloclastite sections that do not have a capping of subaerial lavas are subject to rapid rates of erosion in this region (Andrews and LeMasurier, 1973), and are likely to have been significantly reduced in thickness if they are more than one million years old. The only sections where original thicknesses of hyaloclastite can be determined are the very thin deposits at Holmes Bluff, Bowyer Butte and Patton Bluff. More thorough descriptions of the lithological character of these deposits have been presented previously (LeMasurier, 1972a, 1972b; LeMasurier and Rex, 1982).

The most anomalous aspect of hyaloclastite occurrence in the Hobbs Coast is the scarcity of these deposits in the Ames and Flood ranges. Brandenberger Bluff (Figure 1) is the only significant hyaloclastite deposit in these ranges, which extend from Mt Kauffman south to Mt Andrus (Ames Range) and from Mt Bursey west to Mt Berlin (Flood Range). Most of the other volcanic localities in the Hobbs Coast region, from Mt Petras to the coastal nunataks, are composed entirely, or in large proportion, of hyaloclastite. In order to examine whether the Ames and Flood Range volcanoes represent major eruptive episodes during interglacial periods, the chronology and elevations of these volcanic rocks were studied in some detail. The results are presented below.

## Chronology and Distribution of Subaerial Lavas

In the Hobbs Coast region, the major proportion of volcanic rocks is found in the Ames and Flood Ranges, which include six major trachytic volcanoes that stand between 800 m (Kauffman) and 2200 m (Berlin) above ice level. The outstanding fact about the spatial distribution of hyaloclastites and subaerial lavas is not that subaerial lavas are abundant in the Ames and Flood ranges, which is to be expected, but that hyaloclastites are found mainly in relatively small volcanic centres north and east of these ranges.

Table 2 presents the K-Ar ages of 26 subaerial lavas from the Ames and Flood Ranges, and from a few coastal localities. The elevation of each sample locality above present ice level is also tabulated. The purpose of the latter figure is to help evaluate whether a group of lavas is subaerial because it was erupted during an ice free interval, or whether it is subaerial because it was erupted above the level of the ice sheet. Elevations above sea level cannot be used for this purpose because ice level is the relevant datum for hyaloclastite formation, and ice level has a pronounced curvature in coastal regions (see for example, LeMasurier and Rex, 1982). Where ice level is higher on one side of a volcano than the other, the smaller elevation difference is

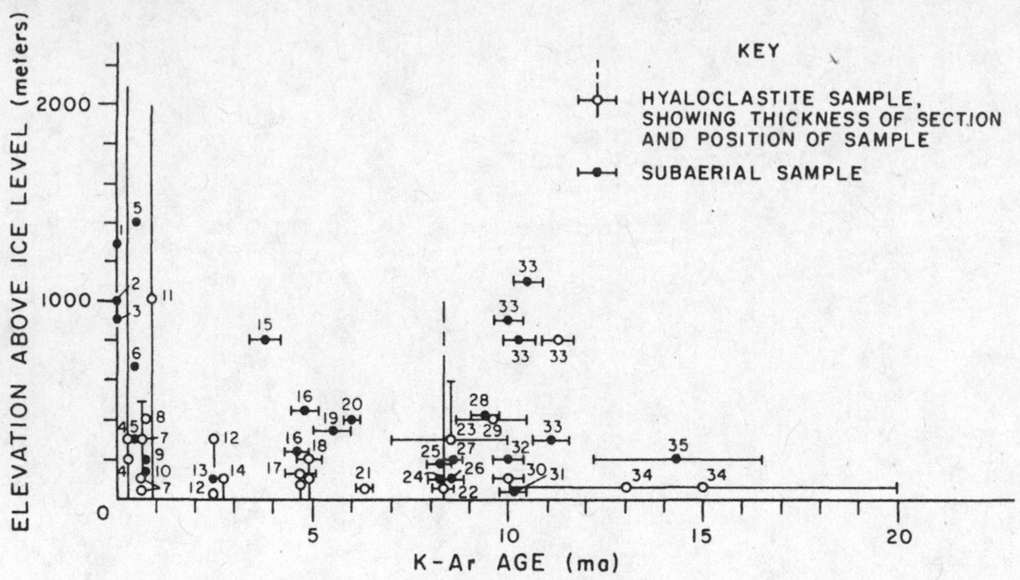


1. MT. BERLIN (23C)
2. MT. BERLIN (31A)
3. MT. ANDRUS (60A)
4. MT. TAKAHE (65B, 65C, 67A), LEMASURIER & REX, 1982
5. TONEY MTN. (75, 76B), LEMASURIER, 1972a
6. MT. BURSEY (28A)
7. SHEPARD ISLAND (8E, 9F, 10D)
8. GRANT ISLAND (11A)
9. MT. BERLIN (7E)
10. MT. BERLIN (14B)
11. MT. MURPHY (62A), LEMASURIER & REX, 1982
12. BRANDENBERGER BLUFF (3C, 4C)
13. MT. BERLIN (67A-35)
14. CRUZEN ISLAND (51c) AND COLEMAN NUNATAK (35E, 35K, 35M)
15. MT. FRAKES, UNPUBLISHED DATA
16. MT. MOULTON (67-1A, 67-2A)
17. SHIBUYA PEAK (38A, 38C)
18. MT. MANTHE, HUDSON MTNS., LEMASURIER & REX, 1982
19. MT. KAUFFMAN (67B-8)
20. MT. BURSEY (27A)
21. HOLMES BLUFF (50A)
22. MT. STEERE, CRARY MTNS., LEMASURIER & REX, 1982
23. JONES MTNS. RUTFORD & OTHERS, 1972
24. HOLMES BLUFF (48B)
25. KOUPEROV PEAK (47B)
26. MT. BURSEY (25A)
27. MT. KOSCIUSKO (40C)
28. MT. BURSEY (67A-21, 29A)
29. BOWYER BUTTE (57D)
30. PATTON BLUFF (37A)
31. MT. BURSEY (24A)
32. MT. KOSCIUSKO (40A)
33. MT. ANDRUS (61C, 44F, 43A, 58D, 59B)
34. TURTLE PEAK, LEMASURIER & REX, 1982
35. MT. ANDRUS (67B-2)

**Figure 3.** Vertical distribution of hyaloclastites and subaerial lavas with respect to the present level of the continental ice sheet. Data are from Table 1 for hyaloclastites and Table 2 for subaerial lavas, unless otherwise noted. Error bars are shown only where errors are more than twice the size of the plotted circle. Dates older than 15 Ma are not shown. Field sample numbers from Tables 1 and 2 are in parentheses.

used. In Figure 3, the elevations of both hyaloclastites and subaerial lavas above present ice level are plotted against their K-Ar ages, to illustrate the vertical distribution of these rocks with respect to their relevant datum. The possible significance of this plot depends on interpretation of the rates of tectonic displacement and the rates of ice level fluctuation in this region, which follow.

## Discussion

*Rates of uplift.* All the information available about the pre-volcanic erosion surface suggests that the differences in elevation of the surface from one nunatak to another have been produced almost entirely by vertical tectonic displacements, with no significant tilting. Peneplain remnants are commonly exposed at elevations between 600 m and 900 m ASL in nunataks along the Hobbs and Ruppert Coasts, and the highest remnant that we know of in all of Marie Byrd Land is the Mt Petras exposure, 2700 m above sea level. The seismic basement contact beneath Toney Mountain (Figure 1) is roughly 3000 m below sea level (Bentley and Clough, 1972) which suggests a total structural relief of 5000-6000 m. If we assume that the peneplain originated near sea level, either as a fluvial surface near the coast or a product of marine planation, then it appears that there has been both uplift and net subsidence, rather than differential uplift alone. The maximum uplift has apparently been roughly 2700 m. If we assume further that fault displacements began when volcanic activity began in this region, roughly 27-28 Ma ago (LeMasurier and Rex, 1982), we can calculate an average rate of uplift of approximately 100 m/million years.

Going a step beyond this approach of regional averaging, it is interesting to note that the age of the volcanic rocks that rest on the very high erosion surface remnant at Mt Petras is 23-25 Ma old (Table 1), nearly the oldest volcanic section in Marie Byrd Land, and that the basal volcanics become younger on successively lower exposures of the surface. This relationship is illustrated in Figure 4, which includes all localities in Marie Byrd Land where dated volcanic rocks rest on the erosion surface. It is reminiscent of the relationship that is commonly found in mountain valleys, where older stream deposits rest on successively higher fluvial terraces, but there is clearly no succession

**TABLE 2: Chronology and Elevations Above Ice Level of Subaerial Lavas**

| Sample No. | Locality | K-Ar age data* %K | Vol. $^{40}Ar$ rad. scc/g $\times 10^{-6}$ | %$^{40}Ar$ rad. | Age (m.a.) | Elev. above ice (m) | Comments |
|---|---|---|---|---|---|---|---|
| | Mt. Berlin (136°W, 76°S) | | | | | | |
| 14B | Mefford Knoll | 4.15 | 0.1009<br>0.1022 | 23.8<br>20.8 | 0.63 ± 0.03 | 125 | |
| 7E | Kraut Rocks | 3.53 | 0.08163<br>0.08929 | 6.9<br>6.4 | 0.62 ± 0.05 | 200 | |
| 23C | Berlin Crater | 3.64 | No radiogenic argon | | <0.1 | 1300 | |
| 31A | Merrem Peak | 4.01 | No radiogenic argon | | <0.1 | 1000 | |
| 67A-35 | Wedemeyer Rocks | 2.33 | 0.2460<br>0.2210 | 40.9<br>48.5 | 2.5 ± 0.2 | 200 | |
| | Mt. Moulton (135°W, 76°S) | | | | | | |
| 67-1A | Edwards Spur | 3.87 | 0.7390<br>0.6580<br>0.7260 | 51.4<br>49.2<br>64.7 | 4.6 ± 0.3 | 200 | |
| 67-2A | Prahl Crags | 4.00 | 0.7040<br>0.8320 | 58.7<br>54.9 | 4.8 ± 0.4 | 450 | |
| | Mt. Bursey (133°W, 76°S) | | | | | | |
| 25A | Starbuck Crater | 3.36 | 1.1084<br>1.1267 | 70.2<br>69.9 | 8.54 ± 0.34 | 100 | |
| 24A | Syrstad Rock | 1.54 | 0.6162<br>0.6294 | 48.5<br>45.1 | 10.4 ± 0.4 | 30 | |
| 67A-7 | Starbuck Crater | 3.93 | 1.240<br>1.640 | 60.9<br>64.3 | 9.0 ± 1.0 | 100 | |
| 67A-21 | Koerner Bluff | 4.25 | 1.604<br>1.594 | 35.6<br>37.9 | 9.42 ± 0.40 | 450 | |
| 29A | 4 km E of Koerner Bluff | 3.43 | 1.2524<br>1.2386 | 73.2<br>69.9 | 9.31 ± 0.37 | 400 | cinder cone |
| 27A | Heaps Rock | 3.95 | 0.9377<br>0.9223 | 69.9<br>72.7 | 6.04 ± 0.24 | 400 | |
| 28A | Hutt Peak | 1.11 | 0.0207<br>0.0165 | 6.1<br>3.8 | 0.43 ± 0.06 | 650 | cinder cone |
| | Mt. Andrus (132°20′W, 75°50′S) | | | | | | |
| 61C | Lind Ridge, W end | 5.09 | 2.1985 | 69.2 | 11.1 ± 0.5 | 300 | feldspar separate |
| 67B-2 | WSW spur | 0.164 | 0.0942 | 12.6 | 14.3 ± 2.2 | 200 | |
| 58D | South Caldera wall | 3.97 | 1.6167<br>1.6465 | 69.1<br>61.3 | 10.5 ± 0.4 | 1100 | |
| 59B | Lind Ridge, centre | 1.71 | 0.6681<br>0.6720 | 69.2<br>59.3 | 10.0 ± 0.4 | 900 | |
| 60A | Lind Ridge, E end | 1.19 | No radiogenic argon | | <0.1 | 900 | cinder cone |
| 44F | NW Flank | 3.69 | 1.6279<br>1.6238 | 47.0<br>51.6 | 11.3 ± 0.4 | 800 | |
| | Mt. Kosciusko (132°15′N, 75°43′S) | | | | | | |
| 40A | West Flank | 3.80 | 1.4701<br>1.5058 | 69.2<br>66.8 | 10.0 ± 0.4 | 200 | |
| 40C | Same as 40A | 1.28 | 0.4308<br>0.4316 | 36.2<br>37.0 | 8.66 ± 0.35 | 200 | cinder cone |
| 67B-6 | Same as 40A | 4.22 | 1.448<br>1.410 | 72.6<br>66.7 | 8.5 ± 0.5 | 200 | |
| 43A | SW Flank | 1.94 | 0.7770<br>0.7788 | 58.1<br>32.9 | 10.3 ± 0.4 | 800 | cinder cone |
| | Mt. Kauffman (132°30′W, 75°35′S) | | | | | | |
| 67B-8 | West Flank | 4.44 | 1.070<br>0.887 | 41.0<br>41.0 | 5.5 ± 0.5 | 350 | |
| | Kouperov Peak (133°45′W, 75°07′S) | | | | | | |
| 47B | South End | 1.33 | 0.4262<br>0.4219 | 39.2<br>28.2 | 8.21 ± 0.33 | 180 | |

*Constants: $\lambda_\beta = 4.72 \times 10^{-10}$ yr$^{-1}$; $\lambda_e = 0.58 \times 10^{-10}$ yr$^{-1}$; $^{40}K/K = 0.0119$ atomic %.
All dated materials are whole rock, unless otherwise noted under comments.

of terrace levels here, and fluvial processes were certainly not involved in producing the relationship in Figure 4.

We interpret Figure 4 to represent the tectonic uplift of fault bounded horsts, in a chronological succession, at a nearly constant rate that we calculate to be 105-122 m/million years. This interpretation assumes, once again, that the erosion surface originated at sea level, and that the volcanics preserved on the erosion surface were extruded when fault displacement began. Presumably lavas extruded along fault lines after uplift had taken place for some time would be able to accumulate only in the adjacent grabens. Based on these assumptions, the linearity of the relationship in Figure 4 implies the following: (1) the average rate of uplift has been essentially the same over the past 25 Ma regardless of whether uplift was continuous, or took place in multiple episodes, and (2) uplift has apparently not yet ceased in the sense that no horst block has yet reached a maximum level at which uplift stops. This kind of relationship is extremely useful because it suggests that one can predict the amount of tectonic uplift of each hyaloclastite section, once its age is known, and then deduce the magnitude of ice level fluctuations that are responsible for the exposure of the section above present ice level.

This approach can be used to interpret the puzzling relationship at Mt Murphy, Toney Mountain and Mt Takahe (LeMasurier, 1972b). Mt Murphy (Figure 1) is 0.8-0.9 Ma old and is composed entirely of subhorizontally bedded hyaloclastite with a well-exposed thickness of 2000 m. Toney Mountain is composed entirely of subaerial lavas, is roughly 0.5 Ma old and also stands 2000 m above ice level. Mt Takahe is too young to be dated by K-Ar with the materials available (<250,000 years), is composed entirely of hyaloclastite, and stands 2100 m above ice level. If our calculated rate of uplift is only grossly accurate, then the amount of tectonic uplift is no more than 110 m for the oldest of these volcanoes, and the exposure of the 2000 m hyaloclastite sections at Mt Murphy and Mt Takahe is caused almost entirely by a fall in ice level. This suggests that the ice level was about 2000 m higher than at present when the bulk of Mt Takahe formed, was near its present level 0.5 Ma ago, and was again 2000 m above its present level when Mt Murphy was formed, just under a million years ago.

In addition to the assumed rate of uplift, an important basis for this inference is the observation that Marie Byrd Land volcanoes which formed beneath the ice sheet are distinctly different from subaerial volcanoes in both form and structure (LeMasurier, 1972b). Those which formed beneath the ice sheet have height/base diameter ratios (h/d) of roughly 0.06 (e.g. Mt Takahe), compared to h/d ratios of about 0.10 for subaerial volcanoes (e.g. Toney Mountain). Internally, the subaerial volcanoes appear to be composed almost entirely of lava flows, while the intraglacial volcanoes are composed mainly of subhorizontally stratified hyaloclastite, as is clearly displayed at Mt Murphy (LeMasurier, 1972b; Andrews and LeMasurier, 1973). During

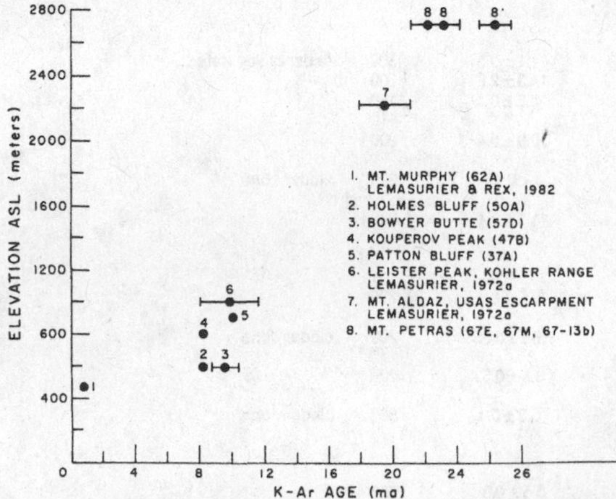


**Figure 4. Elevations of pre-volcanic erosion surface remnants vs. ages of the oldest volcanics resting on the surface. Elevations are estimated from USGS 1:250,000 topographic reconnaissance maps. Dates are from Tables 1 and 2 unless otherwise noted. Field sample numbers are in parentheses. Error bars are shown only where errors are more than twice the size of the plotted circle.**

glacial maxima, the level of the ice sheet must have remained 2000 m higher than its present level, long enough for Mt Murphy and Mt Takahe to have developed their distinctive morphological and structural characteristics. Similarly, the ice must have been near its present level long enough for Toney Mountain to develop its distinctive structure and profile about 0.5 Ma ago. The length of time required for one of these volcanoes to develop the major part of its edifice is not well constrained by the available K-Ar data, but the relationships just described suggest it might have taken only 250,000 to 300,000 years. This seems like a short, even minimal period of time for the growth of a volcano with a volume of 900 km$^3$ (Andrews and LeMasurier, 1973), but it is a long time for a single glacial epoch. The most reasonable statement that can be made at this time is that ice level appears to have fluctuated through a range of roughly 2000 m, with two maxima and two minima (including the present), over the past million years. Further resolution is not possible with the data available.

*Duration of ice sheet instability.* Apparent uniformity of uplift rates over the past 25 Ma suggests that it may be possible to infer how long the large scale fluctuations in ice level have been taking place. The first step in this interpretation is to compare the ages, and elevations above ice level, of hyaloclastites and subaerial lavas, as has been done in Figure 3. Inspection of this diagram does not reveal any time intervals during which subaerial rocks alone were erupted, and no hyaloclastite was formed, as one would expect if there had been long periods of deglaciation in the past. What is most characteristic of the plot is that subaerial lavas and hyaloclastites appear to have been erupted contemporaneously (within the limits of the K-Ar method) on several different occasions in the past, and that in some instances the hyaloclastites now stand higher above ice level than contemporaneous subaerial lavas (e.g. $2.5 \pm 0.2$ Ma; $8.2 \pm 0.5$ Ma).

Occurrences of hyaloclastite standing higher above ice level than contemporary subaerial lavas suggest two possible interpretations: (1) ice level has fluctuated in pre-Quaternary time at a rate that is within the limit of error of K-Ar dating, and/or (2) the hyaloclastites have been tectonically elevated with respect to lavas that lie on stable or subsiding blocks. Tectonic uplift alone, at a rate of 105-122 m/million years could explain the elevation of Brandenberger Bluff hyaloclastites (Table 1) above coeval subaerial lavas at Mt Berlin (Wedemeyer Rock, Table 2), if a fault lies between them, but fluctuating ice levels is an equally viable possibility, within the analytical uncertainty of $\pm 0.2$ Ma. There is a major inconsistency in the record in Upper Miocene time (8-10 Ma), however, that cannot be adequately explained by tectonic displacement. The pre-volcanic erosion surface is overlain by subaerial lavas roughly 8.2 Ma old at Holmes Bluff (Table 1) and at Kouperov Peak (Table 2), each of which now stands 100-200 m above present ice level. However, the data in Figure 4 suggest that the erosion surface in these nunataks was near sea level when these lavas erupted, and has since risen to the present elevations. This implies that there was no ice at sea level on the Marie Byrd Land coast at this time, and hence that the ice level must have been substantially lower than it is today. Similarly, the very thin hyaloclastite section at Bowyer Butte suggests thin ice 9.5 Ma ago. However, the seemingly contemporaneous ($8.3 \pm 0.3$ Ma) hyaloclastite section at Mt Steere is at least 1000 m thick and its base is buried at an unknown depth beneath the ice sheet. This seems to require the existence of a continental ice sheet at this time (LeMasurier and Rex, 1982). The Jones Mountains hyaloclastite section (450 km east of Figure 1) also suggests thick ice along the coast in the interval 7-10 Ma (Rutford et al., 1968). The best explanation for these relationships is that large scale, relatively rapid fluctuations in ice level (or mass of the ice sheet) were a characteristic of the ice sheet in Late Miocene as well as Quaternary time. Over the past 8 Ma the ice sheet has apparently been smaller than it is today during some recessions, much larger than today in major advances, and the rates of these fluctuations seem to have been within the $\pm 0.2$ Ma of K-Ar dating precision.

*Rates of subsidence.* It has been noted above that the seismically determined position of the erosion surface beneath Toney Mountain suggests that grabens in this region represent net subsidence, rather than differential uplift. This raises the question of whether a rate of subsidence can be calculated. A crude estimate of 107 m/Ma can be obtained for Toney Mountain by simply dividing the depth of the erosion surface (approximately 3000 m) by 28 Ma, using the same assumptions as those outlined for determining uplift rates. Another possible method is suggested by the remarkable lack of hyaloclastite in the Ames and Flood Ranges. It seems logical to expect that once grabens and horsts have begun to form, volcanic products erupted along fault controlled vents would accumulate mainly in the grabens, and that these materials would be exposed only if the activity were voluminous enough to build the pile above ice level. Among all the volcanic chains in this region, Toney Mountain most clearly represents a volcanic pile that has accumulated over a subsiding block, and owes its elevation entirely to constructional processes. Hyaloclastites that form in such an environment would occur beneath the subaerial lava cap, and would eventually subside below even the lowest stands of the continental ice sheet. This may explain the scarcity of hyaloclastite in the Ames and Flood Ranges.

Figure 5 was prepared in an effort to determine whether any direct effect of this subsidence could be found other than the seismic data of Bentley and Clough (1972). It is based on the possibility that a relationship between maximum volcano height and crustal thickness, such as the one Thorarinsson (1967) has shown for Iceland, might exist in Marie Byrd Land. Also, volcanoes composed of subaerial lavas are virtually unmodified by erosion in this region, because of extremely slow erosion rates for this rock type (Andrews and LeMasurier, 1973) and constructional summit elevations can be easily determined. The "age" of each volcano is represented by the age of pre-caldera lavas on the caldera rim, because post-caldera cinder cone activity commonly has continued long after the main cone is built. Caldera rim dates were not available for Mt Steere, Mt Frakes, Mt Sidley, Toney Mountain and Mt Takahe, and therefore, the time when the summit elevation of these volcanoes was reached is less certain than for the others. A decline in summit elevation with volcano age is suggested by this diagram, and the decline can be calculated from the graph as an effect of graben subsidence at a rate of roughly 106-138 m/Ma, which compares closely with the rates of uplift of horsts. The graph could also be interpreted as an effect of increasing rates of magma production with time, such as the relationship observed in the Hawaiian chain (Dalrymple et al., 1973). This seems a much less likely possibility for the following reasons: (1) the seismic traverse at Toney Mountain leaves little doubt that subsidence has, in fact, taken place beneath the volcano, and (2) the rates of subsidence are calculated completely independently of the rates of uplift, and yet yield the same figure for these two closely related processes.

*Implications.* It has long been thought that the West Antarctic ice sheet experiences rapid, large-scale advances and retreats (surges), because it is grounded below sea level and because it displays a concave profile over its most rapidly moving segments (Hughes, 1973). The possible effects of this behaviour on world climate and on Cenzoic sea levels have been a topic for lively debate (Wilson, 1964; Mercer, 1978; Matthews and Poore, 1980; Loutit and Kennett, 1981), a full discussion of which is beyond the scope of this paper. It is appropriate to point out, however, that volcanic geology provides evidence that a continental ice sheet was already in existence when volcanism began in Marie Byrd Land in Late Oligocene time, and that the sub-sea level ice filled grabens in this region developed in Late Cenzoic time (LeMasurier and Rex,1982). If the rapid ice level changes described in this paper are in fact a result of grounding below sea level, then the tectonic and volcanic history of this region implies that ice level changes have become larger and more frequent in the latter part of Cenzoic time, as the grabens deepened beneath the ice cover. The possibility that these changes may have been large enough to affect sea level is suggested by the fact that the Cenozoic cycles of sea level change described by Vail et al. (1977) became strikingly more frequent in latest Cenozoic time. Loutit and Kennett (1981) support the widely held view that the Antarctic ice sheet developed mainly in Middle Miocene time (around 14 Ma), and that the Late Cenozoic high frequency sea level cycles result from superimposing glacio-eustatic effects on a longer term cyclic mechanism that has not yet been identified. We would suggest that the longer term cyclic mechanism is glacial, certainly back to Late Oligocene time and perhaps much earlier (Matthews and Poore, 1980; Barron et al., 1981), and that the higher frequency cycles reflect the development of subglacial grabens in West Antarctica. The rate of subsidence of these grabens is therefore of interest not only as a confirmation of uplift rates, but because of its possible glacio-eustatic implications.

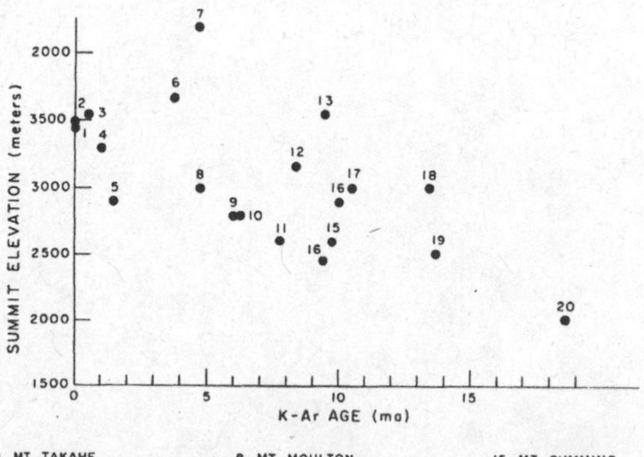


1. MT TAKAHE
2. MT. BERLIN
3. TONEY MOUNTAIN
4. MT. WAESCHE (MAIN Pk.)
5. MT. WAESCHE (CHANG Pk.)
6. MT. FRAKES
7. MT SIDLEY
8. MT. MOULTON
9. MT. BURSEY (MAIN Pk.)
10. MT. HARTIGAN (No CALDERA)
11. MT. HARTIGAN (So. CALDERA)
12. MT. HAMPTON (MAIN Pk.)
13. MT. STEERE
14. MT. BURSEY (KOERNEY BLUFF)
15. MT. CUMMING
16. MT. KOSCIUSKO
17. MT. ANDRUS
18. MT. HAMPTON (WHITNEY Pk)
19. MT. FLINT
20. REYNOLDS RIDGE

**Figure 5. Constructional summit elevations vs. ages of caldera rim lavas (see text for exceptions), for all major central volcanoes in Marie Byrd Land. Elevations are from USGS 1:250,000 topographic reconnaissance maps. Dates are from Tables 1 and 2 unless otherwise noted.**

## Summary and Conclusions

Previous reports have presented evidence that Marie Byrd Land hyaloclastites were produced by subglacial eruptions and that they record a history of continental glaciation in this region from Late Oligocene time to the present. The data presented in this paper bear directly on the magnitude and rate of ice level fluctuations in the past, and on the extent to which tectonic displacements have affected the volcanic record of glacial history. Thus, the new data permit us to develop an aspect of glacial history that was previously very speculative and unconstrained. Our major conclusions are as follows:

(1) Rates of tectonic uplift in Marie Byrd Land can be estimated by a relationship that exists between the elevation above sea level of a pre-volcanic erosion surface, and the age of the oldest volcanic rocks that rest on the surface (Figure 4). A rate of 105-122 m/Ma, that has operated nearly continuously over the past 25 Ma, is suggested by the linearity of the relationship. A rate of subsidence of the erosion surface below sea level can be estimated by independent volcanic and seismic relationships. The rate of roughly 106-138 m/Ma so derived (Figure 5) is consistent with the rate of uplift, and provides a confirmation of the rate of tectonic dispoacement in this region. A corollary of these conclusions is that areas where hyaloclastites are found are commonly those undergoing tectonic uplift, and hyaloclastites are only rarely exposed in subsiding blocks.

(2) Comparisons of the K-Ar ages of hyaloclastites and subaerial lavas do not reveal any interval of time that could be confidently interpreted as a time of deglaciation in Marie Byrd Land. However, the elevations of Quaternary hyaloclastites above ice level are far greater than can be accounted for by 105-122 m of uplift per million years. This leads to the conclusion that, although it has probably not disappeared, the level of the continental ice sheet has fluctuated through a range of 2000 m on at least three occasions within the last one million years.

(3) The data suggest that rapid large-scale changes in size have been a characteristic of the ice sheet in Marie Byrd Land at least since Late Miocene time (8-10 Ma). This inference is based on assumptions that rates of uplift have been uniform and persistent, and that rates of ice level fluctuations have been within the limit of error of K-Ar dating for pre-Quaternary rocks. Thus, pre-Quaternary hyaloclastites that are now found topographically higher than seemingly contemporaneous subaerial lavas can be most easily explained by extrapolation of Quaternary glacial and tectonic behaviour. The data on volcanic rocks older than 10 Ma are not adequate to suggest anything more than what has previously been suggested about the existence of a continental scale ice sheet.

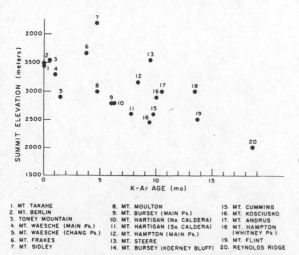

(4) The major implications of the glacial tectonic and volcanic history that has been determined in Marie Byrd Land are: (a) that Cenozoic sea level cycles (and hence sedimentary cycles) are likely to have had significant glacial control at least since Late Oligocene time, and possibly much earlier, and (b) the increasing frequency of sea level cycles in the latest Cenozoic may reflect the development of large-scale fluctuations in the size of the West Antarctic ice sheet, which in turn may be related to the late Cenozoic development of deep subglacial grabens.

*Acknowledgements* This work has been supported by National Science Foundation grants DPP76-04396 and DPP77-27546 administered by the Division of Polar Programmes. We are grateful to J.T. Andrews, L.A. Warner and an anonymous reviewer for reviews and discussion of the manuscript.

## REFERENCES

ANDREWS, J.T. and LeMASURIER, W.E., 1973: Rates of Quaternary glacial erosion and corrie formation, Marie Byrd Land, Antarctica. *Geology, 1,* 75-80.

BARRON, E.J., THOMPSON, S.L. and SCHNEIDER, S.H., 1981: An ice-free Cretaceous? Results from climate model simulations. *Science, 212,* 501-8.

BENTLEY, C.R. and CLOUGH, J.W., 1972: Antarctic subglacial structure from siesmic refraction measurements. *in* Adie, R.J., (ed.). *Antarctic Geology and Geophysics.* Universitetsforlaget, Oslo, 683-91.

COOPER, R.A., LANDIS, C.A., LeMASURIER, W.E. and SPEDEN I.G., 1982: Geologic history and regional patterns in New Zealand and West Antarctica—their paleotectonic and paleogeographic significance; *in* Craddock, C., (ed.). *Antarctic Geoscience,*Madison, Univ. Wisconsin Press, 43-53.

CRADDOCK, C., BASTIEN, T.W. and RUTFORD, R.H., 1964: Geology of the Jones Mountains area; *in*Adie. R.J., (ed.). *Antarctic Geology.* North-Holland, Amsterdam, 171-87.

DALRYMPLE, G.B., SILVER, E.A. and JACKSON, E.D., 1973: Origin of the Hawaiian Islands. *Am. Scientist, 67,* 294-308.

HUGHES, T., 1973: Is the West Antarctic ice sheet disintegrating? *J. Geophys. Res., 78,* 7884-910.

LAUDON, T.S., 1972: Stratigraphy of eastern Ellsworth Land; *in*Adie, R.J., (ed.). *Antarctic Geology and Geophysics.* Universitetsforlaget, Oslo, 215-24.

LeMASURIER, W.E., 1972a: Volcanic record of Cenozoic glacial history of Marie Byrd Land; *in* Adie, R.J. (ed.). *Antarctic Geology and Geophysics.* Universitetsforlaget, Oslo, 251-60.

LeMASURIER, W.E., 1972b: Volcanic record of Antarctic glacial history: implications with regard to Cenozoic sea levels; *in*Price, R.J. and Sugden D.E., (comp.). *Polar Geomorphology.* Spec. Publ. Inst. Brit. Geogr., no. 4, 58-74.

LeMASURIER, W.E., 1978: The Cenozoic West Antarctic System and its associated Volcanic and Structural Features (abs.). Geol. Soc. Am. 1978 Annual Meeting, Abstracts with Programmes, 10, 443.

LeMASURIER, W.E., McINTOSH, W.C. and REX, D.C., 1981: The geology of W.C., 1979: Tillite, glacial striae and hyaloclastite associations on Hobbs Coast, Marie Byrd Land. *U.S. Antarct. J.. 14.* 48-50.

LeMASURIER W.E., MELANDER, O., GRINDLEY, G.W. and McINTOSH, W.C., 1979: Tillite, glacial striae and hyaloclastite associations on Hobbs Coast, Marie Byrd Land. *U.S. Antarct. J., 14,* 48-50.

LeMASURIER, W.E. and REX, D.C., 1982: Volcanic record of Cenozoic glacial history in Marie Byrd Land and western Ellsworth Land: revised chronology and evaluation of tectonic factors; *in*Craddock, C., (ed.). *Antarctic Geoscience,* Madison, Univ. Wisconsin Press, 725-34.

LOUTIT, T.S. and KENNETT, J.P., 1981: New Zealand and Australian Cenozoic sedimentary cycles and global sea-level changes. *Am. Assoc. Petrol. Geol., Bull., 65,* 1586-1601.

MATTHEWS, R.K. and POORE, R.Z., 1980: Tertiary $\delta^{18}O$ record and glacio-eustatic sea level fluctuations. *Geology, 8,* 501-5.

MERCER, J.H., 1978: West Antarctic ice sheet and $CO_2$ greenhouse effect: a threat of disaster. *Nature, 271,* 321-5.

RUTFORD, R.H., CRADDOCK, C. and BASTIEN, T.W., 1968: Late Tertiary glaciation and sea level changes in Antarctica Tertiary sea level fluctuations. *Palaeogeogr., Palaeoclimatol., Palaeoecol., 5,* 15-39.

THORARINSSON, S., 1967: Some problems of volcanism in Iceland. *Geol. Rdsch, 57,* 1-20.

VAIL, P.R., MITCHUM, R.M. Jnr. and THOMPSON, S., III, 1977: Global cycles of relative changes of sea level. *Am. Assoc. Petrol. Geol., Mem., 26,* 83-97.

WILSON, A.T., 1964: Origin of ice ages: an ice shelf theory for Pleistocene glaciation. *Nature, 201,* 147-9.

# SEISMOLOGICAL OBSERVATION ON MOUNT EREBUS, 1980—1981

T. Takanami, *Faculty of Science, Hokkaido University, Sapporo, Japan.*

J. Kienle, *Geophysical Institute, University of Alaska, Fairbanks, Alaska, 99701, U.S.A.*

P. R. Kyle, *Institute of Polar Studies, Ohio State University, Columbus, Ohio, 43210, U.S.A.*

R. R. Dibble, *Department of Geology, Victoria University, Wellington, New Zealand.*

K. Kaminuma and K. Shibuya, *National Institute of Polar Research, 9-10 Kaga I-chome, Itabashi-ku, Tokyo 173, Japan.*

*Abstract* A high sensitivity seismic network was installed on Mt Erebus (3794 m) in Ross Island, Antarctica during the 1980-81 austral summer field season. The network consists of three permanent seismographs with radiotelemetry links to Scott Base. Two stations are on the western flank at Abbott Peak (1973 m) and Hooper Shoulder (1900 m), 10 and 5 km respectively from the summit crater. The third station is on the southern crater rim of Mt Erebus. Each station transmits continuously the output from a single, vertical component, one second period, borehole type seismograph. Data are recorded on a 14-channel slow speed magnetic tape recorder. Around 50 events per day were recorded at each station from late December, 1980 to mid-February 1981. A six-station network with portable data recorders was also occupied around the summit crater from 20th December, 1980 to 6th January, 1981. The hypocentre co-ordinates and the origin time of 94 events out of 131 events could be located by using four or more stations out of nine. Some of the 131 events were accompanied with explosions.

Mount Erebus (3794 m) has erupted intermittently since 1841 when it was discovered by James Ross (1847). At present, it is one of the most active volcanoes in the world. Since December 1972 there has been a lava lake, 100 m in major axis and 60 m in minor axis, in the northern half of the Inner Crater at the summit. Seismological observations at Mt Erebus summit were firstly carried out by a New Zealand party during December 1974. They recorded numerous volcanic earthquakes, and estimated a value of 1.2 for the Gutenberg-Richter parameter b (Kyle et al., 1982).

A project of seismological studies on Mt Erebus was started in co-operation with Japan, New Zealand and the United States in the austral field season of 1980/81, and named International Mt Erebus Seismological Studies (IMESS). Following a preliminary observation in the 1979/80 field season, a permanent seismological network has been established on Mt Erebus by the IMESS. The network consists of three seismographs with radiotelemetry links to Scott Base. This seismological network is operated in one of the wildest and the harshest environments in the world. Data was obtained until July 1981, when the batteries for data sending were flattened.

## NETWORK OF SEISMOLOGICAL OBSERVATION

Three seismographs with radiotelemetry links to Scott Base were installed on Mt Erebus as shown in Figure 1, and a magnetic data recorder was also installed in Scott Base. Stations are on the western flank at Abbott Peak (1793 m), and at Hooper Shoulder (1900 m), 10 km and 5 km from the summit crater respectively, and on the southern crater rim near the summit of Mt Erebus. The stations are 43.6, 35.0 and 39.4 km from Scott Base respectively. Hooper Shoulder Station relays the signals at Abbott Peak to Scott Base. Each station transmits continuously the output from a single, vertical component, one second period, borehole type seismograph. In addition the summit station also telemeters data from two acoustic channels, used to monitor explosions in the crater, and an experimental induction loop, installed by the New Zealand party, for monitoring the flux of electrically charged gases. Data are recorded on a 14-channel slow speed magnetic tape recorder. The signal of one station is also recorded on a visible chart recorder for long term monitoring.

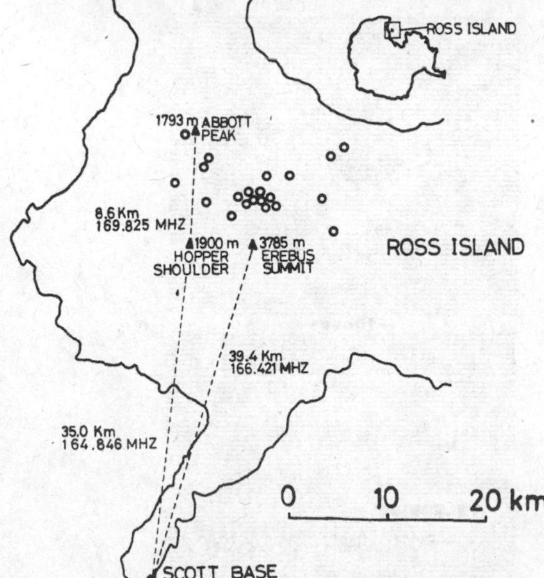


**Figure 1. Locations of seismic stations for radiotelemetry to Scott Base on Mt Erebus, Ross Island, Antarctica, and earthquake locations determined by this radiotelemetry network from 22nd December 1980 to 9th January 1981.**

## NUMBER OF EARTHQUAKES

The observations started on 15th December, 1980 at Abbott Station, on 17th December at Hooper and on 23rd December at the summit. The daily frequency of earthquakes is plotted in Figure 2. The earthquakes in Figure 2 seemed to be near shocks which were counted on playback paper chart recordings at 5 mm/min for the three stations until 17th February 1981. The arrows in Figure 2 are on days when the exact number of events could not be counted because of extreme microseismic activity, so the total number of events may be greater than shown. Magnifications at each station for 1 Hz signals are 25,000 for Mt Erebus summit; and 80,000 for Abbott Peak and Hooper Shoulder. The number of earthquakes for Abbott Peak is 3346 recognisable events in 63 days, 2994 events in 63 days for Hooper Shoulder and 2586 events in 57 days for the summit. Then the mean number of events per day recorded at each station is 53 events for Abbott Peak, 48 events for Hooper Shoulder and 45 events for the summit. Overall, the seismic activity at all stations occurred in swarms with peaks at the beginning of the recording period in late December 1980 and in the middle of February 1981.

Four types of earthquakes were recognised on the monitoring recorder as follows: (1) teleseisms; (2) volcanic earthquakes; (3) icequakes and (4) local earthquakes. Seismograms of a teleseism, and an icequake and a local earthquake are shown in Figures 3 and 4. Teleseisms and local earthquakes have a lower frequency and longer duration than the other types. An example of an event which is difficult to identify, but is either a local earthquake or a volcanic earthquake, is shown with a symbol A in Figure 3. Icequakes are usually characterised by sharp onset, and higher frequency and shorter duration time than natural earthquakes (Kaminuma, 1971; Kaminuma and Haneda, 1979).

Figure 5 gives a typical seismogram of volcanic earthquake and/or an icequake swarm. Symbols V and I show events provisionally identified as volcanic earthquakes and icequakes. There are events which are difficult to be classified as local earthquakes, volcanic earthquakes, or icequakes, on the monitoring recorder of one seismological station.

Figure 6 plotted the daily number of events which were recorded from December 1980 to July 1981 by the long term monitoring

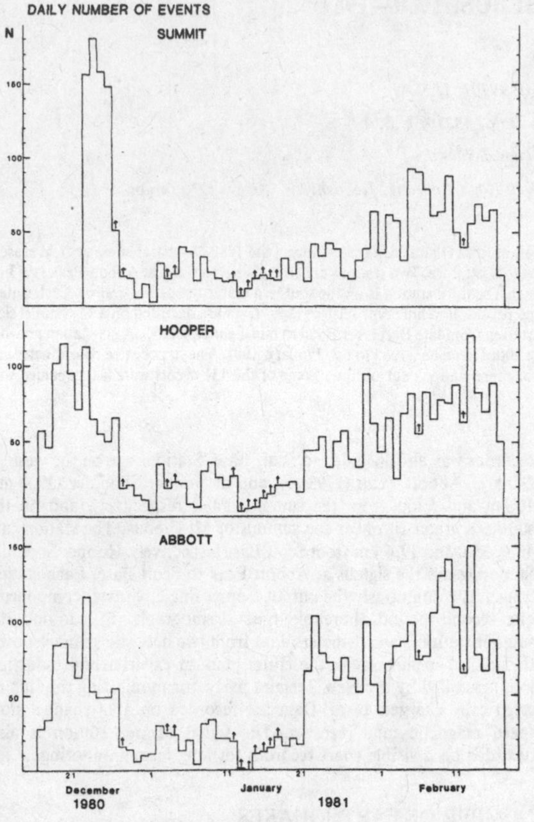


**Figure 2. Daily number of events for the summit, Hooper Shoulder and Abbott stations. Arrows indicate the day that a total of events is more than the number shown in the figure.**

recorder. Remarkable swarm activities with over 20 events per hour were observed in April and July 1981. The events of Figure 6 are both icequakes and earthquakes; most of the events in the swarm are identified as icequakes as shown in Figure 5. Even though the daily number of volcanic earthquakes may be less than that shown in Figure 6, the background numbers of seismic events appear to vary with a periodicity of 2-2.5/month.

## EARTHQUAKE LOCATION

Twenty-one events recorded between 22nd December 1980 and 9th January 1981 were seen on all three stations. S-P times for many of these events ranged from 3 to 6 seconds at the Erebus summit station, some were as low as 1 second and one was high as 10.7 seconds at Abbott Peak. Unfortunately, the velocity structure of Mt Erebus is unknown. The authors' first impression is that the P-wave velocity is extremely low within the main cone, perhaps as low as 1 to 1.5 km/second. With the assumption of such a low velocity, most of the 21 events seen on all three stations were located 5 to 10 km north of the Erebus summit station in the upper Fang Ridge-Fang Glacier region as shown in Figure 1. Some of these could be non-volcanic earthquakes, but the swarmlike character rather suggests a volcanic/tectonic origin. Assuming a higher velocity of 3 km/second would make the epicentres further out by 10 to 30 km.

A six-station network with portable data recorders around the summit crater was operated from 20th December 1980 to 6th January 1981 when the Japanese party stayed at the summit. Then nine seismological stations on Mt Erebus were occupied during this period. The locations of three permanent and six temporary stations are listed in Table 1. For the hypocentre determination of the earthquakes, a P-wave structure was proposed as shown in Figure 7. The Poisson's ratio is estimated to be $\sigma = 0.25$ in the structure. The hypocentre co-ordinates and the origin time based on the P-wave velocity structure model were determined by using more than four stations out of nine. The focal depth was assumed to be under the sea level in the hypocentre determination. The initial value of the focal depth for the hypocentre determination was assumed as 0.5 km, and the computation converged within five iterations. The hypocentre co-ordinates and origin time were determined for 94 out of 131 events. The epicentre location and the vertical profiles are given in Figure 8. The

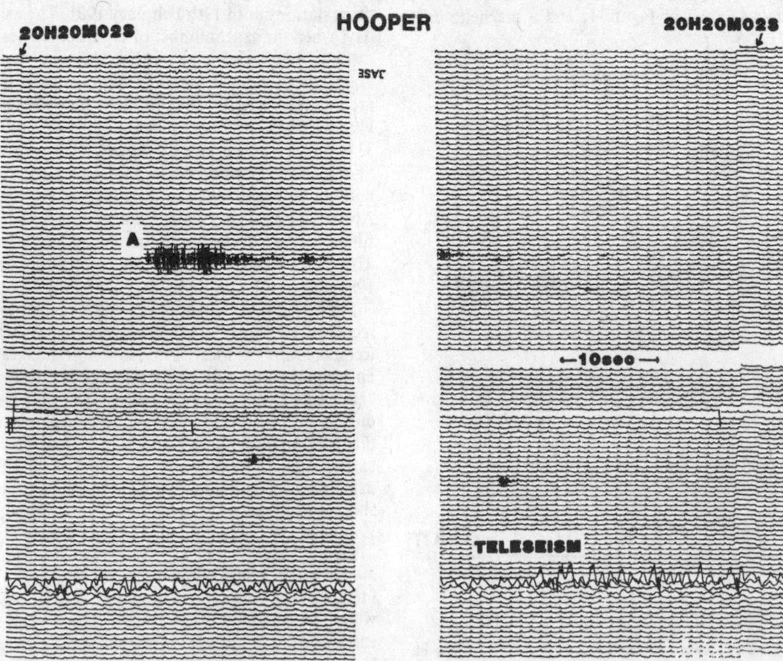


**Figure 3. A typical record of teleseisms on the monitoring seismogram. "A" shows an event which can not be identified as either a local or a volcanic earthquake.**

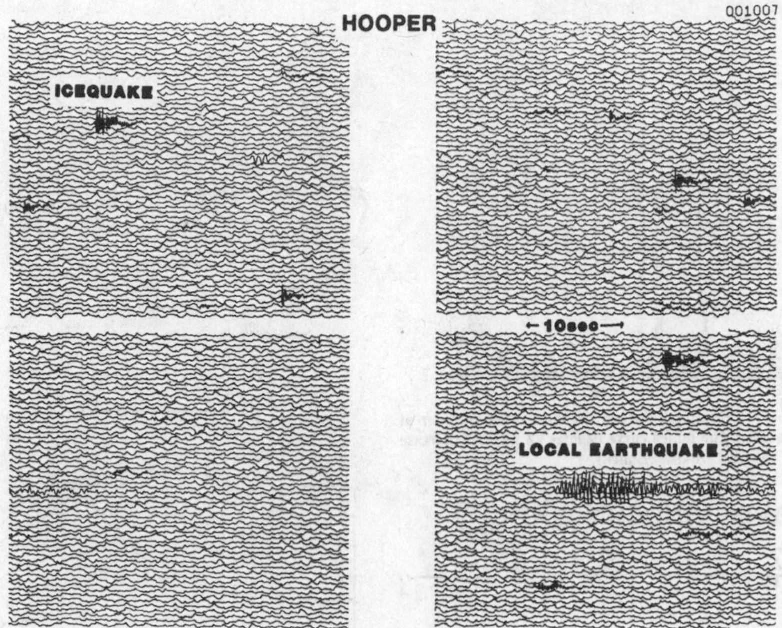


Figure 4. A typical record of local earthquakes and icequakes on the monitoring seismogram.

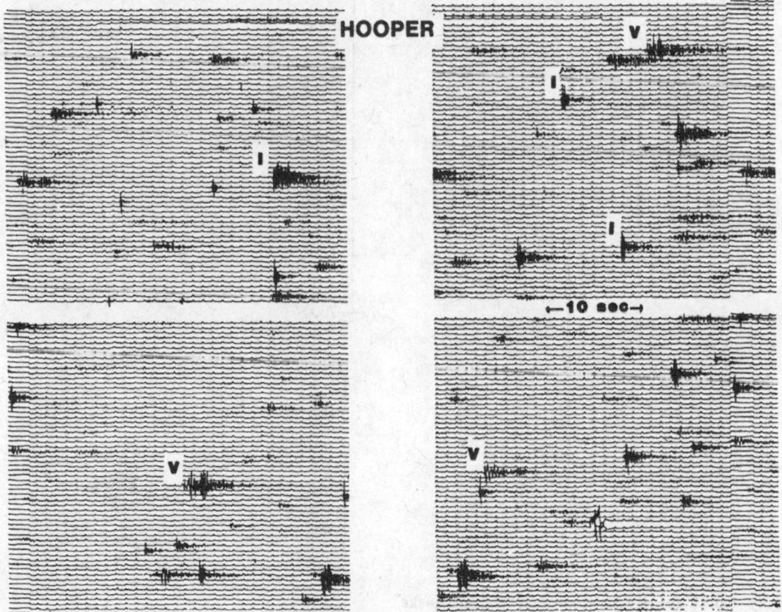


Figure 5. A seismogram of icequake and volcanic earthquake swarms. "I" denotes an icequake and "V" a volcanic earthquake.

focal depth is assumed under the sea level in its computation. The focal depth of most events is converged 1 to 2 km under the sea level. The events with the shallow focal depth might be volcanic earthquakes. Some of them were accompanied by eruptions. Earthquakes in Figure 8 are located over a wider area than those in Figure 1. But the radiotelemetry network comprised only three stations, and the temporary station network was too small to determine the details of earthquake locations throughout Ross Island.

Temporal variations of seismic activity and earthquake locations which were observed in the first season of IMESS were denoted briefly. The IMESS will be continued until the 1983/84 field season. The details of the seismicity in and near Ross Island and the structure of Mt Erebus will be studied in the near future.

**TABLE 1: Locations of Seismic Stations**

| Station | X (km) | Y (km) | Z (m) |
|---|---|---|---|
| Summit | 0.00 | 0.00 | 3780 |
| Abbott | −5.90 | 7.50 | 1793 |
| Hooper | −5.70 | −0.70 | 1900 |
| Summit Network | | | |
| ST1 | 0.03 | 0.89 | 3625 |
| ST2 | −0.18 | 0.17 | 3650 |
| ST3 | −0.08 | 0.17 | 3754 |
| ST4 | −0.46 | −0.10 | 3644 |
| ST5 | −1.16 | 0.52 | 3462 |
| ST6 | 0.38 | 0.45 | 3767 |

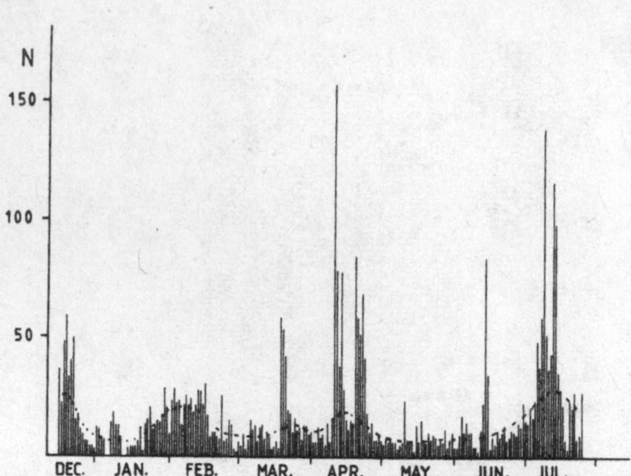


**Figure 6. Daily number of events counted on the monitoring recorder at Hooper Shoulder. The dotted curve denotes an increase-decrease tendency of the background seismicity.**

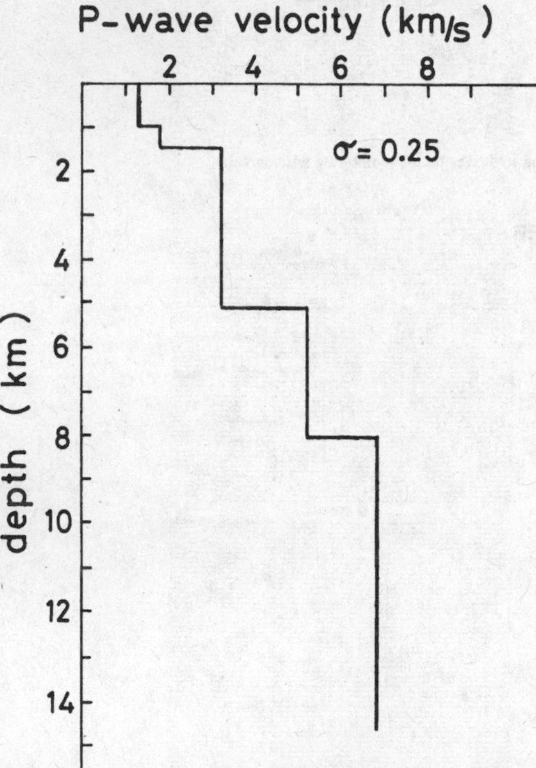


**Figure 7. P-wave velocity model for the determination of earthquake locations.**

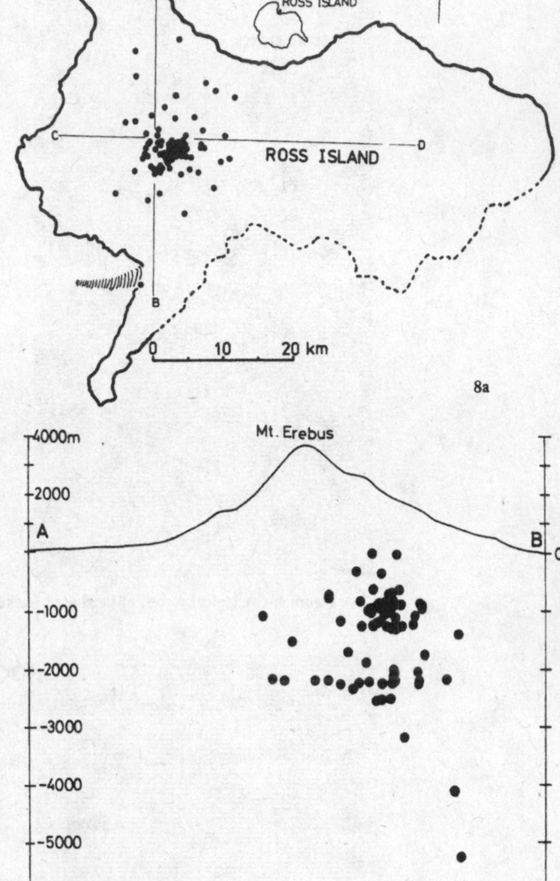


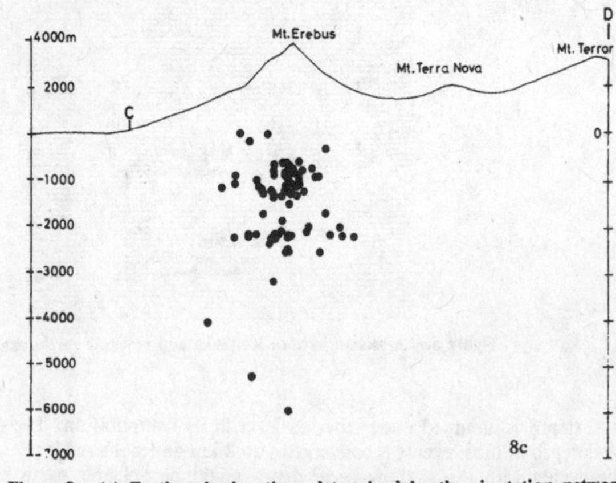


**Figure 8. (a) Earthquake locations determined by the six-station network from 20th December 1980 to 6th January 1981. (b) The vertical projection of the hypocentres along the A-B line. (c) The vertical projection of the hypocentres along the C-D line.**

*Acknowledgements* The authors are grateful to Messrs S. Estes, University of Alaska, K. Terai, National Institute of Polar Research and N. Osada, University of Tokyo for their co-operation in the work on Mt Erebus. The authors are also grateful to Scott Base personnel, and in particular Mr S. Whitfield, who assisted in the establishment of the seismic network. Special mention must go to the VXE-6 helicopter crews who showed considerable patience during the initial setting up of the seismic stations. The work was supported by both National Institute of Polar Research, Japan and National Science Foundation grant DPP7920316.

## REFERENCES

KAMINUMA, K., 1971: Microearthquakes observed at Syowa Station, Antarctica. *Antarctic Rec., 40,* 65-73.

KAMINUMA, K. and HANEDA, T., 1979: Icequakes around Syowa Staion, Antactica. *Antarctic Rec., 65,* 135-48.

KYLE, P.R., DIBBLE, R.R., GIGGENBACH, W.F. and KEYS, J., 1982: Volcanic activity associated with the anorthoclase phonolite lava lake, Mount Erebus, Antarctica; *in* Craddock, C. (ed.) *Antarctic Geoscience.* Univ. Wisconsin Press, Madison, 735-745.

ROSS, J., 1847: *A voyage of discovery and research in the Southern and Antarctic regions.* Vol. 1, John Murray, London.

# THE GEOLOGY OF THE MID-MIOCENE McMURDO VOLCANIC GROUP AT MOUNT MORNING, McMURDO SOUND, ANTARCTICA

P.R. Kyle, *Department of Geoscience, New Mexico Institute of Mining and Technology, Socorro, New Mexico 87801, USA.*

H.L. Muncy, *Mobil Producing, Texas and New Mexico, 9 Greenway Plaza, Houston, Texas 77046, USA.*

**Abstract** Rocks exposed at Gandalf Ridge and near Lake Morning on the northern slopes of Mount Morning represent the earliest documented evidence of volcanism in the Erebus volcanic province of the McMurdo Volcanic Group. K/Ar dating of samples suggests a Mid-Miocene age (14.6 Ma to 18.7 Ma).

**The rocks range in composition from trachyandesite to quartz trachyte and pantellerite. Two different trends, one non-peralkaline and one peralkaline, are recognised. The non-peralkaline trend is represented by a trachyandesite-quartz trachyte sequence and the peralkaline trend by a trachyandesite-comenditic trachyte-comendite-pantellerite sequence. Major and trace element data are consistent with the trachyandesite being derived from an alkali basalt parent by fractional crystallisation of olivine, clinopyroxene, feldspar, opaque oxides and apatite.**

**The volcanism at Mount Morning is believed to have been associated with the fracturing caused by an extensional tectonic regime. This regime is consistent with the widespread alkaline volcanism and the known crustal thinning in the western Ross Sea.**

# Nd AND Sr ISOTOPE GEOCHEMISTRY OF LEUCITE-BEARING LAVAS FROM GAUSSBERG, EAST ANTARCTICA

K.D. Collerson and M.T. McCulloch, *Research School of Earth Sciences, Australian National University, PO Box 4, Canberra, A.C.T., 2601. Australia.*

*Abstract* Gaussberg, an isolated late Pleistocene volcano, is situated on the edge of the Antarctic continent, between Mirny and Davis on the apparent extension of the Kerguelen-Gaussberg ridge. It is composed of potassium rich lavas (leucitite) which contain a varied assemblage of inclusions: (1) cognate leucite-clinopyroxene-magnetite, leucite-clinopyroxene-glass, olivine-clinopyroxene-leucite and leucite-clinopyroxene-phlogopite assemblages; (2) partially fused crustal xenoliths; and (3) rare lherzolites.

Nd and Sr isotopic ratios have been measured on specimens of leucitite from Gaussberg to elucidate aspects of their source characteristics and petrogenesis, and to provide comparative data for analyses from Kerguelen. The leucitite data and data for two cognate nodules are quite uniform and plot slightly to the right of the enriched mantle array $\Sigma_{Nd} = -13.0$ to $-14.9$; $^{87}Sr/^{86}Sr = 0.7092$ to 0.7109). They are significantly more enriched than Nd isotopic data presented previously for Kerguelen by Dosso and Murthy (1980) and White and Hofmann (1982) and lie close to the limit of the field for diopsides from ultramafic nodules in kimberlites (Menzies and Murthy, 1980a). Xenoliths of crustal derivation are extensively contaminated by leucitite and exhibit $\Sigma_{Nd}$ of $-21.7$ to $-24.8$. Rare earth patterns for the three lecitites are extremely fractionated with $La_N/Yb_N$ ratios ranging from 433 to 293.

The Nd and Sr isotopic compositions may be the result of mixing between magma derived from either an enriched or a depleted source and pre-existing continental crust. However, in view of the high Nd and Sr content of the lavas (ca 130 ppm Nd and 1600 to 1900 ppm Sr), and relatively low Nd and Sr contents of identified crustal contaminants (Nd < 50 ppm, Sr ~ 100-1000 ppm), it is difficult to envisage how crustal contamination could uniformly modify Nd and Sr isotopic ratios without resulting in a significant change in either major or trace element chemistry. They are, therefore, interpreted as reflecting derivation of the lavas from a highly enriched mantle source region (with low Sm-Nd and high Rb-Sr ratios) possibly with minor dispersion caused by sea water or crustal $^{87}Sr$ contamination. Although this enrichment may be the result of mantle metasomatic activity, the preferred explanation is that the parent magma for the leucitites was derived by partial melting of diapirs rising from ancient subducted oceanic crust in the deep mantle.

A number of recent models have been devised to explain the structure, geochemistry, geodynamics and evolution of the Earth's mantle (e.g. DePaolo and Wasserburg, 1976a, b; Allègre et al., 1979; O'Nions et al., 1979; Allègre, 1981; Hofmann and White, 1982; Ringwood, 1982). They are based, to a large degree, on inferences about the mantle source characteristics of mid-ocean ridge and oceanic island basalts (MORB's and OIB's). These basalts commonly show an inverse linear correlation between Nd and Sr isotopic ratios and they define the so called mantle array. The majority of OIB's have $^{143}Nd/^{144}Nd$ and $^{87}Sr/^{86}Sr$ ratios which plot in the field of depleted mantle sources, viz. a source region with long-term higher Sm/Nd and lower Rb/Sr than the values estimated for the bulk Earth or primitive mantle (CHUR). Nevertheless, a subordinate number of OIB's and continental hot-spot volcanics exhibit Nd and Sr isotopic compositions which extend the mantle array into the enriched mantle field (O'Nions et al., 1977; Dosso and Murthy, 1980; White and Hofmann, 1982; Hawkesworth and Vollmer, 1979). Although some examples may be the result of crustal contamination (DePaolo, 1981), the cases cited above, as well as data presented by Menzies and Murthy (1980a) for clinopyroxenes from ultramafic nodules in kimberlites, clearly demonstrate the existence of enriched mantle. Therefore, the spectrum of mantle source compositions indicated by the OIB's, continental alkali volcanics and upper mantle clinopyroxenes is a reflection of the presence of gross heterogeneities within the mantle.

In this preliminary paper, we present Sr and Nd isotopic data as well as rare earth element (REE) concentrations for a suite of highly potassic lavas, cognate nodules and crustal xenoliths from Gaussberg volcano, in East Antarctica. The data significantly extend the mantle array within the field of enriched mantle sources. Consequently, these continental lavas are interpreted to have formed by partial melting of such a mantle reservoir. Possible scenarios for the generation of the enriched mantle are explored.

## TECTONIC SETTING

The Kerguelen-Gaussberg ridge in the southern Indian Ocean forms a semicontinuous aseismic zone which has been the locus of hot-spot volcanic activity during the last 27 Ma (Duncan, 1981). Along its ca 2000 km length, this activity has occurred on Kerguelen, McDonald Island, Heard Island, and Gaussberg (Figure 1). Bathymetric and geophysical evidence (Houtz et al., 1977) indicates that the ridge forms a structurally continuous zone which extends from Kerguelen past Heard Island to about 63°S. Between this latitude and Gaussberg on the Antarctic mainland, there is an hiatus in the ridge. This is interpreted to indicate that the Kerguelen-Gaussberg ridge has been the site of intermittent and possibly unrelated, mantle diapiric activity with a single hot-spot causing activity on Kerguelen, McDonald Island and Heard Island and possibly a separate hot-spot causing the volcanism at Gaussberg.

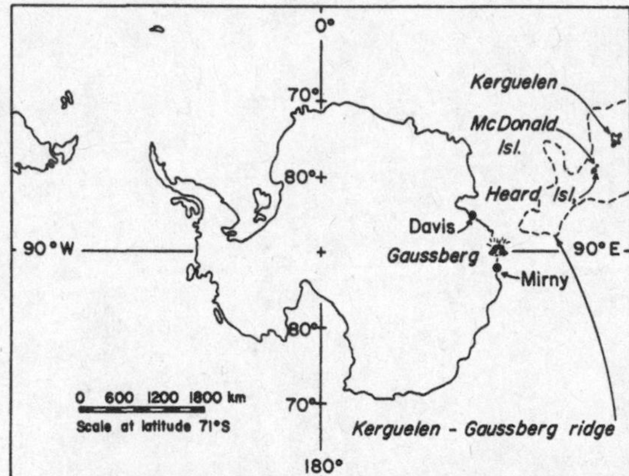


**Figure 1. Locality map showing the positions of Gaussberg, Heard Island, McDonald Island and Kerguelen and the trace of the Kerguelen-Gaussberg Ridge.**

## GENERAL GEOLOGY

### *Field characteristics*

Gaussberg is an impressive ice free ~ 370 m high volcanic landmark, situated at 67°S, 89°E on the East Antarctic shield, between Mirny and Davis. It consists predominantly of leucitite pillow lavas. The pillows are well formed and range in diameter from 10-20 cm to 1 m. They have glassy selvages up to 10 cm wide. Both the fine grained lavas and the glasses are highly vesiculated and some contain amygdules of native sulphur up to 1 cm in diameter. Highly brecciated tops and fronts of individual flows, some with lava tunnels up to 2 m in diameter, are also present. According to Tingey et al. (in press), the physical volcanological features are believed to be the result of subglacial eruption.

### *Geochronology*

Geochronological information about Gaussberg is rather limited and difficult to assess. For example, whole rock K-Ar ages of 20 Ma and 9 Ma reported by Ravich and Krylov (1964) and Soloviev (1972) respectively, are in marked disagreement with mean K-Ar ages of 0.052 ± 0.003 Ma and 0.059 ± 0.002 Ma for leucite separated from two specimens of leucitite (Tingey et al., in press). If a mean age of 0.056 ± 0.005 Ma is accepted as the time of formation of Gaussberg, then it is significantly younger than Kerguelen (27 Ma to 8 Ma: Nougier, 1969; Watkins et al., 1974; Dosso et al., 1979) and was apparently erupting synchronously with volcanic activity on Heard and McDonald Islands (Clarke et al., this volume).

*Petrology*

*The leucitites* The leucitite lavas from Gaussberg range in texture from almost aphyric to hypohyaline and they are commonly extremely vesicular. They contain three dominant microphenocrystal phases; leucite, clinopyroxene, and olivine which range in size up to 1 mm in diameter. Microglomeroporphyritic aggregates of these phases are commonly observed. The groundmass in which these occur is either glass (containing small microlites of phlogopite and leucite) or entirely devitrified to a felted aggregate of opaques, phlogopite, and, possibly, leucite.

*Inclusions* The Gaussberg lavas contain three groups of inclusions: crustal xenoliths, cognate inclusions and mantle nodules. The crustal xenoliths and cognate inclusions are relatively abundant, however the mantle nodules are extremely rare.

Crustal xenoliths are of three types: (1) layered quartzo-feldspathic gneisses; (2) nonfoliated rapakivi granites; and (3) metabasic lithologies. The xenoliths are subrounded to ovoid in shape and occur on a variety of scales up to 2 x 1 m. They are all characterised by beautifully developed partial melting fetures which occur as glassy black to brown veinlets within individual xenoliths, or as fusion rinds on the surface of the xenoliths. All stages in the development of melting and reaction with the host lava can be observed; from grain boundary melting, where the fusion product is a thin film between quartz and feldspar grains, to almost total melting, where spots of residual minerals can be resolved with difficulty in flow banded glassy fusion veins. Microprobe studies (Table 2) have revealed the presence of glass with a range of compositions. Some glass, which is pale in colour, is clearly partial melt derived from the xenolith. The other, dark glass, is geochemically similar to the leucitite. Flow banding, devitrification and mixing between the two compositions produce spectacular textural relationships.

The cognate inclusions range in size from less than 1 cm to 10 cm in diameter. They are composed of varying concentrations of the dominant phenocryst phases observed in the leucitite, viz. leucite, clinopyroxene and magnetite. However, Sheraton and Cundari (1980) also report four phase olivine-leucite-clinopyroxene-phlogopite bearing assemblages. In some examples, the cognate nodules contain a high proportion of volcanic glass which was presumably entrapped and quenched during crystallisation. Ultramafic nodules of mantle derivation are rare in the Gaussberg leucitites. The only known example, an olivine-orthopyroxene-clinopyroxene-spinel assemblage (lherzolite), was described by Sheraton and Cundari (1980).

The cognate nodules of cumulus origin examined in this study are of three types:

(1) Aggregates of leucite and clinopyroxene with intercumulus magnetite;
(2) Aggregates of leucite and clinopyroxene in a hypohyaline matrix; and
(3) Aggregates of leucite and clinopyroxene with intercumulus glass containing laths of phlogopite.

Further details of the petrology, mineral chemistry, and major and trace element geochemistry of the Gaussberg lavas and inclusions are given in Sheraton and Cundari (1980).

## ANALYTICAL METHODS

Prior to crushing, the samples were washed in distilled water in an ultrasonic bath. Crushing was carried out using a jaw crusher and tungsten carbide planetary ball mill until the grain size was less than 200 mesh.

Rb and Sr isotopic ratios and concentrations were determined using a modified version of the isotope dilution technique described by Compston et al. (1977). Replicate analyses of the standard potassium feldspar SRM 607 using a mixed $^{85}$Rb-$^{84}$Sr spike has yielded a mean $^{87}$Rb/$^{86}$Sr ratio of 24.13 ± 0.04 (SE mean) and a mean normalised $^{87}$Sr/$^{86}$Sr of 1.991 ± 0.0008 (SE mean). This is equivalent to an age of 1413 ± 5 Ma, assuming an initial $^{87}$Sr/$^{86}$Sr ($I_{Sr}$) = 0.710. This compares favourably with the analysis reported by the National Bureau of Standards (t = 1412 Ma). Replicate analyses (unspiked) of SRM 987 given an $^{87}$Sr/$^{86}$Sr of 0.71030 ± 0.00004 (2σ). The mean isotopic composition of BCR-1 determined from the analysis of six separate digestions using the mixed spike give an $^{87}$Rb/$^{86}$Sr of 0.4202 ± 21 (SE mean) and an $^{87}$Sr/$^{86}$Sr of 0.70503 ± 2 (2σ).

Sm and Nd isotopic date and concentrations were obtained using a mixed $^{150}$Nd-$^{147}$Sm tracer technique described in McCulloch and Chappell (1982). The Nd isotopic ratios were normalised to $^{146}$Nd/$^{142}$Nd = 0.636151 (Wasserburg et al., 1981) to remove the effects of mass fractionation. The $^{143}$Nd/$^{144}$Nd ratio for BCR-1 determined at ANU is 0.511843 ± 20 (2σ).

Glass compositions were determined by energy dispersive microprobe, following the method of Ware (1981). Rare earth elements were analysed by spark source mass spectrometry using the method described by Taylor and Gorton (1977). Analytical precision and accuracy is typically ca ± 5%.

## RESULTS

### *Sr and Nd geochemistry*

Rb-Sr and Sm-Nd isotopic ratios and concentration data are presented in Table 1 and depicted graphically in Figures 2-4.

***Rb-Sr results*** The leucitites and one of the cognate nodules are characterised by high concentrations of Rb and Sr; viz. 290-442 ppm and 1295-1890 ppm respectively. These ranges are similar to concentration levels reported by Hawkesworth and Vollmer (1979) for leucite bearing lavas from Roccamonfina and Somma-Vesuvius. In Figure 2 it is clear that $^{87}$Sr/$^{86}$Sr ratios for the leucitites are extremely high for mantle derived rocks and define two populations, viz. 0.70923-0.70926 and 0.70972-0.70978. This variation might be caused by some form of source heterogeneity. Alternatively, it might be the result of contamination by radiogenic old crust. However, in view of the fact that the Sr contents of potential contaminants are much lower than measured concentrations in the lavas (Table 1), contamination would appear to be an unlikely cause of the high $^{87}$Sr/$^{86}$Sr ratios. Therefore, although some contamination with disaggregated crustal xenoliths almost certainly occurred, its effect on the $^{87}$Sr/$^{86}$Sr ratio of the lavas would be buffered by the high Sr content of the lavas.

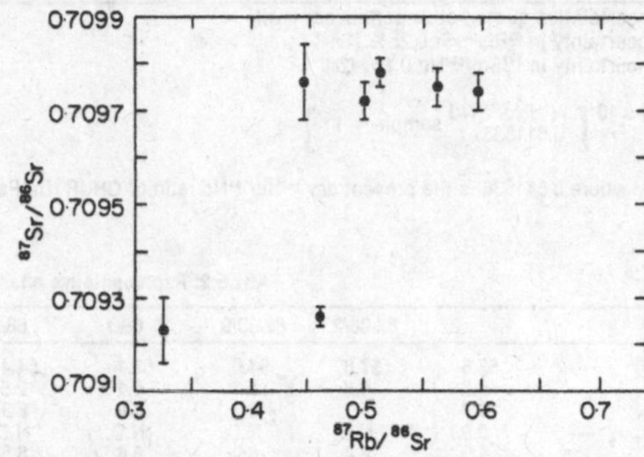


**Figure 2. $^{87}$Sr/$^{86}$Sr versus $^{87}$Rb/$^{86}$Sr diagram showing the distribution of data for leucitite from Gaussberg. The error bars represent 2σ confidence levels.**

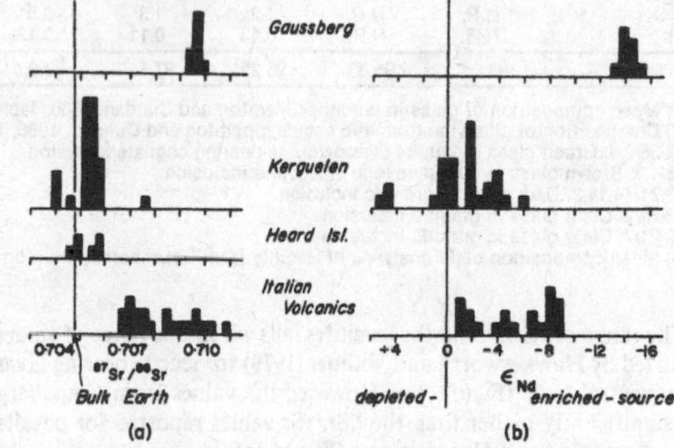


**Figure 3a. Histogram showing initial $^{87}$Sr/$^{86}$Sr ratios for Gaussberg, Heard Island, Kerguelen, and the central Italian leucite bearing volcanics.**

**Figure 3b. Histogram showing source characteristics, as represented in $\Sigma_{Nd}$ units for Gaussberg, Kerguelen, and the central Italian leucite bearing volcanics. Data from Dosso and Murthy (1980); White and Hofmann (1982); Clarke et al. (this volume); Hawkesworth and Vollmer (1979).**

**TABLE 1: Rb-Sr and Sm-Nd concentrations and isotopic ratios for lavas and inclusions from Gaussberg volcano.**

| Sample | Rb (ppm) | Sr (ppm) | $^{87}Rb/^{86}Sr^{(2)}$ | $^{87}Rb/^{86}Sr^{(1)}$ | Sm (ppm) | Nd (ppm) | $^{147}Sm/^{144}Nd^{(3)}$ | $^{147}Sm/^{144}Nd^{(1)}$ | $\Sigma_{Nd}(0)$ |
|---|---|---|---|---|---|---|---|---|---|
| Leucitite:— | | | | | | | | | |
| 79-242 | 289.9 | 1889.5 | 0.325 | 0.70923±7 | 16.35 | 128.7 | 0.0768 | 0.511106±28 | −14.3 |
| 4883 | 303.2 | 1892.8 | 0.463 | 0.70926±2 | 15.85 | 124.1 | 0.0772 | 0.511115±16 | −14.1 |
| 82-27 | 306.7 | 1770.7 | 0.500 | 0.70972±4 | 16.00 | 124.6 | 0.0777 | 0.511170±18 | −13.0 |
| 82-30 | 326.8 | 1581.9 | 0.507 | 0.70974±4 | 15.62 | 122.3 | 0.0773 | 0.511116±18 | −14.1 |
| 4893A | 329.7 | 1694.8 | 0.562 | 0.70975±4 | 16.14 | 128.6 | 0.0759 | 0.511111±20 | −14.2 |
| 79-241 | 301.8 | 1743.8 | 0.443 | 0.70976±8 | 16.12 | 126.4 | 0.0772 | 0.511106±26 | −14.3 |
| 4888 | 311.2 | 1750.4 | 0.514 | 0.70978±3 | 16.23 | 127.5 | 0.0770 | 0.511118±24 | −14.0 |
| Cognate cumulate nodules:— | | | | | | | | | |
| Leucite-clinopyroxene-magnetite-glass | | | | | | | | | |
| 82-38 | 441.6 | 1294.7 | 0.985 | 0.70987±4 | 13.49 | 105.7 | 0.0772 | 0.511073±18 | −14.9 |
| Leucite-clinopyroxene-magnetite | | | | | | | | | |
| 82-35 | 221.8 | 140.4 | 4.562 | 0.71090±5 | 2.31 | 11.36 | 0.1230 | 0.511108±30 | −14.2 |
| Xenoliths of crustal derivation | | | | | | | | | |
| 82-24 | 319.3 | 157.9 | 5.869 | 0.76260±7 | 20.72 | 125.9 | 0.0995 | 0.510574±26 | −24.7 |
| 82-21 | 187.5 | 302.9 | 1.791 | 0.73148±3 | 25.95 | 146.3 | 0.1073 | 0.510569±26 | −24.8 |
| 68 cpx | 16.53 | 35.87 | 1.333 | 0.72746±6 | 6.986 | 24.63 | 0.1716 | 0.510727±30 | −21.7 |

(1) Uncertainties quoted at $2\sigma$ confidence level.
(2) Uncertainty in $^{87}Rb/^{86}\nu Sr$ 0.25% ($1\sigma$)
(3) Uncertainty in $^{147}Sm/^{144}Nd$ 0.1% ($2\sigma$)

$$\Sigma_{Nd}(0) = 10^4 \left[ \frac{(^{143}Nd/^{144}Nd}{0.511836} \text{ sample} - 1 \right]$$

where 0.511836 is the present day $^{143}Nd/^{144}Nd$ ratio of CHUR (De Paolo and Wasserburg: 1976, b 1977).

**TABLE 2: Representative analyses of leucitite and glass from Gaussberg**

| | 1 | 2 | 82-38/2 | 82-38/9 | 68/1 | 68/3 | 82-21/14 | 82-21/11 | 82-21/2 | 82-24/2 | 82-21/7 | 3 |
|---|---|---|---|---|---|---|---|---|---|---|---|---|
| $SiO_2$ | 53.6 | 57.8 | 51.0 | 52.1 | 54.4 | 55.1 | 50.8 | 55.3 | 59.9 | 65.8 | 67.2 | 50.92 |
| $TiO_2$ | 6.2 | 6.8 | 5.1 | 5.1 | 2.5 | 1.9 | 2.3 | 2.2 | 1.3 | 0.11 | 0.3 | 3.42 |
| $Al_2O_3$ | 6.9 | 7.1 | 8.7 | 8.9 | 8.3 | 8.5 | 15.3 | 15.4 | 18.4 | 17.4 | 15.4 | 9.86 |
| $Cr_2O_3$ | 0.03 | N.D. | N.D. | N.D. | N.D. | N.D. | N.D. | N.D. | N.D. | N.D. | N.D. | N.R. |
| FeO | 9.5 | 8.4 | 6.9 | 6.8 | 8.5 | 8.5 | 13.7 | 9.3 | 5.0 | 2.5 | 2.2 | 5.55 |
| MnO | 0.11 | 0.13 | 0.12 | N.D. | N.D. | N.D. | 0.18 | 0.17 | N.D. | N.D. | N.D. | 0.09 |
| MgO | 5.4 | 2.5 | 4.8 | 4.8 | 4.8 | 4.5 | 3.4 | 2.4 | 1.8 | 0.7 | 0.57 | 7.95 |
| CaO | 3.3 | 1.2 | 4.4 | 4.1 | 3.6 | 3.5 | 5.7 | 3.9 | 1.8 | 1.4 | 0.75 | 4.63 |
| $Na_2O$ | 2.6 | 2.3 | 2.4 | 2.7 | 2.5 | 2.2 | 3.1 | 3.5 | 3.6 | 4.3 | 2.9 | 2.65 |
| $K_2O$ | 10.4 | 9.1 | 11.5 | 11.5 | 12.4 | 12.1 | 4.7 | 5.8 | 7.5 | 7.3 | 8.2 | 11.61 |
| $P_2O_5$ | N.R. | N.R. | 1.2 | 1.3 | 0.51 | 0.33 | 0.61 | 0.30 | N.M. | N.M. | N.M. | 1.48 |
| Cl | N.R. | N.R. | 0.13 | 0.11 | 0.12 | 0.07 | 0.25 | 0.21 | 0.17 | N.D. | 0.10 | 0.06 |
| Total | 98.04 | 95.33 | 96.25 | 97.41 | 97.6 | 96.7 | 100.04 | 98.48 | 99.47 | 99.51 | 97.62 | 97.20 |

(1) Mean composition of glass in leucitite (Sheraton and Cundari, 1980; Table 2).
(2) Composition of glass in ultramafic nodule (Sheraton and Cundari, 1980; Table 4).
82-38/2,9: Green glass in leucite-clinopyroxene-bearing cognate inclusion.
68/1,3: Brown glass in clinopyroxene-phlogopite inclusion.
82-21/14,11,2: Dark glass in granitic inclusion.
82-24/2: Clear glass in granitic inclusion.
82-21/7: Clear glass in granitic inclusion.
(3) Mean composition of 11 analyses of leucitite from Gaussberg. (Sheraton and Cundari, 1980; Table 5).

The range of $^{87}Sr/^{86}Sr$ in the leucitites falls within the range of values reported by Hawkesworth and Vollmer (1979) for leucite bearing lavas from central Italy (Figure 3a). However, the values from Gaussberg are significantly higher than the $^{87}Sr/^{86}Sr$ values reported for basalts from Kerguelen and Heard Island (Figure 3a).

*Sm-Nd results* The leucitites exhibit a limited range of variation in both Sm and Nd, viz. 15.6-16.4 ppm and 122.3-128.7 ppm, respectively. Significantly, concentrations in the cognate nodules differ from each other by almost an order-of-magnitude. Sm and Nd values for the glass bearing inclusion are slightly lower than values for the leucitite (13.5 ppm Sm and 105.7 ppm Nd). However, the leucite-clinopyroxene-magnetite bearing nodule has extremely low abundances (2.3 ppm and 11.4 ppm). This indicates that the REE's have been concentrated in the glass, which is presumably a residual liquid. The high concentrations of Sm and Nd in the crustal xenoliths are in marked contrast to Sr concentrations. The crustal xenoliths exhibit virtually the same high levels as the leucitites (Table 1). In view of the petrographic evidence for mixing between glasses of different composition (Table 2), this feature is interpreted to be a direct result of contamination of the crustal xenoliths by the host leucitite. However, prior to this contamination, Sr must have been removed, possibly by clinopyroxene fractionation, from the residual leucitite liquid.

The measured $^{143}Nd/^{144}Nd$ ratios in the leucitites and the cognate nodules exhibit a very restricted range, 0.511073-0.511170. These correspond to a range of $\Sigma_{Nd}(O)$ of from −14.9 to −13.0. In Figure 3b, these data are compared with data from Kerguelen and central Italy. Although the central Italian leucite bearing volcanics described by Hawkesworth and Vollmer (1979) also lie in the enriched mantle field, their sources were significantly less enriched than the data indicate for the Gaussberg sources. Likewise, lavas and plutonic rocks from Kerguelen (Dosso and Murthy, 1980; White and Hofmann, 1982) exhibit a spectrum of $^{143}Nd/^{144}Nd$ ratios. These data clearly demonstrate that the Kerguelen sources ranged from light rare earth (LREE) depleted to enriched ($\Sigma_{Nd}$ + 5 to −5.7). However, the most LREE enriched source for Kerguelen was significantly less enriched than the sources indicated for Gaussberg (Figure 3b).

When plotted in an $\Sigma_{Nd} - \Sigma_{Sr}$ diagram (Figure 4), data for the Gaussberg leucitites and the cognate nodules lie close to the extension of the mantle array into the enriched mantle field. Similar $\Sigma_{Nd}$ values are reported by McCulloch et al. (in press) for highly potassic kimberlites and lamproites from Western Australia. However, they have more radiogenic Sr isotopic compositions and lie to the right of the mantle array (Figure 4). For comparison, also shown are data for clinopyroxene inclusions from kimberlites (Menzies and Murthy, 1980a), whole rock samples from Kerguelen (Dosso and Murthy, 1980; White and Hofmann, 1982), leucite bearing lavas from central Italy (Hawkesworth and Vollmer, 1979) and data for OIB's from the Society Islands, and Samoa and a variety of MORB's (White and Hofmann, 1982).

Crustal inclusions in the lavas have extremely radiogenic $^{87}Sr/^{86}Sr$ and unradiogenic $^{143}Nd/^{144}Nd$ ratios and define a field which lies well to the right of the mantle array and the Gaussberg lava population. This field is typical of Precambrian upper crustal compositions. For comparison, also shown are data for lower and int., nediate Archaean crustal compositions from the Vestfold Block (Collerson et al., this volume).

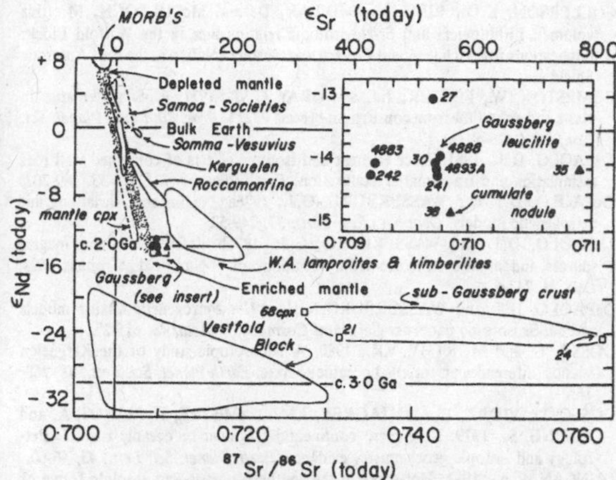


**Figure 4. $\Sigma_{Nd}$ versus $\Sigma_{Sr}$ diagram showing data from Gaussberg relative to the mantle array. For comparison, also shown are data for Roccamonfina and Somma-Vesuvius (Hawkesworth and Vollmer, 1979); Kerguelen (Dosso and Murthy, 1980; White and Hofmann, 1982); Samoa-Society Islands (Societies) and MORB's (White and Hofmann, 1982); mantle diopsides (Menzies and Murthy, 1980a, b); Western Australian kimberlites and lamproites (McCulloch et al., in press); and data for high grade gneisses from the Vestfold Block, Antarctica (Collerson et al., this volume). Details of the mantle array (shown by the stippled ornamentation) were taken from Allegre et al. (1979). Crustal formation ages from DePaolo and Wasserburg (1979) are also shown.**

Model $T^{Nd}_{CHUR}$ and $T^{Sr}_{UR}$ ages (DePaolo and Wasserburg, 1976a, b), calculated for the granitic inclusions in the Gaussberg lavas from the data in Table 1, range from 1973 Ma to 2152 Ma ($T^{Nd}_{CHUR}$) and from 700 Ma to 1093 Ma ($T^{Sr}_{UR}$). In view of the lack of agreement between the Sr and Nd model ages, it is evident that both of these isotopic systems have been disturbed to some extent by geological processes. Nevertheless, they provide clear evidence that the crust through which the Gaussberg magmas passed is Precambrian. Although the precise age of this continental crust cannot be deduced, the $T^{Nd}_{CHUR}$ age of 2152 Ma must be regarded as a minimum estimate. This indicates that early Proterozoic or late Archaean crust occurs at least ~ 550 km east of the Archaean Vestfold Block (cf. Collerson et al., this volume).

The Gaussberg leucitites are extremely uniform in terms of their major element and trace element chemistry (Sheraton and Cundari, 1980) and, as discussed previously, they have extremely high contents of Sr, Nd, and Sm. None of the potential contaminants described above have similar levels of these elements. Even if the Nd and Sr isotopic compositions were modified by the effects of crustal contamination, it is difficult to envisage how undersaturated basic lavas with remarkably uniform geochemical patterns could result, when one considers the volume of crust that would have to be assimilated (cf. DePaolo, 1981).

*Rare earth elements* Rare earth elements determined for three specimens of leucitite, a cognate nodule and a crustal xenolith from Gaussberg are presented in Table 3. These are shown diagrammatically as chondrite normalised patterns in Figure 5. The leucitites are remarkably uniform in terms of their REE geochemistry and show extreme light rare earth enrichment with $La_N/Yb_N$ ratios ranging between 433 and 293. Although these patterns are somewhat similar to those exhibited by the central Italian leucite bearing lavas, they do not show the negative Eu anomaly (Eu/Eu* = 0.68-0.80) described by Hawkesworth and Vollmer (1979) and interpreted to be the result of the LIL and $Eu^{2+}$ enrichment in the source of the Roccamonfina volcanics. However, continental alkali basalts and nephelinites from Ross Island, Antarctica, which are presumably genetically similar lavas, more commonly exhibit positive Eu anomalies (Sun and Hanson, 1975a). The chondrite normalised patterns exhibited by the Gaussberg leucitites (Figure 5) show virtually no anomaly (Eu/Eu* = 1.0-1.1). They are, therefore, interpreted to reflect source, or melt, physical and chemical conditions transitional between those of Roccamonfina and Ross Island.

The rare earth pattern exhibited by the cognate nodule (82-35) does not show the same level of LREE enrichment and is relatively flat with

**TABLE 3: Rare earth element concentration data for lavas and inclusions from Gaussberg**

| | 79-241 | 79-242 | 4883 | 82-35 | 82-21 | 68 cpx | Chondritic REE normalizing factors |
|---|---|---|---|---|---|---|---|
| La | 205 | 248 | 254 | 13.5 | 222 | 5.85 | 0.957 |
| Ce | 415 | 468 | 478 | 27.8 | 440 | 23.1 | 0.957 |
| Pr | 43.5 | 42.0 | 48.7 | 3.36 | 46.5 | 4.0 | 0.137 |
| Nd | 143 | 145 | 160 | 12.4 | 159 | 21.7 | 0.711 |
| Sm | 17.2 | 17.9 | 20.5 | 2.31 | 20.0 | 5.83 | 0.231 |
| Eu | 4.58 | 4.14 | 4.89 | 0.88 | 3.40 | 0.91 | 0.087 |
| Gd | 8.94 | | 10.7 | 2.54 | 20.9 | 5.78 | 0.306 |
| Tb | | | | | 3.13 | 1.05 | 0.058 |
| Dy | | | | 2.85 | 18.0 | 6.70 | 0.381 |
| Ho | 0.53 | 0.55 | 0.69 | 0.55 | 3.39 | 1.51 | 0.0851 |
| Er | 1.07 | | 1.42 | 1.50 | 9.26 | 4.38 | 0.249 |
| Yb | 0.47 | 0.39 | 0.48 | 1.51 | 8.10 | 4.10 | 0.248 |
| $La_N/Yb_N$ | 293 | 433 | 359 | 6.03 | 18.48 | 0.96 | |
| Eu/Eu* | 1.1 | 1.0 | 1.0 | 1.02 | 0.46 | 0.47 | |

Analyst P.E. Oswald-Sealy
Leucitite—79-241, 79-242, 4883
Clinopyroxene-leucite-magnetite cumulate cognate inclusion—82-35
Partially melted granodioritic xenolith—82-21
Clinopyroxene fraction from crustal xenolith—68

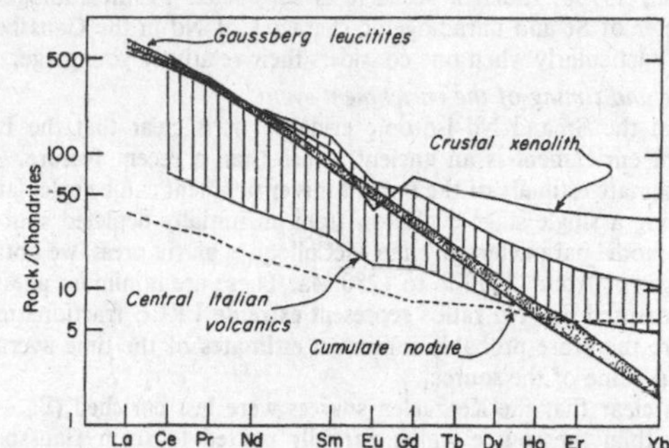


**Figure 5. Chondrite normalised rare earth patterns showing fields for the Gaussberg leucitites and central Italian volcanics (after Hawkesworth and Vollmer, 1979), as well as a glass free cognate nodule and a glass bearing crustal xenolith from Gaussberg.**

a $La_N/Yb_N$ of 6.03 (Figure 5). Apart from an absence of glass and olivine, the mineralogy of this cognate nodule is similar to that of its leucitite host. This observation, together with the previously discussed evidence of contamination between LREE enriched leucitite glass and the crustal xenoliths, clearly demonstrates that the extreme enrichments in the incompatible elements (LIL) are principally taking place in the residual glass. The chondrite normalised REE pattern for the granitic xenolith lies within the range of patterns exhibited by the leucitites between La and Nd, and exhibits similar concentrations of both Nd and Sm. However, the middle and heavy REE's are more enriched, with a pronounced negative Eu anomaly. This portion of the pattern is typically upper crustal (cf. McLennan and Taylor, 1982).

## DISCUSSION

### *Mantle heterogeneity along the Kerguelen-Gaussberg ridge*

Gaussberg leucitites are chemically extremely homogeneous and silica undersaturated (Sheraton and Cundari, 1980), have very high concentrations of Sr, Nd and Sm and plot in a tight field on the extension of the mantle array. Therefore, although crustal xenoliths are present, they apparently had little effect on the overall composition of the magma, or on its Sr and Nd isotopic signature. Consequently, these isotopic parameters are interpreted to reflect aspects of the geochemistry of the subcrustal source of these lavas. This reservoir was clearly enriched in Rb relative to Sr and Nd relative to Sm.

Along the Kerguelen-Gaussberg ridge, it is apparent that volcanic activity occurred synchronously at Gaussberg and Heard Island. Therefore, more than one hot-spot must have been involved. The data presented in this paper show that these hot-spots differed significantly in their levels of enrichment in incompatible elements (Figure 3a). This reflects the presence of grossly inhomogeneous suboceanic and subcontinental mantle in this area, with heterogeneities on a much greater scale, in terms of Rb-Sr and Sm-Nd, than those reported by White and Hofmann (1982) for Kerguelen (cf. Figures 3 and 4).

*Rare earth element and isotopic constraints on the petrogenesis of the Gaussberg leucitites*

A number of petrological models have been devised to account for the formation of such highly potassic melts (see reviews in Sheraton and Cundari, 1981; Kuehner et al., 1981). According to Sheraton and Cundari (1980), the extreme K contents and high K/Na ratios of the Gaussberg leucitites are the result of "small degrees of partial melting of phlogopite rich mantle below the level of amphibole stability". These melts are believed to have experienced a limited amount of high pressure fractionation of garnet and clinopyroxene with subsequent low pressure crystal fractionation of leucite, clinopyroxene, and olivine to produce cognate cumulates.

The REE patterns exhibited by the Gaussberg leucitites could be generated by relatively small degrees of partial melting (~ 1-2%) of a garnet peridotite source with 1-4 times chondritic rare earth abundances (cf. Kay and Gast, 1973). However, such a mechanism would not easily explain (see Hofmann and White, 1982) the levels of Ni, Cr, and V reported for these lavas. A more likely explanation is that they were formed by higher degrees of partial melting of a LIL element enriched source, with higher abundances of the LREE's (cf. Sun and Hanson, 1975b). Such a scenario is supported by the radiogenic character of Sr and unradiogenic character of Nd in the Gaussberg lavas, particularly when one considers their relatively young age.

*Nature and timing of the enrichment event*

From the Sr and Nd isotopic evidence, it is clear that the LIL element enrichment is an ancient rather than a recent feature. An approximate estimate of the time of this enrichment can be calculated assuming a single stage evolution from an initially depleted source. Using model parameters given in McCulloch et al. (in press) we obtain $T^{Nd}_{DM}$ ages of from 1220 Ma to 1280 Ma. These are minimum ages as the measured Sm/Nd ratios represent extreme LREE fractionations and are therefore probably minimum estimates of the time average Sm/Nd value of the source.

It is clear that the Kerguelen sources were less enriched ($\Sigma_{Nd}$ + 5 to − 5) than the source which partially melted to form Gaussberg ($\Sigma_{Nd}$ − 13 to − 15). The nature and origin of these enriched mantle sources is, however, speculative. They could be the result of metasomatism in the mantle (Hawkesworth and Vollmer, 1979; Menzies and Murthy, 1980a, b; Bailey, 1982), or, alternatively, a feature resulting from partial melting of ancient (Precambrian) subducted oceanic crust (cf. Chase, 1981; Hofmann and White, 1982; Ringwood, in press). Such altered subducted oceanic crust would have REE's at ca 10 times chondritic levels, varied Rb-Sr and Sm-Nd ratios, and show varied enrichment in K, Rb, Th and U. The difference in the level of enrichment of the Gaussberg and Kerguelen lavas could reflect the existence of sources of approximately the same age but with different Sm-Nd (and Rb-Sr) ratios or, alternatively, it indicates a real age difference of the sources. Assuming that the ~ 2000 Ma Pb whole rock "isochron" reported by Dosso et al. (1979) dates the time of formation of oceanic crust which, after subduction, melted ~ 1800 Ma later to form Kerguelen (cf. Chase, 1981), then the source region of the Gaussberg hot-spot may be significantly older.

These data provide further confirmation of the extremely heterogeneous character of the suboceanic and subcontinental mantle. Furthermore, they demonstrate that such heterogeneities have existed for extremely long periods of geological time.

*Acknowledgements* Some of the samples analysed in this study were kindly supplied by Dr J.W. Sheraton, Dr J. Parker and Dr I. McDougall. Others were collected by KDC during the 1980/81 summer season. Logistic support was provided by the Australian National Antarctic Research Expeditions. R.S. Taylor and P. Oswald-Sealy kindly provided the REE analyses. Other technical assistance was given by D. Millar and R. Rudowski. Chris Webb assisted in the production of the manuscript.

## REFERENCES

ALLEGRE, C.J., 1981: Chemical geodynamics. *Tectonophysics, 81,* 109-32.

ALLEGRE, C.J., BEN OTHMAN, D., POLVE, M. and RICHARD, P., 1979: The Nd-Sr isotopic correlation in mantle materials and geodynamic consequences. *Phys. Earth Planet. Int., 19,* 293-306.

BAILEY, D.K., 1982: Mantle metasomatism—continuing chemical change within the Earth. *Nature, 296,* 525-30.

CHASE, C.G., 1981: Oceanic island Pb: Two stage histories and mantle evolution. *Earth Planet. Sci. Lett., 52,* 277-84.

CLARKE, I., McDOUGALL, I. and WHITFORD, D.J. (this volume): Volcanic evolution of Heard and McDonald Islands, southern Indian Ocean.

COLLERSON, K.D., REID, E., MILLAR, D. and McCULLOCH, M. (this volume): Lithological and Sr-Nd isotopic relationships in the Vestfold Block: implications for Archaean and Proterozoic crustal evolution in the East Antarctic Shield.

COMPSTON, W., FOSTER, J.J. and GRAY, C.M., 1977: Rb-Sr systematics in clasts and aphanites from consortium breccia 73215. *Proc. 8th Lunar Planet. Sci. Conf., 2,* 2525-49.

DePAOLO, D.J., 1981: Trace element and isotopic effects of combined wall rock assimilation and fractional crystallisation. *Earth Planet. Sci. Lett., 53,* 189-202.

DePAOLO, D.J. and WASSERBURG, G.J., 1976a: Nd isotopic variations and petrogenetic models. *Geophys. Res. Lett., 34,* 249-52.

DePAOLO, D.J. and WASSERBURG, G.J., 1976b: Inferences about magma sources and mantle structure from variations of $^{143}Nd/^{144}Nd$. *Geophys. Res. Lett., 3,* 743-6.

DePAOLO, D.J. and WASSERBURG, G.J., 1979: Petrogenetic mixing models and Nd-Sr isotopic patterns. *Geochim. Cosmochim. Acta, 43,* 615-27.

DOSSO, L. and MURTHY, V.R., 1980: A Nd isotopic study of the Kerguelen islands: inferences on enriched oceanic sources. *Earth Planet. Sci. Lett., 48,* 268-76.

DOSSO, L., VIDAL, P., CANTAGREL, J.M., LAMEYRE, J., MAROT, A. and ZIMINE, S., 1979: Kerguelen: continental fragment or oceanic island?: Petrology and isotopic geochemistry evidence. *Earth Planet. Sci. Lett., 43,* 46-60.

DUNCAN, R.A., 1981: Hot-spots in the southern oceans—an absolute frame of reference for motion of the Gondwana continents. *Tectonophysics, 74,* 29-42.

HAWKESWORTH, C.J. and VOLLMER, R., 1979: Crustal contamination versus enriched mantle: $^{143}Nd/^{144}Nd$ and $^{87}Sr/^{86}Sr$ evidence from the Italian volcanics. *Contrib. Mineral. Petrol., 69,* 151-65.

HOFMANN, A.W. and WHITE, W.S., 1982: Mantle plumes from ancient oceanic crust. *Earth Planet. Sci. Lett., 57,* 421-36.

HOUTZ, R.E., HAYES, D.E. and MARKL, R.G., 1977: Kerguelen plateau bathymetry, sediment distribution and crustal structure. *Mar. Geol., 25,* 95-130.

KAY, R.W. and GAST, P.W., 1973: Rare earth content and origin of alkali rich basalts. *J. Geol., 81,* 653-82.

KUEHNER, S.M., EDGAR, A.D. and ARIMA, M., 1981: Petrogenesis of the ultrapotassic rocks from the Leucite Hills, Wyoming. *Am. Mineral., 66,* 663-77.

McCULLOCH, M.T. and CHAPPELL, B.W., 1982: Nd isotopic characteristics of S- and I-type granites. *Earth Planet. Sci. Lett., 58,* 51-64.

McCULLOCH, M.T., JAQUES, A.L., NELSON, D. and LEWIS, J.D. (in press): Nd and Sr isotopes in kimberlites and lamproites from the West Kimberley, Western Australia; an enriched mantle origin. *Nature.*

McLENNAN, S.M. and TAYLOR, S.R., 1982: Geochemical constraints on the growth of the continental crust. *J. Geol., 90,* 347-61.

MENZIES, M. and MURTHY, R.V., 1980a: Enriched mantle: Nd and Sr isotopes in diopsides from kimberlite nodules. *Nature, 283,* 634-6.

MENZIES, M. and MURTHY, R.V., 1980b: Nd and Sr isotope geochemistry of hydrous mantle nodules and their host alkali basalts: implications for local heterogeneities in metasomatically veined mantle. *Earth Planet. Sci. Lett., 46,* 323-34.

NOUGIER, J., 1969: Contribution a l'étude geologique et geomorphologique desîles Kerguelen (Territoire des Terres Australes et Antarctiques francaises). *CNFRA, 27.*

O'NIONS, R.K., EVENSON, N.M. and HAMILTON, P.J., 1979: Geochemical modelling of mantle differentiation and crustal growth. *J. Geophys. Res., 84,* 6091-101.

O'NIONS, R.K., HAMILTON, P.J. and EVENSON, N.M., 1977: Variations in $^{143}Nd/^{144}Nd$ and $^{87}Sr/^{86}Sr$ ratios in oceanic basalts. *Earth Planet. Sci. Lett., 34,* 13-22.

RAVICH, M.G. and KRYLOV, A.J., 1964: Absolute ages of rocks from East Antarctica; *in* Adie, R.J. (ed.) *Antarctic Geology.* North Holland, Amsterdam, 579-89.

RINGWOOD, A.E., 1982: Phase transformations and differentiation in subducted lithosphere: implications for mantle dynamics, basalt petrogenesis and crustal evolution. *J. Geol., 90,* 611-44.

SHERATON, J.W. and CUNDARI, A., 1980: Leucitites from Gaussberg, Antarctica. *Contrib. Mineral. Petrol., 71,* 417-27.

SOLOVIEV, D.S., 1972: Platform magmatic formation of East Antarctica; *in* Adie, R.J. (ed.) *Antarctic Geology and Geophysics.* Universitetsforlaget, Oslo, 531-38.

SUN, S.S. and HANSON, G.N., 1975a: Origin of Ross Island basanitoids and limitations upon the heterogeneity of mantle sources for alkali basalts and nephelinites. *Contrib. Mineral. Petrol., 52,* 77-106.

SUN, S.S. and HANSON, G.N., 1975b: Evolution of the mantle: geochemical evidence from alkali basalt. *Geology, 3,* 287-302.

TAYLOR, S.R. and GORTON, M.P., 1977: Geochemical application of spark source mass spectrography—III. Element sensitivity, precision and accuracy. *Geochim. Cosmochim. Acta, 41,* 1375-80.

TINGEY, R.J., McDOUGALL, I. and GLEADOW, A.J.W. (in press): The age and mode of formation of Gaussberg, Antarctica. *Geol. Soc. Aust., J.*

WARE, N.G., 1981: Computer programs and calibration with the PIBS technique for quantitative electron probe analysis using a lithium drifted silicon detector. *Computers and Geoscience, 7,* 167-84.

WASSERBURG, G.J., JACOBSEN, S.B., DEPAOLO, D.J., McCULLOCH, M.T. and WEN, T., 1981: Precise determination of Sm/Nd ratios, Sm and Nd isotopic abundances in standard solutions. *Geochim. Cosmochim. Acta, 45,* 2311-23.

WATKINS, N.D., GUNN, B.M., NOUGIER, J. and BAKSI, A.K., 1974: Kerguelen: continental fragment or oceanic island. *Geol. Soc. Am. Bull., 85,* 201-12.

WHITE, W.M. and HOFMANN, A.W., 1982: Sr and Nd isotope geochemistry of oceanic basalts and mantle evolution. *Nature, 296,* 821-6.

# INDEX